DIRECTORY OF

ENGINEERING

AND

ENGINEERING

TECHNOLOGY

UNDERGRADUATE

PROGRAMS

American Society for Engineering Education

11 Dupont Circle, Suite 200

Washington, DC 20036

(202) 986-8500

COPYRIGHT

© 1993 by the American Society for Engineering Education. All rights reserved. No part of this book may be reproduced, stored in a retrieval system, transcribed or transmitted, in any form or by any means—electronic, mechanical, photocopying, recording, or otherwise—except for citations of data for scholarly or reference purposes with full acknowledgment of title, edition, and publisher and written notification to the American Society for Engineering Education prior to use.

PREVIOUS EDITIONS
© 1989, 1990, 1992

Editorial inquiries concerning this book should be addressed to:
    Manager, Publications and Marketing Services
    American Society for Engineering Education
    11 Dupont Circle, Suite 200
    Washington, DC 20036
    USA

ISBN 0-87823-143-9; ISSN 1057-5286
Composition and design by the American Society for Engineering Education
Printed in the United States of America

Foreword

The *ASEE 1993 Directory of Engineering and Engineering Technology Undergraduate Programs* is the most comprehensive resource available for information on baccalaureate programs in both engineering and engineering technology. ASEE began publishing directories of engineering graduate schools in 1967. In its fourth year of publication, ASEE's undergraduate directory has quickly become a vital resource for students, educators, administrators and corporate professionals.

In this year's directory, more than three hundred and fifty colleges have full-page entries. Each of the participating schools has at least one bachelor's (four-year) degree program accredited by the Accreditation Board for Engineering and Technology (ABET) or the Canadian Engineering Accreditation Board (CEAB).* With several new features—a comprehensive index to both engineering and engineering technology degree programs, state-by-state statistical summaries, and an appendix of personnel, the 1993 Directory has been revised to make it even more useful and easy-to-use.

The American Society for Engineering Education is a non-profit organization of individuals and institutions dedicated to improving all aspects of engineering and engineering technology education. For 100 years the Society has played a key role in developing and advocating policies and programs that advance education in engineering and allied branches of science and technology.

Frank L. Huband,
Executive Director

ASEE Publications and Marketing Services Staff

Erin O'Neill, *Manager*
Cliff Cary, *Directories Coordinator*
Cathy Haynie, *Production Designer*
De Thomas, *Advertising Coordinator*
Kirstin Peters, *Publication Sales*
Heidi Gomez, *Publications Assistant*
Sanjai Chandra, *Computer Support*
James Hess, *Computer Support*

*The Accreditation Board for Engineering and Technology (ABET) is the official accrediting authority for engineering and technology programs in the United States. In Canada, the official accrediting body for engineering programs is the Canadian Engineering Accreditation Board (CEAB). These organizations accredit individual engineering and engineering technology programs, not universities, colleges, schools or degrees. For complete information on the accreditation process, you should contact: ABET, 345 East 47th Street, New York, NY 10017-2397; (212)705-7685. For information on accreditation in Canada, contact CEAB, 601-116 Albert Street, Ottawa, ON, K1P 5G3; (613)232-2474.

Table of Contents

Foreword 3

INTRODUCTION
Using the Directory 6
Keys to Acronyms & Abbreviations: *Professional Societies, Honor Societies, Financial Aid Forms, Postal Codes* 10
Alphabetical Index of Participating Institutions 12
Geographical Index of Participating Institutions 14

ENGINEERING SECTION
Participating Engineering Institutions .... TAB
An Introduction to Engineering, by Ray Landis 18
Engineering Disciplines: Descriptions and Career Opportunities, by Don Evans 18
Index to Engineering Degree Programs 20

Comparative Tables–by School
Undergraduate Degrees Awarded–by Category 33
Support Programs for Ethnic Minorities & Women 39
Enrollments and Degrees Awarded 45

Comparative Tables–by Degree Program
All Engineering Disciplines 53
Chemical Engineering 93
Civil Engineering 101
Electrical Engineering 111
Industrial Engineering 125
Mechanical Engineering 133

Institution Profiles 144

ENGINEERING TECHNOLOGY SECTION
Participating Engineering Technology Institutions .... TAB
An Introduction to Engineering Technology, by Stephen Cheshier 426
Engineering Technology Disciplines: Descriptions and Career Opportunities, by Fred Emshousen 426
Index to Engineering Technology Degree Programs 428

Comparative Tables–by School
Undergraduate Degrees Awarded–by Category 431
Support Programs for Ethnic Minorities & Women 433
Enrollments and Degrees Awarded 435

Comparative Table–by Degree Program
All Engineering Technology Disciplines 439

Institution Profiles 448

APPENDIXES & INDEXES
Engineering and Engineering Technology Personnel 527
Index to Engineering and Engineering Technology Personnel 570
ABET/CEAB Accredited Degree Programs 585
Index to ABET/CEAB Accredited Degree Programs 596

Using the Directory

This introduction provides an overview of the various parts of the directory and how to use them. In most cases, you will find additional information within the individual sections. The directory is divided into four sections. (1) *Using the Directory* contains general, introductory, and explanatory information that applies to the entire directory and that affects all participating schools—both engineering and engineering technology. (2) The *Engineering Section* consists of indexes, tables, and institution profiles of engineering schools. (3) Similarly, the *Engineering Technology Section* consists of indexes, tables, and institution profiles of engineering technology schools. To avoid repetition, the constituent parts of both the engineering and the engineering technology sections are discussed together below. (4) Finally, the *Appendixes and Indexes* portion of the directory contains supplementary material about personnel and accredited bachelor's degree programs.

A Note on Participation Criteria

Information throughout this directory was furnished by each of the 356 participating institutions that responded to our survey. Before deciding upon an institution for undergraduate study, students should contact the appropriate offices and consult the degree catalogs at each institution to verify information. Questions about any information provided here should be addressed directly to the respective institutions.

ASEE requires that each participating school offer at least one baccalaureate (usually a four-year bachelor's) engineering or engineering technology degree program accredited by the Accreditation Board for Engineering and Technology (ABET) or the Canadian Engineering Accreditation Board (CEAB). Therefore, institutions that offer baccalaureate engineering or engineering technology degree programs but do not offer at least one *accredited* baccalaureate engineering or engineering technology degree program are not eligible for participation in our directory. Similarly, institutions that offer accredited *associate* degree programs but do not offer at least one accredited *baccalaureate* degree program are not eligible. In short, a qualifying program must be an *accredited baccalaureate* degree program in engineering or engineering technology. For a complete list of these programs, refer to the *Appendix of ABET/CEAB Accredited Degree Programs* in the back of this book.

Introduction

The *Keys to Acronyms & Abbreviations* clarify some of the "shorthand" used to conserve space throughout the directory. Specifically, keys for Professional Societies, Honor Societies, and Financial Aid Forms are for use primarily with the *Institution Profiles,* where these acronyms and abbreviations are most often used. Then, a list of Postal Codes and their corresponding full state and province names is provided to help users locate each school.

The *Alphabetical Index of Participating Institutions* lists all institutions that returned surveys for this directory. In this list, schools appear in alphabetical order, with engineering schools listed first, followed by engineering technology schools. Following the school name, the city and state or province of the main campus are given in parentheses. Next, in brackets, are references to the section of the directory ("E" for Engineering and/or "T" for Engineering Technology) and the institution profile reference number of a school's full-page profile(s).

In the *Geographical Index of Participating Institutions*, schools are grouped by country—either the United States or Canada—and then by state or province. Within each state or province, schools are presented alphabetically by school name. As with the alphabetical index, the city appears in parentheses, followed by section abbreviations ("E" or "T") and institution profile reference number(s).

Engineering & Engineering Technology Sections

The *Engineering and Engineering Technology Sections* are nearly identical in content. Therefore, for brevity, only the *Engineering Section* is described below. Where differences in the *Engineering Technology Section* do occur, these are noted.

The list of *Participating Engineering Institutions* includes 280 engineering schools with at least one baccalaureate engineering degree program accredited by ABET or CEAB. The schools appear in alphabetical order with the institution profile reference number to the left of each school's name. If a school on the engineering list also has an engineering technology entry, that entry number is cross-referenced in brackets. The same is true on the list of 76 engineering technology schools.

To introduce students to engineering, Ray Landis in his article "An Introduction to Engineering" describes the benefits of an engineering education and relates what engineers do, the types of challenges they encounter, and the contributions they make. In the second article—"Engineering Disciplines: Descriptions and Career Opportunities", Don Evans discusses in greater detail five major engineering disciplines—chemical, civil, electrical, industrial, and mechanical engineering—and describes career opportunities available to students in each discipline. The corresponding articles for engineering technology are: "An Introduction to Engineering Technology" by Stephen Cheshier and "Engineering Technology Disciplines: Descriptions and Career Opportunities" by Fred Emshousen.

The index to *Engineering Degree Programs* contains more than 1900 engineering (more than 300 for engineering technology) majors and degree programs at participating institutions. Many but not all of these programs are accredited by ABET or CEAB. To determine which programs are accredited, refer to the *Tables of Degree Programs*, which contain complete information for each degree program. In some cases, similarly titled degree programs are grouped under a broader subject heading. For exact titles of degree programs, refer to the *Tables of Degree Programs* or to the *Institution Profiles*.

COMPARATIVE TABLES

In each of the comparative tables, schools appear in order by the institution profile reference number and the school name, which are in the far left of each row, followed by the state in which the main campus of the college is located. In cases where a school did not provide relevant data, that school has been omitted from the table. At the end of statistical tables, state-by-state totals, as well as grand totals for each country, are provided. In many instances, schools were allowed to clarify or explain their responses with a footnote. Where a school has footnoted their response, a footnote indicator appears in parentheses above and to the right of the data. In cases where no schools in a particular state or province provided relevant data, there will be no totals for that state or province. Following appropriate tables, a separate section devoted to footnotes provides the full text of each footnote.

Comparative Tables—by School

Each table in this section presents a single line of summary data for every school. The first four columns of data in the *Engineering Enrollments and Degrees Awarded* table are the number of students enrolled in all engineering programs by class—freshmen/first-year students, sophomores/second-year students, juniors/third-year students, and seniors/fourth-year and all other students. Then the total number of students enrolled in all engineering degree programs is given by degree level—bachelor's (the sum of the previous four columns), master's, and doctoral. There are no engineering technology doctoral programs. The last three columns represent the number of engineering degrees awarded by type of degree—bachelor's, master's, and doctoral. No doctoral degree is awarded in engineering technology. Unless footnoted otherwise, enrollment figures are for the fall 1991 term and degrees awarded are for the 1991-92 twelve-month academic year.

In the table *Undergraduate Degrees Awarded—by Category*, the figures represent the percentages of bachelor's degree recipients who fall into different categories. The section devoted to students of different ethnic backgrounds and nationalities includes African Americans, Asian (including Pacific Island) Americans, Hispanic Americans, and Native Americans, as well as citizens of other countries. The figure for the column entitled "All Others" was derived by subtracting the above percentages from 100.0%. In most cases, Canadian schools could not provide data in these or corresponding categories; therefore, this information does not appear for most Canadian schools. In addition, information was requested about degree recipients in categories independent of ethnic or national background. These categories are: women, students with self-identified physical or learning disabilities, and students who were older than 25 years of age when they received their degrees. Unless footnoted otherwise, degrees awarded are for the 1991-92, twelve-month academic year.

The table of *Support Programs for Ethnic Minorities and Women* includes information on student chapters of professional societies, as well as on programs that are staffed by the engineering college or by the university. If a university offers a student chapter of the Society of Women Engineers (SWE), the National Society of Black Engineers (NSBE), the Society of Hispanic Professional Engineers (SHPE), or, in the fourth column, any other professional society for historically under represented students, a check mark appears in the corresponding column. Similarly, if a school has its own staffed programs for ethnic minorities or women, a check mark will appears in the appropriate columns. The last two columns indicate (by capital letter) when (Summer, the nine-month Academic year, or Year-round) and for which students (Prematriculation, Freshman, Sophomore, Junior, senioR, and Graduate) at least one program is available.

Comparative Tables—by Degree Program

In each of the engineering degree program tables, the first line of information for an institution contains the institution profile reference number, the school's name, and the state in which the main campus of the engineering college is located. Next, there is an indented line of data for each degree program. The first item on these lines is the name of the degree program, followed by a letter in brackets which indicates whether a school designated this program as belonging to one of six major discipline categories—[H] chemical engineering, [V] civil engineering, [E] electrical engineering, [I] industrial engineering, [M] mechanical engineering, and [O] for all other engineering disciplines. If any footnote was given, the indicator appears next. Subsequent information *for each degree program* appears in the following columns: the full-time equivalent (FTE) of all faculty dedicated to this program (see explanation below*); for undergraduate programs, whether or not the program is accredited by ABET (the Accreditation Board for Engineering and Technology) or CEAB (the Canadian Engineering Accreditation Board), the nominal length and actual average length of the program in years, the time courses are offered (Day, Evening, or Both day and evening), the availability of cooperative education (internship) programs (None, Optional or Required), the percentage of graduates who participated in co-op programs, the number of bachelor's degrees awarded, and the percentage of graduates who were full-time. For graduate programs, the columns contain the number of graduates and the percentage who were full-time in master's programs and, (for engineering schools) doctoral programs. Unless otherwise footnoted, information on degrees is for the 1991-92 twelve-month, academic year; all other information is current.

[*The full-time equivalent (FTE) of part-time faculty is calculated by dividing the total number of course hours taught by all part-time faculty by the full-time teaching load at that university. For example, if 10 part-time professors each teach one 3-hour course (for a total of 30 hours), and the normal course load for a full-time professor is 12 hours, then the FTE of part-time faculty is 30 ÷ 12

= 2.5. The FTE of *all* faculty is the sum of the number of full-time faculty and the FTE of part-time faculty.]

Institution Profiles

Each school's entry will vary according to the completeness of the information provided to ASEE. In instances in which a question was not applicable or information was not available, a section or subsection may have been omitted for that school. Except for enrollments (which are for the fall 1991 term) and degrees awarded (which are for the 1991-92 twelve-month academic year), all information is current as of publication in 1993.

Throughout the institution profiles, respondents had the opportunity to qualify their responses with footnotes. Where they have done so, there will be an indicator in parentheses to the right of the data being footnoted. The corresponding text appears in small type beneath a line at the end of each subsection.

■ INSTITUTION PROFILE

The *Institution Profile* provides the address of the engineering/engineering technology college, the head of the institution (usually the president or the chancellor), and other general information for the institution as a whole. This information includes enrollments for all (not just engineering) students, the type of institution (public, private, or having elements of both), the academic calendar (semesters, quarters, etc.), the setting of the main campus (rural, small town, suburban, or urban), and the nearest city or metropolitan area (including its population and the distance from the main campus). Also indicated are whether courses are offered at one campus or more than one campus, the types of engineering degrees offered (engineering and/or engineering technology), and any other (non-engineering) undergraduate degree-granting colleges, schools, or departments.

■ ENGINEERING ADMISSIONS

Admissions information includes the admissions office address and any admissions requirements or recommendations, including entrance examinations (with minimum score), high school courses (with number of years), requirements for out-of-state residents and for foreign students, and any departmental admissions requirements.

■ FINANCIAL INFORMATION

The *Financial Information* section details expenses and financial aid information. Student expenses are itemized as tuition and fees, college room and board, books and supplies, and other expenses for a full-time, first-year engineering student for the current nine-month academic year (excluding summer terms). If expenses differ for certain groups of students, two sets of expenses are given. Financial aid information includes the financial aid office contact, the types of financial aid offered, forms required to apply for financial aid, and any other financial aid information.

■ ENGINEERING COLLEGE INFORMATION

The *Engineering College Information* portion of a school's entry begins with the head of engineering (usually the dean). The *Undergraduate Engineering Enrollments* section provides enrollment figures by class—first-year students/freshmen, second-year students/sophomores, third-year students/juniors, and all other undergraduate students/seniors, and total undergraduate engineering enrollments for the fall 1991 term.

The section *Undergraduate Degrees Awarded—by Degree Program* gives the name of each degree program or major offered and the number of baccalaureate degrees awarded in each during the 1991-92 twelve-month academic year. Additional information about each program is provided in the *Comparative Tables* section of the directory.

In *Undergraduate Degrees Awarded—by Category*, the figures represent the percentages of bachelor's degree recipients who fall into different categories, as detailed in the *Comparative Tables—by School* section above. Unless footnoted otherwise, degrees awarded are for the 1991-92, twelve-month academic year.

The *Graduate Engineering Enrollments and Degrees Awarded* section provides total enrollments and degrees awarded for master's and (for engineering) doctoral students. Unless otherwise footnoted, enrollments are for the fall 1991 term and degrees awarded are for the 1991-92 twelve-month academic year.

Engineering Student Data includes the number of applicants to the engineering college, as well as the percentage of applicants offered admission for the fall 1991 term. For matriculated engineering students, this section provides the percentage of entering students who were in the top quartile (25%) of their high school graduating class and the average scores for entering students on the SAT and ACT.

Engineering Faculty—by Department gives the number of full-time faculty and the full-time equivalent (FTE) of part-time faculty in each engineering department.

In *Engineering College Description* and *Engineering Degree Programs Description*, each school has the opportunity to describe its programs, exchanges, facilities, joint or double majors, and any other relevant information about its engineering college or specific degrees offered.

A separate *Dual Degree Programs* section has been set aside for discussion of any programs in which a student may obtain simultaneous degrees in engineering and another discipline.

The section on *Transfer Information* addresses two possible methods of transfer—through specific "articulation" agreements with other academic organizations (either colleges and schools within the university or with other universities) and through any other method(s). For each method, schools were asked to provide the number of students who transferred from four-year programs and two-year programs and any special entrance requirements. Next, any residency requirements for transfer students are provided.

The *Graduation Requirements* portion of a school's profile explains the requirements for graduating with an engineering baccalaureate degree.

■ STUDENT PROGRAMS

The first items in this section are lists of engineering *Professional Societies and Honorary Societies* that are available to students. The *Support Programs* section gives information about university-staffed support programs, including whether the program is designed for women, ethnic minorities, or both; when the program is available (summer only, the nine-month academic year, or year round); and at which levels students are eligible (prematriculation, first-year/freshmen, second-year/sophomores, third-year/juniors, all other undergraduates/seniors, and/or graduate students). Next is a list of any student chapter organizations designed specifically for women and/or ethnic minorities. Finally, schools have the opportunity to describe other engineering student support programs at their institution.

Appendixes & Indexes

Both of the appendixes are organized in alphabetical order by institution. Following both is an index of their contents, as detailed below.

ENGINEERING & ENGINEERING TECHNOLOGY PERSONNEL

The appendix of Engineering & Engineering Technology Personnel lists the position, name, title, telephone number, fax number, and Email number for personnel at each institution, including the head of the institution, the head of engineering or engineering technology, up to four associate or assistant deans, department chairs, and the admissions and financial aid contacts. The appendix is divided into engineering and engineering technology subsections, within which institutions are organized in alphabetical order. Following the appendix is an alphabetical index of engineering and engineering technology personnel by last name, then first name.

ABET/CEAB ACCREDITED DEGREE PROGRAMS

This appendix is divided into two parts: engineering programs and engineering technology programs. Each is organized alphabetically by school, with degree programs also listed alphabetically for each school. Each section includes all baccalaureate (usually four-year bachelor's) degree programs accredited by the Accreditation Board for Engineering and Technology (ABET). The engineering section also includes programs accredited by the Canadian Engineering Accreditation Board (CEAB). Following the appendix is an alphabetical index of accredited engineering and engineering technology degree programs.

Keys to Acronyms & Abbreviations

FINANCIAL AID FORMS

ACT	American College Testing form
AFSA	Application for Federal Student Aid
CSS-FAF	College Scholarship Service Financial Aid Form
FAF	Financial Aid Form
FAT	Financial Aid Transcript
FAR	Financial Aid Report
FSA	Federal Student Aid application
FFS	Family Financial Statement
Stafford	Stafford Loan (formerly Guaranteed Student Loan)
IRS	Internal Revenue Service (federal tax return forms)
SAR	Student Aid Report
SDF	Student Data Form
Supplemental	Supplemental Student Loan
Institutional	Institutional application form(s)

PROFESSIONAL SOCIETIES

AAPG	American Association of Petroleum Geologists
AATCC	American Association of Textile Chemists and Colorists
ACerS	The American Ceramic Society
ACI	American Concrete Institute
ACM	Association for Computing Machinery
ACS	American Chemical Society
ACSM	American Congress on Surveying and Mapping
AEG	Association of Engineering Geologists
AFS	American Fisheries Society
AGCA	Associated General Contractors of America
AHS	American Helicopter Society
AIA	American Institute of Architects
AIAA	American Institute of Aeronautics and Astronautics
AIC	The American Institute of Chemists
AIChE	American Institute of Chemical Engineers
AIDD	American Institute for Design and Drafting
AIME	American Institute of Mining, Metallurgical and Petroleum Engineers
AIP	American Institute of Physics
AIPE	American Institute of Plant Engineers
AMS	American Mathematical Society
AMS*	American Meteorological Society
ANS	American Nuclear Society
APS	The American Physical Society
APS*	The American Phytopathological Society
ASA	Acoustical Society of America
ASA*	American Statistical Association
ASAE	American Society of Agricultural Engineers
ASBE	American Society of Body Engineers
ASCE	American Society of Civil Engineers
ASEM	American Society for Engineering Management
ASHRAE	American Society of Heating, Refrigerating & Air-Conditioning Engineers
ASM	American Society for Microbiology
ASM*	Association for Systems Management
ASME	The American Society of Mechanical Engineers
ASNE	American Society of Naval Engineers
ASPRS	American Society for Photogrammetry and Remote Sensing
AWRA	American Water Resources Association
AWS	American Welding Society
BMES	Biomedical Engineering Society
CES	Capstone Engineering Society
CIM	The Canadian Institute of Mining and Metallurgy
CSAE	Canadian Society of Agricultural Engineering
CSCE	Canadian Society for Civil Engineering
CSME	The Canadian Society for Mechanical Engineering
DPMA	Data Processing Management Association
GSA	The Geological Society of America
ICA	International Cartographic Association
IEEE	The Institute of Electrical & Electronics Engineers
IIE	Institute of Industrial Engineers
IES	Illuminating Engineering Society of North America
IES*	Institute of Environmental Sciences
ISA	Instrument Society of America
ITE	Institute of Transportation Engineers
MAA	The Mathematical Association of America
NAHB	National Association of Home Builders
NSAE	National Society of Architectural Engineers
NSPE	National Society of Professional Engineers
ORSA	Operations Research Society of America
OSA	Optical Society of America
SAE	Society of Automotive Engineers
SAME	The Society of American Military Engineers
SAMPE	Society for the Advancement of Material & Process Engineering
SEE	Society of Explosives Engineers
SEG	Society of Exploration Geophysicists
SEM	Society for Experimental Mechanics
SES	Socity of Engineering Science
SES*	Standards Engineering Society
SFPE	Society of Fire Protection Engineers
SIAM	Society for Industrial and Applied Mathematics
SME	Society of Manufacturing Engineers
SME*	Society of Mining Engineers
SNAME	The Society of Naval Architects and Marine Engineers
SPE	Society of Petroleum Engineers
SPE*	Society of Plastics Engineers
SPS	Society of Physics Students
STC	Society for Technical Communication
TAPPI	Technical Association of the Pulp and Paper Industry
TIMS	The Institute of Management Sciences
TMS	The Minerals, Metals, and Materials Society

HONORARY SOCIETIES

AE	Alpha Epsilon
APM	Alpha Pi Mu
CE	Chi Epsilon
EKN	Eta Kappa Nu
KEK	Kappa Eta Kappa
K	Keramos
OCE	Omega Chi Epsilon
PAE	Phi Alpha Epsilon
PKU	Phi Kappa Upsilon
PTS	Pi Tau Sigma
SGT	Sigma Gamma Tau
SPD	Sigma Phi Delta
TAP	Tau Alpha Pi
TBP	Tau Beta Pi
TT	Theta Tau

POSTAL CODES

AB	Alberta
AK	Alaska
AL	Alabama
AR	Arkansas
AZ	Arizona
BC	British Columbia
CA	California
CO	Colorado
CT	Connecticut
DC	District Of Columbia
DE	Delaware
FL	Florida
GA	Georgia
HI	Hawaii
IA	Iowa
ID	Idaho
IL	Illinois
IN	Indiana
KS	Kansas
KY	Kentucky
LA	Louisiana
MA	Massachusetts
MB	Manitoba
MD	Maryland
ME	Maine
MI	Michigan
MN	Minnesota
MO	Missouri
MS	Mississippi
MT	Montana
NB	New Brunswick
NC	North Carolina
ND	North Dakota
NE	Nebraska
NF	Newfoundland
NH	New Hampshire
NJ	New Jersey
NM	New Mexico
NS	Nova Scotia
NV	Nevada
NY	New York
OH	Ohio
OK	Oklahoma
ON	Ontario
OR	Oregon
PA	Pennsylvania
PE	Prince Edward Island
PQ	Quebec
PR	Puerto Rico
RI	Rhode Island
SC	South Carolina
SD	South Dakota
SK	Saskatchewan
TN	Tennessee
TX	Texas
UT	Utah
VA	Virginia
VT	Vermont
WA	Washington
WI	Wisconsin
WV	West Virginia
WY	Wyoming

Alphabetical Index of Participating Institutions

Schools appear in alphabetical order by key word. Following the school name, the city and state or province of the main campus are given in parentheses. Next, in brackets, are references to the section of the directory ("E" for Engineering and/or "T" for Engineering Technology) and the institution profile reference number of a school's full-page profile(s).

A

University of Akron *(Akron, OH)* [E-001] [T-281]
University of Alabama *(Tuscaloosa, AL)* [E-002]
University of Alabama at Birmingham *(Birmingham, AL)* [E-003]
University of Alabama at Huntsville *(Huntsville, AL)* [E-004]
Alabama A&M University *(Normal, AL)* [T-282]
University of Alaska, Fairbanks *(Fairbanks, AK)* [E-005]
University of Alberta *(Edmonton, AB)* [E-006]
Alfred University *(Alfred, NY)* [E-007]
University of Arizona *(Tucson, AZ)* [E-008]
Arizona State University *(Tempe, AZ)* [T-283] [E-009]
University of Arkansas *(Fayetteville, AR)* [E-010]
University of Arkansas at Little Rock *(Little Rock, AR)* [T-284]
Arkansas Tech University *(Russellville, AR)* [E-011]
Auburn University *(Auburn University, AL)* [T-285] [E-012]

B

Baylor University *(Waco, TX)* [E-013]
Boston University *(Boston, MA)* [E-014]
Bradley University *(Peoria, IL)* [E-015] [T-286]
University of Bridgeport *(Bridgeport, CT)* [E-016]
Brigham Young University *(Provo, UT)* [E-017] [T-287]
Bucknell University *(Lewisburg, PA)* [E-018]

C

University of California at Berkeley *(Berkeley, CA)* [E-019]
University of California, Davis *(Davis, CA)* [E-020]
University of California, Irvine *(Irvine, CA)* [E-021]
University of California, Los Angeles *(Los Angeles, CA)* [E-022]
University of California, San Diego *(La Jolla, CA)* [E-023]
University of California, Santa Barbara *(Santa Barbara, CA)* [E-024]
University of California, Santa Cruz *(Santa Cruz, CA)* [E-025]
California Institute of Technology *(Pasadena, CA)* [E-026]
California Polytechnic State University *(San Luis Obispo, CA)* [E-027]
California State University, Chico *(Chico, CA)* [E-028]
California State University, Fresno *(Fresno, CA)* [E-029]
California State University, Long Beach *(Long Beach, CA)* [E-030]
California State University, Los Angeles *(Los Angeles, CA)* [E-031]
California State University, Northridge *(Northridge, CA)* [E-032]
California State University, Sacramento *(Sacramento, CA)* [E-033] [T-288]
California State Polytechnic University, Pomona *(Pomona, CA)* [E-034] [T-289]
Calvin College *(Grand Rapids, MI)* [E-035]
Capitol College *(Laurel, MD)* [T-290]
Carnegie Mellon University *(Pittsburgh, PA)* [E-036]
Case Western Reserve University *(Cleveland, OH)* [E-037]
The Catholic University of America *(Washington, DC)* [E-038]
Central Connecticut State University *(New Britain, CT)* [T-291]
University of Central Florida *(Orlando, FL)* [E-039] [T-292]
Central State University *(Wilberforce, OH)* [E-040]
Christian Brothers University *(Memphis, TN)* [E-041]
University of Cincinnati *(Cincinnati, OH)* [T-293] [E-042]
The Citadel *(Charleston, SC)* [E-043]
City College of the City University of New York *(New York, NY)* [E-044]
City University of New York, College of Staten Island *(Staten Island, NY)* [E-045]
Clarkson University *(Potsdam, NY)* [E-046]
Clemson University *(Clemson, SC)* [E-047]
Cleveland State University *(Cleveland, OH)* [E-048]
University of Colorado at Boulder *(Boulder, CO)* [E-049]
University of Colorado at Colorado Springs *(Colorado Springs, CO)* [E-050]
University of Colorado at Denver *(Denver, CO)* [E-051]
Colorado School of Mines *(Golden, CO)* [E-052]
Colorado State University *(Fort Collins, CO)* [E-053]
Columbia University *(New York, NY)* [E-054]
Concordia University *(Montréal, PQ)* [E-055]
University of Connecticut *(Storrs, CT)* [E-056]
The Cooper Union *(New York, NY)* [E-057]
Cornell University *(Ithaca, NY)* [E-058]

D

Dartmouth College *(Hanover, NH)* [E-059]
University of Dayton *(Dayton, OH)* [E-060] [T-294]
University of Delaware *(Newark, DE)* [E-061]
University of Denver *(Denver, CO)* [E-062]
DeVry Institute of Technology, Chicago *(Chicago, IL)* [T-295]
DeVry Institute of Technology, Columbus *(Columbus, OH)* [T-296]
DeVry Institute of Technology, Atlanta *(Decatur, GA)* [T-297]
DeVry Institute of Technology, DuPage *(Addison, IL)* [T-298]
DeVry Institute of Technology, Dallas *(Irving, TX)* [T-299]
DeVry Institute of Technology, Kansas City *(Kansas City, MO)* [T-300]
DeVry Institute of Technology, Los Angeles *(City of Industry, CA)* [T-301]
DeVry Institute of Technology, Phoenix *(Phoenix, AZ)* [T-302]
University of the District of Columbia *(Washington, DC)* [E-063] [T-303]
Drexel University *(Philadelphia, PA)* [E-064]
Duke University *(Durham, NC)* [E-065]

E

East Tennessee State University *(Johnson City, TN)* [T-304]
École Polytechnique de Montréal *(Montréal, PQ)* [E-066]
Embry-Riddle Aeronautical University - Daytona Beach *(Daytona Beach, FL)* [E-067] [T-305]
Embry-Riddle Aeronautical University, Western Campus *(Prescott, AZ)* [E-068]
University of Evansville *(Evansville, IN)* [E-069]

F

Ferris State University *(Big Rapids, MI)* [T-306] [E-070]
University of Florida *(Gainesville, FL)* [E-071]
Florida Agricultural and Mechanical University *(Tallahassee, FL)* [E-072]
Florida Atlantic University *(Boca Raton, FL)* [E-073]
Florida Institute of Technology *(Melbourne, FL)* [E-074]
Florida International University *(Miami, FL)* [E-075]
Florida State University *(Tallahassee, FL)* [E-076]
Franklin University *(Columbus, OH)* [T-307]

G

Gannon University *(Erie, PA)* [E-078]
George Mason University *(Fairfax, VA)* [E-079]
The George Washington University *(Washington, DC)* [E-080]
University of Georgia *(Athens, GA)* [E-081]
Georgia Institute of Technology *(Atlanta, GA)* [E-082]
GMI Engineering & Management Institute *(Flint, MI)* [E-077]
Gonzaga University *(Spokane, WA)* [E-083]
Grand Valley State University *(Grand Rapids, MI)* [E-084]
Grove City College *(Grove City, PA)* [E-085]

H

University of Hartford *(West Hartford, CT)* [T-308]
Harvard University *(Cambridge, MA)* [E-086]
Harvey Mudd College *(Claremont, CA)* [E-087]
University of Hawaii at Manoa *(Honolulu, HI)* [E-088]
Hofstra University *(Hempstead, NY)* [E-089]
University of Houston *(Houston, TX)* [E-090]
Howard University *(Washington, DC)* [E-091]

I

University of Idaho *(Moscow, ID)* [E-092]
Idaho State University *(Pocatello, ID)* [E-093]
University of Illinois at Chicago *(Chicago, IL)* [E-094]
University of Illinois at Urbana-Champaign *(Urbana, IL)* [E-095]
Illinois Institute of Technology *(Chicago, IL)* [E-096]
Indiana University-Purdue University at Indianapolis *(Indianapolis, IN)* [T-309] [E-097]
University of Iowa *(Iowa City, IA)* [E-098]
Iowa State University *(Ames, IA)* [E-099]

J

The Johns Hopkins University *(Baltimore, MD)* [E-100]

K

University of Kansas *(Lawrence, KS)* [E-101]
Kansas State University *(Manhattan, KS)* [E-102]
University of Kentucky *(Lexington, KY)* [E-103]

L

Lafayette College *(Easton, PA)* [E-104]
Lake Superior State University *(Sault Ste. Marie, MI)* [T-310]
Lamar University *(Beaumont, TX)* [E-105]
Lawrence Technological University *(Southfield, MI)* [E-106]
Lehigh University *(Bethlehem, PA)* [E-108]
LeTourneau University *(Longview, TX)* [E-107]
Louisiana State University *(Baton Rouge, LA)* [E-109]
Louisiana Tech University *(Ruston, LA)* [E-110] [T-311]
University of Louisville *(Louisville, KY)* [E-111]
Loyola College in Maryland *(Baltimore, MD)* [E-112]
Loyola Marymount University *(Los Angeles, CA)* [E-113]

M

University of Maine *(Orono, ME)* [E-114] [T-312]
University of Manitoba *(Winnipeg, MB)* [E-115]
Mankato State University *(Mankato, MN)* [E-116] [T-313]
Marietta College *(Marietta, OH)* [T-117]
Marquette University *(Milwaukee, WI)* [E-118]
University of Maryland Baltimore County *(Baltimore, MD)* [E-119]
University of Maryland, College Park *(College Park, MD)* [E-120]
University of Massachusetts *(Amherst, MA)* [E-121]
University of Massachusetts, Dartmouth *(North Dartmouth, MA)* [E-122] [T-314]
University of Massachusetts, Lowell *(Lowell, MA)* [E-123] [T-315]
Massachusetts Institute of Technology *(Cambridge, MA)* [E-124]
McGill University *(Montréal, PQ)* [E-125]
Memphis State University *(Memphis, TN)* [E-126] [T-316]
Mercer University *(Macon, GA)* [E-127]
Metropolitan State College of Denver *(Denver, CO)* [T-317]
University of Miami *(Coral Gables, FL)* [E-128]
Miami University *(Oxford, OH)* [E-129]
University of Michigan *(Ann Arbor, MI)* [E-130]

University of Michigan-Dearborn *(Dearborn, MI)* [E-131]
Michigan State University *(East Lansing, MI)* [E-132]
Michigan Technological University *(Houghton, MI)* [E-133]
Milwaukee School of Engineering *(Milwaukee, WI)* [E-134] [T-318]
University of Minnesota, Duluth *(Duluth, MN)* [E-135]
University of Minnesota *(Minneapolis, MN)* [E-136]
University of Mississippi *(University, MS)* [E-137]
Mississippi State University *(Mississippi State, MS)* [E-138]
University of Missouri-Columbia *(Columbia, MO)* [E-139]
University of Missouri-Kansas City *(Independence, MO)* [E-140]
University of Missouri-Rolla *(Rolla, MO)* [E-141]
Monmouth College *(West Long Branch, NJ)* [E-142]
Montana College of Mineral Science and Technology *(Butte, MT)* [E-143]
Montana State University *(Bozeman, MT)* [E-144] [T-319]
Morgan State University *(Baltimore, MD)* [E-145]
Murray State University *(Murray, KY)* [T-320]

N

University of Nebraska - Lincoln *(Lincoln, NE)* [E-146]
University of Nevada, Las Vegas *(Las Vegas, NV)* [E-147]
New England College *(Henniker, NH)* [E-148]
University of New Hampshire *(Durham, NH)* [E-149] [T-321]
University of New Haven *(West Haven, CT)* [E-150]
New Jersey Institute of Technology *(Newark, NJ)* [E-151] [T-322]
The University of New Mexico *(Albuquerque, NM)* [E-152]
New Mexico Institute of Mining and Technology *(Socorro, NM)* [E-153]
New Mexico State University *(Las Cruces, NM)* [E-154] [T-323]
University of New Orleans *(New Orleans, LA)* [E-155]
New York Institute of Technology *(Old Westbury, NY)* [E-156]
New York State College of Ceramics at Alfred University *(Alfred, NY)* [E-157]
University of North Carolina-Charlotte *(Charlotte, NC)* [E-158] [T-324]
North Carolina Agricultural and Technical State University *(Greensboro, NC)* [E-159]
North Carolina State University *(Raleigh, NC)* [E-160]
University of North Dakota *(Grand Forks, ND)* [E-161]
North Dakota State University *(Fargo, ND)* [E-162]
Northeastern University *(Boston, MA)* [E-163] [T-325]
Northern Arizona University *(Flagstaff, AZ)* [E-164]
Northern Illinois University *(DeKalb, IL)* [E-165]
Northwestern University *(Evanston, IL)* [E-166]
Norwich University *(Northfield, VT)* [E-167]
University of Notre Dame *(Notre Dame, IN)* [E-168]

O

Oakland University *(Rochester, MI)* [E-169]
Ohio University *(Athens, OH)* [E-170]
Ohio Northern University *(Ada, OH)* [E-171]
Ohio State University *(Columbus, OH)* [E-172]
University of Oklahoma *(Norman, OK)* [E-173]
Oklahoma Christian University of Science and Arts *(Oklahoma City, OK)* [E-174]
Oklahoma State University *(Stillwater, OK)* [E-175] [T-326]
Old Dominion University *(Norfolk, VA)* [E-176] [T-327]
Oregon Institute of Technology *(Klamath Falls, OR)* [T-328]
Oregon State University *(Corvallis, OR)* [E-177]

P

University of the Pacific *(Stockton, CA)* [E-178]
University of Pennsylvania *(Philadelphia, PA)* [E-179]
Pennsylvania State University *(University Park, PA)* [E-180]
Pennsylvania State University at Erie, The Behrend College *(Erie, PA)* [T-329]
Pennsylvania State University - Harrisburg *(Middletown, PA)* [T-330]
Pittsburg State University *(Pittsburg, KS)* [T-331]
University of Pittsburgh *(Pittsburgh, PA)* [E-181]
University of Pittsburgh at Johnstown *(Johnstown, PA)* [T-332]

Point Park College *(Pittsburgh, PA)* [T-333]
Polytechnic University *(Brooklyn, NY)* [E-182]
University of Portland *(Portland, OR)* [E-183]
Portland State University *(Portland, OR)* [E-184]
Prairie View A&M University *(Prairie View, TX)* [T-334] [E-185]
Princeton University *(Princeton, NJ)* [E-186]
University of Puerto Rico, Mayagüez Campus *(Mayagüez, PR)* [E-187]
Purdue University *(West Lafayette, IN)* [E-188] [T-335]

R

Rensselaer Polytechnic Institute *(Troy, NY)* [E-189]
University of Rhode Island *(Kingston, RI)* [E-190]
William Marsh Rice University *(Houston, TX)* [E-191]
University of Rochester *(Rochester, NY)* [E-192]
Rochester Institute of Technology *(Rochester, NY)* [T-336] [E-193]
Rose-Hulman Institute of Technology *(Terre Haute, IN)* [E-194]
Rutgers, The State University *(Piscataway, NJ)* [E-195]

S

Saginaw Valley State University *(University Center, MI)* [E-196]
St. Cloud State University *(St. Cloud, MN)* [E-197] [T-337]
Parks College of Saint Louis University *(Cahokia, IL)* [E-198]
University of San Diego *(San Diego, CA)* [E-199]
San Diego State University *(San Diego, CA)* [E-200]
San Francisco State University *(San Francisco, CA)* [E-201]
San Jose State University *(San Jose, CA)* [E-202]
Santa Clara University *(Santa Clara, CA)* [E-203]
Seattle University *(Seattle, WA)* [E-204]
University of South Alabama *(Mobile, AL)* [E-205]
University of South Carolina *(Columbia, SC)* [E-206]
South Carolina State University *(Orangeburg, SC)* [T-338]
South Dakota School of Mines and Technology *(Rapid City, SD)* [E-207]
South Dakota State University *(Brookings, SD)* [E-208]
University of South Florida *(Tampa, FL)* [E-209]
Southern University and A&M College *(Baton Rouge, LA)* [E-210]
Southern College of Technology *(Marietta, GA)* [T-339]
University of Southern California *(Los Angeles, CA)* [E-211]
University of Southern Colorado *(Pueblo, CO)* [E-212] [T-340]
Southern Illinois University at Carbondale *(Carbondale, IL)* [E-213] [T-341]
Southern Illinois University at Edwardsville *(Edwardsville, IL)* [E-214]
University of Southern Maine *(Gorham, ME)* [E-215]
Southern Methodist University *(Dallas, TX)* [E-216]
University of Southern Mississippi *(Hattiesburg, MS)* [T-342]
Stanford University *(Stanford, CA)* [E-217]
State University of New York at Binghamton *(Binghamton, NY)* [E-218]
State University of New York at Buffalo *(Buffalo, NY)* [E-219]
State University of New York, College of Environmental Science and Forestry *(Syracuse, NY)* [E-220]
State University of New York College of Technology at Farmingdale *(Farmingdale, NY)* [T-343]
State University of New York Maritime College *(Throgs Neck Station, NY)* [E-221]
State University of New York, College at New Paltz *(New Paltz, NY)* [E-222]
State University of New York at Stony Brook *(Stony Brook, NY)* [E-223]
State University of New York Institute of Technology, Utica/Rome *(Utica, NY)* [T-344]
Swarthmore College *(Swarthmore, PA)* [E-224]
Syracuse University *(Syracuse, NY)* [E-225]

T

Technical University of Nova Scotia *(Halifax, NS)* [E-226]
Temple University *(Philadelphia, PA)* [E-227] [T-345]
University of Tennessee, Chattanooga *(Chattanooga, TN)* [E-228]
University of Tennessee, Knoxville *(Knoxville, TN)* [E-229]

University of Tennessee, Martin *(Martin, TN)* [T-346]
Tennessee State University *(Nashville, TN)* [E-230]
Tennessee Technological University *(Cookeville, TN)* [E-231]
University of Texas at Arlington *(Arlington, TX)* [E-232]
University of Texas at Austin *(Austin, TX)* [E-233]
University of Texas at Dallas *(Richardson, TX)* [E-234]
University of Texas at El Paso *(El Paso, TX)* [E-235]
University of Texas at San Antonio *(San Antonio, TX)* [E-236]
Texas A&I University *(Kingsville, TX)* [E-237]
Texas A&M University *(College Station, TX)* [E-238] [T-347]
Texas Tech University *(Lubbock, TX)* [E-239] [T-348]
University of Toledo *(Toledo, OH)* [E-240]
Tri-State University *(Angola, IN)* [E-241] [T-349]
Trinity University *(San Antonio, TX)* [E-242]
Tufts University *(Medford, MA)* [E-243]
Tulane University *(New Orleans, LA)* [E-244]
University of Tulsa *(Tulsa, OK)* [E-245]
Tuskegee University *(Tuskegee, AL)* [E-246]

U

Union College *(Schenectady, NY)* [E-247]
United States Air Force Academy *(USAF Academy, CO)* [E-248]
United States Coast Guard Academy *(New London, CT)* [E-249]
United States Merchant Marine Academy *(Kings Point, NY)* [E-250]
United States Military Academy *(West Point, NY)* [E-251]
United States Naval Academy *(Annapolis, MD)* [E-252]
University of Utah *(Salt Lake City, UT)* [E-253]
Utah State University *(Logan, UT)* [E-254]

V

Valparaiso University *(Valparaiso, IN)* [E-255]
Vanderbilt University *(Nashville, TN)* [E-256]
University of Vermont *(Burlington, VT)* [E-257]
Villanova University *(Villanova, PA)* [E-258]
University of Virginia *(Charlottesville, VA)* [E-259]
Virginia Military Institute *(Lexington, VA)* [E-260]
Virginia Polytechnic Institute and State University *(Blacksburg, VA)* [E-261]
Virginia State University *(Petersburg, VA)* [T-350]

W

University of Washington *(Seattle, WA)* [E-262]
Washington University *(St. Louis, MO)* [E-263]
Washington State University *(Pullman, WA)* [E-264]
Wayne State University *(Detroit, MI)* [E-265]
Webb Institute of Naval Architecture *(Glen Cove, NY)* [E-266]
Weber State University *(Ogden, UT)* [T-351]
Wentworth Institute of Technology *(Boston, MA)* [T-352]
West Virginia University *(Morgantown, WV)* [E-267]
West Virginia Institute of Technology *(Montgomery, WV)* [E-268] [T-353]
Western Carolina University *(Cullowhee, NC)* [T-354]
Western Kentucky University *(Bowling Green, KY)* [T-355]
Western Michigan University *(Kalamazoo, MI)* [E-269] [T-356]
Western New England College *(Springfield, MA)* [E-270]
Wichita State University *(Wichita, KS)* [E-271]
Widener University *(Chester, PA)* [E-272]
Wilkes University *(Wilkes-Barre, PA)* [E-273]
University of Wisconsin-Madison *(Madison, WI)* [E-274]
University of Wisconsin-Milwaukee *(Milwaukee, WI)* [E-275]
University of Wisconsin-Platteville *(Platteville, WI)* [E-276]
Worcester Polytechnic Institute *(Worcester, MA)* [E-277]
Wright State University *(Dayton, OH)* [E-278]
University of Wyoming *(Laramie, WY)* [E-279]

Y

Yale University *(New Haven, CT)* [E-280]

Geographical Index of Participating Institutions

Schools are grouped by country–either the United States or Canada–and then by state or province. Within each state or province, schools are presented alphabetically by school name. As with the alphabetical index, the city appears in parentheses, followed by section abbreviations ("E" or "T") and institution profile reference number(s).

 **UNITED STATES**

ALABAMA

University of Alabama *(Tuscaloosa)* [E-002]
University of Alabama at Birmingham *(Birmingham)* [E-003]
University of Alabama at Huntsville *(Huntsville)* [E-004]
Alabama A&M University *(Normal)* [T-282]
Auburn University *(Auburn University)* [E-012] [T-285]
University of South Alabama *(Mobile)* [E-205]
Tuskegee University *(Tuskegee)* [E-246]

ALASKA

University of Alaska, Fairbanks *(Fairbanks)* [E-005]

ARIZONA

University of Arizona *(Tucson)* [E-008]
Arizona State University *(Tempe)* [E-009] [T-283]
DeVry Institute of Technology, Phoenix *(Phoenix)* [T-302]
Embry-Riddle Aeronautical University, Western Campus *(Prescott)* [E-068]
Northern Arizona University *(Flagstaff)* [E-164]

ARKANSAS

University of Arkansas *(Fayetteville)* [E-010]
University of Arkansas at Little Rock *(Little Rock)* [T-284]
Arkansas Tech University *(Russellville)* [E-011]

CALIFORNIA

University of California at Berkeley *(Berkeley)* [E-019]
University of California, Davis *(Davis)* [E-020]
University of California, Irvine *(Irvine)* [E-021]
University of California, Los Angeles *(Los Angeles)* [E-022]
University of California, San Diego *(La Jolla)* [E-023]
University of California, Santa Barbara *(Santa Barbara)* [E-024]
University of California, Santa Cruz *(Santa Cruz)* [E-025]
California Institute of Technology *(Pasadena)* [E-026]
California Polytechnic State University *(San Luis Obispo)* [E-027]
California State University, Chico *(Chico)* [E-028]
California State University, Fresno *(Fresno)* [E-029]
California State University, Long Beach *(Long Beach)* [E-030]
California State University, Los Angeles *(Los Angeles)* [E-031]
California State University, Northridge *(Northridge)* [E-032]
California State University, Sacramento *(Sacramento)* [E-033] [T-288]
California State Polytechnic University, Pomona *(Pomona)* [E-034] [T-289]
DeVry Institute of Technology, Los Angeles *(City of Industry)* [T-301]
Harvey Mudd College *(Claremont)* [E-087]
Loyola Marymount University *(Los Angeles)* [E-113]
University of the Pacific *(Stockton)* [E-178]
University of San Diego *(San Diego)* [E-199]
San Diego State University *(San Diego)* [E-200]
San Francisco State University *(San Francisco)* [E-201]
San Jose State University *(San Jose)* [E-202]
Santa Clara University *(Santa Clara)* [E-203]
University of Southern California *(Los Angeles)* [E-211]
Stanford University *(Stanford)* [E-217]

COLORADO

University of Colorado at Boulder *(Boulder)* [E-049]
University of Colorado at Colorado Springs *(Colorado Springs)* [E-050]
University of Colorado at Denver *(Denver)* [E-051]
Colorado School of Mines *(Golden)* [E-052]
Colorado State University *(Fort Collins)* [E-053]
University of Denver *(Denver)* [E-062]
Metropolitan State College of Denver *(Denver)* [T-317]
University of Southern Colorado *(Pueblo)* [E-212] [T-340]
United States Air Force Academy *(USAF Academy)* [E-248]

CONNECTICUT

University of Bridgeport *(Bridgeport)* [E-016]
Central Connecticut State University *(New Britain)* [T-291]
University of Connecticut *(Storrs)* [E-056]
University of Hartford *(West Hartford)* [T-308]
University of New Haven *(West Haven)* [E-150]
United States Coast Guard Academy *(New London)* [E-249]
Yale University *(New Haven)* [E-280]

DELAWARE

University of Delaware *(Newark)* [E-061]

DISTRICT OF COLUMBIA

The Catholic University of America *(Washington)* [E-038]
University of the District of Columbia *(Washington)* [E-063] [T-303]
The George Washington University *(Washington)* [E-080]
Howard University *(Washington)* [E-091]

FLORIDA

University of Central Florida *(Orlando)* [E-039] [T-292]
Embry-Riddle Aeronautical University - Daytona Beach *(Daytona Beach)* [E-067] [T-305]
University of Florida *(Gainesville)* [E-071]
Florida Agricultural and Mechanical University *(Tallahassee)* [E-072]
Florida Atlantic University *(Boca Raton)* [E-073]
Florida Institute of Technology *(Melbourne)* [E-074]
Florida International University *(Miami)* [E-075]
Florida State University *(Tallahassee)* [E-076]
University of Miami *(Coral Gables)* [E-128]
University of South Florida *(Tampa)* [E-209]

GEORGIA

DeVry Institute of Technology, Atlanta *(Decatur)* [T-297]
University of Georgia *(Athens)* [E-081]
Georgia Institute of Technology *(Atlanta)* [E-082]
Mercer University *(Macon)* [E-127]
Southern College of Technology *(Marietta)* [T-339]

HAWAII

University of Hawaii at Manoa *(Honolulu)* [E-088]

IDAHO

University of Idaho *(Moscow)* [E-092]
Idaho State University *(Pocatello)* [E-093]

ILLINOIS

Bradley University *(Peoria)* [E-015] [T-286]
DeVry Institute of Technology, Chicago *(Chicago)* [T-295]
DeVry Institute of Technology, DuPage *(Addison)* [T-298]
University of Illinois at Chicago *(Chicago)* [E-094]
University of Illinois at Urbana-Champaign *(Urbana)* [E-095]
Illinois Institute of Technology *(Chicago)* [E-096]
Northern Illinois University *(DeKalb)* [E-165]
Northwestern University *(Evanston)* [E-166]
Parks College of Saint Louis University *(Cahokia)* [E-198]
Southern Illinois University at Carbondale *(Carbondale)* [E-213] [T-341]
Southern Illinois University at Edwardsville *(Edwardsville)* [E-214]

INDIANA

University of Evansville *(Evansville)* [E-069]
Indiana University-Purdue University at Indianapolis *(Indianapolis)* [E-097] [T-309]
University of Notre Dame *(Notre Dame)* [E-168]
Purdue University *(West Lafayette)* [E-188] [T-335]
Rose-Hulman Institute of Technology *(Terre Haute)* [E-194]
Tri-State University *(Angola)* [E-241] [T-349]
Valparaiso University *(Valparaiso)* [E-255]

IOWA

University of Iowa *(Iowa City)* [E-098]
Iowa State University *(Ames)* [E-099]

KANSAS

University of Kansas *(Lawrence)* [E-101]
Kansas State University *(Manhattan)* [E-102]
Pittsburg State University *(Pittsburg)* [T-331]
Wichita State University *(Wichita)* [E-271]

KENTUCKY

University of Kentucky *(Lexington)* [E-103]
University of Louisville *(Louisville)* [E-111]
Murray State University *(Murray)* [T-320]
Western Kentucky University *(Bowling Green)* [T-355]

LOUISIANA

Louisiana State University *(Baton Rouge)* [E-109]
Louisiana Tech University *(Ruston)* [E-110] [T-311]
University of New Orleans *(New Orleans)* [E-155]
Southern University and A&M College *(Baton Rouge)* [E-210]
Tulane University *(New Orleans)* [E-244]

MAINE

University of Maine *(Orono)* [E-114] [T-312]
University of Southern Maine *(Gorham)* [E-215]

MARYLAND

Capitol College *(Laurel)* [T-290]
The Johns Hopkins University *(Baltimore)* [E-100]
Loyola College in Maryland *(Baltimore)* [E-112]
University of Maryland Baltimore County *(Baltimore)* [E-119]
University of Maryland, College Park *(College Park)* [E-120]
Morgan State University *(Baltimore)* [E-145]
United States Naval Academy *(Annapolis)* [E-252]

MASSACHUSETTS

Boston University *(Boston)* [E-014]
Harvard University *(Cambridge)* [E-086]
University of Massachusetts *(Amherst)* [E-121]
University of Massachusetts, Dartmouth *(North Dartmouth)* [E-122] [T-314]
University of Massachusetts, Lowell *(Lowell)* [E-123] [T-315]
Massachusetts Institute of Technology *(Cambridge)* [E-124]
Northeastern University *(Boston)* [E-163] [T-325]
Tufts University *(Medford)* [E-243]
Wentworth Institute of Technology *(Boston)* [T-352]
Western New England College *(Springfield)* [E-270]
Worcester Polytechnic Institute *(Worcester)* [E-277]

MICHIGAN

Calvin College *(Grand Rapids)* [E-035]
Ferris State University *(Big Rapids)* [E-070] [T-306]
GMI Engineering & Management Institute *(Flint)* [E-077]
Grand Valley State University *(Grand Rapids)* [E-084]
Lake Superior State University *(Sault Ste. Marie)* [T-310]
Lawrence Technological University *(Southfield)* [E-106]
University of Michigan *(Ann Arbor)* [E-130]
University of Michigan-Dearborn *(Dearborn)* [E-131]
Michigan State University *(East Lansing)* [E-132]
Michigan Technological University *(Houghton)* [E-133]
Oakland University *(Rochester)* [E-169]
Saginaw Valley State University *(University Center)* [E-196]
Wayne State University *(Detroit)* [E-265]
Western Michigan University *(Kalamazoo)* [E-269] [T-356]

MINNESOTA

Mankato State University *(Mankato)* [E-116] [T-313]
University of Minnesota, Duluth *(Duluth)* [E-135]
University of Minnesota *(Minneapolis)* [E-136]
St. Cloud State University *(St. Cloud)* [E-197] [T-337]

MISSISSIPPI

University of Mississippi *(University)* [E-137]
Mississippi State University *(Mississippi State)* [E-138]
University of Southern Mississippi *(Hattiesburg)* [T-342]

MISSOURI

DeVry Institute of Technology, Kansas City *(Kansas City)* [T-300]
University of Missouri-Columbia *(Columbia)* [E-139]
University of Missouri-Kansas City *(Independence)* [E-140]
University of Missouri-Rolla *(Rolla)* [E-141]
Washington University *(St. Louis)* [E-263]

MONTANA

Montana College of Mineral Science and Technology *(Butte)* [E-143]
Montana State University *(Bozeman)* [E-144] [T-319]

NEBRASKA

University of Nebraska-Lincoln *(Lincoln)* [F-146]

NEVADA

University of Nevada, Las Vegas *(Las Vegas)* [E-147]

NEW HAMPSHIRE

Dartmouth College *(Hanover)* [E-059]
New England College *(Henniker)* [E-148]
University of New Hampshire *(Durham)* [E-149] [T-321]

NEW JERSEY

Monmouth College *(West Long Branch)* [E-142]
New Jersey Institute of Technology *(Newark)* [E-151] [T-322]
Princeton University *(Princeton)* [E-186]
Rutgers, The State University *(Piscataway)* [E-195]

NEW MEXICO

The University of New Mexico *(Albuquerque)* [E-152]
New Mexico Institute of Mining and Technology *(Socorro)* [E-153]
New Mexico State University *(Las Cruces)* [E-154] [T-323]

NEW YORK

Alfred University *(Alfred)* [E-007]
City College of the City University of New York *(New York)* [E-044]
City University of New York, College of Staten Island *(Staten Island)* [E-045]
Clarkson University *(Potsdam)* [E-046]
Columbia University *(New York)* [E-054]
The Cooper Union *(New York)* [E-057]
Cornell University *(Ithaca)* [E-058]
Hofstra University *(Hempstead)* [E-089]
New York Institute of Technology *(Old Westbury)* [E-156]
New York State College of Ceramics at Alfred University *(Alfred)* [E-157]
Polytechnic University *(Brooklyn)* [E-182]
Rensselaer Polytechnic Institute *(Troy)* [E-189]
University of Rochester *(Rochester)* [E-192]
Rochester Institute of Technology *(Rochester)* [E-193] [T-336]
State University of New York at Binghamton *(Binghamton)* [E-218]
State University of New York at Buffalo *(Buffalo)* [E-219]
State University of New York, College of Environmental Science and Forestry *(Syracuse)* [E-220]
State University of New York College of Technology at Farmingdale *(Farmingdale)* [T-343]
State University of New York Maritime College *(Throgs Neck Station)* [E-221]
State University of New York, College at New Paltz *(New Paltz)* [E-222]
State University of New York at Stony Brook *(Stony Brook)* [E-223]
State University of New York Institute of Technology, Utica/Rome *(Utica)* [T-344]
Syracuse University *(Syracuse)* [E-225]
Union College *(Schenectady)* [E-247]
United States Merchant Marine Academy *(Kings Point)* [E-250]
United States Military Academy *(West Point)* [E-251]
Webb Institute of Naval Architecture *(Glen Cove)* [E-266]

NORTH CAROLINA

Duke University *(Durham)* [E-065]
University of North Carolina-Charlotte *(Charlotte)* [E-158] [T-324]
North Carolina Agricultural and Technical State University *(Greensboro)* [E-159]
North Carolina State University *(Raleigh)* [E-160]
Western Carolina University *(Cullowhee)* [T-354]

NORTH DAKOTA

University of North Dakota *(Grand Forks)* [E-161]
North Dakota State University *(Fargo)* [E-162]

OHIO

University of Akron *(Akron)* [E-001] [T-281]
Case Western Reserve University *(Cleveland)* [E-037]
Central State University *(Wilberforce)* [E-040]
University of Cincinnati *(Cincinnati)* [E-042] [T-293]
Cleveland State University *(Cleveland)* [E-048]
University of Dayton *(Dayton)* [E-060] [T-294]
DeVry Institute of Technology, Columbus *(Columbus)* [T-296]
Franklin University *(Columbus)* [T-307]
Marietta College *(Marietta)* [E-117]
Miami University *(Oxford)* [E-129]
Ohio University *(Athens)* [E-170]
Ohio Northern University *(Ada)* [E-171]
Ohio State University *(Columbus)* [E-172]
University of Toledo *(Toledo)* [E-240]
Wright State University *(Dayton)* [E-278]

OKLAHOMA

University of Oklahoma *(Norman)* [E-173]
Oklahoma Christian University of Science and Arts *(Oklahoma City)* [E-174]
Oklahoma State University *(Stillwater)* [E-175] [T-326]
University of Tulsa *(Tulsa)* [E-245]

OREGON

Oregon Institute of Technology *(Klamath Falls)* [T-328]
Oregon State University *(Corvallis)* [E-177]
University of Portland *(Portland)* [E-183]
Portland State University *(Portland)* [E-184]

PENNSYLVANIA

Bucknell University *(Lewisburg)* [E-018]
Carnegie Mellon University *(Pittsburgh)* [E-036]
Drexel University *(Philadelphia)* [E-064]
Gannon University *(Erie)* [E-078]
Grove City College *(Grove City)* [E-085]
Lafayette College *(Easton)* [E-104]
Lehigh University *(Bethlehem)* [E-108]
University of Pennsylvania *(Philadelphia)* [E-179]
Pennsylvania State University *(University Park)* [E-180]
Pennsylvania State University at Erie, The Behrend College *(Erie)* [T-329]
Pennsylvania State University - Harrisburg *(Middletown)* [T-330]
University of Pittsburgh *(Pittsburgh)* [E-181]
University of Pittsburgh at Johnstown *(Johnstown)* [T-332]
Point Park College *(Pittsburgh)* [T-333]
Swarthmore College *(Swarthmore)* [E-224]
Temple University *(Philadelphia)* [E-227] [T-345]
Villanova University *(Villanova)* [E-258]
Widener University *(Chester)* [E-272]
Wilkes University *(Wilkes-Barre)* [E-273]

PUERTO RICO

University of Puerto Rico, Mayagüez Campus *(Mayagüez)* [E-187]

RHODE ISLAND

University of Rhode Island *(Kingston)* [E-190]

SOUTH CAROLINA

The Citadel *(Charleston)* [E-043]
Clemson University *(Clemson)* [E-047]
University of South Carolina *(Columbia)* [E-206]
South Carolina State University *(Orangeburg)* [T-338]

SOUTH DAKOTA

South Dakota School of Mines and Technology *(Rapid City)* [E-207]
South Dakota State University *(Brookings)* [E-208]

TENNESSEE

Christian Brothers University *(Memphis)* [E-041]
East Tennessee State University *(Johnson City)* [T-304]
Memphis State University *(Memphis)* [E-126] [T-316]
University of Tennessee, Chattanooga *(Chattanooga)* [E-228]

Geographical Index

University of Tennessee, Knoxville *(Knoxville)* [E-229]
University of Tennessee, Martin *(Martin)* [T-346]
Tennessee State University *(Nashville)* [E-230]
Tennessee Technological University *(Cookeville)* [E-231]
Vanderbilt University *(Nashville)* [E-256]

TEXAS

Baylor University *(Waco)* [E-013]
DeVry Institute of Technology, Dallas *(Irving)* [T-299]
University of Houston *(Houston)* [E-090]
Lamar University *(Beaumont)* [E-105]
LeTourneau University *(Longview)* [E-107]
Prairie View A&M University *(Prairie View)* [E-185] [T-334]
William Marsh Rice University *(Houston)* [E-191]
Southern Methodist University *(Dallas)* [E-216]
University of Texas at Arlington *(Arlington)* [E-232]
University of Texas at Austin *(Austin)* [E-233]
University of Texas at Dallas *(Richardson)* [E-234]
University of Texas at El Paso *(El Paso)* [E-235]
University of Texas at San Antonio *(San Antonio)* [E-236]
Texas A&I University *(Kingsville)* [E-237]
Texas A&M University *(College Station)* [E-238] [T-347]
Texas Tech University *(Lubbock)* [E-239] [T-348]
Trinity University *(San Antonio)* [E-242]

UTAH

Brigham Young University *(Provo)* [E-017] [T-287]
University of Utah *(Salt Lake City)* [E-253]
Utah State University *(Logan)* [E-254]
Weber State University *(Ogden)* [T-351]

VERMONT

Norwich University *(Northfield)* [E-167]
University of Vermont *(Burlington)* [E-257]

VIRGINIA

George Mason University *(Fairfax)* [E-079]
Old Dominion University *(Norfolk)* [E-176] [T-327]
University of Virginia *(Charlottesville)* [E-259]
Virginia Military Institute *(Lexington)* [E-260]
Virginia Polytechnic Institute and State University *(Blacksburg)* [E-261]
Virginia State University *(Petersburg)* [T-350]

WASHINGTON

Gonzaga University *(Spokane)* [E-083]
Seattle University *(Seattle)* [E-204]
University of Washington *(Seattle)* [E-262]
Washington State University *(Pullman)* [E-264]

WEST VIRGINIA

West Virginia University *(Morgantown)* [E-267]
West Virginia Institute of Technology *(Montgomery)* [E-268] [T-353]

WISCONSIN

Marquette University *(Milwaukee)* [E-118]
Milwaukee School of Engineering *(Milwaukee)* [E-134] [T-318]
University of Wisconsin-Madison *(Madison)* [E-274]
University of Wisconsin-Milwaukee *(Milwaukee)* [E-275]
University of Wisconsin-Platteville *(Platteville)* [E-276]

WYOMING

University of Wyoming *(Laramie)* [E-279]

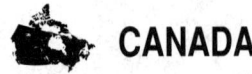

CANADA

ALBERTA

University of Alberta *(Edmonton)* [E-006]

MANITOBA

University of Manitoba *(Winnipeg)* [E-115]

NOVA SCOTIA

Technical University of Nova Scotia *(Halifax)* [E-226]

QUEBEC

Concordia University *(Montréal)* [E-055]
École Polytechnique de Montréal *(Montréal)* [E-066]
McGill University *(Montréal)* [E-125]

Engineering Section

ENGINEERING SECTION
Participating Engineering Institutions .... **TAB**
An Introduction to Engineering, by Ray Landis 18
Engineering Disciplines: Descriptions and Career Opportunities, by Don Evans 18
Index to Engineering Degree Programs 20

Comparative Tables–by School
Undergraduate Degrees Awarded–by Category 33
Support Programs for Ethnic Minorities & Women 39
Enrollments and Degrees Awarded 45

Comparative Tables–by Degree Program
All Engineering Disciplines 53
Chemical Engineering 93
Civil Engineering 101
Electrical Engineering 111
Industrial Engineering 125
Mechanical Engineering 133

Institution Profiles 144

An Introduction to Engineering

An Introduction to Engineering
By Ray Landis—California State University, Los Angeles

What kind of work should I do? This is one of the important and difficult questions of your life. When you make a decision about your career, you will have to consider many questions. Will there be jobs available when I finish my education? Will I be well paid? Will my work challenge me? Will my work make the world a better place? Will my work be respected? The engineering profession offers a resounding YES to all of these questions.

Some of the biggest challenges facing our society today and in the next century call for the intelligent and wise use of technology. Consider the need for renewable energy and controlling pollution, for feeding the worlds population and providing affordable health care. Engineers are being asked to find creative ways to meet these demands.

More and more, engineers are looking at urban development, transportation, and manufacturing enterprises as interconnected systems that affect the quality of life on the planet. Engineers don't design products and systems for each others amazement; they design them for people. And they don't work alone. Tackling complex problems means working with all sorts of other experts—biologists, city planners, lawyers, architects, psychologists, politicians—in the U.S. and around the world.

Engineering has been defined as the art of applying the principles of mathematics and science, experience, judgment, and common sense to make things which benefit people. In other words, engineering is the process of producing a technical product to meet a specific need. Engineers have many different kinds of jobs to choose from. They do research, design, analysis, development, tests, and sales. Some do one or more of these while teaching the next generation of technical experts.

If you are interested in discovering new knowledge, you might consider a career as a research engineer. If you are imaginative and creative, design engineering may be for you. The work of analytical engineers (only about ten percent fall into this category) most closely resembles what you do in your mathematics and science classes. If you like laboratory courses and conducting experiments, test engineering. could be a compatible career for you. If you like to manage projects, look into becoming a development engineer. Sales engineering could be a good choice if you are persuasive and like working with people.

Engineers are also defined by traditional academic areas of study. The five largest of these are chemical, civil, electrical, industrial and mechanical engineering. There are also more specialized areas of engineering, including aerospace, ocean, nuclear, biomedical, petrochemical, and environmental engineering.

Engineers are in great demand. Studies indicate that there will be a shortage of engineers well into the 21st century. Over 1.2 million engineers work in the U.S. today, making engineering second only to teaching as the largest profession.

According to the College Placement Council, the average job offer for *new* engineering graduates in 1991 was $2,783 per month an annual salary of $33,400 and that's just to start! Compare engineers starting salary with those for other fields: chemistry—$2,327; computer science—$2,546; accounting—$2,248; physics—$2,419; humanities—$2,145; health professions—$2,363; business administration—$2,046; mathematics—$2,353; and social sciences—$1,873. Salaries are even higher for those who continue their education and receive a masters or Ph.D. degree in engineering.

Since engineers play a vital role in technological advancement and in solving society's problems, engineering is a highly respected profession. If you have other ambitions—such as medicine, law, or business—engineering offers excellent preparation for graduate study in these areas.

Students who choose engineering as their pre-med major not only have a very good chance of gaining admission to medical school, they have in engineering an excellent fall back career option.

Since it develops the capacity for logical thinking, engineering is good preparation for law school. In particular, there are opportunities in the field of patent law, which calls for an undergraduate engineering degree combined with a law degree.

Another strong combination is an undergraduate degree in engineering and a masters in business administration (MBA). The combination of technical and management training leads to careers in technical management, marketing, or sales—areas with career paths to top management positions.

Many rewards and opportunities will be yours if you graduate in engineering. Through the study of engineering, you will gain a strong technical knowledge base, develop your logical thinking and problem-solving skills, build confidence in your ability, and enhance your self-esteem. Your education will prepare you to gain a position as a practicing engineer or to continue for advanced study in engineering or another field. If you choose employment as an engineer, you will have the opportunity to work on the frontiers of technology and contribute to solving important societal problems. You will be well paid to do work that is both challenging and rewarding. What other field offers comparable benefits?

Engineering Disciplines: Descriptions and Career Opportunities
By Don Evans—Arizona State University

Chemical Engineering
Chemical Engineering occupies a special place among scientific and engineering disciplines by serving as a bridge between chemistry and engineering. This field generally appeals to students who have a strong interest in applying the principles of chemistry to the development of new materials and to the design of processes and equipment to manufacture these new products. Although well trained in chemistry, chemical engineers differ from chemists in that they apply the results of chemical research and discovery by adapting laboratory processes for use in full-scale manufacturing plants. Indeed, chemical engineers play an important role in technology's development since their knowledge is fundamental to processing most materials as well as to lessening the environmental impact of this processing.

Chemical engineers have traditionally played a key role in industries as varied as petroleum, food, artificial fibers, petrochemicals, plastics, ceramics, primary metals, glass, and specialty chemicals. But newer areas such as semiconductors, biotechnology, biomedical engineering, modern materials (composites, superconductors), and environmental engineering all have generated more opportunities for chemical engineers. These factors along with the recent recovery and reported record earnings of the chemical and petroleum industries have created a great demand for chemical engineering graduates.

Civil Engineering

Civil Engineering is most commonly associated with the creation of processes and structures that serve the common good of society. This is generally accomplished through the analysis, planning, design, construction, and maintenance of many types of facilities for government, commerce, and industry. These facilities include: the structural elements and construction techniques of high-rise buildings, factories, schools, airport terminals, rocket launch facilities, and offshore platforms; transportation systems such as bridges, roads and superhighways, runways, tunnels, rail systems, subways, and harbor and river developments; water resources facilities such as dams, canals, and purification plants and delivery systems; and environmental protection facilities such as those for flood control and solid waste and wastewater collection and treatment. Civil engineers are concerned with the impact of their projects on the public and the environment, and they attempt to find technically and economically feasible solutions to society's needs.

Civil engineers engage in technical, administrative, or commercial work with manufacturing companies, construction companies, transportation companies, and power companies. Other opportunities for employment exist in engineering consulting and in the various bureaus of federal, state, and local governments. Civil engineering is one of the better engineering professions from the viewpoint of international travel opportunities or for eventually establishing ones own consulting business.

Electrical Engineering

The professional activities of electrical engineers directly affect the lives of most people every day. Electrical engineers are responsible for designing and developing radio and television broadcast and receiving equipment, high fidelity audio equipment, telephone networks and switching systems, computer systems, electric power generation and distribution systems, and automatic control systems. This latter technology is used in many new airplanes that rely heavily on computers and electronics for stability and control. Electrical engineers are also involved with developing and processing materials such as semiconductors and with miniaturizing electronic components and systems such as the microprocessor chips that permeate nearly every aspect of technology.

Employment opportunities in electrical engineering are extremely varied. Electrical manufacturing companies employ large numbers of electrical engineers for design, testing, research, and sales. Electrical power companies and public utility companies require qualified electrical engineers, as do the companies that control the networks of communications systems. Other opportunities for employment exist with oil companies, railroads, food processing plants, lumber enterprises, biological laboratories, and chemical plants. The aircraft and missile industries use electrical engineers who are familiar with circuit design and employment of flight data computers, servomechanisms, computers, and solid-state devices. There are few industries that do not have at least one electrical engineer on staff.

Industrial Engineering

The industrial engineer is responsible for integrating people, material, capital, energy, and equipment into efficient systems for producing goods and services. Industrial engineers are generally more concerned with the larger picture of management of industries and the production of goods than with detailed development of processes. The realm of industrial engineering includes part of all engineering fields. Generally, industrial engineers work with people and machines, thus it is important that they be educated in both personnel administration and in the relations of people and machines to production.

Opportunities for employment exist in almost every industrial plant and even in many businesses not concerned directly with manufacturing or processing goods. The systems modeling, design, and analysis skills of industrial engineers enable them to improve the efficiency and effectiveness of such diverse systems as manufacturing, health care, criminal justice, finance, and transportation. In many cases industrial engineers have been employed by department stores, insurance companies, consulting companies, and government agencies.

Mechanical Engineering

Mechanical Engineering is a creative discipline that draws upon a number of basic sciences to design the devices, machines, processes, and systems that involve mechanical work and its conversion from, and into, different forms of energy. It includes converting thermal, chemical, and nuclear energy into mechanical energy through engines and power plants; transporting energy via devices like heat exchangers, pipelines, gears, and linkages; and using energy to perform a variety of tasks for the benefit of society, such as in transportation vehicles of all types, manufacturing tools and equipment, and household appliances. Furthermore, since all manufactured products must be constructed of solid materials and because most products contain parts that transmit forces, mechanical engineering is involved in selecting the materials for and ensuring the structural integrity of almost every product on the market.

Mechanical engineering graduates have found employment in almost every type of industry. They have traditionally led the design and development efforts of such products as transportation vehicles, internal combustion and gas turbine engines, refrigeration systems, power plants (conventional, nuclear, and solar), and manufacturing production lines. Manufacturing plants, power-generating stations, public utility companies, transportation companies, airlines, and factories are examples of organizations that need mechanical engineers. In the missile and space industries, mechanical engineers design, develop and test airborne and missile fuel systems, servo-valves, and electro-mechanical control systems. They also work in the building industry to design energy efficient walls, windows, ceilings, and heating and cooling systems.

Index to Engineering Degree Programs

A

Acoustics
090 University of Houston
180 Pennsylvania State University

Administration
(See: Engineering Administration; Water Resources Administration)

Aeronautical & Astronautical Engineering
(See also: Astronautical Engineering)
095 University of Illinois at Urbana-Champaign
124 Massachusetts Institute of Technology
172 Ohio State University
188 Purdue University
217 Stanford University
262 University of Washington

Aeronautical Engineering
(See also: Aerospace Engineering; Airway Science; Astronautical Engineering; Aviation; Aviation Management; Mechanical & Aeronautical Engineering)
026 California Institute of Technology
027 California Polytechnic State University
189 Rensselaer Polytechnic Institute
248 United States Air Force Academy
269 Western Michigan University

Aeronautical Science & Engineering
020 University of California, Davis

Aerospace & Aeronautical Engineering
(See also: Aeronautical Engineering)
067 Embry-Riddle Aeronautical University - Daytona Beach

Aerospace & Mechanical Engineering
(See also: Mechanical Engineering)
168 University of Notre Dame

Aerospace & Ocean Engineering
(See also: Ocean Engineering)
261 Virginia Polytechnic Institute and State University

Aerospace Engineering
(See also: Atmospheric & Space Sciences; Mechanical & Aerospace Engineering; Mechanical, Manufacturing & Aerospace Engineering; Space Operations)
002 University of Alabama
008 University of Arizona
009 Arizona State University
012 Auburn University
014 Boston University
022 University of California, Los Angeles
023 University of California, San Diego
030 California State University, Long Beach
034 California State Polytechnic University, Pomona
037 Case Western Reserve University
039 University of Central Florida
042 University of Cincinnati
049 University of Colorado at Boulder
068 Embry-Riddle Aeronautical University, Western Campus
071 University of Florida
082 Georgia Institute of Technology
096 Illinois Institute of Technology
099 Iowa State University
101 University of Kansas
120 University of Maryland, College Park
130 University of Michigan
138 Mississippi State University
141 University of Missouri-Rolla
160 North Carolina State University
173 University of Oklahoma
175 Oklahoma State University
180 Pennsylvania State University
182 Polytechnic University
198 Parks College of Saint Louis University
202 San Jose State University
211 University of Southern California
219 State University of New York at Buffalo
225 Syracuse University
229 University of Tennessee, Knoxville
232 University of Texas at Arlington
233 University of Texas at Austin
238 Texas A&M University
241 Tri-State University
246 Tuskegee University
252 United States Naval Academy
259 University of Virginia
261 Virginia Polytechnic Institute and State University
267 West Virginia University
271 Wichita State University

Aerospace Engineering & Engineering Mechanics
(See also: Engineering Mechanics; Mechanics)
136 University of Minnesota
200 San Diego State University

Aerospace Science Engineering
130 University of Michigan

Agricultural & Biological Engineering
(See also: Biological Engineering)
058 Cornell University

Agricultural Biosystems Engineering
(See also: Biological Engineering)
008 University of Arizona

Agricultural Engineering
(See also: Biological & Agricultural Engineering; Biological & Irrigation Engineering; Environmental Engineering; Food Engineering; Irrigation; Water Resources Administration)
012 Auburn University
027 California Polytechnic State University
034 California State Polytechnic University, Pomona
053 Colorado State University
071 University of Florida
081 University of Georgia
092 University of Idaho
095 University of Illinois at Urbana-Champaign
099 Iowa State University
102 Kansas State University
103 University of Kentucky
115 University of Manitoba
120 University of Maryland, College Park
132 Michigan State University
136 University of Minnesota
138 Mississippi State University
139 University of Missouri-Columbia
144 Montana State University
146 University of Nebraska - Lincoln
154 New Mexico State University
159 North Carolina Agricultural and Technical State University
162 North Dakota State University
172 Ohio State University
175 Oklahoma State University
180 Pennsylvania State University
188 Purdue University
208 South Dakota State University
226 Technical University of Nova Scotia
229 University of Tennessee, Knoxville
238 Texas A&M University
261 Virginia Polytechnic Institute and State University
264 Washington State University

Agricultural Engineering (Food)
(See also: Food Engineering)
188 Purdue University

Agriculture & Biological Engineering
(See also: Biological Engineering)
047 Clemson University
058 Cornell University

Airway Science
(See also: Aeronautical Engineering)
170 Ohio University

Applied & Engineering Physics
(See also: Engineering Physics; Physics)
052 Colorado School of Mines
058 Cornell University

Applied Engineering
(See also: Engineering; General Engineering)
273 Wilkes University

Applied Mathematics
(See also: Computational & Applied Mathematics; Engineering Mathematics)
026 California Institute of Technology
049 University of Colorado at Boulder
050 University of Colorado at Colorado Springs
051 University of Colorado at Denver
066 École Polytechnique de Montréal
100 The Johns Hopkins University
166 Northwestern University
259 University of Virginia

Applied Mechanics
(See also: Civil Engineering & Applied Mechanics; Engineering Mechanics; Mechanical Engineering & Applied Mechanics; Mechanics; Theoretical & Applied Mechanics)
023 University of California, San Diego
026 California Institute of Technology
130 University of Michigan
200 San Diego State University
259 University of Virginia

Applied Physics
(See also: Applied & Engineering Physics; Electrical Engineering & Applied Physics; Engineering Physics)
023 University of California, San Diego
026 California Institute of Technology
054 Columbia University
100 The Johns Hopkins University
130 University of Michigan
280 Yale University

Applied Science
020 University of California, Davis
023 University of California, San Diego
263 Washington University

Applied Sciences in Engineering
195 Rutgers, The State University

Appropriate Technology
(See also: Technology)
064 Drexel University

Architectural Engineering
(See also: Architectural Studies; Architecture; Art & Architectural History; Civil & Architectural Engineering; Civil, Environmental, & Architectural Engineering; Naval Architecture; Naval Architecture & Marine Engineering)
027 California Polytechnic State University
064 Drexel University
101 University of Kansas
102 Kansas State University
128 University of Miami
134 Milwaukee School of Engineering
159 North Carolina Agricultural and Technical State University
175 Oklahoma State University
180 Pennsylvania State University
230 Tennessee State University
233 University of Texas at Austin

Architectural Studies
264 Washington State University

Architecture
167 Norwich University
264 Washington State University

Arctic Engineering
005 University of Alaska, Fairbanks

Art & Architectural History
243 Tufts University

Astronautical Engineering
(See also: Aeronautical & Astronautical Engineering; Aeronautical Engineering; Aerospace Engineering; Astronautical Engineering)
248 United States Air Force Academy

Atmospheric & Space Sciences
(See also: Aeronautical Engineering; Aerospace Engineering; Astronautical Engineering; Airway Sciences)
130 University of Michigan

Aviation
(See also: Aeronautical Engineering; Airway Science)
172 Ohio State University

Aviation Management
(See also: Engineering Management; Management)
012 Auburn University

B

Biochemical Engineering
(See also: Biological Engineering; Chemical Engineering)
021 University of California, Irvine
066 École Polytechnique de Montréal
090 University of Houston

Bioengineering
(See also: Biological Engineering)
009 Arizona State University
019 University of California at Berkeley
023 University of California, San Diego
047 Clemson University
094 University of Illinois at Chicago
126 Memphis State University
130 University of Michigan
179 University of Pennsylvania
181 University of Pittsburgh
225 Syracuse University
238 Texas A&M University
253 University of Utah
270 Western New England College

Biological & Agricultural Engineering
(See also: Agricultural Engineering)
010 University of Arkansas
020 University of California, Davis
081 University of Georgia
109 Louisiana State University
160 North Carolina State University

Biological & Engineering Science(s)
(See also: Biological Engineering; Engineering Science(s))
263 Washington University

Biological & Irrigation Engineering
(See also: Irrigation Engineering)
254 Utah State University

Biological Engineering
(See also: Agricultural & Biological Engineering; Agricultural Biosystems; Agricultural Engineering Agriculture & Biological Engineering; Biochemical Engineering; Bioengineering; Biomedical Engineering; Bioresource Engineering; Chemical & Biological Engineering)
138 Mississippi State University

Biological System Engineering
146 University of Nebraska - Lincoln

Biomedical Engineering
001 University of Akron
003 University of Alabama at Birmingham
014 Boston University
023 University of California, San Diego
036 Carnegie Mellon University
037 Case Western Reserve University
038 The Catholic University of America
064 Drexel University
065 Duke University
066 École Polytechnique de Montréal
090 University of Houston
098 University of Iowa
099 Iowa State University
100 The Johns Hopkins University
110 Louisiana Tech University
118 Marquette University
128 University of Miami
134 Milwaukee School of Engineering
151 New Jersey Institute of Technology
152 The University of New Mexico
166 Northwestern University
172 Ohio State University
189 Rensselaer Polytechnic Institute
195 Rutgers, The State University
211 University of Southern California
232 University of Texas at Arlington
233 University of Texas at Austin
244 Tulane University
256 Vanderbilt University
259 University of Virginia
277 Worcester Polytechnic Institute
278 Wright State University

Bioresource Engineering
114 University of Maine
195 Rutgers, The State University

Biosystems
(See: Agricultural Biosystems Engineering; Agricultural Engineering; Biological Engineering; Systems)

Building Engineering
(See also: Civil Engineering; Construction Engineering)
055 Concordia University

C

Ceramic Engineering
(See also: Glass)
007 Alfred University
047 Clemson University
082 Georgia Institute of Technology
095 University of Illinois at Urbana-Champaign
099 Iowa State University
141 University of Missouri-Rolla
157 New York State College of Ceramics at Alfred University
172 Ohio State University
195 Rutgers, The State University
262 University of Washington

Ceramic Engineering Science
007 Alfred University
157 New York State College of Ceramics at Alfred University

Chemical & Biochemical Engineering
(See also: Biochemical Engineering)
098 University of Iowa
119 University of Maryland Baltimore County
185 Prairie View A&M University

Chemical & Petroleum Engineering
(See also: Petroleum Engineering)
101 University of Kansas
181 University of Pittsburgh

Chemical Engineering
(See also: Biochemical Engineering; Chemistry; Fuels Engineering; Hazardous Waste Management; Nuclear Engineering; Petroleum Engineering; Textile Chemistry)
001 University of Akron
002 University of Alabama
004 University of Alabama at Huntsville
006 University of Alberta
008 University of Arizona
009 Arizona State University
010 University of Arkansas
012 Auburn University
017 Brigham Young University
018 Bucknell University
019 University of California at Berkeley
020 University of California, Davis
022 University of California, Los Angeles
023 University of California, San Diego
024 University of California, Santa Barbara
026 California Institute of Technology
030 California State University, Long Beach
034 California State Polytechnic University, Pomona
036 Carnegie Mellon University
037 Case Western Reserve University
041 Christian Brothers University
042 University of Cincinnati
044 City College of the City University of New York
046 Clarkson University
047 Clemson University
048 Cleveland State University
049 University of Colorado at Boulder
053 Colorado State University
054 Columbia University
056 University of Connecticut
057 The Cooper Union
058 Cornell University
060 University of Dayton
061 University of Delaware
064 Drexel University
066 École Polytechnique de Montréal
071 University of Florida
072 Florida Agricultural and Mechanical University
074 Florida Institute of Technology
076 Florida State University
082 Georgia Institute of Technology
090 University of Houston
091 Howard University
092 University of Idaho
094 University of Illinois at Chicago
095 University of Illinois at Urbana-Champaign
096 Illinois Institute of Technology
099 Iowa State University
100 The Johns Hopkins University
102 Kansas State University
103 University of Kentucky
104 Lafayette College
105 Lamar University
108 Lehigh University
109 Louisiana State University
110 Louisiana Tech University
111 University of Louisville
114 University of Maine
120 University of Maryland, College Park
121 University of Massachusetts
123 University of Massachusetts, Lowell
124 Massachusetts Institute of Technology
125 McGill University
130 University of Michigan
132 Michigan State University
133 Michigan Technological University
135 University of Minnesota, Duluth
136 University of Minnesota
137 University of Mississippi
138 Mississippi State University
139 University of Missouri-Columbia
141 University of Missouri-Rolla
144 Montana State University
146 University of Nebraska - Lincoln
149 University of New Hampshire
150 University of New Haven
151 New Jersey Institute of Technology
152 The University of New Mexico
154 New Mexico State University
159 North Carolina Agricultural and Technical State University
160 North Carolina State University
161 University of North Dakota
163 Northeastern University
166 Northwestern University
168 University of Notre Dame
170 Ohio University
172 Ohio State University
173 University of Oklahoma
175 Oklahoma State University
177 Oregon State University

179	University of Pennsylvania
180	Pennsylvania State University
181	University of Pittsburgh
182	Polytechnic University
185	Prairie View A&M University
186	Princeton University
187	University of Puerto Rico, Mayagüez Campus
188	Purdue University
189	Rensselaer Polytechnic Institute
190	University of Rhode Island
191	William Marsh Rice University
192	University of Rochester
194	Rose-Hulman Institute of Technology
195	Rutgers, The State University
202	San Jose State University
205	University of South Alabama
206	University of South Carolina
207	South Dakota School of Mines and Technology
209	University of South Florida
211	University of Southern California
217	Stanford University
219	State University of New York at Buffalo
225	Syracuse University
226	Technical University of Nova Scotia
228	University of Tennessee, Chattanooga
229	University of Tennessee, Knoxville
231	Tennessee Technological University
233	University of Texas at Austin
237	Texas A&I University
238	Texas A&M University
239	Texas Tech University
240	University of Toledo
241	Tri-State University
243	Tufts University
244	Tulane University
245	University of Tulsa
246	Tuskegee University
253	University of Utah
256	Vanderbilt University
258	Villanova University
259	University of Virginia
261	Virginia Polytechnic Institute and State University
262	University of Washington
263	Washington University
264	Washington State University
265	Wayne State University
267	West Virginia University
268	West Virginia Institute of Technology
272	Widener University
273	Wilkes University
274	University of Wisconsin-Madison
277	Worcester Polytechnic Institute
279	University of Wyoming
280	Yale University

Chemical Engineering & Materials Science

(See also: Materials Science)
| 020 | University of California, Davis |

Chemical Engineering & Petroleum Refining

(See also: Petroleum Engineering)
| 052 | Colorado School of Mines |

Chemistry

(See also: Chemical Engineering; Petroleum Engineering; Textile Chemistry)
| 052 | Colorado School of Mines |

Civil & Architectural Engineering

(See also: Architectural Engineering)
| 279 | University of Wyoming |

Civil & Construction Engineering

(See also: Construction Engineering)
| 004 | University of Alabama at Huntsville |
| 106 | Lawrence Technological University |

Civil & Environmental Engineering

(See also: Environmental Engineering)
027	California Polytechnic State University
046	Clarkson University
058	Cornell University
074	Florida Institute of Technology
075	Florida International University
090	University of Houston
098	University of Iowa

124	Massachusetts Institute of Technology
130	University of Michigan
147	University of Nevada, Las Vegas
150	University of New Haven
182	Polytechnic University
204	Seattle University
210	Southern University and A&M College
244	Tulane University
254	Utah State University
260	Virginia Military Institute
264	Washington State University
265	Wayne State University

Civil Engineering

(See also: Applied Mechanics; Architectural Engineering; Building Engineering; Construction Engineering; Engineering Science & Mechanics; Environmental Engineering; Land Surveying; Operations Research; Public Works & Civil Engineering)

001	University of Akron
002	University of Alabama
003	University of Alabama at Birmingham
004	University of Alabama at Huntsville
005	University of Alaska, Fairbanks
006	University of Alberta
008	University of Arizona
009	Arizona State University
010	University of Arkansas
012	Auburn University
015	Bradley University
017	Brigham Young University
018	Bucknell University
019	University of California at Berkeley
020	University of California, Davis
021	University of California, Irvine
022	University of California, Los Angeles
026	California Institute of Technology
028	California State University, Chico
029	California State University, Fresno
030	California State University, Long Beach
031	California State University, Los Angeles
033	California State University, Sacramento
034	California State Polytechnic University, Pomona
036	Carnegie Mellon University
037	Case Western Reserve University
038	The Catholic University of America
039	University of Central Florida
041	Christian Brothers University
042	University of Cincinnati
043	The Citadel
044	City College of the City University of New York
047	Clemson University
048	Cleveland State University
051	University of Colorado at Denver
053	Colorado State University
055	Concordia University
056	University of Connecticut
057	The Cooper Union
060	University of Dayton
061	University of Delaware
063	University of the District of Columbia
064	Drexel University
065	Duke University
066	École Polytechnique de Montréal
069	University of Evansville
071	University of Florida
072	Florida Agricultural and Mechanical University
073	Florida Atlantic University
075	Florida International University
076	Florida State University
080	The George Washington University
082	Georgia Institute of Technology
083	Gonzaga University
088	University of Hawaii at Manoa
090	University of Houston
091	Howard University
092	University of Idaho
094	University of Illinois at Chicago
095	University of Illinois at Urbana-Champaign
096	Illinois Institute of Technology
099	Iowa State University
100	The Johns Hopkins University
101	University of Kansas
102	Kansas State University
103	University of Kentucky
104	Lafayette College
105	Lamar University

108	Lehigh University
109	Louisiana State University
110	Louisiana Tech University
111	University of Louisville
113	Loyola Marymount University
114	University of Maine
115	University of Manitoba
118	Marquette University
120	University of Maryland, College Park
121	University of Massachusetts
122	University of Massachusetts, Dartmouth
123	University of Massachusetts, Lowell
126	Memphis State University
128	University of Miami
130	University of Michigan
132	Michigan State University
133	Michigan Technological University
136	University of Minnesota
137	University of Mississippi
138	Mississippi State University
139	University of Missouri-Columbia
140	University of Missouri-Kansas City
141	University of Missouri-Rolla
144	Montana State University
145	Morgan State University
146	University of Nebraska - Lincoln
148	New England College
149	University of New Hampshire
150	University of New Haven
151	New Jersey Institute of Technology
152	The University of New Mexico
154	New Mexico State University
155	University of New Orleans
158	University of North Carolina-Charlotte
159	North Carolina Agricultural and Technical State University
160	North Carolina State University
161	University of North Dakota
162	North Dakota State University
163	Northeastern University
164	Northern Arizona University
166	Northwestern University
167	Norwich University
168	University of Notre Dame
170	Ohio University
171	Ohio Northern University
172	Ohio State University
173	University of Oklahoma
175	Oklahoma State University
176	Old Dominion University
177	Oregon State University
178	University of the Pacific
180	Pennsylvania State University
181	University of Pittsburgh
182	Polytechnic University
183	University of Portland
184	Portland State University
185	Prairie View A&M University
187	University of Puerto Rico, Mayagüez Campus
188	Purdue University
189	Rensselaer Polytechnic Institute
190	University of Rhode Island
191	William Marsh Rice University
194	Rose-Hulman Institute of Technology
195	Rutgers, The State University
200	San Diego State University
201	San Francisco State University
202	San Jose State University
203	Santa Clara University
205	University of South Alabama
206	University of South Carolina
207	South Dakota School of Mines and Technology
208	South Dakota State University
209	University of South Florida
210	Southern University and A&M College
211	University of Southern California
213	Southern Illinois University at Carbondale
214	Southern Illinois University at Edwardsville
216	Southern Methodist University
217	Stanford University
219	State University of New York at Buffalo
225	Syracuse University
226	Technical University of Nova Scotia
227	Temple University
228	University of Tennessee, Chattanooga
229	University of Tennessee, Knoxville
230	Tennessee State University
231	Tennessee Technological University

Index to Engineering Degree Programs

Civil Engineering & Applied Mechanics - Computer Systems Engineering

232 University of Texas at Arlington
233 University of Texas at Austin
235 University of Texas at El Paso
236 University of Texas at San Antonio
237 Texas A&I University
238 Texas A&M University
239 Texas Tech University
240 University of Toledo
241 Tri-State University
243 Tufts University
247 Union College
248 United States Air Force Academy
249 United States Coast Guard Academy
251 United States Military Academy
253 University of Utah
255 Valparaiso University
256 Vanderbilt University
257 University of Vermont
258 Villanova University
259 University of Virginia
261 Virginia Polytechnic Institute and State University
262 University of Washington
263 Washington University
264 Washington State University
267 West Virginia University
268 West Virginia Institute of Technology
272 Widener University
273 Wilkes University
274 University of Wisconsin-Madison
275 University of Wisconsin-Milwaukee
276 University of Wisconsin-Platteville
277 Worcester Polytechnic Institute

Civil Engineering & Applied Mechanics
(See also: Applied Mechanics; Engineering Mechanics; Mechanics)
125 McGill University

Civil Engineering & Engineering Mechanics
(See also: Applied Mechanics; Engineering Mechanics; Mechanics)
054 Columbia University

Civil Engineering & Materials Science
(See also: Materials Science)
020 University of California, Davis

Civil Engineering & Operations Research
(See also: Operations Research)
186 Princeton University

Civil Engineering (Construction)
(See also: Construction Engineering)
160 North Carolina State University

Civil, Environmental, & Architectural Engineering
(See also: Architectural Engineering; Environmental Engineering)
049 University of Colorado at Boulder

Coastal & Oceanographic
(See also: Aerospace & Ocean Engineering; Marine Engineering; Naval Architecture; Ocean Engineering; Oceanography)
071 University of Florida

College Program
(See also: Engineering; General Engineering; Interdisciplinary Engineering)
058 Cornell University

Communications
(See also: Technical Communication; Telecommunications)

Computation & Neural Systems
026 California Institute of Technology

Computational & Applied Mathematics
(See also: Applied Mathematics; Engineering Mathematics; Mathematics)
191 William Marsh Rice University

Computational Engineering
(See also: Applied Mathematics; Engineering Mathematics; Mathematics; Neural Systems)
138 Mississippi State University

Computer & Electrical Engineering
(See also: Electrical Engineering)
135 University of Minnesota, Duluth
188 Purdue University

Computer & Information Science
(See also: Information Systems; Information Technology)
071 University of Florida
137 University of Mississippi
172 Ohio State University

Computer & Systems Engineering
(See also: Systems Engineering)
090 University of Houston
189 Rensselaer Polytechnic Institute

Computer Engineering
(See also: Computer Science; Electrical Engineering; Electrical & Computer Engineering; Electrical Engineering & Computer Science; Electronic & Computer Control System; Engineering Math & Computer Science; Software Systems Engineering; Systems & Computer Science)
006 University of Alberta
008 University of Arizona
012 Auburn University
014 Boston University
016 University of Bridgeport
023 University of California, San Diego
025 University of California, Santa Cruz
027 California Polytechnic State University
028 California State University, Chico
029 California State University, Fresno
033 California State University, Sacramento
037 Case Western Reserve University
039 University of Central Florida
042 University of Cincinnati
055 Concordia University
066 École Polytechnique de Montréal
069 University of Evansville
073 Florida Atlantic University
075 Florida International University
080 The George Washington University
082 Georgia Institute of Technology
092 University of Idaho
094 University of Illinois at Chicago
095 University of Illinois at Urbana-Champaign
099 Iowa State University
109 Louisiana State University
115 University of Manitoba
130 University of Michigan
132 Michigan State University
134 Milwaukee School of Engineering
138 Mississippi State University
139 University of Missouri-Columbia
146 University of Nebraska - Lincoln
151 New Jersey Institute of Technology
152 The University of New Mexico
160 North Carolina State University
168 University of Notre Dame
169 Oakland University
176 Old Dominion University
177 Oregon State University
180 Pennsylvania State University
190 University of Rhode Island
193 Rochester Institute of Technology
202 San Jose State University
203 Santa Clara University
209 University of South Florida
216 Southern Methodist University
225 Syracuse University
233 University of Texas at Austin
238 Texas A&M University
243 Tufts University
253 University of Utah
255 Valparaiso University
258 Villanova University
261 Virginia Polytechnic Institute and State University
267 West Virginia University
278 Wright State University

Computer Engineering & Computer Science
030 California State University, Long Beach
167 Norwich University

Computer, Information, & Control Engineering
(See also: Control Engineering; Information Systems; Information Technology)
130 University of Michigan

Computer Science
(See also: Computer Engineering)
002 University of Alabama
009 Arizona State University
012 Auburn University
018 Bucknell University
022 University of California, Los Angeles
023 University of California, San Diego
024 University of California, Santa Barbara
026 California Institute of Technology
029 California State University, Fresno
032 California State University, Northridge
044 City College of the City University of New York
049 University of Colorado at Boulder
050 University of Colorado at Colorado Springs
054 Columbia University
058 Cornell University
073 Florida Atlantic University
079 George Mason University
080 The George Washington University
092 University of Idaho
094 University of Illinois at Chicago
095 University of Illinois at Urbana-Champaign
100 The Johns Hopkins University
110 Louisiana Tech University
132 Michigan State University
150 University of New Haven
152 The University of New Mexico
158 University of North Carolina-Charlotte
159 North Carolina Agricultural and Technical State University
160 North Carolina State University
166 Northwestern University
169 Oakland University
173 University of Oklahoma
177 Oregon State University
184 Portland State University
186 Princeton University
191 William Marsh Rice University
208 South Dakota State University
211 University of Southern California
216 Southern Methodist University
217 Stanford University
218 State University of New York at Binghamton
235 University of Texas at El Paso
237 Texas A&I University
238 Texas A&M University
239 Texas Tech University
244 Tulane University
253 University of Utah
256 Vanderbilt University
259 University of Virginia
263 Washington University
264 Washington State University
278 Wright State University

Computer Science & Engineering
012 Auburn University
020 University of California, Davis
022 University of California, Los Angeles
023 University of California, San Diego
037 Case Western Reserve University
051 University of Colorado at Denver
056 University of Connecticut
111 University of Louisville
130 University of Michigan
164 Northern Arizona University
179 University of Pennsylvania
209 University of South Florida
240 University of Toledo
262 University of Washington

Computer Science Engineering
038 The Catholic University of America
232 University of Texas at Arlington

Computer Systems Engineering
009 Arizona State University
010 University of Arkansas
269 Western Michigan University

Computing Sciences
042 University of Cincinnati

Construction Engineering
(See also: Building Engineering; Civil & Construction Engineering; Civil Engineering; Civil Engineering (Construction))
099 Iowa State University
152 The University of New Mexico
162 North Dakota State University
263 Washington University

Construction Engineering & Management
(See also: Engineering Management; Management)
130 University of Michigan
177 Oregon State University
188 Purdue University

Construction Engineering Technology
110 Louisiana Tech University

Construction Management
(See also: Engineering Management; Management)
015 Bradley University
029 California State University, Fresno
064 Drexel University
152 The University of New Mexico
160 North Carolina State University
263 Washington University
264 Washington State University

Control Engineering
(See: Computer Engineering; Computer Information, & Control Engineering; Computer Science; Electronic & Computer Control Systems; Systems & Control Engineering)

D

Design
(See: Architectural Engineering; Structural Design)

E

Economics
(See: Engineering-Economic Systems; Mineral Economics)

Electric Power Engineering
156 New York Institute of Technology
189 Rensselaer Polytechnic Institute

Electrical & Computer Engineering
(See also: Computer Engineering)
003 University of Alabama at Birmingham
004 University of Alabama at Huntsville
017 Brigham Young University
020 University of California, Davis
021 University of California, Irvine
024 University of California, Santa Barbara
034 California State Polytechnic University, Pomona
036 Carnegie Mellon University
046 Clarkson University
047 Clemson University
049 University of Colorado at Boulder
050 University of Colorado at Colorado Springs
064 Drexel University
074 Florida Institute of Technology
075 Florida International University
077 GMI Engineering & Management Institute
095 University of Illinois at Urbana-Champaign
096 Illinois Institute of Technology
098 University of Iowa
101 University of Kansas
102 Kansas State University
114 University of Maine
121 University of Massachusetts
122 University of Massachusetts, Dartmouth
128 University of Miami
131 University of Michigan-Dearborn
140 University of Missouri-Kansas City
147 University of Nevada, Las Vegas
154 New Mexico State University
163 Northeastern University
175 Oklahoma State University
178 University of the Pacific
184 Portland State University
187 University of Puerto Rico, Mayagüez Campus
191 William Marsh Rice University
194 Rose-Hulman Institute of Technology
206 University of South Carolina
229 University of Tennessee, Knoxville
233 University of Texas at Austin
254 Utah State University
265 Wayne State University
277 Worcester Polytechnic Institute

Electrical & Electronics Engineering
(See also: Electronics Engineering)
027 California Polytechnic State University
028 California State University, Chico
033 California State University, Sacramento
177 Oregon State University

Electrical Engineering
(See also: Applied Physics; Computer & Electrical Engineering; Computer Engineering; Computer, Information, & Control Engineering; Computer Science; Electric Power Engineering; Electronics Engineering)
001 University of Akron
002 University of Alabama
005 University of Alaska, Fairbanks
006 University of Alberta
007 Alfred University
008 University of Arizona
009 Arizona State University
010 University of Arkansas
012 Auburn University
014 Boston University
015 Bradley University
016 University of Bridgeport
018 Bucknell University
022 University of California, Los Angeles
023 University of California, San Diego
026 California Institute of Technology
029 California State University, Fresno
030 California State University, Long Beach
031 California State University, Los Angeles
034 California State Polytechnic University, Pomona
038 The Catholic University of America
039 University of Central Florida
041 Christian Brothers University
042 University of Cincinnati
043 The Citadel
044 City College of the City University of New York
048 Cleveland State University
051 University of Colorado at Denver
053 Colorado State University
054 Columbia University
055 Concordia University
056 University of Connecticut
057 The Cooper Union
058 Cornell University
060 University of Dayton
061 University of Delaware
062 University of Denver
063 University of the District of Columbia
065 Duke University
066 École Polytechnique de Montréal
068 Embry-Riddle Aeronautical University, Western Campus
069 University of Evansville
071 University of Florida
072 Florida Agricultural and Mechanical University
073 Florida Atlantic University
075 Florida International University
076 Florida State University
078 Gannon University
079 George Mason University
080 The George Washington University
082 Georgia Institute of Technology
083 Gonzaga University
085 Grove City College
088 University of Hawaii at Manoa
089 Hofstra University
090 University of Houston
091 Howard University
092 University of Idaho
094 University of Illinois at Chicago
097 Indiana University-Purdue University at Indianapolis
099 Iowa State University
100 The Johns Hopkins University
103 University of Kentucky
104 Lafayette College
105 Lamar University
106 Lawrence Technological University
109 Louisiana State University
110 Louisiana Tech University
111 University of Louisville
112 Loyola College in Maryland
113 Loyola Marymount University
115 University of Manitoba
116 Mankato State University
118 Marquette University
119 University of Maryland Baltimore County
120 University of Maryland, College Park
123 University of Massachusetts, Lowell
126 Memphis State University
130 University of Michigan
132 Michigan State University
133 Michigan Technological University
134 Milwaukee School of Engineering
136 University of Minnesota
137 University of Mississippi
138 Mississippi State University
139 University of Missouri-Columbia
141 University of Missouri-Rolla
144 Montana State University
145 Morgan State University
146 University of Nebraska - Lincoln
149 University of New Hampshire
150 University of New Haven
151 New Jersey Institute of Technology
152 The University of New Mexico
153 New Mexico Institute of Mining and Technology
155 University of New Orleans
158 University of North Carolina-Charlotte
159 North Carolina Agricultural and Technical State University
160 North Carolina State University
161 University of North Dakota
162 North Dakota State University
164 Northern Arizona University
165 Northern Illinois University
166 Northwestern University
167 Norwich University
168 University of Notre Dame
169 Oakland University
170 Ohio University
171 Ohio Northern University
172 Ohio State University
173 University of Oklahoma
174 Oklahoma Christian University of Science and Arts
176 Old Dominion University
179 University of Pennsylvania
180 Pennsylvania State University
181 University of Pittsburgh
182 Polytechnic University
183 University of Portland
185 Prairie View A&M University
186 Princeton University
188 Purdue University
189 Rensselaer Polytechnic Institute
190 University of Rhode Island
192 University of Rochester
193 Rochester Institute of Technology
195 Rutgers, The State University
196 Saginaw Valley State University
197 St. Cloud State University
198 Parks College of Saint Louis University
199 University of San Diego
200 San Diego State University
201 San Francisco State University
202 San Jose State University
203 Santa Clara University
204 Seattle University
205 University of South Alabama
206 University of South Carolina
207 South Dakota School of Mines and Technology
208 South Dakota State University
209 University of South Florida
210 Southern University and A&M College
211 University of Southern California
213 Southern Illinois University at Carbondale
214 Southern Illinois University at Edwardsville
215 University of Southern Maine
216 Southern Methodist University
217 Stanford University
218 State University of New York at Binghamton
219 State University of New York at Buffalo

Index to Engineering Degree Programs

221 State University of New York Maritime College
222 State University of New York, College at New Paltz
223 State University of New York at Stony Brook
225 Syracuse University
226 Technical University of Nova Scotia
227 Temple University
228 University of Tennessee, Chattanooga
230 Tennessee State University
231 Tennessee Technological University
232 University of Texas at Arlington
234 University of Texas at Dallas
235 University of Texas at El Paso
236 University of Texas at San Antonio
237 Texas A&I University
238 Texas A&M University
239 Texas Tech University
240 University of Toledo
241 Tri-State University
243 Tufts University
244 Tulane University
245 University of Tulsa
246 Tuskegee University
247 Union College
248 United States Air Force Academy
249 United States Coast Guard Academy
251 United States Military Academy
252 United States Naval Academy
253 University of Utah
255 Valparaiso University
256 Vanderbilt University
257 University of Vermont
258 Villanova University
259 University of Virginia
260 Virginia Military Institute
261 Virginia Polytechnic Institute and State University
262 University of Washington
263 Washington University
264 Washington State University
267 West Virginia University
268 West Virginia Institute of Technology
269 Western Michigan University
270 Western New England College
271 Wichita State University
272 Widener University
273 Wilkes University
274 University of Wisconsin-Madison
275 University of Wisconsin-Milwaukee
276 University of Wisconsin-Platteville
278 Wright State University
279 University of Wyoming
280 Yale University

Electrical Engineering & Applied Physics
(See also: Applied Physics; Engineering Physics; Physics)
037 Case Western Reserve University

Electrical Engineering & Computer Science
(See also: Computer Engineering; Computer Science)
019 University of California at Berkeley
094 University of Illinois at Chicago
108 Lehigh University
124 Massachusetts Institute of Technology
264 Washington State University

Electrical Engineering & Materials Science
(See also: Materials Science)
020 University of California, Davis

Electrical Engineering Systems
130 University of Michigan

Electrical Engineering Technology
110 Louisiana Tech University

Electronic & Computer Control System
(See also: Computer Engineering; Control(s) Engineering)
265 Wayne State University

Electronic Engineering Technology
208 South Dakota State University
210 Southern University and A&M College

Electronic Technology
048 Cleveland State University

Electronic(s) Engineering
(See also: Computer Engineering; Computer Science; Control Engineering; Electrical & Electronics Engineering; Electrical Engineering; Microelectronic Engineering; Solid State Electronics)
125 McGill University
142 Monmouth College
210 Southern University and A&M College

Energy
(See: Fuels Engineering; Metallurgical, Mineral, Mining, Natural Gas, Nuclear, Petroleum Engineering)

Energy & Mineral Resources
(See also: Energy; Metallurgical; Mineral; Mining)
233 University of Texas at Austin

Energy Resources
181 University of Pittsburgh

Engineering
(See also: Applied Engineering; Applied Science; Applied Sciences in Engineering; College Program; Engineering Arts; Engineering Interdisciplinary; Engineering Special Studies; General Engineering; Interdisciplinary Engineering)
001 University of Akron
010 University of Arkansas
013 Baylor University
022 University of California, Los Angeles
032 California State University, Northridge
035 Calvin College
052 Colorado School of Mines
057 The Cooper Union
084 Grand Valley State University
087 Harvey Mudd College
089 Hofstra University
093 Idaho State University
104 Lafayette College
107 LeTourneau University
127 Mercer University
130 University of Michigan
133 Michigan Technological University
160 North Carolina State University
224 Swarthmore College
228 University of Tennessee, Chattanooga
238 Texas A&M University
267 West Virginia University

Engineering & Applied Science
(See also: Applied Science)
026 California Institute of Technology

Engineering & Industrial Technology
(See also: Industrial Technology)
030 California State University, Long Beach

Engineering & Public Policy
(See also: Public Policy)
036 Carnegie Mellon University
263 Washington University

Engineering Administration
(See also: Administration)
241 Tri-State University

Engineering Arts
(See also: Engineering)
132 Michigan State University

Engineering Geology
(See also: Geology; Geological Engineering; Geoscience)
064 Drexel University

Engineering Geoscience
(See also: Geology; Geological Engineering; Geoscience)
019 University of California at Berkeley

Engineering Hydrology
(See also: Agricultural Engineering; Biological & Irrigation Engineering; Environmental Engineering; Hydrology; Irrigation; Water Resources Administration)
002 University of Alabama

Engineering Interdisciplinary Studies
(See also: Engineering; Interdisciplinary Studies)
009 Arizona State University

Engineering Management
(See also: Management)
005 University of Alaska, Fairbanks
041 Christian Brothers University
064 Drexel University
080 The George Washington University
090 University of Houston
093 Idaho State University
094 University of Illinois at Chicago
111 University of Louisville
119 University of Maryland Baltimore County
129 Miami University
141 University of Missouri-Rolla
161 University of North Dakota
176 Old Dominion University
178 University of the Pacific
183 University of Portland
184 Portland State University
209 University of South Florida
228 University of Tennessee, Chattanooga
251 United States Military Academy
264 Washington State University
270 Western New England College
273 Wilkes University

Engineering Math & Computer Science
(See also: Mathematics; Computer Engineering/Science)
111 University of Louisville

Engineering Math & Statistics
(See also: Mathematics; Statistics)
019 University of California at Berkeley

Engineering Mathematics
(See also: Mathematics)
008 University of Arizona

Engineering Mechanics
(See also: Mechanics)
002 University of Alabama
008 University of Arizona
042 University of Cincinnati
094 University of Illinois at Chicago
099 Iowa State University
100 The Johns Hopkins University
103 University of Kentucky
138 Mississippi State University
141 University of Missouri-Rolla
146 University of Nebraska - Lincoln
172 Ohio State University
233 University of Texas at Austin
248 United States Air Force Academy
274 University of Wisconsin-Madison

Engineering Operations
(See also: Operations)
099 Iowa State University

Engineering Physics
(See also: Applied Physics; Physics)
006 University of Alberta
008 University of Arizona
015 Bradley University
019 University of California at Berkeley
023 University of California, San Diego
066 École Polytechnique de Montréal
067 Embry-Riddle Aeronautical University - Daytona Beach
094 University of Illinois at Chicago
095 University of Illinois at Urbana-Champaign
101 University of Kansas
114 University of Maine
130 University of Michigan
172 Ohio State University
173 University of Oklahoma
177 Oregon State University
178 University of the Pacific
181 University of Pittsburgh
189 Rensselaer Polytechnic Institute
208 South Dakota State University
219 State University of New York at Buffalo
225 Syracuse University

229 University of Tennessee, Knoxville
239 Texas Tech University
240 University of Toledo
245 University of Tulsa
251 United States Military Academy
259 University of Virginia
278 Wright State University

Engineering Psychology
243 Tufts University

Engineering Science & Mechanics
(See also: Engineering Mechanics; Mechanics)
071 University of Florida
082 Georgia Institute of Technology
119 University of Maryland Baltimore County
180 Pennsylvania State University
229 University of Tennessee, Knoxville
261 Virginia Polytechnic Institute and State University

Engineering Science(s)
009 Arizona State University
023 University of California, San Diego
026 California Institute of Technology
027 California Polytechnic State University
045 City University of New York, College of Staten Island
053 Colorado State University
059 Dartmouth College
064 Drexel University
086 Harvard University
089 Hofstra University
099 Iowa State University
109 Louisiana State University
112 Loyola College in Maryland
137 University of Mississippi
143 Montana College of Mineral Science and Technology
151 New Jersey Institute of Technology
183 University of Portland
189 Rensselaer Polytechnic Institute
213 Southern Illinois University at Carbondale
219 State University of New York at Buffalo
223 State University of New York at Stony Brook
240 University of Toledo
242 Trinity University
248 United States Air Force Academy
256 Vanderbilt University
259 University of Virginia

Engineering Special Studies
(See also: College Program; Engineering; Engineering Interdisciplinary; General Engineering; Interdisciplinary Engineering)
009 Arizona State University

Engineering Technology
(See also: Applied disciplines; Engineering; Technology)
030 California State University, Long Beach
039 University of Central Florida
165 Northern Illinois University
209 University of South Florida
240 University of Toledo

Engineering-Economic Systems
(See also: Economics; Public Policy)
217 Stanford University

Environmental & Resource Engineering
220 State University of New York, College of Environmental Science and Forestry

Environmental Engineering
(See also: Biological & Irrigation Engineering; Civil & Electrical Engineering; Civil, Engineering Hydrology; Environmental, & Architectural Engineering; Forest Engineering; Hazardous Waste Management; Hydrology; Irrigation; Mineral & Environmental Engineering; Paper Science & Engineering; Waste; Water Resources Administration; Wood Products Engineering)
002 University of Alabama
010 University of Arkansas
039 University of Central Florida
042 University of Cincinnati
064 Drexel University
071 University of Florida
082 Georgia Institute of Technology
090 University of Houston
095 University of Illinois at Urbana-Champaign
096 Illinois Institute of Technology

130 University of Michigan
132 Michigan State University
133 Michigan Technological University
143 Montana College of Mineral Science and Technology
151 New Jersey Institute of Technology
160 North Carolina State University
166 Northwestern University
189 Rensselaer Polytechnic Institute
225 Syracuse University
228 University of Tennessee, Chattanooga
237 Texas A&I University
251 United States Military Academy
264 Washington State University
273 Wilkes University

Environmental Engineering Science
026 California Institute of Technology

Environmental Engineering Technology
167 Norwich University

Environmental Health Engineering
(See also: Health)
233 University of Texas at Austin

Environmental Quality Engineering/Science
005 University of Alaska, Fairbanks

Environmental Science
012 Auburn University
042 University of Cincinnati
095 University of Illinois at Urbana-Champaign
100 The Johns Hopkins University
173 University of Oklahoma
191 William Marsh Rice University

Environmental Studies & Environmental Health
(See also: Environmental Health; Health)
243 Tufts University

Environmental Systems Engineering
047 Clemson University

Extractive Metallurgical Engineering
(See also: Metallurgical; Mineral; Mining Engineering)
136 University of Minnesota

F

Facilities Engineering
221 State University of New York Maritime College

Fire Protection Engineering
120 University of Maryland, College Park
277 Worcester Polytechnic Institute

Fluid & Thermal Engineering
(See also: Thermal Engineering)
037 Case Western Reserve University

Food Engineering
(See also: Agricultural Engineering; Agricultural Engineering (Food))
132 Michigan State University

Forest Engineering
(See also: Environmental Engineering; Paper Science & Engineering; Wood Products Engineering)
012 Auburn University
114 University of Maine
220 State University of New York, College of Environmental Science and Forestry

Fuels Engineering
(See also: Energy; Natural Gas; Petroleum)
253 University of Utah

G

Gas
(See: Natural Gas Engineering; Petroleum Engineering)

General Engineering
(See also: Engineering)
011 Arkansas Tech University
014 Boston University
021 University of California, Irvine
062 University of Denver
084 Grand Valley State University
095 University of Illinois at Urbana-Champaign
134 Milwaukee School of Engineering
137 University of Mississippi
175 Oklahoma State University
202 San Jose State University
248 United States Air Force Academy
252 United States Naval Academy

Geological Engineering
(See also: Engineering Geology; Engineering Geoscience; Geology; Geomechanics; Geophysical Engineering; Geosciences; Geotechnical Engineering & Geosciences)
008 University of Arizona
012 Auburn University
052 Colorado School of Mines
066 École Polytechnique de Montréal
115 University of Manitoba
133 Michigan Technological University
136 University of Minnesota
137 University of Mississippi
141 University of Missouri-Rolla
143 Montana College of Mineral Science and Technology
154 New Mexico State University
161 University of North Dakota
173 University of Oklahoma
207 South Dakota School of Mines and Technology
253 University of Utah
274 University of Wisconsin-Madison

Geological Sciences
058 Cornell University

Geology
137 University of Mississippi

Geomechanics
(See also: Mechanics)
192 University of Rochester

Geophysical Engineering
052 Colorado School of Mines
143 Montana College of Mineral Science and Technology

Geosciences
110 Louisiana Tech University

Geotechnical Engineering & Geosciences
094 University of Illinois at Chicago

Glass Engineering Science
(See also: Ceramic Engineering; Ceramic Engineering Sciences)
007 Alfred University
157 New York State College of Ceramics at Alfred University

H

Hazardous Waste Management
(See also: Environmental Engineering; Waste)
265 Wayne State University

Health
(See: Environmental Health Engineering; Environmental Studies & Environmental Health; Health Physics; Industrial Hygiene; Occupational Health & Safety; Radiological Health Engineering)

Health Physics
042 University of Cincinnati
238 Texas A&M University

Human Factors Engineering
(See also: Health; Human Factors Engineering; Man-Machine-Environment Systems; Occupational Health & Safety;

Safety Engineering; Technology & Human Affairs; Work Environment)
278 Wright State University

Hydrology
(See also: Agricultural Engineering; Biological & Irrigation Engineering; Environmental Engineering; Engineering Hydrology; Irrigation; Water Resources Administration)
008 University of Arizona

I

Industrial & Manufacturing Engineering
(See also: Industrial Engineering; Manufacturing Engineering)
190 University of Rhode Island
193 Rochester Institute of Technology
265 Wayne State University

Industrial & Manufacturing Systems Engineering
(See also: Industrial Engineering; Manufacturing Engineering)
077 GMI Engineering & Management Institute
108 Lehigh University

Industrial & Operations Engineering
(See also: Industrial Engineering; Operations Engineering)
130 University of Michigan

Industrial & Systems Engineering
(See also: Industrial Engineering; Systems Engineering)
004 University of Alabama at Huntsville
071 University of Florida
075 Florida International University
131 University of Michigan-Dearborn
170 Ohio University
172 Ohio State University
202 San Jose State University
211 University of Southern California
261 Virginia Polytechnic Institute and State University

Industrial Engineering
(See also: Engineering & Industrial Technology; Engineering Management; Industrial Technology; Management Engineering; Manufacturing Engineering; Operations Research; Operations Research & Industrial Engineering)
002 University of Alabama
008 University of Arizona
009 Arizona State University
010 University of Arkansas
012 Auburn University
015 Bradley University
027 California Polytechnic State University
029 California State University, Fresno
034 California State Polytechnic University, Pomona
037 Case Western Reserve University
039 University of Central Florida
042 University of Cincinnati
047 Clemson University
048 Cleveland State University
064 Drexel University
066 École Polytechnique de Montréal
072 Florida Agricultural and Mechanical University
076 Florida State University
077 GMI Engineering & Management Institute
082 Georgia Institute of Technology
089 Hofstra University
090 University of Houston
094 University of Illinois at Chicago
095 University of Illinois at Urbana-Champaign
098 University of Iowa
099 Iowa State University
102 Kansas State University
105 Lamar University
109 Louisiana State University
110 Louisiana Tech University
111 University of Louisville
115 University of Manitoba
117 Marietta College
118 Marquette University
128 University of Miami
134 Milwaukee School of Engineering
135 University of Minnesota, Duluth
138 Mississippi State University
139 University of Missouri-Columbia
145 Morgan State University
146 University of Nebraska - Lincoln
150 University of New Haven
151 New Jersey Institute of Technology
154 New Mexico State University
156 New York Institute of Technology
159 North Carolina Agricultural and Technical State University
160 North Carolina State University
165 Northern Illinois University
166 Northwestern University
173 University of Oklahoma
177 Oregon State University
180 Pennsylvania State University
181 University of Pittsburgh
182 Polytechnic University
187 University of Puerto Rico, Mayagüez Campus
188 Purdue University
195 Rutgers, The State University
207 South Dakota School of Mines and Technology
209 University of South Florida
212 University of Southern Colorado
214 Southern Illinois University at Edwardsville
218 State University of New York at Binghamton
219 State University of New York at Buffalo
226 Technical University of Nova Scotia
228 University of Tennessee, Chattanooga
229 University of Tennessee, Knoxville
231 Tennessee Technological University
232 University of Texas at Arlington
235 University of Texas at El Paso
237 Texas A&I University
238 Texas A&M University
239 Texas Tech University
240 University of Toledo
262 University of Washington
267 West Virginia University
269 Western Michigan University
270 Western New England College
271 Wichita State University
273 Wilkes University
274 University of Wisconsin-Madison
275 University of Wisconsin-Milwaukee
276 University of Wisconsin-Platteville

Industrial Engineering & Information Systems
(See also: Information)
163 Northeastern University

Industrial Engineering & Management
(See also: Engineering Management; Management)
144 Montana State University
162 North Dakota State University
175 Oklahoma State University
189 Rensselaer Polytechnic Institute
217 Stanford University

Industrial Engineering & Operations Research
(See also: Operations Research)
019 University of California at Berkeley
054 Columbia University
094 University of Illinois at Chicago
121 University of Massachusetts

Industrial Engineering (Furniture Manufacturing)
(See also: Manufacturing Engineering; Wood Products Engineering)
160 North Carolina State University

Industrial Hygiene
(See also: Health)
238 Texas A&M University

Industrial Manufacturing
(See also: Manufacturing Engineering)
209 University of South Florida

Industrial Technology
123 University of Massachusetts, Lowell
170 Ohio University
218 State University of New York at Binghamton
218 State University of New York at Binghamton
237 Texas A&I University
254 Utah State University

Information
(See: Computer & Information Science; Computer, Information, & Control Engineering; Industrial Engineering & Information Systems; Information Systems; Information Technology)

Information Systems
079 George Mason University

Information Technology
079 George Mason University

Integrated Manufacturing Systems Engineering
(See also: Manufacturing Engineering)
160 North Carolina State University

Interdisciplinary Engineering
(See also: Engineering; Engineering Interdisciplinary Studies; General Engineering)
071 University of Florida
097 Indiana University-Purdue University at Indianapolis
169 Oakland University
170 Ohio University
181 University of Pittsburgh
188 Purdue University
192 University of Rochester
238 Texas A&M University

Irrigation
(See also: Agricultural Engineering; Biological & Irrigation Engineering; Environmental Engineering; Engineering Hydrology; Hydrology; Water Resources Administration)

L

Land Surveying & Topography
(See also: Surveying; Topography)
187 University of Puerto Rico, Mayagüez Campus

Land Surveying Engineering
(See also: Surveying; Topography)
187 University of Puerto Rico, Mayagüez Campus
188 Purdue University

M

Macromolecular Science & Engineering
037 Case Western Reserve University
130 University of Michigan

Management
(See: Engineering Management; Industrial Engineering & Management; Management Engineering; Public Policy; Technical Management; Water Resources Administration)

Management Engineering
016 University of Bridgeport
151 New Jersey Institute of Technology

Management of Technology
263 Washington University

Management Science
216 Southern Methodist University

Man-Machine-Environment Systems
(See also: Health; Human Factors Engineering; Occupational Health & Safety; Safety Engineering; Work Environment)
022 University of California, Los Angeles

Manufacturing & Robotics
(See also: Robotics)
152 The University of New Mexico

Manufacturing Engineering
(See also: Industrial & Manufacturing Engineering; Industrial & Manufacturing Systems Engineering; Industrial Engineering (Furniture Manufacturing); Integrated Manufacturing Systems Engineering; Mechanical Engineering & Manufacturing; Mechanical, Manufacturing & Aerospace Engineering)
014 Boston University
015 Bradley University
017 Brigham Young University

019 University of California at Berkeley
022 University of California, Los Angeles
034 California State Polytechnic University, Pomona
036 Carnegie Mellon University
040 Central State University
129 Miami University
151 New Jersey Institute of Technology
197 St. Cloud State University
225 Syracuse University
243 Tufts University
277 Worcester Polytechnic Institute

Manufacturing Systems Engineering
012 Auburn University
073 Florida Atlantic University
077 GMI Engineering & Management Institute
181 University of Pittsburgh
233 University of Texas at Austin
274 University of Wisconsin-Madison

Mapping
(See: Land Surveying; Surveying; Surveying & Mapping)

Marine Engineering
(See also: Naval Architecture; Ocean Engineering; Water)
221 State University of New York Maritime College
250 United States Merchant Marine Academy
252 United States Naval Academy

Marine Engineering Systems
250 United States Merchant Marine Academy

Marine Engineering/Transportation
(See also: Transportation)
250 United States Merchant Marine Academy

Materials & Metallurgical Engineering
(See also: Metallurgical; Mineral; Mining Engineering)
096 Illinois Institute of Technology
153 New Mexico Institute of Mining and Technology

Materials Engineering
(See also: Chemical Engineering & Materials Science; Civil Engineering & Materials Science; Electrical Engineering & Materials Science; Materials Science; Mechanical Engineering & Materials Science; Mechanics & Materials Science; Metallurgical & Materials Engineering; Metallurgical Engineering & Materials Science; Metallurgical, Mining, Mineral & Materials Engineering)
012 Auburn University
022 University of California, Los Angeles
024 University of California, Santa Barbara
027 California Polytechnic State University
042 University of Cincinnati
064 Drexel University
066 École Polytechnique de Montréal
090 University of Houston
120 University of Maryland, College Park
188 Purdue University
189 Rensselaer Polytechnic Institute
202 San Jose State University
273 Wilkes University
275 University of Wisconsin-Milwaukee
277 Worcester Polytechnic Institute

Materials Science
(See also: Materials Engineering)
009 Arizona State University
023 University of California, San Diego
026 California Institute of Technology
042 University of Cincinnati
061 University of Delaware
132 Michigan State University
136 University of Minnesota
229 University of Tennessee, Knoxville
257 University of Vermont

Materials Science & Engineering
(See also: Materials Engineering; Materials Science)
003 University of Alabama at Birmingham
008 University of Arizona
019 University of California at Berkeley
020 University of California, Davis
022 University of California, Los Angeles
036 Carnegie Mellon University
037 Case Western Reserve University

058 Cornell University
071 University of Florida
082 Georgia Institute of Technology
094 University of Illinois at Chicago
100 The Johns Hopkins University
103 University of Kentucky
108 Lehigh University
124 Massachusetts Institute of Technology
130 University of Michigan
133 Michigan Technological University
160 North Carolina State University
166 Northwestern University
168 University of Notre Dame
172 Ohio State University
179 University of Pennsylvania
181 University of Pittsburgh
192 University of Rochester
217 Stanford University
223 State University of New York at Stony Brook
233 University of Texas at Austin
253 University of Utah
256 Vanderbilt University
259 University of Virginia
261 Virginia Polytechnic Institute and State University
263 Washington University
264 Washington State University
265 Wayne State University
274 University of Wisconsin-Madison
278 Wright State University

Mathematical Sciences
(See also: Engineering Mathematics; Mathematics; Statistics)
100 The Johns Hopkins University

Mathematics
(See also: Applied Mathematics; Computational & Neural Systems; Computational & Applied Mathematics; Computational Engineering; Computer Engineering; Computer Science; Engineering Math & Computer Science; Engineering Math & Statistics; Engineering Mathematics; Statistical Sciences; Statistics; Systems Science & Mathematics)
052 Colorado School of Mines

Mechanical & Aeronautical Engineering
(See also: Aeronautical Engineering)
046 Clarkson University
156 New York Institute of Technology

Mechanical & Aerospace Engineering
(See also: Aerospace Engineering; Space)
004 University of Alabama at Huntsville
021 University of California, Irvine
058 Cornell University
074 Florida Institute of Technology
139 University of Missouri-Columbia
140 University of Missouri-Kansas City
175 Oklahoma State University
186 Princeton University
192 University of Rochester
259 University of Virginia

Mechanical Engineering
(See also: Aeronautical Engineering; Aerospace Engineering; Aerospace & Mechanical Engineering; Applied Mechanics; Manufacturing Engineering; Materials Engineering; Mechanics; Nuclear Engineering; Thermal Engineering)
001 University of Akron
002 University of Alabama
003 University of Alabama at Birmingham
004 University of Alabama at Huntsville
005 University of Alaska, Fairbanks
006 University of Alberta
007 Alfred University
008 University of Arizona
009 Arizona State University
010 University of Arkansas
012 Auburn University
014 Boston University
015 Bradley University
016 University of Bridgeport
017 Brigham Young University
018 Bucknell University
019 University of California at Berkeley
020 University of California, Davis
022 University of California, Los Angeles

023 University of California, San Diego
024 University of California, Santa Barbara
026 California Institute of Technology
027 California Polytechnic State University
028 California State University, Chico
029 California State University, Fresno
030 California State University, Long Beach
031 California State University, Los Angeles
033 California State University, Sacramento
034 California State Polytechnic University, Pomona
036 Carnegie Mellon University
037 Case Western Reserve University
038 The Catholic University of America
039 University of Central Florida
041 Christian Brothers University
042 University of Cincinnati
044 City College of the City University of New York
047 Clemson University
048 Cleveland State University
049 University of Colorado at Boulder
051 University of Colorado at Denver
053 Colorado State University
054 Columbia University
055 Concordia University
056 University of Connecticut
057 The Cooper Union
060 University of Dayton
061 University of Delaware
062 University of Denver
063 University of the District of Columbia
064 Drexel University
065 Duke University
066 École Polytechnique de Montréal
069 University of Evansville
071 University of Florida
072 Florida Agricultural and Mechanical University
073 Florida Atlantic University
075 Florida International University
076 Florida State University
077 GMI Engineering & Management Institute
078 Gannon University
080 The George Washington University
082 Georgia Institute of Technology
083 Gonzaga University
085 Grove City College
088 University of Hawaii at Manoa
089 Hofstra University
090 University of Houston
091 Howard University
092 University of Idaho
094 University of Illinois at Chicago
095 University of Illinois at Urbana-Champaign
096 Illinois Institute of Technology
097 Indiana University-Purdue University at Indianapolis
098 University of Iowa
099 Iowa State University
100 The Johns Hopkins University
101 University of Kansas
102 Kansas State University
103 University of Kentucky
104 Lafayette College
105 Lamar University
106 Lawrence Technological University
109 Louisiana State University
110 Louisiana Tech University
111 University of Louisville
113 Loyola Marymount University
114 University of Maine
115 University of Manitoba
116 Mankato State University
118 Marquette University
119 University of Maryland Baltimore County
120 University of Maryland, College Park
121 University of Massachusetts
122 University of Massachusetts, Dartmouth
123 University of Massachusetts, Lowell
124 Massachusetts Institute of Technology
125 McGill University
126 Memphis State University
128 University of Miami
130 University of Michigan
131 University of Michigan-Dearborn
132 Michigan State University
133 Michigan Technological University
134 Milwaukee School of Engineering
136 University of Minnesota
137 University of Mississippi

138 Mississippi State University	252 United States Naval Academy	**Metallurgical Engineering**
141 University of Missouri-Rolla	253 University of Utah	(See also: Energy; Environmental Engineering; Extractive
144 Montana State University	255 Valparaiso University	Metallurgical Engineering; Geological Engineering; Materials &
146 University of Nebraska - Lincoln	256 Vanderbilt University	Metallurgical Engineering; Materials Engineering; Materials
147 University of Nevada, Las Vegas	257 University of Vermont	Science; Metallurgy; Minerals; Mining Engineering; Petroleum
149 University of New Hampshire	258 Villanova University	Engineering)
150 University of New Haven	259 University of Virginia	002 University of Alabama
151 New Jersey Institute of Technology	260 Virginia Military Institute	006 University of Alberta
152 The University of New Mexico	261 Virginia Polytechnic Institute and State University	042 University of Cincinnati
154 New Mexico State University	262 University of Washington	094 University of Illinois at Chicago
155 University of New Orleans	263 Washington University	095 University of Illinois at Urbana-Champaign
159 North Carolina Agricultural and Technical State University	264 Washington State University	099 Iowa State University
160 North Carolina State University	265 Wayne State University	125 McGill University
161 University of North Dakota	267 West Virginia University	141 University of Missouri-Rolla
163 Northeastern University	268 West Virginia Institute of Technology	143 Montana College of Mineral Science and Technology
164 Northern Arizona University	269 Western Michigan University	172 Ohio State University
165 Northern Illinois University	270 Western New England College	181 University of Pittsburgh
166 Northwestern University	271 Wichita State University	207 South Dakota School of Mines and Technology
167 Norwich University	272 Widener University	226 Technical University of Nova Scotia
168 University of Notre Dame	273 Wilkes University	253 University of Utah
169 Oakland University	274 University of Wisconsin-Madison	262 University of Washington
170 Ohio University	275 University of Wisconsin-Milwaukee	274 University of Wisconsin-Madison
171 Ohio Northern University	276 University of Wisconsin-Platteville	
172 Ohio State University	277 Worcester Polytechnic Institute	**Metallurgical Engineering & Material Science**
173 University of Oklahoma	278 Wright State University	036 Carnegie Mellon University
174 Oklahoma Christian University of Science and Arts	279 University of Wyoming	182 Polytechnic University
176 Old Dominion University	280 Yale University	
177 Oregon State University		**Metallurgical, Mining, Mineral & Materials Engineering**
178 University of the Pacific	**Mechanical Engineering & Applied Mechanics**	(See also: Materials Engineering; Materials Science; Minerals;
180 Pennsylvania State University	(See also: Applied Mechanics; Mechanics)	Mining Engineering)
181 University of Pittsburgh	108 Lehigh University	054 Columbia University
182 Polytechnic University	162 North Dakota State University	
183 University of Portland	179 University of Pennsylvania	**Metallurgy**
184 Portland State University		(See also: Metallurgical Engineering)
185 Prairie View A&M University	**Mechanical Engineering & Engineering Science**	056 University of Connecticut
187 University of Puerto Rico, Mayagüez Campus	(See also: Engineering Science)	094 University of Illinois at Chicago
188 Purdue University	158 University of North Carolina-Charlotte	099 Iowa State University
189 Rensselaer Polytechnic Institute		
190 University of Rhode Island	**Mechanical Engineering & Manufacturing**	**Meteorology**
192 University of Rochester	(See also: Manufacturing Engineering)	(See also: Atmospheric & Space Sciences)
193 Rochester Institute of Technology	028 California State University, Chico	130 University of Michigan
194 Rose-Hulman Institute of Technology		
195 Rutgers, The State University	**Mechanical Engineering & Materials Science**	**Microelectronic Engineering**
196 Saginaw Valley State University	(See also: Materials Science)	(See also: Electronics Engineering)
200 San Diego State University	020 University of California, Davis	193 Rochester Institute of Technology
201 San Francisco State University	191 William Marsh Rice University	
202 San Jose State University		**Mineral & Environmental Engineering**
203 Santa Clara University	**Mechanical, Manufacturing & Aerospace Engineering**	(See also: Environmental Engineering)
204 Seattle University	(See also: Manufacturing Engineering; Aerospace Engineering;	153 New Mexico Institute of Mining and Technology
205 University of South Alabama	Space)	
206 University of South Carolina	254 Utah State University	**Mineral Economics**
207 South Dakota School of Mines and Technology		008 University of Arizona
208 South Dakota State University	**Mechanical Technology**	
209 University of South Florida	(See also: Appropriate Technology; Technology)	**Mineral Engineering**
210 Southern University and A&M College	048 Cleveland State University	(See also: Energy & Mineral Resources; Environmental
211 University of Southern California	210 Southern University and A&M College	Engineering; Geological Engineering; Materials Engineering;
213 Southern Illinois University at Carbondale		Metallurgical Engineering; Metallurgical, Mining, Mineral &
214 Southern Illinois University at Edwardsville	**Mechanics**	Materials Engineering; Mining Engineering; Petroleum Engineering)
216 Southern Methodist University	(See also: Aerospace Engineering & Engineering Mechanics;	002 University of Alabama
217 Stanford University	Aerospace Engineering & Mechanics; Applied Mechanics;	019 University of California at Berkeley
218 State University of New York at Binghamton	Engineering Mechanics; Engineering Science & Mechanics;	066 École Polytechnique de Montréal
219 State University of New York at Buffalo	Geomechanics; Materials Engineering; Mechanical Engineering)	
221 State University of New York Maritime College	132 Michigan State University	**Minerals Processing**
223 State University of New York at Stony Brook	176 Old Dominion University	133 Michigan Technological University
225 Syracuse University	189 Rensselaer Polytechnic Institute	
226 Technical University of Nova Scotia	228 University of Tennessee, Chattanooga	**Mining Engineering**
227 Temple University		(See also: Energy; Environmental Engineering; Geological
228 University of Tennessee, Chattanooga	**Mechanics & Materials Science**	Engineering; Materials Engineering; Metallurgical Engineering;
229 University of Tennessee, Knoxville	(See also: Materials Science)	Metallurgical, Mining, Mineral & Materials Engineering)
230 Tennessee State University	195 Rutgers, The State University	006 University of Alberta
231 Tennessee Technological University		008 University of Arizona
232 University of Texas at Arlington	**Medical**	052 Colorado School of Mines
233 University of Texas at Austin	(See: Biomedical Engineering)	066 École Polytechnique de Montréal
235 University of Texas at El Paso		103 University of Kentucky
236 University of Texas at San Antonio	**Metallurgical & Materials Engineering**	125 McGill University
237 Texas A&I University	(See also: Materials Engineering; Materials Science; Minerals;	133 Michigan Technological University
238 Texas A&M University	Mining)	141 University of Missouri-Rolla
239 Texas Tech University	052 Colorado School of Mines	143 Montana College of Mineral Science and Technology
240 University of Toledo	235 University of Texas at El Paso	181 University of Pittsburgh
241 Tri-State University		207 South Dakota School of Mines and Technology
243 Tufts University		213 Southern Illinois University at Carbondale
244 Tulane University		226 Technical University of Nova Scotia
245 University of Tulsa		253 University of Utah
246 Tuskegee University		261 Virginia Polytechnic Institute and State University
247 Union College		
248 United States Air Force Academy		
251 United States Military Academy		

Index to Engineering Degree Programs

N

Natural Gas Engineering
(See also: Energy; Fuels; Petroleum Engineering)
- 237 Texas A&I University

Naval Architecture
(See also: Architectural Engineering; Marine Engineering; Ocean Engineering)
- 019 University of California at Berkeley
- 221 State University of New York Maritime College
- 252 United States Naval Academy

Naval Architecture & Marine Engineering
(See also: Architectural Engineering; Marine Engineering; Ocean Engineering)
- 130 University of Michigan
- 155 University of New Orleans
- 249 United States Coast Guard Academy
- 266 Webb Institute of Naval Architecture

Neural Systems
(See: Computation & Neural Systems; Computer Engineering)

Nuclear Engineering
(See also: Chemical Engineering; Energy; Mining Engineering; Hazardous Waste Management; Mechanical Engineering)
- 008 University of Arizona
- 019 University of California at Berkeley
- 022 University of California, Los Angeles
- 024 University of California, Santa Barbara
- 042 University of Cincinnati
- 066 École Polytechnique de Montréal
- 071 University of Florida
- 082 Georgia Institute of Technology
- 095 University of Illinois at Urbana-Champaign
- 099 Iowa State University
- 102 Kansas State University
- 120 University of Maryland, College Park
- 123 University of Massachusetts, Lowell
- 124 Massachusetts Institute of Technology
- 130 University of Michigan
- 138 Mississippi State University
- 141 University of Missouri-Rolla
- 152 The University of New Mexico
- 160 North Carolina State University
- 172 Ohio State University
- 177 Oregon State University
- 180 Pennsylvania State University
- 188 Purdue University
- 189 Rensselaer Polytechnic Institute
- 229 University of Tennessee, Knoxville
- 238 Texas A&M University
- 253 University of Utah
- 259 University of Virginia
- 274 University of Wisconsin-Madison

Nuclear Engineering & Sciences
- 058 Cornell University
- 071 University of Florida
- 130 University of Michigan
- 189 Rensselaer Polytechnic Institute

O

Occupational Health & Safety
(See also: Health; Human Factors Engineering; Man-Machine-Environment Engineering; Safety Engineering; Work Environment)
- 151 New Jersey Institute of Technology
- 267 West Virginia University

Ocean Engineering
(See also: Aerospace & Ocean Engineering; Coastal & Oceanographic; Marine Engineering; Naval Architecture; Ocean Science; Oceanography)
- 073 Florida Atlantic University
- 074 Florida Institute of Technology
- 124 Massachusetts Institute of Technology
- 190 University of Rhode Island
- 238 Texas A&M University
- 252 United States Naval Academy

Ocean Science
- 130 University of Michigan

Oceanography: Physical
- 130 University of Michigan

Operations
(See: Engineering Operations; Operations Management; Operations Research; Industrial & Operations Engineering; Space Operations)

Operations Management
(See also: Engineering Management; Management)
- 010 University of Arkansas

Operations Research
(See also: Civil Engineering & Operations Research; Industrial Engineering; Industrial Engineering & Operations Research)
- 010 University of Arkansas
- 080 The George Washington University
- 090 University of Houston
- 099 Iowa State University
- 160 North Carolina State University
- 217 Stanford University
- 265 Wayne State University

Operations Research & Engineering
- 058 Cornell University
- 058 Cornell University

Operations Research & Industrial Engineering
(See also: Industrial Engineering)
- 058 Cornell University
- 233 University of Texas at Austin

Operations Research & Management Science
(See also: Engineering Management; Management: Management Science)
- 079 George Mason University

Optical Engineering
- 008 University of Arizona
- 192 University of Rochester

P

Paper & Printing Science & Engineering
(See also: Forest Engineering; Wood Products Engineering)
- 269 Western Michigan University

Paper Science & Engineering
- 129 Miami University
- 220 State University of New York, College of Environmental Science and Forestry

Petroleum Engineering
(See also: Chemical & Petroleum Engineering; Chemical Engineering; Chemical Engineering & Petroleum Refining; Energy; Fuels; Natural Gas; Mining Engineering; Refining)
- 006 University of Alberta
- 019 University of California at Berkeley
- 052 Colorado School of Mines
- 090 University of Houston
- 109 Louisiana State University
- 110 Louisiana Tech University
- 117 Marietta College
- 138 Mississippi State University
- 141 University of Missouri-Rolla
- 143 Montana College of Mineral Science and Technology
- 153 New Mexico Institute of Mining and Technology
- 162 North Dakota State University
- 173 University of Oklahoma
- 181 University of Pittsburgh
- 233 University of Texas at Austin
- 238 Texas A&M University
- 239 Texas Tech University
- 245 University of Tulsa
- 279 University of Wyoming

Physics
(See: Applied & Engineering Physics; Applied Physics; Electrical Engineering & Applied Physics; Engineering Physics; Health Physics)

Plastics Engineering
(See also: Polymer Science & Engineering)
- 123 University of Massachusetts, Lowell

Policy
(See: Public Policy)

Polymer Science & Engineering
(See also: Plastics Engineering)
- 037 Case Western Reserve University

Printing
(See: Paper & Printing Science & Engineering)

Psychology
(See: Engineering Psychology)

Public Policy
(See: Engineering & Public Policy; Engineering-Economic Systems; Public Works; Public Works & Civil Engineering; Technology & Human Affairs; Transportation; Transportation Planning & Engineering; Urban Systems Engineering)

Public Works
(See also: Public Policy)
- 181 University of Pittsburgh

Public Works & Civil Engineering
(See also: Civil Engineering)
- 181 University of Pittsburgh

Q

Quality
(See: Environmental Quality Engineering/Science)

R

Radiological Health Engineering
(See also: Health)
- 238 Texas A&M University

Reliability Engineering
- 008 University of Arizona
- 120 University of Maryland, College Park

Resource(s)
(See: Bioresources Engineering; Energy & Mineral Resources; Energy Resources; Environmental & Resource Engineering; Water Resources Administration)

Robotics
(See: Manufacturing & Robotics)

S

Safety Engineering
(See also: Health; Human Factors Engineering; Man-Machine-Environment Engineering; Occupational Health & Safety; Work Environment)
- 238 Texas A&M University

Science Management
(See also: Management)
- 005 University of Alaska, Fairbanks

Software Systems Engineering
(See also: Computer Engineering)
- 079 George Mason University

Solid State Electronics
(See also: Electronics Engineering)
- 042 University of Cincinnati

Index to Engineering Degree Programs

Space
(See: Aeronautical Engineering; Aerospace Engineering; Atmospheric & Space Sciences; Astronautical Engineering; Mechanical Engineering)

Space Operations
050 University of Colorado at Colorado Springs

Statistical Sciences
079 George Mason University

Statistics
(See also: Engineering Mathematics; Engineering Math & Statistics; Mathematics)
191 William Marsh Rice University

Structural Design
263 Washington University

Structural Engineering
023 University of California, San Diego
263 Washington University

Surveying & Mapping
(See also: Land Surveying & Topography; Land Surveying Engineering)
071 University of Florida
154 New Mexico State University

Surveying Engineering
(See also: Land Surveying & Topography; Land Surveying Engineering)
029 California State University, Fresno
070 Ferris State University
114 University of Maine
172 Ohio State University

Systems
(See: Agricultural Biosystems Engineering; Biological Systems Engineering; Computation & Neural Systems; Computer & Systems Engineering; Computer Systems Engineering; Electrical Engineering Systems; Electronic & Computer Control Systems; Engineering-Economic Systems; Environmental Systems Engineering; Industrial & Manufacturing Systems Engineering; Industrial & Systems Engineering; Industrial Engineering & Information Systems; Information Systems; Integrated Manufacturing Systems Engineering; Man-Machine-Environment Systems; Manufacturing Systems Engineering; Marine Engineering Systems; Neural Systems; Software Systems Engineering; Urban Systems Engineering)

Systems & Computer Science
(See also: Computer Engineering; Computer Science; Systems)
091 Howard University

Systems & Control Engineering
(See also: Control Engineering; Systems)
023 University of California, San Diego
037 Case Western Reserve University

Systems Analysis & Engineering
080 The George Washington University

Systems Engineering
008 University of Arizona
014 Boston University
079 George Mason University
120 University of Maryland, College Park
169 Oakland University
251 United States Military Academy
252 United States Naval Academy
259 University of Virginia

Systems Science
023 University of California, San Diego
218 State University of New York at Binghamton

Systems Science & Engineering
179 University of Pennsylvania

Systems Science & Mathematics
(See also: Mathematics)
263 Washington University

T

Technical Communication
(See also: Communication; Telecommunications)
262 University of Washington

Technical Management
(See also: Management)
100 The Johns Hopkins University

Technology
(See: Applied disciplines; Appropriate Technology; Electronic Engineering Technology; Electronic Technology; Engineering & Industrial Technology; Engineering Technology; Environmental Engineering Technology; Industrial Technology; Information Technology; Management of Technology; Mechanical Technology; Textile Management & Technology)

Technology & Human Affairs
263 Washington University

Telecommunications
(See also: Communications; Technical Communication)
049 University of Colorado at Boulder

Textile Chemistry
(See also: Chemical Engineering; Chemistry)
012 Auburn University
082 Georgia Institute of Technology

Textile Engineering
012 Auburn University
082 Georgia Institute of Technology
160 North Carolina State University

Textile Management & Technology
(See also: Management)
012 Auburn University
082 Georgia Institute of Technology

Textile Science
012 Auburn University

Theoretical & Applied Mechanics
(See also: Applied Mechanics; Mechanics)
058 Cornell University
095 University of Illinois at Urbana-Champaign

Thermal Engineering
(See also: Mechanical Engineering)
228 University of Tennessee, Chattanooga

Topography
(See: Land Surveying & Topography)

Transportation Engineering
(See also: Marine Engineering/Transportation)
189 Rensselaer Polytechnic Institute

Transportation Planning & Engineering
(See also: Public Policy; Urban Systems Engineering)
151 New Jersey Institute of Technology

U

Urban Systems Engineering
(See also: Public Policy; Technology & Human Affairs; Transportation Planning & Engineering)
079 George Mason University

W

Waste
(See: Environmental Engineering; Hazardous Waste Management)

Water Resources Administration
(See also: Agricultural Engineering; Biological & Irrigation Engineering; Environmental Engineering; Engineering Hydrology; Hydrology; Irrigation)
008 University of Arizona

Welding Engineering
107 LeTourneau University
172 Ohio State University

Wood Products Engineering
(See also: Forest Engineering; Industrial Engineering (Furniture Manufacturing))
220 State University of New York, College of Environmental Science and Forestry

Work Environment
(See also: Health; Human Factors Engineering; Man-Machine-Environment Engineering; Occupational Health & Safety; Safety Engineering)
123 University of Massachusetts, Lowell

Ten reasons to choose Cornell Engineering...

1 A broad-based common curriculum for most of the first two years that lets students familiarize themselves with a variety of engineering fields before choosing an area for more-specialized study in their junior and senior years.

2 A wide selection of traditional and nontraditional engineering fields (majors).

3 For students who don't want to pursue an established major, a College Program that lets them design their own curriculum.

4 A combination of elective and required courses that helps students acquire a liberal education along with their technical education. Students may take up to one-third of their courses in the humanities, the social sciences, and the expressive arts in the College of Arts and Sciences and the five other undergraduate colleges at Cornell.

5 Opportunities to take part in research as an undergraduate. (In 1989–1990, the most recent year for which comparisons are available, Cornell's College of Engineering ranked among the top five engineering colleges in the country in total funding for research. The annual engineering research budget exceeds $53 million.)

6 An optional cooperative education program that gives students eight months of full-time work experience in industry and lets them graduate on time with their class.

7 An innovative freshman seminar program and for upper-level students a field-focused Engineering Communications Program that together make writing and speaking skills a natural part of an engineering education.

8 Many attractive combined-degree programs, including dual majors in engineering; dual degrees in engineering and in arts and sciences; a B.S. degree in engineering and an integrated Master of Engineering (M.Eng.) degree; and a B.S. degree in engineering combined with an M.Eng. degree and an M.B.A. degree.

9 Opportunities to study for a semester or a year in another country.

10 Exceptional placement opportunities and—after graduation—membership in a strong and active group of more than 20,000 living alumni of the College of Engineering.

By infusing professional disciplines with the spirit of liberal learning, Cornell is uniquely capable of educating men and women for the modern world.

For more information, contact the College of Engineering Office of Admissions, Cornell University, Carpenter Hall Annex, Ithaca, NY 14853-2201; telephone 607/255-5008.

Undergraduate Degrees Awarded–by Category

Institutions are listed in alphabetical order. All figures below represent the percentages of bachelor's degree recipients who fall into different categories. The section devoted to students of different ethnic backgrounds and nationalities includes African Americans, Asian (including Pacific Island) Americans, Hispanic Americans, and Native Americans, as well as citizens of other countries. The figure for the column "All Others" was derived by subtracting the above percentages from 100.0%. In most most cases, Canadian schools could not provide data in corresponding categories for African Canadians, etc.; therefore, this information does not appear for most Canadian schools. In addition, information was requested about degree recipients in categories independent if ethnic or national background. These categories are: women, students with self-identified physical or learning disabilities, and students who were older than 25 years of age when they received their degrees. Unless footnoted otherwise, degrees awarded are for the 1991-92, twelve-month academic year.

ENGINEERING Institution Profile Reference Number & Name	State/Province	Ethnicity/Nationality						Other Categories		
		African Americans	Asian Americans	Hispanic Amer.	Native Americans	Foreign Citizens	All Others	Women	Students with Disabilities	Over 25 Years of Age
001 University of Akron	OH	0.5	1.3	0.5	0.0	97.7	4.5	13.0	-.-(A)	-.-(A)
002 University of Alabama	AL	8.8	1.1	1.8	-.-	88.3	18.6	19.3	-.-(A)	-.-(A)
003 University of Alabama at Birmingham	AL	5.0	0.0	1.0	0.0	94.0	19.0	12.0	-.-(A)	-.-(A)
004 University of Alabama at Huntsville	AL	5.6	5.6	1.0	1.0	86.8	5.1	19.7	0.0	0.0
005 University of Alaska, Fairbanks	AK	0.0	2.6	5.2	0.0	92.2	2.6	7.8	0.0	31.0
006 University of Alberta	AB	-.-	-.-	-.-	-.-	100.0	-.-	12.3	-.-	-.-
007 Alfred University	NY	11.0	4.0	4.0	1.0	80.0	10.0	22.0	-.-(A)	-.-(A)
008 University of Arizona	AZ	1.0	6.0	7.0	1.0	85.0	16.0	15.0	-.-(A)	-.-(A)
009 Arizona State University	AZ	1.0	10.0	3.0	0.0	86.0	13.0	14.0	-.-	-.-
010 University of Arkansas	AR	3.6	3.2	0.4	0.4	92.4	14.5	6.5		
011 Arkansas Tech University	AR	6.7	0.0	0.0	0.0	93.3	0.0	20.0	0.0	53.4
012 Auburn University	AL	2.9	2.0	0.4	0.1	94.6	0.7	16.7	-.-(A)	-.-(A)
013 Baylor University	TX	0.0	12.0	12.0	0.0	76.0	6.0	24.0	0.0	0.0
014 Boston University	MA	4.3	14.0	2.0	0.0	79.7	17.4	21.7	-.-(A)	-.-(A)
015 Bradley University	IL	0.6	1.0	0.6	0.0	97.8	0.2	11.0	-.-(A)	-.-(A)
016 University of Bridgeport	CT	2.2	5.6	0.0	0.0	92.2	70.0	4.4	0.0	69.0
017 Brigham Young University	UT	-.-	-.-	3.4	1.7	94.9	3.4	3.4	-.-(A)	-.-(1)
018 Bucknell University	PA	1.0	3.0	1.0	0.0	95.0	0.0	15.0(1)	0.0(A)	5.0
019 University of California at Berkeley	CA	2.5	53.7	4.9	0.7	38.2	-.-(A)	20.4	-.-(A)	-.-(A)
020 University of California, Davis	CA	2.5	3.8	7.6	0.3	85.8	0.3	16.4	-.-(A)	-.-(A)
021 University of California, Irvine	CA	1.0	48.0	5.0	1.0	45.0	3.0	17.0	-.-(A)	-.-(A)
022 University of California, Los Angeles	CA	2.7	49.5	7.3	0.2	40.3	5.9(2)	17.6	-.-	-.-
023 University of California, San Diego	CA	1.2	27.8	5.2	0.6	65.2	8.0	17.1	0.0(A)	0.0(A)
024 University of California, Santa Barbara	CA	1.6	18.0	6.6	0.4	73.4	9.8	16.8	-.-(A)	-.-(A)
025 University of California, Santa Cruz	CA	9.0	31.8	4.5	0.0	54.7	-.-	9.0	-.-	-.-
026 California Institute of Technology	CA	1.0	23.0	1.0	1.0	74.0	12.0	13.0	-.-	-.-
027 California Polytechnic State University	CA	1.0	17.0	7.0	1.0	74.0	-.-	13.0	-.-	-.-
028 California State University, Chico	CA	-.-	-.-	-.-	-.-	100.0	-.-	-.-	-.-	-.-
029 California State University, Fresno	CA	2.7	20.0	25.5	1.1	50.7	11.5	8.0	-.-(A)	26.5
030 California State University, Long Beach	CA	3.0	26.0	10.0	0.5	60.5	11.0	14.0	1.0	71.0
031 California State University, Los Angeles	CA	10.5	44.4(2)	7.3	0.0	37.8	12.9	12.9	-.-(A)	-.-(A)
032 California State University, Northridge	CA	7.7(C)	26.0(C)	3.0(C)	0.0(C)	63.3	-.-(A)	15.0(C)	-.-(A)	-.-(A)
033 California State University, Sacramento	CA	4.0	23.0	6.4	-.-	66.6	-.-(A)	14.0	-.-	-.-
034 California State Polytechnic University, Pomona	CA	1.0	47.0	11.0	-.-	41.0	4.0	18.0	-.-	-.-
035 Calvin College	MI	0.0	0.0	0.0	1.9	98.1	13.0	9.3	-.-(A)	9.3
036 Carnegie Mellon University	PA	2.0(C)	5.0(C)	-.-(A)	-.-(A)	93.0	-.-(A)	14.0(C)	0.0	-.-(A)
037 Case Western Reserve University	OH	4.5	14.1	1.5	0.2	79.7	9.1	26.0	0.8	0.7
038 The Catholic University of America	DC	2.2	6.6	6.6	0.0	84.6	17.1	28.4	0.0	0.0
039 University of Central Florida	FL	3.3	8.4	8.3	0.1	79.9	12.0(C)	17.2	0.2	15.0
040 Central State University	OH	75.0	0.0	0.0	0.0	25.0	0.0	50.0	0.0	0.0
041 Christian Brothers University	TN	16.3	3.2	0.0	0.0	80.5	6.5	24.5	-.-(A)	-.-(A)
042 University of Cincinnati	OH	1.4	3.4	0.0	0.0	95.2	0.6	14.2	-.-(A)	-.-(A)
043 The Citadel	SC	-.-	-.-	-.-	-.-	100.0	-.-	4.0	-.-	-.-
044 City College of the City University of New York	NY	23.1	30.1	6.6	0.0	40.2	11.8	11.8	-.-(A)	-.-(A)
045 City University of New York, College of Staten Island	NY	-.-(A)	-.-(A)	-.-(A)	-.-(A)	100.0	-.-(A)	-.-(A)	-.-(A)	-.-(A)

Degrees Awarded—by Category

ENGINEERING

| | | PERCENTAGE OF UNDERGRADUATE DEGREES AWARDED–BY CATEGORY ||||||| |||
| | | Ethnicity/Nationality |||||| Other Categories |||
Institution Profile Reference Number & Name	State/Province	African Americans	Asian Americans	Hispanic Amer.	Native Americans	Foreign Citizens	All Others	Women	Students with Disabilities	Over 25 Years of Age
046 Clarkson University	NY	-.-(1)	-.-(1)	-.-(1)	-.-(1)	100.0	-.-(1)	16.9(1)	-.-(1)	2.9(1)
047 Clemson University	SC	4.0	1.0	1.0	0.0	94.0	12.0	17.0	-.-(A)	-.-(A)
048 Cleveland State University	OH	2.3	2.3	1.5	0.0	93.9	5.9	14.8	-.-(A)	59.0
049 University of Colorado at Boulder	CO	-.-	-.-	-.-	-.-	100.0	-.-	14.0	-.-(A)	-.-(A)
050 University of Colorado at Colorado Springs	CO	4.1	10.2	4.1	0.0	81.6	-.-(A)	24.5	-.-(A)	71.0
051 University of Colorado at Denver	CO	1.3	17.4	4.0	0.0	77.3	7.4	17.4	0.5	69.0
052 Colorado School of Mines	CO	-.-(1)	2.0	3.0	2.0	93.0	22.0	20.0	-.-(1)	-.-(A)
053 Colorado State University	CO	-.-	2.5	3.0	-.-	94.5	-.-(A)	-.-(A)	-.-(A)	-.-(A)
054 Columbia University	NY	3.3	43.9	4.1	-.-	48.7	-.-(A)	12.6	-.-(A)	-.-(A)
055 Concordia University	PQ	-.-	-.-	-.-	-.-	100.0	-.-	-.-	-.-	-.-
056 University of Connecticut	CT	1.2	7.1	2.4	0.0	89.3	1.2	12.7	-.-(A)	-.-(A)
057 The Cooper Union	NY	10.0	40.0	5.0	-.-	45.0	-.-	18.0	-.-	-.-
058 Cornell University	NY	1.5	18.0	3.0	0.0	77.5	9.0	21.0	0.0	0.0
059 Dartmouth College	NH	-.-(A)	7.0	1.7	-.-(A)	91.3	14.0	12.0	-.-(A)	-.-(A)
060 University of Dayton	OH	5.0	-.-	2.0	0.0	93.0	3.0	23.0	-.-	-.-
061 University of Delaware	DE	11.5	7.9	1.1	0.0	79.5	1.1	24.8	-.-(A)	-.-(A)
062 University of Denver	CO	0.0	0.0	0.0	0.0	100.0	37.5	12.5	0.0	6.3
063 University of the District of Columbia	DC	-.-(A)	-.-(A)	-.-(A)	-.-(A)	100.0	-.-(A)	-.-(A)	-.-(A)	-.-(A)
064 Drexel University	PA	5.0	12.0	1.0	0.0	82.0	9.0	14.0	0.2	16.0
065 Duke University	NC	7.8	4.9	-.-	0.0	87.3	2.9	20.6	-.-	-.-
066 École Polytechnique de Montréal	PQ	-.-(A)	-.-(A)	-.-(A)	-.-(A)	100.0	-.-(A)	21.2(5)	-.-(A)	-.-(A)
067 Embry-Riddle Aeronautical University - Daytona Beach	FL	1.0	8.1	2.4	0.0	88.5	-.-(A)	13.7	-.-(A)	-.-(A)
068 Embry-Riddle Aeronautical University, Western Campus	AZ	4.0	6.0	6.0	0.0	84.0	10.0	11.0	0.0	10.0
069 University of Evansville	IN	1.1	0.0	0.0	0.0	98.9	34.8	11.2	0.0	7.8(C)
070 Ferris State University	MI	-.-(A)	-.-(A)	-.-(A)	-.-(A)	100.0	-.-(A)	-.-(B)	-.-(A)	-.-(A)
071 University of Florida	FL	4.3	6.4	9.2	0.0	80.1	5.1	14.4	0.9	-.-(A)
072 Florida Agricultural and Mechanical University	FL	17.7(1)	6.6(1)	5.0(1)	-.-(1)	70.7	-.-(A)	10.5(1)	-.-(A)	30.4(1)
073 Florida Atlantic University	FL	6.8	7.3	10.5	0.0	75.4	12.0	25.7	1.0(C)	10.0(C)
074 Florida Institute of Technology	FL	2.2	2.7	7.6	-.-	87.5	23.9	15.9	-.-	15.0
075 Florida International University	FL	9.2	6.0	57.9	-.-	26.9	15.2	10.9	-.-	31.4
076 Florida State University	FL	17.8(2)	-.-(2)	-.-(2)	-.-(2)	82.2	-.-(A)	10.5(2)	-.-(A)	30.4(2)
077 GMI Engineering & Management Institute	MI	4.0	8.0	2.0	1.0	85.0	5.0	20.0	-.-	-.-
078 Gannon University	PA	2.0	0.0	0.0	0.0	98.0	23.6	11.0	0.0	14.0
079 George Mason University	VA	3.0	26.0	4.3	0.0	66.7	9.0	29.0	-.-(A)	-.-(A)
080 The George Washington University	DC	1.0	6.3	1.8	-.-	90.9	41.4	23.4	-.-(A)	-.-(A)
081 University of Georgia	GA	2.0	0.0	0.0	0.0	98.0	1.0	0.0	0.0	20.0
082 Georgia Institute of Technology	GA	7.4	7.2	3.1	-.-	82.3	2.7	21.2	-.-	-.-(A)
083 Gonzaga University	WA	0.0	-.-	-.-	0.0	100.0	-.-	-.-	-.-	17.0
084 Grand Valley State University	MI	7.8	3.5	1.6	1.0	86.1	0.5	12.5	-.-(A)	-.-(A)
085 Grove City College	PA	0.0	0.0	0.0	0.0	100.0	2.9	8.6	0.0	0.0
086 Harvard University	MA	0.0(A)	0.0(A)	0.0(A)	0.0(A)	100.0	0.0(A)	0.0(A)	0.0(A)	0.0(A)
087 Harvey Mudd College	CA	0.0	25.0	9.0	0.0	66.0	3.0	23.0	0.0	0.0
088 University of Hawaii at Manoa	HI	0.0	84.3	0.6	0.0	15.1	-.-(A)	24.5	-.-(A)	26.4
089 Hofstra University	NY	3.0	5.9	5.9	0.0	85.2	5.9	5.9	-.-	17.6
090 University of Houston	TX	1.6	20.2	6.4	0.0	71.8	17.6	17.6	0.0	0.0
091 Howard University	DC	79.0	0.0(1)	0.0	0.0	21.0	21.0	38.0	0.0	0.0(A)
092 University of Idaho	ID	0.0	2.8	0.0	0.0	97.2	8.5	6.4	-.-(1)	-.-(1)
093 Idaho State University	ID	0.0	0.0	3.0	0.0	97.0	10.0	3.0	3.0	83.0
094 University of Illinois at Chicago	IL	3.4	32.0	6.8	0.6	57.2	2.8	15.8	0.0	12.1
095 University of Illinois at Urbana-Champaign	IL	1.4	13.5	2.0	0.0	83.1	0.7	13.8	-.-(A)	-.-(A)
096 Illinois Institute of Technology	IL	3.0	20.0	5.0	0.1(C)	71.9	7.0	13.0	-.-(A)	-.-(A)
097 Indiana University-Purdue University at Indianapolis	IN	7.0	7.0	1.0	0.0	85.0	6.5	17.0	0.0	0.0
098 University of Iowa	IA	0.0	5.4	1.4	0.0	93.2	6.3	16.3	-.-(A)	-.-(A)
099 Iowa State University	IA	-.-	3.0	1.0	0.0	96.0	9.0	11.0	-.-(A)	-.-(A)
100 The Johns Hopkins University	MD	2.0	17.0	2.0	0.0	79.0	4.0	21.0	-.-(A)	-.-(A)

Degrees Awarded—by Category

ENGINEERING

| | | PERCENTAGE OF UNDERGRADUATE DEGREES AWARDED–BY CATEGORY ||||||||| |
| | | Ethnicity/Nationality |||||| Other Categories ||| |
Institution Profile Reference Number & Name	State/Province	African Americans	Asian Americans	Hispanic Amer.	Native Americans	Foreign Citizens	All Others	Women	Students with Disabilities	Over 25 Years of Age
101 University of Kansas	KS	1.2	5.8	1.2	-.-	91.8	19.0	14.5	-.-(A)	-.-(A)
102 Kansas State University	KS	-.-	1.2	-.-	0.0	98.8	1.7	5.9	0.0	-.-(A)
103 University of Kentucky	KY	1.3	3.9	-.-	-.-	94.8	4.7	10.3	-.-(A)	22.3
104 Lafayette College	PA	0.0	2.6	0.0	0.0	97.4	7.9	25.4	0.0	0.0
105 Lamar University	TX	5.0	0.0	4.0	0.9	90.1	7.0	13.0	0.0	57.0
106 Lawrence Technological University	MI	2.3	2.3	1.2	0.6	93.6	9.2	10.1	-.-(A)	-.-(A)
107 LeTourneau University	TX	2.0	2.0	3.0	0.0	93.0	9.0	6.4	0.0	9.0
108 Lehigh University	PA	2.0	4.0	-.-	0.0	94.0	3.0	15.0	-.-(A)	-.-(A)
109 Louisiana State University	LA	4.9	11.0	3.0	-.-	81.1	13.0	11.0	0.0	39.0
110 Louisiana Tech University	LA	4.0	2.0	0.0	0.0	94.0	3.0	10.0	-.-(A)	-.-(A)
111 University of Louisville	KY	3.4	9.1	-.-	-.-	87.5	5.1(1)	17.7	-.-(A)	-.-(A)
112 Loyola College in Maryland	MD	-.-	-.-	-.-	-.-	100.0	-.-	-.-	-.-	-.-
113 Loyola Marymount University	CA	5.0	15.0	15.0	1.0	64.0	3.0	21.0	1.0	5.0
114 University of Maine	ME	0.0	0.0	0.0	0.0	100.0	5.0	10.0	0.0	15.0
115 University of Manitoba	MB	-.-(A)	-.-(A)	-.-(A)	-.-(A)	100.0	-.-(A)	-.-(A)	-.-(A)	-.-(A)
116 Mankato State University	MN	0.0	10.0	0.0	0.0	90.0	10.0	5.0	1.0	8.0
117 Marietta College	OH	0.0	0.0	0.0	0.0	100.0	25.0	13.0	0.0	13.0
118 Marquette University	WI	2.0	5.0	2.0	0.0	91.0	5.0	17.0	-.-(A)	-.-(A)
119 University of Maryland Baltimore County	MD	15.0	10.0	2.0	0.0	73.0	2.0	21.0	-.-(A)	-.-(A)
120 University of Maryland, College Park	MD	3.7	23.4	2.3	-.-	70.6	4.9	16.5	-.-	-.-(A)
121 University of Massachusetts	MA	2.7	5.4	3.0	-.-	88.9	5.4	18.5	-.-(A)	-.-(A)
122 University of Massachusetts, Dartmouth	MA	-.-(A)	-.-(A)	-.-(A)	-.-(A)	100.0	-.-(A)	-.-(A)	-.-(A)	-.-(A)
123 University of Massachusetts, Lowell	MA	0.0	7.4	0.1	0.0	92.5	4.6	15.5	-.-(A)	-.-(A)
124 Massachusetts Institute of Technology	MA	7.7	19.1	7.2	0.3	65.7	12.1	30.2	-.-(A)	-.-(A)
125 McGill University	PQ	-.-(A)	-.-(A)	-.-(A)	-.-(A)	100.0	4.5	19.6	-.-(A)	-.-(A)
126 Memphis State University	TN	3.5	3.5	0.0	0.0	93.0	-.-(A)	18.8	-.-(A)	5.0
127 Mercer University	GA	14.0	6.0	4.0	2.0	74.0	4.0	22.0	-.-	1.0
128 University of Miami	FL	5.5	2.8	16.6	0.0	75.1	35.0	13.5	0.0	26.4
129 Miami University	OH	0.0	0.0	0.0	0.0	100.0	0.0	15.0	0.0	5.0
130 University of Michigan	MI	3.9	10.2	2.2	0.1	83.6	3.3	20.1	-.-(A)	-.-(A)
131 University of Michigan-Dearborn	MI	0.0	7.0	2.0	0.0	91.0	2.0	21.0	0.0(A)	0.0(A)
132 Michigan State University	MI	4.9	2.6	-.-	-.-	92.5	3.0	20.1	-.-(A)	-.-(A)
133 Michigan Technological University	MI	0.4	1.3	0.4	0.7	97.2	9.1	13.9	-.-(A)	17.8
134 Milwaukee School of Engineering	WI	2.0	3.0	1.0	0.0	94.0	2.0	13.0	-.-(A)	-.-(A)
135 University of Minnesota, Duluth	MN	-.-(A)	-.-(A)	-.-(A)	-.-(A)	100.0	-.-(A)	-.-(A)	-.-(A)	-.-(A)
136 University of Minnesota	MN	-.-	7.3	-.-	-.-	92.7	3.3	14.9	-.-(A)	-.-(A)
137 University of Mississippi	MS	4.8	14.3	1.0	0.0	79.9	0.0	16.7	-.-	-.-(A)
138 Mississippi State University	MS	6.0	3.0	-.-	0.0	91.0	8.0	10.0	-.-(A)	-.-(A)
139 University of Missouri-Columbia	MO	3.0	2.7	0.2	0.0	94.1	7.7	11.0	-.-(A)	-.-(A)
140 University of Missouri-Kansas City	MO	4.0	16.0	3.0	0.0	77.0	7.0	10.0	0.0	9.0
141 University of Missouri-Rolla	MO	1.4	7.6	0.5	0.0	90.5	4.4	17.2	0.0	21.0
142 Monmouth College	NJ	9.5	4.8	0.0	0.0	85.7	19.0	14.3	0.0	-.-(X)
143 Montana College of Mineral Science and Technology	MT	-.-(A)	-.-(A)	-.-(A)	-.-(A)	100.0	-.-(A)	19.0	-.-(A)	-.-(A)
144 Montana State University	MT	-.-(A)	-.-(A)	-.-(A)	-.-(A)	100.0	-.-(A)	12.7	-.-(A)	-.-(A)
145 Morgan State University	MD	92.5	2.9	1.0	0.0	3.6	3.6	34.6	0.0	4.0
146 University of Nebraska - Lincoln	NE	-.-	-.-	-.-	-.-	100.0	10.3	6.7	-.-	-.-
147 University of Nevada, Las Vegas	NV	6.4	6.4	4.3	-.-	82.9	6.4	17.0	-.-(A)	63.8
148 New England College	NH	0.0	0.0	0.0	0.0	100.0	0.0	10.0	0.0	0.0
149 University of New Hampshire	NH	0.0	1.2	-.-	0.0	98.8	-.-	11.3	-.-(A)	-.-(A)
150 University of New Haven	CT	3.0	3.0	0.0	0.0	94.0	3.0	10.0	-.-(A)	40.0(C)
151 New Jersey Institute of Technology	NJ	7.6	14.5	10.4	0.3	67.2	7.3(1)	10.4	-.-(A)	41.4
152 The University of New Mexico	NM	0.5	1.6	15.1	3.2	79.6	7.5	21.0	-.-(A)	-.-(A)
153 New Mexico Institute of Mining and Technology	NM	2.0	3.0	18.0	4.0	73.0	4.0	32.0	-.-(A)	-.-(A)
154 New Mexico State University	NM	0.0	-.-(A)	27.2	1.7	71.1	6.9(3)	16.8	-.-	-.-(A)
155 University of New Orleans	LA	5.5	5.5	5.5	0.9	82.6	10.2	15.7	-.-	-.-

Degrees Awarded—by Category

ENGINEERING

| Institution Profile Reference Number & Name | State/Province | PERCENTAGE OF UNDERGRADUATE DEGREES AWARDED–BY CATEGORY |||||||||
| | | Ethnicity/Nationality |||||| Other Categories |||
		African Americans	Asian Americans	Hispanic Amer.	Native Americans	Foreign Citizens	All Others	Women	Students with Disabilities	Over 25 Years of Age
156 New York Institute of Technology	NY	3.3	5.0	2.3	-.-	89.4	-.-	11.1	-.-	-.-
157 New York State College of Ceramics at Alfred University	NY	0.0	0.8	0.0	0.0	99.2	0.8	11.2	0.0	0.0
158 University of North Carolina-Charlotte	NC	0.7	4.7	0.0	0.7	93.9	6.3	8.7	0.0	45.2
159 North Carolina Agricultural and Technical State University	NC	81.0	1.0	1.0	0.0	17.0	1.0	37.0	-.-(A)	5.0
160 North Carolina State University	NC	6.3	4.7(C)	0.9(C)	0.6(C)	87.5	1.8(C)	17.1	-.-(A)	-.-(A)
161 University of North Dakota	ND	-.-	-.-	-.-	-.-	100.0	-.-	20.0	-.-	-.-
162 North Dakota State University	ND	1.0	3.1	2.0	2.0	91.9	4.9	13.0	1.5	8.5
163 Northeastern University	MA	3.0	5.0	4.0	0.0	88.0	16.5	12.0	2.0	-.-(A)
164 Northern Arizona University	AZ	0.0	4.1	2.7	1.4	91.8	24.3	9.5	0.0	41.9
165 Northern Illinois University	IL	2.9	7.1	1.4	0.0	88.6	8.9	12.9	-.-(A)	27.1
166 Northwestern University	IL	8.2	16.7	1.1	0.0	74.0	8.9	23.4	-.-(A)	-.-(A)
167 Norwich University	VT	1.0	3.2	5.0	0.0	90.8	4.2	7.0	0.0	1.0
168 University of Notre Dame	IN	0.5	3.0	3.0	0.0	93.5	6.4	18.8	0.0	0.0
169 Oakland University	MI	-.-	-.-	-.-	-.-	100.0	-.-	-.-	-.-	-.-
170 Ohio University	OH	2.1	0.0	1.0	0.0	96.9	11.8	14.4	0.0	0.0
171 Ohio Northern University	OH	0.0	1.4	0.0	0.0	98.6	1.4	13.0	0.0	0.0
172 Ohio State University	OH	2.0	3.6	1.7	0.3	92.4	7.5	10.1	0.0(A)	9.0
173 University of Oklahoma	OK	-.-	-.-(A)	-.-	-.-	100.0	-.-(A)	14.5	-.-(A)	-.-(A)
174 Oklahoma Christian University of Science and Arts	OK	0.0	0.0	0.0	0.0	100.0	9.1	0.0	0.0	0.0
175 Oklahoma State University	OK	2.6	2.1	2.1	2.1	91.1	16.1	14.6	-.-(A)	-.-(A)
176 Old Dominion University	VA	6.4	7.6	0.0	-.-	86.0	5.8	21.1	-.-	-.-(A)
177 Oregon State University	OR	-.-(A)	-.-(A)	-.-(A)	-.-(A)	100.0	-.-(A)	12.0	-.-(A)	-.-(A)
178 University of the Pacific	CA	2.4	12.2	4.9	0.0	80.5	43.9	11.0	0.0	0.0
179 University of Pennsylvania	PA	4.8	14.7	3.2	0.0	77.3	13.1	24.3	-.-(A)	2.0
180 Pennsylvania State University	PA	1.3	3.3	0.7	0.0	94.7	1.7	11.8	-.-	8.0
181 University of Pittsburgh	PA	3.7	3.3	0.7	0.0	92.3	0.7	17.9	0.3	10.6
182 Polytechnic University	NY	5.0	38.0	6.0	0.0	51.0	8.0	12.0	0.0	16.0
183 University of Portland	OR	-.-	14.0	-.-	-.-	86.0	25.0	12.0	0.0	-.-(A)
184 Portland State University	OR	1.1	18.5	1.6	0.0	78.8	9.5	14.8	-.-(A)	-.-(A)
185 Prairie View A&M University	TX	87.0	1.0	1.0	0.0	11.0	11.0	30.0	0.0(A)	0.0(A)
186 Princeton University	NJ	5.4	18.9	3.4	0.0	72.3	6.8	18.2	-.-	-.-
187 University of Puerto Rico, Mayagüez Campus	PR	-.-(A)	-.-(A)	97.0	-.-(A)	3.0	3.0	30.0	-.-(A)	-.-(A)
188 Purdue University	IN	3.0	6.4	1.6	0.1	88.9	3.4	19.6	-.-(A)	-.-(A)
189 Rensselaer Polytechnic Institute	NY	3.4	10.0	3.7	-.-	82.9	5.7	18.8	0.0	0.0
190 University of Rhode Island	RI	1.5	6.9	1.5	0.0	90.1	-.-(1)	17.7	-.-(A)	-.-(A)
191 William Marsh Rice University	TX	2.5	8.1	6.2	0.0	83.2	6.2	23.0	-.-(A)	-.-(A)
192 University of Rochester	NY	3.0	6.0	1.0	0.0	90.0	6.0	25.0	0.0	1.0
193 Rochester Institute of Technology	NY	1.4	3.7	-.-	-.-	94.9	5.6	11.2	-.-(A)	9.6
194 Rose-Hulman Institute of Technology	IN	0.0	2.0	-.-	0.0	98.0	3.0	0.0	-.-	0.0
195 Rutgers, The State University	NJ	4.5	16.5	2.7	-.-	76.3	5.6	17.8	-.-	5.9
196 Saginaw Valley State University	MI	2.4	2.4	-.-	-.-	95.2	2.4	4.9	-.-	-.-
197 St. Cloud State University	MN	0.0	0.0	0.0	0.0	100.0	3.0	3.0	-.-(A)	-.-(A)
198 Parks College of Saint Louis University	IL	0.0	0.0	0.0	0.0	100.0	14.0	14.0	0.0	0.0
199 University of San Diego	CA	14.3	14.3	0.0	0.0	71.4	0.0	28.6	0.0	-.-(A)
200 San Diego State University	CA	1.2	28.9	8.1	1.2	60.6	5.6	15.8	-.-(A)	-.-(A)
201 San Francisco State University	CA	7.0	11.0	11.0	0.0	71.0	21.0(2)	16.0	-.-(A)	-.-(A)
202 San Jose State University	CA	2.0	46.0	6.0	1.0	45.0	-.-(A)	-.-(A)	-.-(A)	73.0
203 Santa Clara University	CA	1.1	26.1	8.1	0.1	64.6	8.1	28.0	-.-(A)	-.-(A)
204 Seattle University	WA	-.-	20.0	-.-	-.-	80.0	11.0	13.0	-.-(A)	52.0
205 University of South Alabama	AL	3.6	4.8	1.2	1.2	89.2	17.9	11.9	-.-(A)	-.-(A)
206 University of South Carolina	SC	-.-(C)	-.-(C)	-.-(C)	-.-	100.0	-.-(C)	14.7	-.-(A)	-.-(A)
207 South Dakota School of Mines and Technology	SD	1.0	1.0	1.0	2.0	95.0	14.0	31.0	-.-(A)	28.0
208 South Dakota State University	SD	0.0	0.0	0.0	-.-	100.0	12.0	12.0	-.-(A)	-.-(A)
209 University of South Florida	FL	2.9	10.0	6.7	-.-	80.4	1.7	19.9	-.-	50.0
210 Southern University and A&M College	LA	87.6	0.0	0.0	0.0	12.4	12.4	25.8	0.0	0.0

Degrees Awarded—by Category

ENGINEERING

Institution Profile Reference Number & Name	State/Province	PERCENTAGE OF UNDERGRADUATE DEGREES AWARDED-BY CATEGORY									
		Ethnicity/Nationality						Other Categories		
		African Americans	Asian Americans	Hispanic Amer.	Native Americans	Foreign Citizens	All Others	Women	Students with Disabilities	Over 25 Years of Age
211 University of Southern California	CA	-.-	-.-	-.-	-.-	100.0	-.-	-.-	-.-	-.-
212 University of Southern Colorado	CO	-.-	-.-	-.-	-.-	100.0	14.3	28.6	-.-	14.3
213 Southern Illinois University at Carbondale	IL	2.8	5.1	0.5	0.0	91.6	19.4	6.8	0.5	38.0
214 Southern Illinois University at Edwardsville	IL	7.2	3.6	0.9	1.8	86.5	3.6	18.0	-.-(A)	-.-(A)
215 University of Southern Maine	ME	0.0	0.0	0.0	0.0	100.0	5.0	5.0	0.0(0)	60.0
216 Southern Methodist University	TX	1.8	7.0	3.5	0.0	87.7	8.8	24.6	-.-(A)	-.-(A)
217 Stanford University	CA	9.0	19.9	9.6	1.4	60.1	5.7	21.3	-.-(A)	-.-(A)
218 SUNY at Binghamton	NY	2.0	2.0	1.0	0.0	95.0	4.0	11.0	0.0(A)	0.0(A)
219 SUNY at Buffalo	NY	1.8	6.8	1.0	0.0	90.4	6.0	10.8	-.-	0.0
220 SUNY, College of Environmental Science and Forestry	NY	0.0	2.0	0.0	0.0	98.0	0.0	10.0	0.0	-.-(A)
221 SUNY Maritime College	NY	1.5	3.0	1.5	0.0	94.0	1.5	6.0	0.0	0.0
222 SUNY, College at New Paltz	NY	8.0	4.0	4.0	0.0	84.0	29.0	12.5	0.0	17.0
223 SUNY at Stony Brook	NY	3.8	21.5	4.6	0.0	70.1	6.9	10.0	0.0	-.-(A)
224 Swarthmore College	PA	3.6	0.0	0.0	0.0	96.4	3.6	3.6	0.0	0.0
225 Syracuse University	NY	2.8	6.6	1.1	0.6	88.9	8.8	13.2	-.-(A)	-.-(A)
226 Technical University of Nova Scotia	NS	-.-	-.-	-.-	-.-	100.0	-.-	-.-	-.-	-.-
227 Temple University	PA	10.9	13.4	0.0	0.0	75.7	6.1	13.4	-.-(A)	-.-(A)
228 University of Tennessee, Chattanooga	TN	11.4	2.1	0.5	0.5	85.5	-.-	17.2	-.-	-.-
229 University of Tennessee, Knoxville	TN	5.5	3.2	-.-	-.-	91.3	9.0	15.9	-.-(A)	-.-(A)
230 Tennessee State University	TN	47.2	0.0	0.0	0.0	52.8	12.5	25.0	0.0	20.8
231 Tennessee Technological University	TN	2.0	2.0	1.0	0.0	95.0	0.0	11.6	0.0	0.0
232 University of Texas at Arlington	TX	2.4	16.0	3.3	-.-(A)	78.3	24.5	12.0	-.-(A)	-.-(2)
233 University of Texas at Austin	TX	4.6	14.9	14.3	0.1	66.1	10.0	16.6	-.-(A)	18.7
234 University of Texas at Dallas	TX	0.0	0.0(A)	0.0	0.0	100.0	33.0	-.-	0.0(A)	0.0(A)
235 University of Texas at El Paso	TX	1.9	2.6	69.2	-.-	26.3	2.5	21.7	-.-	-.-
236 University of Texas at San Antonio	TX	0.0	9.8	34.1	0.0	56.1	12.2	9.8	-.-	-.-
237 Texas A&I University	TX	1.0	1.0	28.8	0.0	69.2	35.2	5.6	-.-(A)	-.-(A)
238 Texas A&M University	TX	1.4	7.0	6.5	-.-	85.1	4.7	19.0	-.-	-.-(A)
239 Texas Tech University	TX	-.-	5.0	4.7	-.-	90.3	28.0	9.0	3.5	35.0
240 University of Toledo	OH	1.0	1.0	1.0	0.0	97.0	21.0	13.0	-.-(A)	-.-(A)
241 Tri-State University	IN	1.0	1.0	1.0	0.0	97.0	24.0	9.0	1.0	20.0
242 Trinity University	TX	5.0	0.0	10.0	5.0	80.0	5.0	35.0	0.0	5.0
243 Tufts University	MA	1.7	7.4	1.7	-.-	89.2	7.4	19.4	-.-(A)	-.-(A)
244 Tulane University	LA	8.6	10.0	6.4	0.0	75.0	12.1	24.3	0.0	0.0
245 University of Tulsa	OK	1.1	2.3	2.3	2.3	92.0	34.1	18.8	-.-(A)	-.-(A)
246 Tuskegee University	AL	92.3	0.0	0.0	0.0	7.7	7.7	34.4	0.0	1.0
247 Union College	NY	2.2	3.4	2.2	0.0	92.2	1.1	14.9	-.-(A)	-.-(A)
248 United States Air Force Academy	CO	9.4(2)	4.0(2)	3.1(2)	0.0(2)	83.5	0.6(2)	5.1(2)	0.0(2)	0.9(2)
249 United States Coast Guard Academy	CT	0.0	5.1	1.3	2.5	91.1	2.5	7.6	0.0	0.0
250 United States Merchant Marine Academy	NY	1.1	5.6	0.0	1.1	92.2	0.0	4.5	0.0	1.1
251 United States Military Academy	NY	7.1	5.0	13.9	-.-	74.0	1.3	7.3	0.0	0.0
252 United States Naval Academy	MD	2.7	5.9	4.5	0.5	86.4	2.1	3.7	0.0	0.0
253 University of Utah	UT	0.0	4.3	2.6	0.0	93.1	13.2	11.5	0.0(A)	0.0(A)
254 Utah State University	UT	-.-	4.0	-.-	-.-	96.0	6.0	6.0	0.0(A)	0.0(A)
255 Valparaiso University	IN	0.0	0.0	1.0	0.0	99.0	14.0	22.0	0.0(A)	0.0(A)
256 Vanderbilt University	TN	3.0	4.0	0.0	0.0	93.0	5.0	21.0	0.0	-.-
257 University of Vermont	VT	0.7	3.3	0.0	0.0	96.0	0.7	8.6	-.-(A)	-.-(A)
258 Villanova University	PA	-.-	-.-	-.-	-.-	100.0	-.-	18.2	-.-	-.-
259 University of Virginia	VA	5.6	12.7	1.2	0.0	80.5	2.8	24.5	-.-(A)	1.9
260 Virginia Military Institute	VA	0.0	3.0	1.0	0.0	96.0	2.0	0.0	-.-	7.0
261 Virginia Polytechnic Institute and State University	VA	2.0	8.0	1.0	1.0	88.0	3.0	19.0	1.0	0.0(A)
262 University of Washington	WA	2.0	23.0	2.0	1.0	72.0	3.0	19.0	-.-(A)	-.-(A)
263 Washington University	MO	4.4	8.5	1.2	0.0	85.9	12.5	19.3	-.-	-.-
264 Washington State University	WA	-.-(1)	-.-	-.-(3)	-.-(2)	100.0	-.-(4)	12.0	-.-(A)	-.-(A)
265 Wayne State University	MI	6.1	3.0	1.8	0.0	89.1	20.1	18.3	-.-(A)	-.-(A)

Degrees Awarded—by Category

ENGINEERING

PERCENTAGE OF UNDERGRADUATE DEGREES AWARDED–BY CATEGORY

Institution Profile Reference Number & Name	State/Province	Ethnicity/Nationality						Other Categories		
		African Americans	Asian Americans	Hispanic Amer.	Native Americans	Foreign Citizens	All Others	Women	Students with Disabilities	Over 25 Years of Age
266 Webb Institute of Naval Architecture	NY	0.0	0.0	0.0	0.0	100.0	0.0	25.0	0.0	0.0
267 West Virginia University	WV	1.5	1.8	0.4	0.0	96.3	2.2	10.7	-.-(A)	-.-(A)
268 West Virginia Institute of Technology	WV	-.-	-.-	-.-	-.-	100.0	5.0	1.3	6.0	18.9
269 Western Michigan University	MI	0.0	1.9	-.-	0.0	98.1	18.6	3.8	-.-(A)	-.-(A)
270 Western New England College	MA	-.-	1.3	2.5	-.-	96.2	6.5	15.6	-.-	31.2
271 Wichita State University	KS	0.7	19.0(2)	2.6	1.3	76.4	28.2	10.0	0.0	51.3
272 Widener University	PA	2.5	12.0	1.7	0.0	83.8	17.4	12.0	0.0	15.0
273 Wilkes University	PA	1.0(C)	5.0(C)	1.0(C)	-.-(A)	93.0	3.0(C)	12.0(C)	-.-(C)	8.0(C)
274 University of Wisconsin-Madison	WI	0.5	4.9	0.5	-.-	94.1	14.1	12.3	-.-(A)	-.-(A)
275 University of Wisconsin-Milwaukee	WI	1.8	4.7	2.9	0.0	90.6	2.3	15.8	5.9	35.1
276 University of Wisconsin-Platteville	WI	1.0	3.0	-.-	-.-	96.0	2.0	10.0	0.0	25.0
277 Worcester Polytechnic Institute	MA	0.7	5.2	1.0	0.0	93.1	7.3	21.0	0.2	-.-(A)
278 Wright State University	OH	1.5	6.3	0.0	0.0	92.2	11.7	10.7	-.-(A)	-.-(A)
279 University of Wyoming	WY	0.0	0.0	0.7	0.7	98.6	7.8	10.5	-.-(A)	-.-(A)
280 Yale University	CT	9.0	11.3	4.5	0.0	75.2	11.3	13.6	-.-(2)	0.0

FOOTNOTES: The following footnotes are the same for all schools: (A) Data not available. (B) Data not applicable. (C) Estimated data.

017 Brigham Young University: (1) Average student age at graduation was 25.1

018 Bucknell University: (1) This information is given in whole numbers not percentages.

022 University of California, Los Angeles: (2) Foreign Nationals - students on student (F-1) or exchange (J-1) visas.

031 California State University, Los Angeles: (2) Includes Pacific Islanders.

046 Clarkson University: (1) Figured on Engineering Students only

052 Colorado School of Mines: (1) One graduate in this category.

066 École Polytechnique de Montréal: (5) Figure valid in Fall 1991.

072 Florida Agricultural and Mechanical University: (1) Numbers include both Florida Agricultural and Mechanical University and Florida State University students.

076 Florida State University: (2) Numbers include both FAMU and FSU students. The academic year is Fall 1991, Spring 1992 and Summer 1992.

091 Howard University: (1) There were two (2) Asian Americans.

111 University of Louisville: (1) Computer Science is located in the Engineering College.

151 New Jersey Institute of Technology: (1) International students - not foreign born. NJIT maintains data on a citizen/permanent resident versus non-citizen basis.

154 New Mexico State University: (3) Foreign nationals

190 University of Rhode Island: (1) Based upon overall University baccalaureate graduates

201 San Francisco State University: (2) Foreign students; does not include foreign born legal residents of the U.S.

208 South Dakota State University: (1) Less than 1% were American Indian

232 University of Texas at Arlington: (2) Data entered is for the University as a whole.

248 United States Air Force Academy: (2) Data based on Engineering Majors.

264 Washington State University: (1) Actual % = .006 (2) Actual % = .003 (3) Actual % = .02 (4) Actual % = .04

271 Wichita State University: (2) Permanent residents and citizens.

280 Yale University: (2) Information about disabilities is not available.

Support Programs for Ethnic Minorities & Women

Institutions are listed in alphabetical order. The first items in a school's entry are the institution profile reference number, the school name, and the state or province in which the main campus is located. Next, if a university offers a student chapter organization of the Society of Women Engineers (SWE), the National Society of Black Engineers (NSBE), the Society of Hispanic Professional Engineers (SHPE) or any other professional society for ethnic minorities or women, there will be a check mark (√) in the appropriate column(s). Similarly, if the college or university staffs its own programs for ethnic minorities and/or women, there will be a check mark (√) in the appropriate column(s). The last two columns indicate by capital letter abbreviation during which time periods (**S**ummer, the nine-month **A**cademic Year, or **Y**ear-round) and for students at which levels (**P**rematriculation, **F**reshman, **S**ophomore, **J**unior, senio**R**, and **G**raduate) at least one program is available. For more information about these programs, refer to the *Student Programs* section of the *Institution Profiles*.

ENGINEERING		STUDENT CHAPTER ORGANIZATIONS				COLLEGE/UNIVERSITY-STAFFED PROGRAMS			
								Available:	Level Offered:
Institution Profile Reference Number & Name	State/Province	Society of Women Engineers (SWE)	National Society of Black Engineers (NSBE)	Society of Hispanic Prof. Engineers (SHPE)	Other(s)—See Institution Profiles	For Ethnic Minorities	For Women	Summer / Academic year / Year-round	Prematriculation / Freshman / Sophomore / Junior / senioR / Graduate
001 U of Akron	OH	–	√	–	–	√	–	Y	PFSJRG
002 U of Alabama	AL	√	√	–	–	√	–	Y	PFSJRG
003 U of Alabama at Birmingham	AL	√	√	–	–	√	–	Y	FSJR
004 U of Alabama at Huntsville	AL	√	√	–	–	–	–	–	–
005 U of Alaska, Fairbanks	AK	–	–	–	√	√	√	AY	PFSJRG
006 U of Alberta	AB	–	–	–	–	–	–	–	–
007 Alfred U	NY	√	–	–	√	√	√	Y	PFS
008 U of Arizona	AZ	√	–	√	√	√	√	Y	PFSJR
009 Arizona State U	AZ	√	–	√	√	√	–	A	PFS
010 U of Arkansas	AR	√	√	–	–	√	–	–	PF
011 Arkansas Tech U	AR	–	–	–	–	√	–	Y	FSJRG
012 Auburn U	AL	√	√	–	–	√	√	SAY	PFSJR
013 Baylor U	TX	–	–	–	–	–	–	–	–
014 Boston U	MA	√	√	√	–	√	–	AY	FSJRG
015 Bradley U	IL	–	√	–	√	√	–	Y	PFSJRG
016 U of Bridgeport	CT	√	–	–	–	√	–	Y	PFSJR
017 Brigham Young U	UT	√	–	–	–	–	√	Y	FSJRG
018 Bucknell U	PA	√	√	–	–	√	√	Y	PFSJRG
019 U of California at Berkeley	CA	√	–	–	√	√	√	Y	PFSJRG
020 U of California, Davis	CA	√	√	√	√	√	√	SY	PFSJRG
021 U of California, Irvine	CA	√	√	–	√	√	√	SAY	PFSJR
022 U of California, Los Angeles	CA	√	√	–	√	√	–	Y	PFSJRG
023 U of California, San Diego	CA	√	√	√	√	√	–	AY	PFSJRG
024 U of California, Santa Barbara	CA	√	√	–	–	√	–	Y	PFSJRG
025 U of California, Santa Cruz	CA	–	–	–	–	–	–	–	–
026 California Institute of Technology	CA	√	√	√	–	√	–	S	P
027 California Polytechnic State U	CA	√	√	√	√	√	√	Y	PFSJRG
028 California State U, Chico	CA	√	√	√	√	√	–	Y	PFSJR
029 California State U, Fresno	CA	√	√	√	–	√	–	Y	PFSJRG
030 California State U, Long Beach	CA	√	√	√	–	√	–	Y	PFSJRG
031 California State U, Los Angeles	CA	√	√	√	√	√	–	Y	PFSJR
032 California State U, Northridge	CA	√	√	√	–	√	–	Y	PFSJR
033 California State U, Sacramento	CA	√	√	√	–	√	√	AY	PFSJRG
034 California State Polytechnic U, Pomona	CA	√	√	√	–	√	√	SY	PFSJRG
035 Calvin College	MI	–	–	–	–	√	–	SA	PFSJR
036 Carnegie Mellon U	PA	√	√	√	–	√	√	SY	PFSJRG
037 Case Western Reserve U	OH	√	√	√	–	√	√	AY	PFSJR
038 The Catholic U of America	DC	√	–	√	–	–	–	–	–
039 U of Central Florida	FL	√	√	√	–	√	√	Y	PFSJRG
040 Central State U	OH	–	√	–	√	–	–	–	–

Support Programs

ENGINEERING Institution Profile Reference Number & Name	State/Province	STUDENT CHAPTER ORGANIZATIONS					COLLEGE/UNIVERSITY-STAFFED PROGRAMS			
		Society of Women Engineers (SWE)	National Society of Black Engineers (NSBE)	Society of Hispanic Prof. Engineers (SHPE)	Other(s)—See Institution Profiles	For Ethnic Minorities	For Women	Available: Summer / Academic year / Year-round	Level Offered: Prematriculation Freshman Sophomore Junior SenioR Graduate
041 Christian Brothers U	TN	–	√	–	√	√	√	S	P
042 U of Cincinnati	OH	√	√	–	–	√	√	SY	PF
043 The Citadel	SC	–	–	–	–	–	–	–	–
044 City College of the City U of New York	NY	√	√	–	√	√	–	Y	FSJR
045 City U of New York, College of Staten Island	NY	√	–	–	–	–	–	–	–
046 Clarkson U	NY	√	√	√	√	√	√	A	PFSJRG
047 Clemson U	SC	√	√	–	–	√	–	Y	PFSJR
048 Cleveland State U	OH	√	√	–	√	√	√	A	FSJR
049 U of Colorado at Boulder	CO	√	√	√	√	√	√	Y	PFSJRG
050 U of Colorado at Colorado Springs	CO	√	–	–	–	–	√	Y	PFSJRG
051 U of Colorado at Denver	CO	√	√	√	–	√	√	Y	PFSJRG
052 Colorado School of Mines	CO	√	√	√	√	√	√	SA	PFSJRG
053 Colorado State U	CO	√	√	√	√	√	√	A	FSJR
054 Columbia U	NY	√	√	√	√	√	–	Y	PFSJR
055 Concordia U	PQ	–	–	–	–	–	√	Y	PFSJRG
056 U of Connecticut	CT	√	√	–	–	√	–	Y	PFSJR
057 The Cooper Union	NY	√	√	√	–	√	√	Y	FSJRG
058 Cornell U	NY	√	√	√	–	√	√	Y	PFSJRG
059 Dartmouth College	NH	–	–	–	√	–	–	–	–
060 U of Dayton	OH	√	√	–	–	√	√	Y	PF
061 U of Delaware	DE	√	√	–	–	√	–	SA	PFSJR
062 U of Denver	CO	√	–	√	–	√	√	A	FSJRG
063 U of the District of Columbia	DC	–	–	–	–	–	–	–	–
064 Drexel U	PA	√	–	–	√	√	√	Y	PFSJRG
065 Duke U	NC	√	√	–	–	–	–	–	–
066 École Polytechnique de Montréal	PQ	–	–	–	–	–	–	–	–
067 Embry-Riddle Aeronautical U - Daytona Beach	FL	–	–	–	–	√	√	Y	F
068 Embry-Riddle Aeronautical U, Western Campus	AZ	–	–	–	√	–	–	–	–
069 U of Evansville	IN	√	–	–	–	–	–	–	–
070 Ferris State U	MI	–	–	–	√	√	√	Y	PFSJR
071 U of Florida	FL	√	√	√	–	√	√	Y	PFSJRG
072 Florida Agricultural & Mechanical U	FL	√	√	–	–	√	–	SA	PFS
073 Florida Atlantic U	FL	√	√	–	√	√	√	Y	PFSJRG
074 Florida Institute of Technology	FL	√	–	–	–	–	–	–	–
075 Florida International U	FL	√	√	√	√	√	√	Y	PFSJRG
076 Florida State U	FL	√	√	–	–	√	–	SA	PFS
077 GMI Engineering & Management Institute	MI	√	√	√	–	√	–	–	–
078 Gannon U	PA	–	–	–	–	√	√	Y	PFSJRG
079 George Mason U	VA	√	√	√	–	√	√	Y	FSJRG
080 The George Washington U	DC	√	√	–	√	√	√	Y	PFSJRG
081 U of Georgia	GA	–	–	–	–	√	√	Y	PFSJRG
082 Georgia Institute of Technology	GA	√	√	√	√	√	√	AY	PFSJRG
083 Gonzaga U	WA	√	–	–	–	–	–	–	–
084 Grand Valley State U	MI	–	√	–	√	√	–	Y	PFSJRG
085 Grove City College	PA	√	–	–	–	–	√	A	FSJR
086 Harvard U	MA	–	–	–	√	√	–	–	–
087 Harvey Mudd College	CA	√	–	–	–	√	–	S	P
088 U of Hawaii at Manoa	HI	–	–	–	–	√	–	Y	PFSJR
089 Hofstra U	NY	–	–	–	–	√	–	Y	FSJR
090 U of Houston	TX	√	√	–	√	√	–	Y	PFSJR

Support Programs

ENGINEERING

Institution Profile Reference Number & Name	State/Province	Society of Women Engineers (SWE)	National Society of Black Engineers (NSBE)	Society of Hispanic Prof. Engineers (SHPE)	Other(s)—See Institution Profiles	For Ethnic Minorities	For Women	Summer / Academic year / Year-round	Level Offered: Prematriculation / Freshman / Sophomore / Junior / SenioR / Graduate
091 Howard U	DC	√	√	–	√	√	√	S	P
092 U of Idaho	ID	√	–	–	–	–	√	A	FSJRG
093 Idaho State U	ID	√	–	–	–	–	–	–	–
094 U of Illinois at Chicago	IL	√	√	√	–	√	–	Y	FSJR
095 U of Illinois at Urbana-Champaign	IL	√	√	√	–	√	–	Y	PFSJRG
096 Illinois Institute of Technology	IL	√	√	√	–	√	–	A	FS
097 Indiana U-Purdue U at Indianapolis	IN	–	–	–	√	√	–	Y	PFSJRG
098 U of Iowa	IA	√	–	–	√	√	√	A	FS
099 Iowa State U	IA	√	√	√	√	√	√	A	PFS
100 The Johns Hopkins U	MD	√	√	–	–	√	√	AY	PFSJRG
101 U of Kansas	KS	√	√	√	√	√	–	SY	PFSJRG
102 Kansas State U	KS	√	√	√	√	√	–	A	FSJRG
103 U of Kentucky	KY	√	√	–	–	√	–	Y	PFSJRG
104 Lafayette College	PA	√	√	–	–	√	√	A	FSJR
105 Lamar U	TX	√	√	√	–	√	√	AY	FSJRG
106 Lawrence Technological U	MI	√	√	–	–	√	–	Y	P
107 LeTourneau U	TX	–	–	–	–	√	√	Y	PFSJR
108 Lehigh U	PA	√	√	–	–	√	√	AY	PFSJRG
109 Louisiana State U	LA	√	√	–	–	√	–	Y	PFSJR
110 Louisiana Tech U	LA	–	√	–	–	√	–	Y	PFSJRG
111 U of Louisville	KY	√	√	–	–	√	√	SA	PF
112 Loyola College in Maryland	MD	–	–	–	–	–	–	–	–
113 Loyola Marymount U	CA	√	√	√	–	–	–	–	–
114 U of Maine	ME	√	–	–	–	–	–	–	–
115 U of Manitoba	MB	–	–	–	√	√	√	AY	PFSJRG
116 Mankato State U	MN	√	–	–	–	√	√	Y	FSJRG
117 Marietta College	OH	–	–	–	–	–	–	–	–
118 Marquette U	WI	√	√	–	–	√	–	Y	FSJR
119 U of Maryland Baltimore County	MD	√	√	–	–	√	√	A	FSJRG
120 U of Maryland, College Park	MD	√	√	√	–	√	–	SY	PFSJR
121 U of Massachusetts	MA	√	√	√	–	√	√	Y	PFSJRG
122 U of Massachusetts, Dartmouth	MA	√	–	–	–	√	√	A	FSJR
123 U of Massachusetts, Lowell	MA	√	√	–	–	√	√	A	FSJRG
124 Massachusetts Institute of Technology	MA	√	√	√	√	√	√	SA	PFSJR
125 McGill U	PQ	–	–	–	√	–	–	–	–
126 Memphis State U	TN	√	√	–	–	√	–	S	P
127 Mercer U	GA	√	–	–	√	–	–	–	–
128 U of Miami	FL	√	√	–	–	–	–	–	–
129 Miami U	OH	–	–	–	√	√	√	AY	PFSJRG
130 U of Michigan	MI	√	√	√	√	√	√	Y	PFSJRG
131 U of Michigan-Dearborn	MI	√	√	–	–	√	–	Y	PF
132 Michigan State U	MI	√	√	√	√	√	√	Y	FSJRG
133 Michigan Technological U	MI	√	√	√	√	√	√	SY	PFSJR
134 Milwaukee School of Engineering	WI	√	√	√	√	√	√	SA	PF
135 U of Minnesota, Duluth	MN	–	–	–	√	√	√	AY	PFSJRG
136 U of Minnesota	MN	√	√	–	√	√	–	Y	PFSJRG
137 U of Mississippi	MS	√	√	–	–	√	–	A	FSJRG
138 Mississippi State U	MS	√	√	–	–	√	–	SY	PFSJRG
139 U of Missouri-Columbia	MO	√	√	–	√	√	√	AY	FSJRG
140 U of Missouri-Kansas City	MO	√	√	–	–	√	–	Y	PFSJRG

Support Programs

ENGINEERING Institution Profile Reference Number & Name	State/Province	STUDENT CHAPTER ORGANIZATIONS				COLLEGE/UNIVERSITY-STAFFED PROGRAMS			
		Society of Women Engineers (SWE)	National Society of Black Engineers (NSBE)	Society of Hispanic Prof. Engineers (SHPE)	Other(s)—See Institution Profiles	For Ethnic Minorities	For Women	Available: Summer/Academic year/Year-round	Level Offered: Prematriculation/Freshman/Sophomore/Junior/SenioR/Graduate
141 U of Missouri-Rolla	MO	√	√	√	–	√	√	Y	PFSJRG
142 Monmouth College	NJ	–	–	–	√	√	√	AY	PFSJRG
143 Montana College of Mineral Science & Technology	MT	√	–	–	–	–	–	–	–
144 Montana State U	MT	√	–	–	√	√	√	Y	PFSJRG
145 Morgan State U	MD	√	√	–	√	√	–	SA	PFSJR
146 U of Nebraska - Lincoln	NE	√	√	–	–	–	–	–	–
147 U of Nevada, Las Vegas	NV	√	√	√	√	√	√	Y	PFSJRG
148 New England College	NH	–	–	–	–	–	–	–	–
149 U of New Hampshire	NH	√	–	–	–	√	–	Y	FSJRG
150 U of New Haven	CT	–	–	–	√	–	√	AY	PFSJRG
151 New Jersey Institute of Technology	NJ	√	√	–	√	√	√	AY	PFSJRG
152 The U of New Mexico	NM	√	–	–	√	√	√	Y	FSJR
153 New Mexico Institute of Mining & Technology	NM	√	–	√	√	√	√	Y	FSJR
154 New Mexico State U	NM	√	√	√	√	√	√	Y	PFSJRG
155 U of New Orleans	LA	√	√	√	–	–	–	–	–
156 New York Institute of Technology	NY	√	√	√	–	–	–	–	–
157 New York State College of Ceramics at Alfred U	NY	√	–	–	√	√	–	Y	PFSJR
158 U of North Carolina-Charlotte	NC	√	–	–	–	√	–	SY	PFSJR
159 North Carolina Agricultural & Technical State U	NC	√	–	–	–	√	–	Y	PFSJRG
160 North Carolina State U	NC	√	√	–	–	√	√	SAY	PFS
161 U of North Dakota	ND	√	–	–	–	–	√	Y	PFSJRG
162 North Dakota State U	ND	√	–	–	√	√	–	Y	S
163 Northeastern U	MA	√	√	–	√	√	√	Y	PFSJRG
164 Northern Arizona U	AZ	√	–	√	√	√	–	AY	PFSJRG
165 Northern Illinois U	IL	√	√	√	–	√	–	Y	FSJR
166 Northwestern U	IL	√	√	√	–	√	–	Y	PFSJR
167 Norwich U	VT	–	–	–	√	–	–	–	–
168 U of Notre Dame	IN	√	√	–	–	√	–	Y	FSJR
169 Oakland U	MI	√	√	–	–	√	–	SA	PF
170 Ohio U	OH	√	√	–	√	√	–	S	P
171 Ohio Northern U	OH	√	–	–	–	√	–	Y	PFSJR
172 Ohio State U	OH	√	√	–	–	√	√	Y	PFSJRG
173 U of Oklahoma	OK	√	√	–	–	√	–	Y	PFSJRG
174 Oklahoma Christian U of Science & Arts	OK	–	–	–	–	–	–	–	–
175 Oklahoma State U	OK	√	√	√	√	√	–	SAY	PFSJR
176 Old Dominion U	VA	√	–	–	√	√	√	AY	FSJRG
177 Oregon State U	OR	√	–	–	–	√	–	Y	PFSJRG
178 U of the Pacific	CA	√	√	–	–	√	√	Y	PFSJR
179 U of Pennsylvania	PA	√	√	√	–	√	–	Y	PFSJRG
180 Pennsylvania State U	PA	√	√	√	–	√	√	Y	PFSJRG
181 U of Pittsburgh	PA	√	√	–	–	√	–	Y	PFSJR
182 Polytechnic U	NY	√	√	√	√	–	–	–	–
183 U of Portland	OR	√	–	–	–	√	√	AY	FSJRG
184 Portland State U	OR	√	–	–	√	√	√	Y	P
185 Prairie View A&M U	TX	√	√	√	√	√	–	S	P
186 Princeton U	NJ	√	√	√	–	√	–	Y	SJR
187 U of Puerto Rico, Mayagüez Campus	PR	√	–	–	–	–	–	–	–
188 Purdue U	IN	√	√	√	–	√	√	A	F
189 Rensselaer Polytechnic Institute	NY	√	–	–	–	√	√	Y	PFSJRG
190 U of Rhode Island	RI	√	–	–	–	√	√	A	FSJR

Support Programs

ENGINEERING

Institution Profile Reference Number & Name	State/Province	Society of Women Engineers (SWE)	National Society of Black Engineers (NSBE)	Society of Hispanic Prof. Engineers (SHPE)	Other(s)—See Institution Profiles	For Ethnic Minorities	For Women	Summer / Academic year / Year-round	Prematriculation Freshman Sophomore Junior SenioR Graduate
191 William Marsh Rice U	TX	√	√	–	√	√	√	Y	PFSJRG
192 U of Rochester	NY	√	√	√	–	√	–	SA	PFSJR
193 Rochester Institute of Technology	NY	√	√	√	–	–	–	–	–
194 Rose-Hulman Institute of Technology	IN	–	√	–	√	√	–	Y	FSJRG
195 Rutgers, The State U	NJ	√	√	√	√	√	√	SA	PFSJR
196 Saginaw Valley State U	MI	–	–	–	–	√	√	SAY	PFSJR
197 St. Cloud State U	MN	√	–	–	–	–	–	–	–
198 Parks College of Saint Louis U	IL	–	–	–	√	–	–	–	–
199 U of San Diego	CA	√	–	–	–	√	√	Y	FSJR
200 San Diego State U	CA	√	√	√	√	√	–	Y	PFSJRG
201 San Francisco State U	CA	√	√	√	–	√	√	AY	PFSJR
202 San Jose State U	CA	√	√	√	–	√	√	Y	PFSJRG
203 Santa Clara U	CA	√	√	–	–	√	–	Y	PFSJRG
204 Seattle U	WA	√	–	–	–	–	–	–	–
205 U of South Alabama	AL	√	√	–	–	–	–	–	–
206 U of South Carolina	SC	√	√	–	–	√	–	Y	PFSJR
207 South Dakota School of Mines & Technology	SD	√	–	–	√	√	–	Y	PFSJR
208 South Dakota State U	SD	√	–	–	–	–	–	–	–
209 U of South Florida	FL	√	–	√	–	√	–	Y	FSJR
210 Southern U & A&M College	LA	√	√	–	√	√	√	AY	PFSJR
211 U of Southern California	CA	√	√	√	–	√	–	Y	PFSJRG
212 U of Southern Colorado	CO	√	–	√	–	–	√	A	FSJRG
213 Southern Illinois U at Carbondale	IL	√	√	√	√	√	√	SY	PFSJR
214 Southern Illinois U at Edwardsville	IL	√	√	–	–	√	√	Y	PFSJRG
215 U of Southern Maine	ME	–	–	–	–	–	–	–	–
216 Southern Methodist U	TX	√	√	√	–	√	–	Y	PFSJR
217 Stanford U	CA	√	–	–	√	√	√	AY	PFSJRG
218 State U of New York at Binghamton	NY	√	√	–	√	√	√	A	FSJR
219 State U of New York at Buffalo	NY	√	√	√	–	√	–	AY	FSJR
220 State U of New York, College of Environmental Science & Forestry	NY	–	√	–	√	√	–	A	FSJRG
221 State U of New York Maritime College	NY	√	–	–	–	√	√	Y	PFSJR
222 State U of New York, College at New Paltz	NY	–	√	–	–	–	–	Y	PFSJR
223 State U of New York at Stony Brook	NY	√	–	–	–	√	–	A	PF
224 Swarthmore College	PA	√	–	–	–	√	–	A	PFSJRG
225 Syracuse U	NY	√	–	–	–	√	–	Y	FSJR
226 Technical U of Nova Scotia	NS	–	–	–	–	–	√	–	SJRG
227 Temple U	PA	√	–	–	√	–	–	–	–
228 U of Tennessee, Chattanooga	TN	√	√	–	–	–	–	–	–
229 U of Tennessee, Knoxville	TN	√	√	–	–	√	–	Y	PFSJR
230 Tennessee State U	TN	√	√	–	√	√	√	SAY	PF
231 Tennessee Technological U	TN	√	√	–	–	√	–	Y	PFSJRG
232 U of Texas at Arlington	TX	√	√	√	–	–	–	–	–
233 U of Texas at Austin	TX	√	√	√	√	√	√	Y	PFSJRG
234 U of Texas at Dallas	TX	–	–	–	–	–	–	–	–
235 U of Texas at El Paso	TX	√	–	√	√	√	–	S	P
236 U of Texas at San Antonio	TX	√	–	√	–	–	–	–	–
237 Texas A&I U	TX	√	–	√	–	√	√	Y	PFSJRG
238 Texas A&M U	TX	√	√	–	√	√	√	AY	PFSJRG
239 Texas Tech U	TX	√	√	√	–	√	√	AY	PFSJRG
240 U of Toledo	OH	√	√	–	√	–	–	–	–

Support Programs

ENGINEERING

Institution Profile Reference Number & Name	State/Province	Society of Women Engineers (SWE)	National Society of Black Engineers (NSBE)	Society of Hispanic Prof. Engineers (SHPE)	Other(s)—See Institution Profiles	For Ethnic Minorities	For Women	Available: Summer / Academic year / Year-round	Level Offered: Prematriculation / Freshman / Sophomore / Junior / SenioR / Graduate
241 Tri-State U	IN	√	√	–	√	√	–	Y	PFSJR
242 Trinity U	TX	√	–	–	–	–	√	A	FSJR
243 Tufts U	MA	√	√	–	√	√	√	AY	FSJRG
244 Tulane U	LA	–	√	–	√	√	√	Y	FSJRG
245 U of Tulsa	OK	√	–	–	–	–	–	–	–
246 Tuskegee U	AL	√	√	–	–	√	√	AY	FSJRG
247 Union College	NY	√	√	–	√	√	–	Y	PFSJR
248 United States Air Force Academy	CO	–	–	–	√	√	√	Y	PFSJR
249 United States Coast Guard Academy	CT	√	–	–	–	√	√	S	P
250 United States Merchant Marine Academy	NY	√	–	–	–	–	–	–	–
251 United States Military Academy	NY	√	–	–	–	–	–	–	–
252 United States Naval Academy	MD	√	–	–	–	√	√	Y	FSJR
253 U of Utah	UT	√	√	√	√	√	√	Y	PFSJRG
254 Utah State U	UT	√	–	–	–	√	√	Y	PFSJRG
255 Valparaiso U	IN	√	–	–	–	√	–	Y	FSJRG
256 Vanderbilt U	TN	√	√	–	–	√	–	S	P
257 U of Vermont	VT	√	–	–	–	√	–	SY	PFSJRG
258 Villanova U	PA	√	–	–	√	–	–	–	–
259 U of Virginia	VA	√	√	√	√	√	–	Y	PFSJRG
260 Virginia Military Institute	VA	–	–	–	–	√	–	S	P
261 Virginia Polytechnic Institute & State U	VA	√	√	–	–	√	–	SA	PFSJR
262 U of Washington	WA	√	√	√	√	√	√	AY	PFSJRG
263 Washington U	MO	√	√	–	–	–	–	–	–
264 Washington State U	WA	√	√	√	√	√	√	AY	PFSJRG
265 Wayne State U	MI	√	–	–	√	√	√	Y	PFSJRG
266 Webb Institute of Naval Architecture	NY	–	–	–	–	–	–	–	–
267 West Virginia U	WV	√	√	–	–	√	–	Y	FSJRG
268 West Virginia Institute of Technology	WV	–	–	–	–	–	–	–	–
269 Western Michigan U	MI	√	√	–	√	√	√	Y	PFSJRG
270 Western New England College	MA	√	–	–	–	–	–	–	–
271 Wichita State U	KS	√	–	–	√	√	–	S	P
272 Widener U	PA	√	–	–	–	–	–	–	–
273 Wilkes U	PA	√	–	–	√	√	√	AY	PFSJRG
274 U of Wisconsin-Madison	WI	√	√	√	√	√	–	Y	PFSJRG
275 U of Wisconsin-Milwaukee	WI	√	√	√	√	√	–	Y	PFSJR
276 U of Wisconsin-Platteville	WI	√	–	–	√	√	–	Y	FSJR
277 Worcester Polytechnic Institute	MA	√	–	–	–	√	√	Y	PFSJRG
278 Wright State U	OH	√	√	–	–	√	–	SY	PFSJRG
279 U of Wyoming	WY	√	–	–	–	√	√	Y	FSJRG
280 Yale U	CT	√	√	–	–	√	√	A	FSJRG
Totals:		232	170	93	111	215	127	-	-

Enrollments and Degrees Awarded

Institutions are listed in alphabetical order. The first four columns of data are the number of students enrolled in all engineering programs by class—freshman/first-year students, sophomores/second-year students, juniors/third-year students, and seniors/fourth-year and all other students. Then the total number of students enrolled in all engineering degree programs is given by degree level—bachelor's (the sum of the previous four columns), master's, and doctoral. The last three columns represent the number of engineering degrees awarded by type of degree—bachelor's, master's, and doctoral. Unless footnoted otherwise, enrollment figures are for the fall 1991 term and degrees awarded are for the 1991-92 twelve-month academic year.

ENGINEERING		ENROLLMENTS							DEGREES AWARDED		
		Undergraduate—By Class				Totals—By Level			By Degree		
Institution Profile Reference Number & Name	State/Province	Freshmen/First-year	Sophomores/Second-year	Juniors/Third-year	Seniors/All others	Bachelor's	Master's	Doctoral	Bachelor's	Master's	Doctoral
001 University of Akron	OH	595	316	305	802$^{(1)}$	2,018	320	102	221	78	18
002 University of Alabama	AL	610	346	342	482	1,780	217	94	274	74	14
003 University of Alabama at Birmingham	AL	274	75	108	155	612	115	55	105	43	2
004 University of Alabama at Huntsville	AL	304	239	335	617	1,495	480	138	191	100	11
005 University of Alaska, Fairbanks	AK	90	46	48	104	288	63	4	39	21	--
006 University of Alberta	AB	502	759	544	377	2,182	278	170	371	102	20
007 Alfred University	NY	124	77	114	96	411	181	24	96	21	6
008 University of Arizona	AZ	670	734	653	959	3,016	486	396	498	163	64
009 Arizona State University	AZ	444	459	692	1,289	2,884	1,068	383	460	271	58
010 University of Arkansas	AR	456	329	343	472	1,600	173	49	231	201	10
011 Arkansas Tech University	AR	108	60	37	41	246	--$^{(B)}$	--$^{(B)}$	15	--	--
012 Auburn University	AL	1,132	992	798	876	3,798	448	193	729	143	35
013 Baylor University	TX	100	60	40	30	230	--$^{(2)}$	--$^{(2)}$	25	--	--
014 Boston University	MA	389	285	252	366	1,292	475	71	297	209	13
015 Bradley University	IL	116	129	155	249	649	138	--$^{(B)}$	191	42	--
016 University of Bridgeport	CT	95	53	33	77	258	166	--.	87	29	--
017 Brigham Young University	UT	70	30	66	154	320	13	51	341	64	10
018 Bucknell University	PA	182	155	163	132	632	30	--$^{(B)}$	137	11	--
019 University of California at Berkeley	CA	610$^{(1)}$	477$^{(1)}$	802$^{(1)}$	841$^{(1)}$	2,730	428	1,100	566	355	186
020 University of California, Davis	CA	364	366	522	711	1,963	352	315	293	113	62
021 University of California, Irvine	CA	295	177	247	323	1,042	173	186	172	81	25
022 University of California, Los Angeles	CA	548	436	550$^{(1)}$	660	2,194	190	801	410	207	95
023 University of California, San Diego	CA	336	331	391	785	1,843	164	350	361	164	50
024 University of California, Santa Barbara	CA	319	219	310	285	1,133	144	370	242	122	44
025 University of California, Santa Cruz	CA	43	22	36	41	142	20	29	18	11	2
026 California Institute of Technology	CA	--$^{(1)}$	83	127	110	320	79	343	121	91	62
027 California Polytechnic State University	CA	744	598	699	1,999	4,040	209	--.	498	15	--
028 California State University, Chico	CA	--.	--.	--.	--.	--.	--.	--.	183	3	--
029 California State University, Fresno	CA	--$^{(A)}$	--$^{(A)}$	--$^{(A)}$	1,471$^{(2)}$	1,471	100	--$^{(A)}$	273	9	--
030 California State University, Long Beach	CA	478	307	898	1,888	3,571	1,054	14	474	97	--
031 California State University, Los Angeles	CA	191	113	190	490$^{(1)}$	984	186	--$^{(B)}$	124	33	--
032 California State University, Northridge	CA	294	220	287	762	1,563	484	--$^{(B)}$	261	151	--
033 California State University, Sacramento	CA	224	140	362	803	1,529	273	--$^{(B)}$	242	9	--
034 California State Polytechnic University, Pomona	CA	975	572	735	1,453	3,735	91	--.	484	--	--
035 Calvin College	MI	95	58	69	81	303	--$^{(B)}$	--$^{(B)}$	54	--	--
036 Carnegie Mellon University	PA	379	351	279	264	1,273	231	359	230	124	70
037 Case Western Reserve University	OH	285$^{(1)}$	244	285	307	1,121	620	243	198	190	74
038 The Catholic University of America	DC	62	53	45	68	228	314	65	53	42	6
039 University of Central Florida	FL	565$^{(1)}$	303	399	1,186$^{(2)}$	2,453	775$^{(2)}$	107	413	99	6
040 Central State University	OH	62	18	12	8	100	--.	--.	4	--	--
041 Christian Brothers University	TN	75	70	51	59$^{(1)}$	255	40	--$^{(B)}$	61	10	--
042 University of Cincinnati	OH	472$^{(1)}$	418$^{(1)}$	715$^{(1)}$	352$^{(1)}$	1,957	766$^{(2)}$	326$^{(2)}$	302	192	56
043 The Citadel	SC	146	96	50	65	357	--.	--.	37	--	--
044 City College of the City University of New York	NY	643$^{(1)}$	433	421	486	1,983	593	125	283	170	11
045 City University of New York, College of Staten Island	NY	59	22	29	56	166	--$^{(A)}$	--$^{(A)}$	28	--	--

Enrollments and Degrees Awarded

ENGINEERING

Institution Profile Reference Number & Name	State/Province	ENROLLMENTS Undergraduate—By Class				Totals—By Level			DEGREES AWARDED By Degree		
		Freshmen/First-year	Sophomores/Second-year	Juniors/Third-year	Seniors/All others	Bachelor's	Master's	Doctoral	Bachelor's	Master's	Doctoral
046 Clarkson University	NY	407	335	466	470	1,678	100	98	444	56	18
047 Clemson University	SC	926	757	814	789	3,286	505	153	524	174	34
048 Cleveland State University	OH	151	143	231	373	898	144	38	190	82	7
049 University of Colorado at Boulder	CO	615	553	491	704	2,363	540	445	375	300	50
050 University of Colorado at Colorado Springs	CO	23	21	40	105[2]	189	71	25	76	48	4
051 University of Colorado at Denver	CO	60[1]	77	166	405	708	236	11	149	49	--
052 Colorado School of Mines	CO	629	381	320	384	1,714	--[A]	--[A]	230	96	43
053 Colorado State University	CO	355	259	240	427	1,281	292	175	208	94	39
054 Columbia University	NY	266	219	241	246	972	178[1]	298[1]	235	311	59
055 Concordia University	PQ	--	--	--	--	--	--	--	199	60	11
056 University of Connecticut	CT	266	336	316	358	1,276	292	240	252	91	33
057 The Cooper Union	NY	145	140	120	139	544	67	--	105	18	--
058 Cornell University	NY	788	647	592	608	2,635	337	731	630	409	140
059 Dartmouth College	NH	70[1]	90[1]	78	66	304	52	57	58	23	5
060 University of Dayton	OH	294	216	236	221	967	607	56	193	84	5
061 University of Delaware	DE	263	255	191	268	977	139	125	186	61	46
062 University of Denver	CO	35[1]	15	21[2]	24	95	11	--	16	1	--
063 University of the District of Columbia	DC	95[1]	36[1]	36[1]	70[1]	237	--	--	36	--	--
064 Drexel University	PA	828[1]	670[2]	483	481	2,462	780	77	521	193	35
065 Duke University	NC	245	198	249	198	890	63	191	204	62	26
066 École Polytechnique de Montréal	PQ	1,526[5]	771	649	598	3,544	846[3]	283[5]	613	177	54
067 Embry-Riddle Aeronautical University - Daytona Beach	FL	275	148	129	272	824	36	--[B]	124	1	--
068 Embry-Riddle Aeronautical University, Western Campus	AZ	155	89	75	133	452	--	--	54	--	--
069 University of Evansville	IN	91	48	51	108[1]	298	--[B]	--[B]	84	--	--
070 Ferris State University	MI	6	13	15	37	71	--[A]	--[A]	10	--	--
071 University of Florida	FL	1,230	1,180	1,123	1,025	4,558	1,047	410	530	288	85
072 Florida Agricultural and Mechanical University	FL	485[1]	285[1]	448[1]	543[1]	1,761	126[1]	19[1]	181	49	--
073 Florida Atlantic University	FL	145[1]	82[1]	242[1]	507[2]	976	315[4]	99[1]	191	72	14
074 Florida Institute of Technology	FL	285	305	219	233	1,042	279	49	197	82	7
075 Florida International University	FL	64	131	551	406	1,152	173	11	183	35	--
076 Florida State University	FL	485[1]	285[1]	448[1]	543[1]	1,761	126[1]	19[1]	181	49	--
077 GMI Engineering & Management Institute	MI	535	552	498	609	2,194	49	--	333	--	--
078 Gannon University	PA	44	23	34	48	149	42	--	39	5	--
079 George Mason University	VA	286	214	266	399	1,165	898	293	163	247	13
080 The George Washington University	DC	225	125	125	128	603	1,549[2]	437	111	400	26
081 University of Georgia	GA	23	20	21	33	97	9	4	20	7	1
082 Georgia Institute of Technology	GA	1,760[2]	1,530[2]	1,608[2]	1,888[2]	6,786	1,165	908	1,207	582	129
083 Gonzaga University	WA	95	67	60	86	308	8	--	46	3	--
084 Grand Valley State University	MI	104	73	47	35	259	--[B]	--[B]	23	--	--
085 Grove City College	PA	165	85	57	46	353	--[1]	--[1]	35	--	--
086 Harvard University	MA	--[B]	49	33	39	121	6	49	18	18	10
087 Harvey Mudd College	CA	179	176	123	137	615	6	--	55	6	--
088 University of Hawaii at Manoa	HI	166	180	234	341	921	142	48	159	35	7
089 Hofstra University	NY	55	55	59	102	271	--	--	34	--	--
090 University of Houston	TX	480	399	478	681[1]	2,038	686	278	193	146	25
091 Howard University	DC	214[3]	110	111	157	592	77[2]	13	139	29	6
092 University of Idaho	ID	371	280	265	408	1,324	145	35	165	67	7
093 Idaho State University	ID	130	91	67	80	368	44	8	30	5	--
094 University of Illinois at Chicago	IL	423	389	454	770[1]	2,036	538	278	322	158	38
095 University of Illinois at Urbana-Champaign	IL	1,299	999	1,210	1,630	5,138	872	1,344	1,122	473	249
096 Illinois Institute of Technology	IL	217[1]	172[1]	277[1]	287[2]	953	118[4]	100[5]	240	188	34
097 Indiana University-Purdue University at Indianapolis	IN	52	120	112	276	560	30	--	78	7	--
098 University of Iowa	IA	413	278	271	275	1,237	203	211	226	76	48
099 Iowa State University	IA	1,276	816	810	1,204	4,106	336[1]	401[1]	--	--	--
100 The Johns Hopkins University	MD	225	202	170	168	765	2,344[3]	437[5]	195	776	42

Enrollments and Degrees Awarded

ENGINEERING

			ENROLLMENTS							DEGREES AWARDED		
			Undergraduate—By Class				Totals—By Level			By Degree		
Institution Profile Reference Number & Name	State/Province		Freshmen/First-year	Sophomores/Second-year	Juniors/Third-year	Seniors/All others	Bachelor's	Master's	Doctoral	Bachelor's	Master's	Doctoral
101 University of Kansas	KS		381	245	285	487(1)	1,398	470	95	241	92	18
102 Kansas State University	KS		706	486	515	826	2,533	203	79	336	66	14
103 University of Kentucky	KY		515	322	292	485	1,614	229	131	233	54	21
104 Lafayette College	PA		134	105	97	115	451	--.	--.	114	--	--
105 Lamar University	TX		284	167	169	279(1)	899	206	42	71	--	2
106 Lawrence Technological University	MI		505	493	459	396	1,853	42	--.	346	9	--
107 LeTourneau University	TX		86	47	28	58	219	--.(1)	--.(1)	32	--	--
108 Lehigh University	PA		396	376	363	396	1,531	398	261	359	137	43
109 Louisiana State University	LA		994(1)	409	523	700	2,626	412	224	263	96	18
110 Louisiana Tech University	LA		493(1)	265(1)	225(1)	360(1)	1,343	165(2)	67	174	49	5
111 University of Louisville	KY		471(2)	326(2)	266(2)	238(2)	1,301	487(2)	46(2)	176	124	3
112 Loyola College in Maryland	MD		23	9	6	13	51	289	--.	12	--	--
113 Loyola Marymount University	CA		75	64	57	49	245	150	--.	36	20	--
114 University of Maine	ME		302	223	210	216	951	103	35	163	44	3
115 University of Manitoba	MB		418	325	302	265	1,310	209	131	187	36	13
116 Mankato State University	MN		162	68	61	83	374	--.	--.	38	--	--
117 Marietta College	OH		23	19	11	13	66	--.	--.	8	--	--
118 Marquette University	WI		413	277	379	246	1,315	134	62	257	85	12
119 University of Maryland Baltimore County	MD		197	128	127	211(1)	663	175	74	49	30	--
120 University of Maryland, College Park	MD		625	363	555	1,268	2,811	753	544	585	248	63
121 University of Massachusetts	MA		361	331	399	483(1)	1,574	359	279	297	159	42
122 University of Massachusetts, Dartmouth	MA		157	122	114	146	539	54	--.(B)	109	15	--
123 University of Massachusetts, Lowell	MA		498	380	420	450	1,748	1,052	121	397	247	21
124 Massachusetts Institute of Technology	MA		--.(1)	678	687	666	2,031	1,109	1,114	596	563	239
125 McGill University	PQ		--.(B)	466(1)	786	418	1,670	392	264	297	166	54
126 Memphis State University	TN		203	181	263	513	1,160	188	37	128	50	9
127 Mercer University	GA		117	110	95	139	461	156	--.	45	3	--
128 University of Miami	FL		224	140	132	253	749	151	59	158	52	10
129 Miami University	OH		107	79	62	48	296	15	--.(B)	40	3	--
130 University of Michigan	MI		1,239	1,066	1,106	1,308	4,719	883	916	819	504	238
131 University of Michigan-Dearborn	MI		286	287	316	462	1,351	511	--.(B)	147	65	--
132 Michigan State University	MI		1,119	919	676	861	3,575	372	274	608	148	67
133 Michigan Technological University	MI		1,173	901	1,182	934	4,190	196	94	777	91	24
134 Milwaukee School of Engineering	WI		515	366	266	277	1,424	70	--.	210	12	--
135 University of Minnesota, Duluth	MN		85	79	85	168	417	--.(A)	--.(A)	69	--	--
136 University of Minnesota	MN		540	682	817	1,488	3,527	502	561	657	184	103
137 University of Mississippi	MS		155	84	105	153	497	135	50	84	51	4
138 Mississippi State University	MS		502	398	525	818	2,243	237	69	299	69	14
139 University of Missouri-Columbia	MO		460	340	393	639	1,832	432	157	335	173	15
140 University of Missouri-Kansas City	MO		69	117	178	392	756	79	10	67	19	--
141 University of Missouri-Rolla	MO		790	578	699	1,413	3,480	641	216	566	248	48
142 Monmouth College	NJ		15	17	12	29	73	66	--.(1)	21	10	--
143 Montana College of Mineral Science and Technology	MT		695	321	275	336	1,627	83	--.(B)	95	19	--
144 Montana State University	MT		557	210	229	325(1)	1,321	78	18	181	26	3
145 Morgan State University	MD		226	135	92	81	534	--.(B)	--.(B)	34	--	--
146 University of Nebraska - Lincoln	NE		322	231	341	559	1,453	--.	--.	--	--	--
147 University of Nevada, Las Vegas	NV		595(1)	43	62	119	819	55	3	47	11	--
148 New England College	NH		14	5	8	8	35	--.	--.	8	--	--
149 University of New Hampshire	NH		206	198	166	277	847	149	42	163	49	8
150 University of New Haven	CT		239	215	190	202	846	194	--.(B)	91	113	--
151 New Jersey Institute of Technology	NJ		809	460	609	724	2,602	872(3)	134(4)	316	353	15
152 The University of New Mexico	NM		328(1)	474(2)	236	507	1,545	412(5)	147(7)	182	149	32
153 New Mexico Institute of Mining and Technology	NM		149	82	60	71	362	41	25	30	8	6
154 New Mexico State University	NM		510(1)	375(1)	342(1)	489(1)	1,716	322	97	236	114	14
155 University of New Orleans	LA		433	249	184	301	1,167	130	--.	108	--	--

Enrollments and Degrees Awarded

ENGINEERING

Institution Profile Reference Number & Name	State/Province	ENROLLMENTS Undergraduate—By Class				Totals—By Level			DEGREES AWARDED By Degree		
		Freshmen/First-year	Sophomores/Second-year	Juniors/Third-year	Seniors/All others	Bachelor's	Master's	Doctoral	Bachelor's	Master's	Doctoral
156 New York Institute of Technology	NY	107	106	144	152	509	20	--.	216	--	--
157 New York State College of Ceramics at Alfred University	NY	76	58	64	68	266	39	33	80	14	--
158 University of North Carolina-Charlotte	NC	220	202	276	330	1,028	179	--.	176	54	--
159 North Carolina Agricultural and Technical State University	NC	482	260	208	288[2]	1,238	189	12	193	30	--
160 North Carolina State University	NC	1,491	1,486	1,436	1,586	5,999	687	449	970	269	105
161 University of North Dakota	ND	189	93	143	249	674	37	2	93	15	--
162 North Dakota State University	ND	594	501	408	751	2,254	79	31	282	16	4
163 Northeastern University	MA	298	215[1]	367[1]	417	1,297	1,077	132	423	299	14
164 Northern Arizona University	AZ	288	194	160	212	854	--.	--.	74	--	--
165 Northern Illinois University	IL	254	88	113	155	610	98	--.[B]	120	45	--
166 Northwestern University	IL	350	282	291	299[1]	1,222	326	687	281	125	105
167 Norwich University	VT	111	74	50	48	283	--.	--.	23	--	--
168 University of Notre Dame	IN	368[1]	283	277	241[2]	1,169	61	176	202	51	18
169 Oakland University	MI	208	169	209	278	864	321	46	130	114	8
170 Ohio University	OH	410[1]	275[1]	212[1]	316[1]	1,213	253[1]	41[1]	255	44	7
171 Ohio Northern University	OH	146	76	86	81	389	--.	--.	69	--	--
172 Ohio State University	OH	687	768	912	1,427	3,794	837	769	655	271	85
173 University of Oklahoma	OK	639	440	352	758	2,189	--.[A]	--.[A]	316	131	38
174 Oklahoma Christian University of Science and Arts	OK	54	22	25	38	139	--.[2]	--.[2]	11	--	--
175 Oklahoma State University	OK	416	359	400	406	1,581	493	142	202	151	25
176 Old Dominion University	VA	184[1]	214[1]	257[1]	450[1]	1,105	309[1]	96	171	93	16
177 Oregon State University	OR	506[1]	701[1]	344[1]	496[1]	2,047	340[1]	175[1]	428	134	31
178 University of the Pacific	CA	64	41	111	58	274	6	--.[1]	63	1	--
179 University of Pennsylvania	PA	395	323	330	326	1,374	185	473	200	147	102
180 Pennsylvania State University	PA	2,237	1,649	1,428	1,842	7,156	557	459	1,155	318	92
181 University of Pittsburgh	PA	287	399	372	452	1,510	547	176	301	146	40
182 Polytechnic University	NY	464	284	224	256	1,228	744[1]	193[1]	298	279	27
183 University of Portland	OR	61	36	73	94	264	19	--.[B]	57	9	--
184 Portland State University	OR	181	167	282	565[1]	1,195	335	32	189	82	1
185 Prairie View A&M University	TX	512	143	109	86	850	6	--.	114	--	--
186 Princeton University	NJ	225	203	196	153	777	7	351	148	26	44
187 University of Puerto Rico, Mayagüez Campus	PR	694	667	620	1,831	3,812	166	--.[B]	535	13	--
188 Purdue University	IN	1,704	1,742	1,509	1,490	6,445	972	644	1,248	388	154
189 Rensselaer Polytechnic Institute	NY	838	671	786	628	2,923	475	397	669	338	81
190 University of Rhode Island	RI	224	184	204	279	891	199	103	142	48	18
191 William Marsh Rice University	TX	221	156	156	213	746	120	259	161	76	31
192 University of Rochester	NY	236	204	151	180	771	114[1]	250[2]	149	89	34
193 Rochester Institute of Technology	NY	303	252	290	590	1,435	360	--.	251	50	--
194 Rose-Hulman Institute of Technology	IN	291	267	237	260	1,055	44	--.[B]	227	5	--
195 Rutgers, The State University	NJ	644	628	618	570	2,460	415	342	544	135	51
196 Saginaw Valley State University	MI	75	54	72	116	317	--.[B]	--.[B]	41	--	--
197 St. Cloud State University	MN	79	50	58	74	261	--.	--.	33	--	--
198 Parks College of Saint Louis University	IL	60	64	58	114	296	10	--.	77	--	--
199 University of San Diego	CA	17[1]	8	10	16[2]	51	--.	--.	7	--	--
200 San Diego State University	CA	387	221	451	876	1,935	305	10	322	66	--
201 San Francisco State University	CA	88[1]	76[1]	84[1]	452[1]	700	22	--.[3]	--	--	--
202 San Jose State University	CA	633	418	741	1,389	3,181	847	--.	375	180	--
203 Santa Clara University	CA	200	128	119	146	593	1,169	37	114	95	6
204 Seattle University	WA	62	54	105	120	341	114	--.[B]	97	--	--
205 University of South Alabama	AL	301	175	223	236	935	73	--.[B]	84	8	--
206 University of South Carolina	SC	455	285	259	340	1,339	259	68	227	88	11
207 South Dakota School of Mines and Technology	SD	419	318	212	281	1,230	142	25	187	69	1
208 South Dakota State University	SD	346	208	188	290	1,032	170	--.	194	26	--
209 University of South Florida	FL	313	223	514	1,188	2,238	520	168	386	138	22
210 Southern University and A&M College	LA	372	252	177	279	1,080	--.[A]	--.[A]	134	--	--

Enrollments and Degrees Awarded

ENGINEERING

#	Institution Profile Reference Number & Name	State/Province	Enrollments Undergraduate—By Class Freshmen/First-year	Sophomores/Second-year	Juniors/Third-year	Seniors/All others	Totals—By Level Bachelor's	Master's	Doctoral	Degrees Awarded By Degree Bachelor's	Master's	Doctoral
211	University of Southern California	CA	--.	--.	--.	--.	--	--.	--.	292	490	89
212	University of Southern Colorado	CO	43	18	13	21	95	25	--.(B)	--	3	--
213	Southern Illinois University at Carbondale	IL	227	189	236	435	1,087	157	40	175	59	10
214	Southern Illinois University at Edwardsville	IL	12	87	162	259	520	98	--.(B)	111	33	--
215	University of Southern Maine	ME	45	30	30(1)	18	123	--.	--.	8	--	--
216	Southern Methodist University	TX	--.(1)	57	50	80	187	170(3)	136(5)	72	82	18
217	Stanford University	CA	--.(B)	27	169	368	564	1,284	1,271	252	803	191
218	SUNY at Binghamton	NY	--.(B)	50	120	240	410	188	36	191	139	10
219	SUNY at Buffalo	NY	716	460	409	835	2,420	350	452	482	184	39
220	SUNY, College of Environmental Science and Forestry	NY	35	68	76	67	246	58	31	49	14	--
221	SUNY Maritime College	NY	120	90	80	70	360	--.	--.	59	--	--
222	SUNY, College at New Paltz	NY	29	41	46	62	178	--.(B)	--.(B)	24	--	--
223	SUNY at Stony Brook	NY	172	138	107	301	718	--.(A)	--.(A)	130	49	20
224	Swarthmore College	PA	42	21	29	24	116	--.	--.	18	--	--
225	Syracuse University	NY	252	199	166	249	866	260(2)	218(2)	171	200	43
226	Technical University of Nova Scotia	NS	--.	227	221	178	626	169	75	168	27	14
227	Temple University	PA	30	48	91	139	308	77	13	82	--	--
228	University of Tennessee, Chattanooga	TN	185	215	223	195	818	135	--.	42	22	--
229	University of Tennessee, Knoxville	TN	695	383	328	588(1)	1,994	244	130	340	150	42
230	Tennessee State University	TN	300	167	134	246	847	42(2)	--.(B)	72	11	--
231	Tennessee Technological University	TN	561	288	269	466	1,584	117	52	195	29	7
232	University of Texas at Arlington	TX	304	222	315	1,024(1)	1,865	1,001	263	--	--	--
233	University of Texas at Austin	TX	1,409	907	959	1,559	4,834	1,208	885	687	430	141
234	University of Texas at Dallas	TX	12	11	112	73	208	204	36	42	54	2
235	University of Texas at El Paso	TX	189	256	235	340	1,020	403	15	180	66	--
236	University of Texas at San Antonio	TX	288	205	189	355	1,037	102	--.(A)	82	9	--
237	Texas A&I University	TX	257	130	98	180	665	204	--.(A)	--	--	--
238	Texas A&M University	TX	1,766(1)	1,293(1)	1,569(1)	1,904(1)	6,532	1,369(2)	685(2)	1,046	386	111
239	Texas Tech University	TX	715	432	367	545	2,059	261	120	194	77	14
240	University of Toledo	OH	816	337	455	473	2,081	231	72	287	59	10
241	Tri-State University	IN	113	109	148	115	485	--.(1)	--.(1)	111	--	--
242	Trinity University	TX	120	47	35	39	241	--.(B)	--.(B)	20	--	--
243	Tufts University	MA	176	134	164	205	679	163	47	188	94	13
244	Tulane University	LA	268	183	150	140	741	77	65	130	20	20
245	University of Tulsa	OK	144	109	111	101	465	68	34	85	28	7
246	Tuskegee University	AL	370	176	160	273	979	51	--.	68	13	--
247	Union College	NY	93	70	87	88	338	63	--.(B)	87	33	--
248	United States Air Force Academy	CO	1,131(1)	1,132(1)	1,053(1)	972(1)	4,288	--.(B)	--.(B)	350	--	--
249	United States Coast Guard Academy	CT	152	95	82	64	393	--.	--.	79	--	--
250	United States Merchant Marine Academy	NY	141	149	122	118	530	--.	--.	89	--	--
251	United States Military Academy	NY	--.(1)	--.(1)	376	400	776	--.(2)	--.(2)	381	--	--
252	United States Naval Academy	MD	418	328	399	375	1,520	--.	--.	361	--	--
253	University of Utah	UT	286(1)	214(1)	225(1)	470(1)	1,195	150	213	277	91	44
254	Utah State University	UT	429(1)	241(1)	214(1)	401(A)	1,285	203	96	129	84	15
255	Valparaiso University	IN	96	76	87	107	366	--.	--.	83	--	--
256	Vanderbilt University	TN	318	245	203	241	1,007	139	178	245	60	24
257	University of Vermont	VT	178	137	100	112	527	76	33	104	22	4
258	Villanova University	PA	272	176	200	197	845	373	--.	203	96	--
259	University of Virginia	VA	432	395	333	437	1,597	427	273	323	193	48
260	Virginia Military Institute	VA	131	108	84	110	433	--.	--.	433	--	--
261	Virginia Polytechnic Institute and State University	VA	1,259	1,031	1,057	1,165	4,512	1,632	478	1,111	387	130
262	University of Washington	WA	2(1)	120(1)	563	947	1,632	890	442	603	285	50
263	Washington University	MO	212	216	234	334	996	423	192	231	164	34
264	Washington State University	WA	312	439	360	392	1,503	302	107	505	99	23
265	Wayne State University	MI	424	193	255	517	1,389	921	204	164	300	16

Enrollments and Degrees Awarded

ENGINEERING

Institution Profile Reference Number & Name	State/Province	ENROLLMENTS Undergraduate—By Class Freshmen/First-year	Sophomores/Second-year	Juniors/Third-year	Seniors/All others	Totals—By Level Bachelor's	Master's	Doctoral	DEGREES AWARDED By Degree Bachelor's	Master's	Doctoral
266 Webb Institute of Naval Architecture	NY	23	19	23	16	81	--.	--.	16	--	--
267 West Virginia University	WV	446	316	319	427	1,508	267	110	271	90	21
268 West Virginia Institute of Technology	WV	273	165	159	208	805	16	--.	100	--	--
269 Western Michigan University	MI	273	273	270	397	1,213	123	--.	156	35	--
270 Western New England College	MA	107	100	107	123	437	77	--.	82	45	--
271 Wichita State University	KS	345[1]	306	241	359	1,251	206	88	152	58	10
272 Widener University	PA	75[1]	88[1]	77[1]	95[1]	335	119	--.	115	16	--
273 Wilkes University	PA	55[1]	50[C]	60[C]	45[C]	210	25[C]	--.[B]	66	4	--
274 University of Wisconsin-Madison	WI	836	751	783	1,189	3,559	636	557	539	258	69
275 University of Wisconsin-Milwaukee	WI	176	257	161	433	1,027	178	55	171	50	9
276 University of Wisconsin-Platteville	WI	542	267	259	408	1,476	--.	--.	201	--	--
277 Worcester Polytechnic Institute	MA	496	547[2]	539[3]	442	2,024	--.[1]	--.[1]	393	128	8
278 Wright State University	OH	482[1]	293	286	581	1,642	339	34	204	58	--
279 University of Wyoming	WY	263	161	175	331	930	159	28	155	25	--
280 Yale University	CT	--.[1]	--.[1]	39	47	86	9	144	44	10	25
Totals:		97,926	76,725	82,511	115,712	372,874	73,643	36,081	62,677	25,937	6,089

Enrollments and Degrees Awarded—Footnotes

FOOTNOTES: The following footnotes are the same for all schools: (A) Data not available. (B) Data not applicable. (C) Estimated data.

001 University of Akron: (1) This includes 455 part time students at various stages in degree programs

013 Baylor University: (2) No Graduate Engineering Program

019 University of California at Berkeley: (1) Includes Chemical Engineering. Chemical Engineering is part of the College of Chemistry.

022 University of California, Los Angeles: (1) Juniors: 550 includes 3 students limited status.

026 California Institute of Technology: (1) Majors are chosen in sophomore year.

029 California State University, Fresno: (2) Total number of majors is 1471

031 California State University, Los Angeles: (1) Graduate students pursuing a BS in Engineering total 33.

037 Case Western Reserve University: (1) Freshmen admitted to University and not specific engineering programs.

039 University of Central Florida: (1) Freshmen hrs; Soph. 30-60 hrs; Junior 60-90 hrs; Senior 90 hrs Unclassified students (123) placed in Freshmen no. (2) Includes 214 unclassifieds and post-bacs

040 Central State University: (1) No graduate program available.

041 Christian Brothers University: (1) In addition to the full-time students listed by year, there was a combined total of 36 part-time students enrolled for the academic year 1991-92.

042 University of Cincinnati: (1) College undergraduate programs are 5-years, enrollment was 386 for pre-juniors and 329 for juniors. Enrollment figures for Fall 1992. (2) Enrollment figures are for Fall 1992.

044 City College of the City University of New York: (1) This and the following enrollment figures include full-time students only. In addition, there are 543 part-time students.

050 University of Colorado at Colorado Springs: (2) Includes 50 5th-year seniors.

051 University of Colorado at Denver: (1) Includes all degree programs offered by the College of Engineering (CE, CS, EE, ME, A.Math). (3) PhD programs authorized at the CU-Boulder campus as part of system-wide graduate program.

054 Columbia University: (1) Full time students only; 335 additional part time students.

059 Dartmouth College: (1) Undeclared Majors

062 University of Denver: (1) Includes one part-time student. (2) Includes two part-time students.

063 University of the District of Columbia: (1) 425 part-time students

064 Drexel University: (1) Includes 211 reclassified frosh. (2) Pre-Junior - 416

066 École Polytechnique de Montréal: (1) Fall '91 + Winter '92 + Summer '92 (3) In addition, there are 91 students in D.E.S.S. programs. Figures valid in Fall '91. (4) In addition, 23 students graduated in D.E.S.S. programs. Figures comprise Fall '91 + Winter '92 + Summer '92. (5) Figure valid in Fall 1991.

069 University of Evansville: (1) Includes 5th year co-op students.

072 Florida Agricultural and Mechanical University: (1) Numbers include both Florida Agricultural and Mechanical University and Florida State University students. (2) Numbers include both FAMU and FSU students. Number is for Fall 1991, Spring 1992, and Summer 1992.

073 Florida Atlantic University: (1) Includes Computer Science (2) Includes Computer Science and Second Bachelor's Students (4) Listed number does not include 229 nondegree students enrolled in graduate courses.

076 Florida State University: (1) Numbers include Florida Agricultural and Mechanical University and Florida State University students. (2) Numbers include both FAMU and FSU students. The academic year is Fall 1991, Spring 1992 and Summer 1992.

080 The George Washington University: (2) All campuses included.

082 Georgia Institute of Technology: (2) Includes full-time, part-time, and students on co-op work quarter.

085 Grove City College: (1) No Graduate Degree Programs Offered

086 Harvard University: (1) Engineering Sciences Ph.D. candidates can receive Master's along the way to the Ph.D.

090 University of Houston: (1) There are 300 post-baccalaureate students.

091 Howard University: (2) This figure does not include the 66 part-time students in the master's degree program. (3) The total does not include part-time students in the freshmen through senior levels.

094 University of Illinois at Chicago: (1) Includes 7 nondegree students.

096 Illinois Institute of Technology: (1) Figures are full-time students only. Total part-time undergraduate student enrollment was 343. (2) Includes 5th year students also (all full-time). Total part-time undergraduate student enrollment was 343. (4) Figures are full-time master's students only. Total part-time graduate (master's and doctorate) student enrollment was 877. (5) Figures are full-time doctoral students only. Total part-time graduate (master's and doctoral) student enrollment was 877.

099 Iowa State University: (1) Data available only for fall 1992.

100 The Johns Hopkins University: (1) Plus 31 in part-time programs. (3) Includes 2311 in part-time programs. (4) 672 from part-time programs; 91 from full-time programs. (5) All full-time programs.

101 University of Kansas: (1) This includes level 5 students (454+33=487).

105 Lamar University: (1) Includes 57 post baccalaureate students

107 LeTourneau University: (1) No graduate programs in engineering are offered at LeTourneau University.

109 Louisiana State University: (1) includes pre-engineering program enrollment

110 Louisiana Tech University: (1) Includes Computer Science and Geosciences. (2) Includes Computer Science.

111 University of Louisville: (2) Fall enrollment data only.

119 University of Maryland Baltimore County: (1) Includes all part-time students.

121 University of Massachusetts: (1) Includes post-graduates and non-classified students.

124 Massachusetts Institute of Technology: (1) Students select a major in the sophomore year.

125 McGill University: (1) Three and one half year program following two year junior college. (2) Graduates in 1991-92 academic session.

142 Monmouth College: (1) Program not available

144 Montana State University: (1) Includes second degree students

147 University of Nevada, Las Vegas: (1) Number is pre-engineering students who are generally freshmen.

151 New Jersey Institute of Technology: (3) Additional 7 masters students enrolled in applied chemistry. Additional 138 students enrolled in environmental science. (4) Additional 8 doctoral students enrolled in environmental science.

152 The University of New Mexico: (1) Includes students in associate programs and students working on the freshman year of the curriculum. (2) (5) Includes Computer Science students (98). (6) Includes Computer Science students (19). (7) Includes Computer Science students (4). (8) Includes Computer Science students (2).

154 New Mexico State University: (1) Fall 1992 enrollment (does not include surveying or technology).

156 New York Institute of Technology: (1) Effective 9/92, NYIT is offering the following M.S. programs: Electrical Engineering, Mechanical Engineering and Environmental Technology.

159 North Carolina Agricultural and Technical State University: (1) Graduates of the joint PhD program have degrees conferred from N.C. State University. (2) This includes fifth year Architectural Engineering students.

162 North Dakota State University: (1) Architecture 366, Landscape Architecture 68, Construction Management 92

163 Northeastern University: (1) Cooperative education program is a five-year program. Enrollment in Middler Year (between Sophomore and Junior) was 296.

166 Northwestern University: (1) An additional 40 are fifth year students in cooperative education program.

168 University of Notre Dame: (1) Reported as the number of engineering intents in the Freshman Year of Studies (2) Includes 19 fifth year students

169 Oakland University: (1) Includes Computer Science graduates.

170 Ohio University: (1) Fall 1992-93 Enrollment.

174 Oklahoma Christian University of Science and Arts: (2) There are no engineering graduate programs.

176 Old Dominion University: (1) Fall 1991 data

177 Oregon State University: (1) Includes non-ABET programs in College of Engineering

178 University of the Pacific: (1) School has no doctoral programs.

179 University of Pennsylvania: (1) Includes 36 Bachelor of Applied Science graduates.

182 Polytechnic University: (1) Excludes Computer Science

184 Portland State University: (1) Includes postbaccalaureate undergraduates.

191 William Marsh Rice University: (1) 40 additional students received B.A. degrees. (3) Of these, 46 students received M.S. degrees and 30 received professional master's degrees.

192 University of Rochester: (1) does not include 3-2 students or leave of absences. (2) does not include leave of absence.

198 Parks College of Saint Louis University: (1) Masters in Aerospace Engineering began fall 1991

199 University of San Diego: (1) Freshman enter as undeclared. 31 started in the fall freshman engr preceptorial course, 17 continued in the engineering program in the spring. (2) This figure includes 9 4th yr majors and 7 5th yr graduating seniors (Senior 2) who completed the std 4.5 yr prog in either in the 9th or 10th sem.

201 San Francisco State University: (1) Fall '91 enrollment figures only. (3) No such program at this institution.

215 University of Southern Maine: (1) 8

216 Southern Methodist University: (1) First year students not admitted directly to the engineering school. There were 157 pre-engineering students. (3) Enrollment in applied science master's programs was 227. (4) Applied science master's degrees awarded were 60. (5) Enrollment in applied science doctoral programs was 56. (6) Doctoral degrees awarded in applied sciences was 6. (7) Awarded bachelor's degrees in applied sciences was 15.

Enrollments and Degrees Awarded—Footnotes

223 State University of New York at Stony Brook: (1) SUNY Stony Brook offers B.E. degrees for engineering students.

225 Syracuse University: (2) 682 part-time graduate students not included in this figure. Masters vs. doctoral breakdown not available at this time.

229 University of Tennessee, Knoxville: (1) 256 part-time students not included in totals. (2) Numbers do not include 710 part-time M.S. and PhD students.

230 Tennessee State University: (2) Master of Engineering degree with options in Civil Engineering, Electrical engineering and Mechanical Engineering.

232 University of Texas at Arlington: (1) Number includes seniors as well as non-degreed students enrolled in the College of Engineering.

238 Texas A&M University: (1) CPSC, ENTC and IDIS not included (2) Includes CPSC

241 Tri-State University: (1) No graduate programs

248 United States Air Force Academy: (1) Reflects Fall 92 Enrollment. Data based on all students in the institution. All students must take engineering courses.

251 United States Military Academy: (1) enrollment in a majors program comes at the end of sophomore year (2) USMA is an undergraduate institution

253 University of Utah: (1) Full-time enrollment for autumn qtr. 1991. Additional 448 part-time students (all classes) are not included. (2) An additional 6 master of science degrees granted in computer science. (3) An additional 6 doctoral degrees granted in computer science.

254 Utah State University: (1) Fall Quarter 1991 Data

262 University of Washington: (1) Normally, Freshmen do not register in the College of Engineering. Generally, enter College at Junior standing.

271 Wichita State University: (1) Most freshmen are admitted to University College before they have completed admission requirements for engineering.

272 Widener University: (1) Full-time day students.

273 Wilkes University: (1) First year is common to all engineering majors

277 Worcester Polytechnic Institute: (1) WPI does not separate MS and PhD students in enrollment figures. Total graduate school enrollment for AY 1991-92 is 1057 students. (2) Includes new transfer students. (3) Includes students on co-op.

278 Wright State University: (1) Data in this section includes computer science students.

280 Yale University: (1) Major is not declared until Junior year.

All Engineering Disciplines

Institutions are listed in alphabetical order. The first line in a school's entry contains the institution profile reference number, the school name, and the state or province in which the main campus is located. Subsequent indented lines for each program give the name of the degree program, followed by a major category abbreviation in brackets (H = Chemical, V = Civil, E = Electrical, I = Industrial, M = Mechanical, and O = Other) the full-time equivalent (FTE) of all faculty. For undergraduate programs, the columns represent: accreditation status, the nominal and actual average length of the program, the time that courses are available (Day, Evening or Both), the availability of cooperative (internship) programs (None, Optional, or Required), the percentage of graduates who participated in co-op programs, the number of bachelor's degrees awarded, and the percentage of these graduates who were full-time. For graduate programs, the columns indicate the number of graduates and the percentage who were full-time in master's and then doctoral programs. Unless otherwise footnoted, information on degrees is for the 1991-92 academic year; all other information is current. At the end of the table are (1) a summary of totals by state or province and by country and (2) the full text of all footnotes for this table.

ENGINEERING Institution Profile Reference Number & Name Name of Degree Program	State/Province	FACULTY Full-time equivalent (FTE)	ABET/CEAB accred?	UNDERGRADUATE PROGRAMS							GRADUATE PROGRAMS			
				Length		Time	Co-op		Bachelor's		Master's		Doctoral	
				Nominal length of program in years	Average length of program in years	Day/Eve./Both	None/Opt./Req.	% of graduates in Co-op programs	# of degrees awarded	% of graduates who were full-time	# of degrees awarded	% of graduates who were full-time	# of degrees awarded	% of graduates who were full-time
001 U of Akron	OH													
Biomedical Eng [O]		8.0	—(B)	-.-(B)	-.-(A)	—(B)	—(B)	-.-(B)	—(B)	-.-(B)	7	75.0	1	100.0
Chemical Eng [H]		9.0	Y	4.0	-.-(A)	D	O	69.0	26	87.5(A)	11	67.0	1	100.0
Civil Eng [V]		12.0	Y	4.0	-.-(A)	B	O	57.0	34	75.0(A)	9	40.0	9	80.0
Electrical Eng [E]		18.0	Y	4.0	-.-(A)	B	O	57.0	65	74.8(A)	20	57.0	5	90.0
Eng [O]		-.-	N	4.0	-.-(A)	B	N	-.-(B)	2	-.-(A)	—(B)	-.-(B)	—(B)	-.-(B)
Mechanical Eng [M]		20.0	Y	4.0	-.-(A)	B	O	60.0	94	68.3(A)	31	50.0	2	75.0
002 U of Alabama	AL													
Aerospace Eng [O]		8.2	Y	4.0	4.5	D	O	9.1	33	100.0	5	100.0	—(B)	-.-(B)
Chemical Eng [H]		9.0	Y	4.0	4.5	D	O	24.1	29	100.0	1	100.0	2	100.0
Civil Eng [V]		9.0	Y	4.0	4.5	D	O	11.5	26	100.0	4	100.0	2	100.0
Computer Science [O]		15.0	N(4)	4.0	4.5	D	O	13.3	12	100.0	8	100.0	—.	-.-
Electrical Eng [E](1)		15.0	Y	4.0	4.5	D	O	18.1	83	100.0	18	100.0	1	100.0
Eng Hydrology [O](5)		-.-(6)	N(3)	-.-(A)	-.-(A)	D	O	-.-(A)	—(B)	-.-(B)	0	100.0	0	100.0
Eng Mechanics [O]		10.0	N(3)	-.-(A)	-.-(A)	D	O	-.-(A)	—(B)	-.-(B)	3	100.0	3	100.0
Env'l Eng [O](5)		-.-(6)	N(3)	-.-(A)	-.-(A)	D	O	-.-(A)	—(B)	-.-(B)	5	100.0	—(B)	-.-(B)
Industrial Eng [I]		12.0	Y	4.0	4.5	D	O	20.7	29	100.0	11	100.0	5	100.0
Mechanical Eng [M]		15.0	Y	4.0	4.5	D	O	21.1	57	100.0	10	100.0	3	100.0
Metallurgical Eng [O]		6.8	Y	4.0	4.5	D	O	0.0	2	100.0	5	100.0	3	100.0
Mineral Eng [O](2)		8.0(7)	Y	4.0	4.5	D	O	0.0	3	100.0	4	100.0	—(B)	-.-(B)
003 U of Alabama at Birmingham	AL													
Biomedical Eng [O]		6.0	N(B)	-.-(B)	-.-(B)	D	—(B)	-.-(B)	—(B)	-.-(B)	14	50.0	2	100.0
Civil Eng [V]		8.0	Y	4.0	-.-(A)	B	O	-.-(A)	29	41.0	2	35.0	—(B)	-.-(B)
Electrical & Computer Eng [E]		12.0	Y	4.0	-.-(A)	B	O	-.-(A)	29	48.0	15	30.0	—(B)	-.-(B)
Materials Science & Eng [O]		7.5	Y	4.0	-.-(A)	B	O	-.-(A)	8	73.0	6	0.0	0	100.0
Mechanical Eng [M]		9.0	Y	4.0	-.-(A)	B	O	-.-(A)	39	39.0	6	50.0	—(B)	-.-(B)
004 U of Alabama at Huntsville	AL													
Chemical Eng [H]		4.5	Y	4.0	-.-	B	O	-.-	4	84.0	—.	-.-	—.	-.-
Civil & Construction Eng [V]		-.-	Y	-.-	-.-	B	O	-.-	—.	-.-	—.	-.-(A)	—.	-.-(A)
Civil Eng [V]		4.3	Y	4.0	-.-(A)	B	O	-.-(A)	12	52.9	—.	-.-(A)	—.	-.-(A)
Electrical & Computer Eng [E]		26.9	Y	4.0	-.-(A)	B	O	-.-(A)	116	53.7	49	36.8	5	57.9
Industrial & Systems Eng [I]		13.3	Y	4.0	-.-(A)	B	O	-.-(A)	14	57.4	26	23.8	3	28.1
Mechanical & Aerospace Eng [M]		17.9	Y	4.0	-.-(A)	B	O	-.-(A)	45	64.8	25	35.0	3	59.2
Mechanical Eng [M]		-.-	Y	4.0		B	O		—.					
005 U of Alaska, Fairbanks	AK													
Arctic Eng [V]		-.-	—(B)	-.-(B)	-.-(B)	—(B)	—(B)	-.-(B)	—.	-.-	—(A)	-.-(A)	—(A)	-.-(A)
Civil Eng [V]		10.5	Y	4.0	4.5	D	N	-.-(A)	13	80.0	4	50.0	0	67.0
Electrical Eng [E]		8.0	Y	4.0	4.5	D	N	-.-(A)	13	70.0	4	90.0	—(A)	-.-(A)
Eng Mgmt [O]		-.-	—(B)	-.-(B)	-.-(B)	—(B)	—(B)	-.-(B)	—.	-.-	—(A)	-.-(A)	—(A)	-.-(A)
Env'l Quality Eng [V]		-.-	—(B)	-.-(B)	-.-(B)	—(B)	—(B)	-.-(B)	—.	-.-	—(A)	-.-(A)	—(A)	-.-(A)
Env'l Quality Science [V]		-.-	—(B)	-.-(B)	-.-(B)	—(B)	—(B)	-.-(B)	—.	-.-	—(A)	-.-(A)	—(A)	-.-(A)
Mechanical Eng [M]		8.0	Y	4.0	4.5	D	N	-.-(A)	13	90.0	13	90.0	—(A)	-.-(A)
Science Mgmt [O]		-.-	—(B)	-.-(B)	-.-(B)	—(B)	—(B)	-.-(B)	—.	-.-	—(A)	-.-(A)	—(A)	-.-(A)
006 U of Alberta	AB													
Chemical Eng [H]		18.0	Y	4.0	-.-(A)	D	O	33.0	49	100.0	5	85.0	1	100.0
Civil Eng [V]		34.5	Y	4.0	-.-(A)	D	O	33.0	53	100.0	38	80.5	10	97.0
Computer Eng [O]		-.-(B)	Y	4.0	-.-(A)	D	O	50.0	26	100.0	—.	-.-(B)	—.	-.-(B)
Electrical Eng [E]		36.0(3)	Y	4.0	-.-(A)	D	O	33.0	105	100.0	33	86.4	5	91.0
Eng Physics [O]		-.-(1)	Y	4.0	-.-(A)	D	N	-.-(B)	7	100.0	—.	-.-(B)	—.	-.-(B)
Mechanical Eng [M]		25.0	Y	4.0	-.-(A)	D	O	33.0	92	100.0	12	70.0	2	100.0
Metallurgical Eng [O]		-.-(1)	Y	4.0	-.-(A)	D	O	33.0	7	100.0	—(B)	-.-(B)	—(B)	-.-(B)
Mining Eng [O]		-.-(1)	Y	4.0	-.-(A)	D	O	33.0	6	100.0	—(B)	-.-(B)	—(B)	-.-(B)
Petroleum Eng [O]		-.-(1)	Y	4.0	-.-(A)	D	O	33.0	26	100.0	14(2)	72.0(2)	2(2)	90.0

All Engineering Disciplines

ENGINEERING Institution Profile Reference Number & Name Name of Degree Program	State/Province	FACULTY Full-time equivalent (FTE)	UNDERGRADUATE PROGRAMS							GRADUATE PROGRAMS			
			ABET/CEAB accred?	Length		Time	Co-op		Bachelor's		Master's		Doctoral	
				Nominal length of program in years	Average length of program in years	Day/Eve./Both	None/Opt./Req.	% of graduates in Co-op programs	# of degrees awarded	% of graduates who were full-time	# of degrees awarded	% of graduates who were full-time	# of degrees awarded	% of graduates who were full-time
007 Alfred U	NY													
Ceramic Eng [O]		24.0	Y	4.5	4.5	B	O	90.0	72	95.0	9	80.0	6	100.0
Ceramic Eng Science [O]		24.0	Y	4.5	4.5	B	O	90.0	5	90.0	4	85.0	—.(B)	-.-(B)
Electrical Eng [E]		5.0	Y	4.5	4.5	B	O	90.0	8	100.0	—.(B)	-.-(B)	—.(B)	-.-(B)
Glass Eng Science [O]		24.0	Y	4.5	4.5	B	O	90.0	3	100.0	7	100.0	—.(B)	-.-(B)
Mechanical Eng [M]		5.0	Y	4.5	4.5	B	O	90.0	8	100.0	1	100.0	—.(B)	-.-(B)
008 U of Arizona	AZ													
Aerospace Eng [O](1)		33.0	Y	4.0	-.-	D	O	-.-	53	92.0	4	-.-	1	-.-
Agricultural Biosystems Eng [O]		-.-	Y	4.0	-.-	D	O	-.-	4	85.0	5	-.-	—.	-.-
Chemical Eng [H]		11.0	Y	4.0	-.-	D	O	-.-	22	88.0	2	-.-	1	-.-
Civil Eng [V]		25.0	Y	4.0	-.-	D	O	-.-	47	87.0	15	-.-	14	-.-
Computer Eng [O](2)		-.-	Y	4.0	-.-	D	O	-.-	43	87.0	—.	-.-	—.	-.-
Electrical Eng [E]		48.0	Y	4.0	-.-	D	O	-.-	120	83.0	61	-.-	11	-.-
Eng Mathematics [O]		-.-	N	4.0	-.-	D	O	-.-	11	91.0	—.	-.-	—.	-.-
Eng Mechanics [O](3)		-.-	—	4.0	-.-	D	O	-.-	—.	-.-	3	-.-	—.	-.-
Eng Physics [O]		-.-	N	4.0	-.-	D	O	-.-	8	90.0	—.	-.-	—.	-.-
Geological Eng [O]		12.0	Y	4.0	-.-	D	O	-.-	6	88.0	2	-.-	1	-.-
Hydrology [O]		14.0	—	4.0	-.-	D	O	-.-	11	81.0	10	-.-	4	-.-
Industrial Eng [I]		19.0	Y	4.0	-.-	D	O	-.-	29	88.0	10	-.-	—.	-.-
Materials Science & Eng [O]		21.0	Y	4.0	-.-	D	O	-.-	18	90.0	7	-.-	—.	-.-
Mechanical Eng [M]		33.0	Y	4.0	-.-	D	O	-.-	83	88.0	12	-.-	15	-.-
Mineral Economics [O](4)		-.-	—	4.0	-.-	D	O	-.-	0	-.-	3	-.-	4	-.-
Mining Eng [O](4)		-.-	Y	4.0	-.-	D	O	-.-	4	88.0	3	-.-	3	-.-
Nuclear Eng [O]		9.0	Y	4.0	-.-	D	O	-.-	14	91.0	7	-.-	3	-.-
Optical Eng [O](2)		-.-	N	4.0	-.-	D	O	-.-	2	88.0	—.	-.-	—.	-.-
Reliability Eng [O](5)		-.-	—	-.-	-.-	D	O	-.-	—.	-.-	7	-.-	—.	-.-
Systems Eng [O](5)		-.-	Y	4.0	-.-	D	O	-.-	23	91.0	7	-.-	7	-.-
Water Resources Administration [O](6)		-.-	—	-.-	-.-	D	O	-.-	—.	-.-	5	-.-	—.	-.-
009 Arizona State U	AZ													
Aerospace Eng [O]		40.3	Y	4.0	5.0	D	O	-.-(A)	50	83.0	2	50.0	—.	-.-(B)
Bioengineering [O]		26.0	Y	4.0	5.0	D	O	-.-(A)	22	86.0	5	74.0	—.	-.-(B)
Chemical Eng [H]		26.0	Y	4.0	5.0	D	O	-.-(A)	23	86.0	5	75.0	2	50.0
Civil Eng [V]		18.5	Y	4.0	5.0	D	O	-.-(A)	31	77.0	27	77.0	7	77.0
Computer Science		30.5	N	4.0	4.0	D	O	-.-(A)	48	72.0	64	58.5	14	58.5
Computer Systems Eng [O]		30.5	Y	4.0	5.0	D	O	-.-(A)	25	72.0	—.	-.-	—.	-.-
Electrical Eng [E]		45.0	Y	4.0	5.0	D	O	-.-(A)	142	72.0	101	48.0	13	48.0
Eng Interdisciplinary Studies [O]		-.-	Y	4.0	5.0	D	O	-.-(A)	4	100.0	—.	-.-	—.	-.-
Eng Science(s) [O]		-.-	—	-.-	-.-	-.-	-.-	-.-	—.	-.-	5	40.0	4	50.0
Eng Special Studies [O]		-.-	Y	4.0	5.0	D	O	-.-(A)	7	71.5	—.	-.-	—.	-.-
Industrial Eng [I]		17.5	Y	4.0	5.0	D	O	-.-(A)	25	85.0	31	62.0	10	62.0
Materials Science [O]		26.0	N	4.0	5.0	D	O	-.-(A)	0	-.-(B)	—.	-.-	2	50.0
Mechanical Eng [M]		40.3	Y	4.0	5.0	D	O	-.-(A)	83	83.0	31	66.0	6	66.0
010 U of Arkansas	AR													
Biological & Agricultural Eng [O]		11.0	Y	-.-	-.-	—	—	-.-	2	82.0	1	50.0	—.	100.0
Chemical Eng [H]		13.0	Y	4.0	-.-	D	O	-.-	39	89.0	9	50.0	1	100.0
Civil Eng [V]		11.0	Y	4.0	-.-	D	O	-.-	25	95.9	4	46.1	1	100.0
Computer Systems Eng [O]		11.0	Y	4.0	-.-	D	O	-.-	30	89.4	2	66.7	—.	-.-
Electrical Eng [E]		17.0	Y	4.0	-.-	D	O	-.-	63	90.2	11	68.1	4	-.-
Eng [O]		-.-	—(B)	-.-(B)	-.-(B)	B	—(B)	-.-	—.	-.-	11	0.0	—.	-.-
Env'l Eng [O]		-.-	—(B)	-.-(B)	-.-(B)	D	—(B)	-.-	—.	-.-	1	75.0	—.	-.-
Industrial Eng [I]		-.-	Y	4.0	-.-	D	O	-.-	30	-.-	9	91.7	3	100.0
Mechanical Eng [M]		13.0	Y	4.0	-.-	D	O	-.-	42	90.2	4	31.8	1	100.0
Operations Mgmt [O]		-.-	—(B)	-.-(B)	-.-(B)	E	—(B)	-.-	—.	-.-	149	0.0	—.	-.-
Operations Research [O]		-.-	—(B)	-.-(B)	-.-(B)	D	—(B)	-.-	—.	-.-	0	50.0	—.	-.-
011 Arkansas Tech U	AR													
General Eng [O]		8.0	Y	4.0	4.0	D	N	-.-	15	70.0(C)	—.(B)	-.-(B)	—.(B)	-.-(B)
012 Auburn U	AL													
Aerospace Eng [M]		16.0	Y	4.0	-.-(A)	D	O	12.3	65	88.3(10)	7	69.2(10)	3	70.0(10)
Agricultural Eng [O]		10.0	Y	4.0	-.-(A)	D	O	25.0	4	93.3(10)	2	33.3(10)	1	25.0(10)
Aviation Mgmt [O]		-.-(1)	—(B)	4.0	-.-(A)	D	O	3.8	53	93.6(10)	—.(B)	-.-(B)	—.(B)	-.-(B)
Chemical Eng [H]		15.5	Y	4.0	-.-(A)	D	O	27.7	47	86.0(10)	6	75.9(10)	5	95.2(10)
Civil Eng [V]		20.5	Y	4.0	-.-(A)	D	O	24.4	78	89.2(10)	29	75.0(10)	7	73.7(10)
Computer Eng [E]		-.-(2)	Y	4.0	-.-(A)	D	O	17.0	47	79.2(10)	—.(B)	-.-(B)	—.(B)	-.-(B)
Computer Science [E]		-.-(2)	—(3)	4.0	-.-(A)	D	O	28.6	14	83.6(10)	—.(B)	-.-(B)	—.(B)	-.-(B)
Computer Science & Eng [E]		13.5	—(B)	-.-(B)	-.-(B)	—(B)	—(B)	-.-(B)	—.(B)	-.-(B)	11	18.6(10)	0	36.4(10)
Electrical Eng [E]		28.5	Y	4.0	-.-(A)	D	O	36.8	182	84.1(10)	24	34.4(10)	4	20.7(10)
Env'l Science [O](5)		-.-(5)	—(B)	4.0	-.-(A)	D	O	0.0	12	91.5(10)	—.	-.-	—.(B)	-.-(B)

All Engineering Disciplines

ENGINEERING Institution Profile Reference Number & Name Name of Degree Program	State/Province	FACULTY Full-time equivalent (FTE)	ABET/CEAB accred?	UNDERGRADUATE PROGRAMS					GRADUATE PROGRAMS			
				Length		Time	Co-op	Bachelor's		Master's		Doctoral	
				Nominal length of program in years	Average length of program in years	Day/Eve./Both	None/Opt./Req.	% of graduates in Co-op programs	# of degrees awarded	% of graduates who were full-time	# of degrees awarded	% of graduates who were full-time	# of degrees awarded	% of graduates who were full-time
Forest Eng [O]		-.-(4)	N	4.0	-.-(A)	D	O	0.0	4	91.7(10)	—(B)	-.-(B)	—(B)	-.-(B)
Geological Eng [O](5)		-.-(5)	N	4.0	-.-(A)	D	O	0.0	1	100.0(10)	—(B)	-.-(B)	—(B)	-.-(B)
Industrial Eng [I]		14.0	Y	4.0	-.-(A)	D	O	22.2	54	85.7(10)	20	53.3(10)	6	100.0(10)
Manuf'g Systems Eng [I]		-.-(6)	—(B)	-.-(B)	-.-(B)	—(B)	—(B)	—(B)	—(B)	-.-(B)	2	25.0(10)	—(B)	-.-(B)
Materials Eng [M]		-.-(7)	Y	4.0	-.-(A)	D	O	20.0	15	100.0(10)	15	48.8(10)	3	50.0(10)
Mechanical Eng [M]		27.0	Y	4.0	-.-(A)	D	O	33.6	131	84.3(10)	27	32.9(10)	6	59.3(10)
Textile Chemistry [O]		-.-(8)	—(B)	4.0	-.-(A)	D	O	0.0	6	92.3(10)	—(B)	-.-(B)	—(B)	-.-(B)
Textile Eng [O]		8.0	N	4.0	-.-(A)	D	O	50.0	4	100.0(10)	—(B)	-.-(B)	—(B)	-.-(B)
Textile Mgmt & Tech [O]		-.-(8)	Y	4.0	-.-(A)	D	O	25.0	12	86.2(10)	—(B)	-.-(B)	—(B)	-.-(B)
Textile Science [O](9)		-.-(9)	—(B)	-.-(B)	-.-(B)	—(B)	—(B)	—(B)	—(B)	-.-(B)	0	100.0(10)	—(B)	-.-(B)
013 Baylor U	TX													
Eng [O](1)		10.0	Y	4.0	4.5	D	N	-.-	25	100.0	—(2)	-.-(2)	—(2)	-.-(2)
014 Boston U	MA													
Aerospace Eng [O]		17.0(3)	Y	4.0	4.0	D	O	18.7	56	99.0(1)	9	50.0	1	66.6
Biomedical Eng [O]		21.0	Y	4.0	4.0	D	O	16.8	54	99.0(1)	18	85.0	2	100.0
Computer Eng [O]		33.0(4)	Y	4.0	4.0	D	O	35.5	29	99.0(1)	52	61.2	3(2)	100.0
Electrical Eng [E](2)		33.0(4)	Y	4.0	4.0	D	O	20.0	68	99.0(1)	44	10.0	3(2)	89.0
General Eng [O]		-.-	Y	-.-	-.-	D	O	-.-	-.-	-.-	—.	-.-	—.	-.-
Manuf'g Eng [O]		16.0	Y	4.0	4.0	D	O	17.8	42	99.0(1)	23	53.2	0	100.0
Mechanical Eng [M]		17.0(3)	Y	4.0	4.0	D	O	17.1	38	99.0(1)	23	5.9	1(5)	92.4
Systems Eng [O]		33.0(4)	Y	4.0	4.0	D	O	21.4	10	99.0(1)	40	68.2	3(2)	100.0
015 Bradley U	IL													
Civil Eng [V]		14.0	Y	4.0	4.0(1)	B	O	10.0	21	88.5	10	4.2	—(B)	-.-(B)
Construction Mgmt [O](3)		14.0	N	4.0	4.0(1)	B	O	5.0	15	86.5	—(B)	-.-(B)	—(B)	-.-(B)
Electrical Eng [E](2)		21.3	Y	4.0	4.0(1)	B	O	7.0	42	90.4	13	14.9	—(B)	-.-(B)
Eng Physics [O]		-.-(4)	N	4.0	4.0	B	O	0.0	2	100.0	—(B)	-.-(B)	—(B)	-.-(B)
Industrial Eng [I]		8.5	Y	4.0	4.0(1)	B	O	18.0	34	93.3	6	12.5	—(B)	-.-(B)
Manuf'g Eng [O]		12.5	Y	4.0	4.0(1)	B	O	24.0	21	82.2	0	18.2	—(B)	-.-(B)
Mechanical Eng [M]		12.8	Y	4.0	4.0(1)	B	O	9.0	56	89.2	13	11.6	—(B)	-.-(B)
016 U of Bridgeport	CT													
Computer Eng [O]		7.0	Y	4.0	4.0	B	O	5.0	7	75.0	8	57.5	—.	-.-
Electrical Eng [E]		6.0	Y	4.0	4.0	B	O	5.0	58	77.0	12	41.7	—.	-.-
Mgmt Eng [I]		1.5	N	-.-	-.-	—	—	-.-	0	-.-	—.	-.-	—.	-.-
Mechanical Eng [M]		5.0	Y	4.0	4.0	B	O	5.0	22	72.4	9	31.3	—.	-.-
017 Brigham Young U	UT													
Chemical Eng [H]		13.5	Y	4.0	5.0	B	O	-.-(A)	22	-.-(A)	7	-.-(A)	1	-.-(A)
Civil Eng [V]		16.9	Y	4.0	5.0	B	O	-.-(A)	56	-.-(A)	22	-.-(A)	3	-.-(A)
Electrical & Computer Eng [E]		22.0	Y	4.0	5.0	B	O	-.-(A)	117	-.-(A)	24	-.-(A)	1	-.-(A)
Manuf'g Eng [O]		19.8	Y	4.0	-.-(A)	B	O	-.-(A)	7	-.-(A)	—(A)	-.-(A)	—(A)	-.-(A)
Mechanical Eng [M]		19.7	Y	4.0	5.0	B	O	-.-(A)	139	-.-(A)	11	-.-(A)	5	-.-(A)
018 Bucknell U	PA													
Chemical Eng [H]		8.0(1)	Y	4.0	4.0	D(A)	N(A)	-.-	21	100.0	2	100.0	—(A)	-.-(A)
Civil Eng [V]		8.0(1)	Y	4.0	4.0	D(A)	N(A)	-.-	38	100.0	2	100.0	—(A)	-.-(A)
Computer Science [O]		8.0(1)	N	4.0	4.0	—(A)	N(A)	-.-	13	100.0	—(A)	-.-(A)	—(A)	-.-(A)
Electrical Eng [E]		7.0(1)	Y	4.0	4.0	—(A)	N(A)	-.-	34	100.0	1	100.0	—(A)	-.-(A)
Mechanical Eng [M]		8.0(1)	Y	4.0	4.0	—(A)	N(A)	-.-	31	100.0	6	100.0	—(A)	-.-(A)
019 U of Cal at Berkeley	CA													
Bioengineering [O](1)		-.-	N	4.0	-.-(A)	D	O	9.0	11	100.0	—.	-.-	1	100.0
Chemical Eng [H](2)		20.0	Y	4.0	-.-	D	O	20.0	60	99.0	18	100.0	23	100.0
Civil Eng [V]		43.5	Y	4.0	-.-(A)	D	O	35.0	74	100.0	134	100.0	30	100.0
Electrical Eng & Computer Science [E](1)		78.5	Y	4.0	-.-(A)	D	O	25.0	203	100.0	71	100.0	65	100.0
Eng Geoscience [O]		-.-	N	4.0	-.-(A)	D	O	0.0	0	100.0	—.	-.-	—.	-.-
Eng Math & Statistics [O]		-.-	N	4.0	-.-(A)	D	O	0.0	4	100.0	—.	-.-	—.	-.-
Eng Physics [O]		-.-	N	4.0	-.-(A)	D	O	0.0	6	100.0	—.	-.-	—.	-.-
Industrial Eng & Operations Research [I]		14.0	Y	4.0	-.-(A)	D	O	12.0	34	100.0	32	100.0	12	100.0
Manuf'g Eng [O](3)		-.-	N	4.0	-.-(A)	D	O	0.0	1	100.0	—.	-.-	—.	-.-
Materials Science & Eng [O](1)		23.0	N	4.0	-.-(A)	D	O	19.0	30	100.0	23	100.0	14	100.0
Mechanical Eng [M](1)		42.0	Y	4.0	-.-(A)	D	O	13.0	128	100.0	64	100.0	36	100.0
Mineral Eng [O](4)		-.-	Y	4.0	-.-(A)	D	O	0.0	1	100.0	—.	-.-	—.	-.-
Naval Architecture [O]		3.5	Y	4.0	-.-(A)	D	O	0.0	4	100.0	6	100.0	2	100.0
Nuclear Eng [O](1)		9.0	Y	4.0	-.-(A)	D	O	0.0	9	100.0	7	100.0	3	100.0
Petroleum Eng [O]		-.-	N	4.0	-.-(A)	D	O	0.0	1	100.0	—.	-.-	—.	-.-

All Engineering Disciplines

ENGINEERING Institution Profile Reference Number & Name Name of Degree Program	State/Province	FACULTY Full-time equivalent (FTE)	ABET/CEAB accred?	UNDERGRADUATE PROGRAMS					GRADUATE PROGRAMS			
				Length		Time	Co-op		Bachelor's		Master's		Doctoral	
				Nominal length of program in years	Average length of program in years	Day/Eve./Both	None/Opt./Req.	% of graduates in Co-op programs	# of degrees awarded	% of graduates who were full-time	# of degrees awarded	% of graduates who were full-time	# of degrees awarded	% of graduates who were full-time
020 U of Cal, Davis	CA													
Aeronautical Science & Eng [O][1]		-.-(A)	Y	4.0	-.-(A)	D	O	-.-(A)	5(2)	100.0	—[1]	100.0	—[1]	100.0
Appl Science [O]		11.4	N(4)	-.-(B)	-.-(B)	D	N(B)	-.-(A)	—(B)	-.-(B)	9	100.0	13	100.0
Biological & Agricultural Eng [O]		3.2(3)	Y	4.0	-.-(A)	D	O	-.-(A)	1	100.0	6	100.0	4	100.0
Chemical Eng [H]		11.4	Y	4.0	-.-(A)	D	O	-.-(A)	35	100.0	6	100.0	3	100.0
Chemical Eng & Materials Science [H]		18.4	N	4.0	-.-(A)	D	O(A)	-.-(A)	2	100.0	—[5]	-.-[5]	—[5]	-.-[5]
Civil Eng [V]		25.0	Y	4.0	-.-(A)	D	O	-.-(A)	71	100.0	27	100.0	14	100.0
Civil Eng & Materials Science [V]		32.0	N	4.0	-.-(A)	D	O	-.-(A)	2	100.0	—[6]	-.-[6]	—[6]	-.-[6]
Computer Science & Eng [O]		20.1	Y	4.0	-.-(A)	D	O	-.-(A)	20	100.0	8	100.0	10	100.0
Electrical & Computer Eng [E]		36.6	Y	4.0	-.-(A)	D	O	-.-(A)	91	100.0	30	100.0	5	100.0
Electrical Eng & Materials Science [E]		43.6	N	4.0	-.-(A)	D	O	-.-(A)	4	100.0	—[7]	-.-[7]	—[7]	-.-[7]
Materials Science & Eng [O]		-.-[1]	Y	4.0	-.-(A)	D	O	-.-(A)	5	100.0	—[1]	-.-[1]	—[1]	-.-[1]
Mechanical Eng [M]		37.5	Y	4.0	-.-(A)	D	O	-.-(A)	47	100.0	27	100.0	13	100.0
Mechanical Eng & Materials Science [M]		-.-[1]	N	4.0	-.-(A)	D	O	-.-(A)	10	100.0	—[1]	-.-[1]	—[1]	-.-[1]
021 U of Cal, Irvine	CA													
Biochemical Eng [H][1]		6.0	N(B)	-.-(B)	-.-(B)	D	N	-.-(B)	—(B)	-.-(B)	2	89.0	2	100.0
Civil Eng [V]		17.0	Y	4.0	4.5	D	N	-.-(B)	52	100.0	15	69.0	2	100.0
Electrical & Computer Eng [E]		23.0	Y	4.0	4.5	D	N	-.-(B)	67	100.0	42	77.0	13	100.0
General Eng [O][2]		-.-(3)	N	4.0	4.5	D	N	-.-(B)	0(4)	100.0	—[2]	-.-[2]	—[2]	-.-[2]
Mechanical & Aerospace Eng [M]		22.0	Y	4.0	4.5	D	N	-.-(B)	53	100.0	22	88.0	8	100.0
022 U of Cal, Los Angeles	CA													
Aerospace Eng [O]		8.0	Y	4.0	4.6	—	O	-.-	37	100.0	7	100.0	5	100.0
Chemical Eng [H]		11.0	Y	4.0	4.6	—	O	-.-	28	100.0	8	100.0	3	100.0
Civil Eng [V]		16.0	Y	4.0	4.6	—	O	-.-	62	100.0	29	100.0	11	100.0
Computer Science [O]		29.0	—	4.0	4.6	—	O	-.-	2	100.0	36	100.0	19	100.0
Computer Science & Eng [O]		-.-(6)	Y	4.0	4.6	—	O	-.-	60	100.0	—[5]	100.0	—[5]	100.0
Electrical Eng [E]		35.0	Y	4.0	4.6	—	O	-.-	126	100.0	74	100.0	30	100.0
Eng [O]		-.-(2)	Y	4.0	4.6	—	O	-.-	2	100.0	0	100.0	5	100.0
Man-Machine-Environment Systems [O]		-.-(2)	Y	-.-	-.-	—	—	-.-	—[7]	-.-	2	100.0	1	100.0
Manuf'g Eng [O]		-.-(2)	Y	-.-	-.-	—	—	-.-	—[7]	-.-	8	100.0	—[3]	-.-
Materials Eng [O]		10.0	Y	4.0	4.6	—	O	-.-	13	100.0	—[1]	-.-	—[1]	-.-
Materials Science & Eng [O]		-.-(4)	Y	4.0	-.-	—	—	-.-	—.	-.-	15	100.0	7	100.0
Mechanical Eng [M]		11.0	Y	4.0	4.6	—	O	-.-	80	100.0	25	100.0	13	100.0
Nuclear Eng [O]		8.0	Y	-.-	-.-	—	—	-.-	—[7]	-.-	3	100.0	1	100.0
023 U of Cal, San Diego	CA													
Aerospace Eng [O][1]		36.9(2)	N	-.-	-.-	D	N	-.-	0	0.0	6	100.0	0	100.0
Appl Mechanics [O][1]		36.9(2)	N	4.0	4.7	D	N	-.-	3	100.0	14	100.0	9	100.0
Appl Physics [E]		35.3(2)	N	4.0	4.7	D	N	-.-	1	100.0	22	100.0	9	100.0
Appl Science [O]		-.-	N	-.-	-.-	—	N	-.-	—.	-.-	—.	-.-	—.	-.-
Bioengineering [O][1]		36.9(2)	Y	4.0	4.7	D	N	-.-	16	100.0	15	100.0	3	100.0
Biomedical Eng [O][1]		36.9(2)	N	4.0	4.7	D	N	-.-	21	100.0	—.	-.-	—.	-.-
Chemical Eng [H][1]		36.9(2)	Y	4.0	4.7	D	N	-.-	19	100.0	3	100.0	5	100.0
Computer Eng [O][5]		35.3(2)	N	4.0	4.7	D	N	-.-	31	100.0	2	0.0	—.	-.-
Computer Science [O][3]		28.0	N(4)	4.0	4.7	D	N	-.-	45	100.0	26	100.0	4	100.0
Computer Science & Eng [O][3]		28.0(2)	N(4)	4.0	4.7	D	N	-.-	45	100.0	26	100.0	4	100.0
Electrical Eng [E]		35.3(2)	Y	4.0	4.7	D	N	-.-	75	100.0	39	100.0	15	100.0
Eng Physics [O]		-.-	N	-.-	-.-	D	N	-.-	—.	-.-	—.	-.-	—.	-.-
Eng Science(s) [O]		-.-	N	-.-	-.-	D	N	-.-	—.	-.-	—.	-.-	—.	-.-
Materials Science [O]		-.-	N	-.-(A)	-.-	—(B)	N	-.-	—.	-.-	—.	-.-	—.	-.-
Mechanical Eng [M]		36.9(2)	Y	4.0	4.7	D	N	-.-	84	100.0	9	100.0	0	100.0
Structural Eng [O][1]		36.9(2)	Y	4.0	4.7	D	N	-.-	21	100.0	2	100.0	1	100.0
Systems & Control Eng [O]		-.-	Y	4.0	-.-	D	N	-.-	—.	-.-	—.	-.-	—.	-.-
Systems Science [O]		-.-	N	-.-(B)	-.-	—(B)	N	-.-	—.	-.-	—.	-.-	—.	-.-
024 U of Cal, Santa Barbara	CA													
Chemical Eng [H]		19.3(2)	Y	4.0	4.2	—(B)	N	-.-(B)	21	100.0	2	100.0	3	100.0
Computer Science [O]		16.3	N(4)	4.0	4.2	B	O	-.-(7)	59	100.0	28	100.0	1	100.0
Electrical & Computer Eng [E][6]		31.4	Y	4.0	4.2	—(B)	N	-.-(B)	74	100.0	58	95.0[1]	20	95.0[1]
Materials Eng [O][5]		14.1	—	-.-(B)	-.-(B)	—(B)	N(B)	-.-(B)	—(B)	-.-(B)	6	100.0	12	100.0
Mechanical Eng [M]		21.3	Y	4.0	4.2	—(B)	N	-.-(B)	78	100.0	28	100.0	8	100.0
Nuclear Eng [O]		19.3(2)	Y	4.0	4.2	—(B)	N	-.-(B)	10	100.0	0	100.0	—[3]	-.-[3]
025 U of Cal, Santa Cruz	CA													
Computer Eng [O]		11.0	Y	4.0	4.0	—	N	-.-	18	89.0	11	95.0	2	90.0
026 Cal Inst of Tech	CA													
Aeronautics [O]		10.0	N	-.-(B)	-.-	—(B)	—(B)	-.-	—.	-.-	13	100.0	8	100.0
Appl Mathematics [O]		6.0	N	4.0	-.-	D	—(B)	-.-	2	100.0	1	100.0	5	100.0
Appl Mechanics [O]		6.0	N	-.-(B)	-.-	—(B)	—(B)	-.-	—.	-.-	3	100.0	4	100.0

All Engineering Disciplines

ENGINEERING Institution Profile Reference Number & Name Name of Degree Program	State/Province	FACULTY Full-time equivalent (FTE)	ABET/CEAB accred?	UNDERGRADUATE PROGRAMS Length Nominal length of program in years	Length Average length of program in years	Time Day/Eve./Both	Co-op None/Opt./Req.	Co-op % of graduates in Co-op programs	Bachelor's # of degrees awarded	Bachelor's % of graduates who were full-time	GRADUATE PROGRAMS Master's # of degrees awarded	Master's % of graduates who were full-time	Doctoral # of degrees awarded	Doctoral % of graduates who were full-time
Appl Physics [O]		10.0	N	4.0	-.-	D	—(B)	-.-	15	100.0	7	100.0	7	100.0
Chemical Eng [H]		8.0	Y	4.0	-.-	D	—(B)	-.-	7	100.0	8	100.0	8	100.0
Civil Eng [V]		6.0	N	-.-(B)	-.-	—(B)	—(B)	-.-	—.	-.-	3	100.0	2	100.0
Computation & Neural Systems [O]		1.0	N	-.-(B)	-.-	—(B)	—(B)	-.-	—.	-.-	1	100.0	3	100.0
Computer Science [O]		10.0	N	-.-(B)	-.-	—(B)	—(B)	-.-	—.	-.-	4	100.0	5	100.0
Electrical Eng [E]		15.0	N	4.0	-.-	D	—(B)	-.-	29	100.0	32	100.0	9	100.0
Eng & Appl Science [O]		-.-	Y	4.0	-.-	D	—(B)	-.-	68	100.0	—.	-.-	—.	-.-
Eng Science(s) [O]		1.0	N	-.-(B)	-.-	—(B)	—(B)	-.-	—.	-.-	—.	-.-	1	100.0
Env'l Eng Science [O]		7.0	N	-.-(B)	-.-	—(B)	—(B)	-.-	—.	-.-	9	100.0	4	100.0
Materials Science [O]		3.0	N	-.-(B)	-.-	—(B)	—(B)	-.-	—.	-.-	1	100.0	2	100.0
Mechanical Eng [M]		11.0	N	-.-(B)	-.-	—(B)	—(B)	-.-	—.	-.-	9	100.0	4	100.0
027 Cal Poly State U	CA													
Aeronautical Eng [O]		7.7	Y	4.0	4.9	D	O	-.-	49	-.-	4	-.-	—.	-.-
Agricultural Eng [O]		13.3	Y	4.0	4.7	D	O	-.-	24	-.-	—.	-.-	—.	-.-
Architectural Eng [O]		12.6	Y	4.0	4.7	D	O	-.-	50	-.-	—.	-.-	—.	-.-
Civil & Env'l Eng [V]		16.9	Y	4.0	4.7	D	O	-.-	110	-.-	4	-.-	—.	-.-
Computer Eng [O]		-.-	N(3)	4.0	4.9	D	O	-.-	19	-.-	—.	-.-	—.	-.-
Electrical & Electronics Eng [E]		26.1(2)	Y	4.0	4.8	D	O	-.-	32	-.-	7	-.-	—.	-.-
Eng Science(s) [O]		-.-	N	4.0	-.-	D	O	-.-	7	-.-	—.	-.-	—.	-.-
Industrial Eng [I]		15.8	Y	4.0	5.0	D	O	-.-	44	-.-	—.	-.-	—.	-.-
Materials Eng [O]		-.-	Y	4.0	4.7	D	O	-.-	17	-.-	—.	-.-	—.	-.-
Mechanical Eng [M]		27.2	Y	4.0	5.0	D	O	-.-	146	-.-	—.	-.-	—.	-.-
028 Cal State U, Chico	CA													
Civil Eng [V]		8.7	Y	4.0	-.-(A)	D	O	-.-(A)	34	90.0	—(B)	-.-(B)	—(B)	-.-(B)
Computer Eng [O]		8.7	Y	4.0	-.-(A)	D	O	-.-(A)	34	90.0	—(B)	-.-(B)	—(B)	-.-(B)
Electrical & Electronics Eng [E]		10.3	Y	4.0	-.-(A)	D	O	-.-(A)	35	88.0	3	13.0	—(B)	-.-(B)
Mechanical Eng [M]		8.0	Y	4.0	-.-(A)	D	O	-.-(A)	40	92.0	—(B)	-.-(B)	—(B)	-.-(B)
Mechanical Eng & Manuf'g [M]		8.0	Y	4.0	-.-(A)	D	O	-.-(A)	40	92.0	—(B)	-.-(B)	—(B)	-.-(B)
029 Cal State U, Fresno	CA													
Civil Eng [V]		6.5	Y	4.0	4.5	B(3)	O	3.0	35	70.0	8	25.0	—.	-.-(1)
Computer Eng [O]		-.-	N(4)	4.0	4.5	D	O	-.-	0	93.0	—.	-.-	—.	-.-
Computer Science [O]		7.0	—	4.0	5.0	D	O	2.0	45	85.0	0(4)	60.0	—.	-.-
Construction Mgmt [V]		4.0	—(2)	4.0	4.5	B	O	5.0	40	40.0	—.	-.-	—.	-.-
Electrical Eng [E]		10.0	Y	4.0	4.5	D	O	4.0	76	4.0	—(B)	-.-(B)	—(B)	-.-(B)
Industrial Eng [I]		5.0	Y	4.0	4.5	B	O	10.0	10	4.5	—.	-.-	—.	-.-
Mechanical Eng [M]		7.0	Y	4.0	5.0	D	O	5.0(C)	42	75.0	—.	-.-	—.	-.-
Surveying Eng [O]		5.0	Y	4.0	5.0	B	O	8.0	25	100.0	1	75.0	—.	-.-(1)
030 Cal State U, Long Beach	CA													
Aerospace Eng [O](1)		7.6	—(1)	-.-(1)	-.-(1)	E	N	-.-(B)	—(1)	-.-(1)	2	0.0	—.	-.-(B)
Chemical Eng [H]		5.1	Y	4.0	5.5	B	N	-.-	15	55.0	—.	-.-	—.	-.-
Civil Eng [V]		12.4	Y	4.0	5.5	B	N	-.-	48	55.0	32	10.0	—.	-.-(B)
Computer Eng & Computer Science [O]		18.4	Y	4.0	5.5	B	N	-.-(B)	90	55.0	21	10.0	—.	-.-
Electrical Eng [E]		25.1	Y	4.0	5.5	B	N	-.-	90	55.0	21	10.0	—.	-.-
Eng & Industrial Tech [O]		-.-	N	-.-	-.-	B	N	-.-	—.	-.-	—(2)	-.-(2)	—(2)	-.-(2)
Eng Tech [O](2)		16.1	N	4.0	5.5	B	N	-.-(B)	128	30.0	—.	-.-	—.	-.-
Mechanical Eng [M]		16.4	Y	4.0	5.5	B	N	-.-	103	55.0	21	10.0	—.	-.-
031 Cal State U, Los Angeles	CA													
Civil Eng [V]		9.9	Y	4.0	4.5	B	O	-.-(A)	20	61.0	3	5.0	—(B)	-.-(B)
Electrical Eng [E]		17.3	Y	4.0	4.5	B	O	-.-(A)	87	62.0	26	22.0	—(B)	-.-(B)
Mechanical Eng [M]		10.1	Y	4.0	4.0	B	O	-.-(A)	17	57.0	4	5.0	—(B)	-.-(B)
032 Cal State U, Northridge	CA													
Computer Science [O]		22.0	N(1)	4.0	5.0	B	N	-.-	67	52.0	35	10.0	—(B)	-.-(B)
Eng [O]		50.0	Y	4.0	5.0	B	N	-.-	194	54.0	116	7.0	—(B)	-.-(B)
033 Cal State U, Sacramento	CA													
Civil Eng [V]		20.1	Y	4.0	5.0	B	O	7.0	58	-.-(A)	9	-.-(A)	—(B)	-.-(B)
Computer Eng [O]		5.0	Y	4.0	5.0	B	O	10.0	20	-.-(A)	—(B)	-.-(B)	—(B)	-.-(B)
Electrical & Electronics Eng [E]		15.0	Y	4.0	5.0	B	O	3.3	90	-.-(A)	—(B)	-.-(B)	—(B)	-.-(B)
Mechanical Eng [M]		17.0	Y	4.0	5.0	B	O	4.0	74	-.-(A)	—(B)	-.-(B)	—(B)	-.-(B)
034 Cal State Poly, Pomona	CA													
Aerospace Eng [O]		6.4	Y	4.0	5.5	B	O	-.-	50	78.0	—.	-.-	—.	-.-
Agricultural Eng [O]		3.0	Y	4.0	5.5	B	O	-.-	4	76.0	—.	-.-	—.	-.-
Chemical Eng [H]		8.4	Y	4.0	5.5	B	O	-.-	50	78.0	—.	-.-	—.	-.-

All Engineering Disciplines

ENGINEERING Institution Profile Reference Number & Name Name of Degree Program	State/Province	FACULTY Full-time equivalent (FTE)	UNDERGRADUATE PROGRAMS							GRADUATE PROGRAMS			
			ABET/CEAB accred?	Length		Time	Co-op		Bachelor's		Master's		Doctoral	
				Nominal length of program in years	Average length of program in years	Day/Eve./Both	None/Opt./Req.	% of graduates in Co-op programs	# of degrees awarded	% of graduates who were full-time	# of degrees awarded	% of graduates who were full-time	# of degrees awarded	% of graduates who were full-time
Civil Eng [V]		17.0	Y	4.0	5.5	B	O	-.-	116	81.0	—.	-.-	—.	-.-
Electrical & Computer Eng [E]		38.5	Y	4.0	5.5	—	—	-.-	146	76.0	—.	-.-	—.	-.-
Electrical Eng [E]		-.-	Y	4.0	-.-	B	O	-.-	—.	-.-	—.	-.-	—.	-.-
Industrial Eng [I]		7.0	Y	4.0	5.5	B	O	-.-	35	83.0	—.	-.-	—.	-.-
Manuf'g Eng [O]		6.3	Y	4.0	5.5	B	O	-.-	16	65.0	—.	-.-	—.	-.-
Mechanical Eng [M]		23.8	Y	4.0	5.5	B	O	-.-	67	75.0	—.	-.-	—.	-.-
035 Calvin Coll	MI													
Eng [O]		8.0	Y	4.0	-.-(A)	D	N	-.-(B)	54	-.-(A)	—(B)	-.-(B)	—(B)	-.-(B)
036 Carnegie Mellon U	PA													
Biomedical Eng [O][1]		2.5	N[1]	4.0[1]	4.0[1]	D[1]	N	-.-(1)	0[1]	0.0[1]	1	100.0	0	71.0
Chemical Eng [H]		15.5	Y	4.0	4.0	D	N	-.-(B)	25	99.0	5	37.5	11	98.1
Civil Eng [V]		14.0	Y	4.0	4.0	D	N	-.-(B)	32	100.0	17	95.0	8	94.0
Electrical & Computer Eng [E]		34.5	Y	4.0	4.0	D	N	-.-(B)	98	98.6	50	97.2	31	97.5
Eng & Public Policy [O][2]		8.2	Y	4.0	4.0	D	N	-.-(B)	—(B)	-.-(B)	7	100.0	4	95.5
Manuf'g Eng [O][2]		-.-(B)	N[2]	1.0	1.0	D	N	-.-(B)	—(B)	-.-(B)	11	100.0	—(B)	-.-(B)
Materials Science & Eng [O]		18.5	Y	4.0	4.0	D	O	50.0	13	98.5	11	97.5	7	93.6
Mechanical Eng [M]		23.6	Y	4.0	4.0	D	N	-.-(B)	62	98.9	22	89.0	9	98.0
Metallurgical Eng & Material Science [O]		-.-	Y	4.0	-.-	D	O	-.-	—.	-.-	—.	-.-	—.	-.-
037 Case Western Reserve U	OH													
Aerospace [O]		16.0[3]	N	4.0	4.3	B	O	0.0	1	100.0	—(B)	-.-(B)	—(B)	-.-(B)
Biomedical Eng [O]		15.0	Y	4.0	4.3	B	O	2.0	23	100.0	19	90.0	11	100.0
Chemical Eng [H]		10.0	Y	4.0	4.3	B	O	10.0	26	8.0	10	100.0	4	100.0
Civil Eng [V]		7.0	Y	4.0	4.3	B	O	30.0	7	100.0	14	50.0	—(B)	-.-(B)
Computer Eng [O]		11.0	Y	4.0	4.3	B	O	10.0	23	90.0	20	50.0	6	100.0
Computer Science & Eng [O]		11.0	N	4.0	4.3	B	O	10.0	5	100.0	14	100.0	—(B)	-.-(B)
Electrical Eng & Appl Physics [E]		12.0	Y	4.0	4.3	B	O	30.0	36	80.0	17	50.0	5	100.0
Fluid & Thermal Eng [O]		16.0[4]	Y	4.0	4.3	B	O	10.0	8	100.0	—(B)	-.-(B)	—(B)	-.-(B)
Industrial Eng [I]		8.0[2]	N	4.0	4.3	B	O	30.0	5	60.0	—(B)	-.-(B)	—(B)	-.-(B)
Macromolecular Science & Eng [O][1]		13.0	Y	4.0	4.3	B	O	25.0	4	100.0	21	66.0	10	100.0
Materials Science & Eng [O]		11.0	Y	4.0	4.3	B	O	25.0	4	100.0	21	66.6	10	100.0
Mechanical Eng [M]		16.0	Y	4.0	4.3	B	O	10.0	35	70.0	28	50.0	10	90.0
Polymer Science & Eng [O]		13.0	Y	4.0	4.3	B	O	25.0	4	100.0	8	100.0	12	100.0
Systems & Control Eng [O]		5.0	Y	4.0	4.3	B	O	10.0	17	90.0	18	75.0	6	100.0
038 The Catholic U of America	DC													
Biomedical Eng [O]		5.0	Y	4.0	4.0	D	O	0.0	17	100.0	4	50.0	—.	-.-
Civil Eng [V][1]		11.5	Y	4.0	4.0	D	O[2]	7.5	14	100.0	10	70.0	—.	-.-
Computer Science Eng [O][3]		1.0	N	4.0	4.0	D	O	0.0	2	100.0	—.	-.-	—.	-.-
Computer Science/Eng [E]		-.-	N	4.0	-.-	D	O	-.-	—.	-.-	—.	-.-	—.	-.-
Electrical Eng [E]		9.0	Y	4.0	4.0	D	O	10.0	8	90.0	14	10.0	2	0.0
Mechanical Eng [M]		10.0	Y	4.0	4.0	D	O	10.0	12	93.0	14	7.0	4	100.0
039 U of Central Florida	FL													
Aerospace Eng [O]		7.0	Y	4.0	4.5(C)	B	O	15.0(C)	12	33.5	—(B)	-.-(B)	—(B)	-.-(B)
Civil Eng [V]		11.0	Y	4.0	5.0(C)	B	O	15.0(C)	34	38.4	6	22.0	1	31.0(C)
Computer Eng [O]		12.0	Y	4.0	4.5(C)	B	O	15.0(C)	30	45.0	17	48.0	0	40.0(C)
Electrical Eng [E]		34.0	Y	4.0	4.5(C)	B	O	15.0(C)	152	63.0	43	32.0	2	26.0
Eng Tech [O]		12.0	Y	4.0	5.0(1)	B	O	20.0(C)	96	33.3	—(B)	-.-(B)	—(B)	-.-(B)
Env'l Eng [O]		8.0	Y	4.0	4.5(C)	B	O	15.0	11	68.0	4	41.3	0	40.0(C)
Industrial Eng [I]		17.0	Y	4.0	4.5(C)	B	O	15.0(C)	28	66.7	18	52.0	2	50.0(C)
Mechanical Eng [M]		12.0	Y	4.0	4.5	B	O	15.0(C)	50	68.5	11	16.0	1	20.0(C)
040 Central State U	OH													
Manuf'g [O]		7.3	Y	4.0	4.5(C)	D	N	-.-	4	98.0	—.	-.-	—.	-.-(1)
041 Christian Brothers U	TN													
Chemical Eng [H]		3.0	Y	4.0	4.7	D	N	-.-(B)	9	77.5	—(B)	-.-(B)	—(B)	-.-(B)
Civil Eng [V]		4.0	Y	4.0	4.7	D	N	-.-(B)	14	94.7	—(B)	-.-(B)	—(B)	-.-(B)
Electrical Eng [E]		6.0	Y	4.0	4.7	D	N	-.-(B)	23	86.8	—(B)	-.-(B)	—(B)	-.-(B)
Eng Mgmt [O]		2.5	N	2.0	2.0	E	N	-.-(B)	—(B)	-.-(B)	10	-.-(1)	—(B)	-.-(B)
Mechanical Eng [M]		6.0	Y	4.0	4.7	D	N	-.-(B)	15	86.8	—(B)	-.-(B)	—(B)	-.-(B)
042 U of Cincinnati	OH													
Aerospace Eng [O]		24.0(1)	Y	5.0	5.2	D	R	100.0	42	100.0	49	74.0	7	74.0
Chemical Eng [H]		17.0(1)	Y	5.0	5.2	D	R	100.0	50	100.0	4	96.1	7	81.5
Civil Eng [V]		24.0(1)	Y	5.0	5.2	D	R	100.0	47	100.0	11	72.2	4	81.3
Computer Eng [O]		30.0(1)	Y	5.0	5.2	D	R	100.0	6	100.0	10	59.3	2	95.0
Computing Sciences [O][2]		30.0(1)	—(B)	-.-(B)	-.-(B)	—(B)	—(B)	-.-(B)	—(B)	-.-(B)	—.	-.-	0	100.0
Electrical Eng [E]		30.0(1)	Y	5.0	5.2	D	R	100.0	50	100.0	24	59.3	5	87.2

All Engineering Disciplines

ENGINEERING Institution Profile Reference Number & Name Name of Degree Program	State/Province	FACULTY Full-time equivalent (FTE)	ABET/CEAB accred?	UNDERGRADUATE PROGRAMS Length Nominal length of program in years	Length Average length of program in years	Time Day/Eve./Both	Co-op None/Opt./Req.	Co-op % of graduates in Co-op programs	Bachelor's # of degrees awarded	Bachelor's % of graduates who were full-time	GRADUATE PROGRAMS Master's # of degrees awarded	Master's % of graduates who were full-time	Doctoral # of degrees awarded	Doctoral % of graduates who were full-time
Eng Mechanics [O]		24.0(1)	Y	5.0	5.2	D	R	100.0	18	100.0	5	46.7	2	75.0
Env'l Eng [O](3)		24.0(1)	Y(B)	-.-(B)	-.-(B)	—(B)	—(B)	-.-(B)	—(B)	-.-(B)	21	79.4	6	81.3
Env'l Science [O](2)		24.0(1)	N(B)	-.-(B)	-.-(B)	—(B)	—(B)	-.-(B)	—(B)	-.-(B)	11	58.0	1	50.0
Health Physics [O](2)		30.0(1)	—(B)	-.-(B)	-.-(B)	—(B)	—(B)	-.-(B)	—(B)	-.-(B)	6	87.5	—.	-.-
Industrial Eng [I]		30.0(1)	Y	5.0	5.2	D	R	100.0	31	100.0	14	45.1	4	78.9
Materials Eng [O]		17.0(1)	N(4)	5.0	5.2	D	R	100.0	24	100.0	—(B)	-.-(B)	—(B)	-.-(B)
Materials Science [O](2)		17.0(1)	—(B)	-.-(B)	-.-(B)	—(B)	—(B)	-.-(B)	—(B)	-.-(B)	4	75.0	8	90.0
Mechanical Eng [M]		30.0(1)	Y	5.0	5.2	D	R	100.0	23	100.0	30	58.8	10	78.2
Metallurgical Eng [O](5)		17.0(1)	Y(5)	-.-(5)	-.-(5)	D(5)	R(5)	-.-(5)	—(5)	-.-(5)	2	0.0	0	66.7
Nuclear Eng [O]		30.0(1)	Y	5.0	5.2	D	R	100.0	11	100.0	1	28.6	0	92.0
Solid State Electronics [O](2)		30.0(1)	—(B)	-.-(B)	-.-(B)	—(B)	—(B)	-.-(B)	—(B)	-.-(B)	—.	-.-	—.	-.-
043 The Citadel	SC													
Civil Eng [V]		-.-	Y	4.0	4.0	B	N	-.-	19	30.0	—.	-.-	—.	-.-
Electrical Eng [E]		-.-	Y	4.0	4.0	B	N	-.-	18	30.0	—.	-.-	—.	-.-
044 CUNY-City Coll	NY													
Chemical Eng [H]		10.6	Y	4.0	5.0	D(1)	O	-.-(A)	7	88.8	6	7.5	3	100.0
Civil Eng [V]		16.8	Y	4.0	5.0	D(1)	O	-.-(A)	46	74.3	31	8.4	0	89.4
Computer Science [O]		17.6	N(2)	4.0	5.0	B	O	-.-(A)	54	79.0	81	13.9	—(3)	-.-(3)
Electrical Eng [E]		31.5	Y	4.0	5.0	B	O	-.-(A)	134	74.9	33	24.4	6	100.0
Mechanical Eng [M]		19.3	Y	4.0	5.0	D(1)	O	-.-(A)	42	83.4	19	9.6	2	100.0
045 CUNY-Staten Island	NY													
Eng Science(s) [O]		21.0	Y	4.0	-.-	—	N	-.-	28	-.-	—.	-.-	—.	-.-
046 Clarkson U	NY													
Chemical Eng [H]		15.0	Y	4.0	-.-	D	O	-.-(A)	42	99.9	11	100.0	7	100.0
Civil & Env'l Eng [V]		18.0	Y	4.0	-.-	D	O	-.-(A)	104	99.9	16	100.0	3	100.0
Electrical & Computer Eng [E]		22.0	Y	4.0	-.-	D	O	-.-(A)	144	99.9	15	99.9	2	99.9
Mechanical & Aeronautical Eng [M]		21.5	Y	4.0	-.-	D	O	-.-(A)	154	99.9	14	-.-	6	-.-
047 Clemson U	SC													
Agriculture & Biological Eng [O]		9.0	Y	4.0	-.-	D	O	-.-	9	100.0	4	66.0	2	90.0
Bioengineering [O]		8.0	N	-.-	-.-	D	—	-.-	0	0.0	11	81.0	3	100.0
Ceramic Eng [O]		11.0	Y	4.0	-.-	D	O	-.-	39	96.0	5	78.0	4	81.0
Chemical Eng [H]		12.0	Y	4.0	-.-	D	O	-.-	46	81.0	5	100.0	2	87.0
Civil Eng [V]		18.0	Y	4.0	-.-	D	O	-.-	91	93.0	22	63.0	4	100.0
Electrical & Computer Eng [E]		44.0	Y	4.0	-.-	D	O	-.-	143	87.0	50	82.0	9	90.0
Env'l Systems Eng [O]		10.0	Y	-.-	-.-	D	—	-.-	0	0.0	22	68.0	4	84.0
Industrial Eng [I]		9.0	Y	4.0	-.-	D	O	-.-	58	87.0	11	76.0	1	60.0
Mechanical Eng [M]		31.0	Y	4.0	-.-	D	O	-.-	138	86.0	44	94.0	5	73.0
048 Cleveland State U	OH													
Chemical Eng [H]		9.8	Y	4.0	4.8(1)	D	O	50.0	16	77.6	5	50.0	1	100.0
Civil Eng [V]		9.5	Y	4.0	4.8(1)	D	O	23.5	18	76.6	19	30.0	—.	-.-
Electrical Eng [E](2)		16.5	Y	4.0	5.2(1)	B	O	44.6	48	66.4	24	52.2	2	100.0
Electronic Tech [O]		4.7	N	2.0(3)	4.9(1)	B	O	1.0	29	15.2	—(4)	-.-(4)	—(4)	-.-(4)
Industrial Eng [I]		6.8	Y	4.0	4.3(1)	D	O	42.9	6	74.3	24	53.3	—.	-.-
Mechanical Eng [M]		12.0	Y	4.0	5.9(1)	B	O	22.5	47	57.3	10	23.1	4	100.0
Mechanical Tech [O]		3.0	N	2.0(4)	6.4(1)	E	O	1.0	26	15.7	—(4)	-.-(4)	—(4)	-.-(4)
049 U of Colorado at Boulder	CO													
Aerospace Eng Sciences [O]		47.0	Y	4.0	4.5	D	O	-.-	103	98.1	57	91.9	7	91.9
Appl Mathematics [O]		-.-	N	4.0	4.5	D	O	-.-	27	100.0	—(B)	-.-(B)	—(B)	-.-(B)
Chemical Eng [H]		17.0	Y	4.0	4.5	D	O	-.-	27	95.3	3	90.0	2	90.0
Civil, Env'l, & Architectural Eng [V]		33.0	Y	4.0	4.5	D	O	-.-	55	97.8	43	85.3	8	85.3
Computer Science [O]		34.0	N	4.0	4.5	D	O	-.-	42	94.7	69	81.8	14	81.8
Electrical & Computer Eng [E]		53.0	Y	4.0	4.5	D	O	-.-	73	97.4	73	87.6	15	87.6
Mechanical Eng [M]		22.0	Y	4.0	4.5	D	O	-.-	48	97.9	16	87.5	4	87.5
Telecommunications [O]		-.-	N	-.-	-.-	—	N	-.-	—(B)	-.-(B)	39	66.0	—(B)	-.-(B)
050 U of Colorado-Col Spr	CO													
Appl Mathematics [O]		12.0	N	4.0	5.0	B	O	20.0	9	55.5	2	58.3	—(B)	-.-(B)
Computer Science [O]		14.0	N(1)	4.0	5.0	B	O	20.0	18	49.0	11	43.0	—(B)	-.-(B)
Electrical & Computer Eng [E]		13.8	Y	4.0	5.0	B	O	22.0	49	60.0	24	54.0	4	60.0
Space Operations [O]		1.0	N	2.5	3.5	—(B)	N	-.-(B)	—(B)	-.-(B)	11	65.0	—(B)	-.-(B)
051 U of Colorado at Denver	CO													
Appl Mathematics [O](1)		-.-(2)	N	4.0	4.5	B	O	0.0	5	60.0	—(1)	-.-(1)	—(1)	-.-(1)
Civil Eng [V]		13.4	Y	4.0	4.5	B	O	5.0	20	60.0	15	20.0	—(3)	-.-(3)
Computer Science & Eng [O]		12.8	N(B)	4.0	4.5	B	O	5.0	19	60.0	—(4)	-.-(4)	—(3)	-.-(3)

All Engineering Disciplines

ENGINEERING Institution Profile Reference Number & Name / Name of Degree Program	State/Province	FACULTY Full-time equivalent (FTE)	ABET/CEAB accred?	UNDERGRAD Length Nominal length of program in years	UNDERGRAD Length Average length of program in years	UNDERGRAD Time Day/Eve./Both	UNDERGRAD Co-op None/Opt./Req.	UNDERGRAD Co-op % of graduates in Co-op programs	UNDERGRAD Bachelor's # of degrees awarded	UNDERGRAD Bachelor's % of graduates who were full-time	GRAD Master's # of degrees awarded	GRAD Master's % of graduates who were full-time	GRAD Doctoral # of degrees awarded	GRAD Doctoral % of graduates who were full-time
Electrical Eng [E]		16.1	Y	4.0	4.5	B	O	5.0	69	50.0	31	20.0	—.(3)	-.-(3)
Mechanical Eng [M]		10.1	Y	4.0	4.5	B	O	1.0	36	50.0	3	20.0	—.(3)	-.-(3)
052 Colorado School of Mines	CO													
Appl & Eng Physics [O]		12.0	Y	4.0	4.0	D	O	-.-(A)	16	100.0	5	75.0	4	85.0
Chemical Eng & Petroleum Refining [H]		12.0	Y	4.0	4.0	D	O	-.-(A)	16	100.0	4	80.0	5	90.0
Chemistry [O]		13.0	N	4.0	4.0	D	O	-.-(A)	5	100.0	4	65.0	9	65.0
Eng [O]		17.0	Y	4.0	4.0	D	O	-.-(A)	108	100.0	4	55.0	—.(B)	-.-(B)
Geological Eng [O]		16.0	Y	4.0	4.0	D	O	-.-(A)	15	100.0	29	45.0	5	50.0
Geophysical Eng [O]		9.0	Y	4.0	4.0	D	O	-.-(A)	6	100.0	14	78.0	6	90.0
Mathematics [O]		17.0	N	4.0	4.0	D	O	-.-(A)	17	100.0	13	70.0	1	78.0
Metallurgical & Materials Eng [O]		16.0	Y	4.0	4.0	D	O	-.-(A)	16	100.0	8	74.0	5	85.0
Mining Eng [O]		7.0	Y	4.0	4.0	D	O	-.-(A)	9	100.0	9	79.0	6	100.0
Petroleum Eng [O]		5.0	Y	4.0	4.0	D	O	-.-(A)	22	100.0	6	55.0	2	64.0
053 Colorado State U	CO													
Agricultural Eng [O]		9.0	Y	4.0	4.5	D(A)	N	-.-(A)	3	99.0(C)	9	95.0(C)	3	100.0
Chemical Eng [H]		8.0	Y	4.0	4.5	D(A)	N	-.-(A)	16	99.0(C)	6	95.0(C)	0	95.0(C)
Civil Eng [V]		33.0	Y	4.0	4.5	D(A)	N	-.-(A)	61	93.0	46	74.0	19	40.0
Electrical Eng [E]		19.0	Y	4.0	4.5	D(A)	N	-.-(A)	31	90.0	18	63.0	11	79.0
Eng Science(s) [O]		-.-(B)	Y	4.0	4.5	D(A)	N	-.-(A)	20	96.0	—.(B)	-.-(B)	—.(B)	-.-(B)
Mechanical Eng [M]		23.0	Y	4.0	4.5	D(A)	N	-.-(A)	77	73.0	15	63.0	6	58.0
054 Columbia U	NY													
Appl Physics [O](3)		11.7	N	4.0	4.0	—(B)	N	-.-(B)	16	100.0	15	91.0	1	91.0
Chemical Eng [H](1)		11.0	Y	4.0	4.0	—(B)	N	-.-(B)	27	100.0	17	70.9	2	70.9
Civil Eng & Eng Mechanics [V]		16.0	Y	4.0	4.0	—(B)	N	-.-(B)	15	100.0	31	49.5	2	49.5
Computer Science [O]		24.3	N	4.0	4.0	—(B)	N	-.-(B)	35	100.0	73	73.5	6	73.5
Electrical Eng [E](4)		24.7	Y	4.0	4.0	—(B)	N	-.-(B)	47	100.0	88	63.8	15	63.8
Industrial Eng & Operations Research [I]		11.7	Y(2)	4.0	4.0	—(B)	N	-.-(B)	44	100.0	39	53.5	5	53.5
Mechanical Eng [M]		12.3	Y	4.0	4.0	—(B)	N	-.-(B)	47	100.0	38	80.0	11	80.0
Metallurg'l, Mining, Mineral & Mat'ls Eng [O](5)		15.7	Y	4.0	4.0	—(B)	N	-.-(B)	4	100.0	10	71.4	17	71.4
055 Concordia U	PQ													
Building Eng [O]		-.-	Y	4.0	-.-	B	O	-.-	27	78.0	9	68.0	4	82.0
Civil Eng [V]		-.-	Y	4.0	-.-	B	O	-.-	15	74.0	9	73.0	3	86.0
Computer Eng [O]		-.-	Y	4.0	-.-	B(1)	N	-.-	23	86.0	—.	-.-	—.	-.-
Electrical Eng [E]		-.-	Y	4.0	-.-	B	N	-.-	52	78.0	18	80.0	2	92.0
Mechanical Eng [M]		-.-	Y	4.0	-.-	B	N	-.-	82	81.0	24	87.0	2	83.0
056 U of Conn	CT													
Chemical Eng [H]		14.0	Y	4.0	4.5	D	O	13.0	8	95.0(C)	13(1)	55.0(1)	11(1)	93.0(1)
Civil Eng [V]		23.0	Y	4.0	4.5	D	O	22.0	55	95.0(C)	14	48.0	1(2)	77.0(2)
Computer Science & Eng [O]		19.0	Y(5)	4.0	4.5	D	O	48.0	31	95.0(C)	14	84.0	4	79.0
Electrical Eng [E]		22.0	Y	4.0	4.5	D	O	44.0	80	95.0(C)	22(3)	52.0(3)	2(3)	82.0(3)
Mechanical Eng [M]		21.0	Y	4.0	4.5	D	O	27.0	78	95.0(C)	23(4)	40.0(4)	13	52.0
Metallurgy [O]		10.0	—(B)	-.-(B)	-.-(B)	—(B)	—(B)	-.-(B)	—(B)	-.-(B)	5(6)	37.0(6)	2(6)	87.0(6)
057 The Cooper Union	NY													
Chemical Eng [H]		7.5	Y	4.0	4.0	D	N	-.-	14	100.0	2	100.0	—.	-.-
Civil Eng [V]		8.0	Y	4.0	4.0	D	N	-.-	25	100.0	5	100.0	—.	-.-
Electrical Eng [E]		10.5	Y	4.0	4.0	D	N	-.-	33	100.0	8	100.0	—.	-.-
Eng [O](1)		15.0	N	4.0	4.0	—	—	-.-	10	100.0	—.	-.-	—.	-.-
Mechanical Eng [M]		8.6	Y	4.0	4.0	D	N	-.-	23	100.0	3	100.0	—.	-.-
058 Cornell U	NY													
Agricultural & Biological Eng [O]		-.-	Y	4.0	-.-	D	O	-.-	—.	-.-	—.	-.-	—.	-.-
Agriculture & Biological Eng [O]		22.0	Y	4.0	-.-	—	O	3.0	28	-.-	11	-.-	4	-.-
Appl & Eng Physics [O]		11.5	Y	4.0	-.-	D	O	30.0	30	-.-	23	-.-	18	-.-
Chemical Eng [H]		17.0	Y	4.0	-.-	D	O	32.0	53	-.-	14	-.-	12	-.-
Civil & Env'l Eng [V]		28.3	Y	4.0	-.-	D	O	17.0	57	-.-	58	-.-	12	-.-
Coll Program [O]		13.0	N	4.0	-.-	D	O	0.0	2	-.-	—.	-.-	—.	-.-
Computer Science [O]		25.5	N	4.0	-.-	D	O	38.0	48	-.-	33	-.-	13	-.-
Electrical Eng [E]		42.0	Y	4.0	-.-	D	O	27.0	142	-.-	144	-.-	30	-.-
Geological Sciences [O]		13.2	N	4.0	-.-	D	N	-.-	2	-.-	4	-.-	6	-.-
Materials Science & Eng [O]		14.0	Y	4.0	-.-	D	O	15.0	27	-.-	25	-.-	14	-.-
Mechanical & Aerospace Eng [M]		27.0	Y	4.0	-.-	D	O	29.0	147	-.-	42	-.-	12	-.-
Nuclear Science & Eng [O]		5.0	N	4.0	-.-	D	N	-.-	0	-.-	—.	-.-	—.	-.-
Operations Research & Eng [I]		-.-	Y	4.0	-.-	D	O	-.-	—.	-.-	—.	-.-	—.	-.-
Operations Research & Eng [O]		-.-	Y	4.0	-.-	D	O	-.-	—.	-.-	—.	-.-	—.	-.-
Operations Research & Industrial Eng [O]		19.5	Y	4.0	-.-	—	O	21.0	94	-.-	52	-.-	12	-.-
Theoretical & Appl Mechanics [O]		13.5	N	4.0	-.-	D	N	-.-	0	-.-	3	-.-	7	-.-

All Engineering Disciplines

Institution Profile Reference Number & Name Name of Degree Program	State/Province	FACULTY Full-time equivalent (FTE)	ABET/CEAB accred?	UNDERGRADUATE PROGRAMS							GRADUATE PROGRAMS			
				Length		Time	Co-op		Bachelor's		Master's		Doctoral	
				Nominal length of program in years	Average length of program in years	Day/Eve./Both	None/Opt./Req.	% of graduates in Co-op programs	# of degrees awarded	% of graduates who were full-time	# of degrees awarded	% of graduates who were full-time	# of degrees awarded	% of graduates who were full-time
059 Dartmouth Coll	NH													
Eng Science(s) [O]		30.0	Y	4.0	4.0	D	N	-.-	58	100.0	23	100.0	5	100.0
060 U of Dayton	OH													
Chemical Eng [H]		7.0	Y	4.0	4.0	D	O	35.0	25	97.0	6	40.0	—.	-.-
Civil Eng [V]		10.5	Y	4.0	4.0	D	O	35.0	19	94.0	2	16.0	—.	-.-
Electrical Eng [E]		15.7	Y	4.0	4.0	D	O	35.0	80	89.0	37	12.0	2	46.0
Mechanical Eng [M]		14.8	Y	4.0	4.0	D	O	35.0	69	95.2	39	16.0	3	26.0
061 U of Delaware	DE													
Chemical Eng [H]		21.0	Y	4.0	4.3[1]	B[2]	N	-.-	38	99.0	10	70.0	19	100.0
Civil Eng [V]		18.0	Y	4.0	4.0	B	N	-.-	34	91.4	30	86.2[3]	7	71.2
Electrical Eng [E]		14.6	Y	4.0	4.0	B	N	-.-	52	94.6	8	83.3	8	89.3
Materials Science [O][5]		-.-	N	-.-	-.-	—	—	-.-	—.	-.-	2	100.0	5	96.3
Mechanical Eng [M]		16.0	Y	4.0	4.0	B	N	-.-	62	-.-	11	19.0	7	31.0
062 U of Denver	CO													
Electrical Eng [E]		6.5	Y	4.0	4.0	D	O	18.2	12	3.3	0	100.0	—.	-.-
General Eng [O][1]		13.0	N	4.0	4.0	D	O	-.-	0	-.-	—.	-.-	—.	-.-
Mechanical Eng [M]		7.0	Y	4.0	4.0	D	O	0.0	4	7.7	1	100.0	—.	-.-
063 U of the District of Columbia	DC													
Civil Eng [V]		6.0	Y	4.0	6.0	D	O	-.-[A]	4	-.-[A]	—.	-.-	—.	-.-
Electrical Eng [E]		7.0	Y	4.0	6.0	D	O	-.-[A]	25	-.-[A]	—.	-.-	—.	-.-
Mechanical Eng [M]		5.0	Y	4.0	6.0	D	O	-.-[A]	7	-.-[A]	—.	-.-	—.	-.-
064 Drexel U	PA													
Appropriate Tech [O]		1.0	N	5.0	5.0	B	R	100.0	4	100.0	—.	-.-[B]	—.[B]	-.-[B]
Architectural Eng [O]		6.5	Y	5.0	5.0	D	R	100.0	60	100.0	—.[B]	-.-[B]	—.[B]	-.-[B]
Biomedical Eng [O][2]		13.0	N	-.-[B]	-.-[B]	—[B]	N	-.-[B]	—[B]	-.-[B]	9	38.0	4	100.0
Chemical Eng [H]		10.0	Y	5.0	5.1	B	R	100.0	41	83.0	5	30.0	1	96.0
Civil Eng [V]		14.0	Y	5.0	5.0	B	R	75.0	76	75.0	16	30.0	1	80.0
Construction Mgmt [O]		3.0	N	6.0	6.0[C]	E	N	-.-[B]	8	0.0	—.[B]	-.-[B]	—.[B]	-.-[B]
Electrical & Computer Eng [E]		50.0	Y	5.0	5.3[C]	B	R	100.0	200	95.0	63	15.0	13	90.0
Eng Geology [O][2]		3.0	N	-.-[B]	-.-[B]	—[B]	N	-.-[B]	—[B]	-.-[B]	6	10.0	—.	-.-
Eng Mgmt [O][2]		4.0	N	-.-[B]	-.-[B]	—[B]	N	-.-[B]	—[B]	-.-[B]	27	6.0	—.	-.-
Eng Science(s) [O][1]		-.-	N	4.0	4.0	D	R	90.0	0	100.0	—.	-.-	—.	-.-
Env'l Eng [O][2]		12.0	N	-.-[B]	-.-[B]	—[B]	N	-.-[B]	—[B]	-.-[B]	27	30.0	2	100.0
Industrial Eng [I]		1.0	N	6.0[C]	6.0[C]	E	N	-.-[B]	5	0.0	—.[B]	-.-[B]	—.[B]	-.-[B]
Materials Eng [O]		12.0	Y	5.0	5.0[C]	B	R	85.0	8	80.0	3	100.0	7	100.0
Mechanical Eng [M]		30.0	Y	5.0	5.0	B	R	100.0	119	80.0	37	35.0	7	90.0
065 Duke U	NC													
Biomedical Eng [O]		29.0	Y	4.0	4.0	—[B]	N	-.-	62	100.0	11	100.0	6	100.0
Civil Eng [V]		15.0	Y	4.0	4.0	B[B]	N	-.-	33	100.0	20	100.0	4	100.0
Electrical Eng [E]		18.0	Y	4.0	4.0	—[B]	N	-.-	66	100.0	17	100.0	7	100.0
Mechanical Eng [M]		23.0	Y	4.0	4.0	—[B]	N	-.-	43	100.0	14	100.0	9	100.0
066 École Polytechnique	PQ													
Appl Mathematics [O]		21.0	Y	4.0	4.0[4]	—[A]	—[A]	-.-[A]	—[2]	-.-[3]	8[2]	100.0[3]	5[2]	100.0
Biochemical Eng [O]		6.0	—	-.-[B]	-.-[B]	—[B]	—[B]	-.-[B]	—.	-.-	—.	-.-	—.	-.-
Biomedical Eng [O]		6.0	—	-.-[B]	-.-[B]	—[B]	—[B]	-.-[B]	—.	-.-	10	96.4	4	100.0
Chemical Eng [H]		13.4	Y	4.0	4.5	D	N	-.-	25	97.6	16	96.9	4	100.0
Civil Eng [V]		25.5	Y	4.0	4.5	D	N	-.-	107	97.4	29	88.0	6	100.0
Computer Eng [O]		46.0[5]	Y	4.0	4.5	D	N	-.-	42	97.0	—.[B]	-.-[B]	—.[B]	-.-[B]
Electrical Eng [E]		46.0[5]	Y	4.0	4.5	D	N	-.-	160	96.7	43	87.2	8	100.0
Eng Physics [O]		17.6	Y	4.0	4.5	D	N	-.-	22	96.8	10	100.0	8	100.0
Geological Eng [O]		14.0[6]	Y	4.5	4.5	D	R	100.0	4	87.3	—.[B]	-.-[B]	—.[B]	-.-[B]
Industrial Eng [I]		20.4	Y	4.0	4.5	D	N	-.-	62	96.5	10	71.7	—.[B]	-.-[B]
Materials Eng [O]		12.6	Y	4.0	4.5	D	N	-.-	17	94.3	7	100.0	6	100.0
Mechanical Eng [M][7]		46.2	Y	4.0	4.5	D	N	-.-	167	97.3	30	80.5	6	100.0
Mineral Eng [O]		14.0[6]	—	-.-	-.-	—	—	-.-	—.	-.-	6	100.0	6	100.0
Mining Eng [O]		14.0[6]	Y	4.0	4.5	D	R	100.0	7	81.3	—.[B]	-.-[B]	—.[B]	-.-[B]
Nuclear Eng [O][8]		7.8	—	-.-	-.-	—	—	-.-	—.	-.-	8	100.0	1	100.0
067 Embry-Riddle-Daytona Beach	FL													
Aerospace & Aeronautical Eng [O]		15.0	Y	2.0	3.0	—[A]	O	-.-[A]	115	91.1	1	52.8	—.[B]	-.-[B]
Eng Physics [O]		14.0	N	4.0	4.0	—[A]	O	-.-[A]	9	96.7	—.[B]	-.-[B]	—.[B]	-.-[B]

All Engineering Disciplines

ENGINEERING Institution Profile Reference Number & Name / Name of Degree Program	State/Province	FACULTY Full-time equivalent (FTE)	ABET/CEAB accred?	UNDERGRADUATE PROGRAMS Length Nominal length of program in years	Length Average length of program in years	Time Day/Eve./Both	Co-op None/Opt./Req.	Co-op % of graduates in Co-op programs	Bachelor's # of degrees awarded	Bachelor's % of graduates who were full-time	GRADUATE PROGRAMS Master's # of degrees awarded	Master's % of graduates who were full-time	Doctoral # of degrees awarded	Doctoral % of graduates who were full-time
068 Embry-Riddle-Western Campus	AZ													
Aerospace Eng [O]		9.0	Y	4.0	4.5	B	O	18.0	50	99.0	—.	-.-	—.	-.-
Electrical Eng [E]		5.0	N[1]	4.5	4.5	B	O	50.0	4	100.0	—.	-.-	—.	-.-
069 U of Evansville	IN													
Civil Eng [V]		4.0	N	4.0	4.3	D	O	25.0	4	100.0	—.[B]	-.-[B]	—.[B]	-.-[B]
Computer Eng [O]		6.0[1]	N	4.0	4.2	D	O	0.0[B]	—.[A]	100.0	—.[B]	-.-[B]	—.[B]	-.-[B]
Electrical Eng [E]		7.7	Y	4.0	4.2	D	O	8.2	42	94.0[C]	—.[B]	-.-[B]	—.[B]	-.-[B]
Mechanical Eng [M]		5.2	Y	4.0	4.2	D	O	36.0	38	96.0	—.[B]	-.-[B]	—.[B]	-.-[B]
070 Ferris State U	MI													
Surveying Eng [O]		3.0	Y	4.0	4.2	—.[A]	N	-.-[A]	10	90.0	—.[A]	-.-[A]	—.[A]	-.-[A]
071 U of Florida	FL													
Aerospace Eng [O]		29.0[3]	Y	4.0	4.5	D	O	3.0	46	100.0	15	100.0	4	100.0
Agricultural Eng [O]		33.0	Y	4.0	4.5	D	O	0.0	6	95.0	2	100.0	2	100.0
Chemical Eng [H]		16.5	Y	5.0	5.5	D	O	8.0	31	99.0	3	100.0	4	100.0
Civil Eng [V]		32.0[1]	Y	4.0	4.5	D	O	3.8	69	88.6	39	100.0	4	100.0
Coastal & Oceanographic [O]		9.0	Y[4]	-.-[B]	-.-[B]	D	N	-.-[B]	0	0.0[B]	4	100.0	1	100.0
Computer & Information Science [O]		30.2	Y	4.5	5.2	D	O	19.0	47	76.7	23	100.0	7	100.0
Electrical Eng [E]		46.0	Y	4.0	4.5	D	O	11.0	99	95.0	79	100.0	26	100.0
Eng Science & Mechanics [O]		-.-[3]	Y	4.0	4.5	D	O	3.0	11	100.0	11	100.0	4	100.0
Env'l Eng [O]		17.0	Y	4.0	4.5	D	O	3.0	31	100.0	34	100.0	3	100.0
Industrial & Systems Eng [I][2]		13.3	Y	4.0	4.5	D	O	10.0	41	80.0	17	100.0	3	100.0
Interdisciplinary Eng [O]		-.-[10]	N	4.0	4.5	D	O	0.0	1	100.0	—.[B]	-.-[B]	—.[B]	-.-[B]
Materials Science & Eng [O][7]		24.5	Y[8]	5.0	5.5	D	O	3.0	37	91.0	29	100.0	9	100.0
Mechanical Eng [M]		31.3	Y	4.5	5.0	D	O	3.0	80	98.0	20	100.0	9	100.0
Nuclear Eng [O][9]		11.5	Y	4.0	4.5	D	O	0.0	16	100.0	6	100.0	8	100.0
Nuclear Eng Sciences [O][9]		-.-	N	4.0	4.5	D	O	0.0	4	100.0	6	100.0	1	100.0
Surveying & Mapping [O][5]		-.-	Y	4.0	4.0	D	O	100.0	11	93.3	—.[B]	-.-[B]	—.[B]	-.-[B]
072 Florida A&M U	FL													
Chemical Eng [H]		7.5	Y	4.5	5.0	D	O	-.-[A]	14[1]	-.-[A]	3[1]	-.-[A]	—.	-.-[A]
Civil Eng [V]		10.5	Y	4.5	5.0	D	O	-.-[A]	28[1]	-.-[A]	12[1]	-.-[A]	—.[B]	-.-[B]
Electrical Eng [E]		14.0	Y	4.5	5.0	D	O	-.-[A]	79[1]	-.-[A]	27[1]	-.-[A]	—.[B]	-.-[B]
Industrial Eng [I]		7.0	Y	4.5	5.0	D	O	-.-[A]	8[1]	-.-[A]	—.[B]	-.-[B]	—.[B]	-.-[B]
Mechanical Eng [M]		16.0	Y	4.5	5.0	D	O	-.-[A]	52[1]	-.-[A]	7[1]	-.-[A]	—.[B]	-.-[A]
073 Florida Atlantic U	FL													
Civil Eng [V][1]		3.5	N[2]	-.-[B]	-.-[B]	B[B]	N[B]	-.-[B]	0	-.-[2]	0	42.1	—.[1]	-.-[1]
Computer Eng [O][3]		9.0	N[3]	-.-[3]	-.-[3]	B[B]	—.[A]	-.-[B]	0[3]	-.-[3]	8	39.8	0	83.3
Computer Science [O]		11.0	N[4]	4.0	4.5[5]	B	O	5.0[C]	89	48.0	16	39.8	0[6]	58.8
Electrical Eng [E]		19.5	Y	4.5	5.0[5]	B	O	10.0[C]	61	62.0	12	36.5	3	68.4
Manuf'g Systems Eng [O]		2.5	N[2]	-.-[B]	-.-[B]	—.[B]	—.[B]	-.-[B]	0[2]	-.-[2]	3	45.5	—.[1]	-.-[1]
Mechanical Eng [M]		16.5	Y	4.5	5.0[5]	B	O	10.0[C]	25	66.0	25	38.6	7	72.2
Ocean Eng [O]		14.5	Y	4.5	5.0[5]	B	O	10.0[C]	16	76.3	8	82.8	4	89.5
074 Florida Inst of Tech	FL													
Chemical Eng [H]		5.3	Y	4.0	4.5	D	O	25.0	8	96.7	3	60.0	—.	-.-
Civil & Env'l Eng [V]		6.8	Y	4.0	4.3	D	O	33.3	16	86.4	6	58.8	0	100.0
Electrical & Computer Eng [E]		17.6	Y	4.0	4.5	D	O	33.3	111	86.5	55	45.5	6	20.5
Mechanical & Aerospace Eng [M]		13.9	Y	4.0	4.5	D	O	35.0	40	91.9	11	60.0	1	-.-
Ocean Eng [O]		6.3	Y	4.0	4.3	D	O	5.0	22	96.4	7	65.0	—.	-.-
075 Florida International U	FL													
Civil & Env'l Eng [V]		-.-	Y	-.-	-.-	B	O	-.-	—.	-.-	—.	-.-	—.	-.-
Civil Eng [V]		12.5	Y	4.0	4.5	B	O	10.0	31	77.8	7	100.0	—.	-.-
Computer Eng [O]		6.0	N[1]	4.0	4.5	B	O	100.0	0	74.8	—.[2]	100.0	—.	-.-
Electrical & Computer Eng [E]		-.-	Y	-.-	-.-	B	O	-.-	—.	-.-	—.	-.-	—.	-.-
Electrical Eng [E]		19.5	Y	4.0	4.5	B	O	10.0	84	77.9	18	100.0	0	100.0
Industrial & Systems Eng [I]		10.3	Y	4.0	4.5	B	O	50.0	35	73.1	1	100.0	—.	-.-
Mechanical Eng [M]		16.0	Y	4.0	4.5	B	O	10.0	33	80.1	9	100.0	—.	-.-
076 Florida State U	FL													
Chemical Eng [H]		7.5	Y	4.5	5.0	D	O	-.-[A]	14[1]	-.-[A]	3[1]	-.-[A]	—.	-.-[B]
Civil Eng [V]		10.5	Y	4.5	5.0	D	O	-.-[A]	28[1]	-.-[A]	12[1]	-.-[A]	—.[B]	-.-[B]
Electrical Eng [E]		14.0	Y	4.5	5.0	D	O	-.-[A]	79[1]	-.-[A]	27[1]	-.-[A]	—.[B]	-.-[B]
Industrial Eng [I]		7.0	Y	4.5	5.0	D	O	-.-[A]	8[1]	-.-[A]	—.[B]	-.-[B]	—.[B]	-.-[B]
Mechanical Eng [M]		16.0	Y	4.5	5.0	D	O	-.-[A]	52[1]	-.-[A]	7[1]	-.-[A]	—.	-.-[A]

All Engineering Disciplines

ENGINEERING Institution Profile Reference Number & Name Name of Degree Program	State/Province	FACULTY Full-time equivalent (FTE)	ABET/CEAB accred?	UNDERGRADUATE PROGRAMS					Bachelor's		GRADUATE PROGRAMS			
				Length		Time	Co-op				Master's		Doctoral	
				Nominal length of program in years	Average length of program in years	Day/Eve./Both	None/Opt./Req.	% of graduates in Co-op programs	# of degrees awarded	% of graduates who were full-time	# of degrees awarded	% of graduates who were full-time	# of degrees awarded	% of graduates who were full-time
077 GMI	MI													
Electrical & Computer Eng [E]		18.0	Y	5.0	5.0	E[1]	R	100.0	108	100.0	—.	-.-	—.	-.-
Industrial & Manuf'g Systems Eng [I]		10.0	Y	5.0	5.0	D[1]	R	100.0	26	100.0	—.	-.-	—.	-.-
Industrial Eng [I]		10.0	Y	5.0	5.0	D[1]	R	100.0	26	100.0	—.	-.-	—.	-.-
Manuf'g Systems Eng [O]		10.0	Y	5.0	5.0	D[1]	R	100.0	37	100.0	—.	-.-	—.	-.-
Mechanical Eng [M]		32.0	Y	5.0	5.0	D[1]	R	100.0	136	100.0	—.	-.-	—.	-.-
078 Gannon U	PA													
Electrical Eng [E]		7.0	Y	4.0	4.5	B	O	0.0	23	81.7	3	16.6	—.	-.-
Mechanical Eng [M]		7.0	Y	4.0	4.5	B	O	0.0	16	82.0	2	8.3	—.	-.-
079 George Mason U	VA													
Computer Science [O]		18.2	N[B]	4.0	4.5	B	O	-.-[A]	75	64.0	54	17.0	—[1]	-.-[B]
Electrical Eng [E]		23.3	Y	4.0	4.5	B[6]	O	-.-[A]	82	69.0	34	13.0	—[1]	-.-[B]
Information Systems [O]		12.0	—[A]	-.-[B]	-.-[B]	B	O	-.-[B]	0[2]	-.-[B]	103	14.0	—[1]	-.-[B]
Information Tech [O]		-.-[3]	N[B]	-.-[B]	-.-[B]	B	N[B]	-.-[B]	—[B]	-.-[B]	—[B]	-.-[B]	13	14.0
Operations Research & Mgmt Science [O]		7.5	N[B]	-.-[B]	-.-[B]	B	O[B]	-.-[B]	—[4]	-.-[B]	19	12.0	—[1]	-.-[B]
Software Systems Eng [O]		12.0[5]	—[B]	-.-[B]	-.-[B]	B	—[B]	-.-[B]	—[B]	-.-[B]	13	8.0	—[1]	-.-[B]
Statistical Sciences [O]		7.0	—[B]	-.-[B]	-.-[B]	B	O	-.-[B]	—[B]	-.-[B]	4	33.0	—[1]	-.-[B]
Systems Eng [O]		13.3	N	4.0	4.0	B	O	-.-[B]	4	69.0	20	11.0	—[1]	-.-[B]
Urban Systems Eng [O]		-.-[6]	N	4.0	4.0	B	O	-.-[B]	2	86.0	—.[B]	-.-[B]		
080 The George Washington U	DC													
Civil Eng [V]		10.0	Y	4.0	-.-[A]	B	O	-.-	12	80.0	23	27.0	3	27.0
Computer Eng [O]		6.0	Y	4.0	-.-[A]	D	O	-.-	15	77.0	—[B]	-.-[B]	—[B]	-.-[B]
Computer Science [O]		16.0	N	4.0	-.-[A]	D	O	-.-[A]	8	73.0	54	33.0	7	26.0
Electrical Eng [E]		23.0	Y	4.0	-.-[A]	D	O	-.-[A]	47	79.0	72	20.0	7	23.0
Eng Mgmt [O]		15.0	—[B]	-.-[B]	-.-[A]	D[B]	N	-.-[A]	—[B]	-.-[B]	189	18.0	4	27.0
Mechanical Eng [M]		19.0	Y	4.0	-.-[A]	B	O	-.-[A]	22	92.0	32	35.0	3	37.0
Operations Research [O]		8.0	—[B]	-.-[B]	-.-[A]	D[B]	O[B]	-.-[A]	—[B]	-.-[B]	30	16.0	2	35.0
Systems Analysis & Eng [O]		8.0	Y	4.0	-.-[A]	D	O	-.-[A]	7	74.0	—.[B]	-.-[B]	—[B]	-.-[B]
081 U of Georgia	GA													
Agricultural Eng [O]		-.-	Y	4.0	-.-	D	O	-.-	—.	-.-	—.	-.-	—.	-.-
Biological & Agricultural Eng [O]		28.0	Y	4.0	4.3	D	O	10.0	20	100.0	7	67.0	1	25.0
082 Georgia Tech	GA													
Aerospace Eng [O]		27.3[8]	Y	4.0	4.3	D	O	12.1	65	89.0	52	82.0	20	87.0
Ceramic Eng [O]		12.0[8]	Y	4.0	5.0	D	O	0.0	1	90.0	3	86.0	1	100.0
Chemical Eng [H]		25.0[8]	Y	4.0	4.3	D	O	36.0	72	97.0	8	99.0	8	99.0
Civil Eng [V]		27.0[8]	Y	4.0	4.7	D	O	24.0	116	93.4	54	47.2[4]	3	81.0
Computer Eng [E]		8.0[8]	Y	4.0	4.7	D	O	50.0	14	94.0	—.[B]	-.-[B]	—[B]	-.-[B]
Electrical Eng [E]		72.0[8]	Y	4.0	4.7	D	O	38.0	302	93.0	203	62.0	48	80.0
Eng Science & Mechanics [V]		9.0[8]	Y	4.0	4.5	D	O	57.0	7	95.9	4	90.0	2	66.6
Env'l Eng [V]		8.0[8]	—[2]	-.-[2]	-.-[2]	—[2]	N[2]	-.-[2]	0[2]	0.0[2]	15	54.3[3]	—.[1]	-.-[1]
Industrial Eng [I]		51.7[8]	Y	4.0	4.7	D	O	36.0	253	93.7	117	72.3[7]	16	93.4
Materials Science & Eng [O]		12.0[8]	Y	4.0	5.0	D	O	30.0	12	98.0	3[5]	74.0	3[5]	100.0
Mechanical Eng [M]		47.5[8]	Y	4.0	4.5	D	O	43.0	331	93.9	83	81.0	23	94.0
Nuclear Eng [M]		10.0[8]	Y	4.0	4.4	D	O	43.0	7	94.4	32[6]	34.0	3	83.0
Textile Chemistry [O]		11.0[8]	N	4.0	4.3	D	O	0.0	5	100.0	2[9]	100.0	0[9]	100.0
Textile Eng [O]		11.0[8]	Y	4.0	4.1	D	O	14.0	14	94.0	3[9]	78.0	2[9]	94.0
Textile Mgmt & Tech [O]		11.0[8]	N	4.0	4.2	D	O	38.0	8	93.0	3[9]	64.0	0[9]	75.0
083 Gonzaga U	WA													
Civil Eng [V]		6.0	Y	4.0	-.-	D	N	-.-	12	94.0	—.	-.-	—.	-.-
Electrical Eng [E]		8.0	Y	4.0	-.-	D	N	-.-	22	93.0	1	85.0	—.	-.-
Mechanical Eng [M]		9.0	Y	4.0	-.-	D	N	-.-	12	93.0	2	100.0	—.	-.-
084 Grand Valley State U	MI													
Eng [O]		-.-	Y	4.3	-.-	D	R	-.-[A]	—.	-.-	—.	-.-	—.	-.-
General Eng [O]		9.0	Y	4.3	4.3	D	R	100.0	23	100.0	—.[B]	-.-[B]	—.[B]	-.-[B]
085 Grove City Coll	PA													
Electrical Eng [E]		6.0	Y	4.0	4.1	D	N	-.-	22	100.0	—.[B]	-.-[B]	—.[B]	-.-[B]
Mechanical Eng [M]		6.0	Y	4.0	4.1	D	N	-.-	13	100.0	—.[B]	-.-[B]	—.[B]	-.-[B]
086 Harvard U	MA													
Eng Science(s) [O]		46.2	Y[2]	4.0	-.-	D	N	-.-	18[1]	100.0	18	100.0	10	100.0
087 Harvey Mudd Coll	CA													
Eng [O]		15.0	Y	4.0	4.0	D	N	-.-	55	100.0	6	100.0	—.	-.-

All Engineering Disciplines

ENGINEERING Institution Profile Reference Number & Name Name of Degree Program	State/Province	FACULTY Full-time equivalent (FTE)	ABET/CEAB accred?	UNDERGRADUATE PROGRAMS					Bachelor's		GRADUATE PROGRAMS			
				Length		Time	Co-op				Master's		Doctoral	
				Nominal length of program in years	Average length of program in years	Day/Eve./Both	None/Opt./Req.	% of graduates in Co-op programs	# of degrees awarded	% of graduates who were full-time	# of degrees awarded	% of graduates who were full-time	# of degrees awarded	% of graduates who were full-time
088 U of Hawaii at Manoa	HI													
Civil Eng [V]		19.0	Y	4.0	5.0	D	O	-.-	49	89.0	9	95.0	—.(B)	-.-(B)
Electrical Eng [E]		27.0	Y	4.0	5.0	D	O	-.-	62	87.8	19	74.0	4	88.0
Mechanical Eng [M]		15.0	Y	4.0	5.0	D	O	-.-	48	83.8	7	97.0	3	89.0
089 Hofstra U	NY													
Electrical Eng [E]		5.5	Y	4.0	4.5	B	N	-.-	14	65.4	—.	-.-	—.	-.-
Eng [O](B)		-.-	—	-.-	-.-	—	—	-.-	—.	-.-	—.	-.-	—.	-.-
Eng Science(s) [O]		3.0	Y	4.0	4.5	B	N	-.-	4	69.0	—.	-.-	—.	-.-
Industrial Eng [I]		1.0	N	4.0	4.5	B	N	-.-	3	64.3	—.	-.-	—.	-.-
Mechanical Eng [M]		5.8	Y	4.0	4.5	B	N	-.-	13	77.8	—.	-.-	—.	-.-
090 U of Houston	TX													
Acoustics [O]		-.-	N	-.-	-.-	—	—	-.-	—.	-.-	—.	-.-	—.	-.-
Biochemical Eng [O]		-.-	Y	-.-	-.-	—	—	-.-	—.	-.-	—.	-.-	—.	-.-
Biomedical Eng [O]		-.-	N	-.-	-.-	—	—	-.-	—.	-.-	—.	-.-	—.	-.-
Chemical Eng [H](5)		16.0	Y	4.0(C)	5.0	B	O	38.0	26	73.8	18	12.5	9	91.0
Civil & Env'l Eng [V](1)		14.5	Y	4.0	5.0	B	O	-.-(A)	13	42.7	22	50.5	5	50.5
Civil Eng [V]		-.-	Y	-.-	-.-	—	O	-.-	—.	-.-	—.	-.-	—.	-.-
Computer & Systems Eng [O]		-.-	N	-.-	-.-	—	—	-.-	—.	-.-	—.	-.-	—.	-.-
Electrical Eng [E](2)		28.6	Y	4.0(C)	5.0	B	O	-.-(A)	90	61.0	72	48.0	2	68.0
Eng Mgmt [O]		-.-	N	-.-	-.-	—	—	-.-	—.	-.-	—.	-.-	—.	-.-
Env'l Eng [O]		-.-	N	-.-	-.-	—	—	-.-	—.	-.-	—.	-.-	—.	-.-
Industrial Eng [I](3)		8.0	Y	4.0	5.0	B	O	15.0	13	90.0	20	60.0	5	100.0
Materials Eng [O]		-.-	N	-.-	-.-	—	—	-.-	—.	-.-	—.	-.-	—.	-.-
Mechanical Eng [M](4)		23.3	Y	4.5	-.-(A)	B	O	-.-(A)	51	42.0	14	35.0	4	60.0
Operations Research [O]		-.-	N	-.-	-.-	—	—	-.-	—.	-.-	—.	-.-	—.	-.-
Petroleum Eng [O]		-.-	N	-.-	-.-	—	—	-.-	—.	-.-	—.	-.-	—.	-.-
091 Howard U	DC													
Chemical Eng [H](C)		5.0	Y	4.0	5.0	D	O	20.0	10	100.0	1	90.0	—.(2)	-.-(2)
Civil Eng [V]		12.0	Y	4.0	4.5	D	O	5.0	20	100.0	8	50.0	—.(B)	-.-(B)
Electrical Eng [E](C)		18.0	Y	4.0	5.0	D	O	3.0	56	97.0	15	90.0	4	100.0
Mechanical Eng [M](C)		10.0	Y	4.0	5.0	B	O	0.0	28	85.0	5	75.0	2	80.0
Systems & Computer Sciences [O](C)		1.3	N(3)	4.0	4.0	D	O	-.-	25	-.-	—.(A)	-.-(A)	—.(A)	-.-(A)
092 U of Idaho	ID													
Agricultural Eng [O]		13.5	Y	4.0	6.0	D	O	0.0	7	100.0	6	66.6	1	70.0
Chemical Eng [H]		7.0	Y	4.0	5.0	D	O	-.-	11	94.8	6	59.0	1	64.0
Civil Eng [V]		13.3	Y	4.0	6.0	D	O	10.0	24	93.0	14	64.0	2	100.0
Computer Eng [O]		3.0	N	4.0	5.5	D	O	0.0	6	97.1	5	72.7	—.	-.-
Computer Science [O]		12.7	N	4.0	6.0	D	O	15.0	24	91.7	12	45.0	—.	-.-
Electrical Eng [E]		16.5	Y	4.0	6.0	D	O	2.0	53	93.5	16	67.0	3	46.0
Mechanical Eng [M]		17.0	Y	4.0	5.5	D	O	0.0	40	94.8	8	87.0	0	72.0
093 Idaho State U	ID													
Eng [O]		16.0	Y	4.0	5.0	B(A)	N	-.-(A)	30	83.0	5	25.0	—.	-.-
Eng Mgmt [O]		16.0	N	4.0	5.0	B(A)	N	-.-(A)	0	75.0	—.(A)	-.-(A)	—.(A)	-.-(A)
094 U of Illinois at Chicago	IL													
Bioengineering [O]		3.0(1)	Y	4.0	-.-	D	O	0.0	8	90.0	3	20.0	2	33.0
Chemical Eng [H]		10.0	Y	4.0	-.-	D	O	15.0	13	82.0	6	35.0	5	27.0
Civil Eng [V]		11.0	Y	4.0	-.-	—	—	24.0	17	83.0	10	38.0	1	12.0
Computer Eng [O]		22.0	Y	4.0	-.-	D	O	10.0	39	84.0	—.	-.-	—.	-.-
Computer Science [O]		-.-(2)	N	4.0	-.-	D	O	-.-	0	100.0	—.	-.-	—.	-.-
Electrical Eng [E]		29.0	Y	4.0	-.-	D	O	14.0	138	81.0	—.	-.-	—.	-.-
Electrical Eng & Computer Science [E]		51.0	—	-.-	-.-	—	—	-.-	—.	-.-	81	32.0	13	27.0
Eng Mgmt [O]		-.-(3)	N	4.0	-.-	D	O	0.0	4	71.0	—.	-.-	—.	-.-
Eng Mechanics [O]		6.0	N	-.-	-.-	—	—	-.-	—.	-.-	0	9.0	2	73.0
Eng Physics [O]		-.-(4)	N	4.0	-.-	D	O	0.0	0	82.0	—.	-.-	—.	-.-
Geotechnical Eng & Geosciences [O]		1.0(1)	—	-.-	-.-	—	—	-.-	—.	-.-	—.(5)	-.-	1	88.0
Industrial Eng [I]		6.0	Y	4.0	-.-	D	O	-.-	23	90.0	9	46.0	—.(6)	-.-
Industrial Eng & Operations Research [I]		6.0	—	-.-	-.-	—	—	-.-	—.	-.-	—.(5)	-.-	1	50.0
Materials Science & Eng [O]		4.0	N	4.0	-.-	D	O	-.-	0	57.0	—.	-.-	—.	-.-
Mechanical Eng [M]		24.0	Y	4.0	-.-	D	O	14.0	79	83.0	46	36.0	12	23.0
Metallurgical Eng [O]		4.0	Y	4.0	-.-	D	O	0.0	1	75.0	—.	-.-	—.	-.-
Metallurgy [O]		4.0	—	-.-	-.-	—	—	-.-	—.	-.-	3	45.0	1	41.0
095 Illinois, Urbana-Champaign	IL													
Aeronautical & Astronautical Eng [O]		17.5	Y	4.0	4.5	—(A)	O	13.8	58	99.7	30	74.3	10	76.7
Agricultural Eng [O]		16.0	Y	4.0	4.5	—(A)	O	5.6	18	100.0	9	78.9	10	82.1
Ceramic Eng [O](2)		32.7	Y	4.0	4.5	—(A)	O	9.4	32	100.0	9	100.0	7	93.5

All Engineering Disciplines

ENGINEERING Institution Profile Reference Number & Name Name of Degree Program	State/Province	FACULTY Full-time equivalent (FTE)	ABET/CEAB accred?	UNDERGRADUATE PROGRAMS					GRADUATE PROGRAMS			
				Length		Time	Co-op		Bachelor's		Master's		Doctoral	
				Nominal length of program in years	Average length of program in years	Day/Eve./Both	None/Opt./Req.	% of graduates in Co-op programs	# of degrees awarded	% of graduates who were full-time	# of degrees awarded	% of graduates who were full-time	# of degrees awarded	% of graduates who were full-time
Chemical Eng [H]		10.6	Y	4.0	4.5	—(A)	O	-.-(A)	72	100.0	7	100.0	14	100.0
Civil Eng [V]		55.0	Y	4.0	4.5	—(A)	O	3.0	135	99.2	68	88.9	26	80.2
Computer Eng [O](3)		-.-	Y	4.0	4.5	—(A)	O	10.3	58	100.0	—(B)	-.-(B)	—(B)	-.-(B)
Computer Science [O]		41.5	N	4.0	4.5	—(A)	O	3.1	98	94.0	83	87.6	33	85.0
Electrical & Computer Eng [E]		77.0	Y	4.0	4.5	—(A)	O	7.0	230	99.1	77	92.8	52	93.6
Eng Physics [O]		65.7	N	4.0	4.5	—(A)	O	0.0	19	97.3	52	98.8	34	96.8
Env'l Eng [O](4)		-.-	N	-.-(B)	-.-(B)	—(B)	—(B)	-.-(B)	—(B)	-.-(B)	25	84.4	3	78.8
Env'l Science [O](4)		-.-	N	-.-(B)	-.-	—(B)	—(B)	-.-	—.	-.-	2	84.4	1	78.8
General Eng [O]		19.3	Y	4.0	4.5	—(A)	O	5.7	123	99.8	12	88.2	—(A)	-.-(A)
Industrial Eng [I](1)		-.-	Y	4.0	4.5	—(A)	O	13.0	46	98.0	9	81.0	—(A)	-.-(A)
Mechanical Eng [M]		46.4	Y	4.0	4.5	—(A)	O	15.6	192	98.5	61	100.0	25	93.5
Metallurgical Eng [O](2)		-.-	Y	4.0	4.5	—(A)	O	0.0	13	94.6	10	100.0	18	88.5
Nuclear Eng [O]		13.0	Y	4.0	4.5	—(A)	O	0.0	15	99.1	10	90.9	8	91.5
Theoretical & Appl Mechanics [O](5)		18.2	Y	4.0	4.5	—(A)	O	7.7	13	100.0	9	93.3	8	91.9
096 Illinois Inst of Tech	IL													
Aerospace Eng [O]		21.7	Y	4.0	4.8	B	O	0.0(2)	22	94.8	12(5)	30.0	5(5)	52.0
Chemical Eng [H]		12.7	Y	4.0	4.8	B	O	12.5(2)	24	77.0	23	43.0	11	70.0
Civil Eng [V]		10.1	Y	4.0	4.8	B	O	7.7(2)	13	73.3	21	27.0	2	59.0
Electrical & Computer Eng [E]		33.6	Y	4.0	4.8	B	O	6.1(2)	116	67.1	90	16.0	2	44.0
Env'l Eng [O](3)		7.8	N(4)	4.0	-.-(A)	B	O	-.-(4)	0(4)	80.0	23	23.0	5	78.0
Materials & Metallurgical Eng [O]		8.7	Y	4.0	4.8	B	O	0.0(2)	4	81.8	7	33.0	4	52.0
Mechanical Eng [M]		21.7	Y	4.0	4.8	B	O	3.3(2)	61	68.0	12(5)	30.0	5	52.0
097 Indiana-Purdue, Indianapolis	IN													
Electrical Eng [E]		12.0	Y	4.0	4.0	B	O	0.0	45	53.0	5	0.0	—.	-.-
Interdisciplinary Eng [O]		-.-	N	4.0	4.0	B	O	0.0	1	14.0	—.	-.-	—.	-.-
Mechanical Eng [M]		12.0	Y	4.0	4.0	B	O	0.0	32	57.0	2	33.0	—.	-.-
098 U of Iowa	IA													
Biomedical Eng [O]		8.3	Y	4.0	4.5(C)	D	O	15.0(C)	28	100.0(C)	7	100.0(C)	8	100.0(C)
Chemical & Biochemical Eng [H]		7.6	Y	4.0	4.5(C)	D	O	15.0(C)	20	100.0	3	100.0(C)	1	100.0(C)
Civil & Env'l Eng [V]		15.4	Y	4.0	4.5(C)	D	O	15.0(C)	28	100.0(C)	24	100.0(C)	10	100.0(C)
Electrical & Computer Eng [E]		20.0	Y	4.0	4.5(C)	D	O	15.0(C)	64	100.0(C)	14	100.0(C)	5	100.0(C)
Industrial Eng [I]		7.5	Y	4.0	4.5(C)	D	O	15.0(C)	43	100.0(C)	7	100.0(C)	5	100.0(C)
Mechanical Eng [M]		12.6	Y	4.0	4.5(C)	D	O	15.0(C)	43	100.0(C)	21	100.0(C)	19	100.0(C)
099 Iowa State U	IA													
Aerospace Eng [O]		-.-	Y	4.0	-.-	D	O	-.-	53	-.-	11	-.-	2	-.-
Agricultural Eng [O]		-.-	Y	4.0	-.-	D	O	-.-	6	-.-	4	-.-	3	-.-
Biomedical Eng [O]		-.-	N	-.-(B)	-.-	—(B)	—(B)	-.-	—.	-.-	7	-.-	4	-.-
Ceramic Eng [O]		-.-	Y	4.0	-.-	D	N	-.-	13	-.-	1	-.-	—.	-.-
Chemical Eng [H]		-.-	Y	4.0	-.-	D	O	-.-	43	-.-	4	-.-	8	-.-
Civil Eng [V]		-.-	Y	4.0	-.-	D	O	-.-	60	-.-	39	-.-	1	-.-
Computer Eng [E]		-.-	Y	4.0	-.-	D	O	-.-	48	-.-	10	-.-	1	-.-
Construction Eng [V]		-.-	Y	4.0	-.-	D	O	-.-	30	-.-	—.	-.-	—.	-.-
Electrical Eng [E]		-.-	Y	4.0	-.-	D	O	-.-	167	-.-	27	-.-	4	-.-
Eng Mechanics [O]		-.-	N	-.-(B)	-.-	—(B)	—(B)	-.-	—.	-.-	7	-.-	—.	-.-
Eng Operations [O]		-.-	N	4.0	-.-	D	O	-.-	5	-.-	—.	-.-	—.	-.-
Eng Science(s) [O]		-.-	Y	4.0	-.-	D	O	-.-	3	-.-	—.	-.-	—.	-.-
Industrial Eng [I]		-.-	Y	4.0	-.-	D	O	-.-	86	-.-	11	-.-	3	-.-
Mechanical Eng [M]		-.-	Y	4.0	-.-	D	O	-.-	113	-.-	14	-.-	11	-.-
Metallurgical Eng [O]		-.-	Y	4.0	-.-	D	O	-.-	9	-.-	—.	-.-	—.	-.-
Metallurgy [O]		-.-	N	-.-(B)	-.-	—(B)	—(B)	-.-	—.	-.-	5	-.-	3	-.-
Nuclear Eng [O]		-.-	N	-.-(B)	-.-	—(B)	—(B)	-.-	10	-.-	3	-.-	—.	-.-
Operations Research [I]		-.-	N						—.		—.		—.	
100 The Johns Hopkins U	MD													
Appl Mathematics [O](5)		3.7	N	-.-(6)	-.-(6)	E	—(B)	-.-(B)	0	0.0	29	0.0	—.	-.-
Appl Physics [O](5)		-.-	N	-.-(6)	-.-(6)	E	—(B)	-.-(B)	0	0.0	8	0.0	—.	-.-
Biomedical Eng [O]		18.3	Y	4.0	4.0	D	O	-.-(A)	44	100.0	17	100.0	3	100.0
Chemical Eng [H]		9.7(1)	Y	4.0	4.0	B	O	-.-(A)	44	100.0	7	71.0	4	100.0
Civil Eng [V]		8.7(2)	Y	4.0	4.0	B	O	-.-(A)	11	100.0	9	67.0	1	100.0
Computer Science [O]		37.0(7)	N	4.0	4.0	B	O	-.-(A)	19	100.0	323	2.0	2	100.0
Electrical Eng [E]		36.0(3)	Y	4.0	4.0	B	O	-.-(A)	31	100.0	260	8.0	5	100.0
Eng Mechanics [O](10)		15.3(4)	Y	4.0	4.0	B	O	-.-(A)	16	100.0	13	15.0	3	100.0
Env'l Science & Eng [O]		18.3(8)	N	-.-(B)	-.-(B)	B	N	-.-(B)	0	0.0	18	78.0	7	100.0
Materials Science & Eng [O]		10.3(9)	Y	4.0	4.0	B	O	-.-(A)	4	100.0	11	82.0	11	100.0
Mathematical Sciences [O]		10.7	N	4.0	4.0	D	O	-.-(A)	10	100.0	11	100.0	3	100.0
Mechanical Eng [M](10)		15.3(4)	Y	4.0	4.0	B	O	-.-(A)	16	100.0	13	15.0	3	100.0
Technical Mgmt [O](5)		7.7	N(5)	5.0	-.-(6)	E	—(B)	-.-(A)	0	0.0	57	0.0	—.	-.-

All Engineering Disciplines

ENGINEERING / Institution Profile Reference Number & Name / Name of Degree Program	State/Province	FACULTY Full-time equivalent (FTE)	ABET/CEAB accred?	UNDERGRADUATE PROGRAMS Length Nominal length of program in years	Length Average length of program in years	Time Day/Eve./Both	Co-op None/Opt./Req.	Co-op % of graduates in Co-op programs	Bachelor's # of degrees awarded	Bachelor's % of graduates who were full-time	GRADUATE PROGRAMS Master's # of degrees awarded	Master's % of graduates who were full-time	Doctoral # of degrees awarded	Doctoral % of graduates who were full-time
101 U of Kansas	KS													
Aerospace Eng [O]		10.0	Y	4.0	-.-	D	O	1.0	40	-.-(A)	8	-.-(A)	3	-.-(A)
Architectural Eng [O]		7.0	Y	5.0	-.-	D	O	-.-	32	-.-(A)	6	-.-(A)	—.	-.-(A)
Chemical & Petroleum Eng [H]		13.0	Y	4.0	-.-	D	O	-.-(A)	12	-.-(A)	7	-.-(A)	2	-.-(A)
Civil Eng [V]		25.0	Y	4.0	-.-	D	O	-.-	37	-.-(A)	39	-.-(A)	6	-.-(A)
Electrical & Computer Eng [E]		-.-	Y	4.0	-.-	D	O	-.-	81	-.-(A)	22	-.-(A)	4	-.-(A)
Eng Physics [O]		5.0	Y	4.0	-.-	D	O	-.-	3	-.-(B)	—(B)	-.-(B)	—(B)	-.-(B)
Mechanical Eng [M]		12.0	Y	4.0	-.-	D	O	-.-	36	-.-(A)	10	-.-(A)	3	-.-(A)
102 Kansas State U	KS													
Agricultural Eng [O]		19.0	Y	4.0	4.7	D	O	-.-	3	99.0	2	-.-	2	-.-
Architectural Eng [O]		12.0	Y	5.0	5.5	D	O	-.-	42	95.0	—.	-.-	—.	-.-
Chemical Eng [H]		9.0	Y	4.0	4.7	D	O	-.-	13	98.0	4	-.-	2	-.-
Civil Eng [V]		13.0	Y	4.0	4.7	D	O	-.-	29	98.0	8	-.-	—.	-.-
Electrical & Computer Eng [E]		21.0	Y	4.0	4.7	D	O	-.-	111	96.0	20	-.-	1	-.-
Industrial Eng [I]		10.0	Y	4.0	4.7	D	O	-.-	53	97.0	19	-.-	3	-.-
Mechanical Eng [M]		16.0	Y	4.0	4.7	D	O	-.-	79	96.0	10	-.-	4	-.-
Nuclear Eng [O]		6.0	Y	4.0	4.7	D	O	-.-	6	98.0	3	-.-	2	-.-
103 U of Kentucky	KY													
Agricultural Eng [O]		3.2	Y	4.0	5.0(C)	D	O	-.-(A)	0	97.8	5	-.-(A)	4	75.0(1)
Chemical Eng [H]		12.3	Y	4.0	5.0(C)	D	O	-.-(A)	20	92.4	9	-.-(A)	3	82.9(1)
Civil Eng [V]		20.3	Y	4.0	5.0(C)	D	O	-.-(A)	62	92.5	13	-.-(A)	2	52.2(1)
Electrical Eng [E]		21.3	Y	4.0	5.0(C)	D	O	-.-(A)	93	88.9	5	-.-(A)	4	68.0(1)
Eng Mechanics [O](2)		8.0	—(B)	-.-(B)	-.-(B)	—(B)	—(B)	-.-(B)	—(B)	-.-(B)	4	-.-(1)	1	90.9
Materials Science & Eng [O]		7.0	Y	4.0	5.0(C)	D	O	-.-(A)	7	89.3	4	-.-(A)	2	82.1(1)
Mechanical Eng [M]		19.0	Y	4.0	5.0(C)	D	O	-.-(A)	45	90.9	8	-.-(A)	5	58.3
Mining Eng [O]		10.2	Y	4.0	5.0(C)	D	O	-.-(A)	6	97.8	6	-.-(A)	0	52.0(1)
104 Lafayette Coll	PA													
Chemical Eng [H](1)		8.5	Y	4.0	4.0	D	N	-.-	13	93.0	—.	-.-	—.	-.-
Civil Eng [V](1)		8.5	Y	4.0	4.0	D	N	-.-	25	100.0	—.	-.-	—.	-.-
Electrical Eng [E](1)		9.6	Y	4.0	4.0	D	N	-.-	27	78.0	—.	-.-	—.	-.-
Eng [O](1)		2.0	N	4.0	4.0	B	N	-.-	17(2)	71.0(2)	—.	-.-	—.	-.-
Mechanical Eng [M](1)		8.0	Y	4.0	4.0	B	N	-.-	32	72.0	—.	-.-	—.	-.-
105 Lamar U	TX													
Chemical Eng [H]		5.5	Y	4.0	5.0	D	O	43.0	10	-.-(A)	—(A)	-.-(A)	—.	-.-(A)
Civil Eng [V]		5.3	Y	4.0	5.0	D	O	0.0	6	-.-(A)	—(A)	-.-(A)	2	-.-(A)
Electrical Eng [E]		5.5	Y	4.0	5.0	D	O	74.0	30	-.-(A)	—(A)	-.-(A)	—.	-.-(A)
Industrial Eng [I]		3.5	Y	4.0	4.5	B	O	45.0	5	-.-(A)	—(A)	-.-(A)	—.	-.-(A)
Mechanical Eng [M]		4.8	Y	4.0	5.0	D	O	74.0	20	-.-(A)	—(A)	-.-(A)	—.	-.-(A)
106 Lawrence Tech U	MI													
Civil & Construction Eng [V]		5.8	—(1)	4.0	4.5(C)	B	N	-.-	38(2)	68.3	—.	-.-	—.	-.-
Electrical Eng [E]		17.4	Y	4.0	4.5(C)	B	O	11.3	133	56.2	—.	-.-	—.	-.-
Mechanical Eng [M]		23.4	Y	4.0	4.5(C)	B	O	6.9	175	58.2	9	0.0(3)	—.	-.-
107 LeTourneau U	TX													
Eng [O](1)		8.0	Y	4.0	4.5	D	O	13.0	30	90.0	—(2)	-.-(2)	—(2)	-.-(2)
Welding Eng [O]		1.2	N	4.0	4.5	D	O	0.0	2	100.0	—(2)	-.-(2)	—(2)	-.-(2)
108 Lehigh U	PA													
Chemical Eng [H]		17.0	Y	4.0	4.0	D	O	24.0	45	100.0	21	43.0	13	87.0
Civil Eng [V]		22.0	Y	4.0	4.0	D	O	0.0	62	100.0	21	70.0	9	97.0
Electrical Eng & Computer Science [E](1)		27.0	Y	4.0	4.0	D	O	1.0	79	100.0	38	39.0	6	66.0
Industrial & Manuf'g Systems Eng [I](2)		14.0	Y	4.0	4.0	D	O	0.0(2)	42	100.0	13	50.0	3	80.0
Materials Science & Eng [O]		17.0	Y	4.0	4.0	D	O	1.0	28	100.0	21	80.0	6	99.0
Mechanical Eng & Appl Mechanics [M]		34.0	Y	4.0	4.0	D	O	1.0	103	100.0	23	75.0	6	99.0
109 Louisiana State U	LA													
Biological & Agricultural Eng [O]		12.5	Y	4.0	5.0	D	O	0.0	4	90.0	2	90.0	—(1)	-.-(B)
Chemical Eng [H]		16.5	Y	4.0	5.0	D	O	13.3	22	90.2	9	91.4	5	86.1
Civil Eng [V]		20.8	Y	4.0	5.0	D	O	1.0	36	86.1	12	91.4	4	95.2
Computer Eng [O]		9.0	Y	4.0	5.0	D	O	12.3	22	91.0	—(2)	-.-(B)	—(2)	-.-(B)
Electrical Eng [E]		19.7	Y	4.0	5.0	D	O	6.4	87	85.6	32(1)	92.7	0	91.4
Eng Science(s) [O]		-.-	N	-.-(B)	-.-(B)	D	—(B)	-.-(B)	—(3)	-.-(B)	8	88.9	0	92.3
Industrial Eng [I]		11.5	Y	4.0	5.0	D	O	5.0	23	81.3	21	89.8	—(1)	-.-(B)
Mechanical Eng [M]		20.8	Y	4.0	5.0	D	O	4.8	61	86.9	7	79.1	8	92.3
Petroleum Eng [O]		8.2	Y	4.0	5.0	D	O	0.0	8	87.2	5	89.7	1	100.0

All Engineering Disciplines

ENGINEERING Institution Profile Reference Number & Name Name of Degree Program	State/Province	FACULTY Full-time equivalent (FTE)	ABET/CEAB accred?	UNDERGRADUATE PROGRAMS						GRADUATE PROGRAMS			
				Length Nominal length of program in years	Length Average length of program in years	Time Day/Eve./Both	Co-op None/Opt./Req.	Co-op % of graduates in Co-op programs	Bachelor's # of degrees awarded	Bachelor's % of graduates who were full-time	Master's # of degrees awarded	Master's % of graduates who were full-time	Doctoral # of degrees awarded	Doctoral % of graduates who were full-time
110 Louisiana Tech U	LA													
Biomedical Eng [O]		6.0	Y	4.0	4.7	D	O	-.-(A)	13	-.-(A)	4	-.-(A)	—.	-.-(B)
Chemical Eng [H]		5.0	Y	4.0	4.7	D	O	-.-(A)	19	-.-(A)	4	-.-(A)	1	-.-(A)
Civil Eng [V]		14.0	Y	4.0	4.7	D	O	-.-(A)	19	-.-(A)	2	-.-(A)	1	-.-(A)
Computer Science [O]		6.0	N(1)	4.0	4.7	D	O	-.-(A)	13	-.-(A)	4	-.-(A)	—.	-.-
Construction Eng Tech [O]		3.0	Y	4.0	4.7	D	O	-.-(A)	4	-.-(A)	—.	-.-	—.	-.-(B)
Electrical Eng [E]		11.0	Y	4.0	4.7	D	O	-.-(A)	47	-.-(A)	23	-.-(A)	—.	-.-
Electrical Eng Tech [O]		3.0	Y	4.0	4.0	D	O	-.-(A)	17	-.-(A)	—.	-.-	—.	-.-
Geosciences [O](2)		3.0	N	4.0	4.0	D	O	-.-(A)	2	-.-(A)	—.	-.-	—.	-.-(A)
Industrial Eng [I]		6.0	Y	4.0	4.7	D	O	-.-(A)	4	-.-(A)	4	-.-(A)	1	-.-(A)
Mechanical Eng [M]		13.0	Y	4.0	4.7	D	O	-.-(A)	29	-.-(A)	8	-.-(A)	1	-.-(A)
Petroleum Eng [O](2)		4.0	Y	4.0	4.7	D	O	-.-(A)	7	-.-(A)	—.	-.-(B)	1	-.-(A)
111 U of Louisville	KY													
Chemical Eng [H]		9.0	Y(1)	4.0	4.5	D	R	98.0	24	77.5	11	56.6	0	77.7
Civil Eng [V]		12.0	Y(1)	4.0	4.5	D	R	93.0	22	80.6	15	29.0	—.	-.-
Computer Science & Eng [O]		-.-(2)	N	3.0	3.5	B	N	-.-	0	0.0	0	40.0	1	39.3
Electrical Eng [E]		17.8	Y(1)	4.0	4.5	D	R	98.0	64	66.9	35	23.5	—.	-.-
Eng Mgmt [O](4)		-.-(3)	N	-.-	-.-	—	N	-.-	—.	-.-	2	-.-(A)	—.	-.-
Eng Math & Computer Science [O]		17.0	Y(1)	4.0	4.5	D	R	97.0	16	72.9	22	21.2	—.	-.-
Industrial Eng [I]		9.0	Y(1)	4.0	4.5	D	R	98.0	16	67.1	15	68.4	2	57.1
Mechanical Eng [M]		13.0	Y(1)	4.0	4.5	D	R	92.0	34	72.4	24	45.1	—.	-.-
112 Loyola Coll in Maryland	MD													
Electrical Eng [E]		6.0	N	4.0	4.5	—	N	-.-	2	100.0	—.	-.-	—.	-.-
Eng Science(s) [O]		6.0	Y	4.0	4.0	—	N	-.-	10	100.0	—.	-.-	—.	-.-
113 Loyola Marymount U	CA													
Civil Eng [V]		6.5	Y	4.0	4.2	B	N	-.-	12	100.0	10	0.0	—.	-.-
Electrical Eng [E]		6.5	Y	4.0	4.2	B	N	-.-	12	100.0	5	0.0	—.	-.-
Mechanical Eng [M]		6.5	Y	4.0	4.2	B	N	-.-	12	100.0	5	0.0	—.	-.-
114 U of Maine	ME													
Bio-Resource Eng [O]		-.-	Y	4.0	-.-	D	O	-.-	—.	-.-	—.	-.-	—.	-.-
Chemical Eng [H]		12.0	Y	4.0	4.5	D	N	-.-	30	95.0	4	90.0	—.	100.0
Civil Eng [V]		12.5	Y	4.0	4.5	D	N	-.-	41	94.0	16	100.0	2	100.0
Electrical & Computer Eng [E]		14.5	Y	4.0	4.5	D	N	-.-	39	90.0	12	90.0	—.	-.-(1)
Eng Physics [O]		17.0	Y	4.0	4.5	D	N	-.-	4	100.0	0	100.0	—.(1)	-.-
Forest Eng [O]		3.0	Y	4.0	4.5	D	N	-.-	5	100.0	—.	-.-(2)	—.(1)	-.-
Mechanical Eng [M]		9.0	Y	4.0	4.5	D	N	-.-	33	90.0	—.	100.0	—.(1)	-.-
Surveying Eng [O]		6.0	Y	4.0	4.5	D	N	-.-	11	100.0	12	100.0	1	100.0
115 U of Manitoba	MB													
Agricultural Eng [O]		10.0	Y	4.0	-.-(A)	D	N	-.-(B)	3	-.-(A)	—.	-.-(A)	—.	-.-(A)
Civil Eng [V]		-.-	Y	4.0	-.-(A)	D	O	-.-(A)	48	-.-(B)	14	-.-(A)	8	-.-(A)
Computer Eng [E]		-.-	Y	4.0	-.-(A)	D	O	-.-(A)	17	-.-(A)	—.(1)	-.-(B)	—.(1)	-.-(B)
Electrical Eng [E]		-.-	Y(1)	4.0	-.-(A)	D	O	-.-(A)	50	-.-(A)	11(1)	-.-(A)	5(1)	-.-(A)
Geological Eng [O]		-.-	Y	4.0	-.-(A)	D	N	-.-(B)	—.(A)	-.-(A)	—.	-.-(B)	—.	-.-(B)
Industrial Eng [M]		-.-	Y	4.0	-.-(A)	D	N	-.-(B)	26	-.-(A)	11(2)	-.-(A)	—.(2)	-.-(B)
Mechanical Eng [M]		-.-	Y	4.0	-.-(A)	D	N	-.-(B)	43	-.-(A)	—.(2)	-.-(A)	—.(2)	-.-(B)
116 Mankato State U	MN													
Electrical Eng [E]		7.0	Y	4.0	4.5	D	N	-.-	24	100.0	—.	-.-	—.	-.-
Mechanical Eng [M]		5.0	N(1)	4.0	4.5	D	N	-.-	14	100.0	—.	-.-	—.	-.-
117 Marietta Coll	OH													
Industrial Eng [I]		2.0	N	4.0	4.0	B	N	-.-	3	100.0	—.	-.-	—.	-.-
Petroleum Eng [O]		4.0	Y	4.0	4.5	D	N	-.-(B)	5	100.0	—.	-.-	—.	-.-
118 Marquette U	WI													
Biomedical Eng [O]		12.6	Y	4.0	4.2	B	O	29.0	31	99.0	13	73.0	2	89.0
Civil Eng [V]		11.0	Y	4.0	4.2	B	O	39.0	46	1.0	15	44.0	3	66.0
Electrical Eng [E]		21.2	Y	4.0	4.2	B	O	42.0	98	85.0	42	32.0	3	54.0
Industrial Eng [I]		4.0	Y	4.0	4.2	B	O	25.0	24	98.0	—.(B)	-.-(B)	—.(B)	-.-(B)
Mechanical Eng [M]		15.9	Y	4.0	4.2	B	O	43.0	58	92.0	15	34.0	4	82.0
119 Maryland Baltimore Cty	MD													
Chemical & Biochemical Eng [H]		8.5	Y	4.0	4.5	D	O	42.9	7	86.0	2	67.0	—.	100.0
Electrical Eng [E]		9.0	N(4)	-.-	-.-	B	—	-.-	—.	-.-	15	55.0	—.	87.0
Eng Mgmt [O]		-.-(1)	—	-.-	-.-	—	—	-.-	—.	-.-	2	-.-(2)	—.	-.-(3)
Eng Science & Mechanics [O]		-.-	—	-.-	-.-	—	—	-.-	—.	-.-	2	-.-	—.(3)	-.-(3)
Mechanical Eng [M]		15.0	Y	4.0	4.5	D	O	33.3	42	86.0	9	52.0	—.	88.0

All Engineering Disciplines

ENGINEERING Institution Profile Reference Number & Name Name of Degree Program	State/Province	FACULTY Full-time equivalent (FTE)	ABET/CEAB accred?	UNDERGRADUATE PROGRAMS					Bachelor's		GRADUATE PROGRAMS			
				Length		Time	Co-op				Master's		Doctoral	
				Nominal length of program in years	Average length of program in years	Day/Eve./Both	None/Opt./Req.	% of graduates in Co-op programs	# of degrees awarded	% of graduates who were full-time	# of degrees awarded	% of graduates who were full-time	# of degrees awarded	% of graduates who were full-time
120 U of Maryland, Coll Park	MD													
Aerospace Eng [O]		18.0	Y	4.0	4.5	D	O	12.0	100	10.4	24	64.3	6	72.2
Agricultural Eng [O]		5.0	Y	4.0	4.5	D	O	-.-	3	-.-	3	60.0	0	66.7
Chemical Eng [H]		12.0	Y	4.0	4.5	D	O	16.7	24	88.0	10	80.0	8	80.8
Civil Eng [V]		30.0	Y	4.0	4.5	D	O	1.6	61	85.5	53	38.6	7	45.4
Electrical Eng [E]		65.0	Y	4.0	5.0	D	O	9.9	212	80.9	92	49.2	18	73.2
Fire Protection Eng [O]		7.0	Y	4.0	4.5	D	O	6.2	32	91.0	1	40.0	—.	-.-
Materials Eng [O][3]		6.0	N	-.-[3]	-.-	D	—[3]	-.-	0	-.-	5	35.5	2	56.1
Mechanical Eng [M]		40.0	Y	4.0	4.5	D	O	19.4	144	84.3	44	56.4	17	57.3
Nuclear Eng [O][1]		7.0	Y	4.0	4.5	D	O	-.-	9	91.7	4	42.9	3	58.8
Reliability Eng [O][3]		-.-	N	-.-[3]	-.-	D	—[3]	-.-	0	-.-	8	13.0	2	44.2
Systems Eng [O][3]		-.-	N	-.-[3]	-.-	D	—[3]	-.-	—.	-.-	4	48.1	—.	-.-
121 U of Mass	MA													
Chemical Eng [H]		14.4[4]	Y	4.0	4.6	D	O	36.4	22	96.4	4	83.3	12	69.8
Civil Eng [V]		18.0[4]	Y	4.0	4.6	D	O	18.9	74	98.2	19[1]	67.4	8	40.0
Electrical & Computer Eng [E]		32.2[4]	Y	4.0	4.6	D	O	20.8	72	96.6	73	69.6	15	39.0
Industrial Eng & Operations Research [I]		7.3[4]	Y	4.0	4.6	D	O	36.8	38	96.3	28[2]	43.5	1	30.8
Mechanical Eng [M]		23.7[4]	Y	4.0	4.6	D	O	40.7	91	95.6	35[3]	58.8	6	46.8
122 U of Mass, Dartmouth	MA													
Civil Eng [V]		7.0	Y	4.0	4.0	—	N	-.-	25	100.0	—.	-.-	—.	-.-
Electrical & Computer Eng [E]		17.0	Y	4.0	4.0	—	N	-.-	62	95.0	15	13.5	—.	-.-
Mechanical Eng [M]		10.0	Y	4.0	4.0	—	N	-.-	22	100.0	—.	-.-	—.	-.-
123 U of Mass, Lowell	MA													
Chemical Eng [H]		9.0	Y	4.0	4.5	D	N	-.-	10	100.0	1	100.0	—.	-.-
Civil Eng [V]		19.0	Y	4.0	4.8	D	N	-.-	82	100.0	34	15.0	—.	-.-
Electrical Eng [E]		38.0	Y	4.0	4.7	D	N	-.-	114	100.0	88	33.0	4	100.0
Industrial Tech [I]		5.0	N	4.0	4.9	B	N	-.-	7	100.0	43	8.0	—.	-.-
Mechanical Eng [M]		2.5	Y	4.0	4.7	D	N	-.-	74	100.0	16	50.0	2	25.0
Nuclear Eng [H]		20.5	Y	4.0	4.7	D	N	-.-	74	100.0	16	80.0	2	50.0
Plastics Eng [O]		17.5	Y	4.0	4.7	D	N	-.-	36	100.0	44	50.0	12	75.0
Work Environment [O]		9.0	N	-.-	-.-(A)	D	N	-.-	0	0.0	5	40.0	1	30.0
124 Mass Inst of Tech	MA													
Aeronautics & Astronautics [O]		36.0	Y	4.0	-.-	D	O	-.-[1]	66	100.0	63	100.0	26	100.0
Chemical Eng [H]		32.0	Y	4.0	-.-	D	N	-.-	52	100.0	28	100.0	32	100.0
Civil & Env'l Eng [V]		40.0	Y	4.0	-.-	D	O	0.0	40	100.0	59	100.0	29	100.0
Electrical Eng & Computer Science [E]		115.0	Y	4.0	-.-	D	O	-.-[5]	245	100.0	170	100.0	62	100.0
Materials Science & Eng [O]		36.0	Y	4.0	-.-	D	O	-.-[2]	43	100.0	31	100.0	28	100.0
Mechanical Eng [M]		60.0	Y	4.0	-.-	D	O	-.-[3]	138	100.0	143	100.0	31	100.0
Nuclear Eng [O]		23.0	Y	4.0	-.-	D	O	-.-[4]	7	100.0	28	100.0	20	100.0
Ocean Eng [O]		22.0	Y	4.0	-.-	D	N	-.-	5	100.0	41	100.0	11	100.0
125 McGill U	PQ													
Chemical Eng [H]		20.1	Y	3.5[1]	4.0	D	O[2]	-.-(A)	44	-.-(A)	27	47.6[3]	14	40.5[3]
Civil Eng & Appl Mechanics [V]		20.5	Y	3.5[1]	4.0	D	O[2]	-.-(A)	42	-.-(A)	20	62.2[3]	4	55.2[3]
Electronics Eng [E]		43.3	Y	3.5[1]	4.0[C]	D	O[2]	-.-(A)	100	-.-(A)	36	59.8[3]	10	36.8[3]
Mechanical Eng [M]		24.9	Y	3.5[1]	4.0[C]	D	O[2]	-.-(A)	98	-.-(A)	29	48.4[3]	12	55.8[3]
Metallurgical Eng [O]		18.3	Y	3.5[1]	4.0[C]	D	O[2]	-.-(A)	11	-.-(A)	27[4]	78.9[3]	7[5]	85.2[3]
Mining Eng [O]		18.3	Y	4.0[1]	-.-(A)	D	R	100.0	2	-.-(A)	27[6]	78.9[3]	7[7]	85.2[3]
126 Memphis State U	TN													
Bioengineering [O]		4.0	N	-.-(A)	-.-(A)	D	N	-.-(A)	0	-.-(A)	2	50.0	2	50.0
Civil Eng [V]		15.0	Y	4.0	-.-	D	O	-.-	20	70.0	9	50.0	1	50.0
Electrical Eng [E]		13.0	Y	4.0	5.0	D	O	20.0	54	70.0	27	65.0	3	65.0
Mechanical Eng [M]		13.0	Y	4.0	5.0	D	O	10.0	54	70.0	12	65.0	3	50.0
127 Mercer U	GA													
Eng [O]		-.-(A)	Y	4.0	4.0	B	O	-.-(A)	45	98.0	3	-.-(A)	—.(A)	-.-(A)
128 U of Miami	FL													
Architectural Eng [O]		1.0	Y	4.0	4.0	D	O	1.0	23	92.7	1	3.2	—.	-.-
Biomedical Eng [O]		5.0	N	4.0	4.0	D	O	6.0	0	93.8	12	90.0	5	98.0
Civil Eng [V]		10.0	Y	4.0	4.0	D	O	2.0	23	92.7	11	40.0	2	20.0
Electrical & Computer Eng [E]		10.1	Y	4.0	4.0	D	O	2.0	61	89.9	11	86.0	1	83.0
Industrial Eng [I]		10.5	Y	4.0	4.0	D	O	1.0	23	91.6	10	26.0	0	6.0
Mechanical Eng [M]		7.0	Y	4.0	4.0	D	O	2.0	28	94.4	7	88.0	2	100.0

All Engineering Disciplines

ENGINEERING Institution Profile Reference Number & Name Name of Degree Program	State/Province	FACULTY Full-time equivalent (FTE)	ABET/CEAB accred?	UNDERGRADUATE PROGRAMS						GRADUATE PROGRAMS			
				Length		Time	Co-op		Bachelor's	Master's		Doctoral	
				Nominal length of program in years	Average length of program in years	Day/Eve./Both	None/Opt./Req.	% of graduates in Co-op programs	# of degrees awarded	% of graduates who were full-time	# of degrees awarded	% of graduates who were full-time	# of degrees awarded	% of graduates who were full-time
129 Miami U	OH													
Eng Mgmt [O][1]		5.0	N	4.0	4.0	B	O[2]	-.-[1]	0[1]	100.0	—[B]	-.-[B]	—[B]	-.-[B]
Manuf'g Eng [O]		5.8	Y	4.0	4.0	B	O	-.-[1]	15	96.2	—[B]	-.-[B]	—[B]	-.-[B]
Paper Science & Eng [O]		7.5	N	4.0	4.0	B	O	68.0	25	86.0	3	76.9	—[B]	-.-[B]
130 U of Michigan	MI													
Aerospace Eng [O]		26.6	Y	4.0	4.7	D	O	8.3[7]	107	95.5	34[4]	86.7[5]	17	86.7[5]
Aerospace Science Eng [O][2]		-.-[3]	—	-.-	-.-	D	—	-.-	—.	-.-	1[4]	88.9[5]	—.	-.-
Appl Mechanics [M][2]		-.-[3]	—	-.-	-.-	—	—	-.-	—.	-.-	7[4]	87.5[5]	5	87.5[5]
Appl Physics [E][2]		-.-[3]	—	-.-	-.-	—	—	-.-	—.	-.-	6[4]	-.-[A]	15	-.-[A]
Atmospheric & Space Sciences [O][2]		16.1	—	-.-	-.-	D	—	-.-	—.	-.-	10[4]	94.9[5]	7	94.9[5]
Bioengineering [O][2]		-.-[8]	—	-.-	-.-	—	—	-.-	—.	-.-	24[4]	-.-[A]	10	-.-[A]
Chemical Eng [H]		16.3	Y	4.0	4.7	D	O	8.3[7]	58	96.6	10[4]	88.0[5]	16	88.0[5]
Civil & Env'l Eng [V][1]		26.6	Y	4.0	4.7	D	O	8.3[7]	48	97.7	—.	-.-	—.	-.-
Civil Eng [V][2]		-.-[3]	—	-.-	-.-	—	—	-.-	—.	-.-	21[4]	93.6[5]	20	93.6[5]
Computer Eng [E][1]		-.-[3]	Y	4.0	4.7	D	O	8.3[7]	67	93.6	—.	-.-	—.	-.-
Computer, Information, & Control Eng [E][6]		-.-[3]	—	-.-	-.-	—	—	-.-	—.	-.-	—.	-.-	5	100.0
Computer Science & Eng [E][2]		-.-[3]	—	-.-	-.-	—	—	-.-	—.	-.-	75[4]	80.1[5]	12	80.1[5]
Construction Eng & Mgmt [V][2]		-.-[3]	—	-.-	-.-	—	—	-.-	—.	-.-	15[4]	92.9[5]	—.	-.-
Electrical Eng [E]		102.7	Y	4.0	4.7	D	O	8.3[7]	143	93.3	56[4]	85.8[5]	32	85.8[5]
Electrical Eng Systems [E][2]		-.-[3]	—	-.-	-.-	—	—	-.-	—.	-.-	28[4]	84.0[5]	12	84.0[5]
Eng [O][1]		-.-[8]	N	4.0	4.7	D	O	8.3[7]	7	98.8	—.	-.-	—.	-.-
Eng Physics [O][1]		-.-[3]	N	4.0	4.7	D	O	8.3[7]	4	100.0	—.	-.-	—.	-.-
Env'l Eng [V][2]		-.-[3]	—	-.-	-.-	—	—	-.-	—.	-.-	19[4]	90.0[5]	1	90.0[5]
Industrial & Operations Eng [I]		23.1	Y	4.0	4.7	D	O	8.3[7]	104	96.2	48[4]	88.1[5]	8	88.1[5]
Macromolecular Science & Eng [O][2]		-.-[8]	—	-.-	-.-	—	—	-.-	—.	-.-	2[4]	-.-[A]	6	-.-[A]
Materials Science & Eng [O]		15.9	Y	4.0	4.7	D	O	8.3[7]	17	96.8	15[4]	84.1[5]	7	84.1[5]
Mechanical Eng [M]		62.4	Y	4.0	4.7	D	O	8.3[7]	228	94.2	100[4]	82.3[5]	39	82.3[5]
Meteorology [O][1]		-.-[3]	N	4.0	4.7	D	O	8.3[7]	5	100.0	—.	-.-	—.	-.-
Naval Architecture & Marine Eng [O]		11.8	Y	4.0	4.7	D	O	8.3[7]	18	94.1	21[4]	93.3[5]	8	93.3[5]
Nuclear Eng [O]		11.6	Y	4.0	4.7	D	O	8.3[7]	12	95.5	12[4]	97.2[5]	16	97.2[5]
Nuclear Sciences [O][2]		-.-[3]	—	-.-	-.-	D	—	-.-	—.	-.-	0[4]	91.7[5]	0	91.7[5]
Ocean Science [O][2]		-.-[3]	—	-.-	-.-	D	—	-.-	—.	-.-	—.	-.-	1	100.0
Oceanography: Physical [O]		-.-[3]	N	4.0	4.7	D	O	8.3[7]	1	100.0	—.	-.-	1	100.0
131 U of Michigan-Dearborn	MI													
Electrical & Computer Eng [E]		13.6	Y	4.0	4.5	D	O	0.0[A]	76	65.2	24	4.0	—[B]	-.-[B]
Industrial & Systems Eng [I]		7.5	Y	4.0	4.5	D	O	0.0[A]	13	57.6	15	6.4	—[B]	-.-[B]
Mechanical Eng [M]		15.1	Y	4.0	4.5	D	O	0.0[A]	58	62.6	26	0.0	—[B]	-.-[B]
132 Michigan State U	MI													
Agricultural Eng [O]		22.0	Y	4.0	4.5	D	O	0.0	2	100.0	2	100.0	4	100.0
Chemical Eng [H]		10.0	Y	4.0	4.5	D	O	29.0	62	100.0	13	100.0	9	100.0
Civil Eng [V]		22.0	Y	4.0	4.5	D	O	26.0	94	100.0	21	100.0	10	100.0
Computer Eng [O][1]		26.0[1]	N	4.0	4.5	D	O	5.0	15	100.0	—.	-.-	—.	-.-
Computer Science [O]		23.0	N	4.0	4.5	D	O	18.0	76	100.0	49	100.0	5	100.0
Electrical Eng [E]		26.0	Y	4.0	4.5	D	O	30.0	103	100.0	30	100.0	16	100.0
Eng Arts [O][2]		15.0[2]	N	4.0	4.5	D	O	33.0	65	100.0	—.	-.-	—.	-.-
Env'l Eng [O][3]		22.0[3]	N	-.-	-.-	D	O	0.0	0	0.0	9	100.0	2	100.0
Food Eng [O][4]		22.0[4]	N	4.0	4.5	D	O	12.5	4	100.0	—.	-.-	—.	-.-
Materials Science [O][2]		15.0[2]	Y	4.0	4.5	D	O	6.0	15	100.0	6	100.0	7	100.0
Mechanical Eng [M]		21.0	Y	4.0	4.5	D	O	40.0	168	100.0	12	100.0	6	100.0
Mechanics [O][2]		15.0[2]	N	4.0	4.5	D	O	25.0	4	100.0	6	100.0	8	100.0
133 Michigan Tech U	MI													
Chemical Eng [H]		9.0	Y	4.0	4.3	D	O	12.7	63	97.6	4	100.0	0	87.5
Civil Eng [V]		12.8	Y	4.0	4.3	D	O	18.6	97	92.0	16	97.4	0	100.0
Electrical Eng [E]		26.8	Y	4.0	4.3	D	O	16.7	239	95.8	31	96.2	0	100.0
Eng [O]		25.0[2]	Y	4.0	4.3	D	O	15.8	19	96.9	7	100.0	11	93.0
Env'l Eng [O]		7.0	Y	4.0	4.3	D	O	0.0	29	95.5	—.	-.-	—.	-.-
Geological Eng [O]		2.0	Y	4.0	4.3	D	O	14.3	7	97.3	0	100.0	0	100.0
Materials Science & Eng [O]		15.0	Y	4.0	4.3	D	O	13.8	36	93.8	3	100.0	8	100.0
Mechanical Eng [M]		41.0	Y	4.0	4.3	D	O	15.1	271	95.3	27	98.5	5	93.1
Minerals Processing [O]		3.0	Y	4.0	4.3	D	O	10.0	10	100.0	2	100.0	—[1]	-.-[1]
Mining Eng [O]		4.0	Y	4.0	4.3	D	O	0.0	6	95.3	1	100.0	0	100.0
134 Milwaukee Sch of Eng	WI													
Architectural Eng [O]		7.2	Y	4.0	4.6	D	O[1]	-.-	47	98.0	—.	-.-	—.	-.-
Biomedical Eng [O]		6.0	Y	4.0	4.7	D	N[1]	-.-	6	100.0	—.	-.-	—.	-.-
Computer Eng [O]		12.0	Y	4.0	4.8	D	N[1]	-.-	20	96.0	—.	-.-	—.	-.-

All Engineering Disciplines

ENGINEERING Institution Profile Reference Number & Name Name of Degree Program	State/Province	FACULTY Full-time equivalent (FTE)	ABET/CEAB accred?	UNDERGRADUATE PROGRAMS					Bachelor's		GRADUATE PROGRAMS			
				Length		Time	Co-op				Master's		Doctoral	
				Nominal length of program in years	Average length of program in years	Day/Eve./Both	None/Opt./Req.	% of graduates in Co-op programs	# of degrees awarded	% of graduates who were full-time	# of degrees awarded	% of graduates who were full-time	# of degrees awarded	% of graduates who were full-time
Electrical Eng [E]		14.7	Y	4.0	4.6	D	N[1]	-.-	66	95.0	—.	-.-	—.	-.-
General Eng [O][1]		4.0	N	4.0	4.5	E	N[1]	-.-	0	0.0	12	2.0	—.	-.-
Industrial Eng [I]		6.5	Y	4.0	4.6	D	N[1]	-.-	15	95.0	—.	-.-	—.	-.-
Mechanical Eng [M]		12.5	Y	4.0	4.6	D	N[1]	-.-	56	95.0	—.	-.-	—.	-.-
135 U of Minnesota, Duluth	MN													
Chemical Eng [H]		5.0	Y	4.0	4.6	D	N	-.-	11	100.0	—.	-.-	—.	-.-
Computer & Electrical Eng [E]		9.0	Y	4.0	4.6	D	N	-.-	34	100.0	—.	-.-	—.	-.-
Industrial Eng [I]		6.0	Y	4.0	4.6	D	N	-.-	24	100.0	—.	-.-	—.	-.-
136 U of Minnesota	MN													
Aerospace Eng & Mechanics [O]		22.0[1]	Y	4.0	-.-[A]	D	—	-.-	83	-.-[A]	17	-.-[A]	6	-.-[A]
Agricultural Eng [O]		17.0[1]	Y	4.0	-.-	D	N	-.-	10	-.-[A]	2	-.-[A]	7	-.-[A]
Chemical Eng [H]		31.0[1]	Y	4.0	-.-[A]	D	N[A]	-.-[A]	84	-.-[A]	3	-.-[A]	22	-.-[A]
Civil Eng [V][2]		31.0	Y	4.0	-.-[A]	B	O	-.-[A]	113	-.-[A]	27	-.-[A]	9	-.-[A]
Electrical Eng [E][3]		50.0[1]	Y	4.0	-.-[A]	B	O	-.-[A]	141	-.-[A]	62	-.-[A]	19	-.-[A]
Extractive Metallurgical Eng [O]		6.0	Y	4.0	-.-[A]	D	N	-.-	4	-.-[A]	6	-.-[A]	1	-.-[A]
Geological Eng [O]		5.0[4]	Y	4.0	-.-[A]	D	N	-.-	3	-.-[A]	2	-.-[A]	7	-.-[A]
Materials Science [O]		-.-[1]	Y	4.0	-.-[A]	D	N	-.-[B]	9	-.-[A]	5	-.-[A]	9	-.-[A]
Mechanical Eng [M]		40.0	Y	4.0	-.-[A]	B	O	-.-[A]	210	-.-[A]	60	-.-[A]	23	-.-[A]
137 U of Mississippi	MS													
Chemical Eng [H][1]		5.8	Y	4.0	4.5	D	N	-.-	5	95.3	—.	-.-	—.	-.-
Civil Eng [V][2]		7.0	Y	4.0	4.5	D	N	-.-	18	88.9	—.	-.-	—.	-.-
Computer & Information Science [O][5]		7.0	N[6]	4.0	4.5	D	N	-.-	18	91.9	—.	-.-	—.	-.-
Electrical Eng [E][3]		9.3	Y	4.0	4.5	D	N	-.-	16	95.2	—.	-.-	—.	-.-
Eng Science(s) [O][8]		42.3	—	-.-	-.-	—	N	-.-	0	-.-	50	79.1	4	73.2
General Eng [O][7]		29.3	N	4.0	4.5	D	N	-.-	17	89.2	—.	-.-	—.	-.-
Geological Eng [O][9]		6.0	Y	4.0	4.5	D	N	-.-	2	100.0	—.	-.-	—.	-.-
Geology [O]		3.0	N	4.0	4.5	D	N	-.-	1	100.0	1	100.0	—.	-.-
Mechanical Eng [M][4]		7.2	Y	4.0	4.5	D	N	-.-	7	91.8	—.	-.-	—.	-.-
138 Mississippi State U	MS													
Aerospace Eng [O]		21.0	Y	4.0	4.8	D	O	30.0	35	95.0	11	100.0	4	100.0
Agricultural Eng [O][2]		7.0	Y	4.0	4.8	D	O	30.0	2	100.0	1	100.0	0	100.0
Biological Eng [O]		-.-[1]	Y	4.0	4.8	D	O	30.0	6	95.0	2	100.0	0	100.0
Chemical Eng [H]		10.0	Y	4.0	4.8	D	O	30.0	28	95.0	1	100.0	1	100.0
Civil Eng [V]		12.0	Y	4.0	4.8	D	O	30.0	29	95.0	14	100.0	1	100.0
Computational Eng [O]		-.-[3]	Y[5]	4.0[4]	4.8[4]	D	O[4]	-.-[4]	0[4]	-.-[4]	0	100.0	1	100.0
Computer Eng (Electrical) [E]		-.-[5]	Y	4.0	4.8	D	O	30.0	20	95.0	0	100.0	0	100.0
Electrical Eng [E]		24.0	Y	4.0	4.8	D	O	30.0	75	95.0	22	100.0	4	100.0
Eng Mechanics [O]		-.-[6]	Y	-.-[7]	-.-[7]	D	O[7]	-.-[7]	—[7]	-.-[7]	2	100.0	0	100.0
Industrial Eng [I]		13.0	Y	4.0	4.8	D	O	30.0	46	95.0	6	100.0	1	100.0
Mechanical Eng [M]		16.0	Y	4.0	4.8	D	O	30.0	50	95.0	8	100.0	2	100.0
Nuclear Eng [O][2]		3.0	Y	4.0	4.8	D	O	30.0	4	100.0	0	100.0	0	100.0
Petroleum Eng [O][2]		5.0	Y	4.0	4.8	D	O	0.0	4	95.0	2	100.0	0	100.0
139 U of Missouri-Columbia	MO													
Agricultural Eng [O]		8.0	Y	4.0	4.5	D	O	4.0	4	100.0	3	66.6	1	62.5
Chemical Eng [H]		9.0	Y	4.0	-.-	D	O	4.0	20	95.9	8	55.1	1	57.8
Civil Eng [V][1]		16.0	Y	4.0	-.-	D	O	4.0	46	81.7	19	35.4	—.	12.5
Computer Eng [O]		21.0	Y	4.0	-.-	D	O	4.0	33	92.2	—.	-.-	—.	-.-
Electrical Eng [E][2]		25.0	Y	4.0	-.-	D	O	4.0	106	84.1	96	38.0	7	39.7
Industrial Eng [I]		7.0	Y	4.0	-.-	D	O	4.0	19	96.8	25	62.2	3	66.6
Mechanical & Aerospace Eng [M][3]		23.0	Y	4.0	-.-	D	O	4.0	107	87.0	22	61.1	3	46.4
140 U of Missouri-Kansas City	MO													
Civil Eng [V]		4.0	Y	4.0	5.0	B	N	-.-[2]	13	53.0	2	10.0	—.	-.-
Electrical & Computer Eng [E]		7.0	Y	4.0	5.0	B	N	-.-[2]	31	59.0	14	-.-	—.	-.-
Mechanical & Aerospace Eng [M]		6.0	Y	4.0	5.0	B	N	-.-[2]	23	40.0	3	20.0	0	75.0
141 U of Missouri-Rolla	MO													
Aerospace Eng [O]		7.9	Y	4.0	4.5	D	O	8.3	36	90.0	7	50.0	1	0.0
Ceramic Eng [O]		11.0	Y	4.0	4.5	D	O	14.3	21	90.0	16	50.0	3	0.0
Chemical Eng [H]		12.8	Y	4.0	4.5	D	O	0.0	31	94.0	3	58.0	2	45.0
Civil Eng [V]		16.4	Y	4.0	4.5	D	O	0.0	68	92.0	17	48.0	2	64.0
Electrical Eng [E]		32.0	Y	4.0	4.5	D	O	13.7	145	87.0	68	19.0	6	42.0
Eng Mgmt [O]		17.3	Y	4.0	4.5	D	O	3.3	59	82.0	85	37.0	15	60.0
Eng Mechanics [O][A]		8.0	N	-.-[1]	-.-	—[1]	—[A]	-.-[A]	0	0.0	11	13.0	5	50.0
Geological Eng [O]		6.0	Y	4.0	4.5	D	O	4.8	21	97.0	8	89.0	0	33.0
Mechanical Eng [M]		24.4	Y	4.0	4.5	D	O	12.5	152	87.0	16	43.0	7	45.0
Metallurgical Eng [O]		11.8	Y	4.0	4.5	D	O	5.9	17	91.0	11	29.0	6	40.0

All Engineering Disciplines

ENGINEERING Institution Profile Reference Number & Name Name of Degree Program	State/Province	FACULTY Full-time equivalent (FTE)	ABET/CEAB accred?	UNDERGRADUATE PROGRAMS							GRADUATE PROGRAMS			
				Length		Time	Co-op		Bachelor's		Master's		Doctoral	
				Nominal length of program in years	Average length of program in years	Day/Eve./Both	None/Opt./Req.	% of graduates in Co-op programs	# of degrees awarded	% of graduates who were full-time	# of degrees awarded	% of graduates who were full-time	# of degrees awarded	% of graduates who were full-time
Mining Eng [O]		8.0	Y	4.0	4.5	D	O	0.0	3	93.0	1	100.0	1	60.0
Nuclear Eng [O]		6.0	Y	4.0	4.5	D	O	0.0	9	79.0	4	60.0	0	50.0
Petroleum Eng [O]		4.0	Y	4.0	4.5	D	O	0.0	4	94.0	1	100.0	—.	-.-
142 Monmouth Coll	NJ													
Electronic Eng [E]		5.3	Y	4.0	4.0	B	O(1)	5.0(C)	21	50.0(C)	10	25.0(C)	—.(B)	-.-(B)
143 Montana Coll of Mineral Sci	MT													
Eng Science(s) [O]		7.0	Y	4.5	5.0	D	O	-.-(A)	16	87.3	2	-.-(A)	—.(A)	-.-(A)
Env'l Eng [O]		5.0	Y	4.5	5.0	D	O	-.-(A)	18	96.3	8	-.-(A)	—.(A)	-.-(A)
Geological Eng [O]		5.5	Y	4.5	5.0	D	O	-.-(A)	9	90.0	—.	-.-(A)	—.(A)	-.-(A)
Geophysical Eng [O]		4.0	Y	4.5	5.0	D	O	-.-(A)	0	89.0	—.	-.-(A)	—.(A)	-.-(A)
Metallurgical Eng [O]		6.0	Y	4.5	5.0	D	O	-.-(A)	11	94.5	7	-.-(A)	—.(A)	-.-(A)
Mining Eng [O]		4.0	Y	4.5	5.0	D	O	-.-(A)	24	98.0	1	-.-(A)	—.(A)	-.-(A)
Petroleum Eng [O]		5.0	Y	4.5	5.0	D	O	-.-(A)	17	99.1	1	-.-(A)	—.(A)	-.-(A)
144 Montana State U	MT													
Agricultural Eng [O]		3.0	Y	4.0	4.7	D	N	-.-	0	-.-(1)	—.	-.-	—.	-.-
Chemical Eng [H]		8.0	Y	4.0	4.7	D	N	-.-	27	-.-(1)	6	100.0	1	100.0
Civil Eng [V]		19.4	Y	4.0	4.7	D	N	25.0	32	-.-(1)	2	100.0	0	100.0
Electrical Eng [E]		15.9	Y	4.0	4.7	D	N	-.-	46	-.-(1)	10	100.0	2	100.0
Industrial & Mgmt Eng [I]		6.0	Y	4.0	4.7	D	N	-.-	17	-.-(1)	5	100.0	—.	-.-
Mechanical Eng [M]		13.0	Y	4.0	4.7	D	N	-.-	59	-.-(1)	3	100.0	—.	-.-
145 Morgan State U	MD													
Civil Eng [V]		6.0	Y	4.0	4.5	D	O	0.0	9	94.0	—.	-.-	—.	-.-
Electrical Eng [E]		10.0	Y	4.0	4.5	D	O	0.0	15	91.0	—.	-.-	—.	-.-
Industrial Eng [I]		5.0	Y	4.0	4.5	D	O	0.0	10	93.0	—.	-.-	—.	-.-
146 U of Nebraska - Lincoln	NE													
Agricultural Eng [O]		10.0	Y	4.0	4.5	D	O	-.-	14	-.-	8	-.-	1	-.-
Biological System Eng [O]		10.0	N	4.0	4.5	D	O	-.-	1	-.-	—.	-.-	—.	-.-
Chemical Eng [H]		8.0	Y	4.0	4.5	—	—	-.-	29	-.-	16	-.-	1	-.-
Civil Eng [V]		18.0	Y	4.0	4.5	B	O	-.-	44	-.-	36	-.-	8	-.-
Computer Eng [O]		8.0	N	4.0	4.5	D	O	-.-	47	-.-	30	-.-	2	-.-
Electrical Eng [E]		18.0	Y	4.0	4.5	D	O	-.-	135	-.-	31	-.-	1	-.-
Eng Mechanics [O]		15.0	N	2.0	2.0	D	—	-.-	—.	-.-	11	-.-	—.	-.-
Industrial Eng [I]		10.0	Y	4.0	4.5	D	O	-.-	46	-.-	8	-.-	3	-.-
Mechanical Eng [M]		16.0	Y	4.0	4.5	D	O	-.-	141	-.-	22	-.-	4	-.-
147 U of Nevada, Las Vegas	NV													
Civil & Env'l Eng [V]		9.0	Y	4.0	4.0	B	N	-.-	19	68.2	3	53.0	0	100.0
Electrical & Computer Eng [E]		8.5	Y	4.0	4.0	B	N	-.-	15	72.8	6	26.1	—.(B)	-.-(B)
Mechanical Eng [M]		7.0	Y	4.0	4.0	B	N	-.-	13	70.9	2	53.0	—.(B)	-.-(B)
148 New England Coll	NH													
Civil Eng [V]		4.5	Y	4.0	4.0	D	N	-.-	8	100.0	—.	-.-	—.	-.-
149 U of New Hampshire	NH													
Chemical Eng [H]		7.0	Y	4.0	-.-(A)	D	N	-.-	24	-.-(A)	2	-.-(A)	2	-.-(A)
Civil Eng [V]		10.0	Y	4.0	-.-	D	N	-.-	53	-.-	14	-.-(A)	1	-.-(A)
Electrical Eng [E]		16.0	Y	4.0	-.-(A)	—	N	-.-	42	-.-(A)	22	-.-(A)	3	-.-(A)
Mechanical Eng [M]		14.0	Y	4.0	-.-(A)	D	N	-.-	44	-.-(A)	11	-.-(A)	2	-.-(A)
150 U of New Haven	CT													
Chemical Eng [H]		7.2(1)	N	4.0	4.5	B	O	5.0(C)	2	60.0(C)	—.(B)	-.-(B)	—.(B)	-.-(B)
Civil & Env'l Eng [V]		8.0	N	-.-(B)	-.-(B)	—	—(B)	-.-	—.(B)	-.-(B)	9	20.0	—.(B)	-.-(B)
Civil Eng [V]		8.0	Y	4.0	4.5	B	O	5.0(C)	11	60.0	9	30.0(C)	—.(B)	-.-(B)
Computer Science [O]		12.0	N	4.0	4.5	B	O	10.0	9	60.0(C)	75	40.0(C)	—.(B)	-.-(B)
Electrical Eng [E]		11.0	Y	4.0	4.5(C)	B	O	10.0(C)	32	60.0(A)	—.	30.0(C)	—.(B)	-.-(B)
Industrial Eng [I]		7.0	Y	4.0	4.5(C)	B	O	5.0(A)	4	60.0(C)	14	30.0(C)	—.(B)	-.-(B)
Mechanical Eng [M]		11.0	Y	4.0	4.5(C)	B	O	5.0(A)	33	60.0(C)	6	20.0(C)	—.(B)	-.-(B)
151 New Jersey Inst of Tech	NJ													
Biomedical Eng [O]		3.0(2)	N(1)	-.-(1)	-.-(1)	E(1)	N(1)	-.-(1)	—.(1)	-.-(1)	15	68.9	—.(3)	-.-(3)
Chemical Eng [H]		14.0	Y	4.0	4.5	B	O	14.3	21	80.5	14	41.4	4	81.8
Civil Eng [V]		17.9	Y	4.0	4.5	B	O	4.3	46	72.6	46	20.2	2	80.0
Computer Eng [O]		12.1	N	4.0	4.5	B	O	15.4	13	90.0	—.(4)	-.-(4)	—.(4)	-.-(4)
Electrical Eng [E]		40.7	Y	4.0	4.5	B	O	12.2	123	70.5	99	30.6	4	77.8
Eng Science(s) [O]		3.0(2)	N	4.0	4.5	B	O	0.0	3	81.8	6	29.4	—.(5)	-.-(5)
Env'l Eng [O]		10.9	N(1)	-.-(1)	-.-(1)	E(1)	N(1)	-.-(1)	—.(1)	-.-(1)	41	42.1	—.(3)	-.-(3)
Industrial Eng [I]		9.3	Y	4.0	4.5	B	O	51.7	29	64.0	11	55.6	—.(5)	-.-(5)

All Engineering Disciplines

ENGINEERING Institution Profile Reference Number & Name Name of Degree Program	State/Province	FACULTY Full-time equivalent (FTE)	ABET/CEAB accred?	UNDERGRADUATE PROGRAMS							GRADUATE PROGRAMS			
				Length		Time	Co-op		Bachelor's		Master's		Doctoral	
				Nominal length of program in years	Average length of program in years	Day/Eve./Both	None/Opt./Req.	% of graduates in Co-op programs	# of degrees awarded	% of graduates who were full-time	# of degrees awarded	% of graduates who were full-time	# of degrees awarded	% of graduates who were full-time
Mgmt Eng [O]		7.6	N[1]	-.-[1]	-.-[1]	E[1]	N[1]	-.-[1]	—[1]	-.-[1]	40	12.6	—[3]	-.-[3]
Manuf'g Eng [O]		8.3[2]	N	4.0	4.5	B	O	33.3	3	14.6	25	34.5	—[5]	-.-[5]
Mechanical Eng [M]		31.2	Y	4.0	4.5	B	O	7.7	78	75.2	39	24.2	5	82.1
Occupational Health & Safety [O]		2.0[2]	N[1]	-.-[1]	-.-[1]	E[1]	N[1]	-.-[1]	—[1]	-.-[1]	1	41.7	—[3]	-.-[3]
Transportation Planning & Eng [O]		6.1	N[6]	-.-[6]	-.-[6]	E[6]	N[6]	-.-[6]	—[6]	-.-[6]	16	20.0	0[7]	42.8
152 The U of New Mexico	NM													
Biomedical Eng [O][2]		-.-[2]	N	4.0	5.0	D	O	-.-(A)	2	90.3	—[3]	-.-[3]	—[3]	-.-[3]
Chemical Eng [H]		9.3	Y	4.0	5.0	D	O	-.-(A)	6	88.0	33	100.0[1]	14	100.0[1]
Civil Eng [V]		15.6	Y	4.0	5.0	D	O	-.-(A)	16	82.2	27	100.0[1]	0	100.0[1]
Computer Eng [O]		10.0	Y	4.0	5.0	D	O	-.-(A)	24	68.0	0[4]	100.0[1]	0[4]	100.0[1]
Computer Science [O]		14.8	N[5]	4.0	5.0	D	O	-.-(A)	19	58.4	19	100.0[1]	2	100.0[1]
Construction Eng [O]		1.0	Y	4.0	5.0	D	O	-.-(A)	5	78.9	—[3]	-.-[3]	—[3]	-.-[3]
Construction Mgmt [O]		1.0	N	4.0	5.0	D	O	-.-(A)	1	75.0	—[3]	-.-[3]	—[3]	-.-[3]
Electrical Eng [E]		32.0	Y	4.0	5.0	D	O	-.-(A)	50	70.7	51	100.0[1]	9	100.0[1]
Manuf'g & Robotics [O][6]		2.0	N	4.0	5.0	D	O	-.-(A)	1	42.8	—[3]	-.-[3]	—[3]	-.-[3]
Mechanical Eng [M]		16.8	Y	4.0	5.0	D	O	-.-(A)	55	72.1	16	100.0[1]	3	100.0[1]
Nuclear Eng [O]		7.0	Y	4.0	5.0	D	O	-.-(A)	3	75.0	3	100.0[1]	4	100.0[1]
153 New Mexico Inst of Mining	NM													
Electrical Eng [E]		5.0	N[1]	4.0	4.5	D	O	-.-(A)	1	-.-(A)	—.	-.-(A)		-.-(A)
Materials & Metallurgical Eng [O]		10.0	N[1]	4.0	4.0	D(A)	O	-.-(A)	9	-.-(A)	2	-.-(A)	4	-.-(A)
Mineral & Env'l Eng [O]		8.0	N[1]	4.0	4.5	D	O	-.-(A)	9	-.-(A)	2	-.-(A)	—.	-.-(A)
Petroleum Eng [O]		5.0	Y	4.0	4.0	D(A)	O	-.-(A)	11	-.-(A)	4	-.-(A)	2	-.-(A)
154 New Mexico State U	NM													
Agricultural Eng [O]		3.0	Y	4.0	5.1(C)	D	O	-.-(A)	1	88.0(C)	—[B]	-.-[B]	—[B]	-.-[B]
Chemical Eng [H]		7.7	Y	4.0	5.1(C)	D	O	30.0(C)	17	88.0(C)	12	75.0(C)	1	75.0(C)
Civil Eng [V]		13.2	Y	4.0	5.1(C)	D	O	26.0(C)	45	88.0(C)	11	88.0(C)	3	64.0(C)
Electrical & Computer Eng [E]		35.4	Y[1]	4.0	5.1(C)	D	O	26.0(C)	98	88.0(C)	56	60.0(C)	5	74.0(C)
Geological Eng [O]		3.0	Y	4.0	5.1(C)	D	O	-.-(A)	1	88.0(C)	—[B]	-.-[B]	—[B]	-.-[B]
Industrial Eng [I]		6.5	Y	4.0	5.1(C)	D	O	18.0(C)	12	88.0(C)	29[2]	36.0(C)	0	60.0(C)
Mechanical Eng [M]		18.3	Y	4.0	5.1(C)	D	O	20.0(C)	58	88.0(C)	6	67.0(C)	5	62.0(C)
Surveying & Mapping [O][3]		3.0	N[4]	4.0	5.1(C)	D	O	100.0	4	75.0(C)	—[B]	-.-[B]	—[B]	-.-[B]
155 U of New Orleans	LA													
Civil Eng [V]		5.0	Y	4.0	5.2	D	O	-.-	18	65.0	—.	-.-	—.	-.-
Electrical Eng [E]		10.0	Y	4.0	5.2	B	O	-.-	46	74.0	—.	-.-	—.	-.-
Mechanical Eng [M]		10.0	Y	4.0	5.2	B	O	-.-	27	72.0	—.	-.-	—.	-.-
Naval Architecture & Marine Eng [O]		4.0	Y	4.0	5.2	B	O	-.-	17	77.0	—.	-.-	—.	-.-
156 New York Inst of Tech	NY													
Electric Power Eng [E]		11.1	Y	4.0	4.0	B	O	-.-(A)	111	-.-(A)	—[B]	-.-[B]	—[B]	-.-[B]
Industrial Eng [I]		1.5	N	4.0	4.0	—(A)	O	-.-(A)	14	-.-(A)	—[B]	-.-[B]	—[B]	-.-[B]
Mechanical & Aeronautical Eng [M]		13.5	Y	4.0	4.0	B	O	-.-(A)	91	-.-(A)	—[B]	-.-[B]	—[B]	-.-[B]
157 NY State Coll of Ceramics	NY													
Ceramic Eng [O]		24.0	Y	4.0	4.5	—	O	18.5	71	86.9	7	100.0	—.	-.-
Ceramic Eng Science [O]		24.0	Y	4.0	4.5	—	O	0.0	6	100.0	1	100.0	—.	-.-
Glass Eng Science [O]		24.0	Y	4.0	4.5	—	O	0.0	3	100.0	6	100.0	—.	-.-
158 UNC-Charlotte	NC													
Civil Eng [V]		12.3	Y	4.0	4.5	B	O	26.0	23	86.1	2	27.6	—.	-.-
Computer Science [O]		25.0	N	4.0	4.5	B	O	34.2	50	69.3	26	28.3	—.	-.-
Electrical Eng [E]		21.5	Y	4.0	4.5	B	O	22.2	45	78.9	17	35.8	—.	-.-
Mechanical Eng & Eng Science [M]		15.3	Y	4.0	4.5	B	O	22.4	58	82.9	9	27.5	—.	-.-
159 North Carolina A&T	NC													
Agricultural Eng [O][1]		3.0	Y	4.0	-.-(A)	D	O	-.-(A)	0	-.-(A)	—[B]	-.-[B]	—[B]	-.-[B]
Architectural Eng [O]		7.0	Y	5.0	-.-(A)	D	O	-.-(A)	8	-.-(A)	—.	-.-	—[B]	-.-[B]
Chemical Eng [H]		11.0[1]	Y	4.0	-.-(A)	D	O	-.-(A)	10	-.-(A)	—[B]	-.-[B]	—[B]	-.-[B]
Civil Eng [V]		5.9	Y	4.0	-.-(A)	D	O	-.-(A)	8	-.-(A)	—[B]	-.-[B]	—[B]	-.-[B]
Computer Science [O]		8.0	N	4.0	-.-(A)	D	O	-.-(A)	40	-.-(A)	—[B]	-.-[B]	—[B]	-.-[B]
Electrical Eng [E]		21.0	Y	4.0	-.-(A)	D	O	-.-(A)	65	-.-(A)	21	-.-(A)	—[B]	-.-[B]
Industrial Eng [I]		7.3	Y	4.0	-.-(A)	D	O	-.-(A)	30	-.-(A)	5	-.-(A)	—[B]	-.-[B]
Mechanical Eng [M]		19.1	Y	4.0	-.-(A)	D	O	-.-(A)	32	-.-(A)	4	-.-(A)	—[B]	-.-[B]
160 North Carolina State U	NC													
Aerospace Eng [M]		39.5	Y	4.0	4.5	D	O	20.0	19	89.7	9	82.1	6	62.5
Biological & Agricultural Eng [O]		11.0	Y	4.0	4.5	D	O	20.0	2	90.9	4	33.3	8	54.5
Chemical Eng [H]		16.5	Y	4.0	4.5	D	O	20.0	67	85.5	7	57.9	9	32.6
Civil Eng [V]		36.5	Y	4.0	4.5	D	O	20.0	117	85.6	30	60.9	10	20.8

All Engineering Disciplines

ENGINEERING Institution Profile Reference Number & Name Name of Degree Program	State/Province	FACULTY Full-time equivalent (FTE)	ABET/CEAB accred?	UNDERGRADUATE PROGRAMS							GRADUATE PROGRAMS			
				Length		Time	Co-op		Bachelor's		Master's		Doctoral	
				Nominal length of program in years	Average length of program in years	Day/Eve./Both	None/Opt./Req.	% of graduates in Co-op programs	# of degrees awarded	% of graduates who were full-time	# of degrees awarded	% of graduates who were full-time	# of degrees awarded	% of graduates who were full-time
Civil Eng (Construction) [V]		36.5	Y	4.0	4.5	D	O	20.0	34	85.7	6	60.9	—.	-.-
Computer Eng [E]		54.0	Y	4.0	4.5	D	O	20.0	71	75.4	48	31.0	6	9.1
Computer Science [O]		34.4	N	4.0	4.5	D	O	20.0	107	65.8	31	45.2	—.	21.9
Construction Mgmt [O]		36.5	N	4.0	4.5	D	O	20.0	2	100.0	—.	-.-	—.	-.-
Electrical Eng [E]		54.0	Y	4.0	4.5	D	O	20.0	210	81.3	46	60.6	20	50.0
Eng [O]		5.0(C)	N	4.0	4.5	D	O	0.0	1	100.0	—.	-.-	—.	-.-
Env'l Eng [V](1)		36.5	N	4.0	-.-	D	O	-.-	—.	-.-	—.	-.-	—.	-.-
Industrial Eng [I]		22.5	Y	4.0	4.5	D	O	20.0	85	83.5	20	56.5	6	42.0
Industrial Eng (Furniture Manuf'g) [I]		22.5	N	4.0	4.5	D	O	20.0	8	90.4	—.	-.-	—.	-.-
Integrated Manuf'g Systems Eng [O]		10.0(C)	N	-.-(B)	-.-	D(B)	O(B)	-.-	—.	-.-	19	68.2	—.	-.-
Materials Science & Eng [O]		22.5	Y	4.0	4.5	D	O	20.0	15	88.6	15	54.3	15	28.8
Mechanical Eng [M]		39.5	Y	4.0	4.5	D	O	20.0	216	80.1	18	50.0	18	38.5
Nuclear Eng [O]		11.0	Y	4.0	4.5	D	O	20.0	7	87.7	8	36.8	4	41.2
Operations Research [O]		5.0(C)	N	-.-(B)	-.-	_(B)	_(B)	-.-	—.	-.-	8	55.0	3	46.7
Textile Eng [O]		4.5	Y	4.0	4.5	D	O	20.0	9	95.9	—.	-.-	—.	-.-
161 U of North Dakota	ND													
Chemical Eng [H]		5.0	Y	4.0	5.0	D	O	-.-	13	-.-	2	-.-	—.	-.-
Civil Eng [V]		6.0	Y	4.0	5.0	D	O	-.-	23	-.-	4	-.-	—.	-.-
Electrical Eng [E]		8.0	Y	4.0	5.0	D	O	-.-	27	-.-	7	-.-	—.	-.-
Eng Mgmt [O]		1.0	N	4.0	5.0	D	O	-.-	5	-.-	—.	-.-	—.	-.-
Geological Eng [O]		3.0	Y	4.0	5.0	D	O	-.-	1	-.-	—.(1)	-.-	—.	-.-
Mechanical Eng [M]		10.0	Y	4.0	5.0	D	O	-.-	24	-.-	2	-.-	—.	-.-
162 North Dakota State U	ND													
Agricultural Eng [O](1)		4.0	Y	4.0	4.5	D	O	45.0	13	95.0	2	100.0	0	95.0
Civil Eng [V]		8.0	Y	4.0	4.5	D	O	45.0	56	95.0	2	100.0	0	95.0
Construction Eng [O]		3.0	Y	4.0	4.5	D	O	45.0	6	95.0	1	100.0	1	95.0
Electrical Eng [E](1)		11.0	Y	4.0	4.5	D	O	45.0	104	95.0	8	100.0	2	95.0
Industrial Eng & Mgmt [I](1)		7.0	Y	4.0	4.5	D	O	45.0	40	95.0	2	100.0	0	95.0
Mechanical Eng & Appl Mechanics [M](1)		14.0	Y	4.0	4.5	D	O	-.-	62	95.0	1	100.0	1	100.0
Petroleum Eng [O](2)		3.0	N	4.0	4.5	D	O	45.0	1	100.0	0	100.0	0	95.0
163 Northeastern U	MA													
Chemical Eng [H]		7.0	Y	5.0(1)	5.0	D	O	-.-	25	100.0	13	54.5	1	80.0
Civil Eng [V]		17.0	Y	5.0(1)	5.0	B	O	95.0	78	90.0	42	22.9	1	90.9
Electrical & Computer Eng [E]		50.0	Y	5.0(1)	5.0	B	O	85.0	182	85.0	114	27.6	12	80.3
Industrial Eng & Information Systems [I]		16.0	Y	5.0(1)	5.0	B	O	95.0	28	100.0	84	47.2	0	57.7
Mechanical Eng [M]		22.0	Y	5.0(1)	5.0	B	O	90.0	110	80.0	46	39.6	0	78.9
164 Northern Arizona U	AZ													
Civil Eng [V](1)		6.5	Y	4.0	5.0	D	O	5.0	8	95.0	—.	-.-(B)	—.	-.-(B)
Computer Science & Eng [O]		8.3	Y	4.0	5.0	D	O	5.0	17	95.0	—.	-.-(B)	—.	-.-(B)
Electrical Eng [E](1)		9.0	Y	4.0	5.0	D	O	5.0	28	95.0	—.	-.-(B)	—.	-.-(B)
Mechanical Eng [M](1)		7.3	Y	4.0	5.0	D	O	5.0	21	5.0	—.	-.-(B)	—.	-.-(B)
165 Northern Illinois U	IL													
Electrical Eng [E]		13.0	Y	4.0	4.6	D	O	7.0	41	91.4	16	64.4	—.(B)	-.-(B)
Eng Tech [O]		6.0	N	4.0	4.1	B	O	10.0	40	90.0	15	20.0	—.(B)	-.-(B)
Industrial Eng [I]		6.0	Y	4.0	4.6	D	O	15.0	8	90.0	8	66.0	—.(B)	-.-(B)
Mechanical Eng [M]		11.0	Y	4.0	4.6	D	O	1.0	31	90.0	6	57.0	—.(B)	-.-(B)
166 Northwestern U	IL													
Appl Mathematics [O]		11.5	N	4.0	4.0	D	O	50.0	2	100.0	2	100.0	3	100.0
Biomedical Eng [O]		9.1	Y	4.0	4.0	D	O	0.0	36	100.0	10	98.2	4	100.0
Chemical Eng [H]		15.0	Y	4.0	4.0	D	O	10.5	38	100.0	2	100.0	9	100.0
Civil Eng [V]		23.8	Y	4.0	4.0	D	O	0.0	12	100.0	22	90.4	16	100.0
Computer Science [O]		-.-(1)	N	4.0	4.0	D	O	12.5	24	100.0	26	69.5	23	100.0
Electrical Eng [E]		43.5	Y	4.0	4.0	D	O	6.0	50	100.0	17	91.5	11	100.0
Env'l Eng [O]		-.-(2)	Y	4.0	4.0	D	O	28.6	7	100.0	—.	-.-	—.	-.-
Industrial Eng [I]		16.3	Y	4.0	4.0	D	O	20.0	65	100.0	19	98.0	3	100.0
Materials Science & Eng [O]		17.5	Y	4.0	4.0	D	O	33.3	15	100.0	8	96.8	21	100.0
Mechanical Eng [M]		17.0	Y	4.0	4.0	D	O	15.6	32	100.0	19	94.1	15	100.0
167 Norwich U	VT													
Architecture [O]		7.5	N	5.0	5.0	D	N	-.-	0	-.-	—.	-.-	—.	-.-
Civil Eng [V]		4.0	Y	4.0	4.0	D	O	-.-	23	100.0	—.	-.-	—.	-.-
Computer Eng & Computer Science [O]		2.0	N	4.0	4.0	D	N	-.-	0	100.0	—.	-.-	—.	-.-
Electrical Eng [E]		5.0	Y	4.0	4.0	D	N	-.-	0	100.0	—.	-.-	—.	-.-
Env'l Eng Tech [O]		2.5	Y	4.0	4.0	D	N	-.-	0	100.0	—.	-.-	—.	-.-
Mechanical Eng [M]		5.0	Y	4.0	4.0	D	N	-.-	0	100.0	—.	-.-	—.	-.-

All Engineering Disciplines

ENGINEERING / Institution Profile Reference Number & Name / Name of Degree Program	State/Province	FACULTY Full-time equivalent (FTE)	ABET/CEAB accred?	UNDERGRAD Nominal length of program in years	UNDERGRAD Average length of program in years	Time Day/Eve./Both	Co-op None/Opt./Req.	% of graduates in Co-op programs	Bachelor's # of degrees awarded	Bachelor's % of graduates who were full-time	Master's # of degrees awarded	Master's % of graduates who were full-time	Doctoral # of degrees awarded	Doctoral % of graduates who were full-time
168 U of Notre Dame	IN													
Aerospace & Mechanical Eng [O]		10.0	Y	4.0	4.0	—[B]	N	-.-[B]	43	100.0	3	90.0	2	100.0
Chemical Eng [H]		11.0	Y	4.0	4.0	D	N	-.-	26	100.0	12	100.0	6	100.0
Civil Eng [V]		13.0	Y	4.0	4.0	D	N	-.-	21	100.0	4	100.0	2	100.0
Computer Eng [O]		9.0	N	4.0	4.0	D	N	-.-	0	0.0	0	100.0	0	100.0
Electrical Eng [E]		18.0	Y	4.0	4.0	D	N	-.-	50	100.0	18	90.0	4	85.0
Materials Science & Eng [O]		4.0	Y	4.0	4.0	D	N	-.-	4	100.0	3	100.0	3	100.0
Mechanical Eng [M]		16.0	Y	4.0	4.0	D	N	-.-	58	100.0	11	100.0	1	100.0
169 Oakland U	MI													
Computer Eng [O]		15.0	Y	4.0	4.5	D	O	-.-	18	-.-	36[1]	-.-	—[2]	-.-
Computer Science [O]		15.0	N[3]	4.0	4.5	D	O	-.-	26	-.-	36[1]	-.-	—[2]	-.-
Electrical Eng [E]		16.4[5]	Y	4.0	4.5	D[A]	O	-.-	49	-.-	19[6]	-.-	—[2]	-.-
Interdisciplinary Doctoral Program [O][4]		36.0	—	-.-	-.-	—	—	-.-	—.	-.-	—.	-.-	8	-.-
Mechanical Eng [M]		11.5	Y	4.0	4.5	D	O	-.-	33	-.-	15	-.-	—[2]	-.-
Systems Eng [O][5]		16.4	Y	4.0	4.5	D	O	-.-	4	-.-	8	-.-	—[2]	-.-
170 Ohio U	OH													
Airway Science [O]		2.0	N	4.0	-.-	—	O	0.0	17	-.-	—[B]	-.-	—[B]	-.-
Chemical Eng [H]		10.0	Y	4.0	-.-	—	O	0.0	10	-.-	6	-.-	—.	-.-
Civil Eng [V]		8.0	Y	4.0	-.-	—	O	17.0	18	-.-	5	-.-	—.	-.-
Electrical Eng [E]		21.0	Y	4.0	-.-	—	O	17.0	88	-.-	13	-.-	4	-.-
Industrial & Systems Eng [I]		8.0	Y	4.0	-.-	—	O	20.0	40	-.-	6	-.-	—[B]	-.-
Industrial Tech [O]		8.0	N	4.0	-.-	—	O	-.-	41	-.-	—[B]	-.-	—[B]	-.-
Interdisciplinary Eng [O]		-.-	N	-.-	-.-	—	—	-.-	0	0.0	—.	-.-	3	-.-
Mechanical Eng [M]		9.0	Y	4.0	-.-	—	O	29.0	41	-.-	14	-.-	—.	-.-
171 Ohio Northern U	OH													
Civil Eng [V]		5.0	Y	4.0	4.3	D	N	-.-	17	100.0	—.	-.-	—.	-.-
Electrical Eng [E]		6.0	Y	4.0	4.0	D	O	42.0	19	100.0	—.	-.-	—.	-.-
Mechanical Eng [M]		5.0	Y	4.0	4.0	D	O	39.0	33	100.0	—.	-.-	—.	-.-
172 Ohio State U	OH													
Aeronautical & Astronautical Eng [O]		12.0	Y	4.0	4.5	D	N	-.-	51	94.0	5	100.0	1	100.0
Agricultural Eng [O]		19.0	Y	4.0	4.5	D	O	-.-	16	87.0	4	100.0	5	100.0
Aviation [O]		4.0	N	4.0	4.5	D	N	-.-	17	87.0	—.	-.-	—.	-.-
Biomedical Eng [O]		-.-	N	-.-	-.-	D	O	-.-	0	-.-	8	100.0	1	100.0
Ceramic Eng [O]		-.-	Y	4.0	4.5	D	O	-.-	14	95.0	—.	-.-	—.	-.-
Chemical Eng [H]		12.0	Y	4.0	4.5	D	N	-.-	42	91.0	15	89.0	7	100.0
Civil Eng [V]		23.0	Y	4.0	4.5	D	O	-.-	61	89.0	29	90.0	6	97.7
Computer & Information Science [O]		36.0	N	4.0	4.5	D	O	-.-	87	83.0	0	65.2	0	81.4
Electrical Eng [E]		44.0	Y	4.0	4.5	D	O	-.-	105	77.0	90	100.0	20	100.0
Eng Mechanics [O]		14.0	N	-.-	-.-	D	N	-.-	0	-.-	5	100.0	1	100.0
Eng Physics [O]		-.-	N	4.0	4.5	D	N	-.-	11	85.0	—.	-.-	—.	-.-
Industrial & Systems Eng [I]		20.0	Y	4.0	4.5	D	O	-.-	63	92.0	22	100.0	10	100.0
Materials Science & Eng [O]		21.0	N	-.-	-.-	D	N	-.-	0	0.0	19	100.0	10	100.0
Mechanical Eng [M]		32.0	Y	4.0	4.5	D	O	-.-	134	86.0	53	100.0	12	100.0
Metallurgical Eng [O]		-.-	Y	4.0	4.5	D	O	-.-	16	86.0	—.	-.-	—.	-.-
Nuclear Eng [O]		5.0	N	-.-	-.-	D	N	-.-	0	-.-	3	83.3	6	95.5
Surveying [O]		9.0	Y	4.0	4.5	D	O	-.-	7	80.0	—.	-.-	—.	-.-
Welding Eng [O]		9.0	Y	4.0	4.5	D	O	-.-	31	88.0	18	100.0	6	100.0
173 U of Oklahoma	OK													
Aerospace Eng [O]		4.0	Y	4.0	4.7	B	O	-.-[A]	28	85.3	0	57.1	0	40.0
Chemical Eng [H]		11.0	Y	4.0	4.7	B	O	-.-[A]	28	-.-	8	55.2	3	47.8
Civil Eng [V]		15.0	Y	4.0	4.7	B	O	-.-[A]	21	80.5	24	48.9	3	44.4
Computer Science [O]		8.0	N	4.0	4.7	B	O	-.-[A]	25	-.-[A]	10	-.-[A]	9	-.-[A]
Electrical Eng [E]		16.0	Y	4.0	4.7	B	O	-.-[A]	107	77.8	19	40.2	6	28.6
Eng Physics [O]		3.0	Y	4.0	4.7	B	O	-.-[A]	4	90.4	0	62.5	—.	-.-[2]
Env'l Science [V]		5.0	N	4.0	4.7	B	O	-.-[A]	2	85.1	9	53.2	3	0.0
Geological Eng [O]		2.0	N	4.0	4.7	B	O	-.-[A]	0	66.7	—.	-.-[3]	0	66.7
Industrial Eng [I]		10.0	Y	4.0	4.7	B	O	-.-[A]	33	81.0	25	60.4	6	44.4
Mechanical Eng [M]		10.8	Y	4.0	4.7	B	O	-.-[A]	59	80.7	32	54.0	2	44.8
Petroleum Eng [O]		7.0	Y	4.0	4.7	B	O	-.-[A]	9	87.3	4	57.1	6	30.8
174 Oklahoma Christian	OK													
Electrical Eng [E]		5.0	Y	4.5	4.5	D	N	-.-	6	91.4	—.	-.-	—.	-.-
Mechanical Eng [M]		5.0	Y	4.5	4.5	D	N	-.-	5	91.4	—.	-.-	—.	-.-

All Engineering Disciplines

ENGINEERING Institution Profile Reference Number & Name Name of Degree Program	State/Province	FACULTY Full-time equivalent (FTE)	ABET/CEAB accred?	UNDERGRADUATE PROGRAMS - Length - Nominal length of program in years	Average length of program in years	Time Day/Eve./Both	Co-op None/Opt./Req.	% of graduates in Co-op programs	Bachelor's # of degrees awarded	% of graduates who were full-time	Master's # of degrees awarded	% of graduates who were full-time	Doctoral # of degrees awarded	% of graduates who were full-time
175 Oklahoma State U	OK													
Aerospace [O](3)		20.5	Y	4.0	4.6	D	O	0.0	15	100.0	—(4)	-.-(B)	—(B)	-.-(B)
Agricultural Eng [O]		12.5	Y	4.0	4.6	D	O	0.0	3	100.0	2	100.0	0	100.0
Architectural Eng [O]		16.0	Y	5.0	5.2	D	N	-.-	3	100.0	1	100.0	—(1)	-.-
Chemical Eng [H]		9.5	Y	4.0	4.7	D	O	0.0	27	100.0	14	75.0	4	100.0
Civil Eng [V]		18.0	Y	4.0	4.7	D	O	0.0	17	85.0	34	90.0	5	100.0
Electrical & Computer Eng [E]		19.8	Y	4.0	4.7	D	O	9.8	51	97.0	42	78.0	4	100.0
General Eng [O]		0.3(2)	Y	4.0	4.4	D	O	0.0	0	55.0	—.	-.-	—.	-.-
Industrial Eng & Mgmt [I]		12.5	Y	4.0	4.6	D	O	3.3	30	100.0	28	87.5	5	100.0
Mechanical & Aerospace Eng [M]		20.5	Y	4.0	4.7	D	O	10.7	56	93.0	30	82.5	7	92.8
176 Old Dominion U	VA													
Civil Eng [V]		10.0(1)	Y	4.0	4.5	D	O	-.-(A)	34	88.9	18	27.8	4	69.2
Computer Eng [E]		5.5(1)	Y	4.0	4.5	D	O	-.-(A)	20	82.8	1	0.0	—(B)	-.-(B)
Electrical Eng [E]		12.5(1)	Y	4.0	4.5	D	O	-.-(A)	61	88.1	6	38.5	—.	100.0
Eng Mgmt [O](2)		6.0(1)	N(B)	-.-(B)	-.-(B)	B(B)	—(B)	-.-(B)	—(B)	-.-(B)	52	8.1	3	46.4
Mechanical Eng [M]		21.0(1)	Y	4.0	4.5	D	O	-.-(A)	56	83.0	7	16.7	6	94.5
Mechanics [M]		-.-(3)	N	-.-(B)	-.-(B)	—(B)	—(B)	-.-(B)	—.	-.-	9	5.3	3	94.5
177 Oregon State U	OR													
Chemical Eng [H]		7.0	Y	4.0	4.8(C)	D	O(1)	0.0	21	100.0	10	100.0	1	100.0
Civil Eng [V]		22.0	Y	4.0	4.8(C)	D	N	-.-	64	100.0	31	100.0	4	100.0
Computer Eng [E]		6.0	Y	4.0	4.8(C)	D	O(2)	12.0	19	100.0	—(3)	-.-	—(3)	-.-
Computer Science [O]		15.0	N	4.0	4.0(C)	D	O(1)	0.0	88	100.0	30	100.0	5	100.0
Construction Eng Mgmt [V]		4.0	N	4.0	4.8(C)	D	N	-.-	34	100.0	—(4)	-.-	—(4)	-.-
Electrical & Electronics Eng [E]		22.0	Y	4.0	4.8(C)	D	O(2)	0.0	65	100.0	29(3)	100.0	7(3)	100.0
Eng Physics [O]		3.0	N	4.0	4.8(C)	D	N	-.-	7	100.0	—(4)	-.-	—(3)	-.-
Industrial Eng [I]		11.0	Y	4.0	4.8(C)	D	O(5)	58.0	45	100.0	10	100.0	3	100.0
Mechanical Eng [M]		21.0	Y	4.0	4.8(C)	D	O(1)	0.0	73	100.0	20	100.0	10	100.0
Nuclear Eng [O]		5.0	Y	4.0	4.8(C)	D	N	-.-	12	100.0	4	100.0	1	100.0
178 U of the Pacific	CA													
Civil Eng [V]		5.0	Y	5.0	5.0	D	R	100.0	18	100.0	—(2)	-.-(2)	—(2)	-.-(2)
Electrical & Computer Eng [E]		7.5	Y	5.0	5.0	D	R	56.1	26	100.0	1	20.0	—(3)	-.-(3)
Eng Mgmt [O]		3.0	N	5.0	5.0	D	R	100.0	6	100.0	—(5)	-.-(5)	—(5)	-.-(5)
Eng Physics [O]		4.0	Y	5.0	5.0	D	R	100.0	2	100.0	—(1)	-.-(1)	—(1)	-.-(1)
Mechanical Eng [M]		4.0	Y	5.0	5.0	D	R	95.0	11	95.0	—(4)	-.-(4)	—(4)	-.-(4)
179 U of Penn	PA													
Bioengineering [O]		12.0	Y	4.0	-.-(A)	D	N	-.-(B)	25	100.0	22	87.5	11	83.7
Chemical Eng [H]		11.0	Y	4.0	-.-(A)	D	N	-.-(B)	19	100.0	10	100.0	54	98.1
Computer Science & Eng [O]		23.0	N	4.0	-.-(A)	D	N	-.-(A)	43	100.0	39	55.6	13	82.3
Electrical Eng [E]		20.0	Y	4.0	-.-	D	N	-.-	41	100.0	26	37.1	6	86.9
Materials Science & Eng [O]		16.0	Y	4.0	-.-(A)	D	N	-.-(B)	14	100.0	1	25.0	9	96.6
Mechanical Eng & Appl Mechanics [M]		12.0	Y	4.0	-.-	D	N	-.-	31	100.0	13	36.4	6	84.8
Systems Science & Eng [O]		10.0	Y	4.0	-.-(A)	D	N	-.-(B)	27	100.0	36	17.5	3	69.0
180 Penn State U	PA													
Acoustics [O]		4.0	N	-.-	-.-	—	-.-	-.-	—.	-.-	8	-.-	4	-.-
Aerospace Eng [O]		16.0	Y	4.0	-.-	D	O	16.0	96	-.-	17	-.-	5	-.-
Agricultural Eng [O]		12.0	Y	4.0	-.-	D	O	10.0	19	-.-	3	-.-	3	-.-
Architectural Eng [O]		11.0	Y	5.0	-.-	D	O	12.0	90	-.-	9	-.-	—.	-.-
Chemical Eng [H]		19.0	—	4.0	-.-	—	—	25.0	93	-.-	9	-.-	12	-.-
Civil Eng [V]		28.0	Y	4.0	-.-	D	O	11.0	173	-.-	29	-.-	5	-.-
Computer Eng [O]		12.0	Y	4.0	-.-	D	O	23.0	41	-.-	14	-.-	5	-.-
Electrical Eng [E]		36.0	Y	4.0	-.-	D	O	19.0	240	-.-	47	-.-	15	-.-
Eng Science & Mechanics [O]		30.0	Y	4.0	-.-	D	O	1.0	37	-.-	94	-.-	12	-.-
Industrial Eng [I]		23.0	Y	4.0	-.-	D	O	20.0	132	-.-	34	-.-	5	-.-
Mechanical Eng [M]		41.0	Y	4.0	-.-	—	O	21.0	223	-.-	48	-.-	20	-.-
Nuclear Eng [O]		11.0	Y	4.0	-.-	D	O	1.0	11	-.-	6	-.-	6	-.-
181 U of Pittsburgh	PA													
Bioengineering [O](1)		7.8	—(3)	-.-(3)	-.-(3)	—(3)	—(3)	-.-(3)	—(3)	-.-(3)	0	75.0	—(3)	-.-(3)
Chemical & Petroleum Eng [H](1)		20.3	—(3)	-.-(3)	-.-(3)	—(3)	—(3)	-.-(3)	—(3)	-.-(3)	—.	-.-	—(3)	-.-(3)
Chemical Eng [H]		20.3	Y	4.0	4.0	D	O	20.6	34	94.4	10	66.7	7	71.4
Civil Eng [V]		17.2	Y	4.0	4.0	D	O	12.5	48	95.5	21	33.3	5	66.7
Electrical Eng [E]		21.0	Y	4.0	4.0	D(6)	O	27.8	72	84.4	45	45.5	11	68.8
Energy Resources [O](1)		1.5	—(3)	-.-(3)	-.-(3)	—(3)	—(3)	-.-(3)	—(3)	-.-(3)	7	16.0	—(3)	-.-(3)
Eng Physics [O](4)		3.0	N	4.0	4.0	D	O	0.0	1	100.0	—(4)	-.-(4)	—(4)	-.-(4)
Industrial Eng [I]		14.2	Y	4.0	4.0	D	O	12.2	41	92.5	31	41.0	5	84.2
Interdisciplinary Eng [O](1)		-.-(B)	—(1)	-.-(1)	-.-(1)	—(1)	—(1)	-.-(1)	—(1)	-.-(1)	—(B)	-.-(B)	0	100.0
Manuf'g Systems Eng [O](1)		6.6	—(3)	-.-(3)	-.-(3)	—(3)	—(3)	-.-(3)	—(3)	-.-(3)	9	12.2	—(3)	-.-(3)

All Engineering Disciplines

ENGINEERING Institution Profile Reference Number & Name Name of Degree Program	State/Province	FACULTY Full-time equivalent (FTE)	ABET/CEAB accred?	UNDERGRADUATE PROGRAMS					Bachelor's		GRADUATE PROGRAMS			
				Length		Time	Co-op				Master's		Doctoral	
				Nominal length of program in years	Average length of program in years	Day/Eve./Both	None/Opt./Req.	% of graduates in Co-op programs	# of degrees awarded	% of graduates who were full-time	# of degrees awarded	% of graduates who were full-time	# of degrees awarded	% of graduates who were full-time
Materials Science & Eng [O]		19.0	Y	4.0	4.0	D	O	0.0	7	95.2(5)	3	45.5(5)	2	81.8(5)
Mechanical Eng [M]		18.2	Y	4.0	4.0	D(6)	O	21.1	90	86.9	16	38.0	8	75.7
Metallurgical Eng [O]		19.0	Y	4.0	4.0	D	O	0.0	8	95.2(5)	3	45.5(5)	2	81.8(5)
Mining Eng [O](1)		0.2	—(3)	-.-(3)	-.-(3)	—(3)	—(3)	-.-(3)	—(3)	-.-(3)	1	50.0	—(3)	-.-(3)
Petroleum Eng [O](1)		2.0	—(3)	-.-(3)	-.-(3)	—(3)	—(3)	-.-(3)	—(3)	-.-(3)	—.	-.-	—(3)	-.-(3)
Public Works [O](1)		0.2	—(3)	-.-(3)	-.-(3)	—(3)	—(3)	-.-(3)	—(3)	-.-(3)	—(3)	-.-(3)	—(3)	-.-(3)
Public Works & Civil Eng [O](1)		17.2	—(3)	-.-(3)	-.-(3)	—(3)	—(3)	-.-(3)	—(3)	-.-(3)	—.	-.-	—(3)	-.-(3)
182 Poly U	NY													
Aerospace Eng [O]		10.0	Y	4.0	-.-(1)	D	O	-.-	27	92.0	5	25.0	0	60.0
Chemical Eng [H]		10.0	Y	4.0	-.-(1)	D	O	-.-(B)	11	90.0	5	39.0	1	45.0
Civil & Env'l Eng [V]		-.-	Y	4.0	-.-(1)	—(B)	O	-.-(B)	34	82.0	28(2)	20.0	4	25.0
Civil Eng [V]		23.0	Y	4.0	-.-(1)	D	O	-.-(B)	34	82.0	28(2)	20.0	4	25.0
Electrical Eng [E]		50.0	Y	4.0	-.-(1)	D(B)	O	-.-(B)	129	89.0	179	23.0	14	30.0
Industrial Eng [I]		13.0	Y	4.0	-.-(1)	D(B)	O	-.-(B)	14	80.0	13	37.0	3	11.0
Mechanical Eng [M]		19.0	Y	4.0	-.-(1)	D(B)	O	-.-(B)	49	92.0	21	32.0	1	25.0
Metallurgical Eng & Material Science [O]		6.0	Y	4.0	-.-	D	O	-.-	—.	-.-	—.	-.-	—.	-.-
183 U of Portland	OR													
Civil Eng [V]		4.7	Y	4.0	4.5	D	O	-.-(2)	3	-.-(1)	—.	-.-(3)	—(B)	-.-(B)
Electrical Eng [E](4)		6.0	Y	4.0	4.5	D	O	-.-(2)	37	-.-(1)	7	-.-(3)	—(B)	-.-(B)
Eng Mgmt [O](5)		-.-	N	4.0	4.5	D	O	-.-(2)	2	-.-(1)	—(B)	-.-(B)	—(B)	-.-(B)
Eng Science(s) [O](6)		-.-	N	4.0	4.5	D	O	-.-(2)	0	-.-(1)	—(B)	-.-(B)	—(B)	-.-(B)
Mechanical Eng [M]		7.3	Y	4.0	4.5	D	O	-.-(2)	15	-.-(1)	2	-.-(3)	—(B)	-.-(B)
184 Portland State U	OR													
Civil Eng [V]		8.0	Y	4.0	4.7	B	O	-.-(A)	30	72.0	5	44.2	1	100.0
Computer Science [O]		12.0	N	4.0	4.7	B	O	-.-(A)	62	57.4	15	43.5	—(B)	-.-(B)
Electrical & Computer Eng [E](1)		13.0	Y(2)	4.0	4.7	B	O	-.-(A)	54	66.5	33	51.2	0	60.9
Eng Mgmt [O]		3.0	N	-.-(B)	-.-(B)	—(B)	—(B)	-.-(B)	—(B)	-.-(B)	18	25.0	0	33.0
Mechanical Eng [M]		11.0	Y	4.0	4.7	B	O	-.-(A)	43	69.6	11	54.5	0	100.0
185 Prairie View A&M U	TX													
Chemical & Biochemical Eng [H]		-.-	Y	4.0	-.-	D	O	-.-	—.	-.-	—.	-.-	—.	-.-
Chemical Eng [H]		5.0	N	4.0	4.0	D	O	10.0	13	90.0(C)	—.	-.-	—.	-.-
Civil Eng [V]		5.0	Y	4.0	4.0	D	O	10.0(C)	12	90.0	0	1.0	—.	-.-
Electrical Eng [E]		10.0	Y	4.0	4.0	D	O	20.0	62	90.0	0	1.0	—.	-.-
Mechanical Eng [M]		9.0	Y	4.0	4.0	D	O	15.0	27	90.0	0	1.0	—.	-.-
186 Princeton U	NJ													
Chemical Eng [H]		18.0	Y	4.0	4.0	—	N	-.-	9	100.0	2	100.0	13	100.0
Civil Eng & Operations Research [V]		23.0	Y	4.0	4.0	—	N	-.-	31	100.0	1	100.0	7	100.0
Computer Science [O]		18.0	Y	4.0	4.0	—	N	-.-	20	100.0	1	100.0	8	100.0
Electrical Eng [E]		22.0	Y	4.0	4.0	—	N	-.-	40	100.0	10	100.0	10	100.0
Mechanical & Aerospace Eng [M]		24.0	Y	4.0	4.0	—	N	-.-	48	100.0	12	100.0	6	100.0
187 U of Puerto Rico	PR													
Chemical Eng [H]		19.0	Y	5.0	5.0	B	O	14.0	83	96.0	0	100.0	—(B)	-.-(B)
Civil Eng [V]		28.0	Y	5.0	5.0	B	O	7.0	106	96.0	4	100.0	—(B)	-.-(B)
Electrical & Computer Eng [E]		30.0	Y	5.0	5.0	B	O	19.0	118	97.0	4	100.0	—(B)	-.-(B)
Industrial Eng [I]		18.0	Y	5.0	5.0	B	O	9.0	92	96.0	4	100.0	—(B)	-.-(B)
Land Surveying & Topography [V]		28.0	N	4.0	4.0	—(B)	O	0.0	12	100.0	—(B)	-.-(B)	—(B)	-.-(B)
Land Surveying Eng [O]		-.-	N	-.-	-.-	B	O	-.-	—.	-.-	—.	-.-	—.	-.-
Mechanical Eng [M]		23.5	Y	5.0	5.0	B	O	18.0	124	97.0	1	100.0	—(B)	-.-(B)
188 Purdue U	IN													
Aeronautical & Astronautical Eng [O]		23.0	Y	4.0	4.3	D	O	12.0	146	100.0	27	100.0	21	100.0
Agricultural Eng [O]		19.0	Y	4.0	4.3	D	O	18.0	20	100.0	10	100.0	8	100.0
Agricultural Eng (Food) [O](2)		-.-(2)	Y(2)	4.0(2)	-.-(2)	D(2)	O(2)	-.-(2)	—(2)	-.-(2)	—(2)	-.-(2)	—(2)	-.-(2)
Chemical Eng [H]		23.0	Y	4.0	4.3(1)	D	O	35.0	128	98.0	15	100.0	22	100.0
Civil Eng [V]		58.0	Y	4.0	4.3(1)	D	O	18.0	129	98.0	68	100.0	18	100.0
Computer & Electrical Eng [E](8)		-.-(8)	Y(8)	4.0(8)	-.-(8)	D(8)	O(8)	-.-(8)	—(8)	-.-(8)	—(8)	-.-(8)	—(8)	-.-(8)
Construction Eng & Mgmt [O]		-.-(3)	Y	4.0	4.3	D	N	-.-(4)	35	100.0	—(5)	-.-	—.	-.-
Electrical Eng [E]		74.0	Y	4.0	4.3	D	O	23.0	308	100.0	125	100.0	43	100.0
Industrial Eng [I]		31.0	Y	4.0	4.3	D	O	19.0	195	100.0	60	100.0	11	100.0
Interdisciplinary Eng [O]		-.-(6)	N	4.0	4.3	D	O	9.0	46	100.0	—(7)	-.-	—(7)	-.-
Land Surveying Eng [O]		5.0	Y	4.0	4.3	D	N	-.-	4	100.0	—(7)	-.-	—(7)	-.-
Materials Eng [O]		16.0	Y	4.0	4.3	D	O	18.0	34	100.0	5	100.0	3	100.0
Mechanical Eng [M]		56.0	Y	4.0	4.3	D	O	25.0	194	100.0	72	100.0	26	100.0
Nuclear Eng [O]		8.0	Y	4.0	4.3	D	O	12.0	9	100.0	6	100.0	2	100.0

All Engineering Disciplines

ENGINEERING Institution Profile Reference Number & Name Name of Degree Program	State/Province	FACULTY Full-time equivalent (FTE)	ABET/CEAB accred?	UNDERGRADUATE PROGRAMS							GRADUATE PROGRAMS			
				Length		Time	Co-op		Bachelor's		Master's		Doctoral	
				Nominal length of program in years	Average length of program in years	Day/Eve./Both	None/Opt./Req.	% of graduates in Co-op programs	# of degrees awarded	% of graduates who were full-time	# of degrees awarded	% of graduates who were full-time	# of degrees awarded	% of graduates who were full-time
189 Rensselaer Poly Inst	NY													
Aeronautical Eng [O][1]		9.0	Y	4.0	4.0	D	O	20.0	50	100.0	14	-.-(A)	3	-.-(A)
Biomedical Eng [O]		8.0	Y	4.0	4.0	D	O	15.6	32	100.0	14	-.-(A)	2	-.-(A)
Chemical Eng [H]		15.0	Y	4.0	4.0	D	O	16.2	37	100.0	4	-.-(A)	4	-.-(A)
Civil Eng [V]		12.0	Y	4.0	4.0	D	O	2.6	38	100.0	8	-.-(A)	4	-.-(A)
Computer & Systems Eng [O]		32.0	Y	4.0	4.0	D	O	29.7	37	100.0	40	-.-(A)	9	-.-(A)
Electric Power Eng [E]		6.0	Y	4.0	4.0	D	O	37.5	8	100.0	26	-.-(A)	4	-.-(A)
Electrical Eng [E]		34.0	Y	4.0	4.0	D	O	27.9	147	100.0	42	-.-(A)	17	-.-(A)
Eng Physics [O]		-.-(A)	Y	4.0	4.0	D	O	22.2	9	100.0	3	-.-(A)	—.	—.
Eng Science(s) [O]		1.0	Y	4.0	4.0	D	O	0.0	2	100.0	48	-.-(A)	—.	—.
Env'l Eng [O]		2.0	Y	4.0	4.0	D	O	28.6	21	100.0	7	-.-(A)	—.	—.
Industrial & Mgmt Eng [I]		6.0	Y	4.0	4.0	D	O	32.0	50	100.0	53	-.-(A)	8	-.-(A)
Materials Eng [O]		18.0	Y	4.0	4.0	D	O	29.7	37	100.0	18	-.-(A)	7	-.-(A)
Mechanical Eng [M]		28.0	Y	4.0	4.0	D	O	28.7	188	100.0	55	-.-(A)	16	-.-(A)
Mechanics [O]		8.0	Y	4.0	4.0	D	O	0.0	0	100.0	—.	-.-(A)	—.	-.-(A)
Nuclear Eng [O]		11.6	Y	4.0	4.0	D	O	23.0	13	100.0	5	-.-(A)	—.	—.
Nuclear Eng & Science [O]		11.6	Y	4.0	4.0	D	O	0.0	0	100.0	—.	-.-(A)	7	-.-(A)
Transportation Eng [O]		1.0	Y	4.0	4.0	D	O	0.0	0	100.0	1	-.-(A)	—.	-.-(A)
190 U of Rhode Island	RI													
Chemical Eng [H]		11.0	Y	4.0	4.5	D	N	-.-	—.	90.0	4	60.0	4	60.0
Civil Eng [V]		12.5	Y	4.0	4.5	D	N	-.-	42	90.0	7	40.0	3	40.0
Computer Eng [E][1]		21.0	Y	4.0	4.5	D	N	-.-	11	90.0	—.(B)	-.-(B)	—.(B)	-.-(B)
Electrical Eng [E]		21.0	Y	4.0	4.5	D	N	-.-	36	90.0	14	50.0	6	60.0
Industrial & Manuf'g Eng [I]		7.5	Y	4.0	4.5	B	N	-.-	11	90.0	7	50.0	—.	-.-
Mechanical Eng [M]		18.0	Y	4.0	4.5	D	O	3.0	42	90.0	8	50.0	3	70.0
Ocean Eng [O]		7.0	N	-.-(B)	-.-	D	N(A)	-.-	—.(2)	-.-	8	80.0	2	40.0
191 Rice U	TX													
Chemical Eng [H]		15.0	Y	4.0	4.4	—(B)	N	-.-(B)	15	100.0	6	87.5	9	96.0
Civil Eng [V]		7.4	Y	4.0	4.1	—(B)	N	-.-(B)	13	100.0	5	100.0	2	100.0
Computational & Appl Mathematics [O][7]		11.5	N(1)	4.0	4.3	—(B)	N	-.-(B)	10(1)	100.0(2)	3(3)	75.0(6)	2	97.2
Computer Science [O]		13.6	N(1)	4.0	4.0	—(B)	N	-.-(B)	20(1)	98.8(2)	14	83.3	5	100.0
Electrical & Computer Eng [E]		20.2	Y	4.0	4.2	—(B)	N	-.-(B)	55	98.7	22	53.8	7	96.5
Env'l Science & Eng [O](4)		5.8	N(1)	4.0	4.0	—(B)	N	-.-(B)	2(5)	100.0(2)	9	83.3	1	100.0
Mechanical Eng & Materials Science [M]		17.6	Y	4.0	4.4	—(B)	N	-.-(B)	42	100.0	15	81.8	5	88.4
Statistics [O]		6.2	N(1)	4.0	4.0	—(B)	N	-.-(B)	4(1)	100.0(2)	2(3)	50.0(6)	0	92.9
192 U of Rochester	NY													
Chemical Eng [H]		12.0	Y	4.0	4.0	D	N	-.-	21	93.0	11	14.3	8	93.9
Electrical Eng [E]		16.0	Y	4.0	4.0	D	N	-.-	44	98.0	32	41.7	6	94.6
Geomechanics [O][1]		-.-	N	4.0	4.0	—	N	-.-	4	100.0	—.	-.-	—.	-.-
Interdisciplinary Eng [O](4)		-.-	N	4.0	4.0	D	N	-.-	4	100.0	—.	-.-	—.	-.-
Materials Science & Eng [O](3)		-.-	—	-.-	-.-	—	—	-.-	—.	-.-	—.	-.-	—.	-.-
Mechanical & Aerospace Science [M](5)		-.-	—	-.-	-.-	—	—	-.-	—.	-.-	5	0.0	7	92.3
Mechanical Eng [M](6)		17.0	Y	4.0	4.0	D	N	-.-	42	99.0	—.	-.-	—.	-.-
Optical Eng [O]		14.0	N	4.0	4.0	D	—	8.7(2)	34	98.0	41	81.2	13	98.8
193 Rochester Inst of Tech	NY													
Computer Eng [O]		5.0	Y	5.0	5.0	D	R	100.0	24	94.0	2	34.0	—.	-.-
Electrical Eng [E]		17.5	Y	5.0	-.-	D	R	100.0	93	78.0	30	15.0	—.	-.-
Industrial & Manuf'g Eng [I]		4.5	Y	5.0	5.0	D	R	100.0	25	88.0	6	5.0	—.	-.-
Mechanical Eng [M]		19.0	Y	5.0	5.0	D	R	100.0	78	95.0	6	4.0	—.	-.-
Microelectronic Eng [O]		5.0	Y	5.0	5.0	D	R	100.0	31	95.0	6	61.0	—.	-.-
194 Rose-Hulman	IN													
Chemical Eng [H]		7.0	Y	4.0	4.0	D	N	-.-(A)	32	100.0	2	100.0	—.(B)	-.-(B)
Civil Eng [V]		5.0	Y	4.0	4.0	D	N	-.-	19	100.0	—.	-.-(A)	—.(B)	-.-(B)
Electrical & Computer Eng [E]		15.0	Y	4.0	4.0	D	N	-.-(A)	76	100.0	—.	-.-(A)	—.(B)	-.-(B)
Mechanical Eng [M]		16.0	Y	4.0	4.0	D	N	-.-(A)	100	100.0	3	100.0	—.(B)	-.-(B)
195 Rutgers, The State U	NJ													
Appl Sciences in Eng [O]		-.-	N	4.0	4.5	D	N	-.-	16	96.0	—.	-.-	—.	-.-
Biomedical Eng [O]		8.0	N	-.-	-.-	B	N	-.-	—.	-.-	9	57.0	10	50.0
Bioresource Eng [O]		3.0	Y	4.0	4.5	D	N	-.-	7	100.0	1	60.0	—.	-.-
Ceramic Eng [O]		23.0	Y	4.0	4.5	D	N(A)	-.-(A)	54	57.0	20	61.0	9	81.0
Chemical Eng [H]		13.0	Y	4.0	4.5	D	N(A)	-.-	37	97.0	9	50.0	10	65.0
Civil Eng [V]		10.0	Y	4.0	4.5	D	N	-.-	53	98.0	20	34.0	0	36.0
Electrical Eng [E]		29.0	Y	4.0	4.5	D	N	-.-	136	94.0	39	31.0	10	33.0
Industrial Eng [I]		8.0	Y	4.0	4.5	D	N	-.-	136	94.0	17	50.0	—.	-.-
Mechanical Eng [M]		24.0	Y	4.0	4.5	D	N	-.-	105	97.0	13	58.0	8	60.0
Mechanics & Materials Science [O]		12.0	N	-.-	-.-	B	N	-.-	0	0.0	7	57.0	4	61.0

All Engineering Disciplines

ENGINEERING Institution Profile Reference Number & Name Name of Degree Program	State/Province	FACULTY Full-time equivalent (FTE)	ABET/CEAB accred?	UNDERGRADUATE PROGRAMS						GRADUATE PROGRAMS			
				Length Nominal length of program in years	Length Average length of program in years	Time Day/Eve./Both	None/Opt./Req.	Co-op % of graduates in Co-op programs	Bachelor's # of degrees awarded	Bachelor's % of graduates who were full-time	Master's # of degrees awarded	Master's % of graduates who were full-time	Doctoral # of degrees awarded	Doctoral % of graduates who were full-time
196 **Saginaw Valley State U**	MI													
Electrical Eng [E]		4.1	Y	4.0	4.5	D	O	10.0	20	50.0	—.	-.-	—.	-.-
Mechanical Eng [M]		6.7	Y	4.0	4.5	D	O	10.0	21	50.0	—.	-.-	—.	-.-
197 **St. Cloud State U**	MN													
Electrical Eng [E]		7.0	Y	4.0	5.0(C)	D	N	-.-	30	90.0(C)	—.(B)	-.-(B)	—.(B)	-.-(B)
Manuf'g Eng [O]		4.0	N	4.0	5.0(C)	D	N	-.-(B)	3	90.0(C)	—.(B)	-.-(B)	—.(B)	-.-(B)
198 **Parks Coll of Saint Louis U**	IL													
Aerospace Eng [O]		12.5	Y	4.0	-.-	D	O	0.0	61	98.5	0(1)	65.2	—.(2)	-.-
Electrical Eng [E]		6.0	Y	4.0	-.-	D	O	-.-	16	97.0	—.(3)	-.-	—.	-.-
199 **U of San Diego**	CA													
Electrical Eng [E](1)		6.3	Y	4.5	4.7(2)	B	O	60.0(3)	7	100.0	—.(B)	-.-(B)	—.(B)	-.-(B)
200 **San Diego State U**	CA													
Aerospace Eng & Eng Mechanics [O]		8.6	Y	4.0	5.5	B	O	-.-(A)	50	80.0	8	20.0	—.	-.-(B)
Appl Mechanics [O]		-.-(B)	N	5.0	-.-(A)	B	N	-.-(B)	—(B)	-.-(B)	—.(B)	-.-(B)	0	100.0
Civil Eng [V]		14.2	Y	4.0	5.5	B	O	-.-(A)	72	82.0	10	20.0	—.	-.-(B)
Electrical Eng [E]		18.6	Y	4.0	5.5	B	O	-.-(A)	142	80.0	30	20.0	—.	-.-(B)
Mechanical Eng [M]		15.7	Y	4.0	5.5	B	O	-.-(A)	58	85.0	18	20.0	—.	-.-(B)
201 **San Francisco State U**	CA													
Civil Eng [V]		-.-	Y	4.0	-.-	B	O	-.-	28	-.-	—.	-.-	—.	-.-
Electrical Eng [E]		-.-	Y	4.0	-.-	B	O	-.-	66	-.-	—.	-.-	—.	-.-
Mechanical Eng [M]		-.-	Y	4.0	-.-	B	O	-.-	28	-.-	—.	-.-	—.	-.-
202 **San Jose State U**	CA													
Aerospace Eng [O]		4.3	Y	4.0	-.-(A)	—	—	-.-	40	68.0	0	6.8	—.(A)	-.-(A)
Chemical Eng [H]		4.1	Y	4.0	-.-(A)	D	—	-.-	16	68.0	6	6.8	—.(A)	-.-(A)
Civil Eng [V]		13.2	Y	4.0	-.-(A)	D	—(A)	-.-	49	68.0	27	6.8	—.(A)	-.-(A)
Computer Eng [O]		6.9	Y	4.0	-.-(A)	D	—(A)	-.-	36	68.0	15	6.8	—.(A)	-.-(A)
Electrical Eng [E]		35.5	Y	4.0	-.-(A)	D	—(A)	-.-	133	68.0	86	6.8	—.(A)	-.-(A)
General Eng [O]		-.-	N	4.0	-.-	D	—(A)	-.-	—.	-.-	—.	-.-	—.(A)	-.-(A)
Industrial & Systems Eng [I]		5.0	Y	4.0	-.-(A)	D	—	-.-	34	68.0	20	6.8	—.(A)	-.-(A)
Materials Eng [O]		7.0	Y	4.0	-.-(A)	D	—(A)	-.-	6	68.0	4	6.8	—.(A)	-.-(A)
Mechanical Eng [M]		16.4	Y	4.0	-.-(A)	D	—(A)	-.-	61	68.0	22	6.8	—.(A)	-.-(A)
203 **Santa Clara U**	CA													
Civil Eng [V]		6.0	Y	4.0	4.0	D	O	-.-	25	99.0	2(1)	0.0	—.(B)	-.-(B)
Computer Eng [O]		8.0	Y	4.0	4.0	D	O	-.-(A)	16	98.0	16	5.0	2	-.-(A)
Electrical Eng [E]		13.0	Y	4.0	4.0	D	O	-.-	38	99.0	56	3.5(C)	4	10.0(C)
Mechanical Eng [M]		10.0	Y	4.0	4.0	D	O	-.-(A)	35	98.0	21	2.0	—.	-.-(A)
204 **Seattle U**	WA													
Civil & Env'l Eng [V]		4.0	Y	4.0	4.0	D	N	-.-(B)	14	-.-(A)	—.	-.-	—.	-.-
Electrical Eng [E]		8.0	Y	4.0	4.0	D	N	-.-(B)	52	-.-(A)	—.	-.-	—.	-.-
Mechanical Eng [M]		5.0	Y	4.0	4.0	D	N	-.-(B)	31	-.-(A)	—.	-.-	—.	-.-
205 **U of South Alabama**	AL													
Chemical Eng [H]		-.-	Y	4.0	-.-	B	O	-.-	8	80.0	1	50.0	—.(B)	-.-(B)
Civil Eng [V]		-.-	Y	4.0	-.-	—	O	-.-	7	78.0	—.(B)	-.-(B)	—.(B)	-.-(B)
Electrical Eng [E]		-.-	Y	4.0	-.-	B	O	-.-	43	79.0	4	26.0	—.(B)	-.-(B)
Mechanical Eng [M]		-.-	Y	4.0	-.-	B	O	-.-	26	69.0	3	44.0	—.(B)	-.-(B)
206 **U of South Carolina**	SC													
Chemical Eng [H]		10.0	Y	4.0	4.0	D	O	-.-(A)	12	9.3	4	91.3	2	85.7
Civil Eng [V]		13.0	Y	4.0	4.0	D	O	-.-(A)	15	85.8	16	86.6	2	56.3
Electrical & Computer Eng [E]		20.0	Y	4.0	4.0	D	O(A)	-.-(A)	71	34.2	24	51.9	3	100.0
Electrical Eng [E]		20.0	Y	4.0	4.0	D	O	-.-(A)	71	34.2	24	51.9	3	100.0
Mechanical Eng [M]		15.0	Y	4.0	4.0	D	O	-.-(A)	58	89.9	20	40.0	1	81.3
207 **South Dakota School of Mines**	SD													
Chemical Eng [H]		4.2	Y	4.0	4.5	D	O	-.-(A)	26	-.-(A)	2	-.-(A)	—.(B)	-.-(B)
Civil Eng [V]		14.2	Y	4.0	4.5	D	O	-.-(A)	18	-.-(A)	26	-.-(A)	—.(B)	-.-(B)
Electrical Eng [E]		11.5	Y	4.0	4.5	D	O	-.-(A)	38	-.-(A)	14	-.-(A)	—.(B)	-.-(B)
Geological Eng [O]		4.0	Y	4.0	4.5	D	O	-.-(A)	9	-.-(A)	4	-.-(A)	1	100.0
Industrial Eng [I]		3.0	N(1)	4.0	4.5(C)	D	O	-.-(A)	9	-.-(A)	—.(B)	-.-(B)	—.(B)	-.-(B)
Mechanical Eng [M]		9.0	Y	4.0	4.5	D	O	-.-(A)	69	-.-(A)	15	-.-(A)	—.(B)	-.-(B)
Metallurgical Eng [O]		5.0	Y	4.0	4.5	D	O	-.-(A)	10	-.-(A)	4	-.-(A)	—.(B)	-.-(B)
Mining Eng [O]		5.0	Y	4.0	4.5	D	O	-.-(A)	8	-.-(A)	4	-.-(A)	—.(B)	-.-(B)

All Engineering Disciplines

ENGINEERING Institution Profile Reference Number & Name Name of Degree Program	State/Province	FACULTY Full-time equivalent (FTE)	ABET/CEAB accred?	UNDERGRADUATE PROGRAMS								GRADUATE PROGRAMS			
				Length		Time	Co-op	Bachelor's		Master's		Doctoral	
				Nominal length of program in years	Average length of program in years	Day/Eve./Both	None/Opt./Req.	% of graduates in Co-op programs	# of degrees awarded	% of graduates who were full-time	# of degrees awarded	% of graduates who were full-time	# of degrees awarded	% of graduates who were full-time
208 South Dakota State U	SD													
Agricultural Eng [O]		8.0	Y	4.0	4.5	D	O	1.0	5	100.0	3	100.0	—.	-.-
Civil Eng [V]		11.0	Y	4.0	5.0	D	O	2.0	40	100.0	12	100.0	—.	-.-
Computer Science [O]		7.0	N	4.0	4.0	D	O	1.0	13	100.0	0	100.0	—.	-.-
Electrical Eng [E]		11.0	Y	4.0	4.5	D	O	10.0	67	100.0	6	100.0	—.	-.-
Electronic Eng Tech [O](1)		4.0	N	4.0	4.0	B	O	15.0	32	100.0	—.	-.-	—.	-.-
Eng Physics [O]		9.8	N	4.0	4.5	D	O	0.0	9	100.0	2	100.0	—.	-.-
Mechanical Eng [M]		10.3	Y	4.0	4.5	D	O	1.0	28	100.0	3	100.0	—.	-.-
209 U of South Florida	FL													
Chemical Eng [H]		8.0	Y	4.0	4.5	B	O	10.0	34	95.0	4	100.0	2	100.0
Civil Eng [V]		19.0	Y	4.0	5.5	B	O	25.0	47	60.5	25	40.0	1	33.0
Computer Eng [O]		7.0	Y	4.0	5.5	B	O	10.0	26	63.4(5)	11	22.2(5)	—.	-.-(5)
Computer Science & Eng [O]		8.0	N(4)	4.0	5.5	B	O	10.0	34	60.3(5)	13	36.6(5)	5	48.7(5)
Electrical Eng [E]		17.0	Y	4.0	5.0	B	O	28.0	121	57.0	51	60.0	10	54.0
Eng Mgmt [I]		1.5	N	-.-	-.-	—	—	-.-	—.	-.-	26	2.0(2)	—.	-.-
Eng Tech [O]		4.2	N	4.0	5.5	B	O	25.0	48	20.3(5)	—.	-.-	—.	-.-
Industrial Eng [I]		6.0	Y	4.0	5.0	B	O	10.0	18	50.0	0	100.0	2	100.0
Industrial Manuf'g [I]		1.5(1)	N	-.-	-.-	—	—	-.-	—.	-.-	2	90.0	—.	-.-
Mechanical Eng [M]		9.0	Y	4.3	5.0	B	O	12.0	58	70.0	6	50.0	2	50.0
210 Southern U & A&M Coll	LA													
Civil & Env'l Eng [V]		6.0	Y	4.0	5.2	D	O	25.0	12	12.4	—(A)	-.-(A)	—(A)	-.-(A)
Civil Eng [V]		6.0	Y	-.-	-.-	—	—	-.-	12	12.4	—(A)	-.-(A)	—(A)	-.-(A)
Electrical Eng [E]		10.0	Y	4.0	5.2	D	O	25.0	34	35.1	—(A)	-.-(A)	—(A)	-.-(A)
Electronic Eng Tech [O]		5.0	N	4.0	5.2	D	O	2.0	25	25.8	—(A)	-.-(A)	—(A)	-.-(A)
Electronics Eng [E]		5.0	N	4.0	5.2	D	O	2.0	25	25.8	—(A)	-.-(A)	—(A)	-.-(A)
Mechanical Eng [M]		14.0	Y	4.0	5.2	D	O	25.0	21	21.6	—(A)	-.-(A)	—(A)	-.-(A)
Mechanical Tech [O]		5.0	N	4.0	5.2	D	O	2.0	5	5.2	—(A)	-.-(A)	—(A)	-.-(A)
211 U of Southern Cal	CA													
Aerospace Eng [O]		8.0	Y	4.0	4.0	D	O	-.-(A)	56	100.0	22	44.0	4	64.0
Biomedical Eng [O]		7.0	N	4.0	4.0	D	O	-.-(A)	24	100.0	5	50.0	0	71.0
Chemical Eng [H]		8.0	Y	4.0	4.0	D	O	-.-(A)	8	100.0	12	75.0	5	75.0
Civil Eng [V]		17.0	Y	4.0	4.0	D	O	-.-(A)	15	100.0	18	50.0	9	75.0
Computer Science [O]		25.0	N	4.0	4.0	D	O	-.-(A)	41	100.0	127	48.0	13	62.0
Electrical Eng [E]		21.0	Y	4.0	4.0	D	O	-.-(A)	72	100.0	251	42.0	54	61.0
Industrial & Systems Eng [I]		4.0	Y	4.0	4.0	D	O	-.-(A)	38	100.0	29	57.0	2	39.0
Mechanical Eng [M]		11.0	Y	4.0	4.0	D	O	-.-(A)	38	100.0	26	41.0	2	64.0
212 U of Southern Colorado	CO													
Industrial Eng [I]		4.0	Y(8)	4.0	5.0	D	O	10.0	—.	-.-	3	10.0	—(B)	-.-(B)
213 Southern Illinois-Carbondale	IL													
Civil Eng [V]		11.5	Y	4.0	4.5	D	O	4.0	27	93.0	9	76.0	—(B)	-.-(B)
Electrical Eng [E]		22.5	Y	4.0	4.5	D	O	3.0	97	92.5	27	67.0	—(B)	-.-(B)
Eng Science(s) [O]		-.-	—(B)	-.-(B)	-.-(B)	D	—(B)	-.-(B)	—(B)	-.-(B)	—(B)	-.-(B)	10	75.0
Mechanical Eng [M]		18.0	Y	4.0	4.5	D	—	5.0	50	95.0	16	71.0	—(B)	-.-(B)
Mining Eng [O]		5.0	Y	4.0	4.5	D	O	-.-	1	79.0	7	76.0	—(B)	-.-(B)
214 Southern Illinois-Edwardsville	IL													
Civil Eng [V]		9.0	Y	4.0	-.-(A)	B	O	-.-(A)	31	67.0	6	16.0	—(B)	-.-(B)
Electrical Eng [E]		10.0	Y	4.0	-.-(A)	B	O	-.-(A)	56	66.0	27	16.0	—(B)	-.-(B)
Industrial Eng [I]		3.0	Y	4.0	-.-(A)	D	O	-.-(A)	11	83.0	—(B)	-.-(B)	—(B)	-.-(B)
Mechanical Eng [M]		5.0	N	4.0	-.-(A)	D	O	-.-(A)	13	85.0	—(B)	-.-(B)	—(B)	-.-(B)
215 U of Southern Maine	ME													
Electrical Eng [E]		5.0	Y	4.0	5.5	B	O	20.0	8	35.0	—.	-.-	—.	-.-
216 Southern Methodist U	TX													
Civil Eng [V](1)		5.0	Y	4.0	4.5	D	O	20.0	10	92.3	2	0.0	—.	-.-
Computer Eng [O]		5.0	Y	4.0	4.4	D	O	25.0	8	90.9	—.	-.-	0	66.7
Computer Science [O]		7.0	N	4.0	4.4	D	O	36.4	11	93.8	15	2.9	5	74.3
Electrical Eng [E]		17.0	Y	4.0	5.0	D	O	50.0	20	90.3	37	0.9	9	85.1
Mgmt Science [O]		4.0	N	4.0	4.7	D	O	37.5	8	90.9	21	2.4	2	36.8
Mechanical Eng [M]		8.0	Y	4.0	4.6	D	O	33.3	15	88.2	7	0.0	2	61.9
217 Stanford U	CA													
Aeronautical & Astronautical Eng [O]		17.0	N	4.0	4.0	D	N	-.-	0	100.0	89	64.0	19	100.0
Chemical Eng [H]		10.0	Y	4.0	4.0	D	N	-.-	12	100.0	15	55.0	6	100.0
Civil Eng [V]		26.0	Y	4.0	4.0	D	N	-.-	19	100.0	102	75.0	16	100.0
Computer Science [O]		33.0	N	4.0	4.0	D	N	-.-	46	100.0	93	68.0	19	100.0

All Engineering Disciplines

ENGINEERING Institution Profile Reference Number & Name Name of Degree Program	State/Province	FACULTY Full-time equivalent (FTE)	ABET/CEAB accred?	UNDERGRADUATE PROGRAMS Length Nominal length of program in years	Length Average length of program in years	Time Day/Eve./Both	Co-op None/Opt./Req.	Co-op % of graduates in Co-op programs	Bachelor's # of degrees awarded	Bachelor's % of graduates who were full-time	GRADUATE PROGRAMS Master's # of degrees awarded	Master's % of graduates who were full-time	Doctoral # of degrees awarded	Doctoral % of graduates who were full-time
Electrical Eng [E]		49.0	Y	4.0	4.0	D[B]	N	-.-	63	100.0	182	65.0	68	100.0
Eng-Economic Systems [O]		11.0	N	4.0	4.0	D	N	-.-	0	0.0	78	72.0	8	100.0
Industrial Eng & Eng Mgmt [I]		10.0	Y	4.0	4.0	D	N	-.-	52	100.0	32	78.0	2	100.0
Materials Science & Eng [O]		9.0	N	4.0	4.0	D	N	-.-	1	100.0	35	48.0	15	100.0
Mechanical Eng [M]		32.0	Y	4.0	4.0	D	N	-.-	59	100.0	138	70.0	35	100.0
Operations Research [O]		9.0	N	4.0	4.0	D	N	-.-	0	100.0	39	78.0	3	100.0
218 SUNY-Binghamton	NY													
Computer Science [O]		14.7	Y	4.0	4.0	—[B]	N	-.-[B]	71	75.0	41	55.0	1	50.0
Electrical Eng [E]		18.1	Y	4.0	4.0	—[B]	N	-.-[B]	48	87.0	55	64.0	5	69.0
Industrial Eng [I][1]		-.-[B]	—[B]	-.-[B]	-.-[B]	—[B]	—[B]	-.-[B]	—[B]	-.-[B]	13	40.0	1	88.0[B]
Industrial Tech [O]		-.-[3]	—[B]	4.0	4.0	—[B]	N	-.-[B]	37	28.0	—[B]	-.-[B]	—[B]	-.-[B]
Industrial Tech [I]		-.-	—	-.-	-.-	—	—	-.-	—.	-.-	—.	-.-	—.	-.-
Mechanical Eng [M][2]		14.1	Y	4.0	4.0	—[B]	N	-.-[B]	35	95.0	12	25.0	1	44.0
Systems Science [O][4]		6.6	—[B]	-.-[B]	-.-[B]	—[B]	—[B]	-.-[B]	—[B]	-.-[B]	18	23.0	2	66.0
219 SUNY-Buffalo	NY													
Aerospace Eng [O][1]		-.-	Y	4.0	4.5	D	N	-.-	36	98.0	10	87.5	0	15.0
Chemical Eng [H]		13.0	Y	4.0	4.5	D	N	-.-	21	95.6	10	68.0	9	54.5
Civil Eng [V]		28.0	Y	4.0	4.5	B	N	-.-	65	88.6	34	67.6	9	36.1
Electrical Eng [E]		30.0	Y	4.0	4.5	B	N	-.-	138	86.6	57	63.4	14	36.0
Eng Physics [O][2]		-.-	N	4.0	4.5	D	N	-.-	3	87.5	—.	-.-	—.	-.-
Eng Science(s) [O][2]		-.-	N	4.0	4.5	—	N	-.-	0	0.0	4	57.1	0	66.7
Industrial Eng [I]		11.0	Y	4.0	4.5	D	N	-.-	62	95.7	30	59.7	1	31.9
Mechanical Eng [M]		24.0	Y	4.0	4.5	B	N	-.-	157	91.9	39	56.7	6	25.3
220 SUNY-Env'l Science & Forestry	NY													
Env'l & Resource Eng [O]		26.3	N	-.-[B]	-.-[B]	—[B]	—[B]	-.-[B]	—[B]	-.-[B]	14	50.0	0	50.0
Forest Eng [O]		8.0	Y	4.0	4.2	D	O	5.0	15	96.0	—[B]	-.-[B]	—[B]	-.-[B]
Paper Science & Eng [O]		11.3	N	4.0	4.2	D	N	-.-[B]	14	93.0	—[B]	-.-[B]	—[B]	-.-[B]
Wood Products Eng [O]		9.3	N	4.0	4.2	D	N	-.-[B]	20	92.0	—[B]	-.-[B]	—[B]	-.-[B]
221 SUNY-Maritime	NY													
Electrical Eng [E]		5.0	Y	4.0	4.3	D	N	-.-	28[1]	100.0	—.	-.-	—.	-.-
Facilities Eng [O][2]		6.0[3]	N	4.0	4.3	D	N	-.-	0	100.0	—.	-.-	—.	-.-
Marine Eng [O]		6.0	Y	4.0	4.3	D	—	100.0	16	100.0	—.	-.-	—.	-.-
Mechanical Eng [M]		6.0	N	4.0	4.3	D	N	-.-	3	100.0	—.	-.-	—.	-.-
Naval Architecture [O]		3.0	Y	4.0	4.3	D	N	-.-	12	100.0	—.	-.-	—.	-.-
222 SUNY-New Paltz	NY													
Electrical Eng [E]		7.5	Y	4.0	4.5	—[A]	O	17.0	24	66.7	—[B]	-.-[B]	—[B]	-.-[B]
223 SUNY-Stony Brook	NY													
Electrical Eng [E]		-.-[A]	Y	4.0	-.-[A]	D	N	-.-[A]	67	96.7	27	79.5	6	65.4
Eng Science(s) [O]		-.-[A]	Y	4.0	-.-[A]	D	N	-.-[A]	18	91.3	—.	-.-[A]	—.	-.-[A]
Materials Science & Eng [O]		-.-[A]	—[A]	-.-[A]	-.-[A]	—[A]	—[A]	-.-[A]	—[A]	-.-[A]	5	80.0	5	51.7
Mechanical Eng [M]		-.-[A]	Y	4.0	-.-[A]	D	N	-.-[A]	45	96.9	17	73.0	9	68.3
224 Swarthmore Coll	PA													
Eng [O]		8.3	Y	4.0	4.0	D	N	-.-[B]	18	100.0	—[1]	-.-	—[1]	-.-
225 Syracuse U	NY													
Aerospace Eng [O]		8.0	Y	4.0	4.2	D	O	16.0	31	99.0	5	-.-[A]	2	-.-[A]
Bioengineering [O]		12.0	Y	4.0	4.2	D	O	13.0	16	98.0	—.	-.-[A]	—.	-.-[A]
Chemical Eng [H]		9.0	Y	4.0	4.2	D	O	13.0	8	93.0	8	-.-[A]	3	-.-[A]
Civil Eng [V]		6.0	Y	4.0	4.2	D	O	7.0	15	92.0	6[2]	-.-[A]	—.	-.-[A]
Computer Eng [O]		15.0	Y	4.0	4.2	D	O	25.0	16	97.0	74	-.-[A]	12	-.-[A]
Electrical Eng [E]		23.0	Y	4.0	4.2	D	O	22.0	32	66.0	78	-.-[A]	13	-.-[A]
Eng Physics [O]		1.0	N	4.0	4.2	D	O	0.0	2	100.0	—[B]	-.-[B]	—[B]	-.-[B]
Env'l Eng [O]		6.0	N[3]	4.0	4.2	D	O	0.0	5	92.0	1	-.-[A]	—[B]	-.-[B]
Manuf'g Eng [M]		1.0	Y	4.0	4.2	D	O	50.0	2	99.0	12	-.-[A]	—[B]	-.-[B]
Mechanical Eng [M]		11.0	Y	4.0	4.2	D	O	21.0	44	93.0	16	-.-[A]	13	-.-[A]
226 Technical U of Nova Scotia	NS													
Agricultural Eng [O]		-.-	Y	4.0[2]	5.0	D	O	0.0[3]	4	100.0	1	100.0	2	100.0
Chemical Eng [H]		8.7	Y	4.0[2]	5.0	D	O	90.0	18	100.0	1	100.0	1	100.0
Civil Eng [V]		10.1	Y	4.0[2]	5.0	D	O	0.0[3]	43	100.0	6	83.0	1	100.0
Electrical Eng [E]		14.5	Y	4.0[2]	5.0	D	O	0.0[3]	38	100.0	6	83.0	4	100.0
Industrial Eng [I]		10.1	Y	4.0[2]	5.0	D	O	0.0[3]	19	100.0	2	100.0	1	100.0
Mechanical Eng [M]		15.9	Y	4.0[2]	5.0	D	N	-.-[A]	39	100.0	7	86.0	3	100.0
Metallurgical Eng [O]		7.7	Y	4.0[2]	5.0	D	O	100.0	3	100.0	3	100.0	2	100.0
Mining Eng [O]		7.7	Y	4.0[2]	5.0	D	O	0.0[3]	4	100.0	1	100.0	—[A]	-.-[A]

All Engineering Disciplines

ENGINEERING Institution Profile Reference Number & Name Name of Degree Program	State/Province	FACULTY Full-time equivalent (FTE)	ABET/CEAB accred?	UNDERGRADUATE PROGRAMS					Bachelor's		GRADUATE PROGRAMS			
				Length		Time	Co-op				Master's		Doctoral	
				Nominal length of program in years	Average length of program in years	Day/Eve./Both	None/Opt./Req.	% of graduates in Co-op programs	# of degrees awarded	% of graduates who were full-time	# of degrees awarded	% of graduates who were full-time	# of degrees awarded	% of graduates who were full-time
227 Temple U	PA													
Civil Eng [V]		13.0(C)	Y	4.0	4.0	—(B)	N	-.-(B)	17	84.0	—(A)	-.-(A)	—(B)	-.-(B)
Electrical Eng [E]		16.0	Y	4.0	4.0	—(B)	N	-.-(B)	47	72.2	—(A)	-.-(A)	—.	-.-(A)
Mechanical Eng [M]		15.5	Y	4.0	4.0	—(B)	N	-.-(B)	18	91.2	—(A)	-.-(A)	—(B)	-.-(B)
228 U of Tennessee, Chattanooga	TN													
Chemical Eng [H](1)		-.-(7)	Y	4.0	4.8	B	O	-.-(A)	8	-.-(A)	2	-.-(A)	—.	-.-
Civil Eng [V](2)		-.-(7)	Y	4.0	4.8	B	O	-.-(A)	2	-.-(A)	2	-.-(A)	—.	-.-
Electrical Eng [E](6)		-.-(7)	Y	-.-	-.-	B	O	-.-(A)	—.	-.-	3	-.-(A)	—.	-.-
Eng [O]		27.8(7)	Y	4.0	4.8	B	O	-.-(A)	—.	-.-(A)	—.	-.-	—.	-.-
Eng Mgmt [O]		-.-(7)	N	4.0	4.8	B	O	-.-(A)	4(8)	-.-(A)	9(9)	-.-(A)	—.	-.-
Env'l Eng [O]		-.-(7)	Y	4.0	4.8	B	O	-.-(A)	1	-.-(A)	—.	-.-(A)	—.	-.-
Industrial Eng [I](4)		-.-(7)	Y	4.0	4.8	B	O	-.-(A)	2	-.-(A)	3	-.-(A)	—.	-.-
Mechanical Eng [M](5)		-.-(7)	Y	-.-	-.-	B	O	-.-(A)	—.	-.-	3	-.-(A)	—.	-.-
Mechanics [M](5)		-.-(7)	Y	4.0	4.8	B	O	-.-(A)	13(5)	-.-(A)	—.	-.-	—.	-.-
Thermal Eng [M](5)		-.-(7)	Y	4.0	4.8	B	O	-.-(A)	12	-.-(A)	—.	-.-	—.	-.-
229 U of Tennessee, Knoxville	TN													
Aerospace Eng [O]		6.0	Y	4.0	-.-(A)	D	O	15.4	13	97.0	14	-.-(A)	4	-.-(A)
Agricultural Eng [O]		19.0	Y	4.0	-.-(A)	D	O	-.-(A)	6	97.0	1	-.-(A)	1	-.-(A)
Chemical Eng [H]		13.5	Y	4.0	-.-(A)	D	O	31.4	35	85.4	8	-.-(A)	7	-.-(A)
Civil Eng [V]		19.3	Y	4.0	-.-(A)	D	O	31.4	35	86.7	18	-.-(A)	1	-.-(A)
Electrical & Computer Eng [E]		38.0	Y	4.0	-.-(A)	D	O	26.3	95	85.7	26	-.-(A)	9	-.-(A)
Eng Physics [O]		5.0	N	4.0	-.-(A)	D	O	0.0	2	82.4	—(B)	-.-(B)	—(B)	-.-(B)
Eng Science & Mechanics [O]		19.5	Y	4.0	-.-(A)	D	O	21.4	14	94.1	6	-.-(A)	5	-.-(A)
Industrial Eng [I]		12.0	Y	4.0	-.-(A)	D	O	21.8	69	88.5	47	-.-(A)	—(A)	-.-(A)
Materials Science [O]		13.2	Y	4.0	-.-(A)	D	O	12.5	8	93.0	3	-.-(A)	3	-.-(A)
Mechanical Eng [M]		12.0	Y	4.0	-.-(A)	D	O	37.3	51	87.0	16	-.-(A)	5	-.-(A)
Nuclear Eng [O]		11.3	Y	4.0	-.-(A)	D	O	8.3	12	90.7	11	-.-(A)	7	-.-(A)
230 Tennessee State U	TN													
Architectural Eng [O]		5.0	Y	4.3	5.0	B	O	5.0	10	86.0	—(B)	-.-(B)	—(B)	-.-(B)
Civil Eng [V]		6.0	Y	4.3	5.0	B	O	5.2	12	76.8	5(1)	18.8	—(B)	-.-(B)
Electrical Eng [E]		8.0	Y	4.3	5.0	B	O	7.5	20	79.7	5(1)	44.4	—(B)	-.-(B)
Mechanical Eng [M]		7.0	Y	4.3	5.0	B	O	5.0	30	75.2	1(1)	100.0	—(B)	-.-(B)
231 Tennessee Tech	TN													
Chemical Eng [H]		60.3	Y	4.0	4.0	D	O	10.0	9	100.0	7	100.0	1	100.0
Civil Eng [V]		156.0	Y	4.0	4.0	D	O	10.0	40	100.0	0	100.0	1	100.0
Electrical Eng [E]		204.3	Y	4.0	4.0	D	O	10.0	51	100.0	9	100.0	4	100.0
Industrial Eng [I]		72.0	Y	4.0	4.0	D	O	10.0	23	100.0	4	100.0	0	100.0
Mechanical Eng [M]		228.8	Y	4.0	4.0	D	O	10.0	72	100.0	9	100.0	1	100.0
232 U of Texas at Arlington	TX													
Aerospace Eng [O]		-.-	Y	4.0	-.-	B	O	-.-	18	-.-	12	-.-	—.	-.-
Biomedical Eng [O]		-.-	N	4.0	-.-	B	N	-.-	—.	-.-	10	-.-	2	-.-
Civil Eng [V]		-.-	Y	4.0	-.-	B	O	-.-	20	-.-	28	-.-	—.	-.-
Computer Science Eng [O]		-.-	Y	4.0	-.-	B	O	-.-	71	-.-	67	-.-	4	-.-
Electrical Eng [E]		-.-	Y	4.0	-.-	B	O	-.-	95	-.-	98	-.-	13	-.-
Industrial Eng [I]		-.-	Y	4.0	-.-	B	O	-.-	30	-.-	36	-.-	4	-.-
Mechanical Eng [M]		-.-	Y	4.0	-.-	B	O	-.-	48	-.-	27	-.-	3	-.-
233 U of Texas at Austin	TX													
Aerospace Eng [O]		32.3	Y	4.0	4.7	D	O(1)	22.2	81	98.3	31	84.3	12	88.5
Architectural Eng [O]		-.-(2)	Y	4.0	4.7	D	O(1)	3.2	31	91.4	14	83.3	—(B)	-.-
Biomedical Eng [O]		-.-(3)	N	-.-(B)	-.-(B)	N	-.-(B)	-.-(B)	—(B)	-.-(B)	15	78.0	2	73.3
Chemical Eng [H]		21.2	Y	4.0	4.7	D	O(1)	20.0	85	93.4	14	81.2	19	91.7
Civil Eng [V]		50.6	Y	4.0	4.7	D	O(1)	0.0	59	91.1	87	85.0	24	82.9
Computer Eng [E]		-.-(9)	Y	4.0	4.8	D	O	-.-(9)	0(10)	-.-(9)	—(B)	-.-(B)	—(B)	-.-(B)
Electrical & Computer Eng [E]		64.5	Y	4.0	4.7	D	O(1)	24.2	219	87.8	117	68.5	38	77.2
Energy & Mineral Resources [O]		-.-(5)	N	-.-(B)	-.-(B)	N	-.-(B)	-.-(B)	—(B)	-.-(B)	4	50.0	—(B)	-.-(B)
Eng Mechanics [O]		-.-(4)	N	-.-	-.-	D	N	-.-	—(B)	-.-	7	78.9	4	85.4
Env'l Health Eng [O]		-.-(6)	N	-.-(B)	-.-(B)	D	—(B)	-.-(B)	—(B)	-.-(B)	10	80.0	—(B)	-.-(B)
Manuf'g Systems Eng [O]		-.-(7)	N	-.-(B)	-.-(B)	D	-.-(B)	-.-(B)	—(B)	-.-(B)	13	83.3	—(B)	-.-(B)
Materials Science & Eng [O]		-.-(8)	N	-.-(B)	-.-(B)	D	-.-(B)	-.-(B)	—(B)	-.-(B)	18	88.4	12	89.1
Mechanical Eng [M]		63.6	Y	4.0	4.7	D	O(1)	11.0	181	87.4	74	78.4	21	85.1
Operations Research & Industrial Eng [O]		8.0(1)	N	1.0(A)	-.-(B)	D	—(B)	-.-	—.	-.-	12	90.0	—.	-.-
Petroleum Eng [O]		17.5	Y	4.0	4.8	D	O(1)	0.0	31	97.5	14	79.5	9	87.0
234 U of Texas at Dallas	TX													
Electrical Eng [E]		5.0	Y	-.-	5.0	—	N	-.-	42	-.-	54	-.-	2	-.-

All Engineering Disciplines

ENGINEERING Institution Profile Reference Number & Name Name of Degree Program	State/Province	FACULTY Full-time equivalent (FTE)	ABET/CEAB accred?	UNDERGRADUATE PROGRAMS					Bachelor's		GRADUATE PROGRAMS			
				Length		Time	Co-op				Master's		Doctoral	
				Nominal length of program in years	Average length of program in years	Day/Eve./Both	None/Opt./Req.	% of graduates in Co-op programs	# of degrees awarded	% of graduates who were full-time	# of degrees awarded	% of graduates who were full-time	# of degrees awarded	% of graduates who were full-time
235 U of Texas at El Paso	TX													
Civil Eng [V]		10.3	Y	4.0	-.-	D	O	20.0	28	-.-	15	-.-	—.	-.-
Computer Science [O]		7.5	N(1)	4.0	-.-	D	O	20.0	25	-.-	7	-.-	—.	-.-
Electrical Eng [E]		17.0	Y	4.0	-.-	D	O	20.0	66	-.-	21	-.-	0	100.0
Industrial Eng [I]		5.0	Y	4.0	-.-	D	O	20.0	14	-.-	7	-.-	—.	-.-
Mechanical Eng [M]		9.8	Y	4.0	-.-	D	O	20.0	28	-.-	10	-.-	—.	-.-
Metallurgical & Materials Eng [O]		7.0	Y	4.0	-.-	D	O	20.0	19	-.-	6	-.-	—.	-.-
236 U of Texas at San Antonio	TX													
Civil Eng [V]		7.2	Y	4.0	-.-	B	O	-.-	20	-.-	5	-.-	—.	-.-
Electrical Eng [E]		12.4	Y	4.0	-.-	B	O	-.-	40	-.-	4	-.-	—.	-.-
Mechanical Eng [M]		7.9	Y	4.0	-.-	B	O	-.-	22	-.-	—.	-.-	—.	-.-
237 Texas A&I U	TX													
Chemical Eng [H]		4.0	Y	4.0	-.-	D	N	-.-	12	-.-	4	-.-	—.	-.-
Civil Eng [V]		-.-	Y	4.0	-.-	D	N	-.-	18	-.-	1	-.-	—.	-.-
Computer Science [O]		-.-	N	4.0	-.-	D	N	-.-	3	-.-	4	-.-	—.	-.-
Electrical Eng [E]		-.-	Y	4.0	-.-	D	N	-.-	41	-.-	13	-.-	—.	-.-
Env'l Eng [O](A)		-.-	N	-.-(B)	-.-	B(B)	N	-.-	—.	-.-	1	-.-	—.	-.-
Industrial Eng [I]		-.-	N	-.-(B)	-.-	—(B)	N	-.-	—.	-.-	—.	-.-	—.	-.-
Industrial Tech [O]		-.-	N	4.0	-.-	D	N	-.-	7	-.-	—.	-.-	—.	-.-
Mechanical Eng [M]		-.-	Y	4.0	-.-	D	N	-.-	25	-.-	5	-.-	—.	-.-
Natural Gas Eng [O]		-.-	Y	4.0	-.-	D	N	-.-	19	-.-	1	-.-	—.	-.-
238 Texas A&M U	TX													
Aerospace Eng [O]		15.0	Y	4.0	4.5(C)	D	O	42.0	83	95.0	16	86.0(C)	6	86.0(C)
Agricultural Eng [O]		3.0	Y	4.0	4.5(C)	D	O	0.0	5	95.0(C)	8	73.0(C)	6	73.0(C)
Bioengineering [O](1)		8.0	Y	4.0	4.5(C)	D	O	19.0	36	96.0	6	79.0(C)	3	79.0(C)
Chemical Eng [H]		20.9	Y	4.0	4.5(C)	D	O	42.0	73	91.0	21	88.0(C)	14	88.0(C)
Civil Eng [V]		71.0	Y	4.0	4.5(C)	D	O	14.0	180	94.0	69	77.0(C)	11	77.0(C)
Computer Eng [O](2)		11.0	N(3)	4.0	4.5(C)	D	O	0.0	18	93.0	—.(6)	-.-	—.	-.-
Computer Science [O]		37.0	N	4.0	4.5(C)	D	O	46.0	125	90.0	46	-.-(A)	6	-.-(A)
Electrical Eng [E]		42.0	Y	4.0	4.5(C)	D	O	53.0	216	88.0	76	86.0(C)	21	86.0(C)
Eng [O]		-.-	Y	8.5(C)	-.-	D	N	-.-	—.	-.-	—.	-.-	7	-.-
Health Physics [O](4)		6.0	Y	6.0(C)	-.-	D	O	0.0	—.(7)	-.-	3	-.-	—.	-.-
Industrial Eng [I]		31.2	Y	4.0	4.5(C)	D	O	0.0	100	91.0	38	83.0(C)	5	83.0(C)
Industrial Hygiene [O](4)		3.0	Y	6.0(9)	-.-	D	O	-.-	—.	-.-	—.	-.-	—.	-.-
Interdisciplinary Eng [O](10)		-.-	N	8.5(C)	-.-	D	N	-.-	—.	-.-	1	-.-	2	-.-
Mechanical Eng [M]		60.3	Y	4.0	4.5(A)	D	O	54.0	146	90.0	55	85.0(C)	12	85.0(C)
Nuclear Eng [O]		11.7	Y	4.0	4.5(C)	D	O	13.0	16	93.0	5	84.0(C)	9	84.0(C)
Ocean Eng [O](5)		6.8	Y	4.0	4.5(C)	D	O	0.0	21	95.0	12	76.0(C)	3	76.0(C)
Petroleum Eng [O]		11.8	Y	4.0	4.5(C)	D	O	0.0	25	93.0	25	77.0(C)	6	77.0(C)
Radiological Health Eng [O](4)		6.0	Y	4.0	4.5(C)	D	O	0.0	2	100.0	—.(8)	-.-	—.	-.-
Safety Eng [O](4)		3.0	N	6.0(C)	-.-	D	O	-.-	—.	-.-	5(7)	-.-	—.	-.-
239 Texas Tech U	TX													
Chemical Eng [H]		10.0	Y	4.0	-.-	D	N	-.-	8	96.0	4	84.0	0	89.0
Civil Eng [V]		18.8	Y	4.0	-.-	D	N	-.-	41	91.9	15	83.3	0	91.7
Computer Science [O]		12.3	N	4.0	-.-	D	N	-.-	27	86.1	13	75.5	1	66.7
Electrical Eng [E]		20.6	Y	4.0	-.-	D	N	-.-	25	87.7	9	79.0	5	100.0
Eng Physics [O]		-.-	Y	4.0	-.-	D	N	-.-	—.	-.-	—.	-.-	—.	-.-
Industrial Eng [I]		10.7	Y	4.0	-.-	D	N	-.-	24	91.6	25	81.0	6	79.3
Mechanical Eng [M]		17.2	Y	4.0	-.-	D	N	-.-	62	90.3	10	75.0	2	74.0
Petroleum Eng [O]		4.7	Y	4.0	-.-	D	N	-.-	7	97.6	1	92.3	—.	-.-
240 U of Toledo	OH													
Chemical Eng [H]		8.0	Y	4.0	4.0	D	O	-.-(A)	16	95.0	9	65.0(A)	—.	-.-
Civil Eng [V]		14.0	Y	4.0	4.0	D	O	-.-(A)	25	85.0	9	55.0	—.	-.-
Computer Science & Eng [O]		8.0	Y	4.0	4.0	D	O	-.-(A)	22	80.0	—.	-.-	—.	-.-
Electrical Eng [E]		19.0	Y	4.0	4.0	D	O	-.-(A)	62	78.0	11	50.0	—.	-.-
Eng Physics [O]		-.-(1)	Y	4.0	4.0	D	O	-.-(A)	2	100.0	—.	-.-	—.	-.-
Eng Science(s) [O]		-.-(1)	N	-.-(B)	-.-(B)	—(B)	—(B)	-.-(B)	—(B)	-.-(B)	8	35.0	10	65.0
Eng Tech [O]		5.0	Y	4.0	4.0	B	O	-.-(A)	77	31.0	—.	-.-	—.	-.-
Industrial Eng [I]		8.0	Y	4.0	4.0	D	O	-.-(A)	13	86.0	5	50.0	—.	-.-
Mechanical Eng [M]		17.0	Y	4.0	4.0	D	O	-.-(A)	70	75.0	17	50.0	—.	-.-
241 Tri-State U	IN													
Aerospace Eng [O]		3.0	Y	4.0	4.5	D	O	0.0	5	91.9	—.(3)	-.-(3)	—.(3)	-.-(3)
Chemical Eng [H]		3.0	Y	4.0	4.0	D	O	22.0	9	100.0	—.(3)	-.-(3)	—.(3)	-.-(3)
Civil Eng [V]		4.7	Y	4.0	4.5	D	O	20.0	12	100.0	—.(3)	-.-(3)	—.(3)	-.-(3)

All Engineering Disciplines

ENGINEERING Institution Profile Reference Number & Name Name of Degree Program	State/Province	FACULTY Full-time equivalent (FTE)	ABET/CEAB accred?	UNDERGRADUATE PROGRAMS					Bachelor's		GRADUATE PROGRAMS			
				Length		Time	Co-op				Master's		Doctoral	
				Nominal length of program in years	Average length of program in years	Day/Eve./Both	None/Opt./Req.	% of graduates in Co-op programs	# of degrees awarded	% of graduates who were full-time	# of degrees awarded	% of graduates who were full-time	# of degrees awarded	% of graduates who were full-time
Electrical Eng [E]		7.0	Y	4.0	4.5	D	O	15.0	37	100.0	—.(3)	-.-(3)	—.(3)	-.-(3)
Eng Administration [O](1)		-.-(2)	N	4.0	4.5	D	O	20.0	10	80.0	—.(3)	-.-(3)	—.(3)	-.-(3)
Mechanical Eng [M]		5.0	Y	4.0	4.5	D	O	16.0	38	91.9	—.(3)	-.-(3)	—.(3)	-.-(3)
242 Trinity U	TX													
Eng Science(s) [O]		9.0	Y	4.0	4.3	—	N	-.-	20	100.0	—.	-.-	—.	-.-
243 Tufts U	MA													
Art & Architectural History [V]		5.0	N	4.0	4.0	D	N	-.-(B)	1	100.0	—.(B)	-.-(B)	—.(B)	-.-(B)
Chemical Eng [H]		9.5	Y	4.0	4.0	D	N	-.-(B)	22	100.0	13	92.0	6	66.0
Civil Eng [V]		16.0	Y	4.0	4.0	B	N	-.-(B)	37	100.0	46	30.0	1	50.0
Computer Eng [E]		5.5	Y	4.0	4.0	D	N	-.-(B)	13	100.0	—.(B)	-.-(B)	—.(B)	-.-(B)
Electrical Eng [E]		17.0	Y	4.0	4.0	D	N	-.-(B)	59	100.0	24	53.0	6	100.0
Eng Psychology [O]		4.0	N	4.0	4.0	D	N	-.-(B)	11	100.0	—.(B)	-.-(B)	—.(B)	-.-(B)
Env'l Studies & Env'l Health [V]		6.0	N	4.0	4.0	D	N	-.-(B)	1	100.0	—.(B)	-.-(B)	—.(B)	-.-(B)
Manuf'g Eng [M]		5.0	N	4.0	4.0	D	N	-.-(B)	3	100.0	—.(B)	-.-(B)	—.(B)	-.-(B)
Mechanical Eng [M]		13.0	Y	4.0	4.0	B	N	-.-(B)	41	100.0	11	26.1	0	67.7
244 Tulane U	LA													
Biomedical Eng [O]		9.0	Y	4.0	4.0	B	N	-.-	31	100.0	9	71.0	1	65.0
Chemical Eng [H]		9.0	Y	4.0	4.0	B	N	-.-	16	100.0	0	65.0	4	40.0
Civil & Env'l Eng [V]		8.0	Y	4.0	4.0	B	N	-.-	19	100.0	2	53.0	4	66.0
Computer Science [O]		9.0	N(1)	4.0	4.0	B	N	-.-	18	100.0	1	33.0	6	0.0
Electrical Eng [E]		9.0	Y	4.0	4.0	B	N	-.-	25	100.0	2	87.0	2	20.0
Mechanical Eng [M]		9.0	Y	4.0	4.0	B	N	-.-	21	100.0	6	47.0	3	78.0
245 U of Tulsa	OK													
Chemical Eng [H]		9.0	Y	4.0	4.0	D	N	-.-(B)	10	97.9	10	72.2	1	80.0
Electrical Eng [E]		9.0	Y	4.0	4.0	D	N	-.-(B)	18	90.1	5	50.0	—.(B)	-.-(B)
Eng Physics [O]		7.0	Y	4.0	4.0	D	N	-.-(B)	5	94.4	—.(1)	-.-(1)	—.(2)	-.-(2)
Mechanical Eng [M]		9.0	Y	4.0	4.0	D	N	-.-(3)	33	92.3	3	40.0	0	100.0
Petroleum Eng [O]		9.0	Y	4.0	4.0	D	N	-.-(3)	19	93.1	10	79.1	6	68.4
246 Tuskegee U	AL													
Aerospace Eng [O]		4.0	Y	4.0	5.0	D	O	80.0	6	99.5	—.	-.-	—.	-.-
Chemical Eng [H]		5.0	Y	4.0	5.0	D	O	80.0	2	99.5	—.	-.-	—.	-.-
Electrical Eng [E]		12.0	Y	4.0	5.0	D	O	80.0	43	99.5	4	99.5	—.	-.-
Mechanical Eng [M]		12.0	Y	4.0	5.0	D	O	80.0	17	99.5	9	97.5	—.	-.-
247 Union Coll	NY													
Civil Eng [V]		5.5	Y	4.0	-.-(A)	B	N	-.-(B)	35	90.0	—.(B)	-.-(B)	—.(B)	-.-(B)
Electrical Eng [E]		8.0	Y	4.0	-.-(A)	B	N	-.-(B)	21	77.0	18	39.0	—.(B)	-.-(B)
Mechanical Eng [M]		8.0	Y	4.0	-.-(A)	B	N	-.-(B)	31	81.0	15	11.0	—.(B)	-.-(B)
248 US Air Force Academy	CO													
Aeronautical Eng [O]		26.3	Y	4.0	4.0	D	N	-.-	87	100.0	—.(B)	-.-(B)	—.(B)	-.-(B)
Astronautical Eng [O]		19.4	Y	4.0	4.0	D	N	-.-	42	100.0	—.(B)	-.-(B)	—.(B)	-.-(B)
Civil Eng [V]		20.5	Y	4.0	4.0	D	N	-.-	70	100.0	—.(B)	-.-(B)	—.(B)	-.-(B)
Electrical Eng [E]		27.0	Y	4.0	4.0	D	N	-.-	42	100.0	—.(B)	-.-(B)	—.(B)	-.-(B)
Eng Mechanics [O]		16.4	Y	4.0	4.0	D	N	-.-	42	100.0	—.(B)	-.-(B)	—.(B)	-.-(B)
Eng Science(s) [O]		9.0	Y	4.0	4.0	D	N	-.-	27	100.0	—.(B)	-.-(B)	—.(B)	-.-(B)
General Eng [O]		9.3	N	4.0	4.0	D	N	-.-	31	100.0	—.(B)	-.-(B)	—.(B)	-.-(B)
Mechanical Eng [M]		5.0	Y	4.0	4.0	D	N	-.-	9	100.0	—.(B)	-.-(B)	—.(B)	-.-(B)
249 US Coast Guard Academy	CT													
Civil Eng [V]		8.0	Y	4.0	4.0	D	N	-.-	40	100.0	—.	-.-	—.	-.-
Electrical Eng [E]		8.0	Y	4.0	4.0	D	N	-.-	16	100.0	—.	-.-	—.	-.-
Naval Architecture & Marine Eng [O]		8.0	Y	4.0	4.0	D	N	-.-	23	100.0	—.	-.-	—.	-.-
250 US Merchant Marine Academy	NY													
Marine Eng [O]		21.0(1)	N	4.0	4.0	D	R	100.0	32	100.0	—.	-.-	—.	-.-
Marine Eng Systems [O]		21.0(1)	Y	4.0	4.0	D	R	100.0	47	100.0	—.	-.-	—.	-.-
Marine Eng/Transportation [O]		21.0(1)	N	4.0	4.0	D	R	100.0	10	100.0	—.	-.-	—.	-.-
251 US Military Academy	NY													
Civil Eng [V]		-.-	Y	4.0	4.0	D	N	-.-	59	100.0	—.(B)	-.-(B)	—.(B)	-.-(B)
Electrical Eng [E]		-.-	Y	4.0	4.0	D	N	-.-	40	100.0	—.(B)	-.-(B)	—.(B)	-.-(B)
Eng Mgmt [O]		-.-	Y	4.0	4.0	D	N	-.-	55	100.0	—.(B)	-.-(B)	—.(B)	-.-(B)
Eng Physics [O]		-.-	N	4.0	4.0	D	N	-.-	9	100.0	—.(B)	-.-(B)	—.(B)	-.-(B)
Env'l Eng [O]		-.-	N	4.0	4.0	D	N	-.-	62	100.0	—.(B)	-.-(B)	—.(B)	-.-(B)
Mechanical Eng [M]		-.-	Y	4.0	4.0	D	N	-.-	135	100.0	—.(B)	-.-(B)	—.(B)	-.-(B)
Systems Eng [O]		-.-	N	4.0	4.0	D	N	-.-	21	100.0	—.(B)	-.-(B)	—.(B)	-.-(B)

All Engineering Disciplines

ENGINEERING Institution Profile Reference Number & Name 　Name of Degree Program	State/Province	FACULTY Full-time equivalent (FTE)	ABET/CEAB accred?	UNDERGRADUATE PROGRAMS							GRADUATE PROGRAMS			
				Length Nominal length of program in years	Length Average length of program in years	Time Day/Eve./Both	Co-op None/Opt./Req.	Co-op % of graduates in Co-op programs	Bachelor's # of degrees awarded	Bachelor's % of graduates who were full-time	Master's # of degrees awarded	Master's % of graduates who were full-time	Doctoral # of degrees awarded	Doctoral % of graduates who were full-time
252 US Naval Academy	MD													
Aerospace Eng [O]		14.0	Y	4.0	4.0	D	N	-.-	88	100.0	—.-	-.-	—.-	-.-
Electrical Eng [E]		31.0	Y	4.0	4.0	D	N	-.-	16	100.0	—.-	-.-	—.-	-.-
General Eng [O]		-.-[1]	N	4.0	4.0	D	N	-.-	54	100.0	—.-	-.-	—.-	-.-
Marine Eng [O]		13.0	Y	4.0	4.0	D	N	-.-	20	100.0	—.-	-.-	—.-	-.-
Mechanical Eng [M]		29.0	Y	4.0	4.0	D	N	-.-	56	100.0	—.-	-.-	—.-	-.-
Naval Architecture [O]		11.0	Y	4.0	4.0	D	N	-.-	22	100.0	—.-	-.-	—.-	-.-
Ocean Eng [O]		7.0	Y	4.0	4.0	D	N	-.-	33	100.0	—.-	-.-	—.-	-.-
Systems Eng [O]		31.0	Y	4.0	4.0	D	N	-.-	72	100.0	—.-	-.-	—.-	-.-
253 U of Utah	UT													
Bioengineering [O][2]		5.0	N	-.-[B]	-.-[B]	—[B]	—[B]	-.-[B]	—[B]	-.-[B]	4	-.-[A]	4	-.-[A]
Chemical Eng [H]		15.0[1]	Y	4.0	5.0	D	O	-.-[A]	12	73.0	5	73.0	1	73.0
Civil Eng [V]		9.0	Y	4.0	6.0	D	O	-.-[A]	29	75.0	7	43.0	1	64.0
Computer Eng [O][6]		-.-[7]	N	4.0	5.0	D	—[A]	-.-[B]	20	-.-[A]	—[6]	-.-[B]	—[6]	-.-[B]
Computer Science [O]		19.5	N	4.0	4.0	D	O	-.-[A]	43	-.-[A]	6	100.0	6	100.0
Electrical Eng [E]		15.1	Y	4.0	5.0	D	O	-.-[A]	81	-.-[A]	23	-.-[A]	6	-.-[A]
Fuels Eng [H][2]		15.0[3]	N	-.-[B]	-.-[B]	D[B]	—[B]	-.-[B]	0[2]	-.-[2]	—[4]	-.-[B]	3	-.-[A]
Geological Eng [O][4]		3.5	Y	4.0	4.0	D	O	-.-[A]	1	-.-[A]	—[4]	-.-[B]	—[4]	-.-[B]
Materials Science & Eng [O]		8.5	Y	4.0	5.0	D	O	-.-[A]	11	-.-[A]	16	-.-[A]	6	-.-[A]
Mechanical Eng [M]		15.7	Y	4.0	5.0	D	O	-.-[A]	73	-.-[A]	25	-.-[A]	5	-.-[A]
Metallurgical Eng [O]		10.0	Y	4.0	4.0	D	O	-.-[A]	3	-.-[A]	3	-.-[A]	11	-.-[A]
Mining Eng [O]		10.5	Y	4.0	5.0	D	O	-.-[A]	4	-.-[A]	2	-.-[A]	1	-.-[A]
254 Utah State U	UT													
Biological & Irrigation Eng [O][2]		4.2	Y	4.0	4.6	D	O	0.0[A]	6	90.0	20	52.4	7	36.8
Civil & Env'l Eng [V]		4.1	Y[1]	4.6	4.6	D	O	-.-[A]	6	90.0	20	52.4	7	36.8
Electrical & Computer Eng [E]		12.5	Y	4.0	4.6	D	O	-.-[A]	39	81.3	32	41.1	1	41.7
Industrial Tech [I]		11.2	N[1]	4.0	4.0	D	O	-.-[A]	33	80.7	2	22.2	—.-	-.-
Mechanical, Manuf'g & Aerospace Eng [M]		10.2	Y[1]	4.0	4.6	D	O	0.0[A]	45	88.3	10	55.0	0	25.0
255 Valparaiso U	IN													
Civil Eng [V]		5.0	Y	4.0	4.0	D	O	-.-	21	99.0	—.-	-.-	—.-	-.-
Computer Eng [O]		5.0	Y	4.0	4.0	D	O	-.-	9	99.0	—.-	-.-	—.-	-.-
Electrical Eng [E]		5.5	Y	4.0	4.0	D	O	-.-	24	99.0	—.-	-.-	—.-	-.-
Mechanical Eng [M]		7.3	Y	4.0	4.0	D	O	-.-	29	99.0	—.-	-.-	—.-	-.-
256 Vanderbilt U	TN													
Biomedical Eng [O]		11.0	Y	4.0	-.-	D	N	-.-	30[1]	100.0	7	100.0	3	100.0
Chemical Eng [H]		10.0	Y	4.0	4.0	D	N	-.-[A]	22	100.0	1	100.0	2	100.0
Civil Eng [V]		15.0	Y	4.0	4.3	D	N	-.-	33	100.0	17[2]	85.0	1	90.0
Computer Science [O][6]		10.0	N	4.0	4.0	D	N	-.-	31	100.0	10	73.0	4	83.0
Electrical Eng [E][3]		20.0	Y[4]	4.0	4.0	D[5]	N	-.-	59	100.0	15	84.0	8	81.0
Eng Science(s) [O]		-.-	N	4.0	4.0	D	N	-.-	20	100.0	—.-	-.-	—.-	-.-
Materials Science & Eng [O]		-.-	N	-.-	-.-	—	—	-.-	—.-	-.-	—.-	-.-	—.-	-.-
Mechanical Eng [M]		16.0	Y	4.0	4.0	D	N	-.-	50	100.0	10	100.0	6	100.0
257 U of Vermont	VT													
Civil Eng [V]		8.5	Y	4.0	4.0	D	O	0.0	32	93.0	3	57.1	0	67.0
Electrical Eng [E]		13.5	Y	4.0	4.0	D	O	10.0	43	76.9	14	4.5	0	33.0
Materials Science [O][1]		-.-	N	-.-	-.-	D	N	-.-[B]	—[B]	-.-[B]	3	57.1	4	54.5
Mechanical Eng [M]		10.0	Y	4.0	4.0	D	O	10.0	29	93.9	2	18.2	—.-	-.-
258 Villanova U	PA													
Chemical Eng [H]		8.0	Y	4.0	4.0	—	N	-.-	29	100.0	9	12.0	—.-	-.-
Civil Eng [V]		12.0	Y	4.0	4.0	—	—	-.-	45	100.0	29	18.0	—.-	-.-
Computer Eng [O][1]		20.0	—	4.0	4.0	—	—	-.-	—.-	-.-	—.-	-.-	—.-	-.-
Electrical Eng [E]		20.0	Y	4.0	4.0	—	N	-.-	71	76.0	32	22.0	—.-	-.-
Mechanical Eng [M]		18.0	Y	4.0	4.0	—	—	-.-	58	86.0	26	30.0	—.-	-.-
259 U of Virginia	VA													
Aerospace Eng [M][3]		32.9[4]	Y	4.0	4.1	D	N	-.-	31	100.0	—.-[3]	-.-	—.-[3]	-.-
Appl Mathematics [O]		9.6	N	4.0	4.4	D	N	-.-	13	100.0	2	100.0	4	100.0
Appl Mechanics [V][2]		8.8[1]	—[2]	-.-	-.-	—	—	-.-	—.-	-.-	2	100.0	—[2]	-.-
Biomedical Eng [O][5]		4.1	—[5]	-.-	-.-	—	—	-.-	—.-	-.-	18	83.9	3	93.8
Chemical Eng [H]		8.9	Y	4.0	4.1	D	N	-.-	35	100.0	15	92.7	3	88.9
Civil Eng [V]		8.8[1]	Y	4.0	4.2	D	N	-.-	40	100.0	14	92.7	3	84.6
Computer Science [O]		13.0	N	4.0	4.3	D	N	-.-	26	100.0	24	79.6	2	90.9
Electrical Eng [E]		15.6	Y	4.0	4.1	D	N	-.-	78	100.0	41	84.3	12	90.7
Eng Physics [O][5]		9.9[6]	—[5]	-.-	-.-	—	—	-.-	—.-	-.-	9	81.8	3	100.0
Eng Science(s) [O][3]		9.9[6]	N	4.0	4.3	D	N	-.-	9	100.0	—.-[3]	-.-	—.-[3]	-.-
Materials Science & Eng [O][5]		9.9	—[5]	-.-	-.-	—	—	-.-	—.-	-.-	14	94.1	11	100.0

All Engineering Disciplines

ENGINEERING Institution Profile Reference Number & Name Name of Degree Program	State/Province	FACULTY Full-time equivalent (FTE)	ABET/CEAB accred?	UNDERGRADUATE PROGRAMS							GRADUATE PROGRAMS			
				Length		Time	Co-op		Bachelor's		Master's		Doctoral	
				Nominal length of program in years	Average length of program in years	Day/Eve./Both	None/Opt./Req.	% of graduates in Co-op programs	# of degrees awarded	% of graduates who were full-time	# of degrees awarded	% of graduates who were full-time	# of degrees awarded	% of graduates who were full-time
Mechanical & Aerospace Eng [M](5)		32.9(4)	—(5)	-.-	-.-	—	—	-.-	—.	-.-	27	89.8	4	80.0
Mechanical Eng [M](3)		32.9(4)	Y	4.0	4.3	D	N	-.-	45	100.0	—(3)	-.-	—(3)	-.-
Nuclear Eng [M]		32.9(4)	Y	4.0	4.0	D	N	-.-	3	100.0	5	87.5	3	100.0
Systems Eng [O]		6.1	Y	4.0	4.1	D	N	-.-	43	100.0	22	89.4	0	83.3
260 Virginia Military Inst	VA													
Civil & Env'l Eng [V]		9.0	Y	4.0	-.-	D	N	-.-	175	100.0	—.	-.-	—.	-.-
Electrical Eng [E]		7.0	Y	4.0	-.-	D	N	-.-	84	100.0	—.	-.-	—.	-.-
Mechanical Eng [M]		8.0	Y	4.0	-.-	D	N	-.-	174	100.0	—.	-.-	—.	-.-
261 Virginia Tech	VA													
Aerospace & Ocean Eng [O]		17.0	Y	4.0	4.0	D	O	19.0	86	100.0	16	100.0	17	100.0
Aerospace Eng [O]		-.-	Y	4.0	4.0	D	O	19.0	86	100.0	16	100.0	17	100.0
Agricultural Eng [O]		20.0	Y	4.0	4.0	D	O	0.0	4	100.0	6	100.0	0	100.0
Chemical Eng [H]		11.0	Y	4.0	4.0	D	O	39.0	86	100.0	16	100.0	17	100.0
Civil Eng [V]		38.0	Y	4.0	4.0	D	O	16.0	177	100.0	91	100.0	9	100.0
Computer Eng [O]		8.0	Y	4.0	4.0	D	O	20.0	40	100.0	—.	-.-	—.	-.-
Electrical Eng [E]		44.0	Y	4.0	4.0	D	O	16.0	253	100.0	93	100.0	23	100.0
Eng Science & Mechanics [O]		34.0	Y	4.0	4.0	D	O	14.0	29	100.0	26	100.0	22	100.0
Industrial & Systems Eng [I]		29.0	Y	4.0	4.0	D	O	20.0	103	100.0	68	100.0	10	100.0
Materials Science & Eng [O]		16.0	Y	4.0	4.0	D	O	4.0	25	100.0	9	100.0	—.	-.-
Mechanical Eng [M]		39.0	Y	4.0	4.0	D	O	20.0	212	20.0	44	100.0	15	100.0
Mining Eng [O]		8.0	Y	4.0	4.0	D	O	-.-	10	100.0	2	100.0	—.	-.-
262 U of Washington	WA													
Aeronautical & Astronautical Eng [O]		20.0	Y	4.0	-.-(A)	D	O	-.-(A)	58	86.0	48	-.-(A)	2	-.-(A)
Ceramic Eng [O]		14.0(1)	Y	4.0	-.-(A)	D	O	-.-(A)	13	94.0	—.	-.-(A)	—.	-.-(A)
Chemical Eng [H]		14.0	Y	4.0	-.-(A)	D	O	-.-(A)	52	90.0	9	-.-(A)	10	-.-(A)
Civil Eng [V]		37.0	Y	4.0	-.-(A)	D	O	-.-(A)	117	93.0	79	-.-(A)	13	-.-(A)
Computer Science & Eng [O]		28.3	Y	4.0	-.-(A)	D	O	-.-(A)	17	79.0	18	-.-(A)	3	-.-(A)
Electrical Eng [E]		45.7	Y	4.0	-.-(A)	D	O	-.-(A)	133	80.0	92	-.-(A)	17	-.-(A)
Industrial Eng [I]		7.0	Y	4.0	-.-(A)	D	O	-.-(A)	28	85.0	—(2)	-.-(B)	—(B)	-.-(B)
Mechanical Eng [M]		32.2	Y	4.0	-.-(A)	D	O	-.-(A)	167	82.0	36	-.-(A)	5	-.-(A)
Metallurgical Eng [O]		14.0(3)	Y	4.0	-.-(A)	D	O	-.-(A)	17	87.0	—.	-.-(A)	—.	-.-(A)
Technical Communication [O]		11.5	N	4.0	-.-(A)	D	O	-.-(A)	1	64.0	3	-.-	—.	-.-
263 Washington U	MO													
Appl Science [O](1)		-.-(1)	N	4.0	4.0	D	O	-.-	4	-.-	—(6)	-.-	—(6)	-.-
Biological & Eng Science [O](2)		-.-(2)	N	4.0	4.0	D	O	-.-	1	-.-	—.	-.-	—.	-.-
Biological & Eng Sciences [O](7)		-.-	N	4.0	4.0	D	O	-.-	1	-.-	—(6)	-.-	—(6)	-.-
Chemical Eng [H]		11.2	Y	4.0	4.0	D	O	-.-	24	-.-	6	-.-	14	-.-
Civil Eng [V]		9.0	Y	4.0	4.0	D	O	-.-	21	-.-	3	-.-	2	-.-
Computer Science [O]		16.0	Y	4.0	4.0	D	O	-.-	40	-.-	23	-.-	2	-.-
Construction Eng [O](8)		-.-	N	-.-	-.-	B	O	-.-	—.	-.-	1	-.-	—.	-.-
Construction Mgmt [O](8)		-.-	N	-.-	-.-	B	O	-.-	—.	-.-	14	-.-	—.	-.-
Electrical Eng [E]		21.0	Y	4.0	4.0	D	O	-.-	69	-.-	44	-.-	6	-.-
Eng & Policy [O]		-.-	Y	-.-	-.-	D	O	-.-	—.	-.-	—.	-.-	—.	-.-
Eng & Policy [O]		-.-	Y	-.-	-.-	D	O	-.-	—.	-.-	—.	-.-	—.	-.-
Mgmt of Tech [O](8)		-.-(9)	N	-.-	-.-	B	O	-.-	—.	-.-	4	-.-	—.	-.-
Materials Science & Eng [O](8)		-.-(3)	N	-.-	-.-	B	O	-.-	—.	-.-	4	-.-	—.	-.-
Mechanical Eng [M]		16.5	Y	4.0	4.0	B	O	-.-	61	-.-	51	-.-	7	-.-
Structural Design [O](8)		-.-	N	-.-(A)	-.-	B	O	-.-	—.	-.-	—.	-.-	—.	-.-
Structural Eng [O](8)		-.-(4)	N	-.-	-.-	B	O	-.-	—.	-.-	2	-.-	—.	-.-
Systems Science & Mathematics [O]		10.0	Y	4.0	4.0	D	O	-.-	10	-.-	10	-.-	3	-.-
Tech & Human Affairs [O](8)		-.-(5)	N	-.-	-.-	D	O	-.-	—.	-.-	2	-.-	—.	-.-
264 Washington State U	WA													
Agricultural Eng [O](A)		12.3	Y	4.0	4.5	D	O	-.-(A)	6	100.0	4	100.0	1	100.0
Architectural Studies [O](1)		-.-(6)	N	4.0	4.2	D	O	-.-(A)	42	100.0	—.	-.-	—.	-.-
Architecture [O](1)		22.5	N	5.0	5.2	D	O	-.-(A)	44	100.0	0	100.0	—(B)	-.-(B)
Chemical Eng [H](1)		10.2	Y	4.0	4.5	D	O	-.-	10	100.0	5	100.0	0	100.0
Civil & Env'l Eng [V](1)		24.0	Y	4.0	4.5	D	O	-.-(A)	53	100.0	8	100.0	3	100.0
Civil Eng [V](1)		24.0	Y	4.0	4.5	D	O	-.-(A)	53	100.0	8	100.0	3	100.0
Computer Science [O](1)		-.-(2)	N	4.0	4.5	D	O	-.-(A)	39	100.0	14	100.0	1	100.0
Construction Mgmt [O](1)		-.-(6)	N	5.0	5.2	D	O	-.-(A)	30	100.0	—.	-.-	—.	-.-
Electrical Eng [E]		-.-	Y	4.0(5)	-.-	D(5)	O	-.-	—.	-.-	—.	-.-	—.	-.-
Electrical Eng & Computer Science [E](1)		37.3	Y	4.0	4.5	D	O	-.-(A)	94	100.0	13	100.0	6	100.0
Eng Mgmt [O](3)		2.0	N	-.-(B)	-.-	E	—	-.-	—(B)	-.-(B)	19	-.-	—(B)	-.-(B)
Env'l Eng [O](1)		-.-(5)	Y	-.-	-.-	-.-	—	-.-	—(B)	-.-(B)	5	100.0	—.	-.-
Materials Science & Eng [O](1)		-.-(4)	Y	4.0	4.5	D	O	-.-(A)	13	100.0	10	100.0	3	100.0
Mechanical Eng [M](1)		28.9	Y	4.0	4.5	D	O	-.-(A)	121	100.0	13	100.0	6	100.0

All Engineering Disciplines

ENGINEERING Institution Profile Reference Number & Name Name of Degree Program	State/Province	FACULTY Full-time equivalent (FTE)	ABET/CEAB accred?	UNDERGRADUATE PROGRAMS					GRADUATE PROGRAMS			
				Length		Time	Co-op	Bachelor's		Master's		Doctoral	
				Nominal length of program in years	Average length of program in years	Day/Eve./Both	None/Opt./Req.	% of graduates in Co-op programs	# of degrees awarded	% of graduates who were full-time	# of degrees awarded	% of graduates who were full-time	# of degrees awarded	% of graduates who were full-time
265 Wayne State U	MI													
Chemical Eng [H]		13.0	Y	4.0	4.5	B	O	-.-(A)	18	-.-(A)	15	-.-(A)	3	-.-(A)
Civil & Env'l Eng [V]		8.0	Y	4.0	4.5	B	O	-.-(A)	13	-.-(A)	34	-.-(A)	1	-.-(A)
Electrical & Computer Eng [E]		20.0	Y	4.0	4.5	B	O	-.-(A)	79	-.-(A)	72	-.-(A)	8	-.-(A)
Electronic & Computer Control System [E]		-.-(1)	N	-.-	-.-	—	—	-.-(A)	—.	-.-(B)	37	-.-(A)	—.	-.-(A)
Hazardous Waste Mgmt [H]		-.-(4)	—(2)	-.-	-.-	—	—	-.-	-.-	-.-	16	-.-(A)	—.	-(5)
Industrial & Manuf'g Eng [I]		8.0	Y	4.0	4.5	B	O	-.-(A)	9	-.-(A)	23	-.-(A)	—.	-.-(A)
Materials Science & Eng [O]		7.0	Y	4.0	4.5	B	O	-.-(A)	6	-.-(A)	13	-.-(A)	—.	-.-(A)
Mechanical Eng [M]		21.0	Y	4.0	4.5	B	O	-.-(A)	39	-.-(A)	89	-.-(A)	3	-.-(A)
Operations Research [I]		5.0(3)	—(2)	-.-	-.-	—	—	-.-	—.	-.-	1	-.-(A)	1	-.-
266 Webb Inst	NY													
Naval Architecture & Marine Eng [O]		8.9	Y	4.0	4.0	D	R	100.0	16	100.0	—.	-.-	—.	-.-
267 West Virginia U	WV													
Aerospace Eng [O]		-.-(1)	Y	4.0	4.3	D	O	0.0	29	100.0	3	95.0	1	95.0
Chemical Eng [H]		12.0	Y	4.0	4.5	D	N	-.-	19	100.0	4	65.0	2	90.0
Civil Eng [V]		19.0	Y	4.0	4.0	D	O	0.0	32	100.0	14	18.0	3	30.0
Computer Eng [O]		7.0	Y	4.0	4.0	D	O	5.0	15	100.0	8	90.0	2	90.0
Electrical Eng [E]		11.0	Y	4.0	4.0	D	O	5.0	69	100.0	8	90.0	5	90.0
Eng [O](2)		-.-	—	-.-	-.-	—	—	-.-	—.	-.-	8	75.0	—.	-.-
Industrial Eng [I]		11.0	Y	4.0	4.6	D	O	0.0	42	98.0	11	90.0	6	85.0
Mechanical Eng [M]		28.0	Y	4.0	4.3	D	O	0.0	65	100.0	25	95.0	2	95.0
Occupational Health & Safety [O](2)		-.-	—	-.-	-.-	—	—	-.-	—.	-.-	9	80.0	—.	-.-
268 West Virginia Inst of Tech	WV													
Chemical Eng [H]		4.0	Y	4.0	4.5	D	O	30.0	4	95.0	—.	-.-	—.	-.-
Civil Eng [V]		6.0	Y	4.0	4.5	D	O	25.0	22	90.0	—.	-.-	—.	-.-
Electrical Eng [E]		8.0	Y	4.0	4.5	D	O	35.0	50	95.0	—.	-.-	—.	-.-
Mechanical Eng [M]		5.0	Y	4.0	4.5	D	O	30.0	24	95.0	—.	-.-	—.	-.-
269 Western Michigan U	MI													
Aeronautical Eng [O]		-.-	N	4.0	4.5	D	O	10.0	13	92.5	—.	-.-	—.	-.-
Computer Systems Eng [O]		3.0	Y	4.0	4.5	D	O	10.0	16	98.8	—.	-.-	—.	-.-
Electrical Eng [E]		9.0	Y	4.0	4.5	D	O	10.0	36	90.3	5	10.6	—.	-.-
Industrial Eng [I]		9.0	Y	4.0	4.5	D	O	10.0	17	95.4	8	29.7	—.	-.-
Mechanical Eng [M]		-.-	Y	4.5	4.0	D	O	10.0	68	99.1	16	4.1	—.	-.-
Paper & Printing Science & Eng [O]		5.0	N	4.0	4.5	D	O	10.0	6	97.3	6	83.0	—.	-.-
270 Western New England Coll	MA													
Bioengineering [O]		1.0	N	4.0	4.4	B	N	-.-(B)	4	100.0	—(B)	-.-(B)	—(B)	-.-(B)
Electrical Eng [E]		5.8	Y	4.0	4.4	B	N	-.-(B)	25	63.8	11	-.-(B)	—(B)	-.-(B)
Eng Mgmt [O]		1.0	N	-.-(B)	-.-(B)	E	N	-.-(B)	—(B)	-.-(B)	30	-.-(B)	—(B)	-.-(B)
Industrial Eng [I]		4.1	Y	4.0	4.4	B	N	-.-(B)	13	68.9	—(B)	-.-(B)	—(B)	-.-(B)
Mechanical Eng [M]		6.8	Y	4.0	4.4	B	N	-.-(B)	40	76.2	4	-.-(B)	—(B)	-.-(B)
271 Wichita State U	KS													
Aerospace Eng [O]		12.1	Y	4.0	5.5	B	O	33.0	21	78.8	12	22.2	3	45.5
Electrical Eng [E]		16.0	Y	4.0	5.5	B	O	8.5	82	56.2	15	39.3	4	47.6
Industrial Eng [I]		11.3	Y	4.0	5.5	B	O	6.3	16	63.0	28	35.1	2	40.7
Mechanical Eng [M]		11.5	Y	4.0	5.5	B	O	33.0	33	58.7	3	38.0	1	56.0
272 Widener U	PA													
Chemical Eng [H]		4.2	Y	4.3	4.0	D	O	20.0	10	-.-(A)	1	-.-(A)	—(B)	-.-(B)
Civil Eng [V]		5.5	Y	4.3	4.0	D	O	21.0	19	-.-(A)	1	-.-(A)	—(B)	-.-(B)
Electrical Eng [E]		9.3	Y	4.3	4.0	D	O	36.0	57	-.-(A)	8	-.-(A)	—(B)	-.-(B)
Mechanical Eng [M]		7.2	Y	4.3	4.0	D	O	34.0	29	-.-(A)	6	-.-(A)	—(B)	-.-(B)
273 Wilkes U	PA													
Appl Eng [O](1)		2.0	N	4.0	4.0	B(2)	O	-.-(A)	—(B)	95.0	—(B)	-.-(B)	—(B)	-.-(B)
Electrical Eng [E]		11.5	Y	4.0	4.0	B(2)	O	5.0(C)	45	85.0(C)	4	30.0(C)	—(B)	-.-(B)
Eng Mgmt [O](3)		4.0	N(4)	4.0	4.0	B(2)	O	30.0(C)	8	50.0(C)	—(B)	-.-(B)	—(B)	-.-(B)
Env'l Eng [O]		9.5	N	4.0	4.0	B(2)	O	1.0(C)	10	75.0(C)	—(B)	-.-(B)	—(B)	-.-(B)
Materials Eng [O]		4.0	Y	4.0	4.0	B(2)	O	0.0(C)	3	100.0	—(B)	-.-(B)	—(B)	-.-(B)
Mechanical Eng [M]		5.5	N(4)	4.0	4.0	B(2)	O	1.0(C)	0(5)	95.0(C)	—(B)	-.-(B)	—(B)	-.-(B)
Pre-Chemical Eng [H](6)		-.-(A)	N	-.-(B)	-.-(B)	B	O	-.-(A)	—(B)	-.-(B)	—(B)	-.-(B)	—(B)	-.-(B)
Pre-Civil Eng [V](6)		-.-(A)	N	-.-(B)	-.-(B)	B	O	-.-(A)	—(B)	-.-(B)	—(B)	-.-(B)	—(B)	-.-(B)
Pre-Industrial Eng [I](6)		-.-(A)	N	-.-(B)	-.-(B)	B	O	-.-(A)	—(B)	-.-(B)	—(B)	-.-(B)	—(B)	-.-(B)

All Engineering Disciplines

ENGINEERING Institution Profile Reference Number & Name / Name of Degree Program	State/Province	FACULTY Full-time equivalent (FTE)	ABET/CEAB accred?	UNDERGRADUATE PROGRAMS Length Nominal length of program in years	Length Average length of program in years	Time Day/Eve./Both	Co-op None/Opt./Req.	Co-op % of graduates in Co-op programs	Bachelor's # of degrees awarded	Bachelor's % of graduates who were full-time	GRADUATE PROGRAMS Master's # of degrees awarded	Master's % of graduates who were full-time	Doctoral # of degrees awarded	Doctoral % of graduates who were full-time
274 U of Wisconsin-Madison	WI													
Chemical Eng [H]		19.0	Y	4.0	5.0	D	O	-.-(A)	76	-.-(A)	11	-.-(A)	13	-.-(A)
Civil Eng [V]		31.0	Y	4.0	5.0	D	O	-.-(A)	47	-.-(A)	33	-.-(A)	8	-.-(A)
Electrical Eng [E]		48.5	Y	4.0	5.0	D	O	-.-(A)	159	-.-(A)	69	-.-(A)	12	-.-(A)
Eng Mechanics [O]		14.9	Y	4.0	5.0	D	O	-.-(A)	34	-.-(A)	16	-.-(B)	7	-.-(B)
Geological Eng [O]		3.1	N	4.0	5.0	D	O	-.-(A)	4	-.-(A)	—.-	-.-	-.-	-.-(A)
Industrial Eng [I]		22.5	Y	4.0	5.0	D	O	-.-(A)	54	-.-(A)	37	-.-(A)	9	-.-(A)
Manuf'g Systems Eng [O](2)		-.-(1)	N	1.0	1.0	D	—	-.-	—.-(B)	-.-	19	-.-	—.-	-.-
Materials Science & Eng [O](2)		15.0	Y	-.-	-.-	D	O	-.-	—.-(B)	-.-	9	-.-	—.-	-.-(A)
Mechanical Eng [M]		31.0	Y	4.0	5.0	D	O	-.-(A)	141	-.-(A)	39	-.-(A)	11	-.-(A)
Metallurgical Eng [O]		-.-(1)	N	4.0	5.0	D	O	-.-(A)	11	-.-(A)	7	-.-(A)	2	-.-(A)
Nuclear Eng [O]		12.0	Y	4.0	5.0	D	O	-.-(A)	13	-.-(A)	18	-.-(A)	7	-.-(A)
275 U of Wisconsin-Milwaukee	WI													
Civil Eng [V]		14.3	Y	4.0	5.5	B	O	7.1	42	67.7	13	28.6	3	58.3
Electrical Eng [E]		11.8	Y	4.0	5.5	B	O	10.4	48	72.9	11	42.2	1	85.7
Industrial Eng [I]		8.8	Y	4.0	5.5	B	O	7.1	14	78.6	6	21.1	0	50.0
Materials Eng [O]		7.2	Y	4.0	5.5	B	O	0.0	9	71.4	5	25.0	4	75.0
Mechanical Eng [M]		14.5	Y	4.0	5.5	B	O	8.6	58	72.1	15	28.3	1	63.6
276 U of Wisconsin-Platteville	WI													
Civil Eng [V]		9.7	Y	4.0	4.7	D	O	15.0	47	100.0	—.-	-.-	—.-	-.-
Electrical Eng [E]		9.0	Y	4.0	4.7	D	O	21.0	57	100.0	—.-	-.-	—.-	-.-
Industrial Eng [I]		4.0	Y	4.0	5.0	D	O	27.0	31	100.0	—.-	-.-	—.-	-.-
Mechanical Eng [M]		11.0	Y	4.0	4.7	D	O	39.0	66	100.0	—.-	-.-	—.-	-.-
277 Worcester Poly Inst	MA													
Biomedical Eng [O]		4.0	N(1)	-.-(B)	-.-(B)	B	—(B)	-.-(B)	—.-(B)	-.-(B)	6	-.-(A)	1	-.-(A)
Chemical Eng [H]		10.0	Y	4.0	4.0	D	O	-.-(A)	25	100.0	7	-.-(A)	1	-.-(A)
Civil Eng [V]		11.6	Y	4.0	4.0	D	O	-.-(A)	69	98.5	13	-.-(A)	—.-	-.-(A)
Electrical & Computer Eng [E]		27.3	Y	4.0	4.0	D	O	-.-(A)	112	95.8	57	-.-(A)	5	-.-(A)
Fire Protection Eng [O]		3.0	N(1)	-.-(B)	-.-(B)	B	—(B)	-.-(B)	—.-(B)	-.-(B)	8	-.-(A)	—.-	-.-(A)
Manuf'g Eng [I]		8.0	Y	4.0	4.0	D	O	-.-(A)	22	100.0	4	-.-(A)	—.-	-.-(A)
Materials Eng [O]		6.0	N(1)	-.-(B)	-.-(B)	B	—(B)	-.-(B)	—.-(B)	-.-(B)	8	-.-(B)	—.-	-.-(B)
Mechanical Eng [M]		32.6	Y	4.0	4.0	D	O	-.-(A)	165	98.8	25	-.-(A)	1	-.-(A)
278 Wright State U	OH													
Biomedical Eng [O]		2.8	Y	4.0	-.-	B	O	0.0	12	87.0	0	60.0	—.-	-.-
Computer Eng [O]		7.7	Y	4.0	-.-	B	O	19.4	36	73.7	7	18.5	0	36.4
Computer Science [O]		14.5	Y(1)	4.0	-.-	B	O	27.3	33	60.8	7	14.5	0	52.2
Electrical Eng [E]		12.6	Y	4.0	-.-	B	O	20.0	59	73.0	26	17.6	—.-	-.-
Eng Physics [O]		0.5	Y	4.0	-.-	B	O	33.3	3	72.1	—.-	-.-	—.-	-.-
Human Factors Eng [O]		1.6	N	4.0	-.-	B	O	15.4	13	80.4	5	31.3	—.-	-.-
Materials Science & Eng [O]		3.0	Y	4.0	-.-	B	O	0.0	4	75.0	4	37.5	—.-	-.-
Mechanical Eng [M]		8.0	Y	4.0	-.-	B	O	38.6	44	78.9	9	26.8	—.-	-.-
279 U of Wyoming	WY													
Chemical Eng [H]		7.0	Y	4.0	4.5	D	O	0.0	6	0.0	4	100.0	—.-	-.-
Civil & Architectural Eng [V]		24.0	Y	4.0	5.0	D	O	0.0	54	95.0	11	90.0	—.-	-.-
Electrical Eng [E]		16.2	Y	4.0	5.0	D	O	4.0	45	91.0	4	100.0	0	100.0
Mechanical Eng [M]		12.0	Y	4.0	5.5	D	O	0.0	28	100.0	3	100.0	0	100.0
Petroleum Eng [O]		5.0	Y	4.0	5.0	D	O	5.0	22	10.0	3	10.0	—.-	-.-
280 Yale U	CT													
Appl Physics [O](1)		-.-	N(2)	4.0	4.0	D	N	-.-	4	-.-(2)	—.-	-.-	6	100.0
Chemical Eng [H]		8.0	Y	4.0	4.0	D	N	-.-	—.-	16.0	1	100.0	—.-	-.-
Electrical Eng [E]		15.0	Y	4.0	4.0	D	N	-.-	20	42.0	5	50.0	11	100.0
Mechanical Eng [M]		10.0	Y	4.0	4.0	D	N	-.-	20	42.0	4	75.0	8	100.0

All Engineering Disciplines—Totals

| | | FACULTY | UNDERGRADUATE PROGRAMS | | | | | | GRADUATE PROGRAMS | | | |
| | | | | Length | | Time | Co-op | Bachelor's | | Master's | | Doctoral | |
State/Province/Country	# of schools	Full-time equivalent (FTE)	# accredited programs	Nominal length of program in years	Average length of program in years	Day/Eve./Both	None/Opt./Req.	% of graduates in Co-op programs	# of degrees awarded	% of graduates who were full-time	# of degrees awarded	% of graduates who were full-time	# of degrees awarded	% of graduates who were full-time
AK	1	26.50	3						39		21		-	
AL	6	403.40	37						1,451		381		62	
AR	2	84.00	8						246		201		10	
AZ	4	570.70	28						1,086		434		122	
CA	26	2,459.10	123						6,360		3,122		812	
CO	8	678.60	33						1,404		591		136	
CT	5	249.70	18						553		243		58	
DC	4	205.80	16						339		471		38	
DE	1	69.59	4						186		61		46	
FL	10	860.80	58						2,544		865		144	
GA	3	370.50	15						1,272		592		130	
HI	1	61.00	3						159		35		7	
IA	2	71.40	18						872		219		88	
ID	2	115.00	6						195		72		7	
IL	9	1,085.50	52						2,639		1,123		436	
IN	7	529.40	36						2,033		451		172	
KS	3	228.90	19						729		216		42	
KY	2	179.10	13						409		178		24	
LA	5	326.00	30						809		165		43	
MA	10	1,144.50	49						2,800		1,777		360	
MD	6	582.50	28						1,236		1,054		105	
ME	2	79.00	9						171		44		3	
MI	13	1,109.60	56						3,608		1,266		353	
MN	4	245.00	14						797		184		103	
MO	4	375.30	30						1,199		604		97	
MS	2	227.90	18						383		120		18	
MT	2	101.80	13						276		45		3	
NC	4	739.30	26						1,543		415		131	
ND	2	83.00	11						375		31		4	
NE	1	113.00	6						457		162		20	
NH	3	81.50	6						229		72		13	
NJ	4	406.40	18						1,029		524		110	
NM	3	227.60	15						448		271		52	
NV	1	24.50	3						47		11		-	
NY	25	1,694.70	110						5,197		2,374		488	
OH	13	1,272.60	72						2,626		1,061		262	
OK	4	274.40	24						614		310		70	
OR	3	181.00	13						674		225		32	
PA	15	1,237.10	68						3,575		1,197		382	
PR	1	146.50	5						535		13		-	
RI	1	98.00	6						142		48		18	
SC	3	230.00	15						788		262		45	
SD	2	117.00	11						381		95		1	
TN	7	1,092.50	40						1,083		332		82	
TX	16	1,108.30	78						3,326		1,633		370	
UT	3	260.90	17						747		239		69	
VA	5	672.50	27						2,201		920		207	
VT	2	58.00	7						127		22		4	
WA	4	424.90	23						1,251		387		73	
WI	5	414.90	28						1,378		405		90	
WV	2	111.00	11						371		90		21	
WY	1	64.20	5						155		25		-	
U.S. Totals:	274	23,574.40	1,382						63,094		25,659		5,963	
AB	1	113.50	9						371		102		20	
MB	1	10.00	7						187		36		13	
NS	1	74.70	8						168		27		14	
PQ	3	455.90	22						1,109		403		119	
Canada Totals:	6	654.10	46						1,835		568		166	

FOOTNOTES: The following footnotes are the same for all schools: (A) Data not available. (B) Data not applicable. (C) Estimated data.

002 University of Alabama: (1) Computer Engineering Option (2) The program includes the major specializations of mining engineering and petroleum engineering. (3) Program only available at the graduate level. (4) Accredited by the Computer Science Accreditation Commission of the Computing Sciences Accreditation Board. (5) Administered by the Civil Engineering Department. (6) Faculty for this program included in the count for Civil Engineering. (7) The program includes the major specialization of mining engineering and petroleum engineering.

006 University of Alberta: (1) Mining, Metallurgical and Petroleum FTE total = 17 (2) Totals for all of Metallurgical, Mining and Petroleum Engineering. (3) Total for Computer, Electrical and Engineering Physics.

008 University of Arizona: (1) Faculty are combined with Mechanical Engineering. (2) Faculty are combined with Electrical Engineering. (3) Faculty are combined with Civil Engineering. (4) Faculty are combined with Geological Engineering. (5) Faculty are combined with Industrial Engineering. (6) Faculty are combined with Hydrology.

012 Auburn University: (1) Supported by the faculty of the Department of Aerospace Engineering. (2) Supported by the faculty of the Department of Computer Science and Engineering. (3) Accredited by Computing Sciences Accreditation Board. (4) Supported by faculty of the School of Forestry and the Department of Agricultural Engineering. (5) Interdepartmental program administered by the Department of Civil Engineering. (6) Supported by the faculty of the Departments of Industrial Engineering, Mechanical Engineering, and Management. (7) Supported by the faculty of the Mechanical Engineering Department. (8) Supported by the faculty of the Department of Textile Engineering. (9) Interdepartmental program administered by the Department of Textile Engineering. (10) Based on credit hours taken Fall Quarter 1991.

013 Baylor University: (1) Electrical and Mechanical options (2) No graduate program

014 Boston University: (1) Part time status granted by petition only. (2) Offered through the Electrical, Computer, Systems department. (3) Aerospace/Mechanical faculty (4) Electrical/Computer/Systems faculty (5) Offered through the Aerospace/Mechanical Department.

015 Bradley University: (1) Students typically take the equivalent of nine semesters but finish in 4 years by using summer school and transfer credits. (2) Department offers a computer option within the electrical engineering degree. (3) Program is in Dept. of Civil Engineering and Construction. Bachelor of Science degree with strong business flavor. Accredited by ACCE. (4) Physics and engineering faculty support program.

018 Bucknell University: (1) Data is given in whole numbers.

019 University of California at Berkeley: (1) Double majors are included with department totals. (2) Administered by the College of Chemistry. (3) Administered by Industrial Engineering and Operations Research and Mechanical Engineering. (4) Faculty FTE and graduate degree data included with Materials Science and Engineering.

020 University of California, Davis: (1) Included in Mechanical Engineering (2) Additional 24 students received B.S. in both Mechanical and Aeronautical Science and Engineering. (3) Also supported by faculty in College of Agricultural and Environmental Sciences (4) Graduate Program only (5) Included in Chemical Engineering (6) Included in Civil Engineering (7) Included in Electrical and Computer Engineering

021 University of California, Irvine: (1) This is a graduate program only. (2) This is an undergraduate program only. (3) Supported by faculty FTE included in other programs. (4) This is a new program.

022 University of California, Los Angeles: (1) At the graduate level see Materials Science & Engineering listed below. (2) Faculty in this program belongs to various departments in the School. (3) Not Offered. (4) See Materials Engineering above. (5) At the graduate level see Computer Science listed above. (6) See Computer Science above. (7) Not offered at the B.S. level.

023 University of California, San Diego: (1) A major within the AMES Department. (2) Due to a large amount of shared teaching responsibility, it is impossible to give faculty numbers by program. See specific department for numbers. (3) A major within the CSE Department. (4) Accredited by CSAB. (5) A joint major with CSE and ECE departments.

024 University of California, Santa Barbara: (1) A Limited number of students are sponsored by their employers and are working on their degrees part-time. (2) Faculty number is for the department of Chemical and Nuclear Engineering, thus includes data for both chemical and nuclear engineering faculty. (3) Nuclear Engineering graduate program offers M.S. degrees only. (4) Accredited by the Computer Science Accreditation Commission of the Computing Sciences Accreditation Board. (5) Materials Engineering department offers only graduate programs. (6) The B.S. degree is awarded in Electrical Engineering. (7) Co-op program for computer science started Fall 1992.

027 California Polytechnic State University: (2) FTEF is for Electrical AND Electronic Engineering (3) Accredited by Computer Science Accreditation Board (CSAB)

029 California State University, Fresno: (1) Doctoral program not available (2) American Council for Construction Education (ACCE) (3) Primarily (4) New degree program

030 California State University, Long Beach: (1) The Department offers graduate programs only. (2) The Department offers undergraduate programs only.

032 California State University, Northridge: (1) Accredited by CSAB - Computer Sciences Accrediting Board

036 Carnegie Mellon University: (1) Undergraduate degree not available - Option only. (2) Undergraduate degree not available.

037 Case Western Reserve University: (1) All figures are shared with Polymer Science and Engineering. (2) Department staff is shared with Systems Engineering. (3) Department staff shared with Mechanical Engineering and Fluid and Thermal Engineering. (4) Department staff shared with Mechanical and Aerospace Engineering.

038 The Catholic University of America: (1) The Civil Engineering program includes a concentration in Construction. (2) Student in Construction concentration must participate in a summer internship program arranged by the department. (3) This program is run by the Electrical Engineering department.

039 University of Central Florida: (1) Due to large number of part-time students.

040 Central State University: (1) No graduate program available.

041 Christian Brothers University: (1) This is a part-time, evening program.

042 University of Cincinnati: (1) Of the 6 departments., 5 have more than one program, therefore, faculty provide support in multiple programs. Total FTE for the college is 142. (2) This is a graduate program only and is not ABET accredited. (3) This is a graduate program only and is ABET accredited. (4) After the first graduating Class of 1992, Materials Engineering will be reviewed by ABET for accreditation purposes. (5) Graduating classes of 90 and 91 receive BS in Metallurgical Engineering. Class of 92 and beyond receive BS in Materials Engineering.

044 City College of the City University of New York: (1) Most prerequisite courses can be taken during the evening. All major courses must be taken during the day. (2) The program is accredited by CSAB. (3) During the academic year 1991-1992, the PhD program in Computer Science was not administered by City College and these data are not available.

048 Cleveland State University: (1) includes Co-Op and part-time students (2) Program includes a computer engineering option (3) Students enter with an associates degree (4) No graduate technology program

050 University of Colorado at Colorado Springs: (1) CSAB accredited

051 University of Colorado at Denver: (1) Graduate program is in the College of Liberal Arts & Sciences. (2) A.Math is not a separate department in the College of Engineering; it is a program offered by the College. (3) Coursework available at CU-Denver but Ph.D. is awarded at CU-Boulder through system-wide Graduate School. (4) Coursework is available at CU-Denver, but Masters in Computer Science is awarded by CU-Boulder through the system-wide Graduate School.

054 Columbia University: (1) Includes Bioengineering and Biomechanics. (2) Industrial Engineering is ABET accredited. (3) Includes Applied Mathematics. (4) Includes Solid State Engineering. (5) Henry Krumb School of Mines: includes Metallurgical, Mining, Materials Science, and Mineral Resource Engineering; Mining Engineering ABET accredited.

055 Concordia University: (1) Undergraduate programme only.

056 University of Connecticut: (1) Includes Materials Science and Polymer Science students. (2) Includes Environmental Engineering students. (3) Includes Biological Engineering students. (4) Includes Ocean Engineering students. (5) Accreditation by the Computing Sciences Accreditation Board (CSAB) is anticipated as well. (6) Includes Materials Science students.

057 The Cooper Union: (1) Intended for gifted students who wish flexibility in their engineering program or who intend to continue studies in law, medicine, dentistry, business, etc.

061 University of Delaware: (1) Some students (about 10 %) require an additional term. Nearly all students take courses during the winter term. (2) Senior level technical electives are available in evening hours. (3) sustaining considered as full-time (5) Graduate program

062 University of Denver: (1) The Engineering (General) program began in Fall 1992. The program will be submitted for ABET accreditation at the earliest opportunity.

064 Drexel University: (1) This Program is no longer available. (2) Graduate Program Only.

066 École Polytechnique de Montréal: (2) Figures in this column refers to Fall '91 + Winter '92 + Summer '92. (3) Figures in this column were valid in Fall 1991. (4) This figure is valid for the average student in all our undergraduate programs. (5) This faculty supports Electrical Engineering and Computer Engineering programs. (6) This faculty supports Geological Engineering, Mining Engineering and Mineral Engineering programs. (7) Master's degree in Aeronautics included (8) Includes Energy Engineering

068 Embry-Riddle Aeronautical University, Western Campus: (1) ABET accreditation expected summer 1993

069 University of Evansville: (1) Program supported by EE & CS faculty

071 University of Florida: (1) Civil Engineering and Surveying and Mapping faculty are combined. (2) Industrial Engineering and Systems Engineering options are available (3) Aerospace Engineering and Engineering Science & Mechanics faculty combined (4) A graduate program only; accredited as an engineering-related program (5) Civil Engineering and Surveying and Mapping faculty are combined (7) Program specialities: ceramics, metals, polymers and electronic materials (8) Ceramics, metals and electronic materials specialties are ABET accredited (9) Nuclear Engineering and Nuclear Engineering Sciences faculty are combined (10) Faculty from other departments support this program

072 Florida Agricultural and Mechanical University: (1) Includes FAMU and FSU students. Year is Fall 1991, Spring 1992 and Summer 1992.

073 Florida Atlantic University: (1) Master's program only. (2) No undergraduate programs in civil or manufacturing systems engineering. (3) No bachelor's program. Master's and Ph.D. programs only. (4) Accredited by CSAB (5) Working students often limit their programs to 12 credits per term or less, depending on work schedules. (6) New program, 1991.

075 Florida International University: (1) The Baccalaureate Degree Program in Computer Engineering begin in the Fall of 1990. ABET Accreditation is scheduled for Fall, 1993. (2) The Master's Degree Program in Computer Engineering is expected to graduate its first student sometime during the 1992/93 academic year.

076 Florida State University: (1) Includes FAMU and FSU students. Year is Fall 1992, Spring 1992, and Summer 1992.

077 GMI Engineering & Management Institute: (1) 3360 hours minimum

079 George Mason University: (1) The Information Technology PhD degree program is interdisciplinary and supported by all seven masters programs. (2) Undergraduate courses in Information Systems are offered and eligible for inclusion in other degree programs. (3) The Information Technology PhD degree program is supported by faculty of all six departments within the School of Information Technology and Engineering. (4) Undergraduate courses in this program are offered and eligible for use in other degree programs at the university. (5) This program is supported by faculty of the Information and Software Systems Engineering Department. (6) This program is supported by faculty of the Systems Engineering department.

082 Georgia Institute of Technology: (1) No doctoral program in Environmental Engineering is offered. (2) No undergraduate degree in Environmental Engineering is offered. (3) Includes specialties in Environmental Hydraulics and Water Resources. (4) Includes: Structural Engineering & Mechanics, Construction Management, Geotechnical Engineering, Materials, Transportation, Engineering Computer Graphics. (5) The master degree is "MS in Metallurgical Engineering". The doctoral degree is "PhD with major in Metallurgy". (6) Includes Health Physics. (7) Includes M.S. Operations Research, M.S. Statistics, and M.S. Health Systems. (8) Fall '91 Quarter FTE. (9) Includes: MS in Textile Engineering, Undesignated, Textiles, Textile Chemistry, & Polymers; PhD in Textile & Fiber Engineering, Textile Chemistry, & Polymer tracks.

086 Harvard University: (1) 15 graduates in the A.B. program. (2) The S.B. program is accredited not the A.B. program.

090 University of Houston: (1) BSCE, MCE, MSCE, MSEnvE, PhD (2) BSEE, MEE, MSEE, PhD (3) BSIE, MIE, MSIE, PhD (4) BSME, MME, MSME, PhD (5) BSCHE, MSCHE, MCHE, PhD

091 Howard University: (2) We do not offer the doctorate in Chemical Engineering. (3) Accredited by Computer Sciences Accreditation Board, Inc. (CSAB).

094 University of Illinois at Chicago: (1) Interdisciplinary program. Supported by additional FTE faculty in other departments. (2) Program supported by FTE faculty in Computer Engineering. (3) Program supported by FTE faculty in Industrial Engineering. (4) Program supported by FTE faculty in several engineering departments and by the Department of Physics. (5) Offered as PhD degree only. (6) Offered as M.S. degree only.

095 University of Illinois at Urbana-Champaign: (1) Industrial Engineering is in the Department of Mechanical and Industrial Engineering. (2) Ceramic and Metallurgical Engineering are in the Department of Materials Science and Engineering. (3) Computer Engineering is in the Department of Electrical and Computer Engineering. (4) Environmental Engineering and Environmental Science are in the Department of Civil Engineering. (5) B.S. program in Engineering Mechanics.

096 Illinois Institute of Technology: (2) Reflects percentage of graduates who received certificates of completion in compliance with ABET requirements. (3) Academic year 1991-1992 is first year this program is offered for bachelor's degree. (4) New program. (5) Graduate degrees offered in Mechanical and Aerospace Engineering only.

100 The Johns Hopkins University: (1) Includes .30 FTE in part-time programs. (2) Includes 1.70 FTE in part-time programs. (3) Includes 26.30 FTE in part-time programs. (4) Includes 3.70 FTE in part-time programs. (5) Offered in part-time programs Only. (6) Varies (7) Includes 27.0 FTE in Part-time programs. (8) Includes 3.30 FTE in part-time programs. (9) Includes 1 FTE in part-time programs. (10) Listed as Mechanical Engineering and Engineering Mechanics

103 University of Kentucky: (1) Breakdown of part-time students by degree is not available. Percentage represents total M.S. and Ph.D. enrollment. (2) Graduate degrees programs only

104 Lafayette College: (1) Undergraduate program only. (2) A Bachelor of Arts degree is granted in this program.

106 Lawrence Technological University: (1) Construction Engineering program is ABET accredited. The results of an ABET team evaluation of the Civil Engineering program are awaited. (2) Of this total 26 are Civil Engineering graduates. (3) Program is available only in the evening.

107 LeTourneau University: (1) Electrical and Mechanical Concentrations. (2) No engineering graduate programs at LeTourneau University.

108 Lehigh University: (1) (2) Planned for 1994

109 Louisiana State University: (1) PhD available through Engineering Science (interdisciplinary program). (2) MS and PhD studies under Electrical Engineering (3) Graduate program only

110 Louisiana Tech University: (1) CSAB accredited. (2) Petroleum Engineering and Geosciences Department.

111 University of Louisville: (1) Program is accredited at the master's level. (2) Faculty for this program are located in Electrical and Engineering Math & Computer Science Departments. (3) Faculty located in the Industrial Engineering Department. (4) Master's program only.

114 University of Maine: (1) Ph.D. Program not available. (2) Masters Program Not Available.

115 University of Manitoba: (1) MS and Doctoral degrees are in Computer and Electrical Engineering. These degrees are reported under Electrical Engineering. (2) MS and Doctoral degrees are in Mechanical and Industrial Engineering. These degrees are reported under Industrial Engineering.

116 Mankato State University: (1) ABET accreditation visit scheduled for Fall, 1993.

119 University of Maryland Baltimore County: (1) Joint program with University of Maryland. University College UMBC uses existing UMBC faculty and courses to teach the 12 credits technical electives (2) program is a part-time program designed for engineers who are working full-time (3) MS program Only (4) no B.S.E.E. program

120 University of Maryland, College Park: (1) This program is part of the Bachelor of Science in Engineering Science degree. (3) This program is only offered at the graduate level.

121 University of Massachusetts: (1) Includes 6 M.S. degrees awarded in Environmental Engineering (2) Includes 13 M.S. degrees earned in Engineering Management. Also see footnote under Mechanical Engineering regarding M.S. degrees in Manufacturing. (3) Includes 20 M.S. degrees in Manufacturing Engineering, and inter-departmental program between Mechanical Engineering and Industrial Engineering. (4) Includes department head

124 Massachusetts Institute of Technology: (1) 9.5% of master's degree recipients participated in co-op programs. (2) 58.0% of bachelor's degree recipients participated in co-op programs; 6.4% of master's degree recipients participated in co-op programs (3) 6.9% of master's degree recipients participated in co-op programs. (4) 10.7% of master's recipients participated in co-op programs. (5) 11.0% of bachelor's degree recipients and 34.1% of master's degree recipients participated in co-op programs.

125 McGill University: (1) Following completion of 2 year CEGEP program. International students should add 1 year for pre-engineering studies (2) Internship year program available for students who have less then 45 credits remaining in their academic program (3) Part time figures include candidates who have completed their residence requirements but are working full time on research and thesis requirements (4) Number includes M Engineering graduates in Mining program (5) Number includes Ph D graduates in mining program (6) Number includes M Eng graduates in the Metallurgical Program (7) Number includes Ph D graduates in Metallurgical program

129 Miami University: (1) Program implemented August 1991. (2) Co-op optional for Manufacturing Engineering technical specialty; Co-op required for Paper Science & Engineering technical specialty.

130 University of Michigan: (1) BS and/or BSE program only. (2) Graduate program only. (3) Faculty included in principal department. (4) Includes Professional degrees. (5) All graduate students. (6) Doctoral program only. (7) For whole College. (8) Interdisciplinary program.

132 Michigan State University: (1) This program is administered by the Electrical Engineering Department. (2) This program is administered by the Materials Science and Mechanics Department. (3) This program is administered by the Civil Engineering Department. (4) This program is administered by the Agricultural Engineering Department.

133 Michigan Technological University: (1) Data included in Material Science and Engineering numbers (2) Engineering includes Chemistry, Geology, Geophysics, Applied Geophysics, General Engineering and Doctorate in Engineering.

134 Milwaukee School of Engineering: (1) No co-op program is available, however, 85% of undergraduates participate in an internship program and have relevant work experience prior to graduation.

136 University of Minnesota: (1) Count is number of faculty. We do not have FTE estimates. Chemical Engineering includes Materials Science. (2) Department is Civil and Mineral Engineering. Statistics have been split between Civil and Extractive Metallurgical Eng. (3) Computer Science is a separate department. (4) Five additional faculty contribute to the geo-engineering program from the Civil & Mineral Engineering Dept.

137 University of Mississippi: (1) M.S. and Ph.D. degrees in Engineering Science are offered with emphasis in Chemical Engineering. (2) M.S. and Ph.D. degrees in Engineering Science are offered with emphasis in Civil Engineering. (3) M.S. and Ph.D. degrees in Engineering Science are offered with emphasis in Electrical Engineering. (4) M.S. and Ph.D. degrees in Engineering Science are offered with emphasis in Mechanical Engineering (5) M.S. and Ph.D. degrees in Engineering Science are offered with emphasis in Computer Science. (6) Accredited by Computer Sciences Accreditation Commission of the Computing Sciences Accreditation Board. (7) There are no graduate degree programs in this category. (8) There are no undergraduate degree programs in this category. (9) M.S. and Ph.D. degrees in Engineering Science are offered with emphasis in Geological Engineering.

138 Mississippi State University: (1) Faculty included in Agricultural Engineering (2) Program being phased out. (3) Interdisciplinary program with faculty from many departments. (4) M.S. and Ph.D. programs only. (5) Faculty included with Electrical Engineering. (6) Faculty included in Aerospace Engineering. (7) M.S. program only.

139 University of Missouri-Columbia: (1) Includes statistics from CE/CEP which is located in Kansas City and administered by the Columbia campus. (2) Includes statistics from EE/CEP which is located in Kansas City and administered by the Columbia campus. (3) Includes statistics from ME/CEP which is located in Kansas City and administered by the Columbia campus.

140 University of Missouri-Kansas City: (2) Most students participate in unofficial co-op programs with local firms.

141 University of Missouri-Rolla: (1) Bachelor's Degree not offered.

142 Monmouth College: (1) Part-time only. Maximum of 9 credits applied to degree.

144 Montana State University: (1) Information unavailable but nearly 100%

150 University of New Haven: (1) Chemistry/Chemical Engineering

151 New Jersey Institute of Technology: (1) No undergraduate program. Masters level only. (2) Interdisciplinary program. (3) No doctoral program. Masters level only. (4) Undergraduate program only. No graduate programs offered. (5) No doctoral program. Undergraduate and masters level only. (6) No undergraduate program. Masters and doctoral levels only. (7) New program.

152 The University of New Mexico: (1) We classify all our graduate students as full-time. (2) This program is being phased out. (3) No graduate program. (4) Included with Electrical Engineering. (5) Accredited by the Computer Science Accreditation Board (CSAB). (6) Manufacturing Engineering and Robotics

153 New Mexico Institute of Mining and Technology: (1) Reviewed in 1992

154 New Mexico State University: (1) Accredited as electrical engineering. (2) Includes degrees unearned through off-campus programs at Kirtland and Cannon Air Force bases. (3) Degree is BS in Surveying (4) Accreditation visit scheduled for January 93.

159 North Carolina Agricultural and Technical State University: (1) Some faculty are assigned to common freshman classes.

160 North Carolina State University: (1) The BS in environmental engineering was approved during the 1992 fall semester. As a result, no students are officially enrolled.

161 University of North Dakota: (1) No M.S., M.Engr. degree offered.

162 North Dakota State University: (1) Masters & Ph.D. degree also available. (2) Ph.D. in Engineering with Petroleum Engineering. emphasis also available.

163 Northeastern University: (1) 5 year co-op option; 4 year co-op option; 4-year non-co-op option

164 Northern Arizona University: (1) Includes an emphasis in environmental engineering.

166 Northwestern University: (1) See Electrical Engineering. (2) See Civil Engineering.

169 Oakland University: (1) M.S. Program in Computer Science and Engineering and is within Computer Science and Engineering Department. (2) See Doctoral Program. (3) Program accredited by CSAB. (4) This is an interdisciplinary school wide program in various fields of engineering at O.U. (5) This program is in Electrical and Systems Engineering Department. (6) M.S. program in Electrical and Computer Engineering and is within the Electrical and Systems Engineering Department.

173 University of Oklahoma: (2) There were no doctoral students in this program during this time period. (3) There were no masters students in this program

175 Oklahoma State University: (1) The masters degree is the terminal degree in this discipline. (2) This program draws on the faculty from the other disciplines. (3) Separately accredited option in Mechanical Engineering. (4) Graduate degrees not offered in this program.

176 Old Dominion University: (1) 1991/92 Head count (2) Graduate Programs only (3) See Mechanical Engineering — faculty participate in both programs

177 Oregon State University: (1) New Program 1992-93 (2) New program 1991-92 (3) M.S. & PhD offered in Electrical and Computer Engineering. Students reported under Electrical and Electronics Engineering. (4) No graduate degrees offered (5) Required for the Manufacturing Option

178 University of the Pacific: (1) Engineering Physics program does not offer a master's or doctoral degree. (2) Civil Engineering does not offer a master's or doctoral degree. (3) Electrical Engineering does not offer a doctoral degree. Computer Engineering does not offer a master's or a doctoral degree. (4) Mechanical Engineering does not offer a master's or doctoral degree. (5) Engineering Management does not offer a master's or doctoral degree.

181 University of Pittsburgh: (1) Graduate-level program. (3) Not applicable. This is a masters-level program. (4) Undergraduate-level program. (5) Includes materials science and engineering and metallurgical engineering students. (6) Part-time program is available.

182 Polytechnic University: (1) 4-5 (2) Includes Environmental

183 University of Portland: (1) Data not available by program. Overall undergraduate full-time is 89%. (2) Each year approximately 10 Engineering students participate in co-op programs. The numbers by degree program fluctuate. (3) Data not available by program. Overall graduate full-time is 42%. (4) Includes an Electrical Track and a Computer Track. (5) Interdisciplinary program. (6) A materials science oriented interdisciplinary program.

184 Portland State University: (1) Undergraduate degrees are awarded in either Electrical Engineering or Computer Engineering. (2) Computer Engineering is not ABET accredited.

188 Purdue University: (1) Except for co-op students who require 5.0 years to complete the program. (2) See Agricultural Engineering. (3) Faculty in this program hold appointments in the School of Civil Engineering. (4) Co-op not available in this program. (5) Construction Engineering and Management is an undergraduate program. (6) Faculty for this program reside in faculties of other academic units. (7) Undergraduate program only. (8) See Electrical Engineering.

189 Rensselaer Polytechnic Institute: (1) Faculty counts may overlap in Mechanical Eng., Aeronautical Eng., & Mechanics, in Electrical, Computer, & Systems Eng., and in Nuclear Eng. & Eng. Physics

190 University of Rhode Island: (1) Program offered in department of Electrical Engineering (2) B.S. degree program started in 1991-1992.

191 William Marsh Rice University: (1) The undergraduate degree program leads to a B.A. degree. No B.S. degree program is offered. (2) Percentage of B.A. engineering students who are full-time. (3) The thesis master's degree program leads to an M.A. degree. (4) An undergraduate major in environmental science. or engineering is offered only as a double major with other fields of science. engineering. (5) The undergraduate environmental double major programs lead to B.A. degrees except for a B.S. in Civil Engineering with an Environmental Option. (6) Percentage of M.A. & M.Eng. students who are full-time. (7) As of July 1, 1992 the name of this department was changed from Mathematical Sciences to Computational and Applied Mathematics.

192 University of Rochester: (1) Geomechanics is an undergraduate program. There is no graduate program. (2) Graduate students may participate in a co-op experience (3) Materials science is a graduate program only. Data is merged with Mechanical and Aerospace Science. (4) Interdisciplinary engineering is an undergraduate program only. (5) Mechanical and Aerospace Science is a graduate program only. Materials Science has been merged under this category. (6) Mechanical engineering is an undergraduate program only.

198 Parks College of Saint Louis University: (1) MS program began fall 1991 (2) No Ph.D. program in Engineering (3) No graduate program

199 University of San Diego: (1) The EE program is a nine semester dual B.S./B.A. program accredited by EAC/ABET. (2) Successful full-time majors will be assured completion between the minimum nominal program length of 9 semesters and a maximum of 10 semesters. (3) Typical co-op/internship programs are scheduled for full-time in the summer only, with optional part-time involvement available during regular terms.

203 Santa Clara University: (1) Degree awarded in Engineering Mechanics

207 South Dakota School of Mines and Technology: (1) New program not yet eligible for ABET accreditation.

208 South Dakota State University: (1) Housed in the General Engineering Department. BST degree.

209 University of South Florida: (1) Masters program only. (2) This degree program is for working engineers. (4) C.S.A.B accredited (5) Based on Fall 1991

212 University of Southern Colorado: (8)

216 Southern Methodist University: (1) This degree is not available to newly entering students.

218 State University of New York at Binghamton: (1) Included in Mechanical Engineering Section (2) Includes Industrial Engineering Program (3) Included in Systems Science Program (4) Includes Industrial Technology Program

219 State University of New York at Buffalo: (1) Joint with Mechanical Engineering (2) Joint with Civil

221 State University of New York Maritime College: (1) BE degree awarded. (2) Facilities Engineering is a new program that is not yet accredited. (3) Facilities Engineering faculty are included with Mechanical Engineering.

224 Swarthmore College: (1) Swarthmore does not offer a masters or doctoral program in engineering.

225 Syracuse University: (2) Includes Mechanical/Manufacturing Option (3) Pending.

226 Technical University of Nova Scotia: (2) Three years at TUNS. (3) Just started.

228 University of Tennessee, Chattanooga: (1) BSE with Chemical Concentration. (2) BSE with Civil Concentration. (4) BSE with Industrial Concentration. (5) BSE with Mechanical Concentration. (6) Master's program. (7) All faculty are included under Engineering. (8) BS degree. (9) MS degree.

230 Tennessee State University: (1) An option in Master of Engineering degree program.

233 University of Texas at Austin: (1) Participants in the Co-op Program are away from the campus for three semesters, working for pay and letter grade. (2) Architectural Engineering faculty are included in the numbers for Civil Engineering. (3) Biomedical Engineering faculty are included in faculty for Aerospace, Chemical, Electrical, and Mechanical Engineering. (4) Engineering Mechanics faculty are included with Aerospace Engineering. (5) Energy & Mineral Resources faculty are included with Petroleum Engineering. (6) Environmental Health Engineering faculty are included with Civil Engineering. (7) Manufacturing systems Engineering faculty are included with Chemical, Electrical and Mechanical Engineering and Management in the College of Business. (8) Materials Science & Engineering and Operations Research & Industrial Engineering faculty are included with Mechanical Engineering. (9) Computer Engineering faculty and students are included with Electrical Engineering. (10) The Computer Engineering degree program requires a semester hour total which differs from Electrical Engineering, but the degree conferred is in Electrical Engineering.

235 University of Texas at El Paso: (1) Computer Science program is accredited by CSAB.

238 Texas A&M University: (1) Degree program in Industrial Engineering (2) Has two tracks: ELEN and CPSC (3) Program had its first ABET accreditation review fall 1992. No decision had been made at the time of this compilation of information. (4) Degree Program in Nuclear Engineering (5) Degree program in Civil Engineering (6) CPEN is an undergraduate degree program only (7) A master's degree program (8) An undergraduate degree program (9) Graduate program in Nuclear Engineering (10) Graduate program only

240 University of Toledo: (1) Interdisciplinary degree program.

241 Tri-State University: (1) This program supplements the math, science, and engineering courses of a traditional engineering program with business, management and economics components. (2) The program is based on current engineering and business course offerings. Thus no additional teaching faculty are required. (3) No graduate programs

244 Tulane University: (1) Computer Science accredited by CSAB

245 University of Tulsa: (1) No M.S. program is offered. (2) No doctoral program is offered. (3) Co-op not available.

250 United States Merchant Marine Academy: (1) Same faculty responsible for all three programs.

252 United States Naval Academy: (1) Faculty support is obtained from all engineering departments.

253 University of Utah: (1) Faculty FTE represents support for the undergraduate and graduate chemical engineering programs and the graduate program of fuels engineering. (2) Graduate level program only. (3) Faculty FTE represents support for the undergraduate and graduate programs in chemical engineering and the graduate program in fuels engineering. (4) Geological Engineering is an undergraduate program only. (6) Computer Engineering is an undergraduate program only. (7) Faculty in direct support of Computer Engineering total five (5), but data of full-time equivalent. is not currently available.

254 Utah State University: (1) Currently seeking ABET accreditation to establish Aerospace, Environmental and Civil Engineering as separately accredited degree programs. (2) Previously known as Agricultural and Irrigation Engineering.

256 Vanderbilt University: (1) An additional 14 double majors with Electrical Engineering. (2) Includes EWRE Graduate Degrees. (3) Also offer program in Computer Engineering in cooperation with the Department of Computer Science. (4) Computer Engineering not yet accredited. (5) Occasional evening course. (6) Also offer program in Computer Engineering in cooperation with the Department of Electrical Engineering.

257 University of Vermont: (1) Graduate program only.

258 Villanova University: (1) Program initiated in the Fall of 1993, now accepting applications.

259 University of Virginia: (1) Department of Civil Engineering and Applied Mechanics. (2) Master's degree program. (3) BS degree program. (4) Department of Mechanical, Aerospace and Nuclear Engineering. (5) Graduate degree program. (6) Department of Materials Science and Engineering.

262 University of Washington: (1) Total FTE for Materials Science Department of which Ceramic Engineering is a part. (2) MS in Engineering degree awarded through Inter-Engineering program. 4 degrees awarded 1991-92. (3) Total FTE for Materials Science Department of which Metallurgical Engineering is a part.

All Engineering Disciplines—Footnotes

263 Washington University: (1) The Applied Science major is available through 5 of our 7 departments. (2) Biological and Engineering Science is available through Chemical Eng, Computer Science, Electrical Eng. and Mechanical Eng. (3) Faculty of Materials Science and Engineering come from several of our departments. (4) Faculty are from Civil Engineering. (5) Faculty from the Engineering and Policy Dept. (6) Undergraduate Program Only. (7) Faculty from all departments. (8) Graduate Program only. (9) Faculty from Engineering & Policy

264 Washington State University: (1) Pullman campus (2) See Electrical Engineering for data. (3) Master of Engineering Management degree program. (4) See Mechanical Engineering for data. (5) See Civil Engineering for data. (6) See Architecture for data.

265 Wayne State University: (1) Included in Electrical and Computer Engineering (2) No Undergraduate Program (3) Included in Industrial and Manufacturing Engineering (4) Included in Chemical Engineering (5) Doctoral degree not offered in this area.

267 West Virginia University: (1) Combined with Mechanical Engineering (2) Bachelor Degrees are not available.

273 Wilkes University: (1) Requires a second major or specific concentration other than engineering. (2) There is no distinction between day or evening programs. (3) Requires the specification of a preference area in engineering (i.e. Electrical, Environmental, Materials, or Mechanical), 5-year Engineering Management/MBA. (4) Has not been visited by ABET. (5) New program. First graduating class in 1993. (6) Prepares the student for direct transfer to the third year in another institution.

274 University of Wisconsin-Madison: (1) Same faculty as Mechanical Engineering, IE, and Materials Science and Engineering. (2) Graduate only program.

277 Worcester Polytechnic Institute: (1) Graduate program only.

278 Wright State University: (1) Accredited by the Computing Sciences Accreditation Board.

280 Yale University: (1) Although the Dept. is not an Engineering program it is under the umbrella organization of the Council of Engineering. (2) Applied Physics is not an engineering department.

Chemical Engineering

Institutions are listed in alphabetical order. The first line in a school's entry contains the institution profile reference number, the school name, and the state or province in which the main campus is located. Subsequent indented lines for each program give the name of the degree program, followed by a major category abbreviation in brackets (H = Chemical, V = Civil, E = Electrical, I = Industrial, M = Mechanical, and O = Other) the full-time equivalent (FTE) of all faculty. For undergraduate programs, the columns represent: accreditation status, the nominal and actual average length of the program, the time that courses are available (Day, Evening or Both), the availability of cooperative (internship) programs (None, Optional, or Required), the percentage of graduates who participated in co-op programs, the number of bachelor's degrees awarded, and the percentage of these graduates who were full-time. For graduate programs, the columns indicate the number of graduates and the percentage who were full-time in master's and then doctoral programs. Unless otherwise footnoted, information on degrees is for the 1991-92 academic year; all other information is current. At the end of the table are (1) a summary of totals by state or province and by country and (2) the full text of all footnotes for this table.

ENGINEERING	State/Province	FACULTY Full-time equivalent (FTE)	UNDERGRADUATE PROGRAMS ABET/CEAB accred?	Length Nominal length of program in years	Length Average length of program in years	Time Day/Eve./Both	Co-op None/Opt./Req.	Co-op % of graduates in Co-op programs	Bachelor's # of degrees awarded	Bachelor's % of graduates who were full-time	GRADUATE PROGRAMS Master's # of degrees awarded	Master's % of graduates who were full-time	Doctoral # of degrees awarded	Doctoral % of graduates who were full-time
001 U of Akron Chemical Eng [H]	OH	9.0	Y	4.0	-.-[A]	D	O	69.0	26	87.5[A]	11	67.0	1	100.0
002 U of Alabama Chemical Eng [H]	AL	9.0	Y	4.0	4.5	D	O	24.1	29	100.0	1	100.0	2	100.0
004 U of Alabama at Huntsville Chemical Eng [H]	AL	4.5	Y	4.0	-.-	B	O	-.-	4	84.0	-.-	-.-	-.-	-.-
006 U of Alberta Chemical Eng [H]	AB	18.0	Y	4.0	-.-[A]	D	O	33.0	49	100.0	5	85.0	1	100.0
008 U of Arizona Chemical Eng [H]	AZ	11.0	Y	4.0	-.-	D	O	-.-	22	88.0	2	-.-	1	-.-
009 Arizona State U Chemical Eng [H]	AZ	26.0	Y	4.0	5.0	D	O	-.-[A]	23	86.0	5	75.0	2	50.0
010 U of Arkansas Chemical Eng [H]	AR	13.0	Y	4.0	-.-	D	O	-.-	39	89.0	9	50.0	1	100.0
012 Auburn U Chemical Eng [H]	AL	15.5	Y	4.0	-.-[A]	D	O	27.7	47	86.0[10]	6	75.9[10]	5	95.2[10]
017 Brigham Young U Chemical Eng [H]	UT	13.5	Y	4.0	5.0	B	O	-.-[A]	22	-.-[A]	7	-.-[A]	1	-.-[A]
018 Bucknell U Chemical Eng [H]	PA	8.0[1]	Y	4.0	4.0	D[A]	N[A]	-.-[A]	21	100.0	2	100.0	-.-[A]	-.-[A]
019 U of Cal at Berkeley Chemical Eng [H][2]	CA	20.0	Y	4.0	-.-	D	O	20.0	60	99.0	18	100.0	23	100.0
020 U of Cal, Davis Chemical Eng [H] Chemical Eng & Materials Science [H]	CA	11.4 18.4	Y N	4.0 4.0	-.-[A] -.-[A]	D D	O O[A]	-.-[A] -.-[A]	35 2	100.0 100.0	6 -.-[5]	100.0 -.-[5]	3 -.-[5]	100.0 -.-[5]
021 U of Cal, Irvine Biochemical Eng [H][1]	CA	6.0	N[B]	-.-[B]	-.-[B]	D	N	-.-[B]	-.-[B]	-.-[B]	2	89.0	2	100.0
022 U of Cal, Los Angeles Chemical Eng [H]	CA	11.0	Y	4.0	4.6	--	O	-.-	28	100.0	8	100.0	3	100.0
023 U of Cal, San Diego Chemical Eng [H][1]	CA	36.9[2]	Y	4.0	4.7	D	N	-.-	19	100.0	3	100.0	5	100.0
024 U of Cal, Santa Barbara Chemical Eng [H]	CA	19.3[2]	Y	4.0	4.2	-.-[B]	N	-.-[B]	21	100.0	2	100.0	3	100.0
026 Cal Inst of Tech Chemical Eng [H]	CA	8.0	Y	4.0	-.-	D	-.-[B]	-.-	7	100.0	8	100.0	8	100.0
030 Cal State U, Long Beach Chemical Eng [H]	CA	5.1	Y	4.0	5.5	B	N	-.-	15	55.0	-.-	-.-	-.-	-.-
034 Cal State Poly, Pomona Chemical Eng [H]	CA	8.4	Y	4.0	5.5	B	O	-.-	50	78.0	-.-	-.-	-.-	-.-
036 Carnegie Mellon U Chemical Eng [H]	PA	15.5	Y	4.0	4.0	D	N	-.-[B]	25	99.0	5	37.5	11	98.1
037 Case Western Reserve U Chemical Eng [H]	OH	10.0	Y	4.0	4.3	B	O	10.0	26	8.0	10	100.0	4	100.0
041 Christian Brothers U Chemical Eng [H]	TN	3.0	Y	4.0	4.7	D	N	-.-[B]	9	77.5	-.-[B]	-.-[B]	-.-[B]	-.-[B]

Chemical Engineering

ENGINEERING Institution Profile Reference Number & Name Name of Degree Program	State/Province	FACULTY Full-time equivalent (FTE)	ABET/CEAB accred?	UNDERGRADUATE PROGRAMS							GRADUATE PROGRAMS			
				Length		Time	Co-op		Bachelor's		Master's		Doctoral	
				Nominal length of program in years	Average length of program in years	Day/Eve./Both	None/Opt./Req.	% of graduates in Co-op programs	# of degrees awarded	% of graduates who were full-time	# of degrees awarded	% of graduates who were full-time	# of degrees awarded	% of graduates who were full-time
042 U of Cincinnati Chemical Eng [H]	OH	17.0[1]	Y	5.0	5.2	D	R	100.0	50	100.0	4	96.1	7	81.5
044 CUNY-City Coll Chemical Eng [H]	NY	10.6	Y	4.0	5.0	D[1]	O	-.-[A]	7	88.8	6	7.5	3	100.0
046 Clarkson U Chemical Eng [H]	NY	15.0	Y	4.0	-.-	D	O	-.-[A]	42	99.9	11	100.0	7	100.0
047 Clemson U Chemical Eng [H]	SC	12.0	Y	4.0	-.-	D	O	-.-	46	81.0	5	100.0	2	87.0
048 Cleveland State U Chemical Eng [H]	OH	9.8	Y	4.0	4.8[1]	D	O	50.0	16	77.6	5	50.0	1	100.0
049 U of Colorado at Boulder Chemical Eng [H]	CO	17.0	Y	4.0	4.5	D	O	-.-	27	95.3	3	90.0	2	90.0
052 Colorado School of Mines Chemical Eng & Petroleum Refining [H]	CO	12.0	Y	4.0	4.0	D	O	-.-[A]	16	100.0	4	80.0	5	90.0
053 Colorado State U Chemical Eng [H]	CO	8.0	Y	4.0	4.5	D[A]	N	-.-[A]	16	99.0[C]	6	95.0[C]	0	95.0[C]
054 Columbia U Chemical Eng [H][1]	NY	11.0	Y	4.0	4.0	-.-[B]	N	-.-[B]	27	100.0	17	70.9	2	70.9
056 U of Conn Chemical Eng [H]	CT	14.0	Y	4.0	4.5	D	O	13.0	8	95.0[C]	13[1]	55.0[1]	11[1]	93.0[1]
057 The Cooper Union Chemical Eng [H]	NY	7.5	Y	4.0	4.0	D	N	-.-	14	100.0	2	100.0	-.-	-.-
058 Cornell U Chemical Eng [H]	NY	17.0	Y	4.0	-.-	D	O	32.0	53	-.-	14	-.-	12	-.-
060 U of Dayton Chemical Eng [H]	OH	7.0	Y	4.0	4.0	D	O	35.0	25	97.0	6	40.0	-.-	-.-
061 U of Delaware Chemical Eng [H]	DE	21.0	Y	4.0	4.3[1]	B[2]	N	-.-	38	99.0	10	70.0	19	100.0
064 Drexel U Chemical Eng [H]	PA	10.0	Y	5.0	5.1	B	R	100.0	41	83.0	5	30.0	1	96.0
066 École Polytechnique Chemical Eng [H]	PQ	13.4	Y	4.0	4.5	D	N	-.-	25	97.6	16	96.9	4	100.0
071 U of Florida Chemical Eng [H]	FL	16.5	Y	5.0	5.5	D	O	8.0	31	99.0	3	100.0	4	100.0
072 Florida A&M U Chemical Eng [H]	FL	7.5	Y	4.5	5.0	D	O	-.-[A]	14[1]	-.-[A]	3[1]	-.-[A]	-.-	-.-[A]
074 Florida Inst of Tech Chemical Eng [H]	FL	5.3	Y	4.0	4.5	D	O	25.0	8	96.7	3	60.0	-.-	-.-
076 Florida State U Chemical Eng [H]	FL	7.5	Y	4.5	5.0	D	O	-.-[A]	14[1]	-.-[A]	3[1]	-.-[A]	-.-	-.-[B]
082 Georgia Tech Chemical Eng [H]	GA	25.0[8]	Y	4.0	4.3	D	O	36.0	72	97.0	8	99.0	8	99.0
090 U of Houston Chemical Eng [H][5]	TX	16.0	Y	4.0[C]	5.0	B	O	38.0	26	73.8	18	12.5	9	91.0
091 Howard U Chemical Eng [H][C]	DC	5.0	Y	4.0	5.0	D	O	20.0	10	100.0	1	90.0	-.-[2]	-.-[2]
092 U of Idaho Chemical Eng [H]	ID	7.0	Y	4.0	5.0	D	O	-.-	11	94.8	6	59.0	1	64.0
094 U of Illinois at Chicago Chemical Eng [H]	IL	10.0	Y	4.0	-.-	D	O	15.0	13	82.0	6	35.0	5	27.0
095 Illinois, Urbana-Champaign Chemical Eng [H]	IL	10.6	Y	4.0	4.5	-.-[A]	O	-.-[A]	72	100.0	7	100.0	14	100.0

Chemical Engineering

ENGINEERING Institution Profile Reference Number & Name Name of Degree Program	State/Province	FACULTY Full-time equivalent (FTE)	ABET/CEAB accred?	UNDERGRADUATE PROGRAMS							GRADUATE PROGRAMS			
				Length		Time	Co-op		Bachelor's		Master's		Doctoral	
				Nominal length of program in years	Average length of program in years	Day/Eve./Both	None/Opt./Req.	% of graduates in Co-op programs	# of degrees awarded	% of graduates who were full-time	# of degrees awarded	% of graduates who were full-time	# of degrees awarded	% of graduates who were full-time
096 Illinois Inst of Tech Chemical Eng [H]	IL	12.7	Y	4.0	4.8	B	O	12.5[2]	24	77.0	23	43.0	11	70.0
098 U of Iowa Chemical & Biochemical Eng [H]	IA	7.6	Y	4.0	4.5[C]	D	O	15.0[C]	20	100.0	3	100.0[C]	1	100.0[C]
099 Iowa State U Chemical Eng [H]	IA	-.-	Y	4.0	-.-	D	O	-.-	43	-.-	4	-.-	8	-.-
100 The Johns Hopkins U Chemical Eng [H]	MD	9.7[1]	Y	4.0	4.0	B	O	-.-[A]	44	100.0	7	71.0	4	100.0
101 U of Kansas Chemical & Petroleum Eng [H]	KS	13.0	Y	4.0	-.-	D	O	-.-[A]	12	-.-[A]	7	-.-[A]	2	-.-[A]
102 Kansas State U Chemical Eng [H]	KS	9.0	Y	4.0	4.7	D	O	-.-	13	98.0	4	-.-	2	-.-
103 U of Kentucky Chemical Eng [H]	KY	12.3	Y	4.0	5.0[C]	D	O	-.-[A]	20	92.4	9	-.-[A]	3	82.9[1]
104 Lafayette Coll Chemical Eng [H][1]	PA	8.5	Y	4.0	4.0	D	N	-.-	13	93.0	-.-	-.-	-.-	-.-
105 Lamar U Chemical Eng [H]	TX	5.5	Y	4.0	5.0	D	O	43.0	10	-.-[A]	-.-[A]	-.-[A]	-.-	-.-[A]
108 Lehigh U Chemical Eng [H]	PA	17.0	Y	4.0	4.0	D	O	24.0	45	100.0	21	43.0	13	87.0
109 Louisiana State U Chemical Eng [H]	LA	16.5	Y	4.0	5.0	D	O	13.3	22	90.2	9	91.4	5	86.1
110 Louisiana Tech U Chemical Eng [H]	LA	5.0	Y	4.0	4.7	D	O	-.-[A]	19	-.-[A]	4	-.-[A]	1	-.-[A]
111 U of Louisville Chemical Eng [H]	KY	9.0	Y[1]	4.0	4.5	D	R	98.0	24	77.5	11	56.6	0	77.7
114 U of Maine Chemical Eng [H]	ME	12.0	Y	4.0	4.5	D	N	-.-	30	95.0	4	90.0	-.-	100.0
119 Maryland Baltimore Cty Chemical & Biochemical Eng [H]	MD	8.5	Y	4.0	4.5	D	O	42.9	7	86.0	2	67.0	-.-	100.0
120 U of Maryland, Coll Park Chemical Eng [H]	MD	12.0	Y	4.0	4.5	D	O	16.7	24	88.0	10	80.0	8	80.8
121 U of Mass Chemical Eng [H]	MA	14.4[4]	Y	4.0	4.6	D	O	36.4	22	96.4	4	83.3	12	69.8
123 U of Mass, Lowell Chemical Eng [H] Nuclear Eng [H]	MA	9.0 20.5	Y Y	4.0 4.0	4.5 4.7	D D	N N	-.- -.-	10 74	100.0 100.0	1 16	100.0 80.0	-.- 2	-.- 50.0
124 Mass Inst of Tech Chemical Eng [H]	MA	32.0	Y	4.0	-.-	D	N	-.-	52	100.0	28	100.0	32	100.0
125 McGill U Chemical Eng [H]	PQ	20.1	Y	3.5[1]	4.0	D	O[2]	-.-[A]	44	-.-[A]	27	47.6[3]	14	40.5[3]
130 U of Michigan Chemical Eng [H]	MI	16.3	Y	4.0	4.7	D	O	8.3[7]	58	96.6	10[4]	88.0[5]	16	88.0[5]
132 Michigan State U Chemical Eng [H]	MI	10.0	Y	4.0	4.5	D	O	29.0	62	100.0	13	100.0	9	100.0
133 Michigan Tech U Chemical Eng [H]	MI	9.0	Y	4.0	4.3	D	O	12.7	63	97.6	4	100.0	0	87.5
135 U of Minnesota, Duluth Chemical Eng [H]	MN	5.0	Y	4.0	4.6	D	N	-.-	11	100.0	-.-	-.-	-.-	-.-
136 U of Minnesota Chemical Eng [H]	MN	31.0[1]	Y	4.0	-.-[A]	D	N[A]	-.-[A]	84	-.-[A]	3	-.-[A]	22	-.-[A]
137 U of Mississippi Chemical Eng [H][1]	MS	5.8	Y	4.0	4.5	D	N	-.-	5	95.3	-.-	-.-	-.-	-.-

Chemical Engineering

ENGINEERING Institution Profile Reference Number & Name Name of Degree Program	State/Province	FACULTY Full-time equivalent (FTE)	ABET/CEAB accred?	UNDERGRADUATE PROGRAMS Length Nominal length of program in years	Length Average length of program in years	Time Day/Eve./Both	Co-op None/Opt./Req.	Co-op % of graduates in Co-op programs	Bachelor's # of degrees awarded	Bachelor's % of graduates who were full-time	GRADUATE PROGRAMS Master's # of degrees awarded	Master's % of graduates who were full-time	Doctoral # of degrees awarded	Doctoral % of graduates who were full-time
138 Mississippi State U Chemical Eng [H]	MS	10.0	Y	4.0	4.8	D	O	30.0	28	95.0	1	100.0	1	100.0
139 U of Missouri-Columbia Chemical Eng [H]	MO	9.0	Y	4.0	-.-	D	O	4.0	20	95.9	8	55.1	1	57.8
141 U of Missouri-Rolla Chemical Eng [H]	MO	12.8	Y	4.0	4.5	D	O	0.0	31	94.0	3	58.0	2	45.0
144 Montana State U Chemical Eng [H]	MT	8.0	Y	4.0	4.7	D	N	-.-	27	-.-[1]	6	100.0	1	100.0
146 U of Nebraska - Lincoln Chemical Eng [H]	NE	8.0	Y	4.0	4.5	--	--	-.-	29	-.-	16	-.-	1	-.-
149 U of New Hampshire Chemical Eng [H]	NH	7.0	Y	4.0	-.-[A]	D	N	-.-	24	-.-[A]	2	-.-[A]	2	-.-[A]
150 U of New Haven Chemical Eng [H]	CT	7.2[1]	N	4.0	4.5	B	O	5.0[C]	2	60.0[C]	--[B]	-.-[B]	--[B]	-.-[B]
151 New Jersey Inst of Tech Chemical Eng [H]	NJ	14.0	Y	4.0	4.5	B	O	14.3	21	80.5	14	41.4	4	81.8
152 The U of New Mexico Chemical Eng [H]	NM	9.3	Y	4.0	5.0	D	O	-.-[A]	6	88.0	33	100.0[1]	14	100.0[1]
154 New Mexico State U Chemical Eng [H]	NM	7.7	Y	4.0	5.1[C]	D	O	30.0[C]	17	88.0[C]	12	75.0[C]	1	75.0[C]
159 North Carolina A&T Chemical Eng [H]	NC	11.0[1]	Y	4.0	-.-[A]	D	O	-.-[A]	10	-.-[A]	--[A]	-.-[A]	--[B]	-.-[B]
160 North Carolina State U Chemical Eng [H]	NC	16.5	Y	4.0	4.5	D	O	20.0	67	85.5	7	57.9	9	32.6
161 U of North Dakota Chemical Eng [H]	ND	5.0	Y	4.0	5.0	D	O	-.-	13	-.-	2	-.-	--.	-.-
163 Northeastern U Chemical Eng [H]	MA	7.0	Y	5.0[1]	5.0	D	O	-.-	25	100.0	13	54.5	1	80.0
166 Northwestern U Chemical Eng [H]	IL	15.0	Y	4.0	4.0	D	O	10.5	38	100.0	2	100.0	9	100.0
168 U of Notre Dame Chemical Eng [H]	IN	11.0	Y	4.0	4.0	D	N	-.-	26	100.0	12	100.0	6	100.0
170 Ohio U Chemical Eng [H]	OH	10.0	Y	4.0	-.-	--	O	0.0	10	-.-	6	-.-	--.	-.-
172 Ohio State U Chemical Eng [H]	OH	12.0	Y	4.0	4.5	D	N	-.-	42	91.0	15	89.0	7	100.0
173 U of Oklahoma Chemical Eng [H]	OK	11.0	Y	4.0	4.7	B	O	-.-[A]	28	-.-	8	55.2	3	47.8
175 Oklahoma State U Chemical Eng [H]	OK	9.5	Y	4.0	4.7	D	O	0.0	27	100.0	14	75.0	4	100.0
177 Oregon State U Chemical Eng [H]	OR	7.0	Y	4.0	4.8[C]	D	O[1]	0.0	21	100.0	10	100.0	1	100.0
179 U of Penn Chemical Eng [H]	PA	11.0	Y	4.0	-.-[A]	D	N	-.-[B]	19	100.0	10	100.0	54	98.1
180 Penn State U Chemical Eng [H]	PA	19.0	--	4.0	-.-	--	--	25.0	93	-.-	9	-.-	12	-.-
181 U of Pittsburgh Chemical & Petroleum Eng [H][1] Chemical Eng [H]	PA	20.3 20.3	--[3] Y	-.-[3] 4.0	-.-[3] 4.0	--[3] D	--[3] O	-.-[3] 20.6	-.-[3] 34	-.-[3] 94.4	--. 10	-.- 66.7	--[3] 7	-.-[3] 71.4
182 Poly U Chemical Eng [H]	NY	10.0	Y	4.0	-.-[1]	D	O	-.-[B]	11	90.0	5	39.0	1	45.0
185 Prairie View A&M U Chemical & Biochemical Eng [H]	TX	-.-	Y	4.0	-.-	D	O	-.-	--.	-.-	--.	-.-	--.	-.-

Chemical Engineering

ENGINEERING Institution Profile Reference Number & Name Name of Degree Program	State/Province	FACULTY Full-time equivalent (FTE)	ABET/CEAB accred?	UNDERGRADUATE PROGRAMS							GRADUATE PROGRAMS			
				Length	Length	Time	Co-op	Co-op	Bachelor's	Bachelor's	Master's	Master's	Doctoral	Doctoral
				Nominal length of program in years	Average length of program in years	Day/Eve./Both	None/Opt./Req.	% of graduates in Co-op programs	# of degrees awarded	% of graduates who were full-time	# of degrees awarded	% of graduates who were full-time	# of degrees awarded	% of graduates who were full-time
Chemical Eng [H]		5.0	N	4.0	4.0	D	O	10.0	13	90.0(C)	-.-	-.-	-.-	-.-
186 Princeton U Chemical Eng [H]	NJ	18.0	Y	4.0	4.0	--	N	-.-	9	100.0	2	100.0	13	100.0
187 U of Puerto Rico Chemical Eng [H]	PR	19.0	Y	5.0	5.0	B	O	14.0	83	96.0	0	100.0	-.-(B)	-.-(B)
188 Purdue U Chemical Eng [H]	IN	23.0	Y	4.0	4.3(1)	D	O	35.0	128	98.0	15	100.0	22	100.0
189 Rensselaer Poly Inst Chemical Eng [H]	NY	15.0	Y	4.0	4.0	D	O	16.2	37	100.0	4	-.-(A)	4	-.-(A)
190 U of Rhode Island Chemical Eng [H]	RI	11.0	Y	4.0	4.5	D	N	-.-	-.-	90.0	4	60.0	4	60.0
191 Rice U Chemical Eng [H]	TX	15.0	Y	4.0	4.4	-.-(B)	N	-.-(B)	15	100.0	6	87.5	9	96.0
192 U of Rochester Chemical Eng [H]	NY	12.0	Y	4.0	4.0	D	N	-.-	21	93.0	11	14.3	8	93.9
194 Rose-Hulman Chemical Eng [H]	IN	7.0	Y	4.0	4.0	D	N	-.-(A)	32	100.0	2	100.0	-.-(B)	-.-(B)
195 Rutgers, The State U Chemical Eng [H]	NJ	13.0	Y	4.0	4.5	D	N(A)	-.-	37	97.0	9	50.0	10	65.0
202 San Jose State U Chemical Eng [H]	CA	4.1	Y	4.0	-.-(A)	D	--	-.-	16	68.0	6	6.8	-.-(A)	-.-(A)
205 U of South Alabama Chemical Eng [H]	AL	-.-	Y	4.0	-.-	B	O	-.-	8	80.0	1	50.0	-.-(B)	-.-(B)
206 U of South Carolina Chemical Eng [H]	SC	10.0	Y	4.0	4.0	D	O	-.-(A)	12	9.3	4	91.3	2	85.7
207 South Dakota School of Mines Chemical Eng [H]	SD	4.2	Y	4.0	4.5	D	O	-.-(A)	26	-.-(A)	2	-.-(A)	-.-(B)	-.-(B)
209 U of South Florida Chemical Eng [H]	FL	8.0	Y	4.0	4.5	B	O	10.0	34	95.0	4	100.0	2	100.0
211 U of Southern Cal Chemical Eng [H]	CA	8.0	Y	4.0	4.0	D	O	-.-(A)	8	100.0	12	75.0	5	75.0
217 Stanford U Chemical Eng [H]	CA	10.0	Y	4.0	4.0	D	N	-.-	12	100.0	15	55.0	6	100.0
219 SUNY-Buffalo Chemical Eng [H]	NY	13.0	Y	4.0	4.5	D	N	-.-	21	95.6	10	68.0	9	54.5
225 Syracuse U Chemical Eng [H]	NY	9.0	Y	4.0	4.2	D	O	13.0	8	93.0	8	-.-(A)	3	-.-(A)
226 Technical U of Nova Scotia Chemical Eng [H]	NS	8.7	Y	4.0(2)	5.0	D	O	90.0	18	100.0	1	100.0	1	100.0
228 U of Tennessee, Chattanooga Chemical Eng [H](1)	TN	-.-(7)	Y	4.0	4.8	B	O	-.-(A)	8	-.-(A)	2	-.-(A)	-.-	-.-
229 U of Tennessee, Knoxville Chemical Eng [H]	TN	13.5	Y	4.0	-.-(A)	D	O	31.4	35	85.4	8	-.-(A)	7	-.-(A)
231 Tennessee Tech Chemical Eng [H]	TN	60.3	Y	4.0	4.0	D	O	10.0	9	100.0	7	100.0	1	100.0
233 U of Texas at Austin Chemical Eng [H]	TX	21.2	Y	4.0	4.7	D	O(1)	20.0	85	93.4	14	81.2	19	91.7
237 Texas A&I U Chemical Eng [H]	TX	4.0	Y	4.0	-.-	D	N	-.-	12	-.-	4	-.-	-.-	-.-
238 Texas A&M U Chemical Eng [H]	TX	20.9	Y	4.0	4.5(C)	D	O	42.0	73	91.0	21	88.0(C)	14	88.0(C)
239 Texas Tech U Chemical Eng [H]	TX	10.0	Y	4.0	-.-	D	N	-.-	8	96.0	4	84.0	0	89.0

Chemical Engineering

ENGINEERING Institution Profile Reference Number & Name Name of Degree Program	State/Province	FACULTY Full-time equivalent (FTE)	ABET/CEAB accred?	UNDERGRADUATE PROGRAMS					GRAPHITE			
				Length		Time	Co-op		Bachelor's		Master's		Doctoral	
				Nominal length of program in years	Average length of program in years	Day/Eve./Both	None/Opt./Req.	% of graduates in Co-op programs	# of degrees awarded	% of graduates who were full-time	# of degrees awarded	% of graduates who were full-time	# of degrees awarded	% of graduates who were full-time
240 U of Toledo Chemical Eng [H]	OH	8.0	Y	4.0	4.0	D	O	-.-(A)	16	95.0	9	65.0(A)	-.-	-.-
241 Tri-State U Chemical Eng [H]	IN	3.0	Y	4.0	4.0	D	O	22.0	9	100.0	-.-(3)	-.-(3)	-.-(3)	-.-(3)
243 Tufts U Chemical Eng [H]	MA	9.5	Y	4.0	4.0	D	N	-.-(B)	22	100.0	13	92.0	6	66.0
244 Tulane U Chemical Eng [H]	LA	9.0	Y	4.0	4.0	B	N	-.-	16	100.0	0	65.0	4	40.0
245 U of Tulsa Chemical Eng [H]	OK	9.0	Y	4.0	4.0	D	N	-.-(B)	10	97.9	10	72.2	1	80.0
246 Tuskegee U Chemical Eng [H]	AL	5.0	Y	4.0	5.0	D	O	80.0	2	99.5	-.-	-.-	-.-	-.-
253 U of Utah Chemical Eng [H] Fuels Eng [H](2)	UT	15.0(1) 15.0(3)	Y N	4.0 -.-(B)	5.0 -.-(B)	D D(B)	O -.-(B)	-.-(A) -.-(B)	12 0(2)	73.0 -.-(2)	5 -.-	73.0 -.-(A)	1 3	73.0 -.-(A)
256 Vanderbilt U Chemical Eng [H]	TN	10.0	Y	4.0	4.0	D	N	-.-(A)	22	100.0	1	100.0	2	100.0
258 Villanova U Chemical Eng [H]	PA	8.0	Y	4.0	4.0	--	N	-.-	29	100.0	9	12.0	-.-	-.-
259 U of Virginia Chemical Eng [H]	VA	8.9	Y	4.0	4.1	D	N	-.-	35	100.0	15	92.7	3	88.9
261 Virginia Tech Chemical Eng [H]	VA	11.0	Y	4.0	4.0	D	O	39.0	86	100.0	16	100.0	17	100.0
262 U of Washington Chemical Eng [H]	WA	14.0	Y	4.0	-.-(A)	D	O	-.-(A)	52	90.0	9	-.-(A)	10	-.-(A)
263 Washington U Chemical Eng [H]	MO	11.2	Y	4.0	4.0	D	O	-.-	24	-.-	6	-.-	14	-.-
264 Washington State U Chemical Eng [H](1)	WA	10.2	Y	4.0	4.5	D	O	-.-	10	100.0	5	100.0	0	100.0
265 Wayne State U Chemical Eng [H] Hazardous Waste Mgmt [H]	MI	13.0 -.-(4)	Y -.-(2)	4.0 -.-	4.5 -.-	B --	O -.-	-.-(A) -.-	18 -.-	-.-(A) -.-	15 16	-.-(A) -.-(A)	3 -.-(5)	-.-(A) -.-
267 West Virginia U Chemical Eng [H]	WV	12.0	Y	4.0	4.5	D	N	-.-	19	100.0	4	65.0	2	90.0
268 West Virginia Inst of Tech Chemical Eng [H]	WV	4.0	Y	4.0	4.5	D	O	30.0	4	95.0	-.-	-.-	-.-	-.-
272 Widener U Chemical Eng [H]	PA	4.2	Y	4.3	4.0	D	O	20.0	10	-.-(A)	1	-.-(A)	-.-(B)	-.-(B)
273 Wilkes U Pre-Chemical Eng [H](6)	PA	-.-(A)	N	-.-(B)	-.-(B)	B	O	-.-(A)	-.-(B)	-.-(B)	-.-(B)	-.-(B)	-.-(B)	-.-(B)
274 U of Wisconsin-Madison Chemical Eng [H]	WI	19.0	Y	4.0	5.0	D	O	-.-(A)	76	-.-(A)	11	-.-(A)	13	-.-(A)
277 Worcester Poly Inst Chemical Eng [H]	MA	10.0	Y	4.0	4.0	D	O	-.-(A)	25	100.0	7	-.-(A)	1	-.-(A)
279 U of Wyoming Chemical Eng [H]	WY	7.0	Y	4.0	4.5	D	O	0.0	6	0.0	4	100.0	-.-	-.-
280 Yale U Chemical Eng [H]	CT	8.0	Y	4.0	4.0	D	N	-.-	-.-	16.0	1	100.0	-.-	-.-

Chemical Engineering—Totals

State/Province/Country	FACULTY		UNDERGRADUATE PROGRAMS						GRADUATE PROGRAMS			
				Length	Time	Co-op		Bachelor's	Master's		Doctoral	
	# of schools	Full-time equivalent (FTE)	# accredited programs	Nominal length of program in years	Average length of program in years	Day/Eve./Both	None/Opt./Req.	% of graduates in Co-op programs	# of degrees awarded	% of graduates who were full-time	# of degrees awarded	% of graduates who were full-time	# of degrees awarded	% of graduates who were full-time
AL	5	34.00	5						90		8		7	
AR	1	13.00	1						39		9		1	
AZ	2	37.00	2						45		7		3	
CA	12	166.60	11						273		80		58	
CO	3	37.00	3						59		13		7	
CT	3	29.20	2						10		14		11	
DC	1	5.00	1						10		1		-	
DE	1	21.00	1						38		10		19	
FL	5	44.80	5						101		16		6	
GA	1	25.00	1						72		8		8	
IA	2	7.60	2						63		7		9	
ID	1	7.00	1						11		6		1	
IL	4	48.30	4						147		38		39	
IN	4	44.00	4						195		29		28	
KS	2	22.00	2						25		11		4	
KY	2	21.30	2						44		20		3	
LA	3	30.50	3						57		13		10	
MA	6	102.40	7						230		82		54	
MD	3	30.20	3						75		19		12	
ME	1	12.00	1						30		4		-	
MI	4	48.30	4						201		58		28	
MN	2	36.00	2						95		3		22	
MO	3	33.00	3						75		17		17	
MS	2	15.80	2						33		1		1	
MT	1	8.00	1						27		6		1	
NC	2	27.50	2						77		7		9	
ND	1	5.00	1						13		2		-	
NE	1	8.00	1						29		16		1	
NH	1	7.00	1						24		2		2	
NJ	3	45.00	3						67		25		27	
NM	2	17.00	2						23		45		15	
NY	10	120.10	10						241		88		49	
OH	8	82.80	8						211		66		20	
OK	3	29.50	3						65		32		8	
OR	1	7.00	1						21		10		1	
PA	11	141.80	9						330		72		98	
PR	1	19.00	1						83		-		-	
RI	1	11.00	1						-		4		4	
SC	2	22.00	2						58		9		4	
SD	1	4.20	1						26		2		-	
TN	5	86.80	5						83		18		10	
TX	8	97.59	8						242		67		51	
UT	2	43.50	2						34		12		5	
VA	2	19.90	2						121		31		20	
WA	2	24.20	2						62		14		10	
WI	1	19.00	1						76		11		13	
WV	2	16.00	2						23		4		2	
WY	1	7.00	1						6		4		-	
U.S. Totals:	145	1,739.90	142						3,960		1,021		698	
AB	1	18.00	1						49		5		1	
NS	1	8.69	1						18		1		1	
PQ	2	33.50	2						69		43		18	
Canada Totals:	4	60.20	4						136		49		20	

Chemical Engineering—Footnotes

FOOTNOTES: The following footnotes are the same for all schools: (A) Data not available. (B) Data not applicable. (C) Estimated data.

012 Auburn University: (10) Based on credit hours taken Fall Quarter 1991.

018 Bucknell University: (1) Data is given in whole numbers.

019 University of California at Berkeley: (2) Administered by the College of Chemistry.

020 University of California, Davis: (5) Included in Chemical Engineering

021 University of California, Irvine: (1) This is a graduate program only.

023 University of California, San Diego: (1) A major within the AMES Department. (2) Due to a large amount of shared teaching responsibility, it is impossible to give faculty numbers by program. See specific department for numbers.

024 University of California, Santa Barbara: (2) Faculty number is for the department of Chemical and Nuclear Engineering, thus includes data for both chemical and nuclear engineering faculty.

042 University of Cincinnati: (1) Of the 6 departments., 5 have more than one program, therefore, faculty provide support in multiple programs. Total FTE for the college is 142.

044 City College of the City University of New York: (1) Most prerequisite courses can be taken during the evening. All major courses must be taken during the day.

048 Cleveland State University: (1) includes Co-Op and part-time students

054 Columbia University: (1) Includes Bioengineering and Biomechanics.

056 University of Connecticut: (1) Includes Materials Science and Polymer Science students.

061 University of Delaware: (1) Some students (about 10 %) require an additional term. Nearly all students take courses during the winter term. (2) Senior level technical electives are available in evening hours.

072 Florida Agricultural and Mechanical University: (1) Includes FAMU and FSU students. Year is Fall 1991, Spring 1992 and Summer 1992.

076 Florida State University: (1) Includes FAMU and FSU students. Year is Fall 1992, Spring 1992, and Summer 1992.

082 Georgia Institute of Technology: (8) Fall '91 Quarter FTE.

090 University of Houston: (5) BSCHE, MSCHE, MCHE, PhD

091 Howard University: (2) We do not offer the doctorate in Chemical Engineering.

096 Illinois Institute of Technology: (2) Reflects percentage of graduates who received certificates of completion in compliance with ABET requirements.

100 The Johns Hopkins University: (1) Includes .30 FTE in part-time programs.

103 University of Kentucky: (1) Breakdown of part-time students by degree is not available. Percentage represents total M.S. and Ph.D. enrollment.

104 Lafayette College: (1) Undergraduate program only.

111 University of Louisville: (1) Program is accredited at the master's level.

121 University of Massachusetts: (4) Includes department head

125 McGill University: (1) Following completion of 2 year CEGEP program. International students should add 1 year for pre-engineering studies (2) Internship year program available for students who have less then 45 credits remaining in their academic program (3) Part time figures include candidates who have completed their residence requirements but are working full time on research and thesis requirements

130 University of Michigan: (4) Includes Professional degrees. (5) All graduate students. (7) For whole College.

136 University of Minnesota: (1) Count is number of faculty. We do not have FTE estimates. Chemical Engineering includes Materials Science.

137 University of Mississippi: (1) M.S. and Ph.D. degrees in Engineering Science are offered with emphasis in Chemical Engineering.

144 Montana State University: (1) Information unavailable but nearly 100%

150 University of New Haven: (1) Chemistry/Chemical Engineering

152 The University of New Mexico: (1) We classify all our graduate students as full-time.

159 North Carolina Agricultural and Technical State University: (1) Some faculty are assigned to common freshman classes.

163 Northeastern University: (1) 5 year co-op option; 4 year co-op option; 4-year non-co-op option

177 Oregon State University: (1) New Program 1992-93

181 University of Pittsburgh: (1) Graduate-level program. (3) Not applicable. This is a masters-level program.

182 Polytechnic University: (1) 4-5

188 Purdue University: (1) Except for co-op students who require 5.0 years to complete the program.

226 Technical University of Nova Scotia: (2) Three years at TUNS.

228 University of Tennessee, Chattanooga: (1) BSE with Chemical Concentration. (7) All faculty are included under Engineering.

233 University of Texas at Austin: (1) Participants in the Co-op Program are away from the campus for three semesters, working for pay and letter grade.

241 Tri-State University: (3) No graduate programs

253 University of Utah: (1) Faculty FTE represents support for the undergraduate and graduate chemical engineering programs and the graduate program of fuels engineering. (2) Graduate level program only. (3) Faculty FTE represents support for the undergraduate and graduate programs in chemical engineering and the graduate program in fuels engineering.

264 Washington State University: (1) Pullman campus

265 Wayne State University: (2) No Undergraduate Program (4) Included in Chemical Engineering (5) Doctoral degree not offered in this area.

273 Wilkes University: (6) Prepares the student for direct transfer to the third year in another institution.

Civil Engineering

Institutions are listed in alphabetical order. The first line in a school's entry contains the institution profile reference number, the school name, and the state or province in which the main campus is located. Subsequent indented lines for each program give the name of the degree program, followed by a major category abbreviation in brackets (H = Chemical, V = Civil, E = Electrical, I = Industrial, M = Mechanical, and O = Other) the full-time equivalent (FTE) of all faculty. For undergraduate programs, the columns represent: accreditation status, the nominal and actual average length of the program, the time that courses are available (Day, Evening or Both), the availability of cooperative (internship) programs (None, Optional, or Required), the percentage of graduates who participated in co-op programs, the number of bachelor's degrees awarded, and the percentage of these graduates who were full-time. For graduate programs, the columns indicate the number of graduates and the percentage who were full-time in master's and then doctoral programs. Unless otherwise footnoted, information on degrees is for the 1991-92 academic year; all other information is current. At the end of the table are (1) a summary of totals by state or province and by country and (2) the full text of all footnotes for this table.

ENGINEERING Institution Profile Reference Number & Name Name of Degree Program	State/Province	FACULTY Full-time equivalent (FTE)	ABET/CEAB accred?	UNDERGRADUATE PROGRAMS						Bachelor's		GRADUATE PROGRAMS			
				Length		Time	Co-op				Master's		Doctoral	
				Nominal length of program in years	Average length of program in years	Day/Eve./Both	None/Opt./Req.	% of graduates in Co-op programs	# of degrees awarded	% of graduates who were full-time	# of degrees awarded	% of graduates who were full-time	# of degrees awarded	% of graduates who were full-time
001 U of Akron Civil Eng [V]	OH	12.0	Y	4.0	-.-(A)	B	O	57.0	34	75.0(A)	9	40.0	9	80.0
002 U of Alabama Civil Eng [V]	AL	9.0	Y	4.0	4.5	D	O	11.5	26	100.0	4	100.0	2	100.0
003 U of Alabama at Birmingham Civil Eng [V]	AL	8.0	Y	4.0	-.-(A)	B	O	-.-(A)	29	41.0	2	35.0	-.-(B)	-.-(B)
004 U of Alabama at Huntsville Civil & Construction Eng [V] Civil Eng [V]	AL	-.- 4.3	Y Y	4.0 4.0	-.-(A) -.-(A)	B B	O O	-.-(A) -.-(A)	-.- 12	-.- 52.9	-.- -.-	-.- -.-(A)	-.- -.-	-.- -.-(A)
005 U of Alaska, Fairbanks Arctic Eng [V] Civil Eng [V] Env'l Quality Eng [V] Env'l Quality Science [V]	AK	-.- 10.5 -.- -.-	-.-(B) Y -.-(B) -.-(B)	-.-(B) 4.0 -.-(B) -.-(B)	-.- 4.5 -.- -.-	-.-(B) D -.-(B) -.-(B)	-.-(B) N -.-(B) -.-(B)	-.- -.- -.- -.-	-.- 13 -.- -.-	-.- 80.0 -.- -.-	-.-(A) 4 -.-(A) -.-(A)	-.-(A) 50.0 -.-(A) -.-(A)	-.-(A) 0 -.-(A) -.-(A)	-.-(A) 67.0 -.-(A) -.-(A)
006 U of Alberta Civil Eng [V]	AB	34.5	Y	4.0	-.-(A)	D	O	33.0	53	100.0	38	80.5	10	97.0
008 U of Arizona Civil Eng [V]	AZ	25.0	Y	4.0	-.-	D	O	-.-	47	87.0	15	-.-	14	-.-
009 Arizona State U Civil Eng [V]	AZ	18.5	Y	4.0	5.0	D	O	-.-(A)	31	77.0	27	77.0	7	77.0
010 U of Arkansas Civil Eng [V]	AR	11.0	Y	4.0	-.-	D	O	-.-	25	95.9	4	46.1	1	100.0
012 Auburn U Civil Eng [V]	AL	20.5	Y	4.0	-.-(A)	D	O	24.4	78	89.2(10)	29	75.0(10)	7	73.7(10)
015 Bradley U Civil Eng [V]	IL	14.0	Y	4.0	4.0(1)	B	O	10.0	21	88.5	10	4.2	-.-(B)	-.-(B)
017 Brigham Young U Civil Eng [V]	UT	16.9	Y	4.0	5.0	B	O	-.-(A)	56	-.-(A)	22	-.-(A)	3	-.-(A)
018 Bucknell U Civil Eng [V]	PA	8.0(1)	Y	4.0	4.0	D(A)	N(A)	-.-(A)	38	100.0	2	100.0	-.-	-.-
019 U of Cal at Berkeley Civil Eng [V]	CA	43.5	Y	4.0	-.-(A)	D	O	35.0	74	100.0	134	100.0	30	100.0
020 U of Cal, Davis Civil Eng [V] Civil Eng & Materials Science [V]	CA	25.0 32.0	Y N	4.0 4.0	-.-(A) -.-(A)	D D	O O	-.-(A) -.-(A)	71 2	100.0 100.0	27 -.-(6)	100.0 -.-(6)	14 -.-(6)	100.0 -.-(6)
021 U of Cal, Irvine Civil Eng [V]	CA	17.0	Y	4.0	4.5	D	N	-.-(B)	52	100.0	15	69.0	2	100.0
022 U of Cal, Los Angeles Civil Eng [V]	CA	16.0	Y	4.0	4.6	--	O	-.-	62	100.0	29	100.0	11	100.0
026 Cal Inst of Tech Civil Eng [V]	CA	6.0	N	-.-(B)	-.-	-.-(B)	-.-(B)	-.-	-.-	-.-	3	100.0	2	100.0
027 Cal Poly State U Civil & Env'l Eng [V]	CA	16.9	Y	4.0	4.7	D	O	-.-	110	-.-	4	-.-	-.-	-.-
028 Cal State U, Chico Civil Eng [V]	CA	8.7	Y	4.0	-.-(A)	D	O	-.-(A)	34	90.0	-.-(B)	-.-(B)	-.-(B)	-.-(B)
029 Cal State U, Fresno Civil Eng [V] Construction Mgmt [V]	CA	6.5 4.0	Y -.-(2)	4.0 4.0	4.5 4.5	B(3) B	O O	3.0 5.0	35 40	70.0 40.0	8 -.-	25.0 -.-	-.- -.-	-.-(1) -.-
030 Cal State U, Long Beach Civil Eng [V]	CA	12.4	Y	4.0	5.5	B	N	-.-	48	55.0	32	10.0	-.-	-.-

Civil Engineering

ENGINEERING Institution Profile Reference Number & Name Name of Degree Program	State/Province	FACULTY Full-time equivalent (FTE)	ABET/CEAB accred?	UNDERGRADUATE PROGRAMS						GRADUATE PROGRAMS			
				Length Nominal length of program in years	Length Average length of program in years	Time Day/Eve./Both	Co-op None/Opt./Req.	Co-op % of graduates in Co-op programs	Bachelor's # of degrees awarded	Bachelor's % of graduates who were full-time	Master's # of degrees awarded	Master's % of graduates who were full-time	Doctoral # of degrees awarded	Doctoral % of graduates who were full-time
031 Cal State U, Los Angeles Civil Eng [V]	CA	9.9	Y	4.0	4.5	B	O	-.-(A)	20	61.0	3	5.0	-.-(B)	-.-(B)
033 Cal State U, Sacramento Civil Eng [V]	CA	20.1	Y	4.0	5.0	B	O	7.0	58	-.-(A)	9	-.-(A)	-.-(B)	-.-(B)
034 Cal State Poly, Pomona Civil Eng [V]	CA	17.0	Y	4.0	5.5	B	O	-.-	116	81.0	-.-	-.-	-.-	-.-
036 Carnegie Mellon U Civil Eng [V]	PA	14.0	Y	4.0	4.0	D	N	-.-(B)	32	100.0	17	95.0	8	94.0
037 Case Western Reserve U Civil Eng [V]	OH	7.0	Y	4.0	4.3	B	O	30.0	7	100.0	14	50.0	-.-(B)	-.-(B)
038 The Catholic U of America Civil Eng [V](1)	DC	11.5	Y	4.0	4.0	D	O(2)	7.5	14	100.0	10	70.0	-.-	-.-
039 U of Central Florida Civil Eng [V]	FL	11.0	Y	4.0	5.0(C)	B	O	15.0(C)	34	38.4	6	22.0	1	31.0(C)
041 Christian Brothers U Civil Eng [V]	TN	4.0	Y	4.0	4.7	D	N	-.-(B)	14	94.7	-.-(B)	-.-(B)	-.-(B)	-.-(B)
042 U of Cincinnati Civil Eng [V]	OH	24.0(1)	Y	5.0	5.2	D	R	100.0	47	100.0	11	72.2	4	81.3
043 The Citadel Civil Eng [V]	SC	-.-	Y	4.0	4.0	B	N	-.-	19	30.0	-.-	-.-	-.-	-.-
044 CUNY-City Coll Civil Eng [V]	NY	16.8	Y	4.0	5.0	D(1)	O	-.-(A)	46	74.3	31	8.4	0	89.4
046 Clarkson U Civil & Env'l Eng [V]	NY	18.0	Y	4.0	-.-	D	O	-.-(A)	104	99.9	16	100.0	3	100.0
047 Clemson U Civil Eng [V]	SC	18.0	Y	4.0	-.-	D	O	-.-	91	93.0	22	63.0	4	100.0
048 Cleveland State U Civil Eng [V]	OH	9.5	Y	4.0	4.8(1)	D	O	23.5	18	76.6	19	30.0	-.-	-.-
049 U of Colorado at Boulder Civil, Env'l, & Architectural Eng [V]	CO	33.0	Y	4.0	4.5	D	O	-.-	55	97.8	43	85.3	8	85.3
051 U of Colorado at Denver Civil Eng [V]	CO	13.4	Y	4.0	4.5	B	O	5.0	20	60.0	15	20.0	-.-(3)	-.-(3)
053 Colorado State U Civil Eng [V]	CO	33.0	Y	4.0	4.5	D(A)	N	-.-(A)	61	93.0	46	74.0	19	40.0
054 Columbia U Civil Eng & Eng Mechanics [V]	NY	16.0	Y	4.0	4.0	-.-(B)	N	-.-(B)	15	100.0	31	49.5	2	49.5
055 Concordia U Civil Eng [V]	PQ	-.-	Y	4.0	-.-	B	N	-.-	15	74.0	9	73.0	3	86.0
056 U of Conn Civil Eng [V]	CT	23.0	Y	4.0	4.5	D	O	22.0	55	95.0(C)	14	48.0	1(2)	77.0(2)
057 The Cooper Union Civil Eng [V]	NY	8.0	Y	4.0	4.0	D	N	-.-	25	100.0	5	100.0	-.-	-.-
058 Cornell U Civil & Env'l Eng [V]	NY	28.3	Y	4.0	-.-	D	O	17.0	57	-.-	58	-.-	12	-.-
060 U of Dayton Civil Eng [V]	OH	10.5	Y	4.0	4.0	D	O	35.0	19	94.0	2	16.0	-.-	-.-
061 U of Delaware Civil Eng [V]	DE	18.0	Y	4.0	4.0	B	N	-.-	34	91.4	30	86.2(3)	7	71.2
063 U of the District of Columbia Civil Eng [V]	DC	6.0	Y	4.0	6.0	D	O	-.-(A)	4	-.-(A)	-.-	-.-	-.-	-.-
064 Drexel U Civil Eng [V]	PA	14.0	Y	5.0	5.0	B	R	75.0	76	75.0	16	30.0	1	80.0
065 Duke U Civil Eng [V]	NC	15.0	Y	4.0	4.0	B(B)	N	-.-	33	100.0	20	100.0	4	100.0
066 École Polytechnique Civil Eng [V]	PQ	25.5	Y	4.0	4.5	D	N	-.-	107	97.4	29	88.0	6	100.0

Civil Engineering

ENGINEERING Institution Profile Reference Number & Name Name of Degree Program	State/Province	FACULTY Full-time equivalent (FTE)	ABET/CEAB accred?	UNDERGRADUATE PROGRAMS					Bachelor's		GRADUATE PROGRAMS			
				Length Nominal length of program in years	Average length of program in years	Time Day/Eve./Both	Co-op None/Opt./Req.	% of graduates in Co-op programs	# of degrees awarded	% of graduates who were full-time	Master's # of degrees awarded	% of graduates who were full-time	Doctoral # of degrees awarded	% of graduates who were full-time
069 U of Evansville Civil Eng [V]	IN	4.0	N	4.0	4.3	D	O	25.0	4	100.0	-.-[B]	-.-[B]	-.-[B]	-.-[B]
071 U of Florida Civil Eng [V]	FL	32.0[1]	Y	4.0	4.5	D	O	3.8	69	88.6	39	100.0	4	100.0
072 Florida A&M U Civil Eng [V]	FL	10.5	Y	4.5	5.0	D	O	-.-[A]	28[1]	-.-[A]	12[1]	-.-[A]	-.-[B]	-.-[B]
073 Florida Atlantic U Civil Eng [V][1]	FL	3.5	N[2]	-.-[B]	-.-[B]	B[B]	N[B]	-.-[B]	0	-.-[2]	0	42.1	-.-[1]	-.-[1]
074 Florida Inst of Tech Civil & Env'l Eng [V]	FL	6.8	Y	4.0	4.3	D	O	33.3	16	86.4	6	58.8	0	100.0
075 Florida International U Civil & Env'l Eng [V] Civil Eng [V]	FL	-.- 12.5	Y Y	-.- 4.0	-.- 4.5	B B	O O	-.- 10.0	-.- 31	-.- 77.8	-.- 7	-.- 100.0	-.- -.-	-.- -.-
076 Florida State U Civil Eng [V]	FL	10.5	Y	4.5	5.0	D	O	-.-[A]	28[1]	-.-[A]	12[1]	-.-[A]	-.-[B]	-.-[B]
080 The George Washington U Civil Eng [V]	DC	10.0	Y	4.0	-.-[A]	B	O	-.-	12	80.0	23	27.0	3	27.0
082 Georgia Tech Civil Eng [V] Eng Science & Mechanics [V] Env'l Eng [V]	GA	27.0[8] 9.0[8] 8.0[8]	Y Y -.-[2]	4.0 4.0 -.-[2]	4.7 4.5 -.-[2]	D D -.-[2]	O O N[2]	24.0 57.0 -.-[2]	116 7 0[2]	93.4 95.9 0.0[2]	54 4 15	47.2[4] 90.0 54.3[3]	3 2 -.-	81.0 66.6 -.-[1]
083 Gonzaga U Civil Eng [V]	WA	6.0	Y	4.0	-.-	D	N	-.-	12	94.0	-.-	-.-	-.-	-.-
088 U of Hawaii at Manoa Civil Eng [V]	HI	19.0	Y	4.0	5.0	D	O	-.-	49	89.0	9	95.0	-.-[B]	-.-[B]
090 U of Houston Civil & Env'l Eng [V][1] Civil Eng [V]	TX	14.5 -.-	Y Y	4.0 -.-	5.0 -.-	B --	O O	-.-[A] -.-	13 -.-	42.7 -.-	22 -.-	50.5 -.-	5 -.-	50.5 -.-
091 Howard U Civil Eng [V]	DC	12.0	Y	4.0	4.5	D	O	5.0	20	100.0	8	50.0	-.-[B]	-.-[B]
092 U of Idaho Civil Eng [V]	ID	13.3	Y	4.0	6.0	D	O	10.0	24	93.0	14	64.0	2	100.0
094 U of Illinois at Chicago Civil Eng [V]	IL	11.0	Y	4.0	-.-	--	O	24.0	17	83.0	10	38.0	1	12.0
095 Illinois, Urbana-Champaign Civil Eng [V]	IL	55.0	Y	4.0	4.5	-.-[A]	O	3.0	135	99.2	68	88.9	26	80.2
096 Illinois Inst of Tech Civil Eng [V]	IL	10.1	Y	4.0	4.8	B	O	7.7[2]	13	73.3	21	27.0	2	59.0
098 U of Iowa Civil & Env'l Eng [V]	IA	15.4	Y	4.0	4.5[C]	D	O	15.0[C]	28	100.0[C]	24	100.0[C]	10	100.0[C]
099 Iowa State U Civil Eng [V] Construction Eng [V]	IA	-.- -.-	Y Y	4.0 4.0	-.- -.-	D D	O O	-.- -.-	60 30	-.- -.-	39 -.-	-.- -.-	1 -.-	-.- -.-
100 The Johns Hopkins U Civil Eng [V]	MD	8.7[2]	Y	4.0	4.0	B	O	-.-[A]	11	100.0	9	67.0	1	100.0
101 U of Kansas Civil Eng [V]	KS	25.0	Y	4.0	-.-	D	O	-.-	37	-.-[A]	39	-.-[A]	6	-.-[A]
102 Kansas State U Civil Eng [V]	KS	13.0	Y	4.0	4.7	D	O	-.-	29	98.0	8	-.-	-.-	-.-
103 U of Kentucky Civil Eng [V]	KY	20.3	Y	4.0	5.0[C]	D	O	-.-[A]	62	92.5	13	-.-[A]	2	52.2[1]
104 Lafayette Coll Civil Eng [V][1]	PA	8.5	Y	4.0	4.0	D	N	-.-	25	100.0	-.-	-.-	-.-	-.-
105 Lamar U Civil Eng [V]	TX	5.3	Y	4.0	5.0	B	O	0.0	6	-.-[A]	-.-[A]	-.-[A]	2	-.-[A]
106 Lawrence Tech U Civil & Construction Eng [V]	MI	5.8	-.-[1]	4.0	4.5[C]	B	N	-.-	38[2]	68.3	-.-	-.-	-.-	-.-

Civil Engineering

ENGINEERING Institution Profile Reference Number & Name Name of Degree Program	State/Province	FACULTY Full-time equivalent (FTE)	ABET/CEAB accred?	UNDERGRADUATE PROGRAMS							GRADUATE PROGRAMS			
				Length		Time	Co-op		Bachelor's		Master's		Doctoral	
				Nominal length of program in years	Average length of program in years	Day/Eve./Both	None/Opt./Req.	% of graduates in Co-op programs	# of degrees awarded	% of graduates who were full-time	# of degrees awarded	% of graduates who were full-time	# of degrees awarded	% of graduates who were full-time
108 Lehigh U Civil Eng [V]	PA	22.0	Y	4.0	4.0	D	O	0.0	62	100.0	21	70.0	9	97.0
109 Louisiana State U Civil Eng [V]	LA	20.8	Y	4.0	5.0	D	O	1.0	36	86.1	12	91.4	4	95.2
110 Louisiana Tech U Civil Eng [V]	LA	14.0	Y	4.0	4.7	D	O	-.-(A)	19	-.-(A)	2	-.-(A)	1	-.-(A)
111 U of Louisville Civil Eng [V]	KY	12.0	Y(1)	4.0	4.5	D	R	93.0	22	80.6	15	29.0	--.	-.-
113 Loyola Marymount U Civil Eng [V]	CA	6.5	Y	4.0	4.2	B	N	-.-	12	100.0	10	0.0	--.	-.-
114 U of Maine Civil Eng [V]	ME	12.5	Y	4.0	4.5	D	N	-.-	41	94.0	16	100.0	2	100.0
115 U of Manitoba Civil Eng [V]	MB	-.-	Y	4.0	-.-(A)	D	O	-.-(A)	48	-.-(B)	14	-.-(A)	8	-.-(A)
118 Marquette U Civil Eng [V]	WI	11.0	Y	4.0	4.2	B	O	39.0	46	1.0	15	44.0	3	66.0
120 U of Maryland, Coll Park Civil Eng [V]	MD	30.0	Y	4.0	4.5	D	O	1.6	61	85.5	53	38.6	7	45.4
121 U of Mass Civil Eng [V]	MA	18.0(4)	Y	4.0	4.6	D	O	18.9	74	98.2	19(1)	67.4	8	40.0
122 U of Mass, Dartmouth Civil Eng [V]	MA	7.0	Y	4.0	4.0	--	N	-.-	25	100.0	--.	-.-	--.	-.-
123 U of Mass, Lowell Civil Eng [V]	MA	19.0	Y	4.0	4.8	D	N	-.-	82	100.0	34	15.0	--.	-.-
124 Mass Inst of Tech Civil & Env'l Eng [V]	MA	40.0	Y	4.0	-.-	D	O	0.0	40	100.0	59	100.0	29	100.0
125 McGill U Civil Eng & Appl Mechanics [V]	PQ	20.5	Y	3.5(1)	4.0	D	O(2)	-.-(A)	42	-.-(A)	20	62.2(3)	4	55.2(3)
126 Memphis State U Civil Eng [V]	TN	15.0	Y	4.0	-.-	D	O	-.-	20	70.0	9	50.0	1	50.0
128 U of Miami Civil Eng [V]	FL	10.0	Y	4.0	4.0	D	O	2.0	23	92.7	11	40.0	2	20.0
130 U of Michigan Civil & Env'l Eng [V](1) Civil Eng [V](2) Construction Eng & Mgmt [V](2) Env'l Eng [V](2)	MI	26.6 -.-(3) -.-(3) -.-(3)	Y -- -- --	4.0 -.- -.- -.-	4.7 -.- -.- -.-	D -- -- --	O -- -- --	8.3(7) -- -- --	48 --. --. --.	97.7 -.- -.- -.-	--. 21(4) 15(4) 19(4)	-.- 93.6(5) 92.9(5) 90.0(5)	--. 20 --. 1	-.- 93.6(5) -.- 90.0(5)
132 Michigan State U Civil Eng [V]	MI	22.0	Y	4.0	4.5	D	O	26.0	94	100.0	21	100.0	10	100.0
133 Michigan Tech U Civil Eng [V]	MI	12.8	Y	4.0	4.3	D	O	18.6	97	92.0	16	97.4	0	100.0
136 U of Minnesota Civil Eng [V](2)	MN	31.0	Y	4.0	-.-(A)	B	O	-.-(A)	113	-.-(A)	27	-.-(A)	9	-.-(A)
137 U of Mississippi Civil Eng [V](2)	MS	7.0	Y	4.0	4.5	D	N	-.-	18	88.9	--.	-.-		
138 Mississippi State U Civil Eng [V]	MS	12.0	Y	4.0	4.8	D	O	30.0	29	95.0	14	100.0	1	100.0
139 U of Missouri-Columbia Civil Eng [V](1)	MO	16.0	Y	4.0	-.-	D	O	4.0	46	81.7	19	35.4	--.	12.5
140 U of Missouri-Kansas City Civil Eng [V]	MO	4.0	Y	4.0	5.0	B	N	-.-(2)	13	53.0	2	10.0	--.	-.-
141 U of Missouri-Rolla Civil Eng [V]	MO	16.4	Y	4.0	4.5	D	O	0.0	68	92.0	17	48.0	2	64.0
144 Montana State U Civil Eng [V]	MT	19.4	Y	4.0	4.7	D	O	25.0	32	-.-(1)	2	100.0	0	100.0
145 Morgan State U Civil Eng [V]	MD	6.0	Y	4.0	4.5	D	O	0.0	9	94.0	--.	-.-	--.	-.-

Civil Engineering

ENGINEERING Institution Profile Reference Number & Name Name of Degree Program	State/Province	FACULTY Full-time equivalent (FTE)	ABET/CEAB accred?	UNDERGRADUATE PROGRAMS							GRADUATE PROGRAMS			
				Length		Time	Co-op		Bachelor's		Master's		Doctoral	
				Nominal length of program in years	Average length of program in years	Day/Eve./Both	None/Opt./Req.	% of graduates in Co-op programs	# of degrees awarded	% of graduates who were full-time	# of degrees awarded	% of graduates who were full-time	# of degrees awarded	% of graduates who were full-time
146 U of Nebraska - Lincoln Civil Eng [V]	NE	18.0	Y	4.0	4.5	B	O	-.-	44	-.-	36	-.-	8	-.-
147 U of Nevada, Las Vegas Civil & Env'l Eng [V]	NV	9.0	Y	4.0	4.0	B	N	-.-	19	68.2	3	53.0	0	100.0
148 New England Coll Civil Eng [V]	NH	4.5	Y	4.0	4.0	D	N	-.-	8	100.0	-.-	-.-	-.-	-.-
149 U of New Hampshire Civil Eng [V]	NH	10.0	Y	4.0	-.-	D	N	-.-	53	-.-	14	-.-(A)	1	-.-(A)
150 U of New Haven Civil & Env'l Eng [V] Civil Eng [V]	CT	8.0 8.0	N Y	-.-(B) 4.0	-.-(B) 4.5	-- B	-.-(B) O	-.- 5.0(C)	-.-(B) 11	-.-(B) 60.0	9 9	20.0 30.0(C)	-.-(B) -.-(B)	-.-(B) -.-(B)
151 New Jersey Inst of Tech Civil Eng [V]	NJ	17.9	Y	4.0	4.5	B	O	4.3	46	72.6	46	20.2	2	80.0
152 The U of New Mexico Civil Eng [V]	NM	15.6	Y	4.0	5.0	D	O	-.-(A)	16	82.2	27	100.0(1)	0	100.0(1)
154 New Mexico State U Civil Eng [V]	NM	13.2	Y	4.0	5.1(C)	D	O	26.0(C)	45	88.0(C)	11	88.0(C)	3	64.0(C)
155 U of New Orleans Civil Eng [V]	LA	5.0	Y	4.0	5.2	D	O	-.-	18	65.0	-.-	-.-	-.-	-.-
158 UNC-Charlotte Civil Eng [V]	NC	12.3	Y	4.0	4.5	B	O	26.0	23	86.1	2	27.6	-.-	-.-
159 North Carolina A&T Civil Eng [V]	NC	5.9	Y	4.0	-.-(A)	D	O	-.-(A)	8	-.-(A)	-.-(A)	-.-(A)	-.-(B)	-.-(B)
160 North Carolina State U Civil Eng [V] Civil Eng (Construction) [V] Env'l Eng [V](1)	NC	36.5 36.5 36.5	Y Y N	4.0 4.0 4.0	4.5 4.5 -.-	D D D	O O O	20.0 20.0 -.-	117 34 -.-	85.6 85.7 -.-	30 6 -.-	60.9 60.9 -.-	10 -.- -.-	20.8 -.- -.-
161 U of North Dakota Civil Eng [V]	ND	6.0	Y	4.0	5.0	D	O	-.-	23	-.-	4	-.-	-.-	-.-
162 North Dakota State U Civil Eng [V](1)	ND	8.0	Y	4.0	4.5	D	O	45.0	56	95.0	2	100.0	0	95.0
163 Northeastern U Civil Eng [V]	MA	17.0	Y	5.0(1)	5.0	B	O	95.0	78	90.0	42	22.9	1	90.9
164 Northern Arizona U Civil Eng [V](1)	AZ	6.5	Y	4.0	5.0	D	O	5.0	8	95.0	-.-	-.-(B)	-.-	-.-(B)
166 Northwestern U Civil Eng [V]	IL	23.8	Y	4.0	4.0	D	O	0.0	12	100.0	22	90.4	16	100.0
167 Norwich U Civil Eng [V]	VT	4.0	Y	4.0	4.0	D	O	-.-	23	100.0	-.-	-.-	-.-	-.-
168 U of Notre Dame Civil Eng [V]	IN	13.0	Y	4.0	4.0	D	N	-.-	21	100.0	4	100.0	2	100.0
170 Ohio U Civil Eng [V]	OH	8.0	Y	4.0	-.-	--	O	17.0	18	-.-	5	-.-	-.-	-.-
171 Ohio Northern U Civil Eng [V]	OH	5.0	Y	4.0	4.3	D	N	-.-	17	100.0	-.-	-.-	-.-	-.-
172 Ohio State U Civil Eng [V]	OH	23.0	Y	4.0	4.5	D	O	-.-	61	89.0	29	90.0	6	97.7
173 U of Oklahoma Civil Eng [V] Env'l Science [V]	OK	15.0 5.0	Y N	4.0 4.0	4.7 4.7	B B	O O	-.-(A) -.-(A)	21 2	80.5 85.1	24 9	48.9 53.2	3 3	44.4 0.0
175 Oklahoma State U Civil Eng [V]	OK	18.0	Y	4.0	4.7	D	O	0.0	17	85.0	34	90.0	5	100.0
176 Old Dominion U Civil Eng [V]	VA	10.0(1)	Y	4.0	4.5	D	O	-.-(A)	34	88.9	18	27.8	4	69.2
177 Oregon State U Civil Eng [V] Construction Eng Mgmt [V]	OR	22.0 4.0	Y N	4.0 4.0	4.8(C) 4.8(C)	D D	N N	-.- -.-	64 34	100.0 100.0	31 -.-(4)	100.0 -.-	4 -.-(4)	100.0 -.-

Civil Engineering

ENGINEERING Institution Profile Reference Number & Name Name of Degree Program	State/Province	FACULTY Full-time equivalent (FTE)	UNDERGRADUATE PROGRAMS							GRADUATE PROGRAMS			
			ABET/CEAB accred?	Length		Time	Co-op		Bachelor's		Master's		Doctoral	
				Nominal length of program in years	Average length of program in years	Day/Eve./Both	None/Opt./Req.	% of graduates in Co-op programs	# of degrees awarded	% of graduates who were full-time	# of degrees awarded	% of graduates who were full-time	# of degrees awarded	% of graduates who were full-time
178 U of the Pacific Civil Eng [V]	CA	5.0	Y	5.0	5.0	D	R	100.0	18	100.0	-.-(2)	-.-(2)	-.-(2)	-.-(2)
180 Penn State U Civil Eng [V]	PA	28.0	Y	4.0	-.-	D	O	11.0	173	-.-	29	-.-	5	-.-
181 U of Pittsburgh Civil Eng [V]	PA	17.2	Y	4.0	4.0	D	O	12.5	48	95.5	21	33.3	5	66.7
182 Poly U Civil & Env'l Eng [V] Civil Eng [V]	NY	-.- 23.0	Y Y	4.0 4.0	-.-(1) -.-(1)	--(B) D	O O	-.-(B) -.-	34 34	82.0 82.0	28(2) 28(2)	20.0 20.0	4 4	25.0 25.0
183 U of Portland Civil Eng [V]	OR	4.7	Y	4.0	4.5	D	O	-.-(2)	3	-.-(1)	-.-	-.-(3)	-.-(B)	-.-(B)
184 Portland State U Civil Eng [V]	OR	8.0	Y	4.0	4.7	B	O	-.-(A)	30	72.0	5	44.2	1	100.0
185 Prairie View A&M U Civil Eng [V]	TX	5.0	Y	4.0	4.0	D	O	10.0(C)	12	90.0	0	1.0	-.-	-.-
186 Princeton U Civil Eng & Operations Research [V]	NJ	23.0	Y	4.0	4.0	--	N	-.-	31	100.0	1	100.0	7	100.0
187 U of Puerto Rico Civil Eng [V] Land Surveying & Topography [V]	PR	28.0 28.0	Y N	5.0 4.0	5.0 4.0	B --(B)	O O	7.0 0.0	106 12	96.0 100.0	4 -.-(B)	100.0 -.-(B)	-.-(B) -.-(B)	-.-(B) -.-(B)
188 Purdue U Civil Eng [V]	IN	58.0	Y	4.0	4.3(1)	D	O	18.0	129	98.0	68	100.0	18	100.0
189 Rensselaer Poly Inst Civil Eng [V]	NY	12.0	Y	4.0	4.0	D	O	2.6	38	100.0	8	-.-(A)	4	-.-(A)
190 U of Rhode Island Civil Eng [V]	RI	12.5	Y	4.0	4.5	D	N	-.-	42	90.0	7	40.0	3	40.0
191 Rice U Civil Eng [V]	TX	7.4	Y	4.0	4.1	--(B)	N	-.-(B)	13	100.0	5	100.0	2	100.0
194 Rose-Hulman Civil Eng [V]	IN	5.0	Y	4.0	4.0	D	N	-.-	19	100.0	-.-	-.-(A)	-.-(B)	-.-(B)
195 Rutgers, The State U Civil Eng [V]	NJ	10.0	Y	4.0	4.5	D	N	-.-	53	98.0	20	34.0	0	36.0
200 San Diego State U Civil Eng [V]	CA	14.2	Y	4.0	5.5	B	O	-.-(A)	72	82.0	10	20.0	-.-	-.-(B)
201 San Francisco State U Civil Eng [V]	CA	-.-	Y	4.0	-.-	B	O	-.-	28	-.-	-.-	-.-	-.-	-.-
202 San Jose State U Civil Eng [V]	CA	13.2	Y	4.0	-.-(A)	D	--(A)	-.-	49	68.0	27	6.8	-.-(A)	-.-(A)
203 Santa Clara U Civil Eng [V]	CA	6.0	Y	4.0	4.0	D	O	-.-	25	99.0	2(1)	0.0	-.-(B)	-.-(B)
204 Seattle U Civil & Env'l Eng [V]	WA	4.0	Y	4.0	4.0	D	N	-.-(B)	14	-.-(A)	-.-	-.-	-.-	-.-
205 U of South Alabama Civil Eng [V]	AL	-.-	Y	4.0	-.-	--	O	-.-	7	78.0	-.-(B)	-.-(B)	-.-(B)	-.-(B)
206 U of South Carolina Civil Eng [V]	SC	13.0	Y	4.0	4.0	D	O	-.-(A)	15	85.8	16	86.6	2	56.3
207 South Dakota School of Mines Civil Eng [V]	SD	14.2	Y	4.0	4.5	D	O	-.-(A)	18	-.-(A)	26	-.-(A)	-.-(B)	-.-(B)
208 South Dakota State U Civil Eng [V]	SD	11.0	Y	4.0	5.0	D	O	2.0	40	100.0	12	100.0	-.-	-.-
209 U of South Florida Civil Eng [V]	FL	19.0	Y	4.0	5.5	B	O	25.0	47	60.5	25	40.0	1	33.0
210 Southern U & A&M Coll Civil & Env'l Eng [V] Civil Eng [V]	LA	6.0 6.0	Y Y	4.0 -.-	5.2 -.-	D --	O --	25.0 -.-	12 12	12.4 12.4	-.-(A) -.-(A)	-.-(A) -.-(A)	-.-(A) -.-(A)	-.-(A) -.-(A)
211 U of Southern Cal Civil Eng [V]	CA	17.0	Y	4.0	4.0	D	O	-.-(A)	15	100.0	18	50.0	9	75.0

Civil Engineering

ENGINEERING Institution Profile Reference Number & Name Name of Degree Program	State/Province	FACULTY Full-time equivalent (FTE)	ABET/CEAB accred?	UNDERGRADUATE PROGRAMS						GRADUATE PROGRAMS			
				Length		Time	Co-op		Bachelor's	Master's		Doctoral	
				Nominal length of program in years	Average length of program in years	Day/Eve./Both	None/Opt./Req.	% of graduates in Co-op programs	# of degrees awarded	% of graduates who were full-time	# of degrees awarded	% of graduates who were full-time	# of degrees awarded	% of graduates who were full-time
213 Southern Illinois-Carbondale Civil Eng [V]	IL	11.5	Y	4.0	4.5	D	O	4.0	27	93.0	9	76.0	-.-(B)	-.-(B)
214 Southern Illinois-Edwardsville Civil Eng [V]	IL	9.0	Y	4.0	-.-(A)	B	O	-.-(A)	31	67.0	6	16.0	-.-(B)	-.-(B)
216 Southern Methodist U Civil Eng [V](1)	TX	5.0	Y	4.0	4.5	D	O	20.0	10	92.3	2	0.0	-.-	-.-
217 Stanford U Civil Eng [V]	CA	26.0	Y	4.0	4.0	D	N	-.-	19	100.0	102	75.0	16	100.0
219 SUNY-Buffalo Civil Eng [V]	NY	28.0	Y	4.0	4.5	B	N	-.-	65	88.6	34	67.6	9	36.1
225 Syracuse U Civil Eng [V]	NY	6.0	Y	4.0	4.2	D	O	7.0	15	92.0	6(2)	-.-(A)	-.-	-.-(A)
226 Technical U of Nova Scotia Civil Eng [V]	NS	10.1	Y	4.0(2)	5.0	D	O	0.0(3)	43	100.0	6	83.0	1	100.0
227 Temple U Civil Eng [V]	PA	13.0(C)	Y	4.0	4.0	-.-(B)	N	-.-(B)	17	84.0	-.-(A)	-.-(A)	-.-(B)	-.-(B)
228 U of Tennessee, Chattanooga Civil Eng [V](2)	TN	-.-(7)	Y	4.0	4.8	B	O	-.-(A)	2	-.-(A)	2	-.-(A)	-.-	-.-
229 U of Tennessee, Knoxville Civil Eng [V]	TN	19.3	Y	4.0	-.-(A)	D	O	31.4	35	86.7	18	-.-(A)	1	-.-(A)
230 Tennessee State U Civil Eng [V]	TN	6.0	Y	4.3	5.0	B	O	5.2	12	76.8	5(1)	18.8	-.-(B)	-.-(B)
231 Tennessee Tech Civil Eng [V]	TN	156.0	Y	4.0	4.0	D	O	10.0	40	100.0	0	100.0	1	100.0
232 U of Texas at Arlington Civil Eng [V]	TX	-.-	Y	4.0	-.-	B	O	-.-	20	-.-	28	-.-	-.-	-.-
233 U of Texas at Austin Civil Eng [V]	TX	50.6	Y	4.0	4.7	D	O(1)	0.0	59	91.1	87	85.0	24	82.9
235 U of Texas at El Paso Civil Eng [V]	TX	10.3	Y	4.0	-.-	D	O	20.0	28	-.-	15	-.-	-.-	-.-
236 U of Texas at San Antonio Civil Eng [V]	TX	7.2	Y	4.0	-.-	B	O	-.-	20	-.-	5	-.-	-.-	-.-
237 Texas A&I U Civil Eng [V]	TX	-.-	Y	4.0	-.-	D	N	-.-	18	-.-	1	-.-	-.-	-.-
238 Texas A&M U Civil Eng [V]	TX	71.0	Y	4.0	4.5(C)	D	O	14.0	180	94.0	69	77.0(C)	11	77.0(C)
239 Texas Tech U Civil Eng [V]	TX	18.8	Y	4.0	-.-	D	N	-.-	41	91.9	15	83.3	0	91.7
240 U of Toledo Civil Eng [V]	OH	14.0	Y	4.0	4.0	D	O	-.-(A)	25	85.0	9	55.0	-.-	-.-
241 Tri-State U Civil Eng [V]	IN	4.7	Y	4.0	4.5	D	O	20.0	12	100.0	-.-(3)	-.-(3)	-.-(3)	-.-(3)
243 Tufts U Art & Architectural History [V] Civil Eng [V] Env'l Studies & Env'l Health [V]	MA	5.0 16.0 6.0	N Y N	4.0 4.0 4.0	4.0 4.0 4.0	D B D	N N N	-.-(B) -.-(B) -.-(B)	1 37 1	100.0 100.0 100.0	-.-(B) 46 -.-(B)	-.-(B) 30.0 -.-(B)	-.-(B) 1 -.-(B)	-.-(B) 50.0 -.-(B)
244 Tulane U Civil & Env'l Eng [V]	LA	8.0	Y	4.0	4.0	B	N	-.-	19	100.0	2	53.0	4	66.0
247 Union Coll Civil Eng [V]	NY	5.5	Y	4.0	-.-(A)	B	N	-.-(B)	35	90.0	-.-(B)	-.-(B)	-.-(B)	-.-(B)
248 US Air Force Academy Civil Eng [V]	CO	20.5	Y	4.0	4.0	D	N	-.-	70	100.0	-.-(B)	-.-(B)	-.-(B)	-.-(B)
249 US Coast Guard Academy Civil Eng [V]	CT	8.0	Y	4.0	4.0	D	N	-.-	40	100.0	-.-	-.-	-.-	-.-
251 US Military Academy Civil Eng [V]	NY	-.-	Y	4.0	4.0	D	N	-.-	59	100.0	-.-(B)	-.-(B)	-.-(B)	-.-(B)

Civil Engineering

ENGINEERING Institution Profile Reference Number & Name Name of Degree Program	State/Province	FACULTY Full-time equivalent (FTE)	ABET/CEAB accred?	UNDERGRADUATE PROGRAMS						GRADUATE PROGRAMS			
				Length Nominal length of program in years	Length Average length of program in years	Time Day/Eve./Both	Co-op None/Opt./Req.	Co-op % of graduates in Co-op programs	Bachelor's # of degrees awarded	Bachelor's % of graduates who were full-time	Master's # of degrees awarded	Master's % of graduates who were full-time	Doctoral # of degrees awarded	Doctoral % of graduates who were full-time
253 U of Utah Civil Eng [V]	UT	9.0	Y	4.0	6.0	D	O	-.-(A)	29	75.0	7	43.0	1	64.0
254 Utah State U Civil & Env'l Eng [V]	UT	4.1	Y(1)	4.6	4.6	D	O	-.-(A)	6	90.0	20	52.4	7	36.8
255 Valparaiso U Civil Eng [V]	IN	5.0	Y	4.0	4.0	D	O	-.-	21	99.0	--.	-.-	--.	-.-
256 Vanderbilt U Civil Eng [V]	TN	15.0	Y	4.0	4.3	D	N	-.-	33	100.0	17(2)	85.0	1	90.0
257 U of Vermont Civil Eng [V]	VT	8.5	Y	4.0	4.0	D	O	0.0	32	93.0	3	57.1	0	67.0
258 Villanova U Civil Eng [V]	PA	12.0	Y	4.0	4.0	--	--	-.-	45	100.0	29	18.0	--.	-.-
259 U of Virginia Appl Mechanics [V](2) Civil Eng [V]	VA	8.8(1) 8.8(1)	--(2) Y	-.- 4.0	-.- 4.2	-- D	-- N	-.- -.-	--. 40	-.- 100.0	2 14	100.0 92.7	--(2) 3	-.- 84.6
260 Virginia Military Inst Civil & Env'l Eng [V]	VA	9.0	Y	4.0	-.-	D	N	-.-	175	100.0	--.	-.-	--.	-.-
261 Virginia Tech Civil Eng [V]	VA	38.0	Y	4.0	4.0	D	O	16.0	177	100.0	91	100.0	9	100.0
262 U of Washington Civil Eng [V]	WA	37.0	Y	4.0	-.-(A)	D	O	-.-(A)	117	93.0	79	-.-(A)	13	-.-(A)
263 Washington U Civil Eng [V]	MO	9.0	Y	4.0	4.0	D	O	-.-	21	-.-	3	-.-	2	-.-
264 Washington State U Civil & Env'l Eng [V](1) Civil Eng [V](1)	WA	24.0 24.0	Y Y	4.0 4.0	4.5 4.5	D D	O O	-.-(A) -.-(A)	53 53	100.0 100.0	8 8	100.0 100.0	3 3	100.0 100.0
265 Wayne State U Civil & Env'l Eng [V]	MI	8.0	Y	4.0	4.5	B	O	-.-(A)	13	-.-(A)	34	-.-(A)	1	-.-(A)
267 West Virginia U Civil Eng [V]	WV	19.0	Y	4.0	4.0	D	O	0.0	32	100.0	14	18.0	3	30.0
268 West Virginia Inst of Tech Civil Eng [V]	WV	6.0	Y	4.0	4.5	D	O	25.0	22	90.0	--.	-.-	--.	-.-
272 Widener U Civil Eng [V]	PA	5.5	Y	4.3	4.0	D	O	21.0	19	-.-(A)	1	-.-(A)	--(B)	-.-(B)
273 Wilkes U Pre-Civil Eng [V](6)	PA	-.-(A)	N	-.-(B)	-.-(B)	B	O	-.-(A)	--(B)	-.-(B)	--(B)	-.-(B)	--(B)	-.-(B)
274 U of Wisconsin-Madison Civil Eng [V]	WI	31.0	Y	4.0	5.0	D	O	-.-(A)	47	-.-(A)	33	-.-(A)	8	-.-(A)
275 U of Wisconsin-Milwaukee Civil Eng [V]	WI	14.3	Y	4.0	5.5	B	O	7.1	42	67.7	13	28.6	3	58.3
276 U of Wisconsin-Platteville Civil Eng [V]	WI	9.7	Y	4.0	4.7	D	O	15.0	47	100.0	--.	-.-	--.	-.-
277 Worcester Poly Inst Civil Eng [V]	MA	11.6	Y	4.0	4.0	D	O	-.-(A)	69	98.5	13	-.-(A)	--.	-.-(A)
279 U of Wyoming Civil & Architectural Eng [V]	WY	24.0	Y	4.0	5.0	D	O	0.0	54	95.0	11	90.0	--.	-.-

Civil Engineering—Totals

State/Province/Country	# of schools	Full-time equivalent (FTE)	# accred. programs	Nominal length of program in years	Average length of program in years	Day/Eve./Both	None/Opt./Req.	% of graduates in Co-op programs	# of degrees awarded (Bachelor's)	% of graduates who were full-time	# of degrees awarded (Master's)	% of graduates who were full-time	# of degrees awarded (Doctoral)	% of graduates who were full-time
AK	1	10.50	1						13		4		-	
AL	5	41.80	6						152		35		9	
AR	1	11.00	1						25		4		1	
AZ	3	50.00	3						86		42		21	
CA	20	322.90	19						960		433		84	
CO	4	99.90	4						206		104		27	
CT	3	47.00	3						106		32		1	
DC	4	39.50	4						50		41		3	
DE	1	18.00	1						34		30		7	
FL	9	115.80	9						276		118		8	
GA	1	44.00	2						123		73		5	
HI	1	19.00	1						49		9		-	
IA	2	15.40	3						118		63		11	
ID	1	13.30	1						24		14		2	
IL	7	134.40	7						256		146		45	
IN	6	89.70	5						206		72		20	
KS	2	38.00	2						66		47		6	
KY	2	32.30	2						84		28		2	
LA	5	59.80	6						116		16		9	
MA	7	139.60	7						407		213		39	
MD	3	44.70	3						81		62		8	
ME	1	12.50	1						41		16		2	
MI	5	75.20	4						290		126		32	
MN	1	31.00	1						113		27		9	
MO	4	45.40	4						148		41		4	
MS	2	19.00	2						47		14		1	
MT	1	19.40	1						32		2		-	
NC	4	142.70	5						215		58		14	
ND	2	14.00	2						79		6		-	
NE	1	18.00	1						44		36		8	
NH	2	14.50	2						61		14		1	
NJ	3	50.90	3						130		67		9	
NM	2	28.80	2						61		38		3	
NV	1	9.00	1						19		3		-	
NY	11	161.60	12						527		245		38	
OH	9	113.00	9						246		98		19	
OK	2	38.00	2						40		67		11	
OR	3	38.70	3						131		36		5	
PA	11	142.20	10						535		136		28	
PR	1	56.00	1						118		4		-	
RI	1	12.50	1						42		7		3	
SC	3	31.00	3						125		38		6	
SD	2	25.20	2						58		38		-	
TN	7	215.30	7						156		51		4	
TX	12	195.10	13						420		249		44	
UT	3	30.00	3						91		49		11	
VA	4	74.59	4						426		125		16	
VT	2	12.50	2						55		3		-	
WA	4	95.00	5						249		95		19	
WI	4	66.00	4						182		61		14	
WV	2	25.00	2						54		14		3	
WY	1	24.00	1						54		11		-	
U.S. Totals:	199	3,222.70	203						8,227		3,361		612	
AB	1	34.50	1						53		38		10	
MB	1	-	1						48		14		8	
NS	1	10.10	1						43		6		1	
PQ	3	46.00	3						164		58		13	
Canada Totals:	6	90.59	6						308		116		32	

Civil Engineering—Footnotes

FOOTNOTES: The following footnotes are the same for all schools: (A) Data not available. (B) Data not applicable. (C) Estimated data.

012 Auburn University: (10) Based on credit hours taken Fall Quarter 1991.

015 Bradley University: (1) Students typically take the equivalent of nine semesters but finish in 4 years by using summer school and transfer credits.

018 Bucknell University: (1) Data is given in whole numbers.

020 University of California, Davis: (6) Included in Civil Engineering

029 California State University, Fresno: (1) Doctoral program not available (2) American Council for Construction Education (ACCE) (3) Primarily

038 The Catholic University of America: (1) The Civil Engineering program includes a concentration in Construction. (2) Student in Construction concentration must participate in a summer internship program arranged by the department.

042 University of Cincinnati: (1) Of the 6 departments., 5 have more than one program, therefore, faculty provide support in multiple programs. Total FTE for the college is 142.

044 City College of the City University of New York: (1) Most prerequisite courses can be taken during the evening. All major courses must be taken during the day.

048 Cleveland State University: (1) includes Co-Op and part-time students

051 University of Colorado at Denver: (3) Coursework available at CU-Denver but Ph.D. is awarded at CU-Boulder through system-wide Graduate School.

056 University of Connecticut: (2) Includes Environmental Engineering students.

061 University of Delaware: (3) sustaining considered as full-time

071 University of Florida: (1) Civil Engineering and Surveying and Mapping faculty are combined.

072 Florida Agricultural and Mechanical University: (1) Includes FAMU and FSU students. Year is Fall 1991, Spring 1992 and Summer 1992.

073 Florida Atlantic University: (1) Master's program only. (2) No undergraduate programs in civil or manufacturing systems engineering.

076 Florida State University: (1) Includes FAMU and FSU students. Year is Fall 1992, Spring 1992, and Summer 1992.

082 Georgia Institute of Technology: (1) No doctoral program in Environmental Engineering is offered. (2) No undergraduate degree in Environmental Engineering is offered. (3) Includes specialties in Environmental Hydraulics and Water Resources. (4) Includes: Structural Engineering & Mechanics, Construction Management, Geotechnical Engineering, Materials, Transportation, Engineering Computer Graphics. (8) Fall '91 Quarter FTE.

090 University of Houston: (1) BSCE, MCE, MSCE, MSEnvE, PhD

096 Illinois Institute of Technology: (2) Reflects percentage of graduates who received certificates of completion in compliance with ABET requirements.

100 The Johns Hopkins University: (2) Includes 1.70 FTE in part-time programs.

103 University of Kentucky: (1) Breakdown of part-time students by degree is not available. Percentage represents total M.S. and Ph.D. enrollment.

104 Lafayette College: (1) Undergraduate program only.

106 Lawrence Technological University: (1) Construction Engineering program is ABET accredited. The results of an ABET team evaluation of the Civil Engineering program are awaited. (2) Of this total 26 are Civil Engineering graduates.

111 University of Louisville: (1) Program is accredited at the master's level.

121 University of Massachusetts: (1) Includes 6 M.S. degrees awarded in Environmental Engineering (4) Includes department head

125 McGill University: (1) Following completion of 2 year CEGEP program. International students should add 1 year for pre-engineering studies (2) Internship year program available for students who have less then 45 credits remaining in their academic program (3) Part time figures include candidates who have completed their residence requirements but are working full time on research and thesis requirements

130 University of Michigan: (1) BS and/or BSE program only. (2) Graduate program only. (3) Faculty included in principal department. (4) Includes Professional degrees. (5) All graduate students. (7) For whole College.

136 University of Minnesota: (2) Department is Civil and Mineral Engineering. Statistics have been split between Civil and Extractive Metallurgical Eng.

137 University of Mississippi: (2) M.S. and Ph.D. degrees in Engineering Science are offered with emphasis in Civil Engineering.

139 University of Missouri-Columbia: (1) Includes statistics from CE/CEP which is located in Kansas City and administered by the Columbia campus.

140 University of Missouri-Kansas City: (2) Most students participate in unofficial co-op programs with local firms.

144 Montana State University: (1) Information unavailable but nearly 100%

152 The University of New Mexico: (1) We classify all our graduate students as full-time.

160 North Carolina State University: (1) The BS in environmental engineering was approved during the 1992 fall semester. As a result, no students are officially enrolled.

162 North Dakota State University: (1) Masters & Ph.D. degree also available.

163 Northeastern University: (1) 5 year co-op option; 4 year co-op option; 4-year non-co-op option

164 Northern Arizona University: (1) Includes an emphasis in environmental engineering.

176 Old Dominion University: (1) 1991/92 Head count

177 Oregon State University: (4) No graduate degrees offered

178 University of the Pacific: (2) Civil Engineering does not offer a master's or doctoral degree.

182 Polytechnic University: (1) 4-5 (5) Includes Environmental

183 University of Portland: (1) Data not available by program. Overall undergraduate full-time is 89%. (2) Each year approximately 10 Engineering students participate in co-op programs. The numbers by degree program fluctuate. (3) Data not available by program. Overall graduate full-time is 42%.

188 Purdue University: (1) Except for co-op students who require 5.0 years to complete the program.

203 Santa Clara University: (1) Degree awarded in Engineering Mechanics

216 Southern Methodist University: (1) This degree is not available to newly entering students.

225 Syracuse University: (2) Includes Mechanical/Manufacturing Option

226 Technical University of Nova Scotia: (2) Three years at TUNS. (3) Just started.

228 University of Tennessee, Chattanooga: (2) BSE with Civil Concentration. (7) All faculty are included under Engineering.

230 Tennessee State University: (1) An option in Master of Engineering degree program.

233 University of Texas at Austin: (1) Participants in the Co-op Program are away from the campus for three semesters, working for pay and letter grade.

241 Tri-State University: (3) No graduate programs

254 Utah State University: (1) Currently seeking ABET accreditation to establish Aerospace, Environmental and Civil Engineering as separately accredited degree programs.

256 Vanderbilt University: (2) Includes EWRE Graduate Degrees.

259 University of Virginia: (1) Department of Civil Engineering and Applied Mechanics. (2) Master's degree program.

264 Washington State University: (1) Pullman campus

273 Wilkes University: (6) Prepares the student for direct transfer to the third year in another institution.

Electrical Engineering

Institutions are listed in alphabetical order. The first line in a school's entry contains the institution profile reference number, the school name, and the state or province in which the main campus is located. Subsequent indented lines for each program give the name of the degree program, followed by a major category abbreviation in brackets (H = Chemical, V = Civil, E = Electrical, I = Industrial, M = Mechanical, and O = Other) the full-time equivalent (FTE) of all faculty. For undergraduate programs, the columns represent: accreditation status, the nominal and actual average length of the program, the time that courses are available (Day, Evening or Both), the availability of cooperative (internship) programs (None, Optional, or Required), the percentage of graduates who participated in co-op programs, the number of bachelor's degrees awarded, and the percentage of these graduates who were full-time. For graduate programs, the columns indicate the number of graduates and the percentage who were full-time in master's and then doctoral programs. Unless otherwise footnoted, information on degrees is for the 1991-92 academic year; all other information is current. At the end of the table are (1) a summary of totals by state or province and by country and (2) the full text of all footnotes for this table.

ENGINEERING Institution Profile Reference Number & Name Name of Degree Program	State/Province	FACULTY Full-time equivalent (FTE)	ABET/CEAB accred?	UNDERGRADUATE PROGRAMS							GRADUATE PROGRAMS			
				Length		Time	Co-op		Bachelor's		Master's		Doctoral	
				Nominal length of program in years	Average length of program in years	Day/Eve./Both	None/Opt./Req.	% of graduates in Co-op programs	# of degrees awarded	% of graduates who were full-time	# of degrees awarded	% of graduates who were full-time	# of degrees awarded	% of graduates who were full-time
001 U of Akron Electrical Eng [E]	OH	18.0	Y	4.0	-.-(A)	B	O	57.0	65	74.8(A)	20	57.0	5	90.0
002 U of Alabama Electrical Eng [E][1]	AL	15.0	Y	4.0	4.5	D	O	18.1	83	100.0	18	100.0	1	100.0
003 U of Alabama at Birmingham Electrical & Computer Eng [E]	AL	12.0	Y	4.0	-.-(A)	B	O	-.-(A)	29	48.0	15	30.0	-.-(B)	-.-(B)
004 U of Alabama at Huntsville Electrical & Computer Eng [E]	AL	26.9	Y	4.0	-.-(A)	B	O	-.-(A)	116	53.7	49	36.8	5	57.9
005 U of Alaska, Fairbanks Electrical Eng [E]	AK	8.0	Y	4.0	4.5	D	N	-.-(A)	13	70.0	4	90.0	-.-	-.-
006 U of Alberta Electrical Eng [E]	AB	36.0[3]	Y	4.0	-.-(A)	D	O	33.0	105	100.0	33	86.4	5	91.0
007 Alfred U Electrical Eng [E]	NY	5.0	Y	4.5	4.5	B	O	90.0	8	100.0	-.-(B)	-.-(B)	-.-(B)	-.-(B)
008 U of Arizona Electrical Eng [E]	AZ	48.0	Y	4.0	-.-	D	O	-.-	120	83.0	61	-.-	11	-.-
009 Arizona State U Electrical Eng [E]	AZ	45.0	Y	4.0	5.0	D	O	-.-(A)	142	72.0	101	48.0	13	48.0
010 U of Arkansas Electrical Eng [E]	AR	17.0	Y	4.0	-.-	D	O	-.-	63	90.2	11	68.1	4	-.-
012 Auburn U Computer Eng [E] Computer Science [E] Computer Science & Eng [E] Electrical Eng [E]	AL	-.-[2] -.-[2] 13.5 28.5	Y -.-[3] -.-(B) Y	4.0 4.0 -.-(B) 4.0	-.-(A) -.-(A) -.-(B) -.-(A)	D D -.-(B) D	O O -.-(B) O	17.0 28.6 -.-(B) 36.8	47 14 -.-(B) 182	79.2[10] 83.6[10] -.-(B) 84.1[10]	-.-(B) -.-(B) 11 24	-.-(B) -.-(B) 18.6[10] 34.4[10]	-.-(B) -.-(B) 0 4	-.-(B) -.-(B) 36.4[10] 20.7[10]
014 Boston U Electrical Eng [E][2]	MA	33.0[4]	Y	4.0	4.0	D	O	20.0	68	99.0[1]	44	10.0	3[2]	89.0
015 Bradley U Electrical Eng [E][2]	IL	21.3	Y	4.0	4.0[1]	B	O	7.0	42	90.4	13	14.9	-.-(B)	-.-(B)
016 U of Bridgeport Electrical Eng [E]	CT	6.0	Y	4.0	4.0	B	O	5.0	58	77.0	12	41.7	-.-	-.-
017 Brigham Young U Electrical & Computer Eng [E]	UT	22.0	Y	4.0	5.0	B	O	-.-(A)	117	-.-(A)	24	-.-(A)	1	-.-(A)
018 Bucknell U Electrical Eng [E]	PA	7.0[1]	Y	4.0	4.0	-.-(A)	N(A)	-.-(A)	34	100.0	1	100.0	-.-(A)	-.-(A)
019 U of Cal at Berkeley Electrical Eng & Computer Science [E][1]	CA	78.5	Y	4.0	-.-(A)	D	O	25.0	203	100.0	71	100.0	65	100.0
020 U of Cal, Davis Electrical & Computer Eng [E] Electrical Eng & Materials Science [E]	CA	36.6 43.6	Y N	4.0 4.0	-.-(A) -.-(A)	D D	O O	-.-(A) -.-(A)	91 4	100.0 100.0	30 -.-[7]	100.0 -.-[7]	5 -.-[7]	100.0 -.-[7]
021 U of Cal, Irvine Electrical & Computer Eng [E]	CA	23.0	Y	4.0	4.5	D	N	-.-(B)	67	100.0	42	77.0	13	100.0

Electrical Engineering

ENGINEERING Institution Profile Reference Number & Name Name of Degree Program	State/Province	FACULTY Full-time equivalent (FTE)	ABET/CEAB accred?	UNDERGRADUATE PROGRAMS							GRADUATE PROGRAMS			
				Length		Time	Co-op		Bachelor's		Master's		Doctoral	
				Nominal length of program in years	Average length of program in years	Day/Eve./Both	None/Opt./Req.	% of graduates in Co-op programs	# of degrees awarded	% of graduates who were full-time	# of degrees awarded	% of graduates who were full-time	# of degrees awarded	% of graduates who were full-time
022 U of Cal, Los Angeles Electrical Eng [E]	CA	35.0	Y	4.0	4.6	--	O	-.-	126	100.0	74	100.0	30	100.0
023 U of Cal, San Diego Appl Physics [E] Electrical Eng [E]	CA	35.3(2) 35.3(2)	N Y	4.0 4.0	4.7 4.7	D D	N N	-.- -.-	1 75	100.0 100.0	22 39	100.0 100.0	9 15	100.0 100.0
024 U of Cal, Santa Barbara Electrical & Computer Eng [E](6)	CA	31.4	Y	4.0	4.2	--(B)	N	-.-(B)	74	100.0	58	95.0(1)	20	95.0(1)
026 Cal Inst of Tech Electrical Eng [E]	CA	15.0	N	4.0	-.-	D	-.-(B)	-.-	29	100.0	32	100.0	9	100.0
027 Cal Poly State U Electrical & Electronics Eng [E]	CA	26.1(2)	Y	4.0	4.8	D	O	-.-	32	-.-	7	-.-	-.-	-.-
028 Cal State U, Chico Electrical & Electronics Eng [E]	CA	10.3	Y	4.0	-.-(A)	D	O	-.-(A)	35	88.0	3	13.0	--(B)	-.-(B)
029 Cal State U, Fresno Electrical Eng [E]	CA	10.0	Y	4.0	4.5	D	O	4.0	76	4.0	-.-	-.-	-.-	-.-
030 Cal State U, Long Beach Electrical Eng [E]	CA	25.1	Y	4.0	5.5	B	N	-.-	90	55.0	21	10.0	-.-	-.-
031 Cal State U, Los Angeles Electrical Eng [E]	CA	17.3	Y	4.0	4.5	B	O	-.-(A)	87	62.0	26	22.0	--(B)	-.-(B)
033 Cal State U, Sacramento Electrical & Electronics Eng [E]	CA	15.0	Y	4.0	5.0	B	O	3.3	90	-.-(A)	-.-(B)	-.-(B)	-.-(B)	-.-(B)
034 Cal State Poly, Pomona Electrical & Computer Eng [E] Electrical Eng [E]	CA	38.5 -.-	Y Y	4.0 4.0	5.5 -.-	-- B	-- O	-.- -.-	146 -.-	76.0 -.-	-.- -.-	-.- -.-	-.- -.-	-.- -.-
036 Carnegie Mellon U Electrical & Computer Eng [E]	PA	34.5	Y	4.0	4.0	D	N	-.-(B)	98	98.6	50	97.2	31	97.5
037 Case Western Reserve U Electrical Eng & Appl Physics [E]	OH	12.0	Y	4.0	4.3	B	O	30.0	36	80.0	17	50.0	5	100.0
038 The Catholic U of America Computer Science/Eng [E] Electrical Eng [E]	DC	-.- 9.0	N Y	4.0 4.0	-.- 4.0	D D	O O	-.- 10.0	-.- 8	-.- 90.0	-.- 14	-.- 10.0	-.- 2	-.- 0.0
039 U of Central Florida Electrical Eng [E]	FL	34.0	Y	4.0	4.5(C)	B	O	15.0(C)	152	63.0	43	32.0	2	26.0
041 Christian Brothers U Electrical Eng [E]	TN	6.0	Y	4.0	4.7	D	N	-.-(B)	23	86.8	--(B)	-.-(B)	--(B)	-.-(B)
042 U of Cincinnati Electrical Eng [E]	OH	30.0(1)	Y	5.0	5.2	D	R	100.0	50	100.0	24	59.3	5	87.2
043 The Citadel Electrical Eng [E]	SC	-.-	Y	4.0	4.0	B	N	-.-	18	30.0	-.-	-.-	-.-	-.-
044 CUNY-City Coll Electrical Eng [E]	NY	31.5	Y	4.0	5.0	B	O	-.-(A)	134	74.9	33	24.4	6	100.0
046 Clarkson U Electrical & Computer Eng [E]	NY	22.0	Y	4.0	-.-	D	O	-.-(A)	144	99.9	15	99.9	2	99.9
047 Clemson U Electrical & Computer Eng [E]	SC	44.0	Y	4.0	-.-	D	O	-.-	143	87.0	50	82.0	9	90.0
048 Cleveland State U Electrical Eng [E](2)	OH	16.5	Y	4.0	5.2(1)	B	O	44.6	48	66.4	24	52.2	2	100.0
049 U of Colorado at Boulder Electrical & Computer Eng [E]	CO	53.0	Y	4.0	4.5	D	O	-.-	73	97.4	73	87.6	15	87.6

Electrical Engineering

ENGINEERING Institution Profile Reference Number & Name Name of Degree Program	State/Province	FACULTY Full-time equivalent (FTE)	UNDERGRADUATE PROGRAMS							GRADUATE PROGRAMS			
			ABET/CEAB accred?	Length		Time	Co-op		Bachelor's		Master's		Doctoral	
				Nominal length of program in years	Average length of program in years	Day/Eve./Both	None/Opt./Req.	% of graduates in Co-op programs	# of degrees awarded	% of graduates who were full-time	# of degrees awarded	% of graduates who were full-time	# of degrees awarded	% of graduates who were full-time
050 U of Colorado-Col Spr Electrical & Computer Eng [E]	CO	13.8	Y	4.0	5.0	B	O	22.0	49	60.0	24	54.0	4	60.0
051 U of Colorado at Denver Electrical Eng [E]	CO	16.1	Y	4.0	4.5	B	O	5.0	69	50.0	31	20.0	--.(3)	--.(3)
053 Colorado State U Electrical Eng [E]	CO	19.0	Y	4.0	4.5	D(A)	N	--.(A)	31	90.0	18	63.0	11	79.0
054 Columbia U Electrical Eng [E](4)	NY	24.7	Y	4.0	4.0	--.(B)	N	--.(B)	47	100.0	88	63.8	15	63.8
055 Concordia U Electrical Eng [E]	PQ	-.-	Y	4.0	-.-	B	N	-.-	52	78.0	18	80.0	2	92.0
056 U of Conn Electrical Eng [E]	CT	22.0	Y	4.0	4.5	D	O	44.0	80	95.0(C)	22(3)	52.0(3)	2(3)	82.0(3)
057 The Cooper Union Electrical Eng [E]	NY	10.5	Y	4.0	4.0	D	N	-.-	33	100.0	8	100.0	--.	-.-
058 Cornell U Electrical Eng [E]	NY	42.0	Y	4.0	-.-	D	O	27.0	142	-.-	144	-.-	30	-.-
060 U of Dayton Electrical Eng [E]	OH	15.7	Y	4.0	4.0	D	O	35.0	80	89.0	37	12.0	2	46.0
061 U of Delaware Electrical Eng [E]	DE	14.6	Y	4.0	4.0	B	N	-.-	52	94.6	8	83.3	8	89.3
062 U of Denver Electrical Eng [E]	CO	6.5	Y	4.0	4.0	D	O	18.2	12	3.3	0	100.0	--.	-.-
063 U of the District of Columbia Electrical Eng [E]	DC	7.0	Y	4.0	6.0	D	O	--.(A)	25	--.(A)	--.	-.-		
064 Drexel U Electrical & Computer Eng [E]	PA	50.0	Y	5.0	5.3(C)	B	R	100.0	200	95.0	63	15.0	13	90.0
065 Duke U Electrical Eng [E]	NC	18.0	Y	4.0	4.0	--.(B)	N	-.-	66	100.0	17	100.0	7	100.0
066 École Polytechnique Electrical Eng [E]	PQ	46.0(5)	Y	4.0	4.5	D	N	-.-	160	96.7	43	87.2	8	100.0
068 Embry-Riddle-Western Campus Electrical Eng [E]	AZ	5.0	N(1)	4.5	4.5	B	O	50.0	4	100.0	--.	-.-	--.	-.-
069 U of Evansville Electrical Eng [E]	IN	7.7	Y	4.0	4.2	D	O	8.2	42	94.0(C)	--.(B)	--.(B)	--.(B)	--.(B)
071 U of Florida Electrical Eng [E]	FL	46.0	Y	4.0	4.5	D	O	11.0	99	95.0	79	100.0	26	100.0
072 Florida A&M U Electrical Eng [E]	FL	14.0	Y	4.5	5.0	D	O	--.(A)	79(1)	--.(A)	27(1)	--.(A)	--.(B)	--.(B)
073 Florida Atlantic U Electrical Eng [E]	FL	19.5	Y	4.5	5.0(5)	B	O	10.0(C)	61	62.0	12	36.5	3	68.4
074 Florida Inst of Tech Electrical & Computer Eng [E]	FL	17.6	Y	4.0	4.5	D	O	33.3	111	86.5	55	45.5	6	20.5
075 Florida International U Electrical & Computer Eng [E] Electrical Eng [E]	FL	-.- 19.5	Y Y	-.- 4.0	-.- 4.5	B B	O O	-.- 10.0	--. 84	--. 77.9	--. 18	--. 100.0	--. 0	-.- 100.0
076 Florida State U Electrical Eng [E]	FL	14.0	Y	4.5	5.0	D	O	--.(A)	79(1)	--.(A)	27(1)	--.(A)	--.(B)	--.(B)
077 GMI Electrical & Computer Eng [E]	MI	18.0	Y	5.0	5.0	E(1)	R	100.0	108	100.0	--.	-.-	--.	-.-

Electrical Engineering

ENGINEERING Institution Profile Reference Number & Name Name of Degree Program	State/Province	FACULTY Full-time equivalent (FTE)	UNDERGRADUATE PROGRAMS								GRADUATE PROGRAMS			
			ABET/CEAB accred?	Length		Time	Co-op		Bachelor's		Master's		Doctoral	
				Nominal length of program in years	Average length of program in years	Day/Eve./Both	None/Opt./Req.	% of graduates in Co-op programs	# of degrees awarded	% of graduates who were full-time	# of degrees awarded	% of graduates who were full-time	# of degrees awarded	% of graduates who were full-time
078 Gannon U Electrical Eng [E]	PA	7.0	Y	4.0	4.5	B	O	0.0	23	81.7	3	16.6	-.-	-.-
079 George Mason U Electrical Eng [E]	VA	23.3	Y	4.0	4.5	B(6)	O	-.-(A)	82	69.0	34	13.0	-.-(1)	-.-(B)
080 The George Washington U Electrical Eng [E]	DC	23.0	Y	4.0	-.-(A)	D	O	-.-(A)	47	79.0	72	20.0	7	23.0
082 Georgia Tech Computer Eng [E] Electrical Eng [E]	GA	8.0(8) 72.0(8)	Y Y	4.0 4.0	4.7 4.7	D D	O O	50.0 38.0	14 302	94.0 93.0	-.-(B) 203	-.-(B) 62.0	-.-(B) 48	-.-(B) 80.0
083 Gonzaga U Electrical Eng [E]	WA	8.0	Y	4.0	-.-	D	N	-.-	22	93.0	1	85.0	-.-	-.-
085 Grove City Coll Electrical Eng [E]	PA	6.0	Y	4.0	4.1	D	N	-.-	22	100.0	-.-(B)	-.-(B)	-.-(B)	-.-(B)
088 U of Hawaii at Manoa Electrical Eng [E]	HI	27.0	Y	4.0	5.0	D	O	-.-	62	87.8	19	74.0	4	88.0
089 Hofstra U Electrical Eng [E]	NY	5.5	Y	4.0	4.5	B	N	-.-	14	65.4	-.-	-.-	-.-	-.-
090 U of Houston Electrical Eng [E](2)	TX	28.6	Y	4.0(C)	5.0	B	O	-.-(A)	90	61.0	72	48.0	2	68.0
091 Howard U Electrical Eng [E](C)	DC	18.0	Y	4.0	5.0	D	O	3.0	56	97.0	15	90.0	4	100.0
092 U of Idaho Electrical Eng [E]	ID	16.5	Y	4.0	6.0	D	O	2.0	53	93.5	16	67.0	3	46.0
094 U of Illinois at Chicago Electrical Eng [E] Electrical Eng & Computer Science [E]	IL	29.0 51.0	Y --	4.0 -.-	-.- -.-	D --	O --	14.0 -.-	138 -.-	81.0 -.-	-.- 81	-.- 32.0	-.- 13	-.- 27.0
095 Illinois, Urbana-Champaign Electrical & Computer Eng [E]	IL	77.0	Y	4.0	4.5	-.-(A)	O	7.0	230	99.1	77	92.8	52	93.6
096 Illinois Inst of Tech Electrical & Computer Eng [E]	IL	33.6	Y	4.0	4.8	B	O	6.1(2)	116	67.1	90	16.0	2	44.0
097 Indiana-Purdue, Indianapolis Electrical Eng [E]	IN	12.0	Y	4.0	4.0	B	O	0.0	45	53.0	5	0.0	-.-	-.-
098 U of Iowa Electrical & Computer Eng [E]	IA	20.0	Y	4.0	4.5(C)	D	O	15.0(C)	64	100.0(C)	14	100.0(C)	5	100.0(C)
099 Iowa State U Computer Eng [E] Electrical Eng [E]	IA	-.- -.-	Y Y	4.0 4.0	-.- -.-	D D	O O	-.- -.-	48 167	-.- -.-	10 27	-.- -.-	1 4	-.- -.-
100 The Johns Hopkins U Electrical Eng [E]	MD	36.0(3)	Y	4.0	4.0	B	O	-.-(A)	31	100.0	260	8.0	5	100.0
101 U of Kansas Electrical & Computer Eng [E]	KS	-.-	Y	4.0	-.-	D	-.-	-.-	81	-.-(A)	22	-.-(A)	4	-.-(A)
102 Kansas State U Electrical & Computer Eng [E]	KS	21.0	Y	4.0	4.7	D	O	-.-	111	96.0	20	-.-	1	-.-
103 U of Kentucky Electrical Eng [E]	KY	21.3	Y	4.0	5.0(C)	D	O	-.-(A)	93	88.9	5	-.-(A)	4	68.0(1)
104 Lafayette Coll Electrical Eng [E](1)	PA	9.6	Y	4.0	4.0	B	N	-.-	27	78.0	-.-	-.-	-.-	-.-
105 Lamar U Electrical Eng [E]	TX	5.5	Y	4.0	5.0	D	O	74.0	30	-.-(A)	-.-(A)	-.-(A)	-.-	-.-(A)

Electrical Engineering

ENGINEERING Institution Profile Reference Number & Name Name of Degree Program	State/Province	FACULTY Full-time equivalent (FTE)	ABET/CEAB accred?	UNDERGRADUATE PROGRAMS							GRADUATE PROGRAMS			
				Length		Time	Co-op		Bachelor's		Master's		Doctoral	
				Nominal length of program in years	Average length of program in years	Day/Eve./Both	None/Opt./Req.	% of graduates in Co-op programs	# of degrees awarded	% of graduates who were full-time	# of degrees awarded	% of graduates who were full-time	# of degrees awarded	% of graduates who were full-time
106 Lawrence Tech U Electrical Eng [E]	MI	17.4	Y	4.0	4.5[C]	B	O	11.3	133	56.2	-.-	-.-	-.-	-.-
108 Lehigh U Electrical Eng & Computer Science [E][1]	PA	27.0	Y	4.0	4.0	D	O	1.0	79	100.0	38	39.0	6	66.0
109 Louisiana State U Electrical Eng [E]	LA	19.7	Y	4.0	5.0	D	O	6.4	87	85.6	32[1]	92.7	0	91.4
110 Louisiana Tech U Electrical Eng [E]	LA	11.0	Y	4.0	4.7	D	O	-.-[A]	47	-.-[A]	23	-.-[A]	-.-	-.-[B]
111 U of Louisville Electrical Eng [E]	KY	17.8	Y[1]	4.0	4.5	D	R	98.0	64	66.9	35	23.5	-.-	-.-
112 Loyola Coll in Maryland Electrical Eng [E]	MD	6.0	N	4.0	4.5	--	N	-.-	2	100.0	-.-	-.-	-.-	-.-
113 Loyola Marymount U Electrical Eng [E]	CA	6.5	Y	4.0	4.2	B	N	-.-	12	100.0	5	0.0	-.-	-.-
114 U of Maine Electrical & Computer Eng [E]	ME	14.5	Y	4.0	4.5	D	N	-.-	39	90.0	12	90.0	-.-	-.-[1]
115 U of Manitoba Computer Eng [E] Electrical Eng [E]	MB	-.- -.-	Y Y[1]	4.0 4.0	-.-[A] -.-[A]	D D	O O	-.-[A] -.-[A]	17 50	-.-[A] -.-[A]	--[1] 11[1]	-.-[B] -.-[A]	--[1] 5[1]	-.-[B] -.-[A]
116 Mankato State U Electrical Eng [E]	MN	7.0	Y	4.0	4.5	D	--	-.-	24	100.0	-.-	-.-	-.-	-.-
118 Marquette U Electrical Eng [E]	WI	21.2	Y	4.0	4.2	B	O	42.0	98	85.0	42	32.0	3	54.0
119 Maryland Baltimore Cty Electrical Eng [E]	MD	9.0	N[4]	-.-	-.-	B	--	-.-	-.-	-.-	15	55.0	-.-	87.0
120 U of Maryland, Coll Park Electrical Eng [E]	MD	65.0	Y	4.0	5.0	D	O	9.9	212	80.9	92	49.2	18	73.2
121 U of Mass Electrical & Computer Eng [E]	MA	32.2[4]	Y	4.0	4.6	D	O	20.8	72	96.6	73	69.6	15	39.0
122 U of Mass, Dartmouth Electrical & Computer Eng [E]	MA	17.0	Y	4.0	4.0	--	N	-.-	62	95.0	15	13.5	-.-	-.-
123 U of Mass, Lowell Electrical Eng [E]	MA	38.0	Y	4.0	4.7	D	N	-.-	114	100.0	88	33.0	4	100.0
124 Mass Inst of Tech Electrical Eng & Computer Science [E]	MA	115.0	Y	4.0	-.-	D	O	-.-[5]	245	100.0	170	100.0	62	100.0
125 McGill U Electronics Eng [E]	PQ	43.3	Y	3.5[1]	4.0[C]	D	O[2]	-.-[A]	100	-.-[A]	36	59.8[3]	10	36.8[3]
126 Memphis State U Electrical Eng [E]	TN	13.0	Y	4.0	5.0	D	O	20.0	54	70.0	27	65.0	3	65.0
128 U of Miami Electrical & Computer Eng [E]	FL	10.1	Y	4.0	4.0	D	O	2.0	61	89.9	11	86.0	1	83.0
130 U of Michigan Appl Physics [E][2] Computer Eng [E][1] Computer, Information, & Control Eng [E][6] Computer Science & Eng [E][2] Electrical Eng [E] Electrical Eng Systems [E][2]	MI	-.-[3] -.-[3] -.-[3] -.-[3] 102.7 -.-[3]	-- Y -- -- Y --	-.- 4.0 -.- -.- 4.0 -.-	-.- 4.7 -.- -.- 4.7 -.-	-- D -- -- D --	-- O -- -- O --	-.- 8.3[7] -.- -.- 8.3[7] -.-	-- 67 -- -- 143 --	-.- 93.6 -.- -.- 93.3 -.-	6[4] -.- -.- 75[4] 56[4] 28[4]	-.-[A] -.- -.- 80.1[5] 85.8[5] 84.0[5]	15 -.- 5 12 32 12	-.-[A] -.- 100.0 80.1[5] 85.8[5] 84.0[5]
131 U of Michigan-Dearborn Electrical & Computer Eng [E]	MI	13.6	Y	4.0	4.5	D	O	0.0[A]	76	65.2	24	4.0	-.-[B]	-.-[B]

Electrical Engineering

ENGINEERING Institution Profile Reference Number & Name Name of Degree Program	State/Province	FACULTY Full-time equivalent (FTE)	ABET/CEAB accred?	UNDERGRADUATE PROGRAMS					GRADUATE PROGRAMS			
				Length		Time	Co-op		Bachelor's		Master's		Doctoral	
				Nominal length of program in years	Average length of program in years	Day/Eve./Both	None/Opt./Req.	% of graduates in Co-op programs	# of degrees awarded	% of graduates who were full-time	# of degrees awarded	% of graduates who were full-time	# of degrees awarded	% of graduates who were full-time
132 Michigan State U Electrical Eng [E]	MI	26.0	Y	4.0	4.5	D	O	30.0	103	100.0	30	100.0	16	100.0
133 Michigan Tech U Electrical Eng [E]	MI	26.8	Y	4.0	4.3	D	O	16.7	239	95.8	31	96.2	0	100.0
134 Milwaukee Sch of Eng Electrical Eng [E]	WI	14.7	Y	4.0	4.6	D	N(1)	-.-	66	95.0	-.-	-.-	-.-	-.-
135 U of Minnesota, Duluth Computer & Electrical Eng [E]	MN	9.0	Y	4.0	4.6	D	N	-.-	34	100.0	-.-	-.-	-.-	-.-
136 U of Minnesota Electrical Eng [E](3)	MN	50.0(1)	Y	4.0	-.-(A)	B	O	-.-(A)	141	-.-(A)	62	-.-(A)	19	-.-(A)
137 U of Mississippi Electrical Eng [E](3)	MS	9.3	Y	4.0	4.5	D	N	-.-	16	95.2	-.-	-.-	-.-	-.-
138 Mississippi State U Computer Eng (Electrical) [E] Electrical Eng [E]	MS	-.-(5) 24.0	Y Y	4.0 4.0	4.8 4.8	D D	O O	30.0 30.0	20 75	95.0 95.0	0 22	100.0 100.0	0 4	100.0 100.0
139 U of Missouri-Columbia Electrical Eng [E](2)	MO	25.0	Y	4.0	-.-	D	O	4.0	106	84.1	96	38.0	7	39.7
140 U of Missouri-Kansas City Electrical & Computer Eng [E]	MO	7.0	Y	4.0	5.0	B	N	-.-(2)	31	59.0	14	-.-	-.-	-.-
141 U of Missouri-Rolla Electrical Eng [E]	MO	32.0	Y	4.0	4.5	D	O	13.7	145	87.0	68	19.0	6	42.0
142 Monmouth Coll Electronic Eng [E]	NJ	5.3	Y	4.0	4.0	B	O(1)	5.0(C)	21	50.0(C)	10	25.0(C)	-.-(B)	-.-(B)
144 Montana State U Electrical Eng [E]	MT	15.9	Y	4.0	4.7	D	N	-.-	46	-.-(1)	10	100.0	2	100.0
145 Morgan State U Electrical Eng [E]	MD	10.0	Y	4.0	4.5	D	O	0.0	15	91.0	-.-	-.-	-.-	-.-
146 U of Nebraska - Lincoln Electrical Eng [E]	NE	18.0	Y	4.0	4.5	D	O	-.-	135	-.-	31	-.-	1	-.-
147 U of Nevada, Las Vegas Electrical & Computer Eng [E]	NV	8.5	Y	4.0	4.0	B	N	-.-	15	72.8	6	26.1	-.-(B)	-.-(B)
149 U of New Hampshire Electrical Eng [E]	NH	16.0	Y	4.0	-.-(A)	--	N	-.-	42	-.-(A)	22	-.-(A)	3	-.-(A)
150 U of New Haven Electrical Eng [E]	CT	11.0	Y	4.0	4.5(C)	B	O	10.0(C)	32	60.0(A)	-.-	30.0(C)	-.-(B)	-.-(B)
151 New Jersey Inst of Tech Electrical Eng [E]	NJ	40.7	Y	4.0	4.5	B	O	12.2	123	70.5	99	30.6	4	77.8
152 The U of New Mexico Electrical Eng [E]	NM	32.0	Y	4.0	5.0	D	O	-.-(A)	50	70.7	51	100.0(1)	9	100.0(1)
153 New Mexico Inst of Mining Electrical Eng [E]	NM	5.0	N(1)	4.0	4.5	D	O	-.-(A)	1	-.-(A)	-.-	-.-(A)	-.-	-.-(A)
154 New Mexico State U Electrical & Computer Eng [E]	NM	35.4	Y(1)	4.0	5.1(C)	D	O	26.0(C)	98	88.0(C)	56	60.0(C)	5	74.0(C)
155 U of New Orleans Electrical Eng [E]	LA	10.0	Y	4.0	5.2	B	O	-.-	46	74.0	-.-	-.-	-.-	-.-
156 New York Inst of Tech Electric Power Eng [E]	NY	11.1	Y	4.0	4.0	B	O	-.-(A)	111	-.-(A)	-.-(B)	-.-(B)	-.-(B)	-.-(B)
158 UNC-Charlotte Electrical Eng [E]	NC	21.5	Y	4.0	4.5	B	O	22.2	45	78.9	17	35.8	-.-	-.-

Electrical Engineering

Institution Profile Reference Number & Name / Name of Degree Program	State/Province	FACULTY Full-time equivalent (FTE)	ABET/CEAB accred?	UNDERGRADUATE PROGRAMS Length Nominal length of program in years	Length Average length of program in years	Time Day/Eve./Both	None/Opt./Req.	Co-op % of graduates in Co-op programs	Bachelor's # of degrees awarded	Bachelor's % of graduates who were full-time	GRADUATE PROGRAMS Master's # of degrees awarded	Master's % of graduates who were full-time	Doctoral # of degrees awarded	Doctoral % of graduates who were full-time
159 North Carolina A&T Electrical Eng [E]	NC	21.0	Y	4.0	-.-(A)	D	O	-.-(A)	65	-.-(A)	21	-.-(A)	--.(B)	-.-(B)
160 North Carolina State U Computer Eng [E] Electrical Eng [E]	NC	54.0 54.0	Y Y	4.0 4.0	4.5 4.5	D D	O O	20.0 20.0	71 210	75.4 81.3	48 46	31.0 60.6	6 20	9.1 50.0
161 U of North Dakota Electrical Eng [E]	ND	8.0	Y	4.0	5.0	D	O	-.-	27	-.-	7	-.-	--.	-.-
162 North Dakota State U Electrical Eng [E]$^{(1)}$	ND	11.0	Y	4.0	4.5	D	O	45.0	104	95.0	8	100.0	2	95.0
163 Northeastern U Electrical & Computer Eng [E]	MA	50.0	Y	5.0$^{(1)}$	5.0	B	O	85.0	182	85.0	114	27.6	12	80.3
164 Northern Arizona U Electrical Eng [E]$^{(1)}$	AZ	9.0	Y	4.0	5.0	D	O	5.0	28	95.0	--.(B)		--.	-.-(B)
165 Northern Illinois U Electrical Eng [E]	IL	13.0	Y	4.0	4.6	D	O	7.0	41	91.4	16	64.4	--.(B)	-.-(B)
166 Northwestern U Electrical Eng [E]	IL	43.5	Y	4.0	4.0	D	O	6.0	50	100.0	17	91.5	11	100.0
167 Norwich U Electrical Eng [E]	VT	5.0	Y	4.0	4.0	D	N	-.-	0	100.0	--.	-.-	--.	-.-
168 U of Notre Dame Electrical Eng [E]	IN	18.0	Y	4.0	4.0	D	N	-.-	50	100.0	18	90.0	4	85.0
169 Oakland U Electrical Eng [E]	MI	16.4$^{(5)}$	Y	4.0	4.5	D$^{(A)}$	O	-.-	49	-.-	19$^{(6)}$	-.-	--.(2)	-.-
170 Ohio U Electrical Eng [E]	OH	21.0	Y	4.0	-.-	--	O	17.0	88	-.-	13	-.-	4	-.-
171 Ohio Northern U Electrical Eng [E]	OH	6.0	Y	4.0	4.0	D	O	42.0	19	100.0	--.	-.-	--.	-.-
172 Ohio State U Electrical Eng [E]	OH	44.0	Y	4.0	4.5	D	O	-.-	105	77.0	90	100.0	20	100.0
173 U of Oklahoma Electrical Eng [E]	OK	16.0	Y	4.0	4.7	B	O	-.-(A)	107	77.8	19	40.2	6	28.6
174 Oklahoma Christian Electrical Eng [E]	OK	5.0	Y	4.5	4.5	D	N	-.-	6	91.4	--.	-.-	--.	-.-
175 Oklahoma State U Electrical & Computer Eng [E]	OK	19.8	Y	4.0	4.7	D	O	9.8	51	97.0	42	78.0	4	100.0
176 Old Dominion U Computer Eng [E] Electrical Eng [E]	VA	5.5$^{(1)}$ 12.5$^{(1)}$	Y Y	4.0 4.0	4.5 4.5	D D	O O	-.-(A) -.-(A)	20 61	82.8 88.1	1 6	0.0 38.5	--.(B) --.	-.-(B) 100.0
177 Oregon State U Computer Eng [E] Electrical & Electronics Eng [E]	OR	6.0 22.0	Y Y	4.0 4.0	4.8$^{(C)}$ 4.8$^{(C)}$	D D	O$^{(2)}$ O$^{(2)}$	12.0 0.0	19 65	100.0 100.0	--.(3) 29$^{(3)}$	-.- 100.0	--.(3) 7$^{(3)}$	-.- 100.0
178 U of the Pacific Electrical & Computer Eng [E]	CA	7.5	Y	5.0	5.0	D	R	56.1	26	100.0	1	20.0	--.(3)	-.-(3)
179 U of Penn Electrical Eng [E]	PA	20.0	Y	4.0	-.-	D	N	-.-	41	100.0	26	37.1	6	86.9
180 Penn State U Electrical Eng [E]	PA	36.0	Y	4.0	-.-	D	O	19.0	240	-.-	47	-.-	15	-.-
181 U of Pittsburgh Electrical Eng [E]	PA	21.0	Y	4.0	4.0	D$^{(6)}$	O	27.8	72	84.4	45	45.5	11	68.8

Electrical Engineering

ENGINEERING Institution Profile Reference Number & Name Name of Degree Program	State/Province	FACULTY Full-time equivalent (FTE)	ABET/CEAB accred?	UNDERGRADUATE PROGRAMS Length Nominal	Length Average	Time Day/Eve./Both	Co-op None/Opt./Req.	Co-op % in programs	Bachelor's # degrees	Bachelor's % full-time	Master's # degrees	Master's % full-time	Doctoral # degrees	Doctoral % full-time
182 Poly U Electrical Eng [E]	NY	50.0	Y	4.0	-.-(1)	D(B)	O	-.-(B)	129	89.0	179	23.0	14	30.0
183 U of Portland Electrical Eng [E](4)	OR	6.0	Y	4.0	4.5	D	O	-.-(2)	37	-.-(1)	7	-.-(3)	--.(B)	-.-(B)
184 Portland State U Electrical & Computer Eng [E](1)	OR	13.0	Y(2)	4.0	4.7	B	O	-.-(A)	54	66.5	33	51.2	0	60.9
185 Prairie View A&M U Electrical Eng [E]	TX	10.0	Y	4.0	4.0	D	O	20.0	62	90.0	0	1.0	-.-	-.-
186 Princeton U Electrical Eng [E]	NJ	22.0	Y	4.0	4.0	--	N	-.-	40	100.0	10	100.0	10	100.0
187 U of Puerto Rico Electrical & Computer Eng [E]	PR	30.0	Y	5.0	5.0	B	O	19.0	118	97.0	4	100.0	--.(B)	-.-(B)
188 Purdue U Computer & Electrical Eng [E](8) Electrical Eng [E]	IN	-.-(8) 74.0	Y(8) Y	4.0(8) 4.0	-.-(8) 4.3	D(8) D	O(8) O	-.-(8) 23.0	--(8) 308	--(8) 100.0	--(8) 125	--(8) 100.0	--(8) 43	--(8) 100.0
189 Rensselaer Poly Inst Electric Power Eng [E] Electrical Eng [E]	NY	6.0 34.0	Y Y	4.0 4.0	4.0 4.0	D D	O O	37.5 27.9	8 147	100.0 100.0	26 42	-.-(A) -.-(A)	4 17	-.-(A) -.-(A)
190 U of Rhode Island Computer Eng [E](1) Electrical Eng [E]	RI	21.0 21.0	Y Y	4.0 4.0	4.5 4.5	D D	N N	-.- -.-	11 36	90.0 90.0	--(B) 14	-.-(B) 50.0	--(B) 6	-.-(B) 60.0
191 Rice U Electrical & Computer Eng [E]	TX	20.2	Y	4.0	4.2	--(B)	N	-.-(B)	55	98.7	22	53.8	7	96.5
192 U of Rochester Electrical Eng [E]	NY	16.0	Y	4.0	4.0	D	N	-.-	44	98.0	32	41.7	6	94.6
193 Rochester Inst of Tech Electrical Eng [E]	NY	17.5	Y	5.0	-.-	D	R	100.0	93	78.0	30	15.0	--.	-.-
194 Rose-Hulman Electrical & Computer Eng [E]	IN	15.0	Y	4.0	4.0	D	N	-.-(A)	76	100.0	--.	-.-(A)	--(B)	-.-(B)
195 Rutgers, The State U Electrical Eng [E]	NJ	29.0	Y	4.0	4.5	D	N	-.-	136	94.0	39	31.0	10	33.0
196 Saginaw Valley State U Electrical Eng [E]	MI	4.1	Y	4.0	4.5	D	O	10.0	20	50.0	--.	-.-	--.	-.-
197 St. Cloud State U Electrical Eng [E]	MN	7.0	Y	4.0	5.0(C)	D	N	-.-	30	90.0(C)	--(B)	-.-(B)	--(B)	-.-(B)
198 Parks Coll of Saint Louis U Electrical Eng [E]	IL	6.0	Y	4.0	-.-	D	O	-.-	16	97.0	--(3)	-.-	--.	-.-
199 U of San Diego Electrical Eng [E](1)	CA	6.3	Y	4.5	4.7(2)	B	O	60.0(3)	7	100.0	--(B)	-.-(B)	--(B)	-.-(B)
200 San Diego State U Electrical Eng [E]	CA	18.6	Y	4.0	5.5	B	O	-.-(A)	142	80.0	30	20.0	--.	-.-(B)
201 San Francisco State U Electrical Eng [E]	CA	-.-	Y	4.0	-.-	B	O	-.-	66	-.-	--.	-.-	--.	-.-
202 San Jose State U Electrical Eng [E]	CA	35.5	Y	4.0	-.-(A)	D	--(A)	-.-	133	68.0	86	6.8	--.(A)	-.-(A)
203 Santa Clara U Electrical Eng [E]	CA	13.0	Y	4.0	4.0	D	O	-.-	38	99.0	56	3.5(C)	4	10.0(C)
204 Seattle U Electrical Eng [E]	WA	8.0	Y	4.0	4.0	D	N	-.-(B)	52	-.-(A)	--.	-.-	--.	-.-

Electrical Engineering

ENGINEERING Institution Profile Reference Number & Name Name of Degree Program	State/Province	FACULTY Full-time equivalent (FTE)	ABET/CEAB accred?	UNDERGRADUATE PROGRAMS							GRADUATE PROGRAMS			
				Length		Time	Co-op		Bachelor's		Master's		Doctoral	
				Nominal length of program in years	Average length of program in years	Day/Eve./Both	None/Opt./Req.	% of graduates in Co-op programs	# of degrees awarded	% of graduates who were full-time	# of degrees awarded	% of graduates who were full-time	# of degrees awarded	% of graduates who were full-time
205 U of South Alabama Electrical Eng [E]	AL	-.-	Y	4.0	-.-	B	O	-.-	43	79.0	4	26.0	-.-(B)	-.-(B)
206 U of South Carolina Electrical & Computer Eng [E] Electrical Eng [E]	SC	20.0 20.0	Y Y	4.0 4.0	4.0 4.0	D D	O(A) O	-.-(A) -.-(A)	71 71	34.2 34.2	24 24	51.9 51.9	3 3	100.0 100.0
207 South Dakota School of Mines Electrical Eng [E]	SD	11.5	Y	4.0	4.5	D	O	-.-(A)	38	-.-(A)	14	-.-(A)	-.-(B)	-.-(B)
208 South Dakota State U Electrical Eng [E]	SD	11.0	Y	4.0	4.5	D	O	10.0	67	100.0	6	100.0	-.-	-.-
209 U of South Florida Electrical Eng [E]	FL	17.0	Y	4.0	5.0	B	O	28.0	121	57.0	51	60.0	10	54.0
210 Southern U & A&M Coll Electrical Eng [E] Electronics Eng [E]	LA	10.0 5.0	Y N	4.0 4.0	5.2 5.2	D D	O O	25.0 2.0	34 25	35.1 25.8	-.-(A) -.-(A)	-.-(A) -.-(A)	-.-(A) -.-(A)	-.-(A) -.-(A)
211 U of Southern Cal Electrical Eng [E]	CA	21.0	Y	4.0	4.0	D	O	-.-(A)	72	100.0	251	42.0	54	61.0
213 Southern Illinois-Carbondale Electrical Eng [E]	IL	22.5	Y	4.0	4.5	D	O	3.0	97	92.5	27	67.0	-.-(B)	-.-(B)
214 Southern Illinois-Edwardsville Electrical Eng [E]	IL	10.0	Y	4.0	-.-(A)	B	O	-.-(A)	56	66.0	27	16.0	-.-(B)	-.-(B)
215 U of Southern Maine Electrical Eng [E]	ME	5.0	Y	4.0	5.5	B	O	20.0	8	35.0	-.-	-.-	-.-	-.-
216 Southern Methodist U Electrical Eng [E]	TX	17.0	Y	4.0	5.0	D	O	50.0	20	90.3	37	0.9	9	85.1
217 Stanford U Electrical Eng [E]	CA	49.0	Y	4.0	4.0	D(B)	N	-.-	63	100.0	182	65.0	68	100.0
218 SUNY-Binghamton Electrical Eng [E]	NY	18.1	Y	4.0	4.0	-.-(B)	N	-.-(B)	48	87.0	55	64.0	5	69.0
219 SUNY-Buffalo Electrical Eng [E]	NY	30.0	Y	4.0	4.5	B	N	-.-	138	86.6	57	63.4	14	36.0
221 SUNY-Maritime Electrical Eng [E]	NY	5.0	Y	4.0	4.3	D	N	-.-	28[1]	100.0	-.-	-.-	-.-	-.-
222 SUNY-New Paltz Electrical Eng [E]	NY	7.5	Y	4.0	4.5	-.-(A)	O	17.0	24	66.7	-.-(B)	-.-(B)	-.-(B)	-.-(B)
223 SUNY-Stony Brook Electrical Eng [E]	NY	-.-(A)	Y	4.0	-.-(A)	D	N	-.-(A)	67	96.7	27	79.5	6	65.4
225 Syracuse U Electrical Eng [E]	NY	23.0	Y	4.0	4.2	D	O	22.0	32	66.0	78	-.-(A)	13	-.-(A)
226 Technical U of Nova Scotia Electrical Eng [E]	NS	14.5	Y	4.0[2]	5.0	D	O	0.0[3]	38	100.0	6	83.0	4	100.0
227 Temple U Electrical Eng [E]	PA	16.0	Y	4.0	4.0	-.-(B)	N	-.-(B)	47	72.2	-.-(A)	-.-(A)	-.-	-.-(A)
228 U of Tennessee, Chattanooga Electrical Eng [E][6]	TN	-.-(7)	Y	-.-	-.-	B	O	-.-(A)	-.-	-.-	3	-.-(A)	-.-	-.-
229 U of Tennessee, Knoxville Electrical & Computer Eng [E]	TN	38.0	Y	4.0	-.-(A)	D	O	26.3	95	85.7	26	-.-(A)	9	-.-(A)
230 Tennessee State U Electrical Eng [E]	TN	8.0	Y	4.3	5.0	B	O	7.5	20	79.7	5[1]	44.4	-.-(B)	-.-(B)

Electrical Engineering

Institution Profile Reference Number & Name / Name of Degree Program	State/Province	FACULTY Full-time equivalent (FTE)	ABET/CEAB accred?	UNDERGRADUATE PROGRAMS Length Nominal length of program in years	Length Average length of program in years	Time Day/Eve./Both	Co-op None/Opt./Req.	Co-op % of graduates in Co-op programs	Bachelor's # of degrees awarded	Bachelor's % of graduates who were full-time	GRADUATE PROGRAMS Master's # of degrees awarded	Master's % of graduates who were full-time	Doctoral # of degrees awarded	Doctoral % of graduates who were full-time
231 Tennessee Tech	TN													
Electrical Eng [E]		204.3	Y	4.0	4.0	D	O	10.0	51	100.0	9	100.0	4	100.0
232 U of Texas at Arlington	TX													
Electrical Eng [E]		-.-	Y	4.0	-.-	B	O	-.-	95	-.-	98	-.-	13	-.-
233 U of Texas at Austin	TX													
Computer Eng [E]		-.-[9]	Y	4.0	4.8	D	O	-.-[9]	0[10]	-.-[9]	-.-[B]	-.-[B]	-.-[B]	-.-[B]
Electrical & Computer Eng [E]		64.5	Y	4.0	4.7	D	O[1]	24.2	219	87.8	117	68.5	38	77.2
234 U of Texas at Dallas	TX													
Electrical Eng [E]		5.0	Y	-.-	5.0	--	N	-.-	42	-.-	54	-.-	2	-.-
235 U of Texas at El Paso	TX													
Electrical Eng [E]		17.0	Y	4.0	-.-	D	O	20.0	66	-.-	21	-.-	0	100.0
236 U of Texas at San Antonio	TX													
Electrical Eng [E]		12.4	Y	4.0	-.-	B	O	-.-	40	-.-	4	-.-	-.-	-.-
237 Texas A&I U	TX													
Electrical Eng [E]		-.-	Y	4.0	-.-	D	N	-.-	41	-.-	13	-.-	-.-	-.-
238 Texas A&M U	TX													
Electrical Eng [E]		42.0	Y	4.0	4.5[C]	D	O	53.0	216	88.0	76	86.0[C]	21	86.0[C]
239 Texas Tech U	TX													
Electrical Eng [E]		20.6	Y	4.0	-.-	D	N	-.-	25	87.7	9	79.0	5	100.0
240 U of Toledo	OH													
Electrical Eng [E]		19.0	Y	4.0	4.0	D	O	-.-[A]	62	78.0	11	50.0	-.-	-.-
241 Tri-State U	IN													
Electrical Eng [E]		7.0	Y	4.0	4.5	D	O	15.0	37	100.0	-.-[3]	-.-[3]	-.-[3]	-.-[3]
243 Tufts U	MA													
Computer Eng [E]		5.5	Y	4.0	4.0	D	N	-.-[B]	13	100.0	-.-[B]	-.-[B]	-.-[B]	-.-[B]
Electrical Eng [E]		17.0	Y	4.0	4.0	D	N	-.-[B]	59	100.0	24	53.0	6	100.0
244 Tulane U	LA													
Electrical Eng [E]		9.0	Y	4.0	4.0	B	N	-.-	25	100.0	2	87.0	2	20.0
245 U of Tulsa	OK													
Electrical Eng [E]		9.0	Y	4.0	4.0	D	N	-.-[B]	18	90.1	5	50.0	-.-[B]	-.-[B]
246 Tuskegee U	AL													
Electrical Eng [E]		12.0	Y	4.0	5.0	D	O	80.0	43	99.5	4	99.5	-.-	-.-
247 Union Coll	NY													
Electrical Eng [E]		8.0	Y	4.0	-.-[A]	B	N	-.-[B]	21	77.0	18	39.0	-.-[B]	-.-[B]
248 US Air Force Academy	CO													
Electrical Eng [E]		27.0	Y	4.0	4.0	D	N	-.-	42	100.0	-.-[B]	-.-[B]	-.-[B]	-.-[B]
249 US Coast Guard Academy	CT													
Electrical Eng [E]		8.0	Y	4.0	4.0	D	N	-.-	16	100.0	-.-	-.-	-.-	-.-
251 US Military Academy	NY													
Electrical Eng [E]		-.-	Y	4.0	4.0	D	N	-.-	40	100.0	-.-[B]	-.-[B]	-.-[B]	-.-[B]
252 US Naval Academy	MD													
Electrical Eng [E]		31.0	Y	4.0	4.0	D	N	-.-	16	100.0	-.-	-.-	-.-	-.-
253 U of Utah	UT													
Electrical Eng [E]		15.1	Y	4.0	5.0	D	O	-.-[A]	81	-.-[A]	23	-.-[A]	6	-.-[A]
254 Utah State U	UT													
Electrical & Computer Eng [E]		12.5	Y	4.0	4.6	D	O	-.-[A]	39	81.3	32	41.1	1	41.7
255 Valparaiso U	IN													
Electrical Eng [E]		5.5	Y	4.0	4.0	D	O	-.-	24	99.0	-.-	-.-	-.-	-.-
256 Vanderbilt U	TN													
Electrical Eng [E][3]		20.0	Y[4]	4.0	4.0	D[5]	N	-.-	59	100.0	15	84.0	8	81.0

Electrical Engineering

ENGINEERING Institution Profile Reference Number & Name Name of Degree Program	State/Province	FACULTY Full-time equivalent (FTE)	ABET/CEAB accred?	UNDERGRADUATE PROGRAMS							GRADUATE PROGRAMS			
				Length		Time	Co-op		Bachelor's		Master's		Doctoral	
				Nominal length of program in years	Average length of program in years	Day/Eve./Both	None/Opt./Req.	% of graduates in Co-op programs	# of degrees awarded	% of graduates who were full-time	# of degrees awarded	% of graduates who were full-time	# of degrees awarded	% of graduates who were full-time
257 U of Vermont Electrical Eng [E]	VT	13.5	Y	4.0	4.0	D	O	10.0	43	76.9	14	4.5	0	33.0
258 Villanova U Electrical Eng [E]	PA	20.0	Y	4.0	4.0	--	N	-.-	71	76.0	32	22.0	-.-	-.-
259 U of Virginia Electrical Eng [E]	VA	15.6	Y	4.0	4.1	D	N	-.-	78	100.0	41	84.3	12	90.7
260 Virginia Military Inst Electrical Eng [E]	VA	7.0	Y	4.0	-.-	D	N	-.-	84	100.0	-.-	-.-	-.-	-.-
261 Virginia Tech Electrical Eng [E]	VA	44.0	Y	4.0	4.0	D	O	16.0	253	100.0	93	100.0	23	100.0
262 U of Washington Electrical Eng [E]	WA	45.7	Y	4.0	-.-(A)	D	O	-.-(A)	133	80.0	92	-.-(A)	17	-.-(A)
263 Washington U Electrical Eng [E]	MO	21.0	Y	4.0	4.0	D	O	-.-	69	-.-	44	-.-	6	-.-
264 Washington State U Electrical Eng [E] Electrical Eng & Computer Science [E][1]	WA	-.- 37.3	Y Y	4.0(5) 4.0	-.- 4.5	D(5) D	O O	-.- -.-(A)	-.- 94	-.- 100.0	-.- 13	-.- 100.0	-.- 6	-.- 100.0
265 Wayne State U Electrical & Computer Eng [E] Electronic & Computer Control System [E]	MI	20.0 -.-(1)	Y N	4.0 -.-	4.5 -.-	B --	O --	-.-(A) -.-(A)	79 -.-	-.-(A) -.-(B)	72 37	-.-(A) -.-(A)	8 -.-	-.-(A) -.-(A)
267 West Virginia U Electrical Eng [E]	WV	11.0	Y	4.0	4.0	D	O	5.0	69	100.0	8	90.0	5	90.0
268 West Virginia Inst of Tech Electrical Eng [E]	WV	8.0	Y	4.0	4.5	D	O	35.0	50	95.0	-.-	-.-	-.-	-.-
269 Western Michigan U Electrical Eng [E]	MI	9.0	Y	4.0	4.5	D	O	10.0	36	90.3	5	10.6	-.-	-.-
270 Western New England Coll Electrical Eng [E]	MA	5.8	Y	4.0	4.4	B	N	-.-(B)	25	63.8	11	-.-(B)	-.-(B)	-.-(B)
271 Wichita State U Electrical Eng [E]	KS	16.0	Y	4.0	5.5	B	O	8.5	82	56.2	15	39.3	4	47.6
272 Widener U Electrical Eng [E]	PA	9.3	Y	4.3	4.0	D	O	36.0	57	-.-(A)	8	-.-(A)	-.-(B)	-.-(B)
273 Wilkes U Electrical Eng [E]	PA	11.5	Y	4.0	4.0	B(2)	O	5.0(C)	45	85.0(C)	4	30.0(C)	-.-(B)	-.-(B)
274 U of Wisconsin-Madison Electrical Eng [E]	WI	48.5	Y	4.0	5.0	D	O	-.-(A)	159	-.-(A)	69	-.-(A)	12	-.-(A)
275 U of Wisconsin-Milwaukee Electrical Eng [E]	WI	11.8	Y	4.0	5.5	B	O	10.4	48	72.9	11	42.2	1	85.7
276 U of Wisconsin-Platteville Electrical Eng [E]	WI	9.0	Y	4.0	4.7	D	O	21.0	57	100.0	-.-	-.-	-.-	-.-
277 Worcester Poly Inst Electrical & Computer Eng [E]	MA	27.3	Y	4.0	4.0	D	O	-.-(A)	112	95.8	57	-.-(A)	5	-.-(A)
278 Wright State U Electrical Eng [E]	OH	12.6	Y	4.0	-.-	B	O	20.0	59	73.0	26	17.6	-.-	-.-
279 U of Wyoming Electrical Eng [E]	WY	16.2	Y	4.0	5.0	D	O	4.0	45	91.0	4	100.0	0	100.0
280 Yale U Electrical Eng [E]	CT	15.0	Y	4.0	4.0	D	N	-.-	20	42.0	5	50.0	11	100.0

Electrical Engineering—Totals

		FACULTY	UNDERGRADUATE PROGRAMS						GRADUATE PROGRAMS			
				Length		Time	Co-op	Bachelor's	Master's		Doctoral	
TOTALS State/Province/Country	# of schools	Full-time equivalent (FTE)	# accredited programs	Nominal length of program in years	Average length of program in years	Day/Eve./Both	None/Opt./Req.	% of graduates in Co-op programs	# of degrees awarded	% of graduates who were full-time	# of degrees awarded	% of graduates who were full-time	# of degrees awarded	% of graduates who were full-time

State	# schools	FTE	# accred	Bach #	Master #	Doctoral #
AK	1	8.00	1	13	4	-
AL	6	107.90	7	557	125	10
AR	1	17.00	1	63	11	4
AZ	4	107.00	3	294	162	24
CA	23	633.40	23	1,785	1,036	292
CO	6	135.40	6	276	146	30
CT	5	62.00	5	206	39	13
DC	4	57.00	4	136	101	13
DE	1	14.60	1	52	8	8
FL	9	191.70	10	847	323	48
GA	1	80.00	2	316	203	48
HI	1	27.00	1	62	19	4
IA	2	20.00	3	279	51	10
ID	1	16.50	1	53	16	3
IL	9	306.90	9	786	348	78
IN	7	139.20	8	582	148	47
KS	3	37.00	3	274	57	9
KY	2	39.10	2	157	40	4
LA	5	64.70	5	264	57	2
MA	9	340.80	10	952	596	107
MD	6	157.00	4	276	367	23
ME	2	19.50	2	47	12	-
MI	10	254.00	11	1,053	383	100
MN	4	73.00	4	229	62	19
MO	4	85.00	4	351	222	19
MS	2	33.30	3	111	22	4
MT	1	15.90	1	46	10	2
NC	4	168.50	5	457	149	33
ND	2	19.00	2	131	15	2
NE	1	18.00	1	135	31	1
NH	1	16.00	1	42	22	3
NJ	4	97.00	4	320	158	24
NM	3	72.40	2	149	107	14
NV	1	8.50	1	15	6	-
NY	20	367.40	21	1,452	832	132
OH	10	194.80	10	612	262	43
OK	4	49.80	4	182	66	10
OR	3	47.00	4	175	69	7
PA	14	274.90	14	1,056	317	82
PR	1	30.00	1	118	4	-
RI	1	42.00	2	47	14	6
SC	3	84.00	4	303	98	15
SD	2	22.50	2	105	20	-
TN	7	289.30	7	302	85	24
TX	13	242.80	14	1,001	523	97
UT	3	49.60	3	237	79	8
VA	5	107.90	6	578	175	35
VT	2	18.50	2	43	14	-
WA	4	99.00	5	301	106	23
WI	5	105.20	5	428	122	16
WV	2	19.00	2	119	8	5
WY	1	16.20	1	45	4	-
U.S. Totals:	**245**	**5,502.20**	**257**	**18,420**	**7,854**	**1,501**
AB	1	36.00	1	105	33	5
MB	1	-	2	67	11	5
NS	1	14.50	1	38	6	4
PQ	3	89.30	3	312	97	20
Canada Totals:	**6**	**139.80**	**7**	**522**	**147**	**34**

Electrical Engineering—Footnotes

FOOTNOTES: The following footnotes are the same for all schools: (A) Data not available. (B) Data not applicable. (C) Estimated data.

002 University of Alabama: (1) Computer Engineering Option

006 University of Alberta: (3) Total for Computer, Electrical and Engineering Physics.

012 Auburn University: (2) Supported by the faculty of Computer Science and Engineering. (3) Accredited by Computing Sciences Accreditation Board. (10) Based on credit hours taken Fall Quarter 1991.

014 Boston University: (1) Part time status granted by petition only. (2) Offered through the Electrical, Computer, Systems department. (4) Electrical/Computer/Systems faculty

015 Bradley University: (1) Students typically take the equivalent of nine semesters but finish in 4 years by using summer school and transfer credits. (2) Department offers a computer option within the electrical engineering degree.

018 Bucknell University: (1) Data is given in whole numbers.

019 University of California at Berkeley: (1) Double majors are included with department totals.

020 University of California, Davis: (7) Included in Electrical and Computer Engineering

023 University of California, San Diego: (2) Due to a large amount of shared teaching responsibility, it is impossible to give faculty numbers by program. See specific department for numbers.

024 University of California, Santa Barbara: (1) A Limited number of students are sponsored by their employers and are working on their degrees part-time. (6) The B.S. degree is awarded in Electrical Engineering.

027 California Polytechnic State University: (2) FTEF is for Electrical AND Electronic Engineering

042 University of Cincinnati: (1) Of the 6 departments., 5 have more than one program, therefore, faculty provide support in multiple programs. Total FTE for the college is 142.

048 Cleveland State University: (1) includes Co-Op and part-time students (2) Program includes a computer engineering option

051 University of Colorado at Denver: (3) Coursework available at CU-Denver but Ph.D. is awarded at CU-Boulder through system-wide Graduate School.

054 Columbia University: (4) Includes Solid State Engineering.

056 University of Connecticut: (3) Includes Biological Engineering students.

066 École Polytechnique de Montréal: (5) This faculty supports Electrical Engineering and Computer Engineering programs.

068 Embry-Riddle Aeronautical University, Western Campus: (1) ABET accreditation expected summer 1993

072 Florida Agricultural and Mechanical University: (1) Includes FAMU and FSU students. Year is Fall 1991, Spring 1992 and Summer 1992.

073 Florida Atlantic University: (5) Working students often limit their programs to 12 credits per term or less, depending on work schedules.

076 Florida State University: (1) Includes FAMU and FSU students. Year is Fall 1992, Spring 1992, and Summer 1992.

077 GMI Engineering & Management Institute: (1) 3360 hours minimum

079 George Mason University: (1) The Information Technology PhD degree program is interdisciplinary and supported by all seven masters programs. (6) This program is supported by faculty of the Systems Engineering department.

082 Georgia Institute of Technology: (8) Fall '91 Quarter FTE.

090 University of Houston: (2) BSEE, MEE, MSEE, PhD

096 Illinois Institute of Technology: (2) Reflects percentage of graduates who received certificates of completion in compliance with ABET requirements.

100 The Johns Hopkins University: (3) Includes 26.30 FTE in part-time programs.

103 University of Kentucky: (1) Breakdown of part-time students by degree is not available. Percentage represents total M.S. and Ph.D. enrollment.

104 Lafayette College: (1) Undergraduate program only.

108 Lehigh University: (1)

109 Louisiana State University: (1) PhD available through Eng Science (interdisciplinary program).

111 University of Louisville: (1) Program is accredited at the master's level.

114 University of Maine: (1) Ph.D. Program not available.

115 University of Manitoba: (1) MS and Doctoral degrees are in Computer and Electrical Engineering. These degrees are reported under Electrical Engineering.

119 University of Maryland Baltimore County: (4) no B.S.E.E. program

121 University of Massachusetts: (4) Includes department head

124 Massachusetts Institute of Technology: (5) 11.0% of bachelor's degree recipients and 34.1% of master's degree recipients participated in co-op programs.

125 McGill University: (1) Following completion of 2 year CEGEP program. International students should add 1 year for pre-engineering studies (2) Internship year program available for students who have less then 45 credits remaining in their academic program (3) Part time figures include candidates who have completed their residence requirements but are working full time on research and thesis requirements

130 University of Michigan: (1) BS and/or BSE program only. (2) Graduate program only. (3) Faculty included in principal department. (4) Includes Professional degrees. (5) All graduate students. (6) Doctoral program only. (7) For whole College.

134 Milwaukee School of Engineering: (1) No co-op program is available, however, 85% of undergraduates participate in an internship program and have relevant work experience prior to graduation.

136 University of Minnesota: (1) Count is number of faculty. We do not have FTE estimates. Chemical Engineering includes Materials Science. (3) Computer Science is a separate department.

137 University of Mississippi: (3) M.S. and Ph.D. degrees in Engineering Science are offered with emphasis in Electrical Engineering.

138 Mississippi State University: (5) Faculty included with Electrical Engineering.

139 University of Missouri-Columbia: (2) Includes statistics from EE/CEP which is located in Kansas City and administered by the Columbia campus.

140 University of Missouri-Kansas City: (2) Most students participate in unofficial co-op programs with local firms.

142 Monmouth College: (1) Part-time only. Maximum of 9 credits applied to degree.

144 Montana State University: (1) Information unavailable but nearly 100%

152 The University of New Mexico: (1) We classify all our graduate students as full-time.

153 New Mexico Institute of Mining and Technology: (1) Reviewed in 1992

154 New Mexico State University: (1) Accredited as electrical engineering.

162 North Dakota State University: (1) Masters & Ph.D. degree also available.

163 Northeastern University: (1) 5 year co-op option; 4 year co-op option; 4-year non-co-op option

164 Northern Arizona University: (1) Includes an emphasis in environmental engineering.

169 Oakland University: (2) See Doctoral Program. (5) This program is in Electrical and Systems Engineering Department. (6) M.S. program in Electrical and Computer Engineering and is within the Electrical and Systems Engineering Department.

176 Old Dominion University: (1) 1991/92 Head count

177 Oregon State University: (2) New program 1991-92 (3) M.S. & PhD offered in Electrical and Computer Engineering. Students reported under Electrical and Electronics Engineering.

178 University of the Pacific: (3) Electrical Engineering does not offer a doctoral degree. Computer Engineering does not offer a master's or a doctoral degree.

181 University of Pittsburgh: (6) Part-time program is available.

182 Polytechnic University: (1) 4-5

183 University of Portland: (1) Data not available by program. Overall undergraduate full-time is 89%. (2) Each year approximately 10 Engineering students participate in co-op programs. The numbers by degree program fluctuate. (3) Data not available by program. Overall graduate full-time is 42%. (4) Includes an Electrical Track and a Computer Track.

184 Portland State University: (1) Undergraduate degrees are awarded in either Electrical Engineering or Computer Engineering. (2) Computer Engineering is not ABET accredited.

188 Purdue University: (8) See Electrical Engineering.

190 University of Rhode Island: (1) Program offered in department of Electrical Engineering

198 Parks College of Saint Louis University: (3) No graduate program

199 University of San Diego: (1) The EE program is a nine semester dual B.S./B.A. program accredited by EAC/ABET. (2) Successful full-time majors will be assured completion between the minimum nominal program length of 9 semesters and a maximum of 10 semesters. (3) Typical co-op/internship programs are scheduled for full-time in the summer only, with optional part-time involvement available during regular terms.

221 State University of New York Maritime College: (1) BE degree awarded.

226 Technical University of Nova Scotia: (2) Three years at TUNS. (3) Just started.

228 University of Tennessee, Chattanooga: (6) Master's program. (7) All faculty are included under Engineering.

230 Tennessee State University: (1) An option in Master of Engineering degree program.

233 University of Texas at Austin: (1) Participants in the Co-op Program are away from the campus for three semesters, working for pay and letter grade. (9) Computer Engineering faculty and students are included with Electrical Engineering. (10) The Computer Engineering degree program requires a semester hour total which differs from Electrical Engineering, but the degree conferred is in Electrical Engineering.

241 Tri-State University: (3) No graduate programs

256 Vanderbilt University: (3) Also offer program in Computer Engineering in cooperation with the Department of Computer Science. (4) Computer Engineering not yet accredited. (5) Occasional evening course.

264 Washington State University: (1) Pullman campus (5) See Civil Engineering for data.

265 Wayne State University: (1) Included in Electrical and Computer Engineering

273 Wilkes University: (2) There is no distinction between day or evening programs.

'Hands on' Laboratory Equipment for Teaching

HI-TECH SCIENTIFIC

HI-PLAN 2
STRUCTURES APPARATUS

HI-PLAN 2 apparatus comprises a logical series of over 50 experiments in mechanics of structures from simple equilibrium to indeterminate and elastic/plastic structures. Technicians and/or undergraduates can improve their practical and theoretical knowledge by using computers with HI-PLAN 2 software for immediate analysis of experimental results.

- Full instructions, theory and sample results.
- Dedicated software using Lotus 1-2-3.
- Fully interactive and user friendly operation.

Instead of a "strong floor" for testing real structural members, use our versatile MAGNUS Universal 30 Ton Test Frame as supplied to satisfied university departments world-wide over the past 30 years.

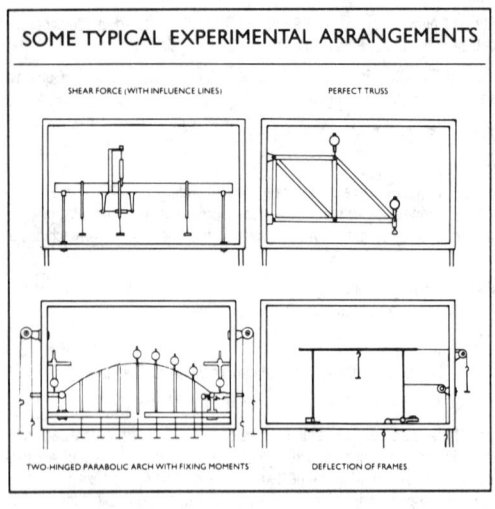

SOME TYPICAL EXPERIMENTAL ARRANGEMENTS
- SHEAR FORCE (WITH INFLUENCE LINES)
- PERFECT TRUSS
- TWO-HINGED PARABOLIC ARCH WITH FIXING MOMENTS
- DEFLECTION OF FRAMES

HI-PLAN 2 NOW INTRODUCES STRUCTURAL DYNAMICS EXPERIMENTS

ENGINEERING EXPERIMENTAL APPARATUS RANGE

There is no alternative to "Hands-On" experience for engineering students. HI-TECH laboratory equipment has been designed by practicing lecturers in technical colleges and universities to demonstrate the principles of engineering and simultaneously to generate confidence in the users. By reducing setting up time and making adjustment of variables straightforward, each experiment is a self contained package that can be completed in a typical laboratory period. The world-wide recognition of the high quality of HI-TECH apparatus comes from over 30 years of good basic design and continual improvement.

- Good instructions.
- Education orientated.
- Nearly 100 experiments.

FRICTION/TRIBOLOGY
MECHANISMS
FORCES

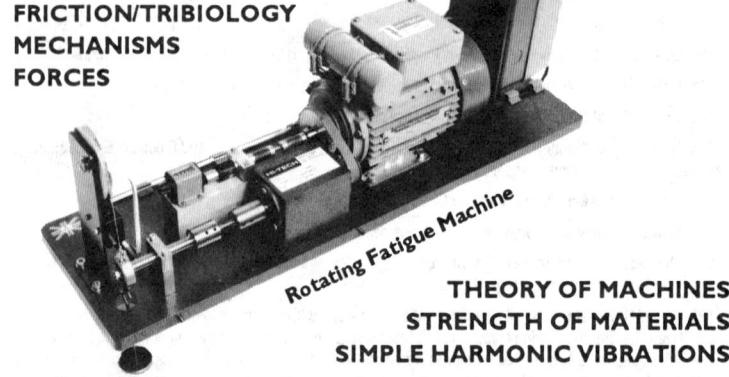
Rotating Fatigue Machine

THEORY OF MACHINES
STRENGTH OF MATERIALS
SIMPLE HARMONIC VIBRATIONS

Call David Mitchell Toll Free on 1 800 334 0724 (USA Only)
Hi-Tech Scientific Ltd., Brunel Road, Salisbury SP2 7PU, United Kingdom.
Telephone: 0722 323643. International +44 722 323643. Facsimile: 0722 412153. International +44 722 412153. Telex: 477877 HITECH G

Industrial Engineering

Institutions are listed in alphabetical order. The first line in a school's entry contains the institution profile reference number, the school name, and the state or province in which the main campus is located. Subsequent indented lines for each program give the name of the degree program, followed by a major category abbreviation in brackets (H = Chemical, V = Civil, E = Electrical, I = Industrial, M = Mechanical, and O = Other) the full-time equivalent (FTE) of all faculty. For undergraduate programs, the columns represent: accreditation status, the nominal and actual average length of the program, the time that courses are available (Day, Evening or Both), the availability of cooperative (internship) programs (None, Optional, or Required), the percentage of graduates who participated in co-op programs, the number of bachelor's degrees awarded, and the percentage of these graduates who were full-time. For graduate programs, the columns indicate the number of graduates and the percentage who were full-time in master's and then doctoral programs. Unless otherwise footnoted, information on degrees is for the 1991-92 academic year; all other information is current. At the end of the table are (1) a summary of totals by state or province and by country and (2) the full text of all footnotes for this table.

ENGINEERING Institution Profile Reference Number & Name / Name of Degree Program	State/Province	FACULTY Full-time equivalent (FTE)	ABET/CEAB accred?	UNDERGRADUATE PROGRAMS Length Nominal length of program in years	Length Average length of program in years	Time Day/Eve./Both	Co-op None/Opt./Req.	Co-op % of graduates in Co-op programs	Bachelor's # of degrees awarded	Bachelor's % of graduates who were full-time	GRADUATE PROGRAMS Master's # of degrees awarded	Master's % of graduates who were full-time	Doctoral # of degrees awarded	Doctoral % of graduates who were full-time
002 U of Alabama / Industrial Eng [I]	AL	12.0	Y	4.0	4.5	D	O	20.7	29	100.0	11	100.0	-.-(B)	-.-(B)
004 U of Alabama at Huntsville / Industrial & Systems Eng [I]	AL	13.3	Y	4.0	-.-(A)	B	O	-.-(A)	14	57.4	26	23.8	3	28.1
008 U of Arizona / Industrial Eng [I]	AZ	19.0	Y	4.0	-.-	D	O	-.-	29	88.0	10	-.-	-.-	-.-
009 Arizona State U / Industrial Eng [I]	AZ	17.5	Y	4.0	5.0	D	O	-.-(A)	25	85.0	31	62.0	10	62.0
010 U of Arkansas / Industrial Eng [I]	AR	-.-	Y	4.0	-.-	D	O	-.-	30	-.-	9	91.7	3	100.0
012 Auburn U / Industrial Eng [I]	AL	14.0	Y	4.0	-.-(A)	D	O	22.2	54	85.7(10)	20	53.3(10)	6	100.0(10)
Manuf'g Systems Eng [I]		-.-(6)	-.-(B)	-.-(B)	-.-(B)	-.-(B)	-.-(B)	-.-(B)	-.-(B)	-.-(B)	2	25.0(10)	-.-(B)	-.-(B)
015 Bradley U / Industrial Eng [I]	IL	8.5	Y	4.0	4.0(1)	B	O	18.0	34	93.3	6	12.5	-.-(B)	-.-(B)
016 U of Bridgeport / Mgmt Eng [I]	CT	1.5	N	-.-	-.-	--	--	-.-	0	-.-	-.-	-.-	-.-	-.-
019 U of Cal at Berkeley / Industrial Eng & Operations Research [I]	CA	14.0	Y	4.0	-.-(A)	D	O	12.0	34	100.0	32	100.0	12	100.0
027 Cal Poly State U / Industrial Eng [I]	CA	15.8	Y	4.0	5.0	D	O	-.-	44	-.-	-.-	-.-	-.-	-.-
029 Cal State U, Fresno / Industrial Eng [I]	CA	5.0	Y	4.0	4.5	B	O	10.0	10	4.5	-.-(B)	-.-(B)	-.-(B)	-.-(B)
034 Cal State Poly, Pomona / Industrial Eng [I]	CA	7.0	Y	4.0	5.5	B	O	-.-	35	83.0	-.-	-.-	-.-	-.-
037 Case Western Reserve U / Industrial Eng [I]	OH	8.0(2)	N	4.0	4.3	B	O	30.0	5	60.0	-.-(B)	-.-(B)	-.-(B)	-.-(B)
039 U of Central Florida / Industrial Eng [I]	FL	17.0	Y	4.0	4.5(C)	B	O	15.0(C)	28	66.7	18	52.0	2	50.0(C)
042 U of Cincinnati / Industrial Eng [I]	OH	30.0(1)	Y	5.0	5.2	D	R	100.0	31	100.0	14	45.1	4	78.9
047 Clemson U / Industrial Eng [I]	SC	9.0	Y	4.0	-.-	D	O	-.-	58	87.0	11	76.0	1	60.0
048 Cleveland State U / Industrial Eng [I]	OH	6.8	Y	4.0	4.3(1)	D	O	42.9	6	74.3	24	53.3	-.-	-.-
054 Columbia U / Industrial Eng & Operations Research [I]	NY	11.7	Y(2)	4.0	4.0	-.-(B)	N	-.-(B)	44	100.0	39	53.5	5	53.5
058 Cornell U / Operations Research & Eng [I]	NY	-.-	Y	4.0	-.-	D	O	-.-	-.-	-.-	-.-	-.-	-.-	-.-
064 Drexel U / Industrial Eng [I]	PA	1.0	N	6.0(C)	6.0(C)	E	N	-.-(B)	5	0.0	-.-(B)	-.-(B)	-.-(B)	-.-(B)

Industrial Engineering

ENGINEERING Institution Profile Reference Number & Name Name of Degree Program	State/Province	FACULTY Full-time equivalent (FTE)	ABET/CEAB accred?	UNDERGRADUATE PROGRAMS					Bachelor's		GRADUATE PROGRAMS			
				Length		Time	Co-op				Master's		Doctoral	
				Nominal length of program in years	Average length of program in years	Day/Eve./Both	None/Opt./Req.	% of graduates in Co-op programs	# of degrees awarded	% of graduates who were full-time	# of degrees awarded	% of graduates who were full-time	# of degrees awarded	% of graduates who were full-time
066 École Polytechnique Industrial Eng [I]	PQ	20.4	Y	4.0	4.5	D	N	-.-	62	96.5	10	71.7	-.-[B]	-.-[B]
071 U of Florida Industrial & Systems Eng [I][2]	FL	13.3	Y	4.0	4.5	D	O	10.0	41	80.0	17	100.0	3	100.0
072 Florida A&M U Industrial Eng [I]	FL	7.0	Y	4.5	5.0	D	O	-.-[A]	8[1]	-.-[A]	-.-[B]	-.-[B]	-.-[B]	-.-[B]
075 Florida International U Industrial & Systems Eng [I]	FL	10.3	Y	4.0	4.5	B	O	50.0	35	73.1	1	100.0	-.-	-.-
076 Florida State U Industrial Eng [I]	FL	7.0	Y	4.5	5.0	D	O	-.-[A]	8[1]	-.-[A]	-.-[B]	-.-[B]	-.-[B]	-.-[B]
077 GMI Industrial & Manuf'g Systems Eng [I] Industrial Eng [I]	MI	10.0 10.0	Y Y	5.0 5.0	5.0 5.0	D[1] D[1]	R R	100.0 100.0	26 26	100.0 100.0	-.- -.-	-.- -.-	-.- -.-	-.- -.-
082 Georgia Tech Industrial Eng [I]	GA	51.7[8]	Y	4.0	4.7	D	O	36.0	253	93.7	117	72.3[7]	16	93.4
089 Hofstra U Industrial Eng [I]	NY	1.0	N	4.0	4.5	B	N	-.-	3	64.3	-.-	-.-	-.-	-.-
090 U of Houston Industrial Eng [I][3]	TX	8.0	Y	4.0	5.0	B	O	15.0	13	90.0	20	60.0	5	100.0
094 U of Illinois at Chicago Industrial Eng [I] Industrial Eng & Operations Research [I]	IL	6.0 6.0	Y --	4.0 -.-	-.- -.-	D --	O --	-.- -.-	23 -.-	90.0 -.-	9 -.-[5]	46.0 -.-	-.-[6] 1	-.- 50.0
095 Illinois, Urbana-Champaign Industrial Eng [I][1]	IL	-.-	Y	4.0	4.5	-.-[A]	O	13.0	46	98.0	9	81.0	-.-[A]	-.-[A]
098 U of Iowa Industrial Eng [I]	IA	7.5	Y	4.0	4.5[C]	D	O	15.0[C]	43	100.0[C]	7	100.0[C]	5	100.0[C]
099 Iowa State U Industrial Eng [I] Operations Research [I]	IA	-.- -.-	Y N	4.0 -.-[B]	-.- -.-	D -.-[B]	O -.-[B]	-.- -.-	86 -.-	-.- -.-	11 -.-	-.- -.-	3 -.-	-.- -.-
102 Kansas State U Industrial Eng [I]	KS	10.0	Y	4.0	4.7	D	O	-.-	53	97.0	19	-.-	3	-.-
105 Lamar U Industrial Eng [I]	TX	3.5	Y	4.0	4.5	B	O	45.0	5	-.-[A]	-.-[A]	-.-[A]	-.-	-.-
108 Lehigh U Industrial & Manuf'g Systems Eng [I][2]	PA	14.0	Y	4.0	4.0	D	O	0.0[2]	42	100.0	13	50.0	3	80.0
109 Louisiana State U Industrial Eng [I]	LA	11.5	Y	4.0	5.0	D	O	5.0	23	81.3	21	89.8	-.-[1]	-.-[B]
110 Louisiana Tech U Industrial Eng [I]	LA	6.0	Y	4.0	4.7	D	O	-.-[A]	4	-.-[A]	4	-.-[A]	1	-.-[A]
111 U of Louisville Industrial Eng [I]	KY	9.0	Y[1]	4.0	4.5	D	R	98.0	16	67.1	15	68.4	2	57.1
117 Marietta Coll Industrial Eng [I]	OH	2.0	N	4.0	4.0	B	N	-.-	3	100.0	-.-	-.-	-.-	-.-
118 Marquette U Industrial Eng [I]	WI	4.0	Y	4.0	4.2	B	O	25.0	24	98.0	-.-[B]	-.-[B]	-.-[B]	-.-[B]
121 U of Mass Industrial Eng & Operations Research [I]	MA	7.3[4]	Y	4.0	4.6	D	O	36.8	38	96.3	28[2]	43.5	1	30.8
123 U of Mass, Lowell Industrial Tech [I]	MA	5.0	N	4.0	4.9	B	N	-.-	7	100.0	43	8.0	-.-	-.-

Industrial Engineering

ENGINEERING Institution Profile Reference Number & Name Name of Degree Program	State/Province	FACULTY Full-time equivalent (FTE)	ABET/CEAB accred?	UNDERGRADUATE PROGRAMS							GRADUATE PROGRAMS			
				Length		Time	Co-op		Bachelor's		Master's		Doctoral	
				Nominal length of program in years	Average length of program in years	Day/Eve./Both	None/Opt./Req.	% of graduates in Co-op programs	# of degrees awarded	% of graduates who were full-time	# of degrees awarded	% of graduates who were full-time	# of degrees awarded	% of graduates who were full-time
128 U of Miami Industrial Eng [I]	FL	10.5	Y	4.0	4.0	D	O	1.0	23	91.6	10	26.0	0	6.0
130 U of Michigan Industrial & Operations Eng [I]	MI	23.1	Y	4.0	4.7	D	O	8.3(7)	104	96.2	48(4)	88.1(5)	8	88.1(5)
131 U of Michigan-Dearborn Industrial & Systems Eng [I]	MI	7.5	Y	4.0	4.5	D	O	0.0(A)	13	57.6	15	6.4	-.-(B)	-.-(B)
134 Milwaukee Sch of Eng Industrial Eng [I]	WI	6.5	Y	4.0	4.6	D	N(1)	-.-	15	95.0	-.-	-.-	-.-	-.-
135 U of Minnesota, Duluth Industrial Eng [I]	MN	6.0	Y	4.0	4.6	D	N	-.-	24	100.0	-.-	-.-	-.-	-.-
138 Mississippi State U Industrial Eng [I]	MS	13.0	Y	4.0	4.8	D	O	30.0	46	95.0	6	100.0	1	100.0
139 U of Missouri-Columbia Industrial Eng [I]	MO	7.0	Y	4.0	-.-	D	O	4.0	19	96.8	25	62.2	3	66.6
144 Montana State U Industrial & Mgmt Eng [I]	MT	6.0	Y	4.0	4.7	D	N	-.-	17	-.-(1)	5	100.0	-.-	-.-
145 Morgan State U Industrial Eng [I]	MD	5.0	Y	4.0	4.5	D	O	0.0	10	93.0	-.-	-.-	-.-	-.-
146 U of Nebraska - Lincoln Industrial Eng [I]	NE	10.0	Y	4.0	4.5	D	O	-.-	46	-.-	8	-.-	3	-.-
150 U of New Haven Industrial Eng [I]	CT	7.0	Y	4.0	4.5(C)	B	O	5.0(A)	4	60.0(C)	14	30.0(C)	-.-(B)	-.-(B)
151 New Jersey Inst of Tech Industrial Eng [I]	NJ	9.3	Y	4.0	4.5	B	O	51.7	29	64.0	11	55.6	-.-(5)	-.-(5)
154 New Mexico State U Industrial Eng [I]	NM	6.5	Y	4.0	5.1(C)	D	O	18.0(C)	12	88.0(C)	29(2)	36.0(C)	0	60.0(C)
156 New York Inst of Tech Industrial Eng [I]	NY	1.5	N	4.0	4.0	-.-(A)	O	-.-(A)	14	-.-(A)	-.-(B)	-.-(B)	-.-(B)	-.-(B)
159 North Carolina A&T Industrial Eng [I]	NC	7.3	Y	4.0	-.-(A)	D	O	-.-(A)	30	-.-(A)	5	-.-(A)	-.-(B)	-.-(B)
160 North Carolina State U Industrial Eng [I] Industrial Eng (Furniture Manuf'g) [I]	NC	22.5 22.5	Y N	4.0 4.0	4.5 4.5	D D	O O	20.0 20.0	85 8	83.5 90.4	20 -.-	56.5 -.-	6 -.-	42.0 -.-
162 North Dakota State U Industrial Eng & Mgmt [I](1)	ND	7.0	Y	4.0	4.5	D	O	45.0	40	95.0	2	100.0	0	95.0
163 Northeastern U Industrial Eng & Information Systems [I]	MA	16.0	Y	5.0(1)	5.0	D	O	95.0	28	100.0	84	47.2	0	57.7
165 Northern Illinois U Industrial Eng [I]	IL	6.0	Y	4.0	4.6	D	O	15.0	8	90.0	8	66.0	-.-(B)	-.-(B)
166 Northwestern U Industrial Eng [I]	IL	16.3	Y	4.0	4.0	D	O	20.0	65	100.0	19	98.0	3	100.0
170 Ohio U Industrial & Systems Eng [I]	OH	8.0	Y	4.0	-.-	--	O	20.0	40	-.-	6	-.-	-.-	-.-
172 Ohio State U Industrial & Systems Eng [I]	OH	20.0	Y	4.0	4.5	D	O	-.-	63	92.0	22	100.0	10	100.0
173 U of Oklahoma Industrial Eng [I]	OK	10.0	Y	4.0	4.7	B	O	-.-(A)	33	81.0	25	60.4	6	44.4
175 Oklahoma State U Industrial Eng & Mgmt [I]	OK	12.5	Y	4.0	4.6	D	O	3.3	30	100.0	28	87.5	5	100.0

Industrial Engineering

ENGINEERING Institution Profile Reference Number & Name Name of Degree Program	State/Province	FACULTY Full-time equivalent (FTE)	ABET/CEAB accred?	UNDERGRADUATE PROGRAMS							GRADUATE PROGRAMS			
				Length		Time	Co-op		Bachelor's		Master's		Doctoral	
				Nominal length of program in years	Average length of program in years	Day/Eve./Both	None/Opt./Req.	% of graduates in Co-op programs	# of degrees awarded	% of graduates who were full-time	# of degrees awarded	% of graduates who were full-time	# of degrees awarded	% of graduates who were full-time
177 Oregon State U Industrial Eng [I]	OR	11.0	Y	4.0	4.8(C)	D	O(5)	58.0	45	100.0	10	100.0	3	100.0
180 Penn State U Industrial Eng [I]	PA	23.0	Y	4.0	-.-	D	O	20.0	132	-.-	34	-.-	5	-.-
181 U of Pittsburgh Industrial Eng [I]	PA	14.2	Y	4.0	4.0	D	O	12.2	41	92.5	31	41.0	5	84.2
182 Poly U Industrial Eng [I]	NY	13.0	Y	4.0	-.-(1)	D(B)	O	-.-(B)	14	80.0	13	37.0	3	11.0
187 U of Puerto Rico Industrial Eng [I]	PR	18.0	Y	5.0	5.0	B	O	9.0	92	96.0	4	100.0	-.-(B)	-.-(B)
188 Purdue U Industrial Eng [I]	IN	31.0	Y	4.0	4.3	D	O	19.0	195	100.0	60	100.0	11	100.0
189 Rensselaer Poly Inst Industrial & Mgmt Eng [I]	NY	6.0	Y	4.0	4.0	D	O	32.0	50	100.0	53	-.-(A)	8	-.-(A)
190 U of Rhode Island Industrial & Manuf'g Eng [I]	RI	7.5	Y	4.0	4.5	B	N	-.-	11	90.0	7	50.0	-.-	-.-
193 Rochester Inst of Tech Industrial & Manuf'g Eng [I]	NY	4.5	Y	5.0	5.0	D	R	100.0	25	88.0	6	5.0	-.-	-.-
195 Rutgers, The State U Industrial Eng [I]	NJ	8.0	Y	4.0	4.5	D	N	-.-	136	94.0	17	50.0	-.-	-.-
202 San Jose State U Industrial & Systems Eng [I]	CA	5.0	Y	4.0	-.-(A)	D	--	-.-	34	68.0	20	6.8	-.-(A)	-.-(A)
207 South Dakota School of Mines Industrial Eng [I]	SD	3.0	N(1)	4.0	4.5(C)	D	O	-.-(A)	9	-.-(A)	-.-(B)	-.-(B)	-.-(B)	-.-(B)
209 U of South Florida Eng Mgmt [I] Industrial Eng [I] Industrial Manuf'g [I]	FL	1.5 6.0 1.5(1)	N Y N	-.- 4.0 -.-	-.- 5.0 -.-	-- B --	-.- O -.-	-.- 10.0 -.-	-.- 18 -.-	-.- 50.0 -.-	26 0 2	2.0(2) 100.0 90.0	-.- 2 -.-	-.- 100.0 -.-
211 U of Southern Cal Industrial & Systems Eng [I]	CA	4.0	Y	4.0	4.0	D	O	-.-(A)	38	100.0	29	57.0	2	39.0
212 U of Southern Colorado Industrial Eng [I]	CO	4.0	Y(8)	4.0	5.0	D	O	10.0	-.-	-.-	3	10.0	-.-(B)	-.-(B)
214 Southern Illinois-Edwardsville Industrial Eng [I]	IL	3.0	Y	4.0	-.-(A)	D	O	-.-(A)	11	83.0	-.-(B)	-.-(B)	-.-(B)	-.-(B)
217 Stanford U Industrial Eng & Eng Mgmt [I]	CA	10.0	Y	4.0	4.0	D	N	-.-	52	100.0	32	78.0	2	100.0
218 SUNY-Binghamton Industrial Eng [I](1) Industrial Tech [I]	NY	-.-(B) -.-	-.-(B) --	-.-(B) -.-	-.-(B) -.-	-.-(B) -.-	-.-(B) -.-	-.-(B) -.-	-.-(B) -.-	-.-(B) -.-	13 -.-	40.0 -.-	1 -.-	88.0 -.-
219 SUNY-Buffalo Industrial Eng [I]	NY	11.0	Y	4.0	4.5	D	N	-.-	62	95.7	30	59.7	1	31.9
226 Technical U of Nova Scotia Industrial Eng [I]	NS	10.1	Y	4.0(2)	5.0	D	O	0.0(3)	19	100.0	2	100.0	1	100.0
228 U of Tennessee, Chattanooga Industrial Eng [I](4)	TN	-.-(7)	Y	4.0	4.8	B	O	-.-(A)	2	-.-(A)	3	-.-(A)	-.-	-.-
229 U of Tennessee, Knoxville Industrial Eng [I]	TN	12.0	Y	4.0	-.-(A)	D	O	21.8	69	88.5	47	-.-(A)	-.-(A)	-.-(A)
231 Tennessee Tech Industrial Eng [I]	TN	72.0	Y	4.0	4.0	D	O	10.0	23	100.0	4	100.0	0	100.0

Industrial Engineering

ENGINEERING Institution Profile Reference Number & Name Name of Degree Program	State/Province	FACULTY Full-time equivalent (FTE)	ABET/CEAB accred?	UNDERGRADUATE PROGRAMS							GRADUATE PROGRAMS			
				Length		Time	Co-op		Bachelor's		Master's		Doctoral	
				Nominal length of program in years	Average length of program in years	Day/Eve./Both	None/Opt./Req.	% of graduates in Co-op programs	# of degrees awarded	% of graduates who were full-time	# of degrees awarded	% of graduates who were full-time	# of degrees awarded	% of graduates who were full-time
232 U of Texas at Arlington Industrial Eng [I]	TX	-.-	Y	4.0	-.-	B	O	-.-	30	-.-	36	-.-	4	-.-
235 U of Texas at El Paso Industrial Eng [I]	TX	5.0	Y	4.0	-.-	D	O	20.0	14	-.-	7	-.-	-.-	-.-
237 Texas A&I U Industrial Eng [I]	TX	-.-	N	-.-(B)	-.-	-.-(B)	N	-.-	-.-	-.-	-.-	-.-	-.-	-.-
238 Texas A&M U Industrial Eng [I]	TX	31.2	Y	4.0	4.5(C)	D	O	0.0	100	91.0	38	83.0(C)	5	83.0(C)
239 Texas Tech U Industrial Eng [I]	TX	10.7	Y	4.0	-.-	D	N	-.-	24	91.6	25	81.0	6	79.3
240 U of Toledo Industrial Eng [I]	OH	8.0	Y	4.0	4.0	D	O	-.-(A)	13	86.0	5	50.0	-.-	-.-
254 Utah State U Industrial Tech [I]	UT	11.2	N(1)	4.0	4.0	D	O	-.-(A)	33	80.7	2	22.2	-.-	-.-
261 Virginia Tech Industrial & Systems Eng [I]	VA	29.0	Y	4.0	4.0	D	O	20.0	103	100.0	68	100.0	10	100.0
262 U of Washington Industrial Eng [I]	WA	7.0	Y	4.0	-.-(A)	D	O	-.-(A)	28	85.0	-.-(2)	-.-(B)	-.-(B)	-.-(B)
265 Wayne State U Industrial & Manuf'g Eng [I] Operations Research [I]	MI	8.0 5.0(3)	Y -.-(2)	4.0 -.-	4.5 -.-	B --	O --	-.-(A) -.-	9 -.-	-.-(A) -.-	23 1	-.-(A) -.-(A)	-.- 1	-.-(A) -.-
267 West Virginia U Industrial Eng [I]	WV	11.0	Y	4.0	4.6	D	O	0.0	42	98.0	11	90.0	6	85.0
269 Western Michigan U Industrial Eng [I]	MI	9.0	Y	4.0	4.5	D	O	10.0	17	95.4	8	29.7	-.-	-.-
270 Western New England Coll Industrial Eng [I]	MA	4.1	Y	4.0	4.4	B	N	-.-(B)	13	68.9	-.-(B)	-.-(B)	-.-(B)	-.-(B)
271 Wichita State U Industrial Eng [I]	KS	11.3	Y	4.0	5.5	B	O	6.3	16	63.0	28	35.1	2	40.7
273 Wilkes U Pre-Industrial Eng [I](6)	PA	-.-(A)	N	-.-(B)	-.-(B)	B	O	-.-(B)	-.-(B)	-.-(B)	-.-(B)	-.-(B)	-.-(B)	-.-(B)
274 U of Wisconsin-Madison Industrial Eng [I]	WI	22.5	Y	4.0	5.0	D	O	-.-(A)	54	-.-(A)	37	-.-(A)	9	-.-(A)
275 U of Wisconsin-Milwaukee Industrial Eng [I]	WI	8.8	Y	4.0	5.5	B	O	7.1	14	78.6	6	21.1	0	50.0
276 U of Wisconsin-Platteville Industrial Eng [I]	WI	4.0	Y	4.0	5.0	D	O	27.0	31	100.0	-.-	-.-	-.-	-.-
277 Worcester Poly Inst Manuf'g Eng [I]	MA	8.0	Y	4.0	4.0	D	O	-.-(A)	22	100.0	4	-.-(A)	-.-	-.-(A)

Industrial Engineering—Totals

State/Province/Country	# of schools	Full-time equivalent (FTE)	# accredited programs	Nominal length of program in years	Average length of program in years	Day/Eve./Both	None/Opt./Req.	% of graduates in Co-op programs	# of degrees awarded (Bachelor's)	% of graduates who were full-time	# of degrees awarded (Master's)	% of graduates who were full-time	# of degrees awarded (Doctoral)	% of graduates who were full-time
AL	3	39.30	3						97		59		9	
AR	1	-	1						30		9		3	
AZ	2	36.50	2						54		41		10	
CA	7	60.80	7						247		113		16	
CO	1	4.00	1						-		3		-	
CT	2	8.50	1						4		14		-	
FL	7	74.09	7						161		74		7	
GA	1	51.70	1						253		117		16	
IA	2	7.50	2						129		18		8	
IL	6	45.80	6						187		51		4	
IN	1	31.00	1						195		60		11	
KS	2	21.30	2						69		47		5	
KY	1	9.00	1						16		15		2	
LA	2	17.50	2						27		25		1	
MA	5	40.40	4						108		159		1	
MD	1	5.00	1						10		-		-	
MI	5	72.59	6						195		95		9	
MN	1	6.00	1						24		-		-	
MO	1	7.00	1						19		25		3	
MS	1	13.00	1						46		6		1	
MT	1	6.00	1						17		5		-	
NC	2	52.30	2						123		25		6	
ND	1	7.00	1						40		2		-	
NE	1	10.00	1						46		8		3	
NJ	2	17.30	2						165		28		-	
NM	1	6.50	1						12		29		-	
NY	9	48.70	6						212		154		18	
OH	7	82.80	5						161		71		14	
OK	2	22.50	2						63		53		11	
OR	1	11.00	1						45		10		3	
PA	5	52.20	3						220		78		13	
PR	1	18.00	1						92		4		-	
RI	1	7.50	1						11		7		-	
SC	1	9.00	1						58		11		1	
SD	1	3.00	-						9		-		-	
TN	3	84.00	3						94		54		-	
TX	7	58.40	6						186		126		20	
UT	1	11.20	-						33		2		-	
VA	1	29.00	1						103		68		10	
WA	1	7.00							28		-		-	
WI	5	45.80	5						138		43		9	
WV	1	11.00	1						42		11		6	
U.S. Totals:	107	1,151.20	96						3,769		1,720		220	
NS	1	10.10	1						19		2		1	
PQ	1	20.40	1						62		10		-	
Canada Totals:	2	30.50	2						81		12		1	

Industrial Engineering—Footnotes

FOOTNOTES: The following footnotes are the same for all schools: (A) Data not available. (B) Data not applicable. (C) Estimated data.

012 Auburn University: (6) Supported by the faculty of the Departments of Industrial Engineering, Mechanical Engineering, and Management. (10) Based on credit hours taken Fall Quarter 1991.

015 Bradley University: (1) Students typically take the equivalent of nine semesters but finish in 4 years by using summer school and transfer credits.

037 Case Western Reserve University: (2) Department staff is shared with Systems Engineering.

042 University of Cincinnati: (1) Of the 6 departments., 5 have more than one program, therefore, faculty provide support in multiple programs. Total FTE for the college is 142.

048 Cleveland State University: (1) includes Co-Op and part-time students

054 Columbia University: (2) Industrial Engineering is ABET accredited.

071 University of Florida: (2) Industrial Engineering and Systems Engineering options are available

072 Florida Agricultural and Mechanical University: (1) Includes FAMU and FSU students. Year is Fall 1991, Spring 1992 and Summer 1992.

076 Florida State University: (1) Includes FAMU and FSU students. Year is Fall 1992, Spring 1992, and Summer 1992.

077 GMI Engineering & Management Institute: (1) 3360 hours minimum

082 Georgia Institute of Technology: (7) Includes M.S. Operations Research, M.S. Statistics, and M.S. Health Systems. (8) Fall '91 Quarter FTE.

090 University of Houston: (3) BSIE, MIE, MSIE, PhD

094 University of Illinois at Chicago: (5) Offered as PhD degree only. (6) Offered as M.S. degree only.

095 University of Illinois at Urbana-Champaign: (1) Industrial Engineering is in the Department of Mechanical and Industrial Engineering.

108 Lehigh University: (2) Planned for 1994

109 Louisiana State University: (1) PhD available through Engineering Science (interdisciplinary program).

111 University of Louisville: (1) Program is accredited at the master's level.

121 University of Massachusetts: (2) Includes 13 M.S. degrees earned in Engineering Management. Also see footnote under Mechanical Engineering regarding M.S. degrees in Manufacturing. (4) Includes department head

130 University of Michigan: (4) Includes Professional degrees. (5) All graduate students. (7) For whole College.

134 Milwaukee School of Engineering: (1) No co-op program is available, however, 85% of undergraduates participate in an internship program and have relevant work experience prior to graduation.

144 Montana State University: (1) Information unavailable but nearly 100%

151 New Jersey Institute of Technology: (5) No doctoral program. Undergraduate and masters level only.

154 New Mexico State University: (2) Includes degrees unearned through off-campus programs at Kirtland and Cannon Air Force bases.

162 North Dakota State University: (1) Masters & Ph.D. degree also available.

163 Northeastern University: (1) 5 year co-op option; 4 year co-op option; 4-year non-co-op option

177 Oregon State University: (5) Required for the Manufacturing Option

182 Polytechnic University: (1) 4-5

207 South Dakota School of Mines and Technology: (1) New program not yet eligible for ABET accreditation.

209 University of South Florida: (1) Masters program only. (2) This degree program is for working engineers.

212 University of Southern Colorado: (8)

218 State University of New York at Binghamton: (1) Included in Mechanical Engineering Section

226 Technical University of Nova Scotia: (2) Three years at TUNS. (3) Just started.

228 University of Tennessee, Chattanooga: (4) BSE with Industrial Concentration. (7) All faculty are included under Engineering.

254 Utah State University: (1) Currently seeking ABET accreditation to establish Aerospace, Environmental and Civil Engineering as separately accredited degree programs.

262 University of Washington: (2) MS in Engineering degree awarded through Inter-Engineering program. 4 degrees awarded 1991-92.

265 Wayne State University: (2) No Undergraduate Program (3) Included in Industrial and Manufacturing Engineering

273 Wilkes University: (6) Prepares the student for direct transfer to the third year in another institution.

1992 ASEE ANNUAL CONFERENCE PROCEEDINGS

"CREATIVITY: EDUCATING WORLD-CLASS ENGINEERS"
HOSTED BY THE UNIVERSITY OF TOLEDO
JUNE 21-25, 1992
TOLEDO, OHIO

*Order the 1992 ASEE Conference Proceedings and take advantage of a 30% savings **only through this offer!***

ASEE member price was $55.00
NOW $38.50

ASEE non-member price was $70.00
NOW $49.00

To order, fill out the ASEE Publications order form in the back of this book, using order number 4183-PU-06-087.

Mechanical Engineering

Institutions are listed in alphabetical order. The first line in a school's entry contains the institution profile reference number, the school name, and the state or province in which the main campus is located. Subsequent indented lines for each program give the name of the degree program, followed by a major category abbreviation in brackets (H = Chemical, V = Civil, E = Electrical, I = Industrial, M = Mechanical, and O = Other) the full-time equivalent (FTE) of all faculty. For undergraduate programs, the columns represent: accreditation status, the nominal and actual average length of the program, the time that courses are available (Day, Evening or Both), the availability of cooperative (internship) programs (None, Optional, or Required), the percentage of graduates who participated in co-op programs, the number of bachelor's degrees awarded, and the percentage of these graduates who were full-time. For graduate programs, the columns indicate the number of graduates and the percentage who were full-time in master's and then doctoral programs. Unless otherwise footnoted, information on degrees is for the 1991-92 academic year; all other information is current. At the end of the table are (1) a summary of totals by state or province and by country and (2) the full text of all footnotes for this table.

ENGINEERING Institution Profile Reference Number & Name / Name of Degree Program	State/Province	FACULTY Full-time equivalent (FTE)	ABET/CEAB accred?	Length Nominal length of program in years	Length Average length of program in years	Time Day/Eve./Both	Co-op None/Opt./Req.	Co-op % of graduates in Co-op programs	Bachelor's # of degrees awarded	Bachelor's % of graduates who were full-time	Master's # of degrees awarded	Master's % of graduates who were full-time	Doctoral # of degrees awarded	Doctoral % of graduates who were full-time
001 U of Akron	OH													
Mechanical Eng [M]		20.0	Y	4.0	-.-(A)	B	O	60.0	94	68.3(A)	31	50.0	2	75.0
002 U of Alabama	AL													
Mechanical Eng [M]		15.0	Y	4.0	4.5	D	O	21.1	57	100.0	10	100.0	3	100.0
003 U of Alabama at Birmingham	AL													
Mechanical Eng [M]		9.0	Y	4.0	-.-(A)	B	O	-.-(A)	39	39.0	6	50.0	-.-(B)	-.-(B)
004 U of Alabama at Huntsville	AL													
Mechanical & Aerospace Eng [M]		17.9	Y	4.0	-.-(A)	B	O	-.-(A)	45	64.8	25	35.0	3	59.2
Mechanical Eng [M]		-.-	Y	4.0	-.-	B	O	-.-	-.-	-.-	-.-	-.-	-.-	-.-
005 U of Alaska, Fairbanks	AK													
Mechanical Eng [M]		8.0	Y	4.0	4.5	D	N	-.-	13	90.0	13	90.0	-.-	-.-
006 U of Alberta	AB													
Mechanical Eng [M]		25.0	Y	4.0	-.-(A)	D	O	33.0	92	100.0	12	70.0	2	100.0
007 Alfred U	NY													
Mechanical Eng [M]		5.0	Y	4.5	4.5	B	O	90.0	8	100.0	1	100.0	-.-(B)	-.-(B)
008 U of Arizona	AZ													
Mechanical Eng [M]		33.0	Y	4.0	-.-	D	O	-.-	83	88.0	12	-.-	15	-.-
009 Arizona State U	AZ													
Mechanical Eng [M]		40.3	Y	4.0	5.0	D	O	-.-(A)	83	83.0	31	66.0	6	66.0
010 U of Arkansas	AR													
Mechanical Eng [M]		13.0	Y	4.0	-.-	D	O	-.-	42	90.2	4	31.8	1	100.0
012 Auburn U	AL													
Aerospace Eng [M]		16.0	Y	4.0	-.-(A)	D	O	12.3	65	88.3(10)	7	69.2(10)	3	70.0(10)
Materials Eng [M]		-.-(7)	Y	4.0	-.-(A)	D	O	20.0	15	100.0(10)	15	48.8(10)	3	50.0(10)
Mechanical Eng [M]		27.0	Y	4.0	-.-(A)	D	O	33.6	131	84.3(10)	27	32.9(10)	6	59.3(10)
014 Boston U	MA													
Mechanical Eng [M]		17.0(3)	Y	4.0	4.0	D	O	17.1	38	99.0(1)	23	5.9	1(5)	92.4
015 Bradley U	IL													
Mechanical Eng [M]		12.8	Y	4.0	4.0(1)	B	O	9.0	56	89.2	13	11.6	-.-(B)	-.-(B)
016 U of Bridgeport	CT													
Mechanical Eng [M]		5.0	Y	4.0	4.0	B	O	5.0	22	72.4	9	31.3	-.-	-.-
017 Brigham Young U	UT													
Mechanical Eng [M]		19.7	Y	4.0	5.0	B	O	-.-(A)	139	-.-(A)	11	-.-(A)	5	-.-(A)
018 Bucknell U	PA													
Mechanical Eng [M]		8.0(1)	Y	4.0	4.0	-.-(A)	N(A)	-.-(A)	31	100.0	6	100.0	-.-(A)	-.-(A)
019 U of Cal at Berkeley	CA													
Mechanical Eng [M](1)		42.0	Y	4.0	-.-(A)	D	O	13.0	128	100.0	64	100.0	36	100.0
020 U of Cal, Davis	CA													
Mechanical Eng [M]		37.5	Y	4.0	-.-(A)	D	O	-.-(A)	47	100.0	27	100.0	13	100.0
Mechanical Eng & Materials Science [M]		-.-(1)	N	4.0	-.-(A)	D	O	-.-(A)	10	100.0	-.-(1)	-.-(1)	-.-(1)	-.-(1)
021 U of Cal, Irvine	CA													
Mechanical & Aerospace Eng [M]		22.0	Y	4.0	4.5	D	N	-.-(B)	53	100.0	22	88.0	8	100.0
022 U of Cal, Los Angeles	CA													
Mechanical Eng [M]		11.0	Y	4.0	4.6	-.-	O	-.-	80	100.0	25	100.0	13	100.0
023 U of Cal, San Diego	CA													
Mechanical Eng [M]		36.9(2)	Y	4.0	4.7	D	N	-.-	84	100.0	9	100.0	0	100.0
024 U of Cal, Santa Barbara	CA													
Mechanical Eng [M]		21.3	Y	4.0	4.2	-.-(B)	N	-.-(B)	78	100.0	28	100.0	8	100.0

Mechanical Engineering

ENGINEERING Institution Profile Reference Number & Name Name of Degree Program	State/Province	FACULTY Full-time equivalent (FTE)	ABET/CEAB accred?	UNDERGRADUATE PROGRAMS					Bachelor's		GRADUATE PROGRAMS			
				Length Nominal length of program in years	Length Average length of program in years	Time Day/Eve./Both	None/Opt./Req.	Co-op % of graduates in Co-op programs	# of degrees awarded	% of graduates who were full-time	Master's # of degrees awarded	Master's % of graduates who were full-time	Doctoral # of degrees awarded	Doctoral % of graduates who were full-time
026 Cal Inst of Tech Mechanical Eng [M]	CA	11.0	N	-.-(B)	-.-	-.-(B)	-.-(B)	-.-	-.-	-.-	9	100.0	4	100.0
027 Cal Poly State U Mechanical Eng [M]	CA	27.2	Y	4.0	5.0	D	O	-.-	146	-.-	-.-	-.-	-.-	-.-
028 Cal State U, Chico Mechanical Eng [M] Mechanical Eng & Manuf'g [M]	CA	8.0 8.0	Y Y	4.0 4.0	-.-(A) -.-(A)	D D	O O	-.-(A) -.-(A)	40 40	92.0 92.0	-.-(B) -.-(B)	-.-(B) -.-(B)	-.-(B) -.-(B)	-.-(B) -.-(B)
029 Cal State U, Fresno Mechanical Eng [M]	CA	7.0	Y	4.0	5.0	D	O	5.0(C)	42	75.0	-.-	-.-	-.-	-.-
030 Cal State U, Long Beach Mechanical Eng [M]	CA	16.4	Y	4.0	5.5	B	N	-.-	103	55.0	21	10.0	-.-	-.-
031 Cal State U, Los Angeles Mechanical Eng [M]	CA	10.1	Y	4.0	4.0	B	O	-.-(A)	17	57.0	4	5.0	-.-(B)	-.-(B)
033 Cal State U, Sacramento Mechanical Eng [M]	CA	17.0	Y	4.0	5.0	B	O	4.0	74	-.-(A)	-.-(B)	-.-(B)	-.-(B)	-.-(B)
034 Cal State Poly, Pomona Mechanical Eng [M]	CA	23.8	Y	4.0	5.5	B	O	-.-	67	75.0	-.-	-.-	-.-	-.-
036 Carnegie Mellon U Mechanical Eng [M]	PA	23.6	Y	4.0	4.0	D	N	-.-(B)	62	98.9	22	89.0	9	98.0
037 Case Western Reserve U Mechanical Eng [M]	OH	16.0	Y	4.0	4.3	B	O	10.0	35	70.0	28	50.0	10	90.0
038 The Catholic U of America Mechanical Eng [M]	DC	10.0	Y	4.0	4.0	D	O	10.0	12	93.0	14	7.0	4	100.0
039 U of Central Florida Mechanical Eng [M]	FL	12.0	Y	4.0	4.5	B	O	15.0(C)	50	68.5	11	16.0	1	20.0(C)
041 Christian Brothers U Mechanical Eng [M]	TN	6.0	Y	4.0	4.7	D	N	-.-(B)	15	86.8	-.-(B)	-.-(B)	-.-(B)	-.-(B)
042 U of Cincinnati Mechanical Eng [M]	OH	30.0(1)	Y	5.0	5.2	D	R	100.0	23	100.0	30	58.8	10	78.2
044 CUNY-City Coll Mechanical Eng [M]	NY	19.3	Y	4.0	5.0	D(1)	O	-.-(A)	42	83.4	19	9.6	2	100.0
046 Clarkson U Mechanical & Aeronautical Eng [M]	NY	21.5	Y	4.0	-.-	D	O	-.-(A)	154	99.9	14	-.-	6	-.-
047 Clemson U Mechanical Eng [M]	SC	31.0	Y	4.0	-.-	D	O	-.-	138	86.0	44	94.0	5	73.0
048 Cleveland State U Mechanical Eng [M]	OH	12.0	Y	4.0	5.9(1)	B	O	22.5	47	57.3	10	23.1	4	100.0
049 U of Colorado at Boulder Mechanical Eng [M]	CO	22.0	Y	4.0	4.5	D	O	-.-	48	97.9	16	87.5	4	87.5
051 U of Colorado at Denver Mechanical Eng [M]	CO	10.1	Y	4.0	4.5	B	O	1.0	36	50.0	3	20.0	-.-(3)	-.-(3)
053 Colorado State U Mechanical Eng [M]	CO	23.0	Y	4.0	4.5	D(A)	N	-.-(A)	77	73.0	15	63.0	6	58.0
054 Columbia U Mechanical Eng [M]	NY	12.3	Y	4.0	4.0	-.-(B)	N	-.-(B)	47	100.0	38	80.0	11	80.0
055 Concordia U Mechanical Eng [M]	PQ	-.-	Y	4.0	-.-	B	N	-.-	82	81.0	24	87.0	2	83.0
056 U of Conn Mechanical Eng [M]	CT	21.0	Y	4.0	4.5	D	O	27.0	78	95.0(C)	23(4)	40.0(4)	13	52.0
057 The Cooper Union Mechanical Eng [M]	NY	8.6	Y	4.0	4.0	D	N	-.-	23	100.0	3	100.0	-.-	-.-
058 Cornell U Mechanical & Aerospace Eng [M]	NY	27.0	Y	4.0	-.-	D	O	29.0	147	-.-	42	-.-	12	-.-
060 U of Dayton Mechanical Eng [M]	OH	14.8	Y	4.0	4.0	D	O	35.0	69	95.2	39	16.0	3	26.0

Mechanical Engineering

ENGINEERING Institution Profile Reference Number & Name Name of Degree Program	State/Province	FACULTY Full-time equivalent (FTE)	ABET/CEAB accred?	UNDERGRADUATE PROGRAMS							GRADUATE PROGRAMS			
				Length		Time	Co-op		Bachelor's		Master's		Doctoral	
				Nominal length of program in years	Average length of program in years	Day/Eve./Both	None/Opt./Req.	% of graduates in Co-op programs	# of degrees awarded	% of graduates who were full-time	# of degrees awarded	% of graduates who were full-time	# of degrees awarded	% of graduates who were full-time
061 U of Delaware Mechanical Eng [M]	DE	16.0	Y	4.0	4.0	B	N	-.-	62	-.-	11	19.0	7	31.0
062 U of Denver Mechanical Eng [M]	CO	7.0	Y	4.0	4.0	D	O	0.0	4	7.7	1	100.0	-.-	-.-
063 U of the District of Columbia Mechanical Eng [M]	DC	5.0	Y	4.0	6.0	D	O	-.-(A)	7	-.-(A)	-.-	-.-	-.-	-.-
064 Drexel U Mechanical Eng [M]	PA	30.0	Y	5.0	5.0	B	R	100.0	119	80.0	37	35.0	7	90.0
065 Duke U Mechanical Eng [M]	NC	23.0	Y	4.0	4.0	-.-(B)	N	-.-	43	100.0	14	100.0	9	100.0
066 École Polytechnique Mechanical Eng [M](7)	PQ	46.2	Y	4.0	4.5	D	N	-.-	167	97.3	30	80.5	6	100.0
069 U of Evansville Mechanical Eng [M]	IN	5.2	Y	4.0	4.2	D	O	36.0	38	96.0	-.-(B)	-.-(B)	-.-(B)	-.-(B)
071 U of Florida Mechanical Eng [M]	FL	31.3	Y	4.5	5.0	D	O	3.0	80	98.0	20	100.0	9	100.0
072 Florida A&M U Mechanical Eng [M]	FL	16.0	Y	4.5	5.0	D	O	-.-(A)	52(1)	-.-(A)	7(1)	-.-(A)	-.-	-.-(A)
073 Florida Atlantic U Mechanical Eng [M]	FL	16.5	Y	4.5	5.0(5)	B	O	10.0(C)	25	66.0	25	38.6	7	72.2
074 Florida Inst of Tech Mechanical & Aerospace Eng [M]	FL	13.9	Y	4.0	4.5	D	O	35.0	40	91.9	11	60.0	1	-.-
075 Florida International U Mechanical Eng [M]	FL	16.0	Y	4.0	4.5	B	O	10.0	33	80.1	9	100.0	-.-	-.-
076 Florida State U Mechanical Eng [M]	FL	16.0	Y	4.5	5.0	D	O	-.-(A)	52(1)	-.-(A)	7(1)	-.-(A)	-.-	-.-(A)
077 GMI Mechanical Eng [M]	MI	32.0	Y	5.0	5.0	D(1)	R	100.0	136	100.0	-.-	-.-	-.-	-.-
078 Gannon U Mechanical Eng [M]	PA	7.0	Y	4.0	4.5	B	O	0.0	16	82.0	2	8.3	-.-	-.-
080 The George Washington U Mechanical Eng [M]	DC	19.0	Y	4.0	-.-(A)	B	O	-.-(A)	22	92.0	32	35.0	3	37.0
082 Georgia Tech Mechanical Eng [M] Nuclear Eng [M]	GA	47.5(8) 10.0(8)	Y Y	4.0 4.0	4.5 4.4	D D	O O	43.0 43.0	331 7	93.9 94.4	83 32(6)	81.0 34.0	23 3	94.0 83.0
083 Gonzaga U Mechanical Eng [M]	WA	9.0	Y	4.0	-.-	D	N	-.-	12	93.0	2	100.0	-.-	-.-
085 Grove City Coll Mechanical Eng [M]	PA	6.0	Y	4.0	4.1	D	N	-.-	13	100.0	-.-(B)	-.-(B)	-.-(B)	-.-(B)
088 U of Hawaii at Manoa Mechanical Eng [M]	HI	15.0	Y	4.0	5.0	D	O	-.-	48	83.8	7	97.0	3	89.0
089 Hofstra U Mechanical Eng [M]	NY	5.8	Y	4.0	4.5	B	N	-.-	13	77.8	-.-	-.-	-.-	-.-
090 U of Houston Mechanical Eng [M](4)	TX	23.3	Y	4.5	-.-(A)	B	O	-.-(A)	51	42.0	14	35.0	4	60.0
091 Howard U Mechanical Eng [M](C)	DC	10.0	Y	4.0	5.0	B	O	0.0	28	85.0	5	75.0	2	80.0
092 U of Idaho Mechanical Eng [M]	ID	17.0	Y	4.0	5.5	D	O	0.0	40	94.8	8	87.0	0	72.0
094 U of Illinois at Chicago Mechanical Eng [M]	IL	24.0	Y	4.0	-.-	D	O	14.0	79	83.0	46	36.0	12	23.0
095 Illinois, Urbana-Champaign Mechanical Eng [M]	IL	46.4	Y	4.0	4.5	-.-(A)	O	15.6	192	98.5	61	100.0	25	93.5
096 Illinois Inst of Tech Mechanical Eng [M]	IL	21.7	Y	4.0	4.8	B	O	3.3(2)	61	68.0	12(5)	30.0	5	52.0

Mechanical Engineering

ENGINEERING Institution Profile Reference Number & Name Name of Degree Program	State/Province	FACULTY Full-time equivalent (FTE)	ABET/CEAB accred?	UNDERGRADUATE PROGRAMS					Bachelor's		GRADUATE PROGRAMS			
				Length		Time	Co-op				Master's		Doctoral	
				Nominal length of program in years	Average length of program in years	Day/Eve./Both	None/Opt./Req.	% of graduates in Co-op programs	# of degrees awarded	% of graduates who were full-time	# of degrees awarded	% of graduates who were full-time	# of degrees awarded	% of graduates who were full-time
097 Indiana-Purdue, Indianapolis Mechanical Eng [M]	IN	12.0	Y	4.0	4.0	B	O	0.0	32	57.0	2	33.0	-.-	-.-
098 U of Iowa Mechanical Eng [M]	IA	12.6	Y	4.0	4.5(C)	D	O	15.0(C)	43	100.0(C)	21	100.0(C)	19	100.0(C)
099 Iowa State U Mechanical Eng [M]	IA	-.-	Y	4.0	-.-	D	O	-.-	113	-.-	14	-.-	11	-.-
100 The Johns Hopkins U Mechanical Eng [M](10)	MD	15.3(4)	Y	4.0	4.0	B	O	-.-(A)	16	100.0	13	15.0	3	100.0
101 U of Kansas Mechanical Eng [M]	KS	12.0	Y	4.0	-.-	D	O	-.-	36	-.-(A)	10	-.-(A)	3	-.-(A)
102 Kansas State U Mechanical Eng [M]	KS	16.0	Y	4.0	4.7	D	O	-.-	79	96.0	10	-.-	4	-.-
103 U of Kentucky Mechanical Eng [M]	KY	19.0	Y	4.0	5.0(C)	D	O	-.-(A)	45	90.9	8	-.-(A)	5	58.3
104 Lafayette Coll Mechanical Eng [M](1)	PA	8.0	Y	4.0	4.0	B	N	-.-	32	72.0	-.-	-.-	-.-	-.-
105 Lamar U Mechanical Eng [M]	TX	4.8	Y	4.0	5.0	D	O	74.0	20	-.-(A)	-.-(A)	-.-(A)	-.-(A)	-.-(A)
106 Lawrence Tech U Mechanical Eng [M]	MI	23.4	Y	4.0	4.5(C)	B	O	6.9	175	58.2	9	0.0(3)	-.-	-.-
108 Lehigh U Mechanical Eng & Appl Mechanics [M]	PA	34.0	Y	4.0	4.0	D	O	1.0	103	100.0	23	75.0	6	99.0
109 Louisiana State U Mechanical Eng [M]	LA	20.8	Y	4.0	5.0	D	O	4.8	61	86.9	7	79.1	8	92.3
110 Louisiana Tech U Mechanical Eng [M]	LA	13.0	Y	4.0	4.7	D	O	-.-(A)	29	-.-(A)	8	-.-(A)	1	-.-(A)
111 U of Louisville Mechanical Eng [M]	KY	13.0	Y(1)	4.0	4.5	D	R	92.0	34	72.4	24	45.1	-.-	-.-
113 Loyola Marymount U Mechanical Eng [M]	CA	6.5	Y	4.0	4.2	B	N	-.-	12	100.0	5	0.0	-.-	-.-
114 U of Maine Mechanical Eng [M]	ME	9.0	Y	4.0	4.5	D	N	-.-	33	90.0	-.-	100.0	-.-(1)	-.-
115 U of Manitoba Industrial Eng [M] Mechanical Eng [M]	MB	-.- -.-	Y Y	4.0 4.0	-.-(A) -.-(A)	D D	N N	-.-(B) -.-(B)	26 43	-.-(A) -.-(A)	11(2) -.-(2)	-.-(A) -.-(A)	-.-(2) -.-(2)	-.-(B) -.-(B)
116 Mankato State U Mechanical Eng [M]	MN	5.0	N(1)	4.0	4.5	D	N	-.-	14	100.0	-.-	-.-	-.-	-.-
118 Marquette U Mechanical Eng [M]	WI	15.9	Y	4.0	4.2	B	O	43.0	58	92.0	15	34.0	4	82.0
119 Maryland Baltimore Cty Mechanical Eng [M]	MD	15.0	Y	4.0	4.5	D	O	33.3	42	86.0	9	52.0	-.-	88.0
120 U of Maryland, Coll Park Mechanical Eng [M]	MD	40.0	Y	4.0	4.5	D	O	19.4	144	84.3	44	56.4	17	57.3
121 U of Mass Mechanical Eng [M]	MA	23.7(4)	Y	4.0	4.6	D	O	40.7	91	95.6	35(3)	58.8	6	46.8
122 U of Mass, Dartmouth Mechanical Eng [M]	MA	10.0	Y	4.0	4.0	--	N	-.-	22	100.0	-.-	-.-	-.-	-.-
123 U of Mass, Lowell Mechanical Eng [M]	MA	2.5	Y	4.0	4.7	D	N	-.-	74	100.0	16	50.0	2	25.0
124 Mass Inst of Tech Mechanical Eng [M]	MA	60.0	Y	4.0	-.-	D	O	-.-(3)	138	100.0	143	100.0	31	100.0
125 McGill U Mechanical Eng [M]	PQ	24.9	Y	3.5(1)	4.0(C)	D	O(2)	-.-(A)	98	-.-(A)	29	48.4(3)	12	55.8(3)
126 Memphis State U Mechanical Eng [M]	TN	13.0	Y	4.0	5.0	D	O	10.0	54	70.0	12	65.0	3	50.0

Mechanical Engineering

Institution Profile Reference Number & Name / Name of Degree Program	State/Province	FACULTY Full-time equivalent (FTE)	ABET/CEAB accred?	UNDERGRADUATE PROGRAMS - Nominal length of program in years	Average length of program in years	Time Day/Eve./Both	Co-op None/Opt./Req.	% of graduates in Co-op programs	Bachelor's # of degrees awarded	% of graduates who were full-time	Master's # of degrees awarded	% of graduates who were full-time	Doctoral # of degrees awarded	% of graduates who were full-time
128 U of Miami Mechanical Eng [M]	FL	7.0	Y	4.0	4.0	D	O	2.0	28	94.4	7	88.0	2	100.0
130 U of Michigan Appl Mechanics [M][2] Mechanical Eng [M]	MI	-.-[3] 62.4	-- Y	-.- 4.0	-.- 4.7	-- D	-- O	-.- 8.3[7]	-.- 228	-.- 94.2	7[4] 100[4]	87.5[5] 82.3[5]	5 39	87.5[5] 82.3[5]
131 U of Michigan-Dearborn Mechanical Eng [M]	MI	15.1	Y	4.0	4.5	D	O	0.0[A]	58	62.6	26	0.0	-.-[B]	-.-[B]
132 Michigan State U Mechanical Eng [M]	MI	21.0	Y	4.0	4.5	D	O	40.0	168	100.0	12	100.0	6	100.0
133 Michigan Tech U Mechanical Eng [M]	MI	41.0	Y	4.0	4.3	D	O	15.1	271	95.3	27	98.5	5	93.1
134 Milwaukee Sch of Eng Mechanical Eng [M]	WI	12.5	Y	4.0	4.6	D	N[1]	-.-	56	95.0	-.-	-.-	-.-	-.-
136 U of Minnesota Mechanical Eng [M]	MN	40.0	Y	4.0	-.-[A]	B	O	-.-[A]	210	-.-[A]	60	-.-[A]	23	-.-[A]
137 U of Mississippi Mechanical Eng [M][4]	MS	7.2	Y	4.0	4.5	D	N	-.-	7	91.8	-.-	-.-	-.-	-.-
138 Mississippi State U Mechanical Eng [M]	MS	16.0	Y	4.0	4.8	D	O	30.0	50	95.0	8	100.0	2	100.0
139 U of Missouri-Columbia Mechanical & Aerospace Eng [M][3]	MO	23.0	Y	4.0	-.-	D	O	4.0	107	87.0	22	61.1	3	46.4
140 U of Missouri-Kansas City Mechanical & Aerospace Eng [M]	MO	6.0	Y	4.0	5.0	B	N	-.-[2]	23	40.0	3	20.0	0	75.0
141 U of Missouri-Rolla Mechanical Eng [M]	MO	24.4	Y	4.0	4.5	D	O	12.5	152	87.0	16	43.0	7	45.0
144 Montana State U Mechanical Eng [M]	MT	13.0	Y	4.0	4.7	D	N	-.-	59	-.-[1]	3	100.0	-.-	-.-
146 U of Nebraska - Lincoln Mechanical Eng [M]	NE	16.0	Y	4.0	4.5	D	O	-.-	141	-.-	22	-.-	4	-.-
147 U of Nevada, Las Vegas Mechanical Eng [M]	NV	7.0	Y	4.0	4.0	B	N	-.-	13	70.9	2	53.0	-.-[B]	-.-[B]
149 U of New Hampshire Mechanical Eng [M]	NH	14.0	Y	4.0	-.-[A]	D	N	-.-	44	-.-[A]	11	-.-[A]	2	-.-[A]
150 U of New Haven Mechanical Eng [M]	CT	11.0	Y	4.0	4.5[C]	B	O	5.0[A]	33	60.0[C]	6	20.0[C]	-.-[B]	-.-[B]
151 New Jersey Inst of Tech Mechanical Eng [M]	NJ	31.2	Y	4.0	4.5	B	O	7.7	78	75.2	39	24.2	5	82.1
152 The U of New Mexico Mechanical Eng [M]	NM	16.8	Y	4.0	5.0	D	O	-.-[A]	55	72.1	16	100.0[1]	3	100.0[1]
154 New Mexico State U Mechanical Eng [M]	NM	18.3	Y	4.0	5.1[C]	D	O	20.0[C]	58	88.0[C]	6	67.0[C]	5	62.0[C]
155 U of New Orleans Mechanical Eng [M]	LA	10.0	Y	4.0	5.2	B	O	-.-	27	72.0	-.-	-.-	-.-	-.-
156 New York Inst of Tech Mechanical & Aeronautical Eng [M]	NY	13.5	Y	4.0	4.0	B	O	-.-[A]	91	-.-[A]	-.-[B]	-.-[B]	-.-[B]	-.-[B]
158 UNC-Charlotte Mechanical Eng & Eng Science [M]	NC	15.3	Y	4.0	4.5	B	O	22.4	58	82.9	9	27.5	-.-	-.-
159 North Carolina A&T Mechanical Eng [M]	NC	19.1	Y	4.0	-.-[A]	D	O	-.-[A]	32	-.-[A]	4	-.-[A]	-.-[B]	-.-[B]
160 North Carolina State U Aerospace Eng [M] Mechanical Eng [M]	NC	39.5 39.5	Y Y	4.0 4.0	4.5 4.5	D D	O O	20.0 20.0	19 216	89.7 80.1	9 18	82.1 50.0	6 18	62.5 38.5
161 U of North Dakota Mechanical Eng [M]	ND	10.0	Y	4.0	5.0	D	O	-.-	24	-.-	2	-.-	-.-	-.-
162 North Dakota State U Mechanical Eng & Appl Mechanics [M][1]	ND	14.0	Y	4.0	4.5	D	O	-.-	62	95.0	1	100.0	1	100.0

Mechanical Engineering

ENGINEERING Institution Profile Reference Number & Name Name of Degree Program	State/Province	FACULTY Full-time equivalent (FTE)	ABET/CEAB accred?	UNDERGRADUATE PROGRAMS					Bachelor's		GRADUATE PROGRAMS			
				Length Nominal length of program in years	Average length of program in years	Time Day/Eve./Both	None/Opt./Req.	Co-op % of graduates in Co-op programs	# of degrees awarded	% of graduates who were full-time	Master's # of degrees awarded	% of graduates who were full-time	Doctoral # of degrees awarded	% of graduates who were full-time
163 Northeastern U Mechanical Eng [M]	MA	22.0	Y	5.0(1)	5.0	B	O	90.0	110	80.0	46	39.6	0	78.9
164 Northern Arizona U Mechanical Eng [M](1)	AZ	7.3	Y	4.0	5.0	D	O	5.0	21	5.0	-.-	-.-(B)	-.-	-.-(B)
165 Northern Illinois U Mechanical Eng [M]	IL	11.0	Y	4.0	4.6	D	O	1.0	31	90.0	6	57.0	-.-(B)	-.-(B)
166 Northwestern U Mechanical Eng [M]	IL	17.0	Y	4.0	4.0	D	O	15.6	32	100.0	19	94.1	15	100.0
167 Norwich U Mechanical Eng [M]	VT	5.0	Y	4.0	4.0	D	N	-.-	0	100.0	-.-	-.-	-.-	-.-
168 U of Notre Dame Mechanical Eng [M]	IN	16.0	Y	4.0	4.0	D	N	-.-	58	100.0	11	100.0	1	100.0
169 Oakland U Mechanical Eng [M]	MI	11.5	Y	4.0	4.5	D	O	-.-	33	-.-	15	-.-	-.-(2)	-.-
170 Ohio U Mechanical Eng [M]	OH	9.0	Y	4.0	-.-	--	O	29.0	41	-.-	14	-.-	-.-	-.-
171 Ohio Northern U Mechanical Eng [M]	OH	5.0	Y	4.0	4.0	D	O	39.0	33	100.0	-.-	-.-	-.-	-.-
172 Ohio State U Mechanical Eng [M]	OH	32.0	Y	4.0	4.5	D	O	-.-	134	86.0	53	100.0	12	100.0
173 U of Oklahoma Mechanical Eng [M]	OK	10.8	Y	4.0	4.7	B	O	-.-(A)	59	80.7	32	54.0	2	44.8
174 Oklahoma Christian Mechanical Eng [M]	OK	5.0	Y	4.5	4.5	D	N	-.-	5	91.4	-.-	-.-	-.-	-.-
175 Oklahoma State U Mechanical & Aerospace Eng [M]	OK	20.5	Y	4.0	4.7	D	O	10.7	56	93.0	30	82.5	7	92.8
176 Old Dominion U Mechanical Eng [M] Mechanics [M]	VA	21.0(1) -.-(3)	Y N	4.0 -.-(B)	4.5 -.-(B)	D -(B)	O -(B)	-.-(A) -.-(B)	56 -.-	83.0 -.-	7 9	16.7 5.3	6 3	94.5 94.5
177 Oregon State U Mechanical Eng [M]	OR	21.0	Y	4.0	4.8(C)	D	O(1)	0.0	73	100.0	20	100.0	10	100.0
178 U of the Pacific Mechanical Eng [M]	CA	4.0	Y	5.0	5.0	D	R	95.0	11	95.0	-.-(4)	-.-(4)	-.-(4)	-.-(4)
179 U of Penn Mechanical Eng & Appl Mechanics [M]	PA	12.0	Y	4.0	-.-	D	N	-.-	31	100.0	13	36.4	6	84.8
180 Penn State U Mechanical Eng [M]	PA	41.0	Y	4.0	-.-	--	O	21.0	223	-.-	48	-.-	20	-.-
181 U of Pittsburgh Mechanical Eng [M]	PA	18.2	Y	4.0	4.0	D(6)	O	21.1	90	86.9	16	38.0	8	75.7
182 Poly U Mechanical Eng [M]	NY	19.0	Y	4.0	-.-(1)	D(B)	O	-.-(B)	49	92.0	21	32.0	1	25.0
183 U of Portland Mechanical Eng [M]	OR	7.3	Y	4.0	4.5	D	O	-.-(2)	15	-.-(1)	2	-.-(3)	-.-(B)	-.-(B)
184 Portland State U Mechanical Eng [M]	OR	11.0	Y	4.0	4.7	B	O	-.-(A)	43	69.6	11	54.5	0	100.0
185 Prairie View A&M U Mechanical Eng [M]	TX	9.0	Y	4.0	4.0	D	O	15.0	27	90.0	0	1.0	-.-	-.-
186 Princeton U Mechanical & Aerospace Eng [M]	NJ	24.0	Y	4.0	4.0	--	N	-.-	48	100.0	12	100.0	6	100.0
187 U of Puerto Rico Mechanical Eng [M]	PR	23.5	Y	5.0	5.0	B	O	18.0	124	97.0	1	100.0	-.-(B)	-.-(B)
188 Purdue U Mechanical Eng [M]	IN	56.0	Y	4.0	4.3	D	O	25.0	194	100.0	72	100.0	26	100.0
189 Rensselaer Poly Inst Mechanical Eng [M]	NY	28.0	Y	4.0	4.0	D	O	28.7	188	100.0	55	-.-(A)	16	-.-(A)

Mechanical Engineering

ENGINEERING Institution Profile Reference Number & Name Name of Degree Program	State/Province	FACULTY Full-time equivalent (FTE)	ABET/CEAB accred?	UNDERGRADUATE PROGRAMS							GRADUATE PROGRAMS			
				Length		Time	Co-op		Bachelor's		Master's		Doctoral	
				Nominal length of program in years	Average length of program in years	Day/Eve./Both	None/Opt./Req.	% of graduates in Co-op programs	# of degrees awarded	% of graduates who were full-time	# of degrees awarded	% of graduates who were full-time	# of degrees awarded	% of graduates who were full-time
190 U of Rhode Island Mechanical Eng [M]	RI	18.0	Y	4.0	4.5	D	O	3.0	42	90.0	8	50.0	3	70.0
191 Rice U Mechanical Eng & Materials Science [M]	TX	17.6	Y	4.0	4.4	--(B)	N	-.-(B)	42	100.0	15	81.8	5	88.4
192 U of Rochester Mechanical & Aerospace Science [M](5) Mechanical Eng [M](6)	NY	-.- 17.0	-- Y	-.- 4.0	-.- 4.0	-- D	-- N	-.- -.-	-.- 42	-.- 99.0	5 --.	0.0 -.-	7 --.	92.3 -.-
193 Rochester Inst of Tech Mechanical Eng [M]	NY	19.0	Y	5.0	5.0	D	R	100.0	78	95.0	6	4.0	--.	-.-
194 Rose-Hulman Mechanical Eng [M]	IN	16.0	Y	4.0	4.0	D	N	-.-(A)	100	100.0	3	100.0	--.(B)	-.-(B)
195 Rutgers, The State U Mechanical Eng [M]	NJ	24.0	Y	4.0	4.5	D	N	-.-	105	97.0	13	58.0	8	60.0
196 Saginaw Valley State U Mechanical Eng [M]	MI	6.7	Y	4.0	4.5	D	O	10.0	21	50.0	--.	-.-	--.	-.-
200 San Diego State U Mechanical Eng [M]	CA	15.7	Y	4.0	5.5	B	O	-.-(A)	58	85.0	18	20.0	--.	-.-(B)
201 San Francisco State U Mechanical Eng [M]	CA	-.-	Y	4.0	-.-	B	O	-.-	28	-.-	--.	-.-	--.	-.-
202 San Jose State U Mechanical Eng [M]	CA	16.4	Y	4.0	-.-(A)	D	--(A)	-.-	61	68.0	22	6.8	--(A)	-.-(A)
203 Santa Clara U Mechanical Eng [M]	CA	10.0	Y	4.0	4.0	D	O	-.-(A)	35	98.0	21	2.0	--.	-.-(A)
204 Seattle U Mechanical Eng [M]	WA	5.0	Y	4.0	4.0	D	N	-.-(B)	31	-.-(A)	--.	-.-	--.	-.-
205 U of South Alabama Mechanical Eng [M]	AL	-.-	Y	4.0	-.-	B	O	-.-	26	69.0	3	44.0	--.(B)	-.-(B)
206 U of South Carolina Mechanical Eng [M]	SC	15.0	Y	4.0	4.0	D	O	-.-(A)	58	89.9	20	40.0	1	81.3
207 South Dakota School of Mines Mechanical Eng [M]	SD	9.0	Y	4.0	4.5	D	O	-.-(A)	69	-.-(A)	15	-.-(A)	--.(B)	-.-(B)
208 South Dakota State U Mechanical Eng [M]	SD	10.3	Y	4.0	4.5	D	O	1.0	28	100.0	3	100.0	--.	-.-
209 U of South Florida Mechanical Eng [M]	FL	9.0	Y	4.3	5.0	B	O	12.0	58	70.0	6	50.0	2	50.0
210 Southern U & A&M Coll Mechanical Eng [M]	LA	14.0	Y	4.0	5.2	D	O	25.0	21	21.6	--(A)	-.-(A)	--(A)	-.-(A)
211 U of Southern Cal Mechanical Eng [M]	CA	11.0	Y	4.0	4.0	D	O	-.-(A)	38	100.0	26	41.0	2	64.0
213 Southern Illinois-Carbondale Mechanical Eng [M]	IL	18.0	Y	4.0	4.5	D	--	5.0	50	95.0	16	71.0	--.(B)	-.-(B)
214 Southern Illinois-Edwardsville Mechanical Eng [M]	IL	5.0	N	4.0	-.-(A)	D	O	-.-(A)	13	85.0	--.(B)	-.-(B)	--.(B)	-.-(B)
216 Southern Methodist U Mechanical Eng [M]	TX	8.0	Y	4.0	4.6	D	O	33.3	15	88.2	7	0.0	2	61.9
217 Stanford U Mechanical Eng [M]	CA	32.0	Y	4.0	4.0	D	N	-.-	59	100.0	138	70.0	35	100.0
218 SUNY-Binghamton Mechanical Eng [M](2)	NY	14.1	Y	4.0	4.0	--(B)	N	-.-(B)	35	95.0	12	25.0	1	44.0
219 SUNY-Buffalo Mechanical Eng [M]	NY	24.0	Y	4.0	4.5	B	N	-.-	157	91.9	39	56.7	6	25.3
221 SUNY-Maritime Mechanical Eng [M]	NY	6.0	N	4.0	4.3	D	N	-.-	3	100.0	--.	-.-	--.	-.-
223 SUNY-Stony Brook Mechanical Eng [M]	NY	-.-(A)	Y	4.0	-.-(A)	D	N	-.-(A)	45	96.9	17	73.0	9	68.3

Mechanical Engineering

ENGINEERING — Institution Profile Reference Number & Name / Name of Degree Program	State/Province	FACULTY Full-time equivalent (FTE)	ABET/CEAB accred?	UNDERGRADUATE PROGRAMS Length Nominal (yrs)	Length Average (yrs)	Time Day/Eve/Both	Co-op None/Opt/Req	Co-op % of grads in Co-op programs	Bachelor's # of degrees awarded	Bachelor's % of grads who were full-time	GRADUATE Master's # of degrees awarded	Master's % of grads who were full-time	Doctoral # of degrees awarded	Doctoral % of grads who were full-time
225 Syracuse U	NY													
Manuf'g Eng [M]		1.0	Y	4.0	4.2	D	O	50.0	2	99.0	12	-.-(A)	-.-(B)	-.-(B)
Mechanical Eng [M]		11.0	Y	4.0	4.2	D	O	21.0	44	93.0	16	-.-(A)	13	-.-(A)
226 Technical U of Nova Scotia	NS													
Mechanical Eng [M]		15.9	Y	4.0(2)	5.0	D	N	-.-(A)	39	100.0	7	86.0	3	100.0
227 Temple U	PA													
Mechanical Eng [M]		15.5	Y	4.0	4.0	--(B)	N	-.-(B)	18	91.2	-.-(A)	-.-(A)	-.-(B)	-.-(B)
228 U of Tennessee, Chattanooga	TN													
Mechanical Eng [M](5)		-.-(7)	Y	-.-	-.-	B	O	-.-(A)	-.-	-.-	3	-.-(A)	-.-	-.-
Mechanics [M](5)		-.-(7)	Y	4.0	4.8	B	O	-.-(A)	13(5)	-.-(A)	-.-	-.-	-.-	-.-
Thermal Eng [M](5)		-.-(7)	Y	4.0	4.8	B	O	-.-(A)	12	-.-(A)	-.-	-.-	-.-	-.-
229 U of Tennessee, Knoxville	TN													
Mechanical Eng [M]		12.0	Y	4.0	-.-(A)	D	O	37.3	51	87.0	16	-.-(A)	5	-.-(A)
230 Tennessee State U	TN													
Mechanical Eng [M]		7.0	Y	4.3	5.0	D	O	5.0	30	75.2	1(1)	100.0	-.-(B)	-.-(B)
231 Tennessee Tech	TN													
Mechanical Eng [M]		228.8	Y	4.0	4.0	D	O	10.0	72	100.0	9	100.0	1	100.0
232 U of Texas at Arlington	TX													
Mechanical Eng [M]		-.-	Y	4.0	-.-	B	O	-.-	48	-.-	27	-.-	3	-.-
233 U of Texas at Austin	TX													
Mechanical Eng [M]		63.6	Y	4.0	4.7	D	O(1)	11.0	181	87.4	74	78.4	21	85.1
235 U of Texas at El Paso	TX													
Mechanical Eng [M]		9.8	Y	4.0	-.-	D	O	20.0	28	-.-	10	-.-	-.-	-.-
236 U of Texas at San Antonio	TX													
Mechanical Eng [M]		7.9	Y	4.0	-.-	B	O	-.-	22	-.-	-.-	-.-	-.-	-.-
237 Texas A&I U	TX													
Mechanical Eng [M]		-.-	Y	4.0	-.-	D	N	-.-	25	-.-	5	-.-	-.-	-.-
238 Texas A&M U	TX													
Mechanical Eng [M]		60.3	Y	4.0	4.5(A)	D	O	54.0	146	90.0	55	85.0(C)	12	85.0(C)
239 Texas Tech U	TX													
Mechanical Eng [M]		17.2	Y	4.0	-.-	D	N	-.-	62	90.3	10	75.0	2	74.0
240 U of Toledo	OH													
Mechanical Eng [M]		17.0	Y	4.0	4.0	D	O	-.-(A)	70	75.0	17	50.0	-.-	-.-
241 Tri-State U	IN													
Mechanical Eng [M]		5.0	Y	4.0	4.5	D	O	16.0	38	91.9	-.-(3)	-.-(3)	-.-(3)	-.-(3)
243 Tufts U	MA													
Manuf'g Eng [M]		5.0	N	4.0	4.0	D	N	-.-(B)	3	100.0	-.-(B)	-.-(B)	-.-(B)	-.-(B)
Mechanical Eng [M]		13.0	Y	4.0	4.0	B	N	-.-(B)	41	100.0	11	26.1	0	67.7
244 Tulane U	LA													
Mechanical Eng [M]		9.0	Y	4.0	4.0	B	N	-.-	21	100.0	6	47.0	3	78.0
245 U of Tulsa	OK													
Mechanical Eng [M]		9.0	Y	4.0	4.0	D	N	-.-(3)	33	92.3	3	40.0	0	100.0
246 Tuskegee U	AL													
Mechanical Eng [M]		12.0	Y	4.0	5.0	D	O	80.0	17	99.5	9	97.5	-.-	-.-
247 Union Coll	NY													
Mechanical Eng [M]		8.0	Y	4.0	-.-(A)	B	N	-.-(B)	31	81.0	15	11.0	-.-(B)	-.-(B)
248 US Air Force Academy	CO													
Mechanical Eng [M]		5.0	Y	4.0	4.0	D	N	-.-	9	100.0	-.-(B)	-.-(B)	-.-(B)	-.-(B)
251 US Military Academy	NY													
Mechanical Eng [M]		-.-	Y	4.0	4.0	D	N	-.-	135	100.0	-.-(B)	-.-(B)	-.-(B)	-.-(B)
252 US Naval Academy	MD													
Mechanical Eng [M]		29.0	Y	4.0	4.0	D	N	-.-	56	100.0	-.-	-.-	-.-	-.-
253 U of Utah	UT													
Mechanical Eng [M]		15.7	Y	4.0	5.0	D	O	-.-(A)	73	-.-(A)	25	-.-(A)	5	-.-(A)
254 Utah State U	UT													
Mechanical, Manuf'g & Aerospace Eng [M]		10.2	Y(1)	4.0	4.6	D	O	0.0(A)	45	88.3	10	55.0	0	25.0

Mechanical Engineering

ENGINEERING Institution Profile Reference Number & Name Name of Degree Program	State/Province	FACULTY Full-time equivalent (FTE)	ABET/CEAB accred?	UNDERGRADUATE PROGRAMS					Bachelor's		GRADUATE PROGRAMS			
				Length		Time	Co-op				Master's		Doctoral	
				Nominal length of program in years	Average length of program in years	Day/Eve./Both	None/Opt./Req.	% of graduates in Co-op programs	# of degrees awarded	% of graduates who were full-time	# of degrees awarded	% of graduates who were full-time	# of degrees awarded	% of graduates who were full-time
255 Valparaiso U Mechanical Eng [M]	IN	7.3	Y	4.0	4.0	D	O	-.-	29	99.0	-.-	-.-	-.-	-.-
256 Vanderbilt U Mechanical Eng [M]	TN	16.0	Y	4.0	4.0	D	N	-.-	50	100.0	10	100.0	6	100.0
257 U of Vermont Mechanical Eng [M]	VT	10.0	Y	4.0	4.0	D	O	10.0	29	93.9	2	18.2	-.-	-.-
258 Villanova U Mechanical Eng [M]	PA	18.0	Y	4.0	4.0	--	--	-.-	58	86.0	26	30.0	-.-	-.-
259 U of Virginia Aerospace Eng [M][3] Mechanical & Aerospace Eng [M][5] Mechanical Eng [M][3] Nuclear Eng [M]	VA	32.9[4] 32.9[4] 32.9[4] 32.9[4]	Y -.-[5] Y Y	4.0 -.- 4.0 4.0	4.1 -.- 4.3 4.0	D -- D D	N -- N N	-.- -.- -.- -.-	31 -.- 45 3	100.0 -.- 100.0 100.0	-.-[3] 27 -.-[3] 5	-.- 89.8 -.- 87.5	-.-[3] 4 -.-[3] 3	-.- 80.0 -.- 100.0
260 Virginia Military Inst Mechanical Eng [M]	VA	8.0	Y	4.0	-.-	D	N	-.-	174	100.0	-.-	-.-	-.-	-.-
261 Virginia Tech Mechanical Eng [M]	VA	39.0	Y	4.0	4.0	D	O	20.0	212	20.0	44	100.0	15	100.0
262 U of Washington Mechanical Eng [M]	WA	32.2	Y	4.0	-.-[A]	D	O	-.-[A]	167	82.0	36	-.-[A]	5	-.-[A]
263 Washington U Mechanical Eng [M]	MO	16.5	Y	4.0	4.0	B	O	-.-	61	-.-	51	-.-	7	-.-
264 Washington State U Mechanical Eng [M][1]	WA	28.9	Y	4.0	4.5	D	O	-.-[A]	121	100.0	13	100.0	6	100.0
265 Wayne State U Mechanical Eng [M]	MI	21.0	Y	4.0	4.5	B	O	-.-[A]	39	-.-[A]	89	-.-[A]	3	-.-[A]
267 West Virginia U Mechanical Eng [M]	WV	28.0	Y	4.0	4.3	D	O	0.0	65	100.0	25	95.0	2	95.0
268 West Virginia Inst of Tech Mechanical Eng [M]	WV	5.0	Y	4.0	4.5	D	O	30.0	24	95.0	-.-	-.-	-.-	-.-
269 Western Michigan U Mechanical Eng [M]	MI	-.-	Y	4.5	4.0	D	O	10.0	68	99.1	16	4.1	-.-	-.-
270 Western New England Coll Mechanical Eng [M]	MA	6.8	Y	4.0	4.4	B	N	-.-[B]	40	76.2	4	-.-[B]	-.-[B]	-.-[B]
271 Wichita State U Mechanical Eng [M]	KS	11.5	Y	4.0	5.5	B	O	33.0	33	58.7	3	38.0	1	56.0
272 Widener U Mechanical Eng [M]	PA	7.2	Y	4.3	4.0	D	O	34.0	29	-.-[A]	6	-.-[A]	-.-[B]	-.-[B]
273 Wilkes U Mechanical Eng [M]	PA	5.5	N[4]	4.0	4.0	B[2]	O	1.0[C]	0[5]	95.0[C]	-.-[B]	-.-[B]	-.-[B]	-.-[B]
274 U of Wisconsin-Madison Mechanical Eng [M]	WI	31.0	Y	4.0	5.0	D	O	-.-[A]	141	-.-[A]	39	-.-[A]	11	-.-[A]
275 U of Wisconsin-Milwaukee Mechanical Eng [M]	WI	14.5	Y	4.0	5.5	B	O	8.6	58	72.1	15	28.3	1	63.6
276 U of Wisconsin-Platteville Mechanical Eng [M]	WI	11.0	Y	4.0	4.7	D	O	39.0	66	100.0	-.-	-.-	-.-	-.-
277 Worcester Poly Inst Mechanical Eng [M]	MA	32.6	Y	4.0	4.0	D	O	-.-[A]	165	98.8	25	-.-[A]	1	-.-[A]
278 Wright State U Mechanical Eng [M]	OH	8.0	Y	4.0	-.-	B	O	38.6	44	78.9	9	26.8	-.-	-.-
279 U of Wyoming Mechanical Eng [M]	WY	12.0	Y	4.0	5.5	D	O	0.0	28	100.0	3	100.0	0	100.0
280 Yale U Mechanical Eng [M]	CT	10.0	Y	4.0	4.0	D	N	-.-	20	42.0	4	75.0	8	100.0

Mechanical Engineering—Totals

State/Province/Country	FACULTY		UNDERGRADUATE PROGRAMS							GRADUATE PROGRAMS			
			Length		Time		Co-op	Bachelor's		Master's		Doctoral	
	# of schools	Full-time equivalent (FTE)	# accredited programs	Nominal length of program in years	Average length of program in years	Day/Eve./Both	None/Opt./Req.	% of graduates in Co-op programs	# of degrees awarded	% of graduates who were full-time	# of degrees awarded	% of graduates who were full-time	# of degrees awarded	% of graduates who were full-time
AK	1	8.00	1						13		13		-	
AL	6	96.90	9						395		102		18	
AR	1	13.00	1						42		4		1	
AZ	3	80.59	3						187		43		21	
CA	22	394.80	22						1,311		439		119	
CO	5	67.09	5						174		35		10	
CT	4	47.00	4						153		42		21	
DC	4	44.00	4						69		51		9	
DE	1	16.00	1						62		11		7	
FL	9	137.70	9						418		103		22	
GA	1	57.50	2						338		115		26	
HI	1	15.00	1						48		7		3	
IA	2	12.60	2						156		35		30	
ID	1	17.00	1						40		8		-	
IL	8	155.90	7						514		173		57	
IN	7	117.50	7						489		88		27	
KS	3	39.50	3						148		23		8	
KY	2	32.00	2						79		32		5	
LA	5	66.80	5						159		21		12	
MA	9	192.60	9						722		303		41	
MD	4	99.30	4						258		66		20	
ME	1	9.00	1						33		-		-	
MI	10	234.10	10						1,197		301		58	
MN	2	45.00	1						224		60		23	
MO	4	69.90	4						343		92		17	
MS	2	23.20	2						57		8		2	
MT	1	13.00	1						59		3		-	
NC	4	136.40	5						368		54		33	
ND	2	24.00	2						86		3		1	
NE	1	16.00	1						141		22		4	
NH	1	14.00	1						44		11		2	
NJ	3	79.20	3						231		64		19	
NM	2	35.10	2						113		22		8	
NV	1	7.00	1						13		2		-	
NY	19	260.10	19						1,334		315		84	
OH	10	163.80	10						590		231		41	
OK	4	45.30	4						153		65		9	
OR	3	39.30	3						131		33		10	
PA	14	234.00	13						825		199		56	
PR	1	23.50	1						124		1		-	
RI	1	18.00	1						42		8		3	
SC	2	46.00	2						196		64		6	
SD	2	19.30	2						97		18		-	
TN	7	282.80	9						297		51		15	
TX	12	221.50	12						667		217		49	
UT	3	45.60	3						257		46		10	
VA	4	199.60	6						521		92		31	
VT	2	15.00	2						29		2		-	
WA	4	75.09	4						331		51		11	
WI	5	84.90	5						379		69		16	
WV	2	33.00	2						89		25		2	
WY	1	12.00	1						28		3		-	
U.S. Totals:	229	4,235.50	235						14,774		3,846		967	
AB	1	25.00	1						92		12		2	
MB	1	-	2						69		11		-	
NS	1	15.90	1						39		7		3	
PQ	3	71.09	3						347		83		20	
Canada Totals:	6	112.00	7						547		113		25	

Mechanical Engineering—Footnotes

FOOTNOTES: The following footnotes are the same for all schools: (A) Data not available. (B) Data not applicable. (C) Estimated data.

012 Auburn University: (7) Supported by the faculty of the Mechanical Engineering Department. (10) Based on credit hours taken Fall Quarter 1991.

014 Boston University: (1) Part time status granted by petition only. (3) Aerospace/Mechanical faculty (5) Offered through the Aerospace/Mechanical Department.

015 Bradley University: (1) Students typically take the equivalent of nine semesters but finish in 4 years by using summer school and transfer credits.

018 Bucknell University: (1) Data is given in whole numbers.

019 University of California at Berkeley: (1) Double majors are included with department totals.

020 University of California, Davis: (1) Included in Mechanical Engineering

023 University of California, San Diego: (2) Due to a large amount of shared teaching responsibility, it is impossible to give faculty numbers by program. See specific department for numbers.

042 University of Cincinnati: (1) Of the 6 departments., 5 have more than one program, therefore, faculty provide support in multiple programs. Total FTE for the college is 142.

044 City College of the City University of New York: (1) Most prerequisite courses can be taken during the evening. All major courses must be taken during the day.

048 Cleveland State University: (1) includes Co-Op and part-time students

051 University of Colorado at Denver: (3) Coursework available at CU-Denver but Ph.D. is awarded at CU-Boulder through system-wide Graduate School.

056 University of Connecticut: (4) Includes Ocean Engineering students.

066 École Polytechnique de Montréal: (7) Master's degree in Aeronautics included

072 Florida Agricultural and Mechanical University: (1) Includes FAMU and FSU students. Year is Fall 1991, Spring 1992 and Summer 1992.

073 Florida Atlantic University: (5) Working students often limit their programs to 12 credits per term or less, depending on work schedules.

076 Florida State University: (1) Includes FAMU and FSU students. Year is Fall 1992, Spring 1992, and Summer 1992.

077 GMI Engineering & Management Institute: (1) 3360 hours minimum

082 Georgia Institute of Technology: (6) Includes Health Physics. (8) Fall '91 Quarter FTE.

090 University of Houston: (4) BSME, MME, MSME, PhD

096 Illinois Institute of Technology: (2) Reflects percentage of graduates who received certificates of completion in compliance with ABET requirements. (5) Graduate degrees offered in Mechanical and Aerospace Engineering only.

100 The Johns Hopkins University: (4) Includes 3.70 FTE in part-time programs. (10) Listed as Mechanical Engineering and Engineering Mechanics

104 Lafayette College: (1) Undergraduate program only.

106 Lawrence Technological University: (3) Program is available only in the evening.

111 University of Louisville: (1) Program is accredited at the master's level.

114 University of Maine: (1) Ph.D. Program not available.

115 University of Manitoba: (2) MS and Doctoral degrees are in Mechanical and Industrial Engineering. These degrees are reported under Industrial Engineering.

116 Mankato State University: (1) ABET accreditation visit scheduled for Fall, 1993.

121 University of Massachusetts: (3) Includes 20 M.S. degrees in Manufacturing Engineering, and inter-departmental program between Mechanical Engineering and Industrial Engineering. (4) Includes department head

124 Massachusetts Institute of Technology: (3) 6.9% of master's degree recipients participated in co-op programs.

125 McGill University: (1) Following completion of 2 year CEGEP program. International students should add 1 year for pre-engineering studies (2) Internship year program available for students who have less then 45 credits remaining in their academic program (3) Part time figures include candidates who have completed their residence requirements but are working full time on research and thesis requirements

130 University of Michigan: (2) Graduate program only. (3) Faculty included in principal department. (4) Includes Professional degrees. (5) All graduate students. (7) For whole College.

134 Milwaukee School of Engineering: (1) No co-op program is available, however, 85% of undergraduates participate in an internship program and have relevant work experience prior to graduation.

137 University of Mississippi: (4) M.S. and Ph.D. degrees in Engineering Science are offered with emphasis in Mechanical Engineering

139 University of Missouri-Columbia: (3) Includes statistics from ME/CEP which is located in Kansas City and administered by the Columbia campus.

140 University of Missouri-Kansas City: (2) Most students participate in unofficial co-op programs with local firms.

144 Montana State University: (1) Information unavailable but nearly 100%

152 The University of New Mexico: (1) We classify all our graduate students as full-time.

162 North Dakota State University: (1) Masters & Ph.D. degree also available.

163 Northeastern University: (1) 5 year co-op option; 4 year co-op option; 4-year non-co-op option

164 Northern Arizona University: (1) Includes an emphasis in environmental engineering.

169 Oakland University: (2) See Doctoral Program.

176 Old Dominion University: (1) 1991/92 Head count (3) See Mechanical Engineering — faculty participate in both programs

177 Oregon State University: (1) New Program 1992-93

178 University of the Pacific: (4) Mechanical Engineering does not offer a master's or doctoral degree.

181 University of Pittsburgh: (6) Part-time program is available.

182 Polytechnic University: (1) 4-5

183 University of Portland: (1) Data not available by program. Overall undergraduate full-time is 89%. (2) Each year approximately 10 Engineering students participate in co-op programs. The numbers by degree program fluctuate. (3) Data not available by program. Overall graduate full-time is 42%.

192 University of Rochester: (5) Mechanical and Aerospace Science is a graduate program only. Materials Science has been merged under this category. (6) Mechanical engineering is an undergraduate program only.

218 State University of New York at Binghamton: (2) Includes Industrial Engineering Program

226 Technical University of Nova Scotia: (2) Three years at TUNS.

228 University of Tennessee, Chattanooga: (5) BSE with Mechanical Concentration. (7) All faculty are included under Engineering.

230 Tennessee State University: (1) An option in Master of Engineering degree program.

233 University of Texas at Austin: (1) Participants in the Co-op Program are away from the campus for three semesters, working for pay and letter grade.

241 Tri-State University: (3) No graduate programs

245 University of Tulsa: (3) Co-op not available.

254 Utah State University: (1) Currently seeking ABET accreditation to establish Aerospace, Environmental and Civil Engineering as separately accredited degree programs.

259 University of Virginia: (3) BS degree program. (4) Department of Mechanical, Aerospace and Nuclear Engineering. (5) Graduate degree program.

264 Washington State University: (1) Pullman campus

273 Wilkes University: (2) There is no distinction between day or evening programs. (4) Has not been visited by ABET. (5) New program. First graduating class in 1993.

001 University of Akron

INSTITUTION PROFILE

HEAD OF THE INSTITUTION
Peggy G. Elliott
Phone: (216)972-7074   **Fax:** (216)972-8652

GENERAL INFORMATION
[All Students—Fall 1991]
Undergraduate enrollment 23,856
Graduate enrollment 4,385
Total institution enrollment 28,241
Type of institution: Public   **Calendar:** Semesters
Location: Akron   **Population:** 250,000
Setting: Urban
Types of engineering degrees: Engineering & Technology
Other degree-granting colleges: Arts & Sciences, Business Administration, Education, Fine & Performing Arts, Law, Nursing, Community and Technical College, Polymer Science and Polymer Engineering, Wayne General and Technical College, University College

ENGINEERING ADMISSIONS

ADMISSIONS OFFICE CONTACT
Martha A. Booth
University of Akron
Admissions Office
Akron, OH 44325-2001
Phone: (216)972-7100

ENGINEERING COLLEGE ADMISSIONS INFORMATION
Admission to the engineering college: After completing second semester of calculus and meeting GPA requirements
Entrance Requirements: SAT, ACT, SAT or ACT required for prospective students 21 and younger, but no minimum score is required
Entrance Recommendations: High School courses—English (4 years), Mathematics (4 years), Science (3 years), Social Science (3 years)
Requirements for foreign students: TOEFL (Score: 500); financial statement; Official documents from foreign schools translated into English.

DEPARTMENTAL ADMISSIONS INFORMATION
Admission to the engineering department: after completion of 30 semester credits and the second semester of calculus subject to restrictions stated below
Additional information: (1) no grade less than C in any calculus course (2) 2.3 GPA established in 3 of the following: (a) all courses, (b) required science courses, (c) required math courses, (d) required engineering courses; (3) no more than six courses repeated

FINANCIAL INFORMATION

ESTIMATED EXPENSES (ACADEMIC YEAR)
[Expenses are for the 1992-93 nine-month academic year.]

State Residents	Undergraduate	Graduate
Tuition and fees	$ 2,841	$ 4,217
College room and board	$ 3,486	$ 3,486
Books and supplies	$ 350	$ 350
Other expenses	$ –	$ –
Total estimated expenses	**$ 6,677**	**$ 8,053**

Out-of-State Residents	Undergraduate	Graduate
Tuition and fees	$ 7,498	$ 7,550
College room and board	$ 3,486	$ 3,486
Books and supplies	$ 350	$ 350
Other expenses	$ –	$ –
Total estimated expenses	**$11,334**	**$ 11,386**

FINANCIAL AID OFFICE CONTACT
Robert D. Hahn, *Director*
University of Akron
Student Financial Aid & Employment
Akron, OH 44325-6211
Phone: (216)972-7032   **Fax:** (216)972-7139

GENERAL FINANCIAL AID INFORMATION
Forms accepted/required: AFSA, FAT, SAR, Institutional, Loan request form
Additional financial aid information: Academic scholarships, Minority scholarships, Performance scholarships, Endowed departmental scholarships, Honors scholarships, Federal and state aid, Presidential scholarships with engineering supplements for full tuition, On campus employment

ENGINEERING COLLEGE INFORMATION

[For additional personnel, refer to the *Appendix*.]

HEAD OF ENGINEERING
Nicholas D. Sylvester
Phone: (216)972-6978   **Fax:** (216)972-5162
EMail: DELUCA@ENGINEER@UAKRON

ENGINEERING COLLEGE ADDRESS
College of Engineering
University of Akron
Akron, OH 44325-3901
Phone: (216)972-7816   **Fax:** (216)972-5162

ENROLLMENTS—BY CLASS
[Numbers are baccalaureate enrollments for the fall 1991 term, unless otherwise footnoted.]
1st-year students/Freshmen 595
2nd-year students/Sophomores 316
3rd-year students/Juniors 305
4th-year students/Seniors .... 802 [1]
Total 2,018

Notes: (1) This includes 455 part time students at various stages in degree programs

NUMBER OF DEGREES AWARDED—BY PROGRAM
[Numbers are engineering baccalaureate degrees awarded during the 1991-1992 academic year, unless otherwise footnoted. For full details about each engineering program, refer to the *Tables of Degree Programs*.]
Biomedical Engineering .... – [B]
Chemical Engineering 26
Civil Engineering 34
Electrical Engineering 65
Engineering 2
Mechanical Engineering 94
Total 221

Notes: (B) Data not applicable.

PERCENTAGE OF DEGREES AWARDED—BY CATEGORY
[Percentages are of all engineering baccalaureate degrees awarded during the 1991-1992 academic year, unless otherwise footnoted.]
African Americans 0.5%
Asian/Pacific Island Americans 1.3%
Hispanic Americans 0.5%
Native Americans .... – %
Foreign Citizens 4.5%
All Others 93.2%
Women 13.0%
Persons with disabilities .... – % [A]
Students over 25 years of age .... – % [A]

Notes: (A) Data not available.

GRADUATE ENROLLMENTS & DEGREES AWARDED
Master's enrollment 320
Master's degrees awarded 78
Doctoral enrollment 102
Doctoral degrees awarded 18

ENGINEERING STUDENT DATA

Applicants to the engineering college
Number of applicants to engineering college 1,002
Percent offered admission 96.0%

Matriculated engineering students
Percentage in top quartile (25%) of High School class 33.3
Average SAT scores: Math—555, Verbal—455, Combined—1010 [1]
Average ACT scores: Math—23, Composite—23 [1]

Notes: (1) SAT or ACT not required for students over 21.

FULL & PART TIME FACULTY—BY DEPARTMENT
[Figures are the head count of full-time faculty and the full-time equivalent (FTE) of part-time faculty for each engineering department or equivalent.]

Department	Full	Part
Chemical Eng	9	–
Civil Eng	12	0.5
Electrical Eng	18	0.5
Biomedical Eng	8	0.3
Mechanical Eng	20	0.8

COLLEGE DESCRIPTION
The College of Engineering at The University of Akron offers ABET accredited degrees in four disciplines. The college also offers an ABET accredited cooperative education program. While the co-op program is optional, it is very popular among students, with over 80% of eligible students participating. Engineering students start their co-op experience during their junior year and receive a full year of work experience. Students are placed in co-op jobs in 15 states. The Chemical Engineering and Mechanical Departments offer, in conjunction with the College of Polymer Science and Polymer Engineering, a unique Polymer Engineering Specialization. All courses required during the first two years of study are available at Wayne College. All college faculty hold the PhD and most are actively engaged in research. The college also offers programs leading to MS and PhD degrees.

DEGREE PROGRAMS DESCRIPTION
The Chemical Engineering department houses a process research center which allows undergraduate students a unique opportunity to participate in current research. Civil engineering students take course in each of Civil Engineering's fundamental areas (structures, environmental, transportation, geotechnical, and hydraulics). The electrical Engineering department also offers engineering computer science classes. Electrical engineering students can specialize through 18 credits of EE electives. Mechanical engineering students complete two senior design projects. Many ME students participate in SAE sponsored competitions (formula one race car, mini-Baja, supermileage vehicle, or cargo aircraft) to satisfy these requirements. The Polymer engineering specialization for Chemical and Mechanical engineering students is very popular with students and co-op employers. Masters' degrees can be completed with or without thesis. The PhD is a college-wide interdisciplinary degree.

TRANSFER INFORMATION
Residency Requirements: Students must complete at least 32 semester hours in residence for a baccalaureate degree with at least 16 semester hours in major.

Transfer without Articulation Agreements
Admission to engineering: after completing 30 semester credits and second semester of calculus and satisfying GPA requirements
Requirements: Minimum established GPA in 3 of following : (a) all courses (b) science courses, (c) calculus courses, (d) engineering courses. Minimum GPA is dependent upon transfer institution: (a) 2.3 from 4 year colleges with ABET accredited engineering programs, (b) 2.5 from 4 year colleges without ABET accredited engineering program, (c) 3.0 from two year colleges

GRADUATION REQUIREMENTS
(1) A minimum of 136 semester credits completed; (2) overall GPA of 2.0; (3) GPA of 2.0 in engineering coursework

STUDENT PROGRAMS

PROFESSIONAL AND HONORARY SOCIETIES
[For key to acronyms, see *Introduction*.]
Professional Societies: AIAA, AIChE, ASCE, ASME, BMES, IEEE, ISA, SAE, Society of Students in Construction (affiliated with AGCA)
Honorary Societies: Eta Kappa Nu, Tau Beta Pi

SUPPORT PROGRAMS
Student Chapter Organizations: National Society of Black Engineers

Minority Engineering Program
For: Ethnic Minorities   **Available:** Year round
Offered: Freshman, Sophomore, Junior, Senior

Precollege Program
For: Ethnic Minorities   **Available:** Year round
Offered: Prematriculation

Black Cultural Center
For: Ethnic Minorities   **Available:** Year round
Offered: Freshman, Sophomore, Junior, Senior, Graduate level

Other Engineering Support Programs: Summer & fall orientation program for freshmen & transfers; career counseling services; international student office; veterans' services; services for students with disabilities; placement services; financial aid office; adult resource center.

002 University of Alabama

INSTITUTION PROFILE

HEAD OF THE INSTITUTION
E. Roger Sayers
Phone: (205)348-5100   Fax: (205)348-8377

GENERAL INFORMATION
[All Students—Fall 1991]
Undergraduate enrollment 15,913
Graduate enrollment 3,383
Total institution enrollment 19,296
Type of institution: Public   Calendar: Semesters
Location: Tuscaloosa   Population: 77,759
Setting: Urban
Types of engineering degrees: Engineering & Technology
Other degree-granting colleges: Arts & Sciences, Business Administration, Communication & Journalism, Education, Law, Nursing, Community Health Sciences, Human Environmental Sciences, Social Work, Library and Information Studies, Graduate School

ENGINEERING ADMISSIONS

ADMISSIONS OFFICE CONTACT
Roy C. Smith
University of Alabama
Box 870132
Tuscaloosa, AL 35487-0132
Phone: (205)348-5666   Fax: (205)348-9046

ENGINEERING COLLEGE ADMISSIONS INFORMATION
Admission to the engineering college: A student applies to the program at the time of application. High school graduates admitted on the basis of acceptable high school records and ACT/SAT scores.
Entrance Requirements: SAT, ACT, The University does not have a minimum score requirement.
Entrance Recommendations: High School courses—English (4 years), College-preparatory Mathematics (3 years), History or Social Science (1 year), Chemistry (1 year)
Requirements for foreign students: TOEFL (Score: 500); financial statement

DEPARTMENTAL ADMISSIONS INFORMATION
Admission to the engineering department: At the time of admission to the institution.

FINANCIAL INFORMATION

ESTIMATED EXPENSES (ACADEMIC YEAR)
[Expenses are for the 1992-93 nine-month academic year.]

State Residents	Undergraduate	Graduate
Tuition and fees	$ 2,068	$ 2,068
College room and board	$ 3,464	$ 3,464
Books and supplies	$ 550	$ 550
Other expenses	$ –(1)	$ –(1)
Total estimated expenses	**$ 6,082**	**$ 6,082**

Out-of-State Residents	Undergraduate	Graduate
Tuition and fees	$ 5,166	$ 5,166
College room and board	$ 3,464	$ 3,464
Books and supplies	$ 550	$ 550
Other expenses	$ –(1)	$ –(1)
Total estimated expenses	**$ 9,180**	**$ 9,180**

Notes: (1) Personal expenses must be considered along with laboratory fees which are $7.50 per credit hour enrolled for each engineering student.

FINANCIAL AID OFFICE CONTACT
Molly M. Lawrence, *Director*
University of Alabama
Box 870162
Tuscaloosa, AL 35487-0162
Phone: (205)348-6756   Fax: (205)348-2989

GENERAL FINANCIAL AID INFORMATION
Forms accepted/required: ACT, FFS, Stafford, SAR
Additional financial aid information: Most financial aid is awarded on the basis of need. However, scholarships and grants totaling nearly $129,000 are available from the College. Engineering students also compete well for University scholarships, e.g., Presidential, Alumni Honors, and National Merit.

ENGINEERING COLLEGE INFORMATION

[For additional personnel, refer to the *Appendix*.]

HEAD OF ENGINEERING
Robert F. Barfield
Phone: (205)348-6405   Fax: (205)348-8573
EMail: BARFIELD@UA1VM.UA.EDU

ENGINEERING COLLEGE ADDRESS
College of Engineering
University of Alabama
Box 870200
Tuscaloosa, AL 35487-0200
Phone: (205)348-6400   Fax: (205)348-8573

ENROLLMENTS—BY CLASS
[Numbers are baccalaureate enrollments for the fall 1991 term, unless otherwise footnoted.]
1st-year students/Freshmen 610
2nd-year students/Sophomores 346
3rd-year students/Juniors 342
4th-year students/Seniors 482
Total 1,780

NUMBER OF DEGREES AWARDED—BY PROGRAM
[Numbers are engineering baccalaureate degrees awarded during the 1991-1992 academic year, unless otherwise footnoted. For full details about each engineering program, refer to the *Tables of Degree Programs*.]

Aerospace Engineering 33
Chemical Engineering 29
Civil Engineering 26
Computer Science 12
Electrical Engineering 83
Engineering Hydrology .... –(B)
Engineering Mechanics .... –(B)
Environmental Engineering .... –(B)
Industrial Engineering 29
Mechanical Engineering 57
Metallurgical Engineering 2
Mineral Engineering 3
Total 274

Notes: (B) Data not applicable.

PERCENTAGE OF DEGREES AWARDED—BY CATEGORY
[Percentages are of all engineering baccalaureate degrees awarded during the 1991-1992 academic year, unless otherwise footnoted.]

African Americans 8.8%
Asian/Pacific Island Americans 1.1%
Hispanic Americans 1.8%
Native Americans .... – %
Foreign Citizens 18.6%
All Others 69.7%
Women 19.3%
Persons with disabilities .... – %(A)
Students over 25 years of age .... – %(A)

Notes: (A) Data not available.

GRADUATE ENROLLMENTS & DEGREES AWARDED
Master's enrollment 217
Master's degrees awarded 74
Doctoral enrollment 94
Doctoral degrees awarded 14

ENGINEERING STUDENT DATA

Applicants to the engineering college
Number of applicants to engineering college 971
Percent offered admission 83.9%

Matriculated engineering students
Percentage in top quartile (25%) of High School class .... –(A)
Average SAT scores: Math—560.8, Verbal—475.3, Combined—1036
Average ACT scores: Math—24.0, Composite—24.7

Notes: (A) Data not available.

FULL & PART TIME FACULTY—BY DEPARTMENT
[Figures are the head count of full-time faculty and the full-time equivalent (FTE) of part-time faculty for each engineering department or equivalent.]

Department	Full	Part
Chemical Eng	9	–
Civil Eng	9	–
Electrical Eng	15	–
Industrial Eng	12	–
Mechanical Eng	15	–
Aerospace Eng	8	–
Computer Science	15	–
Eng Mechanics	10	–
Metallurgical & Materials Eng	6	–
Mineral Eng	8	–
Eng Tech	2	–

COLLEGE DESCRIPTION
The primary objective is to provide students with a quality educational experience which will equip them for rewarding and productive professional careers and for responsible citizenship. Further objectives include the generation of new knowledge through research by its faculty and students and to provide a variety of short courses, conferences, and seminars for practicing engineers, technologists and computer scientists which enables them to stay abreast of the latest advances in their respective fields. The College offers accredited four-year undergraduate programs in engineering, emphasizing engineering fundamentals in the first year, followed by three years of specialized training in one of the eight engineering majors. A four-year accredited undergraduate program in computer science is also available. All programs require courses in the humanities and social studies to satisfy the requirements of the University core curriculum and accreditation criteria of ABET and CSAB.

DEGREE PROGRAMS DESCRIPTION
The B.S. degrees in the various engineering programs prepare the graduate for entry-level professional positions or for graduate study in each of the respective fields. Research-oriented and design-oriented M.S. (thesis and non-thesis), Ed.S. and Ph.D. degree programs are available.

DUAL DEGREE PROGRAMS
Engineering baccalaureate graduates with dual degrees 0
Enrollment requirements: No specific entrance requirements other than those listed above.

TRANSFER INFORMATION
Residency Requirements: A candidate must earn in residence at least one-fourth as many semester hours as the number required for the degree, including nine of the last 18 hours. An engineering student must complete on campus at least one-half of the semester hours required in the discipline in which the degree is awarded.
Transfer without Articulation Agreements
Admission to engineering: At any time
Engineering graduates transferred from ... 4-yr: –   2-yr: –
Requirements: An average of 'C' or higher on all college work attempted is required for admission to The University. If a student has attended more than one institution and earned less than a 'C' average at one of the institutions admission must be approved by the College.

GRADUATION REQUIREMENTS
Bachelor of Science in engineering programs require 130 to 138 semester hours of credit. A student must: 1) complete on this campus at least one-half of the semester hours that are required in the engineering discipline in which a degree is awarded; 2) earn at least a 'C' average on a four-point system on all work attempted and on all work attempted on this campus; 3) earn at least a 'C' average on a four-point system on all work attempted in the student's professional courses and on all work attempted in the student's professional courses on this campus.

STUDENT PROGRAMS

PROFESSIONAL AND HONORARY SOCIETIES
[For key to acronyms, see *Introduction*.]
Professional Societies: ACM, AIAA, AIChE, AIME, ASCE, ASHRAE, ASME, IEEE, IIE, SAE, Alabama Society of Professional Engineers, American Foundrymen's Society
Honorary Societies: Alpha Pi Mu, Chi Epsilon, Eta Kappa Nu, Omega Chi Epsilon, Pi Tau Sigma, Sigma Gamma Tau, Tau Alpha Pi, Tau Beta Pi, Upsilon Pi Epsilon, Alpha Sigma Mu

SUPPORT PROGRAMS
Student Chapter Organizations: Society of Women Engineers, National Society of Black Engineers

Minority Engineering Program
For: Ethnic Minorities   **Available:** Year round
Offered: Prematriculation, Freshman, Sophomore, Junior, Senior, Graduate level

Other Engineering Support Programs: Orientation program for all freshman and transfers; career planning and counseling services; disabled student services; international student services; Engineering Placement Bureau; honors program; residence hall tutorial services; computer services; cooperative education program; honor societies.

003 University of Alabama at Birmingham

INSTITUTION PROFILE

HEAD OF THE INSTITUTION
Charles A. McCallum
Phone: (205)934-4636

GENERAL INFORMATION
[All Students—Fall 1991]
Undergraduate enrollment 612
Graduate enrollment 170
Total institution enrollment 782
Type of institution: Public **Calendar:** Quarters
Location: Birmingham **Population:** 300,000
Setting: Urban
Types of engineering degrees: Engineering
Other degree-granting colleges: Allied Health Sciences, Business Administration, Dentistry, Education, Medicine, Nursing, Arts and Humanities, Natural Sciences and Mathematics, Optometry, Public Health, Social and Behavioral Sciences

ENGINEERING ADMISSIONS

ADMISSIONS OFFICE CONTACT
Norma E. Sorenson
University of Alabama at Birmingham
Office of Student Services
UAB Station
Birmingham, AL 35294-4461
Phone: (205)934-8410 **Fax:** (205)934-8437

ENGINEERING COLLEGE ADMISSIONS INFORMATION
Entrance Requirements: ACT (Score: 21), Or a total mathematics and science reasoning scores of at least 40 is required.
Entrance Recommendations: High School courses—Mathematics (algebra, geometry, trig. & calculus) (4 years), Science (chemistry, physics, biology) (3 years), Typing (1 year), Computer Science (1 year), Applicants who seek admission and have been out of school for a period of time are evaluated on an individual basis.
Requirements for foreign students: UAB admits foreign students who have academic, linguistic and financial abilities to complete their educational objectives. Certain requirements must be met before admission to engineering departments.

DEPARTMENTAL ADMISSIONS INFORMATION
Admission to the engineering department: At the end of the second year.
Additional information: All departments require an overall GPA of 2.0 (on a 4.0 scale), and a mathematics, natural science, and engineering GPA of 2.0, upon the completion of the required core courses.

FINANCIAL INFORMATION

ESTIMATED EXPENSES (ACADEMIC YEAR)
[Expenses are for the 1992-93 nine-month academic year.]

State Residents	Undergraduate	Graduate
Tuition and fees	$ 2,170	$ 2,170
College room and board	$ 5,175	$ 5,175
Books and supplies	$ 750	$ 750
Other expenses	$ 2,700 (1)	$ 2,700 (1)
Total estimated expenses	**$10,795**	**$10,795**
Out-of-State Residents	Undergraduate	Graduate
Tuition and fees	$ 4,092	$ 4,092
College room and board	$ 5,175	$ 5,175
Books and supplies	$ 750	$ 750
Other expenses	$ 2,700 (1)	$ 2,700 (1)
Total estimated expenses	**$12,717**	**$12,717**

Notes: (1) Includes personal and transportation expenses

FINANCIAL AID OFFICE CONTACT
Claude E. McCann, *Director, Financial Aid*
University of Alabama at Birmingham
UAB Station
Birmingham, AL 35294-1150
Phone: (205)934-8223

GENERAL FINANCIAL AID INFORMATION
Forms accepted/required: AFSA, FAT, Stafford, IRS, SAR, Institutional
Additional financial aid information: Students may apply for financial aid if they have been accepted for admission or if they are presently enrolled at UAB. Students applying for financial aid are considered for all programs for which they are eligible. Assistance generally takes the form of a combination of grant, loan and employment.

ENGINEERING COLLEGE INFORMATION

[For additional personnel, refer to the *Appendix*.]

HEAD OF ENGINEERING
Jay Goldman
Phone: (205)934-8400 **Fax:** (205)934-8437
EMail: JGOLDMAN@ENGSYS.ENG.UAB.EDU

ENGINEERING COLLEGE ADDRESS
School of Engineering
University of Alabama at Birmingham
UAB Station
1150 10th Ave. South
Birmingham, AL 35294-4461
Phone: (205)934-8400 **Fax:** (205)934-8437

ENROLLMENTS—BY CLASS
[Numbers are baccalaureate enrollments for the fall 1991 term, unless otherwise footnoted.]
1st-year students/Freshmen 274
2nd-year students/Sophomores 75
3rd-year students/Juniors 108
4th-year students/Seniors 155
Total **612**

NUMBER OF DEGREES AWARDED—BY PROGRAM
[Numbers are engineering baccalaureate degrees awarded during the 1991-1992 academic year, unless otherwise footnoted. For full details about each engineering program, refer to the *Tables of Degree Programs*.]
Biomedical Engineering .... _ (B)
Civil Engineering 29
Electrical & Computer Engineering 29
Materials Science & Engineering 8
Mechanical Engineering 39
Total **105**

Notes: (B) Data not applicable.

PERCENTAGE OF DEGREES AWARDED—BY CATEGORY
[Percentages are of all engineering baccalaureate degrees awarded during the 1991-1992 academic year, unless otherwise footnoted.]
African Americans .... 5.0 %
Asian/Pacific Island Americans .... – %
Hispanic Americans .... 1.0 %
Native Americans .... – %
Foreign Citizens .... 19.0 %
All Others .... 75.0 %
Women .... 12.0 %
Persons with disabilities .... – % (A)
Students over 25 years of age .... – % (A)

Notes: (A) Data not available.

GRADUATE ENROLLMENTS & DEGREES AWARDED
Master's enrollment 115
Master's degrees awarded 43
Doctoral enrollment 55
Doctoral degrees awarded 2

ENGINEERING STUDENT DATA
Applicants to the engineering college
Number of applicants to engineering college 504
Percent offered admission .... 81.0 %
Matriculated engineering students
Percentage in top quartile (25%) of High School class .... – (A)
Average ACT scores: Math—23.5, Composite—23.5

Notes: (A) Data not available.

FULL & PART TIME FACULTY—BY DEPARTMENT
[Figures are the head count of full-time faculty and the full-time equivalent (FTE) of part-time faculty for each engineering department or equivalent.]

Department	Full	Part
Civil Eng	5	3.0
Electrical & Computer Eng	8	4.0
Mechanical Eng	6	3.0
Materials Science & Eng	7	–
Biomedical Eng	5	1.0

COLLEGE DESCRIPTION
The School offers a four-year ABET accredited undergraduate program in four major engineering disciplines - CE, ECE, MSE and ME. The program is well rounded, with special emphasis given to design engineering. All students complete a senior design project. The first two years emphasize English, math, basic sciences and some core engineering sciences. The second two years emphasize engineering science and design in a specific discipline as well as humanities and social sciences. All students receive a broadly based education, as they must satisfy the University Core Curriculum as specified for engineering majors. Graduate opportunities exist in the basic four disciplines plus Biomedical Engineering.

DEGREE PROGRAMS DESCRIPTION
The B.S. degrees in the basic engineering disciplines prepare the graduate to assume an entry-level design engineering position in industry or to pursue graduate study in each of the respective fields or in Biomedical Engineering. The latter program is facilitated by close relationships with the world renown UAB Medical Center. Students who are majoring in one of the many disciplines outside of engineering at UAB may select as their minor an organized program of study in engineering. Graduate study leading to MS and Ph.D. degrees are available to all students who wish additional specialization in engineering design and in engineering research. Student interested in a Biomedical Engineering undergraduate program may elect a specific concentration in CE, ECE, MSE or ME. Students interested in a pre-med program may elect a pre-health option in MSE or ME. Pre-health options satisfy all requirements for admission to any UAB health professional school - Medicine, Dentistry or Optometry.

DUAL DEGREE PROGRAMS
Engineering baccalaureate graduates with dual degrees 0

TRANSFER INFORMATION
Residency Requirements: Students are allowed to transfer a maximum of 68 semester credit hours.
Transfer via Articulation Agreements
Admission to engineering: At any time
Engineering graduates transferred from .. 4-yr: – 2-yr: –
Requirements: Minimum 2.2 GPA (on a 4.0 scale) on all college level work, and a 2.4 GPA on all mathematics, natural sciences, and engineering courses.
Transfer without Articulation Agreements
Admission to engineering: At any time
Engineering graduates transferred from .. 4-yr: – 2-yr: –

GRADUATION REQUIREMENTS
B.S. programs in engineering require 136 semester credit hours. A student must have at least a 2.0 GPA in all hours attempted at UAB and a 2.0 GPA in all natural sciences, mathematics and engineering courses.

STUDENT PROGRAMS

PROFESSIONAL AND HONORARY SOCIETIES
[For key to acronyms, see *Introduction*.]
Professional Societies: ASCE, ASHRAE, ASME, BMES, IEEE, TMS, American Foundrymen's Society (AFS), American Society for Metals (ASM)
Honorary Societies: Eta Kappa Nu, Pi Tau Sigma, Tau Beta Pi, Alpha Sigma Mu

SUPPORT PROGRAMS
Student Chapter Organizations: Society of Women Engineers, National Society of Black Engineers

NSBE Buddy System
For: Ethnic Minorities **Available:** Year round
Offered: Freshman, Sophomore, Junior, Senior
Other Engineering Support Programs: Orientation programs for all freshmen and transfer; individual advising/counseling each term; mentor 'Buddy' programs; peer advisors; tutorial system; School and student newsletters; professional development through student organizations.

004 University of Alabama at Huntsville

■ INSTITUTION PROFILE

HEAD OF THE INSTITUTION
Frank A. Franz
Phone: (205)895-6340 Fax: (205)895-6538

GENERAL INFORMATION
[All Students—Fall 1991]
Undergraduate enrollment 6,376
Graduate enrollment 2,276
Total institution enrollment 8,652
Type of institution: Public Calendar: Other
Location: Huntsville Population: 162,300
Setting: Urban
Types of engineering degrees: Engineering
Other degree-granting colleges: Administrative Science, Liberal Arts, Science, Nursing, Primary Medical Care

■ ENGINEERING ADMISSIONS

ADMISSIONS OFFICE CONTACT
Kenneth O. Thompson
University of Alabama at Huntsville
Engineering Advisement Center
College of Engineering
Huntsville, AL 35899
Phone: (205)895-6877 Fax: (205)895-6758

ENGINEERING COLLEGE ADMISSIONS INFORMATION
Admission to the engineering college: Admitted to pre-engineering then must maintain 2.5/4.0 in the technical subjects and complete certain entry courses.
Entrance Requirements: ACT (Score: 21), High School courses—English (4 years), History or Social Studies (1 year), Algebra (1 year), Geometry (1 year)
Entrance Recommendations: High School courses—College Prep Math (2 years), Physics or Chemistry (1 year), Electives (6 years)
Requirements for foreign students: TOEFL (Score: 500); financial statement

DEPARTMENTAL ADMISSIONS INFORMATION
Admission to the engineering department: The time they have successfully passed a 300 level course in Engineering.

■ FINANCIAL INFORMATION

ESTIMATED EXPENSES (ACADEMIC YEAR)
[Expenses are for the 1992-93 nine-month academic year.]

State Residents	Undergraduate	Graduate
Tuition and fees	$ 2,328	$ 2,238
College room and board	$ 3,075	$ 3,075
Books and supplies	$ 500	$ 500
Other expenses	$ –	$ –
Total estimated expenses	$ 5,903	$ 5,813

Out-of-State Residents	Undergraduate	Graduate
Tuition and fees	$ 4,656	$ 4,476
College room and board	$ 3,075	$ 3,075
Books and supplies	$ 500	$ 500
Other expenses	$ –	$ –
Total estimated expenses	$ 8,231	$ 8,051

FINANCIAL AID OFFICE CONTACT
James B. Gibson, *Director*
University of Alabama at Huntsville
Financial Aid Office
Huntsville, AL 35899
Phone: (205)895-6241

GENERAL FINANCIAL AID INFORMATION
Forms accepted/required: ACT, AFSA, CSS-FAF, FAF, FSA, FFS, Stafford, IRS, SAR, SDF, Supplemental, Institutional
Additional financial aid information: There are scholarships both university wide and college based as well as part-time campus jobs and loans and grants.

■ ENGINEERING COLLEGE INFORMATION

[For additional personnel, refer to the *Appendix*.]

HEAD OF ENGINEERING
Lynn D. Russell
Phone: (205)895-6474 Fax: (205)895-6758

ENGINEERING COLLEGE ADDRESS
College of Engineering
University of Alabama at Huntsville
Huntsville, AL 35899
Phone: (205)895-6474 Fax: (205)895-6758

ENROLLMENTS—BY CLASS
[Numbers are baccalaureate enrollments for the fall 1991 term, unless otherwise footnoted.]
1st-year students/Freshmen 304
2nd-year students/Sophomores 239
3rd-year students/Juniors 335
4th-year students/Seniors 617
Total 1,495

NUMBER OF DEGREES AWARDED—BY PROGRAM
[Numbers are engineering baccalaureate degrees awarded during the 1991-1992 academic year, unless otherwise footnoted. For full details about each engineering program, refer to the *Tables of Degree Programs*.]
Chemical Engineering 4
Civil & Construction Engineering .... –
Civil Engineering 12
Electrical & Computer Engineering 116
Industrial & Systems Engineering 14
Mechanical & Aerospace Engineering 45
Mechanical Engineering .... –
Total 191

PERCENTAGE OF DEGREES AWARDED—BY CATEGORY
[Percentages are of all engineering baccalaureate degrees awarded during the 1991-1992 academic year, unless otherwise footnoted.]
African Americans .... 5.6 %
Asian/Pacific Island Americans .... 5.6 %
Hispanic Americans .... 1.0 %
Native Americans .... 1.0 %
Foreign Citizens .... 5.1 %
All Others .... 81.7 %
Women .... 19.7 %
Persons with disabilities .... – %
Students over 25 years of age .... – %

GRADUATE ENROLLMENTS & DEGREES AWARDED
Master's enrollment 480
Master's degrees awarded 102
Doctoral enrollment 138
Doctoral degrees awarded 11

ENGINEERING STUDENT DATA
Average ACT scores: Math—25.0, Composite—25.0

FULL & PART TIME FACULTY—BY DEPARTMENT
[Figures are the head count of full-time faculty and the full-time equivalent (FTE) of part-time faculty for each engineering department or equivalent.]

Department	Full	Part
Chemical Eng	5	–
Civil Eng	4	–
Electrical & Computer Eng	26	–
Industrial & Systems Eng	11	2.3
Mechanical Eng	20	1.4

COLLEGE DESCRIPTION
The primary strength of the College is the highly qualified faculty. In addition, the entering students are excellent and generally have good academic preparation for entering engineering. The location in the Huntsville high technology community is also a great strength for the engineering program. Huntsville has one of the highest (if not the highest) concentrations per capita of engineers of any city in the country. This concentration of engineers and associated high technology work provides great support for the college. We have several dual degree programs as well as having a campus wide co-op program.

DEGREE PROGRAMS DESCRIPTION
The strong engineering graduate program provides support for the undergraduate program and for the overall engineering unit. The extensive and productive research programs of the college and the university provide many opportunities for development of both faculty and students, and ensure that examples and illustrations given in the courses are the latest and most advanced in the field. We are developing a long range schedule of course offerings in order to provide students better information for planning purposes. Students will be surveyed periodically for input to the schedule. We are initiating new options in optics, propulsion, materials, computers and systems.

DUAL DEGREE PROGRAMS
Engineering baccalaureate graduates with dual degrees 0
Enrollment requirements: Satisfy admission requirements for transfer students.

TRANSFER INFORMATION
Residency Requirements: Student must complete at least one half of the required number of hours at UAH. Maximum number of transfer hours is 65.

Transfer via Articulation Agreements
Admission to engineering: The end of the pre-engineering program, having maintained 2.5/4.0 in technical subjects and taken certain entry courses.
Requirements: Must maintain 2.5/4.0 in technical subjects and complete certain entry courses in pre-engineering.

Transfer without Articulation Agreements
Admission to engineering: The end of the pre-engineering program, having maintained 2.5/4.0 in technical subjects and taken certain entry courses.
Requirements: Must complete pre-engineering courses with 2.5/4.0 in technical subjects.

GRADUATION REQUIREMENTS
Completion of specified number of hours and 2.0 overall as well as in option courses.

■ STUDENT PROGRAMS

PROFESSIONAL AND HONORARY SOCIETIES
[For key to acronyms, see *Introduction*.]
Professional Societies: AIAA, AIChE, ASCE, ASME, IEEE, IIE, SAE
Honorary Societies: Alpha Pi Mu, Eta Kappa Nu, Pi Tau Sigma, Tau Beta Pi

SUPPORT PROGRAMS
Student Chapter Organizations: Society of Women Engineers, National Society of Black Engineers
Other Engineering Support Programs: We have freshman transfer orientation at least three times annually. The College of Engineering has an advisement center for continuous advisement and counseling. We also participate in the Introduction of Minorities to Engineering and Technology (IMET).

005 University of Alaska, Fairbanks

INSTITUTION PROFILE

HEAD OF THE INSTITUTION
Joan K. Wadlow
Phone: (907)474-5213　　　Fax: (907)474-5213

GENERAL INFORMATION
[All Students—Fall 1991]
Undergraduate enrollment 8,260
Graduate enrollment 699
Total institution enrollment 8,959
Type of institution: Public　　Calendar: Semesters
Nearest city: Anchorage　　Population: 250,000
Miles from main campus: 360　　Setting: Small Town
Types of engineering degrees: Engineering
Other degree-granting colleges: Liberal Arts, Agriculture & Land Resources Management, Mineral Engineering, Fisheries and Ocean Sciences, Management

ENGINEERING ADMISSIONS

ADMISSIONS OFFICE CONTACT
Ann Tremarello
University of Alaska, Fairbanks
Registrar, Admissions and Records
Suite 102, Signers' Hall
Fairbanks, AK 99775-0060
Phone: (907)474-7821　　　Fax: (907)474-5379

ENGINEERING COLLEGE ADMISSIONS INFORMATION
Entrance Requirements: SAT, ACT, High School courses—English (3 years), Algebra, Social Science (2 years), Geometry, Physics, Chemistry (1 year), Trigonometry (half year), The test scores with high school GPA are the determining factors.
Entrance Recommendations: High School courses—Foreign Language (2 years), Computers (1 year), Advanced Algebra (half year), Calculus (half year)
Requirements for foreign students: TOEFL (Score: 550); financial statement

DEPARTMENTAL ADMISSIONS INFORMATION
Admission to the engineering department: At the time of admission to the institution.

FINANCIAL INFORMATION

ESTIMATED EXPENSES (ACADEMIC YEAR)
[Expenses are for the 1992-93 nine-month academic year.]

All Students	Undergraduate	Graduate
Tuition and fees	$ 1,508	$ 2,088
College room and board	$ 3,020	$ 3,020
Books and supplies	$ 600	$ 400
Other expenses	$ 522	$ 522
Total estimated expenses	**$ 5,650**	**$ 6,030**

FINANCIAL AID OFFICE CONTACT
Donald E. Scheaffer, *Director of Financial Aid*
University of Alaska, Fairbanks
5th Floor Gruening
Fairbanks, AK 99775-0700
Phone: (907)474-7256　　　Fax: (907)474-7900

GENERAL FINANCIAL AID INFORMATION
Forms accepted/required: ACT, AFSA, CSS-FAF, FAF, FAT, FAR, FSA, FFS, Stafford, IRS, SAR, SDF, All required depending on aid requested.
Additional financial aid information: Need-based scholarships; part-time work on campus; short-term loans; merit-based scholarships; fee/tuition waivers and talent grants; College work study; Alaska Student Loan Program (2 years residency required); Pell Grant

ENGINEERING COLLEGE INFORMATION

[For additional personnel, refer to the *Appendix*.]

HEAD OF ENGINEERING
Frank F. Williams
Phone: (907)474-7330　　　Fax: (907)474-6087
EMail: FFFLW@ACAD3.ALASKA.EDU

ENGINEERING COLLEGE ADDRESS
School of Engineering
University of Alaska, Fairbanks
Room 539, Duckering Building
Fairbanks, AK 99775-0660
Phone: (907)474-7330　　　Fax: (907)474-6087

ENROLLMENTS—BY CLASS
[Numbers are baccalaureate enrollments for the fall 1991 term, unless otherwise footnoted.]
1st-year students/Freshmen 90
2nd-year students/Sophomores 46
3rd-year students/Juniors 48
4th-year students/Seniors 104
Total **288**

NUMBER OF DEGREES AWARDED—BY PROGRAM
[Numbers are engineering baccalaureate degrees awarded during the 1991-1992 academic year, unless otherwise footnoted. For full details about each engineering program, refer to the *Tables of Degree Programs*.]
Arctic Engineering .... –
Civil Engineering 13
Electrical Engineering 13
Engineering Management .... –
Environmental Quality Engineering .... –
Environmental Quality Science .... –
Mechanical Engineering 13
Science Management .... –
Total **39**

PERCENTAGE OF DEGREES AWARDED—BY CATEGORY
[Percentages are of all engineering baccalaureate degrees awarded during the 1991-1992 academic year, unless otherwise footnoted.]
African Americans .... – %
Asian/Pacific Island Americans .... 2.6 %
Hispanic Americans .... 5.2 %
Native Americans .... – %
Foreign Citizens .... 2.6 %
All Others .... 89.6 %
Women .... 7.8 %
Persons with disabilities .... – %
Students over 25 years of age .... 31.0 %

GRADUATE ENROLLMENTS & DEGREES AWARDED
Master's enrollment 63
Master's degrees awarded 12
Doctoral enrollment 4
Doctoral degrees awarded .... –

ENGINEERING STUDENT DATA

Applicants to the engineering college
Number of applicants to engineering college 392
Percent offered admission 63.0%

Matriculated engineering students
Percentage in top quartile (25%) of High School class ...... (A)
Notes: (A) Data not available.

FULL & PART TIME FACULTY—BY DEPARTMENT
[Figures are the head count of full-time faculty and the full-time equivalent (FTE) of part-time faculty for each engineering department or equivalent.]

Department	Full	Part
Civil Eng	10	0.5
Electrical Eng	5	2.0
Mechanical Eng	7	1.0

COLLEGE DESCRIPTION
The School offers four-year undergraduate programs in civil, electrical and mechanical engineering accredited by the Accreditation Board for Engineering and Technology (ABET). Class sizes are small and faculty are readily available to help students outside of regularly scheduled class times. Laboratory access is excellent, as is availability of microcomputers as well as mainframes. Although 'hands-on' or laboratory experience is emphasized, the theoretical aspects of the curriculum are not compromised. Graduates routinely report that they feel exceptionally well prepared to meet the demands of their work as engineers. Many of our graduates have achieved outstanding success at several of the most highly ranked graduate schools in the nation.

DEGREE PROGRAMS DESCRIPTION
Civil engineering at UAF encompasses the following engineering specialty areas: construction, environmental quality, geotechnical, structural, surveying and mapping, transportation, and water resources. Most civil engineering faculty have specialized in northern-regions related research and one of the graduate degree options is arctic engineering. Electrical engineering at UAF provides three areas of emphasis at the undergraduate level: power and automatic control, computer engineering, and electromagnetic communications. Graduate level research is conducted in these specialties, and space physics, adaptive neural networks, digital electronics, instrumentation, and signal processing. Mechanical engineering at UAF includes industrial processes, fluid mechanics, heat transfer, machine design, thermal systems, system dynamics, and study of materials. Graduate research topics have included remote power generation, waste heat recovery, marine propulsion systems, frozen-ground engineering, finite element modeling, corrosion engineering, and plasticity. UAF has recently become a NASA Space Grant Institution. Engineering students participate in rocket payload design and fabrication and see the results launched from the UAF-operated Poker Flat rocket range. Graduate programs in engineering and science management are also offered.

TRANSFER INFORMATION
Residency Requirements: 24 credits in upper-division courses and at least 30 of the last 36 credits for the degree; a minimum of 12 semester credits in each major field and minimum of three semester credits in each minor field.

Transfer without Articulation Agreements
Admission to engineering: Aug 1 for Fall Semester and Dec 1 for Spring Semester
Engineering graduates transferred from . . 4-yr: –　2-yr: –
Requirements: 2.00 GPA, attendance at another accredited institution

GRADUATION REQUIREMENTS
B.S. programs in engineering require 130-133 semester credits, depending on major selected.

STUDENT PROGRAMS

PROFESSIONAL AND HONORARY SOCIETIES
[For key to acronyms, see *Introduction*.]
Professional Societies: ACM, AGCA, AIME, ASCE, ASME, IEEE
Honorary Societies: Tau Beta Pi

SUPPORT PROGRAMS
Student Chapter Organizations: American Indian Science and Engineering Society, Association for Women in Science

American Indian Science and Engineering Society
For: Ethnic Minorities　　**Available:** Academic year
Offered: Prematriculation, Freshman, Sophomore, Junior, Senior, Graduate level

Association for Women in Science
For: Women　　**Available:** Year round
Offered: Freshman, Sophomore, Junior, Senior, Graduate level

Rural Student Services
For: Ethnic Minorities　　**Available:** Academic year
Offered: Prematriculation, Freshman, Sophomore, Junior, Senior, Graduate level

Other Engineering Support Programs: Spring and fall orientation program for all students; academic advising; honors program; computer support group; international student advising; rural student services; veteran's services; placement and career advising; summer sessions; writing center; and tutoring services.

006 University of Alberta

■ INSTITUTION PROFILE

HEAD OF THE INSTITUTION
Paul T. Davenport
Phone: (403)492-3212   **Fax:** (403)492-2726

GENERAL INFORMATION
[All Students—Fall 1991]
Undergraduate enrollment 23,000
Graduate enrollment 3,000
Total institution enrollment 26,000
Type of institution: Public   **Calendar:** Semesters
Location: Edmonton, Alberta   **Population:** 650,000
Setting: Urban
Types of engineering degrees: Engineering

ENGINEERING COLLEGE ADMISSIONS INFORMATION
Entrance Requirements: High School courses—Mathematics (3 years), Chemistry (3 years), Physics (3 years), English (3 years)
Requirements for foreign students: TOEFL (Score: 580)

DEPARTMENTAL ADMISSIONS INFORMATION
Admission to the engineering department: At the end of the first year.
Additional information: Departmental quotas exist. Admission based on first year GPA.

■ FINANCIAL INFORMATION

ESTIMATED EXPENSES (ACADEMIC YEAR)
[Expenses are for the 1992-93 nine-month academic year.]

All Students	Undergraduate	Graduate
Tuition and fees	$ 2,136	$ –
College room and board	$ –	$ –
Books and supplies	$ 750	$ –
Other expenses	$ –	$ –
Total estimated expenses	**$ 2,886**	**$ –**

FINANCIAL AID OFFICE CONTACT
Jiang Liu, *Emergency Aid & Bursary Coordinator*
University of Alberta
Student Financial Aid & Information Center
302G SUB
Edmonton, AB T6G-2J7
Phone: (403)492-3483   **Fax:** (403)492-6701

■ ENGINEERING COLLEGE INFORMATION
[For additional personnel, refer to the *Appendix*.]

HEAD OF ENGINEERING
Fred D. Otto
Phone: (403)492-3596   **Fax:** (403)492-0500
EMail: fotto@vm.ualberta.ca

ENGINEERING COLLEGE ADDRESS
Faculty of Engineering
University of Alberta
5-1 Mechanical Engineering Building
Edmonton, AB T6G-2G8
Phone: (403)492-3320   **Fax:** (403)492-0500

ENROLLMENTS—BY CLASS
[Numbers are baccalaureate enrollments for the fall 1991 term, unless otherwise footnoted.]
1st-year students/Freshmen 502
2nd-year students/Sophomores 759
3rd-year students/Juniors 544
4th-year students/Seniors 377
Total **2,182**

NUMBER OF DEGREES AWARDED—BY PROGRAM
[Numbers are engineering baccalaureate degrees awarded during the 1991-1992 academic year, unless otherwise footnoted. For full details about each engineering program, refer to the *Tables of Degree Programs*.]
Chemical Engineering 49
Civil Engineering 53
Computer Engineering 26
Electrical Engineering 105
Engineering Physics 7
Mechanical Engineering 92
Metallurgical Engineering 7
Mining Engineering 6
Petroleum Engineering 26
Total **371**

PERCENTAGE OF DEGREES AWARDED—BY CATEGORY
[Percentages are of all engineering baccalaureate degrees awarded during the 1991-1992 academic year, unless otherwise footnoted.]
African Canadians .... – %
Asian/Pacific Island Canadians .... – %
Hispanic Canadians .... – %
Native Canadians .... – %
Foreign Citizens .... – %
All Others .... 100 %
Women .... 12.3 %
Persons with disabilities .... – %
Students over 25 years of age .... – %

GRADUATE ENROLLMENTS & DEGREES AWARDED
Master's enrollment 278
Master's degrees awarded 102
Doctoral enrollment 170
Doctoral degrees awarded 20

ENGINEERING STUDENT DATA
Applicants to the engineering college
Number of applicants to engineering college 684
Percent offered admission 93.4%

FULL & PART TIME FACULTY—BY DEPARTMENT
[Figures are the head count of full-time faculty and the full-time equivalent (FTE) of part-time faculty for each engineering department or equivalent.]

Department	Full	Part
Chemical Eng	17	–
Civil Eng	34	–
Electrical Eng	36	–
Mechanical Eng	27	–
Mining, Metallurgical & Petroleum Eng	16	–

DEGREE PROGRAMS DESCRIPTION
Specialization in Computer Process Control is available in the Chemical Engineering Degree.

TRANSFER INFORMATION
Transfer via Articulation Agreements
Admission to engineering: At the end of the freshman year
Engineering graduates transferred from ... 4-yr: – 2-yr: 57
Transfer without Articulation Agreements
Admission to engineering: At any time
Engineering graduates transferred from ... 4-yr: – 2-yr: –

■ STUDENT PROGRAMS

PROFESSIONAL AND HONORARY SOCIETIES
[For key to acronyms, see *Introduction*.]
Professional Societies: ACI, CIM, CSAE, CSCE, CSME, IEEE, SAE, SAMPE, SME*, SPE, Institute of Transportation Engineers (ITE)

SUPPORT PROGRAMS
Other Engineering Support Programs: The Faculty offers an Orientation to Engineering course for all first year students. We have an active student organization and strong involvement with technical societies.

007 Alfred University

■ INSTITUTION PROFILE
HEAD OF THE INSTITUTION
Edward G. Coll Jr.
Phone: (607)871-2101   Fax: (607)871-2339

GENERAL INFORMATION
[All Students—Fall 1991]
Undergraduate enrollment 1,936
Graduate enrollment 322
Total institution enrollment 2,258
Type of institution: Public and Private   **Calendar:** Semesters
Nearest city: Rochester   **Population:** 400,000
Miles from main campus: 70   **Setting:** Small Town
Types of engineering degrees: Engineering
Other degree-granting colleges: Arts & Sciences, Business Administration, Education, Art & Design

■ ENGINEERING ADMISSIONS
ADMISSIONS OFFICE CONTACT
Daniel L. Meyer
Alfred University
Office of Admissions
Alumni Hall
Alfred, NY 14802
Phone: (800)541-9229   Fax: (607)871-2198

ENGINEERING COLLEGE ADMISSIONS INFORMATION
Entrance Requirements: High School courses—English (4 years), Mathematics (4 years), Lab Science (2 years), Social Studies (4 years), Alfred will accept either the SAT or ACT. There are no minimum score requirements.
Entrance Recommendations: High School courses—English (4 years), Mathematics (4 years), Lab Science (3 years), Social Studies (4 years), Alfred will accept either the SAT or ACT. There are no minimum score requirements.
Requirements for foreign students: TOEFL (Score: 550); financial statement; International students must file a summary of their educational experiences by years and a student health record.

DEPARTMENTAL ADMISSIONS INFORMATION
Admission to the engineering department: At the time of admission to the institution.

■ FINANCIAL INFORMATION
ESTIMATED EXPENSES (ACADEMIC YEAR)
[Expenses are for the 1992-93 nine-month academic year.]

State Residents	Undergraduate	Graduate
Tuition and fees	$ 5,785	$ 8,290
College room and board	$ 4,736	$ 4,736
Books and supplies	$ 600	$ 600
Other expenses	$ 400 [1]	$ 400 [1]
Total estimated expenses	**$11,521**	**$14,026**

Out-of-State Residents	Undergraduate	Graduate
Tuition and fees	$ 8,185	$ 8,290
College room and board	$ 4,736	$ 4,736
Books and supplies	$ 600	$ 600
Other expenses	$ 400 [1]	$ 400 [1]
Total estimated expenses	**$13,921**	**$14,026**

Notes: (1) Tuition and fees are $14,998 for mechanical and electrical engineering programs.

FINANCIAL AID OFFICE CONTACT
Earl E. Pierce Jr., *Director of Student Financial Aid*
Alfred University
Alumni Hall
Alfred, NY 14802
Phone: (607)871-2159   Fax: (607)871-2198

GENERAL FINANCIAL AID INFORMATION
Forms accepted/required: CSS-FAF, FAF, FAT, FFS, Stafford, IRS, SAR, Institutional
Additional financial aid information: Alfred awards a full array of financial options including: institutional grants and scholarships, State and Federal grants, work study and loans. Two scholarships of particular interest are the Alfred Award of Merit, which provides full tuition, room and board, and a $500 book stipend for National Merit Finalists; and Alfred's Presidential Scholarship, a $4,000 to $8,000 scholarship for mechanical and electrical engineering, or $500 to $1,500 for ceramic engineering.

■ ENGINEERING COLLEGE INFORMATION
[For additional personnel, refer to the *Appendix*.]
HEAD OF ENGINEERING
Richard W. Ott
Phone: (607)871-2137   Fax: (607)871-2339
EMail: Ott

ENGINEERING COLLEGE ADDRESS
College of Engineering & Professional Studies
Alfred University
Seidlin Engineering Labs
Alfred, NY 14802
Phone: (607)871-2100

ENROLLMENTS—BY CLASS
[Numbers are baccalaureate enrollments for the fall 1991 term, unless otherwise footnoted.]
1st-year students/Freshmen 124
2nd-year students/Sophomores 77
3rd-year students/Juniors 114
4th-year students/Seniors 96
Total 411

NUMBER OF DEGREES AWARDED—BY PROGRAM
[Numbers are engineering baccalaureate degrees awarded during the 1991-1992 academic year, unless otherwise footnoted. For full details about each engineering program, refer to the *Tables of Degree Programs*.]
Ceramic Engineering 72
Ceramic Engineering Science 5
Electrical Engineering 8
Glass Engineering Science 3
Mechanical Engineering 8
Total 96

PERCENTAGE OF DEGREES AWARDED—BY CATEGORY
[Percentages are of all engineering baccalaureate degrees awarded during the 1991-1992 academic year, unless otherwise footnoted.]
African Americans .... 11.0 %
Asian/Pacific Island Americans .... 4.0 %
Hispanic Americans .... 4.0 %
Native Americans .... 1.0 %
Foreign Citizens .... 10.0 %
All Others .... 70.0 %
Women .... 22.0 %
Persons with disabilities .... — % (A)
Students over 25 years of age .... — % (A)

Notes: (A) Data not available.

GRADUATE ENROLLMENTS & DEGREES AWARDED
Master's enrollment 181
Master's degrees awarded 38
Doctoral enrollment 24
Doctoral degrees awarded 6

ENGINEERING STUDENT DATA
Applicants to the engineering college
Number of applicants to engineering college 338
Percent offered admission 83.0%

Matriculated engineering students
Percentage in top quartile (25%) of High School class 80.0
Average SAT scores: Math—630, Verbal—520, Combined—1150
Average ACT scores: Math—27, Composite—28

FULL & PART TIME FACULTY—BY DEPARTMENT
[Figures are the head count of full-time faculty and the full-time equivalent (FTE) of part-time faculty for each engineering department or equivalent.]

Department	Full	Part
Electrical Eng	5	—
Mechanical Eng	5	—
Ceramic Eng & Sciences	24	—

COLLEGE DESCRIPTION
Alfred University is a unique blend of both publicly and privately endowed colleges. Electrical and mechanical engineering comprise the College of Engineering and Professional Studies, which is private. The School of Ceramic Engineering and Sciences is a part of the New York State College of Ceramics, which is public. Alfred's 34 faculty members - all hold the Ph.D. - combine their expertise in teaching with active research effort totaling over $4 million in 1992. Students and faculty have access to over $10 million worth of computers and equipment. The DEC/VAX computer system is campus-wide and available to students 24 hours a day, free of charge. The world's first computerized and automated x-ray diffractometry software was developed at Alfred and is supported by a VAX station 3100. In addition, 90% of engineering students participate in our Co-op program.

DEGREE PROGRAMS DESCRIPTION
Alfred's undergraduate program in Ceramic Engineering is the largest in the country. In the last five years, one-third of all undergraduate degrees were awarded by Alfred University. The University recently received approval to offer a Ph.D. program in glass science, one of only three programs world-wide. Students in all engineering programs are encouraged to complete two internship programs, co-op, or participate in international study programs.

DUAL DEGREE PROGRAMS
Engineering baccalaureate graduates with dual degrees 11

TRANSFER INFORMATION
Residency Requirements: 4 semester residency requirement.
Transfer via Articulation Agreements
Admission to engineering: At any time
Transfer without Articulation Agreements
Admission to engineering: At any time

GRADUATION REQUIREMENTS
All students must maintain a minimum 2.0 GPA.

■ STUDENT PROGRAMS
PROFESSIONAL AND HONORARY SOCIETIES
[For key to acronyms, see *Introduction*.]
Professional Societies: ACerS, ACM, ACS, ASME, IEEE, NSPE, Society of Women Engineers
Honorary Societies: Keramos, Tau Beta Pi

SUPPORT PROGRAMS
Student Chapter Organizations: Society of Women Engineers, Ibero American Student Union, Minorities Educating, Growing & Achieving, Black Student Union - UMOJA, National Career Women's Association

Division of Developmental Studies
For: Women & Ethnic Minorities   **Available:** Year round
Offered: Prematriculation, Freshman, Sophomore

Educational Opportunity Program
For: Women & Ethnic Minorities   **Available:** Year round
Offered: Prematriculation, Freshman, Sophomore

Higher Education Opportunity Program
For: Women & Ethnic Minorities   **Available:** Year round
Offered: Prematriculation, Freshman, Sophomore

Other Engineering Support Programs: All students participate in a three day orientation program designed to assist with their academic and social transition to collegiate life. In addition, a study center provides academic support for the entire campus community; CLASS allows upperclass students to serve as peer counselors and tutors; and the counseling center offers professional assistance on an individual and confidential basis.

008 University of Arizona

INSTITUTION PROFILE

HEAD OF THE INSTITUTION
Manuel T. Pacheco
Phone: (602)621-5511   Fax: (602)621-7475

GENERAL INFORMATION
[All Students—Fall 1991]
Undergraduate enrollment 26,352
Graduate enrollment 7,736
Total institution enrollment **34,088**
Type of institution: Public   Calendar: Semesters
Location: Tucson   Population: 600,000
Setting: Urban
Types of engineering degrees: Engineering
Other degree-granting colleges: Agricultural & Environmental, Allied Health Sciences, Architecture, Arts & Sciences, Business Administration, Communication & Journalism, Education, Fine & Performing Arts, Humanities & Social Sciences, Law, Medicine, Nursing, Pharmacy, Technology & Applied Sciences

ENGINEERING ADMISSIONS

ADMISSIONS OFFICE CONTACT
Jerome A. Lucido
University of Arizona
Office of Admissions
Tucson, AZ 85721-0001
Phone: (602)621-3237   Fax: (602)621-9799

ENGINEERING COLLEGE ADMISSIONS INFORMATION
Entrance Requirements: SAT, ACT, High School courses—English (4 years), Social Studies (2 years), Mathematics (4 years), Laboratory Science (2 years), SAT Minimum Score: 1010 AZ residents/1050 Out-of-State ACT Minimum Score: 23 AZ residents/24 Out-of-State.
Requirements for out-of-state residents: In-state high school applicants must have a class standing in the top 25 percent; or a GPA of 2.75 (3.0 for out-of-state applicants) on a 4.0 scale.
Requirements for foreign students: TOEFL (Score: 500); English placement test and proof of adequate financial support. Admission deadlines: April 1 for Fall. December 1 for Spring. April 1 for Summer. Center for English as a Second Language offers full-time language training.

DEPARTMENTAL ADMISSIONS INFORMATION
Admission to the engineering department: At the time of admission to the institution.
Additional information: Students meet a set of course and GPA requirements to receive advanced standing and enter their junior year. The GPA requirement varies for each department from 2.0 to 2.75 depending on faculty availability. Students must take a writing proficiency examination.

FINANCIAL INFORMATION

ESTIMATED EXPENSES (ACADEMIC YEAR)
[Expenses are for the 1992-93 nine-month academic year.]

State Residents	Undergraduate	Graduate
Tuition and fees	$ 1,540	$ 1,540
College room and board	$ 3,436	$ 3,436
Books and supplies	$ 574	$ 574
Other expenses	$ –	$ –
Total estimated expenses	**$ 5,550**	**$ 5,550**

Out-of-State Residents	Undergraduate	Graduate
Tuition and fees	$ 6,546	$ 6,546
College room and board	$ 3,436	$ 3,436
Books and supplies	$ 574	$ 574
Other expenses	$ –	$ –
Total estimated expenses	**$10,556**	**$10,556**

FINANCIAL AID OFFICE CONTACT
Phyliss K. Bolt-Bannister, *Student Financial Aid Director*
University of Arizona
Administration Building, Room 201B
Tucson, AZ 85721-0001
Phone: (602)621-1643

GENERAL FINANCIAL AID INFORMATION
Forms accepted/required: Free Application for Federal Student Financial Aid
Additional financial aid information: The University provides access to a full range of federal, state and privately donated financial aid funds. Assistance is available based on financial need, academic merit, and program of study. Students should apply as soon as possible after January 1 of the year in which they will enroll.

ENGINEERING COLLEGE INFORMATION
[For additional personnel, refer to the *Appendix*.]

HEAD OF ENGINEERING
Ernest T. Smerdon
Phone: (602)621-6594   Fax: (602)621-2232

ENGINEERING COLLEGE ADDRESS
College of Engineering and Mines
University of Arizona
Geology 134
Tucson, AZ 85721-0001
Phone: (602)621-6032   Fax: (601)621-9995

ENROLLMENTS—BY CLASS
[Numbers are baccalaureate enrollments for the fall 1991 term, unless otherwise footnoted.]
1st-year students/Freshmen 670
2nd-year students/Sophomores 734
3rd-year students/Juniors 653
4th-year students/Seniors 959
Total **3,016**

NUMBER OF DEGREES AWARDED—BY PROGRAM
[Numbers are engineering baccalaureate degrees awarded during the 1991-1992 academic year, unless otherwise footnoted. For full details about each engineering program, refer to the *Tables of Degree Programs*.]
Aerospace Engineering 53
Agricultural Biosystems Engineering 4
Chemical Engineering 22
Civil Engineering 47
Computer Engineering 43
Electrical Engineering 120
Engineering Mathematics 11
Engineering Mechanics .... –
Engineering Physics 8
Geological Engineering 6
Hydrology 11
Industrial Engineering 29
Materials Science & Engineering 18
Mechanical Engineering 83
Mineral Economics .... –
Mining Engineering 4
Nuclear Engineering 14
Optical Engineering 2
Reliability Engineering .... –
Systems Engineering 23
Water Resources Administration .... –
Total **498**

PERCENTAGE OF DEGREES AWARDED—BY CATEGORY
[Percentages are of all engineering baccalaureate degrees awarded during the 1991-1992 academic year, unless otherwise footnoted.]
African Americans 1.0%
Asian/Pacific Island Americans 6.0%
Hispanic Americans 7.0%
Native Americans 1.0%
Foreign Citizens 16.0%
All Others 69.0%
Women 15.0%
Persons with disabilities .... – %(A)
Students over 25 years of age .... – %(A)
Notes: (A) Data not available.

GRADUATE ENROLLMENTS & DEGREES AWARDED
Master's enrollment 486
Master's degrees awarded 164
Doctoral enrollment 396
Doctoral degrees awarded 64

ENGINEERING STUDENT DATA

Applicants to the engineering college
Number of applicants to engineering college 1,958
Percent offered admission 82.0%

Matriculated engineering students
Percentage in top quartile (25%) of High School class 95.0
Average SAT scores: Combined—1078

FULL & PART TIME FACULTY—BY DEPARTMENT
[Figures are the head count of full-time faculty and the full-time equivalent (FTE) of part-time faculty for each engineering department or equivalent.]

Department	Full	Part
Chemical Eng	11	–
Civil Eng & Eng Mechanics	25	–
Electrical & Computer Eng	48	–
Systems & Industrial Eng	19	–
Aerospace & Mechanical Eng	33	–
Agricultural & Biosystems Eng	–	–
Eng Mathematics	–	–
Eng Physics	–	–
Hydrology & Water Resources	14	–
Materials Science & Eng	21	–
Mining & Geological Eng	12	–
Nuclear & Energy Eng	9	–

COLLEGE DESCRIPTION
Engineering students enjoy the availability of the complete spectrum of high quality academic and service programs found only at comprehensive flagship universities. The college offers 4-year undergraduate programs in engineering and in engineering sciences. The freshman year is common to all majors and emphasizes engineering problem-solving, design and computer programming. This is followed by 3 years of specialized training in one of the 17 majors. Co-op is optional in all majors. Average number of years required to actually complete the bachelor's degree: 4.7. As a major partner in a research university, the college research facilities support a large graduate education program. Graduate degrees at the M.S. and Ph.D. levels are offered in 17 majors.

DEGREE PROGRAMS DESCRIPTION
The B.S. degrees in the College prepare graduates for entry-level professional positions or graduate study. In coordination with the B.S. degrees described herein, options are available in Biomedical Engineering, Energy Engineering, Computer Software Engineering, Manufacturing Systems Engineering, and Pre-medical. The M.S. and Ph.D. degrees prepare graduates for research, teaching, and advanced level professional placement. Interdepartmental options at the graduate level include Biomedical Engineering and Energy Systems Engineering.

TRANSFER INFORMATION
Residency Requirements: A minimum of 30 units of credit from the University of Arizona is required for the B.S. degree. It is further required that 18 of the final 30 units be UA credit.

Transfer via Articulation Agreements
Admission to engineering: At any time
Requirements: Transfer programs are articulated with in-state community colleges. Applications and all supporting documents must be received by June 1 for the Fall semester and December 1 for the Spring semester. All transfer credit is reviewed by a faculty advisor to determine applicability toward degree requirements.

Transfer without Articulation Agreements
Admission to engineering: At any time
Requirements: A grade point average of 2.5 is required to transfer into the College.

GRADUATION REQUIREMENTS
Depending on the major selected, B.S. programs in engineering require 127-136 semester hours of specifically designated courses. These are outlined in the University catalog.

STUDENT PROGRAMS

PROFESSIONAL AND HONORARY SOCIETIES
[For key to acronyms, see *Introduction*.]
Professional Societies: AIChE, ANS, ASAE, ASCE, ASME, AWRA, IEEE, IIE, SME*, TMS, Society of Reliability Engineers, American Institute of Aeronautics and Astronautics
Honorary Societies: Eta Kappa Nu, Tau Beta Pi, Alpha Nu Sigma

SUPPORT PROGRAMS
Student Chapter Organizations: Society of Women Engineers, Society of Hispanic Professional Engineers, American Indian Science and Engineering Society, National Society for Black Engineers

Minority Engineering Program
For: Ethnic Minorities   Available: Year round
Offered: Freshman, Sophomore, Junior, Senior

Minority Student Services
For: Ethnic Minorities   Available: Year round
Offered: Freshman, Sophomore, Junior, Senior

Summer Programs
For: Women & Ethnic Minorities   Available: Year round
Offered: Prematriculation

Other Engineering Support Programs: Orientation programs in summer and prior to each semester; Academic Learning Skills Center; academic advising; Student Health Center; Career Services and Placement Center; foreign student office; services for handicapped; Engineering Student Council.

009 Arizona State University

INSTITUTION PROFILE

HEAD OF THE INSTITUTION
Lattie F. Coor
Phone: (602)965-5606   Fax: (602)965-0865

GENERAL INFORMATION
[All Students—Fall 1991]
Undergraduate enrollment 31,426
Graduate enrollment 11,200
Total institution enrollment 42,626

Type of institution: Public   Calendar: Semesters
Nearest city: Phoenix   Population: 983,403
Miles from main campus: 11   Setting: Suburban
Types of engineering degrees: Engineering & Technology
Other degree-granting colleges: Agricultural & Environmental, Architecture, Arts & Sciences, Business Administration, Communication & Journalism, Education, Fine & Performing Arts, Law, Nursing, Extended Education, Public Programs, Social Work, University Honors

ENGINEERING ADMISSIONS

ADMISSIONS OFFICE CONTACT
Susan Clouse
Arizona State University
Undergraduate Admissions
Tempe, AZ 85287-0112
Phone: (602)965-7788   Fax: (602)965-2120

ENGINEERING COLLEGE ADMISSIONS INFORMATION
Entrance Requirements: SAT (Score: 1050), ACT (Score: 23), High School courses—Mathematics (college algebra, geometry, & trig) (3 years), Laboratory Sciences (Physics & Chemistry) (2 years), Computer programming (BASIC) (1 year), Applicants must submit either test scores (SAT of 1050; or ACT of 23) or be in upper 25% of high school class.
Entrance Recommendations: High School courses—Calculus (1 year), Biology (1 year), Computer Programming (1 year)
Requirements for out-of-state residents: Non-resident requirements: upper 25% of high school class; or ACT of 24; or SAT 1050.
Requirements for foreign students: TOEFL (Score: 550); financial statement; Application deadline June 15 for Fall semester, November 15 for Spring semester.

DEPARTMENTAL ADMISSIONS INFORMATION
Admission to the engineering department: At the time of admission to the institution.

FINANCIAL INFORMATION

ESTIMATED EXPENSES (ACADEMIC YEAR)
[Expenses are for the 1992-93 nine-month academic year.]

State Residents	Undergraduate	Graduate
Tuition and fees	$ 1,530	$ 1,530
College room and board	$ 4,110	$ 4,110
Books and supplies	$ 480	$ 480
Other expenses	$ 2,500	$ 2,500[1]
Total estimated expenses	**$ 8,620**	**$ 8,620**

Out-of-State Residents	Undergraduate	Graduate
Tuition and fees	$ 6,940	$ 6,940
College room and board	$ 4,110	$ 4,110
Books and supplies	$ 480	$ 480
Other expenses	$ 2,500	$ 2,500
Total estimated expenses	**$14,030**	**$14,030**

Notes: (1) Personal expenses including travel.

FINANCIAL AID OFFICE CONTACT
Paul Barberini, *Director*
Arizona State University
Student Financial Assistance
Tempe, AZ 85287-0412
Phone: (602)965-4045

GENERAL FINANCIAL AID INFORMATION
Forms accepted/required: FAF, FFS
Additional financial aid information: Scholarships, grants, loans, and employment.

ENGINEERING COLLEGE INFORMATION

[For additional personnel, refer to the *Appendix*.]

HEAD OF ENGINEERING
David C. Chang
Phone: (602)965-1722   Fax: (602)965-2267
EMail: icdcc@asuacad

ENGINEERING COLLEGE ADDRESS
College of Engineering & Applied Sciences
Arizona State University
School of Engineering
Tempe, AZ 85287-5506
Phone: (602)965-3421   Fax: (602)965-2267

ENROLLMENTS—BY CLASS
[Numbers are baccalaureate enrollments for the fall 1991 term, unless otherwise footnoted.]
1st-year students/Freshmen 444
2nd-year students/Sophomores 459
3rd-year students/Juniors 692
4th-year students/Seniors 1,289
Total 2,884

NUMBER OF DEGREES AWARDED—BY PROGRAM
[Numbers are engineering baccalaureate degrees awarded during the 1991-1992 academic year, unless otherwise footnoted. For full details about each engineering program, refer to the *Tables of Degree Programs*.]

Aerospace Engineering 50
Bioengineering 22
Chemical Engineering 23
Civil Engineering 31
Computer Science 48
Computer Systems Engineering 25
Electrical Engineering 142
Engineering Interdisciplinary Studies 4
Engineering Science(s) .... –
Engineering Special Studies 7
Industrial Engineering 25
Materials Science .... –
Mechanical Engineering 83
Total 460

PERCENTAGE OF DEGREES AWARDED—BY CATEGORY
[Percentages are of all engineering baccalaureate degrees awarded during the 1991-1992 academic year, unless otherwise footnoted.]

African Americans 1.0%
Asian/Pacific Island Americans 10.0%
Hispanic Americans 3.0%
Native Americans .... – %
Foreign Citizens 13.0%
All Others 73.0%
Women 14.0%
Persons with disabilities .... – %
Students over 25 years of age .... – %

GRADUATE ENROLLMENTS & DEGREES AWARDED
Master's enrollment 1,068
Master's degrees awarded 271
Doctoral enrollment 383
Doctoral degrees awarded 58

ENGINEERING STUDENT DATA

Applicants to the engineering college
Number of applicants to engineering college 950
Percent offered admission 85.0%

Matriculated engineering students
Percentage in top quartile (25%) of High School class ...... – (A)

Notes: (A) Data not available.

FULL & PART TIME FACULTY—BY DEPARTMENT
[Figures are the head count of full-time faculty and the full-time equivalent (FTE) of part-time faculty for each engineering department or equivalent.]

Department	Full	Part
Chemical, Bio, & Materials Eng	25	1.0
Civil Eng	17	1.5
Electrical Eng	40	5.0
Industrial & Management Systems Eng	15	2.5
Mechanical & Aerospace Eng	35	5.3
Computer Science & Eng	26	4.5

COLLEGE DESCRIPTION

The Bachelor of Science (B.S.) and Bachelor of Science in Engineering (B.S.E.) degrees are made up of three parts: University General Studies, the engineering core, and a major. The courses included in the engineering core serve as basic background material and comprise approximately 33% of the degree program. The majors available are of two types: those associated with a particular department within the School of Engineering and those offered as special and interdisciplinary studies. All curricula are extensions beyond the engineering core and cover a wide variety of subject areas within each field. The Engineering Special and Interdisciplinary Programs are administered by the Office of the Dean and designed for those students whose objectives require more intensity or flexibility than is possible in the traditional departments. About 1/4 of the major credits are referred to as technical electives.

DEGREE PROGRAMS DESCRIPTION

Majors and areas of emphasis leading to the B.S.E. degree are offered by the six engineering departments. Majors of the Engineering Special Programs leading to the B.S.E. degree and the Engineering Interdisciplinary Programs leading to the B.S. degree are non-departmental majors administered by the Office of the Dean. Graduate degrees offered are the M.S., a research oriented degree program requiring a thesis and an oral examination at completion of the program; the M.S.E., a professionally oriented degree program with two options -- one requires a thesis (engineering report or research paper), the other is a non-thesis option; and the Ph.D. degree.

DUAL DEGREE PROGRAMS

Engineering baccalaureate graduates with dual degrees 5
Enrollment requirements: Dual degree students must meet admission requirements and complete all course requirements of both degree programs.

TRANSFER INFORMATION

Residency Requirements: A minimum of 30 semester hours in resident credit courses at ASU required of every candidate for the baccalaureate degree. The final 12 hours immediately preceding graduation must be of resident credit.

Transfer via Articulation Agreements
Admission to engineering: At any time
Requirements: These transfer students must meet the same minimum admission requirements as other transfer students (transfer cum GPA 2.50, and 550 TOEFL if international student).

Transfer without Articulation Agreements
Admission to engineering: At any time
Requirements: These transfer students must meet the minimum transfer requirements of 2.50 cum GPA and TOEFL score of 550 if international student.

GRADUATION REQUIREMENTS

B.S. and B.S.E. degree programs in engineering require 133 semester hours minimum plus University English proficiency requirements. A minimum GPA of 2.00 is required in the overall program, and a minimum GPA of 2.00 is required in all core and major courses. Some programs require "C" minimum grade in selected major courses.

STUDENT PROGRAMS

PROFESSIONAL AND HONORARY SOCIETIES
[For key to acronyms, see *Introduction*.]

Professional Societies: ACS, AIAA, AIChE, ASCE, ASM*, ASME, BMES, IEEE, IEE
Honorary Societies: Alpha Pi Mu, Chi Epsilon, Eta Kappa Nu, Pi Tau Sigma, Tau Beta Pi, Upsilon Pi Epsilon

SUPPORT PROGRAMS
Student Chapter Organizations: Society of Women Engineers, Society of Hispanic Professional Engineers, American Indian Science & Engineering Society, Arizona Council of Black Engineers & Scientists

Minority Engineering Program
For: Ethnic Minorities   **Available:** Academic year
Offered: Prematriculation, Freshman, Sophomore
Other Engineering Support Programs: Orientation for all freshmen and transfers; personal and career counseling services; international students office; veterans' services; disabled students resource office; tutoring services; placement services.

010 University of Arkansas

■ INSTITUTION PROFILE
HEAD OF THE INSTITUTION
Daniel E. Ferritor
Phone: (501)575-4148　　　Fax: (501)575-7575

GENERAL INFORMATION
[All Students—Fall 1991]
Undergraduate enrollment 12,304
Graduate enrollment 2,047
Total institution enrollment 14,351
Type of institution: Public　　**Calendar:** Semesters
Location: Fayetteville, AR　　**Population:** 42,000
Setting: Small Town
Types of engineering degrees: Engineering
Other degree-granting colleges: Architecture, Arts & Sciences, Business Administration, Education, Law, Agriculture & Home Economics, Graduate

ENGINEERING COLLEGE ADMISSIONS INFORMATION
Entrance Requirements: SAT, ACT, High School courses—English (4 years), Social Studies (3 years), Natural Science (min. of 2 yrs. biol,chem,physics) (3 years), Mathematics (3 years), Admission decisions are based on a combination of high school grades and standard admission test scores.
Entrance Recommendations: SAT, ACT, High School courses—English (4 years), Social Studies (3 years), Natural Science (especially chemistry & physics) (4 years), Mathematics (4 years), Some scholarship awards are based upon ACT or SAT test scores.
Requirements for foreign students: TOEFL (Score: 550); For electrical and computer systems engineering students: Minimum composite ACT score of 25 or SAT score of 1000. Minimum score on the Test of Spoken English of 220.

DEPARTMENTAL ADMISSIONS INFORMATION
Admission to the engineering department: At the time of admission to the institution.

■ FINANCIAL INFORMATION
ESTIMATED EXPENSES (ACADEMIC YEAR)
[Expenses are for the 1992-93 nine-month academic year.]

State Residents	Undergraduate	Graduate
Tuition and fees	$ 2,318	$ 2,970
College room and board	$ 3,225	$ 3,225
Books and supplies	$ 600	$ 600
Other expenses	$ –	$ –
Total estimated expenses	**$ 6,143**	**$ 6,795**

Out-of-State Residents	Undergraduate	Graduate
Tuition and fees	$ 5,198	$ 6,030
College room and board	$ 3,225	$ 3,225
Books and supplies	$ 600	$ 600
Other expenses	$ –	$ –
Total estimated expenses	**$ 9,023**	**$ 9,855**

FINANCIAL AID OFFICE CONTACT
Lenthon Clark, *Director of Financial Aid*
University of Arkansas
114 Hunt Hall
Fayetteville, AR 72701
Phone: (501)575-3806　　Fax: (501)575-7575

GENERAL FINANCIAL AID INFORMATION
Forms accepted/required: Need Analysis Form (recognized by Department of Education), University of Arkansas Aid Questionnaire
Additional financial aid information: Need-based scholarships, merit-based scholarships, student loans, College Work-Study Program.

■ ENGINEERING COLLEGE INFORMATION
[For additional personnel, refer to the *Appendix*.]
HEAD OF ENGINEERING
Neil M. Schmitt
Phone: (501)575-3054　　Fax: (501)575-4346
EMail: nms@nschmitt.uark.edu

ENGINEERING COLLEGE ADDRESS
College of Engineering
University of Arkansas
Bell Engineering Center
Fayetteville, AR 72701
Phone: (501)575-3051　　Fax: (501)575-4346

ENROLLMENTS—BY CLASS
[Numbers are baccalaureate enrollments for the fall 1991 term, unless otherwise footnoted.]
1st-year students/Freshmen 456
2nd-year students/Sophomores 329
3rd-year students/Juniors 343
4th-year students/Seniors 472
Total **1,600**

NUMBER OF DEGREES AWARDED—BY PROGRAM
[Numbers are engineering baccalaureate degrees awarded during the 1991-1992 academic year, unless otherwise footnoted. For full details about each engineering program, refer to the *Tables of Degree Programs*.]
Biological & Agricultural Engineering 2
Chemical Engineering 39
Civil Engineering 25
Computer Systems Engineering 30
Electrical Engineering 63
Engineering .... –
Environmental Engineering .... –
Industrial Engineering 30
Mechanical Engineering 42
Operations Management .... –
Operations Research .... –
Total **231**

PERCENTAGE OF DEGREES AWARDED—BY CATEGORY
[Percentages are of all engineering baccalaureate degrees awarded during the 1991-1992 academic year, unless otherwise footnoted.]
African Americans 3.6%
Asian/Pacific Island Americans 3.2%
Hispanic Americans 0.4%
Native Americans 0.4%
Foreign Citizens 14.5%
All Others 77.9%
Women 6.5%
Persons with disabilities .... – %(A)
Students over 25 years of age .... – %(A)

Notes: (A) Data not available.

GRADUATE ENROLLMENTS & DEGREES AWARDED
Master's enrollment 173
Master's degrees awarded 52
Doctoral enrollment 49
Doctoral degrees awarded 12

ENGINEERING STUDENT DATA
Applicants to the engineering college
Number of applicants to engineering college 588
Percent offered admission 100.0%

Matriculated engineering students
Percentage in top quartile (25%) of High School class ...... –(A)
Average ACT scores: Math—25.38, Composite—25.2

Notes: (A) Data not available.

FULL & PART TIME FACULTY—BY DEPARTMENT
[Figures are the head count of full-time faculty and the full-time equivalent (FTE) of part-time faculty for each engineering department or equivalent.]

Department	Full	Part
Chemical Eng	12	1.0
Civil Eng	11	–
Electrical Eng	16	1.0
Industrial Eng	10	1.0
Mechanical Eng	13	1.5
Biological & Agricultural Eng	11	–
Computer Systems Eng	9	2.0

COLLEGE DESCRIPTION
The College of Engineering offers four-year undergraduate programs leading to the Bachelor of Science degree in seven areas. Programs leading to Master of Science degrees and the Doctor of Philosophy degree are also offered. A broad general education is provided, with special emphasis on the physical sciences and their application to the needs of modern civilization. Average number of years required to actually complete the bachelor's degree; 4.5. The freshman curriculum stresses a basic foundation in mathematics, physics, and chemistry. The sophomore, junior, and senior years are spent in a strong concentration on the student's chosen field, with classroom and laboratory work emphasizing practical applications. Electives in the humanities and social sciences aid in providing a well-rounded education. Cooperative Education is an option for all engineering students.

DEGREE PROGRAMS DESCRIPTION
The B.S. degrees in the various engineering disciplines give graduates a solid background in scientific and economic principles and acquaint them with industrial practices in their chosen field. A strong emphasis on practical applications prepares graduates to assume positions of responsibility early in their careers. Advanced study in engineering is encouraged for qualified students. Programs leading to the M.S. and Ph.D degrees are specialized in the selected area of study and prepare students to enter the profession at a level beyond that of baccalaureate graduates.

TRANSFER INFORMATION
Residency Requirements: The full senior year must be completed in residence.

GRADUATION REQUIREMENTS
BS programs in engineering require 132-136 semester hours, depending on the major selected. Every graduate must meet the minimum requirements established by the Engineering Accreditation Commission of the Accreditation Board for Engineering and Technology.

■ STUDENT PROGRAMS
PROFESSIONAL AND HONORARY SOCIETIES
[For key to acronyms, see *Introduction*.]
Professional Societies: AIChE, ANS, ASAE, ASCE, ASME, IEEE, IIE, NSPE, SAE, SAME
Honorary Societies: Alpha Epsilon, Alpha Pi Mu, Chi Epsilon, Eta Kappa Nu, Pi Tau Sigma, Tau Beta Pi, Theta Tau, Alpha Chi Sigma

SUPPORT PROGRAMS
Student Chapter Organizations: Society of Women Engineers, National Society of Black Engineers

Transition-Retention Minority Engineering Programs
For: Ethnic Minorities
Offered: Freshman

Minority Introduction to Engineering
For: Ethnic Minorities
Offered: Prematriculation

Other Engineering Support Programs: Freshman/transfer summer orientation programs, college mentor, mid-semester grade alert for freshmen. Mathematics, Chemistry and English tutors, Academic Development Office, Engineering Writing Center, Mathematics Assistance Center, faculty advisor system within college.

011 Arkansas Tech University

INSTITUTION PROFILE

HEAD OF THE INSTITUTION
Kenneth Kersh
Phone: (501)968-0237

GENERAL INFORMATION
[All Students—Fall 1991]
Undergraduate enrollment 4,200
Graduate enrollment 200
Total institution enrollment 4,400
Type of institution: Public **Calendar:** Semesters
Nearest city: Little Rock, AR **Population:** 195,700
Miles from main campus: 78 **Setting:** Small Town
Types of engineering degrees: Engineering
Other degree-granting colleges: Arts & Sciences, Business Administration, Education, Humanities & Social Sciences, Nursing

ENGINEERING ADMISSIONS

ADMISSIONS OFFICE CONTACT
Tammy Rhodes
Arkansas Tech University
Caraway Building
Russellville, AR 72801-2222
Phone: (501)968-0343

ENGINEERING COLLEGE ADMISSIONS INFORMATION
Entrance Requirements: SAT, ACT (Score: 18), High School courses—Algebra (1 year), Trigonometry
Entrance Recommendations: High School courses—Math (4 years), English (4 years), Physics (1 year), Chemistry (1 year)
Requirements for foreign students: TOEFL (Score: 520); Certified academic records.

DEPARTMENTAL ADMISSIONS INFORMATION
Admission to the engineering department: At the time of admission to the institution.

FINANCIAL INFORMATION

ESTIMATED EXPENSES (ACADEMIC YEAR)
[Expenses are for the 1992-93 nine-month academic year.]

State Residents	Undergraduate	Graduate
Tuition and fees	$ 1,500	$ –
College room and board	$ 2,410	$ –
Books and supplies	$ 450	$ –
Other expenses	$ 60	$ –
Total estimated expenses	$ 4,420	$ –

Out-of-State Residents	Undergraduate	Graduate
Tuition and fees	$ 3,000	$ –
College room and board	$ 2,410	$ –
Books and supplies	$ 450	$ –
Other expenses	$ 120	$ –
Total estimated expenses	$ 5,980	$ –

FINANCIAL AID OFFICE CONTACT
Shirley Goines, *Director of Student Aid*
Arkansas Tech University
Student Center
Russellville, AR 72801-2222
Phone: (501)968-0399

GENERAL FINANCIAL AID INFORMATION
Forms accepted/required: Institutional
Additional financial aid information: Academic scholarships; Leadership scholarships; Arkansas Student Assistance Grant Program; Privately Supported Grant-in-Aid Scholarships.

ENGINEERING COLLEGE INFORMATION

[For additional personnel, refer to the *Appendix*.]

HEAD OF ENGINEERING
Jack Hamm
Phone: (501)968-0353 Fax: (501)968-0677

ENGINEERING COLLEGE ADDRESS
Engineering Department
Arkansas Tech University
Corley Building
Russellville, AR 72801-2222
Phone: (501)968-0663 Fax: (501)968-0677

ENROLLMENTS—BY CLASS
[Numbers are baccalaureate enrollments for the fall 1991 term, unless otherwise footnoted.]
1st-year students/Freshmen 108
2nd-year students/Sophomores 60
3rd-year students/Juniors 37
4th-year students/Seniors 41
Total 246

NUMBER OF DEGREES AWARDED—BY PROGRAM
[Numbers are engineering baccalaureate degrees awarded during the 1991-1992 academic year, unless otherwise footnoted. For full details about each engineering program, refer to the *Tables of Degree Programs*.]
General Engineering 15
Total 15

PERCENTAGE OF DEGREES AWARDED—BY CATEGORY
[Percentages are of all engineering baccalaureate degrees awarded during the 1991-1992 academic year, unless otherwise footnoted.]
African Americans .... 6.7 %
Asian/Pacific Island Americans .... – %
Hispanic Americans .... – %
Native Americans .... – %
Foreign Citizens .... – %
All Others .... 93.3 %
Women .... 20.0 %
Persons with disabilities .... – %
Students over 25 years of age .... 53.4 %

ENGINEERING STUDENT DATA
Applicants to the engineering college
Number of applicants to engineering college 112
Percent offered admission 100.0%

Matriculated engineering students
Percentage in top quartile (25%) of High School class .... – (A)
Average SAT scores: Math—380, Verbal—319, Combined—700
Average ACT scores: Math—21, Composite—21

Notes: (A) Data not available.

FULL & PART TIME FACULTY—BY DEPARTMENT
[Figures are the head count of full-time faculty and the full-time equivalent (FTE) of part-time faculty for each engineering department or equivalent.]

Department	Full	Part
Eng	8	1.0

COLLEGE DESCRIPTION

The Department of Engineering offers a four year program leading to a degree of Bachelor of Science in Engineering. The program is broad in scope and is designed to produce graduates with a professional attitude and approach to problem solving, well educated in the basics of Engineering. The required courses provide a basic foundation in Engineering with an emphasis on computer modeling, computer applications, and digital design. The graduate may elect to attend one of many different graduate schools (electrical, mechanical, industrial, nuclear, mathematics, physics, or business).

DEGREE PROGRAMS DESCRIPTION
Arkansas Tech University engineering graduates are particularly well suited to help their organizations adapt and prosper in the changing environment. The engineering curriculum consists of a dual emphasis on electrical and mechanical engineering. Students take theoretical, design, and laboratory courses in both areas. The dual emphasis requires versatility. At the same time, quality control, reliability, and manufacturing cost are all addressed repeatedly in the design and laboratory courses where modern statistical methods are introduced and practiced.

TRANSFER INFORMATION
Residency Requirements: The last 30 semester hours of work toward a degree the senior year must be done in residence.
Transfer without Articulation Agreements
Admission to engineering: At any time
Engineering graduates transferred from . . 4-yr: –   2-yr: –

STUDENT PROGRAMS

PROFESSIONAL AND HONORARY SOCIETIES
[For key to acronyms, see *Introduction*.]
Professional Societies: NSPE

SUPPORT PROGRAMS
Affirmative Action Director
For: Ethnic Minorities **Available:** Year round
Offered: Freshman, Sophomore, Junior, Senior, Graduate level
Other Engineering Support Programs: Fall orientation program for all freshman and transfer students; academic and career advising; Student Development Center; International Student Office; facilities for disabled students; tutoring programs; placement services.

012 Auburn University

INSTITUTION PROFILE

HEAD OF THE INSTITUTION
William V. Muse
Phone: (205)844-4650   Fax: (205)844-6179

GENERAL INFORMATION
[All Students—Fall 1991]
Undergraduate enrollment 18,985
Graduate enrollment 2,851
Total institution enrollment **21,836**
Type of institution: Public   **Calendar:** Quarters
Nearest city: Montgomery   **Population:** 150,000
Miles from main campus: 60   **Setting:** Small Town
Types of engineering degrees: Engineering & Technology
Other degree-granting colleges: Agricultural & Environmental, Architecture, Business Adm, Education, Nursing, Pharmacy, Technology & Applied Sciences, Forestry, Human Sciences, Liberal Arts, Sciences and Mathematics, Veterinary Medicine

ENGINEERING ADMISSIONS

ADMISSIONS OFFICE CONTACT
Charles F. Reeder
Auburn University
202 Martin Hall
Auburn University, AL 36849-5145
Phone: (204)844-4080   Fax: (205)844-6436

ENGINEERING COLLEGE ADMISSIONS INFORMATION
Entrance Requirements: High School courses—English (4 years), Mathematics (3 years), Science (2 years), Social Studies (3 years), Minimum ACT composite score of 20 (in-state) or 23 (out-of-state) or SAT minimum composite score of 870 (in-state) or 1000 (out-of-state).
Entrance Recommendations: High School courses—Add'l Science (1 year), Add'l Social Studies (1 year), Foreign Language (1 year)
Requirements for out-of-state residents: High school grade point average 2.00 in-state and 3.00 out-of-state.
Requirements for foreign students: TOEFL (Score: 550); financial statement; B average on high school work. 22 on ACT or 1000 on SAT

DEPARTMENTAL ADMISSIONS INFORMATION
Admission to the engineering department: End of the first year.
Additional information: Complete all appropriate freshman courses; earn an overall GPA on all required and approved elective course work as follows: 2.6 for Electrical Engineering; 2.0 for Textile Management and Technology; 2.2 for all other curricula; and recommendation by the Curriculum Admissions Committee.

FINANCIAL INFORMATION

ESTIMATED EXPENSES (ACADEMIC YEAR)
[Expenses are for the 1992-93 nine-month academic year.]

State Residents	Undergraduate	Graduate
Tuition and fees	$ 1,755 [1]	$ 1,755 [1]
College room and board	$ 3,783	$ 3,783
Books and supplies	$ 600	$ 600
Other expenses	$ 1,833 [2]	$ 1,833 [2]
Total estimated expenses	**$ 7,971**	**$ 7,971**

Out-of-State Residents	Undergraduate	Graduate
Tuition and fees	$ 5,265 [1]	$ 5,265 [1]
College room and board	$ 3,783	$ 3,783
Books and supplies	$ 600	$ 600
Other expenses	$ 1,833 [2]	$ 1,833 [2]
Total estimated expenses	**$11,481**	**$11,481**

Notes: (1) Cost is based on 10 to 15 credit hours per quarter.
(2) Miscellaneous expenses (recreation, travel, clothing)

FINANCIAL AID OFFICE CONTACT
Clark Aldridge, *Director*
Auburn University
203 Martin Hall
Auburn University, AL 36849-5144
Phone: (205)844-4723   Fax: (205)844-6085

GENERAL FINANCIAL AID INFORMATION
Forms accepted/required: ACT, FFS, Institutional
Additional financial aid information: Pell Grants, Stafford Loans, PLUS/SLS Loans, Perkins Loan, Supplemental Educational Opportunity Grants, Health Professions Student Loan, Academic Scholarships, College Work-Study, Institutional Employment, Institutional Loans.

ENGINEERING COLLEGE INFORMATION
[For additional personnel, refer to the *Appendix*.]

HEAD OF ENGINEERING
William F. Walker
Phone: (205)844-4326   Fax: (205)844-2672
EMail: wwalker@eng.auburn.edu

ENGINEERING COLLEGE ADDRESS
College of Engineering
Auburn University
108 Ramsay Hall
Auburn University, AL 36849-5330
Phone: (205)844-4326   Fax: (205)844-2672

ENROLLMENTS—BY CLASS
[Numbers are baccalaureate enrollments for the fall 1991 term, unless footnoted.]
1st-year students/Freshmen 1,132
2nd-year students/Sophomores 992
3rd-year students/Juniors 798
4th-year students/Seniors 876
Total **3,798**

NUMBER OF DEGREES AWARDED—BY PROGRAM
[Numbers are engineering baccalaureate degrees awarded during the 1991-1992 academic year, unless otherwise footnoted. For full details about each engineering program, refer to the *Tables of Degree Programs*.]

Aerospace Engineering 65
Agricultural Engineering 4
Aviation Management 53
Chemical Engineering 47
Civil Engineering 78
Computer Engineering 47
Computer Science 14
Computer Science & Engineering .... – [B]
Electrical Engineering 182
Environmental Science 12
Forest Engineering 4
Geological Engineering 1
Industrial Engineering 54
Manufacturing Systems Engineering .... – [B]
Materials Engineering 15
Mechanical Engineering 131
Textile Chemistry 6
Textile Engineering 4
Textile Management & Technology 12
Textile Science .... – [B]
Total **729**

Notes: (B) Data not applicable.

PERCENTAGE OF DEGREES AWARDED—BY CATEGORY
[Percentages are of all engineering baccalaureate degrees awarded during the 1991-1992 academic year, unless otherwise footnoted.]

African Americans 2.9%
Asian/Pacific Island Americans 2.0%
Hispanic Americans 0.4%
Native Americans 0.1%
Foreign Citizens 0.7%
All Others 93.9%
Women 16.7%
Persons with disabilities .... – % [A]
Students over 25 years of age .... – % [A]

Notes: (A) Data not available.

GRADUATE ENROLLMENTS & DEGREES AWARDED
Master's enrollment 448
Master's degrees awarded 143
Doctoral enrollment 193
Doctoral degrees awarded 35

ENGINEERING STUDENT DATA
Applicants to the engineering college
Number of applicants to engineering college .... – [A]
Percent offered admission .... –% [A]

Matriculated engineering students
Percentage in top quartile (25%) of High School class ...... – [A]
Average SAT scores: Math—511.5, Verbal—613.5, Combined—1125.0
Average ACT scores: Math—24.9, Composite—25.0

Notes: (A) Data not available.

FULL & PART TIME FACULTY—BY DEPARTMENT
[Figures are the head count of full-time faculty and the full-time equivalent (FTE) of part-time faculty for each engineering department or equivalent.]

Department	Full	Part
Chemical Eng	15	....
Civil Eng	20	–
Electrical Eng	28	–
Industrial Eng	14	–
Mechanical Eng	27	–
Aerospace Eng	16	–
Agricultural Eng	10	–
Computer Science & Eng	13	0.5
Textile Eng	8	–

COLLEGE DESCRIPTION
Engineering education at Auburn provides in a four-year curriculum both the technical knowledge and the broad general education necessary to equip engineers for their problem-solving challenges. Centered on mathematics and the physical sciences, the curricula also stress the importance of social sciences, humanities, and communication skills. Auburn's engineering programs enable individuals to develop their natural talents and provide knowledge, skills, and understanding that will encourage them to find their places in society & their vocations. Although four academic years are required for each baccalaureate degree, the average student normally requires approximately 14 quarters to complete the particular engineering program. The College has dual-degree arrangements with a number of other colleges and universities. Additionally, there is a very active co-op program available requiring an additional calendar year.

DEGREE PROGRAMS DESCRIPTION
There are 15 baccalaureate degree programs. Additionally, there are 19 Master's degree programs, including both the Master of Science degree and the Professional Master's degree, and 9 Ph.D. programs. Graduate courses for credit and non-credit are offered to off-campus students through the video-tape-based off-campus courses program.

DUAL DEGREE PROGRAMS
Enrollment requirements: Exact placement of transfer students can be determined only upon review of their transcripts by the College of Engineering; a minimum 2.8 grade point average on a 4.0 scale for all programs is required.

TRANSFER INFORMATION
Residency Requirements: Minimum 45 hours in residence in order to receive a bachelor's degree. As a rule, 45 hours must be taken in the final year and in the school or curriculum of graduation.

Transfer via Articulation Agreements
Admission to engineering: any time. Application deadline date: At least three weeks before quarter's opening.
Requirements: Exact placement of transfer students determined only upon review of their transcripts by the College of Engineering; a minimum 2.8 GPA on a 4.0 scale for all programs is required.

Transfer without Articulation Agreements
Admission to engineering: At any time. Application deadline date: At least three weeks before quarter's opening.
Requirements: Exact placement of transfer students can be determined only upon review of transcripts by the College; a minimum 2.8 GPA on a 4.0 scale for all programs is required.

GRADUATION REQUIREMENTS
To earn the bachelor's degree in the College of Engineering, students must complete all subjects in their curriculum, have a minimum GPA of 2.0 in all work attempted at Auburn and have a 2.0 on all courses PASSED in the major. Baccalaureate programs in engineering require 194-210 quarter credits, depending on curriculum selected.

STUDENT PROGRAMS

PROFESSIONAL AND HONORARY SOCIETIES
[For key to acronyms, see *Introduction*.]

Professional Societies: AATCC, ACM, AIAA, AIChE, AMS*, ASAE, ASCE, ASME, IEEE, IIE, Alpha Eta Rho, Technical Association of the Pulp and Paper Industry (TAPPI)
Honorary Societies: Alpha Epsilon, Alpha Pi Mu, Chi Epsilon, Eta Kappa Nu, Omega Chi Epsilon, Pi Tau Sigma, Sigma Gamma Tau, Tau Beta Pi, Phi Psi, Upsilon Pi Epsilon

SUPPORT PROGRAMS
Student Chapter Organizations: Society of Women Engineers, National Society of Black Engineers

Minority Introduction to Engineering (MITE)
For: Women & Ethnic Minorities   **Available:** Summer only
Offered: Prematriculation

E-Day
For: Women & Ethnic Minorities   **Available:** Academic year
Offered: Prematriculation

Engineering Tutorial
Available: Year round
Offered: Freshman, Sophomore, Junior, Senior

University Study Partners
For: Women & Ethnic Minorities   **Available:** Year round
Offered: Freshman, Sophomore, Junior, Senior

Other Engineering Support Programs: There is a summer orientation program for entering Pre-Engineering freshmen. Parents are invited to attend. Counselors are available to students. Engineering tutorial program provides free tutor time for undergraduates in math, chemistry, physics, computer programming, sophomore engineering courses, and most junior courses. University study partners provide help in a broad range of courses.

013 Baylor University

■ INSTITUTION PROFILE

HEAD OF THE INSTITUTION
Herbert H. Reynolds
Phone: (817)755-1311

GENERAL INFORMATION
[All Students—Fall 1991]
Undergraduate enrollment 10,421
Graduate enrollment 1,635
Total institution enrollment **12,056**
Type of institution: Private **Calendar:** Semesters
Location: Waco **Population:** 110,000
Setting: Urban
Types of engineering degrees: Engineering
Other degree-granting colleges: Arts & Sciences, Business Administration, Dentistry, Education, Law, Nursing

■ ENGINEERING ADMISSIONS

ADMISSIONS OFFICE CONTACT
Herman D. Thomas
Baylor University
P.O. Box 97008
Waco, TX 76798-7008
Phone: (817)755-1811

ENGINEERING COLLEGE ADMISSIONS INFORMATION
Entrance Requirements: SAT (Score: 1000), ACT (Score: 21), High School courses—English (4 years), Mathematics (2 years), Natural Science (2 years), Social Science (2 years), Students may submit scores from either SAT or ACT
Entrance Recommendations: High School courses—Trigonometry (1 year), Foreign Language (2 years), Physics (1 year), Chemistry (1 year)
Requirements for foreign students: TOEFL (Score: 540)

DEPARTMENTAL ADMISSIONS INFORMATION
Admission to the engineering department: At the time of admission to the institution.

■ FINANCIAL INFORMATION

ESTIMATED EXPENSES (ACADEMIC YEAR)
[Expenses are for the 1992-93 nine-month academic year.]

All Students	Undergraduate	Graduate
Tuition and fees	$ 6,800	$ —(1)
College room and board	$ 2,200	$ —(1)
Books and supplies	$ 1,000	$ —(1)
Other expenses	$ —	$ —(1)
Total estimated expenses	**$10,000**	**$ —**

Notes: (1) No graduate program in engineering.

FINANCIAL AID OFFICE CONTACT
William J. Dube, *Dean for Academic Scholarships and Student Financial Aid*
Baylor University
P.O. Box 97028
Waco, TX 76798-7028
Phone: (817)755-2611

GENERAL FINANCIAL AID INFORMATION
Forms accepted/required: FAF, Baylor application for Scholarship and Financial Assistance
Additional financial aid information: Participation in federal and state grant and loan programs, merit and/or need based scholarships, College Work Study and jobs on campus.

■ ENGINEERING COLLEGE INFORMATION

[For additional personnel, refer to the *Appendix*.]

HEAD OF ENGINEERING
James D. Bargainer
Phone: (817)755-3871 **Fax:** (817)755-2716
EMail: jim_bargainer@engineering.baylor.edu

ENGINEERING COLLEGE ADDRESS
Department of Engineering and Computer Science
Baylor University
P.O. Box 97356
Waco, TX 76798-7356
Phone: (817)755-3871 **Fax:** (817)755-2716

ENROLLMENTS—BY CLASS
[Numbers are baccalaureate enrollments for the fall 1991 term, unless otherwise footnoted.]
1st-year students/Freshmen 100
2nd-year students/Sophomores 60
3rd-year students/Juniors 40
4th-year students/Seniors 30
Total **230**

NUMBER OF DEGREES AWARDED—BY PROGRAM
[Numbers are engineering baccalaureate degrees awarded during the 1991-1992 academic year, unless otherwise footnoted. For full details about each engineering program, refer to the *Tables of Degree Programs*.]
Engineering 25
Total **25**

PERCENTAGE OF DEGREES AWARDED—BY CATEGORY
[Percentages are of all engineering baccalaureate degrees awarded during the 1991-1992 academic year, unless otherwise footnoted.]
African Americans .... — %
Asian/Pacific Island Americans .... 12.0 %
Hispanic Americans .... 12.0 %
Native Americans .... — %
Foreign Citizens .... 6.0 %
All Others .... 70.0 %
Women .... 24.0 %
Persons with disabilities .... — %
Students over 25 years of age .... — %

ENGINEERING STUDENT DATA

Applicants to the engineering college
Number of applicants to engineering college .... —(A)
Percent offered admission .... —%(A)

Matriculated engineering students
Percentage in top quartile (25%) of High School class .... (A)
Average SAT scores: Math—483, Verbal—594, Combined—1077

Notes: (A) Data not available.

FULL & PART TIME FACULTY—BY DEPARTMENT
[Figures are the head count of full-time faculty and the full-time equivalent (FTE) of part-time faculty for each engineering department or equivalent.]

Department	Full	Part
Division of Eng	10	—

COLLEGE DESCRIPTION
The major goal of the faculty, staff, and administration of Baylor University is to offer an exceptional education in a highly personalized, Christian environment. Class sizes are seldom larger than 30 students with most classes having fewer than 20 students. All classes and laboratories are taught by faculty members who are dedicated to teaching undergraduate students; with research, consulting, and other activities playing only a secondary role. A major strength of the Baylor engineering program is its laboratories which are equipped with state-of-the-art equipment in sufficient quantity to assure the student ready access to all laboratory equipment. The program is a general engineering program which prepares students for life time careers as broadly-based practicing engineers. The student gains depth by selecting either the Electrical or Mechanical option. Students who select the Electrical option may choose to specialize in either electronic systems or computer systems design

DEGREE PROGRAMS DESCRIPTION
The Baylor Program is a general engineering program with a wide range of required engineering core courses which prepares the students for life time careers as a broadly-based practicing engineers. The emphasis is to develop fundamental skills which will have lifelong value. The student gains depth by selecting either the Electrical or the Mechanical option. Students who select the Electrical option then choose to specialize in either electronic systems or computer systems design. Required courses in religion, political science, humanities, and social sciences provide the engineering graduate with a broad-based educational experience. This provided the opportunity for development of many additional activities, aptitudes, and interests.

TRANSFER INFORMATION
Residency Requirements: 60 semester hours at Baylor for bachelor's degree.

Transfer without Articulation Agreements
Admission to engineering: At any time
Engineering graduates transferred from . . 4-yr: 2 2-yr: 3
Requirements: 2.5/4 GPA overall and for last preceding semester. Must be eligible for return to the previous institution.

GRADUATION REQUIREMENTS
A GPA of 2.0/4.0 overall and in major field is required.

■ STUDENT PROGRAMS

PROFESSIONAL AND HONORARY SOCIETIES
[For key to acronyms, see *Introduction*.]
Professional Societies: ASME, IEEE

SUPPORT PROGRAMS
Other Engineering Support Programs: Freshmen/transfer students attend an orientation program during the Summer before entering. Engineering students have an assigned engineering advisor. Aptitude testing is available as well as courses to improve study skills.

014 Boston University

INSTITUTION PROFILE

HEAD OF THE INSTITUTION
John Silber
Phone: (617)353-2200 Fax: (617)353-9764

GENERAL INFORMATION
[All Students—Fall 1991]
Undergraduate enrollment 14,403
Graduate enrollment 14,257
Total institution enrollment **28,660**
Type of institution: Private Calendar: Semesters
Location: Boston Population: 600,000
Setting: Urban
Types of engineering degrees: Engineering
Other degree-granting colleges: Allied Health Sciences, Arts & Sciences, Business Administration, Communication & Journalism, Dentistry, Education, Fine & Performing Arts, Humanities & Social Sciences, Law, Medicine, Public Health, Theology, University Professors Program, Metropolitan College, General Studies

ENGINEERING COLLEGE ADMISSIONS INFORMATION
Entrance Requirements: High School courses—English (4 years), Math (3 years), Science (3 years), Foreign Language/ Social Science (2 years), SAT scores or ACT scores are required. The minimum requirement varies. The average SAT score for entering freshmen in 1991 was: Verbal: 545 Math: 653
Entrance Recommendations: High School courses—Chemistry (1 year), Biology (1 year), Calculus (1 year)
Requirements for foreign students: TOEFL (Score: 550); financial statement

DEPARTMENTAL ADMISSIONS INFORMATION
Admission to the engineering department: At the time of admission to the institution.
Additional information: Engineering students are admitted to programs at the time of admission with the exception of the Engmedic program and Interdisciplinary Studies. Students apply to these two programs in their sophomore year.

FINANCIAL INFORMATION

ESTIMATED EXPENSES (ACADEMIC YEAR)
[Expenses are for the 1992-93 nine-month academic year.]

All Students	Undergraduate	Graduate
Tuition and fees	$16,590	$16,590
College room and board	$7,550	$7,550
Books and supplies	$450	$450
Other expenses	$1,600 [1]	$1,600 [1]
Total estimated expenses	**$26,190**	**$26,190**

Notes: (1) Costs vary with individual circumstances.

FINANCIAL AID OFFICE CONTACT
Barbara Tornow
Boston University
Office of Financial Assistance
881 Commonwealth Ave. 5th floor
Boston, MA 02215
Phone: (617)353-2965 Fax: (617)353-7300

GENERAL FINANCIAL AID INFORMATION
Forms accepted/required: FAF, FAT, FFS, Stafford, IRS, Institutional
Additional financial aid information: The Office of Financial Assistance offers both financial and advisory resources to help in meeting the expenses of Boston University. Financial assistance officers review with students and their families all available means of financing an education, whether through the University or through external funding sources.

ENGINEERING COLLEGE INFORMATION
[For additional personnel, refer to the Appendix.]

HEAD OF ENGINEERING
Charles DeLisi
Phone: (617)353-2800 Fax: (617)353-6322
EMail: in%"delisi@buenga.bu.edu"

ENGINEERING COLLEGE ADDRESS
College of Engineering
Boston University
110 Cummington St. Room 115
Boston, MA 02215
Phone: (617)353-6447 Fax: (617)353-6322

ENROLLMENTS—BY CLASS
[Numbers are baccalaureate enrollments for the fall 1991 term, unless otherwise footnoted.]
1st-year students/Freshmen 389
2nd-year students/Sophomores 285
3rd-year students/Juniors 252
4th-year students/Seniors 366
Total **1,292**

NUMBER OF DEGREES AWARDED—BY PROGRAM
[Numbers are engineering baccalaureate degrees awarded during the 1991-1992 academic year, unless otherwise footnoted. For full details about each engineering program, refer to the Tables of Degree Programs.]
Aerospace Engineering 56
Biomedical Engineering 54
Computer Engineering 29
Electrical Engineering 68
General Engineering .... –
Manufacturing Engineering 42
Mechanical Engineering 38
Systems Engineering 10
Total **297**

PERCENTAGE OF DEGREES AWARDED—BY CATEGORY
[Percentages are of all engineering baccalaureate degrees awarded during the 1991-1992 academic year, unless otherwise footnoted.]
African Americans 4.3%
Asian/Pacific Island Americans 14.0%
Hispanic Americans 2.0%
Native Americans .... – %
Foreign Citizens 17.4%
All Others 62.3%
Women 21.7%
Persons with disabilities .... – % (A)
Students over 25 years of age .... – % (A)

Notes: (A) Data not available.

GRADUATE ENROLLMENTS & DEGREES AWARDED
Master's enrollment 475
Master's degrees awarded 213
Doctoral enrollment 71
Doctoral degrees awarded 6

ENGINEERING STUDENT DATA

Applicants to the engineering college
Number of applicants to engineering college 2,370
Percent offered admission 84.8%

Matriculated engineering students
Percentage in top quartile (25%) of High School class .... – (2)
Average SAT scores: Math—653, Verbal—545, Combined—1198

Notes: (2) Average HS rank of Engineering students accepted = 87%

FULL & PART TIME FACULTY—BY DEPARTMENT
[Figures are the head count of full-time faculty and the full-time equivalent (FTE) of part-time faculty for each engineering department or equivalent.]

Department	Full	Part
Electrical, Computer & Systems Eng	36	–
Aerospace & Mechanical Eng	16	–
Biomedical Eng	21	–
Manufacturing Eng	17	–

COLLEGE DESCRIPTION
The College of Engineering offers the Bachelor of Science degree. Aside from the programs of study in aerospace, biomedical, computer, electrical, manufacturing, mechanical, and systems, an interdisciplinary program allows students to plan their own course of study in such areas as environmental science, space and planetary science, technical management, journalism, and music technology. Through the Engineering/Medical Integrated Curriculum, students majoring in biomedical engineering may begin their medical studies at the Boston University School of Medicine as early as their junior year. The Boston University Collaborative Degree Program (BUCOP) allows students to earn a B.S. from the College of Engineering and a second bachelor's degree from another School or College within the University. The Professional Practice (CO-OP) program offers students a cooperative, paid experience with companies thus giving the students real-world exposure.

DEGREE PROGRAMS DESCRIPTION
AEROSPACE: This program prepares the student for positions in industry and for graduate study in aerospace engineering or in such related fields. BIOMEDICAL: Quantitative engineering methods are used to obtain an understanding of normal and abnormal functions of the human body. COMPUTER: The program provides a strong engineering background centered on digital technology combined with an understanding of the principles and techniques of computer science. ELECTRICAL: Electrical engineering students obtain a strong background in the physical sciences, mathematics, and in computational methods. MANUFACTURING: The courses in manufacturing engineering provide the introduction to technical skills in product development, materials, processes, and production that the manufacturing engineer requires. MECHANICAL: The objective of the B.S. program in Mechanical Engineering is to provide students with the fundamentals of engineering subjects: solid and fluid mechanics, thermodynamics, heat transfer, and energy conversion. SYSTEMS: The basis of the major is formed by electronics, automatic control, data communication, and data processing. INTERDISCIPLINARY: Permits the study of the principles and methods of engineering relevant to many areas.

DUAL DEGREE PROGRAMS
Engineering baccalaureate graduates with dual degrees 1
Enrollment requirements: GPA of 3.0 is required along with core English, math, and science requirements.

TRANSFER INFORMATION
Residency Requirements: 48 credits of upper division Engineering courses are required. Humanities and social sciences have specific requirements also.

Transfer via Articulation Agreements
Admission to engineering: At any time
Engineering graduates transferred from ... 4-yr: 13 2-yr: –
Requirements: GPA of 3.0 is required along with core English, math, and science requirements.

Transfer without Articulation Agreements
Admission to engineering: At any time
Engineering graduates transferred from ... 4-yr: 28 2-yr: –
Requirements: GPA of 3.0 is required along with core English, math and science requirements.

GRADUATION REQUIREMENTS
The B.S. degree requires 130 to 138 semester credits of prescribed curricular courses, depending on the program selected. No more than 16 credits with a grade of D may be applied toward a degree. This graduation requirement applies only to the set of courses presented for graduation, and not to all courses that appear on the transcript. In addition, a cumulative grade point average of 1.7 for all courses taken at Boston University is required.

STUDENT PROGRAMS

PROFESSIONAL AND HONORARY SOCIETIES
[For key to acronyms, see Introduction.]
Professional Societies: AHS, ASME, BMES, IEEE, NSPE, SAE, SME
Honorary Societies: Tau Beta Pi

SUPPORT PROGRAMS
Student Chapter Organizations: Society of Women Engineers, National Society of Black Engineers, Society of Hispanic Professional Engineers

Minority Engineers Society
For: Ethnic Minorities Available: Academic year
Offered: Freshman, Sophomore, Junior, Senior, Graduate level

The King Center
For: Ethnic Minorities Available: Year round
Offered: Freshman, Sophomore, Junior, Senior, Graduate level

Other Engineering Support Programs: The College offers many support services to students beginning with summer and fall orientation programs and the freshmen advising seminar throughout the fall semester. Transfer students are encouraged to come to campus for the Transfer Student Orientation Program. Other support services include a free tutoring office, counseling services, faculty advising, and on-going seminars.

015 Bradley University

■ INSTITUTION PROFILE

HEAD OF THE INSTITUTION
John R. Brazil
Phone: (309)677-3167 **Fax:** (309)677-2330

GENERAL INFORMATION
[All Students—Fall 1991]
Undergraduate enrollment 5,287
Graduate enrollment 946
Total institution enrollment **6,233**
Type of institution: Private **Calendar:** Semesters
Location: Peoria, IL **Population:** 113,000
Setting: Urban
Types of engineering degrees: Engineering & Technology
Other degree-granting colleges: Arts & Sciences, Business Administration, Communication and Fine Arts, Education and Health Sciences

■ ENGINEERING ADMISSIONS

ADMISSIONS OFFICE CONTACT
Gary Bergman
Bradley University
Office of Enrollment Management
100 Swords Hall
Peoria, IL 61625
Phone: (800)447-6460 **Fax:** (309)677-2797

ENGINEERING COLLEGE ADMISSIONS INFORMATION
Entrance Requirements: ACT (Score: 25), High School courses—Algebra (2 years), Plane Geometry (1 year), Trigonometry, Physics (1 year)
Entrance Recommendations: High School courses—Solid Geometry, Chemistry (1 year)

DEPARTMENTAL ADMISSIONS INFORMATION
Admission to the engineering department: At the time of admission to the institution.
Additional information: Admission to IE program requires 1/2 year of graphics. Admission as a transfer student into Civil and Mechanical Engineering programs require a GPA of 2.25 in major courses.

■ FINANCIAL INFORMATION

ESTIMATED EXPENSES (ACADEMIC YEAR)
[Expenses are for the 1992-93 nine-month academic year.]

All Students	Undergraduate	Graduate
Tuition and fees	$ 9,050 (1)	$ 9,050 (1)
College room and board	$ 3,960 (2)	$ 3,960 (2)
Books and supplies	$ 48 (3)	$ 48 (3)
Other expenses	$ 2,000 (4)	$ 2,000 (4)
Total estimated expenses	**$15,058**	**$15,058**

Notes: (1) 12-16 semester hours. Each additional hour-$236. 1-7 hrs, $246/hr; 7 1/2-11 1/2 hrs, $307/hr. $5/hr engineering surcharge fee. (2) Based on double occupancy, 20 meal per week plan. Other meal plans available as well as single occupancy. (3) Activity and health fees. Other fees required for studio art, music, chemistry breakage fees. (4) Books and transportation.

FINANCIAL AID OFFICE CONTACT
David Pardieck, *Director of Financial Aid*
Bradley University
Office of Financial Assistance
14 Swords Hall
Peoria, IL 61625
Phone: (309)677-3089 **Fax:** (309)677-2798

GENERAL FINANCIAL AID INFORMATION
Forms accepted/required: CSS-FAF, FAF, Institutional
Additional financial aid information: There are a number of need based and merit based scholarships available to all students. The university catalog documents these scholarships. In addition the colleges and departments within the colleges have need based and merit based scholarships available. There are also work-study programs available.

■ ENGINEERING COLLEGE INFORMATION

[For additional personnel, refer to the *Appendix*.]
HEAD OF ENGINEERING
John E. Francis
Phone: (309)677-2721 **Fax:** (309)677-3670

ENGINEERING COLLEGE ADDRESS
College of Engineering and Technology
Bradley University
124 Jobst Hall
Peoria, IL 61625
Phone: (309)677-2720 **Fax:** (309)677-3670

ENROLLMENTS—BY CLASS
[Numbers are baccalaureate enrollments for the fall 1991 term, unless otherwise footnoted.]
1st-year students/Freshmen 116
2nd-year students/Sophomores 129
3rd-year students/Juniors 155
4th-year students/Seniors 249
Total **649**

NUMBER OF DEGREES AWARDED—BY PROGRAM
[Numbers are engineering baccalaureate degrees awarded during the 1991-1992 academic year, unless otherwise footnoted. For full details about each engineering program, refer to the *Tables of Degree Programs*.]
Civil Engineering 21
Construction Management 15
Electrical Engineering 42
Engineering Physics 2
Industrial Engineering 34
Manufacturing Engineering 21
Mechanical Engineering 56
Total **191**

PERCENTAGE OF DEGREES AWARDED—BY CATEGORY
[Percentages are of all engineering baccalaureate degrees awarded during the 1991-1992 academic year, unless otherwise footnoted.]
African Americans .... 0.6 %
Asian/Pacific Island Americans .... 1.0 %
Hispanic Americans .... 0.6 %
Native Americans .... – %
Foreign Citizens .... 0.2 %
All Others .... 97.6 %
Women .... 11.0 %
Persons with disabilities .... – % (A)
Students over 25 years of age .... – % (A)
Notes: (A) Data not available.

GRADUATE ENROLLMENTS & DEGREES AWARDED
Master's enrollment 138
Master's degrees awarded 42
Doctoral enrollment .... – (B)
Doctoral degrees awarded .... – (B)
Notes: (B) Data not applicable.

ENGINEERING STUDENT DATA
Applicants to the engineering college
Number of applicants to engineering college 390
Percent offered admission 97.2%

Matriculated engineering students
Percentage in top quartile (25%) of High School class 65.0 (4)
Average SAT scores: Math—580 (2), Verbal—510 (2), Combined—1090 (2)
Average ACT scores: Composite—28 (3)
Notes: (2) University wide data. (3) Composite is for entire College of Engineering and Technology. (4) All university data.

FULL & PART TIME FACULTY—BY DEPARTMENT
[Figures are the head count of full-time faculty and the full-time equivalent (FTE) of part-time faculty for each engineering department or equivalent.]

Department	Full	Part
Civil Eng & Construction	12	1.5
Electrical & Computer Eng & Tech	16	5.3
Industrial Eng	6	2.5
Mechanical Eng	10.5	2.3
Manufacturing	9	3.5

COLLEGE DESCRIPTION
The college has an unclassified engineering program to allow students who are not sure what engineering program they wish to major in to explore various alternatives. The college also has an active cooperative education program. The entering freshman take courses in their major. Most programs begin their laboratory activities no later than their sophomore year. All engineering, technology and construction management programs have a capstone design experience. The cooperative education program is voluntary. The college prides itself in graduating engineers who will be immediately productive for their employers. The faculty of the college emphasize their mentorship role by interacting with students on a one-to-one basis.

DEGREE PROGRAMS DESCRIPTION
The College of Engineering and Technology ascribes to its mission to provide a undergraduate experience which will allow its students to become immediately productive in the work place after graduation and provide the necessary educational background to educate students for further academic work. The faculty are committed to the concept of being a mentor for the students. Close faculty-student interaction is stressed throughout the entire undergraduate experience. The same mentorship approach is used in the graduate programs in the college. The distinctiveness of the college is based on its mentorship approach. The college stresses the laboratory and design experiences throughout the students undergraduate experience.

TRANSFER INFORMATION
Residency Requirements: Minimum of 30 semester hours in residence. 24 of last 30 semester hours must be in residence. Transfer students from junior/community colleges may transfer up to 66 semester hours into a college program.

Transfer without Articulation Agreements
Admission to engineering: At any time
Requirements: (1) 2.5 transfer GPA. (2) 3.00 transfer GPA in EE (3) C or better in math and science. (3) 2.25 transfer GPA in CE and ME courses.

GRADUATION REQUIREMENTS
(1) Minimum 2.00 GPA in College and university courses. (2) Junior level writing course is required. (3) 2.25 GPA in major courses to obtain B.S. in Civil, Industrial, Manufacturing and Mechanical Engineering.

■ STUDENT PROGRAMS

PROFESSIONAL AND HONORARY SOCIETIES
[For key to acronyms, see *Introduction*.]
Professional Societies: AGCA, ASCE, ASME, IEEE, IIE, SAE, SME, American Society of Materials
Honorary Societies: Chi Epsilon, Eta Kappa Nu, Pi Tau Sigma, Sigma Phi Delta, Tau Beta Pi, Beta Tau Epsilon, Alpha Pi Mu

SUPPORT PROGRAMS
Student Chapter Organizations: National Society of Black Engineers, National Technical Association

University Minority Student Services Program
For: Ethnic Minorities **Available:** Year round
Offered: Prematriculation, Freshman, Sophomore, Junior, Senior, Graduate level

Other Engineering Support Programs: Support is provided by a university testing and counseling center, a center for orientation and retention and a summer orientation program during which freshmen are given a math placement test. Tutoring services are provided for core engineering, science and mathematics. A full-time academic counselor is in the college with all students being advised by a full-time faculty member.

016 University of Bridgeport

INSTITUTION PROFILE
HEAD OF THE INSTITUTION
Edward Eigel
Phone: (203)576-4665   Fax: (203)576-4983

GENERAL INFORMATION
[All Students—Fall 1991]
Undergraduate enrollment 1,869
Graduate enrollment 2,035
Total institution enrollment 3,904
Type of institution: Private   **Calendar:** Semesters
Location: Bridgeport   **Population:** 146,000
Setting: Urban
Types of engineering degrees: Engineering
Other degree-granting colleges: Arts & Sciences, Business Administration, Professional Studies, Chiropractic

ENGINEERING ADMISSIONS
ADMISSIONS OFFICE CONTACT
Barbara Maryak
University of Bridgeport
Director of Undergraduate Admissions
Bridgeport, CT 06601
Phone: (203)576-4560   Fax: (203)576-4941

ENGINEERING COLLEGE ADMISSIONS INFORMATION
Entrance Requirements: SAT (Score: 1000), High School courses—Math (4 years), Lab Science (3 years), English (4 years), Social Study (1 year)
Entrance Recommendations: High School courses—Lab Science; Chemistry, and Physics (2 years), Calculus (1 year), Computer Programming (1 year)
Requirements for foreign students: TOEFL (Score: 500); financial statement

DEPARTMENTAL ADMISSIONS INFORMATION
Admission to the engineering department: At the time of admission to the institution.

FINANCIAL INFORMATION
ESTIMATED EXPENSES (ACADEMIC YEAR)
[Expenses are for the 1992-93 nine-month academic year.]

All Students	Undergraduate	Graduate
Tuition and fees	$12,880	$ 5,985
College room and board	$ 6,350	$ 6,350
Books and supplies	$ 450	$ 450
Other expenses	$ 400	$ 400
Total estimated expenses	**$20,080**	**$13,185**

FINANCIAL AID OFFICE CONTACT
Bessie Phakias
University of Bridgeport
Financial Aid Office
Wahlstrom Library, 126 Park Avenue
Bridgeport, CT 06601-2449
Phone: (203)576-4568

GENERAL FINANCIAL AID INFORMATION
Forms accepted/required: CSS-FAF, FAF, FAT, Stafford, IRS, SAR
Additional financial aid information: Need based financial aid plus academic merit and athletic scholarships are available to students from a variety of sources. FAF, IRS, and institutional forms are required.

ENGINEERING COLLEGE INFORMATION
[For additional personnel, refer to the *Appendix*.]
HEAD OF ENGINEERING
Bruce C. Skinner
Phone: (203)576-4111   Fax: (203)576-4766
EMail: skinner@cse.bridgeport.edu

ENGINEERING COLLEGE ADDRESS
School of Engineering
University of Bridgeport
Technology Building
University Avenue
Bridgeport, CT 06601
Phone: (203)576-4111   Fax: (203)576-4766

ENROLLMENTS—BY CLASS
[Numbers are baccalaureate enrollments for the fall 1991 term, unless otherwise footnoted.]
1st-year students/Freshmen 95
2nd-year students/Sophomores 53
3rd-year students/Juniors 33
4th-year students/Seniors 77
Total 258

NUMBER OF DEGREES AWARDED—BY PROGRAM
[Numbers are engineering baccalaureate degrees awarded during the 1991-1992 academic year, unless otherwise footnoted. For full details about each engineering program, refer to the *Tables of Degree Programs*.]
Computer Engineering 7
Electrical Engineering 58
Management Engineering .... –
Mechanical Engineering 22
Total 87

PERCENTAGE OF DEGREES AWARDED—BY CATEGORY
[Percentages are of all engineering baccalaureate degrees awarded during the 1991-1992 academic year, unless otherwise footnoted.]
African Americans 2.2%
Asian/Pacific Island Americans 5.6%
Hispanic Americans .... – %
Native Americans .... – %
Foreign Citizens 70.0%
All Others 22.2%
Women 4.4%
Persons with disabilities .... – %
Students over 25 years of age 69.0%

GRADUATE ENROLLMENTS & DEGREES AWARDED
Master's enrollment 166
Master's degrees awarded 49
Doctoral enrollment .... –
Doctoral degrees awarded .... –

ENGINEERING STUDENT DATA
Applicants to the engineering college
Number of applicants to engineering college 103
Percent offered admission 74.0%

Matriculated engineering students
Percentage in top quartile (25%) of High School class .... – (A)
Average SAT scores: Math—505, Verbal—400

Notes: (A) Data not available.

FULL & PART TIME FACULTY—BY DEPARTMENT
[Figures are the head count of full-time faculty and the full-time equivalent (FTE) of part-time faculty for each engineering department or equivalent.]

Department	Full	Part
Electrical Eng	5	1.0
Mechanical Eng	5	1.0
Computer Science & Eng	6	1.0
Management Eng	1	–

COLLEGE DESCRIPTION
The College of Engineering offers accredited undergraduate degree programs in Computer, Electrical, and Mechanical Engineering. Following a generally common first year, disciplinary specialization begins in the second year and continues throughout the four-year programs. Computer utilization and applications are stressed throughout the entire engineering curricula in a variety of mainframe, PC, and work station environments located in two newly renovated engineering facilities/buildings. Selected social science and humanities programs of study, which fulfill both University Core and ABET requirements, must include a unique senior capstone course. A co-op option is available. Seminar participation, required of all students, may be waived as an equivalent professional experience by co-op students. Many engineering classes are offered in the evening for the benefit of part-time students.

DEGREE PROGRAMS DESCRIPTION
B.S. degree programs in engineering prepare graduates for entry-level professional positions in the government and private sectors, as well as for graduate work in the several disciplines offered by the College. Students in both undergraduate and graduate engineering programs may focus on practice-oriented course offerings or select electives to enhance the engineering science foundations required for additional advanced levels of study.

TRANSFER INFORMATION
Residency Requirements: Students must finish their first 30 credits in residence. In addition a minimum of 18 credits of major area courses must be completed in residence.
Transfer without Articulation Agreements
Admission to engineering: At any time
Engineering graduates transferred from ... 4-yr: 6   2-yr: –

GRADUATION REQUIREMENTS
B.S. degree programs in engineering require 133-136 semester credit hours to be completed with a cumulative grade point average of 2.0 (C) on a 4.0 (A) scale. In addition a grade of C or better is required in the major area courses in each engineering degree program.

STUDENT PROGRAMS
PROFESSIONAL AND HONORARY SOCIETIES
[For key to acronyms, see *Introduction*.]
Professional Societies: ASME, IEEE
Honorary Societies: Eta Kappa Nu, Phi Kappa Phi

SUPPORT PROGRAMS
Student Chapter Organizations: Society of Women Engineers

Engineering For Deserving Youth (EDY)
For: Ethnic Minorities   **Available:** Year round
Offered: Prematriculation, Freshman, Sophomore, Junior, Senior

CPEP (Connecticut Pre Engineering Program)
For: Ethnic Minorities   **Available:** Year round
Offered: Prematriculation

Other Engineering Support Programs: Orientation for all entering freshmen and transfer students. University Life 101, is taken by all freshmen as an in-depth orientation experience. EDY program tutors minority students (20% of the domestic student body). Learning Resources Center provides basic skills assistance, the International Student Office, Career Service and Placement, and services for handicapped students.

017 Brigham Young University

INSTITUTION PROFILE

HEAD OF THE INSTITUTION
Rex E. Lee
Phone: (802)378-2521

GENERAL INFORMATION
[All Students—Fall 1991]
Undergraduate enrollment 26,763
Graduate enrollment 1,786
Total institution enrollment **28,649**
Type of institution: Private   Calendar: Semesters
Nearest city: Salt Lake City   Population: 160,000
Miles from main campus: 45   Setting: Small Town
Types of engineering degrees: Engineering & Technology
Other degree-granting colleges: Education, Fine & Performing Arts, Humanities & Social Sciences, Law, Nursing, Biological & Agricultural, Business, Engineering Sciences & Technology, Physical Education, Family, Home, & Social Sciences

ENGINEERING ADMISSIONS

ADMISSIONS OFFICE CONTACT
Tom M. Gourley
Brigham Young University
Admissions Office
A-153 ASB
Provo, UT 84602
Phone: (801)378-4264   Fax: (801)378-2507

ENGINEERING COLLEGE ADMISSIONS INFORMATION
Admission to the engineering college: Entrance to engineering college with university admission. Formal entrance into the professional programs occurs at the end of sophomore year.
Entrance Requirements: ACT (Score: 25)
Entrance Recommendations: High School courses—Algebra (2 years), Trigonometry/Geometry (1 year), Drafting (1 year), Chemistry and Physics (1 year)
Requirements for foreign students: TOEFL (Score: 500); financial statement

DEPARTMENTAL ADMISSIONS INFORMATION
Admission to the engineering department: At the end of the second year.
Additional information: Completion of specific courses and application to the professional program.

FINANCIAL INFORMATION

ESTIMATED EXPENSES (ACADEMIC YEAR)
[Expenses are for the 1992-93 nine-month academic year.]

All Students	Undergraduate	Graduate
Tuition and fees	$ 2,120 [1]	$ 2,480 [1]
College room and board	$ 3,390	$ 3,390
Books and supplies	$ 630	$ 630
Other expenses	$ 2,020 [2]	$ 2,020 [2]
Total estimated expenses	**$ 8,160**	**$ 8,520**

Notes: (1) Fee for members of the Church of Jesus Christ of Latter-day Saints. Non-members pay $3180 (undergraduate) and $3720 (graduate). (2) Estimated personal and transportation expenses.

FINANCIAL AID OFFICE CONTACT
Norman B. Finlinson, *Director*
Brigham Young University
A-41 ASB
Provo, UT 84602
Phone: (801)378-4104

GENERAL FINANCIAL AID INFORMATION
Forms accepted/required: ACT, FFS, BYU Financial Aid Application
Additional financial aid information: Financial Aid at Brigham Young University comes from various federal, state and university programs in the form of loans, scholarships, and grants.

ENGINEERING COLLEGE INFORMATION

[For additional personnel, refer to the *Appendix*.]

HEAD OF ENGINEERING
L. Douglas Smoot
Phone: (801)378-4326   Fax: (801)378-5705

ENGINEERING COLLEGE ADDRESS
College of Engineering & Technology
Brigham Young University
270 Clyde Building
Provo, UT 84602
Phone: (801)378-4326   Fax: (801)378-5705

ENROLLMENTS—BY CLASS
[Numbers are baccalaureate enrollments for the fall 1991 term, unless otherwise footnoted.]
1st-year students/Freshmen 70
2nd-year students/Sophomores 30
3rd-year students/Juniors 66
4th-year students/Seniors 154
Total **320**

NUMBER OF DEGREES AWARDED—BY PROGRAM
[Numbers are engineering baccalaureate degrees awarded during the 1991-1992 academic year, unless otherwise footnoted. For full details about each engineering program, refer to the *Tables of Degree Programs*.]
Chemical Engineering 22
Civil Engineering 56
Electrical & Computer Engineering 117
Manufacturing Engineering 7
Mechanical Engineering 139
Total **341**

PERCENTAGE OF DEGREES AWARDED—BY CATEGORY
[Percentages are of all engineering baccalaureate degrees awarded during the 1991-1992 academic year, unless otherwise footnoted.]
African Americans .... – %
Asian/Pacific Island Americans .... – %
Hispanic Americans .... 3.4 %
Native Americans .... 1.7 %
Foreign Citizens .... 3.4 %
All Others .... 91.5 %
Women .... 3.4 %
Persons with disabilities .... – % (A)
Students over 25 years of age .... – % (1)

Notes: (A) Data not available. (1) Average student age at graduation was 25.1.

GRADUATE ENROLLMENTS & DEGREES AWARDED
Master's enrollment 13
Master's degrees awarded 18
Doctoral enrollment 51
Doctoral degrees awarded 10

ENGINEERING STUDENT DATA

Applicants to the engineering college
Number of applicants to engineering college .... – (A)
Percent offered admission .... –% (A)

Matriculated engineering students
Percentage in top quartile (25%) of High School class .... – (B)
Average ACT scores: Math—24.9, Composite—26.1

Notes: (A) Data not available. (B) Data not applicable.

FULL & PART TIME FACULTY—BY DEPARTMENT
[Figures are the head count of full-time faculty and the full-time equivalent (FTE) of part-time faculty for each engineering department or equivalent.]

Department	Full	Part
Chemical Eng	12	1.5
Civil Eng	16	–
Electrical & Computer Eng	21	1.0
Mechanical Eng	19	–
Manufacturing Eng & Eng Tech	19	–

COLLEGE DESCRIPTION

The College of Engineering and Technology enrolls some 3,000 students in the five engineering departments. All B.S. degrees are accredited with the Accreditation Board for Engineering and Technology. Each department also offers M.S. and PhD degrees and the college offers a Master of Engineering Management (MEM) degree. Co-op education is available for students wishing work experience. Junior and senior level courses are considered to be professional level and formal acceptance by the major department is required before one may enroll in the courses.

DEGREE PROGRAMS DESCRIPTION
The B.S. degrees offered by each of the five engineering departments are considered entry level degrees. Students possess a fundamental understanding of engineering concepts and are prepared for routine work. Students are advised to seek additional training, preferably the master degree in engineering or some other discipline. A masters degree in engineering may provide advanced professional and/or research preparation. The PhD degree is research oriented and prepares one to enter professional research or university employment.

TRANSFER INFORMATION
Residency Requirements: 30 Semester hours.
Transfer via Articulation Agreements
Admission to engineering: At any time
Requirements: Minimum grade point average of 3.0.
Transfer without Articulation Agreements
Admission to engineering: At any time
Requirements: Minimum grade point average of 3.0

GRADUATION REQUIREMENTS
Engineering programs typically require 135-137 semester hours of credit. Students must maintain a C or better average and can only accumulate 3-6 hours of D credit.

STUDENT PROGRAMS

PROFESSIONAL AND HONORARY SOCIETIES
[For key to acronyms, see *Introduction*.]
Professional Societies: AAPG, ACS, AIAA, AIChE, ASCE, ASME, IEEE, NAHB, SME, Society of Women Engineers
Honorary Societies: Eta Kappa Nu, Tau Beta Pi

SUPPORT PROGRAMS
Student Chapter Organizations: Society of Women Engineers

Society of Women Engineers
For: Women   **Available:** Year round
Offered: Freshman, Sophomore, Junior, Senior, Graduate level
Other Engineering Support Programs: New student orientation.

018 Bucknell University

INSTITUTION PROFILE

HEAD OF THE INSTITUTION
Gary A. Sojka
Phone: (717)524-1511　　　　Fax: (717)524-3760

GENERAL INFORMATION
[All Students—Fall 1991]
Undergraduate enrollment 3,484
Graduate enrollment 226
Total institution enrollment 3,710
Type of institution: Private　　Calendar: Semesters
Nearest city: Harrisburg　　Population: 500,000
Miles from main campus: 60　　Setting: Small Town
Types of engineering degrees: Engineering
Other degree-granting colleges: Arts & Sciences

ENGINEERING ADMISSIONS

ADMISSIONS OFFICE CONTACT
Mark D. Davies
Bucknell University
8 Freas Hall
Lewisburg, PA 17837
Phone: (717)524-1101　　　　Fax: (717)524-3760

ENGINEERING COLLEGE ADMISSIONS INFORMATION
Entrance Requirements: SAT, High School courses—English and Math (4 years), Foreign Language (2 years), Science (3 years), Social Studies (4 years), Bucknell does not have a minimum score required for any of the above tests.
Entrance Recommendations: SAT, High School courses—Chemistry (1 year), Physics (1 year), Math through Trigonometry (4 years), Calculus preferred (1 year), Bucknell does not have a minimum score required for any of the above scores.
Requirements for foreign students: TOEFL (Score: 550)

DEPARTMENTAL ADMISSIONS INFORMATION
Admission to the engineering department: At the time of admission to the institution.

FINANCIAL INFORMATION

ESTIMATED EXPENSES (ACADEMIC YEAR)
[Expenses are for the 1992-93 nine-month academic year.]

All Students	Undergraduate	Graduate
Tuition and fees	$16,670	$ 1,900 [1]
College room and board	$ 4,110	$ 4,110
Books and supplies	$ 1,200	$ 1,200
Other expenses	$ –	$ –
Total estimated expenses	$21,980	$ 7,210

Notes: (1) This amount is per course.

FINANCIAL AID OFFICE CONTACT
Ronald T. Laszewski, *Director of Financial Aid*
Bucknell University
621 St. George Street
Lewisburg, PA 17837
Phone: (717)524-1331　　　　Fax: (717)524-3760

GENERAL FINANCIAL AID INFORMATION
Forms accepted/required: CSS-FAF, FFS, Stafford, IRS, SAR
Additional financial aid information: All scholarships, grants, loans and work study jobs are awarded on a need base.

ENGINEERING COLLEGE INFORMATION
[For additional personnel, refer to the *Appendix*.]

HEAD OF ENGINEERING
Thomas P. Rich
Phone: (717)524-3711　　　　Fax: (717)524-3760

ENGINEERING COLLEGE ADDRESS
College of Engineering
Bucknell University
205 Dana Building
Lewisburg, PA 17837
Phone: (717)524-3711　　　　Fax: (717)524-3760

ENROLLMENTS—BY CLASS
[Numbers are baccalaureate enrollments for the fall 1991 term, unless otherwise footnoted.]
1st-year students/Freshmen 182
2nd-year students/Sophomores 155
3rd-year students/Juniors 163
4th-year students/Seniors 132
Total **632**

NUMBER OF DEGREES AWARDED—BY PROGRAM
[Numbers are engineering baccalaureate degrees awarded during the 1991-1992 academic year, unless otherwise footnoted. For full details about each engineering program, refer to the *Tables of Degree Programs*.]
Chemical Engineering 21
Civil Engineering 38
Computer Science 13
Electrical Engineering 34
Mechanical Engineering 31
Total **137**

PERCENTAGE OF DEGREES AWARDED—BY CATEGORY
[Percentages are of all engineering baccalaureate degrees awarded during the 1991-1992 academic year, unless otherwise footnoted.]
African Americans .... 1.0 %
Asian/Pacific Island Americans .... 3.0 %
Hispanic Americans .... 1.0 %
Native Americans .... – %
Foreign Citizens .... – %
All Others .... 95.0 %
Women .... 15.0 % [1]
Persons with disabilities .... – % [A]
Students over 25 years of age .... 5.0 %

Notes: (A) Data not available. (1) This information is given in whole numbers not percentages.

GRADUATE ENROLLMENTS & DEGREES AWARDED
Master's enrollment 30
Master's degrees awarded 11
Doctoral enrollment .... – [B]
Doctoral degrees awarded .... – [B]

Notes: (B) Data not applicable.

ENGINEERING STUDENT DATA

Applicants to the engineering college
Number of applicants to engineering college .... 1,019 [1]
Percent offered admission 61.0%

Matriculated engineering students
Percentage in top quartile (25%) of High School class .... 67.0 [3]
Average ACT scores: Composite—28

Notes: (1) This information given in a whole number not percentage. (3) This percentage is based on the top 20%.

FULL & PART TIME FACULTY—BY DEPARTMENT
[Figures are the head count of full-time faculty and the full-time equivalent (FTE) of part-time faculty for each engineering department or equivalent.]

Department	Full	Part
Chemical Eng	8	–
Civil Eng	8	–
Electrical Eng	7	–
Mechanical Eng	8	–
Computer Science	8	–

COLLEGE DESCRIPTION
For nearly 100 years Bucknell's Engineering College has held high quality undergraduate education to be its top priority. The result is an intensive engineering curriculum which emphasizes modern laboratory experience in chemical, civil, computer, electrical and mechanical engineering. Students learn from dedicated faculty members in small classes and on independent engineering research and design projects. Typically 20% of engineering students include women. Special attention is given to the freshman experience. The course, Exploring Engineering, is taught to all first semester students. Presented by faculty from all five engineering disciplines, the lectures, laboratories, and design projects in Exploring Engineering help students confirm their specific engineering majors. Well over 90% of all engineering students graduate on time.

DEGREE PROGRAMS DESCRIPTION
Degrees of B.S. in chemical, civil, electrical, and mechanical engineering are accredited by the Engineering Accreditation Commission of the Accreditation Board of Engineering and Technology. A fifth program in computer engineering leads to the degree of B.S. in computer science and engineering. All programs are designed to develop a broad understanding of engineering disciplines, an appreciation of the engineer's individual and professional role in society, and a capacity for lifelong learning. The graduate program leads to the degrees of M.S. in chemical, civil, electrical or mechanical engineering. Each graduate program is individually tailored to meet the needs, preparation, and goals of the student.

DUAL DEGREE PROGRAMS
Engineering baccalaureate graduates with dual degrees 14
Enrollment requirements: Ability to complete the required sequence of courses for both an engineering major and a second major in Bucknell's Bachelor of Arts Program.

TRANSFER INFORMATION
Residency Requirements: 56 credit hours at Bucknell with at least 12 credit hours during the last semester.
Transfer without Articulation Agreements
Admission to engineering: At any time pending space is available and the student has the ability to complete required sequence.
Engineering graduates transferred from ... 4-yr: 6　　2-yr: 5
Requirements: Minimum GPA of 2.5 and minimum of 15 credit hours but no more than 80 credit hours.

GRADUATION REQUIREMENTS
All engineering degree programs require the completion of 34 courses (42 in the combined liberal arts-engineering program) with a GPA of at least 2.0 overall and in engineering.

STUDENT PROGRAMS

PROFESSIONAL AND HONORARY SOCIETIES
[For key to acronyms, see *Introduction*.]
Professional Societies: ACM, AIChE, AMS, ASCE, ASME, IEEE, MAA, NSPE, Society of Women Engineers, Society for Black Engineers
Honorary Societies: Tau Beta Pi, Phi Eta Sigma, Alpha Lambda Delta

SUPPORT PROGRAMS
Student Chapter Organizations: Society of Women Engineers, National Society of Black Engineers

Women's Resource Center
For: Women　　**Available:** Year round
Offered: Prematriculation, Freshman, Sophomore, Junior, Senior, Graduate level

Multicultural Center
For: Women & Ethnic Minorities　　**Available:** Year round
Offered: Prematriculation, Freshman, Sophomore, Junior, Senior, Graduate level

Other Engineering Support Programs: Orientation by Associate Dean; mentoring by faculty especially in EG100; Academic counseling by faculty and deans; off campus studies and international programs; career development center; January Program; summer school; testing services; multicultural center; engineering spirit week.

019 University of California at Berkeley

INSTITUTION PROFILE

HEAD OF THE INSTITUTION
Chang-Lin Tien, *Chancellor*
Phone: (510)642-7464

GENERAL INFORMATION
[All Students—Fall 1991]
Undergraduate enrollment 21,660
Graduate enrollment 8,712
Total institution enrollment **30,372**
Type of institution: Public   **Calendar:** Semesters
Nearest city: San Francisco   **Population:** 700,000
Miles from main campus: 15   **Setting:** Urban
Types of engineering degrees: Engineering
Other degree-granting colleges: Business Administration, Communication & Journalism, Education, Law, Chemistry, Environmental Design, Letters and Science, Natural Resources

ENGINEERING COLLEGE ADMISSIONS INFORMATION
Entrance Requirements: High School courses—U.S. History and Civics or American Government (1 year), English (4 years), Mathematics (4 years), Laboratory Science (1 year), (1) Either the SAT or ACT and (2) Three College Entrance Examination Board achievement tests: English composition, mathematics, and any third test in a different area.
Entrance Recommendations: High School courses—Foreign Language (2 years), Mechanical Drawing (1 year). It is preferred that the third achievement test be in either physics or chemistry.
Requirements for foreign students: TOEFL is required if a regular English course (not ESL) has not been taken in the U.S. Admission is highly competitive; first preference is given to California residents who are citizens or permanent residents of the United States.

DEPARTMENTAL ADMISSIONS INFORMATION
Admission to the engineering department: At the time of admission to the institution.
Additional information: There are no specific departmental requirements; however, admission to the various departments varies with regard to competitiveness. Electrical Engineering and Computer Sciences and the Engineering Science programs are the most competitive, followed by Mechanical Engineering.

FINANCIAL INFORMATION

ESTIMATED EXPENSES (ACADEMIC YEAR)
[Expenses are for the 1992-93 nine-month academic year.]

State Residents	Undergraduate	Graduate
Tuition and fees	$ 3,300	$ 3,370
College room and board	$ 5,730	$ 7,000
Books and supplies	$ 800	$ 500
Other expenses	$ 2,775	$ 2,775
Total estimated expenses	**$12,605**	**$13,645**
Out-of-State Residents	Undergraduate	Graduate
Tuition and fees	$10,999	$11,069
College room and board	$ 5,730	$ 7,000
Books and supplies	$ 800	$ 500
Other expenses	$ 2,775	$ 2,775
Total estimated expenses	**$20,304**	**$21,344**

GENERAL FINANCIAL AID INFORMATION
Forms accepted/required: Additional financial aid information: The Financial Aid Office will mail information about financial aid and scholarships when you check the appropriate box on the application. The types of financial aid available include undergraduate scholarships, federal Pell grants, supplemental Educational Opportunity grants, University loans and grants-in-aid, educational fee grants, Perkins loans, California Insured Student Loans, and work study.

ENGINEERING COLLEGE INFORMATION
[For additional personnel, refer to the *Appendix*.]

HEAD OF ENGINEERING
David A. Hodges
Phone: (510)642-5771

ENGINEERING COLLEGE ADDRESS
College of Engineering
University of California at Berkeley
Student Affairs Office
308 McLaughlin Hall
Berkeley, CA 94720
Phone: (510)642-7594   **Fax:** (510)643-8653

ENROLLMENTS—BY CLASS
[Numbers are baccalaureate enrollments for the fall 1991 term, unless otherwise footnoted.]
1st-year students/Freshmen .... 610 [1]
2nd-year students/Sophomores .... 477 [1]
3rd-year students/Juniors .... 802 [1]
4th-year students/Seniors .... 841 [1]
Total **2,730**

Notes: (1) Includes Chemical Engineering. Chemical Engineering is part of the College of Chemistry.

NUMBER OF DEGREES AWARDED—BY PROGRAM
[Numbers are engineering baccalaureate degrees awarded during the 1991-1992 academic year, unless otherwise footnoted. For full details about each engineering program, refer to the *Tables of Degree Programs*.]
Bioengineering 11
Chemical Engineering 60
Civil Engineering 74
Electrical Engineering & Computer Science 203
Engineering Geoscience .... –
Engineering Math & Statistics 4
Engineering Physics 6
Industrial Engineering & Operations Research 34
Manufacturing Engineering 1
Materials Science & Engineering 30
Mechanical Engineering 128
Mineral Engineering 1
Naval Architecture 4
Nuclear Engineering 9
Petroleum Engineering 1
Total **566**

PERCENTAGE OF DEGREES AWARDED—BY CATEGORY
[Percentages are of all engineering baccalaureate degrees awarded during the 1991-1992 academic year, unless otherwise footnoted.]
African Americans 2.5%
Asian/Pacific Island Americans 53.7%
Hispanic Americans 4.9%
Native Americans 0.7%
Foreign Citizens .... – % (A)
All Others 38.2%
Women 20.4%
Persons with disabilities .... – % (A)
Students over 25 years of age .... – % (A)

Notes: (A) Data not available.

GRADUATE ENROLLMENTS & DEGREES AWARDED
Master's enrollment 428
Master's degrees awarded 355
Doctoral enrollment 1,100
Doctoral degrees awarded 186

ENGINEERING STUDENT DATA
Applicants to the engineering college
Number of applicants to engineering college .... 3,063 [1]
Percent offered admission 33.0%

Matriculated engineering students
Percentage in top quartile (25%) of High School class ... – (A)

Notes: (A) Data not available. (1) For Chemical Engineering: 283 Applicants; 67% offered admission.

FULL & PART TIME FACULTY—BY DEPARTMENT
[Figures are the head count of full-time faculty and the full-time equivalent (FTE) of part-time faculty for each engineering department or equivalent.]

Department	Full	Part
Civil Eng	43	1.5
Electrical Eng & Computer Sciences	74	2.5
Industrial Eng & Operations Research	13	0.5
Mechanical Eng	42	—
Materials Science & Mineral Eng	22	1.0
Naval Architecture & Offshore Eng	3	0.5
Nuclear Eng	8	—
Chemical Eng	20	—

COLLEGE DESCRIPTION
The College of Engineering undergraduate program is a four-year program. There are seven engineering departments, each representing a separate field of engineering. Each has its own faculty, courses, fields of specialization, and curriculum requirements. Some offer more than one undergraduate curriculum, and several offer double-major curricula, allowing a student to earn a B.S. degree in two different fields of engineering. There are also several multidisciplinary curricula which cross both department and college boundaries. A substantial component of humanities and social sciences has become a standard part of engineering curricula. An optional co-op program is also available; it may extend the completion of the student's program to five years. Chemical Engineering is in the College of Chemistry.

DEGREE PROGRAMS DESCRIPTION
The B.S. degrees in the various engineering programs prepare the graduate for entry-level professional positions or for graduate study in each of the respective fields. The B.S. in the Engineering Science programs is intended primarily to prepare students for advanced graduate study. Graduate study in engineering can be carried out in one of three general areas: the M.S. and Ph.D. in Engineering, emphasizing the applications of the natural sciences to the analysis and solution of engineering problems; the M.S. and Ph.D. in Engineering Science; and the M.Eng. and D.Eng. programs of study in professional engineering.

DUAL DEGREE PROGRAMS
Enrollment requirements: Engineering undergraduates who wish to obtain a simultaneous degree in a college or school outside of the College of Engineering should submit a program proposal which: (1) has academic merit; (2) has been approved by the outside college or school; and (3) at the time of submission of the petition should not delay graduation by more than one additional term. Additional information is available from the Engineering Student Affairs Office.

TRANSFER INFORMATION
Residency Requirements: The University specifies that after 90 units have been completed, at least 24 of the remaining units must be completed in residence in the College or School in which the degree is to be taken, in at least two semesters (the one in which the 90 units are exceeded, plus one additional). In addition, the College of Engineering specifies that the final 30 units, comprising 2 consecutive semesters, must be in residence in the College of Engineering.

Transfer via Articulation Agreements
Admission to engineering: Transfers admitted to engineering at the junior level. Application deadlines are November 30 for the Fall semester; July 31 for the Spring. Transfer programs are articulated with community colleges.

Requirements: Requirements vary by major. Most require the following courses: math (2 years--calculus, linear algebra & differential equations, multivariable calculus); physics for scientists and engineers (3 semesters); general chemistry (1 year); fortran; 1 semester each in graphics, mechanics, and materials. GPA varies by major and applicant pool.

Transfer without Articulation Agreements
Admission to engineering: Change of college applicants are evaluated primarily on three factors--completeness of preparation at the level considered (Soph or Jr), grades, and total units including work in progress.

Requirements: Students who have been registered in any college on the UCB campus can transfer into the College of Engineering. only by petitioning for Change of College. Admission to Engineering is highly competitive; petitions are reviewed only once each year--during the spring term. Students who were admitted as transfers to another college are not likely to be accepted for subsequent transfer into Engineering.

GRADUATION REQUIREMENTS
B.S. programs in engineering require a minimum of 120 semester units and a grade-point average of at least 2.00. There are additional College requirements.

STUDENT PROGRAMS

PROFESSIONAL AND HONORARY SOCIETIES
[For key to acronyms, see *Introduction*.]
Professional Societies: AIChE, ANS, ASCE, ASME, IEEE, NSPE, SNAME, SPE, Institute of Industrial Engineers, Material Science and Engineering Association
Honorary Societies: Alpha Pi Mu, Chi Epsilon, Eta Kappa Nu, Pi Tau Sigma, Tau Beta Pi

SUPPORT PROGRAMS
Student Chapter Organizations: Society of Women Engineers, Black Engineering & Science Students Association, Hispanic Engineers & Scientists

College of Chemistry Scholars Program
For: Ethnic Minorities   **Available:** Year round
Offered: Prematriculation, Freshman

Mathematics, Engineering & Science Achievement
For: Ethnic Minorities   **Available:** Year round
Offered: Prematriculation

Minority Engineering Program
For: Ethnic Minorities   **Available:** Year round
Offered: Prematriculation, Freshman, Sophomore, Junior, Senior

Women & Graduate Students of Color in Engineering
For: Women & Ethnic Minorities   **Available:** Year round
Offered: Prematriculation, Freshman, Sophomore, Junior, Senior, Graduate level

Other Engineering Support Programs: New student orientations are offered by CalSO, the College and its student groups. Year-round support is provided by the Minority Engineering Program and College staff. The EJC Buddy Program, tutoring by student honor societies, and faculty advisors provide additional support. The freshman/sophomore seminars allow students to meet the faculty in a small group setting on topics of interest to both.

020 University of California, Davis

INSTITUTION PROFILE
HEAD OF THE INSTITUTION
Theodore L. Hullar
Phone: (916)752-2066 Fax: (916)752-2400

GENERAL INFORMATION
[All Students—Fall 1991]
Undergraduate enrollment 17,898
Graduate enrollment 5,404
Total institution enrollment **23,302**
Type of institution: Public **Calendar:** Quarters
Nearest city: Sacramento **Population:** 385,127
Miles from main campus: 13 **Setting:** Small Town
Types of engineering degrees: Engineering
Other degree-granting colleges: Law, Medicine, Letters and Sciences, Veterinary Medicine, Management, Agricultural and Environmental Sciences

ENGINEERING ADMISSIONS
ADMISSIONS OFFICE CONTACT
Gary Tudor
University of California, Davis
Director of Undergraduate Admissions
175 Mark Hall
Davis, CA 95616-5294
Phone: (916)752-2971 Fax: (916)752-1280

ENGINEERING COLLEGE ADMISSIONS INFORMATION
Admission to the engineering college: Undergraduate students are admitted at the freshman and junior years only.
Entrance Requirements: High School courses—U.S. History (1 year), English (4 years), Mathematics (3 years), Foreign Language (2 years), SAT, CEEB Achievement tests including English Composition, Mathematics I or II and one test from English Literature, Foreign Language, Science or Social Science.
Entrance Recommendations: High School courses—Physics (1 year), Chemistry (1 year), Plane Geometry (1 year), Trigonometry/Analytic Geometry (1 year), American College Test
Requirements for foreign students: International students enrolled in a US High School are eligible to apply for admission at the Freshman level. International transfers at the Junior level who are enrolled at a California Community College or four-year institution are eligible to apply for admission.

DEPARTMENTAL ADMISSIONS INFORMATION
Admission to the engineering department: At the time of admission to the institution.

FINANCIAL INFORMATION
ESTIMATED EXPENSES (ACADEMIC YEAR)
[Expenses are for the 1992-93 nine-month academic year.]

State Residents	Undergraduate	Graduate
Tuition and fees	$ 2,980	$ 3,264
College room and board	$ 5,574	$ 5,584
Books and supplies	$ 737	$ 737
Other expenses	$ 695 [1]	$ 695 [1]
Total estimated expenses	**$ 9,986**	**$10,280**

Out-of-State Residents	Undergraduate	Graduate
Tuition and fees	$10,678	$10,962
College room and board	$ 5,574	$ 5,584
Books and supplies	$ 737	$ 737
Other expenses	$ 695 [1]	$ 695 [1]
Total estimated expenses	**$17,684**	**$17,978**

Notes: (1) Transportation

FINANCIAL AID OFFICE CONTACT
Ronald W. Johnson, *Director*
University of California, Davis
Financial Aid Office
125 North Hall
Davis, CA 95616-5294
Phone: (916)752-2390 Fax: (916)752-7339

GENERAL FINANCIAL AID INFORMATION
Forms accepted/required: ACT, CSS-FAF, FAF, FAT, FAR, FFS, Stafford, IRS, SAR, Supplemental, Scholarship Applications (public/private)
Additional financial aid information: For more information, contact: Financial Aid Office, North Hall, University of California, Davis, CA 95616-5294

ENGINEERING COLLEGE INFORMATION
[For additional personnel, refer to the *Appendix*.]
HEAD OF ENGINEERING
Mohammed S. Ghausi
Phone: (916)752-0554 Fax: (916)752-8058
EMail: msghausi@ucdavis.edu

ENGINEERING COLLEGE ADDRESS
College of Engineering
University of California, Davis
1050 Engineering Unit II
Davis, CA 95616-5294
Phone: (916)752-0553 Fax: (916)752-8058

ENROLLMENTS—BY CLASS
[Numbers are baccalaureate enrollments for the fall 1991 term, unless otherwise footnoted.]
1st-year students/Freshmen 364
2nd-year students/Sophomores 366
3rd-year students/Juniors 522
4th-year students/Seniors 711
Total **1,963**

NUMBER OF DEGREES AWARDED—BY PROGRAM
[Numbers are engineering baccalaureate degrees awarded during the 1991-1992 academic year, unless otherwise footnoted. For full details about each engineering program, refer to the *Tables of Degree Programs*.]
Aeronautical Science & Engineering .... 5 [2]
Applied Science .... – [B]
Biological & Agricultural Engineering 1
Chemical Engineering 35
Chemical Engineering & Materials Science 2
Civil Engineering 71
Civil Engineering & Materials Science 2
Computer Science & Engineering 20
Electrical & Computer Engineering 91
Electrical Engineering & Materials Science 4
Materials Science & Engineering 5
Mechanical Engineering 47
Mechanical Engineering & Materials Science 10
Total **293**

Notes: (B) Data not applicable. (2) Additional 24 students received B.S. in both Mechanical and Aeronautical Science and Engineering.

PERCENTAGE OF DEGREES AWARDED—BY CATEGORY
[Percentages are of all engineering baccalaureate degrees awarded during the 1991-1992 academic year, unless otherwise footnoted.]
African Americans 2.5%
Asian/Pacific Island Americans 3.8%
Hispanic Americans 7.6%
Native Americans 0.3%
Foreign Citizens 0.3%
All Others 85.5%
Women 16.4%
Persons with disabilities .... – % [A]
Students over 25 years of age .... – % [A]

Notes: (A) Data not available.

GRADUATE ENROLLMENTS & DEGREES AWARDED
Master's enrollment 352
Master's degrees awarded 114
Doctoral enrollment 315
Doctoral degrees awarded 53

ENGINEERING STUDENT DATA
Applicants to the engineering college
Number of applicants to engineering college 2,318
Percent offered admission 54.0%

Matriculated engineering students
Percentage in top quartile (25%) of High School class ..... – [A]
Average SAT scores: Math—518, Verbal—643, Combined—1161

Notes: (A) Data not available.

FULL & PART TIME FACULTY—BY DEPARTMENT
[Figures are the head count of full-time faculty and the full-time equivalent (FTE) of part-time faculty for each engineering department or equivalent.]

Department	Full	Part
Chemical Eng	11.1	2.3
Civil Eng	22.1	2.9
Electrical & Computer Eng	34.9	1.8
Mechanical, Aeronautical & Materials Eng	33.9	3.6
Biological & Agricultural Eng	3	0.1
Computer Science & Eng	19	1.1
Applied Science	11.1	0.3

COLLEGE DESCRIPTION
At Davis, the B.S. programs in Engineering are intended to serve as a sound basis for beginning professional practice in engineering design or development, as a general preparation for corporate or government careers, or as a general foundation for pursuing graduate study or continuing education. To these ends, there are thirteen offered, four of which are double majors with Materials Science and Engineering. Additional double major choices may be elected, and a large number of minors may be chosen from the other two undergraduate colleges. Numerous internship opportunities are available through the Internship and Career Center.

DEGREE PROGRAMS DESCRIPTION
The B.S. degrees in the various engineering programs prepare the graduate for a variety of options from an industrial career to advanced study. Graduate study in engineering can be carried out in M.S. or M.Engr. programs which emphasize the scientific and the professional aspects of engineering, respectively. Similarly, further study can be carried out in Ph.D. or D.Engr. programs.

TRANSFER INFORMATION
Residency Requirements: The final 45 units typical of the major must be completed in residence.

Transfer via Articulation Agreements
Admission to engineering: At the end of the sophomore year
Requirements: University of California eligibility. Additionally, in admitting students to 'advanced standing' status, we give priority to Calif. community college students who have completed, with a high GPA, the specific lower-division courses required for admission; each department has its own pattern of these courses.

Transfer without Articulation Agreements
Admission to engineering: At the end of the sophomore year
Requirements: Must meet published course requirements.

GRADUATION REQUIREMENTS
B.S. programs in engineering require 180-214 quarter units, depending on major selected. A full description of requirements can be found in the Engineering Bulletin, a catalog published by the College of Engineering.

STUDENT PROGRAMS
PROFESSIONAL AND HONORARY SOCIETIES
[For key to acronyms, see *Introduction*.]
Professional Societies: AIAA, AIChE, ASAE, ASCE, ASME, IEEE, SAE, TMS
Honorary Societies: Tau Beta Pi

SUPPORT PROGRAMS
Student Chapter Organizations: Society of Women Engineers, National Society of Black Engineers, Society of Hispanic Professional Engineers, American Indian Science and Engineering Society, Pilipino-American Society of Engineers

Center for Women in Engineering
For: Women **Available:** Year round
Offered: Prematriculation, Freshman, Sophomore, Junior, Senior, Graduate level

Minority Engineering Program
For: Ethnic Minorities **Available:** Year round
Offered: Prematriculation, Freshman, Sophomore, Junior, Senior, Graduate level

Minority Opportunities for Research in Engineering
For: Ethnic Minorities **Available:** Year round
Offered: Graduate level

Engineering Summer Residency Program, MESA
For: Ethnic Minorities **Available:** Summer only
Offered: Prematriculation

Other Engineering Support Programs: Summer advising programs are available for incoming freshmen and their parents and for community college transfer students. A fall quarter orientation week program is provided also.

021 University of California, Irvine

INSTITUTION PROFILE

HEAD OF THE INSTITUTION
Dennis Smith
Phone: (714)856-5111 Fax: (714)725-2087

GENERAL INFORMATION
[All Students—Fall 1991]
Undergraduate enrollment 13,811
Graduate enrollment 2,492
Total institution enrollment 16,303
Type of institution: Public **Calendar:** Quarters
Nearest city: Anaheim, California **Population:** 219,000
Miles from main campus: 11 **Setting:** Suburban
Types of engineering degrees: Engineering

Other degree-granting colleges: Business Administration, Education, Fine & Performing Arts, Humanities & Social Sciences, Medicine, Biological Sciences, Physical Sciences, Social Ecology, Information and Computer Science

ENGINEERING ADMISSIONS

ADMISSIONS OFFICE CONTACT
Ann Williams
University of California, Irvine
Office of Admissions
Irvine, CA 92717
Phone: (714)856-6703

ENGINEERING COLLEGE ADMISSIONS INFORMATION
Entrance Requirements: High School courses—History, physics, chemistry (1 year), English, mathematics (4 years); College-preparatory electives (4 years), SAT or ACT required (no required scores); and three College Board Achievement tests required: English composition, mathematics, and one of: literature, foreign language, sciences, or social sciences
Requirements for foreign students: TOEFL (Score: 550); English as a Second Language Placement Test

DEPARTMENTAL ADMISSIONS INFORMATION
Admission to the engineering department: At the time of admission to the institution.
Additional information: Each department has a subcommittee that provides input to the admissions officer. The required grade point average varies from department to department and may fluctuate slightly from year to year.

FINANCIAL INFORMATION

ESTIMATED EXPENSES (ACADEMIC YEAR)
[Expenses are for the 1992-93 nine-month academic year.]

State Residents	Undergraduate	Graduate
Tuition and fees	$ 3,075 [1]	$ 3,642 [1]
College room and board	$ 8,220 [2]	$ 9,250 [2]
Books and supplies	$ 675 [2]	$ 783 [2]
Other expenses	$ —	$ —
Total estimated expenses	**$11,970**	**$13,675**

Out-of-State Residents	Undergraduate	Graduate
Tuition and fees	$10,774 [1]	$11,341 [1]
College room and board	$ 8,220 [2]	$ 9,250 [2]
Books and supplies	$ 675 [2]	$ 783 [2]
Other expenses	$ —	$ —
Total estimated expenses	**$19,669**	**$21,374**

Notes: (1) All tuition and fees are subject to change. (2) Room, board, books, and supplies are estimates based on projected total expenses.

FINANCIAL AID OFFICE CONTACT
Otto W. Reyer, *Director*
University of California, Irvine
Financial Aid
102 Administration
Irvine, CA 92717
Phone: (714)856-6261 Fax: (714)856-4876

GENERAL FINANCIAL AID INFORMATION
Forms accepted/required: Student Aid Application for California (SAAC)
Additional financial aid information: Grants based on need, scholarships based on academic excellence, loans, and work-study programs.

ENGINEERING COLLEGE INFORMATION
[For additional personnel, refer to the *Appendix*.]

HEAD OF ENGINEERING
William A. Sirignano
Phone: (714)856-6002 Fax: (714)856-7966

ENGINEERING COLLEGE ADDRESS
School of Engineering
University of California, Irvine
114 Rockwell Engineering Center
Irvine, CA 92717
Phone: (714)856-4334 Fax: (714)856-7966

ENROLLMENTS—BY CLASS
[Numbers are baccalaureate enrollments for the fall 1991 term, unless otherwise footnoted.]
1st-year students/Freshmen 295
2nd-year students/Sophomores 177
3rd-year students/Juniors 247
4th-year students/Seniors 323
Total 1,042

NUMBER OF DEGREES AWARDED—BY PROGRAM
[Numbers are engineering baccalaureate degrees awarded during the 1991-1992 academic year, unless otherwise footnoted. For full details about each engineering program, refer to the *Tables of Degree Programs*.]
Biochemical Engineering .... — (B)
Civil Engineering 52
Electrical & Computer Engineering 67
General Engineering .... — (4)
Mechanical & Aerospace Engineering 53
Total 172

Notes: (B) Data not applicable. (4) This is a new program.

PERCENTAGE OF DEGREES AWARDED—BY CATEGORY
[Percentages are of all engineering baccalaureate degrees awarded during the 1991-1992 academic year, unless otherwise footnoted.]
African Americans .... 1.0 %
Asian/Pacific Island Americans .... 48.0 %
Hispanic Americans .... 5.0 %
Native Americans .... 1.0 %
Foreign Citizens .... 3.0 %
All Others .... 42.0 %
Women .... 17.0 %
Persons with disabilities .... — % (A)
Students over 25 years of age .... — % (A)

Notes: (A) Data not available.

GRADUATE ENROLLMENTS & DEGREES AWARDED
Master's enrollment 173
Master's degrees awarded 81
Doctoral enrollment 186
Doctoral degrees awarded 25

ENGINEERING STUDENT DATA

Applicants to the engineering college
Number of applicants to engineering college 1,331
Percent offered admission 77.0%

Matriculated engineering students
Percentage in top quartile (25%) of High School class ..... — (A)
Average SAT scores: Math—458, Verbal—612, Combined—1070

Notes: (A) Data not available.

FULL & PART TIME FACULTY—BY DEPARTMENT
[Figures are the head count of full-time faculty and the full-time equivalent (FTE) of part-time faculty for each engineering department or equivalent.]

Department	Full	Part
Biochemical Eng	6	—
Civil Eng	17	4.0
Electrical & Computer Eng	23	4.0
Mechanical & Aerospace Eng	22	1.0

COLLEGE DESCRIPTION
The School of Engineering offers undergraduate and graduate programs of study which emphasize the fundamentals underlying engineering, enabling graduates to continue professional development through formal or informal study. Courses in environmental engineering are offered in Civil Engineering, course in aerospace engineering are offered in Mechanical and Aerospace Engineering, courses in computer engineering are offered in Electrical and Computer Engineering, and undergraduate students with advanced standing may study chemical engineering, computer graphics, or non-traditional engineering programs through the general Engineering program. Undergraduate students may participate in research projects through individual study opportunities. A business minor is offered and students often seek double majors with computer science or biological sciences. A co-op program is not offered.

DEGREE PROGRAMS DESCRIPTION
The B.S. degree is offered in Civil Engineering, Electrical Engineering, general Engineering, and Mechanical Engineering. Civil Engineering offers specializations in environmental engineering, structural engineering, transportation engineering, and water resources engineering. Electrical Engineering has specializations in computer engineering, electrooptics and solid-state devices, power systems, and systems and signal processing. Mechanical Engineering offers specializations in aerospace engineering, combustion and propulsion, heat transfer and fluid mechanics, materials science and engineering, and mechanical systems. The general Engineering program has specializations available in chemical engineering and computer graphics, but students, with the consent of a faculty advisor, may choose any area of special interest. The M.S. and Ph.D. degrees in Engineering are offered in Biochemical Engineering, Civil Engineering, Electrical Engineering, Materials Science and Engineering, and Mechanical Engineering.

DUAL DEGREE PROGRAMS
Enrollment requirements: Varies depending upon major or option.

TRANSFER INFORMATION
Residency Requirements: Credit for the last 36 units of work immediately preceding graduation must be earned in residence at the UCI campus.

Transfer via Articulation Agreements
Admission to engineering: At the end of the sophomore year
Requirements: Completion of required courses (differs with major) in mathematics, physics, chemistry, computing, and lower division writing; GPA must be within cut-off range.

Transfer without Articulation Agreements
Admission to engineering: At the end of the sophomore year
Requirements: Completion of required courses (differs with major) in mathematics, physics, chemistry, computing, and lower division writing; GPA is within cut-off range.

GRADUATION REQUIREMENTS
B.S. programs in engineering require credit for at least 180 units including 24 units in mathematics; completion of three basic engineering courses; completion of four basic science courses with lab; 41 units in breadth courses, including one upper division writing courses; and a GPA of 2.0 or above. Actual course requirements vary between majors.

STUDENT PROGRAMS

PROFESSIONAL AND HONORARY SOCIETIES
[For key to acronyms, see *Introduction*.]
Professional Societies: AGCA, AIAA, ASCE, ASME, IEEE, ITE, SAE, Structural Engineers Association of Southern California (SEAOSC)
Honorary Societies: Chi Epsilon, Eta Kappa Nu, Pi Tau Sigma, Tau Beta Pi

SUPPORT PROGRAMS
Student Chapter Organizations: Society of Women Engineers, National Society of Black Engineers, Mexican-American Engineering Society

Engineering and Computer Science Laboratory-ECSEL
For: Ethnic Minorities **Available:** Academic year
Offered: Freshman, Sophomore, Junior, Senior

Discover UCI
For: Women & Ethnic Minorities **Available:** Summer only
Offered: Prematriculation

Minority Engineering Program
For: Ethnic Minorities **Available:** Year round
Offered: Prematriculation, Freshman, Sophomore, Junior, Senior

Other Engineering Support Programs: Welcome Week, fall orientation program with academic advising; Student-Parent Orientation Program (SPOP), a live-in experience for new students and their parents; Uni-Prep, an intensive five-day program in September for new students; Celebrate UCI, annual spring open house; the Engineering Undergraduate Student Affairs Office has three academic counselors and two part-time peer counselors.

022 University of California, Los Angeles

INSTITUTION PROFILE

HEAD OF THE INSTITUTION
Charles E. Young, *Chancellor*
Phone: (310)825-2121 Fax: (310)206-6030

GENERAL INFORMATION
[All Students—Fall 1991]
Undergraduate enrollment 24,269
Graduate enrollment 8,149
Total institution enrollment 32,418
Type of institution: Public Calendar: Quarters
Location: Los Angeles Population: 3,000,000
Setting: Urban
Types of engineering degrees: Engineering
Other degree-granting colleges: Architecture, Dentistry, Education, Law, Medicine, Nursing, Letters & Science, Arts, Theater, Film, Television, Graduate School of Library & Information Science, Public Health, John E. Anderson Graduate School of Management

ENGINEERING ADMISSIONS

ADMISSIONS OFFICE CONTACT
Stephen E. Jacobsen
University of California, Los Angeles
6412 Boelter Hall
405 Hilgard Avenue
Los Angeles, CA 90024-1600
Phone: (310)825-2941 Fax: (310)825-4061

ENGINEERING COLLEGE ADMISSIONS INFORMATION
Admission to the engineering college: A prospective engineering student applies to the program at the time of application to the university.
Entrance Requirements: High School courses—University A-F requirements, (1) Either (a) ACT, composite score or (b) SAT, total score; (2) Three College Board Achievement Tests (ACH) including: (a) English comp. and (b) mathematics, level 1 or 2 and (c) either English Lit., foreign language, science, or social science.
Requirements for foreign students: A certificate of completion of secondary school with a superior average in academic subjects which would enable them to be admitted to a university in the home country.

DEPARTMENTAL ADMISSIONS INFORMATION
Admission to the engineering department: At the time of admission to the institution.

FINANCIAL INFORMATION

ESTIMATED EXPENSES (ACADEMIC YEAR)
[Expenses are for the 1992-93 nine-month academic year.]

State Residents	Undergraduate	Graduate
Tuition and fees	$ 2,904	$ 3,457
College room and board	$ 5,650 [1]	$ 5,650 [1]
Books and supplies	$ 775	$ 1,020
Other expenses	$ 1,655 [2]	$ 4,140 [2]
Total estimated expenses	**$10,984**	**$14,267**

Out-of-State Residents	Undergraduate	Graduate
Tuition and fees	$10,601	$11,154
College room and board	$ 5,650 [1]	$ 5,650 [1]
Books and supplies	$ 775	$ 1,020
Other expenses	$ 1,655 [2]	$ 4,140
Total estimated expenses	**$18,681**	**$21,964**

Notes: (1) Estimate of on-campus housing costs. (2) Personal and transportation.

FINANCIAL AID OFFICE CONTACT
Lawrence W. Burt
University of California, Los Angeles
Financial Aid Office
A129J Murphy Hall
Los Angeles, CA 90024-1435
Phone: (310)206-0400 Fax: (310)206-1728

GENERAL FINANCIAL AID INFORMATION
Forms accepted/required: Student Aid Application for California (SAAC)
Additional financial aid information: Scholarships, grants, loans, and work-study programs.

ENGINEERING COLLEGE INFORMATION

[For additional personnel, refer to the *Appendix*.]

HEAD OF ENGINEERING
A.R. Frank Wazzan
Phone: (310)206-8245 Fax: (310)206-4061

ENGINEERING COLLEGE ADDRESS
School of Engineering and Applied Science
University of California, Los Angeles
405 Hilgard Avenue
Los Angeles, CA 90024-1600
Phone: (310)825-2757 Fax: (310)825-0761

ENROLLMENTS—BY CLASS
[Numbers are baccalaureate enrollments for the fall 1991 term, unless otherwise footnoted.]
1st-year students/Freshmen 548
2nd-year students/Sophomores 436
3rd-year students/Juniors .... 550 [1]
4th-year students/Seniors 660
Total **2,194**

Notes: (1) Juniors: 550 includes 3 students limited status.

NUMBER OF DEGREES AWARDED—BY PROGRAM
[Numbers are engineering baccalaureate degrees awarded during the 1991-1992 academic year, unless otherwise footnoted. For full details about each engineering program, refer to the *Tables of Degree Programs*.]
Aerospace Engineering 37
Chemical Engineering 28
Civil Engineering 62
Computer Science 2
Computer Science & Engineering 60
Electrical Engineering 126
Engineering 2
Man-Machine-Environment Systems .... — [7]
Manufacturing Engineering .... — [7]
Materials Engineering 13
Materials Science & Engineering .... —
Mechanical Engineering 80
Nuclear Engineering .... — [7]
Total **410**

Notes: (7) Not offered at the B.S. level.

PERCENTAGE OF DEGREES AWARDED—BY CATEGORY
[Percentages are of all engineering baccalaureate degrees awarded during the 1991-1992 academic year, unless otherwise footnoted.]
African Americans 2.7%
Asian/Pacific Island Americans 49.5%
Hispanic Americans 7.3%
Native Americans 0.2%
Foreign Citizens .... 5.9% [2]
All Others 34.4%
Women 17.6%
Persons with disabilities .... — %
Students over 25 years of age .... — %

Notes: (2) Foreign Nationals - students on student (F-1) or exchange (J-1) visas.

GRADUATE ENROLLMENTS & DEGREES AWARDED
Master's enrollment 190
Master's degrees awarded 207
Doctoral enrollment 801
Doctoral degrees awarded 95

ENGINEERING STUDENT DATA
Applicants to the engineering college
Number of applicants to engineering college 3,238
Percent offered admission 41.9%
Average SAT scores: Math—695.4, Verbal—540.6, Combined—1236

FULL & PART TIME FACULTY—BY DEPARTMENT
[Figures are the head count of full-time faculty and the full-time equivalent (FTE) of part-time faculty for each engineering department or equivalent.]

Department	Full	Part
Chemical Eng	11	—
Civil Eng	16	0.4
Electrical Eng	35	2.4
Computer Science Department	29	—
Materials Science & Eng	10	1.2
Mechanical, Aerospace & Nuclear Eng	27	4.3

COLLEGE DESCRIPTION

Students in the School of Engineering and Applied Science may elect one of the eight four-year curricula. The School may nominate exceptionally promising undergraduate students as Departmental Scholars to pursue bachelor's and master's degree programs simultaneously. A Cooperative Education Program is an available plan wherein undergraduate students combine six-month periods of regular employment in private industry or government activities with alternate periods of study. Average number of years required to actually complete the bachelor's degree: 4 2/3 years.

DEGREE PROGRAMS DESCRIPTION

The undergraduate curricula provide students with a firm foundation in engineering and applied science and, prepare graduates for immediate practice of the profession as well as advanced studies. At the graduate level, the School offers programs leading to the degrees of Master of Science, Engineer, and Doctor of Philosophy. The M.S. degree program consists of nine courses of graduate and upper division work in addition to a thesis or a comprehensive examination. The basic program of study for the Ph.D., in most cases, is built around one major field and two minor fields.

TRANSFER INFORMATION

Residency Requirements: Of the last 48 units completed for the Bachelor's degree, 36 must be earned in residence in the School of Engineering and Applied Science on this campus. No more than 16 of the 36 units may be completed in Summer Sessions at UCLA.

Transfer via Articulation Agreements
Admission to engineering: Transfer are admitted at the junior level in the Fall and limited in the Spring. The first day to file undergraduate applications for Fall '93 is 11/1/92. The first day for Spring '93 is 10/1/92.
Engineering graduates transferred from ... 4-yr: 24 2-yr: 99
Requirements: Completion of all the lower division requirements in mathematics, physics and chemistry. Applicants must meet university eligibility requirements and admission to the School is on a competitive basis.

Transfer without Articulation Agreements
Admission to engineering: Same as 'other' for institutions with which we have articulation agreements.
Requirements: Same as Entrance Requirements for institutions with which we have articulation agreements

GRADUATION REQUIREMENTS

B.S. Program in engineering require 180 to 196 quarter units, depending on major selected. At least a final cumulative 2.0 grade point average must be achieved, as well in all upper division University courses offered in satisfaction of the subject and elective requirements of the curriculum. In addition, a 2.0 minimum grade-point average in upper division mathematics, upper division core courses, and the major field is required for graduation.

STUDENT PROGRAMS

PROFESSIONAL AND HONORARY SOCIETIES
[For key to acronyms, see *Introduction*.]
Professional Societies: ACerS, ACM, AIAA, AIChE, ASCE, ASME, BMES, IEEE, American Society of Metals; Engineering Society, University of California (ESUC), Computer Science Undergraduate Assoc. (CSUA); Engr Business Soc (EBS)
Honorary Societies: Eta Kappa Nu, Tau Beta Pi

SUPPORT PROGRAMS
Student Chapter Organizations: Society of Women Engineers, National Society of Black Engineers, Society of Latino Engineers and Scientists, American Indian Science and Engineering Society

Minority Engineering Program (MEP)
For: Ethnic Minorities **Available:** Year round
Offered: Prematriculation, Freshman, Sophomore, Junior, Senior, Graduate level
Other Engineering Support Programs: Orientation programs for all freshmen and transfers in summer, Placement and Career Planning Center, Office of International Students and Scholars, services for students with disabilities.

023 University of California, San Diego

INSTITUTION PROFILE

HEAD OF THE INSTITUTION

Richard C. Atkinson, *Chancellor*
Phone: (619)534-3135 **Fax:** (619)534-6523

GENERAL INFORMATION

[All Students—Fall 1991]

Undergraduate enrollment	14,857
Graduate enrollment	3,384
Total institution enrollment	**18,241**

Type of institution: Public **Calendar:** Quarters
Nearest city: San Diego, California **Population:** 1,139,598
Miles from main campus: 7 **Setting:** Urban
Types of engineering degrees: Engineering
Other degree-granting colleges: Architecture, Arts & Sciences, Humanities & Social Sciences, Medicine, International & Pacific Studies, Scripps Institution of Oceanography

ENGINEERING COLLEGE ADMISSIONS INFORMATION

Admission to the engineering college: Applicants who have demonstrated excellent academic performance prior to being accepted to UCSD are admitted directly into the engineering major.
Entrance Requirements: SAT (Score: 1100), ACT (Score: 26), High School courses—History (1 year), English (4 years), Mathematics (3 years), Laboratory Science (1 year)
Entrance Recommendations: High School courses—Foreign Language (2 years), College Prep Classes (4 years)
Requirements for out-of-state residents: Residents: with SAT of 1600 or ACT of 35, need GPA of 2.78; with GPA of 3.30 or better, eligible with lowest test scores. Non-Residents: GPA of 3.4 or better OR score of 1100 on the SAT or 26 on the ACT. Achievement Tests: 1730 or higher; no area score less than 500.
Requirements for foreign students: TOEFL (Score: 550)

DEPARTMENTAL ADMISSIONS INFORMATION

Admission to the engineering department: Those who have demonstrated excellent performance prior to being admitted to UCSD are admitted directly into the engineering major of their choice.
Additional information: Students not admitted directly into an engineering major must consult the department of their choice and review the requirements necessary to gain admission. In addition, the general-education requirements of UCSD's five undergraduate colleges differ noticeably. Prospective students should review these requirements and take them into account when selecting a college.

FINANCIAL INFORMATION

ESTIMATED EXPENSES (ACADEMIC YEAR)

[Expenses are for the 1992-93 nine-month academic year.]

State Residents	Undergraduate	Graduate
Tuition and fees	$ 2,463	$ –(1)
College room and board	$ 6,152	$ 8,500
Books and supplies	$ 558	$ 500
Other expenses	$ 2,127	$ 2,127
Total estimated expenses	**$11,300**	**$ 11,127**

Out-of-State Residents	Undergraduate	Graduate
Tuition and fees	$ 5,030	$ 11,195
College room and board	$ 6,152	$ 8,500
Books and supplies	$ 558	$ 500
Other expenses	$ 2,127	$ 2,127
Total estimated expenses	**$13,867**	**$ 22,322**

Notes: (1) Graduate students who are residents do not pay tuition and fees.

FINANCIAL AID OFFICE CONTACT

Thomas M. Rutter, *Director*
University of California, San Diego
9500 Gilman Drive
La Jolla, CA 92093-0013
Phone: (619)534-3800 **Fax:** (619)534-5459

GENERAL FINANCIAL AID INFORMATION

Forms accepted/required: ACT, FAF, Student Aid Application for California (SAAC)
Additional financial aid information: UCSD expects students and their families to bear as much of the basic, necessary costs of the student's education as their circumstances will allow. In cases where family resources are insufficient to meet the basic educational costs, the Student Financial Services Office attempts to assist students in obtaining support and financial aid including scholarships, fellowships, grants, loans, and work-study grants.

ENGINEERING COLLEGE INFORMATION

[For additional personnel, refer to the *Appendix*.]

HEAD OF ENGINEERING

M. Lea Rudee
Phone: (619)534-4575 **Fax:** (619)534-4771
EMail: rudee@ucsd.edu

ENGINEERING COLLEGE ADDRESS

Division of Engineering
University of California, San Diego
9500 Gilman Drive
La Jolla, CA 92093-0403
Phone: (619)534-4575 **Fax:** (619)534-4771

ENROLLMENTS—BY CLASS

[Numbers are baccalaureate enrollments for the fall 1991 term, unless footnoted.]

1st-year students/Freshmen	336
2nd-year students/Sophomores	331
3rd-year students/Juniors	391
4th-year students/Seniors	785
Total	**1,843**

NUMBER OF DEGREES AWARDED—BY PROGRAM

[Numbers are engineering baccalaureate degrees awarded during the 1991-1992 academic year, unless otherwise footnoted. For full details about each engineering program, refer to the *Tables of Degree Programs*.]

Aerospace Engineering	–
Applied Mechanics	3
Applied Physics	1
Applied Science	–
Bioengineering	16
Biomedical Engineering	21
Chemical Engineering	19
Computer Engineering	31
Computer Science	45
Computer Science & Engineering	45
Electrical Engineering	75
Engineering Physics	–
Engineering Science(s)	–
Materials Science	–
Mechanical Engineering	84
Structural Engineering	21
Systems & Control Engineering	–
Systems Science	–
Total	**361**

PERCENTAGE OF DEGREES AWARDED—BY CATEGORY

[Percentages are of all engineering baccalaureate degrees awarded during the 1991-1992 academic year, unless otherwise footnoted.]

African Americans	1.2%
Asian/Pacific Island Americans	27.8%
Hispanic Americans	5.2%
Native Americans	0.6%
Foreign Citizens	8.0%
All Others	57.2%
Women	17.1%
Persons with disabilities	– %(A)
Students over 25 years of age	– %(A)

Notes: (A) Data not available.

GRADUATE ENROLLMENTS & DEGREES AWARDED

Master's enrollment	164
Master's degrees awarded	133
Doctoral enrollment	350
Doctoral degrees awarded	39

ENGINEERING STUDENT DATA

Applicants to the engineering college

Number of applicants to engineering college	3,441
Percent offered admission	70.0%

Matriculated engineering students

Percentage in top quartile (25%) of High School class 99.0
Average SAT scores: Math—656, Verbal—519, Combined—1175
Average ACT scores: Composite—26

FULL & PART TIME FACULTY—BY DEPARTMENT

[Figures are the head count of full-time faculty and the full-time equivalent (FTE) of part-time faculty for each engineering department or equivalent.]

Department	Full	Part
Electrical Eng	35	–
Applied Mechanics & Eng Sciences	34	2.9
Computer Science & Eng	28	–

COLLEGE DESCRIPTION

The Division of Engineering at UCSD has experienced dramatic growth and national recognition over the past decade. The number of engineering faculty has doubled, all of whom have outstanding credentials and commitment to teaching and research. The faculty includes ten members of the National Academy of Engineering. A seven-story, $40 million engineering building, the largest building on campus, was completed in fall 1988, and construction of a new building has begun, and should be completed within the next 2 years. Other special facilities and programs include the Center for Magnetic Recording Research, the Charles Lee Powell Structural Systems Laboratory, the Center for Energy Combustion Research, the Institute of Biomedical Engineering, the Institute of Mechanics and Materials, the Center for Highspeed Circuits and Systems, and the San Diego Supercomputer Center. UCSD is situated in beautiful La Jolla overlooking the Pacific Ocean.

DEGREE PROGRAMS DESCRIPTION

The Applied Mechanics & Engineering Sciences (AMES) Department, offers the BA or BS degree in the following areas: aerospace eng., applied mechanics, bioengineering, chemical eng., engineering science, mechanical eng., structural eng., and engineering physics. AMES also offers an MS or PhD in aerospace eng., applied mechanics, applied ocean science, bioengineering, chemical eng., engineering physics, mechanical eng., structural eng., and systems science. The Computer Science & Engineering Department (CSE) offers the BA or BS degree in computer science and computer eng. The graduate program in CSE offers an MS and PhD degree in computer science, computer eng. or related areas. The Electrical and Computer Science (ECE) Department offers BS degree programs in electrical engineering, engineering physics, systems and control engineering, and computer engineering; as well as a BA degree in applied physics and information science. ECE also offers MS and PhD degrees in computer engineering and electrical engineering degrees for applied physics, communication theory and systems, electronic circuits and systems, and intelligent systems-robotics and control.

TRANSFER INFORMATION

Transfer via Articulation Agreements

Admission to engineering: Transfer students must apply no later than at the end of their third quarter of study at UCSD. These students should seek a preliminary appraisal by the department as soon as possible.
Requirements: Requirements for admission to engineering majors are the same for transfer students as for continuing students. Students should be mindful of lower-division prerequisites necessary for admission to upper-division courses. All students interested in engineering should review the UCSD General Catalog for details about admission to the major and graduation requirements.

Transfer without Articulation Agreements

Admission to engineering: Transfer students from any UC who have satisfied lower-division breadth and general education (B/GE) requirements at that campus prior to transfer may consider this requirement satisfied at UCSD.
Requirements: In addition to satisfaction of UC minimum requirements, only transfer students who have completed eighty-four or more transferable quarter-units are considered for admission unless additional standards are met. Priority is given to students transferring from California community colleges.

GRADUATION REQUIREMENTS

Requirements vary between the university's five colleges and Engineering majors. A minimum of 180 units must be completed with at least a 2.0 GPA in major course work, and a minimum GPA of 2.8 overall. Students MUST check with college advisors and departmental academic advisors to insure meeting graduation requirements.

STUDENT PROGRAMS

PROFESSIONAL AND HONORARY SOCIETIES

[For key to acronyms, see *Introduction*.]

Professional Societies: AIAA, AIChE, ASME, IEEE, Bioengineering Society (BES)
Honorary Societies: Tau Beta Pi

SUPPORT PROGRAMS

Student Chapter Organizations: Society of Women Engineers, National Society of Black Engineers, Society of Hispanic Professional Engineers, Mesa Engineering Program (MEP)

Society for Women Engineers
For: Women **Available:** Academic year
Offered: Prematriculation, Freshman, Sophomore, Junior, Senior, Graduate level

National Society of Black Engineers
For: Ethnic Minorities **Available:** Academic year
Offered: Prematriculation, Freshman, Sophomore, Junior, Senior, Graduate level

Society of Hispanic Professional Engineers
For: Ethnic Minorities **Available:** Academic year
Offered: Prematriculation, Freshman, Sophomore, Junior, Senior, Graduate level

Mesa Engineering Program (MEP)
For: Women & Ethnic Minorities **Available:** Year round
Offered: Prematriculation, Freshman, Sophomore, Junior, Senior, Graduate level

Other Engineering Support Programs: Support programs at UCSD consist of an undergraduate college system, intensive orientation programs for new students, Education Abroad Program, Office of Academic Support and Instruction Services (OASIS), Career Services, Disabled Student Services, on-campus housing of freshmen living outside 15-mile radius is nearly always guaranteed.

024 University of California, Santa Barbara

■ INSTITUTION PROFILE

HEAD OF THE INSTITUTION
Barbara S. Uehling
Phone: (805)893-2231   Fax: (805)893-4445

GENERAL INFORMATION
[All Students—Fall 1991]
Undergraduate enrollment 16,176
Graduate enrollment 2,343
Total institution enrollment 18,519
Type of institution: Public   **Calendar:** Quarters
Nearest city: Santa Barbara   **Population:** 76,900
Miles from main campus: 12   **Setting:** Small Town
Types of engineering degrees: Engineering
Other degree-granting colleges: Letters and Science, Creative Studies

■ ENGINEERING ADMISSIONS

ADMISSIONS OFFICE CONTACT
William J. Villa
University of California, Santa Barbara
Office of Admissions
Santa Barbara, CA 93106-2014
Phone: (805)893-2881

ENGINEERING COLLEGE ADMISSIONS INFORMATION
Entrance Requirements: High School courses—English-4, U.S. History-1 (5 years), Algebra-2, plane geometry-1, trigonometry-1/2 (3.5 years), Physics or chemistry, preferably both (1 year), Foreign language-2, college-prep. electives-4 (6 years), SAT or ACT, and three College Board Achievement tests: English comp., math (level 1 or 2), and English literature, science, foreign language, or social science. Minimum scores vary with major desired.
Entrance Recommendations: High School courses—Calculus (1 year), Computer programming, College Board Advanced Placement examinations
Requirements for out-of-state residents: Out-of-state resident admission is based solely upon academic criteria, such as grade point average and test scores. Supplemental criteria, such as extracurricular activities and work experience, are not considered.
Requirements for foreign students: TOEFL (Score: 500); financial statement; College Board Achievement tests in English composition, physics and advanced mathematics - minimum 1730 total, minimum 500 each. SAT examinations - minimum 1100 total, minimum 500 each for math and verbal.

DEPARTMENTAL ADMISSIONS INFORMATION
Admission to the engineering department: At the time of admission to the institution.
Additional information: Computer Science department requires a 2.75 (on a 4-point scale) GPA in pre-major courses to advance to full major. In Electrical Engineering advancement to full major is guaranteed if at end of fifth quarter student has a 2.85 overall GPA, is on schedule in pre-major courses, and will have completed at least 90 units at end of sophomore year; other students may petition for advancement to full major.

■ FINANCIAL INFORMATION

ESTIMATED EXPENSES (ACADEMIC YEAR)
[Expenses are for the 1992-93 nine-month academic year.]

State Residents	Undergraduate	Graduate
Tuition and fees	$ 2,983	$ 3,554
College room and board	$ 5,582 [1]	$ 5,582 [1]
Books and supplies	$ 600	$ 600
Other expenses	$ 132 [2]	$ 132 [2]
Total estimated expenses	**$ 9,297**	**$ 9,868**

Out-of-State Residents	Undergraduate	Graduate
Tuition and fees	$10,682	$11,293
College room and board	$ 5,582 [1]	$ 5,582 [1]
Books and supplies	$ 600	$ 600
Other expenses	$ 132 [2]	$ 132 [2]
Total estimated expenses	**$16,996**	**$17,607**

Notes: (1) Room & board rates vary; amount listed is for on-campus dorm.
(2) Nine-month parking permit.

FINANCIAL AID OFFICE CONTACT
Ron Andrade, *Acting Director of Student Financial Services*
University of California, Santa Barbara
Student Financial Services
Santa Barbara, CA 93106-3180
Phone: (805)893-2432

GENERAL FINANCIAL AID INFORMATION
Forms accepted/required: Financial aid & scholarship section of the UC admission application, Student Aid Application for California
Additional financial aid information: Need-based scholarships; merit-based scholarships; grants; loans; work-study funds; part-time jobs on campus.

■ ENGINEERING COLLEGE INFORMATION

[For additional personnel, refer to the *Appendix*.]
HEAD OF ENGINEERING
Venkatesh Narayanamurti
Phone: (805)893-3141   Fax: (805)893-8124
ENGINEERING COLLEGE ADDRESS
College of Engineering
University of California, Santa Barbara
Santa Barbara, CA 93106-5130
Phone: (805)893-3207   Fax: (805)893-8124

ENROLLMENTS—BY CLASS
[Numbers are baccalaureate enrollments for the fall 1991 term, unless otherwise footnoted.]
1st-year students/Freshmen 319
2nd-year students/Sophomores 219
3rd-year students/Juniors 310
4th-year students/Seniors 285
Total **1,133**

NUMBER OF DEGREES AWARDED—BY PROGRAM
[Numbers are engineering baccalaureate degrees awarded during the 1991-1992 academic year, unless otherwise footnoted. For full details about each engineering program, refer to the *Tables of Degree Programs*.]
Chemical Engineering 21
Computer Science 59
Electrical & Computer Engineering 74
Materials Engineering .... — (B)
Mechanical Engineering 78
Nuclear Engineering 10
Total **242**

Notes: (B) Data not applicable.

PERCENTAGE OF DEGREES AWARDED—BY CATEGORY
[Percentages are of all engineering baccalaureate degrees awarded during the 1991-1992 academic year, unless otherwise footnoted.]
African Americans 1.6%
Asian/Pacific Island Americans 18.0%
Hispanic Americans 6.6%
Native Americans 0.4%
Foreign Citizens 9.8%
All Others 63.6%
Women 16.8%
Persons with disabilities .... — % (A)
Students over 25 years of age .... — % (A)

Notes: (A) Data not available.

GRADUATE ENROLLMENTS & DEGREES AWARDED
Master's enrollment 144
Master's degrees awarded 122
Doctoral enrollment 370
Doctoral degrees awarded 44

ENGINEERING STUDENT DATA
Applicants to the engineering college
Number of applicants to engineering college 1,799
Percent offered admission 74.0%
Matriculated engineering students
Percentage in top quartile (25%) of High School class ... 100.0
Average SAT scores: Math—615, Verbal—470, Combined—1085

FULL & PART TIME FACULTY—BY DEPARTMENT
[Figures are the head count of full-time faculty and the full-time equivalent (FTE) of part-time faculty for each engineering department or equivalent.]

Department	Full	Part
Chemical & Nuclear Eng	18	1.3
Electrical & Computer Eng	29	2.4
Mechanical & Environmental Eng	20	1.3
Materials Eng	14	0.2
Computer Science	13	3.3

COLLEGE DESCRIPTION
The college offers four-year undergraduate programs in engr. and computer science, emphasizing a balance between theory and practical, hands-on engr. The number of engr. faculty has almost doubled in recent years. Laboratory and computer facilities are state-of-the art. Areas of specialization include: ocean engr., materials science, thermal sciences, computer engr., signals and systems, microwaves, solid state, information processing and neural networks, nuclear safety, and chemical process design and development. Research centers include those for: Risk Studies & Safety, Computational Sciences and Engr., Control Engr. and Computation, Information Processing, Macromolecular Science and Engr., Compound Semiconductor Research, High Performance Composites, Ocean Engr., Quantized Electronic Structures, Materials, and Optoelectronic Technology. Double major programs are available; a 5-year B.S./M.S. program in EE is offered; a co-op program in computer science is available.

DEGREE PROGRAMS DESCRIPTION
The B.S. degrees in the various engineering programs prepare the graduate for professional positions in industry, business, or government or for graduate study in engineering or other professional schools such as business, law, or medicine. Graduates may be eligible to pursue a teaching credential. Minimum M.S. degree requirements: Chemical/Nuclear, 30 units plus thesis; Computer Science, 42 units of which 8 may be thesis; Electrical & Computer, 42 units of which 8 may be thesis; Mechanical & Environmental, 42 units of which 12 may be thesis. Non-thesis M.S. degrees require 42 units of course work. No specific graduate course requirements for Ph.D.

DUAL DEGREE PROGRAMS
Enrollment requirements: Agreements are only with other colleges at UCSB. Students must meet admission requirements for both majors.

TRANSFER INFORMATION
Residency Requirements: The residence requirement applies to both freshman and transfer students. Students must complete at least 3 quarters in residence, and at least 35 of the final 45 units must be taken in the college in which the degree is to be awarded. A maximum of 105 quarter units can be transferred from a community college.

Transfer via Articulation Agreements
Admission to engineering: Upon completion or near completion of lower division pre-major requirements. Applicants are admitted for fall quarter only.
Requirements: Completion of pre-major requirements including at least one semester of English comp. & 2-3 general education classes; min. 3.00 GPA overall and in pre-major courses. Required courses include (in quarter units): math (analytical geom. & calculus, including differential equations & linear algebra),24; chemistry (for engrs.) w/lab,12; physics (for engrs.) w/lab,16; FORTRAN,3; electric circuits & devices,4; English comp.,4; general education,8.

Transfer without Articulation Agreements
Admission to engineering: Same as for applicants from schools with which we have articulation agreements.
Requirements: Same as for applicants for schools with which we have articulation agreements.

GRADUATION REQUIREMENTS
B.S. programs in engineering require a minimum of 180 quarter units. Must have at least a 2.0 (C) GPA (1) overall; (2) in major courses; and (3) in upper-division major courses.

■ STUDENT PROGRAMS

PROFESSIONAL AND HONORARY SOCIETIES
[For key to acronyms, see *Introduction*.]
Professional Societies: ACM, AIChE, ANS, ASME, IEEE, SAE, Society of Women Engineers
Honorary Societies: Eta Kappa Nu, Pi Tau Sigma, Tau Beta Pi

SUPPORT PROGRAMS
Student Chapter Organizations: Society of Women Engineers, National Society of Black Engineers, Los Ingenieros (Mex.-Amer. Engr. Society & SHPE)

Women's Center
For: Women   **Available:** Year round
Offered: Freshman, Sophomore, Junior, Senior, Graduate level

Minority Engineering Center
For: Ethnic Minorities   **Available:** Year round
Offered: Prematriculation, Freshman, Sophomore, Junior, Senior, Graduate level

Other Engineering Support Programs: Summer orientation programs for all freshman and transfer students; Campus Learning Assistance Services; career and counseling services; tutorial services; services to disabled students; personal counseling; peer counseling; veterans' services; international students' office.

025 University of California, Santa Cruz

INSTITUTION PROFILE

HEAD OF THE INSTITUTION
Patrick E. Mantey
Phone: (408)459-2158   Fax: (408)459-4829

GENERAL INFORMATION
[All Students—Fall 1991]
Undergraduate enrollment 10,000
Graduate enrollment 900
Total institution enrollment **10,900**

Type of institution: Public   **Calendar:** Quarters
Nearest city: San Jose   **Population:** 900,000
Miles from main campus: 25   **Setting:** Small Town
Types of engineering degrees: Engineering

ENGINEERING ADMISSIONS

ADMISSIONS OFFICE CONTACT
Patrick E. Mantey
University of California, Santa Cruz
225 Applied Sciences Building
Santa Cruz, CA 95064
Phone: (408)459-2320   Fax: (408)459-4829

ENGINEERING COLLEGE ADMISSIONS INFORMATION
Admission to the engineering college: Students may be admitted to major at the time they apply to UCSC. Others may apply on quarterly basis.
Entrance Requirements: SAT, Students are required to take the SAT exam. UCSC does not have a minimum score required for admissions.
Entrance Recommendations: High School courses—4 yrs of math (incl. advanced alg. and trig) (4 years), Chemistry (1 year), Physics (1 year), College Entrance Examination Board's (CEEB) advanced placement examinations in computer science, calculus, chemistry, and English as these examinations may satisfy university requirements.
Requirements for foreign students: TOEFL (Score: 550)

DEPARTMENTAL ADMISSIONS INFORMATION
Admission to the engineering department: Students may be accepted to the major on admission to UCSC and others have the option of applying during the quarter.
Additional information: It is recommended that students take courses in the major to identify their potential within the major. Once they have consulted with the Undergraduate Director and expressed an interest they can formally apply.

FINANCIAL INFORMATION

ESTIMATED EXPENSES (ACADEMIC YEAR)
[Expenses are for the 1992-93 nine-month academic year.]

State Residents	Undergraduate	Graduate
Tuition and fees	$ 3,129	$ 3,620[1]
College room and board	$ 5,931	$ 6,891[2]
Books and supplies	$ 594	$ 744
Other expenses	$ 1,569	$ 492[3]
Total estimated expenses	**$11,223**	**$11,747**

Out-of-State Residents	Undergraduate	Graduate
Tuition and fees	$10,828	$11,309
College room and board	$ 5,931	$ 6,891[4]
Books and supplies	$ 594	$ 744
Other expenses	$ 1,569	$ 492[5]
Total estimated expenses	**$18,922**	**$19,436**

Notes: (1) Fees only. (2) On-campus housing. (3) Transportation. (4) On-campus housing. (5) Transportation.

FINANCIAL AID OFFICE CONTACT
Esperanza L. Nee, *Director of Financial Aid*
University of California, Santa Cruz
201 Hahn Student Services
Santa Cruz, CA 95064
Phone: (408)459-2963

GENERAL FINANCIAL AID INFORMATION
Forms accepted/required: AFSA, CSS-FAF, FAF, FAT, FAR, FSA, Stafford, IRS, SAR, Supplemental
Additional financial aid information: Contact Financial Aid for complete information: (408) 459-2963 Grants, Work-Study Program, Scholarships, Loans

ENGINEERING COLLEGE INFORMATION
[For additional personnel, refer to the *Appendix*.]

HEAD OF ENGINEERING
Patrick E. Mantey
Phone: (408)459-2158   Fax: (408)459-4829
EMail: mantey@cse.ucsc.edu

ENGINEERING COLLEGE ADDRESS
Computer Engineering
University of California, Santa Cruz
1156 High Street
Santa Cruz, CA 95064
Phone: (408)459-4008

ENROLLMENTS—BY CLASS
[Numbers are baccalaureate enrollments for the fall 1991 term, unless otherwise footnoted.]
1st-year students/Freshmen 43
2nd-year students/Sophomores 22
3rd-year students/Juniors 36
4th-year students/Seniors 41
Total **142**

NUMBER OF DEGREES AWARDED—BY PROGRAM
[Numbers are engineering baccalaureate degrees awarded during the 1991-1992 academic year, unless otherwise footnoted. For full details about each engineering program, refer to the *Tables of Degree Programs*.]
Computer Engineering 18
Total **18**

PERCENTAGE OF DEGREES AWARDED—BY CATEGORY
[Percentages are of all engineering baccalaureate degrees awarded during the 1991-1992 academic year, unless otherwise footnoted.]
African Americans .... 9.0 %
Asian/Pacific Island Americans .... 31.8 %
Hispanic Americans .... 4.5 %
Native Americans .... – %
Foreign Citizens .... – %
All Others .... 54.7 %
Women .... 9.0 %
Persons with disabilities .... – %
Students over 25 years of age .... – %

GRADUATE ENROLLMENTS & DEGREES AWARDED
Master's enrollment 20
Master's degrees awarded 11
Doctoral enrollment 29
Doctoral degrees awarded 2

ENGINEERING STUDENT DATA
Applicants to the engineering college
Number of applicants to engineering college 232
Percent offered admission 80.0%

Matriculated engineering students
Percentage in top quartile (25%) of High School class ... 100.0
Average SAT scores: Math—550, Verbal—431, Combined—980

FULL & PART TIME FACULTY—BY DEPARTMENT
[Figures are the head count of full-time faculty and the full-time equivalent (FTE) of part-time faculty for each engineering department or equivalent.]

Department	Full	Part
Computer Eng	11	3.0

COLLEGE DESCRIPTION
This rigorous program in computer engineering offers students exciting possibilities, in part because the campus is so near the electronics industry of Silicon Valley in Santa Clara County. The CE curriculum follows the model of the Computer Society of the Institute of Electrical and Electronics Engineers. The major in computer engineering provides students with excellent training for employment in engineering and a strong background for pursuit of advanced degrees in their chosen areas of specialization. The graduate program in computer engineering offers both the M.S. and Ph.D. degrees.

DEGREE PROGRAMS DESCRIPTION
Undergraduate education at Santa Cruz is organized on a collegiate basis. Eight colleges divide a large university into small academic and social communities. Each college has a distinctive quality derived from its core course and extracurricular programs, its faculty and their academic disciplines, and its architectural style. The small size of the graduate programs at Santa Cruz fosters close working relationships between students and faculty and a high level of interaction among students. Students are encouraged to do independent research and students and faculty from different fields often pursue interdisciplinary research interests. The Santa Cruz campus operates on a quarter-based academic year and uses a narrative evaluation system to record academic performance.

DUAL DEGREE PROGRAMS
Engineering baccalaureate graduates with dual degrees 7
Enrollment requirements: Strong performance in high school algebra, geometry, trigonometry, physics, and chemistry. Excellent SAT and Achievement scores.

TRANSFER INFORMATION
Transfer via Articulation Agreements
Admission to engineering: At time of admission and at any time during quarterly admissions screening.
Engineering graduates transferred from .. 4-yr: –   2-yr: –
Requirements: Junior transfer students are expected to have completed the equivalents of the following courses from the lower-division requirements of this program: Math 11A-B-C, Physics 5A-B-C, 5L-M-N, Chemistry 3A, 3L, and CIS 12A.

Transfer without Articulation Agreements
Admission to engineering: During review process that takes place three times per year.
Engineering graduates transferred from .. 4-yr: 100   2-yr: –
Requirements: The Undergraduate Director reviews a students academic record of the courses they've taken at UCSC. Based on this review a student is admitted to the major or is given some goals to obtain prior to being admitted.

GRADUATION REQUIREMENTS
Students must either complete a senior thesis, score 620 or better on the Computer Science Advanced Graduate Record Examination, score 620 or better on the Engineering Advanced GRE, or take courses 120, 121, 185, and an approved upper-division computer engineering or computer and information sciences course containing a substantial project.

SUPPORT PROGRAMS
Other Engineering Support Programs: Quarterly orientations for all majors. Majors are assigned a faculty adviser within the department. Independent studies and Field studies are available to students to work with a faculty member on a particular project and receive credit.

026 California Institute of Technology

INSTITUTION PROFILE

HEAD OF THE INSTITUTION
Thomas E. Everhart
Phone: (818)356-6301 **Fax:** (818)449-9374

GENERAL INFORMATION
[All Students—Fall 1991]
Undergraduate enrollment 862
Graduate enrollment 1,081
Total institution enrollment 1,943
Type of institution: Private **Calendar:** Quarters
Location: Pasadena **Population:** 125,000
Setting: Urban
Types of engineering degrees: Engineering

ENGINEERING ADMISSIONS

ADMISSIONS OFFICE CONTACT
Carol L. Snow
California Institute of Technology
Admissions Office 1-63
Pasadena, CA 91125
Phone: (818)356-6341 **Fax:** (818)564-8136

ENGINEERING COLLEGE ADMISSIONS INFORMATION
Entrance Requirements: SAT (Score: 1350), High School courses—Chemistry (1 year), English (3 years), Mathematics (4 years), Physics (1 year), Math Level II, English Composition, 1 in biology, chemistry or physics.
Entrance Recommendations: SAT (Score: 1350), High School courses—English (4 years)
Requirements for foreign students: TOEFL (Score: 550)

DEPARTMENTAL ADMISSIONS INFORMATION
Admission to the engineering department: At the end of the first year.

FINANCIAL INFORMATION

ESTIMATED EXPENSES (ACADEMIC YEAR)
[Expenses are for the 1992-93 nine-month academic year.]

All Students	Undergraduate	Graduate
Tuition and fees	$15,160	$14,984
College room and board	$ 5,978	$ 5,697
Books and supplies	$ 660	$ 825
Other expenses	$ 1,275 [1]	$ 4,836 [1]
Total estimated expenses	**$23,073**	**$26,342**

Notes: (1) Expenses include transportation and personal expenses.

FINANCIAL AID OFFICE CONTACT
David J. Levy, *Director of Financial Aid*
California Institute of Technology
Financial Aid Office 12-63
Pasadena, CA 91125
Phone: (818)356-6280

GENERAL FINANCIAL AID INFORMATION
Forms accepted/required: CSS-FAF, FAT, IRS, SAR
Additional financial aid information: Caltech is committed to meeting the full, demonstrated financial need of its students through a variety of grants, scholarships, fellowships, loans, and campus employment opportunities. More information is available in the Caltech catalog and from the Financial Aid Office.

ENGINEERING COLLEGE INFORMATION

[For additional personnel, refer to the *Appendix*.]

HEAD OF ENGINEERING
John H. Seinfeld
Phone: (818)356-4100 **Fax:** (818)585-1729
EMail: john_seinfeld@starbase1.caltech.edu

ENGINEERING COLLEGE ADDRESS
Division of Engineering and Applied Science
California Institute of Technology
1201 E. California Blvd.
Pasadena, CA 91125
Phone: (818)356-4101 **Fax:** (818)585-1729

ENROLLMENTS—BY CLASS
[Numbers are baccalaureate enrollments for the fall 1991 term, unless otherwise footnoted.]
1st-year students/Freshmen .... (1)
2nd-year students/Sophomores 83
3rd-year students/Juniors 127
4th-year students/Seniors 110
Total **320**

Notes: (1) Majors are chosen in sophomore year.

NUMBER OF DEGREES AWARDED—BY PROGRAM
[Numbers are engineering baccalaureate degrees awarded during the 1991-1992 academic year, unless otherwise footnoted. For full details about each engineering program, refer to the *Tables of Degree Programs*.]

Aeronautics .... –
Applied Mathematics 2
Applied Mechanics .... –
Applied Physics 15
Chemical Engineering 7
Civil Engineering .... –
Computation & Neural Systems .... –
Computer Science .... –
Electrical Engineering 29
Engineering & Applied Science 68
Engineering Science(s) .... –
Environmental Engineering Science .... –
Materials Science .... –
Mechanical Engineering .... –
Total **121**

PERCENTAGE OF DEGREES AWARDED—BY CATEGORY
[Percentages are of all engineering baccalaureate degrees awarded during the 1991-1992 academic year, unless otherwise footnoted.]
African Americans 1.0%
Asian/Pacific Island Americans 23.0%
Hispanic Americans 1.0%
Native Americans 1.0%
Foreign Citizens 12.0%
All Others 62.0%
Women 13.0%
Persons with disabilities .... – %
Students over 25 years of age .... – %

GRADUATE ENROLLMENTS & DEGREES AWARDED
Master's enrollment 79
Master's degrees awarded 82
Doctoral enrollment 343
Doctoral degrees awarded 51

ENGINEERING STUDENT DATA

Applicants to the engineering college
Number of applicants to engineering college 929
Percent offered admission 26.0%

Matriculated engineering students
Percentage in top quartile (25%) of High School class ... 100.0
Average SAT scores: Math—760, Verbal—660, Combined—1420

FULL & PART TIME FACULTY—BY DEPARTMENT
[Figures are the head count of full-time faculty and the full-time equivalent (FTE) of part-time faculty for each engineering department or equivalent.]

Department	Full	Part
Chemical Eng	8	–
Civil Eng	6	–
Electrical Eng	15	–
Mechanical Eng	11	–
Applied Mathematics	6	–
Environmental Eng Science	7	–
Graduate Aeronautical Laboratories	10	–
Applied Mechanics	6	–
Applied Physics	10	–
Eng Science	1	–
Materials Science	3	–
Computer Science	10	–
Computation & Neural Systems	1	–

COLLEGE DESCRIPTION
The first year of undergraduate study is the same for all students. At the end of the first year students select an option. Each student is assigned a faculty adviser from a field representing the student's professional interest. In conference with his or her adviser, the student then develops a program of study for the next three years. The program includes Institute-wide requirements in physics, mathematics, and humanities, as well as an additional year of advanced mathematics. Additional requirements vary in the five undergraduate options: Appl. Math., Appl. Phys., Chem. Eng., Elec. Eng., and Eng. Appl. Science. A student may change his or her major interest at any time. Many students decide to prepare for graduate study in a branch of engineering or applied science; others choose to enter professional employment directly after graduation. Undergraduates are encouraged to participate in research activities, an important aspect of the faculty's work, insofar as time per

DEGREE PROGRAMS DESCRIPTION
The BS degree programs in the various fields of engineering and the applied sciences prepare the graduating student for positions in the fields indicated or for graduate study. Programs of graduate study are available within selected areas of the general fields listed. The engineering disciplines offer both MS programs that can be completed within one year and which do not require a thesis and PhD programs. Although Masters Degrees can be earned, the programs in the applied sciences emphasize the PhD program.

DUAL DEGREE PROGRAMS
Enrollment requirements: Students need to be referred by a 3/2 affiliated college.

TRANSFER INFORMATION
Transfer via Articulation Agreements
Admission to engineering: At the end of the freshman year
Engineering graduates transferred from ... 4-yr: 23 2-yr: 1
Requirements: Transfer exams in math, physics and chemistry.

GRADUATION REQUIREMENTS
To qualify for a Bachelor of Science degree a student must obtain passing grades in the courses comprising the Institute requirements, must satisfy the requirements listed under the undergraduate options, and must achieve a grade-point average of not less than 1.9.

STUDENT PROGRAMS

PROFESSIONAL AND HONORARY SOCIETIES
[For key to acronyms, see *Introduction*.]
Professional Societies: AIChE, ASCE, ASME, SIAM
Honorary Societies: Tau Beta Pi

SUPPORT PROGRAMS
Student Chapter Organizations: Society of Women Engineers, National Society of Black Engineers, Society of Hispanic Professional Engineers

Bridge Program
For: Ethnic Minorities **Available:** Summer only
Offered: Prematriculation

Other Engineering Support Programs: SURF (Summer Undergraduate Research Fellowship)

027 California Polytechnic State University

INSTITUTION PROFILE

HEAD OF THE INSTITUTION
Warren J. Baker
Phone: (805)756-6000   Fax: (805)756-1129

GENERAL INFORMATION
[All Students—Fall 1991]
Undergraduate enrollment 13,944
Graduate enrollment 1,153
Total institution enrollment **15,097**
Type of institution: Public   **Calendar:** Quarters
Nearest city: San Francisco   **Population:** 750,000
Miles from main campus: 250   **Setting:** Small Town
Types of engineering degrees: Engineering
Other degree-granting colleges: Business Administration, Agriculture, Liberal Arts, Science & Mathematics, Architecture & Environmental Design

ENGINEERING COLLEGE ADMISSIONS INFORMATION
Entrance Requirements: SAT, High School courses—English (4 years), Mathematics (3 years), US History or US History & Government (1 year), Science with Laboratory (1 year), Index combines SAT & GPA for variable minimum.
Entrance Recommendations: High School courses—Foreign Language (Required) (2 years), Visual & Performing Arts (Required) (1 year), Electives (Required to bring total to 15 unit-years) (3 years)
Requirements for out-of-state residents: 5% Multi Criteria Admission score penalty.
Requirements for foreign students: TOEFL (Score: 500); financial statement

DEPARTMENTAL ADMISSIONS INFORMATION
Admission to the engineering department: At the time of admission to the institution.

FINANCIAL INFORMATION

ESTIMATED EXPENSES (ACADEMIC YEAR)
[Expenses are for the 1992-93 nine-month academic year.]

State Residents	Undergraduate	Graduate
Tuition and fees	$ 2,211	$ 2,211
College room and board	$ 2,860	$ 2,860
Books and supplies	$ 700	$ 700
Other expenses	$ –	$ –
Total estimated expenses	**$ 5,771**	**$ 5,771**

Out-of-State Residents	Undergraduate	Graduate
Tuition and fees	$ 9,400	$ 7,800
College room and board	$ 2,860	$ 2,860
Books and supplies	$ 700	$ 700
Other expenses	$ –	$ –
Total estimated expenses	**$12,960**	**$11,360**

FINANCIAL AID OFFICE CONTACT
Diane L. Ryan, *Director of Financial Aid*
California Polytechnic State University
01-213 Administration
San Luis Obispo, CA 93407
Phone: (805)756-2927

GENERAL FINANCIAL AID INFORMATION
Forms accepted/required: Student Aid Application for California(SAAC)
Additional financial aid information: See Catalog

ENGINEERING COLLEGE INFORMATION
[For additional personnel, refer to the *Appendix*.]

HEAD OF ENGINEERING
Peter Y. Lee
Phone: (805)756-2131   Fax: (805)756-6503
EMail: plee@oasis.calpoly.edu

ENGINEERING COLLEGE ADDRESS
College of Engineering
California Polytechnic State University
13-266
San Luis Obispo, CA 93407
Phone: (805)756-2131   Fax: (805)756-6503

ENROLLMENTS—BY CLASS
[Numbers are baccalaureate enrollments for the fall 1991 term, unless otherwise footnoted.]
1st-year students/Freshmen 744
2nd-year students/Sophomores 598
3rd-year students/Juniors 699
4th-year students/Seniors 1,999
Total **4,040**

NUMBER OF DEGREES AWARDED—BY PROGRAM
[Numbers are engineering baccalaureate degrees awarded during the 1991-1992 academic year, unless otherwise footnoted. For full details about each engineering program, refer to the *Tables of Degree Programs*.]

Aeronautical Engineering 49
Agricultural Engineering 24
Architectural Engineering 50
Civil & Environmental Engineering 110
Computer Engineering 19
Electrical & Electronics Engineering 32
Engineering Science(s) 7
Industrial Engineering 44
Materials Engineering 17
Mechanical Engineering 146
Total **498**

PERCENTAGE OF DEGREES AWARDED—BY CATEGORY
[Percentages are of all engineering baccalaureate degrees awarded during the 1991-1992 academic year, unless otherwise footnoted.]
African Americans .... 1.0 %
Asian/Pacific Island Americans .... 17.0 %
Hispanic Americans .... 7.0 %
Native Americans .... 1.0 %
Foreign Citizens .... – %
All Others .... 74.0 %
Women .... 13.0 %
Persons with disabilities .... – %
Students over 25 years of age .... – %

GRADUATE ENROLLMENTS & DEGREES AWARDED
Master's enrollment 209
Master's degrees awarded 40
Doctoral enrollment .... –
Doctoral degrees awarded .... –

ENGINEERING STUDENT DATA
Applicants to the engineering college
Number of applicants to engineering college 3,255
Percent offered admission 45.0%
Average SAT scores: Math—553, Verbal—680, Combined—1233

FULL & PART TIME FACULTY—BY DEPARTMENT
[Figures are the head count of full-time faculty and the full-time equivalent (FTE) of part-time faculty for each engineering department or equivalent.]

Department	Full	Part
Civil & Environmental Eng	13	5.0
Electronic & Electrical Eng	34	9.0
Industrial Eng	14	5.0
Mechanical Eng	36	2.0
Aeronautical Eng	6	3.0
Computer Eng	5	–
Eng Science	–	–
Materials Eng	7	–
Agricultural Eng	13	2.0
Architectural Eng	9	1.0

COLLEGE DESCRIPTION
The College of Engineering offers programs leading to the Bachelor of Science in ten engineering disciplines and in Computer Science. Bachelor of Science degrees may also be earned in Agricultural Engineering and Architectural Engineering, through their respective colleges. Programs leading to the Master of Science degree are available in Aeronautical Engineering, Civil and Environmental Engineering, Mechanical Engineering, Computer Science and Engineering with specialties in Industrial Engineering and Materials Engineering. A joint MBA/MS Engineering degree program is available with a specialization in Engineering Management. There is a Co-op program and senior projects are required for graduation. Students must declare a major on application for admission. Major changes within the College of Engineering are not difficult to obtain.

DEGREE PROGRAMS DESCRIPTION
The engineering curricula are designed to give the student a practical appreciation of the applied facets of engineering science across discipline lines along with a thorough understanding of the underlying engineering principles. Cal Poly is known for its 'hands on' approach to engineering education - involving substantial laboratory exercises. There are abundant Co-op educational opportunities and students are encouraged to consider Co-op as an integral part of their educational experience at Cal Poly.

TRANSFER INFORMATION
Residency Requirements: 50 quarter units minimum and @ least 30 units of last 40 units counted toward the degree.

Transfer via Articulation Agreements
Admission to engineering: At any time
Requirements: Engineering Calculus Sequence through differential equations, Physics, lower division engineering science and some general education.

Transfer without Articulation Agreements
Admission to engineering: At any time
Requirements: 3.0 Minimum GPA and one quarter @ Cal Poly.

GRADUATION REQUIREMENTS
Major GPA 2.0; Overall GPA 2.0; 30 of last 40 units counting towards degree requirements must be taken in residence. Senior Project required.

STUDENT PROGRAMS

PROFESSIONAL AND HONORARY SOCIETIES
[For key to acronyms, see *Introduction*.]
Professional Societies: ACM, AHS, AIAA, ASAE, ASCE, ASME, IEEE, IIE, NSAE, ORSA
Honorary Societies: Alpha Pi Mu, Chi Epsilon, Pi Tau Sigma, Tau Beta Pi

SUPPORT PROGRAMS
Student Chapter Organizations: Society of Women Engineers, National Society of Black Engineers, Society of Hispanic Professional Engineers, American Indian Engineering Society

Minority Engineering
For: Ethnic Minorities   **Available:** Year round
Offered: Prematriculation, Freshman, Sophomore, Junior, Senior, Graduate level

Women in Engineering
For: Women   **Available:** Year round
Offered: Prematriculation, Freshman, Sophomore, Junior, Senior, Graduate level

Other Engineering Support Programs: Guaranteed Admissions Program With California Community Colleges.

028 California State University, Chico

INSTITUTION PROFILE
HEAD OF THE INSTITUTION
Robin S. Wilson
Phone: (916)898-5201 **Fax:** (916)898-5077

GENERAL INFORMATION
[All Students—Fall 1991]
Undergraduate enrollment 13,458
Graduate enrollment 1,714
Total institution enrollment 15,172
Type of institution: Public **Calendar:** Semesters
Nearest city: Sacramento **Population:** 300,000
Miles from main campus: 90 **Setting:** Small Town
Types of engineering degrees: Engineering
Other degree-granting colleges: Agricultural & Environmental, Business Administration, Communication & Journalism, Education, Natural Sciences, Behavioral and Social Sciences, Humanities and Fine Arts

ENGINEERING ADMISSIONS
ADMISSIONS OFFICE CONTACT
Kenneth C. Edson
California State University, Chico
First and Normal Streets
Chico, CA 95929-0720
Phone: (916)898-6321 **Fax:** (916)898-6824

ENGINEERING COLLEGE ADMISSIONS INFORMATION
Entrance Requirements: High School courses—English (4 years), Mathematics through intermediate algebra (3 years), Science with lab (1 year), Foreign language (2 years), SAT or ACT is required if high school GPA is less than 3.00
Entrance Recommendations: SAT, ACT, High School courses—Physics (1 year), Chemistry (1 year), Mechanical drawing (1 year), Trigonometry & calculus (1 year), Minimum score depends on high school GPA. See catalog and eligibility index.
Requirements for out-of-state residents: Requirements are higher for out-of-state students. See catalog.
Requirements for foreign students: TOEFL (Score: 500); financial statement; 3.0 GPA is required in secondary school.

DEPARTMENTAL ADMISSIONS INFORMATION
Admission to the engineering department: At the time of admission to the institution.

FINANCIAL INFORMATION
ESTIMATED EXPENSES (ACADEMIC YEAR)
[Expenses are for the 1992-93 nine-month academic year.]

State Residents	Undergraduate	Graduate
Tuition and fees	$ 1,096	$ 1,096
College room and board	$ 4,274	$ 4,274
Books and supplies	$ 558	$ 558
Other expenses	$ –	$ –
Total estimated expenses	**$ 5,928**	**$ 5,928**

Out-of-State Residents	Undergraduate	Graduate
Tuition and fees	$ 8,476 [1]	$ 8,476 [1]
College room and board	$ 4,274	$ 4,274
Books and supplies	$ 558	$ 558
Other expenses	$ –	$ –
Total estimated expenses	**$13,308**	**$13,308**

Notes: (1) Includes fee of $246 per unit.

FINANCIAL AID OFFICE CONTACT
David Cook, *Financial Aids Officer*
California State University, Chico
First and Normal Streets
Chico, CA 95929-0705
Phone: (916)898-5065 **Fax:** (916)898-6824

GENERAL FINANCIAL AID INFORMATION
Forms accepted/required: Student Aid Application for California (SAAC)
Additional financial aid information: Scholarships and awards; need-based scholarships; student loans; work-study; program scholarships; EOP grants; Minority Engineering Program scholarships; Cal grants; Pell grants.

ENGINEERING COLLEGE INFORMATION
[For additional personnel, refer to the *Appendix*.]
HEAD OF ENGINEERING
Gary Z. Watters
Phone: (916)898-5963 **Fax:** (916)898-5995
EMail: GWATTERS@OAVAXCSUCHICO.EDU

ENGINEERING COLLEGE ADDRESS
College of Engineering, Computer Science, and Technology
California State University, Chico
First and Normal Streets
Chico, CA 95929-0003
Phone: (916)898-5963 **Fax:** (916)898-5995

NUMBER OF DEGREES AWARDED—BY PROGRAM
[Numbers are engineering baccalaureate degrees awarded during the 1991-1992 academic year, unless otherwise footnoted. For full details about each engineering program, refer to the *Tables of Degree Programs*.]
Civil Engineering 34
Computer Engineering 34
Electrical & Electronics Engineering 35
Mechanical Engineering 40
Mechanical Engineering & Manufacturing 40
Total 183

PERCENTAGE OF DEGREES AWARDED—BY CATEGORY
[Percentages are of all engineering baccalaureate degrees awarded during the 1991-1992 academic year, unless otherwise footnoted.]
African Americans .... – %
Asian/Pacific Island Americans .... – %
Hispanic Americans .... – %
Native Americans .... – %
Foreign Citizens .... – %
All Others .... 100 %
Women .... – %
Persons with disabilities .... – %
Students over 25 years of age .... – %

FULL & PART TIME FACULTY—BY DEPARTMENT
[Figures are the head count of full-time faculty and the full-time equivalent (FTE) of part-time faculty for each engineering department or equivalent.]

Department	Full	Part
Civil Eng	9	–
Electrical/Electronic Eng	8	1.5
Mechanical Eng & Manufacturing	7	1.0
Computer Science & Eng	5	0.3

COLLEGE DESCRIPTION
The College offers a four-year undergraduate program in engineering in four majors. Minors are available in computer science, business, math, physics, and chemistry but no minor is required for graduation. Strong preparation for professional practice as well as graduate education is emphasized. Classes are taught almost exclusively by full-time faculty, many of whom have had significant industrial experience. Undergraduate education is considered important and teachers spend a great deal of time with students. Classes are small and faculty advisers are accessible. Co-op experiences in industry are encouraged and the campus has a co-op office to assist in placement. Student professional organizations have won regional and national recognition.

DEGREE PROGRAMS DESCRIPTION
The BS degrees in engineering prepare students for entry-level positions in engineering practice as well as for graduate school. The programs are practice-oriented and laboratory-intensive with a heavy emphasis on computer applications. The MS programs are offered in Electrical and Mechanical Engineering. These programs also have a strong professional-practice orientation.

TRANSFER INFORMATION
Residency Requirements: Students must complete a minimum of 30 units on the Chico campus, 24 of which must be upper division, 12 of which must be in the major, and 9 of which must be upper division General Education.

Transfer via Articulation Agreements
Admission to engineering: At any time
Engineering graduates transferred from ... 4-yr: 13 2-yr: 82
Requirements: For California residents, good standing at the last university attended and a 2.0 GPA. For non-resident transfers, a 2.4 GPA is required.

Transfer without Articulation Agreements
Admission to engineering: At any time
Engineering graduates transferred from ... 4-yr: – 2-yr: –
Requirements: Same as for articulation agreements.

GRADUATION REQUIREMENTS
B.S. programs in engineering require 132 semester units. Must have 2.00 grade point average in all college courses attempted, all courses at Chico, all courses required for the major, and all department courses. Tests on writing proficiency required before graduation. General Education requirements modified to accommodate high-unit engineering major requirements.

STUDENT PROGRAMS
PROFESSIONAL AND HONORARY SOCIETIES
[For key to acronyms, see *Introduction*.]
Professional Societies: ACM, ASCE, ASME, IEEE, SAMPE, SME, SPE*
Honorary Societies: Eta Kappa Nu, Tau Beta Pi

SUPPORT PROGRAMS
Student Chapter Organizations: Society of Women Engineers, National Society of Black Engineers, Society of Hispanic Professional Engineers, American Indian Science and Engineering Society, Latinos in Technical Careers

Minority Engineering Program
For: Ethnic Minorities **Available:** Year round
Offered: Prematriculation, Freshman, Sophomore, Junior, Senior
Other Engineering Support Programs: Advising and orientation, cooperative education and internships, career counseling, disabled student services, international student office, placement office, veteran's services, math, physics, and chemistry tutoring, student learning center, student writing center, study skills tutoring.

029 California State University, Fresno

INSTITUTION PROFILE
HEAD OF THE INSTITUTION
John D. Welty
Phone: (209)278-2324 Fax: (209)278-4715

GENERAL INFORMATION
[All Students—Fall 1991]
Undergraduate enrollment 15,279
Graduate enrollment 3,623
Total institution enrollment **18,902**
Type of institution: Public Calendar: Semesters
Location: Fresno Population: 350,000
Setting: Urban
Types of engineering degrees: Engineering
Other degree-granting colleges: Business Administration, Education, Natural Sciences, Humanities and Arts, Education, Agricultural Sciences and Technology, Social Sciences

ENGINEERING ADMISSIONS
ADMISSIONS OFFICE CONTACT
Richard Backer
California State University, Fresno
Director of Admissions
5150 N. Maple
Fresno, CA 93740-0057
Phone: (209)278-2191 Fax: (209)278-4812

ENGINEERING COLLEGE ADMISSIONS INFORMATION
Entrance Requirements: SAT, ACT, High School courses—SAT & ACT scores are combined with students high school GPA to determine eligibility index.
Requirements for foreign students: TOEFL (Score: 500)

DEPARTMENTAL ADMISSIONS INFORMATION
Admission to the engineering department: At the time of admission to the institution.

FINANCIAL INFORMATION
ESTIMATED EXPENSES (ACADEMIC YEAR)
[Expenses are for the 1992-93 nine-month academic year.]

State Residents	Undergraduate	Graduate
Tuition and fees	$ 1,486	$ 1,486
College room and board	$ 4,073	$ 4,073
Books and supplies	$ 375	$ 375
Other expenses	$ –	$ –
Total estimated expenses	**$ 5,934**	**$ 5,934**
Out-of-State Residents	Undergraduate	Graduate
Tuition and fees	$ 7,390	$ 7,390 [1]
College room and board	$ 4,073	$ 4,073
Books and supplies	$ 375	$ 375
Other expenses	$ –	$ –
Total estimated expenses	**$11,838**	**$11,838**

Notes: (1) Total fees for student carrying 12 units

FINANCIAL AID OFFICE CONTACT
Joseph W. Heuston, *Director of Financial Aid*
California State University, Fresno
5150 N Maple Ave MS#64
Financial Aid Office
Fresno, CA 93740
Phone: (209)278-6563 Fax: (209)278-7044

GENERAL FINANCIAL AID INFORMATION
Forms accepted/required: AFSA, FAT, Stafford, IRS, SAR, Supplemental, CSU, Fresno Scholarship Application
Additional financial aid information: GRANTS: Pell, BIA, Educational Opportunity, California State University, Cal Grants, State Graduate Fellowships, Bilingual Teacher, Graduate Equity Fellowships FEDERAL LOANS: NDSL, Stafford, SLS, PLUS, Nursing WORK: Federal Work-Study, Campus Student Assistant SCHOLARSHIPS: CSU-Fresno, National Hispanic, Young Black Scholars, Paul Douglas, Other

ENGINEERING COLLEGE INFORMATION
[For additional personnel, refer to the *Appendix*.]
HEAD OF ENGINEERING
Elden K. Shaw
Phone: (209)278-2500 Fax: (209)278-7071

ENGINEERING COLLEGE ADDRESS
School of Engineering
California State University, Fresno
2320 East San Ramon
Fresno, CA 93740-0094
Phone: (209)278-2500 Fax: (209)278-4812

ENROLLMENTS—BY CLASS
[Numbers are baccalaureate enrollments for the fall 1991 term, unless otherwise footnoted.]
1st-year students/Freshmen .... (A)
2nd-year students/Sophomores .... (A)
3rd-year students/Juniors .... (A)
4th-year students/Seniors 1,471 (2)
Total **1,471**

Notes: (A) Data not available. (2) Total number of majors is 1471

NUMBER OF DEGREES AWARDED—BY PROGRAM
[Numbers are engineering baccalaureate degrees awarded during the 1991-1992 academic year, unless otherwise footnoted. For full details about each engineering program, refer to the *Tables of Degree Programs*.]
Civil Engineering 35
Computer Engineering .... –
Computer Science 45
Construction Management 40
Electrical Engineering 76
Industrial Engineering 10
Mechanical Engineering 42
Surveying Engineering 25
Total **273**

PERCENTAGE OF DEGREES AWARDED—BY CATEGORY
[Percentages are of all engineering baccalaureate degrees awarded during the 1991-1992 academic year, unless otherwise footnoted.]
African Americans 2.7%
Asian/Pacific Island Americans 20.0%
Hispanic Americans 25.5%
Native Americans 1.1%
Foreign Citizens 11.5%
All Others 39.2%
Women 8.0%
Persons with disabilities .... – % (A)
Students over 25 years of age 26.5%

Notes: (A) Data not available.

GRADUATE ENROLLMENTS & DEGREES AWARDED
Master's enrollment 100
Master's degrees awarded 8
Doctoral enrollment .... – (A)
Doctoral degrees awarded .... – (A)

Notes: (A) Data not available.

ENGINEERING STUDENT DATA
Applicants to the engineering college
Number of applicants to engineering college 996
Percent offered admission 69.0%

Matriculated engineering students
Percentage in top quartile (25%) of High School class .... – (A)

Notes: (A) Data not available.

FULL & PART TIME FACULTY—BY DEPARTMENT
[Figures are the head count of full-time faculty and the full-time equivalent (FTE) of part-time faculty for each engineering department or equivalent.]

Department	Full	Part
Civil & Surveying Eng	15	–
Electrical & Computer Eng	10	–
Mechanical & Industrial Eng	10	–
Computer Science	7	–

COLLEGE DESCRIPTION
Strong emphasis on undergraduate education.

TRANSFER INFORMATION
Transfer via Articulation Agreements
Admission to engineering: At any time

Transfer without Articulation Agreements
Admission to engineering: At any time
Requirements: 2.0 GPA

STUDENT PROGRAMS
PROFESSIONAL AND HONORARY SOCIETIES
[For key to acronyms, see *Introduction*.]
Professional Societies: ACI, ACM, ACSM, AGCA, ASCE, ASME, ASPRS, IEEE, IIE, California Land Surveyors Association (CLSA), American Public Works Association (APWA)
Honorary Societies: Alpha Pi Mu, Eta Kappa Nu, Tau Beta Pi

SUPPORT PROGRAMS
Student Chapter Organizations: Society of Women Engineers, National Society of Black Engineers, Society of Hispanic Professional Engineers

Minority Engineering Program
For: Ethnic Minorities **Available:** Year round
Offered: Freshman, Sophomore, Junior, Senior, Graduate level

Mathematics Engineering Science Achievement
For: Ethnic Minorities **Available:** Year round
Offered: Prematriculation

030 California State University, Long Beach

INSTITUTION PROFILE

HEAD OF THE INSTITUTION
Curtis L. McCray
Phone: (310)985-4121 Fax: (310)985-5584

GENERAL INFORMATION
[All Students—Fall 1991]
Undergraduate enrollment 23,584
Graduate enrollment 6,487
Total institution enrollment **30,071**
Type of institution: Public Calendar: Semesters
Nearest city: Long Beach Population: 442,000
Miles from main campus: 1 Setting: Urban
Types of engineering degrees: Engineering & Technology
Other degree-granting colleges: Business Administration, Education, Applied Arts & Science, Arts, Liberal Arts, Natural Sciences

ENGINEERING ADMISSIONS

ADMISSIONS OFFICE CONTACT
Fay Denny
California State University, Long Beach
1250 Bellflower Boulevard
Long Beach, CA 90840-0106
Phone: (310)985-5471 Fax: (310)985-8887

ENGINEERING COLLEGE ADMISSIONS INFORMATION
Entrance Requirements: High School courses—English (4 years), Mathematics (3 years), U.S. History (1 year), Foreign Language (2 years), All freshman with an entering GPA below 3.0 (3.6 for non-resident) and any transfer applicants who have fewer than 56 semester or 84 quarter units of transferable college work must submit SAT or ACT test scores.
Entrance Recommendations: High School courses—Visual and Performing Arts (1 year), Electives (3 years), The CSU requires new students to be tested in English and Mathematics after they are admitted. These are not admission tests, but a way to determine a student's preparedness for college work.
Requirements for foreign students: TOEFL (Score: 550); Those who have not attended for at least three years schools at the secondary level or above where English is the principal language of instruction must earn a minimum score of 550 on the Test of English as a Foreign Language (TOEFL) to qualify for admission to the Engineering programs.

DEPARTMENTAL ADMISSIONS INFORMATION
Admission to the engineering department: At the time of admission to the institution.

FINANCIAL INFORMATION

ESTIMATED EXPENSES (ACADEMIC YEAR)
[Expenses are for the 1992-93 nine-month academic year.]

State Residents	Undergraduate	Graduate
Tuition and fees	$ 1,425	$ 1,425
College room and board	$ 6,000	$ 6,000
Books and supplies	$ 800	$ 800
Other expenses	$ 2,500	$ 2,500
Total estimated expenses	**$10,725**	**$10,725**

Out-of-State Residents	Undergraduate	Graduate
Tuition and fees	$ 8,805	$ 8,805
College room and board	$ 6,000	$ 6,000
Books and supplies	$ 800	$ 800
Other expenses	$ 2,500	$ 2,500
Total estimated expenses	**$18,105**	**$18,105**

FINANCIAL AID OFFICE CONTACT
Gloria J. Kapp, *Director*
California State University, Long Beach
Financial Aid
1250 Bellflower Boulevard
Long Beach, CA 90840-0114
Phone: (310)985-8403

GENERAL FINANCIAL AID INFORMATION
Forms accepted/required: AFSA, CSS-FAF, FAF, FAT, FSA, IRS, SAR, SDF, Institutional
Additional financial aid information: Request a Student Aid Application form for California in December, complete it, and mail it to the processing service listed on the form as soon as possible after Jan. 1. Other means of financial aid include: Perkins National Direct Student Loans, College Work Study, California State Educational Opportunity Program Grants. For more information contact: Director of Financial Aid, CSULB, SS/A Bldg. Room 270, 1250 Bellflower Boulevard, Long Beach, CA 90840; (310) 985-4641.

ENGINEERING COLLEGE INFORMATION
[For additional personnel, refer to the *Appendix*.]

HEAD OF ENGINEERING
J. Richard Williams
Phone: (310)985-5123 Fax: (310)985-7561

ENGINEERING COLLEGE ADDRESS
College of Engineering
California State University, Long Beach
1250 Bellflower Boulevard
Long Beach, CA 90840-8306
Phone: (310)985-5121 Fax: (310)985-7561

ENROLLMENTS—BY CLASS
[Numbers are baccalaureate enrollments for the fall 1991 term, unless otherwise footnoted.]
1st-year students/Freshmen 478
2nd-year students/Sophomores 307
3rd-year students/Juniors 898
4th-year students/Seniors 1,888
Total **3,571**

NUMBER OF DEGREES AWARDED—BY PROGRAM
[Numbers are engineering baccalaureate degrees awarded during the 1991-1992 academic year, unless otherwise footnoted. For full details about each engineering program, refer to the *Tables of Degree Programs*.]
Aerospace Engineering .... — (1)
Chemical Engineering 15
Civil Engineering 48
Computer Engineering & Computer Science 90
Electrical Engineering 90
Engineering & Industrial Technology .... —
Engineering Technology 128
Mechanical Engineering 103
Total **474**

Notes: (1) The Department offers graduate programs only.

PERCENTAGE OF DEGREES AWARDED—BY CATEGORY
[Percentages are of all engineering baccalaureate degrees awarded during the 1991-1992 academic year, unless otherwise footnoted.]
African Americans 3.0%
Asian/Pacific Island Americans 26.0%
Hispanic Americans 10.0%
Native Americans 0.5%
Foreign Citizens 11.0%
All Others 49.5%
Women 14.0%
Persons with disabilities 1.0%
Students over 25 years of age 71.0%

GRADUATE ENROLLMENTS & DEGREES AWARDED
Master's enrollment 1,054
Master's degrees awarded 138
Doctoral enrollment 14
Doctoral degrees awarded .... —

ENGINEERING STUDENT DATA
Applicants to the engineering college
Number of applicants to engineering college 992
Percent offered admission 74.0%

Matriculated engineering students
Percentage in top quartile (25%) of High School class 75.0
Average SAT scores: Math—492, Verbal—363, Combined—855
Average ACT scores: Math—25, Composite—20

FULL & PART TIME FACULTY—BY DEPARTMENT
[Figures are the head count of full-time faculty and the full-time equivalent (FTE) of part-time faculty for each engineering department or equivalent.]

Department	Full	Part
Chemical Eng	5	0.1
Civil Eng	10	2.4
Electrical Eng	22	3.1
Mechanical Eng	16	0.4
Aerospace Eng	7	0.6
Computer Eng & Computer Science	16	2.4
Eng Tech	13	3.1

COLLEGE DESCRIPTION
The College of Engineering at California State University, Long Beach, is one the largest engineering colleges in the West. Since it is located in an area with greatest concentration of high-technology industry in the nation, including the largest aerospace firms, it has a strong liaison with industry and the engineering community. With feedback from nine advisory and development councils, curricula are constantly updated and assured of relevancy. The industry executives on these councils also advise on placement opportunities before and after graduation. Because of its location, academic programs aimed at education high-tech professionals are given excellent support by the university. The undergraduate curricula in chemical engineering, civil engineering, computer engineering, electrical engineering, and mechanical engineering are accredited by the Accreditation Board of Engineering and Technology (ABET).

DEGREE PROGRAMS DESCRIPTION
The College of Engineering at California State University, Long Beach provides a broad education for entry into the engineering, computer science, and technology professions. Seven baccalaureate degrees, and six masters' degrees, and the Ph.D. (jointly with The Claremont Graduate School) are offered through its seven departments. Options for specialization and certification, and a dual-degree program are available. Dual-degree programs allow the student to combine an engineering education with one in business, fine arts, humanities, or the sciences. The B.S. programs in chemical engineering, civil engineering, computer engineering, electrical engineering, and mechanical engineering are accredited by the Accreditation Board for Engineering and Technology (ABET).

DUAL DEGREE PROGRAMS
Engineering baccalaureate graduates with dual degrees 2
Enrollment requirements: After three years at the first institution, students transfer to CSULB as juniors to complete the two final years of engineering study. At the end of the first year at CSULB, students who have completed all of the requirements for their first degree are awarded those degrees by the appropriate School or institution. At the end of their fifth year, students who have completed all engineering requirements receive their engineering degrees.

TRANSFER INFORMATION
Residency Requirements: Completion of 30 units in residence at CSULB of which 24 must be upper division and 12 in the major.
Transfer via Articulation Agreements
Admission to engineering: Please see Freshman Admission Requirements.
Engineering graduates transferred from ... 4-yr: 140 2-yr: 330
Requirements: Transfer applicants who have fewer than 56 semester or 84 quarter units of transferable college work must submit scores from the either the Scholastic Aptitude Test of the College Board (SAT) or the American College Test Program (ACT).

GRADUATION REQUIREMENTS
Complete an appropriate number and distribution of units for the degree. Complete 30 units in residence, 24 in upper division, 12 in major. Meet minimal scholarship evaluations: EPT and WPE. Meet mathematical skills evaluation: ELM. Complete University 100 -- The University and Your Future. Complete GE program including requirements in U.S. History, U.S. Constitution and American ideals, and English Composition, and satisfy the GE Themes as specified by the College of Engineering. Complete requirements in major. Receive formal approval by the faculty of the University.

STUDENT PROGRAMS

PROFESSIONAL AND HONORARY SOCIETIES
[For key to acronyms, see *Introduction*.]
Professional Societies: AIAA, AIChE, ASCE, ASME, ASNE, BMES, IEEE, SAE, SAME, SME, Society of Women Engineers (SWE), Association of General Contractors (AGC)
Honorary Societies: Chi Epsilon, Eta Kappa Nu, Pi Tau Sigma, Tau Beta Pi

SUPPORT PROGRAMS
Student Chapter Organizations: Society of Women Engineers, National Society of Black Engineers, Society of Hispanic Professional Engineers

Mexican American Science Engineering Society
For: Ethnic Minorities **Available:** Year round
Offered: Prematriculation, Freshman, Sophomore, Junior, Senior

American Indian Science Engineering Society
For: Ethnic Minorities **Available:** Year round
Offered: Freshman, Sophomore, Junior, Senior, Graduate level

Other Engineering Support Programs: Engineering Problem-Solving Initiative (EPSI) at CSULB provides students opportunities to develop and master 'real-life' problem-solving skills through participation in projects sponsored by industry. In addition to the experience, students also receive stipends.

031 California State University, Los Angeles

INSTITUTION PROFILE

HEAD OF THE INSTITUTION
James M. Rosser
Phone: (213)343-3030   Fax: (213)343-2670

GENERAL INFORMATION
[All Students—Fall 1991]
Undergraduate enrollment 15,607
Graduate enrollment 5,194
Total institution enrollment **20,801**
Type of institution: Public   Calendar: Quarters
Location: Los Angeles   Population: 8,769,900
Setting: Urban
Types of engineering degrees: Engineering
Other degree-granting colleges: Education, Arts & Letters, Health & Human Services, Natural & Social Sciences, Business & Economics

ENGINEERING ADMISSIONS

ADMISSIONS OFFICE CONTACT
Kevin M. Browne
California State University, Los Angeles
Director of Admissions
5151 State University Drive
Los Angeles, CA 90032-8530
Phone: (213)343-3901   Fax: (213)343-2670

ENGINEERING COLLEGE ADMISSIONS INFORMATION
Entrance Requirements: High School courses—English (4 years), Mathematics, Electives (3 years), U.S. History, Science, Visual & Performing Arts (1 year), Foreign Language (2 years), High School GPA times 800 plus combined SAT score must exceed 2800 OR GPA times 200 plus combined ACT score must be greater than 694. In addition, prerequisite High Sch courses must be completed. Special admission categories for some students.
Entrance Recommendations: High School courses—Mathematics (4 years), Chemistry (1 year), Physics (1 year), Mechanical Drawing (CE and ME only) (1 year), High School GPA times 800 plus combined SAT score must exceed 2800 OR GPA times 200 plus combines ACT score must be greater than 694.
Requirements for out-of-state residents: For applicants who neither graduated from a CA high school nor are residents of CA, indexes of either 3402 (SAT) or 822 (ACT) are required.
Requirements for foreign students: TOEFL (Score: 550); financial statement

DEPARTMENTAL ADMISSIONS INFORMATION
Admission to the engineering department: At the time of admission to the institution.

FINANCIAL INFORMATION

ESTIMATED EXPENSES (ACADEMIC YEAR)
[Expenses are for the 1992-93 nine-month academic year.]

State Residents	Undergraduate	Graduate
Tuition and fees	$ 1,428	$ 1,428
College room and board	$ 4,507 (1)	$ 4,507 (1)
Books and supplies	$ 576	$ 576
Other expenses	$ 2,304 (2)	$ 2,304 (2)
Total estimated expenses	**$ 8,815**	**$ 8,815**
Out-of-State Residents	Undergraduate	Graduate
Tuition and fees	$ 1,428 (3)	$ 1,428 (3)
College room and board	$ 4,507 (1)	$ 4,507 (1)
Books and supplies	$ 576	$ 576
Other expenses	$ 2,304 (2)	$ 2,304 (2)
Total estimated expenses	**$ 8,815**	**$ 8,815**

Notes: (1) On campus housing amount. Single, no dependents off campus housing est. $5,673. (2) Transportation estimate $576, misc. personal estimate $1,728. (3) Non residents need to add $164 per unit plus resident fees.

FINANCIAL AID OFFICE CONTACT
Vincent Deanda, *Director*
California State University, Los Angeles
Student Financial Services Office
5151 State University Drive
Los Angeles, CA 90051-8402
Phone: (213)343-3240   Fax: (213)343-2670

GENERAL FINANCIAL AID INFORMATION
Forms accepted/required: ACT, CSS-FAF
Additional financial aid information: Financial aid is primarily need based. Some merit-based scholarships are available. Part-time employment including work-study available on campus. Long-term loans and short-term emergency loans are available. Federal programs include: Pell grant, SEOG, Perkins, Work study, Stafford loan, SLS. State programs include: EOP, State University grant, Graduate Equity Fellowship, CAL grant.

ENGINEERING COLLEGE INFORMATION

[For additional personnel, refer to the *Appendix*.]

HEAD OF ENGINEERING
Raymond B. Landis
Phone: (213)343-4500   Fax: (213)343-4555
EMail: rlandis@calstatela.edu

ENGINEERING COLLEGE ADDRESS
School of Engineering & Technology
California State University, Los Angeles
5151 State University Drive
Los Angeles, CA 90032-8150
Phone: (213)343-3000   Fax: (213)343-2670

ENROLLMENTS—BY CLASS
[Numbers are baccalaureate enrollments for the fall 1991 term, unless otherwise footnoted.]

1st-year students/Freshmen	191
2nd-year students/Sophomores	113
3rd-year students/Juniors	190
4th-year students/Seniors	490 (1)
Total	**984**

Notes: (1) Graduate students pursuing a BS in Engineering total 33.

NUMBER OF DEGREES AWARDED—BY PROGRAM
[Numbers are engineering baccalaureate degrees awarded during the 1991-1992 academic year, unless otherwise footnoted. For full details about each engineering program, refer to the *Tables of Degree Programs*.]

Civil Engineering	20
Electrical Engineering	87
Mechanical Engineering	17
Total	**124**

PERCENTAGE OF DEGREES AWARDED—BY CATEGORY
[Percentages are of all engineering baccalaureate degrees awarded during the 1991-1992 academic year, unless otherwise footnoted.]

African Americans	10.5 %
Asian/Pacific Island Americans	44.4 % (2)
Hispanic Americans	7.3 %
Native Americans	– %
Foreign Citizens	12.9 %
All Others	24.9 %
Women	12.9 %
Persons with disabilities	– % (A)
Students over 25 years of age	– % (A)

Notes: (A) Data not available. (2) Includes Pacific Islanders.

GRADUATE ENROLLMENTS & DEGREES AWARDED

Master's enrollment	186
Master's degrees awarded	33
Doctoral enrollment	– (B)
Doctoral degrees awarded	– (B)

Notes: (B) Data not applicable.

FULL & PART TIME FACULTY—BY DEPARTMENT
[Figures are the head count of full-time faculty and the full-time equivalent (FTE) of part-time faculty for each engineering department or equivalent.]

Department	Full	Part
Civil Eng	9	–
Electrical Eng	15	2.3
Mechanical Eng	9	1.1

COLLEGE DESCRIPTION
The School offers a four-year undergraduate program in engineering. The curriculum emphasizes the fundamentals of mathematics, science, and engineering during the first two years followed by two years of specialized training in either civil, mechanical, or electrical and computer engineering. Faculty are dedicated to undergraduate teaching. Small classes and high level of student/faculty interactions promote a 'small-school' environment. Courses are offered both day and evening, four quarters per year. Excellent physical plant and modern laboratory equipment. Location in highly industrialized technological center results in high level of industry support. Excellent opportunity for part-time and summer employment in industry.

DEGREE PROGRAMS DESCRIPTION
B.S. degree--Curriculum has a strong practical emphasis comprised of extensive engineering design including a required senior design project, 'hands-on' laboratory work, and computer-aided engineering. M.S. degree--Comprehensive exam or thesis option. Forty-five units are required. Students may work on industry sponsored design clinics. Program may be completed by students employed full-time in industry.

TRANSFER INFORMATION
Residency Requirements: Students must complete a minimum of 45 quarter units, including at least 36 upper division units, 18 units in the major, and 12 units in general education courses, in residence at Cal State L.A. for the baccalaureate.

Transfer via Articulation Agreements
Admission to engineering: At the end of the sophomore year
Requirements: Transfer students who were not eligible for admission from high school must complete 84 transferable quarter (56 semester) units with a 2.0 (C) average or better and be in good standing at the last college or university attended.

Transfer without Articulation Agreements
Admission to engineering: At the end of the sophomore year
Requirements: Transfer students who were not eligible for admission from high school must complete 84 transferable quarter (56 semester) units with a 2.0 (C) average or better and be in good standing at the last college or university attended.

GRADUATION REQUIREMENTS
B.S. degree in engineering requires 200-203 quarter units depending on specific major. 2.0 GPA in all college work attempted, all courses attempted at Cal State L.A., and all courses required in major. Passing score on Writing Proficiency Examination. Engineering majors are also subject to special General Education requirements. They must complete several lower division major courses that are also approved for GE credit.

STUDENT PROGRAMS

PROFESSIONAL AND HONORARY SOCIETIES
[For key to acronyms, see *Introduction*.]
Professional Societies: ASCE, ASME, IEEE, SAE, SAMPE, Structural Engineers Association (SEA), Engineering and Technology Student Council
Honorary Societies: Chi Epsilon, Eta Kappa Nu, Pi Tau Sigma, Tau Beta Pi

SUPPORT PROGRAMS
Student Chapter Organizations: Society of Women Engineers, National Society of Black Engineers, Society of Hispanic Professional Engineers, Society of Hispanic Engineering & Science Students, Council for Black Engineering Math & Science Stds

Minority Engineering Program (MEP)
For: Ethnic Minorities   **Available:** Year round
Offered: Prematriculation, Freshman, Sophomore, Junior, Senior

Partnership for Academic Learning & Success (PALS)
For: Ethnic Minorities   **Available:** Year round
Offered: Freshman

Minority Access to Energy Research Careers (MAERC)
For: Ethnic Minorities   **Available:** Year round
Offered: Junior, Senior

Other Engineering Support Programs: Orientation during quarter prior to matriculation for freshman & transfer students; Active network of engineering student organizations; Mandatory academic advisement each quarter; Engineering Student Information Center; Center for Career Planning & Placement; International Student Services; Disabled Student Services; Office of Veterans Services; Women's Resource Center; Learning Resource Center.

032 California State University, Northridge

■ INSTITUTION PROFILE

HEAD OF THE INSTITUTION
Blenda J. Wilson
Phone: (818)885-2121   Fax: (818)885-2254

GENERAL INFORMATION
[All Students—Fall 1991]
Undergraduate enrollment 24,000
Graduate enrollment 6,000
Total institution enrollment 30,000
Type of institution: Public   **Calendar:** Semesters
Location: Los Angeles   **Population:** 4,000,000
Setting: Urban
Types of engineering degrees: Engineering
Other degree-granting colleges: Allied Health Sciences, Arts & Sciences, Business Administration, Communication & Journalism, Education, Fine & Performing Arts, Humanities & Social Sciences

ENGINEERING COLLEGE ADMISSIONS INFORMATION
Admission to the engineering college: Or during the college career, as a "change of major".
Entrance Requirements: SAT, ACT, High School courses—College preparatory program, Eligibility index: combination of high school GPA & SAT or ACT. If GPA is 3.0 or higher, tests are not required.
Entrance Recommendations: High School courses—Strong math and sciences background
Requirements for foreign students: TOEFL (Score: 550)

DEPARTMENTAL ADMISSIONS INFORMATION
Admission to the engineering department: At the time of admission to the institution.

■ FINANCIAL INFORMATION

ESTIMATED EXPENSES (ACADEMIC YEAR)
[Expenses are for the 1992-93 nine-month academic year.]

State Residents	Undergraduate	Graduate
Tuition and fees	$ 770	$ 770
College room and board	$ 5,000	$ 5,000
Books and supplies	$ 300	$ 300
Other expenses	$ —(2)	$ —(2)
Total estimated expenses	**$ 6,070**	**$ 6,070**

Out-of-State Residents	Undergraduate	Graduate
Tuition and fees	$ 770 (1)	$ 770 (1)
College room and board	$ 5,000	$ 5,000
Books and supplies	$ 300	$ 300
Other expenses	$ —(2)	$ —(2)
Total estimated expenses	**$ 6,070**	**$ 6,070**

Notes: (1) Plus $246 per student credit unit (2) Will vary greatly depending on transportation, social, entertainment, general living expenses.

GENERAL FINANCIAL AID INFORMATION
Forms accepted/required: AFSA, FAT, FAR, Stafford, IRS, SAR, Verification forms, Household size forms
Additional financial aid information: Both need-based and merit-based scholarships are available. Also loans, athletic scholarships, part-time jobs on campus, College Work-Study program.

■ ENGINEERING COLLEGE INFORMATION

[For additional personnel, refer to the *Appendix*.]
HEAD OF ENGINEERING
Diane L. Schwartz
Phone: (818)885-4501   Fax: (818)885-2140
EMail: DSCHWARTZ@VAX.CSUN.EDU

ENGINEERING COLLEGE ADDRESS
School of Engineering and Computer Science
California State University, Northridge
18111 Nordhoff St.
P.O. Box 1283 - SECS
Northridge, CA 91328-1283
Phone: (818)885-4501   Fax: (818)885-2140

ENROLLMENTS—BY CLASS
[Numbers are baccalaureate enrollments for the fall 1991 term, unless otherwise footnoted.]
1st-year students/Freshmen 294
2nd-year students/Sophomores 220
3rd-year students/Juniors 287
4th-year students/Seniors 762
Total 1,563

NUMBER OF DEGREES AWARDED—BY PROGRAM
[Numbers are engineering baccalaureate degrees awarded during the 1991-1992 academic year, unless otherwise footnoted. For full details about each engineering program, refer to the *Tables of Degree Programs*.]
Computer Science 67
Engineering 194
Total 261

PERCENTAGE OF DEGREES AWARDED—BY CATEGORY
[Percentages are of all engineering baccalaureate degrees awarded during the 1991-1992 academic year, unless otherwise footnoted.]
African Americans .... 7.7 % (C)
Asian/Pacific Island Americans .... 26.0 % (C)
Hispanic Americans .... 3.0 % (C)
Native Americans .... — % (A)
Foreign Citizens .... — % (C)
All Others .... 63.3 %
Women .... 15.0 % (C)
Persons with disabilities .... — % (A)
Students over 25 years of age .... — % (A)

Notes: (A) Data not available. (C) Estimated data.

GRADUATE ENROLLMENTS & DEGREES AWARDED
Master's enrollment 484
Master's degrees awarded 116
Doctoral enrollment .... — (B)
Doctoral degrees awarded .... — (B)

Notes: (B) Data not applicable.

ENGINEERING STUDENT DATA
Applicants to the engineering college
Number of applicants to engineering college 1,082
Percent offered admission 68.0%
Matriculated engineering students
Percentage in top quartile (25%) of High School class .... (A)
Average SAT scores: Math—477, Verbal—364, Combined—841

Notes: (A) Data not available.

FULL & PART TIME FACULTY—BY DEPARTMENT
[Figures are the head count of full-time faculty and the full-time equivalent (FTE) of part-time faculty for each engineering department or equivalent.]

Department	Full	Part
Civil & Industrial Eng & Applied Mechanics	13	2.0
Electrical & Computer Eng	21	1.5
Mechanical Eng	11	1.5
Computer Science	15	6.0

COLLEGE DESCRIPTION
The School of Engineering and Computer Science offers undergraduate programs in Computer Science and (general) Engineering. The engineering program emphasizes a broad-based background in a number of engineering areas, with students choosing a specialized senior program directed toward their special career objectives. There is a small Honors Co-op program, an SECS Scholarships program, and a strong support network for minorities and women.

DEGREE PROGRAMS DESCRIPTION
The engineering program begins with studies of math, physical sciences, computer science, engineering design, and subjects in humanities and social and behavioral sciences. With the math and science prerequisites completed, the student progresses to the engineering sciences, which bridge the gap between the basic sciences and creative applications. Through the first 3 yrs, all students complete a common core of technical subjects. In the 4th yr. of studies, students pursue senior programs directed toward their special career objectives. The final year strongly emphasizes more advanced concepts of engr. design and includes a capstone experience in which students apply the breadth and depth of their knowledge to solve comprehensive engineering problems. The broad undergraduate program is designed to maximize potential creativity and future professional growth opportunities.

DUAL DEGREE PROGRAMS
Engineering baccalaureate graduates with dual degrees 4

TRANSFER INFORMATION
Residency Requirements: Completion of 30 units in residence at CSUN: 24 of the 30 units must be completed in upper division. 12 of the units must be in the major and 9 of the units must be in general education.

Transfer via Articulation Agreements
Admission to engineering: At any time
Requirements: Transfer students who have completed 56 or more transferable semester units are screened on the basis of college level work. Engineering transfers who have completed English composition and 7 required lower division courses chosen from a specific list of math, science and engr. courses (or their equivalents) with a minimum 2.5 GPA qualify for admission.

Transfer without Articulation Agreements
Admission to engineering: At any time

GRADUATION REQUIREMENTS
The B.S. program in Engineering requires 138 semester units. The B.S. program in Computer Science requires 130 semester units. A GPA of 2.0 is required in all areas: CSUN, overall, major, minor. Meeting of writing skills requirements and completion of general education program.

■ STUDENT PROGRAMS

PROFESSIONAL AND HONORARY SOCIETIES
[For key to acronyms, see *Introduction*.]
Professional Societies: ACM, AIAA, ASCE, ASME, IEEE, SAE, Computer Science Association (CSA), Student Association of Industrial and Manufacturing Engrs. (SAIME)
Honorary Societies: Tau Beta Pi

SUPPORT PROGRAMS
Student Chapter Organizations: Society of Women Engineers, National Society of Black Engineers, Society of Hispanic Professional Engineers, Various campus-wide ethnicity-based groups

Minority Engineering Program (MEP)
For: Ethnic Minorities   **Available:** Year round
Offered: Prematriculation, Freshman, Sophomore, Junior, Senior

Project Preserve
For: Ethnic Minorities   **Available:** Year round
Offered: Freshman, Sophomore, Junior, Senior

Other Engineering Support Programs: The School of Engineering/Computer Science and CSUN offer orientation programs for all freshmen and transfers; academic and career counseling; psychological counseling; study skills center; disabled student program; tutoring programs; placement services; scholarships. SECS programs for minorities (Project Preserve, MESA, Minority Engr. Prog.), also offer a variety of support services.

033 California State University, Sacramento

■ INSTITUTION PROFILE

HEAD OF THE INSTITUTION
Donald R. Gerth
Phone: (916)278-7737 Fax: (916)278-6959

GENERAL INFORMATION
[All Students—Fall 1991]
Undergraduate enrollment 19,406
Graduate enrollment 5,062
Total institution enrollment 24,468
Type of institution: Public Calendar: Semesters
Nearest city: Sacramento Population: 300,000
Miles from main campus: 1 Setting: Urban
Types of engineering degrees: Engineering & Technology
Other degree-granting colleges: Arts & Sciences, Business Administration, Education, Health & Human Services

■ ENGINEERING ADMISSIONS

ADMISSIONS OFFICE CONTACT
Larry Glasmire
California State University, Sacramento
6000 J Street
Sacramento, CA 95819-6048
Phone: (916)278-7111 Fax: (916)278-5443

ENGINEERING COLLEGE ADMISSIONS INFORMATION
Entrance Requirements: SAT (Score: 410), ACT (Score: 10), High School courses—English (4 years), Mathematics (3.5 years), Science with lab (2 years), Foreign Language (2 years)
Entrance Recommendations: High School courses—Mechanical Drawing (1 year), Computer Literacy (1 year)
Requirements for foreign students: TOEFL (Score: 510)

DEPARTMENTAL ADMISSIONS INFORMATION
Admission to the engineering department: At the time of admission to the institution.

■ FINANCIAL INFORMATION

ESTIMATED EXPENSES (ACADEMIC YEAR)
[Expenses are for the 1992-93 nine-month academic year.]

State Residents	Undergraduate	Graduate
Tuition and fees	$ 1,460	$ 1,460
College room and board	$ 4,420	$ 4,420
Books and supplies	$ 525	$ 525
Other expenses	$ 318	$ 318
Total estimated expenses	**$ 6,723**	**$ 6,723**

Out-of-State Residents	Undergraduate	Graduate
Tuition and fees	$ —(1)	$ —(1)
College room and board	$ 4,420	$ 4,420
Books and supplies	$ 525	$ 525
Other expenses	$ 318	$ 318
Total estimated expenses	**$ 5,263**	**$ 5,263**

Notes: (1) $246 per unit plus resident fees.

FINANCIAL AID OFFICE CONTACT
Starla Satchell, *Director of Financial Aid*
California State University, Sacramento
6000 J Street
Sacramento, CA 95819-6044
Phone: (916)278-6554

GENERAL FINANCIAL AID INFORMATION
Forms accepted/required: Institutional
Additional financial aid information: College Work-study, student assistants, a variety of grants, loans, scholarships. To apply a student must fill out a Student Aid Application for California (SAAC). Documents required: Federal Income Tax Return, or Income Certification Form, institutional forms.

■ ENGINEERING COLLEGE INFORMATION
[For additional personnel, refer to the *Appendix*.]

HEAD OF ENGINEERING
Donald H. Gillott
Phone: (916)278-6366 Fax: (916)278-5949
EMail: gillottd@ecs.csus.edu

ENGINEERING COLLEGE ADDRESS
School of Engineering and Computer Science
California State University, Sacramento
6000 J Street
Sacramento, CA 95819-6023
Phone: (916)278-6366 Fax: (916)278-5949

ENROLLMENTS—BY CLASS
[Numbers are baccalaureate enrollments for the fall 1991 term, unless otherwise footnoted.]
1st-year students/Freshmen 224
2nd-year students/Sophomores 140
3rd-year students/Juniors 362
4th-year students/Seniors 803
Total 1,529

NUMBER OF DEGREES AWARDED—BY PROGRAM
[Numbers are engineering baccalaureate degrees awarded during the 1991-1992 academic year, unless otherwise footnoted. For full details about each engineering program, refer to the *Tables of Degree Programs*.]
Civil Engineering 58
Computer Engineering 20
Electrical & Electronics Engineering 90
Mechanical Engineering 74
Total 242

PERCENTAGE OF DEGREES AWARDED—BY CATEGORY
[Percentages are of all engineering baccalaureate degrees awarded during the 1991-1992 academic year, unless otherwise footnoted.]
African Americans .... 4.0 %
Asian/Pacific Island Americans .... 23.0 %
Hispanic Americans .... 6.4 %
Native Americans .... — % (A)
Foreign Citizens .... — % (A)
All Others .... 66.6 %
Women .... 14.0 % (A)
Persons with disabilities .... — % (A)
Students over 25 years of age .... — % (A)

Notes: (A) Data not available.

GRADUATE ENROLLMENTS & DEGREES AWARDED
Master's enrollment 273
Master's degrees awarded 43
Doctoral enrollment .... — (B)
Doctoral degrees awarded .... — (B)

Notes: (B) Data not applicable.

ENGINEERING STUDENT DATA
Applicants to the engineering college
Number of applicants to engineering college 553
Percent offered admission .... 67.5 %

Matriculated engineering students
Percentage in top quartile (25%) of High School class 50.0
Average SAT scores: Math—476, Verbal—402, Combined—878

FULL & PART TIME FACULTY—BY DEPARTMENT
[Figures are the head count of full-time faculty and the full-time equivalent (FTE) of part-time faculty for each engineering department or equivalent.]

Department	Full	Part
Electrical & Electronic Eng	17	2.3
Civil Eng	17	3.1
Mechanical Eng	16	1.0
Computer Science	23	2.4
Biomedical Eng Program	4	—
Construction Management Program	5	—
Mechanical Eng Tech	9	—
Computer Eng	8	—

COLLEGE DESCRIPTION

The School offers a four-year undergraduate program in engineering, emphasizing engineering fundamentals in the first two years, followed by two years of specialized training in 1 of the 4 majors with some liberal studies. Emphasis is on applications through the use of laboratory courses and design projects. There are opportunities to minor in business, computer science, mathematics, geology or physics. Co-op is an option in all programs. During the final two semesters each student completes a senior design project and is encouraged to complete the national standardized Fundamentals of Engineering Exam. Average number of years required to actually complete the bachelor's degree: 5.

DEGREE PROGRAMS DESCRIPTION
The B.S. degrees in the various engineering programs prepare the graduate for entry-level professional positions or for graduate study in each of the respective fields. Graduate study in engineering can be carried out in civil engineering, biomedical engineering, electrical/electronic engineering and mechanical engineering.

TRANSFER INFORMATION
Transfer via Articulation Agreements
Admission to engineering: At any time
Transfer without Articulation Agreements
Admission to engineering: At any time

GRADUATION REQUIREMENTS
B.S. programs in engineering require 132-140 semester credits, depending upon the major selected. An overall grade point average of 2.0 is required for four areas: (1) total courses attempted, (2) CSUS courses attempted, (3) upper division courses applied to the major and minor courses, (4) courses used to complete General Education requirements.

■ STUDENT PROGRAMS

PROFESSIONAL AND HONORARY SOCIETIES
[For key to acronyms, see *Introduction*.]
Professional Societies: ACM, AGCA, AIAA, ASCE, ASME, IEEE, SAE, SAMPE, Society of Women Engineers, IEEE Computer Society
Honorary Societies: Tau Beta Pi, Upsilon Pi Epsilon

SUPPORT PROGRAMS
Student Chapter Organizations: Society of Women Engineers, National Society of Black Engineers, Society of Hispanic Professional Engineers, American Indian Scientist and Engineers

Minority Engineering Program
For: Ethnic Minorities **Available:** Year round
Offered: Prematriculation, Freshman, Sophomore, Junior, Senior, Graduate level

Women's Programs
For: Women **Available:** Academic year
Offered: Prematriculation, Freshman, Sophomore, Junior, Senior, Graduate level

Other Engineering Support Programs: Summer and fall orientation program for all freshmen and transfers; Learning Assessment Program; Career Counseling Services; veteran's services; services for the disabled; placement services; Tutoring Center; Student Health Center; Student Chapters of Engineering Societies; Faculty Advising; International Office; Co-op; Financial Aid; Engineering Career and Placement Service.

California State Polytechnic University, Pomona

INSTITUTION PROFILE
HEAD OF THE INSTITUTION
Bob H. Suzuki
Phone: (909)869-2290

GENERAL INFORMATION
[All Students—Fall 1991]
Undergraduate enrollment 16,525
Graduate enrollment 1,043
Total institution enrollment **17,568**
Type of institution: Public   Calendar: Quarters
Nearest city: Los Angeles   Population: 4,500,000
Miles from main campus: 35   Setting: Urban
Types of engineering degrees: Engineering & Technology
Other degree-granting colleges: Agricultural & Environmental, Architecture, Arts & Sciences, Business Administration, Education, Hotel and Restaurant Management

ENGINEERING ADMISSIONS
ADMISSIONS OFFICE CONTACT
Joseph C. Marshall
California State Polytechnic University, Pomona
Admissions Office
3801 W. Temple Avenue
Pomona, CA 91768
Phone: (909)869-2000

ENGINEERING COLLEGE ADMISSIONS INFORMATION
Entrance Requirements: SAT (Score: 400), ACT (Score: 18), High School courses—CSU Minimum Course Work (1 year), (Units & courses increase through 1995), Eligibility Index requires either the SAT or the ACT.
Entrance Recommendations: SAT (Score: 1000), ACT (Score: 24), High School courses—Mathematics (4 years), Lab Science (2 years)
Requirements for out-of-state residents: Non-resident applicants require an eligibility index of at least 3402, or a high school GPA of at least 3.60.
Requirements for foreign students: TOEFL (Score: 525); financial statement

DEPARTMENTAL ADMISSIONS INFORMATION
Admission to the engineering department: At the time of admission to the institution.
Additional information: Engineering programs typically close early when applications targets are met.

FINANCIAL INFORMATION
ESTIMATED EXPENSES (ACADEMIC YEAR)
[Expenses are for the 1992-93 nine-month academic year.]

State Residents	Undergraduate	Graduate
Tuition and fees	$ 1,342	$ 1,342
College room and board	$ 4,500	$ 4,500
Books and supplies	$ 750	$ 750
Other expenses	$ 954	$ 954
Total estimated expenses	**$ 7,546**	**$ 7,546**

Out-of-State Residents	Undergraduate	Graduate
Tuition and fees	$ 8,300	$ 8,300
College room and board	$ 4,500	$ 4,500
Books and supplies	$ 750	$ 750
Other expenses	$ 954	$ 954
Total estimated expenses	**$14,504**	**$14,504**

FINANCIAL AID OFFICE CONTACT
Al Andino, *Director of Financial Aid*
California State Polytechnic University, Pomona
Office of Financial Aid
3801 W. Temple Avenue
Pomona, CA 91768
Phone: (714)869-3700

GENERAL FINANCIAL AID INFORMATION
Forms accepted/required: CSS-FAF, FAF, IRS
Additional financial aid information: There are numerous opportunities for part-time jobs on campus as well as numerous need-based scholarships.

ENGINEERING COLLEGE INFORMATION
[For additional personnel, refer to the *Appendix*.]
HEAD OF ENGINEERING
Edward C. Hohmann
Phone: (909)869-2600   Fax: (909)869-4370

ENGINEERING COLLEGE ADDRESS
College of Engineering
California State Polytechnic University, Pomona
3801 W. Temple Avenue
Pomona, CA 91768-4066
Phone: (909)869-2600   Fax: (909)869-4370

ENROLLMENTS—BY CLASS
[Numbers are baccalaureate enrollments for the fall 1991 term, unless otherwise footnoted.]
1st-year students/Freshmen 975
2nd-year students/Sophomores 572
3rd-year students/Juniors 735
4th-year students/Seniors 1,453
Total **3,735**

NUMBER OF DEGREES AWARDED—BY PROGRAM
[Numbers are engineering baccalaureate degrees awarded during the 1991-1992 academic year, unless otherwise footnoted. For full details about each engineering program, refer to the *Tables of Degree Programs*.]
Aerospace Engineering 50
Agricultural Engineering 4
Chemical Engineering 50
Civil Engineering 116
Electrical & Computer Engineering 146
Electrical Engineering .... –
Industrial Engineering 35
Manufacturing Engineering 16
Mechanical Engineering 67
Total **484**

PERCENTAGE OF DEGREES AWARDED—BY CATEGORY
[Percentages are of all engineering baccalaureate degrees awarded during the 1991-1992 academic year, unless otherwise footnoted.]
African Americans 1.0%
Asian/Pacific Island Americans 47.0%
Hispanic Americans 11.0%
Native Americans .... – %
Foreign Citizens 4.0%
All Others 37.0%
Women 18.0%
Persons with disabilities .... – %
Students over 25 years of age .... – %

GRADUATE ENROLLMENTS & DEGREES AWARDED
Master's enrollment 91
Master's degrees awarded 28
Doctoral enrollment .... –
Doctoral degrees awarded .... –

ENGINEERING STUDENT DATA
Applicants to the engineering college
Number of applicants to engineering college 2,297
Percent offered admission 65.0%

FULL & PART TIME FACULTY—BY DEPARTMENT
[Figures are the head count of full-time faculty and the full-time equivalent (FTE) of part-time faculty for each engineering department or equivalent.]

Department	Full	Part
Chemical Eng	7	1.4
Civil Eng	14	3.0
Electrical & Computer Eng	35	3.5
Industrial & Manufacturing Eng	13	0.3
Mechanical Eng	20	3.8
Aerospace Eng	5	1.4
Agricultural Eng	3	–

COLLEGE DESCRIPTION
Cal Poly's College of Engineering has a well-earned reputation for helping to meet the technical challenges facing our nation by preparing engineers prepared to contribute significantly to industry upon graduation. The emphasis on a strong theoretical background coordinated with early and significant laboratory experiences continues to make the program unique in engineering education. The college provides study opportunities to over 4400 students in eight accredited engineering and technology curricula and two graduate programs. Each curriculum is designed to give the student both an understanding of the fundamental principles of engineering as an applied science and the practical expertise to apply these principles to actual situations.

DEGREE PROGRAMS DESCRIPTION
Unlike the more traditional engineering curricula which initiate engineering course work in the junior year, Cal Poly's program demands that students take computer programming and engineering orientation courses in the freshman year and that mathematics, basic science and general education courses begin concurrently. Throughout their educational programs, students become adept at using both the university's computing facilities and the college's Computer Aided Engineering Laboratory. Specific features of each curriculum reflect the input of each department's Industry Action Council, composed of leaders in local industry, and each emphasizes laboratory experiences. As a result of this 'learn by doing' environment, graduates of the college continue to be a great demand. The graduate programs build on the undergraduate foundations and are designed to appeal to the employed engineer wishing to advance his understandings of engineering applications.

TRANSFER INFORMATION
Residency Requirements: Total of three quarters in residence, including two immediately preceding graduation.
Transfer without Articulation Agreements
Admission to engineering: At any time
Engineering graduates transferred from ... 4-yr: 50   2-yr: 200
Requirements: Students changing their majors into the College of Engineering must meet the same requirements as transfer applicants from other institutions.

STUDENT PROGRAMS
PROFESSIONAL AND HONORARY SOCIETIES
[For key to acronyms, see *Introduction*.]
Professional Societies: AGCA, AIAA, AIChE, ASCE, ASME, IEEE, IIE, SAE, SME, SPE, American Congress on Surveying and Mapping, Society of Plastics Engineers
Honorary Societies: Alpha Pi Mu, Chi Epsilon, Eta Kappa Nu, Omega Chi Epsilon, Pi Tau Sigma, Sigma Gamma Tau, Tau Alpha Pi, Tau Beta Pi

SUPPORT PROGRAMS
Student Chapter Organizations: Society of Women Engineers, National Society of Black Engineers, Society of Hispanic Professional Engineers, American Indian Science and Engineering Society
Minority Engineering Program
For: Ethnic Minorities   **Available:** Year round
Offered: Prematriculation, Freshman, Sophomore, Junior, Senior, Graduate level
Exploring Engineering
For: Women   **Available:** Summer only
Offered: Prematriculation
Other Engineering Support Programs: All students are strongly urged to participate in the Student Orientation Program conducted by the university prior to the beginning of each quarter.

035 Calvin College

INSTITUTION PROFILE

HEAD OF THE INSTITUTION
Anthony J. Diekema
Phone: (616)957-6100 **Fax:** (616)957-8551

GENERAL INFORMATION
[All Students—Fall 1991]
Undergraduate enrollment 3,819
Graduate enrollment 206
Total institution enrollment 4,025
Type of institution: Private **Calendar:** Other
Location: Grand Rapids **Population:** 190,000
Setting: Suburban
Types of engineering degrees: Engineering
Other degree-granting colleges: Seminary

ENGINEERING ADMISSIONS

ADMISSIONS OFFICE CONTACT
Thomas E. McWhertor
Calvin College
Admissions Office
3201 Burton SE
Grand Rapids, MI 49546-4388
Phone: (800)688-0122 **Fax:** (616)957-8551

ENGINEERING COLLEGE ADMISSIONS INFORMATION
Entrance Requirements: ACT (Score: 20), High School courses—English (3 years), Science (2 years), History/Social Sciences (2 years), Mathematics (2 years)
Entrance Recommendations: SAT (Score: 420), High School courses—English (4 years), Science (2 years), Mathematics (4 years), History/Social Science (3 years)
Requirements for foreign students: TOEFL (Score: 550)

DEPARTMENTAL ADMISSIONS INFORMATION
Admission to the engineering department: At the end of the second year.
Additional information: Completion of the first two years of the model BSE program with a minimum GPA of 2.3 (out of 4.0).

FINANCIAL INFORMATION

ESTIMATED EXPENSES (ACADEMIC YEAR)
[Expenses are for the 1992-93 nine-month academic year.]

All Students	Undergraduate	Graduate
Tuition and fees	$ 8,630	$ –
College room and board	$ 3,520	$ –
Books and supplies	$ 350	$ –
Other expenses	$ –	$ –
Total estimated expenses	$12,500	$ –

FINANCIAL AID OFFICE CONTACT
Wayne K. Hubers, *Director*
Calvin College
Financial Aid Office
3201 Burton SE
Grand Rapids, MI 49546-4388
Phone: (616)957-6134 **Fax:** (616)957-8551

GENERAL FINANCIAL AID INFORMATION
Forms accepted/required: FAF
Additional financial aid information: Need based scholarships; loans; merit-based scholarships; part-time jobs on campus; college work-study programs.

ENGINEERING COLLEGE INFORMATION

[For additional personnel, refer to the *Appendix*.]

HEAD OF ENGINEERING
David Hoekema
Phone: (616)957-6442 **Fax:** (616)957-8551

ENGINEERING COLLEGE ADDRESS
Engineering Department
Calvin College
3201 Burton SE
Grand Rapids, MI 49546-4388
Phone: (616)957-6000 **Fax:** (616)957-8551

ENROLLMENTS—BY CLASS
[Numbers are baccalaureate enrollments for the fall 1991 term, unless otherwise footnoted.]
1st-year students/Freshmen 95
2nd-year students/Sophomores 58
3rd-year students/Juniors 69
4th-year students/Seniors 81
Total 303

NUMBER OF DEGREES AWARDED—BY PROGRAM
[Numbers are engineering baccalaureate degrees awarded during the 1991-1992 academic year, unless otherwise footnoted. For full details about each engineering program, refer to the *Tables of Degree Programs*.]
Engineering 54
Total 54

PERCENTAGE OF DEGREES AWARDED—BY CATEGORY
[Percentages are of all engineering baccalaureate degrees awarded during the 1991-1992 academic year, unless otherwise footnoted.]
African Americans .... – %
Asian/Pacific Island Americans .... – %
Hispanic Americans .... – %
Native Americans 1.9%
Foreign Citizens 13.0%
All Others 85.1%
Women 9.3%
Persons with disabilities .... – % (A)
Students over 25 years of age 9.3%
Notes: (A) Data not available.

FULL & PART TIME FACULTY—BY DEPARTMENT
[Figures are the head count of full-time faculty and the full-time equivalent (FTE) of part-time faculty for each engineering department or equivalent.]

Department	Full	Part
Eng	8	–

COLLEGE DESCRIPTION
The college offers a four year undergraduate program in engineering. A BSE degree with programmed concentrations in civil, electrical, and mechanical engineering is awarded. The first two years including an emphasis on basic science, mathematics, conceptual design, and computer use are a common core for all engineering students. The last two years focus on a particular concentration. A common senior design course sequence emphasizes interdisciplinary projects. Within a Christian liberal arts college context the student is challenged to create technical designs that serve mankind.

DEGREE PROGRAMS DESCRIPTION
The BSE degree prepares the graduate for entry level professional positions or for graduate study in a variety of engineering programs.

TRANSFER INFORMATION
Residency Requirements: All students must complete their last year in residence at Calvin.

Transfer via Articulation Agreements
Admission to engineering: Transfers are admitted to the engineering program up to usually the end of the sophomore year.
Engineering graduates transferred from .. 4-yr: – 2-yr: –
Requirements: The minimum cumulative GPA for students transferring from a 2-year college is 2.5 (out of 4.0).

Transfer without Articulation Agreements
Admission to engineering: Transfers are admitted to the engineering program up to usually the end of the sophomore year.
Engineering graduates transferred from .. 4-yr: 7 2-yr: –
Requirements: The minimum GPA for students transferring from a 4-year institution is 2.0 and from a 2-year college, 2.5.

GRADUATION REQUIREMENTS
The BSE program requires completion of 38 courses with an average GPA of 2.0 (out of 4.0) overall and in the Engineering courses.

STUDENT PROGRAMS

PROFESSIONAL AND HONORARY SOCIETIES
[For key to acronyms, see *Introduction*.]
Professional Societies: ASCE, ASME

SUPPORT PROGRAMS

Entrada
For: Ethnic Minorities **Available:** Summer only
Offered: Prematriculation

Multicultural Student Development
For: Ethnic Minorities **Available:** Academic year
Offered: Freshman, Sophomore, Junior, Senior

Other Engineering Support Programs: Fall orientation program for all freshmen and transfers; learning assessment program; career counseling services; foraying student office; services for blind and handicapped; placement services.

036 Carnegie Mellon University

INSTITUTION PROFILE

HEAD OF THE INSTITUTION
Robert Mehrabian
Phone: (412)268-2200　　　Fax: (412)268-2330

GENERAL INFORMATION
[All Students—Fall 1991]
Undergraduate enrollment 4,273
Graduate enrollment 2,714
Total institution enrollment 7,133
Type of institution: Private　　Calendar: Semesters
Nearest city: Pittsburgh　　Population: 1,336,449
Miles from main campus: 1　　Setting: Urban
Types of engineering degrees: Engineering
Other degree-granting colleges: Fine & Performing Arts, Humanities & Social Sciences, Mellon College of Science, Computer Science, Heinz School of Public Policy & Management, Graduate School of Industrial Administration

ENGINEERING COLLEGE ADMISSIONS INFORMATION
Entrance Requirements: SAT, ACT, High School courses—English (4 years), Mathematics (4 years), Physics (1 year), Chemistry (1 year)
Requirements for foreign students: TOEFL (Score: 575); financial statement; TOEFL minimum score may vary by student.

DEPARTMENTAL ADMISSIONS INFORMATION
Admission to the engineering department: At the end of the first year.

FINANCIAL INFORMATION

ESTIMATED EXPENSES (ACADEMIC YEAR)
[Expenses are for the 1992-93 nine-month academic year.]

All Students	Undergraduate	Graduate
Tuition and fees	$16,100	$16,850
College room and board	$5,210	$8,875 [1]
Books and supplies	$450	$1,075
Other expenses	$1,500 [2]	$2,180 [2]
Total estimated expenses	**$23,260**	**$28,980**

Notes: (1) On-campus room and board not available to graduate students. (2) Includes insurance, transportation, and miscellaneous expenses.

FINANCIAL AID OFFICE CONTACT
Walter C. Cathie, *Associate V.P. for Financial Resources*
Carnegie Mellon University
5000 Forbes Avenue
Pittsburgh, PA 15213
Phone: (412)268-2068　　　Fax: (412)268-7837

GENERAL FINANCIAL AID INFORMATION
Forms accepted/required: FFS, IRS, PHEAA form for PA residents.
Additional financial aid information: Need-based scholarships; merit-based scholarships, part-time jobs on campus; College Work-Study Program.

ENGINEERING COLLEGE INFORMATION
[For additional personnel, refer to the *Appendix*.]

HEAD OF ENGINEERING
Stephen W. Director
Phone: (412)268-2537　　　Fax: (412)268-6421
EMail: director@orion.ece.cmu.edu

ENGINEERING COLLEGE ADDRESS
Carnegie Institute of Technology
Carnegie Mellon University
5000 Forbes Avenue
Pittsburgh, PA 15213
Phone: (412)268-2481　　　Fax: (412)268-6421

ENROLLMENTS—BY CLASS
[Numbers are baccalaureate enrollments for the fall 1991 term, unless otherwise footnoted.]
1st-year students/Freshmen 379
2nd-year students/Sophomores 351
3rd-year students/Juniors 279
4th-year students/Seniors 264
Total **1,273**

NUMBER OF DEGREES AWARDED—BY PROGRAM
[Numbers are engineering baccalaureate degrees awarded during the 1991-1992 academic year, unless otherwise footnoted. For full details about each engineering program, refer to the *Tables of Degree Programs*.]
Biomedical Engineering .... – [1]
Chemical Engineering 25
Civil Engineering 32
Electrical & Computer Engineering 98
Engineering & Public Policy .... – [B]
Manufacturing Engineering .... – [B]
Materials Science & Engineering 13
Mechanical Engineering 62
Metallurgical Engineering & Material Science .... –
Total **230**

Notes: (B) Data not applicable. (1) Undergraduate degree not available - Option only.

PERCENTAGE OF DEGREES AWARDED—BY CATEGORY
[Percentages are of all engineering baccalaureate degrees awarded during the 1991-1992 academic year, unless otherwise footnoted.]
African Americans .... 2.0% [C]
Asian/Pacific Island Americans .... 5.0% [C]
Hispanic Americans .... –% [A]
Native Americans .... –% [A]
Foreign Citizens .... –% [A]
All Others 93.0%
Women .... 14.0% [C]
Persons with disabilities .... –%
Students over 25 years of age .... –% [A]

Notes: (A) Data not available. (C) Estimated data.

GRADUATE ENROLLMENTS & DEGREES AWARDED
Master's enrollment 231
Master's degrees awarded 128
Doctoral enrollment 359
Doctoral degrees awarded 70

ENGINEERING STUDENT DATA
Applicants to the engineering college
Number of applicants to engineering college 2,188
Percent offered admission 67.0%

Matriculated engineering students
Percentage in top quartile (25%) of High School class 98.0
Average SAT scores: Math—690, Verbal—550, Combined—1240

FULL & PART TIME FACULTY—BY DEPARTMENT
[Figures are the head count of full-time faculty and the full-time equivalent (FTE) of part-time faculty for each engineering department or equivalent.]

Department	Full	Part
Chemical Eng	15	–
Civil Eng	16	–
Electrical & Computer Eng	36	–
Mechanical Eng	24	–
Eng & Public Policy	6	–
Materials Science & Eng	18	–
Manufacturing Eng	–	–

COLLEGE DESCRIPTION
Carnegie Institute of Technology has as its continuing goal to maintain excellence in undergraduate and graduate education and research. The degree to which this goal has been achieved is attested to by the demand for its graduates, the success of its alumni, the quality of its faculty, the adoption elsewhere of its innovations, and the national and international recognition it receives in education and research activities. In addition to a 4-year undergraduate program in each department, it is also possible for a student to pursue an interdisciplinary Designated Minor in biomedical eng., electronic materials, engineering design, environmental eng., manufacturing eng., or mechanical behavior of materials, or a double major in engineering and public policy, or to design either double-major or double-degree programs with other non-engineering departments. Co-op is an option in Materials Science and Engineering. Average number of years required to complete the bachelor's degree: 4.

DEGREE PROGRAMS DESCRIPTION
The B.S. degree in the various engineering programs prepare the graduate for entry-level professional positions or for graduate study in each of the respective fields. Graduate study in engineering can be carried out in eight different areas and degrees offered include M.S., M.Eng., and Ph.D., although not every degree is available in each area.

DUAL DEGREE PROGRAMS
Engineering baccalaureate graduates with dual degrees 53
Enrollment requirements: May have department-determined requirements.

TRANSFER INFORMATION
Transfer via Articulation Agreements
Admission to engineering: At any time
Transfer without Articulation Agreements
Admission to engineering: At any time

GRADUATION REQUIREMENTS
B.S. programs in Engineering require a minimum of approximately 390 units of study (approximately 130 semester credits) and a minimum cumulative quality point average of 2.00.

STUDENT PROGRAMS

PROFESSIONAL AND HONORARY SOCIETIES
[For key to acronyms, see *Introduction*.]
Professional Societies: AIChE, ASCE, ASME, IEEE, SAE
Honorary Societies: Tau Beta Pi

SUPPORT PROGRAMS
Student Chapter Organizations: Society of Women Engineers, National Society of Black Engineers, Society of Hispanic Professional Engineers

Minority Academic Advising Service
For: Ethnic Minorities　　Available: Year round
Offered: Prematriculation, Freshman, Sophomore, Junior, Senior, Graduate level

Women Career Guidance Programs
For: Women　　Available: Year round
Offered: Prematriculation, Freshman, Sophomore, Junior, Senior, Graduate level

High School Workshop on Women in Engineering
For: Women　　Available: Summer only
Offered: Prematriculation

Annual Conference on Tech. Opportunities for Women
For: Women　　Available: Summer only
Offered: Prematriculation, Freshman, Sophomore, Junior, Senior, Graduate level

Other Engineering Support Programs: University and College Orientation Program for freshman and transfer students; Foreign Student Orientation Program; Learning Disability Counseling; Handicapped Counseling; Office of Career Services and Placement; Graduate School and Fellowship Advising Program; Carnegie Mellon Action Project.

037 Case Western Reserve University

INSTITUTION PROFILE

HEAD OF THE INSTITUTION
Agnar Pytte
Phone: (216)368-4344 **Fax:** (216)368-5861

GENERAL INFORMATION
[All Students—Fall 1991]
Undergraduate enrollment 1,124
Graduate enrollment 863
Total institution enrollment 1,987
Type of institution: Private **Calendar:** Semesters
Location: Cleveland **Population:** 350,000
Setting: Urban
Types of engineering degrees: Engineering
Other degree-granting colleges: Arts & Sciences, Dentistry, Law, Medicine, Nursing, Management, Applied Social Sciences

ENGINEERING ADMISSIONS

ADMISSIONS OFFICE CONTACT
William T. Conley
Case Western Reserve University
10900 Euclid Ave.
Tomlinson Hall
Cleveland, OH 44106-9920
Phone: (216)368-4450 **Fax:** (216)368-5111

ENGINEERING COLLEGE ADMISSIONS INFORMATION
Entrance Requirements: SAT (Score: 1000), ACT (Score: 27), High School courses—English (4 years), Mathematics (4 years), Physics (1 year), Chemistry (1 year)
Entrance Recommendations: SAT (Score: 1000), ACT (Score: 27), High School courses—Social Studies (2 years), Foreign Language (2 years)
Requirements for foreign students: TOEFL (Score: 550); financial statement

DEPARTMENTAL ADMISSIONS INFORMATION
Admission to the engineering department: At the end of the first year.

FINANCIAL INFORMATION

ESTIMATED EXPENSES (ACADEMIC YEAR)
[Expenses are for the 1992-93 nine-month academic year.]

All Students	Undergraduate	Graduate
Tuition and fees	$14,600	$14,600
College room and board	$ 4,410	$ 4,410
Books and supplies	$ 490	$ 320
Other expenses	$ 1,100	$ 1,100
Total estimated expenses	**$20,600**	**$20,430**

FINANCIAL AID OFFICE CONTACT
Donald Chenelle, *Director of University Financial Aid*
Case Western Reserve University
10900 Euclid Ave
Pardee Hall #129
Cleveland, OH 44106
Phone: (216)368-4530 **Fax:** (216)368-5054

GENERAL FINANCIAL AID INFORMATION
Forms accepted/required: ACT, AFSA, CSS-FAF, FAF, FAT, FFS, Stafford, IRS, SAR, Supplemental, Institutional

ENGINEERING COLLEGE INFORMATION

[For additional personnel, refer to the *Appendix*.]

HEAD OF ENGINEERING
Thomas P. Kicher
Phone: (216)368-4436 **Fax:** (216)368-6939
EMail: Internet:tpk@po.cwru.edu

ENGINEERING COLLEGE ADDRESS
The Case School of Engineering
Case Western Reserve University
10900 Euclid Ave.
Glennan Building
Cleveland, OH 44106-7220
Phone: (216)368-4436 **Fax:** (216)368-6939

ENROLLMENTS—BY CLASS
[Numbers are baccalaureate enrollments for the fall 1991 term, unless otherwise footnoted.]
1st-year students/Freshmen .... 285 [1]
2nd-year students/Sophomores 244
3rd-year students/Juniors 285
4th-year students/Seniors 307
Total **1,121**

Notes: (1) Freshmen admitted to University and not specific engineering programs.

NUMBER OF DEGREES AWARDED—BY PROGRAM

[Numbers are engineering baccalaureate degrees awarded during the 1991-1992 academic year, unless otherwise footnoted. For full details about each engineering program, refer to the *Tables of Degree Programs*.]

Aerospace 1
Biomedical Engineering 23
Chemical Engineering 26
Civil Engineering 7
Computer Engineering 23
Computer Science & Engineering 5
Electrical Engineering & Applied Physics 36
Fluid & Thermal Engineering 8
Industrial Engineering 5
Macromolecular Science & Engineering 4
Materials Science & Engineering 4
Mechanical Engineering 35
Polymer Science & Engineering 4
Systems & Control Engineering 17
Total **198**

PERCENTAGE OF DEGREES AWARDED—BY CATEGORY

[Percentages are of all engineering baccalaureate degrees awarded during the 1991-1992 academic year, unless otherwise footnoted.]

African Americans 4.5%
Asian/Pacific Island Americans 14.1%
Hispanic Americans 1.5%
Native Americans 0.2%
Foreign Citizens 9.1%
All Others 70.6%
Women 26.0%
Persons with disabilities 0.8%
Students over 25 years of age 0.7%

GRADUATE ENROLLMENTS & DEGREES AWARDED

Master's enrollment 620
Master's degrees awarded 155
Doctoral enrollment 243
Doctoral degrees awarded 64

ENGINEERING STUDENT DATA

Applicants to the engineering college
Number of applicants to engineering college 1,400
Percent offered admission 50.0%

Matriculated engineering students
Percentage in top quartile (25%) of High School class 93.0
Average SAT scores: Math—660, Verbal—550, Combined—1210
Average ACT scores: Math—31, Composite—29

FULL & PART TIME FACULTY—BY DEPARTMENT

[Figures are the head count of full-time faculty and the full-time equivalent (FTE) of part-time faculty for each engineering department or equivalent.]

Department	Full	Part
Chemical Eng	10	–
Civil Eng	7	–
Electrical Eng & Applied Physics	12	–
Systems Eng	8	–
Mechanical & Aerospace Eng	15	–
Biomedical Eng	16	–
Computer Eng & Science	11	–
Macromolecular Science	13	–
Material Science & Eng	11	–
Systems Eng	8	–

COLLEGE DESCRIPTION

The Case School of Engineering at Case Western Reserve University presents a unique personality. It is one of the leading research units in the country as determined by research productivity when normalized for the number of full-time faculty. Yet, all of these over 110 full-faculty involved in teaching as well, a sizeable percentage at the under-graduate level. Simply stated, research and consulting are the means by which our faculty practice their profession. The University's compact size and geographical unity also permits considerable interaction between the Case School of Engineering and other University units. Thus double degree programs, such as music and mechanical engineering, are not an un-usual occurrence. A special five year program leading to a M.S. degree in addition to the B.S. is also available as are co-op programs with a large number of leading regional and national companies.

DEGREE PROGRAMS DESCRIPTION

B. S. Level - In addition to the tradition accredited undergraduate degree programs, we have several accredited undergraduate degree programs which are not found at a majority of major engineering schools in the country. These include: biomedical engineering, fluid and thermal engineering science, polymer science and engineering, and systems and control engineering. M.S./Ph.D. level - The same programs described at the B.S. Level are also available at the M.S. and Ph.D. levels. These two are in addition to the more traditional graduate degree programs. The University's proximity to NASA Lewis (located just west of Cleveland presents an opportunity for graduate students in every engineering department. Many graduate students are involved in research supported through Ohio's high-tech Edison Program. The University's medical school and the number of contiguous medical centers provide additional research opportunities in a number of engineering disciplines.

DUAL DEGREE PROGRAMS

Enrollment requirements: The number of transfers was 24, and transfers other than through articulation agreement was 43. Satisfactory grades from college attended, 2.8 or better preferred. Adequate high school courses in Math, Science & English or equivalent.

TRANSFER INFORMATION

Residency Requirements: 60 semester hours

Transfer via Articulation Agreements
Admission to engineering: Binary Program - We have signed agreements with approximately 100 liberal arts institutions that do not offer engineering programs. Students are admitted to the junior year of the engineering program
Engineering graduates transferred from . . 4-yr: 24 2-yr: –
Requirements: Satisfy Case Core requirements as outlined in each articulation agreement. Have 3.0 in Math & Science Laboratory and 3.0 overall.

Transfer without Articulation Agreements
Admission to engineering: At any time
Engineering graduates transferred from . . 4-yr: 12 2-yr: 12
Requirements: Transfer students must meet the same admission requirements as freshman applicants.

GRADUATION REQUIREMENTS

(1) A minimum of 125-137 hours as specified for major. (2) English Composition (3) The Case Core (4) Humanities or social science classes (4 in a sequence) (5) Major requirements

STUDENT PROGRAMS

PROFESSIONAL AND HONORARY SOCIETIES
[For key to acronyms, see *Introduction*.]
Professional Societies: ACerS, ACM, AIAA, AIChE, AIME, ASCE, ASME, BMES, IEEE, SPE*
Honorary Societies: Eta Kappa Nu, Tau Beta Pi, Theta Tau, Mortarboard, Order of Omega

SUPPORT PROGRAMS
Student Chapter Organizations: Society of Women Engineers, National Society of Black Engineers, Society of Hispanic Professional Engineers

Minority Engineering Industrial Opportunity Prog.
For: Women & Ethnic Minorities **Available:** Year round
Offered: Prematriculation, Freshman, Sophomore, Junior, Senior

Society of Women Engineers
For: Women & Ethnic Minorities **Available:** Academic year
Offered: Freshman, Sophomore, Junior, Senior

Minority Scholars Program
For: Women & Ethnic Minorities **Available:** Year round
Offered: Freshman, Sophomore, Junior, Senior

Other Engineering Support Programs: (1) Faculty Advising Program - Faculty Advise Students (2) Pre-Freshman Program - 5 to 6 weeks during the summer, minority students come to school to take courses to get help before taking these same courses in the fall. (3) Walk in Tutoring - Tutoring in the Dorms on Basic Math and Science (4) Free Tutoring Program - Tutoring on a one-to-one basis.

038 The Catholic University of America

INSTITUTION PROFILE
HEAD OF THE INSTITUTION
Patrick Ellis
Phone: (202)319-5100   Fax: (202)319-4441

GENERAL INFORMATION
[All Students—Fall 1991]
Undergraduate enrollment 2,881
Graduate enrollment 3,700
Total institution enrollment **6,581**
Type of institution: Private   **Calendar:** Semesters
Nearest city: Washington, DC   **Population:** 684,000
Miles from main campus: 3   **Setting:** Urban
Types of engineering degrees: Engineering

ENGINEERING ADMISSIONS
ADMISSIONS OFFICE CONTACT
David R. Gibson
The Catholic University of America
620 Michigan Ave., N.E.
Washington, DC 20064
Phone: (202)319-5305   Fax: (202)319-5831

ENGINEERING COLLEGE ADMISSIONS INFORMATION
Admission to the engineering college: A prospective engineering student applies to engineering or to a specific program within engineering at the time of application to the university.
Entrance Requirements: SAT (Score: 1050), High School courses—English (4 years), Science (3 years), Mathematics (4 years), SAT Math Achievement I or II, English, Science
Entrance Recommendations: High School courses—Foreign languages (2 years), History (2 years)
Requirements for foreign students: TOEFL (Score: 500)

DEPARTMENTAL ADMISSIONS INFORMATION
Admission to the engineering department: At the time of admission to the institution.
Additional information: Mathematics through pre-calculus

FINANCIAL INFORMATION
ESTIMATED EXPENSES (ACADEMIC YEAR)
[Expenses are for the 1992-93 nine-month academic year.]

All Students	Undergraduate	Graduate
Tuition and fees	$12,664	$12,664
College room and board	$ 6,000	$ 6,000
Books and supplies	$ 1,000	$ 1,000
Other expenses	$ 200	$ 200
Total estimated expenses	**$19,864**	**$19,864**

FINANCIAL AID OFFICE CONTACT
Doris Torosian, *Director of Financial Aid*
The Catholic University of America
620 Michigan Ave., N.E.
Washington, DC 20064
Phone: (202)319-5307

GENERAL FINANCIAL AID INFORMATION
Forms accepted/required: FAF, Institutional
Additional financial aid information: merit-based scholarships; part-time jobs on campus; college work-study program.

ENGINEERING COLLEGE INFORMATION
[For additional personnel, refer to the *Appendix*.]
HEAD OF ENGINEERING
John J. McCoy
Phone: (202)319-5160   Fax: (202)319-4499

ENGINEERING COLLEGE ADDRESS
School of Engineering
The Catholic University of America
620 Michigan Ave., N.E.
Washington, DC 20064
Phone: (202)319-5200   Fax: (202)319-4499

ENROLLMENTS—BY CLASS
[Numbers are baccalaureate enrollments for the fall 1991 term, unless otherwise footnoted.]
1st-year students/Freshmen 62
2nd-year students/Sophomores 53
3rd-year students/Juniors 45
4th-year students/Seniors 68
Total **228**

NUMBER OF DEGREES AWARDED—BY PROGRAM
[Numbers are engineering baccalaureate degrees awarded during the 1991-1992 academic year, unless otherwise footnoted. For full details about each engineering program, refer to the *Tables of Degree Programs*.]
Biomedical Engineering 17
Civil Engineering 14
Computer Science Engineering 2
Computer Science/Engineering .... –
Electrical Engineering 8
Mechanical Engineering 12
Total **53**

PERCENTAGE OF DEGREES AWARDED—BY CATEGORY
[Percentages are of all engineering baccalaureate degrees awarded during the 1991-1992 academic year, unless otherwise footnoted.]
African Americans 2.2%
Asian/Pacific Island Americans 6.6%
Hispanic Americans 6.6%
Native Americans .... – %
Foreign Citizens 17.1%
All Others 67.5%
Women 28.4%
Persons with disabilities .... – %
Students over 25 years of age .... – %

GRADUATE ENROLLMENTS & DEGREES AWARDED
Master's enrollment 314
Master's degrees awarded 71
Doctoral enrollment 65
Doctoral degrees awarded 6

ENGINEERING STUDENT DATA
Applicants to the engineering college
Number of applicants to engineering college 201
Percent offered admission 90.0%

Matriculated engineering students
Percentage in top quartile (25%) of High School class 45.0
Average SAT scores: Math—642, Verbal—536, Combined—1178

FULL & PART TIME FACULTY—BY DEPARTMENT
[Figures are the head count of full-time faculty and the full-time equivalent (FTE) of part-time faculty for each engineering department or equivalent.]

Department	Full	Part
Civil Eng	7	4.5
Electrical Eng	8	–
Mechanical Eng	12	1.0

COLLEGE DESCRIPTION
At CUA we take pride in being able to offer to students a professional engineering education given within the context of a University rich in the tradition of humanistic values; and given by a strong research-oriented faculty. A student-faculty ratio in undergraduate engineering of 8:1 ensures individual attention for our students. Scholarships, co-op and intern opportunities as well as undergraduate participation in faculty research projects are some additional features of engineering education at CUA.

DEGREE PROGRAMS DESCRIPTION
At the B.S. level, see above. At the Graduate levels our special features are: 1) Diversity of course selection, 2) Non-thesis option for Masters degree programs, 3) Strong interaction with government and industry, 4) Evening hours, 5) Individualized doctoral programs.

DUAL DEGREE PROGRAMS
Engineering baccalaureate graduates with dual degrees 0

TRANSFER INFORMATION
Residency Requirements: One year of residency and completion of the final 30 credits at CUA.
Transfer via Articulation Agreements
Admission to engineering: At any time
Engineering graduates transferred from   ... 4-yr: 2   2-yr: 6
Requirements: Approximately 3.0 average.
Transfer without Articulation Agreements
Admission to engineering: At any time
Engineering graduates transferred from   ... 4-yr: 2   2-yr: 6
Requirements: Approximately 3.0 average.

GRADUATION REQUIREMENTS
A GPA of 2.0 at the bachelors level and a GPA of 3.0 at the Graduate levels.

STUDENT PROGRAMS
PROFESSIONAL AND HONORARY SOCIETIES
[For key to acronyms, see *Introduction*.]
Professional Societies: ASME, IEEE, Society of Women Engineers, Society of Hispanic Professionals
Honorary Societies: Tau Beta Pi

SUPPORT PROGRAMS
Student Chapter Organizations: Society of Women Engineers, Society of Hispanic Professional Engineers
Other Engineering Support Programs: Fall orientation for all freshmen and transfers; counseling services; tutoring services; foreign student office; veterans' services; services for the disabled; placement services.

039 University of Central Florida

INSTITUTION PROFILE

HEAD OF THE INSTITUTION
John C. Hitt
Phone: (407)823-1823   Fax: (407)823-5407

GENERAL INFORMATION
[All Students—Fall 1991]
Undergraduate enrollment 16,640
Graduate enrollment 4,627
Total institution enrollment 21,267
Type of institution: Public   **Calendar:** Semesters
Nearest city: Orlando Metropolitan Area   **Population:** 1,000,000
Miles from main campus: 8   **Setting:** Suburban
Types of engineering degrees: Engineering & Technology
Other degree-granting colleges: Arts & Sciences, Business Administration, Education, Health and Public Affairs

ENGINEERING ADMISSIONS

ADMISSIONS OFFICE CONTACT
John F. Bush
University of Central Florida
4000 University Blvd.
Orlando, FL 32816-0450
Phone: (407)823-3000   Fax: (407)823-5652

ENGINEERING COLLEGE ADMISSIONS INFORMATION
Entrance Requirements: SAT (Score: 1000), ACT (Score: 19), High School courses—Advanced Algebra, Geometry, Trigonometry (3.5 years), Natural Science (3 years), Social Science (3 years), English (4 years), Test of Standard Written English (TSWE), minimum score required: 30
Entrance Recommendations: High School courses—Foreign Language (Required) (2 years), Calculus (1 year), Computer Programming (FORTRAN preferred)
Requirements for foreign students: TOEFL (Score: 550); financial statement

DEPARTMENTAL ADMISSIONS INFORMATION
Admission to the engineering department: At the time of admission to the institution.

FINANCIAL INFORMATION

ESTIMATED EXPENSES (ACADEMIC YEAR)
[Expenses are for the 1992-93 nine-month academic year.]

State Residents	Undergraduate	Graduate
Tuition and fees	$ 1,525 [1]	$ 2,165 [2]
College room and board	$ 3,700 [3]	$ 3,700 [3]
Books and supplies	$ 550 [C]	$ 500 [C]
Other expenses	$ 1,100 [4]	$ 1,100 [4]
Total estimated expenses	**$ 6,875**	**$ 7,465**

Out-of-State Residents	Undergraduate	Graduate
Tuition and fees	$ 5,662 [1]	$ 7,025 [2]
College room and board	$ 3,700 [3]	$ 3,700 [3]
Books and supplies	$ 550 [C]	$ 500 [C]
Other expenses	$ 1,100 [4]	$ 1,100 [4]
Total estimated expenses	**$11,012**	**$12,325**

Notes: (C) Estimated data. (1) Based on 15 hrs Fall and Spring semesters (2) Based on 12 hrs, Spring and Fall semesters (3) Dependent on accommodations and meal plan (4) Estimated transportation costs

FINANCIAL AID OFFICE CONTACT
Mary H. McKinney, *Director of Financial Aid*
University of Central Florida
4000 University Blvd.
Orlando, FL 32816-0450
Phone: (407)823-2827   Fax: (407)823-5652

GENERAL FINANCIAL AID INFORMATION
Forms accepted/required: ACT, CSS-FAF, Institutional
Additional financial aid information: Need-based scholarships; short-term loans; merit-based scholarships; part-time jobs on campus; College Career Work Experience Program.

ENGINEERING COLLEGE INFORMATION

[For additional personnel, refer to the *Appendix*.]

HEAD OF ENGINEERING
Gary E. Whitehouse
Phone: (407)823-2156   Fax: (407)823-5483
EMail: WHITEHSE @ UCF1VM.CC.UCF.EDU

ENGINEERING COLLEGE ADDRESS
College of Engineering
University of Central Florida
4000 University Blvd.
Orlando, FL 32816-0450
Phone: (407)823-2156   Fax: (407)823-5483

ENROLLMENTS—BY CLASS
[Numbers are baccalaureate enrollments for the fall 1991 term, unless otherwise footnoted.]

1st-year students/Freshmen .... 565 [1]
2nd-year students/Sophomores 303
3rd-year students/Juniors 399
4th-year students/Seniors .... 1,186 [2]
Total 2,453

Notes: (1) Freshmen hrs; Soph. 30-60 hrs; Junior 60-90 hrs; Senior 90 hrs Unclassified students (123) placed in Freshmen no. (2) Includes 214 unclassifieds and post-Baccs

NUMBER OF DEGREES AWARDED—BY PROGRAM
[Numbers are engineering baccalaureate degrees awarded during the 1991-1992 academic year, unless otherwise footnoted. For full details about each engineering program, refer to the *Tables of Degree Programs*.]

Aerospace Engineering 12
Civil Engineering 34
Computer Engineering 30
Electrical Engineering 152
Engineering Technology 96
Environmental Engineering 11
Industrial Engineering 28
Mechanical Engineering 50
Total 413

PERCENTAGE OF DEGREES AWARDED—BY CATEGORY
[Percentages are of all engineering baccalaureate degrees awarded during the 1991-1992 academic year, unless otherwise footnoted.]

African Americans 3.3%
Asian/Pacific Island Americans 8.4%
Hispanic Americans 8.3%
Native Americans 0.1%
Foreign Citizens .... 12.0% [C]
All Others 67.9%
Women 17.2%
Persons with disabilities 0.2%
Students over 25 years of age 15.0%

Notes: (C) Estimated data.

GRADUATE ENROLLMENTS & DEGREES AWARDED
Master's enrollment .... 775 [2]
Master's degrees awarded 139
Doctoral enrollment 107
Doctoral degrees awarded 6

Notes: (2) Includes 214 unclassifieds and post-Baccs

ENGINEERING STUDENT DATA

Applicants to the engineering college
Number of applicants to engineering college 1,485
Percent offered admission 13.8%

Matriculated engineering students
Percentage in top quartile (25%) of High School class .... 60.0 [C]
Average SAT scores: Math—596, Verbal—484
Average ACT scores: Composite—25

Notes: (C) Estimated data.

FULL & PART TIME FACULTY—BY DEPARTMENT
[Figures are the head count of full-time faculty and the full-time equivalent (FTE) of part-time faculty for each engineering department or equivalent.]

Department	Full	Part
Civil & Environmental Eng	18	—
Electrical & Computer Eng	46	—
Industrial Eng	18	—
Mechanical & Aerospace Eng	20	—

COLLEGE DESCRIPTION

The University of Central Florida and its College of Engineering are situated in a dynamic, growing region and have kept pace with these developments. The four-year undergraduate engineering curricula emphasize engineering fundamentals enhanced by computer-aided analysis and design. Each engineering program offers an honors in the major for qualifying students, and the college has strong, productive programs to attract and assist minorities and women. Average number of years required to actually complete the bachelor's degree: 4 1/2 (estimated).

DEGREE PROGRAMS DESCRIPTION

The B.S. degrees in the various engineering programs prepare the graduate for entry-level professional positions or for graduate study in each of the respective fields. Graduate study in engineering can be carried out in the general areas of M. S. E. and Ph. D. in engineering, which emphasize the applications of the natural sciences to the analysis and solution of engineering problems.

TRANSFER INFORMATION

Residency Requirements: All students must complete at least two semesters (30 hours) in residence.

Transfer via Articulation Agreements
Admission to engineering: In general and within curriculum, space, and fiscal limitations, admission as a junior to the upper division shall be granted to AA graduates from Florida public institutions.
Engineering graduates transferred from   4-yr: 16   2-yr: 219
Requirements: See the current UCF catalog.

Transfer without Articulation Agreements
Admission to engineering: At any time
Engineering graduates transferred from   4-yr: 25   2-yr: 50
Requirements: All college transfer applicants with fewer than 60 semester hours of acceptable credit must be in good standing and eligible to return to the last institution attended as a degree-seeking student, meet High School unit entrance requirements with at least a 3.0 High School academic GPA and a minimum SAT total score of 1000 or an ACT composite of 23/24 enhanced, and have a B average for all college-level academic courses attempted.

GRADUATION REQUIREMENTS

BS programs in engineering require a minimum of 132 semester hours and a minimum 2.250/4.000 GPA in pre-engineering core, engineering core, and major discipline coursework areas. University requirements include passing the State of Florida College Level Academic Skills Test (CLAST). See the current catalog for additional graduation requirements.

STUDENT PROGRAMS

PROFESSIONAL AND HONORARY SOCIETIES
[For key to acronyms, see *Introduction*.]

Professional Societies: ACM, AIAA, ASCE, ASME, IEEE, IIE, ISA, OSA, SAE, SME, Florida Engineering Society
Honorary Societies: Alpha Pi Mu, Eta Kappa Nu, Pi Tau Sigma, Tau Alpha Pi, Tau Beta Pi

SUPPORT PROGRAMS
Student Chapter Organizations: Society of Women Engineers, National Society of Black Engineers, Society of Hispanic Professional Engineers

Southeastern Consortium for Minorities in Engineering
For: Ethnic Minorities   **Available:** Year round
Offered: Prematriculation

Re-entry Program
For: Women   **Available:** Year round
Offered: Prematriculation, Freshman, Sophomore, Junior, Senior, Graduate level

Other Engineering Support Programs: Orientation and advisement for freshman and transfer students; professional and peer academic advisement; counseling and testing services; academic and career resource and planning services; veterans services; evening/weekend student services.

040 Central State University

INSTITUTION PROFILE
HEAD OF THE INSTITUTION
Arthur E. Thomas
Phone: (513)376-6332 **Fax:** (513)376-6530
GENERAL INFORMATION
[All Students—Fall 1991]
Undergraduate enrollment 3,258
Graduate enrollment 8
Total institution enrollment 3,266
Type of institution: Public **Calendar:** Quarters
Nearest city: Dayton, Ohio **Population:** 205,000
Miles from main campus: 22 **Setting:** Small Town
Types of engineering degrees: Engineering
Other degree-granting colleges: Arts & Sciences, Business Administration, Education, University College (Developmental/Remedial)

ENGINEERING ADMISSIONS
ADMISSIONS OFFICE CONTACT
Robert E. Johnson
Central State University
1400 Brush Row Road
Wilberforce, OH 45384
Phone: (513)376-6348 **Fax:** (513)376-6648

ENGINEERING COLLEGE ADMISSIONS INFORMATION
Admission to the engineering college: Upon certification by the University College that any remedial study requirements have been satisfied.
Entrance Requirements: ACT, High School courses—English (4 years), Mathematics & Science (6 years), Social Sciences (3 years), Foreign Language (2 years), ACT Test is required for admission, but the University has no minimum required score.
Entrance Recommendations: High School courses—Physics (1 year), Chemistry (1 year), Trigonometry/Analytical Geometry (1 year)
Requirements for out-of-state residents: Cumulative GPA of no less than 2.0 or show definite academic improvement.
Requirements for foreign students: TOEFL (Score: 500); financial statement

DEPARTMENTAL ADMISSIONS INFORMATION
Admission to the engineering department: Upon certification by the University College that any remedial study requirements have been satisfied.

FINANCIAL INFORMATION
ESTIMATED EXPENSES (ACADEMIC YEAR)
[Expenses are for the 1992-93 nine-month academic year.]

State Residents	Undergraduate	Graduate
Tuition and fees	$ 2,679	$ –
College room and board	$ 4,293	$ –
Books and supplies	$ 750	$ –
Other expenses	$ 125 (2)	$ – (1)
Total estimated expenses	**$ 7,847**	**$ –**

Out-of-State Residents	Undergraduate	Graduate
Tuition and fees	$ 5,895	$ –
College room and board	$ 4,293	$ –
Books and supplies	$ 750	$ –
Other expenses	$ 125 (2)	$ – (1)
Total estimated expenses	**$11,063**	**$ –**

Notes: (1) No graduate programs available. (2) Includes New Student Orientation Fee: $125 Fall; $25 Winter or Spring.

FINANCIAL AID OFFICE CONTACT
Sunny Terrell, *Executive Director, Financial Aid*
Central State University
1400 Brush Row Road
Wilberforce, OH 45384
Phone: (513)376-6575 **Fax:** (513)376-6530

GENERAL FINANCIAL AID INFORMATION
Forms accepted/required: FAF
Additional financial aid information: Central State University offers financial aid to all eligible students, based upon financial need and academic standing. The FAF Needs Analysis Report is utilized to determine applicant need for the three types of financial aid: grants, loans, employment. Financial aid package may include Pell Grant, Stafford or Plus/SLS Loan, College Work Study, Ohio Instructional Grant, institutional scholarship, etc.

ENGINEERING COLLEGE INFORMATION
[For additional personnel, refer to the *Appendix*.]
HEAD OF ENGINEERING
Melvin A. Johnson Jr.
Phone: (513)376-6324 **Fax:** (513)376-6530
ENGINEERING COLLEGE ADDRESS
College of Arts and Sciences
Central State University
1400 Brush Row Road
Wilberforce, OH 45384
Phone: (513)376-6324 **Fax:** (513)376-6530

ENROLLMENTS—BY CLASS
[Numbers are baccalaureate enrollments for the fall 1991 term, unless otherwise footnoted.]
1st-year students/Freshmen 62
2nd-year students/Sophomores 18
3rd-year students/Juniors 12
4th-year students/Seniors 8
Total 100

NUMBER OF DEGREES AWARDED—BY PROGRAM
[Numbers are engineering baccalaureate degrees awarded during the 1991-1992 academic year, unless otherwise footnoted. For full details about each engineering program, refer to the *Tables of Degree Programs*.]
Manufacturing 4
Total 4

PERCENTAGE OF DEGREES AWARDED—BY CATEGORY
[Percentages are of all engineering baccalaureate degrees awarded during the 1991-1992 academic year, unless otherwise footnoted.]
African Americans .... 75.0 %
Asian/Pacific Island Americans .... – %
Hispanic Americans .... – %
Native Americans .... – %
Foreign Citizens .... – %
All Others .... 25.0 %
Women .... 50.0 %
Persons with disabilities .... – %
Students over 25 years of age .... – %

ENGINEERING STUDENT DATA
Applicants to the engineering college
Number of applicants to engineering college .... 70 (C)
Percent offered admission .... 88.6% (C)

Matriculated engineering students
Percentage in top quartile (25%) of High School class .... – (A)

Notes: (A) Data not available. (C) Estimated data.

FULL- & PART-TIME FACULTY—BY DEPARTMENT
[Figures are the head count of full-time faculty and the full-time equivalent (FTE) of part-time faculty for each engineering department or equivalent.]

Department	Full	Part
Manufacturing Eng	7	0.3

COLLEGE DESCRIPTION
Central State University is situated in a quiet rural setting within a fifty-mile radius of the major metropolitan areas of Columbus, Cincinnati and Dayton. The University is classified as an Historically Black College or University (HBCU) with a long history of meeting the special needs of minorities and students with deficiencies in high school preparation. The Manufacturing Engineering program is one of only ten such programs in the nation with EAC/ABET accreditation at the baccalaureate level. Though no graduate engineering programs are available, the Manufacturing Engineering Department has maintained extensive research involvement and networking with key research consortia, industrial partners and government agencies. Undergraduates may gain valuable engineering experience as research assistants.

DEGREE PROGRAMS DESCRIPTION
Special features: The B.S. degree program in manufacturing engineering at Central State University is one of only ten such programs in the nation with EAC/ABET accreditation. The program follows the guidelines of the Society of Manufacturing Engineers (SME) and has been designed to develop engineers prepared to address industry demands for increasingly sophisticated manufacturing techniques. The curriculum consists of major components of mathematics, chemistry, physics, engineering sciences, and humanities-social sciences, together with the engineering major requirements which can be grouped into the following basic areas: Materials and Processes; Product and Tool Engineering; Computer Control and Manufacturing Systems; Productivity and Quality. The program provides for hands-on experience in the application of CNC, CAD/CAM, robotics, machine vision, computers and microprocessors. Throughout the curriculum, major emphasis is given to the engineering design function. Graduates have gained employment with major companies in the automotive, aerospace and food industries and with government research laboratories. Others have selected jobs with small manufacturing firms or been admitted to graduate engineering programs at major universities.

TRANSFER INFORMATION
Residency Requirements: Must complete a minimum of 45 quarter credit hours at Central State University.

Transfer via Articulation Agreements
Admission to engineering: At any time
Engineering graduates transferred from ... 4-yr: – 2-yr: –
Requirements: In good academic standing at time of transfer or remain out of school for minimum of one quarter.

Transfer without Articulation Agreements
Admission to engineering: At any time
Engineering graduates transferred from ... 4-yr: 1 2-yr: –
Requirements: In good academic standing at the time of transfer or remain out of school for a minimum of one quarter

GRADUATION REQUIREMENTS
Graduation requirements: The B.S. degree in manufacturing engineering requires a total of 219 quarter credit hours with a minimum cumulative GPA of 2.0/4.0. Prior to attaining junior status, all students must take the Collegiate Assessment of Academic Proficiency (CAAP) or any appropriate substitute examination determined by the University. Graduates must pass a Junior English Proficiency (JEP) examination and must take a senior comprehensive exam in manufacturing engineering. A $50 graduation fee is required.

STUDENT PROGRAMS
PROFESSIONAL AND HONORARY SOCIETIES
[For key to acronyms, see *Introduction*.]
Professional Societies: ASME, SME

SUPPORT PROGRAMS
Student Chapter Organizations: National Society of Black Engineers, National Technical Association (NTA)
Other Engineering Support Programs: Freshman orientation programs are provided. The University College has oversight responsibility to remedy deficiencies in high school course work and/or placement examinations scores. Tutorial assistance in math, English and reading is normally available. The Career Services office aids students in securing summer internships and permanent employment upon graduation.

041 Christian Brothers University

INSTITUTION PROFILE

HEAD OF THE INSTITUTION
T. "Brother" Drahmann FSC
Phone: (901)722-0250 Fax: (901)722-0494

GENERAL INFORMATION
[All Students—Fall 1991]
Undergraduate enrollment 1,546
Graduate enrollment 184
Total institution enrollment 1,730

Type of institution: Private **Calendar:** Semesters
Location: Memphis **Population:** 650,000
Setting: Urban
Types of engineering degrees: Engineering
Other degree-granting colleges: Arts, Business, Sciences

ENGINEERING ADMISSIONS

ADMISSIONS OFFICE CONTACT
M. "Brother" Smith FSC
Christian Brothers University
650 East Parkway South
Memphis, TN 38104-5581
Phone: (901)722-0205 Fax: (901)722-0494

ENGINEERING COLLEGE ADMISSIONS INFORMATION
Entrance Requirements: SAT (Score: 800), ACT (Score: 20), High School courses—English (4 years), Mathematics (3 years)
Entrance Recommendations: SAT (Score: 800), ACT (Score: 20), High School courses—Mathematics with geometry, analysis and calculus (4 years), Science with chemistry and physics (3 years)
Requirements for foreign students: TOEFL (Score: 500); financial statement; Transcripts of work done at high school level or above.

DEPARTMENTAL ADMISSIONS INFORMATION
Admission to the engineering department: At the time of admission to the institution.

FINANCIAL INFORMATION

ESTIMATED EXPENSES (ACADEMIC YEAR)
[Expenses are for the 1992-93 nine-month academic year.]

All Students	Undergraduate	Graduate
Tuition and fees	$ 8,390	$ 3,375[1]
College room and board	$ 3,080	—[B]
Books and supplies	$ 650	$ 360
Other expenses	$ 90[2]	—
Total estimated expenses	$12,210	$ 3,735

Notes: (B) Data not applicable. (1) Graduate program is part-time, evening, costing $225 per credit hour. (2) Lab fee of $45 charged for each engineering and science lab.

FINANCIAL AID OFFICE CONTACT
Sandi Mayo, *Director of Financial Aid*
Christian Brothers University
650 East Parkway South
Memphis, TN 38104
Phone: (901)722-0306 Fax: (901)722-0494

GENERAL FINANCIAL AID INFORMATION
Forms accepted/required: ACT, CSS-FAF, FAF, FFS, Institutional
Additional financial aid information: Over two-thirds of the University's student body receives some sort of financial assistance. About half of available aid is in the form of federal and state assistance. Programs included are: Pell Grants, Supplemental Educational Opportunity Grants, National Direct Student Loans, University Work Study, and Guaranteed Student Loans. Tennessee Student Assistance Awards are available to eligible state residents. Other financial aid programs are also available.

ENGINEERING COLLEGE INFORMATION
[For additional personnel, refer to the *Appendix*.]

HEAD OF ENGINEERING
Ray W. Brown
Phone: (901)722-0408 Fax: (901)722-0494

ENGINEERING COLLEGE ADDRESS
School of Engineering
Christian Brothers University
650 East Parkway South
Memphis, TN 38104-5581
Phone: (901)722-0405 Fax: (901)722-0494

ENROLLMENTS—BY CLASS
[Numbers are baccalaureate enrollments for the fall 1991 term, unless otherwise footnoted.]
1st-year students/Freshmen 75
2nd-year students/Sophomores 70
3rd-year students/Juniors 51
4th-year students/Seniors .... 59 [1]
Total 255

Notes: (1) In addition to the full-time students listed by year, there was a combined total of 36 part-time students enrolled for the academic year 1991-92.

NUMBER OF DEGREES AWARDED—BY PROGRAM
[Numbers are engineering baccalaureate degrees awarded during the 1991-1992 academic year, unless otherwise footnoted. For full details about each engineering program, refer to the *Tables of Degree Programs*.]
Chemical Engineering 9
Civil Engineering 14
Electrical Engineering 23
Engineering Management .... —[B]
Mechanical Engineering 15
Total 61

Notes: (B) Data not applicable.

PERCENTAGE OF DEGREES AWARDED—BY CATEGORY
[Percentages are of all engineering baccalaureate degrees awarded during the 1991-1992 academic year, unless otherwise footnoted.]
African Americans 16.3%
Asian/Pacific Island Americans 3.2%
Hispanic Americans .... —%
Native Americans .... —%
Foreign Citizens 6.5%
All Others 74.0%
Women 24.5%
Persons with disabilities .... —% [A]
Students over 25 years of age .... —% [A]

Notes: (A) Data not available.

GRADUATE ENROLLMENTS & DEGREES AWARDED
Master's enrollment 40
Master's degrees awarded 10
Doctoral enrollment .... —[B]
Doctoral degrees awarded .... —[B]

Notes: (B) Data not applicable.

FULL- & PART-TIME FACULTY—BY DEPARTMENT
[Figures are the head count of full-time faculty and the full-time equivalent (FTE) of part-time faculty for each engineering department or equivalent.]

Department	Full	Part
Chemical Eng	3	—
Civil Eng	4	—
Electrical Eng	6	2.0
Mechanical Eng	6	—
Eng Management	1	1.5

COLLEGE DESCRIPTION
The School of Engineering offers four-year undergraduate programs in Chemical, Civil, Electrical and Mechanical engineering as well as a two-year Master's program in Engineering Management. The programs at Christian Brothers University emphasize small classes and individual attention. Students are introduced to engineering design their first year so that they may immediately begin acquiring the skills necessary for successful careers in engineering. Design and engineering problem solving skills continue to be stressed throughout the four years. The process culminates with the Senior Design Project required of all students. Projects are as realistic as possible, some suggested by industrial partners and others resulting from student internships. Students are expected to include the social and moral context as well as technical and economic feasibility in their problem solving design. Average number of years required to actually complete the bachelor's degree: 4.7.

DEGREE PROGRAMS DESCRIPTION
The majors stress the fundamentals of the respective disciplines, using modern tools. Majors are supported by modern laboratories where students perform digital data acquisition and analysis, microprocessor hardware and software design and implementation, environmental analysis, process control and design, materials processing and testing, etc. Modern computer facilities support all four majors with hardware and software for computer aided drafting and design, circuit analysis and design, structural analysis and design, simulation and testing, economic and statistical analysis, spreadsheets and word processing.

DUAL DEGREE PROGRAMS
Engineering baccalaureate graduates with dual degrees 3
Enrollment requirements: Students may register at the 'host' school with the approval of the appropriate academic counselor and a letter from the registrar of the student's 'home' school. Students must have a 2.25 GPA at their 'home' institution. Only students with majors in chemistry, physics or math are eligible to enter engineering programs at the 'host' institution.

TRANSFER INFORMATION
Residency Requirements: 35 of the last 70 hours applied toward the degree and at least one-half of the upper division courses in the major must be earned at Christian Brothers University. Junior and senior level courses from non-EAC/ABET accredited schools will not be transferred.

Transfer without Articulation Agreements
Admission to engineering: Transfer students are admitted to the engineering program through starting junior status.
Requirements: Transfer students must have completed 12 hours or more at another accredited college or university, have a minimum equivalent QPR of 2.50 (on a 4.0 scale), as computed by the Registrar of CBU, and be in good standing at the college of last attendance.

GRADUATION REQUIREMENTS
B.S. programs in engineering require 132 to 135 semester hours, 2.0 GPA in the major courses, and 2.0 GPA overall. M.S. program requires completion of the required 33 semester hours, while maintaining a 3.0 GPA.

STUDENT PROGRAMS

PROFESSIONAL AND HONORARY SOCIETIES
[For key to acronyms, see *Introduction*.]
Professional Societies: AIChE, ASCE, ASME, IEEE, NSPE
Honorary Societies: Tau Beta Pi

SUPPORT PROGRAMS
Student Chapter Organizations: National Society of Black Engineers, Black Student Association

Early Identification Program (EIP)
For: Women & Ethnic Minorities **Available:** Summer only
Offered: Prematriculation

Other Engineering Support Programs: Summer, Fall and Spring Orientation program for all freshmen and transfer students; all engineering students have an assigned engineering faculty advisor; peer counseling; career counseling services; foreign student advising; veterans' counselor; placement services, internships.

042 University of Cincinnati

INSTITUTION PROFILE

HEAD OF THE INSTITUTION
Joseph A. Steger
Phone: (513)556-2201 Fax: (513)556-3010

GENERAL INFORMATION
[All Students—Fall 1991]
Undergraduate enrollment 27,500
Graduate enrollment 7,160
Total institution enrollment 34,660
Type of institution: Public   Calendar: Quarters
Location: Cincinnati   Population: 364,040
Setting: Urban
Types of engineering degrees: Engineering
Other degree-granting colleges: Arts & Sciences, Business Administration, Education, Law, Medicine, Nursing, Pharmacy, Technology & Applied Sciences, Conservatory of Music, Design, Architecture, Art, and Planning, Evening and Continuing Education, University College, Raymond Walters & Clermont Colleges

ENGINEERING ADMISSIONS

ADMISSIONS OFFICE CONTACT
Rudolph Jones
University of Cincinnati
Office of Admissions
One Edwards Center
Cincinnati, OH 45221-0091
Phone: (513)556-1100   Fax: (513)556-1105

ENGINEERING COLLEGE ADMISSIONS INFORMATION
Admission to the engineering college: Admission to College when admitted to University. Additional articulation req.: soc. sci.-2; foreign lan.-2; fine arts-1; additional college prep. subj.-1.
Entrance Requirements: SAT, ACT, High School courses—Mathematics, including trigonometry (4 years), Chemistry (1 year), English (4 years), Physics (1 year), Either SAT or ACT required. Freshman admission criteria - top 30% of high school class.
Requirements for foreign students: TOEFL (Score: 550); TOEFL or the Michigan Test of English Proficiency (min. 85). International applicants must secure their own cooperative employment and submit a letter from the co-op company stating that fact before admission will be considered. Letter of financial support required.

DEPARTMENTAL ADMISSIONS INFORMATION
Admission to the engineering department: At the time of admission to the institution.

FINANCIAL INFORMATION

ESTIMATED EXPENSES (ACADEMIC YEAR)
[Expenses are for the 1992-93 nine-month academic year.]

State Residents	Undergraduate	Graduate
Tuition and fees	$ 3,372	$ 5,247
College room and board	$ 4,431	$ 3,530[1]
Books and supplies	$ 735	$ 820
Other expenses	$ 300[2]	$ –
Total estimated expenses	$ 8,838	$ 9,597

Out-of-State Residents	Undergraduate	Graduate
Tuition and fees	$ 8,049	$ 10,320
College room and board	$ 4,431	$ 3,530[1]
Books and supplies	$ 735	$ 820
Other expenses	$ 300[2]	$ –
Total estimated expenses	$13,515	$14,670

Notes: (1) Graduate student efficiency apartment for 3 quarters. (2) $100 computer fee/academic qtr., 3 academic quarters in freshman and senior years; 2 academic quarters in sophomore, pre-junior and junior years.

FINANCIAL AID OFFICE CONTACT
James Williams, *Director of Student Financial Aid Office*
University of Cincinnati
52 Beecher Hall
Cincinnati, OH 45221-0125
Phone: (513)556-6982

GENERAL FINANCIAL AID INFORMATION
Forms accepted/required: CSS-FAF, FAF, FAR, Stafford, Supplemental
Additional financial aid information: Grants, loans, and work-study are three basic categories of need-based aid; and may be combined as an aid "package". The major grants are: Pell, OIG, and SEOG. The major loans are: Perkins, Stafford, PLUS, SLS and University Loans.

ENGINEERING COLLEGE INFORMATION
[For additional personnel, refer to the *Appendix*.]

HEAD OF ENGINEERING
Constantine N. Papadakis
Phone: (513)556-2933   Fax: (513)556-3626
EMail: CONSTANTINE.PAPADAKIS@UC.EDU

ENGINEERING COLLEGE ADDRESS
College of Engineering
University of Cincinnati
645 Baldwin Hall
Cincinnati, OH 45221-0018
Phone: (513)556-5424   Fax: (513)556-3626

ENROLLMENTS—BY CLASS
[Numbers are baccalaureate enrollments for the fall 1991 term, unless otherwise footnoted.]
1st-year students/Freshmen .... 472[1]
2nd-year students/Sophomores .... 418[1]
3rd-year students/Juniors .... 715[1]
4th-year students/Seniors .... 352[1]
Total 1,957

Notes: (1) College undergraduate programs are 5-years, enrollment was 386 for pre-juniors and 329 for juniors. Enrollment figures for Fall 1992.

NUMBER OF DEGREES AWARDED—BY PROGRAM
[Numbers are engineering baccalaureate degrees awarded during the 1991-1992 academic year, unless otherwise footnoted. For full details about each engineering program, refer to the *Tables of Degree Programs*.]
Aerospace Engineering 42
Chemical Engineering 50
Civil Engineering 47
Computer Engineering .... 6 (B)
Computing Sciences .... – (B)
Electrical Engineering 50
Engineering Mechanics 18
Environmental Engineering .... – (B)
Environmental Science .... – (B)
Health Physics .... – (B)
Industrial Engineering 31
Materials Engineering 24
Materials Science .... – (B)
Mechanical Engineering 23
Metallurgical Engineering .... – (5)
Nuclear Engineering 11
Solid State Electronics .... – (B)
Total 302

Notes: (B) Data not applicable. (5) Graduating classes of 90 and 91 receive BS in Metallurgical Engineering. Class of 92 and beyond receive BS in Materials Engineering.

PERCENTAGE OF DEGREES AWARDED—BY CATEGORY
[Percentages are of all engineering baccalaureate degrees awarded during the 1991-1992 academic year, unless otherwise footnoted.]
African Americans .... 1.4 %
Asian/Pacific Island Americans .... 3.4 %
Hispanic Americans .... – %
Native Americans .... – %
Foreign Citizens .... 0.6 %
All Others .... 94.6 %
Women .... 14.2 %
Persons with disabilities .... – % (A)
Students over 25 years of age .... – % (A)

Notes: (A) Data not available.

GRADUATE ENROLLMENTS & DEGREES AWARDED
Master's enrollment .... 766[2]
Master's degrees awarded 192
Doctoral enrollment .... 326[2]
Doctoral degrees awarded 56

Notes: (2) Enrollment figures are for Fall 1992.

ENGINEERING STUDENT DATA
Applicants to the engineering college
Number of applicants to engineering college .... 1,615[1]
Percent offered admission .... 61.7 %[1]
Matriculated engineering students
Percentage in top quartile (25%) of High School class .... 79.8[2]
Average SAT scores: Math—612[3], Verbal—500[3], Combined—1112[3]
Average ACT scores: Composite—24.0

Notes: (1) Applicant data for Fall 1992. (2) Of the freshmen enrolled Fall 1992, 41.8% ranked in top 10%, and 87% ranked top 33% of their high school graduating class. (3) Enrollment data for Fall 1992.

FULL- & PART-TIME FACULTY—BY DEPARTMENT
[Figures are the head count of full-time faculty and the full-time equivalent (FTE) of part-time faculty for each engineering department or equivalent.]

Department	Full	Part
Chemical Eng	17	–
Civil & Environmental Eng	24.0	–
Electrical & Computer Eng	30	–
Mechanical, Industrial, & Nuclear Eng	30	–
Aerospace Eng & Eng Mechanics	24	–
Materials Science & Eng	17	–
Mechanical, Industrial, & Nuclear Eng	30	–

COLLEGE DESCRIPTION

All Baccalaureate students in the College of Engineering study under the cooperative plan of education. The five-year curricula includes twelve quarters of academic study and up to seven quarters of related practical experience in industry. Upon satisfactory completion of academic and professional practice (co-op) requirements, students are granted a B.S. degree in their major. A pre-medical program and dual degree programs are available. Average number of years required to actually complete the bachelor's degree: 5.2.

DEGREE PROGRAMS DESCRIPTION
Baccalaureate programs are currently accredited by the Accreditation Board for Engineering and Technology (EAC/ABET), except for materials engineering (after first graduating Class of 1992, Materials Engineering will be reviewed by ABET for accreditation purposes). The MS program in environmental engineering is accredited by EAC/ABET.

TRANSFER INFORMATION
Transfer via Articulation Agreements
Admission to engineering: At any time
Engineering graduates transferred from ... 4-yr: –   2-yr: 7
Requirements: As defined by the agreement.
Transfer without Articulation Agreements
Admission to engineering: Any time, up to the beginning of the third year of a five-year program, in order to insure enough co-operative education quarters.
Engineering graduates transferred from ... 4-yr: 60 2-yr: 10
Requirements: 2.70 GPA from 2-year institution; 2.5 GPA from 4-year institution; and appropriate prior course work.

GRADUATION REQUIREMENTS
(1) An overall 2.00 or higher for all course work in the College; (2) A 2.00 or higher in all course work taken in the major; (3) Satisfactory completion of all course requirements in curriculum; (4) Minimum credit hours specified (195-204 quarter credit hours depending on major); (5) Satisfactory completion of all co-op requirements.

STUDENT PROGRAMS

PROFESSIONAL AND HONORARY SOCIETIES
[For key to acronyms, see *Introduction*.]
Professional Societies: AIAA, AIChE, ANS, ASCE, ASME, IEEE, IIE, SME, SPE*, ASHRAE, ITE, SAE
Honorary Societies: Alpha Pi Mu, Chi Epsilon, Eta Kappa Nu, Pi Tau Sigma, Sigma Gamma Tau, Tau Beta Pi, Alpha Nu Sigma, Alpha Sigma Mu

SUPPORT PROGRAMS
Student Chapter Organizations: Society of Women Engineers, National Society of Black Engineers

Summer Institute Program
For: Ethnic Minorities   **Available:** Summer only
Offered: Prematriculation

Corporate Mentoring
For: Ethnic Minorities   **Available:** Year round
Offered: Freshman

Summer Enrichment Program
For: Ethnic Minorities   **Available:** Summer only
Offered: Prematriculation

Cooperative Learning Bridge Program
For: Ethnic Minorities   **Available:** Year round
Offered: Freshman

Society for Women Engineers Outreach Program
For: Women   **Available:** Year round
Offered: Prematriculation

Other Engineering Support Programs: Orientation is offered all freshmen. Introduction to Engineering course is required. A College of Engineering Honors Program and a special program in computer applications for approximately 27 students. Big Brother/Big Sister support program. Freshman tutoring program.

043 The Citadel

INSTITUTION PROFILE

HEAD OF THE INSTITUTION
Claudius E. Watts III
Phone: (803)792-5012 Fax: (803)792-6767

GENERAL INFORMATION
[All Students—Fall 1991]
Undergraduate enrollment 2,185
Graduate enrollment 1,449
Total institution enrollment 3,634

Type of institution: Public Calendar: Semesters
Location: Charleston Population: 500,000
Setting: Urban
Types of engineering degrees: Engineering

ENGINEERING ADMISSIONS

ADMISSIONS OFFICE CONTACT
Wallace I. West
The Citadel
Admission Office
Charleston, SC 29409-0225
Phone: (803)792-5230 Fax: (803)792-7084

ENGINEERING COLLEGE ADMISSIONS INFORMATION
Entrance Requirements: SAT (Score: 800), ACT (Score: 19), High School courses—English (4 years), Math (3 years), Lab Science (2 years), Foreign Language (2 years)
Entrance Recommendations: High School courses—Micro Computing (1 year), Social Studies (2 years), U.S. History (1 year), P.E. or ROTC (1 year)
Requirements for out-of-state residents: Higher SAT and high school GPA.
Requirements for foreign students: TOEFL (Score: 500); financial statement

DEPARTMENTAL ADMISSIONS INFORMATION
Admission to the engineering department: At the time of admission to the institution.

FINANCIAL INFORMATION

FINANCIAL AID OFFICE CONTACT
Hank M. Fuller, *Director of Financial Aid and Scholarships*
The Citadel
Financial Aid and Scholarship Office
Charleston, SC 29409-0225
Phone: (803)792-5187 Fax: (803)792-7084

GENERAL FINANCIAL AID INFORMATION
Forms accepted/required: AFSA, FAT, SAR
Additional financial aid information: Need Board Scholarships Academic Scholarships Federal Grants and Loans College Work Study Part-Time Employment

ENGINEERING COLLEGE ADDRESS
Civil Engineering
The Citadel
171 Moultrie Street
Charleston, SC 29409-0225
Phone: (803)792-5230 Fax: (803)792-7084

ENROLLMENTS—BY CLASS
[Numbers are baccalaureate enrollments for the fall 1991 term, unless otherwise footnoted.]
1st-year students/Freshmen 146
2nd-year students/Sophomores 96
3rd-year students/Juniors 50
4th-year students/Seniors 65
Total 357

NUMBER OF DEGREES AWARDED—BY PROGRAM
[Numbers are engineering baccalaureate degrees awarded during the 1991-1992 academic year, unless otherwise footnoted. For full details about each engineering program, refer to the *Tables of Degree Programs*.]
Civil Engineering 19
Electrical Engineering 18
Total 37

PERCENTAGE OF DEGREES AWARDED—BY CATEGORY
[Percentages are of all engineering baccalaureate degrees awarded during the 1991-1992 academic year, unless otherwise footnoted.]
African Americans .... – %
Asian/Pacific Island Americans .... – %
Hispanic Americans .... – %
Native Americans .... – %
Foreign Citizens .... – %
All Others .... 100 %
Women .... 4.0 %
Persons with disabilities .... – %
Students over 25 years of age .... – %

ENGINEERING STUDENT DATA

Applicants to the engineering college
Number of applicants to engineering college 258
Percent offered admission .... –% (A)

Matriculated engineering students
Percentage in top quartile (25%) of High School class 49.0
Average SAT scores: Math—550, Verbal—467, Combined—1017
Average ACT scores: Math—23, Composite—26

Notes: (A) Data not available.

FULL- & PART-TIME FACULTY—BY DEPARTMENT
[Figures are the head count of full-time faculty and the full-time equivalent (FTE) of part-time faculty for each engineering department or equivalent.]

Department	Full	Part
Civil Eng	9	–
Electrical Eng	8	–

COLLEGE DESCRIPTION
The Citadel provides a quality education through extensive grounding in liberal arts and sciences in a military college atmosphere which challenges students, faculty, and staff to achieve excellence. It emphasizes undergraduate education through a wide range of baccalaureate degree programs in the humanities and social and natural sciences as well as engineering and several other professional fields. The Citadel offers two undergraduate engineering degrees--the Bachelor of Science in Civil Engineering (BSCE) and Bachelor of Science in Electrical Engineering (BSEE). These programs are offered in the day for cadet students only. Both undergraduate programs are offered in the evening for non-cadet students. The engineering facilities include modern classroom, computer and complete laboratories.

DEGREE PROGRAMS DESCRIPTION
The primary missions of the engineering departments are to provide quality study programs which provide the basic educational requirements for the pursuit of a professional career in civil or electrical engineering, to prepare students to pursue post-graduate studies, and to provide an educational background broad enough to meet the requirements of good citizenship, leadership and problem solving. Engineers are people-serving professionals who are major managers of resources as well as technology. The Citadel's engineering programs uniquely strive to achieve special qualification in the management of resources--time, materials, money people and environment--through effective combination of the academic with a disciplined military environment. Consistent with the high ideals of the engineering programs are underpinned by a broad base of ethical knowledge and behavior as well as modern and leading edge technology. The engineering departments function through the linkage of the student, faculty and staff in a special academic community within a quality environment, achieving the intended development of the student through the enriched personal, professional and educational growth of each individual.

TRANSFER INFORMATION
Transfer via Articulation Agreements
Admission to engineering: At the end of the sophomore year
Transfer without Articulation Agreements
Admission to engineering: At any time
Engineering graduates transferred from . . 4-yr: – 2-yr: –
Requirements: Minimum of 18 transferable credits. Minimum of a 2.5 cumulative GPA.

GRADUATION REQUIREMENTS
A student must complete either the Civil and Electrical Engineering major courses of study stated in his catalogue of record and must achieve a minimum grade-point ratio of 2.000 based on all quality points as well as in the major department at The Citadel.

STUDENT PROGRAMS

PROFESSIONAL AND HONORARY SOCIETIES
[For key to acronyms, see *Introduction*.]
Professional Societies: ASCE, IEEE, SAME
Honorary Societies: Tau Beta Pi

SUPPORT PROGRAMS
Other Engineering Support Programs: Freshmen are assigned permanent faculty advisors upon entry. Small class sizes and proven faculty dedication facilitate close and effective association with instructors. Regular meetings with practicing engineers ensure students of all class levels are properly oriented in their study of engineering. Counseling and writing centers and individual extra instruction are available.

044 City College of the City University of New York

INSTITUTION PROFILE

HEAD OF THE INSTITUTION
Augusta Souza-Kappner, *Acting*
Phone: (212)650-7285   Fax: (212)650-7680

GENERAL INFORMATION
[All Students—Fall 1991]
Undergraduate enrollment 11,541
Graduate enrollment 3,242
Total institution enrollment 14,783
Type of institution: Public   Calendar: Semesters
Location: New York   Population: 7,323,000
Setting: Urban
Types of engineering degrees: Engineering
Other degree-granting colleges: Architecture, Arts & Sciences, Education, Nursing

ENGINEERING ADMISSIONS

ADMISSIONS OFFICE CONTACT
Nancy P. Campbell
City College of the City University of New York
Convent Avenue at 138th Street
New York, NY 10031-9198
Phone: (212)650-6448   Fax: (212)650-6417

ENGINEERING COLLEGE ADMISSIONS INFORMATION
Entrance Requirements: Combined SAT score of 900 OR H.S. average of 80 OR top 33% of H.S. class OR GED of 300
Entrance Recommendations: High School courses—Math (4 years), Science (4 years), English (4 years)
Requirements for foreign students: TOEFL (Score: 500)

DEPARTMENTAL ADMISSIONS INFORMATION
Admission to the engineering department: attainment of 2.0 GPA in prerequisite math and science courses.

FINANCIAL INFORMATION

ESTIMATED EXPENSES (ACADEMIC YEAR)
[Expenses are for the 1992-93 nine-month academic year.]

State Residents	Undergraduate	Graduate
Tuition and fees	$ 2,487	$ 3,365
College room and board	$ —[1]	$ —[1]
Books and supplies	$ 500	$ 500
Other expenses	$ 675[2]	$ 675[2]
Total estimated expenses	**$ 3,662**	**$ 4,540**

Out-of-State Residents	Undergraduate	Graduate
Tuition and fees	$ 5,097	$ 5,865
College room and board	$ —[1]	$ —[1]
Books and supplies	$ 500	$ 500
Other expenses	$ 675[2]	$ 675[2]
Total estimated expenses	**$ 6,272**	**$ 7,040**

Notes: (1) CCNY is a commuting college. Estimated expenses, students living at home: $1500. Estimated expenses, self-supporting students: $6930. (2) Estimated expense for transportation.

FINANCIAL AID OFFICE CONTACT
Thelma R. Mason, *Director, Financial Aid*
City College of the City University of New York
Convent Avenue at 138th Street
New York, NY 10031-9198
Phone: (212)650-6656   Fax: (212)650-5829

GENERAL FINANCIAL AID INFORMATION
Forms accepted/required: AFSA, CSS-FAF, FAT, Stafford, SAR, Supplemental
Additional financial aid information: We offer need-based and merit-based scholarships and part-time campus jobs. Merit aid includes: City College Scholarships; General Motors Scholarships; Mellon Minority Undergraduate Fellowships; and Student Aid Association Scholarships. Need-based aid (federal/state) includes Tuition Assistance Program and Aid for Part-Time Study (both for NY State residents); Pell Grants; College Work-Study; Perkins, Stafford, and SLS Loans; Supplemental Educational Opportunity Grants; and Bio-Medical Stipends.

ENGINEERING COLLEGE INFORMATION

[For additional personnel, refer to the *Appendix*.]

HEAD OF ENGINEERING
Charles B. Watkins
Phone: (212)650-5435   Fax: (212)650-5768
EMail: ENGACBW

ENGINEERING COLLEGE ADDRESS
School of Engineering
City College of the City University of New York
Convent Avenue at 138th Street
New York, NY 10031-9198
Phone: (212)650-7000   Fax: (212)650-6417

ENROLLMENTS—BY CLASS
[Numbers are baccalaureate enrollments for the fall 1991 term, unless otherwise footnoted.]
1st-year students/Freshmen .... 643 [1]
2nd-year students/Sophomores 433
3rd-year students/Juniors 421
4th-year students/Seniors 486
Total 1,983

Notes: (1) This and the following enrollment figures include full-time students only. In addition, there are 543 part-time students.

NUMBER OF DEGREES AWARDED—BY PROGRAM
[Numbers are engineering baccalaureate degrees awarded during the 1991-1992 academic year, unless otherwise footnoted. For full details about each engineering program, refer to the *Tables of Degree Programs*.]
Chemical Engineering 7
Civil Engineering 46
Computer Science 54
Electrical Engineering 134
Mechanical Engineering 42
Total 283

PERCENTAGE OF DEGREES AWARDED—BY CATEGORY
[Percentages are of all engineering baccalaureate degrees awarded during the 1991-1992 academic year, unless otherwise footnoted.]
African Americans .... 23.1 %
Asian/Pacific Island Americans .... 30.1 %
Hispanic Americans .... 6.6 %
Native Americans .... — %
Foreign Citizens .... 11.8 %
All Others .... 28.4 %
Women .... 11.8 %
Persons with disabilities .... — % (A)
Students over 25 years of age .... — % (A)

Notes: (A) Data not available.

GRADUATE ENROLLMENTS & DEGREES AWARDED
Master's enrollment 593
Master's degrees awarded 89
Doctoral enrollment 125
Doctoral degrees awarded 11

ENGINEERING STUDENT DATA
Applicants to the engineering college
Number of applicants to engineering college 1,308
Percent offered admission 71.4%

Matriculated engineering students
Percentage in top quartile (25%) of High School class ...... — (A)

Notes: (A) Data not available.

FULL- & PART-TIME FACULTY—BY DEPARTMENT
[Figures are the head count of full-time faculty and the full-time equivalent (FTE) of part-time faculty for each engineering department or equivalent.]

Department	Full	Part
Chemical Eng	9	1.6
Civil Eng	16	1.8
Electrical Eng	29	2.5
Mechanical Eng	18	2.0
Computer Science	15	2.8

COLLEGE DESCRIPTION
The School of Engineering awards BE and ME degrees in Chemical, Civil, Electrical, and Mechanical Engineering and B.S. and M.S. degrees in Computer Science. Through the City University of New York, Ph.D. programs are available in all five disciplines. A $6 million program to acquire new laboratory and computer equipment is nearly complete; recent acquisitions include VAX and DECNET workstations, HP 809 Automatic Network Analyzer, and many others. Each department has its own extensive PC facilities. Externally funded research, in which selected juniors and seniors may participate, typically amounts to $6 million annually. Through Baruch College, a sister CUNY institution, students may obtain an accelerated MBA along with their BE or BS degrees. The School of Engineering reflects City College as a whole in the diversity of student backgrounds and the cultural and recreational advantages of being part of one of the world's great cities.

DEGREE PROGRAMS DESCRIPTION

All BE programs emphasize the fundamentals of the discipline while allowing students to specialize in their area of interest. All programs incorporate a strong design component. Superior students can often work with mentors on research projects, including projects of their own devising; a complete range of modern testing and simulation systems is available for proving out designs. All curricula include a humanities "core" component, considered a model of its kind. We encourage our students to join the professional society of their discipline as well as special-interest societies such as the Society of Women Engineers, etc. (We also sponsor a Concrete Canoe Club, which competes with other colleges in designing, building, and racing concrete canoes.) Courses at the Masters level are offered at night in all disciplines; and in fact, most Masters candidates are part-time students generally taking evening courses. Ph.D. programs are available from the City University of New York in all five School of Engineering curricula; the engineering programs are administered by City College, the computer science program by CUNY. All Ph.D. programs are available at night, although most Ph.D candidates are full-time day students. Many serve as lecturers in their department.

TRANSFER INFORMATION
Residency Requirements: Minimum two semesters at City College plus minimum of 60% of major credits OR 33-36 major credits (depending on degree), whichever is larger. All senior-level courses must be taken at City College.

Transfer via Articulation Agreements
Admission to engineering: At the end of the sophomore year
Engineering graduates transferred from . . . 4-yr: 85 2-yr: 28
Requirements: Minimum grade of C in all courses to be transferred

Transfer without Articulation Agreements
Admission to engineering: At any time
Requirements: Varies according to current GPA, total number of credits completed, and total years of math (calculus and higher) and science (esp. calculus physics) courses completed.

GRADUATION REQUIREMENTS
Minimum cumulative GPA: 2.0. Minimum GPA in major: 2.0. Minimum number of credits: 128. Meet all distribution requirements. Pass English Proficiency (exit writing) exam.

STUDENT PROGRAMS

PROFESSIONAL AND HONORARY SOCIETIES
[For key to acronyms, see *Introduction*.]
Professional Societies: ACM, AIAA, AIChE, ASCE, ASME, IEEE, SAE
Honorary Societies: Chi Epsilon, Eta Kappa Nu, Omega Chi Epsilon, Pi Tau Sigma, Tau Alpha Pi

SUPPORT PROGRAMS
Student Chapter Organizations: Society of Women Engineers, National Society of Black Engineers, Latin American Engineering Society of America, CCAPP (Program for Retention of Engineering Students, Haitian Student Association

PRES (Program for Retention of Engineering Students)
For: Ethnic Minorities   Available: Year round
Offered: Freshman, Sophomore, Junior

Project Preserve
For: Ethnic Minorities   Available: Year round
Offered: Sophomore, Junior, Senior

TRACC (Transfer Retention at City College)
For: Ethnic Minorities   Available: Year round
Offered: Sophomore, Junior, Senior

Other Engineering Support Programs: National Science Research projects at City College, aimed mostly at minority students, include CRCM, CASI, RCMS, CMIPS, and ECSEL. Freshmen and transfer orientation programs are in place; in addition, PRES and CCAPP offer summer math/science coaching for current and future students. Peer counseling and peer tutoring programs are available. AA and similar groups also have campus chapters.

045 City University of New York, College of Staten Island

■ INSTITUTION PROFILE

HEAD OF THE INSTITUTION
Edmond L. Volpe
Phone: (718)390-7940　　　　　　Fax: (718)273-0533

GENERAL INFORMATION
[All Students—Fall 1991]
Undergraduate enrollment 11,000
Graduate enrollment 1,185
Total institution enrollment 12,185

Type of institution: Public　　　　**Calendar:** Semesters
Location: New York City　　　　**Population:** 7,000,000
Setting: Urban
Types of engineering degrees: Engineering

■ ENGINEERING ADMISSIONS

ADMISSIONS OFFICE CONTACT
Panagiotis Razelos
City University of New York, College of Staten Island
130 Stuyvesant Place
Staten Island, NY 10301
Phone: (718)390-7972　　　　　　Fax: (718)273-0533

ENGINEERING COLLEGE ADMISSIONS INFORMATION

DEPARTMENTAL ADMISSIONS INFORMATION
Admission to the engineering department: At the time of admission to the institution.
Additional information: High school average of at least 80 or rank in the top one-third of the graduating class for students currently enrolled in New York high schools.

■ FINANCIAL INFORMATION

ESTIMATED EXPENSES (ACADEMIC YEAR)
[Expenses are for the 1992-93 nine-month academic year.]

State Residents	Undergraduate	Graduate
Tuition and fees	$ 1,225	$ 1,675
College room and board	$ —(A)	$ —(A)
Books and supplies	$ —(A)	$ —(A)
Other expenses	$ —(A)	$ —(A)
Total estimated expenses	**$ 1,225**	**$ 1,675**

Out-of-State Residents	Undergraduate	Graduate
Tuition and fees	$ 2,525	$ 2,925
College room and board	$ —(A)	$ —(A)
Books and supplies	$ —(A)	$ —(A)
Other expenses	$ —(A)	$ —(A)
Total estimated expenses	**$ 2,525**	**$ 2,925**

Notes: (A) Data not available.

FINANCIAL AID OFFICE CONTACT
Sherman Whipkey, *Director of Student Financial Assistance*
City University of New York, College of Staten Island
715 Ocean Terrace
Staten Island, NY 10301
Phone: (718)390-7760

GENERAL FINANCIAL AID INFORMATION
Forms accepted/required: AFSA, Stafford, Supplemental, Institutional
Additional financial aid information: The Office of Financial Assistance administers all aid programs - Federal, State, City and college funded.

■ ENGINEERING COLLEGE INFORMATION

[For additional personnel, refer to the *Appendix*.]

HEAD OF ENGINEERING
Fred R. Naider
Phone: (718)390-7925　　　　　　Fax: (718)273-0533

ENGINEERING COLLEGE ADDRESS
Applied Sciences Department
City University of New York, College of Staten Island
130 Stuyvesant Place
Staten Island, NY 10301
Phone: (718)390-7025　　　　　　Fax: (718)273-0533

ENROLLMENTS—BY CLASS
[Numbers are baccalaureate enrollments for the fall 1991 term, unless otherwise footnoted.]
1st-year students/Freshmen 59
2nd-year students/Sophomores 22
3rd-year students/Juniors 29
4th-year students/Seniors 56
Total **166**

NUMBER OF DEGREES AWARDED—BY PROGRAM
[Numbers are engineering baccalaureate degrees awarded during the 1991-1992 academic year, unless otherwise footnoted. For full details about each engineering program, refer to the *Tables of Degree Programs*.]
Engineering Science(s) 28
Total **28**

PERCENTAGE OF DEGREES AWARDED—BY CATEGORY
[Percentages are of all engineering baccalaureate degrees awarded during the 1991-1992 academic year, unless otherwise footnoted.]
African Americans .... — %(A)
Asian/Pacific Island Americans .... — %(A)
Hispanic Americans .... — %(A)
Native Americans .... — %(A)
Foreign Citizens .... — %(A)
All Others .... 100 %
Women .... — %(A)
Persons with disabilities .... — %(A)
Students over 25 years of age .... — %(A)

Notes: (A) Data not available.

FULL- & PART-TIME FACULTY—BY DEPARTMENT
[Figures are the head count of full-time faculty and the full-time equivalent (FTE) of part-time faculty for each engineering department or equivalent.]

Department	Full	Part
Applied Sciences	19	—

COLLEGE DESCRIPTION
The College of Staten Island offers programs leading to the associate, baccalaureate, and master's degrees, and participates in several doctoral programs in conjunction with the Graduate School of the City University of New York. Admission as an associate candidate is open to anyone with a high school diploma or equivalent; admission to four-year degree programs requires entering freshmen to meet the City University's standards for admission to the senior colleges.

DEGREE PROGRAMS DESCRIPTION
The four-year B.S. program in Engineering Science is offered by the Department of Applied Sciences. Of those baccalaureate programs with significant numbers of graduates, the B.S. program in Engineering Science is by far the most technically demanding in the college. Specifically, the four-year program includes: eighteen (18) credits of mathematics beyond pre-calculus; fifteen (15) credits of chemistry and calculus-based physics; at least two semesters of engineering science courses; and at least one semester of engineering design.

TRANSFER INFORMATION
Transfer via Articulation Agreements
Admission to engineering: At any time
Requirements: The City University policy is that a grade of D is transferable from one branch of CUNY to another, but not accepted from non-CUNY colleges.

Transfer without Articulation Agreements
Admission to engineering: At any time

GRADUATION REQUIREMENTS
2.0 overall average and 2.0 in major

■ STUDENT PROGRAMS

PROFESSIONAL AND HONORARY SOCIETIES
[For key to acronyms, see *Introduction*.]
Professional Societies: ASME, IEEE
Honorary Societies: Tau Alpha Pi

SUPPORT PROGRAMS
Student Chapter Organizations: Society of Women Engineers
Other Engineering Support Programs: An orientation program for all new students provides an introduction to the College, its programs and student life. Each student is assigned an academic adviser and meets with him at least once each semester to discuss the following semester's academic program. An Undergraduate Summer Research Awards Program to provide summer stipends for undergraduates is also available.

046 Clarkson University

INSTITUTION PROFILE
HEAD OF THE INSTITUTION
Richard H. Gallagher
Phone: (315)268-6444 Fax: (315)268-3872

GENERAL INFORMATION
[All Students—Fall 1991]
Undergraduate enrollment 2,846
Graduate enrollment 398
Total institution enrollment 3,244
Type of institution: Private **Calendar:** Semesters
Nearest city: Syracuse **Population:** 164,000
Miles from main campus: 150 **Setting:** Small Town
Types of engineering degrees: Engineering
Other degree-granting colleges: Liberal Studies, Science, Management

ENGINEERING ADMISSIONS
ADMISSIONS OFFICE CONTACT
Robert A. Croot
Clarkson University
Director of Admissions
Holcroft House
Potsdam, NY 13699-5605
Phone: (315)268-6479 Fax: (315)268-5605

ENGINEERING COLLEGE ADMISSIONS INFORMATION
Entrance Requirements: SAT (Score: 1000), High School courses—English (4 years), Mathematics (4 years), Science (3 years), Chemistry & Physics (1 year)
Entrance Recommendations: SAT (Score: 1000)
Requirements for foreign students: TOEFL (Score: 550); financial statement

DEPARTMENTAL ADMISSIONS INFORMATION
Admission to the engineering department: At the time of admission to the institution.
Additional information: All Freshman must maintain a GPA of 2.000

FINANCIAL INFORMATION
ESTIMATED EXPENSES (ACADEMIC YEAR)
[Expenses are for the 1992-93 nine-month academic year.]

All Students	Undergraduate	Graduate
Tuition and fees	$14,510	$ 8,514 [1]
College room and board	$ 5,077	$ — [2]
Books and supplies	$ 600	$ —
Other expenses	$ 1,013	$ —
Total estimated expenses	$21,200	$ 8,514

Notes: (1) Tuition based on taking 9 credit hrs a semester (2) We do not have housing for graduate students

FINANCIAL AID OFFICE CONTACT
Donald T. Mills, *Director of Financial Aid*
Clarkson University
Lewis House
Potsdam, NY 13699-5615
Phone: (315)268-6471

GENERAL FINANCIAL AID INFORMATION
Forms accepted/required: FAF, Institutional
Additional financial aid information: need-based scholarships; short-term loans; merit-based scholarships; part-time jobs on campus; college work-study program.

ENGINEERING COLLEGE INFORMATION
[For additional personnel, refer to the *Appendix*.]
HEAD OF ENGINEERING
William R. Wilcox
Phone: (315)268-6446 Fax: (315)268-3841
EMail: SOE1@CLVM

ENGINEERING COLLEGE ADDRESS
School of Engineering
Clarkson University
Main Street
Potsdam, NY 13699-5700
Phone: (315)268-6446 Fax: (315)268-3841

ENROLLMENTS—BY CLASS
[Numbers are baccalaureate enrollments for the fall 1991 term, unless otherwise footnoted.]
1st-year students/Freshmen 407
2nd-year students/Sophomores 335
3rd-year students/Juniors 466
4th-year students/Seniors 470
Total 1,678

NUMBER OF DEGREES AWARDED—BY PROGRAM
[Numbers are engineering baccalaureate degrees awarded during the 1991-1992 academic year, unless otherwise footnoted. For full details about each engineering program, refer to the *Tables of Degree Programs*.]
Chemical Engineering 42
Civil & Environmental Engineering 104
Electrical & Computer Engineering 144
Mechanical & Aeronautical Engineering 154
Total 444

PERCENTAGE OF DEGREES AWARDED—BY CATEGORY
[Percentages are of all engineering baccalaureate degrees awarded during the 1991-1992 academic year, unless otherwise footnoted.]
African Americans .... — % [1]
Asian/Pacific Island Americans .... — % [1]
Hispanic Americans .... — % [1]
Native Americans .... — % [1]
Foreign Citizens .... — % [1]
All Others .... 100 %
Women .... 16.9% [1]
Persons with disabilities .... — % [1]
Students over 25 years of age .... 2.9% [1]

Notes: (1) Figured on Engineering Students only

GRADUATE ENROLLMENTS & DEGREES AWARDED
Master's enrollment 100
Master's degrees awarded 56
Doctoral enrollment 98
Doctoral degrees awarded 18

ENGINEERING STUDENT DATA
Applicants to the engineering college
Number of applicants to engineering college 1,412
Percent offered admission 93.6%

Matriculated engineering students
Percentage in top quartile (25%) of High School class 72.2
Average SAT scores: Math—620, Verbal—521, Combined—1141

FULL- & PART-TIME FACULTY—BY DEPARTMENT
[Figures are the head count of full-time faculty and the full-time equivalent (FTE) of part-time faculty for each engineering department or equivalent.]

Department	Full	Part
Chemical Eng	15	—
Civil & Environmental Eng	18	—
Electrical & Computer Eng	22	—
Mechanical & Aeronautical Eng	21	—

COLLEGE DESCRIPTION
The Univ. offers 4-year Bach programs in Chem, Civil, Elect. & Comp., Mech. & Aeron. Eng., emphasizing sci. & eng. fundamentals in the 1st year, introductory disc. discipline courses the 2nd year with upper level eng. discipline & elective courses the 3rd & 4th years. Liberal Arts Studies are offered the entire 4 years. The eng. curricula emphasize the importance of the personal computer in all courses taken. Strong business & science colleges at the Univ. make it possible for interested students to combine eng. with elective coursework from these two areas. The Univ. has numerous 2+2 & 3+2 transfer articulation agreements, enabling students to transfer with little or no disruption in obtaining a bach. degree in eng. The SII prog. provides an opportunity for students to gain eng. work experience while completing the B.S. degree. Students are involved in the last year in a design project experience and are urged to sit for the first 8 hrs. of the Prof/Eng/ Lic.

DEGREE PROGRAMS DESCRIPTION
The B.S. degrees in the engineering programs at Clarkson are designed to provide students with an adequate foundation in the scientific, engineering, socio-humanities and management fields to enable them to make significant contributions in their chosen fields while at the same time recognizing their responsibilities to society.

Clarkson offers M.S. and Ph.D. in Chemical Eng.; the M.S. in Civil, Electrical and Mechanical Eng.; and the Ph.D. in Engineering Science.

DUAL DEGREE PROGRAMS
Engineering baccalaureate graduates with dual degrees 0
Enrollment requirements: Preferred 2.75 GPA - GPA below 2.75 may be considered if there are extenuating circumstances and with the approval of the Department Chairman.

TRANSFER INFORMATION
Residency Requirements: Students must complete at least a minimum of two semesters and a maximum of four semesters in residence, are allowed to transfer a maximum of 90 credits. Course requirements could put the transfer student coming in with the 90 credits over 120 credits needed to graduate.

Transfer via Articulation Agreements
Admission to engineering: 2 year college Eng. Science students are admitted at end of second year. 4 year college Eng. Science students are students are admitted at end of third year.
Engineering graduates transferred from ... 4-yr: 46 2-yr: 81
Requirements: Preferred 2.75 GPA - GPA below 2.75 may be considered if there are extenuating circumstances and with the approval of the Department Chairman and recommendation from faculty at previous college.

Transfer without Articulation Agreements
Admission to engineering: 2 year college students are admitted at end of second year. 4 year college students are admitted at end of third year.
Engineering graduates transferred from ... 4-yr: 46 2-yr: 81
Requirements: Preferred 2.75 GPA - GPA below 2.75 may be considered if there are extenuating circumstances and with the approval of the Department chairman. The university may ask for recommendation from the faculty at the previous college.

GRADUATION REQUIREMENTS
B.S. programs in engineering require 120 credit hours, at least 2.000 cumulative index and all departmental course requirements.

STUDENT PROGRAMS
PROFESSIONAL AND HONORARY SOCIETIES
[For key to acronyms, see *Introduction*.]
Professional Societies: ACM, ACS, AIAA, AIChE, ASCE, ASHRAE, ASME, IEEE, SAME, Society of Woman Engineers
Honorary Societies: Chi Epsilon, Eta Kappa Nu, Omega Chi Epsilon, Pi Tau Sigma, Tau Beta Pi

SUPPORT PROGRAMS
Student Chapter Organizations: Society of Women Engineers, National Society of Black Engineers, Society of Hispanic Professional Engineers, Women's Resource Center, American Indian Program, Spectrumm, Minority Student Development Program

Women's Resource Center
For: Women **Available:** Academic year
Offered: Freshman, Sophomore, Junior, Senior

American Indian Program
For: Ethnic Minorities **Available:** Academic year
Offered: Prematriculation, Freshman, Sophomore, Junior, Senior, Graduate level

Spectrum (All Students)
Available: Academic year
Offered: Prematriculation, Freshman, Sophomore, Junior, Senior, Graduate level

Minority Student Development Program
For: Ethnic Minorities **Available:** Academic year
Offered: Prematriculation, Freshman, Sophomore, Junior, Senior, Graduate level

Other Engineering Support Programs: Summer and Fall orientation programs for all freshman and transfers; Learning Assessment program; Career Counseling Services; Foreign Student office; Veterans services; services for blind and handicapped; Placement services. Academic Support Services office; Student Development Center; Writing Laboratory, Education Resources Center.

047 Clemson University

INSTITUTION PROFILE

HEAD OF THE INSTITUTION
A. Max Lennon
Phone: (803)656-3413　　　　　　Fax: (803)656-4676

GENERAL INFORMATION
[All Students—Fall 1991]
Undergraduate enrollment 13,285
Graduate enrollment 4,010
Total institution enrollment **17,295**
Type of institution: Public　　　　Calendar: Semesters
Nearest city: Greenville, SC　　　　Population: 58,000
Miles from main campus: 35　　　　Setting: Small Town
Types of engineering degrees: Engineering
Other degree-granting colleges: Architecture, Education, Nursing, Agricultural Sciences, Sciences, Forest & Recreation Resources, Commerce and Industry, Liberal Arts

ENGINEERING ADMISSIONS

ADMISSIONS OFFICE CONTACT
Michael R. Heintze
Clemson University
105 Sikes Hall
Clemson, SC 29634
Phone: (803)656-2287　　　　　　Fax: (803)656-0622

ENGINEERING COLLEGE ADMISSIONS INFORMATION
Entrance Requirements: SAT, High School courses—English (4 years), Mathematics (3 years), Laboratory Science (2 years), Foreign Language, Social Science, PE, and Other (7 years)
Entrance Recommendations: ACT, High School courses—Additional Mathematics (1 year), Additional Laboratory Science (1 year), Government and Economics (1/2 year each) (1 year)
Requirements for out-of-state residents: Formula (based on high school grades and SAT scores) used to predict academic performance. Requirement is higher for out-of-state students.
Requirements for foreign students: TOEFL (Score: 550)

DEPARTMENTAL ADMISSIONS INFORMATION
Admission to the engineering department: At the end of the first year.

FINANCIAL INFORMATION

ESTIMATED EXPENSES (ACADEMIC YEAR)
[Expenses are for the 1992-93 nine-month academic year.]

State Residents	Undergraduate	Graduate
Tuition and fees	$ 2,310	$ 2,310(1)
College room and board	$ 3,200	$ 3,200
Books and supplies	$ —(B)	$ —(A)
Other expenses	$ 500	$ 575
Total estimated expenses	**$ 6,010**	**$ 6,085**

Out-of-State Residents	Undergraduate	Graduate
Tuition and fees	$ 6,440	$ 6,440(1)
College room and board	$ 3,200	$ 3,200
Books and supplies	$ —(B)	$ —(A)
Other expenses	$ 500	$ 575
Total estimated expenses	**$10,140**	**$10,215**

Notes: (A) Data not available. (B) Data not applicable. (1) Graduate assistants pay $300 per semester.

FINANCIAL AID OFFICE CONTACT
Marvin G. Carmichael, *Director of Financial Aid*
Clemson University
G01 Sikes Hall
Clemson, SC 29634
Phone: (803)656-2280　　　　　　Fax: (803)656-0622

GENERAL FINANCIAL AID INFORMATION
Forms accepted/required: FAF
Additional financial aid information: Clemson offers scholarships, loans, grants and part-time employment. The Financial Aid Office works jointly with the Financial Aid Committee and the Scholarships and Awards Committee.

ENGINEERING COLLEGE INFORMATION
[For additional personnel, refer to the *Appendix*.]

HEAD OF ENGINEERING
Thomas M. Keinath
Phone: (803)656-3202　　　　　　Fax: (803)656-0859
EMail: tom.keinath@eng.clemson.edu

ENGINEERING COLLEGE ADDRESS
College of Engineering
Clemson University
109 Riggs Hall
Box 340901
Clemson, SC 29634-0901
Phone: (803)656-3201　　　　　　Fax: (803)656-0859

ENROLLMENTS—BY CLASS
[Numbers are baccalaureate enrollments for the fall 1991 term, unless otherwise footnoted.]
1st-year students/Freshmen 926
2nd-year students/Sophomores 757
3rd-year students/Juniors 814
4th-year students/Seniors 789
Total **3,286**

NUMBER OF DEGREES AWARDED—BY PROGRAM
[Numbers are engineering baccalaureate degrees awarded during the 1991-1992 academic year, unless otherwise footnoted. For full details about each engineering program, refer to the *Tables of Degree Programs*.]
Agriculture & Biological Engineering 9
Bioengineering .... —
Ceramic Engineering 39
Chemical Engineering 46
Civil Engineering 91
Electrical & Computer Engineering 143
Environmental Systems Engineering .... —
Industrial Engineering 58
Mechanical Engineering 138
Total **524**

PERCENTAGE OF DEGREES AWARDED—BY CATEGORY
[Percentages are of all engineering baccalaureate degrees awarded during the 1991-1992 academic year, unless otherwise footnoted.]
African Americans .... 4.0 %
Asian/Pacific Island Americans .... 1.0 %
Hispanic Americans .... 1.0 %
Native Americans .... — %
Foreign Citizens .... 12.0 %
All Others .... 82.0 %
Women .... 17.0 %
Persons with disabilities .... — % (A)
Students over 25 years of age .... — % (A)

Notes: (A) Data not available.

GRADUATE ENROLLMENTS & DEGREES AWARDED
Master's enrollment 505
Master's degrees awarded 174
Doctoral enrollment 153
Doctoral degrees awarded 34

ENGINEERING STUDENT DATA

Applicants to the engineering college
Number of applicants to engineering college 1,653
Percent offered admission 89.0%

Matriculated engineering students
Percentage in top quartile (25%) of High School class 77.0
Average SAT scores: Math—592, Verbal—485, Combined—1077

FULL- & PART-TIME FACULTY—BY DEPARTMENT
[Figures are the head count of full-time faculty and the full-time equivalent (FTE) of part-time faculty for each engineering department or equivalent.]

Department	Full	Part
Chemical Eng	11	—
Civil Eng	20	—
Electrical & Computer Eng	38	—
Industrial Eng	10	—
Mechanical Eng	29	—
Agricultural & Biological Eng	19	—
Bioengineering	8	—
Ceramic Eng	11	—
Environmental Systems Eng	9	—

COLLEGE DESCRIPTION

The College of Engineering offers bachelor of science degree programs in 9 distinct areas of engineering, 8 of which are accredited by ABET/EAC. The Freshman Engineering program assists students in the selection of an engineering major while providing a strong background in mathematics, science, and computing. Transfer to any engineering major is assured after the freshman program is completed satisfactorily. Cooperative education is available in all engineering majors. The Honors Program allows students to work with faculty on research problems. A summer abroad program, combining academic study in England and travel in Europe, is also available. To meet curricular objectives, effective oral and written communications and computer utilization are integrated throughout each engineering program. Average number of years to complete the bachelor's degree is 4.75.

DEGREE PROGRAMS DESCRIPTION

B.S. degrees are offered by the College of Engineering in nine professional programs: agricultural engineering, ceramic engineering, chemical engineering, civil engineering, computer engineering, electrical engineering, industrial engineering, mechanical engineering, and in one science-based engineering program, engineering analysis. These programs lead to a wide range of career opportunities for graduates, and serve as preparation for further study at the graduate level. The college also offers programs leading to both the M.S. and Ph.D. degrees in its nine degree programs, as well as in bioengineering, engineering mechanics, and environmental systems engineering. A Master of Engineering (M. Engr.) degree is offered in some fields.

DUAL DEGREE PROGRAMS

Engineering baccalaureate graduates with dual degrees 6
Enrollment requirements: Dual degree students must meet all requirements for transfer students. From both 2 and 4 year institutions, requirements are the completion of 30 hours with a GPR 2.50. No 300 level coursework or higher is acceptable from a 2 year institution. Students must have completed 3 years of specified coursework from the sending institution together with a notification of intent to pursue a dual-degree program. Recommendation of advisor suggested.

TRANSFER INFORMATION

Residency Requirements: Candidates for an undergraduate degree must complete at Clemson, a minimum of 30 of the last 36 credits for the degree.

Transfer without Articulation Agreements
Admission to engineering: At any time
Engineering graduates transferred from　　4-yr: 79　　2-yr: 118
Requirements: From both 2 and 4 year institutions, requirements are the completion of 30 hours with a GPR 2.50. No 300 level coursework or higher is acceptable from a 2 year institution.

GRADUATION REQUIREMENTS

Institutional requirements include a 2.0 or higher cumulative grade point ratio on 139 to 144 total semester hours. In addition, candidates for engineering degrees will be required to have a 2.0 or higher cumulative grade-point ratio in all engineering courses taken at Clemson University.

STUDENT PROGRAMS

PROFESSIONAL AND HONORARY SOCIETIES
[For key to acronyms, see *Introduction*.]
Professional Societies: ACerS, ACS, AIChE, ASAE, ASCE, ASME, BMES, IEEE, IIE, NSPE
Honorary Societies: Alpha Pi Mu, Chi Epsilon, Eta Kappa Nu, Keramos, Pi Tau Sigma, Tau Beta Pi

SUPPORT PROGRAMS
Student Chapter Organizations: Society of Women Engineers, National Society of Black Engineers

Program for Engineering Enrichment & Retention
For: Ethnic Minorities　　**Available:** Year round
Offered: Prematriculation, Freshman, Sophomore, Junior, Senior
Other Engineering Support Programs: Summer orientation programs for all freshmen and transfer students entering Clemson in the fall; Freshman Engineering Program for first-year engineering students offers career counseling and the opportunity for identification and correction of any academic difficulties before admission to a degree program.

048 Cleveland State University

INSTITUTION PROFILE
HEAD OF THE INSTITUTION
J. Taylor Sims
Phone: (216)687-3544　　　　　Fax: (216)687-9333
GENERAL INFORMATION
[All Students—Fall 1991]
Undergraduate enrollment 13,518
Graduate enrollment 5,573
Total institution enrollment 19,091
Type of institution: Public　　　　Calendar: Quarters
Location: Cleveland　　　　　　　Population: 659,234
Setting: Urban
Types of engineering degrees: Engineering & Technology
Other degree-granting colleges: Arts & Sciences, Business Administration, Education, Law, Urban Affairs

ENGINEERING ADMISSIONS
ADMISSIONS OFFICE CONTACT
Ruth Ann Moyer
Cleveland State University
Office of Undergraduate Admissions
Euclid Avenue at East 24th Street
Cleveland, OH 44115
Phone: (216)687-3755　　　　　Fax: (216)687-9210
ENGINEERING COLLEGE ADMISSIONS INFORMATION
Entrance Requirements: SAT (Score: 950), ACT (Score: 24), High School courses—Chemistry or Physics (1 year), Mathematics through Trigonometry (4 years), English (4 years)
Requirements for foreign students: TOEFL (Score: 525); financial statement
DEPARTMENTAL ADMISSIONS INFORMATION
Admission to the engineering department: At the end of the first year.

FINANCIAL INFORMATION
ESTIMATED EXPENSES (ACADEMIC YEAR)
[Expenses are for the 1992-93 nine-month academic year.]

State Residents	Undergraduate	Graduate
Tuition and fees	$ 2,976	$ 3,771
College room and board	$ 2,976	$ 2,976
Books and supplies	$ 540	$ 690
Other expenses	$ 725 [1]	$ 725 [1]
Total estimated expenses	**$ 7,217**	**$ 8,162**
Out-of-State Residents	Undergraduate	Graduate
Tuition and fees	$ 5,952	$ 7,542
College room and board	$ 2,976	$ 2,976
Books and supplies	$ 540	$ 690
Other expenses	$ 725 [1]	$ 725 [1]
Total estimated expenses	**$10,193**	**$11,933**

Notes: (1) Transportation and hospitalization
FINANCIAL AID OFFICE CONTACT
William R. Bennett, *Director*
Cleveland State University
Financial Aid
Euclid Avenue at East 24th Street
Cleveland, OH 44115
Phone: (216)687-3764　　　　　Fax: (216)687-9247
GENERAL FINANCIAL AID INFORMATION
Forms accepted/required: FAF, IRS, Institutional, Ohio Instructional Grant
Additional financial aid information: Academic achievement and need-based scholarships; short-term loans; part-time jobs on campus; college work-study program.

ENGINEERING COLLEGE INFORMATION
[For additional personnel, refer to the *Appendix*.]
HEAD OF ENGINEERING
George A. Coulman
Phone: (216)687-2558　　　　　Fax: (216)687-9280
EMail: Coulman@CSVAXD.CSUOHIO.EDU (INTERNET)
ENGINEERING COLLEGE ADDRESS
Fenn College of Engineering
Cleveland State University
Euclid Avenue at East 24th Street
Cleveland, OH 44115
Phone: (216)687-2000　　　　　Fax: (216)687-9366
ENROLLMENTS—BY CLASS
[Numbers are baccalaureate enrollments for the fall 1991 term, unless otherwise footnoted.]
1st-year students/Freshmen 151
2nd-year students/Sophomores 143
3rd-year students/Juniors 231
4th-year students/Seniors 373
Total 898
NUMBER OF DEGREES AWARDED—BY PROGRAM
[Numbers are engineering baccalaureate degrees awarded during the 1991-1992 academic year, unless otherwise footnoted. For full details about each engineering program, refer to the *Tables of Degree Programs*.]
Chemical Engineering 16
Civil Engineering 18
Electrical Engineering 48
Electronic Technology 29
Industrial Engineering 6
Mechanical Engineering 47
Mechanical Technology 26
Total 190
PERCENTAGE OF DEGREES AWARDED—BY CATEGORY
[Percentages are of all engineering baccalaureate degrees awarded during the 1991-1992 academic year, unless otherwise footnoted.]
African Americans 2.3%
Asian/Pacific Island Americans 2.3%
Hispanic Americans 1.5%
Native Americans .... – %
Foreign Citizens 5.9%
All Others 88.0%
Women 14.8%
Persons with disabilities .... – % (A)
Students over 25 years of age 59.0%

Notes: (A) Data not available.
GRADUATE ENROLLMENTS & DEGREES AWARDED
Master's enrollment 144
Master's degrees awarded 64
Doctoral enrollment 38
Doctoral degrees awarded 7
ENGINEERING STUDENT DATA
Applicants to the engineering college
Number of applicants to engineering college 228
Percent offered admission 61.4%
Matriculated engineering students
Percentage in top quartile (25%) of High School class ...... – (A)
Average SAT scores: Math—556, Verbal—456, Combined—1012
Average ACT scores: Composite—24

Notes: (A) Data not available.
FULL- & PART-TIME FACULTY—BY DEPARTMENT
[Figures are the head count of full-time faculty and the full-time equivalent (FTE) of part-time faculty for each engineering department or equivalent.]

Department	Full	Part
Chemical Eng	9.0	0.8
Civil Eng	8.5	1.0
Electrical Eng	15.0	1.5
Industrial Eng	6.5	0.3
Mechanical Eng	11.0	1.0
Tech	6.0	1.7

COLLEGE DESCRIPTION
The Fenn College of Engineering is part of Cleveland State University, which is located in the downtown area of a major city and offers a variety of cultural activities. Since Cleveland is a major industrial center, there are many job opportunities for our graduates. Courses are offered both day and evening, and an optional Co-op Program offers students an opportunity to earn a portion of their expenses as well as apply their course work to on-the-job challenges before graduation. Engineering class sizes are set at 35 students.

DEGREE PROGRAMS DESCRIPTION
All five Engineering programs offer the B.S., M.S., and Doctor of Engineering degrees. All undergraduate programs are offered during the day; Electrical Engineering and Mechanical Engineering, along with Electronic Technology and Mechanical Technology are offered both days and evenings. Electrical Engineering offers an option in Computer Engineering which is offered jointly by the Departments of Electrical Engineering and Computer and Information Science.

TRANSFER INFORMATION
Residency Requirements: A minimum of 36 credits of upper-division course work must be earned in residence at Cleveland State Univ.

Transfer without Articulation Agreements
Admission to engineering: At any time
Engineering graduates transferred from ... 4-yr: 21% 2-yr: 49%
Requirements: Institution must be regionally accredited. Minimum overall GPA of 2.5/4.0 and a minimum GPA of 2.3 in all Mathematics, Science, and Engineering courses. Only grades of C or higher transfer.

GRADUATION REQUIREMENTS
Degree candidates must attain a cumulative grade-point average of at least 2.00 for all work and at least 2.00 for all required courses, including Technical electives, in the combined areas of Engineering, Mathematics, Physics, Computer Science, and Chemistry. Students are limited to two grades of D in Engineering Courses.

STUDENT PROGRAMS
PROFESSIONAL AND HONORARY SOCIETIES
[For key to acronyms, see *Introduction*.]
Professional Societies: ACS, AIAA, AIChE, AIP, AMS, ASCE, ASME, IEEE, IIE, NSPE, Federation of Technology Students
Honorary Societies: Eta Kappa Nu, Tau Beta Pi

SUPPORT PROGRAMS
Student Chapter Organizations: Society of Women Engineers, National Society of Black Engineers, National Technical Association

Minority Tutoring Program
For: Ethnic Minorities　　**Available:** Academic year
Offered: Freshman, Sophomore, Junior, Senior

Society of Women Engineers
For: Women & Ethnic Minorities　　**Available:** Academic year
Offered: Freshman, Sophomore, Junior, Senior

Other Engineering Support Programs: All new students participate in an orientation program to acquaint them with the academic support services. During the summer, skills workshops are available. Career services offer assistance with Co-op, summer jobs, and professional employment. Special programs are available for foreign, handicapped, and women students. Tau Beta Pi runs a tutoring program for all Engineering students.

049 University of Colorado at Boulder

INSTITUTION PROFILE

HEAD OF THE INSTITUTION
Judith E. Albino
Phone: (303)492-6201   Fax: (303)492-6772

GENERAL INFORMATION
[All Students—Fall 1991]
Undergraduate enrollment 20,495
Graduate enrollment 5,076
Total institution enrollment 25,571

Type of institution: Public   Calendar: Semesters
Nearest city: Denver   Population: 1,900,000
Miles from main campus: 21   Setting: Urban
Types of engineering degrees: Engineering
Other degree-granting colleges: Arts & Sciences, Business Administration, Communication & Journalism, Education, Law, Environmental Design, Music

ENGINEERING COLLEGE ADMISSIONS INFORMATION
Entrance Requirements: High School courses—English (4 years), Mathematics (4 years), Natural Science (3 years), Social Science (2 years), There are no absolute minimum ACT or SAT scores, applicants with low scores are reviewed by an Engineering Admissions Committee.
Entrance Recommendations: High School courses—Foreign Language (Required) (2 years)
Requirements for foreign students: TOEFL; financial statement

DEPARTMENTAL ADMISSIONS INFORMATION
Admission to the engineering department: At the time of admission to the institution.

FINANCIAL INFORMATION

ESTIMATED EXPENSES (ACADEMIC YEAR)
[Expenses are for the 1992-93 nine-month academic year.]

State Residents	Undergraduate	Graduate
Tuition and fees	$ 2,962	$ 3,604
College room and board	$ 5,409	$ 5,409
Books and supplies	$ 500	$ 500
Other expenses	$ 900	$ 900
Total estimated expenses	$ 9,771	$ 10,413

Out-of-State Residents	Undergraduate	Graduate
Tuition and fees	$11,828	$ 10,926
College room and board	$ 5,409	$ 5,409
Books and supplies	$ 500	$ 500
Other expenses	$ 900	$ 900
Total estimated expenses	$18,637	$ 17,735

FINANCIAL AID OFFICE CONTACT
Jerry Sullivan, Director, Office of Financial Aid
University of Colorado at Boulder
Campus Box 106
Boulder, CO 80309-0106
Phone: (303)492-5091   Fax: (303)492-0838

GENERAL FINANCIAL AID INFORMATION
Forms accepted/required: ACT, AFSA, CSS-FAF, FAF, FAT, FAR, FSA, FFS, Stafford, IRS, SAR, SDF, United Student Aid Funds Form (preferred--Single File), Free Application for Federal Student Aid
Additional financial aid information: Scholarships, grants, loans, work-study, and hourly student positions are all coordinated by the financial aid office.

ENGINEERING COLLEGE INFORMATION

[For additional personnel, refer to the Appendix.]

HEAD OF ENGINEERING
A. R. Seebass
Phone: (303)492-7006   Fax: (303)492-2199

ENGINEERING COLLEGE ADDRESS
College of Engineering and Applied Science
University of Colorado at Boulder
Campus Box 422
Boulder, CO 80309-0422
Phone: (303)492-5071   Fax: (303)492-2199

ENROLLMENTS—BY CLASS
[Numbers are baccalaureate enrollments for the fall 1991 term, unless otherwise footnoted.]
1st-year students/Freshmen 615
2nd-year students/Sophomores 553
3rd-year students/Juniors 491
4th-year students/Seniors 704
Total 2,363

NUMBER OF DEGREES AWARDED—BY PROGRAM
[Numbers are engineering baccalaureate degrees awarded during the 1991-1992 academic year, unless otherwise footnoted. For full details about each engineering program, refer to the Tables of Degree Programs.]
Aerospace Engineering Sciences 103
Applied Mathematics 27
Chemical Engineering 27
Civil, Environmental, & Architectural Engineering 55
Computer Science 42
Electrical & Computer Engineering 73
Mechanical Engineering 48
Telecommunications .... – (B)
Total 375

Notes: (B) Data not applicable.

PERCENTAGE OF DEGREES AWARDED—BY CATEGORY
[Percentages are of all engineering baccalaureate degrees awarded during the 1991-1992 academic year, unless otherwise footnoted.]
African Americans .... – %
Asian/Pacific Island Americans .... – %
Hispanic Americans .... – %
Native Americans .... – %
Foreign Citizens .... – %
All Others .... 100 %
Women .... 14.0 %
Persons with disabilities .... – % (A)
Students over 25 years of age .... – % (A)

Notes: (A) Data not available.

GRADUATE ENROLLMENTS & DEGREES AWARDED
Master's enrollment 540
Master's degrees awarded 322
Doctoral enrollment 445
Doctoral degrees awarded 50

ENGINEERING STUDENT DATA

Applicants to the engineering college
Number of applicants to engineering college 1,807
Percent offered admission 86.1%

Matriculated engineering students
Percentage in top quartile (25%) of High School class 80.0
Average SAT scores: Math—650, Verbal—510, Combined—1160
Average ACT scores: Math—29, Composite—28

FULL- & PART-TIME FACULTY—BY DEPARTMENT
[Figures are the head count of full-time faculty and the full-time equivalent (FTE) of part-time faculty for each engineering department or equivalent.]

Department	Full	Part
Chemical Eng	14	3.0
Civil, Env'l, & Architectural Eng	27	6.0
Electrical & Computer Eng	43	10.0
Mechanical Eng	17	5.0
Aerospace Eng Sciences	27	20.0
Computer Science	24	10.0

COLLEGE DESCRIPTION
The College offers ten undergraduate majors, all of which require a minimum of four years to complete. Freshmen may elect any of these degree programs or an open option program for their first year. Although admission to the College is selective, the choice of major is open with no enrollment controls. All curricula emphasize engineering and science fundamentals with options for specialization. Students are guided in development of a coherent humanities and social science elective program. The Herbst Humanities Program is offered to students desiring in-depth study of the humanities in a small-group setting. Dual-degree programs are feasible with other curricula on campus. A variety of undergraduate research experiences are available.

DEGREE PROGRAMS DESCRIPTION
The ten Bachelor of Science programs offered by the College prepare the student for professional practice or graduate study in their respective fields. Graduate degrees (M.S. and Ph.D.) in six areas (aerospace, chemical, civil, electrical, and mechanical engineering, plus computer science) are offered. Civil engineering offers an environmental engineering concentration. The Master of Science degree has thesis and non-thesis options. There is a Master of Engineering degree offered primarily to the practicing professional. A Master of Science in Telecommunications is available in an interdisciplinary program.

DUAL DEGREE PROGRAMS
Engineering baccalaureate graduates with dual degrees 14

TRANSFER INFORMATION
Residency Requirements: All students, including transfers, must complete the last 45 semester credit hours in the College of Engineering and Applied Science.

Transfer via Articulation Agreements
Admission to engineering: At the end of the freshman year
Requirements: Required cumulative GPA of 3.00 for community college transfers; 2.75 for transfer from 4-year institution. Prior to acceptance, all transfers must complete one year of college-level Calculus and at least one semester of calculus-based physics and one semester of college chemistry. Chemical engineering applicants should have completed 2 semesters of college chemistry.

Transfer without Articulation Agreements
Admission to engineering: At the end of the freshman year
Requirements: Transfers are admitted to the engineering program up to the middle of the junior year; the last 45 semester credit hours must be earned in the College. The application deadline is April 1. Transfer programs are articulated with all Colorado community colleges. Dual degrees are possible with the Colleges and Schools at UC-Boulder.

GRADUATION REQUIREMENTS
At least 128 semester hours must be completed, the last 45 hours must be completed on the Boulder campus. A cumulative GPA of 2.00 is required in all courses used to fulfill degree requirements and a GPA of 2.00 must be attained in all courses taken in the student's major department. Students who graduated from high school in the spring of 1988 and thereafter must complete any minimum academic preparation standards deficiencies. All students are required to take the Engineer-In-Training examination or the GRE (General and Engineering tests), during their senior year.

STUDENT PROGRAMS

PROFESSIONAL AND HONORARY SOCIETIES
[For key to acronyms, see Introduction.]
Professional Societies: ACM, AGCA, AIAA, AIChE, ASCE, ASME, BMES, IEEE, IES, NSAE
Honorary Societies: Chi Epsilon, Eta Kappa Nu, Omega Chi Epsilon, Pi Tau Sigma, Sigma Gamma Tau, Tau Alpha Pi, Tau Beta Pi

SUPPORT PROGRAMS
Student Chapter Organizations: Society of Women Engineers, National Society of Black Engineers, Society of Hispanic Professional Engineers, Mexican American Engineers & Scientists, American Indian Science and Engineering Society

Women in Engineering Program
For: Women   Available: Year round
Offered: Prematriculation, Freshman, Sophomore, Junior, Senior, Graduate level

Minority Engineering Program
For: Ethnic Minorities   Available: Year round
Offered: Prematriculation, Freshman, Sophomore, Junior, Senior
Other Engineering Support Programs: Fall orientation program for all new freshmen & transfers; career & personal counseling services; a resident hall educational support program; Women in Engineering Program; Minority Engineering Program for underrepresented minority students; Peer Advocate Prog.; student honoraries and service societies; placement services; foreign student office; veterans' services; disabled students' services.

050 University of Colorado at Colorado Springs

INSTITUTION PROFILE

HEAD OF THE INSTITUTION
Judith Albino
Phone: (303)492-6201 Fax: (303)492-6772

GENERAL INFORMATION
[All Students—Fall 1991]
Undergraduate enrollment 4,148
Graduate enrollment 1,603
Total institution enrollment 5,751
Type of institution: Public   Calendar: Semesters
Location: Colorado Springs   Population: 300,000
Setting: Urban
Types of engineering degrees: Engineering
Other degree-granting colleges: Arts & Sciences, Business Administration, Education

ENGINEERING COLLEGE ADMISSIONS INFORMATION
Entrance Requirements: SAT (Score: 1070), ACT (Score: 25), High School courses—English (to include two years composition) (4 years), Mathematics (2 yrs algebra; 1 yr trig or anal geo) (4 years), Natural Science (1 year physics; 1 year chemistry) (3 years), Social Science (Foreign Language) (2 years), Students must be in the upper 40% of their graduation class and have the appropriate background and performance in math and science classes. Engineering Prep Program also available.
Entrance Recommendations: High School courses—Computer Science (2 years), Speech (1 year)
Requirements for foreign students: TOEFL (Score: 550); financial statement

DEPARTMENTAL ADMISSIONS INFORMATION
Admission to the engineering department: At the time of admission to the institution.

FINANCIAL INFORMATION

ESTIMATED EXPENSES (ACADEMIC YEAR)
[Expenses are for the 1992-93 nine-month academic year.]

State Residents	Undergraduate	Graduate
Tuition and fees	$ 1,876	$ 2,146
College room and board	$ —(B)	$ —(B)
Books and supplies	$ 375	$ 375
Other expenses	$ 400	$ 400
Total estimated expenses	$ 2,651	$ 2,921

Out-of-State Residents	Undergraduate	Graduate
Tuition and fees	$ 5,320	$ 5,820
College room and board	$ —(B)	$ —(B)
Books and supplies	$ 375	$ 375
Other expenses	$ 400	$ 400
Total estimated expenses	$ 6,095	$ 6,595

Notes: (B) Data not applicable.

FINANCIAL AID OFFICE CONTACT
Lee Ingalls, *Director of Financial Aid/Student Employment*
University of Colorado at Colorado Springs
1420 Austin Bluffs Parkway
P.O. Box 7150
Colorado Springs, CO 80933-7150
Phone: (719)593-3460   Fax: (719)593-3362

GENERAL FINANCIAL AID INFORMATION
Forms accepted/required: ACT, FAT, FFS, Stafford, SAR, Supplemental, Acceptance by Admissions as a degree student.
Additional financial aid information: Need-based programs: grants (including Pell Grant, SEOG, Colorado Student Grant, Colorado Student Incentive Grant); loans (including Carl Perkins and Stafford); and work study. Non-need based programs include: grants and scholarships (including Colorado Scholars Award, Regents Scholarship, Colorado Diversity Grant, Arnold Scholarship, Graduate Diversity Scholarship and Alumni Scholarship); loans (including PLUS and SLS loans); and no-need work study.

ENGINEERING COLLEGE INFORMATION
[For additional personnel, refer to the *Appendix*.]

HEAD OF ENGINEERING
Pieter A. Frick
Phone: (719)593-3226   Fax: (719)593-3542
EMail: pfrick@wetterhorn.uccs.edu

ENGINEERING COLLEGE ADDRESS
College of Engineering and Applied Science
University of Colorado at Colorado Springs
1420 Austin Bluffs Parkway
P.O. Box 7150
Colorado Springs, CO 80933-7150
Phone: (719)593-3226   Fax: (719)593-3542

ENROLLMENTS—BY CLASS
[Numbers are baccalaureate enrollments for the fall 1991 term, unless otherwise footnoted.]

1st-year students/Freshmen 23
2nd-year students/Sophomores 21
3rd-year students/Juniors 40
4th-year students/Seniors .... 105 [2]
Total 189

Notes: (2) Includes 50 5th-year seniors.

NUMBER OF DEGREES AWARDED—BY PROGRAM
[Numbers are engineering baccalaureate degrees awarded during the 1991-1992 academic year, unless otherwise footnoted. For full details about each engineering program, refer to the *Tables of Degree Programs*.]

Applied Mathematics 9
Computer Science 18
Electrical & Computer Engineering 49
Space Operations .... — [B]
Total 76

Notes: (B) Data not applicable.

PERCENTAGE OF DEGREES AWARDED—BY CATEGORY
[Percentages are of all engineering baccalaureate degrees awarded during the 1991-1992 academic year, unless otherwise footnoted.]

African Americans 4.1%
Asian/Pacific Island Americans 10.2%
Hispanic Americans 4.1%
Native Americans .... — % [A]
Foreign Citizens .... — % [A]
All Others 81.6%
Women 24.5%
Persons with disabilities .... — % [A]
Students over 25 years of age 71.0%

Notes: (A) Data not available.

GRADUATE ENROLLMENTS & DEGREES AWARDED
Master's enrollment 71
Master's degrees awarded 24
Doctoral enrollment 25
Doctoral degrees awarded 4

ENGINEERING STUDENT DATA
Applicants to the engineering college
Number of applicants to engineering college 84
Percent offered admission .... 66.7% [1]

Matriculated engineering students
Percentage in top quartile (25%) of High School class 59.0

Notes: (1) Most of those rejected were offered pre-engineering status.

FULL- & PART-TIME FACULTY—BY DEPARTMENT
[Figures are the head count of full-time faculty and the full-time equivalent (FTE) of part-time faculty for each engineering department or equivalent.]

Department	Full	Part
Electrical & Computer Eng	14	6.0

COLLEGE DESCRIPTION
ABET accredited BSEE with opportunity to specialize in circuit design/computer engineering, physical electronics, or communications/control systems. New facilities with modern state-of-the-art equipment. Relatively small class sizes (generally 45 or less). Free tutoring by student clubs. Sophisticated learning centers to support the learning of math and science. Intern and Co-op programs available usually extending the undergraduate program to five years.

DEGREE PROGRAMS DESCRIPTION
B.S.: The first year of the BSEE curriculum includes courses in mathematics, basic sciences, computer science, composition, and social sciences/humanities. The second year continues with additional courses in these areas, but introduces courses in electrical engineering science. The last two years build on the first two with additional courses in electrical engineering and 15 semester hours of technical electives. The student may choose to specialize in a particular track through the selection of appropriate technical electives. Areas of specialization include electronic circuits, computer-aided design, control systems, communications and signal processing, digital computers, electromagnetic radiation, and microelectronics. In the senior year, participation in a design project is required. M.S. and Ph.D.: Several laboratories specifically available for research have been equipped through research grants, gifts and state funds. These include the Microelectronics Laboratory, the VLSI Design Laboratory, the Microwave/IR Measurements Laboratory, Communications and Signal Processing Laboratory, and the Space and Flight Systems Laboratory.

TRANSFER INFORMATION
Residency Requirements: Last 30 semester hours must be completed after being formally admitted as a degree student.

Transfer via Articulation Agreements
Admission to engineering: At any time
Engineering graduates transferred from ... 4-yr: —   2-yr: 4
Requirements: Must have completed Calculus I and II and one calculus-based physics course with lab. All with grade of 'C' or better and a cumulative GPA of 2.50 or better.

Transfer without Articulation Agreements
Admission to engineering: At any time
Engineering graduates transferred from ... 4-yr: 26   2-yr: 4
Requirements: Must have completed Calculus I and II and one calculus-based physics course with lab. All with grade of 'C' or better and a cumulative GPA of 2.50 or better.

GRADUATION REQUIREMENTS
Cumulative GPA of 2.0 or higher required for graduation. In addition, the ECE Department requires a minimum 2.0 average in the ECE major as well as a minimum 2.0 in both ECE 221 and ECE 241. Of the 15 hours of required technical electives all must be upper division (300 or above) with the exception of CS 245. A minimum of 16 credit hours of design courses are required. All undergraduates are required to take ECE 489, Electrical Engineering Design.

STUDENT PROGRAMS

PROFESSIONAL AND HONORARY SOCIETIES
[For key to acronyms, see *Introduction*.]
Professional Societies: ACM, IEEE, MAA
Honorary Societies: Eta Kappa Nu

SUPPORT PROGRAMS
Student Chapter Organizations: Society of Women Engineers

Society of Women Engineers
For: Women   Available: Year round
Offered: Prematriculation, Freshman, Sophomore, Junior, Senior, Graduate level

Other Engineering Support Programs: Freshman/transfer orientation programs; academic advising and counseling; Math Learning Center; Science Learning Center; Center for Excellence in Writing; Center for Excellence in Oral Communication; Cross Cultural Media Technology Center; University Learning Center; Student Support Services for counseling, testing and placement; Facilitated Mentoring Program.

051 University of Colorado at Denver

INSTITUTION PROFILE

HEAD OF THE INSTITUTION
John C. Buechner, *Chancellor*
Phone: (303)556-2643 Fax: (303)556-2164

GENERAL INFORMATION
[All Students—Fall 1991]
Undergraduate enrollment 6,155
Graduate enrollment 4,934
Total institution enrollment 11,089

Type of institution: Public Calendar: Semesters
Location: Denver Population: 1,800,000
Setting: Urban
Types of engineering degrees: Engineering

Other degree-granting colleges: Architecture, Arts & Sciences, Business Administration, Education, Fine & Performing Arts, Public Affairs

ENGINEERING ADMISSIONS

ADMISSIONS OFFICE CONTACT
Barbara Schneider
University of Colorado at Denver
Executive Director of Admissions
P.O. Box 173364, Campus Box 167
Denver, CO 80217-3364
Phone: (303)556-3287 Fax: (303)556-4838

ENGINEERING COLLEGE ADMISSIONS INFORMATION
Entrance Requirements: SAT (Score: 1000), ACT (Score: 24), High School courses—English (4 years), Math: Alg (2), Geom (1), Trig/Anal. Geometry (1) (4 years), Science: Phys (1), Chem (1), Lab Science (2) (3 years), Other: Foreign Lang (2), Soc. Sci (2), Elective (1) (5 years)

Entrance Recommendations: High School courses—English (4 years), Math: Alg (2), Geom (1), Trig/Anal. Geometry (1) (4 years), Science: Phys (1), Chem (1), Lab Science (2) (3 years), Other: Foreign Lang (2), Soc. Sci (2), Elective (1) (5 years)

Requirements for foreign students: TOEFL (Score: 525); financial statement

DEPARTMENTAL ADMISSIONS INFORMATION
Admission to the engineering department: At the time of admission to the institution.

FINANCIAL INFORMATION

ESTIMATED EXPENSES (ACADEMIC YEAR)
[Expenses are for the 1992-93 nine-month academic year.]

State Residents	Undergraduate	Graduate
Tuition and fees	$ 2,200	$ 1,850[1]
College room and board	$ 3,735[2]	$ 3,735[2]
Books and supplies	$ 500	$ 500
Other expenses	$ 540[3]	$ 540[3]
Total estimated expenses	$ 6,975	$ 6,625

Out-of-State Residents	Undergraduate	Graduate
Tuition and fees	$ 8,400	$ 5,800[4]
College room and board	$ 3,735[2]	$ 3,735[2]
Books and supplies	$ 500	$ 500
Other expenses	$ 540[3]	$ 540[3]
Total estimated expenses	$13,175	$10,575

Notes: (1) $1850 for minimum 5 hours. $3300 for maximum 15 hours. (2) Off-campus housing only. (3) Medical. (4) $5,800 for minimum 5 hours. $9,500 for maximum 15 hours.

FINANCIAL AID OFFICE CONTACT
Elinore Miller, *Director of Financial Aid*
University of Colorado at Denver
Campus Box 125
P.O. Box 173364
Denver, CO 80217-3364
Phone: (303)556-2886 Fax: (303)556-4822

GENERAL FINANCIAL AID INFORMATION
Forms accepted/required: ACT, AFSA, CSS-FAF, FAF, FAT, FSA, FFS, Stafford, IRS, SAR, Institutional

Additional financial aid information: Fellowships, scholarships, grants, student loans, part-time campus employment, and work-study awards. Applicants should submit one of the following 'need-analysis' financial aid forms to be considered for a package of need-based grants, work-study and/or long-term loans.

ENGINEERING COLLEGE INFORMATION

[For additional personnel, refer to the *Appendix*.]

HEAD OF ENGINEERING
Peter E. Jenkins
Phone: (303)556-2870 Fax: (303)556-2511
EMail: PJenkins@cudnvr.denver.colorado.EDU

ENGINEERING COLLEGE ADDRESS
College of Engineering and Applied Science
University of Colorado at Denver
Campus Box 104
P.O. Box 173364
Denver, CO 80217-3364
Phone: (303)556-2870 Fax: (303)556-2511

ENROLLMENTS—BY CLASS
[Numbers are baccalaureate enrollments for the fall 1991 term, unless otherwise footnoted.]
1st-year students/Freshmen .... 60 [1]
2nd-year students/Sophomores 77
3rd-year students/Juniors 166
4th-year students/Seniors 405
Total **708**

Notes: (1) Includes all degree programs offered by the College of Engineering (CE, CS, EE, ME, A.Math).

NUMBER OF DEGREES AWARDED—BY PROGRAM
[Numbers are engineering baccalaureate degrees awarded during the 1991-1992 academic year, unless otherwise footnoted. For full details about each engineering program, refer to the *Tables of Degree Programs*.]
Applied Mathematics 5
Civil Engineering 20
Computer Science & Engineering 19
Electrical Engineering 69
Mechanical Engineering 36
Total **149**

PERCENTAGE OF DEGREES AWARDED—BY CATEGORY
[Percentages are of all engineering baccalaureate degrees awarded during the 1991-1992 academic year, unless otherwise footnoted.]
African Americans 1.3%
Asian/Pacific Island Americans 17.4%
Hispanic Americans 4.0%
Native Americans .... –
Foreign Citizens 7.4%
All Others 69.9%
Women 17.4%
Persons with disabilities 0.5%
Students over 25 years of age 69.0%

GRADUATE ENROLLMENTS & DEGREES AWARDED
Master's enrollment 236
Master's degrees awarded 49
Doctoral enrollment 11
Doctoral degrees awarded .... – [3]

Notes: (3) PhD programs authorized at the CU-Boulder campus as part of system-wide graduate program.

ENGINEERING STUDENT DATA

Applicants to the engineering college
Number of applicants to engineering college 121
Percent offered admission 35.0%

Matriculated engineering students
Percentage in top quartile (25%) of High School class 89.0
Average ACT scores: Math—25, Composite—24

FULL- & PART-TIME FACULTY—BY DEPARTMENT
[Figures are the head count of full-time faculty and the full-time equivalent (FTE) of part-time faculty for each engineering department or equivalent.]

Department	Full	Part
Civil Eng	11	2.4
Electrical Eng	13	3.1
Mechanical Eng	8	2.1
Computer Science & Eng	10	2.8

COLLEGE DESCRIPTION
'Quality Engineering education in the heart of Denver.' For over 75 years the University of Colorado at Denver has provided the Denver metropolitan area with nationally accredited engineering education. This urban campus' strength lies in good faculty-student relationships and informality unique to a university. In recognition of the value of simultaneous professional studies and related employment, engineering classes are scheduled in day and evening hours to accommodate the needs of local employers and working students who attend school part-time. Students may also participate in local co-op education opportunities. A Women & Minorities in Engineering Program addresses the challenges unique to these students. The expertise of the engineering faculty includes soil dynamics, structures, transportation, computer architecture, computer systems, signal processing, CAD, robotics, and biomechanics in additional to traditional areas in civil, electrical, computer, and mechanical engineering

DEGREE PROGRAMS DESCRIPTION
The College maintains a broad-based curriculum in both undergraduate and graduate engineering programs. The four-year undergraduate program stresses fundamental concepts of all aspects of engineering, science, mathematics, computer, and humanities and includes required and elective engineering courses. General education requirements are met by a core curriculum. The required senior design project gives students practical education and teamwork experience. Graduate work leading to the M.S., M.Eng., Ph.D. (in cooperation with CU-Boulder) is offered in Civil, Electrical and Mechanical Engineering, and Computer Science. All graduate courses are held in the late afternoon or evening to accommodate practicing professionals as well as full-time students. Graduate program emphases include: CE: structural, geotechnical, transportation, water resources, and water quality engineering; EE: systems and controls, signal processing, optics and communication, power systems, electronics and semiconductor devices, and fields and radio propagation; CSE: numerical computation, programming languages, software engineering, computer systems, artificial intelligence, and parallel computation; and ME: solid mechanics, fluid mechanics, heat and mass transfer, combustion, materials science, wave propagation and vibrations, and mechanical design.

DUAL DEGREE PROGRAMS
Engineering baccalaureate graduates with dual degrees 1

TRANSFER INFORMATION
Residency Requirements: 30 hours, after admission to the College of Engineering.

Transfer via Articulation Agreements
Admission to engineering: At the end of the freshman year
Engineering graduates transferred from .. 4-yr: – 2-yr: 16
Requirements: Completion of 30 semester hours to include 2 semesters each of Calculus and Physics (Calculus-based); grades of B or higher in Math and Science courses; 2.75 cumulative GPA.

Transfer without Articulation Agreements
Admission to engineering: At any time
Engineering graduates transferred from .. 4-yr: 40 2-yr: 14
Requirements: Completion of 30 hours to include 2 semesters of Calculus and Physics (Calculus-based); grades of C or higher in all courses and a 2.5 GPA.

GRADUATION REQUIREMENTS
B.S. programs in civil, electrical, and mechanical engineering, computer science, and applied mathematics require 128 credit hours (30 credit hour minimum residency at CU-Denver) with a minimum grade point average of 2.0 (C) for all courses attempted and for all major courses. Seniors in the ABET accredited engineering programs are required to take the Fundamentals of Engineering Examination (E.I.T.).

STUDENT PROGRAMS

PROFESSIONAL AND HONORARY SOCIETIES
[For key to acronyms, see *Introduction*.]
Professional Societies: ACM, ASCE, ASME, IEEE
Honorary Societies: Chi Epsilon, Eta Kappa Nu, Pi Tau Sigma, Tau Beta Pi

SUPPORT PROGRAMS
Student Chapter Organizations: Society of Women Engineers, National Society of Black Engineers, Society of Hispanic Professional Engineers

Women and Minorities in Engineering Program
For: Women & Ethnic Minorities Available: Year round
Offered: Prematriculation, Freshman, Sophomore, Junior, Senior, Graduate level

Other Engineering Support Programs: New student orientation. Career Services, Student Services (counseling, tutoring, testing), Internships/Co-Op Education, Financial Aid/Student Employment, Disabled Student Services, International Student Services (counseling, cultural orientation), E.O.P. (Native American, Asian, Black, Hispanic counseling), Women & Minorities in Engineering Program (counseling, mentoring, tutoring).

052 Colorado School of Mines

■ INSTITUTION PROFILE
HEAD OF THE INSTITUTION
George S. Ansell
Phone: (303)273-3280　　　　　　　　Fax: (303)273-3040
GENERAL INFORMATION
[All Students—Fall 1991]
Undergraduate enrollment 1,991
Graduate enrollment 864
Total institution enrollment 2,855
Type of institution: Public　　　**Calendar:** Semesters
Nearest city: Denver, Colorado　**Population:** 1,900,000
Miles from main campus: 20　　　**Setting:** Suburban
Types of engineering degrees: Engineering

■ ENGINEERING ADMISSIONS
ADMISSIONS OFFICE CONTACT
William Young
Colorado School of Mines
Admissions Office
Golden, CO 80401-1887
Phone: (303)273-3220　　　　　　　　Fax: (303)273-3278
ENGINEERING COLLEGE ADMISSIONS INFORMATION
Entrance Requirements: SAT, ACT, High School courses—English (4 years), Mathematics (4 years), Laboratory Science (4 years), Social Studies (4 years)
Requirements for foreign students: TOEFL (Score: 550); financial statement
DEPARTMENTAL ADMISSIONS INFORMATION
Admission to the engineering department: At the time of admission to the institution.

■ FINANCIAL INFORMATION
ESTIMATED EXPENSES (ACADEMIC YEAR)
[Expenses are for the 1992-93 nine-month academic year.]

State Residents	Undergraduate	Graduate
Tuition and fees	$ 3,727	$ 3,727
College room and board	$ 3,500	$ 3,500
Books and supplies	$ 1,000	$ 1,000
Other expenses	$ 720[1]	$ 720
Total estimated expenses	**$ 8,947**	**$ 8,947**

Out-of-State Residents	Undergraduate	Graduate
Tuition and fees	$ 9,999	$ 9,999
College room and board	$ 3,500	$ 3,500
Books and supplies	$ 1,000	$ 1,000
Other expenses	$ 720[1]	$ 720[1]
Total estimated expenses	**$15,219**	**$15,219**

Notes: (1) Health Insurance

FINANCIAL AID OFFICE CONTACT
Roger Koester, *Director of Financial Aid*
Colorado School of Mines
1500 Illinois
Golden, CO 80401-1887
Phone: (303)273-3301　　　　　　　　Fax: (303)273-3278
GENERAL FINANCIAL AID INFORMATION
Forms accepted/required: ACT, CSS-FAF, FAF, FAT, FFS, Stafford, IRS, SAR, Institutional
Additional financial aid information: Federal, state and institutional need-based grants, loans, work study. Academic, athletic, music, minority student merit scholarships.

■ ENGINEERING COLLEGE INFORMATION
[For additional personnel, refer to the *Appendix*.]
HEAD OF ENGINEERING
Franklin D. Schowengerdt
Phone: (303)273-3320　　　　　　　　Fax: (303)273-3040
EMail: FSCHOWEN@DEAN

ENGINEERING COLLEGE ADDRESS
Colorado School of Mines
1500 Illinois
Golden, CO 80401-1887
Phone: (303)273-3000　　　　　　　　Fax: (303)273-3278

ENROLLMENTS—BY CLASS
[Numbers are baccalaureate enrollments for the fall 1991 term, unless otherwise footnoted.]
1st-year students/Freshmen 629
2nd-year students/Sophomores 381
3rd-year students/Juniors 320
4th-year students/Seniors 384
Total 1,714

NUMBER OF DEGREES AWARDED—BY PROGRAM
[Numbers are engineering baccalaureate degrees awarded during the 1991-1992 academic year, unless otherwise footnoted. For full details about each engineering program, refer to the *Tables of Degree Programs*.]
Applied & Engineering Physics 16
Chemical Engineering & Petroleum Refining 16
Chemistry 5
Engineering 108
Geological Engineering 15
Geophysical Engineering 6
Mathematics 17
Metallurgical & Materials Engineering 16
Mining Engineering 9
Petroleum Engineering 22
Total 230

PERCENTAGE OF DEGREES AWARDED—BY CATEGORY
[Percentages are of all engineering baccalaureate degrees awarded during the 1991-1992 academic year, unless otherwise footnoted.]
African Americans .... – %[1]
Asian/Pacific Island Americans .... 2.0 %
Hispanic Americans .... 3.0 %
Native Americans .... 2.0 %
Foreign Citizens .... 22.0 %
All Others .... 71.0 %
Women .... 20.0 %
Persons with disabilities .... – %[1]
Students over 25 years of age .... – %[A]

Notes: (A) Data not available. (1) One graduate in this category.

ENGINEERING STUDENT DATA
Applicants to the engineering college
Number of applicants to engineering college 1,290
Percent offered admission 83.0%
Matriculated engineering students
Percentage in top quartile (25%) of High School class 85.0
Average SAT scores: Math—635, Verbal—520, Combined—1155
Average ACT scores: Math—27, Composite—27

FULL- & PART-TIME FACULTY—BY DEPARTMENT
[Figures are the head count of full-time faculty and the full-time equivalent (FTE) of part-time faculty for each engineering department or equivalent.]

Department	Full	Part
Chemical Eng & Petroleum Refining	9	1.0
Eng	17	2.0
Eng Physics	12	–
Geology & Geological Eng	15	1.0
Geophysical Eng	8	1.0
Metallurgical & Materials Eng	16	–
Mining Eng	7	–
Petroleum Eng	5	–

COLLEGE DESCRIPTION
CSM is an institute of science and engineering with a special focus in the resource areas of energy, minerals and materials. Its primary goal is to engage in education and research which will prepare students and society for responsible stewardship of the earth and its resources. The school specializes in interdisciplinary education and innovative approaches to curriculum. Unusual programs include the Engineering Practices Integrated Course Sequence (EPICS), the McBride Honors Program in Public Affairs for Engineers and an interdisciplinary senior design program. Coop programs are available but not required. A junior-senior summer field session is required in all undergraduate degree programs.

DEGREE PROGRAMS DESCRIPTION
The degree programs at CSM stress quality and interdisciplinary approaches. All of our undergraduate programs require a core course in Environmental Sciences and Engineering Ecology, an 11-semester hour sequence in communications called 'Engineering Practices Integrated Course Sequence (EPICS), a junior-senior summer field session and a capstone design sequence. Our largest department, Engineering, offers a single, non-traditional degree combining aspects of civil, electrical and mechanical. This program serves as a model for interdisciplinary programs across campus. Our graduate programs are also interdisciplinary, but are more specialized. The fastest-growing graduate programs are those in Environmental Sciences and Engineering Ecology, followed closely by those in Materials Science. Both of the programs are interdisciplinary, involving most of the departments at CSM.

TRANSFER INFORMATION
Transfer via Articulation Agreements
Admission to engineering: At any time
Engineering graduates transferred from　...4-yr: 7　2-yr: 5
Requirements: Successful completion of a program specified for that institution.
Transfer without Articulation Agreements
Admission to engineering: At any time
Engineering graduates transferred from　...4-yr: –　2-yr: –
Requirements: Successful completion of specified work at another institution.

GRADUATION REQUIREMENTS
A cumulative GPA which is not less than 2.000

■ STUDENT PROGRAMS
PROFESSIONAL AND HONORARY SOCIETIES
[For key to acronyms, see *Introduction*.]
Professional Societies: AAPG, ACS, AEG, AGCA, AIChE, AIME, ASCE, ASME, GSA, IEEE
Honorary Societies: Tau Beta Pi

SUPPORT PROGRAMS
Student Chapter Organizations: Society of Women Engineers, National Society of Black Engineers, Society of Hispanic Professional Engineers, American Indian Science & Engineering Society, Asian Student Association

Minority Engineering Program
For: Ethnic Minorities　　**Available:** Academic year
Offered: Freshman, Sophomore, Junior, Senior, Graduate level

Summer Bridge Program
For: Women & Ethnic Minorities　**Available:** Summer only
Offered: Freshman

Summer Minority Engineering Training
For: Women & Ethnic Minorities　**Available:** Summer only
Offered: Prematriculation

Tribal Resource Institute for Bus., Eng. and Sci.
For: Ethnic Minorities　　**Available:** Summer only
Offered: Freshman

Other Engineering Support Programs: CSM's Student Development Center provides personal, academic and career counseling for all students. In addition, we offer workshops and seminars on life-skills topics such as time management, substance abuse prevention, assertiveness training, building relationships, communication, stress management, sexual assault prevention and leadership management.

053 Colorado State University

■ INSTITUTION PROFILE

HEAD OF THE INSTITUTION
Albert C. Yates
Phone: (303)491-6211　　　　Fax: (303)491-0501

GENERAL INFORMATION
[All Students—Fall 1991]
Undergraduate enrollment 17,460
Graduate enrollment 3,507
Total institution enrollment 20,967
Type of institution: Public　　　Calendar: Semesters
Nearest city: Denver　　　　　　Population: 1,867,000
Miles from main campus: 70　　　Setting: Urban
Types of engineering degrees: Engineering
Other degree-granting colleges: Humanities & Social Sciences, Agricultural Sciences, Business, Natural Sciences, Natural Resources, Veterinary Medicine and Biomedical Sciences

ENGINEERING COLLEGE ADMISSIONS INFORMATION
Entrance Requirements: SAT (Score: 900), ACT (Score: 22), High School courses—English (4 years), Mathematics (3.5 years), Chemistry/Natural Science (3 years), Social Science (2 years)
Entrance Recommendations: High School courses—English (composition, grammar, literature, speech) (4 years), Mathematics (algebra, geometry, trigonometry) (3.5 years), Chemistry, physics (3 years), Foreign language, mech drawing, or computer prog (1 year)
Requirements for foreign students: TOEFL (Score: 550)

DEPARTMENTAL ADMISSIONS INFORMATION
Admission to the engineering department: At the time of admission to the institution.
Additional information: Mechanical Engineering - ACT 24; SAT 970. Civil Engineering - ACT 23; SAT 930.

■ FINANCIAL INFORMATION

ESTIMATED EXPENSES (ACADEMIC YEAR)
[Expenses are for the 1992-93 nine-month academic year.]

State Residents	Undergraduate	Graduate
Tuition and fees	$ 2,511	$ 2,813
College room and board	$ 3,915	$ 3,915
Books and supplies	$ 500	$ 500
Other expenses	$ 1,224 [1]	$ 1,224 [1]
Total estimated expenses	$ 8,150	$ 8,452
Out-of-State Residents	Undergraduate	Graduate
Tuition and fees	$ 7,677	$ 7,985
College room and board	$ 3,915	$ 3,915
Books and supplies	$ 500	$ 500
Other expenses	$ 1,224 [1]	$ 1,224 [1]
Total estimated expenses	$13,316	$13,624

Notes: (1) Health Insurance and Information Technology Fee

FINANCIAL AID OFFICE CONTACT
G. K. Jacks, *Director*
Colorado State University
Office of Financial Aid
Fort Collins, CO 80523
Phone: (303)491-6321　　　　Fax: (303)491-5010

GENERAL FINANCIAL AID INFORMATION
Forms accepted/required: ACT, FFS
Additional financial aid information: Need-based and merit-based scholarships offered; short-term and long-term loans; hourly employment and work-study employment.

■ ENGINEERING COLLEGE INFORMATION

[For additional personnel, refer to the *Appendix*.]

HEAD OF ENGINEERING
Frank A. Kulacki
Phone: (303)491-6603　　　　Fax: (303)491-5569
EMail: fkulacki%dean%engadmin@vines.colostate.edu

ENGINEERING COLLEGE ADDRESS
College of Engineering
Colorado State University
Engineering Building
Fort Collins, CO 80523
Phone: (303)491-6603　　　　Fax: (303)491-5569

ENROLLMENTS—BY CLASS
[Numbers are baccalaureate enrollments for the fall 1991 term, unless otherwise footnoted.]
1st-year students/Freshmen 355
2nd-year students/Sophomores 259
3rd-year students/Juniors 240
4th-year students/Seniors 427
Total 1,281

NUMBER OF DEGREES AWARDED—BY PROGRAM
[Numbers are engineering baccalaureate degrees awarded during the 1991-1992 academic year, unless otherwise footnoted. For full details about each engineering program, refer to the *Tables of Degree Programs*.]
Agricultural Engineering 3
Chemical Engineering 16
Civil Engineering 61
Electrical Engineering 31
Engineering Science(s) 20
Mechanical Engineering 77
Total 208

PERCENTAGE OF DEGREES AWARDED—BY CATEGORY
[Percentages are of all engineering baccalaureate degrees awarded during the 1991-1992 academic year, unless otherwise footnoted.]
African Americans .... – %
Asian/Pacific Island Americans .... 2.5 %
Hispanic Americans .... 3.0 %
Native Americans .... – %
Foreign Citizens .... – % (A)
All Others .... 94.5 %
Women .... – % (A)
Persons with disabilities .... – % (A)
Students over 25 years of age .... – % (A)

Notes: (A) Data not available.

GRADUATE ENROLLMENTS & DEGREES AWARDED
Master's enrollment 292
Master's degrees awarded 106
Doctoral enrollment 175
Doctoral degrees awarded 47

ENGINEERING STUDENT DATA
Applicants to the engineering college
Number of applicants to engineering college 940
Percent offered admission 81.0%

Matriculated engineering students
Percentage in top quartile (25%) of High School class 66.0
Average SAT scores: Math—605, Verbal—491, Combined—1096
Average ACT scores: Math—27, Composite—26

FULL- & PART-TIME FACULTY—BY DEPARTMENT
[Figures are the head count of full-time faculty and the full-time equivalent (FTE) of part-time faculty for each engineering department or equivalent.]

Department	Full	Part
Agricultural & Chemical Eng	17	–
Civil Eng	33	5.0
Electrical Eng	19	–
Mechanical Eng	23	2.0
Eng Science	–	–
Atmospheric Science	14	1.0

COLLEGE DESCRIPTION
The College of Engineering offers four-year degree programs leading to the B.S. degree in six broad-based fields. Freshmen can also enroll in an open option. The first two years of course work emphasize calculus, physics, chemistry, and basic engineering science to provide the background necessary to take the specialized engineering courses during the junior and senior years. These years focus on applied engineering and design courses within a specific engineering field. Design courses integrate knowledge gained in the engineering curriculum and teach students how to apply that knowledge. In addition, 98% of engineering undergraduate courses are taught by faculty members, not teaching assistants. Students are introduced to the use of computers early in their course work and the Engineering Network Services gives students hands-on experience. A summer research program designed for highly motivated juniors is available.

DEGREE PROGRAMS DESCRIPTION
The B.S. degrees in the various engineering programs prepare the graduate for entry-level professional positions or for graduate study in each of the respective fields. Advanced degree programs (M.S. and Ph.D.) include Atmospheric Science.

TRANSFER INFORMATION
Residency Requirements: A minimum of 32 semester credits must be completed in residence as a major in the College of Engineering.

Transfer via Articulation Agreements
Admission to engineering: At any time
Requirements: Transfers are admitted to the engineering program at any level. Transfer programs are articulated with some community and junior colleges. The university has pre-engineering transfer agreements with selected 4-year institutions within Colorado.

Transfer without Articulation Agreements
Admission to engineering: At any time
Requirements: Completion of first semester of calculus and first semester of calculus-based physics with B's or better. Minimum GPA varies from 2.7 to 3.0 depending upon department and institution from which student is transferring.

GRADUATION REQUIREMENTS
All Bachelor of Science degrees require 128-136 semester credit hours of course work applicable toward specific departmental degree requirements. A minimum overall cumulative GPA of 2.00 and a GPA of 2.00 in departmental courses required; minimum of 16 semester credit hours in humanities and social science course work; and some specific departmental requirements exist.

■ STUDENT PROGRAMS

PROFESSIONAL AND HONORARY SOCIETIES
[For key to acronyms, see *Introduction*.]
Professional Societies: AIAA, AIChE, ASAE, ASCE, ASHRAE, ASME, IEEE, SAE, The Institute of Electrical & Electronics Engineers/Computer Society
Honorary Societies: Alpha Epsilon, Chi Epsilon, Eta Kappa Nu, Omega Chi Epsilon, Pi Tau Sigma, Tau Beta Pi

SUPPORT PROGRAMS
Student Chapter Organizations: Society of Women Engineers, National Society of Black Engineers, Society of Hispanic Professional Engineers, American Indian Society of Engineers and Sciences

The Women and Minority Engineering Program
For: Women & Ethnic Minorities　Available: Academic year
Offered: Freshman, Sophomore, Junior, Senior
Other Engineering Support Programs: Career Counseling and Advising Center available to all students; Freshman Orientation Program; Peer Adviser Program; Placement Services; campus housing-based tutoring program.

054 Columbia University

■ INSTITUTION PROFILE

HEAD OF THE INSTITUTION
Michael I. Sovern
Phone: (212)854-2825　　　　　Fax: (212)854-6466

GENERAL INFORMATION
[All Students—Fall 1991]
Undergraduate enrollment 8,300
Graduate enrollment 10,500
Total institution enrollment **18,800**
Type of institution: Private　　　Calendar: Semesters
Location: New York City　　　　　Population: 8,000,000
Setting: Urban
Types of engineering degrees: Engineering
Other degree-granting colleges: Allied Health Sciences, Architecture, Arts & Sciences, Business Administration, Communication & Journalism, Dentistry, Education, Fine & Performing Arts, Humanities & Social Sciences, Law, Medicine, Nursing, General Studies, International and Public Affairs, Foreign and Comparative Law, Social Work

■ ENGINEERING ADMISSIONS

ADMISSIONS OFFICE CONTACT
Lawrence J. Momo
Columbia University
212 Hamilton Hall
New York, NY 10027-6699
Phone: (212)854-2522　　　　　Fax: (212)854-1209

ENGINEERING COLLEGE ADMISSIONS INFORMATION
Entrance Requirements: SAT, High School courses—Mathematics (preferably through Calculus) (4 years), Physics (1 year), Chemistry (1 year), English (4 years), The American College Testing program will be accepted in lieu of the SAT if all sub-scores are at the 90th percentile or greater. Achievement tests 1) math level I or II, 2)English comp., 3)physics or chemistry.
Entrance Recommendations: High School courses—Foreign Language (3 years), History and Social Studies (4 years)
Requirements for foreign students: TOEFL (Score: 600)

DEPARTMENTAL ADMISSIONS INFORMATION
Admission to the engineering department: At the time of admission to the institution.

■ FINANCIAL INFORMATION

ESTIMATED EXPENSES (ACADEMIC YEAR)
[Expenses are for the 1992-93 nine-month academic year.]

All Students	Undergraduate	Graduate
Tuition and fees	$16,450	$16,450
College room and board	$ 6,010	$ 6,010
Books and supplies	$ 550	$ 550
Other expenses	$ 1,206	$ 1,470
Total estimated expenses	**$24,216**	**$24,480**

FINANCIAL AID OFFICE CONTACT
Deborah B. Pointer, Director of Financial Aid
Columbia University
100 Hamilton Hall
New York, NY 10027-6699
Phone: (212)854-3711　　　　　Fax: (212)854-5353

GENERAL FINANCIAL AID INFORMATION
Forms accepted/required: FAF, IRS, Institutional
Additional financial aid information: A typical financial aid package at Columbia is a combination of an institutional grant which is non-repayable, low interest loans, and college work-study program. The admission process is need-blind, so applying for aid will not be a variable in the selection process. The philosophy of Columbia University is one of need-based financial aid, so financial packages are based on the demonstrated need of the applicant's family and not merit.

■ ENGINEERING COLLEGE INFORMATION
[For additional personnel, refer to the Appendix.]

HEAD OF ENGINEERING
David H. Auston
Phone: (212)854-2993　　　　　Fax: (212)864-0104
EMail: dha3@cunixf.cc.columbia.edu

ENGINEERING COLLEGE ADDRESS
School of Engineering and Applied Science
Columbia University
510 Seeley W. Mudd Building
New York, NY 10027-6699
Phone: (212)854-2993　　　　　Fax: (212)864-0104

ENROLLMENTS—BY CLASS
[Numbers are baccalaureate enrollments for the fall 1991 term, unless otherwise footnoted.]
1st-year students/Freshmen 266
2nd-year students/Sophomores 219
3rd-year students/Juniors 241
4th-year students/Seniors 246
Total **972**

NUMBER OF DEGREES AWARDED—BY PROGRAM
[Numbers are engineering baccalaureate degrees awarded during the 1991-1992 academic year, unless otherwise footnoted. For full details about each engineering program, refer to the Tables of Degree Programs.]
Applied Physics 16
Chemical Engineering 27
Civil Engineering & Engineering Mechanics 15
Computer Science 35
Electrical Engineering 47
Industrial Engineering & Operations Research 44
Mechanical Engineering 47
Metallurgical, Mining, Mineral & Materials Engrg 4
Total **235**

PERCENTAGE OF DEGREES AWARDED—BY CATEGORY
[Percentages are of all engineering baccalaureate degrees awarded during the 1991-1992 academic year, unless otherwise footnoted.]
African Americans 3.3%
Asian/Pacific Island Americans 43.9%
Hispanic Americans 4.1%
Native Americans .... – %
Foreign Citizens .... – % (A)
All Others 48.7%
Women 12.6%
Persons with disabilities .... – % (A)
Students over 25 years of age .... – % (A)
Notes: (A) Data not available.

GRADUATE ENROLLMENTS & DEGREES AWARDED
Master's enrollment 178 (1)
Master's degrees awarded .... – (A)
Doctoral enrollment 298 (1)
Doctoral degrees awarded .... – (A)
Notes: (A) Data not available. (1) Full time students only; 335 additional part time students.

ENGINEERING STUDENT DATA
Applicants to the engineering college
Number of applicants to engineering college 1,423
Percent offered admission 42.0%
Matriculated engineering students
Percentage in top quartile (25%) of High School class ... 100.0
Average SAT scores: Math—719, Verbal—573, Combined—1292

FULL- & PART-TIME FACULTY—BY DEPARTMENT
[Figures are the head count of full-time faculty and the full-time equivalent (FTE) of part-time faculty for each engineering department or equivalent.]

Department	Full	Part
Chemical Eng & Applied Chemistry	8	1.0
Civil Eng & Eng Mechanics	13	3.0
Electrical Eng	21	3.7
Industrial Eng & Operations Research	9	2.7
Mechanical Eng	12	2.3
Applied Physics	10	1.7
Computer Science	19	5.3
Henry Krumb School of Mines	13	2.7

COLLEGE DESCRIPTION
Columbia's School of Engineering and Applied Science provides students with a unique opportunity to benefit from the type of personal attention associated with a small college, while having access to the resources of a comprehensive research university. Undergraduates in the SEAS take classes in the social sciences and humanities through Columbia College providing an undergraduate experience with academic breadth across disciplines, as well as depth, in a chosen field of engineering. A five year liberal arts and engineering program, known as The Combined Plan, offers qualified students the opportunity to earn a BA from over eighty affiliated liberal arts colleges and a BS from Columbia's School of Engineering and Applied Science. Research facilities on campus include the Center for Telecommunications Research established by the National Science Foundation, and a new center for Engineering and Physical Science Research.

DEGREE PROGRAMS DESCRIPTION
Columbia has a long tradition of combining humanistic and technical education in its engineering programs. The four year Bachelor of Science Degree is divided between a two year experience in scientific fundamentals, humanities and social sciences, followed by a program designed for a student's mastery of certain principles and arts central to engineering science. The academic depth of the junior and senior year prepares the student for professional practice upon graduation, while the academic breadth of the program exposes the student to different forms of intellectual inquiry across the University. Specialization, balanced with subjects of broad importance as well as theory and applications are emphasized in the 30 point Master of Science degree. Two doctoral degrees are offered by the University: the Doctor of Engineering Science and the Doctor of Philosophy. The minimum requirements for both are the completion of 60 points of approved graduate study beyond the B.S., the passing of appropriate qualifying exams, and the completion and oral defense of a dissertation based on original research. A Professional degree option is also offered allowing advanced work beyond the M.S., without the required emphasis on research.

DUAL DEGREE PROGRAMS
Engineering baccalaureate graduates with dual degrees 50
Enrollment requirements: Students interested in The Combined Plan can receive a Bachelor's Degree from over 90 liberal arts colleges including Columbia College. Admission to both academic units is necessary. A few qualified undergraduates each year submatriculate into graduate programs at Columbia.

TRANSFER INFORMATION
Transfer via Articulation Agreements
Admission to engineering: At the end of the sophomore year
Engineering graduates transferred from ... 4-yr: 28　2-yr: –
Transfer without Articulation Agreements
Admission to engineering: transfers will submit all application forms, high school and college records by April 1 of their freshman or sophomore year.
Engineering graduates transferred from ... 4-yr: 20　2-yr: 2
Requirements: A transfer student's preparation should be similar in academic content to either the first or second year experience, depending on when the student transfers, as prescribed for Columbia students in the academic bulletin.

GRADUATION REQUIREMENTS
Master of Science degree is granted after the minimum completion of 30 points of approved graduate study extending over at least one academic year and no more than five. A minimum grade-point average of 2.5 is required. A two year masters program is available for undergraduate physical science majors from accredited liberal arts colleges. For departmental requirements of doctoral candidates, please refer to the academic bulletin for the School of Engineering and Applied Science.

■ STUDENT PROGRAMS

PROFESSIONAL AND HONORARY SOCIETIES
[For key to acronyms, see Introduction.]
Professional Societies: ACM, AIChE, AIME, APS, ASCE, ASME, IEEE, IIE, ORSA, SIAM, Nat'l Soc. of Black Eng.; Soc. of Women Eng.; Korean Student Assoc., Society of Hispanic Professional Engineers; Asian-American Society of Engr
Honorary Societies: Alpha Pi Mu, Eta Kappa Nu, Pi Tau Sigma, Tau Beta Pi, Various departmental honor organizations also available.

SUPPORT PROGRAMS
Student Chapter Organizations: Society of Women Engineers, National Society of Black Engineers, Society of Hispanic Professional Engineers, Asian American Society of Engineers
Higher Education Opportunity Program
For: Ethnic Minorities　　**Available:** Year round
Offered: Prematriculation, Freshman, Sophomore, Junior, Senior
Other Engineering Support Programs: Comprehensive programs include, but are not limited to, academic advising by the Office of the Dean of Students, department faculty advising as well as residence advisors in the dorms. A tutoring center, psychological services, and career planning center offer specialized support for students. Less formal support groups are offered by various student organizations.

055 Concordia University

INSTITUTION PROFILE

HEAD OF THE INSTITUTION
Patrick Kenniff
Phone: (514)848-4849   Fax: (514)848-8765

GENERAL INFORMATION
[All Students—Fall 1991]
Undergraduate enrollment 22,401
Graduate enrollment 3,583
Total institution enrollment **25,984**
Type of institution: Private   Calendar: Other
Location: Montréal   Population: 2,000,000
Setting: Urban
Types of engineering degrees: Engineering
Other degree-granting colleges: Arts & Sciences, Business Administration, Communication & Journalism, Education, Fine & Performing Arts, Humanities & Social Sciences, Community and Public Affairs, Science College, The Simone de Beauvoir Institute & Women's Studies

ENGINEERING ADMISSIONS

ADMISSIONS OFFICE CONTACT
Thomas E. Swift
Concordia University
1455 de Maisonneuve Boulevard West
Montréal, PQ H3G-1M8
Phone: (514)848-2668   Fax: (514)848-8631

ENGINEERING COLLEGE ADMISSIONS INFORMATION

Requirements for out-of-state residents: Applicants from Quebec institutions: Diplôme d'études collegiales with 10.10 profile. Comparable qualifications for students from outside Quebec. Where pre-university education is shorter than in Quebec, students enter the Extended Credit Program.
Requirements for foreign students: Applicants whose first language is neither English nor French and who did not take the last 2 years of full-time schooling in an English- or French-language system must demonstrate their proficiency in English for admission purposes and must be tested in English writing skills.

DEPARTMENTAL ADMISSIONS INFORMATION
Admission to the engineering department: At the time of admission to the institution.

FINANCIAL INFORMATION

FINANCIAL AID OFFICE CONTACT
Roger Cote, *Director, Financial Aid*
Concordia University
1455 de Maisonneuve Boulevard West
Montréal, PQ H3G-1M8
Phone: (514)848-3522   Fax: (514)848-3494

GENERAL FINANCIAL AID INFORMATION
Forms accepted/required: Institutional
Additional financial aid information: Available through the Financial Aid Office or the Graduate Awards Office: Concordia Entrance Scholarships; Concordia In-Course Scholarships and Bursaries; awards offered by associations, companies, foundations, societies and clubs; Work Study Programme; Tuition Deferrals; Emergency loans; Quebec loans and bursaries; bursaries to physically disabled students; Canada Scholarships Programme; Undergraduate Awards to women students; NSERC awards.

ENGINEERING COLLEGE INFORMATION
[For additional personnel, refer to the *Appendix*.]

HEAD OF ENGINEERING
M.N.S. Swamy
Phone: (514)848-3060   Fax: (514)848-4509

ENGINEERING COLLEGE ADDRESS
Faculty of Engineering and Computer Science
Concordia University
1455 de Maisonneuve Boulevard West
Montréal, PQ H3G-1M8
Phone: (514)848-2424   Fax: (514)848-3494

NUMBER OF DEGREES AWARDED—BY PROGRAM
[Numbers are engineering baccalaureate degrees awarded during the 1991-1992 academic year, unless otherwise footnoted. For full details about each engineering program, refer to the *Tables of Degree Programs*.]

Building Engineering 27
Civil Engineering 15
Computer Engineering 23
Electrical Engineering 52
Mechanical Engineering 82
Total **199**

PERCENTAGE OF DEGREES AWARDED—BY CATEGORY
[Percentages are of all engineering baccalaureate degrees awarded during the 1991-1992 academic year, unless otherwise footnoted.]

African Canadians .... – %
Asian/Pacific Island Canadians .... – %
Hispanic Canadians .... – %
Native Canadians .... – %
Foreign Citizens .... – %
All Others .... 100 %
Women .... – %
Persons with disabilities .... – %
Students over 25 years of age .... – %

FULL- & PART-TIME FACULTY—BY DEPARTMENT
[Figures are the head count of full-time faculty and the full-time equivalent (FTE) of part-time faculty for each engineering department or equivalent.]

Department	Full	Part
Civil Eng	11	7.8
Electrical & Computer Eng	25	10.8
Mechanical Eng	26	10.5
Centre for Building Studies	16	5.6

COLLEGE DESCRIPTION
Established in 1974 with the merger of Loyola College and Sir George Williams University, Concordia is one of the largest urban universities in Canada. It has long promoted accessibility and innovation as its fundamental characteristics. The Faculty of Engineering and Computer Science offers programmes leading to the degree of Bachelor of Engineering in the fields of Building, Civil, Computer, Electrical, Industrial and Mechanical Engineering. Part-time studies are possible but there is a time limit on the completion of upper level courses. The degree in Building Engineering unique in Canada. The co-operative format is offered for this degree and a combined BEng and MEng programme is available. The Industrial Engineering programme is now an option in Mechanical Engineering. The Faculty offers graduate programmes leading to the degrees of MEng (course-oriented), MEng (Aerospace), MASc (research-oriented) and PhD degrees in Building, Civil, Electrical and Mechanical Engineering.

DEGREE PROGRAMS DESCRIPTION
In addition to the basic engineering sciences, students in the BEng (Building) program learn the fundamentals of building materials, structural analysis and design, building services (acoustical, heating, lighting, air conditioning), economics and project management. Students in the BEng (Civil) program specialize in either Option A (structures and geotechnical) or Option B (transportation and water resources). Students in the BEng (Electrical) program specialize in one of 3 options: Electronics/Communications; Power; Systems. Computer Engineering focuses more directly on the design and use of computer technology for information processing. In the BEng (Mechanical) program 3 options are available: Thermo Fluid and Propulsion; Design and Production; Automation and Control Systems. Industrial Engineering, which was an option of Mechanical Eng. is now a new program. The MEng (Aerospace) degree is offered in co-operation with other Quebec universities.

TRANSFER INFORMATION
Transfer without Articulation Agreements
Admission to engineering: At any time
Engineering graduates transferred from .. 4-yr: 40   2-yr: 160
Requirements: Must hold D.E.C. or equivalent and meet competitive quality requirements due to limited number of spaces available. In Quebec educational system, most students entering freshman year come from 2-year colleges intermediate between high school and university.

GRADUATION REQUIREMENTS
A weighted GPA of at least 2.00 is required for Acceptable Standing in Engineering. Students must be in Acceptable Standing and satisfy the requirements for the program in which they were admitted in order to graduate.

STUDENT PROGRAMS

PROFESSIONAL AND HONORARY SOCIETIES
[For key to acronyms, see *Introduction*.]
Professional Societies: CSCE, CSME, IEEE, SAE, SME, Canadian Aeronautics and Space Institute (CASI), Canadian Society of Industrial Engineers (CSIE)

SUPPORT PROGRAMS
Women in Engineering and Computer Science (WECOS)
For: Women   Available: Year round
Offered: Freshman, Sophomore, Junior, Senior, Graduate level
Office of the Status of Women
For: Women   Available: Year round
Offered: Prematriculation, Freshman, Sophomore, Junior, Senior, Graduate level
Other Engineering Support Programs: Counselling & Development Services offers Educational, Career and Personal Counselling. The Learning Skills and Writing Centre helps students find effective ways of learning. The Group/Workshop Programme includes Assertiveness Training, Study Skills, Exam Anxiety Reduction, Stress Management and Personal Growth groups. Guidance Information Centre: university calendars, employer literature.

056 University of Connecticut

INSTITUTION PROFILE

HEAD OF THE INSTITUTION
Harry J. Hartley
Phone: (203)486-2337 Fax: (203)486-2627

GENERAL INFORMATION
[All Students—Fall 1991]
Undergraduate enrollment 15,472
Graduate enrollment 6,269
Total institution enrollment **21,741**
Type of institution: Public **Calendar:** Semesters
Nearest city: Hartford **Population:** 140,000
Miles from main campus: 35 **Setting:** Small Town
Types of engineering degrees: Engineering
Other degree-granting colleges: Arts & Sciences, Business Administration, Dentistry, Education, Fine & Performing Arts, Law, Medicine, Nursing, Agriculture & Natural Resources, Allied Health Professions, Family Studies, Pharmacy, Social Work

ENGINEERING ADMISSIONS

ADMISSIONS OFFICE CONTACT
Ann L. Huckenbeck
University of Connecticut
Box U-88, Room 5
28 North Eagleville Road
Storrs, CT 06269-3088
Phone: (203)486-3137 Fax: (203)486-1476

ENGINEERING COLLEGE ADMISSIONS INFORMATION
Admission to the engineering college: Admission coincides with admission to the institution. School changes allow students to enter Engineering after matriculating in another school/college.
Entrance Requirements: High School courses—English (4 years), Mathematics (algebra, geometry and trigonometry) (3.5 years), Laboratory Science (physics or chemistry) (2 years), Foreign Language (single language) (2 years), The Scholastic Aptitude Test (SAT) is required of all applicants.
Entrance Recommendations: High School courses—Mathematics (4 years), Laboratory Science (physics and chemistry) (3 years), Foreign Language (single language) (3 years), Advanced Placement tests and Scholastic Aptitude Achievement Tests are recommended.
Requirements for foreign students: The TOEFL and SAT tests are required for foreign applicants in addition to a financial affidavit.

DEPARTMENTAL ADMISSIONS INFORMATION
Admission to the engineering department: A student is admitted to an Engineering major as a freshman. However, declaration of a major need not be done until the sophomore year.

FINANCIAL INFORMATION

ESTIMATED EXPENSES (ACADEMIC YEAR)
[Expenses are for the 1992-93 nine-month academic year.]

State Residents	Undergraduate	Graduate
Tuition and fees	$ 3,902	$ 4,580
College room and board	$ 4,878	$ 5,146
Books and supplies	$ 570	$ 570
Other expenses	$ 1,600	$ 1,600
Total estimated expenses	**$10,950**	**$11,896**

Out-of-State Residents	Undergraduate	Graduate
Tuition and fees	$10,374	$10,798
College room and board	$ 4,878	$ 5,146
Books and supplies	$ 570	$ 570
Other expenses	$ 1,928	$ 1,928
Total estimated expenses	**$17,750**	**$18,442**

FINANCIAL AID OFFICE CONTACT
Veronica G. O'Dette, *Director of Student Financial Aid*
University of Connecticut
U-116, Wilbur Cross Building
233 Glenbrook Road
Storrs, CT 06269-4116
Phone: (203)486-2819

GENERAL FINANCIAL AID INFORMATION
Forms accepted/required: FAF, Stafford, IRS, SAR, Institutional
Additional financial aid information: Financial assistance is available in the form of scholarships, grants, loans, and part-time employment. Student employment is either via the Student Labor Program, open to all students, or the federally funded Work Study Program, open to students with a demonstrated financial need.

ENGINEERING COLLEGE INFORMATION
[For additional personnel, refer to the *Appendix*.]

HEAD OF ENGINEERING
Harold D. Brody
Phone: (203)486-2221 Fax: (203)486-0318

ENGINEERING COLLEGE ADDRESS
School of Engineering
University of Connecticut
U-237
191 Auditorium Road
Storrs, CT 06269-3237
Phone: (203)486-2223 Fax: (203)486-0318

ENROLLMENTS—BY CLASS
[Numbers are baccalaureate enrollments for the fall 1991 term, unless otherwise footnoted.]
1st-year students/Freshmen 266
2nd-year students/Sophomores 336
3rd-year students/Juniors 316
4th-year students/Seniors 358
Total **1,276**

NUMBER OF DEGREES AWARDED—BY PROGRAM
[Numbers are engineering baccalaureate degrees awarded during the 1991-1992 academic year, unless otherwise footnoted. For full details about each engineering program, refer to the *Tables of Degree Programs*.]
Chemical Engineering 8
Civil Engineering 55
Computer Science & Engineering 31
Electrical Engineering 80
Mechanical Engineering 78
Metallurgy .... — (B)
Total **252**

Notes: (B) Data not applicable.

PERCENTAGE OF DEGREES AWARDED—BY CATEGORY
[Percentages are of all engineering baccalaureate degrees awarded during the 1991-1992 academic year, unless otherwise footnoted.]
African Americans 1.2%
Asian/Pacific Island Americans 7.1%
Hispanic Americans 2.4%
Native Americans .... — %
Foreign Citizens 1.2%
All Others 88.1%
Women 12.7%
Persons with disabilities .... — % (A)
Students over 25 years of age .... — % (A)

Notes: (A) Data not available.

GRADUATE ENROLLMENTS & DEGREES AWARDED
Master's enrollment 292
Master's degrees awarded 91
Doctoral enrollment 240
Doctoral degrees awarded 33

ENGINEERING STUDENT DATA
Applicants to the engineering college
Number of applicants to engineering college 1,353
Percent offered admission 54.0%

Matriculated engineering students
Percentage in top quartile (25%) of High School class .... 80.0 (C)
Average SAT scores: Math—627, Verbal—510, Combined—1137

Notes: (C) Estimated data.

FULL- & PART-TIME FACULTY—BY DEPARTMENT
[Figures are the head count of full-time faculty and the full-time equivalent (FTE) of part-time faculty for each engineering department or equivalent.]

Department	Full	Part
Chemical Eng	13	1.5
Civil Eng	22	2.0
Electrical & Systems Eng	19	2.5
Mechanical Eng	17	3.5
Computer Science & Eng	18	2.0
Metallurgy	9	2.5

COLLEGE DESCRIPTION
The School offers a four-year undergraduate program in engineering, emphasizing engineering fundamentals in the first two years, yet allowing for specialized training in each of the five majors. A firm foundation in the liberal arts is achieved by integrating studies in the humanities and social sciences throughout the curriculum. Double majors, combining two of the five majors, are available. A Materials double major is available with the department of Metallurgy. Cooperative Education work experience is an optional program, usually comprising a six-month period and, therefore, results in a delay of graduation by one semester. Study abroad in England or Germany is available for some majors.

DEGREE PROGRAMS DESCRIPTION
The B.S. degree programs are offered in five of the six departments that comprise the School of Engineering, with a Materials double major available at the undergraduate level with the sixth department (Metallurgy). The programs are designed to give sound knowledge of basic principles in mathematics, physics, and chemistry; to offer training in the theory, principles, and practices of engineering; and to present the opportunity to obtain additional instruction and experience in one of the major engineering fields. The graduate degree programs are enhanced by multidisciplinary studies in Biological Engineering, Environmental Engineering, Materials Science, Polymer Science, and Ocean Engineering. Research Centers include the Booth Computer Research Center, the Environmental Research Institute, the Center for Grinding Research and Development, the Advanced Technology Center for Precision Manufacturing, the Photonics Research Center, and the Transportation Institute.

TRANSFER INFORMATION
Residency Requirements: A minimum of thirty (30) credits must be completed at the University.

Transfer via Articulation Agreements
Admission to engineering: At the end of the sophomore year
Requirements: Transfer student applicants to the University must have at least a 2.5/4.0 grade point average at their prior institution. Admission to the Engineering programs normally requires a GPA of 2.8 and above. Students are normally expected to have completed a two year associate degree prior to admission; while those at another four year institution may be admitted earlier.

Transfer without Articulation Agreements
Admission to engineering: At any time
Requirements: Transfer student applicants to the University must have at least a 2.5/4.0 grade point average at their prior institution. Admission to the Engineering programs normally requires a GPA of 2.8 and above. Students are normally expected to have completed a two year associate degree prior to admission; while those at another four year institution may be admitted earlier.

GRADUATION REQUIREMENTS
Students must maintain at least a 2.0/4.0 GPA. in technical classes to be allowed to enter the junior year in the B.S. curriculum. Once admitted to the upper division (normally the junior year), students must maintain at least a 2.0/4.0 in all coursework in order to complete the B.S. degree. Graduate degree requirements vary slightly across all programs; some with thesis work may require more or less coursework than others.

STUDENT PROGRAMS

PROFESSIONAL AND HONORARY SOCIETIES
[For key to acronyms, see *Introduction*.]
Professional Societies: ACM, AIChE, ASCE, ASME, IEEE, TMS
Honorary Societies: Chi Epsilon, Eta Kappa Nu, Omega Chi Epsilon, Pi Tau Sigma, Tau Beta Pi, Upsilon Pi Epsilon

SUPPORT PROGRAMS
Student Chapter Organizations: Society of Women Engineers, National Society of Black Engineers

Minority Engineering Program
For: Ethnic Minorities **Available:** Year round
Offered: Prematriculation, Freshman, Sophomore, Junior, Senior
Other Engineering Support Programs: An Orientation program exists. Support services include Centers for: Counseling, Disabled, International, Women, Afro-American, Puerto Rican/Latin American, and Asian-American students. Summer programs include the Minority Engineering Program's Bridge program and the Center for Academic Programs. Athletes receive support through the Counseling Program for Intercollegiate Athletics.

057 The Cooper Union

■ INSTITUTION PROFILE

HEAD OF THE INSTITUTION
John J. Iselin
Phone: (212)353-4240   Fax: (212)353-4244

GENERAL INFORMATION
[All Students—Fall 1991]
Undergraduate enrollment 1,027
Graduate enrollment 67
Total institution enrollment 1,134
Type of institution: Private   Calendar: Semesters
Location: New York City   Population: 7,500,000
Setting: Urban
Types of engineering degrees: Engineering
Other degree-granting colleges: Architecture, Fine & Performing Arts

■ ENGINEERING ADMISSIONS

ADMISSIONS OFFICE CONTACT
Richard Bory
The Cooper Union
41 Cooper Square
New York, NY 10003
Phone: (212)353-4121   Fax: (212)353-4343

ENGINEERING COLLEGE ADMISSIONS INFORMATION
Entrance Requirements: SAT, High School courses—English (4 years), History-Social Studies (2 years), Mathematics (3.5 years), Physics and Chemistry (1 year), ACT (SAT II) Math and either physics or chemistry.
Requirements for foreign students: TOEFL; financial statement; Must apply from a U.S. home address.

DEPARTMENTAL ADMISSIONS INFORMATION
Admission to the engineering department: At the time of admission to the institution.

■ FINANCIAL INFORMATION

ESTIMATED EXPENSES (ACADEMIC YEAR)
[Expenses are for the 1992-93 nine-month academic year.]

All Students	Undergraduate	Graduate
Tuition and fees	$ 300	$ 350
College room and board	$ 7,000 [2]	$ 7,000 [1]
Books and supplies	$ 400	$ 400
Other expenses	$ –	$ –
Total estimated expenses	$ 7,700	$ 7,750

Notes: (1) Many students commute. However, dormitory facilities are available. (2) Many students commute. However, dormitory facilities are available.

FINANCIAL AID OFFICE CONTACT
Anne-Marie Wiemer-Sumner, Director, Financial Aid & Career Counseling
The Cooper Union
41 Cooper Square
New York, NY 10003
Phone: (212)353-4111   Fax: (212)353-4343

GENERAL FINANCIAL AID INFORMATION
Forms accepted/required: CSS-FAF, FAF, FAT, Stafford, IRS, SAR, Institutional
Additional financial aid information: Financial Aid is based on documented need and includes, grants, loans, part-time campus jobs, assistantships

■ ENGINEERING COLLEGE INFORMATION

[For additional personnel, refer to the *Appendix*.]
HEAD OF ENGINEERING
Eleanor Baum
Phone: (212)353-4285   Fax: (212)353-4341
EMail: baum@green.cooper.edu

ENGINEERING COLLEGE ADDRESS
Albert Nerken School of Engineering
The Cooper Union
51 Astor Place
New York, NY 10003
Phone: (212)353-4285   Fax: (212)353-4341

ENROLLMENTS—BY CLASS
[Numbers are baccalaureate enrollments for the fall 1991 term, unless otherwise footnoted.]
1st-year students/Freshmen 145
2nd-year students/Sophomores 140
3rd-year students/Juniors 120
4th-year students/Seniors 139
Total 544

NUMBER OF DEGREES AWARDED—BY PROGRAM
[Numbers are engineering baccalaureate degrees awarded during the 1991-1992 academic year, unless otherwise footnoted. For full details about each engineering program, refer to the *Tables of Degree Programs*.]
Chemical Engineering 14
Civil Engineering 25
Electrical Engineering 33
Engineering 10
Mechanical Engineering 23
Total 105

PERCENTAGE OF DEGREES AWARDED—BY CATEGORY
[Percentages are of all engineering baccalaureate degrees awarded during the 1991-1992 academic year, unless otherwise footnoted.]
African Americans .... 10.0 %
Asian/Pacific Island Americans .... 40.0 %
Hispanic Americans .... 5.0 %
Native Americans .... – %
Foreign Citizens .... – %
All Others .... 45.0 %
Women .... 18.0 %
Persons with disabilities .... – %
Students over 25 years of age .... – %

GRADUATE ENROLLMENTS & DEGREES AWARDED
Master's enrollment 67
Master's degrees awarded 18
Doctoral enrollment .... –
Doctoral degrees awarded .... –

ENGINEERING STUDENT DATA
Applicants to the engineering college
Number of applicants to engineering college 803
Percent offered admission 33.0%

Matriculated engineering students
Percentage in top quartile (25%) of High School class ... 100.0
Average SAT scores: Math—740, Verbal—610, Combined—1350

FULL- & PART-TIME FACULTY—BY DEPARTMENT
[Figures are the head count of full-time faculty and the full-time equivalent (FTE) of part-time faculty for each engineering department or equivalent.]

Department	Full	Part
Chemical Eng	5	2.5
Civil Eng	5	3.0
Electrical Eng	7	3.5
Mechanical Eng	6	2.6
Eng	12	3.0

COLLEGE DESCRIPTION

Private tuition-free institution. Students enter with extraordinary academic backgrounds. Programs include introduction to engineering design in freshman year. Design projects and participation in research are stressed throughout curricula. Very close interaction with faculty. Exciting urban location permits participation in cultural activities. "New York City is our campus."

DEGREE PROGRAMS DESCRIPTION

All degree programs are highly competitive because of the exceptionally gifted student body. There is strong emphasis on student projects, design and research. This is true both for undergraduate and graduate levels. The degree program in the engineering disciplines are all ABET accredited. Areas of concentration are offered within degree programs in business, medicine, law, environmental engineering, manufacturing engineering, computer engineering, etc. Small class sizes lead to very close interaction with faculty. The unspecified engineering program is intended for gifted students who wish great flexibility in their engineering program or who intend to continue studies in fields such as law, medicine, dentistry, business, etc.

TRANSFER INFORMATION

Residency Requirements: Minimum four semesters full time including two semesters preceding graduation.
Transfer without Articulation Agreements
Admission to engineering: At the end of the freshman year
Engineering graduates transferred from . . 4-yr: 40   2-yr: –
Requirements: Must have completed first year equivalent to that at Cooper Union.

GRADUATION REQUIREMENTS

A minimum of 135 credits; completion of the 55 credit core program consisting of mathematics (17 credits), Physical Science (21 credits), general engineering (5 credits), Humanities and Social Sciences (12 credits); a minimum of 24 credits in the Humanities and Social Sciences (inclusive of the 12 credits in the core program); Satisfaction of the departmental requirements including senior design project; satisfaction of the residence study requirement; a minimum cumulative academic rating (grade point average of 2.0; a minimum 2.0 academic rating for the junior and senior years combined.

■ STUDENT PROGRAMS

PROFESSIONAL AND HONORARY SOCIETIES
[For key to acronyms, see *Introduction*.]
Professional Societies: ACS, AIAA, AIChE, ASCE, ASHRAE, ASME, IEEE, SAE, New York State Water Pollution Control Association, Order of the Engineer
Honorary Societies: Chi Epsilon, Eta Kappa Nu, Pi Tau Sigma, Tau Beta Pi

SUPPORT PROGRAMS
Student Chapter Organizations: Society of Women Engineers, National Society of Black Engineers, Society of Hispanic Professional Engineers

Society of Women Engineers
For: Women   **Available:** Year round
Offered: Freshman, Sophomore, Junior, Senior, Graduate level

National Society of Black Engineers
For: Ethnic Minorities   **Available:** Year round
Offered: Freshman, Sophomore, Junior, Senior, Graduate level

Society of Hispanic Prof. Engineers
For: Ethnic Minorities   **Available:** Year round
Offered: Freshman, Sophomore, Junior, Senior, Graduate level

Other Engineering Support Programs: Orientation program for freshmen run by Student Services Division. Engineering orientation freshman course. Student support coordinator for minority students located in the Dean's office to liaise schoolwide (meetings, counselling, tutoring, etc.).

058 Cornell University

■ INSTITUTION PROFILE
HEAD OF THE INSTITUTION
Frank H. Rhodes
Phone: (607)255-5201 **Fax:** (607)255-9412
GENERAL INFORMATION
[All Students—Fall 1991]
Undergraduate enrollment 13,063
Graduate enrollment 5,905
Total institution enrollment 18,968
Type of institution: Public and Private **Calendar:** Semesters
Nearest city: Syracuse **Population:** 650,000
Miles from main campus: 50 **Setting:** Small Town
Types of engineering degrees: Engineering & Technology
Other degree-granting colleges: Arts & Sciences, Law, Medicine, Agriculture and Life Sciences, Human Ecology, Industrial and Labor Relations, Hotel Administration, Architecture, Art and Planning

■ ENGINEERING ADMISSIONS
ADMISSIONS OFFICE CONTACT
Richard Hale
Cornell University
Director of Admissions
Carpenter Hall Annex
Ithaca, NY 14853
Phone: (607)255-5008 **Fax:** (607)255-9606
ENGINEERING COLLEGE ADMISSIONS INFORMATION
Entrance Requirements: SAT, ACT, High School courses—Chemistry (1 year), Physics (1 year), Mathematics (2 algebra, 1 geometry, 1 pre-calculus) (4 years), English (4 years), SAT or ACT plus 3 achievement tests in Math, English and a Science.
Entrance Recommendations: High School courses—Calculus (1 year), Computer Sciences (1 year), Foreign Language (2 years)
Requirements for foreign students: TOEFL (Score: 600); financial statement
DEPARTMENTAL ADMISSIONS INFORMATION
Admission to the engineering department: At the end of the second year.
Additional information: Affiliation with a department is normally in mid-spring of the sophomore year. Admission requirements vary; some departments require grades of at least C - in each required or elective field course taken prior to formal affiliation with the department.

■ FINANCIAL INFORMATION
ESTIMATED EXPENSES (ACADEMIC YEAR)
[Expenses are for the 1992-93 nine-month academic year.]

All Students	Undergraduate	Graduate
Tuition and fees	$17,276	$17,252
College room and board	$ 5,676	$ _(1)
Books and supplies	$ 480	$ _(1)
Other expenses	$ 1,000	$ _(1)
Total estimated expenses	$24,432	$17,252

Notes: (1) Highly variable
GENERAL FINANCIAL AID INFORMATION
Forms accepted/required: FAF, IRS, Institutional
Additional financial aid information: Need-based scholarships; short-term loans; part-time jobs on campus, and College Work-Study Program, the Cornell tradition.

■ ENGINEERING COLLEGE INFORMATION
[For additional personnel, refer to the *Appendix*.]
HEAD OF ENGINEERING
William B. Streett
Phone: (607)255-9679 **Fax:** (607)255-9606
ENGINEERING COLLEGE ADDRESS
College of Engineering
Cornell University
Carpenter Hall
Ithaca, NY 14853-2201
Phone: (607)255-4326 **Fax:** (607)255-9606

ENROLLMENTS—BY CLASS
[Numbers are baccalaureate enrollments for the fall 1991 term, unless otherwise footnoted.]
1st-year students/Freshmen 788
2nd-year students/Sophomores 647
3rd-year students/Juniors 592
4th-year students/Seniors 608
Total 2,635

NUMBER OF DEGREES AWARDED—BY PROGRAM
[Numbers are engineering baccalaureate degrees awarded during the 1991-1992 academic year, unless otherwise footnoted. For full details about each engineering program, refer to the *Tables of Degree Programs*.]
Agricultural & Biological Engineering .... –
Agriculture & Biological Engineering 28
Applied & Engineering Physics 30
Chemical Engineering 53
Civil & Environmental Engineering 57
College Program 2
Computer Science 48
Electrical Engineering 142
Geological Sciences 2
Materials Science & Engineering 27
Mechanical & Aerospace Engineering 147
Nuclear Science & Engineering .... –
Operations Research & Engineering .... –
Operations Research & Engineering .... –
Operations Research & Industrial Engineering 94
Theoretical & Applied Mechanics .... –
Total 630

PERCENTAGE OF DEGREES AWARDED—BY CATEGORY
[Percentages are of all engineering baccalaureate degrees awarded during the 1991-1992 academic year, unless otherwise footnoted.]
African Americans 1.5%
Asian/Pacific Island Americans 18.0%
Hispanic Americans 3.0%
Native Americans .... – %
Foreign Citizens 9.0%
All Others 68.5%
Women 21.0%
Persons with disabilities .... – %
Students over 25 years of age .... – %

GRADUATE ENROLLMENTS & DEGREES AWARDED
Master's enrollment 337
Master's degrees awarded 414
Doctoral enrollment 731
Doctoral degrees awarded 151

ENGINEERING STUDENT DATA
Applicants to the engineering college
Number of applicants to engineering college 4,327
Percent offered admission 39.5%
Matriculated engineering students
Percentage in top quartile (25%) of High School class 99.0
Average SAT scores: Math—718, Verbal—589, Combined—1307

FULL- & PART-TIME FACULTY—BY DEPARTMENT
[Figures are the head count of full-time faculty and the full-time equivalent (FTE) of part-time faculty for each engineering department or equivalent.]

Department	Full	Part
Chemical Eng	17	–
Electrical Eng	42	–
Agricultural & Biological Eng	22	–
Applied & Eng Physics	10	1.5
Civil & Environmental Eng	28	–
Computer Science	25	–
Geological Sciences	12	1.2
Materials Science & Eng	14	–
Mechanical & Aerospace Eng	27	–
Nuclear Science & Eng	5	–
Operations Research & Eng	19	–
Theoretical & Applied Mechanics	13	–

COLLEGE DESCRIPTION
The college offers a four-year undergraduate program in engineering that emphasizes mathematics, science, and engineering fundamentals in the first two years, followed by two years of specialization in one of 10 upper-class majors or in an individually-designed college program consisting of a major and a minor. Freshmen are required to take two writing-intensive freshman seminars in a university-wide program; they may also elect to take introduction to engineering courses. There is a strong component of course-work in the humanities and social sciences throughout the four years. Dual major programs are available in some fields, along with a 5-year dual degree program with the College of Arts and Sciences. Co-op is available in most fields. The Cornell Abroad Program provides some opportunities for engineering students to spend a semester or year studying abroad.

DEGREE PROGRAMS DESCRIPTION
The B.S. degrees offered in the 10 engineering programs prepare the graduate for entry-level professional positions, for graduate study in the respective fields, and for advanced training in other professional schools such as business, medicine, and law. Graduate programs in 13 areas of engineering include (a) a one-year Master of Engineering degree program for those interested in the practice of engineering (this program includes the M.Eng. degree in Engineering Management), and (b) MS/PhD programs for preparation for careers in research, advanced professional practice or teaching.

TRANSFER INFORMATION
Residency Requirements: 4 semesters minimum
Transfer without Articulation Agreements
Admission to engineering: At the end of the freshman year

GRADUATION REQUIREMENTS
Graduation requirements vary within the various departments. Total credit hours required for the B.S. degree range from 129 to 141 credit hours, which includes 24 credits minimum in humanities and social sciences, and 21 credits of electives.

■ STUDENT PROGRAMS
PROFESSIONAL AND HONORARY SOCIETIES
[For key to acronyms, see *Introduction*.]
Professional Societies: AAPG, AIAA, AIChE, ASAE, ASCE, ASME, IEEE, IIE, ORSA, SAE, Cornell Bioengineering Society, Association of Computer Science Undergraduates
Honorary Societies: Chi Epsilon, Tau Beta Pi, Mu Sigma Tau (Co-op Honorary Society)

SUPPORT PROGRAMS
Student Chapter Organizations: Society of Women Engineers, National Society of Black Engineers, Society of Hispanic Professional Engineers

Engineering Minority Programs Office
For: Ethnic Minorities **Available:** Year round
Offered: Prematriculation, Freshman, Sophomore, Junior, Senior, Graduate level

Women's Program in Engineering
For: Women **Available:** Year round
Offered: Prematriculation, Freshman, Sophomore, Junior, Senior, Graduate level

Other Engineering Support Programs: Summer (optional) and fall orientation for freshmen and transfers; academic advising; peer advising; tutoring programs; learning skills programs; minority programs; career counseling services; services for the disabled; veterans services; placement services.

059 Dartmouth College

INSTITUTION PROFILE
HEAD OF THE INSTITUTION
James O. Freedman
Phone: (603)646-2222 Fax: (603)646-2266

GENERAL INFORMATION
[All Students—Fall 1991]
Undergraduate enrollment 4,266
Graduate enrollment 1,169
Total institution enrollment **5,435**
Type of institution: Private **Calendar:** Quarters
Nearest city: Manchester **Population:** 150,000
Miles from main campus: 75 **Setting:** Small Town
Types of engineering degrees: Engineering
Other degree-granting colleges: Arts & Sciences, Business Administration, Medicine

ENGINEERING ADMISSIONS
ADMISSIONS OFFICE CONTACT
Maria Laskaris
Dartmouth College
Admissions Office
McNutt Hall
Hanover, NH 03755
Phone: (603)646-2875 Fax: (603)646-1216

ENGINEERING COLLEGE ADMISSIONS INFORMATION
Admission to the engineering college: The engineering school admits only post A.B. students. Undergraduates are all in the Arts & Sciences College. Majors are declared at end of sophomore year.
Entrance Requirements: SAT or ACT or 3 achievement tests.
Requirements for foreign students: TOEFL; TOEFL required for non-native speakers of English, but there is no minimum required score. Financial Statement is required if applying for financial aid.

DEPARTMENTAL ADMISSIONS INFORMATION
Admission to the engineering department: At the end of the second year.

FINANCIAL INFORMATION
ESTIMATED EXPENSES (ACADEMIC YEAR)
[Expenses are for the 1992-93 nine-month academic year.]

All Students	Undergraduate	Graduate
Tuition and fees	$17,229	$17,229
College room and board	$5,646	$5,650
Books and supplies	$1,545	$500
Other expenses	$1,400 (2)	$800 (1)
Total estimated expenses	**$25,820**	**$24,179**

Notes: (1) Health insurance fee required unless, student has proof of insurance from another company. (2) Includes activity fee and the purchase of a computer.

FINANCIAL AID OFFICE CONTACT
Virginia S. Hazen, *Director of Financial Aid*
Dartmouth College
Office of Financial Aid
McNutt Hall
Hanover, NH 03755
Phone: (603)646-2451

GENERAL FINANCIAL AID INFORMATION
Forms accepted/required: FAF
Additional financial aid information: Scholarships, work study and loans (all need-based).

ENGINEERING COLLEGE INFORMATION
[For additional personnel, refer to the *Appendix*.]
HEAD OF ENGINEERING
Charles E. Hutchinson
Phone: (603)646-2238 Fax: (603)646-3856

ENGINEERING COLLEGE ADDRESS
Thayer School of Engineering
Dartmouth College
8000 Cummings Hall
Hanover, NH 03755-8000
Phone: (603)646-2230 Fax: (603)646-3856

ENROLLMENTS—BY CLASS
[Numbers are baccalaureate enrollments for the fall 1991 term, unless otherwise footnoted.]
1st-year students/Freshmen 70 (1)
2nd-year students/Sophomores 90 (1)
3rd-year students/Juniors 78
4th-year students/Seniors 66
Total **304**

Notes: (1) Undeclared Majors

NUMBER OF DEGREES AWARDED—BY PROGRAM
[Numbers are engineering baccalaureate degrees awarded during the 1991-1992 academic year, unless otherwise footnoted. For full details about each engineering program, refer to the *Tables of Degree Programs*.]
Engineering Science(s) 58
Total **58**

PERCENTAGE OF DEGREES AWARDED—BY CATEGORY
[Percentages are of all engineering baccalaureate degrees awarded during the 1991-1992 academic year, unless otherwise footnoted.]
African Americans .... – % (A)
Asian/Pacific Island Americans .... 7.0 %
Hispanic Americans .... 1.7 %
Native Americans .... – % (A)
Foreign Citizens .... 14.0 %
All Others .... 77.3 %
Women .... 12.0 %
Persons with disabilities .... – % (A)
Students over 25 years of age .... – % (A)

Notes: (A) Data not available.

GRADUATE ENROLLMENTS & DEGREES AWARDED
Master's enrollment 52
Master's degrees awarded 23
Doctoral enrollment 57
Doctoral degrees awarded 5

FULL- & PART-TIME FACULTY—BY DEPARTMENT
[Figures are the head count of full-time faculty and the full-time equivalent (FTE) of part-time faculty for each engineering department or equivalent.]

Department	Full	Part
Eng Sciences	26	8.0

COLLEGE DESCRIPTION
The Thayer School of Engineering at Dartmouth College, was founded in 1867. It comprises both the undergraduate Department of Engineering Sciences of Dartmouth College and a graduate professional school in engineering. Thayer School is distinguished by its interdisciplinary nature - a school without departments, devoted to the development and teaching of the fundamental scientific bases of engineering, as well as their applications to new and emerging problems in our technological world. Our undergraduate programs emphasize interdisciplinary study in the engineering sciences within the context of a broad-based liberal arts education. At the graduate level we offer degrees through the doctorate, combining scholarship, research, experimentation, and problem solving. Our basic objective is to provide sound technological leadership for the coming decades.

DEGREE PROGRAMS DESCRIPTION
Unique features A.B. in Engineering Sciences 1. Interdisciplinary engineering major 2. Integrated with liberal arts program including foreign language Bachelor of Engineering 1. Post AB program, 1 academic year 2. Design project (2-3 course equivalents) 3. Specialized to meet individual interests 4. Faculty in a single department Master of Science/Doctor of Philosophy 1. Core requirements in math and engineering 2. Oral exam and thesis proposal for Ph.D. 3. Thesis in research or design for both degrees Master of Engineering 1. Engineering design and management 2. Total of 18 courses, no thesis 3. Credit for prior work 4. Four electives may be business courses

DUAL DEGREE PROGRAMS
Engineering baccalaureate graduates with dual degrees 5

TRANSFER INFORMATION
Residency Requirements: 18 Credits
Transfer without Articulation Agreements
Admission to engineering: At any time
Engineering graduates transferred from .. 4-yr: 1 2-yr: 1

GRADUATION REQUIREMENTS
2.0 in the major, 35 courses, 4 courses in each of the divisions (social sciences, science, humanities) outside their major departments, language requirement (3 terms or the equivalent which is 650 in college boards), English 5, Freshmen Seminar and to complete a course in non-western culture.

STUDENT PROGRAMS
PROFESSIONAL AND HONORARY SOCIETIES
[For key to acronyms, see *Introduction*.]
Professional Societies: ASME, Dartmouth Society of Engineers
Honorary Societies: Sigma Xi, Phi Beta Kappa

SUPPORT PROGRAMS
Student Chapter Organizations: Society of Women Engineers, National Society of Black Engineers, AISES-Student Chapter of Amer. Indian Sci&Engineering Soc, WISP (Women in Science Project)
Other Engineering Support Programs: Academic Skills Center: provides intensive academic support for students needing remedial or tutorial assistance. Thayer School Tutoring Program: provides tutoring assistance for students in undergraduate engineering courses.

060 University of Dayton

■ INSTITUTION PROFILE

HEAD OF THE INSTITUTION
Raymond L. Fitz S.M.
Phone: (513)229-4122   Fax: (513)229-3433

GENERAL INFORMATION
[All Students—Fall 1991]
Undergraduate enrollment 6,985
Graduate enrollment 984
Total institution enrollment 7,969
Type of institution: Private   **Calendar:** Trimesters
Location: Dayton   **Population:** 179,000
Setting: Urban
Types of engineering degrees: Engineering
Other degree-granting colleges: Arts & Sciences, Business Administration, Education, Law

■ ENGINEERING ADMISSIONS

ADMISSIONS OFFICE CONTACT
Myron H. Achbach
University of Dayton
300 College Park
St. Mary's Hall Rm. 112
Dayton, OH 45469-1611
Phone: (513)229-4411   Fax: (513)229-3433

ENGINEERING COLLEGE ADMISSIONS INFORMATION
Admission to the engineering college: The end of junior year.
Entrance Requirements: High School courses—English (4 years), Mathematics (4 years), Chemistry (1 year), Physics (1 year)
Entrance Recommendations: High School courses—Additional Academic Units (6 years)
Requirements for foreign students: TOEFL (Score: 500); financial statement; International student applicants must present their academic credentials in official English translation along with their transcripts in the original language. The applicant must also present certification of financial resources available to support an education at the University of Dayton

DEPARTMENTAL ADMISSIONS INFORMATION
Admission to the engineering department: The end of junior year

■ FINANCIAL INFORMATION

ESTIMATED EXPENSES (ACADEMIC YEAR)
[Expenses are for the 1992-93 nine-month academic year.]

All Students	Undergraduate	Graduate
Tuition and fees	$10,810	$ 6,856
College room and board	$ 4,250	$ –
Books and supplies	$ 600	$ 400
Other expenses	$ –	$ –
Total estimated expenses	**$15,660**	**$ 7,256**

FINANCIAL AID OFFICE CONTACT
Joyce J. Wilkins, *Director of Financial Aid*
University of Dayton
300 College Park
St. Mary's Hall Rm. 202
Dayton, OH 45469-1621
Phone: (513)229-4311   Fax: (513)229-3433

GENERAL FINANCIAL AID INFORMATION
Forms accepted/required: ACT, AFSA, CSS-FAF, FAF, FAT, FSA, FFS, Stafford, IRS, SAR, SDF, Supplemental
Additional financial aid information: scholarships, loans, grants, tuition reduction and part-time employment

■ ENGINEERING COLLEGE INFORMATION

[For additional personnel, refer to the *Appendix*.]

HEAD OF ENGINEERING
Joseph Lestingi
Phone: (513)229-2736   Fax: (513)229-2756

ENGINEERING COLLEGE ADDRESS
School of Engineering
University of Dayton
300 College Park
Kettering Labs Rm. 201
Dayton, OH 45469-0228
Phone: (513)229-2736   Fax: (513)229-2756

ENROLLMENTS—BY CLASS
[Numbers are baccalaureate enrollments for the fall 1991 term, unless otherwise footnoted.]
1st-year students/Freshmen 294
2nd-year students/Sophomores 216
3rd-year students/Juniors 236
4th-year students/Seniors 221
Total **967**

NUMBER OF DEGREES AWARDED—BY PROGRAM
[Numbers are engineering baccalaureate degrees awarded during the 1991-1992 academic year, unless otherwise footnoted. For full details about each engineering program, refer to the *Tables of Degree Programs*.]
Chemical Engineering 25
Civil Engineering 19
Electrical Engineering 80
Mechanical Engineering 69
Total **193**

PERCENTAGE OF DEGREES AWARDED—BY CATEGORY
[Percentages are of all engineering baccalaureate degrees awarded during the 1991-1992 academic year, unless otherwise footnoted.]
African Americans .... 5.0 %
Asian/Pacific Island Americans .... – %
Hispanic Americans .... 2.0 %
Native Americans .... – %
Foreign Citizens .... 3.0 %
All Others .... 90.0 %
Women .... 23.0 %
Persons with disabilities .... – %
Students over 25 years of age .... – %

GRADUATE ENROLLMENTS & DEGREES AWARDED
Master's enrollment 607
Master's degrees awarded 174
Doctoral enrollment 56
Doctoral degrees awarded 3

ENGINEERING STUDENT DATA

Applicants to the engineering college
Number of applicants to engineering college 876
Percent offered admission 86.0%

Matriculated engineering students
Percentage in top quartile (25%) of High School class 81.8
Average SAT scores: Math—621, Verbal—524, Combined—1145
Average ACT scores: Composite—27

FULL- & PART-TIME FACULTY—BY DEPARTMENT
[Figures are the head count of full-time faculty and the full-time equivalent (FTE) of part-time faculty for each engineering department or equivalent.]

Department	Full	Part
Chemical Eng	7	–
Civil & Eng Mechanics	9	1.5
Electrical Eng	15	–
Mechanical Eng	13	1.8

COLLEGE DESCRIPTION

The School of Engineering at the University of Dayton prepares men and women for professional careers in engineering and in technology by requiring the completion of a general education sequence of courses in addition to those required in the major. Degrees are offered in chemical, civil, electrical and mechanical engineering, with minors and concentrations such as aerospace, digital systems, and structures available in 14 areas. The undergraduate programs are enhanced by several graduate and doctoral programs and the presence of the University of Dayton Research Institute. Co-op programs are available and add one year to the time required to graduate.

DEGREE PROGRAMS DESCRIPTION

The University of Dayton is committed to the development of the 'complete professional'. This is accomplished by requiring the undergraduate to complete courses with a humanities 'base' and a thematic 'cluster'. Technical and general education courses are organized to progress toward a capstone effort that integrates the educational experience.

TRANSFER INFORMATION

Residency Requirements: 30 Semester Hours.
Transfer without Articulation Agreements
Engineering graduates transferred from   . . . 4-yr: 4   2-yr: –
Requirements: 2.75 Cumulative GPA and B's in Math, Physics and Chemistry

GRADUATION REQUIREMENTS

All prescribed courses outlined in the respective curricula must have been passed with grades of D or better. Although courses may be scheduled in terms other than as listed, all prerequisites and corequisites must be met. All students in the School of Engineering must register under Grade Option 1 for all courses in engineering, mathematics, and science except those offered only under Grade Option 2. The cumulative quality-point average in the student's engineering curriculum must be at least 2.0 (C average).

■ STUDENT PROGRAMS

PROFESSIONAL AND HONORARY SOCIETIES
[For key to acronyms, see *Introduction*.]
Professional Societies: AIAA, AIChE, ASCE, ASME, IEEE, NSPE, Joint Council of Engineers (JCE)
Honorary Societies: Eta Kappa Nu, Pi Tau Sigma, Tau Beta Pi, Tau Nu Kappa

SUPPORT PROGRAMS
Student Chapter Organizations: Society of Women Engineers, National Society of Black Engineers

National Society of Black Engineers
For: Ethnic Minorities   **Available:** Year round
Offered: Freshman

Society of Women Engineers
For: Women   **Available:** Year round
Offered: Prematriculation

Other Engineering Support Programs: Fall orientation program for all freshmen and transfers; Learning Assistance Center; career counseling services; foreign student office; veterans' services; placement services. Students participate in Co-op and Intern programs; Honors and Scholars Programs, complete complement of tutoring services.

061 University of Delaware

INSTITUTION PROFILE

HEAD OF THE INSTITUTION
David P. Roselle
Phone: (302)831-2111

GENERAL INFORMATION
[All Students—Fall 1991]
Undergraduate enrollment 15,248
Graduate enrollment 2,668
Total institution enrollment 20,868
Type of institution: Public and Private Calendar: Semesters
Nearest city: Philadelphia, PA Population: 1,700,000
Miles from main campus: 35 Setting: Small Town
Types of engineering degrees: Engineering
Other degree-granting colleges: Arts & Sciences, Education, Nursing, Business and Economics, Human Resources, Marine Studies, Physical Education, Athletics, Recreation, Urban Affairs and Public Policy

ENGINEERING ADMISSIONS

ADMISSIONS OFFICE CONTACT
N. Bruce Walker
University of Delaware
Dean of Admissions
Newark, DE 19716
Phone: (302)831-8123

ENGINEERING COLLEGE ADMISSIONS INFORMATION
Entrance Requirements: SAT, High School courses—English (4 years), Mathematics - geometry, algebra, pre-calculus (4 years), Sciences (including Chemistry and Physics) (4 years)
Entrance Recommendations: High School courses—Calculus (1 year), Foreign language (2 years), Computer Science (1 year), Social Science and Humanities (4 years), Advanced Placement, CEEB Achievement tests particularly in Mathematics, English, and the Sciences.
Requirements for out-of-state residents: Admission is decided on the basis of a formula which combines the student's GPA, SATM and SATV.
Requirements for foreign students: TOEFL

DEPARTMENTAL ADMISSIONS INFORMATION
Admission to the engineering department: At the time of admission to the institution.

FINANCIAL INFORMATION

ESTIMATED EXPENSES (ACADEMIC YEAR)
[Expenses are for the 1992-93 nine-month academic year.]

State Residents	Undergraduate	Graduate
Tuition and fees	$ 3,390	$ 3,390
College room and board	$ 3,756	$ 3,180
Books and supplies	$ 530	$ 530
Other expenses	$ 367 (2)	$ 352 (1)
Total estimated expenses	$ 8,043	$ 7,452

Out-of-State Residents	Undergraduate	Graduate
Tuition and fees	$ 9,050	$ 9,050
College room and board	$ 3,756	$ 3,180
Books and supplies	$ 530	$ 530
Other expenses	$ 367 (4)	$ 352 (3)
Total estimated expenses	$13,703	$13,112

Notes: (1) Sustaining fee: Master's 60, PhD 80; Health insurance plan varies; Student health fee: 212 (2) Comprehensive student fee 70; Miscellaneous and new student fee: 35; Student health fee: 212; Student Center fee: 50 (3) Same as resident (4) Same as resident

GENERAL FINANCIAL AID INFORMATION
Forms accepted/required: FAF
Additional financial aid information: Need-based grants, loans, employment opportunities; merit-based scholarships.

ENGINEERING COLLEGE INFORMATION

[For additional personnel, refer to the *Appendix*.]

HEAD OF ENGINEERING
Stuart L. Cooper
Phone: (302)831-8017 Fax: (302)831-6751

ENGINEERING COLLEGE ADDRESS
College of Engineering
University of Delaware
135 du Pont Hall
Newark, DE 19716
Phone: (302)831-2401 Fax: (302)831-8179

ENROLLMENTS—BY CLASS
[Numbers are baccalaureate enrollments for the fall 1991 term, unless otherwise footnoted.]
1st-year students/Freshmen 263
2nd-year students/Sophomores 255
3rd-year students/Juniors 191
4th-year students/Seniors 268
Total **977**

NUMBER OF DEGREES AWARDED—BY PROGRAM
[Numbers are engineering baccalaureate degrees awarded during the 1991-1992 academic year, unless otherwise footnoted. For full details about each engineering program, refer to the *Tables of Degree Programs*.]
Chemical Engineering 38
Civil Engineering 34
Electrical Engineering 52
Materials Science .... –
Mechanical Engineering 62
Total **186**

PERCENTAGE OF DEGREES AWARDED—BY CATEGORY
[Percentages are of all engineering baccalaureate degrees awarded during the 1991-1992 academic year, unless otherwise footnoted.]
African Americans 11.5%
Asian/Pacific Island Americans 7.9%
Hispanic Americans 1.1%
Native Americans .... – %
Foreign Citizens 1.1%
All Others 78.4%
Women 24.8%
Persons with disabilities .... – % (A)
Students over 25 years of age .... – % (A)
Notes: (A) Data not available.

GRADUATE ENROLLMENTS & DEGREES AWARDED
Master's enrollment 139
Master's degrees awarded 50
Doctoral enrollment 125
Doctoral degrees awarded 42

ENGINEERING STUDENT DATA
Applicants to the engineering college
Number of applicants to engineering college 1,024
Percent offered admission 77.0%

Matriculated engineering students
Percentage in top quartile (25%) of High School class 73.0 (1)
Average SAT scores: Math—635, Verbal—523, Combined—1158
Notes: (1) top 20%

FULL- & PART-TIME FACULTY—BY DEPARTMENT
[Figures are the head count of full-time faculty and the full-time equivalent (FTE) of part-time faculty for each engineering department or equivalent.]

Department	Full	Part
Chemical Eng	18	–
Civil Eng	18	–
Electrical Eng	15	–
Mechanical Eng	18	–
Materials Science	–	–

COLLEGE DESCRIPTION
Engineering students at the University of Delaware are admitted directly into their chosen engineering major as freshmen. Integrated four-year programs in Chemical, Civil, Electrical, and Mechanical Engineering allow students to develop the basic skills which apply to the broad range of specialties within the chosen major. Students are encouraged to participate in undergraduate research. Six research centers are housed in the College of Engineering: the Center for Applied Coastal Research, the Center for Catalytic Science and Technology, the Center for Composite Materials, the Center for Molecular and Engineering Thermodynamics, the Delaware Transportation Center, and the Orthopedic and Biomechanical Engineering Center.

DEGREE PROGRAMS DESCRIPTION
The College of Engineering offers instruction leading to the degrees of Bachelor of Chemical Engineering, Civil Engineering, Electrical Engineering and Mechanical Engineering. Five-year programs are offered jointly by each of the four engineering majors and the College of Arts and Sciences. Master's degrees are offered in each engineering major as well as a Master of Materials Science and Engineering, and a Master of Applied Sciences in Civil Engineering. Doctor of Philosophy degrees are offered in each engineering major as well as a Doctor of Philosophy in Materials Science and Engineering. Interdisciplinary programs at the graduate level can easily be structured through concentration in the Applied Science degree program areas of biomedical engineering, energy conversion, engineering mechanics, environmental engineering, materials science, and ocean engineering.

DUAL DEGREE PROGRAMS
Engineering baccalaureate graduates with dual degrees 4

TRANSFER INFORMATION
Residency Requirements: Last 30 credit hours must be completed at the University of Delaware.

Transfer without Articulation Agreements
Admission to engineering: At any time
Engineering graduates transferred from .. 4-yr: 18 2-yr: 15
Requirements: Transfer applicants should have an overall B average as well as a B average in math, science, and engineering courses taken during the last year as a full-time student.

GRADUATION REQUIREMENTS
B.S. programs in engineering require 128-131 semester credits, depending on major selected. 2.0 GPA required in major courses as well as overall.

STUDENT PROGRAMS

PROFESSIONAL AND HONORARY SOCIETIES
[For key to acronyms, see *Introduction*.]
Professional Societies: AIChE, ASCE, ASME, IEEE, ITE, NSPE, SAMPE, American Society of Materials
Honorary Societies: Chi Epsilon, Eta Kappa Nu, Pi Tau Sigma, Tau Beta Pi

SUPPORT PROGRAMS
Student Chapter Organizations: Society of Women Engineers, National Society of Black Engineers

Resources to Insure Successful Engineers (RISE)
For: Ethnic Minorities Available: Academic year
Offered: Freshman, Sophomore, Junior, Senior

Summer Academy
For: Ethnic Minorities Available: Summer only
Offered: Prematriculation

Forum to Advance Minority Engineers (FAME)
For: Ethnic Minorities Available: Summer only
Offered: Prematriculation

Other Engineering Support Programs: New Student Orientation, Academic Studies Assistance Program, Academic Advancement Program, Career Planning and Placement, Counseling and Career Services, Math Sciences Teaching and Learning Center, Writing Center, Tutorial Services, University Honors Program.

062 University of Denver

INSTITUTION PROFILE
HEAD OF THE INSTITUTION
Daniel L. Ritchie
Phone: (303)871-2111
GENERAL INFORMATION
[All Students—Fall 1991]
Undergraduate enrollment 2,686
Graduate enrollment 2,993
Total institution enrollment 8,213
Type of institution: Private **Calendar:** Quarters
Location: Denver **Population:** 2,000,000
Setting: Urban
Types of engineering degrees: Engineering
Other degree-granting colleges: Business Administration, Law, Systems Science, Human Services

ENGINEERING ADMISSIONS
ADMISSIONS OFFICE CONTACT
Albert J. Rosa
University of Denver
Engineering Department
Denver, CO 80208
Phone: (303)871-2102 **Fax:** (303)871-4450
ENGINEERING COLLEGE ADMISSIONS INFORMATION
Entrance Requirements: SAT (Score: 1040), ACT (Score: 22), High School courses—Mathematics (2 Algebra, 1 Geometry & Trigonometry) (3 years), Natural Sciences (1 yr of Physics or Chemistry) (2 years), English (4 years), SAT and ACT Scores listed are average scores for DU. They are not requirements.
Entrance Recommendations: High School courses—Advanced Mathematics, Technical Courses
Requirements for foreign students: TOEFL (Score: 570); Foreign students interested in applying for Gradate Teaching Assistantships (GTAs) are required to obtain a score of at least 230 on the Test of Spoken English (TSE).
DEPARTMENTAL ADMISSIONS INFORMATION
Admission to the engineering department: At the time of admission to the institution.

FINANCIAL INFORMATION
ESTIMATED EXPENSES (ACADEMIC YEAR)
[Expenses are for the 1992-93 nine-month academic year.]

All Students	Undergraduate	Graduate
Tuition and fees	$13,740	$13,740
College room and board	$ 4,785 [1]	$ 4,785 [1]
Books and supplies	$ 1,000	$ 1,000
Other expenses	$ –	$ –
Total estimated expenses	**$19,525**	**$19,525**

Notes: (1) This is an average cost. Room and board packages range from $4,098 to $5,472 per academic year.

FINANCIAL AID OFFICE CONTACT
Colleen Hillmeyer, *Director, Financial Aid*
University of Denver
Financial Aid
2199 South University Boulevard
Denver, CO 80208
Phone: (303)871-2681 **Fax:** (303)871-2341
GENERAL FINANCIAL AID INFORMATION
Forms accepted/required: CSS-FAF, FAT, FFS, Stafford, SAR, Supplemental, Institutional
Additional financial aid information: In addition to Federal Financial Aid Programs, the following programs are also offered for engineering students: Work-Study Program; John Kammer Scholarship (based on need and merit, awarded by the engineering department); Colorado Student Incentive Grant (need-based); Colorado Scholars (merit-based); Colorado Student Grant (need-based); DU Educational Grants (need-based).

ENGINEERING COLLEGE INFORMATION
[For additional personnel, refer to the *Appendix*.]
HEAD OF ENGINEERING
John Kice
Phone: (303)871-2693 **Fax:** (303)871-2500
ENGINEERING COLLEGE ADDRESS
Department of Engineering
University of Denver
2390 S. York Street
Denver, CO 80208-0177
Phone: (303)871-2102 **Fax:** (303)871-4450
ENROLLMENTS—BY CLASS
[Numbers are baccalaureate enrollments for the fall 1991 term, unless otherwise footnoted.]
1st-year students/Freshmen .... 35 [1]
2nd-year students/Sophomores 15
3rd-year students/Juniors .... 21 [2]
4th-year students/Seniors 24
Total 95
Notes: (1) Includes one part-time student. (2) Includes two part-time students.
NUMBER OF DEGREES AWARDED—BY PROGRAM
[Numbers are engineering baccalaureate degrees awarded during the 1991-1992 academic year, unless otherwise footnoted. For full details about each engineering program, refer to the *Tables of Degree Programs*.]
Electrical Engineering 12
General Engineering .... –
Mechanical Engineering 4
Total 16
PERCENTAGE OF DEGREES AWARDED—BY CATEGORY
[Percentages are of all engineering baccalaureate degrees awarded during the 1991-1992 academic year, unless otherwise footnoted.]
African Americans .... – %
Asian/Pacific Island Americans .... – %
Hispanic Americans .... – %
Native Americans .... – %
Foreign Citizens .... 37.5 %
All Others .... 62.5 %
Women .... 12.5 %
Persons with disabilities .... – %
Students over 25 years of age .... 6.3 %
GRADUATE ENROLLMENTS & DEGREES AWARDED
Master's enrollment 11
Master's degrees awarded 1
Doctoral enrollment .... –
Doctoral degrees awarded .... –
FULL- & PART-TIME FACULTY—BY DEPARTMENT
[Figures are the head count of full-time faculty and the full-time equivalent (FTE) of part-time faculty for each engineering department or equivalent.]

Department	Full	Part
Electrical Eng	6.5	–
Mechanical Eng	7	–
General Eng	13	–

COLLEGE DESCRIPTION
The University of Denver is a mid-sized university with the benefits of a small liberal arts college curriculum and a large research university program. The engineering curricula capitalizes on both of these benefits and offers an integrated curricula that prepares graduates that are both literate and numerate.

DEGREE PROGRAMS DESCRIPTION
BS Degrees: The Electrical Engineering major offers specializations in communications, computers, robotics and semiconductors. The Mechanical Engineering major offers specializations in materials and robotics. Both degrees can be designed for individualized options through petition. Also, both programs can be taken in an unique integrated program with the Business School, resulting in a BSEE/MBA or BSME/MBA in 5 and 1/3 years. Both degrees are awarded together when the requirements for both degrees have been met. The Engineering (General) Program began in Fall 1992. The program leads to a BSE degree. The program permits students to take electives in a number of specializations which include both traditional engineering areas and non-engineering areas, such as pre-med and pre-law. MS Degrees: The Electrical Engineering master offers specializations in communications and signal processing, systems and controls, electromagnetics, quantum optics, and semiconductors. The Mechanical Engineering master offers specializations in the structure and behavior of materials, and fluid mechanics and heat transfer. An interdisciplinary PhD in Materials Science with the Departments of Physics and Chemistry is administered through the Department of Engineering.
DUAL DEGREE PROGRAMS
Engineering baccalaureate graduates with dual degrees 3
Enrollment requirements: Must pass the GMAT and be accepted to the Graduate Business School.
TRANSFER INFORMATION
Residency Requirements: No more than 138 quarter hours from a 4-year institution or 96 quarter hours from a 2-year institution may be transferred toward a University of Denver Engineering degree.
Transfer via Articulation Agreements
Admission to engineering: Transfers are admitted at the end of their Freshman or Sophomore years.
Engineering graduates transferred from ... 4-yr: – 2-yr: –
Transfer without Articulation Agreements
Admission to engineering: At any time
Engineering graduates transferred from ... 4-yr: – 2-yr: –
GRADUATION REQUIREMENTS
A grade of 'C' or better is required in all required engineering courses.

STUDENT PROGRAMS
PROFESSIONAL AND HONORARY SOCIETIES
[For key to acronyms, see *Introduction*.]
Professional Societies: ASME, IEEE
Honorary Societies: Phi Beta Kappa, Pi Mu Epsilon
SUPPORT PROGRAMS
Student Chapter Organizations: Society of Women Engineers, Society of Hispanic Professional Engineers
Society of Women Engineers
For: Women **Available:** Academic year
Offered: Freshman, Sophomore, Junior, Senior, Graduate level
Society of Hispanic Prof. Engineers
For: Ethnic Minorities **Available:** Academic year
Offered: Freshman, Sophomore, Junior, Senior, Graduate level
Other Engineering Support Programs: FREX 1000 Freshman Experience required of students during their first fall quarter, helps students adjust to college. 'ENGR 0900 Engineering Academy,' required of Freshman students in their second quarter, helps engineering students with the special requirements of engineering programs. A mentor program and tutoring are available free to all freshman and sophomore engineering students.

063 University of the District of Columbia

INSTITUTION PROFILE

HEAD OF THE INSTITUTION
Tilden J. LeMelle
Phone: (202)282-7550   Fax: (202)282-3681

GENERAL INFORMATION
[All Students—Fall 1991]
Undergraduate enrollment 10,380
Graduate enrollment 731
Total institution enrollment 11,990
Type of institution: Public   **Calendar:** Semesters
Location: Washington, DC   **Population:** 638,333
Setting: Urban
Types of engineering degrees: Engineering
Other degree-granting colleges: Business Administration, Education, Life Sciences, Liberal & Fine Arts, Physical Science, Engineering & Technology, University College

ENGINEERING ADMISSIONS

ADMISSIONS OFFICE CONTACT
Sandra B. Dolphin
University of the District of Columbia
4200 Connecticut Ave. N.W.
Washington, DC 20008
Phone: (202)282-3200   Fax: (202)282-3682

ENGINEERING COLLEGE ADMISSIONS INFORMATION
Admission to the engineering college: Completion of specific requirements in Univ Col. Student moves to academic department subject to the major program requirements.
Requirements for foreign students: TOEFL (Score: 500); financial statement

DEPARTMENTAL ADMISSIONS INFORMATION
Admission to the engineering department: Must complete specific requirements of Univ Coll. Then student moves to academic dept subject to major program requirements.

FINANCIAL INFORMATION

ESTIMATED EXPENSES (ACADEMIC YEAR)
[Expenses are for the 1992-93 nine-month academic year.]

State Residents	Undergraduate	Graduate
Tuition and fees	$ 400	$ 760
College room and board	$ –	$ –
Books and supplies	$ _(A)	$ _(A)
Other expenses	$ _(A)	$ _(A)
Total estimated expenses	**$ 400**	**$ 760**

Out-of-State Residents	Undergraduate	Graduate
Tuition and fees	$ 1,480	$ 1,480
College room and board	$ –	$ –
Books and supplies	$ _(A)	$ _(A)
Other expenses	$ _(A)	$ _(A)
Total estimated expenses	**$ 1,480**	**$ 1,480**

Notes: (A) Data not available.

FINANCIAL AID OFFICE CONTACT

Kenneth Howard, *Director of Financial Aid*
University of the District of Columbia
4200 Connecticut Ave N. W.
Washington, DC 20008
Phone: (202)282-3239   Fax: (202)282-3344

GENERAL FINANCIAL AID INFORMATION
Forms accepted/required: ACT, AFSA, CSS-FAF, FAF, FAT, FAR, FSA, FFS, Stafford, IRS, SAR, SDF, Institutional grants, Plus loans
Additional financial aid information: Grants, loans, College work study student employment, job location and development, scholarships

ENGINEERING COLLEGE INFORMATION

[For additional personnel, refer to the *Appendix*.]

HEAD OF ENGINEERING
Philip L. Brach
Phone: (202)282-7427   Fax: (202)282-3677

ENGINEERING COLLEGE ADDRESS
College of Physical Science, Engineering & Technology
University of the District of Columbia
4200 Connecticut Ave. N.W.
Washington, DC 20008
Phone: (202)282-7300   Fax: (202)282-3677

ENROLLMENTS—BY CLASS
[Numbers are baccalaureate enrollments for the fall 1991 term, unless otherwise footnoted.]
1st-year students/Freshmen .... 95 [1]
2nd-year students/Sophomores .... 36 [1]
3rd-year students/Juniors .... 36 [1]
4th-year students/Seniors .... 70 [1]
Total **237**

Notes: (1) 425 part-time students

NUMBER OF DEGREES AWARDED—BY PROGRAM
[Numbers are engineering baccalaureate degrees awarded during the 1991-1992 academic year, unless otherwise footnoted. For full details about each engineering program, refer to the *Tables of Degree Programs*.]
Civil Engineering 4
Electrical Engineering 25
Mechanical Engineering 7
Total **36**

PERCENTAGE OF DEGREES AWARDED—BY CATEGORY
[Percentages are of all engineering baccalaureate degrees awarded during the 1991-1992 academic year, unless otherwise footnoted.]
African Americans .... – % (A)
Asian/Pacific Island Americans .... – % (A)
Hispanic Americans .... – % (A)
Native Americans .... – % (A)
Foreign Citizens .... – % (A)
All Others .... 100 %
Women .... – % (A)
Persons with disabilities .... – % (A)
Students over 25 years of age .... – % (A)

Notes: (A) Data not available.

FULL- & PART-TIME FACULTY—BY DEPARTMENT

[Figures are the head count of full-time faculty and the full-time equivalent (FTE) of part-time faculty for each engineering department or equivalent.]

Department	Full	Part
Civil Eng	6	–
Electrical Eng	7	1.0
Mechanical Eng	5	–

COLLEGE DESCRIPTION
Integration of high tech instrumentation in classroom and laboratory instructional methodology. High quality of laboratories. Benefits of membership in Consortium of Universities in the Washington Metropolitan area.

TRANSFER INFORMATION
Residency Requirements: The last 30 semester hours must be taken at the University of the District of Columbia.
Transfer without Articulation Agreements
Admission to engineering: At any time
Engineering graduates transferred from . . 4-yr: 10   2-yr: 2

GRADUATION REQUIREMENTS
Students are required to have a 2.00 grade point average in all major courses.

STUDENT PROGRAMS

PROFESSIONAL AND HONORARY SOCIETIES
[For key to acronyms, see *Introduction*.]
Professional Societies: AIA, AIP, AMS, APS, ASCE, ASHRAE, ASME, IEEE, NSPE
Honorary Societies: Eta Kappa Nu

SUPPORT PROGRAMS
Other Engineering Support Programs: The University College provides support services such as advising, instructional support, diagnostic testing research and evaluation and special services for the disabled. New student orientation identifies available resources and explains regulations governing student matriculation. The University College administers the Honors Program and English as a Second Language.

064 Drexel University

INSTITUTION PROFILE
HEAD OF THE INSTITUTION
Richard D. Breslin
Phone: (215)895-2100 Fax: (215)895-1714

GENERAL INFORMATION
[All Students—Fall 1991]
Undergraduate enrollment 7,811
Graduate enrollment 3,227
Total institution enrollment 11,038
Type of institution: Private Calendar: Quarters
Location: Philadelphia Population: 1,688,210
Setting: Urban
Types of engineering degrees: Engineering
Other degree-granting colleges: Arts & Sciences, Business Administration, Design Arts, Information Studies, Evening & University College

ENGINEERING ADMISSIONS
ADMISSIONS OFFICE CONTACT
John Russel
Drexel University
Director of Undergraduate Enrollment
32nd & Chestnut Streets
Philadelphia, PA 19104
Phone: (215)895-2400 Fax: (215)895-5939

ENGINEERING COLLEGE ADMISSIONS INFORMATION
Entrance Requirements: SAT (Score: 950), High School courses—Math (2 Algebra, 1 Geometry, 1 Trigonometry) (4 years), English (4 years), Science (2 years)
Entrance Recommendations: High School courses—Chemistry (1 year), Physics (1 year), Achievement tests in English Composition, Math and Science for those who wish to be considered for some academic scholarships.
Requirements for foreign students: TOEFL (Score: 530)

DEPARTMENTAL ADMISSIONS INFORMATION
Admission to the engineering department: At the time of admission to the institution.

FINANCIAL INFORMATION
ESTIMATED EXPENSES (ACADEMIC YEAR)
[Expenses are for the 1992-93 nine-month academic year.]

State Residents	Undergraduate	Graduate
Tuition and fees	$11,635	$13,557
College room and board	$ 5,601	$ 5,601
Books and supplies	$ 500	$ 500
Other expenses	$ —	$ —
Total estimated expenses	**$17,736**	**$19,658**

Out-of-State Residents	Undergraduate	Graduate
Tuition and fees	$11,635	$13,557
College room and board	$ 5,601	$ 5,601
Books and supplies	$ 500	$ 500
Other expenses	$ —	$ —
Total estimated expenses	**$17,736**	**$19,658**

FINANCIAL AID OFFICE CONTACT
Nicholas Flocco, *Director*
Drexel University
Financial Aid
Philadelphia, PA 19104
Phone: (215)895-2537 Fax: (215)895-5939

GENERAL FINANCIAL AID INFORMATION
Forms accepted/required: Institutional, PHEAA Form
Additional financial aid information: Approximately 75% of the freshmen at Drexel enroll with some type of financial aid. Based on need, the award may be in the form of a Drexel grant, a federal loan or grant, part-time work-study employment, or any combination of the above.

ENGINEERING COLLEGE INFORMATION
[For additional personnel, refer to the *Appendix*.]
HEAD OF ENGINEERING
Yatish T. Shah
Phone: (215)895-2210 Fax: (215)895-4929

ENGINEERING COLLEGE ADDRESS
College of Engineering
Drexel University
LeBow Engineering Center
32nd & Chestnut Streets
Philadelphia, PA 19104
Phone: (215)895-2210 Fax: (215)895-4929

ENROLLMENTS—BY CLASS
[Numbers are baccalaureate enrollments for the fall 1991 term, unless otherwise footnoted.]
1st-year students/Freshmen .... 828 [1]
2nd-year students/Sophomores .... 670 [2]
3rd-year students/Juniors 483
4th-year students/Seniors 481
Total 2,462
Notes: (1) Includes 211 reclassified frosh. (2) Pre-Junior - 416

NUMBER OF DEGREES AWARDED—BY PROGRAM
[Numbers are engineering baccalaureate degrees awarded during the 1991-1992 academic year, unless otherwise footnoted. For full details about each engineering program, refer to the *Tables of Degree Programs*.]
Appropriate Technology 4
Architectural Engineering 60
Biomedical Engineering .... — [B]
Chemical Engineering 41
Civil Engineering 76
Construction Management 8
Electrical & Computer Engineering 200
Engineering Geology .... — [B]
Engineering Management .... — [B]
Engineering Science(s) .... —
Environmental Engineering .... — [B]
Industrial Engineering 5
Materials Engineering 8
Mechanical Engineering 119
Total 521
Notes: (B) Data not applicable.

PERCENTAGE OF DEGREES AWARDED—BY CATEGORY
[Percentages are of all engineering baccalaureate degrees awarded during the 1991-1992 academic year, unless otherwise footnoted.]
African Americans 5.0%
Asian/Pacific Island Americans 12.0%
Hispanic Americans 1.0%
Native Americans .... — %
Foreign Citizens 9.0%
All Others 73.0%
Women 14.0%
Persons with disabilities 0.2%
Students over 25 years of age 16.0%

GRADUATE ENROLLMENTS & DEGREES AWARDED
Master's enrollment 780
Master's degrees awarded 307
Doctoral enrollment 77
Doctoral degrees awarded 42

ENGINEERING STUDENT DATA
Applicants to the engineering college
Number of applicants to engineering college 1,404
Percent offered admission 87.0%
Matriculated engineering students
Percentage in top quartile (25%) of High School class 36.0
Average SAT scores: Math—580, Verbal—464, Combined—1043

FULL- & PART-TIME FACULTY—BY DEPARTMENT
[Figures are the head count of full-time faculty and the full-time equivalent (FTE) of part-time faculty for each engineering department or equivalent.]

Department	Full	Part
Chemical Eng	10	2.0
Civil & Architectural Eng	17.0	10.0
Electrical & Computer Eng	49	1.0
Mechanical Eng & Mechanics	28	2.0
Materials Eng	10	2.0

COLLEGE DESCRIPTION
The college offers a 5 year cooperative undergraduate education. Numerous specialty options exist within the seven degree programs as do minor programs. Students participate in alternating six-month classroom and industrial experiences during the middle three years. The senior year includes a very extensive, team-oriented interdisciplinary Senior Design program with some team competitions spanning the range from project identification, proposal writing, project implementation, and reporting to conference presentation. Each department offers a 5 year MS/BS degree program. The college also offers an engineering-based Appropriate Technology undergraduate program. Volunteer students may participate in Drexel's E4 program, a major initiative sponsored by the National Science Foundation and the General Electric Foundation in development of a model curriculum for an integrated learning experience including the basic sciences, engineering, and liberal studies.

DEGREE PROGRAMS DESCRIPTION
The B.S. degrees prepare the graduate for professional positions or for graduate study in the respective fields. All programs are built upon a cooperative education philosophy.

DUAL DEGREE PROGRAMS
Enrollment requirements: Students are admitted at the end of the junior year. These students are in 3-3 programs earning a Bachelor's degree from their previous institution and a cooperative education B.S. degree from Drexel.

TRANSFER INFORMATION
Residency Requirements: At least one-half of the professional courses required for the student's specific program or major must be completed at Drexel. A minimum of 45 credit hours must be completed at Drexel. The senior year must be spent at Drexel.
Transfer via Articulation Agreements
Admission to engineering: Transfer students are admitted to the engineering program at four times during the academic year: Fall, Winter, Spring and Summer.
Engineering graduates transferred from ... 4-yr: – 2-yr: –
Requirements: It is recommended that students complete at least 30 credits of college-level coursework prior to submitting a transfer application. A 2.75 GPA is required. Applicants must submit a high school transcript and transcripts from all post-secondary institutions. Only coursework from regionally accredited institutions is considered for transfer credit.
Transfer without Articulation Agreements
Admission to engineering: Transfer students are admitted to the engineering program at four times during the academic year: Fall, Winter, Spring and Summer.
Engineering graduates transferred from ... 4-yr: – 2-yr: –
Requirements: It is recommended that students complete at least 30 credits of college-level coursework prior to submitting a transfer application. A 2.75 GPA is required. Applicants must submit a high school transcript from all post-secondary institutions. Only coursework from regionally-accredited institutions is considered for transfer credit.

GRADUATION REQUIREMENTS
Completion of 192 quarter hours for the five-year co-op program; completion of specific course requirements for program in which enrolled; GPA no less than 2.0 for all coursework taken at Drexel; completion of no less than 12 months cooperative education.

STUDENT PROGRAMS
PROFESSIONAL AND HONORARY SOCIETIES
[For key to acronyms, see *Introduction*.]
Professional Societies: AIAA, AIChE, AIME, ASCE, ASME, IEEE, NSAE, SAE, SAMPE, SPE*, The Institute of Electrical & Electronics Engineers Computer Society, International Society for Hybrid Microelectronics (ISHM)
Honorary Societies: Chi Epsilon, Eta Kappa Nu, Pi Tau Sigma, Tau Beta Pi

SUPPORT PROGRAMS
Student Chapter Organizations: Society of Women Engineers, Society of Minority Engineers & Scientists, PRIME, PRISM
SUCCESS
For: Women & Ethnic Minorities **Available:** Year round
Offered: Prematriculation, Freshman, Sophomore, Junior, Senior, Graduate level
Women in Engineering Program
For: Women **Available:** Year round
Offered: Prematriculation, Freshman, Sophomore, Junior, Senior, Graduate level
Other Engineering Support Programs: Drexel offers a summer orientation program for all freshmen and transfers; special support programs -- ACT101 and SUCCESS; career counseling services; student support programs for study skills and time management; university life adjustment; Freshmen Center; Co-operative Education Center; and Placement Services Center.

065 Duke University

INSTITUTION PROFILE

HEAD OF THE INSTITUTION
H. Keith H. Brodie
Phone: (919)684-2424

GENERAL INFORMATION
[All Students—Fall 1991]
Undergraduate enrollment 6,071
Graduate enrollment 4,889
Total institution enrollment 11,960
Type of institution: Private   **Calendar:** Semesters
Location: Durham   **Population:** 136,611
Setting: Suburban
Types of engineering degrees: Engineering
Other degree-granting colleges: Arts & Sciences, Business Administration, Law, Medicine, Divinity School, The Environment

ENGINEERING ADMISSIONS

ADMISSIONS OFFICE CONTACT
Christoph O. Guttentag
Duke University
Office of Undergraduate Admissions
2138 Campus Drive Box 90586
Durham, NC 27708-0586
Phone: (919)684-3214

ENGINEERING COLLEGE ADMISSIONS INFORMATION
Entrance Requirements: High School courses—English (4 years), Mathematics (4 years), Natural Science (3 years), Foreign Language (2 years), SAT + 3 Achievement Tests incl. English Composition (with or without essay), Mathematics (Level 1 or 2), + one other
Entrance Recommendations: High School courses—Calculus (1 year), Computers (1 year), Physics (1 year), Standardized tests - recommended: ACT can be substituted for above
Requirements for foreign students: TOEFL and financial statement

DEPARTMENTAL ADMISSIONS INFORMATION
Admission to the engineering department: the time they declare a major.

FINANCIAL INFORMATION

ESTIMATED EXPENSES (ACADEMIC YEAR)
[Expenses are for the 1992-93 nine-month academic year.]

All Students	Undergraduate	Graduate
Tuition and fees	$16,725	$10,370
College room and board	$ 3,975	$ 2,642
Books and supplies	$ 565	$ 1,000
Other expenses	$ 336	$ 2,539
Total estimated expenses	**$21,601**	**$16,551**

FINANCIAL AID OFFICE CONTACT
James A. Belvin Jr., *Director*
Duke University
Financial Aid
2106 Campus Drive, Box 90397
Durham, NC 27708-0397
Phone: (919)684-6225   **Fax:** (919)660-9811

GENERAL FINANCIAL AID INFORMATION
Forms accepted/required: FAF, IRS, Institutional
Additional financial aid information: Need-based scholarships; short-term loans; merit-based scholarships; part-time jobs on campus; College Work-Study Program.

ENGINEERING COLLEGE INFORMATION
[For additional personnel, refer to the *Appendix*.]

HEAD OF ENGINEERING
Earl H. Dowell
Phone: (919)660-5389   **Fax:** (919)684-4860

ENGINEERING COLLEGE ADDRESS
School of Engineering
Duke University
305 Teer Building
Box 90271
Durham, NC 27708-0271
Phone: (919)660-5386   **Fax:** (919)684-4860

ENROLLMENTS—BY CLASS
[Numbers are baccalaureate enrollments for the fall 1991 term, unless otherwise footnoted.]
1st-year students/Freshmen 245
2nd-year students/Sophomores 198
3rd-year students/Juniors 249
4th-year students/Seniors 198
Total **890**

NUMBER OF DEGREES AWARDED—BY PROGRAM
[Numbers are engineering baccalaureate degrees awarded during the 1991-1992 academic year, unless otherwise footnoted. For full details about each engineering program, refer to the *Tables of Degree Programs*.]
Biomedical Engineering 62
Civil Engineering 33
Electrical Engineering 66
Mechanical Engineering 43
Total **204**

PERCENTAGE OF DEGREES AWARDED—BY CATEGORY
[Percentages are of all engineering baccalaureate degrees awarded during the 1991-1992 academic year, unless otherwise footnoted.]
African Americans .... 7.8 %
Asian/Pacific Island Americans .... 4.9 %
Hispanic Americans .... – %
Native Americans .... – %
Foreign Citizens .... 2.9 %
All Others .... 84.4 %
Women .... 20.6 %
Persons with disabilities .... – %
Students over 25 years of age .... – %

GRADUATE ENROLLMENTS & DEGREES AWARDED
Master's enrollment 63
Master's degrees awarded 51
Doctoral enrollment 191
Doctoral degrees awarded 27

ENGINEERING STUDENT DATA
Applicants to the engineering college
Number of applicants to engineering college 2,236
Percent offered admission 28.5%

Matriculated engineering students
Percentage in top quartile (25%) of High School class 99.0
Average SAT scores: Math—730, Verbal—615, Combined—1345

FULL- & PART-TIME FACULTY—BY DEPARTMENT
[Figures are the head count of full-time faculty and the full-time equivalent (FTE) of part-time faculty for each engineering department or equivalent.]

Department	Full	Part
Civil & Environmental Eng	15	–
Electrical Eng	18	–
Mechanical Eng & Materials Science	23	–
Biomedical Eng	29	–

COLLEGE DESCRIPTION
The undergraduate engineering program is a four year program with four engineering departments, each with its own faculty, courses, fields of specialization, and curriculum requirements beyond a set of general requirements. Special academic opportunities include interdisciplinary engineering programs, second majors in any University undergraduate department, certificate programs such as Science Technology, and Human Values, and an International Honors Program. All departments offer a five year BSE/MBA program with the Fuqua School of Business. Independent study is encouraged in the senior year. Overall, emphasis is on engineering education in a liberal arts environment. The educational experience is enhanced by a geographically diverse student body; all 50 states and several foreign countries are represented.

DEGREE PROGRAMS DESCRIPTION
All degree programs offer the B.S.E., M.S., and Ph.D.. M.S. and Ph.D. programs are available also in biochemical engineering, environmental engineering, and materials science. All graduate degree programs are administered through the Graduate School, 127 Allen Bldg. A dual MS/MBA program is offered in conjunction with the Fuqua School of Business. Pre-medical requirements may be met in all B.S.E. degree programs.

DUAL DEGREE PROGRAMS
Engineering baccalaureate graduates with dual degrees 0
Enrollment requirements: Must major in natural science or mathematics; must have 3.0/4.0 GPA overall and 3.3/4.0 in mathematics courses.

TRANSFER INFORMATION
Residency Requirements: Must be resident at Duke for four semesters and earn at least 17 semester course credits (out of 34) at Duke.

Transfer without Articulation Agreements
Admission to engineering: Completion of two or four semesters at fully accredited four year institutions.
Engineering graduates transferred from .. 4-yr: 6   2-yr: –
Requirements: Minimum GPA of 3.0/4.0 at initial institution.

GRADUATION REQUIREMENTS
Of the thirty-four semester courses which fulfill the specified categories in the Bachelor of Science in Engineering degree requirements, thirty-two or their equivalent in number must be passed with grades of P, C-, or better.

STUDENT PROGRAMS

PROFESSIONAL AND HONORARY SOCIETIES
[For key to acronyms, see *Introduction*.]
Professional Societies: ASCE, ASME, BMES, IEEE
Honorary Societies: Chi Epsilon, Eta Kappa Nu, Pi Tau Sigma, Tau Beta Pi

SUPPORT PROGRAMS
Student Chapter Organizations: Society of Women Engineers, National Society of Black Engineers
Other Engineering Support Programs: There are no special engineering student support programs.

066 École Polytechnique de Montréal

■ INSTITUTION PROFILE

HEAD OF THE INSTITUTION
Jean-Paul Gourdeau
Phone: (514)340-4704   Fax: (514)340-3222

GENERAL INFORMATION
[All Students—Fall 1991]
Undergraduate enrollment 3,544
Graduate enrollment 1,220
Total institution enrollment 5,511
Type of institution: Public   **Calendar:** Trimesters
Location: Montréal   **Population:** 2,000,000
Setting: Urban
Types of engineering degrees: Engineering

■ ENGINEERING ADMISSIONS

ADMISSIONS OFFICE CONTACT
Claude Brissette
École Polytechnique de Montréal
Case Postale 6079, Succursale A
Montréal, PQ H3C-3A7
Phone: (514)340-4724   Fax: (514)340-5836

ENGINEERING COLLEGE ADMISSIONS INFORMATION
Entrance Requirements: High School courses—Diplôme d'études collégiales (Québec) (1 year), High school followed by 1 year in university (1 year)
Requirements for out-of-state residents: Because in Québec we have a "cégep" which is equivalent to high school plus a full year, we ask for the equivalent of that degree. So prospective engineering student coming from outside of Québec must respect this criteria.
Requirements for foreign students: A new student, being Canadian or not, must pass an exam to evaluate his or her ability to write or read French language. If he or she fails this test, he or she must pass a special course in French before having gained 45 credits.

DEPARTMENTAL ADMISSIONS INFORMATION
Admission to the engineering department: Best half of students are admitted directly in their program at the admission. The other half are admitted at the end of freshman year.
Additional information: The numbers of places are limited in each program. Presently, these figures are: 130 in Civil eng., 200 in Mechanical eng., 180 in Electrical eng., 55 in Chemical eng., 40 in Materials eng., 25 in Mining eng., 35 in Geological eng., 50 in Eng. Physics, 70 in Industrial eng., 60 in Computer eng. Admission in these programs is based on the scores obtained in the freshman year: the best students have the first choice.

■ FINANCIAL INFORMATION

ESTIMATED EXPENSES (ACADEMIC YEAR)
[Expenses are for the 1992-93 nine-month academic year.]

Residents	Undergraduate	Graduate
Tuition and fees	$ 1,485	$ 1,559 [1]
College room and board	$ 184	$ 178
Books and supplies	$ 500	$ 500
Other expenses	$ —	$ —
Total estimated expenses	$ 2,169	$ 2,237

Non-Residents	Undergraduate	Graduate
Tuition and fees	$ 7,273	$ 7,502
College room and board	$ 184	$ 178
Books and supplies	$ 500	$ 500
Other expenses	$ —	$ —
Total estimated expenses	$ 7,957	$ 8,180

Notes: (1) All figures are in Canadian dollars, they represent a maximum. State resident means "Canadian citizen" or "Canadian resident".

FINANCIAL AID OFFICE CONTACT
France Gaudron, *Responsable de l'aide financière*
École Polytechnique de Montréal
Case Postale 6079, Succursale A
Montréal, PQ H3C-3A7
Phone: (514)340-4842   Fax: (514)340-5836

GENERAL FINANCIAL AID INFORMATION
Forms accepted/required: Forms from Canadian government, Forms from Québec government.
Additional financial aid information: Entry scholarships for undergraduate students; - Scholarships administered by the School (bursaries and other financial awards); - Assistantships for graduate students; - Part-time campus jobs mostly for graduate students.

■ ENGINEERING COLLEGE INFORMATION

[For additional personnel, refer to the *Appendix*.]

HEAD OF ENGINEERING
André Bazergui
Phone: (514)340-4943   Fax: (514)340-4600

ENGINEERING COLLEGE ADDRESS
École Polytechnique de Montréal
Case Postale 6079, Succursale A
Montréal, PQ H3C-3A7
Phone: (514)340-4711   Fax: (514)340-5836

ENROLLMENTS—BY CLASS
[Numbers are baccalaureate enrollments for the fall 1991 term, unless otherwise footnoted.]
1st-year students/Freshmen .... 1,526 [5]
2nd-year students/Sophomores 771
3rd-year students/Juniors 649
4th-year students/Seniors 598
Total 3,544

Notes: (5) Figure valid in Fall 1991.

NUMBER OF DEGREES AWARDED—BY PROGRAM
[Numbers are engineering baccalaureate degrees awarded during the 1991-1992 academic year, unless otherwise footnoted. For full details about each engineering program, refer to the *Tables of Degree Programs*.]

Applied Mathematics .... – [2]
Biochemical Engineering .... –
Biomedical Engineering .... –
Chemical Engineering 25
Civil Engineering 107
Computer Engineering 42
Electrical Engineering 160
Engineering Physics 22
Geological Engineering 4
Industrial Engineering 62
Materials Engineering 17
Mechanical Engineering 167
Mineral Engineering .... –
Mining Engineering 7
Nuclear Engineering .... –
Total 613

Notes: (2) Figures in this column refers to Fall '91 + Winter '92 + Summer '92.

PERCENTAGE OF DEGREES AWARDED—BY CATEGORY
[Percentages are of all engineering baccalaureate degrees awarded during the 1991-1992 academic year, unless otherwise footnoted.]

African Canadians .... – % [A]
Asian/Pacific Island Canadians .... – % [A]
Hispanic Canadians .... – % [A]
Native Canadians .... – % [A]
Foreign Citizens .... – % [A]
All Others .... 100 %
Women .... 21.2 % [5]
Persons with disabilities .... – % [A]
Students over 25 years of age .... – % [A]

Notes: (A) Data not available. (5) Figure valid in Fall 1991.

GRADUATE ENROLLMENTS & DEGREES AWARDED
Master's enrollment .... 846 [3]
Master's degrees awarded .... 147 [4]
Doctoral enrollment .... 283 [5]
Doctoral degrees awarded .... 54 [1]

Notes: (1) Fall '91 + Winter '92 + Summer '92 (3) In addition, there are 91 students in D.E.S.S. programs. Figures valid in Fall '91. (4) In addition, 23 students graduated in D.E.S.S. programs. Figures comprise Fall '91 + Winter '92 + Summer '92. (5) Figure valid in Fall 1991.

ENGINEERING STUDENT DATA
Applicants to the engineering college
Number of applicants to engineering college 2,902
Percent offered admission 68.7%

Matriculated engineering students
Percentage in top quartile (25%) of High School class .... – [A]

Notes: (A) Data not available.

FULL- & PART-TIME FACULTY—BY DEPARTMENT

[Figures are the head count of full-time faculty and the full-time equivalent (FTE) of part-time faculty for each engineering department or equivalent.]

Department	Full	Part
Chemical Eng	12	1.4
Civil Eng	25	–
Electrical Eng	46	–
Industrial Eng	20	–
Mechanical Eng	44	2.2
Metallurgical & materials engineering	11	1.6
Mineral engineering	14	–
Eng Physics	16	1.6
Applied mathematics	21	–
Energy engineering	7	0.8
Biomedical engineering	6	–

COLLEGE DESCRIPTION
École Polytechnique de Montréal is the largest engineering school in Canada. It is affiliated with Université de Montréal, which offers all but non-engineering programs. The school offers 10 four-year undergraduate programs in engineering. Two of which are co-op: Mining engineering, offered in conjunction with McGill University is a 4 year program, and Geological engineering, a 4 year+ 1 trimester program. Each student must complete a design project in the last year, and for all but co-op programs, he or she chooses a "concentration" (30 credits, available in Chemical and Civil engineering programs) or a "specialization" (12 credits). A credited professional internship program is available as an optional basis (8 months or more). All teaching is in French language. Except for a few people (mostly in their last year), all students are full-time students in undergraduate programs. Average number of years actually required to complete the bachelor's degree: 4.5.

DEGREE PROGRAMS DESCRIPTION
École Polytechnique offers 4 graduate programs. (1) The DESS (Diplôme d'études supérieures spécialisées) is a 30 credit program. Part time is permitted; only courses are required (no thesis). Specializations offered are: Ergonomy, project management, as well as all of our other degree programs. (2) The M.Ing (M.Eng). (Maîtrise en ingénierie) is a 45 credit program, full or part time, with project, thesis or internship. It is available in Aeronautics, Applied Mathematics, and all of the other degree programs. (3) The Ph.D. (Philosophiae doctor) is a 90 credit program (98 in Applied Mathematics). Except in special cases, the thesis must be presented in French. A given number of graduate courses are offered in the evening.

TRANSFER INFORMATION
Residency Requirements: A minimum of 60 credits (half of the required 120's) passed at the École Polytechnique are required to get a bachelor degree from the institution.

GRADUATION REQUIREMENTS
In undergraduate programs, the graduation requirements are:- a GPA of at least 1.75/4;- success (minimum grade D) in all mandatory courses (which constitute about 80% of each program);- success in elective courses to total a minimum of 120 credits. For graduate programs, the minimum GPA are: 2.75/4 for D.E.S.S., 3.0/4 for M.Sc.A., M.Ing. and Ph.D. In addition, the thesis must be accepted by a jury.

■ STUDENT PROGRAMS

PROFESSIONAL AND HONORARY SOCIETIES
[For key to acronyms, see *Introduction*.]
Professional Societies: CIM, CSCE, IEEE, SAE, TMS, Institut canadien des ingénieurs, Other Canadian institutes or societies

SUPPORT PROGRAMS
Other Engineering Support Programs: All freshman courses are repeated 2 times a year, and most of them are also available in summer. Workshops in basic scientific courses and in French language are offered. Services: placement, financial aid, exchange programs with foreign universities (mostly in France and US), housing support, health and psychological services.

067 Embry-Riddle Aeronautical University - Daytona Beach

INSTITUTION PROFILE

HEAD OF THE INSTITUTION
Steven Sliwa
Phone: (904)226-6200   Fax: (904)226-6299

GENERAL INFORMATION
[All Students—Fall 1991]
Undergraduate enrollment 4,341
Graduate enrollment 167
Total institution enrollment 4,508
Type of institution: Private   **Calendar:** Semesters
Location: Daytona Beach   **Population:** 370,712
Setting: Urban
Types of engineering degrees: Engineering & Technology

ENGINEERING ADMISSIONS

ADMISSIONS OFFICE CONTACT
Darryl W. Niemeyer
Embry-Riddle Aeronautical University - Daytona Beach
600 S. Clyde Morris Blvd.
Daytona Regional Airport
Daytona Beach, FL 32114-3900
Phone: (904)226-6100   Fax: (904)226-6299

ENGINEERING COLLEGE ADMISSIONS INFORMATION
Entrance Requirements: SAT (Score: 1000), ACT (Score: 24), High School courses—English (4 years), Mathematics (3 years), Sciences (3 years)
Entrance Recommendations: High School courses—Chemistry (1 year), Physics (1 year), Pre-calculus (1 year)
Requirements for foreign students: TOEFL (Score: 500)

DEPARTMENTAL ADMISSIONS INFORMATION
Admission to the engineering department: At the time of admission to the institution.
Additional information: High School GPA = 2.0 and GPA = 2.0 in math and science High School rank = top 50% SAT math = 550 verbal = 450 ACT math = 26 English = 22 Plus, for Engr. Physics majors, High School GPA = 3.0 and GPA = 3.0 in math and science and SAT total = 1100 and ACT comp = 26 and High School rank = top 30%

FINANCIAL INFORMATION

ESTIMATED EXPENSES (ACADEMIC YEAR)
[Expenses are for the 1992-93 nine-month academic year.]

All Students	Undergraduate	Graduate
Tuition and fees	$ 6,200	$ 7,800
College room and board	$ 3,550	—(1)
Books and supplies	$ 800	$ 1,000
Other expenses	$ 750 (2)	$ 750 (2)
Total estimated expenses	$11,300	$ 9,550

Notes: (1) Not available on campus for graduate students. (2) Program and other expenses; does not include transportation and personal expenses.

FINANCIAL AID OFFICE CONTACT
Phillip C. Ledbetter, *Director of Financial Aid*
Embry-Riddle Aeronautical University - Daytona Beach
Embry-Riddle Aeronautical University
600 S. Clyde Morris Blvd.
Daytona Beach, FL 32114-3900
Phone: (904)226-6195

GENERAL FINANCIAL AID INFORMATION
Forms accepted/required: ACT, AFSA, CSS-FAF, FAT, FAR, FFS, Stafford, IRS, SAR, Supplemental, Institutional
Additional financial aid information: Need based scholarships, short-term loans, merit-based scholarships, part-time jobs on campus, college work study program

ENGINEERING COLLEGE INFORMATION

[For additional personnel, refer to the *Appendix*.]

HEAD OF ENGINEERING
Ray Wimberly
Phone: (904)226-6634   Fax: (904)226-6299

ENGINEERING COLLEGE ADDRESS
Engineering Technology Department
Embry-Riddle Aeronautical University - Daytona Beach
600 S. Clyde Morris Blvd.
Daytona Beach Regional Airport
Daytona Beach, FL 32114-3900
Phone: (904)226-6000   Fax: (904)226-6223

ENROLLMENTS—BY CLASS
[Numbers are baccalaureate enrollments for the fall 1991 term, unless otherwise footnoted.]
1st-year students/Freshmen 275
2nd-year students/Sophomores 148
3rd-year students/Juniors 129
4th-year students/Seniors 272
Total **824**

NUMBER OF DEGREES AWARDED—BY PROGRAM
[Numbers are engineering baccalaureate degrees awarded during the 1991-1992 academic year, unless otherwise footnoted. For full details about each engineering program, refer to the *Tables of Degree Programs*.]
Aerospace & Aeronautical Engineering 115
Engineering Physics 9
Total **124**

PERCENTAGE OF DEGREES AWARDED—BY CATEGORY
[Percentages are of all engineering baccalaureate degrees awarded during the 1991-1992 academic year, unless otherwise footnoted.]
African Americans 1.0%
Asian/Pacific Island Americans 8.1%
Hispanic Americans 2.4%
Native Americans .... — %
Foreign Citizens .... — % (A)
All Others 88.5%
Women 13.7%
Persons with disabilities .... — % (A)
Students over 25 years of age .... — % (A)

Notes: (A) Data not available.

GRADUATE ENROLLMENTS & DEGREES AWARDED
Master's enrollment 36
Master's degrees awarded 1
Doctoral enrollment .... — (B)
Doctoral degrees awarded .... — (B)

Notes: (B) Data not applicable.

ENGINEERING STUDENT DATA

Applicants to the engineering college
Number of applicants to engineering college 812
Percent offered admission 77.5%

Matriculated engineering students
Percentage in top quartile (25%) of High School class 41.8
Average SAT scores: Math—600, Verbal—491, Combined—1092
Average ACT scores: Math—27, Composite—26

FULL- & PART-TIME FACULTY—BY DEPARTMENT
[Figures are the head count of full-time faculty and the full-time equivalent (FTE) of part-time faculty for each engineering department or equivalent.]

Department	Full	Part
Electrical Eng	5	–
Aerospace & Aeronautical Eng	24	–
Eng Physics	2	–

COLLEGE DESCRIPTION
Aerospace Engineering: The undergraduate aerospace engineering program provides a thorough grounding in the basics, experience in hands-on laboratories and a practical CAD-based, design-oriented preparation for entry-level engineering careers in the aerospace industry. The department has several wind tunnels and its own airplane. Engineering Physics: The Engineering Physics program has a two semester sequence on designing for spacecraft and orbiting space systems, using CAD/CAM systems. Every student is exposed to hands on experience in the machine shop and is given the opportunity to teach actual physics laboratories. Students explore topics optics, nuclear and atomic physics in open ended design projects in the modern physics laboratory. They have the opportunity to work at the space physics research laboratory at the university, attend field trips, participate in collecting research data, using art electro optical technology and remote sensing equipment.

DEGREE PROGRAMS DESCRIPTION
Embry-Riddle offers the BS degree in Aerospace Engineering at the Daytona Beach and Prescott campuses. This program provides the student the opportunity to acquire aerospace design skills and a broad exposure to theory and modern analysis, measurement, communications, and computational techniques essential for a wide range of entry level engineering positions in the aerospace industry. Embry-Riddle offers a BS degree in Electrical Engineering at the Prescott campus. This program provides the student with the opportunity to acquire a broad background in circuit theory, communication sciences, computers, control systems, electromagnetic fields, energy sources and systems, materials, and electronic devices, as well as specialization in avionics appropriate for entry level engineering positions in the aerospace industry. The BS in Engineering Physics, offered on the Daytona Beach Campus, is designed to develop sufficient depth in both engineering skills and science to produce students who are able to relate basic knowledge to practical problems in engineering. The engineering physicist will have the training of an applied physicist; the ability to attack problems, particularly in the aeronautical and aerospace areas; and the flexibility to extend this basic knowledge to any branch of engineering and science.

TRANSFER INFORMATION
Transfer without Articulation Agreements
Admission to engineering: At any time
Requirements: At least a 2.5 CGPA and at least 12 credit hours completed at another college. Student should be in good academic standing at other college.

GRADUATION REQUIREMENTS
Bachelor of Aerospace Engineering and Engineering Physics programs require 136 semester credit hours, 2.0 CGPA and 2.0 GPA in aerospace and engineering technology courses. Electrical Engineering program requires 135 semester credit hours, 2.0 CGPA and 2.0 GPA in engineering technology courses.

STUDENT PROGRAMS

PROFESSIONAL AND HONORARY SOCIETIES
[For key to acronyms, see *Introduction*.]
Professional Societies: AIAA, SAE, Engineering Student Advisory Council
Honorary Societies: Sigma Gamma Tau, Sigma Phi Delta, Tau Alpha Pi

SUPPORT PROGRAMS

Future Professional Women in Aviation
For: Women   **Available:** Year round
Offered: Freshman

Beta Phi Alpha
For: Women   **Available:** Year round
Offered: Freshman

Kappa Alpha Psi - African American
For: Ethnic Minorities   **Available:** Year round
Offered: Freshman

Other Engineering Support Programs: Orientation program for all freshmen and transfers; career counseling services; academic counseling and tutoring services; international student office; veterans' services; placement services.

068 Embry-Riddle Aeronautical University, Western Campus

■ INSTITUTION PROFILE

HEAD OF THE INSTITUTION
Steven M. Sliwa
Phone: (904)226-6200   Fax: (904)226-6299

GENERAL INFORMATION
[All Students—Fall 1991]
Undergraduate enrollment 1,600
Total institution enrollment 1,600
Type of institution: Private   **Calendar:** Semesters
Nearest city: Phoenix, Arizona   **Population:** 2,300,000
Miles from main campus: 100   **Setting:** Small Town
Types of engineering degrees: Engineering
Other degree-granting colleges: Aeronautical Science, Aviation Business Administration, Aviation Computer Science

ENGINEERING COLLEGE ADMISSIONS INFORMATION
Entrance Recommendations: SAT (Score: 1000), ACT (Score: 48), High School courses—Algebra (2 years), Geometry/Trigonometry (1 year), Physics (1 year), Chemistry
Requirements for foreign students: TOEFL (Score: 500); financial statement

DEPARTMENTAL ADMISSIONS INFORMATION
Admission to the engineering department: At the time of admission to the institution.
Additional information: Recommend math, physics, chemistry, calculus, algebra, geometry, trigonometry and mechanical drawing.

■ FINANCIAL INFORMATION

ESTIMATED EXPENSES (ACADEMIC YEAR)
[Expenses are for the 1992-93 nine-month academic year.]

All Students	Undergraduate	Graduate
Tuition and fees	$ 6,700	$ –
College room and board	$ 3,510	$ –
Books and supplies	$ 700	$ –
Other expenses	$ 250	$ –
Total estimated expenses	$11,160	$ –

FINANCIAL AID OFFICE CONTACT
Dan Lupin, *Director of Financial Aid*
Embry-Riddle Aeronautical University, Western Campus
3200 Willow Creek Road
Prescott, AZ 86301-3720
Phone: (602)776-3762   Fax: (602)445-3184

GENERAL FINANCIAL AID INFORMATION
Forms accepted/required: ACT, AFSA, CSS-FAF, FAF, FAT, FFS, Stafford, IRS, SAR, Supplemental, Institutional
Additional financial aid information: Embry-Riddle offers a full range of financial assistance programs that can help students and their families afford educational costs. Financial assistance programs are typically offered by federal and state governments, these are usually loans, grants and work programs for students who have financial need or who meet certain state residency criteria.

■ ENGINEERING COLLEGE INFORMATION

[For additional personnel, refer to the *Appendix*.]

HEAD OF ENGINEERING
Paul S. Daly
Phone: (602)776-3800   Fax: (602)776-3827

ENGINEERING COLLEGE ADDRESS
Aerospace Engineering & Electrical Engineering
Embry-Riddle Aeronautical University, Western Campus
3200 Willow Creek Road
Prescott, AZ 86301-3720
Phone: (602)776-3728   Fax: (602)776-3740

ENROLLMENTS—BY CLASS
[Numbers are baccalaureate enrollments for the fall 1991 term, unless otherwise footnoted.]
1st-year students/Freshmen 155
2nd-year students/Sophomores 89
3rd-year students/Juniors 75
4th-year students/Seniors 133
Total 452

NUMBER OF DEGREES AWARDED—BY PROGRAM
[Numbers are engineering baccalaureate degrees awarded during the 1991-1992 academic year, unless otherwise footnoted. For full details about each engineering program, refer to the *Tables of Degree Programs*.]
Aerospace Engineering 50
Electrical Engineering 4
Total 54

PERCENTAGE OF DEGREES AWARDED—BY CATEGORY
[Percentages are of all engineering baccalaureate degrees awarded during the 1991-1992 academic year, unless otherwise footnoted.]
African Americans .... 4.0 %
Asian/Pacific Island Americans .... 6.0 %
Hispanic Americans .... 6.0 %
Native Americans .... – %
Foreign Citizens .... 10.0 %
All Others .... 74.0 %
Women .... 11.0 %
Persons with disabilities .... – %
Students over 25 years of age .... 10.0 %

ENGINEERING STUDENT DATA
Applicants to the engineering college
Number of applicants to engineering college 395
Percent offered admission 89.0%

Matriculated engineering students
Percentage in top quartile (25%) of High School class 73.0
Average SAT scores: Math—578, Verbal—489, Combined—1066
Average ACT scores: Math—25, Composite—25

FULL- & PART-TIME FACULTY—BY DEPARTMENT
[Figures are the head count of full-time faculty and the full-time equivalent (FTE) of part-time faculty for each engineering department or equivalent.]

Department	Full	Part
Electrical Eng	5	–
Aerospace Eng	9	2.0

COLLEGE DESCRIPTION
The Aerospace and Electrical Engineering Programs at the Western Campus of Embry-Riddle features a unique combination of scientific theory and engineering applications. Students develop skills and knowledge in small classes and develop a personal relationship with highly qualified and experienced faculty. The focus of the university on aviation and aerospace creates a fertile environment for the exchange of ideas and discussions with fellow students. Cooperative education programs are available and optional. Co-op provides the opportunity to test career interests in jobs directly related to their field of study. A strong career center provides services to increase job opportunities for graduates.

DEGREE PROGRAMS DESCRIPTION
The Electrical Engineering Program will have a new engineering building completed in 1993. Five modern labs support Electrical Engineering through demonstrations, designs and experiments with aerospace electronics applications. The labs are basic circuits/electronic devices, digital circuits and microcomputer applications, aviation communications systems analog/digital control systems and aviation power systems/electronics. The Aerospace Engineering laboratories provide extensive and significant hands-on experience for students. The wind tunnel laboratory contains a research quality subsonic wind tunnel with a 3 x 4 foot test section. A modern supersonic wind tunnel and a shock tube allows investigations of flows with shocks. A materials lab contains the latest equipment to study the properties of materials including heat treatment. The engineering graphics and aircraft design lab have computer-aided design equipment. Aircraft structures and composite labs analyze structural aspects of aerospace vehicles and includes an electron microscope capable of magnifying images 70,000 times.

TRANSFER INFORMATION
Residency Requirements: The last 30 academic credit hours must be completed with Embry-Riddle Aeronautical University for a bachelor's degree.

Transfer via Articulation Agreements
Admission to engineering: At any time
Engineering graduates transferred from ... 4-yr: 2   2-yr: 9
Requirements: 2.5 cumulative GPA and in good academic standing at last college.

Transfer without Articulation Agreements
Admission to engineering: At any time
Engineering graduates transferred from ... 4-yr: 2   2-yr: 4
Requirements: Must meet academic qualifications and degree program capacity must not be filled. A minimum of 12 credit hours must be completed in original degree program before applying.

GRADUATION REQUIREMENTS
2.0 overall. 2.0 in Aerospace Engineering and Electrical Engineering Core Courses.

■ STUDENT PROGRAMS

PROFESSIONAL AND HONORARY SOCIETIES
[For key to acronyms, see *Introduction*.]
Professional Societies: AIAA, IEEE
Honorary Societies: Sigma Gamma Tau

SUPPORT PROGRAMS
Student Chapter Organizations: Women in Engineering (being formed), Women's Support Group
Other Engineering Support Programs: Fall and spring orientation program for all freshmen and transfers; career counseling services; academic counseling and tutoring services; International student counseling; student advisory committee for aerospace engineering (committee is made up of eight students and one faculty advisor); and the newly created Student Success Center.

069 University of Evansville

■ INSTITUTION PROFILE

HEAD OF THE INSTITUTION
James S. Vinson
Phone: (812)479-2151   Fax: (812)479-2320

GENERAL INFORMATION
[All Students—Fall 1991]
Undergraduate enrollment 2,946
Graduate enrollment 125
Total institution enrollment 3,071
Type of institution: Private   **Calendar:** Semesters
Location: Evansville, IN   **Population:** 130,000
Setting: Urban
Types of engineering degrees: Engineering
Other degree-granting colleges: Business Administration, Education, Fine & Performing Arts, Humanities & Social Sciences, Nursing, Harlaxton College-Grantham England, UK

■ ENGINEERING ADMISSIONS

ADMISSIONS OFFICE CONTACT
John Byrd
University of Evansville
1800 Lincoln Ave
Evansville, IN 47722
Phone: (800)423-8633   Fax: (812)479-2320

ENGINEERING COLLEGE ADMISSIONS INFORMATION
Entrance Requirements: SAT (Score: 950), ACT (Score: 23), High School courses—English (4 years), Mathematics (4 years), Chemistry (1 year)
Entrance Recommendations: High School courses—Physics (1 year), Foreign Language (4 years), Calculus (1 year), NEAS - National Engineering Aptitude search by JETS.
Requirements for foreign students: TOEFL (Score: 500); financial statement

DEPARTMENTAL ADMISSIONS INFORMATION
Admission to the engineering department: At the time of admission to the institution.
Additional information: Upper Division Admission requires grades of C or better in all required lower division Math, Science, and Engineering Science courses or a passing grade on the Upper Division Admission test administered by the departments.

■ FINANCIAL INFORMATION

ESTIMATED EXPENSES (ACADEMIC YEAR)
[Expenses are for the 1992-93 nine-month academic year.]

All Students	Undergraduate	Graduate
Tuition and fees	$11,750	$ —
College room and board	$ 3,750	$ —
Books and supplies	$ 500	$ —
Other expenses	$ 1,500 [1]	$ —
Total estimated expenses	$17,500	$ —

Notes: (1) Includes personal computer of $1500

FINANCIAL AID OFFICE CONTACT
Tom Stone, *Director of Financial Aid*
University of Evansville
1800 Lincoln Ave
Evansville, IN 47722
Phone: (812)479-2364   Fax: (812)479-2320

GENERAL FINANCIAL AID INFORMATION
Forms accepted/required: CSS-FAF, SAR, Supplemental, Institutional
Additional financial aid information: Merit Based departmental scholarships based on high school academic record; Need based grants; guaranteed student loans; short term loans; college work study program; part-time jobs off campus;

■ ENGINEERING COLLEGE INFORMATION
[For additional personnel, refer to the *Appendix*.]

HEAD OF ENGINEERING
John R. Tooley
Phone: (812)479-2651   Fax: (812)479-2780
EMail: JACK_TOOLEY%WAYNE-MTS@UM.CC.UMICH.EDU

ENGINEERING COLLEGE ADDRESS
College of Engineering and Computer Science
University of Evansville
1800 Lincoln Ave.
Evansville, IN 47722
Phone: (812)479-2651   Fax: (812)479-2780

ENROLLMENTS—BY CLASS
[Numbers are baccalaureate enrollments for the fall 1991 term, unless otherwise footnoted.]
1st-year students/Freshmen 91
2nd-year students/Sophomores 48
3rd-year students/Juniors 51
4th-year students/Seniors .... 108 [1]
Total 298

Notes: (1) Includes 5th year co-op students.

NUMBER OF DEGREES AWARDED—BY PROGRAM
[Numbers are engineering baccalaureate degrees awarded during the 1991-1992 academic year, unless otherwise footnoted. For full details about each engineering program, refer to the *Tables of Degree Programs*.]
Civil Engineering 4
Computer Engineering .... — (A)
Electrical Engineering 42
Mechanical Engineering 38
Total 84

Notes: (A) Data not available.

PERCENTAGE OF DEGREES AWARDED—BY CATEGORY
[Percentages are of all engineering baccalaureate degrees awarded during the 1991-1992 academic year, unless otherwise footnoted.]
African Americans .... 1.1 %
Asian/Pacific Island Americans .... — %
Hispanic Americans .... — %
Native Americans .... — %
Foreign Citizens .... 34.8 %
All Others .... 64.1 %
Women .... 11.2 %
Persons with disabilities .... — %
Students over 25 years of age .... 7.8 % (C)

Notes: (C) Estimated data.

ENGINEERING STUDENT DATA
Applicants to the engineering college
Number of applicants to engineering college 162
Percent offered admission 91.3%

Matriculated engineering students
Percentage in top quartile (25%) of High School class 93.2
Average SAT scores: Math—572, Verbal—531, Combined—1103

FULL- & PART-TIME FACULTY—BY DEPARTMENT
[Figures are the head count of full-time faculty and the full-time equivalent (FTE) of part-time faculty for each engineering department or equivalent.]

Department	Full	Part
Mechanical & Civil Eng	8	8.5
Electrical Eng & Computer Science	9	9.7
Mechanical & Civil Eng	8	8.5

COLLEGE DESCRIPTION
The College offers four year undergraduate programs in Engineering and Computer Science. A program in Engineering Management is offered by the College in cooperation with the School of Business. Major areas of study are declared after a three semester curriculum; individual project labs are emphasized at all levels; an industrially sponsored individual senior design project is completed during the final two semesters. Incoming students are required to have a personal computer; computer aided engineering is integrated throughout the four years of study. Engineering students participate in the University's rich liberal arts tradition including a unique opportunity to study abroad at Harlaxton College which is the University's campus in England. Optional 5 year Co-op program available for each degree. 4.2 years is the average time to complete the bachelor's degree.

DEGREE PROGRAMS DESCRIPTION
The BS degrees in Engineering prepare the graduate for entry level professional positions and for graduate study. The BS in Engineering Management is intended primarily for entry level positions in technical sales and technical marketing. All undergraduate programs have a practice oriented focus. Computer Engineering combines Computer Science with Electrical Engineering for real time control and signal processing applications. Mechanical Engineering offers specialization in mechanical design, thermo-fluids systems, structural design, or energy systems. Electrical Engineering offers specialization in electronics, electro-optics, electrical power, or digital systems.

TRANSFER INFORMATION
Residency Requirements: At least 63 semester hours must be earned at the University of Evansville including the last 15 hours.
Transfer without Articulation Agreements
Admission to engineering: At any time
Engineering graduates transferred from . . 4-yr: 3   2-yr: 35
Requirements: Evidence of high quality prior academic work; no transfer credit for grades less than C; Not more than 9 additional semester hours may be transferred after matriculation; ABET accreditation required for transfer of engineering courses.

GRADUATION REQUIREMENTS
BS programs in Engineering require 129-136 semester credit hours (depends on major) with a 2.0 over all GPA and a 2.0 minimum GPA in the major courses. A one year foreign language proficiency is required. Writing exams are administered to all incoming students and again as an exit exam at the end of 4 semesters to establish English writing proficiency.

■ STUDENT PROGRAMS

PROFESSIONAL AND HONORARY SOCIETIES
[For key to acronyms, see *Introduction*.]
Professional Societies: ACM, ASCE, ASME, IEEE, NSPE, SAE
Honorary Societies: Eta Kappa Nu, Pi Tau Sigma

SUPPORT PROGRAMS
Student Chapter Organizations: Society of Women Engineers
Other Engineering Support Programs: SOAR (Summer Orientation and Registration) for freshmen and transfers; Placement testing; Career counseling; International student office; Co-op program; Employment placement service.

070 Ferris State University

INSTITUTION PROFILE

HEAD OF THE INSTITUTION
Helen Popovich
Phone: (616)592-2500

GENERAL INFORMATION
[All Students—Fall 1991]
Undergraduate enrollment 12,100
Graduate enrollment 200
Total institution enrollment **12,300**
Type of institution: Public   **Calendar:** Quarters
Nearest city: Grand Rapids   **Population:** 300,000
Miles from main campus: 55   **Setting:** Small Town
Types of engineering degrees: Engineering
Other degree-granting colleges: Allied Health Sciences, Arts & Sciences, Business Administration, Education, Pharmacy, Technology & Applied Sciences, Optometry

ENGINEERING ADMISSIONS

ADMISSIONS OFFICE CONTACT
Duncan Sargent
Ferris State University
901 S. State Street
Big Rapids, MI 49307

ENGINEERING COLLEGE ADMISSIONS INFORMATION
Entrance Requirements: ACT (Score: 20), High School courses—English (3 years), Math (3 years), Physical Science (2 years), Social Science (1 year)
Entrance Recommendations: High School courses—Algebra/Trigonometry/Calculus (4 years), English/Social Science/etc. (4 years), Mechanical Drawing (1 year), Physics/Chemistry (2 years)
Requirements for foreign students: TOEFL (Score: 500); financial statement

DEPARTMENTAL ADMISSIONS INFORMATION
Admission to the engineering department: At the time of admission to the institution.
Additional information: Program admission requirements included high school graduate with a GPA 2.00.

FINANCIAL INFORMATION

ESTIMATED EXPENSES (ACADEMIC YEAR)
[Expenses are for the 1992-93 nine-month academic year.]

State Residents	Undergraduate	Graduate
Tuition and fees	$ 3,000	$ –
College room and board	$ 3,700	$ –
Books and supplies	$ 525	$ –
Other expenses	$ 700	$ –
Total estimated expenses	**$ 7,925**	**$ –**

Out-of-State Residents	Undergraduate	Graduate
Tuition and fees	$ 7,500	$ –
College room and board	$ 3,705	$ –
Books and supplies	$ 525	$ –
Other expenses	$ 700	$ –
Total estimated expenses	**$12,430**	**$ –**

FINANCIAL AID OFFICE CONTACT
Robert Bopp, *Director*
Ferris State University
901 S. State Street
Prakken Building
Big Rapids, MI 49307
Phone: (616)592-2110

GENERAL FINANCIAL AID INFORMATION
Forms accepted/required: FAF, Institutional
Additional financial aid information: Need-based scholarships and loans (local, state and federal), merit-based scholarships, part-time jobs on campus. College Work Study program.

ENGINEERING COLLEGE INFORMATION
[For additional personnel, refer to the *Appendix*.]

HEAD OF ENGINEERING
Joel Galloway
Phone: (616)592-2890   **Fax:** (616)592-2946

ENGINEERING COLLEGE ADDRESS
College of Technology
Ferris State University
901 S. State
Big Rapids, MI 49307
Phone: (616)592-2890   **Fax:** (616)592-2946

ENROLLMENTS—BY CLASS
[Numbers are baccalaureate enrollments for the fall 1991 term, unless otherwise footnoted.]
1st-year students/Freshmen 6
2nd-year students/Sophomores 13
3rd-year students/Juniors 15
4th-year students/Seniors 37
Total **71**

NUMBER OF DEGREES AWARDED—BY PROGRAM
[Numbers are engineering baccalaureate degrees awarded during the 1991-1992 academic year, unless otherwise footnoted. For full details about each engineering program, refer to the *Tables of Degree Programs*.]
Surveying Engineering 10
Total **10**

PERCENTAGE OF DEGREES AWARDED—BY CATEGORY
[Percentages are of all engineering baccalaureate degrees awarded during the 1991-1992 academic year, unless otherwise footnoted.]
African Americans .... – %(A)
Asian/Pacific Island Americans .... – %(A)
Hispanic Americans .... – %(A)
Native Americans .... – %(A)
Foreign Citizens .... – %(A)
All Others .... 100 %
Women .... – %(B)
Persons with disabilities .... – %(A)
Students over 25 years of age .... – %(A)
Notes: (A) Data not available. (B) Data not applicable.

ENGINEERING STUDENT DATA
Applicants to the engineering college
Number of applicants to engineering college 40
Percent offered admission 90.0%
Matriculated engineering students
Percentage in top quartile (25%) of High School class .... – (A)
Notes: (A) Data not available.

FULL- & PART-TIME FACULTY—BY DEPARTMENT
[Figures are the head count of full-time faculty and the full-time equivalent (FTE) of part-time faculty for each engineering department or equivalent.]

Department	Full	Part
Construction Department	28	–

COLLEGE DESCRIPTION
The University's engineering program is housed in the College of Technology. The program is application oriented with heavy emphasis on computer integration and laboratory experience. Faculty are typically hired from business/industry with appropriate professional license and academic preparation. The College places particular emphasis upon the student developing them to their fullest potential and assisting them in career placement. Average number of years required to actually complete the bachelor's degree: 4.1.

DEGREE PROGRAMS DESCRIPTION
The University's four-year degree program in engineering is a blend of theory and application with a strong general education support component.

TRANSFER INFORMATION

Transfer without Articulation Agreements
Admission to engineering: At any time
Engineering graduates transferred from ... 4-yr: 23   2-yr: 27
Requirements: 2.0 college average

GRADUATION REQUIREMENTS
Bachelor of Science in Surveying Engineering requires a minimum of 190 quarter credits, depending on major selected. Overall 2.00 GPA required, 2.00 GPA required in major.

SUPPORT PROGRAMS
Student Chapter Organizations: Women in Technology
Minority Affairs Office
For: Women & Ethnic Minorities   **Available:** Year round
Offered: Prematriculation, Freshman, Sophomore, Junior, Senior
Women in Technology
For: Women   **Available:** Year round
Offered: Prematriculation, Freshman, Sophomore, Junior, Senior
Counseling Services
For: Women & Ethnic Minorities   **Available:** Year round
Offered: Prematriculation, Freshman, Sophomore, Junior, Senior
Other Engineering Support Programs: Summer and Fall orientation program for all freshmen and transfers; Learning Assessment Program; career counseling services, foreign student office, veterans' services, services for blind and handicapped, placement services, academic advising by faculty, educational counseling by counselors.

071 University of Florida

■ INSTITUTION PROFILE

HEAD OF THE INSTITUTION
John V. Lombardi
Phone: (904)392-1311 Fax: (904)392-9506

GENERAL INFORMATION
[All Students—Fall 1991]
Undergraduate enrollment 26,860
Graduate enrollment 7,954
Total institution enrollment 34,814
Type of institution: Public Calendar: Semesters
Location: Gainesville, FL Population: 200,000
Setting: Suburban
Types of engineering degrees: Engineering
Other degree-granting colleges: Agricultural & Environmental, Architecture, Arts & Sciences, Business Administration, Communication & Journalism, Dentistry, Education, Fine & Performing Arts, Law, Medicine, Nursing, Pharmacy, Technology & Applied Sciences, Health & Human Performance, Health Related Professions, Veterinary Medicine

■ ENGINEERING ADMISSIONS

ADMISSIONS OFFICE CONTACT
Barbara T. Fincher
University of Florida
University Registrar
S-222 Criser Hall
Gainesville, FL 32611-2083
Phone: (904)392-1374 Fax: (904)392-3987

ENGINEERING COLLEGE ADMISSIONS INFORMATION
Entrance Requirements: SAT (Score: 840), ACT (Score: 19), High School courses—English (4 years), Math (3 years), Natural Science/Social Science (6 years), Foreign Language (2 years)
Entrance Recommendations: High School courses—Physics (1 year), Chemistry (1 year), Trigonometry, Additional Math
Requirements for out-of-state residents: Entrance requirements vary depending on term & size of applicant pool. Out of state students must meet this minimum index and often need higher requirements since out-of-state beginners are limited to 10% of the class.
Requirements for foreign students: TOEFL (Score: 550); financial statement

DEPARTMENTAL ADMISSIONS INFORMATION
Admission to the engineering department: At the end of the second year.
Additional information: The college has a selective admission policy which is administered through the departments. Specific requirements of this policy vary with programs. Admission advising is available and applicants denied admission to one program may be admissible to another program.

■ FINANCIAL INFORMATION

ESTIMATED EXPENSES (ACADEMIC YEAR)
[Expenses are for the 1992-93 nine-month academic year.]

State Residents	Undergraduate	Graduate
Tuition and fees	$ 1,650	$ 2,530
College room and board	$ 3,950	$ 4,660
Books and supplies	$ 560	$ 560
Other expenses	$ 1,490	$ 1,490
Total estimated expenses	**$ 7,650**	**$ 9,240**

Out-of-State Residents	Undergraduate	Graduate
Tuition and fees	$ 6,410	$ 8,600
College room and board	$ 3,950	$ 4,660
Books and supplies	$ 560	$ 560
Other expenses	$ 1,490	$ 1,490
Total estimated expenses	**$12,410**	**$15,310**

FINANCIAL AID OFFICE CONTACT
Karen L. Fooks, *Director of Financial Aid*
University of Florida
P-113C Peabody Hall
Gainesville, FL 32611-2058
Phone: (904)392-1275 Fax: (904)392-2861

GENERAL FINANCIAL AID INFORMATION
Forms accepted/required: CSS-FAF, FAF, Stafford, Florida Financial Aid Form (FFAF)
Additional financial aid information: All federal and state grant and loan programs; merit and/or need based scholarships; short-term loans; College Work-Study Program. Part-time jobs on campus.

■ ENGINEERING COLLEGE INFORMATION

[For additional personnel, refer to the *Appendix*.]

HEAD OF ENGINEERING
Winfred M. Phillips
Phone: (904)392-6000 Fax: (904)392-9673
EMail: wphil@engnet.ufl.edu

ENGINEERING COLLEGE ADDRESS
College of Engineering
University of Florida
300 Weil Hall
Gainesville, FL 32611-2083
Phone: (904)392-6000 Fax: (904)392-9673

ENROLLMENTS—BY CLASS
[Numbers are baccalaureate enrollments for the fall 1991 term, unless otherwise footnoted.]
1st-year students/Freshmen 1,230
2nd-year students/Sophomores 1,180
3rd-year students/Juniors 1,123
4th-year students/Seniors 1,025
Total 4,558

NUMBER OF DEGREES AWARDED—BY PROGRAM
[Numbers are engineering baccalaureate degrees awarded during the 1991-1992 academic year, unless otherwise footnoted. For full details about each engineering program, refer to the *Tables of Degree Programs*.]
Aerospace Engineering 46
Agricultural Engineering 6
Chemical Engineering 31
Civil Engineering 69
Coastal & Oceanographic .... –
Computer & Information Science 47
Electrical Engineering 99
Engineering Science & Mechanics 11
Environmental Engineering 31
Industrial & Systems Engineering 41
Interdisciplinary Engineering 1
Materials Science & Engineering 37
Mechanical Engineering 80
Nuclear Engineering 16
Nuclear Engineering Sciences 4
Surveying & Mapping 11
Total 530

PERCENTAGE OF DEGREES AWARDED—BY CATEGORY
[Percentages are of all engineering baccalaureate degrees awarded during the 1991-1992 academic year, unless otherwise footnoted.]
African Americans 4.3%
Asian/Pacific Island Americans 6.4%
Hispanic Americans 9.2%
Native Americans .... –%
Foreign Citizens 5.1%
All Others 75.0%
Women 14.4%
Persons with disabilities 0.9%
Students over 25 years of age .... –% (A)

Notes: (A) Data not available.

GRADUATE ENROLLMENTS & DEGREES AWARDED
Master's enrollment 1,047
Master's degrees awarded 302
Doctoral enrollment 410
Doctoral degrees awarded 84

ENGINEERING STUDENT DATA
Applicants to the engineering college
Number of applicants to engineering college .... – (A)
Percent offered admission .... –% (A)

Matriculated engineering students
Percentage in top quartile (25%) of High School class ..... – (1)
Average SAT scores: Math—640, Verbal—520, Combined—1160
Notes: (A) Data not available. (1) Average High School GPA was 3.5/4.0

FULL- & PART-TIME FACULTY—BY DEPARTMENT
[Figures are the head count of full-time faculty and the full-time equivalent (FTE) of part-time faculty for each engineering department or equivalent.]

Department	Full	Part
Chemical Eng	15	1.5
Civil Eng	31	1.0
Electrical Eng	46	–
Industrial & Systems Eng	12	1.3
Mechanical Eng	30	1.5
Aerospace Eng, Mechanics & Eng Science	26	3.0
Agricultural Eng	33	–
Coastal & Oceanographic Eng	9	–
Computer & Information Sciences	27	3.2
Environmental Eng Sciences	17	–
Materials Science & Eng	24	0.5
Nuclear Eng Sciences	10	1.5

COLLEGE DESCRIPTION
The College admits students who have completed or are in the process of completing the pre-engineering curriculum at this University, at community or junior colleges or transferring from other universities. Upper-level coursework takes 2-3 years to complete in any of 17 programs given by 11 undergraduate departments. (Co-op opportunities are available). Extracurricular activities are encouraged within 38 student technical & honor societies. Dual-degree programs & academic minor programs may be arranged. Students pay tuition & fees for a course load suited to individual capabilities and desires. Graduating seniors are encouraged to take the engineer intern exam as a first step toward registration as a professional engineer. Avg # of years required to actually complete the bachelor's degree: 4.5

DEGREE PROGRAMS DESCRIPTION
All programs offer the B.S., M.S., M.E., Engr., and Ph.D. degree. Under Materials Science & Eng are offered ceramic, metals, electronic materials, & polymers specialties. Agricultural Eng offers agrisystems, soil & water, & food & bioprocess engineering. Industrial & Systems Eng offers industrial & systems specialties. Industrial & Systems Eng also offers a concurrent MBA program. A program for a minor in env'l studies (with certificate) is available for students from six programs. Life sciences, biomedical eng and pre-med may be arranged. Coastal & Oceanographic Eng offers graduate degrees only.

DUAL DEGREE PROGRAMS
Engineering baccalaureate graduates with dual degrees 1
Enrollment requirements: Depends on major

TRANSFER INFORMATION
Residency Requirements: Last 30 hours must be in residence.

Transfer via Articulation Agreements
Admission to engineering: transferred from 4 year estimated at 1%; from 2 year at 45%. No of BS EG Graduate transfers from other institutions are estimated.
Engineering graduates transferred from .. 4-yr: 32 2-yr: 178
Requirements: Depends on major, but in general, should have completed pre-engineering math, physics and chemistry

Transfer without Articulation Agreements
Admission to engineering: 4% estimated from 4 year institution, 5% estimated from 2 year institution.
Engineering graduates transferred from .. 4-yr: 59 2-yr: –
Requirements: Depends on major, but should have completed pre-engineering requirements, including math, physics and chemistry

GRADUATION REQUIREMENTS
B.S. programs in engineering require 128-143 credit hours, depending on major selected. Cumulative and upper-level grade point average must be 2.0/greater. Other requirements are detailed in the Undergraduate Catalog.

■ STUDENT PROGRAMS

PROFESSIONAL AND HONORARY SOCIETIES
[For key to acronyms, see *Introduction*.]
Professional Societies: ACM, AIAA, AIChE, ANS, ASAE, ASCE, ASME, IEEE, IIE, NSPE, The American Ceramic Society (ACerS), Society of Environmental Engineers (SEE)
Honorary Societies: Alpha Epsilon, Alpha Pi Mu, Eta Kappa Nu, Keramos, Pi Tau Sigma, Sigma Gamma Tau, Tau Beta Pi, Alpha Nu Sigma, Upsilon Pi Epsilon

SUPPORT PROGRAMS
Student Chapter Organizations: Society of Women Engineers, National Society of Black Engineers, Society of Hispanic Professional Engineers

Minority Affairs Office
For: Women & Ethnic Minorities **Available:** Year round
Offered: Prematriculation, Freshman, Sophomore, Junior, Senior

Tutoring
For: Women & Ethnic Minorities **Available:** Year round
Offered: Prematriculation, Freshman, Sophomore, Junior, Senior

Graduate Minority Programs
For: Ethnic Minorities **Available:** Year round
Offered: Graduate level

Personal Career Counseling
For: Women & Ethnic Minorities **Available:** Year round
Offered: Prematriculation, Freshman, Sophomore, Junior, Senior, Graduate level

Other Engineering Support Programs: Summer, Fall, Spring, orientation program for all freshmen & transfers; academic & career counseling; international student office; services for handicapped; co-operative education opportunities; legal services; tutoring services; honors program for qualifying students, veterans' services; career resource center and placement services; career expos; engineers fair.

072 Florida Agricultural and Mechanical University

INSTITUTION PROFILE

HEAD OF THE INSTITUTION
Frederick S. Humphries
Phone: (904)599-3223 Fax: (904)561-2152

GENERAL INFORMATION
[All Students—Fall 1991]
Undergraduate enrollment 8,308
Graduate enrollment 369
Total institution enrollment **9,243**
Type of institution: Public Calendar: Semesters
Location: Tallahassee Population: 130,000
Setting: Urban
Types of engineering degrees: Engineering & Technology
Other degree-granting colleges: Allied Health Sciences, Architecture, Arts & Sciences, Business Administration, Education, Nursing, Pharmacy, Technology & Applied Sciences, Journalism, Media, and Graphic Arts, General Studies, Graduate Studies, Research and Continuing Ed.

ENGINEERING ADMISSIONS

ADMISSIONS OFFICE CONTACT
Barbara Cox
Florida Agricultural and Mechanical University
Foote Hilyer Administration Center
Tallahassee, FL 32307
Phone: (904)599-3797 Fax: (904)561-2248

ENGINEERING COLLEGE ADMISSIONS INFORMATION
Entrance Requirements: High School courses—English (4 years), Math, Natural Science, Social Science (3 years), Foreign Language (2 years), Additional Academic Electives (4 years), Standard test scores are not required for applicants with a 3.0 (on a 4.0 scale) in the required academic units in high school as well as a satisfactory high school record.
Entrance Recommendations: SAT (Score: 900), ACT (Score: 19), High School courses—Math (4 years), Chemistry, Physics, Computer Programming (1 year), Applicants with less than a High School GPA of 3.0 must have a 2.5 and appropriate standard test score. Sliding scale is used. Exceptions are made by the University committee.
Requirements for foreign students: TOEFL (Score: 500)

DEPARTMENTAL ADMISSIONS INFORMATION
Admission to the engineering department: At the time of admission to the institution.

FINANCIAL INFORMATION

ESTIMATED EXPENSES (ACADEMIC YEAR)
[Expenses are for the 1992-93 nine-month academic year.]

State Residents	Undergraduate	Graduate
Tuition and fees	$ 1,343 [1]	$ 2,552 [1]
College room and board	$ 3,880 [2]	$ 3,880 [2]
Books and supplies	$ 650	$ 700
Other expenses	$ —	$ —
Total estimated expenses	**$ 5,873**	**$ 7,132**

Out-of-State Residents	Undergraduate	Graduate
Tuition and fees	$ 5,152 [1]	$ 8,625 [1]
College room and board	$ 3,880 [2]	$ 3,880 [2]
Books and supplies	$ 650	$ 700
Other expenses	$ —	$ —
Total estimated expenses	**$ 9,682**	**$ 13,205**

Notes: (1) Based on twelve (12) credits per semester for two semester.
(2) Based on most expensive room rate and meal plan.

FINANCIAL AID OFFICE CONTACT
Alton Royal, *Director of Financial Aid*
Florida Agricultural and Mechanical University
Office of Financial Aid
Foote Hilyer Administration Center
Tallahassee, FL 32307
Phone: (904)599-3730 Fax: (904)599-3952

GENERAL FINANCIAL AID INFORMATION
Forms accepted/required: ACT
Additional financial aid information: Pell Grant, Supplemental Educational Opportunity Grant, College Work Study, Perkins Loan, Guaranteed Student Loan, Florida Student Assistant Grant and special scholarships. Some scholarships are awarded by the College of Engineering on a competitive basis.

ENGINEERING COLLEGE INFORMATION
[For additional personnel, refer to the *Appendix*.]

HEAD OF ENGINEERING
Ching-Jen Chen
Phone: (904)487-6100 Fax: (904)487-6486
EMail: CJChen@EVAX.ENG.FSU.EDU

ENGINEERING COLLEGE ADDRESS
FAMU/FSU College of Engineering
Florida Agricultural and Mechanical University
2525 Pottsdamer Rd.
P.O. Box 2175
Tallahassee, FL 32316-2175
Phone: (904)487-6100 Fax: (904)487-6486

ENROLLMENTS—BY CLASS
[Numbers are baccalaureate enrollments for the fall 1991 term, unless otherwise footnoted.]
1st-year students/Freshmen .... 485 [1]
2nd-year students/Sophomores .... 285 [1]
3rd-year students/Juniors .... 448 [1]
4th-year students/Seniors .... 543 [1]
Total **1,761**

Notes: (1) Numbers include both Florida Agricultural and Mechanical University and Florida State University students.

NUMBER OF DEGREES AWARDED—BY PROGRAM
[Numbers are engineering baccalaureate degrees awarded during the 1991-1992 academic year, unless otherwise footnoted. For full details about each engineering program, refer to the *Tables of Degree Programs*.]
Chemical Engineering .... 14 [1]
Civil Engineering .... 28 [1]
Electrical Engineering .... 79 [1]
Industrial Engineering .... 8 [1]
Mechanical Engineering .... 52 [1]
Total **181**

Notes: (1) Includes FAMU and FSU students. Year is Fall 1991, Spring 1992 and Summer 1992.

PERCENTAGE OF DEGREES AWARDED—BY CATEGORY
[Percentages are of all engineering baccalaureate degrees awarded during the 1991-1992 academic year, unless otherwise footnoted.]
African Americans .... 17.7 % [1]
Asian/Pacific Island Americans .... 6.6 % [1]
Hispanic Americans .... 5.0 % [1]
Native Americans .... — % [1]
Foreign Citizens .... — % [A]
All Others .... 70.7 %
Women .... 10.5 % [1]
Persons with disabilities .... — % [A]
Students over 25 years of age .... 30.4 % [1]

Notes: (A) Data not available. (1) Numbers include both Florida Agricultural and Mechanical University and Florida State University students.

GRADUATE ENROLLMENTS & DEGREES AWARDED
Master's enrollment .... 126 [1]
Master's degrees awarded .... 49 [2]
Doctoral enrollment .... 19 [1]
Doctoral degrees awarded .... — [2]

Notes: (1) Numbers include both Florida Agricultural and Mechanical University and Florida State University students. (2) Numbers include both FAMU and FSU students. Number is for Fall 1991, Spring 1992, and Summer 1992.

ENGINEERING STUDENT DATA

Applicants to the engineering college
Number of applicants to engineering college .... — [A]
Percent offered admission .... —% [A]

Matriculated engineering students
Percentage in top quartile (25%) of High School class .... — [A]
Average SAT scores: Combined—914
Average ACT scores: Composite—19.8

Notes: (A) Data not available.

FULL- & PART-TIME FACULTY—BY DEPARTMENT
[Figures are the head count of full-time faculty and the full-time equivalent (FTE) of part-time faculty for each engineering department or equivalent.]

Department	Full	Part
Chemical Eng	7	—
Civil Eng	8	2.5
Electrical Eng	14	—
Industrial Eng	7	—
Mechanical Eng	15	1.0

COLLEGE DESCRIPTION
The FAMU/FSU College of Engineering is a joint program of The Florida Agricultural and Mechanical University (FAMU) and The Florida State University (FSU). Available courses of study lead to the Bachelor and to the Master of Science degree in Chemical, Civil, Electrical, Industrial, and Mechanical Engineering, and to the PhD. in Chemical and Mechanical Engineering. All of the programs are ABET accredited. The primary goals of the FAMU/FSU College of Engineering are to educate engineers of excellence at both the undergraduate and graduate levels, judged by the highest standards in the field and recognized by national peers; to attract and produce greater number of blacks, women and minorities in professional engineering, engineering teaching and research; and to attain national and international recognition of the College through the educational achievements of its faculty and students.

DEGREE PROGRAMS DESCRIPTION
In fall 1992 the College adopted the following admission and retention requirements. 1. Students must achieve a GPA of 2.5 or better in Calculus I, II, Physics I and Chemistry I prior to enrolling in any 2000 level or above engineering course for which any of these courses are prerequisites. A maximum of one repeat of each course is allowed in meeting this requirement. 2. Any student who fails to earn a grade of "C" or better in an engineering course on the third attempt, or withdraws from the course more than twice, is subject to dismissal. 3. Any student who exceeds 30 credit hours of repeated course work is subject to dismissal. 4. Majors must earn a grade of "C" or better in all engineering courses which apply to the degree. This requirement may be waived by the academic dean upon recommendation of the department chair for no more than two engineering courses. 5. Normal policy in the College is to allow two reinstatements. A third reinstatement may be approved in exceptional circumstances. Any student who fails to comply with the agreement approved by the department chair following reinstatement or fails to earn a cumulative GPA of 2.0 or better upon completion of the term is subject to dismissal.

TRANSFER INFORMATION
Residency Requirements: The University requires at least two semesters of residence for any undergraduate degree. If the term of residence is only two semesters, that period must be the student's senior year, provided at least 30 semester hours are earned at FAMU during this period. On the graduate level, a maximum of six credits may be transferred from another institution.

Transfer via Articulation Agreements
Admission to engineering: At any time

Transfer without Articulation Agreements
Admission to engineering: At any time

GRADUATION REQUIREMENTS
Undergraduates: COE: 132-145 credits, complete ABET requirements in Humanities, Social Sciences, sit for FE exam. FAMU: one summer attended, residency requirement met, 'C' average in degree work, maximum of 25% credits by correspondence, etc., complete general education courses, pass Florida College Level Academic Skills Test, Dean's approval. MS: 30-33 credits, 7 year limit, 3.0 GPA each semester, only 6 credits can transfer from other accredited programs, 12 from FAMU. PhD: 45 credits beyond the MS, 3.0 GPA each semester, 5 year limit, transfer credit rule as above.

STUDENT PROGRAMS

PROFESSIONAL AND HONORARY SOCIETIES
[For key to acronyms, see *Introduction*.]
Professional Societies: AIAA, AIChE, ASCE, ASME, IEEE, IIE, SAE, National Society of Black Engineers, Society of Women Engineers
Honorary Societies: Tau Beta Pi

SUPPORT PROGRAMS
Student Chapter Organizations: Society of Women Engineers, National Society of Black Engineers

Engineering Concepts Institute
For: Ethnic Minorities **Available:** Summer only
Offered: Prematriculation

Minority Introduction to Engineering
For: Ethnic Minorities **Available:** Summer only
Offered: Prematriculation

Minority Engineering Retention Program
For: Ethnic Minorities **Available:** Academic year
Offered: Freshman, Sophomore

Other Engineering Support Programs: Summer, Fall, and Spring orientation, academic advising, retention, tutoring, and career counseling programs are utilized to provide support for freshmen, transfer and minority students. Career counseling and placement programs are available for upper classmen. Co-op and internships are also available to qualified students.

073 Florida Atlantic University

INSTITUTION PROFILE

HEAD OF THE INSTITUTION
Anthony J. Catanese
Phone: (407)367-3450   Fax: (407)367-2777

GENERAL INFORMATION
[All Students—Fall 1991]
Undergraduate enrollment 11,401
Graduate enrollment 3,267
Total institution enrollment 14,668
Type of institution: Public   **Calendar:** Semesters
Nearest city: Ft. Lauderdale   **Population:** 172,000
Miles from main campus: 25   **Setting:** Suburban
Types of engineering degrees: Engineering
Other degree-granting colleges: Business Administration, Education, Nursing, Science, Social Science, Arts and Humanities, Liberal Arts, Public Administration

ENGINEERING ADMISSIONS

ADMISSIONS OFFICE CONTACT
Brian Levin-Stankevich
Florida Atlantic University
Director of Admissions
Boca Raton, FL 33431-0991
Phone: (407)367-3040   Fax: (407)367-2758

ENGINEERING COLLEGE ADMISSIONS INFORMATION
Entrance Requirements: High School courses—English (4 years), Mathematics, including trigonometry (3 years), Natural Science, including chemistry (3 years), Social Science (3 years), Minimum qualifications are based on the State University System freshman eligibility scales. Lower GPA requires higher SAT/ACT scores. For example, a 2.0 GPA requires 1050 SAT, while a 2.9 GPA requires 860 SAT.
Entrance Recommendations: High School courses—Foreign Language (2 years), Physics (1 year), College Algebra (1 year), Either SAT or ACT may be taken.
Requirements for foreign students: TOEFL (Score: 550); financial statement; Health insurance is required of all foreign students. Information is available from the Director of Admissions.

DEPARTMENTAL ADMISSIONS INFORMATION
Admission to the engineering department: when program admission requirements are satisfied.
Additional information: Freshmen are admitted directly to departments. Transfer students must have completed calculus and physics with calculus sequences with combined GPAs as follows: EE--2.5/4.0; ME and OE--2.25/4.0 with all attempts counted; CS--2.0/4.0. Students with deficiencies in these areas may be conditionally admitted to the college and admitted to the respective program after removing the deficiencies.

FINANCIAL INFORMATION

ESTIMATED EXPENSES (ACADEMIC YEAR)
[Expenses are for the 1992-93 nine-month academic year.]

State Residents	Undergraduate	Graduate
Tuition and fees	$ 1,400	$ 1,950
College room and board	$ 3,910	$ 3,910
Books and supplies	$ 600	$ 600
Other expenses	$ 2,100 [1]	$ 2,100 [1]
Total estimated expenses	**$ 8,010**	**$ 8,560**

Out-of-State Residents	Undergraduate	Graduate
Tuition and fees	$ 5,200	$ 6,400
College room and board	$ 3,910	$ 3,910
Books and supplies	$ 600	$ 600
Other expenses	$ 2,100 [1]	$ 2,100 [1]
Total estimated expenses	**$11,810**	**$13,010**

Notes: (1) Transportation and personal expenses

FINANCIAL AID OFFICE CONTACT
Olga V. Moas, Director of Student Financial Aid
Florida Atlantic University
500 N.W. 20th Street
Boca Raton, FL 33431-0991
Phone: (407)367-3530   Fax: (407)367-2740

GENERAL FINANCIAL AID INFORMATION
Forms accepted/required: ACT, AFSA, CSS-FAF, FAF, FAT, FAR, FSA, Stafford, IRS, SAR, Supplemental, Institutional
Additional financial aid information: A wide variety of scholarships and loans are available, including student employment opportunities. Scholarship and loan information is available from Student Financial Aid. Student employment and individual departments have information on campus and off-campus employment for students.

ENGINEERING COLLEGE INFORMATION
[For additional personnel, refer to the *Appendix*.]

HEAD OF ENGINEERING
Craig S. Hartley
Phone: (407)367-3400   Fax: (407)367-2659
EMail: HARTLEY@ACC.FAU.EDU

ENGINEERING COLLEGE ADDRESS
College of Engineering
Florida Atlantic University
500 N.W. 20th Street
Boca Raton, FL 33431-0991
Phone: (407)367-3400   Fax: (407)367-2659

ENROLLMENTS—BY CLASS
[Numbers are baccalaureate enrollments for the fall 1991 term, unless otherwise footnoted.]
1st-year students/Freshmen .... 145 [1]
2nd-year students/Sophomores .... 82 [1]
3rd-year students/Juniors .... 242 [1]
4th-year students/Seniors .... 507 [2]
Total **976**

Notes: (1) Includes Computer Science (2) Includes Computer Science and Second Bachelor's Students

NUMBER OF DEGREES AWARDED—BY PROGRAM
[Numbers are engineering baccalaureate degrees awarded during the 1991-1992 academic year, unless otherwise footnoted. For full details about each engineering program, refer to the *Tables of Degree Programs*.]
Civil Engineering .... –
Computer Engineering .... – (3)
Computer Science 89
Electrical Engineering 61
Manufacturing Systems Engineering .... – (2)
Mechanical Engineering 25
Ocean Engineering 16
Total **191**

Notes: (2) No undergraduate programs in civil or manufacturing systems engineering. (3) No bachelor's program. Master's and Ph.D. programs only.

PERCENTAGE OF DEGREES AWARDED—BY CATEGORY
[Percentages are of all engineering baccalaureate degrees awarded during the 1991-1992 academic year, unless otherwise footnoted.]
African Americans .... 6.8 %
Asian/Pacific Island Americans .... 7.3 %
Hispanic Americans .... 10.5 %
Native Americans .... – %
Foreign Citizens .... 12.0 %
All Others .... 63.4 %
Women .... 25.7 %
Persons with disabilities .... 1.0 % (C)
Students over 25 years of age .... 10.0 % (C)

Notes: (C) Estimated data.

GRADUATE ENROLLMENTS & DEGREES AWARDED
Master's enrollment .... 315 [4]
Master's degrees awarded .... 64 [1]
Doctoral enrollment .... 99 [1]
Doctoral degrees awarded 14

Notes: (1) Includes Computer Science (4) Listed number does not include 229 non-degree students enrolled in graduate courses.

ENGINEERING STUDENT DATA
Applicants to the engineering college
Number of applicants to engineering college 302
Percent offered admission 62.3%

Matriculated engineering students
Percentage in top quartile (25%) of High School class – (A)
Average SAT scores: Math—578, Verbal—460, Combined—1038
Average ACT scores: Math—24, Composite—22

Notes: (A) Data not available.

FULL- & PART-TIME FACULTY—BY DEPARTMENT
[Figures are the head count of full-time faculty and the full-time equivalent (FTE) of part-time faculty for each engineering department or equivalent.]

Department	Full	Part
Electrical Eng	16	1.0
Mechanical Eng	19	–
Ocean Eng	18	–
Computer Science & Eng	21	–

COLLEGE DESCRIPTION
Undergraduate engineering and computer science programs at Florida Atlantic University take pride in the personal attention offered to all students. Beginning with the initial orientation and individualized advising session, all students are required to see a faculty advisor prior to registration each term. Faculty take pride in their availability for coursework-related inquiry and experienced faculty teach most entry level courses. All programs have co-op options. All programs have team-oriented senior design projects which emphasize practical application of theoretical knowledge. Unique laboratory opportunities exist in the areas of electromagnetic interference, corrosion, high resolution imaging, robotics, advanced underwater systems, acoustics, chirp sonar, software engineering, materials science and computer aided design.

DEGREE PROGRAMS DESCRIPTION
Bachelors--All College of Engineering bachelors programs offer flexibility in choice of electives to enable broad or specialized programs. Emphasis is placed on team projects, communications skills and engineering ethics throughout the programs. Masters--Programs are very flexible to enable broad or specialized thesis or non-thesis work with interdisciplinary work encouraged. Many courses are offered at remote sites via live, interactive TV and videotape. Ph.D.--Programs are very flexible to enable broad or specialized dissertation work with interdisciplinary work encouraged. Many courses are offered at remote sites via live, interactive TV and videotape. One year of full-time campus residency is required.

DUAL DEGREE PROGRAMS
Engineering baccalaureate graduates with dual degrees 0
Enrollment requirements: Entrance requirements for dual degree programs are the same as for other transfer students.

TRANSFER INFORMATION

Transfer via Articulation Agreements
Admission to engineering: Normally, transfers occur at the end of the sophomore year. Students with 2.0 or higher GPA who meet freshman admission requirements may transfer earlier.
Engineering graduates transferred from .. 4-yr: 25 2-yr: 100
Requirements: Transfer students must have completed calculus and physics with calculus sequences with combined GPAs as follows: EE--2.5/4.0; ME and OE--2.25/4.0 with all attempts counted; CS--2.0/4.0. Students may be conditionally admitted and must satisfy deficiencies within 24 semester hours.

Transfer without Articulation Agreements
Admission to engineering: Normally, transfers occur at the end of the sophomore year. Students may transfer earlier, provided that they have a minimum of 2.0 GPA and meet freshmen admission requirements.
Engineering graduates transferred from .. 4-yr: 5 2-yr: 20
Requirements: Transfer students must have completed calculus and physics with calculus sequences with combined GPAs as follows: EE--2.5/4.0; ME and OE--2.25/4.0 with all attempts counted; CS--2.0/4.0. Conditional admission may be offered with the provision that deficiencies are removed within the first 24 semester hours.

GRADUATION REQUIREMENTS
All programs require an overall 2.0 GPA for graduation. In addition, program requirements are as follows: CS--2.5 GPA in all CS core courses. EE--2.0 in all EE courses with a C or better in each EE core course. ME--C or better in all engineering mechanics, materials science and ME core courses. OE--2.0 GPA in all engineering and science courses attempted at FAU.

STUDENT PROGRAMS

PROFESSIONAL AND HONORARY SOCIETIES
[For key to acronyms, see *Introduction*.]
Professional Societies: ACM, ASME, IEEE, NSPE, SAE, SNAME, Marine Technological Society, National Society of Black Engineers
Honorary Societies: Tau Beta Pi, Upsilon Pi Epsilon

SUPPORT PROGRAMS
Student Chapter Organizations: Society of Women Engineers, National Society of Black Engineers, Florida Atlantic Minorities in Engineering

Minority Student Services
For: Ethnic Minorities   **Available:** Year round
Offered: Prematriculation, Freshman, Sophomore, Junior, Senior, Graduate level

Southeastern Consortium for Minorities in Eng
For: Women & Ethnic Minorities   **Available:** Year round
Offered: Prematriculation

Other Engineering Support Programs: Orientation programs for freshmen and transfer students. Individual academic advising by faculty advisors. Career counseling center.

074 Florida Institute of Technology

INSTITUTION PROFILE
HEAD OF THE INSTITUTION
Lynn E. Weaver
Phone: (407)768-8000 **Fax:** (407)984-8461
GENERAL INFORMATION
[All Students—Fall 1991]
Undergraduate enrollment 2,457
Graduate enrollment 3,369
Total institution enrollment **5,826**
Type of institution: Private **Calendar:** Semesters
Nearest city: Melbourne **Population:** 60,500
Miles from main campus: 1 **Setting:** Urban
Types of engineering degrees: Engineering
Other degree-granting colleges: Aeronautics, Science and Liberal Arts, Psychology, Business

ENGINEERING ADMISSIONS
ADMISSIONS OFFICE CONTACT
Louis T. Levy
Florida Institute of Technology
150 West University Boulevard
Melbourne, FL 32901-6988
Phone: (407)768-8000 **Fax:** (407)723-9468
ENGINEERING COLLEGE ADMISSIONS INFORMATION
Entrance Requirements: SAT (Score: 1000), High School courses—Mathematics (4 years), Chemistry (1 year), English (4 years)
Entrance Recommendations: High School courses—Physics (1 year), Biology (1 year), Additional Mathematics, Including Calculus (2 years), Foreign Language (2 years)
Requirements for foreign students: TOEFL (Score: 550); financial statement
DEPARTMENTAL ADMISSIONS INFORMATION
Admission to the engineering department: At the time of admission to the institution.

FINANCIAL INFORMATION
ESTIMATED EXPENSES (ACADEMIC YEAR)
[Expenses are for the 1992-93 nine-month academic year.]

All Students	Undergraduate	Graduate
Tuition and fees	$11,817	$ 7,182
College room and board	$ 3,723	$ 3,723
Books and supplies	$ 750	$ 750
Other expenses	$ 2,592 [1]	$ 2,592 [1]
Total estimated expenses	**$18,882**	**$14,247**

Notes: (1) Personal expenses and transportation
FINANCIAL AID OFFICE CONTACT
Leonard E. Gude, *Director of Financial Aid & Scholarship*
Florida Institute of Technology
150 W. University Boulevard
Melbourne, FL 32901-6988
Phone: (407)768-8000 **Fax:** (407)984-8461
GENERAL FINANCIAL AID INFORMATION
Forms accepted/required: FAF, Stafford, SAR, SDF
Additional financial aid information: Florida Institute of Technology Grant Program, College Roll Employment available, in addition to all of the standard Grant, Loan, and Work-Study programs normally available. Additionally, Florida residents are eligible for the Florida Tuition Voucher Program and may be eligible for the Florida Tuition Reduction Grant Program.

ENGINEERING COLLEGE INFORMATION
[For additional personnel, refer to the *Appendix*.]
HEAD OF ENGINEERING
Robert L. Sullivan
Phone: (407)768-8000 **Fax:** (407)984-8461
ENGINEERING COLLEGE ADDRESS
College of Engineering
Florida Institute of Technology
150 West University Boulevard
Melbourne, FL 32901-6988
Phone: (407)768-8000 **Fax:** (407)984-8461
ENROLLMENTS—BY CLASS
[Numbers are baccalaureate enrollments for the fall 1991 term, unless otherwise footnoted.]
1st-year students/Freshmen 285
2nd-year students/Sophomores 305
3rd-year students/Juniors 219
4th-year students/Seniors 233
Total **1,042**
NUMBER OF DEGREES AWARDED—BY PROGRAM
[Numbers are engineering baccalaureate degrees awarded during the 1991-1992 academic year, unless otherwise footnoted. For full details about each engineering program, refer to the *Tables of Degree Programs*.]
Chemical Engineering 8
Civil & Environmental Engineering 16
Electrical & Computer Engineering 111
Mechanical & Aerospace Engineering 40
Ocean Engineering 22
Total **197**
PERCENTAGE OF DEGREES AWARDED—BY CATEGORY
[Percentages are of all engineering baccalaureate degrees awarded during the 1991-1992 academic year, unless otherwise footnoted.]
African Americans 2.2%
Asian/Pacific Island Americans 2.7%
Hispanic Americans 7.6%
Native Americans .... – %
Foreign Citizens 23.9%
All Others 63.6%
Women 15.9%
Persons with disabilities .... – %
Students over 25 years of age 15.0%
GRADUATE ENROLLMENTS & DEGREES AWARDED
Master's enrollment 279
Master's degrees awarded 94
Doctoral enrollment 49
Doctoral degrees awarded 7
ENGINEERING STUDENT DATA
Applicants to the engineering college
Number of applicants to engineering college .... 873 [1]
Percent offered admission .... 85.5% [1]
Matriculated engineering students
Percentage in top quartile (25%) of High School class 77.6
Average SAT scores: Math—599.7, Verbal—484.3, Combined—1084.0

Notes: (1) Includes General Engineering Non-degree Program
FULL- & PART-TIME FACULTY—BY DEPARTMENT
[Figures are the head count of full-time faculty and the full-time equivalent (FTE) of part-time faculty for each engineering department or equivalent.]

Department	Full	Part
Chemical Eng	5	–
Civil & Environmental Eng	6	0.8
Electrical & Computer Eng	15	2.6
Mechanical & Aerospace Eng	12	1.9
Oceanography/Ocean Eng & Env'l Science	6	–

COLLEGE DESCRIPTION
The College of Engineering consists of 7 departments offering 4-year undergraduate degree programs in 7 fields of engineering, & Applied Math, Computer Science, Environmental Science & Oceanography. Graduate programs through the doctoral level are offered in most of these areas, & in Engineering Management and Operations Research. In addition, the college maintains several research centers and institutes, including: the Center for Computational Fluid Dynamics, Center for Research on Energy Alternatives, Research Center for Waste Utilization, Center for Electronics Manufacturability, and the Center for the Enhancement of Quality in Engineering Education. These centers serve to encourage & focus collaborative research activities of faculty and students from many different departments within the college, working in the indicated multidisciplinary areas.

DEGREE PROGRAMS DESCRIPTION
B.S. degrees are offered in the eleven fields indicated above, with options available in Computer Science (Business and Scientific). M.S. programs are offered in all of the same fields with options available in Civil Engineering (Construction, Geotechnical, Structures, & Water Resources), Electrical Engineering (Electromagnetics, Microelectronics & Electrooptics, and Systems & Information Processing), Mechanical and Aerospace Engineering (Fluid Dynamics, Heat Transfer & Energy, Combustion & Propulsion, Structures & Materials, Controls & Guidance, Systems & Dynamics), Oceanography (Biological, Chemical, Coastal Zone, Geological, & Physical), and Ocean Engineering (Coastal Processes, Fisheries, Marine Vehicles & Systems, Materials & Structures). Ph.D. programs are offered in all of the same Engineering fields as well as Environmental Science, Applied Mathematics, Computer Science, and Operations Research.

TRANSFER INFORMATION
Residency Requirements: 45 quarter hours, including the final 15 hours prior to graduation.
Transfer via Articulation Agreements
Admission to engineering: At any time
Engineering graduates transferred from ... 4-yr: 50 2-yr: 50
Requirements: There are articulation agreements with most of the Florida State Community College System schools, for the Associate of Arts parallel-program graduates.
Transfer without Articulation Agreements
Admission to engineering: At any time
Engineering graduates transferred from ... 4-yr: 50 2-yr: 50
Requirements: Official copies of transcripts from all previously attended colleges, for those with less than 30 semester hours also must submit official copies of high school transcripts and SAT scores. Grade point average should be at least 2.5 on a 4.0 scale.

STUDENT PROGRAMS
PROFESSIONAL AND HONORARY SOCIETIES
[For key to acronyms, see *Introduction*.]
Professional Societies: AAPG, ACM, ACS, AIAA, AIChE, ASCE, ASME, IEEE, SNAME, STC
Honorary Societies: Chi Epsilon, Eta Kappa Nu, Tau Beta Pi
SUPPORT PROGRAMS
Student Chapter Organizations: Society of Women Engineers
Other Engineering Support Programs: Orientation Week preceding Fall Quarter; Student Success Strategies Program, Individualized Learning Center, Counseling & Psychological Services, International Student Advisor, Campus Ministry, Cooperative Education Program, Office of Career Planning & Placement, and Veterans Affairs Office.

075 Florida International University

INSTITUTION PROFILE

HEAD OF THE INSTITUTION
Modesto A. Maidique
Phone: (305)348-2111 Fax: (305)348-3660

GENERAL INFORMATION
[All Students—Fall 1991]
Undergraduate enrollment 17,500
Graduate enrollment 5,500
Total institution enrollment 23,000
Type of institution: Public Calendar: Semesters
Nearest city: Miami Population: 800,000
Miles from main campus: 3 Setting: Urban
Types of engineering degrees: Engineering
Other degree-granting colleges: Allied Health Sciences, Arts & Sciences, Business Administration, Communication & Journalism, Education, Nursing, Hospitality Management, Computer Science, Public Affairs and Services, Accounting, Design

ENGINEERING ADMISSIONS

ADMISSIONS OFFICE CONTACT
Carmen Brown
Florida International University
Office of Admissions
University Park Campus
Miami, FL 33199
Phone: (305)348-3675 Fax: (305)348-3648

ENGINEERING COLLEGE ADMISSIONS INFORMATION
Admission to the engineering college: The admission policy for freshmen and transfer students are different and the policies vary in each department.
Entrance Requirements: SAT (Score: 1000), ACT (Score: 20), High School courses—English (4 years), Mathematics (3 years), Natural Science/Social Science (6 years), Foreign Language (2 years), Students must first pass the CLAST test before they are formally admitted into the School of Engineering.
Entrance Recommendations: High School courses—Physics (1 year), Chemistry (1 year), Trigonometry/Calculus (1 year), Additional Math (1 year), The State of Florida has developed a test of college level communication and computational skills. The test is called the College Level Academic Skills Test (CLAST).
Requirements for foreign students: TOEFL (Score: 500); financial statement; All foreign students are required to carry medical insurance on a yearly basis.(est. $500/yr.)

DEPARTMENTAL ADMISSIONS INFORMATION
Admission to the engineering department: Admission policy for freshmen and transfer students are different and the policies vary in each department.
Additional information: Students entering FIU with fewer than 48 transfer hours must satisfy all Core Curriculum Requirements while students transferring to FIU with at least 48 hours must satisfy the General Education Requirements and have passed the CLAST test.

FINANCIAL INFORMATION

ESTIMATED EXPENSES (ACADEMIC YEAR)
[Expenses are for the 1992-93 nine-month academic year.]

State Residents	Undergraduate	Graduate
Tuition and fees	$ 1,701[1]	$ 1,920[1]
College room and board	$ 9,900[2]	$ 9,900[2]
Books and supplies	$ 800	$ 800
Other expenses	$ 3,500[3]	$ 3,500[3]
Total estimated expenses	**$15,901**	**$16,120**

Out-of-State Residents	Undergraduate	Graduate
Tuition and fees	$ 6,460[1]	$ 6,511[1]
College room and board	$ 9,900[2]	$ 9,900[2]
Books and supplies	$ 800	$ 800
Other expenses	$ 4,000[4]	$ 4,000[4]
Total estimated expenses	**$21,160**	**$21,211**

Notes: (1) Fees include the Student Health Fee ($27 per semester) and the Athletic Fee ($10 per semester). (2) This amount is estimated at $825 per month to cover room, board, clothing, transportation and incidentals. This cost is for 12 months. (3) All engineering students are greatly encouraged to purchase a personal computer (at minimum a 80386 model) and software. (4) All engineering students are greatly encouraged to purchase a personal computer and software. All international students must carry medical insurance (est. $500)

FINANCIAL AID OFFICE CONTACT
Ana R. Sarasti, Director, Financial Aid
Florida International University
Office of Financial Aid
University Park Campus
Miami, FL 33199
Phone: (305)348-2431 Fax: (305)348-2346

GENERAL FINANCIAL AID INFORMATION
Forms accepted/required: ACT, AFSA, CSS-FAF, FAF, FAT, FAR, FSA, FFS, Stafford, IRS, SAR, SDF
Additional financial aid information: The University adheres to the philosophy that a student is entitled to a college education regardless of his or her financial condition. The Financial Aid Program at the University includes scholarships, grants, loans, and employment.

ENGINEERING COLLEGE INFORMATION

[For additional personnel, refer to the *Appendix*.]
HEAD OF ENGINEERING
Gordon R. Hopkins
Phone: (305)348-2521 Fax: (305)348-3582
EMail: HOPKINS@SERVAX

ENGINEERING COLLEGE ADDRESS
College of Engineering and Design
Florida International University
University Park Campus
Miami, FL 33199
Phone: (305)348-2521 Fax: (305)348-3582

ENROLLMENTS—BY CLASS
[Numbers are baccalaureate enrollments for the fall 1991 term, unless otherwise footnoted.]
1st-year students/Freshmen 64
2nd-year students/Sophomores 131
3rd-year students/Juniors 551
4th-year students/Seniors 406
Total **1,152**

NUMBER OF DEGREES AWARDED—BY PROGRAM
[Numbers are engineering baccalaureate degrees awarded during the 1991-1992 academic year, unless otherwise footnoted. For full details about each engineering program, refer to the *Tables of Degree Programs*.]
Civil & Environmental Engineering .... –
Civil Engineering 31
Computer Engineering .... –
Electrical & Computer Engineering .... –
Electrical Engineering 84
Industrial & Systems Engineering 35
Mechanical Engineering 33
Total **183**

PERCENTAGE OF DEGREES AWARDED—BY CATEGORY
[Percentages are of all engineering baccalaureate degrees awarded during the 1991-1992 academic year, unless otherwise footnoted.]
African Americans 9.2%
Asian/Pacific Island Americans 6.0%
Hispanic Americans 57.9%
Native Americans .... – %
Foreign Citizens 15.2%
All Others 11.7%
Women 10.9%
Persons with disabilities .... – %
Students over 25 years of age 31.4%

GRADUATE ENROLLMENTS & DEGREES AWARDED
Master's enrollment 173
Master's degrees awarded 44
Doctoral enrollment 11
Doctoral degrees awarded .... –

ENGINEERING STUDENT DATA
Applicants to the engineering college
Number of applicants to engineering college 410
Percent offered admission 85.0%
Matriculated engineering students
Percentage in top quartile (25%) of High School class 85.0
Average SAT scores: Math—610, Verbal—480, Combined—1090
Average ACT scores: Math—9, Composite—22

FULL- & PART-TIME FACULTY—BY DEPARTMENT
[Figures are the head count of full-time faculty and the full-time equivalent (FTE) of part-time faculty for each engineering department or equivalent.]

Department	Full	Part
Mechanical Eng	13	3.0
Industrial Eng	10	–
Electrical Eng	19	–
Civil Eng	12	–

COLLEGE DESCRIPTION
The School of Engineering offers baccalaureate degree programs in Civil, Computer, Electrical, Industrial, and Mechanical Engineering; Masters degree programs in Civil, Computer, Electrical, Environmental, Environmental & Urban Systems, Industrial, and Mechanical Engineering; and a doctoral degree program in Electrical Engineering. The various curricula for the School are designed to give the student an education for entry into the profession of engineering. The engineering programs include a strong engineering core foundation designed to prepare the prospective engineer not only with a broad base of fundamental courses in mathematics, sciences, and technical knowledge, but also with a solid cultural background in humanities, social sciences,and English. A full curriculum is offered in both day and evening classes, with many students taking less than a full load while working part-time. Average number of years required to actually complete the bachelor's degree: 4 1/2.

DEGREE PROGRAMS DESCRIPTION
All degree programs offer the B.S. degree. All degree programs offer an M.S. degree. The Civil & Environmental Engineering program offer construction, geotechnical, environmental, structures, surveying, transportation, urban planning, and water resources. The Electrical and Computer Engineering programs offers computers, communication systems, control systems, power systems, and integrated electronics. Industrial Engineering offers simulation and modeling, automation and robotics, and flexible manufacturing systems. Mechanical Engineering offers fluid thermal sciences, mechanics and control of mechanical and dynamic systems, energy systems, heating, ventilation and air conditioning, material sciences, biomechanics and bioengineering, manufacturing methods and computer-aided design.

TRANSFER INFORMATION
Transfer via Articulation Agreements
Admission to engineering: To qualify for admission to the program, transfer students must have satisfied the General Education Requirements and have passed the CLAST test.
Engineering graduates transferred from . . 4-yr: 10 2-yr: 289
Requirements: Students entering FIU with fewer than 48 transfer hours must satisfy all Core Curriculum Requirements while students transferring to FIU with at least 48 hours must satisfy the General Education Requirements. Students must also pass the CLAST test before being eligible for admission to the School of Engineering.

Transfer without Articulation Agreements
Admission to engineering: The same admission policy applies for all transfer students.
Engineering graduates transferred from . . 4-yr: – 2-yr: –
Requirements: The same admission policy applies to all transfer students.

GRADUATION REQUIREMENTS
Obtain the minimum number of credit hours required by the specific program. Complete at least 35 credit hours in the School of Engineering. Attain a minimum GPA of 2.0 in all engineering courses taken at the University. Satisfy the particular requirements for the major and University requirements for graduation. Other requirements are detailed in the Undergraduate Catalog.

STUDENT PROGRAMS

PROFESSIONAL AND HONORARY SOCIETIES
[For key to acronyms, see *Introduction*.]
Professional Societies: ACI, ACS, ACSM, AIA, ASCE, ASME, IEEE, IIE, NSPE, SME
Honorary Societies: Alpha Pi Mu, Pi Tau Sigma, Tau Beta Pi, Alpha Omega Chi, Delta Mu Omega

SUPPORT PROGRAMS
Student Chapter Organizations: Society of Women Engineers, National Society of Black Engineers, Society of Hispanic Professional Engineers, Association of Cuban Engineers

FLorida Action for Minorities in Engineering-FLAME
For: Women & Ethnic Minorities **Available:** Year round
Offered: Prematriculation

MUTEC-Minority UG Training for Energy-rel. Careers
For: Women & Ethnic Minorities **Available:** Year round
Offered: Prematriculation, Freshman, Sophomore, Junior, Senior

PREP - PRe-freshman Enrichment Program
For: Women & Ethnic Minorities **Available:** Year round
Offered: Prematriculation

SABLE-Student Achievement for Black Life Experience
For: Women & Ethnic Minorities **Available:** Year round
Offered: Prematriculation, Freshman, Sophomore, Junior, Senior, Graduate level

Other Engineering Support Programs: Academic Advising Center; Career Planning and Placement; Counseling Services; University Computer Services; Veteran's Services; University Learning/Testing Center; Disabled Student Services; Fall, Spring and Summer Freshmen Orientation Sessions; On-Campus Housing; Instructional Media Services; International Student Services; Minority Student Programs and Services; Student Government Association.

076 Florida State University

INSTITUTION PROFILE

HEAD OF THE INSTITUTION
Dale W. Lick
Phone: (904)644-1085　　　　　　Fax: (904)644-0172

GENERAL INFORMATION
[All Students—Fall 1991]
Undergraduate enrollment 21,300
Graduate enrollment 5,512
Total institution enrollment 28,607

Type of institution: Public　　　　　Calendar: Semesters
Nearest city: Tallahassee　　　　　　Population: 130,000
Miles from main campus: 1　　　　　 Setting: Urban
Types of engineering degrees: Engineering

Other degree-granting colleges: Arts & Sciences, Business Administration, Communication & Journalism, Education, Fine & Performing Arts, Humanities & Social Sciences, Law, Nursing, Library and Information Studies, Motion Picture, Television, Recording Arts, Criminology and Criminal Justice, Human Sciences

ENGINEERING ADMISSIONS

ADMISSIONS OFFICE CONTACT
Peter F. Metarko
Florida State University
Office of Admissions
216B William Johnston Building
Tallahassee, FL 32306
Phone: (904)644-6200　　　　　　Fax: (904)644-6404

ENGINEERING COLLEGE ADMISSIONS INFORMATION
Entrance Requirements: SAT, ACT, High School courses—English (4 years), Math, Natural Science, Social Science (3 years), Foreign Language (2 years), Additional Academic Electives (4 years), Most Florida residents present a High School GPA of 'B' and 25 ACT or 1050 SAT. Non Florida residents are ordinarily held to a higher standard.
Entrance Recommendations: SAT, ACT, High School courses—Math (4 years), Chemistry, Physics, Computer Programming (1 year)
Requirements for foreign students: TOEFL (Score: 550); financial statement

DEPARTMENTAL ADMISSIONS INFORMATION
Admission to the engineering department: At the time of admission to the institution.

FINANCIAL INFORMATION

ESTIMATED EXPENSES (ACADEMIC YEAR)
[Expenses are for the 1992-93 nine-month academic year.]

State Residents	Undergraduate	Graduate
Tuition and fees	$ 1,343 [1]	$ 2,552 [1]
College room and board	$ 3,880 [2]	$ 3,880 [2]
Books and supplies	$ 650	$ 700
Other expenses	$ —	$ —
Total estimated expenses	$ 5,873	$ 7,132

Out-of-State Residents	Undergraduate	Graduate
Tuition and fees	$ 5,152 [1]	$ 8,625 [1]
College room and board	$ 3,880 [2]	$ 3,880 [2]
Books and supplies	$ 650	$ 700
Other expenses	$ —	$ —
Total estimated expenses	$ 9,682	$ 13,205

Notes: (1) Based on twelve (12) credits per semester for two semesters. (2) Based on the most expensive room rate and meal plan.

FINANCIAL AID OFFICE CONTACT
Robert McCloud, *Director of Financial Aid*
Florida State University
Office of Financial Aid
127 Bryan Hall
Tallahassee, FL 32306-1009
Phone: (904)644-5871　　　　　　Fax: (904)644-6404

GENERAL FINANCIAL AID INFORMATION
Forms accepted/required: ACT, AFSA, CSS-FAF, FAF, FAT, FSA, FFS, IRS, SAR, SDF, Institutional, Institutional Verification Form (IVF)
Additional financial aid information: Pell Grant, Supplemental Educational Opportunity Grant, College Work Study, Perkins Loan, Guaranteed Student Loan, Florida Student Assistant Grant and special scholarships. Some scholarships are awarded by the College of Engineering on a competitive basis.

ENGINEERING COLLEGE INFORMATION
[For additional personnel, refer to the *Appendix*.]

HEAD OF ENGINEERING
Ching-Jen Chen
Phone: (904)487-6100　　　　　　Fax: (904)487-6486
EMail: CJChen@EVAX.ENG.FSU.EDU

ENGINEERING COLLEGE ADDRESS
FAMU/FSU College of Engineering
Florida State University
2525 Pottsdamer Rd.
P.O. Box 2175
Tallahassee, FL 32316-2175
Phone: (904)487-6100　　　　　　Fax: (904)487-6486

ENROLLMENTS—BY CLASS
[Numbers are baccalaureate enrollments for the fall 1991 term, unless otherwise footnoted.]

1st-year students/Freshmen	485 [1]
2nd-year students/Sophomores	285 [1]
3rd-year students/Juniors	448 [1]
4th-year students/Seniors	543 [1]
Total	1,761

Notes: (1) Numbers include Florida Agricultural and Mechanical University and Florida State University students.

NUMBER OF DEGREES AWARDED—BY PROGRAM
[Numbers are engineering baccalaureate degrees awarded during the 1991-1992 academic year, unless otherwise footnoted. For full details about each engineering program, refer to the *Tables of Degree Programs*.]

Chemical Engineering	14 [1]
Civil Engineering	28 [1]
Electrical Engineering	79 [1]
Industrial Engineering	8 [1]
Mechanical Engineering	52 [1]
Total	181

Notes: (1) Includes FAMU and FSU students. Year is Fall 1992, Spring 1992, and Summer 1992.

PERCENTAGE OF DEGREES AWARDED—BY CATEGORY
[Percentages are of all engineering baccalaureate degrees awarded during the 1991-1992 academic year, unless otherwise footnoted.]

African Americans	17.8% [2]
Asian/Pacific Island Americans	—% [2]
Hispanic Americans	—% [2]
Native Americans	—% [2]
Foreign Citizens	—% [A]
All Others	82.2%
Women	10.5% [2]
Persons with disabilities	—% [A]
Students over 25 years of age	30.4% [2]

Notes: (A) Data not available. (2) Numbers include both FAMU and FSU students. The academic year is Fall 1991, Spring 1992 and Summer 1992.

GRADUATE ENROLLMENTS & DEGREES AWARDED

Master's enrollment	126 [1]
Master's degrees awarded	49 [2]
Doctoral enrollment	19 [1]
Doctoral degrees awarded	—

Notes: (1) Numbers include Florida Agricultural and Mechanical University and Florida State University students. (2) Numbers include both FAMU and FSU students. The academic year is Fall 1991, Spring 1992 and Summer 1992.

ENGINEERING STUDENT DATA
Applicants to the engineering college
Number of applicants to engineering college .... — [A]
Percent offered admission .... —% [A]
Matriculated engineering students
Percentage in top quartile (25%) of High School class ...... — [A]
Average SAT scores: Combined—1091
Average ACT scores: Composite—25

Notes: (A) Data not available.

FULL- & PART-TIME FACULTY—BY DEPARTMENT
[Figures are the head count of full-time faculty and the full-time equivalent (FTE) of part-time faculty for each engineering department or equivalent.]

Department	Full	Part
Chemical Eng	7	0.5
Civil Eng	8	2.5
Electrical Eng	14	—
Industrial Eng	7	—
Mechanical Eng	14	0.5

COLLEGE DESCRIPTION
The FAMU/FSU College of Engineering is a joint program of The Florida Agricultural and Mechanical University (FAMU) and The Florida State University (FSU). Available courses of study lead to the Bachelor and to the Master of Science degree in Chemical, Civil, Electrical, Industrial, and Mechanical Engineering, and to the PhD. degree in Chemical and Mechanical Engineering. All of the programs are ABET accredited. The primary goals of the FAMU/FSU College of Engineering are to educate engineers of excellence at both the undergraduate and graduate levels, judged by highest standards in the field and recognized by national peers; to attract and produce greater numbers of blacks, women and minorities in professional engineering, engineering teaching and research; to attain national and international recognition of the College through the educational achievements of its faculty and students.

DEGREE PROGRAMS DESCRIPTION
In the fall of 1992 the FAMU/FSU College of Engineering adopted the following admission and retention requirements. 1. Students must achieve a GPA of 2.5 or better in Calculus I,II, Physics I and Chemistry I prior to enrolling in any 2000 level or above engineering course for which any of these courses are prerequisites. A maximum of one repeat of each course is allowed in meeting this requirement. 2. Any student who fails to earn a grade of "C" or better in an engineering course on the third attempt, or withdraws from the course more than twice, is subject to dismissal. 3. Any student who exceeds thirty credit hours of repeated course work is subject to dismissal. 4. Majors must earn a grade of "C" or better in all engineering courses which apply to the degree. This requirement may be waived by the academic dean upon recommendation of the department chair for no more than two engineering courses. 5. Normal policy in the College is to allow two reinstatements. A third reinstatement may be approved in exceptional circumstances. Any student who fails to comply with the agreement approved by the department chair following reinstatement or fails to earn a cumulative GPA of 2.0 or better upon completion of the term is subject to dismissal.

TRANSFER INFORMATION
Residency Requirements: The University requires completion of the last thirty (30) credit hours in residence at this University. Academic Dean can make exceptions for this up to six hours of the final thirty. On the doctoral level: after completing thirty credits of coursework (or being awarded the MS degree), students must be continuously enrolled for at least 24 consecutive months.

Transfer via Articulation Agreements
Admission to engineering: At any time

Transfer without Articulation Agreements
Admission to engineering: At any time

GRADUATION REQUIREMENTS
Undergraduates: COE; 132-145 credits, complete ABET requirements in Humanities, Social Sciences, sit for FE exam. FSU; one summer attended, residency requirement met, 'C' average in degree work, 40 hours 3000 or above, complete liberal studies, pass Florida College Level Academic Skills Test (CLAST), Dean's approval. MS: 30-33 credits, 7 year limit, 3.0 GPA each semester, only 6 credits can transfer from other accredited programs, 12 from FSU. For the PhD: 45 credits beyond the MS, 3.0 GPA each semester, 5 year limit, 24 consecutive months enrollment, credit transfer rule as above.

STUDENT PROGRAMS

PROFESSIONAL AND HONORARY SOCIETIES
[For key to acronyms, see *Introduction*.]

Professional Societies: AIAA, AIChE, ASCE, ASME, IEEE, IIE, SAE, National Society of Black Engineers, Society of Women Engineers
Honorary Societies: Tau Beta Pi

SUPPORT PROGRAMS
Student Chapter Organizations: Society of Women Engineers, National Society of Black Engineers

Engineering Concepts Institute
For: Ethnic Minorities　　　　**Available:** Summer only
Offered: Prematriculation

Minority Introduction to Engineering
For: Ethnic Minorities　　　　**Available:** Summer only
Offered: Prematriculation

Minority Engineering Retention Program
For: Ethnic Minorities　　　　**Available:** Academic year
Offered: Freshman, Sophomore

Other Engineering Support Programs: Summer, Fall, and Spring orientation, academic advising, retention, tutoring, and career counseling programs are utilized to provide support for freshmen, transfer and minority students. Career counseling and placement programs are available for upper classmen. Co-op and internships are also available to qualified students.

077 GMI Engineering & Management Institute

■ INSTITUTION PROFILE

HEAD OF THE INSTITUTION
James E. John
Phone: (313)762-9864   Fax: (313)762-9807

GENERAL INFORMATION
[All Students—Fall 1991]
Undergraduate enrollment 2,367
Graduate enrollment 774
Total institution enrollment 3,141

Type of institution: Private   **Calendar:** Other
Location: Flint   **Population:** 138,192
Setting: Urban
Types of engineering degrees: Engineering
Other degree-granting colleges: Management, Science and Mathematics, Humanities and Social Science

■ ENGINEERING ADMISSIONS

ADMISSIONS OFFICE CONTACT
Kevin A. Pollock
GMI Engineering & Management Institute
1700 West Third Avenue
Flint, MI 48504-4898
Phone: (313)762-7865   Fax: (313)762-9837

ENGINEERING COLLEGE ADMISSIONS INFORMATION
Entrance Requirements: High School courses—English (3 years), Physics or Chemistry (1 year), Algebra (2 years), Geometry (1 year), SAT or ACT
Entrance Recommendations: High School courses—Chemistry, Physics, Foreign Languages, Technical Drawing
Requirements for foreign students: TOEFL (Score: 550); financial statement

DEPARTMENTAL ADMISSIONS INFORMATION
Admission to the engineering department: At the time of admission to the institution.

■ FINANCIAL INFORMATION

ESTIMATED EXPENSES (ACADEMIC YEAR)
[Expenses are for the 1992-93 nine-month academic year.]

All Students	Undergraduate	Graduate
Tuition and fees	$10,150[2]	$10,500[1]
College room and board	$ 2,988[2]	$ 4,482[1]
Books and supplies	$ 500	$ —
Other expenses	$ 2,800	$ —
Total estimated expenses	**$16,438**	**$14,982**

Notes: (1) three terms (2) Two terms

FINANCIAL AID OFFICE CONTACT
Mark J. Delorey, *Director of Financial Aid*
GMI Engineering & Management Institute
1700 West Third Avenue
Flint, MI 48504-4898
Phone: (313)762-7859   Fax: (313)762-9807

GENERAL FINANCIAL AID INFORMATION
Forms accepted/required: Institutional, Application for Federal Student Aid
Additional financial aid information: Merit-based scholarships, need-based grants, part-time jobs on campus, full-time jobs for six months of the year, long-term and short-term loans.

■ ENGINEERING COLLEGE INFORMATION

[For additional personnel, refer to the *Appendix*.]

HEAD OF ENGINEERING
John D. Lorenz
Phone: (313)762-7949   Fax: (313)762-9836

ENGINEERING COLLEGE ADDRESS
School of Engineering
GMI Engineering & Management Institute
1700 West Third Avenue
Flint, MI 48504-4898
Phone: (313)762-9500   Fax: (313)762-9836

ENROLLMENTS—BY CLASS
[Numbers are baccalaureate enrollments for the fall 1991 term, unless otherwise footnoted.]
1st-year students/Freshmen 535
2nd-year students/Sophomores 552
3rd-year students/Juniors 498
4th-year students/Seniors 609
Total **2,194**

NUMBER OF DEGREES AWARDED—BY PROGRAM
[Numbers are engineering baccalaureate degrees awarded during the 1991-1992 academic year, unless otherwise footnoted. For full details about each engineering program, refer to the *Tables of Degree Programs*.]
Electrical & Computer Engineering 108
Industrial & Manufacturing Systems Engineering 26
Industrial Engineering 26
Manufacturing Systems Engineering 37
Mechanical Engineering 136
Total **333**

PERCENTAGE OF DEGREES AWARDED—BY CATEGORY
[Percentages are of all engineering baccalaureate degrees awarded during the 1991-1992 academic year, unless otherwise footnoted.]
African Americans .... 4.0 %
Asian/Pacific Island Americans .... 8.0 %
Hispanic Americans .... 2.0 %
Native Americans .... 1.0 %
Foreign Citizens .... 5.0 %
All Others .... 80.0 %
Women .... 20.0 %
Persons with disabilities .... — %
Students over 25 years of age .... — %

GRADUATE ENROLLMENTS & DEGREES AWARDED
Master's enrollment 49
Master's degrees awarded 7
Doctoral enrollment .... —
Doctoral degrees awarded .... —

ENGINEERING STUDENT DATA
Applicants to the engineering college
Number of applicants to engineering college 1,554
Percent offered admission 69.0%

Matriculated engineering students
Percentage in top quartile (25%) of High School class 9.6
Average SAT scores: Math—630, Verbal—507, Combined—1137
Average ACT scores: Math—27, Composite—26

FULL- & PART-TIME FACULTY—BY DEPARTMENT
[Figures are the head count of full-time faculty and the full-time equivalent (FTE) of part-time faculty for each engineering department or equivalent.]

Department	Full	Part
Electrical & Computer Eng	18	—
Industrial & Manufacturing Systems Eng	10	—
Mechanical Eng	32	—
Manufacturing Systems Eng	10	—

COLLEGE DESCRIPTION

GMI excels in providing its students with both top-notch classroom instruction and career-directed co-op work experience in business and industry. Five years are required to complete the academic program and co-op work experience. Each student must prepare an acceptable thesis to complete the requirements for a bachelor's degree. This is normally accomplished during the fifth year of the program. There are opportunities to minor in humanities and social science, management, applied chemistry, applied mathematics, computer science, and applied optics. Average number of years required to actually complete the bachelor's degree: 5.

DEGREE PROGRAMS DESCRIPTION

The EE program prepares students for a wide spectrum of professional careers focusing on the application of modern electro-science to the design of equipment and systems for processing information and controlling energy. The IE program prepares students for work in a variety of industrial, commercial, and governmental activities engaged in product manufacture or provision of a service. The Mfg. Sys. Engrg. program prepares students in both traditional and advanced manufacturing engineering, with emphasis on the total systems approach to manufacturing operation. The ME program prepares students for careers associated with the design and implementation of mechanical systems and with the conversion, transmission, and utilization of energy. A terminal Master of Science in Engineering degree is offered for engineers in industry. It provides excellent preparation for study at the doctoral level.

TRANSFER INFORMATION

Transfer via Articulation Agreements
Admission to engineering: At the end of the freshman year
Engineering graduates transferred from .. 4-yr: —   2-yr: 4
Requirements: Regular Admissions Requirements plus 3.0 average

Transfer without Articulation Agreements
Admission to engineering: At the end of the freshman year

GRADUATION REQUIREMENTS

Undergraduate thesis required.

■ STUDENT PROGRAMS

PROFESSIONAL AND HONORARY SOCIETIES
[For key to acronyms, see *Introduction*.]
Professional Societies: ASME, IEEE, IIE, SAE, SME
Honorary Societies: Alpha Pi Mu, Eta Kappa Nu, Pi Tau Sigma, Tau Beta Pi, Gamma Mu Iota

SUPPORT PROGRAMS
Student Chapter Organizations: Society of Women Engineers, National Society of Black Engineers, Society of Hispanic Professional Engineers

Office of Minority Affairs
For: Ethnic Minorities

Other Engineering Support Programs: Summer and fall orientation programs for all freshmen and transfers. Freshman orientation required. Extensive academic advising and tutoring, student success workshops and degree counseling by academic departments. Co-op placement office, co-op strategics course for non-employed freshmen. International student office.

078 Gannon University

INSTITUTION PROFILE

HEAD OF THE INSTITUTION
David A. Rubino
Phone: (814)871-5800 Fax: (814)871-7338

GENERAL INFORMATION
[All Students—Fall 1991]
Undergraduate enrollment 3,818
Graduate enrollment 705
Total institution enrollment 4,523
Type of institution: Private   Calendar: Semesters
Location: Erie   Population: 100,000
Setting: Urban
Types of engineering degrees: Engineering & Technology
Other degree-granting colleges: Business Administration, Education, Humanities & Social Sciences, Health Sciences

ENGINEERING ADMISSIONS

ADMISSIONS OFFICE CONTACT
Joyce Scheid-Gilman
Gannon University
Director of Admissions
Erie, PA 16541
Phone: (814)871-7407 Fax: (814)455-6277

ENGINEERING COLLEGE ADMISSIONS INFORMATION
Entrance Requirements: SAT (Score: 850), High School courses—Science (4 years), Mathematics (4 years), English (4 years), Social Sciences and/or Foreign Language (4 years)
Entrance Recommendations: High School courses—Foreign Language (2 years), Pre Calculus (1 year), Calculus (1 year)
Requirements for foreign students: TOEFL; financial statement

DEPARTMENTAL ADMISSIONS INFORMATION
Admission to the engineering department: At the time of admission to the institution.

FINANCIAL INFORMATION

ESTIMATED EXPENSES (ACADEMIC YEAR)
[Expenses are for the 1992-93 nine-month academic year.]

All Students	Undergraduate	Graduate
Tuition and fees	$ 9,680	$ 6,000
College room and board	$ 3,440	$ 3,920
Books and supplies	$ 450	$ 450
Other expenses	$ –	$ –
Total estimated expenses	**$13,570**	**$ 10,370**

FINANCIAL AID OFFICE CONTACT
James A. Treiber, *Director of Financial Aid*
Gannon University
University Square
Erie, PA 16541
Phone: (814)871-7337 Fax: (814)455-6277

GENERAL FINANCIAL AID INFORMATION
Forms accepted/required: PHEAA/Federal Aid Form
Additional financial aid information: Financial aid is generally awarded in the form of a package including scholarship, work, and loan funds. The amount of each type of aid varies according to the University's funds and the student need. During the past year about 80% of Gannon students who applied received financial assistance. Awards ranged from $100 to full cost of tuition, room, and board. The average student received a 32% discount on tuition in addition to state and federal sources.

ENGINEERING COLLEGE INFORMATION
[For additional personnel, refer to the *Appendix*.]

HEAD OF ENGINEERING
William D. Gregory
Phone: (814)871-7616 Fax: (814)455-2631

ENGINEERING COLLEGE ADDRESS
College of Science and Engineering
Gannon University
University Square
Erie, PA 16541
Phone: (814)871-7618 Fax: (814)455-2631

ENROLLMENTS—BY CLASS
[Numbers are baccalaureate enrollments for the fall 1991 term, unless otherwise footnoted.]
1st-year students/Freshmen 44
2nd-year students/Sophomores 23
3rd-year students/Juniors 34
4th-year students/Seniors 48
Total 149

NUMBER OF DEGREES AWARDED—BY PROGRAM
[Numbers are engineering baccalaureate degrees awarded during the 1991-1992 academic year, unless otherwise footnoted. For full details about each engineering program, refer to the *Tables of Degree Programs*.]
Electrical Engineering 23
Mechanical Engineering 16
Total 39

PERCENTAGE OF DEGREES AWARDED—BY CATEGORY
[Percentages are of all engineering baccalaureate degrees awarded during the 1991-1992 academic year, unless otherwise footnoted.]
African Americans .... 2.0 %
Asian/Pacific Island Americans .... – %
Hispanic Americans .... – %
Native Americans .... – %
Foreign Citizens .... 23.6 %
All Others .... 74.4 %
Women .... 11.0 %
Persons with disabilities .... – %
Students over 25 years of age .... 14.0 %

GRADUATE ENROLLMENTS & DEGREES AWARDED
Master's enrollment 42
Master's degrees awarded 5
Doctoral enrollment .... –
Doctoral degrees awarded .... –

ENGINEERING STUDENT DATA
Applicants to the engineering college
Number of applicants to engineering college 122
Percent offered admission 80.0%

Matriculated engineering students
Percentage in top quartile (25%) of High School class 35.0
Average SAT scores: Combined—1004

FULL- & PART-TIME FACULTY—BY DEPARTMENT
[Figures are the head count of full-time faculty and the full-time equivalent (FTE) of part-time faculty for each engineering department or equivalent.]

Department	Full	Part
Electrical Eng	6	2.0
Mechanical Eng	7	2.0

COLLEGE DESCRIPTION
Gannon University is a teaching university with a strong commitment to liberal studies. This core, required of all students, includes literature, theology, philosophy, and history. All students must take at least one writing intensive course per year. Classes are small, even in the first two years. The average class will have 25 students. many upper level courses are limited to 20 students per section. Courses are taught by full-time faculty or adjuncts from industry. There are no graduate assistants teaching courses or laboratories. Both Electrical and Mechanical Engineering have a co-op option. The senior year includes a capstone Engineering Design course. All seniors are required to sign up for the fundamentals of engineering exam.

DEGREE PROGRAMS DESCRIPTION
Gannon offers the professional degrees of BME and BEE at the undergraduate level. The MS in Engineering with a mechanical or electrical emphasis is also available. Electrical Engineering has a very strong laboratory emphasis. It has concentrations in Power Generation and Production, Communications and Control systems, and Computer and Digital Systems. Mechanical Engineering is a broadly based program covering thermal, fluid, and mechanics areas. It has concentrations in Power, Mechanical Design, and Environmental and Fluids Design. The campus is centrally located near extensive manufacturing concerns. Many students obtain part-time employment or serve internships in these industries.

TRANSFER INFORMATION
Transfer without Articulation Agreements
Admission to engineering: At any time
Engineering graduates transferred from ... 4-yr: 6 2-yr: 2
Requirements: Minimum of a 2.0 GPA

GRADUATION REQUIREMENTS
Engineering degrees in Mechanical and Electrical majors require a minimum of 134-139 credits to graduate. All major requirements in the department must be met. The student must also meet all the Liberal Studies requirements of the University. There is also a writing requirement that must be met by taking writing intensive courses. A 2.0 GPA overall and in the major is required. In major courses a student may be required to repeat courses with a D grade.

STUDENT PROGRAMS

PROFESSIONAL AND HONORARY SOCIETIES
[For key to acronyms, see *Introduction*.]
Professional Societies: AMS, ASME, IEEE, NSPE, SAE, SME
Honorary Societies: Eta Kappa Nu, Pi Tau Sigma

SUPPORT PROGRAMS
Educational Opportunity Program
For: Ethnic Minorities **Available:** Year round
Offered: Prematriculation, Freshman, Sophomore, Junior, Senior

Returning to Education Adult Program
For: Women **Available:** Year round
Offered: Prematriculation, Freshman, Sophomore, Junior, Senior, Graduate level

Other Engineering Support Programs: All freshman and transfer students are invited to attend a summer orientation program with their parents. At this time they meet with faculty and plan their first semester courses. There is an office of freshman services that assist the departments. These services include tutoring, writing clinic, freshman grades, and counseling.

079 George Mason University

■ INSTITUTION PROFILE

HEAD OF THE INSTITUTION
George W. Johnson
Phone: (703)993-8700 Fax: (703)993-8707

GENERAL INFORMATION
[All Students—Fall 1991]
Undergraduate enrollment 13,090
Graduate enrollment 7,076
Total institution enrollment 20,829
Type of institution: Public Calendar: Semesters
Nearest city: Washington, DC Population: 617,000
Miles from main campus: 18 Setting: Suburban
Types of engineering degrees: Engineering
Other degree-granting colleges: Arts & Sciences, Business Administration, Nursing, Graduate School of Education, Law

■ ENGINEERING ADMISSIONS

ADMISSIONS OFFICE CONTACT
Patricia Riordan
George Mason University
Office of Admissions
4400 University Drive
Fairfax, VA 22030-4444
Phone: (703)993-2400 Fax: (703)993-2392

ENGINEERING COLLEGE ADMISSIONS INFORMATION
Entrance Requirements: SAT, High School courses—English (4 years), Social studies (3 years), Mathematics (4 years), Laboratory science (2 years)
Entrance Recommendations: High School courses—English (4 years), Social studies (4 years), Mathematics (5 years), Laboratory science (3 years)
Requirements for foreign students: TOEFL (Score: 570); financial statement

DEPARTMENTAL ADMISSIONS INFORMATION
Admission to the engineering department: At the time of admission to the institution.

■ FINANCIAL INFORMATION

ESTIMATED EXPENSES (ACADEMIC YEAR)
[Expenses are for the 1992-93 nine-month academic year.]

State Residents	Undergraduate	Graduate
Tuition and fees	$ 3,372	$ 3,372
College room and board	$ 4,750	$ 4,750
Books and supplies	$ 500	$ 500
Other expenses	$ 1,750 [1]	$ 1,750 [1]
Total estimated expenses	**$10,372**	**$10,372**

Out-of-State Residents	Undergraduate	Graduate
Tuition and fees	$ 8,640	$ 8,640
College room and board	$ 4,750	$ 4,750
Books and supplies	$ 500	$ 500
Other expenses	$ 1,750 [1]	$ 1,750 [1]
Total estimated expenses	**$15,640**	**$15,640**

Notes: (1) Other estimated expenses are for personal and transportation.

FINANCIAL AID OFFICE CONTACT
Jennifer Douglas, *Director*
George Mason University
Finley Building
4400 University Drive
Fairfax, VA 22030-4444
Phone: (703)993-4350

GENERAL FINANCIAL AID INFORMATION
Forms accepted/required: CSS-FAF, Stafford
Additional financial aid information: Need-based and merit-based scholarships; part-time jobs on campus, Perkins Student Loan Program, Supplemental Educational Opportunity Grant, College Work-Study, Pell Grant, Stafford Student Loan Program, PLUS Loan Program, ROTC, Veterans Educational Benefits, Virginia War Orphans Education Program.

■ ENGINEERING COLLEGE INFORMATION
[For additional personnel, refer to the *Appendix*.]

HEAD OF ENGINEERING
Andrew P. Sage
Phone: (703)993-1500 Fax: (703)993-1521
EMail: asage@gmuvax.gmu.edu

ENGINEERING COLLEGE ADDRESS
School of Information Technology and Engineering
George Mason University
4400 University Drive
Room 100, Science & Technology II
Fairfax, VA 22030-4444
Phone: (703)993-1500 Fax: (703)993-1521

ENROLLMENTS—BY CLASS
[Numbers are baccalaureate enrollments for the fall 1991 term, unless otherwise footnoted.]
1st-year students/Freshmen 286
2nd-year students/Sophomores 214
3rd-year students/Juniors 266
4th-year students/Seniors 399
Total **1,165**

NUMBER OF DEGREES AWARDED—BY PROGRAM
[Numbers are engineering baccalaureate degrees awarded during the 1991-1992 academic year, unless otherwise footnoted. For full details about each engineering program, refer to the *Tables of Degree Programs*.]
Computer Science 75
Electrical Engineering 82
Information Systems .... – (2)
Information Technology .... – (B)
Operations Research & Management Science .... – (4)
Software Systems Engineering .... – (B)
Statistical Sciences .... – (B)
Systems Engineering 4
Urban Systems Engineering 2
Total **163**

Notes: (B) Data not applicable. (2) Undergraduate courses in Information Systems are offered and eligible for inclusion in other degree programs. (4) Undergraduate courses in this program are offered and eligible for use in other degree programs at the university.

PERCENTAGE OF DEGREES AWARDED—BY CATEGORY
[Percentages are of all engineering baccalaureate degrees awarded during the 1991-1992 academic year, unless otherwise footnoted.]
African Americans .... 3.0 %
Asian/Pacific Island Americans .... 26.0 %
Hispanic Americans .... 4.3 %
Native Americans .... – %
Foreign Citizens .... 9.0 %
All Others .... 57.7 %
Women .... 29.0 %
Persons with disabilities .... – % (A)
Students over 25 years of age .... – % (A)

Notes: (A) Data not available.

GRADUATE ENROLLMENTS & DEGREES AWARDED
Master's enrollment 898
Master's degrees awarded 247
Doctoral enrollment 293
Doctoral degrees awarded 13

ENGINEERING STUDENT DATA

Applicants to the engineering college
Number of applicants to engineering college 942
Percent offered admission 73.0%

Matriculated engineering students
Percentage in top quartile (25%) of High School class 60.0
Average SAT scores: Math—550, Verbal—400, Combined—950

FULL- & PART-TIME FACULTY—BY DEPARTMENT
[Figures are the head count of full-time faculty and the full-time equivalent (FTE) of part-time faculty for each engineering department or equivalent.]

Department	Full	Part
Electrical & Computer Eng	25	1.4
Systems Eng	11	–
Urban Systems Eng	2	–
Information & Software Systems Eng	12	1.5
Applied & Eng Statistics	7	–
Computer Science	18.5	1.5
Operations Research & Eng	7.5	–

COLLEGE DESCRIPTION
The School of Information Technology and Engineering (SITE) was established in July 1985 to provide modern information technology and engineering programs to meet the needs and expectations of industry and government in these areas of intellectual and professional effort. SITE has established a reputation for excellence. The success of the school is due in large part to on-going close partnerships and interactions between business, government and the University, as well as to the high quality of the faculty. There are approximately 89 full time faculty in the fall of 1992. Eight of these are fellows in at least one professional society. Currently, there are eight endowed chair positions. The faculty have authored more than 50 contemporary textbooks, are editors of 13 professional journals and textbook series, and have managed a number of major conferences. Average number of years required to actually complete the bachelor's degree - 4.5.

DEGREE PROGRAMS DESCRIPTION
The six departments in SITE offer a total of twelve degree programs concentrating on important contemporary technological issues and needs in information technology and engineering. Four bachelor's degree programs are offered: computer science, electrical engineering, systems engineering, and urban systems engineering. The electrical engineering program is ABET accredited. These programs prepare graduates to enter directly into professional employment or to continue studies at the graduate level. Master's programs are offered in computer science, electrical engineering, information systems, operations research and management science, software systems engineering, statistical sciences, and systems engineering. There are options in command, control, communications & intelligence and in urban systems in systems engineering. A single interdisciplinary doctoral program is offered in information technology.

TRANSFER INFORMATION
Residency Requirements: At least one-fourth of the total semester hours presented on the degree application must be completed at the university and must include at least 12 hours of advanced-level courses (numbered 300 or above) in the major program.

Transfer via Articulation Agreements
Admission to engineering: At the end of the sophomore year
Requirements: Must have a 2.5 GPA on a 4.0 scale, or better, and be in good academic standing.

Transfer without Articulation Agreements
Admission to engineering: At any time
Requirements: Must have received a 'B' average in Math courses.

GRADUATION REQUIREMENTS
To qualify for a degree, a student must have been admitted, must have fulfilled all stated requirements for the specific degree, and must have earned a GPA of at least 2.00. A student may not use more than six hours of D grades in the major. The electrical engineering degree requires a total of 133 credit hours.

■ STUDENT PROGRAMS

PROFESSIONAL AND HONORARY SOCIETIES
[For key to acronyms, see *Introduction*.]
Professional Societies: ACM, IEEE, NSPE, Armed Forces Communications and Electronic Association
Honorary Societies: Eta Kappa Nu, Omega Rho, Sigma Xi

SUPPORT PROGRAMS
Student Chapter Organizations: Society of Women Engineers, National Society of Black Engineers, Society of Hispanic Professional Engineers

Minority Student Services
For: Ethnic Minorities **Available:** Year round
Offered: Freshman, Sophomore, Junior, Senior, Graduate level

Disability Support Services
For: Women & Ethnic Minorities **Available:** Year round
Offered: Freshman, Sophomore, Junior, Senior, Graduate level

Other Engineering Support Programs: The Center for New Students, Writing Center, Student Health Center, Health Education Center, Counseling Center (academic and personal), Career Development Center, Cooperative Education, Veteran's Services, Campus Ministry, Office of International Student Services, Student Leadership Development Center, Housing, Mathematics Tutoring Center, Drug Education Center.

080 The George Washington University

■ INSTITUTION PROFILE

HEAD OF THE INSTITUTION
Stephen J. Trachtenberg
Phone: (202)994-6500 Fax: (202)994-0654

GENERAL INFORMATION
[All Students—Fall 1991]
Undergraduate enrollment 6,201
Graduate enrollment 10,091
Total institution enrollment 18,600
Type of institution: Private Calendar: Semesters
Location: Washington Population: 637,000
Setting: Urban
Types of engineering degrees: Engineering
Other degree-granting colleges: Education, Law, Columbian College & Grad. Sch. of Arts & Sciences, Business and Public Management, Medicine and Health Sciences, Business and Public Management, Elliott School of International Affairs

ENGINEERING COLLEGE ADMISSIONS INFORMATION
Entrance Requirements: SAT, ACT, High School courses—English (4 years), Mathematics (4 years), Physics (1 year), Chemistry (1 year)
Entrance Recommendations: SAT (Score: 1100), ACT (Score: 27), High School courses—Pre-calculus (.5 years), Algebra (1 year), History (1 year), Foreign Language (2 years), College Board Achievement Tests - Mathematics and English Composition.
Requirements for foreign students: TOEFL (Score: 550); financial statement; EFL Placement Test waived for students who score above 600 on TOEFL. Financial Statement required of international students.

DEPARTMENTAL ADMISSIONS INFORMATION
Admission to the engineering department: At the time of admission to the institution.

■ FINANCIAL INFORMATION

ESTIMATED EXPENSES (ACADEMIC YEAR)
[Expenses are for the 1992-93 nine-month academic year.]

All Students	Undergraduate	Graduate
Tuition and fees	$15,590	$ 9,892
College room and board	$ 6,310	$ 8,280
Books and supplies	$ 600	$ 600
Other expenses	$ 1,565 [2]	$ 500 [1]
Total estimated expenses	**$24,065**	**$19,272**

Notes: (1) Health Insurance (2) Mandatory fees and personal expenses.

FINANCIAL AID OFFICE CONTACT
Vicki J. Baker
The George Washington University
Student Financial Assistance
2121 Eye Street, NW
Washington, DC 20052
Phone: (202)994-6620 Fax: (202)994-0906

GENERAL FINANCIAL AID INFORMATION
Forms accepted/required: FAF, FAT, Stafford, IRS, Supplemental
Additional financial aid information: Merit-based: graduate and undergraduate, including partial to full tuition, and stipends. Undergraduate merit-based consideration is automatic upon receipt of application. Need-based, including work study. Athletic and performing awards. Graduate teaching assistantships and graduate research assistantships are available.

■ ENGINEERING COLLEGE INFORMATION

[For additional personnel, refer to the Appendix.]

HEAD OF ENGINEERING
Gideon Frieder
Phone: (202)994-6080 Fax: (202)994-4522
EMail: frieder@seas.gwu.edu

ENGINEERING COLLEGE ADDRESS
School of Engineering and Applied Science
The George Washington University
725 23rd Street, N.W.
Washington, DC 20052
Phone: (202)994-6158 Fax: (202)994-4522

ENROLLMENTS—BY CLASS
[Numbers are baccalaureate enrollments for the fall 1991 term, unless otherwise footnoted.]
1st-year students/Freshmen 225
2nd-year students/Sophomores 125
3rd-year students/Juniors 125
4th-year students/Seniors 128
Total **603**

NUMBER OF DEGREES AWARDED—BY PROGRAM
[Numbers are engineering baccalaureate degrees awarded during the 1991-1992 academic year, unless otherwise footnoted. For full details about each engineering program, refer to the Tables of Degree Programs.]
Civil Engineering 12
Computer Engineering 15
Computer Science 8
Electrical Engineering 47
Engineering Management .... – [B]
Mechanical Engineering 22
Operations Research .... – [B]
Systems Analysis & Engineering 7
Total **111**

Notes: (B) Data not applicable.

PERCENTAGE OF DEGREES AWARDED—BY CATEGORY
[Percentages are of all engineering baccalaureate degrees awarded during the 1991-1992 academic year, unless otherwise footnoted.]
African Americans .... 1.0 %
Asian/Pacific Island Americans .... 6.3 %
Hispanic Americans .... 1.8 %
Native Americans .... – %
Foreign Citizens .... 41.4 %
All Others .... 49.5 %
Women .... 23.4 %
Persons with disabilities .... – % [A]
Students over 25 years of age .... – % [A]

Notes: (A) Data not available.

GRADUATE ENROLLMENTS & DEGREES AWARDED
Master's enrollment .... 1,549 [2]
Master's degrees awarded 400
Doctoral enrollment 437
Doctoral degrees awarded 42

Notes: (2) All campuses included.

ENGINEERING STUDENT DATA
Applicants to the engineering college
Number of applicants to engineering college 1,043
Percent offered admission 70.6%

Matriculated engineering students
Percentage in top quartile (25%) of High School class 74.0
Average SAT scores: Math—660, Verbal—565, Combined—1225

FULL- & PART-TIME FACULTY—BY DEPARTMENT
[Figures are the head count of full-time faculty and the full-time equivalent (FTE) of part-time faculty for each engineering department or equivalent.]

Department	Full	Part
Eng Management	15	16.3
Operations Research	8	1.0
Continuing Eng Education	–	–
Civil, Mechanical & Environmental Eng	29	14.3
Electrical Eng & Computer Science	45	17.7

COLLEGE DESCRIPTION
The School offers fully accredited programs leading to undergraduate degrees in civil, computer, electrical (including a pre-med option), and mechanical engineering; systems analysis and engineering; and computer science. Graduate programs are offered in 48 areas of concentration. Emphasis is placed on developing the technical foundation, professional competence and skills at adaptation to remain competitive in the changing career world. Special programs include cooperative education, honors research, Naval ROTC, an accelerated studies program, and programs conducted in cooperation with other schools within the University, as well as programs offered in conjunction with selected liberal arts, community and junior colleges.

DEGREE PROGRAMS DESCRIPTION
A joint BS/JD program with special admission requirements guarantees acceptance in GW's National Law Center with a bachelors degree in any discipline of engineering. Also available are two five-year BS/MS programs: A five-year BS in Systems Analysis and Engineering (SA&E) and MS in Operations Research (OR) and a five-year BS SA&E and MS in the field of Economics. Dual-degree programs have been developed with several liberal arts institutions and pursue a three-year course of study in liberal arts or sciences, then follow a two-year program in engineering at the School of Engineering and Applied Science. Upon successful completion of the program a BA or BS from the first institution and a BS in engineering from GW is awarded.

DUAL DEGREE PROGRAMS
Enrollment requirements: Specified in individual agreements.

TRANSFER INFORMATION
Residency Requirements: A minimum of 30 credit hours including at least 12 hours in the major field must be completed while registered in the school or college from which the degree is sought. Applies to transfers within or from other institutions.

Transfer via Articulation Agreements
Admission to engineering: Transfers are admitted at any time. Application deadline: Fall - June 1, Spring - November 1, Summer April 1.
Requirements: 2.7/4.0 GPA, with good academic standing at all post-secondary institutions previously attended. Only grades of C or better transfer. No strict limit on number of transfer credits but must satisfy residency requirements.

Transfer without Articulation Agreements
Admission to engineering: At any time
Requirements: 2.7/4.0 GPA, with good academic standing at all post-secondary institutions previously attended. Only grades of C or better transfer. No strict limit on number of transfer credits but must satisfy residency requirements.

GRADUATION REQUIREMENTS
B.S. programs require a minimum of 132 semester hours, with a GPA of at least 2.2 in fifth through eighth semester technical courses, and 2.00/4.00 overall for all coursework. Curriculum requirements for specific degrees must be satisfied as outlined in current catalog. Graduate requirements include completion of an approved plan of study, with a minimum of 24 semester hours plus a thesis or 33 hours without a thesis with a minimum GPA of 3.00/4.00; and successful performance on the comprehensive exam. Professional and Doctoral degree requirements are contained in the current catalog.

■ STUDENT PROGRAMS

PROFESSIONAL AND HONORARY SOCIETIES
[For key to acronyms, see Introduction.]
Professional Societies: ACM, ASCE, ASME, IEEE, NSPE, ORSA, Society of Women Engineers, National Society of Black Engineers
Honorary Societies: Eta Kappa Nu, Pi Tau Sigma, Tau Beta Pi, Theta Tau, Omega Rho

SUPPORT PROGRAMS
Student Chapter Organizations: Society of Women Engineers, National Society of Black Engineers, Korean Engineering Student Association

Mentoring incoming women students (School-based)
For: Women Available: Year round
Offered: Prematriculation, Freshman

Multicultural Center (University-based)
For: Ethnic Minorities Available: Year round
Offered: Prematriculation, Freshman, Sophomore, Junior, Senior, Graduate level

Minority Engineering Program
For: Ethnic Minorities Available: Year round
Offered: Prematriculation, Freshman, Sophomore, Junior, Senior, Graduate level

Other Engineering Support Programs: Freshman/transfer orientation 'course,' including mentoring by upperclassmen; free peer tutoring for underclassmen; career services and cooperative education advising; Engineers House; international student services office; full program of Engineers' Week activities; and Tau Beta Pi, Eta Kappa Nu, Theta Tau, and other student societies and organizations.

081 University of Georgia

■ INSTITUTION PROFILE

HEAD OF THE INSTITUTION
Charles B. Knapp
Phone: (706)542-1214 Fax: (706)542-0995

GENERAL INFORMATION
[All Students—Fall 1991]
Undergraduate enrollment 22,961
Graduate enrollment 5,026
Total institution enrollment **27,987**
Type of institution: Public Calendar: Quarters
Nearest city: Atlanta, Georgia Population: 2,800,000
Miles from main campus: 60 Setting: Urban
Types of engineering degrees: Engineering
Other degree-granting colleges: Agricultural & Environmental, Arts & Sciences, Business Administration, Communication & Journalism, Education, Law, Pharmacy, Technology & Applied Sciences, Environmental Design, Family and Consumer Sciences, Forest Resources, Social Work, Veterinary Medicine

■ ENGINEERING ADMISSIONS

ADMISSIONS OFFICE CONTACT
Claire C. Swann
University of Georgia
Undergraduate Admissions Office
114 Academic Building
Athens, GA 30602
Phone: (706)542-8776 Fax: (706)542-1466

ENGINEERING COLLEGE ADMISSIONS INFORMATION
Entrance Requirements: SAT (Score: 950), ACT (Score: 22), High School courses—English (also 2 years of foreign language) (4 years), Math (2 courses in Algebra & 1 in geometry) (3 years), Science (3 years), Social Science (3 years)
Entrance Recommendations: High School courses—Math (trigonometry and pre-calculus) (2 years), Physics (1 year), Computers (1 year)
Requirements for foreign students: TOEFL (Score: 520); financial statement

DEPARTMENTAL ADMISSIONS INFORMATION
Admission to the engineering department: At the time of admission to the institution.

■ FINANCIAL INFORMATION

ESTIMATED EXPENSES (ACADEMIC YEAR)
[Expenses are for the 1992-93 nine-month academic year.]

State Residents	Undergraduate	Graduate
Tuition and fees	$ 2,076	$ 2,076
College room and board	$ 2,898	$ 2,898
Books and supplies	$ 500	$ 500
Other expenses	$ –	$ –
Total estimated expenses	**$ 5,474**	**$ 5,474**

Out-of-State Residents	Undergraduate	Graduate
Tuition and fees	$ 5,520	$ 5,520
College room and board	$ 2,898	$ 2,898
Books and supplies	$ 500	$ 500
Other expenses	$ –	$ –
Total estimated expenses	**$ 8,918**	**$ 8,918**

FINANCIAL AID OFFICE CONTACT
Ray Tripp, *Director*
University of Georgia
Financial Aid Office
220 Academic Building
Athens, GA 30602
Phone: (706)542-8208

GENERAL FINANCIAL AID INFORMATION
Forms accepted/required: Additional financial aid information: The Financial Aid Office handles all types of financial aid. There are various types of grants and loans offered through this office. Also, in addition to this, the College of Agricultural and Environmental Sciences and the Department of Biological and Agricultural Engineering have separate scholarships available that are selected within the College and Department.

■ ENGINEERING COLLEGE INFORMATION

[For additional personnel, refer to the *Appendix*.]

HEAD OF ENGINEERING
William P. Flatt
Phone: (706)542-3924 Fax: (706)542-0803

ENGINEERING COLLEGE ADDRESS
Biological and Agricultural Engineering
University of Georgia
Agriculture Drive
Driftmier Engineering Center
Athens, GA 30602-4435
Phone: (706)542-1653 Fax: (706)542-8806

ENROLLMENTS—BY CLASS
[Numbers are baccalaureate enrollments for the fall 1991 term, unless otherwise footnoted.]
1st-year students/Freshmen 23
2nd-year students/Sophomores 20
3rd-year students/Juniors 21
4th-year students/Seniors 33
Total **97**

NUMBER OF DEGREES AWARDED—BY PROGRAM
[Numbers are engineering baccalaureate degrees awarded during the 1991-1992 academic year, unless otherwise footnoted. For full details about each engineering program, refer to the *Tables of Degree Programs*.]
Agricultural Engineering . . . . –
Biological & Agricultural Engineering 20
Total **20**

PERCENTAGE OF DEGREES AWARDED—BY CATEGORY
[Percentages are of all engineering baccalaureate degrees awarded during the 1991-1992 academic year, unless otherwise footnoted.]
African Americans 2.0%
Asian/Pacific Island Americans . . . . – %
Hispanic Americans . . . . – %
Native Americans . . . . – %
Foreign Citizens 1.0%
All Others 97.0%
Women . . . . – %
Persons with disabilities . . . . – %
Students over 25 years of age 20.0%

GRADUATE ENROLLMENTS & DEGREES AWARDED
Master's enrollment 9
Master's degrees awarded 7
Doctoral enrollment 4
Doctoral degrees awarded 1

ENGINEERING STUDENT DATA

Applicants to the engineering college
Number of applicants to engineering college . . . . – (A)
Percent offered admission . . . . –% (A)

Matriculated engineering students
Percentage in top quartile (25%) of High School class . . . . . . – (A)
Average SAT scores: Math—558, Verbal—482, Combined—1040

Notes: (A) Data not available.

FULL- & PART-TIME FACULTY—BY DEPARTMENT
[Figures are the head count of full-time faculty and the full-time equivalent (FTE) of part-time faculty for each engineering department or equivalent.]

Department	Full	Part
Biological & Agricultural Eng	33	–

COLLEGE DESCRIPTION
The undergraduate agricultural engineering program is a four-year general engineering program. It provides a well-balanced education in the humanities, social sciences, basic sciences, engineering sciences, and engineering design. A 25 quarter credit hour option is available during the senior year in one of the following areas: 1) Mechanical systems, 2) Electrical systems, 3) Processing, 4) Water and soil management and 5) Structural design. During the senior year each student completes a senior design sequence project and is required to complete the national standardized Fundamentals of Engineering Exam. The full curriculum is offered in a day program with some students taking less than a full load while working part time or participating in the co-op program. An optional co-op program in available for sophomores, juniors, and seniors. Average number of years required to actually complete the bachelor's degree: 4.33.

DEGREE PROGRAMS DESCRIPTION
The B.S. degree program in agricultural engineering prepares the graduate for entry level professional positions in agri-business and allied industries and agencies. The M.S. degree program in Agricultural Engineering emphasizes advanced study in science and engineering and the methods and practices of engineering research (thesis required). The Ph.D. degree program in biological and agricultural engineering provides unique graduate study integrating the biological sciences with engineering.

TRANSFER INFORMATION
Residency Requirements: In order to be awarded a baccalaureate degree, students must earn a minimum of 90 quarter credit hours in residence.

Transfer via Articulation Agreements
Admission to engineering: At any time
Engineering graduates transferred from . . 4-yr: 4 2-yr: 5
Requirements: Must have a 2.5 GPA

Transfer without Articulation Agreements
Admission to engineering: At any time
Engineering graduates transferred from . . 4-yr: – 2-yr: –
Requirements: Must have a 2.5 GPA

GRADUATION REQUIREMENTS
Must have a 2.0 cumulative GPA and take the Engineer in Training Exam as the senior exit exam.

■ STUDENT PROGRAMS

PROFESSIONAL AND HONORARY SOCIETIES
[For key to acronyms, see *Introduction*.]
Professional Societies: ASAE

SUPPORT PROGRAMS

Minority Services & Programs
For: Women & Ethnic Minorities Available: Year round
Offered: Prematriculation, Freshman, Sophomore, Junior, Senior, Graduate level

Other Engineering Support Programs: Freshmen/transfer orientation program Summer intern program

082 Georgia Institute of Technology

INSTITUTION PROFILE

HEAD OF THE INSTITUTION
John P. Crecine
Phone: (404)894-5051   Fax: (404)853-9163

GENERAL INFORMATION
[All Students—Fall 1991]
Undergraduate enrollment 9,487
Graduate enrollment 3,327
Total institution enrollment **12,814**
Type of institution: Public   Calendar: Quarters
Location: Atlanta   Population: 2,800,000
Setting: Urban
Types of engineering degrees: Engineering
Other degree-granting colleges: Computing, Management, Policy, and International Affairs, Sciences, Architecture

ENGINEERING ADMISSIONS

ADMISSIONS OFFICE CONTACT
Deborah D. Smith
Georgia Institute of Technology
Office of Admissions
225 North Avenue
Atlanta, GA 30332-0320
Phone: (404)894-4154   Fax: (404)894-9511

ENGINEERING COLLEGE ADMISSIONS INFORMATION
Admission to the engineering college: The decision is based on evaluation of high school grades, courses taken, and SAT/ACT.
Entrance Requirements: High School courses—English (4 years), Math (4 years), Social Sciences (3 years), Foreign Language; Laboratory Science (2 years), The SAT or ACT is required but there are no minimum scores for the SAT or ACT for admission. SAT is preferred. ACT is accepted.
Entrance Recommendations: High School courses—American History, World History, Economics, Govt. (1 year), Chemistry, Plane Geometry (1 year), Trigonometry, Advanced Algebra (min. 1/2 year)
Requirements for out-of-state residents: Because GT is state supported, it admits all qualified state residents first. Out-of-state students are usually required to have higher SAT scores.
Requirements for foreign students: TOEFL (Score: 550); financial statement; Financial Aid is not available for international students. Transcripts must be translated to English.

DEPARTMENTAL ADMISSIONS INFORMATION
Admission to the engineering department: At the time of admission to the institution.

FINANCIAL INFORMATION

ESTIMATED EXPENSES (ACADEMIC YEAR)
[Expenses are for the 1992-93 nine-month academic year.]

State Residents	Undergraduate	Graduate
Tuition and fees	$ 2,202 [1]	$ 2,202 [1]
College room and board	$ 4,308	$ 4,308
Books and supplies	$ 795	$ 795
Other expenses	$ 1,164	$ 1,164
Total estimated expenses	**$ 8,469**	**$ 8,469**

Out-of-State Residents	Undergraduate	Graduate
Tuition and fees	$ 6,531 [1]	$ 6,531 [1]
College room and board	$ 4,308	$ 4,308
Books and supplies	$ 795	$ 795
Other expenses	$ 1,164	$ 1,164
Total estimated expenses	**$12,798**	**$12,798**

Notes: (1) A student holding a 1/3 time or greater research or teaching assistantship only pays reduced fees of approximately $163.00 per quarter.

FINANCIAL AID OFFICE CONTACT
Curley M. Williams, Director
Georgia Institute of Technology
Office of Student Financial Plan. & Services
225 North Avenue
Atlanta, GA 30332-0460
Phone: (404)894-4160   Fax: (404)853-9396

GENERAL FINANCIAL AID INFORMATION
Forms accepted/required: FAT, SAR, Free Application for Federal Student Aid preferred; CSS-FAF acceptable, Georgia Tech Financial Aid Application
Additional financial aid information: Merit and need based scholarships; Short-term and emergency loans; Participates in the Pell Grant Program, the Supplemental Educational Opportunity Grant (SEOG), the College Work-Study Program, the Perkins Loan Program, and the Stafford Loan Program.

ENGINEERING COLLEGE INFORMATION
[For additional personnel, refer to the *Appendix*.]

HEAD OF ENGINEERING
John A. White
Phone: (404)894-3350   Fax: (404)853-0168
EMail: jwhite@gatech.edu

ENGINEERING COLLEGE ADDRESS
College of Engineering
Georgia Institute of Technology
225 North Avenue
Atlanta, GA 30332-0360
Phone: (404)894-3350   Fax: (404)853-0168

ENROLLMENTS—BY CLASS
[Numbers are baccalaureate enrollments for the fall 1991 term, unless footnoted.]
1st-year students/Freshmen .... 1,760 [2]
2nd-year students/Sophomores .... 1,530 [2]
3rd-year students/Juniors .... 1,608 [2]
4th-year students/Seniors .... 1,888 [2]
Total **6,786**

Notes: (2) Includes full-time, part-time, and students on co-op work quarter.

NUMBER OF DEGREES AWARDED—BY PROGRAM
[Numbers are engineering baccalaureate degrees awarded during the 1991-1992 academic year, unless otherwise footnoted. For full details about each engineering program, refer to the *Tables of Degree Programs*.]

Aerospace Engineering 65
Ceramic Engineering 1
Chemical Engineering 72
Civil Engineering 116
Computer Engineering 14
Electrical Engineering 302
Engineering Science & Mechanics 7
Environmental Engineering .... — [2]
Industrial Engineering 253
Materials Science & Engineering 12
Mechanical Engineering 331
Nuclear Engineering 7
Textile Chemistry 5
Textile Engineering 14
Textile Management & Technology 8
Total **1,207**

Notes: (2) No undergraduate degree in Environmental Engineering is offered.

PERCENTAGE OF DEGREES AWARDED—BY CATEGORY
[Percentages are of all engineering baccalaureate degrees awarded during the 1991-1992 academic year, unless otherwise footnoted.]
African Americans 7.4%
Asian/Pacific Island Americans 7.2%
Hispanic Americans 3.1%
Native Americans .... – %
Foreign Citizens 2.7%
All Others 79.6%
Women 21.2%
Persons with disabilities .... – %
Students over 25 years of age .... – % (A)

Notes: (A) Data not available.

GRADUATE ENROLLMENTS & DEGREES AWARDED
Master's enrollment 1,165
Master's degrees awarded 579
Doctoral enrollment 908
Doctoral degrees awarded 129

ENGINEERING STUDENT DATA
Applicants to the engineering college
Number of applicants to engineering college 4,622
Percent offered admission 66.0%
Matriculated engineering students
Percentage in top quartile (25%) of High School class 99.9
Average SAT scores: Math—663, Verbal—556, Combined—1219
Average ACT scores: Math—32

FULL- & PART-TIME FACULTY—BY DEPARTMENT
[Figures are the head count of full-time faculty and the full-time equivalent (FTE) of part-time faculty for each engineering department or equivalent.]

Department	Full	Part
Chemical Eng	25	–
Civil Eng (incl Env'l Eng & Eng Sci Mech)	34	2.0
Electrical Eng (incl Computer Eng)	83	1.0
Industrial & Systems Eng	51	0.7
Mechanical Eng (incl Nuclear Eng)	56	1.5
Aerospace Eng	27	0.3
Materials Science & Eng (incl Ceramics)	12	–
Textile & Fiber Eng (incl Polymer & Textile Chem)	11	–

COLLEGE DESCRIPTION
Ga Tech's College of Engineering offers both the resources of a major technological university and a location in the heart of cosmopolitan Atlanta. The engineering education programs offer a diverse student body and faculty committed to excellence in teaching and research. In 1990-91, Ga Tech's College of Engineering ranked first in total engineering degrees awarded, first in engineering degrees awarded to underrepresented minorities, and first in baccalaureate degrees to women. In 1992, Ga Tech was the first majority institution to receive the NSF Institutional Achievement Award for increased minority participation in engineering, mathematics, and the sciences. The College recently occupied its new Manufacturing Research Center supporting a major, multi-disciplinary research effort in modern manufacturing systems and technologies. In 1996, Ga Tech will serve as the site of the Olympic Village, a unique opportunity for international collaboration for all students, staff, and faculty.

DEGREE PROGRAMS DESCRIPTION
Most of the B.S. level programs are also available as cooperative education programs. Most of the M.S. programs are offered as both designated and undesignated degrees for incoming students with or without ABET accredited bachelor's degrees, respectively.

DUAL DEGREE PROGRAMS
Engineering baccalaureate graduates with dual degrees 56
Enrollment requirements: Applicants must meet all requirements for transfer students.

TRANSFER INFORMATION
Residency Requirements: Candidates for an undergraduate degree must complete in residence the final fifty quarter hours of credit required for the degree.

Transfer via Articulation Agreements
Admission to engineering: Regents Engineering Transfer Program which is available at several state schools.
Engineering graduates transferred from ... 4-yr: 12   2-yr: –
Requirements: Must meet all requirements for transfer students.

Transfer without Articulation Agreements
Admission to engineering: Applicants must meet all requirements for transfer students.
Engineering graduates transferred from ... 4-yr: –   2-yr: –
Requirements: Transfer students must have 45 quarter hours or 30 semester hours with a minimum of 2.7 GPA for residents and a 3.0 GPA for non-residents. Transfer students must have 2 courses of calculus, chemistry and English.

GRADUATION REQUIREMENTS
The Institute requires at least a 2.0 average for graduation. Some programs require a 2.0 or better in major courses or a subset of technical coursework & C or better in all lower division math courses.

STUDENT PROGRAMS

PROFESSIONAL AND HONORARY SOCIETIES
[For key to acronyms, see *Introduction*.]
Professional Societies: AATCC, AIAA, AIChE, AIME, ANS, ASCE, ASME, IEEE, IIE, ORSA, American Ceramic Society, American Chemical Society
Honorary Societies: Alpha Pi Mu, Chi Epsilon, Eta Kappa Nu, Keramos, Omega Chi Epsilon, Pi Tau Sigma, Sigma Gamma Tau, Tau Beta Pi, Phi Kappa Phi, Alpha Nu Sigma

SUPPORT PROGRAMS
Student Chapter Organizations: Society of Women Engineers, National Society of Black Engineers, Society of Hispanic Professional Engineers, Black Graduate Student Association, Georgia Tech African-American Association, 15 various int'l support organizations.

Office of Minority Educational Development
For: Ethnic Minorities   **Available:** Year round
Offered: Prematriculation, Freshman, Sophomore, Junior, Senior, Graduate level

Minority Program
For: Ethnic Minorities   **Available:** Year round
Offered: Freshman, Sophomore, Junior, Senior, Graduate level

Women's Programs
For: Women   **Available:** Academic year
Offered: Freshman, Sophomore, Junior, Senior, Graduate level

Students of Tech Expand your Potential (STEP)
For: Ethnic Minorities   **Available:** Academic year
Offered: Freshman, Sophomore, Junior, Senior, Graduate level

Other Engineering Support Programs: Orientation programs are held during the summer and at the beginning of each quarter for all freshmen and transfers. Other support programs include: Career planning and placement services, counseling, Asset Program (pairs faculty with freshmen to assist in adjustment to Ga Tech), Freshman Experience (incoming support groups and services to assist in the transition to college life).

083 Gonzaga University

INSTITUTION PROFILE

HEAD OF THE INSTITUTION
Bernard J. Coughlin
Phone: (509)328-4220   Fax: (509)484-2818

GENERAL INFORMATION
[All Students—Fall 1991]
Undergraduate enrollment 2,951
Graduate enrollment 1,792
Total institution enrollment 4,743
Type of institution: Private   **Calendar:** Semesters
Location: Spokane   **Population:** 171,000
Setting: Urban
Types of engineering degrees: Engineering
Other degree-granting colleges: Arts & Sciences, Business Administration, Education, Law, Professional Studies

ENGINEERING ADMISSIONS

ADMISSIONS OFFICE CONTACT
Phillip Ballinger
Gonzaga University
502 E Boone
Spokane, WA 99258
Phone: (509)328-4220   Fax: (509)484-2818

ENGINEERING COLLEGE ADMISSIONS INFORMATION
Entrance Requirements: SAT (Score: 1000), ACT (Score: 22), High School courses—English (4 years), Mathematics (3 years), Foreign Language (2 years), History (1 year)
Requirements for foreign students: TOEFL (Score: 520); financial statement

DEPARTMENTAL ADMISSIONS INFORMATION
Admission to the engineering department: At the time of admission to the institution.

FINANCIAL INFORMATION

ESTIMATED EXPENSES (ACADEMIC YEAR)
[Expenses are for the 1992-93 nine-month academic year.]

All Students	Undergraduate	Graduate
Tuition and fees	$11,600	$ 9,100
College room and board	$ 4,100	$ 6,100
Books and supplies	$ 600	$ 850
Other expenses	$ 2,300	$ 2,300
Total estimated expenses	**$18,600**	**$18,350**

FINANCIAL AID OFFICE CONTACT
Nancy E. Ryan, *Associate Director, Financial Aid*
Gonzaga University
502 E. Boone
Spokane, WA 99258
Phone: (509)328-4220   Fax: (509)484-2818

GENERAL FINANCIAL AID INFORMATION
Forms accepted/required: CSS-FAF, IRS, SAR
Additional financial aid information: Need based scholarships, merit scholarships, grant programs, work study jobs, loans.

ENGINEERING COLLEGE INFORMATION

[For additional personnel, refer to the *Appendix*.]

HEAD OF ENGINEERING
Zia A. Yamayee
Phone: (509)328-4220   Fax: (509)484-2818

ENGINEERING COLLEGE ADDRESS
School of Engineering
Gonzaga University
502 E. Boone Avenue
Spokane, WA 99258
Phone: (509)328-4220   Fax: (509)484-5871

ENROLLMENTS—BY CLASS
[Numbers are baccalaureate enrollments for the fall 1991 term, unless otherwise footnoted.]
1st-year students/Freshmen 95
2nd-year students/Sophomores 67
3rd-year students/Juniors 60
4th-year students/Seniors 86
Total 308

NUMBER OF DEGREES AWARDED—BY PROGRAM
[Numbers are engineering baccalaureate degrees awarded during the 1991-1992 academic year, unless otherwise footnoted. For full details about each engineering program, refer to the *Tables of Degree Programs*.]
Civil Engineering 12
Electrical Engineering 22
Mechanical Engineering 12
Total 46

PERCENTAGE OF DEGREES AWARDED—BY CATEGORY
[Percentages are of all engineering baccalaureate degrees awarded during the 1991-1992 academic year, unless otherwise footnoted.]
African Americans .... — %
Asian/Pacific Island Americans .... — %
Hispanic Americans .... — %
Native Americans .... — %
Foreign Citizens .... — %
All Others .... 100.0 %
Women .... — %
Persons with disabilities .... — %
Students over 25 years of age .... 17.0 %

GRADUATE ENROLLMENTS & DEGREES AWARDED
Master's enrollment 8
Master's degrees awarded 1
Doctoral enrollment .... —
Doctoral degrees awarded .... —

ENGINEERING STUDENT DATA

Applicants to the engineering college
Number of applicants to engineering college 291
Percent offered admission .... —% [1]

Matriculated engineering students
Percentage in top quartile (25%) of High School class 67.0
Average SAT scores: Math—514, Verbal—468, Combined—982 [2]
Average ACT scores: Composite—24

Notes: (1) Students with low GPAs and/or scores, if admissible are placed in Arts & Sciences. (2) 1/4 of all students take ACT. SAT + ACT average = 1010

FULL- & PART-TIME FACULTY—BY DEPARTMENT
[Figures are the head count of full-time faculty and the full-time equivalent (FTE) of part-time faculty for each engineering department or equivalent.]

Department	Full	Part
Civil Eng	5	—
Electrical Eng	7	—
Mechanical Eng	6	—

COLLEGE DESCRIPTION

The school offers three ABET-accredited undergraduate degree programs in civil engineering, electrical engineering, and mechanical engineering which emphasize strong and rigorous technical curricula, combined with a broad liberal arts education. Computer-aided design/computer-aided engineering is available to the student. The curriculum prepares graduates to become competent practicing professionals; it prepares the students for graduate school and ultimately for research and/or teaching in academia or industry. The classes also challenge the intellect of the students and help them learn the value of analytical and logical thinking. There are opportunities to minor in computer science, mathematics or physics. A senior design project in the last semester is required along with the recommendation that the national standardized Fundamentals of Engineering exam be taken. Average number of years required to actually complete the bachelor's degree: 4.5.

DEGREE PROGRAMS DESCRIPTION

Four-year Bachelor of Science degrees in civil, electrical and mechanical engineering are offered. The goal of the undergraduate programs is to provide an engineering education that prepares the student with a baccalaureate degree to be a professional engineer. The programs provide a base for graduate study and for lifelong learning in support of evolving career objectives, which include being informed, effective, and responsible participants in the engineering profession and society.

TRANSFER INFORMATION

Residency Requirements: Transfer students must complete at least 30 semester hours in residence. Students may not transfer more than sixty-four semester hour or ninety-six quarter hour credits from a two-year college.

Transfer via Articulation Agreements
Admission to engineering: At any time
Engineering graduates transferred from   4-yr: 5   2-yr: 9
Requirements: 2.5 GPA from a two year degree program; 2.0 GPA from a four year institution.

Transfer without Articulation Agreements
Admission to engineering: At any time
Engineering graduates transferred from   4-yr: —   2-yr: —
Requirements: 2.5 GPA from a two year degree program; 2.0 GPA from a four year institution.

GRADUATION REQUIREMENTS

Fulfillment of the general degree requirements of the University including the University core curriculum. Completion of the common core courses in the School of Engineering. Attainment of an average cumulative grade point of 2.0 in all engineering course work.

STUDENT PROGRAMS

PROFESSIONAL AND HONORARY SOCIETIES
[For key to acronyms, see *Introduction*.]
Professional Societies: ASCE, ASME, IEEE
Honorary Societies: Delta Tau Sigma, Alpha Sigma Nu, Jesuit Honor Society

SUPPORT PROGRAMS
Student Chapter Organizations: Society of Women Engineers
Other Engineering Support Programs: Students are advised by engineering faculty on course requirements and electives. The admissions and student life departments provide freshmen/transfer student orientation programs.

084 Grand Valley State University

INSTITUTION PROFILE
HEAD OF THE INSTITUTION
Arend D. Lubbers
Phone: (616)895-2182 **Fax:** (616)895-3503

GENERAL INFORMATION
[All Students—Fall 1991]
Undergraduate enrollment 10,527
Graduate enrollment 2,340
Total institution enrollment **12,867**
Type of institution: Public **Calendar:** Semesters
Nearest city: Grand Rapids **Population:** 186,530
Miles from main campus: 12 **Setting:** Urban
Types of engineering degrees: Engineering
Other degree-granting colleges: Allied Health Sciences, Arts & Sciences, Business Administration, Communication & Journalism, Education, Fine & Performing Arts, Humanities & Social Sciences, Nursing

ENGINEERING ADMISSIONS
ADMISSIONS OFFICE CONTACT
Paul D. Plotkowski
Grand Valley State University
301 West Fulton, Suite 617
Grand Rapids, MI 49504-6495
Phone: (616)771-6750 **Fax:** (616)771-6642

ENGINEERING COLLEGE ADMISSIONS INFORMATION
Entrance Requirements: ACT (Score: 19), High School courses—Mathematics (3 years), English (4 years), Laboratory Science (2 years), Non-Lab Science (1 year), The freshman class average ACT composite is over 23 and the average high school GPA is 3.26. The middle 50% of the freshman class scored between 20 & 25 on the ACT.
Entrance Recommendations: SAT (Score: 810), ACT (Score: 19), High School courses—Mathematics (4 years), English (4 years), Laboratory Science (3 years)
Requirements for foreign students: TOEFL (Score: 550); financial statement

DEPARTMENTAL ADMISSIONS INFORMATION
Admission to the engineering department: At the end of the second year.
Additional information: Secondary admission is on a competitive basis. Requirements for application are: 1) Completion of Fundamentals of Engineering course sequence (math, phys, chem, English, fr. & soph. engineering) with overall GPA of 2.5 and a C or better in all fundamentals courses.

FINANCIAL INFORMATION
ESTIMATED EXPENSES (ACADEMIC YEAR)
[Expenses are for the 1992-93 nine-month academic year.]

State Residents	Undergraduate	Graduate
Tuition and fees	$ 2,588	$ –
College room and board	$ –[1]	$ –
Books and supplies	$ 400	$ –
Other expenses	$ 90[2]	$ –
Total estimated expenses	**$ 3,078**	**$ –**

Out-of-State Residents	Undergraduate	Graduate
Tuition and fees	$ 5,870	$ –
College room and board	$ –[1]	$ –
Books and supplies	$ 400	$ –
Other expenses	$ 90[2]	$ –
Total estimated expenses	**$ 6,360**	**$ –**

Notes: (1) There are a variety of room and board options available at a wide range of expense. (2) Lab fees are assigned for courses which require special equipment. This figure represents three such courses per year.

FINANCIAL AID OFFICE CONTACT
Ken Fridsma, *Director of Financial Aid*
Grand Valley State University
One Campus Drive
Siedman House
Allendale, MI 49401
Phone: (616)895-3234 **Fax:** (616)895-3180

GENERAL FINANCIAL AID INFORMATION
Forms accepted/required: FAF
Additional financial aid information: Extensive need based financial aid is available through participation in all federal, state and campus based programs. In addition, Grand Valley has a very generous package of scholarships based upon performance which are not need based.

ENGINEERING COLLEGE INFORMATION
[For additional personnel, refer to the *Appendix*.]

HEAD OF ENGINEERING
Paul D. Plotkowski, *Director*
Phone: (616)771-6750 **Fax:** (616)771-6642
EMail: plotkowp@gvsu.edu

ENGINEERING COLLEGE ADDRESS
School of Engineering
Grand Valley State University
301 West Fulton
Suite 617
Grand Rapids, MI 49504-6495
Phone: (616)771-6750 **Fax:** (616)771-6642

ENROLLMENTS—BY CLASS
[Numbers are baccalaureate enrollments for the fall 1991 term, unless otherwise footnoted.]
1st-year students/Freshmen 104
2nd-year students/Sophomores 73
3rd-year students/Juniors 47
4th-year students/Seniors 35
Total **259**

NUMBER OF DEGREES AWARDED—BY PROGRAM
[Numbers are engineering baccalaureate degrees awarded during the 1991-1992 academic year, unless otherwise footnoted. For full details about each engineering program, refer to the *Tables of Degree Programs*.]
Engineering .... –
General Engineering 23
Total **23**

PERCENTAGE OF DEGREES AWARDED—BY CATEGORY
[Percentages are of all engineering baccalaureate degrees awarded during the 1991-1992 academic year, unless otherwise footnoted.]
African Americans 7.8%
Asian/Pacific Island Americans 3.5%
Hispanic Americans 1.6%
Native Americans 1.0%
Foreign Citizens 0.5%
All Others 85.6%
Women 12.5%
Persons with disabilities .... –%[A]
Students over 25 years of age .... –%[A]

Notes: (A) Data not available.

ENGINEERING STUDENT DATA
Applicants to the engineering college
Number of applicants to engineering college .... –[B]
Percent offered admission .... –%[B]

Matriculated engineering students
Percentage in top quartile (25%) of High School class ...... –[A]
Average ACT scores: Math—21.8[1], Composite—23.1[1]

Notes: (A) Data not available. (B) Data not applicable. (1) This data reflects the total freshman population at Grand Valley State University.

FULL- & PART-TIME FACULTY—BY DEPARTMENT
[Figures are the head count of full-time faculty and the full-time equivalent (FTE) of part-time faculty for each engineering department or equivalent.]

Department	Full	Part
School of Eng	9	1.0

COLLEGE DESCRIPTION
The engineering program at GVSU is very practical and teaching oriented. Admission to major standing prior to the junior year is competitive. Upon secondary admission, the students begin the mandatory co-op sequence. The program is designed to be completed in 4 years plus one additional summer term. The class sizes are small (by design) and the program is housed in a three year old $30 million facility.

DEGREE PROGRAMS DESCRIPTION
The Engineering Program at GVSU is accredited by ABET as a general engineering program with mechanical and electrical emphasis areas. Both programs are very practical in flavor. All courses are taught by faculty (no teaching assistants) and the program requires 1 year of co-operative education experience.

TRANSFER INFORMATION
Residency Requirements: Transfer students must complete the last 30 hours at GVSU. The engineering residence requirement is 24 credits of engineering courses at the 300 level or above at GVSU including the capstone senior project.

Transfer via Articulation Agreements
Admission to engineering: At any time
Engineering graduates transferred from 4-yr: 2 2-yr: 1
Requirements: Students must complete the fundamentals sequence and at least 8 credits of engineering work at GVSU prior to application for competitive admission to the school of engineering.

Transfer without Articulation Agreements
Admission to engineering: At any time
Engineering graduates transferred from ..... 4-yr: – 2-yr: –
Requirements: Students must complete the engineering fundamentals course sequence and at least 8 credits of engineering courses at GVSU prior to application for competitive admission to the school of engineering.

GRADUATION REQUIREMENTS
The student is required to have a C or better in all required engineering courses and an overall GPA of 2.5 or better. Successful completion of the co-op program is required.

STUDENT PROGRAMS
PROFESSIONAL AND HONORARY SOCIETIES
[For key to acronyms, see *Introduction*.]
Professional Societies: ACM, ACS, AIP, ASME, SPS
Honorary Societies: Phi Kappa Upsilon

SUPPORT PROGRAMS
Student Chapter Organizations: National Society of Black Engineers, Black Student Union, El Renacimiento, Minority Engineering Committee

Multicultural Center
For: Ethnic Minorities **Available:** Year round
Offered: Prematriculation, Freshman, Sophomore, Junior, Senior, Graduate level

Excel
For: Ethnic Minorities **Available:** Year round
Offered: Prematriculation, Freshman, Sophomore, Junior, Senior

Other Engineering Support Programs: GVSU has an extensive freshman/transfer student orientation program. In addition, careful student advising is coordinated by both professional counselors and faculty.

085 Grove City College

INSTITUTION PROFILE

HEAD OF THE INSTITUTION
Jerry H. Combee
Phone: (412)458-2500 **Fax:** (412)458-2190

GENERAL INFORMATION
[All Students—Fall 1991]
Undergraduate enrollment 2,169
Total institution enrollment 2,169
Type of institution: Private **Calendar:** Semesters
Nearest city: Pittsburgh **Population:** 2,056,705
Miles from main campus: 60 **Setting:** Small Town
Types of engineering degrees: Engineering
Other degree-granting colleges: Arts & Sciences, Business Administration, Communication & Journalism, Education, Fine & Performing Arts, Humanities & Social Sciences

ENGINEERING ADMISSIONS

ADMISSIONS OFFICE CONTACT
Jeffrey C. Mincey
Grove City College
Office of Admissions
100 Campus Drive
Grove City, PA 16127-2104
Phone: (412)458-2100 **Fax:** (412)458-2190

ENGINEERING COLLEGE ADMISSIONS INFORMATION
Entrance Requirements: High School courses—Algebra (2 years), Geometry (1 year), Trigonometry (1 year), Applicants must take either SAT or ACT test. No minimum score required.
Entrance Recommendations: High School courses—Pre-calculus (1 year), Physics (1 year), Chemistry (1 year), English (4 years)
Requirements for foreign students: Applicant must take TOEFL test. No minimum score required.

DEPARTMENTAL ADMISSIONS INFORMATION
Admission to the engineering department: At the time of admission to the institution.

FINANCIAL INFORMATION

ESTIMATED EXPENSES (ACADEMIC YEAR)
[Expenses are for the 1992-93 nine-month academic year.]

All Students	Undergraduate	Graduate
Tuition and fees	$ 5,490	$ —
College room and board	$ 2,780	$ —
Books and supplies	$ 400	$ —
Other expenses	$ —	$ —
Total estimated expenses	**$ 8,670**	**$ —**

FINANCIAL AID OFFICE CONTACT
Anne P. Bowne, *Director of Financial Aid*
Grove City College
100 Campus Drive
Grove City, PA 16127-2104
Phone: (412)458-2163 **Fax:** (412)458-2190

GENERAL FINANCIAL AID INFORMATION
Forms accepted/required: CSS-FAF, FAT, Institutional
Additional financial aid information: Trustee Academic Scholarships, College Scholarships (based on both academics and need), Student Loans, on campus employment.

ENGINEERING COLLEGE INFORMATION

[For additional personnel, refer to the *Appendix*.]

HEAD OF ENGINEERING
Joseph F. Goncz Jr.
Phone: (412)458-2033 **Fax:** (412)458-2190

ENGINEERING COLLEGE ADDRESS
Department of Engineering
Grove City College
100 Campus Drive
Grove City, PA 16127-2104
Phone: (412)458-2000 **Fax:** (412)458-2190

ENROLLMENTS—BY CLASS
[Numbers are baccalaureate enrollments for the fall 1991 term, unless otherwise footnoted.]
1st-year students/Freshmen 165
2nd-year students/Sophomores 85
3rd-year students/Juniors 57
4th-year students/Seniors 46
Total 353

NUMBER OF DEGREES AWARDED—BY PROGRAM
[Numbers are engineering baccalaureate degrees awarded during the 1991-1992 academic year, unless otherwise footnoted. For full details about each engineering program, refer to the *Tables of Degree Programs*.]
Electrical Engineering 22
Mechanical Engineering 13
Total 35

PERCENTAGE OF DEGREES AWARDED—BY CATEGORY
[Percentages are of all engineering baccalaureate degrees awarded during the 1991-1992 academic year, unless otherwise footnoted.]
African Americans .... — %
Asian/Pacific Island Americans .... — %
Hispanic Americans .... — %
Native Americans .... — %
Foreign Citizens .... 2.9 %
All Others .... 97.1 %
Women .... 8.6 %
Persons with disabilities .... — %
Students over 25 years of age .... — %

ENGINEERING STUDENT DATA
Applicants to the engineering college
Number of applicants to engineering college 409
Percent offered admission .... 71.0 %
Matriculated engineering students
Percentage in top quartile (25%) of High School class .... 74.0 [1]
Average SAT scores: Math—615, Verbal—505, Combined—1120
Average ACT scores: Composite—26

Notes: (1) Percentage in top 20% of high school class.

FULL- & PART-TIME FACULTY—BY DEPARTMENT
[Figures are the head count of full-time faculty and the full-time equivalent (FTE) of part-time faculty for each engineering department or equivalent.]

Department	Full	Part
Department of Eng	6	—

COLLEGE DESCRIPTION

Grove City is an independent Christian college of liberal arts and sciences. From its founding days the College has endeavored to give young people the best in liberal and scientific education at the lowest possible cost, and in keeping with this historic policy still maintains one of the lowest tuitions of any independent, quality college. The greatest strength of the engineering program is a single-minded focus on quality undergraduate education. Student class size is small and the style of instruction is essentially "one-on-one," with faculty grading all assignments, and extensive student/faculty interaction. Engineering graduates are well grounded in fundamental engineering concepts and are well prepared to enter the engineering field as well as the civic and cultural life of society.

DEGREE PROGRAMS DESCRIPTION

The B.S. degree programs in electrical and mechanical engineering prepare students for graduate study and entry level positions in the engineering profession. The teaching of engineering design is a program strength. The small senior-class size allows the departments to offer the Senior Experience in Design program, which spans numerous courses in the senior year. The SED programs require groups of students to work on significant design problems, simultaneously promoting interaction with industry and coordination of design work across several courses. Several student design teams have placed first and second in national design competitions.

TRANSFER INFORMATION

Residency Requirements: Students must live on campus or commute from home.
Transfer without Articulation Agreements
Admission to engineering: At any time
Engineering graduates transferred from .. 4-yr: 2 2-yr: 1
Requirements: Verification of honorable dismissal and acceptable grades.

GRADUATION REQUIREMENTS

B.S. programs in Engineering require 136 semester hours. A minimum 2.0 GPA overall and in the major is required for graduation.

STUDENT PROGRAMS

PROFESSIONAL AND HONORARY SOCIETIES
[For key to acronyms, see *Introduction*.]
Professional Societies: ASME, IEEE, NSPE, SAE, SPS

SUPPORT PROGRAMS
Student Chapter Organizations: Society of Women Engineers
Association For Women Students
For: Women **Available:** Academic year
Offered: Freshman, Sophomore, Junior, Senior
Other Engineering Support Programs: Summer and Fall Orientation Programs for Freshman; Career Development and Placement Center; Scholar Lecture Series; Guest Artist Series; Alcohol and Drug Awareness Program; Tutoring and Study Skill Course.

086 Harvard University

■ INSTITUTION PROFILE
HEAD OF THE INSTITUTION
Neil L. Rudenstine
Phone: (617)495-1502

GENERAL INFORMATION
[All Students—Fall 1991]
Undergraduate enrollment	6,678
Graduate enrollment	3,368
Total institution enrollment	**10,246**

Type of institution: Private   **Calendar:** Semesters
Nearest city: Boston   **Population:** 600,000
Miles from main campus: 4   **Setting:** Urban
Types of engineering degrees: Engineering
Other degree-granting colleges: Business Administration, Law, Medicine, Design, Government

■ ENGINEERING ADMISSIONS
ADMISSIONS OFFICE CONTACT
Marlyn M. Lewis
Harvard University
8 Garden Street
Cambridge, MA 02138
Phone: (617)495-5339

ENGINEERING COLLEGE ADMISSIONS INFORMATION
Entrance Requirements: SAT, ACT, Applicants must submit either their SAT or ACT scores and any three achievement tests.
Requirements for foreign students: TOEFL

DEPARTMENTAL ADMISSIONS INFORMATION
Admission to the engineering department: At the end of the first year.

■ FINANCIAL INFORMATION
ESTIMATED EXPENSES (ACADEMIC YEAR)
[Expenses are for the 1992-93 nine-month academic year.]
All Students	Undergraduate	Graduate
Tuition and fees	$17,054	$17,054
College room and board	$5,840	$5,770
Books and supplies	$1,736	$775
Other expenses	$1,220	$ –
Total estimated expenses	**$25,850**	**$23,599**

FINANCIAL AID OFFICE CONTACT
James S. Miller, *Director of Financial Aid*
Harvard University
8 Garden Street
Cambridge, MA 02138
Phone: (617)495-1580

GENERAL FINANCIAL AID INFORMATION
Forms accepted/required: FAF, Institutional
Additional financial aid information: All financial aid awards are based on financial need as determined by the committee on financial aid. We offer no aid based on academic excellence, musical or athletic talent, or other criteria separate from financial need.

■ ENGINEERING COLLEGE INFORMATION
[For additional personnel, refer to the *Appendix*.]
HEAD OF ENGINEERING
Paul C. Martin
Phone: (617)495-5829   Fax: (617)495-9837

ENGINEERING COLLEGE ADDRESS
Division of Applied Sciences
Harvard University
29 Oxford Street
Cambridge, MA 02138
Phone: (617)495-5829   Fax: (617)495-9837

ENROLLMENTS—BY CLASS
[Numbers are baccalaureate enrollments for the fall 1991 term, unless otherwise footnoted.]
1st-year students/Freshmen	0 [B]
2nd-year students/Sophomores	49
3rd-year students/Juniors	33
4th-year students/Seniors	39
Total	**121**

Notes: [B] Data not applicable.

NUMBER OF DEGREES AWARDED—BY PROGRAM
[Numbers are engineering baccalaureate degrees awarded during the 1991-1992 academic year, unless otherwise footnoted. For full details about each engineering program, refer to the *Tables of Degree Programs*.]
Engineering Science(s)	18 [1]
Total	**18**

Notes: [1] 15 graduates in the A.B. program.

PERCENTAGE OF DEGREES AWARDED—BY CATEGORY
[Percentages are of all engineering baccalaureate degrees awarded during the 1991-1992 academic year, unless otherwise footnoted.]
African Americans	– % [A]
Asian/Pacific Island Americans	– % [A]
Hispanic Americans	– % [A]
Native Americans	– % [A]
Foreign Citizens	– % [A]
All Others	100.0 %
Women	– % [A]
Persons with disabilities	– % [A]
Students over 25 years of age	– % [A]

Notes: [A] Data not available.

GRADUATE ENROLLMENTS & DEGREES AWARDED
Master's enrollment	6
Master's degrees awarded	18 [1]
Doctoral enrollment	49
Doctoral degrees awarded	10

Notes: [1] Engineering Sciences Ph.D. candidates can receive Master's along the way to the Ph.D.

ENGINEERING STUDENT DATA
Applicants to the engineering college
Number of applicants to engineering college	1,430
Percent offered admission	14.0%

Matriculated engineering students
Percentage in top quartile (25%) of High School class	99.0

FULL- & PART-TIME FACULTY—BY DEPARTMENT
[Figures are the head count of full-time faculty and the full-time equivalent (FTE) of part-time faculty for each engineering department or equivalent.]
Department	Full	Part
Division of Applied Sciences	51	2.4

COLLEGE DESCRIPTION
Engineering students at Harvard receive a thorough and versatile engineering education in a stimulating liberal arts and science environment. The program includes an education in mathematics and the basic sciences and the engineering sciences, challenging design group problems, and individual design projects. Career objectives of majors include engineering practice, research and business, law, and government with a large technical component. Most students complete programs in four years. 20% of the candidates for the S.B. degree take 5 years, supplementing the required program with a more extensive range of electives. Under an exchange agreement appropriate courses at the Massachusetts Institute of Technology may be taken for full credit at no additional cost.

DEGREE PROGRAMS DESCRIPTION
Harvard offers programs that lead to A.B., S.B., S.M., M.E., and Ph.D. degrees in Engineering Sciences. The requirement for the various degrees are flexible, allowing students to design programs that meet varied career objectives. The A.B. program, in which undergraduates can take full advantage of Harvard's diverse offerings, has two major options: one broadly-based option has a substantial component in mechanics and materials; the other focuses on electrical, systems, and computer engineering. The S.B. program, which meets requirements set by ABET, requires more course work in engineering, including challenging group and individual design projects in the junior and senior years.

TRANSFER INFORMATION
Residency Requirements: Harvard requires two years' residence.
Transfer via Articulation Agreements
Admission to engineering: At the end of the freshman year
Engineering graduates transferred from   . . . 4-yr: –   2-yr: –

GRADUATION REQUIREMENTS
Sixteen full (or 32 half) semester courses. During freshman year, students must take a half course in expository writing. Knowledge of one foreign language - achievement test score of 560 or better or an AP3, 560 on Harvard Placement test or passing a year of foreign language is required. Also, must pass a two-part quantitative reasoning requirement (written or on-site computer exam). Semester courses must be from the core curriculum.

■ STUDENT PROGRAMS
PROFESSIONAL AND HONORARY SOCIETIES
[For key to acronyms, see *Introduction*.]
Professional Societies: IIE
Honorary Societies: Phi Beta Kappa

SUPPORT PROGRAMS
Student Chapter Organizations: Harvard Society of Engineers, Black Scientists and Engineers

American Indians at Harvard
For: Ethnic Minorities

Asian American Association
For: Ethnic Minorities

Black Students' Association
For: Ethnic Minorities

La Organizacion
For: Ethnic Minorities

087 Harvey Mudd College

■ INSTITUTION PROFILE

HEAD OF THE INSTITUTION
Henry E. Riggs
Phone: (714)621-8120

GENERAL INFORMATION
[All Students—Fall 1991]
Undergraduate enrollment 616
Graduate enrollment 13
Total institution enrollment **629**
Type of institution: Private   **Calendar:** Semesters
Nearest city: Los Angeles   **Population:** 3,000,000
Miles from main campus: 35   **Setting:** Suburban
Types of engineering degrees: Engineering

■ ENGINEERING ADMISSIONS

ADMISSIONS OFFICE CONTACT
Patricia Coleman
Harvey Mudd College
301 E. 12th Street, Kinston Hall
Claremont, CA 91711-5990
Phone: (714)621-8011   **Fax:** (714)621-8360

ENGINEERING COLLEGE ADMISSIONS INFORMATION
Admission to the engineering college: Beginning of second semester of Sophomore year.
Entrance Requirements: SAT (Score: 1100), High School courses—English (4 years), Mathematics (4 years), Chemistry and Physics (1 year), History (1 year)
Requirements for foreign students: TOEFL (Score: 600); financial statement

DEPARTMENTAL ADMISSIONS INFORMATION
Admission to the engineering department: Middle of sophomore year.

■ FINANCIAL INFORMATION

ESTIMATED EXPENSES (ACADEMIC YEAR)
[Expenses are for the 1992-93 nine-month academic year.]

All Students	Undergraduate	Graduate
Tuition and fees	$13,480	$ –
College room and board	$ 5,580	$ –
Books and supplies	$ 450	$ –
Other expenses	$ 385	$ –
Total estimated expenses	**$19,895**	**$ –**

FINANCIAL AID OFFICE CONTACT
Noe Ortiz, *Director of Financial Aid*
Harvey Mudd College
301 E. 12th St
Claremont, CA 91711-5628
Phone: (714)621-8055

GENERAL FINANCIAL AID INFORMATION
Forms accepted/required: FAF
Additional financial aid information: Need-based scholarships, loans, part-time campus jobs.

■ ENGINEERING COLLEGE INFORMATION

[For additional personnel, refer to the *Appendix*.]

HEAD OF ENGINEERING
John I. Molinder
Phone: (714)621-8019   **Fax:** (714)621-8465
EMail: JMOLINDER@HMCVAX.CLAREMONT.EDU

ENGINEERING COLLEGE ADDRESS
Engineering Department
Harvey Mudd College
301 E. 12th Street
Parsons Building, Room 369
Claremont, CA 91711-5990
Phone: (714)621-8019   **Fax:** (714)621-8465

ENROLLMENTS—BY CLASS
[Numbers are baccalaureate enrollments for the fall 1991 term, unless otherwise footnoted.]
1st-year students/Freshmen 179
2nd-year students/Sophomores 176
3rd-year students/Juniors 123
4th-year students/Seniors 137
Total **615**

NUMBER OF DEGREES AWARDED—BY PROGRAM
[Numbers are engineering baccalaureate degrees awarded during the 1991-1992 academic year, unless otherwise footnoted. For full details about each engineering program, refer to the *Tables of Degree Programs*.]
Engineering 55
Total **55**

PERCENTAGE OF DEGREES AWARDED—BY CATEGORY
[Percentages are of all engineering baccalaureate degrees awarded during the 1991-1992 academic year, unless otherwise footnoted.]
African Americans .... – %
Asian/Pacific Island Americans .... 25.0 %
Hispanic Americans .... 9.0 %
Native Americans .... – %
Foreign Citizens .... 3.0 %
All Others .... 63.0 %
Women .... 23.0 %
Persons with disabilities .... – %
Students over 25 years of age .... – %

GRADUATE ENROLLMENTS & DEGREES AWARDED
Master's enrollment 6
Master's degrees awarded 6
Doctoral enrollment .... –
Doctoral degrees awarded .... –

ENGINEERING STUDENT DATA
Applicants to the engineering college
Number of applicants to engineering college 541
Percent offered admission 51.0%

Matriculated engineering students
Percentage in top quartile (25%) of High School class ... 100.0
Average SAT scores: Math—580, Verbal—700

FULL- & PART-TIME FACULTY—BY DEPARTMENT
[Figures are the head count of full-time faculty and the full-time equivalent (FTE) of part-time faculty for each engineering department or equivalent.]

Department	Full	Part
Eng	15	–

COLLEGE DESCRIPTION

Harvey Mudd College, one of the six Claremont Colleges, is a private, residential institution located in a small town setting; the academic year is based on a two-semester system. The school offers a Bachelor of Science Degree in biology, chemistry, engineering, mathematics, and physics, with one-third of the curriculum in the technical core, one-third in the major, and one-third in humanities and social sciences. The engineering program is offered one one campus and includes required industry-sponsored clinic projects: it does not have extension centers. The master's degree in engineering, a fifth year program, is open only to Harvey Mudd students. Total institutional enrollment is 616 undergraduates and 13 graduates.

DEGREE PROGRAMS DESCRIPTION

The engineering program provides a broad-based knowledge and experience in synthesis and analysis. It is designed to prepare students for both professional practice and advanced study in various engineering specialties. Leading to an unspecialized bachelor's or master's degree, the program emphasizes an interdisciplinary approach to problem solving. Professional experience that draws on this broad knowledge base is provides by requiring students to undertake challenging design problems in the Engineering Clinic, working on industry-sponsored projects. Engineering majors may choose to emphasize a particular engineering specialty by appropriate choice of elective courses and clinic projects. Students may apply for the Master of Engineering fifth year program during their third year.

TRANSFER INFORMATION

Residency Requirements: Students must complete the last four semesters in residence.

Transfer without Articulation Agreements
Admission to engineering: At the end of the sophomore year
Engineering graduates transferred from .. 4-yr: 10 2-yr: –
Requirements: Good standing at institution previously attended; two references from a mathematics, science, or engineering teacher at candidate's college; three Achievement Test scores (see freshman requirements); transcripts; course descriptions; essay.

GRADUATION REQUIREMENTS

B.S. program in engineering requires 128 semester units with a minimum GPA of 2.0 in the major and CUM.

■ STUDENT PROGRAMS

PROFESSIONAL AND HONORARY SOCIETIES
[For key to acronyms, see *Introduction*.]
Professional Societies: ASME, IEEE

SUPPORT PROGRAMS
Student Chapter Organizations: Society of Women Engineers

NSF funded Freshman Institute
For: Ethnic Minorities   **Available:** Summer only
Offered: Prematriculation

Other Engineering Support Programs: Professional societies student chapters; fall orientation program; career planning and placement services.

088 University of Hawaii at Manoa

INSTITUTION PROFILE

HEAD OF THE INSTITUTION
Yuen C. Paul
Phone: (808)956-5280

GENERAL INFORMATION
[All Students—Fall 1991]
Undergraduate enrollment 13,017
Graduate enrollment 6,299
Total institution enrollment **19,316**
Type of institution: Public   **Calendar:** Semesters
Location: Honolulu   **Population:** 600,000
Setting: Urban
Types of engineering degrees: Engineering
Other degree-granting colleges: Architecture, Arts & Sciences, Business Administration, Education, Law, Medicine, Nursing, Hawaiian, Asian and Pacific Studies, Tropical Agriculture and Human Resources, Public Health, Library and Information Studies, Travel Industry Management

ENGINEERING ADMISSIONS

ADMISSIONS OFFICE CONTACT
Deane H. Kihara
University of Hawaii at Manoa
2540 Dole Street #250
Honolulu, HI 96822
Phone: (808)956-8404   **Fax:** (808)956-2291

ENGINEERING COLLEGE ADMISSIONS INFORMATION
Entrance Requirements: SAT, High School courses—English (4 years), Mathematics (4 years), Physics (1 year), Chemistry (1 year)
Entrance Recommendations: High School courses—Foreign Language (2 years), Engineering Drawing (1 year)
Requirements for foreign students: TOEFL; financial statement

DEPARTMENTAL ADMISSIONS INFORMATION
Admission to the engineering department: At the time of admission to the institution.

FINANCIAL INFORMATION

ESTIMATED EXPENSES (ACADEMIC YEAR)
[Expenses are for the 1992-93 nine-month academic year.]

State Residents	Undergraduate	Graduate
Tuition and fees	$ 1,440	$ 1,817
College room and board	$ 3,400	$ 3,400
Books and supplies	$ 650	$ 650
Other expenses	$ 1,345(A)	$ 1,345(1)
Total estimated expenses	**$ 6,835**	**$ 7,212**

Out-of-State Residents	Undergraduate	Graduate
Tuition and fees	$ 4,340	$ 5,337
College room and board	$ 3,100	$ 3,100
Books and supplies	$ 650	$ 650
Other expenses	$ 1,345(A)	$ 1,345(1)
Total estimated expenses	**$ 9,435**	**$ 10,432**

Notes: (A) Data not available. (1) Health insurance and personal expenses

FINANCIAL AID OFFICE CONTACT
Annabelle C. Fong, *Director, Financial Aid Office*
University of Hawaii at Manoa
2442 Campus Road
Honolulu, HI 96822
Phone: (808)956-7251   **Fax:** (808)956-5076

GENERAL FINANCIAL AID INFORMATION
Forms accepted/required: CSS-FAF, Institutional
Additional financial aid information: Need-based scholarships, merit-based scholarships; part-time jobs on campus; college work-study program.

ENGINEERING COLLEGE INFORMATION
[For additional personnel, refer to the *Appendix*.]

HEAD OF ENGINEERING
Reginald H. Young
Phone: (808)956-7727   **Fax:** (808)956-2291
EMail: young

ENGINEERING COLLEGE ADDRESS
College of Engineering
University of Hawaii at Manoa
2540 Dole Street #240
Honolulu, HI 96822
Phone: (808)956-7727   **Fax:** (808)956-2291

ENROLLMENTS—BY CLASS
[Numbers are baccalaureate enrollments for the fall 1991 term, unless otherwise footnoted.]
1st-year students/Freshmen 166
2nd-year students/Sophomores 180
3rd-year students/Juniors 234
4th-year students/Seniors 341
Total **921**

NUMBER OF DEGREES AWARDED—BY PROGRAM
[Numbers are engineering baccalaureate degrees awarded during the 1991-1992 academic year, unless otherwise footnoted. For full details about each engineering program, refer to the *Tables of Degree Programs*.]
Civil Engineering 49
Electrical Engineering 62
Mechanical Engineering 48
Total **159**

PERCENTAGE OF DEGREES AWARDED—BY CATEGORY
[Percentages are of all engineering baccalaureate degrees awarded during the 1991-1992 academic year, unless otherwise footnoted.]
African Americans .... – %
Asian/Pacific Island Americans 84.3%
Hispanic Americans 0.6%
Native Americans .... – %
Foreign Citizens .... – % (A)
All Others 15.1%
Women 24.5%
Persons with disabilities .... – % (A)
Students over 25 years of age 26.4%
Notes: (A) Data not available.

GRADUATE ENROLLMENTS & DEGREES AWARDED
Master's enrollment 142
Master's degrees awarded 35
Doctoral enrollment 48
Doctoral degrees awarded 7

ENGINEERING STUDENT DATA
Applicants to the engineering college
Number of applicants to engineering college 291
Percent offered admission 89.0%
Matriculated engineering students
Percentage in top quartile (25%) of High School class 71.0
Average SAT scores: Math—610, Verbal—450, Combined—1060

FULL- & PART-TIME FACULTY—BY DEPARTMENT
[Figures are the head count of full-time faculty and the full-time equivalent (FTE) of part-time faculty for each engineering department or equivalent.]

Department	Full	Part
Civil Eng	19	–
Electrical Eng	27	–
Mechanical Eng	15	–

COLLEGE DESCRIPTION
The programs of study in all engineering curricula provide a fundamental science-oriented education with coverage of communications, the humanities and social sciences; the basic physical sciences; and include both theoretical and experimental course work designed to enable our graduates to meet the challenges of our technology-oriented society. Cooperative education job assignments are options in all programs.

DEGREE PROGRAMS DESCRIPTION
The B.S. degrees in the three curricula prepare the graduate for entry-level positions in the engineering community or for graduate study. The College also offers advanced graduate studies at both the M.S. and Ph.D. levels.

TRANSFER INFORMATION
Residency Requirements: Students must earn a minimum of 30 semester hours in residence and are allowed to transfer a maximum of 60 semester hours.
Transfer via Articulation Agreements
Admission to engineering: At any time
Engineering graduates transferred from . . . 4-yr: 15   2-yr: 17
Requirements: Minimum GPA of 3.0. Completion of English Composition, two semesters of Calculus, University Physics with lab, General Chemistry with lab.
Transfer without Articulation Agreements
Admission to engineering: At any time
Engineering graduates transferred from . . . 4-yr: 9   2-yr: 2
Requirements: Minimum GPA of 3.0. Completion of English Composition, two semesters of Calculus, University Physics with lab, General Chemistry with lab.

GRADUATION REQUIREMENTS
B.S. programs in engineering require approximately 142 semester hours, including two years of a foreign language. GPA in the major courses must be at least 2.0.

STUDENT PROGRAMS

PROFESSIONAL AND HONORARY SOCIETIES
[For key to acronyms, see *Introduction*.]
Professional Societies: ASCE, ASHRAE, ASME, IEEE
Honorary Societies: Chi Epsilon, Eta Kappa Nu, Pi Tau Sigma

SUPPORT PROGRAMS
College Opportunities Program
For: Ethnic Minorities   **Available:** Year round
Offered: Prematriculation, Freshman, Sophomore

Operation Manong
For: Ethnic Minorities   **Available:** Year round
Offered: Freshman, Sophomore, Junior, Senior

Other Engineering Support Programs: Fall orientation program for all freshmen and transfer students; foreign student office; veterans services; placement center.

089 Hofstra University

■ INSTITUTION PROFILE
HEAD OF THE INSTITUTION
James M. Shuart
Phone: (516)463-6800 **Fax:** (516)564-4296

GENERAL INFORMATION
[All Students—Fall 1991]
Undergraduate enrollment 7,733
Graduate enrollment 4,276
Total institution enrollment 12,009
Type of institution: Private **Calendar:** Semesters
Nearest city: New York City **Population:** 8,000,000
Miles from main campus: 25 **Setting:** Suburban
Types of engineering degrees: Engineering
Other degree-granting colleges: Arts & Sciences, Business Administration, Education, Law, New College

■ ENGINEERING ADMISSIONS
ADMISSIONS OFFICE CONTACT
Joan I. Mohr
Hofstra University
Holland House 100
Hempstead, NY 11550-1090
Phone: (516)463-6700 **Fax:** (516)564-4296

ENGINEERING COLLEGE ADMISSIONS INFORMATION
Entrance Requirements: SAT, ACT, High School courses—English (4 years), Language (2 years), Mathematics (4 years), Physics and Chemistry (1 year)
Entrance Recommendations: High School courses—Computers (1 year), Mechanical Drawing (1 year), SAT Math Achievement
Requirements for foreign students: TOEFL (Score: 500)

DEPARTMENTAL ADMISSIONS INFORMATION
Admission to the engineering department: At the time of admission to the institution.

■ FINANCIAL INFORMATION
ESTIMATED EXPENSES (ACADEMIC YEAR)
[Expenses are for the 1992-93 nine-month academic year.]

All Students	Undergraduate	Graduate
Tuition and fees	$10,490	$ –
College room and board	$ 5,230	$ –
Books and supplies	$ 720	$ –
Other expenses	$ –	$ –
Total estimated expenses	$16,440	$ –

FINANCIAL AID OFFICE CONTACT
Jean A. Belmont, *Director*
Hofstra University
Financial Aid & Academic Records
126
Hempstead, NY 11550-1090
Phone: (516)463-6680 **Fax:** (516)564-4291

GENERAL FINANCIAL AID INFORMATION
Forms accepted/required: CSS-FAF
Additional financial aid information: Hofstra makes financial aid available to many students in the form of scholarships, grants, loans and jobs.

■ ENGINEERING COLLEGE INFORMATION
[For additional personnel, refer to the *Appendix*.]
HEAD OF ENGINEERING
David M. Rooney
Phone: (516)463-5545 **Fax:** (516)564-4296

ENGINEERING COLLEGE ADDRESS
Department of Engineering
Hofstra University
104A Weed Hall
133 Hofstra University
Hempstead, NY 11550-1090
Phone: (516)463-5544 **Fax:** (516)564-4296

ENROLLMENTS—BY CLASS
[Numbers are baccalaureate enrollments for the fall 1991 term, unless otherwise footnoted.]
1st-year students/Freshmen 55
2nd-year students/Sophomores 55
3rd-year students/Juniors 59
4th-year students/Seniors 102
Total **271**

NUMBER OF DEGREES AWARDED—BY PROGRAM
[Numbers are engineering baccalaureate degrees awarded during the 1991-1992 academic year, unless otherwise footnoted. For full details about each engineering program, refer to the *Tables of Degree Programs*.]
Electrical Engineering 14
Engineering . . . . –
Engineering Science(s) 4
Industrial Engineering 3
Mechanical Engineering 13
Total **34**

PERCENTAGE OF DEGREES AWARDED—BY CATEGORY
[Percentages are of all engineering baccalaureate degrees awarded during the 1991-1992 academic year, unless otherwise footnoted.]
African Americans . . . . 3.0 %
Asian/Pacific Island Americans . . . . 5.9 %
Hispanic Americans . . . . 5.9 %
Native Americans . . . . – %
Foreign Citizens . . . . 5.9 %
All Others . . . . 79.3 %
Women . . . . 5.9 %
Persons with disabilities . . . . – %
Students over 25 years of age . . . . 17.6 %

ENGINEERING STUDENT DATA
Applicants to the engineering college
Number of applicants to engineering college 368
Percent offered admission . . . . 75.0 %
Matriculated engineering students
Percentage in top quartile (25%) of High School class 60.0
Average SAT scores: Math—551, Verbal—463, Combined—1014

FULL- & PART-TIME FACULTY—BY DEPARTMENT
[Figures are the head count of full-time faculty and the full-time equivalent (FTE) of part-time faculty for each engineering department or equivalent.]

Department	Full	Part
Department of Eng	14	1.5

COLLEGE DESCRIPTION
The Department of Engineering is located within the College of Liberal Arts and Sciences. All the engineering programs are administered by the department and lead to either a Bachelor of Engineering or Bachelor of Science degree. There is also a Bachelor of Arts program for students seeking to combine a business and engineering background. The first two years of all curricula are virtually identical, with an emphasis on mathematics, the natural sciences and a strong component of the liberal arts. Each student then chooses a major and works closely with an academic adviser in determining the specialization within that program. At more advanced levels, classes tend to be small and opportunities for individual project work and research are readily available. All programs are offered on a full-time and part-time basis, with the average time to complete the bachelor's degree being four and one-half years for full-time students.

DEGREE PROGRAMS DESCRIPTION
The Department of Engineering emphasizes small classes and extensive interaction between students and full-time faculty, in a curriculum that offers a balance between analytical, laboratory, and design courses. The Engineering Science degree combines the elements of both mechanical and electrical engineering among its common courses, and then allows the student to specialize (with 18 semester hours of courses and an independent design project) in one of three currently growing fields: biomedical, environmental or structural engineering. The Mechanical Engineering degree also permits concentrations in thermal systems, applied mechanics or in aerospace engineering. It includes 15 credits of technical electives and a choice of two out of three design courses. The Electrical Engineering degree is somewhat more structured, with 9 credits of electives and a three-credit individual design project. There is, however, a Computer Engineering option which replaces about 18 credits of electrical courses with an equal amount of computer science credits.

DUAL DEGREE PROGRAMS
Engineering baccalaureate graduates with dual degrees 1

TRANSFER INFORMATION
Residency Requirements: At least 30 semester hours in residence, as well as meeting core course requirements in the College of Liberal Arts and Sciences.
Transfer without Articulation Agreements
Admission to engineering: Transfer students must complete at least 24 credits of college work or meet freshman application criteria. No more than 69 credits from two-year engineering programs can be transferred.
Engineering graduates transferred from . . 4-yr: 5 2-yr: 5

GRADUATION REQUIREMENTS
A student must satisfy the University core course requirements in science, mathematics and social science. In addition, the student must have a 2.0 GPA in the major and in a selected subset of engineering courses.

■ STUDENT PROGRAMS
PROFESSIONAL AND HONORARY SOCIETIES
[For key to acronyms, see *Introduction*.]
Professional Societies: AIAA, ASCE, ASME, IEEE, IIE, NSPE, Society of Women Engineers, National Society of Black Engineers
Honorary Societies: Theta Tau

SUPPORT PROGRAMS
Student Chapter Organizations: Society of Women Engineers, National Society of Black Engineers

CSTEP
For: Ethnic Minorities **Available:** Year round
Offered: Freshman, Sophomore, Junior, Senior
Other Engineering Support Programs: Students may visit Hofstra and stay overnight during the summer for orientation. At this time Engineering faculty address the students and a variety of student life issues are discussed. The University provides a no charge tutorial service for all students requesting it.

090 University of Houston

INSTITUTION PROFILE

HEAD OF THE INSTITUTION
James H. Pickering
Phone: (713)743-8820   Fax: (713)743-8838

GENERAL INFORMATION
[All Students—Fall 1991]
Undergraduate enrollment 25,514
Graduate enrollment 8,093
Total institution enrollment 33,607
Type of institution: Public   Calendar: Semesters
Location: Houston   Population: 2,400,000
Setting: Urban
Types of engineering degrees: Engineering
Other degree-granting colleges: Architecture, Business Administration, Education, Fine & Performing Arts, Humanities & Social Sciences, Law, Pharmacy, Technology & Applied Sciences, Hotel & Restaurant Management, Optometry, Natural Science & Mathematics, Social Work, Technology

ENGINEERING COLLEGE ADMISSIONS INFORMATION
Entrance Requirements: SAT (Score: 850), ACT (Score: 21), High School courses—English (4 years), Chemistry (1 year), Physics (1 year), Mathematics (4 years)
Entrance Recommendations: High School courses—Mechanical Drawing (1 year)
Requirements for foreign students: TOEFL (Score: 550); financial statement

DEPARTMENTAL ADMISSIONS INFORMATION
Admission to the engineering department: At the time of admission to the institution.

FINANCIAL INFORMATION

ESTIMATED EXPENSES (ACADEMIC YEAR)
[Expenses are for the 1992-93 nine-month academic year.]

State Residents	Undergraduate	Graduate
Tuition and fees	$ 1,550	$ 1,800
College room and board	$ 4,000	$ 2,500
Books and supplies	$ 700	$ 900
Other expenses	$ –(A)	$ –(A)
Total estimated expenses	**$ 6,250**	**$ 5,200**

Out-of-State Residents	Undergraduate	Graduate
Tuition and fees	$ 6,200	$ 4,500
College room and board	$ 4,000	$ 2,500
Books and supplies	$ 700	$ 900
Other expenses	$ –(A)	$ –(A)
Total estimated expenses	**$10,900**	**$ 7,900**

Notes: (A) Data not available.

FINANCIAL AID OFFICE CONTACT
Robert Sheridan, *Director of Scholarships & Financial Aids*
University of Houston
4800 Calhoun Rd
Houston, TX 77204-2160
Phone: (713)743-1010

GENERAL FINANCIAL AID INFORMATION
Forms accepted/required: ACT, AFSA, FAF, FAT, FAR, FSA, FFS, Stafford, IRS, SAR, SDF, Supplemental
Additional financial aid information: The Office of Scholarships and Financial Aid provides assistance to students in obtaining scholarships, grants, loans, and equipment. FFS, IRS, and a completed SDF are required.

ENGINEERING COLLEGE INFORMATION
[For additional personnel, refer to the *Appendix*.]

HEAD OF ENGINEERING
Roger Eichhorn
Phone: (713)743-4200   Fax: (713)743-4214
EMail: Eichhorn@uh.edu

ENGINEERING COLLEGE ADDRESS
Cullen College Of Engineering
University of Houston
4800 Calhoun Rd
Houston, TX 77204-4814
Phone: (713)743-4200   Fax: (713)743-4214

ENROLLMENTS—BY CLASS
[Numbers are baccalaureate enrollments for the fall 1991 term, unless otherwise footnoted.]
1st-year students/Freshmen 480
2nd-year students/Sophomores 399
3rd-year students/Juniors 478
4th-year students/Seniors .... 681 [1]
Total 2,038
Notes: (1) There are 300 post-baccalaureate students.

NUMBER OF DEGREES AWARDED—BY PROGRAM
[Numbers are engineering baccalaureate degrees awarded during the 1991-1992 academic year, unless otherwise footnoted. For full details about each engineering program, refer to the *Tables of Degree Programs*.]
Acoustics .... –
Biochemical Engineering .... –
Biomedical Engineering .... –
Chemical Engineering 26
Civil & Environmental Engineering 13
Civil Engineering .... –
Computer & Systems Engineering .... –
Electrical Engineering 90
Engineering Management .... –
Environmental Engineering .... –
Industrial Engineering 13
Materials Engineering .... –
Mechanical Engineering 51
Operations Research .... –
Petroleum Engineering .... –
Total 193

PERCENTAGE OF DEGREES AWARDED—BY CATEGORY
[Percentages are of all engineering baccalaureate degrees awarded during the 1991-1992 academic year, unless otherwise footnoted.]
African Americans .... 1.6 %
Asian/Pacific Island Americans .... 20.2 %
Hispanic Americans .... 6.4 %
Native Americans .... – %
Foreign Citizens .... 17.6 %
All Others .... 54.2 %
Women .... 17.6 %
Persons with disabilities .... – %
Students over 25 years of age .... – %

GRADUATE ENROLLMENTS & DEGREES AWARDED
Master's enrollment 686
Master's degrees awarded 181
Doctoral enrollment 278
Doctoral degrees awarded 25

ENGINEERING STUDENT DATA

Applicants to the engineering college
Number of applicants to engineering college 3,011
Percent offered admission .... 24.9 %

Matriculated engineering students
Percentage in top quartile (25%) of High School class .... 52.5 [1]
Average SAT scores: Math—522, Verbal—458, Combined—980
Notes: (1) Fall 90 data given. Fall 91 data not yet available.

FULL- & PART-TIME FACULTY—BY DEPARTMENT
[Figures are the head count of full-time faculty and the full-time equivalent (FTE) of part-time faculty for each engineering department or equivalent.]

Department	Full	Part
Chemical Eng	16	–
Civil & Environmental Eng	14	0.5
Electrical Eng	28	–
Mechanical Eng	22	1.0
Industrial Eng	8	–

COLLEGE DESCRIPTION
The college offers four-year undergraduate programs in five traditional engineering disciplines. Mathematics and sciences are emphasized in the first three semesters, followed by five semesters of more specialized training in one of the engineering majors. A comprehensive core curriculum, including literature, social science and cultural heritage courses, provides a background in general knowledge. Through the core curriculum and engineering courses, the college strives to prepare students for their role as productive members of society by providing a comprehensive education that includes significant practical experience. Emphasis in the engineering program is placed on instilling a thorough understanding of fundamental concepts, supplying laboratory experience and incorporating the use of computer-aided engineering.

DEGREE PROGRAMS DESCRIPTION
Each individual department has its unique specializations. Chemical Engineering specializes in process engineering, process control, biotechnology, electronic materials, environmental & petroleum engineering. Civil and Environmental Engineering emphasizes geotechnical, environmental, structural designs, & hydrosystem engineering. Electrical Engineering has four degree plan options: Electromagnetics and Solid State Electronics; Computer Engineering; Controls, Power and Communications; and a General EE option. Industrial Engineering prepares students for manufacturing & management system, knowledge based methodologies, & ergonomics engineering. Mechanical Engineering emphasizes fluid mechanics, thermodynamics, material science, & mechanical design engineering.

TRANSFER INFORMATION
Residency Requirements: Students must complete last 30 hours in residence. Must have a cumulative grade point average of 2.0 and in major courses for graduation.

Transfer via Articulation Agreements
Admission to engineering: Dual enrollment can be admitted after the first semester.
Engineering graduates transferred from ... 4-yr: –   2-yr: –
Requirements: Applicants who have 15 or 29 semester hours of college credits must have a 2.5 cumulative grade point average. Those with 30 or more semester hours require a 2.0 cumulative grade point average.

Transfer without Articulation Agreements
Admission to engineering: At any time
Engineering graduates transferred from ... 4-yr: –   2-yr: –
Requirements: Applicants must be in good academic standing

GRADUATION REQUIREMENTS
All students must maintain C- or better in Science, Math, & Engineering courses. Major courses require a 2.0 grade point average. All students need a cumulative grade point average of 2.0 for graduation.

STUDENT PROGRAMS

PROFESSIONAL AND HONORARY SOCIETIES
[For key to acronyms, see *Introduction*.]
Professional Societies: AIChE, ASCE, ASME, IEEE, IIE
Honorary Societies: Alpha Pi Mu, Chi Epsilon, Eta Kappa Nu, Omega Chi Epsilon, Pi Tau Sigma, Tau Beta Pi

SUPPORT PROGRAMS
Student Chapter Organizations: Society of Women Engineers, National Society of Black Engineers, Mexican American Engineers & Scientists, Program For Minority Engineering Students

Program for Minority Engineering Students
For: Ethnic Minorities   **Available:** Year round
Offered: Prematriculation, Freshman, Sophomore, Junior, Senior
Other Engineering Support Programs: Orientation program for all Freshmen & all transfers; Career Counseling Services; Foreign Students Office; Veterans' Services for Blind and Handicapped; Placement Services; Learning support Services.

091 Howard University

INSTITUTION PROFILE

HEAD OF THE INSTITUTION
Franklyn G. Jenifer
Phone: (202)806-2500 Fax: (202)806-5960

GENERAL INFORMATION
[All Students—Fall 1991]
Undergraduate enrollment 8,153
Graduate enrollment 2,782
Total institution enrollment **10,935**

Type of institution: Private		**Calendar:** Semesters	
Nearest city: Washington, DC		**Population:** 640,000	
Miles from main campus: 1		**Setting:** Urban	

Types of engineering degrees: Engineering
Other degree-granting colleges: Allied Health Sciences, Architecture, Arts & Sciences, Business Administration, Communication & Journalism, Dentistry, Education, Fine & Performing Arts, Law, Medicine, Nursing, Pharmacy, Technology & Applied Sciences, Divinity, Human Ecology, Social Work, Liberal Arts, Continuing Education

ENGINEERING ADMISSIONS

ADMISSIONS OFFICE CONTACT
Emmett R. Griffin Jr.
Howard University
Director, Office of Admissions
2400 Sixth Street, N.W., Room 110
Washington, DC 20059
Phone: (202)806-2700 Fax: (202)806-4465

ENGINEERING COLLEGE ADMISSIONS INFORMATION
Entrance Requirements: SAT (Score: 1000), ACT (Score: 24), High School courses—English (3 years), Mathematics (algebra 2; geometry 1; trig 1/2) (3 years), Science (2 years), History or Civics (1 year)
Entrance Recommendations: SAT (Score: 950), ACT (Score: 24), High School courses—Pre-Calculus (1 year), Calculus (1 year), ACH-M is recommended.
Requirements for foreign students: TOEFL (Score: 600); financial statement

DEPARTMENTAL ADMISSIONS INFORMATION
Admission to the engineering department: For Mechanical Engineering, students may be admitted during the freshman and sophomore years, or when admitted to the institution.
Additional information: For Electrical Engineering, a high score in Mathematics is needed, preferably in algebra and trigonometry. Also student must have a minimum of 2.0 GPA.

FINANCIAL INFORMATION

ESTIMATED EXPENSES (ACADEMIC YEAR)
[Expenses are for the 1992-93 nine-month academic year.]

All Students	Undergraduate	Graduate
Tuition and fees	$ 7,005	$ 7,945
College room and board	$ 3,927	$ 4,410
Books and supplies	$ 600	$ 600
Other expenses	$ 2,430	$ 3,217
Total estimated expenses	**$13,962**	**$16,172**

FINANCIAL AID OFFICE CONTACT
Adrienne W. Price, *Director, Financial Aid and Student Employment*
Howard University
2400 Sixth Street, N.W.
Administration Building, Room 211
Washington, DC 20059
Phone: (202)806-2800

GENERAL FINANCIAL AID INFORMATION
Forms accepted/required: CSS-FAF, IRS, SAR, Pell Grant
Additional financial aid information: Academic Scholarships Need-based Scholarships College Work-Study Program (CWS) Loans Endowment Funds Gifts and Grants Trustee Scholarship and Fellowships Howard University Student Employment Program (HUSEP)

ENGINEERING COLLEGE INFORMATION

[For additional personnel, refer to the *Appendix*.]

HEAD OF ENGINEERING
M. Lucius Walker Jr.
Phone: (202)806-6565 Fax: (202)462-1810
EMail: walker@echo.eng.umd.edu

ENGINEERING COLLEGE ADDRESS
School of Engineering
Howard University
2300 Sixth Street, N.W.
L.K. Downing Hall, Room 1016
Washington, DC 20059
Phone: (202)806-6565 Fax: (202)462-1810

ENROLLMENTS—BY CLASS
[Numbers are baccalaureate enrollments for the fall 1991 term, unless otherwise footnoted.]
1st-year students/Freshmen .... 214 [3]
2nd-year students/Sophomores 110
3rd-year students/Juniors 111
4th-year students/Seniors 157
Total **592**

Notes: (3) The total does not include part-time students in the freshmen through senior levels.

NUMBER OF DEGREES AWARDED—BY PROGRAM
[Numbers are engineering baccalaureate degrees awarded during the 1991-1992 academic year, unless otherwise footnoted. For full details about each engineering program, refer to the *Tables of Degree Programs*.]

Chemical Engineering 10
Civil Engineering 20
Electrical Engineering 56
Mechanical Engineering 28
Systems & Computer Sciences 25
Total **139**

PERCENTAGE OF DEGREES AWARDED—BY CATEGORY
[Percentages are of all engineering baccalaureate degrees awarded during the 1991-1992 academic year, unless otherwise footnoted.]

African Americans .... 79.0 %
Asian/Pacific Island Americans .... – % [1]
Hispanic Americans .... – %
Native Americans .... – %
Foreign Citizens .... 21.0 %
All Others .... – %
Women .... 38.0 %
Persons with disabilities .... – %
Students over 25 years of age .... – % [A]

Notes: (A) Data not available. (1) There were two (2) Asian Americans.

GRADUATE ENROLLMENTS & DEGREES AWARDED
Master's enrollment .... 77 [2]
Master's degrees awarded .... 53 [C]
Doctoral enrollment 13
Doctoral degrees awarded .... 11 [C]

Notes: (C) Estimated data. (2) This figure does not include the 66 part-time students in the master's degree program.

ENGINEERING STUDENT DATA
Applicants to the engineering college
Number of applicants to engineering college 728
Percent offered admission 54.0%

Matriculated engineering students
Percentage in top quartile (25%) of High School class .... – [A]
Average SAT scores: Math—556, Verbal—439, Combined—995
Average ACT scores: Math—20, Composite—20

Notes: (A) Data not available.

FULL- & PART-TIME FACULTY—BY DEPARTMENT
[Figures are the head count of full-time faculty and the full-time equivalent (FTE) of part-time faculty for each engineering department or equivalent.]

Department	Full	Part
Chemical Eng	5	2.0
Civil Eng	13	7.0
Electrical Eng	20	3.0
Mechanical Eng	11	5.0
Systems & Computer Science	11	2.0

COLLEGE DESCRIPTION
Nurturing atmosphere typical of HBCUs. Strong student services unit which offers PREFACE (a pre-freshman bridge program), tutoring, and Co-op. Member of WMACU. Domestic and foreign exchange programs.

DEGREE PROGRAMS DESCRIPTION
Initial emphasis in all departments is on the fundamentals of engineering education. Depending on the department enrolled, the student is able to take elective courses in engineering, related sciences, business, liberal arts, and Afro-American studies. Thus, although attention is given to specialization, instruction is focused primarily on fundamentals and the application of these fundamentals to engineering analyses and design problems which are essential to professional growth and life-long learning in this era of rapid advancement in technological and scientific knowledge. In addition, each department in the School of Engineering has specialized graduate and undergraduate areas of concentration. Contact the department for further information on their specific areas.

TRANSFER INFORMATION
Transfer via Articulation Agreements
Admission to engineering: Currently, Howard has an agreement with Virginia Union. The student may apply for admission after completing their freshman year.
Engineering graduates transferred from .. 4-yr: 128 2-yr: 42
Requirements: Cumulative GPA of at least 2.5 Strong math and science background. Twelve (12) semester hours (or 18 quarter hours) of course work

Transfer without Articulation Agreements
Admission to engineering: Normally students are admitted in the fall semester of each academic year. A limited number of students are permitted to enroll in the spring and summer semesters.
Requirements: Cumulative GPA of at least 2.5; strong math and science background; twelve (12) semester credit hours (or 18 quarter hours) of course work. NOTE: Transfer students on probation or suspension at their current or previous college/university are not eligible for admission consideration. Also, students must be in residence for the last 30 semester hours prior to graduation.

GRADUATION REQUIREMENTS
The student must satisfy all entrance and department or degree course requirements; have a cumulative GPA of at least 2.0 for all course work and 2.0 in the major engineering field; achieve a grade of C or higher in at least 5/6ths of the graduation credits; complete in residence the last 30 credits in the degree curriculum; be recommended by an Engineering faculty member; and complete at least 18 credits in the social science or humanities areas. Students receiving two bachelor degrees must also satisfy requirements for both and complete at least 30 credits in differing curricula.

STUDENT PROGRAMS

PROFESSIONAL AND HONORARY SOCIETIES
[For key to acronyms, see *Introduction*.]
Professional Societies: AIChE, ASCE, ASME, IEEE, NSBE; SWE; Engr Student Council; SSCS; NOBCCHE;, Howard Engineer Magazine; NIPLA; and Student Pugwash
Honorary Societies: Omega Chi Epsilon, Tau Beta Pi, Sigma Xi, Sigma Xi

SUPPORT PROGRAMS
Student Chapter Organizations: Society of Women Engineers, National Society of Black Engineers, ECSEL Ambassadors, PREFACE, AFOTEC, NOAA and ASCE

PREFACE
For: Women & Ethnic Minorities   **Available:** Summer only
Offered: Prematriculation

AFOTEC
For: Women & Ethnic Minorities   **Available:** Summer only
Offered: Prematriculation

NOAA
For: Women & Ethnic Minorities   **Available:** Summer only
Offered: Prematriculation

Other Engineering Support Programs: New Student Orientation Counseling and the Engineering Coalition of Schools for Excellence in Education and Leadership which is geared to increase enrollment and recruitment of minorities and women into the engineering field. Also, in Electrical Engineering (EE) are the EE Mentor Program and the Summer Outreach Program.

092 University of Idaho

INSTITUTION PROFILE

HEAD OF THE INSTITUTION
Elizabeth A. Zinser
Phone: (208)885-6365   Fax: (208)885-6558

GENERAL INFORMATION
[All Students—Fall 1991]
Undergraduate enrollment 7,693
Graduate enrollment 1,799
Total institution enrollment **9,492**
Type of institution: Public   Calendar: Semesters
Nearest city: Spokane   Population: 175,000
Miles from main campus: 100   Setting: Small Town
Types of engineering degrees: Engineering
Other degree-granting colleges: Education, Law, Business & Economics, Forestry, Wildlife & Range Sciences, Letters & Science, Mines & Earth Resources, Art & Architecture, Agriculture

ENGINEERING ADMISSIONS

ADMISSIONS OFFICE CONTACT
Weldon R. Tovey
University of Idaho
College of Engineering
Moscow, ID 83843
Phone: (208)885-6479   Fax: (208)885-6645

ENGINEERING COLLEGE ADMISSIONS INFORMATION
Entrance Requirements: SAT (Score: 830), ACT (Score: 19), High School courses—Mathematics at Algebra level or higher (4 years), Natural Science (including Chemistry & Physics) (3 years), English (4 years), Speech (one semester) (1 year), Either the SAT or ACT test is required but not both.
Requirements for foreign students: TOEFL (Score: 550); financial statement

DEPARTMENTAL ADMISSIONS INFORMATION
Admission to the engineering department: At the time of admission to the institution.

FINANCIAL INFORMATION

ESTIMATED EXPENSES (ACADEMIC YEAR)
[Expenses are for the 1992-93 nine-month academic year.]

State Residents	Undergraduate	Graduate
Tuition and fees	$ 1,296	$ 1,728
College room and board	$ 3,084	$ 3,084
Books and supplies	$ 770	$ 770
Other expenses	$ 100	$ —
Total estimated expenses	**$ 5,250**	**$ 5,582**
Out-of-State Residents	Undergraduate	Graduate
Tuition and fees	$ 4,196	$ 4,628
College room and board	$ 3,084	$ 3,084
Books and supplies	$ 770	$ 770
Other expenses	$ 100	$ —
Total estimated expenses	**$ 8,150**	**$ 8,482**

FINANCIAL AID OFFICE CONTACT
Daniel D. Davenport, *Director*
University of Idaho
Financial Aid Office
Moscow, ID 83843
Phone: (208)885-6312

GENERAL FINANCIAL AID INFORMATION
Forms accepted/required: FAF, Stafford, Institutional, UI Scholarship & Financial Aid Application
Additional financial aid information: Part-time employment, scholarships, State Student Incentive Grants, Perkins National Direct Student Loans, Stafford Guaranteed Student Loans, Parent Loans for Undergraduate Study, Supplemental Loans to Assist Students, and Pell and Supplemental Educational Opportunity Grants.

ENGINEERING COLLEGE INFORMATION
[For additional personnel, refer to the *Appendix*.]

HEAD OF ENGINEERING
Richard T. Jacobsen
Phone: (208)885-6479   Fax: (208)885-6645

ENGINEERING COLLEGE ADDRESS
College of Engineering
University of Idaho
Moscow, ID 83843
Phone: (208)885-6479   Fax: (208)885-6645

ENROLLMENTS—BY CLASS
[Numbers are baccalaureate enrollments for the fall 1991 term, unless otherwise footnoted.]
1st-year students/Freshmen 371
2nd-year students/Sophomores 280
3rd-year students/Juniors 265
4th-year students/Seniors 408
Total **1,324**

NUMBER OF DEGREES AWARDED—BY PROGRAM
[Numbers are engineering baccalaureate degrees awarded during the 1991-1992 academic year, unless otherwise footnoted. For full details about each engineering program, refer to the *Tables of Degree Programs*.]
Agricultural Engineering 7
Chemical Engineering 11
Civil Engineering 24
Computer Engineering 6
Computer Science 24
Electrical Engineering 53
Mechanical Engineering 40
Total **165**

PERCENTAGE OF DEGREES AWARDED—BY CATEGORY
[Percentages are of all engineering baccalaureate degrees awarded during the 1991-1992 academic year, unless otherwise footnoted.]
African Americans .... — %
Asian/Pacific Island Americans 2.8%
Hispanic Americans .... — %
Native Americans .... — %
Foreign Citizens 8.5%
All Others 88.7%
Women 6.4%
Persons with disabilities .... — %
Students over 25 years of age .... — %

GRADUATE ENROLLMENTS & DEGREES AWARDED
Master's enrollment 145
Master's degrees awarded 55
Doctoral enrollment 35
Doctoral degrees awarded 7

ENGINEERING STUDENT DATA
Applicants to the engineering college
Number of applicants to engineering college 511
Percent offered admission 99.2%
Matriculated engineering students
Percentage in top quartile (25%) of High School class ... 65.3
Average SAT scores: Math—592, Verbal—485, Combined—1077
Average ACT scores: Composite—24.4

FULL- & PART-TIME FACULTY—BY DEPARTMENT
[Figures are the head count of full-time faculty and the full-time equivalent (FTE) of part-time faculty for each engineering department or equivalent.]

Department	Full	Part
Chemical Eng	7	—
Civil Eng	12	1.3
Electrical Eng	16	—
Mechanical Eng	17	—
Computer Science	12	—

COLLEGE DESCRIPTION
The College of Engineering is the home of a National Center for Applied Transportation Technology, funded by the Federal Department of Transportation, and the Microelectronics Research Center, which is one of the national Space Engineering Research Centers supported by NASA. A large number of firms and agencies from throughout the country send interviewers to the University of Idaho campus each year seeking to hire Idaho engineering students. The size of the college is near the median of engineering colleges in the country. It is not so large that the importance of the student as an individual is lost; it is large enough to support the faculty and facilities needed for top-quality education.

DEGREE PROGRAMS DESCRIPTION
Baccalaureate degree programs are based on a solid foundation of mathematics, science and engineering science, which has resulted in extraordinarily high performance on the Fundamentals of Engineering examination. Each curriculum is designed to prepare students for both entry level professional positions and/or for graduate school. Intensive capstone design classes, coupled with a practice-oriented faculty, give the graduates a pragmatic background. Both thesis and non-thesis master's degrees are offered, as well as the PhD.

TRANSFER INFORMATION
Residency Requirements: After a student has completed 88 credits, he or she must complete a minimum of 32 credits in University of Idaho courses taken for regular resident credit.
Transfer without Articulation Agreements
Admission to engineering: At any time
Requirements: Minimum GPA of 2.8 for out-of-state transfer students.

GRADUATION REQUIREMENTS
2.00 GPA minimum for all programs; 2.00 GPA minimum in upper division engineering classes for Civil Engineering. No grade lower than "C" for Mechanical Engineering.

STUDENT PROGRAMS

PROFESSIONAL AND HONORARY SOCIETIES
[For key to acronyms, see *Introduction*.]
Professional Societies: ACM, AIChE, ASAE, ASCE, ASME, IEEE
Honorary Societies: Tau Beta Pi

SUPPORT PROGRAMS
Student Chapter Organizations: Society of Women Engineers
Society of Women Engineers
For: Women   **Available:** Academic year
Offered: Freshman, Sophomore, Junior, Senior, Graduate level

093 Idaho State University

■ INSTITUTION PROFILE
HEAD OF THE INSTITUTION
Richard L. Bowen
Phone: (208)236-3440 Fax: (208)236-4000

GENERAL INFORMATION
[All Students—Fall 1991]
Undergraduate enrollment 8,768
Graduate enrollment 1,987
Total institution enrollment 10,755
Type of institution: Public Calendar: Semesters
Nearest city: Salt Lake City, UT Population: 1,500,000
Miles from main campus: 170 Setting: Small Town
Types of engineering degrees: Engineering
Other degree-granting colleges: Allied Health Sciences, Arts & Sciences, Business Administration, Communication & Journalism, Education, Fine & Performing Arts, Humanities & Social Sciences, Nursing, Pharmacy, Technology & Applied Sciences

■ ENGINEERING ADMISSIONS
ADMISSIONS OFFICE CONTACT
V. Charyulu
Idaho State University
P.O. Box 8060
Pocatello, ID 83209-8060
Phone: (208)236-2902 Fax: (208)236-4538

ENGINEERING COLLEGE ADMISSIONS INFORMATION
Entrance Requirements: SAT, ACT, High School courses—English (4 years), Math (3 years), Social Sciences (2 years), Humanities (2 years), There are no minimum score requirements for ACT or SAT tests.
Entrance Recommendations: High School courses—Computer Science (1 year), Mechanical Drawing (1 year), Speech (1 year)
Requirements for out-of-state residents: Graduation from an accredited high school and a 2.0 GPA in core subjects.
Requirements for foreign students: TOEFL (Score: 500); financial statement

DEPARTMENTAL ADMISSIONS INFORMATION
Admission to the engineering department: At the time of admission to the institution.

■ FINANCIAL INFORMATION
ESTIMATED EXPENSES (ACADEMIC YEAR)
[Expenses are for the 1992-93 nine-month academic year.]

State Residents	Undergraduate	Graduate
Tuition and fees	$ 1,292	$ 1,700
College room and board	$ 2,600	$ 2,600
Books and supplies	$ 600	$ 600
Other expenses	$ 160 [1]	$ 160 [1]
Total estimated expenses	$ 4,652	$ 5,060

Out-of-State Residents	Undergraduate	Graduate
Tuition and fees	$ 3,942	$ 4,350
College room and board	$ 2,600	$ 2,600
Books and supplies	$ 600	$ 600
Other expenses	$ 160 [1]	$ 160 [1]
Total estimated expenses	$ 7,302	$ 7,710

Notes: (1) Health Insurance

FINANCIAL AID OFFICE CONTACT
Doug Severs, *Director*
Idaho State University
P.O. Box 8077
Pocatello, ID 83209-8077
Phone: (208)236-2756 Fax: (208)236-4000

GENERAL FINANCIAL AID INFORMATION
Forms accepted/required: CSS-FAF, Institutional
Additional financial aid information: Need-based scholarships; short-term loans; merit-based scholarships; part-time jobs on campus; College Work Study Program, Perkins student loans, special non-resident waivers, St. student incentive grants, Pell grants; Stafford student loans, parent loans for students, supplemental loans for students.

■ ENGINEERING COLLEGE INFORMATION
[For additional personnel, refer to the *Appendix*.]
HEAD OF ENGINEERING
V. Charyulu
Phone: (206)236-2902 Fax: (208)236-4538

ENGINEERING COLLEGE ADDRESS
College of Engineering
Idaho State University
P.O. Box 8060
Pocatello, ID 83209-8060
Phone: (208)236-2902 Fax: (208)236-4538

ENROLLMENTS—BY CLASS
[Numbers are baccalaureate enrollments for the fall 1991 term, unless otherwise footnoted.]
1st-year students/Freshmen 130
2nd-year students/Sophomores 91
3rd-year students/Juniors 67
4th-year students/Seniors 80
Total 368

NUMBER OF DEGREES AWARDED—BY PROGRAM
[Numbers are engineering baccalaureate degrees awarded during the 1991-1992 academic year, unless otherwise footnoted. For full details about each engineering program, refer to the *Tables of Degree Programs*.]
Engineering 30
Engineering Management .... –
Total 30

PERCENTAGE OF DEGREES AWARDED—BY CATEGORY
[Percentages are of all engineering baccalaureate degrees awarded during the 1991-1992 academic year, unless otherwise footnoted.]
African Americans .... – %
Asian/Pacific Island Americans .... – %
Hispanic Americans .... 3.0 %
Native Americans .... – %
Foreign Citizens .... 10.0 %
All Others .... 87.0 %
Women .... 3.0 %
Persons with disabilities .... 3.0 %
Students over 25 years of age .... 83.0 %

GRADUATE ENROLLMENTS & DEGREES AWARDED
Master's enrollment 44
Master's degrees awarded 5
Doctoral enrollment 8
Doctoral degrees awarded .... –

ENGINEERING STUDENT DATA
Applicants to the engineering college
Number of applicants to engineering college .... – (A)
Percent offered admission .... –% (A)

Matriculated engineering students
Percentage in top quartile (25%) of High School class .... – (A)
Average ACT scores: Math—22.7, Composite—22.9

Notes: (A) Data not available.

■ FULL- & PART-TIME FACULTY—BY DEPARTMENT
[Figures are the head count of full-time faculty and the full-time equivalent (FTE) of part-time faculty for each engineering department or equivalent.]

Department	Full	Part
Eng	16	–

COLLEGE DESCRIPTION
Four-year undergraduate program in engineering with specialization in 2 of 6 possible areas including: structures, geotechnics, measurement and control, digital systems, thermal-fluids and nuclear power. Four-year undergraduate program in engineering management. Students must satisfy standard university educational goals with an additional requirement of 3 credit-hours of advanced level humanities or social sciences. During the last two semesters each completes a senior design project and is also expected to complete the Fundamentals of Engineering exam. Co-op program is not available. Graduate programs include: M.S. in Measurement and Control, M.S. and Ph.D. in Nuclear Science and Engineering. Students may pursue the study of Hazardous Waste Management as options under both M.S. programs. Training reactor available for coursework and research.

DEGREE PROGRAMS DESCRIPTION
The B.S. program in Engineering prepares the graduate for entry-level professional positions or for graduate study in a variety of engineering fields. The emphasis is development of design competency in the selected field of engineering. The College of Engineering and the College of Business also jointly offer a program leading to a BS in Engineering Management or a B.S. in Engineering Management with Nuclear Engineering emphasis. Graduate programs include: M.S. in Measurement and Control, M.S. and PhD. in Nuclear Science and Engineering. Students may pursue the study of Hazardous Waste Management as options under both M.S. programs.

DUAL DEGREE PROGRAMS
Engineering baccalaureate graduates with dual degrees 1

TRANSFER INFORMATION
Residency Requirements: Of the final 40 credits applied to meet graduation requirements, 32 must be taken in residence at ISU. At least 16 upper division credits directly applicable to the major area of study and/or the degree sought must be taken while in residence on ISU campus. At least 128 credits are required for graduation with a bachelor's degree.

Transfer via Articulation Agreements
Admission to engineering: At any time

Transfer without Articulation Agreements
Admission to engineering: At any time

GRADUATION REQUIREMENTS
B.S. programs in engineering require 128 semester credits and 2.0 GPA or better Engineering courses for graduation.

■ STUDENT PROGRAMS
PROFESSIONAL AND HONORARY SOCIETIES
[For key to acronyms, see *Introduction*.]
Professional Societies: ANS, ASME, IEEE, ISA, NSPE, Society of Women Engineers

SUPPORT PROGRAMS
Student Chapter Organizations: Society of Women Engineers
Other Engineering Support Programs: Fall and spring semester orientation program for all engineering students; academic skills center; career counseling services; foreign student offices; veteran's services.

094 University of Illinois at Chicago

■ INSTITUTION PROFILE

HEAD OF THE INSTITUTION
Stanley O. Ikenberry
Phone: (312)996-8800

GENERAL INFORMATION
[All Students—Fall 1991]
Undergraduate enrollment 15,837
Graduate enrollment 6,114
Total institution enrollment 24,208
Type of institution: Public **Calendar:** Semesters
Location: Chicago **Population:** 2,783,726
Setting: Urban
Types of engineering degrees: Engineering
Other degree-granting colleges: Allied Health Sciences, Architecture, Arts & Sciences, Business Administration, Dentistry, Education, Medicine, Nursing, Jane Addams College of Social Work, Kinesiology, Pharmacy

ENGINEERING COLLEGE ADMISSIONS INFORMATION
Entrance Requirements: ACT, High School courses—English (4 years), Mathematics (3.5 years), Lab Sciences (2 years), Social Studies (1) & Free Electives (2) (3 years)
Entrance Recommendations: High School courses—English (4 years), Mathematics (4 years), Lab Sciences (3 years), Social Studies (2) & Free Electives (2) (4 years)
Requirements for out-of-state residents: Admission is based on combination of Composite ACT Score and High School Percentile Rank referred to as Selection Index. SI is higher for out-of-state residents.
Requirements for foreign students: TOEFL (Score: 480); Foreign Students must have minimum of 3.5 GPA (out of 5.0 grade point average) for College review.

DEPARTMENTAL ADMISSIONS INFORMATION
Admission to the engineering department: At the time of admission to the institution.

■ FINANCIAL INFORMATION

ESTIMATED EXPENSES (ACADEMIC YEAR)
[Expenses are for the 1992-93 nine-month academic year.]

State Residents	Undergraduate	Graduate
Tuition and fees	$ 3,700 (1)	$ 4,198
College room and board	$ 4,500 (2)	$ 4,500 (2)
Books and supplies	$ 700	$ 700
Other expenses	$ 4,000 (3)	$ 4,000 (3)
Total estimated expenses	**$12,900**	**$13,398**

Out-of-State Residents	Undergraduate	Graduate
Tuition and fees	$ 8,410 (1)	$ 9,700
College room and board	$ 4,500 (2)	$ 4,500 (2)
Books and supplies	$ 700	$ 700
Other expenses	$ 4,000 (3)	$ 4,000 (3)
Total estimated expenses	**$17,610**	**$18,900**

Notes: (1) These tuition rates are for JR/SEN/Nondegree. Rate for FR/SOPH is 3,442 (resident) and 7,636 (nonresident). (2) This cost is an average. R/B rates vary depending on type of accommodations and meal plans. (3) Estimated costs for transportation, medical care, personal items, clothing and incidentals.

GENERAL FINANCIAL AID INFORMATION
Forms accepted/required: Single File Forms (additional forms may be required).
Additional financial aid information: University grants; University tuition waivers; student-to-student grants; supplemental educational opportunity grants; Pell grants; Illinois State Monetary awards; Illinois Veterans grants; College work-study; regular student employment.

■ ENGINEERING COLLEGE INFORMATION

[For additional personnel, refer to the *Appendix*.]
HEAD OF ENGINEERING
Paul M. Chung
Phone: (312)996-2400 **Fax:** (312)996-8664

ENGINEERING COLLEGE ADDRESS
College of Engineering (M/C 159)
University of Illinois at Chicago
851 South Morgan Street
123 Science and Engineering Offices
Chicago, IL 60607
Phone: (312)996-3463 **Fax:** (312)996-8664

ENROLLMENTS—BY CLASS
[Numbers are baccalaureate enrollments for the fall 1991 term, unless otherwise footnoted.]
1st-year students/Freshmen 423
2nd-year students/Sophomores 389
3rd-year students/Juniors 454
4th-year students/Seniors 770 (1)
Total **2,036**
Notes: (1) Includes 7 nondegree students.

NUMBER OF DEGREES AWARDED—BY PROGRAM
[Numbers are engineering baccalaureate degrees awarded during the 1991-1992 academic year, unless otherwise footnoted. For full details about each engineering program, refer to the *Tables of Degree Programs*.]
Bioengineering 8
Chemical Engineering 13
Civil Engineering 17
Computer Engineering 39
Computer Science .... –
Electrical Engineering 138
Electrical Engineering & Computer Science .... –
Engineering Management 4
Engineering Mechanics .... –
Engineering Physics .... –
Geotechnical Engineering & Geosciences .... –
Industrial Engineering 23
Industrial Engineering & Operations Research .... –
Materials Science & Engineering .... –
Mechanical Engineering 79
Metallurgical Engineering 1
Metallurgy .... –
Total **322**

PERCENTAGE OF DEGREES AWARDED—BY CATEGORY
[Percentages are of all engineering baccalaureate degrees awarded during the 1991-1992 academic year, unless otherwise footnoted.]
African Americans 3.4%
Asian/Pacific Island Americans 32.0%
Hispanic Americans 6.8%
Native Americans 0.6%
Foreign Citizens 2.8%
All Others 54.4%
Women 15.8%
Persons with disabilities .... – %
Students over 25 years of age 12.1%

GRADUATE ENROLLMENTS & DEGREES AWARDED
Master's enrollment 538
Master's degrees awarded 158
Doctoral enrollment 278
Doctoral degrees awarded 38

ENGINEERING STUDENT DATA
Applicants to the engineering college
Number of applicants to engineering college 1,024
Percent offered admission 66.3%
Matriculated engineering students
Percentage in top quartile (25%) of High School class 65.8
Average ACT scores: Math—24.7, Composite—23.4

FULL- & PART-TIME FACULTY—BY DEPARTMENT
[Figures are the head count of full-time faculty and the full-time equivalent (FTE) of part-time faculty for each engineering department or equivalent.]

Department	Full	Part
Chemical Eng	10	–
Civil Eng, Mechanics, & Metallurgy	15	–
Electrical Eng & Computer Science	51	–
Bioengineering Program	3	–
Mechanical Eng	30	–

COLLEGE DESCRIPTION
The College offers Bachelor of Science degrees in twelve engineering disciplines. One of the programs is jointly administered with the Business College. All programs emphasize engineering fundamentals, liberal arts content and design projects. Early involvement in faculty research activities is encouraged. Co-op education is an option in all of the majors. Average number of years required to actually complete the bachelor's degree: 4.0.

DEGREE PROGRAMS DESCRIPTION
The B.S. degree in the College of Engineering prepares the student for entry-level professional positions or for graduate study in each of the respective fields. The College offers programs leading to the M.S. and Ph.D degrees in Bioengineering, Chemical Engineering, Civil Engineering, Electrical Engineering and Computer Science, Engineering Mechanics, Industrial Engineering, Mechanical Engineering, and Metallurgy. There is also a PhD program in Geotechnical Engineering and Geosciences.

TRANSFER INFORMATION
Residency Requirements: Students transferring from a 2-year institution with Junior standing must complete the last 60 semester hours at the University of Illinois at Chicago. Students transferring from a 4-year institution must complete the last 30 hours at the University of Illinois at Chicago.
Transfer via Articulation Agreements
Admission to engineering: At the end of the freshman year
Engineering graduates transferred from ... 4-yr: 55 2-yr: 110
Requirements: Students must have a 3.5 GPA (out of 5.0 grade point average) for admission.
Transfer without Articulation Agreements
Admission to engineering: At the end of the freshman year
Requirements: Intercollege transfer students must complete 30 semester hours in another college before they can transfer to the College of Engineering. They must have a 3.25 UIC GPA and a 3.50 transfer GPA if they have work from another institution.

GRADUATION REQUIREMENTS
All undergraduate majors require a minimum of 128 semester hours for the baccalaureate degree. An overall GPA of 3.0 (out of 5.0 grade points) is required. A 3.0 GPA is also required of all work in the major field.

■ STUDENT PROGRAMS

PROFESSIONAL AND HONORARY SOCIETIES
[For key to acronyms, see *Introduction*.]
Professional Societies: AIAA, AIChE, ASCE, ASHRAE, ASME, BMES, IEEE, IIE, ITE, SAE, Society of Manufacturing Engineers (SME), Association for Computing Machinery (ACM)
Honorary Societies: Eta Kappa Nu, Tau Beta Pi

SUPPORT PROGRAMS
Student Chapter Organizations: Society of Women Engineers, National Society of Black Engineers, Society of Hispanic Professional Engineers

Minority Engineering Recruitment & Retention Prog.
For: Ethnic Minorities **Available:** Year round
Offered: Freshman, Sophomore, Junior, Senior

Other Engineering Support Programs: Summer orientation programs for freshmen and transfer students; mandatory orientation course required of all students in the first semester of attendance; mandatory academic advisement every semester. An 'Engineering Solution Center' advises undergraduate students on where to seek assistance on various problems.

095 University of Illinois at Urbana-Champaign

■ INSTITUTION PROFILE

HEAD OF THE INSTITUTION

Stanley O. Ikenberry
Phone: (217)333-3070 **Fax:** (217)333-3072

GENERAL INFORMATION

[All Students—Fall 1991]

Undergraduate enrollment	25,846
Graduate enrollment	9,038
Total institution enrollment	**34,884**

Type of institution: Public **Calendar:** Semesters
Nearest city: Urbana-Champaign **Population:** 100,000
Miles from main campus: 1 **Setting:** Urban
Types of engineering degrees: Engineering

Other degree-granting colleges: Agricultural & Environmental, Business Administration, Communication & Journalism, Education, Fine & Performing Arts, Applied Life Studies, Institute of Aviation, Liberal Arts and Sciences, Medicine, Social Work

ENGINEERING COLLEGE ADMISSIONS INFORMATION

Entrance Requirements: SAT, ACT, High School courses—English (4 years), Mathematics, including trigonometry (3.5 years), Science, excluding general science (2 years), Foreign Language (2 years), Freshmen Admission Requirements: Starting with the spring semester 1993, three years of one foreign language will be required for admission to the College of Engineering.
Entrance Recommendations: SAT, ACT, High School courses—English (4 years), Mathematics (including trigonometry) (4 years), Social Sciences (2 years), Science (excluding General Science) (2 years), Advanced Placement Examinations
Requirements for foreign students: TOEFL (Score: 520); financial statement

DEPARTMENTAL ADMISSIONS INFORMATION

Admission to the engineering department: At the time of admission to the institution.
Additional information: Guideline - ACT:C (HSPR) - 23(99); 24(97); 25(95); 26(92); 27(90); 28(88); 29(86); 30(84); 31(82); 32(80); 33(78); 34(75).

■ FINANCIAL INFORMATION

ESTIMATED EXPENSES (ACADEMIC YEAR)

[Expenses are for the 1992-93 nine-month academic year.]

State Residents	Undergraduate	Graduate
Tuition and fees	$ 3,598 (2)	$ 4,238
College room and board	$ 4,058	$ 4,058
Books and supplies	$ 490	$ 490
Other expenses	$ 1,952 (3)	$ 1,952 (1)
Total estimated expenses	**$10,098**	**$10,738**

Out-of-State Residents	Undergraduate	Graduate
Tuition and fees	$ 7,850 (5)	$ 9,770
College room and board	$ 4,058	$ 4,058
Books and supplies	$ 490	$ 490
Other expenses	$ 1,952 (6)	$ 1,952 (4)
Total estimated expenses	**$14,350**	**$16,270**

Notes: (1) Travel allowance (400) and personal expenses (1552). (2) Juniors and seniors, state residents: tuition and fees 3858. (3) Travel allowance (400) and personal expenses (1552). (4) Travel allowance (400) and personal expenses (1552). (5) Juniors and seniors, non-residents: tuition and fees 8630. (6) Travel allowance (400) and personal expenses (1552).

FINANCIAL AID OFFICE CONTACT

Orlo Austin, *Director*
University of Illinois at Urbana-Champaign
Office of Student Financial Aid
610 East John Street
Champaign, IL 61820
Phone: (217)333-0100

GENERAL FINANCIAL AID INFORMATION

Forms accepted/required: ACT, AFSA, CSS-FAF, FAF, FAT, FSA, FFS, Stafford, IRS, SAR, Supplemental, Institutional
Additional financial aid information: Scholarships, grants, loans, and student employment are administered by the Office of Student Financial Aid. Undergraduate students who are Illinois residents applying for financial assistance must also apply to the Illinois State Scholarship Commission for aid. Available also, the federal grant program - Basic Education Opportunity Program.

■ ENGINEERING COLLEGE INFORMATION

[For additional personnel, refer to the *Appendix*.]

HEAD OF ENGINEERING

William R. Schowalter
Phone: (217)333-2150 **Fax:** (217)244-7705
EMail: schow@ux1.cso.uiuc.edu

ENGINEERING COLLEGE ADDRESS

College of Engineering
University of Illinois at Urbana-Champaign
106 Engineering Hall
1308 West Green Street
Urbana, IL 61801
Phone: (217)333-2150 **Fax:** (217)244-7705

ENROLLMENTS—BY CLASS

[Numbers are baccalaureate enrollments for the fall 1991 term, unless otherwise footnoted.]

1st-year students/Freshmen	1,299
2nd-year students/Sophomores	999
3rd-year students/Juniors	1,210
4th-year students/Seniors	1,630
Total	**5,138**

NUMBER OF DEGREES AWARDED—BY PROGRAM

[Numbers are engineering baccalaureate degrees awarded during the 1991-1992 academic year, unless otherwise footnoted. For full details about each engineering program, refer to the *Tables of Degree Programs*.]

Aeronautical & Astronautical Engineering	58
Agricultural Engineering	18
Ceramic Engineering	32
Chemical Engineering	72
Civil Engineering	135
Computer Engineering	58
Computer Science	98
Electrical & Computer Engineering	230
Engineering Physics	19
Environmental Engineering	– (B)
Environmental Science	–
General Engineering	123
Industrial Engineering	46
Mechanical Engineering	192
Metallurgical Engineering	13
Nuclear Engineering	15
Theoretical & Applied Mechanics	13
Total	**1,122**

Notes: (B) Data not applicable.

PERCENTAGE OF DEGREES AWARDED—BY CATEGORY

[Percentages are of all engineering baccalaureate degrees awarded during the 1991-1992 academic year, unless otherwise footnoted.]

African Americans	1.4%
Asian/Pacific Island Americans	13.5%
Hispanic Americans	2.0%
Native Americans	– %
Foreign Citizens	0.7%
All Others	82.4%
Women	13.8%
Persons with disabilities	– % (A)
Students over 25 years of age	– % (A)

Notes: (A) Data not available.

GRADUATE ENROLLMENTS & DEGREES AWARDED

Master's enrollment	872
Master's degrees awarded	473
Doctoral enrollment	1,344
Doctoral degrees awarded	249

ENGINEERING STUDENT DATA

Applicants to the engineering college

Number of applicants to engineering college	2,762
Percent offered admission	77.7%

Matriculated engineering students

Percentage in top quartile (25%) of High School class 97.6
Average ACT scores: Math—30.4, Composite—29.2

FULL- & PART-TIME FACULTY—BY DEPARTMENT

[Figures are the head count of full-time faculty and the full-time equivalent (FTE) of part-time faculty for each engineering department or equivalent.]

Department	Full	Part
Chemical Eng	11	–
Civil Eng	55	–
Electrical & Computer Eng	77	–
Aeronautical & Astronautical Eng	17	–
Agricultural Eng	16	–
Computer Science	41	–
General Eng	19	–
Materials Science & Eng	33	–
Nuclear Eng	13	–
Physics Eng	66	–
Theoretical & Applied Mech. Eng	18	–
Mechanical & Industrial Eng	46	–

COLLEGE DESCRIPTION

The College offers a four-year undergraduate program in engineering, emphasizing engineering fundamentals in the first two years, followed by two years of specialized instruction in one of the 14 majors, with some social sciences and humanities in all four years. Emphasis is on engineering analysis, synthesis, computer-aided engineering, and hands-on lab experience. Each engineering department has its own faculty, courses, fields of specialization, and curriculum requirements. Five-year double-degrees include: combined Engineering/Liberal Arts and Sciences program and the combined Agricultural Engineering/Agricultural program. The five-year Cooperative Engineering Education program allows students to gain professional experience in their field by alternating periods of work with study. Average number of years required to actually complete the bachelor's degree: 8.6 semesters.

DEGREE PROGRAMS DESCRIPTION

The College of Engineering prepares men and women for professional careers in engineering and for responsible technical and semi-technical positions in industry, commerce, education, and government. The engineering curricula, though widely varied and specialized, are built on a general foundation of scientific theory applicable to many different fields. Work in the classroom and laboratory is brought into sharper focus by practical problems that the student solves by methods similar to those of practicing engineers. While each student pursues a curriculum chosen to meet his or her own career goals, all students take certain common courses. Basic courses in mathematics, chemistry, physics, rhetoric, and computer science are required in the first two years. Although the curricula are progressively specialized in the third and fourth years, each student is required to take some courses outside his or her chosen field.

DUAL DEGREE PROGRAMS

Engineering baccalaureate graduates with dual degrees 23
Enrollment requirements: LAS/ENGR. Each student must file an approved program with the engineering college office and with the liberal arts & sciences college office. Most combinations of engineering and liberal arts curricula may be completed in ten semesters, provided that the student does not have deficiencies in the entrance requirements of either college.

TRANSFER INFORMATION

Residency Requirements: Students transferring from a 2-year institution must complete their last 60 hours on this campus. Students transferring from a 4-year institution must complete their last 30 hours on this campus.

Transfer via Articulation Agreements

Admission to engineering: At the end of the sophomore year
Engineering graduates transferred from . . 4-yr: 121 2-yr: 161
Requirements: (1.) You must complete 60 or more semester hours at an accredited institution; (2.) you must complete basic college mathematics (through calculus), chemistry (one year), and a general physics sequence (one or one and a half years) and the English composition (rhetoric requirement); (3.) variable grade point average - depending on supply and demand.

Transfer without Articulation Agreements

Admission to engineering: At the end of the sophomore year

GRADUATION REQUIREMENTS

B.S. programs in engineering require 122-134 semester credits, depending on major selected. A five-year program in cooperative education is available in all college curricula. A five-year program permits a student to earn a B.S. degree in a field of engineering and a B.A. or a B.S. degree from the College of Liberal Arts & Sciences.

■ STUDENT PROGRAMS

PROFESSIONAL AND HONORARY SOCIETIES

[For key to acronyms, see *Introduction*.]

Professional Societies: ACerS, ACM, AIAA, AIChE, ANS, ASAE, ASCE, ASME, IEEE, IIE, Society of Automotive Engineers (SAE), Society of Manufacturing Engineers (SME)
Honorary Societies: Alpha Epsilon, Alpha Pi Mu, Chi Epsilon, Eta Kappa Nu, Keramos, Pi Tau Sigma, Sigma Gamma Tau, Tau Beta Pi, Alpha Nu Sigma, Alpha Sigma Mu, Gamma Epsilon, Phi Kappa Phi, Phi Lambda Upsilon

SUPPORT PROGRAMS

Student Chapter Organizations: Society of Women Engineers, National Society of Black Engineers, Society of Hispanic Professional Engineers

Association of Minority Students in Engineering

For: Ethnic Minorities **Available:** Year round
Offered: Prematriculation, Freshman, Sophomore, Junior, Senior, Graduate level

Other Engineering Support Programs: An all-college James Scholar Honors program; International Program Option; Bioengineering; Manufacturing; five-year Engineering-Liberal Arts and Sciences program; Transfer Student Visitation Day; rehabilitation and handicapped services; Engineering Placement Office; Engineers Open House.

096 Illinois Institute of Technology

■ INSTITUTION PROFILE

HEAD OF THE INSTITUTION
Lewis Collens
Phone: (312)567-5198   Fax: (312)567-3004

GENERAL INFORMATION
[All Students—Fall 1991]
Undergraduate enrollment 2,458
Graduate enrollment 4,111
Total institution enrollment 6,569
Type of institution: Private   Calendar: Semesters
Location: Chicago   Population: 2,783,720
Setting: Urban
Types of engineering degrees: Engineering
Other degree-granting colleges: Architecture, Business Administration, Law, Liberal Arts

■ ENGINEERING ADMISSIONS

ADMISSIONS OFFICE CONTACT
William Black
Illinois Institute of Technology
10 West 33rd Street
Chicago, IL 60616
Phone: (312)567-3025   Fax: (312)567-8828

ENGINEERING COLLEGE ADMISSIONS INFORMATION
Entrance Requirements: SAT, ACT, High School courses—English (4 years), History (1 year), Math (3.5 years), Lab Science (2 years)
Entrance Recommendations: High School courses—Physics (1 year), Chemistry (1 year)
Requirements for foreign students: TOEFL (Score: 550)

DEPARTMENTAL ADMISSIONS INFORMATION
Admission to the engineering department: At the time of admission to the institution.

■ FINANCIAL INFORMATION

ESTIMATED EXPENSES (ACADEMIC YEAR)
[Expenses are for the 1992-93 nine-month academic year.]

All Students	Undergraduate	Graduate
Tuition and fees	$13,070	$13,460
College room and board	$4,610	$4,610
Books and supplies	$500	$500
Other expenses	$800	$800
Total estimated expenses	**$18,980**	**$19,370**

FINANCIAL AID OFFICE CONTACT
Walter J. O'Neill, *Director of Financial Aid*
Illinois Institute of Technology
3300 South Federal
Student Finance Center
Chicago, IL 60616
Phone: (312)567-3033   Fax: (312)567-3302

GENERAL FINANCIAL AID INFORMATION
Forms accepted/required: CSS-FAF, FAT, IRS, SAR, Institutional
Additional financial aid information: The Student Finance Center administers a comprehensive financial aid program consisting of loans, work, scholarships, and grants for undergraduate students.

■ ENGINEERING COLLEGE INFORMATION
[For additional personnel, refer to the *Appendix*.]

HEAD OF ENGINEERING
Stephen M. Copley
Phone: (312)567-3009   Fax: (312)567-5205

ENGINEERING COLLEGE ADDRESS
Armour College of Engineering and Science
Illinois Institute of Technology
10 West 32nd Street
Chicago, IL 60616
Phone: (312)567-3000   Fax: (312)567-5205

ENROLLMENTS—BY CLASS
[Numbers are baccalaureate enrollments for the fall 1991 term, unless otherwise footnoted.]

1st-year students/Freshmen .... 217 [1]
2nd-year students/Sophomores .... 172 [1]
3rd-year students/Juniors .... 277 [1]
4th-year students/Seniors .... 287 [2]
Total **953**

Notes: (1) Figures are full-time students only. Total part-time undergraduate student enrollment was 343. (2) Includes 5th year students also (all full-time). Total part-time undergraduate student enrollment was 343.

NUMBER OF DEGREES AWARDED—BY PROGRAM
[Numbers are engineering baccalaureate degrees awarded during the 1991-1992 academic year, unless otherwise footnoted. For full details about each engineering program, refer to the *Tables of Degree Programs*.]

Aerospace Engineering 22
Chemical Engineering 24
Civil Engineering 13
Electrical & Computer Engineering 116
Environmental Engineering .... – [4]
Materials & Metallurgical Engineering 4
Mechanical Engineering 61
Total **240**

Notes: (4) New program.

PERCENTAGE OF DEGREES AWARDED—BY CATEGORY
[Percentages are of all engineering baccalaureate degrees awarded during the 1991-1992 academic year, unless otherwise footnoted.]

African Americans 3.0%
Asian/Pacific Island Americans 20.0%
Hispanic Americans 5.0%
Native Americans .... 0.1% [C]
Foreign Citizens 7.0%
All Others 64.9%
Women 13.0%
Persons with disabilities .... – % [A]
Students over 25 years of age .... – % [A]

Notes: (A) Data not available. (C) Estimated data.

GRADUATE ENROLLMENTS & DEGREES AWARDED
Master's enrollment .... 118 [4]
Master's degrees awarded 173
Doctoral enrollment .... 100 [5]
Doctoral degrees awarded 29

Notes: (4) Figures are full-time master's students only. Total part-time graduate (master's and doctorate) student enrollment was 877. (5) Figures are full-time doctoral students only. Total part-time graduate (master's and doctoral) student enrollment was 877.

ENGINEERING STUDENT DATA

Applicants to the engineering college
Number of applicants to engineering college 763
Percent offered admission 84.0%

Matriculated engineering students
Percentage in top quartile (25%) of High School class 95.0
Average SAT scores: Math—570, Verbal—440, Combined—1010
Average ACT scores: Math—25, Composite—23

FULL- & PART-TIME FACULTY—BY DEPARTMENT
[Figures are the head count of full-time faculty and the full-time equivalent (FTE) of part-time faculty for each engineering department or equivalent.]

Department	Full	Part
Chemical Eng	11	1.7
Civil Eng	8	2.1
Electrical & Computer Eng	24	9.6
Mechanical & Aerospace Eng	19	2.7
Metallurgical & Materials Eng	7	1.7
Environmental Eng	7	–

COLLEGE DESCRIPTION
The Armour College engineering programs feature special emphasis in quality, creativity, ethics and leadership across the curriculum. There are several five-year double degree programs available, as well as six-year BS/JD and seven-year BS/MD degrees. The BS degree in Environmental Engineering is new and offers expanded options in environmental studies. Most engineering disciplines offer professional specializations for students interested in focusing in specific areas and minors for those seeking breadth. The Fluid Dynamics Research Center on campus is the site for the National Diagnostics Facility, the world's largest university wind tunnel. There is no limit on the number of courses in which a student may enroll, and summer school is free.

DEGREE PROGRAMS DESCRIPTION
All B.S. curricula in Armour College of Engineering and Science prepare the student for both entry-level professional positions and for study at the post-baccalaureate level. The 5-year co-op program integrates college studies with working experience in industry, business and government. Undergraduates have many opportunities to take part in faculty research projects as well. In addition to B.S. degrees, all departments offer M.S. and Ph.D. degrees. Five-year BS/MS, seven-year BS/MD, and six-year BS/JD degrees are also offered, and the college offers two interdisciplinary master's programs--Manufacturing Engineering and Computer Engineering.

TRANSFER INFORMATION
Residency Requirements: Minimum of 45 credits
Transfer via Articulation Agreements
Admission to engineering: At any time
Engineering graduates transferred from ... 4-yr: 46 2-yr: 103
Requirements: 3.0 GPA (4.0 scale)
Transfer without Articulation Agreements
Admission to engineering: At any time
Requirements: 3.0 GPA on 4.0 scale

GRADUATION REQUIREMENTS
To graduate, students in all undergraduate curricula must complete: Departmental curriculum; credit hour requirements as appropriate (minimum 126 hours); General Education requirements; final 45 hours must be completed in residence; a minimum cumulative GPA of 2.0 and minimum 2.0 in major department courses; completion within eight calendar years (full-time), twelve (part-time).

■ STUDENT PROGRAMS

PROFESSIONAL AND HONORARY SOCIETIES
[For key to acronyms, see *Introduction*.]
Professional Societies: ACM, AIA, AIAA, AIChE, ASCE, ASME, IEEE, SAE
Honorary Societies: Chi Epsilon, Eta Kappa Nu, Pi Tau Sigma, Tau Beta Pi, Phi Eta Sigma, Phi Lambda Upsilon

SUPPORT PROGRAMS
Student Chapter Organizations: Society of Women Engineers, National Society of Black Engineers, Society of Hispanic Professional Engineers

Academic Challenge for Excellence in Scholastics
For: Ethnic Minorities   **Available:** Academic year
Offered: Freshman, Sophomore

Other Engineering Support Programs: Freshmen participate in Project HAWK, small classes taught in computer classrooms by faculty from the students' chosen fields. HAWK faculty serve as advisors and mentors, helping students explore opportunities and develop goals. The Educational Technology Center (ETC) is a multi-media resource center providing tutoring, testing and grading, remediation, and enrichment materials for IIT students.

097 Indiana University-Purdue University at Indianapolis

INSTITUTION PROFILE
HEAD OF THE INSTITUTION
Gerald L. Bepko, *Chancellor*
Phone: (317)274-4615   Fax: (317)274-4615

GENERAL INFORMATION
[All Students—Fall 1991]
Undergraduate enrollment 2,341
Graduate enrollment 30
Total institution enrollment 2,371
Type of institution: Public   **Calendar:** Semesters
Location: Indianapolis   **Population:** 1,000,000
Setting: Urban
Types of engineering degrees: Engineering & Technology
Other degree-granting colleges: Allied Health Sciences, Business Administration, Communication & Journalism, Dentistry, Education, Law, Medicine, Nursing, Social Work, Liberal Arts, Science, Continuing Studies, Public and Environmental Affairs

ENGINEERING ADMISSIONS
ADMISSIONS OFFICE CONTACT
Christine Fitzpatrick
Indiana University-Purdue University at Indianapolis
799 West Michigan Street
Indianapolis, IN 46202-5160
Phone: (317)274-0804   Fax: (317)274-4567

ENGINEERING COLLEGE ADMISSIONS INFORMATION
Admission to the engineering college: After completion of: Chemistry - 1 year; Laboratory Science - 1 year; History of Social Studies - 1 year.
Entrance Requirements: SAT (Score: 900), ACT, High School courses—Algebra (2.5 years), Geometry (1 year), Trigonometry, English (3 years), SAT: Verbal--400, Mathematics--500. ACT: English--21, Mathematics--23.
Entrance Recommendations: High School courses—Advanced mathematics (1 year), Computers (1 year), Foreign language (1 year), Physical Science (1 year)
Requirements for foreign students: TOEFL (Score: 550); financial statement

DEPARTMENTAL ADMISSIONS INFORMATION
Admission to the engineering department: At the time of admission to the institution.
Additional information: All engineering departments require the student to achieve a 2.00 (on a 4-point scale) GPA in the freshman year. Entrance requirements: Minimum 2.0 GPA required. Non-resident transfer students are required to have a 2.5 GPA.

FINANCIAL INFORMATION
ESTIMATED EXPENSES (ACADEMIC YEAR)
[Expenses are for the 1992-93 nine-month academic year.]

All Students	Undergraduate	Graduate
Tuition and fees	$ 2,500	$ 2,500
College room and board	$ —(B)	$ —(B)
Books and supplies	$ 1,000	$ 1,000
Other expenses	$ —(B)	$ —(B)
Total estimated expenses	$ 3,500	$ 3,500

Notes: (B) Data not applicable.

FINANCIAL AID OFFICE CONTACT
Natala Hart
Indiana University-Purdue University at Indianapolis
Office of Scholarships & Financial Aids
425 University Boulevard
Indianapolis, IN 46202-5145
Phone: (317)274-4162

GENERAL FINANCIAL AID INFORMATION
Forms accepted/required: FFS, IRS, Institutional
Additional financial aid information: Need-based scholarships; short-term loans; merit-based scholarships; part-time jobs on campus; College Work-Study Program.

ENGINEERING COLLEGE INFORMATION
[For additional personnel, refer to the *Appendix*.]
HEAD OF ENGINEERING
R. Bruce Renda
Phone: (317)274-0800   Fax: (317)274-0832

ENGINEERING COLLEGE ADDRESS
School of Engineering and Technology
Indiana University-Purdue University at Indianapolis
799 West Michigan Street
Indianapolis, IN 46202-5160
Phone: (317)274-2533   Fax: (317)274-4567

ENROLLMENTS—BY CLASS
[Numbers are baccalaureate enrollments for the fall 1991 term, unless otherwise footnoted.]
1st-year students/Freshmen 52
2nd-year students/Sophomores 120
3rd-year students/Juniors 112
4th-year students/Seniors 276
Total **560**

NUMBER OF DEGREES AWARDED—BY PROGRAM
[Numbers are engineering baccalaureate degrees awarded during the 1991-1992 academic year, unless otherwise footnoted. For full details about each engineering program, refer to the *Tables of Degree Programs*.]
Electrical Engineering 45
Interdisciplinary Engineering 1
Mechanical Engineering 32
Total **78**

PERCENTAGE OF DEGREES AWARDED—BY CATEGORY
[Percentages are of all engineering baccalaureate degrees awarded during the 1991-1992 academic year, unless otherwise footnoted.]
African Americans .... 7.0 %
Asian/Pacific Island Americans .... 7.0 %
Hispanic Americans .... 1.0 %
Native Americans .... — %
Foreign Citizens .... 6.5 %
All Others .... 78.5 %
Women .... 17.0 %
Persons with disabilities .... — %
Students over 25 years of age .... — %

GRADUATE ENROLLMENTS & DEGREES AWARDED
Master's enrollment 30
Master's degrees awarded .... —
Doctoral enrollment .... —
Doctoral degrees awarded .... —

ENGINEERING STUDENT DATA
Applicants to the engineering college
Number of applicants to engineering college 286
Percent offered admission 58.7%

Matriculated engineering students
Percentage in top quartile (25%) of High School class .... — (A)
Average SAT scores: Math—583, Verbal—468, Combined—1051

Notes: (A) Data not available.

FULL- & PART-TIME FACULTY—BY DEPARTMENT
[Figures are the head count of full-time faculty and the full-time equivalent (FTE) of part-time faculty for each engineering department or equivalent.]

Department	Full	Part
Electrical Eng	13	—
Mechanical Eng	12	—

COLLEGE DESCRIPTION
The School of Engineering and Technology is a state institution located in a very large urban area. Undergraduate engineering programs emphasize both fundamental and creative engineering skills to prepare students for employment in industry. Many opportunities for co-op or other part-time positions are obtained in the electronics and automotive industries surrounding Indianapolis. Classes are offered during both day and evening in order to provide opportunity for the full-time employee. The School also supports the Computer Network Center, Robotics and Automation Laboratory, Computational Biomechanics Laboratory and a Computational Fluid Dynamics Laboratory that provide additional opportunities for both student and faculty research. Average number of years required to actually complete the bachelor's degree: 5

DEGREE PROGRAMS DESCRIPTION
Undergraduate engineering programs in both electrical and mechanical engineering have strong engineering science and engineering design components and lead, respectively, to the B.S.E.E. or B.S.M.E. degree. An interdisciplinary engineering program is also offered leading to the B.S.E. degree. A special engineering management option is available within the interdisciplinary program. The Freshman Engineering Program is the program for all beginning engineering students, including recent high school graduates, transfer students and second degree students.

TRANSFER INFORMATION
Residency Requirements: In-state students must have a minimum cumulative grade point average of 2.0 on a 4.0 scale. Out-of-state students must have a minimum cumulative grade point average 2.5 on a 4.0 scale.

GRADUATION REQUIREMENTS
All baccalaureate programs require at least 127 semester hours, depending on major selected and a minimum 2.0 degree grade point average. Students are also required to achieve a minimum 2.0 grade average for all required engineering courses.

STUDENT PROGRAMS
PROFESSIONAL AND HONORARY SOCIETIES
[For key to acronyms, see *Introduction*.]
Professional Societies: ACM, AIAA, ASME, IEEE, NSPE, SAE, Society of Women Engineers (SWE), National Society of Black Engineers (NSBE)
Honorary Societies: Tau Alpha Pi

SUPPORT PROGRAMS
Student Chapter Organizations: Minority Engineering Advancement Program, National Society of Black Engineers

Minority Engineering Advancement Program
For: Ethnic Minorities   **Available:** Year round
Offered: Prematriculation, Freshman, Sophomore, Junior, Senior, Graduate level

IUPUI Office of Minority Student Services
For: Ethnic Minorities   **Available:** Year round
Offered: Freshman, Sophomore, Junior, Senior, Graduate level

Other Engineering Support Programs: Summer and Fall orientation program for all freshmen and transfer; English, mathematics, and reading placement testing program; career counseling service; international student office; disabled student services office; placement services.

098 University of Iowa

INSTITUTION PROFILE
HEAD OF THE INSTITUTION
Hunter R. Rawlings III
Phone: (319)335-3549

GENERAL INFORMATION
[All Students—Fall 1991]
Undergraduate enrollment	20,957
Graduate enrollment	6,506
Total institution enrollment	**27,463**

Type of institution: Public **Calendar:** Semesters
Nearest city: Cedar Rapids **Population:** 130,000
Miles from main campus: 35 **Setting:** Urban
Types of engineering degrees: Engineering
Other degree-granting colleges: Arts & Sciences, Business Administration, Dentistry, Education, Law, Medicine, Nursing, Pharmacy, Technology & Applied Sciences, Graduate

ENGINEERING ADMISSIONS
ADMISSIONS OFFICE CONTACT
Michael Barron
University of Iowa
Director of Admissions
Iowa City, IA 52242-1527
Phone: (319)335-1566

ENGINEERING COLLEGE ADMISSIONS INFORMATION
Entrance Requirements: High School courses—English/language arts (4 years), Mathematics (4 years), Physical Sciences (minimum 1 yr each of chem & phys) (3 years), Foreign language (plus 2 yrs of social studies) (2 years), Minimum ACT math and composite subscores of 25 each.
Entrance Recommendations: High School courses—Computer programming
Requirements for out-of-state residents: Non-residents must rank in upper 30 percentile & earn ACT math and comp subscores of 26 or above, or comparable SAT scores. Meeting minimums does not guarantee admission.
Requirements for foreign students: TOEFL (Score: 530); financial statement

DEPARTMENTAL ADMISSIONS INFORMATION
Admission to the engineering department: At the time of admission to the institution.
Additional information: Admission standards for biomedical engineering are more selective.

FINANCIAL INFORMATION
ESTIMATED EXPENSES (ACADEMIC YEAR)
[Expenses are for the 1992-93 nine-month academic year.]

State Residents	Undergraduate	Graduate
Tuition and fees	$ 2,088	$ 2,478
College room and board	$ 3,206	$ 3,206
Books and supplies	$ 580	$ 580
Other expenses	$ 260	$ 260
Total estimated expenses	**$ 6,134**	**$ 6,524**

Out-of-State Residents	Undergraduate	Graduate
Tuition and fees	$ 7,052	$ 7,350
College room and board	$ 3,206	$ 3,206
Books and supplies	$ 580	$ 580
Other expenses	$ 260	$ 260
Total estimated expenses	**$11,098**	**$11,396**

FINANCIAL AID OFFICE CONTACT
Mark S. Warner, *Director of Student Financial Aid*
University of Iowa
Office of Student Financial Aid
Iowa City, IA 52242-1527
Phone: (319)335-1450

GENERAL FINANCIAL AID INFORMATION
Forms accepted/required: FAF, FFS
Additional financial aid information: Need-based grants, loans, and college work-study; merit-based scholarships; part-time jobs on and off campus.

ENGINEERING COLLEGE INFORMATION
[For additional personnel, refer to the *Appendix*.]
HEAD OF ENGINEERING
Richard K. Miller
Phone: (319)335-5766
ENGINEERING COLLEGE ADDRESS
College of Engineering
University of Iowa
Iowa City, IA 52242-1527
Phone: (312)335-5763

ENROLLMENTS—BY CLASS
[Numbers are baccalaureate enrollments for the fall 1991 term, unless otherwise footnoted.]
1st-year students/Freshmen	413
2nd-year students/Sophomores	278
3rd-year students/Juniors	271
4th-year students/Seniors	275
Total	**1,237**

NUMBER OF DEGREES AWARDED—BY PROGRAM
[Numbers are engineering baccalaureate degrees awarded during the 1991-1992 academic year, unless otherwise footnoted. For full details about each engineering program, refer to the *Tables of Degree Programs*.]
Biomedical Engineering	28
Chemical & Biochemical Engineering	20
Civil & Environmental Engineering	28
Electrical & Computer Engineering	64
Industrial Engineering	43
Mechanical Engineering	43
Total	**226**

PERCENTAGE OF DEGREES AWARDED—BY CATEGORY
[Percentages are of all engineering baccalaureate degrees awarded during the 1991-1992 academic year, unless otherwise footnoted.]
African Americans	– %
Asian/Pacific Island Americans	5.4 %
Hispanic Americans	1.4 %
Native Americans	– %
Foreign Citizens	6.3 %
All Others	86.9 %
Women	16.3 %
Persons with disabilities	– % (A)
Students over 25 years of age	– % (A)

Notes: (A) Data not available.

GRADUATE ENROLLMENTS & DEGREES AWARDED
Master's enrollment	203
Master's degrees awarded	76
Doctoral enrollment	211
Doctoral degrees awarded	48

ENGINEERING STUDENT DATA
Applicants to the engineering college
Number of applicants to engineering college	933
Percent offered admission	81.0%

Matriculated engineering students
Percentage in top quartile (25%) of High School class 80.0
Average ACT scores: Math—28, Composite—27

FULL- & PART-TIME FACULTY—BY DEPARTMENT
[Figures are the head count of full-time faculty and the full-time equivalent (FTE) of part-time faculty for each engineering department or equivalent.]

Department	Full	Part
Chemical & Biochemical Eng	8	–
Civil & Environmental Eng	18	–
Electrical & Computer Eng	20	–
Industrial Eng	7	–
Mechanical Eng	13	–
Biomedical Eng	9	–

COLLEGE DESCRIPTION
The College is medium sized and located on a campus with a strong tradition in liberal arts, the human health sciences, and the professions. Engineering students constitute 6% of all undergrads on campus. Hence, opportunities abound for engineering students who wish to pursue interests beyond a major in engineering. For example, dual-degree programs with over 50 majors in liberal arts are available as well as minors in any field in liberal arts or in business. Because the College utilizes an extensive common core curriculum, students may postpone making a choice of engineering major or may change engineering majors during the first three semesters, usually without loss of time or credit. Coop education is elective and adds a calendar year to time to graduation.

DEGREE PROGRAMS DESCRIPTION
Degrees with honors are available to qualified undergraduate students who complete an honors project. A state-of-the-art, interactive computer network with HP/Apollo engineering workstations is available for students at all levels, from freshmen through PhD students. In addition to the research laboratories affiliated with the six graduate programs described above, special purpose research facilities are also available through two research units of the College, the Iowa Institute of Hydraulic Research (IIHR) and the Center for Computer Aided Engineering (CCAD). The IIHR is an international leader in hydraulic engineering and fluid mechanics and has 5 buildings with special purpose water/wind tunnels, a refrigerated towing tank, model test basins, etc. The CCAD, which specializes in the simulation and design optimization of mechanical systems, has extensive computer facilities that include a supercomputer, a special purpose graphics facility, and a full-scale vehicle driving simulator.

DUAL DEGREE PROGRAMS
Engineering baccalaureate graduates with dual degrees 3
Enrollment requirements: Students must meet the admission requirements for liberal arts and engineering colleges.

TRANSFER INFORMATION
Residency Requirements: Students must be enrolled in the College at least the last 30 semester hours, or 45 of last 60, or a total of 90 semester hours. The maximum number of credits from a two-year college that may be applied towards a BSE degree is 64 semester hours.

Transfer without Articulation Agreements
Admission to engineering: After completing a minimum of 24 semester hours of credits to include at least 1 calculus and 1 chemistry course.
Engineering graduates transferred from 4-yr: 12 2-yr: 36
Requirements: Minimum transfer GPA of 2.7/4.0 for Iowa residents and 3.1/4.0 for non-residents. A rolling admission procedure is used. Transfer applicants are held to same high school course requirements as freshmen.

GRADUATION REQUIREMENTS
Each BSE degree program requires: 128 semester hours of credit in a specific curriculum; a grade of C- or better for engineering calculus I and II; and a cumulative grade point average of at least 2.00. For specific MS and PhD degree requirements for each of the 6 graduate programs, please consult the general university catalog.

STUDENT PROGRAMS
PROFESSIONAL AND HONORARY SOCIETIES
[For key to acronyms, see *Introduction*.]
Professional Societies: AIChE, ASCE, ASME, BMES, IEEE, IIE, SAE, Society of Computer Simulation
Honorary Societies: Alpha Pi Mu, Chi Epsilon, Eta Kappa Nu, Omega Chi Epsilon, Pi Tau Sigma, Tau Beta Pi, Theta Tau, Alpha Eta Mu Beta

SUPPORT PROGRAMS
Student Chapter Organizations: Society of Women Engineers, Multi-Ethnic Student Association

Peer Mentoring for Minorities and Women
For: Women & Ethnic Minorities **Available:** Academic year
Offered: Freshman

Free Tutorial Program
For: Women & Ethnic Minorities **Available:** Academic year
Offered: Freshman, Sophomore

Other Engineering Support Programs: The College offers freshmen/transfer orientation programs; free tutoring for all freshmen and sophomore core courses; a peer assistant program using upper class students to assist with counseling and advising all engineering freshmen; an intervention program for new students; a merit-based scholarship program; a short-term loan program; and an employment placement service for graduating students.

099 Iowa State University

INSTITUTION PROFILE

HEAD OF THE INSTITUTION
Martin C. Jischke
Phone: (515)294-2042

GENERAL INFORMATION
[All Students—Fall 1991]
Undergraduate enrollment 20,855
Graduate enrollment 4,395
Total institution enrollment 25,250

Type of institution: Public **Calendar:** Semesters
Nearest city: Des Moines **Population:** 250,000
Miles from main campus: 40 **Setting:** Urban
Types of engineering degrees: Engineering
Other degree-granting colleges: Business Administration, Education, Family and Consumer Sciences, Veterinary Medicine, Liberal Arts and Sciences, Agriculture, Design

ENGINEERING ADMISSIONS

ADMISSIONS OFFICE CONTACT
Karsten Smedal
Iowa State University
Director of Admissions
314 Alumni Hall
Ames, IA 50011
Phone: (515)294-0815 **Fax:** (515)294-1088

ENGINEERING COLLEGE ADMISSIONS INFORMATION
Entrance Requirements: High School courses—English (4 years), Mathematics (3 years), Science (3 years), Social Studies (2 years)
Entrance Recommendations: High School courses—Algebra (2 years), Geometry/Trigonometry (1 year), Physics (1 year), Chemistry (1 year)
Requirements for foreign students: TOEFL (Score: 500); financial statement; All the curricula in the College have enrollment limitations.

DEPARTMENTAL ADMISSIONS INFORMATION
Admission to the engineering department: All students are considered to be in a pre-professional program. Basic program requirements must be met before proceeding to a professional program.

FINANCIAL INFORMATION

ESTIMATED EXPENSES (ACADEMIC YEAR)
[Expenses are for the 1992-93 nine-month academic year.]

State Residents	Undergraduate	Graduate
Tuition and fees	$ 2,088	$ 2,478[1]
College room and board	$ 3,044	$ 2,960
Books and supplies	$ 500	$ 300[2]
Other expenses	$ 200[3]	$ 200[3]
Total estimated expenses	$ 5,832	$ 5,938

Out-of-State Residents	Undergraduate	Graduate
Tuition and fees	$ 6,856	$ 7,148
College room and board	$ 3,044	$ 2,960
Books and supplies	$ 500	$ 300[2]
Other expenses	$ 200[3]	$ 200[3]
Total estimated expenses	$10,600	$10,608

Notes: (1) Amount shown is for 1992-1993 academic year. (2) Cost may vary by degree program. (3) A computer fee of $100 per semester is required of any student who is majoring in any engineering program.

FINANCIAL AID OFFICE CONTACT
Earl Dowling, *Director*
Iowa State University
Student Financial Aid Office
12 Beardshear Hall
Ames, IA 50011
Phone: (515)294-2223 **Fax:** (515)294-0907

GENERAL FINANCIAL AID INFORMATION
Forms accepted/required: FFS, Institutional
Additional financial aid information: Grants, scholarships, loans, and part-time jobs are available and based on merit, financial need, or their demonstrated talent.

ENGINEERING COLLEGE INFORMATION
[For additional personnel, refer to the *Appendix*.]

HEAD OF ENGINEERING
David T. Kao
Phone: (515)294-5933 **Fax:** (515)294-9273

ENGINEERING COLLEGE ADDRESS
College of Engineering
Iowa State University
104 Marston Hall
Ames, IA 50011-2150
Phone: (515)294-5933 **Fax:** (515)294-9273

ENROLLMENTS—BY CLASS
[Numbers are baccalaureate enrollments for the fall 1991 term, unless otherwise footnoted.]
1st-year students/Freshmen 1,276
2nd-year students/Sophomores 816
3rd-year students/Juniors 810
4th-year students/Seniors 1,204
Total 4,106

NUMBER OF DEGREES AWARDED—BY PROGRAM
[Numbers are engineering baccalaureate degrees awarded during the 1991-1992 academic year, unless otherwise footnoted. For full details about each engineering program, refer to the *Tables of Degree Programs*.]
Aerospace Engineering 53
Agricultural Engineering 6
Biomedical Engineering .... –
Ceramic Engineering 13
Chemical Engineering 43
Civil Engineering 60
Computer Engineering 48
Construction Engineering 30
Electrical Engineering 167
Engineering Mechanics .... –
Engineering Operations 5
Engineering Science(s) 3
Industrial Engineering 86
Mechanical Engineering 113
Metallurgical Engineering 9
Metallurgy .... –
Nuclear Engineering 10
Operations Research .... –
Total 646

PERCENTAGE OF DEGREES AWARDED—BY CATEGORY
[Percentages are of all engineering baccalaureate degrees awarded during the 1991-1992 academic year, unless otherwise footnoted.]
African Americans .... – %
Asian/Pacific Island Americans 3.0%
Hispanic Americans 1.0%
Native Americans .... – %
Foreign Citizens 9.0%
All Others 87.0%
Women 11.0%
Persons with disabilities .... – % (A)
Students over 25 years of age .... – % (A)

Notes: (A) Data not available.

GRADUATE ENROLLMENTS & DEGREES AWARDED
Master's enrollment .... 336 [1]
Master's degrees awarded 126
Doctoral enrollment .... 401 [1]
Doctoral degrees awarded 41

Notes: (1) Data available only for fall 1992.

ENGINEERING STUDENT DATA

Applicants to the engineering college
Number of applicants to engineering college 2,608
Percent offered admission 54.0%

Matriculated engineering students
Percentage in top quartile (25%) of High School class 64.5
Average ACT scores: Composite—26

FULL- & PART-TIME FACULTY—BY DEPARTMENT
[Figures are the head count of full-time faculty and the full-time equivalent (FTE) of part-time faculty for each engineering department or equivalent.]

Department	Full	Part
Chemical Eng	15	–
Civil & Construction Eng	32	1.3
Electrical Eng & Computer Eng	43	1.0
Industrial & Manufacturing Systems Eng	16	–
Mechanical Eng	34	–
Agricultural & Biosystems Eng	13	–
Biomedical Eng	–	–
Materials Science & Eng	13	–
Aerospace Eng & Eng Mechanics	36	1.3

COLLEGE DESCRIPTION

The College offers four-year undergraduate programs with 8 departments, 1 division, and 13 curricula. All curricula are divided into two phases: a basic and a professional program. The basic program consists mainly of subjects fundamental and common to all curricula. During this phase, lower-level students make use of the extensive computer facilities available in the College. Upper-level courses emphasize the use of modern laboratory facilities. Double degree programs are available. Five-year co-op programs are available college-wide. International exchange programs are available in some curricula. A concurrent program leading to a B.S. in engineering and an M.S. in business is available.

DEGREE PROGRAMS DESCRIPTION
The College has 8 departments and offers B.S. degrees in 13 curricula, M.E. degrees in 8 curricula, M.S. degrees in 14 curricula, and Ph.D. degrees in 14 curricula. In addition to the 8 departments, the College has a Division of Engineering Fundamentals and Multidisciplinary Design. This unit has responsibility for the integration and coordination of common cross-College areas such as the basic program.

DUAL DEGREE PROGRAMS
Engineering baccalaureate graduates with dual degrees 3
Enrollment requirements: Must have 30 more credits than the curriculum that requires the most credits.

TRANSFER INFORMATION
Residency Requirements: Students must complete at least the last 32 credits in residence.

Transfer via Articulation Agreements
Admission to engineering: At any time
Requirements: Figures for the number of engineering graduates transferring from 4 year institutions and community colleges are not available. Minimum 2.0 GPA. Less than 24 credits college work, admission will be based on high school records only and student must rank in top half of class or certain ACT scores will be considered.

Transfer without Articulation Agreements
Admission to engineering: At any time
Engineering graduates transferred from .. 4-yr: – 2-yr: –
Requirements: Figures for number of engineering graduates transferring from 4 year institutions and community colleges are not available. Minimum 2.0 GPA. Less than 24 credits college work, admission will be based on high school records only and student must rank in top half of class or certain ACT scores will be considered.

GRADUATION REQUIREMENTS
In order to graduate in a professional engineering curriculum, a student must have a minimum GPA of 2.00 in a department-designated group of 200-level and above courses. These courses will total not less than 24 nor more than 48 semester credits.

STUDENT PROGRAMS

PROFESSIONAL AND HONORARY SOCIETIES
[For key to acronyms, see *Introduction*.]
Professional Societies: ACerS, AIAA, AIChE, ASAE, ASCE, ASME, IEEE, IIE, SAE, SAME
Honorary Societies: Alpha Epsilon, Alpha Pi Mu, Chi Epsilon, Eta Kappa Nu, Keramos, Omega Chi Epsilon, Pi Tau Sigma, Sigma Gamma Tau, Tau Beta Pi, Sigma Lambda Chi

SUPPORT PROGRAMS
Student Chapter Organizations: Society of Women Engineers, National Society of Black Engineers, Society of Hispanic Professional Engineers, American Indian Science and Engineering Society, Asian American Engineering Society

Minority Engineering Program
For: Women & Ethnic Minorities **Available:** Academic year
Offered: Prematriculation, Freshman, Sophomore

Academic Excellence Workshop
For: Women & Ethnic Minorities **Available:** Academic year
Offered: Freshman, Sophomore

Other Engineering Support Programs: Comprehensive freshmen/transfer orientation programs during summer, fall and spring; placement services for summer, coop, and permanent employment; university tutoring program; university services for adult and handicapped students; college directed individual academic and career advising, course registration and scheduling assistance, undergraduate mentoring, and tutoring.

100 The Johns Hopkins University

■ INSTITUTION PROFILE

HEAD OF THE INSTITUTION
William C. Richardson
Phone: (410)516-8068   **Fax:** (410)516-7075

GENERAL INFORMATION
[All Students—Fall 1991]
Undergraduate enrollment 6,014
Graduate enrollment 9,217
Total institution enrollment 15,231
Type of institution: Private   **Calendar:** Semesters
Setting: Urban
Types of engineering degrees: Engineering
Other degree-granting colleges: Arts & Sciences, Medicine, Nursing, Hygiene and Public Health, Advanced International Studies, Continuing Studies, Peabody Institute

■ ENGINEERING ADMISSIONS

ADMISSIONS OFFICE CONTACT
Richard M. Fuller
The Johns Hopkins University
Admissions; 140 Garland Hall
Charles & 34th Streets
Baltimore, MD 21218
Phone: (410)516-8171   **Fax:** (410)516-5200

ENGINEERING COLLEGE ADMISSIONS INFORMATION
Entrance Requirements: SAT, High School courses—Mathematics (4 years), Laboratory Science, incl. Chemistry and Physics (3 years), English (4 years), Three achievement tests or the ACT alone.
Entrance Recommendations: High School courses—Foreign Language (2 years), Computing (1 year), History (1 year), SAT math achievement; math advanced placement
Requirements for foreign students: TOEFL (Score: 600); financial statement

DEPARTMENTAL ADMISSIONS INFORMATION
Admission to the engineering department: At the time of admission to the institution.

■ FINANCIAL INFORMATION

ESTIMATED EXPENSES (ACADEMIC YEAR)
[Expenses are for the 1992-93 nine-month academic year.]

All Students	Undergraduate	Graduate
Tuition and fees	$17,000 [1]	$16,750 [1]
College room and board	$ 6,300 [4]	$ 8,700 [2]
Books and supplies	$ 1,280	$ 1,750
Other expenses	$ — [5]	$ 500 [3]
Total estimated expenses	**$24,580**	**$27,700**

Notes: (1) $420 Matriculation Fee charged first-time students only. (2) Twelve-months. (3) Transportation. (4) $1500 for commuter. (5) Varies $0-$1200 depending on distance. Two annual round trips considered.

FINANCIAL AID OFFICE CONTACT
Ellen Frishberg, *Director, Student Financial Services*
The Johns Hopkins University
146 Garland Hall
Charles and 34th Streets
Baltimore, MD 21218
Phone: (410)516-8028   **Fax:** (410)516-6025

GENERAL FINANCIAL AID INFORMATION
Forms accepted/required: FAF, FAT, IRS, Institutional, If applicable, divorce/separated statement, If applicable, business/farm supplement
Additional financial aid information: Financial aid, ordinarily a combination of grants, low-interest loans, and work-study jobs, is awarded to students of academic promise who demonstrate financial need. Aid is awarded on an annual basis.

■ ENGINEERING COLLEGE INFORMATION

[For additional personnel, refer to the *Appendix*.]

HEAD OF ENGINEERING
Don P. Giddens
Phone: (410)516-8350   **Fax:** (410)516-8627
EMail: dgiddens@jhuvms.bitnet

ENGINEERING COLLEGE ADDRESS
G.W.C. Whiting School of Engineering
The Johns Hopkins University
Charles & 34th Streets
Baltimore, MD 21218
Phone: (410)516-8350

ENROLLMENTS—BY CLASS
[Numbers are baccalaureate enrollments for the fall 1991 term, unless otherwise footnoted.]
1st-year students/Freshmen 225
2nd-year students/Sophomores 202
3rd-year students/Juniors 170
4th-year students/Seniors 168
Total 765

NUMBER OF DEGREES AWARDED—BY PROGRAM
[Numbers are engineering baccalaureate degrees awarded during the 1991-1992 academic year, unless otherwise footnoted. For full details about each engineering program, refer to the *Tables of Degree Programs*.]
Applied Mathematics .... –
Applied Physics .... –
Biomedical Engineering 44
Chemical Engineering 44
Civil Engineering 11
Computer Science 19
Electrical Engineering 31
Engineering Mechanics 16
Environmental Science & Engineering .... –
Materials Science & Engineering 4
Mathematical Sciences 10
Mechanical Engineering 16
Technical Management .... –
Total 195

PERCENTAGE OF DEGREES AWARDED—BY CATEGORY
[Percentages are of all engineering baccalaureate degrees awarded during the 1991-1992 academic year, unless otherwise footnoted.]
African Americans .... 2.0 %
Asian/Pacific Island Americans .... 17.0 %
Hispanic Americans .... 2.0 %
Native Americans .... – %
Foreign Citizens .... 4.0 %
All Others .... 75.0 %
Women .... 21.0 %
Persons with disabilities .... – % (A)
Students over 25 years of age .... – % (A)

Notes: (A) Data not available.

GRADUATE ENROLLMENTS & DEGREES AWARDED
Master's enrollment .... 2,344 [3]
Master's degrees awarded .... 763 [4]
Doctoral enrollment .... 437 [5]
Doctoral degrees awarded 39

Notes: (3) Includes 2311 in part-time programs. (4) 672 from part-time programs; 91 from full-time programs. (5) All full-time programs.

ENGINEERING STUDENT DATA

Applicants to the engineering college
Number of applicants to engineering college 1,792
Percent offered admission 55.0%

Matriculated engineering students
Percentage in top quartile (25%) of High School class 92.0
Average SAT scores: Math—590, Verbal—705, Combined—1295

FULL- & PART-TIME FACULTY—BY DEPARTMENT
[Figures are the head count of full-time faculty and the full-time equivalent (FTE) of part-time faculty for each engineering department or equivalent.]

Department	Full	Part
Chemical Eng	8	1.7
Electrical & Computer Eng	15	21.0
Civil Eng	7	3.7
Mechanical Eng	11	4.3
Biomedical Eng	14	3.3
Computer Science	10	27.0
Geography & Environmental Eng	13	5.3
Materials Science & Eng	7	3.3
Mathematical Sciences	9	1.7

COLLEGE DESCRIPTION

The Whiting School, committed to quality education in the setting of a premier research university, offers four-year undergraduate programs in nine departments, an honors bachelors/masters program, and graduate programs leading to the masters and Ph.D. The part-time master's degree programs in engineering and applied science are the largest in the country. The school maintains close ties with Arts & Sciences, Medicine, the Applied Physics Laboratory and industry in the Baltimore-Washington area. The undergraduate student body ranks among the top nationally. The student/faculty ratio is about 8; classes are small with significant opportunity for individual faculty/student interaction, particularly independent study and research projects.

DEGREE PROGRAMS DESCRIPTION

Bachelors degree programs offered provide students with a solid, fundamental background necessary to begin a professional career or study for an advanced degree. B.S. degrees are awarded in all full-time programs, and B.A. degrees in some. Full-time graduate study involving coursework and individual research can be pursued for the M.A., the M.S.E. and the Ph.D. Part-time, professionally-oriented masters programs are available at three sites.

TRANSFER INFORMATION

Residency Requirements: Transfer students must complete at least 60 credits at Johns Hopkins and fulfill a two-year residence requirement.

Transfer via Articulation Agreements
Admission to engineering: At the end of the sophomore year
Engineering graduates transferred from . . . 4-yr: 1   2-yr: –
Requirements: Applications are reviewed individually.

Transfer without Articulation Agreements
Admission to engineering: Generally, at end of sophomore year, but can be at end of freshman year.
Engineering graduates transferred from . . . 4-yr: 7   2-yr: –
Requirements: Generally a 3.0 GPA is required; applications are reviewed individually.

GRADUATION REQUIREMENTS

Bachelor degrees in engineering require 120-128 semester credits, depending on the major. Specific requirements are imposed by individual departments. The University requires a minimum GPA of 2.00 for graduation.

■ STUDENT PROGRAMS

PROFESSIONAL AND HONORARY SOCIETIES
[For key to acronyms, see *Introduction*.]
Professional Societies: AIChE, AIME, ASCE, ASME, BMES, IEEE, ORSA, TMS
Honorary Societies: Eta Kappa Nu, Tau Beta Pi

SUPPORT PROGRAMS
Student Chapter Organizations: Society of Women Engineers, National Society of Black Engineers

Office of Multi Cultural Student Affairs
For: Ethnic Minorities   **Available:** Year round
Offered: Freshman, Sophomore, Junior, Senior, Graduate level

Summer Scholars Program
For: Ethnic Minorities   **Available:** Academic year
Offered: Prematriculation

Women's Center
For: Women   **Available:** Year round
Offered: Freshman, Sophomore, Junior, Senior, Graduate level

Black Student Union
For: Ethnic Minorities   **Available:** Year round
Offered: Freshman, Sophomore, Junior, Senior, Graduate level

Other Engineering Support Programs: Orientation program for new undergraduates; undergraduate affairs office; academic advising; study skills program; student societies, SWE and HoMES; health, counseling, and psychiatric services; career counseling and placement; student employment; foreign student services; cultural and social programs; faculty-student interaction program.

101 University of Kansas

■ INSTITUTION PROFILE

HEAD OF THE INSTITUTION
Gene A. Budig
Phone: (913)864-3131 **Fax:** (913)864-4120

GENERAL INFORMATION
[All Students—Fall 1991]
Undergraduate enrollment 19,428
Graduate enrollment 6,667
Total institution enrollment 26,661
Type of institution: Public **Calendar:** Semesters
Nearest city: Kansas City **Population:** 1,000,000
Miles from main campus: 30 **Setting:** Small Town
Types of engineering degrees: Engineering
Other degree-granting colleges: Allied Health Sciences, Architecture, Arts & Sciences, Business Administration, Communication & Journalism, Education, Fine & Performing Arts, Law, Medicine, Nursing, Social Welfare, Graduate School, Pharmacy

■ ENGINEERING ADMISSIONS

ADMISSIONS OFFICE CONTACT
Deborah B. Castrop
University of Kansas
Office of Admissions
126 Strong Hall
Lawrence, KS 66045
Phone: (913)864-3911 **Fax:** (913)864-5230

ENGINEERING COLLEGE ADMISSIONS INFORMATION
Entrance Requirements: ACT, High School courses—Algebra (2 years), Trigonometry (1 year), MATH 22
Entrance Recommendations: High School courses—Chemistry (1 year), Physics (1 year), English (4 years), Additional Math (2 years)
Requirements for out-of-state residents: All admissions are on a selective basis.
Requirements for foreign students: Students from foreign institutions are not accepted directly into the School of Engineering but may apply for transfer after at least one semester in the College of Liberal Arts & Sciences or in some other U.S. institution.

DEPARTMENTAL ADMISSIONS INFORMATION
Admission to the engineering department: At the time of admission to the institution.
Additional information: Same as freshmen or transfer admission requirements

■ FINANCIAL INFORMATION

ESTIMATED EXPENSES (ACADEMIC YEAR)
[Expenses are for the 1992-93 nine-month academic year.]

State Residents	Undergraduate	Graduate
Tuition and fees	$ 1,800	$ 2,200
College room and board	$ 3,000	$ 3,000
Books and supplies	$ 500	$ 500
Other expenses	$ —(1)	$ —(1)
Total estimated expenses	**$ 5,300**	**$ 5,700**

Out-of-State Residents	Undergraduate	Graduate
Tuition and fees	$ 6,000	$ 6,400
College room and board	$ 3,000	$ 3,000
Books and supplies	$ 500	$ 500
Other expenses	$ —(1)	$ —(1)
Total estimated expenses	**$ 9,500**	**$ 9,900**

Notes: (1) ID card $5, $15 per credit hour is charged for all engineering courses

FINANCIAL AID OFFICE CONTACT
Diane DelBuono, *Director of Financial Aid*
University of Kansas
22 Strong Hall
Lawrence, KS 66045
Phone: (913)864-4700

GENERAL FINANCIAL AID INFORMATION
Forms accepted/required: ACT, FFS, SDF
Additional financial aid information: scholarships, loans, part-time employment and work-study

■ ENGINEERING COLLEGE INFORMATION

[For additional personnel, refer to the *Appendix*.]

HEAD OF ENGINEERING
Carl E. Locke Jr.
Phone: (913)864-3881 **Fax:** (913)864-3199

ENGINEERING COLLEGE ADDRESS
School of Engineering
University of Kansas
4010 Learned Hall
Lawrence, KS 66045-2210
Phone: (913)864-3881 **Fax:** (913)864-3199

ENROLLMENTS—BY CLASS
[Numbers are baccalaureate enrollments for the fall 1991 term, unless otherwise footnoted.]
1st-year students/Freshmen 381
2nd-year students/Sophomores 245
3rd-year students/Juniors 285
4th-year students/Seniors 487 (1)
Total 1,398

Notes: (1) This includes level 5 students (454+33=487).

NUMBER OF DEGREES AWARDED—BY PROGRAM
[Numbers are engineering baccalaureate degrees awarded during the 1991-1992 academic year, unless otherwise footnoted. For full details about each engineering program, refer to the *Tables of Degree Programs*.]
Aerospace Engineering 40
Architectural Engineering 32
Chemical & Petroleum Engineering 12
Civil Engineering 37
Electrical & Computer Engineering 81
Engineering Physics 3
Mechanical Engineering 36
Total 241

PERCENTAGE OF DEGREES AWARDED—BY CATEGORY
[Percentages are of all engineering baccalaureate degrees awarded during the 1991-1992 academic year, unless otherwise footnoted.]
African Americans 1.2%
Asian/Pacific Island Americans 5.8%
Hispanic Americans 1.2%
Native Americans .... — %
Foreign Citizens 19.0%
All Others 72.8%
Women 14.5%
Persons with disabilities .... — % (A)
Students over 25 years of age .... — % (A)

Notes: (A) Data not available.

GRADUATE ENROLLMENTS & DEGREES AWARDED
Master's enrollment 470
Master's degrees awarded 112
Doctoral enrollment 95
Doctoral degrees awarded 18

FULL- & PART-TIME FACULTY—BY DEPARTMENT
[Figures are the head count of full-time faculty and the full-time equivalent (FTE) of part-time faculty for each engineering department or equivalent.]

Department	Full	Part
Chemical & Petroleum Eng	10	—
Civil Eng	20	—
Electrical & Computer Eng	20	—
Mechanical Eng	12	—
Aerospace Eng	7	—
Architectural Eng	5	—
Eng Physics	1	—
Eng Management	2	—

COLLEGE DESCRIPTION
The School of Engineering is located in Learned Hall on the main campus. An engineering library, Spahr Hall, adjoins Learned Hall. The school also conducts engineering research programs in Nichols Hall, the Space Technology Center on west campus. Much of the research in engineering is administered through the non-profit Center for Research, Inc. CRINC helps the university and the School of Engineering attract sponsored research in engineering and related disciplines. Combined programs are available for students interested in pursuing degrees in both business and civil or mechanical engineering, chemical engineering/bio-medical, chemical engineering/environmental and chemical/pre-medical. Most undergraduate courses are taught by full-time faculty. The School of Engineering Career Services Center helps students seeking permanent employment and undergraduates seeking career-related summer employment.

DEGREE PROGRAMS DESCRIPTION
Each of the nine undergraduate degrees include courses in 6 general areas of study: basic sciences, communications, humanities and social sciences, basic engineering sciences, specialized engineering sciences, and engineering design. Appropriate laboratory experience that combines elements of theory and practice is included in each student's program, together with extensive computer-based experience. Through proper planning with their advisors, students may delay choosing specific fields of engineering until they have completed several semesters of college work. Four different graduate degrees can be earned in engineering. The M.S. degree can be earned in Architectural, Aerospace, Civil, Environmental Health Sciences, Water Resources Science, Chemical, Petroleum, Electrical, Engineering Management, Mechanical, Water Resources Engineering, and Environmental Health Engineering. The M.E. degree can be earned in Aerospace. The D.E. degree can be earned in Aerospace, Civil, and Mechanical Engineering. The Ph.D. degree is available in Aerospace, Chemical, Civil, Electrical, Environmental Health Engineering, Environmental Health Science, Mechanical, and Petroleum Engineering.

DUAL DEGREE PROGRAMS
Engineering baccalaureate graduates with dual degrees 6
Enrollment requirements: For a dual degree the student must meet the entrance requirements of both schools.

TRANSFER INFORMATION
Residency Requirements: Students must complete their last 30 hours in residence at KU.

Transfer via Articulation Agreements
Admission to engineering: At any time
Requirements: Transfer students should have overall grade point averages of at least 3.0 with A's & B's in math and science courses.

Transfer without Articulation Agreements
Admission to engineering: At any time
Requirements: Transfer students should have overall grade point averages of at least 3.0 with A's & B's in math and science courses.

GRADUATION REQUIREMENTS
In addition to course requirements, all engineering students must meet the following grade point average requirements: a 2.0 overall KU GPA, a 2.0 GPA in all courses applied to the degree, and a 2.0 GPA in all courses taken in the School of Engineering.

■ STUDENT PROGRAMS

PROFESSIONAL AND HONORARY SOCIETIES
[For key to acronyms, see *Introduction*.]
Professional Societies: AIAA, AIChE, ASCE, ASME, IEEE, IES, ITE, NSAE, NSPE, SAE, The Society of American Military Engineers (SAME), Society of Petroleum Engineers (SPE)
Honorary Societies: Chi Epsilon, Eta Kappa Nu, Phi Alpha Epsilon, Pi Tau Sigma, Sigma Gamma Tau, Tau Beta Pi, Theta Tau

SUPPORT PROGRAMS
Student Chapter Organizations: Society of Women Engineers, National Society of Black Engineers, Society of Hispanic Professional Engineers, American Indian Science and Engineering Society

Minority Engineering Programs
For: Ethnic Minorities **Available:** Year round
Offered: Freshman, Sophomore, Junior, Senior, Graduate level

Early Entry Summer Program For Minorities
For: Ethnic Minorities **Available:** Summer only
Offered: Prematriculation

MACESA Summer Program
For: Ethnic Minorities **Available:** Summer only
Offered: Prematriculation

Other Engineering Support Programs: Summer orientation and enrollment, Renewable Engineering scholarships, Free math and physics tutoring

102 Kansas State University

INSTITUTION PROFILE

HEAD OF THE INSTITUTION
Jon Wefald
Phone: (913)532-6221 Fax: (913)532-7639

GENERAL INFORMATION
[All Students—Fall 1991]
Undergraduate enrollment 17,105
Graduate enrollment 3,607
Total institution enrollment 20,712
Type of institution: Public **Calendar:** Semesters
Nearest city: Topeka **Population:** 115,266
Miles from main campus: 55 **Setting:** Small Town
Types of engineering degrees: Engineering & Technology
Other degree-granting colleges: Architecture, Arts & Sciences, Business Administration, Education, Veterinary Medicine, Human Ecology, Agriculture, Technology

ENGINEERING ADMISSIONS

ADMISSIONS OFFICE CONTACT
Richard N. Elkins
Kansas State University
119 Anderson Hall
Manhattan, KS 66506-0102
Phone: (913)532-6250 Fax: (913)532-5632

ENGINEERING COLLEGE ADMISSIONS INFORMATION
Entrance Requirements: High School courses—Algebra (2 years), Geometry (1 year), Trigonometry, The ACT is required for student counseling, but not for admission.
Entrance Recommendations: High School courses—English (4 years), Chemistry (1 year), Physics (1 year), Calculus (1 year)
Requirements for out-of-state residents: Kansas applicants: high school diploma Out-of-state applicants: 3.0 GPA or ACT composite of 21 or higher
Requirements for foreign students: TOEFL (Score: 550); All appropriate immigration standards and requirements must be met.

DEPARTMENTAL ADMISSIONS INFORMATION
Admission to the engineering department: At the time of admission to the institution.

FINANCIAL INFORMATION

ESTIMATED EXPENSES (ACADEMIC YEAR)
[Expenses are for the 1992-93 nine-month academic year.]

State Residents	Undergraduate	Graduate
Tuition and fees	$ 2,040 [1]	$ 2,418 [1]
College room and board	$ 2,840	$ 2,840
Books and supplies	$ 500	$ 500
Other expenses	$ –	$ –
Total estimated expenses	$ 5,380	$ 5,758

Out-of-State Residents	Undergraduate	Graduate
Tuition and fees	$ 6,212 [1]	$ 6,638 [1]
College room and board	$ 2,840	$ 2,840
Books and supplies	$ 500	$ 500
Other expenses	$ –	$ –
Total estimated expenses	$ 9,552	$ 9,978

Notes: (1) Includes $100/semester Engineering Student Fee.

FINANCIAL AID OFFICE CONTACT
Larry Moeder, Director
Kansas State University
104 Fairchild Hall
Manhattan, KS 66506-1104
Phone: (913)532-6420 Fax: (913)532-5632

GENERAL FINANCIAL AID INFORMATION
Forms accepted/required: FFS, IRS, Institutional
Additional financial aid information: Need-based and merit-based scholarships; loan programs with Perkins, Guaranteed Student Loans and short-term; Supplemental Educational Opportunity Grants; numerous part-time jobs.

ENGINEERING COLLEGE INFORMATION
[For additional personnel, refer to the *Appendix*.]

HEAD OF ENGINEERING
Donald E. Rathbone
Phone: (913)532-5590 Fax: (913)532-7810
EMail: DEANENGR@KSUVM.KSU.EDU

ENGINEERING COLLEGE ADDRESS
College of Engineering
Kansas State University
Durland Hall
Manhattan, KS 66506
Phone: (913)532-5590 Fax: (913)532-7810

ENROLLMENTS—BY CLASS
[Numbers are baccalaureate enrollments for the fall 1991 term, unless otherwise footnoted.]
1st-year students/Freshmen 706
2nd-year students/Sophomores 486
3rd-year students/Juniors 515
4th-year students/Seniors 826
Total 2,533

NUMBER OF DEGREES AWARDED—BY PROGRAM
[Numbers are engineering baccalaureate degrees awarded during the 1991-1992 academic year, unless otherwise footnoted. For full details about each engineering program, refer to the *Tables of Degree Programs*.]
Agricultural Engineering 3
Architectural Engineering 42
Chemical Engineering 13
Civil Engineering 29
Electrical & Computer Engineering 111
Industrial Engineering 53
Mechanical Engineering 79
Nuclear Engineering 6
Total 336

PERCENTAGE OF DEGREES AWARDED—BY CATEGORY
[Percentages are of all engineering baccalaureate degrees awarded during the 1991-1992 academic year, unless otherwise footnoted.]
African Americans .... – %
Asian/Pacific Island Americans .... 1.2 %
Hispanic Americans .... – %
Native Americans .... – %
Foreign Citizens .... 1.7 %
All Others .... 97.1 %
Women .... 5.9 %
Persons with disabilities .... – %
Students over 25 years of age .... – % (A)

Notes: (A) Data not available.

GRADUATE ENROLLMENTS & DEGREES AWARDED
Master's enrollment 203
Master's degrees awarded 65
Doctoral enrollment 79
Doctoral degrees awarded 14

ENGINEERING STUDENT DATA

Applicants to the engineering college
Number of applicants to engineering college 1,179
Percent offered admission .... 84.6 %

Matriculated engineering students
Percentage in top quartile (25%) of High School class 90.0
Average ACT scores: Math—26.0, Composite—25.6

FULL- & PART-TIME FACULTY—BY DEPARTMENT
[Figures are the head count of full-time faculty and the full-time equivalent (FTE) of part-time faculty for each engineering department or equivalent.]

Department	Full	Part
Chemical Eng	9	–
Civil Eng	13	2.2
Electrical & Computer Eng	21	2.1
Industrial Eng	10	–
Mechanical Eng	16	1.3
Architectural Eng & Construction Science	12	1.2
Nuclear Eng	6	–
Agricultural Eng	19	–

COLLEGE DESCRIPTION

The largest and most comprehensive engineering college in Kansas. B.S. degrees are offered in ten engineering fields plus construction science and engineering technology. Most engineering students have full-time academic programs and are involved in a wide variety of university and college organizations and activities. Annual Leadership Institute is available. Co-op is an option; industrial experience in summer is encouraged. All faculty are involved in major graduate/research programs, a primary mission of the college is undergraduate professional education. Average number of years required to actually complete the bachelor's degree: 4.7.

DEGREE PROGRAMS DESCRIPTION

Building on a solid foundation of physical science, mathematics and engineering science, the engineering curricula emphasize analysis and design in the last two years. Computer applications are used extensively in appropriate areas. The curricula also include a variety of humanities and social science electives distributed throughout all four years. The B.S. degree programs in engineering prepare the graduate for entry-level professional positions or for graduate study. In addition to engineering, with the selection of appropriate electives, graduates may pursue graduate degrees in business, law or medicine.

TRANSFER INFORMATION

Residency Requirements: Courses in the major field shall be taken in residence unless an exception is granted by the major department. At least 20 of the last 30 credits for the degree must be in residence with at least 30 resident hours total.

Transfer via Articulation Agreements
Admission to engineering: Transfers may be admitted prior to the start of any academic term; applications should be received at least two months prior to the start of the term.
Requirements: 2.0 (C) average in previous academic work is the minimum, but a higher GPA is generally required.

Transfer without Articulation Agreements
Admission to engineering: Transfers may be admitted prior to the start of any academic term; applications should be received at least two months prior to the start of the term.
Requirements: 2.0 (C) average in previous academic work is the minimum, but a higher GPA is generally required.

GRADUATION REQUIREMENTS

Credits for B.S. degrees: Agr. Eng. - 135; Arch. Eng. - 162 (5-year program); Chem. Eng. - 134; Civil Eng. - 134; Comp Eng. - 135; Elec. Eng. 135; Ind. Eng. - 133; Mfg. Eng. - 133; Mech. Eng. - 135; Nucl. Eng. - 132; Constr. Sci. - 133; Eng. Tech. -126

STUDENT PROGRAMS

PROFESSIONAL AND HONORARY SOCIETIES
[For key to acronyms, see *Introduction*.]
Professional Societies: AIChE, ANS, ASAE, ASCE, ASHRAE, ASME, IEEE, IIE, NSAE, NSPE, Society of Manufacturing Engineers, Society of Automotive Engineers
Honorary Societies: Alpha Epsilon, Alpha Pi Mu, Chi Epsilon, Eta Kappa Nu, Omega Chi Epsilon, Phi Alpha Epsilon, Pi Tau Sigma, Tau Alpha Pi, Tau Beta Pi, Sigma Lambda Chi, Alpha Nu Sigma

SUPPORT PROGRAMS
Student Chapter Organizations: Society of Women Engineers, National Society of Black Engineers, Society of Hispanic Professional Engineers, American Indian Science & Engineering Society

Minority Engineering Program
For: Ethnic Minorities **Available:** Academic year
Offered: Freshman, Sophomore, Junior, Senior, Graduate level
Other Engineering Support Programs: Summer and fall orientation programs for all new freshmen and transfers; academic assistance seminars; veterans' services; handicapped services, counseling services; foreign student office; career planning and placement center. Free tutoring available for introductory math and science classes.

103 University of Kentucky

■ INSTITUTION PROFILE

HEAD OF THE INSTITUTION
Charles T. Wethington Jr.
Phone: (606)257-1704 Fax: (606)257-5640

GENERAL INFORMATION
[All Students—Fall 1991]
Undergraduate enrollment 17,657
Graduate enrollment 6,475
Total institution enrollment **24,132**

Type of institution: Public **Calendar:** Semesters
Nearest city: Cincinnati, OH **Population:** 355,000
Miles from main campus: 75 **Setting:** Urban
Types of engineering degrees: Engineering
Other degree-granting colleges: Agricultural & Environmental, Allied Health Sciences, Architecture, Arts & Sciences, Business Administration, Communication & Journalism, Dentistry, Education, Fine & Performing Arts, Humanities & Social Sciences, Law, Medicine, Nursing, Pharmacy, Technology & Applied Sciences, Library and Information Science, Human Environmental Sciences

■ ENGINEERING ADMISSIONS

ADMISSIONS OFFICE CONTACT
Joseph L. Fink III
University of Kentucky
Office of Admissions
Lexington, KY 40506-0054
Phone: (606)257-2000 Fax: (606)257-3823

ENGINEERING COLLEGE ADMISSIONS INFORMATION
Admission to the engineering college: A prospective engineering student is admitted into pre-engineering within the College at the time of admission to the institution.
Entrance Requirements: High School courses—English (4 years), Math including Algebra I & II and Geometry (3 years), Science chosen from Biology, Chemistry & Physics (2 years), Social Studies (incl. Amer. History & World Civ.) (2 years), ACT Assessment (or SAT) scores are required. Although there is no minimum score requirement, the recommended is 21 for ACT or the SAT equivalent.
Entrance Recommendations: High School courses—Foreign Language (2 years), Computers (1 year), Mechanical Drawing (1 year), Fourth year of Math (1 year)
Requirements for foreign students: TOEFL (Score: 525); financial statement

DEPARTMENTAL ADMISSIONS INFORMATION
Admission to the engineering department: Admission to engineering standing in any department requires the completion of at least 50 semester hrs applicable towards the degree with a minimum cum GPA of 2.25.
Additional information: Students must have at least a 2.25 GPA in required English, chemistry, physics and calculus courses. Higher grade point averages and specific program requirements may be enforced for admission into some departments.

■ FINANCIAL INFORMATION

ESTIMATED EXPENSES (ACADEMIC YEAR)
[Expenses are for the 1992-93 nine-month academic year.]

State Residents	Undergraduate	Graduate
Tuition and fees	$ 1,998 (3)	$ 2,158 (3)
College room and board	$ 2,952 (2)	$ 3,069 (1)
Books and supplies	$ 500 (C)	$ 500 (C)
Other expenses	$ –	$ –
Total estimated expenses	**$ 5,450**	**$ 5,727**

Out-of-State Residents	Undergraduate	Graduate
Tuition and fees	$ 5,358 (3)	$ 5,838 (3)
College room and board	$ 2,952 (2)	$ 3,069 (1)
Books and supplies	$ 500 (C)	$ 500 (C)
Other expenses	$ –	$ –
Total estimated expenses	**$ 8,810**	**$ 9,407**

Notes: (C) Estimated data. (1) 1 bedroom, single occupancy @ $341/month. Includes basic furnishings and utilities. Does not include board or telephone. Available 12 months per year. (2) All undergraduate residence halls for academic year. Includes 2 meals, 5 days (Monday-Friday). (3) Additional fees for 4-week and 8-week summer sessions.

FINANCIAL AID OFFICE CONTACT
David H. Stockham, *Director*
University of Kentucky
Office of Student Financial Aid
Lexington, KY 40506-3172
Phone: (606)257-3172

GENERAL FINANCIAL AID INFORMATION
Forms accepted/required: FAF, IRS, Institutional, A separate application is required for merit-based scholarships.
Additional financial aid information: Need-based scholarships; short-term loans; merit-based scholarships; part-time jobs on campus; College Work-Study programs.

■ ENGINEERING COLLEGE INFORMATION
[For additional personnel, refer to the *Appendix*.]

HEAD OF ENGINEERING
Thomas W. Lester
Phone: (606)257-1687 Fax: (606)258-4922
EMail: thomas.lester@ukwang.uky.edu

ENGINEERING COLLEGE ADDRESS
College of Engineering
University of Kentucky
177 Anderson Hall
Lexington, KY 40506-0046
Phone: (606)257-1687 Fax: (606)258-4922

ENROLLMENTS—BY CLASS
[Numbers are baccalaureate enrollments for the fall 1991 term, unless otherwise footnoted.]
1st-year students/Freshmen 515
2nd-year students/Sophomores 322
3rd-year students/Juniors 292
4th-year students/Seniors 485
Total **1,614**

NUMBER OF DEGREES AWARDED—BY PROGRAM
[Numbers are engineering baccalaureate degrees awarded during the 1991-1992 academic year, unless otherwise footnoted. For full details about each engineering program, refer to the *Tables of Degree Programs*.]
Agricultural Engineering .... –
Chemical Engineering 20
Civil Engineering 62
Electrical Engineering 93
Engineering Mechanics .... – (B)
Materials Science & Engineering 7
Mechanical Engineering 45
Mining Engineering 6
Total **233**

Notes: (B) Data not applicable.

PERCENTAGE OF DEGREES AWARDED—BY CATEGORY
[Percentages are of all engineering baccalaureate degrees awarded during the 1991-1992 academic year, unless otherwise footnoted.]
African Americans 1.3%
Asian/Pacific Island Americans 3.9%
Hispanic Americans .... – %
Native Americans .... – %
Foreign Citizens 4.7%
All Others 90.1%
Women 10.3%
Persons with disabilities .... – % (A)
Students over 25 years of age 22.3%

Notes: (A) Data not available.

GRADUATE ENROLLMENTS & DEGREES AWARDED
Master's enrollment 229
Master's degrees awarded 54
Doctoral enrollment 131
Doctoral degrees awarded 21

ENGINEERING STUDENT DATA
Applicants to the engineering college
Number of applicants to engineering college 798
Percent offered admission 88.2%

Matriculated engineering students
Percentage in top quartile (25%) of High School class ...... – (A)
Average SAT scores: Combined—1015 (1)
Average ACT scores: Composite—26

Notes: (A) Data not available. (1) Average SAT scores are for ALL freshmen. Separate statistics are not maintained for Engineering freshmen.

FULL- & PART-TIME FACULTY—BY DEPARTMENT
[Figures are the head count of full-time faculty and the full-time equivalent (FTE) of part-time faculty for each engineering department or equivalent.]

Department	Full	Part
Chemical Eng	12	–
Civil Eng	18	1.3
Electrical Eng	21	–
Mechanical Eng	21	–
Agricultural Eng	22	–
Eng Mechanics	8	–
Materials Science & Eng	7	–
Mining Eng	10	–

COLLEGE DESCRIPTION
The College offers four-year undergraduate programs in engineering, emphasizing basic science and engineering fundamentals in the first two years, followed by two years of more specialized training in one of our seven majors, with some liberal arts studies throughout all four years. Emphasis is on a balanced program in order to prepare each student either for an entry level professional position or for graduate study. Co-op programs are available in all departments and would normally extend a program to five years. Transfer programs are articulated with regional colleges and universities.

DEGREE PROGRAMS DESCRIPTION
The B.S. degrees in the various engineering programs prepare the graduate for entry-level professional positions or for graduate study in each of the respective fields. M.S. and Ph.D. degrees, emphasizing the applications of the natural sciences to the analysis and solution of engineering problems, are available in all eight program areas. An MCE and MMinE degree, both emphasizing professional engineering practice, are also available in civil engineering and mining engineering respectively.

TRANSFER INFORMATION
Residency Requirements: Students must complete 30 of the last 36 credit hours of their program from the University of Kentucky.

Transfer via Articulation Agreements
Admission to engineering: At any time
Engineering graduates transferred from   4-yr: 69   2-yr: 52

Transfer without Articulation Agreements
Admission to engineering: At any time
Engineering graduates transferred from   4-yr: 28   2-yr: 5

GRADUATION REQUIREMENTS
B.S. programs in engineering require 129-141 semester credits depending on the major selected. A cumulative GPA of 2.0 is required. Some departments also have additional graduation requirements.

■ STUDENT PROGRAMS

PROFESSIONAL AND HONORARY SOCIETIES
[For key to acronyms, see *Introduction*.]

Professional Societies: AIAA, AIChE, ASAE, ASCE, ASME, IEEE, NSPE, SME, SME*, American Society of Metals, Kentucky Society of Professional Engineers
Honorary Societies: Chi Epsilon, Eta Kappa Nu, Omega Chi Epsilon, Pi Tau Sigma, Tau Beta Pi, Alpha Sigma Mu

SUPPORT PROGRAMS
Student Chapter Organizations: Society of Women Engineers, National Society of Black Engineers

Women in Engineering
For: Women **Available:** Year round
Offered: Prematriculation, Freshman, Sophomore, Junior, Senior, Graduate level

Personal and career counseling
For: Women & Ethnic Minorities **Available:** Year round
Offered: Prematriculation, Freshman, Sophomore, Junior, Senior, Graduate level

Other Engineering Support Programs: Summer and fall orientation programs for all freshmen and transfers; learning assessment program; academic advising; placement and career counseling services; foreign student office; veterans' services; services for the blind and handicapped.

104 Lafayette College

INSTITUTION PROFILE
HEAD OF THE INSTITUTION
Robert I. Rotberg
Phone: (215)250-5200   Fax: (215)250-0157

GENERAL INFORMATION
[All Students—Fall 1991]
Undergraduate enrollment 2,225
Total institution enrollment 2,225
Type of institution: Private   Calendar: Semesters
Nearest city: Allentown   Population: 105,000
Miles from main campus: 15   Setting: Suburban
Types of engineering degrees: Engineering
Other degree-granting colleges: Social Sciences Division, Natural Sciences Division, Humanities Division

ENGINEERING ADMISSIONS
ADMISSIONS OFFICE CONTACT
G. Gary Ripple
Lafayette College
Director of Admissions
118 Markle Hall
Easton, PA 18042-1770
Phone: (215)250-5110   Fax: (215)250-9850

ENGINEERING COLLEGE ADMISSIONS INFORMATION
Entrance Requirements: SAT, ACT, High School courses—English (4 years), Chemistry and Physics (1 year), Foreign Language (2 years), Mathematics (4 years), SAT; achievement tests (3) including English composition with or without the essay component, and preferably a sequence of math and science options.
Entrance Recommendations: High School courses—Additional academic subjects (5 years), Calculus or pre-calculus (1 year), Computers (1 year), Engineering Drawing (5 years), ACT in lieu of SAT
Requirements for foreign students: TOEFL; financial statement; TOEFL, if secondary school language of instruction is other than English. Otherwise SAT required.

DEPARTMENTAL ADMISSIONS INFORMATION
Admission to the engineering department: At the time of admission to the institution.

FINANCIAL INFORMATION
ESTIMATED EXPENSES (ACADEMIC YEAR)
[Expenses are for the 1992-93 nine-month academic year.]

All Students	Undergraduate	Graduate
Tuition and fees	$16,795	$ –
College room and board	$ 5,190	$ –
Books and supplies	$ 550	$ –
Other expenses	$ 1,000	$ –
Total estimated expenses	$23,535	$ –

FINANCIAL AID OFFICE CONTACT
Barry W. McCarty, *Director of Student Financial Aid*
Lafayette College
107 Markle Hall
Easton, PA 18042-1777
Phone: (215)250-5055   Fax: (215)250-9850

GENERAL FINANCIAL AID INFORMATION
Forms accepted/required: FAF, IRS
Additional financial aid information: Need-based scholarships (80% of accepted freshmen with need are aided to full need); aid awards include: scholarships, low interest loans, campus employment; college also provides fully subsidized parent loans.

ENGINEERING COLLEGE INFORMATION
[For additional personnel, refer to the *Appendix*.]
HEAD OF ENGINEERING
Michael A. Paolino
Phone: (215)250-5403   Fax: (215)250-0351
EMail: PM#0@Lafayacs

ENGINEERING COLLEGE ADDRESS
Engineering Division
Lafayette College
Dana Hall of Engineering
Easton, PA 18042-1775
Phone: (215)250-5403   Fax: (215)250-0351

ENROLLMENTS—BY CLASS
[Numbers are baccalaureate enrollments for the fall 1991 term, unless otherwise footnoted.]
1st-year students/Freshmen 134
2nd-year students/Sophomores 105
3rd-year students/Juniors 97
4th-year students/Seniors 115
Total 451

NUMBER OF DEGREES AWARDED—BY PROGRAM
[Numbers are engineering baccalaureate degrees awarded during the 1991-1992 academic year, unless otherwise footnoted. For full details about each engineering program, refer to the *Tables of Degree Programs*.]
Chemical Engineering 13
Civil Engineering 25
Electrical Engineering 27
Engineering .... 17 [2]
Mechanical Engineering 32
Total 114

Notes: (2) A Bachelor of Arts degree is granted in this program.

PERCENTAGE OF DEGREES AWARDED—BY CATEGORY
[Percentages are of all engineering baccalaureate degrees awarded during the 1991-1992 academic year, unless otherwise footnoted.]
African Americans .... – %
Asian/Pacific Island Americans .... 2.6 %
Hispanic Americans .... – %
Native Americans .... – %
Foreign Citizens .... 7.9 %
All Others .... 89.5 %
Women .... 25.4 %
Persons with disabilities .... – %
Students over 25 years of age .... – %

ENGINEERING STUDENT DATA
Applicants to the engineering college
Number of applicants to engineering college .... 4,040 [1]
Percent offered admission .... 58.0% [2]

Matriculated engineering students
Percentage in top quartile (25%) of High School class 85.0
Average SAT scores: Math—624, Verbal—549, Combined—1173

Notes: (1) Applies to the entire first year class. (2) Applies to the entire first year class.

FULL- & PART-TIME FACULTY—BY DEPARTMENT
[Figures are the head count of full-time faculty and the full-time equivalent (FTE) of part-time faculty for each engineering department or equivalent.]

Department	Full	Part
Chemical Eng	8	0.5
Eng	2	–
Civil Eng	8	0.5
Electrical Eng	9	0.6
Mechanical Eng	7	1.0

COLLEGE DESCRIPTION
The college offers four-year engineering programs in an undergraduate environment. The student faculty ratio is 12:1. Equipment and laboratories are state-of-the-art and include modern CADD facilities. A common freshman year for all engineering students is followed by three years of study in a specific discipline. There is a strong laboratory component in all programs. Computer applications are integrated into the four years of study. Engineering students also select courses from the arts, humanities, sciences and social sciences. With additional course work, or advanced placement, students can complete the engineering degree with a minor, outside engineering, in four years. A one semester study abroad and five-year, two-degree programs are available. A Bachelor of Arts in engineering is also offered. This program permits flexibility in designing a program in engineering. Average number of years required to actually complete the bachelor's degree: 4.

DEGREE PROGRAMS DESCRIPTION
The B.S. degrees in the various engineering programs prepare the graduate for entry into the profession or for graduate study in each of the respective fields. Qualified students have the opportunity to work closely with a faculty member on an undergraduate research project which can lead to independent study and an honors thesis in the senior year. Several opportunities are available for practical experience through summer industrial employment, summer student-faculty research, study abroad, and cooperative projects with industry. Graduates of the B.A. engineering program go on to successful careers in manufacturing, marketing, and finance; some graduates also continue study in professional schools.

DUAL DEGREE PROGRAMS
Engineering baccalaureate graduates with dual degrees 4

TRANSFER INFORMATION
Residency Requirements: Students must spend a minimum of two years in residence.

Transfer without Articulation Agreements
Admission to engineering: At the end of the freshman year
Engineering graduates transferred from ... 4-yr: 25   2-yr: 3

GRADUATION REQUIREMENTS
B.S. engineering programs require 131-132 semester credits.

STUDENT PROGRAMS
PROFESSIONAL AND HONORARY SOCIETIES
[For key to acronyms, see *Introduction*.]
Professional Societies: AIChE, ASCE, ASME, IEEE, Leonardo Society
Honorary Societies: Eta Kappa Nu, Pi Tau Sigma, Tau Beta Pi

SUPPORT PROGRAMS
Student Chapter Organizations: Society of Women Engineers, National Society of Black Engineers

Association of Black Collegians
For: Ethnic Minorities   **Available:** Academic year
Offered: Freshman, Sophomore, Junior, Senior

Association of Lafayette Women
For: Women   **Available:** Academic year
Offered: Freshman, Sophomore, Junior, Senior

Student Alliance of Latinos of South America
For: Ethnic Minorities   **Available:** Academic year
Offered: Freshman, Sophomore, Junior, Senior

East Asian Club
For: Ethnic Minorities   **Available:** Academic year
Offered: Freshman, Sophomore, Junior, Senior

Other Engineering Support Programs: Summer orientation program for all freshmen and transfers; career counseling services; foreign student services; facilities for the handicapped; placement services; academic tutors.

105 Lamar University

INSTITUTION PROFILE

HEAD OF THE INSTITUTION
Brock Brentlinger, *Interim*
Phone: (409)880-8401   Fax: (409)880-8404

GENERAL INFORMATION
[All Students—Fall 1991]
Undergraduate enrollment 8,586
Graduate enrollment 760
Total institution enrollment 9,346

Type of institution: Public   Calendar: Semesters
Nearest city: Houston, TX   Population: 2,000,000
Miles from main campus: 80   Setting: Urban
Types of engineering degrees: Engineering
Other degree-granting colleges: Arts & Sciences, Fine Arts and Communication, Business, Education and Human Development, Engineering

ENGINEERING ADMISSIONS

ADMISSIONS OFFICE CONTACT
James Rush
Lamar University
P.O. Box 10009
Beaumont, TX 77710
Phone: (409)880-8353

ENGINEERING COLLEGE ADMISSIONS INFORMATION
Admission to the engineering college: A provisional engineering student is accepted to an engineering department after completing 51 hours of the program with at least a 2.25 grade point average.
Entrance Requirements: SAT (Score: 900), ACT (Score: 21), High School courses—Mathematics (4 years), English (4 years), Natural Science (2 years), Foreign Language (1 year), Texas Academic Skills Program
Entrance Recommendations: High School courses—Mathematics (4 years), English (4 years), Natural Science (2 years), Foreign Language (1 year)
Requirements for foreign students: TOEFL (Score: 500); financial statement

DEPARTMENTAL ADMISSIONS INFORMATION
Admission to the engineering department: A provisional engineering student is accepted to an engineering department after completing 51 hours of the program with at least a 2.25 grade point average.
Additional information: Completion of 51 semester hours of the Common Program for all pre-engineering majors with a GPA of 2.25 or more on all required courses. A minimum of 45 semester hours (at least 25 semester hours in engineering at the 300 and 400 level) must be earned after admission to the department.

FINANCIAL INFORMATION

ESTIMATED EXPENSES (ACADEMIC YEAR)
[Expenses are for the 1992-93 nine-month academic year.]

State Residents	Undergraduate	Graduate
Tuition and fees	$ 553	$ 553
College room and board	$ 1,514	$ 1,514
Books and supplies	$ 300	$ 300
Other expenses	$ –	$ –
Total estimated expenses	**$ 2,367**	**$ 2,367**

Out-of-State Residents	Undergraduate	Graduate
Tuition and fees	$ 1,801	$ 1,801
College room and board	$ 1,514	$ 1,514
Books and supplies	$ 300	$ 300
Other expenses	$ –	$ –
Total estimated expenses	**$ 3,615**	**$ 3,615**

FINANCIAL AID OFFICE CONTACT
Ralynn Castete, *Director, Student Financial Aid*
Lamar University
P.O. Box 10042
Beaumont, TX 77710
Phone: (409)880-8450

GENERAL FINANCIAL AID INFORMATION
Forms accepted/required: FAF, FFS
Additional financial aid information: Financial assistance in the form of scholarships, grants, loans and employment is available to a number of qualified students.

ENGINEERING COLLEGE INFORMATION
[For additional personnel, refer to the *Appendix*.]

HEAD OF ENGINEERING
Fred M. Young
Phone: (409)880-8741   Fax: (409)880-8121
EMail: (fred@lub001.lamar.educ)

ENGINEERING COLLEGE ADDRESS
College of Engineering
Lamar University
P.O. Box 10057
Beaumont, TX 77710
Phone: (409)880-8741   Fax: (409)880-8404

ENROLLMENTS—BY CLASS
[Numbers are baccalaureate enrollments for the fall 1991 term, unless otherwise footnoted.]

1st-year students/Freshmen 284
2nd-year students/Sophomores 167
3rd-year students/Juniors 169
4th-year students/Seniors .... 279 [1]
Total **899**

Notes: (1) Includes 57 post baccalaureate students

NUMBER OF DEGREES AWARDED—BY PROGRAM
[Numbers are engineering baccalaureate degrees awarded during the 1991-1992 academic year, unless otherwise footnoted. For full details about each engineering program, refer to the *Tables of Degree Programs*.]

Chemical Engineering 10
Civil Engineering 6
Electrical Engineering 30
Industrial Engineering 5
Mechanical Engineering 20
Total **71**

PERCENTAGE OF DEGREES AWARDED—BY CATEGORY
[Percentages are of all engineering baccalaureate degrees awarded during the 1991-1992 academic year, unless otherwise footnoted.]

African Americans .... 5.0 %
Asian/Pacific Island Americans .... – %
Hispanic Americans .... 4.0 %
Native Americans .... 0.9 %
Foreign Citizens .... 7.0 %
All Others .... 83.1 %
Women .... 13.0 %
Persons with disabilities .... – %
Students over 25 years of age .... 57.0 %

GRADUATE ENROLLMENTS & DEGREES AWARDED
Master's enrollment 206
Master's degrees awarded 49
Doctoral enrollment 42
Doctoral degrees awarded 3

ENGINEERING STUDENT DATA
Applicants to the engineering college
Number of applicants to engineering college .... – (A)
Percent offered admission .... –% (A)

Matriculated engineering students
Percentage in top quartile (25%) of High School class ...... – (A)
Average SAT scores: Math—504, Verbal—460, Combined—964

Notes: (A) Data not available.

FULL- & PART-TIME FACULTY—BY DEPARTMENT
[Figures are the head count of full-time faculty and the full-time equivalent (FTE) of part-time faculty for each engineering department or equivalent.]

Department	Full	Part
Chemical Eng	5	–
Civil Eng	4	–
Electrical Eng	4.5	1.5
Industrial Eng	3	–
Mechanical Eng	4	–
Mathematics	17	–
Computer Science	9	–

COLLEGE DESCRIPTION
The college offers a four-year undergraduate program in Chemical, Civil, Electrical, Industrial and Mechanical Engineering. The first three semesters are common to all of the programs for entering freshmen. Entering freshmen and transfer students are considered provisional majors. Admission to one of the disciplines requires completion of at least 51 semester hours of the common program with a minimum GPA of 2.25 on all required courses. The college also offers a four-year baccalaureate degree in industrial technology. The graduate program offers Master of Engineering, Master of Engineering Science, Master of Engineering Management, Master of Environmental Studies, Master of Environmental Engineering and Doctor of Engineering. Co-op is an option in all four-year programs. Average number of years required to actually complete the bachelor's degree: 5.

DEGREE PROGRAMS DESCRIPTION
The BS degree in each of the engineering programs prepares the graduate for entry-level professional positions or for graduate study. Graduate study in engineering prepares the graduate to analyze and solve complex engineering problems.

TRANSFER INFORMATION
Residency Requirements: 30 semester hours in residence at Lamar University with at least 24 semester hours earned after attaining Senior classification.

Transfer via Articulation Agreements
Admission to engineering: At the end of the sophomore year
Engineering graduates transferred from . . 4-yr: –   2-yr: –
Requirements: The College of Engineering has agreements between several Texas Junior Colleges that allow students to transfer into the College after successfully completing an approved 2 + 2 plan.

Transfer without Articulation Agreements
Admission to engineering: Transfer into our provisional program until all requirements to transfer into the department have been met.
Engineering graduates transferred from . . 4-yr: –   2-yr: –
Requirements: Students must have at least a 2.0 grade point average.

GRADUATION REQUIREMENTS
B.S. programs in engineering require 137-144 semester hours credit, depending on major selected.

STUDENT PROGRAMS

PROFESSIONAL AND HONORARY SOCIETIES
[For key to acronyms, see *Introduction*.]
Professional Societies: ACM, AIChE, AMS, ASCE, ASME, IEEE, IIE
Honorary Societies: Alpha Pi Mu, Chi Epsilon, Eta Kappa Nu, Omega Chi Epsilon, Pi Tau Sigma, Tau Beta Pi, Pi Mu Epsilon, Upsilon Pi Epsilon

SUPPORT PROGRAMS
Student Chapter Organizations: Society of Women Engineers, National Society of Black Engineers, Society of Hispanic Professional Engineers

Society of Women Engineers
For: Women   **Available:** Academic year
Offered: Freshman, Sophomore, Junior, Senior, Graduate level

National Society of Black Engineers
For: Ethnic Minorities   **Available:** Year round
Offered: Freshman, Sophomore, Junior, Senior, Graduate level

Society of Hispanic Prof. Engineers
For: Ethnic Minorities   **Available:** Academic year
Offered: Freshman, Sophomore, Junior, Senior, Graduate level

Other Engineering Support Programs: Summer and fall orientation program for all freshmen and transfers; Learning Skills Program; career counseling and placement center; international student office; veteran affairs office; services for handicapped students, and cooperative education office.

106 Lawrence Technological University

INSTITUTION PROFILE
HEAD OF THE INSTITUTION
Richard E. Marburger
Phone: (313)356-0200 **Fax:** (313)356-6458
GENERAL INFORMATION
[All Students—Fall 1991]
Undergraduate enrollment 4,586
Graduate enrollment 300
Total institution enrollment **4,886**
Type of institution: Private **Calendar:** Quarters
Location: Southfield **Population:** 75,728
Setting: Suburban
Types of engineering degrees: Engineering & Technology
Other degree-granting colleges: Architecture, Arts & Sciences, Management

ENGINEERING ADMISSIONS
ADMISSIONS OFFICE CONTACT
Timothy R. Kennedy
Lawrence Technological University
21000 W.10 Mile Road
Southfield, MI 48075-1058
Phone: (313)356-0200 **Fax:** (313)356-6458
ENGINEERING COLLEGE ADMISSIONS INFORMATION
Entrance Requirements: ACT, High School courses—English (4 years), Math (4 years), Natural Science (4 years), Social Science (3 years), ACT results are required of students entering as freshmen. Test scores are for counseling and will not be used for the admission decision.
Requirements for foreign students: TOEFL (Score: 550); financial statement
DEPARTMENTAL ADMISSIONS INFORMATION
Admission to the engineering department: At the time of admission to the institution.

FINANCIAL INFORMATION
ESTIMATED EXPENSES (ACADEMIC YEAR)
[Expenses are for the 1992-93 nine-month academic year.]

All Students	Undergraduate	Graduate
Tuition and fees	$ 7,422 [3]	$ 4,416 [1]
College room and board	$ 1,944 [4]	$ — [2]
Books and supplies	$ 540	$ 360
Other expenses	$ —	$ —
Total estimated expenses	**$ 9,906**	**$ 4,776**

Notes: (1) Graduate students are not full-time students. Tuition indicated is for 8 credit hours per term. (2) Graduate students do not reside on campus. (3) International student tuition is 35% higher. (4) Furnished apartments with kitchen for four students.

FINANCIAL AID OFFICE CONTACT
Paul F. Kinder, *Director of Financial Aid & Veterans Affairs*
Lawrence Technological University
21000 West Ten Mile Road
Southfield, MI 48075-1058
Phone: (313)356-0200 **Fax:** (313)356-6458
GENERAL FINANCIAL AID INFORMATION
Forms accepted/required: Additional financial aid information: Need-based scholarships and grants; loans, merit-based scholarships; part-time jobs on campus; college work-study program.

ENGINEERING COLLEGE INFORMATION
[For additional personnel, refer to the *Appendix*.]
HEAD OF ENGINEERING
Joseph B. Olivieri
Phone: (313)356-0200 **Fax:** (313)356-6458
EMail: Bitnet: OLIVIERI@LTUVAX

ENGINEERING COLLEGE ADDRESS
College of Engineering
Lawrence Technological University
21000 W. Ten Mile Road
Southfield, MI 48075-1058
Phone: (313)356-0200 **Fax:** (313)356-6458
ENROLLMENTS—BY CLASS
[Numbers are baccalaureate enrollments for the fall 1991 term, unless otherwise footnoted.]
1st-year students/Freshmen 505
2nd-year students/Sophomores 493
3rd-year students/Juniors 459
4th-year students/Seniors 396
Total **1,853**
NUMBER OF DEGREES AWARDED—BY PROGRAM
[Numbers are engineering baccalaureate degrees awarded during the 1991-1992 academic year, unless otherwise footnoted. For full details about each engineering program, refer to the *Tables of Degree Programs*.]
Civil & Construction Engineering .... 38 [2]
Electrical Engineering 133
Mechanical Engineering 175
Total **346**
Notes: (2) Of this total 26 are Civil Engineering graduates.
PERCENTAGE OF DEGREES AWARDED—BY CATEGORY
[Percentages are of all engineering baccalaureate degrees awarded during the 1991-1992 academic year, unless otherwise footnoted.]
African Americans 2.3%
Asian/Pacific Island Americans 2.3%
Hispanic Americans 1.2%
Native Americans 0.6%
Foreign Citizens 9.2%
All Others 84.4%
Women 10.1%
Persons with disabilities .... — % [A]
Students over 25 years of age .... — % [A]
Notes: (A) Data not available.
GRADUATE ENROLLMENTS & DEGREES AWARDED
Master's enrollment 42
Master's degrees awarded 9
Doctoral enrollment .... —
Doctoral degrees awarded .... —
ENGINEERING STUDENT DATA
Applicants to the engineering college
Number of applicants to engineering college 486
Percent offered admission 85.2%
Matriculated engineering students
Percentage in top quartile (25%) of High School class 38.6
Average ACT scores: Math—23, Composite—22
FULL- & PART-TIME FACULTY—BY DEPARTMENT
[Figures are the head count of full-time faculty and the full-time equivalent (FTE) of part-time faculty for each engineering department or equivalent.]

Department	Full	Part
Civil Eng	4	1.8
Electrical Eng	13	4.4
Mechanical Eng	15	8.4

COLLEGE DESCRIPTION
The College offers a four-year undergraduate program in Engineering, emphasizing engineering fundamentals in the first 2 years, followed by 2 years of specialized training in one of 3 majors, with some liberal arts studies throughout all four years. An evening program leading to a degree in 6 years is available with many classes taught by faculty from industry. Computer-aided engineering is emphasized in all programs. Co-op is an option. Implementation of the University's motto of 'Theory and Practice' is a practical philosophy in our instruction and is culminated in the senior year project. Engineering seniors also are required to complete the National Standard Fundamentals of Engineering Exam for graduation.

DEGREE PROGRAMS DESCRIPTION
Lawrence Tech's undergraduate engineering program includes both theoretical and practical dimensions - consistent with the motto of the University, 'Theory and Practice.' The faculty consists of engineers with academic and professional credentials as well as significant industrial experience. Many engineering faculty are concurrently working with industry which insures that the program reflects a strong real-world influence. Lawrence Tech's Master of Engineering in Manufacturing Systems (MEMS) is designed for working professionals who are graduates of an ABET-accredited undergraduate engineering program and who have at least one year of experience in industry. The MEMS program emphasizes the vital relationships and interplay between manufacturing, engineering, research, suppliers, marketing, sales and management. The program is designed to help students understand the systematic relationships that pervade the modern manufacturing process. The combination of Lawrence Tech's practical orientation, academic experience, resources in this specialized field, and convenient accessibility is unrivaled in the area.

TRANSFER INFORMATION
Residency Requirements: Students must complete the last 42 quarter hours at LTU, 20 of those hours must be in the student's major.
Transfer via Articulation Agreements
Admission to engineering: At the end of the sophomore year
Engineering graduates transferred from ... 4-yr: – 2-yr: 115
Requirements: Minimum 2.0 GPA in pre-engineering curriculum.
Transfer without Articulation Agreements
Admission to engineering: At any time
Engineering graduates transferred from ... 4-yr: 90 2-yr: 30
Requirements: Minimum 2.0 GPA.
GRADUATION REQUIREMENTS
Undergraduate students must attain a 2.00 GPA in all credit hours earned at LTU as well as a 2.00 GPA in their major field. Graduate students must complete all required coursework within seven years of matriculation and maintain a minimum 3.0 cumulative GPA.

STUDENT PROGRAMS
PROFESSIONAL AND HONORARY SOCIETIES
[For key to acronyms, see *Introduction*.]
Professional Societies: AIAA, ASCE, ASHRAE, ASME, IEEE, NSPE, SAE, SME, Engineering Society of Detroit, Ham Radio Club
Honorary Societies: Eta Kappa Nu, Pi Tau Sigma, Tau Beta Pi, Theta Tau, Psi Delta Epsilon (LTU Honor Society for Civil Engineering Students)
SUPPORT PROGRAMS
Student Chapter Organizations: Society of Women Engineers, National Society of Black Engineers
DAPCEP (Detroit Area Pre-College Engineering Prog)
For: Ethnic Minorities **Available:** Year round
Offered: Prematriculation
Other Engineering Support Programs: Foreign student services; Veterans services; Placement Services; Freshmen Seminar (study skills) course; Engineering Faculty Academic Counseling Services; Engineering Honor Societies Tutoring; Dean of Students Office counseling services.

107 LeTourneau University

■ INSTITUTION PROFILE
HEAD OF THE INSTITUTION
Alvin O. Austin
Phone: (903)753-0231   **Fax:** (903)237-2730
GENERAL INFORMATION
[All Students—Fall 1991]
Undergraduate enrollment 841
Total institution enrollment 841
Type of institution: Private   **Calendar:** Semesters
Nearest city: Dallas-Ft. Worth   **Population:** 3,700,000
Miles from main campus: 120   **Setting:** Small Town
Types of engineering degrees: Engineering & Technology
Other degree-granting colleges: Business Administration, Aviation, Liberal Arts, Mathematics and Computer Science, Natural Sciences and Physical Education, Secondary Education

■ ENGINEERING ADMISSIONS
ADMISSIONS OFFICE CONTACT
Howard G. Wilson
LeTourneau University
2100 S. Mobberly
P.O. Box 7001
Longview, TX 75607-7001
Phone: (903)753-0231   **Fax:** (903)237-2732
ENGINEERING COLLEGE ADMISSIONS INFORMATION
Entrance Requirements: SAT (Score: 800), ACT (Score: 18), High School courses—English, Journalism, or Speech (4 years), Social Science (2 years), Mathematics (exclusive of general math) (2 years), Natural Science (1 year)
Entrance Recommendations: High School courses—Mathematics (4 years), Chemistry & Physics (2 years)
Requirements for foreign students: TOEFL (Score: 500); financial statement; Deposit of $5,000 (American)
DEPARTMENTAL ADMISSIONS INFORMATION
Admission to the engineering department: At the time of admission to the institution.
Additional information: To remain in the engineering program, a student must maintain a cumulative GPA of 2.10 for all engineering courses taken at LeTourneau University after 12 credits in engineering courses are completed at LeTourneau. In addition, a minimum grade of 'C' is required in designated courses.

■ FINANCIAL INFORMATION
ESTIMATED EXPENSES (ACADEMIC YEAR)
[Expenses are for the 1992-93 nine-month academic year.]

All Students	Undergraduate	Graduate
Tuition and fees	$ 7,800	$ –
College room and board	$ 3,800	$ –
Books and supplies	$ 600	$ –
Other expenses	$ 200 [1]	$ –
Total estimated expenses	**$12,400**	**$ –**

Notes: (1) General fee covering admission to student activities and maintenance and supervision of recreational facilities. Also includes lab fees.

FINANCIAL AID OFFICE CONTACT
Willard Rusk
LeTourneau University
Office of Financial Aid
2100 S. Mobberly, P.O. Box 7001
Longview, TX 75607-7001
Phone: (903)753-0231   **Fax:** (903)237-2732

GENERAL FINANCIAL AID INFORMATION
Forms accepted/required: FAF, FFS
Additional financial aid information: LeTourneau University participates in all Federal Title IV Programs. Texas has special grant and loan programs. LeTourneau has a full range of institutional scholarships, grants and work opportunities.

■ ENGINEERING COLLEGE INFORMATION
[For additional personnel, refer to the *Appendix*.]
HEAD OF ENGINEERING
John R. Busch
Phone: (903)753-0231   **Fax:** (903)237-2732
ENGINEERING COLLEGE ADDRESS
Division of Engineering & Engineering Technology
LeTourneau University
2100 S. Mobberly
P.O. Box 7001
Longview, TX 75607-7001
Phone: (903)753-0231   **Fax:** (903)237-2732
ENROLLMENTS—BY CLASS
[Numbers are baccalaureate enrollments for the fall 1991 term, unless otherwise footnoted.]
1st-year students/Freshmen 86
2nd-year students/Sophomores 47
3rd-year students/Juniors 28
4th-year students/Seniors 58
Total **219**
NUMBER OF DEGREES AWARDED—BY PROGRAM
[Numbers are engineering baccalaureate degrees awarded during the 1991-1992 academic year, unless otherwise footnoted. For full details about each engineering program, refer to the *Tables of Degree Programs*.]
Engineering 30
Welding Engineering 2
Total **32**
PERCENTAGE OF DEGREES AWARDED—BY CATEGORY
[Percentages are of all engineering baccalaureate degrees awarded during the 1991-1992 academic year, unless otherwise footnoted.]
African Americans 2.0%
Asian/Pacific Island Americans 2.0%
Hispanic Americans 3.0%
Native Americans .... – %
Foreign Citizens 9.0%
All Others 84.0%
Women 6.4%
Persons with disabilities .... – %
Students over 25 years of age 9.0%
ENGINEERING STUDENT DATA
Applicants to the engineering college
Number of applicants to engineering college 120
Percent offered admission 90.0%
Matriculated engineering students
Percentage in top quartile (25%) of High School class ...... – (A)
Average SAT scores: Math—600[C], Verbal—480[C], Combined—1080
Average ACT scores: Math—28[C], Composite—25
Notes: (A) Data not available. (C) Estimated data.
FULL- & PART-TIME FACULTY—BY DEPARTMENT
[Figures are the head count of full-time faculty and the full-time equivalent (FTE) of part-time faculty for each engineering department or equivalent.]

Department	Full	Part
Electrical Eng	4	–
Mechanical Eng	4	–
Welding Eng	–	1.2

■ COLLEGE DESCRIPTION
LeTourneau University is a co-educational, interdenominational Christian university built on a tradition of engineering and engineering technology. A general engineering program with electrical and mechanical concentrations is offered which provides a well-rounded engineering education intended to prepare students for a wide range of opportunities in industry and graduate school. Engineering design is emphasized and integrated into the entire curriculum as well as communications skills. All students must complete a two-semester capstone design sequence in either the mechanical or electrical area. All graduating seniors are strongly encouraged to take the national standardized Fundamentals of Engineering Examination. A separate degree in Welding Engineering is offered, as well as a bachelor's program in engineering technology with electrical, mechanical and welding concentrations. Students in all programs may participate in a co-op program which offers many different job opportunities.
DEGREE PROGRAMS DESCRIPTION
Degrees in the programs listed prepare graduates for entry into the engineering profession and/or graduate study.
DUAL DEGREE PROGRAMS
Engineering baccalaureate graduates with dual degrees 0
TRANSFER INFORMATION
Residency Requirements: A minimum of one year in residence at LeTourneau University, including the last 30 semester credit hours for a degree, is required for graduation. In addition, at least 12 semester credit hours must be in the engineering major.
Transfer without Articulation Agreements
Admission to engineering: At any time
Engineering graduates transferred from . . 4-yr: 3   2-yr: 3
GRADUATION REQUIREMENTS
Credit requirement for B.S. in engineering is 137 credit hours. Overall GPA must be 2.0/4.0.

■ STUDENT PROGRAMS
PROFESSIONAL AND HONORARY SOCIETIES
[For key to acronyms, see *Introduction*.]
Professional Societies: ASME, AWS, IEEE, SAE
Honorary Societies: Epsilon Eta Sigma (Local Engineering Honorary Society), Gold Key (University-wide Honorary Society)
SUPPORT PROGRAMS
International Students Organization
Available: Year round
Offered: Prematriculation, Freshman, Sophomore, Junior, Senior
Advising and counseling for women and minorities
For: Women & Ethnic Minorities   **Available:** Year round
Offered: Prematriculation, Freshman, Sophomore, Junior, Senior
Other Engineering Support Programs: Counseling and pre-advising for prospective students; orientation program for all entering freshmen and transfer students; mathematics readiness testing; English reading and readiness testing; career and counseling services; tutoring services; career placement services for graduates and alumni.

108 Lehigh University

INSTITUTION PROFILE
HEAD OF THE INSTITUTION
Peter W. Likins
Phone: (215)758-3155 Fax: (215)758-5402
GENERAL INFORMATION
[All Students—Fall 1991]
Undergraduate enrollment 4,489
Graduate enrollment 2,067
Total institution enrollment **6,556**
Type of institution: Private Calendar: Semesters
Nearest city: Philadelphia Population: 4,000,000
Miles from main campus: 60 Setting: Small Town
Types of engineering degrees: Engineering
Other degree-granting colleges: Arts & Sciences, Education, Business and Economics

ENGINEERING ADMISSIONS
ADMISSIONS OFFICE CONTACT
Patricia G. Boig
Lehigh University
Office of Admissions
27 Memorial Dr. W., Alum Mem Bldg #27
Bethlehem, PA 18015-3045
Phone: (215)758-3100 Fax: (215)758-4361
ENGINEERING COLLEGE ADMISSIONS INFORMATION
Entrance Requirements: High School courses—College preparatory math (4 years), English (4 years), Laboratory science (chemistry included) (2 years), Foreign language (2 years), The SAT test is required, however, there is no minimum score requirement. The ACT may be substituted with permission of the Admissions Office. Achievement tests for English, Math 1 or 2, and either Chemistry or Physics are required.
Entrance Recommendations: High School courses—Physical science (including chem. and physics) (2 years), Lehigh recognizes advanced placement scores in the admissions process.
Requirements for foreign students: TOEFL is required for foreign students, however, no minimum scores are required.
DEPARTMENTAL ADMISSIONS INFORMATION
Admission to the engineering department: At the end of the first year.

FINANCIAL INFORMATION
ESTIMATED EXPENSES (ACADEMIC YEAR)
[Expenses are for the 1992-93 nine-month academic year.]

All Students	Undergraduate	Graduate
Tuition and fees	$15,650	$11,790
College room and board	$4,940	$2,870
Books and supplies	$600	$600
Other expenses	$250	$24
Total estimated expenses	**$21,440**	**$15,284**

FINANCIAL AID OFFICE CONTACT
William E. Stanford, Director, Financial Aid
Lehigh University
218 W. Packer Ave, Room 209
Bldg # 194
Bethlehem, PA 18015
Phone: (215)758-3181 Fax: (215)758-4361
GENERAL FINANCIAL AID INFORMATION
Forms accepted/required: FAF, Institutional
Additional financial aid information: Financial aid is need based and GUARANTEED to qualified U.S. citizens and permanent residents.

ENGINEERING COLLEGE INFORMATION
[For additional personnel, refer to the Appendix.]
HEAD OF ENGINEERING
Sunder H. Advani
Phone: (215)758-5308 Fax: (215)758-5623

ENGINEERING COLLEGE ADDRESS
College of Engineering and Applied Science
Lehigh University
19 Memorial Drive West
Packard Lab #19
Bethlehem, PA 18015-3045
Phone: (215)758-4025 Fax: (215)758-5623
ENROLLMENTS—BY CLASS
[Numbers are baccalaureate enrollments for the fall 1991 term, unless otherwise footnoted.]
1st-year students/Freshmen 396
2nd-year students/Sophomores 376
3rd-year students/Juniors 363
4th-year students/Seniors 396
Total **1,531**
NUMBER OF DEGREES AWARDED—BY PROGRAM
[Numbers are engineering baccalaureate degrees awarded during the 1991-1992 academic year, unless otherwise footnoted. For full details about each engineering program, refer to the Tables of Degree Programs.]
Chemical Engineering 45
Civil Engineering 62
Electrical Engineering & Computer Science 79
Industrial & Manufacturing Systems Engineering 42
Materials Science & Engineering 28
Mechanical Engineering & Applied Mechanics 103
Total **359**
PERCENTAGE OF DEGREES AWARDED—BY CATEGORY
[Percentages are of all engineering baccalaureate degrees awarded during the 1991-1992 academic year, unless otherwise footnoted.]
African Americans .... 2.0 %
Asian/Pacific Island Americans .... 4.0 %
Hispanic Americans .... – %
Native Americans .... – %
Foreign Citizens .... 3.0 %
All Others .... 91.0 %
Women .... 15.0 %
Persons with disabilities .... – % (A)
Students over 25 years of age .... – % (A)
Notes: (A) Data not available.
GRADUATE ENROLLMENTS & DEGREES AWARDED
Master's enrollment 398
Master's degrees awarded 153
Doctoral enrollment 261
Doctoral degrees awarded 43
ENGINEERING STUDENT DATA
Applicants to the engineering college
Number of applicants to engineering college 1,579
Percent offered admission .... 82.0 %
Matriculated engineering students
Percentage in top quartile (25%) of High School class 86.0
Average SAT scores: Math—653, Verbal—511
FULL- & PART-TIME FACULTY—BY DEPARTMENT
[Figures are the head count of full-time faculty and the full-time equivalent (FTE) of part-time faculty for each engineering department or equivalent.]

Department	Full	Part
Chemical Eng	17	–
Civil Eng	22	–
Computer Science & Electrical Eng	27	–
Industrial Eng	14	–
Mechanical Eng	34	–
Materials Science & Eng	17	–

COLLEGE DESCRIPTION
To fulfill varied and changing roles of scientists and engineers, students must learn to appreciate not only the power of technology but also its impact on people. The academic programs of this college are designed to insure that the student achieves technical proficiency and an awareness of the broader scientific and human values under which technologies are developed and applied. Students learn the concepts of creative and disciplined problem-solving that can later be applied to their professional and personal lives. Special initiatives for undergraduates include, for example, individual and team research projects under faculty guidance. Programs in arts/engineering; bioengineering; environmental engineering; and science, technology and society are also available. The six departments boast ratios of students to faculty that are among the lowest in the U.S.
DEGREE PROGRAMS DESCRIPTION
The undergraduate student in the engineering college has the opportunity to participate in individual projects with faculty. Students may enter a five-year dual degree program in arts engineering. Also, the Science Technology and Society program is an interdisciplinary program drawing coursework from the engineering, arts, and business colleges. The goal of the program is to explore the relationships between science and technological advancement and their effect on the quality of human life. In addition to studying for degrees in academic departments, undergraduates may earn degrees in special interdisciplinary programs, such as polymer science, engineering physics, biochemistry and chemistry.
DUAL DEGREE PROGRAMS
Engineering baccalaureate graduates with dual degrees 7
TRANSFER INFORMATION
Transfer without Articulation Agreements
Admission to engineering: Students in good standing in other academic programs at Lehigh may petition to enter the engineering college.
Engineering graduates transferred from ... 4-yr: 27 2-yr: –
Requirements: January and August, students who have attended other colleges are admitted with advanced standing. High school requirements must be met, and a 3.2 general average is required to transfer in. To enter in the Spring semester, deadline for application is Nov. 1; for Fall, April 1. In order to receive a Lehigh degree, the last 30 credit hours must be taken at Lehigh.
GRADUATION REQUIREMENTS
Credit requirements vary from department to department although the range is between 130-136 for graduation. A 2.0 GPA is required.

STUDENT PROGRAMS
PROFESSIONAL AND HONORARY SOCIETIES
[For key to acronyms, see Introduction.]
Professional Societies: ACM, ACS, AIChE, ASCE, ASME, IEEE, IIE, SAE, SME, SPE*
Honorary Societies: Alpha Pi Mu, Chi Epsilon, Eta Kappa Nu, Pi Tau Sigma, Tau Beta Pi
SUPPORT PROGRAMS
Student Chapter Organizations: Society of Women Engineers, National Society of Black Engineers, Society of Hispanic Professional Engineers

Challenge for Success
For: Ethnic Minorities **Available:** Year round
Offered: Prematriculation, Freshman, Sophomore, Junior

Engineering Study Group
For: Ethnic Minorities **Available:** Academic year
Offered: Freshman

Women in Science and Engineering (WISE)
For: Women **Available:** Year round
Offered: Freshman, Sophomore, Junior, Senior, Graduate level
Other Engineering Support Programs: The engineering college has an orientation program for freshmen. Faculty advisors are assigned to all students. Tutoring is available. Career education and development is available through Career Services. A program for international students is available. Student health services and counseling are free of charge. Internships and co-op programs are available through the departments.

109 Louisiana State University

■ INSTITUTION PROFILE
HEAD OF THE INSTITUTION
William E. Davis
Phone: (504)388-6977   Fax: (504)388-5982

GENERAL INFORMATION
[All Students—Fall 1991]
Undergraduate enrollment 21,200
Graduate enrollment 5,400
Total institution enrollment 26,600

Type of institution: Public   **Calendar:** Semesters
Location: Baton Rouge, LA   **Population:** 250,000
Setting: Suburban
Types of engineering degrees: Engineering
Other degree-granting colleges: Arts & Sciences, Business Administration, Education, Basic Sciences, Design, Music, General College, Agriculture

■ ENGINEERING ADMISSIONS
ADMISSIONS OFFICE CONTACT
Lisa B. Harris
Louisiana State University
Office of Admissions
110 T. Boyd Hall
Baton Rouge, LA 70803
Phone: (504)388-1175   Fax: (504)388-5991

ENGINEERING COLLEGE ADMISSIONS INFORMATION
Admission to the engineering college: Students are admitted from Junior Division (pre-engineering) upon completion of 24 hours with a GPA 2.0, and are eligible to schedule Calculus I.
Entrance Requirements: High School courses—English (4 years) and Social Studies (7 years), College Preparatory Mathematics (3 years), Biology, Chemistry, and Physics (3 years), Foreign Language (2 yrs) and Preparatory Electives (4 years), ACT is preferred by the university, but SAT can be used.
Entrance Recommendations: ACT, High School courses—Pre-Calculus and Calculus (1 year), Physics (1 year), Chemistry (1 year), Freshmen are required to submit scores on the ACT or SAT, and this information will be used with other data for admission and placement.
Requirements for foreign students: TOEFL (Score: 500); financial statement

DEPARTMENTAL ADMISSIONS INFORMATION
Admission to the engineering department: Admission to a program is identical to admission to the College.

■ FINANCIAL INFORMATION
ESTIMATED EXPENSES (ACADEMIC YEAR)
[Expenses are for the 1992-93 nine-month academic year.]

State Residents	Undergraduate	Graduate
Tuition and fees	$ 2,170	$ 2,170
College room and board	$ 3,200	$ 3,200
Books and supplies	$ 400	$ 400
Other expenses	$ –	$ –
Total estimated expenses	**$ 5,770**	**$ 5,770**

Out-of-State Residents	Undergraduate	Graduate
Tuition and fees	$ 5,470	$ 5,470
College room and board	$ 3,200	$ 3,200
Books and supplies	$ 400	$ 400
Other expenses	$ –	$ –
Total estimated expenses	**$ 9,070**	**$ 9,070**

FINANCIAL AID OFFICE CONTACT
Ester M. Hill
Louisiana State University
Office of Student Aid and Scholarships
Room 202 Himes Hall
Baton Rouge, LA 70803
Phone: (504)388-3103   Fax: (504)388-6300

GENERAL FINANCIAL AID INFORMATION
Forms accepted/required: ACT, FFS, Institutional
Additional financial aid information: Pell Grants, Supplemental Educational Opportunity Grants, State Student Incentive Grants, College Work Study, Perkins Loans, Guaranteed Student Loans, Parent Loans for Undergraduates, Supplemental Loans, Tuition Waver Scholarships for in-state and out-of-state residents, Non-Resident fee wavers.

■ ENGINEERING COLLEGE INFORMATION
[For additional personnel, refer to the *Appendix*.]
HEAD OF ENGINEERING
Edward McLaughlin
Phone: (504)388-5701   Fax: (504)388-5990

ENGINEERING COLLEGE ADDRESS
College of Engineering
Louisiana State University
Rm 3304 CEBA Building
Baton Rouge, LA 70803
Phone: (504)388-5731   Fax: (504)388-5990

ENROLLMENTS—BY CLASS
[Numbers are baccalaureate enrollments for the fall 1991 term, unless otherwise footnoted.]

1st-year students/Freshmen	994 [1]
2nd-year students/Sophomores	409
3rd-year students/Juniors	523
4th-year students/Seniors	700
Total	**2,626**

Notes: (1) includes pre-engineering program enrollment

NUMBER OF DEGREES AWARDED—BY PROGRAM
[Numbers are engineering baccalaureate degrees awarded during the 1991-1992 academic year, unless otherwise footnoted. For full details about each engineering program, refer to the *Tables of Degree Programs*.]

Biological & Agricultural Engineering	4
Chemical Engineering	22
Civil Engineering	36
Computer Engineering	22
Electrical Engineering	87
Engineering Science(s)	– [3]
Industrial Engineering	23
Mechanical Engineering	61
Petroleum Engineering	8
Total	**263**

Notes: (3) Graduate program only

PERCENTAGE OF DEGREES AWARDED—BY CATEGORY
[Percentages are of all engineering baccalaureate degrees awarded during the 1991-1992 academic year, unless otherwise footnoted.]

African Americans	4.9%
Asian/Pacific Island Americans	11.0%
Hispanic Americans	3.0%
Native Americans	– %
Foreign Citizens	13.0%
All Others	68.1%
Women	11.0%
Persons with disabilities	– %
Students over 25 years of age	39.0%

GRADUATE ENROLLMENTS & DEGREES AWARDED

Master's enrollment	412
Master's degrees awarded	98
Doctoral enrollment	224
Doctoral degrees awarded	20

ENGINEERING STUDENT DATA
Applicants to the engineering college

Number of applicants to engineering college	– (A)
Percent offered admission	90.0% (C)

Matriculated engineering students
Percentage in top quartile (25%) of High School class .... – (A)
Average ACT scores: Math—24.9, Composite—24.6
Notes: (A) Data not available. (C) Estimated data.

FULL- & PART-TIME FACULTY—BY DEPARTMENT
[Figures are the head count of full-time faculty and the full-time equivalent (FTE) of part-time faculty for each engineering department or equivalent.]

Department	Full	Part
Chemical Eng	19	2.0
Civil Eng	17	2.8
Electrical & Computer Eng	29	0.5
Industrial & Manufacturing Systems Eng	12	0.5
Mechanical Eng	20	1.8
Biological & Agricultural Eng	11	1.5
Petroleum Eng	8	0.5

COLLEGE DESCRIPTION
The College offers a four-year undergraduate degree in eight programs from seven departments. All degrees require fundamental and specialized training in engineering, as well as some liberal arts studies (the University General Education requirements) throughout the four years. There are opportunities to minor in several areas, including computer science, Environmental Engineering, and others. Co-op is available is most programs.

DEGREE PROGRAMS DESCRIPTION
B.S. degrees prepare graduates for professional positions or graduate studies with programs in biological and agricultural, chemical, civil, computer, electrical, industrial, mechanical, and petroleum engineering. M.S. programs are available in the above areas as well as the interdisciplinary Master of Engineering (an application-oriented degree for students in preparation for engr. practice beyond the B.S. degree in engr.), or in interdisciplinary areas such as materials or environmental engr. The M.S. in Engr. Science, offered for students with B.S. degrees in pure or applied science, emphasizes interdisciplinary studies. Ph.D. degrees are awarded in chemical, civil, electrical, mechanical, petroleum, and engr. science.

DUAL DEGREE PROGRAMS
Engineering baccalaureate graduates with dual degrees 1
Enrollment requirements: Determined by Department involved

TRANSFER INFORMATION
Residency Requirements: A student must successfully complete 30 hours in college residence, including 15 hours of required engineering courses, or approved technical electives at the 3000 or 4000 level. Nine hours of these must be at the 4000 level in the major department.

Transfer via Articulation Agreements
Admission to engineering: At any time
Requirements: Transfer students are required to have a 2.0 GPA, and their transcripts are evaluated by the Dean's office to determine admissibility. Application deadlines are July 1, December 1, and May 1 for fall, spring, and summer.

Transfer without Articulation Agreements
Admission to engineering: At any time
Requirements: Transfer students are required to have a 2.0 GPA, and their transcripts are evaluated by the Dean's office to determine admissibility. Application deadlines are July 1, December 1, and May 1 for fall, spring, and summer.

GRADUATION REQUIREMENTS
Complete established curriculum with 2.0 GPA for all work taken at LSU and at US institutions; 2.0 GPA on all courses in major department at LSU and at US institutions; meet residency requirements; demonstrate English, Math and Physics proficiency; checkout in the semester prior to graduation. LSU employs a grading system whereby all efforts are used to compute GPA.

■ STUDENT PROGRAMS
PROFESSIONAL AND HONORARY SOCIETIES
[For key to acronyms, see *Introduction*.]
Professional Societies: AIAA, AIChE, ASAE, ASCE, ASME, IEEE, IIE, SAME, SPE
Honorary Societies: Eta Kappa Nu, Pi Tau Sigma, Tau Beta Pi, Chi Epsilon, Pi Epsilon Tau

SUPPORT PROGRAMS
Student Chapter Organizations: Society of Women Engineers, National Society of Black Engineers, Society of Hispanic Professional Engineers

Minority Engineering Program
For: Ethnic Minorities   **Available:** Year round
Offered: Prematriculation, Freshman, Sophomore, Junior, Senior
Other Engineering Support Programs: The University conducts a 2 day 'Spring Testing' for entering freshmen with ACT 25, as well as 2 day Freshmen Orientation sessions (pre-enrollment counseling) during the summer. The College has 2 Academic Counselors, and a MEP Coordinator for advising of existing, entering and transfer students. Recruitment activities are coordinated by the Recruiting Coordinator.

110 Louisiana Tech University

INSTITUTION PROFILE
HEAD OF THE INSTITUTION
Daniel D. Reneau
Phone: (318)257-3785　　　　　　Fax: (318)257-2928
GENERAL INFORMATION
[All Students—Fall 1991]
Undergraduate enrollment 9,283
Graduate enrollment 1,097
Total institution enrollment **10,380**
Type of institution: Public　　**Calendar:** Quarters
Nearest city: Shreveport　　**Population:** 198,525
Miles from main campus: 72　　**Setting:** Small Town
Types of engineering degrees: Engineering & Technology
Other degree-granting colleges: Arts & Sciences, Education, Human Ecology, Life Sciences, Administration and Business

ENGINEERING ADMISSIONS
ADMISSIONS OFFICE CONTACT
Karen T. Akin
Louisiana Tech University
Box 3178
Wyly Tower 221
Ruston, LA 71272
Phone: (318)257-3036　　　　　　Fax: (318)257-2499
ENGINEERING COLLEGE ADMISSIONS INFORMATION
Admission to the engineering college: Good academic standing and at least a 2.2 GPA based on hours attempted in the Freshman Engineering Curriculum.
Entrance Requirements: High School courses—English (4 years), Mathematics (3 years), Social Studies, Science (3 years), Electives (4.5 years), For Fall of 1993 an ACT composite of 22, or a 2.0 GPA on admission courses, or rank in the upper 50% of class is required.
Entrance Recommendations: High School courses—English (4 years), Algebra (2 years), Plane Geometry (1 year), Trigonometry, Chemistry, Physics (1 year)
Requirements for foreign students: TOEFL (Score: 500); financial statement
DEPARTMENTAL ADMISSIONS INFORMATION
Admission to the engineering department: Good academic standing and at least a 2.2 GPA based on hours attempted in the Freshman Curriculum.

FINANCIAL INFORMATION
ESTIMATED EXPENSES (ACADEMIC YEAR)
[Expenses are for the 1992-93 nine-month academic year.]

State Residents	Undergraduate	Graduate
Tuition and fees	$ 2,118	$ 2,118
College room and board	$ 2,115	$ 2,115
Books and supplies	$ 600	$ 600
Other expenses	$ 111 [1]	$ 111 [1]
Total estimated expenses	**$ 4,944**	**$ 4,944**

Out-of-State Residents	Undergraduate	Graduate
Tuition and fees	$ 3,273	$ 3,273
College room and board	$ 2,115	$ 2,115
Books and supplies	$ 600	$ 600
Other expenses	$ 111 [1]	$ 111 [1]
Total estimated expenses	**$ 6,099**	**$ 6,099**

Notes: (1) Engineering Fee.

FINANCIAL AID OFFICE CONTACT
Etienna R. Winzer, *Director, Division of Financial Aid*
Louisiana Tech University
P.O. Box 7925 T.S.
Ruston, LA 71272
Phone: (318)257-2641
GENERAL FINANCIAL AID INFORMATION
Forms accepted/required: SAR, SDF, Institutional
Additional financial aid information: Need-based scholarships; Pell Grant; Supplemental Educational Opportunity Grant; National Direct Student Loan; College Work-Study; Stafford Loan; State Student Incentive Grant; Louisiana College Tuition Plan are all available.

ENGINEERING COLLEGE INFORMATION
[For additional personnel, refer to the *Appendix*.]
HEAD OF ENGINEERING
Barry A. Benedict
Phone: (318)257-4647　　　　　　Fax: (318)257-2562
EMail: benedict@engr.latech.edu
ENGINEERING COLLEGE ADDRESS
College of Engineering
Louisiana Tech University
P.O. Box 10348 T.S.
Ruston, LA 71272
Phone: (318)257-4647　　　　　　Fax: (318)257-2562
ENROLLMENTS—BY CLASS
[Numbers are baccalaureate enrollments for the fall 1991 term, unless otherwise footnoted.]
1st-year students/Freshmen .... 493 [1]
2nd-year students/Sophomores .... 265 [1]
3rd-year students/Juniors .... 225 [1]
4th-year students/Seniors .... 360 [1]
Total **1,343**
Notes: (1) Includes Computer Science and Geosciences.

NUMBER OF DEGREES AWARDED—BY PROGRAM
[Numbers are engineering baccalaureate degrees awarded during the 1991-1992 academic year, unless otherwise footnoted. For full details about each engineering program, refer to the *Tables of Degree Programs*.]
Biomedical Engineering 13
Chemical Engineering 19
Civil Engineering 19
Computer Science 13
Construction Engineering Technology 4
Electrical Engineering 47
Electrical Engineering Technology 17
Geosciences 2
Industrial Engineering 4
Mechanical Engineering 29
Petroleum Engineering 7
Total **174**

PERCENTAGE OF DEGREES AWARDED—BY CATEGORY
[Percentages are of all engineering baccalaureate degrees awarded during the 1991-1992 academic year, unless otherwise footnoted.]
African Americans 4.0%
Asian/Pacific Island Americans 2.0%
Hispanic Americans .... – %
Native Americans .... – %
Foreign Citizens 3.0%
All Others 91.0%
Women 10.0%
Persons with disabilities .... – % (A)
Students over 25 years of age .... – % (A)
Notes: (A) Data not available.

GRADUATE ENROLLMENTS & DEGREES AWARDED
Master's enrollment .... 165 [2]
Master's degrees awarded .... 49 [2]
Doctoral enrollment 67
Doctoral degrees awarded 6
Notes: (2) Includes Computer Science.

ENGINEERING STUDENT DATA
Applicants to the engineering college
Number of applicants to engineering college .... – (A)
Percent offered admission .... –% (A)
Matriculated engineering students
Percentage in top quartile (25%) of High School class .... – (A)
Average ACT scores: Math—23.8, Composite—24.1
Notes: (A) Data not available.

FULL- & PART-TIME FACULTY—BY DEPARTMENT
[Figures are the head count of full-time faculty and the full-time equivalent (FTE) of part-time faculty for each engineering department or equivalent.]

Department	Full	Part
Chemical Eng	5	–
Civil Eng	14	–
Electrical Eng	11	–
Industrial Eng	6	–
Mechanical Eng	13	–
Biomedical Eng	6	–
Computer Science	7	–
Petroleum Eng & Geosciences	7	–

COLLEGE DESCRIPTION
The college offers a four-year undergraduate program in engineering as well as graduate training leading to the M.S., D.E. and Ph.D. degrees. All engineering students follow a common freshman curriculum. The college consists of seven departments and eleven programs (including two technology programs). All engineering and technology programs are accredited by ABET and each curriculum meets state general education requirements. The curricula provide a strong foundation in engineering fundamentals with liberal arts studies incorporated throughout the program. Development of communication skills is stressed. Computers are thoroughly integrated throughout the curricula. Average number of years required to actually complete the bachelor's degree: 4.7

DEGREE PROGRAMS DESCRIPTION
The B.S. degrees in the various engineering programs prepare the graduate for entry-level professional positions or for graduate study in each of the respective fields. Graduate study in engineering can be carried out in one of the following areas: the M.S. in all engineering departments as well as computer science, manufacturing systems engineering, and operations research; the practice oriented D. Engr; and the Ph.D. in Biomedical Engineering.

TRANSFER INFORMATION
Residency Requirements: Three-fourths of the hours required for graduation must have been completed in college residence.
Transfer via Articulation Agreements
Admission to engineering: At any time
Engineering graduates transferred from ... 4-yr: – 2-yr: –
Requirements: Students must have an overall GPA of at least 2.0 out of 4.0 in all transfer courses. Transfer students are subject to the same requirements for admission as all other students.
Transfer without Articulation Agreements
Admission to engineering: At any time
Engineering graduates transferred from ... 4-yr: – 2-yr: –
Requirements: Students must have an overall GPA of at least 2.0 out of 4.0 in all transfer courses. Transfer students are subject to the same requirements for admission as all other students.

GRADUATION REQUIREMENTS
B.S. programs in the College of Engineering require 137-140 semester credit hours, depending on the major selected. A minimum overall GPA of 2.0 is required for graduation.

STUDENT PROGRAMS
PROFESSIONAL AND HONORARY SOCIETIES
[For key to acronyms, see *Introduction*.]
Professional Societies: AIChE, ASCE, ASME, BMES, IEEE, IIE, ITE, SPE, Society of Automotive Engineers (SAE), Associated General Contractors (AGC)
Honorary Societies: Alpha Pi Mu, Chi Epsilon, Eta Kappa Nu, Omega Chi Epsilon, Pi Tau Sigma, Tau Alpha Pi, Tau Beta Pi, Alpha Eta Mu Beta, Upsilon Pi Epsilon, Pi Epsilon Tau

SUPPORT PROGRAMS
Student Chapter Organizations: National Society of Black Engineers

Minority Engineering Program
For: Ethnic Minorities　　**Available:** Year round
Offered: Prematriculation, Freshman, Sophomore, Junior, Senior, Graduate level

Minority Scholarship Program
For: Ethnic Minorities　　**Available:** Year round
Offered: Prematriculation, Freshman, Sophomore, Junior, Senior, Graduate level

Counseling Center
For: Women & Ethnic Minorities　　**Available:** Year round
Offered: Prematriculation, Freshman, Sophomore, Junior, Senior, Graduate level

Career Planning and Placement Center
For: Women & Ethnic Minorities　　**Available:** Year round
Offered: Prematriculation, Freshman, Sophomore, Junior, Senior, Graduate level

Other Engineering Support Programs: Five summer orientation sessions for all freshmen; Engineering Tutoring and Supplemental Instruction; Counseling Center offering career, academic, and personal counseling; Career Planning & Placement Center; International Student Office; Financial Aid Office; Services for Handicapped; and Engineering Scholarship program.

111 University of Louisville

INSTITUTION PROFILE

HEAD OF THE INSTITUTION
Donald C. Swain
Phone: (502)588-5420　　　　　　　　　Fax: (502)588-5682

GENERAL INFORMATION
[All Students—Fall 1991]
Undergraduate enrollment 16,977
Graduate enrollment 3,809
Total institution enrollment 20,876
Type of institution: Public　　　　Calendar: Semesters
Location: Louisville　　　　　　　　Population: 664,937
Setting: Urban
Types of engineering degrees: Engineering
Other degree-granting colleges: Allied Health Sciences, Arts & Sciences, Dentistry, Education, Law, Medicine, Nursing, Business and Public Administration, Music

ENGINEERING ADMISSIONS

ADMISSIONS OFFICE CONTACT
Donald L. Cole
University of Louisville
Academic Services
Speed Scientific School
Louisville, KY 40292-0001
Phone: (502)588-6194　　　　　　　　Fax: (502)588-7033

ENGINEERING COLLEGE ADMISSIONS INFORMATION
Entrance Requirements: ACT (Score: 23), High School courses—Physics (1 year), Mathematics (Alg. I, Alg. II, Geometry, Adv. Math) (4 years), Chemistry (1 year), English (4 years)
Entrance Recommendations: High School courses—Computer, Graphics
Requirements for foreign students: TOEFL (Score: 550); IESL advanced level in College of Arts & Sciences

DEPARTMENTAL ADMISSIONS INFORMATION
Admission to the engineering department: The middle of the sophomore year.
Additional information: Electrical Eng., 2.25 min. GPA; Engrg. Math & Comp. Science, 2.25 min. GPA; Mechanical Eng., 2.25 min.; other departments 2.0 min. GPA.

FINANCIAL INFORMATION

ESTIMATED EXPENSES (ACADEMIC YEAR)
[Expenses are for the 1992-93 nine-month academic year.]

State Residents	Undergraduate	Graduate
Tuition and fees	$ 2,510	$ 2,723
College room and board	$ 4,630	$ 4,630
Books and supplies	$ 681	$ 681
Other expenses	$ 2,996 [1]	$ 2,996 [1]
Total estimated expenses	**$10,817**	**$11,030**

Out-of-State Residents	Undergraduate	Graduate
Tuition and fees	$ 6,995	$ 7,636
College room and board	$ 4,630	$ 4,630
Books and supplies	$ 681	$ 681
Other expenses	$ 2,996 [1]	$ 2,996 [1]
Total estimated expenses	**$15,302**	**$15,943**

Notes: (1) Travel & Personal Expenses.

FINANCIAL AID OFFICE CONTACT
Gilbert B. Tanner, *Director*
University of Louisville
University Center, 100
Louisville, KY 40292-0001
Phone: (588)502-5511　　　　　　　　Fax: (502)588-0182

GENERAL FINANCIAL AID INFORMATION
Forms accepted/required: CSS-FAF, Stafford, SAR, Kentucky Higher Education Assistance Authority (KHEAA)
Additional financial aid information: The University of Louisville offers a wide variety of student financial aid programs provided by federal, state, university and private funds. These programs include grants, loans, employment opportunities and scholarships, which may be awarded in varying proportions to help meet the student's unique financial needs.

ENGINEERING COLLEGE INFORMATION

[For additional personnel, refer to the *Appendix*.]

HEAD OF ENGINEERING
Thomas R. Hanley
Phone: (502)588-6281　　　　　　　　Fax: (502)588-7033
EMail: TRHANL01@ULKYVM.LOUISVILLE.EDU

ENGINEERING COLLEGE ADDRESS
Speed Scientific School
University of Louisville
Third and Eastern Parkway
Louisville, KY 40292-0001
Phone: (502)588-6281　　　　　　　　Fax: (502)588-7033

ENROLLMENTS—BY CLASS
[Numbers are baccalaureate enrollments for the fall 1991 term, unless otherwise footnoted.]

1st-year students/Freshmen .... 471 [2]
2nd-year students/Sophomores .... 326 [2]
3rd-year students/Juniors .... 266 [2]
4th-year students/Seniors .... 238 [2]
Total 1,301

Notes: (2) Fall enrollment data only.

NUMBER OF DEGREES AWARDED—BY PROGRAM
[Numbers are engineering baccalaureate degrees awarded during the 1991-1992 academic year, unless otherwise footnoted. For full details about each engineering program, refer to the *Tables of Degree Programs*.]

Chemical Engineering 24
Civil Engineering 22
Computer Science & Engineering .... –
Electrical Engineering 64
Engineering Management .... –
Engineering Math & Computer Science 16
Industrial Engineering 16
Mechanical Engineering 34
Total 176

PERCENTAGE OF DEGREES AWARDED—BY CATEGORY
[Percentages are of all engineering baccalaureate degrees awarded during the 1991-1992 academic year, unless otherwise footnoted.]

African Americans .... 3.4 %
Asian/Pacific Island Americans .... 9.1 %
Hispanic Americans .... – %
Native Americans .... – %
Foreign Citizens .... 5.1 % [1]
All Others .... 82.4 %
Women .... 17.7 %
Persons with disabilities .... – % [A]
Students over 25 years of age .... – % [A]

Notes: (A) Data not available. (1) Computer Science is located in the Engineering College.

GRADUATE ENROLLMENTS & DEGREES AWARDED
Master's enrollment .... 487 [2]
Master's degrees awarded 124
Doctoral enrollment .... 46 [2]
Doctoral degrees awarded 3

Notes: (2) Fall enrollment data only.

ENGINEERING STUDENT DATA
Applicants to the engineering college
Number of applicants to engineering college 488
Percent offered admission 70.7%

Matriculated engineering students
Percentage in top quartile (25%) of High School class 73.0
Average ACT scores: Math—25, Composite—25

FULL- & PART-TIME FACULTY—BY DEPARTMENT
[Figures are the head count of full-time faculty and the full-time equivalent (FTE) of part-time faculty for each engineering department or equivalent.]

Department	Full	Part
Chemical Eng	9	–
Civil Eng	12	–
Electrical Eng	18	–
Industrial Eng	9	–
Mechanical Eng	13	–
Eng Math & Computer Science	19	–

COLLEGE DESCRIPTION

The University of Louisville is a state-supported urban university located in Kentucky's largest metropolitan area. The Speed Scientific School was established in 1924 as the engineering college. With an enrollment of 8.5% of the University's total enrollment, the School offers the benefits of a small college having access to the resources of a large university. The courses in the degree programs are taught by 85 faculty through 6 departments and include 3 alternating semesters of co-op work experience for all students, with job placement by a professional staff. Interdisciplinary research focus areas in addition to departmental research are bio/food engineering, manufacturing engineering, materials engineering, environmental engineering, and computer science & engineering.

DEGREE PROGRAMS DESCRIPTION

The five-year professional program of studies leading to the Master of Engineering degree has been adopted by the faculty as the program considered best-suited to meet the future requirements of the engineering profession and society. This integrated program of academic studies and meaningful co-op work experiences provides students with the opportunity to observe and participate in the practice of engineering. Required courses in the humanities and social sciences give students the social, ethical, and ecological awareness needed in their profession. Engineering communications instruction assists students in oral and written presentations of technical material. Engineering design experience within the departmental curricula builds upon the fundamental concepts of mathematics, basic sciences, the humanities and social sciences, engineering topics, and communications skills. Graduates of this program have been provided with the resources and advanced instruction necessary to practice engineering with consideration for ethical, legal, ecological, and technical principles.

TRANSFER INFORMATION

Residency Requirements: Students must complete at least 30 of the last 36 semester hours in residence in the Speed Scientific School.

Transfer via Articulation Agreements
Admission to engineering: At any time
Requirements: A 3.0 GPA is needed.

Transfer without Articulation Agreements
Admission to engineering: At any time
Requirements: A 3.0 GPA is needed.

GRADUATION REQUIREMENTS

For Bachelor of Science in Engineering Science 133-139 semester hours; for Master of Engineering degree 160-162 semester hours. Specific programs requirements vary.

STUDENT PROGRAMS

PROFESSIONAL AND HONORARY SOCIETIES
[For key to acronyms, see *Introduction*.]
Professional Societies: ACM, AIChE, ASCE, ASHRAE, ASME, IEEE, IIE, NSPE, SME
Honorary Societies: Alpha Pi Mu, Chi Epsilon, Eta Kappa Nu, Omega Chi Epsilon, Pi Tau Sigma, Tau Beta Pi, Sigma Xi

SUPPORT PROGRAMS
Student Chapter Organizations: Society of Women Engineers, National Society of Black Engineers

Summer Enrichment Programs
For: Women & Ethnic Minorities　　**Available:** Summer only
Offered: Prematriculation

Faculty Mentors
For: Women & Ethnic Minorities　　**Available:** Academic year
Offered: Freshman

Minority Student Orientation
For: Women & Ethnic Minorities　　**Available:** Academic year
Offered: Freshman

Other Engineering Support Programs: Tutoring and Academic Counseling

112 Loyola College in Maryland

INSTITUTION PROFILE
HEAD OF THE INSTITUTION
Joseph A. Sellenger S.J.
Phone: (410)617-2201
GENERAL INFORMATION
[All Students—Fall 1991]
Undergraduate enrollment 3,330
Graduate enrollment 2,919
Total institution enrollment 6,249
Type of institution: Private   **Calendar:** Semesters
Location: Baltimore   **Population:** 787,000
Setting: Urban
Types of engineering degrees: Engineering

ENGINEERING ADMISSIONS
ADMISSIONS OFFICE CONTACT
Robert D. Shelton
Loyola College in Maryland
EE&ES Dept
Baltimore, MD 21210
Phone: (410)617-2852   **Fax:** (410)617-5123
ENGINEERING COLLEGE ADMISSIONS INFORMATION
Entrance Requirements: SAT, High School courses—English (4 years), Math (3 years), Science (2 years), History
Entrance Recommendations: High School courses—Foreign Language (2 years)
Requirements for foreign students: TOEFL

FINANCIAL INFORMATION
ESTIMATED EXPENSES (ACADEMIC YEAR)
[Expenses are for the 1992-93 nine-month academic year.]

All Students	Undergraduate	Graduate
Tuition and fees	$ 5,550	$ –
College room and board	$ 5,800	$ –
Books and supplies	$ 550	$ –
Other expenses	$ –	$ –
Total estimated expenses	**$11,900**	**$ –**

FINANCIAL AID OFFICE CONTACT
Mark Lindenmeyer
Loyola College in Maryland
Financial Aid Office
4501 N. Charles Street
Baltimore, MD 21210
Phone: (410)617-5000

GENERAL FINANCIAL AID INFORMATION
Forms accepted/required: FAF
Additional financial aid information: A wide variety of financial aid is available. About 60% of Loyola students receive some type of financial aid. ROTC scholarships are especially supportive at Loyola.

ENGINEERING COLLEGE INFORMATION
[For additional personnel, refer to the *Appendix*.]
HEAD OF ENGINEERING
David Roswell
Phone: (410)617-2563
ENGINEERING COLLEGE ADDRESS
Electrical Engineering and Engineering Science
Loyola College in Maryland
4501 North Charles Street
Baltimore, MD 21210-2699
Phone: (410)617-2464   **Fax:** (410)617-5123
ENROLLMENTS—BY CLASS
[Numbers are baccalaureate enrollments for the fall 1991 term, unless otherwise footnoted.]
1st-year students/Freshmen 23
2nd-year students/Sophomores 9
3rd-year students/Juniors 6
4th-year students/Seniors 13
Total **51**

NUMBER OF DEGREES AWARDED—BY PROGRAM
[Numbers are engineering baccalaureate degrees awarded during the 1991-1992 academic year, unless otherwise footnoted. For full details about each engineering program, refer to the *Tables of Degree Programs*.]
Electrical Engineering 2
Engineering Science(s) 10
Total **12**

PERCENTAGE OF DEGREES AWARDED—BY CATEGORY
[Percentages are of all engineering baccalaureate degrees awarded during the 1991-1992 academic year, unless otherwise footnoted.]
African Americans .... – %
Asian/Pacific Island Americans .... – %
Hispanic Americans .... – %
Native Americans .... – %
Foreign Citizens .... – %
All Others .... 100 %
Women .... – %
Persons with disabilities .... – %
Students over 25 years of age .... – %

GRADUATE ENROLLMENTS & DEGREES AWARDED
Master's enrollment 289
Master's degrees awarded 45
Doctoral enrollment .... –
Doctoral degrees awarded .... –

FULL- & PART-TIME FACULTY—BY DEPARTMENT
[Figures are the head count of full-time faculty and the full-time equivalent (FTE) of part-time faculty for each engineering department or equivalent.]

Department	Full	Part
Electrical Eng	8	–
Electrical Eng & Eng Science	6	–

COLLEGE DESCRIPTION
Loyola College is a medium-sized, very selective, Catholic liberal arts college, under the aegis of the Society of Jesus in collaboration with the Sisters of Mercy, serving Maryland and the mid-Atlantic region. It is committed to education in the traditional arts and sciences and tomorrow's business and technology. It is one of 28 Jesuit colleges in the U.S. that uphold the 450 year tradition of a strong liberal arts core while also providing education in contemporary engineering innovation.

DEGREE PROGRAMS DESCRIPTION
The engineering science program permits specialization in materials science, electronics, and computer engineering. The electrical engineering program has tracks in computer engineering and in communications and signal processing. Both programs contain a substantial liberal arts component, which provides more breadth than many engineering programs.

STUDENT PROGRAMS
PROFESSIONAL AND HONORARY SOCIETIES
[For key to acronyms, see *Introduction*.]
Professional Societies: IEEE

113 Loyola Marymount University

INSTITUTION PROFILE

HEAD OF THE INSTITUTION
Thomas P. O'Malley
Phone: (310)338-2775 Fax: (310)338-2766

GENERAL INFORMATION
[All Students—Fall 1991]
Undergraduate enrollment 4,000
Graduate enrollment 1,000
Total institution enrollment **5,000**
Type of institution: Private Calendar: Semesters
Nearest city: Los Angeles Population: 3,000,000
Miles from main campus: 10 Setting: Urban
Types of engineering degrees: Engineering
Other degree-granting colleges: Business Administration, Liberal Arts, Communication and Fine Arts

ENGINEERING ADMISSIONS

ADMISSIONS OFFICE CONTACT
Matthew X. Fissinger
Loyola Marymount University
Loyola Boulevard at West 80th Street
Los Angeles, CA 90045
Phone: (310)338-2750 Fax: (310)338-2797

ENGINEERING COLLEGE ADMISSIONS INFORMATION
Entrance Requirements: SAT, ACT, High School courses—English (4 years), Mathematics (4 years), Chemistry (1 year), Physics (1 year)
Entrance Recommendations: SAT, High School courses—Foreign Language (2 years), Social Sciences (3 years), Academic Electives (3 years), SAT Math Achievement, AP Math
Requirements for foreign students: TOEFL (Score: 550); financial statement

DEPARTMENTAL ADMISSIONS INFORMATION
Admission to the engineering department: At the time of admission to the institution.

FINANCIAL INFORMATION

ESTIMATED EXPENSES (ACADEMIC YEAR)
[Expenses are for the 1992-93 nine-month academic year.]

All Students	Undergraduate	Graduate
Tuition and fees	$12,200	$ 5,000
College room and board	$ 5,800	$ –
Books and supplies	$ 600	$ 500
Other expenses	$ –	$ –
Total estimated expenses	**$18,600**	**$ 5,500**

FINANCIAL AID OFFICE CONTACT
Donna Palmer, *Director of Financial Aid*
Loyola Marymount University
Loyola Boulevard at West 80th Street
Los Angeles, CA 90045
Phone: (310)338-2753

GENERAL FINANCIAL AID INFORMATION
Forms accepted/required: Student Aid Application for California, Application for Admission to LMU
Additional financial aid information: Need-based scholarships. Ability-based scholarships. Leadership-based scholarships. Part-time campus employment.

ENGINEERING COLLEGE INFORMATION
[For additional personnel, refer to the *Appendix*.]

HEAD OF ENGINEERING
Gerald S. Jakubowski
Phone: (310)338-2834 Fax: (310)338-7339

ENGINEERING COLLEGE ADDRESS
College of Science and Engineering
Loyola Marymount University
Loyola Boulevard at West 80th Street
Los Angeles, CA 90045
Phone: (310)338-2834 Fax: (310)338-7339

ENROLLMENTS—BY CLASS
[Numbers are baccalaureate enrollments for the fall 1991 term, unless otherwise footnoted.]
1st-year students/Freshmen 75
2nd-year students/Sophomores 64
3rd-year students/Juniors 57
4th-year students/Seniors 49
Total **245**

NUMBER OF DEGREES AWARDED—BY PROGRAM
[Numbers are engineering baccalaureate degrees awarded during the 1991-1992 academic year, unless otherwise footnoted. For full details about each engineering program, refer to the *Tables of Degree Programs*.]
Civil Engineering 12
Electrical Engineering 12
Mechanical Engineering 12
Total **36**

PERCENTAGE OF DEGREES AWARDED—BY CATEGORY
[Percentages are of all engineering baccalaureate degrees awarded during the 1991-1992 academic year, unless otherwise footnoted.]
African Americans 5.0%
Asian/Pacific Island Americans 15.0%
Hispanic Americans 15.0%
Native Americans 1.0%
Foreign Citizens 3.0%
All Others 61.0%
Women 21.0%
Persons with disabilities 1.0%
Students over 25 years of age 5.0%

GRADUATE ENROLLMENTS & DEGREES AWARDED
Master's enrollment 150
Master's degrees awarded 20
Doctoral enrollment .... –
Doctoral degrees awarded .... –

ENGINEERING STUDENT DATA

Applicants to the engineering college
Number of applicants to engineering college 320
Percent offered admission 60.0%

Matriculated engineering students
Percentage in top quartile (25%) of High School class .... –(A)
Average SAT scores: Math—575, Verbal—500, Combined—1075

Notes: (A) Data not available.

FULL- & PART-TIME FACULTY—BY DEPARTMENT
[Figures are the head count of full-time faculty and the full-time equivalent (FTE) of part-time faculty for each engineering department or equivalent.]

Department	Full	Part
Civil Eng	5	4.0
Electrical Eng	5	10.0
Mechanical Eng	5	9.0

COLLEGE DESCRIPTION
The College offers a 4-year undergraduate program in the basic majors of Civil, Mechanical, and Electrical Engineering. Engineering fundamentals are emphasized the first year and a half, followed by two and a half years of specialized training in one of the three majors. A strong liberal arts/humanities core is required. Engineering design is emphasized during the final year of each student's program. All courses are taught by faculty and personalized academic advisement is emphasized. Professionalism is fostered through the student sections of national professional engineering societies and through the student chapter of Tau Beta Pi, the National Engineering Honor Society. Average number of years required to actually complete the bachelor's degree is 4.2.

DEGREE PROGRAMS DESCRIPTION
A Bachelor's of Science degree prepares the graduate for entry into industry as well as graduate school. All three engineering programs are accredited by ABET. The strong liberal arts/humanities core helps us to produce a well rounded engineer who is adept at interacting and communicating with others. The Master's of Science in Engineering programs, offering advanced study in specialized areas, are designed for part-time students.

TRANSFER INFORMATION
Residency Requirements: The last 30 semester hours of academic work must be completed at Loyola Marymount University.
Transfer via Articulation Agreements
Admission to engineering: At any time
Engineering graduates transferred from . . 4-yr: –   2-yr: 2
Requirements: At least a 2.7 cumulative average on a 4.0 scale-rolling admission.
Transfer without Articulation Agreements
Admission to engineering: At any time
Engineering graduates transferred from . . 4-yr: 2   2-yr: 10
Requirements: At least a 2.7 cumulative average on a 4.0 scale-rolling admission.

GRADUATION REQUIREMENTS
A minimum of 138 semester hours including core and program requirements; a GPA of 2.0(C) for the requirements of the major and for all work completed at Loyola Marymount University.

STUDENT PROGRAMS

PROFESSIONAL AND HONORARY SOCIETIES
[For key to acronyms, see *Introduction*.]
Professional Societies: ACM, ASCE, ASME, IEEE, SAE
Honorary Societies: Tau Beta Pi

SUPPORT PROGRAMS
Student Chapter Organizations: Society of Women Engineers, National Society of Black Engineers, Society of Hispanic Professional Engineers
Other Engineering Support Programs: Orientation programs for all freshmen and transfer students; Learning Resource Center; Career Counseling and Placement; Psychological Counseling; Campus Ministry; Health Center; and strong academic counseling system.

114 University of Maine

INSTITUTION PROFILE
HEAD OF THE INSTITUTION
Frederick E. Hutchinson
Phone: (207)581-1512   Fax: (207)581-1517

GENERAL INFORMATION
[All Students—Fall 1991]
Undergraduate enrollment 10,727
Graduate enrollment 2,077
Total institution enrollment 12,804
Type of institution: Public   **Calendar:** Semesters
Nearest city: Bangor   **Population:** 35,000
Miles from main campus: 10   **Setting:** Small Town
Types of engineering degrees: Engineering & Technology
Other degree-granting colleges: Business Administration, Education, Nursing, Applied Sciences & Agriculture, Arts and Humanities, Forest Resources, Sciences, Social and Behavioral Sciences

ENGINEERING ADMISSIONS
ADMISSIONS OFFICE CONTACT
William J. Munsey
University of Maine
115 Chadbourne Hall
Orono, ME 04469-5713
Phone: (207)581-1561   Fax: (207)581-1556

ENGINEERING COLLEGE ADMISSIONS INFORMATION
Entrance Requirements: SAT (Score: 950), High School courses—English (4 years), Chemistry (1 year), Physics (1 year), Mathematics (incl 2 yrs Algebra, 1 geom, 1/2 trig) (4 years)
Entrance Recommendations: High School courses—Foreign Language (2 years), Computer Science (1 year), Fine Arts (1 year)
Requirements for foreign students: TOEFL (Score: 500); financial statement

DEPARTMENTAL ADMISSIONS INFORMATION
Admission to the engineering department: At the time of admission to the institution.

FINANCIAL INFORMATION
ESTIMATED EXPENSES (ACADEMIC YEAR)
[Expenses are for the 1992-93 nine-month academic year.]

State Residents	Undergraduate	Graduate
Tuition and fees	$ 3,231	$ 2,088
College room and board	$ 4,362	$ 4,362
Books and supplies	$ 500	$ 500
Other expenses	$ 500	$ 500
Total estimated expenses	**$ 8,593**	**$ 7,450**

Out-of-State Residents	Undergraduate	Graduate
Tuition and fees	$ 9,691	$ 5,904
College room and board	$ 4,362	$ 4,362
Books and supplies	$ 500	$ 500
Other expenses	$ 500	$ 500
Total estimated expenses	**$15,053**	**$11,266**

FINANCIAL AID OFFICE CONTACT
Peggy L. Crawford, *Director*
University of Maine
Wingate Hall
Orono, ME 04469-5781
Phone: (207)581-1324

GENERAL FINANCIAL AID INFORMATION
Forms accepted/required: FAF, FAT, Stafford, IRS, SAR
Additional financial aid information: Part-time jobs, University Scholarships

ENGINEERING COLLEGE INFORMATION
[For additional personnel, refer to the *Appendix*.]
HEAD OF ENGINEERING
Norman Smith
Phone: (207)581-2216   Fax: (207)581-2220

ENGINEERING COLLEGE ADDRESS
College of Engineering
University of Maine
101 Barrows Hall
Orono, ME 04469-5708
Phone: (207)581-2216   Fax: (207)581-2220

ENROLLMENTS—BY CLASS
[Numbers are baccalaureate enrollments for the fall 1991 term, unless otherwise footnoted.]
1st-year students/Freshmen 302
2nd-year students/Sophomores 223
3rd-year students/Juniors 210
4th-year students/Seniors 216
Total **951**

NUMBER OF DEGREES AWARDED—BY PROGRAM
[Numbers are engineering baccalaureate degrees awarded during the 1991-1992 academic year, unless otherwise footnoted. For full details about each engineering program, refer to the *Tables of Degree Programs*.]
Bio-Resource Engineering .... –
Chemical Engineering 30
Civil Engineering 41
Electrical & Computer Engineering 39
Engineering Physics 4
Forest Engineering 5
Mechanical Engineering 33
Surveying Engineering 11
Total **163**

PERCENTAGE OF DEGREES AWARDED—BY CATEGORY
[Percentages are of all engineering baccalaureate degrees awarded during the 1991-1992 academic year, unless otherwise footnoted.]
African Americans .... – %
Asian/Pacific Island Americans .... – %
Hispanic Americans .... – %
Native Americans .... – %
Foreign Citizens .... 5.0 %
All Others .... 95.0 %
Women .... 10.0 %
Persons with disabilities .... – %
Students over 25 years of age .... 15.0 %

GRADUATE ENROLLMENTS & DEGREES AWARDED
Master's enrollment 103
Master's degrees awarded 49
Doctoral enrollment 35
Doctoral degrees awarded 4

ENGINEERING STUDENT DATA
Applicants to the engineering college
Number of applicants to engineering college 626
Percent offered admission 73.0%

Matriculated engineering students
Percentage in top quartile (25%) of High School class 60.0
Average SAT scores: Math—604, Verbal—501, Combined—1105

FULL- & PART-TIME FACULTY—BY DEPARTMENT
[Figures are the head count of full-time faculty and the full-time equivalent (FTE) of part-time faculty for each engineering department or equivalent.]

Department	Full	Part
Chemical Eng	12	–
Civil Eng	12	–
Electrical & Computer Eng	14	–
Mechanical Eng	9	–
Surveying Eng	6	–
Eng Physics	17	–
Bio-Resource Eng	8	–
Forest Eng	3	–

COLLEGE DESCRIPTION
The College offers a wide range of four year undergraduate programs in Engineering (Chemical, Civil, Computer, Electrical, Eng. Physics, Bio-Resource, Forest Engineering, Mechanical, and Surveying). Students can select a department on entry or may remain as 'undeclared' for up to two years. There are many courses common to all majors in the freshman year to allow easy transfer between programs. Most programs have a co-op option with Chemical Engineering having the largest percentage of co-op students. All programs have a design project, some programs begin the project during the junior year. All students are encouraged to take the Engineering-In-Training Exam. There are opportunities for work with faculty throughout the students' stay.

DEGREE PROGRAMS DESCRIPTION
The B.S. degrees in the various engineering programs prepare the graduate for entry-level professional positions or for graduate study in each of the respective fields. Graduate study in engineering can be carried out in one of two general areas: the M.S. and Ph.D. in engineering, emphasizing the applications of the natural sciences to the analysis and solution of engineering problems.

DUAL DEGREE PROGRAMS
Engineering baccalaureate graduates with dual degrees 1

TRANSFER INFORMATION
Residency Requirements: A minimum of 30 credits of Engineering courses.
Transfer without Articulation Agreements
Admission to engineering: At any time
Engineering graduates transferred from 4-yr: 34   2-yr: 10
Requirements: Transfers must have a 2.5/4.0 GPA with at least C's in math and sciences.

GRADUATION REQUIREMENTS
A cumulative average of not less than 2.0. Passing grades in all required courses. Total credit hours required varies (128-134 semester hours).

STUDENT PROGRAMS
PROFESSIONAL AND HONORARY SOCIETIES
[For key to acronyms, see *Introduction*.]
Professional Societies: ACSM, AIChE, ASAE, ASCE, ASME, IEEE, SPE*, SPS, TAPPI
Honorary Societies: Chi Epsilon, Eta Kappa Nu, Tau Alpha Pi, Tau Beta Pi

SUPPORT PROGRAMS
Student Chapter Organizations: Society of Women Engineers
Other Engineering Support Programs: Summer and fall orientation program for all freshmen and transfers. Pre first semester testing in math and English. Career counseling, foreign student office, counseling and Health Center.

115 University of Manitoba

■ INSTITUTION PROFILE
HEAD OF THE INSTITUTION
Arnold Naimark
Phone: (204)474-9345 Fax: (204)275-1160

GENERAL INFORMATION
[All Students—Fall 1991]
Undergraduate enrollment 21,922
Graduate enrollment 3,201
Total institution enrollment **25,513**
Type of institution: Public Calendar: Semesters
Nearest city: Winnipeg Population: 660,000
Miles from main campus: 6 Setting: Suburban
Types of engineering degrees: Engineering
Other degree-granting colleges: Agricultural & Environmental, Architecture, Arts & Sciences, Business Administration, Dentistry, Education, Fine & Performing Arts, Humanities & Social Sciences, Law, Medicine, Nursing, Pharmacy, Technology & Applied Sciences

■ ENGINEERING ADMISSIONS
ADMISSIONS OFFICE CONTACT
Shari Campbell
University of Manitoba
Admissions Office
424 Admissions Office
Winnipeg, MB R3T-2N2
Phone: (204)474-8813 Fax: (204)275-6534

ENGINEERING COLLEGE ADMISSIONS INFORMATION
Entrance Requirements: High School courses—Chemistry 300 (3 years), Mathematics 300 (3 years), Physics 300 (3 years)
Entrance Recommendations: High School courses—English 300 (3 years)
Requirements for foreign students: TOEFL (Score: 550); TOEFL (Min. 550) for those whose native tongue is not English. Minimum GPA 3.50, or 82% high school average.

DEPARTMENTAL ADMISSIONS INFORMATION
Admission to the engineering department: At the time of admission to the institution.

■ FINANCIAL INFORMATION
ESTIMATED EXPENSES (ACADEMIC YEAR)
[Expenses are for the 1992-93 nine-month academic year.]

All Students	Undergraduate	Graduate
Tuition and fees	$ 2,650	$ 2,260
College room and board	$ 5,340	$ 5,340
Books and supplies	$ 990 (C)	$ 990
Other expenses	$ _ (A)	$ _ (A)
Total estimated expenses	$ 8,980	$ 8,590

Notes: (A) Data not available. (C) Estimated data.

FINANCIAL AID OFFICE CONTACT
Peter Dueck
University of Manitoba
Financial Aids and Awards Office
422 University Centre
Winnipeg, MB R3T-2N2
Phone: (204)474-9261 Fax: (204)275-6534

GENERAL FINANCIAL AID INFORMATION
Forms accepted/required: Student Aid, Financial Aid
Additional financial aid information: Scholarships, bursaries, loans; also on-campus employment (contact Office of Special Programs and Advocacy, 473 University Centre).

■ ENGINEERING COLLEGE INFORMATION
[For additional personnel, refer to the *Appendix*.]
HEAD OF ENGINEERING
Garland E. Laliberte
Phone: (204)474-9806 Fax: (204)275-3773
EMail: BITNET LALIBER@ UOFMCC

ENGINEERING COLLEGE ADDRESS
Faculty of Engineering
University of Manitoba
Office of the Dean
350 Engineering Building
Winnipeg, MB R3T-2N2
Phone: (204)474-9806 Fax: (204)275-3773

ENROLLMENTS—BY CLASS
[Numbers are baccalaureate enrollments for the fall 1991 term, unless otherwise footnoted.]
1st-year students/Freshmen 418
2nd-year students/Sophomores 325
3rd-year students/Juniors 302
4th-year students/Seniors 265
Total **1,310**

NUMBER OF DEGREES AWARDED—BY PROGRAM
[Numbers are engineering baccalaureate degrees awarded during the 1991-1992 academic year, unless otherwise footnoted. For full details about each engineering program, refer to the *Tables of Degree Programs*.]
Agricultural Engineering 3
Civil Engineering 48
Computer Engineering 17
Electrical Engineering 50
Geological Engineering .... – (A)
Industrial Engineering 26
Mechanical Engineering 43
Total **187**

Notes: (A) Data not available.

PERCENTAGE OF DEGREES AWARDED—BY CATEGORY
[Percentages are of all engineering baccalaureate degrees awarded during the 1991-1992 academic year, unless otherwise footnoted.]
African Canadians .... – % (A)
Asian/Pacific Island Canadians .... – % (A)
Hispanic Canadians .... – % (A)
Native Canadians .... – % (A)
Foreign Citizens .... – % (A)
All Others .... 100 %
Women .... – % (A)
Persons with disabilities .... – % (A)
Students over 25 years of age .... – % (A)

Notes: (A) Data not available.

GRADUATE ENROLLMENTS & DEGREES AWARDED
Master's enrollment 209
Master's degrees awarded 40
Doctoral enrollment 131
Doctoral degrees awarded 14

ENGINEERING STUDENT DATA
Applicants to the engineering college
Number of applicants to engineering college 1,018
Percent offered admission 50.0%

Matriculated engineering students
Percentage in top quartile (25%) of High School class 24.5

FULL- & PART-TIME FACULTY—BY DEPARTMENT
[Figures are the head count of full-time faculty and the full-time equivalent (FTE) of part-time faculty for each engineering department or equivalent.]

Department	Full	Part
Civil Eng	22	1.2
Agricultural Eng	10	–
Electrical & Computer Eng	27	1.3
Geological Eng	4	0.1
Mechanical & Industrial Eng	22	1.6

COLLEGE DESCRIPTION
The Faculty offers a four-year undergraduate program in engineering, emphasizing science and mathematics fundamentals in the common first year, followed by three years of design-oriented content in one of seven disciplines. Humanities and social science electives and technical electives are required in each program. During the final year, each student (singly, or with a partner) must write a graduation thesis on a design project or a theoretical study. Students in Civil Engineering may spend an additional year in a supervised industrial setting. These students do not write a graduation thesis. An Industrial Internship Program featuring a 16-month supervised industrial work experience after third year is available for students in Electrical and Computer Engineering. Students take an average of 4.7 years of academic study (excluding co-operative and internship work experience) to complete the four-year bachelor's degree program.

DEGREE PROGRAMS DESCRIPTION
B.Sc. degrees in seven engineering programs following first year (except Industrial Engineering which follows second year Mechanical Engineering) prepare the graduate for entry-level professional positions, or for graduate study in the respective fields. Co-operative education currently available in Civil Engineering, Computer Engineering and Electrical Engineering, but will be introduced in other programs in due course. Postgraduate studies may be carried out in each of the departments, leading to M.Sc., M.Eng. (aimed at practicing engineers), or Ph.D. degrees. Within each department, several areas of specialization are available for postgraduate studies, as well as for elective course selection in the undergraduate programs.

TRANSFER INFORMATION
Residency Requirements: Students must complete at least fifty percent of a full-time program in the Faculty of Engineering.
Transfer without Articulation Agreements
Admission to engineering: Transfer students must have completed Mathematics, Physics and Chemistry 300 or their equivalents. Transfer credit may be given for courses successfully completed elsewhere.
Engineering graduates transferred from .. 4-yr: – 2-yr: 1
Requirements: Specific requirements vary depending on demand, but generally an average grade of at least C+ is necessary for eligibility.

GRADUATION REQUIREMENTS
C-grade or better is required in each course except that D-grades count as a pass after first year and if both sessional and cumulative grade point averages are C or better; Min. C+ sessional grade point average is required in first year, Min. C sessional and cumulative grade point averages are required in senior years. Min. C session grade point average is required in final session. Min. 50% of full program is required to be taken in the Faculty of Engineering.

■ STUDENT PROGRAMS
PROFESSIONAL AND HONORARY SOCIETIES
[For key to acronyms, see *Introduction*.]
Professional Societies: ASAE, CIM, CSAE, CSCE, CSME, IEEE, SAE
Honorary Societies: Sigma Phi Delta, University of Manitoba Industrial Engineering Society, University of Manitoba Geological Engineering Society

SUPPORT PROGRAMS
Student Chapter Organizations: Native Students Office

Access Program (Women in Science & Engineering)
For: Women Available: Academic year
Offered: Prematriculation

Engineering Access Program
For: Ethnic Minorities Available: Year round
Offered: Prematriculation, Freshman, Sophomore, Junior, Senior, Graduate level

Other Engineering Support Programs: Orientation session for first year students on registration day; Counseling Service, Special Programs and Advocacy (Student Advocate, Chaplain's Association, Disabled Student Services, Learning Resource Center, On-Campus Part-time Employment Service); Ombudsman; International Centre for Students; On- and Off-Campus Accommodation Assistance.

116 Mankato State University

INSTITUTION PROFILE
HEAD OF THE INSTITUTION
Richard R. Rush
Phone: (507)389-1111 **Fax:** (507)389-5859
GENERAL INFORMATION
[All Students—Fall 1991]
Undergraduate enrollment 13,100
Graduate enrollment 1,675
Total institution enrollment **14,775**
Type of institution: Public **Calendar:** Quarters
Nearest city: Minneapolis-St. Paul **Population:** 2,500,000
Miles from main campus: 70 **Setting:** Urban
Types of engineering degrees: Engineering & Technology
Other degree-granting colleges: Education, Nursing, Arts and Humanities, Health and Human Performance, Natural Sciences, Math and Home Economics, Social and Behavioral Sciences, Business

ENGINEERING ADMISSIONS
ADMISSIONS OFFICE CONTACT
Jack Parkins
Mankato State University
Admissions Office
MSU Box 55, P.O. Box 8400
Mankato, MN 56002-8400
Phone: (507)389-1822 **Fax:** (507)389-1040
ENGINEERING COLLEGE ADMISSIONS INFORMATION
Entrance Requirements: ACT (Score: 21)
Entrance Recommendations: High School courses—Algebra (2 years), Geometry (1 year), Physics and Chemistry (1 year), College Algebra and Trigonometry (1 year)
Requirements for foreign students: TOEFL (Score: 500); financial statement; If from a non-English speaking country, successful completion of Level 109 in an English Language School; or a Michigan English Proficiency Test Score of 80. Past academic work documents must be officially notarized.
DEPARTMENTAL ADMISSIONS INFORMATION
Admission to the engineering department: At the end of the second year.
Additional information: Pre-engineering requirements must be met. Must have a cumulative GPA of 2.5 for all science, math and engineering courses. Admission to the engineering programs is selective and subject to approval of the Engineering Academic Standards Committee.

FINANCIAL INFORMATION
ESTIMATED EXPENSES (ACADEMIC YEAR)
[Expenses are for the 1992-93 nine-month academic year.]

State Residents	Undergraduate	Graduate
Tuition and fees	$ 2,392	$ 1,700[1]
College room and board	$ 2,620	$ 2,620
Books and supplies	$ 400	$ 500
Other expenses	$ –	$ –
Total estimated expenses	**$ 5,412**	**$ 4,820**

Out-of-State Residents	Undergraduate	Graduate
Tuition and fees	$ 4,382	$ 2,370[1]
College room and board	$ 2,620	$ 2,620
Books and supplies	$ 400	$ 500
Other expenses	$ –	$ –
Total estimated expenses	**$ 7,402**	**$ 5,490**

Notes: (1) Based on 8 credits per quarter for 3 quarters.
FINANCIAL AID OFFICE CONTACT
Sandra Loerts, *Director, Student Financial Aid Office*
Mankato State University
109 Wigley Administration Building
MSU Box 37
Mankato, MN 56002-8400
Phone: (507)389-1185 **Fax:** (507)389-5114

GENERAL FINANCIAL AID INFORMATION
Forms accepted/required: FFS
Additional financial aid information: Allis Foundation Scholarships, University Scholarships, Pell Grants, Minnesota State Grants, Supplemental Educational Opportunity Grants, Stafford Loans, Perkins Loans, and 27 College of Physics, Engineering and Technology academic scholarships.

ENGINEERING COLLEGE INFORMATION
[For additional personnel, refer to the *Appendix*.]
HEAD OF ENGINEERING
Robert J. Herickhoff
Phone: (507)389-1521 **Fax:** (507)389-1095
ENGINEERING COLLEGE ADDRESS
College of Physics, Engineering, and Technology
Mankato State University
Trafton Science Center
MSU Box 3
Mankato, MN 56002-8400
Phone: (507)389-1521 **Fax:** (507)389-1095
ENROLLMENTS—BY CLASS
[Numbers are baccalaureate enrollments for the fall 1991 term, unless otherwise footnoted.]
1st-year students/Freshmen 162
2nd-year students/Sophomores 68
3rd-year students/Juniors 61
4th-year students/Seniors 83
Total **374**
NUMBER OF DEGREES AWARDED—BY PROGRAM
[Numbers are engineering baccalaureate degrees awarded during the 1991-1992 academic year, unless otherwise footnoted. For full details about each engineering program, refer to the *Tables of Degree Programs*.]
Electrical Engineering 24
Mechanical Engineering 14
Total **38**
PERCENTAGE OF DEGREES AWARDED—BY CATEGORY
[Percentages are of all engineering baccalaureate degrees awarded during the 1991-1992 academic year, unless otherwise footnoted.]
African Americans .... – %
Asian/Pacific Island Americans .... 10.0 %
Hispanic Americans .... – %
Native Americans .... – %
Foreign Citizens .... 10.0 %
All Others .... 80.0 %
Women .... 5.0 %
Persons with disabilities .... 1.0 %
Students over 25 years of age .... 8.0 %
ENGINEERING STUDENT DATA
Applicants to the engineering college
Number of applicants to engineering college 69
Percent offered admission .... 100.0 %
Matriculated engineering students
Percentage in top quartile (25%) of High School class 80.0
FULL- & PART-TIME FACULTY—BY DEPARTMENT
[Figures are the head count of full-time faculty and the full-time equivalent (FTE) of part-time faculty for each engineering department or equivalent.]

Department	Full	Part
Electrical Eng	7	–
Mechanical Eng	5	–

COLLEGE DESCRIPTION
MSU's EE and ME programs represent an outgrowth of an extended electronics option for physics majors. Started in 1986, the programs now offer a fully accredited degree in EE with accreditation in ME now in progress. The programs offer areas of emphasis in semiconductors, VLSI design, controls, instrumentation, communications, kinematics, machine design, robotics, industrial automation, and computer engineering. Students enjoy access to over $9,000,000 in equipment including complete cleanroom facilities, advanced analytical instrumentation, pilot lines, numerical machining, and state-of-the-art facilities for communications design and analysis. The programs work cooperatively with the Community Colleges throughout Minnesota to assure ease of transition from these institutions into the MSU environment through a joint admission agreement.
DEGREE PROGRAMS DESCRIPTION
Mankato State University offers B.S. degree programs in 2 engineering disciplines: electrical engineering and mechanical engineering. These programs are distinguished by excellent state-of-the-art equipment and laboratory facilities. Student projects are encouraged and supported. Such projects have won numerous regional and national awards. Most of the faculty hold Ph.D.s and have had extensive industry experience. Close liaison with industry has greatly benefited the engineering programs!
DUAL DEGREE PROGRAMS
Engineering baccalaureate graduates with dual degrees 0
TRANSFER INFORMATION
Residency Requirements: Transfers must complete at least 45 hours at Mankato State University during the last 2 years prior to graduation.
Transfer via Articulation Agreements
Admission to engineering: At the end of the sophomore year
Engineering graduates transferred from ... 4-yr: – 2-yr: 30
Requirements: There is a joint admission agreement with the Minnesota Community College System. Course equivalency determines transfer acceptance of courses.
Transfer without Articulation Agreements
Admission to engineering: At any time
Engineering graduates transferred from ... 4-yr: – 2-yr: –
Requirements: Admission requirements must be met. Applicants screened by Engineering Academic Standards Committee for admission to the junior level program.

STUDENT PROGRAMS
PROFESSIONAL AND HONORARY SOCIETIES
[For key to acronyms, see *Introduction*.]
Professional Societies: ACS, AIP, APS, ASME, IEEE, SAE, SME, SPS
Honorary Societies: Phi Kappa Phi
SUPPORT PROGRAMS
Student Chapter Organizations: Society of Women Engineers
Cultural Diversity Program
For: Women & Ethnic Minorities **Available:** Year round
Offered: Freshman, Sophomore, Junior, Senior, Graduate level
Women's Studies
For: Women **Available:** Year round
Offered: Freshman, Sophomore, Junior, Senior, Graduate level
Other Engineering Support Programs: MSU offers a variety of support and services to incoming students. At the freshman level, MSU offers an introductory engineering sequence. In addition, the college supports dormitory programs for engineering students. These programs include the availability of computer facilities, laboratory facilities, and faculty advisors in selected locations. Several student organizations are active.

117 Marietta College

INSTITUTION PROFILE

HEAD OF THE INSTITUTION
Patrick D. McDonough
Phone: (614)374-4701 Fax: (614)374-4896

GENERAL INFORMATION
[All Students—Fall 1991]
Undergraduate enrollment 1,317
Graduate enrollment 62
Total institution enrollment 1,379
Type of institution: Private **Calendar:** Semesters
Nearest city: Columbus, Ohio **Population:** 1,385,000
Miles from main campus: 100 **Setting:** Small Town
Types of engineering degrees: Engineering
Other degree-granting colleges: Arts & Sciences, Business Administration, Communication & Journalism, Education, Fine & Performing Arts, Humanities & Social Sciences, Sports Medicine

ENGINEERING ADMISSIONS

ADMISSIONS OFFICE CONTACT
Dennis R. DePerro
Marietta College
215 Fifth Street
Marietta, OH 45750-4005
Phone: (800)331-7896 Fax: (614)374-4896

ENGINEERING COLLEGE ADMISSIONS INFORMATION
Entrance Requirements: High School courses—English (4 years), Mathematics (3 years), Science (3 years), Social Studies (3 years), Either the SAT or ACT is required. Composite scores of 1000 SAT or 23 ACT are recommended, but weighed with other academic indicators.
Entrance Recommendations: High School courses—Foreign Language (2 years)
Requirements for foreign students: TOEFL (Score: 525); financial statement

DEPARTMENTAL ADMISSIONS INFORMATION
Admission to the engineering department: At the time of admission to the institution.

FINANCIAL INFORMATION

ESTIMATED EXPENSES (ACADEMIC YEAR)
[Expenses are for the 1992-93 nine-month academic year.]

All Students	Undergraduate	Graduate
Tuition and fees	$12,370	$ –
College room and board	$ 3,530 [2]	$ –
Books and supplies	$ 500	$ –
Other expenses	$ 290 [1]	$ –
Total estimated expenses	$16,690	$ –

Notes: (1) Includes $25 application fee, $20 motor vehicle registration, and $100 residence hall damage deposit. (2) $825 for meal plan; $940 for room.

FINANCIAL AID OFFICE CONTACT
James M. Bauer, *Associate Dean and Director of Financial Aid*
Marietta College
215 Fifth Street
Marietta, OH 45750
Phone: (800)331-2709 Fax: (614)374-4896

GENERAL FINANCIAL AID INFORMATION
Forms accepted/required: FAF
Additional financial aid information: Our financial aid programs are designed to make a Marietta College education accessible to those who, for financial reasons, might otherwise be unable to attend. The underlying assumption of financial aid, with the exception of talent or merit based scholarships, is that the family bears the responsibility to meet College costs to the extent it is able. Financial assistance from College, federal, and state sources is used to bridge the gap between the family effort and College costs.

ENGINEERING COLLEGE INFORMATION

[For additional personnel, refer to the *Appendix*.]

HEAD OF ENGINEERING
Robert W. Chase
Phone: (614)374-4776 Fax: (614)374-4896

ENGINEERING COLLEGE ADDRESS
Petroleum, Industrial, and Binary Engineering
Marietta College
215 Fifth Street
Marietta, OH 45750-4617
Phone: (614)374-4775 Fax: (614)374-4896

ENROLLMENTS—BY CLASS
[Numbers are baccalaureate enrollments for the fall 1991 term, unless otherwise footnoted.]
1st-year students/Freshmen 23
2nd-year students/Sophomores 19
3rd-year students/Juniors 11
4th-year students/Seniors 13
Total 66

NUMBER OF DEGREES AWARDED—BY PROGRAM
[Numbers are engineering baccalaureate degrees awarded during the 1991-1992 academic year, unless otherwise footnoted. For full details about each engineering program, refer to the *Tables of Degree Programs*.]
Industrial Engineering 3
Petroleum Engineering 5
Total 8

PERCENTAGE OF DEGREES AWARDED—BY CATEGORY
[Percentages are of all engineering baccalaureate degrees awarded during the 1991-1992 academic year, unless otherwise footnoted.]
African Americans .... – %
Asian/Pacific Island Americans .... – %
Hispanic Americans .... – %
Native Americans .... – %
Foreign Citizens .... 25.0 %
All Others .... 75.0 %
Women .... 13.0 %
Persons with disabilities .... – %
Students over 25 years of age .... 13.0 %

ENGINEERING STUDENT DATA
Applicants to the engineering college
Number of applicants to engineering college 56
Percent offered admission 66.0%
Matriculated engineering students
Percentage in top quartile (25%) of High School class 80.0
Average SAT scores: Combined—1045
Average ACT scores: Composite—24.5

FULL- & PART-TIME FACULTY—BY DEPARTMENT
[Figures are the head count of full-time faculty and the full-time equivalent (FTE) of part-time faculty for each engineering department or equivalent.]

Department	Full	Part
Industrial Eng	–	2.0
Petroleum Eng	4	–
Binary Eng Programs	1	–

COLLEGE DESCRIPTION
Marietta College, ranked as the number one regional liberal arts college in the Midwest in the 1991 issue of U.S. News and World Report, is the only private liberal arts college in the U.S. to offer a B.S. degree in petroleum engineering. The petroleum engineering program is accredited by EAC of ABET and was ranked eleventh among all U.S. petroleum engineering programs in the latest issue of the Gourman Report. Marietta is the only regional liberal arts college in the country to have an engineering program ranked in the top twenty in any engineering discipline. Marietta College offers no graduate engineering degrees. Faculty are totally devoted to the undergraduate program and class sizes are typically small.

DEGREE PROGRAMS DESCRIPTION

As well as degrees in Petroleum Engineering and Industrial Engineering, Marietta College offers courses in Binary Engineering. Courses in Binary Engineering may be applied toward a Bachelor of Arts degree from Marietta College or transferred to another university to be applied toward a Bachelor of Science degree.

DUAL DEGREE PROGRAMS
Engineering baccalaureate graduates with dual degrees 0

TRANSFER INFORMATION
Transfer via Articulation Agreements
Admission to engineering: At any time
Engineering graduates transferred from .. 4-yr: – 2-yr: –
Requirements: All transfer students must furnish the same credentials as freshman applicants. Transfer credits will be determined by the Registrar in consultation with academic departments. Credit will be allowed for courses completed with a grade of C or higher.

Transfer without Articulation Agreements
Admission to engineering: At any time
Engineering graduates transferred from .. 4-yr: – 2-yr: 1
Requirements: All transfer applicants must furnish the same credentials as freshman applicants. Transfer credits will be determined by the Registrar in conjunction with the academic departments. Credit will be allowed for courses completed with a grade of C or higher.

GRADUATION REQUIREMENTS
Students majoring in engineering must have a 2.0 in their major to graduate. Students majoring in petroleum engineering must also take the Fundamentals of Engineering Examination as a requirement for the degree of B.S. in Petroleum Engineering.

STUDENT PROGRAMS

PROFESSIONAL AND HONORARY SOCIETIES
[For key to acronyms, see *Introduction*.]
Professional Societies: IIE, SPE
Honorary Societies: Pi Epsilon Tau

SUPPORT PROGRAMS
Other Engineering Support Programs: Marietta College seeks to develop strong communication skills and critical thinking skills in its students beginning with a freshman experience course tied to a truly unique general education curriculum in the liberal arts. Engineering majors take an introductory course in their discipline during the freshman year as well.

118 Marquette University

INSTITUTION PROFILE
HEAD OF THE INSTITUTION
Albert J. DiUlio S.J.
Phone: (414)288-7223

GENERAL INFORMATION
[All Students—Fall 1991]
Undergraduate enrollment	8,400
Graduate enrollment	3,375
Total institution enrollment	11,775

Type of institution: Private **Calendar:** Semesters
Location: Milwaukee, WI **Population:** 687,000
Setting: Urban
Types of engineering degrees: Engineering
Other degree-granting colleges: Arts & Sciences, Business Administration, Communication & Journalism, Dentistry, Education, Fine & Performing Arts, Law, Nursing

ENGINEERING ADMISSIONS
ADMISSIONS OFFICE CONTACT
Michael T. Istwan
Marquette University
Office of Admissions
1217 W. Wisconsin Avenue
Milwaukee, WI 53233
Phone: (414)288-7302 **Fax:** (414)288-3764

ENGINEERING COLLEGE ADMISSIONS INFORMATION
Entrance Requirements: SAT, ACT, High School courses—Algebra (1 year), Geometry (1 year), Advanced Algebra (1 year)
Entrance Recommendations: SAT (Score: 1000), ACT (Score: 23), High School courses—Mathematics/Science (4 years), Chemistry or Physics (1 year), Biology (Biomedical Engineering) (1 year)
Requirements for foreign students: TOEFL (Score: 550); financial statement

DEPARTMENTAL ADMISSIONS INFORMATION
Admission to the engineering department: At the time of admission to the institution.

FINANCIAL INFORMATION
ESTIMATED EXPENSES (ACADEMIC YEAR)
[Expenses are for the 1992-93 nine-month academic year.]

All Students	Undergraduate	Graduate
Tuition and fees	$10,400	$ 335[(1)]
College room and board	$ 4,250	$ 4,250
Books and supplies	$ 600	$ 600[(C)]
Other expenses	$ 1,350[(C)]	$ 1,350[(C)]
Total estimated expenses	$16,600	$ 6,535

Notes: (C) Estimated data. (1) per semester credit hour.

FINANCIAL AID OFFICE CONTACT
Daniel L. Goyette, *Director of Student Financial Aid*
Marquette University
1212 W. Wisconsin Avenue
Milwaukee, WI 53233
Phone: (414)288-7390

GENERAL FINANCIAL AID INFORMATION
Forms accepted/required: FAF
Additional financial aid information: Academic Scholarships, Need-Based Scholarships, College Work Study Program, Regular Student Employment.

ENGINEERING COLLEGE INFORMATION
[For additional personnel, refer to the *Appendix*.]
HEAD OF ENGINEERING
Robert L. Reid
Phone: (414)288-6720 **Fax:** (414)288-7082

ENGINEERING COLLEGE ADDRESS
College of Engineering
Marquette University
1515 W. Wisconsin Avenue
Milwaukee, WI 53233
Phone: (414)288-7079 **Fax:** (414)288-7082

ENROLLMENTS—BY CLASS
[Numbers are baccalaureate enrollments for the fall 1991 term, unless otherwise footnoted.]

1st-year students/Freshmen	413
2nd-year students/Sophomores	277
3rd-year students/Juniors	379
4th-year students/Seniors	246
Total	1,315

NUMBER OF DEGREES AWARDED—BY PROGRAM
[Numbers are engineering baccalaureate degrees awarded during the 1991-1992 academic year, unless otherwise footnoted. For full details about each engineering program, refer to the *Tables of Degree Programs*.]

Biomedical Engineering	31
Civil Engineering	46
Electrical Engineering	98
Industrial Engineering	24
Mechanical Engineering	58
Total	257

PERCENTAGE OF DEGREES AWARDED—BY CATEGORY
[Percentages are of all engineering baccalaureate degrees awarded during the 1991-1992 academic year, unless otherwise footnoted.]

African Americans	2.0%
Asian/Pacific Island Americans	5.0%
Hispanic Americans	2.0%
Native Americans	– %
Foreign Citizens	5.0%
All Others	86.0%
Women	17.0%
Persons with disabilities	– % [(A)]
Students over 25 years of age	– % [(A)]

Notes: (A) Data not available.

GRADUATE ENROLLMENTS & DEGREES AWARDED
Master's enrollment	134
Master's degrees awarded	85
Doctoral enrollment	62
Doctoral degrees awarded	12

ENGINEERING STUDENT DATA
Applicants to the engineering college
Number of applicants to engineering college	998
Percent offered admission	84.5%

Matriculated engineering students
Percentage in top quartile (25%) of High School class 70.0
Average SAT scores: Math—597, Verbal—498, Combined—1095
Average ACT scores: Math—27, Composite—26

FULL- & PART-TIME FACULTY—BY DEPARTMENT
[Figures are the head count of full-time faculty and the full-time equivalent (FTE) of part-time faculty for each engineering department or equivalent.]

Department	Full	Part
Civil & Environmental Eng	11	... –
Electrical & Computer Eng	21	... 0.2
Industrial Eng	4	–
Mechanical Eng	15	... 0.9
Biomedical Eng	12	... 0.6

COLLEGE DESCRIPTION
Marquette University is Wisconsin's largest independent institution of higher learning, and is the nations largest religiously affiliated college of engineering. Founded in 1881, Marquette is a co-educational, Jesuit, Catholic, Urban institution. The College of Engineering was established in 1908. All engineering degree programs are available as an engineering Minor. The college also has an established Minor program in Business Administration and Mathematics. As pioneers in Engineering Cooperative Education, Marquette offers an optional Co-op experience in all degree programs. All engineering programs have a strong Liberal Arts component.

DEGREE PROGRAMS DESCRIPTION
The College of Engineering offers a traditional bachelor's degree program which is pursued through the academic departments of Biomedical Engineering, Civil and Environmental Engineering, Electrical and Computer Engineering, Mechanical and Industrial Engineering. M.S. degree programs are available in Biomedical, Civil, Electrical, and Mechanical Engineering. Because major advances in contemporary engineering are fueled by shared knowledge from subfields and diverse disciplines, the College of Engineering established a College-Wide Ph.D. Program to develop a synergistic relationship between its four departments and to emphasize multi-disciplinary research and teaching approaches.

DUAL DEGREE PROGRAMS
Engineering baccalaureate graduates with dual degrees 0
Enrollment requirements: Minimum overall GPA =2.500 with GPA in mathematics/sciences courses = 2.500.

TRANSFER INFORMATION
Residency Requirements: Minimum 32 semester hours of upper division course work in the major.
Transfer via Articulation Agreements
Admission to engineering: At any time
Engineering graduates transferred from . . . 4-yr: 4   2-yr: 9
Requirements: Entrance requirements vary depending on institution/s of origin.
Transfer without Articulation Agreements
Admission to engineering: At any time
Engineering graduates transferred from . . . 4-yr: 3   2-yr: –
Requirements: Minimum overall GPA 2.500 with grades = 2.500 in mathematics/sciences.

GRADUATION REQUIREMENTS
Minimum overall GPA =2.000 with a minimum GPA = 2.000 in the engineering major.

STUDENT PROGRAMS
PROFESSIONAL AND HONORARY SOCIETIES
[For key to acronyms, see *Introduction*.]
Professional Societies: ASCE, ASME, BMES, IEEE, IIE, ITE, SAE, SME, Society of Photo-Optical Instrumentation Engineers (SPIE), Society of Women Engineers (SWE)
Honorary Societies: Chi Epsilon, Eta Kappa Nu, Pi Tau Sigma, Sigma Phi Delta, Tau Beta Pi, Alpha Eta Mu Beta, Alpha Omega Epsilon

SUPPORT PROGRAMS
Student Chapter Organizations: Society of Women Engineers, National Society of Black Engineers

Research Careers for Minority Scholars (NSF-RCMS)
For: Ethnic Minorities **Available:** Year round
Offered: Freshman, Sophomore, Junior, Senior
Other Engineering Support Programs: Young Scholars Program for High School Juniors. Hands-on Mini-Courses for High School Seniors.

119 University of Maryland Baltimore County

■ INSTITUTION PROFILE

HEAD OF THE INSTITUTION
Freeman A. Hrabowski III
Phone: (410)455-2274　　　　　Fax: (410)455-1210

GENERAL INFORMATION
[All Students—Fall 1991]
Undergraduate enrollment 8,929
Graduate enrollment 1,439
Total institution enrollment **10,368**

Type of institution: Public　　　　Calendar: Semesters
Nearest city: Baltimore　　　　　　Population: 800,000
Miles from main campus: 5　　　　Setting: Suburban
Types of engineering degrees: Engineering
Other degree-granting colleges: Arts & Sciences, Nursing, Social Work

■ ENGINEERING ADMISSIONS

ADMISSIONS OFFICE CONTACT
Mindy A. Hand
University of Maryland Baltimore County
5401 Wilkens Avenue
Rm 102 Academic Services/Theatre
Baltimore, MD 21228-5398
Phone: (410)455-2291　　　　　Fax: (410)455-1094

ENGINEERING COLLEGE ADMISSIONS INFORMATION
Admission to the engineering college: Freshmen applicants are admitted with 580 Math SAT score and 1100 total score or 3.0 GPA in high school. Others follow the transfer criteria listed separately.
Entrance Requirements: SAT (Score: 1100), High School courses—English (4 years), Social Science/History (3 years), Lab Based Science (2 years), Math (Algebra & Geometry & Foreign Language Two) (3 years), 1100 minimum with minimum 580 Math
Entrance Recommendations: High School courses—Trigonometry (1 year), Calculus I (1 year), Physics I (1 year), Chemistry I (1 year)
Requirements for foreign students: TOEFL (Score: 550); financial statement; 3.5 cumulative GPA for pre-engr & transfer students who are not eligible for admission to the COE at time of acceptance into the university.

DEPARTMENTAL ADMISSIONS INFORMATION
Admission to the engineering department: Admission depends on GPA and required courses.

■ FINANCIAL INFORMATION

ESTIMATED EXPENSES (ACADEMIC YEAR)
[Expenses are for the 1992-93 nine-month academic year.]

State Residents	Undergraduate	Graduate
Tuition and fees	$ 3,076 [1]	$ 3,780
College room and board	$ 4,482 [2]	$ 4,482
Books and supplies	$ 500	$ 500
Other expenses	$ 27 [3]	$ 27
Total estimated expenses	**$ 8,085**	**$ 8,789**

Out-of-State Residents	Undergraduate	Graduate
Tuition and fees	$ 8,302 [1]	$ 6,300
College room and board	$ 4,482 [2]	$ 4,482
Books and supplies	$ 500	$ 500
Other expenses	$ 27 [3]	$ 27
Total estimated expenses	**$13,311**	**$11,309**

Notes: (1) fee information based on enrolling in 10 credits per semester (2) depends on meal plan and living arrangement selected (traditional hall or apartment) (3) optional health plan

FINANCIAL AID OFFICE CONTACT
Thomas R. Taylor Jr., *Director of Financial Aid*
University of Maryland Baltimore County
5401 Wilkens Avenue
Office of Financial Aid
Baltimore, MD 21228-5398
Phone: (410)455-2387　　　　　Fax: (410)455-1094

GENERAL FINANCIAL AID INFORMATION
Forms accepted/required: FAF, Financial Aid Transcripts from previous instit. must be submitted.
Additional financial aid information: UMBC Administers and Coordinates a variety of federal state and institutional student aid programs including grants, scholarships, loans and paid employment

■ ENGINEERING COLLEGE INFORMATION

[For additional personnel, refer to the *Appendix*.]

HEAD OF ENGINEERING
Duane F. Bruley
Phone: (410)455-3714　　　　　Fax: (410)455-3559
EMail: bruley@umbc2.umbc.edu

ENGINEERING COLLEGE ADDRESS
College of Engineering
University of Maryland Baltimore County
5401 Wilkens Avenue
Room 314 Engr/Computer Science Bldg
Baltimore, MD 21228-5398
Phone: (410)455-3270　　　　　Fax: (410)455-3559

ENROLLMENTS—BY CLASS
[Numbers are baccalaureate enrollments for the fall 1991 term, unless otherwise footnoted.]
1st-year students/Freshmen 197
2nd-year students/Sophomores 128
3rd-year students/Juniors 127
4th-year students/Seniors .... 211 [1]
Total **663**

Notes: (1) Includes all part-time students.

NUMBER OF DEGREES AWARDED—BY PROGRAM
[Numbers are engineering baccalaureate degrees awarded during the 1991-1992 academic year, unless otherwise footnoted. For full details about each engineering program, refer to the *Tables of Degree Programs*.]
Chemical & Biochemical Engineering 7
Electrical Engineering .... –
Engineering Management .... –
Engineering Science & Mechanics .... –
Mechanical Engineering 42
Total **49**

PERCENTAGE OF DEGREES AWARDED—BY CATEGORY
[Percentages are of all engineering baccalaureate degrees awarded during the 1991-1992 academic year, unless otherwise footnoted.]
African Americans .... 15.0 %
Asian/Pacific Island Americans .... 10.0 %
Hispanic Americans .... 2.0 %
Native Americans .... – %
Foreign Citizens .... 2.0 %
All Others .... 71.0 %
Women .... 21.0 %
Persons with disabilities .... – % [A]
Students over 25 years of age .... – % [A]

Notes: (A) Data not available.

GRADUATE ENROLLMENTS & DEGREES AWARDED
Master's enrollment 175
Master's degrees awarded 28
Doctoral enrollment 74
Doctoral degrees awarded 27

ENGINEERING STUDENT DATA

Applicants to the engineering college
Number of applicants to engineering college .... 62 [1]
Percent offered admission .... 100.0% [2]

Matriculated engineering students
Percentage in top quartile (25%) of High School class ...... [A]
Average SAT scores: Math—613, Verbal—481, Combined—1094

Notes: (A) Data not available. (1) does not include pre-engineering students (2) does not include pre engineering students

■ FULL- & PART-TIME FACULTY—BY DEPARTMENT

[Figures are the head count of full-time faculty and the full-time equivalent (FTE) of part-time faculty for each engineering department or equivalent.]

Department	Full	Part
Chemical & Biochemical Eng	8.5	–
Electrical Eng	9	–
Mechanical Eng	15	–
Eng Management	–	–

COLLEGE DESCRIPTION
UMBC offers programs leading to a B.S. in Engineering with options in chemical and biochemical engineering and in mechanical engineering. A pre-engineering program is also available for students interested in the following additional disciplines of engineering: aerospace, civil, and electrical. At the graduate level, UMBC offers M.S. and Ph.D. programs in electrical, chemical and biochemical and mechanical engineering. It also offers jointly with the University of Maryland University College a program leading to the M.S. in Engineering Management degree. Graduate programs in other engineering disciplines have been approved, and plans are under way to offer these at UMBC in the near future.

DEGREE PROGRAMS DESCRIPTION
UMBC recently opened a $26 million engineering/computer science building which contains state-of-the-art engineering instruction and research equipment. Engineering research facilities are located in the Technology Research Center. Research focus areas include biomedical, bioprocessing, biotechnology, photonics, manufacturing, robotics and signal processing. The UMBC campus is located in a suburban environment, just outside the Baltimore beltway. It is ideally located in view of downtown Baltimore's revitalized Inner Harbor, Baltimore/Washington International Airport and is only 30 miles north of the nation's capital. As a mid-sized university of 10,000 students together with 1,400 faculty and staff, UMBC has the programs and facilities of a larger institution while retaining the flexibility and the personal nature of a much smaller campus.

TRANSFER INFORMATION
Residency Requirements: students must do their final 30 credits at UMBC. Only 67 credits from a community college may be used toward and engineering degree at UMBC.

Transfer without Articulation Agreements
Admission to engineering: transfer students min. cum. GPA: 3.0 MD resident, 3.2 Out-of-State, 3.5 Intl. Complete CHEM 101, CHEM 102, CHEM 102L, MATH 151, MATH 152, PHYS 121 or equiv. with min. grade of "C". Completion of 28 crs.

GRADUATION REQUIREMENTS
Overall GPA of at least 2.0 and a grade of "C" or better in all engineering courses. Students must satisfy 30 credits of general education requirements. These include Engl 393, Technical Writing and an upper level course in each of two tracks in social science, arts and humanities, or language.

■ STUDENT PROGRAMS

PROFESSIONAL AND HONORARY SOCIETIES
[For key to acronyms, see *Introduction*.]
Professional Societies: AIChE, ASME, IEEE, SAE
Honorary Societies: Tau Beta Pi

SUPPORT PROGRAMS
Student Chapter Organizations: Society of Women Engineers, National Society of Black Engineers

Society of Women Engineers
For: Women　　　　**Available:** Academic year
Offered: Freshman, Sophomore, Junior, Senior, Graduate level

National Society of Black Engineers
For: Ethnic Minorities
Offered: Freshman, Sophomore, Junior, Senior, Graduate level
Other Engineering Support Programs: Orientation program for new undergraduates, undergraduate affairs office academic advising; student societies; health, counseling and psychiatric services; career development and placement; cooperative education.

120 University of Maryland, College Park

■ INSTITUTION PROFILE

HEAD OF THE INSTITUTION
William E. Kirwan
Phone: (301)405-5803

GENERAL INFORMATION
[All Students—Fall 1991]
Undergraduate enrollment 25,361
Graduate enrollment 9,262
Total institution enrollment 34,623
Type of institution: Public   Calendar: Semesters
Nearest city: Washington, DC   Population: 650,000
Miles from main campus: 10   Setting: Suburban
Types of engineering degrees: Engineering
Other degree-granting colleges: Agricultural & Environmental, Architecture, Business Administration, Communication & Journalism, Education, Arts & Humanities, Behavioral & Social Sciences, Computer, Mathematics & Physical Sciences, Life Sciences, Health & Human Performance

■ ENGINEERING ADMISSIONS

ADMISSIONS OFFICE CONTACT
Linda Clement
University of Maryland, College Park
0101 Mitchell Building
College Park, MD 20742-5235
Phone: (301)314-8385

ENGINEERING COLLEGE ADMISSIONS INFORMATION
Admission to the engineering college: For admission as new freshmen, students must have a 550 on the math portion of the SAT and either an overall SAT score of 1100 or a high school GPA of 3.0.
Entrance Requirements: SAT (Score: 1100), High School courses—English (4 years), Chemistry (1 year), Physics (1 year), Mathematics (4 years), SAT is required with a minimum 550 required on the math portion.
Entrance Recommendations: High School courses—Foreign Language (2 years), History/Social sciences (3 years), Computers (1 year)
Requirements for foreign students: Non-native speakers of English must submit a TOEFL score unless they have received a degree from a tertiary level institution in the U.S.

DEPARTMENTAL ADMISSIONS INFORMATION
Admission to the engineering department: Admission to the College of Engineering with the exception of the Aerospace Engineering Program.
Additional information: The Aerospace department requires a 3.5 high school GPA for direct admission to the department as a freshman. Other students are admitted to pre-Aerospace and must complete 13 prerequisite courses with a 2.5 GPA.

■ FINANCIAL INFORMATION

ESTIMATED EXPENSES (ACADEMIC YEAR)
[Expenses are for the 1992-93 nine-month academic year.]

State Residents	Undergraduate	Graduate
Tuition and fees	$ 2,778	$ 3,280
College room and board	$ 2,500	$ —
Books and supplies	$ 500	$ 1,000
Other expenses	$ —	$ —
Total estimated expenses	**$ 5,778**	**$ 4,280**

Out-of-State Residents	Undergraduate	Graduate
Tuition and fees	$ 8,382	$ 5,880
College room and board	$ 2,500	$ —
Books and supplies	$ 500	$ 1,000
Other expenses	$ —	$ —
Total estimated expenses	**$11,382**	**$ 6,880**

FINANCIAL AID OFFICE CONTACT
Ulysses Glee Jr., *Director*
University of Maryland, College Park
Lee Building
College Park, MD 20742-5211
Phone: (301)314-8313

GENERAL FINANCIAL AID INFORMATION
Forms accepted/required: FAF
Additional financial aid information: Financial aid is available through scholarships (merit and need-based); low interest loans, long-term loans; grants; work-study employment; part-time employment.

■ ENGINEERING COLLEGE INFORMATION

[For additional personnel, refer to the *Appendix*.]
HEAD OF ENGINEERING
George E. Dieter
Phone: (301)405-3868   Fax: (301)314-9867

ENGINEERING COLLEGE ADDRESS
College of Engineering
University of Maryland, College Park
1137 Engineering Classroom Building
College Park, MD 20742-3011
Phone: (301)405-3868   Fax: (301)314-9867

ENROLLMENTS—BY CLASS
[Numbers are baccalaureate enrollments for the fall 1991 term, unless otherwise footnoted.]
1st-year students/Freshmen 625
2nd-year students/Sophomores 363
3rd-year students/Juniors 555
4th-year students/Seniors 1,268
Total **2,811**

NUMBER OF DEGREES AWARDED—BY PROGRAM
[Numbers are engineering baccalaureate degrees awarded during the 1991-1992 academic year, unless otherwise footnoted. For full details about each engineering program, refer to the *Tables of Degree Programs*.]
Aerospace Engineering 100
Agricultural Engineering 3
Chemical Engineering 24
Civil Engineering 61
Electrical Engineering 212
Fire Protection Engineering 32
Materials Engineering .... —
Mechanical Engineering 144
Nuclear Engineering 9
Reliability Engineering .... —
Systems Engineering .... —
Total **585**

PERCENTAGE OF DEGREES AWARDED—BY CATEGORY
[Percentages are of all engineering baccalaureate degrees awarded during the 1991-1992 academic year, unless otherwise footnoted.]
African Americans 3.7%
Asian/Pacific Island Americans 23.4%
Hispanic Americans 2.3%
Native Americans .... — %
Foreign Citizens 4.9%
All Others 65.7%
Women 16.5%
Persons with disabilities .... — %
Students over 25 years of age .... — % (A)
Notes: (A) Data not available.

GRADUATE ENROLLMENTS & DEGREES AWARDED
Master's enrollment 753
Master's degrees awarded 248
Doctoral enrollment 544
Doctoral degrees awarded 63

ENGINEERING STUDENT DATA
Applicants to the engineering college
Number of applicants to engineering college .... — (A)
Percent offered admission .... —% (A)
Matriculated engineering students
Percentage in top quartile (25%) of High School class ...... — (A)
Average SAT scores: Math—644, Verbal—514, Combined—1158
Notes: (A) Data not available.

FULL- & PART-TIME FACULTY—BY DEPARTMENT
[Figures are the head count of full-time faculty and the full-time equivalent (FTE) of part-time faculty for each engineering department or equivalent.]

Department	Full	Part
Chemical Eng	12	1.0
Civil Eng	26	3.0
Electrical Eng	59	15.0
Mechanical Eng	45	15.0
Fire Protection Eng	6	3.0
Materials & Nuclear Eng	17	3.0
Agricultural Eng	5	—
Aerospace Eng	13	14.0

COLLEGE DESCRIPTION
The B.S. degree is offered in 8 engineering departments plus an undesignated major in which a student may combine study in two engineering fields or in an engineering field and a non-engineering field. Bachelors degrees are normally completed in 4 to 5 years and include 1-1/2 years of general science, math and engineering science courses. The remaining semesters are more specialized within the student's major with some liberal arts studies throughout all four years. Dual degree programs are available between UMCP and 17 colleges and universities in which a student earns undergraduate degrees from both institutions. An optional cooperative education program is available in which students work the equivalent of 50 weeks (40 hrs/wk) during alternating semesters in school and at the job. International opportunities are available for students to study either German or Japanese at UMCP and then spend a semester working in engineering companies in the respective countries.

DEGREE PROGRAMS DESCRIPTION
The Bachelor of Science degree is awarded in eight fields and prepares the graduate for both entry-level professional positions and/or graduate school. The University of Maryland offers the only accredited Fire Protection Engineering program in the nation. Advanced study in engineering, including thesis and non-thesis Master of Science degrees and Ph.D. degrees, is available in eleven areas of speciality.

DUAL DEGREE PROGRAMS
Engineering baccalaureate graduates with dual degrees 5
Enrollment requirements: Students must be transferring from one of the liberal arts institutions with which the College of Engineering has an agreement.

TRANSFER INFORMATION
Residency Requirements: Students must do their final 30 credits in residence at UMCP. Only 65 credits from a community college may be used toward a degree in Engineering at UMCP.
Transfer via Articulation Agreements
Admission to engineering: Articulation agreements with Maryland community colleges. Admission is competitive. Students can apply upon completing 28 semester credits and specific courses.
Engineering graduates transferred from . . . 4-yr: 195 2-yr: 292
Requirements: Students must have 28 semester credits, Freshman English, Calculus I & II, Physics I, and Chemistry I & II, plus a competitive GPA.
Transfer without Articulation Agreements
Admission to engineering: Admission is competitive. Students can apply upon completing 28 semester credits and specific courses.
Engineering graduates transferred from . . . 4-yr: 195 2-yr: 292
Requirements: Students must have completed 28 semester credits including Freshman English, Calculus I & II, Physics I, and Chemistry I & II with a competitive GPA.

GRADUATION REQUIREMENTS
Students must have a 'C' or better in all Engineering courses. Most degree programs require 130 to 134 credits. Students must have an overall GPA of 2.0 to graduate.

■ STUDENT PROGRAMS

PROFESSIONAL AND HONORARY SOCIETIES
[For key to acronyms, see *Introduction*.]
Professional Societies: AHS, AIAA, AIChE, ANS, ASCE, ASME, IEEE, SAE, SAME, SFPE
Honorary Societies: Alpha Epsilon, Chi Epsilon, Eta Kappa Nu, Omega Chi Epsilon, Pi Tau Sigma, Sigma Gamma Tau, Tau Beta Pi, Salamander, Alpha Nu Sigma

SUPPORT PROGRAMS
Student Chapter Organizations: Society of Women Engineers, National Society of Black Engineers, Society of Hispanic Professional Engineers

Center for Minorities in Science & Engineering
For: Ethnic Minorities   **Available:** Year round
Offered: Freshman, Sophomore, Junior, Senior

BRIDGE Program
For: Ethnic Minorities   **Available:** Year round
Offered: Prematriculation, Freshman, Sophomore, Junior, Senior

Summer Program for Women and Minorities
For: Women & Ethnic Minorities **Available:** Summer only
Offered: Prematriculation

Other Engineering Support Programs: Freshman and Transfer students attend orientation upon admittance to the University of Maryland. The College of Engineering also teaches an optional one credit course for ongoing orientation and introduction to engineering majors. Other support services include: academic advising, tutoring, individual and group counseling, and financial aid.

121 University of Massachusetts

■ INSTITUTION PROFILE

HEAD OF THE INSTITUTION
Michael K. Hooker
Phone: (617)287-7000

GENERAL INFORMATION
[All Students—Fall 1991]
Undergraduate enrollment 17,212
Graduate enrollment 5,816
Total institution enrollment 23,028

Type of institution: Public Calendar: Semesters
Nearest city: Springfield, MA Population: 157,000
Miles from main campus: 25 Setting: Small Town
Types of engineering degrees: Engineering
Other degree-granting colleges: Agricultural & Environmental, Allied Health Sciences, Arts & Sciences, Business Administration, Education, Fine & Performing Arts, Humanities & Social Sciences, Nursing, Physical Education

■ ENGINEERING ADMISSIONS

ADMISSIONS OFFICE CONTACT
Timm R. Rinehart
University of Massachusetts
Admission's Office
Whitmore Administration Building
Amherst, MA 01003
Phone: (413)545-0222

ENGINEERING COLLEGE ADMISSIONS INFORMATION
Entrance Requirements: SAT (Score: 500), High School courses—English (4 years), Mathematics and Science (4 years), Social Science (2 years), Foreign Language (2 years), The University has an Admissions Eligibility Index that relates class rank to combined SAT score. For ex., if a student has a class rank in the upper 15%, the minimum SAT score required is 500; if class rank is upper 80%, 1150 is minimum required.
Entrance Recommendations: High School courses—Physics with lab (1 year), Chemistry with lab (1 year)
Requirements for foreign students: TOEFL (Score: 550)

DEPARTMENTAL ADMISSIONS INFORMATION
Admission to the engineering department: At the time of admission to the institution.
Additional information: In general, the criteria for admission to a major department is a C or better in the freshmen engineering core courses.

■ FINANCIAL INFORMATION

ESTIMATED EXPENSES (ACADEMIC YEAR)
[Expenses are for the 1992-93 nine-month academic year.]

State Residents	Undergraduate	Graduate
Tuition and fees	$ 5,362	$ 5,455
College room and board	$ 3,693	$ 3,693
Books and supplies	$ 600	$ —(B)
Other expenses	$ —	$ —
Total estimated expenses	$ 9,655	$ 9,148

Out-of-State Residents	Undergraduate	Graduate
Tuition and fees	$11,462	$11,020
College room and board	$ 3,693	$ 3,693
Books and supplies	$ 600	$ —(B)
Other expenses	$ —	$ —
Total estimated expenses	$15,755	$14,713

Notes: (B) Data not applicable.

FINANCIAL AID OFFICE CONTACT
Burt Batty, *Director*
University of Massachusetts
255 Whitmore Administration Building
Amherst, MA 01003
Phone: (413)545-0801

GENERAL FINANCIAL AID INFORMATION
Forms accepted/required: CSS-FAF, IRS
Additional financial aid information: State, federal, campus-based grants and private scholarships are awarded on the basis of a variety of criteria; college work-study, loans; part-time campus jobs also available.

■ ENGINEERING COLLEGE INFORMATION
[For additional personnel, refer to the *Appendix*.]

HEAD OF ENGINEERING
Keith R. Carver
Phone: (413)545-0300 Fax: (413)545-0724

ENGINEERING COLLEGE ADDRESS
College of Engineering
University of Massachusetts
125 Marston Hall
Amherst, MA 01003
Phone: (413)545-0300 Fax: (413)545-0724

ENROLLMENTS—BY CLASS
[Numbers are baccalaureate enrollments for the fall 1991 term, unless otherwise footnoted.]
1st-year students/Freshmen 361
2nd-year students/Sophomores 331
3rd-year students/Juniors 399
4th-year students/Seniors 483 (1)
Total 1,574

Notes: (1) Includes post-graduates and non-classified students.

NUMBER OF DEGREES AWARDED—BY PROGRAM
[Numbers are engineering baccalaureate degrees awarded during the 1991-1992 academic year, unless otherwise footnoted. For full details about each engineering program, refer to the *Tables of Degree Programs*.]
Chemical Engineering 22
Civil Engineering 74
Electrical & Computer Engineering 72
Industrial Engineering & Operations Research 38
Mechanical Engineering 91
Total 297

PERCENTAGE OF DEGREES AWARDED—BY CATEGORY
[Percentages are of all engineering baccalaureate degrees awarded during the 1991-1992 academic year, unless otherwise footnoted.]
African Americans 2.7%
Asian/Pacific Island Americans 5.4%
Hispanic Americans 3.0%
Native Americans .... — %
Foreign Citizens 5.4%
All Others 83.5%
Women 18.5%
Persons with disabilities .... — %(A)
Students over 25 years of age .... — %(A)

Notes: (A) Data not available.

GRADUATE ENROLLMENTS & DEGREES AWARDED
Master's enrollment 359
Master's degrees awarded 159
Doctoral enrollment 279
Doctoral degrees awarded 42

ENGINEERING STUDENT DATA
Applicants to the engineering college
Number of applicants to engineering college 1,460
Percent offered admission 89.5%

Matriculated engineering students
Percentage in top quartile (25%) of High School class 50.0
Average SAT scores: Math—599, Verbal—476, Combined—1075

FULL- & PART-TIME FACULTY—BY DEPARTMENT
[Figures are the head count of full-time faculty and the full-time equivalent (FTE) of part-time faculty for each engineering department or equivalent.]

Department	Full	Part
Chemical Eng	13	1.4
Civil Eng	18	—
Electrical Eng	32	—
Industrial Eng	6	1.3
Mechanical Eng	23	—

COLLEGE DESCRIPTION
The College offers a four-year undergraduate program in engineering - a freshmen curriculum common to all engineering majors, followed by three years of specialized training in one of the six majors, with approximately 20% of the courses in liberal arts. A co-op program is available which consists of one 8-month co-op followed by an optional summer co-op. Co-op students take between 5 and 5 1/2 years to earn their degrees. This past Spring the Gunness Engineering Student Center was completed; it provides students a pleasant atmosphere in which to study and to hold meetings.

DEGREE PROGRAMS DESCRIPTION
The following designates the areas of strength in the College's five programs: Chemical - polymer engineering, catalysis & reaction engineering, geotechnical engineering, transportation, structural engineering & mechanics; Electrical & Computer Engineering - microwave electronics, controls & communication systems, computer engineering; Industrial Engineering & Operations Research - human factors, operations research, and production; Mechanical Engineering - manufacturing, materials, thermo-fluids, design controls.

DUAL DEGREE PROGRAMS
Engineering baccalaureate graduates with dual degrees 5

TRANSFER INFORMATION
Residency Requirements: Students must complete at least 45 graduation credits while at residence at the University.

Transfer via Articulation Agreements
Admission to engineering: At the end of the sophomore year
Requirements: It is recommended that transfer applicants have completed courses in general physics, chemistry, calculus, and computer science.

Transfer without Articulation Agreements
Admission to engineering: At any time
Requirements: It is recommended that transfer applicants have completed courses in general physics, chemistry, calculus, and computer science.

GRADUATION REQUIREMENTS
2.0 in all major courses and 2.0 overall grade point average required for graduation.

■ STUDENT PROGRAMS

PROFESSIONAL AND HONORARY SOCIETIES
[For key to acronyms, see *Introduction*.]
Professional Societies: AIChE, ASCE, ASME, IEEE, IIE
Honorary Societies: Alpha Pi Mu, Chi Epsilon, Eta Kappa Nu, Pi Tau Sigma, Tau Beta Pi

SUPPORT PROGRAMS
Student Chapter Organizations: Society of Women Engineers, National Society of Black Engineers, Society of Hispanic Professional Engineers

Minority Engineering Program
For: Ethnic Minorities Available: Year round
Offered: Prematriculation, Freshman, Sophomore, Junior, Senior, Graduate level

Women in Engineering Program
For: Women Available: Year round
Offered: Prematriculation, Freshman, Sophomore, Junior, Senior, Graduate level

Other Engineering Support Programs: A summer orientation program for all freshmen and transfer students is offered by the New Students Program and the College. The following offices provide academic and/or counseling services: Disability Service Office, Mather Career Center (Career Planning Placement and Co-op Education Programs), Counseling and Academic Development Center, and Veteran's Assistance and Counseling Services.

122 University of Massachusetts, Dartmouth

■ INSTITUTION PROFILE
HEAD OF THE INSTITUTION
Joseph C. Deck, *Chancellor*
Phone: (508)999-8004 Fax: (508)999-8860

GENERAL INFORMATION
[All Students—Fall 1991]
Undergraduate enrollment 5,049
Graduate enrollment 223
Total institution enrollment 5,272

Type of institution: Public **Calendar:** Semesters
Nearest city: New Bedford **Population:** 95,000
Miles from main campus: 9 **Setting:** Suburban
Types of engineering degrees: Engineering & Technology
Other degree-granting colleges: Arts & Sciences, Nursing, Visual & Performing Arts, Business & Industry

■ ENGINEERING ADMISSIONS
ADMISSIONS OFFICE CONTACT
Raymond Barrows
University of Massachusetts, Dartmouth
Old Westport Road
North Dartmouth, MA 02747-2300
Phone: (508)999-8605 Fax: (508)999-8901

ENGINEERING COLLEGE ADMISSIONS INFORMATION
Entrance Requirements: High School courses—English (4 years), Foreign language (2 years), Social Science (2 years), Mathematics (including trigonometry) (3.5 years)
Requirements for foreign students: TOEFL

DEPARTMENTAL ADMISSIONS INFORMATION
Admission to the engineering department: At the time of admission to the institution.

■ FINANCIAL INFORMATION
ESTIMATED EXPENSES (ACADEMIC YEAR)
[Expenses are for the 1992-93 nine-month academic year.]

State Residents	Undergraduate	Graduate
Tuition and fees	$ 3,453	$ 3,808
College room and board	$ 4,300	$ 4,300
Books and supplies	$ 500	$ 500
Other expenses	$ —(B)	$ —(B)
Total estimated expenses	**$ 8,253**	**$ 8,608**

Out-of-State Residents	Undergraduate	Graduate
Tuition and fees	$ 8,339	$ 8,339
College room and board	$ 4,300	$ 4,300
Books and supplies	$ 500	$ 500
Other expenses	$ —(B)	$ —(A)
Total estimated expenses	**$13,139**	**$13,139**

Notes: (A) Data not available. (B) Data not applicable.

FINANCIAL AID OFFICE CONTACT
Gerald S. Coutinho, *Director*
University of Massachusetts, Dartmouth
Old Westport Road
North Dartmouth, MA 02747-2300
Phone: (508)999-8632 Fax: (508)999-8901

GENERAL FINANCIAL AID INFORMATION
Forms accepted/required: FAF
Additional financial aid information: The usual range of financial aid is available, through federal, state and university sources, as described in the University's Admissions Bulletin.

■ ENGINEERING COLLEGE INFORMATION
[For additional personnel, refer to the *Appendix*.]
HEAD OF ENGINEERING
L. Bryce Andersen
Phone: (508)999-8539 Fax: (508)999-8485

ENGINEERING COLLEGE ADDRESS
College of Engineering
University of Massachusetts, Dartmouth
Old Westport Road
North Dartmouth, MA 02747-2300
Phone: (508)999-8539 Fax: (508)999-8485

ENROLLMENTS—BY CLASS
[Numbers are baccalaureate enrollments for the fall 1991 term, unless otherwise footnoted.]
1st-year students/Freshmen 157
2nd-year students/Sophomores 122
3rd-year students/Juniors 114
4th-year students/Seniors 146
Total 539

NUMBER OF DEGREES AWARDED—BY PROGRAM
[Numbers are engineering baccalaureate degrees awarded during the 1991-1992 academic year, unless otherwise footnoted. For full details about each engineering program, refer to the *Tables of Degree Programs*.]
Civil Engineering 25
Electrical & Computer Engineering 62
Mechanical Engineering 22
Total 109

PERCENTAGE OF DEGREES AWARDED—BY CATEGORY
[Percentages are of all engineering baccalaureate degrees awarded during the 1991-1992 academic year, unless otherwise footnoted.]
African Americans .... — % (A)
Asian/Pacific Island Americans .... — % (A)
Hispanic Americans .... — % (A)
Native Americans .... — % (A)
Foreign Citizens .... — % (A)
All Others .... 100 %
Women .... — % (A)
Persons with disabilities .... — % (A)
Students over 25 years of age .... — % (A)
Notes: (A) Data not available.

GRADUATE ENROLLMENTS & DEGREES AWARDED
Master's enrollment 54
Master's degrees awarded 15
Doctoral enrollment .... —(B)
Doctoral degrees awarded .... —(B)
Notes: (B) Data not applicable.

ENGINEERING STUDENT DATA
Applicants to the engineering college
Number of applicants to engineering college 360
Percent offered admission 65.8%

Matriculated engineering students
Percentage in top quartile (25%) of High School class .... —(A)
Average SAT scores: Math—446, Verbal—574, Combined—1020
Notes: (A) Data not available.

FULL- & PART-TIME FACULTY—BY DEPARTMENT
[Figures are the head count of full-time faculty and the full-time equivalent (FTE) of part-time faculty for each engineering department or equivalent.]

Department	Full	Part
Civil Eng	7	—
Electrical & Computer Eng	17	0.8
Mechanical Eng	10	0.4

COLLEGE DESCRIPTION
The College of Engineering emphasizes undergraduate education in the setting of a small yet diversified university. The modern 700 acre suburban campus includes residence halls for 2200 students. Engineering laboratory facilities are modern and equipment is state-of-the-art. Engineering freshman need not declare a major at the time of admissions. Any of the four majors may be selected at any time during the freshman year. Two majors in engineering technology are also available for students who prefer more applications-oriented programs. Freshmen with strong high school preparation can complete the B.S. program in 4 years. For a variety of reasons some students choose to extend their programs over 4 1/2 years or 5 years. Such an extension provides more time for in-depth study, part-time employment or active participation in student organizations or athletics.

DEGREE PROGRAMS DESCRIPTION
The B.S. programs prepare the graduate for entry-level professional positions or for graduate study. The M.S. in electrical engineering provides a variety of advanced courses for students who qualify for graduate study.

TRANSFER INFORMATION
Transfer via Articulation Agreements
Admission to engineering: At any time
Transfer without Articulation Agreements
Admission to engineering: At any time

GRADUATION REQUIREMENTS
B.S. programs require 131 to 134 semester credits, depending on major.

■ STUDENT PROGRAMS
PROFESSIONAL AND HONORARY SOCIETIES
[For key to acronyms, see *Introduction*.]
Professional Societies: ASCE, ASME, IEEE, NSPE, SAE, SME
Honorary Societies: Eta Kappa Nu, Tau Alpha Pi

SUPPORT PROGRAMS
Student Chapter Organizations: Society of Women Engineers

College Now
For: Women & Ethnic Minorities **Available:** Academic year
Offered: Freshman, Sophomore, Junior, Senior

Start Program
For: Women & Ethnic Minorities **Available:** Academic year
Offered: Freshman

Other Engineering Support Programs: Orientations are held for freshmen and transfer students. An optional summer course is available to strengthen mathematics and physics background for freshmen. Tutoring is available for all courses.

123 University of Massachusetts, Lowell

■ INSTITUTION PROFILE
HEAD OF THE INSTITUTION
William Hogan
Phone: (508)934-2201

GENERAL INFORMATION
[All Students—Fall 1991]
Undergraduate enrollment 3,016
Graduate enrollment 1,173
Total institution enrollment **4,189**
Type of institution: Public **Calendar:** Semesters
Nearest city: Lowell, Massachusetts **Population:** 70,000
Miles from main campus: 1/2 **Setting:** Urban
Types of engineering degrees: Engineering & Technology
Other degree-granting colleges: Arts & Sciences, Education, Health Professions, Management Science, Fine Arts

■ ENGINEERING ADMISSIONS
ADMISSIONS OFFICE CONTACT
Sharon L. Quigley
University of Massachusetts, Lowell
One University Ave.
Lowell, MA 01854
Phone: (508)934-2570 Fax: (508)452-1445

ENGINEERING COLLEGE ADMISSIONS INFORMATION
Entrance Requirements: SAT (Score: 900), High School courses—English (4 years), Mathematics (3.5 years), Social Science/History (3 years), Natural/Physical Science (with labs) (2 years)
Entrance Recommendations: ACT (Score: 18), High School courses—Calculus (1 year), Foreign language (1 year)
Requirements for foreign students: TOEFL (Score: 550)

DEPARTMENTAL ADMISSIONS INFORMATION
Admission to the engineering department: At the end of the first year.
Additional information: A cumulative average of 2.5 or greater guarantees entry to department of choice. Less than 2.5 requires approval of Department Head.

■ FINANCIAL INFORMATION
ESTIMATED EXPENSES (ACADEMIC YEAR)
[Expenses are for the 1992-93 nine-month academic year.]

State Residents	Undergraduate	Graduate
Tuition and fees	$ 4,337	$ 4,691
College room and board	$ 3,862	$ 3,862
Books and supplies	$ 500	$ 500
Other expenses	$ 500	$ 500
Total estimated expenses	**$ 9,199**	**$ 9,553**

Out-of-State Residents	Undergraduate	Graduate
Tuition and fees	$ 9,059	$ 9,059
College room and board	$ 3,862	$ 3,862
Books and supplies	$ 500	$ 500
Other expenses	$ 500	$ 500
Total estimated expenses	**$13,921**	**$13,921**

FINANCIAL AID OFFICE CONTACT
Walter Costello, *Director of Financial Aid*
University of Massachusetts, Lowell
McGauvran Student Center South Campus
Lowell, MA 01854
Phone: (508)934-4223

GENERAL FINANCIAL AID INFORMATION
Forms accepted/required: CSS-FAF, IRS, Institutional

■ ENGINEERING COLLEGE INFORMATION
[For additional personnel, refer to the *Appendix*.]
HEAD OF ENGINEERING
Aldo Crugnola
Phone: (508)934-2575 Fax: (508)452-1445

ENGINEERING COLLEGE ADDRESS
James B. Francis College of Engineering
University of Massachusetts, Lowell
One University Avenue
Lowell, MA 01854
Phone: (508)934-2570 Fax: (508)452-1445

ENROLLMENTS—BY CLASS
[Numbers are baccalaureate enrollments for the fall 1991 term, unless otherwise footnoted.]
1st-year students/Freshmen 498
2nd-year students/Sophomores 380
3rd-year students/Juniors 420
4th-year students/Seniors 450
Total **1,748**

NUMBER OF DEGREES AWARDED—BY PROGRAM
[Numbers are engineering baccalaureate degrees awarded during the 1991-1992 academic year, unless otherwise footnoted. For full details about each engineering program, refer to the *Tables of Degree Programs*.]
Chemical Engineering 10
Civil Engineering 82
Electrical Engineering 114
Industrial Technology 7
Mechanical Engineering 74
Nuclear Engineering 74
Plastics Engineering 36
Work Environment .... –
Total **397**

PERCENTAGE OF DEGREES AWARDED—BY CATEGORY
[Percentages are of all engineering baccalaureate degrees awarded during the 1991-1992 academic year, unless otherwise footnoted.]
African Americans .... – %
Asian/Pacific Island Americans 7.4%
Hispanic Americans 0.1%
Native Americans .... – %
Foreign Citizens 4.6%
All Others 87.9%
Women 15.5%
Persons with disabilities .... – % (A)
Students over 25 years of age .... – % (A)

Notes: (A) Data not available.

GRADUATE ENROLLMENTS & DEGREES AWARDED
Master's enrollment 1,052
Master's degrees awarded 234
Doctoral enrollment 121
Doctoral degrees awarded 17

ENGINEERING STUDENT DATA
Applicants to the engineering college
Number of applicants to engineering college 962
Percent offered admission 82.4%

Matriculated engineering students
Percentage in top quartile (25%) of High School class 28.0
Average SAT scores: Math—560, Verbal—500, Combined—1060

FULL- & PART-TIME FACULTY—BY DEPARTMENT
[Figures are the head count of full-time faculty and the full-time equivalent (FTE) of part-time faculty for each engineering department or equivalent.]

Department	Full	Part
Chemical & Nuclear Eng	14	–
Civil Eng	15	4.0
Electrical Eng	35	3.0
Industrial Eng	4	1.0
Mechanical Eng	20	0.5
Plastics Eng	14	3.5
Work Environment	9	–
Eng Tech	6	20.0

COLLEGE DESCRIPTION
The College offers a Track II program - an alternative 5 year program which allows a phased introduction to the rigorous engineering curriculum. There is an optional cooperative education program. Also, there are dual degree programs with other colleges (3+2) as well as articulation agreements with community colleges that result in (2+2) transfer programs. For talented students, there is a combined BS/MS in engineering program (5 year plan).

DEGREE PROGRAMS DESCRIPTION
The Chemical & Nuclear Engineering Department offers a masters level program in Materials. Also, PhD options with the Physics Department in Energy Engineering. The Civil Engineering Department offers a masters level program in Environmental Studies and a PhD option in Environmental Studies with the Department of Chemistry. The Electrical Engineering Department offers masters level programs in both Computer and Systems Engineering, and a Doctor of Engineering program in Electrical Engineering. The Mechanical Engineering Department offers a Doctor of Engineering program in Mechanical Engineering and a PhD option in Engineering Mechanics in conjunction with the Physics Department. The Plastics Engineering Department offers a Doctor of Engineering program in Plastics Engineering and a PhD in Polymer Science/Plastics Engineering is offered.

DUAL DEGREE PROGRAMS
Engineering baccalaureate graduates with dual degrees 1
Enrollment requirements: Specific course grades and a cum equal to or greater than 2.5

TRANSFER INFORMATION
Residency Requirements: A minimum of 30 credits in residence.

Transfer via Articulation Agreements
Admission to engineering: At any time
Engineering graduates transferred from   .. 4-yr: 50   2-yr: 54
Requirements: A cum of 2.5 or greater from the transferring institution.

Transfer without Articulation Agreements
Admission to engineering: At any time
Engineering graduates transferred from   .. 4-yr: 41   2-yr: –
Requirements: Specific grades in selected courses at the transferring institution along with a minimum cum equal to 2.5.

GRADUATION REQUIREMENTS
A cumulative GPA of 2.00 in the major, as well as an overall cumulative GPA of 2.00. Also, a minimum 30 semester hours residency requirement and 120 credit hours total.

■ STUDENT PROGRAMS
PROFESSIONAL AND HONORARY SOCIETIES
[For key to acronyms, see *Introduction*.]
Professional Societies: ACS, AIChE, ASCE, ASME, IEEE, SAE, SPE*, TAPPI
Honorary Societies: Chi Epsilon, Eta Kappa Nu, Omega Chi Epsilon, Pi Tau Sigma, Tau Beta Pi

SUPPORT PROGRAMS
Student Chapter Organizations: Society of Women Engineers, National Society of Black Engineers

Equal Opportunity Program
For: Ethnic Minorities **Available:** Academic year
Offered: Freshman, Sophomore, Junior, Senior

Office of Minorities Services
For: Women & Ethnic Minorities **Available:** Academic year
Offered: Freshman, Sophomore, Junior, Senior, Graduate level

Other Engineering Support Programs: A large peer tutoring and advising program is available through our Centers for Learning and Student Services (CLASS) oriented to helping freshman and transfer students achieve academic success. CLASS also provides counseling, and study skill programs.

124 Massachusetts Institute of Technology

INSTITUTION PROFILE

HEAD OF THE INSTITUTION
Charles M. Vest
Phone: (617)253-0148   Fax: (617)253-3124

GENERAL INFORMATION
[All Students—Fall 1991]
Undergraduate enrollment 4,325
Graduate enrollment 5,216
Total institution enrollment 9,541

Type of institution: Private   Calendar: Semesters
Nearest city: Boston   Population: 574,283
Miles from main campus: 1   Setting: Urban
Types of engineering degrees: Engineering
Other degree-granting colleges: Humanities & Social Sciences, Science, Sloan School of Management, Architecture & Planning, Health Sciences & Technology

ENGINEERING ADMISSIONS

ADMISSIONS OFFICE CONTACT
Michael C. Behnke
Massachusetts Institute of Technology
Director of Admissions
77 Massachusetts Avenue, Room 3-108
Cambridge, MA 02139
Phone: (617)258-5515   Fax: (617)253-8000

ENGINEERING COLLEGE ADMISSIONS INFORMATION
Entrance Requirements: SAT, ACT, Competitive SAT Scores: 650 on SAT-M, 600 on SAT-V; Competitive ACT Score: 30 or higher
Entrance Recommendations: High School courses—Mathematics (through Trigonometry) (4 years), Science (4 years), English (4 years), Foreign Language (2+ yrs. history, soc. studies) (2 years), Plus 3 achievement tests: Math, Science, English or History
Requirements for foreign students: TOEFL (Score: 570); Standardized tests plus math and science achievement tests.

DEPARTMENTAL ADMISSIONS INFORMATION
Admission to the engineering department: At the end of the first year.

FINANCIAL INFORMATION

ESTIMATED EXPENSES (ACADEMIC YEAR)
[Expenses are for the 1992-93 nine-month academic year.]

All Students	Undergraduate	Graduate
Tuition and fees	$16,900	$16,900
College room and board	$5,330 [1]	$9,000 [1]
Books and supplies	$620	$650
Other expenses	$1,400	$3,125
Total estimated expenses	**$24,250**	**$29,675**

Notes: (1) Average

FINANCIAL AID OFFICE CONTACT
Stanley G. Hudson, *Director of Student Financial Aid*
Massachusetts Institute of Technology
77 Massachusetts Avenue
Room 5-119
Cambridge, MA 02139
Phone: (617)253-4971   Fax: (617)258-8301

GENERAL FINANCIAL AID INFORMATION
Forms accepted/required: IRS, Institutional, CSS-FAF Packet which includes: Free application for Federal Stud. Aid, and the Financial Aid Form (FAF)
Additional financial aid information: Financial aid at MIT typically consists of grants, long-term student loans, and campus employment. Grants are outright scholarships that do not have to be repaid. Loans come from three sources: the Perkins Loan Program, the Guaranteed Student Loan Program, and MIT's Technology Loan Fund. These loans carry below-market interest rates and don't require repayment until after graduation. Employment is available on campus, and you can often find a position within your area of academic interest.

ENGINEERING COLLEGE INFORMATION
[For additional personnel, refer to the *Appendix*.]

HEAD OF ENGINEERING
Joel Moses
Phone: (617)253-3292   Fax: (617)253-8549
EMail: Moses@LArch.Lcs.MIT.Edu

ENGINEERING COLLEGE ADDRESS
Office of the Dean of Engineering
Massachusetts Institute of Technology
77 Massachusetts Avenue, Rm. 1-207
Cambridge, MA 02139
Phone: (617)253-3291   Fax: (617)253-8549

ENROLLMENTS—BY CLASS
[Numbers are baccalaureate enrollments for the fall 1991 term, unless otherwise footnoted.]
1st-year students/Freshmen .... 0 [1]
2nd-year students/Sophomores 678
3rd-year students/Juniors 687
4th-year students/Seniors 666
Total 2,031
Notes: (1) Students select a major in the sophomore year.

NUMBER OF DEGREES AWARDED—BY PROGRAM
[Numbers are engineering baccalaureate degrees awarded during the 1991-1992 academic year, unless otherwise footnoted. For full details about each engineering program, refer to the *Tables of Degree Programs*.]
Aeronautics & Astronautics 66
Chemical Engineering 52
Civil & Environmental Engineering 40
Electrical Engineering & Computer Science 245
Materials Science & Engineering 43
Mechanical Engineering 138
Nuclear Engineering 7
Ocean Engineering 5
Total 596

PERCENTAGE OF DEGREES AWARDED—BY CATEGORY
[Percentages are of all engineering baccalaureate degrees awarded during the 1991-1992 academic year, unless otherwise footnoted.]
African Americans 7.7%
Asian/Pacific Island Americans 19.1%
Hispanic Americans 7.2%
Native Americans 0.3%
Foreign Citizens 12.1%
All Others 53.6%
Women 30.2%
Persons with disabilities .... – % [A]
Students over 25 years of age .... – % [A]
Notes: (A) Data not available.

GRADUATE ENROLLMENTS & DEGREES AWARDED
Master's enrollment 1,109
Master's degrees awarded 606
Doctoral enrollment 1,114
Doctoral degrees awarded 239

ENGINEERING STUDENT DATA
Applicants to the engineering college
Number of applicants to engineering college ...... 6,481 [1]
Percent offered admission 31.0%

Matriculated engineering students
Percentage in top quartile (25%) of High School class ... 100.0
Average SAT scores: Math—735, Verbal—618, Combined—1353
Average ACT scores: Math—32, Composite—31

Notes: (1) We do not admit students to a major. The data are for all freshmen, 60% of whom will become engineering majors.

FULL- & PART-TIME FACULTY—BY DEPARTMENT
[Figures are the head count of full-time faculty and the full-time equivalent (FTE) of part-time faculty for each engineering department or equivalent.]

Department	Full	Part
Chemical Eng	32	–
Civil & Environmental Eng	40	–
Electrical Eng & Computer Science	115	–
Mechanical Eng	60	–
Ocean Eng	22	–
Aeronautics & Astronautics	36	–
Materials Science & Eng	36	–
Nuclear Eng	23	–

COLLEGE DESCRIPTION
The MIT School of Engineering offers undergraduate and graduate degrees in Aeronautics & Astronautics, Civil Engineering, Chemical Engineering, Electrical Engineering and Computer Science, Materials Science and Engineering, Mechanical Engineering, Nuclear Engineering, and Ocean Engineering. There is an Engineering Internship Program which combines traditional on-campus academic programs with off-campus work experience in industry and government. There are graduate programs available in interdisciplinary fields such as Technology and Policy, Biomedical Engineering, Environmental Studies, and Management of Technology.

DEGREE PROGRAMS DESCRIPTION
MIT undergraduate programs are designed to assist students in developing the capabilities, understanding, and maturity necessary in order to meet the challenges of today's society. MIT undergraduate students are encouraged to develop both a basic knowledge and a continuing interest in a particular field of study. The Undergraduate Research Opportunities Program which allows undergraduates to work with faculty on ongoing research is an exciting component of the undergraduate curricula. There is, in addition, an Engineering Internship Program which combines on-campus education with off-campus work experience. Each January there is a short Independent Activities Period offering workshops, short courses, seminars, and field trips. In addition to S.M., Ph.D., and Sc.D. degrees, MIT graduate students may pursue professional Engineer's degrees.

DUAL DEGREE PROGRAMS
Engineering baccalaureate graduates with dual degrees 1
Enrollment requirements: Transfer applicants must provide standardized test score data (see freshman section).

TRANSFER INFORMATION
Residency Requirements: Must attend the Institute at least 3 academic terms including the term of graduation.
Transfer without Articulation Agreements
Admission to engineering: Students who have completed a minimum of one year and a maximum of 2.5 years with high standing at a recognized college or engineering school may be considered for transfer admission.
Engineering graduates transferred from ... 4-yr: –   2-yr: –
Requirements: Transfer applicants must provide standardized test score data (see freshman section).

GRADUATION REQUIREMENTS
The Institute expects undergraduate students to complete the requirements for a bachelor's degree within four years. The typical undergraduate would carry a course load of 45 to 48 units per semester. An undergraduate may carry a minimum of 36 units of credit with a term rating of 3.0 or more (on a scale of 5.0). Graduate students are considered to be making satisfactory progress as long as their cumulative grade point average exceeds 4.0 and the number of terms of enrollment does not exceed the designated limit (five for an M.S. candidate, ten for a Ph.D. or Sc.D. candidate).

STUDENT PROGRAMS

PROFESSIONAL AND HONORARY SOCIETIES
[For key to acronyms, see *Introduction*.]
Professional Societies: AIAA, AIChE, ASCE, ASEM, ASME, IEEE, NSPE, SAE, SME
Honorary Societies: Chi Epsilon, Eta Kappa Nu, Pi Tau Sigma, Sigma Gamma Tau, Tau Beta Pi

SUPPORT PROGRAMS
Student Chapter Organizations: Society of Women Engineers, National Society of Black Engineers, Society of Hispanic Professional Engineers, American Indian Science & Engineering Society

Program XL
For: Women & Ethnic Minorities   **Available:** Academic year
Offered: Freshman

Project Interphase
For: Ethnic Minorities   **Available:** Summer only
Offered: Prematriculation

Second Summer Program
For: Ethnic Minorities   **Available:** Summer only
Offered: Freshman

Tutorial Services
For: Women & Ethnic Minorities   **Available:** Academic year
Offered: Freshman, Sophomore, Junior, Senior

Other Engineering Support Programs: Each student at MIT has a faculty advisor. The Office of the Dean for Undergrad Education & Student Affairs offers counseling services and additional services for international students, minorities and handicapped students. Faculty & graduate residents and tutors are available in Institute houses. The Religious Counselors, Student Financial Aid and Career Services provide additional resources.

125 McGill University

■ INSTITUTION PROFILE

HEAD OF THE INSTITUTION
David L. Johnston
Phone: (514)398-4180   Fax: (514)398-7379

GENERAL INFORMATION
[All Students—Fall 1991]
Undergraduate enrollment 14,181
Graduate enrollment 5,364
Total institution enrollment 19,545

Type of institution: Public   Calendar: Semesters
Location: Montréal   Population: 3,000,000
Setting: Urban
Types of engineering degrees: Engineering

■ ENGINEERING ADMISSIONS

ADMISSIONS OFFICE CONTACT
Mariela Johansen
McGill University
845 Sherbrooke St.W.
James Administration Bldg.
Montréal, PQ H3A-2T5
Phone: (514)398-3672   Fax: (514)398-4193

ENGINEERING COLLEGE ADMISSIONS INFORMATION
Entrance Requirements: SAT (Score: 1100), High School courses—Mathematics (3 years), Physics (2 years), Chemistry (2 years)
Requirements for out-of-state residents: Residents from Quebec have completed two years of junior college and judged on the basis of courses completed. High school applicants from outside the province are judged on the basis of high school grades with emphasis on math, phys & chem marks.
Requirements for foreign students: Government requires statement of financial independence prior to granting of student visa.

DEPARTMENTAL ADMISSIONS INFORMATION
Admission to the engineering department: At the time of admission to the institution.
Additional information: Programs in electrical engineering, computer engineering and mechanical engineering are limited enrollment programs. As a consequence the required averages in math, physics and chemistry are higher than in other programs.

■ FINANCIAL INFORMATION

ESTIMATED EXPENSES (ACADEMIC YEAR)
[Expenses are for the 1992-93 nine-month academic year.]

All Students	Undergraduate	Graduate
Tuition and fees	$ 7,800 [1]	$ 2,000 [1]
College room and board	$ 6,000	$ 6,000
Books and supplies	$ 600	$ 600
Other expenses	$ 300 [2]	— [B]
Total estimated expenses	$14,700	$ 8,600

Notes: (B) Data not applicable. (1) Based upon 30 credit course load plus student services & student society fees & course material charges.
(2) Compulsory Health & Accident Insurance for International Students.

FINANCIAL AID OFFICE CONTACT
Judy Stymest, *Director of Student Aid Office*
McGill University
3637 Peel Street
Montréal, PQ H3A-1X1
Phone: (514)398-6013

GENERAL FINANCIAL AID INFORMATION
Forms accepted/required: FAF, FFS, Institutional
Additional financial aid information: Need based bursaries

■ ENGINEERING COLLEGE INFORMATION

[For additional personnel, refer to the *Appendix*.]

HEAD OF ENGINEERING
Pierre R. Bélanger
Phone: (514)398-7251   Fax: (514)398-7379
EMail: pierre@eng1.lan.McGill.CA

ENGINEERING COLLEGE ADDRESS
Faculty of Engineering
McGill University
817 Sherbrooke St. W.
Macdonald Engineering Building
Montréal, PQ H3A-2K6
Phone: (514)398-7257   Fax: (514)398-7379

ENROLLMENTS—BY CLASS
[Numbers are baccalaureate enrollments for the fall 1991 term, unless otherwise footnoted.]
1st-year students/Freshmen .... — [B]
2nd-year students/Sophomores .... 466 [1]
3rd-year students/Juniors 786
4th-year students/Seniors 418
Total 1,670

Notes: (B) Data not applicable. (1) Three and one half year program following two year junior college.

NUMBER OF DEGREES AWARDED—BY PROGRAM
[Numbers are engineering baccalaureate degrees awarded during the 1991-1992 academic year, unless otherwise footnoted. For full details about each engineering program, refer to the *Tables of Degree Programs*.]
Chemical Engineering 44
Civil Engineering & Applied Mechanics 42
Electronics Engineering 100
Mechanical Engineering 98
Metallurgical Engineering 11
Mining Engineering 2
Total 297

PERCENTAGE OF DEGREES AWARDED—BY CATEGORY
[Percentages are of all engineering baccalaureate degrees awarded during the 1991-1992 academic year, unless otherwise footnoted.]
African Canadians .... — % [A]
Asian/Pacific Island Canadians .... — % [A]
Hispanic Canadians .... — % [A]
Native Canadians .... — % [A]
Foreign Citizens .... 4.5 %
All Others .... 95.5 %
Women .... 19.6 %
Persons with disabilities .... — % [A]
Students over 25 years of age .... — % [A]

Notes: (A) Data not available.

■ GRADUATE ENROLLMENTS & DEGREES AWARDED

Master's enrollment 392
Master's degrees awarded .... 139 [2]
Doctoral enrollment 264
Doctoral degrees awarded .... 47 [2]

Notes: (2) Graduates in 1991-92 academic session.

■ ENGINEERING STUDENT DATA

Applicants to the engineering college
Number of applicants to engineering college 1,607
Percent offered admission 48.2%

Matriculated engineering students
Percentage in top quartile (25%) of High School class .... — [A]

Notes: (A) Data not available.

■ FULL- & PART-TIME FACULTY—BY DEPARTMENT

[Figures are the head count of full-time faculty and the full-time equivalent (FTE) of part-time faculty for each engineering department or equivalent.]

Department	Full	Part
Chemical Eng	16.5	3.6
Civil Eng	16.4	4.1
Electrical Eng	34	9.3
Mechanical Eng	21.5	3.4
Mining And Metallurgical Eng	16	2.3

COLLEGE DESCRIPTION
McGill has been rated as the top university in Canada in surveys conducted by Mcleans Magazine.

TRANSFER INFORMATION
Residency Requirements: Maximum of 45 credits granted for previous studies. Students must complete remaining credits, typically 60 - 63, at McGill.

Transfer without Articulation Agreements
Admission to engineering: At any time
Requirements: B average in relevant courses. Minim CGPA of 2.7 is required. In limited enrollment programs CGPA requirement is higher; 2.8 in mechanical and 3.0 in electrical and computer engineering.

GRADUATION REQUIREMENTS
Completion of required courses and technical electives with a minimum CGPA of 2.0

■ STUDENT PROGRAMS

PROFESSIONAL AND HONORARY SOCIETIES
[For key to acronyms, see *Introduction*.]
Professional Societies: AIChE, AIME, CIM, CSCE, CSME, IEEE, SAE
Honorary Societies: Scarlet Key

SUPPORT PROGRAMS
Student Chapter Organizations: POWE (Power of Women in Engineering), WISE (Women in Science and Engineering)

126 Memphis State University

INSTITUTION PROFILE

HEAD OF THE INSTITUTION
V. Lane Rawlins
Phone: (901)678-2234   Fax: (901)678-2000

GENERAL INFORMATION
[All Students—Fall 1991]
Undergraduate enrollment 15,767
Graduate enrollment 3,451
Total institution enrollment 19,218
Type of institution: Public   **Calendar:** Semesters
Location: Memphis, Tennessee   **Population:** 610,337
Setting: Urban
Types of engineering degrees: Engineering & Technology
Other degree-granting colleges: Arts & Sciences, Business Administration, Communication & Journalism, Education, Fine & Performing Arts, Law, Nursing, University College

ENGINEERING ADMISSIONS

ADMISSIONS OFFICE CONTACT
Carol Ferguson
Memphis State University
Corner of Central and Zach Curlin
Engineering Science Room 201
Memphis, TN 38152
Phone: (901)678-2171   Fax: (901)678-4180

ENGINEERING COLLEGE ADMISSIONS INFORMATION
Entrance Requirements: ACT (Score: 18), High School courses—English (4 years), Science (4 years), Math (4 years)
Entrance Recommendations: ACT (Score: 18), High School courses—Trigonometry (1 year), Calculus (1 year), Physics (1 year), Chemistry (1 year)
Requirements for foreign students: TOEFL (Score: 550); F-1 and J-1 student Visa Health Certificate

DEPARTMENTAL ADMISSIONS INFORMATION
Admission to the engineering department: At the time of admission to the institution.

FINANCIAL INFORMATION

ESTIMATED EXPENSES (ACADEMIC YEAR)
[Expenses are for the 1992-93 nine-month academic year.]

State Residents	Undergraduate	Graduate
Tuition and fees	$ 828	$ 991
College room and board	$ 700	$ 700
Books and supplies	$ 250	$ 250
Other expenses	$ —(A)	$ —(A)
Total estimated expenses	**$ 1,778**	**$ 1,941**

Out-of-State Residents	Undergraduate	Graduate
Tuition and fees	$ 2,490	$ 2,673
College room and board	$ 700	$ 700
Books and supplies	$ 250	$ 250
Other expenses	$ —(A)	$ —(A)
Total estimated expenses	**$ 3,440**	**$ 3,623**

Notes: (A) Data not available.

GENERAL FINANCIAL AID INFORMATION
Forms accepted/required: ACT
Additional financial aid information: Pell Grants, Supplemental Educational Opportunity Grant, and Tennessee Student Assistance Award Program.

ENGINEERING COLLEGE INFORMATION

[For additional personnel, refer to the *Appendix*.]

HEAD OF ENGINEERING
John D. Ray
Phone: (901)678-2171   Fax: (901)678-4180

ENGINEERING COLLEGE ADDRESS
Herff College of Engineering
Memphis State University
Corner of Central and Zach Curlin
Engineering Science Building Room 201
Memphis, TN 38152
Phone: (901)678-2171   Fax: (901)678-4180

ENROLLMENTS—BY CLASS
[Numbers are baccalaureate enrollments for the fall 1991 term, unless otherwise footnoted.]
1st-year students/Freshmen 203
2nd-year students/Sophomores 181
3rd-year students/Juniors 263
4th-year students/Seniors 513
Total **1,160**

NUMBER OF DEGREES AWARDED—BY PROGRAM
[Numbers are engineering baccalaureate degrees awarded during the 1991-1992 academic year, unless otherwise footnoted. For full details about each engineering program, refer to the *Tables of Degree Programs*.]
Bioengineering .... –
Civil Engineering 20
Electrical Engineering 54
Mechanical Engineering 54
Total **128**

PERCENTAGE OF DEGREES AWARDED—BY CATEGORY
[Percentages are of all engineering baccalaureate degrees awarded during the 1991-1992 academic year, unless otherwise footnoted.]
African Americans 3.5%
Asian/Pacific Island Americans 3.5%
Hispanic Americans .... – %
Native Americans .... – %
Foreign Citizens .... – %(A)
All Others 93.0%
Women 18.8%
Persons with disabilities .... – %(A)
Students over 25 years of age 5.0%

Notes: (A) Data not available.

GRADUATE ENROLLMENTS & DEGREES AWARDED
Master's enrollment 188
Master's degrees awarded 43
Doctoral enrollment 37
Doctoral degrees awarded 6

ENGINEERING STUDENT DATA
Average ACT scores: Math—21.3, Composite—22.1

FULL- & PART-TIME FACULTY—BY DEPARTMENT
[Figures are the head count of full-time faculty and the full-time equivalent (FTE) of part-time faculty for each engineering department or equivalent.]

Department	Full	Part
Civil Eng	15	1.0
Electrical Eng	13	1.0
Mechanical Eng	13	1.0
Biomedical Eng	5	–

COLLEGE DESCRIPTION

The college's function is to serve the educational and research needs of the industrial community, the metropolitan area, the state and the nation. It accomplishes this function by providing: (1) undergraduate professional education in the principal fields of engineering, (2) undergraduate education in both technical and educational technology, (3) graduate education in all areas, (4) a program of continuing education for the engineering and technological practitioners of the area, (5) assistance in the solution of the industrial problems through utilization of physical facilities and the professional talents of the faculty and students, (6) a forum for the interchange of ideas and experiences among members of the industrial community through conferences, institutes and short courses and (7) an increase in the accumulation of knowledge in special fields of interest by a continuing program of study and research.

DEGREE PROGRAMS DESCRIPTION
Our College strives to give a quality education to students seeking an Engineering degree.

TRANSFER INFORMATION
Residency Requirements: Thirty of the last 33 hours must be completed on campus
Transfer via Articulation Agreements
Admission to engineering: At any time
Engineering graduates transferred from ... 4-yr: 10   2-yr: 20
Transfer without Articulation Agreements
Admission to engineering: At any time
Engineering graduates transferred from ... 4-yr: –   2-yr: –

GRADUATION REQUIREMENTS
2.0 GPA in courses in the major, 132 semester hours; 2.0 overall University GPA.

STUDENT PROGRAMS

PROFESSIONAL AND HONORARY SOCIETIES
[For key to acronyms, see *Introduction*.]
Professional Societies: ASCE, ASME, IEEE, SAE
Honorary Societies: Pi Tau Sigma, Tau Alpha Pi, Tau Beta Pi

SUPPORT PROGRAMS
Student Chapter Organizations: Society of Women Engineers, National Society of Black Engineers

Early Scholars
For: Ethnic Minorities   **Available:** Summer only
Offered: Prematriculation

Other Engineering Support Programs: Summer Orientation Programs for Incoming Freshman and Transfer students. Engineering Learning Center which provides free tutoring programs for engineering students.

127 Mercer University

■ INSTITUTION PROFILE
HEAD OF THE INSTITUTION
R. Kirby Godsey
Phone: (912)752-2500 **Fax:** (912)752-2108

GENERAL INFORMATION
[All Students—Fall 1991]
Undergraduate enrollment 3,959
Graduate enrollment 1,323
Total institution enrollment **5,282**
Type of institution: Private **Calendar:** Quarters
Location: Macon, GA **Population:** 150,000
Setting: Urban
Types of engineering degrees: Engineering
Other degree-granting colleges: Arts & Sciences, Business Administration, Law, Medicine, Pharmacy, Technology & Applied Sciences, University College

■ ENGINEERING ADMISSIONS
ADMISSIONS OFFICE CONTACT
Lea Weissenburger
Mercer University
Office of Undergraduate Admissions
1400 Coleman Avenue
Macon, GA 31207
Phone: (912)752-2650 **Fax:** (912)752-2828

ENGINEERING COLLEGE ADMISSIONS INFORMATION
Entrance Requirements: SAT (Score: 1000), High School courses—Mathematics (Algebra, Geometry, Trigonometry) (3 years), Physical sciences (Physics, Chemistry) (2 years), A nomograph of SAT scores and high school GPA is used to determine admission, therefore it is unfair to state a minimum SAT score.
Entrance Recommendations: High School courses—Calculus (1 year), Biology (1 year), Computers (1 year), English, Humanities, and Social Sciences (4 years)
Requirements for foreign students: TOEFL (Score: 550)

DEPARTMENTAL ADMISSIONS INFORMATION
Admission to the engineering department: At the time of admission to the institution.

■ FINANCIAL INFORMATION
ESTIMATED EXPENSES (ACADEMIC YEAR)
[Expenses are for the 1992-93 nine-month academic year.]

All Students	Undergraduate	Graduate
Tuition and fees	$10,287	$ –
College room and board	$ 3,900 [1]	$ –
Books and supplies	$ 800 [2]	$ –
Other expenses	$ –	$ –
Total estimated expenses	**$14,987**	**$ –**

Notes: (1) Includes cost of 7 day meal ticket. Other meal ticket programs are available at varying cost. (2) Cost will vary among students depending on professors and courses selected.

FINANCIAL AID OFFICE CONTACT
Pam Anderson, *Office Manager*
Mercer University
Office of Financial Aid
1400 Coleman Avenue
Macon, GA 31207
Phone: (912)752-2670 **Fax:** (912)752-2313

GENERAL FINANCIAL AID INFORMATION
Forms accepted/required: FAF
Additional financial aid information: College Work Study Program, Perkins Loans, Stafford Loans, Pell Grant, SEOG, SIG, Need-based Scholarships. Georgia Tuition Equalization Grant is available to Georgia residents. Tuition reductions are available to residents of Bibb and six surrounding counties.

■ ENGINEERING COLLEGE INFORMATION
[For additional personnel, refer to the *Appendix*.]
HEAD OF ENGINEERING
Carroll B. Gambrell
Phone: (912)752-2377 **Fax:** (912)752-2166

ENGINEERING COLLEGE ADDRESS
School of Engineering
Mercer University
1400 Coleman Avenue
Macon, GA 31207
Phone: (912)752-2343 **Fax:** (912)752-2166

ENROLLMENTS—BY CLASS
[Numbers are baccalaureate enrollments for the fall 1991 term, unless otherwise footnoted.]
1st-year students/Freshmen 117
2nd-year students/Sophomores 110
3rd-year students/Juniors 95
4th-year students/Seniors 139
Total **461**

NUMBER OF DEGREES AWARDED—BY PROGRAM
[Numbers are engineering baccalaureate degrees awarded during the 1991-1992 academic year, unless otherwise footnoted. For full details about each engineering program, refer to the *Tables of Degree Programs*.]
Engineering 45
Total **45**

PERCENTAGE OF DEGREES AWARDED—BY CATEGORY
[Percentages are of all engineering baccalaureate degrees awarded during the 1991-1992 academic year, unless otherwise footnoted.]
African Americans .... 14.0 %
Asian/Pacific Island Americans .... 6.0 %
Hispanic Americans .... 4.0 %
Native Americans .... 2.0 %
Foreign Citizens .... 4.0 %
All Others .... 70.0 %
Women .... 22.0 %
Persons with disabilities .... – %
Students over 25 years of age .... 1.0 %

GRADUATE ENROLLMENTS & DEGREES AWARDED
Master's enrollment 156
Master's degrees awarded 4
Doctoral enrollment .... –
Doctoral degrees awarded .... –

ENGINEERING STUDENT DATA
Average SAT scores: Math—580, Verbal—468, Combined—1048

FULL- & PART-TIME FACULTY—BY DEPARTMENT
[Figures are the head count of full-time faculty and the full-time equivalent (FTE) of part-time faculty for each engineering department or equivalent.]

Department	Full	Part
Electrical & Computer Eng	8	–
Biomedical & Environmental Eng	2	–
Computer & Information Systems	1	1.3
Eng Core	5	–
Industrial & Systems Eng	4	1.1
Mechanical & Aerospace Eng	4	2.2
Technical Communications	1	–

COLLEGE DESCRIPTION
The hallmarks of Mercer School of Engineering include: offering a broad-based engineering curriculum, placing emphasis on the liberal arts, teaching in small classes where students and professors interact, and utilizing computer technology. The School offers a four-year undergraduate program, leading to the Bachelor of Science in Engineering degree, with specializations in Biomedical, Electrical, Industrial, and Mechanical Engineering. Minors and second majors are available in liberal arts and business. Freshmen complete the freshman engineering sequence culminating in a freshman design project. During the senior year, each student completes a senior design project and formally presents the project in a written document as well as in an oral presentation to faculty and peers. The co-op program requires four quarters of work experience, and requires approximately one extra year to complete the B.S.E. degree.

DEGREE PROGRAMS DESCRIPTION
The curriculum leading to the BSE degree does not develop any of the psychomotor skills, but rather provides the preparation for development, planning, design, research, graduate work, and with certain electives, for operation, production, testing, maintenance, and management. The Biomedical Engineering specialization has the option to meet all pre-medical requirements. The graduate programs are designed to accommodate part-time students who are working full time, in the areas of Biomedical, Electrical, and Mechanical Engineering, Engineering Management, Computer and Information Systems, and Technical Management. A minimum of 48 credit hours are required for the MSE degree, including research, thesis, or special project credit. Dual degrees are articulated with liberal arts colleges including Wesleyan College and Presbyterian College.

DUAL DEGREE PROGRAMS
Engineering baccalaureate graduates with dual degrees 1
Enrollment requirements: A Dual-Degree student must be in good standing with the previous school he/she attended and have a GPA of greater than 2.0

TRANSFER INFORMATION
Residency Requirements: The last year of academic work (45 quarter hours) must be done in residence at Mercer.
Transfer via Articulation Agreements
Admission to engineering: At any time
Engineering graduates transferred from .. 4-yr: 12 2-yr: 8
Requirements: To transfer into the School of Engineering, a student must have an overall GPA of greater than 2.0 and be in good academic standing.
Transfer without Articulation Agreements
Admission to engineering: At any time
Engineering graduates transferred from .. 4-yr: 4 2-yr: –
Requirements: Transfer students must be in good standing at their previous institution and have at least a 2.0 GPA.

GRADUATION REQUIREMENTS
The B.S.E. degree requires a minimum of 204 quarter hours with a cumulative GPA of 2.0. Engineering students must earn a grade of C or better in the mathematics and physics courses listed in the first two years of the program of study of the engineering specialization. Each department has additional graduation requirements.

■ STUDENT PROGRAMS
PROFESSIONAL AND HONORARY SOCIETIES
[For key to acronyms, see *Introduction*.]
Professional Societies: ACM, ACS, ASME, BMES, IEEE, IIE, MAA, SAE, STC, Association of Mercer Engineers
Honorary Societies: Tau Beta Pi, Phi Kappa Phi, Phi Eta Sigma

SUPPORT PROGRAMS
Student Chapter Organizations: Society of Women Engineers, Organization of Black Students (OBS)
Other Engineering Support Programs: The Learning Skills Center helps interested Mercer students reach his or her full academic potential by stressing individual development in learning. The Center provides peer tutors to assist students in writing as well as in major subject areas. Many services are available on a drop-in basis, and there is no charge.

128 University of Miami

■ INSTITUTION PROFILE

HEAD OF THE INSTITUTION
Edward T. Foote
Phone: (305)284-5155 Fax: (305)284-3768

GENERAL INFORMATION
[All Students—Fall 1991]
Undergraduate enrollment 8,914
Graduate enrollment 3,276
Total institution enrollment **14,245**

Type of institution: Private	Calendar: Semesters
Nearest city: Miami	Population: 1,000,000
Miles from main campus: 13	Setting: Suburban

Types of engineering degrees: Engineering
Other degree-granting colleges: Architecture, Arts & Sciences, Business Administration, Communication & Journalism, Education, Law, Medicine, Nursing, Rosenstiel School of Marine & Atmospheric Science, Graduate School of International Studies, Music

■ ENGINEERING ADMISSIONS

ADMISSIONS OFFICE CONTACT
Martina S. Hahn
University of Miami
P.O. Box 248294
1251 Memorial Drive
Coral Gables, FL 33124-0620
Phone: (305)284-2404 Fax: (305)284-4792

ENGINEERING COLLEGE ADMISSIONS INFORMATION
Entrance Requirements: SAT (Score: 1000), ACT (Score: 24), High School courses—English (4 years), Mathematics (3 years), Natural & Social Sciences (3 years), Foreign Language (2 years)
Entrance Recommendations: High School courses—Pre-Calculus (1 year), Physics (1 year)
Requirements for foreign students: TOEFL (Score: 550)

DEPARTMENTAL ADMISSIONS INFORMATION
Admission to the engineering department: At the time of admission to the institution.
Additional information: Biomedical Engineering - To be in good academic competition students should have 1000 SAT, 3.0 GPA average and three years math and science.

■ FINANCIAL INFORMATION

ESTIMATED EXPENSES (ACADEMIC YEAR)
[Expenses are for the 1992-93 nine-month academic year.]

All Students	Undergraduate	Graduate
Tuition and fees	$11,082	$15,050
College room and board	$ 5,930	$ 5,930
Books and supplies	$ 716	$ 556
Other expenses	$ 2,596 (1)	$ 2,596 (2)
Total estimated expenses	**$20,324**	**$24,132**

Notes: (1) Included are personal expenses and medical insurance.
(2) Included are personal expenses and medical insurance.

FINANCIAL AID OFFICE CONTACT
Martin J. Carney, *Director, Financial Assistance Services*
University of Miami
P.O. Box 248187
1204 Dickinson Drive, Bldg. 37K
Coral Gables, FL 33124-5240
Phone: (305)284-5212 Fax: (305)284-4082

GENERAL FINANCIAL AID INFORMATION
Forms accepted/required: ACT, CSS-FAF, FAT, Stafford, IRS, SAR, Supplemental, Institutional
Additional financial aid information: There are need-based grants, academic scholarships, loans and work-study available to qualified students.

■ ENGINEERING COLLEGE INFORMATION
[For additional personnel, refer to the *Appendix*.]

HEAD OF ENGINEERING
Martin Becker
Phone: (305)284-2404 Fax: (305)284-4792

ENGINEERING COLLEGE ADDRESS
College of Engineering
University of Miami
P.O. Box 248296
1251 Memorial Drive
Coral Gables, FL 33124-0620
Phone: (305)284-2404 Fax: (305)284-4792

ENROLLMENTS—BY CLASS
[Numbers are baccalaureate enrollments for the fall 1991 term, unless otherwise footnoted.]
1st-year students/Freshmen 224
2nd-year students/Sophomores 140
3rd-year students/Juniors 132
4th-year students/Seniors 253
Total **749**

NUMBER OF DEGREES AWARDED—BY PROGRAM
[Numbers are engineering baccalaureate degrees awarded during the 1991-1992 academic year, unless otherwise footnoted. For full details about each engineering program, refer to the *Tables of Degree Programs*.]
Architectural Engineering 23
Biomedical Engineering .... –
Civil Engineering 23
Electrical & Computer Engineering 61
Industrial Engineering 23
Mechanical Engineering 28
Total **158**

PERCENTAGE OF DEGREES AWARDED—BY CATEGORY
[Percentages are of all engineering baccalaureate degrees awarded during the 1991-1992 academic year, unless otherwise footnoted.]
African Americans 5.5%
Asian/Pacific Island Americans 2.8%
Hispanic Americans 16.6%
Native Americans .... – %
Foreign Citizens 35.0%
All Others 40.1%
Women 13.5%
Persons with disabilities .... – %
Students over 25 years of age 26.4%

GRADUATE ENROLLMENTS & DEGREES AWARDED
Master's enrollment 151
Master's degrees awarded 45
Doctoral enrollment 59
Doctoral degrees awarded 10

ENGINEERING STUDENT DATA
Applicants to the engineering college
Number of applicants to engineering college 830
Percent offered admission 75.0%

Matriculated engineering students
Percentage in top quartile (25%) of High School class 70.0
Average SAT scores: Math—612, Verbal—512, Combined—1124
Average ACT scores: Composite—26

FULL- & PART-TIME FACULTY—BY DEPARTMENT
[Figures are the head count of full-time faculty and the full-time equivalent (FTE) of part-time faculty for each engineering department or equivalent.]

Department	Full	Part
Civil Eng	12	1.0
Electrical Eng	17	1.6
Industrial Eng	10	–
Mechanical Eng	13	1.2
Biomedical Eng	5	–
Architectural Eng	15	1.0

COLLEGE DESCRIPTION

The College of Engineering (CoE) is located in the urban community of Coral Gables, and is situated at the crossroads of the Americas. Established in 1947, the College is housed in the J. Neville McArthur Building which recently completed a $2 million addition to accommodate its growing research and academic activities. With the exception of Engineering Science, which is structured for students whose goal is a graduate degree, all undergraduate programs are EAC/ABET-accredited. The CoE is unique in that it offers an architectural engineering program plus a BS-MD program, jointly sponsored by the College of Engineering and the University of Miami School of Medicine. Further, certain undergraduate and graduate degree programs are subsidized by the State of Florida for its resident students.

DEGREE PROGRAMS DESCRIPTION

Architectural Engineering - One of the few accredited undergraduate programs in the nation. A master's degree is also offered. BS-MD program - This program earns the student admission to the University of Miami Medical School upon admittance. The successful student earns an engineering degree in addition to completing his/her requirements for medical school. Engineering Science (BSES) - For those students whose goal is to earn a graduate degree. The program allows the student to structure a curriculum to fit his/her special needs. Options are available in biomedical and ocean engineering for selected students.

TRANSFER INFORMATION

Transfer via Articulation Agreements
Admission to engineering: At any time
Requirements: 2.5 GPA

Transfer without Articulation Agreements
Admission to engineering: At any time
Requirements: 2.5 GPA

GRADUATION REQUIREMENTS

2.0 overall GPA, 2.0 GPA at the University of Miami, 2.0 GPA in Professional Studies Biomedical Engineering - 125 credits needed for graduation, Architectural Engineering - 127 credits needed for graduation, Civil Engineering - 127 credits needed for graduation, Electrical Engineering - 123 credits needed for graduation, Computer Engineering - 125 credits needed for graduation, Industrial Engineering - 125 credits needed for graduation, Mechanical Engineering - 125 credits needed for graduation

■ STUDENT PROGRAMS

PROFESSIONAL AND HONORARY SOCIETIES
[For key to acronyms, see *Introduction*.]
Professional Societies: AIAA, ASCE, ASHRAE, ASME, IEEE, IIE, SAE, SME, Florida Engineering Society
Honorary Societies: Alpha Pi Mu, Chi Epsilon, Eta Kappa Nu, Phi Alpha Epsilon, Pi Tau Sigma, Tau Beta Pi

SUPPORT PROGRAMS
Student Chapter Organizations: Society of Women Engineers, National Society of Black Engineers
Other Engineering Support Programs: Engineering Advisory Board Adopt-A-Freshman

129 Miami University

INSTITUTION PROFILE

HEAD OF THE INSTITUTION
Paul G. Risser
Phone: (513)529-2345   Fax: (513)529-2121

GENERAL INFORMATION
[All Students—Fall 1991]
Undergraduate enrollment 14,385
Graduate enrollment 1,351
Total institution enrollment 15,736
Type of institution: Public   Calendar: Semesters
Nearest city: Cincinnati   Population: 1,401,500
Miles from main campus: 50   Setting: Small Town
Types of engineering degrees: Engineering
Other degree-granting colleges: Arts & Sciences, Business Administration, Education, Interdisciplinary Studies, Fine Arts

ENGINEERING ADMISSIONS

ADMISSIONS OFFICE CONTACT
Errol A. Gundler
Miami University
School of Applied Science
123 Kreger Hall
Oxford, OH 45056
Phone: (513)529-4036   Fax: (513)529-3841

ENGINEERING COLLEGE ADMISSIONS INFORMATION
Entrance Requirements: SAT, ACT, High School courses—English (4 years), Math (including Algebra II) (3 years), Natural Science-including a physical & bio science (3 years), Social Science (3 years), No minimum score required.
Entrance Recommendations: SAT, ACT, High School courses—Chemistry (1 year), Physics (1 year), Mathematics (4 years), Computer Programming (1 year), No minimum score required.
Requirements for foreign students: TOEFL (Score: 530); financial statement

DEPARTMENTAL ADMISSIONS INFORMATION
Admission to the engineering department: At the time of admission to the institution.

FINANCIAL INFORMATION

ESTIMATED EXPENSES (ACADEMIC YEAR)
[Expenses are for the 1992-93 nine-month academic year.]

State Residents	Undergraduate	Graduate
Tuition and fees	$ 4,096	$ 4,286
College room and board	$ 3,620	$ 5,768
Books and supplies	$ 480	$ 686
Other expenses	$ –	$ –
Total estimated expenses	$ 8,196	$ 10,740

Out-of-State Residents	Undergraduate	Graduate
Tuition and fees	$ 8,686	$ 8,876
College room and board	$ 3,620	$ 5,768
Books and supplies	$ 480	$ 686
Other expenses	$ –	$ –
Total estimated expenses	$12,786	$ 15,330

FINANCIAL AID OFFICE CONTACT
Diane Stemper, Director of Student Financial Aid
Miami University
Edwards House
Oxford, OH 45056
Phone: (513)529-4734   Fax: (513)529-3841

GENERAL FINANCIAL AID INFORMATION
Forms accepted/required: FAF, IRS
Additional financial aid information: Merit-based scholarships, need-based scholarships, minority scholarships, scholarships for women pursuing engineering, federal and state grants, federal and private loans, part-time campus work, college work-study program.

ENGINEERING COLLEGE INFORMATION

[For additional personnel, refer to the Appendix.]

HEAD OF ENGINEERING
David C. Haddad
Phone: (513)529-4036   Fax: (513)529-3841

ENGINEERING COLLEGE ADDRESS
School of Applied Science
Miami University
123 Kreger Hall
Oxford, OH 45056
Phone: (513)529-4036   Fax: (513)529-3841

ENROLLMENTS—BY CLASS
[Numbers are baccalaureate enrollments for the fall 1991 term, unless otherwise footnoted.]
1st-year students/Freshmen 107
2nd-year students/Sophomores 79
3rd-year students/Juniors 62
4th-year students/Seniors 48
Total 296

NUMBER OF DEGREES AWARDED—BY PROGRAM
[Numbers are engineering baccalaureate degrees awarded during the 1991-1992 academic year, unless otherwise footnoted. For full details about each engineering program, refer to the Tables of Degree Programs.]
Engineering Management .... – (1)
Manufacturing Engineering 15
Paper Science & Engineering 25
Total 40

Notes: (1) Program implemented August 1991.

PERCENTAGE OF DEGREES AWARDED—BY CATEGORY
[Percentages are of all engineering baccalaureate degrees awarded during the 1991-1992 academic year, unless otherwise footnoted.]
African Americans .... – %
Asian/Pacific Island Americans .... – %
Hispanic Americans .... – %
Native Americans .... – %
Foreign Citizens .... – %
All Others .... 100.0 %
Women .... 15.0 %
Persons with disabilities .... – %
Students over 25 years of age .... 5.0 %

GRADUATE ENROLLMENTS & DEGREES AWARDED
Master's enrollment 15
Master's degrees awarded 3
Doctoral enrollment .... – (B)
Doctoral degrees awarded .... – (B)

Notes: (B) Data not applicable.

ENGINEERING STUDENT DATA

Applicants to the engineering college
Number of applicants to engineering college 239
Percent offered admission 86.0%

Matriculated engineering students
Percentage in top quartile (25%) of High School class 83.8
Average SAT scores: Math—613, Verbal—495, Combined—1108
Average ACT scores: Math—26, Composite—26

FULL- & PART-TIME FACULTY—BY DEPARTMENT
[Figures are the head count of full-time faculty and the full-time equivalent (FTE) of part-time faculty for each engineering department or equivalent.]

Department	Full	Part
Manufacturing Eng	5	1.0
Paper Science & Eng	7	–
Eng Management	–	–

COLLEGE DESCRIPTION
The mission of the School of Applied Science is to build upon Miami's strong leadership development and liberal education tradition to create selected engineering programs which are needed by society. The School offers course work which leads to B.S. degrees in Manufacturing Engineering and Paper Science and Engineering and minors in electrical engineering, manufacturing engineering, and mechanical engineering. The School offers an interdisciplinary program with the school of business in Engineering Management. These programs develop the professional competencies in students that are necessary to analyze, synthesize, and solve the complex problems inherent in a modern, dynamic society. To this end, the programs are built on the fundamental knowledge and methods of mathematics, systems analysis, the physical science, communications, the social science, the life sciences and the humanities. Co-op and summer intern programs are available to all engineering students.

DEGREE PROGRAMS DESCRIPTION
Engineering Management: This is an interdisciplinary program which combines engineering, business, science, mathematics, and liberal arts. Graduates are prepared to fill positions in quality control, technical sales/service, and production supervision. Manufacturing engineering: Prepare students for careers in the rapidly expanding fields of manufacturing engineering. Manufacturing Engineering deals with designing, developing, and controlling the manufacturing process. Graduates typically work as manufacturing engineers in areas such as product and process design, quality control, and computer-aided manufacturing. Paper Science and Engineering: This program provides a broad scientific education preparing students for professional entry-level positions in the pulp and paper or allied industries. Students learn to apply scientific and engineering principles to solve industry problems by following a course sequence emphasizing chemistry, chemical engineering, and paper engineering.

TRANSFER INFORMATION
Residency Requirements: All students must completed at least 32 hours from Miami University, including 12 of the final 20 hours.

Transfer via Articulation Agreements
Admission to engineering: Transfer students are admitted to the engineering programs when they have completed the Associate of Science degrees from selected Community Colleges.
Engineering graduates transferred from .. 4-yr: –   2-yr: 2
Requirements: Students from the selected Community Colleges must have received the Associate of Science degree and attained a GPA of 2.5.

Transfer without Articulation Agreements
Admission to engineering: At any time
Engineering graduates transferred from .. 4-yr: 4   2-yr: 4
Requirements: The number of transfer students that can be offered admission each semester is limited. Those students who have better academic records are given priority. In recent years, most students with a 2.75 or better GPA have been accepted for admission. Students also must have completed the equivalent of 20 semester hours of college work. Course acceptance is done on a course by course basis.

GRADUATION REQUIREMENTS
B.S. degree programs in engineering require a minimum of 135 semester credit hours. A minimum graduation index of 2.0 (4.0=A) is required for graduation.

STUDENT PROGRAMS

PROFESSIONAL AND HONORARY SOCIETIES
[For key to acronyms, see Introduction.]
Professional Societies: ACM, SME, Technical Association of Pulp and Paper Industry (TAPPI)
Honorary Societies: Phi Kappa Phi, Omega Rho

SUPPORT PROGRAMS
Student Chapter Organizations: National Technical Association

Minority Professional Leadership Program (MPLP)
For: Ethnic Minorities   Available: Year round
Offered: Prematriculation, Freshman, Sophomore, Junior, Senior

Academic Enhancement Program (AEP)
For: Ethnic Minorities   Available: Academic year
Offered: Freshman, Sophomore

Miami Academic Achievers Program (MAAP)
For: Women & Ethnic Minorities   Available: Academic year
Offered: Freshman, Sophomore, Junior, Senior

Center for Black Culture and Learning
For: Ethnic Minorities   Available: Year round
Offered: Freshman, Sophomore, Junior, Senior, Graduate level

Other Engineering Support Programs: Summer and fall orientation programs for all freshmen; upper-class mentor system for incoming freshmen; a faculty advisor for each engineering student starting the first semester of the freshmen year; Learning Assistance Center; tutoring services, psychological counseling, International Students Office, handicap services, learning disabled services, placement services, community service office.

130 University of Michigan

INSTITUTION PROFILE

HEAD OF THE INSTITUTION
James J. Duderstadt
Phone: (313)764-6270

GENERAL INFORMATION
[All Students—Fall 1991]
Undergraduate enrollment 23,121
Graduate enrollment 13,422
Total institution enrollment **36,543**

Type of institution: Public **Calendar:** Semesters
Nearest city: Detroit, Michigan **Population:** 1,000,000
Miles from main campus: 35 **Setting:** Small Town
Types of engineering degrees: Engineering
Other degree-granting colleges: Architecture, Business Administration, Dentistry, Education, Law, Medicine, Nursing, Pharmacy, Technology & Applied Sciences, Natural Resources, Social Work, Literature, Science, and the Arts, Information and Library Science, Public Health

ENGINEERING ADMISSIONS

ADMISSIONS OFFICE CONTACT
Theodore L. Spencer
University of Michigan
Office of Admissions
1220 Student Activities Building
Ann Arbor, MI 48109-1316
Phone: (313)764-7433

ENGINEERING COLLEGE ADMISSIONS INFORMATION
Entrance Requirements: SAT, ACT, High School courses—English (3 years), Mathematics (3 years), Laboratory Science (1 year), Academic Electives (4 years), SAT or ACT required
Entrance Recommendations: High School courses—English (4 years), Mathematics (4 years), Laboratory Science, Chemistry and Physics (2 years), Foreign Language as Academic Electives (2 years)
Requirements for out-of-state residents: Out-of-State applicants must have somewhat higher GPA's and SAT/ACT scores.
Requirements for foreign students: TOEFL (Score: 550); TOEFL score of 550 or above or a Michigan Test (MELAB) score of 80 or above is required.

DEPARTMENTAL ADMISSIONS INFORMATION
Admission to the engineering department: The end of freshman year, or at the latest the end of sophomore year.
Additional information: Mechanical Engineering and Applied Mechanics has a set of specific admissions criteria for sophomores.

FINANCIAL INFORMATION

ESTIMATED EXPENSES (ACADEMIC YEAR)
[Expenses are for the 1992-93 nine-month academic year.]

State Residents	Undergraduate	Graduate
Tuition and fees	$ 4,528 [2]	$ 7,930
College room and board	$ 4,855	$ 4,855
Books and supplies	$ 460	$ 575
Other expenses	$ 1,501 [1]	$ 1,501 [1]
Total estimated expenses	**$11,344**	**$14,861**

Out-of-State Residents	Undergraduate	Graduate
Tuition and fees	$14,204 [3]	$15,748
College room and board	$ 4,855	$ 4,855
Books and supplies	$ 460	$ 575
Other expenses	$ 1,501 [1]	$ 1,501 [1]
Total estimated expenses	**$21,020**	**$22,679**

Notes: (1) Travel and miscellaneous. (2) Lower division; upper division $5334. (3) Lower division; upper division $15,674.

GENERAL FINANCIAL AID INFORMATION
Forms accepted/required: Institutional
Additional financial aid information: Need and merit-based scholarships, work-study, and loans are available.

ENGINEERING COLLEGE INFORMATION
[For additional personnel, refer to the *Appendix*.]

HEAD OF ENGINEERING
Peter M. Banks
Phone: (313)764-8475 **Fax:** (313)763-9487
EMail: Peter_Banks@um.cc.umich.edu

ENGINEERING COLLEGE ADDRESS
College of Engineering
University of Michigan
2309 EECS Building
1301 Beal Avenue
Ann Arbor, MI 48109-2116
Phone: (313)764-8470 **Fax:** (313)763-9487

ENROLLMENTS—BY CLASS
[Numbers are baccalaureate enrollments for the fall 1991 term, unless otherwise footnoted.]
1st-year students/Freshmen 1,239
2nd-year students/Sophomores 1,066
3rd-year students/Juniors 1,106
4th-year students/Seniors 1,308
Total **4,719**

NUMBER OF DEGREES AWARDED—BY PROGRAM
[Numbers are engineering baccalaureate degrees awarded during the 1991-1992 academic year, unless otherwise footnoted. For full details about each engineering program, refer to the *Tables of Degree Programs*.]

Aerospace Engineering 107
Aerospace Science Engineering .... –
Applied Mechanics .... –
Applied Physics .... –
Atmospheric & Space Sciences .... –
Bioengineering .... –
Chemical Engineering 58
Civil & Environmental Engineering 48
Civil Engineering .... –
Computer Engineering 67
Computer Science & Engineering .... –
Computer, Information, & Control Engineering ...... –
Construction Engineering & Management .... –
Electrical Engineering 143
Electrical Engineering Systems .... –
Engineering 7
Engineering Physics 4
Environmental Engineering .... –
Industrial & Operations Engineering 104
Macromolecular Science & Engineering .... –
Materials Science & Engineering 17
Mechanical Engineering 228
Meteorology 5
Naval Architecture & Marine Engineering 18
Nuclear Engineering 12
Nuclear Sciences .... –
Ocean Science .... –
Oceanography: Physical 1
Total **819**

PERCENTAGE OF DEGREES AWARDED—BY CATEGORY
[Percentages are of all engineering baccalaureate degrees awarded during the 1991-1992 academic year, unless otherwise footnoted.]
African Americans 3.9%
Asian/Pacific Island Americans 10.2%
Hispanic Americans 2.2%
Native Americans 0.1%
Foreign Citizens 3.3%
All Others 80.3%
Women 20.1%
Persons with disabilities .... – % (A)
Students over 25 years of age .... – % (A)

Notes: (A) Data not available.

GRADUATE ENROLLMENTS & DEGREES AWARDED
Master's enrollment 883
Master's degrees awarded 504
Doctoral enrollment 916
Doctoral degrees awarded 238

ENGINEERING STUDENT DATA
Applicants to the engineering college
Number of applicants to engineering college 3,693
Percent offered admission 75.1%

Matriculated engineering students
Percentage in top quartile (25%) of High School class 97.0
Average SAT scores: Math—666, Verbal—539, Combined—1205
Average ACT scores: Math—29.1, Composite—28.0

FULL- & PART-TIME FACULTY—BY DEPARTMENT
[Figures are the head count of full-time faculty and the full-time equivalent (FTE) of part-time faculty for each engineering department or equivalent.]

Department	Full	Part
Chemical Eng	19	–
Civil & Environmental Eng	26	1.6
Electrical Eng & Computer Science	93	–
Industrial & Operations Eng	23	1.3
Mechanical Eng & Applied Mechanics	46	1.5
Aerospace Eng	25	0.5
Atmospheric, Oceanic & Space Sciences	19	3.3
Materials Science & Eng	18	0.1
Naval Architecture & Marine Eng	16	0.6
Nuclear Eng	15	–

COLLEGE DESCRIPTION
The undergraduate engineering program at Michigan is housed within the context of a research institution. The program seeks to prepare students for the future 20 to 30 years from the present. Seventy percent of our entering freshmen will graduate from the College of Engineering. Of these, 50 percent will graduate in four years. The rest take an additional term or summer half term. The College has dual degree programs with Music, Art, Architecture, and Literature, Science and the Arts. The Co-Op experience is available after the sophomore year. Completion of the degree including the Co-Op experience generally takes 9 semesters in residence.

DEGREE PROGRAMS DESCRIPTION
For complete degree program information, please obtain a copy of the Bulletin.

TRANSFER INFORMATION
Residency Requirements: A minimum of 30 credits earned at Michigan (Ann Arbor) is necessary for any undergraduate degree.

Transfer via Articulation Agreements
Admission to engineering: At the end of the sophomore year
Requirements: The equivalent of the first two years of the Michigan engineering program must be completed before transferring. Grade point requirements vary from program to program.

Transfer without Articulation Agreements
Admission to engineering: Three-two dual degree programs with several liberal arts colleges are articulated.
Engineering graduates transferred from . . . 4-yr: – 2-yr: –
Requirements: The equivalent of the first two years of the Michigan engineering program must be completed before transferring. Grade point requirements vary from program to program.

STUDENT PROGRAMS

PROFESSIONAL AND HONORARY SOCIETIES
[For key to acronyms, see *Introduction*.]
Professional Societies: AIAA, AIChE, ANS, ASCE, ASME, IEEE, IIE, ORSA, SAE, SNAME
Honorary Societies: Alpha Pi Mu, Chi Epsilon, Eta Kappa Nu, Pi Tau Sigma, Sigma Gamma Tau, Tau Beta Pi, Alpha Sigma Mu

SUPPORT PROGRAMS
Student Chapter Organizations: Society of Women Engineers, National Society of Black Engineers, Society of Hispanic Professional Engineers, American Indian Science and Engineering Society, Society of Minority Engineering Students

Minority Engineering Program Office
For: Ethnic Minorities **Available:** Year round
Offered: Prematriculation, Freshman, Sophomore, Junior, Senior, Graduate level

Pilot Women's Program
For: Women **Available:** Year round
Offered: Prematriculation, Freshman, Sophomore, Junior, Senior, Graduate level

Other Engineering Support Programs: Orientation programs for freshmen and transfer students and career guidance and academic counseling for freshmen. Presentations for students to explore the opportunities in many areas of engineering in the College. Financial support is provided based on performance and need. Program advisors from department faculty provide academic counseling.

131 University of Michigan-Dearborn

■ INSTITUTION PROFILE

HEAD OF THE INSTITUTION
Bernard W. Klein
Phone: (313)593-5500　　　　Fax: (313)593-5452

GENERAL INFORMATION
[All Students—Fall 1991]
Undergraduate enrollment 7,006
Graduate enrollment 1,067
Total institution enrollment **8,073**
Type of institution: Public　　**Calendar:** Semesters
Nearest city: Detroit　　**Population:** 1,000,000
Miles from main campus: 10　　**Setting:** Suburban
Types of engineering degrees: Engineering
Other degree-granting colleges: Arts & Sciences, Business Administration, Education

■ ENGINEERING ADMISSIONS

ADMISSIONS OFFICE CONTACT
Carol Mack
University of Michigan-Dearborn
4901 Evergreen Rd.
Dearborn, MI 48128-1491
Phone: (313)593-5199　　　　Fax: (313)593-5452

ENGINEERING COLLEGE ADMISSIONS INFORMATION
Entrance Requirements: SAT (Score: 1050), ACT (Score: 22), High School courses—Science (2 years), Mathematics (4 years), Basic computer programming (1 year)
Entrance Recommendations: High School courses—English (4 years), Physics (1 year), Chemistry (1 year), Structured computer programming (1 year), National Engineering Aptitude Search Test
Requirements for foreign students: TOEFL (Score: 550); financial statement

DEPARTMENTAL ADMISSIONS INFORMATION
Admission to the engineering department: At the time of admission to the institution.

■ FINANCIAL INFORMATION

ESTIMATED EXPENSES (ACADEMIC YEAR)
[Expenses are for the 1992-93 nine-month academic year.]

State Residents	Undergraduate	Graduate
Tuition and fees	$ 2,900	$ 3,400
College room and board	$ 4,000	$ 4,000
Books and supplies	$ 500	$ 500
Other expenses	$ 1,000	$ 1,000
Total estimated expenses	**$ 8,400**	**$ 8,900**

Out-of-State Residents	Undergraduate	Graduate
Tuition and fees	$ 9,350	$ 10,960
College room and board	$ 4,000	$ 4,000
Books and supplies	$ 500	$ 500
Other expenses	$ 1,000	$ 1,000
Total estimated expenses	**$14,850**	**$ 16,460**

FINANCIAL AID OFFICE CONTACT
John A. Mason, *Director, Office of Financial Aid*
University of Michigan-Dearborn
4901 Evergreen Rd.
Dearborn, MI 48128-1491
Phone: (313)593-5300　　　　Fax: (313)593-5452

GENERAL FINANCIAL AID INFORMATION
Forms accepted/required: ACT, AFSA, CSS-FAF, FAT, FFS, Stafford, SAR, Supplemental
Additional financial aid information: Financial Aid consists of the following four types of assistance: scholarships, grants, loans and employment. With the exception of some scholarships, most financial assistance through the Office of Financial Aid is awarded on the basis of financial need.

■ ENGINEERING COLLEGE INFORMATION
[For additional personnel, refer to the *Appendix*.]

HEAD OF ENGINEERING
Subrata Sengupta
Phone: (313)593-5290　　　　Fax: (313)593-9967

ENGINEERING COLLEGE ADDRESS
School of Engineering
University of Michigan-Dearborn
4901 Evergreen Rd.
Dearborn, MI 48128-1491
Phone: (313)593-5510　　　　Fax: (313)593-9967

ENROLLMENTS—BY CLASS
[Numbers are baccalaureate enrollments for the fall 1991 term, unless otherwise footnoted.]
1st-year students/Freshmen 286
2nd-year students/Sophomores 287
3rd-year students/Juniors 316
4th-year students/Seniors 462
Total **1,351**

NUMBER OF DEGREES AWARDED—BY PROGRAM
[Numbers are engineering baccalaureate degrees awarded during the 1991-1992 academic year, unless otherwise footnoted. For full details about each engineering program, refer to the *Tables of Degree Programs*.]
Electrical & Computer Engineering 76
Industrial & Systems Engineering 13
Mechanical Engineering 58
Total **147**

PERCENTAGE OF DEGREES AWARDED—BY CATEGORY
[Percentages are of all engineering baccalaureate degrees awarded during the 1991-1992 academic year, unless otherwise footnoted.]
African Americans . . . . – %
Asian/Pacific Island Americans . . . . 7.0 %
Hispanic Americans . . . . 2.0 %
Native Americans . . . . – %
Foreign Citizens . . . . 2.0 %
All Others . . . . 89.0 %
Women . . . . 21.0 %
Persons with disabilities . . . . – % (A)
Students over 25 years of age . . . . – % (A)
Notes: (A) Data not available.

GRADUATE ENROLLMENTS & DEGREES AWARDED
Master's enrollment 511
Master's degrees awarded 98
Doctoral enrollment . . . . – (B)
Doctoral degrees awarded . . . . – (B)
Notes: (B) Data not applicable.

ENGINEERING STUDENT DATA
Applicants to the engineering college
Number of applicants to engineering college 463
Percent offered admission . . . . 33.0 %

Matriculated engineering students
Percentage in top quartile (25%) of High School class 65.0
Average SAT scores: Math—570, Verbal—470
Average ACT scores: Math—24, Composite—24

FULL- & PART-TIME FACULTY—BY DEPARTMENT
[Figures are the head count of full-time faculty and the full-time equivalent (FTE) of part-time faculty for each engineering department or equivalent.]

Department	Full	Part
Electrical & Computer Eng	12	1.7
Industrial & Systems Eng	8	1.9
Mechanical Eng	13	2.0

COLLEGE DESCRIPTION
The School of Engineering at The University of Michigan-Dearborn offers programs leading to Bachelor of Science in Engineering Degrees in electrical, industrial and systems, and mechanical. These degrees are accredited by the Accreditation Board for Engineering and Technology (ABET) A Bachelor of Science in Engineering Mathematics can also be earned in conjunction with a principal degree in electrical, industrial and systems, or mechanical. A Bachelor of Science degree in Computer and Information Science, with option in Computer Science or Information Systems is also offered by the School. The School of Engineering offers two modes of study: conventional programs, requiring only the completion of academic course requirements, and cooperative internship programs, in which students complete the academic subject requirements plus a series of professional internships.

DEGREE PROGRAMS DESCRIPTION
Undergraduate degree programs: B.S. (Electrical Engineering); B.S. (Industrial and Systems Engineering); B.S. (Mechanical Engineering). Graduate degree programs: M.S.E. (Electrical Engineering); M.S.E. (Industrial Engineering); M.S.E. (Manufacturing Engineering); M.S.E. (Mechanical Engineering).

DUAL DEGREE PROGRAMS
Engineering baccalaureate graduates with dual degrees 5

TRANSFER INFORMATION
Transfer via Articulation Agreements
Admission to engineering: At any time
Engineering graduates transferred from . . 4-yr: 20% 2-yr: 40%
Requirements: Transfer students are required to have a 2.75 cumulative grade point average as well as a 2.75 math and 2.75 science grade point average.

Transfer without Articulation Agreements
Admission to engineering: At any time
Engineering graduates transferred from . . 4-yr: – 2-yr: –
Requirements: Students are required to have a 2.75 cumulative grade point average as well as a 2.75 math and a 2.75 science grade point average.

GRADUATION REQUIREMENTS
Admitted to a degree program in the school. Must complete the required and elective courses and credit hours in a given degree program with a grade of C or better (undergraduates) and B or better (graduate). Minimum grade average of C or better in engineering courses for undergraduate and B or better for graduate. At least 30 credit hours of upper level course work for undergraduates. Others as specified in appropriate announcements.

■ STUDENT PROGRAMS

PROFESSIONAL AND HONORARY SOCIETIES
[For key to acronyms, see *Introduction*.]
Professional Societies: ASME, IEEE, IIE, NSPE, SAE
Honorary Societies: Alpha Pi Mu, Eta Kappa Nu, Tau Beta Pi, Theta Tau

SUPPORT PROGRAMS
Student Chapter Organizations: Society of Women Engineers, National Society of Black Engineers

Program for Academic Support
For: Ethnic Minorities　　**Available:** Year round
Offered: Prematriculation, Freshman

Other Engineering Support Programs: Prior to registration each term all students are mailed a computer generated academic progress report indicating academic progress indicating all degree requirements, classes completed at UM-D with grades earned, classes transferred, current elections, and petitions processed. Tau Beta Pi members offer tutoring in mathematics, chemistry, physics and engineering courses.

132 Michigan State University

■ INSTITUTION PROFILE
HEAD OF THE INSTITUTION
Gordon Guyer
Phone: (517)355-6560 Fax: (517)355-4670
GENERAL INFORMATION
[All Students—Fall 1991]
Undergraduate enrollment 33,684
Graduate enrollment 8,404
Total institution enrollment 42,088
Type of institution: Public Calendar: Semesters
Nearest city: Lansing Population: 135,000
Miles from main campus: 2 Setting: Small Town
Types of engineering degrees: Engineering
Other degree-granting colleges: Agricultural & Environmental, Business Administration, Communication & Journalism, Education, Humanities & Social Sciences, Nursing, Veterinary Medicine, Human Ecology, James Madison, Natural Science, Arts and Letters

■ ENGINEERING ADMISSIONS
ADMISSIONS OFFICE CONTACT
William Turner
Michigan State University
250 Administration Building
East Lansing, MI 48824-1046
Phone: (517)355-8332 Fax: (517)336-2069
ENGINEERING COLLEGE ADMISSIONS INFORMATION
Entrance Requirements: SAT, ACT, High School courses—Mathematics (3.5 years), English (4 years), Science (2 years), Social Science (2 years)
Entrance Recommendations: PSAT, High School courses—Foreign Language (3 years), Mathematics, Science (1 year)
Requirements for foreign students: TOEFL; financial statement
DEPARTMENTAL ADMISSIONS INFORMATION
Admission to the engineering department: At the end of the second year.
Additional information: There are specific course and grade point requirements that must be met by the end of the sophomore year for admission to all majors.

■ FINANCIAL INFORMATION
ESTIMATED EXPENSES (ACADEMIC YEAR)
[Expenses are for the 1992-93 nine-month academic year.]

State Residents	Undergraduate	Graduate
Tuition and fees	$ 3,477 [3]	$ 2,812 [1]
College room and board	$ 3,568	$ 3,568
Books and supplies	$ 504	$ 504
Other expenses	$ 459 [2]	$ 459 [2]
Total estimated expenses	**$ 8,008**	**$ 7,343**

Out-of-State Residents	Undergraduate	Graduate
Tuition and fees	$ 9,248 [3]	$ 5,696 [1]
College room and board	$ 3,568	$ 3,568
Books and supplies	$ 504	$ 504
Other expenses	$ 459 [2]	$ 459 [2]
Total estimated expenses	**$13,779**	**$ 10,227**

Notes: (1) Based on 8 credits per semester. (2) Includes miscellaneous fees and taxes. (3) Based on 15 credits per semester.

GENERAL FINANCIAL AID INFORMATION
Forms accepted/required: ACT, AFSA, CSS-FAF, FSA, Stafford, Institutional
Additional financial aid information: Need based scholarships, short term loans, scholarships, assistantships, jobs and work-study.

■ ENGINEERING COLLEGE INFORMATION
[For additional personnel, refer to the Appendix.]
HEAD OF ENGINEERING
Theodore A. Bickart
Phone: (517)355-5113 Fax: (517)355-2288
EMail: BICKART@egr.msu.edu
ENGINEERING COLLEGE ADDRESS
College of Engineering
Michigan State University
104 Engineering Building
East Lansing, MI 48824-1226
Phone: (517)355-5120 Fax: (517)336-1356

ENROLLMENTS—BY CLASS
[Numbers are baccalaureate enrollments for the fall 1991 term, unless otherwise footnoted.]
1st-year students/Freshmen 1,119
2nd-year students/Sophomores 919
3rd-year students/Juniors 676
4th-year students/Seniors 861
Total 3,575

NUMBER OF DEGREES AWARDED—BY PROGRAM
[Numbers are engineering baccalaureate degrees awarded during the 1991-1992 academic year, unless otherwise footnoted. For full details about each engineering program, refer to the Tables of Degree Programs.]
Agricultural Engineering 2
Chemical Engineering 62
Civil Engineering 94
Computer Engineering 15
Computer Science 76
Electrical Engineering 103
Engineering Arts 65
Environmental Engineering .... –
Food Engineering 4
Materials Science 15
Mechanical Engineering 168
Mechanics 4
Total 608

PERCENTAGE OF DEGREES AWARDED—BY CATEGORY
[Percentages are of all engineering baccalaureate degrees awarded during the 1991-1992 academic year, unless otherwise footnoted.]
African Americans .... 4.9 %
Asian/Pacific Island Americans .... 2.6 %
Hispanic Americans .... – %
Native Americans .... – %
Foreign Citizens .... 3.0 %
All Others .... 89.5 %
Women .... 20.1 %
Persons with disabilities .... – % [A]
Students over 25 years of age .... – % [A]
Notes: (A) Data not available.

GRADUATE ENROLLMENTS & DEGREES AWARDED
Master's enrollment 372
Master's degrees awarded 148
Doctoral enrollment 274
Doctoral degrees awarded 67

ENGINEERING STUDENT DATA
Applicants to the engineering college
Number of applicants to engineering college 2,600
Percent offered admission .... –% [A]
Matriculated engineering students
Percentage in top quartile (25%) of High School class 82.0
Average SAT scores: Math—600, Verbal—560, Combined—1160
Average ACT scores: Math—26, Composite—24 [2]
Notes: (A) Data not available. (2) Test scores apply to students admitted as pre-engineering freshmen.

FULL- & PART-TIME FACULTY—BY DEPARTMENT
[Figures are the head count of full-time faculty and the full-time equivalent (FTE) of part-time faculty for each engineering department or equivalent.]

Department	Full	Part
Agricultural Eng	22	... –
Chemical Eng	10	... 2.0
Civil Eng	22	... 1.2
Computer Science	23	... 1.1
Electrical Eng	26	... –
Mechanical Eng	21	... 1.0
Materials Science & Mechanics	15	... 2.0

COLLEGE DESCRIPTION
Special programs include: a second B.S. degree combining Engineering for International Service; a option in Biomedical Engineering that can be supplemental to any of the several undergraduate majors; extensive honors opportunities including independent study, honors courses, honors projects and undergraduate assistantships; extensive international opportunities including a one year exchange with the University of Surrey, England, one semester in Aachen, Germany and one semester in Kaiserslautern, Germany; the C.I.C. consortium provides numerous opportunities for national and international exchanges; computing resources are extensive and include 24 hour access to personal computers, Unix-based engineering workstations, laser printers, plotting, and software for many engineering applications. Cooperative Engineering Education is a five year option program that is available in all majors in the college.

DEGREE PROGRAMS DESCRIPTION
Special programs include: a second B.S. degree combining Engineering for International Service; an option in Biomedical Engineering that can be supplemental to any of the several undergraduate majors; extensive honors opportunities including independent study, honors courses, honors projects and undergraduate assistantships; extensive international opportunity including a one year exchange with the University of Surrey, England, one semester Aachen, Germany and one semester in Kaiserslautern, Germany; the C.I.C. consortium provides numerous opportunities for national and international exchanges; comp

DUAL DEGREE PROGRAMS
Engineering baccalaureate graduates with dual degrees 5
Enrollment requirements: Admissions is based on completion of a specific set of courses in mathematics, chemistry, physics, and computer science taken during the first two years plus a combined/cumulative technical grade point average. The course requirements are the same for all majors; however, the grade point averages vary according to major.

TRANSFER INFORMATION
Residency Requirements: Complete one year's work, normally the year of graduation, earning at least 30 credits in courses given by Michigan State University.
Transfer via Articulation Agreements
Admission to engineering: At the end of each term.
Requirements: A minimum grade point average of 3.0 for consideration; the grade point average for acceptance is determined by the number of applications for each major.
Transfer without Articulation Agreements
Admission to engineering: At the end of each term.
Requirements: Requirements are the same as above.

■ STUDENT PROGRAMS
PROFESSIONAL AND HONORARY SOCIETIES
[For key to acronyms, see Introduction.]
Professional Societies: ACerS, ACM, AIAA, AIChE, ASAE, ASCE, ASME, IEEE, SAE, SPE*, ASM International (ASM), Society of Engineering Science (SES)
Honorary Societies: Alpha Epsilon, Chi Epsilon, Eta Kappa Nu, Omega Chi Epsilon, Phi Alpha Epsilon, Pi Tau Sigma, Tau Beta Pi, Alpha Sigma Mu

SUPPORT PROGRAMS
Student Chapter Organizations: Society of Women Engineers, National Society of Black Engineers, Society of Hispanic Professional Engineers, American Indian Science and Engineering Society

Society of Women Engineers
For: Women & Ethnic Minorities **Available:** Year round
Offered: Freshman, Sophomore, Junior, Senior, Graduate level

National Society of Black Engineers
For: Ethnic Minorities **Available:** Year round
Offered: Freshman, Sophomore, Junior, Senior, Graduate level

Society of Hispanic Professional Engineers
For: Ethnic Minorities **Available:** Year round
Offered: Freshman, Sophomore, Junior, Senior, Graduate level

American Indian Science and Engineering Society
For: Ethnic Minorities **Available:** Year round
Offered: Freshman, Sophomore, Junior, Senior, Graduate level

Other Engineering Support Programs: Support programs for engineering students include tutorial assistance provided by honorary organizations, required participation in summer academic orientation programs for all students, and excellent counseling and student life programs within the University.

133 Michigan Technological University

INSTITUTION PROFILE

HEAD OF THE INSTITUTION
Curtis J. Tompkins
Phone: (906)487-2200 Fax: (906)487-2935

GENERAL INFORMATION
[All Students—Fall 1991]
Undergraduate enrollment 6,360
Graduate enrollment 601
Total institution enrollment 6,961
Type of institution: Public Calendar: Quarters
Location: Houghton, MI Population: 10,000
Setting: Small Town
Types of engineering degrees: Engineering
Other degree-granting colleges: Arts & Sciences, Business Administration, Technology, Forestry

ENGINEERING ADMISSIONS

ADMISSIONS OFFICE CONTACT
Joseph Galetto
Michigan Technological University
1400 Townsend Drive
Houghton, MI 49931-1295
Phone: (906)487-2335 Fax: (906)487-2245

ENGINEERING COLLEGE ADMISSIONS INFORMATION
Entrance Requirements: High School courses—Algebra (1 year), Chemistry or Physics (1 year), Trigonometry, English (3 years)
Entrance Recommendations: SAT (Score: 1050), ACT (Score: 25), High School courses—Math (4 years), English (4 years), Chemistry, Foreign Language (2 years), Physics (1 year), In general, students should also rank in the upper quartile of their graduating class.
Requirements for foreign students: TOEFL (Score: 500); financial statement; Students should have a minimum of 52 on TOEFL Sec. 1, and a Declaration and Certification of Finances.

DEPARTMENTAL ADMISSIONS INFORMATION
Admission to the engineering department: Admission to engineering departments may be elected by the student upon admission or after matriculation.
Additional information: Students may transfer into an engineering department after matriculation on a space available basis, preference given to the more academically able.

FINANCIAL INFORMATION

ESTIMATED EXPENSES (ACADEMIC YEAR)
[Expenses are for the 1992-93 nine-month academic year.]

State Residents	Undergraduate	Graduate
Tuition and fees	$ 3,249	$ 2,700
College room and board	$ 3,604	$ 3,604
Books and supplies	$ 600	$ 600
Other expenses	$ 400 [1]	$ 400 [1]
Total estimated expenses	$ 7,853	$ 7,304

Out-of-State Residents	Undergraduate	Graduate
Tuition and fees	$ 7,326	$ 6,098
College room and board	$ 3,604	$ 3,604
Books and supplies	$ 600	$ 600
Other expenses	$ 400 [1]	$ 400 [1]
Total estimated expenses	$11,930	$10,702

Notes: (1) Other expenses estimate is for laboratory fees

FINANCIAL AID OFFICE CONTACT
Tim T. Malette, *Director of Financial Aid*
Michigan Technological University
1400 Townsend Drive
Houghton, MI 49931
Phone: (906)487-2622 Fax: (906)487-2398

GENERAL FINANCIAL AID INFORMATION
Forms accepted/required: CSS-FAF, FAF, FAT, FFS, IRS, SAR
Additional financial aid information: Need-based scholarships; merit-based scholarships; campus-based employment; College Work Study; short term loans.

ENGINEERING COLLEGE INFORMATION

[For additional personnel, refer to the *Appendix*.]

HEAD OF ENGINEERING
Vernon B. Watwood, *Interim*
Phone: (906)487-2005 Fax: (906)487-2782
EMail: vbwatwoo@mtu.edu

ENGINEERING COLLEGE ADDRESS
College of Engineering
Michigan Technological University
1400 Townsend Drive
Houghton, MI 49931-1295
Phone: (906)487-2005 Fax: (906)487-2782

ENROLLMENTS—BY CLASS
[Numbers are baccalaureate enrollments for the fall 1991 term, unless otherwise footnoted.]
1st-year students/Freshmen 1,173
2nd-year students/Sophomores 901
3rd-year students/Juniors 1,182
4th-year students/Seniors 934
Total 4,190

NUMBER OF DEGREES AWARDED—BY PROGRAM
[Numbers are engineering baccalaureate degrees awarded during the 1991-1992 academic year, unless otherwise footnoted. For full details about each engineering program, refer to the *Tables of Degree Programs*.]
Chemical Engineering 63
Civil Engineering 97
Electrical Engineering 239
Engineering 19
Environmental Engineering 29
Geological Engineering 7
Materials Science & Engineering 36
Mechanical Engineering 271
Minerals Processing 10
Mining Engineering 6
Total 777

PERCENTAGE OF DEGREES AWARDED—BY CATEGORY
[Percentages are of all engineering baccalaureate degrees awarded during the 1991-1992 academic year, unless otherwise footnoted.]
African Americans 0.4%
Asian/Pacific Island Americans 1.3%
Hispanic Americans 0.4%
Native Americans 0.7%
Foreign Citizens 9.1%
All Others 88.1%
Women 13.9%
Persons with disabilities .... – % (A)
Students over 25 years of age 17.8%

Notes: (A) Data not available.

GRADUATE ENROLLMENTS & DEGREES AWARDED
Master's enrollment 196
Master's degrees awarded 84
Doctoral enrollment 94
Doctoral degrees awarded 20

ENGINEERING STUDENT DATA

Applicants to the engineering college
Number of applicants to engineering college 2,044
Percent offered admission 96.6%

Matriculated engineering students
Percentage in top quartile (25%) of High School class 77.4
Average SAT scores: Math—598, Verbal—484, Combined—1082
Average ACT scores: Math—26, Composite—25

FULL- & PART-TIME FACULTY—BY DEPARTMENT
[Figures are the head count of full-time faculty and the full-time equivalent (FTE) of part-time faculty for each engineering department or equivalent.]

Department	Full	Part
Chemical Eng	9	–
Civil & Environmental Eng	19	–
Electrical Eng	24	2.8
Chemistry	16	–
Geological Eng, Geology, & Geophysics	9	1.0
Mining Eng	4	–
Metallurgical & Materials Eng	18	–
Mechanical Eng	41	–
General Eng	1	–

COLLEGE DESCRIPTION

The College offers 12 programs leading to B.S., 12 to M.S. and 9 to the Ph.D. Many B.S. programs offer several specializations within the discipline. Each program is designed to prepare graduates for entry directly into industrial positions or to continue study at the graduate level. There are opportunities to earn second bachelor degrees in a fifth year with several other disciplines including business, computer science, some physical and life sciences and mathematics and to prepare for secondary school teacher certification.

DEGREE PROGRAMS DESCRIPTION
B.S. degrees are offered in 10 engineering specialties and 3 physical sciences to prepare graduates for entry level professional positions or for graduate study. The B.S. in Engineering allows qualified students to organize interdisciplinary specialties not achievable in most programs. M.S. and Ph.D. programs are conducted in several disciplines and in interdisciplinary combinations under the Ph.D. in Engineering.

DUAL DEGREE PROGRAMS
Enrollment requirements: Students may pursue dual baccalaureate degrees concurrently or at different times.

TRANSFER INFORMATION
Residency Requirements: Last 36 credits prior to graduation must be taken in residence.

Transfer via Articulation Agreements
Admission to engineering: At the end of the sophomore year
Engineering graduates transferred from . . 4-yr: 12 2-yr: –
Requirements: Student must meet MTU internal transfer requirements and be recommended by their institution.

Transfer without Articulation Agreements
Admission to engineering: At any time
Engineering graduates transferred from . . 4-yr: 86 2-yr: 93

GRADUATION REQUIREMENTS
B.S. programs in engineering require a minimum of 196 quarter credits meeting general education and ABET accreditation standards. Graduating students must achieve 2.00 GPA overall and departmental minimum.

STUDENT PROGRAMS

PROFESSIONAL AND HONORARY SOCIETIES
[For key to acronyms, see *Introduction*.]
Professional Societies: ACS, AIChE, AIME, ASCE, ASHRAE, ASME, IEEE, NSPE, SEG, SME, American Society for Metals (ASM), Society of Women Engineers (SWE)
Honorary Societies: Eta Kappa Nu, Pi Tau Sigma, Tau Beta Pi, Phi Lambda Upsilon, Alpha Sigma MU

SUPPORT PROGRAMS
Student Chapter Organizations: Society of Women Engineers, National Society of Black Engineers, American Indian Science & Engineering Society, Hispanic Student Organization

Minority College Access Program (MCAP)
For: Ethnic Minorities Available: Summer only
Offered: Prematriculation

Minorities in Engineering Workshop (MIE)
For: Ethnic Minorities Available: Summer only
Offered: Prematriculation

Wade McCree Incentive Scholarship
For: Ethnic Minorities Available: Year round
Offered: Prematriculation, Freshman, Sophomore, Junior, Senior

WIE Minority Participation
For: Women Available: Summer only
Offered: Prematriculation

Other Engineering Support Programs: Summer and Fall orientation program for all new students. Career counseling programs, personal counseling, International Student office, International Exchange programs for work and/or study, veterans' services, placement services.

134 Milwaukee School of Engineering

■ INSTITUTION PROFILE
HEAD OF THE INSTITUTION
Hermann Viets
Phone: (414)277-7100   Fax: (414)277-7468

GENERAL INFORMATION
[All Students—Fall 1991]
Undergraduate enrollment 2,775
Graduate enrollment 391
Total institution enrollment 3,166
Type of institution: Private   **Calendar:** Quarters
Location: Milwaukee   **Population:** 1,400,000
Setting: Urban
Types of engineering degrees: Engineering & Technology

■ ENGINEERING ADMISSIONS
ADMISSIONS OFFICE CONTACT
T. O. Smith
Milwaukee School of Engineering
1025 N. Broadway
Milwaukee, WI 53202-3109
Phone: (800)332-6763   Fax: (414)277-7475

ENGINEERING COLLEGE ADMISSIONS INFORMATION
Entrance Requirements: ACT, High School courses—English (4 years), Chemistry (1 year), Physics (1 year), Mathematics (Algebra 2 Geometry 1 Pre-Calculus 1) (4 years)
Entrance Recommendations: SAT, High School courses—Foreign Language (2 years), Computers (1 year), Mechanical Drawing (1 year)
Requirements for foreign students: TOEFL (Score: 500); financial statement

DEPARTMENTAL ADMISSIONS INFORMATION
Admission to the engineering department: At the time of admission to the institution.

■ FINANCIAL INFORMATION
ESTIMATED EXPENSES (ACADEMIC YEAR)
[Expenses are for the 1992-93 nine-month academic year.]

All Students	Undergraduate	Graduate
Tuition and fees	$ 9,960	$ 7,020[1]
College room and board	$ 3,045	$ 3,045
Books and supplies	$ 1,000	$ 1,000
Other expenses	$ 1,920	$ 1,920
Total estimated expenses	**$15,925**	**$12,985**

Notes: (1) Based on 9 credits per quarter.

FINANCIAL AID OFFICE CONTACT
Susan Hebert, *Director of Financial Aid*
Milwaukee School of Engineering
1025 N. Broadway
Milwaukee, WI 53202-3109
Phone: (414)277-7224   Fax: (414)277-7450

GENERAL FINANCIAL AID INFORMATION
Forms accepted/required: ACT, AFSA, CSS-FAF, FAT, FSA, FFS, Stafford, SAR, Supplemental, Institutional, As required
Additional financial aid information: Academic and need-based scholarships; Federal and State grant programs; Federal and institutional loan programs; College work-study program.

■ ENGINEERING COLLEGE INFORMATION
[For additional personnel, refer to the *Appendix*.]

HEAD OF ENGINEERING
Thomas W. Davis
Phone: (414)277-7324   Fax: (414)277-7470
EMail: davis@kirk.msoe.edu

ENGINEERING COLLEGE ADDRESS
Milwaukee School of Engineering
1025 N. Broadway
Milwaukee, WI 53202-3109
Phone: (414)277-7300   Fax: (414)277-7477

ENROLLMENTS—BY CLASS
[Numbers are baccalaureate enrollments for the fall 1991 term, unless otherwise footnoted.]
1st-year students/Freshmen 515
2nd-year students/Sophomores 366
3rd-year students/Juniors 266
4th-year students/Seniors 277
Total 1,424

NUMBER OF DEGREES AWARDED—BY PROGRAM
[Numbers are engineering baccalaureate degrees awarded during the 1991-1992 academic year, unless otherwise footnoted. For full details about each engineering program, refer to the *Tables of Degree Programs*.]
Architectural Engineering 47
Biomedical Engineering 6
Computer Engineering 20
Electrical Engineering 66
General Engineering .... –
Industrial Engineering 15
Mechanical Engineering 56
Total 210

PERCENTAGE OF DEGREES AWARDED—BY CATEGORY
[Percentages are of all engineering baccalaureate degrees awarded during the 1991-1992 academic year, unless otherwise footnoted.]
African Americans .... 2.0 %
Asian/Pacific Island Americans .... 3.0 %
Hispanic Americans .... 1.0 %
Native Americans .... – %
Foreign Citizens .... 2.0 %
All Others .... 92.0 %
Women .... 13.0 %
Persons with disabilities .... – % (A)
Students over 25 years of age .... – % (A)

Notes: (A) Data not available.

GRADUATE ENROLLMENTS & DEGREES AWARDED
Master's enrollment 70
Master's degrees awarded 12
Doctoral enrollment .... –
Doctoral degrees awarded .... –

ENGINEERING STUDENT DATA
Applicants to the engineering college
Number of applicants to engineering college 884
Percent offered admission .... 90.0 %

Matriculated engineering students
Percentage in top quartile (25%) of High School class 72.0
Average SAT scores: Math—570, Verbal—460, Combined—1030
Average ACT scores: Math—25.60, Composite—24.87

FULL- & PART-TIME FACULTY—BY DEPARTMENT
[Figures are the head count of full-time faculty and the full-time equivalent (FTE) of part-time faculty for each engineering department or equivalent.]

Department	Full	Part
Electrical Eng	32	11.5
Industrial Eng	22	8.0
Mechanical Eng	22	8.0
Computer Eng	32	11.5
Biomedical Eng	32	11.5
Architectural Eng	6	1.9

COLLEGE DESCRIPTION
MSOE offers four-year degree programs in six engineering disciplines. In each program students complete laboratory intensive courses in their field of specialization plus numerous cross disciplinary courses in other engineering fields, and appropriate course work in humanities, social science, business and communication. There are opportunities for minors or dual degrees available in business and management, and technical communication. Undergraduates have the opportunity to gain additional experience by conducting industrially sponsored applied research through the college's Applied Technology Center and research consortia in fluid power, biomedical systems, rapid prototyping and other areas. Several foreign work/study programs are available. The college also offers an interdisciplinary Master of Science degree program in engineering. Average number of years required to actually complete the bachelor's degree: 4.6

DEGREE PROGRAMS DESCRIPTION
The degree programs prepare students for entry level professional positions and for continued graduate study. Courses and laboratory sessions are taught by qualified faculty with current industrial experience in their field of specialty. Laboratories are industrially sponsored and utilize full size industrial quality equipment. Course sequencing and course development are guided by 18 business and industry advisory committees. Students may pursue advanced study at the Master's level in engineering or engineering management. These programs are industrially oriented and multidisciplinary. Students in the Master of Science degree program in engineering may pursue specialities in materials, fluid power, computers, electronics and controls, manufacturing and imaging.

TRANSFER INFORMATION
Residency Requirements: Students must complete at least 50% of required courses at MSOE for a degree from MSOE.

Transfer via Articulation Agreements
Admission to engineering: At any time
Engineering graduates transferred from   . . . 4-yr: –   2-yr: –
Requirements: Official transcripts of all high school, college and university courses must be submitted. A grade of 'C' or better is required for a course to be considered for transfer. A financial aid transcript from the Financial Aid Office of any previously attended college must be submitted to the Financial Aid Office. Completion of minimum course work as specified in the Academic Catalog.

Transfer without Articulation Agreements
Admission to engineering: At any time
Engineering graduates transferred from   . . . 4-yr: 82   2-yr: 115
Requirements: Evaluated on an individual basis.

GRADUATION REQUIREMENTS
B.S. programs in engineering require 203 to 208 quarter credits, depending on major selected. A 2.0 cumulative grade point average and a 2.0 major grade point average is required.

■ STUDENT PROGRAMS
PROFESSIONAL AND HONORARY SOCIETIES
[For key to acronyms, see *Introduction*.]
Professional Societies: AAPG, ACM, AGCA, AIA, ASCE, ASME, IEEE, IIE, SAE, SME
Honorary Societies: Eta Kappa Nu, Kappa Eta Kappa, Tau Alpha Pi, Tau Omega Mu, Circle K

SUPPORT PROGRAMS
Student Chapter Organizations: Society of Women Engineers, National Society of Black Engineers, Society of Hispanic Professional Engineers, Society of International Students

Upward Bound
For: Ethnic Minorities   **Available:** Summer only
Offered: Prematriculation

Catalyst for Future Success
For: Ethnic Minorities   **Available:** Summer only
Offered: Prematriculation

IDEAS Program
For: Women   **Available:** Academic year
Offered: Freshman

Other Engineering Support Programs: Fall orientation program for all freshmen and transfer students; New student orientation course (1 quarter) for all freshmen; First year mentoring program; Learning Resource Center; Personal and career counseling services; Foreign student office; Veteran's services; Services for handicapped students; Placement services; Student support services; Student Life activities.

135 University of Minnesota, Duluth

■ INSTITUTION PROFILE
HEAD OF THE INSTITUTION
Lawrence A. Ianni
Phone: (218)726-7106   Fax: (218)726-6535

GENERAL INFORMATION
[All Students—Fall 1991]
Undergraduate enrollment 7,411
Graduate enrollment 376
Total institution enrollment **7,787**

Type of institution: Public   **Calendar:** Quarters
Nearest city: Minneapolis-St. Paul   **Population:** 350,000
Miles from main campus: 150   **Setting:** Urban
Types of engineering degrees: Engineering
Other degree-granting colleges: Business Administration, Education, Fine & Performing Arts, Medicine, Liberal Arts

■ ENGINEERING ADMISSIONS
ADMISSIONS OFFICE CONTACT
Gerald R. Allen
University of Minnesota, Duluth
10 University Drive
Duluth, MN 55812-2496
Phone: (218)726-7500   Fax: (218)726-6389

ENGINEERING COLLEGE ADMISSIONS INFORMATION
Entrance Requirements: ACT (Score: 19), High School courses—English (4 years), Social Studies including Amer History (2 years), Mathematics (3 years), Science (3 years)
Entrance Recommendations: ACT (Score: 19), High School courses—English (4 years), Social Studies including Amer History (2 years), Mathematics (3 years), Science (3 years)
Requirements for foreign students: TOEFL (Score: 500); financial statement

DEPARTMENTAL ADMISSIONS INFORMATION
Admission to the engineering department: At the end of the second year.
Additional information: Satisfactory completion of lower division requirements. Admission is competitive, based on cumulative grade point average and space availability.

■ FINANCIAL INFORMATION
ESTIMATED EXPENSES (ACADEMIC YEAR)
[Expenses are for the 1992-93 nine-month academic year.]

State Residents	Undergraduate	Graduate
Tuition and fees	$ 3,045	$ –
College room and board	$ 3,300	$ –
Books and supplies	$ 700	$ –
Other expenses	$ –	$ –
Total estimated expenses	**$ 7,045**	**$ –**

Out-of-State Residents	Undergraduate	Graduate
Tuition and fees	$ 8,421	$ –
College room and board	$ 3,300	$ –
Books and supplies	$ 700	$ –
Other expenses	$ –	$ –
Total estimated expenses	**$12,421**	**$ –**

FINANCIAL AID OFFICE CONTACT
Nicholas F. Whelihan, *Director*
University of Minnesota, Duluth
10 University Drive
Duluth, MN 55812-2496
Phone: (218)726-7500   Fax: (218)726-8787

GENERAL FINANCIAL AID INFORMATION
Forms accepted/required: FAT, FFS, Stafford, SAR, Supplemental
Additional financial aid information: Academic Scholarships, Perkins Loans, Robert Stafford Loans, Supplemental Education, State and Federal College work study, Pell and State grants.

■ ENGINEERING COLLEGE INFORMATION
[For additional personnel, refer to the *Appendix*.]

HEAD OF ENGINEERING
Sabra S. Anderson
Phone: (218)726-7201   Fax: (218)726-6360
EMail: SANDERSO@UB.D.UMN.EDU

ENGINEERING COLLEGE ADDRESS
College of Science and Engineering
University of Minnesota, Duluth
10 University Drive
140 Engineering Building
Duluth, MN 55812-2496
Phone: (218)726-7201   Fax: (218)726-6360

ENROLLMENTS—BY CLASS
[Numbers are baccalaureate enrollments for the fall 1991 term, unless otherwise footnoted.]
1st-year students/Freshmen 85
2nd-year students/Sophomores 79
3rd-year students/Juniors 85
4th-year students/Seniors 168
Total **417**

NUMBER OF DEGREES AWARDED—BY PROGRAM
[Numbers are engineering baccalaureate degrees awarded during the 1991-1992 academic year, unless otherwise footnoted. For full details about each engineering program, refer to the *Tables of Degree Programs*.]
Chemical Engineering 11
Computer & Electrical Engineering 34
Industrial Engineering 24
Total **69**

PERCENTAGE OF DEGREES AWARDED—BY CATEGORY
[Percentages are of all engineering baccalaureate degrees awarded during the 1991-1992 academic year, unless otherwise footnoted.]
African Americans .... – % (A)
Asian/Pacific Island Americans .... – % (A)
Hispanic Americans .... – % (A)
Native Americans .... – % (A)
Foreign Citizens .... – % (A)
All Others .... 100 %
Women .... – % (A)
Persons with disabilities .... – % (A)
Students over 25 years of age .... – % (A)
Notes: (A) Data not available.

FULL- & PART-TIME FACULTY—BY DEPARTMENT
[Figures are the head count of full-time faculty and the full-time equivalent (FTE) of part-time faculty for each engineering department or equivalent.]

Department	Full	Part
Chemical Eng	5	1.0
Industrial Eng	6	1.0
Computer Eng	9	1.0

COLLEGE DESCRIPTION
College features a balance between theoretical and practical experience as well as sensitivity to social and environmental impact. There are many opportunities for undergraduate research projects, internships, and relevant work experiences. The college also offers a degree in Computer Science and many students graduate with double majors in Computer Science and Computer Engineering. The college is initiating a co-op program.

DEGREE PROGRAMS DESCRIPTION
Internships and relevant work experiences are featured in all of the Bachelor's degrees in engineering. A balance of theoretical and practical experience, as well as a sensitivity to social and environmental impact are observed. The Industrial Engineering program has a significant manufacturing and computer bias.

TRANSFER INFORMATION
Residency Requirements: Minimum of 45 resident quarter credits.
Transfer via Articulation Agreements
Admission to engineering: At the end of the sophomore year
Engineering graduates transferred from .. 4-yr: –   2-yr: 4
Requirements: Completion of lower division requirements; admission competitive and based on GPA and space availability.
Transfer without Articulation Agreements
Admission to engineering: At any time
Engineering graduates transferred from .. 4-yr: –   2-yr: 2
Requirements: Good academic standing (2.00 or above); listed as pre-engineering students until other requirements are met.

GRADUATION REQUIREMENTS
Completion of required number of credits. Cumulative GPA within major, minor and overall of 2.00 (C) or above. Filing of degree requirement form. Compliance with general regulations governing the grant of degrees.

■ STUDENT PROGRAMS
PROFESSIONAL AND HONORARY SOCIETIES
[For key to acronyms, see *Introduction*.]
Professional Societies: ACM, ACS, AIChE, AIME, AIP, APS, GSA, IEEE, IIE
Honorary Societies: Tau Beta Pi

SUPPORT PROGRAMS
Student Chapter Organizations: Women in Science and Engineering, American Indian Science and Engineering Society

Women in Science and Engineering
For: Women   **Available:** Academic year
Offered: Freshman, Sophomore, Junior, Senior, Graduate level

Women Mentoring Women
For: Women & Ethnic Minorities   **Available:** Academic year
Offered: Freshman

American Indian Learning Resource Center
For: Ethnic Minorities   **Available:** Year round
Offered: Prematriculation, Freshman, Sophomore, Junior, Senior, Graduate level

Afro-American Student Service Coordinator
For: Ethnic Minorities   **Available:** Year round
Offered: Prematriculation, Freshman, Sophomore, Junior, Senior, Graduate level

Other Engineering Support Programs: Freshman/transfer student orientation/registration program; individual academic advising program; Career and Placement Services; department career advisers; Student Health & Counseling Center.

136 University of Minnesota

INSTITUTION PROFILE

HEAD OF THE INSTITUTION
Nils Hasselmo
Phone: (612)626-1616   Fax: (612)625-3875

GENERAL INFORMATION
[All Students—Fall 1991]
Undergraduate enrollment 29,514
Graduate enrollment 8,505
Total institution enrollment **38,019**
Type of institution: Public   Calendar: Quarters
Location: Minneapolis-St. Paul   Population: 1,900,000
Setting: Urban
Types of engineering degrees: Engineering
Other degree-granting colleges: Agricultural & Environmental, Allied Health Sciences, Architecture, Arts & Sciences, Business Administration, Communication & Journalism, Dentistry, Education, Fine & Performing Arts, Humanities & Social Sciences, Law, Medicine, Nursing, Pharmacy, Technology & Applied Sciences, Veterinary Medicine, Architecture & Landscape Architecture, Public Health, Liberal Arts, Natural Sciences

ENGINEERING COLLEGE ADMISSIONS INFORMATION
Admission to the engineering college: Students can enter IT as freshmen, sophomores or juniors.
Entrance Requirements: High School courses—English (4 years), Foreign Language (2 years), Mathematics (4 years), History (2 years), ACT required for residents of MN, IA, ND, SD, WI. ACT, SAT or PSAT for residents of other states or countries. ACT Math = 26; SAT Math = 550; PSAT Math = 55.
Entrance Recommendations: High School courses—Biology (required) (1 year), Chemistry (required) (1 year), Physics (required) (1 year)
Requirements for out-of-state residents: Resident admission: Minnesota, North Dakota, South Dakota, Wisconsin, and Manitoba. All others must meet higher standards.
Requirements for foreign students: TOEFL (Score: 550); financial statement

DEPARTMENTAL ADMISSIONS INFORMATION
Admission to the engineering department: At the end of the second year.
Additional information: Complete lower division courses; Grade Point Average of at least 2.3 to 2.7 depending on major. (Scale of 0 - 4.0)

FINANCIAL INFORMATION

ESTIMATED EXPENSES (ACADEMIC YEAR)
[Expenses are for the 1992-93 nine-month academic year.]

All Students	Undergraduate	Graduate
Tuition and fees	$ 2,772 [1]	$ 3,380
College room and board	$ 3,400	$ 3,400
Books and supplies	$ 700	$ 800
Other expenses	$ —[A]	$ —[A]
Total estimated expenses	**$ 6,872**	**$ 7,580**

Notes: (A) Data not available. (1) Data are for lower division. Upper division annual cost is $3192 per year.

GENERAL FINANCIAL AID INFORMATION
Forms accepted/required: ACT, Institutional
Additional financial aid information: Need-based scholarships, student loans, work study, merit scholarships.

ENGINEERING COLLEGE INFORMATION
[For additional personnel, refer to the *Appendix*.]

HEAD OF ENGINEERING
Gordon S. Beavers
Phone: (612)624-2006   Fax: (612)624-2841
EMail: beavers@mailbox.mail.umn.edu

ENGINEERING COLLEGE ADDRESS
Institute of Technology Office of Student Affairs
University of Minnesota
106 Lind Hall
207 Church St. SE
Minneapolis, MN 55455-0193
Phone: (612)624-8504   Fax: (612)626-0261

ENROLLMENTS—BY CLASS
[Numbers are baccalaureate enrollments for the fall 1991 term, unless otherwise footnoted.]
1st-year students/Freshmen 540
2nd-year students/Sophomores 682
3rd-year students/Juniors 817
4th-year students/Seniors 1,488
Total **3,527**

NUMBER OF DEGREES AWARDED—BY PROGRAM
[Numbers are engineering baccalaureate degrees awarded during the 1991-1992 academic year, unless otherwise footnoted. For full details about each engineering program, refer to the *Tables of Degree Programs*.]
Aerospace Engineering & Mechanics 83
Agricultural Engineering 10
Chemical Engineering 84
Civil Engineering 113
Electrical Engineering 141
Extractive Metallurgical Engineering 4
Geological Engineering 3
Materials Science 9
Mechanical Engineering 210
Total **657**

PERCENTAGE OF DEGREES AWARDED—BY CATEGORY
[Percentages are of all engineering baccalaureate degrees awarded during the 1991-1992 academic year, unless otherwise footnoted.]
African Americans .... – %
Asian/Pacific Island Americans .... 7.3 %
Hispanic Americans .... – %
Native Americans .... – %
Foreign Citizens .... 3.3 %
All Others .... 89.4 %
Women .... 14.9 %
Persons with disabilities .... – % (A)
Students over 25 years of age .... – % (A)
Notes: (A) Data not available.

GRADUATE ENROLLMENTS & DEGREES AWARDED
Master's enrollment 502
Master's degrees awarded 184
Doctoral enrollment 561
Doctoral degrees awarded 103

ENGINEERING STUDENT DATA
Applicants to the engineering college
Number of applicants to engineering college 1,572
Percent offered admission .... 60.0 %
Matriculated engineering students
Percentage in top quartile (25%) of High School class 98.7
Average SAT scores: Math—646.6, Verbal—507.8, Combined—1154.4
Average ACT scores: Math—28.0, Composite—26.6

FULL- & PART-TIME FACULTY—BY DEPARTMENT
[Figures are the head count of full-time faculty and the full-time equivalent (FTE) of part-time faculty for each engineering department or equivalent.]

Department	Full	Part
Chemical Eng & Materials Science	31	–
Civil & Mineral Eng	37	–
Electrical Eng	50	–
Mechanical Eng	42	–
Aerospace Eng & Mechanics	22	–
Agricultural Eng	17	–
Computer Science	29	–

COLLEGE DESCRIPTION
The Institute of Technology offers four-year undergraduate programs in engineering, computer science, mathematics, chemistry, physics, astronomy and earth sciences. The engineering programs emphasize a thorough grounding in basic sciences and engineering fundamentals during the first two years, with specialized education in the major during the last two years. Liberal education courses are taken throughout the program. Students may obtain an additional BA degree through the College of Liberal Arts, or they may augment their engineering degrees with minors in liberal arts or management. Co-op programs are available in some majors. Each student completes a senior engineering design project. Honors degrees are available.

DEGREE PROGRAMS DESCRIPTION
The various engineering bachelor's degrees prepare the graduate for entry level professional positions or for graduate study. Graduate study in engineering can lead to the M.S., the Master of Engineering, or the Ph.D. Interdisciplinary programs in management of technology are available through the Center for the Development of Technological Leadership, 107 Lind Hall, 207 Church St. SE, Minneapolis, MN 55455.

DUAL DEGREE PROGRAMS
Engineering baccalaureate graduates with dual degrees 6
Enrollment requirements: May transfer with 3 or more years of academic work, including all appropriate course work. GPA: 3.0 for EE, ME, Aero; 2.8 for others.

TRANSFER INFORMATION
Residency Requirements: 45 quarter credits after admission, 30 of which are taken during the senior year.
Transfer via Articulation Agreements
Admission to engineering: At the end of the sophomore year
Engineering graduates transferred from ... 4-yr: 120 2-yr: –
Requirements: Completion of freshman and sophomore courses; high Grade Point Average.
Transfer without Articulation Agreements
Admission to engineering: At the end of the sophomore year
Engineering graduates transferred from ... 4-yr: – 2-yr: 200
Requirements: Completion of freshman and sophomore pre-engineering courses; high GPA.

STUDENT PROGRAMS

PROFESSIONAL AND HONORARY SOCIETIES
[For key to acronyms, see *Introduction*.]
Professional Societies: AIAA, AIChE, AIME, ASAE, ASCE, ASME, IEEE, IIE, SAE, SPS, Environmental Engineering Society
Honorary Societies: Alpha Epsilon, Chi Epsilon, Eta Kappa Nu, Kappa Eta Kappa, Pi Tau Sigma, Sigma Gamma Tau, Tau Beta Pi, Theta Tau, Alpha Chi Sigma, Alpha Sigma Kappa, Plumb Bob, Triangle, IT Student Honors Group

SUPPORT PROGRAMS
Student Chapter Organizations: Society of Women Engineers, National Society of Black Engineers, American Indian Science and Engineering Society, MINIT (Minorities in IT), Minn. African Am Society of Engineers (MAASE)
Project Technology Power
For: Ethnic Minorities   Available: Year round
Offered: Prematriculation, Freshman, Sophomore, Junior, Senior, Graduate level
Asian/Pacific Amer, African Amer Learning Resource Centers
For: Ethnic Minorities   Available: Year round
Offered: Freshman, Sophomore, Junior, Senior, Graduate level
American Indian, Chicano/Latino Learning Resources Centers
For: Ethnic Minorities   Available: Year round
Offered: Prematriculation, Freshman, Sophomore, Junior, Senior, Graduate level
Other Engineering Support Programs: Freshman-sophomore honors program; two-day orientation for new students; faculty advising for 'teams' of all freshmen; career counseling courses; extensive tutorial program; office for students with disabilities; placement office; office for international students.

137 University of Mississippi

■ INSTITUTION PROFILE

HEAD OF THE INSTITUTION
R. Gerald Turner, *Chancellor*
Phone: (601)232-7111   Fax: (601)232-5935

GENERAL INFORMATION
[All Students—Fall 1991]
Undergraduate enrollment 8,791
Graduate enrollment 1,741
Total institution enrollment 11,033

Type of institution: Public   **Calendar:** Semesters
Nearest city: Memphis, Tennessee   **Population:** 1,000,000
Miles from main campus: 70   **Setting:** Small Town
Types of engineering degrees: Engineering

Other degree-granting colleges: Allied Health Sciences, Arts & Sciences, Business Administration, Dentistry, Education, Law, Medicine, Nursing, Pharmacy, Technology & Applied Sciences, Accountancy

■ ENGINEERING ADMISSIONS

ADMISSIONS OFFICE CONTACT
Allie M. Smith
University of Mississippi
101 Carrier Hall
University, MS 38677
Phone: (601)232-7407   Fax: (601)232-7796

ENGINEERING COLLEGE ADMISSIONS INFORMATION

Entrance Requirements: ACT (Score: 20), High School courses—English (4 years), Mathematics: 2 algebra, 1 geometry, 1 trigonometry (4 years), Natural Sciences: chemistry, physics, biology (3 years), Social sciences (must include US History. & Am. Gov.) (2.5 years)
Entrance Recommendations: High School courses—Foreign language (2 years), Computer (1 year)
Requirements for foreign students: TOEFL (Score: 550); financial statement; High school diploma with B average or better.

DEPARTMENTAL ADMISSIONS INFORMATION
Admission to the engineering department: At the time of admission to the institution.

■ FINANCIAL INFORMATION

ESTIMATED EXPENSES (ACADEMIC YEAR)
[Expenses are for the 1992-93 nine-month academic year.]

State Residents	Undergraduate	Graduate
Tuition and fees	$ 2,435	$ 2,435
College room and board	$ 3,370	$ 4,460
Books and supplies	$ 460	$ 760
Other expenses	$ –	$ –
Total estimated expenses	$ 6,265	$ 7,655

Out-of-State Residents	Undergraduate	Graduate
Tuition and fees	$ 4,395	$ 4,395
College room and board	$ 3,370	$ 4,460
Books and supplies	$ 460	$ 760
Other expenses	$ –	$ –
Total estimated expenses	$ 8,225	$ 9,615

FINANCIAL AID OFFICE CONTACT
Thomas G. Hood, *Director of Financial Aid*
University of Mississippi
25 Old Chemistry Building
University, MS 38677
Phone: (601)232-7175   Fax: (601)234-8155

GENERAL FINANCIAL AID INFORMATION
Forms accepted/required: FAF, Institutional
Additional financial aid information: Need-based scholarships; short-term loans; long-term loans; federally, state, and foundation funded grants; out-of-state tuition grants; service scholarships; ROTC scholarships, merit-based scholarships; part-time jobs on campus; College Work-Study program.

■ ENGINEERING COLLEGE INFORMATION

[For additional personnel, refer to the *Appendix*.]

HEAD OF ENGINEERING
Allie M. Smith
Phone: (601)232-7407   Fax: (601)232-7796
EMail: ENAS@UMSVM

ENGINEERING COLLEGE ADDRESS
School of Engineering
University of Mississippi
Carrier Hall
University, MS 38677
Phone: (601)232-7407   Fax: (601)232-7796

ENROLLMENTS—BY CLASS
[Numbers are baccalaureate enrollments for the fall 1991 term, unless otherwise footnoted.]
1st-year students/Freshmen 155
2nd-year students/Sophomores 84
3rd-year students/Juniors 105
4th-year students/Seniors 153
Total 497

NUMBER OF DEGREES AWARDED—BY PROGRAM
[Numbers are engineering baccalaureate degrees awarded during the 1991-1992 academic year, unless otherwise footnoted. For full details about each engineering program, refer to the *Tables of Degree Programs*.]
Chemical Engineering 5
Civil Engineering 18
Computer & Information Science 18
Electrical Engineering 16
Engineering Science(s) .... –
General Engineering 17
Geological Engineering 2
Geology 1
Mechanical Engineering 7
Total 84

PERCENTAGE OF DEGREES AWARDED—BY CATEGORY
[Percentages are of all engineering baccalaureate degrees awarded during the 1991-1992 academic year, unless otherwise footnoted.]
African Americans .... 4.8 %
Asian/Pacific Island Americans .... 14.3 %
Hispanic Americans .... 1.0 %
Native Americans .... – %
Foreign Citizens .... – %
All Others .... 79.9 %
Women .... 16.7 %
Persons with disabilities .... – %
Students over 25 years of age .... – % (A)

Notes: (A) Data not available.

GRADUATE ENROLLMENTS & DEGREES AWARDED
Master's enrollment 135
Master's degrees awarded 51
Doctoral enrollment 50
Doctoral degrees awarded 4

ENGINEERING STUDENT DATA
Applicants to the engineering college
Number of applicants to engineering college 482
Percent offered admission 60.0%
Average ACT scores: Composite—24.7

FULL- & PART-TIME FACULTY—BY DEPARTMENT
[Figures are the head count of full-time faculty and the full-time equivalent (FTE) of part-time faculty for each engineering department or equivalent.]

Department	Full	Part
Chemical Eng	5	0.8
Civil Eng	7	–
Electrical Eng	9	0.3
Mechanical Eng	7	0.2
Geology & Geological Eng	6	–
Computer & Information Science	7	–

COLLEGE DESCRIPTION
The school offers four-year undergraduate programs of study which stress the engineering sciences and are based on fundamental concepts and principles of the natural sciences and mathematics. The accredited programs are in the five basic and broad engineering fields—chemical, civil, electrical, geological, and mechanical—as well as in the field of computer science. The school also offers a broad and flexible Bachelor of Engineering degree which is designed to provide students the opportunity to gain an understanding of scientific and engineering knowledge which will enhance their career objectives in such areas as engineering, science, medicine, law, telecommunications, military, business, etc. This degree program also has many option paths available in engineering's areas, such as astronautical engineering and computer engineering.

DEGREE PROGRAMS DESCRIPTION
The B.S. degrees in the various engineering programs, computer science, and geology prepare the graduate for entry-level professional positions in practice, research, or graduate study in each of the respective fields. The Bachelor of Engineering degree provides students the opportunity to gain an understanding of scientific and engineering knowledge which will enhance their career objectives in such areas as engineering, science, medicine, law, telecommunications, military, business, etc. Graduate study can be carried out in all areas of engineering science at both the M.S. and Ph.D. levels with emphasis in the following general areas: chemical, civil, electrical, geological, and mechanical engineering, computer science, environmental, geology, and computational science and hydroscience. An M.S. program of study is also available in geology.

DUAL DEGREE PROGRAMS
Enrollment requirements: Minimum of 2.00 GPA on all course work taken at other approved colleges. International transfer students must have an overall B average or better on all courses taken at other approved colleges.

TRANSFER INFORMATION
Residency Requirements: At least one year of work in residence is required, including principally senior work for the baccalaureate degrees and totaling not less than 30 semester hours in engineering courses.

Transfer via Articulation Agreements
Admission to engineering: At any time
Requirements: Min. 2.00 GPA on course work taken at other approved colleges. International transfer students must have an overall B average or better on all courses taken at other approved colleges.

Transfer without Articulation Agreements
Admission to engineering: At any time
Requirements: Min. 2.00 GPA on course work taken at other approved colleges. International transfer students must have an overall B average or better on all courses taken at other approved colleges.

GRADUATION REQUIREMENTS
Undergraduate degree programs in engineering require 127-138 semester credits, depending on major selected. In addition to the number of semester hours of prescribed courses, candidates for degrees must have earned a grade point average of 2.00, on a 4.00 system, on all courses submitted in fulfillment of degree requirements, in all engineering courses, and in the designated major department courses.

■ STUDENT PROGRAMS

PROFESSIONAL AND HONORARY SOCIETIES
[For key to acronyms, see *Introduction*.]
Professional Societies: ACM, ACS, AEG, AIAA, AIChE, ASCE, ASME, GSA, IEEE, NSPE
Honorary Societies: Chi Epsilon, Eta Kappa Nu, Sigma Gamma Tau, Tau Beta Pi, Phi Kappa Phi

SUPPORT PROGRAMS
Student Chapter Organizations: Society of Women Engineers, National Society of Black Engineers

Black Student Union
For: Ethnic Minorities   **Available:** Academic year
Offered: Freshman, Sophomore, Junior, Senior, Graduate level

Other Engineering Support Programs: Summer, fall, and spring orientation programs for all new undergraduate students; Learning Center; tutoring; Center for Women's Studies; personal and career counseling; foreign student office; veterans' services; services for blind and handicapped; placement services.

138 Mississippi State University

INSTITUTION PROFILE

HEAD OF THE INSTITUTION
Donald W. Zacharias
Phone: (601)325-3221 **Fax:** (601)325-8028

GENERAL INFORMATION
[All Students—Fall 1991]
Undergraduate enrollment 11,657
Graduate enrollment 2,210
Total institution enrollment 13,867
Type of institution: Public **Calendar:** Semesters
Nearest city: Jackson, Mississippi **Population:** 202,000
Miles from main campus: 100 **Setting:** Small Town
Types of engineering degrees: Engineering
Other degree-granting colleges: Architecture, Arts & Sciences, Education, Agriculture and Home Economics, Business and Industry, Forest Resources, Veterinary Medicine

ENGINEERING ADMISSIONS

ADMISSIONS OFFICE CONTACT
Jerry B. Inmon
Mississippi State University
Director of Admissions
P.O. Box 5268
Mississippi State, MS 39762-9700
Phone: (601)325-2224 **Fax:** (601)325-1846

ENGINEERING COLLEGE ADMISSIONS INFORMATION
Entrance Requirements: High School courses—English (4 years), Mathematics (4 years), Science (3 years), Social studies and/or foreign languages (2 years), The applicant must submit an official American College Test (ACT) score of 18 or more. Non-resident applicants may submit a Scholastic Aptitude Test (SAT) score of at least 710 in lieu of ACT.
Entrance Recommendations: High School courses—Electives (2 years), Mechanical drawing/computer aided design (1 year), Keyboarding
Requirements for foreign students: TOEFL (Score: 550); financial statement

DEPARTMENTAL ADMISSIONS INFORMATION
Admission to the engineering department: At the time of admission to the institution.

FINANCIAL INFORMATION

ESTIMATED EXPENSES (ACADEMIC YEAR)
[Expenses are for the 1992-93 nine-month academic year.]

State Residents	Undergraduate	Graduate
Tuition and fees	$ 2,223	$ 2,456
College room and board	$ 3,400	$ 3,400
Books and supplies	$ 600	$ 700
Other expenses	$ –	$ –
Total estimated expenses	**$ 6,223**	**$ 6,556**

Out-of-State Residents	Undergraduate	Graduate
Tuition and fees	$ 3,686	$ 4,415
College room and board	$ 3,400	$ 3,400
Books and supplies	$ 600	$ 700
Other expenses	$ –	$ –
Total estimated expenses	**$ 7,686**	**$ 8,515**

FINANCIAL AID OFFICE CONTACT
Audrey S. Lambert, *Director of Student Financial Aid and Scholarships*
Mississippi State University
P.O. Box 6238
106 Magruder
Mississippi State, MS 39762
Phone: (601)325-2450

GENERAL FINANCIAL AID INFORMATION
Forms accepted/required: FFS, IRS, Institutional
Additional financial aid information: Need-based grants; short term loans; merit-based scholarships; part-time campus employment; College Work-Study Program.

ENGINEERING COLLEGE INFORMATION
[For additional personnel, refer to the *Appendix*.]

HEAD OF ENGINEERING
Robert A. Altenkirch
Phone: (601)325-2269 **Fax:** (601)325-8573
EMail: raa@de.msstate.edu

ENGINEERING COLLEGE ADDRESS
College of Engineering
Mississippi State University
P.O. Drawer DE
106 McCain Engineering Building
Mississippi State, MS 39762-5635
Phone: (601)325-2269 **Fax:** (601)325-8573

ENROLLMENTS—BY CLASS
[Numbers are baccalaureate enrollments for the fall 1991 term, unless otherwise footnoted.]
1st-year students/Freshmen 502
2nd-year students/Sophomores 398
3rd-year students/Juniors 525
4th-year students/Seniors 818
Total **2,243**

NUMBER OF DEGREES AWARDED—BY PROGRAM
[Numbers are engineering baccalaureate degrees awarded during the 1991-1992 academic year, unless otherwise footnoted. For full details about each engineering program, refer to the *Tables of Degree Programs*.]
Aerospace Engineering 35
Agricultural Engineering 2
Biological Engineering 6
Chemical Engineering 28
Civil Engineering 29
Computational Engineering .... – (4)
Computer Engineering (Electrical) 20
Electrical Engineering 75
Engineering Mechanics .... – (7)
Industrial Engineering 46
Mechanical Engineering 50
Nuclear Engineering 4
Petroleum Engineering 4
Total **299**

Notes: (4) M.S. and Ph.D. programs only. (7) M.S. program only.

PERCENTAGE OF DEGREES AWARDED—BY CATEGORY
[Percentages are of all engineering baccalaureate degrees awarded during the 1991-1992 academic year, unless otherwise footnoted.]
African Americans 6.0%
Asian/Pacific Island Americans 3.0%
Hispanic Americans .... – %
Native Americans .... – %
Foreign Citizens 8.0%
All Others 83.0%
Women 10.0%
Persons with disabilities .... – % (A)
Students over 25 years of age .... – % (A)

Notes: (A) Data not available.

GRADUATE ENROLLMENTS & DEGREES AWARDED
Master's enrollment 237
Master's degrees awarded 69
Doctoral enrollment 69
Doctoral degrees awarded 14

ENGINEERING STUDENT DATA
Applicants to the engineering college
Number of applicants to engineering college 1,095
Percent offered admission 85.0%
Matriculated engineering students
Percentage in top quartile (25%) of High School class 90.0
Average ACT scores: Math—27, Composite—25.52

FULL- & PART-TIME FACULTY—BY DEPARTMENT
[Figures are the head count of full-time faculty and the full-time equivalent (FTE) of part-time faculty for each engineering department or equivalent.]

Department	Full	Part
Chemical Eng	10	–
Civil Eng	12	–
Electrical & Computer Eng	24	–
Industrial Eng	13	1.0
Mechanical & Nuclear Eng	19	–
Agricultural & Biological Eng	7	–
Aerospace Eng	21	–
Petroleum Eng	4	–

COLLEGE DESCRIPTION
The college offers eleven, four-year engineering undergraduate curricula, all of which are accredited by the Accreditation Board for Engineering and Technology. The programs in Aerospace Engineering, Petroleum Engineering and Nuclear Engineering are being phased out. No new admissions were allowed for fall semester 1992. A large but optional cooperative education program extends the time required to graduate to five years. An undergraduate core curriculum of 45 semester hours provides a liberal education in writing, mathematics, sciences, humanities, social sciences, and behavioral sciences. As home to one of eighteen National Science Foundation Engineering Research Centers, the college provides state-of-the-art research experiences for both undergraduate and graduate students. Annual external funding for research exceeds $100,000 per faculty member.

DEGREE PROGRAMS DESCRIPTION
The B.S. degrees in the various engineering programs prepare graduates for entry-level professional positions and for graduate study in many fields. Graduate study in engineering can be pursued in one of two general areas: the M.S. and Ph.D. in engineering, research degrees that emphasize the application of mathematics and the natural sciences to the analysis and design of engineering systems; and the M.Eng. program of study in professional engineering.

DUAL DEGREE PROGRAMS
Engineering baccalaureate graduates with dual degrees 5

TRANSFER INFORMATION
Residency Requirements: No more than half of the semester hours required in a curriculum may be transferred from a community/junior college. For all students, at least 32 semester hours of upper level courses must be completed in residence.

Transfer via Articulation Agreements
Admission to engineering: At any time
Engineering graduates transferred from ... 4-yr: 20 2-yr: 120
Requirements: Minimum 2.0/4.0 grade point average is required for admission.

Transfer without Articulation Agreements
Admission to engineering: At any time
Requirements: Minimum 2.0/4.0 grade point average required for admission.

GRADUATION REQUIREMENTS
The B.S. degree requires 139 semester hours of appropriate course work, including a 45-hour core curriculum and 32 hours of upper level courses completed in residence. A minimum 2.0/4.0 grade point average is required on all college work, on all courses attempted at MSU, and on all courses attempted at MSU that apply to the curriculum.

STUDENT PROGRAMS

PROFESSIONAL AND HONORARY SOCIETIES
[For key to acronyms, see *Introduction*.]
Professional Societies: AIAA, AIChE, ANS, ASAE, ASCE, ASME, IEEE, IIE, NSPE, SPE, Society of Manufacturing Engineers, National Biomedical Engineering Society
Honorary Societies: Alpha Pi Mu, Chi Epsilon, Eta Kappa Nu, Omega Chi Epsilon, Pi Tau Sigma, Sigma Gamma Tau, Tau Beta Pi, Gamma Sigma Delta, Pi Epsilon Tau and Alpha Nu Sigma

SUPPORT PROGRAMS
Student Chapter Organizations: Society of Women Engineers, National Society of Black Engineers

University Familiarization/Minorities Engineering
For: Ethnic Minorities **Available:** Summer only
Offered: Prematriculation

Mississippi Alliance for Minority Participation
For: Ethnic Minorities **Available:** Year round
Offered: Prematriculation, Freshman, Sophomore, Junior, Senior, Graduate level

Southeastern Consortium for Minorities in Engr.
For: Ethnic Minorities **Available:** Year round
Offered: Prematriculation

Other Engineering Support Programs: There are summer and fall orientation programs for all new students, a learning skills center, counseling center, career services center, college-wide computing laboratories and network support, and student support services for handicapped students.

139 University of Missouri-Columbia

■ INSTITUTION PROFILE

HEAD OF THE INSTITUTION
Charles Kiesler
Phone: (314)882-3387

GENERAL INFORMATION
[All Students—Fall 1991]
Undergraduate enrollment 18,446
Graduate enrollment 6,214
Total institution enrollment **24,660**
Type of institution: Public Calendar: Semesters
Nearest city: St. Louis, Missouri Population: 452,801
Miles from main campus: 125 Setting: Urban
Types of engineering degrees: Engineering
Other degree-granting colleges: Arts & Sciences, Business Administration, Education, Law, Medicine, Nursing, Agriculture, Journalism, Human & Environmental Sciences, Library & Informational Science, Veterinary Medicine

ENGINEERING COLLEGE ADMISSIONS INFORMATION
Admission to the engineering college: Students with lower ACT scores initially admitted to Arts/Science pre-engineering program
Entrance Requirements: ACT (Score: 24.0), High School courses—English (4 years), Math (3.5 years), Science with laboratory (2 years), Minimum Math score required is 22.0
Entrance Recommendations: High School courses—As much additional math/science as possible
Requirements for foreign students: Missouri English Language Test

DEPARTMENTAL ADMISSIONS INFORMATION
Admission to the engineering department: At the time of admission to the institution.

■ FINANCIAL INFORMATION

ESTIMATED EXPENSES (ACADEMIC YEAR)
[Expenses are for the 1992-93 nine-month academic year.]

State Residents	Undergraduate	Graduate
Tuition and fees	$ 3,468	$ 3,527
College room and board	$ 3,146	$ 5,793
Books and supplies	$ 503	$ 734
Other expenses	$ 1,494 [1]	$ 4,549 [1]
Total estimated expenses	**$ 8,611**	**$14,603**

Out-of-State Residents	Undergraduate	Graduate
Tuition and fees	$ 8,004	$ 6,914
College room and board	$ 3,146	$ 5,793
Books and supplies	$ 503	$ 734
Other expenses	$ 1,494 [1]	$ 4,549 [1]
Total estimated expenses	**$13,147**	**$17,990**

Notes: (1) Personal expenses and transportation.

FINANCIAL AID OFFICE CONTACT
Joseph M. Camille, *Director*
University of Missouri-Columbia
11 Jesse Hall
Columbia, MO 65211
Phone: (314)882-3795

GENERAL FINANCIAL AID INFORMATION
Forms accepted/required: FAF, FFS
Additional financial aid information: FFS or FAF required for federal aid. College and institution scholarships available.

■ ENGINEERING COLLEGE INFORMATION

[For additional personnel, refer to the *Appendix*.]

HEAD OF ENGINEERING
Anthony L. Hines
Phone: (314)882-4378 Fax: (314)882-2490

ENGINEERING COLLEGE ADDRESS
College of Engineering
University of Missouri-Columbia
W1025 Engineering Complex
Columbia, MO 65211
Phone: (314)882-4375 Fax: (314)882-2490

ENROLLMENTS—BY CLASS
[Numbers are baccalaureate enrollments for the fall 1991 term, unless otherwise footnoted.]
1st-year students/Freshmen 460
2nd-year students/Sophomores 340
3rd-year students/Juniors 393
4th-year students/Seniors 639
Total **1,832**

NUMBER OF DEGREES AWARDED—BY PROGRAM
[Numbers are engineering baccalaureate degrees awarded during the 1991-1992 academic year, unless otherwise footnoted. For full details about each engineering program, refer to the *Tables of Degree Programs*.]
Agricultural Engineering 4
Chemical Engineering 20
Civil Engineering 46
Computer Engineering 33
Electrical Engineering 106
Industrial Engineering 19
Mechanical & Aerospace Engineering 107
Total **335**

PERCENTAGE OF DEGREES AWARDED—BY CATEGORY
[Percentages are of all engineering baccalaureate degrees awarded during the 1991-1992 academic year, unless otherwise footnoted.]
African Americans 3.0%
Asian/Pacific Island Americans 2.7%
Hispanic Americans 0.2%
Native Americans .... – %
Foreign Citizens 7.7%
All Others 86.4%
Women 11.0%
Persons with disabilities .... – % (A)
Students over 25 years of age .... – % (A)

Notes: (A) Data not available.

GRADUATE ENROLLMENTS & DEGREES AWARDED
Master's enrollment 432
Master's degrees awarded 185
Doctoral enrollment 157
Doctoral degrees awarded 17

ENGINEERING STUDENT DATA

Applicants to the engineering college
Number of applicants to engineering college 732
Percent offered admission 69.0%

Matriculated engineering students
Percentage in top quartile (25%) of High School class .... 62.2 [1]
Average ACT scores: Math—27.3, Composite—27.0

Notes: (1) F'91 data

FULL- & PART-TIME FACULTY—BY DEPARTMENT
[Figures are the head count of full-time faculty and the full-time equivalent (FTE) of part-time faculty for each engineering department or equivalent.]

Department	Full	Part
Chemical Eng	9	–
Civil Eng	12	–
Electrical & Computer Eng	21	–
Industrial Eng	6	–
Mechanical & Aerospace Eng	18	–
Nuclear Eng	6	–
Agricultural Eng	8	–
Civil Eng (Kansas City)	4	–
Electrical Eng (Kansas City)	4	–
Mechanical Eng (Kansas City)	5	–

COLLEGE DESCRIPTION
A course in civil engineering was taught on the Columbia campus in 1849, giving the University of Missouri the distinction of being the first institution west of the Mississippi River to offer engineering education. At Mizzou, engineering offers a liberalized, 4-year 126-hour curriculum. Electives allow the student to choose from many interdisciplinary opportunities. Bachelor, master and doctoral degrees in engineering are available in agricultural, chemical, civil, electrical, industrial, and mechanical; MS and PhD, nuclear; and BS, computer. In addition to its programs on the Columbia campus, the College also administers undergraduate and graduate programs in civil, electrical and mechanical engineering in Kansas City. Average number of years to complete the bachelor's degree: 4.5.

DEGREE PROGRAMS DESCRIPTION
BS degrees: Agricultural, Chemical, Civil, Computer, Electrical, Industrial & Mechanical Engineering. BS option: Food Engineering option available in Agricultural Engineering; Bio-Chemical Engineering option and Chemical-Environmental option available in Chemical Engineering. BS minors available with Math, Computer Science and various other Arts/Science departments. Dual BS degree possibilities available for EE/Physics, ME/Physics, etc. MS/PhD degrees: Agricultural, Chemical, Civil, Electrical, Industrial, Mechanical & Aerospace, and Nuclear Engineering.

TRANSFER INFORMATION
Residency Requirements: At least 30 upper-level credit hours must be completed. A minimum of 21 of the 30 credit hours must be upper-level engineering courses. A student transferring from another UM campus must complete at least 15 hours in residence on campus where degree program is located. 12 of the 15 hours must be in engineering and approved by department awarding degree.

Transfer via Articulation Agreements
Admission to engineering: Transferring students must have a GPA of C (2.0/4.0) in all college-level course work attempted at previous institutions; freshmen must meet Engineering freshman admission requirements.

GRADUATION REQUIREMENTS
126-hour curriculum. Students must earn a grade point average of C (2.0 on 4.0 scale) in all courses required for graduation. All departments require a minimum GPA of 2.0 in the engineering courses with some departmental variations as to whether all engineering courses or those in the major are considered.

■ STUDENT PROGRAMS

PROFESSIONAL AND HONORARY SOCIETIES
[For key to acronyms, see *Introduction*.]
Professional Societies: AIAA, AIChE, ANS, ASAE, ASCE, ASME, IEEE, IIE, NSPE, SAE, American Society for Metals Inter., Society of Manufacturing Engineers
Honorary Societies: Alpha Pi Mu, Chi Epsilon, Eta Kappa Nu, Omega Chi Epsilon, Pi Tau Sigma, Tau Beta Pi

SUPPORT PROGRAMS
Student Chapter Organizations: Society of Women Engineers, National Society of Black Engineers, Minorities Engineering Program, Women's Center

Minorities Engineering Program
For: Ethnic Minorities Available: Academic year
Offered: Freshman, Sophomore, Junior, Senior

Campus Women's Center
For: Women Available: Year round
Offered: Freshman, Sophomore, Junior, Senior, Graduate level
Other Engineering Support Programs: College of Engineering students are assigned a faculty adviser to assist them in reaching academic and professional goals. Students are encouraged to meet with their advisers as often as needed. The Career Services Office helps graduating seniors with employment. Co-op opportunities are also available.

140 University of Missouri-Kansas City

■ INSTITUTION PROFILE
HEAD OF THE INSTITUTION
George A. Russell
Phone: (314)882-2011 **Fax:** (314)884-4204
GENERAL INFORMATION
[All Students—Fall 1991]
Undergraduate enrollment 6,253
Graduate enrollment 3,728
Total institution enrollment 11,159
Type of institution: Public **Calendar:** Semesters
Nearest city: Kansas City MO **Population:** 450,000
Miles from main campus: 3 **Setting:** Urban
Types of engineering degrees: Engineering
Other degree-granting colleges: Allied Health Sciences, Business Administration, Communication & Journalism, Dentistry, Education, Fine & Performing Arts, Humanities & Social Sciences, Law, Medicine, Nursing, Pharmacy, Technology & Applied Sciences, Biological Sciences, Conservatory of Music

■ ENGINEERING ADMISSIONS
ADMISSIONS OFFICE CONTACT
Lorraine F. O'Brien
University of Missouri-Kansas City
600 W Mechanic
Independence, MO 64050-1799
Phone: (816)235-1250 **Fax:** (816)235-1260
ENGINEERING COLLEGE ADMISSIONS INFORMATION
Entrance Requirements: ACT (Score: 24), High School courses—English (4 years), Mathematics (3 years), Social Studies and Fine Arts (3 years), Sciences (2 years)
Entrance Recommendations: High School courses—Additional Mathematics (1 year), Additional Science (1 year), Additional Social Studies or Fine Arts (1 year)
Requirements for foreign students: TOEFL (Score: 500); financial statement; Satisfactory evaluations in oral and written communications through the Michigan Test.
DEPARTMENTAL ADMISSIONS INFORMATION
Admission to the engineering department: At the time of admission to the institution.

■ FINANCIAL INFORMATION
ESTIMATED EXPENSES (ACADEMIC YEAR)
[Expenses are for the 1992-93 nine-month academic year.]

All Students	Undergraduate	Graduate
Tuition and fees	$ 1,488	$ 1,310
College room and board	$ 3,095	$ 3,095
Books and supplies	$ 350	$ 350
Other expenses	$ 100	$ 100
Total estimated expenses	**$ 5,033**	**$ 4,855**

FINANCIAL AID OFFICE CONTACT
Buford B. Baber, *Director, Financial Aid*
University of Missouri-Kansas City
5100 Rockhill Road
Kansas City, MO 64110
Phone: (816)235-1154

GENERAL FINANCIAL AID INFORMATION
Forms accepted/required: ACT, Institutional, Family Financial Statement
Additional financial aid information: Please inquire.

■ ENGINEERING COLLEGE INFORMATION
[For additional personnel, refer to the *Appendix*.]
HEAD OF ENGINEERING
Anthony L. Hines
Phone: (314)882-4378 **Fax:** (314)882-0397
ENGINEERING COLLEGE ADDRESS
Engineering Programs
University of Missouri-Kansas City
600 West Mechanic St.
Independence, MO 64050-1799
Phone: (816)235-1250 **Fax:** (816)235-1260
ENROLLMENTS—BY CLASS
[Numbers are baccalaureate enrollments for the fall 1991 term, unless otherwise footnoted.]
1st-year students/Freshmen 69
2nd-year students/Sophomores 117
3rd-year students/Juniors 178
4th-year students/Seniors 392
Total 756
NUMBER OF DEGREES AWARDED—BY PROGRAM
[Numbers are engineering baccalaureate degrees awarded during the 1991-1992 academic year, unless otherwise footnoted. For full details about each engineering program, refer to the *Tables of Degree Programs*.]
Civil Engineering 13
Electrical & Computer Engineering 31
Mechanical & Aerospace Engineering 23
Total 67
PERCENTAGE OF DEGREES AWARDED—BY CATEGORY
[Percentages are of all engineering baccalaureate degrees awarded during the 1991-1992 academic year, unless otherwise footnoted.]
African Americans 4.0%
Asian/Pacific Island Americans 16.0%
Hispanic Americans 3.0%
Native Americans .... – %
Foreign Citizens 7.0%
All Others 70.0%
Women 10.0%
Persons with disabilities .... – %
Students over 25 years of age 9.0%
GRADUATE ENROLLMENTS & DEGREES AWARDED
Master's enrollment 79
Master's degrees awarded 19
Doctoral enrollment 10
Doctoral degrees awarded .... –
ENGINEERING STUDENT DATA
Applicants to the engineering college
Number of applicants to engineering college 69
Percent offered admission 71.0%
Matriculated engineering students
Percentage in top quartile (25%) of High School class 80.0
Average ACT scores: Math—27, Composite—27

FULL- & PART-TIME FACULTY—BY DEPARTMENT
[Figures are the head count of full-time faculty and the full-time equivalent (FTE) of part-time faculty for each engineering department or equivalent.]

Department	Full	Part
Civil Eng	4	–
Electrical & Computer Eng	7	–
Mechanical & Aerospace Eng	6	–

COLLEGE DESCRIPTION
Programs are offered in civil, electrical, and mechanical engineering. All are ABET-accredited. The programs serve an urban population, and many students hold full-time or part-time jobs in engineering firms. There is a preparatory pre-engineering program for entering students not yet fully qualified. An honors program allows seniors to include some graduate work in their curricula.
DEGREE PROGRAMS DESCRIPTION
All three departments offer BS, MS, and PhD programs.
TRANSFER INFORMATION
Residency Requirements: 30 hours minimum in residence with 21 hours of upper level engineering course work.
Transfer via Articulation Agreements
Admission to engineering: At any time
Engineering graduates transferred from ...4-yr: – 2-yr: 117
Requirements: 2.0 GPA in transferable work and in good academic standing (international students must have a 2.5 GPA).
Transfer without Articulation Agreements
Admission to engineering: At any time
Engineering graduates transferred from ...4-yr: 5 2-yr: 20
Requirements: 2.0 GPA in transferable work and in good academic standing (international students must have a 2.5 GPA).
GRADUATION REQUIREMENTS
A GPA of 2.0 is required for graduation.

■ STUDENT PROGRAMS
PROFESSIONAL AND HONORARY SOCIETIES
[For key to acronyms, see *Introduction*.]
Professional Societies: ASCE, ASHRAE, ASME, IEEE
Honorary Societies: Eta Kappa Nu, Pi Tau Sigma, Tau Beta Pi
SUPPORT PROGRAMS
Student Chapter Organizations: Society of Women Engineers, National Society of Black Engineers
Minority Student Affairs Office
For: Ethnic Minorities **Available:** Year round
Offered: Prematriculation, Freshman, Sophomore, Junior, Senior, Graduate level
Other Engineering Support Programs: The College of Arts and Sciences administers the Pre-Engineering Program for marginally qualified freshman students.

141 University of Missouri-Rolla

■ INSTITUTION PROFILE

HEAD OF THE INSTITUTION
John T. Park
Phone: (314)341-4114 Fax: (314)341-6306

GENERAL INFORMATION
[All Students—Fall 1991]
Undergraduate enrollment 4,434
Graduate enrollment 1,223
Total institution enrollment 5,657
Type of institution: Public Calendar: Semesters
Nearest city: St. Louis, Missouri Population: 400,000
Miles from main campus: 102 Setting: Small Town
Types of engineering degrees: Engineering
Other degree-granting colleges: Arts & Sciences

■ ENGINEERING ADMISSIONS

ADMISSIONS OFFICE CONTACT
David J. Allen
University of Missouri-Rolla
102 Parker
Rolla, MO 65401-0249
Phone: (800)522-0938 Fax: (314)341-6308

ENGINEERING COLLEGE ADMISSIONS INFORMATION
Entrance Requirements: High School courses—English (4 years), Mathematics (3 years), Science (2 years), Social Studies (2 years). Students must submit a test score for either ACT, SAT or SCAT. The sum of the students high school class rank percentile and aptitude test percentile must be 120 or higher.
Entrance Recommendations: High School courses—English (4 years), Mathematics (4 years), Chemistry and Physics (2 years), Foreign Language (2 years)
Requirements for foreign students: TOEFL (Score: 550); financial statement

DEPARTMENTAL ADMISSIONS INFORMATION
Admission to the engineering department: At the end of the first year.
Additional information: The minimum requirement for admission to a department is a cumulative grade point average 2.00, except for aerospace engineering, electrical engineering, and mechanical engineering, each of which require a 2.25 GPA.

■ FINANCIAL INFORMATION

ESTIMATED EXPENSES (ACADEMIC YEAR)
[Expenses are for the 1992-93 nine-month academic year.]

State Residents	Undergraduate	Graduate
Tuition and fees	$ 2,771 [2]	$ 2,266 [1]
College room and board	$ 3,450	$ 3,450
Books and supplies	$ 600	$ 600
Other expenses	$ –	$ –
Total estimated expenses	**$ 6,821**	**$ 6,316**

Out-of-State Residents	Undergraduate	Graduate
Tuition and fees	$ 7,300 [2]	$ 5,600 [1]
College room and board	$ 3,450	$ 3,450
Books and supplies	$ 600	$ 600
Other expenses	$ –	$ –
Total estimated expenses	**$11,350**	**$ 9,650**

Notes: (1) Based on 9 credit hours for each of two semesters (2) Based on 14 credit hours for each of two semesters

FINANCIAL AID OFFICE CONTACT
Robert W. Whites, *Associate Director*
University of Missouri-Rolla
Admissions and Financial Aid
Parker Hall
Rolla, MO 65401-0249
Phone: (314)341-4282 Fax: (314)341-4082

GENERAL FINANCIAL AID INFORMATION
Forms accepted/required: FFS, Free Application for Federal Student Aid (FAFSA)
Additional financial aid information: A variety of programs is available to assist students in financing their education. These programs include scholarships, grants, loans, and work-study. Some scholarships are awarded on the basis of need while others are not.

■ ENGINEERING COLLEGE INFORMATION

[For additional personnel, refer to the *Appendix*.]
HEAD OF ENGINEERING
Robert L. Davis
Phone: (314)341-4151 Fax: (314)341-4979

ENGINEERING COLLEGE ADDRESS
School of Engineering, School of Mines and Metallurgy
University of Missouri-Rolla
12th & Pine Streets
Rolla, MO 65401-0249
Phone: (314)341-4111

ENROLLMENTS—BY CLASS
[Numbers are baccalaureate enrollments for the fall 1991 term, unless otherwise footnoted.]
1st-year students/Freshmen 790
2nd-year students/Sophomores 578
3rd-year students/Juniors 699
4th-year students/Seniors 1,413
Total 3,480

NUMBER OF DEGREES AWARDED—BY PROGRAM
[Numbers are engineering baccalaureate degrees awarded during the 1991-1992 academic year, unless otherwise footnoted. For full details about each engineering program, refer to the *Tables of Degree Programs*.]
Aerospace Engineering 36
Ceramic Engineering 21
Chemical Engineering 31
Civil Engineering 68
Electrical Engineering 145
Engineering Management 59
Engineering Mechanics .... –
Geological Engineering 21
Mechanical Engineering 152
Metallurgical Engineering 17
Mining Engineering 3
Nuclear Engineering 9
Petroleum Engineering 4
Total 566

PERCENTAGE OF DEGREES AWARDED—BY CATEGORY
[Percentages are of all engineering baccalaureate degrees awarded during the 1991-1992 academic year, unless otherwise footnoted.]
African Americans 1.4%
Asian/Pacific Island Americans 7.6%
Hispanic Americans 0.5%
Native Americans .... – %
Foreign Citizens 4.4%
All Others 86.1%
Women 17.2%
Persons with disabilities .... – % (A)
Students over 25 years of age 21.0%

Notes: (A) Data not available.

GRADUATE ENROLLMENTS & DEGREES AWARDED
Master's enrollment 641
Master's degrees awarded 248
Doctoral enrollment 216
Doctoral degrees awarded 48

ENGINEERING STUDENT DATA
Applicants to the engineering college
Number of applicants to engineering college 1,992
Percent offered admission 98.0%
Matriculated engineering students
Percentage in top quartile (25%) of High School class 77.0
Average SAT scores: Math—624, Verbal—521, Combined—1146
Average ACT scores: Math—27.3, Composite—27.2

FULL- & PART-TIME FACULTY—BY DEPARTMENT
[Figures are the head count of full-time faculty and the full-time equivalent (FTE) of part-time faculty for each engineering department or equivalent.]

Department	Full	Part
Chemical Eng	12	.... –
Civil Eng	15	.... 1.4
Electrical Eng	30	.... 2.0
Basic Eng	9	.... 1.3
Ceramic Eng	11	.... –
Geological & Petroleum Eng	10	.... –
Metallurgical Eng	10	.... 1.8
Nuclear Eng	6	.... –
Eng Management	16	.... 1.4
Mechanical & Aerospace Eng & Eng Mechanics	36	.... 4.3
Mining Eng	8	.... –

COLLEGE DESCRIPTION
The University of Missouri-Rolla is nationally known for the excellent quality of its undergraduate programs. Because of this reputation, especially among employers of engineers, the programs attract excellent students. The Mission of the institution is to serve as the leading center in the State of Missouri for education in engineering and related sciences and the university has a long tradition for doing so. Throughout its history of 120 years, the campus has focused its activities on science and engineering and on the arts and sciences program that complement the engineering offerings. A co-op program which normally takes 5 years to complete is available. All of the engineering programs are accredited by the Engineering Accreditation Commission of the Accrediting Board for Engineering and Technology.

DEGREE PROGRAMS DESCRIPTION
The University of Missouri-Rolla, founded in 1870 as the University of Missouri School of Mines and Metallurgy, currently has an enrollment of about 5200 students with about 300 faculty. In addition to the twelve fields of engineering, there are also degree programs in six fields of science and math and seven areas of humanities and social sciences. The campus has a well-defined and accepted mission to continue to be Missouri's technological university. The engineering programs are accredited and enjoy an excellent reputation for producing quality graduates, particularly among employers of engineers. The campus size is considered by many to be ideal, yet the engineering programs are large enough to be a major source of engineering manpower in the United States. The size of the engineering program is such that there is flexibility in scheduling courses yet the class sizes are relatively small, allowing for individual attention. There are many very active student organizations on campus - professional, service, recreational and social. Students have an opportunity to participate in campus and community service activities and at the same time develop leadership skills and learn how to be effective team members, two important traits for a successful career.

TRANSFER INFORMATION
Residency Requirements: The last 60 semester hours of course work required for a degree must be completed in residence on campus. With prior approval up to 15 of the last 60 hours may be transferred from another institution.

Transfer via Articulation Agreements
Admission to engineering: At the end of the sophomore year
Engineering graduates transferred from .. 4-yr: 89 2-yr: 121
Requirements: A transfer student must have completed at least 24 semester hours of college level work with a cumulative grade point average of 2.50 or better for admission to an engineering department except, electrical engineering, which requires a 2.75 GPA.

Transfer without Articulation Agreements
Admission to engineering: At any time
Engineering graduates transferred from .. 4-yr: 120 2-yr: –
Requirements: Entrance requirements are the same as for students entering through the provisions of an articulation agreement.

GRADUATION REQUIREMENTS
A cumulative grade point average of at least 2.00 (on a scale of 4 grade points per credit hour for an A) is required for graduation. The cumulative GPA is calculated using all courses taken for college credit for which a letter grade is given. Both grades are used in the case of repeated courses. In addition the GPA must be at least 2.00 in all courses taken at UMR and GPA must be at least 2.00 in all courses taken in the major discipline.

■ STUDENT PROGRAMS

PROFESSIONAL AND HONORARY SOCIETIES
[For key to acronyms, see *Introduction*.]
Professional Societies: AIAA, AIChE, ANS, ASCE, ASEM, ASME, IEEE, SAE, SME*, SPE, Society of Metallurgical Engineers
Honorary Societies: Chi Epsilon, Eta Kappa Nu, Keramos, Omega Chi Epsilon, Pi Tau Sigma, Sigma Gamma Tau, Tau Beta Pi, Theta Tau, Alpha Nu Sigma, Alpha Sigma Mu

SUPPORT PROGRAMS
Student Chapter Organizations: Society of Women Engineers, National Society of Black Engineers, Society of Hispanic Professional Engineers

Minority Engineering Program
For: Ethnic Minorities **Available:** Year round
Offered: Prematriculation, Freshman, Sophomore, Junior, Senior, Graduate level

Women In Engineering
For: Women **Available:** Year round
Offered: Prematriculation, Freshman, Sophomore, Junior, Senior, Graduate level

Other Engineering Support Programs: All 1st year students enroll in the Freshman Engineering Program, take a common core of courses, receive focused counseling and information on engineering careers and campus resources available to them. The Transfer Assistance Program includes a model engineering transfer program, and counselors who work with students before and after transferring to campus.

142 Monmouth College

■ INSTITUTION PROFILE
HEAD OF THE INSTITUTION
Samuel H. Magill
Phone: (908)571-3402   Fax: (908)571-3570

GENERAL INFORMATION
[All Students—Fall 1991]
Undergraduate enrollment 2,826
Graduate enrollment 1,118
Total institution enrollment **3,944**

Type of institution: Private **Calendar:** Semesters
Nearest city: New York **Population:** 8,000,000
Miles from main campus: 45 **Setting:** Small Town
Types of engineering degrees: Engineering
Other degree-granting colleges: Arts & Sciences, Business Administration

■ ENGINEERING ADMISSIONS
ADMISSIONS OFFICE CONTACT
Barry W. Ward
Monmouth College
Director of Undergraduate Admissions
West Long Branch, NJ 07764-1898
Phone: (908)571-3456   Fax: (908)571-3629

ENGINEERING COLLEGE ADMISSIONS INFORMATION
Entrance Requirements: SAT (Score: 950), High School courses—Algebra (2 years), Trigonometry, Geometry, Science (2 years)
Entrance Recommendations: High School courses—Chemistry (1 year), Physics (1 year)
Requirements for foreign students: TOEFL (Score: 525); financial statement; Present proof of completion of secondary school. Foreign language documents must be translated into English and notarized. SAT is recommended.

DEPARTMENTAL ADMISSIONS INFORMATION
Admission to the engineering department: At the time of admission to the institution.

■ FINANCIAL INFORMATION
ESTIMATED EXPENSES (ACADEMIC YEAR)
[Expenses are for the 1992-93 nine-month academic year.]

All Students	Undergraduate	Graduate
Tuition and fees	$11,160	$ 6,534[(1)]
College room and board	$ 4,800[(C)]	$ 4,800[(C)]
Books and supplies	$ 800	$ 500
Other expenses	$ –	$ –
Total estimated expenses	**$16,760**	**$11,834**

Notes: (C) Estimated data. (1) 18 credits at $338 per credit plus comprehensive college fee.

FINANCIAL AID OFFICE CONTACT
Hank Mackiewicz, *Dean of Financial Aid*
Monmouth College
West Long Branch, NJ 07764-1898
Phone: (908)571-3463   Fax: (908)571-3629

GENERAL FINANCIAL AID INFORMATION
Forms accepted/required: FAF, IRS, Scholarship Application, Personal Data Form (PDF)
Additional financial aid information: Monmouth College believes that no qualified student should be denied an educational opportunity due to a lack of financial resources. The College utilizes institutional, federal, and state resources to help meet the cost of attendance. There are academic merit awards, athletic grants, and other forms of scholarships available. College work-study and on-campus employment is also available.

■ ENGINEERING COLLEGE INFORMATION
[For additional personnel, refer to the *Appendix*.]
HEAD OF ENGINEERING
Richard A. Kuntz
Phone: (908)571-3409   Fax: (908)571-3523
EMail: kuntz@monmouth.edu

ENGINEERING COLLEGE ADDRESS
School of Information Sciences and Technology
Monmouth College
Electronic Engineering Department
Cedar Avenue
West Long Branch, NJ 07764-1898
Phone: (908)571-3400   Fax: (908)571-3693

ENROLLMENTS—BY CLASS
[Numbers are baccalaureate enrollments for the fall 1991 term, unless otherwise footnoted.]
1st-year students/Freshmen 15
2nd-year students/Sophomores 17
3rd-year students/Juniors 12
4th-year students/Seniors 29
Total **73**

NUMBER OF DEGREES AWARDED—BY PROGRAM
[Numbers are engineering baccalaureate degrees awarded during the 1991-1992 academic year, unless otherwise footnoted. For full details about each engineering program, refer to the *Tables of Degree Programs*.]
Electronic Engineering 21
Total **21**

PERCENTAGE OF DEGREES AWARDED—BY CATEGORY
[Percentages are of all engineering baccalaureate degrees awarded during the 1991-1992 academic year, unless otherwise footnoted.]
African Americans .... 9.5 %
Asian/Pacific Island Americans .... 4.8 %
Hispanic Americans .... – %
Native Americans .... – %
Foreign Citizens .... 19.0 %
All Others .... 66.7 %
Women .... 14.3 %
Persons with disabilities .... – %
Students over 25 years of age .... – %

GRADUATE ENROLLMENTS & DEGREES AWARDED
Master's enrollment 66
Master's degrees awarded 10
Doctoral enrollment .... –[(1)]
Doctoral degrees awarded .... –[(1)]

Notes: (1) Program not available.

ENGINEERING STUDENT DATA
Applicants to the engineering college
Number of applicants to engineering college 39
Percent offered admission 69.2%
Matriculated engineering students
Percentage in top quartile (25%) of High School class ...... –[(A)]
Average SAT scores: Math—540, Verbal—430

Notes: (A) Data not available.

FULL- & PART-TIME FACULTY—BY DEPARTMENT
[Figures are the head count of full-time faculty and the full-time equivalent (FTE) of part-time faculty for each engineering department or equivalent.]

Department	Full	Part
Electronic Eng	5	0.3

COLLEGE DESCRIPTION
Monmouth College is an independent, comprehensive institution with 45 undergraduate and graduate degree programs and concentrations offered in three schools: Arts and Sciences, Business Administration, and Information Sciences and Technology. The College is nationally recognized for its innovative Leadership and Social Responsibility program, and for its student support services. The student-faculty ratio is 16 - 1, with an average class size of 21. About 1000 students reside on campus. The Electronic Engineering program is ABET accredited, and is supported by excellent laboratories and computer facilities. In addition to general purpose EE laboratories, specialized laboratories include those for image processing, microprocessor development, and microwave engineering. Part-time co-op educational experience is available.

DEGREE PROGRAMS DESCRIPTION
The B.S. in Electronic Engineering at Monmouth College prepares the graduate for entry-level positions in industry and for graduate studies. The curriculum infuses professional study with a liberal education; students take more than one-quarter of their courses in humanities and social sciences. All of the graduates are required to demonstrate effective writing skills. The student studies in depth those technical subjects which are essential to the design of computer, communication, and control systems. The M.S. degree has concentrations offered in communications/signal processing, computer engineering, electron devices/optical electronics, and military electronics. Graduate courses from other disciplines (computer science, mathematics, software engineering) may be applied to M.S. degree requirements.

TRANSFER INFORMATION
Residency Requirements: Completion of a minimum of 32 credits at Monmouth with at least 16 credits in Electronic Engineering.
Transfer via Articulation Agreements
Admission to engineering: At the end of the sophomore year
Engineering graduates transferred from ... 4-yr: – 2-yr: –
Requirements: Completion of courses specified in transfer agreement and good academic standing.
Transfer without Articulation Agreements
Admission to engineering: At any time
Engineering graduates transferred from ... 4-yr: 2 2-yr: 4
Requirements: A grade-point-average of 2.5 (out of 4.0) for previous college work.

GRADUATION REQUIREMENTS
For the B.S. degree, the required course work must be completed with a cumulative grade-point-average of at least a 2.00 (C-average), with at least a 2.10 average for courses in the major. In addition, all candidates for the degree must pass a writing proficiency examination. For the M.S. degree, all course work must be completed with at least a 3.00 (B-average).

■ STUDENT PROGRAMS
PROFESSIONAL AND HONORARY SOCIETIES
[For key to acronyms, see *Introduction*.]
Professional Societies: AIP, AMS, IEEE, SPS
Honorary Societies: Eta Kappa Nu, Mu Kappa Sigma

SUPPORT PROGRAMS
Student Chapter Organizations: AASU/African-American Student Union, Chinese Student Association, Latin American Club

Life and Career Advising Center
For: Women & Ethnic Minorities **Available:** Year round
Offered: Freshman, Sophomore, Junior, Senior, Graduate level

International Student and Scholar Services Office
For: Women & Ethnic Minorities **Available:** Year round
Offered: Prematriculation, Freshman, Sophomore, Junior, Senior, Graduate level

PATH/Program for Advancement Through High School
For: Women & Ethnic Minorities **Available:** Academic year
Offered: Prematriculation

Women's Studies Program
For: Women **Available:** Academic year
Offered: Freshman, Sophomore, Junior, Senior

Other Engineering Support Programs: Advising services including academic, career, and psychological counseling are centralized in the Life and Career Advising Center, with additional academic advising support offered to freshman. The Office of Handicapped Student Affairs serves students with special needs. A transfer counselor assists all transfer students with their educational planning and personal concerns.

143 Montana College of Mineral Science and Technology

INSTITUTION PROFILE

HEAD OF THE INSTITUTION
Lindsay D. Norman
Phone: (406)496-4129 Fax: (406)496-4133

GENERAL INFORMATION
[All Students—Fall 1991]
Undergraduate enrollment 1,903
Graduate enrollment 82
Total institution enrollment 1,985
Type of institution: Public Calendar: Semesters
Location: Butte, MT Population: 40,000
Setting: Small Town
Types of engineering degrees: Engineering

ENGINEERING ADMISSIONS

ADMISSIONS OFFICE CONTACT
Ed Johnson
Montana College of Mineral Science and Technology
Butte, MT 59701
Phone: (406)496-4178 Fax: (406)496-4133

ENGINEERING COLLEGE ADMISSIONS INFORMATION
Entrance Requirements: ACT, High School courses—English (4 years), Mathematics (Algebra I,II, Geometry) (3 years), Social Studies (3 years), Laboratory Science (2 years)
Entrance Recommendations: High School courses—Foreign Language (2 years), Computer Science (2 years), Visual and Performing Arts (2 years), Vocational Education Units (2 years)

DEPARTMENTAL ADMISSIONS INFORMATION
Admission to the engineering department: At the time of admission to the institution.

FINANCIAL INFORMATION

ESTIMATED EXPENSES (ACADEMIC YEAR)
[Expenses are for the 1992-93 nine-month academic year.]

State Residents	Undergraduate	Graduate
Tuition and fees	$ 717	$ 665
College room and board	$ 1,500	$ 1,500
Books and supplies	$ 400	$ 300
Other expenses	$ –	$ –
Total estimated expenses	**$ 2,617**	**$ 2,465**
Out-of-State Residents	Undergraduate	Graduate
Tuition and fees	$ 1,943	$ 1,780
College room and board	$ 1,500	$ 1,500
Books and supplies	$ 400	$ 300
Other expenses	$ –	$ –
Total estimated expenses	**$ 3,843**	**$ 3,580**

FINANCIAL AID OFFICE CONTACT
Frank Kondelis, *Director of Financial Aid*
Montana College of Mineral Science and Technology
Butte, MT 59701
Phone: (406)496-4212 Fax: (406)496-4133

GENERAL FINANCIAL AID INFORMATION
Forms accepted/required: AFSA, Institutional
Additional financial aid information: There are a large number of need-based scholarships and part time work opportunities available.

ENGINEERING COLLEGE INFORMATION

[For additional personnel, refer to the *Appendix*.]

HEAD OF ENGINEERING
Thomas Waring
Phone: (406)496-4127 Fax: (406)496-4133

ENROLLMENTS—BY CLASS
[Numbers are baccalaureate enrollments for the fall 1991 term, unless otherwise footnoted.]
1st-year students/Freshmen 695
2nd-year students/Sophomores 321
3rd-year students/Juniors 275
4th-year students/Seniors 336
Total 1,627

NUMBER OF DEGREES AWARDED—BY PROGRAM
[Numbers are engineering baccalaureate degrees awarded during the 1991-1992 academic year, unless otherwise footnoted. For full details about each engineering program, refer to the *Tables of Degree Programs*.]
Engineering Science(s) 16
Environmental Engineering 18
Geological Engineering 9
Geophysical Engineering .... –
Metallurgical Engineering 11
Mining Engineering 24
Petroleum Engineering 17
Total 95

PERCENTAGE OF DEGREES AWARDED—BY CATEGORY
[Percentages are of all engineering baccalaureate degrees awarded during the 1991-1992 academic year, unless otherwise footnoted.]
African Americans .... – % (A)
Asian/Pacific Island Americans .... – % (A)
Hispanic Americans .... – % (A)
Native Americans .... – % (A)
Foreign Citizens .... – % (A)
All Others .... 100 %
Women .... 19.0 %
Persons with disabilities .... – % (A)
Students over 25 years of age .... – % (A)

Notes: (A) Data not available.

GRADUATE ENROLLMENTS & DEGREES AWARDED
Master's enrollment 83
Master's degrees awarded 19
Doctoral enrollment .... – (B)
Doctoral degrees awarded .... – (B)

Notes: (B) Data not applicable.

ENGINEERING STUDENT DATA

Applicants to the engineering college
Number of applicants to engineering college .... – (A)
Percent offered admission .... –% (A)

Matriculated engineering students
Percentage in top quartile (25%) of High School class .... – (A)
Average SAT scores: Math—551, Verbal—477
Average ACT scores: Math—22, Composite—22

Notes: (A) Data not available.

FULL- & PART-TIME FACULTY—BY DEPARTMENT

[Figures are the head count of full-time faculty and the full-time equivalent (FTE) of part-time faculty for each engineering department or equivalent.]

Department	Full	Part
Petroleum Eng	5	–
Environmental Eng	5	–
Metallurgical Eng	7	–
Geophysical Eng	4	–
Geological Eng	5	0.5
Mining Eng	4	–

COLLEGE DESCRIPTION
Montana Tech is a fully accredited mineral science and engineering college, recently noted as "America's finest small science and engineering college". Academic programs are known among employers as practical and graduates enjoy near 100% placement and multiple job offers

DEGREE PROGRAMS DESCRIPTION
In addition to engineering degree programs, Tech offers programs in computer science, chemistry, mathematics, geochemistry, mineral economics and society and technology.

TRANSFER INFORMATION
Transfer via Articulation Agreements
Admission to engineering: At the end of the sophomore year
Transfer without Articulation Agreements
Admission to engineering: At any time

STUDENT PROGRAMS

PROFESSIONAL AND HONORARY SOCIETIES
[For key to acronyms, see *Introduction*.]
Professional Societies: ACM, AIME, AWS, SME*, SPE
Honorary Societies: Tau Beta Pi, Theta Tau

SUPPORT PROGRAMS
Student Chapter Organizations: Society of Women Engineers
Other Engineering Support Programs: Tech maintains an extensive counseling program available to all students. In addition the Learning Resource Center offers individualized tutoring and self-paced instructional programs.

144 Montana State University

INSTITUTION PROFILE

HEAD OF THE INSTITUTION
Michael P. Malone
Phone: (406)994-2341 Fax: (406)994-1893

GENERAL INFORMATION
[All Students—Fall 1991]
Undergraduate enrollment 9,313
Graduate enrollment 798
Total institution enrollment **10,111**
Type of institution: Public **Calendar:** Semesters
Nearest city: Billings **Population:** 80,000
Miles from main campus: 140 **Setting:** Small Town
Types of engineering degrees: Engineering & Technology
Other degree-granting colleges: Agricultural & Environmental, Architecture, Arts & Sciences, Business Administration, Education, Fine & Performing Arts, Nursing

ENGINEERING ADMISSIONS

ADMISSIONS OFFICE CONTACT
Rhonda Duffus
Montana State University
120 Hamilton Hall
Bozeman, MT 59717
Phone: (406)994-2452 Fax: (406)994-1923

ENGINEERING COLLEGE ADMISSIONS INFORMATION
Entrance Requirements: High School courses—English (4 years), Mathematics (3 years), Social Studies (3 years), Laboratory Science (2 years), ACT or SAT
Entrance Recommendations: SAT (Score: 800), ACT (Score: 20), High School courses—Foreign language (2 years)
Requirements for foreign students: TOEFL (Score: 525); financial statement; Evaluation of academic credentials by ECE

DEPARTMENTAL ADMISSIONS INFORMATION
Admission to the engineering department: At the time of admission to the institution.

FINANCIAL INFORMATION

ESTIMATED EXPENSES (ACADEMIC YEAR)
[Expenses are for the 1992-93 nine-month academic year.]

State Residents	Undergraduate	Graduate
Tuition and fees	$ 1,850	$ 1,850
College room and board	$ 3,500	$ 3,500
Books and supplies	$ 550	$ 550
Other expenses	$ 2,300[1]	$ 2,300[1]
Total estimated expenses	**$ 8,200**	**$ 8,200**

Out-of-State Residents	Undergraduate	Graduate
Tuition and fees	$ 5,550	$ 5,550
College room and board	$ 3,500	$ 3,500
Books and supplies	$ 550	$ 550
Other expenses	$ 2,300[1]	$ 2,300[1]
Total estimated expenses	**$11,900**	**$11,900**

Notes: (1) Personal/Transportation

FINANCIAL AID OFFICE CONTACT
James R. Craig, *Director, Financial Aid Services*
Montana State University
135 Strand Union Building
Bozeman, MT 59717
Phone: (406)994-2845 Fax: (406)994-5488

GENERAL FINANCIAL AID INFORMATION
Forms accepted/required: AFSA, Stafford, SAR
Additional financial aid information: Students applying for financial assistance are considered for all aid programs for which they are eligible. Assistance is offered in the form of grants, scholarships, long term loans, and work opportunities. Priority consideration is given students who apply by March 1 of each year.

ENGINEERING COLLEGE INFORMATION

[For additional personnel, refer to the *Appendix*.]
HEAD OF ENGINEERING
David F. Gibson
Phone: (406)994-2272 Fax: (406)994-6098
EMail: ADEDG@MTSUNIX1.BITNET

ENGINEERING COLLEGE ADDRESS
College of Engineering
Montana State University
212 Roberts Hall
Bozeman, MT 59717-0382
Phone: (406)994-2272 Fax: (406)994-6098

ENROLLMENTS—BY CLASS
[Numbers are baccalaureate enrollments for the fall 1991 term, unless otherwise footnoted.]
1st-year students/Freshmen 557
2nd-year students/Sophomores 210
3rd-year students/Juniors 229
4th-year students/Seniors .... 325 [1]
Total **1,321**
Notes: (1) Includes second degree students

NUMBER OF DEGREES AWARDED—BY PROGRAM
[Numbers are engineering baccalaureate degrees awarded during the 1991-1992 academic year, unless otherwise footnoted. For full details about each engineering program, refer to the *Tables of Degree Programs*.]
Agricultural Engineering .... –
Chemical Engineering 27
Civil Engineering 32
Electrical Engineering 46
Industrial & Management Engineering 17
Mechanical Engineering 59
Total **181**

PERCENTAGE OF DEGREES AWARDED—BY CATEGORY
[Percentages are of all engineering baccalaureate degrees awarded during the 1991-1992 academic year, unless otherwise footnoted.]
African Americans .... – %[A]
Asian/Pacific Island Americans .... – %[A]
Hispanic Americans .... – %[A]
Native Americans .... – %[A]
Foreign Citizens .... – %[A]
All Others .... 100 %
Women .... 12.7 %
Persons with disabilities .... – %[A]
Students over 25 years of age .... – %[A]
Notes: (A) Data not available.

GRADUATE ENROLLMENTS & DEGREES AWARDED
Master's enrollment 78
Master's degrees awarded 28
Doctoral enrollment 18
Doctoral degrees awarded 3

ENGINEERING STUDENT DATA
Applicants to the engineering college
Number of applicants to engineering college 879
Percent offered admission 88.1%

Matriculated engineering students
Percentage in top quartile (25%) of High School class .... –[A]
Average SAT scores: Math—592[1], Verbal—495[1], Combined—1087[1]
Average ACT scores: Composite—24.5[1]
Notes: (A) Data not available. (1) Include engineering technology and computer science

FULL- & PART-TIME FACULTY—BY DEPARTMENT
[Figures are the head count of full-time faculty and the full-time equivalent (FTE) of part-time faculty for each engineering department or equivalent.]

Department	Full	Part
Chemical Eng	8	–
Civil & Agricultural Eng	19	3.7
Electrical Eng	16	1.0
Industrial & Management Eng	7	0.9
Mechanical Eng	11	1.6

COLLEGE DESCRIPTION

The College of Engineering is the largest college on the university campus. It offers four-year undergraduate programs in six areas of engineering, as well as three engineering technology programs and computer science. Programs in engineering emphasize hands-on laboratory experiences and culminate with an engineering design capstone project. Seniors are urged to take the Fundamentals of Engineering exam and have traditionally passed the exam at a rate far in excess of the national average. State-of-the-art computer capabilities exist throughout the College and University. Engineering coursework is complemented by a University Core Curriculum in the humanities, social sciences, and fine arts. Optional co-op programs are available in some of the engineering curricula. An Engineering Research Center in environmental biotechnology provides research opportunities for students. The average number of years required to complete an engineering degree is approximately 4.5 years.

DEGREE PROGRAMS DESCRIPTION

The B.S. degrees in the various engineering programs prepare the graduate for entry-level professional positions of for graduate study in each of the respective fields. The M.S. degree is available at Montana State University in each of the engineering areas, and the Ph.D. degree is offered in four areas of engineering - Chemical, Civil, Electrical, and Mechanical.

DUAL DEGREE PROGRAMS

Engineering baccalaureate graduates with dual degrees 2

TRANSFER INFORMATION

Residency Requirements: A minimum of 30 resident credits are required and a minimum of 23 of the last 30 credits earned to meet the graduation requirement must be resident credits.

Transfer via Articulation Agreements
Admission to engineering: At any time
Engineering graduates transferred from ... 4-yr: 50 2-yr: 10
Requirements: In-state residents must be admissible to their former school. Out-of-state residents must have a 2.0 transferable GPA.

GRADUATION REQUIREMENTS

Engineering students are required to maintain a 2.0 GPA to be eligible for graduation. In addition, a student must receive a C grade or better in all courses required for graduation.

STUDENT PROGRAMS

PROFESSIONAL AND HONORARY SOCIETIES
[For key to acronyms, see *Introduction*.]
Professional Societies: ACM, AGCA, AIChE, ASAE, ASCE, ASME, IEEE, IIE, NSPE
Honorary Societies: Alpha Pi Mu, Chi Epsilon, Eta Kappa Nu, Pi Tau Sigma, Tau Beta Pi

SUPPORT PROGRAMS
Student Chapter Organizations: Society of Women Engineers, American Indian Science and Engineering Society

Women's Center
For: Women **Available:** Year round
Offered: Freshman, Sophomore, Junior, Senior, Graduate level

Center for Native American Studies
For: Ethnic Minorities **Available:** Year round
Offered: Prematriculation, Freshman, Sophomore, Junior, Senior, Graduate level

Other Engineering Support Programs: Programs for support of engineering students include freshman/transfer orientation programs; programs for students-over-the-traditional-age; programs for students returning to school after an extended period, as well as those needing special academic assistance. Other programs are conducted by Disabled Student Services, Career Services, Veterans Affairs, and International Students.

145 Morgan State University

■ INSTITUTION PROFILE
HEAD OF THE INSTITUTION
Earl S. Richardson
Phone: (410)319-3200　　　　　　Fax: (410)319-3107

GENERAL INFORMATION
[All Students—Fall 1991]
Undergraduate enrollment 4,914
Graduate enrollment 488
Total institution enrollment 5,402
Type of institution: Public　　**Calendar:** Semesters
Location: Baltimore　　　　　**Population:** 740,000
Setting: Urban
Types of engineering degrees: Engineering

■ ENGINEERING ADMISSIONS
ADMISSIONS OFFICE CONTACT
Chelseia Harold-Miller
Morgan State University
Cold Spring La. & Hillen Rd.
Baltimore, MD 21239
Phone: (410)319-3000

ENGINEERING COLLEGE ADMISSIONS INFORMATION
Entrance Requirements: SAT (Score: 750), ACT (Score: 18)
Entrance Recommendations: SAT (Score: 750), ACT (Score: 18), High School courses—Mathematics (4 years), Science (3 years), English (4 years)
Requirements for foreign students: TOEFL (Score: 550)

DEPARTMENTAL ADMISSIONS INFORMATION
Admission to the engineering department: At the time of admission to the institution.

■ FINANCIAL INFORMATION
ESTIMATED EXPENSES (ACADEMIC YEAR)
[Expenses are for the 1992-93 nine-month academic year.]

State Residents	Undergraduate	Graduate
Tuition and fees	$ 2,438	$ –
College room and board	$ 4,640	$ –
Books and supplies	$ 1,000	$ –
Other expenses	$ –	$ –
Total estimated expenses	$ 8,078	$ –

Out-of-State Residents	Undergraduate	Graduate
Tuition and fees	$ 4,780	$ –
College room and board	$ 4,640	$ –
Books and supplies	$ 1,000	$ –
Other expenses	$ –	$ –
Total estimated expenses	$10,420	$ –

FINANCIAL AID OFFICE CONTACT
Reginald T. Cureton, *Director of Financial Aid*
Morgan State University
Cold Spring La. & Hillen Rd.
Baltimore, MD 21239
Phone: (410)319-3170　　　　　Fax: (410)319-3852

GENERAL FINANCIAL AID INFORMATION
Forms accepted/required: FAF
Additional financial aid information: Outstanding students are considered for a variety of scholarships. Maryland residents may apply for State Scholarships, Senatorial Scholarships, and House of Delegates Scholarships. Four-year ROTC scholarships are offered by the U.S. Army. About 80% of Morgan's students receive some form of financial assistance. The University also helps students find employment on campus and in the surrounding community.

■ ENGINEERING COLLEGE INFORMATION
[For additional personnel, refer to the *Appendix*.]
HEAD OF ENGINEERING
Eugene M. Deloatch
Phone: (410)319-3231　　　　　Fax: (410)319-3843
EMail: deloatch@echo.eng.umd.edu

ENGINEERING COLLEGE ADDRESS
Clarence M. Mitchell, Jr. School of Engineering
Morgan State University
5200 Perring Parkway
Baltimore, MD 21239
Phone: (410)319-3231　　　　　Fax: (410)319-3843

ENROLLMENTS—BY CLASS
[Numbers are baccalaureate enrollments for the fall 1991 term, unless otherwise footnoted.]
1st-year students/Freshmen 226
2nd-year students/Sophomores 135
3rd-year students/Juniors 92
4th-year students/Seniors 81
Total **534**

NUMBER OF DEGREES AWARDED—BY PROGRAM
[Numbers are engineering baccalaureate degrees awarded during the 1991-1992 academic year, unless otherwise footnoted. For full details about each engineering program, refer to the *Tables of Degree Programs*.]
Civil Engineering 9
Electrical Engineering 15
Industrial Engineering 10
Total **34**

PERCENTAGE OF DEGREES AWARDED—BY CATEGORY
[Percentages are of all engineering baccalaureate degrees awarded during the 1991-1992 academic year, unless otherwise footnoted.]
African Americans .... 92.5 %
Asian/Pacific Island Americans .... 2.9 %
Hispanic Americans .... 1.0 %
Native Americans .... – %
Foreign Citizens .... 3.6 %
All Others .... – %
Women .... 34.6 %
Persons with disabilities .... – %
Students over 25 years of age .... 4.0 %

ENGINEERING STUDENT DATA
Applicants to the engineering college
Number of applicants to engineering college 400
Percent offered admission 60.0%

Matriculated engineering students
Percentage in top quartile (25%) of High School class 55.0
Average SAT scores: Math—480, Verbal—420, Combined—900

FULL- & PART-TIME FACULTY—BY DEPARTMENT
[Figures are the head count of full-time faculty and the full-time equivalent (FTE) of part-time faculty for each engineering department or equivalent.]

Department	Full	Part
Civil Eng	6	1.0
Electrical Eng	10	1.2
Industrial Eng	5	1.7

COLLEGE DESCRIPTION
Morgan State University has the unique distinction of being designated as Maryland's public urban university. Prior to the opening of the School of Engineering there were no publicly supported degree-granting engineering programs in Maryland's greater Baltimore region. Its goal is to be of service to local government and its agencies, federal laboratories, private industry, and the professional development needs of individuals. The population of minorities in Maryland/Morgan is considerably greater than that of the nation as a whole. Morgan has formal dual degree engineering program arrangements with 4-year institutions. Articulated programs exist with community colleges. Transfers to the engineering programs are encouraged.

DEGREE PROGRAMS DESCRIPTION
The School of Engineering offers programs leading to the Baccalaureate degrees in electrical, civil, and industrial engineering. The goal in each of these areas is to establish programs of the highest quality. Quality is measured by the preparation and commitment of the faculty and students, the vision of the administration, and the adequacy of the support facilities. Based on national standards in each of the areas mentioned above, the Morgan School of Engineering measures well.

DUAL DEGREE PROGRAMS
Engineering baccalaureate graduates with dual degrees 0
Enrollment requirements: Exist in formal agreements on file with specific institutions may vary with each institution.

TRANSFER INFORMATION
Residency Requirements: Must be resident for a minimum of one year and complete a minimum of 34 credit hours of required or advanced courses.

Transfer without Articulation Agreements
Admission to engineering: At any time
Requirements: Be in good standing at the home institution.

GRADUATION REQUIREMENTS
Graduate must have a minimum of 2.0 in his/her major courses and a 2.0 GPA overall.

■ STUDENT PROGRAMS
PROFESSIONAL AND HONORARY SOCIETIES
[For key to acronyms, see *Introduction*.]
Professional Societies: ASCE, IEEE, IIE, ORSA, SAE

SUPPORT PROGRAMS
Student Chapter Organizations: Society of Women Engineers, National Society of Black Engineers, ECSEL

Engineering Enrichment Program
For: Ethnic Minorities　　**Available:** Summer only
Offered: Prematriculation

Summer Catch-up Program
For: Ethnic Minorities　　**Available:** Summer only
Offered: Freshman, Sophomore, Junior, Senior

Engineering Careers Exploration Program
For: Ethnic Minorities　　**Available:** Summer only
Offered: Prematriculation

Project No Fail
For: Ethnic Minorities　　**Available:** Academic year
Offered: Freshman, Sophomore, Junior, Senior

Other Engineering Support Programs: All new (non-transfer) students are initially enrolled in the Freshman Engineering Program and must complete a one credit course entitled, Introduction to Engineering. Students are designated as freshmen engineers until all required first-year courses are successfully completed. Tutors are available to all students on request. Pre-college outreach programs are available to high schools.

146 University of Nebraska - Lincoln

INSTITUTION PROFILE

HEAD OF THE INSTITUTION
Graham B. Spanier
Phone: (402)472-2116 Fax: (402)472-5110

GENERAL INFORMATION
[All Students—Fall 1991]
Total institution enrollment 22,500
Type of institution: Public Calendar: Semesters
Location: Lincoln, NE Population: 192,000
Setting: Urban
Types of engineering degrees: Engineering & Technology
Other degree-granting colleges: Agricultural & Environmental, Architecture, Arts & Sciences, Business Administration, Communication & Journalism, Dentistry, Education, Law

ENGINEERING ADMISSIONS

ADMISSIONS OFFICE CONTACT
John E. Beacon
University of Nebraska - Lincoln
14th and R Streets
Lincoln, NE 68588-0411
Phone: (402)472-3620 Fax: (402)472-3603

ENGINEERING COLLEGE ADMISSIONS INFORMATION
Entrance Requirements: High School courses—Math (3.5 years), Science (2 years), English (4 years)
Entrance Recommendations: SAT (Score: 1025), ACT (Score: 24), High School courses—
Requirements for foreign students: TOEFL (Score: 550); financial statement Additional information: Depending on GPA, but minimum is 2.5.

FINANCIAL INFORMATION

ESTIMATED EXPENSES (ACADEMIC YEAR)
[Expenses are for the 1992-93 nine-month academic year.]

State Residents	Undergraduate	Graduate
Tuition and fees	$ 2,187[1]	$ 1,806[1]
College room and board	$ 3,260	$ 3,260
Books and supplies	$ 500	$ 300
Other expenses	$ 200	$ 200
Total estimated expenses	**$ 6,147**	**$ 5,566**

Out-of-State Residents	Undergraduate	Graduate
Tuition and fees	$ 5,367[1]	$ 3,669[1]
College room and board	$ 3,260	$ 3,260
Books and supplies	$ 500	$ 300
Other expenses	$ 200	$ 200
Total estimated expenses	**$ 9,327**	**$ 7,429**

Notes: (1) Based on 9 credit hour load for graduate students and 15 credit hour load for undergraduates.

FINANCIAL AID OFFICE CONTACT
John E. Beacon, *Director*
University of Nebraska - Lincoln
Scholarships & Financial Aid
14th and R Streets
Lincoln, NE 68588
Phone: (402)472-2030 Fax: (402)472-3603

GENERAL FINANCIAL AID INFORMATION
Forms accepted/required: ACT, AFSA, CSS-FAF, FAF, FAT, FSA, FFS, Stafford, IRS, SAR, Supplemental
Additional financial aid information: Work study available in many departments.

ENGINEERING COLLEGE INFORMATION
[For additional personnel, refer to the *Appendix*.]

HEAD OF ENGINEERING
Stanley R. Liberty
Phone: (402)472-3181 Fax: (402)472-7792
EMail: sliberty@unl.edu

ENGINEERING COLLEGE ADDRESS
Engineering & Technology
University of Nebraska - Lincoln
14th and R Streets
Lincoln, NE 68588-0501
Phone: (402)472-3620 Fax: (402)472-7792

ENROLLMENTS—BY CLASS
[Numbers are baccalaureate enrollments for the fall 1991 term, unless otherwise footnoted.]
1st-year students/Freshmen 322
2nd-year students/Sophomores 231
3rd-year students/Juniors 341
4th-year students/Seniors 559
Total **1,453**

NUMBER OF DEGREES AWARDED—BY PROGRAM
[Numbers are engineering baccalaureate degrees awarded during the 1991-1992 academic year, unless otherwise footnoted. For full details about each engineering program, refer to the *Tables of Degree Programs*.]
Agricultural Engineering 14
Biological System Engineering 1
Chemical Engineering 29
Civil Engineering 44
Computer Engineering 47
Electrical Engineering 135
Engineering Mechanics .... –
Industrial Engineering 46
Mechanical Engineering 141
Total **457**

PERCENTAGE OF DEGREES AWARDED—BY CATEGORY
[Percentages are of all engineering baccalaureate degrees awarded during the 1991-1992 academic year, unless otherwise footnoted.]
African Americans .... – %
Asian/Pacific Island Americans .... – %
Hispanic Americans .... – %
Native Americans .... – %
Foreign Citizens .... 10.3 %
All Others .... 89.7 %
Women .... 6.7 %
Persons with disabilities .... – %
Students over 25 years of age .... – %

GRADUATE ENROLLMENTS & DEGREES AWARDED
Master's enrollment .... –
Master's degrees awarded 106
Doctoral enrollment .... –
Doctoral degrees awarded 12

ENGINEERING STUDENT DATA
Average ACT scores: Composite—27

FULL- & PART-TIME FACULTY—BY DEPARTMENT
[Figures are the head count of full-time faculty and the full-time equivalent (FTE) of part-time faculty for each engineering department or equivalent.]

Department	Full	Part
Chemical Eng	8	–
Civil Eng	18	–
Electrical Eng	18	–
Industrial & Management Systems Eng	10	–
Mechanical Eng	16	–
Agricultural Eng	10	–
Biological Systems Eng	10	–
Computer Eng	8	–

COLLEGE DESCRIPTION
The College offers programs on both the Lincoln and Omaha campuses. Undergraduate programs in engineering, computer science, construction management, and engineering technology are offered. Students are able to participate in Co-op and internship programs, a protege program and numerous co-curricular activities. The Co-op program adds 9 to 18 months to the regular degree programs. Typical students graduate from regular degree programs in 4.5 years. International exchange programs are available.

DEGREE PROGRAMS DESCRIPTION
The College offers Bachelor of Science degree programs in each of the following engineering fields: Agricultural Engineering, Biological Systems Engineering, Chemical Engineering, Civil Engineering, Computer Engineering, Electrical Engineering, Industrial Engineering and Mechanical Engineering. Master degrees are offered in all of the above plus Engineering Mechanics and Manufacturing. The College offers a unified PhD program with specialities following the BS programs.

TRANSFER INFORMATION
Residency Requirements: 30 of the last 36 hours must be taken at UNL.

Transfer without Articulation Agreements
Admission to engineering: At any time
Engineering graduates transferred from ... 4-yr: – 2-yr: –
Requirements: GPA of 2.5 for Nebraska natives; GPA of 3.0 for non-residents.

GRADUATION REQUIREMENTS
Overall GPA must be at least 2.50 in Engineering and Technology.

STUDENT PROGRAMS

PROFESSIONAL AND HONORARY SOCIETIES
[For key to acronyms, see *Introduction*.]
Professional Societies: ACM, AIChE, ASAE, ASCE, IEEE, IIE, ITE, SAE, SAME
Honorary Societies: Alpha Epsilon, Alpha Pi Mu, Chi Epsilon, Eta Kappa Nu, Pi Tau Sigma, Tau Beta Pi

SUPPORT PROGRAMS
Student Chapter Organizations: Society of Women Engineers, National Society of Black Engineers
Other Engineering Support Programs: Red Letter Days, New Student Enrollment Days and Big Red Welcome prior to enrollment. Adviser Advantage and introductory talks about various engineering and technology majors. Protege, Intern and Co-op programs. Academic Success Center, Counseling Center, Office of Multicultural Affairs, Foreign Student Office, Handicapped Services, Placement Office, tutoring programs, other services.

147 University of Nevada, Las Vegas

INSTITUTION PROFILE
HEAD OF THE INSTITUTION
Robert C. Maxson
Phone: (702)895-3201 Fax: (702)895-1088

GENERAL INFORMATION
[All Students—Fall 1991]
Undergraduate enrollment 11,021
Graduate enrollment 1,584
Total institution enrollment **12,605**
Type of institution: Public Calendar: Semesters
Location: Las Vegas, NV Population: 860,000
Setting: Urban
Types of engineering degrees: Engineering
Other degree-granting colleges: Allied Health Sciences, Business Administration, Education, Fine & Performing Arts, Human Performance and Development, Hotel Administration, Liberal Arts, Science and Mathematics

ENGINEERING ADMISSIONS
ADMISSIONS OFFICE CONTACT
Larry Mason
University of Nevada, Las Vegas
Director of Admissions
4505 Maryland Parkway, Box 451021
Las Vegas, NV 89154-1021
Phone: (702)895-3443 Fax: (702)895-3850

ENGINEERING COLLEGE ADMISSIONS INFORMATION
Entrance Requirements: SAT (Score: 475), ACT (Score: 21), High School courses—English (4 years), Mathematics (3 years), Social Science (3 years), Natural Science (3 years), One or the other is required; used for placement only.
Entrance Recommendations: SAT (Score: 475), ACT (Score: 21), High School courses—Computer
Requirements for foreign students: TOEFL (Score: 525); financial statement

DEPARTMENTAL ADMISSIONS INFORMATION
Admission to the engineering department: In order to become a full-fledged major a student must satisfy the prerequisites for General Chemistry I, Calculus I, and English 101 (composition & rhetoric).
Additional information: Credit or registration in English I, Calculus I, and General Chemistry I.

FINANCIAL INFORMATION
ESTIMATED EXPENSES (ACADEMIC YEAR)
[Expenses are for the 1992-93 nine-month academic year.]

State Residents	Undergraduate	Graduate
Tuition and fees	$ 1,332	$ 1,395
College room and board	$ 4,780	$ 4,780
Books and supplies	$ 300 (C)	$ 400 (C)
Other expenses	$ 20 (1)	$ 20 (1)
Total estimated expenses	**$ 6,432**	**$ 6,595**

Out-of-State Residents	Undergraduate	Graduate
Tuition and fees	$ 5,382	$ 5,445
College room and board	$ 4,780	$ 4,780
Books and supplies	$ 400 (C)	$ 400 (C)
Other expenses	$ 20 (1)	$ 20 (1)
Total estimated expenses	**$10,582**	**$10,645**

Notes: (C) Estimated data. (1) Non-refundable application fee.

FINANCIAL AID OFFICE CONTACT
Judy Belanger, *Director*
University of Nevada, Las Vegas
Student Financial Services
4505 Maryland Parkway
Las Vegas, NV 89154
Phone: (702)895-3695 Fax: (702)895-1353

GENERAL FINANCIAL AID INFORMATION
Forms accepted/required: ACT, CSS-FAF, FFS, Institutional, USAF Single File, SAAC Cal Grant
Additional financial aid information: The university provides a wide variety of financial resources for qualified students. Aids such as loans, grants, scholarships, student employment, and grants-in-aid are awarded to help financially needy students, to recognize achievement and special talent, and/or to reward service to the university and the community.

ENGINEERING COLLEGE INFORMATION
[For additional personnel, refer to the *Appendix*.]
HEAD OF ENGINEERING
William R. Wells
Phone: (702)895-3699 Fax: (702)895-4059
EMail: wcube@unlv.edu

ENGINEERING COLLEGE ADDRESS
Howard R. Hughes College of Engineering
University of Nevada, Las Vegas
4505 Maryland Parkway
Box 454005
Las Vegas, NV 89154-4005
Phone: (702)895-3699 Fax: (702)895-4059

ENROLLMENTS—BY CLASS
[Numbers are baccalaureate enrollments for the fall 1991 term, unless otherwise footnoted.]
1st-year students/Freshmen 595 (1)
2nd-year students/Sophomores 43
3rd-year students/Juniors 62
4th-year students/Seniors 119
Total **819**

Notes: (1) Number is pre-engineering students who are generally freshmen.

NUMBER OF DEGREES AWARDED—BY PROGRAM
[Numbers are engineering baccalaureate degrees awarded during the 1991-1992 academic year, unless otherwise footnoted. For full details about each engineering program, refer to the *Tables of Degree Programs*.]
Civil & Environmental Engineering 19
Electrical & Computer Engineering 15
Mechanical Engineering 13
Total **47**

PERCENTAGE OF DEGREES AWARDED—BY CATEGORY
[Percentages are of all engineering baccalaureate degrees awarded during the 1991-1992 academic year, unless otherwise footnoted.]
African Americans .... 6.4 %
Asian/Pacific Island Americans .... 6.4 %
Hispanic Americans .... 4.3 %
Native Americans .... – %
Foreign Citizens .... 6.4 %
All Others .... 76.5 %
Women .... 17.0 %
Persons with disabilities .... – % (A)
Students over 25 years of age .... 63.8 %

Notes: (A) Data not available.

GRADUATE ENROLLMENTS & DEGREES AWARDED
Master's enrollment 55
Master's degrees awarded 10
Doctoral enrollment 3
Doctoral degrees awarded .... –

FULL- & PART-TIME FACULTY—BY DEPARTMENT
[Figures are the head count of full-time faculty and the full-time equivalent (FTE) of part-time faculty for each engineering department or equivalent.]

Department	Full	Part
Civil & Environmental Eng	10	–
Electrical & Computer Eng	11	–
Mechanical Eng	9	–

COLLEGE DESCRIPTION
The college offers course work leading to the B.S. degree in six disciplines. The curricula provide the student with a broad base founded on liberal arts and basic sciences within a framework of 46 credits of university core requirements. The programs are complemented by courses in engineering science, analysis, and design with emphasis on problem solving, computer applications, and laboratory experience. The disciplines were selected because of their importance in this area and to the profession as a whole. The physical plant is one of the nation's best with modern, well-equipped engineering laboratories and classrooms. Extensive computer resources are made available to students and faculty throughout the entire study program. Senior design projects are required and numerous opportunities exist for advanced undergraduates to work on research projects with faculty. The average number of years required to actually complete the bachelor's degree is 5 years.

DEGREE PROGRAMS DESCRIPTION
The B.S. degree programs in the various disciplines prepare the graduate for entry level professional positions or for graduate study in each field. These programs prepare graduates for the Fundamentals of Engineering examination and registration as Professional Engineers. M.S. programs emphasize advanced engineering practice and research within specific areas of engineering and computer science.

DUAL DEGREE PROGRAMS
Engineering baccalaureate graduates with dual degrees 0
Enrollment requirements: The degree must be the same as the degree of the first major. The addition of a second major must be completed prior to undertaking the last 30 semester credits of work required for the degree sought.

TRANSFER INFORMATION
Residency Requirements: A candidate for the bachelor's degree must complete the last 30 semester credits in uninterrupted resident credit as a major in the college from which the degree is expected.

Transfer via Articulation Agreements
Admission to engineering: At any time
Engineering graduates transferred from .. 4-yr: 17 2-yr: 20
Requirements: Student must transfer from an accredited university and 15 semester hours of transferable credit must have been completed with an overall GPA of 2.00 on a 4.00 scale.

Transfer without Articulation Agreements
Admission to engineering: At any time
Engineering graduates transferred from .. 4-yr: – 2-yr: –
Requirements: Minimum 2.3 GPA from transfer students. Transfer students admitted on probation must sign an advising agreement for the upcoming semester.

GRADUATION REQUIREMENTS
Candidates for degrees in engineering programs are required to complete all university General Education Core Requirements in addition to the college and departmental requirements. Students must complete a two-semester sequence in either social science or humanities, maintain a GPA of 2.00 or higher in all immediate prerequisites of all engineering courses, and complete one of the engineering curricula in the college.

STUDENT PROGRAMS
PROFESSIONAL AND HONORARY SOCIETIES
[For key to acronyms, see *Introduction*.]
Professional Societies: ACM, AGCA, AIA, AIAA, ASCE, ASHRAE, ASME, IEEE, ITE, NSPE
Honorary Societies: Engineering Honor Society, Upsilon Pi Epsilon

SUPPORT PROGRAMS
Student Chapter Organizations: Society of Women Engineers, National Society of Black Engineers, Society of Hispanic Professional Engineers, American Indian Science & Engineering Society

Minority Engineering Program
For: Ethnic Minorities **Available:** Year round
Offered: Prematriculation, Freshman, Sophomore, Junior, Senior, Graduate level
Other Engineering Support Programs: The Student Development Center offers new student orientation designed to assist in the transition to college by serving as an introduction to the life of the university. The college offers a separate orientation program to engineering students as a part of the university program.

148 New England College

INSTITUTION PROFILE

HEAD OF THE INSTITUTION
William R. O'Connell Jr.
Phone: (603)428-2222   Fax: (603)428-7230

GENERAL INFORMATION
[All Students—Fall 1991]
Undergraduate enrollment 1,209
Graduate enrollment 48
Total institution enrollment **1,257**
Type of institution: Private   **Calendar:** Semesters
Nearest city: Boston   **Population:** 574,000
Miles from main campus: 90   **Setting:** Small Town
Types of engineering degrees: Engineering

ENGINEERING ADMISSIONS

ADMISSIONS OFFICE CONTACT
John F. Spaulding
New England College
Davis House Admissions Office
26 Bridge Street
Henniker, NH 03242-3299
Phone: (603)428-2223   Fax: (603)428-7230

ENGINEERING COLLEGE ADMISSIONS INFORMATION
Entrance Requirements: High School courses—English (4 years), Mathematics (3 years), General Science (1 year), Laboratory Science (2 years)
Entrance Recommendations: High School courses—Trigonometry/Pre-Calculus (1 year), Physics (1 year), Chemistry (1 year)

DEPARTMENTAL ADMISSIONS INFORMATION
Admission to the engineering department: At the time of admission to the institution.
Additional information: Students must have the science and math credentials adequate to meet the demands of the curriculum. Students may be admitted to Engineering conditionally. They may take some or all of the recommended freshman courses. They must meet with their Advisor at the end of the semester to discuss their progress. At that time, they may be admitted to the Civil Engineering Program.

FINANCIAL INFORMATION

ESTIMATED EXPENSES (ACADEMIC YEAR)
[Expenses are for the 1992-93 nine-month academic year.]

All Students	Undergraduate	Graduate
Tuition and fees	$11,990	$ –
College room and board	$ 4,970	$ –
Books and supplies	$ 400	$ –
Other expenses	$ 700	$ –
Total estimated expenses	**$18,060**	**$ –**

FINANCIAL AID OFFICE CONTACT
Sandy Schneider, *Director*
New England College
Administration Building
7 Main Street
Henniker, NH 03242-3299
Phone: (603)428-2211   Fax: (603)428-7230

GENERAL FINANCIAL AID INFORMATION
Forms accepted/required: CSS-FAF, FAT, Stafford, IRS, SAR, Institutional
Additional financial aid information: Need based financial aid comprised of grants, loans and work-study programs. All returning students are eligible to be considered for academic excellence awards.

ENGINEERING COLLEGE INFORMATION
[For additional personnel, refer to the *Appendix*.]

HEAD OF ENGINEERING
Donald G. Blanchard
Phone: (603)428-2245   Fax: (603)428-7230

ENGINEERING COLLEGE ADDRESS
Engineering Program
New England College
New Science Building
Depot Hill Road
Henniker, NH 03242-000
Phone: (603)428-2211   Fax: (603)428-7230

ENROLLMENTS—BY CLASS
[Numbers are baccalaureate enrollments for the fall 1991 term, unless otherwise footnoted.]
1st-year students/Freshmen 14
2nd-year students/Sophomores 5
3rd-year students/Juniors 8
4th-year students/Seniors 8
Total **35**

NUMBER OF DEGREES AWARDED—BY PROGRAM
[Numbers are engineering baccalaureate degrees awarded during the 1991-1992 academic year, unless otherwise footnoted. For full details about each engineering program, refer to the *Tables of Degree Programs*.]
Civil Engineering 8
Total **8**

PERCENTAGE OF DEGREES AWARDED—BY CATEGORY
[Percentages are of all engineering baccalaureate degrees awarded during the 1991-1992 academic year, unless otherwise footnoted.]
African Americans .... – %
Asian/Pacific Island Americans .... – %
Hispanic Americans .... – %
Native Americans .... – %
Foreign Citizens .... – %
All Others .... 100.0 %
Women .... 10.0 %
Persons with disabilities .... – %
Students over 25 years of age .... – %

ENGINEERING STUDENT DATA
Applicants to the engineering college
Number of applicants to engineering college 31
Percent offered admission 45.0%

Matriculated engineering students
Percentage in top quartile (25%) of High School class 16.6
Average SAT scores: Math—500, Verbal—370, Combined—870

FULL- & PART-TIME FACULTY—BY DEPARTMENT
[Figures are the head count of full-time faculty and the full-time equivalent (FTE) of part-time faculty for each engineering department or equivalent.]

Department	Full	Part
Civil Eng	4	–

COLLEGE DESCRIPTION
The Civil Engineering Program at New England College is uniquely established within the setting of a small liberal arts college. Students enjoy close contact with professors and hands-on experience with actual engineering problems. The personal atmosphere and small classes enable students to learn how to use high precision instruments to gather data and draw plans using computer-assisted drafting systems. This practical experience gives students the skills needed to identify and analyze problems and design workable solutions. The Program has gained an international reputation for excellence, and its graduates are recruited by many private firms and government agencies, because its students are trained with an emphasis on practical applications.

DEGREE PROGRAMS DESCRIPTION
A Bachelor of Science in Civil Engineering from New England College permits a certain amount of specialization in the fields of structures, geotechnical, transportation, water resources, urban planning and design; an environmental engineering tract is being developed. A total of 16 credits of general education, as defined in ABET criteria, must be completed as shown in the curriculum sequence. The College's focus on international education perspective exposes students to critical and creative thinking opportunities that help develop effective communication skills and responsible academic commitments. The normal College requirements for writing courses must be satisfied by all students; the junior year writing activity is incorporated into the engineering courses. The senior year professional seminar program and capstone engineering design project are trademarks of the Program. The Program also maintains a close association with the College's Business and Environmental Sciences Programs.

TRANSFER INFORMATION
Residency Requirements: To graduate from New England College with a BSCE degree, students must complete 60 credits from New England College or 30 credits in the senior year, and no less than 12 credits within the major. In addition, engineering majors must complete 9 credits of engineering science and 12 credits of engineering design courses.

Transfer via Articulation Agreements
Admission to engineering: At the end of the sophomore year
Engineering graduates transferred from ..... 4-yr: –   2-yr: –

Transfer without Articulation Agreements
Admission to engineering: At any time
Engineering graduates transferred from ..... 4-yr: 2   2-yr: –

GRADUATION REQUIREMENTS
The Civil Engineering major is designed to exceed the minimum curricular criteria established by ABET. A total of 136 credits is required for a BSCE, including: 16 credits of general education, 33 credits of mathematics and sciences, 38 credits of engineering science, and 19 credits of engineering design. The remaining credits are engineering and general electives. Only engineering courses with a C- or better are accepted for graduation requirements.

STUDENT PROGRAMS

PROFESSIONAL AND HONORARY SOCIETIES
[For key to acronyms, see *Introduction*.]
Professional Societies: ASCE
Honorary Societies: Phi Tau Beta Honor Society, Red and Blue Society

SUPPORT PROGRAMS
Other Engineering Support Programs: All freshman and transfer students are enrolled in the orientation program where contacts with faculty and student chapter members are made on a bi-weekly basis. All students are assigned a Faculty Advisor, and there is a full-time Academic Advising Center available for student referrals when necessary.

149 University of New Hampshire

■ INSTITUTION PROFILE
HEAD OF THE INSTITUTION
Dale F. Nitzschke
Phone: (603)862-2450   Fax: (603)862-3060
GENERAL INFORMATION
[All Students—Fall 1991]
Undergraduate enrollment 10,704
Graduate enrollment 1,553
Total institution enrollment 12,257
Type of institution: Public   **Calendar:** Semesters
Nearest city: Dover, NH   **Population:** 26,000
Miles from main campus: 4   **Setting:** Small Town
Types of engineering degrees: Engineering & Technology
Other degree-granting colleges: Liberal Arts, Life Science and Agriculture, Business and Economics, Health and Human Services

■ ENGINEERING ADMISSIONS
ADMISSIONS OFFICE CONTACT
David W. Kraus
University of New Hampshire
Grant House
Durham, NH 03824
Phone: (603)862-1360
ENGINEERING COLLEGE ADMISSIONS INFORMATION
Entrance Requirements: High School courses—English (4 years), Mathematics(Including trigonometry) (4 years), Science (including physics and/or chemistry) (3 years)
Entrance Recommendations: SAT, High School courses—Social science (2 years), Foreign language (3 years)
Requirements for foreign students: TOEFL (Score: 550)
DEPARTMENTAL ADMISSIONS INFORMATION
Admission to the engineering department: At the time of admission to the institution.

■ FINANCIAL INFORMATION
ESTIMATED EXPENSES (ACADEMIC YEAR)
[Expenses are for the 1992-93 nine-month academic year.]

State Residents	Undergraduate	Graduate
Tuition and fees	$ 3,290	$ 4,057
College room and board	$ 3,600	$ 3,600
Books and supplies	$ 250	$ 250
Other expenses	$ –	$ –
Total estimated expenses	**$ 7,140**	**$ 7,907**

Out-of-State Residents	Undergraduate	Graduate
Tuition and fees	$ 9,840	$ 10,530
College room and board	$ 3,600	$ 3,600
Books and supplies	$ 250	$ 250
Other expenses	$ –	$ –
Total estimated expenses	**$13,690**	**$ 14,380**

FINANCIAL AID OFFICE CONTACT
Richard Craig, *Director, Financial Aid*
University of New Hampshire
Stoke Hall
Durham, NH 03824
Phone: (603)862-3600
GENERAL FINANCIAL AID INFORMATION
Forms accepted/required: FAF, IRS
Additional financial aid information: Aid is available from University grants and scholarships, Pell Grant Programs, UNH loan funds, Perkins Loans, Higher Education Act Loans and college work-study programs.

■ ENGINEERING COLLEGE INFORMATION
[For additional personnel, refer to the *Appendix*.]
HEAD OF ENGINEERING
Otis J. Sproul
Phone: (603)862-1781   Fax: (603)862-2486
ENGINEERING COLLEGE ADDRESS
College of Engineering and Physical Sciences
University of New Hampshire
Kingsbury Hall
Durham, NH 03824
Phone: (603)862-1234   Fax: (603)862-2486
ENROLLMENTS—BY CLASS
[Numbers are baccalaureate enrollments for the fall 1991 term, unless otherwise footnoted.]
1st-year students/Freshmen 206
2nd-year students/Sophomores 198
3rd-year students/Juniors 166
4th-year students/Seniors 277
Total **847**

NUMBER OF DEGREES AWARDED—BY PROGRAM
[Numbers are engineering baccalaureate degrees awarded during the 1991-1992 academic year, unless otherwise footnoted. For full details about each engineering program, refer to the *Tables of Degree Programs*.]
Chemical Engineering 24
Civil Engineering 53
Electrical Engineering 42
Mechanical Engineering 44
Total **163**

PERCENTAGE OF DEGREES AWARDED—BY CATEGORY
[Percentages are of all engineering baccalaureate degrees awarded during the 1991-1992 academic year, unless otherwise footnoted.]
African Americans .... – %
Asian/Pacific Island Americans .... 1.2 %
Hispanic Americans .... – %
Native Americans .... – %
Foreign Citizens .... – %
All Others .... 98.8 %
Women .... 11.3 %
Persons with disabilities .... – % (A)
Students over 25 years of age .... – % (A)
Notes: (A) Data not available.

GRADUATE ENROLLMENTS & DEGREES AWARDED
Master's enrollment 149
Master's degrees awarded 53
Doctoral enrollment 42
Doctoral degrees awarded 9
ENGINEERING STUDENT DATA
Applicants to the engineering college
Number of applicants to engineering college 580
Percent offered admission 82.0%
Matriculated engineering students
Percentage in top quartile (25%) of High School class 70.0
Average SAT scores: Combined—1098

FULL- & PART-TIME FACULTY—BY DEPARTMENT
[Figures are the head count of full-time faculty and the full-time equivalent (FTE) of part-time faculty for each engineering department or equivalent.]

Department	Full	Part
Chemical Eng	7	–
Civil Eng	10	–
Electrical & Computer Eng	16	–
Mechanical Eng	14	–

COLLEGE DESCRIPTION
The undergraduate degree is given at the end of four years of study and includes the University's general education requirements as well as the engineering departments' major requirements. The laboratory portion of the coursework is a particularly strong aspect of the student's education. The University and the engineering departments have an Honors Program. An informal coo-op program, arranged primarily in the summer, is available on an individual basis. A cooperative engineering/business program awards the MBA and the BS in engineering degree at the end of five years. Interdisciplinary minors are available in environmental engineering, hydrology, materials science and ocean engineering. An exchange program with the Technical University of Budapest, Budapest, Hungary exists for engineering students in their junior year. Graduate programs through the doctoral level are available in all departments.

DEGREE PROGRAMS DESCRIPTION
Engineering programs prepare students for productive employment in their engineering areas and for entry into graduate school. The faculty operate with five broad educational objectives under which the student should: (1) be able to reason; (2) understand the appropriate professional subject matter; (3) possess a broad education in the basic sciences; (4) have an understanding and concern for life and humanity; and (5) appreciate that education is a life long process.

DUAL DEGREE PROGRAMS
Engineering baccalaureate graduates with dual degrees 0
Enrollment requirements: Students must complete 32 credits beyond those required for the first degree.

TRANSFER INFORMATION
Residency Requirements: All students are required to complete their last 32 credits in residence.

Transfer without Articulation Agreements
Admission to engineering: At any time
Engineering graduates transferred from   4-yr: 9   2-yr: 4

GRADUATION REQUIREMENTS
Students must achieve the following minimum number of credits: Chemical Engineering 129; Civil Engineering 133; Electrical Engineering 128; Mechanical Engineering 128. The Departments of Civil Mechanical and Electrical Engineering require a 2.00 grade point average in their engineering courses for graduation.

■ STUDENT PROGRAMS
PROFESSIONAL AND HONORARY SOCIETIES
[For key to acronyms, see *Introduction*.]
Professional Societies: ACM, ACS, AIChE, AMS, ASCE, ASME, IEEE, IES, TAPPI
Honorary Societies: Tau Beta Pi, Phi Kappa Phi

SUPPORT PROGRAMS
Student Chapter Organizations: Society of Women Engineers

Office of Multicultural Student Affairs
For: Ethnic Minorities   **Available:** Year round
Offered: Freshman, Sophomore, Junior, Senior, Graduate level
Other Engineering Support Programs: A June orientation is expected of all entering freshmen plus an orientation in September prior to the start of the fall semester. All students are assigned academic advisors in their major department. The University operates centers to assist in the development of study habits, career planning and placement, as well as personal counseling.

150 University of New Haven

INSTITUTION PROFILE
HEAD OF THE INSTITUTION
Lawrence J. DeNardis
Phone: (203)932-7275 Fax: (203)937-0756
GENERAL INFORMATION
[All Students—Fall 1991]
Undergraduate enrollment 3,399
Graduate enrollment 2,500
Total institution enrollment 5,899
Type of institution: Private **Calendar:** Semesters
Nearest city: New Haven **Population:** 130,000
Miles from main campus: 3 **Setting:** Urban
Types of engineering degrees: Engineering
Other degree-granting colleges: Arts & Sciences, Business Administration, Hotel/Restaurant/Tourism, Public Safety and Professional Studies, Graduate School

ENGINEERING ADMISSIONS
ADMISSIONS OFFICE CONTACT
Steven Briggs
University of New Haven
Office of Admissions
300 Orange Avenue
West Haven, CT 06516
Phone: (203)932-7319 Fax: (203)933-5610
ENGINEERING COLLEGE ADMISSIONS INFORMATION
Admission to the engineering college: Upon satisfying Entry-Level Engineering requirements
Entrance Requirements: SAT, High School courses—English (4 years), Algebra (2 years), Geometry/Trigonometry (1 year), Science (including one year of physics) (2 years)
Entrance Recommendations: Mathematics Achievement
Requirements for foreign students: TOEFL (Score: 500); financial statement
DEPARTMENTAL ADMISSIONS INFORMATION
Admission to the engineering department: Upon completion of the Entry Level Engineering Program (usually at the end of freshman year). Transfer students must complete 12 credit hours.
Additional information: Each degree program may require up to 6 hours of additional credits within the major to satisfy the Entry Level Engineering Program requirement for that degree.

FINANCIAL INFORMATION
ESTIMATED EXPENSES (ACADEMIC YEAR)
[Expenses are for the 1992-93 nine-month academic year.]

All Students	Undergraduate	Graduate
Tuition and fees	$ 9,820	$ 7,875
College room and board	$ 2,480	$ 9,000
Books and supplies	$ _(A)	$ _(A)
Other expenses	$ _(A)	$ _(A)
Total estimated expenses	$12,300	$16,875

Notes: (A) Data not available.
FINANCIAL AID OFFICE CONTACT
Jane C. Sangeloty, *Director of Financial Aid*
University of New Haven
300 Orange Avenue
West Haven, CT 06516
Phone: (203)932-7312
GENERAL FINANCIAL AID INFORMATION
Forms accepted/required: CSS-FAF, FAT, IRS, Institutional
Additional financial aid information: The University of New Haven offers a comprehensive financial aid program. Students receive assistance in the form of grants, scholarships, student loans and part-time employment. Funds are available from federal, state, institutional and private sources. Approximately 60% of the University's full-time undergraduate students receive some form of financial assistance.

ENGINEERING COLLEGE INFORMATION
[For additional personnel, refer to the *Appendix*.]
HEAD OF ENGINEERING
M. J. Kenig
Phone: (203)932-7168 Fax: (203)932-7394
ENGINEERING COLLEGE ADDRESS
School of Engineering
University of New Haven
Buckman Hall
300 Orange Avenue
West Haven, CT 06516
Phone: (203)932-7000 Fax: (203)932-7394
ENROLLMENTS—BY CLASS
[Numbers are baccalaureate enrollments for the fall 1991 term, unless otherwise footnoted.]
1st-year students/Freshmen 239
2nd-year students/Sophomores 215
3rd-year students/Juniors 190
4th-year students/Seniors 202
Total 846
NUMBER OF DEGREES AWARDED—BY PROGRAM
[Numbers are engineering baccalaureate degrees awarded during the 1991-1992 academic year, unless otherwise footnoted. For full details about each engineering program, refer to the *Tables of Degree Programs*.]
Chemical Engineering 2
Civil & Environmental Engineering .... _(B)
Civil Engineering 11
Computer Science 9
Electrical Engineering 32
Industrial Engineering 4
Mechanical Engineering 33
Total 91
Notes: (B) Data not applicable.
PERCENTAGE OF DEGREES AWARDED—BY CATEGORY
[Percentages are of all engineering baccalaureate degrees awarded during the 1991-1992 academic year, unless otherwise footnoted.]
African Americans .... 3.0 %
Asian/Pacific Island Americans .... 3.0 %
Hispanic Americans .... – %
Native Americans .... – %
Foreign Citizens .... 3.0 %
All Others .... 91.0 %
Women .... 10.0 %
Persons with disabilities .... – %(A)
Students over 25 years of age .... 40.0 %(C)
Notes: (A) Data not available. (C) Estimated data.
GRADUATE ENROLLMENTS & DEGREES AWARDED
Master's enrollment 194
Master's degrees awarded 30
Doctoral enrollment .... _(B)
Doctoral degrees awarded .... _(B)
Notes: (B) Data not applicable.
FULL- & PART-TIME FACULTY—BY DEPARTMENT
[Figures are the head count of full-time faculty and the full-time equivalent (FTE) of part-time faculty for each engineering department or equivalent.]

Department	Full	Part
Chemistry/Chemical Eng	5	2.2
Civil & Environmental Eng	6	2.0
Electrical & Computer Eng	9	1.0
Industrial Eng	4	3.0
Mechanical Eng	10	1.0
Computer Science	8	4.0

COLLEGE DESCRIPTION
The University of New Haven is a private co-educational institution located in a suburban setting. The academic year follows the semester system. Degrees are offered in the Schools of Engineering, Arts and Sciences, Business, Hotel/Tourism, Public Safety/Professional Studies, Graduate School. The engineering degree programs are offered on the main campus on a full and part time basis and at the Groton campus on a part time basis. Co-op opportunities are available in all degree programs offered by the School of Engineering. All programs are nominally four year programs but most students take between four and five years of full time study to complete degree requirements.

DEGREE PROGRAMS DESCRIPTION
All new students are admitted to the ENTRY LEVEL ENGINEERING PROGRAM (ELEP) which has a common freshman year. Individual engineering degree programs admit students to the PROFESSIONAL LEVEL ENGINEERING PROGRAM (PLEP) upon satisfactory completion of the ELEP core. Transfer students who have satisfactorily completed the ELEP core are provisionally admitted to take twelve credit hours before they can be admitted to PLEP. All programs provide a solid foundation in mathematics, computer and physical sciences, engineering science and the humanities and social sciences. Laboratory study and utilization of the Engineering Computer Center are integrated into each program and serve to develop skills essential for the application of theoretical and empirical modeling, analysis, and design. The Co-op option is available in all programs. Engineering classes are small averaging about 20 students. Most courses are scheduled both day and evening hours providing students additional scheduling flexibility.

TRANSFER INFORMATION
Residency Requirements: Degree candidates must complete a minimum of 30 semester hours of work in residence at the University of New Haven.
Transfer without Articulation Agreements
Admission to engineering: Transfers are admitted to the entry level engineering program up to the end of their junior year in other programs. Transfer credits are uniformly evaluated for all degree programs.
Requirements: Admission is into the Entry Level Program. Matriculation into the Professional Level Programs is dependent upon successfully meeting Entry Level Program requirements and successfully completing 12 credit hours in the School of Engineering.

GRADUATION REQUIREMENTS
Students must meet degree requirements for all programs as described in the University Bulletin. These include meeting the University Core requirements, design credit requirements and writing proficiency requirements. All students must pass a writing proficiency examination after completion of 57 credit hours.

STUDENT PROGRAMS
PROFESSIONAL AND HONORARY SOCIETIES
[For key to acronyms, see *Introduction*.]
Professional Societies: ASCE, ASME, IEEE, IIE, ORSA, SME, TIMS
Honorary Societies: Alpha Epsilon, Chi Epsilon, Eta Kappa Nu, Pi Tau Sigma, The Order of the Engineer
SUPPORT PROGRAMS
Student Chapter Organizations: Mentor program for freshman women, Women's Health Center, Women's studies course offerings, Office of Minority Student Affairs

Mentor program for freshmen women
For: Women **Available:** Academic year
Offered: Prematriculation, Freshman

Women's Health Center
For: Women **Available:** Year round
Offered: Prematriculation, Freshman, Sophomore, Junior, Senior, Graduate level

Women's studies course offering
For: Women **Available:** Academic year
Offered: Freshman, Sophomore, Junior, Senior

Other Engineering Support Programs: Orientation program for all freshmen and transfers; Freshmen Advising and Freshmen Experience Seminar; Center for Learning Resources provides tutoring and study skills workshops; Career Development Office; Counseling Center; International Student Office; Minority Student Office; Disabled Student Services. Advisement for new students occurs within the Entry Level Engineering Program.

151 New Jersey Institute of Technology

INSTITUTION PROFILE

HEAD OF THE INSTITUTION
Saul K. Fenster
Phone: (201)596-3101   Fax: (201)624-2541

GENERAL INFORMATION
[All Students—Fall 1991]
Undergraduate enrollment 4,876
Graduate enrollment 2,521
Total institution enrollment 7,397
Type of institution: Public   Calendar: Semesters
Location: Newark   Population: 275,221
Setting: Urban
Types of engineering degrees: Engineering & Technology
Other degree-granting colleges: Science and Liberal Arts, Architecture, Industrial Management

ENGINEERING ADMISSIONS

ADMISSIONS OFFICE CONTACT
William Anderson
New Jersey Institute of Technology
University Heights
Newark, NJ 07102-9938
Phone: (201)596-3300   Fax: (201)802-1854

ENGINEERING COLLEGE ADMISSIONS INFORMATION
Entrance Requirements: SAT, High School courses—English (4 years), Mathematics incl algebra, geometry & trigonometry (3.5 years), Lab Sciences (chemistry and physics preferred) (2 years), Also mathematics achievement test level I or II
Requirements for foreign students: TOEFL (Score: 520); financial statement

DEPARTMENTAL ADMISSIONS INFORMATION
Admission to the engineering department: At the time of admission to the institution.

FINANCIAL INFORMATION

ESTIMATED EXPENSES (ACADEMIC YEAR)
[Expenses are for the 1992-93 nine-month academic year.]

State Residents	Undergraduate	Graduate
Tuition and fees	$ 4,524	$ 6,136
College room and board	$ 4,980 [1]	$ 4,980 [1]
Books and supplies	$ 750	$ 750
Other expenses	$ –	$ –
Total estimated expenses	**$10,254**	**$11,866**

Out-of-State Residents	Undergraduate	Graduate
Tuition and fees	$ 8,642	$ 8,808
College room and board	$ 4,980 [1]	$ 4,980 [1]
Books and supplies	$ 750	$ 750
Other expenses	$ –	$ –
Total estimated expenses	**$14,372**	**$14,538**

Notes: (1) Average cost.

FINANCIAL AID OFFICE CONTACT
Mary Hurdle, *Director of Financial Aid*
New Jersey Institute of Technology
University Heights
Newark, NJ 07102-9938
Phone: (201)596-3479

GENERAL FINANCIAL AID INFORMATION
Forms accepted/required: CSS-FAF, FAF, FAT, Stafford, IRS, SAR, Supplemental, Institutional, New Jersey Financial Aid Form (NJFAF)
Additional financial aid information: Financial Aid offered includes: Pell Grants, SEOG'S, Slate Grants/Scholarships, Academic Merit Scholarships, Private Scholarships, Stafford Loans, Perkins Loans, plus/SLS Loans, Institutional Fund Scholarships, Grants, Loans, and part time campus employment.

ENGINEERING COLLEGE INFORMATION
[For additional personnel, refer to the *Appendix*.]

HEAD OF ENGINEERING
George Pincus
Phone: (201)596-3213   Fax: (201)596-2316
EMail: Pincus@Admin1.NJIT.Edu

ENGINEERING COLLEGE ADDRESS
Newark College of Engineering
New Jersey Institute of Technology
University Heights
323 M.L. King Blvd.
Newark, NJ 07102-9938
Phone: (201)596-3000   Fax: (201)596-2316

ENROLLMENTS—BY CLASS
[Numbers are baccalaureate enrollments for the fall 1991 term, unless otherwise footnoted.]

1st-year students/Freshmen	809
2nd-year students/Sophomores	460
3rd-year students/Juniors	609
4th-year students/Seniors	724
Total	**2,602**

NUMBER OF DEGREES AWARDED—BY PROGRAM
[Numbers are engineering baccalaureate degrees awarded during the 1991-1992 academic year, unless otherwise footnoted. For full details about each engineering program, refer to the *Tables of Degree Programs*.]

Biomedical Engineering	– (1)
Chemical Engineering	21
Civil Engineering	46
Computer Engineering	13
Electrical Engineering	123
Engineering Science(s)	3
Environmental Engineering	– (1)
Industrial Engineering	29
Management Engineering	– (1)
Manufacturing Engineering	3
Mechanical Engineering	78
Occupational Health & Safety	– (1)
Transportation Planning & Engineering	– (6)
Total	**316**

Notes: (1) No undergraduate program. Masters level only. (6) No undergraduate program. Masters and doctoral levels only.

PERCENTAGE OF DEGREES AWARDED—BY CATEGORY
[Percentages are of all engineering baccalaureate degrees awarded during the 1991-1992 academic year, unless otherwise footnoted.]

African Americans	7.6 %
Asian/Pacific Island Americans	14.5 %
Hispanic Americans	10.4 %
Native Americans	0.3 %
Foreign Citizens	7.3 % (1)
All Others	59.9 %
Women	10.4 %
Persons with disabilities	– % (A)
Students over 25 years of age	41.4 %

Notes: (A) Data not available. (1) International students - not foreign born. NJIT maintains data on a citizen/permanent resident versus non-citizen basis.

GRADUATE ENROLLMENTS & DEGREES AWARDED

Master's enrollment	872 (3)
Master's degrees awarded	353
Doctoral enrollment	134 (4)
Doctoral degrees awarded	15

Notes: (3) Additional 7 masters students enrolled in applied chemistry. Additional 138 students enrolled in environmental science. (4) Additional 8 doctoral students enrolled in environmental science.

ENGINEERING STUDENT DATA

Applicants to the engineering college
Number of applicants to engineering college 958
Percent offered admission 68.1%

Matriculated engineering students
Percentage in top quartile (25%) of High School class 49.8
Average SAT scores: Math—602, Verbal—470, Combined—1072

FULL- & PART-TIME FACULTY—BY DEPARTMENT
[Figures are the head count of full-time faculty and the full-time equivalent (FTE) of part-time faculty for each engineering department or equivalent.]

Department	Full	Part
Chemical Eng, Chemistry & Env'l Science	30	1.8
Civil & Environmental Eng	26	1.2
Electrical & Computer Eng	38	4.7
Mechanical & Industrial Eng	42	2.5
Eng Tech	12	7.5

COLLEGE DESCRIPTION
Co-op is available - 2 full time work periods. Participation in Co-op may extend the time required to complete the degree program by one year. Double major, dual BS degree, BS/MS, BS/MD, and BS/DMD options available. The program includes courses in the basic sciences and the humanities and social sciences. Thus the overall program provides students with an education designed to permit students to make major contributions in many areas. More than 30 major research and public service centers on campus including center for manufacturing systems and hazardous substance management center.

DEGREE PROGRAMS DESCRIPTION
Bachelors level-all engineering majors pursue a common first year. Each curriculum is sufficiently broad to permit a grad to enter the engineering profession immediately or continue on to advanced study. Master's degree programs are available in Applied Science, Biomedical Engineering, Computer Engineering, Chemical Engineering, Chemistry, Civil Engineering, Elect. Engineering, Engineering Science, Envir. Engineering, Envir. Science, Ind. Engineering, Mgmt Engineering, Manuf. Engineering, Materials Science and Engineering, Mech. Engineering, Occupational Safety and Health Engineering, and Transportation. Doctoral Programs and the Degree of Engineer are offered in Chemical, Civil, Electrical, and Mech. Engineering. NJIT encourages interdisciplinary approaches to study and research. A number of graduate Programs - particularly Biomedical Engineering, Manuf. Engineering, Envir. Science, and Engineering and Applied Science-Draw upon the instructional and lab resources of two or more academic depts. In addition, a program in cooperative education enables graduate students to combine related professional work experiences with their advanced studies.

DUAL DEGREE PROGRAMS
Engineering baccalaureate graduates with dual degrees 0
Enrollment requirements: Written approval to undertake this curriculum must be obtained from the Department(s) involved and the Dean(s) of the appropriate college(s).

TRANSFER INFORMATION
Residency Requirements: Students must complete at least 33 credits approved by the Department of their major study

Transfer via Articulation Agreements
Admission to engineering: In the freshman, sophomore, or junior year.
Engineering graduates transferred from .. 4-yr: –   2-yr: 129
Requirements: 2.50 GPA or better recommended. Transcripts of all attempted post secondary School of Work. Applicants who have earned fewer than 30 college credits may be asked to submit a secondary School of Transcript and/or Standardized Test Scores. A rolling admission procedure is used.

Transfer without Articulation Agreements
Admission to engineering: In the freshman, sophomore, or junior year.
Engineering graduates transferred from .. 4-yr: 15   2-yr: 2
Requirements: 2.50 GPA or better recommended. Transcripts of all attempted post- secondary School of Work. Applicants who have earned fewer than 30 college credits may be asked to submit a secondary School of Transcript and/or Standardized Test Scores (SAT OR ACT). A rolling admission procedure is used.

GRADUATION REQUIREMENTS
Bachelor of Science Programs in Engineering require 139-143 credits (including 2 credits of Physical Education) depending on major selected. Students must attain a grade point average of 2.0 in all the courses listed in the catalog as being required in the third and fourth years of the appropriate curriculum.

STUDENT PROGRAMS

PROFESSIONAL AND HONORARY SOCIETIES
[For key to acronyms, see *Introduction*.]
Professional Societies: ACS, AIChE, ASCE, ASME, IEEE, IIE, ITE, SAE, SME, Society of Women Engineers (SWE)
Honorary Societies: Alpha Pi Mu, Chi Epsilon, Eta Kappa Nu, Omega Chi Epsilon, Pi Tau Sigma, Tau Alpha Pi, Tau Beta Pi, Phi Eta Sigma

SUPPORT PROGRAMS
Student Chapter Organizations: Society of Women Engineers, National Society of Black Engineers, Black Association of Student Engineers (BASE), Hispanic Organization of Students in Tech (HOST)

Educational Opportunity Program
For: Ethnic Minorities   **Available:** Year round
Offered: Prematriculation, Freshman, Sophomore, Junior, Senior, Graduate level

Women in Engineering, Science and Technology
For: Women   **Available:** Year round
Offered: Prematriculation, Freshman, Sophomore, Junior, Senior

Women Students Over 25
For: Women   **Available:** Academic year
Offered: Freshman, Sophomore, Junior, Senior, Graduate level

Big Sister/Little Sister
For: Women   **Available:** Academic year
Offered: Freshman, Sophomore, Junior, Senior

Other Engineering Support Programs: Freshman orientation, Transfer student orientation, educational opportunity fund summer program to strengthen a student's background, personal counseling, academic counseling, career counseling, English as a second language (ESL) Courses, tutoring program, career services, International student services, health services, and services to physically challenged and learning disabled students.

152 The University of New Mexico

■ INSTITUTION PROFILE
HEAD OF THE INSTITUTION
Richard E. Peck
Phone: (505)277-2626 Fax: (505)277-5965
GENERAL INFORMATION
[All Students—Fall 1991]
Undergraduate enrollment 20,069
Graduate enrollment 4,940
Total institution enrollment **25,009**
Type of institution: Public Calendar: Semesters
Location: Albuquerque Population: 500,000
Setting: Urban
Types of engineering degrees: Engineering
Other degree-granting colleges: Allied Health Sciences, Architecture, Arts & Sciences, Business Administration, Education, Fine & Performing Arts, Law, Medicine, Nursing, Pharmacy, Technology & Applied Sciences, Dental Programs

■ ENGINEERING ADMISSIONS
ADMISSIONS OFFICE CONTACT
Cynthia Stuart
The University of New Mexico
Admissions Office
Albuquerque, NM 87131
Phone: (505)277-2446 Fax: (505)277-6686
ENGINEERING COLLEGE ADMISSIONS INFORMATION
Admission to the engineering college: Completion of a minimum of 26 hours: must include 18 hours from calculus, chemistry, physics, engineering; 2.2-3.0 GPA, depending on major.
Entrance Requirements: SAT (Score: 660), ACT (Score: 17), High School courses—English (4 years), Single language other than English (2 years), Mathematics (3 years), Natural Sciences and Social Sciences (2 years)
Entrance Recommendations: ACT (Score: 17), High School courses—English (4 years), Single language other than English (2 years), Mathematics (3 years), Natural Sciences and Social Sciences (2 years)
Requirements for foreign students: TOEFL (Score: 550); financial statement; Official certified transcripts
DEPARTMENTAL ADMISSIONS INFORMATION
Admission to the engineering department: See above.

■ FINANCIAL INFORMATION
ESTIMATED EXPENSES (ACADEMIC YEAR)
[Expenses are for the 1992-93 nine-month academic year.]

State Residents	Undergraduate	Graduate
Tuition and fees	$ 1,656	$ 1,862[1]
College room and board	$ 3,021	$ 3,021
Books and supplies	$ 600	$ 600
Other expenses	$ —[A]	$ —[A]
Total estimated expenses	**$ 5,277**	**$ 5,483**

Out-of-State Residents	Undergraduate	Graduate
Tuition and fees	$ 5,880	$ 6,092[1]
College room and board	$ 3,021	$ 3,021
Books and supplies	$ 600	$ 600
Other expenses	$ —[A]	$ —[A]
Total estimated expenses	**$ 9,501**	**$ 9,713**

Notes: (A) Data not available. (1) Includes a $16.00 Graduate Student Association fee.

FINANCIAL AID OFFICE CONTACT
John E. Whiteside, *Director*
The University of New Mexico
Mesa Vista Hall North
Albuquerque, NM 87131
Phone: (505)277-2041
GENERAL FINANCIAL AID INFORMATION
Forms accepted/required: AFSA, FAT, IRS, SAR, Supplemental, Institutional
Additional financial aid information: Need-based scholarships; merit-based scholarships; part-time and work-study jobs on and off campus; and federal and state funded loans and grants.

■ ENGINEERING COLLEGE INFORMATION
[For additional personnel, refer to the *Appendix*.]
HEAD OF ENGINEERING
James E. Thompson
Phone: (505)277-5521 Fax: (505)277-0813
ENGINEERING COLLEGE ADDRESS
College of Engineering
The University of New Mexico
Farris Engineering Center
Albuquerque, NM 87131
Phone: (505)277-5521 Fax: (505)277-0813
ENROLLMENTS—BY CLASS
[Numbers are baccalaureate enrollments for the fall 1991 term, unless otherwise footnoted.]
1st-year students/Freshmen .... 328 [1]
2nd-year students/Sophomores .... 474 [2]
3rd-year students/Juniors 236
4th-year students/Seniors 507
Total **1,545**

Notes: (1) Includes students in associate programs and students working on the freshman year of the curriculum. (2) Includes students in associate programs and students working on the freshman year of the curriculum.

NUMBER OF DEGREES AWARDED—BY PROGRAM
[Numbers are engineering baccalaureate degrees awarded during the 1991-1992 academic year, unless otherwise footnoted. For full details about each engineering program, refer to the *Tables of Degree Programs*.]
Biomedical Engineering 2
Chemical Engineering 6
Civil Engineering 16
Computer Engineering 24
Computer Science 19
Construction Engineering 5
Construction Management 1
Electrical Engineering 50
Manufacturing & Robotics 1
Mechanical Engineering 55
Nuclear Engineering 3
Total **182**

PERCENTAGE OF DEGREES AWARDED—BY CATEGORY
[Percentages are of all engineering baccalaureate degrees awarded during the 1991-1992 academic year, unless otherwise footnoted.]
African Americans 0.5%
Asian/Pacific Island Americans 1.6%
Hispanic Americans 15.1%
Native Americans 3.2%
Foreign Citizens 7.5%
All Others 72.1%
Women 21.0%
Persons with disabilities .... — % [A]
Students over 25 years of age .... — % [A]

Notes: (A) Data not available.

GRADUATE ENROLLMENTS & DEGREES AWARDED
Master's enrollment .... 412 [5]
Master's degrees awarded .... 120 [6]
Doctoral enrollment .... 147 [7]
Doctoral degrees awarded .... 18 [8]

Notes: (5) Includes Computer Science students (98). (6) Includes Computer Science students (19). (7) Includes Computer Science students (4). (8) Includes Computer Science students (2).

ENGINEERING STUDENT DATA
Applicants to the engineering college
Number of applicants to engineering college 350
Percent offered admission 90.0%
Matriculated engineering students
Percentage in top quartile (25%) of High School class .... — [A]

Notes: (A) Data not available.

FULL- & PART-TIME FACULTY—BY DEPARTMENT
[Figures are the head count of full-time faculty and the full-time equivalent (FTE) of part-time faculty for each engineering department or equivalent.]

Department	Full	Part
Chemical & Nuclear Eng	14	2.3
Civil Eng	17	0.6
Electrical & Computer Eng	40	2.0
Mechanical Eng	18	0.8
Computer Science	14	0.8

COLLEGE DESCRIPTION
The College offers four-year accredited undergraduate programs in chemical, civil, computer, construction, electrical, mechanical and nuclear engineering and computer science. In addition, the College offers more flexible four-year degrees in manufacturing engineering and robotics and construction management. Co-op programs are available in all fields. The college offers graduate study at the MS and PhD levels in chemical, civil, electrical, mechanical, and nuclear engineering and in computer science. The College offers a certificate through the Waste-Management Education and Research Consortium. This is the nation's first consortium created to develop resources and address waste management issues.

DEGREE PROGRAMS DESCRIPTION
The College of Engineering at the University of New Mexico advances, interprets, and disseminates knowledge through teaching, research, and service. The College performs these activities in ways that attract high quality students and faculty to produce the highest quality graduates. The College enjoys three distinct advantages: (1) New Mexico's cultural and ethnic diversity, (2) location in the commercial, industrial, and population center of the State, and (3) proximity to federal laboratories. The College conducts basic and applied research in engineering and scientific disciplines to advance the knowledge base and to strengthen the teaching program. Through its research activities the College also seeks to improve and protect the natural and built environments, and to achieve the economic goals of the University, the State, and the Nation. The College recognizes the importance of multi-disciplinary approaches to research and seeks to enhance research programs through collaboration with other disciplines both within and outside the University.

TRANSFER INFORMATION
Residency Requirements: A minimum of 30 semester hours of credit, exclusive of extension and correspondence (independent study) credit, must be earned at UNM. Of these 30 semester hours in residence, 15 semester hours must be earned after the candidate has accumulated 92 hours of earned semester hour credit; these 15 hours, however, do not necessarily have to be the last hours of a degree program.
Transfer via Articulation Agreements
Admission to engineering: See transfer via articulation agreements.
Transfer without Articulation Agreements
Admission to engineering: Same as for articulated agreements.

GRADUATION REQUIREMENTS
Each candidate for degree must have at least a 2.0 grade point average on work taken at the University of New Mexico which is counted toward graduation. Among the credits presented for graduation not more than 9 credit hours shall be D. BS and BE programs require 130 to 136 semester hours, depending on major.

■ STUDENT PROGRAMS
PROFESSIONAL AND HONORARY SOCIETIES
[For key to acronyms, see *Introduction*.]
Professional Societies: ACM, AGCA, AIAA, AIChE, ANS, ASCE, ASME, IEEE, SAMPE, IEEE-Computer Division, Students for the Exploration and Development of Space
Honorary Societies: Chi Epsilon, Eta Kappa Nu, Pi Tau Sigma, Tau Beta Pi

SUPPORT PROGRAMS
Student Chapter Organizations: Society of Women Engineers, Hispanic Engineering Organization, American Indian Science and Engineering Society

Minority Engineering Programs
For: Women & Ethnic Minorities **Available:** Year round
Offered: Freshman, Sophomore, Junior, Senior

NASA Training Project
For: Ethnic Minorities **Available:** Year round
Offered: Freshman, Sophomore, Junior, Senior

Native American Programs, College of Engineering
For: Ethnic Minorities **Available:** Year round
Offered: Freshman, Sophomore, Junior, Senior

Other Engineering Support Programs: Summer orientation program encouraged for all incoming students; comprehensive tutoring available to all students in freshman and sophomore courses; career counseling services; foreign students office; veterans services; services for handicapped students; placement services; advising services available year round.

153 New Mexico Institute of Mining and Technology

INSTITUTION PROFILE

HEAD OF THE INSTITUTION
Laurence H. Lattman
Phone: (505)835-5600 **Fax:** (505)835-6329

GENERAL INFORMATION
[All Students—Fall 1991]
Undergraduate enrollment 1,278
Graduate enrollment 329
Total institution enrollment 1,607
Type of institution: Public **Calendar:** Semesters
Nearest city: Albuquerque **Population:** 750,000
Miles from main campus: 75 **Setting:** Small Town
Types of engineering degrees: Engineering

ENGINEERING ADMISSIONS

ADMISSIONS OFFICE CONTACT
Louise Chamberlin
New Mexico Institute of Mining and Technology
Campus Station
Socorro, NM 87801
Phone: (800)428-8324 **Fax:** (505)835-6329

ENGINEERING COLLEGE ADMISSIONS INFORMATION
Entrance Requirements: ACT (Score: 21), High School courses—English (4 years), Science (2 years), Math (3 years), Social Science (3 years)
Entrance Recommendations: High School courses—Biology (2 years), Chemistry (2 years), Physics (2 years), Algebra, Trigonometry and Intro. to Calculus (2 years)
Requirements for foreign students: TOEFL (Score: 540)

DEPARTMENTAL ADMISSIONS INFORMATION
Admission to the engineering department: At the time of admission to the institution.

FINANCIAL INFORMATION

ESTIMATED EXPENSES (ACADEMIC YEAR)
[Expenses are for the 1992-93 nine-month academic year.]

State Residents	Undergraduate	Graduate
Tuition and fees	$ 1,781	$ 1,794
College room and board	$ 1,990	$ 1,990
Books and supplies	$ 1,200	$ 1,200
Other expenses	$ –	$ –
Total estimated expenses	**$ 4,971**	**$ 4,984**

Out-of-State Residents	Undergraduate	Graduate
Tuition and fees	$ 5,481	$ 5,572
College room and board	$ 1,990	$ 1,990
Books and supplies	$ 1,200	$ 1,200
Other expenses	$ –	$ –
Total estimated expenses	**$ 8,671**	**$ 8,762**

FINANCIAL AID OFFICE CONTACT
Anne Hansen, *Director, Financial Aid*
New Mexico Institute of Mining and Technology
Campus Station
Socorro, NM 87801
Phone: (505)835-5333 **Fax:** (505)835-6329

GENERAL FINANCIAL AID INFORMATION
Forms accepted/required: FAF, FFS, Stafford, SAR, Institutional
Additional financial aid information: Grants, loans, college work study, or campus employment. Need-based scholarships/grants. Merit-based scholarships.

ENGINEERING COLLEGE INFORMATION
[For additional personnel, refer to the *Appendix*.]

HEAD OF ENGINEERING
Carl J. Popp, *Vice President for Academic Affairs*
Phone: (505)835-5227 **Fax:** (505)835-6329

ENGINEERING COLLEGE ADDRESS
New Mexico Institute of Mining and Technology
Campus Station
Socorro, NM 87801
Phone: (505)835-5227 **Fax:** (505)835-6329

ENROLLMENTS—BY CLASS
[Numbers are baccalaureate enrollments for the fall 1991 term, unless otherwise footnoted.]
1st-year students/Freshmen 149
2nd-year students/Sophomores 82
3rd-year students/Juniors 60
4th-year students/Seniors 71
Total **362**

NUMBER OF DEGREES AWARDED—BY PROGRAM
[Numbers are engineering baccalaureate degrees awarded during the 1991-1992 academic year, unless otherwise footnoted. For full details about each engineering program, refer to the *Tables of Degree Programs*.]
Electrical Engineering 1
Materials & Metallurgical Engineering 9
Mineral & Environmental Engineering 9
Petroleum Engineering 11
Total **30**

PERCENTAGE OF DEGREES AWARDED—BY CATEGORY
[Percentages are of all engineering baccalaureate degrees awarded during the 1991-1992 academic year, unless otherwise footnoted.]
African Americans 2.0%
Asian/Pacific Island Americans 3.0%
Hispanic Americans 18.0%
Native Americans 4.0%
Foreign Citizens 4.0%
All Others 69.0%
Women 32.0%
Persons with disabilities .... –% (A)
Students over 25 years of age .... –% (A)

Notes: (A) Data not available.

GRADUATE ENROLLMENTS & DEGREES AWARDED
Master's enrollment 41
Master's degrees awarded 9
Doctoral enrollment 25
Doctoral degrees awarded 6

ENGINEERING STUDENT DATA
Applicants to the engineering college
Number of applicants to engineering college .... – (A)
Percent offered admission .... –% (A)

Matriculated engineering students
Percentage in top quartile (25%) of High School class ... 62.0
Average ACT scores: Math—24.4, Composite—24.9

Notes: (A) Data not available.

FULL- & PART-TIME FACULTY—BY DEPARTMENT
[Figures are the head count of full-time faculty and the full-time equivalent (FTE) of part-time faculty for each engineering department or equivalent.]

Department	Full	Part
Electrical Eng	5	–
Mineral & Environmental Eng	8	–
Materials & Metallurgical Eng	9	–
Petroleum Eng	6	–

COLLEGE DESCRIPTION
The engineering programs are organized under four departments (Electrical Engineering, Material and Metallurgical Engineering, Mining, Geological and Environmental Engineering and Petroleum Engineering) with the exception of General Engineering which is supervised by a faculty committee. The major emphasis is in programs related to natural resource engineering. All of the full-time engineering faculty possess Ph.D. degrees and most have had considerable industrial experience. The average number of years required to actually complete the bachelor's degree is 4.5.

DEGREE PROGRAMS DESCRIPTION
The B.S. degree programs prepare students to enter directly into professional positions or graduate school. The B.S. in engineering science is designed to prepare students for graduate work or to allow them to design special options. B.B. degrees are offered in engineering science, electrical engineering, environmental engineering, geological engineering, materials engineering, metallurgical engineering, mining engineering and petroleum engineering. M.S. degrees are offered in materials, minerals and petroleum engineering and Ph.D. degrees are offered in materials and petroleum engineering.

TRANSFER INFORMATION
Residency Requirements: Minimum of 30 required hours of coursework must be taken to graduate at New Mexico Institute of Mining and Technology. Must be the last 30 hours of coursework.

Transfer via Articulation Agreements
Admission to engineering: Transfer students should have taken 3 semesters of Calculus, calculus-based physics, chemistry, humanities courses, biology or geology in order to transfer as a junior-level student.
Requirements: 2.0 GPA. A rolling admissions procedure is used.

STUDENT PROGRAMS

PROFESSIONAL AND HONORARY SOCIETIES
[For key to acronyms, see *Introduction*.]
Professional Societies: AAPG, ACM, AEG, IEEE, NSPE, SPE, SPS, STC
Honorary Societies: Tau Beta Pi

SUPPORT PROGRAMS
Student Chapter Organizations: Society of Women Engineers, Society of Hispanic Professional Engineers, Minority Scholars Support Program (MSSP), NAMES/MIMES, NM Alliance for Science Education

Academic Support Assistance Program
For: Women & Ethnic Minorities **Available:** Year round
Offered: Freshman, Sophomore, Junior, Senior

Other Engineering Support Programs: Tutoring and counseling are available. Orientation for new students is comprehensive and designed to acclimate new students quickly. Academic advising is performed by engineering faculty in departments in which students intend to major.

154 New Mexico State University

INSTITUTION PROFILE

HEAD OF THE INSTITUTION
James E. Halligan
Phone: (505)646-2035 Fax: (505)646-6334

GENERAL INFORMATION
[All Students—Fall 1991]
Undergraduate enrollment 12,957
Graduate enrollment 2,543
Total institution enrollment 15,500

Type of institution: Public **Calendar:** Semesters
Location: Las Cruces, New Mexico **Population:** 62,126
Setting: Small Town
Types of engineering degrees: Engineering & Technology
Other degree-granting colleges: Arts & Sciences, Education, Agriculture & Home Economics, Human & Community Services (includes nursing), Business Administration & Economics

ENGINEERING ADMISSIONS

ADMISSIONS OFFICE CONTACT
Bill Bruner
New Mexico State University
Box 30001, Dept. 3A
Las Cruces, NM 88003-0001
Phone: (505)646-3121 Fax: (505)646-6330

ENGINEERING COLLEGE ADMISSIONS INFORMATION
Admission to the engineering college: Admitted regularly if their English ACT score is 16 or greater and their performance on a local math exam is satisfactory. Provisional admission is possible.
Entrance Requirements: High School courses—English (4 years), Science (beyond General Science) (2 years), Mathematics (3 years), Foreign language or fine arts (1 year), Regular admission requires a HS GPA of 2.0 & Enhanced ACT of 20 or a HS GPA of 2.5 or an Enhanced ACT of 21. Provisional admission is possible for students not meeting these standards.
Entrance Recommendations: ACT (Score: 20), High School courses—Mathematics (4 years), Chemistry and physics (1 year each) (2 years), Foreign language (2 years), Computers & technical drawing (1 yr each) (2 years)
Requirements for foreign students: TOEFL (Score: 500); financial statement

DEPARTMENTAL ADMISSIONS INFORMATION
Admission to the engineering department: Regular admission requires Enhanced English ACT of 16 or greater and a satisfactory score on a local math exam. Provisional admission is allowed.
Additional information: Students not qualifying for regular admission will be admitted into the General Engineering program. Such students have two years to complete any deficiencies with a minimum GPA. For Ag E, C E, Geol E and M E, the minimum GPA is a 2.1; for I E, a 2.3; and for E E & Chem E the minimum is a 2.5.

FINANCIAL INFORMATION

ESTIMATED EXPENSES (ACADEMIC YEAR)
[Expenses are for the 1992-93 nine-month academic year.]

State Residents	Undergraduate	Graduate
Tuition and fees	$ 1,756	$ 1,876
College room and board	$ 2,668 [1]	$ 2,668 [1]
Books and supplies	$ 575	$ 580
Other expenses	$ –	$ –
Total estimated expenses	$ 4,999	$ 5,124

Out-of-State Residents	Undergraduate	Graduate
Tuition and fees	$ 5,686	$ 5,806
College room and board	$ 2,668 [1]	$ 2,668 [1]
Books and supplies	$ 575	$ 580
Other expenses	$ –	$ –
Total estimated expenses	$ 8,929	$ 9,054

Notes: (1) Board cost is for 12 meals per week.

FINANCIAL AID OFFICE CONTACT
Greeley Myers, *Director of Financial Aid*
New Mexico State University
Box 30001, Dept. 5100
Las Cruces, NM 88003-0001
Phone: (505)646-4105 Fax: (505)646-6330

GENERAL FINANCIAL AID INFORMATION
Forms accepted/required: FAT, Stafford, IRS, SAR, Supplemental, United Student Aid Fund Single File Application, New Mexico Supplemental Information Form
Additional financial aid information: Academic scholarships (including special scholarships for out of state students) from the University, the College of Engineering, and Air Force & Army ROTC. Need based financial aid includes Pell Grant, SEOG, Perkins DSL, and work study. Minority students are also eligible for Designated Academic Opportunity Scholarships (DAOS) and National Action Council for Minorities (NACME) Scholarships. Non-need based work study is also available.

ENGINEERING COLLEGE INFORMATION

[For additional personnel, refer to the *Appendix*.]

HEAD OF ENGINEERING
J. Derald Morgan
Phone: (505)646-2914 Fax: (505)646-3549

ENGINEERING COLLEGE ADDRESS
College of Engineering
New Mexico State University
Box 30001, Dept. 3449
Las Cruces, NM 88003-0001
Phone: (505)646-2912 Fax: (505)646-3549

ENROLLMENTS—BY CLASS
[Numbers are baccalaureate enrollments for the fall 1991 term, unless otherwise footnoted.]
1st-year students/Freshmen .... 510 [1]
2nd-year students/Sophomores .... 375 [1]
3rd-year students/Juniors .... 342 [1]
4th-year students/Seniors .... 489 [1]
Total 1,716

Notes: (1) Fall 1992 enrollment (does not include surveying or technology).

NUMBER OF DEGREES AWARDED—BY PROGRAM
[Numbers are engineering baccalaureate degrees awarded during the 1991-1992 academic year, unless otherwise footnoted. For full details about each engineering program, refer to the *Tables of Degree Programs*.]
Agricultural Engineering 1
Chemical Engineering 17
Civil Engineering 45
Electrical & Computer Engineering 98
Geological Engineering 1
Industrial Engineering 12
Mechanical Engineering 58
Surveying & Mapping 4
Total 236

PERCENTAGE OF DEGREES AWARDED—BY CATEGORY
[Percentages are of all engineering baccalaureate degrees awarded during the 1991-1992 academic year, unless otherwise footnoted.]
African Americans .... – %
Asian/Pacific Island Americans .... – % (A)
Hispanic Americans .... 27.2 %
Native Americans .... 1.7 %
Foreign Citizens .... 6.9 % (3)
All Others .... 64.2 %
Women .... 16.8 %
Persons with disabilities .... – % (A)
Students over 25 years of age .... – % (A)

Notes: (A) Data not available. (3) Foreign nationals

GRADUATE ENROLLMENTS & DEGREES AWARDED
Master's enrollment 322
Master's degrees awarded 114
Doctoral enrollment 97
Doctoral degrees awarded 14

ENGINEERING STUDENT DATA

Applicants to the engineering college
Number of applicants to engineering college .... 576 [1]
Percent offered admission 85.0%

Matriculated engineering students
Percentage in top quartile (25%) of High School class ... – (A)
Average ACT scores: Math—26.3 [2], Composite—25.5

Notes: (A) Data not available. (1) Includes applicants for engineering technology and surveying. (2) Average ACT scores for students admitted in the regular admission status.

FULL- & PART-TIME FACULTY—BY DEPARTMENT
[Figures are the head count of full-time faculty and the full-time equivalent (FTE) of part-time faculty for each engineering department or equivalent.]

Department	Full	Part
Chemical Eng	7	–
Civil Eng	13	–
Electrical Eng	33	2.4
Industrial Eng	6	–
Mechanical Eng	17	1.3
Surveying	3	–
Agricultural Eng	3	–
Geological Eng	3	–

COLLEGE DESCRIPTION
The College offers four year programs in engineering, engineering technology, and surveying. Freshman students are integrated into the College of Engineering. Students with adequate mathematics skills are immediately enrolled in a departmental FORTRAN or C course taught by engineering faculty. The College offers an Environmental minor. Co-op education is strongly encouraged and is available to all majors. All students are required to meet the University's General Education requirements which ensures a well rounded education. The College has a combined BA/BS program with Western New Mexico University of Silver City.

DEGREE PROGRAMS DESCRIPTION
The BS degrees in the various engineering programs prepare the graduate for entry-level professional positions or for graduate study. The degree programs require foundation courses in mathematics, science, and engineering science plus specialized courses in engineering science and design. All degree programs also require written and oral communications courses and General Education requirements. Within this general framework, students are provided maximum flexibility to shape their curriculum. The College of Engineering offers a unique Environmental Minor which requires students to take at least two courses by interactive television from other universities. New Mexico State University has an unusually diverse student body. Students from Indian reservations, ranches, farms, Hispanic communities, metropolitan Albuquerque and El Paso, and Los Alamos study, work and play with students from many other states and foreign countries. The New Mexico State University College of Engineering includes the Particle Astrophysics Laboratory, the New Mexico Space Grant Consortium, the Waste Management Education and Research Consortium (WERC), the Teaching Factory, the Bridge Inspection Program and the NASA Space Telemetering Center. All offer research opportunities and employment to both graduate and undergraduate students.

DUAL DEGREE PROGRAMS
Engineering baccalaureate graduates with dual degrees 0

TRANSFER INFORMATION
Residency Requirements: Students must complete the last 30 hours in residence.

Transfer without Articulation Agreements
Admission to engineering: Transfer students are placed in the General Engineering status. Regular admission requires establishing a 2.1 (AgE,CE,GeolE,ME) or a 2.3 (IE) or a 2.5 (ChE & EE).

GRADUATION REQUIREMENTS
Agricultural, chemical, civil, geological and mechanical engineering require a 2.0 GPA in departmental courses. The University requires an overall 2.0 GPA. The last 30 hours of coursework must be taken at NMSU. The University also requires 55 hours of upper division coursework.

STUDENT PROGRAMS

PROFESSIONAL AND HONORARY SOCIETIES
[For key to acronyms, see *Introduction*.]
Professional Societies: ACSM, AGCA, AIAA, AIChE, ASAE, ASCE, ASME, IEEE, IIE, SAE
Honorary Societies: Alpha Pi Mu, Chi Epsilon, Eta Kappa Nu, Omega Chi Epsilon, Pi Tau Sigma, Tau Alpha Pi, Tau Beta Pi

SUPPORT PROGRAMS
Student Chapter Organizations: Society of Women Engineers, National Society of Black Engineers, Society of Hispanic Professional Engineers, American Indians in Science and Engineering

American Indian Programs
For: Ethnic Minorities **Available:** Year round
Offered: Prematriculation, Freshman, Sophomore, Junior, Senior, Graduate level

Black Programs
For: Ethnic Minorities **Available:** Year round
Offered: Prematriculation, Freshman, Sophomore, Junior, Senior, Graduate level

Chicano Programs
For: Ethnic Minorities **Available:** Year round
Offered: Prematriculation, Freshman, Sophomore, Junior, Senior, Graduate level

Women's Center
For: Women **Available:** Year round
Offered: Freshman, Sophomore, Junior, Senior, Graduate level
Other Engineering Support Programs: Summer orientation programs for new students, fall and spring orientation courses, Center for Counseling and Student Development, Center for Learning Assistance, federally funded student and peer counseling, veterans services, and free student organization sponsored tutoring. The College also provides student study halls in every department.

155 University of New Orleans

■ INSTITUTION PROFILE

HEAD OF THE INSTITUTION
Gregory M. O'Brien
Phone: (504)286-6201   Fax: (504)286-6872

GENERAL INFORMATION
[All Students—Fall 1991]
Undergraduate enrollment 12,435
Graduate enrollment 3,649
Total institution enrollment 16,084
Type of institution: Public   Calendar: Semesters
Location: New Orleans   Population: 1,000,000
Setting: Urban
Types of engineering degrees: Engineering
Other degree-granting colleges: Business Administration, Education, Liberal Arts, Sciences, Urban & Public Affairs

■ ENGINEERING ADMISSIONS

ADMISSIONS OFFICE CONTACT
Roslyn Sheley
University of New Orleans
Admissions Office
New Orleans, LA 70148
Phone: (504)286-6000

ENGINEERING COLLEGE ADMISSIONS INFORMATION
Entrance Requirements: SAT (Score: 810), ACT (Score: 20), High School courses—English (4 years), Mathematics - Algebra I, II, Geometry (3 years), Natural Sciences - Biology & 2 Science Electives (3 years), Social Sciences (3 years)
Requirements for foreign students: TOEFL (Score: 500); financial statement; Proper Immigration and Naturalization Service documents are required, i.e., I-20, passport, student visa.

DEPARTMENTAL ADMISSIONS INFORMATION
Admission to the engineering department: Min of 12 sem hrs with a 3.0 avg, 18 sem hrs with a 2.5 avg, or 24 sem hrs with a 2.2 avg. A 2.5 avg. is required for all transfer students.

■ FINANCIAL INFORMATION

ESTIMATED EXPENSES (ACADEMIC YEAR)
[Expenses are for the 1992-93 nine-month academic year.]

State Residents	Undergraduate	Graduate
Tuition and fees	$ 1,181	$ 1,181
College room and board	$ –	$ –
Books and supplies	$ 510	$ 510
Other expenses	$ –	$ –
Total estimated expenses	**$ 1,691**	**$ 1,691**

Out-of-State Residents	Undergraduate	Graduate
Tuition and fees	$ 2,577	$ 2,577
College room and board	$ –	$ –
Books and supplies	$ 510	$ 510
Other expenses	$ –	$ –
Total estimated expenses	**$ 3,087**	**$ 3,087**

FINANCIAL AID OFFICE CONTACT
Avon Dennis, *Director*
University of New Orleans
Student Financial Aid Office
New Orleans, LA 70148
Phone: (504)286-6603   Fax: (504)286-7393

GENERAL FINANCIAL AID INFORMATION
Forms accepted/required: For more information, contact: Student Financial Aid Office at UNO.
Additional financial aid information: All forms of financial aid - grants, loans, scholarships, and part-time campus work are handled through the Student Financial Aid Office, which will evaluate each student's need for aid in accord with the financial aid policy. Recommendations will be made on what aid is available and what aid the student is eligible to receive.

■ ENGINEERING COLLEGE INFORMATION
[For additional personnel, refer to the *Appendix*.]

HEAD OF ENGINEERING
John N. Crisp
Phone: (504)286-6327   Fax: (504)286-7413

ENGINEERING COLLEGE ADDRESS
College of Engineering
University of New Orleans
Lakefront Campus
New Orleans, LA 70148
Phone: (504)286-6000   Fax: (504)286-7413

ENROLLMENTS—BY CLASS
[Numbers are baccalaureate enrollments for the fall 1991 term, unless otherwise footnoted.]
1st-year students/Freshmen 433
2nd-year students/Sophomores 249
3rd-year students/Juniors 184
4th-year students/Seniors 301
Total 1,167

NUMBER OF DEGREES AWARDED—BY PROGRAM
[Numbers are engineering baccalaureate degrees awarded during the 1991-1992 academic year, unless otherwise footnoted. For full details about each engineering program, refer to the *Tables of Degree Programs*.]
Civil Engineering 18
Electrical Engineering 46
Mechanical Engineering 27
Naval Architecture & Marine Engineering 17
Total 108

PERCENTAGE OF DEGREES AWARDED—BY CATEGORY
[Percentages are of all engineering baccalaureate degrees awarded during the 1991-1992 academic year, unless otherwise footnoted.]
African Americans 5.5%
Asian/Pacific Island Americans 5.5%
Hispanic Americans 5.5%
Native Americans 0.9%
Foreign Citizens 10.2%
All Others 72.4%
Women 15.7%
Persons with disabilities .... – %
Students over 25 years of age .... – %

GRADUATE ENROLLMENTS & DEGREES AWARDED
Master's enrollment 130
Master's degrees awarded 33
Doctoral enrollment .... –
Doctoral degrees awarded .... –

FULL- & PART-TIME FACULTY—BY DEPARTMENT
[Figures are the head count of full-time faculty and the full-time equivalent (FTE) of part-time faculty for each engineering department or equivalent.]

Department	Full	Part
Civil Eng	8	2.5
Electrical Eng	11	5.5
Mechanical Eng	10	1.3
Naval Architecture & Marine Eng	4	–

COLLEGE DESCRIPTION
The College of Engineering offers undergraduate degree programs in civil engineering, electrical engineering, mechanical engineering, and naval architecture and marine engineering. Dual degree programs are available in conjunction with three other local universities, resulting in both a physics degree and an engineering degree obtained in five years. An active five year co-op education program (work/study) is currently active, resulting in a B.S. degree. A consortium of regional states allows students from nearby states to enroll in unique programs without paying out-of-state tuition. Major research efforts exist in the areas of urban waste management and environmental protection.

DEGREE PROGRAMS DESCRIPTION
The undergraduate degree program in engineering provides a broad engineering education in preparation for: a) Professional employment, mainly as civil, electrical, mechanical engineering, or marine engineers, or as naval architects, in design, development, production, operation, and sales, or b) Graduate study in the various fields of engineering and the physical sciences. Emphasis is placed on fundamentals in the basic fields of civil, electrical, and mechanical engineering, as well as naval architecture and marine engineering, followed by applications in the areas of engineering design and planning.

DUAL DEGREE PROGRAMS
Engineering baccalaureate graduates with dual degrees 0

TRANSFER INFORMATION
Transfer via Articulation Agreements
Admission to engineering: Students attend the cooperating univ. for 3 years, majoring in physics, and then transferring to UNO for two additional years. Students are awarded the two degrees upon completion of the 5 year program.
Engineering graduates transferred from .. 4-yr: –   2-yr: –
Transfer without Articulation Agreements
Admission to engineering: At any time
Engineering graduates transferred from .. 4-yr: –   2-yr: –
Requirements: Of the 30 semester hours transfer students must complete at UNO (residency requirement), a minimum of 15 semester hours must be senior level work in the student's major field. A student cannot receive junior or senior level degree credit for a freshman/sophomore course taken at another college or university. Junior and senior level engineering courses will be accepted if completed at an institution that confers baccalaureate engineering degrees

GRADUATION REQUIREMENTS
The B.S. in CE, EE, ME, NA&ME: (1) Completion of a program of study in civil engineering, electrical engineering, mechanical engineering, or naval architecture and marine engineering; (2) a C average in all work attempted, a C average in the semesters containing the last 60 hours of courses offered for the degree, and a C average in engineering courses; (3) approval of all electives; (4) completion of a program of general studies.

■ STUDENT PROGRAMS

PROFESSIONAL AND HONORARY SOCIETIES
[For key to acronyms, see *Introduction*.]
Professional Societies: ASCE, ASME, AWRA, AWS, IEEE, ISA, SAME, SNAME
Honorary Societies: Eta Kappa Nu, Pi Tau Sigma, Tau Beta Pi

SUPPORT PROGRAMS
Student Chapter Organizations: Society of Women Engineers, National Society of Black Engineers, Society of Hispanic Professional Engineers
Other Engineering Support Programs: Louisiana Engineering Society (LES)

154 New Mexico State University

INSTITUTION PROFILE

HEAD OF THE INSTITUTION
James E. Halligan
Phone: (505)646-2035 Fax: (505)646-6334

GENERAL INFORMATION
[All Students—Fall 1991]
Undergraduate enrollment 12,957
Graduate enrollment 2,543
Total institution enrollment 15,500
Type of institution: Public Calendar: Semesters
Location: Las Cruces, New Mexico Population: 62,126
Setting: Small Town
Types of engineering degrees: Engineering & Technology
Other degree-granting colleges: Arts & Sciences, Education, Agriculture & Home Economics, Human & Community Services (includes nursing), Business Administration & Economics

ENGINEERING ADMISSIONS

ADMISSIONS OFFICE CONTACT
Bill Bruner
New Mexico State University
Box 30001, Dept. 3A
Las Cruces, NM 88003-0001
Phone: (505)646-3121 Fax: (505)646-6330

ENGINEERING COLLEGE ADMISSIONS INFORMATION
Admission to the engineering college: Admitted regularly if their English ACT score is 16 or greater and their performance on a local math exam is satisfactory. Provisional admission is possible.
Entrance Requirements: High School courses—English (4 years), Science (beyond General Science) (2 years), Mathematics (3 years), Foreign language or fine arts (1 year), Regular admission requires a HS GPA of 2.0 & Enhanced ACT of 20 or a HS GPA of 2.5 or an Enhanced ACT of 21. Provisional admission is possible for students not meeting these standards.
Entrance Recommendations: ACT (Score: 20), High School courses—Mathematics (4 years), Chemistry and physics (1 year each) (2 years), Foreign language (2 years), Computers & technical drawing (1 yr each) (2 years)
Requirements for foreign students: TOEFL (Score: 500); financial statement

DEPARTMENTAL ADMISSIONS INFORMATION
Admission to the engineering department: Regular admission requires Enhanced English ACT of 16 or greater and a satisfactory score on a local math exam. Provisional admission is allowed.
Additional information: Students not qualifying for regular admission will be admitted into the General Engineering program. Such students have two years to complete any deficiencies with a minimum GPA. For Ag E, C E, Geol E, and M E, the minimum GPA is a 2.1; for I E, a 2.3; and for E E & Chem E the minimum is a 2.5.

FINANCIAL INFORMATION

ESTIMATED EXPENSES (ACADEMIC YEAR)
[Expenses are for the 1992-93 nine-month academic year.]

State Residents	Undergraduate	Graduate
Tuition and fees	$ 1,756	$ 1,876
College room and board	$ 2,668 [1]	$ 2,668 [1]
Books and supplies	$ 575	$ 580
Other expenses	$ –	$ –
Total estimated expenses	$ 4,999	$ 5,124

Out-of-State Residents	Undergraduate	Graduate
Tuition and fees	$ 5,686	$ 5,806
College room and board	$ 2,668 [1]	$ 2,668 [1]
Books and supplies	$ 575	$ 580
Other expenses	$ –	$ –
Total estimated expenses	$ 8,929	$ 9,054

Notes: (1) Board cost is for 12 meals per week.

FINANCIAL AID OFFICE CONTACT
Greeley Myers, *Director of Financial Aid*
New Mexico State University
Box 30001, Dept. 5100
Las Cruces, NM 88003-0001
Phone: (505)646-4105 Fax: (505)646-6330

GENERAL FINANCIAL AID INFORMATION
Forms accepted/required: FAT, Stafford, IRS, SAR, Supplemental, United Student Aid Fund Single File Application, New Mexico Supplemental Information Form
Additional financial aid information: Academic scholarships (including special scholarships for out of state students) from the University, the College of Engineering, and Air Force & Army ROTC. Need based financial aid includes Pell Grant, SEOG, Perkins DSL, and work study. Minority students are also eligible for Designated Academic Opportunity Scholarships (DAOS) and National Action Council for Minorities (NACME) Scholarships. Non-need based work study is also available.

ENGINEERING COLLEGE INFORMATION

[For additional personnel, refer to the *Appendix*.]

HEAD OF ENGINEERING
J. Derald Morgan
Phone: (505)646-2914 Fax: (505)646-3549

ENGINEERING COLLEGE ADDRESS
College of Engineering
New Mexico State University
Box 30001, Dept. 3449
Las Cruces, NM 88003-0001
Phone: (505)646-2912 Fax: (505)646-3549

ENROLLMENTS—BY CLASS
[Numbers are baccalaureate enrollments for the fall 1991 term, unless otherwise footnoted.]

1st-year students/Freshmen .... 510 [1]
2nd-year students/Sophomores .... 375 [1]
3rd-year students/Juniors .... 342 [1]
4th-year students/Seniors .... 489 [1]
Total 1,716

Notes: (1) Fall 1992 enrollment (does not include surveying or technology).

NUMBER OF DEGREES AWARDED—BY PROGRAM
[Numbers are engineering baccalaureate degrees awarded during the 1991-1992 academic year, unless otherwise footnoted. For full details about each engineering program, refer to the *Tables of Degree Programs*.]

Agricultural Engineering 1
Chemical Engineering 17
Civil Engineering 45
Electrical & Computer Engineering 98
Geological Engineering 1
Industrial Engineering 12
Mechanical Engineering 58
Surveying & Mapping 4
Total 236

PERCENTAGE OF DEGREES AWARDED—BY CATEGORY
[Percentages are of all engineering baccalaureate degrees awarded during the 1991-1992 academic year, unless otherwise footnoted.]

African Americans .... – % (A)
Asian/Pacific Island Americans .... – % (A)
Hispanic Americans .... 27.2 %
Native Americans .... 1.7 %
Foreign Citizens .... 6.9 % (3)
All Others .... 64.2 %
Women .... 16.8 %
Persons with disabilities .... – % (A)
Students over 25 years of age .... – % (A)

Notes: (A) Data not available. (3) Foreign nationals

GRADUATE ENROLLMENTS & DEGREES AWARDED
Master's enrollment 322
Master's degrees awarded 114
Doctoral enrollment 97
Doctoral degrees awarded 14

ENGINEERING STUDENT DATA
Applicants to the engineering college
Number of applicants to engineering college ...... 576 [1]
Percent offered admission .... 85.0 %

Matriculated engineering students
Percentage in top quartile (25%) of High School class ..... – (A)
Average ACT scores: Math—26.3 [2], Composite—25.5

Notes: (A) Data not available. (1) Includes applicants for engineering technology and surveying. (2) Average ACT scores for students admitted in the regular engineering status.

FULL- & PART-TIME FACULTY—BY DEPARTMENT
[Figures are the head count of full-time faculty and the full-time equivalent (FTE) of part-time faculty for each engineering department or equivalent.]

Department	Full	Part
Chemical Eng	7	–
Civil Eng	13	–
Electrical Eng	33	2.4
Industrial Eng	6	–
Mechanical Eng	17	1.3
Surveying	3	–
Agricultural Eng	3	–
Geological Eng	3	–

COLLEGE DESCRIPTION
The College offers four year programs in engineering, engineering technology, and surveying. Freshman students are integrated into the College of Engineering. Students with adequate mathematics skills are immediately enrolled in a departmental FORTRAN or C course taught by engineering faculty. The College offers an Environmental minor. Co-op education is strongly encouraged and is available to all majors. All students are required to meet the University's General Education requirements which ensures a well rounded education. The College has a combined BA/BS program with Western New Mexico University of Silver City.

DEGREE PROGRAMS DESCRIPTION
The BS degrees in the various engineering programs prepare the graduate for entry-level professional positions or for graduate study. The degree programs require foundation courses in mathematics, science, and engineering science plus specialized courses in engineering science and design. All degree programs also require written and oral communications courses and General Education requirements. Within this general framework, students are provided maximum flexibility to shape their curriculum. The College of Engineering offers a unique Environmental Minor which requires students to take at least two courses by interactive television from other universities. New Mexico State University has an unusually diverse student body. Students from Indian reservations, ranches, farms, Hispanic communities, metropolitan Albuquerque and El Paso, and Los Alamos study, work and play with students from many other states and foreign countries. The New Mexico State University College of Engineering includes the Particle Astrophysics Laboratory, the New Mexico Space Grant Consortium, the Waste Management Education and Research Consortium (WERC), the Teaching Factory, the Bridge Inspection Program and the NASA Space Telemetering Center. All offer research opportunities and employment to both graduate and undergraduate students.

DUAL DEGREE PROGRAMS
Engineering baccalaureate graduates with dual degrees 0

TRANSFER INFORMATION
Residency Requirements: Students must complete the last 30 hours in residence.
Transfer without Articulation Agreements
Admission to engineering: Transfer students are placed in the General Engineering status. Regular admission requires establishing a 2.1 (AgE,CE,GeolE,ME) or a 2.3 (IE) or a 2.5 (ChE & EE).

GRADUATION REQUIREMENTS
Agricultural, chemical, civil, geological and mechanical engineering require a 2.0 GPA in departmental courses. The University requires an overall 2.0 GPA. The last 30 hours of coursework must be taken at NMSU. The University also requires 55 hours of upper division coursework.

STUDENT PROGRAMS

PROFESSIONAL AND HONORARY SOCIETIES
[For key to acronyms, see *Introduction*.]
Professional Societies: ACSM, AGCA, AIAA, AIChE, ASAE, ASCE, ASME, IEEE, IIE, SAE
Honorary Societies: Alpha Pi Mu, Chi Epsilon, Eta Kappa Nu, Omega Chi Epsilon, Pi Tau Sigma, Tau Alpha Pi, Tau Beta Pi

SUPPORT PROGRAMS
Student Chapter Organizations: Society of Women Engineers, National Society of Black Engineers, Society of Hispanic Professional Engineers, American Indians in Science and Engineering

American Indian Programs
For: Ethnic Minorities **Available:** Year round
Offered: Prematriculation, Freshman, Sophomore, Junior, Senior, Graduate level

Black Programs
For: Ethnic Minorities **Available:** Year round
Offered: Prematriculation, Freshman, Sophomore, Junior, Senior, Graduate level

Chicano Programs
For: Ethnic Minorities **Available:** Year round
Offered: Prematriculation, Freshman, Sophomore, Junior, Senior, Graduate level

Women's Center
For: Women **Available:** Year round
Offered: Freshman, Sophomore, Junior, Senior, Graduate level

Other Engineering Support Programs: Summer orientation programs for new students, fall and spring orientation courses, Center for Counseling and Student Development, Center for Learning Assistance, federally funded student and peer counseling, veterans services, and free student organization sponsored tutoring. The College also provides student study halls in every department.

155 University of New Orleans

■ INSTITUTION PROFILE

HEAD OF THE INSTITUTION
Gregory M. O'Brien
Phone: (504)286-6201   Fax: (504)286-6872

GENERAL INFORMATION
[All Students—Fall 1991]
Undergraduate enrollment 12,435
Graduate enrollment 3,649
Total institution enrollment **16,084**
Type of institution: Public   **Calendar:** Semesters
Location: New Orleans   **Population:** 1,000,000
Setting: Urban
Types of engineering degrees: Engineering
Other degree-granting colleges: Business Administration, Education, Liberal Arts, Sciences, Urban & Public Affairs

■ ENGINEERING ADMISSIONS

ADMISSIONS OFFICE CONTACT
Roslyn Sheley
University of New Orleans
Admissions Office
New Orleans, LA 70148
Phone: (504)286-6000

ENGINEERING COLLEGE ADMISSIONS INFORMATION
Entrance Requirements: SAT (Score: 810), ACT (Score: 20), High School courses—English (4 years), Mathematics - Algebra I, II, Geometry (3 years), Natural Sciences - Biology & 2 Science Electives (3 years), Social Sciences (3 years)
Requirements for foreign students: TOEFL (Score: 500); financial statement; Proper Immigration and Naturalization Service documents are required, i.e., I-20, passport, student visa.

DEPARTMENTAL ADMISSIONS INFORMATION
Admission to the engineering department: Min of 12 sem hrs with a 3.0 avg, 18 sem hrs with a 2.5 avg, or 24 sem hrs with a 2.2 avg. A 2.5 avg. is required for all transfer students.

■ FINANCIAL INFORMATION

ESTIMATED EXPENSES (ACADEMIC YEAR)
[Expenses are for the 1992-93 nine-month academic year.]

State Residents	Undergraduate	Graduate
Tuition and fees	$ 1,181	$ 1,181
College room and board	$ –	$ –
Books and supplies	$ 510	$ 510
Other expenses	$ –	$ –
Total estimated expenses	**$ 1,691**	**$ 1,691**
Out-of-State Residents	Undergraduate	Graduate
Tuition and fees	$ 2,577	$ 2,577
College room and board	$ –	$ –
Books and supplies	$ 510	$ 510
Other expenses	$ –	$ –
Total estimated expenses	**$ 3,087**	**$ 3,087**

FINANCIAL AID OFFICE CONTACT
Avon Dennis, *Director*
University of New Orleans
Student Financial Aid Office
New Orleans, LA 70148
Phone: (504)286-6603   Fax: (504)286-7393

GENERAL FINANCIAL AID INFORMATION
Forms accepted/required: For more information, contact: Student Financial Aid Office at UNO.
Additional financial aid information: All forms of financial aid - grants, loans, scholarships, and part-time campus work are handled through the Student Financial Aid Office, which will evaluate each student's need for aid in accord with the financial aid policy. Recommendations will be made on what aid is available and what aid the student is eligible to receive.

■ ENGINEERING COLLEGE INFORMATION
[For additional personnel, refer to the *Appendix*.]

HEAD OF ENGINEERING
John N. Crisp
Phone: (504)286-6327   Fax: (504)286-7413

ENGINEERING COLLEGE ADDRESS
College of Engineering
University of New Orleans
Lakefront Campus
New Orleans, LA 70148
Phone: (504)286-6000   Fax: (504)286-7413

ENROLLMENTS—BY CLASS
[Numbers are baccalaureate enrollments for the fall 1991 term, unless otherwise footnoted.]
1st-year students/Freshmen 433
2nd-year students/Sophomores 249
3rd-year students/Juniors 184
4th-year students/Seniors 301
Total **1,167**

NUMBER OF DEGREES AWARDED—BY PROGRAM
[Numbers are engineering baccalaureate degrees awarded during the 1991-1992 academic year, unless otherwise footnoted. For full details about each engineering program, refer to the *Tables of Degree Programs*.]
Civil Engineering 18
Electrical Engineering 46
Mechanical Engineering 27
Naval Architecture & Marine Engineering 17
Total **108**

PERCENTAGE OF DEGREES AWARDED—BY CATEGORY
[Percentages are of all engineering baccalaureate degrees awarded during the 1991-1992 academic year, unless otherwise footnoted.]
African Americans 5.5%
Asian/Pacific Island Americans 5.5%
Hispanic Americans 5.5%
Native Americans 0.9%
Foreign Citizens 10.2%
All Others 72.4%
Women 15.7%
Persons with disabilities .... – %
Students over 25 years of age .... – %

GRADUATE ENROLLMENTS & DEGREES AWARDED
Master's enrollment 130
Master's degrees awarded 33
Doctoral enrollment .... –
Doctoral degrees awarded .... –

FULL- & PART-TIME FACULTY—BY DEPARTMENT
[Figures are the head count of full-time faculty and the full-time equivalent (FTE) of part-time faculty for each engineering department or equivalent.]

Department	Full	Part
Civil Eng	8	2.5
Electrical Eng	11	5.5
Mechanical Eng	10	1.3
Naval Architecture & Marine Eng	4	–

COLLEGE DESCRIPTION
The College of Engineering offers undergraduate degree programs in civil engineering, electrical engineering, mechanical engineering, and naval architecture and marine engineering. Dual degree programs are available in conjunction with three other local universities, resulting in both a physics degree and an engineering degree obtained in five years. An active five year co-op education program (work/study) is currently active, resulting in a B.S. degree. A consortium of regional states allows students from nearby states to enroll in unique programs without paying out-of-state tuition. Major research efforts exist in the areas of urban waste management and environmental protection.

DEGREE PROGRAMS DESCRIPTION
The undergraduate degree program in engineering provides a broad engineering education in preparation for: a) Professional employment, mainly as civil, electrical, mechanical engineering, or marine engineers, or as naval architects, in design, development, production, operation, and sales, or b) Graduate study in the various fields of engineering and the physical sciences. Emphasis is placed on fundamentals in the basic fields of civil, electrical, and mechanical engineering, as well as naval architecture and marine engineering, followed by applications in the areas of engineering design and planning.

DUAL DEGREE PROGRAMS
Engineering baccalaureate graduates with dual degrees 0

TRANSFER INFORMATION
Transfer via Articulation Agreements
Admission to engineering: Students attend the cooperating univ. for 3 years, majoring in physics, and then transferring to UNO for two additional years. Students are awarded the two degrees upon completion of the 5 year program.
Engineering graduates transferred from .. 4-yr: –   2-yr: –
Transfer without Articulation Agreements
Admission to engineering: At any time
Engineering graduates transferred from .. 4-yr: –   2-yr: –
Requirements: Of the 30 semester hours transfer students must complete at UNO (residency requirement), a minimum of 15 semester hours must be senior level work in the student's major field. A student cannot receive junior or senior level degree credit for a freshman/sophomore course taken at another college or university. Junior and senior level engineering courses will be accepted if completed at an institution that confers baccalaureate engineering degrees

GRADUATION REQUIREMENTS
The B.S. in CE, EE, ME, NA&ME: (1) Completion of a program of study in civil engineering, electrical engineering, mechanical engineering, or naval architecture and marine engineering; (2) a C average in all work attempted, a C average in the semesters containing the last 60 hours of courses offered for the degree, and a C average in engineering courses; (3) approval of all electives; (4) completion of a program of general studies.

■ STUDENT PROGRAMS

PROFESSIONAL AND HONORARY SOCIETIES
[For key to acronyms, see *Introduction*.]
Professional Societies: ASCE, ASME, AWRA, AWS, IEEE, ISA, SAME, SNAME
Honorary Societies: Eta Kappa Nu, Pi Tau Sigma, Tau Beta Pi

SUPPORT PROGRAMS
Student Chapter Organizations: Society of Women Engineers, National Society of Black Engineers, Society of Hispanic Professional Engineers
Other Engineering Support Programs: Louisiana Engineering Society (LES)

156 New York Institute of Technology

■ INSTITUTION PROFILE
HEAD OF THE INSTITUTION
Matthew Schure
Phone: (516)686-7650 Fax: (516)686-6830

GENERAL INFORMATION
[All Students—Fall 1991]
Undergraduate enrollment 9,000
Graduate enrollment 2,525
Total institution enrollment 11,525
Type of institution: Private Calendar: Semesters
Nearest city: New York City Population: 8,000,000
Miles from main campus: 30 Setting: Suburban
Types of engineering degrees: Engineering & Technology
Other degree-granting colleges: Architecture, Arts & Sciences, Business Administration, Education, Humanities & Social Sciences, New York College of Osteopathic Medicine

■ ENGINEERING ADMISSIONS
ADMISSIONS OFFICE CONTACT
Arthur Lambert
New York Institute of Technology
Admissions/Gerry House
Old Westbury, NY 11568
Phone: (516)686-7520 Fax: (516)626-0419

ENGINEERING COLLEGE ADMISSIONS INFORMATION
Entrance Requirements: SAT (Score: 900)
Requirements for foreign students: TOEFL (Score: 450); financial statement

DEPARTMENTAL ADMISSIONS INFORMATION
Admission to the engineering department: At the time of admission to the institution.
Additional information: The minimum SAT score must be 500 in Math and 400 in Verbal. Engineering students should also have adequate mathematics preparation to permit entry into Calculus I.

■ FINANCIAL INFORMATION
ESTIMATED EXPENSES (ACADEMIC YEAR)
[Expenses are for the 1992-93 nine-month academic year.]

All Students	Undergraduate	Graduate
Tuition and fees	$ 8,950	$ 285[1]
College room and board	$ 5,200	$ 5,200
Books and supplies	$ 500	$ 225
Other expenses	$ –	$ –
Total estimated expenses	**$14,650**	**$ 5,710**

Notes: (1) per credit

FINANCIAL AID OFFICE CONTACT
Doreen Meyers, *Director of Financial Aid*
New York Institute of Technology
Dorothy Schure Campus
Old Westbury, NY 11586
Phone: (516)686-7680 Fax: (516)626-2627

GENERAL FINANCIAL AID INFORMATION
Forms accepted/required: FAF, IRS
Additional financial aid information: Honor and Challenge Scholarship: freshmen-SAT 1100/ACT no less than 27. Up to $4640. Transfer Scholarships: 60 credits or more from accredited colleges. Up to $4640. Public Sector Scholarship: spouse, parent, student employed in public sector. 20% of tuition. Government Service Scholarship Program: undergraduate adult students employed in government or civil service. Up to $4640. College work-study program and student aid jobs available.

■ ENGINEERING COLLEGE INFORMATION
[For additional personnel, refer to the *Appendix*.]
HEAD OF ENGINEERING
Heskia Heskiaoff
Phone: (516)686-7931 Fax: (516)626-0673

ENGINEERING COLLEGE ADDRESS
School of Engineering & Technology
New York Institute of Technology
Dorothy Schure Campus
Old Westbury, NY 11568
Phone: (516)686-7906 Fax: (516)626-0673

ENROLLMENTS—BY CLASS
[Numbers are baccalaureate enrollments for the fall 1991 term, unless otherwise footnoted.]
1st-year students/Freshmen 107
2nd-year students/Sophomores 106
3rd-year students/Juniors 144
4th-year students/Seniors 152
Total 509

NUMBER OF DEGREES AWARDED—BY PROGRAM
[Numbers are engineering baccalaureate degrees awarded during the 1991-1992 academic year, unless otherwise footnoted. For full details about each engineering program, refer to the *Tables of Degree Programs*.]
Electric Power Engineering 111
Industrial Engineering 14
Mechanical & Aeronautical Engineering 91
Total 216

PERCENTAGE OF DEGREES AWARDED—BY CATEGORY
[Percentages are of all engineering baccalaureate degrees awarded during the 1991-1992 academic year, unless otherwise footnoted.]
African Americans 3.3%
Asian/Pacific Island Americans 5.0%
Hispanic Americans 2.3%
Native Americans .... – %
Foreign Citizens .... – %
All Others 89.4%
Women 11.1%
Persons with disabilities .... – %
Students over 25 years of age .... – %

GRADUATE ENROLLMENTS & DEGREES AWARDED
Master's enrollment 20
Master's degrees awarded .... –[1]
Doctoral enrollment .... –
Doctoral degrees awarded .... –

Notes: (1) Effective 9/92, NYIT is offering the following M.S. programs: Electrical Engineering, Mechanical Engineering and Environmental Technology.

ENGINEERING STUDENT DATA
Applicants to the engineering college
Number of applicants to engineering college 180

FULL- & PART-TIME FACULTY—BY DEPARTMENT
[Figures are the head count of full-time faculty and the full-time equivalent (FTE) of part-time faculty for each engineering department or equivalent.]

Department	Full	Part
Electrical Eng	13	4.0
Mechanical Eng	8	6.6
Industrial Eng	1	1.0

COLLEGE DESCRIPTION
New York Institute of Technology emphasizes career-oriented education. Its open-access philosophy is reflected in a modest tuition/rate and entrance policies which provide people the opportunity to demonstrate their ability to perform in college. Applicants who do not satisfy the engineering entrance requirements may enroll in our pre-engineering program and transfer into engineering once they have satisfied the transfer requirements. NYIT has an optional co-op program which provides students with one or two placements of a semester (plus an optional summer) each. Opportunities for students to spend time in Europe are available through collaborative arrangements with two universities in Yugoslavia.

DEGREE PROGRAMS DESCRIPTION
There is heavy design emphasis in our engineering curricula. Project courses develop the skills needed in industry. A culminating capstone course requires students to work in teams, integrating skills in engineering, computer applications, communications, project management, and related areas. The projects are reflective of real-world problems; many of them are defined by companies. Students are required to make verbal presentations, produce documentation which is professional in content and presentation, plan and schedule their work to comply deadlines, consider costs and marketing factors, and to take into account legal, safety, and environmental issues. The important role of computers in engineering is reflected in our programs. The use of computers in engineering systems and the use of computers in engineering analysis and design are stressed. Computer aided design and analysis, simulation and modeling, programming and numerical analysis are among the areas included in the curricula.

TRANSFER INFORMATION
Residency Requirements: Engineering students must complete at least 30 credits in residency at New York Institute of Technology - 15 of which must be in their major.
Transfer without Articulation Agreements
Admission to engineering: At any time
Requirements: Transferring students must have a minimum cumulative average of 2.0 and must have completed at least 12 credits of required advanced Mathematics, Physics, Computer Science and engineering with a minimum average 2.5 in these courses.

GRADUATION REQUIREMENTS
Engineering students must have a minimum average of 2.00 in all major courses and must take a minimum of 15 credits.

■ STUDENT PROGRAMS
PROFESSIONAL AND HONORARY SOCIETIES
[For key to acronyms, see *Introduction*.]
Professional Societies: ACM, AIAA, ASHRAE, ASME, IEEE, SAE, SME
Honorary Societies: Eta Kappa Nu, Tau Alpha Pi

SUPPORT PROGRAMS
Student Chapter Organizations: Society of Women Engineers, National Society of Black Engineers, Society of Hispanic Professional Engineers
Other Engineering Support Programs: Entering students are given placement exams in math and English. Based on these test results, students are placed into the first course in the sequence or into one of a sequence of non-credit remedial courses. In addition, all students can get extra help in any subject area through our tutoring program.

157 New York State College of Ceramics at Alfred University

■ INSTITUTION PROFILE
HEAD OF THE INSTITUTION
Edward G. Coll Jr.
Phone: (607)871-2101   Fax: (607)871-2339

GENERAL INFORMATION
[All Students—Fall 1991]
Undergraduate enrollment 188,3.5
Graduate enrollment 32,7.5
Total institution enrollment 2,211
Type of institution: Public   Calendar: Semesters
Nearest city: Rochester, NY   Population: 152,000
Miles from main campus: 65   Setting: Small Town
Types of engineering degrees: Engineering
Other degree-granting colleges: Liberal Arts & Sciences, Business, Engineering & Professional Studies, Art & Design

■ ENGINEERING ADMISSIONS
ADMISSIONS OFFICE CONTACT
Daniel L. Meyer
New York State College of Ceramics at Alfred University
Office of Admissions
P.O. Box 765
Alfred, NY 14802
Phone: (607)871-2115   Fax: (607)871-2198

ENGINEERING COLLEGE ADMISSIONS INFORMATION
Entrance Requirements: SAT, ACT, High School courses—English (4 years), Mathematics (4 years), Laboratory Science including Chemistry and Physics (2 years), Social Studies and History (2 years), SAT or ACT may be used. Advanced Placement given in appropriate cases to qualified students.
Entrance Recommendations: High School courses—Computer Science (1 year), Foreign Language (2 years), Calculus (1 year)
Requirements for foreign students: TOEFL (Score: 550); financial statement

DEPARTMENTAL ADMISSIONS INFORMATION
Admission to the engineering department: At the time of admission to the institution.

■ FINANCIAL INFORMATION
ESTIMATED EXPENSES (ACADEMIC YEAR)
[Expenses are for the 1992-93 nine-month academic year.]

State Residents	Undergraduate	Graduate
Tuition and fees	$ 5,585	$ 8,090
College room and board	$ 4,736	$ 6,000
Books and supplies	$ 500	$ 500
Other expenses	$ 200 [1]	$ 200 [1]
Total estimated expenses	$11,021	$14,790

Out-of-State Residents	Undergraduate	Graduate
Tuition and fees	$ 7,985	$ 8,090
College room and board	$ 4,736	$ 6,000
Books and supplies	$ 500	$ 500
Other expenses	$ 200 [1]	$ 200 [1]
Total estimated expenses	$13,421	$14,790

Notes: (1) Student Service Fee

FINANCIAL AID OFFICE CONTACT
Earl E. Pierce, *Director of Financial Aid*
New York State College of Ceramics at Alfred University
Office of Student Financial Aid
Alfred, NY 14802
Phone: (607)871-2159   Fax: (607)871-2198

GENERAL FINANCIAL AID INFORMATION
Forms accepted/required: CSS-FAF, FAF, FAT, Stafford, IRS, SAR, Supplemental, Institutional
Additional financial aid information: Grants are available on merit and/or need basis. Applicants are required to complete the FAF form and the Alfred University Financial Aid form. Financial planning manual available on request from the Director of Financial Aid, 607/871-2159.

■ ENGINEERING COLLEGE INFORMATION
[For additional personnel, refer to the *Appendix*.]

HEAD OF ENGINEERING
Alastair N. Cormack
Phone: (607)871-2422   Fax: (607)871-2305
EMail: Cormack@Ceramics.alfred.edu

ENGINEERING COLLEGE ADDRESS
School of Ceramic Engineering and Sciences
New York State College of Ceramics at Alfred University
McMahon Engineering Building
Alfred, NY 14802
Phone: (607)871-2448   Fax: (607)871-2305

ENROLLMENTS—BY CLASS
[Numbers are baccalaureate enrollments for the fall 1991 term, unless otherwise footnoted.]
1st-year students/Freshmen 76
2nd-year students/Sophomores 58
3rd-year students/Juniors 64
4th-year students/Seniors 68
Total 266

NUMBER OF DEGREES AWARDED—BY PROGRAM
[Numbers are engineering baccalaureate degrees awarded during the 1991-1992 academic year, unless otherwise footnoted. For full details about each engineering program, refer to the *Tables of Degree Programs*.]
Ceramic Engineering 71
Ceramic Engineering Science 6
Glass Engineering Science 3
Total 80

PERCENTAGE OF DEGREES AWARDED—BY CATEGORY
[Percentages are of all engineering baccalaureate degrees awarded during the 1991-1992 academic year, unless otherwise footnoted.]
African Americans .... – %
Asian/Pacific Island Americans 0.8%
Hispanic Americans .... – %
Native Americans .... – %
Foreign Citizens 0.8%
All Others 98.4%
Women 11.2%
Persons with disabilities .... – %
Students over 25 years of age .... – %

GRADUATE ENROLLMENTS & DEGREES AWARDED
Master's enrollment 39
Master's degrees awarded 20
Doctoral enrollment 33
Doctoral degrees awarded 6

ENGINEERING STUDENT DATA
Applicants to the engineering college
Number of applicants to engineering college 146
Percent offered admission 93.2%

Matriculated engineering students
Percentage in top quartile (25%) of High School class 84.0
Average SAT scores: Math—643, Verbal—535, Combined—1178
Average ACT scores: Math—26, Composite—26

FULL- & PART-TIME FACULTY—BY DEPARTMENT
[Figures are the head count of full-time faculty and the full-time equivalent (FTE) of part-time faculty for each engineering department or equivalent.]

Department	Full	Part
Ceramic Eng & Sciences	24	–

COLLEGE DESCRIPTION
Alfred U. has educated 1/3 of all the ceramic engineers in the US since the program's founding in 1900. Generally considered as one of the best in the world, the program reflects its excellence in many ways. First, it has superb undergraduate teaching labs and specialized research labs. Premier among these is a new $250,000 lab where students and faculty study the thermo-mechanical properties of technical ceramics for severe environments, and high-stress applications. Second, graduates and undergraduates benefit from ongoing, critical research in the field of ceramics which is carried out in 7 research-related facilities. Among these are the Center for Advanced Ceramic Technology, a leader in the study of high performance structural, electro-optical, and superconducting ceramics, and the Center for Glass Research. Third, the Scholes Library has the most extensive collection of books and research materials on ceramic engineering to found anywhere.

DEGREE PROGRAMS DESCRIPTION
The B.S. degree programs prepare graduates for entry-level positions in industry or for graduate study. Ceramic engineers study inorganic, substances such as oxides, carbides, and nitrides. Ceramics are vital to many areas including: communications, medicine, space, metal processing and computer technology. Ceramic engineers often choose one area of specialization, such as automotive ceramics, electro-ceramics, high-temperature materials, or glasses. Ceramic scientists work at the cutting edge of the field, developing new ideas, processes, and products. If you are interested in a career in R&D, or graduate study leading to teaching or research careers, this may be best for you. The studies concentrate on the basic chemistry and physics of materials to understand the important connections between structure and properties. This curriculum offers preparation for graduate school or industrial work stressing research and development. The glass engineering science program is concerned with developing uses of glass in eyewear, tableware, light bulbs, windows, television, fiber optics, structural glass and heat resistant glass. You may concentrate on the engineering challenges of glass production or on the more scientific aspects of glass.

DUAL DEGREE PROGRAMS
Engineering baccalaureate graduates with dual degrees 5

TRANSFER INFORMATION
Residency Requirements: Transfer students must be in residence at Alfred University at least during the senior year. Students who have earned all but 8 or fewer of the credits required for graduation may be permitted by the Dean to complete degree requirements elsewhere.

Transfer via Articulation Agreements
Admission to engineering: Transfer student's credentials are evaluated on a course by course basis.
Engineering graduates transferred from .. 4-yr: –   2-yr: 1
Requirements: All transfer students who apply to one of our engineering programs must meet prerequisite math and science course work to be considered for admission.

Transfer without Articulation Agreements
Admission to engineering: At any time
Engineering graduates transferred from .. 4-yr: –   2-yr: 1
Requirements: All transfer students who apply to one of our engineering programs must meet prerequisite math and science course work to be considered for admission.

■ STUDENT PROGRAMS
PROFESSIONAL AND HONORARY SOCIETIES
[For key to acronyms, see *Introduction*.]
Professional Societies: ACerS, NSPE, Keramos
Honorary Societies: Keramos, Tau Beta Pi

SUPPORT PROGRAMS
Student Chapter Organizations: Society of Women Engineers, UMOJA, IASU, MEGA, Peer Mentor Program

Educational Opportunity Program
For: Ethnic Minorities   **Available:** Year round
Offered: Prematriculation, Freshman, Sophomore, Junior, Senior

Division of Developmental Studies
For: Ethnic Minorities   **Available:** Year round
Offered: Prematriculation, Freshman, Sophomore, Junior, Senior

Other Engineering Support Programs: E.O.P.: Enables students from all ethnic backgrounds whose economic and educational circumstances have placed limitations on their college opportunities to further their education. D.D.S.: Academic assistance program to strengthen skills and help the student succeed with the E.O.P. M.E.G.A.: Student organization promoting multi-cultural awareness through educational and social activities.

158 University of North Carolina-Charlotte

■ INSTITUTION PROFILE
HEAD OF THE INSTITUTION
James H. Woodward
Phone: (704)547-2201　　　　Fax: (704)547-2144

GENERAL INFORMATION
[All Students—Fall 1991]
Undergraduate enrollment 13,087
Graduate enrollment 2,276
Total institution enrollment **15,363**
Type of institution: Public　　　Calendar: Semesters
Nearest city: Charlotte　　　　　Population: 417,000
Miles from main campus: 1　　　　Setting: Urban
Types of engineering degrees: Engineering & Technology
Other degree-granting colleges: Architecture, Arts & Sciences, Business Administration, Education, Nursing

■ ENGINEERING ADMISSIONS
ADMISSIONS OFFICE CONTACT
Kathi M. Baucom
University of North Carolina-Charlotte
Director of Admissions
Charlotte, NC 28223
Phone: (704)547-2213　　　　Fax: (704)547-2144

ENGINEERING COLLEGE ADMISSIONS INFORMATION
Entrance Requirements: SAT (Score: 900), ACT (Score: 24), High School courses—English (4 years), Mathematics (3 years), Social Studies (2 years), Science (3 years)
Entrance Recommendations: High School courses—Algebra (2 years), Trigonometry (1 year), Geometry (1 year), Advanced Math or Calculus (1 year)
Requirements for foreign students: TOEFL (Score: 500); financial statement

DEPARTMENTAL ADMISSIONS INFORMATION
Admission to the engineering department: At the time of admission to the institution.
Additional information: All applicants must present acceptable grades in algebra, trigonometry, plane and solid geometry, or the equivalent in integrated courses. A student satisfying SAT MATH 500, SAT VERBAL 400, HS RANK 1/2 or SAT MATH 500, SAT VERBAL 350, PGA 2.5, HS RANK 1/2 may be admitted directly to the lower division of a department (first two years of specified courses). A student not admitted to the lower division but having SAT MATH 450, SAT VERBAL 350, HS RANK 1/2, will be admitted to the Freshman Engineering Program.

■ FINANCIAL INFORMATION
ESTIMATED EXPENSES (ACADEMIC YEAR)
[Expenses are for the 1992-93 nine-month academic year.]

State Residents	Undergraduate	Graduate
Tuition and fees	$ 1,191	$ 1,191
College room and board	$ 3,126	$ 3,126
Books and supplies	$ 600	$ 500
Other expenses	$ 125	$ 125
Total estimated expenses	**$ 5,042**	**$ 4,942**

Out-of-State Residents	Undergraduate	Graduate
Tuition and fees	$ 6,863	$ 6,863
College room and board	$ 3,126	$ 3,126
Books and supplies	$ 600	$ 500
Other expenses	$ 125	$ 125
Total estimated expenses	**$10,714**	**$10,614**

FINANCIAL AID OFFICE CONTACT
Curtis R. Whalen, *Director of Student Financial Aid*
University of North Carolina-Charlotte
Charlotte, NC 28223
Phone: (704)547-2461　　　　Fax: (704)547-2144

GENERAL FINANCIAL AID INFORMATION
Forms accepted/required: FAF
Additional financial aid information: Scholarships, grants, loans, part-time employment; College Work-Study Program. Required forms: Financial Aid Form of the College Scholarship Service.

■ ENGINEERING COLLEGE INFORMATION
[For additional personnel, refer to the *Appendix*.]
HEAD OF ENGINEERING
Robert D. Snyder
Phone: (704)547-2301　　　　Fax: (704)547-2352
ENGINEERING COLLEGE ADDRESS
College of Engineering
University of North Carolina-Charlotte
Highway 49
Smith Engineering Building
Charlotte, NC 28223
Phone: (704)547-2301　　　　Fax: (704)547-2352

ENROLLMENTS—BY CLASS
[Numbers are baccalaureate enrollments for the fall 1991 term, unless otherwise footnoted.]
1st-year students/Freshmen 220
2nd-year students/Sophomores 202
3rd-year students/Juniors 276
4th-year students/Seniors 330
Total **1,028**

NUMBER OF DEGREES AWARDED—BY PROGRAM
[Numbers are engineering baccalaureate degrees awarded during the 1991-1992 academic year, unless otherwise footnoted. For full details about each engineering program, refer to the *Tables of Degree Programs*.]
Civil Engineering 23
Computer Science 50
Electrical Engineering 45
Mechanical Engineering & Engineering Science 58
Total **176**

PERCENTAGE OF DEGREES AWARDED—BY CATEGORY
[Percentages are of all engineering baccalaureate degrees awarded during the 1991-1992 academic year, unless otherwise footnoted.]
African Americans .... 0.7 %
Asian/Pacific Island Americans .... 4.7 %
Hispanic Americans .... – %
Native Americans .... 0.7 %
Foreign Citizens .... 6.3 %
All Others .... 87.6 %
Women .... 8.7 %
Persons with disabilities .... – %
Students over 25 years of age .... 45.2 %

GRADUATE ENROLLMENTS & DEGREES AWARDED
Master's enrollment 179
Master's degrees awarded 29
Doctoral enrollment .... –
Doctoral degrees awarded .... –

ENGINEERING STUDENT DATA
Applicants to the engineering college
Number of applicants to engineering college 495
Percent offered admission .... 80.0 %
Matriculated engineering students
Percentage in top quartile (25%) of High School class 88.7
Average SAT scores: Math—568, Verbal—477, Combined—1046.5

FULL- & PART-TIME FACULTY—BY DEPARTMENT
[Figures are the head count of full-time faculty and the full-time equivalent (FTE) of part-time faculty for each engineering department or equivalent.]

Department	Full	Part
Civil Eng	12	–
Electrical Eng	20	1.5
Mechanical Eng	15	–
Computer Science	21	4.0

COLLEGE DESCRIPTION
The College offers a four-year undergraduate program in engineering, emphasizing engineering fundamentals in the first year, followed by three years of specialized education in one of the three majors, with some liberal arts studies throughout all four years. There are opportunities to minor in a number of other areas. The College offers a dual-degree program (math/engineering) with Johnson C. Smith University in Charlotte, NC, and Belmont Abbey College in Belmont, NC. Approved pre-engineering programs at two- and four-year institutions throughout North Carolina provide opportunities for transfer into engineering programs. Co-op is an option in all programs. During the last semester most students complete senior design projects. Students are encouraged to complete the national standardized Fundamentals of Engineering exam. Average number of years required to actually complete the bachelor's degree: approximately 4.5.

DEGREE PROGRAMS DESCRIPTION
The bachelor's degrees in the various engineering programs prepare graduates for careers as professional engineers and/or for graduate study in each of the respective fields. Masters degrees provide an opportunity for graduate-level engineering education both to improve on-the-job skills and to provide an advanced degree program for career development and/or additional graduate study beyond the masters. Doctoral programs are offered through a cooperative arrangement with North Carolina State University at Raleigh.

DUAL DEGREE PROGRAMS
Engineering baccalaureate graduates with dual degrees 1
Enrollment requirements: Minimum 2.50 GPA

TRANSFER INFORMATION
Residency Requirements: Students must complete the last 30 semester hours of credit toward the degree in residence as well as the last 12 semester hours of the work in the major field.
Transfer via Articulation Agreements
Admission to engineering: At any time
Engineering graduates transferred from ... 4-yr: 20　2-yr: 10
Requirements: Minimum 2.50 GPA
Transfer without Articulation Agreements
Admission to engineering: At any time
Engineering graduates transferred from ... 4-yr: 23　2-yr: 8
Requirements: Minimum 2.50 GPA

GRADUATION REQUIREMENTS
Bachelor's programs in engineering require 132 semester credits. Co-op students must complete at least three work periods.

■ STUDENT PROGRAMS
PROFESSIONAL AND HONORARY SOCIETIES
[For key to acronyms, see *Introduction*.]
Professional Societies: ACM, AGCA, ASCE, ASHRAE, ASME, IEEE, ITE, NSPE, SAE, SME, Society of Women Engineers (SWE)
Honorary Societies: Tau Beta Pi

SUPPORT PROGRAMS
Student Chapter Organizations: Society of Women Engineers

University Transitional Opportunities Prog (UTOP)
For: Ethnic Minorities　　**Available:** Summer only
Offered: Prematriculation

Student Advising for Freshman Excellence (SAFE)
For: Ethnic Minorities　　**Available:** Year round
Offered: Freshman, Sophomore

Student Support Services (SSS)
For: Ethnic Minorities　　**Available:** Year round
Offered: Freshman, Sophomore, Junior, Senior

Other Engineering Support Programs: Summer and fall orientation program for all freshmen and transfers; Learning Center; tutorial services; supplemental instruction; Disabled Student Services; Writing Resources Center; Counseling Center; Center for Student Employment and Career Services; Center for International Studies.

159 North Carolina Agricultural and Technical State University

INSTITUTION PROFILE

HEAD OF THE INSTITUTION
Edward B. Fort
Phone: (919)334-7940

GENERAL INFORMATION
[All Students—Fall 1991]
Undergraduate enrollment 6,344
Graduate enrollment 775
Total institution enrollment 7,119
Type of institution: Public   Calendar: Semesters
Location: Greensboro, NC   Population: 196,000
Setting: Urban
Types of engineering degrees: Engineering
Other degree-granting colleges: Arts & Sciences, Education, Nursing, Business and Economics, Agriculture, Technology, Graduate School

ENGINEERING ADMISSIONS

ADMISSIONS OFFICE CONTACT
John Smith
North Carolina Agricultural and Technical State University
Room 101, Dowdy Building
1601 E. Market Street
Greensboro, NC 27411
Phone: (919)334-7946

ENGINEERING COLLEGE ADMISSIONS INFORMATION
Entrance Requirements: High School courses—English (4 years), Mathematics (3 years), Science (3 years), Social Studies (2 years), SAT required. In-state applicants: Minimum GPA/SAT ranges from 2.3/1000 to 3.4/750. Out-of-state: Minimum GPA/SAT ranges from 2.5/1000 to 3.4/800.
Entrance Recommendations: High School courses—Foreign Language (2 years), Advanced Mathematics (1 year)
Requirements for out-of-state residents: Profiles relating minimum HS GPA with minimum SAT scores are used for admission criteria. Different profiles are used for in-state and for out-of-state applicants.
Requirements for foreign students: In addition to satisfying all requirements to the school, must have a satisfactory score on the TOEFL and be eligible for an F-1 visa.

DEPARTMENTAL ADMISSIONS INFORMATION
Admission to the engineering department: At the time of admission to the institution.

FINANCIAL INFORMATION

ESTIMATED EXPENSES (ACADEMIC YEAR)
[Expenses are for the 1992-93 nine-month academic year.]

State Residents	Undergraduate	Graduate
Tuition and fees	$ 1,270	$ 1,270
College room and board	$ 3,040	$ 3,040
Books and supplies	$ 350	$ 350
Other expenses	$ 5 [1]	$ 5 [1]
Total estimated expenses	$ 4,665	$ 4,665

Out-of-State Residents	Undergraduate	Graduate
Tuition and fees	$ 6,942	$ 6,942
College room and board	$ 3,040	$ 3,040
Books and supplies	$ 350	$ 350
Other expenses	$ 5 [1]	$ 5 [1]
Total estimated expenses	$10,337	$10,337

Notes: (1) Chemistry Laboratory Fee

FINANCIAL AID OFFICE CONTACT
Delores S. Davis, *Director of Financial Aid*
North Carolina Agricultural and Technical State University
Room 102, Dowdy Building
1601 E. Market Street
Greensboro, NC 27411
Phone: (919)334-7973

GENERAL FINANCIAL AID INFORMATION
Forms accepted/required: FAF
Additional financial aid information: Both need-based and merit-based scholarships are available. Short-term loans, part-time campus jobs, and College Work-Study programs are available.

ENGINEERING COLLEGE INFORMATION
[For additional personnel, refer to the *Appendix*.]

HEAD OF ENGINEERING
Harold L. Martin
Phone: (919)334-7589   Fax: (919)334-7540
EMail: hlm@vanity.ncat.edu

ENGINEERING COLLEGE ADDRESS
School of Engineering
North Carolina Agricultural and Technical State University
Room 651, McNair Hall
1601 E. Market Street
Greensboro, NC 27411
Phone: (919)334-7589   Fax: (919)334-7540

ENROLLMENTS—BY CLASS
[Numbers are baccalaureate enrollments for the fall 1991 term, unless otherwise footnoted.]
1st-year students/Freshmen 482
2nd-year students/Sophomores 260
3rd-year students/Juniors 208
4th-year students/Seniors .... 288 [2]
Total 1,238
Notes: (2) This includes fifth year Architectural Engineering students.

NUMBER OF DEGREES AWARDED—BY PROGRAM
[Numbers are engineering baccalaureate degrees awarded during the 1991-1992 academic year, unless otherwise footnoted. For full details about each engineering program, refer to the *Tables of Degree Programs*.]
Agricultural Engineering .... –
Architectural Engineering 8
Chemical Engineering 10
Civil Engineering 8
Computer Science 40
Electrical Engineering 65
Industrial Engineering 30
Mechanical Engineering 32
Total 193

PERCENTAGE OF DEGREES AWARDED—BY CATEGORY
[Percentages are of all engineering baccalaureate degrees awarded during the 1991-1992 academic year, unless otherwise footnoted.]
African Americans 81.0%
Asian/Pacific Island Americans 1.0%
Hispanic Americans 1.0%
Native Americans .... – %
Foreign Citizens 1.0%
All Others 16.0%
Women 37.0%
Persons with disabilities .... – % [A]
Students over 25 years of age 5.0%
Notes: (A) Data not available.

GRADUATE ENROLLMENTS & DEGREES AWARDED
Master's enrollment 189
Master's degrees awarded 53
Doctoral enrollment 12
Doctoral degrees awarded .... – [1]
Notes: (1) Graduates of the joint PhD program have degrees conferred from N.C. State University.

ENGINEERING STUDENT DATA
Applicants to the engineering college
Number of applicants to engineering college .... – [A]
Percent offered admission .... –% [A]

Matriculated engineering students
Percentage in top quartile (25%) of High School class .. – [A]
Average SAT scores: Combined—935
Notes: (A) Data not available.

FULL- & PART-TIME FACULTY—BY DEPARTMENT
[Figures are the head count of full-time faculty and the full-time equivalent (FTE) of part-time faculty for each engineering department or equivalent.]

Department	Full	Part
Chemical Eng	11	–
Civil Eng	5	0.9
Electrical Eng	21	–
Industrial Eng	7	0.3
Mechanical Eng	17	2.1
Architectural Eng	7	–
Agricultural Eng	3	–
Computer Science	8	–

COLLEGE DESCRIPTION
The school offers four-year undergraduate programs in Chemical, Civil, Electrical, Industrial, and Mechanical Engineering. Architectural Engineering is offered in a five-year undergraduate program. The School of Engineering offers a four-year joint program with the School of Agriculture leading to an undergraduate degree in Agricultural Engineering. Engineering fundamentals are covered in the first two years, with specific major course emphasis placed in the remainder of the course of study. The school offers a dual-degree program with Bennett College. A co-op program is supported by the University through the Office of Cooperative Education. The school also offers degrees in Computer Science.

DEGREE PROGRAMS DESCRIPTION
The School of Engineering grants B.S. degrees in Architectural, Chemical, Civil, Electrical, Industrial, and Mechanical Engineering as well as Computer Science. It jointly with the School of Agriculture grants a B.S. in Agricultural Engineering. The School of Engineering grants M.S. degrees in in Architectural, Electrical, Industrial, and Mechanical Engineering. An M.S. degree in Engineering, with options in Chemical and in Civil Engineering, is granted as well. An M.S. in Agricultural Science is offered jointly with the School of Agriculture. Ph.D. degrees are available in most Engineering disciplines through an inter-institutional program between North Carolina State and North Carolina A&T State Universities. A dual-degree program is offered with Bennett College.

DUAL DEGREE PROGRAMS
Engineering baccalaureate graduates with dual degrees 0

TRANSFER INFORMATION
Residency Requirements: The student must complete at least three semesters in full-time residence at the University, including the two semesters immediately preceding the the period when the degree requirements are completed. At least half of the credits in the student's major field must be taken at the University.

Transfer via Articulation Agreements
Admission to engineering: At any time
Requirements: At previous school, transfer student must not have been on social or academic probation, must have had a 2.5 of 4.0 GPA, and must not have been suspended or dropped. If transferring from other than an accredited four-year program in engineering, a 3.0 is required. If less than 30 earned credits are transferred, then student must meet all requirements for freshman admission.

Transfer without Articulation Agreements
Admission to engineering: At any time
Engineering graduates transferred from .. 4-yr: –   2-yr: 83
Requirements: Same requirements as for students transferring from programs where an articulation agreement exists.

GRADUATION REQUIREMENTS
B.S. programs in engineering require 130-160 credit hours depending upon the major selected. A GPA of 2.0 in all courses and in the student's major field of study must be achieved. A required senior exam must be taken by all engineering students as a graduation requirement.

STUDENT PROGRAMS

PROFESSIONAL AND HONORARY SOCIETIES
[For key to acronyms, see *Introduction*.]
Professional Societies: ACM, AIA, AIChE, ASAE, ASCE, ASME, IEEE, IIE, NSAE, NSPE, American Society of Heating, Refrigerating, & Air Conditioning Engineers
Honorary Societies: Alpha Pi Mu, Eta Kappa Nu, Pi Tau Sigma, Tau Beta Pi, Phi Gamma Kappa, Alpha Delta Epsilon

SUPPORT PROGRAMS
Student Chapter Organizations: Society of Women Engineers, National Society of Black Engineers

Office of Veteran and Handicapped Student Affairs
For: Ethnic Minorities   **Available:** Year round
Offered: Prematriculation, Freshman, Sophomore, Junior, Senior, Graduate level

Office of Counseling Services
For: Ethnic Minorities   **Available:** Year round
Offered: Prematriculation, Freshman, Sophomore, Junior, Senior, Graduate level

Other Engineering Support Programs: Freshman Advisement/Learning Assistance Center; Orientation for freshmen and transfer students; Foreign Student Advisor; Tutoring Program; Placement Center; Co-op Education Program.

160 North Carolina State University

■ INSTITUTION PROFILE

HEAD OF THE INSTITUTION
Larry K. Monteith
Phone: (919)515-2191

GENERAL INFORMATION
[All Students—Fall 1991]
Undergraduate enrollment 18,285
Graduate enrollment 8,871
Total institution enrollment **27,156**
Type of institution: Public   **Calendar:** Semesters
Location: Raleigh   **Population:** 215,000
Setting: Urban
Types of engineering degrees: Engineering
Other degree-granting colleges: Agricultural & Environmental, Business Administration, Education, Humanities & Social Sciences, Design, Forest Resources, Physical and Mathematical Sciences, Textiles, Veterinary Medicine

■ ENGINEERING ADMISSIONS

ADMISSIONS OFFICE CONTACT
George R. Dixon
North Carolina State University
Director of Admissions
Box 7103-NCSU
Raleigh, NC 27695-7103
Phone: (919)515-2434   **Fax:** (919)515-5039

ENGINEERING COLLEGE ADMISSIONS INFORMATION
Entrance Requirements: High School courses—English (4 years), Mathematics inc. Algebra I & II and Geometry (3 years), Science inc. 1 life science and 1 physical science (3 years), Social sciences (including 1 U.S. History) (2 years), Admissions is based on an evaluation of the high school record, including the level and difficulty of courses completed, over-all grade point average, class rank, and SAT or ACT scores. Extracurricular activities are given some consideration.
Entrance Recommendations: High School courses—Foreign language (2 years), Score on SAT Mathematics Achievement Test Level II determines eligibility for placement in first calculus course.
Requirements for foreign students: TOEFL (Score: 550); financial statement

DEPARTMENTAL ADMISSIONS INFORMATION
Admission to the engineering department: At the end of the first year.
Additional information: Eligibility for admission to degree programs requires completion of freshman year courses. Admission to degree programs is based upon NCSU grade point average and time to complete required courses. In most cases, a 2.9 over-all grade point average should be sufficient for admission to first-choice program.

■ FINANCIAL INFORMATION

ESTIMATED EXPENSES (ACADEMIC YEAR)
[Expenses are for the 1992-93 nine-month academic year.]

State Residents	Undergraduate	Graduate
Tuition and fees	$ 1,318	$ 1,324
College room and board	$ 3,350	$ 3,350
Books and supplies	$ 500	$ 500
Other expenses	$ 1,250 [1]	$ 1,250 [1]
Total estimated expenses	**$ 6,418**	**$ 6,424**

Out-of-State Residents	Undergraduate	Graduate
Tuition and fees	$ 7,902	$ 7,908
College room and board	$ 3,350	$ 3,350
Books and supplies	$ 500	$ 500
Other expenses	$ 1,250 [1]	$ 1,250 [1]
Total estimated expenses	**$13,002**	**$13,008**

Notes: (1) Personal expenses and transportation

FINANCIAL AID OFFICE CONTACT
Julia E. Rice, *Director*
North Carolina State University
Office of Financial Aid
Box 7302
Raleigh, NC 27695-7302
Phone: (919)515-2421

GENERAL FINANCIAL AID INFORMATION
Forms accepted/required: FAF
Additional financial aid information: Need-based scholarships; Pell Grants; Supplemental Educational Opportunity Grants; NC Student Incentive Grants; long-term low interest loans; emergency short-term loans; part-time work-study jobs on campus; merit-based scholarships.

■ ENGINEERING COLLEGE INFORMATION

[For additional personnel, refer to the *Appendix*.]

HEAD OF ENGINEERING
Wilbur L. Meier
Phone: (919)515-2311   **Fax:** (919)515-2463
EMail: wmeier@eos.ncsu.edu

ENGINEERING COLLEGE ADDRESS
College of Engineering
North Carolina State University
Box 7901-NCSU
Raleigh, NC 27695-7901
Phone: (919)515-2311   **Fax:** (919)515-2463

ENROLLMENTS—BY CLASS
[Numbers are baccalaureate enrollments for the fall 1991 term, unless otherwise footnoted.]
1st-year students/Freshmen 1,491
2nd-year students/Sophomores 1,486
3rd-year students/Juniors 1,436
4th-year students/Seniors 1,586
Total **5,999**

NUMBER OF DEGREES AWARDED—BY PROGRAM
[Numbers are engineering baccalaureate degrees awarded during the 1991-1992 academic year, unless otherwise footnoted. For full details about each engineering program, refer to the *Tables of Degree Programs*.]
Aerospace Engineering 19
Biological & Agricultural Engineering 2
Chemical Engineering 67
Civil Engineering 117
Civil Engineering (Construction) 34
Computer Engineering 71
Computer Science 107
Construction Management 2
Electrical Engineering 210
Engineering 1
Environmental Engineering .... –
Industrial Engineering 85
Industrial Engineering (Furniture Manufacturing) .. 8
Integrated Manufacturing Systems Engineering ... –
Materials Science & Engineering 15
Mechanical Engineering 216
Nuclear Engineering 7
Operations Research .... –
Textile Engineering 9
Total **970**

PERCENTAGE OF DEGREES AWARDED—BY CATEGORY
[Percentages are of all engineering baccalaureate degrees awarded during the 1991-1992 academic year, unless otherwise footnoted.]
African Americans .... 6.3 %
Asian/Pacific Island Americans .... 4.7 % (C)
Hispanic Americans .... 0.9 % (C)
Native Americans .... 0.6 % (C)
Foreign Citizens .... 1.8 % (C)
All Others .... 85.7 %
Women .... 17.1 %
Persons with disabilities .... – % (A)
Students over 25 years of age .... – % (A)

Notes: (A) Data not available. (C) Estimated data.

GRADUATE ENROLLMENTS & DEGREES AWARDED
Master's enrollment 687
Master's degrees awarded 295
Doctoral enrollment 449
Doctoral degrees awarded 97

ENGINEERING STUDENT DATA

Applicants to the engineering college
Number of applicants to engineering college 2,695
Percent offered admission 77.2%

Matriculated engineering students
Percentage in top quartile (25%) of High School class .... 70.0 (C)
Average SAT scores: Math—606, Verbal—500, Combined—1106

Notes: (C) Estimated data.

FULL- & PART-TIME FACULTY—BY DEPARTMENT
[Figures are the head count of full-time faculty and the full-time equivalent (FTE) of part-time faculty for each engineering department or equivalent.]

Department	Full	Part
Chemical Eng	16	16.5
Civil Eng	35	36.5
Electrical & Computer Eng	45	54.0
Industrial Eng	23	22.5
Mechanical & Aerospace Eng	44	39.5
Biological & Agricultural Eng	11	11.0
Computer Science	39	34.4
Materials Science & Eng	24	22.5
Nuclear Eng	11	11.0
Textile Eng, Chemistry, & Science	9	4.5

COLLEGE DESCRIPTION
The College offers sixteen undergraduate programs in engineering. Students may elect to combine these four-year baccalaureate programs with a second, dual-degree major, or a minor option in over sixty areas, and they can participate in the co-op program. Co-op participation normally adds at least one year to the time required to earn a B. S. degree. New freshmen students are enrolled in the Engineering Undesignated program and they must complete freshman-level course requirements in order to apply for admission to degree programs. The Engineering Scholars program stimulates growth and personal development for outstanding students who are admitted as freshmen. The Benjamin Franklin Scholars program is available to students who wish to earn dual degrees in engineering and the humanities or social sciences. The College of Engineering also offers degrees in computer science.

DEGREE PROGRAMS DESCRIPTION
The baccalaureate program provides preparation for entry into industry, government, business or private practice, as well as graduate school. Post-baccalaureate professional degrees in chemical, civil, electrical, industrial, materials science and engineering, and nuclear engineering. A Master of Technology for International Development and a Master of Integrated Manufacturing System Engineering are offered, as are M.S. and Ph.D. degrees in computer science and operations research. Doctoral degrees are offered in all engineering curricula.

DUAL DEGREE PROGRAMS
Engineering baccalaureate graduates with dual degrees 44

TRANSFER INFORMATION
Residency Requirements: B.S. candidates must earn at least 48 of their last 60 hours of credit at NCSU while enrolled as degree candidates.

Transfer via Articulation Agreements
Admission to engineering: At the end of the sophomore year
Engineering graduates transferred from   ... 4-yr: 31   2-yr: 10

Transfer without Articulation Agreements
Admission to engineering: At any time
Engineering graduates transferred from   ... 4-yr: 160   2-yr: 80
Requirements: Students must have completed 28 semester credit hours or 42 quarter hours with a minimum GPA of 2.8. College-level courses in English and mathematics must have been completed with a grade of 'C' or better.

GRADUATION REQUIREMENTS
B.S. degree programs in engineering require 120-135 semester credits, depending upon major selected. Student's over-all GPA must be at least 2.0; the GPA for all major courses must be at least 2.0, or students must earn a grade of 'C' or better in all required major courses.

■ STUDENT PROGRAMS

PROFESSIONAL AND HONORARY SOCIETIES
[For key to acronyms, see *Introduction*.]
Professional Societies: ACM, AIAA, AIChE, ANS, ASAE, ASCE, ASME, IEEE, IIE, SPE*, Textile Engineering Society, NCSU Furniture Club
Honorary Societies: Alpha Epsilon, Alpha Pi Mu, Chi Epsilon, Eta Kappa Nu, Pi Tau Sigma, Sigma Gamma Tau, Tau Beta Pi, Theta Tau, Upsilon Pi Epsilon, Alpha Sigma Mu

SUPPORT PROGRAMS
Student Chapter Organizations: Society of Women Engineers, National Society of Black Engineers

African - American Symposium
For: Ethnic Minorities   **Available:** Summer only
Offered: Prematriculation

Summer Transition Program
For: Ethnic Minorities   **Available:** Summer only
Offered: Prematriculation

Peer Mentor Program
For: Ethnic Minorities   **Available:** Academic year
Offered: Freshman

College of Engineering Tutelage Program
For: Women & Ethnic Minorities   **Available:** Year round
Offered: Freshman, Sophomore

Other Engineering Support Programs: Summer and fall orientation for freshmen and transfer students; the Summer Transition program; career and personal counseling services; International Student Office; Engineering Scholars program; North Carolina State Fellows program; Study Abroad Office; Handicapped Student Services; Academic Skills Program; the College of Engineering Tutelage program; the First-Year Experience.

161 University of North Dakota

■ INSTITUTION PROFILE

HEAD OF THE INSTITUTION
Kendall R. Baker
Phone: (701)777-2121 Fax: (701)777-3866

GENERAL INFORMATION
[All Students—Fall 1991]
Undergraduate enrollment 10,111
Graduate enrollment 1,829
Total institution enrollment 11,940
Type of institution: Public   Calendar: Semesters
Nearest city: Winnipeg, Manitoba   Population: 500,000
Miles from main campus: 150   Setting: Small Town
Types of engineering degrees: Engineering
Other degree-granting colleges: Arts & Sciences, Business Administration, Communication & Journalism, Education, Fine & Performing Arts, Humanities & Social Sciences, Law, Medicine, Nursing, Aviation

■ ENGINEERING ADMISSIONS

ADMISSIONS OFFICE CONTACT
Donna M. Bruce
University of North Dakota
Box 8070
Grand Forks, ND 58202-8070
Phone: (701)777-3821 Fax: (701)777-2696

ENGINEERING COLLEGE ADMISSIONS INFORMATION
Admission to the engineering college: By a formal admission process, students are admitted to a professional engr. degree program.
Entrance Requirements: High School courses—1993 English Written & Oral Communication (4 years), 1993 Math (Alg. I and above) (3 years), 1993 Lab Science (3 years), 1993 Social Science (3 years)
Entrance Recommendations: High School courses—English (4 years), Math (3 years), Science (3 years), Social Science (3 years)

DEPARTMENTAL ADMISSIONS INFORMATION
Admission to the engineering department: A student who has completed 24 credit hours may transfer from Univ. College to the SEM.
Additional information: Students must have a minimum GPA of 2.0. A formal admission process is necessary for students to enter a professional engineering degree program, and only those admitted students will be eligible to graduate with an engineering degree. All of the professional engineering degree programs require the following for admission: (A) minimum grade of C must be earned in foundation courses and (B) an additional science course prescribed by each admitting department. At least four engineering science courses or acceptable equivalents prescribed by each admitting department.

■ FINANCIAL INFORMATION

ESTIMATED EXPENSES (ACADEMIC YEAR)
[Expenses are for the 1992-93 nine-month academic year.]

State Residents	Undergraduate	Graduate
Tuition and fees	$ 2,145	$ 2,355
College room and board	$ 2,455	$ 2,455
Books and supplies	$ 400	$ 300
Other expenses	$ 1,600 (1)	$ –
Total estimated expenses	$ 6,600	$ 5,110

Out-of-State Residents	Undergraduate	Graduate
Tuition and fees	$ 5,253	$ 5,616
College room and board	$ 2,455	$ 2,455
Books and supplies	$ 400	$ 300
Other expenses	$ 1,600	$ –
Total estimated expenses	$ 9,708	$ 8,371

Notes: (1) Personal expenses. Special low tuition/fee structure for contiguous states and provinces.

FINANCIAL AID OFFICE CONTACT
Mark Brickson, *Director*
University of North Dakota
Box 7205 University Station
Grand Forks, ND 58202-7205
Phone: (701)777-3121

GENERAL FINANCIAL AID INFORMATION
Forms accepted/required: ACT, CSS-FAF, FAT, FAR, FSA, FFS, Stafford, IRS, SAR, SDF, Supplemental, Institutional
Additional financial aid information: Four different types of financial aid are offered:(1) employment, (2) loans, (3) scholarships, and (4) grants.

■ ENGINEERING COLLEGE INFORMATION

[For additional personnel, refer to the *Appendix*.]

HEAD OF ENGINEERING
Mogens Henriksen
Phone: (701)777-3412 Fax: (701)777-4838
EMail: henrik@eng.und.nodak.edu

ENGINEERING COLLEGE ADDRESS
School of Engineering and Mines
University of North Dakota
Box 8155 University Station
Grand Forks, ND 58202-8155
Phone: (701)777-3411 Fax: (701)777-4838

ENROLLMENTS—BY CLASS
[Numbers are baccalaureate enrollments for the fall 1991 term, unless otherwise footnoted.]
1st-year students/Freshmen 189
2nd-year students/Sophomores 93
3rd-year students/Juniors 143
4th-year students/Seniors 249
Total 674

NUMBER OF DEGREES AWARDED—BY PROGRAM
[Numbers are engineering baccalaureate degrees awarded during the 1991-1992 academic year, unless otherwise footnoted. For full details about each engineering program, refer to the *Tables of Degree Programs*.]
Chemical Engineering 13
Civil Engineering 23
Electrical Engineering 27
Engineering Management 5
Geological Engineering 1
Mechanical Engineering 24
Total 93

PERCENTAGE OF DEGREES AWARDED—BY CATEGORY
[Percentages are of all engineering baccalaureate degrees awarded during the 1991-1992 academic year, unless otherwise footnoted.]
African Americans .... – %
Asian/Pacific Island Americans .... – %
Hispanic Americans .... – %
Native Americans .... – %
Foreign Citizens .... – %
All Others .... 100 %
Women .... 20.0 %
Persons with disabilities .... – %
Students over 25 years of age .... – %

GRADUATE ENROLLMENTS & DEGREES AWARDED
Master's enrollment 37
Master's degrees awarded 15
Doctoral enrollment 2
Doctoral degrees awarded .... –

ENGINEERING STUDENT DATA
Applicants to the engineering college
Number of applicants to engineering college 177
Percent offered admission 86.0%
Average SAT scores: Math—639, Verbal—520
Average ACT scores: Math—25.3, Composite—25.0

FULL- & PART-TIME FACULTY—BY DEPARTMENT
[Figures are the head count of full-time faculty and the full-time equivalent (FTE) of part-time faculty for each engineering department or equivalent.]

Department	Full	Part
Chemical Eng	5	–
Civil Eng	6	–
Electrical Eng	8	–
Mechanical Eng	10	–
Geology & Geological Eng	11	–
Eng Management	1	–
Eng Physics	3	–

COLLEGE DESCRIPTION

The School of Engineering and Mines offers programs in the classical fields of engineering. Although graduate programs through the doctorate are offered, undergraduate engineering education remains its main mission. The programs are small as a result of enrollment management which limits each program to a suitable number of students. All major technical societies are present; the SWE Chapter has been recognized as the Best Chapter in the Nation four times in the past eleven years. The graduates are recruited regionally and nationally. The school has supportive alumni who form an effective network, assisting by enhancing facilities and helping students to find professional positions in industry and government. Laboratories are modern and well equipped. The faculty members are strong teachers; many are also active scholars. The graduate program has a special focus in energy and environment. The school collaborates in many interdisciplinary research projects with other units on campus

DEGREE PROGRAMS DESCRIPTION
All undergraduate engineering programs are design oriented. The favorable student to faculty ratio ensures close interaction between students and faculty members. Many undergraduate students find part-time employment as research associates in the engineering departments as well as in the other research units on campus. Many students also take advantage of the co-op opportunities offered through the University Co-op office. Each undergraduate program has curricular breath and appropriate depth. The graduate-level activities focus on energy and environment. The school offers M.S. degrees in chemical, civil, electrical and mechanical engineering and a Ph.D. in energy engineering. The chemical and mechanical engineering graduate programs include catalysis, combustion, thermal sciences, fluid mechanics, and modeling. The civil engineering program offers strong environmental systems courses; civil and mechanical engineering jointly offer depth in solid mechanics. The electrical engineering graduate emphasis is on machines and power electronics. Each academic unit is research active; interaction with the Energy and Environmental Research Center, a unit of the University of North Dakota with a research volume in excess of $12 million per year, offer clinical opportunities for many students, thus enhancing the educational experience

TRANSFER INFORMATION
Transfer via Articulation Agreements
Admission to engineering: At the end of the sophomore year

Transfer without Articulation Agreements
Admission to engineering: At the end of the sophomore year
Engineering graduates transferred from .. 4-yr: – 2-yr: –

GRADUATION REQUIREMENTS
B.S. programs in engineering require 135-147 semester credits, depending on program.

■ STUDENT PROGRAMS

PROFESSIONAL AND HONORARY SOCIETIES
[For key to acronyms, see *Introduction*.]
Professional Societies: AIChE, AIME, ASCE, ASEM, ASME, IEEE, NSPE, SME*, SPE, Society of Women Engineers
Honorary Societies: Eta Kappa Nu, Tau Beta Pi, Sigma Gamma Epsilon

SUPPORT PROGRAMS
Student Chapter Organizations: Society of Women Engineers

Women in Engineering
For: Women **Available:** Year round
Offered: Prematriculation, Freshman, Sophomore, Junior, Senior, Graduate level

Other Engineering Support Programs: UND holds orientation programs for new students (freshman and transfer students) each semester. Emphasis is on acquainting students with people, programs and resources at UND and the surrounding community. Career Counseling Services provides individual counseling, career testing and information for students who are in the process of choosing a career field or selecting an academic major.

162 North Dakota State University

INSTITUTION PROFILE
HEAD OF THE INSTITUTION
J. L. Ozbun
Phone: (701)237-7211   Fax: (701)237-7050
GENERAL INFORMATION
[All Students—Fall 1991]
Undergraduate enrollment 2,264
Graduate enrollment 120
Total institution enrollment **2,384**
Type of institution: Public   **Calendar:** Semesters
Location: Fargo, ND   **Population:** 65,000
Setting: Urban
Types of engineering degrees: Engineering & Technology
Other degree-granting colleges: Agricultural & Environmental, Business Administration, Humanities & Social Sciences, Pharmacy, Technology & Applied Sciences, Science-Math, Human Development & Education, University Studies

ENGINEERING ADMISSIONS
ADMISSIONS OFFICE CONTACT
Robert A. Preloger
North Dakota State University
Ceres Hall 124
Fargo, ND 58105
Phone: (701)237-8643
ENGINEERING COLLEGE ADMISSIONS INFORMATION
Admission to the engineering college: Transfer and new students can be admitted to individual units.
Entrance Requirements: ACT (Score: 24), High School courses—English (4 years), Mathematics (3 years), Lab Science (3 years), Social Science (3 years)
Entrance Recommendations: ACT (Score: 22), High School courses—Graphics, Physics, Chemistry
Requirements for out-of-state residents: Out-of-state transfer student must have 2.5 GPA.
Requirements for foreign students: TOEFL (Score: 525); financial statement; Must be able to support themselves by showing 10,000 bank deposit or other evidence. TOEFL passing score required.
DEPARTMENTAL ADMISSIONS INFORMATION
Admission to the engineering department: Admitted to University & College of Engineering at the beginning of freshman year.
Additional information: Some Engineering departments have special requirements such as: Mechanical and Electrical Engineering require GPA of 2.85, ACT score of 22, and upper 25% of high school graduating class.

FINANCIAL INFORMATION
ESTIMATED EXPENSES (ACADEMIC YEAR)
[Expenses are for the 1992-93 nine-month academic year.]

State Residents	Undergraduate	Graduate
Tuition and fees	$ 1,860[1]	$ 1,121
College room and board	$ 2,482[2]	$ 2,482
Books and supplies	$ 172[2]	$ 280
Other expenses	$ 50	$ 50
Total estimated expenses	**$ 4,564**	**$ 3,933**
Out-of-State Residents	Undergraduate	Graduate
Tuition and fees	$ 4,968	$ 2,858
College room and board	$ 2,482	$ 2,482
Books and supplies	$ 172	$ 280
Other expenses	$ 50	$ 50
Total estimated expenses	**$ 7,672**	**$ 5,670**

Notes: (1) $2,328 for MN residents. (2) Same for Minnesota residents.
FINANCIAL AID OFFICE CONTACT
Wayne K. Tesmer, *Financial Aid Director*
North Dakota State University
Ceres Hall 202
Fargo, ND 58105
Phone: (701)237-7538
GENERAL FINANCIAL AID INFORMATION
Forms accepted/required: ACT, AFSA, CSS-FAF, FAF, FAT, FAR, FSA, FFS, Stafford, IRS, SAR, SDF, Supplemental, Institutional

ENGINEERING COLLEGE INFORMATION
[For additional personnel, refer to the *Appendix*.]
HEAD OF ENGINEERING
Joseph Stanislao
Phone: (701)237-7494   Fax: (701)237-7195

ENGINEERING COLLEGE ADDRESS
College of Engineering and Architecture
North Dakota State University
P.O. Box 5285, NDSU Station
Fargo, ND 58105
Phone: (701)237-7494   Fax: (701)237-7195
ENROLLMENTS—BY CLASS
[Numbers are baccalaureate enrollments for the fall 1991 term, unless otherwise footnoted.]
1st-year students/Freshmen 594
2nd-year students/Sophomores 501
3rd-year students/Juniors 408
4th-year students/Seniors 751
Total **2,254**

NUMBER OF DEGREES AWARDED—BY PROGRAM
[Numbers are engineering baccalaureate degrees awarded during the 1991-1992 academic year, unless otherwise footnoted. For full details about each engineering program, refer to the *Tables of Degree Programs*.]
Agricultural Engineering 13
Civil Engineering 56
Construction Engineering 6
Electrical Engineering 104
Industrial Engineering & Management 40
Mechanical Engineering & Applied Mechanics 62
Petroleum Engineering 1
Total **282**

PERCENTAGE OF DEGREES AWARDED—BY CATEGORY
[Percentages are of all engineering baccalaureate degrees awarded during the 1991-1992 academic year, unless otherwise footnoted.]
African Americans 1.0%
Asian/Pacific Island Americans 3.1%
Hispanic Americans 2.0%
Native Americans 2.0%
Foreign Citizens 4.9%
All Others 87.0%
Women 13.0%
Persons with disabilities 1.5%
Students over 25 years of age 8.5%

GRADUATE ENROLLMENTS & DEGREES AWARDED
Master's enrollment 79
Master's degrees awarded 30
Doctoral enrollment 31
Doctoral degrees awarded 4

ENGINEERING STUDENT DATA
Applicants to the engineering college
Number of applicants to engineering college .... 950[1]
Percent offered admission 85.0%
Matriculated engineering students
Percentage in top quartile (25%) of High School class 85.5
Average ACT scores: Math—23.0, Composite—24.5

Notes: (1) Excludes 402 applicants for Architecture & Landscape Arch.

FULL- & PART-TIME FACULTY—BY DEPARTMENT
[Figures are the head count of full-time faculty and the full-time equivalent (FTE) of part-time faculty for each engineering department or equivalent.]

Department	Full	Part
Civil Eng & Construction	11	1.0
Electrical & Electronics Eng	19	2.0
Industrial Eng	7	–
Mechanical Eng	14	3.0
Agricultural Eng	4	4.0
Eng Science	4	1.0
Aero-Manufacturing Eng Tech	2	1.0
Aerospace Studies	4	–
Military Science	5	–

COLLEGE DESCRIPTION
College depts. are administered by a dept chair with assigned faculty and staff. Students are assigned to a dept. for counseling according to their field of study. The general program provides preparatory counseling and studies for entering students. A five-year professional degree completes the programs in architecture and landscape architecture. Specialized study options are available in bioengineering, sanitary engr., environ. planning, industrial management, plus system, transportation and manufacturing engineering. The Master of Science in engineering fields and the Master of Architecture degrees are available. The Doctor of Philosophy in Engr. degree is a single doctoral program administered by the Graduate School and the College of Engr. & Arch. The Robert Perkins Engr. Computer Center provides a powerful tool for research and industrial development.

DEGREE PROGRAMS DESCRIPTION
Agric. Engr: combines biology, principles of engr., & engr. design to improve production & processing methods for agric. commodities. Civil Engr: math & sci. background in planning, design, constr., maint., & operation of our civilization's large, permanent engr. projects. Constr. Engr. & Mgmnt: combines engr., constr., business, & management educ. for work in the constr. industry. Elec. & Electronics Engr: background for work in electronics & electrical systems. Options in communications; digital, control, & power systems; microwaves, computer engr., & bioengr. Industrial Engr: specialized knowledge & skills educ. in math & physical/social sci with engr. analysis & design principles & methods to specify, predict & eval. results to be obtained from integrated systems of people, materials, equip. & energy. Engr. Physics: engr., math & physics educ. for employment in optics, electronics, reactor engr., biophysics, radiology sci., & meteorology. Petr. Engr: engr., math & phys. sci. educ. emphasizing chemistry & geology, with petr. engr. courses in field delineation & eval., reservoir engr., well stimulation, etc. Mechan. Engr: math and basic & applied science educ. providing background in thermodynamic properties, energy conversion & conservation, and concept development and improvement of mechanical

DUAL DEGREE PROGRAMS
Engineering baccalaureate graduates with dual degrees 113
Enrollment requirements: 2nd degree must have been in residence for 1 semester, taking 12 credits or more during the time period.

TRANSFER INFORMATION
Transfer via Articulation Agreements
Admission to engineering: Admitted at any time. Application Deadline: Rolling basis.
Engineering graduates transferred from   ... 4-yr: 10% 2-yr: 12%
Requirements: 2.50 GPA required if we have previous students. 3.00 GPA for situations we are not familiar with.
Transfer without Articulation Agreements
Admission to engineering: Admitted to engineering program at any time. Application deadline: Rolling basis.
Engineering graduates transferred from   ... 4-yr: 15% 2-yr: 20%

GRADUATION REQUIREMENTS
Undergraduates must maintain 2.0 GPA minimum and meet all ABET accreditation requirements. The graduate programs are customized to the individual graduate student.

STUDENT PROGRAMS
PROFESSIONAL AND HONORARY SOCIETIES
[For key to acronyms, see *Introduction*.]
Professional Societies: AGCA, AIA, ASAE, ASCE, ASME, IEEE, IIE, ITE, SME, SPE, Biomedical Engineering Society (BMES), American Water Resources Association (AWRA)
Honorary Societies: Alpha Pi Mu, Chi Epsilon, Eta Kappa Nu, Omega Chi Epsilon, Pi Tau Sigma, Sigma Phi Delta, Tau Alpha Pi, Tau Beta Pi, Phi Kappa Phi, Sigma Lambda Chi

SUPPORT PROGRAMS
Student Chapter Organizations: Society of Women Engineers, E & A Council

Native American Science & Engr. Society (NASES)
For: Ethnic Minorities
Offered: Sophomore

Upward-Bound for Veterans & Native Americans
For: Ethnic Minorities   **Available:** Year round
Offered: Sophomore

Other Engineering Support Programs: Honorary Societies provide academic help in professional and service courses free of charge. The general program of the College provides special counseling and program preparation for students who have not yet determined the professional field they wish to study. This program endeavors to place entering students in the most suitable degree-granting program of the college.

163 Northeastern University

INSTITUTION PROFILE

HEAD OF THE INSTITUTION
John A. Curry
Phone: (617)437-2101 Fax: (617)437-5015

GENERAL INFORMATION
[All Students—Fall 1991]
Undergraduate enrollment 23,590
Graduate enrollment 5,292
Total institution enrollment 31,779
Type of institution: Private Calendar: Quarters
Location: Boston Population: 578,000
Setting: Urban
Types of engineering degrees: Engineering & Technology
Other degree-granting colleges: Allied Health Sciences, Arts & Sciences, Business Administration, Law, Nursing, Pharmacy, Technology & Applied Sciences, Computer Science, Criminal Justice

ENGINEERING ADMISSIONS

ADMISSIONS OFFICE CONTACT
Michael F. Clifford
Northeastern University
Undergraduate Admissions
139 Richards Hall
Boston, MA 02115
Phone: (617)437-2200 Fax: (617)437-8780

ENGINEERING COLLEGE ADMISSIONS INFORMATION
Entrance Requirements: SAT (Score: 950), High School courses—Mathematics (4 years), Chemistry (1 year), Physics (1 year)
Entrance Recommendations: High School courses—Additional science courses (1 year)
Requirements for foreign students: TOEFL (Score: 550)

DEPARTMENTAL ADMISSIONS INFORMATION
Admission to the engineering department: At the time of admission to the institution.

FINANCIAL INFORMATION

ESTIMATED EXPENSES (ACADEMIC YEAR)
[Expenses are for the 1992-93 nine-month academic year.]

All Students	Undergraduate	Graduate
Tuition and fees	$11,490 [1]	$ 7,950
College room and board	$ 6,000 [2]	$ 8,700
Books and supplies	$ –	$ 700
Other expenses	$ 795 [3]	$ 600
Total estimated expenses	**$18,285**	**$17,950**

Notes: (1) Tuition for three quarters of freshman year. (2) Varies according to accommodation and meal plan selected. (3) Includes health service fee; other fees, such as parking, are optional

FINANCIAL AID OFFICE CONTACT
Jean C. Eddy, *Dean of Financial Aid*
Northeastern University
356 Richards Hall
360 Huntington Avenue
Boston, MA 02115
Phone: (617)437-3190 Fax: (617)437-8623

GENERAL FINANCIAL AID INFORMATION
Forms accepted/required: FAF
Additional financial aid information: Financial aid includes state assistance programs, federal programs and University and other private scholarships. Academic performance-based scholarships are available.

ENGINEERING COLLEGE INFORMATION
[For additional personnel, refer to the *Appendix*.]

HEAD OF ENGINEERING
Paul H. King
Phone: (617)437-2152 Fax: (617)437-8504

ENGINEERING COLLEGE ADDRESS
College of Engineering
Northeastern University
Snell Engineering Center
360 Huntington Avenue
Boston, MA 02115
Phone: (617)437-2152 Fax: (617)437-2501

ENROLLMENTS—BY CLASS
[Numbers are baccalaureate enrollments for the fall 1991 term, unless otherwise footnoted.]
1st-year students/Freshmen 298
2nd-year students/Sophomores .... 215 [1]
3rd-year students/Juniors .... 367 [1]
4th-year students/Seniors 417
Total 1,297

Notes: (1) Cooperative education program is a five-year program. Enrollment in Middler Year (between Sophomore and Junior) was 296.

NUMBER OF DEGREES AWARDED—BY PROGRAM
[Numbers are engineering baccalaureate degrees awarded during the 1991-1992 academic year, unless otherwise footnoted. For full details about each engineering program, refer to the *Tables of Degree Programs*.]
Chemical Engineering 25
Civil Engineering 78
Electrical & Computer Engineering 182
Industrial Engineering & Information Systems 28
Mechanical Engineering 110
Total 423

PERCENTAGE OF DEGREES AWARDED—BY CATEGORY
[Percentages are of all engineering baccalaureate degrees awarded during the 1991-1992 academic year, unless otherwise footnoted.]
African Americans .... 3.0 %
Asian/Pacific Island Americans .... 5.0 %
Hispanic Americans .... 4.0 %
Native Americans .... – %
Foreign Citizens .... 16.5 %
All Others .... 71.5 %
Women .... 12.0 %
Persons with disabilities .... 2.0 %
Students over 25 years of age .... – % [A]

Notes: (A) Data not available.

GRADUATE ENROLLMENTS & DEGREES AWARDED
Master's enrollment 1,077
Master's degrees awarded 312
Doctoral enrollment 132
Doctoral degrees awarded 14

ENGINEERING STUDENT DATA
Applicants to the engineering college
Number of applicants to engineering college 1,503
Percent offered admission 65.3%
Matriculated engineering students
Percentage in top quartile (25%) of High School class 80.0
Average SAT scores: Math—620, Verbal—480, Combined—1100

FULL- & PART-TIME FACULTY—BY DEPARTMENT
[Figures are the head count of full-time faculty and the full-time equivalent (FTE) of part-time faculty for each engineering department or equivalent.]

Department	Full	Part
Chemical Eng	7	....
Civil Eng	16	1.5
Electrical & Computer Eng	49	1.0
Industrial Eng & Information Systems	16	–
Mechanical Eng	22	1.0

COLLEGE DESCRIPTION
The College offers four- and five-year cooperative education undergraduate programs in engineering, with majors available in chemical, civil, electrical and computer, industrial and mechanical engineering. Northeastern University is the world leader in cooperative education, and the College of Engineering has developed a high quality undergraduate program that integrates classroom learning in basic engineering science and engineering design with practical industrial experience that prepares graduates for professional practice or for graduate school. The program is sufficiently large to make it possible to offer a wide variety of elective courses. The Departments of Electrical and Computer Engineering, Industrial Engineering and Information Systems, and Mechanical Engineering offer programs leading to both the bachelor's and master's degrees in five years for students with outstanding academic records.

DEGREE PROGRAMS DESCRIPTION
The B.S. degrees with co-op experience are designed to prepare students for entry-level professional positions or for graduate study in their chosen fields. The College offers a broad range of graduate programs leading to the Master of Science, Engineer Degree, Doctor of Engineering (Chemical Engineering) and Ph.D.

TRANSFER INFORMATION
Residency Requirements: Students transferring from another college or university are not eligible to receive the bachelor of science degree until they have completed at least 48 quarter hours at Northeastern University immediately preceding graduation.
Transfer via Articulation Agreements
Admission to engineering: At any time
Engineering graduates transferred from .. 4-yr: 10 2-yr: 23
Requirements: Good academic standing and a minimum grade point average of 2.0 from accredited four-year institutions and 2.6 from accredited two-year institutions.
Transfer without Articulation Agreements
Admission to engineering: At any time
Engineering graduates transferred from .. 4-yr: 117 2-yr: 18
Requirements: Good academic standing. Decision is based on review of entire academic record.

GRADUATION REQUIREMENTS
Candidates for graduation must complete all of the prescribed work in the curriculum in which they seek to qualify, with no academic deficiencies in required course work. Degree requirements will be based upon the year of graduation as determined by the date of entry or re-entry into the College of Engineering.

STUDENT PROGRAMS

PROFESSIONAL AND HONORARY SOCIETIES
[For key to acronyms, see *Introduction*.]
Professional Societies: AIChE, ASCE, ASME, IEEE, IIE, SAE, Engineering Student Council
Honorary Societies: Alpha Pi Mu, Chi Epsilon, Eta Kappa Nu, Omega Chi Epsilon, Pi Tau Sigma, Tau Alpha Pi, Tau Beta Pi

SUPPORT PROGRAMS
Student Chapter Organizations: Society of Women Engineers, Black Engineering Student Society
NUPRIME
For: Ethnic Minorities Available: Year round
Offered: Prematriculation, Freshman, Sophomore, Junior, Senior
Women in Engineering
For: Women Available: Year round
Offered: Prematriculation, Freshman, Sophomore, Junior, Senior, Graduate level

Other Engineering Support Programs: Comprehensive academic support is provided by the Office of Engineering Student Services and departmental academic advisors. Special orientation programs and monitoring to allow early warning of academic problems are provided for freshmen. Peer and graduate student tutors are available as are the resources of the University's Counseling and Testing Service.

164 Northern Arizona University

INSTITUTION PROFILE
HEAD OF THE INSTITUTION
Eugene M. Hughes
Phone: (602)523-3232   Fax: (602)523-4230

GENERAL INFORMATION
[All Students—Fall 1991]
Undergraduate enrollment 14,056
Graduate enrollment 4,435
Total institution enrollment 18,491
Type of institution: Public   **Calendar:** Semesters
Location: Flagstaff   **Population:** 46,000
Setting: Small Town
Types of engineering degrees: Engineering & Technology
Other degree-granting colleges: Allied Health Sciences, Arts & Sciences, Business Administration, Communication & Journalism, Education, Fine & Performing Arts, Humanities & Social Sciences, Forestry, Hotel and Restaurant Management

ENGINEERING ADMISSIONS
ADMISSIONS OFFICE CONTACT
Molly S. Carder
Northern Arizona University
Director of Admissions
Box 04084
Flagstaff, AZ 86011-4084
Phone: (602)523-6002   Fax: (602)523-2220

ENGINEERING COLLEGE ADMISSIONS INFORMATION
Admission to the engineering college: Admission upon completion of 8 specified pre-engineering courses with C or better, overall GPA of 2.5 on at least 12 hours at NAU.
Entrance Requirements: SAT (Score: 930), ACT (Score: 22), High School courses—English (4 years), Mathematics (3 years), Laboratory Science (2 years), Social Science (including American History) (2 years), High school GPA of 3.00 or above or graduation in the upper 25% of high school class also qualifies for unconditional admission. Students not achieving these scores may receive conditional admission on an individual basis.
Entrance Recommendations: High School courses—Algebra, Geometry (3 years), Trigonometry, Pre-calculus (1 year), Chemistry (1 year), Physics (1 year), Advanced placement tests may give college credit.
Requirements for out-of-state residents: For unconditional admission, minimum ACT score for non-residents is 24; minimum SAT is 1010.
Requirements for foreign students: TOEFL (Score: 500); financial statement

DEPARTMENTAL ADMISSIONS INFORMATION
Admission to the engineering department: Students may declare an engineering major upon admission to the university, but must be admitted to an engineering program to take sophomore courses.
Additional information: Admission is based on completion of specified pre-engineering courses with grades of C or above and establishing a cumulative GPA of 2.50 or above on at least 12 hours in residence.

FINANCIAL INFORMATION
ESTIMATED EXPENSES (ACADEMIC YEAR)
[Expenses are for the 1992-93 nine-month academic year.]

State Residents	Undergraduate	Graduate
Tuition and fees	$ 1,590	$ –
College room and board	$ 2,699	$ –
Books and supplies	$ 500	$ –
Other expenses	$ –(A)	$ –
Total estimated expenses	**$ 4,789**	**$ –**

Out-of-State Residents	Undergraduate	Graduate
Tuition and fees	$ 6,242	$ –
College room and board	$ 2,699	$ –
Books and supplies	$ 500	$ –
Other expenses	$ –(A)	$ –
Total estimated expenses	**$ 9,441**	**$ –**

Notes: (A) Data not available.

FINANCIAL AID OFFICE CONTACT
James D. Pritchard, *Director of Financial Aid*
Northern Arizona University
Financial Aid Office
Box 04108
Flagstaff, AZ 86011-4108
Phone: (602)523-4951

GENERAL FINANCIAL AID INFORMATION
Forms accepted/required: FFS, United Student Aid Funds 'Singleform'
Additional financial aid information: Work Study program, part time jobs on campus, federal and state grants, merit and need-based scholarships

ENGINEERING COLLEGE INFORMATION
[For additional personnel, refer to the *Appendix*.]
HEAD OF ENGINEERING
Spencer L. Brinkerhoff
Phone: (602)523-2880   Fax: (602)523-2300
ENGINEERING COLLEGE ADDRESS
College of Engineering and Technology
Northern Arizona University
Box 15600
Flagstaff, AZ 86011-1560
Phone: (602)523-5252   Fax: (602)523-2300

ENROLLMENTS—BY CLASS
[Numbers are baccalaureate enrollments for the fall 1991 term, unless otherwise footnoted.]
1st-year students/Freshmen 288
2nd-year students/Sophomores 194
3rd-year students/Juniors 160
4th-year students/Seniors 212
Total **854**

NUMBER OF DEGREES AWARDED—BY PROGRAM
[Numbers are engineering baccalaureate degrees awarded during the 1991-1992 academic year, unless otherwise footnoted. For full details about each engineering program, refer to the *Tables of Degree Programs*.]
Civil Engineering 8
Computer Science & Engineering 17
Electrical Engineering 28
Mechanical Engineering 21
Total **74**

PERCENTAGE OF DEGREES AWARDED—BY CATEGORY
[Percentages are of all engineering baccalaureate degrees awarded during the 1991-1992 academic year, unless otherwise footnoted.]
African Americans .... – %
Asian/Pacific Island Americans 4.1%
Hispanic Americans 2.7%
Native Americans 1.4%
Foreign Citizens 24.3%
All Others 67.5%
Women 9.5%
Persons with disabilities .... – %
Students over 25 years of age 41.9%

FULL- & PART-TIME FACULTY—BY DEPARTMENT
[Figures are the head count of full-time faculty and the full-time equivalent (FTE) of part-time faculty for each engineering department or equivalent.]

Department	Full	Part
Civil Eng	6	0.5
Electrical Eng	9	–
Mechanical Eng	7	0.2
Computer Science & Eng	8	–

COLLEGE DESCRIPTION
Dedicated to undergraduate instruction, the College of Engineering and Technology places great emphasis on student-faculty interaction. Virtually all lecture and laboratory classes, at all academic levels, are taught by full-time faculty members. Most faculty hold PhD degrees and have industrial experience. The emphasis is on engineering design and all students must complete a team oriented, capstone design project.

DEGREE PROGRAMS DESCRIPTION
Programs leading to the B.S. in Engineering are offered with majors in Civil, Electrical, and Mechanical Engineering. The programs are built upon a strong foundation of basic science and mathematics, supplemented by an extensive core of general engineering topics. Individual degree programs are characterized by a requirement that students study all the major sub-fields of their disciplines and a curriculum structure that builds increasing competence in design, capped by a required design project. The B.S. in Computer Science and Engineering program shares these characteristics and is further distinguished by a requirement that all students receive a thorough grounding in the topics traditionally associated with computer science and those associated with computer engineering.

TRANSFER INFORMATION
Residency Requirements: The University requires at least 30 hours in residence. In addition, the College requires transfer students to complete at least 21 hours of required upper division courses (exclusive of liberal studies) at NAU, including at least 15 hours of upper division courses from the College of Engineering and Technology

Transfer via Articulation Agreements
Admission to engineering: At any time
Engineering graduates transferred from ... 4-yr: 5   2-yr: 14
Requirements: Transfer students with fewer than 12 hours must meet freshman admission standards. Transfer students with 12 or more hours must have a minimum GPA of 2.00 (AZ residents) or 2.50 (non-residents) for unconditional admission. Transfer students must satisfy engineering admission requirements to take courses at the sophomore level.

Transfer without Articulation Agreements
Admission to engineering: At any time
Engineering graduates transferred from ... 4-yr: 2   2-yr: 4

GRADUATION REQUIREMENTS
(1) Completion of a specified program of study totaling 136 semester hours, with a minimum cumulative grade point average of 2.00. (2) Minimum cumulative grade point average of 2.25 in engineering and computer science & engineering classes. (3) No more than two grades of D in mathematics, science, engineering and computer science & engineering courses.

STUDENT PROGRAMS
PROFESSIONAL AND HONORARY SOCIETIES
[For key to acronyms, see *Introduction*.]
Professional Societies: ACM, AIAA, ASCE, ASME, IEEE, SAE
Honorary Societies: Tau Alpha Pi, Tau Beta Pi, Phi Kappa Phi

SUPPORT PROGRAMS
Student Chapter Organizations: Society of Women Engineers, Society of Hispanic Professional Engineers, American Indian Science and Engineering Society

Native American Engineering Education Program
For: Ethnic Minorities   **Available:** Year round
Offered: Prematriculation, Freshman, Sophomore, Junior, Senior

Hispanic Engineering Education Program
For: Ethnic Minorities   **Available:** Year round
Offered: Prematriculation, Freshman, Sophomore, Junior, Senior

University Multicultural Center
For: Ethnic Minorities   **Available:** Academic year
Offered: Freshman, Sophomore, Junior, Senior

Other Engineering Support Programs: The University offers tutoring and special assistance throughout the Learning Assistance Center. Tau Beta Pi volunteers conduct engineering-specific tutoring sessions.

165 Northern Illinois University

■ INSTITUTION PROFILE

HEAD OF THE INSTITUTION
John E. LaTourette
Phone: (815)753-9501 **Fax:** (815)753-8686

GENERAL INFORMATION
[All Students—Fall 1991]
Undergraduate enrollment 18,220
Graduate enrollment 6,675
Total institution enrollment **24,895**
Type of institution: Public   **Calendar:** Semesters
Nearest city: Chicago   **Population:** 3,000,000
Miles from main campus: 60   **Setting:** Small Town
Types of engineering degrees: Engineering & Technology
Other degree-granting colleges: Arts & Sciences, Education, Law, Professional Studies, Continuing Education, Graduate School, Visual & Performing Arts, Business

■ ENGINEERING ADMISSIONS

ADMISSIONS OFFICE CONTACT
Joy M. Pauschke
Northern Illinois University
Associate Dean
College of Engineering & Eng. Technology
DeKalb, IL 60115
Phone: (815)753-1442 **Fax:** (815)753-0362

ENGINEERING COLLEGE ADMISSIONS INFORMATION
Entrance Requirements: ACT (Score: 19), High School courses—English (3 years), College preparatory math (2 years), Physical or biological sciences (2 years), Social Science (2 years), Minimum ACT scores are 19 or rank in upper 1/2 of class or 23 and rank in upper 2/3 of class.
Entrance Recommendations: High School courses—English (4 years), College preparatory math (3 years), Physical or biological sciences (3 years)
Requirements for foreign students: TOEFL (Score: 550); financial statement; International students with fewer than 45 semester hours of college credit must rank in the equivalent of the upper one-third of their high school graduating class. Must submit letters of recommendation, all transcripts or national examination scores with English translation.

DEPARTMENTAL ADMISSIONS INFORMATION
Admission to the engineering department: End of third semester
Additional information: 1. Completion of minimum of 41 semester hours of course work with minimum 2.30 GPA on a 4.00 scale. 2. Completion of the following courses with a minimum of 2.30 GPA English 103, English 104, Calculus 1 and 2, Engineering Mechanics 1, and General Physics. 3. An applicant must meet deadlines as stated in undergraduate catalog. Admission is competitive and will be based primarily on the GPA in the prescribed courses and secondarily on overall GPA earned, whether in work taken at NIU or transferred from other institutions.

■ FINANCIAL INFORMATION

ESTIMATED EXPENSES (ACADEMIC YEAR)
[Expenses are for the 1992-93 nine-month academic year.]

State Residents	Undergraduate	Graduate
Tuition and fees	$ 2,647 [1]	$ 2,677
College room and board	$ 2,870	$ 2,870
Books and supplies	$ 520	$ 520
Other expenses	$ 1,460 [2]	$ 1,460 [2]
Total estimated expenses	**$ 7,497**	**$ 7,527**
Out-of-State Residents	Undergraduate	Graduate
Tuition and fees	$ 6,247	$ 6,324
College room and board	$ 2,870	$ 2,870
Books and supplies	$ 520	$ 520
Other expenses	$ 1,460 [2]	$ 1,460 [2]
Total estimated expenses	**$11,097**	**$11,174**

Notes: (1) Tuition is assessed on a graduated scale. Full time = twelve hours ($2647), 13 hrs. = $2894; 14 hrs. = $2960; 15 hrs. = $3026; 16 hrs. = $3092.
(2) Transportation and personal expenses.

FINANCIAL AID OFFICE CONTACT
Jerry D. Augsburger, *Director, Student Financial Aid*
Northern Illinois University
Swen Parson 245
Student Financial Aid
DeKalb, IL 60115
Phone: (815)753-1300

GENERAL FINANCIAL AID INFORMATION
Forms accepted/required: FAF, FAT, Institutional
Additional financial aid information: Scholarships: Gift assistance usually based on academic achievement, major, and/or special ability. Grants: Gift assistance usually based on financial need. Loans: Funds to be repaid with interest. Employment: Earnings from a part-time job on- or off-campus.

■ ENGINEERING COLLEGE INFORMATION

[For additional personnel, refer to the *Appendix*.]

HEAD OF ENGINEERING
Romualdas Kasuba
Phone: (815)753-1283 **Fax:** (815)753-1310
EMail: kasuba@ceet.niu.edu

ENGINEERING COLLEGE ADDRESS
College of Engineering and Engineering Technology
Northern Illinois University
Engineering Building
DeKalb, IL 60115
Phone: (815)753-1442 **Fax:** (815)753-0362

ENROLLMENTS—BY CLASS
[Numbers are baccalaureate enrollments for the fall 1991 term, unless otherwise footnoted.]
1st-year students/Freshmen 254
2nd-year students/Sophomores 88
3rd-year students/Juniors 113
4th-year students/Seniors 155
Total **610**

NUMBER OF DEGREES AWARDED—BY PROGRAM
[Numbers are engineering baccalaureate degrees awarded during the 1991-1992 academic year, unless otherwise footnoted. For full details about each engineering program, refer to the *Tables of Degree Programs*.]
Electrical Engineering 41
Engineering Technology 40
Industrial Engineering 8
Mechanical Engineering 31
Total **120**

PERCENTAGE OF DEGREES AWARDED—BY CATEGORY
[Percentages are of all engineering baccalaureate degrees awarded during the 1991-1992 academic year, unless otherwise footnoted.]
African Americans .... 2.9 %
Asian/Pacific Island Americans .... 7.1 %
Hispanic Americans .... 1.4 %
Native Americans .... – %
Foreign Citizens .... 8.9 %
All Others .... 79.7 %
Women .... 12.9 %
Persons with disabilities .... – % (A)
Students over 25 years of age .... 27.1 %

Notes: (A) Data not available.

GRADUATE ENROLLMENTS & DEGREES AWARDED
Master's enrollment 98
Master's degrees awarded 30
Doctoral enrollment .... – (B)
Doctoral degrees awarded .... – (B)

Notes: (B) Data not applicable.

ENGINEERING STUDENT DATA
Applicants to the engineering college
Number of applicants to engineering college 623
Percent offered admission 76.9%
Matriculated engineering students
Percentage in top quartile (25%) of High School class 43.5
Average ACT scores: Composite—24

FULL- & PART-TIME FACULTY—BY DEPARTMENT

[Figures are the head count of full-time faculty and the full-time equivalent (FTE) of part-time faculty for each engineering department or equivalent.]

Department	Full	Part
Electrical Eng	13	–
Industrial Eng	6	–
Mechanical Eng	10	1.0
Tech	18	1.0

COLLEGE DESCRIPTION
The College of Engineering and Engineering Technology, established in 1985, has four departments: Electrical Engineering, Industrial Engineering, Mechanical Engineering, and Technology. All departments offer B.S. and M.S. programs. Because of its youth, the college focuses on the future by incorporating the latest in engineering education philosophies and relevant technological changes into its programs. Majority of the engineering faculty have recent full-time industrial experience which is reflected in the professional portions of the programs. The college has a three-semester pre-engineering program and a limited admission policy into the undergraduate engineering programs. All graduating engineering seniors must take the Engineer In Training or equivalent exam. The cooperative education program is available but not required.

DEGREE PROGRAMS DESCRIPTION
The offered undergraduate programs provide training for entry level professional positions or for graduate work at NIU or other institutions. The Master of Science degree programs are offered in electrical engineering, industrial engineering, mechanical engineering, and industrial management through the Technology Department in cooperation with the College of Business. Numerous companies and research laboratories in the Northern Illinois region cooperate with the college in providing widely ranging senior design and graduate projects.

TRANSFER INFORMATION
Residency Requirements: Candidates for degrees from Northern must earn their last 30 hours of credit in course work offered by Northern Illinois University.

Transfer via Articulation Agreements
Admission to engineering: At any time
Engineering graduates transferred from   4-yr: 8   2-yr: 34
Transfer without Articulation Agreements
Admission to engineering: At any time
GRADUATION REQUIREMENTS
All graduating engineering students must take the Engineering In Training or an equivalent exam.

■ STUDENT PROGRAMS

PROFESSIONAL AND HONORARY SOCIETIES
[For key to acronyms, see *Introduction*.]
Professional Societies: ASHRAE, ASME, IEEE, IIE, SME, SPE*, International Society of Hybrid Microelectronics (ISHM)
Honorary Societies: Eta Kappa Nu

SUPPORT PROGRAMS
Student Chapter Organizations: Society of Women Engineers, National Society of Black Engineers, Society of Hispanic Professional Engineers

Educational Services and Programs
For: Ethnic Minorities   **Available:** Year round
Offered: Freshman, Sophomore, Junior, Senior
Other Engineering Support Programs: ICPS 101 - University Experience - first semester freshman year course - section offered for engineering and technology majors. - summer orientation program part of university program. - mandatory advising - college tutors

166 Northwestern University

INSTITUTION PROFILE

HEAD OF THE INSTITUTION
Arnold R. Weber
Phone: (708)491-7456 Fax: (708)491-8406

GENERAL INFORMATION
[All Students—Fall 1991]
Undergraduate enrollment 7,458
Graduate enrollment 4,417
Total institution enrollment 11,924
Type of institution: Private Calendar: Quarters
Nearest city: Chicago Population: 3,500,000
Miles from main campus: 12 Setting: Suburban
Types of engineering degrees: Engineering
Other degree-granting colleges: Arts & Sciences, Dentistry, Law, Medicine, Journalism, Music, Speech, Education and Social Policy

ENGINEERING ADMISSIONS

ADMISSIONS OFFICE CONTACT
Carol Lunkenheimer
Northwestern University
Office of Admission
1801 Hinman Ave.
Evanston, IL 60204-3060
Phone: (708)491-7271 Fax: (708)467-7317

ENGINEERING COLLEGE ADMISSIONS INFORMATION
Entrance Requirements: High School courses—Mathematics (4 years), Laboratory Science (2 years), SAT or ACT
Entrance Recommendations: High School courses—English (4 years), Foreign Language (3 years), History/Social Studies (3 years), Electives (2 years), CEEB Achievement Tests in English Composition, Math level I or II, and either Physics or Chemistry
Requirements for foreign students: TOEFL (Score: 600); financial statement; outstanding academic record

DEPARTMENTAL ADMISSIONS INFORMATION
Admission to the engineering department: At the time of admission to the institution.

FINANCIAL INFORMATION

ESTIMATED EXPENSES (ACADEMIC YEAR)
[Expenses are for the 1992-93 nine-month academic year.]

All Students	Undergraduate	Graduate
Tuition and fees	$15,075	$15,075
College room and board	$ 6,138	$ —(A)
Books and supplies	$ 651 (A)	$ —(A)
Other expenses	$ —(A)	$ —(A)
Total estimated expenses	**$21,864**	**$15,075**

Notes: (A) Data not available.

FINANCIAL AID OFFICE CONTACT
Carolyn Lindley, *Director of Financial Aid*
Northwestern University
Office of Financial Aid
1801 Hinman Ave.
Evanston, IL 60204-3060
Phone: (708)491-7400

GENERAL FINANCIAL AID INFORMATION
Forms accepted/required: CSS-FAF, Institutional
Additional financial aid information: Need based financial aid, may include grant, loan and part-time employment in any of several possible combinations; industry sponsored merit based scholarships; College Work-Study Program (CWS).

ENGINEERING COLLEGE INFORMATION
[For additional personnel, refer to the *Appendix*.]

HEAD OF ENGINEERING
Jerome B. Cohen
Phone: (708)491-5220 Fax: (708)491-8539
EMail: jbc@ccadmin.tech.nwu.edu

ENGINEERING COLLEGE ADDRESS
McCormick School of Engineering and Applied Science
Northwestern University
2145 Sheridan Rd.
Evanston, IL 60208-3102
Phone: (708)491-7379 Fax: (708)491-8539

ENROLLMENTS—BY CLASS
[Numbers are baccalaureate enrollments for the fall 1991 term, unless otherwise footnoted.]
1st-year students/Freshmen 350
2nd-year students/Sophomores 282
3rd-year students/Juniors 291
4th-year students/Seniors 299 (1)
Total 1,222

Notes: (1) An additional 40 are fifth year students in cooperative education program.

NUMBER OF DEGREES AWARDED—BY PROGRAM
[Numbers are engineering baccalaureate degrees awarded during the 1991-1992 academic year, unless otherwise footnoted. For full details about each engineering program, refer to the *Tables of Degree Programs*.]
Applied Mathematics 2
Biomedical Engineering 36
Chemical Engineering 38
Civil Engineering 12
Computer Science 24
Electrical Engineering 50
Environmental Engineering 7
Industrial Engineering 65
Materials Science & Engineering 15
Mechanical Engineering 32
Total 281

PERCENTAGE OF DEGREES AWARDED—BY CATEGORY
[Percentages are of all engineering baccalaureate degrees awarded during the 1991-1992 academic year, unless otherwise footnoted.]
African Americans .... 8.2 %
Asian/Pacific Island Americans .... 16.7 %
Hispanic Americans .... 1.1 %
Native Americans .... — %
Foreign Citizens .... 8.9 %
All Others .... 65.1 %
Women .... 23.4 %
Persons with disabilities .... — % (A)
Students over 25 years of age .... — % (A)

Notes: (A) Data not available.

GRADUATE ENROLLMENTS & DEGREES AWARDED
Master's enrollment 326
Master's degrees awarded 204
Doctoral enrollment 687
Doctoral degrees awarded 109

ENGINEERING STUDENT DATA

Applicants to the engineering college
Number of applicants to engineering college 1,862
Percent offered admission 62.6%

Matriculated engineering students
Percentage in top quartile (25%) of High School class 97.3
Average SAT scores: Math—565, Verbal—692, Combined—1257
Average ACT scores: Composite—29.1

FULL- & PART-TIME FACULTY—BY DEPARTMENT
[Figures are the head count of full-time faculty and the full-time equivalent (FTE) of part-time faculty for each engineering department or equivalent.]

Department	Full	Part
Chemical Eng	15	—
Civil Eng	22	1.8
Electrical Eng & Computer Science	43	—
Industrial Eng	14	2.3
Mechanical Eng	17	—
Biomedical Eng	8	1.1
Eng Science & Applied Mathematics	11	—
Materials Science & Eng	17	0.5

COLLEGE DESCRIPTION

The undergraduate curriculum is intended to provide the maximum flexibility consistent with the development of a sound professional preparation. The basic structure presented in all fields of engineering prescribes a distribution of studies: 38% in basic math, science and eng. science; 33% in major field courses; and 29% in social sciences, humanities and unrestricted electives. Flexibility enables honors programs in which students with outstanding credentials can be admitted directly from high school to undergraduate/graduate studies include the Honors Medical Program - U.G. Eng.:Medicine; Honors Research Program - U.G. Eng./Grad. School; and Honors Eng./Management Program - U.G. Eng./Management. Other combined programs of study lead to the award of dual degrees: BS-DDS(Biomedical Eng./Dental School); BS/BA (Eng./Arts and Science), BS-BM (Eng./Music) and a combined bachelor's program between engineering and journalism. Co-op is an option in all fields of engineering study.

DEGREE PROGRAMS DESCRIPTION

All B.S. curricula in the McCormick School follow a common pattern and a common goal to prepare for both entry-level professional positions and for study at the post-baccalaureate level. The five year Co-op program adds a more practice oriented dimension to this educational experience. All departments offer not only the B.S. degree, also the M.S. and Ph.D. Although some M.S. degrees may not require a thesis, most M.S. degrees and all Ph.D. are strongly research oriented and require a thesis based on substantive original investigation. Two programs are offered at the Masters level; Master of Engineering Management and Master of Manufacturing Engineering. In addition, a Masters of Manufacturing Management is offered jointly with the Kellogg Graduate School of Management.

DUAL DEGREE PROGRAMS

Engineering baccalaureate graduates with dual degrees 8

TRANSFER INFORMATION

Residency Requirements: Two years residence requirement for all students.

Transfer without Articulation Agreements
Admission to engineering: At the end of the freshman year
Engineering graduates transferred from ... 4-yr: 16 2-yr: 2
Requirements: completed freshman year, maintained a satisfactory scholastic record (a competitive average is B)

GRADUATION REQUIREMENTS

All B.S. degree programs in engineering require 48 quarter course units (192 quarter hours) with a grade point average of C or better in all work presented for the degree.

STUDENT PROGRAMS

PROFESSIONAL AND HONORARY SOCIETIES
[For key to acronyms, see *Introduction*.]
Professional Societies: AIChE, ASCE, ASME, BMES, IEEE, IIE, SAE, American Society for Metals
Honorary Societies: Eta Kappa Nu, Omega Chi Epsilon, Pi Tau Sigma, Tau Beta Pi, Phi Lambda Upsilon, Sigma Xi

SUPPORT PROGRAMS
Student Chapter Organizations: Society of Women Engineers, National Society of Black Engineers, Society of Hispanic Professional Engineers

Minority Engineering Opportunity Program
For: Ethnic Minorities **Available:** Year round
Offered: Prematriculation, Freshman, Sophomore, Junior, Senior

Office of African-American Student Affairs
For: Ethnic Minorities **Available:** Year round
Offered: Freshman, Sophomore, Junior, Senior

Other Engineering Support Programs: New Student Week orientation for all freshman and transfer students with option of summer component for those who can attend, freshman program, tutoring program, counseling offices in engineering, University offices for counseling and guidance, foreign students, veterans, and placement.

167 Norwich University

INSTITUTION PROFILE

HEAD OF THE INSTITUTION
Richard W. Schneider
Phone: (802)485-2065 Fax: (802)485-2580

GENERAL INFORMATION
[All Students—Fall 1991]
Undergraduate enrollment 1,450
Total institution enrollment 1,450
Type of institution: Private **Calendar:** Semesters
Nearest city: Burlington, Vermont **Population:** 40,000
Miles from main campus: 50 **Setting:** Small Town
Types of engineering degrees: Engineering & Technology
Other degree-granting colleges: Architecture, Business Administration, Communication & Journalism, Education, Fine & Performing Arts, Humanities & Social Sciences, Nursing

ENGINEERING ADMISSIONS

ADMISSIONS OFFICE CONTACT
Frank E. Griffis
Norwich University
Main Street
Northfield, VT 05663
Phone: (802)485-2000 Fax: (802)485-2580

ENGINEERING COLLEGE ADMISSIONS INFORMATION
Entrance Requirements: SAT, ACT, High School courses—English (4 years), Laboratory (3 years), Mathematics (4 years)
Entrance Recommendations: High School courses—Advanced mathematics (1 year)

DEPARTMENTAL ADMISSIONS INFORMATION
Admission to the engineering department: At the time of admission to the institution.

FINANCIAL INFORMATION

ESTIMATED EXPENSES (ACADEMIC YEAR)
[Expenses are for the 1992-93 nine-month academic year.]

All Students	Undergraduate	Graduate
Tuition and fees	$12,214	$ –
College room and board	$ 4,776	$ –
Books and supplies	$ 400	$ –
Other expenses	$ –	$ –
Total estimated expenses	$17,390	$ –

FINANCIAL AID OFFICE CONTACT
Karen Waring, *Financial Aid Director*
Norwich University
Financial Aid Office
Northfield, VT 05663
Phone: (802)485-2015 Fax: (802)485-2580

GENERAL FINANCIAL AID INFORMATION
Forms accepted/required: FAF
Additional financial aid information: Student financial aid is based upon need and qualified students are admitted without regard to need. Student aid packages include work-study, loans and grants.

ENGINEERING COLLEGE INFORMATION
[For additional personnel, refer to the *Appendix*.]

HEAD OF ENGINEERING
Eugene Sevi
Phone: (802)485-2275 Fax: (802)485-2580
EMail: Sevi

ENGINEERING COLLEGE ADDRESS
Engineering, Architecture and Technology Division
Norwich University
Main Street
Northfield, VT 05663
Phone: (802)485-2256 Fax: (802)485-2580

ENROLLMENTS—BY CLASS
[Numbers are baccalaureate enrollments for the fall 1991 term, unless otherwise footnoted.]
1st-year students/Freshmen 111
2nd-year students/Sophomores 74
3rd-year students/Juniors 50
4th-year students/Seniors 48
Total 283

NUMBER OF DEGREES AWARDED—BY PROGRAM
[Numbers are engineering baccalaureate degrees awarded during the 1991-1992 academic year, unless otherwise footnoted. For full details about each engineering program, refer to the *Tables of Degree Programs*.]
Architecture .... –
Civil Engineering 23
Computer Engineering & Computer Science .... –
Electrical Engineering .... –
Environmental Engineering Technology .... –
Mechanical Engineering .... –
Total 23

PERCENTAGE OF DEGREES AWARDED—BY CATEGORY
[Percentages are of all engineering baccalaureate degrees awarded during the 1991-1992 academic year, unless otherwise footnoted.]
African Americans 1.0%
Asian/Pacific Island Americans 3.2%
Hispanic Americans 5.0%
Native Americans .... – %
Foreign Citizens 4.2%
All Others 86.6%
Women 7.0%
Persons with disabilities .... – %
Students over 25 years of age 1.0%

ENGINEERING STUDENT DATA
Applicants to the engineering college
Number of applicants to engineering college 498
Percent offered admission 85.0%

Matriculated engineering students
Percentage in top quartile (25%) of High School class 31.0
Average SAT scores: Math—780, Verbal—420, Combined—1200
Average ACT scores: Math—18.5, Composite—22

FULL- & PART-TIME FACULTY—BY DEPARTMENT
[Figures are the head count of full-time faculty and the full-time equivalent (FTE) of part-time faculty for each engineering department or equivalent.]

Department	Full	Part
Civil Eng	4	–
Electrical & Computer Eng	7	–
Mechanical Eng	5	–
Environmental Eng Tech	2.5	–
Architecture	6	2.5

COLLEGE DESCRIPTION
Norwich University has two campuses. One is a military college and the other is a traditional civilian college. At each college leadership training is emphasized and students can select from a Peace Corps and ROTC programs. Norwich offers small classes and individual attention with a faculty that is selected for its professionalism, ability to communicate and desire to teach. Norwich is the first school in the country to offer engineering education and continues that tradition with excellent placement of its graduates in industry, government and graduate school.

DEGREE PROGRAMS DESCRIPTION
The Division of Engineering, Architecture and Technology is comprised of the Departments of: Civil Engineering, Computer Science and Engineering, Electrical Engineering, Engineering Technology (Environmental), Mechanical Engineering, and Architecture. The Head of the Division of Engineering, Architecture and Technology is Professor Eugene A. Sevi, M.S. Degrees Offered: Bachelor of Science in Civil Engineering, Computer Science, Computer Engineering, Electrical Engineering, Mechanical Engineering, Engineering Technology (Environmental), and Bachelor of Architecture. Course Offerings: All Engineering and Technology courses are offered on the Military College campus while all Architecture courses are offered on the Vermont College campus. Freshman and sophomore years are common to all engineering students with the exception of one course in the second semester of the sophomore year. Junior and senior year listings are found under Engineering and Technology curricula headings. A student who has completed all degree requirements except for attaining a 2.00 average must take at least 50 percent of all subsequent course work in technical material (subject to approval by the Division Head).

TRANSFER INFORMATION
Residency Requirements: There are no formal residency requirements, but students must complete their last two years at Norwich University.

Transfer via Articulation Agreements
Admission to engineering: At any time
Engineering graduates transferred from ... 4-yr: 2 2-yr: 4
Transfer without Articulation Agreements
Admission to engineering: At any time
Engineering graduates transferred from ... 4-yr: – 2-yr: 1

GRADUATION REQUIREMENTS
Students are required to have a 2.00 GPA to graduate.

STUDENT PROGRAMS

PROFESSIONAL AND HONORARY SOCIETIES
[For key to acronyms, see *Introduction*.]
Professional Societies: ASCE, ASME, IEEE, SAME, Air and Waste Management Association
Honorary Societies: Chi Epsilon, Eta Kappa Nu, Tau Alpha Pi, Tau Beta Pi

SUPPORT PROGRAMS
Student Chapter Organizations: Norwich treats all people equally.
Other Engineering Support Programs: Norwich's small classes allow for personal instruction and individual assistance. The teaching faculty can often make the difference for a student who is experiencing difficulty.

168 University of Notre Dame

■ INSTITUTION PROFILE
HEAD OF THE INSTITUTION
Edward A. Malloy
Phone: (219)239-6755   Fax: (219)239-7428

GENERAL INFORMATION
[All Students—Fall 1991]
Undergraduate enrollment 7,623
Graduate enrollment 2,462
Total institution enrollment 10,085
Type of institution: Private   Calendar: Semesters
Nearest city: South Bend, IN   Population: 100,000
Miles from main campus: 2   Setting: Suburban
Types of engineering degrees: Engineering
Other degree-granting colleges: Business Administration, Law, Arts and Letters, Science, Freshman Year of Studies

■ ENGINEERING ADMISSIONS
ADMISSIONS OFFICE CONTACT
Kevin M. Rooney
University of Notre Dame
Director of Admissions
Notre Dame, IN 46556-5602
Phone: (219)239-7505

ENGINEERING COLLEGE ADMISSIONS INFORMATION
Admission to the engineering college: Students are admitted to Notre Dame in the Freshman Year of Studies. They enter the College of Engineering as sophomores.
Entrance Requirements: SAT, High School courses—English (4 years), Algebra and geometry (3 years), Calculus or pre-calculus (1 year), Foreign Language (2 years), SAT is required but no minimum score is specified.
Entrance Recommendations: High School courses—History (1 year), Chemistry (1 year), Physics (1 year), Additional English, history, language or science (3 years)
Requirements for foreign students: TOEFL; financial statement; TOEFL is required but no minimum score is specified.

DEPARTMENTAL ADMISSIONS INFORMATION
Admission to the engineering department: At the end of the first year.

■ FINANCIAL INFORMATION
ESTIMATED EXPENSES (ACADEMIC YEAR)
[Expenses are for the 1992-93 nine-month academic year.]

All Students	Undergraduate	Graduate
Tuition and fees	$14,650	$14,530
College room and board	$3,790	$3,900
Books and supplies	$550	$500
Other expenses	$1,660(C)	$1,570(C)
Total estimated expenses	$20,650	$20,500

Notes: (C) Estimated data.

FINANCIAL AID OFFICE CONTACT
Joseph A. Russo, *Director of Financial Aid*
University of Notre Dame
Administration Building
Notre Dame, IN 46556-5602
Phone: (219)239-6436

GENERAL FINANCIAL AID INFORMATION
Forms accepted/required: FAF, IRS
Additional financial aid information: Need based scholarships and other forms of financial aid including loans and work study.

■ ENGINEERING COLLEGE INFORMATION
[For additional personnel, refer to the *Appendix*.]
HEAD OF ENGINEERING
Anthony N. Michel
Phone: (219)239-5534   Fax: (219)239-8007

ENGINEERING COLLEGE ADDRESS
College of Engineering
University of Notre Dame
Fitzpatrick Hall of Engineering
Notre Dame, IN 46556-5637
Phone: (219)239-5530   Fax: (219)239-8007

ENROLLMENTS—BY CLASS
[Numbers are baccalaureate enrollments for the fall 1991 term, unless otherwise footnoted.]
1st-year students/Freshmen .... 368 [1]
2nd-year students/Sophomores 283
3rd-year students/Juniors 277
4th-year students/Seniors .... 241 [2]
Total 1,169

Notes: (1) Reported as the number of engineering intents in the Freshman Year of Studies (2) Includes 19 fifth year students

NUMBER OF DEGREES AWARDED—BY PROGRAM
[Numbers are engineering baccalaureate degrees awarded during the 1991-1992 academic year, unless otherwise footnoted. For full details about each engineering program, refer to the *Tables of Degree Programs*.]
Aerospace & Mechanical Engineering 43
Chemical Engineering 26
Civil Engineering 21
Computer Engineering .... –
Electrical Engineering 50
Materials Science & Engineering 4
Mechanical Engineering 58
Total 202

PERCENTAGE OF DEGREES AWARDED—BY CATEGORY
[Percentages are of all engineering baccalaureate degrees awarded during the 1991-1992 academic year, unless otherwise footnoted.]
African Americans 0.5%
Asian/Pacific Island Americans 3.0%
Hispanic Americans 3.0%
Native Americans .... – %
Foreign Citizens 6.4%
All Others 87.1%
Women 18.8%
Persons with disabilities .... – %
Students over 25 years of age .... – %

GRADUATE ENROLLMENTS & DEGREES AWARDED
Master's enrollment 61
Master's degrees awarded 51
Doctoral enrollment 176
Doctoral degrees awarded 18

ENGINEERING STUDENT DATA
Applicants to the engineering college
Number of applicants to engineering college 861
Percent offered admission 42.7%
Matriculated engineering students
Percentage in top quartile (25%) of High School class 95.0
Average SAT scores: Math—665, Verbal—566, Combined—1231

FULL- & PART-TIME FACULTY—BY DEPARTMENT
[Figures are the head count of full-time faculty and the full-time equivalent (FTE) of part-time faculty for each engineering department or equivalent.]

Department	Full	Part
Chemical Eng	11	–
Civil Eng & Geological Sciences	18	0.5
Electrical Eng	23	1.0
Aerospace & Mechanical Eng	27	1.0
Computer Science & Eng	9	1.5
School of Architecture	12	3.0

COLLEGE DESCRIPTION
Engineering programs were initiated at Notre Dame in 1873, the first under Catholic auspices in the United States. Notre Dame seeks to provide quality engineering education which emphasizes the fundamentals of mathematics, science and engineering science in a highly residential and value centered environment. Programs are available in which a student may earn a Bachelor of Arts and a Bachelor of Science in an engineering discipline in a five year course of study. Foreign study opportunities are available in this program. Students in four year engineering programs may study in London in a summer session. A five year BS engineering and MBA program is available to those who qualify and are accepted.

DEGREE PROGRAMS DESCRIPTION
All students must complete course distribution requirements for the bachelor's degree. These include courses in English, history, social science, literature or fine arts, philosophy and theology. Graduate programs emphasize study at the doctoral level.

DUAL DEGREE PROGRAMS
Engineering baccalaureate graduates with dual degrees 19

TRANSFER INFORMATION
Residency Requirements: All students must be in residence a minimum of two years, including the last year.
Transfer via Articulation Agreements
Admission to engineering: At the end of the sophomore year
Engineering graduates transferred from ... 4-yr: 11   2-yr: 2
Transfer without Articulation Agreements
Admission to engineering: At the end of the freshman year
Engineering graduates transferred from ... 4-yr: 2   2-yr: –

GRADUATION REQUIREMENTS
To be eligible for a degree, each student must complete the specified curricula and have a minimum cumulative grade point average of 2.00 on a 4 point scale. The last year of study must be in residence.

■ STUDENT PROGRAMS
PROFESSIONAL AND HONORARY SOCIETIES
[For key to acronyms, see *Introduction*.]
Professional Societies: AIA, AIAA, AIChE, ASCE, ASME, IEEE, SAME, TMS
Honorary Societies: Chi Epsilon, Eta Kappa Nu, Pi Tau Sigma, Sigma Gamma Tau, Tau Beta Pi, Tau Sigma Delta, Alpha Sigma Mu

SUPPORT PROGRAMS
Student Chapter Organizations: Society of Women Engineers, National Society of Black Engineers
Minority Engineering Program
For: Ethnic Minorities   **Available:** Year round
Offered: Freshman, Sophomore, Junior, Senior
Other Engineering Support Programs: Undergraduate students in the College of Engineering publish a quarterly magazine, 'The Technical Review'. Students are represented on the Engineering College Council as well as on the University Academic Council. The Joint Engineering Council is the umbrella student organization who sponsors a career fair each year.

169 Oakland University

■ INSTITUTION PROFILE

HEAD OF THE INSTITUTION
Sandra P. Parckard
Phone: (313)370-3500 Fax: (313)370-3504

GENERAL INFORMATION
[All Students—Fall 1991]
Undergraduate enrollment 10,016
Graduate enrollment 2,514
Total institution enrollment **12,530**
Type of institution: Public   Calendar: Semesters
Nearest city: Rochester   Population: 60,000
Miles from main campus: 4   Setting: Suburban
Types of engineering degrees: Engineering
Other degree-granting colleges: Allied Health Sciences, Arts & Sciences, Business Administration, Communication & Journalism, Education, Fine & Performing Arts, Humanities & Social Sciences, Nursing

■ ENGINEERING ADMISSIONS

ADMISSIONS OFFICE CONTACT
J. W. Rose
Oakland University
Director of Admissions and Scholarships
Rochester, MI 48309-4401
Phone: (313)370-3360 Fax: (313)370-2286

ENGINEERING COLLEGE ADMISSIONS INFORMATION
Entrance Requirements: ACT, High School courses—Mathematics (4 years), English (4 years), Physical or Life Sciences (3 years)
Entrance Recommendations: SAT, ACT, High School courses—Drafting, Computer Programming, Machine Shop Practice
Requirements for foreign students: TOEFL; financial statement

DEPARTMENTAL ADMISSIONS INFORMATION
Admission to the engineering department: admission to the institution. Transfer into a major of choice, after admission into the engineering or computer science, is generally available.

■ FINANCIAL INFORMATION

ESTIMATED EXPENSES (ACADEMIC YEAR)
[Expenses are for the 1992-93 nine-month academic year.]

State Residents	Undergraduate	Graduate
Tuition and fees	$ 2,456	$ 3,576
College room and board	$ 3,715	$ 3,715
Books and supplies	$ 400	$ 400
Other expenses	$ 400[1]	$ 400[1]
Total estimated expenses	**$ 6,971**	**$ 8,091**

Out-of-State Residents	Undergraduate	Graduate
Tuition and fees	$ 7,488	$ 7,920
College room and board	$ 3,715	$ 3,715
Books and supplies	$ 400	$ 400
Other expenses	$ 400[1]	$ 400[1]
Total estimated expenses	**$12,003**	**$12,435**

Notes: (1) Course and other fees.

FINANCIAL AID OFFICE CONTACT
Lee Anderson, *Director of Financial Aid Programs*
Oakland University
161 North Foundation
Rochester, MI 48309-4401
Phone: (313)370-3370 Fax: (313)370-2286

GENERAL FINANCIAL AID INFORMATION
Forms accepted/required: FAF, Stafford, SAR
Additional financial aid information: Need-based financial aid; short-term loans; merit-based scholarships; part-time jobs on campus; College Work-Study Program, Pell Grants, Perkins Loan Program, and other University and State programs. Required forms: FAF, IRS, institutional form.

■ ENGINEERING COLLEGE INFORMATION

[For additional personnel, refer to the *Appendix*.]

HEAD OF ENGINEERING
Howard R. Whitt
Phone: (313)370-2217 Fax: (313)370-4261
EMail: witt@argo.acs.oakland.edu

ENGINEERING COLLEGE ADDRESS
School of Engineering and Computer Science
Oakland University
Dodge Hall of Engineering
Rochester, MI 48309-4401
Phone: (313)370-2217 Fax: (313)370-4261

ENROLLMENTS—BY CLASS
[Numbers are baccalaureate enrollments for the fall 1991 term, unless otherwise footnoted.]
1st-year students/Freshmen 208
2nd-year students/Sophomores 169
3rd-year students/Juniors 209
4th-year students/Seniors 278
Total 864

NUMBER OF DEGREES AWARDED—BY PROGRAM
[Numbers are engineering baccalaureate degrees awarded during the 1991-1992 academic year, unless otherwise footnoted. For full details about each engineering program, refer to the *Tables of Degree Programs*.]
Computer Engineering 18
Computer Science 26
Electrical Engineering 49
Interdisciplinary Doctoral Program .... –
Mechanical Engineering 33
Systems Engineering 4
Total 130

PERCENTAGE OF DEGREES AWARDED—BY CATEGORY
[Percentages are of all engineering baccalaureate degrees awarded during the 1991-1992 academic year, unless otherwise footnoted.]
African Americans .... – %
Asian/Pacific Island Americans .... – %
Hispanic Americans .... – %
Native Americans .... – %
Foreign Citizens .... – %
All Others .... 100 %
Women .... – %
Persons with disabilities .... – %
Students over 25 years of age .... – %

GRADUATE ENROLLMENTS & DEGREES AWARDED
Master's enrollment 321
Master's degrees awarded 78
Doctoral enrollment 46
Doctoral degrees awarded 8

ENGINEERING STUDENT DATA
Average ACT scores: Composite—25

FULL- & PART-TIME FACULTY—BY DEPARTMENT
[Figures are the head count of full-time faculty and the full-time equivalent (FTE) of part-time faculty for each engineering department or equivalent.]

Department	Full	Part
Mechanical Eng	10	1.5
Computer Science & Eng	12	3.0
Electrical & Systems Eng	14	2.4

COLLEGE DESCRIPTION
The school offers four year undergraduate programs in computer, electrical, mechanical and systems engineering, emphasizing math, science, and engineering fundamentals in the first year, followed by about three years of specialized training, with some liberal arts studies throughout all four years. Emphasis is on computer-aided engineering. The program includes a comprehensive engineering core. Most courses have a laboratory associated with them providing a 'real world', hands-on experience to the students. Also, engineering design projects are required in a number of courses. The school also offers a B.S. degree in computer science and B.S. degrees in engineering physics and chemistry jointly with the College of Arts and Sciences. Co-op programs are available in all majors and will take approximately an additional year to complete the degree. The school also has strong commitment to graduate study and research

DEGREE PROGRAMS DESCRIPTION
The B.S. in engineering with a major in Mechanical, Electrical, Computer or Systems engineering and B.S. in computer science program prepare the graduate for entry-level professional positions or for graduate study in each of the respective fields. Masters study can be carried out in one of four areas--Computer Science and Engineering, Electrical and Computer Engineering, Mechanical Engineering, and Systems Engineering. The Ph.D. in Systems Engineering involves a blending of various disciplines and is well suited for studies in areas of robotics, electronics, communications, mechanics, manufacturing systems, fluid and thermal systems, dynamic systems and control, computer hardware and software systems, artificial intelligence and expert systems. The school is committed to providing a high quality education at the undergraduate level as well as masters and doctoral levels.

TRANSFER INFORMATION
Residency Requirements: A student must complete at least 32 semester credit hours at Oakland University of 300 or above level courses including at least 24 credits in engineering subjects required for the major, of which 16 credits must be in design. Not more than 62 credits may be transferred from a community or junior college. Meet the university standard in English composition.

Transfer without Articulation Agreements
Admission to engineering: Transfers are admitted to the engineer and computer science programs at any level. The application deadlines are July 15 and Nov. 15. Transfer programs are articulated with community colleges
Requirements: A cumulative grade point average of 3.0. Student should show preparation in calculus and college physics.

GRADUATION REQUIREMENTS
B.S. programs in engineering requires 128 semester credits. B.S. program in computer science requires 128 semester credits. M.S. requires 32 credit hours. Thesis option available. Ph.D. requires 56 credits of course work beyond bachelors. 24 credits of doctoral dissertation.

■ STUDENT PROGRAMS

PROFESSIONAL AND HONORARY SOCIETIES
[For key to acronyms, see *Introduction*.]
Professional Societies: ACM, AIAA, ASME, IEEE, NSPE, SAE, SME
Honorary Societies: Eta Kappa Nu, Tau Beta Pi, Theta Tau

SUPPORT PROGRAMS
Student Chapter Organizations: Society of Women Engineers, National Society of Black Engineers

Academic Opportunity Program
For: Ethnic Minorities   Available: Summer only
Offered: Freshman

Upward Bound
For: Ethnic Minorities   Available: Summer only
Offered: Prematriculation

Detroit Area Pre-College Engineering Prog.(DAPCEP)
For: Ethnic Minorities   Available: Academic year
Offered: Prematriculation

Detroit Area Pre-College Engineering Prog.(DAPCEP)
For: Ethnic Minorities   Available: Summer only
Offered: Prematriculation

Other Engineering Support Programs: Oakland University has mandatory freshman and transfer orientation programs for all undergraduate students including engineering and computer science students. The Academic Skills Center administers seminars on study-skills, time management and tutoring for specific courses. Professional student organizations in the school provide additional educational experiences and career development.

170 Ohio University

INSTITUTION PROFILE

HEAD OF THE INSTITUTION
Charles Ping
Phone: (614)593-1804 **Fax:** (614)593-9196

GENERAL INFORMATION
[All Students—Fall 1991]
Undergraduate enrollment 15,323
Graduate enrollment 2,529
Total institution enrollment **18,248**
Type of institution: Public **Calendar:** Quarters
Nearest city: Columbus **Population:** 632,910
Miles from main campus: 75 **Setting:** Small Town
Types of engineering degrees: Engineering
Other degree-granting colleges: Southern Campus, Lancaster Campus, Eastern Campus, Chillicothe Campus

ENGINEERING COLLEGE ADMISSIONS INFORMATION
Entrance Requirements: SAT (Score: 1000), ACT (Score: 24), High School courses—Mathematics (4 years), English (4 years), Chemistry (1 year), Physics (1 year), Must be in the top 30% of their graduating class.
Entrance Recommendations: SAT (Score: 1000), ACT (Score: 24), High School courses—Mathematics (4 years), English (4 years), Chemistry (1 year), Physics (1 year)
Requirements for foreign students: financial statement

DEPARTMENTAL ADMISSIONS INFORMATION
Admission to the engineering department: At the time of admission to the institution.
Additional information: Students meeting academic requirements given in section 13 are accepted for admission within the constraints of the department's capability to provide a quality education. Those not meeting freshmen admission requirements can enroll in the pre-engineering major code in the University College. After satisfactorily completing a group of specified mathematics and science courses, students may transfer into one of the College's programs within the constraints of the department's capability to provide quality education.

FINANCIAL INFORMATION

ESTIMATED EXPENSES (ACADEMIC YEAR)
[Expenses are for the 1992-93 nine-month academic year.]

State Residents	Undergraduate	Graduate
Tuition and fees	$ 3,900	$ 4,329
College room and board	$ 4,017	$ 4,017
Books and supplies	$ 500	$ 500
Other expenses	$ —(1)	$ —(1)
Total estimated expenses	$ 8,417	$ 8,846

Out-of-State Residents	Undergraduate	Graduate
Tuition and fees	$ 7,563	$ 7,992
College room and board	$ 4,017	$ 4,017
Books and supplies	$ 500	$ 500
Other expenses	$ —(1)	$ —(1)
Total estimated expenses	$12,080	$12,509

Notes: (1) Optional Insurance is $316. per year.

FINANCIAL AID OFFICE CONTACT
Carolyn Sabatino, *Director of Financial Aid*
Ohio University
020 Chubb
Athens, OH 45701
Phone: (614)593-4141 **Fax:** (614)593-4140

GENERAL FINANCIAL AID INFORMATION
Forms accepted/required: CSS-FAF
Additional financial aid information: Scholarships, grants, loans and various need-based financial aids are available including: Ohio Instructional Grant, Pell Grant, Stafford student Loan, College work-study, Perkins Loan, and the Supplemental Education Opportunity Grant.

ENGINEERING COLLEGE INFORMATION
[For additional personnel, refer to the *Appendix*.]

HEAD OF ENGINEERING
T. Richard Robe
Phone: (614)593-1479 **Fax:** (614)593-0659

ENGINEERING COLLEGE ADDRESS
College of Engineering and Technology
Ohio University
Stocker Center
Athens, OH 45701-2979
Phone: (614)593-1474 **Fax:** (614)593-0659

ENROLLMENTS—BY CLASS
[Numbers are baccalaureate enrollments for the fall 1991 term, unless otherwise footnoted.]
1st-year students/Freshmen .... 410 [1]
2nd-year students/Sophomores .... 275 [1]
3rd-year students/Juniors .... 212 [1]
4th-year students/Seniors .... 316 [1]
Total **1,213**

Notes: (1) Fall 1992-93 Enrollment.

NUMBER OF DEGREES AWARDED—BY PROGRAM
[Numbers are engineering baccalaureate degrees awarded during the 1991-1992 academic year, unless otherwise footnoted. For full details about each engineering program, refer to the *Tables of Degree Programs*.]
Airway Science 17
Chemical Engineering 10
Civil Engineering 18
Electrical Engineering 88
Industrial & Systems Engineering 40
Industrial Technology 41
Interdisciplinary Engineering .... –
Mechanical Engineering 41
Total **255**

PERCENTAGE OF DEGREES AWARDED—BY CATEGORY
[Percentages are of all engineering baccalaureate degrees awarded during the 1991-1992 academic year, unless otherwise footnoted.]
African Americans .... 2.1 %
Asian/Pacific Island Americans .... – %
Hispanic Americans .... 1.0 %
Native Americans .... – %
Foreign Citizens .... 11.8 %
All Others .... 85.1 %
Women .... 14.4 %
Persons with disabilities .... – %
Students over 25 years of age .... – %

GRADUATE ENROLLMENTS & DEGREES AWARDED
Master's enrollment .... 253 [1]
Master's degrees awarded 44
Doctoral enrollment .... 41 [1]
Doctoral degrees awarded 7

Notes: (1) Fall 1992-93 Enrollment.

ENGINEERING STUDENT DATA
Applicants to the engineering college
Number of applicants to engineering college 1,088
Percent offered admission 83.3%
Matriculated engineering students
Percentage in top quartile (25%) of High School class 64.1
Average SAT scores: Math—568, Verbal—471, Combined—1039
Average ACT scores: Math—24.6, Composite—24.2

FULL- & PART-TIME FACULTY—BY DEPARTMENT
[Figures are the head count of full-time faculty and the full-time equivalent (FTE) of part-time faculty for each engineering department or equivalent.]

Department	Full	Part
Chemical Eng	10	–
Civil Eng	8	–
Electrical Eng	21	1.8
Industrial & Systems Eng	8	1.2
Mechanical Eng	9	1.2

COLLEGE DESCRIPTION
Ohio University, the first institution in the old Northwest territory, was chartered in 1804. All departments in the College of Engineering are housed in a modern facility providing well equipped laboratories, classrooms and offices. The College is the recipient of a multi-million dollar endowment from the late C. Paul Stocker and his wife Beth. Currently endowments totaling over $15 million provide many benefits to the College's programs. Cooperative education programs are offered in all departments. Several pre-college programs provide unique opportunities for students from the 6th grade through 12th grade. The Pre-Engineering Program for Minorities (PEP), which is offered in summer, provides a transition program from high school to college. Seven centers enhance inter-departmental cooperation in the academic and research programs.

DEGREE PROGRAMS DESCRIPTION
The Bachelor of Science Degree is offered in all departments in the College. The Masters of Science degree is offered in all the engineering programs: chemical, civil, electrical, industrial and systems, and mechanical. The Ph.D. degree is offered in chemical and electrical engineering. An interdisciplinary Ph.D. degree is offered with focus areas in geotechnical and environmental, materials processing and intelligent systems. In addition to the engineering degree programs a Bachelor of Science is offered in both Industrial Technology and Airway Science. Seven centers within the College provide a forum for interdepartmental research and education. Endowments in excess of $15 million provide unique benefits to the programs within the College.

DUAL DEGREE PROGRAMS
Engineering baccalaureate graduates with dual degrees 0000

TRANSFER INFORMATION
Residency Requirements: For students completing less that 96 credit hours at Ohio University the residence requirements is the final year (3 quarters) with 48 hours of credit. For students completing 96 or more quarter hours at Ohio University, the final quarter shall be in residence.

Transfer via Articulation Agreements
Admission to engineering: Transfer from other colleges within Ohio University is based on students' academic records.
Engineering graduates transferred from ... 4-yr: – 2-yr: –
Requirements: 2.50 GPA from an ABET Accredited school or 3.00 from a two year community college.

Transfer without Articulation Agreements
Admission to engineering: Admission based on qualifications.
Engineering graduates transferred from ... 4-yr: – 2-yr: –
Requirements: Transfer articulation exists with several Ohio Community Colleges and two year technical schools.

GRADUATION REQUIREMENTS
A 2.00 or above is required in all courses required for graduation. A 2.00 or above in all courses in engineering and technology is required for graduation. A 2.00 or above in all major courses is required for graduation.

STUDENT PROGRAMS

PROFESSIONAL AND HONORARY SOCIETIES
[For key to acronyms, see *Introduction*.]
Professional Societies: AIChE, ASCE, ASME, IEEE, IIE, SME
Honorary Societies: Alpha Pi Mu, Eta Kappa Nu, Tau Beta Pi

SUPPORT PROGRAMS
Student Chapter Organizations: Society of Women Engineers, National Society of Black Engineers, Society of Black Engineers

Pre-Engineering Program for Minorities
For: Ethnic Minorities **Available:** Summer only
Offered: Prematriculation

Other Engineering Support Programs: New students attend a 2-day Pre-College program during the summer prior to fall enrollment. The purpose of the program is for students to be evaluated and advised and to be given information about the University's special programs. Final Fall schedules are developed and academic advisors are assigned. During the academic year students meet with their advisors each quarter for advising.

171 Ohio Northern University

INSTITUTION PROFILE

HEAD OF THE INSTITUTION
DeBow Freed
Phone: (419)772-2031   Fax: (419)772-1932

GENERAL INFORMATION
[All Students—Fall 1991]
Undergraduate enrollment 2,343
Graduate enrollment 448
Total institution enrollment 2,791
Type of institution: Private   **Calendar:** Quarters
Nearest city: Lima   **Population:** 80,000
Miles from main campus: 15   **Setting:** Small Town
Types of engineering degrees: Engineering
Other degree-granting colleges: Arts & Sciences, Business Administration, Law, Pharmacy, Technology & Applied Sciences

ENGINEERING ADMISSIONS

ADMISSIONS OFFICE CONTACT
Karen Condeni
Ohio Northern University
Office of Admissions
Ada, OH 45810
Phone: (419)772-2260   Fax: (419)772-2313

ENGINEERING COLLEGE ADMISSIONS INFORMATION
Entrance Requirements: SAT (Score: 1000), ACT (Score: 20), High School courses—Algebra (2 years), Physics (1 year), Geometry (1 year), Pre-calculus (1 year)
Entrance Recommendations: High School courses—Foreign language (3 years), Calculus (1 year), Intro to computers & CAD (1 year), Chemistry (1 year)
Requirements for foreign students: TOEFL (Score: 500); financial statement

DEPARTMENTAL ADMISSIONS INFORMATION
Admission to the engineering department: At the time of admission to the institution.

FINANCIAL INFORMATION

ESTIMATED EXPENSES (ACADEMIC YEAR)
[Expenses are for the 1992-93 nine-month academic year.]

All Students	Undergraduate	Graduate
Tuition and fees	$14,550	$ –
College room and board	$ 3,615	$ –
Books and supplies	$ 675	$ –
Other expenses	$ 885	$ –
Total estimated expenses	$19,725	$ –

FINANCIAL AID OFFICE CONTACT
Karen Condeni, *Vice President/Dean, Financial Aid*
Ohio Northern University
Office of Financial Aid
Ada, OH 45810
Phone: (419)772-2260   Fax: (419)772-2313

GENERAL FINANCIAL AID INFORMATION
Forms accepted/required: ACT, AFSA, CSS-FAF, FAF, FAT, Stafford, IRS, SAR, SDF, Supplemental, Institutional
Additional financial aid information: Presidential Scholarship, Achievement Award, Honor Scholarship, Dean's Scholarship, Ohio Academic Scholarship, ONU Grants, Methodist Grants, Sibling Grants, Meth. Ministerial Deduction, Pell Grants, SEOG, Ohio Choice Grant, Campus Employment, Work/Study (CWS) Perkins Loan, Stafford Loan (GSL)

ENGINEERING COLLEGE INFORMATION
[For additional personnel, refer to the *Appendix*.]

HEAD OF ENGINEERING
Bruce E. Burton
Phone: (419)772-2372   Fax: (419)772-2404
EMail: bburton@newton.onu.edu

ENGINEERING COLLEGE ADDRESS
T.J. Smull College of Engineering
Ohio Northern University
Ada, OH 45810
Phone: (419)772-2372   Fax: (419)772-2404

ENROLLMENTS—BY CLASS
[Numbers are baccalaureate enrollments for the fall 1991 term, unless otherwise footnoted.]
1st-year students/Freshmen 146
2nd-year students/Sophomores 76
3rd-year students/Juniors 86
4th-year students/Seniors 81
Total 389

NUMBER OF DEGREES AWARDED—BY PROGRAM
[Numbers are engineering baccalaureate degrees awarded during the 1991-1992 academic year, unless otherwise footnoted. For full details about each engineering program, refer to the *Tables of Degree Programs*.]
Civil Engineering 17
Electrical Engineering 19
Mechanical Engineering 33
Total 69

PERCENTAGE OF DEGREES AWARDED—BY CATEGORY
[Percentages are of all engineering baccalaureate degrees awarded during the 1991-1992 academic year, unless otherwise footnoted.]
African Americans .... – %
Asian/Pacific Island Americans .... 1.4 %
Hispanic Americans .... – %
Native Americans .... – %
Foreign Citizens .... 1.4 %
All Others .... 97.2 %
Women .... 13.0 %
Persons with disabilities .... – %
Students over 25 years of age .... – %

ENGINEERING STUDENT DATA
Applicants to the engineering college
Number of applicants to engineering college 361
Percent offered admission 80.0%

Matriculated engineering students
Percentage in top quartile (25%) of High School class 71.0
Average ACT scores: Math—25.5, Composite—24.5

FULL- & PART-TIME FACULTY—BY DEPARTMENT
[Figures are the head count of full-time faculty and the full-time equivalent (FTE) of part-time faculty for each engineering department or equivalent.]

Department	Full	Part
Civil Eng	5	–
Electrical Eng	5	–
Mechanical Eng	5	–

COLLEGE DESCRIPTION
Emphasis is on teaching. Small classes--20 to 30. Open door policy. High emphasis placed on laboratory. Computers readily available in many locations. Many interdisciplinary programs are available including: business, computer science, environmental, foreign language, pre-law, and pre-med. Extracurricular activities are encouraged. Excellent five-year co-op program, both domestic and international.

DEGREE PROGRAMS DESCRIPTION
ONU's engineering college offers a B.S. degree in civil, electrical, and mechanical engineering.

DUAL DEGREE PROGRAMS
Engineering baccalaureate graduates with dual degrees 0

TRANSFER INFORMATION
Residency Requirements: One-year residence.
Transfer via Articulation Agreements
Admission to engineering: At any time
Engineering graduates transferred from .. 4-yr: –   2-yr: –
Requirements: Must have a 2.0 grade point average with appropriate math and science courses.

Transfer without Articulation Agreements
Admission to engineering: At any time
Engineering graduates transferred from .. 4-yr: 9   2-yr: –
Requirements: Must have a 2.0 grade point average with appropriate math and science courses.

GRADUATION REQUIREMENTS
A 2.0 grade point average in all engineering courses and a 2.0 grade point average overall.

STUDENT PROGRAMS

PROFESSIONAL AND HONORARY SOCIETIES
[For key to acronyms, see *Introduction*.]
Professional Societies: ASCE, ASME, IEEE, NSPE
Honorary Societies: Tau Beta Pi

SUPPORT PROGRAMS
Student Chapter Organizations: Society of Women Engineers
Multicultural Development Office
For: Ethnic Minorities   **Available:** Year round
Offered: Prematriculation, Freshman, Sophomore, Junior, Senior
Other Engineering Support Programs: Four two-day freshman summer orientation programs, one transfer summer orientation program, one international student orientation program, and one-week (summer) engineering math refresher course.

172 Ohio State University

INSTITUTION PROFILE

HEAD OF THE INSTITUTION
E. Gordon Gee
Phone: (614)292-2424 **Fax:** (614)292-1231

GENERAL INFORMATION
[All Students—Fall 1991]
Undergraduate enrollment 40,785
Graduate enrollment 13,528
Total institution enrollment **54,313**
Type of institution: Public **Calendar:** Quarters
Nearest city: Columbus **Population:** 63,300
Miles from main campus: 1 **Setting:** Urban
Types of engineering degrees: Engineering
Other degree-granting colleges: Agricultural & Environmental, Allied Health Sciences, Architecture, Arts & Sciences, Business Administration, Communication & Journalism, Dentistry, Education, Fine & Performing Arts, Humanities & Social Sciences, Law, Medicine, Nursing, Pharmacy, Technology & Applied Sciences, Social Work, Mathematical & Physical Sciences

ENGINEERING ADMISSIONS

ADMISSIONS OFFICE CONTACT
James J. Mager
Ohio State University
1800 Cannon Drive
Columbus, OH 43210-1230
Phone: (614)292-3980 **Fax:** (614)292-4818

ENGINEERING COLLEGE ADMISSIONS INFORMATION
Admission to the engineering college: The time of Admission to the institution with minimum test math scores of 25 ACT/550 SAT; or at end of freshman year with minimum GPA of 2.0/4.0.
Entrance Requirements: High School courses—English (4 years), Mathematics (3 years), Science, Foreign Language, Social Science (2 years), Arts (1 year)
Entrance Recommendations: PSAT, SAT, ACT, High School courses—Mathematics (4 years), Science, Foreign Language, Social Science (3 years), Arts (1 year)
Requirements for foreign students: TOEFL (Score: 550); financial statement

DEPARTMENTAL ADMISSIONS INFORMATION
Admission to the engineering department: At the end of the first year.
Additional information: Acceptance to an engineering major based on the cumulative point-hour ratio after completion of program specified pre-major courses. Students that have completed the pre-major courses and have a CPHR of 3.00 or above are assured of acceptance into the major of their choice.

FINANCIAL INFORMATION

ESTIMATED EXPENSES (ACADEMIC YEAR)
[Expenses are for the 1992-93 nine-month academic year.]

State Residents	Undergraduate	Graduate
Tuition and fees	$ 2,799	$ 3,966
College room and board	$ 4,074	$ 4,074
Books and supplies	$ 375	$ 450
Other expenses	$ 2,100	$ 2,100
Total estimated expenses	**$ 9,348**	**$ 10,590**

Out-of-State Residents	Undergraduate	Graduate
Tuition and fees	$ 8,292	$ 10,278
College room and board	$ 4,074	$ 4,074
Books and supplies	$ 375	$ 450
Other expenses	$ 2,100	$ 2,100
Total estimated expenses	**$14,841**	**$ 16,902**

FINANCIAL AID OFFICE CONTACT
Mary B. Haldane, *Director of Student Financial Aid*
Ohio State University
1800 Cannon Drive
Columbus, OH 43210-1230
Phone: (614)292-0300 **Fax:** (614)292-9264

GENERAL FINANCIAL AID INFORMATION
Forms accepted/required: FAF, FAT, Institutional
Additional financial aid information: Scholarships; Grants; Cooperative Housing; Self Help Programs (Loans, employment)

ENGINEERING COLLEGE INFORMATION

[For additional personnel, refer to the *Appendix*.]

HEAD OF ENGINEERING
Jose B. Cruz Jr.
Phone: (614)292-2836 **Fax:** (614)292-9021

ENGINEERING COLLEGE ADDRESS
College of Engineering
Ohio State University
2070 Neil Avenue
Columbus, OH 43210-1275
Phone: (614)292-2651 **Fax:** (614)292-9021

ENROLLMENTS—BY CLASS
[Numbers are baccalaureate enrollments for the fall 1991 term, unless otherwise footnoted.]
1st-year students/Freshmen 687
2nd-year students/Sophomores 768
3rd-year students/Juniors 912
4th-year students/Seniors 1,427
Total **3,794**

NUMBER OF DEGREES AWARDED—BY PROGRAM
[Numbers are engineering baccalaureate degrees awarded during the 1991-1992 academic year, unless otherwise footnoted. For full details about each engineering program, refer to the *Tables of Degree Programs*.]
Aeronautical & Astronautical Engineering 51
Agricultural Engineering 16
Aviation 17
Biomedical Engineering .... –
Ceramic Engineering 14
Chemical Engineering 42
Civil Engineering 61
Computer & Information Science 87
Electrical Engineering 105
Engineering Mechanics .... –
Engineering Physics 11
Industrial & Systems Engineering 63
Materials Science & Engineering .... –
Mechanical Engineering 134
Metallurgical Engineering 16
Nuclear Engineering .... –
Surveying 7
Welding Engineering 31
Total **655**

PERCENTAGE OF DEGREES AWARDED—BY CATEGORY
[Percentages are of all engineering baccalaureate degrees awarded during the 1991-1992 academic year, unless otherwise footnoted.]
African Americans 2.0%
Asian/Pacific Island Americans 3.6%
Hispanic Americans 1.7%
Native Americans 0.3%
Foreign Citizens 7.5%
All Others 84.9%
Women 10.1%
Persons with disabilities .... –%
Students over 25 years of age 9.0%

GRADUATE ENROLLMENTS & DEGREES AWARDED
Master's enrollment 837
Master's degrees awarded 284
Doctoral enrollment 769
Doctoral degrees awarded 100

ENGINEERING STUDENT DATA
Applicants to the engineering college
Number of applicants to engineering college 2,332
Percent offered admission 88.0%

Matriculated engineering students
Percentage in top quartile (25%) of High School class 59.0
Average SAT scores: Math—589, Verbal—490, Combined—1079
Average ACT scores: Math—25.5, Composite—24.9

FULL- & PART-TIME FACULTY—BY DEPARTMENT
[Figures are the head count of full-time faculty and the full-time equivalent (FTE) of part-time faculty for each engineering department or equivalent.]

Department	Full	Part
Chemical Eng	12	–
Civil Eng	23	–
Electrical Eng	44	–
Industrial Eng	20	–
Mechanical Eng	32	–
Aeronautical & Astronautical Eng	12	–
Agricultural Eng	19	–
Aviation	4	–
Computer & Information Science	36	–
Eng Mechanics	14	–
Eng Physics	–	–
Materials Science & Eng	21	–
Geodetic Science & Surveying	9	–
Welding Eng	9	–
Eng Graphics	12	–

COLLEGE DESCRIPTION
The College of Engineering has over 300 full-time PhD faculty and a student-faculty ratio of 17:1. The faculty are world class scholars that administer a $49 million research budget but also have a strong commitment to the class-room. Selective admission ensures high quality; one student in five ranked in the top 5 percent of their high school class. Special programs support international study, job placement, minority assistance, women's programs, honors enrichment, and freshmen tutoring. Over 30 engineering student organizations are available.

TRANSFER INFORMATION
Transfer without Articulation Agreements
Admission to engineering: At any time
Engineering graduates transferred from ... 4-yr: – 2-yr: –
Requirements: All transfer applications are reviewed by departmental admissions committees for admission decisions. The factors considered are grade point average, courses taken, and rigor and applicability of course content. Transfer students admitted into Engineering as pre-majors will need to satisfy the specific academic requirements of the desired major in competition with continuing Ohio State engineering pre-majors.

GRADUATION REQUIREMENTS
(1) 2.0 GPA in the major courses. (2) 2.0 GPA overall grade point average requirement.

STUDENT PROGRAMS

PROFESSIONAL AND HONORARY SOCIETIES
[For key to acronyms, see *Introduction*.]
Professional Societies: ACerS, ACS, AIA, AIAA, ASAE, ASCE, ASME, IEEE, IIE, NSPE
Honorary Societies: Alpha Pi Mu, Chi Epsilon, Eta Kappa Nu, Keramos, Pi Tau Sigma, Sigma Gamma Tau, Tau Beta Pi, Theta Tau, TEXNIKOI, Triangle

SUPPORT PROGRAMS
Student Chapter Organizations: Society of Women Engineers, National Society of Black Engineers

Women in Engineering
For: Women **Available:** Year round
Offered: Prematriculation, Freshman, Sophomore, Junior, Senior, Graduate level

Minority Engineering Program
For: Ethnic Minorities **Available:** Year round
Offered: Prematriculation, Freshman, Sophomore, Junior, Senior, Graduate level

Other Engineering Support Programs: Special support activities include orientation programs for new freshmen and transfer students; a freshman tutoring program; elective courses focusing on career guidance and issues for women in engineering careers; and an engineering speaking society that visits high schools to discuss engineering careers.

173 University of Oklahoma

INSTITUTION PROFILE

HEAD OF THE INSTITUTION
Richard L. Van Horn
Phone: (405)325-3916   Fax: (405)325-7605

GENERAL INFORMATION
[All Students—Fall 1991]
Undergraduate enrollment 14,685
Graduate enrollment 4,279
Total institution enrollment 18,964

Type of institution: Public   Calendar: Semesters
Nearest city: Oklahoma City, OK   Population: 750,000
Miles from main campus: 20   Setting: Urban
Types of engineering degrees: Engineering
Other degree-granting colleges: Allied Health Sciences, Architecture, Arts & Sciences, Business Administration, Dentistry, Education, Fine & Performing Arts, Law, Medicine, Nursing, Pharmacy, Technology & Applied Sciences

ENGINEERING ADMISSIONS

ADMISSIONS OFFICE CONTACT
Marc S. Borish
University of Oklahoma
Director of Admissions
Norman, OK 73019-0631
Phone: (405)325-2251   Fax: (405)325-7047

ENGINEERING COLLEGE ADMISSIONS INFORMATION
Admission to the engineering college: Must complete 24 hours of college credit with a grade of C or better in each course.
Entrance Requirements: ACT (Score: 22), High School courses—English (4 years), College Prep Math (Alg I, Geom, Alg II or higher (3 years), Lab Science (2 years), History (including one year American History) (2 years), ACT or SAT scores are required; however, there is no minimum score necessary for College of Engineering. University requires a 21 ACT or 950 SAT for admission to the University for freshmen.
Entrance Recommendations: SAT (Score: 950), ACT (Score: 21), High School courses—College Prep Math (4 years), Physics (1 year), Chemistry (2 years), Foreign Language (2 years), ACT or SAT scores are required; however, there is no minimum score necessary for College of Engineering. University requires a 21 ACT or 950 SAT for admission to the University for freshmen.
Requirements for foreign students: TOEFL minimum of 550 is a University, not College of Engineering, requirement, and the score must be no more than 2 years old.

DEPARTMENTAL ADMISSIONS INFORMATION
Admission to the engineering department: Students are admitted to their respective schools/depts concurrently with admission to the CoE.
Additional information: Departmental admission requirements are the same as those for admission to the College of Engineering.

FINANCIAL INFORMATION

ESTIMATED EXPENSES (ACADEMIC YEAR)
[Expenses are for the 1992-93 nine-month academic year.]

State Residents	Undergraduate	Graduate
Tuition and fees	$ 1,165 [2]	$ 1,258 [1]
College room and board	$ 3,358	$ 3,358
Books and supplies	$ 670	$ 670
Other expenses	$ 200 [3]	$ 200 [3]
Total estimated expenses	**$ 5,393**	**$ 5,486**

Out-of-State Residents	Undergraduate	Graduate
Tuition and fees	$ 4,565 [2]	$ 3,747 [1]
College room and board	$ 3,358	$ 3,358
Books and supplies	$ 670	$ 670
Other expenses	$ 200 [3]	$ 200 [3]
Total estimated expenses	**$ 8,793**	**$ 7,975**

Notes: (1) based on 9 hours/semester (2) based on 15 hours/semester (3) computer fees

FINANCIAL AID OFFICE CONTACT
Mary Mowdy, *Interim Director of Financial Aid*
University of Oklahoma
Hester-Robertson Hall
Norman, OK 73019-0631
Phone: (405)325-4521   Fax: (405)325-7608

GENERAL FINANCIAL AID INFORMATION
Forms accepted/required: FFS
Additional financial aid information: need-based scholarships; short-term loans; student loans; merit-based scholarships; grants; part-time jobs on campus; work-study program. For engineering scholarships, contact the appropriate department or the Associate Dean for Academic Programs, 202 W. Boyd, Room 107, Norman, OK 73019.

ENGINEERING COLLEGE INFORMATION

[For additional personnel, refer to the *Appendix*.]

HEAD OF ENGINEERING
Billy L. Crynes
Phone: (405)325-2621   Fax: (405)325-7508
EMail: crynes@mailhost.ecn.uoknor.edu

ENGINEERING COLLEGE ADDRESS
College of Engineering
University of Oklahoma
202 West Boyd
CEC 107
Norman, OK 73019-0631
Phone: (405)325-2621   Fax: (405)325-7508

ENROLLMENTS—BY CLASS
[Numbers are baccalaureate enrollments for the fall 1991 term, unless otherwise footnoted.]
1st-year students/Freshmen 639
2nd-year students/Sophomores 440
3rd-year students/Juniors 352
4th-year students/Seniors 758
Total **2,189**

NUMBER OF DEGREES AWARDED—BY PROGRAM
[Numbers are engineering baccalaureate degrees awarded during the 1991-1992 academic year, unless otherwise footnoted. For full details about each engineering program, refer to the *Tables of Degree Programs*.]

Aerospace Engineering 28
Chemical Engineering 28
Civil Engineering 21
Computer Science 25
Electrical Engineering 107
Engineering Physics 4
Environmental Science 2
Geological Engineering .... –
Industrial Engineering 33
Mechanical Engineering 59
Petroleum Engineering 9
Total **316**

PERCENTAGE OF DEGREES AWARDED—BY CATEGORY
[Percentages are of all engineering baccalaureate degrees awarded during the 1991-1992 academic year, unless otherwise footnoted.]
African Americans .... – %
Asian/Pacific Island Americans .... – % (A)
Hispanic Americans .... – %
Native Americans .... – %
Foreign Citizens .... – % (A)
All Others .... 100 %
Women .... 14.5 %
Persons with disabilities .... – % (A)
Students over 25 years of age .... – % (A)

Notes: (A) Data not available.

GRADUATE ENROLLMENTS & DEGREES AWARDED
Master's enrollment .... – (A)
Master's degrees awarded 134
Doctoral enrollment .... – (A)
Doctoral degrees awarded 39

Notes: (A) Data not available.

ENGINEERING STUDENT DATA

Applicants to the engineering college
Number of applicants to engineering college .... – (A)
Percent offered admission .... –% (A)

Matriculated engineering students
Percentage in top quartile (25%) of High School class .... – (A)
Average SAT scores: Combined—1235
Average ACT scores: Composite—26.5 [1]

Notes: (A) Data not available. (1) based on 326 students reporting scores

FULL- & PART-TIME FACULTY—BY DEPARTMENT
[Figures are the head count of full-time faculty and the full-time equivalent (FTE) of part-time faculty for each engineering department or equivalent.]

Department	Full	Part
Chemical Eng	11	0.3
Civil Eng & Environmental Science	16	–
Electrical Eng	21	2.0
Industrial Eng	9	0.8
Aerospace & Mechanical Eng	15	0.8
Eng Physics	3	–
Petroleum & Geological Eng	8	2.0
Computer Science	9	–

COLLEGE DESCRIPTION

Instruction in professional engineering was first given at the University of Oklahoma in 1899. The college now offers four-year undergraduate degrees in 12 engineering fields, computer science, and environmental science. Graduate programs are also available in 12 engineering and related fields. An optional co-op program is available which combines a sequence of academic study and engineering employment. The time required to complete an engineering degree program as a co-op student is minimally 4 years and 9 months for the usual eight semester program where every summer will be spent either in school or on work assignments.

DEGREE PROGRAMS DESCRIPTION

The B.S. degrees in the various engineering programs qualify the graduate for a professional position or for graduate study in a particular field. A student desiring a graduate degree can elect to work towards a Masters of Science, Doctor of Philosophy, or Doctor of Engineering. Some programs offer a non-thesis Master's option. For more information about graduate degrees, please contact the Graduate College, Buchanan Hall, Room 313, University of Oklahoma, Norman, OK 73019 or the specific department.

DUAL DEGREE PROGRAMS

Enrollment requirements: Students may receive a dual degree upon completion of all requirements for both degrees, provided that the work includes at least 30 add'l hrs of upper-division engineering, applied science & elective courses appropriate to the field of the second degree. Admission req's for the CoE are the same for all students.

TRANSFER INFORMATION

Residency Requirements: Students must spend 2 semesters or the equivalent in residence, with at least one semester enrolled as a CoE student. They must complete 36 of their last 48 curricular semester hrs in residence, with 24 of the 36 being in their major field.

GRADUATION REQUIREMENTS

B.S. programs in engineering require 128-139 semester hours, depending on the major selected. Up to 64 hours of credit earned at a junior college may be accepted toward B.S. engineering degrees; students must earn a C (2.00) or better GPA at the University of Oklahoma in combined college work in the following areas: major, curriculum, and all coursework.

STUDENT PROGRAMS

PROFESSIONAL AND HONORARY SOCIETIES
[For key to acronyms, see *Introduction*.]
Professional Societies: ACM, ACS, AIAA, AIChE, ASCE, IIE, ORSA, SME, SPE*, American Society of Mechanical Engineers, National Society of Black Engineers
Honorary Societies: Alpha Pi Mu, Chi Epsilon, Eta Kappa Nu, Pi Tau Sigma, Sigma Gamma Tau, Tau Beta Pi, American Indian Science and Engineering Society, Society of Hispanic Professional Engineers

SUPPORT PROGRAMS
Student Chapter Organizations: Society of Women Engineers, National Society of Black Engineers

Minority Engineering Programs
For: Ethnic Minorities   **Available:** Year round
Offered: Prematriculation, Freshman, Sophomore, Junior, Senior, Graduate level

Other Engineering Support Programs: The College of Engineering offers student advising in its Koch Undergraduate Student Advising Center.

174 Oklahoma Christian University of Science and Arts

■ INSTITUTION PROFILE

HEAD OF THE INSTITUTION
J. Terry Johnson
Phone: (405)425-5100　　　　　Fax: (405)425-5316

GENERAL INFORMATION
[All Students—Fall 1991]
Undergraduate enrollment 1,652
Graduate enrollment 38
Total institution enrollment 1,690

Type of institution: Private　　　Calendar: Trimesters
Nearest city: Oklahoma City　　　Population: 400,000
Miles from main campus: 10　　　Setting: Suburban
Types of engineering degrees: Engineering
Other degree-granting colleges: Business Administration, Education, Biblical Studies, Liberal Arts

■ ENGINEERING ADMISSIONS

ADMISSIONS OFFICE CONTACT
Bob Rowley
Oklahoma Christian University of Science and Arts
P.O. Box 11000
2501 E. Memorial Road
Oklahoma City, OK 73136-1100
Phone: (405)425-5054　　　　　Fax: (405)425-5316

ENGINEERING COLLEGE ADMISSIONS INFORMATION
Entrance Requirements: SAT, ACT
Entrance Recommendations: SAT, ACT (Score: 20), High School courses—Math (3 years), Science (3 years)
Requirements for foreign students: TOEFL (Score: 550); financial statement

DEPARTMENTAL ADMISSIONS INFORMATION
Admission to the engineering department: Students must apply for admission into engineering after they have completed at least 45 hours and before enrolling in a junior level course.
Additional information: Students must apply for admission into engineering after they have completed at least 45 hours and before enrolling in a junior level course.

■ FINANCIAL INFORMATION

ESTIMATED EXPENSES (ACADEMIC YEAR)
[Expenses are for the 1992-93 nine-month academic year.]

All Students	Undergraduate	Graduate
Tuition and fees	$ 5,080	$ –
College room and board	$ 2,840	$ –
Books and supplies	$ –	$ –
Other expenses	$ 200 [1]	$ –
Total estimated expenses	$ 8,120	$ –

Notes: (1) Mandatory fees

FINANCIAL AID OFFICE CONTACT
Andy Carpenter, *Director of Financial Aid*
Oklahoma Christian University of Science and Arts
P.O. Box 11000
2501 E. Memorial Road
Oklahoma City, OK 73136-1100
Phone: (405)425-5105　　　　　Fax: (405)425-5316

GENERAL FINANCIAL AID INFORMATION
Forms accepted/required: ACT, AFSA, FAT, FSA, FFS, Stafford, IRS, SAR, SDF, Supplemental, Institutional
Additional financial aid information: Pell Grants, Supplemental grants, Work-study, Stafford Loans, Supplemental Loans for independent students.

■ ENGINEERING COLLEGE INFORMATION

[For additional personnel, refer to the *Appendix*.]

HEAD OF ENGINEERING
W. Joe Watson
Phone: (405)425-5426　　　　　Fax: (405)425-5316

ENGINEERING COLLEGE ADDRESS
College of Science and Engineering
Oklahoma Christian University of Science and Arts
P.O. Box 11000
2501 E. Memorial Road
Oklahoma City, OK 73136-1100
Phone: (405)425-5400　　　　　Fax: (405)425-5316

ENROLLMENTS—BY CLASS
[Numbers are baccalaureate enrollments for the fall 1991 term, unless otherwise footnoted.]
1st-year students/Freshmen 54
2nd-year students/Sophomores 22
3rd-year students/Juniors 25
4th-year students/Seniors 38
Total 139

NUMBER OF DEGREES AWARDED—BY PROGRAM
[Numbers are engineering baccalaureate degrees awarded during the 1991-1992 academic year, unless otherwise footnoted. For full details about each engineering program, refer to the *Tables of Degree Programs*.]
Electrical Engineering 6
Mechanical Engineering 5
Total 11

PERCENTAGE OF DEGREES AWARDED—BY CATEGORY
[Percentages are of all engineering baccalaureate degrees awarded during the 1991-1992 academic year, unless otherwise footnoted.]
African Americans .... – %
Asian/Pacific Island Americans .... – %
Hispanic Americans .... – %
Native Americans .... – %
Foreign Citizens .... 9.1 %
All Others .... 90.9 %
Women .... – %
Persons with disabilities .... – %
Students over 25 years of age .... – %

ENGINEERING STUDENT DATA
Applicants to the engineering college
Number of applicants to engineering college 101
Percent offered admission .... 100.0% [1]

Matriculated engineering students
Percentage in top quartile (25%) of High School class .... –

Notes: (1) OC is an open admission institution.

FULL- & PART-TIME FACULTY—BY DEPARTMENT
[Figures are the head count of full-time faculty and the full-time equivalent (FTE) of part-time faculty for each engineering department or equivalent.]

Department	Full	Part
Electrical Eng	5	–
Mechanical Eng	5	–

COLLEGE DESCRIPTION
The Department of Engineering offers baccalaureate degree programs in electrical and mechanical engineering. Both programs are built on strong fundamental and theoretical engineering foundations as well as strong emphases on the practice of engineering, including engineering analysis, synthesis, and design; heavy emphasis is placed on the development of effective communications skills, on the liberal arts, and their place in engineering. The priority of both programs is the undergraduate engineering student and his or her engineering education (no graduate programs are offered).

DEGREE PROGRAMS DESCRIPTION
The degree programs offer the B. S. degree. In Electrical Engineering, four areas of emphasis are offered - communications, computers, controls, and electronics, thus allowing the students to select and study those areas in greater depth, building upon the earlier courses required in all of these areas. In Mechanical Engineering several upper division optional courses of study are provided - aerodynamic design, vibrations, and composite engineering materials. Seniors in both programs complete a year-long, systems design project course in which they work in interdisciplinary teams to apply the studied engineering courses and design methodologies to solution of an engineering problem (which may be industry or faculty sponsored). Summer internships are strongly encouraged for the practical experience gained and for enhancement of the overall preparation for an engineering career. As part of a liberal arts university, the engineering programs place great emphasis on engineering within the liberal arts context, providing the graduates with additional perspectives of values, societal awareness, and community and professional involvement.

DUAL DEGREE PROGRAMS
Engineering baccalaureate graduates with dual degrees 0
Enrollment requirements: Same as described for transfer students.

TRANSFER INFORMATION
Residency Requirements: A graduation requirement of all OC students is that at least one-half, or 22 hours (whichever is smaller) of the work in a major/minor program must be completed at OC. In majors containing a specialization (as engineering does), at least one-half of the work required by that specialization must be completed at OC.

Transfer via Articulation Agreements
Admission to engineering: At any time
Engineering graduates transferred from ... 4-yr: –　2-yr: –
Requirements: Transfer credits with grades less than C in the physical sciences, mathematics, computer science, engineering science and engineering will not count toward graduation, with the exception of instances where a D occurred earlier in a course sequence, followed by a C or better later in that sequence, if the grades for all courses that are specified prerequisites for courses to be taken at OC are C or higher.

Transfer without Articulation Agreements
Admission to engineering: Same as for departmental admission
Engineering graduates transferred from ... 4-yr: 2　2-yr: 1
Requirements: Same as described for schools with articulation agreements.

GRADUATION REQUIREMENTS
Beyond the university's requirement of an overall GPA of not less than 2.0 (of 4.0), Engineering requires at least a 2.4 GPA in all upper division (junior/senior) engineering courses and a minimum grade of C (2.0 of 4.0) in each upper division course that is a prerequisite for other courses in the degree plan; also, beyond the university's general education requirements, Engineering requires courses in technical communications, engineering economics, and upper division humanities.

■ STUDENT PROGRAMS

PROFESSIONAL AND HONORARY SOCIETIES
[For key to acronyms, see *Introduction*.]
Professional Societies: ASHRAE, ASME, IEEE

SUPPORT PROGRAMS
Other Engineering Support Programs: Freshman are offered a summer orientation; engineering freshmen and transfers with less than 32 hours take an introduction to engineering. Engineering transfer students enroll in a leveling course on design methodologies; all students have access to tutors and remedial courses. Academic, inter-personal, and placement counseling are strongly emphasized aspects of the student-faculty interaction.

175 Oklahoma State University

■ INSTITUTION PROFILE

HEAD OF THE INSTITUTION
John R. Campbell
Phone: (405)744-6386 Fax: (405)744-6285

GENERAL INFORMATION
[All Students—Fall 1991]
Undergraduate enrollment 14,835
Graduate enrollment 3,810
Total institution enrollment **19,476**
Type of institution: Public Calendar: Semesters
Nearest city: Oklahoma City Population: 403,000
Miles from main campus: 65 Setting: Small Town
Types of engineering degrees: Engineering & Technology
Other degree-granting colleges: Agricultural & Environmental, Arts & Sciences, Business Administration, Education, Human and Environmental Science, Veterinary Medicine

■ ENGINEERING ADMISSIONS

ADMISSIONS OFFICE CONTACT
Larry D. Zirkle
Oklahoma State University
Director of Student Services
EN 101
Stillwater, OK 74078
Phone: (405)744-5276 Fax: (495)744-6187

ENGINEERING COLLEGE ADMISSIONS INFORMATION
Entrance Requirements: ACT (Score: 22), High School courses—English (4 years), History (must include American History) (2 years), Laboratory Science (2 years), Mathematics (Algebra I and above) (2 years)
Entrance Recommendations: High School courses—Select from computer science, foreign language, (4 years), Economics, geography, government, speech, SAT equivalent accepted in lieu of ACT score.
Requirements for foreign students: TOEFL (Score: 500); financial statement

DEPARTMENTAL ADMISSIONS INFORMATION
Admission to the engineering department: At the end of the second year.
Additional information: An overall GPA of 2.30; a GPA of at least 2.5 in mathematics, physical science, and English, and required engineering science courses. Non-resident transfer students must have at least a 2.70 GPA, including a 2.50 or better in math, science and English.

■ FINANCIAL INFORMATION

ESTIMATED EXPENSES (ACADEMIC YEAR)
[Expenses are for the 1992-93 nine-month academic year.]

State Residents	Undergraduate	Graduate
Tuition and fees	$ 1,644	$ 2,064
College room and board	$ 2,960	$ 2,960
Books and supplies	$ 536	$ 576
Other expenses	$ 2,374	$ 2,374
Total estimated expenses	$ 7,514	$ 7,974

Out-of-State Residents	Undergraduate	Graduate
Tuition and fees	$ 4,622 [1]	$ 5,742 [1]
College room and board	$ 2,960	$ 2,960
Books and supplies	$ 536	$ 576
Other expenses	$ 2,374	$ 2,374
Total estimated expenses	$10,492	$11,652

Notes: (1) Children of non-resident alumni treated as resident students.

FINANCIAL AID OFFICE CONTACT
Charles W. Bruce, *Director of Financial Aids*
Oklahoma State University
Hanner Hall
Stillwater, OK 74078
Phone: (405)744-6604 Fax: (405)744-6438

GENERAL FINANCIAL AID INFORMATION
Forms accepted/required: ACT, CSS-FAF, FAF, FAR, FFS, Stafford, IRS, SDF, Institutional
Additional financial aid information: The full spectrum of financial aids are available

■ ENGINEERING COLLEGE INFORMATION

[For additional personnel, refer to the *Appendix*.]

HEAD OF ENGINEERING
Karl N. Reid
Phone: (405)744-5140 Fax: (405)744-7545
EMail: kreid@master.ceat.okstate.edu

ENGINEERING COLLEGE ADDRESS
College of Engineering, Architecture & Technology
Oklahoma State University
Stillwater, OK 74078
Phone: (405)744-5140 Fax: (405)744-6187

ENROLLMENTS—BY CLASS
[Numbers are baccalaureate enrollments for the fall 1991 term, unless otherwise footnoted.]
1st-year students/Freshmen 416
2nd-year students/Sophomores 359
3rd-year students/Juniors 400
4th-year students/Seniors 406
Total **1,581**

NUMBER OF DEGREES AWARDED—BY PROGRAM
[Numbers are engineering baccalaureate degrees awarded during the 1991-1992 academic year, unless otherwise footnoted. For full details about each engineering program, refer to the *Tables of Degree Programs*.]
Aerospace 15
Agricultural Engineering 3
Architectural Engineering 3
Chemical Engineering 27
Civil Engineering 17
Electrical & Computer Engineering 51
General Engineering .... –
Industrial Engineering & Management 30
Mechanical & Aerospace Engineering 56
Total **202**

PERCENTAGE OF DEGREES AWARDED—BY CATEGORY
[Percentages are of all engineering baccalaureate degrees awarded during the 1991-1992 academic year, unless otherwise footnoted.]
African Americans .... 2.6 %
Asian/Pacific Island Americans .... 2.1 %
Hispanic Americans .... 2.1 %
Native Americans .... 2.1 %
Foreign Citizens .... 16.1 %
All Others .... 75.0 %
Women .... 14.6 %
Persons with disabilities .... – % (A)
Students over 25 years of age .... – % (A)

Notes: (A) Data not available.

GRADUATE ENROLLMENTS & DEGREES AWARDED
Master's enrollment 493
Master's degrees awarded 178
Doctoral enrollment 142
Doctoral degrees awarded 25

ENGINEERING STUDENT DATA
Applicants to the engineering college
Number of applicants to engineering college .... – (A)
Percent offered admission .... –% (A)

Matriculated engineering students
Percentage in top quartile (25%) of High School class ..... – (1)
Average ACT scores: Math—28, Composite—26.0

Notes: (A) Data not available. (1) Percentage of freshmen ranked in the top 10% of their high school class: 41.6%

FULL- & PART-TIME FACULTY—BY DEPARTMENT
[Figures are the head count of full-time faculty and the full-time equivalent (FTE) of part-time faculty for each engineering department or equivalent.]

Department	Full	Part
Chemical Eng	9	0.5
Civil Eng	18	–
Electrical & Computer Eng	19	0.8
Industrial Eng & Management	12	0.5
Mechanical & Aerospace Eng	20	0.5
Agricultural Eng	11	1.5
Architectural Eng	16	–
General Eng	–	–

COLLEGE DESCRIPTION
Engineering, Technology, and Architecture are all combined in one college. Students may choose from this spectrum at admission or at a later stage of their academic program. Co-op is available as an option for all engineering and technology students, and adds one year to the length of the program. Graduate degree programs are offered at masters and doctoral levels for all engineering programs, and at the masters level for architectural engineering.

DEGREE PROGRAMS DESCRIPTION
The Professional School concept is used for the bachelors programs. Students are admitted to Pre-engineering for the first two years, and then if qualified, admitted to the upper division for coursework in the major. A 76 hour common curriculum, including general education, characterizes each of the programs. Masters level programs include both the MS and Master of (designated) Engineering. The latter is a practice-oriented degree usually including an internship instead of a thesis. PhD programs stress research, such that the proper preparation is the MS degree. The dissertation normally extends to 18-24 semester hours out of a total of 90+ hours, or 60+ hours beyond a masters degree.

TRANSFER INFORMATION
Residency Requirements: At least half the major requirements (not less than 30 hours) must be earned at OSU over two semesters, a semester and a summer, or over three summers.
Transfer without Articulation Agreements
Admission to engineering: At any time
Requirements: Transfers from other colleges at OSU are treated the same as transfers from other institutions.

GRADUATION REQUIREMENTS
The GPA required for graduation is 2.0 overall, and 2.0 in the major. A 'C' or better is required in each course that is a prerequisite for a course in the major.

■ STUDENT PROGRAMS

PROFESSIONAL AND HONORARY SOCIETIES
[For key to acronyms, see *Introduction*.]
Professional Societies: ACM, AIA, AIAA, AIChE, ASAE, ASCE, ASME, IEEE, IIE
Honorary Societies: Alpha Epsilon, Alpha Pi Mu, Chi Epsilon, Eta Kappa Nu, Omega Chi Epsilon, Pi Tau Sigma, Tau Beta Pi

SUPPORT PROGRAMS
Student Chapter Organizations: Society of Women Engineers, National Society of Black Engineers, Society of Hispanic Professional Engineers, American Indian Science and Engineering Society, Society of Asian American Engineers

Council of Partners
For: Ethnic Minorities **Available:** Year round
Offered: Prematriculation, Freshman, Sophomore, Junior, Senior

HUES (Housing for Underrepresented eng. Stud.)
For: Ethnic Minorities **Available:** Academic year
Offered: Freshman, Sophomore, Junior

ESAAP (Summer Academic Achievement Program)
For: Ethnic Minorities **Available:** Summer only
Offered: Freshman

Minority Special Orientation
For: Ethnic Minorities **Available:** Academic year
Offered: Freshman

Other Engineering Support Programs: Freshman Orientation (Alpha) each fall is a three-day program to familiarize new students with the campus, student organizations, library, etc.

176 Old Dominion University

■ INSTITUTION PROFILE
HEAD OF THE INSTITUTION
James V. Koch
Phone: (804)683-3159　　　　　　　　Fax: (804)683-5679
GENERAL INFORMATION
[All Students—Fall 1991]
Undergraduate enrollment 11,624
Graduate enrollment 5,062
Total institution enrollment **16,686**
Type of institution: Public　　　**Calendar:** Semesters
Location: Norfolk　　　　　　　**Population:** 261,229
Setting: Urban
Types of engineering degrees: Engineering & Technology
Other degree-granting colleges: Education, Arts and Letters, Business and Public Administration, Health Sciences, Sciences

■ ENGINEERING ADMISSIONS
ADMISSIONS OFFICE CONTACT
Angela N. Boyd
Old Dominion University
Office of Admissions
Norfolk, VA 23529-0050
Phone: (804)683-3637　　　　　　　Fax: (804)683-3647
ENGINEERING COLLEGE ADMISSIONS INFORMATION
Entrance Requirements: SAT (Score: 850)
Entrance Recommendations: High School courses—English (4 years), Mathematics (4 years), Natural/Computer Sciences (4 years), Foreign language (3 years), Advanced Placement
Requirements for foreign students: TOEFL (Score: 550); financial statement
DEPARTMENTAL ADMISSIONS INFORMATION
Admission to the engineering department: At the time of admission to the institution.

■ FINANCIAL INFORMATION
ESTIMATED EXPENSES (ACADEMIC YEAR)
[Expenses are for the 1992-93 nine-month academic year.]

State Residents	Undergraduate	Graduate
Tuition and fees	$ 3,978	$ 2,714
College room and board	$ 4,450	$ 4,450
Books and supplies	$ 550	$ 400
Other expenses	$ –(A)	$ –(A)
Total estimated expenses	**$ 8,978**	**$ 7,564**

Out-of-State Residents	Undergraduate	Graduate
Tuition and fees	$ 9,792	$ 6,818
College room and board	$ 4,450	$ 4,450
Books and supplies	$ 550	$ 400
Other expenses	$ –(A)	$ –(A)
Total estimated expenses	**$14,792**	**$11,668**

Notes: (A) Data not available.
FINANCIAL AID OFFICE CONTACT
Helga A. Greenfield, *Director*
Old Dominion University
Financial Aid Office
Norfolk, VA 23529-0052
Phone: (804)683-3683　　　　　　　Fax: (804)683-5357
GENERAL FINANCIAL AID INFORMATION
Forms accepted/required: FAF, Stafford, Institutional, Free Application for Federal Student Financial Aid (FAFSA)
Additional financial aid information: Need based scholarships; merit based scholarships; grants; loans; part-time employment on campus; work study programs.

■ ENGINEERING COLLEGE INFORMATION
[For additional personnel, refer to the *Appendix*.]
HEAD OF ENGINEERING
Ernest J. Cross Jr.
Phone: (804)683-3787　　　　　　　Fax: (804)683-4898

ENGINEERING COLLEGE ADDRESS
College of Engineering and Technology
Old Dominion University
Kaufman-Duckworth Hall, Room 101
Norfolk, VA 23529-0236
Phone: (804)683-3787　　　　　　　Fax: (804)683-4898
ENROLLMENTS—BY CLASS
[Numbers are baccalaureate enrollments for the fall 1991 term, unless otherwise footnoted.]
1st-year students/Freshmen .... 184 [1]
2nd-year students/Sophomores .... 214 [1]
3rd-year students/Juniors .... 257 [1]
4th-year students/Seniors .... 450 [1]
Total **1,105**

Notes: (1) Fall 1991 data
NUMBER OF DEGREES AWARDED—BY PROGRAM
[Numbers are engineering baccalaureate degrees awarded during the 1991-1992 academic year, unless otherwise footnoted. For full details about each engineering program, refer to the *Tables of Degree Programs*.]
Civil Engineering 34
Computer Engineering 20
Electrical Engineering 61
Engineering Management .... – [B]
Mechanical Engineering 56
Mechanics .... –
Total **171**

Notes: (B) Data not applicable.
PERCENTAGE OF DEGREES AWARDED—BY CATEGORY
[Percentages are of all engineering baccalaureate degrees awarded during the 1991-1992 academic year, unless otherwise footnoted.]
African Americans .... 6.4 %
Asian/Pacific Island Americans .... 7.6 %
Hispanic Americans .... – %
Native Americans .... – %
Foreign Citizens .... 5.8 %
All Others .... 80.2 %
Women .... 21.1 %
Persons with disabilities .... – %
Students over 25 years of age .... – % [A]

Notes: (A) Data not available.
GRADUATE ENROLLMENTS & DEGREES AWARDED
Master's enrollment .... 309 [1]
Master's degrees awarded 93
Doctoral enrollment 96
Doctoral degrees awarded 16

Notes: (1) Fall 1991 data
ENGINEERING STUDENT DATA
Applicants to the engineering college
Number of applicants to engineering college .... 566 [1]
Percent offered admission .... 65.0% [1]
Matriculated engineering students
Percentage in top quartile (25%) of High School class 44.0
Average SAT scores: Math—550, Verbal—440, Combined—990

Notes: (1) Fall 1991 data
FULL- & PART-TIME FACULTY—BY DEPARTMENT
[Figures are the head count of full-time faculty and the full-time equivalent (FTE) of part-time faculty for each engineering department or equivalent.]

Department	Full	Part
Civil & Environmental Eng	10	–
Electrical & Computer Eng	18	–
Mechanical Eng & Mechanics	22	–
Eng Management	8	2.8

COLLEGE DESCRIPTION
The University offers four year undergraduate programs in engineering and engineering technology as well as engineering degrees at the master's and doctoral level. Following a common first year emphasizing math, science and engineering fundamentals, the engineering undergraduates specialize in one of the four major areas. The computer and experimental laboratory facilities are well equipped with state of the art equipment. Co-operative education (a five year program) is available in all departments as well as opportunities to minor in several areas. The College offers a dual degree program for selected engineering students to earn a B.A. degree. Capstone courses are required in each major and students are encouraged to complete the national standardized Fundamentals of Engineering exam during their last year. Average number of years required to actually complete the bachelor's degree is 4.5.
DEGREE PROGRAMS DESCRIPTION
The B.S. degrees in the College's engineering programs prepare the graduate for entry level professional positions or for graduate study in various engineering disciplines. The Master of Engineering (non-thesis), Master of Science (thesis) and Ph.D. degree program are offered in each engineering department. All of the undergraduate degree programs in Engineering are ABET accredited.
DUAL DEGREE PROGRAMS
Engineering baccalaureate graduates with dual degrees 1
Enrollment requirements: Any engineering program and any program in the College of Arts and Letters (a five year program) limited to National Merit Semi-finalist caliber.
TRANSFER INFORMATION
Residency Requirements: Student must earn a minimum of 30 semester credits at the university.
Transfer via Articulation Agreements
Admission to engineering: A range of articulation agreements including 3 + 2, 2 + 2, and transfers are admitted at any time.
Engineering graduates transferred from ... 4-yr: –　2-yr: 178
Requirements: Must be in good academic standing at previous institution. Only grades of C or better transfer.
Transfer without Articulation Agreements
Admission to engineering: At any time
Requirements: Must be in good academic standing at previous institution. Only grades of C or better transfer.
GRADUATION REQUIREMENTS
B.S. programs in engineering require 135-138 semester hours, depending on major; exit writing exam; 2.00 GPA in major and overall.

■ STUDENT PROGRAMS
PROFESSIONAL AND HONORARY SOCIETIES
[For key to acronyms, see *Introduction*.]
Professional Societies: ACI, AGCA, AIAA, AIPE, ASCE, ASHRAE, ASME, IEEE, SAE, SME
Honorary Societies: Chi Epsilon, Eta Kappa Nu, Pi Tau Sigma, Tau Alpha Pi, Tau Beta Pi, Theta Tau
SUPPORT PROGRAMS
Student Chapter Organizations: Society of Women Engineers, Fellowship of Minority Engineers and Scientists, Affiliated with Nat'l Society of Black Engineers

Women's Center
For: Women　　　　　　　**Available:** Year round
Offered: Freshman, Sophomore, Junior, Senior, Graduate level

Under-Represented Minorities in Engr Programs
For: Ethnic Minorities　　　**Available:** Academic year
Offered: Freshman, Sophomore, Junior, Senior

Other Engineering Support Programs: The University conducts a summer orientation program for freshmen. During the program all testing, advising and registration is completed.

177 Oregon State University

■ INSTITUTION PROFILE

HEAD OF THE INSTITUTION
John V. Byrne
Phone: (503)737-4133 Fax: (503)737-3033

GENERAL INFORMATION
[All Students—Fall 1991]
Undergraduate enrollment 11,000
Graduate enrollment 3,000
Total institution enrollment **14,000**
Type of institution: Public Calendar: Quarters
Nearest city: Eugene, Oregon Population: 100,000
Miles from main campus: 45 Setting: Small Town
Types of engineering degrees: Engineering

Other degree-granting colleges: Agricultural & Environmental, Business Administration, Fine & Performing Arts, Humanities & Social Sciences, Pharmacy, Technology & Applied Sciences, Forestry, Health and Human Performance, Home Economics and Education, Oceanography, Science

■ ENGINEERING ADMISSIONS

ADMISSIONS OFFICE CONTACT
Kay Conrad
Oregon State University
Office of Admissions
Administrative Services B104
Corvallis, OR 97331-2106
Phone: (503)737-4411 Fax: (503)737-2400

ENGINEERING COLLEGE ADMISSIONS INFORMATION
Entrance Requirements: High School courses—English (4 years), Mathematics (3 years), Social Studies (3 years), Science (2 years), SAT or ACT required for placement purposes. No minimum score required unless high school GPA is less than 3.0.
Entrance Recommendations: High School courses—Chemistry (1 year), Physics (1 year), Trigonometry/Pre-Calculus (1 year)
Requirements for foreign students: TOEFL (Score: 550); financial statement

DEPARTMENTAL ADMISSIONS INFORMATION
Admission to the engineering department: At the time of admission to the institution.

■ FINANCIAL INFORMATION

ESTIMATED EXPENSES (ACADEMIC YEAR)
[Expenses are for the 1992-93 nine-month academic year.]

State Residents	Undergraduate	Graduate
Tuition and fees	$ 2,600	$ 3,780
College room and board	$ 3,600	$ 3,600
Books and supplies	$ 500	$ 600
Other expenses	$ –	$ –
Total estimated expenses	**$ 6,700**	**$ 7,980**

Out-of-State Residents	Undergraduate	Graduate
Tuition and fees	$ 6,900	$ 5,800
College room and board	$ 3,600	$ 3,600
Books and supplies	$ 500	$ 600
Other expenses	$ –	$ –
Total estimated expenses	**$11,000**	**$10,000**

FINANCIAL AID OFFICE CONTACT
Keith McCreight, *Director, Financial Aid Office*
Oregon State University
Administrative Services A218
Corvallis, OR 97331-2120
Phone: (503)737-2241

GENERAL FINANCIAL AID INFORMATION
Forms accepted/required: FAF
Additional financial aid information: Scholarships, grants, loans, and part-time employment are available.

■ ENGINEERING COLLEGE INFORMATION

[For additional personnel, refer to the *Appendix*.]

HEAD OF ENGINEERING
S John T. Owen
Phone: (503)737-4525 Fax: (503)737-3462
EMail: owensj@ccmail.orst.edu

ENGINEERING COLLEGE ADDRESS
College of Engineering
Oregon State University
Covell Hall 101
Corvallis, OR 97331-2409
Phone: (503)737-4525 Fax: (503)737-3462

ENROLLMENTS—BY CLASS
[Numbers are baccalaureate enrollments for the fall 1991 term, unless otherwise footnoted.]
1st-year students/Freshmen .... 506 [1]
2nd-year students/Sophomores .... 701 [1]
3rd-year students/Juniors .... 344 [1]
4th-year students/Seniors .... 496 [1]
Total **2,047**

Notes: (1) Includes non-ABET programs in College of Engineering

NUMBER OF DEGREES AWARDED—BY PROGRAM
[Numbers are engineering baccalaureate degrees awarded during the 1991-1992 academic year, unless otherwise footnoted. For full details about each engineering program, refer to the *Tables of Degree Programs*.]
Chemical Engineering 21
Civil Engineering 64
Computer Engineering 19
Computer Science 88
Construction Engineering Management 34
Electrical & Electronics Engineering 65
Engineering Physics 7
Industrial Engineering 45
Mechanical Engineering 73
Nuclear Engineering 12
Total **428**

PERCENTAGE OF DEGREES AWARDED—BY CATEGORY
[Percentages are of all engineering baccalaureate degrees awarded during the 1991-1992 academic year, unless otherwise footnoted.]
African Americans .... – % [A]
Asian/Pacific Island Americans .... – % [A]
Hispanic Americans .... – % [A]
Native Americans .... – % [A]
Foreign Citizens .... – % [A]
All Others .... 100 %
Women .... 12.0 %
Persons with disabilities .... – % [A]
Students over 25 years of age .... – % [A]

Notes: (A) Data not available.

GRADUATE ENROLLMENTS & DEGREES AWARDED
Master's enrollment .... 340 [1]
Master's degrees awarded .... 137 [1]
Doctoral enrollment .... 175 [1]
Doctoral degrees awarded .... 31 [1]

Notes: (1) Includes non-ABET programs in College of Engineering

ENGINEERING STUDENT DATA
Applicants to the engineering college
Number of applicants to engineering college 1,304
Percent offered admission 84.0%
Matriculated engineering students
Percentage in top quartile (25%) of High School class .... – [A]
Average SAT scores: Combined—1040 [C]

Notes: (A) Data not available. (C) Estimated data.

FULL- & PART-TIME FACULTY—BY DEPARTMENT
[Figures are the head count of full-time faculty and the full-time equivalent (FTE) of part-time faculty for each engineering department or equivalent.]

Department	Full	Part
Chemical Eng	7	–
Civil Eng	26	–
Electrical & Computer Eng	28	–
Industrial & Manufacturing Eng	11	–
Mechanical Eng	21	–
Computer Science	15	–
Eng Physics	8	–
Nuclear Eng	5	–

COLLEGE DESCRIPTION
Since its origin in 1893, the college has graduated over 20,000 students and has established a solid reputation for high quality engineering education. Undergraduate programs are organized into two years of Pre-Professional study and two years of Professional study. Students may take the Pre-Professional program in the College or at one of many transfer schools that offer an equivalent curriculum. Admission to a Professional Program is competitive and based upon the Pre-Professional academic record. Two six month co-op experiences are required for the Manufacturing Option in the Industrial Engineering program. An optional co-op program is available to students in Chemical Engineering, Computer Engineering, Electrical and Electronics Engineering, and Computer Science. Co-op students require five years to complete a degree.

DEGREE PROGRAMS DESCRIPTION
The curriculum leading to the B.S. degree prepares students for entry level professional positions in engineering or computer science. Advancement to higher professional levels requires additional knowledge gained through experience or graduate study plus experience gained through professional practice. B.S. degree graduates are also prepared for advanced studies at the M.S. and Ph.D. levels.

TRANSFER INFORMATION
Residency Requirements: Must complete last 45 credits at OSU.

Transfer without Articulation Agreements
Admission to engineering: At any time
Requirements: Transfer students must complete 36 transferable credits (including English composition and math) with a GPA of 2.25.

GRADUATION REQUIREMENTS
Graduation requires a 2.25 GPA in required courses.

■ STUDENT PROGRAMS

PROFESSIONAL AND HONORARY SOCIETIES
[For key to acronyms, see *Introduction*.]
Professional Societies: ACM, AGCA, AIChE, ANS, ASAE, ASCE, ASME, IEEE, IIE, SME
Honorary Societies: Eta Kappa Nu, Pi Tau Sigma, Tau Beta Pi

SUPPORT PROGRAMS
Student Chapter Organizations: Society of Women Engineers

Educational Opportunities Program
For: Ethnic Minorities Available: Year round
Offered: Prematriculation, Freshman, Sophomore, Junior, Senior, Graduate level

Black Student Union
For: Ethnic Minorities Available: Year round
Offered: Prematriculation, Freshman, Sophomore, Junior, Senior, Graduate level

Hispanic Student Union
For: Ethnic Minorities Available: Year round
Offered: Prematriculation, Freshman, Sophomore, Junior, Senior, Graduate level

Native American Student Association
For: Ethnic Minorities Available: Year round
Offered: Prematriculation, Freshman, Sophomore, Junior, Senior, Graduate level

Other Engineering Support Programs: A General Engineering program is available for Freshmen who have not decided on a major but who are interested in engineering. There is a special interest dormitory for women in engineering and science. Tau Beta Pi offers free tutoring for lower division students.

178 University of the Pacific

■ INSTITUTION PROFILE
HEAD OF THE INSTITUTION
Bill Atchley
Phone: (209)946-2222 Fax: (209)946-2652
GENERAL INFORMATION
[All Students—Fall 1991]
Undergraduate enrollment 3,492
Graduate enrollment 503
Total institution enrollment **3,995**
Type of institution: Private **Calendar:** Semesters
Nearest city: Sacramento **Population:** 500,000
Miles from main campus: 40 **Setting:** Suburban
Types of engineering degrees: Engineering
Other degree-granting colleges: Dentistry, Law

■ ENGINEERING ADMISSIONS
ADMISSIONS OFFICE CONTACT
Edward L. Schoenberg
University of the Pacific
3601 Pacific Avenue
Stockton, CA 95211
Phone: (209)946-2211
ENGINEERING COLLEGE ADMISSIONS INFORMATION
Entrance Recommendations: SAT (Score: 990), High School courses—English (4 years), Math (4 years), Foreign Language (1 year), U.S. History/Government (1 year)
Requirements for foreign students: TOEFL (Score: 475)
DEPARTMENTAL ADMISSIONS INFORMATION
Admission to the engineering department: At the time of admission to the institution.

■ FINANCIAL INFORMATION
ESTIMATED EXPENSES (ACADEMIC YEAR)
[Expenses are for the 1992-93 nine-month academic year.]

All Students	Undergraduate	Graduate
Tuition and fees	$14,660	$14,660
College room and board	$5,300	$5,300
Books and supplies	$750	$500
Other expenses	$330	$330
Total estimated expenses	**$21,040**	**$20,790**

FINANCIAL AID OFFICE CONTACT
Lynn Fox, Director
University of the Pacific
3601 Pacific Avenue
Stockton, CA 95211
Phone: (209)946-2421
GENERAL FINANCIAL AID INFORMATION
Forms accepted/required: FAF, FAT, FFS, IRS, SAR, SDF
Additional financial aid information: Scholarships, grants, loans and job opportunities.

■ ENGINEERING COLLEGE INFORMATION
[For additional personnel, refer to the *Appendix*.]
HEAD OF ENGINEERING
Ashland Brown
Phone: (209)946-3091 Fax: (209)946-3086
ENGINEERING COLLEGE ADDRESS
School of Engineering
University of the Pacific
3601 Pacific Avenue
Stockton, CA 95211
Phone: (209)946-3091 Fax: (209)946-3086

ENROLLMENTS—BY CLASS
[Numbers are baccalaureate enrollments for the fall 1991 term, unless otherwise footnoted.]
1st-year students/Freshmen 64
2nd-year students/Sophomores 41
3rd-year students/Juniors 111
4th-year students/Seniors 58
Total **274**

NUMBER OF DEGREES AWARDED—BY PROGRAM
[Numbers are engineering baccalaureate degrees awarded during the 1991-1992 academic year, unless otherwise footnoted. For full details about each engineering program, refer to the *Tables of Degree Programs*.]
Civil Engineering 18
Electrical & Computer Engineering 26
Engineering Management 6
Engineering Physics 2
Mechanical Engineering 11
Total **63**

PERCENTAGE OF DEGREES AWARDED—BY CATEGORY
[Percentages are of all engineering baccalaureate degrees awarded during the 1991-1992 academic year, unless otherwise footnoted.]
African Americans 2.4%
Asian/Pacific Island Americans 12.2%
Hispanic Americans 4.9%
Native Americans .... – %
Foreign Citizens 43.9%
All Others 36.6%
Women 11.0%
Persons with disabilities .... – %
Students over 25 years of age .... – %

GRADUATE ENROLLMENTS & DEGREES AWARDED
Master's enrollment 6
Master's degrees awarded 1
Doctoral enrollment .... – (1)
Doctoral degrees awarded .... – (1)

Notes: (1) School has no doctoral programs.

ENGINEERING STUDENT DATA
Applicants to the engineering college
Number of applicants to engineering college 188
Percent offered admission 85.0%

Matriculated engineering students
Percentage in top quartile (25%) of High School class 53.0
Average SAT scores: Math—588, Verbal—436, Combined—1024

FULL- & PART-TIME FACULTY—BY DEPARTMENT
[Figures are the head count of full-time faculty and the full-time equivalent (FTE) of part-time faculty for each engineering department or equivalent.]

Department	Full	Part
Civil Eng	5	–
Electrical & Computer Eng	7	0.5
Mechanical Eng	4	–
Eng Physics	4	–
Eng Management	3	–

COLLEGE DESCRIPTION
The School of Engineering provides a broad undergraduate program leading to a Bachelor of Science degree in five areas of study. The engineering course work provides the specialized education serving as the foundation for the professional career, and it is reinforced with one year of full-time employment through the mandatory Cooperative Education Program. Under the Co-op plan, students alternate two paid work experiences with periods of study. Thus, the Co-op Program requires a five year course of study. The University's five year guarantee ensures that the student will get the classes needed to graduate at the time scheduled in the program. Four of the engineering programs are accredited by ABET: Civil, Computer, Electrical, Mechanical and Engineering Physics. An undergraduate degree in Engineering Management is also offered. All engineering students have the opportunity to participate in the International Engineering Minor program.

DEGREE PROGRAMS DESCRIPTION
Civil, BS Through use of upper division electives students concentrate coursework in several areas including engineering, fluid mechanics, engineering mechanics and materials and environmental and geotechnical engineering. Electrical,BS,MS Communications, energy conversion and control systems, digital systems and microcomputers. Mechanical,BS, Energy systems or thermal sciences, mechanical systems or applied mechanics. Computer Engineering, BS, Computer design, application and design of related hardware and software. Combines the necessary elements of electrical engineering and computer science. Engineering Physics, BS, Provides broad training in physics and math and basic training in engineering and design. Engineering Management, BS, Combines study in engineering and selected courses in economics and business administration.

DUAL DEGREE PROGRAMS
Engineering baccalaureate graduates with dual degrees 3

TRANSFER INFORMATION
Residency Requirements: A transfer student must take at least 16 of the last 32 units in the School of Engineering with a minimum of 12 of the 16 units from courses in his/her discipline.

Transfer via Articulation Agreements
Admission to engineering: At any time
Engineering graduates transferred from ... 4-yr: 4 2-yr: 18
Requirements: Minimum GPA of 2.0 and a maximum of 64 units.

Transfer without Articulation Agreements
Admission to engineering: At any time
Engineering graduates transferred from ... 4-yr: 4 2-yr: 18
Requirements: A minimum GPA of 2.0 and a maximum of 64 units accepted from two year community college.

GRADUATION REQUIREMENTS
Unit requirements depends on major selected. All students must have a minimum GPA of 2.0 for all coursework (cumulative) counting towards the degree, all coursework taken at UOP, and all engineering coursework. In addition, all U.S. citizens must complete 50 weeks of co-operative education credits.

■ STUDENT PROGRAMS
PROFESSIONAL AND HONORARY SOCIETIES
[For key to acronyms, see *Introduction*.]
Professional Societies: ASCE, ASEM, ASME, IEEE
Honorary Societies: Eta Kappa Nu, Tau Beta Pi
SUPPORT PROGRAMS
Student Chapter Organizations: Society of Women Engineers, National Society of Black Engineers, Society of Hispanic Professional Engineers

MEP Minority Engineering Program
For: Women & Ethnic Minorities **Available:** Year round
Offered: Prematriculation, Freshman, Sophomore, Junior, Senior

CIP Community Involvement Program
For: Women & Ethnic Minorities **Available:** Year round
Offered: Prematriculation, Freshman, Sophomore, Junior, Senior
Other Engineering Support Programs: Fall and Spring Student-to-Student Advising Programs for freshman and transfer students. School of Engrg. Co-op Placement; Academic Skills Center; Tau Beta Pi Tutoring Center; Professional Student Chapters for each of the six majors; Counseling Services, Foreign Student Services; Office of Student Life.

179 University of Pennsylvania

INSTITUTION PROFILE

HEAD OF THE INSTITUTION
Sheldon Hackney
Phone: (215)898-7221   Fax: (215)898-9659

GENERAL INFORMATION
[All Students—Fall 1991]
Undergraduate enrollment 11,311
Graduate enrollment 10,863
Total institution enrollment **22,174**

Type of institution: Private   Calendar: Semesters
Location: Philadelphia   Population: 1,585,000
Setting: Urban
Types of engineering degrees: Engineering
Other degree-granting colleges: Arts & Sciences, Business Administration, Communication & Journalism, Dentistry, Education, Fine & Performing Arts, Law, Medicine, Nursing, Veterinary Medicine, Social Work

ENGINEERING ADMISSIONS

ADMISSIONS OFFICE CONTACT
Willis J. Stetson
University of Pennsylvania
1 College Hall
Philadelphia, PA 19104-6376
Phone: (215)898-7507   Fax: (215)898-9670

ENGINEERING COLLEGE ADMISSIONS INFORMATION
Entrance Requirements: SAT, Most successful applicants have taken the most rigorous program available at their school. SAT verbal and math tests are required, plus three achievement tests including English Composition and Mathematics.
Entrance Recommendations: High School courses—Mathematics (4 years), Science (4 years), SAT test scores are preferred but ACT may be taken instead.
Requirements for foreign students: TOEFL

DEPARTMENTAL ADMISSIONS INFORMATION
Admission to the engineering department: Most are admitted at admission to institution; curriculum deferred students defer decision to end of first year.

FINANCIAL INFORMATION

ESTIMATED EXPENSES (ACADEMIC YEAR)
[Expenses are for the 1992-93 nine-month academic year.]

All Students	Undergraduate	Graduate
Tuition and fees	$16,838	$17,834
College room and board	$6,330	$6,330
Books and supplies	$480	$480
Other expenses	$1,202	$1,202
Total estimated expenses	**$24,850**	**$25,846**

FINANCIAL AID OFFICE CONTACT
William M. Schilling, *Director of Student Financial Aid*
University of Pennsylvania
212 Franklin Building
Philadelphia, PA 19104-6270
Phone: (215)898-6784   Fax: (215)573-5428

GENERAL FINANCIAL AID INFORMATION
Forms accepted/required: FAF, IRS
Additional financial aid information: All financial aid is need-based and includes grants, loans, and work-study jobs.

ENGINEERING COLLEGE INFORMATION
[For additional personnel, refer to the *Appendix*.]

HEAD OF ENGINEERING
Gregory C. Farrington
Phone: (215)898-7244   Fax: (215)573-2018
EMail: farringt@eniac.seas.upenn.edu

ENGINEERING COLLEGE ADDRESS
School of Engineering and Applied Science
University of Pennsylvania
109 Towne Building
Philadelphia, PA 19104-6391
Phone: (215)898-7246   Fax: (215)898-1130

ENROLLMENTS—BY CLASS
[Numbers are baccalaureate enrollments for the fall 1991 term, unless otherwise footnoted.]
1st-year students/Freshmen 395
2nd-year students/Sophomores 323
3rd-year students/Juniors 330
4th-year students/Seniors 326
Total **1,374**

NUMBER OF DEGREES AWARDED—BY PROGRAM
[Numbers are engineering baccalaureate degrees awarded during the 1991-1992 academic year, unless otherwise footnoted. For full details about each engineering program, refer to the *Tables of Degree Programs*.]
Bioengineering 25
Chemical Engineering 19
Computer Science & Engineering 43
Electrical Engineering 41
Materials Science & Engineering 14
Mechanical Engineering & Applied Mechanics 31
Systems Science & Engineering 27
Total **200**

PERCENTAGE OF DEGREES AWARDED—BY CATEGORY
[Percentages are of all engineering baccalaureate degrees awarded during the 1991-1992 academic year, unless otherwise footnoted.]
African Americans 4.8%
Asian/Pacific Island Americans 14.7%
Hispanic Americans 3.2%
Native Americans .... – %
Foreign Citizens 13.1%
All Others 64.2%
Women 24.3%
Persons with disabilities .... – % (A)
Students over 25 years of age 2.0%

Notes: (A) Data not available.

GRADUATE ENROLLMENTS & DEGREES AWARDED
Master's enrollment 185
Master's degrees awarded 158
Doctoral enrollment 473
Doctoral degrees awarded 54

ENGINEERING STUDENT DATA
Applicants to the engineering college
Number of applicants to engineering college 1,716
Percent offered admission 53.4%

Matriculated engineering students
Percentage in top quartile (25%) of High School class ... 100.0
Average SAT scores: Math—706, Verbal—580, Combined—1286

FULL- & PART-TIME FACULTY—BY DEPARTMENT
[Figures are the head count of full-time faculty and the full-time equivalent (FTE) of part-time faculty for each engineering department or equivalent.]

Department	Full	Part
Chemical Eng	11	–
Electrical Eng	20	–
Mechanical Eng & Applied Mechanics	12	–
Materials Science & Eng	16	–
Bioengineering	12	–
Systems	10	–
Computer & Information Science	23	–

COLLEGE DESCRIPTION
Engineering at Penn is a relatively small engineering school (with about 1400 undergraduates and 105 faculty) in the heart of a major research institution with a full complement of graduate, professional, and other schools. Thus, academic opportunities abound, and the engineering curricula have been designed to be sufficiently flexible to allow students to take advantage of these opportunities. For example, 28 percent of our students are also pursuing a degree from one of the other undergraduate schools, mainly the Wharton School of Business, and School of Arts and Sciences. We offer only a full-time, four year curriculum. A co-op program is not available; however, the Career Planning and Placement Service is very effective in helping students find summer positions and internships.

DEGREE PROGRAMS DESCRIPTION
At the B.S. level, over one-quarter of our students are dual degree students, earning an engineering degree and another University of Penn degree. Included in this group are those students in the Management and Technology program who earn degrees from SEAS and the Wharton School of Business. Many research opportunities are available for undergraduates and a number submatriculate each year into our graduate programs. The University of Pennsylvania offers programs leading to the degrees of Master of Science in Engineering, Master of Science in Transportation, and Doctor of Philosophy. A number of special programs are available. The objective of the Executive Master of Science in Engineering Program is to provide a new, rigorous educational program which gives its graduates the intellectual base to exercise creative leadership in the development and commercialization of emerging technologies. Joint masters level programs of study are available between engineering and The Wharton School of Business, the School of Medicine, and the Departments of Architecture and City and Regional Planning within the Graduate School of Fine Arts.

DUAL DEGREE PROGRAMS
Engineering baccalaureate graduates with dual degrees 64

TRANSFER INFORMATION
Residency Requirements: Two years of study at the University of Pennsylvania.

Transfer without Articulation Agreements
Admission to engineering: At any time
Engineering graduates transferred from   .. 4-yr: 20   2-yr: –

Requirements: Students transferring after 1 year of college, 1 course in chemistry, 1 course in physics, 2 courses in calculus and 2 courses in the social sciences and humanities. After 2 years of college, 4 courses in math, 2 courses in physics, 1 course in chemistry, 3-4 courses in social sciences and humanities and several engineering courses.

GRADUATION REQUIREMENTS
Cumulative GPA must be 2.00 or higher plus all failures and incompletes must be cleared.

STUDENT PROGRAMS

PROFESSIONAL AND HONORARY SOCIETIES
[For key to acronyms, see *Introduction*.]
Professional Societies: AIChE, ASCE, ASME, IEEE, Bioengineering Society, Society of Systems Engineers
Honorary Societies: Eta Kappa Nu, Pi Tau Sigma, Tau Beta Pi, Hexagon Senior Society

SUPPORT PROGRAMS
Student Chapter Organizations: Society of Women Engineers, National Society of Black Engineers, Society of Hispanic Professional Engineers

Engineering Minority Programs
For: Ethnic Minorities   **Available:** Year round
Offered: Prematriculation, Freshman, Sophomore, Junior, Senior, Graduate level

Other Engineering Support Programs: A Pre-Freshman program; SEAS freshman coaching program involving collaborative working groups; Tutoring Center; Tau Beta Pi Tutoring specifically for engineering students; William Penn Scholars program, a collaborative working group and mentoring program for minority students; Hexagon Society and Career Planning programs for choosing majors and careers.

180 Pennsylvania State University

INSTITUTION PROFILE

HEAD OF THE INSTITUTION
Joab L. Thomas
Phone: (814)865-7611 Fax: (814)863-4631

GENERAL INFORMATION
[All Students—Fall 1991]
Undergraduate enrollment 59,883
Graduate enrollment 10,714
Total institution enrollment **70,597**

Type of institution: Public **Calendar:** Semesters
Nearest city: Harrisburg **Population:** 52,376
Miles from main campus: 89 **Setting:** Urban
Types of engineering degrees: Engineering & Technology
Other degree-granting colleges: Architecture, Business Administration, Communication & Journalism, Education, Agricultural Sciences, Earth & Mineral Sciences, Science, Health & Human Development, Liberal Arts

ENGINEERING ADMISSIONS

ADMISSIONS OFFICE CONTACT
Anna M. Griswold
Pennsylvania State University
201 Shields Building
University Park, PA 16802-1400
Phone: (814)865-5471 Fax: (814)863-7590

ENGINEERING COLLEGE ADMISSIONS INFORMATION
Entrance Requirements: High School courses—English (4 years), Math (3 years), Arts/Humanities/Social Sciences/foreign languages (5 years), Science (3 years), SAT or ACT.
Entrance Recommendations: High School courses—English (4 years), Math (at least one year of calculus) (3 years), Arts/Humanities/Social Sciences/foreign languages (5 years), Science (at least 1 yr of chem & 1 yr of physics) (3 years)
Requirements for out-of-state residents: Freshman admission decisions are based on required Carnegie units, final grades in academic subjects beginning with ninth grade, verbal and math scores of the SAT or ACT, and class rank for applicants who have taken AP and/or honors courses.
Requirements for foreign students: financial statement

DEPARTMENTAL ADMISSIONS INFORMATION
Admission to the engineering department: At the end of the second year.
Additional information: To enter any major in the College of Engineering, the following courses must be completed with a grade of C or better: Math 140, 141; Chem 12, 14; Physics 201. Also, students with a GPA of 3.00 or higher will receive their major of choice. Those students with a GPA below a 3.0 will be assigned to a major based on indicated preferences and their GPA ranking. (Enrollment controls are used in the College of Engineering.)

FINANCIAL INFORMATION

ESTIMATED EXPENSES (ACADEMIC YEAR)
[Expenses are for the 1992-93 nine-month academic year.]

State Residents	Undergraduate	Graduate
Tuition and fees	$ 5,018[3]	$ 5,558[1]
College room and board	$ 3,790[4]	$ 3,850[2]
Books and supplies	$ 464[C]	$ 464[C]
Other expenses	$ –	$ –
Total estimated expenses	**$ 9,272**	**$ 9,872**

Out-of-State Residents	Undergraduate	Graduate
Tuition and fees	$10,044[3]	$10,646[7]
College room and board	$ 3,790[4]	$ 3,850[2]
Books and supplies	$ 464[C]	$ 464[C]
Other expenses	$ –	$ –
Total estimated expenses	**$14,298**	**$14,960**

Notes: (C) Estimated data. (1) Tuition = $2554; surcharge = $200; computer fee = $35 (2) Double room with meal plan. (3) Includes tuition, tuition surcharge, and computer fee. (4) This is for a double room plus a meal plan which ranges from $870-$1040 per semester. (7) Tuition $5,088; surcharge $200; Computer fee $35

FINANCIAL AID OFFICE CONTACT
Anna M. Griswold, *Director, Student Aid*
Pennsylvania State University
314 Shields Building
University Park, PA 16802-1400
Phone: (814)865-6301 Fax: (814)863-0322

GENERAL FINANCIAL AID INFORMATION
Forms accepted/required: ACT, AFSA, CSS-FAF, FAF, FAT, Stafford, IRS, SAR, Supplemental, Institutional
Additional financial aid information: Financial Aid opportunities include need-based and merit-based scholarships, and part-time (wage payroll and work-study) employment.

ENGINEERING COLLEGE INFORMATION

[For additional personnel, refer to the *Appendix*.]

HEAD OF ENGINEERING
David N. Wormley
Phone: (814)865-7537 Fax: (814)863-4749
EMail: CHW1@PSUADMIN

ENGINEERING COLLEGE ADDRESS
College of Engineering
Pennsylvania State University
101 Hammond Building
University Park, PA 16802-1400
Phone: (814)865-7537 Fax: (814)863-4749

ENROLLMENTS—BY CLASS
[Numbers are baccalaureate enrollments for the fall 1991 term, unless otherwise footnoted.]
1st-year students/Freshmen 2,237
2nd-year students/Sophomores 1,649
3rd-year students/Juniors 1,428
4th-year students/Seniors 1,842
Total **7,156**

NUMBER OF DEGREES AWARDED—BY PROGRAM
[Numbers are engineering baccalaureate degrees awarded during the 1991-1992 academic year, unless otherwise footnoted. For full details about each engineering program, refer to the *Tables of Degree Programs*.]
Acoustics .... –
Aerospace Engineering 96
Agricultural Engineering 19
Architectural Engineering 90
Chemical Engineering 93
Civil Engineering 173
Computer Engineering 41
Electrical Engineering 240
Engineering Science & Mechanics 37
Industrial Engineering 132
Mechanical Engineering 223
Nuclear Engineering 11
Total **1,155**

PERCENTAGE OF DEGREES AWARDED—BY CATEGORY
[Percentages are of all engineering baccalaureate degrees awarded during the 1991-1992 academic year, unless otherwise footnoted.]
African Americans 1.3%
Asian/Pacific Island Americans 3.3%
Hispanic Americans 0.7%
Native Americans .... – %
Foreign Citizens 1.7%
All Others 93.0%
Women 11.8%
Persons with disabilities .... – %
Students over 25 years of age 8.0%

GRADUATE ENROLLMENTS & DEGREES AWARDED
Master's enrollment 557
Master's degrees awarded 392
Doctoral enrollment 459
Doctoral degrees awarded 137

ENGINEERING STUDENT DATA
Applicants to the engineering college
Number of applicants to engineering college 4,854
Percent offered admission 94.0%
Average SAT scores: Math—590, Verbal—476, Combined—1066

FULL- & PART-TIME FACULTY—BY DEPARTMENT
[Figures are the head count of full-time faculty and the full-time equivalent (FTE) of part-time faculty for each engineering department or equivalent.]

Department	Full	Part
Chemical Eng	19	1.0
Civil & Environmental Eng	28	2.0
Electrical & Computer Eng	48	9.0
Industrial & Management Systems Eng	23	1.0
Mechanical Eng	41	1.3
Aerospace Eng	16	0.5
Agricultural & Biological Eng	12	–
Architectural Eng	11	1.6
Eng Science & Mechanics	30	2.0
Nuclear Eng	11	–
Mining Eng	11	–
Ceramic Science & Eng	14	–
Metals Science & Eng	11	–
Petroleum & Natural Gas Eng	6	–
Bioengineering	5	–

COLLEGE DESCRIPTION
The mission of the College of Engineering is to provide high-quality education in engineering and engineering technology, to conduct leading research and scholarship, and to provide service to the profession, the Commonwealth, and the nation. Enrichment opportunities include: (a) University Scholars Program--honors course work is completed each year and a senior thesis is required; (b) study-abroad program; (c) dual-degree program with the College of Liberal Arts; (d) minors in bioeng, chem eng, eng mech, STS (sci, tech & soc); (e) student teaching intern program--undergrads, working with faculty, have the opportunity to experience teaching and associated responsibilities; (f) Project WISE (Workplace Integration Skills for Engineers); (g) co-op program--in all majors & requires one year of participation with 2-3 assignments. The College is a member of the Engineering Coalition of Schools for Excellence in Education and Leadership and the National Space Grant College Program.

DEGREE PROGRAMS DESCRIPTION
All departments offer the B.S., M.S., and Ph.D. degrees. Architectural Engineering offers a Bachelor of Architectural Engineering in a five-year program. Acoustics and Bioengineering do not offer a B.S. degree, although an undergraduate minor is available in Bioengineering. The Civil Engineering Department offers an Environmental Engineering option at the undergraduate level and an M.S., M.Eng., and Ph.D. in Environmental Engineering. The M.Eng. degree is not offered by Aerospace, Computer, Electrical, Ceramic Science, and Metals Engineering. A B.S. in Engineering with Mechanical or Electrical options is offered by the Behrend College in Erie. Four programs, Ceramic Science and Engineering, Metals Science and Engineering, Mining Engineering, and Petroleum and Natural Gas Engineering are offered by the College of Earth and Mineral Sciences.

DUAL DEGREE PROGRAMS
Engineering baccalaureate graduates with dual degrees 4
Enrollment requirements: Students spend 3 years in the College of Liberal Arts and transfer to engineering to complete their final two years. They will earn a B.A. and a B.S.

TRANSFER INFORMATION
Residency Requirements: At least 36 of the last 60 credits must be in courses offered by the University or in cooperative degree programs that have been established by formal agreement and approved by the University Faculty Senate.

Transfer via Articulation Agreements
Admission to engineering: Approximately 23 schools participate in a 3-2 program with our College. The students spend the first 3 years at a 3-2 participating school and complete the final 2 years at Penn State.
Engineering graduates transferred from ... 4-yr: 60 2-yr: 11
Requirements: A minimum GPA of 3.0 is required for admission to any of our engineering programs.

Transfer without Articulation Agreements
Admission to engineering: At any time
Requirements: A minimum GPA of 3.50 is required for admission to civil, computer, and mechanical engineering.

GRADUATION REQUIREMENTS
A student must have obtained at least a 2.00 cumulative GPA and a 2.00 GPA in the major in order to receive a B.S. degree in the College of Engineering.

STUDENT PROGRAMS

PROFESSIONAL AND HONORARY SOCIETIES
[For key to acronyms, see *Introduction*.]
Professional Societies: AHS, AIAA, AIChE, ANS, ASCE, ASME, IEEE, IIE, SAE, SES
Honorary Societies: Alpha Pi Mu, Chi Epsilon, Eta Kappa Nu, Omega Chi Epsilon, Phi Alpha Epsilon, Pi Tau Sigma, Sigma Gamma Tau, Tau Beta Pi

SUPPORT PROGRAMS
Student Chapter Organizations: Society of Women Engineers, National Society of Black Engineers, Society of Hispanic Professional Engineers

Minority Engineering Program
For: Ethnic Minorities **Available:** Year round
Offered: Prematriculation, Freshman, Sophomore, Junior, Senior, Graduate level

Women in Engineering Program
For: Women **Available:** Year round
Offered: Prematriculation, Freshman, Sophomore, Junior, Senior, Graduate level

Other Engineering Support Programs: Orientations are conducted before each semester for all freshmen and transfer students. Students have access to Career Development and Placement Services, Academic Assistance Program, Center for Women Students, Returning Adult Student Center, Developmental Year Program, Veterans Assistance Office, International Student Office, & Counseling Center.

181 University of Pittsburgh

INSTITUTION PROFILE

HEAD OF THE INSTITUTION
John D. O'Connor
Phone: (412)624-4200 Fax: (412)624-1150

GENERAL INFORMATION
[All Students—Fall 1991]
Undergraduate enrollment 18,250
Graduate enrollment 9,723
Total institution enrollment 27,973

Type of institution: Private **Calendar:** Trimesters
Location: Pittsburgh **Population:** 2,242,798
Setting: Urban
Types of engineering degrees: Engineering

Other degree-granting colleges: Allied Health Sciences, Arts & Sciences, Business Administration, Dentistry, Education, Law, Medicine, Nursing, Pharmacy, Technology & Applied Sciences, Social Work, Library and Information Sciences, Public and International Affairs, Public Health, General Studies

ENGINEERING ADMISSIONS

ADMISSIONS OFFICE CONTACT
Betsy A. Porter
University of Pittsburgh
Director of Admissions
2nd Floor, Bruce Hall
Pittsburgh, PA 15260
Phone: (412)624-7488 Fax: (412)648-8815

ENGINEERING COLLEGE ADMISSIONS INFORMATION
Entrance Requirements: SAT (Score: 1100), High School courses—English, mathematics (4 years), Chemistry, physics, history (1 year), Academic electives (4.5 years)
Entrance Recommendations: ACT (Score: 27), High School courses—Computers (1 year)
Requirements for foreign students: TOEFL (Score: 500); financial statement

DEPARTMENTAL ADMISSIONS INFORMATION
Admission to the engineering department: At the end of the first year.
Additional information: 2.00 (on a 4-point scale) at end of freshman engineering program; 2.50 for regional campus relocatees.

FINANCIAL INFORMATION

ESTIMATED EXPENSES (ACADEMIC YEAR)
[Expenses are for the 1992-93 nine-month academic year.]

State Residents	Undergraduate	Graduate
Tuition and fees	$ 5,900	$ 7,390
College room and board	$ 4,250	$ –
Books and supplies	$ 600	$ 600
Other expenses	$ 376 [1]	$ 326 [1]
Total estimated expenses	**$11,126**	**$ 8,316**

Out-of-State Residents	Undergraduate	Graduate
Tuition and fees	$12,710	$14,780
College room and board	$ 4,250	$ –
Books and supplies	$ 600	$ 600
Other expenses	$ 376 [1]	$ 326 [1]
Total estimated expenses	**$17,936**	**$15,706**

Notes: (1) Computer fee, health fee, and student activity fee.

FINANCIAL AID OFFICE CONTACT
Betsy A. Porter, *Director of Admissions and Financial Aid*
University of Pittsburgh
2nd Floor, Bruce Hall
Pittsburgh, PA 15260
Phone: (412)624-7488 Fax: (412)648-8815

GENERAL FINANCIAL AID INFORMATION
Forms accepted/required: FAF, FAT, Stafford, IRS, SAR, Supplemental, Pennsylvania State Grant application for PA residents., Out-of-state students check with their state agencies.
Additional financial aid information: Freshmen must submit the FAF. PA residents must complete PHEAA application for state grant eligibility. Student and parent tax returns are also required. Students are reviewed for both grants, need-based scholarships, and self-help aid (workstudy and loans) following submission of aid forms. Merit-based scholarships are part of the admission process and do not require a separate application.

ENGINEERING COLLEGE INFORMATION
[For additional personnel, refer to the *Appendix*.]

HEAD OF ENGINEERING
Charles A. Sorber
Phone: (412)624-9800 Fax: (412)624-1108

ENGINEERING COLLEGE ADDRESS
School of Engineering
University of Pittsburgh
253 Benedum Engineering Hall
Pittsburgh, PA 15261
Phone: (412)624-9800 Fax: (412)624-1108

ENROLLMENTS—BY CLASS
[Numbers are baccalaureate enrollments for the fall 1991 term, unless otherwise footnoted.]
1st-year students/Freshmen 287
2nd-year students/Sophomores 399
3rd-year students/Juniors 372
4th-year students/Seniors 452
Total **1,510**

NUMBER OF DEGREES AWARDED—BY PROGRAM
[Numbers are engineering baccalaureate degrees awarded during the 1991-1992 academic year, unless otherwise footnoted. For full details about each engineering program, refer to the *Tables of Degree Programs*.]

Bioengineering .... – (3)
Chemical & Petroleum Engineering .... – (3)
Chemical Engineering 34
Civil Engineering 48
Electrical Engineering 72
Energy Resources .... – (3)
Engineering Physics 1
Industrial Engineering 41
Interdisciplinary Engineering .... – (1)
Manufacturing Systems Engineering .... – (3)
Materials Science & Engineering 7
Mechanical Engineering 90
Metallurgical Engineering 8
Mining Engineering .... – (3)
Petroleum Engineering .... – (3)
Public Works .... – (3)
Public Works & Civil Engineering .... – (3)
Total **301**

Notes: (1) Graduate-level program. (3) Not applicable. This is a masters-level program.

PERCENTAGE OF DEGREES AWARDED—BY CATEGORY
[Percentages are of all engineering baccalaureate degrees awarded during the 1991-1992 academic year, unless otherwise footnoted.]
African Americans .... 3.7 %
Asian/Pacific Island Americans .... 3.3 %
Hispanic Americans .... 0.7 %
Native Americans .... – %
Foreign Citizens .... 0.7 %
All Others .... 91.6 %
Women .... 17.9 %
Persons with disabilities .... 0.3 %
Students over 25 years of age .... 10.6 %

GRADUATE ENROLLMENTS & DEGREES AWARDED
Master's enrollment 547
Master's degrees awarded 146
Doctoral enrollment 176
Doctoral degrees awarded 40

ENGINEERING STUDENT DATA

Applicants to the engineering college
Number of applicants to engineering college 799
Percent offered admission .... 73.0 %

Matriculated engineering students
Percentage in top quartile (25%) of High School class 77.6
Average SAT scores: Math—611, Verbal—501, Combined—1112

FULL- & PART-TIME FACULTY—BY DEPARTMENT
[Figures are the head count of full-time faculty and the full-time equivalent (FTE) of part-time faculty for each engineering department or equivalent.]

Department	Full	Part
Civil Eng	17	0.2
Chemical & Petroleum Eng	20	0.3
Electrical Eng	21	–
Industrial Eng	14	0.2
Mechanical Eng	18	0.2
Materials Science & Eng	19	–
Freshman Program & Career Development	–	–

COLLEGE DESCRIPTION
The School offers four-year undergraduate programs in engineering, emphasizing engineering and science fundamentals in the first year, followed by three years of specialized education in chemical, civil, electrical, engineering-physics, industrial, mechanical, materials science, or metallurgical engineering, with some required and elective liberal arts courses. The School also offers a five-year, dual-degree program leading to a B.A. or B.S. in liberal arts and a B.S. in engineering. Co-op is an option in all programs. Average number of years required to actually complete the bachelor's degree: 4.

DEGREE PROGRAMS DESCRIPTION
Full-time B.S. degree programs are offered in chemical, civil, electrical, engineering physics, industrial, materials science, mechanical and metallurgical engineering, with electrical and mechanical engineering also available for part-time students. Special programs can be structured based upon individual student interest and ability. A Co-Op Program is available. At the graduate level, the School offers programs leading to the M.S. and Ph.D. degrees in chemical, civil, electrical, industrial, mechanical, and materials science and metallurgical engineering. M.S. degree programs in bioengineering, manufacturing systems, mining, and petroleum engineering, and in energy resources are also available, as are professional master's programs in energy resources and public works administration. In addition, specialization in environmental systems management, operations research, engineering management, ceramics, and polymer engineering are offered. An Interdisciplinary Ph.D. program is available.

DUAL DEGREE PROGRAMS
Engineering baccalaureate graduates with dual degrees 4

TRANSFER INFORMATION
Residency Requirements: The work of the senior year (a minimum of 26 credits) must be completed while in residence at the School of Engineering, University of Pittsburgh.

Transfer via Articulation Agreements
Admission to engineering: At any time
Engineering graduates transferred from .. 4-yr: – 2-yr: –

Requirements: Upon completing the course work as listed in the articulation agreement with a 2.80 QPA or higher, students are guaranteed placement in the department of their choice. The articulation agreement was put in to effect in June 1991.

Transfer without Articulation Agreements
Admission to engineering: At any time
Engineering graduates transferred from .. 4-yr: 106 2-yr: 12

Requirements: 2.50 or higher on a 4.0 scale at the institution previously attended.

GRADUATION REQUIREMENTS
A B.S. student must have completed the required courses and earned the total number of credits required by the program in which the student is enrolled and have a minimum QPA. of 2.0 for (a) all courses completed at the University of Pittsburgh and (b) all departmental courses. Advanced-standing credits partially fulfill course requirements, but grades and credits earned in such courses are not included in the QPA. calculations. A minimum of 3.0 QPA. is required for graduate students.

STUDENT PROGRAMS

PROFESSIONAL AND HONORARY SOCIETIES
[For key to acronyms, see *Introduction*.]

Professional Societies: AIChE, ASCE, ASME, IEEE, IIE, NSPE, SAE, TMS, Society of Women Engineers (SWE), National Society of Black Engineers (NSBE)
Honorary Societies: Alpha Pi Mu, Chi Epsilon, Eta Kappa Nu, Omega Chi Epsilon, Pi Tau Sigma, Tau Beta Pi, Omega Rho

SUPPORT PROGRAMS
Student Chapter Organizations: Society of Women Engineers, National Society of Black Engineers

Pitt Engineering Impact Program (PEI-P)
For: Ethnic Minorities **Available:** Year round
Offered: Prematriculation, Freshman, Sophomore, Junior, Senior
Other Engineering Support Programs: Summer orientation program for all freshmen; Learning Skills Center; career counseling services; international services office; disabled student services/veterans' services; placement services; psychological services.

182 Polytechnic University

INSTITUTION PROFILE

HEAD OF THE INSTITUTION
George Bugliarello
Phone: (718)260-3500   Fax: (718)260-3755

GENERAL INFORMATION
[All Students—Fall 1991]
Undergraduate enrollment 1,541
Graduate enrollment 1,926
Total institution enrollment 3,467

Type of institution: Private   **Calendar:** Semesters
Location: New York   **Population:** 7,500,000
Setting: Urban
Types of engineering degrees: Engineering

ENGINEERING ADMISSIONS

ADMISSIONS OFFICE CONTACT
Ellen Hartigan
Polytechnic University
Six MetroTech Center
Brooklyn, NY 11201
Phone: (718)260-3100   Fax: (718)260-3136

ENGINEERING COLLEGE ADMISSIONS INFORMATION
Entrance Requirements: SAT (Score: 1000), ACT (Score: 24), High School courses—English (4 years), Math (4 years), Science (4 years), Other courses required for graduation in NY State, Minimum SAT 1000 combined score ACT 24 and Composite TOEFL 500 (international students only)
Entrance Recommendations: Three Achievement Tests: English, Mathematics I or II and Physics or Chemistry
Requirements for foreign students: TOEFL (Score: 500); Interviews are strongly recommended, but may be required of some candidates for admission. Certification of finances required.

DEPARTMENTAL ADMISSIONS INFORMATION
Admission to the engineering department: There are no special admission requirements.

FINANCIAL INFORMATION

ESTIMATED EXPENSES (ACADEMIC YEAR)
[Expenses are for the 1992-93 nine-month academic year.]

All Students	Undergraduate	Graduate
Tuition and fees	$15,000	$15,000
College room and board	$5,000	$5,000
Books and supplies	$600	$416
Other expenses	$1,858	$1,858
Total estimated expenses	$22,458	$22,274

FINANCIAL AID OFFICE CONTACT
Veronica Lukas, *Director of Financial Aid*
Polytechnic University
Six MetroTech Center
Brooklyn, NY 11201
Phone: (718)260-3300   Fax: (718)260-3136

GENERAL FINANCIAL AID INFORMATION
Forms accepted/required: ACT, FAF, Stafford, Institutional, Free Application for Federal Student Aid (1993-94)
Additional financial aid information: There are several types of financial aid available at the institution. They are as follows: merit-based scholarships, need-based grants, loans and on-campus employment.

ENGINEERING COLLEGE INFORMATION

[For additional personnel, refer to the *Appendix*.]

HEAD OF ENGINEERING
Roger P. Roess
Phone: (718)260-3550   Fax: (718)260-3136

ENGINEERING COLLEGE ADDRESS
Engineering Division
Polytechnic University
Six MetroTech Center
Brooklyn, NY 11201
Phone: (718)260-3600   Fax: (718)260-3136

ENROLLMENTS—BY CLASS
[Numbers are baccalaureate enrollments for the fall 1991 term, unless otherwise footnoted.]
1st-year students/Freshmen 464
2nd-year students/Sophomores 284
3rd-year students/Juniors 224
4th-year students/Seniors 256
Total 1,228

NUMBER OF DEGREES AWARDED—BY PROGRAM
[Numbers are engineering baccalaureate degrees awarded during the 1991-1992 academic year, unless otherwise footnoted. For full details about each engineering program, refer to the *Tables of Degree Programs*.]
Aerospace Engineering 27
Chemical Engineering 11
Civil & Environmental Engineering 34
Civil Engineering 34
Electrical Engineering 129
Industrial Engineering 14
Mechanical Engineering 49
Metallurgical Engineering & Material Science .... –
Total 298

PERCENTAGE OF DEGREES AWARDED—BY CATEGORY
[Percentages are of all engineering baccalaureate degrees awarded during the 1991-1992 academic year, unless otherwise footnoted.]
African Americans 5.0%
Asian/Pacific Island Americans 38.0%
Hispanic Americans 6.0%
Native Americans .... – %
Foreign Citizens 8.0%
All Others 43.0%
Women 12.0%
Persons with disabilities .... – %
Students over 25 years of age 16.0%

GRADUATE ENROLLMENTS & DEGREES AWARDED
Master's enrollment .... 744 [1]
Master's degrees awarded 386
Doctoral enrollment .... 193 [1]
Doctoral degrees awarded 41

Notes: (1) Excludes Computer Science

ENGINEERING STUDENT DATA
Applicants to the engineering college
Number of applicants to engineering college 1,162
Percent offered admission 68.0%

Matriculated engineering students
Percentage in top quartile (25%) of High School class 73.0
Average SAT scores: Math—480, Verbal—630, Combined—1110

FULL- & PART-TIME FACULTY—BY DEPARTMENT

[Figures are the head count of full-time faculty and the full-time equivalent (FTE) of part-time faculty for each engineering department or equivalent.]

Department	Full	Part
Chemical Eng	8	–
Civil Eng	16	5.7
Electrical Eng	39	9.0
Industrial Eng	4	3.0
Mechanical Eng	11	2.7
Aerospace Eng	7	1.0
Metallurgy & Materials Science	7	0.5

COLLEGE DESCRIPTION
Polytechnic is fully focused on programs relating to science and technology. Over 95% of the student body is enrolled in engineering and science programs. Thus, Polytechnic has the opportunity to focus all areas of its curricula, including mathematics, management and the liberal arts on the technological world around us. Polytechnic offers a Co-op option for all of its majors, but does not require it. Co-op students generally earn their bachelors degree in five years.

TRANSFER INFORMATION
Residency Requirements: All transfers (from within the school and from other institutions) must satisfy the university residence requirement.

Transfer without Articulation Agreements
Admission to engineering: At any time
Requirements: GPA required - 2.75 from accredited colleges or universities.

STUDENT PROGRAMS

PROFESSIONAL AND HONORARY SOCIETIES
[For key to acronyms, see *Introduction*.]
Professional Societies: ACM, AIAA, AIChE, ASCE, ASME, IEEE, IIE, SAE
Honorary Societies: Chi Epsilon, Eta Kappa Nu, Omega Chi Epsilon, Pi Tau Sigma, Sigma Gamma Tau, Tau Beta Pi

SUPPORT PROGRAMS
Student Chapter Organizations: Society of Women Engineers, National Society of Black Engineers, Society of Hispanic Professional Engineers, Higher Education Opportunity Program (HEOP)
Other Engineering Support Programs: Polytechnic maintains a number of additional student support programs including: unified freshman advising program; a required freshman orientation course lasting one semester; drop-in learning/tutoring centers on both campuses; an Office of Special services offering programmed tutoring, counseling, and related services.

183 University of Portland

■ INSTITUTION PROFILE
HEAD OF THE INSTITUTION
David T. Tyson
Phone: (503)283-7101 Fax: (503)283-7399

GENERAL INFORMATION
[All Students—Fall 1991]
Undergraduate enrollment 2,222
Graduate enrollment 424
Total institution enrollment 2,646

Type of institution: Private **Calendar:** Semesters
Location: Portland, OR **Population:** 450,000
Setting: Suburban
Types of engineering degrees: Engineering
Other degree-granting colleges: Arts & Sciences, Business Administration, Education, Nursing

■ ENGINEERING ADMISSIONS
ADMISSIONS OFFICE CONTACT
Daniel B. Reilly
University of Portland
5000 N. Willamette Boulevard
Portland, OR 97203-5798
Phone: (503)283-7147 Fax: (503)283-7399

ENGINEERING COLLEGE ADMISSIONS INFORMATION
Entrance Requirements: High School courses—Mathematics (3 years), English (3 years), SAT or ACT required.
Entrance Recommendations: High School courses—Mathematics (4 years), English (4 years), Physical and life science (3 years), Computer programming (1 year)
Requirements for foreign students: TOEFL (Score: 500)

DEPARTMENTAL ADMISSIONS INFORMATION
Admission to the engineering department: The student may select (or change to) a specific Engineering program at admission, or at any later time.

■ FINANCIAL INFORMATION
ESTIMATED EXPENSES (ACADEMIC YEAR)
[Expenses are for the 1992-93 nine-month academic year.]

All Students	Undergraduate	Graduate
Tuition and fees	$10,300	$ 8,200
College room and board	$ 3,790	$ 3,790
Books and supplies	$ 600	$ 600
Other expenses	$ 80	$ —
Total estimated expenses	$14,770	$12,590

FINANCIAL AID OFFICE CONTACT
Rita A. Lambert, *Director of Financial Aid*
University of Portland
5000 N. Willamette Boulevard
Portland, OR 97203-5798
Phone: (503)283-7311 Fax: (503)283-7399

GENERAL FINANCIAL AID INFORMATION
Forms accepted/required: CSS-FAF, Institutional, Free Application for Federal Student Aid (FAFSA)
Additional financial aid information: Need-based scholarships; merit-based scholarships; loans; part-time jobs on campus; College Work-Study Program.

■ ENGINEERING COLLEGE INFORMATION
[For additional personnel, refer to the *Appendix*.]
HEAD OF ENGINEERING
Thomas J. Nelson
Phone: (503)283-7314 Fax: (503)283-7399

ENGINEERING COLLEGE ADDRESS
Multnomah School of Engineering
University of Portland
5000 N. Willamette Boulevard
Portland, OR 97203-5798
Phone: (503)283-7314 Fax: (503)283-7399

ENROLLMENTS—BY CLASS
[Numbers are baccalaureate enrollments for the fall 1991 term, unless otherwise footnoted.]
1st-year students/Freshmen 61
2nd-year students/Sophomores 36
3rd-year students/Juniors 73
4th-year students/Seniors 94
Total 264

NUMBER OF DEGREES AWARDED—BY PROGRAM
[Numbers are engineering baccalaureate degrees awarded during the 1991-1992 academic year, unless otherwise footnoted. For full details about each engineering program, refer to the *Tables of Degree Programs*.]
Civil Engineering 3
Electrical Engineering 37
Engineering Management 2
Engineering Science(s) .... —
Mechanical Engineering 15
Total 57

PERCENTAGE OF DEGREES AWARDED—BY CATEGORY
[Percentages are of all engineering baccalaureate degrees awarded during the 1991-1992 academic year, unless otherwise footnoted.]
African Americans .... — %
Asian/Pacific Island Americans .... 14.0 %
Hispanic Americans .... — %
Native Americans .... — %
Foreign Citizens .... 25.0 %
All Others .... 61.0 %
Women .... 12.0 %
Persons with disabilities .... — %
Students over 25 years of age .... — % (A)

Notes: (A) Data not available.

GRADUATE ENROLLMENTS & DEGREES AWARDED
Master's enrollment 19
Master's degrees awarded 9
Doctoral enrollment .... — (B)
Doctoral degrees awarded .... — (B)

Notes: (B) Data not applicable.

ENGINEERING STUDENT DATA
Applicants to the engineering college
Number of applicants to engineering college 249
Percent offered admission 78.7%

Matriculated engineering students
Percentage in top quartile (25%) of High School class 79.9
Average SAT scores: Math—597, Verbal—478, Combined—1075

FULL- & PART-TIME FACULTY—BY DEPARTMENT
[Figures are the head count of full-time faculty and the full-time equivalent (FTE) of part-time faculty for each engineering department or equivalent.]

Department	Full	Part
Civil Eng	4	0.7
Electrical Eng	5	1.0
Mechanical Eng	7	0.3

COLLEGE DESCRIPTION
The University of Portland emphasizes undergraduate education. Teaching excellence is our primary goal. We offer careful counseling, individual attention to the student, and faculty accessibility. The Civil, Electrical, and Mechanical Engineering programs are accredited by EAC/ABET. All engineering classes are taught by professors: we do not use teaching assistants. Students are admitted directly into the Engineering School and they can major in any of our degree programs without further qualification. Superior computing and laboratory facilities supplement the theoretical material; easy access to these for undergraduate students is school policy. A broad liberal arts core complements the technical material, and technical writing is stressed. While we highlight the traditional student and campus residentiality, we also welcome the commuter and adult student.

DEGREE PROGRAMS DESCRIPTION
The B.S. degrees in the various engineering programs prepare the graduate for entry-level professional positions or for graduate study in each of the respective fields. The M.S. degree programs are aimed particularly at employed engineers who need additional coursework in a specific area of interest. Master's thesis is optional.

TRANSFER INFORMATION
Residency Requirements: The last 30 semester credits must be completed at the University. Seventy-five percent of the upper division courses in the major must be completed at the University.
Transfer via Articulation Agreements
Admission to engineering: At the end of the sophomore year
Engineering graduates transferred from .. 4-yr: — 2-yr: 3
Transfer without Articulation Agreements
Admission to engineering: At any time

GRADUATION REQUIREMENTS
2.0 GPA in all courses taken at the University of Portland, and 2.0 in the major field.

■ STUDENT PROGRAMS
PROFESSIONAL AND HONORARY SOCIETIES
[For key to acronyms, see *Introduction*.]
Professional Societies: ACM, ASCE, ASME, IEEE, ITE, NSPE, SAE
Honorary Societies: Eta Kappa Nu

SUPPORT PROGRAMS
Student Chapter Organizations: Society of Women Engineers

Minority Student Counselor
For: Ethnic Minorities **Available:** Academic year
Offered: Freshman, Sophomore, Junior, Senior, Graduate level

Family Away from Home
For: Ethnic Minorities **Available:** Year round
Offered: Freshman, Sophomore, Junior, Senior, Graduate level

Special Women's Counseling Group
For: Women **Available:** Academic year
Offered: Freshman, Sophomore, Junior, Senior, Graduate level

Other Engineering Support Programs: Summer preregistration for freshmen and transfers; Fall orientation program for freshmen and transfers; small-group meetings with freshmen; counseling and consulting services including test anxiety; career counseling services; placement services; international student office.

184 Portland State University

INSTITUTION PROFILE
HEAD OF THE INSTITUTION
Judith A. Ramaley
Phone: (503)725-4411 **Fax:** (503)725-4499

GENERAL INFORMATION
[All Students—Fall 1991]
Undergraduate enrollment 10,502
Graduate enrollment 3,783
Total institution enrollment **14,285**
Type of institution: Public **Calendar:** Quarters
Location: Portland Metropolitan **Population:** 1,535,400
Setting: Urban
Types of engineering degrees: Engineering
Other degree-granting colleges: Business Administration, Education, Fine & Performing Arts, Graduate School of Social Work, Urban and Public Affairs, Liberal Arts & Sciences, Extended Studies

ENGINEERING ADMISSIONS
ADMISSIONS OFFICE CONTACT
H. Chik M. Erzurumlu
Portland State University
P.O. Box 751
Portland, OR 97207-0751
Phone: (503)725-4631 **Fax:** (503)725-4298

ENGINEERING COLLEGE ADMISSIONS INFORMATION
Entrance Requirements: SAT, ACT, High School courses—English (4 years), Chemistry (1 year), Physics (1 year), Mathematics (4 years)
Entrance Recommendations: High School courses—Foreign Language (2 years), Computers (1 year), Mechanical Drawing (1 year)
Requirements for foreign students: TOEFL (Score: 525); financial statement

DEPARTMENTAL ADMISSIONS INFORMATION
Admission to the engineering department: At the end of the second year.
Additional information: The engineering departments require a minimum GPA of 2.25 (on a 4-point scale) in all engineering course work at the end of the sophomore year; all grades must be C or better.

FINANCIAL INFORMATION
ESTIMATED EXPENSES (ACADEMIC YEAR)
[Expenses are for the 1992-93 nine-month academic year.]

State Residents	Undergraduate	Graduate
Tuition and fees	$ 2,658	$ 3,633
College room and board	$ 4,168 [1]	$ 4,168 [1]
Books and supplies	$ 750	$ 750
Other expenses	$ 1,755 [2]	$ 1,755 [2]
Total estimated expenses	**$ 9,331**	**$ 10,306**

Out-of-State Residents	Undergraduate	Graduate
Tuition and fees	$ 6,939	$ 5,799
College room and board	$ 4,168 [1]	$ 4,168 [1]
Books and supplies	$ 750	$ 750
Other expenses	$ 1,755 [2]	$ 1,755 [2]
Total estimated expenses	**$13,612**	**$ 12,472**

Notes: (1) Room and board is offered through the PSS Residence Life Program. (2) Includes transportation and miscellaneous personal expenses.

FINANCIAL AID OFFICE CONTACT
John E. Anderson, *Director of Financial Aid*
Portland State University
P.O. Box 751
Portland, OR 97207-0751
Phone: (503)725-3461 **Fax:** (503)725-4882

GENERAL FINANCIAL AID INFORMATION
Forms accepted/required: CSS-FAF, Stafford, IRS, Supplemental, Institutional
Additional financial aid information: Departmental and school scholarships; university need/merit-based scholarships; grants; long- and short-term loans; part-time jobs on campus; College Work-Study Program.

ENGINEERING COLLEGE INFORMATION
[For additional personnel, refer to the *Appendix*.]
HEAD OF ENGINEERING
H. Chik M. Erzurumlu
Phone: (503)725-4631 **Fax:** (503)725-4298
EMail: bfhe@eas.pdx.edu

ENGINEERING COLLEGE ADDRESS
School of Engineering and Applied Science
Portland State University
P.O. Box 751
Portland, OR 97207-0751
Phone: (503)725-4631 **Fax:** (503)725-4298

ENROLLMENTS—BY CLASS
[Numbers are baccalaureate enrollments for the fall 1991 term, unless otherwise footnoted.]
1st-year students/Freshmen 181
2nd-year students/Sophomores 167
3rd-year students/Juniors 282
4th-year students/Seniors .... 565 [1]
Total **1,195**

Notes: (1) Includes post-baccalaureate undergraduates.

NUMBER OF DEGREES AWARDED—BY PROGRAM
[Numbers are engineering baccalaureate degrees awarded during the 1991-1992 academic year, unless otherwise footnoted. For full details about each engineering program, refer to the *Tables of Degree Programs*.]
Civil Engineering 30
Computer Science 62
Electrical & Computer Engineering 54
Engineering Management .... – [B]
Mechanical Engineering 43
Total **189**

Notes: (B) Data not applicable.

PERCENTAGE OF DEGREES AWARDED—BY CATEGORY
[Percentages are of all engineering baccalaureate degrees awarded during the 1991-1992 academic year, unless otherwise footnoted.]
African Americans 1.1%
Asian/Pacific Island Americans 18.5%
Hispanic Americans 1.6%
Native Americans .... – %
Foreign Citizens 9.5%
All Others 69.3%
Women 14.8%
Persons with disabilities .... – % [A]
Students over 25 years of age .... – % [A]

Notes: (A) Data not available.

GRADUATE ENROLLMENTS & DEGREES AWARDED
Master's enrollment 335
Master's degrees awarded 82
Doctoral enrollment 32
Doctoral degrees awarded .... –

ENGINEERING STUDENT DATA
Applicants to the engineering college
Number of applicants to engineering college 185
Percent offered admission 93.0%

Matriculated engineering students
Percentage in top quartile (25%) of High School class ...... – [A]

Notes: (A) Data not available.

FULL- & PART-TIME FACULTY—BY DEPARTMENT
[Figures are the head count of full-time faculty and the full-time equivalent (FTE) of part-time faculty for each engineering department or equivalent.]

Department	Full	Part
Civil Eng	8	0.4
Electrical Eng	13	0.1
Mechanical Eng	11	0.8
Computer Science	12	1.4
Eng Management Graduate Program	3	–

COLLEGE DESCRIPTION
The undergraduate offerings of the School of Engineering and Applied Science consist of four degrees in engineering and one degree in computer science. Each is a four-year program with its own faculty, courses, fields of specialization, and curriculum requirements. All programs incorporate computer-aided design techniques using state-of-the-art computing resources. Co-op education credits may be earned at local companies and used to meet undergraduate degree requirements. Graduating seniors are encouraged to complete the national Fundamentals of Engineering examination. Average number of years required to actually complete the bachelor's degree: 4 2/3.

DEGREE PROGRAMS DESCRIPTION
The B.S. degrees in the various engineering programs and computer science prepare the graduate for entry-level professional positions or for graduate study in each of the respective fields. The M.S. degrees in these fields permit students to concentrate in various specialty areas in either the thesis project, or non-thesis (course work only) option. The M.S. degree in Engineering Management is designed for technical professionals who require management capabilities as well as technical preparation. The Ph.D. degree can be earned in Electrical and Computer Engineering, in Civil Engineering, Mechanical Engineering, and Engineering Management as an option in the Systems Science Program, and in Civil Engineering as an option in the Environmental Sciences and Resources Program.

TRANSFER INFORMATION
Residency Requirements: After admission to the University, 45 of the final 60 credits or 165 of the total credits must be taken in residence.

Transfer via Articulation Agreements
Admission to engineering: At the end of the sophomore year
Engineering graduates transferred from ... 4-yr: – 2-yr: –
Requirements: See "Departmental Admission Requirements".

Transfer without Articulation Agreements
Admission to engineering: At the end of the sophomore year
Engineering graduates transferred from ... 4-yr: – 2-yr: –
Requirements: See "Departmental Admission Requirements" above.

GRADUATION REQUIREMENTS
B.S. programs in engineering and computer science require 186-204 quarter credits, depending on major selected.

STUDENT PROGRAMS
PROFESSIONAL AND HONORARY SOCIETIES
[For key to acronyms, see *Introduction*.]
Professional Societies: ACM, ASCE, ASME, IEEE, NSPE, SAE
Honorary Societies: Eta Kappa Nu, Tau Beta Pi

SUPPORT PROGRAMS
Student Chapter Organizations: Society of Women Engineers, National Society of Black Engineers, American Indian Science and Engineering Society

Oregon-MESA
For: Women & Ethnic Minorities **Available:** Year round
Offered: Prematriculation

Other Engineering Support Programs: Summer orientation and advising program for all freshmen and transfers; student advising/referral center; counseling and testing services; health and legal services; services for handicapped students, international students, and veterans; placement services; and School of Engineering academic advising for entering and lower division students.

185 Prairie View A&M University

INSTITUTION PROFILE
HEAD OF THE INSTITUTION
Julius W. Becton
Phone: (409)857-2111 Fax: (409)857-3928

GENERAL INFORMATION
[All Students—Fall 1991]
Undergraduate enrollment 5,130
Graduate enrollment 530
Total institution enrollment 5,660

Type of institution: Public   **Calendar:** Semesters
Nearest city: Houston, Texas   **Population:** 2,000,000
Miles from main campus: 45   **Setting:** Small Town
Types of engineering degrees: Engineering

Other degree-granting colleges: Education, Nursing, Applied Science and Engineering Technology, Arts and Sciences, Benjamin Banneker, Business

ENGINEERING ADMISSIONS
ADMISSIONS OFFICE CONTACT
Robert F. Ford
Prairie View A&M University
P.O. Box 2610
Prairie View, TX 77446
Phone: (409)857-2618 Fax: (409)857-4956

ENGINEERING COLLEGE ADMISSIONS INFORMATION
Entrance Requirements: High School courses—Physical Science 1 unit (1 year), Geometry 1 unit Trig- Preferred (1 year), English (but not required) (1 year), Algebra (2 units) (1 year)
Entrance Recommendations: SAT (Score: 950), ACT (Score: 21), High School courses—Physical Science 2 units (1 year), Algebra & Trigonometry (2 units) (1 year), English (3 units) (1 year)

DEPARTMENTAL ADMISSIONS INFORMATION
Admission to the engineering department: At the time of admission to the institution.
Additional information: If a student does not meet the recommended scores of 950 SAT or 21 ACT, he or she will have to take and pass the Calculus Readiness Test; if the student does not pass the Calculus Readiness Test, they will have to enter the pre-engineering program.

FINANCIAL INFORMATION
ESTIMATED EXPENSES (ACADEMIC YEAR)
[Expenses are for the 1992-93 nine-month academic year.]

State Residents	Undergraduate	Graduate
Tuition and fees	$ 509	$ 421
College room and board	$ 1,700	$ –
Books and supplies	$ 300	$ 150
Other expenses	$ –	$ –
Total estimated expenses	**$ 2,509**	**$ 571**

Out-of-State Residents	Undergraduate	Graduate
Tuition and fees	$ 2,209	$ 1,663
College room and board	$ 1,700	$ –
Books and supplies	$ 300	$ 150
Other expenses	$ –	$ –
Total estimated expenses	**$ 4,209**	**$ 1,813**

FINANCIAL AID OFFICE CONTACT
Advergus D. James, *Director*
Prairie View A&M University
P.O. Box C
Prairie View, TX 77446
Phone: (409)857-2424

GENERAL FINANCIAL AID INFORMATION
Forms accepted/required: ACT, AFSA, CSS-FAF, FAF, FAT, FAR, FSA, FFS, Stafford, IRS, SAR, SDF, Non-Filing Tax Statement, Statement of Educational Purpose and Default Form

ENGINEERING COLLEGE INFORMATION
[For additional personnel, refer to the *Appendix*.]

HEAD OF ENGINEERING
John Foster
Phone: (409)857-2212 Fax: (409)857-2222

ENGINEERING COLLEGE ADDRESS
College of Engineering and Architecture
Prairie View A&M University
P. O. Box 397
Prairie View, TX 77446-0397
Phone: (409)857-2211 Fax: (409)857-2222

ENROLLMENTS—BY CLASS
[Numbers are baccalaureate enrollments for the fall 1991 term, unless otherwise footnoted.]
1st-year students/Freshmen 512
2nd-year students/Sophomores 143
3rd-year students/Juniors 109
4th-year students/Seniors 86
Total **850**

NUMBER OF DEGREES AWARDED—BY PROGRAM
[Numbers are engineering baccalaureate degrees awarded during the 1991-1992 academic year, unless otherwise footnoted. For full details about each engineering program, refer to the *Tables of Degree Programs*.]
Chemical & Biochemical Engineering .... –
Chemical Engineering 13
Civil Engineering 12
Electrical Engineering 62
Mechanical Engineering 27
Total **114**

PERCENTAGE OF DEGREES AWARDED—BY CATEGORY
[Percentages are of all engineering baccalaureate degrees awarded during the 1991-1992 academic year, unless otherwise footnoted.]
African Americans .... 87.0 %
Asian/Pacific Island Americans .... 1.0 %
Hispanic Americans .... 1.0 %
Native Americans .... – %
Foreign Citizens .... 11.0 %
All Others .... – %
Women .... 30.0 %
Persons with disabilities .... – % (A)
Students over 25 years of age .... – % (A)

Notes: (A) Data not available.

GRADUATE ENROLLMENTS & DEGREES AWARDED
Master's enrollment 6
Master's degrees awarded .... –
Doctoral enrollment .... –
Doctoral degrees awarded .... –

ENGINEERING STUDENT DATA
Applicants to the engineering college
Number of applicants to engineering college .... – (A)

Matriculated engineering students
Percentage in top quartile (25%) of High School class 30.0
Average SAT scores: Combined—950
Average ACT scores: Composite—19

Notes: (A) Data not available.

FULL- & PART-TIME FACULTY—BY DEPARTMENT
[Figures are the head count of full-time faculty and the full-time equivalent (FTE) of part-time faculty for each engineering department or equivalent.]

Department	Full	Part
Chemical Eng	3	2.0
Civil Eng	6	2.0
Electrical Eng	8	2.0
Mechanical Eng	6	2.0
Architecture	6	3.0
Graduate Program	6	–

COLLEGE DESCRIPTION
The central mission is to strive to provide for the dual pursuit of general and professional education. The fostering of professional ethics, the development of analytical and conceptual skills in problem solving, and to produce Graduates who are equipped for leadership roles in the technological society. The primary goal objective is to provide professional training of the highest quality which, when supplemented by postgraduate practical experience or graduate studies, will enable the student to practice engineering with a high level of competence within the broadest spectrum of responsibility to society. Co-op programs and internship programs are offered but not required for graduating.

DEGREE PROGRAMS DESCRIPTION
B.S. Degree, M.S. Degree; Joint Ph.D. Degree.

TRANSFER INFORMATION
Transfer via Articulation Agreements
Admission to engineering: At any time
Transfer without Articulation Agreements
Admission to engineering: At any time

GRADUATION REQUIREMENTS
Students must graduate with a GPA of 2.00 and have completed all required courses within their department.

STUDENT PROGRAMS
PROFESSIONAL AND HONORARY SOCIETIES
[For key to acronyms, see *Introduction*.]
Professional Societies: AIChE, ASCE, ASME, IEEE
Honorary Societies: Chi Epsilon, Eta Kappa Nu, Pi Tau Sigma, Tau Beta Pi

SUPPORT PROGRAMS
Student Chapter Organizations: Society of Women Engineers, National Society of Black Engineers, Society of Hispanic Professional Engineers, Tau Beta Pi, American Society of Mechanical Engineers, Institute for Electrical and Electronic Engineers, American Society of Civil Engineers

Minority Introduction to Engineering (MITE)
For: Ethnic Minorities   **Available:** Summer only
Offered: Prematriculation

Engineering Concepts Institute (ECI)
For: Ethnic Minorities   **Available:** Summer only
Offered: Prematriculation

Other Engineering Support Programs: Freshmen Advisory Council Departmental Counselors for Transfer Students Office for Recruitment and Special Summer Programs.

186 Princeton University

INSTITUTION PROFILE
HEAD OF THE INSTITUTION
Harold T. Shapiro
Phone: (609)258-6100   Fax: (609)258-1294
GENERAL INFORMATION
[All Students—Fall 1991]
Undergraduate enrollment 4,525
Graduate enrollment 1,913
Total institution enrollment **6,438**
Type of institution: Private   **Calendar:** Semesters
Nearest city: Trenton   **Population:** 100,000
Miles from main campus: 11   **Setting:** Suburban
Types of engineering degrees: Engineering
Other degree-granting colleges: Architecture, Arts & Sciences, Humanities & Social Sciences, Woodrow Wilson School

ENGINEERING COLLEGE ADMISSIONS INFORMATION
Admission to the engineering college: By degree transfer
Entrance Requirements: SAT, High School courses—Math (4 years), Chemistry (1 year), Physics (1 year), Achievement tests, including one in math and one in either chemistry or physics.
Requirements for foreign students: TOEFL; financial statement

DEPARTMENTAL ADMISSIONS INFORMATION
Admission to the engineering department: At the end of the first year.

FINANCIAL INFORMATION
ESTIMATED EXPENSES (ACADEMIC YEAR)
[Expenses are for the 1992-93 nine-month academic year.]

All Students	Undergraduate	Graduate
Tuition and fees	$17,750	$17,850
College room and board	$5,517	$4,175[1]
Books and supplies	$600	$ –
Other expenses	$ –[1]	$ –[1]
Total estimated expenses	**$23,867**	**$22,025**

Notes: (1) Estimated aver. for grad. room fee and limited board contract in the Graduate College. No estimate is provided for travel and incidental living expenses.

FINANCIAL AID OFFICE CONTACT
Don M. Betterton, *Director, Undergraduate Financial Aid*
Princeton University
205 West College
Princeton, NJ 08544-5263
Phone: (609)258-3330   Fax: (609)258-2853
GENERAL FINANCIAL AID INFORMATION
Forms accepted/required: CSS-FAF
Additional financial aid information: Based on financial need.

ENGINEERING COLLEGE INFORMATION
[For additional personnel, refer to the *Appendix*.]
HEAD OF ENGINEERING
James Wei
Phone: (609)258-2260   Fax: (609)258-6744
EMail: jameswei@pucc
ENGINEERING COLLEGE ADDRESS
School of Engineering
Princeton University
Olden Street - Engineering Quadrangle
Princeton, NJ 08544-5263
Phone: (609)258-3000   Fax: (609)258-6744

ENROLLMENTS—BY CLASS
[Numbers are baccalaureate enrollments for the fall 1991 term, unless otherwise footnoted.]
1st-year students/Freshmen 225
2nd-year students/Sophomores 203
3rd-year students/Juniors 196
4th-year students/Seniors 153
Total **777**

NUMBER OF DEGREES AWARDED—BY PROGRAM
[Numbers are engineering baccalaureate degrees awarded during the 1991-1992 academic year, unless otherwise footnoted. For full details about each engineering program, refer to the *Tables of Degree Programs*.]
Chemical Engineering 9
Civil Engineering & Operations Research 31
Computer Science 20
Electrical Engineering 40
Mechanical & Aerospace Engineering 48
Total **148**

PERCENTAGE OF DEGREES AWARDED—BY CATEGORY
[Percentages are of all engineering baccalaureate degrees awarded during the 1991-1992 academic year, unless otherwise footnoted.]
African Americans 5.4%
Asian/Pacific Island Americans 18.9%
Hispanic Americans 3.4%
Native Americans .... – %
Foreign Citizens 6.8%
All Others 65.5%
Women 18.2%
Persons with disabilities .... – %
Students over 25 years of age .... – %

GRADUATE ENROLLMENTS & DEGREES AWARDED
Master's enrollment 7
Master's degrees awarded 26
Doctoral enrollment 351
Doctoral degrees awarded 44

ENGINEERING STUDENT DATA
Applicants to the engineering college
Number of applicants to engineering college 2,297
Percent offered admission 15.4%
Matriculated engineering students
Percentage in top quartile (25%) of High School class 98.0
Average SAT scores: Math—728, Verbal—638, Combined—1366

FULL- & PART-TIME FACULTY—BY DEPARTMENT
[Figures are the head count of full-time faculty and the full-time equivalent (FTE) of part-time faculty for each engineering department or equivalent.]

Department	Full	Part
Chemical Eng	18	–
Civil Eng & Operations Research	23	–
Electrical Eng	22	–
Mechanical & Aerospace Eng	24	–
Computer Science	18	–

COLLEGE DESCRIPTION
4-yr. engineering curricula are offered through 5 academic departments; 5 interdepartmental programs confer certificates of proficiency at graduation. 7 ABET-accredited programs are available. In all engineering curricula independent work, with faculty guidance, is encouraged for upper-class students as a means of fostering critical thinking, ingenuity, and creativity. Writing skills and breadth of knowledge beyond technical areas are among the goals of a Princeton engineering education. Students may also combine an engineering curriculum with study in depth in other fields such as environmental studies, public & international affairs, materials science, applied and computational mathematics, cognitive studies, musical performance, women's studies, creative writing, or foreign languages and cultures. Advanced standing, allowing graduation after 6 or 7 semesters, is an option for students entering with adequate advanced placement.

DEGREE PROGRAMS DESCRIPTION
Princeton stresses the creative application of mathematical and scientific principles to the solution of everyday problems in our society. While pursuing their majors, students are engaged in independent research projects supervised by faculty members. A Princeton engineering education also fosters clear writing and well-seasoned argumentation, familiarity with important issues in the humanities and social sciences, and breadth of knowledge beyond technical areas.

TRANSFER INFORMATION
Residency Requirements: Minimum 2 years.
Transfer without Articulation Agreements
Admission to engineering: At any time up to the end of the sophomore year.
Engineering graduates transferred from ....4-yr: –   2-yr: –

GRADUATION REQUIREMENTS
36 courses are required for the BSE degree. These must include at least 7 courses in the humanities and social sciences. A course satisfying the University writing requirement must be included; so must a course leading to computer proficiency. Course requirements in physics (2), chemistry (1), and mathematics (4) may be reduced by advanced placement. Other requirements, including a core of 8 upper-class major courses, are set by departments for different programs of concentration; a C average in the departmental core is required for graduation.

STUDENT PROGRAMS
PROFESSIONAL AND HONORARY SOCIETIES
[For key to acronyms, see *Introduction*.]
Professional Societies: AIAA, AIChE, ASCE, ASME, IEEE, Society of Hispanic Professional Engineers, Society of Women Engineers
Honorary Societies: Tau Beta Pi, Sigma Xi
SUPPORT PROGRAMS
Student Chapter Organizations: Society of Women Engineers, National Society of Black Engineers, Society of Hispanic Professional Engineers

PROMES
For: Ethnic Minorities   **Available:** Year round
Offered: Sophomore, Junior, Senior
Other Engineering Support Programs: Summer program for selected pre-matriculating first-year students; Orientation Program for first year students; First-Year Study Halls for physics, chemistry, and math; Individual Tutoring Program; Interactors Program providing junior and senior level mentors for first-year students; First-Year Faculty Advising Program; Sophomore Study Halls for introductory departmental offerings.

187 University of Puerto Rico, Mayagüez Campus

■ INSTITUTION PROFILE

HEAD OF THE INSTITUTION
Alejandro M. Ruiz
Phone: (809)265-3878 Fax: (809)834-3031

GENERAL INFORMATION
[All Students—Fall 1991]
Undergraduate enrollment 3,812
Graduate enrollment 166
Total institution enrollment 3,978
Type of institution: Public **Calendar:** Semesters
Location: Mayagüez **Population:** 90,000
Setting: Urban
Types of engineering degrees: Engineering
Other degree-granting colleges: Agricultural & Environmental, Arts & Sciences, Business Administration, Fine & Performing Arts, Humanities & Social Sciences, Nursing

■ ENGINEERING ADMISSIONS

ADMISSIONS OFFICE CONTACT
Jorge Ortiz
University of Puerto Rico, Mayagüez Campus
Associate Dean of Engineering
Post Box 5000
Mayagüez, PR 00681-5000
Phone: (809)832-4040 Fax: (809)833-1190

ENGINEERING COLLEGE ADMISSIONS INFORMATION
Entrance Requirements: SAT (Score: 488), (SAT Spanish Version)
Entrance Recommendations: SAT (Score: 488), High School courses—Mathematics, Physics, Chemistry, SAT Spanish Version
Requirements for foreign students: Total full-time tuition and fees are $3500 at graduate level and $2600 at undergraduate level.

DEPARTMENTAL ADMISSIONS INFORMATION
Admission to the engineering department: At the time of admission to the institution.
Additional information: There are General Admission Index (GAI) requirements which are based on SAT test scores (Spanish version) and high school GPA.

■ FINANCIAL INFORMATION

ESTIMATED EXPENSES (ACADEMIC YEAR)
[Expenses are for the 1992-93 nine-month academic year.]

All Students	Undergraduate	Graduate
Tuition and fees	$ 1,458	$ 2,000
College room and board	$ 1,000	$ 1,000
Books and supplies	$ 800	$ 1,000
Other expenses	$ 600	$ 600
Total estimated expenses	**$ 3,858**	**$ 4,600**

FINANCIAL AID OFFICE CONTACT
Pedro J. Aubret, *Director*
University of Puerto Rico, Mayagüez Campus
Post Box 5000, College Station
Mayagüez, PR 00681-5000
Phone: (809)832-4040

GENERAL FINANCIAL AID INFORMATION
Forms accepted/required: AFSA, FAT, Stafford, IRS, SAR, SDF, Institutional
Additional financial aid information: There are basic federal financial aid and loan programs and work-study and part-time job programs available on the campus.

■ ENGINEERING COLLEGE INFORMATION

[For additional personnel, refer to the *Appendix*.]

HEAD OF ENGINEERING
José F. Lluch
Phone: (809)833-1121 Fax: (809)833-1190
EMail: J_Lluch @ Rumac.UPR.CLU.Edu

ENGINEERING COLLEGE ADDRESS
College of Engineering
University of Puerto Rico, Mayagüez Campus
Post Box 5000
Mayagüez, PR 00681-5000
Phone: (809)832-4040 Fax: (809)832-0119

ENROLLMENTS—BY CLASS
[Numbers are baccalaureate enrollments for the fall 1991 term, unless otherwise footnoted.]
1st-year students/Freshmen 694
2nd-year students/Sophomores 667
3rd-year students/Juniors 620
4th-year students/Seniors 1,831
Total **3,812**

NUMBER OF DEGREES AWARDED—BY PROGRAM
[Numbers are engineering baccalaureate degrees awarded during the 1991-1992 academic year, unless otherwise footnoted. For full details about each engineering program, refer to the *Tables of Degree Programs*.]
Chemical Engineering 83
Civil Engineering 106
Electrical & Computer Engineering 118
Industrial Engineering 92
Land Surveying & Topography 12
Land Surveying Engineering .... –
Mechanical Engineering 124
Total **535**

PERCENTAGE OF DEGREES AWARDED—BY CATEGORY
[Percentages are of all engineering baccalaureate degrees awarded during the 1991-1992 academic year, unless otherwise footnoted.]
African Americans .... – % (A)
Asian/Pacific Island Americans .... – % (A)
Hispanic Americans .... 97.0 %
Native Americans .... – %
Foreign Citizens .... 3.0 %
All Others .... – %
Women .... 30.0 %
Persons with disabilities .... – % (A)
Students over 25 years of age .... – % (A)
Notes: (A) Data not available.

GRADUATE ENROLLMENTS & DEGREES AWARDED
Master's enrollment 166
Master's degrees awarded 13
Doctoral enrollment .... – (B)
Doctoral degrees awarded .... – (B)
Notes: (B) Data not applicable.

■ ENGINEERING STUDENT DATA

Applicants to the engineering college
Number of applicants to engineering college 1,014
Percent offered admission 82.0%

Matriculated engineering students
Percentage in top quartile (25%) of High School class ... 100.0

FULL- & PART-TIME FACULTY—BY DEPARTMENT
[Figures are the head count of full-time faculty and the full-time equivalent (FTE) of part-time faculty for each engineering department or equivalent.]

Department	Full	Part
Chemical Eng	19	–
Civil Eng	28	–
Electrical Eng	30	–
Industrial Eng	18	1.6
Mechanical Eng	23	–
General Eng	33	1.7

COLLEGE DESCRIPTION
College of Engineering offers a co-op course involving co-op experience with industry or government. This experience accounts for 3-6 credits of free electives.

DEGREE PROGRAMS DESCRIPTION
At the undergraduate level in the Civil Engineering Department, the student may graduate with a Bachelor of Science in Surveying and Topography. In the Electrical Engineering program, the student may graduate in Electrical or Computer Engineering. At the master's level in the Department of Industrial Engineering, the student may graduate with a Master of Engineering degree in Management Systems Engineering.

TRANSFER INFORMATION
Transfer without Articulation Agreements
Admission to engineering: At any time
Engineering graduates transferred from .. 4-yr: 12 2-yr: –
Requirements: Students coming from other colleges of the UPR system must have 12 credit hours in science, math and engineering with a GPA of at least 3.00. They must have completed at least 75% of the courses for which they registered. External transfer students must have completed 48 credit hours with 3.00 or better and must have approved 75% of the courses in math and sciences.

GRADUATION REQUIREMENTS
The grade point average required for graduation is 2.00. In addition, graduates must have earned a grade point average of 2.00 in the courses taken in their major fields.

■ STUDENT PROGRAMS

PROFESSIONAL AND HONORARY SOCIETIES
[For key to acronyms, see *Introduction*.]
Professional Societies: AIChE, ASAE, ASCE, ASME, IEEE, IIE, SME
Honorary Societies: Alpha Pi Mu, Tau Beta Pi

SUPPORT PROGRAMS
Student Chapter Organizations: Society of Women Engineers
Other Engineering Support Programs: College of Engineering has a counseling program for academic purposes at all year levels.

188 Purdue University

■ INSTITUTION PROFILE

HEAD OF THE INSTITUTION
Steven C. Beering
Phone: (317)494-9708

GENERAL INFORMATION
[All Students—Fall 1991]
Undergraduate enrollment 29,663
Graduate enrollment 6,204
Total institution enrollment 38,068
Type of institution: Public **Calendar:** Semesters
Nearest city: Indianapolis **Population:** 700,000
Miles from main campus: 60 **Setting:** Small Town
Types of engineering degrees: Engineering
Other degree-granting colleges: Agricultural & Environmental, Education, Humanities & Social Sciences, Nursing, Pharmacy, Technology & Applied Sciences, Science, Veterinary Medicine, Consumer and Family Sciences, Management

ENGINEERING COLLEGE ADMISSIONS INFORMATION
Admission to the engineering college: All students are admitted into the freshman engineering program. Upon completion of the program requirements, students move to a professional program.
Entrance Requirements: High School courses—Algebra (3 sem) Geometry & Trigonometry (3 sem) (3 years), Chemistry (1 year), Lab Science (1 year), English (3 years), SAT verbal 400, Math 500 or ACT Engl 21, Math 23
Entrance Recommendations: High School courses—Computer programming & Keyboarding (1 year), Mathematics (4 years), English (4 years), Physics (1 year)
Requirements for out-of-state residents: Must meet subject matter and quality requirements. The most qualified applicants will be selected for 600 available spaces.

DEPARTMENTAL ADMISSIONS INFORMATION
Admission to the engineering department: At the time of admission to the institution.

■ FINANCIAL INFORMATION

ESTIMATED EXPENSES (ACADEMIC YEAR)
[Expenses are for the 1992-93 nine-month academic year.]

State Residents	Undergraduate	Graduate
Tuition and fees	$ 2,520[1]	$ 2,520[1]
College room and board	$ 3,610[2]	$ 3,610[2]
Books and supplies	$ 500	$ 600
Other expenses	$ 1,270[3]	$ 1,270[3]
Total estimated expenses	**$ 7,900**	**$ 8,000**

Out-of-State Residents	Undergraduate	Graduate
Tuition and fees	$ 8,192[1]	$ 8,192[1]
College room and board	$ 3,610[2]	$ 3,610[2]
Books and supplies	$ 500	$ 600
Other expenses	$ 1,270[3]	$ 1,270[3]
Total estimated expenses	**$13,572**	**$13,672**

Notes: (1) Rates are subject to change without published notice. (2) Average for a single student. (3) Miscellaneous and travel.

FINANCIAL AID OFFICE CONTACT
Joyce Hall, *Director of Financial Aid*
Purdue University
Division of Financial Aid
Schleman Hall of Student Services
West Lafayette, IN 47907
Phone: (317)494-5050

GENERAL FINANCIAL AID INFORMATION
Forms accepted/required: FAF, IRS, Institutional
Additional financial aid information: Need-based scholarships; short-term loans; merit-based scholarships; athletic scholarships; part-time jobs on campus; College Work-Study Program.

■ ENGINEERING COLLEGE INFORMATION

[For additional personnel, refer to the *Appendix*.]

HEAD OF ENGINEERING
Henry T. Yang
Phone: (317)494-5346

ENGINEERING COLLEGE ADDRESS
Schools of Engineering
Purdue University
1280 Engineering Admin. Bldg., Rm. 101
West Lafayette, IN 47907-1280
Phone: (317)494-5345 **Fax:** (317)494-9321

ENROLLMENTS—BY CLASS
[Numbers are baccalaureate enrollments for the fall 1991 term, unless otherwise footnoted.]
1st-year students/Freshmen 1,704
2nd-year students/Sophomores 1,742
3rd-year students/Juniors 1,509
4th-year students/Seniors 1,490
Total **6,445**

NUMBER OF DEGREES AWARDED—BY PROGRAM
[Numbers are engineering baccalaureate degrees awarded during the 1991-1992 academic year, unless otherwise footnoted. For full details about each engineering program, refer to the *Tables of Degree Programs*.]
Aeronautical & Astronautical Engineering 146
Agricultural Engineering 20
Agricultural Engineering (Food) .... –[2]
Chemical Engineering 128
Civil Engineering 129
Computer & Electrical Engineering .... –[8]
Construction Engineering & Management 35
Electrical Engineering 308
Industrial Engineering 195
Interdisciplinary Engineering 46
Land Surveying Engineering 4
Materials Engineering 34
Mechanical Engineering 194
Nuclear Engineering 9
Total **1,248**

Notes: (2) See Agricultural Engineering. (8) See Electrical Engineering.

PERCENTAGE OF DEGREES AWARDED—BY CATEGORY
[Percentages are of all engineering baccalaureate degrees awarded during the 1991-1992 academic year, unless otherwise footnoted.]
African Americans .... 3.0 %
Asian/Pacific Island Americans .... 6.4 %
Hispanic Americans .... 1.6 %
Native Americans .... 0.1 %
Foreign Citizens .... 3.4 %
All Others .... 85.5 %
Women .... 19.6 %
Persons with disabilities .... – %[A]
Students over 25 years of age .... – %[A]

Notes: (A) Data not available.

GRADUATE ENROLLMENTS & DEGREES AWARDED
Master's enrollment 972
Master's degrees awarded 411
Doctoral enrollment 644
Doctoral degrees awarded 154

ENGINEERING STUDENT DATA
Applicants to the engineering college
Number of applicants to engineering college 5,825
Percent offered admission 84.0%
Matriculated engineering students
Percentage in top quartile (25%) of High School class 90.0
Average SAT scores: Math—633, Verbal—506, Combined—1139

FULL- & PART-TIME FACULTY—BY DEPARTMENT
[Figures are the head count of full-time faculty and the full-time equivalent (FTE) of part-time faculty for each engineering department or equivalent.]

Department	Full	Part
Aeronautics & Astronautics	23	–
Agricultural Eng	19	–
Chemical Eng	23	–
Civil Eng	58	–
Construction Eng & Management	–	–
Electrical Eng	74	–
Freshman Eng	7	–
Industrial Eng	31	–
Interdisciplinary Eng Studies	1	–
Materials Eng	16	–
Mechanical Eng	56	–
Nuclear Eng	8	–

COLLEGE DESCRIPTION
The Schools of Engineering offer course work leading to the B.S. degree in twelve disciplines. The curricula provide the student with a broad base on which to build an engineering career. They are founded on the basic sciences complemented by courses in the engineering sciences, analysis and design that emphasize synthesis, computer applications and problem solving. Extensive laboratory and computer experience is provided via modern facilities in all basic areas of each discipline. For selected students, co-op is an option in ten of the twelve degree programs. The Department of Freshman Engineering is the entry point for all beginning students. Qualified students are admitted to the professional engineering programs upon satisfactory completion of the pre-engineering program requirements. Average number of years required to actually complete the bachelor's degree for students not participating in the co-op program is 4.3.

DEGREE PROGRAMS DESCRIPTION
The B.S. degrees in the various engineering programs prepare the graduate for entry-level professional positions or for graduate study in each of the respective fields. Advanced degree programs (M.S. and Ph.D.) are available as indicated elsewhere.

DUAL DEGREE PROGRAMS
Engineering baccalaureate graduates with dual degrees 6

TRANSFER INFORMATION
Residency Requirements: Students must complete the senior year in residence. Some engineering departments have a maximum number of engineering courses which can be transferred.
Transfer without Articulation Agreements
Admission to engineering: At the end of the freshman year
Requirements: Certain academic requirements must be met. Some programs may close because of space limitations.

GRADUATION REQUIREMENTS
B.S. programs in engineering require 124-133 semester credit hours, depending upon major selected. A minimum graduation index of 4.0 (6.0=A) is required for graduation.

■ STUDENT PROGRAMS

PROFESSIONAL AND HONORARY SOCIETIES
[For key to acronyms, see *Introduction*.]
Professional Societies: ACerS, ACSM, AIAA, AIChE, AIME, ANS, ASAE, ASCE, ASME, IEEE, Institute of Industrial Engineers, National Society of Professional Engineers
Honorary Societies: Alpha Epsilon, Alpha Pi Mu, Chi Epsilon, Eta Kappa Nu, Omega Chi Epsilon, Pi Tau Sigma, Sigma Gamma Tau, Tau Beta Pi, Theta Tau, Omega Rho

SUPPORT PROGRAMS
Student Chapter Organizations: Society of Women Engineers, National Society of Black Engineers, Society of Hispanic Professional Engineers

Minority Engineering Program
For: Ethnic Minorities **Available:** Academic year
Offered: Freshman

Women in Engineering
For: Women **Available:** Academic year
Offered: Freshman

Other Engineering Support Programs: Summer and fall orientation program for all freshmen and transfers; academic and career counseling; psychological counseling; study skills center; international student office; disabled student program; tutoring programs; placement services.

189 Rensselaer Polytechnic Institute

■ INSTITUTION PROFILE
HEAD OF THE INSTITUTION
Roland W. Schmitt
Phone: (518)276-6211   Fax: (518)276-8702
GENERAL INFORMATION
[All Students—Fall 1991]
Undergraduate enrollment 4,398
Graduate enrollment 2,259
Total institution enrollment 6,839
Type of institution: Private   Calendar: Semesters
Nearest city: Albany   Population: 100,000
Miles from main campus: 5   Setting: Urban
Types of engineering degrees: Engineering
Other degree-granting colleges: Architecture, Humanities & Social Sciences, Science, Management

■ ENGINEERING ADMISSIONS
ADMISSIONS OFFICE CONTACT
Conrad Sharrow
Rensselaer Polytechnic Institute
110 Eighth Street
Troy, NY 12180-3590
Phone: (518)276-6216   Fax: (518)276-4072
ENGINEERING COLLEGE ADMISSIONS INFORMATION
Entrance Requirements: SAT, ACT, High School courses—English (4 years), Social Studies (3 years), Science (3 years), Math (4 years), SAT or ACT required (Minimum SAT Math score 600, minimum ACT Math score 28).
Entrance Recommendations: Achievement: English Composition, Math, Chemistry or Physics
Requirements for foreign students: TOEFL (Score: 550); financial statement
DEPARTMENTAL ADMISSIONS INFORMATION
Admission to the engineering department: Students with the appropriate background may be admitted at any time.

■ FINANCIAL INFORMATION
ESTIMATED EXPENSES (ACADEMIC YEAR)
[Expenses are for the 1992-93 nine-month academic year.]

All Students	Undergraduate	Graduate
Tuition and fees	$16,402	$14,790
College room and board	$ 5,468	$ 5,700
Books and supplies	$ 500	$ 1,190
Other expenses	$ 630	$ 825
Total estimated expenses	**$23,000**	**$22,505**

FINANCIAL AID OFFICE CONTACT
James H. Stevenson, *Director of Financial Aid*
Rensselaer Polytechnic Institute
110 Eighth Street
Troy, NY 12180-3590
Phone: (518)276-6813   Fax: (518)276-4072
GENERAL FINANCIAL AID INFORMATION
Forms accepted/required: FAF, Parent and Student Income Tax Return required
Additional financial aid information: Need-based scholarships; loans; part-time jobs on campus; College Work-Study Program. Financial aid awarded based on need. Need is met.

■ ENGINEERING COLLEGE INFORMATION
[For additional personnel, refer to the *Appendix*.]
HEAD OF ENGINEERING
James M. Tien
Phone: (518)276-6486   Fax: (518)276-8788
EMail: JMTIEN@RPITSMTS.Bitnet
ENGINEERING COLLEGE ADDRESS
School of Engineering
Rensselaer Polytechnic Institute
110 Eighth Street
Troy, NY 12180-3590
Phone: (518)276-6203   Fax: (518)276-8788

ENROLLMENTS—BY CLASS
[Numbers are baccalaureate enrollments for the fall 1991 term, unless otherwise footnoted.]
1st-year students/Freshmen 838
2nd-year students/Sophomores 671
3rd-year students/Juniors 786
4th-year students/Seniors 628
Total **2,923**

NUMBER OF DEGREES AWARDED—BY PROGRAM
[Numbers are engineering baccalaureate degrees awarded during the 1991-1992 academic year, unless otherwise footnoted. For full details about each engineering program, refer to the *Tables of Degree Programs*.]
Aeronautical Engineering 50
Biomedical Engineering 32
Chemical Engineering 37
Civil Engineering 38
Computer & Systems Engineering 37
Electric Power Engineering 8
Electrical Engineering 147
Engineering Physics 9
Engineering Science(s) 2
Environmental Engineering 21
Industrial & Management Engineering 50
Materials Engineering 37
Mechanical Engineering 188
Mechanics .... –
Nuclear Engineering 13
Nuclear Engineering & Science .... –
Transportation Engineering .... –
Total **669**

PERCENTAGE OF DEGREES AWARDED—BY CATEGORY
[Percentages are of all engineering baccalaureate degrees awarded during the 1991-1992 academic year, unless otherwise footnoted.]
African Americans 3.4%
Asian/Pacific Island Americans 10.0%
Hispanic Americans 3.7%
Native Americans .... – %
Foreign Citizens 5.7%
All Others 77.2%
Women 18.8%
Persons with disabilities .... – %
Students over 25 years of age .... – %

GRADUATE ENROLLMENTS & DEGREES AWARDED
Master's enrollment 475
Master's degrees awarded 310
Doctoral enrollment 397
Doctoral degrees awarded 73

ENGINEERING STUDENT DATA
Applicants to the engineering college
Number of applicants to engineering college 3,169
Percent offered admission 84.0%

Matriculated engineering students
Percentage in top quartile (25%) of High School class 85.3
Average SAT scores: Math—671, Verbal—545, Combined—1216

FULL- & PART-TIME FACULTY—BY DEPARTMENT
[Figures are the head count of full-time faculty and the full-time equivalent (FTE) of part-time faculty for each engineering department or equivalent.]

Department	Full	Part
Chemical Eng	15	–
Civil & Environmental Eng	14	2.8
Electrical, Computer, & Systems Eng	40	5.6
Decision Sciences & Eng Systems	23	–
Mechanical Eng, Aeronautical Eng, & Mechanics	34	3.3
Biomedical Eng	8	–
Electric Power Eng	6	–
Materials Eng	18	2.7
Nuclear Eng & Eng Physics	9	2.6

COLLEGE DESCRIPTION
The School offers a four-year undergraduate program in engineering emphasizing engineering fundamentals in the first two years, followed by two years of specialized training in one of the fourteen majors, with some liberal arts studies throughout all four years. Emphasis is on computer-aided engineering. There are opportunities to minor in management, computer science, humanities or social sciences. Co-op is an option in all programs. During the senior year each student completes a senior design project and is also urged to take the national standardized Fundamentals of Engineering exam. Average number of years required to actually complete the bachelor's degree: 4.

DEGREE PROGRAMS DESCRIPTION
The B.S. degrees in the various engineering programs prepare the graduate for entry-level professional positions or for graduate study in each of the respective fields. The B.S. in engineering science is intended primarily for students who do not intend to pursue engineering careers. Graduate study in engineering can be carried out in the M.E., D.Eng., or the M.S. and Ph.D. programs.

DUAL DEGREE PROGRAMS
Enrollment requirements: Same as above.

TRANSFER INFORMATION
Residency Requirements: Two years with the last 30 credits being completed on the Rensselaer campus.
Transfer via Articulation Agreements
Admission to engineering: At the end of the third/junior year from four-year liberal arts colleges
Requirements: A 3.0 GPA in all engineering fields except: aeronautical & mechanical engineering require a 3.2 GPA and computer & systems engineering require a 3.3 GPA. A letter of recommendation from a faculty member, appropriate department head or dean is required.
Transfer without Articulation Agreements
Admission to engineering: At any time
Requirements: A 3.0 GPA in all engineering fields except: aeronautical & mechanical engineering require a 3.2 GPA and computer & systems engineering require a 3.3 GPA.

GRADUATION REQUIREMENTS
A GPA of 1.8 is required for graduation. The B.S. programs in engineering require 130-38 semester credits, depending on major selected.

■ STUDENT PROGRAMS
PROFESSIONAL AND HONORARY SOCIETIES
[For key to acronyms, see *Introduction*.]
Professional Societies: ACerS, AIAA, AIChE, ANS, ASCE, ASME, BMES, IEEE, SAE, SME, Society for the Advancement of Material and Process Engineering, Society of Manufacturing Engineers
Honorary Societies: Alpha Pi Mu, Chi Epsilon, Eta Kappa Nu, Pi Tau Sigma, Sigma Gamma Tau, Tau Beta Pi, Alpha Sigma Mu, Epsilon Delta Sigma

SUPPORT PROGRAMS
Student Chapter Organizations: Society of Women Engineers, National Society of Black Engineers, Society of Hispanic Professional Engineers

Office of Minority Student Affairs
For: Women & Ethnic Minorities   **Available:** Year round
Offered: Prematriculation, Freshman, Sophomore, Junior, Senior, Graduate level

Other Engineering Support Programs: Summer and fall orientation program for all freshmen and transfers; Learning Center; career counseling services; foreign student office; services for blind and handicapped; placement services; counseling center.

190 University of Rhode Island

■ INSTITUTION PROFILE
HEAD OF THE INSTITUTION
Robert L. Carothers
Phone: (401)792-2444   **Fax:** (401)792-7149
GENERAL INFORMATION
[All Students—Fall 1991]
Undergraduate enrollment 10,820
Graduate enrollment 2,446
Total institution enrollment 13,266
Type of institution: Public   **Calendar:** Semesters
Nearest city: Providence   **Population:** 260,000
Miles from main campus: 25   **Setting:** Small Town
Types of engineering degrees: Engineering
Other degree-granting colleges: Arts & Sciences, Business Administration, Nursing, Human Science and Services, Resource Development, Pharmacy, Graduate School of Oceanography, Continuing Education

■ ENGINEERING ADMISSIONS
ADMISSIONS OFFICE CONTACT
Catherine L. Zeiser
University of Rhode Island
Office of Admissions
Green Hall
Kingston, RI 02881-0807
Phone: (401)792-7100   **Fax:** (401)792-2002
ENGINEERING COLLEGE ADMISSIONS INFORMATION
Admission to the engineering college: Students are first admitted to the University College. They are transferred to Engineering after completing 30 credits and earned at least a 2.0 QPA.
Entrance Requirements: SAT (Score: 1000), High School courses—English (4 years), Mathematics (4 years), Physics or natural science (2 years), Social science (2 years)
Entrance Recommendations: SAT (Score: 1000), High School courses—Foreign language (2 years)
Requirements for foreign students: TOEFL (Score: 500); financial statement
DEPARTMENTAL ADMISSIONS INFORMATION
Admission to the engineering department: At the end of the first year.
Additional information: Students are accepted to the Departments (College of Engineering) after they complete the freshman year requirements and earned at least a 2.0 overall QPA and a 2.0 cum on the required math, sciences, and engineering courses.

■ FINANCIAL INFORMATION
ESTIMATED EXPENSES (ACADEMIC YEAR)
[Expenses are for the 1992-93 nine-month academic year.]

All Students	Undergraduate	Graduate
Tuition and fees	$ 8,833	$ 9,481
College room and board	$ 5,200	$ 5,625 [1]
Books and supplies	$ 450	$ 500
Other expenses	$ 1,275 [2]	$ 950 [2]
Total estimated expenses	**$15,758**	**$ 16,556**

Notes: (1) On-campus room and board are not available for graduate students. Estimated cost is for off-campus rent and food expenses. (2) Other expenses are estimates for travel and incidentals and include weekend meals for undergraduates with room and board.

FINANCIAL AID OFFICE CONTACT
Horace J. Amaral Jr., *Assistant Dean of Student Financial Aid*
University of Rhode Island
Roosevelt Hall
Kingston, RI 02881
Phone: (401)792-2314
GENERAL FINANCIAL AID INFORMATION
Forms accepted/required: FAF
Additional financial aid information: The financial aid programs are designed to serve students from the widest possible range of society and all students are encouraged to apply. In most cases, financial aid will be awarded in a 'package' of grants, loans, and student employment opportunities. The purpose is to assist students in meeting their costs.

■ ENGINEERING COLLEGE INFORMATION
[For additional personnel, refer to the *Appendix*.]
HEAD OF ENGINEERING
Thomas J. Kim, *Interim Dean*
Phone: (401)792-2186   **Fax:** (401)782-1066
ENGINEERING COLLEGE ADDRESS
College of Engineering
University of Rhode Island
Bliss Hall
Kingston, RI 02881-0805
Phone: (401)792-2186   **Fax:** (401)782-1066
ENROLLMENTS—BY CLASS
[Numbers are baccalaureate enrollments for the fall 1991 term, unless otherwise footnoted.]
1st-year students/Freshmen 224
2nd-year students/Sophomores 184
3rd-year students/Juniors 204
4th-year students/Seniors 279
Total 891
NUMBER OF DEGREES AWARDED—BY PROGRAM
[Numbers are engineering baccalaureate degrees awarded during the 1991-1992 academic year, unless otherwise footnoted. For full details about each engineering program, refer to the *Tables of Degree Programs*.]
Chemical Engineering .... –
Civil Engineering 42
Computer Engineering 11
Electrical Engineering 36
Industrial & Manufacturing Engineering 11
Mechanical Engineering 42
Ocean Engineering .... – [2]
Total 142
Notes: (2) B.S. degree program started in 1991-1992.
PERCENTAGE OF DEGREES AWARDED—BY CATEGORY
[Percentages are of all engineering baccalaureate degrees awarded during the 1991-1992 academic year, unless otherwise footnoted.]
African Americans 1.5%
Asian/Pacific Island Americans 6.9%
Hispanic Americans 1.5%
Native Americans .... – %
Foreign Citizens .... – % [1]
All Others 90.1%
Women 17.7%
Persons with disabilities .... – % [A]
Students over 25 years of age .... – % [A]
Notes: (A) Data not available. (1) Based upon overall University baccalaureate graduates
GRADUATE ENROLLMENTS & DEGREES AWARDED
Master's enrollment 199
Master's degrees awarded 48
Doctoral enrollment 103
Doctoral degrees awarded 18
ENGINEERING STUDENT DATA
Applicants to the engineering college
Number of applicants to engineering college 684
Percent offered admission 85.0%
Matriculated engineering students
Percentage in top quartile (25%) of High School class 78.0
Average SAT scores: Math—595, Verbal—477, Combined—1072
FULL- & PART-TIME FACULTY—BY DEPARTMENT
[Figures are the head count of full-time faculty and the full-time equivalent (FTE) of part-time faculty for each engineering department or equivalent.]

Department	Full	Part
Chemical Eng	9	2.0
Civil & Environmental Eng	11	2.5
Electrical Eng	20	1.0
Industrial & Manufacturing Eng	5	–
Mechanical Eng & Applied Mechanics	18	–
Ocean Eng	7	–

COLLEGE DESCRIPTION
The goal of the college is to stimulate the students to become creative, responsible engineers, aware of the social implications of their work, and flexible enough to adjust to the rapid changes taking place in all branches of engineering. Engineers from all fields are heavily involved in the solution of technological and socio-technological problems. The International Engineering Program provides students with unique opportunities to prepare for their engineering careers and spend one semester abroad in a professional internship. The five-year program will lead to two degrees, BS in Engineering and BA in German. A new engineering design program, which emphasizes the integration of design throughout the curriculum, has been developed. Students are assigned a design project in their sophomore year. Average number of years required to actually complete the bachelor's degree: 4.5. A French International Engineering program is being developed to be offered in the Fall 1993.

DEGREE PROGRAMS DESCRIPTION
All of the engineering curricula are based on an intense study of mathematics and the basic sciences, and of the engineering sciences common to all branches of the profession. On this base is built the in-depth study of the important principles and concepts of each separate discipline. These principles are applied to the understanding and solution of problems of current interest and importance in the field. Each curriculum is designed to provide the knowledge and ability necessary for practice as a professional engineer, or for successful graduate study. Graduate study in engineering leading to M.S. and Ph.D. degrees can be carried out in all six departments.

DUAL DEGREE PROGRAMS
Engineering baccalaureate graduates with dual degrees 1
TRANSFER INFORMATION
Residency Requirements: The work of the senior year shall be taken at the University of Rhode Island. Exceptions must be approved by the Dean.
Transfer via Articulation Agreements
Admission to engineering: At any time
Engineering graduates transferred from ... 4-yr: 2   2-yr: 60
Requirements: Most successful applicants offer a GPA above 2.40.
Transfer without Articulation Agreements
Admission to engineering: At any time
Engineering graduates transferred from ... 4-yr: 5   2-yr: 11
Requirements: Most successful applicants after a GPA above 2.40.
GRADUATION REQUIREMENTS
To graduate, a student must have completed the work for the curriculum in which he or she is enrolled and must have earned at least a 2.00 quality point average. In addition, he/she also has to maintain a minimum 2.00 average in all mathematics, sciences, and engineering courses.

■ STUDENT PROGRAMS
PROFESSIONAL AND HONORARY SOCIETIES
[For key to acronyms, see *Introduction*.]
Professional Societies: ACM, AIChE, ASCE, ASME, IEEE, IIE, SME
Honorary Societies: Chi Epsilon, Pi Tau Sigma, Tau Beta Pi
SUPPORT PROGRAMS
Student Chapter Organizations: Society of Women Engineers, National Society of Black Engineers
Women Study Program
For: Women & Ethnic Minorities   **Available:** Academic year
Offered: Freshman, Sophomore, Junior, Senior
African and Afro American Studies
For: Women & Ethnic Minorities   **Available:** Academic year
Offered: Freshman, Sophomore, Junior, Senior
Other Engineering Support Programs: Summer and fall orientation program for all freshmen and transfers; learning assessment program; Career counseling services; foreign student services; services for handicapped.

191 William Marsh Rice University

INSTITUTION PROFILE

HEAD OF THE INSTITUTION
George E. Rupp
Phone: (713)527-4041 Fax: (713)285-5271

GENERAL INFORMATION
[All Students—Fall 1991]
Undergraduate enrollment 2,667
Graduate enrollment 1,382
Total institution enrollment **4,291**

Type of institution: Private Calendar: Semesters
Location: Houston Population: 2,000,000
Setting: Urban
Types of engineering degrees: Engineering
Other degree-granting colleges: Architecture, Music, Humanities, Social Sciences, Natural Sciences, Graduate School of Administration

ENGINEERING ADMISSIONS

ADMISSIONS OFFICE CONTACT
Ron W. Moss
William Marsh Rice University
Director of Admissions
P.O. Box 1892
Houston, TX 77251-1892
Phone: (713)527-4036 Fax: (713)285-5271

ENGINEERING COLLEGE ADMISSIONS INFORMATION
Entrance Requirements: SAT, High School courses—English (4 years), Math-incl. either trigonometry or elementary anal. (3 years), Chemistry (1 year), Physics (1 year). Although the SAT test is required, there is no required minimum score. Three achievement tests are required and must include: English composition (with or without essay); Math I, II or IIc; and chemistry or physics.
Requirements for foreign students: TOEFL (Score: 550); financial statement

DEPARTMENTAL ADMISSIONS INFORMATION
Admission to the engineering department: At the end of the second year.

FINANCIAL INFORMATION

ESTIMATED EXPENSES (ACADEMIC YEAR)
[Expenses are for the 1992-93 nine-month academic year.]

All Students	Undergraduate	Graduate
Tuition and fees	$ 8,833	$ 9,481
College room and board	$ 5,200 (1)	$ 5,625 (1)
Books and supplies	$ 450	$ 500
Other expenses	$ 1,275 (2)	$ 950 (2)
Total estimated expenses	**$15,758**	**$ 16,556**

Notes: (1) On-campus room and board are not available for graduate students. Estimated cost is for off-campus rent and food expense. (2) Other expenses are estimates for travel and incidentals, and include weekend meals for undergraduates with room and board.

FINANCIAL AID OFFICE CONTACT
G. D. Hunt, Director, Office of Financial Aid
William Marsh Rice University
P.O. Box 1892
Houston, TX 77251-1892
Phone: (723)527-4958 Fax: (713)285-5322

GENERAL FINANCIAL AID INFORMATION
Forms accepted/required: FAF, Rice Financial Aid Application
Additional financial aid information: Grants; low-interest loans; campus work opportunities; merit-based scholarships; need-based scholarships. Students in M.S. and Ph.D. programs often receive tuition waivers and fellowship or research assistant stipends.

ENGINEERING COLLEGE INFORMATION
[For additional personnel, refer to the Appendix.]

HEAD OF ENGINEERING
Michael M. Carroll
Phone: (713)527-4009 Fax: (713)285-5300
EMail: DENG@ricevm1.rice.edu

ENGINEERING COLLEGE ADDRESS
George R. Brown School of Engineering
William Marsh Rice University
P.O. Box 1892
Houston, TX 77251-1892
Phone: (713)527-4009 Fax: (713)285-5300

ENROLLMENTS—BY CLASS
[Numbers are baccalaureate enrollments for the fall 1991 term, unless otherwise footnoted.]
1st-year students/Freshmen 221
2nd-year students/Sophomores 156
3rd-year students/Juniors 156
4th-year students/Seniors 213
Total **746**

NUMBER OF DEGREES AWARDED—BY PROGRAM
[Numbers are engineering baccalaureate degrees awarded during the 1991-1992 academic year, unless otherwise footnoted. For full details about each engineering program, refer to the Tables of Degree Programs.]
Chemical Engineering 15
Civil Engineering 13
Computational & Applied Mathematics 10 (1)
Computer Science 20 (1)
Electrical & Computer Engineering 55
Environmental Science & Engineering 2 (5)
Mechanical Engineering & Materials Science 42
Statistics 4 (1)
Total **161**

Notes: (1) The undergraduate degree program leads to a B.A. degree. No B.S. degree program is offered. (5) The undergraduate environmental double major programs lead to B.A. degrees except for a B.S. in Civil Engineering with an Environmental Option.

PERCENTAGE OF DEGREES AWARDED—BY CATEGORY
[Percentages are of all engineering baccalaureate degrees awarded during the 1991-1992 academic year, unless otherwise footnoted.]
African Americans 2.5%
Asian/Pacific Island Americans 8.1%
Hispanic Americans 6.2%
Native Americans .... – %
Foreign Citizens 6.2%
All Others 77.0%
Women 23.0%
Persons with disabilities .... – % (A)
Students over 25 years of age .... – % (A)

Notes: (A) Data not available.

GRADUATE ENROLLMENTS & DEGREES AWARDED
Master's enrollment 120
Master's degrees awarded 76 (3)
Doctoral enrollment 259
Doctoral degrees awarded 31

Notes: (3) Of these, 46 students received M.S. degrees and 30 received professional master's degrees.

ENGINEERING STUDENT DATA

Applicants to the engineering college
Number of applicants to engineering college 1,814
Percent offered admission 19.7%

Matriculated engineering students
Percentage in top quartile (25%) of High School class 80.0 (4)

Notes: (4) All entering engineering students, except those for whom no rank was reported, were in the top 25% of their high school class.

FULL- & PART-TIME FACULTY—BY DEPARTMENT
[Figures are the head count of full-time faculty and the full-time equivalent (FTE) of part-time faculty for each engineering department or equivalent.]

Department	Full	Part
Chemical Eng	15	–
Civil Eng	7	–
Electrical & Computer Eng	18	2.2
Mechanical Eng & Materials Science	16	1.6
Computer Science	12	1.6
Environmental Science & Eng	5	0.8
Computational & Applied Mathematics	10	1.5
Statistics	5	1.2

COLLEGE DESCRIPTION

Undergraduates may major in any of five engineering fields plus computational and applied mathematics (formerly mathematical sciences), computer science, or statistics. These four-year programs lead to either the B.A. or B.S. degree and may qualify students for further study leading to a fifth-year professional master's degree, a Master of Science degree, or a Doctor of Philosophy degree. Students are required to take foundation courses and distribution courses in areas outside their major interest. Students are encouraged to take a double major or a coherent minor to meet these requirements and to balance their education between science and engineering and the liberal arts. Rice does not offer co-op programs. Rice admits bright students and graduates well-educated engineers.

DEGREE PROGRAMS DESCRIPTION

Curricula in engineering at Rice University lead to either B.A. or B.S. degrees in the fields of chemical engineering, civil engineering, electrical and computer engineering, mechanical engineering, and in materials science and engineering. In computational and applied mathematics, computer science, environmental science and engineering, and statistics curricula lead to the B.A. degree. These curricula may also be used as part of integrated five-year programs that lead to professional master's degrees in each of the above fields. They also prepare the student for graduate study toward either the Master of Science, the Master of Arts, or the Doctor of Philosophy degrees or for entry into the engineering profession.

DUAL DEGREE PROGRAMS

Engineering baccalaureate graduates with dual degrees 24

TRANSFER INFORMATION

Residency Requirements: Transfer students must be registered in residence at Rice for at least four full semesters during the fall or spring terms and must complete not less than 60 semester hours for a Rice degree.

Transfer without Articulation Agreements
Admission to engineering: Transfers are admitted to the engineering program at any level after the freshman year.
Engineering graduates transferred from .. 4-yr: 11 2-yr: –
Requirements: SAT, application, transcripts, evaluations, and interview required. Minimum 3.2 (4.0 scale) GPA. recommended on all college work. Special emphasis given to performance at the college level.

GRADUATION REQUIREMENTS

A student taking the B.A. program is required to pass a minimum of 120 semester hours. The major department may specify no more than 80 semester hours for the major. A student following a B.S. program in engineering must pass a minimum of 134 semester hours (137 for chemical engineering). No department may specify more than 92 semester hours (104 for chemical engineering) for the B.S. degree. All students must complete their degree requirements with a minimum GPA of 1.67 in all Rice courses and a minimum GPA of 2.00 for those courses taken in fulfillment of their major requirements.

STUDENT PROGRAMS

PROFESSIONAL AND HONORARY SOCIETIES
[For key to acronyms, see Introduction.]
Professional Societies: ACM, AIChE, ASCE, ASME, IEEE, American Society of Metallurgists
Honorary Societies: Eta Kappa Nu, Tau Beta Pi

SUPPORT PROGRAMS
Student Chapter Organizations: Society of Women Engineers, National Society of Black Engineers, Mexican American Engineering Association

Office of Minority Affairs
For: Women & Ethnic Minorities **Available:** Year round
Offered: Prematriculation, Freshman, Sophomore, Junior, Senior, Graduate level
Other Engineering Support Programs: Orientation Week for freshmen and transfer students; Career Planning and Placement Office; Office of Student Advising; International Student Office.

192 University of Rochester

INSTITUTION PROFILE
HEAD OF THE INSTITUTION
G. Dennis O'Brien
Phone: (716)275-8356   Fax: (716)256-2473

GENERAL INFORMATION
[All Students—Fall 1991]
Undergraduate enrollment	4,822
Graduate enrollment	2,673
Total institution enrollment	**7,495**

Type of institution: Private   Calendar: Semesters
Nearest city: Rochester   Population: 250,000
Miles from main campus: 2   Setting: Urban
Types of engineering degrees: Engineering
Other degree-granting colleges: Arts & Sciences, Business Administration, Dentistry, Education, Medicine, Nursing, Eastman School of Music

ENGINEERING ADMISSIONS
ADMISSIONS OFFICE CONTACT
Wayne A. Locust
University of Rochester
Undergraduate Admissions
Meliora Hall
Rochester, NY 14627
Phone: (716)275-3221   Fax: (716)461-4595

ENGINEERING COLLEGE ADMISSIONS INFORMATION
Admission to the engineering college: Initially admitted to Arts & Science, intended engineering majors have faculty advisors, enroll in engineering courses as freshmen, enter at sophomore year end.
Entrance Requirements: High School courses—Mathematics (4 years), Laboratory Science (2 years), The SAT or ACT is required. Students are evaluated on the strength of their application.
Entrance Recommendations: High School courses—English (4 years), Social Studies (4 years), Foreign Language (2 years), Physics (1 year), Students are urged to submit the CEEB Achievement exams in the areas of English, mathematics, foreign language and science.
Requirements for foreign students: TOEFL; Foreign students must submit their SAT or ACT scores. The TOEFL score is required; no minimum score is needed but at least a 600 is preferred. A financial statement is necessary for financial aid consideration.

DEPARTMENTAL ADMISSIONS INFORMATION
Admission to the engineering department: At the end of the second year.
Additional information: Chemical Engineering requires a 2.15 GPA in all chemistry and chemical engineering courses taken in the first two years; EE requires a 2.3 GPA minimum in four EE courses and a 2.0 GPA in all other courses; ME requires at least a 2.0 in all ME courses and an overall 2.0 GPA;Optics requires an overall GPA of 2.0, a 2.0 in physics and math sequences and a 'C' or better in optics courses taken in the first two years; Geomechanics requires an overall 2.0 GPA; the Interdisciplinary program requires a 2.7 GPA, the BA in engineering science requires an overall 2.3 GPA for admission.

FINANCIAL INFORMATION
ESTIMATED EXPENSES (ACADEMIC YEAR)
[Expenses are for the 1992-93 nine-month academic year.]

All Students	Undergraduate	Graduate
Tuition and fees	$16,060	$16,800
College room and board	$ 6,015	$ 6,500
Books and supplies	$ 500	$ 500
Other expenses	$ 394 (1)	$ — (2)
Total estimated expenses	**$22,969**	**$23,800**

Notes: (1) student related fees such as activity fee and health fees. personal expenses are not included in the totals. (2) personal expenses have not been included in these figures.

FINANCIAL AID OFFICE CONTACT
Ryan Williams, *Director of Financial Aid*
University of Rochester
Meliora Hall
Rochester, NY 14627
Phone: (716)275-3226   Fax: (716)461-4595

GENERAL FINANCIAL AID INFORMATION
Forms accepted/required: FAF
Additional financial aid information: A strong program of financial aid provides scholarships, grants, loans and part-time employment. Selection is based on financial need; however, after need has been established, the aid package is based on merit. Prospective engineering students are eligible for University scholarships including National Merit, Urban League, Alumni, Bausch & Lomb, and Xerox. Eastman Kodak and General Motors award scholarships to engineering students amounting to the cost of tuition for undergraduate study.

ENGINEERING COLLEGE INFORMATION
[For additional personnel, refer to the *Appendix*.]
HEAD OF ENGINEERING
Bruce W. Arden
Phone: (716)275-4151   Fax: (716)461-4735
EMail: arden@ee.rochester.edu

ENGINEERING COLLEGE ADDRESS
College of Engineering & Applied Science
University of Rochester
401 Dewey Hall
Rochester, NY 14627
Phone: (716)275-4151   Fax: (716)461-4735

ENROLLMENTS—BY CLASS
[Numbers are baccalaureate enrollments for the fall 1991 term, unless otherwise footnoted.]
1st-year students/Freshmen	236
2nd-year students/Sophomores	204
3rd-year students/Juniors	151
4th-year students/Seniors	180
Total	**771**

NUMBER OF DEGREES AWARDED—BY PROGRAM
[Numbers are engineering baccalaureate degrees awarded during the 1991-1992 academic year, unless otherwise footnoted. For full details about each engineering program, refer to the *Tables of Degree Programs*.]
Chemical Engineering	21
Electrical Engineering	44
Geomechanics	4
Interdisciplinary Engineering	4
Materials Science & Engineering	–
Mechanical & Aerospace Science	–
Mechanical Engineering	42
Optical Engineering	34
Total	**149**

PERCENTAGE OF DEGREES AWARDED—BY CATEGORY
[Percentages are of all engineering baccalaureate degrees awarded during the 1991-1992 academic year, unless otherwise footnoted.]
African Americans	3.0 %
Asian/Pacific Island Americans	6.0 %
Hispanic Americans	1.0 %
Native Americans	– %
Foreign Citizens	6.0 %
All Others	84.0 %
Women	25.0 %
Persons with disabilities	– %
Students over 25 years of age	1.0 %

GRADUATE ENROLLMENTS & DEGREES AWARDED
Master's enrollment	114 (1)
Master's degrees awarded	82
Doctoral enrollment	250 (2)
Doctoral degrees awarded	27

Notes: (1) does not include 3-2 students or leave of absences. (2) does not include leave of absence.

ENGINEERING STUDENT DATA
Applicants to the engineering college
Number of applicants to engineering college	1,224
Percent offered admission	64.0%

Matriculated engineering students
Percentage in top quartile (25%) of High School class	72.0

FULL- & PART-TIME FACULTY—BY DEPARTMENT
[Figures are the head count of full-time faculty and the full-time equivalent (FTE) of part-time faculty for each engineering department or equivalent.]

Department	Full	Part
Chemical Eng	12	0.8
Electrical Eng	16	1.0
Mechanical Eng	17	0.8
The Institute of Optics	14	2.2

COLLEGE DESCRIPTION
The College of Engineering & Applied Science offers BS degrees in ChE, EE, ME, Engineering and Applied Science (an interdisciplinary program), Geomechanics and Optics. ChE, EE, and ME are ABET accredited. Other opportunities include a BS, MS, & Ph.D. certificate in Biomedical Engrg, a certificate in management studies and a BA in Engrg Science. A five-year enriched program of study towards a BS/BA degree is available. Rochester's Take Five program allows selected engineering undergraduates a fifth year of courses tuition free. A five year program leading to the BS/MS in Electrical Engrg or Optics is available. The ChE, EE, and ME departments, and the Institute of Optics offer degrees at all levels. The College research centers include the New York State Center for Advanced Optical Technology, The Laboratory for Laser Energetics, the Center for Optics Manufacturing, the Rochester Center for Biomedical Ultrasound and the Center for Opto-Electronics and Imaging.

DEGREE PROGRAMS DESCRIPTION
The University of Rochester's engineering programs are based on the fundamentals of science and engineering. They exist in and benefit from a strong liberal arts environment. These programs prepare students equally to assume positions as practicing engineers in industry or to enter graduate and professional schools for further education.

DUAL DEGREE PROGRAMS
Engineering baccalaureate graduates with dual degrees 4
Enrollment requirements: Students who are accepted for dual degrees must complete the requirements for degree conferral in both colleges.

TRANSFER INFORMATION
Residency Requirements: Students must be enrolled in residence in the College, completing at least 12 hours each semester, for one academic year. This requirement may be waived under some circumstances.

Transfer without Articulation Agreements
Admission to engineering: At the end of the sophomore year
Engineering graduates transferred from . . . 4-yr: 10   2-yr: 5
Requirements: Departmentally determined; at least 2.0 GPA (some depts. are higher), and completion of first two year curriculum as shown in University Bulletin. Students may be conditionally admitted lacking some requirements.

GRADUATION REQUIREMENTS
B.S. engineering programs require between 128-132 semester hours depending on major; distribution requirements in liberal arts and engineering must be satisfied. One year residency required for undergraduates, a cum average of at least 2.0 for all coursework taken for credit and an average of at least 2.0 in courses specified by the department of concentration. M.S. engineering programs require between 30-32 hours. Students may elect to take a comprehensive exam in their subject or prepare a thesis. Ph.D. programs require 90 semester hours which includes successful defense.

STUDENT PROGRAMS
PROFESSIONAL AND HONORARY SOCIETIES
[For key to acronyms, see *Introduction*.]
Professional Societies: AIChE, ASME, IEEE, OSA
Honorary Societies: Tau Beta Pi

SUPPORT PROGRAMS
Student Chapter Organizations: Society of Women Engineers, National Society of Black Engineers, Society of Hispanic Professional Engineers

Early Connection Orientation
For: Ethnic Minorities   **Available:** Summer only
Offered: Prematriculation

Faculty Mentor Program
For: Ethnic Minorities   **Available:** Academic year
Offered: Freshman

Minority Peer Counseling Program
For: Ethnic Minorities   **Available:** Academic year
Offered: Freshman

Higher Education Opportunity Program
For: Ethnic Minorities   **Available:** Academic year
Offered: Freshman, Sophomore, Junior, Senior

Other Engineering Support Programs: Early Connection Orientation is a six week program designed to assist freshmen in making the transition from high school to college. Summer orientation programs are held for all new engineering students. Each engineering department offers auvising sessions prior to fall and spring registration periods. Tau Beta Pi members offer free tutoring services to engineering students.

193 Rochester Institute of Technology

INSTITUTION PROFILE

HEAD OF THE INSTITUTION
Albert J. Simone
Phone: (716)475-2396   Fax: (716)475-5700

GENERAL INFORMATION
[All Students—Fall 1991]
Undergraduate enrollment 11,150
Graduate enrollment 1,868
Total institution enrollment 13,018
Type of institution: Private   **Calendar:** Quarters
Nearest city: Rochester   **Population:** 500,000
Miles from main campus: 3   **Setting:** Suburban
Types of engineering degrees: Engineering & Technology
Other degree-granting colleges: Agricultural & Environmental, Business Administration, Liberal Arts, Science, Imaging Arts and Sciences, Continuing Education, National Technical Institute of the Deaf

ENGINEERING ADMISSIONS

ADMISSIONS OFFICE CONTACT
Daniel Shelley
Rochester Institute of Technology
Director of Admissions
P.O. Box 9887
Rochester, NY 14623-0887
Phone: (716)475-6631   Fax: (716)475-5476

ENGINEERING COLLEGE ADMISSIONS INFORMATION
Entrance Requirements: SAT, ACT, High School courses—English (4 years), Mathematics (including algebra, geometry, trig) (4 years), Physics (1 year), Chemistry (1 year), RIT considers the student's entire record. A low SAT or ACT test score would not necessarily disqualify a student for admission.
Entrance Recommendations: High School courses—Additional mathematics, College credit possible for advanced placement tests.
Requirements for foreign students: TOEFL (Score: 525); financial statement

DEPARTMENTAL ADMISSIONS INFORMATION
Admission to the engineering department: At the time of admission to the institution.

FINANCIAL INFORMATION

ESTIMATED EXPENSES (ACADEMIC YEAR)
[Expenses are for the 1992-93 nine-month academic year.]

All Students	Undergraduate	Graduate
Tuition and fees	$12,525	$13,536
College room and board	$ 5,285 [1]	$ 5,286 [1]
Books and supplies	$ 500	$ 500
Other expenses	$ 195	$ 100
Total estimated expenses	**$18,505**	**$19,422**

Notes: (1) includes 14 meal plus plan

FINANCIAL AID OFFICE CONTACT
Verna J. Hazen, *Director*
Rochester Institute of Technology
Bausch & Lomb Building
P.O. Box 9887
Rochester, NY 14623-0887
Phone: (716)475-5520   Fax: (716)475-7270

GENERAL FINANCIAL AID INFORMATION
Forms accepted/required: ACT, AFSA, CSS-FAF
Additional financial aid information: Need-based scholarships, merit-based scholarships, college work study, part-time jobs on campus, grants and entitlements also available. Financial aid awards approximately 50 million dollars annually in grants, loans, jobs. RIT awards merit and need based scholarships and participates in all applicable federal and state programs.

ENGINEERING COLLEGE INFORMATION

[For additional personnel, refer to the *Appendix*.]

HEAD OF ENGINEERING
Paul E. Petersen
Phone: (716)475-2146   Fax: (716)475-6879
EMail: PEPEEE

ENGINEERING COLLEGE ADDRESS
College of Engineering
Rochester Institute of Technology
P.O. Box 9887
Rochester, NY 14623-0887
Phone: (716)475-2145   Fax: (715)475-6879

ENROLLMENTS—BY CLASS
[Numbers are baccalaureate enrollments for the fall 1991 term, unless otherwise footnoted.]
1st-year students/Freshmen 303
2nd-year students/Sophomores 252
3rd-year students/Juniors 290
4th-year students/Seniors 590
Total **1,435**

NUMBER OF DEGREES AWARDED—BY PROGRAM
[Numbers are engineering baccalaureate degrees awarded during the 1991-1992 academic year, unless otherwise footnoted. For full details about each engineering program, refer to the *Tables of Degree Programs*.]
Computer Engineering 24
Electrical Engineering 93
Industrial & Manufacturing Engineering 25
Mechanical Engineering 78
Microelectronic Engineering 31
Total **251**

PERCENTAGE OF DEGREES AWARDED—BY CATEGORY
[Percentages are of all engineering baccalaureate degrees awarded during the 1991-1992 academic year, unless otherwise footnoted.]
African Americans 1.4%
Asian/Pacific Island Americans 3.7%
Hispanic Americans .... – %
Native Americans .... – %
Foreign Citizens 5.6%
All Others 89.3%
Women 11.2%
Persons with disabilities .... – % (A)
Students over 25 years of age 9.6%

Notes: (A) Data not available.

GRADUATE ENROLLMENTS & DEGREES AWARDED
Master's enrollment 360
Master's degrees awarded 112
Doctoral enrollment .... –
Doctoral degrees awarded .... –

ENGINEERING STUDENT DATA

Applicants to the engineering college
Number of applicants to engineering college 1,296
Percent offered admission 89.0%

Matriculated engineering students
Percentage in top quartile (25%) of High School class .... 58.8 [5]
Average ACT scores: Composite—25 [4]

Notes: (4) Composite based on mean (5) Top 20%

FULL- & PART-TIME FACULTY—BY DEPARTMENT
[Figures are the head count of full-time faculty and the full-time equivalent (FTE) of part-time faculty for each engineering department or equivalent.]

Department	Full	Part
Electrical Eng	22	–
Industrial & Manufacturing Eng	7	–
Mechanical Eng	21	–
Computer Eng	5.5	–
Microelectronic Eng	6.5	–

COLLEGE DESCRIPTION

The College of Engineering offers undergraduate engineering as a five year program which includes five quarters of cooperative education. The curricula is strong in fundamentals which leads to specialization in the fourth and fifth years. The Mechanical, Electrical, and Computer Engineering programs offer the combined BS/MS sequence.

DEGREE PROGRAMS DESCRIPTION
The five programs offered by the College of Engineering are planned to prepare students to fit into present day industrial and community life and to lay a foundation for graduate work in specialized fields. Each is a five year program with cooperative education as part of the requirement. The M.S. degree is offered in electrical, mechanical and computer engineering and a joint M.S. degree with the College of Science in materials science and engineering. There are also M.E. degrees with options in electrical, mechanical and microelectronic manufacturing.

DUAL DEGREE PROGRAMS
Engineering baccalaureate graduates with dual degrees 0

TRANSFER INFORMATION
Residency Requirements: Program cumulative GPA of at least 2.0 and a principal field of study GPA of at least 2.0. Students must successfully complete a minimum of 45 credit hours in the College granting the degree.

Transfer via Articulation Agreements
Admission to engineering: At the end of the sophomore year
Requirements: 2.80 GPA

Transfer without Articulation Agreements
Admission to engineering: At any time
Engineering graduates transferred from   .. 4-yr: 16   2-yr: 68

GRADUATION REQUIREMENTS
B.S. programs in engineering require a minimum of 180 quarter credits. Program cumulative GPA of at least 2.0; principal field of study GPA of at least 2.0.

STUDENT PROGRAMS

PROFESSIONAL AND HONORARY SOCIETIES
[For key to acronyms, see *Introduction*.]
Professional Societies: ASME, IEEE, IIE, SAE, SME, Society of Photographic & Instrumentation Engineers (SPIE)
Honorary Societies: Eta Kappa Nu, Pi Tau Sigma, Tau Beta Pi

SUPPORT PROGRAMS
Student Chapter Organizations: Society of Women Engineers, National Society of Black Engineers, Society of Hispanic Professional Engineers, HEOP, CAP?
Other Engineering Support Programs: Freshmen/transfer orientation programs are offered twice during the summer before the fall quarter for freshmen and once for transfers. A third orientation is held in the fall before classes begin. Entering freshmen who do not meet the requirements of the calculus exam are required to take a special six-hour calculus and analytic geometry course.

194 Rose-Hulman Institute of Technology

INSTITUTION PROFILE

HEAD OF THE INSTITUTION
Samuel F. Hulbert
Phone: (812)877-8202 Fax: (812)877-9925

GENERAL INFORMATION
[All Students—Fall 1991]
Undergraduate enrollment 1,294
Graduate enrollment 81
Total institution enrollment 1,375
Type of institution: Private   Calendar: Quarters
Nearest city: Terre Haute, Indiana   Population: 60,000
Miles from main campus: 1   Setting: Suburban
Types of engineering degrees: Engineering

ENGINEERING ADMISSIONS

ADMISSIONS OFFICE CONTACT
Chuck Howard
Rose-Hulman Institute of Technology
5500 Wabash Avenue
Terre Haute, IN 47803-3999
Phone: (812)877-1511 Fax: (812)877-3198

ENGINEERING COLLEGE ADMISSIONS INFORMATION
Entrance Requirements: SAT (Score: 1000), ACT, High School courses—Mathematics (4 years), Physics (1 year), Chemistry (1 year), English (4 years), Students are required to take either the SAT or the ACT. The minimum score for SAT is 400 verbal, 550 math, 1000 total. For the ACT the minimum score is 20 English, and 27 math.
Entrance Recommendations: High School courses—Mechanical Drawing (1 year), Foreign Language (2 years), Computer Programming (1 year), Speech
Requirements for foreign students: TOEFL (Score: 550)

DEPARTMENTAL ADMISSIONS INFORMATION
Admission to the engineering department: An applicant is admitted to Rose-Hulman rather than to particular degree programs.

FINANCIAL INFORMATION

ESTIMATED EXPENSES (ACADEMIC YEAR)
[Expenses are for the 1992-93 nine-month academic year.]

All Students	Undergraduate	Graduate
Tuition and fees	$11,670	$11,670
College room and board	$ 3,295 [2]	$ 8,865 [1]
Books and supplies	$ 500	$ 500
Other expenses	$ 600	$ 600
Total estimated expenses	**$16,065**	**$21,635**

Notes: (1) & Misc. (2) & Misc.

FINANCIAL AID OFFICE CONTACT
R. Paul Steward, *Director of Financial Aid*
Rose-Hulman Institute of Technology
5500 Wabash Avenue
Terre Haute, IN 47803-3999
Phone: (812)877-1511 Fax: (812)877-9925

GENERAL FINANCIAL AID INFORMATION
Forms accepted/required: Free Application for Federal Student Aid (FAFSA)
Additional financial aid information: Scholarship (merit-based), grants (Institutional, State, and Federal), Work-study, and loans, free application for Federal Student Aid (FAFSA).

ENGINEERING COLLEGE INFORMATION
[For additional personnel, refer to the *Appendix*.]

HEAD OF ENGINEERING
James R. Eifert
Phone: (812)877-8222 Fax: (812)877-9925
EMail: Eifert@NEXTWORK.Rose-Hulman.EDU'

ENGINEERING COLLEGE ADDRESS
Rose-Hulman Institute of Technology
5500 Wabash Avenue
Terre Haute, IN 47803-3999
Phone: (812)877-1511 Fax: (812)877-3198

ENROLLMENTS—BY CLASS
[Numbers are baccalaureate enrollments for the fall 1991 term, unless otherwise footnoted.]
1st-year students/Freshmen 291
2nd-year students/Sophomores 267
3rd-year students/Juniors 237
4th-year students/Seniors 260
Total **1,055**

NUMBER OF DEGREES AWARDED—BY PROGRAM
[Numbers are engineering baccalaureate degrees awarded during the 1991-1992 academic year, unless otherwise footnoted. For full details about each engineering program, refer to the *Tables of Degree Programs*.]
Chemical Engineering 32
Civil Engineering 19
Electrical & Computer Engineering 76
Mechanical Engineering 100
Total **227**

PERCENTAGE OF DEGREES AWARDED—BY CATEGORY
[Percentages are of all engineering baccalaureate degrees awarded during the 1991-1992 academic year, unless otherwise footnoted.]
African Americans .... – %
Asian/Pacific Island Americans .... 2.0 %
Hispanic Americans .... – %
Native Americans .... – %
Foreign Citizens .... 3.0 %
All Others .... 95.0 %
Women .... – %
Persons with disabilities .... – %
Students over 25 years of age .... – %

GRADUATE ENROLLMENTS & DEGREES AWARDED
Master's enrollment 44
Master's degrees awarded 15
Doctoral enrollment .... – [B]
Doctoral degrees awarded .... – [B]

Notes: (B) Data not applicable.

ENGINEERING STUDENT DATA
Applicants to the engineering college
Number of applicants to engineering college 3,190
Percent offered admission 60.0%

Matriculated engineering students
Percentage in top quartile (25%) of High School class 94.0
Average SAT scores: Math—680, Verbal—540, Combined—1220
Average ACT scores: Math—31, Composite—31

FULL- & PART-TIME FACULTY—BY DEPARTMENT
[Figures are the head count of full-time faculty and the full-time equivalent (FTE) of part-time faculty for each engineering department or equivalent.]

Department	Full	Part
Chemical Eng	7	–
Civil Eng	5	–
Electrical & Computer Eng	15	–
Mechanical Eng	16	–

COLLEGE DESCRIPTION
Rose-Hulman Institute of Technology is committed to excellence in the liberal education of scientists and engineers. This commitment leads to an emphasis on high quality instruction in the classroom and laboratory and to an array of co-curricular activities directed toward the development of the whole person. The Institute's commitment to personalized instruction and mentoring by the faculty has helped to create a campus environment of collegiality among the faculty and students. A strong emphasis is placed upon the development of the communication skills of the students and a rigorous humanities and social science program. The Institution emphasis in these latter areas strives to ensure that Rose-Hulman graduates are not only technically competent but are aware of the place of their technical endeavors in the larger societal issues. *In the Fall of 1995 Rose-Hulman will admit women upper-class transfer and freshmen students to its undergraduate programs.

DEGREE PROGRAMS DESCRIPTION
The B.S. degrees in the various engineering programs prepare the graduate for entry-level professional positions or for graduate study in each of the respective fields. Students may also earn a Technical Translator's Certificate in Russian or German. Japanese is also available. M.S. degree programs are available as indicated.

TRANSFER INFORMATION
Residency Requirements: There is a minimum residence requirement of two years with all required courses of senior year taken at Rose-Hulman.

Transfer without Articulation Agreements
Admission to engineering: At the end of the sophomore year
Engineering graduates transferred from . . . 4-yr: 14 2-yr: 2
Requirements: Dependent upon grades, courses taken, and recommendation.

GRADUATION REQUIREMENTS
195 minimum quarter hours with 'C' average.

STUDENT PROGRAMS

PROFESSIONAL AND HONORARY SOCIETIES
[For key to acronyms, see *Introduction*.]
Professional Societies: AIChE, ASCE, ASME, IEEE
Honorary Societies: Eta Kappa Nu, Omega Chi Epsilon, Pi Tau Sigma, Tau Beta Pi, Alpha Lambda Delta

SUPPORT PROGRAMS
Student Chapter Organizations: National Society of Black Engineers, International Student Advisor

International Student Advisor
For: Ethnic Minorities   **Available:** Year round
Offered: Freshman, Sophomore, Junior, Senior, Graduate level
Other Engineering Support Programs: Rose-Hulman has a Fall orientation program for all freshmen for academic credit; career counseling; learning center (peer tutoring); psychological counseling; placement services.

195 Rutgers, The State University

INSTITUTION PROFILE

HEAD OF THE INSTITUTION
Francis L. Lawrence
Phone: (908)932-7454   Fax: (908)932-8060

GENERAL INFORMATION
[All Students—Fall 1991]
Undergraduate enrollment 35,588
Graduate enrollment 13,105
Total institution enrollment 48,693

Type of institution: Public   **Calendar:** Semesters
Nearest city: New Brunswick, NJ   **Population:** 41,442
Miles from main campus: 5   **Setting:** Suburban
Types of engineering degrees: Engineering
Other degree-granting colleges: Agricultural & Environmental, Arts & Sciences, Communication & Journalism, Education, Fine & Performing Arts, Humanities & Social Sciences, Law, Nursing, Pharmacy, Technology & Applied Sciences, Rutgers, Douglass, Cook, Livingston

ENGINEERING ADMISSIONS

ADMISSIONS OFFICE CONTACT
M. E. Mitchell
Rutgers, The State University
Administrative Services Building
P.O. Box 2101
New Brunswick, NJ 08903-2101
Phone: (908)932-3770   Fax: (908)932-0237

ENGINEERING COLLEGE ADMISSIONS INFORMATION
Entrance Requirements: SAT, High School courses—English (4 years), Math (4 years), Chemistry (1 year), Physics (1 year)
Entrance Recommendations: High School courses—Computer Programming (1 year), Foreign Language (2 years)
Requirements for foreign students: TOEFL (Score: 550); financial statement

DEPARTMENTAL ADMISSIONS INFORMATION
Admission to the engineering department: At the end of the first year.

FINANCIAL INFORMATION

ESTIMATED EXPENSES (ACADEMIC YEAR)
[Expenses are for the 1992-93 nine-month academic year.]

State Residents	Undergraduate	Graduate
Tuition and fees	$ 4,398	$ 5,031
College room and board	$ 4,334	$ 4,800
Books and supplies	$ 800	$ 600
Other expenses	$ –	$ –
Total estimated expenses	**$ 9,532**	**$ 10,431**

Out-of-State Residents	Undergraduate	Graduate
Tuition and fees	$ 8,135	$ 7,189
College room and board	$ 4,334	$ 4,800
Books and supplies	$ 800	$ 600
Other expenses	$ –	$ –
Total estimated expenses	**$13,269**	**$ 12,589**

FINANCIAL AID OFFICE CONTACT
Carl H. Buck, *Director of Financial Aid*
Rutgers, The State University
Records Hall
Room 140
New Brunswick, NJ 08903
Phone: (908)932-7755

GENERAL FINANCIAL AID INFORMATION
Forms accepted/required: FAF, Stafford, IRS, Institutional, New Jersey Financial Aid form (NJFAF)
Additional financial aid information: Need-based scholarships; short-term loans; merit-based scholarships; part-time employment on campus; college work-study program.

ENGINEERING COLLEGE INFORMATION
[For additional personnel, refer to the *Appendix*.]

HEAD OF ENGINEERING
Ellis H. Dill
Phone: (908)932-2214   Fax: (908)932-5313

ENGINEERING COLLEGE ADDRESS
College of Engineering
Rutgers, The State University
P.O. Box 909
Piscataway, NJ 08855-0909
Phone: (908)932-2214   Fax: (908)932-5313

ENROLLMENTS—BY CLASS
[Numbers are baccalaureate enrollments for the fall 1991 term, unless otherwise footnoted.]
1st-year students/Freshmen 644
2nd-year students/Sophomores 628
3rd-year students/Juniors 618
4th-year students/Seniors 570
Total **2,460**

NUMBER OF DEGREES AWARDED—BY PROGRAM
[Numbers are engineering baccalaureate degrees awarded during the 1991-1992 academic year, unless otherwise footnoted. For full details about each engineering program, refer to the *Tables of Degree Programs*.]
Applied Sciences in Engineering 16
Biomedical Engineering .... –
Bioresource Engineering 7
Ceramic Engineering 54
Chemical Engineering 37
Civil Engineering 53
Electrical Engineering 136
Industrial Engineering 136
Mechanical Engineering 105
Mechanics & Materials Science .... –
Total **544**

PERCENTAGE OF DEGREES AWARDED—BY CATEGORY
[Percentages are of all engineering baccalaureate degrees awarded during the 1991-1992 academic year, unless otherwise footnoted.]
African Americans 4.5%
Asian/Pacific Island Americans 16.5%
Hispanic Americans 2.7%
Native Americans .... – %
Foreign Citizens 5.6%
All Others 70.7%
Women 17.8%
Persons with disabilities .... – %
Students over 25 years of age 5.9%

GRADUATE ENROLLMENTS & DEGREES AWARDED
Master's enrollment 415
Master's degrees awarded 132
Doctoral enrollment 342
Doctoral degrees awarded 51

ENGINEERING STUDENT DATA
Applicants to the engineering college
Number of applicants to engineering college 2,737
Percent offered admission 73.7%

Matriculated engineering students
Percentage in top quartile (25%) of High School class 74.5
Average SAT scores: Math—640, Verbal—499, Combined—1139

FULL- & PART-TIME FACULTY—BY DEPARTMENT
[Figures are the head count of full-time faculty and the full-time equivalent (FTE) of part-time faculty for each engineering department or equivalent.]

Department	Full	Part
Chemical & Biochemical Eng	13	–
Civil & Environmental Eng	10	1.6
Electrical & Computer Eng	29	2.7
Industrial Eng	8	–
Mechanical & Aerospace Eng	24	2.1
Ceramics	23	–
Biomedical Eng	8	1.4
Mechanics & Materials Science	12	1.0

COLLEGE DESCRIPTION
Four-year undergraduate curricula leading to the degree of Bachelor of Science are offered in seven engineering disciplines. In addition, a flexible four-year program in applied sciences in engineering is administered by an interdepartmental committee. Several curricula offer concentrations within the fundamental engineering disciplines including aerospace, biochemical, biomedical, computer, environmental, and packaging engineering. All curricula emphasize fundamental principles and are sufficiently comprehensive to form a foundation for advanced study in engineering or other fields. Five-year dual-degree programs are offered which lead to a B.A. or an M.B.A. in addition to the B.S. in engineering. A B.S./M.D. program is available to highly qualified students.

DEGREE PROGRAMS DESCRIPTION
The B.S. degrees in the various engineering programs prepare the graduate for entry-level professional positions, for graduate study in each of the respective fields, or for advanced study in fields such as law, medicine, or business administration. The B.S. in engineering science is intended primarily for advanced graduate study. All departments which offer B.S. degrees in engineering also offer strong, research-oriented graduate programs leading to the M.S. and Ph.D. degrees in engineering. Two additional departments, Biomedical Engineering and Mechanics and Materials Science, offer graduate programs only but offer undergraduate courses available to all majors.

DUAL DEGREE PROGRAMS
Engineering baccalaureate graduates with dual degrees 27
Enrollment requirements: Students complete first two years (including chemistry, calculus, and physics) at Rutgers, Douglass, Livingston or Cook Colleges with a GPA. of approximately 2.5 or better.

TRANSFER INFORMATION
Residency Requirements: Students must complete at least 30 of their final 42 credits at Rutgers University.

Transfer via Articulation Agreements
Admission to engineering: At the end of the sophomore year
Engineering graduates transferred from .. 4-yr: – 2-yr: 15
Requirements: A.S. Degree in Engineering Science, GPA. greater than 3.0.

Transfer without Articulation Agreements
Admission to engineering: At any time
Engineering graduates transferred from .. 4-yr: 41 2-yr: 54
Requirements: Should have completed one year of calculus, chemistry, and physics. Overall GPA. of 3.0 or better.

GRADUATION REQUIREMENTS
B.S. programs in engineering require 131-138 semester credits, depending on major. Common core includes 7 credits-chemistry, 16 credits-math, 12 credits-physics, 3 credits-computer programming, 3 credits-engineering mechanics, and 18 credits-humanities/social science electives. Minimum overall GPA. is 1.800; minimum GPA. in major is 2.000.

STUDENT PROGRAMS

PROFESSIONAL AND HONORARY SOCIETIES
[For key to acronyms, see *Introduction*.]
Professional Societies: ACerS, AIAA, AIChE, ASAE, ASCE, ASME, IEEE, IIE, SAE, SME, Biomedical Engineering Society, Optical Society of America
Honorary Societies: Alpha Pi Mu, Chi Epsilon, Eta Kappa Nu, Keramos, Pi Tau Sigma, Tau Beta Pi

SUPPORT PROGRAMS
Student Chapter Organizations: Society of Women Engineers, National Society of Black Engineers, Society of Hispanic Professional Engineers, Minority Engineering Educational Task, Rutgers Women in Math, Science and Engineering

Summer EOF Program
For: Women & Ethnic Minorities   **Available:** Summer only
Offered: Prematriculation

Educational Opportunity Fund Program
For: Women & Ethnic Minorities   **Available:** Academic year
Offered: Freshman, Sophomore, Junior, Senior

Summer Engineering Institute
For: Ethnic Minorities   **Available:** Summer only
Offered: Prematriculation, Freshman

Other Engineering Support Programs: Summer and fall orientation program for all engineering freshmen and transfers; summer pre-freshmen program for eligible students; personal and career counseling services; foreign student office; veterans' services; services for handicapped; placement services.

196 Saginaw Valley State University

■ INSTITUTION PROFILE
HEAD OF THE INSTITUTION
Eric R. Gilbertson
Phone: (517)790-4041 **Fax:** (517)790-1314
GENERAL INFORMATION
[All Students—Fall 1991]
Undergraduate enrollment 5,726
Graduate enrollment 1,143
Total institution enrollment **6,869**
Type of institution: Public **Calendar:** Semesters
Nearest city: Saginaw **Population:** 90,000
Miles from main campus: 2 **Setting:** Suburban
Types of engineering degrees: Engineering
Other degree-granting colleges: Education, Arts and Behavioral Sciences, Nursing and Allied Health Sciences, Business and Management

■ ENGINEERING ADMISSIONS
ADMISSIONS OFFICE CONTACT
Thomas E. Kullgren
Saginaw Valley State University
2250 Pierce Road
Pioneer Hall 208
University Center, MI 48710
Phone: (517)790-4144 **Fax:** (517)790-2717
ENGINEERING COLLEGE ADMISSIONS INFORMATION
Entrance Requirements: ACT (Score: 14)
Entrance Recommendations: High School courses—English (4 years), Mathematics (3 years), Science (3 years), Social Science (3 years)
Requirements for foreign students: TOEFL (Score: 525)
DEPARTMENTAL ADMISSIONS INFORMATION
Admission to the engineering department: At the time of admission to the institution.

■ FINANCIAL INFORMATION
ESTIMATED EXPENSES (ACADEMIC YEAR)
[Expenses are for the 1992-93 nine-month academic year.]

State Residents	Undergraduate	Graduate
Tuition and fees	$ 2,270	$ –
College room and board	$ 3,200	$ –
Books and supplies	$ 500	$ –
Other expenses	$ –	$ –
Total estimated expenses	**$ 5,970**	**$ –**

Out-of-State Residents	Undergraduate	Graduate
Tuition and fees	$ 4,650	$ –
College room and board	$ 3,200	$ –
Books and supplies	$ 500	$ –
Other expenses	$ –	$ –
Total estimated expenses	**$ 8,350**	**$ –**

FINANCIAL AID OFFICE CONTACT
William L. Healy, *Director, Scholarship & Financial Aid*
Saginaw Valley State University
2250 Pierce Road
University Center, MI 48710
Phone: (517)790-4106
GENERAL FINANCIAL AID INFORMATION
Forms accepted/required: CSS-FAF, Stafford, Institutional
Additional financial aid information: Need-based scholarships, Merit scholarships, Part-time campus jobs, loans, grants

■ ENGINEERING COLLEGE INFORMATION
[For additional personnel, refer to the *Appendix*.]
HEAD OF ENGINEERING
Thomas E. Kullgren
Phone: (517)790-4141 **Fax:** (517)790-2717
ENGINEERING COLLEGE ADDRESS
Engineering & Technology
Saginaw Valley State University
2250 Pierce Road
Pioneer Hall 208
University Center, MI 48710
Phone: (517)790-4144 **Fax:** (517)790-2717
ENROLLMENTS—BY CLASS
[Numbers are baccalaureate enrollments for the fall 1991 term, unless otherwise footnoted.]
1st-year students/Freshmen 75
2nd-year students/Sophomores 54
3rd-year students/Juniors 72
4th-year students/Seniors 116
Total **317**
NUMBER OF DEGREES AWARDED—BY PROGRAM
[Numbers are engineering baccalaureate degrees awarded during the 1991-1992 academic year, unless otherwise footnoted. For full details about each engineering program, refer to the *Tables of Degree Programs*.]
Electrical Engineering 20
Mechanical Engineering 21
Total **41**
PERCENTAGE OF DEGREES AWARDED—BY CATEGORY
[Percentages are of all engineering baccalaureate degrees awarded during the 1991-1992 academic year, unless otherwise footnoted.]
African Americans .... 2.4 %
Asian/Pacific Island Americans .... 2.4 %
Hispanic Americans .... – %
Native Americans .... – %
Foreign Citizens .... 2.4 %
All Others .... 92.8 %
Women .... 4.9 %
Persons with disabilities .... – %
Students over 25 years of age .... – %
ENGINEERING STUDENT DATA
Applicants to the engineering college
Number of applicants to engineering college 100
Percent offered admission 75.0%
Matriculated engineering students
Percentage in top quartile (25%) of High School class ...... – (B)
Average ACT scores: Math—19.4, Composite—20.2
Notes: (B) Data not applicable.
FULL- & PART-TIME FACULTY—BY DEPARTMENT
[Figures are the head count of full-time faculty and the full-time equivalent (FTE) of part-time faculty for each engineering department or equivalent.]

Department	Full	Part
Electrical Eng	5	–
Mechanical Eng	6	–

COLLEGE DESCRIPTION
Engineering at SVSU is a hands-on experience led almost exclusively by full-time faculty in classes averaging 14 students. Facilities and laboratory equipment are modern and first class. Generous endowments support laboratories and scholarships. Students can participate in overseas exchange programs, the NASA Space Grant Program, a vibrant optional co-op with major regional manufacturers and undergraduate research and testing laboratory employment. Students are part of a major technology development and transfer program in environmental science and engineering. A system complete with full-time staff and foundation funding has been constructed to deal with engineering student pipeline issues.
DEGREE PROGRAMS DESCRIPTION
Students in both Electrical and Mechanical Engineering receive a strong hands-on experience in a variety of modern laboratories using state-of-the art equipment. Design and research experiences are important parts of the program and a number of students are accepted by graduate schools for advanced study. The vast majority of courses are taught by full-time faculty and class sizes average 14. The University is located in a heavily industrialized region with a strong base of employment for engineers. Electrical Engineering Department offers an undergraduate program of foundation courses and courses specializing in computer engineering, robotics, digital controls, communications and power systems. The Mechanical Engineering Department offers an undergraduate program of foundation courses and specialized courses in areas such as finite elements, fracture and fatigue and vibrations.
TRANSFER INFORMATION
Residency Requirements: May only transfer in grades of C or better; must complete at least 31 hours in residence.
Transfer via Articulation Agreements
Admission to engineering: At any time
Engineering graduates transferred from ... 4-yr: 20 2-yr: 70
Requirements: Transfer grades of C or better only. 62 hours maximum transfer from a 2-year college. 93 hours maximum from a 4-year college.
Transfer without Articulation Agreements
Admission to engineering: At any time
Engineering graduates transferred from ... 4-yr: 20 2-yr: 70
Requirements: Transfer grades of C or better only. 62 hours maximum transfer from a 2-year college. 93 hours maximum from a 4 year college.

■ STUDENT PROGRAMS
PROFESSIONAL AND HONORARY SOCIETIES
[For key to acronyms, see *Introduction*.]
Professional Societies: ACM, ACS, AMS, ASME, IEEE, MAA, Society of Women Engineers
SUPPORT PROGRAMS
MI Space Grant Consortium (NASA)
For: Women & Ethnic Minorities **Available:** Year round
Offered: Freshman, Sophomore, Junior, Senior
Mid MI Minority Pre-Engineering Program (M3PEP)
For: Women & Ethnic Minorities **Available:** Summer only
Offered: Freshman
Project Get-Set
For: Women & Ethnic Minorities **Available:** Academic year
Offered: Prematriculation
Other Engineering Support Programs: Freshmen concepts and careers course, faculty advising program, junior year overseas program, co-op program, endowed scholarships, employment in an on-campus testing lab, undergraduate research on-campus testing lab, undergraduate research, learning resources center w/tutoring.

197 St. Cloud State University

INSTITUTION PROFILE

HEAD OF THE INSTITUTION
Robert O. Bess
Phone: (612)255-2122

GENERAL INFORMATION
[All Students—Fall 1991]
Undergraduate enrollment 14,955
Graduate enrollment 1,370
Total institution enrollment 16,325
Type of institution: Public **Calendar:** Quarters
Nearest city: Minneapolis **Population:** 300,000
Miles from main campus: 65 **Setting:** Urban
Types of engineering degrees: Engineering & Technology
Other degree-granting colleges: Business Administration, Education, Social Sciences, Fine Arts and Humanities

ENGINEERING ADMISSIONS

ADMISSIONS OFFICE CONTACT
Bruce W. Ellis
St. Cloud State University
ECC 211
720 South Fourth Avenue
St. Cloud, MN 56301-4498
Phone: (612)255-3252 **Fax:** (612)654-5127

ENGINEERING COLLEGE ADMISSIONS INFORMATION
Entrance Requirements: ACT (Score: 25), Students must satisfy entrance requirements with either a 25 score on the ACT or by being in the upper 50% of their high school graduating class.
Entrance Recommendations: High School courses—Mathematics (4 years), Physics (1 year), Chemistry (1 year)
Requirements for foreign students: TOEFL (Score: 475); financial statement; Students must carry health insurance (offered by the school).

DEPARTMENTAL ADMISSIONS INFORMATION
Admission to the engineering department: At the end of the second year.
Additional information: To be considered for admission, the student shall have completed all lower division pre-engineering classes with an honor point ratio (HPR) of 2.5 or better as well as an HPR of 2.5 or better in all classes. All attempts are counted in establishing the HPR.

FINANCIAL INFORMATION

ESTIMATED EXPENSES (ACADEMIC YEAR)
[Expenses are for the 1992-93 nine-month academic year.]

State Residents	Undergraduate	Graduate
Tuition and fees	$ 1,967	$ 2,721
College room and board	$ 2,295	$ 2,295
Books and supplies	$ 345	$ 345
Other expenses	$ –	$ –
Total estimated expenses	**$ 4,607**	**$ 5,361**

Out-of-State Residents	Undergraduate	Graduate
Tuition and fees	$ 3,215[1]	$ 3,825[1]
College room and board	$ 2,295	$ 2,295
Books and supplies	$ 345	$ 345
Other expenses	$ –	$ –
Total estimated expenses	**$ 5,855**	**$ 6,465**

Notes: (1) Tuition reciprocity agreements exist with Wisconsin, North Dakota, and South Dakota.

FINANCIAL AID OFFICE CONTACT
Frank Loncorich, *Director of Financial Aid*
St. Cloud State University
S.C.S.U. AS 106
720 South Fourth Avenue
St. Cloud, MN 56301-4498
Phone: (612)255-2047

GENERAL FINANCIAL AID INFORMATION
Forms accepted/required: FFS

ENGINEERING COLLEGE INFORMATION

[For additional personnel, refer to the *Appendix*.]

HEAD OF ENGINEERING
G. Richard Hogan
Phone: (612)255-2192 **Fax:** (612)255-4262

ENGINEERING COLLEGE ADDRESS
College of Science and Technology
St. Cloud State University
720 South Fourth Avenue
St. Cloud, MN 56301-4498
Phone: (612)255-2192 **Fax:** (612)255-4262

ENROLLMENTS—BY CLASS
[Numbers are baccalaureate enrollments for the fall 1991 term, unless otherwise footnoted.]
1st-year students/Freshmen 79
2nd-year students/Sophomores 50
3rd-year students/Juniors 58
4th-year students/Seniors 74
Total **261**

NUMBER OF DEGREES AWARDED—BY PROGRAM
[Numbers are engineering baccalaureate degrees awarded during the 1991-1992 academic year, unless otherwise footnoted. For full details about each engineering program, refer to the *Tables of Degree Programs*.]
Electrical Engineering 30
Manufacturing Engineering 3
Total **33**

PERCENTAGE OF DEGREES AWARDED—BY CATEGORY
[Percentages are of all engineering baccalaureate degrees awarded during the 1991-1992 academic year, unless otherwise footnoted.]
African Americans .... – %
Asian/Pacific Island Americans .... – %
Hispanic Americans .... – %
Native Americans .... – %
Foreign Citizens .... 3.0 %
All Others .... 97.0 %
Women .... 3.0 %
Persons with disabilities .... – % (A)
Students over 25 years of age .... – % (A)

Notes: (A) Data not available.

ENGINEERING STUDENT DATA

Applicants to the engineering college
Number of applicants to engineering college 72
Percent offered admission 99.0%

Matriculated engineering students
Percentage in top quartile (25%) of High School class .... – (A)
Average ACT scores: Composite—22[1]

Notes: (A) Data not available. (1) As reported by ACT for all new freshmen only

FULL- & PART-TIME FACULTY—BY DEPARTMENT

[Figures are the head count of full-time faculty and the full-time equivalent (FTE) of part-time faculty for each engineering department or equivalent.]

Department	Full	Part
Electrical Eng	7	–
Manufacturing Eng	4	–

COLLEGE DESCRIPTION
The St. Cloud State University College of Science and Technology houses nine departments which include the majors in electrical engineering, manufacturing engineering, engineering technology (general and manufacturing), photographic engineering technology (administrative and technological), and photographic science and instrumentation. The manufacturing engineering major is one of about a dozen programs and the photographic engineering technology is one of two in the country. We strongly encourage full time internships/co-op programs with a length of 3 to 6 months. A common double major selected by students is electrical engineering and physics/electro-optics. Our student SWE chapter hosts a seventh and eighth grade girls engineering program annually.

DEGREE PROGRAMS DESCRIPTION
The Bachelor of Science major in Electrical Engineering emphasizes analog and digital electronics, computers, communications, controls and computer aided design. The course work is designed to equip the student with the knowledge and skills necessary to work in research and design positions in the high technology electronics, information processing and computer industries. This program is also suitable as a preparation for graduate study in electrical engineering, computer engineering and computer science. The Bachelor of Science major in Manufacturing Engineering is designed to provide students with a thorough knowledge of the entire manufacturing process, from design through production and management. This involves integrated systems design, the development of manufacturing processes and effective communication with workers and management. The Manufacturing Engineering program equips the student with skills to work in the areas of design, manufacturing, production, quality control, and engineering management. The program is also suitable as preparation for graduate study in manufacturing engineering, industrial engineering, and engineering management.

TRANSFER INFORMATION
Residency Requirements: To be eligible for graduation a student must be in residence at least three quarters and must have earned at least 45 quarter hours of credit in residence out of the last 96 quarter hours. Students transferring from two year institutions must earn at least 96 quarter credits at a four year institution in addition to the credits from their two year institution.

GRADUATION REQUIREMENTS
(1) Engineering graduates must have within their 54 credit general education program at least 24 credits of humanities and social sciences that emphasize cultural values and include advanced level classes. (2) Students must maintain at least a 2.0 HPR each quarter or be placed in a probationary or disqualified status.

STUDENT PROGRAMS

PROFESSIONAL AND HONORARY SOCIETIES
[For key to acronyms, see *Introduction*.]
Professional Societies: ACM, ACS, AIP, IEEE, SME
Honorary Societies: Eta Kappa Nu, Kappa Eta Kappa

SUPPORT PROGRAMS
Student Chapter Organizations: Society of Women Engineers
Other Engineering Support Programs: All pre-engineering students are assigned a faculty adviser with whom they meet quarterly for course selection and advisement purposes. Summer orientation is held for all freshman and transfer students.

198 Parks College of Saint Louis University

■ INSTITUTION PROFILE
HEAD OF THE INSTITUTION
Lawrence H. Biondi S.J.
Phone: (314)658-2471   Fax: (314)658-7105
GENERAL INFORMATION
[All Students—Fall 1991]
Undergraduate enrollment 1,127
Graduate enrollment 10
Total institution enrollment **1,137**
Type of institution: Private   **Calendar:** Semesters
Nearest city: St. Louis, Missouri   **Population:** 1,000,000
Miles from main campus: 4   **Setting:** Small Town
Types of engineering degrees: Engineering & Technology
Other degree-granting colleges: Allied Health Sciences, Arts & Sciences, Business Administration, Communication & Journalism, Dentistry, Education, Fine & Performing Arts, Humanities & Social Sciences, Law, Medicine, Nursing, Public Health

■ ENGINEERING ADMISSIONS
ADMISSIONS OFFICE CONTACT
Sarah Nandor
Parks College of Saint Louis University
Highway 157 & Falling Springs Road
Cahokia, IL 62206-1430
Phone: (618)337-7575   Fax: (618)332-6802
ENGINEERING COLLEGE ADMISSIONS INFORMATION
Entrance Requirements: High School courses—Mathematics (4 years), English (3 years), Science (3 years), Social Science (2 years)
Entrance Recommendations: SAT, ACT, High School courses—English (4 years), Science (4 years), Social Science (3 years), Foreign Language (2 years)
Requirements for foreign students: financial statement
DEPARTMENTAL ADMISSIONS INFORMATION
Admission to the engineering department: At the time of admission to the institution.
Additional information: English placement exam required. Proficiency examination in Pre-calculus Mathematics required.

■ FINANCIAL INFORMATION
ESTIMATED EXPENSES (ACADEMIC YEAR)
[Expenses are for the 1992-93 nine-month academic year.]

All Students	Undergraduate	Graduate
Tuition and fees	$ 7,720 [2]	$ 7,560 [1]
College room and board	$ 3,750	$ 3,400
Books and supplies	$ 225	$ 500
Other expenses	$ –	$ –
Total estimated expenses	**$11,695**	**$11,460**

Notes: (1) based on 9 credit hours/semester (2) based on 12-18 credit hours/semester
FINANCIAL AID OFFICE CONTACT
Rachel M. Phillipone, *Financial Aid Coordinator*
Parks College of Saint Louis University
Highway 157 & Falling Springs Road
Cahokia, IL 62206-1430
Phone: (618)337-7575   Fax: (618)332-6802
GENERAL FINANCIAL AID INFORMATION
Forms accepted/required: FAF, FFS, IRS, Institutional, FAFSA
Additional financial aid information: Need based scholarships, loans, Grants, College work/study

■ ENGINEERING COLLEGE INFORMATION
[For additional personnel, refer to the *Appendix*.]
HEAD OF ENGINEERING
Peggy Baty
Phone: (618)337-7575   Fax: (618)332-6802
ENGINEERING COLLEGE ADDRESS
Parks College of Saint Louis University
Highway 157 & Falling Springs Road
Cahokia, IL 62206-1430
Phone: (618)337-7575   Fax: (618)332-6802
ENROLLMENTS—BY CLASS
[Numbers are baccalaureate enrollments for the fall 1991 term, unless otherwise footnoted.]
1st-year students/Freshmen 60
2nd-year students/Sophomores 64
3rd-year students/Juniors 58
4th-year students/Seniors 114
Total **296**
NUMBER OF DEGREES AWARDED—BY PROGRAM
[Numbers are engineering baccalaureate degrees awarded during the 1991-1992 academic year, unless otherwise footnoted. For full details about each engineering program, refer to the *Tables of Degree Programs*.]
Aerospace Engineering 61
Electrical Engineering 16
Total **77**
PERCENTAGE OF DEGREES AWARDED—BY CATEGORY
[Percentages are of all engineering baccalaureate degrees awarded during the 1991-1992 academic year, unless otherwise footnoted.]
African Americans .... – %
Asian/Pacific Island Americans .... – %
Hispanic Americans .... – %
Native Americans .... – %
Foreign Citizens .... 14.0 %
All Others .... 86.0 %
Women .... 14.0 %
Persons with disabilities .... – %
Students over 25 years of age .... – %
GRADUATE ENROLLMENTS & DEGREES AWARDED
Master's enrollment 10
Master's degrees awarded .... – [1]
Doctoral enrollment .... –
Doctoral degrees awarded .... –
Notes: (1) Masters in Aerospace Engineering began fall 1991
ENGINEERING STUDENT DATA
Applicants to the engineering college
Number of applicants to engineering college 220
Percent offered admission .... 75.0 %
Average SAT scores: Math—540, Verbal—430, Combined—970
Average ACT scores: Math—24, Composite—23
FULL- & PART-TIME FACULTY—BY DEPARTMENT
[Figures are the head count of full-time faculty and the full-time equivalent (FTE) of part-time faculty for each engineering department or equivalent.]

Department	Full	Part
Electrical Eng	6	–
Aerospace Eng	11	1.5

COLLEGE DESCRIPTION
Saint Louis University, founded in 1818, is the oldest University west of the Mississippi. The University is a Catholic Jesuit institution. Parks College, established in 1927, is one of the four campuses of the University and has as its emphasis, excellence in aerospace education. The Aerospace Engineering and Electrical Engineering programs are ABET accredited. Ten other bachelor degree programs are offered on the Parks campus. A Masters of Science in Aerospace Engineering began in the fall of 1991. An optional cooperative educational program allows students to gain industrial experience. In conjunction with the Physics department at the University, a 3-2 program is available which results in the B.A. in Physics and B.S. in Aerospace Engineering.
DEGREE PROGRAMS DESCRIPTION
The ABET accredited Aerospace and Electrical Engineering curricula prepare its graduates for entry into productive engineering positions involving analysis, design, and development of either aerospace vehicles or electrical and electronic systems, respectively. The Aerospace curriculum contains course work in such areas at aerodynamics, structural design, propulsion, flight mechanics, and astrodynamics. Electives, such as orbital mechanics or flight simulation provide provide opportunities for greater depth of study. The Electrical Engineering program is directed toward sequential development of course work to provide both depth and breadth. Instruction is provided in such areas as electrical and electronic circuits, signals and systems, control systems, communication systems, electromagnetics, digital systems, micro processors, and energy conversions. This is followed by specialized courses in integrated circuit design, digital signal processing, filter design, microwaves, radar systems, spacecraft communications or image processing. Both curricula include a two - semester design sequence to provide a meaningful engineering design experience. Modern well equipped laboratories provide the students with hands on experience. An evening graduate program leading to the M.S. in Aerospace Engineering is available.
DUAL DEGREE PROGRAMS
Engineering baccalaureate graduates with dual degrees 1
Enrollment requirements: A 3-2 program is available with the Physics Department of the University. The student can receive the B.A. in Physics and B.S. in Aerospace Engineering
TRANSFER INFORMATION
Residency Requirements: Transfer students must complete the final 30 hours at Parks or on the Frost Campus of the University.
Transfer via Articulation Agreements
Requirements: A 2.5 GPA
Transfer without Articulation Agreements
Admission to engineering: At any time
Requirements: A 2.5 GPA
GRADUATION REQUIREMENTS
An overall GPA of 2.0 out of 4.0 is required.

■ STUDENT PROGRAMS
PROFESSIONAL AND HONORARY SOCIETIES
[For key to acronyms, see *Introduction*.]
Professional Societies: AHS, AIAA, IEEE, SAE
Honorary Societies: Alpha Sigma Nu, Alpha Chi
SUPPORT PROGRAMS
Student Chapter Organizations: Black Student Alliance
Other Engineering Support Programs: Professional Development (0 credit) required of all freshmen; Enrichment Center provides academic counseling

199 University of San Diego

INSTITUTION PROFILE

HEAD OF THE INSTITUTION
Author E. Hughes
Phone: (619)260-4520 Fax: (619)260-6833

GENERAL INFORMATION
[All Students—Fall 1991]
Undergraduate enrollment 3,901
Graduate enrollment 2,140
Total institution enrollment 6,041
Type of institution: Private Calendar: Semesters
Nearest city: San Diego Population: 1,000,000
Miles from main campus: 6 Setting: Urban
Types of engineering degrees: Engineering
Other degree-granting colleges: Arts & Sciences, Business Administration, Education, Law, Nursing, Graduate & Continuing Education

ENGINEERING ADMISSIONS

ADMISSIONS OFFICE CONTACT
Warren W. Muller
University of San Diego
Serra Hall, Rm 203
5998 Alcala Park
San Diego, CA 92110-2492
Phone: (619)260-4506 Fax: (619)260-6836

ENGINEERING COLLEGE ADMISSIONS INFORMATION
Admission to the engineering college: After beginning of the first semester of attendance.
Entrance Requirements: SAT, While no minimum scores are stated, 450 minimum each is desired in verbal and math; avg scores were 485, 533 respectively.
Entrance Recommendations: SAT, ACT, High School courses—English (4 years), Mathematics (4 years), Social Sciences (4 years), Natural Sciences (4 years), For residents, SAT is required, non-residents may submit either SAT or ACT. High school performance in academic subjects given the highest consideration.
Requirements for out-of-state residents: Out-of-state residents must submit either SAT &/or ACT scores; in-state residents must submit SAT scores.
Requirements for foreign students: TOEFL (Score: 550); financial statement; SAT or ACT scores required. Must fund support from non-university sources. Avg UG score: 580. A minimum of 600 is required for graduate students.

DEPARTMENTAL ADMISSIONS INFORMATION
Admission to the engineering department: After beginning of the first semester in attendance for freshman; or upon transfer with more than 24 units.
Additional information: Strongly recommended prior preparation of college algebra, plane trig, freshman composition and two to four years high school foreign language. Also one of following: Upper 25% of graduating class; 3.0/4.0 in college prep courses; minimum SAT/ACT of 450/19 verbal, 500/18 math, 1050/23 overall.

FINANCIAL INFORMATION

ESTIMATED EXPENSES (ACADEMIC YEAR)
[Expenses are for the 1992-93 nine-month academic year.]

All Students	Undergraduate	Graduate
Tuition and fees	$12,240	$10,400
College room and board	$ 5,710	$ 6,560
Books and supplies	$ 600	$ –
Other expenses	$ 2,430 [1]	$ –
Total estimated expenses	**$20,980**	**$16,960**

Notes: (1) For on-campus students, includes transportation ($660) and personal ($1770) expenses.

FINANCIAL AID OFFICE CONTACT
Judith Lewis-Logue, *Director, Financial Aid*
University of San Diego
Serra Hall Rm 202
5998 Alcala Park
San Diego, CA 92110-2492
Phone: (619)260-4514

GENERAL FINANCIAL AID INFORMATION
Forms accepted/required: ACT, CSS-FAF, FAF, FFS, Stafford, SAR, Supplemental
Additional financial aid information: Merit-based scholarships, need-based scholarships, grants (federal, state, USD), loans (federal, USD), federal work study, student staff positions, off-campus employment.

ENGINEERING COLLEGE INFORMATION

[For additional personnel, refer to the *Appendix*.]

HEAD OF ENGINEERING
Thomas A. Kanneman
Phone: (619)260-4628 Fax: (619)260-4619
EMail: t_kanneman@usdcsd.acusd.edu

ENGINEERING COLLEGE ADDRESS
Engineering
University of San Diego
Loma Hall Rm 230
5998 Alcala Park
San Diego, CA 92110-2492
Phone: (619)260-4627 Fax: (619)260-4619

ENROLLMENTS—BY CLASS
[Numbers are baccalaureate enrollments for the fall 1991 term, unless otherwise footnoted.]
1st-year students/Freshmen .... 17 [1]
2nd-year students/Sophomores 8
3rd-year students/Juniors 10
4th-year students/Seniors .... 16 [2]
Total 51

Notes: (1) Freshman enter as undeclared. 31 started in the fall freshman engineering preceptorial course, 17 continued in the engineering program in the spring. (2) This figure includes 9 4th yr majors and 7 5th yr graduating seniors (Senior 2) who completed the standard 4.5 yr program in either in the 9th or 10th semester.

NUMBER OF DEGREES AWARDED—BY PROGRAM
[Numbers are engineering baccalaureate degrees awarded during the 1991-1992 academic year, unless otherwise footnoted. For full details about each engineering program, refer to the *Tables of Degree Programs*.]
Electrical Engineering 7
Total 7

PERCENTAGE OF DEGREES AWARDED—BY CATEGORY
[Percentages are of all engineering baccalaureate degrees awarded during the 1991-1992 academic year, unless otherwise footnoted.]
African Americans 14.3%
Asian/Pacific Island Americans 14.3%
Hispanic Americans .... – %
Native Americans .... – %
Foreign Citizens .... – %
All Others 71.4%
Women 28.6%
Persons with disabilities .... – %
Students over 25 years of age .... – % [A]

Notes: (A) Data not available.

ENGINEERING STUDENT DATA
Applicants to the engineering college
Number of applicants to engineering college .... 2,970 [5]
Percent offered admission .... 78.0% [6]

Matriculated engineering students
Percentage in top quartile (25%) of High School class .... [A]
Average SAT scores: Math—536, Verbal—482, Combined—1018

Notes: (A) Data not available. (5) Freshman are not admitted to a major upon application or entry to USD. Figure is for freshman applicants to USD. (6) See footnote 5. Of 31 freshman in the fall semester Engineering preceptorial course, 17 continued in the spring engineering course.

FULL- & PART-TIME FACULTY—BY DEPARTMENT
[Figures are the head count of full-time faculty and the full-time equivalent (FTE) of part-time faculty for each engineering department or equivalent.]

Department	Full	Part
Electrical Eng	6	0.3

COLLEGE DESCRIPTION

Engineering was initiated at USD in fall 1986, starting with EE. The EE program is accredited by EAC/ABET, covering the initial graduating class of 1991. Engineering moved into a new building with expanded facilities in June, 1992. A nine semester dual BS/BA program is offered in engineering. Co-op/internship programs are available, but not required, and usually involve full-time summer periods and/or part-time programs. Most students are full-time students and will complete (i.e. required courses are available) the program in the minimum time (9 semester) or or at most in 10 semesters.

DEGREE PROGRAMS DESCRIPTION

The EE program is housed in new facilities in the newly completed Loam Hall with three modern computer-based instrumentation and systems laboratories, with a fourth reserved for future development. The program is an electronics and computer engineering oriented program with intensive computer-aided engineering and design laboratory components. The program offers a balanced coverage of topics in basic engineering, electronic devices, circuits and systems, digital systems, electronic communications, control systems, signal processing, instrumentation and computer-aided testing. Additional emphasis is given over typical engineering programs in non-technical subjects and in oral and written communication preparation as is suggested by the dual BS/BA designation. Third semester foreign language proficiency is required.

TRANSFER INFORMATION

Residency Requirements: Graduates must earn a minimum of the final 30 semester units of credit in residence at USD or if not, 54 of the last 60 units completed at USD. Since the engineering program is a 9 to 10 semester 153 credit program, some exceptions are possible, based on the merits of the individual case.

Transfer via Articulation Agreements
Admission to engineering: Transfers typically enter at any time up to the beginning of 6th semester of the 9 to 10 semester program depending on preparation.
Engineering graduates transferred from . . 4-yr: 3   2-yr: –
Requirements: A 2.7/4.0 minimum transfer GPA required by USD; 3.0 minimum recommended for engineering.

Transfer without Articulation Agreements
Admission to engineering: Same as for articulated agreements.
Engineering graduates transferred from . . 4-yr: –   2-yr: –
Requirements: Same as for articulated agreements. Overall USD had 114 4yr and 165 2yr transfer graduates.

GRADUATION REQUIREMENTS

USD requires at least 124 semester units credit, of which 48 must be upper-division (UD). The engineering degree requires between 144 and 153 units, of which 54-63 must be UD. The lower range depends on the amount of AP, CLEP, and proficiency exam credit, as well as variations in applying transfer credit. In addition to completion of the general education program, graduates must have a 2.0 minimum GPA overall in all college courses, and a GPA 2.0 or better with no less that C- grades in the total UD units required for the major. The final 30 units must be completed at USD, or if not, 54 of the last 60

STUDENT PROGRAMS

PROFESSIONAL AND HONORARY SOCIETIES
[For key to acronyms, see *Introduction*.]
Professional Societies: IEEE
Honorary Societies: Delta Epsilon Sigma, Sigma Psi

SUPPORT PROGRAMS
Student Chapter Organizations: Society of Women Engineers

Educational Opportunity Program
For: Women & Ethnic Minorities **Available:** Year round
Offered: Freshman, Sophomore, Junior, Senior

Other Engineering Support Programs: Fall new student orientation week; Spring orientation (2 days); Counseling Center & Academic Services; Career Services; Student Health Services; International Student Services; Educational Opportunity Programs.

200 San Diego State University

INSTITUTION PROFILE
HEAD OF THE INSTITUTION
Thomas B. Day
Phone: (619)594-5201   **Fax:** (619)594-5642

GENERAL INFORMATION
[All Students—Fall 1991]
Undergraduate enrollment 26,134
Graduate enrollment 6,817
Total institution enrollment **32,951**
Type of institution: Public   **Calendar:** Semesters
Nearest city: San Diego   **Population:** 1,000,000
Miles from main campus: 6   **Setting:** Urban
Types of engineering degrees: Engineering
Other degree-granting colleges: Business Administration, Education, Health and Human Services, Arts and Letters, Sciences, Professional Studies and Fine Arts

ENGINEERING COLLEGE ADMISSIONS INFORMATION
Entrance Requirements: High School courses—English/Foreign Language (6 years), Mathematics (3 years), US History/Visual and Performing Arts (2 years), Science (1 year), Have a qualifiable eligibility index (e.i). The e.i. is the combination of your high school GPA and your score on either the ACT or SAT test. Your GPA is based on grades earned during your first three years of high school (excluding p.e.).
Entrance Recommendations: SAT, ACT, The SAT or ACT minimum score required for admission depends on your high school GPA
Requirements for foreign students: TOEFL (Score: 550)

DEPARTMENTAL ADMISSIONS INFORMATION
Admission to the engineering department: At the time of admission to the institution.

FINANCIAL INFORMATION
ESTIMATED EXPENSES (ACADEMIC YEAR)
[Expenses are for the 1992-93 nine-month academic year.]

State Residents	Undergraduate	Graduate
Tuition and fees	$ 1,500	$ 1,500
College room and board	$ 6,420 [2]	$ 6,420 [2]
Books and supplies	$ 558	$ 558
Other expenses	$ 2,460 [3]	$ 2,460 [3]
Total estimated expenses	**$10,938**	**$10,938**

Out-of-State Residents	Undergraduate	Graduate
Tuition and fees	$ 1,500 [1]	$ 1,500 [1]
College room and board	$ 6,420 [2]	$ 6,420 [2]
Books and supplies	$ 558	$ 558
Other expenses	$ 2,460 [3]	$ 2,460 [3]
Total estimated expenses	**$10,938**	**$10,938**

Notes: (1) Plus $246 per unit attempted (2) Assumes living off-campus (3) Transportation and personal

FINANCIAL AID OFFICE CONTACT
William D. Boyd, *Director of Financial Aid*
San Diego State University
Student Service Annex
San Diego, CA 92182-0763
Phone: (619)594-6323   **Fax:** (619)594-5642

GENERAL FINANCIAL AID INFORMATION
Forms accepted/required: FAF, IRS, Institutional
Additional financial aid information: Need-based scholarships; loans; merit-based scholarships; athletic scholarships; part-time jobs on campus; college work-study program.

ENGINEERING COLLEGE INFORMATION
[For additional personnel, refer to the *Appendix*.]
HEAD OF ENGINEERING
George T. Craig
Phone: (619)594-6061   **Fax:** (619)594-6005
EMail: gcraig@sciences.sdsu.edu

ENGINEERING COLLEGE ADDRESS
College of Engineering
San Diego State University
5300 Campanile Drive
San Diego, CA 92182-0416
Phone: (619)594-6061   **Fax:** (619)594-6005

ENROLLMENTS—BY CLASS
[Numbers are baccalaureate enrollments for the fall 1991 term, unless otherwise footnoted.]
1st-year students/Freshmen 387
2nd-year students/Sophomores 221
3rd-year students/Juniors 451
4th-year students/Seniors 876
Total **1,935**

NUMBER OF DEGREES AWARDED—BY PROGRAM
[Numbers are engineering baccalaureate degrees awarded during the 1991-1992 academic year, unless otherwise footnoted. For full details about each engineering program, refer to the *Tables of Degree Programs*.]
Aerospace Engineering & Engineering Mechanics 50
Applied Mechanics .... – [B]
Civil Engineering 72
Electrical Engineering 142
Mechanical Engineering 58
Total **322**

Notes: (B) Data not applicable.

PERCENTAGE OF DEGREES AWARDED—BY CATEGORY
[Percentages are of all engineering baccalaureate degrees awarded during the 1991-1992 academic year, unless otherwise footnoted.]
African Americans 1.2%
Asian/Pacific Island Americans 28.9%
Hispanic Americans 8.1%
Native Americans 1.2%
Foreign Citizens 5.6%
All Others 55.0%
Women 15.8%
Persons with disabilities .... – % [A]
Students over 25 years of age .... – % [A]

Notes: (A) Data not available.

GRADUATE ENROLLMENTS & DEGREES AWARDED
Master's enrollment 305
Master's degrees awarded 66
Doctoral enrollment 10
Doctoral degrees awarded .... –

ENGINEERING STUDENT DATA
Applicants to the engineering college
Number of applicants to engineering college 1,284
Percent offered admission 81.0%

Matriculated engineering students
Percentage in top quartile (25%) of High School class ...... – [A]
Average SAT scores: Math—474, Verbal—422, Combined—896
Average ACT scores: Math—20.8, Composite—20.2

Notes: (A) Data not available.

FULL- & PART-TIME FACULTY—BY DEPARTMENT
[Figures are the head count of full-time faculty and the full-time equivalent (FTE) of part-time faculty for each engineering department or equivalent.]

Department	Full	Part
Civil Eng	13	2.6
Electrical Eng	17	3.0
Mechanical Eng	8	5.1
Aerospace Eng & Eng Mechanics	9	0.4
Eng Sciences/Applied Mechanics	1	0.5

COLLEGE DESCRIPTION
The objective of the engineering program at San Diego State is to provide the intellectual and physical environment best calculated to encourage students to develop their capacities toward a successful career in the profession of engineering, knowing the need for engineers to maintain a professional proficiency in a rapidly changing technology and advancing state of the art. Moreover, the effective development and application of technology depends on responsible judgments by professionals cognizant of the total needs of society and how technology affects people. Thus, the engineering graduate should have the academic background necessary for personal and professional growth.

DEGREE PROGRAMS DESCRIPTION
B.S. Program - Design oriented programs with strong ties to the San Diego regional professional community. All programs contain both depth and breadth in areas of specialization. M.S. Program - Program accommodates the part-time student by offering early morning as well as late afternoon, evening classes. Masters students involved with externally funded faculty directed research programs. Ph.D. Program - Offers a Ph.D. degree in Applied Mechanics. The program is offered jointly with the University of California, San Diego (UCSD). Students may enter the program after completion of either B.S. or M.S. degree. Students spend one year in residence on UCSD campus taking coursework and the remainder of their time at San Diego State University taking coursework and conducting their research.

TRANSFER INFORMATION
Residency Requirements: To qualify for a bachelor's degree, each of the following must be completed at this institution: A. A minimum of 30 units total, of which at least 24 units must be in upper division courses. B. At least half of the upper division units required for the major, unless waived by the department. C. At least nine units in upper division General Education courses.

Transfer without Articulation Agreements
Admission to engineering: At any time
Engineering graduates transferred from ... 4-yr: 30 2-yr: 193

GRADUATION REQUIREMENTS
Requires 2.0 (C average) in major, 2.0 overall, and 2.0 at San Diego State University.

STUDENT PROGRAMS
PROFESSIONAL AND HONORARY SOCIETIES
[For key to acronyms, see *Introduction*.]
Professional Societies: AGCA, AIAA, ASCE, ASHRAE, ASME, IEEE, SAE, SPE*, Association for Energy Engineers (AEE)
Honorary Societies: Chi Epsilon, Eta Kappa Nu, Pi Tau Sigma, Sigma Gamma Tau, Tau Beta Pi

SUPPORT PROGRAMS
Student Chapter Organizations: Society of Women Engineers, National Society of Black Engineers, Society of Hispanic Professional Engineers, Vietnamese Engineers Society

Minority Engineering Program
For: Ethnic Minorities   **Available:** Year round
Offered: Prematriculation, Freshman, Sophomore, Junior, Senior, Graduate level

201 San Francisco State University

■ INSTITUTION PROFILE

HEAD OF THE INSTITUTION
Robert A. Corrigan
Phone: (415)338-1381

GENERAL INFORMATION
[All Students—Fall 1991]
Undergraduate enrollment 20,492
Graduate enrollment 7,419
Total institution enrollment 27,911
Type of institution: Public **Calendar:** Semesters
Location: San Francisco **Population:** 750,000
Setting: Urban
Types of engineering degrees: Engineering
Other degree-granting colleges: Business Administration, Education, Fine & Performing Arts, Behavioral and Social Sciences, Ethnic Studies, Health, Physical Education, recreation, Humanities, Science

ENGINEERING COLLEGE ADMISSIONS INFORMATION
Entrance Requirements: SAT, ACT, High School courses—English (4 years), Mathematics (3 years), Lab Science (1 year), US History (1 year)
Entrance Recommendations: High School courses—Mathematics (4 years), Lab Sciences (3 years), Foreign Language (2 years), Visual and Performing Arts (1 year)
Requirements for out-of-state residents: In-state GPA and SAT composite minimum of 2800, or GPA and ACT composite minimum of 694 is required. Corresponding figures for Out-of-State are 3402 and 842 respectively.
Requirements for foreign students: TOEFL (Score: 500)

DEPARTMENTAL ADMISSIONS INFORMATION
Admission to the engineering department: At the time of admission to the institution.

■ FINANCIAL INFORMATION

FINANCIAL AID OFFICE CONTACT
Jeffrey S. Baker, *Director, Financial Aids*
San Francisco State University
1600 Holloway Avenue
San Francisco, CA 94132
Phone: (415)338-2437

GENERAL FINANCIAL AID INFORMATION
Forms accepted/required: Student Aid Application for California (SAAC)
Additional financial aid information: Pell Grant, Supplemental Educational Opportunity Grant, State University Grant, California State Educational Opportunity Program Grant, College Work-Study Program, Perkins Loan Program, Stafford Loan Program, Cal Grant A, Cal Grant B, Various internal and external scholarships, part-time jobs. Contact Financial Aid Office for current information.

■ ENGINEERING COLLEGE INFORMATION
[For additional personnel, refer to the *Appendix*.]

HEAD OF ENGINEERING
V. V. Krishnan
Phone: (415)338-1174 Fax: (415)338-0525

ENGINEERING COLLEGE ADDRESS
Division of Engineering
San Francisco State University
1600 Holloway Avenue
San Francisco, CA 94132
Phone: (415)338-1174 Fax: (415)338-0525

ENROLLMENTS—BY CLASS
[Numbers are baccalaureate enrollments for the fall 1991 term, unless otherwise footnoted.]
1st-year students/Freshmen .... 88 [1]
2nd-year students/Sophomores .... 76 [1]
3rd-year students/Juniors .... 84 [1]
4th-year students/Seniors .... 452 [1]
Total 700

Notes: (1) Fall '91 enrollment figures only.

NUMBER OF DEGREES AWARDED—BY PROGRAM
[Numbers are engineering baccalaureate degrees awarded during the 1991-1992 academic year, unless otherwise footnoted. For full details about each engineering program, refer to the *Tables of Degree Programs*.]
Civil Engineering 28
Electrical Engineering 66
Mechanical Engineering 28
Total 122

PERCENTAGE OF DEGREES AWARDED—BY CATEGORY
[Percentages are of all engineering baccalaureate degrees awarded during the 1991-1992 academic year, unless otherwise footnoted.]
African Americans .... 7.0 %
Asian/Pacific Island Americans .... 11.0 %
Hispanic Americans .... 11.0 %
Native Americans .... – %
Foreign Citizens .... 21.0 % [2]
All Others .... 50.0 %
Women .... 16.0 %
Persons with disabilities .... – % [A]
Students over 25 years of age .... – % [A]

Notes: (A) Data not available. (2) Foreign students; does not include foreign born legal residents of the U.S.

GRADUATE ENROLLMENTS & DEGREES AWARDED
Master's enrollment 22
Master's degrees awarded .... –
Doctoral enrollment .... – [3]
Doctoral degrees awarded .... – [3]

Notes: (3) No such program at this institution.

ENGINEERING STUDENT DATA
Applicants to the engineering college
Number of applicants to engineering college 693
Percent offered admission 40.0%

Matriculated engineering students
Percentage in top quartile (25%) of High School class .... – [A]

Notes: (A) Data not available.

FULL- & PART-TIME FACULTY—BY DEPARTMENT
[Figures are the head count of full-time faculty and the full-time equivalent (FTE) of part-time faculty for each engineering department or equivalent.]

Department	Full	Part
Civil Eng	4	1.0
Electrical Eng	8	2.0
Mechanical Eng	7	1.0

COLLEGE DESCRIPTION
Excellence in teaching the individual is the primary mission and feature. With equal emphasis on practice and theory, SFSU offers a quality education leading to a successful career in engineering. Unique features include emphasis on projects, teamwork, and development of communications skills necessary for success. Support networks have been established to meet the changing needs of today's student. Programs such as co-op, Design Center, Minority Engineering, Women in Engineering are specifically designed to assist engineering students. In addition, students and graduates have opportunities to participate in the most dynamic working environment - the San Francisco Bay Area.

DEGREE PROGRAMS DESCRIPTION
Each degree program is designed to give students a solid background in mathematics, sciences and engineering fundamentals. In addition to a broad and comprehensive set of required subjects, students may choose the following emphasis areas: B.S. in CE - structural, geotechnical, environmental, and construction management. B.S. in EE - communications, computers, electronics, power systems, and robotics and control. B.S. in ME - machine design, thermal-fluid systems, and robotics and control. M.S. in engineering: Flexible program which can be tailored to the individual needs of students. Selected specializations within the traditional areas of Civil, Electrical and Mechanical Engineering.

TRANSFER INFORMATION
Residency Requirements: 30 semester units must be earned in residence. 24 of these units shall be in upper division courses.

Transfer via Articulation Agreements
Admission to engineering: At any time
Requirements: Grade point average of 2.0 (C) or better and is in good standing at the last college or university attended.

Transfer without Articulation Agreements
Admission to engineering: At any time
Requirements: Grade point average of 2.0 or better and is in good standing at the last college or university attended.

GRADUATION REQUIREMENTS
Each student must complete all requirements in one of the approved engineering curricula and satisfy all other university graduation requirements. GPA overall: 2.0 (C), GPA in major: 2.0 (C).

■ STUDENT PROGRAMS

PROFESSIONAL AND HONORARY SOCIETIES
[For key to acronyms, see *Introduction*.]
Professional Societies: ASCE, ASME, IEEE, ISA

SUPPORT PROGRAMS
Student Chapter Organizations: Society of Women Engineers, National Society of Black Engineers, Society of Hispanic Professional Engineers

Minority Engineering Program (MEP)
For: Ethnic Minorities **Available:** Year round
Offered: Prematriculation, Freshman, Sophomore, Junior, Senior

Women in Engineering Program (WEP)
For: Women **Available:** Academic year
Offered: Prematriculation, Freshman, Sophomore, Junior, Senior

Other Engineering Support Programs: New students orientation program Mandatory academic advising program Scholarship programs Counseling and Psychological Services Advising Center

202 San Jose State University

■ INSTITUTION PROFILE

HEAD OF THE INSTITUTION
Handel Evans
Phone: (408)924-2400

GENERAL INFORMATION
[All Students—Fall 1991]
Undergraduate enrollment 23,642
Graduate enrollment 6,419
Total institution enrollment **30,061**
Type of institution: Public **Calendar:** Semesters
Location: San Jose **Population:** 750,000
Setting: Urban
Types of engineering degrees: Engineering
Other degree-granting colleges: Business Administration, Education, Fine & Performing Arts, Humanities & Social Sciences, Science, Social Sciences, Social Work, Applied Sciences and Arts

ENGINEERING COLLEGE ADMISSIONS INFORMATION
Admission to the engineering college: the time of application to the University. Applications for Fall semester should be submitted in November and applications for Spring semester in August.
Entrance Requirements: High School courses—English (4 years), Mathematics (3 years), U.S. History and Government (1 year), Foreign Language (2 years), SAT/ACT if high school GPA is less than 3.00. TOEFL may be required for some students.
Entrance Recommendations: High School courses—Computers (1 year), Advanced math and science (1 year)
Requirements for out-of-state residents: The College of Engineering only accepts California State residents.
Requirements for foreign students: TOEFL (Score: 550); Currently, non-residents of California are not eligible for admission to undergraduate engineering programs.

DEPARTMENTAL ADMISSIONS INFORMATION
Admission to the engineering department: At the time of admission to the institution.
Additional information: There are currently no special departmental requirements. Currently, all applicants to undergraduate engineering programs must be California residents.

■ FINANCIAL INFORMATION

ESTIMATED EXPENSES (ACADEMIC YEAR)
[Expenses are for the 1992-93 nine-month academic year.]

All Students	Undergraduate	Graduate
Tuition and fees	$ 1,556	$ 1,556[1]
College room and board	$ 4,564	$ 4,564
Books and supplies	$ 694	$ 694
Other expenses	$ 1,368	$ 1,368
Total estimated expenses	**$ 8,182**	**$ 8,182**

Notes: (1) All figures are based on ONE academic year

FINANCIAL AID OFFICE CONTACT
Donald Ryan, *Financial Aid Director*
San Jose State University
Financial Aid Student Services
San Jose, CA 96192-0036
Phone: (408)924-1676

GENERAL FINANCIAL AID INFORMATION
Forms accepted/required: FAF
Additional financial aid information: Many financial aid packages are available through the University Financial Aid Office.

■ ENGINEERING COLLEGE INFORMATION

[For additional personnel, refer to the *Appendix*.]
HEAD OF ENGINEERING
Jay D. Pinson
Phone: (408)924-3800 **Fax:** (408)924-3818

ENGINEERING COLLEGE ADDRESS
College of Engineering
San Jose State University
One Washington Square
San Jose, CA 95192-0080
Phone: (408)924-3800 **Fax:** (408)924-3818

ENROLLMENTS—BY CLASS
[Numbers are baccalaureate enrollments for the fall 1991 term, unless otherwise footnoted.]
1st-year students/Freshmen 633
2nd-year students/Sophomores 418
3rd-year students/Juniors 741
4th-year students/Seniors 1,389
Total **3,181**

NUMBER OF DEGREES AWARDED—BY PROGRAM
[Numbers are engineering baccalaureate degrees awarded during the 1991-1992 academic year, unless otherwise footnoted. For full details about each engineering program, refer to the *Tables of Degree Programs*.]
Aerospace Engineering 40
Chemical Engineering 16
Civil Engineering 49
Computer Engineering 36
Electrical Engineering 133
General Engineering .... –
Industrial & Systems Engineering 34
Materials Engineering 6
Mechanical Engineering 61
Total **375**

PERCENTAGE OF DEGREES AWARDED—BY CATEGORY
[Percentages are of all engineering baccalaureate degrees awarded during the 1991-1992 academic year, unless otherwise footnoted.]
African Americans .... 2.0 %
Asian/Pacific Island Americans .... 46.0 %
Hispanic Americans .... 6.0 %
Native Americans .... 1.0 %
Foreign Citizens .... – % (A)
All Others .... 45.0 %
Women .... – % (A)
Persons with disabilities .... – % (A)
Students over 25 years of age .... 73.0 %

Notes: (A) Data not available.

GRADUATE ENROLLMENTS & DEGREES AWARDED
Master's enrollment 847
Master's degrees awarded 213
Doctoral enrollment .... –
Doctoral degrees awarded .... –

ENGINEERING STUDENT DATA
Applicants to the engineering college
Number of applicants to engineering college 1,082
Percent offered admission .... –% (1)

Matriculated engineering students
Percentage in top quartile (25%) of High School class .... – (A)

Notes: (A) Data not available. (1) Only students ranked in top 1/3 of H.S. class apply for SJSU

FULL- & PART-TIME FACULTY—BY DEPARTMENT
[Figures are the head count of full-time faculty and the full-time equivalent (FTE) of part-time faculty for each engineering department or equivalent.]

Department	Full	Part
Chemical Eng	4	–
Civil Eng	13	–
Electrical Eng	31	3.5
Industrial & Systems Eng	4	1.0
Mechanical Eng	13	3.4
Aerospace Eng	4	–
Computer Eng	5	2.1
Materials Eng	5	1.3
General Eng	4	2.1

COLLEGE DESCRIPTION
The College of Engineering is located in the heart of 'Silicon Valley' and is the leading supplier of engineers to this high tech area. One of the major strengths of the engineering program at San Jose State is the emphasis that all departments place on laboratory and lecture-laboratory courses. The facility gives engineering students access to an increasing number of hands-on laboratory and lecture-laboratory courses. All the laboratories in the building are linked by a network to provide increased student access to the outstanding computer facilities available in the College of Engineering.

DEGREE PROGRAMS DESCRIPTION
The B.S. degree in the various engineering programs prepares the graduate for entry-level professional positions or for graduate study. Graduate study is M.S. degree only. M.S. in General Engineering is special interdisciplinary program involving several departments and the program requires a project. All M.S. programs in engineering emphasize engineering design and solutions to engineering problems.

TRANSFER INFORMATION
Residency Requirements: Maximum of 70 semester units can be transferred from a community college. Must complete at least 24 upper division semester units in School of Engineering and Science with a minimum of 15 units in the major department.
Transfer via Articulation Agreements
Admission to engineering: At the end of the sophomore year
Engineering graduates transferred from ... 4-yr: 23 2-yr: 243
Requirements: Currently all applicants for undergraduate engineering programs must be California residents.
Transfer without Articulation Agreements
Admission to engineering: At any time
Engineering graduates transferred from ... 4-yr: 23 2-yr: 243

GRADUATION REQUIREMENTS
B.S. programs in engineering require 129-140 semester units, depending on major selected.

■ STUDENT PROGRAMS

PROFESSIONAL AND HONORARY SOCIETIES
[For key to acronyms, see *Introduction*.]
Professional Societies: ACM, AIAA, AIChE, ASCE, ASME, IEEE, IIE, IES*, SAMPE, SME
Honorary Societies: Chi Epsilon, Eta Kappa Nu, Pi Tau Sigma, Tau Beta Pi

SUPPORT PROGRAMS
Student Chapter Organizations: Society of Women Engineers, National Society of Black Engineers, Society of Hispanic Professional Engineers

Minority Engineering Program
For: Women & Ethnic Minorities **Available:** Year round
Offered: Prematriculation, Freshman, Sophomore, Junior, Senior, Graduate level

Other Engineering Support Programs: Fall and Spring orientation program for all freshman and transfers: career counseling services: placement services: disabled student services.

203 Santa Clara University

INSTITUTION PROFILE
HEAD OF THE INSTITUTION
Paul L. Locatelli S.J.
Phone: (408)554-4100
GENERAL INFORMATION
[All Students—Fall 1991]
Undergraduate enrollment 4,079
Graduate enrollment 3,723
Total institution enrollment 7,802
Type of institution: Private **Calendar:** Quarters
Nearest city: San Jose **Population:** 800,000
Miles from main campus: 2 **Setting:** Urban
Types of engineering degrees: Engineering
Other degree-granting colleges: Arts & Sciences, Business Administration, Education, Law, Counseling Psychology and Education

ENGINEERING ADMISSIONS
ADMISSIONS OFFICE CONTACT
Daniel J. Saracino
Santa Clara University
455 El Camino Real
Santa Clara, CA 95053
Phone: (408)554-4700 **Fax:** (408)554-5255
ENGINEERING COLLEGE ADMISSIONS INFORMATION
Entrance Requirements: SAT, High School courses—High School Diploma required, college preparatory (4 years)
Entrance Recommendations: High School courses—Algebra, Laboratory Science (2 years), English (4 years), History, Plane Geometry,(Trigonometry 1/2 yr.) (1 year), Foreign Language (Elective 2-1/2 yrs.) (3 years)
Requirements for foreign students: TOEFL (Score: 550)
DEPARTMENTAL ADMISSIONS INFORMATION
Admission to the engineering department: At the time of admission to the institution.

FINANCIAL INFORMATION
ESTIMATED EXPENSES (ACADEMIC YEAR)
[Expenses are for the 1992-93 nine-month academic year.]

All Students	Undergraduate	Graduate
Tuition and fees	$12,150	$ 7,632
College room and board	$ 5,556	$ 5,184
Books and supplies	$ 576	$ 576
Other expenses	$ 2,304	$ 2,412[1]
Total estimated expenses	**$20,586**	**$15,804**

Notes: (1) Other expenses include transportation and personal expenses.
FINANCIAL AID OFFICE CONTACT
Rita LeBarre, *Director of Financial Aid*
Santa Clara University
Santa Clara, CA 95053
Phone: (408)554-4505 **Fax:** (408)554-6926
GENERAL FINANCIAL AID INFORMATION
Forms accepted/required: FAF, SAAC
Additional financial aid information: Santa Clara University offers a broadly based program of financial assistance for students including scholarships, grants, loans, and part-time on-campus employment. Before receiving any financial aid a student must complete a "SAAC" (Student Aid Application California). "FAF" is required for out-of-state students.

ENGINEERING COLLEGE INFORMATION
[For additional personnel, refer to the *Appendix*.]
HEAD OF ENGINEERING
Terry E. Shoup
Phone: (408)554-4600 **Fax:** (408)554-5474
EMail: tshoup@scu.bitnet
ENGINEERING COLLEGE ADDRESS
School of Engineering
Santa Clara University
455 El Camino Real
Santa Clara, CA 95053
Phone: (408)554-4600 **Fax:** (408)554-5474
ENROLLMENTS—BY CLASS
[Numbers are baccalaureate enrollments for the fall 1991 term, unless otherwise footnoted.]
1st-year students/Freshmen 200
2nd-year students/Sophomores 128
3rd-year students/Juniors 119
4th-year students/Seniors 146
Total 593
NUMBER OF DEGREES AWARDED—BY PROGRAM
[Numbers are engineering baccalaureate degrees awarded during the 1991-1992 academic year, unless otherwise footnoted. For full details about each engineering program, refer to the *Tables of Degree Programs*.]
Civil Engineering 25
Computer Engineering 16
Electrical Engineering 38
Mechanical Engineering 35
Total 114
PERCENTAGE OF DEGREES AWARDED—BY CATEGORY
[Percentages are of all engineering baccalaureate degrees awarded during the 1991-1992 academic year, unless otherwise footnoted.]
African Americans .... 1.1 %
Asian/Pacific Island Americans .... 26.1 %
Hispanic Americans .... 8.1 %
Native Americans .... 0.1 %
Foreign Citizens .... 8.1 %
All Others .... 56.5 %
Women .... 28.0 %
Persons with disabilities .... – %(A)
Students over 25 years of age .... – %(A)
Notes: (A) Data not available.
GRADUATE ENROLLMENTS & DEGREES AWARDED
Master's enrollment 1,169
Master's degrees awarded 299
Doctoral enrollment 37
Doctoral degrees awarded 6
ENGINEERING STUDENT DATA
Applicants to the engineering college
Number of applicants to engineering college 403
Percent offered admission 87.3%
Matriculated engineering students
Percentage in top quartile (25%) of High School class .... –(A)
Average SAT scores: Math—579, Verbal—503, Combined—1082
Notes: (A) Data not available.
FULL- & PART-TIME FACULTY—BY DEPARTMENT
[Figures are the head count of full-time faculty and the full-time equivalent (FTE) of part-time faculty for each engineering department or equivalent.]

Department	Full	Part
Civil Eng	6	–
Electrical Eng	13	–
Mechanical Eng	10	–
Computer Eng	8	–

COLLEGE DESCRIPTION
The undergraduate engineering program is a four-year program. The program for the Bachelor of Science degree includes courses that fulfill the goals of the University curriculum; that is, education of the "whole person" in a program which imparts a common body of knowledge to all students, develops the capacity for critical judgment and clear expression, and encourages a desire for life-long learning. The degree program also includes courses in the School of Engineering curriculum and the engineering departmental major curriculum. The college also has computing laboratory facilities as well as several specialized laboratories. A cooperative education program is available but not required.

DEGREE PROGRAMS DESCRIPTION
Besides the Master's degrees described in Section 18, M.S. degrees are offered in Applied Mathematics, Engineering, Engineering Management, and Engineering Mechanics. Engineer's Degrees are offered in Electrical Engineering and Mechanical Engineering. Three Ph.D. programs are offered: Electrical Engineering, Computer Engineering and Mechanical Engineering. A minor in engineering at the undergraduate level is also offered.

DUAL DEGREE PROGRAMS
Engineering baccalaureate graduates with dual degrees 0

TRANSFER INFORMATION
Residency Requirements: Total minimum units required for graduation--197.
Transfer via Articulation Agreements
Admission to engineering: At any time
Requirements: College transcripts required (min. 3.0 GPA). High school transcripts and SAT scores required for applicants with fewer than 30 semester./45 qtr. hours of transferable credit. Maximum of 90 credits transferable from a 2-yr. institution. Required college courses: calculus, chem. and physics.
Transfer without Articulation Agreements
Admission to engineering: At any time

GRADUATION REQUIREMENTS
The program leading to a degree of Bachelor of Science consists of a minimum of 197 units with a GPA of at least a C (2.0) in all course work at the University and in the student's engineering courses. In addition to fulfilling the University Curriculum requirements, students have specific departmental requirements for each major.

STUDENT PROGRAMS
PROFESSIONAL AND HONORARY SOCIETIES
[For key to acronyms, see *Introduction*.]
Professional Societies: AIAA, ASCE, ASME, IEEE, SAE, Institute of Electrical & Electronic Engineers/Computer Science, Association of General Contractors (AGC)
Honorary Societies: Eta Kappa Nu, Pi Tau Sigma, Tau Beta Pi, Upsilon Pi Epsilon
SUPPORT PROGRAMS
Student Chapter Organizations: Society of Women Engineers, National Society of Black Engineers

Society of Hispanic/American Scientists & Engineers
For: Ethnic Minorities **Available:** Year round
Offered: Prematriculation, Freshman, Sophomore, Junior, Senior, Graduate level
Other Engineering Support Programs: Student support programs include freshman and transfer student orientations, academic assistance (tutorial programs), personal counseling, and career counseling.

204 Seattle University

INSTITUTION PROFILE
HEAD OF THE INSTITUTION
William J. Sullivan S.J.
Phone: (206)296-1891 Fax: (206)296-2163

GENERAL INFORMATION
[All Students—Fall 1991]
Undergraduate enrollment 3,089
Graduate enrollment 1,676
Total institution enrollment **4,765**
Type of institution: Private   Calendar: Quarters
Nearest city: Seattle   Population: 500,000
Miles from main campus: 1   Setting: Urban
Types of engineering degrees: Engineering
Other degree-granting colleges: Arts & Sciences, Business Administration, Education, Nursing, Matteo Ricci College, Graduate School

ENGINEERING ADMISSIONS
ADMISSIONS OFFICE CONTACT
Lee Gerig
Seattle University
Dean of Admissions
Seattle, WA 98122
Phone: (206)296-5800

ENGINEERING COLLEGE ADMISSIONS INFORMATION
Entrance Requirements: SAT (Score: 850), ACT (Score: 19), High School courses—High school mathematics, pref. incl. trigonometry (3 years), Laboratory science (2 years), English (4 years), Social Studies (2 years)
Entrance Recommendations: High School courses—Academic Electives (5 years)
Requirements for foreign students: TOEFL (Score: 520)

DEPARTMENTAL ADMISSIONS INFORMATION
Admission to the engineering department: At the time of admission to the institution.

FINANCIAL INFORMATION
ESTIMATED EXPENSES (ACADEMIC YEAR)
[Expenses are for the 1992-93 nine-month academic year.]

All Students	Undergraduate	Graduate
Tuition and fees	$11,520	$ 332[1]
College room and board	$ 4,479	—[B]
Books and supplies	$ 2,919	—[B]
Other expenses	$ 55[2]	—[B]
Total estimated expenses	**$18,973**	**332**

Notes: (B) Data not applicable. (1) Graduate program is part-time. $332 is cost per credit at 1992-93 rates. (2) Lab fees.

FINANCIAL AID OFFICE CONTACT
James White, Director of Financial Aid
Seattle University
Financial Aid Office
Seattle, WA 98122
Phone: (206)296-5840 Fax: (206)296-2163

GENERAL FINANCIAL AID INFORMATION
Forms accepted/required: FAF
Additional financial aid information: Need based scholarships and grants; part-time jobs on campus and work study program.

ENGINEERING COLLEGE INFORMATION
[For additional personnel, refer to the *Appendix*.]
HEAD OF ENGINEERING
Kathleen Mailer
Phone: (206)296-5500 Fax: (206)296-2179
EMail: kmailer@seattleu.edu

ENGINEERING COLLEGE ADDRESS
School of Science and Engineering
Seattle University
Broadway and Madison
Seattle, WA 98122
Phone: (206)296-5500 Fax: (206)296-2179

ENROLLMENTS—BY CLASS
[Numbers are baccalaureate enrollments for the fall 1991 term, unless otherwise footnoted.]
1st-year students/Freshmen 62
2nd-year students/Sophomores 54
3rd-year students/Juniors 105
4th-year students/Seniors 120
Total **341**

NUMBER OF DEGREES AWARDED—BY PROGRAM
[Numbers are engineering baccalaureate degrees awarded during the 1991-1992 academic year, unless otherwise footnoted. For full details about each engineering program, refer to the *Tables of Degree Programs*.]
Civil & Environmental Engineering 14
Electrical Engineering 52
Mechanical Engineering 31
Total **97**

PERCENTAGE OF DEGREES AWARDED—BY CATEGORY
[Percentages are of all engineering baccalaureate degrees awarded during the 1991-1992 academic year, unless otherwise footnoted.]
African Americans .... — %
Asian/Pacific Island Americans .... 20.0 %
Hispanic Americans .... — %
Native Americans .... — %
Foreign Citizens .... 11.0 %
All Others .... 69.0 %
Women .... 13.0 %
Persons with disabilities .... — %[A]
Students over 25 years of age .... 52.0 %

Notes: (A) Data not available.

GRADUATE ENROLLMENTS & DEGREES AWARDED
Master's enrollment 114
Master's degrees awarded 50
Doctoral enrollment .... —[B]
Doctoral degrees awarded .... —[B]

Notes: (B) Data not applicable.

ENGINEERING STUDENT DATA
Applicants to the engineering college
Number of applicants to engineering college 171
Percent offered admission .... 89.0 %

Matriculated engineering students
Percentage in top quartile (25%) of High School class .... —[A]
Average SAT scores: Math—527[1], Verbal—463[1], Combined—990[1]

Notes: (A) Data not available. (1) Includes all freshmen, not just engineers.

FULL- & PART-TIME FACULTY—BY DEPARTMENT
[Figures are the head count of full-time faculty and the full-time equivalent (FTE) of part-time faculty for each engineering department or equivalent.]

Department	Full	Part
Civil & Environmental Eng	4	—
Electrical Eng	8	—
Mechanical Eng	5	—

COLLEGE DESCRIPTION
The Thomas J. Bannan Center for Science and Engineering, dedicated in early January, 1990, provides Seattle University with the finest undergraduate science and engineering facilities available in the Northwest. The Center includes the Engineering Building which was completed in 1987 and contains 76,000 square feet of state-of-the-art laboratories, classrooms, and faculty offices. In addition, the Building also houses the innovative Engineering Design Center, one of only two such centers in the country.

DEGREE PROGRAMS DESCRIPTION
The engineering common studies program is essentially standard across the Departments of Civil and Environmental, Electrical, and Mechanical Engineering; the capstone design sequence is multi-disciplinary in character and thus cuts across departmental lines.

TRANSFER INFORMATION
Residency Requirements: Must complete the final 45 credits in residence.
Transfer without Articulation Agreements
Admission to engineering: To change major from one academic department to one of the engineering programs, student must first have completed 25 credits in residence, 15 of which must be in science & engineering courses, with a 2.5 GPA.
Requirements: After 25 credits, 15 of which are in science and engineering courses, with a 2.5 GPA or better. Students must apply 30 days before the start of the quarter.

GRADUATION REQUIREMENTS
Students must achieve an overall GPA of 2.5, as well as a major GPA of 2.5, by the time of graduation. All undergraduate engineering students are required to take the Washington State Engineer-in-Training (EIT) examination before being granted their degree.

STUDENT PROGRAMS
PROFESSIONAL AND HONORARY SOCIETIES
[For key to acronyms, see *Introduction*.]
Professional Societies: ASCE, ASME, IEEE
Honorary Societies: Tau Beta Pi

SUPPORT PROGRAMS
Student Chapter Organizations: Society of Women Engineers
Other Engineering Support Programs: Summer and fall orientation programs for all freshmen and transfers; academic and career counseling; psychological counseling; learning skills center; international student office; tutoring programs; career planning and placement seminars.

205 University of South Alabama

■ INSTITUTION PROFILE

HEAD OF THE INSTITUTION
Frederick P. Whiddon
Phone: (205)460-6111　　　　　Fax: (205)460-7541

GENERAL INFORMATION
[All Students—Fall 1991]
Undergraduate enrollment 10,534
Graduate enrollment 1,506
Total institution enrollment **12,506**
Type of institution: Public　　　Calendar: Quarters
Location: Mobile　　　　　　　Population: 196,278
Setting: Suburban
Types of engineering degrees: Engineering
Other degree-granting colleges: Allied Health Sciences, Arts & Sciences, Business Administration, Education, Medicine, Nursing, Continuing Education and Special Programs, Computer and Information Sciences

■ ENGINEERING ADMISSIONS

ADMISSIONS OFFICE CONTACT
J. David Stearns
University of South Alabama
307 University Blvd.
Mobile, AL 36688
Phone: (205)460-6494

ENGINEERING COLLEGE ADMISSIONS INFORMATION
Entrance Requirements: SAT (Score: 800), ACT (Score: 19)
Entrance Recommendations: High School courses—Mathematics (4 years), English (4 years), Physics, Chemistry, or Biology (2 years), History, Social Sciences, or Foreign Language (2 years)
Requirements for foreign students: TOEFL (Score: 500); financial statement

DEPARTMENTAL ADMISSIONS INFORMATION
Admission to the engineering department: At the time of admission to the institution.

■ FINANCIAL INFORMATION

ESTIMATED EXPENSES (ACADEMIC YEAR)
[Expenses are for the 1992-93 nine-month academic year.]

State Residents	Undergraduate	Graduate
Tuition and fees	$ 2,295	$ 1,383
College room and board	$ 2,895	$ 2,895
Books and supplies	$ 480	$ 480
Other expenses	$ –	$ –
Total estimated expenses	**$ 5,670**	**$ 4,758**

Out-of-State Residents	Undergraduate	Graduate
Tuition and fees	$ 2,895	$ 1,983
College room and board	$ 2,895	$ 2,895
Books and supplies	$ 480	$ 480
Other expenses	$ –	$ –
Total estimated expenses	**$ 6,270**	**$ 5,358**

FINANCIAL AID OFFICE CONTACT
Grady L. Collins, *Director*
University of South Alabama
Administration Bldg. Room 260
Mobile, AL 36688
Phone: (205)460-6231

GENERAL FINANCIAL AID INFORMATION
Forms accepted/required: ACT, FAT, FFS, Stafford, IRS
Additional financial aid information: In addition to academic and performance scholarships, various scholarships are available to deserving students i.e. foreign students, children of veterans, etc. Employment is also available for students through the College Work Study Program, the Student Assistant Program and Cooperative Education.

■ ENGINEERING COLLEGE INFORMATION
[For additional personnel, refer to the *Appendix*.]

HEAD OF ENGINEERING
David T. Hayhurst
Phone: (205)460-6140　　　　　Fax: (205)460-6343

ENGINEERING COLLEGE ADDRESS
College of Engineering
University of South Alabama
307 University Boulevard
Mobile, AL 36688
Phone: (205)460-6140　　　　　Fax: (205)460-6343

ENROLLMENTS—BY CLASS
[Numbers are baccalaureate enrollments for the fall 1991 term, unless otherwise footnoted.]
1st-year students/Freshmen 301
2nd-year students/Sophomores 175
3rd-year students/Juniors 223
4th-year students/Seniors 236
Total **935**

NUMBER OF DEGREES AWARDED—BY PROGRAM
[Numbers are engineering baccalaureate degrees awarded during the 1991-1992 academic year, unless otherwise footnoted. For full details about each engineering program, refer to the *Tables of Degree Programs*.]
Chemical Engineering 8
Civil Engineering 7
Electrical Engineering 43
Mechanical Engineering 26
Total **84**

PERCENTAGE OF DEGREES AWARDED—BY CATEGORY
[Percentages are of all engineering baccalaureate degrees awarded during the 1991-1992 academic year, unless otherwise footnoted.]
African Americans 3.6%
Asian/Pacific Island Americans 4.8%
Hispanic Americans 1.2%
Native Americans 1.2%
Foreign Citizens 17.9%
All Others 71.3%
Women 11.9%
Persons with disabilities .... – % (A)
Students over 25 years of age .... – % (A)

Notes: (A) Data not available.

GRADUATE ENROLLMENTS & DEGREES AWARDED
Master's enrollment 73
Master's degrees awarded 8
Doctoral enrollment .... – (B)
Doctoral degrees awarded .... – (B)

Notes: (B) Data not applicable.

■ ENGINEERING STUDENT DATA

Applicants to the engineering college
Number of applicants to engineering college 298
Percent offered admission 93.7%

Matriculated engineering students
Percentage in top quartile (25%) of High School class ..... – (A)
Average SAT scores: Math—500, Verbal—583, Combined—1083
Average ACT scores: Math—23.3, Composite—23.3

Notes: (A) Data not available.

FULL- & PART-TIME FACULTY—BY DEPARTMENT
[Figures are the head count of full-time faculty and the full-time equivalent (FTE) of part-time faculty for each engineering department or equivalent.]

Department	Full	Part
Chemical Eng	5	–
Civil Eng	6	–
Electrical Eng	10	–
Mechanical Eng	8	–

COLLEGE DESCRIPTION
Dual degrees are permitted with approval of each degree granting department. Co-op programs (non-ABET accredited) are available. Minimum graduation time is five years with co-op.

DEGREE PROGRAMS DESCRIPTION
No lecture or laboratory courses taught by graduate students.

DUAL DEGREE PROGRAMS
Engineering baccalaureate graduates with dual degrees 0

TRANSFER INFORMATION
Residency Requirements: A candidate for graduation must complete a minimum of 48 quarter hours of credit at the University of South Alabama in upper-division coursework (Jr. and Sr. levels). The College of Engineering requires that a minimum of 24 quarter hours of this credit be completed in the student's major department.

Transfer without Articulation Agreements
Admission to engineering: At any time
Engineering graduates transferred from　..　4-yr: 44　2-yr: 64

GRADUATION REQUIREMENTS
Minimum GPA of 2.00 in major and GPA of 2.00 overall. GPA calculations include all courses taken. Students failing to earn GPA of 2.00 in major may graduate provided a minimum grade of C is earned in all major courses counted toward the degree.

■ STUDENT PROGRAMS

PROFESSIONAL AND HONORARY SOCIETIES
[For key to acronyms, see *Introduction*.]
Professional Societies: AAPG, ACM, AIChE, ASCE, ASHRAE, ASME, IEEE
Honorary Societies: Eta Kappa Nu, Pi Tau Sigma, Tau Beta Pi

SUPPORT PROGRAMS
Student Chapter Organizations: Society of Women Engineers, National Society of Black Engineers
Other Engineering Support Programs: No specific support programs.

206 University of South Carolina

■ INSTITUTION PROFILE
HEAD OF THE INSTITUTION
John M. Palms
Phone: (803)777-2001   Fax: (803)777-9480

GENERAL INFORMATION
[All Students—Fall 1991]
Undergraduate enrollment 32,888
Graduate enrollment 8,871
Total institution enrollment 41,759
Type of institution: Public   **Calendar:** Semesters
Location: Columbia   **Population:** 410,000
Setting: Urban
Types of engineering degrees: Engineering
Other degree-granting colleges: Business Administration, Education, Humanities & Social Sciences, Law, Medicine, Nursing, Pharmacy, Technology & Applied Sciences, Applied Professional Sciences, Criminal Justice, Health, Science and Mathematics, South Carolina Honors College

ENGINEERING COLLEGE ADMISSIONS INFORMATION
Entrance Requirements: SAT (Score: 900), ACT (Score: 22), High School courses—English (4 years), Mathematics (3 years), Laboratory Science (2 years), Social Studies (3 years)
Entrance Recommendations: High School courses—Foreign Language (2 years), Algebra (2 years), Geometry/Trigonometry (2 years), Chemistry or Physics (2 years)
Requirements for foreign students: TOEFL (Score: 550); financial statement

DEPARTMENTAL ADMISSIONS INFORMATION
Admission to the engineering department: At the time of admission to the institution.

■ FINANCIAL INFORMATION
ESTIMATED EXPENSES (ACADEMIC YEAR)
[Expenses are for the 1992-93 nine-month academic year.]

State Residents	Undergraduate	Graduate
Tuition and fees	$ 1,409	$ 1,474
College room and board	$ 3,210	$ 3,210
Books and supplies	495	495
Other expenses	$ 2,986 (1)	$ 2,986 (1)
Total estimated expenses	**$ 8,100**	**$ 8,165**

Out-of-State Residents	Undergraduate	Graduate
Tuition and fees	$ 3,523	$ 2,948
College room and board	$ 3,210	$ 3,210
Books and supplies	495	495
Other expenses	$ 2,986 (1)	$ 2,986 (1)
Total estimated expenses	**$10,214**	**$ 9,639**

Notes: (1) Matriculation & Application Fees $25 Each; Travel $795; Personal and Miscellaneous $2,206.

FINANCIAL AID OFFICE CONTACT
Denise Wellman
University of South Carolina
Student Financial Aid & Scholarships
1714 College Street
Columbia, SC 29208
Phone: (803)777-3215   Fax: (803)777-0941

GENERAL FINANCIAL AID INFORMATION
Forms accepted/required: FFS, Stafford, IRS
Additional financial aid information: Academic scholarships and awards: need-based grants, long- and short-term loans, College Work-Study Program, part-time jobs on campus, and veterans benefits.

■ ENGINEERING COLLEGE INFORMATION
[For additional personnel, refer to the *Appendix*.]
HEAD OF ENGINEERING
W. K. Humphries
Phone: (803)777-4259   Fax: (803)777-9597

ENGINEERING COLLEGE ADDRESS
College of Engineering
University of South Carolina
Swearingen Engineering Center
Columbia, SC 29208
Phone: (803)777-4177   Fax: (803)777-9597

ENROLLMENTS—BY CLASS
[Numbers are baccalaureate enrollments for the fall 1991 term, unless otherwise footnoted.]
1st-year students/Freshmen 455
2nd-year students/Sophomores 285
3rd-year students/Juniors 259
4th-year students/Seniors 340
Total 1,339

NUMBER OF DEGREES AWARDED—BY PROGRAM
[Numbers are engineering baccalaureate degrees awarded during the 1991-1992 academic year, unless otherwise footnoted. For full details about each engineering program, refer to the *Tables of Degree Programs*.]
Chemical Engineering 12
Civil Engineering 15
Electrical & Computer Engineering 71
Electrical Engineering 71
Mechanical Engineering 58
Total 227

PERCENTAGE OF DEGREES AWARDED—BY CATEGORY
[Percentages are of all engineering baccalaureate degrees awarded during the 1991-1992 academic year, unless otherwise footnoted.]
African Americans .... – % (C)
Asian/Pacific Island Americans .... – % (C)
Hispanic Americans .... – % (C)
Native Americans .... – %
Foreign Citizens .... – % (C)
All Others .... 100 %
Women .... 14.7 %
Persons with disabilities .... – % (A)
Students over 25 years of age .... – % (A)

Notes: (A) Data not available. (C) Estimated data.

GRADUATE ENROLLMENTS & DEGREES AWARDED
Master's enrollment 259
Master's degrees awarded 66
Doctoral enrollment 68
Doctoral degrees awarded 11

ENGINEERING STUDENT DATA
Applicants to the engineering college
Number of applicants to engineering college 760
Percent offered admission 88.8%
Matriculated engineering students
Percentage in top quartile (25%) of High School class 54.8
Average SAT scores: Math—545, Verbal—440, Combined—985

FULL- & PART-TIME FACULTY—BY DEPARTMENT
[Figures are the head count of full-time faculty and the full-time equivalent (FTE) of part-time faculty for each engineering department or equivalent.]

Department	Full	Part
Chemical Eng	10	–
Civil Eng	13	1.0
Electrical Eng	20	–
Mechanical Eng	15	1.0

■ COLLEGE DESCRIPTION
The College offers a four-year undergraduate program in engineering, emphasizing engineering fundamentals in the first year, followed by three years of specialized training in one of the four majors, with some liberal arts studies throughout all four years. Students may take the first two years in an evening program, but must complete the degree during the day. Evening students take a part-time load while usually working full time. Students may also take some courses on the system campuses, but must complete the degree on the Columbia campus. Co-op is an option in all undergraduate programs.

DEGREE PROGRAMS DESCRIPTION
All degree programs offer the BS, MS, ME, and PhD degrees. Under BS are offered Chemical, Civil, Electrical/Computer and Mechanical Engineering. The BS degree programs prepare the graduate for entry-level professional positions or for graduate study in each of the respective fields. The College offers graduate programs leading to the degrees of Master of Engineering, Master of Science, and Doctor of Philosophy in Chemical, Civil, Electrical and Computer, and Mechanical Engineering. Graduate studies in the College of Engineering involve the student in a balanced program emphasizing either design or research.

TRANSFER INFORMATION
Residency Requirements: Students must complete at least 60 semester hours of credit in residence.
Transfer via Articulation Agreements
Admission to engineering: At the end of the sophomore year
Engineering graduates transferred from ... 4-yr: –   2-yr: –
Requirements: Minimum 3.0 GPA on 30 semester hours at two-year institutions. SAT scores must be submitted.
Transfer without Articulation Agreements
Admission to engineering: At any time
Engineering graduates transferred from ... 4-yr: 49   2-yr: 21
Requirements: Minimum 2.5 GPA on 30 semester hours at four-year institutions. SAT scores must be submitted.

GRADUATION REQUIREMENTS
Degree requirements for the M.S., M.E. and Ph.D. vary according to the program. For the B.S. degree in engineering in the various majors, a cumulative GPA of 2.0 is required for graduation, as well as a 2.0 GPA in both major courses and all-engineering courses.

■ STUDENT PROGRAMS
PROFESSIONAL AND HONORARY SOCIETIES
[For key to acronyms, see *Introduction*.]
Professional Societies: AIChE, ASCE, ASHRAE, ASME, IEEE
Honorary Societies: Chi Epsilon, Eta Kappa Nu, Omega Chi Epsilon, Pi Tau Sigma, Tau Beta Pi

SUPPORT PROGRAMS
Student Chapter Organizations: Society of Women Engineers, National Society of Black Engineers
Minority Engineering Programs
For: Ethnic Minorities   **Available:** Year round
Offered: Prematriculation, Freshman, Sophomore, Junior, Senior
Other Engineering Support Programs: University services available to engineering students include summer and fall orientation programs for all freshmen and transfers, Academic Skills Center, Career Planning and Placement, Disabled Student Services, Veterans Affairs, personal counseling, Women's Center, and tutoring services.

207 South Dakota School of Mines and Technology

■ INSTITUTION PROFILE

HEAD OF THE INSTITUTION
Richard J. Gowen
Phone: (605)394-2411 Fax: (605)394-6131

GENERAL INFORMATION
[All Students—Fall 1991]
Undergraduate enrollment 2,132
Graduate enrollment 318
Total institution enrollment 2,450
Type of institution: Public Calendar: Semesters
Nearest city: Denver Population: 500,000
Miles from main campus: 350 Setting: Urban
Types of engineering degrees: Engineering

■ ENGINEERING ADMISSIONS

ADMISSIONS OFFICE CONTACT
Gary A. Bjordal
South Dakota School of Mines and Technology
Admissions Office
501 E. Saint Joseph Street
Rapid City, SD 57701-3995
Phone: (605)394-2400 Fax: (605)394-6131

ENGINEERING COLLEGE ADMISSIONS INFORMATION
Entrance Requirements: SAT (Score: 930), ACT (Score: 23), High School courses—English (4 years), Mathematics and Laboratory Science (5 years), Social Science/Social Studies (3 years), Computer Science and Fine Arts (1 year), Either SAT or ACT required, ACT preferred.
Entrance Recommendations: ACT (Score: 23), High School courses—Mathematics (4 years), Laboratory Science (4 years)
Requirements for foreign students: TOEFL (Score: 530); financial statement; Complete secondary and college academic transcripts translated into English. Written recommendations from two instructors or professors familiar with the academic performance and capabilities of the applicant.

DEPARTMENTAL ADMISSIONS INFORMATION
Admission to the engineering department: At the time of admission to the institution.

■ FINANCIAL INFORMATION

ESTIMATED EXPENSES (ACADEMIC YEAR)
[Expenses are for the 1992-93 nine-month academic year.]

State Residents	Undergraduate	Graduate
Tuition and fees	$ 2,509	$ 4,247
College room and board	$ 2,478	$ 2,478
Books and supplies	$ 475	$ 650
Other expenses	$ –	$ –
Total estimated expenses	**$ 5,462**	**$ 7,375**
Out-of-State Residents	Undergraduate	Graduate
Tuition and fees	$ 4,460	$ 5,170
College room and board	$ 2,478	$ 2,478
Books and supplies	$ 475	$ 650
Other expenses	$ –	$ –
Total estimated expenses	**$ 7,413**	**$ 8,298**

FINANCIAL AID OFFICE CONTACT
Sharon K. Colombe, *Director of Financial Aid*
South Dakota School of Mines and Technology
Financial Aid Office
501 E. Saint Joseph Street
Rapid City, SD 57701-3995
Phone: (605)394-2274 Fax: (605)394-6131

GENERAL FINANCIAL AID INFORMATION
Forms accepted/required: CSS-FAF, FFS, Educational Assistance Corp. Application for Student Aid (preferred)
Additional financial aid information: Pell Grant Supplemental Educational Opportunity Grant Perkins Loan College Work Study Stafford Loan State Student Incentive Grant

■ ENGINEERING COLLEGE INFORMATION

[For additional personnel, refer to the *Appendix*.]
HEAD OF ENGINEERING
William L. Hughes
Phone: (605)394-2256 Fax: (605)394-6131

ENGINEERING COLLEGE ADDRESS
South Dakota School of Mines and Technology
501 East Saint Joseph Street
Rapid City, SD 57701-3995
Phone: (605)394-2400 Fax: (605)394-6131

ENROLLMENTS—BY CLASS
[Numbers are baccalaureate enrollments for the fall 1991 term, unless otherwise footnoted.]
1st-year students/Freshmen 419
2nd-year students/Sophomores 318
3rd-year students/Juniors 212
4th-year students/Seniors 281
Total **1,230**

NUMBER OF DEGREES AWARDED—BY PROGRAM
[Numbers are engineering baccalaureate degrees awarded during the 1991-1992 academic year, unless otherwise footnoted. For full details about each engineering program, refer to the *Tables of Degree Programs*.]
Chemical Engineering 26
Civil Engineering 18
Electrical Engineering 38
Geological Engineering 9
Industrial Engineering 9
Mechanical Engineering 69
Metallurgical Engineering 10
Mining Engineering 8
Total **187**

PERCENTAGE OF DEGREES AWARDED—BY CATEGORY
[Percentages are of all engineering baccalaureate degrees awarded during the 1991-1992 academic year, unless otherwise footnoted.]
African Americans 1.0%
Asian/Pacific Island Americans 1.0%
Hispanic Americans 1.0%
Native Americans 2.0%
Foreign Citizens 14.0%
All Others 81.0%
Women 31.0%
Persons with disabilities .... – % (A)
Students over 25 years of age 28.0%
Notes: (A) Data not available.

GRADUATE ENROLLMENTS & DEGREES AWARDED
Master's enrollment 142
Master's degrees awarded 79
Doctoral enrollment 25
Doctoral degrees awarded 2

ENGINEERING STUDENT DATA
Applicants to the engineering college
Number of applicants to engineering college 1,081
Percent offered admission 95.0%
Matriculated engineering students
Percentage in top quartile (25%) of High School class 53.8
Average ACT scores: Math—27, Composite—24

FULL- & PART-TIME FACULTY—BY DEPARTMENT
[Figures are the head count of full-time faculty and the full-time equivalent (FTE) of part-time faculty for each engineering department or equivalent.]

Department	Full	Part
Chemical Eng	4	–
Civil Eng	11	–
Electrical Eng	11	–
Industrial Eng	3	–
Mechanical Eng	8	–
Geological Eng	4	–
Metallurgical Eng	5	–
Mining Eng	4	1.0

COLLEGE DESCRIPTION

South Dakota School of Mines and Technology is one of only a very few moderately-sized, affordably-priced universities that specialize in the areas of engineering and science. This high degree of specialization enhances SDSM&T's national and international reputation for excellence, makes our graduates highly sought by industry, allows us to furnish our classrooms and laboratories with the latest in high-tech, state-of-the-art scientific equipment and computer facilities and provides a greater amount of fellowship among our students, making it easy for students to study together in a 'group effort'.

DEGREE PROGRAMS DESCRIPTION

B.S. Level: The extraordinarily small class size allows for a great degree of personalized attention and gives the student opportunities to pursue individualized research projects at a level usually not available to students at larger universities until they are graduate students.

TRANSFER INFORMATION

Residency Requirements: Must complete final 30 semester credit hours at SDSM&T to receive degree.
Transfer without Articulation Agreements
Admission to engineering: At any time
Requirements: Must be in good academic standing at all colleges and universities attended previously. Recommended cumulative grade point average of 3.0 ('B').

GRADUATION REQUIREMENTS

(1) Mathematics-16 credit hours, mathematics sequence begins with Math 123, Calculus I. (2) Laboratory Science-minimum of 16 credit hours, Chem 112/113, Physics 119 and 219 required. (3) Humanities and Social Science-16 credit hours, minimum of 6 credit hours in each area. (4) Free Elective vary with the individual department Students must attain a 2.0 ('C') cumulative grade point average. The final 30 credit hours must be completed at this institution.

■ STUDENT PROGRAMS

PROFESSIONAL AND HONORARY SOCIETIES
[For key to acronyms, see *Introduction*.]
Professional Societies: ACS, AIChE, ASCE, ASME, IEEE, SAME, SPS, American Society of Metallurgists International/Minerals,Metals,Materials Soc., Association for Computing Machinery
Honorary Societies: Eta Kappa Nu, Tau Beta Pi, Theta Tau, Alpha Chi Sigma, Alpha Sigma Mu

SUPPORT PROGRAMS
Student Chapter Organizations: Society of Women Engineers, American Indian Science and Engineering Society

Scientific Knowledge for Indian Learning's Leadership
For: Ethnic Minorities **Available:** Year round
Offered: Prematriculation, Freshman, Sophomore, Junior, Senior
Other Engineering Support Programs: Tech Learning Centre-provides free tutoring services Dean of Students Office-counseling services

208 South Dakota State University

■ INSTITUTION PROFILE
HEAD OF THE INSTITUTION
Robert T. Wagner
Phone: (605)688-4111   **Fax:** (605)688-5822
GENERAL INFORMATION
[All Students—Fall 1991]
Undergraduate enrollment 7,520
Graduate enrollment 1,030
Total institution enrollment **8,550**
Type of institution: Public   **Calendar:** Semesters
Nearest city: Sioux Falls, SD   **Population:** 108,000
Miles from main campus: 58   **Setting:** Small Town
Types of engineering degrees: Engineering & Technology
Other degree-granting colleges: Arts & Sciences, Education, Nursing, Agricultural & Biological Sciences, Home Economics, General Registration, Pharmacy

■ ENGINEERING ADMISSIONS
ADMISSIONS OFFICE CONTACT
Tracy Welsh
South Dakota State University
Admissions Office
Administration Building Box 2201
Brookings, SD 57007-1998
Phone: (605)688-4121   **Fax:** (605)688-6384
ENGINEERING COLLEGE ADMISSIONS INFORMATION
Entrance Requirements: ACT (Score: 22), High School courses—English (4 years), Social Science (3 years), Laboratory Science & Math (Algebra or higher) (2 years), Computer Science
Entrance Recommendations: ACT (Score: 22), High School courses—Algebra (2 years), Geometry (1 year), Graphics (1 year), Physics (1 year)
Requirements for foreign students: TOEFL (Score: 500); financial statement
DEPARTMENTAL ADMISSIONS INFORMATION
Admission to the engineering department: At the time of admission to the institution.

■ FINANCIAL INFORMATION
ESTIMATED EXPENSES (ACADEMIC YEAR)
[Expenses are for the 1992-93 nine-month academic year.]

State Residents	Undergraduate	Graduate
Tuition and fees	$ 2,146 (4)	$ 1,640 (1)
College room and board	$ 1,828 (2)	$ 1,828 (2)
Books and supplies	$ 600 (C)	$ 320 (C)
Other expenses	$ — (3)	$ — (3)
Total estimated expenses	**$ 4,574**	**$ 3,788**
Out-of-State Residents	Undergraduate	Graduate
Tuition and fees	$ 3,213 (4)	$ 2,805 (1)
College room and board	$ 1,828 (2)	$ 1,828 (2)
Books and supplies	$ 600 (C)	$ 320 (C)
Other expenses	$ — (3)	$ — (3)
Total estimated expenses	**$ 5,641**	**$ 4,953**

Notes: (C) Estimated data. (1) 9 credit hours for semester (2) double occupancy (3) Engineering Education Fee per credit $10.50 and Engineering/Science Lab Fees (per course) $15.00 (4) 16 credit hours per semester

FINANCIAL AID OFFICE CONTACT
Jay Larsen, *Financial Aid Director*
South Dakota State University
Financial Aid Office
ADMIN 106 Box 2201
Brookings, SD 57007-0198
Phone: (605)688-4695   **Fax:** (605)688-5822

GENERAL FINANCIAL AID INFORMATION
Forms accepted/required: ACT, AFSA, CSS-FAF, FAF, FAT, FAR, FSA, FFS, Stafford, IRS, SAR, Supplemental, Education Assistance Corporation
Additional financial aid information: Scholarships, State Incentive Grants, Pell Grants, Supplemental Educational Opportunity Grants, Perkins Loans, Work Study Program, Stafford Student Loans, PLUS and SLS Loans, Governmental Agency Student Financial Aid, Student Employment, Veterans Assistance, Aid to Members of SD National Guard

■ ENGINEERING COLLEGE INFORMATION
[For additional personnel, refer to the *Appendix*.]
HEAD OF ENGINEERING
Duane E. Sander
Phone: (605)688-4161   **Fax:** (605)688-5878
EMail: DYERB@ENGSDSTATE.EDU
ENGINEERING COLLEGE ADDRESS
College of Engineering
South Dakota State University
Box 2219
Crothers Engineering Hall 201
Brookings, SD 57007-0096
Phone: (605)688-4161   **Fax:** (605)688-5878
ENROLLMENTS—BY CLASS
[Numbers are baccalaureate enrollments for the fall 1991 term, unless otherwise footnoted.]
1st-year students/Freshmen 346
2nd-year students/Sophomores 208
3rd-year students/Juniors 188
4th-year students/Seniors 290
Total **1,032**
NUMBER OF DEGREES AWARDED—BY PROGRAM
[Numbers are engineering baccalaureate degrees awarded during the 1991-1992 academic year, unless otherwise footnoted. For full details about each engineering program, refer to the *Tables of Degree Programs*.]
Agricultural Engineering 5
Civil Engineering 40
Computer Science 13
Electrical Engineering 67
Electronic Engineering Technology 32
Engineering Physics 9
Mechanical Engineering 28
Total **194**
PERCENTAGE OF DEGREES AWARDED—BY CATEGORY
[Percentages are of all engineering baccalaureate degrees awarded during the 1991-1992 academic year, unless otherwise footnoted.]
African Americans .... — %
Asian/Pacific Island Americans .... — %
Hispanic Americans .... — %
Native Americans .... — % (1)
Foreign Citizens .... 12.0 %
All Others .... 88.0 %
Women .... 12.0 %
Persons with disabilities .... — % (A)
Students over 25 years of age .... — % (A)
Notes: (A) Data not available. (1) Less than 1% were American Indian
GRADUATE ENROLLMENTS & DEGREES AWARDED
Master's enrollment 170
Master's degrees awarded 35
Doctoral enrollment .... —
Doctoral degrees awarded .... —

ENGINEERING STUDENT DATA
Applicants to the engineering college
Number of applicants to engineering college 380
Percent offered admission 98.2%
Matriculated engineering students
Percentage in top quartile (25%) of High School class 49.1
Average ACT scores: Composite—23.1
FULL- & PART-TIME FACULTY—BY DEPARTMENT
[Figures are the head count of full-time faculty and the full-time equivalent (FTE) of part-time faculty for each engineering department or equivalent.]

Department	Full	Part
Civil Eng	11	—
Electrical Eng	11	—
Mechanical Eng	9	1.3
Agricultural Eng	8	—
Computer Science	6	1.0
General Eng	10	—
Eng Physics	9	—

COLLEGE DESCRIPTION
There are special programs in promotional techniques for engineers, industrial management, engineering physics, double-major programs in Mechanical and Electrical Engineering and Engineering Physics and student originated research projects. Informal co-op and internship programs have been established with local companies. Degree completion is extended to 4 1/2 to 5 years in co-op programs.
DEGREE PROGRAMS DESCRIPTION
The baccalaureate programs emphasize practical application of engineering principles. The master's degree programs provide for concentration in an engineering area such as CE, EE, ME, AE and Engineering Physics. In addition, interdisciplinary and management oriented programs are encouraged.
TRANSFER INFORMATION
Residency Requirements: Successful completion of at least 32 hours at South Dakota State University with a minimum of 20 credit hours of junior and senior (300-400) level courses. Credits earned by examination are not counted as resident credit unless an exception has been made because of special program features.
Transfer without Articulation Agreements
Admission to engineering: At any time
Requirements: Transfer students are eligible for admission if they meet the following (1) Have a cumulative grade point average of 2.0 (2) Are in good standing with their most recently attended school.
GRADUATION REQUIREMENTS
Each department has specific GPA requirements for designated course sequences. Minimum GPA for graduation is 2.0 overall for courses required for degree.

■ STUDENT PROGRAMS
PROFESSIONAL AND HONORARY SOCIETIES
[For key to acronyms, see *Introduction*.]
Professional Societies: ACM, ASAE, ASCE, ASHRAE, ASME, IEEE, NSPE, SAME, SPS, National Council of Teachers in Mathematics (student chapter)
Honorary Societies: Alpha Epsilon, Chi Epsilon, Eta Kappa Nu, Pi Tau Sigma, Sigma Phi Delta, Tau Beta Pi, Sigma Pi Sigma, Pi Mu Epsilon
SUPPORT PROGRAMS
Student Chapter Organizations: Society of Women Engineers
Other Engineering Support Programs: An Engineering Orientation class is provided for fall and spring semesters to provide academic guidance and career information to new engineering students. The Career and Academic Planning Center provides career guidance and academic skills assistance. Native American Advisor available university-wide.

209 University of South Florida

■ INSTITUTION PROFILE

HEAD OF THE INSTITUTION
Francis T. Borkowski
Phone: (813)974-2791   Fax: (813)974-5530

GENERAL INFORMATION
[All Students—Fall 1991]
Undergraduate enrollment 22,252
Graduate enrollment 5,266
Total institution enrollment 33,280
Type of institution: Public   Calendar: Semesters
Location: Tampa   Population: 280,000
Setting: Urban
Types of engineering degrees: Engineering
Other degree-granting colleges: Architecture, Arts & Sciences, Business Administration, Education, Fine & Performing Arts, Medicine, Nursing, Public Health, Architecture

■ ENGINEERING ADMISSIONS

ADMISSIONS OFFICE CONTACT
George R. Card
University of South Florida
College of Engineering
Tampa, FL 33620-5350
Phone: (813)974-2684   Fax: (813)974-5094

ENGINEERING COLLEGE ADMISSIONS INFORMATION
Entrance Requirements: SAT (Score: 1050), ACT (Score: 24), High School courses—English (4) + Foreign Language (2) (6 years), Mathematics (3 years), Natural Science (3 years), Social Science (3 years)
Entrance Recommendations: High School courses—Algebra I (1 year), Algebra II (1 year), Geometry (1 year), Trigonometry (1 year)
Requirements for foreign students: TOEFL (Score: 550); financial statement

DEPARTMENTAL ADMISSIONS INFORMATION
Admission to the engineering department: At the time of admission to the institution.
Additional information: Completion of the following courses with either: (1) a 'C' or better in each course on first attempt, or (2) a GPA in these courses of 2.2 with a final grade of 'C' or better in each course. Eng. Orientation; Statics; Thermodynamics I; FORTRAN for Eng.; Eng. Statistics I; Intro. Elec. Syst. I.

■ FINANCIAL INFORMATION

ESTIMATED EXPENSES (ACADEMIC YEAR)
[Expenses are for the 1992-93 nine-month academic year.]

State Residents	Undergraduate	Graduate
Tuition and fees	$ 1,410	$ 1,630
College room and board	$ 3,260	$ 3,260
Books and supplies	$ 300	$ 425
Other expenses	$ —	$ —
Total estimated expenses	**$ 4,970**	**$ 5,315**

Out-of-State Residents	Undergraduate	Graduate
Tuition and fees	$ 4,710	$ 5,273
College room and board	$ 3,260	$ 3,260
Books and supplies	$ 300	$ 425
Other expenses	$ —	$ —
Total estimated expenses	**$ 8,270**	**$ 8,958**

FINANCIAL AID OFFICE CONTACT
Gwyndolyn Francis, *Director of Financial Aid*
University of South Florida
4202 E. Fowler Ave., SVC 1102
Tampa, FL 33620-5350
Phone: (813)974-4700   Fax: (813)974-5144

GENERAL FINANCIAL AID INFORMATION
Forms accepted/required: ACT, AFSA, CSS-FAF, FFS
Additional financial aid information: Financial aid is available in the form of scholarships, grants, part-time employment, and low interest loans. Temporary deferments of registration fees, and short-term, interest free loans are available to students whose aid is delayed in delivery.

■ ENGINEERING COLLEGE INFORMATION
[For additional personnel, refer to the *Appendix*.]

HEAD OF ENGINEERING
Michael G. Kovac
Phone: (813)974-3780   Fax: (813)974-5094
EMail: kovac@ec.usf.edu

ENGINEERING COLLEGE ADDRESS
College of Engineering
University of South Florida
4202 Fowler Avenue
Tampa, FL 33620-5350
Phone: (813)974-3780   Fax: (813)974-5094

ENROLLMENTS—BY CLASS
[Numbers are baccalaureate enrollments for the fall 1991 term, unless otherwise footnoted.]
1st-year students/Freshmen 313
2nd-year students/Sophomores 223
3rd-year students/Juniors 514
4th-year students/Seniors 1,188
Total 2,238

NUMBER OF DEGREES AWARDED—BY PROGRAM
[Numbers are engineering baccalaureate degrees awarded during the 1991-1992 academic year, unless otherwise footnoted. For full details about each engineering program, refer to the *Tables of Degree Programs*.]
Chemical Engineering 34
Civil Engineering 47
Computer Engineering 26
Computer Science & Engineering 34
Electrical Engineering 121
Engineering Management .... –
Engineering Technology 48
Industrial Engineering 18
Industrial Manufacturing .... –
Mechanical Engineering 58
Total 386

PERCENTAGE OF DEGREES AWARDED—BY CATEGORY
[Percentages are of all engineering baccalaureate degrees awarded during the 1991-1992 academic year, unless otherwise footnoted.]
African Americans 2.9%
Asian/Pacific Island Americans 10.0%
Hispanic Americans 6.7%
Native Americans .... – %
Foreign Citizens 1.7%
All Others 78.7%
Women 19.9%
Persons with disabilities .... – %
Students over 25 years of age 50.0%

GRADUATE ENROLLMENTS & DEGREES AWARDED
Master's enrollment 520
Master's degrees awarded 138
Doctoral enrollment 168
Doctoral degrees awarded 22

ENGINEERING STUDENT DATA
Applicants to the engineering college
Number of applicants to engineering college 635
Percent offered admission 53.8%

Matriculated engineering students
Percentage in top quartile (25%) of High School class ...... –(A)
Average SAT scores: Math—580, Verbal—500, Combined—1080
Average ACT scores: Math—26, Composite—24

Notes: (A) Data not available.

FULL- & PART-TIME FACULTY—BY DEPARTMENT
[Figures are the head count of full-time faculty and the full-time equivalent (FTE) of part-time faculty for each engineering department or equivalent.]

Department	Full	Part
Chemical Eng	8	–
Civil Eng	17	2.5
Electrical Eng	25	3.4
Industrial Eng	8	1.9
Mechanical Eng	8	–
Computer Science & Eng	14	1.3

COLLEGE DESCRIPTION

The mission of the USF College of Engineering encompasses several major themes. At the undergraduate level there is a strong commitment to prepare students for careers in industry, government, and private practice and for advanced studies in professional schools of engineering, science, law, business, and medicine. At the graduate level the commitment is to prepare students to make major contributions to the advancement of engineering by conducting research of national significance and prominence. The College emphasizes close liaison with industry and government to provide students with the skills and perspectives needed to assume technological leadership. Average number of years required to actually complete the bachelor's degree: 5. Co-op is available for students in all programs. Civil and Mechanical seniors are required to take the Fundamentals of Engineering Examination and all other seniors are strongly encouraged to take the F.E. Examination.

DEGREE PROGRAMS DESCRIPTION

The B.S. degrees in the various engineering programs prepare the graduate for entry-level professional positions and for graduate study in each of the respective fields. The B.S. in the engineering science program is intended primarily to prepare students for advanced graduate study. Graduate study in engineering can be carried out in engineering oriented or engineering science oriented programs at the Master and Doctoral levels.

TRANSFER INFORMATION

Residency Requirements: Students transferring from Junior and Community College must complete a minimum of 60 hrs. Students transferring from upper level institutions must complete a minimum of 30 hours.

Transfer via Articulation Agreements
Admission to engineering: At the end of the sophomore year
Engineering graduates transferred from .. 4-yr: – 2-yr: 155
Requirements: Transfer students must have completed with a 2.5 GPA at acceptable schools, the equivalent of USF courses which follow; one year of Engr. Calculus, one year of Gen. Chemistry. Grades of D in above courses are not accepted.

Transfer without Articulation Agreements
Admission to engineering: At any time
Engineering graduates transferred from .. 4-yr: 127 2-yr: 50
Requirements: Transfer students must have a 2.5 GPA at acceptable schools, the equivalent of USF courses which follow; one year of Engr. Calculus, one year of Gen. Chemistry. Grades of D in above courses are not accepted.

GRADUATION REQUIREMENTS

All engineering programs require successful completion of 136 approved semester credit hours. All Civil and Mechanical seniors must take Fundamentals of Engineering Exam.

■ STUDENT PROGRAMS

PROFESSIONAL AND HONORARY SOCIETIES
[For key to acronyms, see *Introduction*.]
Professional Societies: ACM, AIAA, AIChE, ASCE, ASME, IEEE, IIE, ITE, NSPE, Human Factors Society, Water Environmental Federation
Honorary Societies: Alpha Pi Mu, Chi Epsilon, Pi Tau Sigma, Tau Beta Pi

SUPPORT PROGRAMS
Student Chapter Organizations: Society of Women Engineers, Society of Hispanic Professional Engineers, Society for the Advancement of Minorities in Engr Sciences

Project Thrust
For: Ethnic Minorities   **Available:** Year round
Offered: Freshman, Sophomore, Junior, Senior

Minority Engineering Program
For: Ethnic Minorities   **Available:** Year round
Offered: Freshman, Sophomore, Junior, Senior

Other Engineering Support Programs: Summer, Fall, and Spring Orientation and advising for all new freshman and transfer students; Career and personal counseling services; foreign student office; veterans services; blind and handicap services; placement and cooperative education services.

210 Southern University and A&M College

INSTITUTION PROFILE

HEAD OF THE INSTITUTION
Marvin L. Yates
Phone: (504)771-5020 Fax: (504)771-2026

GENERAL INFORMATION
[All Students—Fall 1991]
Undergraduate enrollment 9,338
Graduate enrollment 1,137
Total institution enrollment 10,475
Type of institution: Public **Calendar:** Semesters
Nearest city: New Orleans **Population:** 94,440
Miles from main campus: 85 **Setting:** Urban
Types of engineering degrees: Engineering & Technology
Other degree-granting colleges: Agricultural & Environmental, Allied Health Sciences, Architecture, Arts & Sciences, Business Administration, Communication & Journalism, Education, Fine & Performing Arts, Humanities & Social Sciences, Law, Nursing

ENGINEERING ADMISSIONS

ADMISSIONS OFFICE CONTACT
Henry J. Bellaire
Southern University and A&M College
Office of Admissions
Southern Branch P.O.
Baton Rouge, LA 70813
Phone: (504)771-5124

ENGINEERING COLLEGE ADMISSIONS INFORMATION
Entrance Requirements: High School courses—Math (2 years), Science (2 years), English (4 years), There is no minimum score required for admission. Southern University has an open admissions policy. All high school graduates and students who have successfully completed the General Education Development (GED) Test are eligible for admission.
Entrance Recommendations: SAT (Score: 630), ACT (Score: 16), High School courses—Algebra, Plane geometry, trigonometry, calculus (4 years), English (4 years), History (1 year), Chemistry, Physics, General Science, Biology (4 years), ACT, SAT, and TOEFL. No minimum score required. Open admission policy.
Requirements for foreign students: TOEFL (Score: 400); financial statement; Application must be completed, submitted and received by the Admissions Office at least 90 days prior to the anticipated date of registration if the applicant is still in his country.

DEPARTMENTAL ADMISSIONS INFORMATION
Admission to the engineering department: At the end of the first year.
Additional information: Upon completion of freshman year requirements, admission to the College of Engineering is based on the completion of 24 or more semester hours of credit with a 2.0 or better GPA, and credit for or eligibility to schedule analytical geometry and calculus for curricula leading to the bachelor of science degrees in civil, electrical, or mechanical engineering, or college algebra for curricula leading to the bachelor of science degree in engineering technology.

FINANCIAL INFORMATION

ESTIMATED EXPENSES (ACADEMIC YEAR)
[Expenses are for the 1992-93 nine-month academic year.]

All Students	Undergraduate	Graduate
Tuition and fees	$ 794	$ 803
College room and board	$ 1,429	$ 1,429
Books and supplies	$ 300	$ 900
Other expenses	$ 17	$ 17
Total estimated expenses	**$ 2,540**	**$ 3,149**

FINANCIAL AID OFFICE CONTACT
Cynthia L. Tarver, *Director*
Southern University and A&M College
114 F.G. Clark Activity Center
Southern Branch P.O.
Baton Rouge, LA 70813
Phone: (504)771-2796

GENERAL FINANCIAL AID INFORMATION
Forms accepted/required: ACT, AFSA, CSS-FAF, FAF, FAT, FSA, FFS, Stafford, IRS, SAR, SDF, Supplemental
Additional financial aid information: Academic Scholarships and Student Work study Program.

ENGINEERING COLLEGE INFORMATION
[For additional personnel, refer to the *Appendix*.]

HEAD OF ENGINEERING
Trent V. Montgomery
Phone: (504)771-5290 Fax: (504)771-2072
EMail: trent@subrvm.bitnet

ENGINEERING COLLEGE ADDRESS
College of Engineering
Southern University and A&M College
Southern Branch P.O.
Baton Rouge, LA 70813
Phone: (504)771-5290 Fax: (504)771-2072

ENROLLMENTS—BY CLASS
[Numbers are baccalaureate enrollments for the fall 1991 term, unless otherwise footnoted.]
1st-year students/Freshmen 372
2nd-year students/Sophomores 252
3rd-year students/Juniors 177
4th-year students/Seniors 279
Total 1,080

NUMBER OF DEGREES AWARDED—BY PROGRAM
[Numbers are engineering baccalaureate degrees awarded during the 1991-1992 academic year, unless otherwise footnoted. For full details about each engineering program, refer to the *Tables of Degree Programs*.]
Civil & Environmental Engineering 12
Civil Engineering 12
Electrical Engineering 34
Electronic Engineering Technology 25
Electronics Engineering 25
Mechanical Engineering 21
Mechanical Technology 5
Total 134

PERCENTAGE OF DEGREES AWARDED—BY CATEGORY
[Percentages are of all engineering baccalaureate degrees awarded during the 1991-1992 academic year, unless otherwise footnoted.]
African Americans .... 87.6 %
Asian/Pacific Island Americans .... – %
Hispanic Americans .... – %
Native Americans .... – %
Foreign Citizens .... 12.4 %
All Others .... – %
Women .... 25.8 %
Persons with disabilities .... – %
Students over 25 years of age .... – %

ENGINEERING STUDENT DATA
Applicants to the engineering college
Number of applicants to engineering college 105
Percent offered admission 100.0%

Matriculated engineering students
Percentage in top quartile (25%) of High School class .... – (A)

Notes: (A) Data not available.

FULL- & PART-TIME FACULTY—BY DEPARTMENT
[Figures are the head count of full-time faculty and the full-time equivalent (FTE) of part-time faculty for each engineering department or equivalent.]

Department	Full	Part
Civil Eng	6	1.0
Electrical Eng	9	1.0
Mechanical Eng	14	5.0
Division of Tech	10	3.0

COLLEGE DESCRIPTION
The College assumes a social mission to educate and encourage minority students to select engineering as their major. Evidence of success in this area is measured by the amount of funds spent on the Engineering Summer Institute, the Retention Program, and the Dual-degree Programs. Approximately 200 pre-college/elementary to high-school students are affected by the ESI Program. Dual-degree Programs have been established with Alcorn and Jackson State University of Mississippi and Xavier University of New Orleans, LA. The College is a member of the Synthesis Engineering Coalition sponsored by the National Science Foundation. The Coalition plans to thoroughly restructure undergraduate engineering education.

DEGREE PROGRAMS DESCRIPTION
BSCE - Curriculum is designed to give students a thorough knowledge of basic science and suitable training in applying fundamental principles, in the analysis, design, and maintenance of civil engineering works. BSEE - To educate students in the methods of design, application, and analysis of electrical systems. BSME - designed to give students a preparation which leads to a thorough understanding of how to apply laws of basic science while simultaneously stimulating the development of creative thinking, professional attitude, and economic judgment. BS EET/MET - The program centers on electronic circuits and their applications in various technical fields. The fundamental basic theories of engineering, science, and mathematics essential to understanding and competence in application, form an important part of the program.

DUAL DEGREE PROGRAMS
Engineering baccalaureate graduates with dual degrees 0
Enrollment requirements: An official and detailed transcript showing entrance credits evidence of good standing and scholarship must be submitted to the Office of Admissions.

TRANSFER INFORMATION
Residency Requirements: A candidate for the Baccalaureate degree must complete the last thirty semester hours of studies in residence at Southern University and that period of residence shall not be less than one semester and one summer session, or less than three summer sessions, during either of which periods the candidate is enrolled in the college in which the degree is offered.

Transfer via Articulation Agreements
Admission to engineering: At the end of the sophomore year
Engineering graduates transferred from ... 4-yr: – 2-yr: –

Transfer without Articulation Agreements
Admission to engineering: Maximum transfer hours = 93 Contact the Dean of Engineering or Associate Dean of Engineering. 504-771-3798.
Engineering graduates transferred from ... 4-yr: 13 2-yr: 1
Requirements: An official and detailed transcript showing entrance credits, evidence of good standing and scholarship must be submitted to the Office of Admissions.

GRADUATION REQUIREMENTS
Graduation Requirements: 1. Attainment of the overall grade point average of 2.00. 2. Attainment of a grade point average of 2.00 in courses taken in the College of Engineering and the major area of study. 3. Attainment of a 'C' or better in each course in the major area presented to fulfill credit hour requirement in the major. 4. Passage of the departmental comprehensive examination and the university-wide writing proficiency examination. 5. The attainment of an acceptable score on the GRE examination.

STUDENT PROGRAMS

PROFESSIONAL AND HONORARY SOCIETIES
[For key to acronyms, see *Introduction*.]
Professional Societies: ASCE, ASME, IEEE
Honorary Societies: Chi Epsilon, Eta Kappa Nu, Pi Tau Sigma

SUPPORT PROGRAMS
Student Chapter Organizations: Society of Women Engineers, National Society of Black Engineers, American Society of Civil Engineers, American Society of Mechanical Engineers, The Engineering Technology Club

Engineering Recruitment Program
For: Women & Ethnic Minorities **Available:** Year round
Offered: Prematriculation, Freshman, Sophomore

Engineering Retention Program
For: Women & Ethnic Minorities **Available:** Academic year
Offered: Freshman, Sophomore, Junior, Senior

EIT Program
For: Women & Ethnic Minorities **Available:** Academic year
Offered: Junior, Senior

Other Engineering Support Programs: Other support programs for engineering students include the following: 1. Tutorial Program 2. Engineering Summer Institute 3. Honors Program 4. Student Development Program 5. Co-op

211 University of Southern California

■ INSTITUTION PROFILE

HEAD OF THE INSTITUTION
Steven B. Sample
Phone: (213)740-2111

GENERAL INFORMATION
[All Students—Fall 1991]
Undergraduate enrollment 13,500
Graduate enrollment 14,000
Total institution enrollment **27,500**
Type of institution: Private **Calendar:** Semesters
Nearest city: Los Angeles **Population:** 8,000,000
Miles from main campus: 2 **Setting:** Urban
Types of engineering degrees: Engineering
Other degree-granting colleges: Allied Health Sciences, Architecture, Arts & Sciences, Business Administration, Communication & Journalism, Dentistry, Education, Fine & Performing Arts, Humanities & Social Sciences, Law, Medicine, Nursing, Pharmacy, Technology & Applied Sciences, Cinema-Television, Gerontology, Public Administration, Social Work, Urban and Regional Planning

■ ENGINEERING ADMISSIONS

ADMISSIONS OFFICE CONTACT
Clarke T. Howatt
University of Southern California
School of Engineering
Olin Hall, Rm. 106, University Park
Los Angeles, CA 90089-1455
Phone: (213)740-4530 Fax: (213)740-4530

ENGINEERING COLLEGE ADMISSIONS INFORMATION
Entrance Requirements: SAT (Score: 1000), ACT (Score: 25), High School courses—Natural Sciences/Social Sciences (1 year), Humanities/English/Foreign Language/Math (1 year), Three electives - Computers Science/Drama/Speech (1 year), Electives cont'd - Fine Arts/Music/Journalism (1 year), Minimum of 550 Math SAT or 27 ACT
Entrance Recommendations: ACT (Score: 25), High School courses—Chemistry (1 year), Physics (1 year), Math (including analysis and pre-calculus) (4 years)
Requirements for foreign students: financial statement

DEPARTMENTAL ADMISSIONS INFORMATION
Admission to the engineering department: At the time of admission to the institution.

■ FINANCIAL INFORMATION

ESTIMATED EXPENSES (ACADEMIC YEAR)
[Expenses are for the 1992-93 nine-month academic year.]

All Students	Undergraduate	Graduate
Tuition and fees	$16,020	$10,870
College room and board	$6,040	$7,182
Books and supplies	$600	$962
Other expenses	$2,210	$3,090
Total estimated expenses	**$24,870**	**$22,104**

FINANCIAL AID OFFICE CONTACT
Catherine Thomas, *Director, Financial Aid*
University of Southern California
Student Administrative Services Rm. 328
University Park
Los Angeles, CA 90089-0914
Phone: (212)740-5444

GENERAL FINANCIAL AID INFORMATION
Forms accepted/required: ACT, CSS-FAF, FAF, Stafford, IRS, SAR, Institutional, Free Application for Federal Student Aid (FAFSA)
Additional financial aid information: Need-based financial assistance; merit scholarships and research awards; alternative financing.

■ ENGINEERING COLLEGE INFORMATION

[For additional personnel, refer to the *Appendix*.]

HEAD OF ENGINEERING
Leonard M. Silverman
Phone: (213)740-0884 Fax: (213)740-8493

ENGINEERING COLLEGE ADDRESS
School of Engineering
University of Southern California
Olin Hall, Rm. 106
University Park
Los Angeles, CA 90089-1455
Phone: (213)740-4530 Fax: (213)740-8690

NUMBER OF DEGREES AWARDED—BY PROGRAM
[Numbers are engineering baccalaureate degrees awarded during the 1991-1992 academic year, unless otherwise footnoted. For full details about each engineering program, refer to the *Tables of Degree Programs*.]

Aerospace Engineering 56
Biomedical Engineering 24
Chemical Engineering 8
Civil Engineering 15
Computer Science 41
Electrical Engineering 72
Industrial & Systems Engineering 38
Mechanical Engineering 38
Total **292**

PERCENTAGE OF DEGREES AWARDED—BY CATEGORY
[Percentages are of all engineering baccalaureate degrees awarded during the 1991-1992 academic year, unless otherwise footnoted.]

African Americans .... – %
Asian/Pacific Island Americans .... – %
Hispanic Americans .... – %
Native Americans .... – %
Foreign Citizens .... – %
All Others .... 100 %
Women .... – %
Persons with disabilities .... – %
Students over 25 years of age .... – %

ENGINEERING STUDENT DATA

Applicants to the engineering college
Number of applicants to engineering college 1,702
Percent offered admission 70.0%

Matriculated engineering students
Percentage in top quartile (25%) of High School class .... –(A)
Average SAT scores: Math—650, Verbal—513, Combined—1163
Notes: (A) Data not available.

FULL- & PART-TIME FACULTY—BY DEPARTMENT
[Figures are the head count of full-time faculty and the full-time equivalent (FTE) of part-time faculty for each engineering department or equivalent.]

Department	Full	Part
Chemical Eng	8	2.0
Civil Eng	17	3.0
Mechanical Eng	11	3.0
Industrial Eng	4	5.0
Electrical Eng	21	–
Aerospace Eng	13	2.0
Biomedical Eng	6	2.0
Computer Science	25	3.0

COLLEGE DESCRIPTION
The School of Engineering offers 4 yr degree programs in AE, BME, CHE, CE, CECS (Computer Eng/Computer Sci), CSCI, EE, ENE, ISE, & ME. Dual major programs are available between BME & EE, BME & ME, combined majors between CE & ARCH, EE & Recording Arts; minors in Environmental Eng. and PTE, Business and most departments in the College of Letters, Arts and Science. A Three-Two program is offered between specific Liberal Arts College around the country and the School of Engineering. Co-op is an option in all engineering departments. In addition to Semester Abroad Programs within the University, Engineering also offers an Overseas Summer Program. Important aspects of our program are: 1) Engineering classes begin in the freshmen year; 2) most undergraduate classes are taught by full-time faculty; 3) large numbers of undergraduates (including freshmen) participate in research with faculty.

DEGREE PROGRAMS DESCRIPTION
The B.S. degree programs in the various departments prepare the graduate for entry-level professional positions or for graduate study. Optional participation in the Co-op program enhances the on-campus curriculum. Students in the B.S. degree programs begin their engineering coursework in their freshmen year. Graduate programs in engineering lead to the M.S., Engineer and Ph.D. degrees. Programs at the Master's and Engineer level emphasize both theoretical concepts and the analysis and solution of complex engineering problems. The Ph.D requires an original dissertation demonstrating independent research and a scholarly result.

DUAL DEGREE PROGRAMS
Engineering baccalaureate graduates with dual degrees 10

TRANSFER INFORMATION
Residency Requirements: Must complete last 48 units of degree at USC

Transfer via Articulation Agreements
Admission to engineering: Considered transfer students after completion of 30 or more transferable semester units (45 quarter units).
Engineering graduates transferred from . . 4-yr: – 2-yr: –
Requirements: 30 or more transferable semester units (45 quarter units) from a fully accredited college or university with a minimum of 3.0 GPA. Dual degree programs are available with several liberal arts colleges nationwide.

Transfer without Articulation Agreements
Admission to engineering: Must have achieved a 3.0 GPA on transferable work. Lower GPA's are considered by some departments.
Engineering graduates transferred from . . 4-yr: – 2-yr: –
Requirements: Minimum 3.0 GPA on coursework completed.

GRADUATION REQUIREMENTS
Bachelor's degree programs require 128-135 semester units depending on the major chosen with an overall 2.0 cumulative GPA and a 2.0 cumulative GPA in the major.

■ STUDENT PROGRAMS

PROFESSIONAL AND HONORARY SOCIETIES
[For key to acronyms, see *Introduction*.]
Professional Societies: ACM, AIAA, AIChE, AIME, ASCE, ASME, BMES, IEEE, IIE, National Society of Black Engineers, Society of Hispanic Professional Engineers
Honorary Societies: Alpha Pi Mu, Chi Epsilon, Eta Kappa Nu, Omega Chi Epsilon, Pi Tau Sigma, Sigma Gamma Tau, Sigma Phi Delta, Tau Beta Pi, Omega Rho, Pi Epsilon Tau

SUPPORT PROGRAMS
Student Chapter Organizations: Society of Women Engineers, National Society of Black Engineers, Society of Hispanic Professional Engineers

Minority Engineering Program
For: Ethnic Minorities **Available:** Year round
Offered: Prematriculation, Freshman, Sophomore, Junior, Senior, Graduate level
Other Engineering Support Programs: On-campus and out-of-state orientation programs; Engineering Tutoring Program; University Learning Center; Minority Bridge Program; University Career Development Center; Academic counseling; University Counseling Center; University Office for Disabled Students

212 University of Southern Colorado

INSTITUTION PROFILE
HEAD OF THE INSTITUTION
Robert C. Shirley
Phone: (719)549-2306　　　　　Fax: (719)549-2219

GENERAL INFORMATION
[All Students—Fall 1991]
Undergraduate enrollment 4,337
Graduate enrollment 151
Total institution enrollment 4,488
Type of institution: Public　　Calendar: Semesters
Location: Pueblo　　Population: 100,000
Setting: Urban
Types of engineering degrees: Engineering & Technology
Other degree-granting colleges: Humanities & Social Sciences, Science and Mathematics, Center for Teaching and Learning, Business

ENGINEERING ADMISSIONS
ADMISSIONS OFFICE CONTACT
Margie Wade
University of Southern Colorado
Dean of Admissions & Enrollment Services
2200 Bonforte Blvd.
Pueblo, CO 81001-4901
Phone: (719)549-2461　　　　　Fax: (719)549-2219

ENGINEERING COLLEGE ADMISSIONS INFORMATION
Entrance Requirements: SAT, ACT, ACT or SAT is required. Admission is based on an index which combines test scores and high school grade point average.
Entrance Recommendations: High School courses—English (4 years), Algebra (2 years), Trigonometry (1 year), Physics (1 year)
Requirements for foreign students: TOEFL (Score: 500); financial statement

DEPARTMENTAL ADMISSIONS INFORMATION
Admission to the engineering department: At the time of admission to the institution.

FINANCIAL INFORMATION
ESTIMATED EXPENSES (ACADEMIC YEAR)
[Expenses are for the 1992-93 nine-month academic year.]

State Residents	Undergraduate	Graduate
Tuition and fees	$ 1,726	$ 1,726
College room and board	$ 3,536 [1]	$ 3,536 [1]
Books and supplies	$ 650 [C]	$ 650 [C]
Other expenses	$ — [2]	$ — [2]
Total estimated expenses	**$ 5,912**	**$ 5,912**

Out-of-State Residents	Undergraduate	Graduate
Tuition and fees	$ 5,958	$ 5,958
College room and board	$ 3,536 [1]	$ 3,536 [1]
Books and supplies	$ 650 [C]	$ 650 [C]
Other expenses	$ — [2]	$ — [2]
Total estimated expenses	**$10,144**	**$10,144**

Notes: (C) Estimated data. (1) Based on double room and 19-meal plan. (2) Special fees vary from $2 to $95.

FINANCIAL AID OFFICE CONTACT
Gina T. Mestas, *Director of Financial Aid*
University of Southern Colorado
2200 Bonforte Blvd.
Pueblo, CO 81001-4901
Phone: (719)549-2753　　　　　Fax: (719)549-2938

GENERAL FINANCIAL AID INFORMATION
Forms accepted/required: ACT, AFSA, CSS-FAF, FAT, FSA, FFS, Stafford, IRS, SAR, Supplemental, Institutional, Comprehensive Financial Aid Report (CFAR), Financial Aid Form Need Analysis Report (FAFNAR)
Additional financial aid information: Need-based Scholarships College Work-Study Program Student Loan Programs Private Scholarships USC President's Scholarships

ENGINEERING COLLEGE INFORMATION
[For additional personnel, refer to the *Appendix*.]
HEAD OF ENGINEERING
Ray L. Sisson
Phone: (719)549-2696　　　　　Fax: (719)549-2519
EMail: SISSON@COMET.USCOLO.EDU

ENGINEERING COLLEGE ADDRESS
College of Applied Science and Engineering Technology
University of Southern Colorado
2200 Bonforte Blvd.
Pueblo, CO 81001-4901
Phone: (719)549-2100　　　　　Fax: (719)549-2461

ENROLLMENTS—BY CLASS
[Numbers are baccalaureate enrollments for the fall 1991 term, unless otherwise footnoted.]
1st-year students/Freshmen 43
2nd-year students/Sophomores 18
3rd-year students/Juniors 13
4th-year students/Seniors 21
Total **95**

NUMBER OF DEGREES AWARDED—BY PROGRAM
[Numbers are engineering baccalaureate degrees awarded during the 1991-1992 academic year, unless otherwise footnoted. For full details about each engineering program, refer to the *Tables of Degree Programs*.]
Industrial Engineering .... —
Total .... **—**

PERCENTAGE OF DEGREES AWARDED—BY CATEGORY
[Percentages are of all engineering baccalaureate degrees awarded during the 1991-1992 academic year, unless otherwise footnoted.]
African Americans .... — %
Asian/Pacific Island Americans .... — %
Hispanic Americans .... — %
Native Americans .... — %
Foreign Citizens .... 14.3 %
All Others .... 85.7 %
Women .... 28.6 %
Persons with disabilities .... — %
Students over 25 years of age .... 14.3 %

GRADUATE ENROLLMENTS & DEGREES AWARDED
Master's enrollment 25
Master's degrees awarded 3
Doctoral enrollment .... — [B]
Doctoral degrees awarded .... — [B]

Notes: (B) Data not applicable.

ENGINEERING STUDENT DATA
Applicants to the engineering college
Number of applicants to engineering college 61
Percent offered admission 95.0%

Matriculated engineering students
Percentage in top quartile (25%) of High School class .... — [A]

Notes: (A) Data not available.

FULL- & PART-TIME FACULTY—BY DEPARTMENT
[Figures are the head count of full-time faculty and the full-time equivalent (FTE) of part-time faculty for each engineering department or equivalent.]

Department	Full	Part
Industrial & Systems Eng	4	—

COLLEGE DESCRIPTION
The industrial engineering major leads to a bachelor of science in industrial engineering (BSIEN) degree. The department also provides courses for the first two years of other engineering disciplines for potential transfer students, courses for engineering options in chemistry and physics, and a master of science in systems engineering (MS) degree. A minor in industrial engineering is under development.

DEGREE PROGRAMS DESCRIPTION
The fundamental goal of the engineering department is to provide students with high quality instruction in industrial engineering which is broad-based and strongly rooted in mathematics, physical science and engineering science. Graduates of this program will be well-trained industrial engineers who can make immediate contributions to the operation of high-tech industries, service organizations, and traditional manufacturers. Opportunities for students to interact with real world problems are available through co-operative work experiences and the senior engineering design course. The master of systems engineering is closely related to industrial engineering.

TRANSFER INFORMATION
Residency Requirements: A minimum of 30 semester hours of resident instruction as approved by the engineering department must be earned in residence at USC.

Transfer via Articulation Agreements
Admission to engineering: At the end of the sophomore year
Engineering graduates transferred from . . . 4-yr: — 2-yr: 3
Requirements: Students who complete satisfactorily the list of equivalent courses with a cumulative grade point average of 2.5 or above. Block transfer agreements are in place with Colorado State University, Fort Lewis College, and Pikes Peak Community College.

Transfer without Articulation Agreements
Admission to engineering: At any time
Engineering graduates transferred from . . . 4-yr: 6 2-yr: 3
Requirements: Students transferring into industrial engineering from other universities and/or departments must have earned a minimum overall 2.5 grade point average. Course equivalences are evaluated on an individual basis.

GRADUATION REQUIREMENTS
A cumulative average of at least 2.00 in all industrial engineering core courses is required for graduation. Each senior engineering student is required to take the Fundamentals of Engineering Exam (EIT) as prescribed by the Colorado State Board of Registration for Professional Engineers.

STUDENT PROGRAMS
PROFESSIONAL AND HONORARY SOCIETIES
[For key to acronyms, see *Introduction*.]
Professional Societies: ASCE, ASEM, ASME, IEEE, IIE, ORSA, SME, TIMS, Society of Women Engineers

SUPPORT PROGRAMS
Student Chapter Organizations: Society of Women Engineers, Society of Hispanic Professional Engineers

Women's Resource Center
For: Women　　**Available:** Academic year
Offered: Freshman, Sophomore, Junior, Senior, Graduate level
Other Engineering Support Programs: Indian Summer Orientation Programs (freshmen and transfers); Counseling and Career Services

213 Southern Illinois University at Carbondale

■ INSTITUTION PROFILE

HEAD OF THE INSTITUTION
John C. Guyon
Phone: (618)453-2341   Fax: (618)453-5362

GENERAL INFORMATION
[All Students—Fall 1991]
Undergraduate enrollment 20,339
Graduate enrollment 4,427
Total institution enrollment 24,766
Type of institution: Public   **Calendar:** Semesters
Nearest city: St. Louis, Missouri   **Population:** 2,500,000
Miles from main campus: 100   **Setting:** Small Town
Types of engineering degrees: Engineering & Technology
Other degree-granting colleges: Business Administration, Education, Law, Medicine, Agriculture, Communications and Fine Arts, Liberal Arts, Science, Technical Careers

■ ENGINEERING ADMISSIONS

ADMISSIONS OFFICE CONTACT
Jerre C. Pfaff
Southern Illinois University at Carbondale
Admissions and Records
Carbondale, IL 62901
Phone: (618)453-4321   Fax: (618)453-3250

ENGINEERING COLLEGE ADMISSIONS INFORMATION
Entrance Requirements: High School courses—English (4 years), Mathematics (3.5 years), Science (3 years), Social Studies (3 years), Beginning freshmen and transfers with less than 26 semester hours must have an ACT of 23 or greater if they are in the top half of their high school class and 19 or greater if they are in the top quarter of their high school class.
Entrance Recommendations: High School courses—English (4 years), Chemistry (1 year), Physics (1 year), Mathematics (3.5 years)
Requirements for foreign students: TOEFL (Score: 520); financial statement

DEPARTMENTAL ADMISSIONS INFORMATION
Admission to the engineering department: At the time of admission to the institution.

■ FINANCIAL INFORMATION

ESTIMATED EXPENSES (ACADEMIC YEAR)
[Expenses are for the 1992-93 nine-month academic year.]

State Residents	Undergraduate	Graduate
Tuition and fees	$ 3,006	$ 3,006
College room and board	$ 3,024	$ 3,024
Books and supplies	$ 550	$ 650
Other expenses	$ 2,600	$ 3,000
Total estimated expenses	**$ 9,180**	**$ 9,680**

Out-of-State Residents	Undergraduate	Graduate
Tuition and fees	$ 7,406	$ 7,406
College room and board	$ 3,024	$ 3,024
Books and supplies	$ 550	$ 650
Other expenses	$ 2,600	$ 3,000
Total estimated expenses	**$13,580**	**$14,080**

FINANCIAL AID OFFICE CONTACT
Pamela A. Britton, *Director, Financial Aid*
Southern Illinois University at Carbondale
Financial Aid
Carbondale, IL 62901
Phone: (618)453-4334

GENERAL FINANCIAL AID INFORMATION
Forms accepted/required: ACT, AFSA, CSS-FAF, FFS, PHEA
Additional financial aid information: Scholarships and awards are based on scholastic achievements of new freshmen and Illinois community college transfers; federal and state grant programs; and on-campus student work jobs.

■ ENGINEERING COLLEGE INFORMATION

[For additional personnel, refer to the *Appendix*.]

HEAD OF ENGINEERING
Juh W. Chen
Phone: (618)453-4321   Fax: (618)453-4235

ENGINEERING COLLEGE ADDRESS
College of Engineering
Southern Illinois University at Carbondale
Carbondale, IL 62901
Phone: (618)453-4321   Fax: (618)453-4235

ENROLLMENTS—BY CLASS
[Numbers are baccalaureate enrollments for the fall 1991 term, unless otherwise footnoted.]
1st-year students/Freshmen 227
2nd-year students/Sophomores 189
3rd-year students/Juniors 236
4th-year students/Seniors 435
Total 1,087

NUMBER OF DEGREES AWARDED—BY PROGRAM
[Numbers are engineering baccalaureate degrees awarded during the 1991-1992 academic year, unless otherwise footnoted. For full details about each engineering program, refer to the *Tables of Degree Programs*.]
Civil Engineering 27
Electrical Engineering 97
Engineering Science(s) .... – (B)
Mechanical Engineering 50
Mining Engineering 1
Total 175
Notes: (B) Data not applicable.

PERCENTAGE OF DEGREES AWARDED—BY CATEGORY
[Percentages are of all engineering baccalaureate degrees awarded during the 1991-1992 academic year, unless otherwise footnoted.]
African Americans .... 2.8 %
Asian/Pacific Island Americans .... 5.1 %
Hispanic Americans .... 0.5 %
Native Americans .... – %
Foreign Citizens .... 19.4 %
All Others .... 72.2 %
Women .... 6.8 %
Persons with disabilities .... 0.5 %
Students over 25 years of age .... 38.0 %

GRADUATE ENROLLMENTS & DEGREES AWARDED
Master's enrollment 157
Master's degrees awarded 59
Doctoral enrollment 40
Doctoral degrees awarded 10

ENGINEERING STUDENT DATA
Applicants to the engineering college
Number of applicants to engineering college 557
Percent offered admission .... 89.0 %

Matriculated engineering students
Percentage in top quartile (25%) of High School class 45.0
Average ACT scores: Math—24.2, Composite—23.5

FULL- & PART-TIME FACULTY—BY DEPARTMENT
[Figures are the head count of full-time faculty and the full-time equivalent (FTE) of part-time faculty for each engineering department or equivalent.]

Department	Full	Part
Civil Eng	11	–
Electrical Eng	21	1.5
Mechanical Eng	17	1.0
Mining Eng	5	–

COLLEGE DESCRIPTION
The College offers bachelor's and master's degree programs in four engineering disciplines-civil engineering, electrical engineering, mechanical engineering and mining engineering as well as a doctor of philosophy degree in engineering science with options to specialize in electrical systems, engineering mechanics or fossil energy. All of the undergraduate programs share a common core of courses in general education, mathematics, basic sciences, engineering science and engineering design. Each department of the college maintains well equipped laboratories for instruction and research. The college maintains excellent computer facilities for student use. Students have the opportunity to participate in eight professional engineering societies. Opportunities are provided for students to take the engineer-in-training exam as a first step toward becoming a registered professional engineer.

DEGREE PROGRAMS DESCRIPTION
The B.S. degrees in the various engineering programs prepare the graduate for entry-level professional positions or for graduate study in each of the respective fields. The engineering programs share a common core of general education, mathematics, basic sciences, engineering science and engineering design. Master's degrees are available in civil, electrical, mechanical and mining engineering. The Ph.D. program in engineering science has three areas of concentration, electrical, systems, fossil energy and mechanics.

TRANSFER INFORMATION
Residency Requirements: A student must complete 90 semester hours of credit at Southern Illinois University at Carbondale or complete sixty senior institution hours, and the last thirty consecutive semester hours are taken at Southern Illinois University at Carbondale.
Transfer via Articulation Agreements
Admission to engineering: At any time
Engineering graduates transferred from   . . 4-yr: –   2-yr: –
Requirements: Transfer students are admitted to the engineering programs at any level. The application deadline is at least 30 days prior to the beginning of classes. Transfer students with more than 26 semester hours of course work must have a 2.4 GPA to be admitted directly into engineering.
Transfer without Articulation Agreements
Admission to engineering: At any time
Engineering graduates transferred from   . . 4-yr: –   2-yr: –
Requirements: Students from other academic units have the same entrance requirements as students from institutions with articulation agreements.

GRADUATION REQUIREMENTS
The B.S. degree programs in engineering require 31 semester hours of general education; 18-24 semester hours of basic science; 17 semester hours of mathematics and 63-68 semester hours of engineering courses depending upon the specific major. Total semester hours 133-135. A GPA of 2.0 out of a possible 4.0 is required in the major (engineering) as well as overall.

■ STUDENT PROGRAMS

PROFESSIONAL AND HONORARY SOCIETIES
[For key to acronyms, see *Introduction*.]
Professional Societies: AEG, ASCE, ASHRAE, ASME, IEEE, SME, SME*
Honorary Societies: Tau Beta Pi

SUPPORT PROGRAMS
Student Chapter Organizations: Society of Women Engineers, National Society of Black Engineers, Society of Hispanic Professional Engineers, Blacks in Engineering and Allied Technologies

Minority Engineering Program
For: Ethnic Minorities   **Available:** Year round
Offered: Prematriculation, Freshman, Sophomore, Junior, Senior

Women in Engineering
For: Women   **Available:** Summer only
Offered: Prematriculation

Minority Engineering Summer Program
For: Ethnic Minorities   **Available:** Summer only
Offered: Prematriculation

Minority Bridge Program
For: Ethnic Minorities   **Available:** Summer only
Offered: Freshman

Other Engineering Support Programs: Comprehensive university orientation programs for all new students and parents; academic advisement provided by trained advisors in the unit; career development center; testing services; counseling center; services for students with disabilities; international programs and services; and university placement center.

214 Southern Illinois University at Edwardsville

INSTITUTION PROFILE
HEAD OF THE INSTITUTION
Earl E. Lazerson
Phone: (618)692-2475   Fax: (618)692-2270

GENERAL INFORMATION
[All Students—Fall 1991]

Undergraduate enrollment	8,961
Graduate enrollment	2,847
Total institution enrollment	**11,808**

Type of institution: Public   **Calendar:** Quarters
Nearest city: St. Louis, MO   **Population:** 1,007,800
Miles from main campus: 20   **Setting:** Small Town
Types of engineering degrees: Engineering
Other degree-granting colleges: Business Administration, Dentistry, Education, Nursing, Humanities, Social Sciences, Fine Arts & Communication, Sciences

ENGINEERING ADMISSIONS
ADMISSIONS OFFICE CONTACT
Harlan H. Bengtson
Southern Illinois University at Edwardsville
School of Engineering
SIUE Box 1800
Edwardsville, IL 62026-1800
Phone: (618)692-2534   Fax: (618)692-2555

ENGINEERING COLLEGE ADMISSIONS INFORMATION
Admission to the engineering college: Most are admitted in the sophomore year.
Entrance Requirements: High School courses—Mathematics (3 years), Science (2 years), English (4 years), Social Studies (3 years), The sum of the ACT composite national percentile score and the percentile rank from the high school graduating class must be at least 100.
Requirements for foreign students: TOEFL (Score: 550); financial statement

DEPARTMENTAL ADMISSIONS INFORMATION
Admission to the engineering department: At the end of the second year.
Additional information: Completion of a specified lower division engineering program with a GPA of at least 2.0/4.0. Space limitations may raise the minimum allowable GPA for some programs.

FINANCIAL INFORMATION
ESTIMATED EXPENSES (ACADEMIC YEAR)
[Expenses are for the 1992-93 nine-month academic year.]

State Residents	Undergraduate	Graduate
Tuition and fees	$ 2,158	$ 2,187
College room and board	$ 3,000	$ 3,000
Books and supplies	$ 150	$ 400
Other expenses	$ 1,500	$ 1,500
Total estimated expenses	**$ 6,808**	**$ 7,087**

Out-of-State Residents	Undergraduate	Graduate
Tuition and fees	$ 5,610	$ 5,884
College room and board	$ 3,000	$ 3,000
Books and supplies	$ 150	$ 400
Other expenses	$ 1,500	$ 1,500
Total estimated expenses	**$10,260**	**$10,784**

FINANCIAL AID OFFICE CONTACT
William D. Burns, *Acting Director*
Southern Illinois University at Edwardsville
Student Work & Financial Aid Office
Rendleman Bldg., Room 2308
Edwardsville, IL 62026-1060
Phone: (618)692-3880

GENERAL FINANCIAL AID INFORMATION
Forms accepted/required: Singlefile form, United Student Aid Funds
Additional financial aid information: Title IV financial aid, need-based scholarships and part-time jobs are available.

ENGINEERING COLLEGE INFORMATION
[For additional personnel, refer to the *Appendix*.]
HEAD OF ENGINEERING
Colby V. Ardis
Phone: (618)692-2541   Fax: (618)692-3374

ENGINEERING COLLEGE ADDRESS
School of Engineering
Southern Illinois University at Edwardsville
University Park
SIUE Box 1804
Edwardsville, IL 62026-1804
Phone: (618)692-2541   Fax: (618)692-3374

ENROLLMENTS—BY CLASS
[Numbers are baccalaureate enrollments for the fall 1991 term, unless otherwise footnoted.]

1st-year students/Freshmen	12
2nd-year students/Sophomores	87
3rd-year students/Juniors	162
4th-year students/Seniors	259
Total	**520**

NUMBER OF DEGREES AWARDED—BY PROGRAM
[Numbers are engineering baccalaureate degrees awarded during the 1991-1992 academic year, unless otherwise footnoted. For full details about each engineering program, refer to the *Tables of Degree Programs*.]

Civil Engineering	31
Electrical Engineering	56
Industrial Engineering	11
Mechanical Engineering	13
Total	**111**

PERCENTAGE OF DEGREES AWARDED—BY CATEGORY
[Percentages are of all engineering baccalaureate degrees awarded during the 1991-1992 academic year, unless otherwise footnoted.]

African Americans	7.2%
Asian/Pacific Island Americans	3.6%
Hispanic Americans	0.9%
Native Americans	1.8%
Foreign Citizens	3.6%
All Others	82.9%
Women	18.0%
Persons with disabilities	— % (A)
Students over 25 years of age	— % (A)

Notes: (A) Data not available.

GRADUATE ENROLLMENTS & DEGREES AWARDED

Master's enrollment	98
Master's degrees awarded	33
Doctoral enrollment	— (B)
Doctoral degrees awarded	— (B)

Notes: (B) Data not applicable.

FULL- & PART-TIME FACULTY—BY DEPARTMENT
[Figures are the head count of full-time faculty and the full-time equivalent (FTE) of part-time faculty for each engineering department or equivalent.]

Department	Full	Part
Civil Eng	9	—
Electrical Eng	10	—
Industrial Eng	3	—
Mechanical Eng	5	—

COLLEGE DESCRIPTION
The School of Engineering, along with Southern Illinois University at Edwardsville in general, places great emphasis on 'recognized excellence in education.' This is done through undergraduate and Master's degree programs. Co-op programs are available for all undergraduate degrees, and typically extend the time requirement for the degree to five years.

DEGREE PROGRAMS DESCRIPTION
The Civil Engineering B.S. program allows specialization in Environmental, Structural, or Transportation Engineering. The first three years of the program are available in evening offerings. The Electrical Engineering B.S. program has a strong laboratory component including the areas of signal processing, computers, control systems, and image processing. The entire program is available in evening offerings. The Industrial Engineering B.S. program has a manufacturing emphasis and includes laboratories equipped for instruction in the areas of robotics and computer integrated manufacturing. The first two years of the program are available in evening offerings. The Mechanical Engineering B.S. program includes emphasis on dynamics and control, fluid mechanics and thermodynamics, and materials. The first two years of the program are available in evening offerings. The M.S. degree program in Civil Engineering is designed as a part-time program for engineers in the area who are employed full-time. It is possible to specialize in either environmental or structural engineering. The M.S. degree program in Electrical Engineering has good research capabilities in computers, controls, and signal and image processing. The program is available in an evening format to serve the needs of engineers who are working full-t

TRANSFER INFORMATION
Residency Requirements: At least 48 quarter hours of credit must be earned at SIUE. At least 96 quarter hours of credit must be earned at a four-year college or university.

Transfer via Articulation Agreements
Admission to engineering: At any time
Requirements: For entry to the junior year, a specified lower division engineering program must be completed with a GPA of at least 2.0/4.0. Space limitations may raise the minimum allowable GPA for some programs.

Transfer without Articulation Agreements
Admission to engineering: At any time
Requirements: GPA of at least 2.25/4.0 in the lower division engineering courses if they are not Illinois residents.

GRADUATION REQUIREMENTS
Overall GPA of at least 2.0/4.0. GPA in major courses of at least 2.0/4.0. GPA in major courses numbered above 300 of at least 2.0/4.0.

STUDENT PROGRAMS
PROFESSIONAL AND HONORARY SOCIETIES
[For key to acronyms, see *Introduction*.]
Professional Societies: AGCA, ASCE, IEEE, IIE
Honorary Societies: Eta Kappa Nu, Sigma Lambda Chi, CE Honor Society

SUPPORT PROGRAMS
Student Chapter Organizations: Society of Women Engineers, National Society of Black Engineers

Minority Engineering Program
For: Women & Ethnic Minorities   **Available:** Year round
Offered: Prematriculation, Freshman, Sophomore, Junior, Senior, Graduate level

Other Engineering Support Programs: University-wide orientation programs are available for new freshmen and new transfer students. All engineering majors have an engineering faculty member as an academic advisor. A wide variety of counseling and other student support services are available to all SIUE students.

215 University of Southern Maine

■ INSTITUTION PROFILE

HEAD OF THE INSTITUTION
Richard Pattenaude
Phone: (207)780-4480 Fax: (207)780-4549

GENERAL INFORMATION
[All Students—Fall 1991]
Undergraduate enrollment 8,000
Graduate enrollment 2,000
Total institution enrollment **10,000**

Type of institution: Public **Calendar:** Semesters
Nearest city: Portland **Population:** 75,000
Miles from main campus: 10 **Setting:** Suburban
Types of engineering degrees: Engineering
Other degree-granting colleges: Arts & Sciences, Business Administration, Education, Law, Nursing

■ ENGINEERING ADMISSIONS

ADMISSIONS OFFICE CONTACT
Daniel Palubniak
University of Southern Maine
37 College Avenue
Gorham, ME 04038
Phone: (207)780-4970

ENGINEERING COLLEGE ADMISSIONS INFORMATION
Entrance Recommendations: High School courses—Algebra (2 years), Trigonometry (1 year), Physics (1 year), Chemistry (1 year)

DEPARTMENTAL ADMISSIONS INFORMATION
Admission to the engineering department: At the time of admission to the institution.
Additional information: Freshman are strongly encouraged to have a background in physics, chemistry, algebra and trigonometry. Remediation is available.

■ FINANCIAL INFORMATION

ESTIMATED EXPENSES (ACADEMIC YEAR)
[Expenses are for the 1992-93 nine-month academic year.]

State Residents	Undergraduate	Graduate
Tuition and fees	$ 2,136	$ 2,160
College room and board	$ 2,400	$ 2,400
Books and supplies	$ 400	$ 400
Other expenses	$ –	$ –
Total estimated expenses	**$ 4,936**	**$ 4,960**
Out-of-State Residents	**Undergraduate**	**Graduate**
Tuition and fees	$ 6,048	$ 6,096
College room and board	$ 2,400	$ 2,400
Books and supplies	$ 400	$ 400
Other expenses	$ –	$ –
Total estimated expenses	**$ 8,848**	**$ 8,896**

FINANCIAL AID OFFICE CONTACT
Helen F. Parker, *Senior Associate Director*
University of Southern Maine
37 College Avenue
Gorham, ME 04038
Phone: (207)780-5250

GENERAL FINANCIAL AID INFORMATION
Forms accepted/required: ACT, FSA, Stafford, Supplemental
Additional financial aid information: Work study, scholarships etc. are available.

■ ENGINEERING COLLEGE INFORMATION
[For additional personnel, refer to the *Appendix*.]

HEAD OF ENGINEERING
Brian C. Hodgkin
Phone: (207)780-5582 Fax: (207)780-5129

ENGINEERING COLLEGE ADDRESS
School of Applied Science
University of Southern Maine
37 College Avenue
Gorham, ME 04038
Phone: (207)780-5584 Fax: (207)780-5129

ENROLLMENTS—BY CLASS
[Numbers are baccalaureate enrollments for the fall 1991 term, unless otherwise footnoted.]
1st-year students/Freshmen 45
2nd-year students/Sophomores 30
3rd-year students/Juniors 30 (1)
4th-year students/Seniors 18
Total **123**

Notes: (1) 8

NUMBER OF DEGREES AWARDED—BY PROGRAM
[Numbers are engineering baccalaureate degrees awarded during the 1991-1992 academic year, unless otherwise footnoted. For full details about each engineering program, refer to the *Tables of Degree Programs*.]
Electrical Engineering 8
Total **8**

PERCENTAGE OF DEGREES AWARDED—BY CATEGORY
[Percentages are of all engineering baccalaureate degrees awarded during the 1991-1992 academic year, unless otherwise footnoted.]
African Americans .... – %
Asian/Pacific Island Americans .... – %
Hispanic Americans .... – %
Native Americans .... – %
Foreign Citizens .... 5.0 %
All Others .... 95.0 %
Women .... 5.0 %
Persons with disabilities .... – %
Students over 25 years of age .... 60.0 %

ENGINEERING STUDENT DATA
Applicants to the engineering college
Number of applicants to engineering college 55
Percent offered admission .... 70.0 %

Matriculated engineering students
Percentage in top quartile (25%) of High School class 40.0
Average SAT scores: Math—550, Verbal—510, Combined—1060

FULL- & PART-TIME FACULTY—BY DEPARTMENT
[Figures are the head count of full-time faculty and the full-time equivalent (FTE) of part-time faculty for each engineering department or equivalent.]

Department	Full	Part
Electrical Eng	5	0.5

COLLEGE DESCRIPTION
The School of Applied Science of the University of Southern Maine has baccalaureate programs in electrical engineering, computer science, industrial technology and vocational education and master's programs in applied immunology and computer science. The majority of the students are employed either full or part-time. The average age of the students is 27. Co-op opportunities exist in local and regional industrial firms. Double majors for electrical engineering students are available in mathematics and computer science. A minor is available in chemistry.

DEGREE PROGRAMS DESCRIPTION
The University of Southern Maine grants both B.A. and B.S. degrees in a number of areas and M.A. and M.S. degrees in selected areas. Over half of the undergraduate students are non-traditional and over three quarters of the graduate students are non-traditional.

TRANSFER INFORMATION
Transfer via Articulation Agreements
Admission to engineering: At any time
Transfer without Articulation Agreements
Admission to engineering: At any time
Engineering graduates transferred from   . . 4-yr: –   2-yr: –

GRADUATION REQUIREMENTS
A 2.00 overall average and a 2.00 in the major, based on a 4.00 system are required for graduation.

■ STUDENT PROGRAMS

PROFESSIONAL AND HONORARY SOCIETIES
[For key to acronyms, see *Introduction*.]
Professional Societies: ACS, IEEE
Honorary Societies: Alpha Epsilon

SUPPORT PROGRAMS
Other Engineering Support Programs: Engineering students are initially counseled by the professional advising staff. Upon matriculation in the engineering program the are assigned a faculty advisor with whom they meet periodically throughout their time in the program.

216 Southern Methodist University

INSTITUTION PROFILE

HEAD OF THE INSTITUTION
A. Kenneth Pye
Phone: (214)768-3300   Fax: (214)768-4138

GENERAL INFORMATION
[All Students—Fall 1991]
Undergraduate enrollment 5,500
Graduate enrollment 3,300
Total institution enrollment 8,800
Type of institution: Private   **Calendar:** Semesters
Nearest city: Dallas, TX   **Population:** 1,000,000
Miles from main campus: 5   **Setting:** Suburban
Types of engineering degrees: Engineering
Other degree-granting colleges: Dedman College (Arts, Humanities and Sciences), Cox School of Business, Meadows School of the Arts, Law, Theology

ENGINEERING COLLEGE ADMISSIONS INFORMATION
Admission to the engineering college: Complete 24 semester hours and achieve a 2.0 or higher GPA in 5 courses consisting of first year English, a year of calculus and one semester of science.
Entrance Requirements: High School courses—Mathematics (4 years), Chemistry (1 year), Physics (1 year), Foreign Language (2 years), Submission of SAT or ACT required; SAT preferred.
Entrance Recommendations: High School courses—English (4 years), Biology (1 year), Social Studies (2 years), Computer Science or Programming (1 year)
Requirements for foreign students: TOEFL (Score: 550); financial statement

DEPARTMENTAL ADMISSIONS INFORMATION
Admission to the engineering department: Same as for Engineering College.

FINANCIAL INFORMATION

ESTIMATED EXPENSES (ACADEMIC YEAR)
[Expenses are for the 1992-93 nine-month academic year.]

All Students	Undergraduate	Graduate
Tuition and fees	$12,688 [2]	$12,688 [1]
College room and board	$ 4,896	$ 4,896
Books and supplies	$ 1,000	$ 1,000
Other expenses	$ — [A]	$ — [A]
Total estimated expenses	**$18,584**	**$18,584**

Notes: (A) Data not available. (1) Assumes 12 semester hour load. (2) Covers 12 to 18 semester hour load.

GENERAL FINANCIAL AID INFORMATION
Forms accepted/required: Institutional, For all need-based aid:, Financial Aid Form filed with College Scholarship Service
Additional financial aid information: Undergraduate financial aid available includes honors scholarships, need-based grants, loans and part-time on-campus employment.

ENGINEERING COLLEGE INFORMATION

[For additional personnel, refer to the *Appendix*.]
HEAD OF ENGINEERING
André G. Vacroux
Phone: (214)768-3050   Fax: (214)768-3845
EMail: dorle@seas.smu.edu

ENGINEERING COLLEGE ADDRESS
School of Engineering & Applied Science
Southern Methodist University
115 Caruth Hall
3145 Dyer Street
Dallas, TX 75275-0335
Phone: (214)768-3100   Fax: (214)768-3883

ENROLLMENTS—BY CLASS

[Numbers are baccalaureate enrollments for the fall 1991 term, unless otherwise footnoted.]
1st-year students/Freshmen .... [1]
2nd-year students/Sophomores 57
3rd-year students/Juniors 50
4th-year students/Seniors 80
Total **187**

Notes: (1) First year students not admitted directly to the engineering school. There were 157 pre-engineering students.

NUMBER OF DEGREES AWARDED—BY PROGRAM

[Numbers are engineering baccalaureate degrees awarded during the 1991-1992 academic year, unless otherwise footnoted. For full details about each engineering program, refer to the *Tables of Degree Programs*.]
Civil Engineering 10
Computer Engineering 8
Computer Science 11
Electrical Engineering 20
Management Science 8
Mechanical Engineering 15
Total **72**

PERCENTAGE OF DEGREES AWARDED—BY CATEGORY

[Percentages are of all engineering baccalaureate degrees awarded during the 1991-1992 academic year, unless otherwise footnoted.]
African Americans 1.8%
Asian/Pacific Island Americans 7.0%
Hispanic Americans 3.5%
Native Americans .... — %
Foreign Citizens 8.8%
All Others 78.9%
Women 24.6%
Persons with disabilities .... — % [A]
Students over 25 years of age .... — % [A]

Notes: (A) Data not available.

GRADUATE ENROLLMENTS & DEGREES AWARDED

Master's enrollment .... 170 [3]
Master's degrees awarded .... 67 [4]
Doctoral enrollment .... 136 [5]
Doctoral degrees awarded .... 13 [6]

Notes: (3) Enrollment in applied science master's programs was 227. (4) Applied science master's degrees awarded were 60. (5) Enrollment in applied science doctoral programs was 56. (6) Doctoral degrees awarded in applied sciences was 6.

ENGINEERING STUDENT DATA

Applicants to the engineering college
Number of applicants to engineering college .... 533 [1]
Percent offered admission 64.0%

Matriculated engineering students
Percentage in top quartile (25%) of High School class 75.3
Average SAT scores: Math—615, Verbal—520, Combined—1135

Notes: (1) Data are for Fall 1991.

FULL- & PART-TIME FACULTY—BY DEPARTMENT

[Figures are the head count of full-time faculty and the full-time equivalent (FTE) of part-time faculty for each engineering department or equivalent.]

Department	Full	Part
Electrical Eng	18	1.0
Computer Science & Eng	13	1.0
Mechanical Eng	11	1.4

COLLEGE DESCRIPTION

The School of Engineering and Applied Science offers traditional engineering programs and also supports flexible specializations in conjunction with several liberal arts and science emphases. Options are offered in biomedical engineering and pre-med to technically oriented students. SMU's co-op program takes advantage of the proximity of the approximately 800 high tech corporations in the Dallas-Fort Worth metroplex. For instance, a student co-oping at the nearby Superconducting Super Collider could earn a dual degree (EE + physics) requiring only three more courses than the standard BSEE. Small classes are taught by full-time faculty; professors are accessible outside of class; undergraduate research opportunities are available. SMU students come from every state in the Union and as many as 70 foreign countries. Study abroad programs are also available.

DEGREE PROGRAMS DESCRIPTION

Computer Engineering BSCpE, MSCpE, PhD; Computer Science BS, MS, PhD; Electrical Engineering BSEE, MSEE ,PhD; Mechanical Engineering BSME, MSME, PhD; Management Science BS; Engineering Management MSEM, DEng; Operations Research MS, PhD; Telecommunications Systems Management MS; Materials Science & Engineering MS; Hazardous & Waste Materials Management MS.

DUAL DEGREE PROGRAMS

Engineering baccalaureate graduates with dual degrees 4
Enrollment requirements: Honors standing.

TRANSFER INFORMATION

Residency Requirements: 60 semester hours must be earned in residence, including 30 in the major department or interdisciplinary program; of the last 60 semester hours, 45 must be taken in residence.

Transfer via Articulation Agreements
Admission to engineering: At any time
Engineering graduates transferred from ... 4-yr: –   2-yr: 2
Requirements: Evaluation of transcript.

Transfer without Articulation Agreements
Admission to engineering: At any time
Engineering graduates transferred from ... 4-yr: 8   2-yr: –
Requirements: Evaluation of transcript.

GRADUATION REQUIREMENTS

Completion of: (1) the required semester credit hours; (2) GPAs of at least 2.0 (a) overall, (b) in all courses taken at SMU, and (c) in the major field courses; and (3) engineering design requirements.

STUDENT PROGRAMS

PROFESSIONAL AND HONORARY SOCIETIES
[For key to acronyms, see *Introduction*.]
Professional Societies: ACM, ASEM, ASME, IEEE, ORSA, TIMS
Honorary Societies: Eta Kappa Nu, Pi Tau Sigma, Tau Beta Pi, Theta Tau

SUPPORT PROGRAMS
Student Chapter Organizations: Society of Women Engineers, National Society of Black Engineers, Society of Hispanic Professional Engineers

Minority Co-op Program
For: Ethnic Minorities   **Available:** Year round
Offered: Prematriculation, Freshman, Sophomore, Junior, Senior

Intercultural Education & Minority Student Serv.
For: Ethnic Minorities   **Available:** Year round
Offered: Freshman, Sophomore, Junior, Senior

Learning Enhancement Center
For: Ethnic Minorities   **Available:** Year round
Offered: Prematriculation, Freshman, Sophomore, Junior, Senior

Other Engineering Support Programs: Support programs and services include freshman mentoring, Learning Enhancement Center academic skills training and tutorial services, Tau Beta Pi tutoring and faculty advising.

217 Stanford University

INSTITUTION PROFILE

HEAD OF THE INSTITUTION
Gerhard R. Casper
Phone: (415)723-2481 Fax: (415)725-6847

GENERAL INFORMATION
[All Students—Fall 1991]
Undergraduate enrollment 6,564
Graduate enrollment 7,329
Total institution enrollment **13,893**
Type of institution: Private Calendar: Quarters
Nearest city: San Francisco Population: 750,000
Miles from main campus: 35 Setting: Suburban
Types of engineering degrees: Engineering
Other degree-granting colleges: Business Administration, Education, Humanities & Social Sciences, Law, Medicine, Earth Sciences

ENGINEERING ADMISSIONS

ADMISSIONS OFFICE CONTACT
James M. Montoya
Stanford University
Office of Undergraduate Admissions
Old Union
Stanford, CA 94035-3005
Phone: (415)723-2091

ENGINEERING COLLEGE ADMISSIONS INFORMATION
Admission to the engineering college: Freshman are admitted to the university, not to the Engineering school. All majors are open. There are no special admissions requirements for engineering.
Entrance Requirements: All applicants must complete either the SAT or the ACT.
Entrance Recommendations: Three of the College Board Achievement Tests are recommended, one of which preferably should be English Composition.
Requirements for foreign students: TOEFL strongly recommended if applicant has lived in an English-speaking country for fewer than five years, or if English is not the applicant's primary language.

DEPARTMENTAL ADMISSIONS INFORMATION
Admission to the engineering department: Students declare majors by the end of the second year. All majors are open to all undergrads. There are no special admissions requirements for engineering.

FINANCIAL INFORMATION

ESTIMATED EXPENSES (ACADEMIC YEAR)
[Expenses are for the 1992-93 nine-month academic year.]

All Students	Undergraduate	Graduate
Tuition and fees	$16,536	$17,706
College room and board	$ 6,313	$ 7,980
Books and supplies	$ 792	$ 948
Other expenses	$ 1,278	$ 855
Total estimated expenses	**$24,919**	**$27,489**

FINANCIAL AID OFFICE CONTACT
Robert P. Huff, *Director of Financial Aids*
Stanford University
Old Union Room 214
Stanford, CA 94305-3021
Phone: (415)723-3058

GENERAL FINANCIAL AID INFORMATION
Forms accepted/required: FAF, FAT, IRS, Student Aid Application For California (California residents), Stanford Entering Student Financial Aid Supplemental Application
Additional financial aid information: Term-Time Job Eligibility; Federal - College Work Study (based on exceptional financial need); Stanford - Department funded (based on need) Student Loans: Federal - Perkins; State/Federal - Stafford, SLS; Stanford: University long term, Stanford Supplemental Student Loan Program Scholarships and Grants: Stanford - endowment, current gifts, operating budget; State - Cal Grant A, Cal Grant B; Federal - SEOG, Pell grant, BIA

ENGINEERING COLLEGE INFORMATION
[For additional personnel, refer to the *Appendix*.]

HEAD OF ENGINEERING
James F. Gibbons
Phone: (415)723-3938 Fax: (415)723-5599
EMail: Gibbons@Sierra.Stanford.Edu

ENGINEERING COLLEGE ADDRESS
School of Engineering
Stanford University
Terman Engineering Center
Stanford, CA 94305-4027
Phone: (415)723-3935 Fax: (415)723-5599

ENROLLMENTS—BY CLASS
[Numbers are baccalaureate enrollments for the fall 1991 term, unless otherwise footnoted.]
1st-year students/Freshmen .... (B)
2nd-year students/Sophomores 27
3rd-year students/Juniors 169
4th-year students/Seniors 368
Total **564**
Notes: (B) Data not applicable.

NUMBER OF DEGREES AWARDED—BY PROGRAM
[Numbers are engineering baccalaureate degrees awarded during the 1991-1992 academic year, unless otherwise footnoted. For full details about each engineering program, refer to the *Tables of Degree Programs*.]
Aeronautical & Astronautical Engineering .... –
Chemical Engineering 12
Civil Engineering 19
Computer Science 46
Electrical Engineering 63
Engineering-Economic Systems .... –
Industrial Engineering & Engineering Management 52
Materials Science & Engineering 1
Mechanical Engineering 59
Operations Research .... –
Total **252**

PERCENTAGE OF DEGREES AWARDED—BY CATEGORY
[Percentages are of all engineering baccalaureate degrees awarded during the 1991-1992 academic year, unless otherwise footnoted.]
African Americans .... 9.0 %
Asian/Pacific Island Americans .... 19.9 %
Hispanic Americans .... 9.6 %
Native Americans .... 1.4 %
Foreign Citizens .... 5.7 %
All Others .... 54.4 %
Women .... 21.3 %
Persons with disabilities .... – % (A)
Students over 25 years of age .... – % (A)
Notes: (A) Data not available.

GRADUATE ENROLLMENTS & DEGREES AWARDED
Master's enrollment 1,284
Master's degrees awarded 818
Doctoral enrollment 1,271
Doctoral degrees awarded 222

ENGINEERING STUDENT DATA
Applicants to the engineering college
Number of applicants to engineering college .... – (3)
Percent offered admission .... –% (3)
Matriculated engineering students
Percentage in top quartile (25%) of High School class 85.0 (4)
Notes: (3) Students are accepted to the University, not to individual schools.
(4) Data was not available for 14% of the freshman class.

FULL- & PART-TIME FACULTY—BY DEPARTMENT
[Figures are the head count of full-time faculty and the full-time equivalent (FTE) of part-time faculty for each engineering department or equivalent.]

Department	Full	Part
Chemical Eng	10	–
Civil Eng	26	1.0
Electrical Eng	49	–
Industrial Eng & Eng Management	10	–
Mechanical Eng	32	1.0
Computer Science	33	2.0
Materials Science & Eng	9	–
Aeronautics & Astronautics	17	–
Eng-Economic Systems	11	–
Operations Research	9	–

COLLEGE DESCRIPTION
Stanford's engineering education emphasizes fundamentals, embedded in a rich liberal arts undergraduate curriculum. Curricula are notable for their flexibility and breadth. In addition to traditional engineering majors, students may enroll in interdisciplinary majors, or they may design their own majors. Engineering students have the opportunity to study in overseas programs, including two programs in Berlin and Kyoto especially designed for engineers, and to explore other disciplines through honors programs and double majors. The University offers the resources of a faculty who are both teachers and leading scholars, and research opportunities abound. The student body at Stanford is quite diverse, creating a rich intellectual and social environment.

DEGREE PROGRAMS DESCRIPTION
B.S. programs in engineering are characterized by flexibility and an emphasis on fundamentals. The Coterminal Program allows students to complete both B.S. and M.S. degrees coterminally after five years. M.S. programs are professionally oriented and may be completed in one academic year. They are course-based and do not require theses. There is an extraordinary diversity of Ph.D. programs, and interdisciplinary work is encouraged.

DUAL DEGREE PROGRAMS
Enrollment requirements: Stanford does have a 3/2 program leading to an A.B. from a liberal arts institution and a B.S. from Stanford. However, 3/2 candidates must apply as regular transfer applicants and are expected to meet the same admissions standards as all other transfer applicants.

TRANSFER INFORMATION
Residency Requirements: A maximum of 90 quarter units may be transferred (out of a total of 180 for graduation).
Transfer without Articulation Agreements
Admission to engineering: Transfer students are admitted to the university. They may declare any major, including engineering.
Engineering graduates transferred from   . . 4-yr: –   2-yr: –

GRADUATION REQUIREMENTS
All students must satisfy four general requirements: the Writing Requirement, the Distribution Requirements, the Foreign Language Requirement, and Major requirements. Engineering graduates must achieve a 2.0 GPA for all engineering coursework (engineering fundamentals and engineering depth).

STUDENT PROGRAMS

PROFESSIONAL AND HONORARY SOCIETIES
[For key to acronyms, see *Introduction*.]
Professional Societies: AGCA, AIAA, AIChE, ASCE, ASME, IEEE, IIE, ORSA
Honorary Societies: Tau Beta Pi

SUPPORT PROGRAMS
Student Chapter Organizations: Society of Women Engineers, Society of Black Scientists and Engineers, Society of Chicano/Latino Engineers and Scientists, American Indian Science and Engineering Society

Science, Engineering, Math Achievement Program
For: Ethnic Minorities **Available:** Academic year
Offered: Freshman, Sophomore

Minority Women in Science and Engineering
For: Women & Ethnic Minorities **Available:** Academic year
Offered: Freshman, Sophomore, Junior, Senior, Graduate level

Pre-College Science and Math Program
For: Ethnic Minorities **Available:** Year round
Offered: Prematriculation

Academic Support Tutorials
For: Ethnic Minorities **Available:** Year round
Offered: Freshman, Sophomore, Junior, Senior, Graduate level

Other Engineering Support Programs: Students find support from a variety of sources on campus, including the Undergraduate Advising Center, the Center for Teaching and Learning, and Counseling & Psychological Services. Tau Beta Pi, the engineering honor society, provides free tutoring for Engineering Fundamentals coursework.

218 State University of New York at Binghamton

■ INSTITUTION PROFILE
HEAD OF THE INSTITUTION
Lois B. DeFleur
Phone: (607)777-2131   Fax: (607)777-4000

GENERAL INFORMATION
[All Students—Fall 1991]
Undergraduate enrollment 8,928
Graduate enrollment 2,955
Total institution enrollment 11,883
Type of institution: Public   Calendar: Semesters
Nearest city: Binghamton   Population: 80,000
Miles from main campus: 5   Setting: Suburban
Types of engineering degrees: Engineering
Other degree-granting colleges: Arts & Sciences, Business Administration, Education, Nursing

■ ENGINEERING ADMISSIONS
ADMISSIONS OFFICE CONTACT
Michael F. McGoff
State University of New York at Binghamton
The Watson School of Engineering
Binghamton, NY 13902-6000
Phone: (607)777-6203   Fax: (607)777-4822

ENGINEERING COLLEGE ADMISSIONS INFORMATION
Admission to the engineering college: Engineering Programs - Upper Division Only

DEPARTMENTAL ADMISSIONS INFORMATION
Admission to the engineering department: Junior level status
Additional information: Applicants are admitted to Junior level status in the Watson School after having completed the A.S. in Engineering Science or equivalent. Students are accepted directly into either the department of Electrical Engineering or Mechanical Engineering.

■ FINANCIAL INFORMATION
ESTIMATED EXPENSES (ACADEMIC YEAR)
[Expenses are for the 1992-93 nine-month academic year.]

State Residents	Undergraduate	Graduate
Tuition and fees	$ 2,957	$ 4,199
College room and board	$ 4,760	$ 5,580
Books and supplies	$ 600	$ 600
Other expenses	$ 1,103	$ 3,311
Total estimated expenses	**$ 9,420**	**$13,690**
Out-of-State Residents	Undergraduate	Graduate
Tuition and fees	$ 6,857	$ 7,515
College room and board	$ 4,760	$ 5,580
Books and supplies	$ 600	$ 600
Other expenses	$ 1,303	$ 3,315
Total estimated expenses	**$13,520**	**$17,010**

FINANCIAL AID OFFICE CONTACT
Christina M. Knickerbocker, *Director*
State University of New York at Binghamton
Student Financial Aid and Employment
P.O. Box 6000
Binghamton, NY 13902-6000
Phone: (607)777-2428

GENERAL FINANCIAL AID INFORMATION
Forms accepted/required: FAF, FAT
Additional financial aid information: All Title IV campus-based programs are offered: supplemental educational opportunity grants (SEOG), Perkins Loans, Stafford Loans, College Work-study. Other assistance is available through Pell Grants, Supplemental Loans to Students (SLS), Parent Loans (PLUS), New York State Tuition Assistance Program (TAP) and various University grants.

■ ENGINEERING COLLEGE INFORMATION
[For additional personnel, refer to the *Appendix*.]

HEAD OF ENGINEERING
Lyle D. Feisel
Phone: (607)777-2871   Fax: (607)777-4822
EMail: Feisel@BINGTJW

ENGINEERING COLLEGE ADDRESS
The Watson School of Engineering and Applied Science
State University of New York at Binghamton
P.O. Box 6000
Binghamton, NY 13902-6000
Phone: (607)777-6203   Fax: (607)777-4822

ENROLLMENTS—BY CLASS
[Numbers are baccalaureate enrollments for the fall 1991 term, unless otherwise footnoted.]
1st-year students/Freshmen .... 0 (B)
2nd-year students/Sophomores 50
3rd-year students/Juniors 120
4th-year students/Seniors 240
Total 410
Notes: (B) Data not applicable.

NUMBER OF DEGREES AWARDED—BY PROGRAM
[Numbers are engineering baccalaureate degrees awarded during the 1991-1992 academic year, unless otherwise footnoted. For full details about each engineering program, refer to the *Tables of Degree Programs*.]
Computer Science 71
Electrical Engineering 48
Industrial Engineering .... – (B)
Industrial Technology 37
Industrial Technology .... –
Mechanical Engineering 35
Systems Science .... – (B)
Total 191
Notes: (B) Data not applicable.

PERCENTAGE OF DEGREES AWARDED—BY CATEGORY
[Percentages are of all engineering baccalaureate degrees awarded during the 1991-1992 academic year, unless otherwise footnoted.]
African Americans .... 2.0 %
Asian/Pacific Island Americans .... 2.0 %
Hispanic Americans .... 1.0 %
Native Americans .... – %
Foreign Citizens .... 4.0 %
All Others .... 91.0 %
Women .... 11.0 %
Persons with disabilities .... – % (A)
Students over 25 years of age .... – % (A)
Notes: (A) Data not available.

GRADUATE ENROLLMENTS & DEGREES AWARDED
Master's enrollment 188
Master's degrees awarded 80
Doctoral enrollment 36
Doctoral degrees awarded 7

FULL- & PART-TIME FACULTY—BY DEPARTMENT
[Figures are the head count of full-time faculty and the full-time equivalent (FTE) of part-time faculty for each engineering department or equivalent.]

Department	Full	Part
Electrical Eng	17	1.1
Mechanical & Industrial Eng	14	–
Computer Science	13	1.7
Systems Science	6	–

COLLEGE DESCRIPTION
Engineering programs have small classes. Individual attention is the norm. Laboratory facilities are excellent. Strong connection with the local high-tech industry. Superior quality of faculty research in electronics packaging recently recognized by the National Science Foundation when it funded a major Cooperative Research Center at the Watson School. Upper division engineering program students come from all over New York State. The School has active 2/2 and 3/2 programs with many other colleges. The HPEO program (Harpur Pre-Engineering Option) enables students to attend SUNY Binghamton's prestigious Harpur College for their first two years of study. Watson has both 3/2 and 2/2 programs with Harpur. There is an NSF funded graduate Engineering Education partnership with Morgan State University. No traditional co-op program, but many students are employed part-time and during summer in high-tech firms.

DEGREE PROGRAMS DESCRIPTION
Master of Science Electrical Engineering: Minimum of 30 credit hours. Four courses required in the selected field of specialization. Two electrical engineering courses outside the selected field of specialization. One technical elective which can be from another department. Thesis option with seminar and oral examination requirements, and preparation of a publishable paper, or non-thesis option with one additional course in each category above. Mechanical Engineering: Minimum of 30 credit hours. Three core courses in specified areas. Thesis option requires a total of eight courses and oral examination of thesis content. Non-thesis option requires one additional course plus a technical project and oral examination of project content. Industrial Engineering: minimum of 31 credit hours. Six core courses are required. Thesis option requires a total of eight courses and oral examination of thesis content. Non-thesis option requires a total of ten courses and a technical project.

DUAL DEGREE PROGRAMS
Engineering baccalaureate graduates with dual degrees 6
Enrollment requirements: Coursework equivalent to an Associate of Science in Engineering Science prior to entering junior level engineering curricula.

TRANSFER INFORMATION
Residency Requirements: Minimum of 30 credits completed at State University of New York at Binghamton, 15 of which must be Watson School credits.

Transfer via Articulation Agreements
Admission to engineering: At the end of the sophomore year Engineering graduates transferred from . . . 4-yr: 37   2-yr: 47
Requirements: Associate of Science degree in Engineering Science or equivalent. (The 1991-92 figures for BS engineering graduates from 4-year institutions includes 25 from Binghamton's Arts and Sciences College's pre-engineering option.)

Transfer without Articulation Agreements
Admission to engineering: There is a pre-engineering option with Harpur College of Arts and Sciences at SUNY-Binghamton.
Requirements: Associate of Science degree in Engineering Science or equivalent.

GRADUATION REQUIREMENTS
To receive a degree from the Watson School an undergraduate must successfully complete all degree requirements with a cumulative grade point average of 2.0. A minimum of 30 credit hours (15 of which must be in Watson School courses) must be completed at the State University of New York at Binghamton to meet the school's residency requirement.

■ STUDENT PROGRAMS
PROFESSIONAL AND HONORARY SOCIETIES
[For key to acronyms, see *Introduction*.]
Professional Societies: ACM, ASME, IEEE, SAE, SME, Society of Women Engineers, National Society of Black Engineers
Honorary Societies: Eta Kappa Nu, Tau Beta Pi, Upsilon Pi Epsilon

SUPPORT PROGRAMS
Student Chapter Organizations: Society of Women Engineers, National Society of Black Engineers, Collegiate Science and Technology Entry Program

Collegiate Science & Technology Entry Program
For: Ethnic Minorities   Available: Academic year
Offered: Freshman, Sophomore, Junior, Senior

National Society of Black Engineers
For: Ethnic Minorities   Available: Academic year
Offered: Freshman, Sophomore, Junior, Senior

Society of Women Engineers
For: Women   Available: Academic year
Offered: Freshman, Sophomore, Junior, Senior

Other Engineering Support Programs: PREP time introduces new EE/ME juniors to their program and peers. CSTEP provides tutoring, academic counseling, industry visits, guest speakers and a resource room. Eligible students are economically disadvantaged and/or Black, Hispanic, Native American or Native Alaskan. NSBE provides academic and social support plus career counseling for its members.

219 State University of New York at Buffalo

■ INSTITUTION PROFILE
HEAD OF THE INSTITUTION
William R. Greiner
Phone: (716)645-2901 Fax: (716)645-3728

GENERAL INFORMATION
[All Students—Fall 1991]
Undergraduate enrollment 17,263
Graduate enrollment 8,752
Total institution enrollment **26,015**
Type of institution: Public **Calendar:** Semesters
Nearest city: Buffalo **Population:** 339,000
Miles from main campus: 3 **Setting:** Urban
Types of engineering degrees: Engineering
Other degree-granting colleges: Millard Fillmore College (Evening Division)

■ ENGINEERING ADMISSIONS
ADMISSIONS OFFICE CONTACT
Kevin Durkin
State University of New York at Buffalo
Director of Admissions
Hayes Annex A
Buffalo, NY 14214
Phone: (716)829-2333 Fax: (716)829-3902

ENGINEERING COLLEGE ADMISSIONS INFORMATION
Entrance Requirements: SAT, ACT, High School courses—College preparatory mathematics (3 years), College preparatory science (3 years), English (4 years), Either the SAT or the ACT test scores may be submitted. Admission is determined based upon the combination of SAT or ACT score, class rank and class average.
Entrance Recommendations: High School courses—4th year of mathematics (4 years), Foreign language (3 years), Mechanical drawing (1 year), SAT math achievement Advanced Placement math
Requirements for foreign students: TOEFL (Score: 550); financial statement; One year in US.

DEPARTMENTAL ADMISSIONS INFORMATION
Admission to the engineering department: At the end of the second year.
Additional information: Departmental admission occurs during sophomore year and is based on average in technical courses. Minimum 2.0 GPA in technical and engineering courses required with some programs requiring higher average.

■ FINANCIAL INFORMATION
ESTIMATED EXPENSES (ACADEMIC YEAR)
[Expenses are for the 1992-93 nine-month academic year.]

State Residents	Undergraduate	Graduate
Tuition and fees	$ 3,070	$ 4,305
College room and board	$ 4,277	$ 4,277
Books and supplies	$ 683	$ 683
Other expenses	$ 1,785 [1]	$ 1,785 [1]
Total estimated expenses	**$ 9,815**	**$11,050**
Out-of-State Residents	Undergraduate	Graduate
Tuition and fees	$ 6,970	$ 7,621
College room and board	$ 4,277	$ 4,277
Books and supplies	$ 683	$ 683
Other expenses	$ 1,785 [1]	$ 1,785 [1]
Total estimated expenses	**$13,715**	**$14,366**

Notes: (1) Estimate personal expenses and transportation.

FINANCIAL AID OFFICE CONTACT
Michael D. Randall, *Director*
State University of New York at Buffalo
Office of Financial Aid
Hayes Annex C
Buffalo, NY 14214
Phone: (716)829-3724 Fax: (716)829-2022

GENERAL FINANCIAL AID INFORMATION
Forms accepted/required: AFSA, Verification documents as requested, Separate TAP application
Additional financial aid information: Perkins Loans, college work study, need-based scholarships, Pell, TAP and student loans.

■ ENGINEERING COLLEGE INFORMATION
[For additional personnel, refer to the *Appendix*.]
HEAD OF ENGINEERING
George C. Lee
Phone: (716)645-2771 Fax: (716)645-2495
EMail: FEAGCLEE@ubvms.cc.buffalo.edu

ENGINEERING COLLEGE ADDRESS
School of Engineering and Applied Sciences
State University of New York at Buffalo
410 Bonner Hall
Buffalo, NY 14260-1900
Phone: (716)645-2774 Fax: (716)645-2495

ENROLLMENTS—BY CLASS
[Numbers are baccalaureate enrollments for the fall 1991 term, unless otherwise footnoted.]
1st-year students/Freshmen 716
2nd-year students/Sophomores 460
3rd-year students/Juniors 409
4th-year students/Seniors 835
Total **2,420**

NUMBER OF DEGREES AWARDED—BY PROGRAM
[Numbers are engineering baccalaureate degrees awarded during the 1991-1992 academic year, unless otherwise footnoted. For full details about each engineering program, refer to the *Tables of Degree Programs*.]
Aerospace Engineering 36
Chemical Engineering 21
Civil Engineering 65
Electrical Engineering 138
Engineering Physics 3
Engineering Science(s) .... —
Industrial Engineering 62
Mechanical Engineering 157
Total **482**

PERCENTAGE OF DEGREES AWARDED—BY CATEGORY
[Percentages are of all engineering baccalaureate degrees awarded during the 1991-1992 academic year, unless otherwise footnoted.]
African Americans 1.8%
Asian/Pacific Island Americans 6.8%
Hispanic Americans 1.0%
Native Americans .... — %
Foreign Citizens 6.0%
All Others 84.4%
Women 10.8%
Persons with disabilities .... — %
Students over 25 years of age .... — %

GRADUATE ENROLLMENTS & DEGREES AWARDED
Master's enrollment 350
Master's degrees awarded 184
Doctoral enrollment 452
Doctoral degrees awarded 39

ENGINEERING STUDENT DATA
Applicants to the engineering college
Number of applicants to engineering college 2,536
Percent offered admission 59.6%

Matriculated engineering students
Percentage in top quartile (25%) of High School class 91.0
Average SAT scores: Math—622, Verbal—499, Combined—1121
Average ACT scores: Composite—26

FULL- & PART-TIME FACULTY—BY DEPARTMENT
[Figures are the head count of full-time faculty and the full-time equivalent (FTE) of part-time faculty for each engineering department or equivalent.]

Department	Full	Part
Chemical Eng	13	0.3
Civil Eng	28	1.0
Electrical & Computer Eng	30	1.4
Industrial Eng	11	0.4
Mechanical & Aerospace Eng	24	1.4

COLLEGE DESCRIPTION
The University at Buffalo has the principal engineering school in the SUNY system. It offers six four-year ABET accredited undergraduate programs plus an Engineering Physics program (jointly offered with Physics). The first year of each of the BS programs is common, with most specialization occurring in the third and fourth years. Particularly noteworthy are the design orientation of each of the programs and the high degree of computer integration. Minors are available with many other programs. Many students transfer from community colleges throughout the State and from four-year liberal arts colleges (as part of articulated 3/2 programs). On average, the BS degree program is completed in 4.5 years. Outstanding facilities are available on the University's new Amherst Campus.

DEGREE PROGRAMS DESCRIPTION
The BS degree programs provide qualifications both for immediate employment and for subsequent graduate study. Graduate study is available leading to degrees of Master of Engineering, Master of Science and Doctor of Philosophy. Graduates find employment nationally and internationally in major corporations. Several internationally known research centers provide exceptional opportunities for advanced study.

TRANSFER INFORMATION
Residency Requirements: Students must complete at least 32 credit hours in residence.
Transfer via Articulation Agreements
Admission to engineering: Transfer programs are articulated with Engineering Science programs in community colleges. Dual degrees are articulated on a 3-2 basis with four year liberal arts colleges.
Engineering graduates transferred from . . 4-yr: 29 2-yr: 70
Transfer without Articulation Agreements
Admission to engineering: At any time
Engineering graduates transferred from . . 4-yr: 40 2-yr: 42

GRADUATION REQUIREMENTS
BS programs in engineering require 132-140 semester credits, depending on major; minimum 2.0 overall and engineering averages required.

■ STUDENT PROGRAMS
PROFESSIONAL AND HONORARY SOCIETIES
[For key to acronyms, see *Introduction*.]
Professional Societies: AIAA, AIChE, ASCE, ASME, BMES, IEEE, IIE, NSPE, SAE, SME, Society of Women Engineers (SWE), National Society of Black Engineers (NSBE)
Honorary Societies: Alpha Pi Mu, Chi Epsilon, Eta Kappa Nu, Omega Chi Epsilon, Pi Tau Sigma, Sigma Gamma Tau, Tau Beta Pi, Omega Rho

SUPPORT PROGRAMS
Student Chapter Organizations: Society of Women Engineers, National Society of Black Engineers, Society of Hispanic Professional Engineers

Engineering minority affairs staff
For: Ethnic Minorities **Available:** Year round
Offered: Freshman, Sophomore, Junior, Senior

Special tutoring services
For: Ethnic Minorities **Available:** Academic year
Offered: Freshman, Sophomore

Other Engineering Support Programs: Summer orientation program for all freshmen and transfers; separate engineering advisement staff; career planning and placement services; personal counseling service; veterans affairs office; office for handicapped services; office for international education.

220 State University of New York, College of Environmental Science & Forestry

■ INSTITUTION PROFILE
HEAD OF THE INSTITUTION
Ross S. Whaley
Phone: (315)470-6681   Fax: (315)470-6932
GENERAL INFORMATION
[All Students—Fall 1991]
Undergraduate enrollment 1,197
Graduate enrollment 661
Total institution enrollment **1,858**
Type of institution: Public   **Calendar:** Semesters
Location: Syracuse, NY   **Population:** 163,860
Setting: Urban
Types of engineering degrees: Engineering
Other degree-granting colleges: Landscape Architecture, Chemistry, Environmental and Forest Biology, Forestry, Environmental Studies

■ ENGINEERING ADMISSIONS
ADMISSIONS OFFICE CONTACT
Dennis O. Stratton
State University of New York, College of Environmental Science and Forestry
Director of Admissions
106 Bray Hall
Syracuse, NY 13210-2779
Phone: (315)470-6600   Fax: (315)470-6933
ENGINEERING COLLEGE ADMISSIONS INFORMATION
Entrance Requirements: SAT (Score: 1050), High School courses—Mathematics (4 years), Science (4 years), SAT or ACT required.
Entrance Recommendations: High School courses—Chemistry (1 year), Physics (1 year)
Requirements for foreign students: TOEFL (Score: 550); 1.) Essay 2.) Teacher Recommendation
DEPARTMENTAL ADMISSIONS INFORMATION
Admission to the engineering department: Admitted to FEG and PSE as freshmen, sophomores or juniors. Admitted to WPE as juniors only.
Additional information: Each department requires specific coursework for transfer students.

■ FINANCIAL INFORMATION
ESTIMATED EXPENSES (ACADEMIC YEAR)
[Expenses are for the 1992-93 nine-month academic year.]

State Residents	Undergraduate	Graduate
Tuition and fees	$ 2,937	$ 4,243
College room and board	$ 5,920	$ 5,920
Books and supplies	$ 600	$ 600
Other expenses	$ 650	$ 1,090
Total estimated expenses	**$10,107**	**$11,853**

Out-of-State Residents	Undergraduate	Graduate
Tuition and fees	$ 6,837	$ 7,559
College room and board	$ 5,920	$ 5,920
Books and supplies	$ 600	$ 600
Other expenses	$ 850	$ 1,290
Total estimated expenses	**$14,207**	**$15,369**

FINANCIAL AID OFFICE CONTACT
John E. View, *Director of Financial Aid and EOP Program*
State University of New York, College of Environmental Science and Forestry
115 Bray Hall
1 Forestry Drive
Syracuse, NY 13210
Phone: (315)470-6670   Fax: (315)470-6933
GENERAL FINANCIAL AID INFORMATION
Forms accepted/required: ESF College Aid Application, FAFSA - Free Application for Federal Student Aid
Additional financial aid information: 1.) ESF - College Aid 2.) Work Study

■ ENGINEERING COLLEGE INFORMATION
[For additional personnel, refer to the *Appendix*.]
HEAD OF ENGINEERING
Robert H. Brock, *Director*
Phone: (315)470-6633   Fax: (315)470-6958
ENGINEERING COLLEGE ADDRESS
Division of Engineering
State University of New York, College of Environmental Science and Forestry
312 Bray Hall
1 Forestry Drive
Syracuse, NY 13210-2778
Phone: (315)470-6633   Fax: (315)470-6958
ENROLLMENTS—BY CLASS
[Numbers are baccalaureate enrollments for the fall 1991 term, unless otherwise footnoted.]
1st-year students/Freshmen 35
2nd-year students/Sophomores 68
3rd-year students/Juniors 76
4th-year students/Seniors 67
Total **246**
NUMBER OF DEGREES AWARDED—BY PROGRAM
[Numbers are engineering baccalaureate degrees awarded during the 1991-1992 academic year, unless otherwise footnoted. For full details about each engineering program, refer to the *Tables of Degree Programs*.]
Environmental & Resource Engineering .... – (B)
Forest Engineering 15
Paper Science & Engineering 14
Wood Products Engineering 20
Total **49**
Notes: (B) Data not applicable.
PERCENTAGE OF DEGREES AWARDED—BY CATEGORY
[Percentages are of all engineering baccalaureate degrees awarded during the 1991-1992 academic year, unless otherwise footnoted.]
African Americans .... – %
Asian/Pacific Island Americans .... 2.0 %
Hispanic Americans .... – %
Native Americans .... – %
Foreign Citizens .... – %
All Others .... 98.0 %
Women .... 10.0 %
Persons with disabilities .... – %
Students over 25 years of age .... – % (A)
Notes: (A) Data not available.
GRADUATE ENROLLMENTS & DEGREES AWARDED
Master's enrollment 58
Master's degrees awarded 14
Doctoral enrollment 31
Doctoral degrees awarded .... –
ENGINEERING STUDENT DATA
Applicants to the engineering college
Number of applicants to engineering college 96
Percent offered admission .... 31.0 %
Matriculated engineering students
Percentage in top quartile (25%) of High School class 81.0
Average SAT scores: Math—623, Verbal—511, Combined—1134
FULL- & PART-TIME FACULTY—BY DEPARTMENT
[Figures are the head count of full-time faculty and the full-time equivalent (FTE) of part-time faculty for each engineering department or equivalent.]

Department	Full	Part
Forest Eng	8	–
Paper Science & Eng	11	–
Wood Products Eng	8	–

COLLEGE DESCRIPTION
The Div. of Eng. coordinates the activities of the individually administered Faculties of Forest Eng., Paper Sci. & Eng., and Wood Prod. Eng. Programs under the Div. of Eng. enjoy strong financial support from the Empire State Paper Research Assoc., Inc. and the Syracuse Pulp & Paper Foundation. These strong industrial ties provide access to information, technical expertise & equipment. The Wood Prod. Eng. Program has excellent facilities for modern microscopy in the N.C. Brown Center for Ultrastructure Studies. In addition, a Tropical Timber Information Center provides identification of wood samples & information about tropical woods for both general characteristics & technical properties. Special strength exists in the Forest Eng. Program in the area of natural resource sensing, measurement & analysis. A broad range of faculty expertise is available in this area in addition to specialized laboratories and instrumentation.

DEGREE PROGRAMS DESCRIPTION
The Division of Engineering at ESF has three distinct degree programs at the undergraduate (BS) level which are Forest Eng., Paper Science & Eng., & Wood Products Eng. Forest Eng. instruction focuses on: locating & quantifying resources; designing harvesting, conveyance & transportation systems & networks for water & timber; designing structures, facilities & pollution abatement systems; & engineering planning for the development of sites & regions for multiple use. Paper Science & Eng. gives a broad base of study to prepare people for professional positions in the pulp & paper industry. It provides education in the physical sciences & chemical engineering, with specific emphasis on those aspects of these disciplines which relate to the manufacture of pulp and paper. The Wood Products Eng. Program prepares students for a wide variety of professional occupations in heavy construction or in the use of wood as a material. These interests are presented in two curriculum options: Construction and Wood Science & Technology. The Division of Engineering has a graduate program (MS & PhD) in Environmental & Resource Engineering (ERE). ERE is concerned with the application of science and engineering to the conservation, restoration, holistic development, and improved utilization of the natural environment and its forest-related r

TRANSFER INFORMATION
Transfer via Articulation Agreements
Admission to engineering: Generally at the end of the freshman or sophomore year but a few transfer in the middle of the freshman or sophomore year. Wood Prod. Eng. only accepts junior transfers.
Engineering graduates transferred from   . . . 4-yr: 22   2-yr: 41
Requirements: The student must have satisfied the articulation agreement in place between their institution and ESF.
Transfer without Articulation Agreements
Admission to engineering: If space accommodates.
Engineering graduates transferred from   . . . 4-yr: 4   2-yr: 7
Requirements: The student should have a 2.3 - 2.5 minimum cumulative GPA. Varies with department.
GRADUATION REQUIREMENTS
B.S. programs require an overall grade point average of at least 2.0.

■ STUDENT PROGRAMS
PROFESSIONAL AND HONORARY SOCIETIES
[For key to acronyms, see *Introduction*.]
Professional Societies: ACM, ACS, ACSM, AGCA, ASAE, ASPRS, AWRA, TAPPI, Paper Industries Management Association
Honorary Societies: Alpha Xi Sigma
SUPPORT PROGRAMS
Student Chapter Organizations: National Society of Black Engineers, Society of Hispanic Professional Engineers, Baobab Society, Association for Women in Science
Tutoring for Non-EOP Minority Students
For: Ethnic Minorities   **Available:** Academic year
Offered: Freshman, Sophomore, Junior, Senior, Graduate level
Other Engineering Support Programs: The Forest Engineering Club, Papyrus Club and Wood Engineers Club provide support for engineering students. Freshman/transfer orientation courses/programs are required. Academic and counseling programs are available from student services.

221 State University of New York Maritime College

INSTITUTION PROFILE

HEAD OF THE INSTITUTION
Floyd H. Miller
Phone: (718)409-7270 **Fax:** (718)409-7392

GENERAL INFORMATION
[All Students—Fall 1991]
Undergraduate enrollment 722
Graduate enrollment 177
Total institution enrollment **899**
Type of institution: Public **Calendar:** Other
Nearest city: New York City **Population:** 7,000,000
Miles from main campus: 10 **Setting:** Suburban
Types of engineering degrees: Engineering

ENGINEERING ADMISSIONS

ADMISSIONS OFFICE CONTACT
Peter Cooney
State University of New York Maritime College
Director of Admissions
Fort Schuyler
Throgs Neck Station, NY 10465-4198
Phone: (718)409-7222 **Fax:** (718)409-7392

ENGINEERING COLLEGE ADMISSIONS INFORMATION
Entrance Requirements: SAT (Score: 950), High School courses—English (4 years), Physics or Chemistry (1 year), Math (3 years)
Entrance Recommendations: PSAT (Score: 950), SAT (Score: 950), ACT (Score: 22), High School courses—Math 12 (1 year), Additional Science (1 year), Mechanical Drawing (1 year)
Requirements for foreign students: TOEFL (Score: 500); financial statement

DEPARTMENTAL ADMISSIONS INFORMATION
Admission to the engineering department: At the time of admission to the institution.

FINANCIAL INFORMATION

ESTIMATED EXPENSES (ACADEMIC YEAR)
[Expenses are for the 1992-93 nine-month academic year.]

State Residents	Undergraduate	Graduate
Tuition and fees	$ 2,650	$ –
College room and board	$ 4,344	$ –
Books and supplies	$ 600	$ –
Other expenses	$ 2,110	$ –
Total estimated expenses	**$ 9,704**	**$ –**
Out-of-State Residents	Undergraduate	Graduate
Tuition and fees	$ 6,556	$ –
College room and board	$ 4,344	$ –
Books and supplies	$ 600	$ –
Other expenses	$ 2,110	$ –
Total estimated expenses	**$13,610**	**$ –**

FINANCIAL AID OFFICE CONTACT
Howard L. English, *Director of Financial Aid*
State University of New York Maritime College
Fort Schuyler
Throgs Neck Station
Bronx, NY 10465-4198
Phone: (718)409-7277 **Fax:** (718)409-7392

GENERAL FINANCIAL AID INFORMATION
Forms accepted/required: FAF
Additional financial aid information: The college has the full range of Federal and State Financial Aid programs as well as private scholarship funds.

ENGINEERING COLLEGE INFORMATION

[For additional personnel, refer to the *Appendix*.]

HEAD OF ENGINEERING
Jose Femenia, *Chairman*
Phone: (718)409-7411 **Fax:** (718)409-7421

ENGINEERING COLLEGE ADDRESS
Engineering Department
State University of New York Maritime College
Fort Schuyler
Throgs Neck Station, NY 10465-4198
Phone: (718)409-7200 **Fax:** (718)409-7392

ENROLLMENTS—BY CLASS
[Numbers are baccalaureate enrollments for the fall 1991 term, unless otherwise footnoted.]
1st-year students/Freshmen 120
2nd-year students/Sophomores 90
3rd-year students/Juniors 80
4th-year students/Seniors 70
Total **360**

NUMBER OF DEGREES AWARDED—BY PROGRAM
[Numbers are engineering baccalaureate degrees awarded during the 1991-1992 academic year, unless otherwise footnoted. For full details about each engineering program, refer to the *Tables of Degree Programs*.]
Electrical Engineering .... 28 [1]
Facilities Engineering .... –
Marine Engineering 16
Mechanical Engineering 3
Naval Architecture 12
Total **59**
Notes: (1) BE degree awarded.

PERCENTAGE OF DEGREES AWARDED—BY CATEGORY
[Percentages are of all engineering baccalaureate degrees awarded during the 1991-1992 academic year, unless otherwise footnoted.]
African Americans 1.5%
Asian/Pacific Island Americans 3.0%
Hispanic Americans 1.5%
Native Americans .... – %
Foreign Citizens 1.5%
All Others 92.5%
Women 6.0%
Persons with disabilities .... – %
Students over 25 years of age .... – %

ENGINEERING STUDENT DATA

Applicants to the engineering college
Number of applicants to engineering college 475
Percent offered admission 70.0%

Matriculated engineering students
Percentage in top quartile (25%) of High School class 45.0
Average SAT scores: Math—550, Verbal—450, Combined—1000

FULL- & PART-TIME FACULTY—BY DEPARTMENT
[Figures are the head count of full-time faculty and the full-time equivalent (FTE) of part-time faculty for each engineering department or equivalent.]

Department	Full	Part
Electrical Eng	5	–
Mechanical Eng	6	–
Marine Eng	6	–
Naval Architecture	3	–
Facilities Eng	6	–

COLLEGE DESCRIPTION
Maritime College is a unique institution having a long history of serving the marine industry and of adapting its curricula to serve the academic and professional needs of students and graduates. The principal strength of the engr. unit comes from the synergistic combination of a diverse faculty covering various academic disciplines and prof. backgrounds, exceptional teaching facilities, & labs & curricula that meld design oriented 4/year engr. programs & a fifth for license preparation. The academic design oriented phase of the programs occurs during the 4 conventional 30 week academic years & the major part of the Merchant Marine license training program is concentrated during the three 11 week summer sea terms aboard the colleges training ship.

DEGREE PROGRAMS DESCRIPTION
The Bachelor of Engineering degree in various engineering programs prepares the graduate for entry-level professional positions either ashore in the practice of engineering or at-sea as a licensed officer in the Merchant Marine or Armed Forces. The Bachelor of Science degree engineering program combines a non-design oriented engineering program with a strong general education in humanities and management. Graduates of the BE programs are well prepared for graduate studies in engineering.

TRANSFER INFORMATION
Residency Requirements: Due to Federal requirement for Merchant Marine license all students transferring in from non-maritime type schools must spend 2.5 yrs in residence at the college.

Transfer via Articulation Agreements
Admission to engineering: At the end of the sophomore year
Engineering graduates transferred from . . 4-yr: 5 2-yr: 4
Requirements: Students must complete Math through diff. equa., 12 credits calculus based physics, 6 credits chem, 3 credits comp sci, 12 credits Eng/hum, 9 credits engr sciences, 4 credits phys ed. Transfer students must demonstrate proficiency in college level work.

Transfer without Articulation Agreements
Admission to engineering: At any time
Engineering graduates transferred from . . 4-yr: 5 2-yr: 4
Requirements: Transfer applicants must demonstrate proficiency in college level work and a potential for completing the engr. courses required of Maritime College Engineering students.

GRADUATION REQUIREMENTS
For graduation students must complete the course of study as published in the college catalog with a minimum cumulative grade point average (CGPA) of 2.0/4.0.

STUDENT PROGRAMS

PROFESSIONAL AND HONORARY SOCIETIES
[For key to acronyms, see *Introduction*.]
Professional Societies: ANS, ASME, IEEE, SNAME
Honorary Societies: Order of the Engineer

SUPPORT PROGRAMS
Student Chapter Organizations: Society of Women Engineers

Student Support Services
For: Women & Ethnic Minorities **Available:** Year round
Offered: Prematriculation, Freshman, Sophomore, Junior, Senior
Other Engineering Support Programs: The Student Support Services provides personal, academic/career counseling/tutoring aid to disabled cadets and Seminars to broaden skills and academic performance. Association of Women Merchant Marine Officers (Student Chapter) AWMMO: Was recently formed and is unique among Maritime Academies in the nation, designed to guide female cadets for careers in the maritime industry.

222 State University of New York, College at New Paltz

INSTITUTION PROFILE

HEAD OF THE INSTITUTION
Alice Chandler
Phone: (914)257-3288

GENERAL INFORMATION
[All Students—Fall 1991]
Undergraduate enrollment 6,619
Graduate enrollment 1,856
Total institution enrollment 8,475
Type of institution: Public Calendar: Semesters
Nearest city: Poughkeepsie Population: 28,844
Miles from main campus: 15 Setting: Small Town
Types of engineering degrees: Engineering
Other degree-granting colleges: Arts & Sciences, Education, Fine & Performing Arts

ENGINEERING ADMISSIONS

ADMISSIONS OFFICE CONTACT
Robert J. Seaman
State University of New York, College at New Paltz
Dean of Admissions, HAB 405
75 S. Manheim Blvd.
New Paltz, NY 12561-2499
Phone: (914)257-3200

ENGINEERING COLLEGE ADMISSIONS INFORMATION
Entrance Requirements: Admission to The College is offered on a competitive basis with the primary emphasis on academic achievement. Prospective students must take either the SAT or ACT.
Requirements for foreign students: TOEFL (Score: 525); financial statement

DEPARTMENTAL ADMISSIONS INFORMATION
Admission to the engineering department: The time when the student has completed the pre-engineering coursework with a GPA. of 2.5 or better.
Additional information: The student must complete the pre-engineering coursework with a GPA. of 2.5 or better. This coursework consists of Freshman Composition 1 & 2, General Chemistry 1, Calculus 1,2,3, & 4, General Physics 1 & 2, Introduction to Computing, Engineering Graphics, and Introduction to Engineering Science.

FINANCIAL INFORMATION

ESTIMATED EXPENSES (ACADEMIC YEAR)
[Expenses are for the 1992-93 nine-month academic year.]

State Residents	Undergraduate	Graduate
Tuition and fees	$ 2,921	$ 4,227
College room and board	$ 4,390	$ 4,390
Books and supplies	$ 590	$ 590
Other expenses	$ 1,600[1]	$ 1,600[1]
Total estimated expenses	$ 9,501	$ 10,807

Out-of-State Residents	Undergraduate	Graduate
Tuition and fees	$ 6,821	$ 7,533
College room and board	$ 4,390	$ 4,390
Books and supplies	$ 590	$ 590
Other expenses	$ 1,600[2]	$ 1,600[2]
Total estimated expenses	$ 13,401	$ 14,113

Notes: (1) Optional Student Health Insurance 72.00 (2) Mandatory Health Fee 308.00

FINANCIAL AID OFFICE CONTACT
Daniel Sistarenik, Director of Financial Aid Programs
State University of New York, College at New Paltz
HAB 603A
75 S. Manheim Blvd.
New Paltz, NY 12561-2499
Phone: (914)257-3250

GENERAL FINANCIAL AID INFORMATION
Forms accepted/required: FAF, IRS form
Additional financial aid information: Need-based grants, loans, part-time jobs on campus, College Work study Program.

ENGINEERING COLLEGE INFORMATION
[For additional personnel, refer to the Appendix.]

HEAD OF ENGINEERING
Owen Hill
Phone: (914)257-3720 Fax: (914)257-3009

ENGINEERING COLLEGE ADDRESS
School of Engineering and Business Administration
State University of New York, College at New Paltz
75 S. Manheim Blvd.
New Paltz, NY 12561-2499
Phone: (914)257-3720 Fax: (914)257-3009

ENROLLMENTS—BY CLASS
[Numbers are baccalaureate enrollments for the fall 1991 term, unless otherwise footnoted.]
1st-year students/Freshmen 29
2nd-year students/Sophomores 41
3rd-year students/Juniors 46
4th-year students/Seniors 62
Total 178

NUMBER OF DEGREES AWARDED—BY PROGRAM
[Numbers are engineering baccalaureate degrees awarded during the 1991-1992 academic year, unless otherwise footnoted. For full details about each engineering program, refer to the Tables of Degree Programs.]
Electrical Engineering 24
Total 24

PERCENTAGE OF DEGREES AWARDED—BY CATEGORY
[Percentages are of all engineering baccalaureate degrees awarded during the 1991-1992 academic year, unless otherwise footnoted.]
African Americans .... 8.0 %
Asian/Pacific Island Americans .... 4.0 %
Hispanic Americans .... 4.0 %
Native Americans .... – %
Foreign Citizens .... 29.0 %
All Others .... 55.0 %
Women .... 12.5 %
Persons with disabilities .... – %
Students over 25 years of age .... 17.0 %

ENGINEERING STUDENT DATA
Applicants to the engineering college
Number of applicants to engineering college 364
Percent offered admission 45.0%
Matriculated engineering students
Percentage in top quartile (25%) of High School class 45.0
Average SAT scores: Math—587, Verbal—456, Combined—1043

FULL- & PART-TIME FACULTY—BY DEPARTMENT
[Figures are the head count of full-time faculty and the full-time equivalent (FTE) of part-time faculty for each engineering department or equivalent.]

Department	Full	Part
Electrical Eng	7	0.5

COLLEGE DESCRIPTION
The school offers a four year undergraduate program leading to the B.S. in Electrical Engineering. The first three semesters constitute a pre-engineering curriculum emphasizing basic mathematics and science. The remaining five semesters include core electrical engineering coursework and specialized senior electives. The four areas of specialization are Microelectronics, including VLSI design; Fields & Waves, including microwaves; Systems Engineering, including signal processing and control; and Computer Engineering. A co-op office aids students in locating positions in industry. During the senior year students are expected to undertake a major design project under the guidance of a member of the faculty.

DEGREE PROGRAMS DESCRIPTION
The pre-engineering curriculum which can be completed during the sophomore year provides a solid foundation in mathematics and basic sciences. The core electrical engineering courses, which are taken by all students, cover the basic areas of electrical engineering from electronics to fields and waves to linear systems. Senior electives and associated high level laboratories build on the basic knowledge and skills mastered in earlier coursework. The undergraduate experience culminates in a senior design project encompassing two semesters in which the student works under the guidance of a faculty member to identify, formulate, and solve a substantial design problem. In most cases the senior laboratories, which are available for senior design, are outfitted with sophisticated, industry grade equipment providing the student with valuable high level experience. In depth study can be undertaken in the areas of fields & waves, computer engineering, microelectronics, and systems engineering.

DUAL DEGREE PROGRAMS
Engineering baccalaureate graduates with dual degrees 0
Enrollment requirements: Completion of specific coursework with a 2.5 GPA. or better overall.

TRANSFER INFORMATION
Residency Requirements: Students must complete 30 credits in residence.
Transfer via Articulation Agreements
Admission to engineering: At the end of the sophomore year
Engineering graduates transferred from   ... 4-yr: –   2-yr: 1
Requirements: The student must earn an A.S. in Engineering Science with a C or better in engineering pre-requisite coursework.
Transfer without Articulation Agreements
Admission to engineering: The time when all of the pre-engineering coursework is completed with a 2.5 GPA. or better.
Engineering graduates transferred from   ... 4-yr: 9   2-yr: 1
Requirements: Overall GPA. of 2.5 or better and strong math background.

GRADUATION REQUIREMENTS
Students must have completed the pre-engineering coursework with a GPA. of 2.5 or better. Seniors must satisfactorily complete a major design project in the senior year.

STUDENT PROGRAMS

PROFESSIONAL AND HONORARY SOCIETIES
[For key to acronyms, see Introduction.]
Professional Societies: IEEE
Honorary Societies: College-wide Honors Program

SUPPORT PROGRAMS
Student Chapter Organizations: National Society of Black Engineers

Collegiate Science & Technology Entry Program
For: Ethnic Minorities   Available: Year round
Offered: Prematriculation, Freshman, Sophomore, Junior, Senior
Other Engineering Support Programs: Freshman and transfer orientation sessions held each summer and during intercession involve personnel from Academic Advising, Admissions, and the Department of Electrical Engineering. C-STEP (Collegiate Science & Technology Entry Program), provides a pre-freshman summer program and academic support and enrichment throughout the ensuing years for economically underprivileged and minority students.

223 State University of New York at Stony Brook

■ INSTITUTION PROFILE

HEAD OF THE INSTITUTION
John H. Marburger
Phone: (516)632-6265　　　　Fax: (516)632-6252

GENERAL INFORMATION
[All Students—Fall 1991]
Undergraduate enrollment 11,440
Graduate enrollment 6,192
Total institution enrollment **17,632**
Type of institution: Public　　**Calendar:** Semesters
Nearest city: New York City　　**Population:** 7,000,000
Miles from main campus: 60　　**Setting:** Suburban
Types of engineering degrees: Engineering
Other degree-granting colleges: Arts & Sciences, Dentistry, Medicine, Nursing, Allied Health Professions, Policy Analysis and Public Management, Social Welfare

■ ENGINEERING ADMISSIONS

ADMISSIONS OFFICE CONTACT
GiGi Lamens
State University of New York at Stony Brook
Undergraduate Admissions Office
Administration Bldg., Room 118
Stony Brook, NY 11794-2200
Phone: (516)632-6868　　　　Fax: (516)632-9027

ENGINEERING COLLEGE ADMISSIONS INFORMATION
Admission to the engineering college: Students not admitted to the college at the time they enter the university may apply after one semester of physics and calculus.
Entrance Requirements: SAT, ACT, High School courses—English (4 years), Mathematics (4 years), Physics (1 year), Chemistry (1 year)
Entrance Recommendations: SAT, High School courses—Computers (1 year), Social Studies (3 years), Foreign Language (2 years), Other Science (1 year)
Requirements for foreign students: TOEFL (Score: 550); financial statement

DEPARTMENTAL ADMISSIONS INFORMATION
Admission to the engineering department: Students not admitted to the college at the time they enter the university may apply after completing at least one semester of physics and calculus.
Additional information: Admission is competitive and on a space-available basis.

■ FINANCIAL INFORMATION

ESTIMATED EXPENSES (ACADEMIC YEAR)
[Expenses are for the 1992-93 nine-month academic year.]

State Residents	Undergraduate	Graduate
Tuition and fees	$ 2,929	$ 4,162
College room and board	$ 4,413	—(1)
Books and supplies	$ 800	—(A)
Other expenses	$ —(A)	$ —(A)
Total estimated expenses	**$ 8,142**	**$ 4,162**

Out-of-State Residents	Undergraduate	Graduate
Tuition and fees	$ 6,829	$ 7,478
College room and board	$ 4,413	—(1)
Books and supplies	$ 800	—(A)
Other expenses	$ —(A)	$ —(A)
Total estimated expenses	**$12,042**	**$ 7,478**

Notes: (A) Data not available. (1) Varies depending on apartment.

FINANCIAL AID OFFICE CONTACT
Sherwood Johnson, *Director, Financial Aid*
State University of New York at Stony Brook
Financial Aid Office
230 Administration Bldg.
Stony Brook, NY 11794-0851
Phone: (516)632-6840

GENERAL FINANCIAL AID INFORMATION
Forms accepted/required: FAF, Stony Brook Institutional Application
Additional financial aid information: Need-based scholarship, loan and tuition assistance programs; work-study programs; NYS Regents Scholarships; private scholarships (merit and need).

■ ENGINEERING COLLEGE INFORMATION

[For additional personnel, refer to the *Appendix*.]

HEAD OF ENGINEERING
Yacov Shamash
Phone: (516)632-8380　　　　Fax: (516)632-8205
EMail: yshamash@ccmail.sunysb.edu

ENGINEERING COLLEGE ADDRESS
College of Engineering and Applied Sciences
State University of New York at Stony Brook
Engineering Building, Room 100
Stony Brook, NY 11794-2200
Phone: (516)632-8380　　　　Fax: (516)632-8205

ENROLLMENTS—BY CLASS
[Numbers are baccalaureate enrollments for the fall 1991 term, unless otherwise footnoted.]
1st-year students/Freshmen 172
2nd-year students/Sophomores 138
3rd-year students/Juniors 107
4th-year students/Seniors 301
Total **718**

NUMBER OF DEGREES AWARDED—BY PROGRAM
[Numbers are engineering baccalaureate degrees awarded during the 1991-1992 academic year, unless otherwise footnoted. For full details about each engineering program, refer to the *Tables of Degree Programs*.]
Electrical Engineering 67
Engineering Science(s) 18
Materials Science & Engineering .... —(A)
Mechanical Engineering 45
Total **130**

Notes: (A) Data not available.

PERCENTAGE OF DEGREES AWARDED—BY CATEGORY
[Percentages are of all engineering baccalaureate degrees awarded during the 1991-1992 academic year, unless otherwise footnoted.]
African Americans .... 3.8 %
Asian/Pacific Island Americans .... 21.5 %
Hispanic Americans .... 4.6 %
Native Americans .... — %
Foreign Citizens .... 6.9 %
All Others .... 63.2 %
Women .... 10.0 %
Persons with disabilities .... — %
Students over 25 years of age .... — %(A)

Notes: (A) Data not available.

GRADUATE ENROLLMENTS & DEGREES AWARDED
Master's enrollment .... —(A)
Master's degrees awarded 49
Doctoral enrollment .... —(A)
Doctoral degrees awarded 20

Notes: (A) Data not available.

FULL- & PART-TIME FACULTY—BY DEPARTMENT
[Figures are the head count of full-time faculty and the full-time equivalent (FTE) of part-time faculty for each engineering department or equivalent.]

Department	Full	Part
Electrical Eng	19	7.5
Mechanical Eng	15	7.5
Materials Science & Eng	11	–

COLLEGE DESCRIPTION
Four-year B.E. programs are designed to enable graduates to develop with technology and to contribute to society in the general sense. Courses for all freshmen include physics, calculus, chemistry, computer science and English composition, with specialization beginning in the second year. At least three courses in arts and humanities and at least three others in behavioral and social sciences are required, and an upper-level writing requirement must be met. Considerable 'hands-on' experience is provided via required lab courses, and computers are widely used in design and for general problem solving. During senior year, an extensive design project is completed, and the opportunity exists for participation in faculty research. A student may opt for a second non-engineering major or up to three non-engineering minors. A double baccalaureate with the College of Arts and Sciences is also possible. Average number of years required to actually complete the bachelor's degree: 4.7.

DEGREE PROGRAMS DESCRIPTION
The B.E. degrees in the various engineering programs prepare the graduate for entry-level professional positions or for graduate study in various engineering fields. The engineering science major may be used to concentrate in materials science or to provide a general engineering education. The electrical engineering major may be used to concentrate in computer engineering. The M.S. and Ph.D. degrees are awarded in three areas: electrical engineering, mechanical engineering, and materials science and engineering.

TRANSFER INFORMATION
Residency Requirements: Beginning with the 58th credit, at least 36 credits must be earned at Stony Brook. At least seven engineering core courses and/or approved technical elective courses offered by the College must be completed at Stony Brook, at least five of which must be offered by the department of the student's major.

Transfer via Articulation Agreements
Admission to engineering: At any time
Transfer without Articulation Agreements
Admission to engineering: At any time

GRADUATION REQUIREMENTS
B.E. programs require a minimum of 128 credits, a 2.00 GPA overall at Stony Brook, and grades of 2.00 or better in certain courses in engineering and related fields.

■ STUDENT PROGRAMS

PROFESSIONAL AND HONORARY SOCIETIES
[For key to acronyms, see *Introduction*.]
Professional Societies: ASME, IEEE, SAE, SAMPE, Society of Women Engineers, The Computer Science Society
Honorary Societies: Eta Kappa Nu, Tau Beta Pi

SUPPORT PROGRAMS
Student Chapter Organizations: Society of Women Engineers, National Society of Black Engineers, Society of Hispanic Professional Engineers

Five-year B.E. or B.S. Degree Program
For: Ethnic Minorities　　**Available:** Academic year
Offered: Prematriculation, Freshman

Other Engineering Support Programs: Summer, fall and winter orientation programs for all new students; Math Learning Center; Writing Clinic; Career Development Office; Disabled Student Services; Veterans Affairs Office; Foreign Student Office; University Counseling Center; free tutoring by membership of Tau Beta Pi.

224 Swarthmore College

INSTITUTION PROFILE
HEAD OF THE INSTITUTION
Alfred H. Bloom
Phone: (215)328-8314 **Fax:** (215)328-8673
GENERAL INFORMATION
[All Students—Fall 1991]
Undergraduate enrollment 1,350
Total institution enrollment 1,350
Type of institution: Private **Calendar:** Semesters
Nearest city: Philadelphia, PA **Population:** 160,000
Miles from main campus: 10 **Setting:** Suburban
Types of engineering degrees: Engineering

ENGINEERING ADMISSIONS
ADMISSIONS OFFICE CONTACT
Robert A. Barr
Swarthmore College
500 College Avenue
Swarthmore, PA 19081-1397
Phone: (215)328-8308 **Fax:** (215)328-8673
ENGINEERING COLLEGE ADMISSIONS INFORMATION
Entrance Requirements: High School courses—History, social studies, & literature (1 year), Mathematics (4 years), Physics/Chemistry (1 year)
Entrance Recommendations: High School courses—Physics (1 year), Pre-calculus (1 year), Calculus (1 year), Foreign language (4 years)
DEPARTMENTAL ADMISSIONS INFORMATION
Admission to the engineering department: At the end of the second year.
Additional information: An average grade of C or better in all courses taken in the freshman and sophomore years and an average grade of C or better in all the introductory engineering, mathematics and science courses.

FINANCIAL INFORMATION
ESTIMATED EXPENSES (ACADEMIC YEAR)
[Expenses are for the 1992-93 nine-month academic year.]

All Students	Undergraduate	Graduate
Tuition and fees	$17,460	$ –
College room and board	$ 5,844	$ –
Books and supplies	$ 1,000	$ –
Other expenses	$ –[B]	$ –
Total estimated expenses	**$24,304**	**$ –**

Notes: (B) Data not applicable.
FINANCIAL AID OFFICE CONTACT
Laura Talbot, *Director of Financial Aid*
Swarthmore College
500 College Avenue
Swarthmore, PA 19081-1397
Phone: (215)328-8358 **Fax:** (215)328-8673
GENERAL FINANCIAL AID INFORMATION
Forms accepted/required: FAF, Institutional
Additional financial aid information: Most aid is need-based and is usually a combination of scholarship, loan, and student employment. Forty-nine percent of the student body currently receives financial aid.

ENGINEERING COLLEGE INFORMATION
[For additional personnel, refer to the *Appendix*.]
HEAD OF ENGINEERING
Frederick L. Orthlieb
Phone: (215)328-8080 **Fax:** (215)328-8082
EMail: forthli1@swarthmore.edu

ENGINEERING COLLEGE ADDRESS
Department of Engineering
Swarthmore College
500 College Avenue
Swarthmore, PA 19081-1397
Phone: (215)328-8071 **Fax:** (215)328-8082
ENROLLMENTS—BY CLASS
[Numbers are baccalaureate enrollments for the fall 1991 term, unless otherwise footnoted.]
1st-year students/Freshmen 42
2nd-year students/Sophomores 21
3rd-year students/Juniors 29
4th-year students/Seniors 24
Total 116
NUMBER OF DEGREES AWARDED—BY PROGRAM
[Numbers are engineering baccalaureate degrees awarded during the 1991-1992 academic year, unless otherwise footnoted. For full details about each engineering program, refer to the *Tables of Degree Programs*.]
Engineering 18
Total **18**
PERCENTAGE OF DEGREES AWARDED—BY CATEGORY
[Percentages are of all engineering baccalaureate degrees awarded during the 1991-1992 academic year, unless otherwise footnoted.]
African Americans 3.6%
Asian/Pacific Island Americans .... – %
Hispanic Americans .... – %
Native Americans .... – %
Foreign Citizens 3.6%
All Others 92.8%
Women 3.6%
Persons with disabilities .... – %
Students over 25 years of age .... – %
ENGINEERING STUDENT DATA
Applicants to the engineering college
Number of applicants to engineering college .... 417 [1]
Percent offered admission 27.8%
Matriculated engineering students
Percentage in top quartile (25%) of High School class 96.4
Average SAT scores: Math—706, Verbal—628, Combined—1334
Notes: (1) Includes 118 who listed engineering as one of two majors.
FULL- & PART-TIME FACULTY—BY DEPARTMENT
[Figures are the head count of full-time faculty and the full-time equivalent (FTE) of part-time faculty for each engineering department or equivalent.]

Department	Full	Part
Eng	8	–

COLLEGE DESCRIPTION
The department, one of eighteen at the College, offers a four-year undergraduate program leading to the Bachelor of Science in Engineering. The major requires four courses in mathematics, four courses in the sciences, and twelve in engineering. Students have the opportunity to take twelve courses in the humanities or social sciences, and to fashion a major or minor in an academic area other than engineering. Typically, one-third of the graduating class has double majors. Of the twelve engineering courses, six are core courses intended to provide breadth in the student's background, while six are engineering electives in which the student can explore areas of special interest, including the traditional areas of mechanical, electrical, civil, and computer engineering. During the last semester each student completes a senior design project and formally presents the results in a written document as well as an oral presentation to faculty and peers.
DEGREE PROGRAMS DESCRIPTION
Four courses in mathematics and four in the sciences are required for the engineering degree. Within the department, the following core courses are required of all students: Mechanics, Physical Systems Analysis I and II, Experimentation for Engineering Design, Thermofluid Mechanics, and Engineering Design. The first four courses are normally taken in the freshman and sophomore years. Six elective courses in engineering complete the student's degree requirements in the department. College-wide requirements include distribution and language requirements. In addition, each student must have twenty courses outside of the major department in order to graduate.
DUAL DEGREE PROGRAMS
Engineering baccalaureate graduates with dual degrees 5
Enrollment requirements: A "B" average is required in all of the student's courses.
TRANSFER INFORMATION
Transfer without Articulation Agreements
Admission to engineering: At the end of the sophomore year
Engineering graduates transferred from ... 4-yr: – 2-yr: –
Requirements: The engineering department has not admitted transfers above the third semester; all engineering transfers have come from ABET accredited programs with average grades above B and have been awarded appropriate numbers of Swarthmore course credits based on careful review course syllabi and transcripts by the department chair.
GRADUATION REQUIREMENTS
Satisfaction or completion of: 1) 32 courses with an average grade of C or better; 2) 12 courses with an average grade of C or better within the engineering department; 3) distribution requirements with at least 20 courses outside of the engineering department; 4) foreign language requirement; 5) the senior design project (which meets the College-wide requirement for a comprehensive examination).

STUDENT PROGRAMS
PROFESSIONAL AND HONORARY SOCIETIES
[For key to acronyms, see *Introduction*.]
Professional Societies: ASCE, ASME, IEEE, Society of Women Engineers
Honorary Societies: Tau Beta Pi
SUPPORT PROGRAMS
Student Chapter Organizations: Society of Women Engineers
Black Cultural Center
For: Ethnic Minorities **Available:** Academic year
Offered: Prematriculation, Freshman, Sophomore, Junior, Senior, Graduate level
Alice Paul Women's Center
For: Women **Available:** Academic year
Offered: Prematriculation, Freshman, Sophomore, Junior, Senior, Graduate level
Other Engineering Support Programs: Prospective engineering students are assigned faculty advisers from the department upon admission. The College's Career Planning and Placement Office plays a major role in supporting engineering students. The Student Orientation Committee includes programs for freshmen interested in becoming engineering students in the fall and spring semesters of the freshman year.

225 Syracuse University

INSTITUTION PROFILE

HEAD OF THE INSTITUTION
Kenneth A. Shaw, *Chancellor*
Phone: (315)443-2235 Fax: (315)443-3503

GENERAL INFORMATION
[All Students—Fall 1991]
Undergraduate enrollment 13,780
Graduate enrollment 7,126
Total institution enrollment **20,906**
Type of institution: Private Calendar: Semesters
Location: Syracuse Population: 400,000
Setting: Urban
Types of engineering degrees: Engineering
Other degree-granting colleges: Architecture, Arts & Sciences, Communication & Journalism, Education, Fine & Performing Arts, Humanities & Social Sciences, Law, Nursing, Social Work, Human Development, Information Studies, Management

ENGINEERING ADMISSIONS

ADMISSIONS OFFICE CONTACT
David C. Smith
Syracuse University
Admissions
201 Tolley Administration Building
Syracuse, NY 13244-1100
Phone: (315)443-3611

ENGINEERING COLLEGE ADMISSIONS INFORMATION
Entrance Requirements: SAT, ACT, High School courses—English (4 years), Social Studies (3 years), Mathematics and Sciences (4 years), Foreign Language (2 years), SAT & ACT minimum scores not available.
Entrance Recommendations: Personal interview recommended.
Requirements for foreign students: TOEFL (Score: 525); financial statement

DEPARTMENTAL ADMISSIONS INFORMATION
Admission to the engineering department: At the time of admission to the institution.

FINANCIAL INFORMATION

ESTIMATED EXPENSES (ACADEMIC YEAR)
[Expenses are for the 1992-93 nine-month academic year.]

All Students	Undergraduate	Graduate
Tuition and fees	$14,112	$10,049
College room and board	$ 5,920	$ 5,920
Books and supplies	$ 560	$ 985
Other expenses	$ –	$ –
Total estimated expenses	$20,592	$16,954

FINANCIAL AID OFFICE CONTACT
Christopher Walsh, *Director of Financial Aid*
Syracuse University
201 Tolley Administration Building
Syracuse, NY 13244-1100
Phone: (315)449-1513 Fax: (315)443-1531

GENERAL FINANCIAL AID INFORMATION
Forms accepted/required: FAF
Additional financial aid information: Merit scholarships; need-based scholarships; student loans; College Work-Study; part-time jobs on campus.

ENGINEERING COLLEGE INFORMATION

[For additional personnel, refer to the *Appendix*.]

HEAD OF ENGINEERING
Steven C. Chamberlain
Phone: (315)443-4341 Fax: (315)443-4936
EMail: Steve_Chamberlain@ISR.SYR.EDU

ENGINEERING COLLEGE ADDRESS
L.C. Smith College of Engineering and Computer Science
Syracuse University
223 Link Hall
Syracuse, NY 13244-1240
Phone: (315)443-2545 Fax: (315)443-4936

ENROLLMENTS—BY CLASS

[Numbers are baccalaureate enrollments for the fall 1991 term, unless otherwise footnoted.]
1st-year students/Freshmen 252
2nd-year students/Sophomores 199
3rd-year students/Juniors 166
4th-year students/Seniors 249
Total **866**

NUMBER OF DEGREES AWARDED—BY PROGRAM

[Numbers are engineering baccalaureate degrees awarded during the 1991-1992 academic year, unless otherwise footnoted. For full details about each engineering program, refer to the *Tables of Degree Programs*.]

Aerospace Engineering 31
Bioengineering 16
Chemical Engineering 8
Civil Engineering 15
Computer Engineering 16
Electrical Engineering 32
Engineering Physics 2
Environmental Engineering 5
Manufacturing Engineering 2
Mechanical Engineering 44
Total **171**

PERCENTAGE OF DEGREES AWARDED—BY CATEGORY

[Percentages are of all engineering baccalaureate degrees awarded during the 1991-1992 academic year, unless otherwise footnoted.]

African Americans 2.8%
Asian/Pacific Island Americans 6.6%
Hispanic Americans 1.1%
Native Americans 0.6%
Foreign Citizens 8.8%
All Others 80.1%
Women 13.2%
Persons with disabilities .... – % (A)
Students over 25 years of age .... – % (A)

Notes: (A) Data not available.

GRADUATE ENROLLMENTS & DEGREES AWARDED

Master's enrollment 260 (2)
Master's degrees awarded 31
Doctoral enrollment 218 (2)
Doctoral degrees awarded 196

Notes: (2) 682 part-time graduate students not included in this figure. Masters vs. doctoral breakdown not available at this time.

ENGINEERING STUDENT DATA

Applicants to the engineering college
Number of applicants to engineering college 1,221
Percent offered admission 85.5%

Matriculated engineering students
Percentage in top quartile (25%) of High School class .... 75.0 (A)
Average SAT scores: Math—599.2, Verbal—473.6, Combined—1072.8

Notes: (A) Data not available.

FULL- & PART-TIME FACULTY—BY DEPARTMENT

[Figures are the head count of full-time faculty and the full-time equivalent (FTE) of part-time faculty for each engineering department or equivalent.]

Department	Full	Part
Chemical Eng & Materials Science	9	–
Civil & Environmental Eng	6	–
Electrical & Computer Eng	23	–
Mechanical, Aerospace & Manufacturing Eng	–	–
Bioengineering	12	–
Aerospace Eng	8	–
Computer Eng	15	–
Eng Physics	1	–
Environmental Eng	6	–
Manufacturing Eng	1	–
Mechanical Eng	11	–

COLLEGE DESCRIPTION

All Engineering Programs can be completed in four years. Each program emphasizes engineering and science fundamentals; the development of problem solving and design skills; applications of computers throughout the curriculum; and a broad introduction to the social sciences and humanities. A capstone design course or project is required in the senior year. A coop program requiring a total of one year of work experience in several segments before graduation is available for all disciplines. A five-year combined-degree program, leading to a B.S. in Arts and Sciences and a B.S. in Engineering is also available. Students may participate in a Junior Year Abroad program in London.

DEGREE PROGRAMS DESCRIPTION

The B.S. degrees prepare students for entry-level professional positions or for graduate study in the appropriate discipline. Students completing the B.S. in Bioengineering have been very successful in obtaining admission to medical school. Both thesis and non-thesis options exist in many of the M.S. programs. All Ph.D programs require a dissertation and lead to high level research activities in industry and government. Ph.D degrees also lead to faculty positions in engineering schools and colleges.

DUAL DEGREE PROGRAMS

Engineering baccalaureate graduates with dual degrees 5
Enrollment requirements: The College of Engineering offers a combined degree with the College of Arts and Sciences. The two degree programs must be unrelated disciplines that lead to two different career objectives and normally must be distinguished by different degree labels (degree types). Students must fully meet requirements of both degrees and complete at least 30 credits beyond the normal requirements for one degree (usually 120 credits).

TRANSFER INFORMATION

Residency Requirements: Undergraduates must take at least 30 credit hours of coursework at Syracuse University in order to be granted its degree.

Transfer via Articulation Agreements
Admission to engineering: Admission at the end of third year.
Engineering graduates transferred from .. 4-yr: 3 2-yr: –

Requirements: Completion of three years of engineering science. Grade point average of at least 2.50/4.00. Students must have a minimum grade point average of 2.500/4.000 and a grade of "C" or better in Engineering, Math, and Science courses.

Transfer without Articulation Agreements
Admission to engineering: Must complete at least one year of pre-engineering studies.
Engineering graduates transferred from .. 4-yr: 19 2-yr: 21

Requirements: Students must have a minimum grade point average of 2.500/4.000 and a grade of 'C' or better in Engineering, Math, and Science courses.

GRADUATION REQUIREMENTS

B.S. Programs in engineering require from 132 to 140 credit hours, depending on major. At least a 2.000 GPA. is required with at least a 2.000 GPA. in mathematics, science, and engineering courses.

STUDENT PROGRAMS

PROFESSIONAL AND HONORARY SOCIETIES
[For key to acronyms, see *Introduction*.]

Professional Societies: ACM, AIAA, AIChE, ASCE, ASME, BMES, IEEE, SAE, National Society of Black Engineers (NSBE), Society of Women Engineers (SWE)
Honorary Societies: Chi Epsilon, Eta Kappa Nu, Pi Tau Sigma, Sigma Gamma Tau, Tau Beta Pi

SUPPORT PROGRAMS
Student Chapter Organizations: Society of Women Engineers, National Society of Black Engineers, Society of Hispanic Professional Engineers

Minority Engineering Program
For: Ethnic Minorities **Available:** Year round
Offered: Freshman, Sophomore, Junior, Senior

Other Engineering Support Programs: Summer registration for all freshmen; fall orientation program for all freshmen and transfers; academic advising services; foreign student office; veteran's office; tutoring and personal counseling services; services for blind and handicapped; placement services.

226 Technical University of Nova Scotia

■ INSTITUTION PROFILE

HEAD OF THE INSTITUTION
Peter F. Adams
Phone: (902)420-7644 Fax: (902)420-7551

GENERAL INFORMATION
[All Students—Fall 1991]
Undergraduate enrollment 995
Graduate enrollment 340
Total institution enrollment 1,335
Type of institution: Public Calendar: Trimesters
Location: Halifax, Nova Scotia Population: 250,000
Setting: Urban
Types of engineering degrees: Engineering
Other degree-granting colleges: Architecture, Computer Science

■ ENGINEERING ADMISSIONS

ADMISSIONS OFFICE CONTACT
Lamont Pelletier
Technical University of Nova Scotia
1360 Barrington Street
P.O. Box 1000
Halifax, NS B3J-2X4
Phone: (902)420-7624 Fax: (902)420-7551

ENGINEERING COLLEGE ADMISSIONS INFORMATION
Admission to the engineering college: Students are admitted after a two year engineering program at eight Associated Universities.

DEPARTMENTAL ADMISSIONS INFORMATION
Admission to the engineering department: At the time of admission to the institution.

■ FINANCIAL INFORMATION

ESTIMATED EXPENSES (ACADEMIC YEAR)
[Expenses are for the 1992-93 nine-month academic year.]

Residents	Undergraduate	Graduate
Tuition and fees	$ 1,303 (1)	$ 1,185 (1)
College room and board	$ 2,174 (1)	$ 2,476 (1)
Books and supplies	$ –	$ –
Other expenses	$ –	$ –
Total estimated expenses	**$ 3,477**	**$ 3,661**

Non-Residents	Undergraduate	Graduate
Tuition and fees	$ 1,303	$ 1,185
College room and board	$ 2,174	$ 2,476
Books and supplies	$ –	$ –
Other expenses	$ 2,350 (2)	$ 2,350 (2)
Total estimated expenses	**$ 5,827**	**$ 6,011**

Notes: (1) Tuition and fees are per term. (2) Foreign Student Differential fee per academic year.

FINANCIAL AID OFFICE CONTACT
William F. Caley, *Associate Dean, Undergraduate*
Technical University of Nova Scotia
1360 Barrington Street
P.O. Box 1000
Halifax, NS B3J-2X4
Phone: (902)420-7608 Fax: (902)420-7551

GENERAL FINANCIAL AID INFORMATION
Forms accepted/required: Institutional
Additional financial aid information: Scholarships, Canada Student Loans & Bursary.

■ ENGINEERING COLLEGE INFORMATION
[For additional personnel, refer to the *Appendix*.]

HEAD OF ENGINEERING
Donald A. Roy
Phone: (902)420-7600 Fax: (902)420-7551

ENGINEERING COLLEGE ADDRESS
Faculty of Engineering
Technical University of Nova Scotia
1360 Barrington Street
P.O. Box 1000
Halifax, NS B3J-2X4
Phone: (902)420-7500 Fax: (902)420-7551

ENROLLMENTS—BY CLASS
[Numbers are baccalaureate enrollments for the fall 1991 term, unless otherwise footnoted.]
1st-year students/Freshmen .... –
2nd-year students/Sophomores 227
3rd-year students/Juniors 221
4th-year students/Seniors 178
Total **626**

NUMBER OF DEGREES AWARDED—BY PROGRAM
[Numbers are engineering baccalaureate degrees awarded during the 1991-1992 academic year, unless otherwise footnoted. For full details about each engineering program, refer to the *Tables of Degree Programs*.]
Agricultural Engineering 4
Chemical Engineering 18
Civil Engineering 43
Electrical Engineering 38
Industrial Engineering 19
Mechanical Engineering 39
Metallurgical Engineering 3
Mining Engineering 4
Total **168**

PERCENTAGE OF DEGREES AWARDED—BY CATEGORY
[Percentages are of all engineering baccalaureate degrees awarded during the 1991-1992 academic year, unless otherwise footnoted.]
African Canadians .... – %
Asian/Pacific Island Canadians .... – %
Hispanic Canadians .... – %
Native Canadians .... – %
Foreign Citizens .... – %
All Others .... 100 %
Women .... – %
Persons with disabilities .... – %
Students over 25 years of age .... – %

GRADUATE ENROLLMENTS & DEGREES AWARDED
Master's enrollment 169
Master's degrees awarded 36
Doctoral enrollment 75
Doctoral degrees awarded 14

FULL- & PART-TIME FACULTY—BY DEPARTMENT
[Figures are the head count of full-time faculty and the full-time equivalent (FTE) of part-time faculty for each engineering department or equivalent.]

Department	Full	Part
Chemical Eng	6	–
Civil Eng	10	–
Electrical Eng	12	–
Industrial Eng	8	–
Mechanical Eng	12	–
Agricultural Eng	6	–
Mining & Metallurgical Eng	10	–
Applied Mathematics	5	–
Food Science & Tech	3	–

COLLEGE DESCRIPTION

The Faculty of Engineering has seven engineering departments, one service department (the Department of Applied Mathematics) and one applied science department (the Department of Food Science and Technology). Each engineering department, with the exception of one, deals with one undergraduate discipline and is responsible for the degree program in that discipline. The Department of Mining and Metallurgical Engineering administers degree programs in the disciplines of Mining Engineering and Metallurgical Engineering. Engineering disciplines offering co-operative programs schedule work periods in industry at various times of the year. This sequencing may vary according to the discipline.

DEGREE PROGRAMS DESCRIPTION

The engineering program at the Technical University of Nova Scotia is of five years duration for students who have completed senior matriculation (Nova Scotia Grade XII) including mathematics, physics, and chemistry. Each Associated University establishes its own entrance requirements. The program at each Associated University contains courses fulfilling the minimum entrance requirements established by the Senate of the Technical University of Nova Scotia. Students who complete the applied science or engineering program at an Associated University receive a Certificate or Diploma are admitted to the program in Civil, Electrical, Mechanical, Mining, Agricultural, Chemical, Metallurgical or Industrial at TUNS without examination. Students who have completed equivalent university studies elsewhere may also be admitted.

TRANSFER INFORMATION

Residency Requirements: Minimum residence requirement is fifty percent of TUNS portion of the program which must include final year.
Transfer via Articulation Agreements
Admission to engineering: At the end of the sophomore year
Transfer without Articulation Agreements
Admission to engineering: At any time

GRADUATION REQUIREMENTS

Sessional Numerical Average of at least 60 percent per each session.

■ STUDENT PROGRAMS

PROFESSIONAL AND HONORARY SOCIETIES
[For key to acronyms, see *Introduction*.]
Professional Societies: CIM, CSAE, CSCE, CSME, IEEE, IIE

SUPPORT PROGRAMS
TUNS Women on Campus
For: Women
Offered: Sophomore, Junior, Senior, Graduate level

227 Temple University

INSTITUTION PROFILE

HEAD OF THE INSTITUTION
Peter J. Liacouras
Phone: (215)787-7405

GENERAL INFORMATION
[All Students—Fall 1991]
Undergraduate enrollment 22,000
Graduate enrollment 10,000
Total institution enrollment 32,000
Type of institution: Public and Private **Calendar:** Semesters
Location: Philadelphia **Population:** 1,600,000
Setting: Urban
Types of engineering degrees: Engineering & Technology
Other degree-granting colleges: Allied Health Sciences, Arts & Sciences, Business Administration, Communication & Journalism, Dentistry, Education, Fine & Performing Arts, Law, Medicine, Nursing, Pharmacy, Technology & Applied Sciences, Social Administration, Landscape Architecture & Horticulture, Health, Physical Education, Recreation & Dance

ENGINEERING COLLEGE ADMISSIONS INFORMATION
Admission to the engineering college: Minimum of 600 for math SAT.; 550 minimum for verbal SAT.; upper three percentiles in high school class
Entrance Requirements: SAT (Score: 1100), High School courses—English (4 years), Mathematics (preferably through calculus) (4 years), Foreign Language (2 years), Laboratory Science (2 years)
Entrance Recommendations: High School courses—Computer Science (1 year), Advanced Placement Tests
Requirements for foreign students: TOEFL (Score: 550)

DEPARTMENTAL ADMISSIONS INFORMATION
Admission to the engineering department: Same as freshman engineering admission requirements.

FINANCIAL INFORMATION

ESTIMATED EXPENSES (ACADEMIC YEAR)
[Expenses are for the 1992-93 nine-month academic year.]

State Residents	Undergraduate	Graduate
Tuition and fees	$ 4,983	$ 4,356[1]
College room and board	$ 5,540	$ _(A)
Books and supplies	$ 800[C]	$ _(A)
Other expenses	$ _(A)	$ _(A)
Total estimated expenses	$11,323	$ 4,356

Out-of-State Residents	Undergraduate	Graduate
Tuition and fees	$ 9,197	$ 5,490[1]
College room and board	$ 5,540	$ _(A)
Books and supplies	$ 800[C]	$ _(A)
Other expenses	$ _(A)	$ _(A)
Total estimated expenses	$15,537	$ 5,490

Notes: (A) Data not available. (C) Estimated data. (1) Based on 9 credit hours per semester.

FINANCIAL AID OFFICE CONTACT
John Morris, *Director*
Temple University
2nd Floor, Conwell Hall
Broad & Montgomery Streets
Philadelphia, PA 19122
Phone: (215)787-1492

GENERAL FINANCIAL AID INFORMATION
Forms accepted/required: State PHEAA form
Additional financial aid information: University, state, and federal low-interest loans, awards based on merit, work-study positions, ROTC scholarships.

ENGINEERING COLLEGE INFORMATION
[For additional personnel, refer to the *Appendix*.]

HEAD OF ENGINEERING
Charles K. Alexander
Phone: (215)787-7959 **Fax:** (215)787-6936

ENGINEERING COLLEGE ADDRESS
College of Engineering, Computer Sciences and Architecture
Temple University
12th and Norris Streets
Philadelphia, PA 19122
Phone: (215)787-7800 **Fax:** (215)787-6936

ENROLLMENTS—BY CLASS
[Numbers are baccalaureate enrollments for the fall 1991 term, unless otherwise footnoted.]
1st-year students/Freshmen 30
2nd-year students/Sophomores 48
3rd-year students/Juniors 91
4th-year students/Seniors 139
Total 308

NUMBER OF DEGREES AWARDED—BY PROGRAM
[Numbers are engineering baccalaureate degrees awarded during the 1991-1992 academic year, unless otherwise footnoted. For full details about each engineering program, refer to the *Tables of Degree Programs*.]
Civil Engineering 17
Electrical Engineering 47
Mechanical Engineering 18
Total 82

PERCENTAGE OF DEGREES AWARDED—BY CATEGORY
[Percentages are of all engineering baccalaureate degrees awarded during the 1991-1992 academic year, unless otherwise footnoted.]
African Americans .... 10.9 %
Asian/Pacific Island Americans .... 13.4 %
Hispanic Americans .... – %
Native Americans .... – %
Foreign Citizens .... 6.1 %
All Others .... 69.6 %
Women .... 13.4 %
Persons with disabilities .... – % (A)
Students over 25 years of age .... – % (A)

Notes: (A) Data not available.

GRADUATE ENROLLMENTS & DEGREES AWARDED
Master's enrollment 77
Master's degrees awarded 12
Doctoral enrollment 13
Doctoral degrees awarded .... –

ENGINEERING STUDENT DATA
Applicants to the engineering college
Number of applicants to engineering college .... _ (A)
Percent offered admission .... _% (A)

Matriculated engineering students
Percentage in top quartile (25%) of High School class .... 75.0 (C)
Average SAT scores: Math—600[C], Verbal—550[C], Combined—1150[C]

Notes: (A) Data not available. (C) Estimated data.

FULL- & PART-TIME FACULTY—BY DEPARTMENT
[Figures are the head count of full-time faculty and the full-time equivalent (FTE) of part-time faculty for each engineering department or equivalent.]

Department	Full	Part
Civil Eng	10	3.0
Electrical Eng	16	–
Mechanical Eng	12	3.5

COLLEGE DESCRIPTION
The college offers a four-year B.S.E. program in civil, electrical, and mechanical engineering. The first year is dominated by required core courses. Fundamentals of engineering are emphasized during the second year, with specialized courses offered years three and four. A design project is required during the fourth year. Minors are available in mathematics and physics.

DEGREE PROGRAMS DESCRIPTION
The B.S.E. degrees prepare the student for graduate school or for entry-level engineering positions. M.S.E. and Ph.D. degrees in engineering are offered in some engineering areas.

TRANSFER INFORMATION
Residency Requirements: Students must be in residence at Temple University for the last 30 credits prior to receiving their degree.

Transfer via Articulation Agreements
Admission to engineering: At any time
Requirements: Transfer students from a non-ABET-accredited engineering program or non-engineering major must have completed at least 15 semester hours with a GPA of 3.0. Transfers from an ABET-approved program must have a a 2.3 GPA.

Transfer without Articulation Agreements
Admission to engineering: At any time
Requirements: Same as entrance requirements for units with articulation agreements.

GRADUATION REQUIREMENTS
B.S.E. candidates must successfully complete a minimum of 124 semester hours. of courses with a minimum GPA. of 2.3 for engineering and major courses.

STUDENT PROGRAMS

PROFESSIONAL AND HONORARY SOCIETIES
[For key to acronyms, see *Introduction*.]
Professional Societies: ACI, ACM, ASCE, ASME, IEEE, SAE, American Society of Metals (ASM), Pennsylvania Society of Land Surveyors (PSLS)
Honorary Societies: Eta Kappa Nu

SUPPORT PROGRAMS
Student Chapter Organizations: Society of Women Engineers, Minority Engineering Student Association (MESA)
Other Engineering Support Programs: Orientation program for freshman and transfer students; free tutoring center in college; career counseling and job placement; veterans, international, and handicapped student office; academic counseling office; psychological services are university services.

228 University of Tennessee, Chattanooga

■ INSTITUTION PROFILE
HEAD OF THE INSTITUTION
Frederick W. Obear
Phone: (615)755-4141 Fax: (615)756-5559
GENERAL INFORMATION
[All Students—Fall 1991]
Undergraduate enrollment 6,840
Graduate enrollment 1,048
Total institution enrollment **7,888**
Type of institution: Public   Calendar: Semesters
Location: Chattanooga   Population: 275,000
Setting: Urban
Types of engineering degrees: Engineering
Other degree-granting colleges: Arts & Sciences, Business Administration, Education, Humanities & Social Sciences, Nursing, Arts & Sciences, Business Administration, Education, Health & Human Services, Nursing

■ ENGINEERING ADMISSIONS
ADMISSIONS OFFICE CONTACT
Patsy Reynolds
University of Tennessee, Chattanooga
Admissions Office
615 McCallie Avenue
Chattanooga, TN 37403-2598
Phone: (615)755-4662
ENGINEERING COLLEGE ADMISSIONS INFORMATION
Entrance Requirements: SAT, ACT, High School courses—Algebra (2 years), Geometry (1 year), English (4 years), Lab science (2 years), GPA 2.75 and ACT 16 (640 SAT) or GPA 2.00 and ACT 21 (900 SAT) and 13 required high school units.
Entrance Recommendations: High School courses—Trigonometry (1 year), Pre-calculus (1 year), Chemistry (1 year), Physics (1 year)
DEPARTMENTAL ADMISSIONS INFORMATION
Admission to the engineering department: At the time of admission to the institution.
Additional information: Special admission requirement: Math ACT 25 or greater

■ FINANCIAL INFORMATION
ESTIMATED EXPENSES (ACADEMIC YEAR)
[Expenses are for the 1992-93 nine-month academic year.]

State Residents	Undergraduate	Graduate
Tuition and fees	$ 1,670	$ 2,158
College room and board	$ 3,428	$ 3,428
Books and supplies	$ 600	$ 600
Other expenses	$ 2,510	$ 2,510[C]
Total estimated expenses	**$ 8,208**	**$ 8,696**

Out-of-State Residents	Undergraduate	Graduate
Tuition and fees	$ 5,202	$ 5,758
College room and board	$ 3,428	$ 3,428
Books and supplies	$ 600	$ 600
Other expenses	$ 2,510	$ 2,510[C]
Total estimated expenses	**$11,740**	**$12,296**

Notes: (C) Estimated data.

FINANCIAL AID OFFICE CONTACT
Ray P. Fox, *Dean, Admissions & Records*
University of Tennessee, Chattanooga
615 McCallie Avenue
Chattanooga, TN 37403-2598
Phone: (615)755-4511
GENERAL FINANCIAL AID INFORMATION
Forms accepted/required: ACT, FAT, FAR, FFS, IRS, SAR, Institutional

■ ENGINEERING COLLEGE INFORMATION
[For additional personnel, refer to the *Appendix*.]
HEAD OF ENGINEERING
Ronald B. Cox
Phone: (615)755-4121 Fax: (615)755-5229
ENGINEERING COLLEGE ADDRESS
School of Engineering
University of Tennessee, Chattanooga
615 McCallie Avenue
Chattanooga, TN 37403-2598
Phone: (615)755-4121 Fax: (615)755-5229
ENROLLMENTS—BY CLASS
[Numbers are baccalaureate enrollments for the fall 1991 term, unless otherwise footnoted.]
1st-year students/Freshmen 185
2nd-year students/Sophomores 215
3rd-year students/Juniors 223
4th-year students/Seniors 195
Total **818**
NUMBER OF DEGREES AWARDED—BY PROGRAM
[Numbers are engineering baccalaureate degrees awarded during the 1991-1992 academic year, unless otherwise footnoted. For full details about each engineering program, refer to the *Tables of Degree Programs*.]
Chemical Engineering 8
Civil Engineering 2
Electrical Engineering .... –
Engineering .... –
Engineering Management .... 4 [8]
Environmental Engineering 1
Industrial Engineering 2
Mechanical Engineering .... –
Mechanics .... 13 [5]
Thermal Engineering 12
Total **42**

Notes: (5) BSE with Mechanical Concentration. (8) BS degree.
PERCENTAGE OF DEGREES AWARDED—BY CATEGORY
[Percentages are of all engineering baccalaureate degrees awarded during the 1991-1992 academic year, unless otherwise footnoted.]
African Americans 11.4%
Asian/Pacific Island Americans 2.1%
Hispanic Americans 0.5%
Native Americans 0.5%
Foreign Citizens .... – %
All Others 85.5%
Women 17.2%
Persons with disabilities .... – %
Students over 25 years of age .... – %

■ GRADUATE ENROLLMENTS & DEGREES AWARDED
Master's enrollment 135
Master's degrees awarded 16
Doctoral enrollment .... –
Doctoral degrees awarded .... –

■ ENGINEERING STUDENT DATA
Applicants to the engineering college
Number of applicants to engineering college 120
Percent offered admission 88.0%
Matriculated engineering students
Percentage in top quartile (25%) of High School class .... – [A]
Average ACT scores: Math—20.0, Composite—22.19
Notes: (A) Data not available.
FULL- & PART-TIME FACULTY—BY DEPARTMENT
[Figures are the head count of full-time faculty and the full-time equivalent (FTE) of part-time faculty for each engineering department or equivalent.]

Department	Full	Part
Chemical Eng	5	–
Civil Eng	2	–
Electrical Eng	7	1.0
Industrial/Engr Management/Manufacturing	4	0.5
Mechanical Eng	7	1.5

COLLEGE DESCRIPTION
Co-op program (involves 15% of students). Comprehensive, interdisciplinary design experience. Year-long senior design project. Emphasis on interdisciplinary engineering--design, total quality. Faculty - 90% Ph.D., 90% Registered Professional Engineers. Graduates comfortable with project management assignments and proficient in working in teams.
TRANSFER INFORMATION
Residency Requirements: The final 30 semester hours must be completed in residence on this campus.
Transfer via Articulation Agreements
Admission to engineering: At any time
Engineering graduates transferred from ... 4-yr: 21 2-yr: 18
Requirements: Must meet all UTC continuation standards
Transfer without Articulation Agreements
Admission to engineering: At any time
Requirements: Must meet all UTC continuation standards

■ STUDENT PROGRAMS
PROFESSIONAL AND HONORARY SOCIETIES
[For key to acronyms, see *Introduction*.]
Professional Societies: ACM, AIChE, ASCE, ASEM, ASHRAE, ASME, DPMA, IEEE, IIE, NSPE, Association of General Contractors, Tennessee Society of Professional Engineers
Honorary Societies: Tau Beta Pi, Upsilon Pi Epsilon
SUPPORT PROGRAMS
Student Chapter Organizations: Society of Women Engineers, National Society of Black Engineers
Other Engineering Support Programs: Freshman Engineering and Advising Program (Dr. Mike Jones, Director); Freshman Seminar Series (Programs on careers, ethics, professionalism); Senior Seminar Series (Programs on careers, ethics, professionalism).

229 University of Tennessee, Knoxville

INSTITUTION PROFILE

HEAD OF THE INSTITUTION
Joseph E. Johnson
Phone: (615)974-2241　　　　Fax: (615)974-3753

GENERAL INFORMATION
[All Students—Fall 1991]
Undergraduate enrollment 19,342
Graduate enrollment 6,191
Total institution enrollment **25,533**
Type of institution: Public　　**Calendar:** Semesters
Nearest city: Knoxville, TN　　**Population:** 300,000
Miles from main campus: 1　　**Setting:** Urban
Types of engineering degrees: Engineering
Other degree-granting colleges: Agricultural & Environmental, Architecture, Arts & Sciences, Business Administration, Communication & Journalism, Education, Law, Nursing

ENGINEERING ADMISSIONS

ADMISSIONS OFFICE CONTACT
Gordon E. Stanley
University of Tennessee, Knoxville
320 Student Services Building
Knoxville, TN 37996-0200
Phone: (615)974-2184　　　　Fax: (615)974-6435

ENGINEERING COLLEGE ADMISSIONS INFORMATION
Entrance Requirements: High School courses—English (4 years), Mathematics (3 years), Natural Science (2 years), Foreign Language (2 years), SAT and ACT scores are used in combination with high school GPA to determine admission, there is no definite minimum.
Entrance Recommendations: High School courses—Trigonometry or Advanced Mathematics (1 year), American History (1 year), World History or geography (1 year)
Requirements for foreign students: TOEFL (Score: 525); Foreign applications are reviewed by the Freshman Engineering Advising Center to determine admission.

DEPARTMENTAL ADMISSIONS INFORMATION
Admission to the engineering department: At the end of the first year.
Additional information: Departmental admission requirements vary with departments. Enrollment demand is the main criterion for establishing departmental admission.

FINANCIAL INFORMATION

ESTIMATED EXPENSES (ACADEMIC YEAR)
[Expenses are for the 1992-93 nine-month academic year.]

State Residents	Undergraduate	Graduate
Tuition and fees	$ 1,898	$ 2,308
College room and board	$ 3,200	$ 3,200
Books and supplies	$ 1,800	$ 1,600
Other expenses	$ —(A)	$ —(A)
Total estimated expenses	**$ 6,898**	**$ 7,108**

Out-of-State Residents	Undergraduate	Graduate
Tuition and fees	$ 5,498	$ 5,708
College room and board	$ 3,200	$ 3,200
Books and supplies	$ 1,800	$ 1,600
Other expenses	$ —(A)	$ —(A)
Total estimated expenses	**$10,498**	**$10,508**

Notes: (A) Data not available.

FINANCIAL AID OFFICE CONTACT
John E. Mays, *Director of Financial Aid*
University of Tennessee, Knoxville
115 Student Services Building
Knoxville, TN 37996-0210
Phone: (615)974-3131

GENERAL FINANCIAL AID INFORMATION
Forms accepted/required: Financial Aid Office should be contacted for information about forms.
Additional financial aid information: Financial aid is available throughout the Campus Financial Aid Office and through the College of Engineering Office of Associate Dean for Academic Affairs.

ENGINEERING COLLEGE INFORMATION
[For additional personnel, refer to the *Appendix*.]

HEAD OF ENGINEERING
Jerry E. Stoneking
Phone: (615)974-5321　　　　Fax: (615)974-2669
EMail: BITNET:Stonekin:UTK VAX

ENGINEERING COLLEGE ADDRESS
College of Engineering
University of Tennessee, Knoxville
124 Perkins Hall
Knoxville, TN 37996-2000
Phone: (615)974-5321　　　　Fax: (615)974-2669

ENROLLMENTS—BY CLASS
[Numbers are baccalaureate enrollments for the fall 1991 term, unless otherwise footnoted.]
1st-year students/Freshmen 695
2nd-year students/Sophomores 383
3rd-year students/Juniors 328
4th-year students/Seniors .... 588 [1]
Total **1,994**

Notes: (1) 256 part-time students not included in totals.

NUMBER OF DEGREES AWARDED—BY PROGRAM
[Numbers are engineering baccalaureate degrees awarded during the 1991-1992 academic year, unless otherwise footnoted. For full details about each engineering program, refer to the *Tables of Degree Programs*.]
Aerospace Engineering 13
Agricultural Engineering 6
Chemical Engineering 35
Civil Engineering 35
Electrical & Computer Engineering 95
Engineering Physics 2
Engineering Science & Mechanics 14
Industrial Engineering 69
Materials Science 8
Mechanical Engineering 51
Nuclear Engineering 12
Total **340**

PERCENTAGE OF DEGREES AWARDED—BY CATEGORY
[Percentages are of all engineering baccalaureate degrees awarded during the 1991-1992 academic year, unless otherwise footnoted.]
African Americans 5.5%
Asian/Pacific Island Americans 3.2%
Hispanic Americans .... —%
Native Americans .... —%
Foreign Citizens 9.0%
All Others 82.3%
Women 15.9%
Persons with disabilities .... —% (A)
Students over 25 years of age .... —% (A)

Notes: (A) Data not available.

GRADUATE ENROLLMENTS & DEGREES AWARDED
Master's enrollment 244
Master's degrees awarded 193
Doctoral enrollment 130
Doctoral degrees awarded .... 47 [2]

Notes: (2) Numbers do not include 710 part-time M.S. and PhD students.

ENGINEERING STUDENT DATA

Applicants to the engineering college
Number of applicants to engineering college 968
Percent offered admission .... —% (A)

Matriculated engineering students
Percentage in top quartile (25%) of High School class .... — (A)
Average SAT scores: Math—597.1, Combined—1072.2
Average ACT scores: Math—25.2, Composite—25

Notes: (A) Data not available.

FULL- & PART-TIME FACULTY—BY DEPARTMENT
[Figures are the head count of full-time faculty and the full-time equivalent (FTE) of part-time faculty for each engineering department or equivalent.]

Department	Full	Part
Chemical Eng	13	–
Civil Eng	18	1.3
Electrical Eng	38	–
Industrial Eng	12	–
Materials Science & Eng	10	3.2
Eng Science & Mechanics	19	–
Nuclear Eng	11	–
Mechanical & Aerospace Eng	18	–
Agricultural Eng	19	–
Eng Physics	5	–

COLLEGE DESCRIPTION
The University of Tennessee, Knoxville, established in 1794, is a public, Land Grant Institution located in an urban setting. The UTK engineering program is the fourth oldest in the United States. The College provides orientation sessions for all new students, operates a Freshman Advising Center, offers a required Freshman Seminar, offers a Cooperative Engineering option, and operates several workstation based computer laboratories. The College has a very close working relationship with the College of Business Administration including a minor in Business Administration for engineering majors. The College has a nationally recognized Minority Engineering Program supported by industry with a $11,000 stipend in addition to co-op earnings.

DEGREE PROGRAMS DESCRIPTION
Some undergraduate degree programs have a progression requirement. The Undergraduate Catalog for the University of Tennessee, Knoxville should be consulted for detailed information on progression requirements.

TRANSFER INFORMATION
Residency Requirements: The last 30 semester hours of course work for the degree must be taken at the University of Tennessee, Knoxville.

Transfer via Articulation Agreements
Admission to engineering: At any time
Engineering graduates transferred from　　..　4-yr: –　　2-yr: –
Requirements: Admission of transfer students is determined by each department.

Transfer without Articulation Agreements
Admission to engineering: At any time
Engineering graduates transferred from　　..　4-yr: –　　2-yr: –
Requirements: Transfer eligibility is determined by each department based on academic record and enrollment demand in the department.

GRADUATION REQUIREMENTS
An overall 2.00 GPA is required for graduation. Some departments have minimum major GPA requirements, consult university catalog.

STUDENT PROGRAMS

PROFESSIONAL AND HONORARY SOCIETIES
[For key to acronyms, see *Introduction*.]
Professional Societies: AIAA, AIChE, ANS, ASAE, ASCE, ASME, BMES, IEEE, IIE, SAE, National Society of Black Engineers, American Society for Metals
Honorary Societies: Alpha Pi Mu, Chi Epsilon, Eta Kappa Nu, Pi Tau Sigma, Tau Beta Pi

SUPPORT PROGRAMS
Student Chapter Organizations: Society of Women Engineers, National Society of Black Engineers

Minority Engineering Scholarship Program
For: Ethnic Minorities　　**Available:** Year round
Offered: Prematriculation, Freshman, Sophomore, Junior, Senior
Other Engineering Support Programs: The College provides an orientation program for all new students, both freshman and transfers. The advising of all freshman is done centrally in the Freshman Engineering Advising Center. The Minority Engineering Scholarship Program provides a two week orientation program prior to the beginning of fall semester for all new students entering MESP.

230 Tennessee State University

INSTITUTION PROFILE

HEAD OF THE INSTITUTION
James A. Hefner
Phone: (615)320-3432 Fax: (615)320-3376

GENERAL INFORMATION
[All Students—Fall 1991]
Undergraduate enrollment 6,605
Graduate enrollment 986
Total institution enrollment **7,591**
Type of institution: Public Calendar: Semesters
Location: Nashville Population: 780,000
Setting: Urban
Types of engineering degrees: Engineering
Other degree-granting colleges: Arts & Sciences, Education, Nursing, Agriculture and Home Economics, Allied Health Professionals, Business, Graduate Studies and Research

ENGINEERING ADMISSIONS

ADMISSIONS OFFICE CONTACT
Erskine R. Vanderbilt
Tennessee State University
3500 John A. Merritt Boulevard
Nashville, TN 37209-1561
Phone: (615)320-3420 Fax: (615)320-3114

ENGINEERING COLLEGE ADMISSIONS INFORMATION
Entrance Requirements: SAT (Score: 720), ACT (Score: 19), High School courses—English (4 years), Algebra (2 years), Geometry or Advanced Mathematics (1 year), Natural/Physical Science with lab. (2 years)
Entrance Recommendations: High School courses—Social sciences (1 year), Visual and/or Performing Arts (1 year), United States History (1 year), A Single Foreign Language (2 years)
Requirements for foreign students: TOEFL (Score: 500); financial statement

DEPARTMENTAL ADMISSIONS INFORMATION
Admission to the engineering department: At the time of admission to the institution.
Additional information: All students are required to pass the Engineering Entrance Examination (EEE) in physics, chemistry, and calculus by the end of the sophomore year, in order to take junior level engineering courses. Eligibility to take this test is a GPA of 2.5 in the above courses and a cumulative GPA of 2.5 in all courses. A score of 75% is required to pass EEE. EEE test is given five times a year, and a student must pass this test within three attempts. Tutorial sessions are provided by student organizations to prepare students for the test.

FINANCIAL INFORMATION

ESTIMATED EXPENSES (ACADEMIC YEAR)
[Expenses are for the 1992-93 nine-month academic year.]

State Residents	Undergraduate	Graduate
Tuition and fees	$ 1,632	$ 2,098
College room and board	$ 2,660	$ 2,660
Books and supplies	$ 1,000	$ 800
Other expenses	$ —(1)	$ —(1)
Total estimated expenses	**$ 5,292**	**$ 5,558**

Out-of-State Residents	Undergraduate	Graduate
Tuition and fees	$ 5,234	$ 5,700
College room and board	$ 2,660	$ 2,660
Books and supplies	$ 1,000	$ 800
Other expenses	$ —(1)	$ —(1)
Total estimated expenses	**$ 8,894**	**$ 9,160**

Notes: (1) A one time orientation fee of $20 and in addition a fee of $25 for foreign students.

FINANCIAL AID OFFICE CONTACT
Wilson Lee, *Director of Financial Aid*
Tennessee State University
3500 John A. Merritt Boulevard
Nashville, TN 37209-1561
Phone: (615)320-3440 Fax: (615)320-3114

GENERAL FINANCIAL AID INFORMATION
Forms accepted/required: FAF, Institutional, College of Engineering and Technology Scholarship Form
Additional financial aid information: A number of grants, scholarships and loans from federal, state, and private sources are available to students. Also, employment is possible through the Work-Study Program, College Work-Aid Program, and the Academic Work Scholarship Program. In addition engineering students may also qualify for work aid scholarship as tutors and graders. Merit based scholarships exist for minority and non-minority students. Undergraduate research assistantships are available to outstanding juniors and seniors.

ENGINEERING COLLEGE INFORMATION

[For additional personnel, refer to the *Appendix*.]

HEAD OF ENGINEERING
Decatur B. Rogers
Phone: (615)320-3550 Fax: (615)320-3554

ENGINEERING COLLEGE ADDRESS
College of Engineering & Technology
Tennessee State University
3500 John A. Merritt Boulevard
Nashville, TN 37209-1561
Phone: (615)320-3131 Fax: (615)320-3114

ENROLLMENTS—BY CLASS
[Numbers are baccalaureate enrollments for the fall 1991 term, unless otherwise footnoted.]
1st-year students/Freshmen 300
2nd-year students/Sophomores 167
3rd-year students/Juniors 134
4th-year students/Seniors 246
Total **847**

NUMBER OF DEGREES AWARDED—BY PROGRAM
[Numbers are engineering baccalaureate degrees awarded during the 1991-1992 academic year, unless otherwise footnoted. For full details about each engineering program, refer to the *Tables of Degree Programs*.]
Architectural Engineering 10
Civil Engineering 12
Electrical Engineering 20
Mechanical Engineering 30
Total **72**

PERCENTAGE OF DEGREES AWARDED—BY CATEGORY
[Percentages are of all engineering baccalaureate degrees awarded during the 1991-1992 academic year, unless otherwise footnoted.]
African Americans .... 47.2 %
Asian/Pacific Island Americans .... – %
Hispanic Americans .... – %
Native Americans .... – %
Foreign Citizens .... 12.5 %
All Others .... 40.3 %
Women .... 25.0 %
Persons with disabilities .... – %
Students over 25 years of age .... 20.8 %

GRADUATE ENROLLMENTS & DEGREES AWARDED
Master's enrollment 42 (2)
Master's degrees awarded 11
Doctoral enrollment .... – (B)
Doctoral degrees awarded .... – (B)

Notes: (B) Data not applicable. (2) Master of Engineering degree with options in Civil Engineering, Electrical engineering and Mechanical Engineering.

ENGINEERING STUDENT DATA
Applicants to the engineering college
Number of applicants to engineering college 750
Percent offered admission 70.5%

Matriculated engineering students
Percentage in top quartile (25%) of High School class 60.0
Average ACT scores: Math—21, Composite—20.8

FULL- & PART-TIME FACULTY—BY DEPARTMENT
[Figures are the head count of full-time faculty and the full-time equivalent (FTE) of part-time faculty for each engineering department or equivalent.]

Department	Full	Part
Civil Eng	6	–
Electrical Eng	8	0.1
Mechanical Eng	7	0.5
Architectural Eng	5	–

COLLEGE DESCRIPTION
The School of Engineering and Technology is located in the greatest area of industrial expansion in the Southeast--within 30 miles of Nissan Motor Manufacturing Corporation and General Motors' Saturn Manufacturing plant. The average engineering class size is about 25 students. Design and computer usage is implemented throughout the curricula as well as the use of communication skills. Our laboratories are being equipped with leading edge technology equipment in Robotics, Computer-aided-design and Computer integrated manufacturing, to name a few. Engineering faculty have obtained doctoral degrees from 15 different renowned universities, and they also teach all the engineering laboratories. About 33% of the faculty are engaged in DoD, DOE, and NASA funded research projects, and they employ undergraduate and graduate students to assist them in research activities. Summer employment and Co-op opportunities are available to Students. About 20% of our graduates pursue graduate studies

DEGREE PROGRAMS DESCRIPTION
The College of Engineering and Technology offers four ABET accredited engineering degree programs; Architectural, Civil, Electrical, and Mechanical Engineering. The Architectural Engineering program develops expertise in the areas of architectural design, structural design, environmental systems design and construction management. The Civil Engineering department develops expertise in structural design, transportation engineering, water resource management and geotechnical engineering. Communications, electromagnetics, power engineering, computer engineering, control systems, circuit analysis and design expertise are developed in Electrical Engineering program, while the Mechanical Engineering program develops expertise in robotics, computer-aided design, mechanical systems design, thermal systems design, manufacturing science and materials processing. Our students present design projects to panel of judges from industry and participate actively in the student contests. The College is in the process of introducing Computer Engineering and Manufacturing Engineering concentrations in Electrical and Mechanical Engineering programs respectively. A Master of Science degree in Computer and Information Systems Engineering (CISE) is planned for Fall 1993.

TRANSFER INFORMATION
Residency Requirements: A transfer student must spend at least one academic year in residence at the university, and while in residence earn not less than 30 semester hours of credit with a minimum average of 'c' (2.0). Students transferring from a junior or community college must complete a minimum of 60 hours of credit for the bachelor's degree at TSU.

Transfer via Articulation Agreements
Admission to engineering: At the end of the sophomore year
Engineering graduates transferred from ... 4-yr: 10 2-yr: 12
Requirements: Students must have a GPA of 2.5 or better and must pass engineering entrance examination prior to taking junior level engineering courses.

Transfer without Articulation Agreements
Admission to engineering: At any time
Engineering graduates transferred from ... 4-yr: 37 2-yr: 32
Requirements: Students are admitted to engineering program after they have passed the engineering entrance examination. (EEE) A student must have a GPA of 2.5 in physics, chemistry and calculus courses and overall GPA of 2.5 or better to take EEE test.

GRADUATION REQUIREMENTS
Students are not permitted to accumulate more than two 'D' grades in all required engineering, mathematics, physics, and chemistry course(s). Students must repeat the courses in which a grade of 'D' is received the very next time the course(s) is offered. All engineering students are required to take ACT-COMP and EIT examinations one semester prior to graduation.

STUDENT PROGRAMS

PROFESSIONAL AND HONORARY SOCIETIES
[For key to acronyms, see *Introduction*.]
Professional Societies: ASCE, ASHRAE, ASME, IEEE, NSAE
Honorary Societies: Eta Kappa Nu, Pi Tau Sigma

SUPPORT PROGRAMS
Student Chapter Organizations: Society of Women Engineers, National Society of Black Engineers, Engineering, Math, and Science Saturday Academy, Minority Introduction to Engineering, Engineering Concept Institute, Freshman Mentor Program

Engineering, Math and Science Saturday Academy
For: Women & Ethnic Minorities Available: Year round
Offered: Prematriculation

Minority Introduction To Engineering
For: Ethnic Minorities Available: Summer only
Offered: Prematriculation

Engineering Concept Institute
For: Ethnic Minorities Available: Summer only
Offered: Prematriculation

Freshman Mentor Program
For: Ethnic Minorities Available: Academic year
Offered: Freshman

Other Engineering Support Programs: The school has a PLATO-based tutorial center using IBM PC/70s for engineering freshmen & sophomores. Courses on algebra, geometry, trigonometry, calculus, physics, chemistry, statics and dynamics are available. Student tutors are provided for all freshman engineering courses and they hold tutorial sessions at least twice a week. All faculty also set aside 10 hours per week for student advisement

231 Tennessee Technological University

■ INSTITUTION PROFILE

HEAD OF THE INSTITUTION
Angelo A. Volpe
Phone: (615)372-3241   Fax: (615)372-3898

GENERAL INFORMATION
[All Students—Fall 1991]
Undergraduate enrollment 7,261
Graduate enrollment 899
Total institution enrollment 8,160
Type of institution: Public   **Calendar:** Semesters
Nearest city: Nashville   **Population:** 1,000,000
Miles from main campus: 85   **Setting:** Small Town
Types of engineering degrees: Engineering
Other degree-granting colleges: Arts & Sciences, Business Administration, Education, Nursing, Agriculture & Home Economics

■ ENGINEERING ADMISSIONS

ADMISSIONS OFFICE CONTACT
James C. Perry
Tennessee Technological University
Box 5006
North Dixie Avenue
Cookeville, TN 38505
Phone: (615)372-3888   Fax: (615)372-3898

ENGINEERING COLLEGE ADMISSIONS INFORMATION
Entrance Requirements: ACT (Score: 20), High School courses—Mathematics (3 years), Science (2 years), English (4 years)
Entrance Recommendations: High School courses—Mathematics (4 years), Science (4 years)
Requirements for foreign students: TOEFL (Score: 500); financial statement

DEPARTMENTAL ADMISSIONS INFORMATION
Admission to the engineering department: Completion of 30 semester hours with a 2.5 QPA including at least a C grade in Calculus I.

■ FINANCIAL INFORMATION

ESTIMATED EXPENSES (ACADEMIC YEAR)
[Expenses are for the 1992-93 nine-month academic year.]

State Residents	Undergraduate	Graduate
Tuition and fees	$ 1,634	$ 8,553
College room and board	$ 2,712	$ 6,093
Books and supplies	$ 660	$ 990
Other expenses	$ –	$ –
Total estimated expenses	**$ 5,006**	**$ 15,636**

Out-of-State Residents	Undergraduate	Graduate
Tuition and fees	$ 5,236	$ 8,553
College room and board	$ 2,712	$ 6,093
Books and supplies	$ 660	$ 990
Other expenses	$ –	$ –
Total estimated expenses	**$ 8,608**	**$ 15,636**

FINANCIAL AID OFFICE CONTACT
Raymond L. Holbrook, *Director, Financial Aid*
Tennessee Technological University
Box 5076
North Dixie Avenue
Cookeville, TN 38505
Phone: (615)372-3073   Fax: (615)372-6138

GENERAL FINANCIAL AID INFORMATION
Forms accepted/required: ACT, AFSA, FAF, FAT, FSA, FFS, Stafford, SAR, Institutional
Additional financial aid information: Merit-based scholarships; Need-based scholarships; Part-time campus jobs; Short-term loans

■ ENGINEERING COLLEGE INFORMATION
[For additional personnel, refer to the *Appendix*.]

HEAD OF ENGINEERING
George M. Swisher
Phone: (615)372-3172   Fax: (615)372-6172
EMail: GMS8735@TNTECH.EDU

ENGINEERING COLLEGE ADDRESS
College of Engineering
Tennessee Technological University
Box 5005
Peachtree Street
Cookeville, TN 38505
Phone: (615)372-3172   Fax: (615)372-6172

ENROLLMENTS—BY CLASS
[Numbers are baccalaureate enrollments for the fall 1991 term, unless otherwise footnoted.]

1st-year students/Freshmen 561
2nd-year students/Sophomores 288
3rd-year students/Juniors 269
4th-year students/Seniors 466
Total **1,584**

NUMBER OF DEGREES AWARDED—BY PROGRAM
[Numbers are engineering baccalaureate degrees awarded during the 1991-1992 academic year, unless otherwise footnoted. For full details about each engineering program, refer to the *Tables of Degree Programs*.]

Chemical Engineering 9
Civil Engineering 40
Electrical Engineering 51
Industrial Engineering 23
Mechanical Engineering 72
Total **195**

PERCENTAGE OF DEGREES AWARDED—BY CATEGORY
[Percentages are of all engineering baccalaureate degrees awarded during the 1991-1992 academic year, unless otherwise footnoted.]

African Americans .... 2.0 %
Asian/Pacific Island Americans .... 2.0 %
Hispanic Americans .... 1.0 %
Native Americans .... – %
Foreign Citizens .... – %
All Others .... 95.0 %
Women .... 11.6 %
Persons with disabilities .... – %
Students over 25 years of age .... – %

GRADUATE ENROLLMENTS & DEGREES AWARDED
Master's enrollment 117
Master's degrees awarded 30
Doctoral enrollment 52
Doctoral degrees awarded 7

ENGINEERING STUDENT DATA
Applicants to the engineering college
Number of applicants to engineering college 622
Percent offered admission 88.0%

Matriculated engineering students
Percentage in top quartile (25%) of High School class .... – (A)
Average ACT scores: Math—25, Composite—24

Notes: (A) Data not available.

FULL- & PART-TIME FACULTY—BY DEPARTMENT
[Figures are the head count of full-time faculty and the full-time equivalent (FTE) of part-time faculty for each engineering department or equivalent.]

Department	Full	Part
Chemical Eng	5	0.3
Civil Eng	13	–
Electrical Eng	17	0.3
Industrial Eng	6	–
Mechanical Eng	19	0.8

COLLEGE DESCRIPTION
The College offers a four-year undergraduate program in Engineering and Industrial Technology as well as MS and PhD programs. All freshman engineering students enter the College through the Basic Engineering Department with transfer to one of the degree-granting departments after meeting certain specific requirements. Emphasis during the first year is given to computer-aided engineering using the Sun Sparcstation I's for graphics and programming. Co-op is an option in all degrees emphasizing a year-on/year-off program. Students complete a senior design project and are expected to complete the national standardized Fundamentals of Engineering exam as well as the ACT comp exam. All programs are accredited by ABET or NAIT.

DEGREE PROGRAMS DESCRIPTION
The BS degrees in the various engineering programs prepare the graduate for entry-level professional positions or for graduate study in each of the respective fields. The program highlights use of Sun Sparcstation I computers for freshman graphics. Graduate study in engineering can be carried out in the MS or PhD program emphasizing the applications of the natural sciences to the analysis and solution of engineering problems.

TRANSFER INFORMATION
Residency Requirements: Two semesters in the junior and senior years, including the last semester, must be in residence.

Transfer via Articulation Agreements
Admission to engineering: At any time

Transfer without Articulation Agreements
Admission to engineering: At any time

GRADUATION REQUIREMENTS
Student must have 2.00 (C) QPA in four different areas: All coursework at TTU; all coursework in major at TTU; all coursework for all colleges/universities; all coursework in major for all colleges/universities

■ STUDENT PROGRAMS

PROFESSIONAL AND HONORARY SOCIETIES
[For key to acronyms, see *Introduction*.]
Professional Societies: AIChE, ASCE, ASHRAE, ASME, AWS, IEEE, IIE, SAME, SME
Honorary Societies: Chi Epsilon, Eta Kappa Nu, Pi Tau Sigma, Tau Beta Pi, Theta Tau

SUPPORT PROGRAMS
Student Chapter Organizations: Society of Women Engineers, National Society of Black Engineers

Minority Engineering
For: Ethnic Minorities   **Available:** Year round
Offered: Prematriculation, Freshman, Sophomore, Junior, Senior, Graduate level

Other Engineering Support Programs: UNIV 101, University Life, is an introduction to university life with particular emphasis on factors relating to college success. Additionally, there is a fall orientation program; career counseling; international student affairs office; placement services; and other normal services.

232 University of Texas at Arlington

■ INSTITUTION PROFILE
HEAD OF THE INSTITUTION
Wendell H. Nedderman
Phone: (817)273-2101 Fax: (817)794-5656

GENERAL INFORMATION
[All Students—Fall 1991]
Undergraduate enrollment 20,404
Graduate enrollment 4,321
Total institution enrollment 24,725
Type of institution: Public Calendar: Semesters
Location: Arlington Population: 250,000
Setting: Urban
Types of engineering degrees: Engineering
Other degree-granting colleges: Architecture, Business Administration, Nursing, Liberal Arts, Science, Social Work, Urban & Public Affairs

■ ENGINEERING ADMISSIONS
ADMISSIONS OFFICE CONTACT
Zack Prince
University of Texas at Arlington
703 Monroe
Arlington, TX 76019
Phone: (817)273-3565

ENGINEERING COLLEGE ADMISSIONS INFORMATION
Entrance Requirements: SAT (Score: 1100), ACT (Score: 28), High School courses—English (4 years), Algebra and Geometry (2 years), Social Science (3 years), Science (2 years)
Entrance Recommendations: High School courses—Trigonometry (1 year), Calculus (1 year), Physics or Chemistry (1 year), Foreign Language (a single modern) (2 years)
Requirements for foreign students: financial statement

DEPARTMENTAL ADMISSIONS INFORMATION
Admission to the engineering department: At the time of admission to the institution.
Additional information: Students must resent a SAT score of 1100 or higher or a composite ACT score of 28 or higher. In addition, score at least 550 on the math and 450 on the verbal portions of the SAT or score at least 27 on the math and 23 on the verbal portions of the ACT. Students must have completed the prerequisites to enroll in the appropriate engineering math and science courses. They also must present two high school units of single foreign language.

■ FINANCIAL INFORMATION
ESTIMATED EXPENSES (ACADEMIC YEAR)
[Expenses are for the 1992-93 nine-month academic year.]

State Residents	Undergraduate	Graduate
Tuition and fees	$ 1,566	$ 2,430
College room and board	$ 4,500	$ 4,500
Books and supplies	$ 600	$ 550
Other expenses	$ —(B)	$ —(B)
Total estimated expenses	**$ 6,666**	**$ 7,480**

Out-of-State Residents	Undergraduate	Graduate
Tuition and fees	$ 6,534	$ 6,894
College room and board	$ 4,500	$ 4,500
Books and supplies	$ 550	$ 600
Other expenses	$ —(B)	$ —(B)
Total estimated expenses	**$11,584**	**$11,994**

Notes: (B) Data not applicable.

FINANCIAL AID OFFICE CONTACT
Judy Walker, *Director*
University of Texas at Arlington
800 South Cooper
703 Monroe
Arlington, TX 76019
Phone: (817)273-3561 Fax: (817)273-5555

GENERAL FINANCIAL AID INFORMATION
Forms accepted/required: ACT, AFSA, CSS-FAF, FAF, FAT, FFS, Stafford, IRS, SAR, SDF, Supplemental, Institutional, Military Discharge Paper, Driver's License
Additional financial aid information: Types available: Gift Aid Grants Work Study Student Loans Scholarships

■ ENGINEERING COLLEGE INFORMATION
[For additional personnel, refer to the *Appendix*.]
HEAD OF ENGINEERING
John H. McElroy
Phone: (817)273-2571 Fax: (817)273-2548

ENGINEERING COLLEGE ADDRESS
College of Engineering
University of Texas at Arlington
416 Yates
Arlington, TX 76019
Phone: (817)273-2571 Fax: (817)273-2548

ENROLLMENTS—BY CLASS
[Numbers are baccalaureate enrollments for the fall 1991 term, unless otherwise footnoted.]
1st-year students/Freshmen 304
2nd-year students/Sophomores 222
3rd-year students/Juniors 315
4th-year students/Seniors 1,024 (1)
Total 1,865
Notes: (1) Number includes seniors as well as non-degreed students enrolled in the College of Engineering.

NUMBER OF DEGREES AWARDED—BY PROGRAM
[Numbers are engineering baccalaureate degrees awarded during the 1991-1992 academic year, unless otherwise footnoted. For full details about each engineering program, refer to the *Tables of Degree Programs*.]
Aerospace Engineering 18
Biomedical Engineering .... –
Civil Engineering 20
Computer Science Engineering 71
Electrical Engineering 95
Industrial Engineering 30
Mechanical Engineering 48
Total 282

PERCENTAGE OF DEGREES AWARDED—BY CATEGORY
[Percentages are of all engineering baccalaureate degrees awarded during the 1991-1992 academic year, unless otherwise footnoted.]
African Americans .... 2.4 %
Asian/Pacific Island Americans .... 16.0 %
Hispanic Americans .... 3.3 %
Native Americans .... – % (A)
Foreign Citizens .... 24.5 %
All Others .... 53.8 %
Women .... 12.0 %
Persons with disabilities .... – % (A)
Students over 25 years of age .... – % (2)
Notes: (A) Data not available. (2) Data entered is for the University as a whole.

GRADUATE ENROLLMENTS & DEGREES AWARDED
Master's enrollment 1,001
Master's degrees awarded 282
Doctoral enrollment 263
Doctoral degrees awarded 27

ENGINEERING STUDENT DATA
Applicants to the engineering college
Number of applicants to engineering college .... – (A)
Percent offered admission .... –% (A)

Matriculated engineering students
Percentage in top quartile (25%) of High School class ..... – (A)
Average SAT scores: Math—636, Verbal—521, Combined—1157
Average ACT scores: Math—28, Composite—27
Notes: (A) Data not available.

FULL- & PART-TIME FACULTY—BY DEPARTMENT
[Figures are the head count of full-time faculty and the full-time equivalent (FTE) of part-time faculty for each engineering department or equivalent.]

Department	Full	Part
Civil Eng	14	2.3
Electrical Eng	23	4.4
Industrial Eng	8	–
Mechanical Eng	15	2.1
Aerospace Eng	11	1.2
Computer Science Eng	15	8.4

COLLEGE DESCRIPTION
Engineering has been defined as 'The art of directing the great sources of power in nature, for the use and convenience of man.' Engineering is a rewarding and satisfying career for those women and men with the talent and determination to meet the challenges of rapidly changing technologies and complex societal problems in urgent need of solutions. The College of Engineering provides an opportunity for study in several of the branes of engineering under the guidance of an excellent faculty. Baccalaureate degree programs are offered in aerospace, civil, computer science, electrical, industrial, and mechanical engineering. These programs are accredited by the Accreditation Board for Engineering and Technology (ABET). Graduate degrees, both masters and doctoral, are offered in each of these disciplines, and in other areas of specialization: biomedical, computer science, engineering mechanics, manufacturing, and materials science.

DEGREE PROGRAMS DESCRIPTION
The B.S. program consists of courses offered both during the day and at night. The M.S. program consists of both a research track (24 semester hours plus thesis) and a non research track (36 semester hours plus a comprehensive examination). The Ph.D. programs normally require research projects in areas supported by sponsored research projects. Courses for all graduate degrees are offered both day and night. Most graduate theory courses are offered over closed circuit television to over 30 industrial sites in the Dallas Fort Worth metroplex.

TRANSFER INFORMATION
Residency Requirements: Each candidate for a degree must complete and receive credit in residence for at least 30 semester hours. Successful completion of a course of study prescribed by the major department, including a minimum of 36 advanced hours.

Transfer via Articulation Agreements
Admission to engineering: At any time
Engineering graduates transferred from ... 4-yr: 6 2-yr: 75
Requirements: Transfer students must meet the same requirements for admission as students entering without transfer credit (SAT or ACT score, foreign language, etc.) and minimum 2.5 out of 4.0 GPA on all courses applicable to the engineering degree and have a three calculation GPA on transfer credits sufficient to enter in good academic standing into the College of Engineering.

Transfer without Articulation Agreements
Admission to engineering: At any time
Engineering graduates transferred from ... 4-yr: 78 2-yr: 25
Requirements: Same as listed above

GRADUATION REQUIREMENTS
The degree requirements for each engineering program are consistent with the criteria established by ABET. The College determines the acceptability of lower level courses transferred from community or junior colleges in the region by formal coordination with each of these institutions and ensuring that they set the same standards as the establish course syllabi consistent with those used by UTA. Additionally, the college monitors the GPA performance of transfer students from those institutions from which we receive the majority of our transfer students.

■ STUDENT PROGRAMS
PROFESSIONAL AND HONORARY SOCIETIES
[For key to acronyms, see *Introduction*.]
Professional Societies: ACM, AHS, AIAA, ASCE, ASHRAE, ASME, IEEE, IIE, SAE, American Production & Inventory Control Society, American Institute of Industrial Engineers
Honorary Societies: Alpha Pi Mu, Chi Epsilon, Eta Kappa Nu, Pi Tau Sigma, Sigma Gamma Tau, Tau Beta Pi, Upsilon Pi Epsilon, Pi Sigma Pi

SUPPORT PROGRAMS
Student Chapter Organizations: Society of Women Engineers, National Society of Black Engineers, Society of Hispanic Professional Engineers
Other Engineering Support Programs: The College of Engineering has installed a counseling and probation program to more closely track the progress of engineering students and to identify those needing academic assistance by the end of their first enrollment. The College Probation rules requires students that are doing poorly in math and science courses to take appropriate remediation prior to continuing their degree program.

233 University of Texas at Austin

■ INSTITUTION PROFILE

HEAD OF THE INSTITUTION
Robert M. Berdahl
Phone: (512)471-1232　　　Fax: (512)471-8102

GENERAL INFORMATION
[All Students—Fall 1991]
Undergraduate enrollment 35,911
Graduate enrollment 13,342
Total institution enrollment **49,253**

Type of institution: Public　　**Calendar:** Semesters
Location: Austin, Texas　　**Population:** 465,622
Setting: Urban
Types of engineering degrees: Engineering
Other degree-granting colleges: Architecture, Arts & Sciences, Business Administration, Communication & Journalism, Education, Fine & Performing Arts, Law, Nursing, Pharmacy, Technology & Applied Sciences, Liberal Arts, Library and Information Science, Natural Sciences, Public Affairs, Social Work

■ ENGINEERING ADMISSIONS

ADMISSIONS OFFICE CONTACT
Shirley F. Binder
University of Texas at Austin
Main Building 7
Austin, TX 78712-1159
Phone: (512)471-1711　　　Fax: (512)471-3529

ENGINEERING COLLEGE ADMISSIONS INFORMATION
Entrance Requirements: SAT, ACT, High School courses—English (4 years), Foreign Language (2 years), Mathematics at the level of Algebra I or higher (3 years), Science (2 years), See below.
Entrance Recommendations: High School courses—Mathematics beyond minimum required (1 year), Chemistry (beyond minimum required) (1 year), Physics beyond minimum required (1 year), Two achievement tests: English Composition and Math Level I.
Requirements for out-of-state residents: In-State: Top 10% of high school class or rank 11-25% plus SAT 1000 or ACT 24; rank 26-50% plus SAT 1100 or ACT 26; Non-Texas residents: top quarter of high school class AND SAT 1100 or ACT 26.
Requirements for foreign students: TOEFL (Score: 550); financial statement

DEPARTMENTAL ADMISSIONS INFORMATION
Admission to department: At the time of admission to the institution.
Additional information: Admission to the aerospace, chemical, electrical, and mechanical programs is limited, meaning that an unspecified higher SAT or ACT score may be required.

■ FINANCIAL INFORMATION

ESTIMATED EXPENSES (ACADEMIC YEAR)
[Expenses are for the 1992-93 nine-month academic year.]

State Residents	Undergraduate	Graduate
Tuition and fees	$ 1,372 (4)	$ 1,366 (1)
College room and board	$ 3,920 (2)	$ 3,920 (2)
Books and supplies	$ 600	$ 700
Other expenses	$ 2,260 (5)	$ 2,650 (3)
Total estimated expenses	**$ 8,152**	**$ 8,636**

Out-of-State Residents	Undergraduate	Graduate
Tuition and fees	$ 5,512 (4)	$ 3,598 (1)
College room and board	$ 3,920 (2)	$ 3,920 (2)
Books and supplies	$ 600	$ 700
Other expenses	$ 2,260 (5)	$ 2,650 (3)
Total estimated expenses	**$12,292**	**$10,868**

Notes: (1) Nine semester hours (2) On campus. Off campus: 4500. (3) Transportation: 800; Miscellaneous: 1850. (4) Fifteen semester hours. (5) Transportation: 800; Miscellaneous: 1460.

FINANCIAL AID OFFICE CONTACT
Patricia S. Harris, *Director*
University of Texas at Austin
Student Financial Services
Austin, TX 78713-7758
Phone: (512)471-4001　　　Fax: (512)475-6296

GENERAL FINANCIAL AID INFORMATION
Forms accepted/required: FSA, Stafford, SAR, Supplemental, Institutional, Single File Form (USA Funds)
Additional financial aid information: Federal, state, and private sources of financial aid include part-time (work-study) employment, grants, and loans. Need and merit based scholarships exist for minority and non-minority students.

■ ENGINEERING COLLEGE INFORMATION

[For additional personnel, refer to the *Appendix*.]

HEAD OF ENGINEERING
Herbert H. Woodson
Phone: (512)471-1136　　　Fax: (512)471-3955
EMail: H_Woodson.Dean_of_Eng@Engdeangate.CE.UTexas.EDU

ENGINEERING COLLEGE ADDRESS
College of Engineering
University of Texas at Austin
Cockrell Hall 10.3
Austin, TX 78712-1080
Phone: (512)471-1166　　　Fax: (512)471-3955

ENROLLMENTS—BY CLASS
[Numbers are baccalaureate enrollments for the fall 1991 term, unless footnoted.]
1st-year students/Freshmen 1,409
2nd-year students/Sophomores 907
3rd-year students/Juniors 959
4th-year students/Seniors 1,559
Total **4,834**

NUMBER OF DEGREES AWARDED—BY PROGRAM
[Numbers are engineering baccalaureate degrees awarded during the 1991-1992 academic year, unless otherwise footnoted. For full details about each engineering program, refer to the *Tables of Degree Programs*.]

Aerospace Engineering 81
Architectural Engineering 31
Biomedical Engineering .... – (B)
Chemical Engineering 85
Civil Engineering 59
Computer Engineering .... – (10)
Electrical & Computer Engineering 219
Energy & Mineral Resources .... – (B)
Engineering Mechanics .... – (B)
Environmental Health Engineering .... – (B)
Manufacturing Systems Engineering .... – (B)
Materials Science & Engineering .... – (B)
Mechanical Engineering 181
Operations Research & Industrial Engineering .... –
Petroleum Engineering 31
Total **687**

Notes: (B) Data not applicable. (10) The Computer Engineering degree program requires a semester hour total which differs from Electrical Engineering, but the degree conferred is in Electrical Engineering.

PERCENTAGE OF DEGREES AWARDED—BY CATEGORY
[Percentages are of all engineering baccalaureate degrees awarded during the 1991-1992 academic year, unless otherwise footnoted.]
African Americans 4.6%
Asian/Pacific Island Americans 14.9%
Hispanic Americans 14.3%
Native Americans 0.1%
Foreign Citizens 10.0%
All Others 56.1%
Women 16.6%
Persons with disabilities .... – % (A)
Students over 25 years of age 18.7%

Notes: (A) Data not available.

GRADUATE ENROLLMENTS & DEGREES AWARDED
Master's enrollment 1,208
Master's degrees awarded 413
Doctoral enrollment 885
Doctoral degrees awarded 141

ENGINEERING STUDENT DATA
Applicants to the engineering college
Number of applicants to engineering college 2,325
Percent offered admission 69.4%

Matriculated engineering students
Percentage in top quartile (25%) of High School class 87.9
Average SAT scores: Math—633, Verbal—507, Combined—1139 (1)
Average ACT scores: Math—27.1 (2), Composite—25.6 (2)

Notes: (1) Applies to 1038 of 1080 new freshmen. (2) Applies to 358 of 1080 new freshmen.

FULL- & PART-TIME FACULTY—BY DEPARTMENT
[Figures are the head count of full-time faculty and the full-time equivalent (FTE) of part-time faculty for each engineering department or equivalent.]

Department	Full	Part
Chemical Eng	19	2.2
Civil Eng	47	3.6
Electrical & Computer Eng	58	6.5
Mechanical Eng	58	5.5
Aerospace Eng & Eng Mechanics	30	2.3
Petroleum Eng	16	1.5

COLLEGE DESCRIPTION

There are eight undergraduate degree programs, each of which offers from two to eight areas of specialization, totaling 39 option areas. Dual degree programs exist for qualified students in two areas. One combines Engineering & the College of Liberal Arts Plan II Honors program to allow students to earn both a B.S. in engineering and a B.A. degree simultaneously. A six-year plan offered jointly with the School of Architecture allows students to earn both a B.S. in engineering and a bachelor of architecture. Co-op programs are available after completion of prerequisite courses with at least a 2.5 GPA. Utilizing two or three semesters for co-op employment usually lengthens the degree program by approximately six months.

DEGREE PROGRAMS DESCRIPTION

The 7 B.S. programs in engineering are oriented toward application & design, preparing students for careers as professional engineers or for graduate study. The 13 M.S. in engineering degree programs offer advanced analysis & solutions to engineering problems. The one M.A. program offers the same for problems in energy & mineral resources. All master's programs are available with thesis, report, or no thesis & no report options. The 9 doctoral programs offer the most advanced opportunities for research & solution of engineering problems preparatory to a career of teaching &/or research.

DUAL DEGREE PROGRAMS

Enrollment requirements: Applicants to the architectural engineering/architecture dual degree programs must apply for admission to the School of Architecture according to the procedures and deadlines established by the school. Admission to the Engineering/Plan II Dual degree program requires two separate applications: one to the University for the College of Engineering and one to the Plan II Honors Program.

TRANSFER INFORMATION

Residency Requirements: Must complete in residence at least two long-session semesters, and at least thirty semester hours of work counted toward the degree in the major field. At least the last twenty-four semester hours of technical courses counted toward an engineering degree must be taken while the student is registered as an undergraduate engineering major at the University.

Transfer via Articulation Agreements
Admission to engineering: During sophomore year
Engineering graduates transferred from ... 4-yr: 187　2-yr: 205
Requirements: Applicant must have completed a minimum of thirty semesters of transferable credit with a grade point average of at least 3.0. Additional criteria may be utilized for aerospace, chemical, electrical, and mechanical engineering.

Transfer without Articulation Agreements
Admission to engineering: Applicants must have completed a minimum of thirty semester hours of transferable credit.
Engineering graduates transferred from ... 4-yr: 80　2-yr: 90
Requirements: GPA of at least 3.0 in transferable credit. Additional criteria may be needed for aerospace, chem, elec, & mechanical.

GRADUATION REQUIREMENTS

GPA of at least 2.00 in all courses and in all major courses taken at the University. Hours required: Aerospace Eng: 130; Architectural Eng: 129; Chemical Eng & Petroleum Eng: 131; Civil Eng: 124; Electrical Eng: 129; Computer Eng: 132; & Mechanical Eng: 135.

■ STUDENT PROGRAMS

PROFESSIONAL AND HONORARY SOCIETIES

[For key to acronyms, see *Introduction*.]
Professional Societies: AIAA, AIChE, ASCE, ASME, IEEE, NSAE, NSPE, SPE, Theta Tau
Honorary Soc: Chi Epsilon, Eta Kappa Nu, Omega Chi Epsilon, Pi Tau Sigma, Sigma Gamma Tau, Tau Beta Pi, Pi Epsilon Tau

SUPPORT PROGRAMS

Student Chapter Organizations: Society of Women Engineers, National Society of Black Engineers, Society of Hispanic Professional Engineers, Pi Sigma Pi

Equal Opportunity in Engineering
For: Ethnic Minorities　　**Available:** Year round
Offered: Prematriculation, Freshman, Sophomore, Junior, Senior, Graduate level

Women in Engineering Program
For: Women　　**Available:** Year round
Offered: Prematriculation, Freshman, Sophomore, Junior, Senior, Graduate level

Other Engineering Support Programs: An orientation program is provided before each session for new freshmen and transfer students. The Student Affairs Office (SAO) provides both academic and non-academic counseling for undergraduates. The SAO offers a faculty-student mentor program for freshmen. Summer bridge programs are provided for minority students. Course work & career advising is provided by faculty.

234 University of Texas at Dallas

■ INSTITUTION PROFILE

HEAD OF THE INSTITUTION
Robert H. Rutford
Phone: (214)690-2201   Fax: (214)690-2237

GENERAL INFORMATION
[All Students—Fall 1991]
Undergraduate enrollment 5,026
Graduate enrollment 3,954
Total institution enrollment 8,980
Type of institution: Public   **Calendar:** Semesters
Nearest city: Dallas   **Population:** 1,006,000
Miles from main campus: 18   **Setting:** Urban
Types of engineering degrees: Engineering
Other degree-granting colleges: Business Administration, Fine Arts, Natural Sciences, Social Sciences, Interdisciplinary, Arts & Humanities

■ ENGINEERING ADMISSIONS

ADMISSIONS OFFICE CONTACT
Jackie Beitler
University of Texas at Dallas
P.O. Box 830688
2601 North Floyd Road
Richardson, TX 75083-0688
Phone: (214)690-2976   Fax: (214)690-2813

ENGINEERING COLLEGE ADMISSIONS INFORMATION
Entrance Requirements: SAT (Score: 1100), ACT (Score: 27), High School courses—Language Arts (4 years), Math (3.5 years), Foreign Language (2 years), Science and Social Sciences (3 years), TASP
Entrance Recommendations: High School courses—Algebra (2 years), Plane Geometry (1 year), Trigonometry (5 years), Chemistry and Physics (1 year)
Requirements for out-of-state residents: SAT--In State, 1100; ACT--In State, 27; SAT--Non Res, 1200; ACT--Non Res, 29.
Requirements for foreign students: TOEFL (Score: 550); Application fee for foreign students - $75.00

DEPARTMENTAL ADMISSIONS INFORMATION
Admission to the engineering department: At the time of admission to the institution.
Additional information: Must meet general admission requirements

■ FINANCIAL INFORMATION

ESTIMATED EXPENSES (ACADEMIC YEAR)
[Expenses are for the 1992-93 nine-month academic year.]

State Residents	Undergraduate	Graduate
Tuition and fees	$ 1,094	$ 1,094
College room and board	$ 4,248 [1]	$ 4,248 [1]
Books and supplies	$ 356	$ 356
Other expenses	$ 2,437	$ 2,437
Total estimated expenses	**$ 8,135**	**$ 8,135**

Out-of-State Residents	Undergraduate	Graduate
Tuition and fees	$ 3,038	$ 3,038
College room and board	$ 4,248 [1]	$ 4,248 [1]
Books and supplies	$ 356	$ 356
Other expenses	$ 2,437	$ 2,437
Total estimated expenses	**$10,079**	**$10,079**

Notes: (1) No university room and board available on campus. Privately owned/operated apartments.

FINANCIAL AID OFFICE CONTACT
Michael O'Rear, *Director*
University of Texas at Dallas
P.O. Box 830688
2601 North Floyd
Richardson, TX 75083-0688
Phone: (214)690-2941   Fax: (214)690-2947

GENERAL FINANCIAL AID INFORMATION
Forms accepted/required: FAF, FAR, Institutional
Additional financial aid information: Need based, academic achievement, part time campus employment, work study.

■ ENGINEERING COLLEGE INFORMATION

[For additional personnel, refer to the *Appendix*.]

HEAD OF ENGINEERING
Blake E. Cherrington
Phone: (214)690-2974   Fax: (214)690-2813
EMail: cher@utdallas.edu

ENGINEERING COLLEGE ADDRESS
Electrical Engineering
University of Texas at Dallas
P.O. Box 830688
2601 North Floyd Road
Richardson, TX 75083-0688
Phone: (214)690-2000   Fax: (214)690-2813

ENROLLMENTS—BY CLASS
[Numbers are baccalaureate enrollments for the fall 1991 term, unless otherwise footnoted.]
1st-year students/Freshmen 12
2nd-year students/Sophomores 11
3rd-year students/Juniors 112
4th-year students/Seniors 73
Total **208**

NUMBER OF DEGREES AWARDED—BY PROGRAM
[Numbers are engineering baccalaureate degrees awarded during the 1991-1992 academic year, unless otherwise footnoted. For full details about each engineering program, refer to the *Tables of Degree Programs*.]
Electrical Engineering 42
Total **42**

PERCENTAGE OF DEGREES AWARDED—BY CATEGORY
[Percentages are of all engineering baccalaureate degrees awarded during the 1991-1992 academic year, unless otherwise footnoted.]
African Americans .... – % (A)
Asian/Pacific Island Americans .... – % (A)
Hispanic Americans .... – %
Native Americans .... – %
Foreign Citizens .... 33.0 %
All Others .... 67.0 %
Women .... – %
Persons with disabilities .... – % (A)
Students over 25 years of age .... – % (A)
Notes: (A) Data not available.

GRADUATE ENROLLMENTS & DEGREES AWARDED
Master's enrollment 204
Master's degrees awarded 53
Doctoral enrollment 36
Doctoral degrees awarded 2

ENGINEERING STUDENT DATA

Applicants to the engineering college
Number of applicants to engineering college 54
Percent offered admission 67.0%

Matriculated engineering students
Percentage in top quartile (25%) of High School class 65.0
Average SAT scores: Math—530, Verbal—506, Combined—1036

FULL- & PART-TIME FACULTY—BY DEPARTMENT
[Figures are the head count of full-time faculty and the full-time equivalent (FTE) of part-time faculty for each engineering department or equivalent.]

Department	Full	Part
Electrical Eng	18	–

COLLEGE DESCRIPTION
The Erik Jonsson School of Engineering and Computer Science provides undergraduate degree preparation for professional practice as an engineer or computer scientist and offers a strong foundation for graduate study in these fields. The programs of the school concentrate on electronic information processing devices and technologies that are involved with the acquisition, interpretation, transmission, and utilization of information. The Electrical Engineering program emphasizes study in microelectronics and telecommunications.

DEGREE PROGRAMS DESCRIPTION
The BSEE program emphasizes study in microelectronics and telecommunications. Available concentration courses in the MSEE program are communications and signal processing; digital systems; solid state devices and circuits; optical devices, materials, and systems; as well as in physics and mathematics. Doctoral level research opportunities emphasize communications systems; optical communications; digital signal processing; digital computer systems and networks; solid state devices and circuits; electronic materials and processing; optical materials devices, and systems; lasers and photonics; vision systems for manufacturing; and CAD systems for robotics and manufacturing.

TRANSFER INFORMATION
Transfer via Articulation Agreements
Admission to engineering: Completion of Required freshman and sophomore courses.
Requirements: No grade lower than "C" 2.5 GPA.
Transfer without Articulation Agreements
Admission to engineering: Completion of Freshman and Sophomore courses, no grade lower than "C", 2.5 GPA

GRADUATION REQUIREMENTS
2.0 in major; 2.0 overall.

■ STUDENT PROGRAMS

PROFESSIONAL AND HONORARY SOCIETIES
[For key to acronyms, see *Introduction*.]
Professional Societies: IEEE

SUPPORT PROGRAMS
Student Chapter Organizations: Society of Women Engineers

235 University of Texas at El Paso

■ INSTITUTION PROFILE
HEAD OF THE INSTITUTION
Diana S. Natalicio
Phone: (915)747-5555
GENERAL INFORMATION
[All Students—Fall 1991]
Undergraduate enrollment 14,599
Graduate enrollment 2,610
Total institution enrollment **17,209**
Type of institution: Public   Calendar: Semesters
Location: El Paso, Texas   Population: 591,610
Setting: Urban
Types of engineering degrees: Engineering
Other degree-granting colleges: Business Administration, Education, Nursing, Science, Liberal Arts

■ ENGINEERING ADMISSIONS
ADMISSIONS OFFICE CONTACT
Stephen Riter
University of Texas at El Paso
Dean College of Engineering
El Paso, TX 79968
Phone: (915)747-5460   Fax: (915)747-5616
ENGINEERING COLLEGE ADMISSIONS INFORMATION
Entrance Requirements: SAT (Score: 700), ACT (Score: 18), High School courses—Algebra (2 years), Geometry (1 year), Trigonometry, English (4 years)
Requirements for foreign students: TOEFL (Score: 500)
DEPARTMENTAL ADMISSIONS INFORMATION
Admission to the engineering department: At the time of admission to the institution.

■ FINANCIAL INFORMATION
ESTIMATED EXPENSES (ACADEMIC YEAR)
[Expenses are for the 1992-93 nine-month academic year.]

State Residents	Undergraduate	Graduate
Tuition and fees	$ 1,114	$ 1,422
College room and board	$ 4,000	$ 4,000
Books and supplies	$ 500	$ 500
Other expenses	$ –	$ –
Total estimated expenses	**$ 5,614**	**$ 5,922**
Out-of-State Residents	Undergraduate	Graduate
Tuition and fees	$ 4,426	$ 3,902
College room and board	$ 4,000	$ 4,000
Books and supplies	$ 500	$ 500
Other expenses	$ –	$ –
Total estimated expenses	**$ 8,926**	**$ 8,402**

FINANCIAL AID OFFICE CONTACT
Beto Lopez, *Director Recruitment and Scholarships*
University of Texas at El Paso
Academic Services Building, 101
El Paso, TX 79968
Phone: (915)747-5896
GENERAL FINANCIAL AID INFORMATION
Forms accepted/required: ACT, CSS-FAF, General Application Form (local)
Additional financial aid information: Numerous need based and merit awards are available.

■ ENGINEERING COLLEGE INFORMATION
[For additional personnel, refer to the *Appendix*.]
HEAD OF ENGINEERING
Stephen Riter
Phone: (915)747-5460   Fax: (915)747-5616
ENGINEERING COLLEGE ADDRESS
College of Engineering
University of Texas at El Paso
El Paso, TX 79968
Phone: (915)747-5460   Fax: (915)747-5616
ENROLLMENTS—BY CLASS
[Numbers are baccalaureate enrollments for the fall 1991 term, unless otherwise footnoted.]
1st-year students/Freshmen 189
2nd-year students/Sophomores 256
3rd-year students/Juniors 235
4th-year students/Seniors 340
Total **1,020**

NUMBER OF DEGREES AWARDED—BY PROGRAM
[Numbers are engineering baccalaureate degrees awarded during the 1991-1992 academic year, unless otherwise footnoted. For full details about each engineering program, refer to the *Tables of Degree Programs*.]
Civil Engineering 28
Computer Science 25
Electrical Engineering 66
Industrial Engineering 14
Mechanical Engineering 28
Metallurgical & Materials Engineering 19
Total **180**

PERCENTAGE OF DEGREES AWARDED—BY CATEGORY
[Percentages are of all engineering baccalaureate degrees awarded during the 1991-1992 academic year, unless otherwise footnoted.]
African Americans .... 1.9 %
Asian/Pacific Island Americans .... 2.6 %
Hispanic Americans .... 69.2 %
Native Americans .... – %
Foreign Citizens .... 2.5 %
All Others .... 23.8 %
Women .... 21.7 %
Persons with disabilities .... – %
Students over 25 years of age .... – %

GRADUATE ENROLLMENTS & DEGREES AWARDED
Master's enrollment 403
Master's degrees awarded 87
Doctoral enrollment 15
Doctoral degrees awarded .... –

ENGINEERING STUDENT DATA
Applicants to the engineering college
Number of applicants to engineering college 360
Percent offered admission 100.0%
Matriculated engineering students
Percentage in top quartile (25%) of High School class 41.0
Average SAT scores: Math—504, Verbal—404, Combined—908
Average ACT scores: Math—18.4, Composite—18.6

FULL- & PART-TIME FACULTY—BY DEPARTMENT
[Figures are the head count of full-time faculty and the full-time equivalent (FTE) of part-time faculty for each engineering department or equivalent.]

Department	Full	Part
Civil Eng	10	0.3
Electrical Eng	15	0.5
Industrial Eng	5	–
Mechanical Eng	9	0.8
Metallurgical & Materials Eng	7	1.0
Computer Science	6	1.5

COLLEGE DESCRIPTION
The College offers four year undergraduate programs in engineering and computer science. The Freshman year program includes an introduction to engineering designed to orient students to the practice of engineering and help select a major. Each curriculum stresses hands on laboratory experiences and culminates in a significant senior design project. Participation in Coop is encouraged. Numerous opportunities exist for undergraduate participation in research projects. A joint B.S./M.S. in five years is available.

DEGREE PROGRAMS DESCRIPTION
Baccalaureate degree programs in engineering are intended to prepare students for productive lifetime careers as engineers through balanced offering of courses in general studies, basic mathematics, the physical sciences, engineering design. Both thesis and non-thesis M.S. degrees are offered in all of the undergraduate disciplines plus one in Manufacturing Engineering. A Ph.D. is offered in Electrical Engineering and Metallurgical and Materials Engineering.

TRANSFER INFORMATION
Transfer without Articulation Agreements
Admission to engineering: At any time
Requirements: A minimum overall GPA of 2.5 for all institutions attended is required. International students must meet the additional requirement of an overall minimum GPA of 3.0 in mathematics, chemistry, physics, and engineering for all institutions attended.

GRADUATION REQUIREMENTS
Degree programs in engineering require 132 to 136 hours; a 2.0 overall grade point average as well as a 2.0 average in the major.

■ STUDENT PROGRAMS
PROFESSIONAL AND HONORARY SOCIETIES
[For key to acronyms, see *Introduction*.]
Professional Societies: ACM, AGCA, ASCE, ASME, IEEE, IIE, SAE, American Foundrymen's Society, The Metallurgical Society
Honorary Societies: Alpha Pi Mu, Chi Epsilon, Eta Kappa Nu, Pi Tau Sigma, Tau Beta Pi, Upsilon Pi Upsilon, Alpha Sigma Mu

SUPPORT PROGRAMS
Student Chapter Organizations: Society of Women Engineers, Society of Hispanic Professional Engineers, Mexican American Engineering and Science Society

Project 'Success'
For: Ethnic Minorities   **Available:** Summer only
Offered: Prematriculation
Other Engineering Support Programs: Summer and Fall orientation for all Freshmen and transfers; study skills and tutorial program; international student program; career counseling and placement office; Engineering offers summer bridge program to quickly integrate students into engineering, correct prior educational deficiencies and increase the likelihood of succeeding in engineering.

236 University of Texas at San Antonio

INSTITUTION PROFILE
HEAD OF THE INSTITUTION
Samuel A. Kirkpatrick
Phone: (210)691-4101　　　　　　Fax: (210)691-4655

GENERAL INFORMATION
[All Students—Fall 1991]
Undergraduate enrollment 13,849
Graduate enrollment 1,910
Total institution enrollment 15,759
Type of institution: Public　　　**Calendar:** Semesters
Location: San Antonio　　　**Population:** 1,300,000
Setting: Urban
Types of engineering degrees: Engineering
Other degree-granting colleges: Business, Fine Arts and Humanities, Social and Behavioral Sciences

ENGINEERING COLLEGE ADMISSIONS INFORMATION
Entrance Requirements: SAT (Score: 700), ACT (Score: 18), High School courses—English (4 years)
Entrance Recommendations: High School courses—English (4 years), Mathematics (4 years), Natural Science (2 years), Social Science (2 years)
Requirements for foreign students: TOEFL (Score: 550)

DEPARTMENTAL ADMISSIONS INFORMATION
Admission to the engineering department: At the time of admission to the institution.

FINANCIAL INFORMATION
ESTIMATED EXPENSES (ACADEMIC YEAR)
[Expenses are for the 1992-93 nine-month academic year.]

State Residents	Undergraduate	Graduate
Tuition and fees	$ 600[1]	$ 400[1]
College room and board	$ —	$ —
Books and supplies	$ —	$ —
Other expenses	$ —	$ —
Total estimated expenses	**$ 600**	**$ 400**

Out-of-State Residents	Undergraduate	Graduate
Tuition and fees	$ 2,200[1]	$ 1,325[1]
College room and board	$ —	$ —
Books and supplies	$ —	$ —
Other expenses	$ —	$ —
Total estimated expenses	**$ 2,200**	**$ 1,325**

Notes: (1) Tuition for 9 semester hours, including fees.

GENERAL FINANCIAL AID INFORMATION
Forms accepted/required: Additional financial aid information: For information, contact: Office of Student Financial Aid.

ENGINEERING COLLEGE INFORMATION
[For additional personnel, refer to the *Appendix*.]
HEAD OF ENGINEERING
James H. Tracey
Phone: (210)691-4450　　　　　　Fax: (210)691-4445
EMail: jtracey@lonestar

ENGINEERING COLLEGE ADDRESS
College of Sciences and Engineering
University of Texas at San Antonio
6900 North Loop 1604 West
San Antonio, TX 78249-0661
Phone: (210)691-4490　　　　　　Fax: (210)691-5589

ENROLLMENTS—BY CLASS
[Numbers are baccalaureate enrollments for the fall 1991 term, unless otherwise footnoted.]
1st-year students/Freshmen 288
2nd-year students/Sophomores 205
3rd-year students/Juniors 189
4th-year students/Seniors 355
Total **1,037**

NUMBER OF DEGREES AWARDED—BY PROGRAM
[Numbers are engineering baccalaureate degrees awarded during the 1991-1992 academic year, unless otherwise footnoted. For full details about each engineering program, refer to the *Tables of Degree Programs*.]
Civil Engineering 20
Electrical Engineering 40
Mechanical Engineering 22
Total **82**

PERCENTAGE OF DEGREES AWARDED—BY CATEGORY
[Percentages are of all engineering baccalaureate degrees awarded during the 1991-1992 academic year, unless otherwise footnoted.]
African Americans .... — %
Asian/Pacific Island Americans .... 9.8 %
Hispanic Americans .... 34.1 %
Native Americans .... — %
Foreign Citizens .... 12.2 %
All Others .... 43.9 %
Women .... 9.8 %
Persons with disabilities .... — %
Students over 25 years of age .... — %

GRADUATE ENROLLMENTS & DEGREES AWARDED
Master's enrollment 102
Master's degrees awarded 9
Doctoral enrollment .... —
Doctoral degrees awarded .... —

ENGINEERING STUDENT DATA
Matriculated engineering students
Percentage in top quartile (25%) of High School class 47.0
Average SAT scores: Combined—952
Average ACT scores: Composite—21

FULL- & PART-TIME FACULTY—BY DEPARTMENT
[Figures are the head count of full-time faculty and the full-time equivalent (FTE) of part-time faculty for each engineering department or equivalent.]

Department	Full	Part
Division of Eng	20	7.5
Civil Eng	5	2.2
Electrical Eng	9	3.4
Mechanical Eng	6	1.9

COLLEGE DESCRIPTION
Aggressive young engineering program in new and modern facilities in large metropolitan area.

DEGREE PROGRAMS DESCRIPTION
The B.S. degrees in the various engineering programs prepare the graduate for entry-level professional positions and graduate study in civil, electrical and mechanical engineering.

TRANSFER INFORMATION
Residency Requirements: A minimum of 30 semester credit hours of UTSA courses must be completed; 24 of the last 30 must be completed in residence; of the minimum 39 upper-division hours required, 18 must be earned in UTSA courses; and at least six hours of upper-division UTSA course work in the major must be completed.

Transfer via Articulation Agreements
Admission to engineering: At any time
Engineering graduates transferred from . . . 4-yr: 5　　2-yr: 30

Transfer without Articulation Agreements
Admission to engineering: At any time
Engineering graduates transferred from . . . 4-yr: 17　　2-yr: 11

GRADUATION REQUIREMENTS
A student must earn a grade of C or better in all prerequisite courses and must satisfy current ABET requirements. Minimum hours for graduation is 134 hours.

STUDENT PROGRAMS
PROFESSIONAL AND HONORARY SOCIETIES
[For key to acronyms, see *Introduction*.]
Professional Societies: ASCE, ASHRAE, ASME, IEEE, SAE

SUPPORT PROGRAMS
Student Chapter Organizations: Society of Women Engineers, Society of Hispanic Professional Engineers
Other Engineering Support Programs: Orientation program for all freshmen and transfers; career planning and placement services; services for blind and handicapped.

237 Texas A&I University

■ INSTITUTION PROFILE

HEAD OF THE INSTITUTION
Manuel L. Ibanez
Phone: (512)595-3207 **Fax:** (512)595-3218

GENERAL INFORMATION
[All Students—Fall 1991]
Undergraduate enrollment 4,874
Graduate enrollment 1,063
Total institution enrollment **5,937**
Type of institution: Public **Calendar:** Semesters
Nearest city: Corpus Christi, Texas **Population:** 250,000
Miles from main campus: 40 **Setting:** Small Town
Types of engineering degrees: Engineering
Other degree-granting colleges: Arts & Sciences, Business Administration, Education, Agriculture and Home Economics

■ ENGINEERING ADMISSIONS

ADMISSIONS OFFICE CONTACT
Ruth T. Fletcher
Texas A&I University
Campus Box 105
Kingsville, TX 78363
Phone: (512)595-2811

ENGINEERING COLLEGE ADMISSIONS INFORMATION
Entrance Requirements: SAT (Score: 850), ACT (Score: 21), High School courses—English (4 years), Math (4 years), Physics (1 year), Chemistry (1 year)
Entrance Recommendations: High School courses—Calculus (1 year), Computer Science
Requirements for foreign students: TOEFL (Score: 525); financial statement

DEPARTMENTAL ADMISSIONS INFORMATION
Admission to the engineering department: At the time of admission to the institution.

■ FINANCIAL INFORMATION

ESTIMATED EXPENSES (ACADEMIC YEAR)
[Expenses are for the 1992-93 nine-month academic year.]

State Residents	Undergraduate	Graduate
Tuition and fees	$ 1,362 (2)	$ 978 (1)
College room and board	$ 2,816	$ 2,816
Books and supplies	$ 900 (C)	$ 800 (C)
Other expenses	– (A)	– (A)
Total estimated expenses	**$ 5,078**	**$ 4,594**

Out-of-State Residents	Undergraduate	Graduate
Tuition and fees	$ 5,106 (2)	$ 3,474 (1)
College room and board	$ 2,816	$ 2,816
Books and supplies	$ 900 (C)	$ 800 (C)
Other expenses	– (A)	– (A)
Total estimated expenses	**$ 8,822**	**$ 7,090**

Notes: (A) Data not available. (C) Estimated data. (1) 12 semester hours per semester (2) 18 semester hours per semester

FINANCIAL AID OFFICE CONTACT
Auturo Pecos, *Director, Financial Aid*
Texas A&I University
Campus Box 115
Kingsville, TX 78363
Phone: (512)595-3911

GENERAL FINANCIAL AID INFORMATION
Forms accepted/required: CSS-FAF, FAF, FAT, Stafford, IRS, SAR, Supplemental, Institutional
Additional financial aid information: work study, need-based scholarships and part time jobs

■ ENGINEERING COLLEGE INFORMATION

[For additional personnel, refer to the *Appendix*.]

HEAD OF ENGINEERING
Phil V. Compton
Phone: (512)595-2001 **Fax:** (512)595-2106
EMail: KFPVC00@TAIMVS1

ENGINEERING COLLEGE ADDRESS
College of Engineering
Texas A&I University
Campus Box 188
Kingsville, TX 78362
Phone: (512)595-2001 **Fax:** (512)595-2106

ENROLLMENTS—BY CLASS
[Numbers are baccalaureate enrollments for the fall 1991 term, unless otherwise footnoted.]
1st-year students/Freshmen 257
2nd-year students/Sophomores 130
3rd-year students/Juniors 98
4th-year students/Seniors 180
Total **665**

NUMBER OF DEGREES AWARDED—BY PROGRAM
[Numbers are engineering baccalaureate degrees awarded during the 1991-1992 academic year, unless otherwise footnoted. For full details about each engineering program, refer to the *Tables of Degree Programs*.]
Chemical Engineering 12
Civil Engineering 18
Computer Science 3
Electrical Engineering 41
Environmental Engineering .... –
Industrial Engineering .... –
Industrial Technology 7
Mechanical Engineering 25
Natural Gas Engineering 19
Total **125**

PERCENTAGE OF DEGREES AWARDED—BY CATEGORY
[Percentages are of all engineering baccalaureate degrees awarded during the 1991-1992 academic year, unless otherwise footnoted.]
African Americans .... 1.0 %
Asian/Pacific Island Americans .... 1.0 %
Hispanic Americans .... 28.8 %
Native Americans .... – %
Foreign Citizens .... 35.2 %
All Others .... 34.0 %
Women .... 5.6 %
Persons with disabilities .... – % (A)
Students over 25 years of age .... – % (A)

Notes: (A) Data not available.

GRADUATE ENROLLMENTS & DEGREES AWARDED
Master's enrollment 204
Master's degrees awarded 29
Doctoral enrollment .... – (A)
Doctoral degrees awarded .... – (A)

Notes: (A) Data not available.

FULL- & PART-TIME FACULTY—BY DEPARTMENT
[Figures are the head count of full-time faculty and the full-time equivalent (FTE) of part-time faculty for each engineering department or equivalent.]

Department	Full	Part
Chemical Eng & Natural Gas Eng	7	0.5
Civil Eng	5	–
Electrical Eng & Computer Science	12	–
Mechanical Eng & Industrial Eng	10	–
Environmental Eng	4	–
Industrial Tech	–	–

COLLEGE DESCRIPTION

The faculty has a blend of industrial and academic experience which provides our students with a well-rounded educational experience. The small class size of our engineering classes allows for individual attention by faculty with students. Well-equipped laboratories and computer facilities are available for students. A study center and tutorial support are available to assist freshmen and sophomores.

DEGREE PROGRAMS DESCRIPTION

The B.S. level engineering programs are designed to give the student an understanding of the fundamental principles underlying engineering science and engineering practice. Each curriculum contains basic courses to develop a solid foundation in mathematics, chemistry and physics, and includes a general background in humanities and social sciences. Building on this background, the engineering science courses provide application of basic principles and the analysis of engineering systems. The engineering design component of the curriculum in each area provides the engineering student with methods and techniques for the solution of technological problems of society. The College offers programs of study leading to both the Master of Science and the Master of Engineering degrees. The Master of Engineering degree is a special program intended to prepare students for professional careers in engineering and provide the opportunity for advanced studies to practicing engineers. Students who intend to continue academic work toward a Doctor of Philosophy degree are urged to follow the Master of Science degree with a major in engineering.

TRANSFER INFORMATION

Residency Requirements: last 36 hours must be in residence

Transfer via Articulation Agreements
Admission to engineering: At any time
Requirements: Overall grade point average of 2.5/4.0

Transfer without Articulation Agreements
Admission to engineering: At any time
Engineering graduates transferred from . . 4-yr: – 2-yr: –
Requirements: Overall grade point average of a 2.5/4.0

GRADUATION REQUIREMENTS

The degree of bachelor of science with an engineering major is granted to the candidate fulfilling one of the prescribed curricula in the University catalog. In addition to general requirements for graduation (given in the Texas A&I University catalog), the candidate must maintain a grade point average of 2.0 on (1) all coursework attempted, (2) all coursework attempted at A&I, (3) all mathematics and natural science courses specified for the degree, and (4) all engineering courses in the major specified for the degree.

■ STUDENT PROGRAMS

PROFESSIONAL AND HONORARY SOCIETIES
[For key to acronyms, see *Introduction*.]
Professional Societies: ACM, AIChE, ASCE, ASHRAE, ASME, IEEE, SME, SPE
Honorary Societies: Eta Kappa Nu, Pi Tau Sigma, Tau Beta Pi

SUPPORT PROGRAMS
Student Chapter Organizations: Society of Women Engineers, Society of Hispanic Professional Engineers

Women and Minority Engineering Program
For: Women & Ethnic Minorities **Available:** Year round
Offered: Prematriculation, Freshman, Sophomore, Junior, Senior, Graduate level

Other Engineering Support Programs: Support programs for freshmen include academic and personal counseling, study center, tutorial support, and clustering of students in common freshmen classes.

238 Texas A&M University

INSTITUTION PROFILE
HEAD OF THE INSTITUTION
William H. Mobley
Phone: (409)845-2217   Fax: (409)845-5027

GENERAL INFORMATION
[All Students—Fall 1991]
Undergraduate enrollment 33,024
Graduate enrollment 7,973
Total institution enrollment 40,997
Type of institution: Public   **Calendar:** Semesters
Nearest city: Houston   **Population:** 2,782,414
Miles from main campus: 97   **Setting:** Suburban
Types of engineering degrees: Engineering & Technology
Other degree-granting colleges: Architecture, Business Administration, Education, Medicine, Geosciences and Maritime Studies, Agriculture and Life Sciences, Liberal Arts, Science, Veterinary Medicine

ENGINEERING ADMISSIONS
ADMISSIONS OFFICE CONTACT
Gary R. Engelgau
Texas A&M University
Office of Admissions
College Station, TX 77843-0100
Phone: (409)845-1040   Fax: (409)845-0727

ENGINEERING COLLEGE ADMISSIONS INFORMATION
Entrance Requirements: High School courses—English (4 years), Mathematics (3.5 years), Science (2 years), Social Studies (2.5 years), Top 10%: TX resident--no min; non-resident--1100 SAT or 27 ACT. 1st qt: TX--1000 SAT or 24 ACT; non-res--1100 SAT or 27 ACT. 2nd qt: TX--1100 SAT or 27 ACT; non-res--not eligible. 3rd qt: TX--1200 SAT or 29 ACT; non-res--not eligible.
Entrance Recommendations: High School courses—Trigonometry, Advanced math, Foreign language (2 years), Computer Science (1 year), ECAT and MAT (Level I and II)
Requirements for foreign students: TOEFL (Score: 550); financial statement; Written and oral English proficiency tested locally before registration

DEPARTMENTAL ADMISSIONS INFORMATION
Admission to the engineering department: At the time of admission to the institution.
Additional information: There are no special departmental requirements at the freshman level. Admission to engineering courses at the sophomore or higher levels is on the basis of GPA and on a common body of knowledge.

FINANCIAL INFORMATION
ESTIMATED EXPENSES (ACADEMIC YEAR)
[Expenses are for the 1992-93 nine-month academic year.]

State Residents	Undergraduate	Graduate
Tuition and fees	$ 1,526 [3]	$ 1,238 [1]
College room and board	$ 3,738 [C]	$ 4,200 [2]
Books and supplies	$ 600 [C]	$ 1,000 [C]
Other expenses	$ 1,914 [C]	$ 2,974 [C]
Total estimated expenses	$ 7,778	$ 9,412

Out-of-State Residents	Undergraduate	Graduate
Tuition and fees	$ 5,606 [3]	$ 5,488 [1]
College room and board	$ 3,738 [C]	$ 4,200 [2]
Books and supplies	$ 600 [C]	$ 1,000 [C]
Other expenses	$ 1,914 [C]	$ 2,974 [C]
Total estimated expenses	$11,858	$ 13,662

Notes: (C) Estimated data. (1) Based on nine credit hours per semester (2) Grad housing is available in the University-owned apartments (3) Based on twelve credit hours per semester

FINANCIAL AID OFFICE CONTACT
Donald L. Engelage, *Director*
Texas A&M University
Student Financial Aid Office
College Station, TX 77843-1252
Phone: (409)845-3236

GENERAL FINANCIAL AID INFORMATION
Forms accepted/required: FAF, FFS, SAR
Additional financial aid information: Merit-based scholarships, awards, fellowships and grants; short-term loans; part-time campus jobs; work-study program; valedictorian tuition exemption; and need-based scholarships.

ENGINEERING COLLEGE INFORMATION
[For additional personnel, refer to the *Appendix*.]
HEAD OF ENGINEERING
Kenneth L. Peddicord
Phone: (409)845-7203   Fax: (409)845-8986
EMail: KLP7201

ENGINEERING COLLEGE ADDRESS
College of Engineering
Texas A&M University
301 Wisenbaker Engineering Research Ctr.
College Station, TX 77843-3126
Phone: (409)845-1321   Fax: (409)845-8986

ENROLLMENTS—BY CLASS
[Numbers are baccalaureate enrollments for the fall 1991 term, unless footnoted.]
1st-year students/Freshmen .... 1,766 [1]
2nd-year students/Sophomores .... 1,293 [1]
3rd-year students/Juniors .... 1,569 [1]
4th-year students/Seniors .... 1,904 [1]
Total 6,532

Notes: (1) CPSC, ENTC and IDIS not included

NUMBER OF DEGREES AWARDED—BY PROGRAM
[Numbers are engineering baccalaureate degrees awarded during the 1991-1992 academic year, unless otherwise footnoted. For full details about each engineering program, refer to the *Tables of Degree Programs*.]

Aerospace Engineering 83
Agricultural Engineering 5
Bioengineering 36
Chemical Engineering 73
Civil Engineering 180
Computer Engineering 18
Computer Science 125
Electrical Engineering 216
Engineering .... – [7]
Health Physics .... –
Industrial Engineering 100
Industrial Hygiene .... –
Interdisciplinary Engineering .... –
Mechanical Engineering 146
Nuclear Engineering 16
Ocean Engineering 21
Petroleum Engineering 25
Radiological Health Engineering 2
Safety Engineering .... –
Total 1,046

Notes: (7) A master's degree program

PERCENTAGE OF DEGREES AWARDED—BY CATEGORY
[Percentages are of all engineering baccalaureate degrees awarded during the 1991-1992 academic year, unless otherwise footnoted.]
African Americans 1.4%
Asian/Pacific Island Americans 7.0%
Hispanic Americans 6.5%
Native Americans .... – %
Foreign Citizens 4.7%
All Others 80.4%
Women 19.0%
Persons with disabilities .... – % (A)
Students over 25 years of age .... – %

Notes: (A) Data not available.

GRADUATE ENROLLMENTS & DEGREES AWARDED
Master's enrollment .... 1,369 [2]
Master's degrees awarded .... 386 [2]
Doctoral enrollment .... 685 [2]
Doctoral degrees awarded .... 111 [2]

Notes: (2) Includes CPSC

ENGINEERING STUDENT DATA
Applicants to the engineering college
Number of applicants to engineering college 3,182
Percent offered admission 85.0%

Matriculated engineering students
Percentage in top quartile (25%) of High School class .... 78.2 [1]
Average SAT scores: Math—623.3, Verbal—513.4, Combined—1137
Average ACT scores: Math—27, Composite—26

Notes: (1) Percentage represents the entire university

FULL- & PART-TIME FACULTY—BY DEPARTMENT
[Figures are the head count of full-time faculty and the full-time equivalent (FTE) of part-time faculty for each engineering department or equivalent.]

Department	Full	Part
Chemical Eng	20	–
Civil Eng	63.6	7.4
Electrical Eng	42	–
Industrial Eng	31.2	–
Mechanical Eng	60	–
Aerospace Eng	15	–
Computer Science	32	5.0
Nuclear Eng	8	3.7
Petroleum Eng	11	–
Eng Tech	26	1.8
Agricultural Eng	2	1.0

COLLEGE DESCRIPTION
The college offers a four-year program in engineering, emphasizing engineering fundamentals in the common freshman year, followed by three years of specialized education leading to one of fourteen majors. This program combines the fundamentals of a University Core Curriculum with a specialized technical education in the College of Engineering, preparing the student for industrial work or graduate study. Cooperative education is an option in the undergraduate program. Honors classes, Engineering Scholars Program, and undergraduate research projects are available to qualified students.

DEGREE PROGRAMS DESCRIPTION
The B.S. degree prepares students both for entry-level professional careers or graduate school; the M.S. & M.E. provide more advanced preparation, either for industry or PhD level study; the PhD prepares students for a research career in industry or academics; the Doctor of Engineering (DE) prepares students for advanced careers in engineering & management & the Interdisciplinary Engineering PhD (ITDE) prepares students for research careers in technical areas.

DUAL DEGREE PROGRAMS
Engineering baccalaureate graduates with dual degrees 1

TRANSFER INFORMATION
Residency Requirements: A candidate for a baccalaureate degree must enroll in and complete at least 30 of the last 36 hours immediately preceding graduation in residence at this institution.

Transfer via Articulation Agreements
Admission to engineering: At any time
Engineering graduates transferred from ... 4-yr: 15   2-yr: 18
Requirements: 30 hrs or less-3.0 GPR and requirements for entering freshmen; 31-45 hrs, 3.0 GPR; 46-60 hrs, 2.5 GPR; 61+ hrs, 2.0 GPR. Transfer students, regardless of transfer hours, are admitted with a Lower Division classification and must meet the standards and criteria for admission to a major degree sequence (good academic standing and credit for specific courses).

Transfer without Articulation Agreements
Admission to engineering: At any time
Engineering graduates transferred from ... 4-yr: 40   2-yr: 46
Requirements: Same as stated in the previous question.

GRADUATION REQUIREMENTS
BS programs in engineering require 134-136 semester credits, depending on selected major. The undergraduate must complete with a minimum 2.0 GPR of all undergraduate course work attempted at TAMU.

STUDENT PROGRAMS
PROFESSIONAL AND HONORARY SOCIETIES
[For key to acronyms, see *Introduction*.]
Professional Societies: ACM, AIAA, AIChE, ANS, ASCE, ASME, IEEE, IIE, SME, SPE, Texas Society of Professional Engineers (TSPE), Institute of Transportation Engineers (ITE)
Honorary Societies: Alpha Pi Mu, Chi Epsilon, Eta Kappa Nu, Omega Chi Epsilon, Pi Tau Sigma, Sigma Gamma Tau, Tau Alpha Pi, Tau Beta Pi, Alpha Nu Sigma, Upsilon Pi Epsilon

SUPPORT PROGRAMS
Student Chapter Organizations: Society of Women Engineers, National Society of Black Engineers, Mexican Americans in Engineering and Science, Committee on Awareness (CAMAC, NAACP, & BAC)

Minority Engr. Prog. (African Amer.,Hisp, & Nat. Am.)
For: Ethnic Minorities   **Available:** Year round
Offered: Prematriculation, Freshman, Sophomore, Junior, Senior, Graduate level

Multicultural Ctr.(FACES, I CARE, & EXCELL)
For: Ethnic Minorities   **Available:** Year round
Offered: Prematriculation, Freshman, Sophomore, Junior, Senior, Graduate level

Minority Enrichment & Dev. through Acad.& Leadership Skills
For: Ethnic Minorities   **Available:** Academic year
Offered: Freshman, Sophomore, Junior

Other Engineering Support Programs: Summer, fall and spring orientation programs for all freshmen and transfers; services for handicapped and veterans; alcohol and drugs education and prevention services; international student services; counseling services; career planning and placement services.

239 Texas Tech University

INSTITUTION PROFILE

HEAD OF THE INSTITUTION
Robert W. Lawless
Phone: (806)742-2121 Fax: (806)742-2138

GENERAL INFORMATION
[All Students—Fall 1991]
Undergraduate enrollment 20,287
Graduate enrollment 3,807
Total institution enrollment **24,094**
Type of institution: Public Calendar: Semesters
Nearest city: Amarillo Population: 150,000
Miles from main campus: 120 Setting: Urban
Types of engineering degrees: Engineering & Technology
Other degree-granting colleges: Agricultural & Environmental, Allied Health Sciences, Architecture, Arts & Sciences, Business Administration, Education, Law, Medicine, Nursing, Home Economics

ENGINEERING ADMISSIONS

ADMISSIONS OFFICE CONTACT
Marty Grassel
Texas Tech University
Box 45005
Lubbock, TX 79409-5005
Phone: (806)742-1482 Fax: (806)742-2007

ENGINEERING COLLEGE ADMISSIONS INFORMATION
Entrance Requirements: SAT (Score: 900), ACT (Score: 22), High School courses—English (4 years), Chemistry (1 year), Physics (1 year), Math (geometry, trigonometry, & algebra II) (3 years), Texas Academic Skills Program (TASP) Test
Requirements for foreign students: TOEFL (Score: 550); financial statement; Should apply one year in advance.

DEPARTMENTAL ADMISSIONS INFORMATION
Admission to the engineering department: At the time of admission to the institution.
Additional information: A minimum Grade Point Average of 2.0 is required by all departments.

FINANCIAL INFORMATION

ESTIMATED EXPENSES (ACADEMIC YEAR)
[Expenses are for the 1992-93 nine-month academic year.]

State Residents	Undergraduate	Graduate
Tuition and fees	$ 1,230	$ 1,230[1]
College room and board	$ 3,600	$ 3,600[2]
Books and supplies	$ 700	$ 700
Other expenses	$ –	$ –[3]
Total estimated expenses	**$ 5,530**	**$ 5,530**

Out-of-State Residents	Undergraduate	Graduate
Tuition and fees	$ 4,500	$ 4,500
College room and board	$ 3,600	$ 3,600
Books and supplies	$ 700	$ 700
Other expenses	$ –	$ –
Total estimated expenses	**$ 8,800**	**$ 8,800**

Notes: (1) Tuition fee is based on a 15 hour courseload. (2) Air conditioned dormitories may be slightly higher. (3) Students are required to purchase their own drawing equipment, calculators, etc.

FINANCIAL AID OFFICE CONTACT
Ronny Barnes
Texas Tech University
Office of Financial Aids for Students
Box 45011
Lubbock, TX 79409-5011
Phone: (806)742-3681

GENERAL FINANCIAL AID INFORMATION
Forms accepted/required: AFSA, CSS-FAF, FAF, FAT, FAR, FSA, FFS, Stafford, IRS, SAR, SDF, Supplemental
Additional financial aid information: Financial aid available includes need-based and merit-based scholarships, loans, grants, part-time jobs on campus, college work-study program.

ENGINEERING COLLEGE INFORMATION

[For additional personnel, refer to the *Appendix*.]

HEAD OF ENGINEERING
Mason H. Somerville
Phone: (806)742-3451 Fax: (806)742-3493

ENGINEERING COLLEGE ADDRESS
College of Engineering
Texas Tech University
Box 43103
Lubbock, TX 79409-3103
Phone: (806)742-3451 Fax: (806)742-3493

ENROLLMENTS—BY CLASS
[Numbers are baccalaureate enrollments for the fall 1991 term, unless otherwise footnoted.]
1st-year students/Freshmen 715
2nd-year students/Sophomores 432
3rd-year students/Juniors 367
4th-year students/Seniors 545
Total **2,059**

NUMBER OF DEGREES AWARDED—BY PROGRAM
[Numbers are engineering baccalaureate degrees awarded during the 1991-1992 academic year, unless otherwise footnoted. For full details about each engineering program, refer to the *Tables of Degree Programs*.]
Chemical Engineering 8
Civil Engineering 41
Computer Science 27
Electrical Engineering 25
Engineering Physics .... –
Industrial Engineering 24
Mechanical Engineering 62
Petroleum Engineering 7
Total **194**

PERCENTAGE OF DEGREES AWARDED—BY CATEGORY
[Percentages are of all engineering baccalaureate degrees awarded during the 1991-1992 academic year, unless otherwise footnoted.]
African Americans .... – %
Asian/Pacific Island Americans .... 5.0 %
Hispanic Americans .... 4.7 %
Native Americans .... – %
Foreign Citizens .... 28.0 %
All Others .... 62.3 %
Women .... 9.0 %
Persons with disabilities .... 3.5 %
Students over 25 years of age .... 35.0 %

GRADUATE ENROLLMENTS & DEGREES AWARDED
Master's enrollment 261
Master's degrees awarded 86
Doctoral enrollment 120
Doctoral degrees awarded 15

ENGINEERING STUDENT DATA
Applicants to the engineering college
Number of applicants to engineering college 882
Percent offered admission 73.3%
Matriculated engineering students
Percentage in top quartile (25%) of High School class 37.0
Average SAT scores: Math—468, Verbal—439, Combined—907
Average ACT scores: Composite—24

FULL- & PART-TIME FACULTY—BY DEPARTMENT
[Figures are the head count of full-time faculty and the full-time equivalent (FTE) of part-time faculty for each engineering department or equivalent.]

Department	Full	Part
Chemical Eng	10	–
Civil Eng	17	1.8
Electrical Eng	19	1.6
Industrial Eng	10	0.7
Mechanical Eng	15	2.2
Petroleum Eng	3	1.7
Computer Science	8	0.8
Eng Tech	8	1.7

COLLEGE DESCRIPTION
Our primary goal is to educate students at the undergraduate and graduate levels who are exceptionally prepared to fill leadership roles as professionals cognizant of technology and its economic and political role in the world. We strive to produce graduates who are: (1) Equipped with technical competence and problem solving ability; (2) Able to communicate and work well with others; (3) Sensitive to the needs of society; and (4) Well-educated in the humanities, as well as the engineering disciplines. All students acquire a background in mathematics, chemistry, and physics before going into a particular specialty. Students who were not computer literate upon entering the University become so during the course of their graduate studies.

DEGREE PROGRAMS DESCRIPTION
The B.S. degrees prepare the graduate to join more than 15,000 other College of Engineering graduates from Texas Tech University who have filled professional positions in industry or government or who have entered graduate schools from coast to coast. The B.A. degree, which provides special emphasis on the application of scientific discovery and the effect of technology upon society, is available for students who do not want to practice engineering. A dual degree in Architecture/Civil Engineering leads to options beyond either of the individual programs. Graduate study in engineering can be pursued at the Master's level, which requires about 1.5 years beyond the B.S. degree, or the doctoral level, which leads to the Ph.D. after 3 or more years of study.

DUAL DEGREE PROGRAMS
Engineering baccalaureate graduates with dual degrees 3

TRANSFER INFORMATION
Residency Requirements: A student must complete their last 30 hours in residence and are allowed a maximum of 66 transfer hours from a Junior College.

Transfer without Articulation Agreements
Admission to engineering: At any time
Engineering graduates transferred from . . 4-yr: 31 2-yr: 63
Requirements: Must have a 2.0 GPA to transfer.

GRADUATION REQUIREMENTS
The B.S. programs in engineering require 132-139 semester hours, depending on major, with a minimum 2.0 Grade Point Average.

STUDENT PROGRAMS

PROFESSIONAL AND HONORARY SOCIETIES
[For key to acronyms, see *Introduction*.]
Professional Societies: ACI, ACM, AGCA, AIChE, ASCE, ASME, IEEE, IIE, NSPE, SPE
Honorary Societies: Alpha Epsilon, Alpha Pi Mu, Chi Epsilon, Eta Kappa Nu, Omega Chi Epsilon, Pi Tau Sigma, Tau Alpha Pi, Tau Beta Pi, Upsilon Pi Epsilon, Pi Epsilon Tau

SUPPORT PROGRAMS
Student Chapter Organizations: Society of Women Engineers, National Society of Black Engineers, Society of Hispanic Professional Engineers, Minority Engineering Program, Society of Hispanic Professional Engineers

Minority Engineering Program
For: Women & Ethnic Minorities Available: Year round
Offered: Prematriculation, Freshman, Sophomore, Junior, Senior, Graduate level

National Society of Black Engineers
For: Women & Ethnic Minorities Available: Year round
Offered: Prematriculation, Freshman, Sophomore, Junior, Senior, Graduate level

Society of Women Engineers
For: Women & Ethnic Minorities Available: Academic year
Offered: Freshman, Sophomore, Junior, Senior, Graduate level

Society of Hispanic Professional Engineers
For: Women & Ethnic Minorities Available: Academic year
Offered: Freshman, Sophomore, Junior, Senior, Graduate level

Other Engineering Support Programs: The University hosts several Summer Orientation Sessions for both entering and transfer students. Support services available include: Engineering Communications Center, a Faculty Mentoring Program, the Minority Engineering Program, Programs for Academic Support Services, Inc.(PASS), Career Planning and Placement Center.

240 University of Toledo

INSTITUTION PROFILE
HEAD OF THE INSTITUTION
Frank E. Horton
Phone: (419)537-2211
GENERAL INFORMATION
[All Students—Fall 1991]
Undergraduate enrollment 21,620
Graduate enrollment 3,349
Total institution enrollment **24,969**
Type of institution: Public **Calendar:** Quarters
Location: Toledo **Population:** 650,000
Setting: Urban
Types of engineering degrees: Engineering & Technology
Other degree-granting colleges: Arts & Sciences, Business Administration, Law, Pharmacy, Technology & Applied Sciences, University College, Univ. Community & Tech. Educational & Allied Prof., Continuing Education

ENGINEERING ADMISSIONS
ADMISSIONS OFFICE CONTACT
R. Eastop
University of Toledo
Admission Services
Toledo, OH 43606-3390
Phone: (419)537-2696
ENGINEERING COLLEGE ADMISSIONS INFORMATION
Entrance Requirements: SAT (Score: 900), ACT (Score: 21), High School courses—English (4 years), Math (4 years), Chemistry (1 year)
Entrance Recommendations: High School courses—Natural science (chemistry, physics, biology) (3 years), Social sciences (such as history, civics) (2.5 years), Computer science (1 year), Advance Placement Math, Chemistry Placement Exam
Requirements for foreign students: TOEFL (Score: 500)
DEPARTMENTAL ADMISSIONS INFORMATION
Admission to the engineering department: At the time of admission to the institution.
Additional information: All new B.S. students in the College of Engineering (freshmen or transfer, excluding the B.E.T. students) admitted into a Pre-Engineering Program. This allows students to enroll in calculus, chemistry, engineering graphics and computer programming to demonstrate potential success in the engineering program. All students transfer into an engineering discipline before 75 credit hours of attempted work. All departments require a 2.0 accumulate GPA (4 pt. scale); 2.3 for Computer Science and 2.5 for non-residents in Electrical Engineering.

FINANCIAL INFORMATION
ESTIMATED EXPENSES (ACADEMIC YEAR)
[Expenses are for the 1992-93 nine-month academic year.]

State Residents	Undergraduate	Graduate
Tuition and fees	$ 3,072	$ 4,341
College room and board	$ 2,929	$ 2,929
Books and supplies	$ 500	$ 600
Other expenses	$ 474	$ 474
Total estimated expenses	**$ 6,975**	**$ 8,344**

Out-of-State Residents	Undergraduate	Graduate
Tuition and fees	$ 7,377	$ 8,646
College room and board	$ 2,929	$ 2,929
Books and supplies	$ 500	$ 600
Other expenses	$ 474	$ 474
Total estimated expenses	**$11,280**	**$12,649**

FINANCIAL AID OFFICE CONTACT
Richard Lasko, *Director of Financial Aid*
University of Toledo
Toledo, OH 43606-3390
Phone: (419)537-2056
GENERAL FINANCIAL AID INFORMATION
Forms accepted/required: FAF, Institutional
Additional financial aid information: There are various types of financial aid available such as need-based scholarships; work-study employment; part-time campus jobs; guaranteed student loans; Perkins loans; Ohio Instructional Grants; Pell Grants; plus special engineering scholarships and awards.

ENGINEERING COLLEGE INFORMATION
[For additional personnel, refer to the *Appendix*.]
HEAD OF ENGINEERING
Edward Lumsdaine
Phone: (419)537-2707
ENGINEERING COLLEGE ADDRESS
College of Engineering
University of Toledo
Toledo, OH 43606-3390
Phone: (419)537-2707 Fax: (419)537-2805
ENROLLMENTS—BY CLASS
[Numbers are baccalaureate enrollments for the fall 1991 term, unless otherwise footnoted.]
1st-year students/Freshmen 816
2nd-year students/Sophomores 337
3rd-year students/Juniors 455
4th-year students/Seniors 473
Total **2,081**
NUMBER OF DEGREES AWARDED—BY PROGRAM
[Numbers are engineering baccalaureate degrees awarded during the 1991-1992 academic year, unless otherwise footnoted. For full details about each engineering program, refer to the *Tables of Degree Programs*.]
Chemical Engineering 16
Civil Engineering 25
Computer Science & Engineering 22
Electrical Engineering 62
Engineering Physics 2
Engineering Science(s) .... _ (B)
Engineering Technology 77
Industrial Engineering 13
Mechanical Engineering 70
Total **287**
Notes: (B) Data not applicable.
PERCENTAGE OF DEGREES AWARDED—BY CATEGORY
[Percentages are of all engineering baccalaureate degrees awarded during the 1991-1992 academic year, unless otherwise footnoted.]
African Americans 1.0%
Asian/Pacific Island Americans 1.0%
Hispanic Americans 1.0%
Native Americans .... _ %
Foreign Citizens 21.0%
All Others 76.0%
Women 13.0%
Persons with disabilities .... _ % (A)
Students over 25 years of age .... _ % (A)
Notes: (A) Data not available.
GRADUATE ENROLLMENTS & DEGREES AWARDED
Master's enrollment 231
Master's degrees awarded 57
Doctoral enrollment 72
Doctoral degrees awarded 10
ENGINEERING STUDENT DATA
Applicants to the engineering college
Number of applicants to engineering college 850
Percent offered admission 89.0%
Matriculated engineering students
Percentage in top quartile (25%) of High School class 50.0
Average SAT scores: Combined—1050
Average ACT scores: Math—24, Composite—24
FULL- & PART-TIME FACULTY—BY DEPARTMENT
[Figures are the head count of full-time faculty and the full-time equivalent (FTE) of part-time faculty for each engineering department or equivalent.]

Department	Full	Part
Chemical Eng	9	–
Civil Eng	7	–
Electrical Eng	19	–
Computer Science & Eng	7	–
Industrial Eng	8	–
Mechanical Eng	8	–
Eng Tech	5	12.0
Eng Physics	–	–

COLLEGE DESCRIPTION
The College offers a four-year undergraduate program in engineering, emphasizing engineering fundamentals followed by a specialized training in one of seven undergraduate majors, six master's degree majors, and four major areas of study in the doctoral degree program. A bachelor of engineering technology is also available. Emphasis is placed on computer-aided engineering. Computer facilities are modern with students having ample access. Students may pursue a dual degree; co-op is an option in most programs. Creative thinking skills and problem solving techniques are introduced to all entering engineering freshmen and during the last undergraduate quarter, students complete a senior design project. All students are encouraged to complete the Fundamentals of Engineering examination. The faculty is known for dedication to teaching excellence, scholarly activities and continued research. The average number of years required to actually complete the B.S. degree is four years.

DEGREE PROGRAMS DESCRIPTION
The B.S. degrees in various engineering programs prepare the graduate for entry-level professional positions or for the graduate study in each of the respective fields. The B.S. degree in Engineering Physics is intended primarily to prepare students for advanced graduate study. A Bachelor of Engineering Technology is offered in four areas: electronics engineering technology, mechanical engineering technology, manufacturing engineering technology, and construction technology. The Master of Science degree is offered in Chemical, Civil, Electrical, Industrial, and Mechanical Engineering as well as Engineering Science. The Doctor of Philosophy degree in Engineering Science is offered in four interdisciplinary areas: chemical and biological transport, engineering mechanics, electronics and energy, and systems theory and engineering.

TRANSFER INFORMATION
Residency Requirements: B.S. transfer students must earn at least 50 hours of undergraduate credit in residence, with 20 credits in major area. Full-time students must take their last quarter in residence; part-time, their last 20 hours in residence.
Transfer via Articulation Agreements
Admission to engineering: Transcripts are required at least four weeks prior to the beginning of a given quarter.
GRADUATION REQUIREMENTS
B.S. transfer students must earn at least 50 hours of undergraduate credit in residence, with 20 credits in major area. Full-time students must take their last quarter in residence; part-time, their last 20 hours in residence.

STUDENT PROGRAMS
PROFESSIONAL AND HONORARY SOCIETIES
[For key to acronyms, see *Introduction*.]
Professional Societies: ACM, AIChE, ASCE, ASME, IEEE, IIE, SAE, Ohio Society of Professional Engineers (OSPE), Society for the Advancement of Management (SAM)
Honorary Societies: Alpha Pi Mu, Eta Kappa Nu, Pi Tau Sigma, Tau Alpha Pi, Tau Beta Pi, Civil Engineering Honor Society (UTCEHS), Phi Sigma Rho
SUPPORT PROGRAMS
Student Chapter Organizations: Society of Women Engineers, National Society of Black Engineers, PREP Program (for aspiring minority students), Toledo EXCEL Program, Minority Mentorship Program, Minority Orientation Program
Other Engineering Support Programs: SOAR, Tutoring Services, Pre-Engineering, Engineering Orientation Course, and Creativity & Problem Solving Course.

241 Tri-State University

■ INSTITUTION PROFILE

HEAD OF THE INSTITUTION
William G. Meyers
Phone: (219)665-4187 Fax: (219)665-4292

GENERAL INFORMATION
[All Students—Fall 1991]
Undergraduate enrollment 1,060
Total institution enrollment **1,060**
Type of institution: Private **Calendar:** Quarters
Nearest city: Fort Wayne, Indiana **Population:** 200,000
Miles from main campus: 45 **Setting:** Small Town
Types of engineering degrees: Engineering & Technology
Other degree-granting colleges: Arts & Sciences, Business Administration

■ ENGINEERING ADMISSIONS

ADMISSIONS OFFICE CONTACT
Walter Lilley
Tri-State University
Admissions Office
Angola, IN 46703-0307
Phone: (219)665-4139 Fax: (219)665-4292

ENGINEERING COLLEGE ADMISSIONS INFORMATION
Entrance Requirements: SAT, ACT, High School courses—Algebra (2 years), Science and Social Studies (2 years), Chemistry, Physics, Geometry, Trigonometry (1 year), English (4 years), Students are required to take EITHER the SAT or ACT. Additionally, all freshman are given mathematics and English placement exams during orientation.
Entrance Recommendations: High School courses—Calculus or Pre-Calculus (1 year), Computer Fundamentals (1 year), Mechanical Drawing (1 year)
Requirements for foreign students: financial statement; English proficiency verification is required. TOEFL is recommended. Mathematics and English placement examinations are given during orientation.

DEPARTMENTAL ADMISSIONS INFORMATION
Admission to the engineering department: At the time of admission to the institution.

■ FINANCIAL INFORMATION

ESTIMATED EXPENSES (ACADEMIC YEAR)
[Expenses are for the 1992-93 nine-month academic year.]

All Students	Undergraduate	Graduate
Tuition and fees	$10,500 [1]	$ — [3]
College room and board	$ 3,900	$ — [3]
Books and supplies	$ 500	$ — [3]
Other expenses	$ 500 [2]	$ — [3]
Total estimated expenses	**$15,400**	**$ —**

Notes: (1) Non-engineering full-time tuition is approximately $9476. Tuition for a quarter credit hour is $214 for engineering and $184 for all other schools. (2) Personal Expenses (3) No graduate programs

FINANCIAL AID OFFICE CONTACT
Susan Stroh, *Financial Aid Director*
Tri-State University
Financial Aid Office
Angola, IN 46703-0307
Phone: (219)665-4174 Fax: (219)665-4292

GENERAL FINANCIAL AID INFORMATION
Forms accepted/required: FAF, Stafford, Supplemental, Institutional, Perkins Loan
Additional financial aid information: Scholarships (need and merit based), Grants (Institutional, State and Federal), Campus Employment, Work Study, Athletic Awards, Loans

■ ENGINEERING COLLEGE INFORMATION
[For additional personnel, refer to the *Appendix*.]

HEAD OF ENGINEERING
William G. Meyers
Phone: (219)665-4187 Fax: (219)665-4292

ENGINEERING COLLEGE ADDRESS
School of Engineering
Tri-State University
300 South Darling Street
Angola, IN 46703-0307
Phone: (219)665-4100 Fax: (219)665-4292

ENROLLMENTS—BY CLASS
[Numbers are baccalaureate enrollments for the fall 1991 term, unless otherwise footnoted.]
1st-year students/Freshmen 113
2nd-year students/Sophomores 109
3rd-year students/Juniors 148
4th-year students/Seniors 115
Total **485**

NUMBER OF DEGREES AWARDED—BY PROGRAM
[Numbers are engineering baccalaureate degrees awarded during the 1991-1992 academic year, unless otherwise footnoted. For full details about each engineering program, refer to the *Tables of Degree Programs*.]
Aerospace Engineering 5
Chemical Engineering 9
Civil Engineering 12
Electrical Engineering 37
Engineering Administration 10
Mechanical Engineering 38
Total **111**

PERCENTAGE OF DEGREES AWARDED—BY CATEGORY
[Percentages are of all engineering baccalaureate degrees awarded during the 1991-1992 academic year, unless otherwise footnoted.]
African Americans .... 1.0 %
Asian/Pacific Island Americans .... 1.0 %
Hispanic Americans .... 1.0 %
Native Americans .... — %
Foreign Citizens .... 24.0 %
All Others .... 73.0 %
Women .... 9.0 %
Persons with disabilities .... 1.0 %
Students over 25 years of age .... 20.0 %

ENGINEERING STUDENT DATA
Applicants to the engineering college
Number of applicants to engineering college 565
Percent offered admission 95.0%
Matriculated engineering students
Percentage in top quartile (25%) of High School class 60.0
Average SAT scores: Math—443, Verbal—568, Combined—1011
Average ACT scores: Math—25, Composite—24

FULL- & PART-TIME FACULTY—BY DEPARTMENT
[Figures are the head count of full-time faculty and the full-time equivalent (FTE) of part-time faculty for each engineering department or equivalent.]

Department	Full	Part
Chemical Eng	3	—
Civil Eng	4	—
Electrical Eng	7	—
Mechanical Eng	5	—
Aerospace Eng	3	—
Eng Administration	—	—

COLLEGE DESCRIPTION
Tri-State University provides a unique university experience to students through small classes and a faculty committed to quality, exclusively undergraduate, education. The School of Business and the School of Arts and Sciences work closely with the School of Engineering to provide students with opportunities to expand their programs of study in these areas. The emphasis is on teaching and hands-on laboratory instruction. All classes and laboratories are presented and directed by experienced faculty. Questions are welcomed through the faculty open-door policy. A Co-op experience is encouraged and requires approximately one additional year for program completion. A competitive 3 month Japanese Internship program is also available.

DEGREE PROGRAMS DESCRIPTION
Five ABET accredited engineering programs and an Engineering Administration program prepare students for entry level professional practice positions or for graduate study in business or their respective technical field. Communication skills and a commitment to humanities and social sciences is evident throughout each program.

DUAL DEGREE PROGRAMS
Engineering baccalaureate graduates with dual degrees 0

TRANSFER INFORMATION
Residency Requirements: A minimum of 45 quarter credit hours must be earned at Tri-State University to be eligible for a baccalaureate degree.

Transfer via Articulation Agreements
Admission to engineering: At any time
Engineering graduates transferred from . . 4-yr: 20 2-yr: 30
Requirements: Satisfactory academic performance at the previous institution.

Transfer without Articulation Agreements
Admission to engineering: At any time
Requirements: Satisfactory academic performance at the previous institution.

GRADUATION REQUIREMENTS
206 quarter credit hours must be completed with an average overall grade point of at least 2.0/4.0. Additionally, a student must satisfy program course requirements.

■ STUDENT PROGRAMS

PROFESSIONAL AND HONORARY SOCIETIES
[For key to acronyms, see *Introduction*.]
Professional Societies: ACM, AIAA, AIChE, AIDD, ASCE, ASME, IEEE, SAE, SME, National Society of Minority Students, Society of Women Engineers
Honorary Societies: Chi Epsilon, Eta Kappa Nu, Omega Chi Epsilon, Pi Tau Sigma, Sigma Gamma Tau, Sigma Phi Delta, Tau Beta Pi, Phi Eta Sigma, Skull and Bones

SUPPORT PROGRAMS
Student Chapter Organizations: Society of Women Engineers, National Society of Black Engineers, National Society of Minority Students, Malaysian Student Association, International Student Association, Muslim Student Association

International Students Office
For: Ethnic Minorities **Available:** Year round
Offered: Prematriculation, Freshman, Sophomore, Junior, Senior

English Language Center
For: Ethnic Minorities **Available:** Year round
Offered: Prematriculation, Freshman, Sophomore, Junior, Senior

Other Engineering Support Programs: The Charter Counseling Center, a Health Services Center and the Gettig Fitness and Wellness Center are available for students as needed. The Career Center provides assistance in locating co-op and full-time employment. An orientation program emphasizing academic success and professional development is required for engineering freshman. Non-credit preparatory courses are also available.

242 Trinity University

INSTITUTION PROFILE
HEAD OF THE INSTITUTION
Ronald K. Calgaard
Phone: (210)736-8401 Fax: (210)736-8400

GENERAL INFORMATION
[All Students—Fall 1991]
Undergraduate enrollment 2,168
Graduate enrollment 116
Total institution enrollment 2,284
Type of institution: Private Calendar: Semesters
Location: San Antonio Population: 900,000
Setting: Urban
Types of engineering degrees: Engineering

ENGINEERING ADMISSIONS
ADMISSIONS OFFICE CONTACT
Beth Allen
Trinity University
Office of Admissions
715 Stadium Drive
San Antonio, TX 78212
Phone: (210)736-7207 Fax: (210)736-8164

ENGINEERING COLLEGE ADMISSIONS INFORMATION
Entrance Requirements: SAT, ACT, High School courses—English (4 years), College prep. mathematics incl. trig. or pre-calc. (3 years), Laboratory science (2 years), Social studies (2); foreign language (2) (4 years), Trinity accepts SAT or ACT test scores and has no published minimum.
Entrance Recommendations: SAT, ACT, High School courses—Foreign language (3 years), Computer programming (1 year), Calculus (1 year), Trinity has no published minimum.
Requirements for foreign students: TOEFL; financial statement

DEPARTMENTAL ADMISSIONS INFORMATION
Admission to the engineering department: At the end of the second year.
Additional information: Mathematics 311, 312, 323 with an average of 2.0 or better; Physics 311/111, 312/112 and Chemistry 318/118 with an average of 2.0 or better. Engineering Science 301, 302, 313, 314, 320/120 and 115 with an average of 2.0 or better; a grade of C or better in ES 313, 314, 320/120 and approval by the department. Although admission to the major is at the end of the sophomore year, course requirements begin with the first semester of the freshman year.

FINANCIAL INFORMATION
ESTIMATED EXPENSES (ACADEMIC YEAR)
[Expenses are for the 1992-93 nine-month academic year.]

All Students	Undergraduate	Graduate
Tuition and fees	$11,060	$ –
College room and board	$ 4,640	$ –
Books and supplies	$ 400	$ –
Other expenses	$ 35 (1)	$ –
Total estimated expenses	**$16,135**	**$ –**

Notes: (1) These expenses are: Residence Hall damage fee and Association of Residence Hall programming fee.

FINANCIAL AID OFFICE CONTACT
Estelle Frerichs, *Director of Financial Aid*
Trinity University
715 Stadium Drive
San Antonio, TX 78212
Phone: (210)736-8315

GENERAL FINANCIAL AID INFORMATION
Forms accepted/required: ACT, AFSA, FFS, IRS, SDF, Institutional
Additional financial aid information: There are need based scholarships, merit-based scholarships, part-time campus jobs, and a College Work Study Program. In addition, there are NSPE-based scholarships for entering students and Swalm Fund scholarships for fifth year study.

ENGINEERING COLLEGE INFORMATION
[For additional personnel, refer to the *Appendix*.]
HEAD OF ENGINEERING
John S. Dickey
Phone: (210)736-7414 Fax: (210)736-7229

ENGINEERING COLLEGE ADDRESS
Engineering Science Department
Trinity University
715 Stadium Drive
San Antonio, TX 78212
Phone: (210)736-7511 Fax: (210)736-7569

ENROLLMENTS—BY CLASS
[Numbers are baccalaureate enrollments for the fall 1991 term, unless otherwise footnoted.]
1st-year students/Freshmen 120
2nd-year students/Sophomores 47
3rd-year students/Juniors 35
4th-year students/Seniors 39
Total **241**

NUMBER OF DEGREES AWARDED—BY PROGRAM
[Numbers are engineering baccalaureate degrees awarded during the 1991-1992 academic year, unless otherwise footnoted. For full details about each engineering program, refer to the *Tables of Degree Programs*.]
Engineering Science(s) 20
Total **20**

PERCENTAGE OF DEGREES AWARDED—BY CATEGORY
[Percentages are of all engineering baccalaureate degrees awarded during the 1991-1992 academic year, unless otherwise footnoted.]
African Americans 5.0%
Asian/Pacific Island Americans .... – %
Hispanic Americans 10.0%
Native Americans 5.0%
Foreign Citizens 5.0%
All Others 75.0%
Women 35.0%
Persons with disabilities .... – %
Students over 25 years of age 5.0%

ENGINEERING STUDENT DATA
Applicants to the engineering college
Number of applicants to engineering college .... – (B)
Percent offered admission .... –% (B)
Matriculated engineering students
Percentage in top quartile (25%) of High School class .... – (A)
Average SAT scores: Math—677, Verbal—597, Combined—1273
Average ACT scores: Math—31, Composite—29
Notes: (A) Data not available. (B) Data not applicable.

FULL- & PART-TIME FACULTY—BY DEPARTMENT
[Figures are the head count of full-time faculty and the full-time equivalent (FTE) of part-time faculty for each engineering department or equivalent.]

Department	Full	Part
Eng Science	9	–

COLLEGE DESCRIPTION
Trinity University is an independent, nonsectarian, co-educational university in the tradition of the liberal arts and sciences. The academic year is based on a semester system. The school offers one engineering degree, the Bachelor of Science in Engineering Science. Trinity's undergraduate Engineering Science Program is accredited by the Engineering Accreditation Commission of the Accreditation Board for Engineering and Technology (EAC/ABET). Major academic divisions include: Sciences, Mathematics and Engineering, Behavioral Sciences and Business and Administrative Studies, and Humanities and Art.

DEGREE PROGRAMS DESCRIPTION
The goal of the Engineering Science Department at Trinity University is to develop high quality engineering graduates who have the ability to think critically and creatively and who possess a broad background in the engineering sciences. Trinity's program is one of only 16 EAC/ABET accredited programs of this type in the country. Special features include: small classes and laboratories, close contact with full time faculty, and a capstone design experience for seniors in which the student applies his/her coursework in the actual design, construction and testing of an engineering device. The Bachelor of Science program in Engineering Science requires 129 semester credits.

TRANSFER INFORMATION
Transfer without Articulation Agreements
Admission to engineering: At any time
Engineering graduates transferred from . . . 4-yr: – 2-yr: –
Requirements: Trinity University may evaluate and accept credit earned at other accredited educational institutions. Each course is evaluated separately to determine if it can apply toward a Trinity degree.

GRADUATION REQUIREMENTS
At least 60 semester hours, including the last 30, must be completed in residence to complete a baccalaureate degree. Thirty of the 60 hours must be advanced hours, including at least 15 hours in the major. In order to qualify for the Bachelor of Science in Engineering Science, students must complete the University's Common Curriculum, have a GPA of at least 2.0 overall, and a GPA of at least 2.0 in the major.

STUDENT PROGRAMS
PROFESSIONAL AND HONORARY SOCIETIES
[For key to acronyms, see *Introduction*.]
Professional Societies: AIChE, ASME, IEEE, NSPE, SAE, Trinity Engineers: A student run engineering association, Society of Women Engineers
Honorary Societies: Phi Beta Kappa

SUPPORT PROGRAMS
Student Chapter Organizations: Society of Women Engineers
Society of Women Engineers
For: Women **Available:** Academic year
Offered: Freshman, Sophomore, Junior, Senior
Other Engineering Support Programs: The small size of the engineering program permits the majority of engineering students to receive academic advising from faculty in the Department. The University provides many high quality support programs through the Office of Counseling and Career Services and through student-run groups.

243 Tufts University

INSTITUTION PROFILE

HEAD OF THE INSTITUTION
John DiBiaggio
Phone: (617)627-3300

GENERAL INFORMATION
[All Students—Fall 1991]
Undergraduate enrollment	4,400
Graduate enrollment	1,184
Total institution enrollment	5,584

Type of institution: Private **Calendar:** Semesters
Nearest city: Boston **Population:** 620,000
Miles from main campus: 7 **Setting:** Suburban
Types of engineering degrees: Engineering

Other degree-granting colleges: Arts & Sciences, Dentistry, Medicine, Law and Diplomacy, Nutrition, Graduate Biomedical Sciences, Veterinary

ENGINEERING ADMISSIONS

ADMISSIONS OFFICE CONTACT
David D. Cuttino
Tufts University
Undergraduate Admissions
Bendetson Hall
Medford, MA 02155
Phone: (617)628-5000 **Fax:** (617)627-3860

ENGINEERING COLLEGE ADMISSIONS INFORMATION
Entrance Requirements: SAT, English, Physics or Chemistry, and Math Achievement Tests
Entrance Recommendations: High School courses—English (4 years), Math (4 years), Science (4 years), History (2 years)
Requirements for foreign students: TOEFL (Score: 550); financial statement

DEPARTMENTAL ADMISSIONS INFORMATION
Admission to the engineering department: At the end of the first year.

FINANCIAL INFORMATION

ESTIMATED EXPENSES (ACADEMIC YEAR)
[Expenses are for the 1992-93 nine-month academic year.]

All Students	Undergraduate	Graduate
Tuition and fees	$17,897	$17,970
College room and board	$ 5,443	$ 5,443
Books and supplies	$ 1,250	$ 1,250
Other expenses	$ 447	$ 315
Total estimated expenses	**$25,037**	**$24,978**

FINANCIAL AID OFFICE CONTACT
William F. Eastwood, *Director of Financial Aid*
Tufts University
128 Professors Row
Medford, MA 02155
Phone: (617)628-5000

GENERAL FINANCIAL AID INFORMATION
Forms accepted/required: FAF, Tufts Aid Application
Additional financial aid information: Tufts University students are eligible for a full range of financial aid in the form of university, state and federal grants, long-term Perkins (formerly NDSL) and Stafford (GSL) loans, Supplementary Educational Opportunity Grants (SEOG), Pell grants, Tufts National Merit Scholarships, as well as federally subsidized (CWSP) campus employment.

ENGINEERING COLLEGE INFORMATION
[For additional personnel, refer to the *Appendix*.]

HEAD OF ENGINEERING
Frederick C. Nelson
Phone: (617)627-3237 **Fax:** (617)627-3819

ENGINEERING COLLEGE ADDRESS
College of Engineering
Tufts University
200 College Avenue
105 Anderson Hall
Medford, MA 02155
Phone: (617)628-5000 **Fax:** (617)627-3819

ENROLLMENTS—BY CLASS
[Numbers are baccalaureate enrollments for the fall 1991 term, unless otherwise footnoted.]

1st-year students/Freshmen	176
2nd-year students/Sophomores	134
3rd-year students/Juniors	164
4th-year students/Seniors	205
Total	**679**

NUMBER OF DEGREES AWARDED—BY PROGRAM
[Numbers are engineering baccalaureate degrees awarded during the 1991-1992 academic year, unless otherwise footnoted. For full details about each engineering program, refer to the *Tables of Degree Programs*.]

Art & Architectural History	1
Chemical Engineering	22
Civil Engineering	37
Computer Engineering	13
Electrical Engineering	59
Engineering Psychology	11
Environmental Studies & Environmental Health	1
Manufacturing Engineering	3
Mechanical Engineering	41
Total	**188**

PERCENTAGE OF DEGREES AWARDED—BY CATEGORY
[Percentages are of all engineering baccalaureate degrees awarded during the 1991-1992 academic year, unless otherwise footnoted.]

African Americans	1.7%
Asian/Pacific Island Americans	7.4%
Hispanic Americans	1.7%
Native Americans	— %
Foreign Citizens	7.4%
All Others	81.8%
Women	19.4%
Persons with disabilities	— % (A)
Students over 25 years of age	— % (A)

Notes: (A) Data not available.

GRADUATE ENROLLMENTS & DEGREES AWARDED
Master's enrollment	163
Master's degrees awarded	96
Doctoral enrollment	47
Doctoral degrees awarded	15

ENGINEERING STUDENT DATA
Applicants to the engineering college
Number of applicants to engineering college	1,006
Percent offered admission	57.0%

Matriculated engineering students
Percentage in top quartile (25%) of High School class 92.0
Average SAT scores: Math—684, Verbal—562, Combined—1246

FULL- & PART-TIME FACULTY—BY DEPARTMENT
[Figures are the head count of full-time faculty and the full-time equivalent (FTE) of part-time faculty for each engineering department or equivalent.]

Department	Full	Part
Chemical Eng	8	1.5
Civil Eng	11	5.0
Electrical Eng	14	3.0
Mechanical Eng	12	1.0

COLLEGE DESCRIPTION
The goal of the College of Engineering is to provide an education that produces graduates who are effective in solving current engineering problems and are well prepared for continuous learning about new engineering principles; that fosters the attitudes and skills that will allow them to increase the nation's store of technological knowledge; and that places them in the cultural, economic, environmental, and ethical contexts in which they will practice their profession. The College of Engineering offers eight bachelor of science degrees; BS in Chemical, Civil, Electrical (including Computer Engineering option) and Mechanical; bachelor of science in engineering science; bachelor of science in engineering (including Computer Science and Engineering, and Manufacturing Engineering); bachelor of science in engineering physics and bachelor of science (including Engineering Psychology). A Minor in Engineering Management is also available for students graduating with an engineering degree.

DEGREE PROGRAMS DESCRIPTION
The Bachelor of Science degree in the engineering programs prepare students for entry-level professional positions or graduate study. The five engineering departments offer both M.S. and Ph.D. degrees. The College offers a combined five-year Liberal Arts/Engineering program in which students earn both Engineering and Liberal Arts degrees. Alternatively an undergraduate may earn one degree with a second major with the consent of the appropriate departments. A combined Bachelor's and Master's Degrees Program is available. Participation in Tufts Programs Abroad is also possible.

TRANSFER INFORMATION
Residency Requirements: Transfer students must have a minimum of 2 years residence. The last two semesters must be in residence at Tufts.

Transfer without Articulation Agreements
Admission to engineering: Students are admitted at the end of freshman year, in January of their sophomore year, and at the end of the sophomore year.

Requirements: Admission is competitive based on the assessment of college and secondary school records, recommendations and the history of extracurricular and personal involvement.

GRADUATION REQUIREMENTS
Four years of full time study totaling 38 courses of which grades of C- or better are required in two-thirds of the courses submitted for the degree. (For students who transfer to Tufts, grades of C- or better must be earned in not less than two-thirds of the courses taken at Tufts.) In addition it is expected that grades of C- or better will be earned in at least seventy-five percent of the courses taken in the department of concentration.

STUDENT PROGRAMS

PROFESSIONAL AND HONORARY SOCIETIES
[For key to acronyms, see *Introduction*.]
Professional Societies: ACerS, ACS, AIAA, AIChE, ASCE, ASME, IEEE, SME, SPE*, International Society of Hybrid Microelectronics, The Catalysis Society of America
Honorary Societies: Eta Kappa Nu, Tau Beta Pi, Sigma Xi

SUPPORT PROGRAMS
Student Chapter Organizations: Society of Women Engineers, National Society of Black Engineers, Personal and career counseling

The Society of Women Engineers
For: Women **Available:** Academic year
Offered: Freshman, Sophomore, Junior, Senior, Graduate level

The National Society of Black Engineers
For: Ethnic Minorities **Available:** Year round
Offered: Freshman, Sophomore, Junior, Senior, Graduate level

Other Engineering Support Programs: Summer and fall orientation program for freshmen and transfer students; career counseling services; foreign student office; services for handicapped; placement services; individual faculty advisors.

244 Tulane University

INSTITUTION PROFILE

HEAD OF THE INSTITUTION
Eamon M. Kelly
Phone: (504)865-5201   **Fax:** (504)865-5202

GENERAL INFORMATION
[All Students—Fall 1991]
Undergraduate enrollment 7,084
Graduate enrollment 4,403
Total institution enrollment **11,487**
Type of institution: Private   **Calendar:** Semesters
Location: New Orleans, LA   **Population:** 500,000
Setting: Urban
Types of engineering degrees: Engineering
Other degree-granting colleges: Allied Health Sciences, Architecture, Arts & Sciences, Business Administration, Fine & Performing Arts, Humanities & Social Sciences, Law, Medicine, Social Work, Architecture

ENGINEERING COLLEGE ADMISSIONS INFORMATION
Entrance Requirements: High School courses—English (4 years), Math (4 years), Laboratory Science (3 years), Social Studies (with emphasis on history) (2 years), Minimum SAT scores: Verbal 450 and Math 500; Minimum ACT scores: Composite 23 and Math 25.
Entrance Recommendations: High School courses—Calculus (1 year), Physics (1 year), Computer (1 year), College Board Achievement tests in English Composition, Physics, Math
Requirements for foreign students: TOEFL (Score: 550); financial statement; High school transcript with translation and description of grading system. ESLAT for Puerto Rican applicants

DEPARTMENTAL ADMISSIONS INFORMATION
Admission to the engineering department: At the time of admission to the institution.

FINANCIAL INFORMATION

ESTIMATED EXPENSES (ACADEMIC YEAR)
[Expenses are for the 1992-93 nine-month academic year.]

All Students	Undergraduate	Graduate
Tuition and fees	$18,185	$18,185
College room and board	$ 5,660	$ 5,660
Books and supplies	$ 350	$ 450
Other expenses	$ 800 [1]	$ 3,210 [1]
Total estimated expenses	**$24,995**	**$27,505**

Notes: (1) Includes travel and miscellaneous expenses.

FINANCIAL AID OFFICE CONTACT
Thomas P. Lovett, *Director of Financial Aid*
Tulane University
Financial Aid
New Orleans, LA 70118
Phone: (504)865-5723

GENERAL FINANCIAL AID INFORMATION
Forms accepted/required: CSS-FAF, FAF, IRS
Additional financial aid information: Aid is available based on need as well as academic merit through scholarships, grants, loans and campus employment: College work-study Program. Required forms: FAF and IRS for undergraduate and GAPSFAS for graduate students.

ENGINEERING COLLEGE INFORMATION
[For additional personnel, refer to the *Appendix*.]

HEAD OF ENGINEERING
William C. Van Buskirk
Phone: (504)865-5766   **Fax:** (504)862-8747

ENGINEERING COLLEGE ADDRESS
School of Engineering
Tulane University
6823 St. Charles Avenue
New Orleans, LA 70118
Phone: (504)865-5764   **Fax:** (504)862-8747

ENROLLMENTS—BY CLASS
[Numbers are baccalaureate enrollments for the fall 1991 term, unless otherwise footnoted.]
1st-year students/Freshmen 268
2nd-year students/Sophomores 183
3rd-year students/Juniors 150
4th-year students/Seniors 140
Total **741**

NUMBER OF DEGREES AWARDED—BY PROGRAM
[Numbers are engineering baccalaureate degrees awarded during the 1991-1992 academic year, unless otherwise footnoted. For full details about each engineering program, refer to the *Tables of Degree Programs*.]
Biomedical Engineering 31
Chemical Engineering 16
Civil & Environmental Engineering 19
Computer Science 18
Electrical Engineering 25
Mechanical Engineering 21
Total **130**

PERCENTAGE OF DEGREES AWARDED—BY CATEGORY
[Percentages are of all engineering baccalaureate degrees awarded during the 1991-1992 academic year, unless otherwise footnoted.]
African Americans 8.6%
Asian/Pacific Island Americans 10.0%
Hispanic Americans 6.4%
Native Americans .... – %
Foreign Citizens 12.1%
All Others 62.9%
Women 24.3%
Persons with disabilities .... – %
Students over 25 years of age .... – %

GRADUATE ENROLLMENTS & DEGREES AWARDED
Master's enrollment 77
Master's degrees awarded 22
Doctoral enrollment 65
Doctoral degrees awarded 20

ENGINEERING STUDENT DATA
Applicants to the engineering college
Number of applicants to engineering college 1,133
Percent offered admission 80.0%

Matriculated engineering students
Percentage in top quartile (25%) of High School class 91.0
Average SAT scores: Math—649, Verbal—571, Combined—1220
Average ACT scores: Composite—28

FULL- & PART-TIME FACULTY—BY DEPARTMENT
[Figures are the head count of full-time faculty and the full-time equivalent (FTE) of part-time faculty for each engineering department or equivalent.]

Department	Full	Part
Chemical Eng	9	–
Civil & Environmental Eng	8	–
Electrical Eng	9	–
Mechanical Eng	9	–
Biomedical Eng	9	–
Computer Science	9	–

COLLEGE DESCRIPTION
School of Engineering students pursue undergraduate and advanced degrees in biomedical, chemical, civil, computer, electrical, and mechanical engineering. There is also a flexible interdisciplinary degree program at the undergraduate level leading to a BS degree in Engineering. In addition, students can earn degrees in computer science. All undergraduate curricula are four years in length, with some liberal arts requirements. In conjunction with Tulane's A.B. Freeman School of Business, students may earn a B.S. in Engineering and an MBA in five years, rather than the traditional six. A B.S./M.S. program allows a student to complete the B.S. and M.S. program in five years with continuing financial aid. A program is also offered in Premedical Studies in Engineering. Average number of years required to actually complete the BS degree is four.

DEGREE PROGRAMS DESCRIPTION
The undergraduate Computer Science program leads to the B.S. in Computer Science. All other undergraduate programs lead to the B.S. in Engineering (Biomedical, Chemical, Civil, Computer, Electrical, Mechanical Engineering, and Engineering Science). The Graduate Division of the School of Engineering offers the M.Engr. and D.Engr. degrees in Biomedical, Chemical, Civil, Computer Science, Electrical and Mechanical Engineering. Programs leading to the M.S. and Ph.D. in these fields are administered by the Graduate School of Tulane University.

DUAL DEGREE PROGRAMS
Engineering baccalaureate graduates with dual degrees 2

TRANSFER INFORMATION
Residency Requirements: Minimum of one year residency with all senior courses taken at Tulane.

Transfer without Articulation Agreements
Admission to engineering: At any time
Engineering graduates transferred from ... 4-yr: –   2-yr: –
Requirements: At least a 2.5 GPA on a 4.0 scale and strength in Math & Science.

GRADUATION REQUIREMENTS
Minimum 2.0 overall GPA and 2.0 in major. Last two semesters of coursework toward the degree must be completed at Tulane.

STUDENT PROGRAMS

PROFESSIONAL AND HONORARY SOCIETIES
[For key to acronyms, see *Introduction*.]
Professional Societies: ACM, AIChE, ASCE, ASHRAE, ASME, BMES, IEEE, TES-LES Student Chapter (Tulane Eng. Society/Louisiana Eng. Society)
Honorary Societies: Eta Kappa Nu, Omega Chi Epsilon, Pi Tau Sigma, Tau Beta Pi, Alpha Eta Mu Beta, Upsilon Pi Epsilon

SUPPORT PROGRAMS
Student Chapter Organizations: National Society of Black Engineers, Office of Multi-cultural Affairs, Delta Sigma Theta Sorority

Educational Resource Center
For: Women & Ethnic Minorities   **Available:** Year round
Offered: Freshman, Sophomore, Junior, Senior, Graduate level

Career Services Center
For: Women & Ethnic Minorities   **Available:** Year round
Offered: Freshman, Sophomore, Junior, Senior, Graduate level

Other Engineering Support Programs: Fall and Spring Orientation for Freshman and Transfers; Educational Resource Center; Pre-professional Advising; Career Services Center; Counseling and Testing; and Student Affairs.

245 University of Tulsa

INSTITUTION PROFILE

HEAD OF THE INSTITUTION
Robert H. Donaldson
Phone: (918)631-2305 Fax: (918)631-2247

GENERAL INFORMATION
[All Students—Fall 1991]
Undergraduate enrollment 3,463
Graduate enrollment 1,659
Total institution enrollment 5,122
Type of institution: Private Calendar: Semesters
Location: Tulsa Population: 500,000
Setting: Urban
Types of engineering degrees: Engineering
Other degree-granting colleges: Arts & Sciences, Business Administration, Law

ENGINEERING ADMISSIONS

ADMISSIONS OFFICE CONTACT
John C. Corso
University of Tulsa
Office of Admission
600 South College Avenue
Tulsa, OK 74104-3189
Phone: (918)631-2307

ENGINEERING COLLEGE ADMISSIONS INFORMATION
Entrance Requirements: Either SAT or ACT is acceptable. There are no minimum score requirements.
Entrance Recommendations: High School courses—College Preparatory English (4 years), College Prep. math, including algebra II & Trig. (4 years), College Preparatory social studies (3 years), College Preparatory sciences (4 years), SAT or ACT. There are no minimum score requirements.
Requirements for foreign students: TOEFL (Score: 500); financial statement; English proficiency is not used in determining the admissibility of students to the university. Proficiency is required, however, before a student will be allowed to enroll in academic courses.

DEPARTMENTAL ADMISSIONS INFORMATION
Admission to the engineering department: At the time of admission to the institution.
Additional information: In addition to meeting to meeting the university's admission requirements, students are expected to have high school credit for four semesters of algebra, one, semester of trigonometry, two semesters of plane geometry, six semesters of English, and two semesters of chemistry.

FINANCIAL INFORMATION

ESTIMATED EXPENSES (ACADEMIC YEAR)
[Expenses are for the 1992-93 nine-month academic year.]

All Students	Undergraduate	Graduate
Tuition and fees	$ 9,380 [3]	$ 6,840 [1]
College room and board	$ 3,775	$ 3,775
Books and supplies	$ 750	$ 480
Other expenses	$ –	$ –
Total estimated expenses	**$13,905**	**$11,095**

Notes: (1) 9 hours per semester (3) 12 - 18 hours per semester

FINANCIAL AID OFFICE CONTACT
David L. Gruen, *Director*
University of Tulsa
Student Financial Services
600 South College Avenue
Tulsa, OK 74104-3189
Phone: (918)631-2526

GENERAL FINANCIAL AID INFORMATION
Forms accepted/required: ACT, CSS-FAF, FAF, FFS, SDF, Institutional
Additional financial aid information: Grants, loans, College Work-Study Program, and both need-based and merit-based scholarships are available.

ENGINEERING COLLEGE INFORMATION

[For additional personnel, refer to the *Appendix*.]

HEAD OF ENGINEERING
Lewis M. Duncan
Phone: (918)631-2288 Fax: (918)631-2286
EMail: LDuncan@vax1.utulsa.edu

ENGINEERING COLLEGE ADDRESS
College of Engineering and Applied Sciences
University of Tulsa
600 South College Avenue
Tulsa, OK 74104-3189
Phone: (918)631-2000 Fax: (918)631-2247

ENROLLMENTS—BY CLASS
[Numbers are baccalaureate enrollments for the fall 1991 term, unless otherwise footnoted.]
1st-year students/Freshmen 144
2nd-year students/Sophomores 109
3rd-year students/Juniors 111
4th-year students/Seniors 101
Total **465**

NUMBER OF DEGREES AWARDED—BY PROGRAM
[Numbers are engineering baccalaureate degrees awarded during the 1991-1992 academic year, unless otherwise footnoted. For full details about each engineering program, refer to the *Tables of Degree Programs*.]
Chemical Engineering 10
Electrical Engineering 18
Engineering Physics 5
Mechanical Engineering 33
Petroleum Engineering 19
Total **85**

PERCENTAGE OF DEGREES AWARDED—BY CATEGORY
[Percentages are of all engineering baccalaureate degrees awarded during the 1991-1992 academic year, unless otherwise footnoted.]
African Americans .... 1.1 %
Asian/Pacific Island Americans .... 2.3 %
Hispanic Americans .... 2.3 %
Native Americans .... 2.3 %
Foreign Citizens .... 34.1 %
All Others .... 57.9 %
Women .... 18.8 %
Persons with disabilities .... – % [A]
Students over 25 years of age .... – % [A]

Notes: (A) Data not available.

GRADUATE ENROLLMENTS & DEGREES AWARDED
Master's enrollment 68
Master's degrees awarded 28
Doctoral enrollment 34
Doctoral degrees awarded 7

ENGINEERING STUDENT DATA

Applicants to the engineering college
Number of applicants to engineering college 377
Percent offered admission 95.0%

Matriculated engineering students
Percentage in top quartile (25%) of High School class .... 55.0 [1]
Average SAT scores: Math—516[1], Verbal—597[1], Combined—1113[1]
Average ACT scores: Math—27.0[1], Composite—26.3[1]

Notes: (1) Data reflects all students in the College of Engineering and Applied Sciences.

FULL- & PART-TIME FACULTY—BY DEPARTMENT
[Figures are the head count of full-time faculty and the full-time equivalent (FTE) of part-time faculty for each engineering department or equivalent.]

Department	Full	Part
Chemical Eng	9	–
Electrical Eng	9	–
Mechanical Eng	9	–
Eng Physics	7	–
Petroleum Eng	9	–

COLLEGE DESCRIPTION

The College of Engineering and Applied Sciences offers curricula designed to provide a liberal education in both the humanities and the sciences, so that students gain insight into the roles of the engineer and scientist within the complex interactions of American society. Emphasis is upon developing the students' curiosity and creativity rather than mere indoctrination in current industrial technologies. Dual degree and double major programs are available. Freshmen entering the college with exceptional records will be considered for the Honors Program which includes honors courses, a scholarship, a Junior Colloquium, and a Senior Project.

DEGREE PROGRAMS DESCRIPTION
Classes are small, with a student/faculty ratio of fifteen-to-one. All courses are taught by faculty, not graduate assistants. Students complete the Tulsa Curriculum, which includes social sciences, humanities, and three courses in writing, thus making them well-rounded and marketable to employers upon graduation.

TRANSFER INFORMATION
Residency Requirements: The last 30 hours of course work must be complete at The University of Tulsa in order to earn a degree.
Transfer via Articulation Agreements
Admission to engineering: At any time
Requirements: An applicant who has previously attended other colleges or universities should be in good standing at the institution last attended. A 2.5 grade point average is recommended for consideration for admission.

Transfer without Articulation Agreements
Admission to engineering: At any time
Requirements: An applicant who has previously attended other colleges or universities should be in good standing at the institution last attended. A 2.5 grade point average is recommended for consideration for admission.

GRADUATION REQUIREMENTS
A student must complete all major courses in the prescribed curriculum with at least a 2.0 grade point average. An overall grade point average of 2.0 is also required.

STUDENT PROGRAMS

PROFESSIONAL AND HONORARY SOCIETIES
[For key to acronyms, see *Introduction*.]
Professional Societies: ACS, AIChE, ASHRAE, ASME, IEEE, SAE, SPE, SPS
Honorary Societies: Eta Kappa Nu, Tau Beta Pi, Sigma Pi Sigma

SUPPORT PROGRAMS
Student Chapter Organizations: Society of Women Engineers
Other Engineering Support Programs: Support programs include freshman orientation, career planning and personal development, counseling and psychological services, Computer Resource Center, and the Writing Center. A wide variety of student organizations, fraternities, sororities and extracurricular activities are also available to students. The Student Advocacy Center is available to students with special educational needs.

246 Tuskegee University

INSTITUTION PROFILE

HEAD OF THE INSTITUTION
Benjamin F. Payton
Phone: (205)727-8501   Fax: (205)727-5276

GENERAL INFORMATION
[All Students—Fall 1991]
Undergraduate enrollment 3,525
Graduate enrollment 177
Total institution enrollment 3,702
Type of institution: Private   **Calendar:** Semesters
Nearest city: Montgomery   **Population:** 200,000
Miles from main campus: 40   **Setting:** Suburban
Types of engineering degrees: Engineering
Other degree-granting colleges: Arts & Sciences, Education, Veterinary Medicine, Agriculture and Home Economics, Nursing and Allied Health, Business

ENGINEERING ADMISSIONS

ADMISSIONS OFFICE CONTACT
Lee Young
Tuskegee University
Office of Admissions
Tuskegee, AL 36088
Phone: (205)727-8390

ENGINEERING COLLEGE ADMISSIONS INFORMATION
Entrance Requirements: SAT (Score: 800), ACT (Score: 19), High School courses—English (3 years), Mathematics (4 years), Physical Science (1 year), Social Sciences (3 years)
Entrance Recommendations: High School courses—Physics (2 years), Chemistry (2 years), English (1 year), History (2 years), Math level I or II
Requirements for foreign students: TOEFL (Score: 540)

DEPARTMENTAL ADMISSIONS INFORMATION
Admission to the engineering department: At the time of admission to the institution.

FINANCIAL INFORMATION

ESTIMATED EXPENSES (ACADEMIC YEAR)
[Expenses are for the 1992-93 nine-month academic year.]

All Students	Undergraduate	Graduate
Tuition and fees	$ 6,535	$ 6,535
College room and board	$ 3,270	$ 3,270
Books and supplies	$ 940	$ 940
Other expenses	$ 375	$ 375
Total estimated expenses	$11,120	$11,120

FINANCIAL AID OFFICE CONTACT
Barbara T. Blair, *Director of Financial Aid*
Tuskegee University
Old Administration Building
Tuskegee, AL 36088
Phone: (205)727-8201

GENERAL FINANCIAL AID INFORMATION
Forms accepted/required: FAF, Appropriate Income Tax Form, Need Analysis

ENGINEERING COLLEGE INFORMATION
[For additional personnel, refer to the *Appendix*.]

HEAD OF ENGINEERING
Shaik Jeelani
Phone: (205)727-8355   Fax: (205)727-8090

ENGINEERING COLLEGE ADDRESS
School of Engineering and Architecture
Tuskegee University
Tuskegee, AL 36088
Phone: (205)727-8355   Fax: (205)727-8090

ENROLLMENTS—BY CLASS
[Numbers are baccalaureate enrollments for the fall 1991 term, unless otherwise footnoted.]
1st-year students/Freshmen 370
2nd-year students/Sophomores 176
3rd-year students/Juniors 160
4th-year students/Seniors 273
Total **979**

NUMBER OF DEGREES AWARDED—BY PROGRAM
[Numbers are engineering baccalaureate degrees awarded during the 1991-1992 academic year, unless otherwise footnoted. For full details about each engineering program, refer to the *Tables of Degree Programs*.]
Aerospace Engineering 6
Chemical Engineering 2
Electrical Engineering 43
Mechanical Engineering 17
Total **68**

PERCENTAGE OF DEGREES AWARDED—BY CATEGORY
[Percentages are of all engineering baccalaureate degrees awarded during the 1991-1992 academic year, unless otherwise footnoted.]
African Americans .... 92.3 %
Asian/Pacific Island Americans .... – %
Hispanic Americans .... – %
Native Americans .... – %
Foreign Citizens .... 7.7 %
All Others .... – %
Women .... 34.4 %
Persons with disabilities .... – %
Students over 25 years of age .... 1.0 %

GRADUATE ENROLLMENTS & DEGREES AWARDED
Master's enrollment 51
Master's degrees awarded 13
Doctoral enrollment .... –
Doctoral degrees awarded .... –

ENGINEERING STUDENT DATA
Applicants to the engineering college
Number of applicants to engineering college 1,081
Percent offered admission 59.0%

Matriculated engineering students
Percentage in top quartile (25%) of High School class 14.0
Average SAT scores: Math—520, Verbal—340, Combined—860
Average ACT scores: Composite—19

FULL- & PART-TIME FACULTY—BY DEPARTMENT
[Figures are the head count of full-time faculty and the full-time equivalent (FTE) of part-time faculty for each engineering department or equivalent.]

Department	Full	Part
Chemical Eng	5	–
Electrical Eng	12	12.0
Mechanical Eng	11	1.0
Aerospace Science Eng	3	1.0

COLLEGE DESCRIPTION
The school of Engineering and Architecture offers course work leading to the B.S. degree in four disciplines (Aerospace, Chemical, Electrical and Mechanical Eng.). The curricula provide the student with basic science, math, and humanities as well as a fundamental knowledge of the specific engineering discipline. Course work in the engineering dept. is designed to provide a clear understanding of engineering science, analysis and design, computer applications and laboratory exploration. The program is further strengthened by courses in the social sciences, ethics and economics, which provide a foundation for a successful career as an engineer in a complex society.

DEGREE PROGRAMS DESCRIPTION
Degree programs offered in the school of Engineering and Architecture are: Bachelor of Science in Aerospace Science, Chemical, Electrical and Mechanical Engineering; Bachelor of Arts in Architectural Science, Bachelor of Architecture, and Bachelor of Science in Construction Science and Management.

DUAL DEGREE PROGRAMS
Engineering baccalaureate graduates with dual degrees 6

TRANSFER INFORMATION
Residency Requirements: The student must have been a full-time resident for at least two semesters, and must be enrolled during the semester in which the degree is to be awarded. The student will not be permitted to take courses at another institution during the semester the degree is awarded.

Transfer without Articulation Agreements
Admission to engineering: beginning of each semester
Engineering graduates transferred from ..... 4-yr: 9   2-yr: –
Requirements: Students who wish to enter from other colleges or universities must be eligible to re-enter the institution last attended and must furnish official transcript and course descriptions of work done in all institutions previously attended. Credits will be given toward graduation for those courses with a grade if "C" or better and which are similar to those in the curriculum.

GRADUATION REQUIREMENTS
Students are required to complete all courses in the departmental curriculum with at least a 2.00 GPA overall. A grade of "C" or above is required for certain courses such as English and Reading, as well as all required courses in mathematics, Science, and Engineering. An English proficiency examination is also required before graduation.

STUDENT PROGRAMS

PROFESSIONAL AND HONORARY SOCIETIES
[For key to acronyms, see *Introduction*.]
Professional Societies: ACS, AGCA, AIA, AIAA, AIChE, ASME, IEEE, SAME
Honorary Societies: Eta Kappa Nu, Pi Tau Sigma, CHE Honor Society, Sigma Lambda Chi

SUPPORT PROGRAMS
Student Chapter Organizations: Society of Women Engineers, National Society of Black Engineers

Counseling Center
For: Women & Ethnic Minorities   **Available:** Academic year
Offered: Freshman, Sophomore, Junior, Senior, Graduate level

Career Development and Placement Services
For: Women & Ethnic Minorities   **Available:** Year round
Offered: Sophomore, Junior, Senior, Graduate level

Student Life and Development
For: Women & Ethnic Minorities   **Available:** Year round
Offered: Freshman, Sophomore, Junior, Senior, Graduate level

Office of International Programs
For: Ethnic Minorities   **Available:** Year round
Offered: Freshman, Sophomore, Junior, Senior, Graduate level
Other Engineering Support Programs: Freshman Accelerated Start-up Training for Retention in the Engineering Curricula (FASTREC). Reading and math tutorials, Engineering peer tutorials. Minority Introduction to Engineering (MITE). Cooperative Education.

247 Union College

■ INSTITUTION PROFILE
HEAD OF THE INSTITUTION
Roger H. Hull
Phone: (518)370-6101
GENERAL INFORMATION
[All Students—Fall 1991]
Undergraduate enrollment 2,000
Graduate enrollment 200
Total institution enrollment **2,200**
Type of institution: Private **Calendar:** Trimesters
Nearest city: Albany, New York **Population:** 150,000
Miles from main campus: 15 **Setting:** Suburban
Types of engineering degrees: Engineering

■ ENGINEERING ADMISSIONS
ADMISSIONS OFFICE CONTACT
Daniel M. Lundquist
Union College
Becker Hall
Schenectady, NY 12308-2311
Phone: (518)370-6112
ENGINEERING COLLEGE ADMISSIONS INFORMATION
Entrance Requirements: High School courses—Mathematics (3 years), English (4 years), Physics (1 year), Chemistry (1 year), Union College requires the ACT OR 3 Achievement tests (including English Composition). The SAT is optional.
Entrance Recommendations: High School courses—Mathematics (4 years)
Requirements for foreign students: TOEFL (Score: 550); financial statement
DEPARTMENTAL ADMISSIONS INFORMATION
Admission to the engineering department: At the time of admission to the institution.

■ FINANCIAL INFORMATION
ESTIMATED EXPENSES (ACADEMIC YEAR)
[Expenses are for the 1992-93 nine-month academic year.]

All Students	Undergraduate	Graduate
Tuition and fees	$16,748	$ 8,365
College room and board	$ 5,722	$ —(B)
Books and supplies	$ 450	$ 450
Other expenses	$ 1,430	$ 1,430
Total estimated expenses	**$24,350**	**$10,245**

Notes: (B) Data not applicable.
FINANCIAL AID OFFICE CONTACT
Michael S. Brown, *Director of Financial Aid*
Union College
Becker Hall
Schenectady, NY 12308-2311
Phone: (518)370-6123
GENERAL FINANCIAL AID INFORMATION
Forms accepted/required: FAF, IRS, Institutional
Additional financial aid information: Financial aid includes need based scholarships, on campus jobs, a limited number of loans from the College's own loan funds and from the Perkins Loan/NDSL Fund which the College administers, and CAUSE awards (forgivable loans for public service).

■ ENGINEERING COLLEGE INFORMATION
[For additional personnel, refer to the *Appendix*.]
HEAD OF ENGINEERING
Lawrence J. Hollander
Phone: (518)370-6038 **Fax:** (518)370-6789
EMail: HOLLANDL@GAR.UNION.EDU

ENGINEERING COLLEGE ADDRESS
Division of Engineering and Applied Science
Union College
807 Union Street
Schenectady, NY 12308-2311
Phone: (518)370-6000 **Fax:** (518)370-6789
ENROLLMENTS—BY CLASS
[Numbers are baccalaureate enrollments for the fall 1991 term, unless otherwise footnoted.]
1st-year students/Freshmen 93
2nd-year students/Sophomores 70
3rd-year students/Juniors 87
4th-year students/Seniors 88
Total **338**
NUMBER OF DEGREES AWARDED—BY PROGRAM
[Numbers are engineering baccalaureate degrees awarded during the 1991-1992 academic year, unless otherwise footnoted. For full details about each engineering program, refer to the *Tables of Degree Programs*.]
Civil Engineering 35
Electrical Engineering 21
Mechanical Engineering 31
Total **87**
PERCENTAGE OF DEGREES AWARDED—BY CATEGORY
[Percentages are of all engineering baccalaureate degrees awarded during the 1991-1992 academic year, unless otherwise footnoted.]
African Americans 2.2%
Asian/Pacific Island Americans 3.4%
Hispanic Americans 2.2%
Native Americans .... — %
Foreign Citizens 1.1%
All Others 91.1%
Women 14.9%
Persons with disabilities .... — % (A)
Students over 25 years of age .... — % (A)
Notes: (A) Data not available.
GRADUATE ENROLLMENTS & DEGREES AWARDED
Master's enrollment 63
Master's degrees awarded 33
Doctoral enrollment .... — (B)
Doctoral degrees awarded .... — (B)
Notes: (B) Data not applicable.
ENGINEERING STUDENT DATA
Applicants to the engineering college
Number of applicants to engineering college 410
Percent offered admission 60.0%
Matriculated engineering students
Percentage in top quartile (25%) of High School class 84.0
Average ACT scores: Composite—27
FULL- & PART-TIME FACULTY—BY DEPARTMENT
[Figures are the head count of full-time faculty and the full-time equivalent (FTE) of part-time faculty for each engineering department or equivalent.]

Department	Full	Part
Civil Eng	5	—
Electrical Eng	7	1.3
Mechanical Eng	7	1.0

COLLEGE DESCRIPTION
Union is an independent, primarily undergraduate, liberal arts and engineering college. Perhaps the most distinctive feature of the academic program is the existence within the liberal arts framework of a strong engineering program. EAC/ABET accredited degrees are offered in Civil, Electrical and Mechanical Engineering, both day and evening. The academic year is divided into three ten-week terms. There are approximately 2000 students on the campus, 400 of whom are engineers. The male/female ratio among the students is approximately 50-50. Approximately 3000 applications are received seeking entrance to the freshman class of 510 students. In 1995, the College will be celebrating its bicentennial and the sesquicentennial of engineering. Seniors are required to complete capstone design projects and are encouraged to complete the national standardized Fundamentals of Engineering examination. The laboratories are well-equipped and the computer facilities are excellent.
DEGREE PROGRAMS DESCRIPTION
Master's degrees are offered in Electrical Engineering and Mechanical Engineering, part-time or full-time. Ten courses are required. A thesis is optional and may count as two of the ten courses.
DUAL DEGREE PROGRAMS
Engineering baccalaureate graduates with dual degrees 3
TRANSFER INFORMATION
Residency Requirements: Normally, at least two years of residency is required.
Transfer via Articulation Agreements
Admission to engineering: At the end of the sophomore year
Requirements: Usually a cumulative average of 3.0 or better is required for transfer students. In making its decisions, the Admissions Committee leans heavily upon the work completed and upon the recommendation of the appropriate official at the college presently attended.
Transfer without Articulation Agreements
Admission to engineering: Transfers are admitted to the day school at the end of the freshmen year. They are admitted to the evening division at any time.
Requirements: Usually a cumulative average of 3.0 or better is required for transfer students. In making its decisions, the Admissions Committee leans heavily upon the work completed and upon the recommendation of the appropriate official at the college presently attended.
GRADUATION REQUIREMENTS
The baccalaureate engineering programs require the completion of 39 courses. The student must attain a minimum cumulative index of 1.80 overall and 2.00 in the major.

■ STUDENT PROGRAMS
PROFESSIONAL AND HONORARY SOCIETIES
[For key to acronyms, see *Introduction*.]
Professional Societies: ACI, ACM, ACS, ASCE, ASME, IEEE, SPS
Honorary Societies: Chi Epsilon, Eta Kappa Nu, Pi Tau Sigma, Tau Beta Pi, Sigma Xi
SUPPORT PROGRAMS
Student Chapter Organizations: Society of Women Engineers, National Society of Black Engineers, Asian Student Union, African and Latino Alliance of Students, Women's Union
Academic Opportunity Program
For: Ethnic Minorities **Available:** Year round
Offered: Prematriculation, Freshman, Sophomore, Junior, Senior
Other Engineering Support Programs: Fall orientation program for all entering students; Academic Opportunity Program (Higher Education Opportunity Program) for low-income students; the Writing Center; the Career Development Center; the Counseling Center; Health Services.

248 United States Air Force Academy

INSTITUTION PROFILE
HEAD OF THE INSTITUTION
Bradley C. Hosmer
Phone: (719)472-4140 **Fax:** (719)472-4146
GENERAL INFORMATION
[All Students—Fall 1991]
Undergraduate enrollment 4,288
Total institution enrollment 4,288
Type of institution: Public **Calendar:** Semesters
Nearest city: Colorado Springs **Population:** 350,000
Miles from main campus: 7 **Setting:** Suburban
Types of engineering degrees: Engineering
Other degree-granting colleges: Basic Sciences, Humanities, Social Sciences

ENGINEERING ADMISSIONS
ADMISSIONS OFFICE CONTACT
Rolland R. Stoneman
United States Air Force Academy
Cadet Admissions
HQ USAFA/RRS
USAF Academy, CO 80840-5651
Phone: (719)472-2520 **Fax:** (719)472-3647
ENGINEERING COLLEGE ADMISSIONS INFORMATION
Admission to the engineering college: Students are admitted to a specific engineering program at any time between their freshman year and the end of their sophomore year.
Entrance Recommendations: SAT (Score: 1100), ACT (Score: 25), High School courses—Mathematics (through calculus) (4 years), English (4 years), Lab Sciences (including physics & chemistry) (4 years), Modern Foreign Language (2 years)
Requirements for foreign students: Foreign students are admitted on a case-by-case basis according to agreements worked out with the respective foreign country.
DEPARTMENTAL ADMISSIONS INFORMATION
Admission to the engineering department: Students are admitted to an engineering program at any time between their freshman year and the end of their sophomore year.

FINANCIAL INFORMATION
ESTIMATED EXPENSES (ACADEMIC YEAR)
[Expenses are for the 1992-93 nine-month academic year.]

All Students	Undergraduate	Graduate
Tuition and fees	$ –[1]	$ –
College room and board	$ –	$ –
Books and supplies	$ –	$ –
Other expenses	$ –	$ –
Total estimated expenses	$ –	$ –

Notes: (1) All students receive a full scholarship.

ENGINEERING COLLEGE INFORMATION
[For additional personnel, refer to the *Appendix*.]
HEAD OF ENGINEERING
Cary A. Fisher
Phone: (719)472-2531 **Fax:** (719)472-2944
EMail: dfemmail@falconnet.usafa.af.mil
ENGINEERING COLLEGE ADDRESS
Engineering Division
United States Air Force Academy
HQ USAFA/DF
USAF Academy, CO 80840-5701
Phone: (719)472-2531 **Fax:** (719)472-2944

ENROLLMENTS—BY CLASS
[Numbers are baccalaureate enrollments for the fall 1991 term, unless otherwise footnoted.]
1st-year students/Freshmen .... 1,131 [1]
2nd-year students/Sophomores .... 1,132 [1]
3rd-year students/Juniors .... 1,053 [1]
4th-year students/Seniors .... 972 [1]
Total 4,288
Notes: (1) Reflects Fall 92 Enrollment. Data based on all students in the institution. All students must take engineering courses.

NUMBER OF DEGREES AWARDED—BY PROGRAM
[Numbers are engineering baccalaureate degrees awarded during the 1991-1992 academic year, unless otherwise footnoted. For full details about each engineering program, refer to the *Tables of Degree Programs*.]
Aeronautical Engineering 87
Astronautical Engineering 42
Civil Engineering 70
Electrical Engineering 42
Engineering Mechanics 42
Engineering Science(s) 27
General Engineering 31
Mechanical Engineering 9
Total 350

PERCENTAGE OF DEGREES AWARDED—BY CATEGORY
[Percentages are of all engineering baccalaureate degrees awarded during the 1991-1992 academic year, unless otherwise footnoted.]
African Americans .... 9.4% [2]
Asian/Pacific Island Americans .... 4.0% [2]
Hispanic Americans .... 3.1% [2]
Native Americans .... –% [2]
Foreign Citizens .... 0.6% [2]
All Others 82.9%
Women .... 5.1% [2]
Persons with disabilities .... –% [2]
Students over 25 years of age .... 0.9% [2]
Notes: (2) Data based on Engineering Majors.

ENGINEERING STUDENT DATA
Applicants to the engineering college
Number of applicants to engineering college .... – [1]
Percent offered admission .... –% [1]
Matriculated engineering students
Percentage in top quartile (25%) of High School class 94.5
Average SAT scores: Math—659[2], Verbal—567[2], Combined—1226[2]
Average ACT scores: Math—29.1[2], Composite—28.9[2]
Notes: (1) Freshmen do not directly apply to nor are directly accepted into Engineering Programs. (2) Reflects Fall 92 Freshmen.

FULL- & PART-TIME FACULTY—BY DEPARTMENT
[Figures are the head count of full-time faculty and the full-time equivalent (FTE) of part-time faculty for each engineering department or equivalent.]

Department	Full	Part
Civil Eng	22	–
Electrical Eng	27	–
Eng Mechanics	27	–
Astronautics	25	–
Aeronautics	32	–

COLLEGE DESCRIPTION
The United States Air Force Academy graduates officers to serve in the United States Air Force. Cadets are provided a broad education in the basic sciences, engineering, humanities and social sciences. Academic studies are integrated with military, aviation and athletic training to prepare cadets for careers in the Air Force. The engineering programs build on this broad core to prepare cadets for careers as engineers. The Academy is unique in the extent of its facilities for undergraduate engineering education. Extensive lab facilities allow the integration of classroom theory with the physical world. Fabrication facilities allow for the production of designs. Work stations, a computer network, and a PC for each cadet prepare the engineering student for effectiveness in the CAD/CAM environment.

DEGREE PROGRAMS DESCRIPTION
All cadets graduate with a bachelor of science degree. Only those cadets who graduate with an accredited degree have that degree printed on their diploma. Each degree program has a set of required majors courses to provide breadth and depth within the specific engineering discipline. Beyond this students may specialize by choosing from a variety of design options and sequences. Aeronautical Engineering students may focus on airframes or engines. Astronautical Engineering students choose an emphasis in astrodynamics, spacecraft systems, or controls. Civil Engineering students may specialize in structures, environmental, geotechnical or general practices. Electrical Engineering students may elect a systems, communications, computer engineering or controls sequence. In Engineering Mechanics the student selects from options in dynamics, materials or structures. Engineering Sciences majors have the flexibility to select among eight tracks in areas such as aeronautics, space systems, astrodynamics, control systems, structures, materials, mechanical engineering, and controls/structures. The General Engineering program is not ABET accredited but does offer the cadet maximum flexibility in structuring their academic program.

DUAL DEGREE PROGRAMS
Engineering baccalaureate graduates with dual degrees 5

TRANSFER INFORMATION
Residency Requirements: Must complete a full four year program.
Transfer without Articulation Agreements
Admission to engineering: Students may transfer credits but must be appointed to the Academy and complete a full four year program. Students may be admitted to an engineering program up to the end of their sophomore year.
Engineering graduates transferred from ... 4-yr: –   2-yr: –

STUDENT PROGRAMS
PROFESSIONAL AND HONORARY SOCIETIES
[For key to acronyms, see *Introduction*.]
Professional Societies: AIAA, AIP, ASCE, ASME, IEEE, SAME
Honorary Societies: Sigma Gamma Tau, USAFA has a distinguished graduate program.
SUPPORT PROGRAMS
Student Chapter Organizations: Way of Life Committee, Los Padrinos
Cadet Counseling Center (supports all cadets)
For: Women & Ethnic Minorities **Available:** Year round
Offered: Prematriculation, Freshman, Sophomore, Junior, Senior
Other Engineering Support Programs: All incoming freshmen have the opportunity to tour the Air Force Academy prior to the start of their freshman year. Each cadet is assigned an academic advisor to help them in course selections throughout their four years at the Academy.

249 United States Coast Guard Academy

■ INSTITUTION PROFILE
HEAD OF THE INSTITUTION
Thomas T. Matteson
Phone: (203)444-8285 **Fax:** (203)444-8288
GENERAL INFORMATION
[All Students—Fall 1991]
Undergraduate enrollment 976
Total institution enrollment 976
Type of institution: Public **Calendar:** Semesters
Nearest city: New London, CT **Population:** 29,000
Miles from main campus: 2 **Setting:** Small Town
Types of engineering degrees: Engineering

■ ENGINEERING ADMISSIONS
ADMISSIONS OFFICE CONTACT
Robert W. Thorne
United States Coast Guard Academy
15 Mohegan Avenue
New London, CT 06320-4195
Phone: (203)444-8500 **Fax:** (203)444-8288
ENGINEERING COLLEGE ADMISSIONS INFORMATION
Admission to the engineering college: A prospective engineering student applies to a specific engineering major at the end of the sophomore year.
Entrance Requirements: SAT (Score: 950), ACT, High School courses—Algebra (2 years), Geometry (1 year), English (3 years)
Entrance Recommendations: High School courses—Mathematics (4 years), English (4 years)
Requirements for foreign students: Applicants are evaluated based on high school standings, standardized test results and leadership potential. Also must be 17-22 as of 1 July of summer entering; U.S. citizenship required; must be unmarried, must pass physical.
DEPARTMENTAL ADMISSIONS INFORMATION
Admission to the engineering department: A prospective engineering student applies to a specific engineering major at the end of the sophomore year.
Additional information: Engineering department applicants must have a 2.00/4.00 average for all math, science, computer science and engineering courses taken prior to the beginning of the junior year. A grade of C or better must be achieved in Statics/Strengths of Materials (Civil and Marine Engineers), Dynamics (Marine Engineers), and Electrical Engineering I and II (Electrical Engineers).

■ FINANCIAL INFORMATION
ESTIMATED EXPENSES (ACADEMIC YEAR)
[Expenses are for the 1992-93 nine-month academic year.]

All Students	Undergraduate	Graduate
Tuition and fees	$ —(1)	$ —
College room and board	$ —	$ —
Books and supplies	$ —	$ —
Other expenses	$ —	$ —
Total estimated expenses	$ —	$ —

Notes: (1) Tuition is free. $1500 deposit is required at the beginning of the 4-year undergraduate program for initial issue of uniform and equipment.
GENERAL FINANCIAL AID INFORMATION
Forms accepted/required: Additional financial aid information: All students are paid one-half of an Ensign's base pay.

■ ENGINEERING COLLEGE INFORMATION
[For additional personnel, refer to the *Appendix*.]
HEAD OF ENGINEERING
John C. Maxham
Phone: (203)444-8547 **Fax:** (203)444-8546
ENGINEERING COLLEGE ADDRESS
Department of Engineering
United States Coast Guard Academy
15 Mohegan Avenue
New London, CT 06320-4195
Phone: (203)444-8547 **Fax:** (203)444-8546
ENROLLMENTS—BY CLASS
[Numbers are baccalaureate enrollments for the fall 1991 term, unless otherwise footnoted.]
1st-year students/Freshmen 152
2nd-year students/Sophomores 95
3rd-year students/Juniors 82
4th-year students/Seniors 64
Total 393
NUMBER OF DEGREES AWARDED—BY PROGRAM
[Numbers are engineering baccalaureate degrees awarded during the 1991-1992 academic year, unless otherwise footnoted. For full details about each engineering program, refer to the *Tables of Degree Programs*.]
Civil Engineering 40
Electrical Engineering 16
Naval Architecture & Marine Engineering 23
Total 79
PERCENTAGE OF DEGREES AWARDED—BY CATEGORY
[Percentages are of all engineering baccalaureate degrees awarded during the 1991-1992 academic year, unless otherwise footnoted.]
African Americans .... — %
Asian/Pacific Island Americans .... 5.1 %
Hispanic Americans .... 1.3 %
Native Americans .... 2.5 %
Foreign Citizens .... 2.5 %
All Others .... 88.6 %
Women .... 7.6 %
Persons with disabilities .... — %
Students over 25 years of age .... — %
ENGINEERING STUDENT DATA
Applicants to the engineering college
Number of applicants to engineering college 4,924
Percent offered admission .... 9.8 %
Matriculated engineering students
Percentage in top quartile (25%) of High School class 96.1
Average SAT scores: Math—638, Verbal—544, Combined—1182
Average ACT scores: Math—27
FULL- & PART-TIME FACULTY—BY DEPARTMENT
[Figures are the head count of full-time faculty and the full-time equivalent (FTE) of part-time faculty for each engineering department or equivalent.]

Department	Full	Part
Civil Eng	8	—
Electrical Eng	8	—
Naval Architecture & Marine Eng	8	—

COLLEGE DESCRIPTION
The Coast Guard Academy is a military academy located in a small city setting; the academic year is based on a semester system. The school offers degrees in civil, electrical, and naval architecture & marine engineering. Other academic divisions include: Humanities, Computer Science, Mathematics, Economics and Management, and Science. The engineering program is offered on campus; it does not have extension centers. Graduating students earn a Bachelor of Science degree and a Coast Guard Ensign Commission in four years. Major area course work commences in the sophomore year, culminating in a senior capstone design project. The academic program includes standard college level humanities, science and mathematics core courses. Professional development courses such as navigation and leadership provide special skills required of a Coast Guard officer.
DEGREE PROGRAMS DESCRIPTION
The three B.S. programs prepare graduates for entry level technical positions within and outside the Coast Guard as well as further graduate work. All programs focus on practical applications as emphasized by the capstone design project in each discipline. Many senior design projects solve Coast Guard and other real world engineering problems.
GRADUATION REQUIREMENTS
Pass all core courses; earn a minimum of 126.0 credits excluding physical education; accumulate at least 90.0 credits of course work with C or better grades exclusive of physical education; complete courses and obtain a 2.00 average in all required upper division courses in a major; be in residence at the Academy for at least four academic years; pass required physical education courses and meet minimum swimming standards; maintain a high sense of integrity.

■ STUDENT PROGRAMS
PROFESSIONAL AND HONORARY SOCIETIES
[For key to acronyms, see *Introduction*.]
Professional Societies: ASCE, IEEE, SNAME, Society of Women Engineers (SWE), Armed Forces Communications & Electronics Association (AFCEA)
SUPPORT PROGRAMS
Student Chapter Organizations: Society of Women Engineers
Minority Introduction to Engineering (MITE)
For: Ethnic Minorities **Available:** Summer only
Offered: Prematriculation
Academy Introduction Mission (AIM)
For: Women & Ethnic Minorities **Available:** Summer only
Offered: Prematriculation
Other Engineering Support Programs: Psychological testing and counseling; academic counseling; academic assistance program available Sunday through Friday nights, mandatory for freshmen in academic difficulty; Cadet Writing Center; Efficient Reading Program; summer calculus programs; freshmen are issued Macintosh SE computers with internal hard disk and floppy disk drives. Each engineering program has a computer aided design lab.

250 United States Merchant Marine Academy

■ INSTITUTION PROFILE
HEAD OF THE INSTITUTION
Paul L. Krinsky
Phone: (516)773-5348 Fax: (516)773-5390
GENERAL INFORMATION
[All Students—Fall 1991]
Undergraduate enrollment 968
Total institution enrollment 968
Type of institution: Public Calendar: Quarters
Nearest city: New York Population: 7,500,000
Miles from main campus: 15 Setting: Suburban
Types of engineering degrees: Engineering
ENGINEERING COLLEGE ADMISSIONS INFORMATION
Admission to the engineering college: Students enter the engineering programs at the end of the second quarter of the freshman year.
Entrance Requirements: SAT (Score: 950), High School courses—English (3 years), Mathematics (algebra, geometry, trigonometry) (3 years), Chemistry or Physics, with laboratory (1 year)
Entrance Recommendations: High School courses—Mathematics (4 years), Chemistry (1 year), Physics (1 year)
Requirements for foreign students: TOEFL
DEPARTMENTAL ADMISSIONS INFORMATION
Admission to the engineering department: Students enter the engineering programs at the end of the second quarter of the freshman year.

■ FINANCIAL INFORMATION
ESTIMATED EXPENSES (ACADEMIC YEAR)
[Expenses are for the 1992-93 nine-month academic year.]

All Students	Undergraduate	Graduate
Tuition and fees	$ –[1]	$ –
College room and board	$ –[1]	$ –
Books and supplies	$ –[1]	$ –
Other expenses	$ 3,450[2]	$ –
Total estimated expenses	**$ 3,450**	**$ –**

Notes: (1) Tuition, room and board, fees, books, and many other expenses are paid for by the Government. (2) Estimated deposit to cover initial issue of uniforms and personal computer.

FINANCIAL AID OFFICE CONTACT
Frances Ferguson, *Financial Aid Officer*
United States Merchant Marine Academy
Admissions Office
Kings Point, NY 11024
Phone: (516)773-5391 Fax: (516)773-5509
GENERAL FINANCIAL AID INFORMATION
Forms accepted/required: Additional financial aid information: Students requiring financial aid to obtain the required deposit can seek assistance from programs such as the Guaranteed Student Loan (GLS), the Parental Loan for Undergraduate Students (PLUS), and Auxiliary Loans to Assist Students (ALAS).

■ ENGINEERING COLLEGE INFORMATION
[For additional personnel, refer to the *Appendix*.]
HEAD OF ENGINEERING
James A. Harbach
Phone: (516)773-5471 Fax: (516)773-5534
ENGINEERING COLLEGE ADDRESS
Department of Engineering
United States Merchant Marine Academy
Kings Point, NY 11024-1699
Phone: (516)773-5471 Fax: (516)773-5534
ENROLLMENTS—BY CLASS
[Numbers are baccalaureate enrollments for the fall 1991 term, unless otherwise footnoted.]
1st-year students/Freshmen 141
2nd-year students/Sophomores 149
3rd-year students/Juniors 122
4th-year students/Seniors 118
Total 530
NUMBER OF DEGREES AWARDED—BY PROGRAM
[Numbers are engineering baccalaureate degrees awarded during the 1991-1992 academic year, unless otherwise footnoted. For full details about each engineering program, refer to the *Tables of Degree Programs*.]
Marine Engineering 32
Marine Engineering Systems 47
Marine Engineering/Transportation 10
Total 89
PERCENTAGE OF DEGREES AWARDED—BY CATEGORY
[Percentages are of all engineering baccalaureate degrees awarded during the 1991-1992 academic year, unless otherwise footnoted.]
African Americans 1.1%
Asian/Pacific Island Americans 5.6%
Hispanic Americans .... – %
Native Americans 1.1%
Foreign Citizens .... – %
All Others 92.2%
Women 4.5%
Persons with disabilities .... – %
Students over 25 years of age 1.1%
ENGINEERING STUDENT DATA
Applicants to the engineering college
Number of applicants to engineering college .... 1,805[1]
Percent offered admission .... 15.6%[1]
Matriculated engineering students
Percentage in top quartile (25%) of High School class 75.0
Average SAT scores: Math—598, Verbal—543, Combined—1141

Notes: (1) Based on total nominations to the Academy.

COLLEGE DESCRIPTION
The U.S. Merchant Marine Academy is one of the five federal service academies. Tuition, fees, room, board, books, and many other costs are paid for by the government. Entrance to the Academy is through Congressional nomination. The Engineering Department offers programs in Marine Engineering, Marine Engineering Systems, and Marine Engineering/Marine Transportation. All programs are four year, including two five month cooperative work/study periods aboard United States flag commercial ships. In addition to receiving a Bachelor of Science degree, all engineering students complete the requirements for a United States Coast Guard license as a Third Assistant Engineer and a commission as Ensign, USNR.

DEGREE PROGRAMS DESCRIPTION
The engineering programs at the U.S. Merchant Marine Academy are a unique combination of traditional undergraduate engineering education combined with training in the operation of power systems. All students complete two five month cooperative work/study periods aboard several U.S. flag commercial ships. Graduates are qualified for a variety of professional positions in the maritime industry including service at sea as a licensed engineering officer; as an active-duty officer in the Navy, Army, Air Force, Marines, or Coast Guard; or as an engineer in the maritime industry ashore.

GRADUATION REQUIREMENTS
B.S. programs require the completion of approximately 225 quarter credit hours. A minimum average of 2.0 (A=4.0) is required for graduation. In addition, students in the Marine Engineering Systems program must have a minimum average of 2.0 in their major. All engineering students must complete the requirements for a United States Coast Guard license as Third Assistant Engineer and a commission as Ensign, USNR.

■ STUDENT PROGRAMS
PROFESSIONAL AND HONORARY SOCIETIES
[For key to acronyms, see *Introduction*.]
Professional Societies: ASNE, SNAME
SUPPORT PROGRAMS
Student Chapter Organizations: Society of Women Engineers
Other Engineering Support Programs: Freshman advisor program; counseling on personal, career and academic matters.

251 United States Military Academy

■ INSTITUTION PROFILE

HEAD OF THE INSTITUTION
Howard D. Graves
Phone: (914)938-2610   Fax: (914)938-3214

GENERAL INFORMATION
[All Students—Fall 1991]
Undergraduate enrollment 4,386
Total institution enrollment 4,386
Type of institution: Public **Calendar:** Semesters
Nearest city: New York City **Population:** 7,500,000
Miles from main campus: 60 **Setting:** Small Town
Types of engineering degrees: Engineering

■ ENGINEERING ADMISSIONS

ADMISSIONS OFFICE CONTACT
Pierce A. Rushton Jr.
United States Military Academy
606 Thayer Rd
West Point, NY 10996-9902
Phone: (914)938-4041   Fax: (914)938-3021

ENGINEERING COLLEGE ADMISSIONS INFORMATION
Entrance Recommendations: SAT, ACT, High School courses—Math (4 years), English (4 years), Chemistry (1 year), Modern Foreign Language (2 years)
Requirements for foreign students: Foreign students must be nominated by agreement between the United States and their country

DEPARTMENTAL ADMISSIONS INFORMATION
Admission to the engineering department: At the end of the second year.

■ FINANCIAL INFORMATION

ESTIMATED EXPENSES (ACADEMIC YEAR)
[Expenses are for the 1992-93 nine-month academic year.]

All Students	Undergraduate	Graduate
Tuition and fees	$ —[1]	$ —
College room and board	$ —[1]	$ —
Books and supplies	$ —[2]	$ —
Other expenses	$ 1,500[2]	$ —
Total estimated expenses	**$ 1,500**	**$ —**

Notes: (1) Tuition and room & board are paid for by the government (2) An initial deposit of $1500 is required to help defray the initial expenses for cadet uniforms, personal computers and supplies

■ ENGINEERING COLLEGE INFORMATION

[For additional personnel, refer to the *Appendix*.]

HEAD OF ENGINEERING
Gerald E. Galloway Jr.
Phone: (914)938-2000   Fax: (914)938-5438

ENGINEERING COLLEGE ADDRESS
United States Military Academy
606 Thayer Rd
West Point, NY 10996-5000
Phone: (914)938-4011   Fax: (914)938-3021

ENROLLMENTS—BY CLASS
[Numbers are baccalaureate enrollments for the fall 1991 term, unless otherwise footnoted.]
1st-year students/Freshmen .... (1)
2nd-year students/Sophomores .... (1)
3rd-year students/Juniors 376
4th-year students/Seniors 400
Total 776

Notes: (1) enrollment in a majors program comes at the end of sophomore year

NUMBER OF DEGREES AWARDED—BY PROGRAM
[Numbers are engineering baccalaureate degrees awarded during the 1991-1992 academic year, unless otherwise footnoted. For full details about each engineering program, refer to the *Tables of Degree Programs*.]
Civil Engineering 59
Electrical Engineering 40
Engineering Management 55
Engineering Physics 9
Environmental Engineering 62
Mechanical Engineering 135
Systems Engineering 21
Total 381

PERCENTAGE OF DEGREES AWARDED—BY CATEGORY
[Percentages are of all engineering baccalaureate degrees awarded during the 1991-1992 academic year, unless otherwise footnoted.]
African Americans 7.1%
Asian/Pacific Island Americans 5.0%
Hispanic Americans 13.9%
Native Americans .... — %
Foreign Citizens 1.3%
All Others 72.7%
Women 7.3%
Persons with disabilities .... — %
Students over 25 years of age .... — %

ENGINEERING STUDENT DATA
Applicants to the engineering college
Number of applicants to engineering college 12,249
Percent offered admission 10.0%

Matriculated engineering students
Percentage in top quartile (25%) of High School class .... 84.7 [1]
Average SAT scores: Math—648, Verbal—559, Combined—1207
Average ACT scores: Math—28.3, Composite—27
Notes: (1) Percentage of students in top 20% of high school class.

FULL- & PART-TIME FACULTY—BY DEPARTMENT
[Figures are the head count of full-time faculty and the full-time equivalent (FTE) of part-time faculty for each engineering department or equivalent.]

Department	Full	Part
Civil & Mechanical Eng	43	—
Electrical Eng	39	—
Geography & Environmental Eng	25	—
Systems Eng	38	—
Physics	35	—

COLLEGE DESCRIPTION
The U.S. Military Academy provides a foundation in mathematics, basic sciences, engineering sciences, humanities, behavior sciences, and social sciences. The 31-course core curriculum represents the essential broad base of knowledge necessary for success as a commissioned officer. Cadets may enter any field of study or major without restriction.

DEGREE PROGRAMS DESCRIPTION
All cadets receive a bachelor of science degree upon graduation and are commissioned as officers to serve in the United States Army.

GRADUATION REQUIREMENTS
Cadets must successfully complete the baseline requirement of a field of study of 40 academic courses; successfully complete or validate each course in the core curriculum; complete eight semesters of physical education; complete four military science intensive sessions; and achieve a cumulative grade point average of at least 2.0.

■ STUDENT PROGRAMS

PROFESSIONAL AND HONORARY SOCIETIES
[For key to acronyms, see *Introduction*.]
Professional Societies: ACM, AHS, AIAA, ASCE, ASME, IEEE, SAE, SAME
Honorary Societies: Eta Kappa Nu, Phi Kappa Phi (disciplinary)

SUPPORT PROGRAMS
Student Chapter Organizations: Society of Women Engineers

252 United States Naval Academy

■ INSTITUTION PROFILE
HEAD OF THE INSTITUTION
Thomas C. Lynch
Phone: (410)267-2202 **Fax:** (410)267-2303
GENERAL INFORMATION
[All Students—Fall 1991]
Undergraduate enrollment 4,300
Total institution enrollment 4,300
Type of institution: Public **Calendar:** Semesters
Location: Annapolis, MD **Population:** 35,000
Setting: Suburban
Types of engineering degrees: Engineering
Other degree-granting colleges: Mathematics and Science, Humanities and Social Sciences

■ ENGINEERING ADMISSIONS
ADMISSIONS OFFICE CONTACT
John W. Renard
United States Naval Academy
Candidate Guidance Office
Annapolis, MD 21402-5018
Phone: (410)267-4361 **Fax:** (410)267-4348
ENGINEERING COLLEGE ADMISSIONS INFORMATION
Entrance Requirements: SAT or ACT
Entrance Recommendations: High School courses—Mathematics (through trigonometry) (4 years), Chemistry (1 year), English (4 years), Modern Foreign Language (2 years), SAT or ACT
Requirements for foreign students: TOEFL; Must be a resident and officially nominated by their country.
DEPARTMENTAL ADMISSIONS INFORMATION
Admission to the engineering department: At the end of the first year.

■ FINANCIAL INFORMATION
ESTIMATED EXPENSES (ACADEMIC YEAR)
[Expenses are for the 1992-93 nine-month academic year.]

All Students	Undergraduate	Graduate
Tuition and fees	$ —	$ —
College room and board	$ —	$ —
Books and supplies	$ —	$ —
Other expenses	$ 1,500 [1]	$ —
Total estimated expenses	$ 1,500	$ —

Notes: (1) All candidates entering the Naval Academy are required to deposit $1500 as partial payment for uniforms and supplies, including a personal computer.

GENERAL FINANCIAL AID INFORMATION
Forms accepted/required: Additional financial aid information: A $3,400 interest-free loan from the federal government is advanced to entering midshipmen to help defray first-year costs not covered by the deposit.

■ ENGINEERING COLLEGE INFORMATION
[For additional personnel, refer to the *Appendix*.]
HEAD OF ENGINEERING
Philip F. Grasser
Phone: (410)267-2391 **Fax:** (410)267-2591
EMail: E05205@n1.usna.navy.mil

ENGINEERING COLLEGE ADDRESS
Division of Engineering and Weapons
United States Naval Academy
590 Holloway Road
Annapolis, MD 21402-5042
Phone: (410)267-2391 **Fax:** (410)267-2591
ENROLLMENTS—BY CLASS
[Numbers are baccalaureate enrollments for the fall 1991 term, unless otherwise footnoted.]
1st-year students/Freshmen 418
2nd-year students/Sophomores 328
3rd-year students/Juniors 399
4th-year students/Seniors 375
Total 1,520
NUMBER OF DEGREES AWARDED—BY PROGRAM
[Numbers are engineering baccalaureate degrees awarded during the 1991-1992 academic year, unless otherwise footnoted. For full details about each engineering program, refer to the *Tables of Degree Programs*.]
Aerospace Engineering 88
Electrical Engineering 16
General Engineering 54
Marine Engineering 20
Mechanical Engineering 56
Naval Architecture 22
Ocean Engineering 33
Systems Engineering 72
Total 361
PERCENTAGE OF DEGREES AWARDED—BY CATEGORY
[Percentages are of all engineering baccalaureate degrees awarded during the 1991-1992 academic year, unless otherwise footnoted.]
African Americans .... 2.7 %
Asian/Pacific Island Americans .... 5.9 %
Hispanic Americans .... 4.5 %
Native Americans .... 0.5 %
Foreign Citizens .... 2.1 %
All Others .... 84.3 %
Women .... 3.7 %
Persons with disabilities .... — %
Students over 25 years of age .... — %
ENGINEERING STUDENT DATA
Applicants to the engineering college
Number of applicants to engineering college 11,588
Percent offered admission .... 12.0 %
Matriculated engineering students
Percentage in top quartile (25%) of High School class 85.1
Average SAT scores: Math—678, Verbal—576, Combined—1254
Average ACT scores: Math—31, Composite—25
FULL- & PART-TIME FACULTY—BY DEPARTMENT
[Figures are the head count of full-time faculty and the full-time equivalent (FTE) of part-time faculty for each engineering department or equivalent.]

Department	Full	Part
Aerospace Eng	14	—
Mechanical Eng	29	—
Electrical Eng	31	—
Weapons & Systems Eng	31	—
Naval Architecture, Ocean, & Marine Eng	31	—

COLLEGE DESCRIPTION
The Naval Academy offers eight majors in engineering. Seven of these majors are accredited by the Engineering Accreditation Commission of the Accreditation Board of Engineering and Technology. Each engineering curriculum offers an integrated program of study, utilizing a balance between classroom theory, laboratory work and practical application. The Naval Academy's laboratory facilities are some of the most advanced and extensive in the country. During the final semester, each engineering student completes a senior design project and is also expected to take the Engineer-in-Training (EIT) examination. Average number of years required to complete the bachelor's degree: 4.
DEGREE PROGRAMS DESCRIPTION
Bachelor of science degrees specifying a major field are awarded to midshipmen upon graduation. They are commissioned as ensigns in the U.S. Navy or second lieutenants in the U.S. Marine Corps and begin at least six years of service as commissioned officers. Postgraduate education is encouraged, and midshipmen with outstanding academic records can compete for a number of scholarships for postgraduate school right after graduation from the Naval Academy or after an initial operational assignment.
GRADUATION REQUIREMENTS
Complete or validate the required professional, general distribution, and majors courses specified in the chosen major; achieve a final cumulative quality point rating (CQPR) of at least 2.0; meet required standards in professional studies and at-sea training; meet required standards of military performance, conduct, honor, and physical education, and accept a commission in the Navy or Marine Corps, unless one is not offered due to physical disqualification.

■ STUDENT PROGRAMS
PROFESSIONAL AND HONORARY SOCIETIES
[For key to acronyms, see *Introduction*.]
Professional Societies: AIAA, ANS, ASME, ASNE, IEEE, SAE, SNAME
Honorary Societies: Pi Tau Sigma, Tau Beta Pi
SUPPORT PROGRAMS
Student Chapter Organizations: Society of Women Engineers
Major selection counseling
For: Women & Ethnic Minorities **Available:** Year round
Offered: Freshman
Personal counseling
For: Women & Ethnic Minorities **Available:** Year round
Offered: Freshman, Sophomore, Junior, Senior
Service selection counseling
For: Women & Ethnic Minorities **Available:** Year round
Offered: Junior, Senior

253 University of Utah

INSTITUTION PROFILE

HEAD OF THE INSTITUTION
Arthur K. Smith
Phone: (801)581-5701   Fax: (801)581-6892

GENERAL INFORMATION
[All Students—Fall 1991]
Undergraduate enrollment 21,018
Graduate enrollment 4,563
Total institution enrollment **25,581**
Type of institution: Public   **Calendar:** Quarters
Location: Salt Lake City   **Population:** 700,000
Setting: Urban
Types of engineering degrees: Engineering
Other degree-granting colleges: Allied Health Sci, Architecture, Business Adm, Education, Fine & Performing Arts, Humanities, Law, Medicine, Nursing, Pharmacy, Technology & Appl Sciences, Mines & Earth Sciences, Liberal Arts, Social & Behavioral Science, Science

ENGINEERING ADMISSIONS

ADMISSIONS OFFICE CONTACT
J. Stayner Landward
University of Utah
Director of Admission
250 Student Services Building
Salt Lake City, UT 84112
Phone: (801)581-7281   Fax: (801)585-3034

ENGINEERING COLLEGE ADMISSIONS INFORMATION
Admission to the engineering college: the level of pre-engineering upon admissions with intermediate status following freshman year for Engineering; at general admissions for Mines.
Entrance Requirements: High School courses—English (4 years), Math (one must be algebra), biology, & chemistry (2 years), Foreign language (2 years), History (1 year), Combination of ACT/SAT scores with high school GPA creates admission index number upon which admissions is based.
Entrance Recommendations: High School courses—Plane Geometry, Algebra, Trigonometry (3.5 years), Physics (1 year), Chemistry (1 year), Technical Communications (1 year)
Requirements for foreign students: TOEFL (Score: 500); financial statement; International Student Application for Admission, health statement, authorized English record of academic experience.

DEPARTMENTAL ADMISSIONS INFORMATION
Admission to the engineering department: upon successful completion of pre-engr. courses (usually end of freshman year) for Engineering; at entrance for College of Mines & Earth Sciences.
Additional information: Department admission requirements and restrictions are program specific. In general, admissions and continued progress in engineering programs are based on the student's performance in course work. Students are encouraged to contact the department of interest for detailed information.

FINANCIAL INFORMATION

ESTIMATED EXPENSES (ACADEMIC YEAR)
[Expenses are for the 1992-93 nine-month academic year.]

State Residents	Undergraduate	Graduate
Tuition and fees	$ 2,105 [3]	$ 1,589 [1]
College room and board	$ 4,210	$ 4,210
Books and supplies	$ 640	$ 640
Other expenses	$ 1,592 [2]	$ 1,592 [2]
Total estimated expenses	**$ 8,547**	**$ 8,031**

Out-of-State Residents	Undergraduate	Graduate
Tuition and fees	$ 6,075 [3]	$ 4,515 [1]
College room and board	$ 4,210	$ 4,210
Books and supplies	$ 640	$ 640
Other expenses	$ 1,592 [2]	$ 1,592 [2]
Total estimated expenses	**$12,517**	**$10,957**

Notes: (1) Based on 3 quarters of 9 credit hour graduate registration. (2) Estimated travel and other living expenses. (3) Based on 3 quarters of 15 credit hour undergraduate registration.

FINANCIAL AID OFFICE CONTACT
Harold R. Weight, *Director of Financial Aid*
University of Utah
105 Student Services Building
Salt Lake City, UT 84112
Phone: (801)581-6211   Fax: (801)585-3034

GENERAL FINANCIAL AID INFORMATION
Forms accepted/required: AFSA, FAT, IRS, SAR, Institutional
Additional financial aid information: Need-based and/or merit-based scholarships; short-term low-interest loans; need-based no-repayment grants; part-time jobs on campus; work-study program.

ENGINEERING COLLEGE INFORMATION

[For additional personnel, refer to the *Appendix*.]

HEAD OF ENGINEERING
David W. Pershing
Phone: (801)581-6911   Fax: (801)581-8692
EMail: deandave@dean.eng.utah.edu

ENGINEERING COLLEGE ADDRESS
College of Engineering
University of Utah
2202 MEB
Salt Lake City, UT 84112
Phone: (801)581-6911   Fax: (801)581-8692

OTHER ENGINEERING ADDRESS
College of Mines and Earth Sciences
University of Utah
209 WBB
Salt Lake City, UT 84112

ENROLLMENTS—BY CLASS
[Numbers are baccalaureate enrollments for the fall 1991 term, unless footnoted.]
1st-year students/Freshmen .... 286 [1]
2nd-year students/Sophomores .... 214 [1]
3rd-year students/Juniors .... 225 [1]
4th-year students/Seniors .... 470 [1]
Total **1,195**

Notes: (1) Full-time enrollment for autumn qtr. 1991. Additional 448 part-time students (all classes) are not included.

NUMBER OF DEGREES AWARDED—BY PROGRAM
[Numbers are engineering baccalaureate degrees awarded during the 1991-1992 academic year, unless otherwise footnoted. For full details about each engineering program, refer to the *Tables of Degree Programs*.]
Bioengineering .... – [B]
Chemical Engineering 12
Civil Engineering 29
Computer Engineering 20
Computer Science 43
Electrical Engineering 81
Fuels Engineering .... – [2]
Geological Engineering 1
Materials Science & Engineering 11
Mechanical Engineering 73
Metallurgical Engineering 3
Mining Engineering 4
Total **277**

Notes: (B) Data not applicable. (2) Graduate level program only.

PERCENTAGE OF DEGREES AWARDED—BY CATEGORY
[Percentages are of all engineering baccalaureate degrees awarded during the 1991-1992 academic year, unless otherwise footnoted.]
African Americans .... – %
Asian/Pacific Island Americans .... 4.3 %
Hispanic Americans .... 2.6 %
Native Americans .... – %
Foreign Citizens .... 13.2 %
All Others .... 79.9 %
Women .... 11.5 %
Persons with disabilities .... – % [A]
Students over 25 years of age .... – % [A]

Notes: (A) Data not available.

GRADUATE ENROLLMENTS & DEGREES AWARDED
Master's enrollment 150
Master's degrees awarded .... 87 [2]
Doctoral enrollment 213
Doctoral degrees awarded .... 38 [3]

Notes: (2) An additional 6 master of science degrees granted in computer science. (3) An additional 6 doctoral degrees granted in computer science.

ENGINEERING STUDENT DATA
Applicants to the engineering college
Number of applicants to engineering college .... – [A]
Percent offered admission .... –% [A]
Matriculated engineering students
Percentage in top quartile (25%) of High School class .... – [A]
Average ACT scores: Math—21.5, Composite—22.8

Notes: (A) Data not available.

FULL- & PART-TIME FACULTY—BY DEPARTMENT
[Figures are the head count of full-time faculty and the full-time equivalent (FTE) of part-time faculty for each engineering department or equivalent.]

Department	Full	Part
Chemical & Fuels Eng	13	2.0
Civil Eng	9	–
Electrical Eng	14	1.1
Mechanical Eng	14	1.7
Computer Science	18	–
Metallurgical Eng	9	1.0
Mining Eng	5	5.5
Geological Eng	1	2.5
Materials Science & Eng	7	1.1
Bioengineering	2	3.0

COLLEGE DESCRIPTION
Engineering disciplines at the University are housed in two colleges: the College of Engineering and the College of Mines & Earth Sciences. Several programs provide direct support to undergraduates: the Engineering Clinic provides students with the opportunity to work on specific industrial problems with faculty and industrial representatives; the Mentoring Program links entering freshmen one-on-one with a faculty member; the Undergraduate Research Opportunity Plan provides undergraduates with the opportunity to work in a research laboratory; and the Cooperative Education Program integrates student's academic and career interests with a work experience.

DEGREE PROGRAMS DESCRIPTION
Bioengineering and Fuels Engineering are graduate degree programs only. Computer Engineering is an undergraduate program only. All other programs within the College of Engineering offer B.S., M.S., M.E. (non-thesis M.S.), and Ph.D. degrees. Electrical Engineering offers an E.E. (non-thesis Ph.D) degree. In the College of Mines and Earth Sciences all programs offer B.S., M.E., M.S., and Ph.D. degrees.

TRANSFER INFORMATION
Residency Requirements: Last 45 quarter hours of B.S. degree must be at the University of Utah.
Transfer via Articulation Agreements
Admission to engineering: Some depts accept students up to the end of the soph year; others accept students at any time, as long as the final 45 qtr. hours are completed in residence.
Requirements: Must satisfy general admission requirements set by the University. Status is reviewed on a case-by-case basis.
Transfer without Articulation Agreements
Admission to engineering: Some departments will accept students up to end of sophomore year. Others will accept any time as long as 45 quarter hours are completed at the University of Utah.
Requirements: Students must satisfy general admission requirements set by the University of Utah.

GRADUATION REQUIREMENTS
University over-all GPA must be greater than 2.00. Each department has different GPA requirements for course-groupings (e.g. first 2 years and major course work often have different GPA requirements). Chemical, Civil, Mechanical, and Mining Engineering require students to pass the Fundamentals of Engineering (FE) exam.

STUDENT PROGRAMS

PROFESSIONAL AND HONORARY SOCIETIES
[For key to acronyms, see *Introduction*.]
Professional Societies: ACM, AIChE, ANS, ASCE, ASME, BMES, IEEE, NSPE, SME, TMS, Society of Mining Engineers, The American Ceramic Society
Honorary Societies: Chi Epsilon, Eta Kappa Nu, Pi Tau Sigma, Tau Beta Pi, Alpha Nu Sigma

SUPPORT PROGRAMS
Student Chapter Organizations: Society of Women Engineers, National Society of Black Engineers, Society of Hispanic Professional Engineers, American Indian Science & Engineering Society, Minority Engineering Students Organization, Today's Black Engineers

Minority Engineering Program
For: Women & Ethnic Minorities   **Available:** Year round
Offered: Prematriculation, Freshman, Sophomore, Junior, Senior, Graduate level

Women's Resource Center
For: Women   **Available:** Year round
Offered: Prematriculation, Freshman, Sophomore, Junior, Senior, Graduate level

Center for Ethnic Student Affairs
For: Ethnic Minorities   **Available:** Year round
Offered: Prematriculation, Freshman, Sophomore, Junior, Senior, Graduate level

Minority Health Science Center
For: Ethnic Minorities   **Available:** Year round
Offered: Prematriculation, Freshman, Sophomore, Junior, Senior, Graduate level

Other Engineering Support Programs: One-half day orientation & registration during the summer for all freshman; academic counseling & tutoring services; special engineering seminars; international, women & veteran services; professional societies; financial aid and scholarship; placement and career guidance service; faculty mentoring for freshman students.

254 Utah State University

■ INSTITUTION PROFILE
HEAD OF THE INSTITUTION
Emert George
Phone: (801)750-1157 Fax: (801)750-3880

GENERAL INFORMATION
[All Students—Fall 1991]
Undergraduate enrollment 12,431
Graduate enrollment 2,687
Total institution enrollment **15,118**

Type of institution: Public Calendar: Quarters
Nearest city: Salt Lake City Population: 165,000
Miles from main campus: 85 Setting: Small Town
Types of engineering degrees: Engineering & Technology
Other degree-granting colleges: Agricultural & Environmental, Arts & Sciences, Business Administration, Education, Humanities & Social Sciences, Family Life, Natural Resources, Science

■ ENGINEERING ADMISSIONS
ADMISSIONS OFFICE CONTACT
J. Rodney Clark
Utah State University
Director of Admissions
Logan, UT 84322-1600
Phone: (801)750-1106 Fax: (801)750-2644

ENGINEERING COLLEGE ADMISSIONS INFORMATION
Admission to the engineering college: Students are admitted on pre-professional status as freshman and are admitted to the professional program after successfully completing the sophomore year.
Entrance Requirements: ACT, High School courses—English (4 years), Chemistry (1 year), Physics (1 year), Mathematics (including algebra, geometry, trig.) (4 years), Admission is formula based on a combination of high school GPA and ACT composite score.
Entrance Recommendations: High School courses—Computers (1 year), Mechanical Drawing (1 year)
Requirements for foreign students: TOEFL (Score: 500); financial statement

DEPARTMENTAL ADMISSIONS INFORMATION
Admission to the engineering department: At the time of admission to the institution.
Additional information: At the end of the sophomore year, students apply for the Professional Program in their major. Minimum GPA to apply is 2.3 in engineering/math/science courses. However, there is not always room for all applicants in a requested department.

■ FINANCIAL INFORMATION
ESTIMATED EXPENSES (ACADEMIC YEAR)
[Expenses are for the 1992-93 nine-month academic year.]

State Residents	Undergraduate	Graduate
Tuition and fees	$ 1,554	$ 918
College room and board	$ 3,240	$ 3,240
Books and supplies	$ 390	$ 390
Other expenses	$ –	$ –
Total estimated expenses	**$ 5,184**	**$ 4,548**

Out-of-State Residents	Undergraduate	Graduate
Tuition and fees	$ 4,164	$ 2,553
College room and board	$ 3,240	$ 3,240
Books and supplies	$ 390	$ 390
Other expenses	$ –	$ –
Total estimated expenses	**$ 7,794**	**$ 6,183**

FINANCIAL AID OFFICE CONTACT
Vicki L. Atkinson, *Director of Financial Aid*
Utah State University
Logan, UT 84322-1800
Phone: (801)750-1022 Fax: (801)750-2571

GENERAL FINANCIAL AID INFORMATION
Forms accepted/required: FFS, Institutional
Additional financial aid information: Need-based scholarships and grants: short-term loans; Stafford & Perkins loans: State grant money: merit-based scholarships: part-time jobs on campus: College work-study program.

■ ENGINEERING COLLEGE INFORMATION
[For additional personnel, refer to the *Appendix*.]
HEAD OF ENGINEERING
A. Bruce Bishop
Phone: (801)750-2776 Fax: (801)750-1248

ENGINEERING COLLEGE ADDRESS
College of Engineering
Utah State University
D.F. Petersen Engineering Building
Logan, UT 84322-4100
Phone: (801)750-2775 Fax: (801)750-1248

ENROLLMENTS—BY CLASS
[Numbers are baccalaureate enrollments for the fall 1991 term, unless otherwise footnoted.]
1st-year students/Freshmen 429 (1)
2nd-year students/Sophomores 241 (1)
3rd-year students/Juniors 214 (1)
4th-year students/Seniors .... 401 (A)
Total **1,285**
Notes: (A) Data not available. (1) Fall Quarter 1991 Data

NUMBER OF DEGREES AWARDED—BY PROGRAM
[Numbers are engineering baccalaureate degrees awarded during the 1991-1992 academic year, unless otherwise footnoted. For full details about each engineering program, refer to the *Tables of Degree Programs*.]
Biological & Irrigation Engineering 6
Civil & Environmental Engineering 6
Electrical & Computer Engineering 39
Industrial Technology 33
Mechanical, Manufacturing & Aerospace Engineering 45
Total **129**

PERCENTAGE OF DEGREES AWARDED—BY CATEGORY
[Percentages are of all engineering baccalaureate degrees awarded during the 1991-1992 academic year, unless otherwise footnoted.]
African Americans .... – %
Asian/Pacific Island Americans .... 4.0 %
Hispanic Americans .... – %
Native Americans .... – %
Foreign Citizens .... 6.0 %
All Others .... 90.0 %
Women .... 6.0 %
Persons with disabilities .... – % (A)
Students over 25 years of age .... – % (A)
Notes: (A) Data not available.

GRADUATE ENROLLMENTS & DEGREES AWARDED
Master's enrollment 203
Master's degrees awarded 94
Doctoral enrollment 96
Doctoral degrees awarded 16

FULL- & PART-TIME FACULTY—BY DEPARTMENT
[Figures are the head count of full-time faculty and the full-time equivalent (FTE) of part-time faculty for each engineering department or equivalent.]

Department	Full	Part
Civil & Environmental Eng	29	–
Electrical Eng	21	–
Industrial Tech & Education	13	–
Mechanical & Aerospace Eng	14	1.0
Biological & Irrigation Eng	17	–

COLLEGE DESCRIPTION
The college offers four-year undergraduate programs in engineering, emphasizing engineering fundamentals in the first two years. The first two years also serve as a proving ground, and performance there qualifies students for admission to a Professional Engineering Program in one of the six majors. Liberal Arts courses are scattered throughout the four years. All programs have a strong computer emphasis. Minors are not required but are often pursued in math, physics, computer science and business. Co-op experience is encouraged but not required. All majors complete a senior design project. Majors in Civil, Manufacturing, Mechanical, and Aerospace Engineering must pass the national standardized Fundamentals of Engineering exam. The average student takes about 4.6 years to actually complete the bachelor's degree.

DEGREE PROGRAMS DESCRIPTION
The BS degrees in the various engineering programs prepare the graduate for entry-level professional positions or for graduate study in each of the respective fields. Graduate study can be carried out in one of two general areas: the MS and PhD in engineering, emphasizing the applications of the sciences to the analysis and solution of engineering problems. Students pursuing the MS degree may select either the thesis or the non-thesis option. The thesis option emphasizes research methods and laboratory skills. The non-thesis option emphasizes design and/or engineering practice.

DUAL DEGREE PROGRAMS
Engineering baccalaureate graduates with dual degrees 7

TRANSFER INFORMATION
Residency Requirements: Students must complete at least 45 credits in residence at USU's Logan campus or designated residence areas. No more than 120 credits of transfer credit from a junior college will be accepted toward graduation.
Transfer via Articulation Agreements
Admission to engineering: At any time
Engineering graduates transferred from 4-yr: 18 2-yr: 38
Requirements: Criteria include resources available in the requested department and the transfer GPA performance in technical coursework along with an evaluation of the program of the former college or university.
Transfer without Articulation Agreements
Admission to engineering: At any time
Engineering graduates transferred from ...4-yr: – 2-yr: –
Requirements: Criteria include resources available in the requested department and the transfer GPA in technical course work along with an evaluation of the student's former program.

GRADUATION REQUIREMENTS
B.S. programs in engineering require 195-196 quarter credits, depending on major selected. The student's overall GPA and the engineering/math/science GPA must be 2.0 or higher. Civil, Manufacturing, and Mechanical & Aerospace Engineering majors must pass the Fundamentals of Engineering Exam. The MS degree requires a minimum of 45 to 51 quarter hours of credit past a BS engineering degree. The actual number of credits depends on the particular program. The PhD degree requires a research dissertation and normally takes two years beyond the MS.

■ STUDENT PROGRAMS
PROFESSIONAL AND HONORARY SOCIETIES
[For key to acronyms, see *Introduction*.]
Professional Societies: AIAA, ASAE, ASCE, ASME, AWRA, AWS, IEEE, SAE, SME, Society of Women Engineers
Honorary Societies: Chi Epsilon, Tau Beta Pi

SUPPORT PROGRAMS
Student Chapter Organizations: Society of Women Engineers
Minority Student Affairs
For: Ethnic Minorities **Available:** Year round
Offered: Prematriculation, Freshman, Sophomore, Junior, Senior, Graduate level
Women's Center
For: Women **Available:** Year round
Offered: Prematriculation, Freshman, Sophomore, Junior, Senior, Graduate level

Other Engineering Support Programs: Summer and fall orientation for all freshmen and transfers: Freshman Orientation course; Alcohol & Drug Abuse Education/Prevention; Career Development; Career Placement; Cooperative Education; Counseling Center; Disabled Student Center; Helpline; Honors Program; International Student Office; Learning Assistance Center; and Veterans Affairs.

255 Valparaiso University

■ INSTITUTION PROFILE

HEAD OF THE INSTITUTION
Alan F. Harre
Phone: (219)464-5115 Fax: (219)464-5381

GENERAL INFORMATION
[All Students—Fall 1991]
Undergraduate enrollment 3,549
Graduate enrollment 208
Total institution enrollment 3,757
Type of institution: Private Calendar: Semesters
Nearest city: Chicago Population: 6,000,000
Miles from main campus: 60 Setting: Small Town
Types of engineering degrees: Engineering
Other degree-granting colleges: Arts & Sciences, Business Administration, Law, Nursing, Christ College (Honors)

■ ENGINEERING ADMISSIONS

ADMISSIONS OFFICE CONTACT
Karen Foust
Valparaiso University
Valparaiso, IN 46383
Phone: (219)464-5011 Fax: (219)464-5381

ENGINEERING COLLEGE ADMISSIONS INFORMATION
Entrance Requirements: High School courses—Mathematics-algebra, geometry, trigonometry (3.5 years), Chemistry or Physics (1 year)
Entrance Recommendations: High School courses—English (4 years), Mathematics (4 years), Social Studies/Foreign Language (2 years), Laboratory Sciences (2 years)
Requirements for foreign students: TOEFL (Score: 550); financial statement

DEPARTMENTAL ADMISSIONS INFORMATION
Admission to the engineering department: Student is admitted to department at two-thirds point in first semester of freshman year contingent upon an academically successful first semester.
Additional information: 1. A cumulative grade point average (GPA) of 2.000 for all required courses. 2. A cumulative GPA of 2.000 in General Engineering and applicable mathematics and chemistry courses. 3. A cumulative GPA of 2.000 in General Education courses.

■ FINANCIAL INFORMATION

ESTIMATED EXPENSES (ACADEMIC YEAR)
[Expenses are for the 1992-93 nine-month academic year.]

All Students	Undergraduate	Graduate
Tuition and fees	$10,910	$ —
College room and board	$ 3,940	$ —
Books and supplies	$ 500	$ —
Other expenses	$ 400	$ —
Total estimated expenses	$15,750	$ —

FINANCIAL AID OFFICE CONTACT
Katherine E. Wehling, Director
Valparaiso University
Union Street
Kretzmann Hall
Valparaiso, IN 46383-6259
Phone: (219)464-5014 Fax: (219)464-5381

GENERAL FINANCIAL AID INFORMATION
Forms accepted/required: AFSA, CSS-FAF, FAT, FAR, FSA, Stafford, IRS, SAR, SDF, Supplemental, Institutional, Perkins
Additional financial aid information: Founders Scholarships, Board of Directors Scholarships, Honors Scholarships, Presidential Scholarships, University Scholarships, Lutheran Presidents' Scholarships, Indiana Scholarships, Alumni Heritage Awards, Diversity Awards, Leadership Awards, Martin Luther Awards, Lutheran High School Principal's Awards, Athletic Grants, Music Grants, University Grants, Federal Grants, State Grants.

■ ENGINEERING COLLEGE INFORMATION

[For additional personnel, refer to the *Appendix*.]

HEAD OF ENGINEERING
Stuart G. Walesh
Phone: (219)464-5121 Fax: (219)464-5381

ENGINEERING COLLEGE ADDRESS
College of Engineering
Valparaiso University
Union Street
Gellerson Room 105
Valparaiso, IN 46383
Phone: (219)464-5121 Fax: (219)464-5381

ENROLLMENTS—BY CLASS
[Numbers are baccalaureate enrollments for the fall 1991 term, unless otherwise footnoted.]
1st-year students/Freshmen 96
2nd-year students/Sophomores 76
3rd-year students/Juniors 87
4th-year students/Seniors 107
Total 366

NUMBER OF DEGREES AWARDED—BY PROGRAM
[Numbers are engineering baccalaureate degrees awarded during the 1991-1992 academic year, unless otherwise footnoted. For full details about each engineering program, refer to the *Tables of Degree Programs*.]
Civil Engineering 21
Computer Engineering 9
Electrical Engineering 24
Mechanical Engineering 29
Total 83

PERCENTAGE OF DEGREES AWARDED—BY CATEGORY
[Percentages are of all engineering baccalaureate degrees awarded during the 1991-1992 academic year, unless otherwise footnoted.]
African Americans .... — %
Asian/Pacific Island Americans .... — %
Hispanic Americans .... 1.0 %
Native Americans .... — %
Foreign Citizens .... 14.0 %
All Others .... 85.0 %
Women .... 22.0 %
Persons with disabilities .... — % (A)
Students over 25 years of age .... — % (A)

Notes: (A) Data not available.

ENGINEERING STUDENT DATA
Applicants to the engineering college
Number of applicants to engineering college .... — (A)
Percent offered admission .... —% (A)

Matriculated engineering students
Percentage in top quartile (25%) of High School class ...... — (A)
Average SAT scores: Math—505, Verbal—635, Combined—1140
Average ACT scores: Composite—27

Notes: (A) Data not available.

FULL- & PART-TIME FACULTY—BY DEPARTMENT
[Figures are the head count of full-time faculty and the full-time equivalent (FTE) of part-time faculty for each engineering department or equivalent.]

Department	Full	Part
Civil Eng	5	—
Electrical & Computer Eng	8	—
Mechanical Eng	7	—

COLLEGE DESCRIPTION
Small lectures and laboratories with hands-on laboratory experience. Experienced faculty members present all lectures and personally direct all laboratories. Computer use in curriculum includes drafting, analysis, design, manufacturing, data acquisition, testing, and simulation. Optional Cooperative Education program (5 year)—3 summers and 2 semesters as a para-professional in an engineering organization. Optional double degree program (5 year)—B.S. in engineering and B.S. or B.A. in another college. Optional undergraduate research. Engineering major with business minor (4.5 years). International study—Cambridge; Japan; Paris; Peoples Republic of China; Puebla; Reutligen and Tübingen, Germany.

DEGREE PROGRAMS DESCRIPTION
The College of Engineering is a demanding, exclusively undergraduate engineering college operating in the Christian tradition of Valparaiso University; led by readily available faculty having a varied academic and professional practice background; and offering a comprehensive curriculum that stresses the laboratory and requires professional practice experience. Bachelor's degrees may be earned in civil, computer, electrical, and mechanical engineering. The curriculum is comprehensive in that about one-fourth of the credits are earned in mathematics and basic sciences, one-fourth in engineering analysis and design, one-fourth in engineering sciences, and one-fourth in humanities and social sciences. Approximately forty percent of the credits required for graduation in engineering are taken in one of more of the other colleges at the University. All engineering students participate in at least one of two available professional practice experiences which are the Independent Study Project and Cooperative Education.

DUAL DEGREE PROGRAMS
Engineering baccalaureate graduates with dual degrees 1

TRANSFER INFORMATION
Transfer without Articulation Agreements
Admission to engineering: At any time
Requirements: Urged to communicate with the chair of the department in which they hope to major prior to formally applying for admission to obtain a preliminary assessment of the duration of their course of study. The transfer student can apply a maximum of 3 credit hours in theology. With more than 68 credit hours or advanced standing, students need only to complete one theology course.

GRADUATION REQUIREMENTS
A cumulative GPA of 2.000 in General Engineering courses, mathematics, and chemistry, as well as in the specific engineering major and in General Education courses applicable to the degree. Each engineering degree requires 133-137 credits for graduation.

■ STUDENT PROGRAMS

PROFESSIONAL AND HONORARY SOCIETIES
[For key to acronyms, see *Introduction*.]
Professional Societies: ASCE, ASME, IEEE, Computer Engineering Society, The Society of Women Engineers (SWE)
Honorary Societies: Tau Beta Pi

SUPPORT PROGRAMS
Student Chapter Organizations: Society of Women Engineers

Minority Programs Office
For: Ethnic Minorities Available: Year round
Offered: Freshman, Sophomore, Junior, Senior, Graduate level
Other Engineering Support Programs: The Exploring Engineering course is taken by all first year students and helps them confirm engineering, select a major, and acquire certain computation tools and techniques. Christ College is an honors program for students who would like an enriched program in the humanities as well as satisfy engineering education requirements.

256 Vanderbilt University

■ INSTITUTION PROFILE
HEAD OF THE INSTITUTION
Joe B. Wyatt
Phone: (615)322-2168
GENERAL INFORMATION
[All Students—Fall 1991]
Undergraduate enrollment 5,547
Graduate enrollment 4,034
Total institution enrollment 9,581
Type of institution: Private **Calendar:** Semesters
Nearest city: Nashville, TN **Population:** 532,000
Miles from main campus: 1 **Setting:** Urban
Types of engineering degrees: Engineering
Other degree-granting colleges: Arts & Sciences, Business Administration, Education, Law, Medicine, Nursing, Blair School of Music, Divinity School, Graduate School

■ ENGINEERING ADMISSIONS
ADMISSIONS OFFICE CONTACT
Neill F. Sanders
Vanderbilt University
Undergraduate Admissions
2305 West End Avenue
Nashville, TN 37203-1725
Phone: (615)322-2561
ENGINEERING COLLEGE ADMISSIONS INFORMATION
Entrance Recommendations: SAT, ACT, High School courses—Mathematics (4 years), Chemistry (1 year), Physics (1 year), SAT or ACT Three Achievement Tests (Math, English, & Physics or other Science)
Requirements for foreign students: TOEFL (Score: 500); financial statement
DEPARTMENTAL ADMISSIONS INFORMATION
Admission to the engineering department: At the time of admission to the institution.

■ FINANCIAL INFORMATION
ESTIMATED EXPENSES (ACADEMIC YEAR)
[Expenses are for the 1992-93 nine-month academic year.]

All Students	Undergraduate	Graduate
Tuition and fees	$15,455	$ 5,994
College room and board	$ 5,420	$ 5,460
Books and supplies	$ 575	$ 450
Other expenses	$ 826	$ –
Total estimated expenses	**$22,276**	**$11,904**

FINANCIAL AID OFFICE CONTACT
David Mohning, *Director, Financial Aid*
Vanderbilt University
2309 West End Avenue
Nashville, TN 37203-1725
Phone: (615)322-3591
GENERAL FINANCIAL AID INFORMATION
Forms accepted/required: CSS-FAF, FAF, Stafford, IRS, SAR, Supplemental, Institutional
Additional financial aid information: Federal, State and Private Scholarships are available Need-based scholarships Honor scholarships Student Loans Student Employment

■ ENGINEERING COLLEGE INFORMATION
[For additional personnel, refer to the *Appendix*.]
HEAD OF ENGINEERING
Edward A. Parrish
Phone: (615)322-2762 Fax: (615)343-8006
EMail: eap@vuse.vanderbilt.edu

ENGINEERING COLLEGE ADDRESS
School of Engineering
Vanderbilt University
400 24th Avenue South
Nashville, TN 37235
Phone: (615)322-2762 Fax: (615)343-8006

ENROLLMENTS—BY CLASS
[Numbers are baccalaureate enrollments for the fall 1991 term, unless otherwise footnoted.]
1st-year students/Freshmen 318
2nd-year students/Sophomores 245
3rd-year students/Juniors 203
4th-year students/Seniors 241
Total **1,007**

NUMBER OF DEGREES AWARDED—BY PROGRAM
[Numbers are engineering baccalaureate degrees awarded during the 1991-1992 academic year, unless otherwise footnoted. For full details about each engineering program, refer to the *Tables of Degree Programs*.]
Biomedical Engineering .... 30 [1]
Chemical Engineering 22
Civil Engineering 33
Computer Science 31
Electrical Engineering 59
Engineering Science(s) 20
Materials Science & Engineering .... –
Mechanical Engineering 50
Total **245**

Notes: (1) An additional 14 double majors with Electrical Engineering.

PERCENTAGE OF DEGREES AWARDED—BY CATEGORY
[Percentages are of all engineering baccalaureate degrees awarded during the 1991-1992 academic year, unless otherwise footnoted.]
African Americans .... 3.0 %
Asian/Pacific Island Americans .... 4.0 %
Hispanic Americans .... – %
Native Americans .... – %
Foreign Citizens .... 5.0 %
All Others .... 88.0 %
Women .... 21.0 %
Persons with disabilities .... – %
Students over 25 years of age .... – %

GRADUATE ENROLLMENTS & DEGREES AWARDED
Master's enrollment 139
Master's degrees awarded 68
Doctoral enrollment 178
Doctoral degrees awarded 21

ENGINEERING STUDENT DATA
Applicants to the engineering college
Number of applicants to engineering college 1,173
Percent offered admission 76.0%
Matriculated engineering students
Percentage in top quartile (25%) of High School class 16.0
Average SAT scores: Math—651, Verbal—533, Combined—1184
Average ACT scores: Composite—27

FULL- & PART-TIME FACULTY—BY DEPARTMENT
[Figures are the head count of full-time faculty and the full-time equivalent (FTE) of part-time faculty for each engineering department or equivalent.]

Department	Full	Part
Chemical Eng	10	–
Civil & Environmental Eng	15	–
Electrical Eng	22	–
Mechanical Eng	16	–
Computer Science	10	–
Materials Science & Eng	9	–
Biomedical Eng	11	–

COLLEGE DESCRIPTION
Small enough for the student to be known by faculty and to know faculty. Large enough to be comprehensive. There are no special Humanities/Social Science courses for engineers. They take the same classes in those subjects as Arts and Science Majors. 40% of graduates have two or more majors. Second majors may be in an Arts and Science discipline. Junior year abroad programs.

DEGREE PROGRAMS DESCRIPTION
We offer Bachelor of Science, Bachelor of Engineering, Master of Engineering, Master of Science, and Doctor of Philosophy. The Bachelor of Engineering degree is offered in programs accredited by ABET. Master of Science degree requires a thesis. Master of Engineering (Professional Degree) requires a project.

DUAL DEGREE PROGRAMS
Engineering baccalaureate graduates with dual degrees 0
Enrollment requirements: 3.00 Grade Point Average

TRANSFER INFORMATION
Residency Requirements: 60 hours and 4 semesters
Transfer via Articulation Agreements
Admission to engineering: Enter at the end of the junior year.
Requirements: 3.0 Grade Point Average
Transfer without Articulation Agreements
Admission to engineering: Transfers admitted at the end of any semester after the freshman year. Only transfers from outside the university are included in the statistics in this survey.
Engineering graduates transferred from ... 4-yr: 17   2-yr: 2
Requirements: 3.0 Grade Point Average

GRADUATION REQUIREMENTS
Total number of hours specific to each department. 2.00 GPA Overall 2.00 GPA in Engineering Courses 2.00 GPA in Major Courses

■ STUDENT PROGRAMS
PROFESSIONAL AND HONORARY SOCIETIES
[For key to acronyms, see *Introduction*.]
Professional Societies: ACM, AIChE, ASCE, ASME, BMES, IEEE, American Society for Metals (ASM)
Honorary Societies: Chi Epsilon, Eta Kappa Nu, Pi Tau Sigma, Tau Beta Pi, Alpha Sigma Mu, Sigma Xi

SUPPORT PROGRAMS
Student Chapter Organizations: Society of Women Engineers, National Society of Black Engineers

Minority Summer Research Program
For: Ethnic Minorities **Available:** Summer only
Offered: Prematriculation
Other Engineering Support Programs: Summer Academic Orientation Program, Matching Gears Program (Undergraduate advisors for freshmen), PAVE (Preparatory Academics for Vanderbilt Engineers), Fall Orientation Program Vucept (Upper class students who serve as guides for new freshmen during orientation)

257 University of Vermont

INSTITUTION PROFILE

HEAD OF THE INSTITUTION
Thomas P. Salmon
Phone: (802)656-3186   Fax: (802)656-8432

GENERAL INFORMATION
[All Students—Fall 1991]
Undergraduate enrollment 8,103
Graduate enrollment 1,198
Total institution enrollment 9,301
Type of institution: Public   Calendar: Semesters
Nearest city: Montréal   Population: 1,015,420
Miles from main campus: 100   Setting: Small Town
Types of engineering degrees: Engineering
Other degree-granting colleges: Allied Health Sciences, Arts & Sciences, Business Administration, Medicine, Nursing, Agriculture & Life Sciences, Education & Social Services, Natural Resources

ENGINEERING ADMISSIONS

ADMISSIONS OFFICE CONTACT
Susan A. Wertheimer
University of Vermont
Undergraduate Admissions
Clement House, 194 South Prospect St.
Burlington, VT 05405-0156
Phone: (802)656-3370   Fax: (802)656-8611

ENGINEERING COLLEGE ADMISSIONS INFORMATION
Admission to the engineering college: Admission to the institution, or when student in another unit has completed minimum of 30 credits with 2.25 cumulative grade point average.
Entrance Requirements: SAT, High School courses—Mathematics (4 years), Physics (1 year), Chemistry (1 year), ACT accepted in place of SAT. First-year students take Calculus Readiness Test at Orientation.
Entrance Recommendations: Achievement tests in Mathematics, Physics and Chemistry are helpful, but not required.
Requirements for out-of-state residents: Priority given to all qualified Vermonters. Non-residents accepted on a competitive, space-available basis.
Requirements for foreign students: TOEFL (Score: 550); financial statement

DEPARTMENTAL ADMISSIONS INFORMATION
Admission to the engineering department: Admission to the institution. Re-evaluated at end of Sophomore year: 2.0 cumulative GPA required at that time.

FINANCIAL INFORMATION

ESTIMATED EXPENSES (ACADEMIC YEAR)
[Expenses are for the 1992-93 nine-month academic year.]

State Residents	Undergraduate	Graduate
Tuition and fees	$ 6,166	$ 6,244
College room and board	$ 4,266	$ _(A)
Books and supplies	$ 465	$ _(A)
Other expenses	$ _(A)	$ _(A)
Total estimated expenses	$10,897	$ 6,244

Out-of-State Residents	Undergraduate	Graduate
Tuition and fees	$14,766	$14,860
College room and board	$ 4,266	$ _(A)
Books and supplies	$ 465	$ _(A)
Other expenses	$ _(A)	$ _(A)
Total estimated expenses	$19,497	$14,860

Notes: (A) Data not available.

FINANCIAL AID OFFICE CONTACT
Donald M. Honeman, *Director*
University of Vermont
Financial Aid Office
330 Waterman
Burlington, VT 05405-0156
Phone: (802)656-3156

GENERAL FINANCIAL AID INFORMATION
Forms accepted/required: CSS-FAF, FAT, Stafford, IRS, SAR
Additional financial aid information: Aid is most often awarded in combinations or 'packages' of various types of aid. Almost all awards include some loan or work-study for which the student states a preference on the admissions and financial aid applications. International students should address inquiries to the Office of International Educational Services.

ENGINEERING COLLEGE INFORMATION
[For additional personnel, refer to the *Appendix*.]

HEAD OF ENGINEERING
George F. Pinder
Phone: (802)656-3390   Fax: (802)656-8802
EMail: pinder@uvm.edu

ENGINEERING COLLEGE ADDRESS
College of Engineering and Mathematics
University of Vermont
109 Votey
Burlington, VT 05405-0156
Phone: (802)656-3390   Fax: (802)656-8802

ENROLLMENTS—BY CLASS
[Numbers are baccalaureate enrollments for the fall 1991 term, unless otherwise footnoted.]
1st-year students/Freshmen 178
2nd-year students/Sophomores 137
3rd-year students/Juniors 100
4th-year students/Seniors 112
Total 527

NUMBER OF DEGREES AWARDED—BY PROGRAM
[Numbers are engineering baccalaureate degrees awarded during the 1991-1992 academic year, unless otherwise footnoted. For full details about each engineering program, refer to the *Tables of Degree Programs*.]
Civil Engineering 32
Electrical Engineering 43
Materials Science .... _(B)
Mechanical Engineering 29
Total 104

Notes: (B) Data not applicable.

PERCENTAGE OF DEGREES AWARDED—BY CATEGORY
[Percentages are of all engineering baccalaureate degrees awarded during the 1991-1992 academic year, unless otherwise footnoted.]
African Americans 0.7%
Asian/Pacific Island Americans 3.3%
Hispanic Americans .... – %
Native Americans .... – %
Foreign Citizens 0.7%
All Others 95.3%
Women 8.6%
Persons with disabilities .... – %(A)
Students over 25 years of age .... – %(A)

Notes: (A) Data not available.

GRADUATE ENROLLMENTS & DEGREES AWARDED
Master's enrollment 76
Master's degrees awarded 29
Doctoral enrollment 33
Doctoral degrees awarded 4

ENGINEERING STUDENT DATA
Applicants to the engineering college
Number of applicants to engineering college 534
Percent offered admission 91.4%

FULL- & PART-TIME FACULTY—BY DEPARTMENT
[Figures are the head count of full-time faculty and the full-time equivalent (FTE) of part-time faculty for each engineering department or equivalent.]

Department	Full	Part
Civil Eng	8	–
Electrical Eng	13	–
Mechanical Eng	10	–

COLLEGE DESCRIPTION
Engineering education at UVM combines the study of mathematics and the physical, life and engineering sciences with application to the analysis and design of equipment, processes and complete systems. The breadth and flexibility of the engineering programs provide a sound background for engineering practice in private or public domains, for graduate study in engineering and science, and for further professional study in such fields as business, law or medicine. At least 18 credit hours in the humanities and social sciences are required to broaden the student's understanding of humankind. Engineering students may extend their undergraduate curriculum beyond the typical four years. A Co-op Program enables students to apply their learning to a full-time, paid position in a business setting related to their major. Over 90% of UVM students taking the Fundamentals in Engineering Test (FET) pass it the first time, compared to a national average of 38%.

DEGREE PROGRAMS DESCRIPTION
The Electrical Engineering program includes areas of concentration in plasma physics, electro-optics, cryoelectronics, signal processing, and control systems theory and applications. Options: computer, biomedical, or pre-medical. Research is under way to develop new techniques that will enable electrical information to travel faster and more efficiently. Environmental engineering, soil mechanics, hydraulics and hydrology, structural engineering and transportation engineering are are all part of the Civil Engineering program. Options in environmental engineering and general civil engineering are offered. Special emphasis on structural dynamics, finite element methods, groundwater remediation and aerosols. A complete toxicology experimental laboratory is maintained for the study of airborne contaminants. To develop more efficient manufacturing processes, to design better machinery, and to improve the production and transmission of energy, are among the goals of the Mechanical Engineering program. Options: general, biomedical, manufacturing and pre-medical. Special emphasis in areas of biomechanics, design, fluids, materials, manufacturing, combustion and vibrations. Laboratory equipment in all important areas is available. The Materials Science program is available to graduate students.

TRANSFER INFORMATION
Residency Requirements: Must complete 30 of final 45 credits in College of Engineering & Mathematics at University of Vermont.
Transfer without Articulation Agreements
Admission to engineering: After 30 credit hours with minimum 2.25 grade point average.
Requirements: A 2.25 grade point average is required for transfers from other UVM units.

GRADUATION REQUIREMENTS
To receive a degree in a major, students must have a minimum cumulative average of 2.0, and must complete 30 of the last 45 hours of credit in residence at UVM as matriculated students. Additional degree requirements are specified for each major.

STUDENT PROGRAMS

PROFESSIONAL AND HONORARY SOCIETIES
[For key to acronyms, see *Introduction*.]
Professional Societies: ACM, ASCE, ASEM, ASME, IEEE, MAA
Honorary Societies: Chi Epsilon, Tau Beta Pi, Phi Eta Sigma, Upsilon Pi Epsilon

SUPPORT PROGRAMS
Student Chapter Organizations: Society of Women Engineers

Center for Cultural Pluralism
For: Ethnic Minorities   **Available:** Year round
Offered: Freshman, Sophomore, Junior, Senior, Graduate level

Summer Minority Program
For: Ethnic Minorities   **Available:** Summer only
Offered: Prematriculation

Office of Multicultural Affairs
For: Ethnic Minorities   **Available:** Year round
Offered: Prematriculation, Freshman, Sophomore, Junior, Senior, Graduate level

Research Apprentice Program
For: Ethnic Minorities   **Available:** Summer only
Offered: Prematriculation

Other Engineering Support Programs: The Learning Co-op provides academic help to students in most course-related subjects. Resource Center peer advisors augment faculty advising in engineering, mathematics and business administration, and also offer information on tutors, textbook exchange, internships, etc. Help/study sessions are provided at the departmental level for many introductory mathematics courses.

258 Villanova University

■ INSTITUTION PROFILE

HEAD OF THE INSTITUTION
Edmund J. Dobbin
Phone: (215)645-4510   Fax: (215)645-4514

GENERAL INFORMATION
[All Students—Fall 1991]
Undergraduate enrollment 6,354
Graduate enrollment 3,484
Total institution enrollment 9,838
Type of institution: Private   **Calendar:** Semesters
Nearest city: Philadelphia   **Population:** 1,585,000
Miles from main campus: 12   **Setting:** Suburban
Types of engineering degrees: Engineering
Other degree-granting colleges: Arts & Sciences, Law, Nursing, Commerce and Finance

ENGINEERING COLLEGE ADMISSIONS INFORMATION
Admission to the engineering college: the time of admission to the institution.
Entrance Requirements: SAT (Score: 950), High School courses—English (4 years), Chemistry (1 year), Physics (1 year), Mathematics (4 years)
Entrance Recommendations: ACT, High School courses—Foreign Language (1 year), History (1 year), Social Studies (1 year), Computer Science (1 year), Advanced Placement in mathematics, chemistry, English, physics, biology, computer science, American history, classics, European history, modern languages, and political science.
Requirements for foreign students: TOEFL (Score: 550); financial statement

DEPARTMENTAL ADMISSIONS INFORMATION
Admission to the engineering department: The time of admission to the institution. Students declare their engineering major at the end of freshman year.

■ FINANCIAL INFORMATION

ESTIMATED EXPENSES (ACADEMIC YEAR)
[Expenses are for the 1992-93 nine-month academic year.]

All Students	Undergraduate	Graduate
Tuition and fees	$14,050	$10,616
College room and board	$ 6,080	$ –
Books and supplies	$ 500	$ 300
Other expenses	$ –	$ –
Total estimated expenses	**$20,630**	**$10,916**

FINANCIAL AID OFFICE CONTACT
George Walter, *Financial Aid Officer*
Villanova University
Villanova, PA 19085
Phone: (215)645-4010   Fax: (215)645-7599

GENERAL FINANCIAL AID INFORMATION
Forms accepted/required: FAF, IRS, Pennsylvania State Grant, Federal Student Aid Application
Additional financial aid information: Non-need based Presidential scholarships; need-based institutional and state/federal grants and loans; part-time campus work study program: Pell Grant, Supplemental Educational Opportunity Grant, College Work Study, Perkins Loan, Stratford Student Loan, Parents' Loan for Undergraduate Students.

■ ENGINEERING COLLEGE INFORMATION

[For additional personnel, refer to the *Appendix*.]
HEAD OF ENGINEERING
Robert D. Lynch
Phone: (215)645-4940   Fax: (215)645-4941
EMail: RDLYNCH@VUVAXCOM

ENGINEERING COLLEGE ADDRESS
College of Engineering
Villanova University
Villanova, PA 19085-1681
Phone: (215)645-4940   Fax: (215)645-4941

ENROLLMENTS—BY CLASS
[Numbers are baccalaureate enrollments for the fall 1991 term, unless otherwise footnoted.]
1st-year students/Freshmen 272
2nd-year students/Sophomores 176
3rd-year students/Juniors 200
4th-year students/Seniors 197
Total 845

NUMBER OF DEGREES AWARDED—BY PROGRAM
[Numbers are engineering baccalaureate degrees awarded during the 1991-1992 academic year, unless otherwise footnoted. For full details about each engineering program, refer to the *Tables of Degree Programs*.]
Chemical Engineering 29
Civil Engineering 45
Computer Engineering .... –
Electrical Engineering 71
Mechanical Engineering 58
Total 203

PERCENTAGE OF DEGREES AWARDED—BY CATEGORY
[Percentages are of all engineering baccalaureate degrees awarded during the 1991-1992 academic year, unless otherwise footnoted.]
African Americans .... – %
Asian/Pacific Island Americans .... – %
Hispanic Americans .... – %
Native Americans .... – %
Foreign Citizens .... – %
All Others .... 100 %
Women .... 18.2 %
Persons with disabilities .... – %
Students over 25 years of age .... – %

GRADUATE ENROLLMENTS & DEGREES AWARDED
Master's enrollment 373
Master's degrees awarded 108
Doctoral enrollment .... –
Doctoral degrees awarded .... –

ENGINEERING STUDENT DATA
Applicants to the engineering college
Number of applicants to engineering college 868
Percent offered admission 91.0%
Matriculated engineering students
Percentage in top quartile (25%) of High School class ... 70.0
Average SAT scores: Math—632.4, Verbal—525.6, Combined—1157.9

FULL- & PART-TIME FACULTY—BY DEPARTMENT
[Figures are the head count of full-time faculty and the full-time equivalent (FTE) of part-time faculty for each engineering department or equivalent.]

Department	Full	Part
Chemical Eng	8	1.0
Civil Eng	12	1.0
Electrical Eng	20	3.8
Mechanical Eng	18	1.5
Computer Eng	20	3.8

COLLEGE DESCRIPTION
Villanova's undergraduate engineering program is characterized by a personalized approach to technical education in a humanistic context. The undergraduate curriculum stresses fundamentals and design experience. The design process is central to the College's goal of educating technically competent engineers who have developed creative problem-solving skills. All engineering freshmen are assigned to a team that analyzes an engineering problem, and designs a solution. Senior design projects serve as capstone learning experiences that prepare graduates to enter the profession with real-world design experience. Technical courses are complemented by courses in the humanities and social sciences, that engender a values-centered approach to the ethical complexities of the engineering profession. Number of years typically required to complete the bachelor's degree: 4.0.

DEGREE PROGRAMS DESCRIPTION
In the early years the BChE program emphasizes the basic principles whose understanding is required of all chemical engineers. Later courses utilize these principles to develop skill in chemical engineering design. The B.C.E. program prepares men and women to confidently enter the profession as well as to continue into graduate level studies. Emphasis is placed on recognizing societal needs and creative problem solving to meet those needs. The curriculum develops the concept of design coupled with engineering judgment. During the first three years all students in the B.E.E. program take essentially the same core program. This provides a firm foundation for a senior year consisting of a series of electives in specialized areas (1) communications and signal processing, (2) systems and control, (3) computer engineering, (4) electronics, (5) electromagnetics and applied physics, and (6) digital systems, and an in-depth, two-term design project. The first two years of the B.M.E. program are devoted to laying a foundation of mathematics, physical science, and the general engineering sciences. The junior and senior years are devoted primarily to career elective courses which permit a minor concentration in CAD/CAM and simulation, materials and solid mechanics, thermo-fluid engineering, business, or arts and science.

DUAL DEGREE PROGRAMS
Engineering baccalaureate graduates with dual degrees 0
Enrollment requirements: Entrance requirements are the same as those for regular admission from high school.

TRANSFER INFORMATION
Residency Requirements: Specific residence requirements for graduation of all transfer students: The senior year and at least half of all major courses must be taken at Villanova University.
Transfer without Articulation Agreements
Admission to engineering: the end of the first semester of the freshman year.
Engineering graduates transferred from ... 4-yr: 10   2-yr: 5

GRADUATION REQUIREMENTS
Undergraduate engineering students must successfully complete all of the courses required for the engineering degree sought, with a cumulative overall and technical grade point average of at least 2.00. At least half of all the engineering courses and, normally, the senior year of an engineering bachelor's degree program, must be taken at Villanova University. The number of credits for a Bachelor of Engineering degree vary according to the student's major; for a B.Ch.E. 146 semester credits, B.C.E. 139 semester credits, B.E.E. 142 semester credits, and B.M.E. 130 semester credits.

■ STUDENT PROGRAMS

PROFESSIONAL AND HONORARY SOCIETIES
[For key to acronyms, see *Introduction*.]
Professional Societies: AIChE, ASCE, ASME, IEEE, Society of Women Engineers (SWE), The Institute of Transportation Engineers
Honorary Societies: Chi Epsilon, Eta Kappa Nu, Pi Tau Sigma, Tau Beta Pi

SUPPORT PROGRAMS
Student Chapter Organizations: Society of Women Engineers, VUMES, V.U. Minority Engineers and Scientists
Other Engineering Support Programs: University-wide support services: Academic Advancement Program with summer preparation and academic assistance for disadvantaged or underprepared students; study skills and career counseling; an international student advisor; job placement; tutoring, group study sessions, and summer and year-long internship in industry. Participation in solar-powered racing team.

259 University of Virginia

INSTITUTION PROFILE

HEAD OF THE INSTITUTION
John T. Casteen III
Phone: (804)924-3337   Fax: (804)924-3792

GENERAL INFORMATION
[All Students—Fall 1991]
Undergraduate enrollment 11,400
Graduate enrollment 6,200
Total institution enrollment 17,600
Type of institution: Public   **Calendar:** Semesters
Location: Charlottesville, VA   **Population:** 90,000
Setting: Small Town
Types of engineering degrees: Engineering
Other degree-granting colleges: Architecture, Arts & Sciences, Business Administration, Education, Law, Medicine, Nursing

ENGINEERING ADMISSIONS

ADMISSIONS OFFICE CONTACT
William J. Thurneck
University of Virginia
Thornton Hall
McCormick Road
Charlottesville, VA 22903-2442
Phone: (804)924-3155   Fax: (804)982-2734

ENGINEERING COLLEGE ADMISSIONS INFORMATION
Entrance Requirements: SAT, High School courses—Mathematics (4 years), English (4 years), Science (including Chemistry and Physics) (3 years), Social Studies (1 year), Achievement Tests: Mathematics, English Composition, and a third test, either Science, Foreign Language, or History.
Entrance Recommendations: High School courses—Foreign Language (2 years), Science Achievement Test recommended as third test.
Requirements for foreign students: TOEFL (Score: 600); financial statement

DEPARTMENTAL ADMISSIONS INFORMATION
Admission to the engineering department: At the end of the second year.
Additional information: Admission to a specific engineering program may be limited if over-subscription occurs.

FINANCIAL INFORMATION

ESTIMATED EXPENSES (ACADEMIC YEAR)
[Expenses are for the 1992-93 nine-month academic year.]

State Residents	Undergraduate	Graduate
Tuition and fees	$ 3,900	$ 3,900
College room and board	$ 3,850	$ 6,250
Books and supplies	$ 500	$ 500
Other expenses	$ 1,000 [1]	$ 2,260 [1]
Total estimated expenses	**$ 9,250**	**$12,910**

Out-of-State Residents	Undergraduate	Graduate
Tuition and fees	$10,836	$10,836
College room and board	$ 3,850	$ 6,250
Books and supplies	$ 500	$ 500
Other expenses	$ 1,000 [1]	$ 2,260 [1]
Total estimated expenses	**$16,186**	**$19,846**

Notes: (1) Personal Expenses.

FINANCIAL AID OFFICE CONTACT
Wayne M. Sparks, *Director of Financial Aid to Students*
University of Virginia
Miller Hall
P.O. Box 9021
Charlottesville, VA 22906
Phone: (804)924-3725

GENERAL FINANCIAL AID INFORMATION
Forms accepted/required: FAF
Additional financial aid information: Financial aid is normally a combination of gift assistance, loan funds, and/or employment.

ENGINEERING COLLEGE INFORMATION
[For additional personnel, refer to the *Appendix*.]

HEAD OF ENGINEERING
Edgar A. Starke Jr.
Phone: (804)924-3593   Fax: (804)982-2734
EMail: EAS1O@virginia.edu

ENGINEERING COLLEGE ADDRESS
School of Engineering and Applied Science
University of Virginia
Thornton Hall
McCormick Road
Charlottesville, VA 22903-2442
Phone: (804)924-3072   Fax: (804)982-2734

ENROLLMENTS—BY CLASS
[Numbers are baccalaureate enrollments for the fall 1991 term, unless otherwise footnoted.]
1st-year students/Freshmen 432
2nd-year students/Sophomores 395
3rd-year students/Juniors 333
4th-year students/Seniors 437
Total **1,597**

NUMBER OF DEGREES AWARDED—BY PROGRAM
[Numbers are engineering baccalaureate degrees awarded during the 1991-1992 academic year, unless otherwise footnoted. For full details about each engineering program, refer to the *Tables of Degree Programs*.]
Aerospace Engineering 31
Applied Mathematics 13
Applied Mechanics .... –
Biomedical Engineering .... –
Chemical Engineering 35
Civil Engineering 40
Computer Science 26
Electrical Engineering 78
Engineering Physics .... –
Engineering Science(s) 9
Materials Science & Engineering .... –
Mechanical & Aerospace Engineering .... –
Mechanical Engineering 45
Nuclear Engineering 3
Systems Engineering 43
Total **323**

PERCENTAGE OF DEGREES AWARDED—BY CATEGORY
[Percentages are of all engineering baccalaureate degrees awarded during the 1991-1992 academic year, unless otherwise footnoted.]
African Americans 5.6%
Asian/Pacific Island Americans 12.7%
Hispanic Americans 1.2%
Native Americans .... – %
Foreign Citizens 2.8%
All Others 77.7%
Women 24.5%
Persons with disabilities .... – % (A)
Students over 25 years of age 1.9%

Notes: (A) Data not available.

GRADUATE ENROLLMENTS & DEGREES AWARDED
Master's enrollment 427
Master's degrees awarded 193
Doctoral enrollment 273
Doctoral degrees awarded 48

ENGINEERING STUDENT DATA
Applicants to the engineering college
Number of applicants to engineering college 2,719
Percent offered admission 34.5%

Matriculated engineering students
Percentage in top quartile (25%) of High School class 97.0
Average SAT scores: Math—694, Verbal—573, Combined—1267

FULL- & PART-TIME FACULTY—BY DEPARTMENT
[Figures are the head count of full-time faculty and the full-time equivalent (FTE) of part-time faculty for each engineering department or equivalent.]

Department	Full	Part
Chemical Eng	11	0.5
Civil Eng	13	1.3
Electrical Eng	20	0.8
Mechanical, Aerospace & Nuclear Eng	35	3.0
Applied Mathematics	11	–
Biomedical Eng	6	–
Computer Science	19	1.3
Humanities	9	1.3
Materials Science & Eng	18	0.5
Systems Eng	10	–

COLLEGE DESCRIPTION
The School of Engineering and Applied Science is the first school of technical studies founded within a multi-discipline state university. It offers bachelor of science degrees in aerospace, chemical, civil, electrical, mechanical, nuclear and systems engineering as well as applied mathematics, computer science and engineering science. The School has its own Department of Applied Mathematics and Division of Humanities. Each student must write a senior thesis. This is done under the joint direction of a technical faculty advisor and a faculty member of the Humanities Division. Thirty first-year students are selected each year to participate in the Rodman Scholars Program, which consists of a special program of studies and activities. The School strongly encourages each student to obtain engineering/technical experience via summer employment after the third year. The School's Office of Career Planning and Placement is available to provide assistance in obtaining such positions.

DEGREE PROGRAMS DESCRIPTION
All undergraduates are required to prepare a thesis as part of the undergraduate degree requirement. A faculty member from the Humanities Division of the School, along with a faculty advisor in the selected technical area, oversee the preparation and presentation of this undergraduate thesis. The Accelerated Bachelor's/Master's Degree Program is a five year program in which exceptional undergraduates can obtain both the BS and the MS in a period of five years. In this program, students prepare a Master's thesis which also satisfies the undergraduate thesis requirement. In addition to the Master's of Science degree which requires a Master's thesis, a non-thesis Master's program is available which includes additional course work in place of the thesis.

TRANSFER INFORMATION
Residency Requirements: Minimum of two semesters in residence; must be registered in residence during the semester in which degree is awarded.
Transfer without Articulation Agreements
Engineering graduates transferred from   .. 4-yr: 28   2-yr: 12

GRADUATION REQUIREMENTS
Overall grade point average of 2.0 on a 4.0 scale required for all undergraduate degrees, 3.0 for all graduate degrees.

STUDENT PROGRAMS

PROFESSIONAL AND HONORARY SOCIETIES
[For key to acronyms, see *Introduction*.]
Professional Societies: ACM, AIAA, AIChE, ANS, ASCE, ASEM, ASHRAE, ASME, IEEE, Trigon
Honorary Societies: Chi Epsilon, Eta Kappa Nu, Pi Tau Sigma, Sigma Gamma Tau, Tau Beta Pi, Theta Tau, Omega Rho, Omicron Xi

SUPPORT PROGRAMS
Student Chapter Organizations: Society of Women Engineers, National Society of Black Engineers, Society of Hispanic Professional Engineers, Graduate Society of Black Engineers.
Office of Minority Programs
For: Ethnic Minorities   **Available:** Year round
Offered: Prematriculation, Freshman, Sophomore, Junior, Senior, Graduate level
Other Engineering Support Programs: The School or University offers the following support programs: orientation for first-year students; tutorial services; student, residential and faculty advisors; Office of Career Planning and Placement; University Counseling Center; Ombudsman Services; Disability Services; Learning Needs and Evaluation Center; Student's Legal Services; Student Health Center; and several others.

260 Virginia Military Institute

■ INSTITUTION PROFILE
HEAD OF THE INSTITUTION
John W. Knapp
Phone: (703)464-7311 **Fax:** (703)464-7660
GENERAL INFORMATION
[All Students—Fall 1991]
Undergraduate enrollment 1,281
Total institution enrollment 1,281
Type of institution: Public **Calendar:** Semesters
Nearest city: Roanoke, VA **Population:** 100,000
Miles from main campus: 50 **Setting:** Small Town
Types of engineering degrees: Engineering
Other degree-granting colleges: Sciences, Liberal Arts

■ ENGINEERING ADMISSIONS
ADMISSIONS OFFICE CONTACT
Daniel A. Troppoli
Virginia Military Institute
Route 11 North
Lexington, VA 24450-0304
Phone: (703)464-7211 **Fax:** (703)464-7746
ENGINEERING COLLEGE ADMISSIONS INFORMATION
Entrance Requirements: SAT (Score: 1000), ACT (Score: 23), High School courses—English (4 years), Algebra (2 years), Geometry (1 year)
Entrance Recommendations: SAT (Score: 1000), ACT (Score: 23), High School courses—Advanced Mathematics (1 year), Social Studies (3 years), Laboratory Science (3 years), Foreign Language (3 years)
Requirements for foreign students: TOEFL (Score: 500); financial statement
DEPARTMENTAL ADMISSIONS INFORMATION
Admission to the engineering department: At the time of admission to the institution.

■ FINANCIAL INFORMATION
ESTIMATED EXPENSES (ACADEMIC YEAR)
[Expenses are for the 1992-93 nine-month academic year.]

State Residents	Undergraduate	Graduate
Tuition and fees	$ 4,640	$ –
College room and board	$ 3,690	$ –
Books and supplies	$ 500	$ –
Other expenses	$ –	$ –
Total estimated expenses	**$ 8,830**	**$ –**

Out-of-State Residents	Undergraduate	Graduate
Tuition and fees	$10,290	$ –
College room and board	$ 3,690	$ –
Books and supplies	$ 500	$ –
Other expenses	$ –	$ –
Total estimated expenses	**$14,480**	**$ –**

FINANCIAL AID OFFICE CONTACT
Timothy P. Golden, *Director of Financial Aid*
Virginia Military Institute
Route 11, North
Lexington, VA 24450-0304
Phone: (703)464-7208
GENERAL FINANCIAL AID INFORMATION
Forms accepted/required: CSS-FAF, FAF, FAR, FFS, Stafford, IRS, Institutional
Additional financial aid information: Scholarships and Grants (need and non-need based); the Institute Scholars Program; College Work-Study; Loans; Special Funds.

■ ENGINEERING COLLEGE INFORMATION
[For additional personnel, refer to the *Appendix*.]
HEAD OF ENGINEERING
Richard S. Trandel
Phone: (703)464-7308 **Fax:** (703)464-7663
ENGINEERING COLLEGE ADDRESS
Engineering Division
Virginia Military Institute
Route 11 North
Lexington, VA 24450-0304
Phone: (703)464-7308 **Fax:** (703)464-7663
ENROLLMENTS—BY CLASS
[Numbers are baccalaureate enrollments for the fall 1991 term, unless otherwise footnoted.]
1st-year students/Freshmen 131
2nd-year students/Sophomores 108
3rd-year students/Juniors 84
4th-year students/Seniors 110
Total 433
NUMBER OF DEGREES AWARDED—BY PROGRAM
[Numbers are engineering baccalaureate degrees awarded during the 1991-1992 academic year, unless otherwise footnoted. For full details about each engineering program, refer to the *Tables of Degree Programs*.]
Civil & Environmental Engineering 175
Electrical Engineering 84
Mechanical Engineering 174
Total 433
PERCENTAGE OF DEGREES AWARDED—BY CATEGORY
[Percentages are of all engineering baccalaureate degrees awarded during the 1991-1992 academic year, unless otherwise footnoted.]
African Americans .... – %
Asian/Pacific Island Americans .... 3.0 %
Hispanic Americans .... 1.0 %
Native Americans .... – %
Foreign Citizens .... 2.0 %
All Others .... 94.0 %
Women .... – %
Persons with disabilities .... – %
Students over 25 years of age .... 7.0 %
ENGINEERING STUDENT DATA
Applicants to the engineering college
Number of applicants to engineering college 313
Percent offered admission .... 87.0 %
Matriculated engineering students
Percentage in top quartile (25%) of High School class 40.0
Average SAT scores: Math—560, Verbal—485, Combined—1045
Average ACT scores: Composite—22
FULL- & PART-TIME FACULTY—BY DEPARTMENT
[Figures are the head count of full-time faculty and the full-time equivalent (FTE) of part-time faculty for each engineering department or equivalent.]

Department	Full	Part
Civil & Environmental Eng	9	–
Electrical Eng	7	–
Mechanical Eng	8	–

COLLEGE DESCRIPTION
All engineering students begin their studies with science and mathematics to form a strong groundwork for more specialized courses. Statics, dynamics, and solid mechanics further prepare cadets for each year's increasingly advanced courses. The VMI engineering curricula require substantial amounts of lab work and extensive use of computers. In addition to technical classes, the engineering student studies literature, history, psychology, language and humanities courses. Cadets are encouraged to take the Fundamentals of Engineering Examination during their first class (senior) year to prepare them to become professional engineers in the future. Departments sponsor student membership in professional societies (civil engineering, ASCE; electrical engineering, IEEE; mechanical engineering, ASME). Participation in professional activities is emphasized. Average number of years required to actually complete the bachelor's degree: 4.3.
DEGREE PROGRAMS DESCRIPTION
The B.S. degrees in the various engineering programs prepare the graduate for entry-level professional positions, service in the Armed Forces, or for graduate study.
TRANSFER INFORMATION
Residency Requirements: At least two years (four semesters) of residence at VMI are required regardless of the number of course credit approved for transfer.
Transfer without Articulation Agreements
Admission to engineering: At the end of the freshman year
Engineering graduates transferred from . . . 4-yr: 2   2-yr: 3
Requirements: All applicants, must be unmarried males not more than 22 years of age on matriculation. At least two years of residence at VMI are required. Transfer applicants must show at least a 'C' (or 2.0 on a 4.0 scale) cumulative quality point average on all courses attempted. They must be in good standing with respect to their academic and conduct records and eligible to return to the college, which must be accredited.
GRADUATION REQUIREMENTS
An overall GPA of 2.0 is required in all courses. Each engineering program requires 144 semester hours.

■ STUDENT PROGRAMS
PROFESSIONAL AND HONORARY SOCIETIES
[For key to acronyms, see *Introduction*.]
Professional Societies: ASCE, ASME, IEEE, NSPE, SAME
Honorary Societies: Eta Kappa Nu, Tau Beta Pi
SUPPORT PROGRAMS
The Minority Retention Program
For: Ethnic Minorities **Available:** Summer only
Offered: Prematriculation
Other Engineering Support Programs: Summer orientation program for some freshmen, career counseling services, placement services, and tutoring.

261 Virginia Polytechnic Institute and State University

■ INSTITUTION PROFILE

HEAD OF THE INSTITUTION
James D. McComas
Phone: (703)231-6231   Fax: (703)231-4264

GENERAL INFORMATION
[All Students—Fall 1991]
Undergraduate enrollment 19,308
Graduate enrollment 4,604
Total institution enrollment 23,912
Type of institution: Public   Calendar: Semesters
Nearest city: Roanoke   Population: 95,000
Miles from main campus: 45   Setting: Small Town
Types of engineering degrees: Engineering
Other degree-granting colleges: Arts & Sciences, Business Administration, Education, Human Resources, Veterinary Medicine, Agricultural & Life Sciences, Architecture & Urban Studies, Forestry

■ ENGINEERING ADMISSIONS

ADMISSIONS OFFICE CONTACT
David R. Bousquet
Virginia Polytechnic Institute and State University
104 Burruss Hall
Blacksburg, VA 24061-0202
Phone: (703)341-6267   Fax: (703)231-3242

ENGINEERING COLLEGE ADMISSIONS INFORMATION
Entrance Requirements: SAT, High School courses—English (4 years), Math (4 years), Laboratory Science (3 years), Social Science (2 years)
Entrance Recommendations: High School courses—Foreign or classical language (2 years), Computer Science (1 year)
Requirements for foreign students: TOEFL (Score: 550)

DEPARTMENTAL ADMISSIONS INFORMATION
Admission to the engineering department: At the end of the first year.
Additional information: Students transfer into a degree-granting department after completing the required first-year courses and earn a grade-point-average of 1.9.

■ FINANCIAL INFORMATION

ESTIMATED EXPENSES (ACADEMIC YEAR)
[Expenses are for the 1992-93 nine-month academic year.]

State Residents	Undergraduate	Graduate
Tuition and fees	$ 3,538	$ 4,084
College room and board	$ 1,194	$ 1,456
Books and supplies	$ 500	$ 500
Other expenses	$ 2,000 [1]	$ —
Total estimated expenses	**$ 7,232**	**$ 6,040**

Out-of-State Residents	Undergraduate	Graduate
Tuition and fees	$ 8,986	$ 5,794
College room and board	$ 1,194	$ 1,456
Books and supplies	$ 500	$ 500
Other expenses	$ 2,000 [2]	$ —
Total estimated expenses	**$12,680**	**$ 7,750**

Notes: (1) All freshmen and transfer students are required to purchase a personal computer. (2) All freshmen and transfer students are required to purchase a personal computer.

FINANCIAL AID OFFICE CONTACT
Anne H. Clarke, *Director*
Virginia Polytechnic Institute and State University
222 Burruss Hall
Blacksburg, VA 24060-0222
Phone: (703)231-5179   Fax: (703)231-9139

GENERAL FINANCIAL AID INFORMATION
Forms accepted/required: FAF, IRS, Institutional
Additional financial aid information: Nine major state and federal programs; merit-based scholarships; work-study program.

■ ENGINEERING COLLEGE INFORMATION

[For additional personnel, refer to the *Appendix*.]

HEAD OF ENGINEERING
Wayne Clough
Phone: (703)231-6641   Fax: (703)231-7248
EMail: GWC @ VTVM1.CC.VT.EDU

ENGINEERING COLLEGE ADDRESS
College of Engineering
Virginia Polytechnic Institute and State University
333 Norris Hall
Blacksburg, VA 24061-0217
Phone: (703)231-6641   Fax: (703)231-7248

ENROLLMENTS—BY CLASS
[Numbers are baccalaureate enrollments for the fall 1991 term, unless otherwise footnoted.]
1st-year students/Freshmen 1,259
2nd-year students/Sophomores 1,031
3rd-year students/Juniors 1,057
4th-year students/Seniors 1,165
Total **4,512**

NUMBER OF DEGREES AWARDED—BY PROGRAM
[Numbers are engineering baccalaureate degrees awarded during the 1991-1992 academic year, unless otherwise footnoted. For full details about each engineering program, refer to the *Tables of Degree Programs*.]
Aerospace & Ocean Engineering 86
Aerospace Engineering 86
Agricultural Engineering 4
Chemical Engineering 86
Civil Engineering 177
Computer Engineering 40
Electrical Engineering 253
Engineering Science & Mechanics 29
Industrial & Systems Engineering 103
Materials Science & Engineering 25
Mechanical Engineering 212
Mining Engineering 10
Total **1,111**

PERCENTAGE OF DEGREES AWARDED—BY CATEGORY
[Percentages are of all engineering baccalaureate degrees awarded during the 1991-1992 academic year, unless otherwise footnoted.]
African Americans 2.0%
Asian/Pacific Island Americans 8.0%
Hispanic Americans 1.0%
Native Americans 1.0%
Foreign Citizens 3.0%
All Others 85.0%
Women 19.0%
Persons with disabilities 1.0%
Students over 25 years of age .... — % (A)
Notes: (A) Data not available.

GRADUATE ENROLLMENTS & DEGREES AWARDED
Master's enrollment 1,632
Master's degrees awarded 399
Doctoral enrollment 478
Doctoral degrees awarded 121

ENGINEERING STUDENT DATA

Applicants to the engineering college
Number of applicants to engineering college 4,903
Percent offered admission 56.0%

Matriculated engineering students
Percentage in top quartile (25%) of High School class 92.0
Average SAT scores: Math—660, Verbal—544, Combined—1204

FULL- & PART-TIME FACULTY—BY DEPARTMENT
[Figures are the head count of full-time faculty and the full-time equivalent (FTE) of part-time faculty for each engineering department or equivalent.]

Department	Full	Part
Chemical Eng	11	—
Civil Eng	38	—
Electrical Eng/Computer Eng	52	—
Industrial & Systems Eng	29	—
Mechanical Eng	39	—
Aerospace & Ocean Eng	17	—
Agricultural Eng	20	—
Eng Science & Mechanics	34	—
Materials Science & Eng	16	—
Mining Eng	8	—

COLLEGE DESCRIPTION
The College offers a four-year program in engineering (five years for co-op students) plus Masters and Ph.D programs in 12 majors. All undergraduate students begin study with one year of engineering fundamentals before transferring to a selected program. The College is a leader in the integration of personal computers into the curriculum, and all students are required to own a PC and to use it in many courses. Students have access to outstanding computing and experimental facilities. Cooperation with industry and government makes possible undergraduate student participation in funded research projects and exciting design projects, such as design and construction of the 80-foot long National Aerospace Plane mock-up for the 1989 Paris Air Show and winning entries in SAE racing competitions and the GM Solar Powered Car competition. A majority of students complete the B. S. degree in 4 years (5 for co-op students).

DEGREE PROGRAMS DESCRIPTION
Many engineering students enroll in dual and second degree programs with second degrees in other engineering fields or in areas in the Humanities, Sciences, Math or Social Sciences. A second degree requires a minimum of thirty additional credits of study. Students within one semester of earning the B.S. degree may, upon application and acceptance to graduate school, enroll for graduate credit, taking graduate courses concurrently with senior level B.S. coursework.

TRANSFER INFORMATION
Residency Requirements: The senior year, with a minimum of 30 hours, must be completed in residence, or 30 of the last 40 hours must be completed in residence. Only approved elective courses taken in absentia are transferred to complete requirements.

Transfer without Articulation Agreements
Admission to engineering: At any time
Engineering graduates transferred from .. 4-yr: 21   2-yr: 53
Requirements: College level calculus and chemistry required. Students who have not completed the equivalent of the Virginia Tech freshman year, will be admitted in General Engineering rather than a degree-granting program. Transfer students must own a personal computer meeting minimum standards.

GRADUATION REQUIREMENTS
Overall and in-major courses must average 2.0.

■ STUDENT PROGRAMS

PROFESSIONAL AND HONORARY SOCIETIES
[For key to acronyms, see *Introduction*.]
Professional Societies: AIAA, AIChE, ASAE, ASCE, ASME, IEEE, IIE, SAE, SNAME
Honorary Societies: Alpha Epsilon, Alpha Pi Mu, Chi Epsilon, Eta Kappa Nu, Omega Chi Epsilon, Pi Tau Sigma, Sigma Gamma Tau, Tau Beta Pi, Alpha Sigma Mu

SUPPORT PROGRAMS
Student Chapter Organizations: Society of Women Engineers, National Society of Black Engineers

Minority Center
For: Ethnic Minorities   **Available:** Academic year
Offered: Freshman, Sophomore, Junior, Senior

BEST
For: Ethnic Minorities   **Available:** Academic year
Offered: Freshman

VA TECH Academic Success Program
For: Ethnic Minorities   **Available:** Academic year
Offered: Freshman, Sophomore

Summer Transition Program
For: Ethnic Minorities   **Available:** Summer only
Offered: Prematriculation

Other Engineering Support Programs: Summer and fall orientation program for all freshmen and transfer students; career counseling services; International Student Office; Veterans Affairs; services for blind and handicapped; placement services; study center (tutoring mostly freshmen).

262 University of Washington

INSTITUTION PROFILE

HEAD OF THE INSTITUTION
William P. Gerberding
Phone: (206)543-5010   Fax: (206)543-3951

GENERAL INFORMATION
[All Students—Fall 1991]
Undergraduate enrollment 25,092
Graduate enrollment 9,177
Total institution enrollment **34,269**
Type of institution: Public   Calendar: Quarters
Location: Seattle   Population: 500,000
Setting: Urban
Types of engineering degrees: Engineering
Other degree-granting colleges: Architecture, Arts & Sciences, Business Administration, Dentistry, Education, Law, Medicine, Nursing, Pharmacy, Technology & Applied Sciences, Public Health and Community Medicine, Social Work, Public Affairs, Forest Resources, Ocean and Fishery Sciences

ENGINEERING ADMISSIONS

ADMISSIONS OFFICE CONTACT
William J. Heideger
University of Washington
Engineering Advising Center, FH-18
Seattle, WA 98195
Phone: (206)543-1770   Fax: (206)685-0666

ENGINEERING COLLEGE ADMISSIONS INFORMATION
Admission to the engineering college: Upon completion of departmental prerequisite requirements (normally about the end of 2nd year).
Entrance Requirements: College of Engineering does not have standard test requirement. University requires verbal & quantitative composite score from SAT or ACT. No minimum score required.
Entrance Recommendations: High School courses—Mathematics (4 years), Lab Science (2 years), College of Engineering does not have standard test recommendation. University requires verbal & quantitative composite score from SAT or ACT. No minimum score required.
Requirements for out-of-state residents: Priority must be given to in-state residents.
Requirements for foreign students: TOEFL (Score: 580); financial statement; College of Engineering does not have special requirements.

DEPARTMENTAL ADMISSIONS INFORMATION
Admission to the engineering department: Upon completion of departmental prerequisite requirements.

FINANCIAL INFORMATION

ESTIMATED EXPENSES (ACADEMIC YEAR)
[Expenses are for the 1992-93 nine-month academic year.]

State Residents	Undergraduate	Graduate
Tuition and fees	$ 2,274	$ 3,537
College room and board	$ 5,274	$ 5,274
Books and supplies	$ 633	$ 633
Other expenses	$ 2,275 [1]	$ 2,275 [1]
Total estimated expenses	**$10,456**	**$11,719**

Out-of-State Residents	Undergraduate	Graduate
Tuition and fees	$ 6,345	$ 8,850
College room and board	$ 5,274	$ 5,274
Books and supplies	$ 633	$ 633
Other expenses	$ 2,275 [1]	$ 2,275 [1]
Total estimated expenses	**$14,527**	**$17,032**

Notes: (1) Personal expenses and transportation

FINANCIAL AID OFFICE CONTACT
Eric Godfrey, *Assistant Vice President and Director*
University of Washington
105 Schmitz, PE-20
Seattle, WA 98195
Phone: (206)543-6101

GENERAL FINANCIAL AID INFORMATION
Forms accepted/required: FAF
Additional financial aid information: Grants, departmental scholarships, private scholarships, work study, loans.

ENGINEERING COLLEGE INFORMATION
[For additional personnel, refer to the *Appendix*.]

HEAD OF ENGINEERING
J. R. Bowen
Phone: (206)543-0340   Fax: (206)685-0666
EMail: bowen@engr.washington.edu

ENGINEERING COLLEGE ADDRESS
College of Engineering
University of Washington
Office of the Dean
371 Loew Hall, Mail Stop FH-10
Seattle, WA 98195
Phone: (206)543-0340   Fax: (206)685-0666

ENROLLMENTS—BY CLASS
[Numbers are baccalaureate enrollments for the fall 1991 term, unless otherwise footnoted.]
1st-year students/Freshmen .... 2 [1]
2nd-year students/Sophomores .... 120 [1]
3rd-year students/Juniors 563
4th-year students/Seniors 947
Total **1,632**

Notes: (1) Normally, Freshmen do not register in the College of Engineering. Generally, enter College at Junior standing.

NUMBER OF DEGREES AWARDED—BY PROGRAM
[Numbers are engineering baccalaureate degrees awarded during the 1991-1992 academic year, unless otherwise footnoted. For full details about each engineering program, refer to the *Tables of Degree Programs*.]
Aeronautical & Astronautical Engineering 58
Ceramic Engineering 13
Chemical Engineering 52
Civil Engineering 117
Computer Science & Engineering 17
Electrical Engineering 133
Industrial Engineering 28
Mechanical Engineering 167
Metallurgical Engineering 17
Technical Communication 1
Total **603**

PERCENTAGE OF DEGREES AWARDED—BY CATEGORY
[Percentages are of all engineering baccalaureate degrees awarded during the 1991-1992 academic year, unless otherwise footnoted.]
African Americans .... 2.0 %
Asian/Pacific Island Americans .... 23.0 %
Hispanic Americans .... 2.0 %
Native Americans .... 1.0 %
Foreign Citizens .... 3.0 %
All Others .... 69.0 %
Women .... 19.0 %
Persons with disabilities .... – % (A)
Students over 25 years of age .... – % (A)

Notes: (A) Data not available.

GRADUATE ENROLLMENTS & DEGREES AWARDED
Master's enrollment 890
Master's degrees awarded 307
Doctoral enrollment 442
Doctoral degrees awarded 61

FULL- & PART-TIME FACULTY—BY DEPARTMENT
[Figures are the head count of full-time faculty and the full-time equivalent (FTE) of part-time faculty for each engineering department or equivalent.]

Department	Full	Part
Chemical Eng	14	–
Civil Eng	37	–
Electrical Eng	44	1.7
Industrial Eng	7	–
Mechanical Eng	31	1.2
Aeronautics & Astronautics	18	2.0
Materials Science & Eng	14	–
Computer Science & Eng	26	2.3
Technical Communication	11	–

COLLEGE DESCRIPTION
The undergraduate engineering program is a four-year program which emphasizes mathematics, science and engineering fundamentals in the first two years, followed by specialized training in one of the ten departmental degree programs. There are opportunities to develop an area of specialization within individual departments, as well as a B.S.E. interdisciplinary engineering studies program for students whose interests are not adequately served by one of the existing degrees. Computer applications to engineering problems are utilized in all programs and a distribution of humanities and social science courses is included through the four years. A Co-op placement is an option for all students and each student participates in a senior design project.

DEGREE PROGRAMS DESCRIPTION
The B.S. degrees in the various engineering programs prepare the graduate for the professional practice of engineering or for additional study at the graduate level. The B.S.E. in the bioengineering program is intended primarily to prepare students for medical school. Graduate study in engineering may lead to either the M.S. or Ph.D. in one of the engineering departments or the M.S.E., M. Eng., or M.S. in the College without designation of a specific major.

TRANSFER INFORMATION
Residency Requirements: Transfer students from 2 yr schools must complete a minimum of 96 credits at the University; transfer students from 4 yr institutions must complete 45 credits at the University. All transfer students must complete their final year in residence.
Transfer without Articulation Agreements
Admission to engineering: Upon completion of departmental prerequisite requirements.

GRADUATION REQUIREMENTS
192 credits minimum is required within certain areas of study and some specific courses within certain areas.

STUDENT PROGRAMS

PROFESSIONAL AND HONORARY SOCIETIES
[For key to acronyms, see *Introduction*.]
Professional Societies: ACerS, AIAA, AIChE, AIME, ASCE, ASME, BMES, IEEE, IIE, STC, Institute of Transportation Engineers (ITE), Society of Manufacturing Engineers (SME)
Honorary Societies: Chi Epsilon, Keramos, Pi Tau Sigma, Tau Beta Pi

SUPPORT PROGRAMS
Student Chapter Organizations: Society of Women Engineers, National Society of Black Engineers, Society of Hispanic Professional Engineers, American Indian Science and Engineering Society

Minority Science and Engineering Program
For: Ethnic Minorities   Available: Academic year
Offered: Prematriculation, Freshman, Sophomore, Junior, Senior

Minority Internship Program
For: Ethnic Minorities   Available: Academic year
Offered: Freshman, Sophomore, Junior, Senior

Minority Graduate Prog. for Engineering & Science
For: Ethnic Minorities   Available: Year round
Offered: Graduate level

Women in Engineering
For: Women & Ethnic Minorities   Available: Year round
Offered: Prematriculation, Freshman, Sophomore, Junior, Senior, Graduate level

Other Engineering Support Programs: Engineering Advising Center provides academic counseling, freshman / transfer student orientation, dept. prerequisite guidance, and a career planning course. Career Services arranges summer internships and Engrg. Cooperative Educ. offers 6-month industrial placements. MESA (Math, Engrg. Sci Achievement) works with secondary school students to increase the number of minorities in these areas.

263 Washington University

INSTITUTION PROFILE

HEAD OF THE INSTITUTION
William H. Danforth
Phone: (314)935-5100

GENERAL INFORMATION
[All Students—Fall 1991]
Undergraduate enrollment 4,896
Graduate enrollment 4,533
Total institution enrollment **9,429**
Type of institution: Private **Calendar:** Semesters
Nearest city: St. Louis **Population:** 1,000,000
Miles from main campus: 1 **Setting:** Suburban
Types of engineering degrees: Engineering & Technology
Other degree-granting colleges: Architecture, Arts & Sciences, Business Administration, Fine & Performing Arts, Law, Medicine, Social Work

ENGINEERING ADMISSIONS

ADMISSIONS OFFICE CONTACT
William L. Marsden
Washington University
Campus Box 1164
Engineering Student Services
St. Louis, MO 63130
Phone: (314)935-6130 **Fax:** (314)935-4434

ENGINEERING COLLEGE ADMISSIONS INFORMATION
Entrance Requirements: High School courses—English (4 years), Mathematics (4 years), Chemistry (1 year), Physics (1 year), SAT and/or ACT
Entrance Recommendations: High School courses—Calculus (1 year), Foreign Languages (2 years), History (3 years), Social Sciences (3 years), College Board Achievement Tests TOEFL for all students for whom English is a second language
Requirements for foreign students: University's International Student Application, SAT, TOEFL, and verification of the existence of funds in United States dollars for tuition and living expenses.

DEPARTMENTAL ADMISSIONS INFORMATION
Admission to the engineering department: Engineering students admitted as freshmen select a specific engineering program any time. Transfer students are admitted to an engineering program

FINANCIAL INFORMATION

ESTIMATED EXPENSES (ACADEMIC YEAR)
[Expenses are for the 1992-93 nine-month academic year.]

All Students	Undergraduate	Graduate
Tuition and fees	$ 8,375 [3]	$ 8,375 [1]
College room and board	$ 5,394	$ 5,394
Books and supplies	$ 728	$ 728
Other expenses	$ 1,578 [2]	$ 1,578 [2]
Total estimated expenses	**$16,075**	**$16,075**

Notes: (1) $700 per unit. Estimate is for 12 units each semester. (2) Insurance, recreation, laundry, clothing, incidentals (3) This figure is true for any student caring between 12 and 21 credits. A fee of $700 per credit is charged any student who registers for over 21 credits.

FINANCIAL AID OFFICE CONTACT
Dennis J. Martin, *Assistant Provost and Director of Financial Aid*
Washington University
Campus Box 1041
One Brookings Drive
St. Louis, MO 63130
Phone: (314)935-5900

GENERAL FINANCIAL AID INFORMATION
Forms accepted/required: FAF, FAT, Stafford, IRS, SAR, Supplemental, Institutional, Divorced - Separated Parent's Statement, if applicable
Additional financial aid information: Need-based scholarships and grants; merit-based scholarships; long-term subsidized loans; Cost Stabilization Plan; part-time jobs on campus; College Work-Study Program

ENGINEERING COLLEGE INFORMATION

[For additional personnel, refer to the *Appendix*.]

HEAD OF ENGINEERING
Christopher I. Byrnes
Phone: (314)935-6166 **Fax:** (314)935-4434

ENGINEERING COLLEGE ADDRESS
School of Engineering and Applied Science
Washington University
1 Brookings Drive
Campus Box 1100
St. Louis, MO 63130
Phone: (314)935-6100 **Fax:** (314)935-4434

ENROLLMENTS—BY CLASS
[Numbers are baccalaureate enrollments for the fall 1991 term, unless otherwise footnoted.]
1st-year students/Freshmen 212
2nd-year students/Sophomores 216
3rd-year students/Juniors 234
4th-year students/Seniors 334
Total **996**

NUMBER OF DEGREES AWARDED—BY PROGRAM
[Numbers are engineering baccalaureate degrees awarded during the 1991-1992 academic year, unless otherwise footnoted. For full details about each engineering program, refer to the *Tables of Degree Programs*.]

Applied Science 4
Biological & Engineering Science 1
Biological & Engineering Sciences 1
Chemical Engineering 24
Civil Engineering 21
Computer Science 40
Construction Engineering .... –
Construction Management .... –
Electrical Engineering 69
Engineering & Policy .... –
Engineering & Policy .... –
Management of Technology .... –
Materials Science & Engineering .... –
Mechanical Engineering 61
Structural Design .... –
Structural Engineering .... –
Systems Science & Mathematics 10
Technology & Human Affairs .... –
Total **231**

PERCENTAGE OF DEGREES AWARDED—BY CATEGORY
[Percentages are of all engineering baccalaureate degrees awarded during the 1991-1992 academic year, unless otherwise footnoted.]
African Americans 4.4%
Asian/Pacific Island Americans 8.5%
Hispanic Americans 1.2%
Native Americans .... – %
Foreign Citizens 12.5%
All Others 73.4%
Women 19.3%
Persons with disabilities .... – %
Students over 25 years of age .... – %

GRADUATE ENROLLMENTS & DEGREES AWARDED
Master's enrollment 423
Master's degrees awarded 168
Doctoral enrollment 192
Doctoral degrees awarded 36

ENGINEERING STUDENT DATA
Applicants to the engineering college
Number of applicants to engineering college 1,538
Percent offered admission 73.1%

Matriculated engineering students
Percentage in top quartile (25%) of High School class 97.8
Average SAT scores: Math—690, Verbal—561, Combined—1251
Average ACT scores: Math—30.8, Composite—29.5

FULL- & PART-TIME FACULTY—BY DEPARTMENT

[Figures are the head count of full-time faculty and the full-time equivalent (FTE) of part-time faculty for each engineering department or equivalent.]

Department	Full	Part
Chemical Eng	10	–
Civil Eng	6	–
Electrical Eng	22	–
Mechanical Eng	14	0.5
Computer Science	14	–
Eng & Policy	5	1.0
Systems Science & Mathematics	8	0.5

COLLEGE DESCRIPTION
The School offers 4-year undergraduate programs in engineering; 5-year programs combining engineering bachelors and masters degrees and engineering bachelors and the M.B.A.; and a 7-year program combining selected engineering bachelors and the M. of Architecture. Double engineering major programs are available - B.S.E.E./B.S. Computer Science is most common. The School offers minors in computing, engineering and public policy, and structures. Minors are also available in architecture, business and most areas in arts and sciences. Co-op is available - normally extends undergraduate program to 5 years. The school has an exchange program (Junior year) with the Institut National des Sciences Applique'es (INSA) in Lyon, France.

DEGREE PROGRAMS DESCRIPTION
B.S.: 1. Opportunity for undergraduates to participate in on-going faculty research. 2. Student receives individual attention throughout their undergraduate program and have the opportunity for close interaction with faculty. 3. Almost all undergraduate engineering courses are taught by professors. 4. Engineering and Public Policy degree is one of only two such programs accredited by ABET. 5. Double degree programs available in Computer Engineering, Environmental Resources and Process Control.

DUAL DEGREE PROGRAMS
Engineering baccalaureate graduates with dual degrees 42
Enrollment requirements: Completion of specific courses; overall 3.0/4.0 or better.

TRANSFER INFORMATION
Residency Requirements: Completion of at least 30 units of upper level (junior or higher) engineering courses from our engineering school.

Transfer via Articulation Agreements
Admission to engineering: Dual Degree: at end of their 3rd year.
Requirements: Minimum of 3.0/4.0 cum GPA.

Transfer without Articulation Agreements
Admission to engineering: At any time
Requirements: Minimum of a 3.0/4.0 cum GPA.

GRADUATION REQUIREMENTS
No more than 1/4 of all courses taken on campus may have grades below 'C'. No more than 1/4 of all courses in the major may have grades below 'C'.

STUDENT PROGRAMS

PROFESSIONAL AND HONORARY SOCIETIES
[For key to acronyms, see *Introduction*.]
Professional Societies: ACM, AIAA, AIChE, ASCE, ASHRAE, ASME, IEEE, SAE, SIAM, SME
Honorary Societies: Eta Kappa Nu, Pi Tau Sigma, Tau Beta Pi, Civil Engineering Honor Society

SUPPORT PROGRAMS
Student Chapter Organizations: Society of Women Engineers, National Society of Black Engineers
Other Engineering Support Programs: Extensive orientation program for all university freshman and transfer students. Peer Advising program for all new engineering freshman. Freshman Seminar course for all engineering freshman. Free tutoring for all engineering students in all courses.

264 Washington State University

INSTITUTION PROFILE

HEAD OF THE INSTITUTION
Samuel H. Smith
Phone: (509)335-6666 **Fax:** (509)335-0103

GENERAL INFORMATION
[All Students—Fall 1991]
Undergraduate enrollment 14,893
Graduate enrollment 2,953
Total institution enrollment **17,846**

Type of institution: Public **Calendar:** Semesters
Nearest city: Spokane, Washington **Population:** 170,200
Miles from main campus: 80 **Setting:** Small Town
Types of engineering degrees: Engineering
Other degree-granting colleges: Arts & Sciences, Education, Agriculture and Home Economics, Veterinary Medicine, Business and Economics, Pharmacy, Nursing

ENGINEERING COLLEGE ADMISSIONS INFORMATION
Entrance Requirements: SAT, ACT, High School courses—Foreign Language (2 years) English (4 years), Mathematics (Alg., Geom., Alg. II/Trig) (3 years), Science (including one year of lab science) (2 years), Social Science (3 years), Admissions is based on a combination of test scores (ACT or SAT) and grade point average.
Entrance Recommendations: High School courses—Science (3 years), Laboratory Science (2 years), Mathematics (all the school has to offer), Fine, Visual, & Performing Arts (required) (1 year), Math Placement and Writing Placement tests are administered prior to registration for the first term.
Requirements for foreign students: TOEFL (Score: 520); financial statement

DEPARTMENTAL ADMISSIONS INFORMATION
Admission to the engineering department: At the end of the second year.
Additional information: Students are admitted after completing required core courses with a satisfactory grade point average.

FINANCIAL INFORMATION

ESTIMATED EXPENSES (ACADEMIC YEAR)
[Expenses are for the 1992-93 nine-month academic year.]

State Residents	Undergraduate	Graduate
Tuition and fees	$ 2,254	$ 3,538
College room and board	$ 3,860	$ 4,310
Books and supplies	$ 733	$ 600
Other expenses	$ –	$ 600
Total estimated expenses	**$ 6,847**	**$ 9,048**

Out-of-State Residents	Undergraduate	Graduate
Tuition and fees	$ 6,345	$ 8,850
College room and board	$ 3,860	$ 4,310
Books and supplies	$ 733	$ 600
Other expenses	$ –	$ 600
Total estimated expenses	**$10,938**	**$14,360**

GENERAL FINANCIAL AID INFORMATION
Forms accepted/required: FAF
Additional financial aid information: Merit and need-based scholarships; minority scholarships; short term loans; Perkins Loan; Stafford Loan; part-time jobs on campus; work-study; Pell Grant; Washington State Need Grant

ENGINEERING COLLEGE INFORMATION
[For additional personnel, refer to the *Appendix*.]

HEAD OF ENGINEERING
Reid C. Miller
Phone: (509)335-5593 **Fax:** (509)335-9608

ENGINEERING COLLEGE ADDRESS
College of Engineering and Architecture
Washington State University
Pullman, WA 99164-2714
Phone: (509)335-1584 **Fax:** (509)335-9608

ENROLLMENTS—BY CLASS
[Numbers are baccalaureate enrollments for the fall 1991 term, unless otherwise footnoted.]
1st-year students/Freshmen 312
2nd-year students/Sophomores 439
3rd-year students/Juniors 360
4th-year students/Seniors 392
Total **1,503**

NUMBER OF DEGREES AWARDED—BY PROGRAM
[Numbers are engineering baccalaureate degrees awarded during the 1991-1992 academic year, unless otherwise footnoted. For full details about each engineering program, refer to the *Tables of Degree Programs*.]
Agricultural Engineering 6
Architectural Studies 42
Architecture 44
Chemical Engineering 10
Civil & Environmental Engineering 53
Civil Engineering 53
Computer Science 39
Construction Management 30
Electrical Engineering .... –
Electrical Engineering & Computer Science 94
Engineering Management .... –(B)
Environmental Engineering .... –(B)
Materials Science & Engineering 13
Mechanical Engineering 121
Total **505**

Notes: (B) Data not applicable.

PERCENTAGE OF DEGREES AWARDED—BY CATEGORY
[Percentages are of all engineering baccalaureate degrees awarded during the 1991-1992 academic year, unless otherwise footnoted.]
African Americans .... – %(1)
Asian/Pacific Island Americans .... – %
Hispanic Americans .... – %(3)
Native Americans .... – %(2)
Foreign Citizens .... – %(4)
All Others .... 100 %
Women .... 12.0 %
Persons with disabilities .... – %(A)
Students over 25 years of age .... – %(A)

Notes: (A) Data not available. (1) Actual % = .006 (2) Actual % = .003 (3) Actual % = .02 (4) Actual % = .04

GRADUATE ENROLLMENTS & DEGREES AWARDED
Master's enrollment 302
Master's degrees awarded 91
Doctoral enrollment 107
Doctoral degrees awarded 20

ENGINEERING STUDENT DATA
Applicants to the engineering college
Number of applicants to engineering college 538
Percent offered admission 79.0%

FULL- & PART-TIME FACULTY—BY DEPARTMENT
[Figures are the head count of full-time faculty and the full-time equivalent (FTE) of part-time faculty for each engineering department or equivalent.]

Department	Full	Part
Chemical Eng	9	10.1
Civil & Environmental Eng	21	25.7
Electrical Eng & Computer Science	36	39.2
Mechanical & Materials Eng	24	26.8
Biological Systems Eng	10	10.4
Architecture	19	20.6

COLLEGE DESCRIPTION

WSU in Pullman is a residential campus that includes the College of Engineering and Architecture with 1970 undergraduate students and 119 faculty. The college is small enough to offer personal attention, yet large enough to offer a wide range of academic options. In 1990, the college launched an NSF Research Experience for Undergraduates (REU) Program to provide a 10-week quality summer research opportunity for students. WSU has an exchange agreement for their engineering students for a one or two-semester enrollment at either The Technical University of Denmark or The Engineering Academy of Denmark. Through the Professional Experience Program in WSU's Office of Career Services, students gain work experience on summer or semester internships.

DEGREE PROGRAMS DESCRIPTION

WSU undergraduate engineering programs emphasize hands-on experience and integration of social issues into engineering problem solving. Students select from topical areas of coherence in general education to develop depth and integration of humanities, social sciences, and sciences with their engineering coursework. WSU engineering graduates typically perform very well in engineering fundamentals examinations and are sought by employers. Interdisciplinary programs of study in engineering science leading to the degrees of Master of Science in Engineering or Doctor of Philosophy provide flexibility in designing programs of study tailored to the specialized needs of students with non-traditional backgrounds. There is also an interdisciplinary Ph.D. program in Materials Science which accepts students from the sciences and engineering who wish to pursue graduate research in materials engineering.

TRANSFER INFORMATION

Residency Requirements: Community college transfer students must complete 60 semester credits at WSU. Four-year college transfer students must complete 30 semester credits at WSU.

Transfer without Articulation Agreements
Admission to engineering: Students are considered for admission to engineering programs when they have completed a required set of core courses. A fixed number is admitted based on GPA and other factors.

STUDENT PROGRAMS

PROFESSIONAL AND HONORARY SOCIETIES
[For key to acronyms, see *Introduction*.]
Professional Societies: ACM, AGCA, AIA, AIChE, ASAE, ASCE, ASME, IEEE, American Society of Metals (ASM/AIME), Air and Waste Management Association (AWMA)
Honorary Societies: Tau Beta Pi

SUPPORT PROGRAMS
Student Chapter Organizations: Society of Women Engineers, National Society of Black Engineers, Society of Hispanic Professional Engineers, American Indian Science and Engineering Society

Women Engineering Program
For: Women **Available:** Year round
Offered: Freshman, Sophomore, Junior, Senior, Graduate level

Minority Engineering Program
For: Ethnic Minorities **Available:** Year round
Offered: Freshman, Sophomore, Junior, Senior, Graduate level

MESA--Mathematics Engineering Science Achievement
For: Ethnic Minorities **Available:** Academic year
Offered: Prematriculation

Other Engineering Support Programs: Summer Orientation Program; Fall Engineering Orientation; Student Advising and Learning Center; Writing Lab; Science Learning and Instructional Center; Career Services and Placement Office; Professional Experience Program; Counseling and Testing; Joint Program for Career Development; University Honors; Office of International Education; Computing Service Center; WEP and MEP Summer Bridge Workshop

265 Wayne State University

■ INSTITUTION PROFILE

HEAD OF THE INSTITUTION
David W. Adamany
Phone: (313)577-2230 Fax: (313)577-3200

GENERAL INFORMATION
[All Students—Fall 1991]
Undergraduate enrollment 20,811
Graduate enrollment 13,193
Total institution enrollment **34,004**

Type of institution: Public **Calendar:** Semesters
Location: Detroit **Population:** 1,000,000
Setting: Urban
Types of engineering degrees: Engineering & Technology
Other degree-granting colleges: Allied Health Sciences, Arts & Sciences, Business Administration, Education, Fine & Performing Arts, Law, Medicine, Nursing, Pharmacy, Technology & Applied Sciences, Social Work, Urban, Labor and Metropolitan Affairs, Lifelong Learning

■ ENGINEERING ADMISSIONS

ADMISSIONS OFFICE CONTACT
H. Allan Knappenberger
Wayne State University
College of Engineering
Detroit, MI 48202
Phone: (313)577-3040 Fax: (313)577-5300

ENGINEERING COLLEGE ADMISSIONS INFORMATION
Entrance Requirements: High School courses—English (4 years), Math (4 years), Social Sciences (2 years), Physics and Chemistry (2 years)
Entrance Recommendations: ACT (Score: 21), High School courses—Biology (1 year), Additional Social Science, History or Language (3 years), Fine Arts (2 years)
Requirements for foreign students: TOEFL (Score: 550); financial statement; Test of written English. A minimum score of 5.5 is required on the WE.

DEPARTMENTAL ADMISSIONS INFORMATION
Admission to the engineering department: At the time of admission to the institution.

■ FINANCIAL INFORMATION

ESTIMATED EXPENSES (ACADEMIC YEAR)
[Expenses are for the 1992-93 nine-month academic year.]

State Residents	Undergraduate	Graduate
Tuition and fees	$ 2,934 [3]	$ 2,948 [1]
College room and board	$ 5,306 [2]	$ 5,306 [2]
Books and supplies	$ 469	$ 469
Other expenses	$ 2,632 [4]	$ 2,632 [4]
Total estimated expenses	**$11,341**	**$11,355**

Out-of-State Residents	Undergraduate	Graduate
Tuition and fees	$ 6,474 [3]	$ 6,275 [1]
College room and board	$ 5,306 [2]	$ 5,306 [2]
Books and supplies	$ 469	$ 469
Other expenses	$ 2,632 [4]	$ 2,632 [4]
Total estimated expenses	**$14,881**	**$14,682**

Notes: (1) 24 credit hours (2) Estimate. There are no on campus facilities. (3) Average based on 32 hours of credit. (4) Transportation and other miscellaneous expenses.

FINANCIAL AID OFFICE CONTACT
Judith L. Florian, *Director*
Wayne State University
3 West HNJ-SSC
Detroit, MI 48202
Phone: (313)577-3378 Fax: (313)577-6648

GENERAL FINANCIAL AID INFORMATION
Forms accepted/required: CSS-FAF, FAF, FAT, FFS, SAR, The Wayne State Application for Financial Aid
Additional financial aid information: A wide range of private scholarships, grants, loans and employment are available to students in the College.

■ ENGINEERING COLLEGE INFORMATION

[For additional personnel, refer to the *Appendix*.]
HEAD OF ENGINEERING
Fred W. Beaufait
Phone: (313)577-3775 Fax: (313)577-5300
EMail: internet: beaufait@dbo.eng.wayne.edu

ENGINEERING COLLEGE ADDRESS
College of Engineering
Wayne State University
5050 Anthony Wayne Drive
Detroit, MI 48202
Phone: (313)577-3780 Fax: (313)577-5300

ENROLLMENTS—BY CLASS
[Numbers are baccalaureate enrollments for the fall 1991 term, unless otherwise footnoted.]
1st-year students/Freshmen 424
2nd-year students/Sophomores 193
3rd-year students/Juniors 255
4th-year students/Seniors 517
Total **1,389**

NUMBER OF DEGREES AWARDED—BY PROGRAM
[Numbers are engineering baccalaureate degrees awarded during the 1991-1992 academic year, unless otherwise footnoted. For full details about each engineering program, refer to the *Tables of Degree Programs*.]
Chemical Engineering 18
Civil & Environmental Engineering 13
Electrical & Computer Engineering 79
Electronic & Computer Control System .... –
Hazardous Waste Management .... –
Industrial & Manufacturing Engineering 9
Materials Science & Engineering 6
Mechanical Engineering 39
Operations Research .... –
Total **164**

PERCENTAGE OF DEGREES AWARDED—BY CATEGORY
[Percentages are of all engineering baccalaureate degrees awarded during the 1991-1992 academic year, unless otherwise footnoted.]
African Americans .... 6.1 %
Asian/Pacific Island Americans .... 3.0 %
Hispanic Americans .... 1.8 %
Native Americans .... – %
Foreign Citizens .... 20.1 %
All Others .... 69.0 %
Women .... 18.3 %
Persons with disabilities .... – % (A)
Students over 25 years of age .... – % (A)

Notes: (A) Data not available.

GRADUATE ENROLLMENTS & DEGREES AWARDED
Master's enrollment 921
Master's degrees awarded 300
Doctoral enrollment 204
Doctoral degrees awarded 16

ENGINEERING STUDENT DATA
Applicants to the engineering college
Number of applicants to engineering college 477
Percent offered admission 78.0%

Matriculated engineering students
Percentage in top quartile (25%) of High School class .... – (A)

Notes: (A) Data not available.

FULL- & PART-TIME FACULTY—BY DEPARTMENT
[Figures are the head count of full-time faculty and the full-time equivalent (FTE) of part-time faculty for each engineering department or equivalent.]

Department	Full	Part
Chemical Eng	13	–
Civil & Environmental Eng	8	–
Electrical & Computer Eng	20	2.5
Industrial & Manufacturing Eng	8	–
Mechanical Eng	21	1.2
Materials Science & Eng	7	–

COLLEGE DESCRIPTION
The College is organized into a Division of Engineering Technology and Departments of Chemical Engineering, Civil and Environmental Engineering, Electrical and Computer Engineering, Industrial and Manufacturing Engineering, Materials Science and Engineering and Mechanical Engineering. The Engineering College facilities include 215,500 square feet of classroom, office and laboratory space. During the 1986-87 academic year, the College occupied a completely redesigned, extensively refurbished and equipped laboratory facility and a new three story office addition. Research is conducted in all six academic departments in the College as well as in the Bioengineering Research Center and the Center for Automotive Research. In the 1991-92 fiscal year the 78 faculty in the College received 99 externally funded research awards totaling $4,643,099. Average number of years required to actually complete the bachelor's degree: 4.5 years.

DEGREE PROGRAMS DESCRIPTION
The College offers ABET (i.e. Accreditation Board for Engineering and Technology) accredited educational programs leading to the B.S. degree in six engineering disciplines (i.e. Chemical, Civil, Electrical, Industrial, Materials Science and Engineering, and Mechanical Engineering. A co-op program is available as an option in all six programs in engineering. Programs leading to the M.S. and Ph.D. degrees are offered in the above six disciplines as well as in Computer Engineering and in Operations Research. An M.S. degree program is offered in Electronic and Computer Control Systems. The Technology Division offers programs leading to the B.S. degree in Electrical and Electronics, Mechanical, and Manufacturing and Industrial Engineering Technology.

TRANSFER INFORMATION
Residency Requirements: A minimum of thirty credits must be earned at Wayne State University. No more than sixty-four credits can be transferred from a two-year institution.

Transfer via Articulation Agreements
Admission to engineering: At any time
Requirements: Transfer students with a h.p.a. of 2.0 or above, from an accredited institution, who have completed the MAT 201-204 sequence with a grade 'C' or better will be admitted to a professional program, subject to departmental criteria. Students who do not meet these requirements may be admitted to the pre-professional program.

Transfer without Articulation Agreements
Admission to engineering: At any time
Requirements: See above requirements.

GRADUATION REQUIREMENTS
B.S. programs in engineering require the completion of 136-140 semester credits depending on major selected. A minimum 2.0 h.p.a. in the total residence credits and a minimum 2.0 h.p.a. in the total work taken in the department of specialization is required.

■ STUDENT PROGRAMS

PROFESSIONAL AND HONORARY SOCIETIES
[For key to acronyms, see *Introduction*.]
Professional Societies: AIChE, AIME, ASCE, ASME, IEEE, IIE, SAE, SME, American Society of Metallurgists, Engineering Society of Detroit, Student Chapter
Honorary Societies: Chi Epsilon, Eta Kappa Nu, Pi Tau Sigma, Tau Alpha Pi, Tau Beta Pi, Theta Tau, Sigma Xi

SUPPORT PROGRAMS
Student Chapter Organizations: Society of Women Engineers, Assoc. of Black Engineers and Applied Scientists

Women's Resource Center
For: Women

Engineering Help Center
For: Ethnic Minorities **Available:** Year round
Offered: Prematriculation, Freshman, Sophomore, Junior, Senior, Graduate level

Other Engineering Support Programs: A wide range of University wide services are available. These include: freshmen/transfer orientation programs, legal aid, personal psychological development, career planning and development, reading and study skills, academic skills, testing and evaluation, health, recreational, housing, physically handicapped, religious and placement services and a tutoring service.

266 Webb Institute of Naval Architecture

■ INSTITUTION PROFILE

HEAD OF THE INSTITUTION
James J. Conti
Phone: (516)671-2277 **Fax:** (516)674-9838

GENERAL INFORMATION
[All Students—Fall 1991]
Undergraduate enrollment 81
Total institution enrollment 81
Type of institution: Private **Calendar:** Semesters
Nearest city: New York **Population:** 7,000,000
Miles from main campus: 26 **Setting:** Suburban
Types of engineering degrees: Engineering

■ ENGINEERING ADMISSIONS

ADMISSIONS OFFICE CONTACT
William G. Murray
Webb Institute of Naval Architecture
Crescent Beach Rd.
Glen Cove, NY 11542-1398
Phone: (516)671-2213 **Fax:** (516)674-9838

ENGINEERING COLLEGE ADMISSIONS INFORMATION
Entrance Requirements: SAT, High School courses—Physics (1 year), Chemistry (1 year), College Preparatory Mathematics (4 years), English (4 years), In addition to the SAT, achievement tests in English Composition, Physics or Chemistry, and Mathematics Level I or II are required. No official minimum score on the SAT is required, but students who score below 1250 are seldom admitted.
Entrance Recommendations: High School courses—Calculus (1 year), Mechanical Drawing (1 year)
Requirements for foreign students: Webb Institute admits only U.S. citizens.

DEPARTMENTAL ADMISSIONS INFORMATION
Admission to the engineering department: At the time of admission to the institution.

■ FINANCIAL INFORMATION

ESTIMATED EXPENSES (ACADEMIC YEAR)
[Expenses are for the 1992-93 nine-month academic year.]

All Students	Undergraduate	Graduate
Tuition and fees	$ –(1)	$ –
College room and board	$ 4,800	$ –
Books and supplies	$ 650	$ –
Other expenses	$ –	$ –
Total estimated expenses	**$ 5,450**	**$ –**

Notes: (1) Each student receives a scholarship that covers tuition and fees.

FINANCIAL AID OFFICE CONTACT
William G. Murray, *Director of Admissions*
Webb Institute of Naval Architecture
Crescent Beach Rd.
Glen Cove, NY 11542-1398
Phone: (516)671-2213 **Fax:** (516)674-9838

GENERAL FINANCIAL AID INFORMATION
Forms accepted/required: CSS-FAF
Additional financial aid information: Students may qualify for Pell Grant, Stafford Student Loans and local financial aid programs.

■ ENGINEERING COLLEGE INFORMATION

[For additional personnel, refer to the *Appendix*.]

HEAD OF ENGINEERING
Jacques B. Hadler
Phone: (516)671-2215 **Fax:** (516)674-9838

ENGINEERING COLLEGE ADDRESS
Webb Institute of Naval Architecture
Crescent Beach Rd.
Glen Cove, NY 11542-1398
Phone: (516)671-2277 **Fax:** (516)674-9838

ENROLLMENTS—BY CLASS
[Numbers are baccalaureate enrollments for the fall 1991 term, unless otherwise footnoted.]
1st-year students/Freshmen 23
2nd-year students/Sophomores 19
3rd-year students/Juniors 23
4th-year students/Seniors 16
Total 81

NUMBER OF DEGREES AWARDED—BY PROGRAM
[Numbers are engineering baccalaureate degrees awarded during the 1991-1992 academic year, unless otherwise footnoted. For full details about each engineering program, refer to the *Tables of Degree Programs*.]
Naval Architecture & Marine Engineering 16
Total 16

PERCENTAGE OF DEGREES AWARDED—BY CATEGORY
[Percentages are of all engineering baccalaureate degrees awarded during the 1991-1992 academic year, unless otherwise footnoted.]
African Americans .... – %
Asian/Pacific Island Americans .... – %
Hispanic Americans .... – %
Native Americans .... – %
Foreign Citizens .... – %
All Others .... 100.0 %
Women .... 25.0 %
Persons with disabilities .... – %
Students over 25 years of age .... – %

ENGINEERING STUDENT DATA
Applicants to the engineering college
Number of applicants to engineering college 88
Percent offered admission 33.0%
Matriculated engineering students
Percentage in top quartile (25%) of High School class ... 100.0
Average SAT scores: Math—716, Verbal—602, Combined—1318

COLLEGE DESCRIPTION
Webb Institute of Naval Architecture is a very small, highly selective private college with a single degree program - naval architecture and marine engineering. The pace and depth of the course work is appropriate to the gifted student body. Each student completes 8 weeks of work experience each academic year. Each senior completes a thesis as a requirement for graduation. Students pay for room and board and buy their own books, but each student receives a scholarship that covers tuition and fees.

DEGREE PROGRAMS DESCRIPTION
Webb Institute offers a single degree program leading to the B.S. in Naval Architecture and Marine Engineering. The program consists of 146.5 semester credit hours of academic work, including a thesis, and approximately 32 weeks of work experience. Classes are small, and students and faculty have a very close working relationship.

TRANSFER INFORMATION
Residency Requirements: Webb Institute does not grant transfer credit to students transferring from another institution. All enrolled students are required to reside on campus.

GRADUATION REQUIREMENTS
A cumulative grade average of 70 (on a 100 point scale) is required for graduation.

SUPPORT PROGRAMS
Other Engineering Support Programs: Students participate in local sections of the Society of Naval Architects and Marine Engineers and the Institute of Marine Engineers.

267 West Virginia University

INSTITUTION PROFILE
HEAD OF THE INSTITUTION
Neil S. Bucklew
Phone: (304)293-5531 Fax: (304)293-7417
GENERAL INFORMATION
[All Students—Fall 1991]
Undergraduate enrollment 16,282
Graduate enrollment 6,178
Total institution enrollment **22,460**
Type of institution: Public   Calendar: Semesters
Nearest city: Pittsburgh   Population: 500,000
Miles from main campus: 80   Setting: Urban
Types of engineering degrees: Engineering
Other degree-granting colleges: Arts & Sciences, Business Administration, Communication & Journalism, Dentistry, Education, Fine & Performing Arts, Law, Medicine, Nursing, Pharmacy, Technology & Applied Sciences, Agriculture & Forestry, Mineral & Energy Resources, Physical Education, Social Work

ENGINEERING ADMISSIONS
ADMISSIONS OFFICE CONTACT
Glenn Carter
West Virginia University
Office of Admissions & Records
P.O. Box 6009
Morgantown, WV 26506-6009
Phone: (304)293-2124 Fax: (304)293-3080
ENGINEERING COLLEGE ADMISSIONS INFORMATION
Entrance Requirements: SAT (Score: 800), ACT (Score: 21), High School courses—Algebra (2 years), Plane Geometry (1 year), Trigonometry / Advanced Math (1 year)
Requirements for out-of-state residents: In-State Residents: ACT math score of 25 (SAT 500) or high school grade point average of at least 3.0 plus an ACT math score of at least 21 (SAT 450). Out-of-State Residents: ACT math score of 25 (SAT 500) or higher.
Requirements for foreign students: TOEFL (Score: 550); financial statement; Graduate Record Exam (PhD = 90%) (MS = 80%).
DEPARTMENTAL ADMISSIONS INFORMATION
Admission to the engineering department: At the end of the first year.
Additional information: Only general engineering students who have a GPA of 2.0 or better are eligible for admission to a program. Students who have passed five courses: Chemistry 15, English 1, Engineering 1, Engineering 2, and Mathematics 15 will be evaluated for admission to a program based upon their overall GPA, with particular attention to the grades in the above five courses plus all other math, chemistry, and physics courses taken.

FINANCIAL INFORMATION
ESTIMATED EXPENSES (ACADEMIC YEAR)
[Expenses are for the 1992-93 nine-month academic year.]

State Residents	Undergraduate	Graduate
Tuition and fees	$ 1,950	$ 2,050
College room and board	$ 1,500	$ 1,500
Books and supplies	$ 750	$ 750
Other expenses	$ 2,150 [1]	$ 2,150 [1]
Total estimated expenses	**$ 6,350**	**$ 6,450**
Out-of-State Residents	Undergraduate	Graduate
Tuition and fees	$ 5,300	$ 5,550
College room and board	$ 1,500	$ 1,500
Books and supplies	$ 850	$ 850
Other expenses	$ 2,150 [1]	$ 2,150 [1]
Total estimated expenses	**$ 9,800**	**$ 10,050**

Notes: (1) Transportation and personal expenses
FINANCIAL AID OFFICE CONTACT
Neil E. Bolyard, *Director*
West Virginia University
Student Financial Aid Office
Morgantown, WV 26506
Phone: (304)293-5242 Fax: (304)293-4890
GENERAL FINANCIAL AID INFORMATION
Forms accepted/required: FAF
Additional financial aid information: Need-based scholarships; merit-based scholarships; short-term loans; part-time jobs on campus; College Work-Study Program.

ENGINEERING COLLEGE INFORMATION
[For additional personnel, refer to the *Appendix*.]
HEAD OF ENGINEERING
Robert M. Desmond
Phone: (304)293-4821 Fax: (304)293-5024
EMail: DESMOND@FACULTY.COE.WVU.WVNET.EDU
ENGINEERING COLLEGE ADDRESS
College of Engineering
West Virginia University
P.O. Box 6101
Morgantown, WV 26506-6101
Phone: (304)293-4821 Fax: (304)293-5024
ENROLLMENTS—BY CLASS
[Numbers are baccalaureate enrollments for the fall 1991 term, unless otherwise footnoted.]
1st-year students/Freshmen 446
2nd-year students/Sophomores 316
3rd-year students/Juniors 319
4th-year students/Seniors 427
Total **1,508**
NUMBER OF DEGREES AWARDED—BY PROGRAM
[Numbers are engineering baccalaureate degrees awarded during the 1991-1992 academic year, unless otherwise footnoted. For full details about each engineering program, refer to the *Tables of Degree Programs*.]
Aerospace Engineering 29
Chemical Engineering 19
Civil Engineering 32
Computer Engineering 15
Electrical Engineering 69
Engineering .... –
Industrial Engineering 42
Mechanical Engineering 65
Occupational Health & Safety .... –
Total **271**
PERCENTAGE OF DEGREES AWARDED—BY CATEGORY
[Percentages are of all engineering baccalaureate degrees awarded during the 1991-1992 academic year, unless otherwise footnoted.]
African Americans .... 1.5 %
Asian/Pacific Island Americans .... 1.8 %
Hispanic Americans .... 0.4 %
Native Americans .... – %
Foreign Citizens .... 2.2 %
All Others .... 94.1 %
Women .... 10.7 %
Persons with disabilities .... – % (A)
Students over 25 years of age .... – % (A)
Notes: (A) Data not available.
GRADUATE ENROLLMENTS & DEGREES AWARDED
Master's enrollment 267
Master's degrees awarded 78
Doctoral enrollment 110
Doctoral degrees awarded 21
ENGINEERING STUDENT DATA
Applicants to the engineering college
Number of applicants to engineering college 740
Percent offered admission 100.0%
Matriculated engineering students
Percentage in top quartile (25%) of High School class ..... – (A)
Average SAT scores: Math—583, Combined—1029
Average ACT scores: Math—25, Composite—25
Notes: (A) Data not available.
FULL- & PART-TIME FACULTY—BY DEPARTMENT
[Figures are the head count of full-time faculty and the full-time equivalent (FTE) of part-time faculty for each engineering department or equivalent.]

Department	Full	Part
Chemical Eng	12	–
Civil Eng	19	–
Electrical & Computer Eng	17	–
Industrial Eng	11	1.5
Mechanical & Aerospace Eng	28	3.8

COLLEGE DESCRIPTION
The college offers four-year ABET accredited undergraduate programs in engineering, emphasizing engineering fundamentals in the first year, followed by three years of specialized training in one of the seven majors, with some liberal arts studies throughout all four years. Emphasis is on decision-making incorporating the use of computers starting in the freshman year. The college permits a dual-degree program for engineering students (e.g., Electrical Engineering and Computer Science). During the last semester in most programs, each student, working in teams, sometimes interdisciplinary, completes a senior design project.
DEGREE PROGRAMS DESCRIPTION
The B.S. degrees in the various engineering programs prepare the graduate for entry-level professional positions or for graduate study in each of the respective fields. Graduate study in engineering can be carried out in one of three general areas, including the M.S. and Ph.D. in engineering, emphasizing the applications of the natural sciences to the analysis and solution of engineering problems.
TRANSFER INFORMATION
Residency Requirements: Transfer students must complete a total of 90 hours or at least the last 30 hours of work in residence at WVU. Students are required to earn up to 15 hours in the major field regardless of the number of hours or the nature of the courses transferred.
Transfer via Articulation Agreements
Admission to engineering: At the end of the sophomore year
Engineering graduates transferred from ... 4-yr: 3   2-yr: 8
Requirements: Any student who is a resident of West Virginia, has been admitted to a sending institution, and meets the College of Engineering admission requirements, and maintains a grade point average of 2.0 or higher during four semesters (64 credit hours) at a sending institution will be assured admission into a baccalaureate program in the College, provided the student has satisfactorily completed all prerequisite courses.
Transfer without Articulation Agreements
Admission to engineering: At any time
Engineering graduates transferred from ... 4-yr: 56   2-yr: –
Requirements: Transfer students applying for admission to the College must have completed Math 15 and 16 and Chemistry 15 and 16 or Physics 11 and 12 (or their equivalent) with an overall 2.5 GPA and a 2.5 GPA in math and science.
GRADUATION REQUIREMENTS
B.S. programs in engineering require 136 to 141 credits, depending on the major selected, and a GPA of 2.0 or better for all engineering courses attempted.

STUDENT PROGRAMS
PROFESSIONAL AND HONORARY SOCIETIES
[For key to acronyms, see *Introduction*.]
Professional Societies: AIAA, AIChE, ASCE, ASME, IEEE, IIE, SAE
Honorary Societies: Alpha Pi Mu, Chi Epsilon, Eta Kappa Nu, Omega Chi Epsilon, Pi Tau Sigma, Sigma Gamma Tau, Tau Beta Pi
SUPPORT PROGRAMS
Student Chapter Organizations: Society of Women Engineers, National Society of Black Engineers
Center for Black Culture
For: Ethnic Minorities   **Available:** Year round
Offered: Freshman, Sophomore, Junior, Senior, Graduate level
Other Engineering Support Programs: Summer and Fall Orientation Program for all freshmen and transfer students; Career Services Center; Counseling and Psychological Services Center; Disability Services; Office of Student Life; Math Learning Center; and International Student Office.

268 West Virginia Institute of Technology

INSTITUTION PROFILE
HEAD OF THE INSTITUTION
John C. Carrier
Phone: (304)442-3146 Fax: (304)442-3059
GENERAL INFORMATION
[All Students—Fall 1991]
Undergraduate enrollment 2,984
Graduate enrollment 16
Total institution enrollment 3,000
Type of institution: Public Calendar: Semesters
Nearest city: Charleston, WV Population: 50,000
Miles from main campus: 30 Setting: Small Town
Types of engineering degrees: Engineering
Other degree-granting colleges: Arts and Sciences, Business and Economics, Community and Technical College

ENGINEERING ADMISSIONS
ADMISSIONS OFFICE CONTACT
Robert P. Scholl
West Virginia Institute of Technology
Registrar/Admissions Office
Montgomery, WV 25136
Phone: (304)442-3151 Fax: (304)442-3059
ENGINEERING COLLEGE ADMISSIONS INFORMATION
Entrance Requirements: ACT (Score: 21), High School courses—College Preparatory Mathematics (4 years), English (4 years), Social Studies (3 years), Laboratory Sciences (2 years)
Requirements for foreign students: TOEFL (Score: 500); financial statement
DEPARTMENTAL ADMISSIONS INFORMATION
Admission to the engineering department: At the time of admission to the institution.

FINANCIAL INFORMATION
ESTIMATED EXPENSES (ACADEMIC YEAR)
[Expenses are for the 1992-93 nine-month academic year.]

State Residents	Undergraduate	Graduate
Tuition and fees	$ 1,832	$ 1,976
College room and board	$ 3,752	$ 3,752
Books and supplies	$ 600	$ 600
Other expenses	$ –	$ –
Total estimated expenses	**$ 6,184**	**$ 6,328**

Out-of-State Residents	Undergraduate	Graduate
Tuition and fees	$ 4,002	$ 4,578
College room and board	$ 3,752	$ 3,752
Books and supplies	$ 600	$ 600
Other expenses	$ –	$ –
Total estimated expenses	**$ 8,354**	**$ 8,930**

FINANCIAL AID OFFICE CONTACT
Nina M. Morton, *Director of Financial Aid*
West Virginia Institute of Technology
Financial Aid Office
Montgomery, WV 25136
Phone: (304)442-3032 Fax: (304)442-3059
GENERAL FINANCIAL AID INFORMATION
Forms accepted/required: AFSA, FAF, Institutional
Additional financial aid information: Pell Grant WV Higher Education Grant Supplemental Educational Opportunity Grant College Work Study Perkins Loan Stafford Loans Scholarships

ENGINEERING COLLEGE INFORMATION
[For additional personnel, refer to the *Appendix*.]
HEAD OF ENGINEERING
Ernest E. Nester
Phone: (304)442-3161 Fax: (304)442-3307
ENGINEERING COLLEGE ADDRESS
Leonard C. Nelson College of Engineering
West Virginia Institute of Technology
Dean's Office
Montgomery, WV 25136
Phone: (304)442-3161 Fax: (304)442-3307
ENROLLMENTS—BY CLASS
[Numbers are baccalaureate enrollments for the fall 1991 term, unless otherwise footnoted.]
1st-year students/Freshmen 273
2nd-year students/Sophomores 165
3rd-year students/Juniors 159
4th-year students/Seniors 208
Total 805
NUMBER OF DEGREES AWARDED—BY PROGRAM
[Numbers are engineering baccalaureate degrees awarded during the 1991-1992 academic year, unless otherwise footnoted. For full details about each engineering program, refer to the *Tables of Degree Programs*.]
Chemical Engineering 4
Civil Engineering 22
Electrical Engineering 50
Mechanical Engineering 24
Total 100
PERCENTAGE OF DEGREES AWARDED—BY CATEGORY
[Percentages are of all engineering baccalaureate degrees awarded during the 1991-1992 academic year, unless otherwise footnoted.]
African Americans .... – %
Asian/Pacific Island Americans .... – %
Hispanic Americans .... – %
Native Americans .... – %
Foreign Citizens .... 5.0 %
All Others .... 95.0 %
Women .... 1.3 %
Persons with disabilities .... 6.0 %
Students over 25 years of age .... 18.9 %
GRADUATE ENROLLMENTS & DEGREES AWARDED
Master's enrollment 16
Master's degrees awarded 11
Doctoral enrollment .... –
Doctoral degrees awarded .... –
FULL- & PART-TIME FACULTY—BY DEPARTMENT
[Figures are the head count of full-time faculty and the full-time equivalent (FTE) of part-time faculty for each engineering department or equivalent.]

Department	Full	Part
Chemical Eng	4	–
Civil Eng	6	–
Electrical Eng	7	–
Mechanical Eng	5	–

COLLEGE DESCRIPTION
The primary mission is to provide high quality, accredited undergraduate education in the engineering and computer science disciplines. These programs are intended for WV residents who have the necessary background and others who meet the admission requirements. A secondary mission is to provide graduate engineering education in a small number of specific areas of interest to local industry. A CO-OP program is available.

DEGREE PROGRAMS DESCRIPTION
While not mandatory, WVIT encourages its students to participate in our CO-OP program. This program of alternate periods of work and college requires five years to complete, but the rewards in experience and knowledge are well worth the additional time. Some of the CO-OP students have been able to pay for their education after the first year.
TRANSFER INFORMATION
Residency Requirements: All students must complete 30 of their last 36 hours at WVIT.
Transfer via Articulation Agreements
Admission to engineering: At the end of the sophomore year
Engineering graduates transferred from ... 4-yr: 15 2-yr: 10
Requirements: Students who met College of Engineering admission requirements and are in good academic standing at the sending institution are automatically admitted.
Transfer without Articulation Agreements
Admission to engineering: At any time
Engineering graduates transferred from ... 4-yr: 10 2-yr: 10
Requirements: Must have completed College Algebra and Trigonometry with grades of C or better and have a cumulative grade point average of at least 2.5.
GRADUATION REQUIREMENTS
Overall GPA of at least 2.0 in all courses attempted Overall GPA of at least 2.0 in all courses taken at WVIT Overall GPA of at least 2.0 in all professional courses taken at WVIT Overall GPA of at least 2.0 in all professional courses

STUDENT PROGRAMS
PROFESSIONAL AND HONORARY SOCIETIES
[For key to acronyms, see *Introduction*.]
Professional Societies: ACM, AIChE, ASCE, ASME, IEEE, ISA
Honorary Societies: Eta Kappa Nu, Pi Tau Sigma, Tau Beta Pi

269 Western Michigan University

INSTITUTION PROFILE

HEAD OF THE INSTITUTION
Diether H. Haenicke
Phone: (616)387-2351　　　　Fax: (616)387-2355

GENERAL INFORMATION
[All Students—Fall 1991]
Undergraduate enrollment 20,130
Graduate enrollment 3,928
Total institution enrollment **24,058**
Type of institution: Public　　Calendar: Semesters
Nearest city: Grand Rapids　　Population: 181,843
Miles from main campus: 50　　Setting: Urban
Types of engineering degrees: Engineering & Technology
Other degree-granting colleges: Arts & Sciences, Education, Haworth College of Business, Health & Human Services, Lee Honors College, Fine Arts

ENGINEERING ADMISSIONS

ADMISSIONS OFFICE CONTACT
Stanley E. Henderson
Western Michigan University
Director of Undergraduate Admissions
2240 Seibert Administration Building
Kalamazoo, MI 49008
Phone: (616)387-2000　　　　Fax: (616)387-2355

ENGINEERING COLLEGE ADMISSIONS INFORMATION
Admission to the engineering college: Beginning students are admitted to pre-engineering. They apply to a specific engineering program at the end of the sophomore year.
Entrance Requirements: ACT
Entrance Recommendations: High School courses—Mathematics and English (each) (4 years), Physical or Life Sciences (3 years), Social Sciences (2 years), Basic Computer Course (1 year)
Requirements for foreign students: TOEFL (Score: 550); MTELP (85%); TOEFL score can be 500 if student takes remedial English class first semester.

DEPARTMENTAL ADMISSIONS INFORMATION
Admission to the engineering department: At the end of the second year.
Additional information: A grade of 'C' or better is required in the following: Calculus I, II, III (12 sem hrs); gen chem (4 sem hrs); calculus-based physics (8 sem hrs); freshman writing (3 sem hrs); hum/soc sci (6-8 sem hrs); comp progr (1-3 sem hrs); 10-12 sem hrs of sci and engr sci depending on engineering program.

FINANCIAL INFORMATION

ESTIMATED EXPENSES (ACADEMIC YEAR)
[Expenses are for the 1992-93 nine-month academic year.]

State Residents	Undergraduate	Graduate
Tuition and fees	$ 2,905 [2]	$ 2,412 [1]
College room and board	$ 3,830	$ 3,830
Books and supplies	$ 470	$ 470
Other expenses	$ –	$ –
Total estimated expenses	**$ 7,205**	**$ 6,712**

Out-of-State Residents	Undergraduate	Graduate
Tuition and fees	$ 6,795 [3]	$ 5,373 [1]
College room and board	$ 3,830	$ 3,830
Books and supplies	$ 470	$ 470
Other expenses	$ –	$ –
Total estimated expenses	**$11,095**	**$ 9,673**

Notes: (1) Based on 9 hrs per sem as full-time (2) Based on 15 hrs per sem as full-time; figure shown is average of tuition/fees for lower (fresh/soph) and upper (jr/sr) level--lower is 2780, upper is 3030. (3) Based on 15 hrs per sem as full-time; figure shown is average of tuition/fees for lower (fresh/soph) and upper (jr/sr) level--lower is 6490, upper is 7100.

FINANCIAL AID OFFICE CONTACT
John A. Kundel, *Director of Student Financial Aid*
Western Michigan University
Student Financial Aid Office
3306 Student Services Building
Kalamazoo, MI 49008
Phone: (616)387-6000　　　　Fax: (616)387-6989

GENERAL FINANCIAL AID INFORMATION
Forms accepted/required: FSA, FFS, Institutional
Additional financial aid information: Need-based scholarships and grants; short-term loans; merit-based scholarships; part-time jobs on campus; student employment referral; College Work-Study program.

ENGINEERING COLLEGE INFORMATION

[For additional personnel, refer to the *Appendix*.]

HEAD OF ENGINEERING
Leonard R. Lamberson
Phone: (616)387-4017　　　　Fax: (616)387-4024
EMail: LAMBERSON@GW.WMICH.EDU

ENGINEERING COLLEGE ADDRESS
College of Engineering and Applied Sciences
Western Michigan University
2022 Kohrman Hall
Kalamazoo, MI 49008-5062
Phone: (616)387-4017　　　　Fax: (616)387-4024

ENROLLMENTS—BY CLASS
[Numbers are baccalaureate enrollments for the fall 1991 term, unless otherwise footnoted.]
1st-year students/Freshmen 273
2nd-year students/Sophomores 273
3rd-year students/Juniors 270
4th-year students/Seniors 397
Total **1,213**

NUMBER OF DEGREES AWARDED—BY PROGRAM
[Numbers are engineering baccalaureate degrees awarded during the 1991-1992 academic year, unless otherwise footnoted. For full details about each engineering program, refer to the *Tables of Degree Programs*.]
Aeronautical Engineering 13
Computer Systems Engineering 16
Electrical Engineering 36
Industrial Engineering 17
Mechanical Engineering 68
Paper & Printing Science & Engineering 6
Total **156**

PERCENTAGE OF DEGREES AWARDED—BY CATEGORY
[Percentages are of all engineering baccalaureate degrees awarded during the 1991-1992 academic year, unless otherwise footnoted.]
African Americans .... – %
Asian/Pacific Island Americans .... 1.9 %
Hispanic Americans .... – %
Native Americans .... – %
Foreign Citizens .... 18.6 %
All Others .... 79.5 %
Women .... 3.8 %
Persons with disabilities .... – % (A)
Students over 25 years of age .... – % (A)

Notes: (A) Data not available.

GRADUATE ENROLLMENTS & DEGREES AWARDED
Master's enrollment 123
Master's degrees awarded 34
Doctoral enrollment .... –
Doctoral degrees awarded .... –

ENGINEERING STUDENT DATA
Applicants to the engineering college
Number of applicants to engineering college .... 706 [1]
Percent offered admission .... 84.0 %
Matriculated engineering students
Percentage in top quartile (25%) of High School class 35.0
Average ACT scores: Math—23.8, Composite—23.1

Notes: (1) Information given here will be for the Fall 1991 term

FULL- & PART-TIME FACULTY—BY DEPARTMENT
[Figures are the head count of full-time faculty and the full-time equivalent (FTE) of part-time faculty for each engineering department or equivalent.]

Department	Full	Part
Electrical Eng	12	–
Industrial Eng	9	3.6
Mechanical & Aeronautical Eng	17	–
Paper & Printing Science & Eng	5	1.1

COLLEGE DESCRIPTION
The College offers four-year undergraduate programs in engineering, engineering technology and related technical management and support areas. A pre-engineering program stresses engineering fundamentals in the first two years and specialized education in a specific area is provided during the last two. General education in the humanities and social sciences is distributed throughout the four years. Computer-aided engineering is emphasized at all levels, as are communication skills. Co-op and other field experience opportunities are optional. During the last semester, students complete a senior design project, usually involving a real problem in an area industry. Graduate programs attract both full-time students and part-time students who are engineers working in industry throughout west Michigan. Average number of years required to actually complete the Bachelor's degree is four and a half.

DEGREE PROGRAMS DESCRIPTION

The B.S.E. degrees prepare the graduate for entry-level professional positions or for graduate study in each of the respective fields. In the undergraduate programs, emphasis is placed on linking theory to practice. The college offers M.S.E. degrees in Electrical Engineering, Mechanical Engineering and Industrial Engineering, and M.S. degrees in Operations Research, Engineering Management, and Paper Science and Engineering.

TRANSFER INFORMATION
Residency Requirements: Half of the program must be from a four-year accredited institution, 30 semester hours from WMU.
Transfer without Articulation Agreements
Admission to engineering: At any time

GRADUATION REQUIREMENTS
B.S.E. programs require 130 to 136 semester credits, depending on major selected.

STUDENT PROGRAMS

PROFESSIONAL AND HONORARY SOCIETIES
[For key to acronyms, see *Introduction*.]
Professional Societies: ASME, IEEE, IIE, NSPE, SAE, SME, SPE*, TAPPI, American Production & Inventory Control Society (APICS)
Honorary Societies: Alpha Pi Mu, Pi Tau Sigma, Tau Beta Pi, Theta Tau, Phi Kappa Phi

SUPPORT PROGRAMS
Student Chapter Organizations: Society of Women Engineers, National Society of Black Engineers, Pre-college intro programs to engineering, WMU

Martin Luther King Program
For: Ethnic Minorities　　**Available:** Year round
Offered: Freshman

Minority Student Services
For: Ethnic Minorities　　**Available:** Year round
Offered: Prematriculation, Freshman, Sophomore, Junior, Senior, Graduate level

Center for Women's Services
For: Women　　**Available:** Year round
Offered: Prematriculation, Freshman, Sophomore, Junior, Senior, Graduate level

Other Engineering Support Programs: Freshman and transfer orientation; placement services; international student services; and handicapped student services.

270 Western New England College

INSTITUTION PROFILE
HEAD OF THE INSTITUTION
Beverly W. Miller
Phone: (413)782-1243   Fax: (413)782-1746
GENERAL INFORMATION
[All Students—Fall 1991]
Undergraduate enrollment 3,090
Graduate enrollment 1,731
Total institution enrollment **4,821**
Type of institution: Private   **Calendar:** Semesters
Location: Springfield   **Population:** 150,000
Setting: Urban
Types of engineering degrees: Engineering
Other degree-granting colleges: Arts & Sciences, Business Administration, Law

ENGINEERING ADMISSIONS
ADMISSIONS OFFICE CONTACT
Lori-Ann Paterwic
Western New England College
1215 Wilbraham Road
Springfield, MA 01119
Phone: (413)782-3111   Fax: (413)782-1746
ENGINEERING COLLEGE ADMISSIONS INFORMATION
Entrance Requirements: SAT (Score: 900), High School courses—English (4 years), Mathematics (4 years), U.S. History (1 year), Physics and/or Chemistry (preferably both) (1 year), No cut-off, average SAT scores 900-1000+, Based on 90-91 Freshman Class scores. Results of the SAT or ACT examinations should be forwarded to the Admissions Office.
Entrance Recommendations: Results of the SAT or ACT examinations should be forwarded to the Admissions Office.
Requirements for foreign students: TOEFL (Score: 500); financial statement; SAT are not required. Supplementary Application Form to be completed only if student is enrolled in a post-secondary institution during the current semester.
DEPARTMENTAL ADMISSIONS INFORMATION
Admission to the engineering department: At the time of admission to the institution.

FINANCIAL INFORMATION
ESTIMATED EXPENSES (ACADEMIC YEAR)
[Expenses are for the 1992-93 nine-month academic year.]

All Students	Undergraduate	Graduate
Tuition and fees	$ 8,656	$ –(1)
College room and board	$ 5,200	$ –
Books and supplies	$ 500	$ –
Other expenses	$ 500	$ –
Total estimated expenses	**$14,856**	**$ –**

Notes: (1) Graduate are charged $256 per credit hour - not a flat rate.
FINANCIAL AID OFFICE CONTACT
Kathy M. Chambers, *Director of Financial Aid*
Western New England College
1215 Wilbraham Road
Springfield, MA 01119
Phone: (413)782-3111   Fax: (413)796-2008
GENERAL FINANCIAL AID INFORMATION
Forms accepted/required: CSS-FAF, Institutional, Free Application for Federal Student Aid
Additional financial aid information: Perkins, Stafford, SLS, PLUS, Pell, SEOG, State grants, CWS, various scholarships, Tuition Assistance Grant, Tuition Remission

ENGINEERING COLLEGE INFORMATION
[For additional personnel, refer to the *Appendix*.]
HEAD OF ENGINEERING
Stephen C. Crist
Phone: (413)782-1273   Fax: (413)782-1746
ENGINEERING COLLEGE ADDRESS
School of Engineering
Western New England College
1215 Wilbraham Road
Springfield, MA 01119
Phone: (413)782-1272   Fax: (413)782-1746
ENROLLMENTS—BY CLASS
[Numbers are baccalaureate enrollments for the fall 1991 term, unless otherwise footnoted.]
1st-year students/Freshmen 107
2nd-year students/Sophomores 100
3rd-year students/Juniors 107
4th-year students/Seniors 123
Total **437**
NUMBER OF DEGREES AWARDED—BY PROGRAM
[Numbers are engineering baccalaureate degrees awarded during the 1991-1992 academic year, unless otherwise footnoted. For full details about each engineering program, refer to the *Tables of Degree Programs*.]
Bioengineering 4
Electrical Engineering 25
Engineering Management .... –(B)
Industrial Engineering 13
Mechanical Engineering 40
Total **82**
Notes: (B) Data not applicable.
PERCENTAGE OF DEGREES AWARDED—BY CATEGORY
[Percentages are of all engineering baccalaureate degrees awarded during the 1991-1992 academic year, unless otherwise footnoted.]
African Americans .... – %
Asian/Pacific Island Americans 1.3%
Hispanic Americans 2.5%
Native Americans .... – %
Foreign Citizens 6.5%
All Others 89.7%
Women 15.6%
Persons with disabilities .... – %
Students over 25 years of age 31.2%
GRADUATE ENROLLMENTS & DEGREES AWARDED
Master's enrollment 77
Master's degrees awarded 45
Doctoral enrollment .... –
Doctoral degrees awarded .... –
ENGINEERING STUDENT DATA
Applicants to the engineering college
Number of applicants to engineering college 283
Percent offered admission 86.0%
Matriculated engineering students
Percentage in top quartile (25%) of High School class 30.0
Average SAT scores: Math—510, Verbal—420, Combined—930
FULL- & PART-TIME FACULTY—BY DEPARTMENT
[Figures are the head count of full-time faculty and the full-time equivalent (FTE) of part-time faculty for each engineering department or equivalent.]

Department	Full	Part
Electrical Eng	6	–
Industrial & Manufacturing Eng	4	1.1
Mechanical Eng	6	–

COLLEGE DESCRIPTION
The Engineering School at WNEC offers personalized laboratory intensive programs that prepare each student for a profession. The programs are designed to serve the needs of students seeking a degree on either a full-time or part-time basis. The College is committed to excellence in undergraduate teaching. As a result, classes are small and instruction is provided by Full-time faculty. Graduate programs are offered only on a part-time basis to serve the working professional.

DEGREE PROGRAMS DESCRIPTION
The EE department offers two B.S. options: Electrical and Computer. Both options are accredited by ABET under Electrical Engineering. The ME department also offers two B.S. options: Mechanical and Manufacturing. These are both ABET accredited ME programs. The IE department offers B.S. options in Industrial and Manufacturing engineering, both accredited as IE programs. A Bachelor's of Engineering with a concentration in Bioengineering is also offered. All undergraduate programs are offered on campus for full-time day and part-time evening students. M.S. programs are offered in the evening for part time students in Electrical Engineering, Mechanical Engineering, and Engineering Management.

TRANSFER INFORMATION
Residency Requirements: Not more than 70 credits are acceptable in transfer from two-year colleges, nor more than 90 credits from four-year colleges and universities (including any applicable two-year college credits).
Transfer via Articulation Agreements
Admission to engineering: At the end of the sophomore year
Requirements: Associate degree graduates who have followed the prescribed programs of study at these specific institutes are allowed the opportunity to complete requirements for baccalaureate degrees in two years at WNEC. Prospective transfer students are encouraged to submit applications to the Admissions Office well in advance of the desired entrance date. Midyear candidates should apply for admission before Jan. 1.
Transfer without Articulation Agreements
Admission to engineering: At any time
GRADUATION REQUIREMENTS
Attain a minimum GPA of 2.0, and a 2.0 minimum for courses in the major.

STUDENT PROGRAMS
PROFESSIONAL AND HONORARY SOCIETIES
[For key to acronyms, see *Introduction*.]
Professional Societies: ASME, IEEE, IIE, Engineering Medicine and Biology Society (EMBS), American Society for Quality Control (ASQC)
Honorary Societies: Tau Beta Pi
SUPPORT PROGRAMS
Student Chapter Organizations: Society of Women Engineers
Other Engineering Support Programs: SOAR (Summer Orientation & Registration) programs are for new students held in June/July. Each new student & his/her parents are invited for a 2-day program that includes placement testing, registration, and information sessions. Freshmen take a College Success Skills course specifically for engineers. Other support programs are Inst. wide, including professional counseling, peer advising, & placement.

271 Wichita State University

■ INSTITUTION PROFILE

HEAD OF THE INSTITUTION
Warren B. Armstrong
Phone: (316)689-3001 Fax: (316)689-3093

GENERAL INFORMATION
[All Students—Fall 1991]
Undergraduate enrollment 13,100
Graduate enrollment 2,679
Total institution enrollment **15,779**
Type of institution: Public Calendar: Semesters
Location: Wichita Population: 300,000
Setting: Urban
Types of engineering degrees: Engineering
Other degree-granting colleges: Allied Health Sciences, Arts & Sciences, Business Administration, Education, Fine & Performing Arts

■ ENGINEERING ADMISSIONS

ADMISSIONS OFFICE CONTACT
Rita Abent
Wichita State University
111 Jardine Hall
Wichita, KS 67260-0124
Phone: (800)362-2594 Fax: (316)689-3795

ENGINEERING COLLEGE ADMISSIONS INFORMATION
Admission to the engineering college: Completion of 24 hours with 2.0 or higher (on a 4-point scale) including 'C' or better in Calculus I, Physics I or Chemistry, English I, II and Speech.
Entrance Recommendations: SAT, ACT, High School courses—English (4 years), Mathematics (algebra, geometry, & pre-calculus) (4 years), Physics (1 year), Chemistry (1 year); ACT or SAT is strongly recommended for placement purposes; in addition, Advanced Placement math & Advanced Placement chemistry are recommended.
Requirements for out-of-state residents: All Kansas high school graduates are admitted to the University under the State's Open Admissions law. Out-of-state students must rank in the upper one-half of class, or have a 2.0 on a 4-point scale, or present acceptable ACT or SAT test scores.
Requirements for foreign students: TOEFL (Score: 530); financial statement

DEPARTMENTAL ADMISSIONS INFORMATION
Admission to the engineering department: Admission to Engineering College with a declared major.

■ FINANCIAL INFORMATION

ESTIMATED EXPENSES (ACADEMIC YEAR)
[Expenses are for the 1992-93 nine-month academic year.]

State Residents	Undergraduate	Graduate
Tuition and fees	$ 1,500 [2]	$ 1,352 [1]
College room and board	$ 3,005	$ 3,005
Books and supplies	$ 450	$ 450
Other expenses	$ — [3]	$ — [3]
Total estimated expenses	**$ 4,955**	**$ 4,807**

Out-of-State Residents	Undergraduate	Graduate
Tuition and fees	$ 4,837 [2]	$ 3,883 [1]
College room and board	$ 3,005	$ 3,005
Books and supplies	$ 450	$ 450
Other expenses	$ — [3]	$ — [3]
Total estimated expenses	**$ 8,292**	**$ 7,338**

Notes: (1) Based on 9 hrs per sem as full-time for 2 semesters (2) Based on 12 hrs per sem as full-time for 2 semesters (3) Engineering Equipment Fee $15 per credit hour for engineering courses

FINANCIAL AID OFFICE CONTACT
Larry G. Rector, Director, Student Financial Planning & Assistance
Wichita State University
223 Grace Wilkie Hall
Box 24
Wichita, KS 67260-0024
Phone: (316)689-3430

GENERAL FINANCIAL AID INFORMATION
Forms accepted/required: FFS, IRS, Institutional
Additional financial aid information: Need-based scholarships; short-term loans; merit-based scholarships; part-time jobs on campus; University Work-Study Program.

■ ENGINEERING COLLEGE INFORMATION

[For additional personnel, refer to the *Appendix*.]

HEAD OF ENGINEERING
William J. Wilhelm
Phone: (316)689-3400 Fax: (316)689-3853

ENGINEERING COLLEGE ADDRESS
College of Engineering
Wichita State University
1845 Fairmount
Wichita, KS 67260-0044
Phone: (316)689-3400 Fax: (316)689-3853

ENROLLMENTS—BY CLASS
[Numbers are baccalaureate enrollments for the fall 1991 term, unless otherwise footnoted.]
1st-year students/Freshmen .... 345 [1]
2nd-year students/Sophomores 306
3rd-year students/Juniors 241
4th-year students/Seniors 359
Total **1,251**

Notes: (1) Most freshmen are admitted to University College before they have completed admission requirements for engineering.

NUMBER OF DEGREES AWARDED—BY PROGRAM
[Numbers are engineering baccalaureate degrees awarded during the 1991-1992 academic year, unless otherwise footnoted. For full details about each engineering program, refer to the *Tables of Degree Programs*.]
Aerospace Engineering 21
Electrical Engineering 82
Industrial Engineering 16
Mechanical Engineering 33
Total **152**

PERCENTAGE OF DEGREES AWARDED—BY CATEGORY
[Percentages are of all engineering baccalaureate degrees awarded during the 1991-1992 academic year, unless otherwise footnoted.]
African Americans 0.7%
Asian/Pacific Island Americans .... 19.0% [2]
Hispanic Americans 2.6%
Native Americans 1.3%
Foreign Citizens 28.2%
All Others 48.2%
Women 10.0%
Persons with disabilities .... – %
Students over 25 years of age 51.3%

Notes: (2) Permanent residents and citizens.

GRADUATE ENROLLMENTS & DEGREES AWARDED
Master's enrollment 206
Master's degrees awarded 58
Doctoral enrollment 88
Doctoral degrees awarded 10

ENGINEERING STUDENT DATA
Applicants to the engineering college
Number of applicants to engineering college 996
Percent offered admission 75.0%

Matriculated engineering students
Percentage in top quartile (25%) of High School class ...... – [A]
Average ACT scores: Composite—24

Notes: (A) Data not available.

FULL- & PART-TIME FACULTY—BY DEPARTMENT
[Figures are the head count of full-time faculty and the full-time equivalent (FTE) of part-time faculty for each engineering department or equivalent.]

Department	Full	Part
Electrical Eng	14	2.0
Industrial Eng	11	–
Mechanical Eng	11	0.5
Aerospace Eng	12	0.1

COLLEGE DESCRIPTION

The College offers a four-year undergraduate program in four departments, emphasizing engineering science, design, communication skills, and liberal education, as well as a specialized curriculum for each major. Evening course sections accommodate working and part-time students. The curriculum emphasizes computer use. Students complete a senior design course. A strong co-operative engineering education program allows students to gain a variety of industrial work experiences, and the University's location in the State's manufacturing center encourages frequent interaction with practicing engineers. The time needed to complete the program with co-op is about six years.

DEGREE PROGRAMS DESCRIPTION
The B.S. degrees offered by the four departments prepare the graduate for entry-level professional positions or for graduate study. An M.S. degree and Ph.D. degree are offered by each department. Co-operative research programs with local industry provide opportunities to investigate manufacturing and aviation-related problems with state-of-the-art equipment.

TRANSFER INFORMATION
Residency Requirements: Transfer students must complete at least 30 semester hours of course credit at Wichita State. Also, at least 24 of the last 30 hours or 50 of the last 60 hours must be completed at Wichita State.

Transfer via Articulation Agreements
Admission to engineering: All transfers must complete the same requirements for admission to engineering. Kansas community college associate degrees are articulated with the University's general education curriculum.

Transfer without Articulation Agreements
Admission to engineering: If they meet the admission requirements for the College of Engineering, as specified above.

GRADUATION REQUIREMENTS
B.S. programs in engineering require 132-135 semester hours, depending on major, and a minimum of a 2.00 GPA in major and overall work.

■ STUDENT PROGRAMS

PROFESSIONAL AND HONORARY SOCIETIES
[For key to acronyms, see *Introduction*.]
Professional Societies: AIAA, ASHRAE, ASME, IEEE, IIE, NSPE, SAE, SME
Honorary Societies: Alpha Pi Mu, Eta Kappa Nu, Pi Tau Sigma, Sigma Gamma Tau, Tau Beta Pi

SUPPORT PROGRAMS
Student Chapter Organizations: Society of Women Engineers, Minority Engineering Student Organization
MACESA
For: Ethnic Minorities Available: Summer only
Offered: Prematriculation
Other Engineering Support Programs: Orientation programs for freshmen/transfers/international students; career counseling /placement services; international program office; veterans' services; handicapped services; pre-school; math tutoring and writing lab; intensive English program; personal counseling/testing services; co-op education services; the College tutoring program for mathematics, physics, and basic engineering courses.

272 Widener University

INSTITUTION PROFILE

HEAD OF THE INSTITUTION
Robert J. Bruce
Phone: (215)499-4100   Fax: (215)876-9751

GENERAL INFORMATION
[All Students—Fall 1991]
Undergraduate enrollment 4,590
Graduate enrollment 4,264
Total institution enrollment 8,854
Type of institution: Private   **Calendar:** Semesters
Nearest city: Philadelphia   **Population:** 2,000,000
Miles from main campus: 10   **Setting:** Suburban
Types of engineering degrees: Engineering
Other degree-granting colleges: Agricultural & Environmental, Arts & Sciences, Business Administration, Education, Fine & Performing Arts, Humanities & Social Sciences, Law, Nursing, Hotel and Restaurant Management, Physical Therapy, Management, Institute for Clinical Graduate Psychology, Social Work

ENGINEERING ADMISSIONS

ADMISSIONS OFFICE CONTACT
Michael Mahoney
Widener University
One University Place
Chester, PA 19013
Phone: (215)499-4126   Fax: (215)876-9751

ENGINEERING COLLEGE ADMISSIONS INFORMATION
Entrance Requirements: SAT (Score: 1050), High School courses—Physics (1 year), Chemistry (1 year), Math (3 years), English (4 years)
Entrance Recommendations: SAT (Score: 1050), High School courses—Physics (1 year), Chemistry (1 year), Math (3 years), English (4 years)
Requirements for foreign students: TOEFL (Score: 500); financial statement

DEPARTMENTAL ADMISSIONS INFORMATION
Admission to the engineering department: At the time of admission to the institution.

FINANCIAL INFORMATION

ESTIMATED EXPENSES (ACADEMIC YEAR)
[Expenses are for the 1992-93 nine-month academic year.]

All Students	Undergraduate	Graduate
Tuition and fees	$11,795 (4)	$ 370 (1)
College room and board	$ 5,000	$ 5,100 (2)
Books and supplies	$ 500 (5)	$ 400
Other expenses	$ —	$ 1,000 (3)
Total estimated expenses	$17,295	$ 6,870

Notes: (1) Per Credit - Graduate courses are all three credits. (2) Dormitory (3) travel, supplies, etc. (4) includes engineering fee and application fee. (5) For full year

FINANCIAL AID OFFICE CONTACT
Ethel Desmarais, *Director*
Widener University
One University Place
Chester, PA 19013
Phone: (215)499-4171   Fax: (215)876-9751

GENERAL FINANCIAL AID INFORMATION
Forms accepted/required: CSS-FAF, FAF, FAT, Stafford, IRS, SAR, Supplemental, Institutional, PHEAA FORM
Additional financial aid information: All Federal Loans, some Institutional Grants and Aids, Presidential Academic Achievement, Institutional Work Study.

ENGINEERING COLLEGE INFORMATION
[For additional personnel, refer to the *Appendix*.]

HEAD OF ENGINEERING
Thomas G. McWilliams
Phone: (215)499-4036   Fax: (215)499-4059

ENGINEERING COLLEGE ADDRESS
School of Engineering
Widener University
One University Place
Chester, PA 19013
Phone: (215)499-4036   Fax: (215)499-4059

ENROLLMENTS—BY CLASS
[Numbers are baccalaureate enrollments for the fall 1991 term, unless otherwise footnoted.]
1st-year students/Freshmen 75 (1)
2nd-year students/Sophomores 88 (1)
3rd-year students/Juniors 77 (1)
4th-year students/Seniors 95 (1)
Total 335

Notes: (1) Full-time day students.

NUMBER OF DEGREES AWARDED—BY PROGRAM
[Numbers are engineering baccalaureate degrees awarded during the 1991-1992 academic year, unless otherwise footnoted. For full details about each engineering program, refer to the *Tables of Degree Programs*.]
Chemical Engineering 10
Civil Engineering 19
Electrical Engineering 57
Mechanical Engineering 29
Total 115

PERCENTAGE OF DEGREES AWARDED—BY CATEGORY
[Percentages are of all engineering baccalaureate degrees awarded during the 1991-1992 academic year, unless otherwise footnoted.]
African Americans . . . . 2.5 %
Asian/Pacific Island Americans . . . . 12.0 %
Hispanic Americans . . . . 1.7 %
Native Americans . . . . — %
Foreign Citizens . . . . 17.4 %
All Others . . . . 66.4 %
Women . . . . 12.0 %
Persons with disabilities . . . . — %
Students over 25 years of age . . . . 15.0 %

GRADUATE ENROLLMENTS & DEGREES AWARDED
Master's enrollment 119
Master's degrees awarded 41
Doctoral enrollment . . . . —
Doctoral degrees awarded . . . . —

ENGINEERING STUDENT DATA
Applicants to the engineering college
Number of applicants to engineering college 320
Percent offered admission 92.0%

Matriculated engineering students
Percentage in top quartile (25%) of High School class 50.0
Average SAT scores: Math—560, Verbal—459, Combined—1018

FULL- & PART-TIME FACULTY—BY DEPARTMENT
[Figures are the head count of full-time faculty and the full-time equivalent (FTE) of part-time faculty for each engineering department or equivalent.]

Department	Full	Part
Chemical Eng	4	0.5
Civil Eng	5	1.2
Electrical Eng	9	1.2
Mechanical Eng	8	—
Graduate Programs & Research	—	—

COLLEGE DESCRIPTION
Students are offered the advantage of small classes within a diversified university. The small class size and academic advising system promote a close working relationship with the faculty. Classes are taught by faculty not by graduate assistants. An optional four year cooperative education program is offered. Students gain 12 months of full time engineering job experience (one 4 and one 8 month period) and receive their B.S.E. degree in 4 years. Students may enter as undecided majors. A dual major program is available in electrical engineering/physics. Minors are also available. A 3 yr. B.S.E. accelerated degree program and a 4 year combined B.S.E./M.E. are available. Senior engineering majors take a senior projects course.

DEGREE PROGRAMS DESCRIPTION
Bachelor of Science in Engineering with majors in Chemical, Civil, Electrical and Mechanical. Master of Engineering in Chemical (w/ Environmental Option), Civil (w/Enviro. Option), Computer Software and Telecommunication, Electrical, Engineering Management (w/Enviro. Option),Mechanical, Dual M.E./M.B.A..

DUAL DEGREE PROGRAMS
Engineering baccalaureate graduates with dual degrees 1

TRANSFER INFORMATION
Residency Requirements: Must complete final 45 credits at Widener.
Transfer via Articulation Agreements
Admission to engineering: Following completion of requirements as stated in the articulation agreements.
Requirements: GPA of 2.0 and a minimum grade of C in Engineering Courses
Transfer without Articulation Agreements
Admission to engineering: At any time with a minimum of 9 credits.
Requirements: Must have a C or better.

GRADUATION REQUIREMENTS
(A) Minimum 2.0 GPA. (B) Maximum 2 D's in the FR/SO technical courses excluding lab courses. (C) Maximum 2 D's in junior technical courses excluding lab courses. (D) Maximum 2 D's in senior technical courses excluding lab courses. (E) Minimum 12 social science credits. (F) Minimum 12 humanities credits.

STUDENT PROGRAMS

PROFESSIONAL AND HONORARY SOCIETIES
[For key to acronyms, see *Introduction*.]
Professional Societies: AAPG, ACM, ACS, AIChE, ASCE, ASME, IEEE, NSPE, SAE, SPS, Society of Women Engineers
Honorary Societies: Tau Beta Pi

SUPPORT PROGRAMS
Student Chapter Organizations: Society of Women Engineers
Other Engineering Support Programs: There are numerous programs and services at Widener University intended to help students achieve academic success. Many of these are under the authority of the dean of Freshman Studies and assistant provost for Academic Support Services.

273 Wilkes University

INSTITUTION PROFILE

HEAD OF THE INSTITUTION
Christopher N. Breiseth
Phone: (717)831-4000   Fax: (717)824-2934

GENERAL INFORMATION
[All Students—Fall 1991]
Undergraduate enrollment 2,425
Graduate enrollment 675
Total institution enrollment 3,100
Type of institution: Private   **Calendar:** Semesters
Nearest city: Philadelphia   **Population:** 1,700,000
Miles from main campus: 110   **Setting:** Urban
Types of engineering degrees: Engineering
Other degree-granting colleges: Business, Society and Public Policy, Liberal Arts and Human Sciences

ENGINEERING ADMISSIONS

ADMISSIONS OFFICE CONTACT
Emory P. Guffrovich Jr.
Wilkes University
Chase Hall
South River Street
Wilkes-Barre, PA 18766
Phone: (800)945-5378   Fax: (717)829-2434

ENGINEERING COLLEGE ADMISSIONS INFORMATION
Entrance Requirements: SAT (Score: 890), High School courses—Laboratory Sciences (1 year), English (4 years), Geometry (1 year), Pre-Calculus (1 year)
Entrance Recommendations: SAT, ACT, High School courses—Computers (1 year), History (1 year), Chemistry & Physics (1 year), Calculus (1 year), MPT Math Placement Test
Requirements for foreign students: TOEFL (Score: 500); financial statement

DEPARTMENTAL ADMISSIONS INFORMATION
Admission to the engineering department: At the time of admission to the institution.
Additional information: A strong math and science background is required.

FINANCIAL INFORMATION

ESTIMATED EXPENSES (ACADEMIC YEAR)
[Expenses are for the 1992-93 nine-month academic year.]

All Students	Undergraduate	Graduate
Tuition and fees	$ 9,868	$ 335 (4)
College room and board	$ 4,500 (1)	$ _ (B)
Books and supplies	$ 800	$ _ (3)
Other expenses	$ _ (2)	$ _ (2)
Total estimated expenses	$15,168	$ 335

Notes: (B) Data not applicable. (1) Maximum meal plan (2) All campus activities, including sports, are free for full time students. (3) Proportional to the number of credits. Full time students should expect a cost of $1000 per year. (4) Graduate tuition is based on $335 per credit. Full time student carries a minimum of 9 credits per semester.

FINANCIAL AID OFFICE CONTACT
Rachael L. Lohman, *Director of Financial Aid*
Wilkes University
Sturdevant Hall
170 South Franklin Street
Wilkes-Barre, PA 18766
Phone: (717)824-4651   Fax: (717)829-6662

GENERAL FINANCIAL AID INFORMATION
Forms accepted/required: Stafford, Institutional, Free Application for Federal Student Aid (was called AFSA)
Additional financial aid information: Full tuition trustee scholarships; Carpenter engineering scholarships; various academic and need-based scholarships and loans; work study and student assistantships.

ENGINEERING COLLEGE INFORMATION
[For additional personnel, refer to the *Appendix*.]

HEAD OF ENGINEERING
Umid R. Nejib
Phone: (717)831-4800   Fax: (717)829-2434
EMail: nejib @ wilkesl.wilkes.edu

ENGINEERING COLLEGE ADDRESS
School of Science and Engineering
Wilkes University
Stark Learning Center
South River Street
Wilkes-Barre, PA 18766
Phone: (800)945-5378   Fax: (717)829-2434

ENROLLMENTS—BY CLASS
[Numbers are baccalaureate enrollments for the fall 1991 term, unless otherwise footnoted.]
1st-year students/Freshmen 55 (1)
2nd-year students/Sophomores .... 50 (C)
3rd-year students/Juniors .... 60 (C)
4th-year students/Seniors .... 45 (C)
Total 210
Notes: (C) Estimated data. (1) First year is common to all engineering majors

NUMBER OF DEGREES AWARDED—BY PROGRAM
[Numbers are engineering baccalaureate degrees awarded during the 1991-1992 academic year, unless otherwise footnoted. For full details about each engineering program, refer to the *Tables of Degree Programs*.]
Applied Engineering .... _ (B)
Electrical Engineering 45
Engineering Management 8
Environmental Engineering 10
Materials Engineering 3
Mechanical Engineering .... _ (5)
Pre-Chemical Engineering .... _ (B)
Pre-Civil Engineering .... _ (B)
Pre-Industrial Engineering .... _ (B)
Total 66
Notes: (B) Data not applicable. (5) New program. First graduating class in 1993.

PERCENTAGE OF DEGREES AWARDED—BY CATEGORY
[Percentages are of all engineering baccalaureate degrees awarded during the 1991-1992 academic year, unless otherwise footnoted.]
African Americans .... 1.0% (C)
Asian/Pacific Island Americans .... 5.0% (C)
Hispanic Americans .... 1.0% (C)
Native Americans .... _ % (A)
Foreign Citizens .... 3.0% (C)
All Others 90.0%
Women .... 12.0% (C)
Persons with disabilities .... _ % (C)
Students over 25 years of age .... 8.0% (C)
Notes: (A) Data not available. (C) Estimated data.

GRADUATE ENROLLMENTS & DEGREES AWARDED
Master's enrollment .... 25 (C)
Master's degrees awarded .... 5 (C)
Doctoral enrollment .... _ (B)
Doctoral degrees awarded .... _ (B)
Notes: (B) Data not applicable. (C) Estimated data.

ENGINEERING STUDENT DATA
Applicants to the engineering college
Number of applicants to engineering college 135
Percent offered admission 86.6%

Matriculated engineering students
Percentage in top quartile (25%) of High School class 49.0
Average SAT scores: Math—570, Verbal—470, Combined—1040

FULL- & PART-TIME FACULTY—BY DEPARTMENT
[Figures are the head count of full-time faculty and the full-time equivalent (FTE) of part-time faculty for each engineering department or equivalent.]

Department	Full	Part
GeoEnvironmental Eng	7	2.5
Electrical & Computer Eng	10	2.5
Mechanical & Materials Eng	7	1.5

COLLEGE DESCRIPTION

The school offers a wide variety of innovative engineering programs with a common first year followed by three years of specialized education in one of the majors with emphasis on writing and communications skills. The educational environment is based on the notion that the relationship between the faculty and the student is similar to that of the craftsman and the apprentice. It provides the students access to state-of-the-art laboratories on a daily basis as part of the required curriculum. The student has ample opportunity to interact seriously with industry. Emphasis is placed on computer aided engineering all the way through prototype. A senior project is a requirement. Wilkes offers a unique B.A. degree in applied and engineering sciences and a five-year B.S./M.S. program. All degree programs are also available in the evening. Matriculated students are permitted to take up to 18 credits per semester without incurring additional tuition cost.

DEGREE PROGRAMS DESCRIPTION
The B.S. degree in the various engineering programs prepare the graduate for entry-level professional positions or for graduate study. The B.A. degree in applied and engineering sciences prepares the student for careers in the health sciences, law, social services, planning, and international development. The M.S. degree in electrical engineering emphasizes problem solving and system integration. The completion of a research-oriented thesis based on real life problems is required.

DUAL DEGREE PROGRAMS
Engineering baccalaureate graduates with dual degrees 2
Enrollment requirements: Similar guidelines as matriculation.

TRANSFER INFORMATION
Residency Requirements: Students must complete at least 30 credits at Wilkes University half of which must be in the major field.

Transfer via Articulation Agreements
Admission to engineering: At the end of the freshman year
Engineering graduates transferred from 4-yr: 2   2-yr: 15
Requirements: Transfer must possess a minimum of 2.0 GPA. Each course must have a 'C' or better to be transferred. All engineering courses are evaluated by the respective engineering department.

Transfer without Articulation Agreements
Admission to engineering: At the end of the freshman year
Engineering graduates transferred from 4-yr: 3   2-yr: 5
Requirements: Same as Articulation transfers

GRADUATION REQUIREMENTS
B.S. programs in engineering require 134-137 semester credits, depending on major selected. B. A. in applied and Engineering Sciences requires 124 semester credits. The 5 years B.S. Engineering Management/MBA emphasizes industrial problems and the integration of new technologies and services.

STUDENT PROGRAMS

PROFESSIONAL AND HONORARY SOCIETIES
[For key to acronyms, see *Introduction*.]
Professional Societies: ACerS, ACS, AIPE, ASHRAE, ASME, IEEE, MAA, International Society of Hybrid Microelectronics (ISHM), Pennsylvania Society of Professional Engineers (PSPE)
Honorary Societies: Eta Kappa Nu, Sigma Xi, Sigma Pi Sigma

SUPPORT PROGRAMS
Student Chapter Organizations: Society of Women Engineers, Engineering Club

Multicultural Organization
For: Ethnic Minorities   **Available:** Year round
Offered: Freshman, Sophomore, Junior, Senior, Graduate level

Minority Recruiting
For: Women & Ethnic Minorities   **Available:** Academic year
Offered: Prematriculation, Freshman, Sophomore, Junior, Senior

Other Engineering Support Programs: Summer orientation programs for all freshmen; cooperative education program and free tutorial services; five student clubs and nine student chapters of professional organizations; personal and career counseling services; international student affairs; veteran services; services for the blind and handicapped; and placement services.

274 University of Wisconsin-Madison

INSTITUTION PROFILE

HEAD OF THE INSTITUTION
Katharine Lyall, *Acting*
Phone: (608)262-4048

GENERAL INFORMATION
[All Students—Fall 1991]
Undergraduate enrollment	28,900
Graduate enrollment	10,145
Total institution enrollment	**43,196**

Type of institution: Public   **Calendar:** Semesters
Location: Madison   **Population:** 190,100
Setting: Urban
Types of engineering degrees: Engineering
Other degree-granting colleges: Agricultural & Environmental, Arts & Sciences, Business Administration, Education, Law, Medicine, Nursing, Pharmacy, Technology & Applied Sciences, Family Resources and Consumer Science, Veterinary Medicine

ENGINEERING ADMISSIONS

ADMISSIONS OFFICE CONTACT
Keith White
University of Wisconsin-Madison
140 Peterson Bldg, 750 University Avenue
Madison, WI 53706-1490
Phone: (608)262-3961

ENGINEERING COLLEGE ADMISSIONS INFORMATION
Entrance Requirements: SAT, ACT, High School courses—English (4 years), Mathematics - Algebra, Geometry, 3rd year of Math (3 years), History and Social Studies (2 years), Natural Science (2 years), In-state students must submit ACT test results. Out-of-state students can submit ACT or SAT test results. Students are not admitted solely on the basis of their standard test results.
Entrance Recommendations: SAT, ACT, High School courses—Single Foreign Language or Computer Science & Statistics (2 years), Fine Arts, Communication Arts (2 years), Math (4 years), Physics, Chemistry, Computer Science (each) (1 year)
Requirements for out-of-state residents: In order to be admitted from out-of-state, students must be in the top 10% of their graduating high school class. The normal criteria for in-state students is to graduate in the top 40% of the high school class.
Requirements for foreign students: TOEFL; Foreign students are admitted on a case-by-case basis.

DEPARTMENTAL ADMISSIONS INFORMATION
Admission to the engineering department: English requirement, 24 or more degree cr., 17 or more cr. in calculus, statistics, chemistry, comp. science, statics or physics with a GPA of at least 2.50.
Additional information: When applications exceed programs capacity cumulative grade point averages and other factors.

FINANCIAL INFORMATION

ESTIMATED EXPENSES (ACADEMIC YEAR)
[Expenses are for the 1992-93 nine-month academic year.]

State Residents	Undergraduate	Graduate
Tuition and fees	$ 2,346	$ 3,239
College room and board	$ 3,850	$ 4,180
Books and supplies	$ 510	$ 510
Other expenses	$ 1,410	$ 1,880
Total estimated expenses	**$ 8,116**	**$ 9,809**

Out-of-State Residents	Undergraduate	Graduate
Tuition and fees	$ 7,840	$ 9,825
College room and board	$ 3,850	$ 4,180
Books and supplies	$ 510	$ 510
Other expenses	$ 1,410	$ 1,880
Total estimated expenses	**$13,610**	**$16,395**

GENERAL FINANCIAL AID INFORMATION
Forms accepted/required: FAF, FAT, UW-Financial Aid Form
Additional financial aid information: Grants and work-study employment; short-term loans; department and college scholarships.

ENGINEERING COLLEGE INFORMATION

[For additional personnel, refer to the *Appendix*.]

HEAD OF ENGINEERING
John G. Bollinger
Phone: (608)262-3482   Fax: (608)262-6400
EMail: boleng@engr.wisc.edu

ENGINEERING COLLEGE ADDRESS
College of Engineering
University of Wisconsin-Madison
1415 Johnson Drive
Madison, WI 53706-1691
Phone: (608)262-3484   Fax: (608)262-6400

ENROLLMENTS—BY CLASS
[Numbers are baccalaureate enrollments for the fall 1991 term, unless otherwise footnoted.]

1st-year students/Freshmen	836
2nd-year students/Sophomores	751
3rd-year students/Juniors	783
4th-year students/Seniors	1,189
Total	**3,559**

NUMBER OF DEGREES AWARDED—BY PROGRAM
[Numbers are engineering baccalaureate degrees awarded during the 1991-1992 academic year, unless otherwise footnoted. For full details about each engineering program, refer to the *Tables of Degree Programs*.]

Chemical Engineering	76
Civil Engineering	47
Electrical Engineering	159
Engineering Mechanics	34
Geological Engineering	4
Industrial Engineering	54
Manufacturing Systems Engineering	– (B)
Materials Science & Engineering	– (B)
Mechanical Engineering	141
Metallurgical Engineering	11
Nuclear Engineering	13
Total	**539**

Notes: (B) Data not applicable.

PERCENTAGE OF DEGREES AWARDED—BY CATEGORY
[Percentages are of all engineering baccalaureate degrees awarded during the 1991-1992 academic year, unless otherwise footnoted.]

African Americans	0.5%
Asian/Pacific Island Americans	4.9%
Hispanic Americans	0.5%
Native Americans	– %
Foreign Citizens	14.1%
All Others	80.0%
Women	12.3%
Persons with disabilities	– % (A)
Students over 25 years of age	– % (A)

Notes: (A) Data not available.

GRADUATE ENROLLMENTS & DEGREES AWARDED

Master's enrollment	636
Master's degrees awarded	258
Doctoral enrollment	557
Doctoral degrees awarded	75

ENGINEERING STUDENT DATA
Applicants to the engineering college
Number of applicants to engineering college	1,952
Percent offered admission	79.0%

Matriculated engineering students
Percentage in top quartile (25%) of High School class 81.0
Average SAT scores: Math—637, Verbal—511, Combined—1148
Average ACT scores: Math—27.4, Composite—26.5

FULL- & PART-TIME FACULTY—BY DEPARTMENT
[Figures are the head count of full-time faculty and the full-time equivalent (FTE) of part-time faculty for each engineering department or equivalent.]

Department	Full	Part
Chemical Eng	19	–
Civil & Environmental Eng	32	–
Electrical & Computer Eng	48	–
Industrial Eng	22	–
Mechanical Eng	31	–
Eng Mechanics	15	–
Nuclear Eng & Eng Physics	12	–
Material Science & Eng	17	–
Eng Professional Development	19	–

COLLEGE DESCRIPTION
One of our greatest strengths is our faculty, and the undergraduate degree programs enable students to meet and work with members of a world-class faculty. Organizations such as fourteen consortia and 29 research centers, and and physical facilities such as 108 named laboratories also contribute to quality. It is possible for undergraduate students to work with faculty in any of these organizations and laboratories. The opportunities for highly motivated undergraduates to participate in special programs are almost unlimited on the Madison campus. Undergraduate engineering students may declare second majors in the College of Letters and Science, and many do. A very active co-operative education program is available; increasing numbers of students and engineering employers are finding co-op to be superior to the more traditional methods of gaining employment and employees. The Madison campus has seven dual-degree agreements with smaller campuses, both private and public.

DEGREE PROGRAMS DESCRIPTION
A number of undergraduate degree programs have options: the computer engineering option is available in the electrical engineering degree program; astronautics is an option in engineering mechanics; and 1) surveying and 2) construction engineering and management are options in the civil engineering program. Certificate programs are available in 1) technical communication and 2) technical Japanese. A wide range of opportunities are available at the graduate level. Large formal organizations with physical facilities exist in the manufacturing systems engineering and materials science areas. Research centers and consortia from applied superconductivity to Wisconsin power electronics are very active. Professional development degrees are available.

DUAL DEGREE PROGRAMS
Engineering baccalaureate graduates with dual degrees 0
Enrollment requirements: English requirement, 24 or more degree credits, 17 or more credits in calculus (including second semester), statistics, chemistry, computer science, statics or physics with a grade point average of at least 2.50. Cumulative GPA of 3.0 or higher. Other requirements are unique to each agreement.

TRANSFER INFORMATION
Residency Requirements: 30 credits in residence in the College of Engineering. 15 credits in the degree-granting department, last 2 semesters full-time (12 or more credits) in residence. Some programs have additional requirements.

Transfer via Articulation Agreements
Admission to engineering: English requirement, 24 or more deg. cr., 17 or more cr. in calculus, statistics, chemistry, comp. science, statics or physics with a GPA of 2.50.
Engineering graduates transferred from   . . . 4-yr: –   2-yr: –
Requirements: English requirement, 24 or more degree credits, 17 or more credits in calculus (including second semester), statistics, chemistry, computer science, statics or physics with a grade point average of at least 2.50.

Transfer without Articulation Agreements
Admission to engineering: English requirement, 24 or more degree cr., 17 or more cr. in calculus, statistics, chemistry, comp. science, statics or physics with a GPA of at least 2.50.
Engineering graduates transferred from   . . . 4-yr: –   2-yr: –
Requirements: English requirement, 24 or more degree crs., 17 or more credits in calculus (including second semester), statistics, chemistry, computer science, statics or physics with a grade point average of at least 2.50.

GRADUATION REQUIREMENTS
A variation of GPA known as PCR (point credit ratio) must exceed or equal 2.000 for sessions and semester containing the last 60 credits, and for all courses taken in the student's department. At least 30 credits, including 15 credits from the student's department, must be taken while a student of the College of Engineering. GPAs for the last semester and combined last two semesters equal or exceed 2.000.

STUDENT PROGRAMS

PROFESSIONAL AND HONORARY SOCIETIES
[For key to acronyms, see *Introduction*.]
Professional Societies: AIAA, AIChE, ANS, ASCE, ASME, AWS, IEEE, IIE, SAE, SME, Wisconsin Black Engineering Student Society, Society of Women Engineers (SWE)
Honorary Societies: Alpha Pi Mu, Chi Epsilon, Eta Kappa Nu, Kappa Eta Kappa, Pi Tau Sigma, Tau Beta Pi, Theta Tau, Alpha Chi Sigma, Alpha Sigma Mu

SUPPORT PROGRAMS
Student Chapter Organizations: Society of Women Engineers, National Society of Black Engineers, Society of Hispanic Professional Engineers, American Indian Science and Engineering Society

College of Engineering-Office of Minority Affairs
For: Ethnic Minorities   **Available:** Year round
Offered: Prematriculation, Freshman, Sophomore, Junior, Senior, Graduate level
Other Engineering Support Programs: SOAR Program (Summer Orientation and Registration) for all students enrolling in engineering. Program for new freshman and transfer includes dinner with faculty and staff and a day and a half orientation and registration for the next semester's courses. The College also has a 1 credit course (EPD 100) for students who are undecided on their careers.

275 University of Wisconsin-Milwaukee

■ INSTITUTION PROFILE
HEAD OF THE INSTITUTION
John H. Schroeder, *Chancellor*
Phone: (414)229-4331   Fax: (414)229-4553
GENERAL INFORMATION
[All Students—Fall 1991]
Undergraduate enrollment 20,557
Graduate enrollment 4,899
Total institution enrollment **25,456**
Type of institution: Public   **Calendar:** Semesters
Location: Milwaukee   **Population:** 961,800
Setting: Urban
Types of engineering degrees: Engineering
Other degree-granting colleges: Allied Health Sciences, Architecture, Arts & Sciences, Business Administration, Education, Fine & Performing Arts, Nursing, Social Welfare, Library and Information Science, Graduate School

■ ENGINEERING ADMISSIONS
ADMISSIONS OFFICE CONTACT
Beth L. Weckmueller
University of Wisconsin-Milwaukee
Office of Admissions
P.O. Box 749
Milwaukee, WI 53201-0749
Phone: (414)229-3800   Fax: (414)229-6940
ENGINEERING COLLEGE ADMISSIONS INFORMATION
Entrance Requirements: ACT, High School courses—English (4 years), Mathematics (algebra, geometry, trigonometry) (3 years), History and Social Sciences (3 years), Chemistry (1 year), and other Natural Sciences (2 years), SAT scores rather than ACT may be submitted by out-of-state applicants. ACT required of in-state applicants but admission may not be denied solely on basis of test scores.
Entrance Recommendations: High School courses—Advanced Mathematics (1 year), Natural Sciences (physics, biology) (1 year), Computer Programming
Requirements for foreign students: TOEFL (Score: 550); financial statement
DEPARTMENTAL ADMISSIONS INFORMATION
Admission to the engineering department: Completion of 48 credits countable, including 30 credits required, for the degree.
Additional information: A GPA (for both the 48 countable and 30 required credits) of 2.40 for mechanical and 2.25 for civil, electrical, industrial and materials engineering.

■ FINANCIAL INFORMATION
ESTIMATED EXPENSES (ACADEMIC YEAR)
[Expenses are for the 1992-93 nine-month academic year.]

State Residents	Undergraduate	Graduate
Tuition and fees	$ 2,393	$ 3,286
College room and board	$ 4,394	$ 4,394
Books and supplies	$ 600	$ 600
Other expenses	$ –	$ –
Total estimated expenses	**$ 7,387**	**$ 8,280**

Out-of-State Residents	Undergraduate	Graduate
Tuition and fees	$ 7,679	$ 9,872
College room and board	$ 4,394	$ 4,394
Books and supplies	$ 600	$ 600
Other expenses	$ –	$ –
Total estimated expenses	**$12,673**	**$14,866**

FINANCIAL AID OFFICE CONTACT
Mary E. Roggeman, *Director, Department of Financial Aid*
University of Wisconsin-Milwaukee
P.O. Box 469
Milwaukee, WI 53201-0469
Phone: (414)229-4541
GENERAL FINANCIAL AID INFORMATION
Forms accepted/required: ACT, FAF, FAT, FFS, Stafford, IRS, SAR, Institutional
Additional financial aid information: Need-based scholarships, merit-based scholarships, grants, loans, College-Work-Study, part-time employment on and off campus.

■ ENGINEERING COLLEGE INFORMATION
[For additional personnel, refer to the *Appendix*.]
HEAD OF ENGINEERING
Charles F. James Jr.
Phone: (414)229-4126   Fax: (414)229-6958
ENGINEERING COLLEGE ADDRESS
College of Engineering and Applied Science
University of Wisconsin-Milwaukee
P.O. Box 784
Milwaukee, WI 53201-0784
Phone: (414)229-4768   Fax: (414)229-6958
ENROLLMENTS—BY CLASS
[Numbers are baccalaureate enrollments for the fall 1991 term, unless otherwise footnoted.]
1st-year students/Freshmen 176
2nd-year students/Sophomores 257
3rd-year students/Juniors 161
4th-year students/Seniors 433
Total **1,027**
NUMBER OF DEGREES AWARDED—BY PROGRAM
[Numbers are engineering baccalaureate degrees awarded during the 1991-1992 academic year, unless otherwise footnoted. For full details about each engineering program, refer to the *Tables of Degree Programs*.]
Civil Engineering 42
Electrical Engineering 48
Industrial Engineering 14
Materials Engineering 9
Mechanical Engineering 58
Total **171**
PERCENTAGE OF DEGREES AWARDED—BY CATEGORY
[Percentages are of all engineering baccalaureate degrees awarded during the 1991-1992 academic year, unless otherwise footnoted.]
African Americans 1.8%
Asian/Pacific Island Americans 4.7%
Hispanic Americans 2.9%
Native Americans .... – %
Foreign Citizens 2.3%
All Others 88.3%
Women 15.8%
Persons with disabilities 5.9%
Students over 25 years of age 35.1%
GRADUATE ENROLLMENTS & DEGREES AWARDED
Master's enrollment 178
Master's degrees awarded 47
Doctoral enrollment 55
Doctoral degrees awarded 14
ENGINEERING STUDENT DATA
Applicants to the engineering college
Number of applicants to engineering college 466
Percent offered admission 75.5%
Matriculated engineering students
Percentage in top quartile (25%) of High School class 45.0
Average ACT scores: Math—24.3, Composite—23.3
FULL- & PART-TIME FACULTY—BY DEPARTMENT
[Figures are the head count of full-time faculty and the full-time equivalent (FTE) of part-time faculty for each engineering department or equivalent.]

Department	Full	Part
Civil Eng & Mechanics	14	0.3
Electrical Eng	10	1.8
Industrial & Systems Eng	8	0.8
Mechanical Eng	13	1.5
Materials Eng	7	0.2

COLLEGE DESCRIPTION
The College offers four-year undergraduate programs in five engineering disciplines and computer science. The engineering curricula have a broad base in the basic sciences and engineering fundamentals. On completion of the pre-engineering program, qualified students are admitted to an engineering major (normally at the end of the sophomore year). The programs provide a balance between analysis and design. Each major requires laboratory and computer experience, and a cap-stone design project. Co-op is an option in all majors, and adds an additional three semesters to the time required to complete the program. Students may obtain double-majors. Articulation agreements permit students to take the first two years (pre-engineering) at UW-Green Bay, UW-Parkside, or UW Center-Waukesha County.

DEGREE PROGRAMS DESCRIPTION
The B.S. degrees in the engineering programs prepare the graduate for entry-level professional positions or for graduate study in each of the respective fields. Some degree programs include a range of emphases: Civil Engineering offers emphases in geotechnical engineering, structural engineering, environmental and water resources engineering, and transportation and municipal engineering; Mechanical Engineering offers emphases in energy engineering, mechanical design, and general mechanical engineering.
TRANSFER INFORMATION
Residency Requirements: Transfer students must complete at least 24 of the last 30 degree credits in residence.
Transfer via Articulation Agreements
Admission to engineering: At the end of the sophomore year
Engineering graduates transferred from .. 4-yr: 25   2-yr: 9
Requirements: Entrance requirements are the same as those for admission to the College of Engineering and Applied Science at UW-Milwaukee. Number of transfers given above (and below) are those who transferred during the 1991-92 academic year.
Transfer without Articulation Agreements
Admission to engineering: At any time
Engineering graduates transferred from .. 4-yr: 24   2-yr: 20
Requirements: Students with more than 12 and less than 30 transfer credits are admitted to pre-engineering if their GPA is 2.0 or higher, if they would have been admissible as freshmen, and they have completed first semester calculus. Students with more than 30 transfer credits should have a GPA of at least 2.25. Students with more than 48 credits are considered for admission directly into an engineering major.
GRADUATION REQUIREMENTS
B.S. programs in engineering require 131-134 semester credit hours, depending on the major selected. Students must maintain an overall GPA of at least 2.0 on all work attempted at the University, at least 2.0 in all courses at the junior level or above in the major department, and at least 2.0 in all courses taken in the College.

■ STUDENT PROGRAMS
PROFESSIONAL AND HONORARY SOCIETIES
[For key to acronyms, see *Introduction*.]
Professional Societies: AIAA, ASCE, ASME, IEEE, IIE, ITE, SAE, SME, TMS, Society of Women Engineers, National Society of Black Engineers
Honorary Societies: Pi Tau Sigma, Tau Beta Pi, Theta Tau, Triangle Fraternity
SUPPORT PROGRAMS
Student Chapter Organizations: Society of Women Engineers, National Society of Black Engineers, Society of Hispanic Professional Engineers, American Indian Science and Engineering Society

Multicultural Engineering Program
For: Women & Ethnic Minorities   **Available:** Year round
Offered: Prematriculation, Freshman, Sophomore, Junior, Senior

American Indian Student Services
For: Ethnic Minorities   **Available:** Year round
Offered: Prematriculation, Freshman, Sophomore, Junior, Senior

African American Student Academic Services
For: Ethnic Minorities   **Available:** Year round
Offered: Prematriculation, Freshman, Sophomore, Junior, Senior

Hispanic Student Academic Services
For: Ethnic Minorities   **Available:** Year round
Offered: Prematriculation, Freshman, Sophomore, Junior, Senior

Other Engineering Support Programs: Summer and spring orientation programs for freshmen and transfer students; academic and career counseling; student health services; disabled student services; day care center; computing services; employment placement services; tutoring; learning skills program; placement and proficiency testing.

276 University of Wisconsin-Platteville

INSTITUTION PROFILE

HEAD OF THE INSTITUTION
Lee A. Halgren, *Interim*
Phone: (608)342-1234

GENERAL INFORMATION
[All Students—Fall 1991]
Undergraduate enrollment 5,001
Graduate enrollment 213
Total institution enrollment 5,214

Type of institution: Public　　　Calendar: Semesters
Nearest city: Madison, WI　　　Population: 190,000
Miles from main campus: 70　　　Setting: Small Town
Types of engineering degrees: Engineering

Other degree-granting colleges: Arts & Sciences, Education, Agriculture, Business, Industry, and Communication

ENGINEERING ADMISSIONS

ADMISSIONS OFFICE CONTACT
Richard R. Schumacher
University of Wisconsin-Platteville
1 University Plaza
Platteville, WI 53818-3099
Phone: (608)342-1125　　　Fax: (608)342-1122

ENGINEERING COLLEGE ADMISSIONS INFORMATION
Entrance Requirements: ACT (Score: 21), High School courses—English (4 years), Mathematics (algebra and above) (2 years), Natural Science (min. 1 from biol., phys., chem.) (2 years), Social Sciences (3 years)
Entrance Recommendations: High School courses—Mathematics (4 years), Chemistry (1 year), Physics (1 year), Computer Science (1 year)
Requirements for foreign students: TOEFL (Score: 550); financial statement

DEPARTMENTAL ADMISSIONS INFORMATION
Admission to the engineering department: To enter an engrg. prog. from Gen. Engrg. a student must complete 6 core courses, have a Core GPA of 2.1-Civil; 2.0-Ind.; 2.3-Elec., 2.4-Mech.

FINANCIAL INFORMATION

ESTIMATED EXPENSES (ACADEMIC YEAR)
[Expenses are for the 1992-93 nine-month academic year.]

State Residents	Undergraduate	Graduate
Tuition and fees	$ 1,956	$ 2,302
College room and board	$ 2,410	$ 2,410
Books and supplies	$ 300	$ 300
Other expenses	$ —	$ —
Total estimated expenses	**$ 4,666**	**$ 5,012**

Out-of-State Residents	Undergraduate	Graduate
Tuition and fees	$ 5,810 [2]	$ 6,836 [1]
College room and board	$ 2,410	$ 2,410
Books and supplies	$ 300	$ 300
Other expenses	$ —	$ —
Total estimated expenses	**$ 8,520**	**$ 9,546**

Notes: (1) Minnesota Residents-$2330 (2) Minnesota Residents-$2258

FINANCIAL AID OFFICE CONTACT
Elizabeth Tucker, *Director of Financial Aids*
University of Wisconsin-Platteville
1 University Plaza
Platteville, WI 53818-3099
Phone: (608)342-1836

GENERAL FINANCIAL AID INFORMATION
Forms accepted/required: FAF, FFS, Stafford, Institutional, Free Application for Federal Student Aid
Additional financial aid information: State and federal need-based grants and loans; need and merit-based scholarships; short-term loans; student employment, including internships, co-ops, regular and work study.

ENGINEERING COLLEGE INFORMATION
[For additional personnel, refer to the *Appendix*.]

HEAD OF ENGINEERING
Ronald A. Yeske
Phone: (608)342-1561　　　Fax: (608)342-1566

ENGINEERING COLLEGE ADDRESS
College of Engineering
University of Wisconsin-Platteville
1 University Plaza
Platteville, WI 53818-3099
Phone: (608)342-1561　　　Fax: (608)342-1566

ENROLLMENTS—BY CLASS
[Numbers are baccalaureate enrollments for the fall 1991 term, unless otherwise footnoted.]
1st-year students/Freshmen 542
2nd-year students/Sophomores 267
3rd-year students/Juniors 259
4th-year students/Seniors 408
Total 1,476

NUMBER OF DEGREES AWARDED—BY PROGRAM
[Numbers are engineering baccalaureate degrees awarded during the 1991-1992 academic year, unless otherwise footnoted. For full details about each engineering program, refer to the *Tables of Degree Programs*.]
Civil Engineering 47
Electrical Engineering 57
Industrial Engineering 31
Mechanical Engineering 66
Total 201

PERCENTAGE OF DEGREES AWARDED—BY CATEGORY
[Percentages are of all engineering baccalaureate degrees awarded during the 1991-1992 academic year, unless otherwise footnoted.]
African Americans .... 1.0 %
Asian/Pacific Island Americans .... 3.0 %
Hispanic Americans .... — %
Native Americans .... — %
Foreign Citizens .... 2.0 %
All Others .... 94.0 %
Women .... 10.0 %
Persons with disabilities .... — %
Students over 25 years of age .... 25.0 %

ENGINEERING STUDENT DATA
Applicants to the engineering college
Number of applicants to engineering college 762
Percent offered admission 55.0%

Matriculated engineering students
Percentage in top quartile (25%) of High School class .... — (A)
Average ACT scores: Composite—22.1 (A)

Notes: (A) Data not available.

FULL- & PART-TIME FACULTY—BY DEPARTMENT
[Figures are the head count of full-time faculty and the full-time equivalent (FTE) of part-time faculty for each engineering department or equivalent.]

Department	Full	Part
Civil Eng	9	0.7
Electrical Eng	9	—
Industrial Eng	4	—
Mechanical Eng	11	—
General Eng	5	1.0

COLLEGE DESCRIPTION
The UW-Platteville undergraduate program in engineering emphasizes the basic sciences, reinforced with the engineering sciences, followed by an integrated sequence of engineering analysis and design. The college offers a four year undergraduate program in engineering, emphasizing engineering fundamentals in the first year, followed by three years of specialized training in one of four majors (civil, electrical, industrial, and mechanical) plus required courses in the humanities and social sciences. Typical minors include mathematics and computer science. Co-op is an option for all four degree programs. During the last semester, each student completes a senior design project and is encouraged to write the EIT exam.

DEGREE PROGRAMS DESCRIPTION
The BS degrees in the four engineering majors (civil, electrical, industrial, and mechanical) prepare graduates for entry-level professional positions. No graduate programs are presently offered. The four degree programs include a range of emphases: civil offers emphases in construction management, environmental engineering, geotechnical engineering, surface mining, structural design and transportation engineering. Electrical offers emphases in communications and electronics, computers, controls, and power and energy. Industrial and mechanical engineering do not offer any formal areas of emphasis. Students are well-prepared in the basics.

TRANSFER INFORMATION
Residency Requirements: Students must complete at least 32 credits in residence, with 23 of the last 32 being in residence.
Transfer via Articulation Agreements
Admission to engineering: At any time
Requirements: The entrance requirements for students in the joint enrollment program are the same as for freshmen entering UW-Platteville.
Transfer without Articulation Agreements
Admission to engineering: At any time
Engineering graduates transferred from ... 4-yr: 48　2-yr: 30
Requirements: 2.0 GPA from UW Center schools, 2.5 GPA for WI residents at 4 year institutions.

GRADUATION REQUIREMENTS
BS programs in engineering require 135-138 credits, depending on the major selected. Each program has specified courses which must be completed with a minimum grade.

STUDENT PROGRAMS

PROFESSIONAL AND HONORARY SOCIETIES
[For key to acronyms, see *Introduction*.]
Professional Societies: AIME, ASCE, ASHRAE, ASME, IEEE, IIE, SAE, SME
Honorary Societies: Alpha Pi Mu, Chi Epsilon, Tau Beta Pi, Theta Tau, Eta Lambda Epsilon

SUPPORT PROGRAMS
Student Chapter Organizations: Society of Women Engineers, International Student Club, Multicultural Center, Women's Center

Multicultural Services
For: Ethnic Minorities　　　Available: Year round
Offered: Freshman, Sophomore, Junior, Senior

Other Engineering Support Programs: Other programs include summer and fall orientation programs for all new freshmen and transfer students; career and personal counseling services; study skill seminars; placement services; tutoring; special assistance for the handicapped and disadvantaged; peer advisor program; reading clinic; math and science learning center; and writing lab.

277 Worcester Polytechnic Institute

INSTITUTION PROFILE
HEAD OF THE INSTITUTION
Jon C. Strauss
Phone: (508)831-5200 **Fax:** (508)831-5753
GENERAL INFORMATION
[All Students—Fall 1991]
Undergraduate enrollment 2,845
Graduate enrollment 1,057
Total institution enrollment **3,902**
Type of institution: Private **Calendar:** Other
Nearest city: Worcester **Population:** 170,000
Miles from main campus: 1 **Setting:** Urban
Types of engineering degrees: Engineering

ENGINEERING ADMISSIONS
ADMISSIONS OFFICE CONTACT
Robert G. Voss
Worcester Polytechnic Institute
100 Institute Road
Worcester, MA 01609-2280
Phone: (508)831-5286 **Fax:** (508)831-5483
ENGINEERING COLLEGE ADMISSIONS INFORMATION
Entrance Requirements: High School courses—English (4 years), Mathematics (4 years), Physics (1 year), Chemistry (1 year), The SAT and 3 Achievement Tests or the ACT are required.
Entrance Recommendations: High School courses—Biology (1 year), Foreign Language (2 years), Social Studies (2 years)
Requirements for foreign students: TOEFL (Score: 550); financial statement
DEPARTMENTAL ADMISSIONS INFORMATION
Admission to the engineering department: At the time of admission to the institution.

FINANCIAL INFORMATION
ESTIMATED EXPENSES (ACADEMIC YEAR)
[Expenses are for the 1992-93 nine-month academic year.]

All Students	Undergraduate	Graduate
Tuition and fees	$14,555	$ 8,280
College room and board	$ 4,820	$ 5,780
Books and supplies	$ 490	$ 794
Other expenses	$ 870	$ 2,025
Total estimated expenses	**$20,735**	**$16,879**

FINANCIAL AID OFFICE CONTACT
Michael J. Curley, *Director of Financial Aid*
Worcester Polytechnic Institute
100 Institute Road
Worcester, MA 01609-2280
Phone: (508)831-5469 **Fax:** (508)831-5753
GENERAL FINANCIAL AID INFORMATION
Forms accepted/required: CSS-FAF, IRS, SAR, Institutional
Additional financial aid information: Need based scholarships and loans, part-time jobs on campus, College Work-Study Program and auxiliary and parental loans are available. National Merit Scholarship and NSPE sponsor.

ENGINEERING COLLEGE INFORMATION
[For additional personnel, refer to the *Appendix*.]
HEAD OF ENGINEERING
Francis C. Lutz
Phone: (508)831-5404 **Fax:** (508)831-5774
EMail: FCLUTZ@JAKE.WPI.EDU
ENGINEERING COLLEGE ADDRESS
Departments of Engineering
Worcester Polytechnic Institute
100 Institute Road
Worcester, MA 01609-2280
Phone: (508)831-5000 **Fax:** (508)831-5483
ENROLLMENTS—BY CLASS
[Numbers are baccalaureate enrollments for the fall 1991 term, unless otherwise footnoted.]
1st-year students/Freshmen 496
2nd-year students/Sophomores .... 547 [2]
3rd-year students/Juniors .... 539 [3]
4th-year students/Seniors 442
Total **2,024**
Notes: (2) Includes new transfer students. (3) Includes students on co-op.

NUMBER OF DEGREES AWARDED—BY PROGRAM
[Numbers are engineering baccalaureate degrees awarded during the 1991-1992 academic year, unless otherwise footnoted. For full details about each engineering program, refer to the *Tables of Degree Programs*.]
Biomedical Engineering .... – [B]
Chemical Engineering 25
Civil Engineering 69
Electrical & Computer Engineering 112
Fire Protection Engineering .... – [B]
Manufacturing Engineering 22
Materials Engineering .... – [B]
Mechanical Engineering 165
Total **393**
Notes: (B) Data not applicable.

PERCENTAGE OF DEGREES AWARDED—BY CATEGORY
[Percentages are of all engineering baccalaureate degrees awarded during the 1991-1992 academic year, unless otherwise footnoted.]
African Americans 0.7%
Asian/Pacific Island Americans 5.2%
Hispanic Americans 1.0%
Native Americans .... – %
Foreign Citizens 7.3%
All Others 85.8%
Women 21.0%
Persons with disabilities 0.2%
Students over 25 years of age .... – % [A]
Notes: (A) Data not available.

GRADUATE ENROLLMENTS & DEGREES AWARDED
Master's enrollment .... – [1]
Master's degrees awarded 120
Doctoral enrollment .... – [1]
Doctoral degrees awarded 8
Notes: (1) WPI does not separate MS and PhD students in enrollment figures. Total graduate school enrollment for AY 1991-92 is 1057 students.

ENGINEERING STUDENT DATA
Applicants to the engineering college
Number of applicants to engineering college .... 1,968 [2]
Percent offered admission 77.0%
Matriculated engineering students
Percentage in top quartile (25%) of High School class 92.0
Average SAT scores: Math—660[4], Verbal—540[4]
Average ACT scores: Composite—29
Notes: (2) Entering Fall 1992 (4) Mean.

FULL- & PART-TIME FACULTY—BY DEPARTMENT
[Figures are the head count of full-time faculty and the full-time equivalent (FTE) of part-time faculty for each engineering department or equivalent.]

Department	Full	Part
Chemical Eng	10	–
Civil Eng	11	0.6
Electrical Eng	25	2.3
Manufacturing Eng	8	–
Mechanical Eng	31	1.6
Biomedical Eng	4	–
Fire Protection Eng	3	–
Materials Eng	6	–

COLLEGE DESCRIPTION
WPI offers an undergraduate program, the WPI Plan, emphasizing a new approach to education. The academic year is based on four seven-week terms. Degree requirements include: the Major Qualifying Project which demonstrates the ability to solve problems in a chosen field; a Humanities Sufficiency to help develop skills through theoretical courses and independent study; and an Interactive Qualifying Project which investigates the relationship between technology and society. Co-op programs are optional and are available. Overseas exchange programs include sites in Germany, Sweden, Scotland and Ireland. Project Centers are located in London, Venice, Bangkok, Washington D.C., Puerto Rico, and San Francisco. Exceptional facilities are available to the students for their experimental laboratory activities. Combined BS/MS programs are available as well as double majors.

DEGREE PROGRAMS DESCRIPTION
The B.S. degree programs prepare graduates for entry-level professional employment or graduate study in their fields. Graduate study in engineering can be carried out at the M.S. and/or Ph.D. level. resources available to students range from CAD/CAM laboratories and computers to electron microscopes. Students may work with New England educational institutions which include teaching hospitals, research centers, industries, museums and cultural centers, thus staying at the forefront of their field. WPI offers the only graduate level Fire Protection Engineering program in the U.S. An M.B.A. is offered. Combined B.S./M.S. and B.S./M.B.A. programs are available to qualified students who elect to study for both degrees and WPI. Taking the Fundamentals of Engineering Exam is encouraged.

DUAL DEGREE PROGRAMS
Engineering baccalaureate graduates with dual degrees 9
TRANSFER INFORMATION
Residency Requirements: Two years of full-time study
Transfer via Articulation Agreements
Admission to engineering: At the end of the sophomore year
Requirements: In general, a 3.0 GPA is required.
Transfer without Articulation Agreements
Admission to engineering: At the end of the sophomore year
Requirements: In general, a 3.0 GPA is expected.
GRADUATION REQUIREMENTS
B.S. programs require completion of a Humanities Sufficiency, an Interactive Qualifying Project, a Major Qualifying Project and Departmental Distribution Requirements. In addition, students combining their requirements into a coherent program of international study, including an international IQP conducted abroad, are eligible for the designation "International Scholar." All requirements are detailed in the Undergraduate Catalog.

STUDENT PROGRAMS
PROFESSIONAL AND HONORARY SOCIETIES
[For key to acronyms, see *Introduction*.]
Professional Societies: ACM, AIAA, AIChE, ANS, ASCE, ASME, IEEE, SAE, SFPE, SME
Honorary Societies: Chi Epsilon, Eta Kappa Nu, Pi Tau Sigma, Tau Beta Pi, Phi Lambda Upsilon, Upsilon Pi Epsilon
SUPPORT PROGRAMS
Student Chapter Organizations: Society of Women Engineers
Office for Multi-Cultural Affairs
For: Women & Ethnic Minorities **Available:** Year round
Offered: Prematriculation, Freshman, Sophomore, Junior, Senior, Graduate level
Student Counseling Center
For: Women & Ethnic Minorities **Available:** Year round
Offered: Freshman, Sophomore, Junior, Senior, Graduate level
Office of Student Affairs
For: Women & Ethnic Minorities **Available:** Year round
Offered: Freshman, Sophomore, Junior, Senior, Graduate level
Office of Graduate and Career Plans
For: Women & Ethnic Minorities **Available:** Year round
Offered: Freshman, Sophomore, Junior, Senior, Graduate level
Other Engineering Support Programs: WPI offers fall orientation programs for freshmen and transfers; a summer program in science and math; career and counseling services; both residential and overseas project centers; cooperative education; a combined BS/MS program; a Math and Science peer tutoring program; an extensive Major Selection program; Manufacturing Engineering Applications Center; and a Writing Resource Center.

278 Wright State University

■ INSTITUTION PROFILE
HEAD OF THE INSTITUTION
Paige E. Mulhollan
Phone: (513)873-2312　　　　　　Fax: (513)873-2421
GENERAL INFORMATION
[All Students—Fall 1991]
Undergraduate enrollment 13,831
Graduate enrollment 3,930
Total institution enrollment **17,761**
Type of institution: Public　　　**Calendar:** Quarters
Nearest city: Dayton　　　　　　**Population:** 182,000
Miles from main campus: 10　　　**Setting:** Suburban
Types of engineering degrees: Engineering
Other degree-granting colleges: Business Administration, Education, Science and Mathematics, Liberal Arts, Nursing, Medicine, Professional Psychology

■ ENGINEERING ADMISSIONS
ADMISSIONS OFFICE CONTACT
Kenneth A. Davenport
Wright State University
Dayton, OH 45435-0001
Phone: (513)873-2211　　　　　　Fax: (513)873-3301
ENGINEERING COLLEGE ADMISSIONS INFORMATION
Admission to the engineering college: The end of the freshman year after meeting college/department entrance requirements.
Entrance Requirements: High School courses—English (4 years), Mathematics including Algebra I and II (3 years), Science (3 years), Social sciences (including two units of history) (3 years), ACT or SAT scores are required for admission.
Entrance Recommendations: High School courses—Foreign language (required) (2 years), Arts (required) (1 year), Chemistry (1 year), Trigonometry (1 year)
Requirements for out-of-state residents: Ohio residents must have graduated from an accredited high school or have passed an equivalency test (GED). Out-of-state residents must present evidence of above-average ability to do college work and meet the regular admission requirements.
Requirements for foreign students: TOEFL (Score: 500); Financial certification; complete academic record; descriptive syllabus of courses taken from foreign universities.
DEPARTMENTAL ADMISSIONS INFORMATION
Admission to the engineering department: The end of the freshman year after meeting college/department entrance requirements.
Additional information: Admission to a department major requires a 2.25 GPA (on a 4-point scale); completion of 45 quarter credits of college level work; and, completion of specified core courses with a C or better in each course.

■ FINANCIAL INFORMATION
ESTIMATED EXPENSES (ACADEMIC YEAR)
[Expenses are for the 1992-93 nine-month academic year.]

State Residents	Undergraduate	Graduate
Tuition and fees	$ 2,949	$ 3,723
College room and board	$ 4,353	$ 4,353
Books and supplies	$ 660	$ 660
Other expenses	$ 1,731	$ 1,731[1]
Total estimated expenses	**$ 9,693**	**$10,467**
Out-of-State Residents	Undergraduate	Graduate
Tuition and fees	$ 5,883	$ 6,657
College room and board	$ 4,353	$ 4,353
Books and supplies	$ 660	$ 660
Other expenses	$ 1,731	$ 1,731
Total estimated expenses	**$12,627**	**$13,401**

Notes: (1) For all on-campus students, the estimated transportation allowance is $639, and the estimated allowance for personal expenses is $1,092.
FINANCIAL AID OFFICE CONTACT
David R. Darr, *Director of Financial Aid*
Wright State University
Dayton, OH 45435-0001
Phone: (513)873-2321　　　　　　Fax: (513)873-3301
GENERAL FINANCIAL AID INFORMATION
Forms accepted/required: Institutional, Free Application for Federal Student Aid (FAFSA)
Additional financial aid information: University offers merit-based scholarships, need-based scholarships, and need-based federal and state financial aid. Part-time campus employment also is available.

■ ENGINEERING COLLEGE INFORMATION
[For additional personnel, refer to the *Appendix*.]
HEAD OF ENGINEERING
James E. Brandeberry
Phone: (513)873-5007　　　　　　Fax: (513)873-5009
EMail: jbrand@matrix.cs.wright.edu
ENGINEERING COLLEGE ADDRESS
College of Engineering and Computer Science
Wright State University
Dayton, OH 45435-0001
Phone: (513)873-5001　　　　　　Fax: (513)873-5009
ENROLLMENTS—BY CLASS
[Numbers are baccalaureate enrollments for the fall 1991 term, unless otherwise footnoted.]
1st-year students/Freshmen .... 482[1]
2nd-year students/Sophomores 293
3rd-year students/Juniors 286
4th-year students/Seniors 581
Total **1,642**
Notes: (1) Data in this section includes computer science students.
NUMBER OF DEGREES AWARDED—BY PROGRAM
[Numbers are engineering baccalaureate degrees awarded during the 1991-1992 academic year, unless otherwise footnoted. For full details about each engineering program, refer to the *Tables of Degree Programs*.]
Biomedical Engineering 12
Computer Engineering 36
Computer Science 33
Electrical Engineering 59
Engineering Physics 3
Human Factors Engineering 13
Materials Science & Engineering 4
Mechanical Engineering 44
Total **204**
PERCENTAGE OF DEGREES AWARDED—BY CATEGORY
[Percentages are of all engineering baccalaureate degrees awarded during the 1991-1992 academic year, unless otherwise footnoted.]
African Americans 1.5%
Asian/Pacific Island Americans 6.3%
Hispanic Americans .... – %
Native Americans .... – %
Foreign Citizens 11.7%
All Others 80.5%
Women 10.7%
Persons with disabilities .... – %[A]
Students over 25 years of age .... – %[A]
Notes: (A) Data not available.
GRADUATE ENROLLMENTS & DEGREES AWARDED
Master's enrollment 339
Master's degrees awarded 73
Doctoral enrollment 34
Doctoral degrees awarded .... –
ENGINEERING STUDENT DATA
Applicants to the engineering college
Number of applicants to engineering college .... 529[1]
Percent offered admission 94.1%
Matriculated engineering students
Percentage in top quartile (25%) of High School class 47.0
Average SAT scores: Math—524, Verbal—449, Combined—973
Average ACT scores: Math—23, Composite—23
Notes: (1) Data in this section includes computer science students.
FULL- & PART-TIME FACULTY—BY DEPARTMENT
[Figures are the head count of full-time faculty and the full-time equivalent (FTE) of part-time faculty for each engineering department or equivalent.]

Department	Full	Part
Electrical Eng	18	1.0
Mechanical & Materials Eng	12	5.0
Biomedical & Human Factors Eng	8	1.0
Computer Science & Eng	23	4.0

COLLEGE DESCRIPTION
The college offers four-year undergraduate programs in engineering and computer science. The first two years emphasize fundamentals, followed by two years of specialized education in one of eight majors. Emphasis is placed on computer-aided engineering and design. A dual major program is available with the Department of Mathematics and Statistics. Honors programs are available in all departments. Co-op is an option in all undergraduate programs. Most courses are offered in both day and evening to allow employed students to complete their studies on a part-time basis. Senior students are prepared and encouraged to take the Fundamentals of Engineering examination.
DEGREE PROGRAMS DESCRIPTION
The undergraduate degree programs prepare students for professional positions or graduate study in each of the respective fields. The M.S. program in systems engineering is broad in scope, offering the students the opportunity to concentrate in biomedical, electrical, human factors, materials science, or mechanical core areas. The M.S. program in computer science, the M.S. program in computer engineering and the Ph.D. program in computer science and engineering include concentration of study in specific areas of computer science and engineering. Programmatic strength lies in the unique blend of faculty expertise, in the marriage of theory with software and hardware design, and in the laboratory facilities available to the program.
TRANSFER INFORMATION
Residency Requirements: Students must complete all requirements in their degree program. A minimum of 45 quarter credit hours must be earned at Wright State, 30 of which must be in courses numbered 300 or above. At least 15 of the last 45 hours of credit must be taken at Wright State.
Transfer via Articulation Agreements
Admission to engineering: Transfer programs are articulated at the freshman and soph level with local community colleges. Transfer students are admitted to a department major when they meet college/department entrance requirements.
Transfer without Articulation Agreements
Admission to engineering: Transfer students are admitted to a department major when they meet college/department entrance requirements.
Requirements: International students should apply for admission four months prior to the quarter in which they wish to enroll.
GRADUATION REQUIREMENTS
Undergraduate degree programs require 201-207 quarter credits depending on the major selected. Students must earn at least a minimum cumulative grade point average of 2.0 and at least a minimum cumulative grade point average of 2.0 in their major; complete the General Education requirements; and, fulfill the university's residence requirements.

■ STUDENT PROGRAMS
PROFESSIONAL AND HONORARY SOCIETIES
[For key to acronyms, see *Introduction*.]
Professional Societies: ACM, AIAA, ASME, BMES, IEEE, NSPE, SAE, SAME, SPS, American Society of Metals International, Human Factors Engineering Society
Honorary Societies: Tau Beta Pi, Sigma Pi Sigma, Alpha Sigma Mu
SUPPORT PROGRAMS
Student Chapter Organizations: Society of Women Engineers, National Society of Black Engineers
WRIGHT STEPP Pre-Engineering Summer Program
For: Ethnic Minorities　　　**Available:** Summer only
Offered: Prematriculation
Horizons in Science and Engineering
For: Ethnic Minorities　　　**Available:** Summer only
Offered: Prematriculation
Bolinga Cultural Resources Center
For: Ethnic Minorities　　　**Available:** Year round
Offered: Freshman, Sophomore, Junior, Senior, Graduate level
Other Engineering Support Programs: Orientation and welcome programs; freshmen orientation course and mentoring program; counseling and peer support services; handicapped student services; learning-disabled services; tutoring services; developmental education program; veteran's affairs; career services; cooperative education; International Student Office; student employment; Army and Air Force ROTC; International Exchange Program.

279 University of Wyoming

INSTITUTION PROFILE

HEAD OF THE INSTITUTION
Terry P. Roark
Phone: (307)766-4121   Fax: (307)766-2271

GENERAL INFORMATION
[All Students—Fall 1991]
Undergraduate enrollment 9,444
Graduate enrollment 3,212
Total institution enrollment **12,656**

Type of institution: Public	**Calendar:** Semesters
Nearest city: Cheyenne, WY	**Population:** 55,000
Miles from main campus: 45	**Setting:** Small Town

Types of engineering degrees: Engineering
Other degree-granting colleges: Arts & Sciences, Education, Law, Agriculture, Health Sciences, Commerce & Industry

ENGINEERING ADMISSIONS

ADMISSIONS OFFICE CONTACT
Richard A. Davis
University of Wyoming
Office of Admissions
Box 3435
Laramie, WY 82071-3435
Phone: (307)766-5160   Fax: (307)766-4042

ENGINEERING COLLEGE ADMISSIONS INFORMATION
Entrance Recommendations: SAT, ACT, High School courses—Math (3 years), Science (2 years), Minimum scores not applicable.
Requirements for out-of-state residents: Graduation from an accredited high school.
Requirements for foreign students: TOEFL (Score: 525); financial statement

DEPARTMENTAL ADMISSIONS INFORMATION
Admission to the engineering department: At the time of admission to the institution.

FINANCIAL INFORMATION

ESTIMATED EXPENSES (ACADEMIC YEAR)
[Expenses are for the 1992-93 nine-month academic year.]

State Residents	Undergraduate	Graduate
Tuition and fees	$ 1,430	$ 1,722
College room and board	$ 4,009	$ 4,009
Books and supplies	$ 960	$ 960
Other expenses	$ 1,945 [1]	$ 1,945 [1]
Total estimated expenses	**$ 8,344**	**$ 8,636**

Out-of-State Residents	Undergraduate	Graduate
Tuition and fees	$ 2,251	$ 2,397
College room and board	$ 4,009	$ 4,009
Books and supplies	$ 960	$ 960
Other expenses	$ 1,945 [1]	$ 1,945 [1]
Total estimated expenses	**$ 9,165**	**$ 9,311**

Notes: (1) Personal expenses.

FINANCIAL AID OFFICE CONTACT
John F. Nutter, *Director*
University of Wyoming
Office of Financial Aid
Box 3335
Laramie, WY 82071-3335
Phone: (307)766-2118   Fax: (307)766-4042

GENERAL FINANCIAL AID INFORMATION
Forms accepted/required: FAF, FFS, UW Application for Financial Aid
Additional financial aid information: Student loans, work-study, scholarships.

ENGINEERING COLLEGE INFORMATION
[For additional personnel, refer to the *Appendix*.]

HEAD OF ENGINEERING
Samuel D. Hakes
Phone: (307)766-4253   Fax: (307)766-4444

ENGINEERING COLLEGE ADDRESS
College of Engineering
University of Wyoming
Box 3295
Laramie, WY 82071-3295
Phone: (307)766-4253   Fax: (307)766-4444

ENROLLMENTS—BY CLASS
[Numbers are baccalaureate enrollments for the fall 1991 term, unless otherwise footnoted.]
1st-year students/Freshmen 263
2nd-year students/Sophomores 161
3rd-year students/Juniors 175
4th-year students/Seniors 331
Total **930**

NUMBER OF DEGREES AWARDED—BY PROGRAM
[Numbers are engineering baccalaureate degrees awarded during the 1991-1992 academic year, unless otherwise footnoted. For full details about each engineering program, refer to the *Tables of Degree Programs*.]
Chemical Engineering 6
Civil & Architectural Engineering 54
Electrical Engineering 45
Mechanical Engineering 28
Petroleum Engineering 22
Total **155**

PERCENTAGE OF DEGREES AWARDED—BY CATEGORY
[Percentages are of all engineering baccalaureate degrees awarded during the 1991-1992 academic year, unless otherwise footnoted.]
African Americans .... – %
Asian/Pacific Island Americans .... – %
Hispanic Americans .... 0.7 %
Native Americans .... 0.7 %
Foreign Citizens .... 7.8 %
All Others .... 90.8 %
Women .... 10.5 %
Persons with disabilities .... – % (A)
Students over 25 years of age .... – % (A)

Notes: (A) Data not available.

GRADUATE ENROLLMENTS & DEGREES AWARDED
Master's enrollment 159
Master's degrees awarded 34
Doctoral enrollment 28
Doctoral degrees awarded 2

ENGINEERING STUDENT DATA
Applicants to the engineering college
Number of applicants to engineering college .... – (A)
Percent offered admission .... –% (A)
Matriculated engineering students
Percentage in top quartile (25%) of High School class 62.3
Average ACT scores: Math—25.3, Composite—24.5

Notes: (A) Data not available.

FULL- & PART-TIME FACULTY—BY DEPARTMENT
[Figures are the head count of full-time faculty and the full-time equivalent (FTE) of part-time faculty for each engineering department or equivalent.]

Department	Full	Part
Chemical Eng	7	–
Civil & Architectural Eng	24	–
Electrical Eng	16	1.2
Mechanical Eng	12	–
Petroleum Eng	6	0.5

COLLEGE DESCRIPTION
The College of Engineering requires that all full-time undergraduate students complete at least 12 semester hours of course work applicable to an engineering degree and have a GPA of 2.0 or better each semester. The required credit hours can be completed in a four-year program, but because of the rigorous nature of some courses involved, students may require additional time to complete degree requirements. Co-op programs are available but not required.

DEGREE PROGRAMS DESCRIPTION
All degree programs offer the B.S., M.S. and Ph.D. degrees.

TRANSFER INFORMATION
Transfer without Articulation Agreements
Admission to engineering: At any time
Engineering graduates transferred from . . 4-yr: –   2-yr: –

GRADUATION REQUIREMENTS
A cumulative GPA of 2.00 is required by the University of Wyoming for all bachelor's degrees. In addition, Engineering students must earn a cumulative grade point average of 2.00 in all Engineering courses taken at UW. A grade of 'C' or better must be earned in all courses which are prerequisite to any required engineering science course. B.S. programs in engineering require 132 credit hours. Other requirements are listed in the Undergraduate Bulletin.

STUDENT PROGRAMS

PROFESSIONAL AND HONORARY SOCIETIES
[For key to acronyms, see *Introduction*.]
Professional Societies: AIChE, ASAE, ASCE, ASME, IEEE, SAE, SPE, Society of Women Engineers (SWE)
Honorary Societies: Tau Beta Pi

SUPPORT PROGRAMS
Student Chapter Organizations: Society of Women Engineers

Minority Affairs Office
For: Women & Ethnic Minorities   **Available:** Year round
Offered: Freshman, Sophomore, Junior, Senior, Graduate level

Women's Center
For: Women   **Available:** Year round
Offered: Freshman, Sophomore, Junior, Senior, Graduate level

Other Engineering Support Programs: During the summer, the university offers both freshmen and transfer orientation programs on campus. Counseling is offered through the Office of Academic Advising.

280 Yale University

INSTITUTION PROFILE
HEAD OF THE INSTITUTION
Howard R. Lamar
Phone: (203)432-2550 Fax: (203)432-7105

GENERAL INFORMATION
[All Students—Fall 1991]
Undergraduate enrollment	5,295
Graduate enrollment	5,598
Total institution enrollment	**10,893**

Type of institution: Private **Calendar:** Semesters
Location: New Haven **Population:** 130,000
Setting: Urban
Types of engineering degrees: Engineering
Other degree-granting colleges: Agricultural & Environmental, Architecture, Arts & Sciences, Business Administration, Fine & Performing Arts, Law, Medicine, Nursing, Divinity, Forestry, EPH, Music, Art

ENGINEERING ADMISSIONS
ADMISSIONS OFFICE CONTACT
Richard H. Shaw Jr.
Yale University
1502 Yale Station
149 Elm Street
New Haven, CT 06520
Phone: (203)432-1900 Fax: (203)432-7329

ENGINEERING COLLEGE ADMISSIONS INFORMATION
Entrance Requirements: SAT, ACT, We require the SAT and 3 Achievement Tests. Students may substitute the ACT.
Entrance Recommendations: High School courses—Mathematics (3 years), Chemistry (1 year), Physics (1 year)
Requirements for foreign students: TOEFL; We require a preliminary application form. A financial statement is required from students requesting financial aid.

DEPARTMENTAL ADMISSIONS INFORMATION
Admission to the engineering department: At the end of the second year.
Additional information: All Yale students must complete specific core course requirements. If a student plans to be in an accredited program in engineering, he or she must take math and either physics or chemistry in the freshman year.

FINANCIAL INFORMATION
ESTIMATED EXPENSES (ACADEMIC YEAR)
[Expenses are for the 1992-93 nine-month academic year.]

All Students	Undergraduate	Graduate
Tuition and fees	$17,500	$15,920[1]
College room and board	$6,200	$7,065[1]
Books and supplies	$1,920[2]	$900[1]
Other expenses	$–	$2,955[2]
Total estimated expenses	**$25,620**	**$26,840**

Notes: (1) Expenses are for students enrolled in the Graduate School of Arts and Sciences. Expenses vary for students enrolled in the University's professional Schools (2) Includes hospitalization, personal expenses, and transportation.

FINANCIAL AID OFFICE CONTACT
Jacqueline Foster, *Director of Financial Aid*
Yale University
143 Elm Street
Box 2170 Yale Station
New Haven, CT 06520
Phone: (203)432-0360

GENERAL FINANCIAL AID INFORMATION
Forms accepted/required: AFSA, CSS-FAF, FAF, FAT, Stafford, SAR, Institutional, Parent and student tax returns, CSS Business/Farm &/or Divorced/Separated Parents Statement
Additional financial aid information: All scholarships are based on demonstrated financial need. However, any student at Yale may work part-time as a student aid, regardless of financial need.

ENGINEERING COLLEGE INFORMATION
[For additional personnel, refer to the *Appendix*.]
HEAD OF ENGINEERING
Gary L. Haller, *Chairman*
Phone: (203)432-4220 Fax: (203)432-2797

ENGINEERING COLLEGE ADDRESS
Council of Engineering
Yale University
P.O. Box 1968 Yale Station
10 Hillhouse Avenue
New Haven, CT 06520-1968
Phone: (203)432-4220 Fax: (203)432-2797

ENROLLMENTS—BY CLASS
[Numbers are baccalaureate enrollments for the fall 1991 term, unless otherwise footnoted.]
1st-year students/Freshmen	(1)
2nd-year students/Sophomores	(1)
3rd-year students/Juniors	39
4th-year students/Seniors	47
Total	**86**

Notes: (1) Major is not declared until Junior year.

NUMBER OF DEGREES AWARDED—BY PROGRAM
[Numbers are engineering baccalaureate degrees awarded during the 1991-1992 academic year, unless otherwise footnoted. For full details about each engineering program, refer to the *Tables of Degree Programs*.]
Applied Physics	4
Chemical Engineering	–
Electrical Engineering	20
Mechanical Engineering	20
Total	**44**

PERCENTAGE OF DEGREES AWARDED—BY CATEGORY
[Percentages are of all engineering baccalaureate degrees awarded during the 1991-1992 academic year, unless otherwise footnoted.]
African Americans	9.0%
Asian/Pacific Island Americans	11.3%
Hispanic Americans	4.5%
Native Americans	–%
Foreign Citizens	11.3%
All Others	63.9%
Women	13.6%
Persons with disabilities	–%[2]
Students over 25 years of age	–%

Notes: (2) Information about disabilities is not available.

GRADUATE ENROLLMENTS & DEGREES AWARDED
Master's enrollment	9
Master's degrees awarded	7
Doctoral enrollment	144
Doctoral degrees awarded	25

ENGINEERING STUDENT DATA
Applicants to the engineering college
Number of applicants to engineering college	875
Percent offered admission	21.1%

Average SAT scores: Math—695[2], Verbal—650[2], Combined—1345[2]
Average ACT scores: Composite—30

Notes: (2) SAT scores are averages of ranges for the middle 50% of our entering class. These ranges are: Math 650-740; Verbal 600-700; Combined 1250-1440.

FULL- & PART-TIME FACULTY—BY DEPARTMENT
[Figures are the head count of full-time faculty and the full-time equivalent (FTE) of part-time faculty for each engineering department or equivalent.]
Department	Full	Part
Chemical Eng	8	2.0
Electrical Eng	15	2.0
Mechanical Eng	10	3.0
Applied Physics	13	–

COLLEGE DESCRIPTION
Engineering is taught in a number of programs offered by departments within the College rather than in a school distinct from the arts and sciences. Thus students following a degree program in engineering have ready access to a wide range of courses in the arts and sciences and the opportunity to explore other disciplines while pursuing their specialization. At the same time, students interested in engineering but not intending to specialize in it have the opportunity to become acquainted with some aspects of modern technology. Engineering programs are offered in the departments of Applied Physics, Chemical Engineering, Electrical Engineering, and Mechanical Engineering. Curricula in the programs range from those accredited by the Accreditation Board for Engineering and Technology (ABET), to those with a lesser technical content, allowing students freedom to include focuses of a non-technical nature in their studies. In addition, a related major in Applied Mathematics is available.

DEGREE PROGRAMS DESCRIPTION
The distribution requirements for the Bachelor's Degree are intended to assure that by graduation all students are competent in a foreign language at the intermediate level and that their work, including their studies in their major programs, has been grounded in a sound acquaintance with a variety of fields of inquiry and approaches to knowledge. Yale does not require prescribed courses in specific subjects, but instead encourages undergraduates to design programs that best reflect their own intellectual interests, that open the maximum range of intellectual opportunities through which those interests can be expanded, and that direct their studies to that most elusive of goals, a liberal education.

TRANSFER INFORMATION
Residency Requirements: Students transferring to Yale from other universities must be enrolled at Yale for at least two years.
Transfer without Articulation Agreements
Admission to engineering: Students are ineligible for transfer admission to Yale if they have completed more than two years of college study at another institution.
Engineering graduates transferred from ... 4-yr: – 2-yr: –
Requirements: SAT or ACT. The transfer program is very small. Last year, 626 students applied for admission; 35 were offered admission.

STUDENT PROGRAMS
PROFESSIONAL AND HONORARY SOCIETIES
[For key to acronyms, see *Introduction*.]
Professional Societies: ASME, IEEE, Yale Science & Engineering Association
Honorary Societies: Tau Beta Pi, Sigma Xi

SUPPORT PROGRAMS
Student Chapter Organizations: Society of Women Engineers, National Society of Black Engineers

Women In Science
For: Women **Available:** Academic year
Offered: Freshman, Sophomore, Junior, Senior, Graduate level

Academic Mentorship Program in the Sciences
For: Ethnic Minorities
Offered: Freshman, Sophomore, Junior, Senior, Graduate level

Other Engineering Support Programs: A freshman orientation meeting for all those expressing an interest in Engineering and Applied Science; math/science tutors in the residential colleges; freshman counselors; 12 college deans; faculty advisor assigned to each freshman; ethnic counselors; directors of undergraduate studies; three-day freshman orientation program.

Educating the Technological Leaders of the 21st Century

The School of Engineering and Applied Science (SEAS) at SMU educates the "complete engineer" with in-depth technological programs and a broad foundation in the humanities and social sciences. Other significant educational advantages include:

Small classes taught by faculty. Students, even at the first-year level, learn from full-time faculty and senior professors — not teaching assistants in large lecture halls.

Emphasis on quality programs. We focus our resources on the fields in engineering and applied science that are most in demand — computer engineering, electrical engineering, mechanical engineering, computer science, and management science.

Exciting specializations and other options. For instance, we offer one of the best biomedical specialization programs in the nation. Ninety-nine percent of all SEAS students with a biomedical specialization who have applied to medical school have been accepted. Students also can choose double majors such as our recently added dual degree program in electrical engineering and physics.

An option to earn while you learn. Students in the SEAS cooperative education program and the minority co-op program can earn approximately $45,000, with full participation, while they live on campus.

Opportunities for women to excel. Our enrollment of women engineering students is about twice the national average, and women engineers are leaders throughout SEAS and SMU.

The Dallas advantage. Because SMU is located in the heart of one of the nation's top three high-tech centers, SEAS offers undergraduates an added dimension that many other schools cannot. For instance, we already have begun an extensive collaboration with the recently established Superconducting Super Collider (SSC) Laboratory, only a 30-minute drive from campus.

For more information, call: Southern Methodist University Office of Undergraduate Admission   Toll-free: 1-800-323-0672 or (214) 768-2058

Engineering Technology Section

ENGINEERING TECHNOLOGY SECTION
Participating Engineering Technology Institutions .... **TAB**
An Introduction to Engineering Technology, by Stephen Cheshier 426
Engineering Technology Disciplines: Descriptions and Career Opportunities, by Fred Emshousen 426
Index to Engineering Technology Degree Programs 428

Comparative Tables–by School
Undergraduate Degrees Awarded–by Category 431
Support Programs for Ethnic Minorities & Women 433
Enrollments and Degrees Awarded 435

Comparative Table–by Degree Program
All Engineering Technology Disciplines 439

Institution Profiles 448

An Introduction to Engineering Technology

An Introduction to Engineering Technology
By Stephen R. Cheshier—Southern College of Technology

The field of engineering technology came into its own after WWII, as many veterans who had received technical training took advantage of the G. I. Bill to continue their education to the associates degree level. As a result, many of todays engineering technology colleges were either established or extended at that time.

After Sputnik in 1957, a stronger emphasis was placed on preparation for research and on integrating more math and science into engineering curricula. In many research universities, this necessitated a decreased emphasis on the arts and on practical elements of engineering as they applied to industrial preparation. Seeing this developing void, many of the stronger two-year engineering technology colleges added two more years of advanced applied technical work, thus establishing the baccalaureate engineering technology degree (the first of which was accredited in the late 1960s by ECPD–now TAC/ABET). Most also preserved their strong associates programs.

The baccalaureate degree programs grew both in quantity and quality during the next several decades, such that today there are 315 TAC/ABET-accredited BS/BET programs in over 90 disciplines in about 110 colleges and universities. At the associates degree level, there are 450 TAC/ABET-accredited programs in about 120 disciplines in approximately 160 colleges and universities. Many others exist that have not sought TAC/ABET accreditation.

The most popular program is electrical/electronics, followed by mechanical/manufacturing, and civil engineering technologies. These three categories account for 85 percent of the graduates nationally.

Engineering technology programs are characterized by their focus on application and practice, and by their approximately 50/50 mix of theory and laboratory experience. Typically, the faculty hold technical masters degrees (usually in engineering, but occasionally in engineering technology), and many are registered professional engineers.

Students recruited into engineering technology programs usually come from the top half of their graduating class, rather than from the top 5-10 percent exclusively (as is the case in some of the major engineering programs). Women and minorities enroll at about the same percentages as in engineering programs.

Graduates are recruited by most major technological companies in the U.S. They are employed across the technological spectrum, but are better suited to areas that deal with application, implementation, and production, as opposed to the conceptual design and research functions. The technical sales and customer services fields also account for many placements.

A definition recently adopted by ASEE's Engineering Technology Council illustrates the programs educational emphasis and the most fitting placements for their graduates:

Engineering technology is the profession in which a knowledge of mathematics and natural sciences gained by higher education, experience, and practice is devoted primarily to the implementation and extension of existing technology for the benefit of humanity. Engineering technology education focuses primarily on the applied aspects of science and engineering aimed at preparing graduates for practice in that portion of the technological spectrum closest to the product improvement, manufacturing, and engineering operational functions.

Engineering Technology Disciplines: Descriptions and Career Opportunities
By Fred Emshousen—Purdue University

Civil and Construction Engineering Technology
Civil and construction engineering technology stresses production rather than design, and management practices rather than skills. It encompasses the entire construction process—managing people, machines, and money in the civil engineering and construction industry. All areas of commerce, transportation, water systems, manufacturing, utilities, and housing require facilities developed by civil engineers and civil and construction engineering technologists.

New housing required by the expanding population, the need for new commercial facilities, and the modernization of existing facilities should create many new employment opportunities in the construction industry. Civil and construction technologists work in the office, the field, or both. They are often project managers, estimators, schedulers, and cost technologists and work for all types of contractors–residential, building, industrial, highway, bridge, mechanical, electrical, and specialty.

Electrical and Electronic Engineering Technology
Electrical and electronic engineering technology combines engineering knowledge and methods with technical skills to support engineering activities, emphasizing practical application rather than theory and design. Areas of concentration include automated manufacturing and test operations, industrial controls, microcomputer-related hardware applications, instrumentation, linear analog integrated circuits, power generation and distribution, digital communications, and biomedical electronic applications.

The continual development and application of new electronic devices and systems ensures a strong demand for electrical and electronic technologists. Major industrial and business concerns across the nation have interesting and challenging positions available in aeronautics, medical electronics, computers, broadcasting, telecommunications, factory automation, and robotics. Graduates may work within the engineering team in applied design, product development, manufacturing, production, or technical operations. The demand for electrical and electronic technologists is expected to grow by 50% by 1995, more than three times the average rate of all occupations.

Mechanical Engineering Technology
Mechanical engineering technology is concerned with the implementation of current mechanical engineering practices. Mechanical engineering technologists apply basic knowledge of mathematics and engineering science to the solution of design problems and to the operation and testing of engineering and mechanical systems. Individuals can enter the field with either a two-year associates degree or a four-year baccalaureate degree.

The growing demand for modern and complex industrial machinery, machine tools, and computer-controlled processes makes the employment outlook excellent for mechanical engineering technologists. Positions are available in power generation, manufacturing operations, project planning, production supervision, plant operations, quality assurance, reliability testing, and technical sales and services. The occupational growth rate for mechanical engineering technologists is expected to be about 37% - more than twice the projected growth of 15% for all occupations.

Manufacturing Engineering Technology

Manufacturing engineering technology is concerned with the many facets of making products. Manufacturing technologists are key people in the process that takes the design engineers concept and specifications and helps to translate them into actual production of manufactured goods at the lowest possible cost. Manufacturing technologists direct and coordinate manufacturing processes in industrial plants. Technical areas of work encompass planning, production methods, fabrication, assembly, materials handling, scheduling, quality assurance and a wide range of related manufacturing activities.

Due to the rapidly changing nature of modern plant manufacturing, the manufacturing technologist can expect to find diverse and challenging opportunities in both large and small industries. They can work in such areas as automated testing or product improvement. As technology grows more complex and competition becomes more keen, the demand for manufacturing technologists will continue to grow and to provide excellent job opportunities and salaries.

Index to Engineering Technology Degree Programs

A

Aeronautical Engineering Technology
(See also: Aircraft Engineering Technology; Aviation; Avionics)
283 Arizona State University
343 State University of New York College of Technology at Farmingdale

Aerospace Maintenance Engineering Technology
325 Northeastern University

Air Conditioning
(See: HVACR Engineering Technology)

Aircraft Engineering Technology
305 Embry-Riddle Aeronautical University - Daytona Beach

Aircraft Maintenance Engineering Technology
(See also: Aeronautical Engineering Technology; Aviation; Avionics)
356 Western Michigan University
343 State University of New York College of Technology at Farmingdale

Apparel Engineering Technology
(See also: Textile Engineering Technology)
339 Southern College of Technology

Architectural Engineering Technology
(See also: Construction/Architectural Engineering Technology)
293 University of Cincinnati
308 University of Hartford
316 Memphis State University
342 University of Southern Mississippi
352 Wentworth Institute of Technology

Automated Manufacturing Technology
281 University of Akron

Automated Systems Engineering Technology
310 Lake Superior State University

Automotive Engineering Technology
313 Mankato State University
343 State University of New York College of Technology at Farmingdale
356 Western Michigan University

Aviation Technology & Operations
(See also: Aeronautical Engineering Technology; Aircraft Engineering Technology; Avionics Engineering Technology)
356 Western Michigan University

Avionics Engineering Technology
(See also: Aeronautical Engineering Technology; Aircraft Engineering Technology; Aviation Engineering Technology)
305 Embry-Riddle Aeronautical University - Daytona Beach

B

Biomedical Engineering Technology
304 East Tennessee State University
343 State University of New York College of Technology at Farmingdale

Building Construction Technology
(See also: Civil Engineering Technology; Construction Engineering Technology)
352 Wentworth Institute of Technology

C

Chemical Process Technology
294 University of Dayton

Chemical Technology
293 University of Cincinnati

Civil & Construction Engineering Technology
(See also: Building Construction Technology; Construction Engineering Technology)
291 Central Connecticut State University
320 Murray State University
345 Temple University

Civil Engineering Technology
(See also: Building Construction Technology; Construction Engineering Technology; Surveying)
282 Alabama A&M University
315 University of Massachusetts, Lowell
317 Metropolitan State College of Denver
320 Murray State University
323 New Mexico State University
324 University of North Carolina-Charlotte
327 Old Dominion University
328 Oregon Institute of Technology
332 University of Pittsburgh at Johnstown
333 Point Park College
336 Rochester Institute of Technology
338 South Carolina State University
339 Southern College of Technology
340 University of Southern Colorado
341 Southern Illinois University at Carbondale
343 State University of New York College of Technology at Farmingdale
346 University of Tennessee, Martin
352 Wentworth Institute of Technology
355 Western Kentucky University

Civil & Surveying Engineering Technology
(See also: Surveying Engineering Technology)
353 West Virginia Institute of Technology

Communications
(See: Telecommunications Engineering Technology)

Computer Aided Drafting & Design Technology
(See also: Computing Graphics Technology; Design; Drafting)
349 Tri-State University

Computer Engineering Technology
(See also: Drafting & Computer Drafting Technology; Electrical & Computer Engineering Technology; Electronic(s) & Computer Technology)
284 University of Arkansas at Little Rock
290 Capitol College
292 University of Central Florida
308 University of Hartford
310 Lake Superior State University
316 Memphis State University
322 New Jersey Institute of Technology
325 Northeastern University
334 Prairie View A&M University
336 Rochester Institute of Technology
339 Southern College of Technology
342 University of Southern Mississippi
352 Wentworth Institute of Technology

Computer Integrated Manufacturing Technology
(See also: Manufacturing Engineering Technology)
309 Indiana University-Purdue University at Indianapolis
335 Purdue University

Computer Science
334 Prairie View A&M University

Computer Systems Technology
328 Oregon Institute of Technology
344 State University of New York Institute of Technology, Utica/Rome

Computing Graphics Technology
(See also: Computer Aided Design & Drafting Technology; Design; Drafting)
343 State University of New York College of Technology at Farmingdale

Construction & Contracting Engineering Technology
(See also: Contracting)
322 New Jersey Institute of Technology

Construction Engineering Technology
(See also: Building Construction Technology)
284 University of Arkansas at Little Rock
286 Bradley University
303 University of the District of Columbia
304 East Tennessee State University
311 Louisiana Tech University
319 Montana State University
331 Pittsburg State University
342 University of Southern Mississippi
348 Texas Tech University

Construction Management Technology
(See also: Management)
288 California State University, Sacramento
293 University of Cincinnati
312 University of Maine
326 Oklahoma State University
331 Pittsburg State University

Construction Science & Management
(See also: Management)
356 Western Michigan University

Construction/Architectural Engineering Technology
(See also: Architectural Engineering Technology)
309 Indiana University-Purdue University at Indianapolis
343 State University of New York College of Technology at Farmingdale

Contracting
(See: Construction & Contracting Engineering Technology; Construction Engineering Technology)

D

Design Engineering Technology
(See also: Computer Aided Drafting & Design Technology; Computing Graphics Technology; Drafting; Engineering Design Graphics; Engineering Graphics; Industrial Design; Mechanical Design Technology; Product Design Engineering Technology; Structural Design & Construction Engineering Technology)
292 University of Central Florida

Drafting & Computer Drafting Technology
(See also: Design; Drafting)
281 University of Akron

Drafting & Design Engineering Technology
(See also: Computer Aided Drafting & Design Technology; Computing Graphics Technology; Design Engineering Technology)
282 Alabama A&M University
310 Lake Superior State University
349 Tri-State University
353 West Virginia Institute of Technology

E

Education
(See: Technology Education)

Electrical & Computer Engineering Technology
(See also: Computer Engineering Technology)
320 Murray State University
330 Pennsylvania State University - Harrisburg

Electrical Engineering Technology
(See also: Computer Engineering Technology; Electronic(s) Engineering Technology)
282 Alabama A&M University
286 Bradley University
293 University of Cincinnati
309 Indiana University-Purdue University at Indianapolis

311 Louisiana Tech University
312 University of Maine
314 University of Massachusetts, Dartmouth
316 Memphis State University
320 Murray State University
321 University of New Hampshire
322 New Jersey Institute of Technology
324 University of North Carolina-Charlotte
325 Northeastern University
327 Old Dominion University
329 Pennsylvania State University at Erie, The Behrend College
332 University of Pittsburgh at Johnstown
333 Point Park College
334 Prairie View A&M University
335 Purdue University
336 Rochester Institute of Technology
338 South Carolina State University
339 Southern College of Technology
341 Southern Illinois University at Carbondale
343 State University of New York College of Technology at Farmingdale
344 State University of New York Institute of Technology, Utica/Rome
345 Temple University
346 University of Tennessee, Martin
353 West Virginia Institute of Technology
355 Western Kentucky University

Electrical/Electronic(s) Engineering Technology
287 Brigham Young University
290 Capitol College
306 Ferris State University
310 Lake Superior State University
318 Milwaukee School of Engineering
319 Montana State University
328 Oregon Institute of Technology
348 Texas Tech University

Electromechanical Engineering Technology
(See also: Electrical Engineering Technology; Mechanical Engineering Technology)
355 Western Kentucky University

Electromechanical Systems Engineering Technology
(See also: Electrical Engineering Technology; Mechanical Engineering Technology; Systems)
303 University of the District of Columbia

Electronic(s) Engineering Technology
(See also: Electrical Engineering Technology; Electrical/Electronics Engineering Technology)
281 University of Akron
283 Arizona State University
284 University of Arkansas at Little Rock
287 Brigham Young University
292 University of Central Florida
294 University of Dayton
295 DeVry Institute of Technology, Chicago
296 DeVry Institute of Technology, Columbus
297 DeVry Institute of Technology, Atlanta
298 DeVry Institute of Technology, DuPage
299 DeVry Institute of Technology, Dallas
300 DeVry Institute of Technology, Kansas City
301 DeVry Institute of Technology, Los Angeles
302 DeVry Institute of Technology, Phoenix
304 East Tennessee State University
307 Franklin University
308 University of Hartford
313 Mankato State University
315 University of Massachusetts, Lowell
317 Metropolitan State College of Denver
323 New Mexico State University
331 Pittsburg State University
340 University of Southern Colorado
342 University of Southern Mississippi
347 Texas A&M University
350 Virginia State University
351 Weber State University
352 Wentworth Institute of Technology
353 West Virginia Institute of Technology
354 Western Carolina University

Electronic(s) & Computer Technology
(See also: Computer Engineering Technology; Electrical Engineering Technology)
326 Oklahoma State University

Engineering Design Graphics
(See also: Design; Drafting; Graphics)
304 East Tennessee State University

Engineering Graphics
(See also: Design; Drafting; Graphics)
356 Western Michigan University

Engineering Metallurgy
356 Western Michigan University

Engineering Science
343 State University of New York College of Technology at Farmingdale

Engineering Technology
(See also: General Engineering Technology)
289 California State Polytechnic University, Pomona
342 University of Southern Mississippi
345 Temple University
347 Texas A&M University
353 West Virginia Institute of Technology

Environmental Engineering Technology
294 University of Dayton
320 Murray State University
330 Pennsylvania State University - Harrisburg
345 Temple University

F

Fire Protection & Safety Technology
326 Oklahoma State University

G

General Engineering Technology
(See also: Engineering Technology)
326 Oklahoma State University
337 St. Cloud State University

Graphics
(See: Computing Graphics Technology; Design; Drafting; Engineering Design Graphics; Engineering Graphics)

H

HVACR Engineering Technology
(Heating, Ventilation, Air Conditioning, and Refrigeration Engineering Technology)
306 Ferris State University

Heating
(See: HVACR Engineering Technology)

I

Imaging Engineering Technology
337 St. Cloud State University

Imaging Science
337 St. Cloud State University

Industrial Design
(See also: Design; Drafting; Graphics)
356 Western Michigan University

Industrial Engineering Technology
282 Alabama A&M University
294 University of Dayton
304 East Tennessee State University
309 Indiana University-Purdue University at Indianapolis
338 South Carolina State University
339 Southern College of Technology
342 University of Southern Mississippi
344 State University of New York Institute of Technology, Utica/Rome
353 West Virginia Institute of Technology

Information Systems Engineering Technology
(See also: Systems)
292 University of Central Florida

L

Land
(See: Surveying & Land Information Systems)

Laser Optical Engineering Technology
328 Oregon Institute of Technology

M

Maintenance
(See: Aerospace Maintenance Engineering Technology; Aircraft Maintenance Engineering Technology)

Management
(See: Construction Management Technology; Construction Science & Management; Textile Management & Technology)

Manufacturing Engineering Technology
(See also: Automated Manufacturing Technology; Computer Integrated Manufacturing Technology)
281 University of Akron
283 Arizona State University
284 University of Arkansas at Little Rock
286 Bradley University
287 Brigham Young University
291 Central Connecticut State University
294 University of Dayton
304 East Tennessee State University
306 Ferris State University
313 Mankato State University
316 Memphis State University
318 Milwaukee School of Engineering
320 Murray State University
322 New Jersey Institute of Technology
324 University of North Carolina-Charlotte
325 Northeastern University
326 Oklahoma State University
328 Oregon Institute of Technology
331 Pittsburg State University
336 Rochester Institute of Technology
337 St. Cloud State University
342 University of Southern Mississippi
343 State University of New York College of Technology at Farmingdale
347 Texas A&M University
349 Tri-State University
351 Weber State University
352 Wentworth Institute of Technology
354 Western Carolina University
356 Western Michigan University

Manufacturing Systems
(See also: Systems)
341 Southern Illinois University at Carbondale

Mapping
(See: Surveying & Mapping)

Mechanical Design Technology
(See also: Design; Drafting; Graphics)
326 Oklahoma State University

Index to Engineering Technology Degree Programs

Mechanical Engineering Technology
(See also: Electromechanical Engineering Technology; Electromechanical Systems Engineering Technology)
- 281 University of Akron
- 282 Alabama A&M University
- 284 University of Arkansas at Little Rock
- 288 California State University, Sacramento
- 293 University of Cincinnati
- 294 University of Dayton
- 307 Franklin University
- 309 Indiana University-Purdue University at Indianapolis
- 310 Lake Superior State University
- 312 University of Maine
- 314 University of Massachusetts, Dartmouth
- 315 University of Massachusetts, Lowell
- 317 Metropolitan State College of Denver
- 318 Milwaukee School of Engineering
- 319 Montana State University
- 321 University of New Hampshire
- 322 New Jersey Institute of Technology
- 323 New Mexico State University
- 324 University of North Carolina-Charlotte
- 325 Northeastern University
- 327 Old Dominion University
- 328 Oregon Institute of Technology
- 329 Pennsylvania State University at Erie, The Behrend College
- 330 Pennsylvania State University - Harrisburg
- 331 Pittsburg State University
- 332 University of Pittsburgh at Johnstown
- 333 Point Park College
- 334 Prairie View A&M University
- 335 Purdue University
- 336 Rochester Institute of Technology
- 338 South Carolina State University
- 339 Southern College of Technology
- 340 University of Southern Colorado
- 341 Southern Illinois University at Carbondale
- 342 University of Southern Mississippi
- 343 State University of New York College of Technology at Farmingdale
- 344 State University of New York Institute of Technology, Utica/Rome
- 345 Temple University
- 346 University of Tennessee, Martin
- 347 Texas A&M University
- 348 Texas Tech University
- 350 Virginia State University
- 351 Weber State University
- 352 Wentworth Institute of Technology
- 353 West Virginia Institute of Technology
- 355 Western Kentucky University

Mechanical Power Technology
- 326 Oklahoma State University

Mechanical-Structural Engineering Technology
(See also: Structural Design & Construction Engineering Technology)
- 325 Northeastern University

Medical
(See: Biomedical Engineering Technology)

Metallurgy
(See: Engineering Metallurgy)

O

Operations Engineering Technology
(See also: Aviation Technology & Operations; Technical Operations)
- 292 University of Central Florida

Optical
(See: Laser Optical Engineering Technology)

P

Plastics Engineering Technology
- 306 Ferris State University
- 329 Pennsylvania State University at Erie, The Behrend College
- 331 Pittsburg State University

Product Design Engineering Technology
(See also: Design)
- 306 Ferris State University

R

Refrigeration
(See: HVACR Engineering Technology)

S

Space
(See: Aerospace)

Structural Design & Construction Engineering Technology
(See also: Construction Engineering Technology; Design; Mechanical-Structural Engineering Technology)
- 330 Pennsylvania State University - Harrisburg

Surveying & Construction Technology
(See also: Civil Engineering Technology; Construction Engineering Technology)
- 281 University of Akron

Surveying & Land Information Systems
- 284 University of Arkansas at Little Rock

Surveying & Mapping
- 304 East Tennessee State University

Surveying Engineering Technology
(See also: Civil & Surveying Engineering Technology; Civil Engineering Technology; Construction Engineering Technology)
- 306 Ferris State University
- 322 New Jersey Institute of Technology
- 328 Oregon Institute of Technology
- 353 West Virginia Institute of Technology

Systems
(See: Automated Systems Engineering Technology; Computer Systems Technology; Electromechanical Systems Engineering Technology; Information Systems Engineering Technology; Manufacturing Systems; Surveying & Land Information Systems)

T

Technical Operations
(See also: Operations)
- 327 Old Dominion University

Technical Studies
- 353 West Virginia Institute of Technology

Technology Education
- 353 West Virginia Institute of Technology

Telecommunications Engineering Technology
- 290 Capitol College
- 336 Rochester Institute of Technology

Textile Engineering Technology
(See also: Apparel Engineering Technology)
- 339 Southern College of Technology

Textile Management & Technology
- 285 Auburn University

V

Ventilation
(See: HVACR Engineering Technology)

W

Welding Engineering Technology
- 306 Ferris State University

Undergraduate Degrees Awarded–by Category

Institutions are listed in alphabetical order. All figures below represent the percentages of bachelor's degree recipients who fall into different categories. The section devoted to students of different ethnic backgrounds and nationalities includes African Americans, Asian (including Pacific Island) Americans, Hispanic Americans, and Native Americans, as well as citizens of other countries. The figure for the column "All Others" was derived by subtracting the above percentages from 100.0%. In most most cases, Canadian schools could not provide data in corresponding categories for African Canadians, etc.; therefore, this information does not appear for most Canadian schools. In addition, information was requested about degree recipients in categories independent if ethnic or national background. These categories are: women, students with self-identified physical or learning disabilities, and students who were older than 25 years of age when they received their degrees. Unless footnoted otherwise, degrees awarded are for the 1991-92, twelve-month academic year.

ENGINEERING TECHNOLOGY

PERCENTAGE OF UNDERGRADUATE DEGREES AWARDED–BY CATEGORY

	Institution Profile Reference Number & Name	State/Province	Ethnicity/Nationality						Other Categories		
			African Americans	Asian Americans	Hispanic Amer.	Native Americans	Foreign Citizens	All Others	Women	Students with Disabilities	Over 25 Years of Age
281	University of Akron	OH	2.0	0.0	0.0	1.0	97.0	1.0	2.0	-.-(A)	-.-(A)
282	Alabama A&M University	AL	75.0	0.0	0.0	0.0	25.0	25.0	30.0	0.0	2.0
283	Arizona State University	AZ	0.0	6.4	3.2	1.0	89.4	9.8	3.0	-.-	-.-
284	University of Arkansas at Little Rock	AR	3.1	6.2	0.0	0.0	90.7	0.0	6.2	-.-(A)	47.9
285	Auburn University	AL	8.3	0.0	0.0	0.0	91.7	0.0	58.3	-.-(A)	-.-(A)
286	Bradley University	IL	5.0	5.0	1.0	0.0	89.0	2.0	6.0	-.-(A)	-.-(A)
287	Brigham Young University	UT	-.-(A)	-.-(A)	3.4	1.7	94.9	3.4	3.4	-.-(A)	-.-(1)
288	California State University, Sacramento	CA	2.8	11.9	10.6	2.3	72.4	0.5	6.0	-.-(A)	-.-(A)
289	California State Polytechnic University, Pomona	CA	3.0	43.0	17.0	-.-	37.0	36.0	8.0	-.-	-.-
290	Capitol College	MD	18.0	12.0	2.0	1.0	67.0	17.0	10.0	1.0	68.0
291	Central Connecticut State University	CT	6.0	1.0	0.0	0.0	93.0	0.0	6.0	0.4	-.-
292	University of Central Florida	FL	11.5	6.1	10.8	0.0	71.6	5.4(C)	33.9	0.2(C)	33.0(C)
293	University of Cincinnati	OH	1.6	3.5	-.-	0.0	94.9	0.0(A)	6.5	0.0(A)	0.0(A)
294	University of Dayton	OH	6.9	3.9	0.0	0.0	89.2	3.9	20.6	-.-(A)	-.-(A)
295	DeVry Institute of Technology, Chicago	IL	13.6	22.7	15.6	0.6	47.5	3.2	7.8	-.-(A)	4.8
296	DeVry Institute of Technology, Columbus	OH	7.7	3.2	-.-	0.0	89.1	5.0	3.2	-.-(A)	28.1(A)
297	DeVry Institute of Technology, Atlanta	GA	32.0	8.2	2.5	0.0	57.3	2.5	13.1	-.-(A)	36.9
298	DeVry Institute of Technology, DuPage	IL	2.9	9.6	4.8	0.0	82.7	0.0	0.0	-.-(A)	33.7
299	DeVry Institute of Technology, Dallas	TX	8.9	12.2	12.2	0.0	66.7	6.7	7.8	-.-(A)	60.0
300	DeVry Institute of Technology, Kansas City	MO	1.9	7.4	0.0	0.0	90.7	1.9	3.7	-.-(A)	51.9
301	DeVry Institute of Technology, Los Angeles	CA	11.0	30.1	21.9	1.4	35.6	2.7	1.4	-.-(A)	56.2
302	DeVry Institute of Technology, Phoenix	AZ	2.2	14.2	4.9	0.0	78.7	7.7	2.2	-.-(A)	48.1
303	University of the District of Columbia	DC	-.-(A)	-.-(A)	-.-(A)	-.-(A)	100.0	-.-(A)	-.-(A)	-.-(A)	-.-(A)
304	East Tennessee State University	TN	2.2	0.0	0.0	0.0	97.8	0.0	8.9	0.0	20.0
305	Embry-Riddle Aeronautical University - Daytona Beach	FL	4.4	11.5	12.7	0.0	71.4	-.-(A)	18.9	-.-(A)	-.-(A)
306	Ferris State University	MI	-.-(A)	-.-(A)	-.-(A)	-.-(A)	100.0	-.-(A)	-.-(A)	-.-(A)	-.-(A)
307	Franklin University	OH	-.-(1)	-.-(1)	-.-(1)	-.-(1)	100.0	-.-(1)	8.5	-.-(1)	-.-(1)
308	University of Hartford	CT	15.0	2.0	5.6	0.0	77.4	0.0	7.3	0.0(A)	0.0(A)
309	Indiana University-Purdue University at Indianapolis	IN	6.0	2.0	0.8	0.0	91.2	0.2	20.0	-.-(B)	-.-(B)
310	Lake Superior State University	MI	-.-	-.-	-.-	-.-	100.0	-.-	-.-	-.-	-.-
311	Louisiana Tech University	LA	33.0	0.0	0.0	0.0	67.0	0.0	0.1	0.0	0.0
312	University of Maine	ME	0.0	0.0	0.0	4.0	96.0	0.0	4.0	0.0	27.0
313	Mankato State University	MN	0.0	5.0	0.0	0.0	95.0	1.0	1.0	0.5	8.0
314	University of Massachusetts, Dartmouth	MA	-.-(A)	-.-(A)	-.-(A)	-.-(A)	100.0	-.-(A)	-.-(A)	-.-(A)	-.-(A)
315	University of Massachusetts, Lowell	MA	-.-(A)	-.-(A)	-.-(A)	-.-(A)	100.0	-.-(A)	-.-(A)	-.-(A)	-.-(A)
316	Memphis State University	TN	21.0	-.-(A)	-.-(A)	-.-(A)	79.0	-.-(A)	18.0	-.-(A)	12.0
317	Metropolitan State College of Denver	CO	-.-(A)	-.-(A)	-.-(A)	-.-(A)	100.0	-.-(A)	-.-(A)	-.-(A)	-.-(A)
318	Milwaukee School of Engineering	WI	2.0	2.0	1.0	1.0	94.0	1.0	10.0	-.-(A)	-.-(A)
319	Montana State University	MT	-.-(A)	-.-(A)	-.-(A)	-.-(A)	100.0	-.-(A)	-.-(A)	1.3	-.-(A)
320	Murray State University	KY	-.-(A)	-.-(A)	-.-(A)	-.-(A)	100.0	-.-(A)	-.-(A)	5.0	-.-(A)
321	University of New Hampshire	NH	0.0	0.0	0.0	0.0	100.0	0.0	2.2	0.0	-.-(A)
322	New Jersey Institute of Technology	NJ	8.5	7.6	6.8	0.0	77.1	2.5(2)	3.4	-.-(A)	76.3
323	New Mexico State University	NM	1.0	2.0	35.0	4.0	58.0	2.0	11.0	3.0	60.0
324	University of North Carolina-Charlotte	NC	5.2	3.5	0.9	0.9	89.5	3.5	5.2	0.0	69.3
325	Northeastern University	MA	-.-	-.-	-.-	-.-	100.0	-.-	-.-	-.-	-.-

Degrees Awarded—by Category

ENGINEERING TECHNOLOGY

| Institution Profile Reference Number & Name | State/Province | PERCENTAGE OF UNDERGRADUATE DEGREES AWARDED–BY CATEGORY |||||||||
| | | Ethnicity/Nationality |||||| Other Categories |||
		African Americans	Asian Americans	Hispanic Amer.	Native Americans	Foreign Citizens	All Others	Women	Students with Disabilities	Over 25 Years of Age
326 Oklahoma State University	OK	0.0	2.3	0.0	0.8	96.9	14.7	4.7	-.-(A)	-.-(A)
327 Old Dominion University	VA	4.5	7.4	-.-	0.0	88.1	2.2	12.6	0.0	-.-(A)
328 Oregon Institute of Technology	OR	0.0	5.5	1.5	1.0	92.0	2.0	5.5	-.-(A)	-.-(A)
329 Pennsylvania State University at Erie, The Behrend College	PA	-.-	-.-	-.-	-.-	100.0	-.-	-.-	-.-	81.0
330 Pennsylvania State University - Harrisburg	PA	2.4	-.-	-.-	-.-	97.6	-.-	2.0	-.-	30.1
331 Pittsburg State University	KS	1.3	1.3	1.3	0.0	96.1	11.5	3.8	-.-(A)	33.3
332 University of Pittsburgh at Johnstown	PA	-.-	-.-	-.-	-.-	100.0	-.-	10.8	-.-	-.-
333 Point Park College	PA	2.0	-.-	-.-	-.-	98.0	-.-(A)	4.4	-.-(A)	-.-(A)
334 Prairie View A&M University	TX	85.0	2.0	2.0	2.0	9.0	9.0	36.6	0.0	0.0
335 Purdue University	IN	-.-	-.-	1.0	-.-	99.0	1.2	2.0	-.-(A)	12.0
336 Rochester Institute of Technology	NY	1.0	3.0	0.1	0.3	95.6	1.0	4.0	0.0	30.0
337 St. Cloud State University	MN	0.0	10.0	0.0	0.0	90.0	0.0	35.0	0.0	25.0
338 South Carolina State University	SC	95.2	1.6	1.6	0.0	1.6	1.6	12.0	0.0	21.9
339 Southern College of Technology	GA	16.8	2.9	1.5	0.2	78.6	7.5	13.2	0.2	45.0
340 University of Southern Colorado	CO	2.4	2.4	30.6	1.9	62.7	7.3	7.1	-.-(A)	-.-(A)
341 Southern Illinois University at Carbondale	IL	5.7	1.9	1.9	0.0	90.5	9.6	7.6	0.0(B)	40.0
342 University of Southern Mississippi	MS	21.9	3.1	1.1	-.-	73.9	-.-(A)	15.4	-.-(A)	-.-(A)
343 SUNY College of Technology at Farmingdale	NY	-.-	-.-	-.-	-.-	100.0	-.-	-.-	-.-	-.-
344 SUNY Institute of Technology, Utica/Rome	NY	4.0	3.0	3.0	0.0	90.0	2.0	4.0	-.-(A)	-.-(A)
345 Temple University	PA	12.0	7.4	2.7	0.0	77.9	8.3	11.1	-.-(A)	-.-(A)
346 University of Tennessee, Martin	TN	5.0	0.0	0.0	0.0	95.0	15.0	10.0	0.0	25.0
347 Texas A&M University	TX	6.3	4.5	10.0	0.0	79.2	-.-	1.8	0.0(A)	41.0
348 Texas Tech University	TX	4.3	4.3	12.4	0.0	79.0	1.4	-.-	6.2	36.7
349 Tri-State University	IN	-.-(1)	-.-(1)	-.-(1)	-.-(1)	100.0	-.-(1)	-.-(1)	-.-(1)	-.-(1)
350 Virginia State University	VA	66.0	5.0	0.0	0.0	29.0	5.0	22.0	0.0	33.0
351 Weber State University	UT	1.0	3.7	2.0	0.6	92.7	2.3	8.1	-.-(A)	-.-(A)
352 Wentworth Institute of Technology	MA	7.9	5.0	2.9	0.2	84.0	6.1	11.4	-.-(A)	-.-(A)
353 West Virginia Institute of Technology	WV	7.0	3.0	3.0	3.0	84.0	3.0	33.0	1.0	-.-
354 Western Carolina University	NC	3.8	3.0	0.0	0.0	93.2	-.-	7.6	-.-	-.-
355 Western Kentucky University	KY	0.0	0.0	0.0	0.0	100.0	2.3	2.3	0.0	10.0
356 Western Michigan University	MI	3.1	-.-	-.-	-.-	96.9	3.1	9.9	-.-(A)	-.-(A)

FOOTNOTES: The following footnotes are the same for all schools: (A) Data not available. (B) Data not applicable. (C) Estimated data.

287 **Brigham Young University:** (1) Average student age at graduation was 25.1 years.

307 **Franklin University:** (1) Minority enrollment = 16.5%. Specific data not available.

322 **New Jersey Institute of Technology:** (2) International students - not foreign born. NJIT maintains data on a citizen/permanent resident versus non-citizen basis.

349 **Tri-State University:** (1) The baccalaureate engineering technology program is new and as such has not yet produced a graduate.

Support Programs for Ethnic Minorities & Women

Institutions are listed in alphabetical order. The first items in a school's entry are the institution profile reference number, the school name, and the state or province in which the main campus is located. Next, if a university offers a student chapter organization of the Society of Women Engineers (SWE), the National Society of Black Engineers (NSBE), the Society of Hispanic Professional Engineers (SHPE) or any other professional society for ethnic minorities or women, there will be a check mark (√) in the appropriate column(s). Similarly, if the college or university staffs its own programs for ethnic minorities and/or women, there will be a check mark (√) in the appropriate column(s). The last two columns indicate by capital letter abbreviation during which time periods (**S**ummer, the nine-month **A**cademic Year, or **Y**ear-round) and for students at which levels (**P**rematriculation, **F**reshman, **S**ophomore, **J**unior, senio**R**, and **G**raduate) at least one program is available. For more information about these programs, refer to the *Student Programs* section of the *Institution Profiles*.

ENGINEERING TECHNOLOGY										
		STUDENT CHAPTER ORGANIZATIONS				COLLEGE/UNIVERSITY-STAFFED PROGRAMS				
								Available:		Level Offered:
Institution Profile Reference Number & Name	State/Province	Society of Women Engineers (SWE)	National Society of Black Engineers (NSBE)	Society of Hispanic Prof. Engineers (SHPE)	Other(s)—See Institution Profiles	For Ethnic Minorities	For Women	Summer Academic year Year-round	Prematriculation Freshman Sophomore Junior SenioR Graduate	
281 U of Akron	OH	–	√	–	–	√	√	Y	PFSJRG	
282 Alabama A&M U	AL	–	√	–	√	–	–	–	–	
283 Arizona State U	AZ	√	–	√	√	√	–	A	PFS	
284 U of Arkansas at Little Rock	AR	√	–	–	√	√	√	AY	FSJRG	
285 Auburn U	AL	√	√	–	–	√	√	SAY	PFSJR	
286 Bradley U	IL	–	√	–	√	√	–	A	FSJR	
287 Brigham Young U	UT	√	–	–	–	–	–	–	–	
288 California State U, Sacramento	CA	√	√	√	√	√	√	AY	PFSJRG	
289 California State Polytechnic U, Pomona	CA	√	√	√	–	√	√	SY	PFSJRG	
290 Capitol College	MD	√	√	–	–	–	–	–	–	
291 Central Connecticut State U	CT	–	–	–	√	√	√	AY	PFSJRG	
292 U of Central Florida	FL	–	–	–	√	√	√	Y	PFSJRG	
293 U of Cincinnati	OH	√	–	–	√	√	√	Y	FSJR	
294 U of Dayton	OH	√	√	–	√	√	√	SAY	PFSJRG	
295 DeVry Institute of Technology, Chicago	IL	–	–	√	√	√	√	Y	FSJR	
296 DeVry Institute of Technology, Columbus	OH	√	–	–	–	√	√	Y	FSJR	
297 DeVry Institute of Technology, Atlanta	GA	√	√	–	–	√	√	Y	FSJR	
298 DeVry Institute of Technology, DuPage	IL	–	–	–	√	√	√	Y	FSJR	
299 DeVry Institute of Technology, Dallas	TX	√	–	√	–	√	√	Y	FSJR	
300 DeVry Institute of Technology, Kansas City	MO	–	–	–	–	√	√	Y	FSJR	
301 DeVry Institute of Technology, Los Angeles	CA	–	√	√	–	√	√	Y	FSJRG	
302 DeVry Institute of Technology, Phoenix	AZ	–	–	–	√	√	√	Y	FSJR	
303 U of the District of Columbia	DC	–	–	–	–	–	–	–	–	
304 East Tennessee State U	TN	–	–	–	√	–	–	–	–	
305 Embry-Riddle Aeronautical U - Daytona Beach	FL	–	–	–	√	√	√	Y	F	
306 Ferris State U	MI	–	–	–	√	√	√	Y	PFSJR	
307 Franklin U	OH	–	–	–	–	–	–	–	–	
308 U of Hartford	CT	√	–	–	√	–	√	A	FSJR	
309 Indiana U-Purdue U at Indianapolis	IN	√	√	–	√	√	–	Y	PFSJRG	
310 Lake Superior State U	MI	√	–	–	–	–	–	–	–	
311 Louisiana Tech U	LA	–	√	–	–	√	√	Y	PFSJRG	
312 U of Maine	ME	√	–	–	–	–	–	–	–	
313 Mankato State U	MN	√	–	–	–	√	√	Y	FSJRG	
314 U of Massachusetts, Dartmouth	MA	√	–	–	–	√	√	A	F	
315 U of Massachusetts, Lowell	MA	–	√	–	–	√	√	A	FSJRG	
316 Memphis State U	TN	√	√	–	–	√	–	S	P	
317 Metropolitan State College of Denver	CO	–	–	–	–	–	–	–	–	
318 Milwaukee School of Engineering	WI	√	√	√	√	√	√	SA	PF	
319 Montana State U	MT	√	–	–	√	√	√	Y	PFSJRG	
320 Murray State U	KY	–	–	–	–	–	–	–	–	

Support Programs

ENGINEERING TECHNOLOGY

Institution Profile Reference Number & Name	State/Province	Society of Women Engineers (SWE)	National Society of Black Engineers (NSBE)	Society of Hispanic Prof. Engineers (SHPE)	Other(s)—See Institution Profiles	For Ethnic Minorities	For Women	Summer / Academic year / Year-round	Prematriculation / Freshman / Sophomore / Junior / SenioR / Graduate
321 U of New Hampshire	NH	√	–	–	–	√	–	Y	FSJRG
322 New Jersey Institute of Technology	NJ	–	–	–	–	√	√	AY	PJR
323 New Mexico State U	NM	√	√	√	√	√	√	Y	PFSJRG
324 U of North Carolina-Charlotte	NC	√	–	–	–	√	–	SY	PFSJR
325 Northeastern U	MA	√	–	–	√	√	√	Y	PFSJR
326 Oklahoma State U	OK	√	√	√	√	√	–	SA	PFSJ
327 Old Dominion U	VA	√	–	–	√	√	√	AY	FSJRG
328 Oregon Institute of Technology	OR	√	–	–	–	–	–	–	–
329 Pennsylvania State U at Erie, The Behrend College	PA	–	√	–	√	–	–	–	–
330 Pennsylvania State U - Harrisburg	PA	–	–	–	–	√	–	Y	JRG
331 Pittsburg State U	KS	√	–	–	–	√	√	AY	PFSJRG
332 U of Pittsburgh at Johnstown	PA	–	–	–	–	√	√	Y	FSJR
333 Point Park College	PA	–	–	–	–	–	–	–	–
334 Prairie View A&M U	TX	√	√	–	√	√	–	A	FS
335 Purdue U	IN	√	–	–	–	√	–	SA	PFSJR
336 Rochester Institute of Technology	NY	√	√	–	–	–	–	–	–
337 St. Cloud State U	MN	√	–	–	√	√	√	Y	PFSJRG
338 South Carolina State U	SC	–	√	–	–	√	√	Y	FSJRG
339 Southern College of Technology	GA	–	√	–	√	√	–	Y	FSJR
340 U of Southern Colorado	CO	√	–	√	–	√	√	Y	PFSJRG
341 Southern Illinois U at Carbondale	IL	√	√	√	√	√	–	S	P
342 U of Southern Mississippi	MS	√	√	–	–	–	–	–	–
343 State U of New York College of Technology at Farmingdale	NY	–	–	–	–	–	–	–	–
344 State U of New York Institute of Technology, Utica/Rome	NY	√	–	–	–	√	√	A	JR
345 Temple U	PA	√	–	–	√	–	–	–	–
346 U of Tennessee, Martin	TN	√	–	–	–	√	–	Y	PFSJRG
347 Texas A&M U	TX	√	√	–	√	–	–	–	–
348 Texas Tech U	TX	√	√	√	√	√	–	Y	FS
349 Tri-State U	IN	√	√	–	√	√	–	Y	PFSJR
350 Virginia State U	VA	–	–	–	–	–	–	–	–
351 Weber State U	UT	–	–	–	–	–	–	–	–
352 Wentworth Institute of Technology	MA	√	√	–	–	–	√	Y	FSJR
353 West Virginia Institute of Technology	WV	–	–	–	–	√	√	A	FSJR
354 Western Carolina U	NC	–	–	–	–	√	–	A	–
355 Western Kentucky U	KY	–	–	–	–	√	–	Y	PFSJRG
356 Western Michigan U	MI	√	√	–	√	√	√	Y	PFSJRG
Totals:		46	29	13	33	54	39	-	-

Enrollments and Degrees Awarded

Institutions are listed in alphabetical order. The first four columns of data are the number of students enrolled in all engineering technology programs by class—freshman/first-year students, sophomores/second-year students, juniors/third-year students, and seniors/fourth-year and all other students. Then the total number of students enrolled in all engineering technology degree programs is given by degree level—bachelor's (the sum of the previous four columns) and master's. The last three columns represent the number of engineering technology degrees awarded by type of degree—bachelor's and master's. Unless footnoted otherwise, enrollment figures are for the fall 1991 term and degrees awarded are for the 1991-92 twelve-month academic year.

	ENGINEERING TECHNOLOGY Institution Profile Reference Number & Name	State/Province	ENROLLMENTS Undergraduate—By Class				ENROLLMENTS Totals—By Level		DEGREES AWARDED By Degree	
			Freshmen/First-year	Sophomores/Second-year	Juniors/Third-year	Seniors/All others	Bachelor's	Master's	Bachelor's	Master's
281	University of Akron	OH	267[1]	182[1]	206[1]	274[1]	929	--.[B]	99	--
282	Alabama A&M University	AL	350	200	150	75	775	--.[B]	46	--
283	Arizona State University	AZ	20	27	81	208	336	166	93	15
284	University of Arkansas at Little Rock	AR	--.[A]	83	94	164	341	--.[B]	44	--
285	Auburn University	AL	--.[B]	9	9	12	30	--.[B]	12	--
286	Bradley University	IL	3	11	50	86	150	--.[B]	54	--
287	Brigham Young University	UT	70	30	66	154	320	13	59	11
288	California State University, Sacramento	CA	13	10	66	129	218	--.[B]	33	--
289	California State Polytechnic University, Pomona	CA	93	110	216	437	856	--.	128	--
290	Capitol College	MD	80	110	180	143	513	--.	73	--
291	Central Connecticut State University	CT	62	40	76	129	307	--.[B]	130	--
292	University of Central Florida	FL	--.[1]	--.[1]	46[2]	295[2]	341	--.[3]	96	--
293	University of Cincinnati	OH	250	185	138	68	641	--.	94	--
294	University of Dayton	OH	116	81	117	151	465	--.[B]	102	--
295	DeVry Institute of Technology, Chicago	IL	293	147	134	184	758	--.[1]	154	--
296	DeVry Institute of Technology, Columbus	OH	367	133	101	199	800	--.[1]	221	--
297	DeVry Institute of Technology, Atlanta	GA	330	134	80	102	646	--.[1]	122	--
298	DeVry Institute of Technology, DuPage	IL	232	109	95	117	553	--.[1]	104	--
299	DeVry Institute of Technology, Dallas	TX	249	90	93	127	559	--.[2]	90	--
300	DeVry Institute of Technology, Kansas City	MO	155	61	33	67	316	--.[1]	54	--
301	DeVry Institute of Technology, Los Angeles	CA	208	67	57	96	428	--.[1]	73	--
302	DeVry Institute of Technology, Phoenix	AZ	381	204	99	170	854	--.[1]	183	--
303	University of the District of Columbia	DC	188[1]	18[2]	5[3]	11[4]	222	--.	16	--
304	East Tennessee State University	TN	140	100	83	140	463	48	89	28
305	Embry-Riddle Aeronautical University - Daytona Beach	FL	98	49	47	91	285	36	58	--
306	Ferris State University	MI	--	--	196	292	488	--.[A]	143	--
307	Franklin University	OH	--.[A]	--.[A]	--.[A]	--.[A]	--	--.[2]	48	--
308	University of Hartford	CT	62	38	67	56	223	--.	55	--
309	Indiana University-Purdue University at Indianapolis	IN	155	897	145	306	1,503	--.[B]	108	--
310	Lake Superior State University	MI	67	88	59	64	278	--.	55	--
311	Louisiana Tech University	LA	16	16	17	36	85	--.[B]	21	--
312	University of Maine	ME	116	105	83	93	397	--.	54	--
313	Mankato State University	MN	61	53	61	94	269	--.[1]	44	--
314	University of Massachusetts, Dartmouth	MA	80	61	46	41	228	--.[B]	44	--
315	University of Massachusetts, Lowell	MA	--.[A]	--.[A]	--.[A]	--.[A]	--	--.	56	--
316	Memphis State University	TN	49	54	103	217	423	14	--	--
317	Metropolitan State College of Denver	CO	--.[A]	--.[A]	--.[A]	--.[A]	--	--.	64	--
318	Milwaukee School of Engineering	WI	368	272	213	115	968	--.	81	--
319	Montana State University	MT	47	77	95	119[1]	338	--.	73	--
320	Murray State University	KY	122	77	69	96	364	51	41	--
321	University of New Hampshire	NH	--.[B]	--.[B]	43	66	109	--.[B]	23	--
322	New Jersey Institute of Technology	NJ	--.[1]	--.[1]	291	223	514	--.[3]	118	--
323	New Mexico State University	NM	49	58	72	138	317	--.	79	--
324	University of North Carolina-Charlotte	NC	--	--	227	183	410	--.	112	--
325	Northeastern University	MA	134	98[1]	94[1]	94	420	--.	150	--

Enrollments and Degrees Awarded

ENGINEERING TECHNOLOGY

Institution Profile Reference Number & Name	State/Province	ENROLLMENTS Undergraduate—By Class				Totals—By Level		DEGREES AWARDED By Degree	
		Freshmen/First-year	Sophomores/Second-year	Juniors/Third-year	Seniors/All others	Bachelor's	Master's	Bachelor's	Master's
326 Oklahoma State University	OK	65	101	167	303	636	--.[1]	113	--
327 Old Dominion University	VA	68[1]	69[1]	155[1]	276[1]	568	--.[B]	135	--
328 Oregon Institute of Technology	OR	318	219	208	371	1,116	--.[B]	167	--
329 Pennsylvania State University at Erie, The Behrend College	PA	77	105	94	102	378	--.	45	--
330 Pennsylvania State University - Harrisburg	PA	--.[2]	--.[2]	252	325	577	--.	288	--
331 Pittsburg State University	KS	95	73	66	146[1]	380	--.[B]	77	--
332 University of Pittsburgh at Johnstown	PA	86	98	81	69	334	--.	59	--
333 Point Park College	PA	154	42	147	222	565	--.[B]	83	--
334 Prairie View A&M University	TX	142	55	24	19	240	--.	40	--
335 Purdue University	IN	326	394	180	225	1,125	--.[B]	237	--
336 Rochester Institute of Technology	NY	152	164	617	423	1,356	--.[B]	305	--
337 St. Cloud State University	MN	30	26	29	44	129	--.	25	--
338 South Carolina State University	SC	151	123	60	123	457	--.[B]	60	--
339 Southern College of Technology	GA	701	600	623	635	2,559	53	337	--
340 University of Southern Colorado	CO	73	45	50	84	252	--.[B]	44	--
341 Southern Illinois University at Carbondale	IL	12	33	53	130	228	--.[B]	52	7
342 University of Southern Mississippi	MS	76	66	122	183	447	34	95	4
343 State University of New York College of Technology at Farmingdale	NY	402	305[1]	55	55[2]	817	--.[B]	38	--
344 State University of New York Institute of Technology, Utica/Rome	NY	--	--	203	157	360	--.	74	--
345 Temple University	PA	204	136	107	175	622	--.[B]	107	--
346 University of Tennessee, Martin	TN	73	32	16	34	155	--.[2]	20	--
347 Texas A&M University	TX	34	74	209	482	799	--.[B]	222	--
348 Texas Tech University	TX	37	41	44	88	210	--.	42	--
349 Tri-State University	IN	20	15	1	--	36	--.[3]	--	--
350 Virginia State University	VA	69	29	17	35	150	--.[B]	18	--
351 Weber State University	UT	190	113	114	286	703	--.[1]	114	--
352 Wentworth Institute of Technology	MA	875	765	567[1]	301[1]	2,508	--.[3]	349	--
353 West Virginia Institute of Technology	WV	124	177	15	53	369	--.	125	--
354 Western Carolina University	NC	20	15	47	50	132	--.[B]	27	--
355 Western Kentucky University	KY	84	56	59	97	296	--.	40	--
356 Western Michigan University	MI	163	236	253	345	997	--.[1]	131	--
Totals:		10,112	8,291	8,838	11,700	38,941	415	7,205	50

Enrollments and Degrees Awarded

FOOTNOTES: The following footnotes are the same for all schools: (A) Data not available. (B) Data not applicable. (C) Estimated data.

281 University of Akron: (1) Headcount - includes PT & FT students.

292 University of Central Florida: (1) The Engineering Technology program does not admit freshmen. Students must complete AA or AS before being admitted to the Engineering Technology program. (2) 0 SH freshman 30 SH sophomore 60 SH junior 90 SH senior. (3) The Engineering Technology program does not offer the master's degree.

295 DeVry Institute of Technology, Chicago: (1) No graduate degree offered

296 DeVry Institute of Technology, Columbus: (1) No graduate degree offered

297 DeVry Institute of Technology, Atlanta: (1) No graduate degree offered

298 DeVry Institute of Technology, DuPage: (1) No graduate degree offered

299 DeVry Institute of Technology, Dallas: (2) No graduate degree offered

300 DeVry Institute of Technology, Kansas City: (1) No graduate degree offered

301 DeVry Institute of Technology, Los Angeles: (1) No graduate degree offered

302 DeVry Institute of Technology, Phoenix: (1) No graduate degree offered

303 University of the District of Columbia: (1) 31 full-time students, 157 part-time students. (2) 18 full-time, 170 part-time. (3) 5 full-time, 41 part-time. (4) 11 full-time, 30 part-time.

307 Franklin University: (2) No graduate programs to date.

313 Mankato State University: (1) Physics

319 Montana State University: (1) Includes second degree students.

322 New Jersey Institute of Technology: (1) Engineering Technology is an upper division program only. (3) No graduate level program offered.

325 Northeastern University: (1) Program is a five-year program; enrollment in Middler Year (between sophomore and junior) was 103

326 Oklahoma State University: (1) Graduate degree programs not offered.

327 Old Dominion University: (1) Fall 1991 data

330 Pennsylvania State University - Harrisburg: (2) This campus offers Junior and Senior courses only.

331 Pittsburg State University: (1) Four students are classified as "special" and are not counted above.

343 State University of New York College of Technology at Farmingdale: (1) Freshman and Sophomore part time students equal 155 (2) Junior and Senior part time students equal 72

346 University of Tennessee, Martin: (2) No graduate programs taught in Engineering Technology.

349 Tri-State University: (2) Seven students graduated in 1991-92 with Associate Degrees in Engineering Technology. (3) No graduate programs.

351 Weber State University: (1) There are no Engineering programs at WSU, only engineering technology.

352 Wentworth Institute of Technology: (1) Includes students on cooperative work semester (2) 527 associate degree eng. tech. program graduates in various major fields. (3) Master's degree is not awarded at Wentworth.

356 Western Michigan University: (1) No master's program available in Engineering Technology at this time.

The American Society for Engineering Education

recognizes the support of its corporate members and their contribution to the advancement of engineering and engineering technology education

Aluminum Corporation of America
American Society of Mechanical Engineers
Armfield, Inc.
Autodesk
CH2M HILL, Inc.
Cad Key
Chevron Research & Technology Company
Corning Glass Works
Dow Chemical Company
E.I. du Pont de Nemours & Company
Eastman Kodak Company
Edison Electric Institute
Ford Motor Company
GTE Telephone Operations
HEMAR Insurance Corporation of America
Hewlett-Packard Company
Hughes Aircraft Company
IBM Corporation
Lawrence Livermore National Laboratories

Los Angeles Department of Water & Power
McGraw-Hill, Inc.
Microsoft Corporation
Motorola, Inc.
NAMEPA
NASA Langley Research Center
Pacific Gas & Electric Company
Procter & Gamble Company
Professional Publications, Inc.
Saunders College Publishing/HBJ
Siemens Aktiengesellschaft
Southern California Gas Company
Structural Dynamics Research Corporation
T.R.W. Space & Technology Group
Texas Instruments, Inc.
United Technologies Corporation
Westinghouse Electric Company
John Wiley & Sons, Inc.

For more information on ways your own company can contribute to the future of engineering education, call:

Dr. Woodrow Leake
Deputy Executive Director
American Society for Engineering Education
(202) 986-8510

All Engineering Technology Disciplines

Institutions are listed in alphabetical order. The first line in a school's entry contains the institution profile reference number, the school name, and the state or province in which the main campus is located. Subsequent indented lines for each program give the name of the degree program, followed by a major category abbreviation in brackets (H = Chemical, V = Civil, E = Electrical, I = Industrial, M = Mechanical, and O = Other) the full-time equivalent (FTE) of all faculty. For undergraduate programs, the columns represent: accreditation status, the nominal and actual average length of the program, the time that courses are available (Day, Evening or Both), the availability of cooperative (internship) programs (None, Optional, or Required), the percentage of graduates who participated in co-op programs, the number of bachelor's degrees awarded, and the percentage of these graduates who were full-time. For graduate programs, the columns indicate the number of graduates and the percentage who were full-time in master's and then doctoral programs. Unless otherwise footnoted, information on degrees is for the 1991-92 academic year; all other information is current. At the end of the table are (1) a summary of totals by state or province and by country and (2) the full text of all footnotes for this table.

ENGINEERING TECHNOLOGY	State/Province	FACULTY Full-time equivalent (FTE)	UNDERGRADUATE PROGRAMS							GRADUATE PROGRAMS Master's	
Institution Profile Reference Number & Name / Name of Degree Program			ABET/CEAB accred?	Nominal length of program in years	Average length of program in years	Day/Eve./Both	None/Opt./Req.	% of graduates in Co-op programs	# of degrees awarded	% of graduates who were full-time	# of degrees awarded	% of graduates who were full-time
281 U of Akron	OH											
Automated Manuf'g Tech [O]		0.6(2)	N	4.0	4.5	B	O	0.0	3	22.7	-.-(B)	-.-(B)
Drafting & Computer Drafting Tech [O](3)		3.7	N	2.0	2.5	D	O	15.8	-.-(B)	-.-(B)	-.-(B)	-.-(B)
Electronic Tech [O]		2.3(2)	Y(1)	4.0	4.5	B	O	1.5	58	40.1	-.-(B)	-.-(B)
Electronic Tech [O](3)		7.4	Y(1)	2.0	2.5	B	O	1.7	-.-(B)	-.-(B)	-.-(B)	-.-(B)
Manuf'g Tech [O](3)		2.0	N	2.0	2.5	B	O	0.0	-.-(B)	-.-(B)	-.-(B)	-.-(B)
Mechanical Tech [O](3)		4.4	Y(1)	2.0	2.5	B	O	22.7	-.-(B)	-.-(B)	-.-(B)	-.-(B)
Mechanical Tech [O](4)		2.9(2)	Y(1)	4.0	4.5	B	O	0.0	38	34.4	-.-(B)	-.-(B)
Surveying & Construction Tech [O](3)		3.1	Y(1)	2.0	2.5	B	O	0.0	-.-(B)	-.-(B)	-.-(B)	-.-(B)
282 Alabama A&M U	AL											
Civil Eng [O]		10.0	Y	4.0	4.0	B	O	15.0	3	97.0	-.-	-.-
Drafting & Design Tech [O]		3.0	Y	4.0	4.0	B	O	10.0	2	95.0	-.-	-.-
Electrical Eng [O]		10.0	Y	4.0	4.0	B	O	10.0	31	95.0	-.-	-.-
Industrial Tech [O]		5.0	N	4.0	4.0	B	O	1.0	1	1.0	-.-	-.-
Mechanical Eng [O]		10.0	N	4.0	4.0	B	O	25.0	9	95.0	-.-	-.-
283 Arizona State U	AZ											
Aeronautical Eng Tech [O]		9.0	Y	4.0	4.0	D	O	-.-	18	78.2	2	50.0
Electronic Eng Tech [O]		11.0	Y	4.0	4.0	D	O	-.-	43	64.0	8	42.0
Manuf'g Eng Tech [O]		17.0	Y	4.0	4.0	D	O	-.-	32	62.0	5	42.0
284 U of Arkansas at Little Rock	AR											
Computer Eng Tech [O]		3.5	Y	4.0	-.-(A)	B	O	-.-(A)	3	53.8	-.-(B)	-.-(B)
Construction Eng Tech [O]		4.0	Y	4.0	-.-(A)	B	O	-.-(A)	11	62.0	-.-(B)	-.-(B)
Electronic(s) Eng Tech [O]		5.0	Y	4.0	-.-(A)	B	O	-.-(A)	10	46.5	-.-(B)	-.-(B)
Manuf'g Eng Tech [O]		3.0	Y	4.0	-.-(A)	B	O	-.-(A)	1	58.3	-.-(B)	-.-(B)
Mechanical Eng Tech [O]		5.0	Y	4.0	-.-(A)	B	O	-.-(A)	14	56.3	-.-(B)	-.-(B)
Surveying & Land Information Systems [O]		2.0	N	4.0	-.-(A)	B	O	-.-(A)	5	55.6	-.-(B)	-.-(B)
285 Auburn U	AL											
Textile Mgmt & Tech [O]		8.0(1)	Y	4.0	-.-(A)	D	O	25.0	12	86.2(2)	-.-(B)	-.-(B)
286 Bradley U	IL											
Construction [O]		-.-	N	4.0	-.-	B	O	-.-	-.-	-.-	-.-(B)	-.-(B)
Electrical Eng Tech [O]		21.3	Y	4.0	4.0	B	O	-.-	23	85.7	-.-(B)	-.-(B)
Manuf'g Tech [O]		12.5	Y	4.0	4.0	B	O	-.-	31	45.7	-.-(B)	-.-(B)
287 Brigham Young U	UT											
Electrical/Electronic(s) Eng Tech [O]		-.-(2)	Y	4.0	-.-	-.-(A)	O	-.-	-.-	-.-	-.-	-.-
Electronic Eng Tech [E]		-.-(2)	Y	4.0	5.0	B	O	-.-(A)	25	-.-(A)	-.-(B)	-.-(B)
Manuf'g Eng [O]		-.-(2)	N(1)	4.0	5.0	B	O	-.-(A)	7	-.-(A)	-.-	-.-(A)
Manuf'g Eng Tech [O]		19.7(2)	N(1)	4.0	5.0	B	O	-.-(A)	27	-.-(A)	11	-.-(A)
288 Cal State U, Sacramento	CA											
Construction Mgmt Tech [O]		5.5	Y	4.0	5.0	B	O	-.-	16	-.-	-.-	-.-
Mechanical Eng Tech [O]		9.3	Y	2.0	3.0	B	O	5.9	17	-.-	-.-	-.-
289 Cal State Poly, Pomona	CA											
Eng Tech [O]		12.7	Y	4.0	5.5	B	O	-.-	128	58.0	-.-	-.-
290 Capitol Coll	MD											
Computer Eng Tech [O]		4.0	Y	4.0	4.5	B	O	25.0	8	61.0	-.-(B)	-.-(B)
Electrical/Electronic(s) Eng Tech [O]		4.0	Y	4.0	4.5	B	O	28.0	63	37.0	-.-(B)	-.-(B)
Telecommunications Eng Tech [O]		3.0	N	4.0	4.5	B	O	0.0	2	39.0	-.-(B)	-.-(B)
291 Central Conn State U	CT											
Civil & Construction Tech [V](1)		4.5	Y	4.0	5.0	B	O	15.0	130(3)	-.-(A)	-.-	-.-
Manuf'g Tech [O](1)		4.5	Y(2)	4.0	5.0	B	O	22.0	-.-(4)	-.-(A)	-.-	-.-

All Engineering Technology Disciplines

ENGINEERING TECHNOLOGY Institution Profile Reference Number & Name Name of Degree Program	State/Province	FACULTY Full-time equivalent (FTE)	ABET/CEAB accred?	UNDERGRADUATE PROGRAMS Length Nominal length of program in years	Length Average length of program in years	Time Day/Eve./Both	Co-op None/Opt./Req.	Co-op % of graduates in Co-op programs	Bachelor's # of degrees awarded	Bachelor's % of graduates who were full-time	GRADUATE PROGRAMS Master's # of degrees awarded	Master's % of graduates who were full-time
292 U of Central Florida	FL											
Computer Eng Tech [O]		2.2	Y	4.0	5.0(1)	B	O	20.0(C)	11	33.0(C)	-.-(2)	-.-(2)
Design Eng Tech [O]		2.2	Y	4.0	5.0(1)	B	O	20.0(C)	29	33.0(C)	-.-(2)	-.-(2)
Electronic(s) Eng Tech [O]		3.7	Y	4.0	5.0(1)	B	O	20.0(C)	30	33.0(C)	-.-(2)	-.-(2)
Information Systems Eng Tech [O]		1.7	Y	4.0	5.0(1)	B	O	20.0(C)	11	33.0(C)	-.-(2)	-.-(2)
Operations Eng Tech [O]		2.2	Y	4.0	5.0(1)	B	O	20.0(C)	15	33.0(C)	-.-(2)	-.-(2)
293 U of Cincinnati	OH											
Architectural Eng Tech [O](1)		10.6	Y	5.0(2)	5.0	B	R	100.0(6)	13	96.3	-.-	-.-
Chemical Tech [O](3)		3.8	Y	2.0(3)	2.0	B	-.-(4)	100.0	0	0.0	-.-	-.-
Construction Mgmt Tech [O](5)		10.6	Y	5.0(2)	5.0	B	R	100.0	29	88.8	-.-	-.-
Electrical Eng Tech [O]		10.0	Y	4.0	4.0	B	-.-(7)	54.0	23	51.1	-.-	-.-
Mechanical Eng Tech [O](8)		15.1	Y	4.0	4.0	B	O(9)	74.0	29	48.3	-.-	-.-
294 U of Dayton	OH											
Chemical Process Tech [O]		1.5	N	4.0	4.3	D	O	100.0	3	96.6	-.-(B)	-.-(B)
Electronic Eng Tech [O]		6.3	Y	4.0	4.3	B	O	42.8	21	63.2	-.-(B)	-.-(B)
Env'l Eng Tech [O]		1.5	N	4.0	4.3	D	O	20.0	10	96.1	-.-(B)	-.-(B)
Industrial Eng Tech [O]		2.6	Y	4.0	4.3	B	O	33.3	30	72.7	-.-(B)	-.-(B)
Manuf'g Eng Tech [O]		3.4	Y	4.0	4.3	B	O	33.3	3	68.8	-.-(B)	-.-(B)
Mechanical Eng Tech [O]		3.3	Y	4.0	4.3	B	O	28.6	35	69.9	-.-(B)	-.-(B)
295 DeVry Inst-Chicago	IL											
Electronic Eng Tech [O]		14.5	Y	3.0	4.0	D	O	-.-(A)	154	88.3	-.-(1)	-.-(1)
Electronic(s) Eng Tech [O]		-.-	Y	3.0	-.-	D	O	-.-	-.-	-.-	-.-	-.-
296 DeVry Inst-Columbus	OH											
Electronic Eng Tech [O]		15.0	Y	3.0	4.0	D	O	-.-(A)	221	93.6	-.-(1)	-.-(1)
Electronic(s) Eng Tech [O]		-.-	Y	3.0	-.-	D	O	-.-	-.-	-.-	-.-	-.-
297 DeVry Inst-Atlanta	GA											
Electronic Eng Tech [O]		11.0	Y	3.0	4.0	D	O	-.-(A)	122	94.7	-.-(1)	-.-(1)
298 DeVry Inst-DuPage	IL											
Electronic Eng Tech [O]		12.5	Y	3.0	4.0	D	O	10.0	104	90.6	-.-(1)	-.-(1)
Electronic(s) Eng Tech [O]		-.-	Y	3.0	-.-	D	O	-.-	-.-	-.-	-.-	-.-
299 DeVry Inst-Dallas	TX											
Electronic Eng Tech [O]		11.5	Y	3.0	4.0	D	N	-.-(A)	90	87.5	-.-(1)	-.-(1)
300 DeVry Inst-Kansas City	MO											
Electronic Eng Tech [O]		9.5	Y	3.0	4.0	D	O	-.-(A)	54	92.1	-.-(1)	-.-(1)
Electronic(s) Eng Tech [O]		-.-	Y	3.0	-.-	D	N	-.-	-.-	-.-	-.-	-.-
301 DeVry Inst-Los Angeles	CA											
Electronic Eng Tech [O]		9.5	Y	3.0	4.0	D	N	-.-(A)	73	89.7	-.-(1)	-.-(1)
Electronic(s) Eng Tech [O]		-.-	Y	3.0	-.-	D	N	-.-	-.-	-.-	-.-	-.-
302 DeVry Inst-Phoenix	AZ											
Electronic Eng Tech [O]		-.-	Y	3.0	4.0	D	O	-.-(A)	183	90.5	-.-(1)	-.-(1)
Electronic(s) Eng Tech [O]		-.-	Y	3.0	-.-	D	O	-.-	-.-	-.-	-.-	-.-
303 U of the District of Columbia	DC											
Construction Eng Tech [O]		2.0	Y	4.0	6.0(1)	E	O	-.-(A)	10	7.0	-.-	-.-
Electromechanical Systems Eng Tech [O]		2.0	Y	4.0	6.0	E	N	-.-(B)	6	9.0	-.-	-.-
304 East Tennessee State U	TN											
Biomedical Eng Tech [O]		2.0	N	4.0	4.5	D	O	-.-(A)	0	75.0	-.-	-.-
Construction Eng Tech [O]		3.0	Y	4.0	4.5	D	O	5.0	18	75.0	18	0.0
Electronics Eng [O]		3.0	N	4.0	4.5	D	O	5.0	27	90.0	6	50.0
Eng Design Graphics [O]		2.3	N	4.0	4.5	D	O	5.0	17	80.0	-.-	-.-
Industrial Tech [O]		1.3	N	4.0	4.5	D	O	0.0	13	75.0	-.-	-.-
Manuf'g Eng Tech [O]		3.0	Y	4.0	4.5	D	O	10.0	12	85.0	4	50.0
Surveying & Mapping [O]		2.0	N	4.0	4.5	D	O	2.5	2	80.0	-.-	-.-
305 Embry-Riddle-Daytona Beach	FL											
Aircraft Eng Tech [O]		7.0	Y	4.0	4.0	-.-(A)	-.-(A)	-.-(A)	35	88.1	-.-(B)	-.-(B)
Avionics Eng Tech/Avionics Tech [O]		11.0	N	4.0	4.0	-.-(A)	-.-(A)	-.-(A)	23	76.1	-.-(B)	-.-(B)
306 Ferris State U	MI											
Electrical/Electronic(s) Eng Tech [O](A)		4.0(A)	N(A)	2.0	2.1	-.-(A)	N(A)	-.-(A)	18(A)	80.0(A)	-.-(A)	-.-(A)
HVACR Eng Tech [O]		2.0(A)	N	2.0	2.1	--	N(A)	-.-(A)	13	90.0	-.-(A)	-.-(A)

All Engineering Technology Disciplines

ENGINEERING TECHNOLOGY Institution Profile Reference Number & Name Name of Degree Program	State/Province	FACULTY Full-time equivalent (FTE)	ABET/CEAB accred?	UNDERGRADUATE PROGRAMS							GRADUATE PROGRAMS	
				Length		Time	Co-op		Bachelor's		Master's	
				Nominal length of program in years	Average length of program in years	Day/Eve./Both	None/Opt./Req.	% of graduates in Co-op programs	# of degrees awarded	% of graduates who were full-time	# of degrees awarded	% of graduates who were full-time
Manuf'g Eng Tech [O]		6.0	N	2.0	2.1	B	N(A)	-.-	26	70.0	-.-(A)	-.-(A)
Plastics Eng Tech [O]		6.0	N	2.0	2.1	B	N(A)	-.-	48	90.0	-.-(A)	-.-(A)
Product Design Eng Tech [O]		3.0	N	2.0	2.1	B	N(A)	-.-	18	80.0	-.-	-.-(A)
Surveying Eng [O]		3.0	Y	4.0(A)	4.2	--	N	-.-	5	90.0	-.-(A)	-.-(A)
Welding Eng Tech [O]		3.0	N	2.0	-.-	--	N(A)	-.-	15	100.0		
307 Franklin U	OH											
Electronics Eng Tech [O]		5.0	Y	4.0	4.0	B	O	-.-(A)	25	22.0	-.-(B)	-.-(B)
Mechanical Eng Tech [O]		5.0	Y	4.0	4.0	B	O	-.-(A)	23	19.0	-.-(B)	-.-(B)
308 U of Hartford	CT											
Architectural Eng Tech [O]		5.0	N	4.0	4.0	B	O	-.-(A)	0(1)	89.0	-.-	-.-
Computer Eng Tech [O](2)		4.4	N	2.0	2.0	B	N	-.-	0	0.0	-.-	-.-
Electronic Eng Tech [O]		9.0	Y	4.0	4.2	B	O	5.0	55	71.0	-.-	-.-
309 Indiana-Purdue, Indianapolis	IN											
Computer Integrated Manuf'g Tech [O]		3.0	N	4.0	4.0	B	O	-.-(A)	8	41.0	-.-(B)	-.-(B)
Construction/Architectural Eng Tech [O]		7.0	N	4.0	4.0	B	O	-.-(A)	21	38.0	-.-(B)	-.-(B)
Electrical Eng Tech [O]		8.0	Y	4.0	4.0	B	O	-.-(A)	39	28.0	-.-(B)	-.-(B)
Industrial Eng Tech [O]		1.0	Y	4.0	4.0	B	O	-.-(A)	3	13.0	-.-(B)	-.-(B)
Mechanical Eng Tech [O]		8.0	Y	4.0	4.0	B	O	-.-(A)	37	34.0	-.-(B)	-.-(B)
310 Lake Superior State U	MI											
Automated Systems Eng Tech [O]		-.-	Y	4.0	-.-	D	O	-.-	11	-.-	-.-	-.-
Computer Eng Tech [O]		1.0	Y	1.0	-.-	D	N	-.-	-.-	-.-	-.-	-.-
Drafting & Design Eng Tech [O]		-.-	Y	2.0	-.-	D	N	-.-	-.-	-.-	-.-	-.-
Electrical/Electronic(s) Eng Tech [O]		4.7	Y	4.0	-.-	D	O	-.-	30	-.-	-.-	-.-
Mechanical Eng Tech [O]		5.6	Y	4.0	-.-	D	O	-.-	14	-.-	-.-	-.-
311 Louisiana Tech U	LA											
Construction Eng Tech [O]		3.0	Y	4.0	4.7	D	O	0.0	4	100.0	-.-(1)	-.-(1)
Electrical Eng Tech [O]		3.0	Y	4.0	4.7	D	O	0.0	17	100.0	-.-(1)	-.-(1)
312 U of Maine	ME											
Construction Mgmt Tech [O]		-.-	Y	4.0	4.5	D	O	37.0	8	98.0	-.-	-.-
Electrical Eng Tech [O]		-.-	Y	4.0	4.5	D	O	37.0	21	98.0	-.-	-.-
Mechanical Eng Tech [O]		-.-	Y	4.0	4.5	D	O	37.0	25	98.0	-.-	-.-
313 Mankato State U	MN											
Automotive Eng Tech [O]		2.0	N	4.0	4.0	B	N	-.-	10	90.0	-.-	-.-
Electronic Eng Tech [O]		3.0	Y	4.0	4.0	B	N	-.-	28	84.0	-.-	-.-
Manuf'g Eng Tech [O]		3.0	Y	4.0	4.0	B	N	-.-	6	80.0	-.-	-.-
314 U of Mass, Dartmouth	MA											
Electrical Eng Tech [O]		4.0	Y	4.0	4.5	--	N	-.-	30	90.0	-.-	-.-
Mechanical Eng Tech [O]		10.0	Y	4.0	4.5	--	N	-.-	14	90.0	-.-	-.-
315 U of Mass, Lowell	MA											
Civil Eng Tech [O]		5.0	Y	8.0	8.0	E	N	-.-	6	-.-(A)	-.-	-.-
Electronic(s) Eng Tech [O]		11.0	Y	8.0	8.0	E	N	-.-	29	-.-(A)	-.-	-.-
Mechanical Eng Tech [O]		10.0	Y	8.0	8.0	E	N	-.-	21	-.-(A)	-.-	-.-
316 Memphis State U	TN											
Architectural Eng Tech [O]		-.-	Y	4.0	-.-	B	O	-.-	15	-.-	-.-	-.-
Computer Eng Tech [O]		-.-	Y	4.0	-.-	B	O	-.-	16	-.-	-.-	-.-
Electrical Eng Tech [O]		-.-	Y	4.0	-.-	B	O	-.-	27	-.-	-.-	-.-
Manuf'g Eng Tech [O]		-.-	Y	4.0	-.-	B	O	-.-	14	-.-	-.-	-.-
317 Metropolitan State-Denver	CO											
Civil Eng Tech [O]		6.3	Y	4.0	-.-(A)	B	O	-.-(A)	11	-.-(B)	-.-(B)	-.-(B)
Electronic(s) Eng Tech [O]		9.4	Y	4.0	-.-(A)	B	O	-.-(A)	38	-.-(A)	-.-(B)	-.-(B)
Mechanical Eng Tech [O]		4.5	Y	4.0	-.-(A)	B	O	-.-(A)	15	-.-(A)	-.-(B)	-.-(B)
318 Milwaukee Sch of Eng	WI											
Electrical/Electronic(s) Eng Tech [O]		14.2	Y	4.0	6.2	B	N(1)	-.-	44	40.0	-.-	-.-
Manuf'g Eng Tech [O]		5.2	N	4.0	6.2	B	N(1)	-.-	7	12.0	-.-	-.-
Mechanical Eng Tech [O]		11.0	Y	4.0	6.2	B	N(1)	-.-	30	38.0	-.-	-.-
319 Montana State U	MT											
Construction Eng Tech [O]		3.8	Y	4.0	4.7	D	O	50.0	28	-.-(1)	-.-	-.-
Electrical/Electronic(s) Eng Tech [O]		3.7	Y	4.0	4.7	D	N	-.-	13	-.-(1)	-.-	-.-
Mechanical Eng Tech [O]		5.0	Y	4.0	4.7	D	N	-.-	32	-.-(1)	-.-	-.-

All Engineering Technology Disciplines

ENGINEERING TECHNOLOGY Institution Profile Reference Number & Name Name of Degree Program	State/Province	FACULTY Full-time equivalent (FTE)	UNDERGRADUATE PROGRAMS							GRADUATE PROGRAMS	
			ABET/CEAB accred?	Length		Time	Co-op		Bachelor's	Master's	
				Nominal length of program in years	Average length of program in years	Day/Eve./Both	None/Opt./Req.	% of graduates in Co-op programs	# of degrees awarded	% of graduates who were full-time	# of degrees awarded	% of graduates who were full-time
320 Murray State U	KY											
Civil Eng Tech [O]		3.7	Y	4.0	4.0	B	R	100.0	10	90.0	-.-(B)	-.-(B)
Civil/Construction Eng Tech [O]		1.7	Y	4.0	4.0	B	R	100.0	3	90.0	-.-(B)	-.-(B)
Electrical Eng Tech [O]		3.5	Y	4.0	4.0	D	O	-.-(B)	10	90.0	-.-(B)	-.-(B)
Electrical/Computer Eng Tech [O]		1.5	Y	4.0	4.0	B	O	-.-(A)	1	90.0	-.-(B)	-.-(B)
Env'l Eng Tech [O]		1.5	N	4.0	4.0	B	R	100.0	6	90.0	-.-(B)	-.-(B)
Manuf'g Eng Tech [O]		4.7	Y	4.0	4.0	B	O	-.-(A)	11	90.0	-.-(B)	-.-(B)
321 U of New Hampshire	NH											
Electrical Eng Tech [O]		2.0	Y	2.0	-.-(A)	-.-(B)	N	-.-(B)	10	40.0	-.-(B)	-.-(B)
Mechanical Eng Tech [O]		2.0	Y	2.0	-.-(A)	-.-(B)	N	-.-	13	40.0	-.-(B)	-.-(B)
322 New Jersey Inst of Tech	NJ											
Computer Eng Tech [O]		-.-(1)	N	2.0	3.0	B	O	16.7	6	30.0	-.-(2)	-.-(2)
Construction & Contracting Eng Tech [O]		-.-(1)	Y	2.0	3.0	B	O	18.7	32	33.9	-.-(2)	-.-(2)
Electrical Eng Tech [O]		-.-(1)	Y	2.0	3.0	B	O	15.7	51	34.8	-.-(2)	-.-(2)
Manuf'g Eng Tech [O]		-.-(1)	Y	2.0	3.0	B	O	14.3	7	24.2	-.-(2)	-.-(2)
Mechanical Eng Tech [O]		-.-(1)	Y	2.0	3.0	B	O	19.0	21	30.0	-.-(2)	-.-(2)
Surveying Eng Tech [O]		-.-(1)	N	2.0	3.0	B	O	0.0	1	10.0	-.-(2)	-.-(2)
323 New Mexico State U	NM											
Civil Eng Tech [O]		3.0	Y	4.0	4.8	D	O	15.0(C)	11	75.0(C)	-.-(B)	-.-(B)
Electronic(s) Eng Tech [O]		7.0	Y	4.0	4.8	D	O	15.0(C)	55	75.0(C)	-.-(B)	-.-(B)
Mechanical Eng Tech [O]		4.0	Y	4.0	4.8	D	O	15.0(C)	13	75.0(C)	-.-(B)	-.-(B)
324 UNC-Charlotte	NC											
Civil Eng Tech [O]		4.1	Y	2.0	2.5	B	O	10.0	20	79.2	-.-	-.-
Electrical Eng Tech [O]		8.1	Y	2.0	2.5	B	O	11.8	59	56.9	-.-	-.-
Manuf'g Eng Tech [O]		3.0	N	2.0	2.5	B	O	66.6	6	57.9	-.-	-.-
Mechanical Eng Tech [O]		7.1	Y	2.0	2.5	B	O	29.6	27	48.7	-.-	-.-
325 Northeastern U	MA											
Aerospace Maintenance Eng Tech [O]		2.0	N	3.5	3.0	B	R	100.0	4	100.0	-.-	-.-
Computer Tech [O]		8.0	N	5.0	5.0	B	R	100.0	18	100.0	-.-	-.-
Electrical Eng Tech [O]		10.5	Y	5.0	5.0	B	R	100.0	76	100.0	-.-	-.-
Manuf'g Eng Tech [O]		2.0	N	7.0	7.0	E	N	-.-	1	0.0	-.-	-.-
Mechanical Eng Tech [O]		7.0	Y	5.0	5.0	B	R	100.0	47	100.0	-.-	-.-
Mechanical-Structural Eng Tech [O]		2.0	Y	7.0	7.0	E	N	-.-	4	0.0	-.-	-.-
326 Oklahoma State U	OK											
Construction Mgmt Tech [O]		3.0	Y	4.0	4.2	D	O	0.0	8	-.-(2)	-.-(1)	-.-
Electronic(s) & Computer Tech [O]		7.0	Y	4.0	4.2	D	O	11.7	35	-.-(2)	-.-(1)	-.-
Fire Protection & Safety Tech [O]		4.0	Y	4.0	4.2	D	O	0.0	27	-.-(2)	-.-(1)	-.-
General Tech [O](2)		-.-(3)	N	4.0	4.2	D	O	0.0	1	-.-(2)	-.-(1)	-.-
Manuf'g Eng Tech [O]		7.0	Y	4.0	4.2	D	O	0.0(2)	4	-.-(2)	-.-(1)	-.-
Mechanical Design Tech [O]		7.0	Y	4.0	4.2	D	O	7.2	29	-.-(2)	-.-(1)	-.-
Mechanical Power Tech [O]		5.5	Y	4.0	4.2	D	O	11.1	9	-.-(1)	-.-(2)	-.-
327 Old Dominion U	VA											
Civil Eng Tech [O]		4.0	Y	4.0	4.5	B	O	-.-(A)	31	92.7	-.-(B)	-.-(B)
Electrical Eng Tech [O]		8.0	Y	4.0	4.5	B	O	-.-(A)	53	70.7	-.-(B)	-.-(B)
Mechanical Eng Tech [O]		7.0	Y	4.0	4.5	B	O	-.-(A)	38	74.6	-.-(B)	-.-(B)
Technical Operations [O]		1.0	N	4.0	4.5	B	O	-.-(A)	13	77.8	-.-(B)	-.-(B)
328 Oregon Inst of Tech	OR											
Civil Eng Tech [O]		6.0	Y	4.0	4.7	D	O	-.-(A)	30	90.2	-.-(B)	-.-(B)
Computer Systems Tech [O]		9.0	Y	4.0	4.7	D	O	-.-(A)	20	70.2	-.-(B)	-.-(B)
Electrical/Electronic(s) Eng Tech [O]		12.5	Y	4.0	4.7	D	O	-.-(A)	37	71.4	-.-(B)	-.-(B)
Laser Optical Eng Tech [O]		3.0	Y	4.0	4.7	D	O	-.-(A)	15	72.1	-.-(B)	-.-(B)
Manuf'g Eng Tech [O]		6.0	Y	4.0	4.7	D	O	-.-(A)	14	75.8	-.-(B)	-.-(B)
Mechanical Eng Tech [O]		8.0	Y	4.0	4.7	D	O	-.-(A)	41	87.4	-.-(B)	-.-(B)
Surveying [O]		3.0	Y	4.0	4.7	D	O	-.-(A)	10	87.0	-.-(B)	-.-(B)
329 Penn State-Erie	PA											
Electrical Eng [O]		5.2	Y	4.0	4.0	D	N	-.-	15	87.0	-.-	-.-
Mechanical Eng [O]		9.7	Y	4.0	4.0	D	N	-.-	21	99.0	-.-	-.-
Plastics Eng [O]		3.2	N	4.0	4.0	B	N	-.-	9	100.0	-.-	-.-
330 Penn State U - Harrisburg	PA											
Electrical & Computer Eng Tech [O]		12.0	Y	4.0	4.4	B	N	-.-(B)	68	88.0	-.-	-.-
Enviromental Eng Tech [O]		5.0	Y	4.0	4.1	B	N	-.-(B)	137	-.-(A)	-.-	-.-

All Engineering Technology Disciplines

ENGINEERING TECHNOLOGY Institution Profile Reference Number & Name Name of Degree Program	State/Province	FACULTY Full-time equivalent (FTE)	ABET/CEAB accred?	UNDERGRADUATE PROGRAMS					Bachelor's		GRADUATE PROGRAMS	
				Length		Time	Co-op				Master's	
				Nominal length of program in years	Average length of program in years	Day/Eve./Both	None/Opt./Req.	% of graduates in Co-op programs	# of degrees awarded	% of graduates who were full-time	# of degrees awarded	% of graduates who were full-time
Mechanical Eng Tech [O]		10.0	Y	4.0	4.1	B	N	-.-(B)	53	89.0	-.-	-.-
Structural Design & Construction Eng Tech [O]		6.0	Y	4.0	4.5	B	N	-.-(B)	30	90.0	-.-	-.-
331 Pittsburg State U	KS											
Construction Eng Tech [O]		3.0	Y	4.0	4.0	D	O	-.-(A)	5	88.1	-.-(B)	-.-(B)
Construction Mgmt Tech [O]		3.0	N	4.0	4.0	D	O	-.-(A)	8	84.8	-.-(B)	-.-(B)
Electronic(s) Eng Tech [O]		4.2	Y	4.0	4.0	D	O	-.-(A)	21	85.9	-.-(B)	-.-(B)
Manuf'g Eng Tech [O]		3.0	Y	4.0	4.0	D	O	-.-(A)	7	62.1	-.-(B)	-.-(B)
Mechanical Eng Tech [O]		4.0	Y	4.0	4.0	D	O	-.-(A)	8	88.7	-.-(B)	-.-(B)
Plastics Eng Tech [O]		4.2	Y	4.0	4.0	D	O	-.-(A)	28	95.3	-.-(B)	-.-(B)
332 U of Pittsburgh at Johnstown	PA											
Civil Eng Tech [O]		6.0	Y	9.0	4.0	D	N	-.-	19	100.0	-.-	-.-
Electrical Eng Tech [O]		6.0	Y	9.0	4.0	D	N	-.-	23	100.0	-.-	-.-
Mechanical Eng Tech [O]		6.0	Y	9.0	4.0	D	N	-.-	17	100.0	-.-	-.-
333 Point Park Coll	PA											
Civil Eng Tech [O]		2.8	Y	3.0	4.0	-.-(A)	O	-.-(A)	12	22.7	-.-(B)	-.-(B)
Electrical Eng Tech [O]		4.0	Y	3.0	4.0	-.-(A)	O	-.-(A)	50	14.5	-.-(B)	-.-(B)
Mechanical Eng Tech [O]		3.0	Y	3.0	4.0	-.-(A)	O	-.-(A)	21	11.3	-.-(B)	-.-(B)
334 Prairie View A&M U	TX											
Computer Eng Tech [O]		7.0(1)	Y	4.0	4.0	D	O	-.-(A)	14(C)	85.0(C)	-.-	-.-
Computer Science [O]		8.2	Y	4.0	4.0	D	O	-.-(A)	-.-	85.0(C)	-.-	-.-
Electrical Eng Tech [O]		-.-(1)	Y	4.0	4.0	D	O	-.-(A)	13(C)	85.0(C)	-.-	-.-
Mechanical Eng Tech [O]		-.-(1)	N	4.0	4.0	D	O	-.-(A)	13(C)	85.0(C)	-.-	-.-
335 Purdue U	IN											
Computer Integrated Manuf'g Tech [O]		6.0	N	4.0	4.0	D	O	-.-(A)	25	-.-(A)	-.-	-.-
Electrical Eng Tech [O]		24.0	Y	4.0	4.0	D	O	-.-(A)	111	-.-(A)	-.-	-.-
Mechanical Eng Tech [O]		16.0	Y	4.0	4.0	D	O	-.-(A)	101	-.-(A)	-.-	-.-
336 Rochester Inst of Tech	NY											
Civil Eng Tech [O]		5.0	Y	5.0	5.0	D	R	100.0	52	100.0	-.-(B)	-.-(B)
Computer Eng Tech [O](1)		4.0	Y	5.0	5.0	D	R	100.0	22	100.0	-.-(B)	-.-(B)
Electrical Eng Tech [O]		7.0	Y	5.0	5.0	B	R	100.0	112	85.0	-.-(B)	-.-(B)
Manuf'g Eng Tech [O]		4.0	Y	5.0	5.0	B	R	100.0	41	90.0	-.-(B)	-.-(B)
Mechanical Eng Tech [O]		7.0	Y	5.0	5.0	B	R	100.0	75	80.0	-.-(B)	-.-(B)
Telecommunications Eng Tech [O](2)		3.0	N	5.0	5.0	B	R	100.0	3	100.0	-.-	-.-
337 St. Cloud State U	MN											
Eng Tech (General) [O]		3.0	N	4.0	4.5	B	O(1)	67.0(1)	2	100.0	-.-	-.-
Imaging Eng Tech [O]		3.0	N(2)	4.0	4.5	B	O	0.0	7	100.0	-.-	-.-
Imaging Science [O]		2.0	N	4.0	4.5	B	O	0.0	1	100.0	-.-	-.-
Manuf'g Eng Tech [O]		4.0	Y	4.0	4.5	B(1)	O(1)	40.0(1)	15	90.0	-.-	-.-
338 South Carolina State U	SC											
Civil Eng Tech [O]		4.0	Y	4.0	5.0	D	O	10.0	15	15.4	-.-	-.-
Electrical Eng Tech [O]		5.0	Y	4.0	5.0	D	O	21.0	31	60.3	-.-	-.-
Industrial Eng Tech [O]		3.0	Y	4.0	5.0	D	O	10.0	7	7.7	-.-	-.-
Mechanical Eng Tech [O]		4.0	Y	4.0	5.0	D	O	15.0	7	16.6	-.-	-.-
339 Southern Coll of Tech	GA											
Apparel Eng Tech [O]		1.5	Y	4.0	5.0	B	O	0.0	2	79.0	-.-(B)	-.-(B)
Civil Eng Tech [O]		10.4	Y	4.0	5.0	B	O	28.0	46	62.0	-.-(B)	-.-(B)
Computer Eng Tech [O]		5.0	Y	4.0	5.0	B	O	27.0	15	64.0	-.-(B)	-.-(B)
Electrical Eng Tech [O]		15.7	Y	4.0	5.0	B	O	15.0	118	62.0	-.-(B)	-.-(B)
Industrial Eng Tech [O]		14.3	Y	4.0	5.0	B	O	25.0	80	59.0	-.-(B)	-.-(B)
Mechanical Eng Tech [O]		14.4	Y	4.0	5.0	B	O	18.0	72	59.0	-.-(B)	-.-(B)
Textile Eng Tech [O]		1.5	Y	4.0	5.0	B	O	0.0	4	84.0	-.-(B)	-.-(B)
340 U of Southern Colorado	CO											
Civil Eng Tech [O]		4.3	Y	4.0	4.5(C)	D	O	5.0(C)	17	85.0(C)	-.-(B)	-.-(B)
Electronic(s) Eng Tech [O]		3.5	Y	4.0	4.5(C)	D	O	5.0(C)	12	90.0(C)	-.-(B)	-.-(B)
Mechanical Eng Tech [O]		4.5	Y	4.0	4.5(C)	D	O	5.0(C)	15	85.0(B)	-.-(B)	-.-(B)
341 Southern Illinois-Carbondale	IL											
Civil Eng Tech [O]		3.0	Y	4.0	4.5	D	O	3.0	8	87.0	-.-(B)	-.-(B)
Electrical Eng Tech [O]		4.0	Y	4.0	4.5	D	O	2.0	24	94.0	-.-(B)	-.-(B)
Manuf'g Systems [O]		-.-(B)	-.-(B)	-.-(B)	-.-(B)	D	-.-(B)	-.-(B)	-.-(B)	-.-(B)	7	64.0
Mechanical Eng Tech [O]		3.0	Y	4.0	4.5	D	O	2.0	20	91.0	-.-(B)	-.-(B)

All Engineering Technology Disciplines

ENGINEERING TECHNOLOGY Institution Profile Reference Number & Name Name of Degree Program	State/Province	FACULTY Full-time equivalent (FTE)	ABET/CEAB accred?	UNDERGRADUATE PROGRAMS					GRADUATE PROGRAMS	
				Length		Time	Co-op	Bachelor's		Master's	
				Nominal length of program in years	Average length of program in years	Day/Eve./Both	None/Opt./Req.	% of graduates in Co-op programs	# of degrees awarded	% of graduates who were full-time	# of degrees awarded	% of graduates who were full-time

342 U of Southern Mississippi	MS											
Architectural Eng Tech [O]		2.5	Y	4.0	-.-	D	O	-.-	13	-.-	-.-	-.-
Computer Eng Tech [O]		2.5	Y	4.0	-.-	D	O	-.-	7	-.-	-.-	-.-
Construction Eng Tech [O]		2.5	Y	4.0	-.-	D	O	-.-	16	-.-	-.-	-.-
Electronic Eng Tech [O]		4	Y	-.-	-.-	--	--	-.-	36	-.-	-.-	-.-
Eng Tech [O][1]		-.-(A)	--(B)	-.-(B)	3.0	--(A)	--(A)	-.-(A)	--(B)	-.-(B)	0(2)	-.-(A)
Industrial Eng Tech [O]		2.5	Y	4.0	-.-	D	O	-.-	15	-.-	-.-	-.-
Manuf'g Tech [O]		1.5	N	-.-	-.-	--	--	-.-	--(B)	-.-	4	-.-
Mechanical Eng Tech [O]		2.5	Y	4.0	-.-	D	O	-.-	8	-.-	-.-	-.-
343 SUNY-Farmingdale	NY											
Aerospace Tech [O]		6.3	N	2.0	2.0	D	N	-.-	-.-	-.-	-.-	-.-
Aircraft Maintenance Tech [O]		1.0	N	2.0	3.0	B	N	-.-	-.-	-.-	-.-	-.-
Automotive Eng Tech [O]		5.0	Y	2.0	2.5	D	N	-.-	-.-	-.-	-.-	-.-
Biomedical Eng Tech [O]		1.0	Y	2.0	2.0	D	N	-.-	-.-	-.-	-.-	-.-
Civil Eng Tech [O]		5.0	Y	2.0	2.5	D	N	-.-	-.-	-.-	-.-	-.-
Computing Graphics Tech [O]		1.5	N	2.0	2.0	B	N	-.-	-.-	-.-	-.-	-.-
Construction/Architectural Eng Tech [O]		7.0	Y	2.0	2.5	B	N	-.-	-.-	-.-	-.-	-.-
Electrical Eng Tech [O]		15.3	Y	4.0	4.0	B	N	-.-	31	70.0	-.-	-.-
Eng Science [O]		2.0	N	2.0	2.0	B	N	-.-	-.-	-.-	-.-	-.-
Manuf'g Eng Tech [O]		3.0	Y	2.0	2.0	B	N	-.-	7	25.0	-.-	-.-
Mechanical Eng Tech [O]		5.0	Y	2.0	2.5	B	N	-.-	-.-	-.-	-.-	-.-
344 SUNY-Utica/Rome	NY											
Computer Systems Tech [O]		1.0	Y	2.0	2.0	B	N	-.-	36	-.-(A)	--(B)	-.-(B)
Electrical Eng Tech [O]		8.3	Y	2.0	2.0	B	N	-.-	7(1)	-.-(A)	--(B)	-.-(B)
Industrial Eng Tech [O]		2.3	Y	2.0	2.0	B	N	-.-(A)	11	-.-(A)	--(B)	-.-(B)
Mechanical Eng Tech [O]		4.6	Y	2.0	2.0	B	N	-.-	20	-.-(A)	--(B)	-.-(B)
345 Temple U	PA											
Civil & Construction Eng [O]		13.0	Y	4.0	4.0	--(A)	O	-.-(A)	26	74.8	--(B)	-.-(B)
Civil/Construction Eng Tech [O]		-.-	Y	4.0	-.-	--(A)	O	-.-(A)	-.-	-.-	--(B)	-.-(B)
Electrical Eng Tech [O]		16.0	Y	4.0	4.0	--(A)	O	-.-(A)	37	64.5	--(B)	-.-(B)
Eng Tech [O]		15.5	N	4.0	4.0	--(A)	O	-.-(A)	2	64.2	--(B)	-.-(B)
Env'l Eng Tech [O]		13.0	N	4.0	4.0	--(A)	O	-.-(A)	19	77.7	--(B)	-.-(B)
Mechanical Eng Tech [O]		15.5	Y	4.0	4.0	--(A)	O	-.-(A)	23	65.4	--(B)	-.-(B)
346 U of Tennessee, Martin	TN											
Civil Eng Tech [O]		3.0	Y	4.0	-.-(A)	D	O	-.-(A)	2	89.9	--(1)	-.-(1)
Electrical Eng Tech [O]		3.0	Y	4.0	-.-(B)	D	O	-.-(A)	11	86.9	--(1)	-.-(1)
Mechanical Eng Tech [O]		3.0	Y	4.0	-.-(A)	D	O	-.-(A)	7	87.5	--(1)	-.-(1)
347 Texas A&M U	TX											
Electronic Eng Tech [O]		9.0(2)	Y	4.0	5.0	D	O	4.0	56(2)	0.0(A)	--(A)	-.-(A)
Eng Tech [O][1]		22.0	Y	4.0	5.0	D	O	10.0	111	0.0(A)	--(B)	-.-(B)
Manuf'g Eng Tech [O]		19.0(2)	Y	4.0	5.0	D	O	0.0	19(2)	0.0(A)	--(B)	-.-(B)
Mechanical Eng Tech [O]		36.0(2)	Y	4.0	5.0	D	O	6.0	36(2)	0.0(A)	--(B)	-.-(B)
348 Texas Tech U	TX											
Construction Eng Tech [O]		3.5	Y	4.0	4.5	D	O	-.-	8	-.-	-.-	-.-
Electrical/Electronic(s) Eng Tech [O]		3.5	Y	4.0	4.5	D	O	-.-	18	-.-	-.-	-.-
Mechanical Eng Tech [O]		3.2	Y	4.0	4.5	D	O	-.-	16	-.-	-.-	-.-
349 Tri-State U	IN											
Computer Aided Drafting & Design Tech [O]		2.5	N	4.0	4.8	D	O	0.0	0(3)	100.0	--(2)	-.-(2)
Drafting & Design Tech [O]		2.5	Y	2.0	2.5	D	O	2.0	--(1)	-.-(1)	--(2)	-.-(2)
Manuf'g Tech [O]		2.0	N	2.0	2.5	D	O	2.0	0(1)	0.0(1)	--(2)	-.-(2)
350 Virginia State U	VA											
Electronic Eng Tech [O]		3.0	Y	4.0	4.5	B	O	10.0	9	90.0	--(1)	-.-(1)
Mechanical Eng [O]		3.0	Y	4.0	4.5	B	O	10.0	9	80.0	--(1)	-.-
351 Weber State U	UT											
Electronic Eng Tech [O]		11.2	Y	4.0	5.0(C)	B	O	-.-(A)	87	-.-(A)	--(B)	-.-(B)
Manuf'g Eng Tech [O]		11.0	Y	4.0	5.0	D	O	-.-(A)	23	-.-(A)	--(B)	-.-(B)
Mechanical Eng Tech [O]		5.0	Y	4.0	5.0	D	O	-.-(A)	4	-.-(A)	--(B)	-.-(B)
352 Wentworth Inst of Tech	MA											
Architectural Eng Tech [O]		26.3	Y	4.0	4.5	B(2)	R	88.0	63	87.0	--(1)	-.-(1)
Building Construction Tech [O]		9.0	N	4.0	4.5	B(2)	R	95.0	42	95.0	--(1)	-.-(1)
Civil Eng Tech [O]		7.0	Y	4.0	4.5	D	R	100.0	55	92.0	--(1)	-.-(1)
Computer Eng Tech [O]		17.0	Y	4.0	4.5	D	R	100.0	19	97.0	--(1)	-.-(1)

All Engineering Technology Disciplines

ENGINEERING TECHNOLOGY Institution Profile Reference Number & Name Name of Degree Program	State/Province	FACULTY Full-time equivalent (FTE)	ABET/CEAB accred?	UNDERGRADUATE PROGRAMS							GRADUATE PROGRAMS	
				Length		Time	Co-op		Bachelor's		Master's	
				Nominal length of program in years	Average length of program in years	Day/Eve./Both	None/Opt./Req.	% of graduates in Co-op programs	# of degrees awarded	% of graduates who were full-time	# of degrees awarded	% of graduates who were full-time
Electronic(s) Eng Tech [O]		17.0	Y	4.0	4.5	B(B)	R	75.0	99	82.0	--.(1)	-.-(1)
Manuf'g Eng Tech [O]		17.4	Y	4.0	4.5	D	R	100.0	17	100.0	--.(1)	-.-(1)
Mechanical Eng Tech [O]		17.4	Y	4.0	4.5	B	R	65.0	54	85.0	--.(1)	-.-(1)
353 West Virginia Inst of Tech	WV											
Civil/Surveying Eng Tech [O](A)		2.0	N	2.0	2.5	D	O	-.-	6	100.0	--.	-.-
Drafting & Design Eng Tech [O](1)		2.0	Y	2.0	2.5	D	O	-.-	18	100.0	--.	-.-
Electrical Eng Tech [O](A)		3.0	Y	2.0	2.5	D	O	-.-	23	100.0	--.	-.-
Electronic(s) Eng Tech [O]		2.0	Y	2.0	2.0	D	O	-.-	30	100.0	--.	-.-
Eng Tech [O]		-.-	Y	2.0	2.0	D	O	-.-	6	100.0	--.	-.-
Industrial Tech [O]		1.0	N	2.0	2.0	D	O	-.-	16	100.0	--.	-.-
Mechanical Eng Tech [O](A)		2.0	Y	2.0	2.5	D	O	-.-	12	100.0	--.	-.-
Surveying Eng Tech [O](1)		1.0	N	2.0	2.0	D	O	-.-	3	100.0	--.	-.-
Technical Studies [O](1)		-.-	N	2.0	2.5	D	O	-.-	7	100.0	--.	-.-
Tech Education [O]		2.0	N	2.0	-.-	D	O	-.-	4	100.0	--.	-.-
354 Western Carolina U	NC											
Electronic(s) Eng Tech [O]		3.0	N	4.0	4.0	--	--	10.0	11	62.0	--.	-.-
Manuf'g Eng Tech [O]		3.7	Y	4.0	4.0	--	O	3.0	16	46.7	--.	-.-
355 Western Kentucky U	KY											
Civil Eng Tech [O]		3.1	Y	4.0	4.7	B	O	-.-(A)	10	-.-(A)	--.(B)	-.-(B)
Electrical Eng Tech [O]		4.0	Y	4.0	4.7	B	O	-.-(A)	16	-.-(A)	--.(B)	-.-(B)
Electromechanical Eng Tech [O]		-.-(B)	N	4.0	4.7	B	R	100.0	3	-.-(A)	--.(B)	-.-(B)
Mechanical Eng Tech [O]		3.0	Y	4.0	4.7	B	O	-.-(A)	11	-.-(A)	--.(B)	-.-(B)
356 Western Michigan U	MI											
Aircraft Maintenance Eng Tech [O]		-.-	N	4.5	4.5	D	O	1.0	0	95.0	--.	-.-
Automotive Eng Tech [O]		1.0	N	4.0	4.5	D	O	-.-	6	95.0	--.	-.-
Aviation Tech & Operations [O]		-.-	N	4.0	4.5	D	O	1.0	55	98.0	--.	-.-
Construction Science & Mgmt [O]		2.3	N	4.0	4.5	D	O	-.-	21	90.0	--.	-.-
Eng Graphics [O]		-.-	N	4.0	4.5	D	O	-.-	30	95.0	--.	-.-
Eng Metallurgy [O]		-.-	N	4.0	4.5	D	O	-.-	0	100.0	--.	-.-
Industrial Design [O]		-.-	N	4.0	4.5	D	O	-.-	9	95.0	--.	-.-
Manuf'g Eng Tech [O]		-.-	Y	4.0	4.5	D	O	-.-	10	90.0	--.	-.-

All Engineering Technology Disciplines—Totals

TOTALS

State/Province/Country	# of schools	FACULTY Full-time equivalent (FTE)	UNDERGRADUATE PROGRAMS ABET/CEAB accred?	Length Nominal length of program in years	Length Average length of program in years	Time Day/Eve./Both	None/Opt./Req.	Co-op % of graduates in Co-op programs	Bachelor's # of degrees awarded	Bachelor's % of graduates who were full-time	GRADUATE PROGRAMS Master's # of degrees awarded	Master's % of graduates who were full-time
AL	2	46.00	4						58		-	-
AR	1	22.50	5						44		-	-
AZ	2	37.00	5						276		15	-
CA	3	37.00	5						234		-	-
CO	2	32.50	6						108		-	-
CT	2	27.40	3						185		-	-
DC	1	4.00	2						16		-	-
FL	2	30.00	6						154		-	-
GA	2	73.80	8						459		-	-
IL	4	70.80	9						364		7	-
IN	3	80.00	6						345		-	-
KS	1	21.40	5						77		-	-
KY	2	26.70	8						81		-	-
LA	1	6.00	2						21		-	-
MA	4	182.60	14						599		-	-
MD	1	11.00	2						73		-	-
ME	1	-	3						54		-	-
MI	3	41.60	7						329		-	-
MN	2	20.00	3						69		-	-
MO	1	9.50	2						54		-	-
MS	1	18.00	6						95		4	-
MT	1	12.50	3						73		-	-
NC	2	29.00	4						139		-	-
NH	1	4.00	2						23		-	-
NJ	1	-	4						118		-	-
NM	1	14.00	3						79		-	-
NY	3	98.30	16						417		-	-
OH	5	120.10	18						564		-	-
OK	1	33.50	6						113		-	-
OR	1	47.50	7						167		-	-
PA	5	151.90	16						582		-	-
SC	1	16.00	4						60		-	-
TN	3	25.60	9						181		28	-
TX	4	122.90	11						394		-	-
UT	2	46.90	5						173		11	-
VA	2	26.00	5						153		-	-
WI	1	30.40	2						81		-	-
WV	1	15.00	5						125		-	-
U.S. Totals:	76	1,591.40	231						7,137		65	-

All Engineering Technology Disciplines—Footnotes

FOOTNOTES: The following footnotes are the same for all schools: (A) Data not available. (B) Data not applicable. (C) Estimated data.

281 University of Akron: (1) Accredited by TAC/ABET. (2) FTE faculty figures represent last 2 yrs. of program. First 2 yrs. are included in associate program figures. (3) AAS degree. (4) BSMT degree.

285 Auburn University: (1) Total FTE for Textile Engineering program. FTE for Textile Management and Technology cannot be differentiated. (2) Based on credit hours taken Fall Quarter 91.

287 Brigham Young University: (1) New program currently under consideration by EAC/ABET for accreditation. (2) All faculty are listed under Manufacturing Engineering Technology.

291 Central Connecticut State University: (1) BSET degree. No graduate degrees offered. (2) Technology Accreditation Commission. (3) Total for both programs. (4) Degrees are reported under Civil and Construction Technology.

292 University of Central Florida: (1) Due primarily to large percentage of students who work part time, go to school part time. (2) The Engineering Technology programs do not offer the master's degree.

293 University of Cincinnati: (1) AAS program is in Architectural Technology; BS program is in Architectural Engineering Technology. (2) BS program is 5 years (including co-op); AAS program is 2 yrs (including co-op). (3) Presently, the Chemical Technology is an AAS degree program; a BS program is planned for implementation in 1994. (4) Presently there is not a BS degree offered. Co-op is a requirement for AAS program. (5) AAS program is called Civil and Construction Engineering Technology; BS program is called Construction Management and is a "plus 2" program. (6) 100% of BS graduates co-oped. 92% of AAS graduates co-oped. (7) BS program has no co-op requirement; co-op is a requirement for AAS program. (8) Department offers an AAS degree in Manufacturing Engineering Technology with BS planned for 1993. (9) BS program has co-op as an option; co-op is a requirement for the AAS program.

295 DeVry Institute of Technology, Chicago: (1) No graduate degree offered

296 DeVry Institute of Technology, Columbus: (1) Graduate Degree not offered

297 DeVry Institute of Technology, Atlanta: (1) No graduate degree offered

298 DeVry Institute of Technology, DuPage: (1) No graduate degree offered

299 DeVry Institute of Technology, Dallas: (1) No graduate degree offered

300 DeVry Institute of Technology, Kansas City: (1) No graduate degree offered

301 DeVry Institute of Technology, Los Angeles: (1) No graduate degree offered

302 DeVry Institute of Technology, Phoenix: (1) No graduate degree offered

303 University of the District of Columbia: (1) Most students are part-time. All courses are offered in p.m.

308 University of Hartford: (1) AET Program instituted in 1991. (2) The CET program is a two year program which awards a AAS.

311 Louisiana Tech University: (1) Graduate degrees not available for technology.

318 Milwaukee School of Engineering: (1) No co-op program is available, however, 85% of undergraduates participate in an internship program and have relevant work experience prior to graduation.

319 Montana State University: (1) Information unavailable but near 100%

322 New Jersey Institute of Technology: (1) No separate faculty for this program. The Engineering Technology Department as a whole has 17.2 faculty FTEs. (2) No graduate level program offered.

326 Oklahoma State University: (1) Graduate programs not offered in Technology (2) Approximately 83% of all engineering technology students are full time. Data not broken down by program. (3) This program draws on coursework in other programs

334 Prairie View A&M University: (1) Faculty figures for the Computer, Electrical, and Mechanical Engineering Technology programs are reported together under Computer Engineering Technology.

336 Rochester Institute of Technology: (1) Awards both AAS & BS. (2) New Program; includes Technical and Management options.

337 St. Cloud State University: (1) The system in place is internships. (2) This program is in process of being accredited.

342 University of Southern Mississippi: (1) Emphasis: Construction Management, Electronics Sys., CADD/CAM/CIM, Manufacturing Sys., Production & Inventory, Quality Control, & Industrial Technology. (2) Master's degree program only. Because this is a new program, there are no graduates yet.

344 State University of New York Institute of Technology, Utica/Rome: (1) Admit upper level students only.

346 University of Tennessee, Martin: (1) No graduate programs in Engineering Technology.

347 Texas A&M University: (1) One Engineering Technology Degree for all three separately accredited programs - Electronics, Mechanical & Manufacturing. (2) This data is included in the total for Engineering Technology Degree Program.

349 Tri-State University: (1) Associate degree program (2) No graduate programs (3) The baccalaureate engineering technology program is new and as such has not yet produced a graduate.

350 Virginia State University: (1) There is no Graduate Program in Engineering Technology at this Institution.

352 Wentworth Institute of Technology: (1) Master's degree is not awarded at Wentworth. (2) Program offered during day and on weekend.

353 West Virginia Institute of Technology: (1) Bachelor of Science in Engineering Technology-Surveying.

281 University of Akron

■ INSTITUTION PROFILE
HEAD OF THE INSTITUTION
Peggy G. Elliott
Phone: (216)972-7074
GENERAL INFORMATION
[All Students—Fall 1991]
Undergraduate enrollment 25,390
Graduate enrollment 4,389
Total institution enrollment 29,779
Type of institution: Public **Calendar:** Semesters
Location: Akron **Population:** 223,019
Setting: Urban
Types of engineering degrees: Engineering Technology
Other degree-granting colleges: Arts & Sciences, Business Administration, Education, Fine & Performing Arts, Law, Nursing, Polymer Science, Wayne General and Technical College

■ ENGINEERING TECHNOLOGY ADMISSIONS
ADMISSIONS OFFICE CONTACT
Martha A. Booth
University of Akron
Admissions Office
Akron, OH 44325-2001
Phone: (216)972-7100
TECHNOLOGY COLLEGE ADMISSIONS INFORMATION
Admission to the engineering technology college: Immediately following or during freshmen orientation period. (Graduation from an accredited high school or an earned GED is prerequisite.)
Entrance Requirements: SAT, ACT, SAT or ACT required for those under the age of 21.
Entrance Recommendations: High School courses—English (4 years), Mathematics (3 years), Science (3 years), Social Science (3 years)
Requirements for foreign students: TOEFL (Score: 500); financial statement; Official documents (high school &/or college transcripts) from foreign schools translated into English.
DEPARTMENTAL ADMISSIONS INFORMATION
Admission to the engineering technology department: Engineering technology students are admitted immediately following or during the freshmen orientation period.
Additional information: Technology students are required to maintain a minimum 2.0 GPA (on a 4.0 scale) in the math, science, and basic major courses in the first two years.

■ FINANCIAL INFORMATION
ESTIMATED EXPENSES (ACADEMIC YEAR)
[Expenses are for the 1992-93 nine-month academic year.]

State Residents	Undergraduate	Graduate
Tuition and fees	$ 2,841	$ —
College room and board	$ 3,486	$ —
Books and supplies	$ 500	$ —
Other expenses	$ 80 [1]	$ —
Total estimated expenses	$ 6,907	$ —
Out-of-State Residents	**Undergraduate**	**Graduate**
Tuition and fees	$ 8,177 [2]	$ —
College room and board	$ 3,486	$ —
Books and supplies	$ 500	$ —
Other expenses	$ 80 [1]	$ —
Total estimated expenses	$12,243	$ —

Notes: (1) Course lab fee estimate. (2) $255.56 per credit hr. for out-of-state residents. Assume 16 hr. load per semester.
FINANCIAL AID OFFICE CONTACT
Robert D. Hahn, *Director*
University of Akron
Student Financial Aid & Employment
Akron, OH 44325-6211
Phone: (216)972-7032 **Fax:** (216)972-7139
GENERAL FINANCIAL AID INFORMATION
Forms accepted/required: AFSA, FAF, Stafford, Supplemental, Institutional, OIG form, PLUS (Parent Loan) Application.
Additional financial aid information: Need-based state and federal grants, merit-based scholarships, part-time jobs, educational bank loans (some based on financial need, some not), and the College Work Study Program are available.

■ TECHNOLOGY COLLEGE INFORMATION
[For additional personnel, refer to the *Appendix*.]
HEAD OF ENGINEERING TECHNOLOGY
Frederick J. Sturm
Phone: (216)972-7220 **Fax:** (216)972-5300
ENGINEERING TECHNOLOGY COLLEGE ADDRESS
Engineering & Science Technology Division
University of Akron
Community & Technical College
Akron, OH 44325-6104
Phone: (216)972-7051 **Fax:** (216)972-6990
TECHNOLOGY ENROLLMENTS—BY CLASS
[Numbers are technology baccalaureate enrollments for the fall 1991 term, unless otherwise footnoted.]
1st-year students/Freshmen .... 267 [1]
2nd-year students/Sophomores .... 182 [1]
3rd-year students/Juniors .... 206 [1]
4th-year students/Seniors .... 274 [1]
Total 929
Notes: (1) Head count - includes PT & FT students.
NUMBER OF DEGREES AWARDED—BY PROGRAM
[Numbers are engineering technology baccalaureate degrees awarded during the 1991-1992 academic year. For full details about each engineering technology program, refer to the *Table of Degree Programs*.]
Automated Manufacturing Technology 3
Drafting & Computer Drafting Technology .... — [B]
Electronic Technology 58
Electronic Technology .... — [B]
Manufacturing Technology .... — [B]
Mechanical Technology .... — [B]
Mechanical Technology 38
Surveying & Construction Technology .... — [B]
Total 99
Notes: (B) Data not applicable.
PERCENTAGE OF DEGREES AWARDED—BY CATEGORY
[Percentages are of all engineering technology baccalaureate degrees awarded during the 1991-1992 academic year, unless otherwise footnoted.]
African Americans .... 2.0 %
Asian/Pacific Island Americans .... — %
Hispanic Americans .... — %
Native Americans .... 1.0 %
Foreign Citizens .... 1.0 %
All Others .... 96.0 %
Women .... 2.0 %
Persons with disabilities .... — % [A]
Students over 25 years of age .... — % [A]
Notes: (A) Data not available.
ENGINEERING TECHNOLOGY STUDENT DATA
Applicants to the engineering technology college
Number of applicants to engineering technology college ... 108
Percent offered admission .... 96.3 %
Matriculated engineering technology students
Percentage in top quartile (25%) of High School class .... 80.6 [2]
Average SAT scores: Math—477[1] Verbal—383[1]
Average ACT scores: Math—18.7[1]
Notes: (1) Data not avail. for all students. SAT/ACT required only for those under 21. Scores shown are extrapolated based on weighted avg of students accepted. (2) High school rank required only for those under the age of 25.
FULL- & PART-TIME FACULTY—BY DEPARTMENT
[Figures are the head count of full-time faculty and the full-time equivalent (FTE) of part-time faculty for each engineering technology department or equivalent.]

Department	Full	Part
Surveying & Construction Tech	1	0.8
Electronic Tech	8	2.3
Manufacturing & Automated Manufacturing Tech	2	0.8
Mechanical Tech	5	1.9
Drafting & Computer Drafting Tech	2	1.6

TECHNOLOGY COLLEGE DESCRIPTION
The Engineering & Science Technology Division of the Community & Technical College offers three 2+2 programs (Bachelor of Science in Electronic Technology, Bachelor of Science in Mechanical Technology, and Bachelor of Science in Automated Mfg. Tech.) Graduates of the Associate of Applied Science program in Surveying & Construction Technology may continue on for a Bachelor of Construction Technology degree (3 additional yrs., including a required co-op); graduates of the Associate of Applied Science programs in Drafting or Surveying & Construction Technology (Surveying Option only) may continue on for a Bachelor of Science in Geography/Cartography. Co-op programs are available for all programs (required only in the BCT program). Length of program would be extended by the number of co-op terms elected by the student.
TECHNOLOGY DEGREE PROGRAMS DESCRIPTION
The C & T College's 'plus-two' programs lead toward a Bachelor of Science degree in Electronic Technology, Mechanical Technology, or Automated Manufacturing Technology, with an emphasis in industrial application of theoretical concepts. The Electronic program offers studies in circuit analysis, digital and linear electronics, communications, computer programming, and control systems. The Mechanical program offers an orientation in mechanical design, CNC machinery, and production/supervisory topics. The Automated Manufacturing program provides studies in CAD, CAM, CIM, and the techniques of managing the automated manufacturing system. The College also awards Associate of Applied Science degrees in Manufacturing, Drafting & Computer Drafting, and Surveying & Construction Technology; a certificate program is available in Drafting & Computer Drafting Technology.
DUAL DEGREE PROGRAMS
Technology baccalaureate graduates with dual degrees 2
Enrollment requirements: Completion of the appropriate associate degree.
TRANSFER INFORMATION
Residency Requirements: Students must complete at least 32 semester hours in residence for a baccalaureate degree or 16 semester hours for an associate degree program.
Transfer via Articulation Agreements
Admission to engineering technology: At any time
Technology graduates transferred from ... 4-yr: — 2-yr: 15
Requirements: Minimum 2.00 GPA. A rolling admission procedure is used.
Transfer without Articulation Agreements
Admission to engineering technology: At any time
Technology graduates transferred from ... 4-yr: 10 2-yr: 20
Requirements: Minimum 2.00 GPA. A rolling admission procedure is used.
GRADUATION REQUIREMENTS
Bachelor of Science degree programs require 132-137 semester credit hours; Associate of Applied Science degree programs require 64-71 semester credit hours, depending on major. An overall GPA of 2.00 or above is required for graduation. Students must complete the last 16 hrs. for the associate of applied science and last 32 hrs. for the baccalaureate degree. If they wish to transfer coursework in during this time, they must obtain special permission.

■ STUDENT PROGRAMS
PROFESSIONAL AND HONORARY SOCIETIES
[For key to acronyms, see *Introduction*.]
Professional Societies: ACSM, ASME, IEEE, Society of Students in Construction (affiliated with AGCA)
Honorary Societies: Tau Alpha Pi
SUPPORT PROGRAMS
Student Chapter Organizations: National Society of Black Engineers

Black Cultural Center
For: Ethnic Minorities **Available:** Year round
Offered: Freshman, Sophomore, Junior, Senior, Graduate level

Adult Resource Center
For: Women **Available:** Year round
Offered: Prematriculation, Freshman, Sophomore, Junior, Senior, Graduate level

Other Engineering Support Programs: Summer & fall orientation prog. for freshmen & transfers; career counseling services.; foreign student office; veterans' services.; services. for handicapped; placement services. Updated list of scholarships and one-on-one counseling is avail. through the Adult Resource Center. National Society of Black Engineers, contact the Black Cultural Center in East Hall, room 206 (peer counseling available also.)

282 Alabama A&M University

INSTITUTION PROFILE
HEAD OF THE INSTITUTION
David Henson
Phone: (205)851-5230

GENERAL INFORMATION
[All Students—Fall 1991]
Undergraduate enrollment 3,500
Graduate enrollment 1,700
Total institution enrollment **5,200**
Type of institution: Public **Calendar:** Semesters
Location: Huntsville **Population:** 169,000
Setting: Suburban
Types of engineering degrees: Engineering Technology
Other degree-granting colleges: Arts & Sciences, Business Administration, Education, Agriculture and Home Economics

ENGINEERING TECHNOLOGY ADMISSIONS
ADMISSIONS OFFICE CONTACT
James O. Heyward
Alabama A&M University
P.O. Box 908
Normal, AL 35762-908
Phone: (205)851-5245

TECHNOLOGY COLLEGE ADMISSIONS INFORMATION
Entrance Requirements: SAT, ACT
Entrance Recommendations: SAT, ACT
Requirements for foreign students: TOEFL; financial statement

DEPARTMENTAL ADMISSIONS INFORMATION
Admission to the engineering technology department: At the time of admission to the institution.

FINANCIAL INFORMATION
ESTIMATED EXPENSES (ACADEMIC YEAR)
[Expenses are for the 1992-93 nine-month academic year.]

State Residents	Undergraduate	Graduate
Tuition and fees	$ 1,500	$ 1,400(C)
College room and board	$ 1,725	$ —(A)
Books and supplies	$ 300	$ 800
Other expenses	$ 500	$ —(A)
Total estimated expenses	**$ 4,025**	**$ 2,200**

Out-of-State Residents	Undergraduate	Graduate
Tuition and fees	$ 2,500	$ 1,800(B)
College room and board	$ 4,500	$ —(B)
Books and supplies	$ 400	$ 400
Other expenses	$ 500	$ —(A)
Total estimated expenses	**$ 7,900**	**$ 2,200**

Notes: (A) Data not available. (B) Data not applicable. (C) Estimated data.

FINANCIAL AID OFFICE CONTACT
Percy N. Lanier, *Director*
Alabama A&M University
P.O. Box 907
Normal, AL 35763-907
Phone: (205)851-5400

GENERAL FINANCIAL AID INFORMATION
Forms accepted/required: Stafford, IRS
Additional financial aid information: Bi-weekly jobs, Work-study jobs (Federal Funds), Academic Scholarships.

TECHNOLOGY COLLEGE INFORMATION
[For additional personnel, refer to the *Appendix*.]
HEAD OF ENGINEERING TECHNOLOGY
Arthur J. Bond
Phone: (205)851-5560 **Fax:** (205)851-5580

ENGINEERING TECHNOLOGY COLLEGE ADDRESS
School of Engineering & Technology
Alabama A&M University
P.O. Box 1148
Normal, AL 35762-1148
Phone: (205)851-5560 **Fax:** (205)851-5580

TECHNOLOGY ENROLLMENTS—BY CLASS
[Numbers are technology baccalaureate enrollments for the fall 1991 term, unless otherwise footnoted.]
1st-year students/Freshmen 350
2nd-year students/Sophomores 200
3rd-year students/Juniors 150
4th-year students/Seniors 75
Total **775**

NUMBER OF DEGREES AWARDED—BY PROGRAM
[Numbers are engineering technology baccalaureate degrees awarded during the 1991-1992 academic year, unless otherwise footnoted. For full details about each engineering technology program, refer to the *Table of Degree Programs*.]
Civil Engineering 3
Drafting & Design Technology 2
Electrical Engineering 31
Industrial Technology 1
Mechanical Engineering 9
Total **46**

PERCENTAGE OF DEGREES AWARDED—BY CATEGORY
[Percentages are of all engineering technology baccalaureate degrees awarded during the 1991-1992 academic year, unless otherwise footnoted.]
African Americans .... 75.0 %
Asian/Pacific Island Americans .... – %
Hispanic Americans .... – %
Native Americans .... – %
Foreign Citizens .... 25.0 %
All Others .... – %
Women .... 30.0 %
Persons with disabilities .... – %
Students over 25 years of age .... 2.0 %

FULL- & PART-TIME FACULTY—BY DEPARTMENT
[Figures are the head count of full-time faculty and the full-time equivalent (FTE) of part-time faculty for each engineering technology department or equivalent.]

Department	Full	Part
Eng Tech	10	–

DUAL DEGREE PROGRAMS
Technology baccalaureate graduates with dual degrees 0

TRANSFER INFORMATION
Transfer without Articulation Agreements
Admission to engineering technology: At any time
Technology graduates transferred from ... 4-yr: 2   2-yr: 10

GRADUATION REQUIREMENTS
GPA requirement in major courses is 2.00. GPA overall grade point average is 2.00.

STUDENT PROGRAMS
PROFESSIONAL AND HONORARY SOCIETIES
[For key to acronyms, see *Introduction*.]
Professional Societies: ASCE
Honorary Societies: Sigma Tau Epsilon

SUPPORT PROGRAMS
Student Chapter Organizations: National Society of Black Engineers, American Society for Civil Engineering, ISEE

283 Arizona State University

INSTITUTION PROFILE

HEAD OF THE INSTITUTION
Lattie F. Coor
Phone: (602)965-5606 **Fax:** (602)965-0865

GENERAL INFORMATION
[All Students—Fall 1991]

Undergraduate enrollment	31,426
Graduate enrollment	11,200
Total institution enrollment	**42,626**

Type of institution: Public **Calendar:** Semesters
Nearest city: Phoenix **Population:** 983,403
Miles from main campus: 11 **Setting:** Suburban
Types of engineering degrees: Engineering & Technology
Other degree-granting colleges: Agricultural & Environmental, Architecture, Arts & Sciences, Business Administration, Communication & Journalism, Education, Fine & Performing Arts, Law, Nursing, Public Programs, Social Work, University Honors

ENGINEERING TECHNOLOGY ADMISSIONS

ADMISSIONS OFFICE CONTACT
Susan Clouse
Arizona State University
Undergraduate Admissions
Tempe, AZ 85287-0112
Phone: (602)965-7788 **Fax:** (602)965-2120

TECHNOLOGY COLLEGE ADMISSIONS INFORMATION
Entrance Requirements: SAT (Score: 930), ACT (Score: 22), High School courses—Mathematics (Algebra I & II; geometry (3 years), Applicants (Arizona residents) must submit either test scores (SAT 930; or ACT 22) or rank in top 50% of high school class.
Entrance Recommendations: High School courses—Chemistry (1 year), Physics (1 year)
Requirements for out-of-state residents: Non-resident: Rank in top 25% of high school class; or ACT of 24; or SAT of 1010.
Requirements for foreign students: TOEFL (Score: 500); financial statement; Application deadline: June 15 for Fall semester, November 15 for Spring semester.

DEPARTMENTAL ADMISSIONS INFORMATION
Admission to the engineering technology department: At the time of admission to the institution.

FINANCIAL INFORMATION

ESTIMATED EXPENSES (ACADEMIC YEAR)
[Expenses are for the 1992-93 nine-month academic year.]

State Residents	Undergraduate	Graduate
Tuition and fees	$ 1,530	$ 1,530
College room and board	$ 4,110	$ 4,110
Books and supplies	$ 480	$ 480
Other expenses	$ 2,500	$ 2,500[1]
Total estimated expenses	**$ 8,620**	**$ 8,620**

Out-of-State Residents	Undergraduate	Graduate
Tuition and fees	$ 6,940	$ 6,940
College room and board	$ 4,110	$ 4,110
Books and supplies	$ 480	$ 480
Other expenses	$ 2,500	$ 2,500
Total estimated expenses	**$14,030**	**$14,030**

Notes: (1) Personal expenses including travel.

FINANCIAL AID OFFICE CONTACT
Paul Barberini, *Director*
Arizona State University
Student Financial Assistance
Tempe, AZ 85287-0412
Phone: (602)965-4045

GENERAL FINANCIAL AID INFORMATION
Forms accepted/required: FAF, FFS
Additional financial aid information: Scholarships, grants, loans, & employment.

TECHNOLOGY COLLEGE INFORMATION
[For additional personnel, refer to the *Appendix*.]

HEAD OF ENGINEERING TECHNOLOGY
David C. Chang
Phone: (602)965-1722 **Fax:** (602)965-2267
EMail: icdcc@asuacad

ENGINEERING TECHNOLOGY COLLEGE ADDRESS
College of Engineering & Applied Sciences
Arizona State University
School of Technology
Tempe, AZ 85287-6506
Phone: (602)965-3874 **Fax:** (602)965-5089

TECHNOLOGY ENROLLMENTS—BY CLASS
[Numbers are technology baccalaureate enrollments for the fall 1991 term, unless otherwise footnoted.]

1st-year students/Freshmen	20
2nd-year students/Sophomores	27
3rd-year students/Juniors	81
4th-year students/Seniors	208
Total	**336**

NUMBER OF DEGREES AWARDED—BY PROGRAM
[Numbers are engineering technology baccalaureate degrees awarded during the 1991-1992 academic year, unless otherwise footnoted. For full details about each engineering technology program, refer to the *Table of Degree Programs*.]

Aeronautical Engineering Technology	18
Electronic Engineering Technology	43
Manufacturing Engineering Technology	32
Total	**93**

PERCENTAGE OF DEGREES AWARDED—BY CATEGORY
[Percentages are of all engineering technology baccalaureate degrees awarded during the 1991-1992 academic year, unless otherwise footnoted.]

African Americans	– %
Asian/Pacific Island Americans	6.4 %
Hispanic Americans	3.2 %
Native Americans	1.0 %
Foreign Citizens	9.8 %
All Others	79.6 %
Women	3.0 %
Persons with disabilities	– %
Students over 25 years of age	– %

GRADUATE ENROLLMENTS & DEGREES AWARDED

Master's enrollment	166
Master's degrees awarded	15
Doctoral enrollment	–
Doctoral degrees awarded	–

ENGINEERING TECHNOLOGY STUDENT DATA

Applicants to the engineering technology college

Number of applicants to engineering technology college	85
Percent offered admission	92.0%

Matriculated engineering technology students

Percentage in top quartile (25%) of High School class	– (A)

Notes: (A) Data not available.

FULL- & PART-TIME FACULTY—BY DEPARTMENT
[Figures are the head count of full-time faculty and the full-time equivalent (FTE) of part-time faculty for each engineering technology department or equivalent.]

Department	Full	Part
Aeronautical Tech	9	1.8
Electronics & Computer Tech	11	–
Manufacturing & Industrial Tech	17	3.3

TECHNOLOGY COLLEGE DESCRIPTION
Four year Bachelor of Science degree programs are offered in aeronautical engineering technology, electronics engineering technology, and manufacturing engineering technology. These programs are made up of university General Studies, engineering technology core, and major courses. The technology graduate contributes an applications orientation to complement the engineer's more theoretical concepts. A student will be educated to render practical decisions with safety and economy in mind; install and operate technical systems; develop or improve a product; to revise systems; and provide customer support when needed. Co-op is an option; students who choose this program ideally complete 12 months employment in addition to the four-year undergraduate program.

TECHNOLOGY DEGREE PROGRAMS DESCRIPTION
Majors and areas of emphasis are offered leading to the B.S. degree. These programs allow the student to earn a degree in a technological field which stresses theory reinforced by laboratory application. A Master of Technology degree with a major in Technology is also offered. This program is designed for flexibility, permitting the student to select a combination of courses in technology and supporting areas to meet individual career goals. A practicum or applied project is required.

DUAL DEGREE PROGRAMS
Technology baccalaureate graduates with dual degrees 5
Enrollment requirements: Dual degree students must meet admission requirements and complete all course requirements of both degree programs.

TRANSFER INFORMATION
Residency Requirements: A minimum of 30 semester hours earned in resident credit courses at ASU is required of every candidate for the baccalaureate degree. The final 12 hours immediately preceding graduation must be of resident credit.

Transfer via Articulation Agreements
Admission to engineering technology: At any time
Requirements: Arizona Residents: Minimum cum GPA 2.25; Non-residents: Minimum cum GPA 2.50.

Transfer without Articulation Agreements
Admission to engineering technology: At any time
Requirements: These transfer students must meet the minimum cum GPA requirements (2.25, Arizona residents; 2.50, non-residents) and TOEFL score of 500 if international student.

GRADUATION REQUIREMENTS
B.S. degree programs in engineering technology require 126 semester hours minimum plus University English proficiency requirements. A minimum cum GPA of 2.00 is required in the overall program.

STUDENT PROGRAMS

PROFESSIONAL AND HONORARY SOCIETIES
[For key to acronyms, see *Introduction*.]
Professional Societies: AIAA, ASME, IEEE
Honorary Societies: Tau Alpha Pi

SUPPORT PROGRAMS
Student Chapter Organizations: Society of Women Engineers, Society of Hispanic Professional Engineers, American Indian Science & Engineering Society, Arizona Council of Black Engineers & Scientists

Minority Engineering Program
For: Ethnic Minorities **Available:** Academic year
Offered: Prematriculation, Freshman, Sophomore
Other Engineering Support Programs: Orientation program for all freshmen and transfers; personal and career counseling services; international students; veteran's services; disabled students resource office; tutoring services; and placement services.

284 University of Arkansas at Little Rock

INSTITUTION PROFILE
HEAD OF THE INSTITUTION
Alan B. Sugg
Phone: (501)686-2505 Fax: (501)686-2507
GENERAL INFORMATION
[All Students—Fall 1991]
Undergraduate enrollment 10,569
Graduate enrollment 1,455
Total institution enrollment 12,419
Type of institution: Public **Calendar:** Semesters
Location: Little Rock, AR **Population:** 176,798
Setting: Urban
Types of engineering degrees: Engineering Technology
Other degree-granting colleges: Business Administration, Education, Law, Arts, Humanities, & Social Sciences, Professional & Public Affairs, University College

ENGINEERING TECHNOLOGY ADMISSIONS
ADMISSIONS OFFICE CONTACT
Sue Pines
University of Arkansas at Little Rock
Director of Admissions
2801 South University
Little Rock, AR 72204-1099
Phone: (501)569-3127
TECHNOLOGY COLLEGE ADMISSIONS INFORMATION
Admission to the engineering technology college: the time they are eligible to enroll in ENGL 1311 and MATH 1302.
Entrance Requirements: SAT (Score: 750), ACT (Score: 19)
Entrance Recommendations: SAT (Score: 750), ACT (Score: 19), High School courses—Mathematics (Algebra I & II, Plane Geometry, Trig) (4 years), Science: Physics, Chemistry (2 years), Computers (1 year), Drafting (1 year)
Requirements for foreign students: TOEFL (min. 500 points) or Michigan English Language Assessment Battery (min. 85 percent) and financial statement.
DEPARTMENTAL ADMISSIONS INFORMATION
Admission to the engineering technology department: A prospective student applies at the time of application to the college and is admitted on the basis of acceptable high school records and ACT or SAT scores.
Additional information: All entering students are initially admitted to the University College. They are accepted by the department only after they have demonstrated competency in both mathematics and English.

FINANCIAL INFORMATION
ESTIMATED EXPENSES (ACADEMIC YEAR)
[Expenses are for the 1992-93 nine-month academic year.]

State Residents	Undergraduate	Graduate
Tuition and fees	$ 2,000	$ –
College room and board	$ 2,550	$ –
Books and supplies	$ 600	$ –
Other expenses	$ –	$ –
Total estimated expenses	**$ 5,150**	**$ –**

Out-of-State Residents	Undergraduate	Graduate
Tuition and fees	$ 4,800	$ –
College room and board	$ 2,559	$ –
Books and supplies	$ 600	$ –
Other expenses	$ –	$ –
Total estimated expenses	**$ 7,959**	**$ –**

GENERAL FINANCIAL AID INFORMATION
Forms accepted/required: ACT, IRS, Institutional Forms
Additional financial aid information: Merit-Based Scholarships; Need-Based Scholarships; Engineering Technology Scholarships for academic excellence; Short-term Loans; Federal Grants; College Work-study Programs. Required Forms: FFS, IRS, and Institutional Forms.

TECHNOLOGY COLLEGE INFORMATION
[For additional personnel, refer to the *Appendix*.]
HEAD OF ENGINEERING TECHNOLOGY
Charles A. Stevens
Phone: (501)569-3150 Fax: (501)569-8020
EMail: CASTEVENS%EIVAX@UALR.EDU
ENGINEERING TECHNOLOGY COLLEGE ADDRESS
Department of Engineering Technology
University of Arkansas at Little Rock
2801 South University
Little Rock, AR 72204-1099
Phone: (501)569-8200 Fax: (501)569-8020
TECHNOLOGY ENROLLMENTS—BY CLASS
[Numbers are technology baccalaureate enrollments for the fall 1991 term, unless otherwise footnoted.]
1st-year students/Freshmen .... (A)
2nd-year students/Sophomores 83
3rd-year students/Juniors 94
4th-year students/Seniors 164
Total 341
Notes: (A) Data not available.
NUMBER OF DEGREES AWARDED—BY PROGRAM
[Numbers are engineering technology baccalaureate degrees awarded during the 1991-1992 academic year, unless otherwise footnoted. For full details about each engineering technology program, refer to the *Table of Degree Programs*.]
Computer Engineering Technology 3
Construction Engineering Technology 11
Electronic(s) Engineering Technology 10
Manufacturing Engineering Technology 1
Mechanical Engineering Technology 14
Surveying & Land Information Systems 5
Total 44
PERCENTAGE OF DEGREES AWARDED—BY CATEGORY
[Percentages are of all engineering technology baccalaureate degrees awarded during the 1991-1992 academic year, unless otherwise footnoted.]
African Americans .... 3.1 %
Asian/Pacific Island Americans .... 6.2 %
Hispanic Americans .... – %
Native Americans .... – %
Foreign Citizens .... – %
All Others .... 90.7 %
Women .... 6.2 %
Persons with disabilities .... – % (A)
Students over 25 years of age .... 47.9 %
Notes: (A) Data not available.
FULL- & PART-TIME FACULTY—BY DEPARTMENT
[Figures are the head count of full-time faculty and the full-time equivalent (FTE) of part-time faculty for each engineering technology department or equivalent.]

Department	Full	Part
Electronics Eng Tech	3	2.0
Manufacturing Eng Tech	2	1.0
Mechanical Eng Tech	3	2.0
Computer Eng Tech	3	–
Construction Eng Tech	3	1.0
Surveying	1	1.0

TECHNOLOGY COLLEGE DESCRIPTION
Situated in the industrial hub of the state, the engineering technology program enjoys a modern building with state-of-the-art laboratory facilities. The program has seen continuous growth since its inception in 1977. It offers both associate and baccalaureate degrees in construction, electronics and mechanical engineering technology on a 2 + 2 format. Also baccalaureate degrees are available in computer and manufacturing engineering technology and surveying. Both day and evening classes are available to accommodate full-time and part-time working students. The program curricula emphasize extensive laboratory work supplemented by a strong theoretical background and substantial computer experience. Senior design projects are often carried out in coalition with local industries to provide real-life experience to the graduates. The seniors are encouraged to take the EIT exam. Co-op program is optional, but encouraged. Double-major option with other disciplines is available.

TECHNOLOGY DEGREE PROGRAMS DESCRIPTION
The B.S. degree program in engineering technology emphasizes the operational and applied aspects of engineering principles along with practical design to prepare graduates for entry-level professional positions or for graduate study. Special features of the program include excellent computing facilities, a 13 million dollar new building with state-of-the-art laboratory equipment and a close working relationship with the industrial community to keep the curricula tuned to satisfy industry's needs. Each of the five engineering technology programs awards the B.S. degree and is accredited by the Technology Accreditation Commission (TAC) of the Accreditation Board for Engineering and Technology (ABET). The associate degree programs in construction, electronics, and mechanical engineering technology are also TAC/ABET accredited.

TRANSFER INFORMATION
Residency Requirements: Transfer students must earn a minimum of 30 consecutive hours in residence, not including credit by examination.
Transfer via Articulation Agreements
Admission to engineering technology: At any time
Technology graduates transferred from .....4-yr: – 2-yr: –
Requirements: Students who have attempted nine hours or less at another college or university must meet the admission requirements for entering freshmen. Transfer credit will be given only for grades of C or better earned in regionally accredited institutions.
Transfer without Articulation Agreements
Admission to engineering technology: At any time
Technology graduates transferred from .....4-yr: – 2-yr: –
Requirements: Students who have attempted nine hours or less at another college or university must meet the admission requirements for entering freshmen. Transfer credit will be given only for grades of C or better earned in regionally accredited institutions.

GRADUATION REQUIREMENTS
Bachelor of Science in Engineering Technology requires 132 to 134 semester credit hours, depending on majors. The University also requires the students to pass a proficiency examination in written English. Associate degree programs require 64 to 66 semester credits depending on majors. A minimum of C average is required on all work attempted at the University and in the major for graduation.

STUDENT PROGRAMS
PROFESSIONAL AND HONORARY SOCIETIES
[For key to acronyms, see *Introduction*.]
Professional Societies: ACSM, AGCA, ASHRAE, ASME, IEEE, SAME, SME, Society of Women Engineers
Honorary Societies: Tau Alpha Pi, Phi Kappa Phi
SUPPORT PROGRAMS
Student Chapter Organizations: Society of Women Engineers, International Students' Club, Assoc. of Minority Educational Needs & Development, Multi-Cultural Diversity Society, Non-Traditional Age Student Club

Minority Incentive Grants
For: Ethnic Minorities **Available:** Academic year
Offered: Freshman, Sophomore, Junior, Senior, Graduate level
Minority Affairs through Dean of Students' Office
For: Ethnic Minorities **Available:** Year round
Offered: Freshman, Sophomore, Junior, Senior, Graduate level
Career Counseling and Placement
For: Women & Ethnic Minorities **Available:** Year round
Offered: Freshman, Sophomore, Junior, Senior, Graduate level
Other Engineering Support Programs: Summer and Fall orientation program for all freshmen and transfers; skills assessment testing program; individual academic advising by faculty advisors; Dean of Students' office; counseling and career planning services; veterans' services; services for disabled students; foreign student office and Intensive English Language program.

285 Auburn University

INSTITUTION PROFILE

HEAD OF THE INSTITUTION
William V. Muse
Phone: (205)844-4650   Fax: (205)844-6179

GENERAL INFORMATION
[All Students—Fall 1991]
Undergraduate enrollment 18,985
Graduate enrollment 2,851
Total institution enrollment 21,836

Type of institution: Public   **Calendar:** Quarters
Nearest city: Montgomery   **Population:** 150,000
Miles from main campus: 60   **Setting:** Small Town
Types of engineering degrees: Engineering & Technology
Other degree-granting colleges: Agricultural & Environmental, Architecture, Business Administration, Education, Nursing, Pharmacy, Technology & Applied Sciences, Forestry, Human Sciences, Liberal Arts, Sciences and Mathematics, Veterinary Medicine

ENGINEERING TECHNOLOGY ADMISSIONS

ADMISSIONS OFFICE CONTACT
Charles F. Reeder
Auburn University
202 Martin Hall
Auburn University, AL 36849-5145
Phone: (205)844-4080   Fax: (205)844-6436

TECHNOLOGY COLLEGE ADMISSIONS INFORMATION
Entrance Requirements: High School courses—English (4 years), Mathematics (3 years), Science (2 years), Social Studies (3 years), Minimum ACT composite score of 20 (in-state) or 23 (out-of-state) or SAT minimum composite score of 870 (in-state) or 1000 (out-of-state).
Entrance Recommendations: High School courses—Additional Science (1 year), Additional Social Studies (1 year), Foreign Language (1 year)
Requirements for out-of-state residents: High school grade point average 2.00 in-state and 3.00 out-of-state.
Requirements for foreign students: TOEFL (Score: 550); financial statement; B average on high school work. 22 on ACT or 1000 on SAT is required.

DEPARTMENTAL ADMISSIONS INFORMATION
Admission to the engineering technology department: At the end of the first year.
Additional information: Complete all appropriate freshman courses; earn an overall grade point average of 2.0 on all required and approved elective course work; and recommendation by the Curriculum Admissions Committee.

FINANCIAL INFORMATION

ESTIMATED EXPENSES (ACADEMIC YEAR)
[Expenses are for the 1992-93 nine-month academic year.]

State Residents	Undergraduate	Graduate
Tuition and fees	$ 1,755[1]	$ 1,755[1]
College room and board	$ 3,783	$ 3,783
Books and supplies	$ 600	$ 600
Other expenses	$ 1,833[2]	$ 1,833[2]
Total estimated expenses	**$ 7,971**	**$ 7,971**

Out-of-State Residents	Undergraduate	Graduate
Tuition and fees	$ 5,265[1]	$ 5,265[1]
College room and board	$ 3,783	$ 3,783
Books and supplies	$ 600	$ 600
Other expenses	$ 1,833[2]	$ 1,833[2]
Total estimated expenses	**$11,481**	**$11,481**

Notes: (1) Cost is based on 10 to 15 credit hours per quarter.
(2) Miscellaneous expenses (recreation, travel, clothing).

FINANCIAL AID OFFICE CONTACT
Clark Aldridge, *Director*
Auburn University
203 Martin Hall
Auburn University, AL 36849-5144
Phone: (205)844-4723   Fax: (205)844-6085

GENERAL FINANCIAL AID INFORMATION

Forms accepted/required: ACT, FFS, Institutional
Additional financial aid information: Pell Grants, Stafford Loans, PLUS/SLS Loans, Perkins Loan, Supplemental Educational Opportunity Grants, Health Professions Student Loan, Academic Scholarships, College Work-Study, Institutional Employment, Institutional Loans.

TECHNOLOGY COLLEGE INFORMATION

[For additional personnel, refer to the *Appendix*.]

HEAD OF ENGINEERING TECHNOLOGY
William F. Walker
Phone: (205)844-4326   Fax: (205)844-2672
EMail: wwalker@eng.auburn.edu

ENGINEERING TECHNOLOGY COLLEGE ADDRESS
College of Engineering
Auburn University
108 Ramsay Hall
Auburn University, AL 36849-5330
Phone: (205)844-4326   Fax: (205)844-2672

TECHNOLOGY ENROLLMENTS—BY CLASS
[Numbers are technology baccalaureate enrollments for the fall 1991 term, unless otherwise footnoted.]
1st-year students/Freshmen .... (B)
2nd-year students/Sophomores 9
3rd-year students/Juniors 9
4th-year students/Seniors 12
Total 30

Notes: (B) Data not applicable.

NUMBER OF DEGREES AWARDED—BY PROGRAM
[Numbers are engineering technology baccalaureate degrees awarded during the 1991-1992 academic year, unless otherwise footnoted. For full details about each engineering technology program, refer to the *Table of Degree Programs*.]
Textile Management & Technology 12
Total 12

PERCENTAGE OF DEGREES AWARDED—BY CATEGORY
[Percentages are of all engineering technology baccalaureate degrees awarded during the 1991-1992 academic year, unless otherwise footnoted.]
African Americans 8.3%
Asian/Pacific Island Americans .... – %
Hispanic Americans .... – %
Native Americans .... – %
Foreign Citizens .... – %
All Others 91.7%
Women 58.3%
Persons with disabilities .... – % (A)
Students over 25 years of age .... – % (A)

Notes: (A) Data not available.

ENGINEERING TECHNOLOGY STUDENT DATA
Applicants to the engineering technology college
Number of applicants to engineering technology college ...... – (A)
Percent offered admission .... –% (A)

Matriculated engineering technology students
Percentage in top quartile (25%) of High School class .... – (A)
Average ACT scores: Math—26.3, Composite—26.3

Notes: (A) Data not available.

FULL- & PART-TIME FACULTY—BY DEPARTMENT
[Figures are the head count of full-time faculty and the full-time equivalent (FTE) of part-time faculty for each engineering technology department or equivalent.]

Department	Full	Part
Textile Eng	8	–

TECHNOLOGY COLLEGE DESCRIPTION
Textile Management and Technology prepares graduates for careers in textiles by providing a background in management, marketing, finance, and computer science, along with textile production and processes. Although four academic years are required for the baccalaureate degree, the average student normally requires approximately 14 quarters to complete the technology program. The College of Engineering has dual-degree arrangements with a number of other colleges and universities. An optional co-op program is available requiring one additional calendar year.

TECHNOLOGY DEGREE PROGRAMS DESCRIPTION

There is one engineering technology program and 16 baccalaureate degree engineering programs. Additionally, there are 20 Master's degree engineering programs, including both the Master of Science degree and the Professional Master's degree. Nine Ph.D. programs are offered. Graduate level courses for credit and non-credit are offered to the off-campus students through the videotape-based off-campus courses program.

DUAL DEGREE PROGRAMS
Enrollment requirements: Exact placement of transfer students can be determined only upon review of their transcripts by the College of Engineering; a minimum 2.8 grade point average on a 4.0 scale for all programs is required.

TRANSFER INFORMATION
Residency Requirements: Minimum of forty-five hours must be earned in residence in order to receive the bachelor's degree. As a general rule 45 hours must be taken in the final year and in the school or curriculum of graduation.

Transfer via Articulation Agreements
Admission to engineering technology: any time. Application deadline date: At least three weeks before quarter's opening.
Requirements: Exact placement of transfer students can be determined only upon review of their transcripts by the College of Engineering; a minimum 2.8 grade point average on a 4.0 scale for all programs is required.

Transfer without Articulation Agreements
Admission to engineering technology: any time. Application deadline date: At least three weeks before quarter's opening.
Requirements: Exact placement of transfer students can be determined only upon review of their transcripts by the College of Engineering; a minimum 2.8 grade point average on a 4.0 scale for all programs is required.

GRADUATION REQUIREMENTS
To earn the bachelor's degree in the College of Engineering, students must complete all of the subjects in their curriculum, have a minimum grade point average of 2.0 in all work attempted at Auburn University and have a 2.0 on all courses PASSED in the major at Auburn. The Textile Management and Technology Program requires 196 quarter hours.

STUDENT PROGRAMS

PROFESSIONAL AND HONORARY SOCIETIES
[For key to acronyms, see *Introduction*.]
Professional Societies: AATCC
Honorary Societies: Phi Psi

SUPPORT PROGRAMS
Student Chapter Organizations: Society of Women Engineers, National Society of Black Engineers

Minority Introduction to Engineering (MITE)
For: Women & Ethnic Minorities   **Available:** Summer only
Offered: Prematriculation

E-Day
For: Women & Ethnic Minorities   **Available:** Academic year
Offered: Prematriculation

Engineering Tutorial
For: Women & Ethnic Minorities   **Available:** Year round
Offered: Freshman, Sophomore, Junior, Senior

University Study Partners
For: Women & Ethnic Minorities   **Available:** Year round
Offered: Freshman, Sophomore, Junior, Senior

Other Engineering Support Programs: There is a summer orientation program for entering Pre-Engineering freshmen. Parents are invited to attend. Counselors are available to students. Engineering tutorial program provides free tutor time for undergraduates in math, chemistry, physics, computer programming, sophomore engineering courses, and most junior courses. University study partners provide help in a broad range of courses.

286 Bradley University

■ INSTITUTION PROFILE
HEAD OF THE INSTITUTION
John R. Brazil
Phone: (309)677-3167 Fax: (309)677-2330

GENERAL INFORMATION
[All Students—Fall 1991]
Undergraduate enrollment 5,287
Graduate enrollment 946
Total institution enrollment 6,233
Type of institution: Private **Calendar:** Semesters
Location: Peoria, IL **Population:** 113,000
Setting: Urban
Types of engineering degrees: Engineering & Technology
Other degree-granting colleges: Arts & Sciences, Business Administration, Communication and Fine Arts, Education and Health Sciences

■ ENGINEERING TECHNOLOGY ADMISSIONS
ADMISSIONS OFFICE CONTACT
Gary Bergman
Bradley University
Office of Enrollment Management
100 Swords Hall
Peoria, IL 61625
Phone: (800)447-6460 Fax: (309)677-2797

TECHNOLOGY COLLEGE ADMISSIONS INFORMATION
Entrance Requirements: ACT (Score: 22), High School courses—Algebra (1 year), Plane Geometry (1 year)

DEPARTMENTAL ADMISSIONS INFORMATION
Admission to the engineering technology department: At the time of admission to the institution.

■ FINANCIAL INFORMATION
ESTIMATED EXPENSES (ACADEMIC YEAR)
[Expenses are for the 1992-93 nine-month academic year.]

All Students	Undergraduate	Graduate
Tuition and fees	$ 9,050 [1]	$ 9,050 [1]
College room and board	$ 3,960 [2]	$ 3,960 [2]
Books and supplies	$ 48 [3]	$ 48 [3]
Other expenses	$ 2,000 [4]	$ 2,000 [4]
Total estimated expenses	**$15,058**	**$15,058**

Notes: (1) Based on 12-16 hrs. 1-7 hrs, $246/hr; 7 1/2-11 1/2 hrs, $307/hr; over 16 hrs, $236/hr over 16; $5/hr surcharge on engineering and technology courses. (2) Based on double occupancy, 20 meals per week. Other meal plans available as well as single occupancy. (3) Health and activity fees. (4) Books and transportation.

FINANCIAL AID OFFICE CONTACT
David Pardieck, *Director of Financial Aid*
Bradley University
Office of Financial Assistance
14 Swords Hall
Peoria, IL 61625
Phone: (309)677-3089 Fax: (309)677-2798

GENERAL FINANCIAL AID INFORMATION
Forms accepted/required: CSS-FAF, FAF, Institutional
Additional financial aid information: Bradley University has a number of need based and merit based awards available to students. These are listed in the University catalog. In addition, the various colleges and department within the university also have available need and merit based awards.

■ TECHNOLOGY COLLEGE INFORMATION
[For additional personnel, refer to the *Appendix*.]

HEAD OF ENGINEERING TECHNOLOGY
John E. Francis
Phone: (309)677-2720 Fax: (309)677-3670

ENGINEERING TECHNOLOGY COLLEGE ADDRESS
College of Engineering and Technology
Bradley University
124 Jobst Hall
Peoria, IL 61625
Phone: (309)677-2720 Fax: (309)677-3670

TECHNOLOGY ENROLLMENTS—BY CLASS
[Numbers are technology baccalaureate enrollments for the fall 1991 term, unless otherwise footnoted.]
1st-year students/Freshmen 3
2nd-year students/Sophomores 11
3rd-year students/Juniors 50
4th-year students/Seniors 86
Total **150**

NUMBER OF DEGREES AWARDED—BY PROGRAM
[Numbers are engineering technology baccalaureate degrees awarded during the 1991-1992 academic year, unless otherwise footnoted. For full details about each engineering technology program, refer to the *Table of Degree Programs*.]
Construction .... –
Electrical Engineering Technology 23
Manufacturing Technology 31
Total **54**

PERCENTAGE OF DEGREES AWARDED—BY CATEGORY
[Percentages are of all engineering technology baccalaureate degrees awarded during the 1991-1992 academic year, unless otherwise footnoted.]
African Americans .... 5.0 %
Asian/Pacific Island Americans .... 5.0 %
Hispanic Americans .... 1.0 %
Native Americans .... – %
Foreign Citizens .... 2.0 %
All Others .... 87.0 %
Women .... 6.0 %
Persons with disabilities .... – % [A]
Students over 25 years of age .... – % [A]

Notes: (A) Data not available.

ENGINEERING TECHNOLOGY STUDENT DATA
Applicants to the engineering technology college
Number of applicants to engineering technology college 55
Percent offered admission 96.4%

Matriculated engineering technology students
Percentage in top quartile (25%) of High School class .... 65.0 [1]
Average SAT scores: Math—580[1], Verbal—510[1], Combined—1090[1]
Average ACT scores: Composite—28[2]

Notes: (1) University wide data. (2) Composite is for entire College of Engineering and Technology.

■ FULL- & PART-TIME FACULTY—BY DEPARTMENT
[Figures are the head count of full-time faculty and the full-time equivalent (FTE) of part-time faculty for each engineering technology department or equivalent.]

Department	Full	Part
Electrical Eng Tech	16	5.3
Manufacturing Tech	9	3.5

TECHNOLOGY COLLEGE DESCRIPTION
All technology programs are housed in departments with parallel engineering programs. The technology classes are taught by full-time engineering faculty. In addition, there is a comprehensive cooperative education program available for the technology students. The technology programs are rich with laboratory and other hands-on experiences. The faculty of the college emphasize their mentorship role by interacting with students on a one-to-one basis. The programs have capstone project experience.

TECHNOLOGY DEGREE PROGRAMS DESCRIPTION
(1) The Manufacturing Technology undergraduate programs have two options-one in Manufacturing Processes and one in Manufacturing Design. Both of these programs are accredited by ABET. (2) The Electrical Engineering Technology program is unique in that it emphasizes hands-on laboratory experiences throughout the entire curriculum.

TRANSFER INFORMATION
Residency Requirements: (1) Minimum of 30 semester hours on campus. (2) Maximum of 66 semester hours of transfer credit from a junior college may be applied to a program. (3) 24 of last 30 semester hours must be in residence.

Transfer without Articulation Agreements
Admission to engineering technology: At any time
Requirements: (1) 2.5 transfer grade point average (2) C or better in mathematics and science courses

GRADUATION REQUIREMENTS
To graduate from the technology or construction programs the student must have achieved an overall grade point average of 2.0 on a 4.0 scale both on a university wide basis and on a college basis.

■ STUDENT PROGRAMS
PROFESSIONAL AND HONORARY SOCIETIES
[For key to acronyms, see *Introduction*.]
Professional Societies: AGCA, ASCE, ASME, IEEE, IIE, SAE, SME, American Society of Materials
Honorary Societies: Chi Epsilon, Eta Kappa Nu, Pi Tau Sigma, Sigma Phi Delta, Tau Beta Pi, Beta Tau Epsilon, Sigma Lambda Chi

SUPPORT PROGRAMS
Student Chapter Organizations: National Society of Black Engineers, University Minority Services Program

National Technical Association
For: Ethnic Minorities **Available:** Academic year
Offered: Freshman, Sophomore, Junior, Senior
Other Engineering Support Programs: Support is provided by a university testing and counseling center, a center for orientation and retention and a summer orientation program during which freshmen are given a math placement test. Tutoring services are provided for core engineering, science and mathematics. A full-time academic counselor is in the college with all students being advised by a full-time faculty member.

287 Brigham Young University

■ INSTITUTION PROFILE
HEAD OF THE INSTITUTION
Rex E. Lee
Phone: (801)378-2521
GENERAL INFORMATION
[All Students—Fall 1991]
Undergraduate enrollment 26,863
Graduate enrollment 1,786
Total institution enrollment **28,649**
Type of institution: Private **Calendar:** Semesters
Nearest city: Salt Lake City **Population:** 160,000
Miles from main campus: 45 **Setting:** Small Town
Types of engineering degrees: Engineering & Technology
Other degree-granting colleges: Education, Fine & Performing Arts, Humanities & Social Sciences, Law, Nursing, Biological & Agricultural, Business, Engineering Sciences & Technology, Physical Education, Family, Home & Social Sciences

■ ENGINEERING TECHNOLOGY ADMISSIONS
ADMISSIONS OFFICE CONTACT
Tom M. Gourley
Brigham Young University
Admissions Office
A-153 ASB
Provo, UT 84602
Phone: (801)378-2507 **Fax:** (801)378-4264
TECHNOLOGY COLLEGE ADMISSIONS INFORMATION
Admission to the engineering technology college: Entry with University admission. Entrance to Prof. Program at end of Sophomore year.
Entrance Requirements: ACT (Score: 25)
Entrance Recommendations: High School courses—Algebra (2 years), Trigonometry/Geometry (1 year), Drafting (1 year), Chemistry and Physics (1 year)
Requirements for foreign students: TOEFL (Score: 500)
DEPARTMENTAL ADMISSIONS INFORMATION
Admission to the engineering technology department: At the end of the second year.
Additional information: Completion of specific courses and application to Professional Program.

■ FINANCIAL INFORMATION
ESTIMATED EXPENSES (ACADEMIC YEAR)
[Expenses are for the 1992-93 nine-month academic year.]

All Students	Undergraduate	Graduate
Tuition and fees	$ 2,120 [1]	$ 2,480 [1]
College room and board	$ 3,390	$ 3,390
Books and supplies	$ 630	$ 630
Other expenses	$ 2,020 [2]	$ 2,020 [2]
Total estimated expenses	**$ 8,160**	**$ 8,520**

Notes: (1) Fee for members of the Church of Jesus Christ of Latter-day Saints. Non-member pay $3180 (undergraduate) and $3270 (graduate). (2) Estimated personal and transportation expenses.

FINANCIAL AID OFFICE CONTACT
Norman B. Finlinson, *Director*
Brigham Young University
A-41 ASB
Provo, UT 84602
Phone: (801)378-4104
GENERAL FINANCIAL AID INFORMATION
Forms accepted/required: ACT, FAT, FFS, BYU Financial Aid Application
Additional financial aid information: Direct inquiries and application for financial aid to University Financial Aid Office. Department offers many student positions as Teaching Assistants, Paper Graders, and Maintenance Assistants.

■ TECHNOLOGY COLLEGE INFORMATION
[For additional personnel, refer to the *Appendix*.]
HEAD OF ENGINEERING TECHNOLOGY
L. Douglas Smoot
Phone: (801)378-4326 **Fax:** (801)378-5705
ENGINEERING TECHNOLOGY COLLEGE ADDRESS
College of Engineering & Technology
Brigham Young University
270 Clyde Building
Provo, UT 84602
Phone: (801)378-4326 **Fax:** (801)378-6300
TECHNOLOGY ENROLLMENTS—BY CLASS
[Numbers are technology baccalaureate enrollments for the fall 1991 term, unless otherwise footnoted.]
1st-year students/Freshmen 70
2nd-year students/Sophomores 30
3rd-year students/Juniors 66
4th-year students/Seniors 154
Total **320**

NUMBER OF DEGREES AWARDED—BY PROGRAM
[Numbers are engineering technology baccalaureate degrees awarded during the 1991-1992 academic year, unless otherwise footnoted. For full details about each engineering technology program, refer to the *Table of Degree Programs*.]
Electrical/Electronic(s) Engineering Technology .... –
Electronic Engineering Technology 25
Manufacturing Engineering 7
Manufacturing Engineering Technology 27
Total **59**

PERCENTAGE OF DEGREES AWARDED—BY CATEGORY
[Percentages are of all engineering technology baccalaureate degrees awarded during the 1991-1992 academic year, unless otherwise footnoted.]
African Americans .... – % [A]
Asian/Pacific Island Americans .... – % [A]
Hispanic Americans .... 3.4 %
Native Americans .... 1.7 %
Foreign Citizens .... 3.4 %
All Others .... 91.5 %
Women .... 3.4 %
Persons with disabilities .... – % [A]
Students over 25 years of age .... – % [1]

Notes: (A) Data not available. (1) Average student age at graduation was 25.1 years.

GRADUATE ENROLLMENTS & DEGREES AWARDED
Master's enrollment 13
Master's degrees awarded 18
Doctoral enrollment .... –
Doctoral degrees awarded .... –

ENGINEERING TECHNOLOGY STUDENT DATA
Applicants to the engineering technology college
Number of applicants to engineering technology college ..... – [A]
Percent offered admission .... –% [A]

Matriculated engineering technology students
Percentage in top quartile (25%) of High School class ...... – [A]
Average ACT scores: Math—24.9, Composite—26.1

Notes: (A) Data not available.

FULL- & PART-TIME FACULTY—BY DEPARTMENT
[Figures are the head count of full-time faculty and the full-time equivalent (FTE) of part-time faculty for each engineering technology department or equivalent.]

Department	Full	Part
Manufacturing Eng & Eng Tech	18	–
Manufacturing Eng	19	–
Manufacturing Eng & Eng Tech	19	–

TECHNOLOGY COLLEGE DESCRIPTION
The College of Engineering & Technology enrolls some 3000 students in the 5 engineering and engineering technology departments. All BS degrees are accredited with the Accreditation Board for Engineering and Technology. Each department also offers MS and PhD degrees, and the College offers a Master of Technology Management (MTM) degree. Coop education is available for students wishing work experience. Junior & senior level courses are considered to be professional level, and formal acceptance by the major department is required before one may enroll in the courses.

TECHNOLOGY DEGREE PROGRAMS DESCRIPTION
BS - Electronics Engineering Technology - Students interested in applying engineering principles and supervising people in areas of design, development, production, construction, operations, sales, and management. Provides training in communications, digital electronics, circuit analysis, instrumentation, real-time programming, and computer-aided process control & testing. BS - Manufacturing Engineering Technology & Manufacturing Engineering - Closely related, with much shared between them. Principle difference is that the engineer is more responsible for research & development of the tools & processes used to produce quality products within competitive cost constraints. The engineering technology graduate is more responsible for the selection and deployment of tools and processes for cost-effective manufacturing. Both must understand the relationships between diverse manufacturing functions. Differences in the curricula are in the depth of mathematical rigor & in the scope of projects assigned.

TRANSFER INFORMATION
Residency Requirements: 30 Semester Hours
Transfer via Articulation Agreements
Admission to engineering technology: At any time
Technology graduates transferred from ... 4-yr: 7   2-yr: 15
Requirements: Minimum GPA of 3.0
Transfer without Articulation Agreements
Admission to engineering technology: At any time
Technology graduates transferred from ... 4-yr: 5   2-yr: 10
Requirements: Minimum GPA of 3.0
GRADUATION REQUIREMENTS
Consult with Department for allowance of 'D' credit.

■ STUDENT PROGRAMS
PROFESSIONAL AND HONORARY SOCIETIES
[For key to acronyms, see *Introduction*.]
Professional Societies: ASME, IEEE, SAMPE, SME, SPE*, National Computer Graphics Association (NCGA), A Materials Society (AMS)
Honorary Societies: Tau Beta Pi, Phi Kappa Phi, Sigma Xi
SUPPORT PROGRAMS
Student Chapter Organizations: Society of Women Engineers
Other Engineering Support Programs: New Student Orientation

288 California State University, Sacramento

■ INSTITUTION PROFILE
HEAD OF THE INSTITUTION
Donald R. Gerth
Phone: (916)278-7737 **Fax:** (916)278-6959
GENERAL INFORMATION
[All Students—Fall 1991]
Undergraduate enrollment 19,406
Graduate enrollment 5,062
Total institution enrollment **24,468**
Type of institution: Public **Calendar:** Semesters
Nearest city: Sacramento **Population:** 3,000,000
Miles from main campus: 1 **Setting:** Urban
Types of engineering degrees: Engineering & Technology
Other degree-granting colleges: Arts & Sciences, Business Administration, Education, Health and Human Services

■ ENGINEERING TECHNOLOGY ADMISSIONS
ADMISSIONS OFFICE CONTACT
Larry Glasmire
California State University, Sacramento
Admissions and Records
6000 J Street
Sacramento, CA 95819-6048
Phone: (916)278-7111 **Fax:** (916)278-5443
TECHNOLOGY COLLEGE ADMISSIONS INFORMATION
Entrance Requirements: SAT (Score: 410), ACT (Score: 10), High School courses—Math (3 years), English (4 years), METs should contact a counselor
Entrance Recommendations: High School courses—CM (geometry, trigonometry, chemistry, physics), Mechanical drawing, METs (knowledge of mechanical drawing and, computer literacy)
Requirements for foreign students: TOEFL (Score: 510)
DEPARTMENTAL ADMISSIONS INFORMATION
Admission to the engineering technology department: At the time of admission to the institution.

■ FINANCIAL INFORMATION
ESTIMATED EXPENSES (ACADEMIC YEAR)
[Expenses are for the 1992-93 nine-month academic year.]

State Residents	Undergraduate	Graduate
Tuition and fees	$ 1,460	$ 1,460
College room and board	$ 4,420	$ 4,420
Books and supplies	$ 525	$ 525
Other expenses	$ 318	$ 318
Total estimated expenses	**$ 6,723**	**$ 6,723**
Out-of-State Residents	Undergraduate	Graduate
Tuition and fees	$ —(1)	$ —(1)
College room and board	$ 4,420	$ 4,420
Books and supplies	$ 525	$ 525
Other expenses	$ 318	$ 318
Total estimated expenses	**$ 5,263**	**$ 5,263**

Notes: (1) $246 per unit plus resident fees.
FINANCIAL AID OFFICE CONTACT
Starla Satchell, *Director of Financial Aid*
California State University, Sacramento
Financial Aid Services
6000 J Street
Sacramento, CA 95819-6044
Phone: (916)278-6554
GENERAL FINANCIAL AID INFORMATION
Forms accepted/required: Institutional
Additional financial aid information: College Work-study, student assistants, a variety of grants, loans, scholarships. To apply a student must fill out a Student Aid Application for California (SAAC). Documents required: Federal Income Tax Return, or Income Certification Form, Institutional Forms.

■ TECHNOLOGY COLLEGE INFORMATION
[For additional personnel, refer to the *Appendix*.]
HEAD OF ENGINEERING TECHNOLOGY
Donald H. Gillott
Phone: (916)278-6366 **Fax:** (916)278-5949
ENGINEERING TECHNOLOGY COLLEGE ADDRESS
Engineering Technology Programs
California State University, Sacramento
6000 J Street
Sacramento, CA 95819-6023
Phone: (916)278-6616 **Fax:** (916)278-5949
TECHNOLOGY ENROLLMENTS—BY CLASS
[Numbers are technology baccalaureate enrollments for the fall 1991 term, unless otherwise footnoted.]
1st-year students/Freshmen 13
2nd-year students/Sophomores 10
3rd-year students/Juniors 66
4th-year students/Seniors 129
Total **218**
NUMBER OF DEGREES AWARDED—BY PROGRAM
[Numbers are engineering technology baccalaureate degrees awarded during the 1991-1992 academic year, unless otherwise footnoted. For full details about each engineering technology program, refer to the *Table of Degree Programs*.]
Construction Management Technology 16
Mechanical Engineering Technology 17
Total **33**
PERCENTAGE OF DEGREES AWARDED—BY CATEGORY
[Percentages are of all engineering technology baccalaureate degrees awarded during the 1991-1992 academic year, unless otherwise footnoted.]
African Americans 2.8%
Asian/Pacific Island Americans 11.9%
Hispanic Americans 10.6%
Native Americans 2.3%
Foreign Citizens 0.5%
All Others 71.9%
Women 6.0%
Persons with disabilities .... — %(A)
Students over 25 years of age .... — %(A)
Notes: (A) Data not available.
ENGINEERING TECHNOLOGY STUDENT DATA
Applicants to the engineering technology college
Number of applicants to engineering technology college 44
Percent offered admission 70.5%
Matriculated engineering technology students
Percentage in top quartile (25%) of High School class 50.0
Average SAT scores: Math—476, Verbal—402, Combined—878
FULL- & PART-TIME FACULTY—BY DEPARTMENT
[Figures are the head count of full-time faculty and the full-time equivalent (FTE) of part-time faculty for each engineering technology department or equivalent.]

Department	Full	Part
Mechanical Eng Tech	9	0.3
Construction Management	5	0.5

TECHNOLOGY COLLEGE DESCRIPTION
The Construction Management option in Engineering Technology is fully accredited by both the Technical Accreditation Commission of ABET and the American Council for Construction Education. It prepares students for managerial positions with contractors and other organizations involved in the construction process. The Mechanical Engineering Technology program at CSUS is the only CSU MET program in Northern California. It is accredited by the Technical Accreditation Commission of the ABET. The MET program provides university level preparation for a professional career in the technological spectrum between the more theoretically oriented mechanical engineer and the more applied technician.

TECHNOLOGY DEGREE PROGRAMS DESCRIPTION
The Construction Management curriculum consists of 3 distinctive but inter-related components. They are engineering fundamentals, construction management, and business administration. The entire sequence of courses in the CM component utilizes the functional approach as a framework for studying the management of the construction process. The emphasis in Mechanical Engineering Technology is placed on applied practice rather than analytical methods. The technical electives in the program include courses in material science, electrical installation, computer-aided design, computer-aided manufacturing, heating-vent-air condition and quality control. Business electives can be selected from a variety of management and organizational behavior and environment.
TRANSFER INFORMATION
Residency Requirements: Students must complete at least 30 of the total units on campus excluding extension credits. At least 24 of these must be upper division units in the major.
Transfer via Articulation Agreements
Admission to engineering technology: CM students are admitted in the Spring and MET students are admitted in the Fall
Transfer without Articulation Agreements
Admission to engineering technology: CM students are admitted in the Spring and MET students are admitted in the Fall
GRADUATION REQUIREMENTS
Construction Management: The program consists of 59 units in the pre-major (lower division) and 60 units in the major (upper division); the BS with the Business minor requires a total of 140 units. Mechanical Engineering Technology: The program consists of 42 units in the pre-major (lower division) and 57 units in the major (upper division). The BS requires a total of 132 units.

■ STUDENT PROGRAMS
PROFESSIONAL AND HONORARY SOCIETIES
[For key to acronyms, see *Introduction*.]
Professional Societies: AGCA, ASME, MAA, SAE, SAMPE
Honorary Societies: Tau Beta Pi
SUPPORT PROGRAMS
Student Chapter Organizations: Society of Women Engineers, National Society of Black Engineers, Society of Hispanic Professional Engineers, American Indian Scientist and Engineers
Minority Engineering Program
For: Women & Ethnic Minorities **Available:** Year round
Offered: Prematriculation, Freshman, Sophomore, Junior, Senior
Women's Programs
For: Women & Ethnic Minorities **Available:** Academic year
Offered: Prematriculation, Freshman, Sophomore, Junior, Senior, Graduate level
Other Engineering Support Programs: Spring and Fall orientation and advising for all freshmen and transfers. Career counseling, services for disabled, placement services, foreign student office, tutoring, veterans service.

California State Polytechnic University, Pomona

INSTITUTION PROFILE

HEAD OF THE INSTITUTION
Bob H. Suzuki
Phone: (909)869-2290

GENERAL INFORMATION
[All Students—Fall 1991]
Undergraduate enrollment 16,525
Graduate enrollment 1,043
Total institution enrollment **17,568**
Type of institution: Public   **Calendar:** Quarters
Nearest city: Los Angeles   **Population:** 4,500,000
Miles from main campus: 35   **Setting:** Urban
Types of engineering degrees: Engineering & Technology
Other degree-granting colleges: Agricultural & Environmental, Architecture, Arts & Sciences, Business Administration, Education, Hotel and Restaurant Management

ENGINEERING TECHNOLOGY ADMISSIONS

ADMISSIONS OFFICE CONTACT
Joseph C. Marshall
California State Polytechnic University, Pomona
Admissions Office
3801 W. Temple Avenue
Pomona, CA 91768
Phone: (909)869-2000

TECHNOLOGY COLLEGE ADMISSIONS INFORMATION
Entrance Requirements: SAT (Score: 400), ACT (Score: 18), High School courses—CSU Minimum Course Work (1 year), (Units & courses increase through 1995), Eligibility Index requires either the SAT or the ACT.
Entrance Recommendations: SAT (Score: 1000), ACT (Score: 24), High School courses—Mathematics (4 years), Lab Science (2 years)
Requirements for out-of-state residents: Non-resident applicants require an eligibility index of at least 3402, or a high school GPA of at least 3.60.
Requirements for foreign students: TOEFL (Score: 525); financial statement

DEPARTMENTAL ADMISSIONS INFORMATION
Admission to the engineering technology department: At the time of admission to the institution.

FINANCIAL INFORMATION

ESTIMATED EXPENSES (ACADEMIC YEAR)
[Expenses are for the 1992-93 nine-month academic year.]

State Residents	Undergraduate	Graduate
Tuition and fees	$ 1,342	$ 1,342
College room and board	$ 4,500	$ 4,500
Books and supplies	$ 750	$ 750
Other expenses	$ 954	$ 954
Total estimated expenses	**$ 7,546**	**$ 7,546**

Out-of-State Residents	Undergraduate	Graduate
Tuition and fees	$ 8,300	$ 8,300
College room and board	$ 4,500	$ 4,500
Books and supplies	$ 750	$ 750
Other expenses	$ 954	$ 954
Total estimated expenses	**$14,504**	**$14,504**

FINANCIAL AID OFFICE CONTACT
Al Andino, *Director of Financial Aid*
California State Polytechnic University, Pomona
Office of Financial Aid
3801 W. Temple Avenue
Pomona, CA 91768
Phone: (909)869-3700

GENERAL FINANCIAL AID INFORMATION
Forms accepted/required: CSS-FAF, FAF, IRS
Additional financial aid information: There are numerous opportunities for part-time jobs on campus as well as numerous need-based scholarships.

TECHNOLOGY COLLEGE INFORMATION

[For additional personnel, refer to the *Appendix*.]
HEAD OF ENGINEERING TECHNOLOGY
Edward C. Hohmann
Phone: (909)869-2600   **Fax:** (909)869-4370
ENGINEERING TECHNOLOGY COLLEGE ADDRESS
College of Engineering
California State Polytechnic University, Pomona
3801 W. Temple Avenue
Pomona, CA 91768-4066
Phone: (909)869-2600   **Fax:** (909)869-4370

TECHNOLOGY ENROLLMENTS—BY CLASS
[Numbers are technology baccalaureate enrollments for the fall 1991 term, unless otherwise footnoted.]
1st-year students/Freshmen 93
2nd-year students/Sophomores 110
3rd-year students/Juniors 216
4th-year students/Seniors 437
Total **856**

NUMBER OF DEGREES AWARDED—BY PROGRAM
[Numbers are engineering technology baccalaureate degrees awarded during the 1991-1992 academic year, unless otherwise footnoted. For full details about each engineering technology program, refer to the *Table of Degree Programs*.]
Engineering Technology 128
Total **128**

PERCENTAGE OF DEGREES AWARDED—BY CATEGORY
[Percentages are of all engineering technology baccalaureate degrees awarded during the 1991-1992 academic year, unless otherwise footnoted.]
African Americans .... 3.0 %
Asian/Pacific Island Americans .... 43.0 %
Hispanic Americans .... 17.0 %
Native Americans .... — %
Foreign Citizens .... 36.0 %
All Others .... 1.0 %
Women .... 8.0 %
Persons with disabilities .... — %
Students over 25 years of age .... — %

ENGINEERING TECHNOLOGY STUDENT DATA
Applicants to the engineering technology college
Number of applicants to engineering technology college ... 184
Percent offered admission 66.0%

FULL- & PART-TIME FACULTY—BY DEPARTMENT
[Figures are the head count of full-time faculty and the full-time equivalent (FTE) of part-time faculty for each engineering technology department or equivalent.]

Department	Full	Part
Eng Tech	10	2.7

TECHNOLOGY COLLEGE DESCRIPTION

Cal Poly's College of Engineering has a well-earned reputation for helping to meet the technical challenges facing our nation by preparing engineers prepared to contribute significantly to industry upon graduation. The emphasis on a strong theoretical background coordinated with early and significant laboratory experience continues to make the program unique in engineering education. The college provides study opportunities to over 4400 students in eight accredited engineering and technology and two graduate programs. Each curriculum is designed to give the student both an understanding of the fundamental principles of engineering as an applied science and the practical expertise to apply these principles to actual situations.

TECHNOLOGY DEGREE PROGRAMS DESCRIPTION
Unlike the more traditional engineering curricula which initiate engineering course work in the junior year, Cal Poly's program demands that students take computer programming and engineering orientation courses in the freshman year and that mathematics, basic science and general education begin concurrently. Throughout their educational programs, students become adept at using both the university's computing facilities and the college's Computer Aided Engineering Laboratory. Specific features of the curricula reflect the input of the program's Industry Action Council, composed of leaders of local industry, and emphasizes laboratory experiences. As a result of this 'learn by doing' environment, graduates of the college continue to be in great demand.

TRANSFER INFORMATION
Residency Requirements: Total of three quarters in residence, including two immediately preceding graduation.

Transfer without Articulation Agreements
Admission to engineering technology: At any time
Technology graduates transferred from ... 4-yr: 10   2-yr: 70

STUDENT PROGRAMS

PROFESSIONAL AND HONORARY SOCIETIES
[For key to acronyms, see *Introduction*.]

Professional Societies: AGCA, AIAA, AIChE, ASCE, ASME, IEEE, IIE, SAE, SME, SPE, American Congress on Surveying and Mapping, Society of Plastics Engineers
Honorary Societies: Alpha Pi Mu, Chi Epsilon, Eta Kappa Nu, Omega Chi Epsilon, Pi Tau Sigma, Sigma Gamma Tau, Tau Alpha Pi, Tau Beta Pi

SUPPORT PROGRAMS
Student Chapter Organizations: Society of Women Engineers, National Society of Black Engineers, Society of Hispanic Professional Engineers

Minority Engineering Program
For: Ethnic Minorities   **Available:** Year round
Offered: Prematriculation, Freshman, Sophomore, Junior, Senior, Graduate level

Exploring Engineering
For: Women   **Available:** Summer only
Offered: Prematriculation

Other Engineering Support Programs: All students are strongly urged to participate in the Student Orientation Program conducted by the university prior to the beginning of each quarter.

290 Capitol College

■ INSTITUTION PROFILE
HEAD OF THE INSTITUTION
G. William Troxler
Phone: (301)953-0060   Fax: (301)953-3876

GENERAL INFORMATION
[All Students—Fall 1991]
Undergraduate enrollment 700
Graduate enrollment 100
Total institution enrollment **800**
Type of institution: Private   **Calendar:** Semesters
Nearest city: Washington   **Population:** 700,000
Miles from main campus: 25   **Setting:** Suburban
Types of engineering degrees: Engineering & Technology

■ ENGINEERING TECHNOLOGY ADMISSIONS
ADMISSIONS OFFICE CONTACT
Anthony G. Miller
Capitol College
11301 Springfield Road
Laurel, MD 20708
Phone: (301)953-0060   Fax: (301)953-3876

TECHNOLOGY COLLEGE ADMISSIONS INFORMATION
Entrance Requirements: SAT, High School courses—Math (3 years), English (4 years), Laboratory Science (2 years), Social Science (2 years), SAT or ACT Tests required - no minimum score. Applicants evaluated on an individual basis.
Entrance Recommendations: SAT
Requirements for foreign students: TOEFL (Score: 500); financial statement

DEPARTMENTAL ADMISSIONS INFORMATION
Admission to the engineering technology department: At the time of admission to the institution.

■ FINANCIAL INFORMATION
ESTIMATED EXPENSES (ACADEMIC YEAR)
[Expenses are for the 1992-93 nine-month academic year.]

All Students	Undergraduate	Graduate
Tuition and fees	$ 7,434	$ 300[1]
College room and board	$ 2,600	$ —[B]
Books and supplies	$ 600	$ —[B]
Other expenses	$ —[B]	$ —[B]
Total estimated expenses	**$10,634**	**$ 300**

Notes: (B) Data not applicable. (1) per credit

FINANCIAL AID OFFICE CONTACT
Sheila Sauls-White, *Director*
Capitol College
Financial Aid
11301 Springfield Road
Laurel, MD 20708
Phone: (301)953-0060   Fax: (301)953-3876

GENERAL FINANCIAL AID INFORMATION
Forms accepted/required: CSS-FAF, FAT, Stafford, IRS, SAR, Supplemental
Additional financial aid information: All forms of federal financial aid. Scholarships.

■ TECHNOLOGY COLLEGE INFORMATION
[For additional personnel, refer to the *Appendix*.]
HEAD OF ENGINEERING TECHNOLOGY
Earl E. Gottsman
Phone: (301)953-0060   Fax: (301)953-3876

ENGINEERING TECHNOLOGY COLLEGE ADDRESS
College of Engineering
Capitol College
11301 Springfield Road
Laurel, MD 20708
Phone: (301)953-0060   Fax: (301)953-3876

TECHNOLOGY ENROLLMENTS—BY CLASS
[Numbers are technology baccalaureate enrollments for the fall 1991 term, unless otherwise footnoted.]
1st-year students/Freshmen 80
2nd-year students/Sophomores 110
3rd-year students/Juniors 180
4th-year students/Seniors 143
Total **513**

NUMBER OF DEGREES AWARDED—BY PROGRAM
[Numbers are engineering technology baccalaureate degrees awarded during the 1991-1992 academic year, unless otherwise footnoted. For full details about each engineering technology program, refer to the *Table of Degree Programs*.]
Computer Engineering Technology 8
Electrical/Electronic(s) Engineering Technology 63
Telecommunications Engineering Technology 2
Total **73**

PERCENTAGE OF DEGREES AWARDED—BY CATEGORY
[Percentages are of all engineering technology baccalaureate degrees awarded during the 1991-1992 academic year, unless otherwise footnoted.]
African Americans .... 18.0 %
Asian/Pacific Island Americans .... 12.0 %
Hispanic Americans .... 2.0 %
Native Americans .... 1.0 %
Foreign Citizens .... 17.0 %
All Others .... 50.0 %
Women .... 10.0 %
Persons with disabilities .... 1.0 %
Students over 25 years of age .... 68.0 %

ENGINEERING TECHNOLOGY STUDENT DATA
Applicants to the engineering technology college
Number of applicants to engineering technology college ... 100
Percent offered admission .... 70.0 %
Matriculated engineering technology students
Percentage in top quartile (25%) of High School class ...... –
Average SAT scores: Math—430, Verbal—410, Combined—840

FULL- & PART-TIME FACULTY—BY DEPARTMENT
[Figures are the head count of full-time faculty and the full-time equivalent (FTE) of part-time faculty for each engineering technology department or equivalent.]

Department	Full	Part
Electrical/Electronic(s) Eng Tech	9	2.5
Computer & Mathematics Department	4	1.0

TECHNOLOGY COLLEGE DESCRIPTION
Capitol College is a private, non-profit college which offers practical educational experiences that enable graduates to advance, manage and communicate changes in the information age. Teaching at CC focuses on students. The student/faculty ratio of 14.1 encourages individual attention and personalized instruction. The curriculum is designed to challenge students, yet give them the opportunity to succeed. The academic programs are supported by laboratories dedicated to engineering/electronics, computers, telecommunications, physics and chemistry. In addition to basic test equipment, the engineering/electronics lab contains advanced prototying and design equipment. The College's open-lab policy makes these facilities available to students more than 72 hrs/week. Approximately 10% of the student body participates in the cooperative education program to gain valuable work experience in their field of study.

TECHNOLOGY DEGREE PROGRAMS DESCRIPTION
Engineering, Engineering Technology, Computers and Telecommunications. B.S. degree programs are all laboratory intensive, allowing students to gain practical knowledge along with theory gained in the classroom. Extensive laboratory facilities available to students over 72 hrs/wk. Working in conjunction with industry, programs are constantly updated to reflect changing technology. Students interested in the field of Optoengineering Engineering Technology are offered the opportunity to take a number of courses offered in this field. Career preparation is a key part of the Capitol College experience. Graduates are in great demand by business and industry.

TRANSFER INFORMATION
Residency Requirements: Must complete 40 credits at Capitol College - 20 in the major of choice. Can transfer a maximum of 70 credits from a 2-yr. school.
Transfer via Articulation Agreements
Admission to engineering technology: At any time
Technology graduates transferred from ... 4-yr: –   2-yr: 65
Requirements: Each course must have a grade of 2.0 to transfer.
Transfer without Articulation Agreements
Admission to engineering technology: At any time
Technology graduates transferred from ... 4-yr: 8   2-yr: 36
Requirements: Same as previous question.

GRADUATION REQUIREMENTS
Must have 2.0 GPA and have completed the required number of credits in the major field of study.

■ STUDENT PROGRAMS
PROFESSIONAL AND HONORARY SOCIETIES
[For key to acronyms, see *Introduction*.]
Professional Societies: IEEE
Honorary Societies: Tau Alpha Pi, Alpha Chi
SUPPORT PROGRAMS
Student Chapter Organizations: Society of Women Engineers, National Society of Black Engineers
Other Engineering Support Programs: The College offers an orientation program for both Freshman and transfer students. The College operates a Tutoring Resource Center which provides students with academic assistance in all academic areas.

291 Central Connecticut State University

■ INSTITUTION PROFILE

HEAD OF THE INSTITUTION
John W. Shumaker
Phone: (203)827-7203 **Fax:** (203)827-7033

GENERAL INFORMATION
[All Students—Fall 1991]
Undergraduate enrollment 11,219
Graduate enrollment 2,565
Total institution enrollment 13,784
Type of institution: Public **Calendar:** Semesters
Nearest city: Hartford, Connecticut **Population:** 135,000
Miles from main campus: 9 **Setting:** Suburban
Types of engineering degrees: Engineering Technology
Other degree-granting colleges: Arts & Sciences, Business Administration, Communication & Journalism, Education, Nursing

■ ENGINEERING TECHNOLOGY ADMISSIONS

ADMISSIONS OFFICE CONTACT
Hakim-Salahu A. Din
Central Connecticut State University
1615 Stanley Street
New Britain, CT 06050
Phone: (203)827-7543 **Fax:** (203)827-7220

TECHNOLOGY COLLEGE ADMISSIONS INFORMATION
Entrance Requirements: SAT, High School courses—English (4 years), Mathematics (3 years)
Entrance Recommendations: SAT, High School courses—English (4 years), Mathematics (3 years), Science (2 years), Foreign Language (3 years)
Requirements for foreign students: TOEFL (Score: 500); financial statement

DEPARTMENTAL ADMISSIONS INFORMATION
Admission to the engineering technology department: At the time of admission to the institution.

■ FINANCIAL INFORMATION

ESTIMATED EXPENSES (ACADEMIC YEAR)
[Expenses are for the 1992-93 nine-month academic year.]

State Residents	Undergraduate	Graduate
Tuition and fees	$ 2,804	$ 3,175
College room and board	$ 4,154	$ –
Books and supplies	$ 500	$ –
Other expenses	$ 483	$ 8,812
Total estimated expenses	**$ 7,941**	**$ 11,987**

Out-of-State Residents	Undergraduate	Graduate
Tuition and fees	$ 7,160	$ 7,511
College room and board	$ 4,154	$ –
Books and supplies	$ 500	$ –
Other expenses	$ 483	$ 13,148
Total estimated expenses	**$12,297**	**$ 20,659**

FINANCIAL AID OFFICE CONTACT
John Taylor, *Director of Financial Aid*
Central Connecticut State University
1615 Stanley Street
New Britain, CT 06050 **Fax:** (203)827-7200

GENERAL FINANCIAL AID INFORMATION
Forms accepted/required: ACT, FAF
Additional financial aid information: Need based on scholastic scholarships, athletic, work study, graduate assistantships.

■ TECHNOLOGY COLLEGE INFORMATION

[For additional personnel, refer to the *Appendix*.]
HEAD OF ENGINEERING TECHNOLOGY
John R. Wright
Phone: (203)827-7378 **Fax:** (203)827-7982

ENGINEERING TECHNOLOGY COLLEGE ADDRESS
Engineering Technology
Central Connecticut State University
1615 Stanley Street
New Britain, CT 06050
Phone: (203)827-7370 **Fax:** (203)827-7982

TECHNOLOGY ENROLLMENTS—BY CLASS
[Numbers are technology baccalaureate enrollments for the fall 1991 term, unless otherwise footnoted.]
1st-year students/Freshmen 62
2nd-year students/Sophomores 40
3rd-year students/Juniors 76
4th-year students/Seniors 129
Total **307**

NUMBER OF DEGREES AWARDED—BY PROGRAM
[Numbers are engineering technology baccalaureate degrees awarded during the 1991-1992 academic year, unless otherwise footnoted. For full details about each engineering technology program, refer to the *Table of Degree Programs*.]
Civil & Construction Technology 130 (3)
Manufacturing Technology .... – (4)
Total **130**

Notes: (3) Total for both programs. (4) Degrees are reported under Civil and Construction Technology.

PERCENTAGE OF DEGREES AWARDED—BY CATEGORY
[Percentages are of all engineering technology baccalaureate degrees awarded during the 1991-1992 academic year, unless otherwise footnoted.]
African Americans 6.0%
Asian/Pacific Island Americans 1.0%
Hispanic Americans .... – %
Native Americans .... – %
Foreign Citizens .... – %
All Others 93.0%
Women 6.0%
Persons with disabilities 0.4%
Students over 25 years of age .... – %

ENGINEERING TECHNOLOGY STUDENT DATA
Applicants to the engineering technology college
Number of applicants to engineering technology college 96
Percent offered admission 60.0%
Average SAT scores: Math—518, Verbal—422, Combined—940

FULL- & PART-TIME FACULTY—BY DEPARTMENT
[Figures are the head count of full-time faculty and the full-time equivalent (FTE) of part-time faculty for each engineering technology department or equivalent.]

Department	Full	Part
Civil & Construction Eng Tech	4.0	–
Manufacturing Eng Tech	5	–
Industrial Systems Eng Tech	5	–

TECHNOLOGY COLLEGE DESCRIPTION

Under the School of Technology, the Engineering Technology Department offers a Bachelor of Science Engineering Technology degree with specializations in Construction, Industrial Systems or Manufacturing Engineering Technology. Each degree program is a 130 credit hours, with a General Education requirement of 62 hours, which includes mathematics, computer science, science, communications, English, humanities/social science and approved Engineering Technology courses.

TECHNOLOGY DEGREE PROGRAMS DESCRIPTION

The Engineering Technology degree at CCSU is an analytical degree program based on a curriculum designed to give a student actual laboratory and field application experience. Though one of the largest schools of technology in New England, classes are still kept at the 18-25 student level. This allows for personal contact and interaction between instructors and is one of the major advantages of the program. All classes are taught by professors. Graduate students are utilized to provide open lab times and individual instruction opportunities.

TRANSFER INFORMATION

Residency Requirements: 45 credits "In Residence" 15 credits in major; 9 credits in minor or concentration
Transfer via Articulation Agreements
Admission to engineering technology: At any time
Requirements: Minimum 2.0/4.0 cumulative GPA
Transfer without Articulation Agreements
Admission to engineering technology: At any time
Requirements: Minimum 2.0/4.0 cumulative GPA

■ STUDENT PROGRAMS

PROFESSIONAL AND HONORARY SOCIETIES
[For key to acronyms, see *Introduction*.]
Professional Societies: AGCA, ASCE, IES, SME
Honorary Societies: Epsilon Pi Tau

SUPPORT PROGRAMS
Student Chapter Organizations: ASME - Minority Grant, ASUP - Minority Grant

Minority Mentoring
For: Ethnic Minorities **Available:** Academic year
Offered: Prematriculation, Freshman, Sophomore, Junior, Senior

Employment Opportunities Program
For: Ethnic Minorities **Available:** Year round
Offered: Freshman, Sophomore, Junior, Senior, Graduate level

Department of Higher Education Minority Grant
For: Women & Ethnic Minorities **Available:** Academic year

Teacher Ed. Grant- Minorities from Hft. Grad. Prg.
For: Women & Ethnic Minorities **Available:** Academic year
Offered: Graduate level

292 University of Central Florida

■ INSTITUTION PROFILE
HEAD OF THE INSTITUTION
John C. Hitt
Phone: (407)823-1823   **Fax:** (407)823-5407
GENERAL INFORMATION
[All Students—Fall 1991]
Undergraduate enrollment 16,640
Graduate enrollment 4,627
Total institution enrollment 21,267
Type of institution: Public   **Calendar:** Semesters
Nearest city: Orlando Metropolitan Area   **Population:** 1,000,000
Miles from main campus: 8   **Setting:** Suburban
Types of engineering degrees: Engineering & Technology
Other degree-granting colleges: Arts & Sciences, Business Administration, Education, Health and Public Affairs

■ ENGINEERING TECHNOLOGY ADMISSIONS
ADMISSIONS OFFICE CONTACT
John F. Bush
University of Central Florida
4000 University Blvd.
Orlando, FL 32816-0450
Phone: (407)823-3000   **Fax:** (407)823-5652
TECHNOLOGY COLLEGE ADMISSIONS INFORMATION
Entrance Requirements: SAT (Score: 1000), ACT (Score: 19), Freshmen are not admitted to the Engineering Technology program. Students must complete the AA or AS degree before being admitted to the Engineering Technology program.
Requirements for foreign students: TOEFL (Score: 550); financial statement
DEPARTMENTAL ADMISSIONS INFORMATION
Admission to the engineering technology department: At the end of the second year.
Additional information: Engineering Technology requires that a student have an AA or AS degree to be admitted to the program.

■ FINANCIAL INFORMATION
ESTIMATED EXPENSES (ACADEMIC YEAR)
[Expenses are for the 1992-93 nine-month academic year.]

State Residents	Undergraduate	Graduate
Tuition and fees	$ 1,525 [5]	$ 2,165 [1]
College room and board	$ 3,700 [2]	$ 3,700 [2]
Books and supplies	$ 550 [3]	$ 500 [3]
Other expenses	$ 1,100 [4]	$ 1,100 [4]
Total estimated expenses	**$ 6,875**	**$ 7,465**

Out-of-State Residents	Undergraduate	Graduate
Tuition and fees	$ 5,662 [5]	$ 7,025 [1]
College room and board	$ 3,700 [2]	$ 3,700 [2]
Books and supplies	$ 550 [3]	$ 500 [3]
Other expenses	$ 1,100 [4]	$ 1,100 [4]
Total estimated expenses	**$11,012**	**$12,325**

Notes: (1) Based on 12 SH, Fall and Spring Semesters. (2) Dependent on accommodations and meal plan. (3) Dependent on the number of SH in which the student is enrolled. (4) Estimated transportation costs. (5) Based on 15 SH Fall and Spring Semesters.

FINANCIAL AID OFFICE CONTACT
Mary H. McKinney, *Director of Financial Aid*
University of Central Florida
4000 University Blvd.
Orlando, FL 32816-0450
Phone: (407)823-2827   **Fax:** (407)823-5652

GENERAL FINANCIAL AID INFORMATION
Forms accepted/required: ACT, CSS-FAF, Institutional
Additional financial aid information: Need-based scholarships; short-term loans; merit-based scholarships; part-time jobs on campus; College Career Work Experience Program.

■ TECHNOLOGY COLLEGE INFORMATION
[For additional personnel, refer to the *Appendix*.]
HEAD OF ENGINEERING TECHNOLOGY
Gary E. Whitehouse
Phone: (407)823-2156   **Fax:** (407)823-5483
EMail: WHITEHSE @ UCF1VM.CC.UCF.EDU
ENGINEERING TECHNOLOGY COLLEGE ADDRESS
College of Engineering
University of Central Florida
4000 University Blvd.
Orlando, FL 32816-0450
Phone: (407)823-2268   **Fax:** (407)823-5483
TECHNOLOGY ENROLLMENTS—BY CLASS
[Numbers are technology baccalaureate enrollments for the fall 1991 term, unless otherwise footnoted.]
1st-year students/Freshmen .... 0 [1]
2nd-year students/Sophomores .... 0 [1]
3rd-year students/Juniors .... 46 [2]
4th-year students/Seniors .... 295 [2]
Total 341

Notes: (1) The Engineering Technology program does not admit freshmen. Students must complete AA or AS before being admitted to the Engineering Technology program. (2) 0 SH freshman 30 SH sophomore 60 SH junior 90 SH senior.

NUMBER OF DEGREES AWARDED—BY PROGRAM
[Numbers are engineering technology baccalaureate degrees awarded during the 1991-1992 academic year, unless otherwise footnoted. For full details about each engineering technology program, refer to the *Table of Degree Programs*.]
Computer Engineering Technology 11
Design Engineering Technology 29
Electronic(s) Engineering Technology 30
Information Systems Engineering Technology 11
Operations Engineering Technology 15
Total 96

PERCENTAGE OF DEGREES AWARDED—BY CATEGORY
[Percentages are of all engineering technology baccalaureate degrees awarded during the 1991-1992 academic year, unless otherwise footnoted.]
African Americans 11.5%
Asian/Pacific Island Americans 6.1%
Hispanic Americans 10.8%
Native Americans .... – %
Foreign Citizens .... 5.4% [C]
All Others 66.2%
Women 33.9%
Persons with disabilities .... 0.2% [C]
Students over 25 years of age .... 33.0% [C]

Notes: (C) Estimated data.

FULL- & PART-TIME FACULTY—BY DEPARTMENT
[Figures are the head count of full-time faculty and the full-time equivalent (FTE) of part-time faculty for each engineering technology department or equivalent.]

Department	Full	Part
Eng Tech	12	–

TECHNOLOGY COLLEGE DESCRIPTION
The Bachelor's of Science in Engineering Technology is a plus-two program which accepts course credits from ABET accredited and non-ABET accepted Associate Degree programs. Five options within the degree are offered: Design Engineering Technology, Computer Engineering Technology, Electronics Engineering Technology, Information Systems Engineering Technology, and Operations Engineering Technology. All Engineering Technology students are assigned a faculty advisor for all academic counseling and guidance. All faculty in the department serve as advisors for students within their programs of study. The five options are offered during the day and evening hours by the full-time faculty.

TECHNOLOGY DEGREE PROGRAMS DESCRIPTION
The upper division Bachelor of Science in Engineering Technology (BSET) program at the University of Central Florida is designed to advance the engineering technician to the engineering technologist level. Graduates of baccalaureate programs are termed Engineering Technologists. The Engineering Technologist is primarily involved in application of existing technologies and devices to the solution of routine engineering problems.

DUAL DEGREE PROGRAMS
Technology baccalaureate graduates with dual degrees 0

TRANSFER INFORMATION
Residency Requirements: Students must complete at least 2 semesters (30 SH) in residence.
Transfer via Articulation Agreements
Admission to engineering technology: At the end of the sophomore year
Technology graduates transferred from ... 4-yr: 14   2-yr: 59
Requirements: Engineering Technology requires that a student have an AA or an AS degree with a minimum GPA of 2.0.
Transfer without Articulation Agreements
Admission to engineering technology: At the end of the sophomore year
Technology graduates transferred from ... 4-yr: 7   2-yr: 14
Requirements: Consist of possession of an AA or AS degree with a minimum GPA of 2.0.

GRADUATION REQUIREMENTS
BSET programs require 128 semester credits and a 2.00/4.00 GPA requirement in the engineering technology core and major discipline coursework. University requirements include passing the State of Florida College Level Academic Skills Test (CLAST).

■ STUDENT PROGRAMS
PROFESSIONAL AND HONORARY SOCIETIES
[For key to acronyms, see *Introduction*.]
Professional Societies: American Society for Engineering Technology (ASET)
Honorary Societies: Tau Alpha Pi
SUPPORT PROGRAMS
Student Chapter Organizations: ASET, Tau Alpha Pi
Southeastern Consortium for Minorities in Engineering
For: Ethnic Minorities   **Available:** Year round
Offered: Prematriculation

Re-entry Program
For: Women   **Available:** Year round
Offered: Prematriculation, Freshman, Sophomore, Junior, Senior, Graduate level
Other Engineering Support Programs: Orientation for transfer students; academic advisement; counseling and testing services. Career resource and planning services; veterans' services; evening/weekend student services.

293 University of Cincinnati

INSTITUTION PROFILE

HEAD OF THE INSTITUTION
Joseph A. Steger
Phone: (513)556-2201　　　　Fax: (513)556-3010

GENERAL INFORMATION
[All Students—Fall 1991]
Undergraduate enrollment 1,516
Total institution enrollment 1,516
Type of institution: Public　　Calendar: Quarters
Location: Cincinnati　　Population: 364,000
Setting: Urban
Types of engineering degrees: Engineering Technology
Other degree-granting colleges: Architecture, Arts & Sciences, Business Administration, Education, Law, Medicine, Nursing, Pharmacy, Technology & Applied Sciences, Clermont - Branch Campus, Raymond Walters - Branch Campus, Evening & Continuing Education, University College, College Conservatory of Music

ENGINEERING TECHNOLOGY ADMISSIONS

ADMISSIONS OFFICE CONTACT
Cheryll A. Dunn
University of Cincinnati
2220 Victory Parkway
Cincinnati, OH 45206-2822
Phone: (513)556-6561　　　　Fax: (513)556-4599

TECHNOLOGY COLLEGE ADMISSIONS INFORMATION
Entrance Requirements: SAT (Score: 450), ACT (Score: 18), High School courses—English (4 years), Math (3 years), Physics (1 year), Chemistry (1 year), On the SAT the 450 is the Math score. On the ACT the 18 represents Math Verbal and Composite combined scores.
Entrance Recommendations: High School courses—Social Science (2 years), Foreign Language (2 years), Fine Arts (1 year)
Requirements for foreign students: Applicants whose native language is not English are required to pass the TOEFL. For more information write to: Office of Admissions, Room 100 French Hall, M.L. #91, University of Cincinnati, Cincinnati, OH 45221-0091 U.S.A.

DEPARTMENTAL ADMISSIONS INFORMATION
Admission to the engineering technology department: At the time of admission to the institution.
Additional information: The Construction Science Department requires students be in the upper half of their respective high school graduation classes. Applicants must have achieved a minimum score of 550 or 22 on the math sections of the SAT or ACT respectively. Transfer students must meet these requirements and have achieved a 2.5 cumulative GPA for previous college work.

FINANCIAL INFORMATION

ESTIMATED EXPENSES (ACADEMIC YEAR)
[Expenses are for the 1992-93 nine-month academic year.]

State Residents	Undergraduate	Graduate
Tuition and fees	$ 3,372	$ –
College room and board	$ 4,236 (1)	$ –
Books and supplies	$ 600 (C)	$ –
Other expenses	$ –	$ –
Total estimated expenses	**$ 8,208**	**$ –**

Out-of-State Residents	Undergraduate	Graduate
Tuition and fees	$ 8,049	$ –
College room and board	$ 4,236 (1)	$ –
Books and supplies	$ 600 (C)	$ –
Other expenses	$ –	$ –
Total estimated expenses	**$12,885**	**$ –**

Notes: (C) Estimated data. (1) This is for the "10-meal" plan. A 14-meal per week plan is $4371 and a 19-meal plan is $4431.

GENERAL FINANCIAL AID INFORMATION
Forms accepted/required: AFSA, FAF, Ohio Instructional Grant
Additional financial aid information: Students are encouraged to apply for financial aid by completing an FAF Form, available in late November. Deadline is March 1st. Form includes Supplemental Education Opportunity Grant, Pell Grant, Perkins Loan, College Work Study & Guaranteed Student Loan. Undergraduate Ohio residents should also apply for the Ohio Instructional Grant.

TECHNOLOGY COLLEGE INFORMATION
[For additional personnel, refer to the *Appendix*.]

HEAD OF ENGINEERING TECHNOLOGY
Fritz J. Kryman
Phone: (513)556-6556　　　　Fax: (513)556-4599

ENGINEERING TECHNOLOGY COLLEGE ADDRESS
OMI College of Applied Science
University of Cincinnati
2220 Victory Parkway
Cincinnati, OH 45206-2822
Phone: (513)556-6567　　　　Fax: (513)556-4599

TECHNOLOGY ENROLLMENTS—BY CLASS
[Numbers are technology baccalaureate enrollments for the fall 1991 term, unless otherwise footnoted.]
1st-year students/Freshmen 250
2nd-year students/Sophomores 185
3rd-year students/Juniors 138
4th-year students/Seniors 68
Total 641

NUMBER OF DEGREES AWARDED—BY PROGRAM
[Numbers are engineering technology baccalaureate degrees awarded during the 1991-1992 academic year, unless otherwise footnoted. For full details about each engineering technology program, refer to the *Table of Degree Programs*.]
Architectural Engineering Technology 13
Chemical Technology .... –
Construction Management Technology 29
Electrical Engineering Technology 23
Mechanical Engineering Technology 29
Total **94**

PERCENTAGE OF DEGREES AWARDED—BY CATEGORY
[Percentages are of all engineering technology baccalaureate degrees awarded during the 1991-1992 academic year, unless otherwise footnoted.]
African Americans .... 1.6 %
Asian/Pacific Island Americans .... 3.5 %
Hispanic Americans .... – %
Native Americans .... – %
Foreign Citizens .... – % (A)
All Others .... 94.9 %
Women .... 6.5 %
Persons with disabilities .... – % (A)
Students over 25 years of age .... – % (A)

Notes: (A) Data not available.

ENGINEERING TECHNOLOGY STUDENT DATA
Applicants to the engineering technology college
Number of applicants to engineering technology college ... 542
Percent offered admission 78.6%

Matriculated engineering technology students
Percentage in top quartile (25%) of High School class 39.5
Average SAT scores: Math—516, Verbal—409
Average ACT scores: Composite—21

FULL- & PART-TIME FACULTY—BY DEPARTMENT
[Figures are the head count of full-time faculty and the full-time equivalent (FTE) of part-time faculty for each engineering technology department or equivalent.]

Department	Full	Part
Construction Science	8	2.6
Electrical Eng Tech	6	4.0
Mechanical Eng Tech	7	8.1
Chemical Tech	3	–
Mathematics, Physics, & Computing Tech	8	5.5
Humanities/Social Sciences/Communication	5	3.5

TECHNOLOGY COLLEGE DESCRIPTION
The OMI College of Applied Science (founded in 1828 as the Ohio Mechanics Institute) is one of 17 colleges of the University of Cincinnati and offers both Associate and Baccalaureate degree programs in engineering technology. In June, 1989, the college moved to its new location overlooking the Ohio River two miles from the main campus of the University of Cincinnati. The new campus offers additional laboratory space and high bay areas for specialized courses. The College is self-contained in that, with its own Humanities/Social Science Department and Math/Physics/Computing Department, traditional support courses in these disciplines are carefully integrated into the engineering technology curricula. All associate level programs have a co-op requirement and many of the baccalaureate programs also have co-op (either as a requirement or as an option). Each department provides computing support for its students and the college-wide computing facility is a hallmark for the College.

TECHNOLOGY DEGREE PROGRAMS DESCRIPTION
All BS degree programs require a senior design project. The unique relationships among the College's departments, its advisory boards, and local and regional business and industry have allowed many students to publish their results and even to obtain patents for their designs.

TRANSFER INFORMATION
Residency Requirements: A transfer student must fulfill a minimum of 45 credit hours of departmentally approved coursework at the College to be eligible for a degree from the College.

Transfer via Articulation Agreements
Admission to engineering technology: One agreement can transfer at the end of the freshman year. All other agreements transfer at the end of the sophomore year.
Technology graduates transferred from ... 4-yr: 81　2-yr: –
Requirements: See transfer requirements.

Transfer without Articulation Agreements
Admission to engineering technology: At any time
Requirements: 1) Students must be in good standing according to standards of this college and transfer college. Preference given to 2.5 or better cumulative GPA. 2) Only courses with grade C or better transfer. 3) Only courses from regionally accredited institutions. 4) Credit given only for courses with comparable equivalent in program entered; demonstrated proficiency may be required. 5) Transfer credits granted to matriculated students only.

GRADUATION REQUIREMENTS
Students must fulfill the requirements of the program from which they expect to graduate, have at least a 2.0 cumulative GPA and pass the College English exit exam. For a student whose registration is continuous, requirements for graduation are those published for the year the student first registered. When registration is not continuous, requirements are those applicable at the time of final continuous registration.

STUDENT PROGRAMS

PROFESSIONAL AND HONORARY SOCIETIES
[For key to acronyms, see *Introduction*.]
Professional Societies: AIA, ASME, IEEE, SME, Society of Women Engineers, Student National Technical Association
Honorary Societies: Tau Alpha Pi, Scepter Society

SUPPORT PROGRAMS
Student Chapter Organizations: Society of Women Engineers, Student National Technical Association

Women and Minority Scholarship Program
For: Women & Ethnic Minorities　**Available:** Year round
Offered: Freshman

Herschede Scholarship Program
For: Ethnic Minorities　**Available:** Year round
Offered: Freshman, Sophomore, Junior, Senior

Other Engineering Support Programs: Four days in summer are for student/parent orientation. The Student Development and Counseling Office provides ongoing guidance as do each student's academic advisor. A special Summer Preparatory Program provides an intense exposure to math, English and physics for the students needing review or accepted conditionally. A one-year Technical Access Program is a 'bridge' for underprepared students

294 University of Dayton

■ INSTITUTION PROFILE

HEAD OF THE INSTITUTION
Raymond L. Fitz
Phone: (513)229-4122   Fax: (513)229-4545

GENERAL INFORMATION
[All Students—Fall 1991]
Undergraduate enrollment 6,985
Graduate enrollment 4,188
Total institution enrollment 11,173
Type of institution: Private   **Calendar:** Semesters
Location: Dayton   **Population:** 179,000
Setting: Urban
Types of engineering degrees: Engineering & Technology
Other degree-granting colleges: Arts & Sciences, Business Administration, Education, Law

■ ENGINEERING TECHNOLOGY ADMISSIONS

ADMISSIONS OFFICE CONTACT
Myron H. Achbach
University of Dayton
300 College Park
Dayton, OH 45469-1611
Phone: (513)229-4411   Fax: (513)229-4545

TECHNOLOGY COLLEGE ADMISSIONS INFORMATION
Entrance Requirements: High School courses—English (4 years), Algebra (2 years), Geometry (1 year), Social Science (2 years), Either ACT or SAT is required.
Entrance Recommendations: High School courses—Physics (1 year), Chemistry (1 year), Trigonometry (1 year), Computers (1 year), Both SAT and ACT are recommended.
Requirements for foreign students: TOEFL (Score: 500); financial statement; International student applicants must present their academic credentials in official English translation along with their transcripts in the original language. The applicant must also present certification of financial resources available to support an education at the University of Dayton.

DEPARTMENTAL ADMISSIONS INFORMATION
Admission to the engineering technology department: Admission to the Institution. Undeclared status is available.

■ FINANCIAL INFORMATION

ESTIMATED EXPENSES (ACADEMIC YEAR)
[Expenses are for the 1992-93 nine-month academic year.]

All Students	Undergraduate	Graduate
Tuition and fees	$10,810	$ –
College room and board	$ 4,250	$ –
Books and supplies	$ 600	$ –
Other expenses	$ –(1)	$ –
Total estimated expenses	$15,660	$ –

Notes: (1) Personal expenses vary.

GENERAL FINANCIAL AID INFORMATION
Forms accepted/required: FAF, FAT, SAR, Institutional
Additional financial aid information: Need-based scholarships; short-term loans; merit-based scholarships; part-time jobs on campus; College Work Study Program.

■ TECHNOLOGY COLLEGE INFORMATION
[For additional personnel, refer to the *Appendix*.]

HEAD OF ENGINEERING TECHNOLOGY
Robert L. Mott
Phone: (513)229-2333   Fax: (513)229-2756

ENGINEERING TECHNOLOGY COLLEGE ADDRESS
Engineering Technology Division
University of Dayton
300 College Park
Dayton, OH 45469-0244
Phone: (513)229-2333   Fax: (513)229-2756

TECHNOLOGY ENROLLMENTS—BY CLASS
[Numbers are technology baccalaureate enrollments for the fall 1991 term, unless otherwise footnoted.]
1st-year students/Freshmen 116
2nd-year students/Sophomores 81
3rd-year students/Juniors 117
4th-year students/Seniors 151
Total 465

NUMBER OF DEGREES AWARDED—BY PROGRAM
[Numbers are engineering technology baccalaureate degrees awarded during the 1991-1992 academic year, unless otherwise footnoted. For full details about each engineering technology program, refer to the *Table of Degree Programs*.]
Chemical Process Technology 3
Electronic Engineering Technology 21
Environmental Engineering Technology 10
Industrial Engineering Technology 30
Manufacturing Engineering Technology 3
Mechanical Engineering Technology 35
Total 102

PERCENTAGE OF DEGREES AWARDED—BY CATEGORY
[Percentages are of all engineering technology baccalaureate degrees awarded during the 1991-1992 academic year, unless otherwise footnoted.]
African Americans 6.9%
Asian/Pacific Island Americans 3.9%
Hispanic Americans .... – %
Native Americans .... – %
Foreign Citizens 3.9%
All Others 85.3%
Women 20.6%
Persons with disabilities .... – % (A)
Students over 25 years of age .... – % (A)

Notes: (A) Data not available.

ENGINEERING TECHNOLOGY STUDENT DATA
Applicants to the engineering technology college
Number of applicants to engineering technology college ... 149
Percent offered admission 88.0%

Matriculated engineering technology students
Percentage in top quartile (25%) of High School class 45.0
Average SAT scores: Math—538, Verbal—459, Combined—997
Average ACT scores: Composite—22

FULL- & PART-TIME FACULTY—BY DEPARTMENT
[Figures are the head count of full-time faculty and the full-time equivalent (FTE) of part-time faculty for each engineering technology department or equivalent.]

Department	Full	Part
Electronic Eng Tech	6	–
Industrial Eng Tech	2	1.1
Manufacturing Eng Tech	3	–
Eng Tech Service Courses	3	1.4
Mechanical Eng Tech	3	–
Chemical Process Tech	1	–
Environmental Eng Tech	1	–

TECHNOLOGY COLLEGE DESCRIPTION
Our goal is to prepare the Complete Professional who can pursue a meaningful professional career with concern for the human condition and for moral and ethical values. All of our six programs stress the APPLICATION of technical knowledge and are 4-years long, leading to the B.S. degree. Technical electives allow flexibility for the student with interest in a related field to have a minor in that field. Our placement record is excellent and our graduates report sustained professional advancement in both salary and responsibility. The Engineering Technology Division is a unit of the School of Engineering with its own faculty, courses and curricula. The University of Dayton has offered bachelor degree engineering technology programs since 1965 and this experience has attracted a strong faculty with a commitment to teaching and advising. Co-op program is available. Average number of years required to actually complete the bachelor's degree: 4.30.

TECHNOLOGY DEGREE PROGRAMS DESCRIPTION
The University's four year engineering technology programs emphasize application of engineering principles and are designed to provide excellent preparation in the major field and breadth in both technical and non-technical areas so that the graduate may work effectively with persons of varied educational backgrounds. All programs award the Bachelor of Science degree.

DUAL DEGREE PROGRAMS
Technology baccalaureate graduates with dual degrees 1

TRANSFER INFORMATION
Residency Requirements: Transfer students must take their last 30 credits at the University of Dayton. Students transferring from a two year institution must earn at least 60 credits at U.D.

Transfer via Articulation Agreements
Admission to engineering technology: At the end of the sophomore year
Technology graduates transferred from   ... 4-yr: –   2-yr: –
Requirements: Minimum - 2.0 GPA.

Transfer without Articulation Agreements
Admission to engineering technology: At any time
Technology graduates transferred from   ... 4-yr: –   2-yr: –
Requirements: Minimum 2.0 GPA.

GRADUATION REQUIREMENTS
Student must pass all prescribed courses and have a GPA of 2.0 on a 4.0 scale. The Bachelor of Science in Engineering Technology requires 129 to 131 semester credit hours depending upon the major selected.

■ STUDENT PROGRAMS

PROFESSIONAL AND HONORARY SOCIETIES
[For key to acronyms, see *Introduction*.]
Professional Societies: ACS, AIChE, ASM*, ASME, IEEE, IIE, SAE, SME, Hospital Information Management Systems Society (HIMSS), Association of Environmental Professionals (AEP)
Honorary Societies: Tau Alpha Pi, Tau Nu Kappa

SUPPORT PROGRAMS
Student Chapter Organizations: Society of Women Engineers, National Society of Black Engineers, Office of Minority Student Affairs, Women in Engineering Program, Women in Graduate Engineering Studies (WINGES)

Society of Women Engineers
For: Women   **Available:** Year round
Offered: Prematriculation, Freshman, Sophomore, Junior, Senior, Graduate level

National Society of Black Engineers
For: Ethnic Minorities   **Available:** Academic year
Offered: Prematriculation, Freshman, Sophomore, Junior, Senior

Office of Minority Student Affairs
For: Ethnic Minorities   **Available:** Year round
Offered: Freshman, Sophomore, Junior, Senior, Graduate level

Women in Engineering Program
For: Women   **Available:** Summer only
Offered: Prematriculation

Other Engineering Support Programs: Fall orientation program for all freshmen and transfers; Learning Assistance Center; career counseling services; foreign student office; veterans' services; placement services. Students participate in Co-op and Intern programs; Honors and Scholars Programs, complete complement of tutoring services.

295 DeVry Institute of Technology, Chicago

■ INSTITUTION PROFILE

HEAD OF THE INSTITUTION
E. Arthur Stunard
Phone: (312)929-8500 Fax: (312)348-1780

GENERAL INFORMATION
[All Students—Fall 1991]
Undergraduate enrollment 3,331
Total institution enrollment 3,331
Type of institution: Private **Calendar:** Semesters
Location: Chicago **Population:** 2,783,726
Setting: Suburban
Types of engineering degrees: Engineering Technology
Other degree-granting colleges: Accounting, Business Operations, Computer Information Systems, Electronics Technician (AAS), Telecommunications Management

■ ENGINEERING TECHNOLOGY ADMISSIONS

ADMISSIONS OFFICE CONTACT
Richard L. Yaconis
DeVry Institute of Technology, Chicago
3300 North Campbell Avenue
Chicago, IL 60618-5994
Phone: (312)929-6550 Fax: (312)348-1780

TECHNOLOGY COLLEGE ADMISSIONS INFORMATION
Admission to the engineering technology college: A student applies to the program after meeting with a DeVry representative during senior year of high school or after completing high school or equivalent.
Entrance Requirements: Entrance examinations in math and algebra must be taken if the applicant has not taken a recommended standardized test, or has failed to achieve the required score on the standardized test.
Entrance Recommendations: SAT (Score: 480), ACT (Score: 20), High School courses—Algebra/Advanced Algebra (1 year), Trigonometry (1 year), Calculus (1 year), Physics (1 year), WPCT may also be taken. A score of 20 on the math section of the ACT or 480 on the math section of the SAT or WPCT is required.
Requirements for foreign students: TOEFL (Score: 500); Provide a certified copy of the applicant's high school (or college, if applicable) transcript (in English); a notarized statement of financial support or a certified government sponsor letter; proof of English language proficiency. A Form I-20AB will be provided to the applicant.

DEPARTMENTAL ADMISSIONS INFORMATION
Admission to the engineering technology department: At the time of admission to the institution.
Additional information: Prospective students who wish to complete an application must either visit the institute and meet with a DeVry admissions representative, or meet with a DeVry field representative assigned to the student's geographical area.

■ FINANCIAL INFORMATION

ESTIMATED EXPENSES (ACADEMIC YEAR)
[Expenses are for the 1992-93 nine-month academic year.]

All Students	Undergraduate	Graduate
Tuition and fees	$ 5,249	$ —(1)
College room and board	$ 3,326	$ —(1)
Books and supplies	$ 500	$ —(1)
Other expenses	$ 4,006 (2)	$ —(1)
Total estimated expenses	**$13,081**	**$ —**

Notes: (1) No graduate degree offered (2) Transportation $2,095 Personal/Misc. $1,911

FINANCIAL AID OFFICE CONTACT
Betty Glenn, *Director of Financial Aid*
DeVry Institute of Technology, Chicago
3300 North Campbell Avenue
Chicago, IL 60618-5994
Phone: (312)929-8500 Fax: (312)348-1780

GENERAL FINANCIAL AID INFORMATION
Forms accepted/required: AFSA, CSS-FAF, FAF, FAT, FSA, FFS, Stafford, IRS, SAR, Supplemental, Institutional
Additional financial aid information: Pell Grants; Supplement Educational Opportunity Grants; Perkins Loans; College Work-Study; Guaranteed Student Loan Programs: Stafford, SLS, PLUS; EDUCARD Plan-monthly payment plan agreement; DeVry Scholarship program. Preferred method of application: College Scholarship Service's Financial Aid Form.

■ TECHNOLOGY COLLEGE INFORMATION

[For additional personnel, refer to the *Appendix*.]

HEAD OF ENGINEERING TECHNOLOGY
Eugene E. Miller
Phone: (312)929-8500 Fax: (312)348-1780

ENGINEERING TECHNOLOGY COLLEGE ADDRESS
Electronics Engineering Technology
DeVry Institute of Technology, Chicago
3300 North Campbell Avenue
Chicago, IL 60618-5994
Phone: (312)929-8500 Fax: (312)348-1780

TECHNOLOGY ENROLLMENTS—BY CLASS
[Numbers are technology baccalaureate enrollments for the fall 1991 term, unless otherwise footnoted.]
1st-year students/Freshmen 293
2nd-year students/Sophomores 147
3rd-year students/Juniors 134
4th-year students/Seniors 184
Total 758

NUMBER OF DEGREES AWARDED—BY PROGRAM
[Numbers are engineering technology baccalaureate degrees awarded during the 1991-1992 academic year, unless otherwise footnoted. For full details about each engineering technology program, refer to the *Table of Degree Programs*.]
Electronic Engineering Technology 154
Electronic(s) Engineering Technology .... –
Total **154**

PERCENTAGE OF DEGREES AWARDED—BY CATEGORY
[Percentages are of all engineering technology baccalaureate degrees awarded during the 1991-1992 academic year, unless otherwise footnoted.]
African Americans 13.6%
Asian/Pacific Island Americans 22.7%
Hispanic Americans 15.6%
Native Americans 0.6%
Foreign Citizens 3.2%
All Others 44.3%
Women 7.8%
Persons with disabilities .... – % (A)
Students over 25 years of age 4.8%

Notes: (A) Data not available.

FULL- & PART-TIME FACULTY—BY DEPARTMENT
[Figures are the head count of full-time faculty and the full-time equivalent (FTE) of part-time faculty for each engineering technology department or equivalent.]

Department	Full	Part
Electronics Eng Tech	12	2.5

TECHNOLOGY COLLEGE DESCRIPTION
The Institute's baccalaureate degree program, Electronics Engineering Technology, offers the capability to refine and produce a design idea - to make the design concept into a real product. All of the skills and knowledge acquired throughout the program are brought together in a senior project, as students gain valuable industry experience by actually designing and building a microprocessor-based system. Laboratories are available at scheduled hours during the school day, evenings, and Saturdays. Electronics laboratory facilities include student work spaces, equipped with oscilloscopes, signal generators, multimeters and power supplies. These workstations are utilized for basic experiments in electronic circuits. Advanced laboratories are equipped to complement course work in digital circuits, digital computers, microprocessors, communication systems, industrial electronics and control systems. A physics laboratory offers an additional range of equipment.

TECHNOLOGY DEGREE PROGRAMS DESCRIPTION
The Institute's baccalaureate degree program, Electronics Engineering Technology, is designed to meet the need for technical specialists who can aggressively confront and solve complex problems in areas as diverse as space communication systems, local area networks and artificial intelligence.

TRANSFER INFORMATION
Residency Requirements: Candidates for a degree who have transferred from a non-DeVry school must complete more than thirty-five percent of the required credit hours in a program at a DeVry Institute to successfully meet the program requirements.

Transfer without Articulation Agreements
Admission to engineering technology: At any time
Requirements: Entrance requirements are the same for transfer students as they are for new students, unless the applicant possesses an accredited degree which waives the entrance requirements.

GRADUATION REQUIREMENTS
To graduate from any program, a student must maintain a Cumulative Grade Point Average (CGPA) of not less than 2.00 and satisfactorily complete all current curriculum requirements or the equivalent. Students who complete their work in any program with a CGPA of 3.50 or higher are graduated with honors and are so recognized at the graduation ceremony.

■ STUDENT PROGRAMS

PROFESSIONAL AND HONORARY SOCIETIES
[For key to acronyms, see *Introduction*.]
Professional Societies: DPMA, IEEE, ISA
Honorary Societies: Tau Alpha Pi

SUPPORT PROGRAMS
Student Chapter Organizations: Society of Hispanic Professional Engineers, Afro-American Student Organization, National Technical Association, Indo-Pakistani Club, Phillipino Club

Academic Support Center
For: Women & Ethnic Minorities **Available:** Year round
Offered: Freshman, Sophomore, Junior, Senior

Learning Resource Center
For: Women & Ethnic Minorities **Available:** Year round
Offered: Freshman, Sophomore, Junior, Senior

Other Engineering Support Programs: New students take a Student Orientation course for credit to get them acclimated to the college environment. Other support services include: career counseling services; graduate placement services; tutorial services; housing/part-time employment services.

296 DeVry Institute of Technology, Columbus

INSTITUTION PROFILE
HEAD OF THE INSTITUTION
Richard A. Czerniak
Phone: (614)253-7291   **Fax:** (614)252-4108
GENERAL INFORMATION
[All Students—Fall 1991]
Undergraduate enrollment 2,745
Total institution enrollment 2,745
Type of institution: Private   **Calendar:** Semesters
Location: Columbus   **Population:** 632,910
Setting: Urban
Types of engineering degrees: Engineering Technology
Other degree-granting colleges: Accounting, Business Operations, Computer Information Systems, Electronics Technician (AAS)

ENGINEERING TECHNOLOGY ADMISSIONS
ADMISSIONS OFFICE CONTACT
Richard Rodman
DeVry Institute of Technology, Columbus
1350 Alum Creek Drive
Columbus, OH 43209-2764
Phone: (614)253-1525   **Fax:** (614)252-4108
TECHNOLOGY COLLEGE ADMISSIONS INFORMATION
Admission to the engineering technology college: A student applies to the program after meeting with a DeVry representative during senior year of high school or after completing high school or equivalent.
Entrance Requirements: Entrance examinations in math and algebra must be taken if the applicant has not taken a recommended standardized test, or has failed to achieve the required score on the standardized test.
Entrance Recommendations: SAT (Score: 480), ACT (Score: 20), High School courses—Algebra/Advanced Algebra (1 year), Trigonometry (1 year), Calculus (1 year), Physics (1 year), WPCT may also be taken. A score of 20 on the math section of the ACT or 480 on the math section of the SAT or WPCT is required.
Requirements for foreign students: TOEFL (Score: 500); Provide a certified copy of the applicant's high school (or college, if applicable) transcript (in English); a notarized statement of financial support or a certified government sponsor letter; proof of English language proficiency. A Form I-20AB will be provided by the Institute.
DEPARTMENTAL ADMISSIONS INFORMATION
Admission to the engineering technology department: At the time of admission to the institution.
Additional information: Prospective students who wish to complete an application must either visit the institute and meet with a DeVry admissions representative, or meet with a DeVry field representative assigned to the student's geographical area.

FINANCIAL INFORMATION
ESTIMATED EXPENSES (ACADEMIC YEAR)
[Expenses are for the 1992-93 nine-month academic year.]

All Students	Undergraduate	Graduate
Tuition and fees	$ 5,249	$ —[2]
College room and board	$ 2,923	$ —[2]
Books and supplies	$ 500	$ —[2]
Other expenses	$ 3,958 [1]	$ —[2]
Total estimated expenses	**$12,630**	**$ —**

Notes: (1) Transportation $2,047 Personal/Miscellaneous $1,911 (2) No graduate degree offered

FINANCIAL AID OFFICE CONTACT
Cynthia A. Price, *Director of Financial Aid*
DeVry Institute of Technology, Columbus
1350 Alum Creek Drive
Columbus, OH 43209-2764
Phone: (614)253-7291   **Fax:** (614)252-4108
GENERAL FINANCIAL AID INFORMATION
Forms accepted/required: AFSA, CSS-FAF, FAF, FAT, FSA, FFS, Stafford, IRS, SAR, Supplemental, Institutional, Ohio Instructional Grant Application (OIG), Statement of Educational Purpose/Title IV Certification
Additional financial aid information: Pell Grants; Supplement Educational Opportunity Grants; Perkins Loans; College Work-Study; Guaranteed Student Loan Programs: Stafford, SLS, PLUS; EDUCARD Plan-monthly payment plan agreement; DeVry Scholarship program; Ohio Instructional Grant. Preferred method of application: College Scholarship Services Financial Aid Form.

TECHNOLOGY COLLEGE INFORMATION
[For additional personnel, refer to the *Appendix*.]
HEAD OF ENGINEERING TECHNOLOGY
Edward A. Wilson
Phone: (614)253-7291   **Fax:** (614)252-4108
ENGINEERING TECHNOLOGY COLLEGE ADDRESS
Electronics Engineering Technology
DeVry Institute of Technology, Columbus
1350 Alum Creek Drive
Columbus, OH 43209-2764
Phone: (614)253-7291   **Fax:** (614)252-4108
TECHNOLOGY ENROLLMENTS—BY CLASS
[Numbers are technology baccalaureate enrollments for the fall 1991 term, unless otherwise footnoted.]
1st-year students/Freshmen 367
2nd-year students/Sophomores 133
3rd-year students/Juniors 101
4th-year students/Seniors 199
Total 800
NUMBER OF DEGREES AWARDED—BY PROGRAM
[Numbers are engineering technology baccalaureate degrees awarded during the 1991-1992 academic year, unless otherwise footnoted. For full details about each engineering technology program, refer to the *Table of Degree Programs*.]
Electronic Engineering Technology 221
Electronic(s) Engineering Technology .... —
Total 221
PERCENTAGE OF DEGREES AWARDED—BY CATEGORY
[Percentages are of all engineering technology baccalaureate degrees awarded during the 1991-1992 academic year, unless otherwise footnoted.]
African Americans 7.7%
Asian/Pacific Island Americans 3.2%
Hispanic Americans .... —%
Native Americans .... —%
Foreign Citizens 5.0%
All Others 84.1%
Women 3.2%
Persons with disabilities .... —% (A)
Students over 25 years of age .... 28.1% (A)
Notes: (A) Data not available.
FULL- & PART-TIME FACULTY—BY DEPARTMENT
[Figures are the head count of full-time faculty and the full-time equivalent (FTE) of part-time faculty for each engineering technology department or equivalent.]

Department	Full	Part
Electronics Eng Tech	14	1.0

TECHNOLOGY COLLEGE DESCRIPTION
The Institute's baccalaureate degree program, Electronics Engineering Technology, offers the capability to refine and produce a design idea - to make the design concept into a real product. All of the skills and knowledge acquired throughout the program are brought together in a senior project, as students gain valuable industry experience by actually designing and building a microprocessor-based system. Laboratories are available at scheduled hours during the school day, evenings, and Saturdays. Electronics laboratory facilities include student work spaces, equipped with oscilloscopes, signal generators, multimeters and power supplies. These workstations are utilized for basic experiments in electronic circuits. Advanced laboratories are equipped to complement course work in digital circuits, digital computers, microprocessors, communication systems, industrial electronics and control systems. A physics laboratory offers an additional range of equipment.
TECHNOLOGY DEGREE PROGRAMS DESCRIPTION
The Institute's baccalaureate degree program, Electronics Engineering Technology, is designed to meet the need for technical specialists who can aggressively confront and solve complex problems in areas as diverse as space communication systems, local area networks and artificial intelligence.
TRANSFER INFORMATION
Residency Requirements: Candidates for a degree who have transferred from a non-DeVry school must complete more than thirty-five percent of the required credit hours in a program at a DeVry Institute to successfully meet the program requirements.
Transfer without Articulation Agreements
Admission to engineering technology: At any time
Requirements: Entrance requirements are the same for transfer students as they are for new students, unless the applicant possesses an accredited degree which waives the entrance requirements.
GRADUATION REQUIREMENTS
To graduate from any program, a student must maintain a Cumulative Grade Point Average (CGPA) of not less than 2.00 and satisfactorily complete all current curriculum requirements or the equivalent. Students who complete their work in any program with a CGPA of 3.50 or higher are graduated with honors and are so recognized at the graduation ceremony.

STUDENT PROGRAMS
PROFESSIONAL AND HONORARY SOCIETIES
[For key to acronyms, see *Introduction*.]
Professional Societies: DPMA, IEEE
Honorary Societies: Tau Alpha Pi
SUPPORT PROGRAMS
Student Chapter Organizations: Society of Women Engineers
Academic Support Center
For: Women & Ethnic Minorities   **Available:** Year round
Offered: Freshman, Sophomore, Junior, Senior
Learning Resource Center
For: Women & Ethnic Minorities   **Available:** Year round
Offered: Freshman, Sophomore, Junior, Senior
Other Engineering Support Programs: New students take a Student Orientation course for credit to get them acclimated to the college environment. Other support services include: career counseling services; graduate placement services; tutorial services; housing/part-time employment services; financial aid services; and co-op services.

297 DeVry Institute of Technology, Atlanta

■ INSTITUTION PROFILE

HEAD OF THE INSTITUTION
Ron Bush
Phone: (404)292-7900   Fax: (404)292-2321

GENERAL INFORMATION
[All Students—Fall 1991]
Undergraduate enrollment 3,133
Total institution enrollment 3,133

Type of institution: Private	**Calendar:** Semesters
Nearest city: Atlanta	**Population:** 394,017
Miles from main campus: 10	**Setting:** Suburban

Types of engineering degrees: Engineering Technology
Other degree-granting colleges: Accounting, Business Operations, Computer Information Systems, Electronic Technician (AAS)

■ ENGINEERING TECHNOLOGY ADMISSIONS

ADMISSIONS OFFICE CONTACT
Susan Hirst
DeVry Institute of Technology, Atlanta
250 N. Arcadia
Decatur, GA 30030-2198
Phone: (404)292-2645   Fax: (404)292-2321

TECHNOLOGY COLLEGE ADMISSIONS INFORMATION

Admission to the engineering technology college: A student applies to the program after meeting with a DeVry representative during senior year of high school or after completing high school or equivalent.

Entrance Requirements: Entrance examinations in math and algebra must be taken if the applicant has not taken a recommended standardized test, or has failed to achieve the required score on the standardized test.

Entrance Recommendations: SAT (Score: 480), ACT (Score: 20), High School courses—Algebra/Advanced Algebra (1 year), Trigonometry (1 year), Calculus (1 year), Physics (1 year), WPCT may also be taken. A score of 20 on the math section of the ACT or 480 on the math section of the SAT or WPCT is required.

Requirements for foreign students: TOEFL (Score: 500); Provide the Inst. with a certified copy of the applicant's H.S. (or college, if applicable) transcript (in English); a notarized statement of financial support or a certified government sponsor letter; proof of Eng. language proficiency. A Form I-20AB will be provided to the applicant by the Ins

DEPARTMENTAL ADMISSIONS INFORMATION

Admission to the engineering technology department: At the time of admission to the institution.

Additional information: Prospective students who wish to complete an application must either visit the Institute and meet with a DeVry admissions representative, or meet with a DeVry field representative assigned to the student's geographical area.

■ FINANCIAL INFORMATION

ESTIMATED EXPENSES (ACADEMIC YEAR)
[Expenses are for the 1992-93 nine-month academic year.]

All Students	Undergraduate	Graduate
Tuition and fees	$ 5,249	$ _(1)
College room and board	$ 3,227	$ _(1)
Books and supplies	$ 500	$ _(1)
Other expenses	$ 4,137 (2)	$ _(1)
Total estimated expenses	**$13,113**	**$ –**

Notes: (1) No graduate degree offered (2) Transportation $2,226 Personal/Misc. $1,911

FINANCIAL AID OFFICE CONTACT
Loretta Franklin, *Dean of Student Finance*
DeVry Institute of Technology, Atlanta
250 N. Arcadia Ave.
Decatur, GA 30030-2198
Phone: (404)292-7900   Fax: (404)292-2321

GENERAL FINANCIAL AID INFORMATION
Forms accepted/required: AFSA, CSS-FAF, FAF, FAT, FSA, FFS, Stafford, IRS, SAR, Supplemental, Institutional

Additional financial aid information: Pell Grants; Supplemental Educational Opportunity Grants; Perkins Loans; College Work-Study; Guaranteed Student Loan Programs: Stafford, SLS, PLUS; EDUCARD Plan - monthly payment plan agreement; DeVry Scholarship program. Preferred method of application: College Scholarship Service's Financial Aid Form.

■ TECHNOLOGY COLLEGE INFORMATION

[For additional personnel, refer to the *Appendix*.]

HEAD OF ENGINEERING TECHNOLOGY
Jack Griffin
Phone: (404)292-7900   Fax: (404)292-2321

ENGINEERING TECHNOLOGY COLLEGE ADDRESS
Electronics Engineering Technology
DeVry Institute of Technology, Atlanta
250 N. Arcadia
Decatur, GA 30030-2198
Phone: (404)292-7900   Fax: (404)292-2321

TECHNOLOGY ENROLLMENTS—BY CLASS
[Numbers are technology baccalaureate enrollments for the fall 1991 term, unless otherwise footnoted.]

1st-year students/Freshmen 330
2nd-year students/Sophomores 134
3rd-year students/Juniors 80
4th-year students/Seniors 102
Total **646**

NUMBER OF DEGREES AWARDED—BY PROGRAM
[Numbers are engineering technology baccalaureate degrees awarded during the 1991-1992 academic year, unless otherwise footnoted. For full details about each engineering technology program, refer to the *Table of Degree Programs*.]

Electronic Engineering Technology 122
Total **122**

PERCENTAGE OF DEGREES AWARDED—BY CATEGORY
[Percentages are of all engineering technology baccalaureate degrees awarded during the 1991-1992 academic year, unless otherwise footnoted.]

African Americans .... 32.0 %
Asian/Pacific Island Americans .... 8.2 %
Hispanic Americans .... 2.5 %
Native Americans .... – %
Foreign Citizens .... 2.5 %
All Others .... 54.8 %
Women .... 13.1 %
Persons with disabilities .... – % (A)
Students over 25 years of age .... 36.9 %

Notes: (A) Data not available.

FULL- & PART-TIME FACULTY—BY DEPARTMENT
[Figures are the head count of full-time faculty and the full-time equivalent (FTE) of part-time faculty for each engineering technology department or equivalent.]

Department	Full	Part
Electronic Eng Tech	11	–

TECHNOLOGY COLLEGE DESCRIPTION
The Institute's baccalaureate degree program, Electr. Eng. Tech., offers the capability to refine and produce a design idea - to make the design concept into a real product. All of the skills and knowledge acquired throughout the program are brought together in a senior project, as students gain valuable industry experience by actually designing and building a microprocessor-based system. Laboratories are available at scheduled hours during the school day, evenings, and Saturdays. Electronics laboratory facilities include student work spaces equipped with oscilloscopes, signal generators, multimeters and power supplies. These workstations are utilized for basic experiments in electronic circuits. Advanced laboratories are equipped to complement course work in digital circuits, digital computers, microprocessors, communication systems, industrial electronics and control systems. A physics laboratory offers an additional range of equipment.

TECHNOLOGY DEGREE PROGRAMS DESCRIPTION
The Institute's baccalaureate degree program, Electronics Engineering Technology, is designed to meet the need for technical specialists who can aggressively confront and solve complex problems in areas as diverse as space communication systems, local area networks and artificial intelligence.

TRANSFER INFORMATION
Residency Requirements: Candidates for a degree who have transferred from a non-DeVry school must complete more than thirty-five percent of the required credit hours in a program at a DeVry Institute to successfully meet the program requirements.

Transfer via Articulation Agreements
Admission to engineering technology: At any time
Requirements: Entrance requirements are the same for transfer students as they are for new students, unless the applicant possesses an accredited degree which waives the entrance requirements.

Transfer without Articulation Agreements
Admission to engineering technology: At any time
Requirements: Entrance requirements are the same for transfer students as they are for new students, unless the applicant possesses an accredited degree which waives the entrance requirements.

GRADUATION REQUIREMENTS
To graduate from any program, a student must maintain a Cumulative Grade Point Average (CGPA) of not less than 2.00 and satisfactorily complete all current curriculum requirements or the equivalent. Students who complete their work in any program with a CGPA of 3.50 or higher are graduated with honors and are so recognized at the graduation ceremony.

■ STUDENT PROGRAMS

PROFESSIONAL AND HONORARY SOCIETIES
[For key to acronyms, see *Introduction*.]
Professional Societies: DPMA, IEEE
Honorary Societies: Tau Alpha Pi

SUPPORT PROGRAMS
Student Chapter Organizations: Society of Women Engineers, National Society of Black Engineers

Academic Support Center
For: Women & Ethnic Minorities   **Available:** Year round
Offered: Freshman, Sophomore, Junior, Senior

Learning Resource Center
For: Women & Ethnic Minorities   **Available:** Year round
Offered: Freshman, Sophomore, Junior, Senior

Mentor Program
For: Women & Ethnic Minorities   **Available:** Year round
Offered: Freshman

Other Engineering Support Programs: New students take a Student Orientation course for credit to get them acclimated to the college environment. Other support services include: English as a Second Language; career counseling services; graduate placement services; tutorial services; housing/part-time employment services; student advisor services.

298 DeVry Institute of Technology, DuPage

INSTITUTION PROFILE
HEAD OF THE INSTITUTION
Jerry R. Dill
Phone: (708)953-1300 **Fax:** (708)953-1236
GENERAL INFORMATION
[All Students—Fall 1991]
Undergraduate enrollment 2,564
Total institution enrollment 2,564
Type of institution: Private **Calendar:** Semesters
Nearest city: Chicago **Population:** 2,783,726
Miles from main campus: 20 **Setting:** Suburban
Types of engineering degrees: Engineering Technology
Other degree-granting colleges: Accounting, Business Operations, Computer Information Systems, Electronics Technician (AAS), Telecommunications Management

ENGINEERING TECHNOLOGY ADMISSIONS
ADMISSIONS OFFICE CONTACT
Virginia Mechnig
DeVry Institute of Technology, DuPage
1221 North Swift Road
Addison, IL 60101-5614
Phone: (708)953-2000 **Fax:** (708)953-1236
TECHNOLOGY COLLEGE ADMISSIONS INFORMATION
Admission to the engineering technology college: A student applies to the program after meeting with a DeVry representative during senior year of high school or after completing high school or equivalent.
Entrance Requirements: Entrance examinations in math and algebra must be taken if the applicant has not taken a recommended standardized test, or has failed to achieve the required score on the standardized test.
Entrance Recommendations: SAT (Score: 480), ACT (Score: 20), High School courses—Algebra/Advanced Algebra (1 year), Trigonometry (1 year), Calculus (1 year), Physics (1 year), WPCT may also be taken. A score of 20 on the math section of the ACT or 480 on the math section of the SAT or WPCT is required.
Requirements for foreign students: TOEFL (Score: 500); Provide a certified copy of the applicant's high school (or college, if applicable) transcript (in English); a notarized statement of financial support or a certified government sponsor letter; proof of English language proficiency. A Form I-20AB will be provided to the applicant.
DEPARTMENTAL ADMISSIONS INFORMATION
Admission to the engineering technology department: At the time of admission to the institution.
Additional information: Prospective students who wish to complete an application must either visit the institute and meet with a DeVry admissions representative, or meet with a DeVry field representative assigned to the student's geographical area.

FINANCIAL INFORMATION
ESTIMATED EXPENSES (ACADEMIC YEAR)
[Expenses are for the 1992-93 nine-month academic year.]

All Students	Undergraduate	Graduate
Tuition and fees	$ 5,249	$ _(1)
College room and board	$ 3,326	$ _(1)
Books and supplies	$ 500	$ _(1)
Other expenses	$ 4,006 (2)	$ _(1)
Total estimated expenses	$13,081	$ —

Notes: (1) No graduate degree offered (2) Transportation $2,095 Personal/Misc. $1,911

FINANCIAL AID OFFICE CONTACT
Diane Battistella, *Director of Financial Aid*
DeVry Institute of Technology, DuPage
1221 North Swift Road
Addison, IL 60101-5614
Phone: (708)953-1300 **Fax:** (708)953-1236
GENERAL FINANCIAL AID INFORMATION
Forms accepted/required: AFSA, CSS-FAF, FAF, FAT, FSA, FFS, Stafford, IRS, SAR, Supplemental, Institutional
Additional financial aid information: Pell Grants; Supplement Educational Opportunity Grants; Perkins Loans; College Work-Study; Guaranteed Student Loan Programs: Stafford, SLS, PLUS; EDUCARD Plan-monthly payment plan agreement; DeVry Scholarship program. Preferred method of application: College Scholarship Service's Financial Aid Form.

TECHNOLOGY COLLEGE INFORMATION
[For additional personnel, refer to the *Appendix*.]
HEAD OF ENGINEERING TECHNOLOGY
Sylvia Washington
Phone: (708)953-1300 **Fax:** (708)953-1236
ENGINEERING TECHNOLOGY COLLEGE ADDRESS
Electronics Engineering Technology
DeVry Institute of Technology, DuPage
1221 North Swift Road
Addison, IL 60101-5614
Phone: (708)953-1300 **Fax:** (708)953-1236
TECHNOLOGY ENROLLMENTS—BY CLASS
[Numbers are technology baccalaureate enrollments for the fall 1991 term, unless otherwise footnoted.]
1st-year students/Freshmen 232
2nd-year students/Sophomores 109
3rd-year students/Juniors 95
4th-year students/Seniors 117
Total 553
NUMBER OF DEGREES AWARDED—BY PROGRAM
[Numbers are engineering technology baccalaureate degrees awarded during the 1991-1992 academic year, unless otherwise footnoted. For full details about each engineering technology program, refer to the *Table of Degree Programs*.]
Electronic Engineering Technology 104
Electronic(s) Engineering Technology .... —
Total 104
PERCENTAGE OF DEGREES AWARDED—BY CATEGORY
[Percentages are of all engineering technology baccalaureate degrees awarded during the 1991-1992 academic year, unless otherwise footnoted.]
African Americans 2.9%
Asian/Pacific Island Americans 9.6%
Hispanic Americans 4.8%
Native Americans .... — %
Foreign Citizens .... — %
All Others 82.7%
Women .... — %
Persons with disabilities .... — % (A)
Students over 25 years of age 33.7%
Notes: (A) Data not available.
FULL- & PART-TIME FACULTY—BY DEPARTMENT
[Figures are the head count of full-time faculty and the full-time equivalent (FTE) of part-time faculty for each engineering technology department or equivalent.]

Department	Full	Part
Electronics Eng Tech	12	0.5

TECHNOLOGY COLLEGE DESCRIPTION
The Institute's baccalaureate degree program, Electronics Engineering Technology, offers the capability to refine and produce a design idea - to make the design concept into a real product. All of the skills and knowledge acquired throughout the program are brought together in a senior project, as students gain valuable industry experience by actually designing and building a microprocessor-based system. Laboratories are available at scheduled hours during the school day, evenings, and Saturdays. Electronics laboratory facilities include student work spaces, equipped with oscilloscopes, signal generators, multimeters and power supplies. These workstations are utilized for basic experiments in electronic circuits. Advanced laboratories are equipped to complement course work in digital circuits, digital computers, microprocessors, communication systems, industrial electronics and control systems. A physics laboratory offers an additional range of equipment.
TECHNOLOGY DEGREE PROGRAMS DESCRIPTION
The Institute's baccalaureate degree program, Electronics Engineering Technology, is designed to meet the need for technical specialists who can aggressively confront and solve complex problems in areas as diverse as space communication systems, local area networks and artificial intelligence.
TRANSFER INFORMATION
Residency Requirements: Candidates for a degree who have transferred from a non-DeVry school must complete more than thirty-five percent of the required credit hours in a program at a DeVry Institute to successfully meet the program requirements.
Transfer via Articulation Agreements
Admission to engineering technology: At any time
Requirements: Entrance requirements are the same for transfer students as they are for new students, unless the applicant possesses an accredited degree which waives the entrance requirements.
Transfer without Articulation Agreements
Admission to engineering technology: At any time
Requirements: Entrance requirements are the same for transfer students as they are for new students, unless the applicant possesses an accredited degree which waives the entrance requirements.
GRADUATION REQUIREMENTS
To graduate from any program, a student must maintain a Cumulative Grade Point Average (CGPA) of not less than 2.00 and satisfactorily complete all current curriculum requirements or the equivalent. Students who complete their work in any program with a CGPA of 3.50 or higher are graduated with honors and are so recognized at the graduation ceremony.

STUDENT PROGRAMS
PROFESSIONAL AND HONORARY SOCIETIES
[For key to acronyms, see *Introduction*.]
Professional Societies: DPMA, IEEE, ISA
Honorary Societies: Tau Alpha Pi
SUPPORT PROGRAMS
Student Chapter Organizations: Assn. for the Advancement of Hispanic Students, Black Leadership Assn. Standing Together, Gamma Phi Theta

Academic Skills Center
For: Women & Ethnic Minorities **Available:** Year round
Offered: Freshman, Sophomore, Junior, Senior

Learning Resource Center
For: Women & Ethnic Minorities **Available:** Year round
Offered: Freshman, Sophomore, Junior, Senior

English as a Second Language
For: Ethnic Minorities **Available:** Year round
Offered: Freshman, Sophomore, Junior, Senior

Other Engineering Support Programs: New students take a Student Orientation course for credit to get them acclimated to college environment. Other support services include: career counseling services; graduate placement services; tutorial services; housing/part-time employment services.

299 DeVry Institute of Technology, Dallas

■ INSTITUTION PROFILE
HEAD OF THE INSTITUTION
Frank Cannon
Phone: (214)258-6767　　　　　　　　Fax: (214)659-1748

GENERAL INFORMATION
[All Students—Fall 1991]
Undergraduate enrollment 2,240
Total institution enrollment 2,240

Type of institution: Private　　**Calendar:** Semesters
Nearest city: Dallas　　**Population:** 1,006,877
Miles from main campus: 10　　**Setting:** Suburban
Types of engineering degrees: Engineering Technology
Other degree-granting colleges: Accounting, Business Operations, Computer Information Systems, Electronic Technician (AAS)

■ ENGINEERING TECHNOLOGY ADMISSIONS
ADMISSIONS OFFICE CONTACT
Danny Millan
DeVry Institute of Technology, Dallas
4250 N. Beltline Road
Irving, TX 75038-4299
Phone: (214)258-6330　　　　　　　　Fax: (214)659-1748

TECHNOLOGY COLLEGE ADMISSIONS INFORMATION
Admission to the engineering technology college: A student applies to the program after meeting with a DeVry representative during senior year of high school or after completing high school or equivalent.
Entrance Requirements: Entrance examinations in math and algebra must be taken if the applicant has not taken a recommended standardized test, or has failed to achieve the required score on the standardized test.
Entrance Recommendations: SAT (Score: 480), ACT (Score: 20), High School courses—Algebra/Advanced Algebra (1 year), Trigonometry (1 year), Calculus (1 year), Physics (1 year), WPCT may also be taken. A score of 20 on the math section of the ACT or 480 on the math section of the SAT or WPCT is required.
Requirements for foreign students: TOEFL (Score: 500); Provide Institute with a certified copy of the applicant's H.S. (or college, if applicable) transcript (in Eng); a notarized statement of financial support or a cert. government sponsor letter; proof of Eng. lang. proficiency. A Form I-20AB will be provided to the applicant by the Institute.

DEPARTMENTAL ADMISSIONS INFORMATION
Admission to the engineering technology department: At the time of admission to the institution.
Additional information: Prospective students who wish to complete an application must either visit the Institute and meet with a DeVry admissions representative, or meet with a DeVry field representative assigned to the student's geographical area.

■ FINANCIAL INFORMATION
ESTIMATED EXPENSES (ACADEMIC YEAR)
[Expenses are for the 1992-93 nine-month academic year.]

All Students	Undergraduate	Graduate
Tuition and fees	$ 5,249	$ —[1]
College room and board	$ 3,534	$ —[1]
Books and supplies	$ 500	$ —[1]
Other expenses	$ 4,399 [2]	$ —[1]
Total estimated expenses	**$13,682**	**$ —**

Notes: (1) No graduate degree offered (2) Transportation $2,488 Personal/Miscellaneous $1,911

FINANCIAL AID OFFICE CONTACT
Laura Myers, *Director of Financial Aid*
DeVry Institute of Technology, Dallas
4250 North Beltline Road
Irving, TX 75038-4299
Phone: (214)258-6767　　　　　　　　Fax: (214)659-1748

GENERAL FINANCIAL AID INFORMATION
Forms accepted/required: AFSA, CSS-FAF, FAF, FAT, FSA, FFS, Stafford, IRS, SAR, Supplemental, Institutional, Verification Worksheet
Additional financial aid information: Pell Grants; Supplemental Educational Opportunity Grants; Perkins Loans; College Work-Study; Guaranteed Student Loan Programs: Stafford, SLS, PLUS; EDUCARD Plan-monthly payment plan agreement; DeVry Scholarship program. Preferred method of application: College Scholarship Service's Financial Aid Form.

■ TECHNOLOGY COLLEGE INFORMATION
[For additional personnel, refer to the *Appendix*.]

HEAD OF ENGINEERING TECHNOLOGY
Charles Reck
Phone: (214)258-6767　　　　　　　　Fax: (214)659-1748

ENGINEERING TECHNOLOGY COLLEGE ADDRESS
Electronics Engineering Technology
DeVry Institute of Technology, Dallas
4250 N. Beltline Road
Irving, TX 75038-4299
Phone: (214)258-6767　　　　　　　　Fax: (214)659-1748

TECHNOLOGY ENROLLMENTS—BY CLASS
[Numbers are technology baccalaureate enrollments for the fall 1991 term, unless otherwise footnoted.]
1st-year students/Freshmen 249
2nd-year students/Sophomores 90
3rd-year students/Juniors 93
4th-year students/Seniors 127
Total **559**

NUMBER OF DEGREES AWARDED—BY PROGRAM
[Numbers are engineering technology baccalaureate degrees awarded during the 1991-1992 academic year, unless otherwise footnoted. For full details about each engineering technology program, refer to the *Table of Degree Programs*.]
Electronic Engineering Technology 90
Total **90**

PERCENTAGE OF DEGREES AWARDED—BY CATEGORY
[Percentages are of all engineering technology baccalaureate degrees awarded during the 1991-1992 academic year, unless otherwise footnoted.]
African Americans 8.9%
Asian/Pacific Island Americans 12.2%
Hispanic Americans 12.2%
Native Americans .... — %
Foreign Citizens 6.7%
All Others 60.0%
Women 7.8%
Persons with disabilities .... — %[A]
Students over 25 years of age 60.0%

Notes: (A) Data not available.

FULL- & PART-TIME FACULTY—BY DEPARTMENT
[Figures are the head count of full-time faculty and the full-time equivalent (FTE) of part-time faculty for each engineering technology department or equivalent.]

Department	Full	Part
Electronic Eng Tech	11	0.5

TECHNOLOGY COLLEGE DESCRIPTION
The Institute's baccalaureate program, Electronics Engineering Technology, offers the capability to refine and produce a design idea - to make the design concept into a real product. All of the skills and knowledge acquired throughout the program are brought together in a senior project, as students gain valuable industry experience by actually designing and building a microprocessor-based system. Laboratories are available at scheduled hours during the school day, evenings, and Saturdays. Electronics laboratory facilities include student work spaces, equipped with oscilloscopes, signal generators, multimeters and power supplies. These workstations are utilized for basic experiments in electronic circuits. Advanced laboratories are equipped to complement course work in digital circuits, digital computers, microprocessors, communication systems, industrial electronics and control systems. A physics laboratory offers an additional range of equipment.

TECHNOLOGY DEGREE PROGRAMS DESCRIPTION
The Institute's baccalaureate degree program, Electronics Engineering Technology, is designed to meet the need for technical specialists who can aggressively confront and solve complex problems in areas as diverse as space communication systems, local area networks and artificial intelligence.

TRANSFER INFORMATION
Residency Requirements: Candidates for a degree who have transferred from a non-DeVry school must complete more than thirty-five percent of the required credit hours in a program at a DeVry Institute to successfully meet the program requirements.

Transfer without Articulation Agreements
Admission to engineering technology: At any time
Requirements: Entrance requirements are the same for transfer students as they are for new students, unless the applicant possesses an accredited degree which waives the entrance requirements.

GRADUATION REQUIREMENTS
To graduate from any program, a student must maintain a Cumulative Grade Point Average (CGPA) of not less than 2.00 and satisfactorily complete all current curriculum requirements or the equivalent. Students who complete their work in any program with a CGPA of 3.50 or higher are graduated with honors and are so recognized at the graduation ceremony.

■ STUDENT PROGRAMS
PROFESSIONAL AND HONORARY SOCIETIES
[For key to acronyms, see *Introduction*.]
Professional Societies: DPMA, IEEE
Honorary Societies: Tau Alpha Pi

SUPPORT PROGRAMS
Student Chapter Organizations: Society of Women Engineers, Society of Hispanic Professional Engineers

Academic Support Center
For: Women & Ethnic Minorities　　**Available:** Year round
Offered: Freshman, Sophomore, Junior, Senior

Learning Resource Center
For: Women & Ethnic Minorities　　**Available:** Year round
Offered: Freshman, Sophomore, Junior, Senior

Other Engineering Support Programs: New students take a Student Orientation course for credit to get them acclimated to the college environment. Other support services include: Basic Skills Tutoring; English as a Second Language (ESL); career counseling services; graduate placement services; coursework tutorial services; housing/part-time employment services.

300 DeVry Institute of Technology, Kansas City

■ INSTITUTION PROFILE
HEAD OF THE INSTITUTION
C. Robert LeValley
Phone: (816)941-0430 **Fax:** (816)941-0896
GENERAL INFORMATION
[All Students—Fall 1991]
Undergraduate enrollment 1,814
Total institution enrollment 1,814
Type of institution: Private **Calendar:** Semesters
Location: Kansas City, MO **Population:** 435,146
Setting: Suburban
Types of engineering degrees: Engineering Technology
Other degree-granting colleges: Accounting, Business Operations, Computer Information Systems, Electronics Technician (AAS), Telecommunications Management

■ ENGINEERING TECHNOLOGY ADMISSIONS
ADMISSIONS OFFICE CONTACT
Gale Dykes-Grimmett
DeVry Institute of Technology, Kansas City
11224 Holmes Road
Kansas City, MO 64131-3626
Phone: (816)941-2810 **Fax:** (816)941-0896
TECHNOLOGY COLLEGE ADMISSIONS INFORMATION
Admission to the engineering technology college: A student applies to the program after meeting with a DeVry representative during senior year of high school or after completing high school or equivalent.
Entrance Requirements: Entrance examinations in math and algebra must be taken if the applicant has not taken a recommended standardized test, or has failed to achieve the required score on the standardized test.
Entrance Recommendations: SAT (Score: 480), ACT (Score: 20), High School courses—Algebra/Advanced Algebra (1 year), Trigonometry (1 year), Calculus (1 year), Physics (1 year), WPCT may also be taken. A score of 20 on the math section of the ACT or 480 on the math section of the SAT or WPCT is required.
Requirements for foreign students: TOEFL (Score: 500); Provide a certified copy of the applicant's high school (or college, if applicable) transcript (in English); a notarized statement of financial support or a certified government sponsor letter; proof of English language proficiency. A Form I-20AB will be provided to the applicant.
DEPARTMENTAL ADMISSIONS INFORMATION
Admission to the engineering technology department: At the time of admission to the institution.
Additional information: Prospective students who wish to complete an application must either visit the institute and meet with a DeVry admissions representative, or meet with a DeVry field representative assigned to the student's geographical area.

■ FINANCIAL INFORMATION
ESTIMATED EXPENSES (ACADEMIC YEAR)
[Expenses are for the 1992-93 nine-month academic year.]

All Students	Undergraduate	Graduate
Tuition and fees	$ 5,249	$ —[1]
College room and board	$ 2,827	$ —[1]
Books and supplies	$ 500	$ —[1]
Other expenses	$ 4,129 [2]	$ —[1]
Total estimated expenses	**$12,705**	**$ —**

Notes: (1) No graduate degree offered (2) Transportation $2,218 Personal/Misc. $1,911

FINANCIAL AID OFFICE CONTACT
Sharon Lamson, *Director of Financial Aid*
DeVry Institute of Technology, Kansas City
11224 Holmes Road
Kansas City, MO 64131-3626
Phone: (816)941-0430 **Fax:** (816)941-0896
GENERAL FINANCIAL AID INFORMATION
Forms accepted/required: AFSA, CSS-FAF, FAF, FAT, FSA, FFS, Stafford, IRS, SAR, Supplemental, Institutional
Additional financial aid information: Pell Grants; Supplemental Educational Opportunity Grants; Perkins Loans; College Work-Study; Guaranteed Student Loan Programs: Stafford, SLS, PLUS; EDUCARD Plan-monthly payment plan agreement; DeVry Scholarship programs. Preferred method of application: College Scholarship Service's Financial Aid Form.

■ TECHNOLOGY COLLEGE INFORMATION
[For additional personnel, refer to the *Appendix*.]
HEAD OF ENGINEERING TECHNOLOGY
Paul Stephanchick
Phone: (816)941-0430 **Fax:** (816)941-0896
ENGINEERING TECHNOLOGY COLLEGE ADDRESS
Electronics Engineering Technology
DeVry Institute of Technology, Kansas City
11224 Holmes Road
Kansas City, MO 64131-3626
Phone: (816)941-0430 **Fax:** (816)941-0896
TECHNOLOGY ENROLLMENTS—BY CLASS
[Numbers are technology baccalaureate enrollments for the fall 1991 term, unless otherwise footnoted.]
1st-year students/Freshmen 155
2nd-year students/Sophomores 61
3rd-year students/Juniors 33
4th-year students/Seniors 67
Total 316
NUMBER OF DEGREES AWARDED—BY PROGRAM
[Numbers are engineering technology baccalaureate degrees awarded during the 1991-1992 academic year, unless otherwise footnoted. For full details about each engineering technology program, refer to the *Table of Degree Programs*.]
Electronic Engineering Technology 54
Electronic(s) Engineering Technology .... –
Total 54
PERCENTAGE OF DEGREES AWARDED—BY CATEGORY
[Percentages are of all engineering technology baccalaureate degrees awarded during the 1991-1992 academic year, unless otherwise footnoted.]
African Americans 1.9%
Asian/Pacific Island Americans 7.4%
Hispanic Americans .... – %
Native Americans .... – %
Foreign Citizens 1.9%
All Others 88.8%
Women 3.7%
Persons with disabilities .... – % (A)
Students over 25 years of age 51.9%
Notes: (A) Data not available.
FULL- & PART-TIME FACULTY—BY DEPARTMENT
[Figures are the head count of full-time faculty and the full-time equivalent (FTE) of part-time faculty for each engineering technology department or equivalent.]

Department	Full	Part
Electronics Eng Tech	9	0.5

TECHNOLOGY COLLEGE DESCRIPTION
The institute's baccalaureate degree program, Electronics Engineering Technology, offers the capability to refine and produce a design idea - to make the design concept into a real product. All of the skills and knowledge acquired throughout the program are brought together in a senior project, as students gain valuable industry experience by actually designing and building a microprocessor-based system. Laboratories are available at scheduled hours during the school day, evenings, and Saturdays. Electronics laboratory facilities include student work spaces, equipped with oscilloscopes, signal generators, multimeters and power supplies. These workstations are utilized for basic experiments in electronic circuits. Advanced laboratories are equipped to complement course work in digital circuits, digital computers, microprocessors, communication systems, industrial electronics, and control systems. A physics laboratory offers an additional range of equipment.
TECHNOLOGY DEGREE PROGRAMS DESCRIPTION
The Institute's baccalaureate degree program, Electronics Engineering Technology, is designed to meet the need for technical specialists who can aggressively confront and solve complex problems in areas as diverse as space communication systems, local area networks and artificial intelligence.
TRANSFER INFORMATION
Residency Requirements: Candidates for a degree who have transferred from a non-DeVry school must complete more than thirty-five percent of the required credit hours in a program at DeVry Institute to successfully meet the program requirements.
Transfer without Articulation Agreements
Admission to engineering technology: At any time
Requirements: Entrance requirements are the same for transfer students as they are for new students, unless the applicant possesses an accredited degree which waived the entrance requirements.
GRADUATION REQUIREMENTS
To graduate from any program, a student must maintain a Cumulative Grade Point Average (CGPA) of not less than 2.00 and satisfactorily complete all current curriculum requirements or the equivalent. Students who complete their work in any program with a CGPA of 3.50 or higher are graduated with honors and are so recognized at the graduation ceremony.

■ STUDENT PROGRAMS
PROFESSIONAL AND HONORARY SOCIETIES
[For key to acronyms, see *Introduction*.]
Professional Societies: DPMA, IEEE
Honorary Societies: Tau Alpha Pi
SUPPORT PROGRAMS
Academic Support Center
For: Women & Ethnic Minorities **Available:** Year round
Offered: Freshman, Sophomore, Junior, Senior
Learning Resource Center
For: Women & Ethnic Minorities **Available:** Year round
Offered: Freshman, Sophomore, Junior, Senior
Other Engineering Support Programs: New students take a Student Orientation course for credit to get them acclimated to the college environment. Other support services include: career counseling services; graduate placement services; tutorial services; housing/part-time employment services; student advisor services.

301 DeVry Institute of Technology, Los Angeles

■ INSTITUTION PROFILE

HEAD OF THE INSTITUTION
David Moore
Phone: (213)699-9927 **Fax:** (213)692-6272

GENERAL INFORMATION
[All Students—Fall 1991]
Undergraduate enrollment 1,884
Total institution enrollment 1,884
Type of institution: Private **Calendar:** Semesters
Nearest city: Los Angeles **Population:** 3,485,398
Miles from main campus: 19 **Setting:** Urban
Types of engineering degrees: Engineering Technology
Other degree-granting colleges: Accounting, Business Operations, Computer Information Systems, Electronics Technician (AAS), Telecommunications Management

■ ENGINEERING TECHNOLOGY ADMISSIONS

ADMISSIONS OFFICE CONTACT
Keith E. Paridy
DeVry Institute of Technology, Los Angeles
12801 Crossroads Parkway South
City of Industry, CA 91746-3495
Phone: (213)692-0551 **Fax:** (213)692-6272

TECHNOLOGY COLLEGE ADMISSIONS INFORMATION
Admission to the engineering technology college: A student applies to the program after meeting with a DeVry representative during senior year of high school or after completing high school or equivalent.
Entrance Requirements: Entrance examinations in math and algebra must be taken if the applicant has not taken a recommended standardized test, or has failed to achieve the required score on the standardized test.
Entrance Recommendations: SAT (Score: 480), ACT (Score: 20), High School courses—Algebra/Advanced Algebra (1 year), Trigonometry (1 year), Calculus (1 year), Physics (1 year), WPCT may also be taken. A score of 20 on the math section of the ACT or 480 on the math section of the SAT or WPCT.
Requirements for foreign students: TOEFL (Score: 500); Provide a certified copy of the applicant's high school (or college, if applicable) transcript (in English): a notarized statement of financial support or certified government sponsor letter; proof of English language proficiency. A Form I-20AB will be provided to the applicant.

DEPARTMENTAL ADMISSIONS INFORMATION
Admission to the engineering technology department: At the time of admission to the institution.
Additional information: Prospective students who wish to complete an application must either visit the institute and meet with a DeVry admissions representative, or meet with a DeVry field representative assigned to the student's geographical area.

■ FINANCIAL INFORMATION

ESTIMATED EXPENSES (ACADEMIC YEAR)
[Expenses are for the 1992-93 nine-month academic year.]

All Students	Undergraduate	Graduate
Tuition and fees	$ 5,249	$ —[1]
College room and board	$ 4,298	$ —[1]
Books and supplies	$ 500	$ —[1]
Other expenses	$ 4,231 [2]	$ —[1]
Total estimated expenses	**$14,278**	**$ —**

Notes: (1) No graduate degree offered (2) Transportation 2,320 Personal/Misc. 1,911

FINANCIAL AID OFFICE CONTACT
Lynn Taylor, *Director of Financial Aid*
DeVry Institute of Technology, Los Angeles
12801 Crossroads Parkway South
City of Industry, CA 91746-3495
Phone: (213)699-9927 **Fax:** (213)692-6272

GENERAL FINANCIAL AID INFORMATION
Forms accepted/required: AFSA, CSS-FAF, FAF, FAT, FSA, FFS, Stafford, IRS, SAR, Supplemental, Institutional, Cal Grant A, B, and C.
Additional financial aid information: Pell Grants; Supplement Educational Opportunity Grants; Perkins Loans; College Work-Study; Guaranteed Student Loan Programs: Stafford, SLS, PLUS; EDUCARD Plan-monthly payment plan agreement; DeVry Scholarship program. Preferred method of application: College Scholarship Service's Financial Aid Form.

■ TECHNOLOGY COLLEGE INFORMATION

[For additional personnel, refer to the *Appendix*.]
HEAD OF ENGINEERING TECHNOLOGY
Iraj Borbor
Phone: (213)699-9927 **Fax:** (213)692-6272

ENGINEERING TECHNOLOGY COLLEGE ADDRESS
Electronics Engineering Technology
DeVry Institute of Technology, Los Angeles
12801 Crossroads Parkway South
City of Industry, CA 91746-3495
Phone: (213)699-9927 **Fax:** (213)692-6272

TECHNOLOGY ENROLLMENTS—BY CLASS
[Numbers are technology baccalaureate enrollments for the fall 1991 term, unless otherwise footnoted.]
1st-year students/Freshmen 208
2nd-year students/Sophomores 67
3rd-year students/Juniors 57
4th-year students/Seniors 96
Total **428**

NUMBER OF DEGREES AWARDED—BY PROGRAM
[Numbers are engineering technology baccalaureate degrees awarded during the 1991-1992 academic year, unless otherwise footnoted. For full details about each engineering technology program, refer to the *Table of Degree Programs*.]
Electronic Engineering Technology 73
Electronic(s) Engineering Technology .... —
Total **73**

PERCENTAGE OF DEGREES AWARDED—BY CATEGORY
[Percentages are of all engineering technology baccalaureate degrees awarded during the 1991-1992 academic year, unless otherwise footnoted.]
African Americans .... 11.0 %
Asian/Pacific Island Americans .... 30.1 %
Hispanic Americans .... 21.9 %
Native Americans .... 1.4 %
Foreign Citizens .... 2.7 %
All Others .... 32.9 %
Women .... 1.4 %
Persons with disabilities .... — % (A)
Students over 25 years of age .... 56.2 %

Notes: (A) Data not available.

FULL- & PART-TIME FACULTY—BY DEPARTMENT
[Figures are the head count of full-time faculty and the full-time equivalent (FTE) of part-time faculty for each engineering technology department or equivalent.]

Department	Full	Part
Electronics Eng Tech	9	0.5

TECHNOLOGY COLLEGE DESCRIPTION
The Institute's baccalaureate degree program, Electronics Engineering Technology, offers the capability to refine and produce a design idea - to make the design concept into a real product. All of the skills and knowledge acquired throughout the program are brought together in a senior project, as students gain valuable industry experience by actually designing and building a microprocessor-based system. Laboratories are available at scheduled hours during the school day, evenings, and Saturdays. Electronics laboratory facilities include student work spaces, equipped with oscilloscopes, signal generators, multimeters and power supplies. These workstations are utilized for basic experiments in electronic circuits. Advanced laboratories are equipped to complement course work in digital circuits, digital computers, microprocessors, communication systems, industrial electronics and control systems. A physics laboratory offers an additional range of equipment.

TECHNOLOGY DEGREE PROGRAMS DESCRIPTION
The Institute's baccalaureate degree program, Electronics Engineering Technology, is designed to meet the need for technical specialists who can aggressively confront and solve complex problems in areas as diverse as space communication systems, local area networks and artificial intelligence.

TRANSFER INFORMATION
Residency Requirements: Candidates for a degree who have transferred from a non-DeVry school must complete more than thirty-five percent of the required credit hours in a program at a DeVry Institute to successfully meet the program requirements.

Transfer via Articulation Agreements
Admission to engineering technology: At any time
Requirements: Entrance requirements are the same for transfer students as they are for new students, unless the applicant possesses an accredited degree which waives the entrance requirements.

Transfer without Articulation Agreements
Admission to engineering technology: At any time
Requirements: Entrance requirements are the same for transfer students as they are for new students, unless the applicant possesses an accredited degree which waives the entrance requirements.

GRADUATION REQUIREMENTS
To graduate from any program, a student must maintain a Cumulative Grade Point Average (CGPA) of not less than 2.00 and satisfactory complete all current curriculum requirements or the equivalent. Students who complete their work in any program with a CGPA of 3.50 or higher are graduated with honors and are so recognized at the graduation ceremony.

■ STUDENT PROGRAMS

PROFESSIONAL AND HONORARY SOCIETIES
[For key to acronyms, see *Introduction*.]
Professional Societies: DPMA, IEEE
Honorary Societies: Tau Alpha Pi

SUPPORT PROGRAMS
Student Chapter Organizations: National Society of Black Engineers, Society of Hispanic Professional Engineers

Academic Support Center
For: Women & Ethnic Minorities **Available:** Year round
Offered: Freshman, Sophomore, Junior, Senior

Learning Resource Center
For: Women & Ethnic Minorities **Available:** Year round
Offered: Freshman, Sophomore, Junior, Senior, Graduate level
Other Engineering Support Programs: New students take a Student Orientation course for credit to get them acclimated to the college environment. Other support services include: Focus English; career counseling services; graduate placement services; tutorial services; housing/part-time employment services; student advisor services; technical review hour.

302 DeVry Institute of Technology, Phoenix

INSTITUTION PROFILE
HEAD OF THE INSTITUTION
James A. Dugan
Phone: (602)870-9222   Fax: (602)870-1209

GENERAL INFORMATION
[All Students—Fall 1991]
Undergraduate enrollment 2,719
Total institution enrollment 2,719
Type of institution: Private   **Calendar:** Semesters
Location: Phoenix   **Population:** 983,403
Setting: Suburban
Types of engineering degrees: Engineering Technology
Other degree-granting colleges: Accounting, Business Operations, Computer Information Systems, Electronics Technician (AAS)

ENGINEERING TECHNOLOGY ADMISSIONS
ADMISSIONS OFFICE CONTACT
Kim Galetti
DeVry Institute of Technology, Phoenix
2149 West Dunlap Avenue
Phoenix, AZ 85021-2995
Phone: (602)870-9201   Fax: (602)870-1209

TECHNOLOGY COLLEGE ADMISSIONS INFORMATION
Admission to the engineering technology college: A student applies to the program after meeting with a DeVry representative during senior year of high school or after completing high school or equivalent.
Entrance Requirements: Entrance examinations in math and algebra must be taken if the applicant has not taken a recommended standardized test, or has failed to achieve the required score on the standardized test.
Entrance Recommendations: SAT (Score: 480), ACT (Score: 20), High School courses—Algebra/Advanced Algebra (1 year), Trigonometry (1 year), Calculus (1 year), Physics (1 year), WPCT may also be taken. A score of 20 on the math section of the ACT or 480 on the math section of the SAT or WPCT is required.
Requirements for foreign students: TOEFL (Score: 500); Provide a certified copy of the applicant's high school (or college, if applicable) transcript (in English); a notarized statement of financial support or certified government sponsor letter; proof of English language proficiency. A Form I-20AB will be provided to the applicant.

DEPARTMENTAL ADMISSIONS INFORMATION
Admission to the engineering technology department: At the time of admission to the institution.
Additional information: Prospective students who wish to complete an application must either visit the Institute and meet with a DeVry admissions representative, or meet with a DeVry field representative assigned to the student's geographical area.

FINANCIAL INFORMATION
ESTIMATED EXPENSES (ACADEMIC YEAR)
[Expenses are for the 1992-93 nine-month academic year.]

All Students	Undergraduate	Graduate
Tuition and fees	$ 5,249	$ _(1)
College room and board	$ 4,107	$ _(1)
Books and supplies	$ 500	$ _(1)
Other expenses	$ 4,267 (2)	$ _(1)
Total estimated expenses	**$14,123**	**$ —**

Notes: (1) No graduate degree offered (2) Transportation $2,060 Personal/Miscellaneous $1,911

FINANCIAL AID OFFICE CONTACT
Scott Morrison, *Director of Financial Aid*
DeVry Institute of Technology, Phoenix
2149 West Dunlap Avenue
Phoenix, AZ 85021-2995
Phone: (602)870-9222   Fax: (602)870-1209

GENERAL FINANCIAL AID INFORMATION
Forms accepted/required: AFSA, CSS-FAF, FAF, FAT, FSA, FFS, Stafford, IRS, SAR, Supplemental, Institutional
Additional financial aid information: Pell Grants; Supplement Educational Opportunity Grants; Perkins Loans; College Work-Study; Guaranteed Student Loan Programs: Stafford, SLS, PLUS; EDUCARD Plan-monthly payment plan agreement; DeVry Scholarship program. Preferred method of application: College Scholarship Service's Financial Aid Form.

TECHNOLOGY COLLEGE INFORMATION
[For additional personnel, refer to the *Appendix*.]
HEAD OF ENGINEERING TECHNOLOGY
Michael Thomas
Phone: (602)870-9222   Fax: (602)870-1209

ENGINEERING TECHNOLOGY COLLEGE ADDRESS
Electronics Engineering Technology
DeVry Institute of Technology, Phoenix
2149 West Dunlap Avenue
Phoenix, AZ 85021-2995
Phone: (602)870-9222   Fax: (602)870-1209

TECHNOLOGY ENROLLMENTS—BY CLASS
[Numbers are technology baccalaureate enrollments for the fall 1991 term, unless otherwise footnoted.]
1st-year students/Freshmen 381
2nd-year students/Sophomores 204
3rd-year students/Juniors 99
4th-year students/Seniors 170
Total 854

NUMBER OF DEGREES AWARDED—BY PROGRAM
[Numbers are engineering technology baccalaureate degrees awarded during the 1991-1992 academic year, unless otherwise footnoted. For full details about each engineering technology program, refer to the *Table of Degree Programs*.]
Electronic Engineering Technology 183
Electronic(s) Engineering Technology .... –
Total 183

PERCENTAGE OF DEGREES AWARDED—BY CATEGORY
[Percentages are of all engineering technology baccalaureate degrees awarded during the 1991-1992 academic year, unless otherwise footnoted.]
African Americans .... 2.2 %
Asian/Pacific Island Americans .... 14.2 %
Hispanic Americans .... 4.9 %
Native Americans .... – %
Foreign Citizens .... 7.7 %
All Others .... 71.0 %
Women .... 2.2 %
Persons with disabilities .... – % (A)
Students over 25 years of age .... 48.1 %

Notes: (A) Data not available.

FULL- & PART-TIME FACULTY—BY DEPARTMENT
[Figures are the head count of full-time faculty and the full-time equivalent (FTE) of part-time faculty for each engineering technology department or equivalent.]

Department	Full	Part
Electronics Eng Tech	14	1.5

TECHNOLOGY COLLEGE DESCRIPTION
The Institute's baccalaureate degree program, Electronics Engineering Technology, offers the capability to refine and produce a design idea - to make the design concept into a real product. All of the skills and knowledge acquired throughout the program are brought together in a senior project, as students gain valuable industry experience by actually designing and building a microprocessor-based system. Laboratories are available at scheduled hours during the school day, evenings, and Saturdays. Electronics laboratory facilities include student work spaces, equipped with oscilloscopes, signal generators, multimeters and power supplies. These workstations are utilized for basic experiments in electronic circuits. Advanced laboratories are equipped to complement course work in digital circuits, digital computers, microprocessors, communication systems, industrial electronics and control systems. A physics laboratory offers an additional range of equipment.

TECHNOLOGY DEGREE PROGRAMS DESCRIPTION
The Institute's baccalaureate degree program, Electronics Engineering Technology, is designed to meet the need for technical specialists who can aggressively confront and solve complex problems in areas as diverse as space communications systems, local area networks and artificial intelligence.

TRANSFER INFORMATION
Residency Requirements: Candidates for a degree who have transferred from a non-DeVry school must complete more than thirty-five percent of the required credit hours in a program at a DeVry Institute to successfully meet the program requirements.

Transfer without Articulation Agreements
Admission to engineering technology: At any time
Requirements: Entrance requirements are the same for transfer students as they are for new students, unless the applicant possesses an accredited degree which waives the entrance requirements.

GRADUATION REQUIREMENTS
To graduate from any program, a student must maintain a Cumulative Grade Point Average (CGPA) of not less than 2.00 and satisfactorily complete all current curriculum requirements or the equivalent. Students who complete their work in any program with a CGPA of 3.50 or higher are graduated with honors and are so recognized at the graduation ceremony.

STUDENT PROGRAMS
PROFESSIONAL AND HONORARY SOCIETIES
[For key to acronyms, see *Introduction*.]
Professional Societies: DPMA, IEEE
Honorary Societies: Tau Alpha Pi

SUPPORT PROGRAMS
Student Chapter Organizations: Society of Hispanic Professional Engineers, Arizona Council of Black Engineers, Women in Business and Technology

Academic Support Center
For: Women & Ethnic Minorities   **Available:** Year round
Offered: Freshman, Sophomore, Junior, Senior

Learning Resource Center
For: Women & Ethnic Minorities   **Available:** Year round
Offered: Freshman, Sophomore, Junior, Senior

Other Engineering Support Programs: New students take a Student Orientation course for credit to get them acclimated to the college environment. Other support services include: English prep and reading improvement classes; academic counseling; career counseling services; graduate placement services; tutorial services; housing/part-time employment services.

303 University of the District of Columbia

■ INSTITUTION PROFILE

HEAD OF THE INSTITUTION
Tilden J. LeMelle
Phone: (202)282-7550 Fax: (202)282-3681

GENERAL INFORMATION
[All Students—Fall 1991]
Undergraduate enrollment 10,380
Graduate enrollment 710
Total institution enrollment **11,990**
Type of institution: Public **Calendar:** Semesters
Location: Washington, DC **Population:** 638,333
Setting: Urban
Types of engineering degrees: Engineering Technology
Other degree-granting colleges: Business Administration, Education, Life Sciences, Liberal and Fine Arts, University College, Physical Sciences & Technology

■ ENGINEERING TECHNOLOGY ADMISSIONS

ADMISSIONS OFFICE CONTACT
Sandra B. Dolphin
University of the District of Columbia
4200 Connecticut Ave. N.W.
Washington, DC 20008
Phone: (202)282-7300 Fax: (202)282-3679

TECHNOLOGY COLLEGE ADMISSIONS INFORMATION
Admission to the engineering technology college: Upon completion of specific requirements to exit University College, students move into the academic department subject to the major program requirements.
Requirements for foreign students: TOEFL (Score: 500); financial statement

DEPARTMENTAL ADMISSIONS INFORMATION
Admission to the engineering technology department: Completion of specific University College courses is required first.
Additional information: This is a "2+2" program.

■ FINANCIAL INFORMATION

ESTIMATED EXPENSES (ACADEMIC YEAR)
[Expenses are for the 1992-93 nine-month academic year.]

State Residents	Undergraduate	Graduate
Tuition and fees	$ 400	$ 760
College room and board	$ —	$ —
Books and supplies	$ —(A)	$ —(A)
Other expenses	$ —(A)	$ —(A)
Total estimated expenses	**$ 400**	**$ 760**

Out-of-State Residents	Undergraduate	Graduate
Tuition and fees	$ 1,480	$ 1,480
College room and board	$ —	$ —
Books and supplies	$ —(A)	$ —(A)
Other expenses	$ —(A)	$ —(A)
Total estimated expenses	**$ 1,480**	**$ 1,480**

Notes: (A) Data not available.

■ FINANCIAL AID OFFICE CONTACT
Kenneth Howard, *Director of Financial Aid*
University of the District of Columbia
4200 Connecticut Ave N. W.
Washington, DC 20008
Phone: (202)282-3239 Fax: (202)282-3344

GENERAL FINANCIAL AID INFORMATION
Forms accepted/required: AFSA, CSS-FAF, FAF, FAT, FAR, FSA, FFS, Stafford, IRS, SAR, SDF, Supplemental, Plus Loan, Institutional Grants
Additional financial aid information: Grants, loans, College work-study, student employment, job location and development, scholarships

■ TECHNOLOGY COLLEGE INFORMATION
[For additional personnel, refer to the *Appendix*.]

HEAD OF ENGINEERING TECHNOLOGY
Philip L. Brach
Phone: (202)282-7427 Fax: (202)282-3677

ENGINEERING TECHNOLOGY COLLEGE ADDRESS
College of Physical Science, Engineering & Technology
University of the District of Columbia
4200 Connecticut Ave. N.W.
Washington, DC 20008
Phone: (202)282-7300 Fax: (202)282-3677

TECHNOLOGY ENROLLMENTS—BY CLASS
[Numbers are technology baccalaureate enrollments for the fall 1991 term, unless otherwise footnoted.]
1st-year students/Freshmen .... 188 [1]
2nd-year students/Sophomores .... 18 [2]
3rd-year students/Juniors .... 5 [3]
4th-year students/Seniors .... 11 [4]
Total **222**

Notes: (1) 31 full-time students, 157 part-time students. (2) 18 full-time, 170 part-time. (3) 5 full-time, 41 part-time. (4) 11 full-time, 30 part-time.

NUMBER OF DEGREES AWARDED—BY PROGRAM
[Numbers are engineering technology baccalaureate degrees awarded during the 1991-1992 academic year, unless otherwise footnoted. For full details about each engineering technology program, refer to the *Table of Degree Programs*.]
Construction Engineering Technology 10
Electromechanical Systems Engineering Technology 6
Total **16**

PERCENTAGE OF DEGREES AWARDED—BY CATEGORY
[Percentages are of all engineering technology baccalaureate degrees awarded during the 1991-1992 academic year, unless otherwise footnoted.]
African Americans .... — % (A)
Asian/Pacific Island Americans .... — % (A)
Hispanic Americans .... — % (A)
Native Americans .... — % (A)
Foreign Citizens .... — % (A)
All Others .... 100 %
Women .... — % (A)
Persons with disabilities .... — % (A)
Students over 25 years of age .... — % (A)

Notes: (A) Data not available.

■ FULL- & PART-TIME FACULTY—BY DEPARTMENT
[Figures are the head count of full-time faculty and the full-time equivalent (FTE) of part-time faculty for each engineering technology department or equivalent.]

Department	Full	Part
Architectural & Civil Eng Tech	7	2.0
Electrical & Mechanical Eng Tech	8	—

TECHNOLOGY COLLEGE DESCRIPTION
Integration of high tech instrumentation in classroom and laboratory instructional methodology. High quality of laboratories. Benefits of membership in Consortium of Universities in the Washington Metropolitan area.

TRANSFER INFORMATION
Residency Requirements: The last 30 semester credit hours must be completed at UDC.

GRADUATION REQUIREMENTS
A grade of "c" or better is required for all departmental courses, in all required math courses and in all technical electives.

■ STUDENT PROGRAMS

PROFESSIONAL AND HONORARY SOCIETIES
[For key to acronyms, see *Introduction*.]
Professional Societies: AIA, IEEE
Honorary Societies: Tau Alpha Pi

SUPPORT PROGRAMS
Other Engineering Support Programs: The University College provides support services such as advising, instructional support, diagnostic testing research and evaluation and special services for the disabled. New student orientation identifies available resources and explains regulations governing student matriculation. The University College administers the Honors Program and English as a Second Language.

304 East Tennessee State University

INSTITUTION PROFILE
HEAD OF THE INSTITUTION
Roy S. Nicks
Phone: (615)929-4211 **Fax:** (615)929-4004
GENERAL INFORMATION
[All Students—Fall 1991]
Undergraduate enrollment 9,992
Graduate enrollment 1,719
Total institution enrollment 11,711
Type of institution: Public **Calendar:** Semesters
Nearest city: Knoxville, Tennessee **Population:** 320,000
Miles from main campus: 100 **Setting:** Urban
Types of engineering degrees: Engineering Technology
Other degree-granting colleges: Allied Health Sciences, Arts & Sciences, Business Administration, Education, Medicine, Nursing, Applied Science and Technology

ENGINEERING TECHNOLOGY ADMISSIONS
ADMISSIONS OFFICE CONTACT
Nancy L. Dishner
East Tennessee State University
P.O. Box 70,733
106 Dossett Hall
Johnson City, TN 37614-0733
Phone: (615)929-4213 **Fax:** (615)929-5770
TECHNOLOGY COLLEGE ADMISSIONS INFORMATION
Entrance Requirements: SAT (Score: 720), ACT (Score: 19), High School courses—English (4 years), Algebra I&II, Natural/Physical Science--(each) (2 years), Geometry, Social Studies, US History--(each) (1 year), Single Foreign Language (2 years), HS-GPA 2.3 or higher on 4.0 scale & ACT score 19 or higher or SAT score 720. Regulated admission: HS GPA 2.0-2.3-ACT score 17 or 18. Can take no more than 12 sem. hrs. during first semester. Adult over 21 required to take the AAPP test.
Entrance Recommendations: SAT (Score: 720), ACT (Score: 19), High School courses—Visual or Performing Arts (required) (1 year), ACT or Sat (under 21 years of age.) Over 21 and not taking ACT or SAT must complete the AAPP test battery.
Requirements for foreign students: TOEFL (Score: 500); financial statement; Academic standards in the home country must be met first.
DEPARTMENTAL ADMISSIONS INFORMATION
Admission to the engineering technology department: At the time of admission to the institution.

FINANCIAL INFORMATION
ESTIMATED EXPENSES (ACADEMIC YEAR)
[Expenses are for the 1992-93 nine-month academic year.]

State Residents	Undergraduate	Graduate
Tuition and fees	$ 1,382	$ 1,884
College room and board	$ 1,260	$ 1,350
Books and supplies	$ 400	$ 400
Other expenses	$ 300	$ 300
Total estimated expenses	**$ 3,342**	**$ 3,934**

Out-of-State Residents	Undergraduate	Graduate
Tuition and fees	$ 3,204	$ 3,364
College room and board	$ 1,260	$ 1,350
Books and supplies	$ 400	$ 400
Other expenses	$ 300	$ 300
Total estimated expenses	**$ 5,164**	**$ 5,414**

FINANCIAL AID OFFICE CONTACT
Linda J. Clemons, *Director, Office of Financial Aid*
East Tennessee State University
P.O. Box 70,722, ETSU
Johnson City, TN 37614-0722
Phone: (615)929-4313 **Fax:** (615)929-5770
GENERAL FINANCIAL AID INFORMATION
Forms accepted/required: ACT, AFSA, FAT, FSA, FFS, Stafford, IRS, SAR, Institutional
Additional financial aid information: PELL, SEOG, NDSL, GSL, SLS, Student work programs- (CWSP, RSWP, WSP), TSAC Awards, Voc. Rehab., Private Awards, Performing Grants (Athletic/Academic), Graduate Assistantships, ROTC Scholarships, Minority Scholarships

TECHNOLOGY COLLEGE INFORMATION
[For additional personnel, refer to the *Appendix*.]
HEAD OF ENGINEERING TECHNOLOGY
James A. Hales
Phone: (615)929-4465 **Fax:** (615)929-6435
EMail: A31HALES@ETSU
ENGINEERING TECHNOLOGY COLLEGE ADDRESS
Department of Technology
East Tennessee State University
P.O. Box 70,552
Dossett Drive
Johnson City, TN 37614-0552
Phone: (615)929-4310 **Fax:** (615)929-6435
TECHNOLOGY ENROLLMENTS—BY CLASS
[Numbers are technology baccalaureate enrollments for the fall 1991 term, unless otherwise footnoted.]
1st-year students/Freshmen 140
2nd-year students/Sophomores 100
3rd-year students/Juniors 83
4th-year students/Seniors 140
Total **463**
NUMBER OF DEGREES AWARDED—BY PROGRAM
[Numbers are engineering technology baccalaureate degrees awarded during the 1991-1992 academic year, unless otherwise footnoted. For full details about each engineering technology program, refer to the *Table of Degree Programs*.]
Biomedical Engineering Technology .... –
Construction Engineering Technology 18
Electronics Engineering 27
Engineering Design Graphics 17
Industrial Technology 13
Manufacturing Engineering Technology 12
Surveying & Mapping 2
Total **89**
PERCENTAGE OF DEGREES AWARDED—BY CATEGORY
[Percentages are of all engineering technology baccalaureate degrees awarded during the 1991-1992 academic year, unless otherwise footnoted.]
African Americans .... 2.2 %
Asian/Pacific Island Americans .... – %
Hispanic Americans .... – %
Native Americans .... – %
Foreign Citizens .... – %
All Others .... 97.8 %
Women .... 8.9 %
Persons with disabilities .... – %
Students over 25 years of age .... 20.0 %
GRADUATE ENROLLMENTS & DEGREES AWARDED
Master's enrollment 48
Master's degrees awarded 9

Doctoral enrollment .... –
Doctoral degrees awarded .... –
FULL- & PART-TIME FACULTY—BY DEPARTMENT
[Figures are the head count of full-time faculty and the full-time equivalent (FTE) of part-time faculty for each engineering technology department or equivalent.]

Department	Full	Part
Electronics Eng Tech	3	–
Manufacturing Eng Tech	3	–
Biomedical Eng Tech	2	–
Construction Eng Tech	3	–
Eng Design Graphics Tech	2	–
Industrial Tech	1	–
Surveying & Mapping	2	–

TECHNOLOGY COLLEGE DESCRIPTION
The department of technology is a member of the IBM-CIM in Higher Education Alliance providing extensive computer and related laboratory equipment throughout the department. Co-op programs are available for all program options and usually add one year to the degree program. A student/faculty exchange program is available with sister institutions in Shanghai and Beijing, China.
TECHNOLOGY DEGREE PROGRAMS DESCRIPTION
The B.S. degree program has 15 well equipped laboratories available to support the application aspects of course requirements. The department has a partnership with the IBM Corporation, Siemans Industrial Automation Inc., and the Lucky Computer Company which has resulted in a high level of computerization including an IBM 9370, and dozens of 386 and 486 PC's and PLC's. The department faculty are connected to the campus token ring network. Approximately 80% of the faculty hold doctoral degrees. The graduate program is designed to be interdisciplinary allowing students to utilize courses from any other graduate department that are relative to their professional program of study. Cooperative education opportunities exist for both undergraduate and graduate students.
TRANSFER INFORMATION
Residency Requirements: Minimum of 35 semester hours including last full semester.
Transfer via Articulation Agreements
Admission to engineering technology: At any time
Technology graduates transferred from ... 4-yr: 6 2-yr: 43
GRADUATION REQUIREMENTS
Same as University requirements

STUDENT PROGRAMS
PROFESSIONAL AND HONORARY SOCIETIES
[For key to acronyms, see *Introduction*.]
Professional Societies: ACM, ACSM, AGCA, IEEE, IIE, NAHB, SME
Honorary Societies: Epsilon Pi Tau
SUPPORT PROGRAMS
Student Chapter Organizations: Black Affairs Association
Other Engineering Support Programs: University offers: Four summer freshmen/transfer orientation programs during year; department of technology is responsible for freshmen/transfer orientations June through August; and adult orientations, fall, spring, summer. Department conducts a career awareness program in summer designed to reduce the effects of sex role stereotyping in engineering and technology for young women.

305 Embry-Riddle Aeronautical University - Daytona Beach

■ INSTITUTION PROFILE

HEAD OF THE INSTITUTION
Steven Silwa
Phone: (905)226-6200   Fax: (904)226-6299

GENERAL INFORMATION
[All Students—Fall 1991]
Undergraduate enrollment 4,341
Graduate enrollment 167
Total institution enrollment 4,508
Type of institution: Private   Calendar: Semesters
Location: Daytona Beach   Population: 370,712
Setting: Urban
Types of engineering degrees: Engineering & Technology

■ ENGINEERING TECHNOLOGY ADMISSIONS

ADMISSIONS OFFICE CONTACT
Darryl W. Niemeyer
Embry-Riddle Aeronautical University - Daytona Beach
600 S. Clyde Morris Blvd.
Daytona Regional Airport
Daytona Beach, FL 32114-3900
Phone: (904)226-6100   Fax: (904)226-6299

TECHNOLOGY COLLEGE ADMISSIONS INFORMATION
Entrance Requirements: SAT (Score: 1000), ACT (Score: 24), High School courses—English (4 years), Laboratory (3 years), Sciences (3 years)
Entrance Recommendations: High School courses—Chemistry (1 year), Physics (1 year), Pre-calculus (1 year)
Requirements for foreign students: TOEFL (Score: 500)

DEPARTMENTAL ADMISSIONS INFORMATION
Admission to the engineering technology department: At the time of admission to the institution.

■ FINANCIAL INFORMATION

ESTIMATED EXPENSES (ACADEMIC YEAR)
[Expenses are for the 1992-93 nine-month academic year.]

All Students	Undergraduate	Graduate
Tuition and fees	$ 6,200	$ 7,800
College room and board	$ 3,550	$ —[1]
Books and supplies	$ 800	$ 1,000
Other expenses	$ 750[2]	$ 750[2]
Total estimated expenses	$11,300	$ 9,550

Notes: (1) Not available on campus for graduate students. (2) Program and other expenses does not include transportation and other personal expenses.

FINANCIAL AID OFFICE CONTACT
Phillip C. Ledbetter, *Director of Financial Aid*
Embry-Riddle Aeronautical University - Daytona Beach
600 S. Clyde Morris Blvd.
Daytona Beach Regional Airport
Daytona Beach, FL 32114-3900
Phone: (904)226-6195

GENERAL FINANCIAL AID INFORMATION
Forms accepted/required: ACT, AFSA, CSS-FAF, FAT, FAR, FFS, Stafford, IRS, SAR, Supplemental, Institutional
Additional financial aid information: Need based scholarships, short-term loans, merit-based scholarships, part-time jobs on campus, college work study program.

■ TECHNOLOGY COLLEGE INFORMATION

[For additional personnel, refer to the *Appendix*.]

HEAD OF ENGINEERING TECHNOLOGY
William Martin
Phone: (904)226-6821   Fax: (904)226-6011

ENGINEERING TECHNOLOGY COLLEGE ADDRESS
Engineering Technology Department
Embry-Riddle Aeronautical University - Daytona Beach
600 S. Clyde Morris Blvd.
Daytona Beach Regional Airport
Daytona Beach, FL 32114-3900
Phone: (904)226-6000   Fax: (904)226-6223

TECHNOLOGY ENROLLMENTS—BY CLASS
[Numbers are technology baccalaureate enrollments for the fall 1991 term, unless otherwise footnoted.]
1st-year students/Freshmen 98
2nd-year students/Sophomores 49
3rd-year students/Juniors 47
4th-year students/Seniors 91
Total 285

NUMBER OF DEGREES AWARDED—BY PROGRAM
[Numbers are engineering technology baccalaureate degrees awarded during the 1991-1992 academic year, unless otherwise footnoted. For full details about each engineering technology program, refer to the *Table of Degree Programs*.]
Aircraft Engineering Technology 35
Avionics Engineering Tech/Avionics Technology 23
Total 58

PERCENTAGE OF DEGREES AWARDED—BY CATEGORY
[Percentages are of all engineering technology baccalaureate degrees awarded during the 1991-1992 academic year, unless otherwise footnoted.]
African Americans 4.4%
Asian/Pacific Island Americans 11.5%
Hispanic Americans 12.7%
Native Americans .... — %
Foreign Citizens .... — % (A)
All Others 71.4%
Women 18.9%
Persons with disabilities .... — % (A)
Students over 25 years of age .... — % (A)

Notes: (A) Data not available.

GRADUATE ENROLLMENTS & DEGREES AWARDED
Master's enrollment 36
Master's degrees awarded 1
Doctoral enrollment .... — (A)
Doctoral degrees awarded .... — (A)

Notes: (A) Data not available.

ENGINEERING TECHNOLOGY STUDENT DATA
Applicants to the engineering technology college
Number of applicants to engineering technology college ... 117
Percent offered admission 71.2%

Matriculated engineering technology students
Percentage in top quartile (25%) of High School class 11.5
Average SAT scores: Math—527, Verbal—460, Combined—987
Average ACT scores: Math—23, Composite—23

FULL- & PART-TIME FACULTY—BY DEPARTMENT
[Figures are the head count of full-time faculty and the full-time equivalent (FTE) of part-time faculty for each engineering technology department or equivalent.]

Department	Full	Part
Civil Eng	4	—
Electrical Eng	4	—
Mechanical Eng	6	—
Environmental Eng Tech	2.5	—
Computer Science & Computer Eng	3	—
Architecture	5	—

TECHNOLOGY COLLEGE DESCRIPTION
Aviation/Aerospace is emphasized. Programs provide extensive hands-on laboratory experiences including work in wind tunnel, non-destructive testing (NDT), electronics/avionics laboratories, and an FAA-approved repair station. This department also offers a BS program in Aviation Technology which allows a student to combine general education courses with two out of three options. Those options are Flight Technology, Avionics Technology, and Aviation Maintenance Technology.

TECHNOLOGY DEGREE PROGRAMS DESCRIPTION
AIRCRAFT ENGINEERING TECHNOLOGY: Embry-Riddle offers the Bachelor of Science degree in Aircraft Engineering Technology at the Daytona Beach Campus. The ACET program is designed to provide the student with a solid foundation in math and the natural sciences as well as a broad exposure to technology courses that address the application of scientific and engineering principles. The program provides a strong background in such areas as Applied Aerodynamics, Structural and Systems Analysis, Aircraft Performance and Design as well as Quality Assurance, Testing and other disciplines that are necessary for wide variety of careers in the Aviation Industry AVIONICS ENGINEERING TECHNOLOGY: The Avionics Engineering Technology program prepares individuals for challenging careers in aviation high technology as avionic technologists and technicians. The program provide an understanding of electronics theory, avionics system design analysis, logistic support and strong foundation in general education. The University also offers a program in Avionics Technology that leads to an Associate in Science degree. Avionics Technology that leads to an Associate in Science degree. Avionics Technology may also be selected as an area of concentration in various degree programs.

GRADUATION REQUIREMENTS
Bachelor of Aircraft Engineering Technology program requires 130 semester credit hours, 2.0 CGPA and 2.0 GPA in engineering technology courses. Bachelor of Avionics Engineering Technology program requires 129 semester credit hours, 2.0 CGPA and 2.0 GPA in engineering technology courses. Associate of avionics Technology program requires 78 semester credit hours, 2.0 CGPA and 2.0 GPA in engineering technology courses.

■ STUDENT PROGRAMS

PROFESSIONAL AND HONORARY SOCIETIES
[For key to acronyms, see *Introduction*.]
Professional Societies: IEEE, SAE, Engineering Student Advisory Council
Honorary Societies: Sigma Gamma Tau, Sigma Phi Delta, Tau Alpha Pi, Avionics Engineering Technology Honor Society Norwich treats all people equally.

SUPPORT PROGRAMS

Beta Phi Alpha
For: Women   **Available:** Year round
Offered: Freshman

Future Professional Women in Aviation
For: Women   **Available:** Year round
Offered: Freshman

Kappa Alpha Psi - African American
For: Ethnic Minorities   **Available:** Year round
Offered: Freshman

American Institute of Aeronautics and Astronautics
For: Women & Ethnic Minorities   **Available:** Year round
Offered: Freshman

Other Engineering Support Programs: Orientation program for all freshmen and transfers; career counseling services; academic counseling and tutoring services; international student office; veterans' services; placement services.

306 Ferris State University

INSTITUTION PROFILE
HEAD OF THE INSTITUTION
Helen Popovich
Phone: (616)592-2500

GENERAL INFORMATION
[All Students—Fall 1991]
Undergraduate enrollment 12,100
Graduate enrollment 200
Total institution enrollment 12,300
Type of institution: Public   **Calendar:** Quarters
Nearest city: Grand Rapids   **Population:** 300,000
Miles from main campus: 55   **Setting:** Small Town
Types of engineering degrees: Engineering & Technology
Other degree-granting colleges: Allied Health Sciences, Arts & Sciences, Business Administration, Education, Pharmacy, Technology & Applied Sciences, Optometry

ENGINEERING TECHNOLOGY ADMISSIONS
ADMISSIONS OFFICE CONTACT
Duncan Sargent
Ferris State University
901 S. State Street
Big Rapids, MI 49307-2295
Phone: (616)592-2100

TECHNOLOGY COLLEGE ADMISSIONS INFORMATION
Entrance Requirements: ACT (Score: 20), High School courses—English (3 years), Math (3 years), Physical Science (2 years), Social Science (1 year)
Entrance Recommendations: High School courses—Algebra, Trigonometry, Calculus (4 years), English, Social Science, etc. (4 years), Mechanical Drawing (1 year), Physics, Chemistry (2 years)
Requirements for foreign students: TOEFL (Score: 500); financial statement

DEPARTMENTAL ADMISSIONS INFORMATION
Admission to the engineering technology department: At the end of the second year.
Additional information: Program admission requirements included Mathematics GPA 2.50 (on a 4.00 scale) and a GPA 2.75 in all Freshmen/Sophomore major coursework and a cumulative 2.50 GPA.

FINANCIAL INFORMATION
ESTIMATED EXPENSES (ACADEMIC YEAR)
[Expenses are for the 1992-93 nine-month academic year.]

All Students	Undergraduate	Graduate
Tuition and fees	$ 3,000 (A)	$ –
College room and board	$ 3,700 (A)	$ –
Books and supplies	$ 525 (A)	$ –
Other expenses	$ 700 (A)	$ –
Total estimated expenses	**$ 7,925**	**$ –**

Notes: (A) Data not available.

FINANCIAL AID OFFICE CONTACT
Robert Bopp, *Director*
Ferris State University
901 S. State Street
Prakken Building
Big Rapids, MI 49307
Phone: (616)592-2110

GENERAL FINANCIAL AID INFORMATION
Forms accepted/required: FAF, Institutional
Additional financial aid information: Need-based scholarships and loans (local, state and federal), merit-based scholarships, part-time jobs on campus. College Work Study program.

TECHNOLOGY COLLEGE INFORMATION
[For additional personnel, refer to the *Appendix*.]
HEAD OF ENGINEERING TECHNOLOGY
Joel Galloway
Phone: (616)592-2898   Fax: (616)592-2946

ENGINEERING TECHNOLOGY COLLEGE ADDRESS
College of Technology
Ferris State University
901 S. State Street
Big Rapids, MI 49307
Phone: (616)592-2890   Fax: (616)592-2946

TECHNOLOGY ENROLLMENTS—BY CLASS
[Numbers are technology baccalaureate enrollments for the fall 1991 term, unless otherwise footnoted.]
1st-year students/Freshmen 0
2nd-year students/Sophomores 0
3rd-year students/Juniors 196
4th-year students/Seniors 292
Total 488

NUMBER OF DEGREES AWARDED—BY PROGRAM
[Numbers are engineering technology baccalaureate degrees awarded during the 1991-1992 academic year, unless otherwise footnoted. For full details about each engineering technology program, refer to the *Table of Degree Programs*.]
Electrical/Electronic(s) Engineering Technology .... 18 (A)
HVACR Engineering Technology 13
Manufacturing Engineering Technology 26
Plastics Engineering Technology 48
Product Design Engineering Technology 18
Surveying Engineering 5
Welding Engineering Technology 15
Total 143

Notes: (A) Data not available.

PERCENTAGE OF DEGREES AWARDED—BY CATEGORY
[Percentages are of all engineering technology baccalaureate degrees awarded during the 1991-1992 academic year, unless otherwise footnoted.]
African Americans .... – % (A)
Asian/Pacific Island Americans .... – % (A)
Hispanic Americans .... – % (A)
Native Americans .... – % (A)
Foreign Citizens .... – % (A)
All Others .... 100 %
Women .... – % (A)
Persons with disabilities .... – % (A)
Students over 25 years of age .... – % (A)

Notes: (A) Data not available.

ENGINEERING TECHNOLOGY STUDENT DATA
Applicants to the engineering technology college
Number of applicants to engineering technology college ... 270 (A)
Percent offered admission .... 80.0% (A)

Matriculated engineering technology students
Percentage in top quartile (25%) of High School class ...... – (A)

Notes: (A) Data not available.

FULL- & PART-TIME FACULTY—BY DEPARTMENT
[Figures are the head count of full-time faculty and the full-time equivalent (FTE) of part-time faculty for each engineering technology department or equivalent.]

Department	Full	Part
Construction Department	28	–
Electrical/Electronics Eng Tech	12	–
Manufacturing Eng Technologies	37	–

TECHNOLOGY COLLEGE DESCRIPTION
The University's engineering technology programs are housed in the College of Technology. The curricula are designed around a 2 + 2 laddered concept with the first two years culminating as an exit point with the Associate in Applied Science degree. All engineering technology programs require one ten-week industrial internship. Programs are application oriented with heavy emphasis on computer integration and laboratory experience. Faculty are typically hired from business/industry with appropriate professional license and academic preparation. The College places particular emphasis upon the student, developing them to their fullest potential and assisting them in career placement. Opportunities exist for students and faculty to become involved in applied research and development via the College's Technology Transfer Center. Average number of years required to actually complete the bachelor's degree: 4.1.

TECHNOLOGY DEGREE PROGRAMS DESCRIPTION
The University's four-year degree programs in engineering technology are a blend of theory and application with a strong general education support component. All engineering technology programs that award the BS degree have a junior year level entry point. The College of Technology also awards Associate in Applied Science degrees for several programs in Automotive, Graphic Arts, Construction Technology, Design, Electronics, and Manufacturing Technology.

DUAL DEGREE PROGRAMS
Technology baccalaureate graduates with dual degrees 0
Enrollment requirements: same requirements as engineering technology degree programs

TRANSFER INFORMATION
Residency Requirements: 45 quarter credits minimum (generally senior year)
Transfer via Articulation Agreements
Admission to engineering technology: At the end of the sophomore year
Technology graduates transferred from ... 4-yr: 10   2-yr: 30
Requirements: Transfers are admitted to the technology program as Junior level and application deadline is 2/15 for Fall. Minimum 2.50 cumulative GPA and 2.75 in major.

GRADUATION REQUIREMENTS
Bachelor of Science in Engineering Technology programs require a minimum of 190 quarter credits, depending on major selected. Overall 2.00 GPA required, 2.00 GPA required in major.

STUDENT PROGRAMS
PROFESSIONAL AND HONORARY SOCIETIES
[For key to acronyms, see *Introduction*.]
Professional Societies: ACSM, AGCA, AIA, ASHRAE, AWS, IEEE, ISA, SAE, SME, SPE*, National Association of Home Builders (NAHB)

SUPPORT PROGRAMS
Student Chapter Organizations: Women in Technology

Minority Affairs Office
For: Women & Ethnic Minorities   **Available:** Year round
Offered: Prematriculation, Freshman, Sophomore, Junior, Senior

Women in Technology
For: Women   **Available:** Year round
Offered: Prematriculation, Freshman, Sophomore, Junior, Senior

Counseling Services
For: Women & Ethnic Minorities   **Available:** Year round
Offered: Prematriculation, Freshman, Sophomore, Junior, Senior

Other Engineering Support Programs: Summer and Fall orientation program for all freshmen and transfers; Learning Assessment Program; career counseling services, foreign student office, veterans' services, services for blind and handicapped, placement services, academic advising by faculty, educational counseling by counselors.

307 Franklin University

■ INSTITUTION PROFILE
HEAD OF THE INSTITUTION
Paul J. Otte
Phone: (614)341-6234   Fax: (614)221-7723
GENERAL INFORMATION
[All Students—Fall 1991]
Undergraduate enrollment 3,960
Total institution enrollment **3,960**
Type of institution: Private   **Calendar:** Trimesters
Location: Columbus   **Population:** 620,000
Setting: Urban
Types of engineering degrees: Engineering Technology
Other degree-granting colleges: Arts & Sciences

■ ENGINEERING TECHNOLOGY ADMISSIONS
ADMISSIONS OFFICE CONTACT
Linda M. Steele
Franklin University
201 South Grant Avenue
Columbus, OH 43215-5399
Phone: (614)224-6237   Fax: (614)224-7723
TECHNOLOGY COLLEGE ADMISSIONS INFORMATION
Entrance Requirements: University placement tests in math, English, reading per University requirements.
Entrance Recommendations: ACT, High School courses—Mathematics (4 years), Physics (1 year), Drafting (1 year), ACT may be substituted in-lieu of placement tests.
Requirements for foreign students: TOEFL (Score: 550)
DEPARTMENTAL ADMISSIONS INFORMATION
Admission to the engineering technology department: At the time of admission to the institution.

■ FINANCIAL INFORMATION
ESTIMATED EXPENSES (ACADEMIC YEAR)
[Expenses are for the 1992-93 nine-month academic year.]

All Students	Undergraduate	Graduate
Tuition and fees	$ 5,760	$ –
College room and board	–	$ –
Books and supplies	$ 450	$ –
Other expenses	$ –	$ –
Total estimated expenses	**$ 6,210**	**$ –**

FINANCIAL AID OFFICE CONTACT
Katherine M. Faye, *Director of Financial Aid*
Franklin University
201 South Grant Avenue
Columbus, OH 43215-5399
Phone: (614)224-6237   Fax: (614)221-7723

GENERAL FINANCIAL AID INFORMATION
Forms accepted/required: AFSA, FAF, FAT, Stafford, IRS, SAR, Supplemental, Institutional
Additional financial aid information: Scholarships, college work study, and university part-time employment.

■ TECHNOLOGY COLLEGE INFORMATION
[For additional personnel, refer to the *Appendix*.]
HEAD OF ENGINEERING TECHNOLOGY
Gena R. Proulx
Phone: (614)341-6351   Fax: (614)221-7723
ENGINEERING TECHNOLOGY COLLEGE ADDRESS
College of Business and Technology
Franklin University
201 South Grant Avenue
Columbus, OH 43215-5399
Phone: (614)224-6337   Fax: (614)221-7723
TECHNOLOGY ENROLLMENTS—BY CLASS
[Numbers are technology baccalaureate enrollments for the fall 1991 term, unless otherwise footnoted.]
1st-year students/Freshmen .... (A)
2nd-year students/Sophomores .... (A)
3rd-year students/Juniors .... (A)
4th-year students/Seniors .... (A)
Total **0**
Notes: (A) Data not available.
NUMBER OF DEGREES AWARDED—BY PROGRAM
[Numbers are engineering technology baccalaureate degrees awarded during the 1991-1992 academic year, unless otherwise footnoted. For full details about each engineering technology program, refer to the *Table of Degree Programs*.]
Electronics Engineering Technology 25
Mechanical Engineering Technology 23
Total **48**
PERCENTAGE OF DEGREES AWARDED—BY CATEGORY
[Percentages are of all engineering technology baccalaureate degrees awarded during the 1991-1992 academic year, unless otherwise footnoted.]
African Americans .... – % (1)
Asian/Pacific Island Americans .... – % (1)
Hispanic Americans .... – % (1)
Native Americans .... – % (1)
Foreign Citizens .... – % (1)
All Others .... 100 %
Women .... 8.5 %
Persons with disabilities .... – % (1)
Students over 25 years of age .... – % (1)
Notes: (1) Minority enrollment = 16.5%. Specific data not available.

■ FULL- & PART-TIME FACULTY—BY DEPARTMENT
[Figures are the head count of full-time faculty and the full-time equivalent (FTE) of part-time faculty for each engineering technology department or equivalent.]

Department	Full	Part
Electronics Eng Tech	2	3.0
Mechanical Eng Tech	2	3.0

TECHNOLOGY COLLEGE DESCRIPTION
The Engineering Technology programs are designed to prepare students for technical careers in industry and business. Students are provided with a sound theoretical and practical background coupled with the mathematical skills required to understand, develop, and use theories. In both major programs, the overall goal is to provide students with a strong foundation in engineering fundamentals and principles. Supplemental laboratory work experiences enhance classroom theory to prepare graduates for work in such areas as product design and application, estimating, laboratory evaluation and measurement, and system servicing and installation. The Engineering Technology college facilities include well-equipped labs for both programs.

TRANSFER INFORMATION
Residency Requirements: Must earn 16 credit hours of major area courses at the 300 or 400 level for the bachelor's degree. 30 of the last 45 credit hours must be earned at Franklin University.
Transfer without Articulation Agreements
Admission to engineering technology: At any time
GRADUATION REQUIREMENTS
Minimum GPA of 2.0 in major area and minimum GPA of 2.0 in overall coursework. Must complete residence requirements. Must complete payment of tuition and fees.

■ STUDENT PROGRAMS
PROFESSIONAL AND HONORARY SOCIETIES
[For key to acronyms, see *Introduction*.]
Professional Societies: ACM, ASME, IEEE
Honorary Societies: Tau Pi Phi
SUPPORT PROGRAMS
Other Engineering Support Programs: A professional academic advisor supports the faculty teaching in the technology areas. For all new first-year and transfer students, an orientation session is conducted each trimester. On a weekly basis throughout a trimester, seminars are conducted by the Advisement Center Staff on topics related to Math and Speech Anxiety, Stress Management, and Writing Effective Research Papers, and more.

308 University of Hartford

■ INSTITUTION PROFILE
HEAD OF THE INSTITUTION
Humphrey Tonkin
Phone: (203)768-4417 Fax: (203)768-5417

GENERAL INFORMATION
[All Students—Fall 1991]
Undergraduate enrollment 5,604
Graduate enrollment 1,982
Total institution enrollment 7,586
Type of institution: Private **Calendar:** Semesters
Nearest city: Hartford **Population:** 140,000
Miles from main campus: 4 **Setting:** Suburban
Types of engineering degrees: Engineering & Technology
Other degree-granting colleges: Allied Health Sciences, Arts & Sciences, Business Administration, Education, Nursing, Music, Basic Studies, Hartford Art School, Hartford College for Women, Hartt School of Music

■ ENGINEERING TECHNOLOGY ADMISSIONS
ADMISSIONS OFFICE CONTACT
Richard P. Mills Jr.
University of Hartford
200 Bloomfield Avenue
West Hartford, CT 06117
Phone: (800)766-4024 Fax: (203)768-5074

TECHNOLOGY COLLEGE ADMISSIONS INFORMATION
Entrance Requirements: SAT (Score: 800), High School courses—English (4 years), Mathematics (Algebra, Geometry, Trig, recommended) (2.5 years), Laboratory Science (Physics preferred) (1 year), Social Studies (1 year)
Entrance Recommendations: High School courses—Calculus (1 year), Geometry (1 year), Trigonometry (1 year), Electronics/Computers (2 years), English Composition Achievement Test. Math Level I Achievement Exam or Math Level II Achievement Exam.
Requirements for foreign students: TOEFL (Score: 550); There are no additional requirements for foreign students.

DEPARTMENTAL ADMISSIONS INFORMATION
Admission to the engineering technology department: At the time of admission to the institution.

■ FINANCIAL INFORMATION
ESTIMATED EXPENSES (ACADEMIC YEAR)
[Expenses are for the 1992-93 nine-month academic year.]

All Students	Undergraduate	Graduate
Tuition and fees	$13,600	$ –
College room and board	$ 5,922	$ –
Books and supplies	$ 500	$ –
Other expenses	$ 60 [1]	$ –
Total estimated expenses	**$20,082**	**$ –**

Notes: (1) Registration Fee

FINANCIAL AID OFFICE CONTACT
Daniel E. Small, *Director of Student Financial Assistance*
University of Hartford
200 Bloomfield Avenue
West Hartford, CT 06117
Phone: (203)768-4904 Fax: (203)768-4961

GENERAL FINANCIAL AID INFORMATION
Forms accepted/required: FAF
Additional financial aid information: Need based scholarships; short term loans; merit based scholarships; part-time jobs on campus; college work-study programs; standard state and federal support.

■ TECHNOLOGY COLLEGE INFORMATION
[For additional personnel, refer to the *Appendix*.]
HEAD OF ENGINEERING TECHNOLOGY
Alan J. Hadad
Phone: (203)768-4308 Fax: (203)768-5074

ENGINEERING TECHNOLOGY COLLEGE ADDRESS
Ward College of Technology
University of Hartford
200 Bloomfield Avenue
West Hartford, CT 06117
Phone: (203)768-4112 Fax: (203)768-5074

TECHNOLOGY ENROLLMENTS—BY CLASS
[Numbers are technology baccalaureate enrollments for the fall 1991 term, unless otherwise footnoted.]
1st-year students/Freshmen 62
2nd-year students/Sophomores 38
3rd-year students/Juniors 67
4th-year students/Seniors 56
Total 223

NUMBER OF DEGREES AWARDED—BY PROGRAM
[Numbers are engineering technology baccalaureate degrees awarded during the 1991-1992 academic year, unless otherwise footnoted. For full details about each engineering technology program, refer to the *Table of Degree Programs*.]
Architectural Engineering Technology .... – [1]
Computer Engineering Technology .... –
Electronic Engineering Technology 55
Total 55

Notes: (1) AET Program instituted in 1991.

PERCENTAGE OF DEGREES AWARDED—BY CATEGORY
[Percentages are of all engineering technology baccalaureate degrees awarded during the 1991-1992 academic year, unless otherwise footnoted.]
African Americans .... 15.0 %
Asian/Pacific Island Americans .... 2.0 %
Hispanic Americans .... 5.6 %
Native Americans .... – %
Foreign Citizens .... – %
All Others .... 77.4 %
Women .... 7.3 %
Persons with disabilities .... – % (A)
Students over 25 years of age .... – % (A)

Notes: (A) Data not available.

ENGINEERING TECHNOLOGY STUDENT DATA
Applicants to the engineering technology college
Number of applicants to engineering technology college ... 103
Percent offered admission 91.0%
Matriculated engineering technology students
Percentage in top quartile (25%) of High School class .... 30.0 (A)
Average SAT scores: Math—426, Verbal—368, Combined—794

Notes: (A) Data not available.

FULL- & PART-TIME FACULTY—BY DEPARTMENT
[Figures are the head count of full-time faculty and the full-time equivalent (FTE) of part-time faculty for each engineering technology department or equivalent.]

Department	Full	Part
Electronic & Computer Eng Tech	9	1.4
Architectural Eng Tech	2	2.0
Mathematics & Technical Communications	3	1.0

TECHNOLOGY COLLEGE DESCRIPTION
The College offers four-year Architectural and Electronic Engineering Technology programs which are laboratory and studio intensive. Co-op is offered starting in the sophomore year. A four-and-a-half year program is available for those who have the ability to complete the four year program but lack the educational preparation. The College places particular emphasis on seeking out students with academic promise who face cultural, economic, or personal barriers and assisting them in reaching their potential. The College also offers two-year programs in Computer Engineering Technology and Electronic Engineering Technology.

TECHNOLOGY DEGREE PROGRAMS DESCRIPTION
The College's four-year programs leading to either a Bachelor of Science degree in Architectural Engineering Technology or Electronic Engineering Technology stress the application of theoretical concepts in a laboratory or studio environment. The College also awards Associate degrees (two year) in Computer Engineering Technology and Electronic Engineering Technology. All of these programs are structured to provide a proper foundation in mathematics and the basic sciences, while ensuring extensive hands-on laboratory/studio instruction. There is also a strong component of general education with particular emphasis upon communication skills. Both the Electronic Engineering Technology and the Architectural Engineering Technology programs are the only four-year programs in these disciplines offered in Connecticut. The Electronic Engineering Technology programs are accredited by ABET.

TRANSFER INFORMATION
Residency Requirements: The last thirty (30) credits for BS degrees must be taken in residence.
Transfer via Articulation Agreements
Admission to engineering technology: At the end of the sophomore year
Technology graduates transferred from ... 4-yr: 2   2-yr: 14
Transfer without Articulation Agreements
Admission to engineering technology: At any time
Technology graduates transferred from ... 4-yr: 5   2-yr: 4

GRADUATION REQUIREMENTS
BSAET program requires a minimum of 136 credits for graduation.
BSEET program requires a minimum of 127 credits for graduation.
AASCET program requires a minimum of 68 credits for graduation.
AASEET program requires a minimum of 67 credits for graduation.
The last thirty (30) credits of BS programs must be taken in residence. For all programs of study a minimum overall GPA of 2.00 is required for graduation. In addition a GPA of 2.00 is required in the major.

■ STUDENT PROGRAMS
PROFESSIONAL AND HONORARY SOCIETIES
[For key to acronyms, see *Introduction*.]
Professional Societies: IEEE, ISA, American Institute of Architectural Students (AIAS)
Honorary Societies: Tau Alpha Pi, Tau Beta Pi

SUPPORT PROGRAMS
Student Chapter Organizations: Society of Women Engineers, Women of Ward

Women of Ward
For: Women **Available:** Academic year
Offered: Freshman, Sophomore, Junior, Senior
Other Engineering Support Programs: COLLEGE-WIDE: Gateway program, Aetna/Ward Career Ladder program, Open Lab program and Peer Tutor Program, Ward College Student Council. UNIVERSITY-WIDE: Orientation for all incoming students, career counseling services, foreign student office, disabled student services, career placement services, faculty advising system, learning skills center, personal counseling services and health services.

309 Indiana University-Purdue University at Indianapolis

INSTITUTION PROFILE

HEAD OF THE INSTITUTION
Gerald L. Bepko, *Chancellor*
Phone: (317)274-4417   Fax: (317)274-4615

GENERAL INFORMATION
[All Students—Fall 1991]
Undergraduate enrollment 2,341
Graduate enrollment 30
Total institution enrollment **2,371**
Type of institution: Public   Calendar: Semesters
Location: Indianapolis   Population: 1,000,000
Setting: Urban
Types of engineering degrees: Engineering & Technology
Other degree-granting colleges: Allied Health Sciences, Business Administration, Dentistry, Education, Law, Medicine, Nursing, Social Work, Liberal Arts, Science, Physical Education, Public and Environmental Affairs

ENGINEERING TECHNOLOGY ADMISSIONS

ADMISSIONS OFFICE CONTACT
Christine Fitzpatrick
Indiana University-Purdue University at Indianapolis
799 West Michigan Street
Indianapolis, IN 46202-5160
Phone: (317)274-0804   Fax: (317)274-4567

TECHNOLOGY COLLEGE ADMISSIONS INFORMATION
Admission to the engineering technology college: After completion of Chemistry-1 year, Laboratory Science-1 year, History of Social Studies-1 year.
Entrance Requirements: SAT (Score: 750), ACT (Score: 16), High School courses—English (3 years), Algebra (1 year), Geometry (1 year), Laboratory Science (1 year), SAT: Verbal--400, Mathematics--500. ACT: English--21, Mathematics--23.
Entrance Recommendations: High School courses—Algebra (1 year), Trigonometry (1 year), Computers (1 year), Physical Science (1 year)
Requirements for foreign students: TOEFL (Score: 550); financial statement

DEPARTMENTAL ADMISSIONS INFORMATION
Admission to the engineering technology department: At the time of admission to the institution.

FINANCIAL INFORMATION

ESTIMATED EXPENSES (ACADEMIC YEAR)
[Expenses are for the 1992-93 nine-month academic year.]

All Students	Undergraduate	Graduate
Tuition and fees	$ 2,500	$ 2,500
College room and board	$ _(B)	$ _(B)
Books and supplies	$ 1,000	$ 1,000
Other expenses	$ _(B)	$ _(B)
Total estimated expenses	**$ 3,500**	**$ 3,500**

Notes: (B) Data not applicable.

FINANCIAL AID OFFICE CONTACT
Natala Hart
Indiana University-Purdue University at Indianapolis
Office of Scholarships & Financial Aids
425 University Boulevard
Indianapolis, IN 46202-5145
Phone: (317)274-4162

GENERAL FINANCIAL AID INFORMATION
Forms accepted/required: FFS, IRS, Institutional
Additional financial aid information: Need-based scholarships: short-term loans: merit-based scholarships: part-time jobs on campus: College Work-Study Program. Required forms: FFS, IRS, institutional form

TECHNOLOGY COLLEGE INFORMATION
[For additional personnel, refer to the *Appendix*.]

HEAD OF ENGINEERING TECHNOLOGY
R. Bruce Renda
Phone: (317)274-0800   Fax: (317)274-0832

ENGINEERING TECHNOLOGY COLLEGE ADDRESS
School of Engineering and Technology
Indiana University-Purdue University at Indianapolis
799 West Michigan Street
Indianapolis, IN 46202-5160
Phone: (317)274-2533   Fax: (317)274-4567

TECHNOLOGY ENROLLMENTS—BY CLASS
[Numbers are technology baccalaureate enrollments for the fall 1991 term, unless otherwise footnoted.]
1st-year students/Freshmen 155
2nd-year students/Sophomores 897
3rd-year students/Juniors 145
4th-year students/Seniors 306
Total **1,503**

NUMBER OF DEGREES AWARDED—BY PROGRAM
[Numbers are engineering technology baccalaureate degrees awarded during the 1991-1992 academic year, unless otherwise footnoted. For full details about each engineering technology program, refer to the *Table of Degree Programs*.]
Computer Integrated Manufacturing Technology 8
Construction/Architectural Engineering Technology 21
Electrical Engineering Technology 39
Industrial Engineering Technology 3
Mechanical Engineering Technology 37
Total **108**

PERCENTAGE OF DEGREES AWARDED—BY CATEGORY
[Percentages are of all engineering technology baccalaureate degrees awarded during the 1991-1992 academic year, unless otherwise footnoted.]
African Americans .... 6.0 %
Asian/Pacific Island Americans .... 2.0 %
Hispanic Americans .... 0.8 %
Native Americans .... – %
Foreign Citizens .... 0.2 %
All Others .... 91.0 %
Women .... 20.0 %
Persons with disabilities .... – % (B)
Students over 25 years of age .... – % (B)

Notes: (B) Data not applicable.

ENGINEERING TECHNOLOGY STUDENT DATA
Applicants to the engineering technology college
Number of applicants to engineering technology college ... 120
Percent offered admission 80.0%

Matriculated engineering technology students
Percentage in top quartile (25%) of High School class 20.0
Average SAT scores: Math—490, Verbal—416, Combined—906

FULL- & PART-TIME FACULTY—BY DEPARTMENT
[Figures are the head count of full-time faculty and the full-time equivalent (FTE) of part-time faculty for each engineering technology department or equivalent.]

Department	Full	Part
Construction Tech	7	1.6
Electrical Eng Tech	7	1.5
Manufacturing Tech	10	2.8

TECHNOLOGY COLLEGE DESCRIPTION
The school's four-year engineering technology programs follow the traditional 2 + 2 model, located in a very large urban area. The school allows the student to pursue either a full-time or part-time plan of study. An optional cooperative education program is provided although opportunities for full-time or part-time employment while attending school are abundant. The school also offers two-year programs in engineering technology. Average number of years required to actually complete the bachelor's degree: 4.

TECHNOLOGY DEGREE PROGRAMS DESCRIPTION
The school's four-year programs in engineering technology emphasize practical application of theoretical concepts for industrial employment. All engineering technology program graduates earn the B.S. degree. In addition to engineering technology, the school also offers B.S. degree programs in Computer Technology and Supervision.

TRANSFER INFORMATION
Residency Requirements: Students must complete a minimum of 32 semester hours in either the Indiana University or Purdue University system.

Transfer via Articulation Agreements
Admission to engineering technology: At any time
Technology graduates transferred from ... 4-yr: –   2-yr: –

Transfer without Articulation Agreements
Admission to engineering technology: At any time
Technology graduates transferred from ... 4-yr: –   2-yr: –

GRADUATION REQUIREMENTS
All baccalaureate programs require 128-142 semester hours, depending on major selected and a minimum 2.0 grade point average

STUDENT PROGRAMS

PROFESSIONAL AND HONORARY SOCIETIES
[For key to acronyms, see *Introduction*.]
Professional Societies: ACM, AIAA, ASME, IEEE, NSPE, SAE, SME, National Society of Black Engineers (NSBE), Society of Women Engineers (SWE)

SUPPORT PROGRAMS
Student Chapter Organizations: Society of Women Engineers, National Society of Black Engineers, Minority Engineering Advancement Program

Minority Engineering Advancement Program
For: Ethnic Minorities   **Available:** Year round
Offered: Prematriculation, Freshman, Sophomore, Junior, Senior, Graduate level

IUPUI Office of Minority Student Services
For: Ethnic Minorities   **Available:** Year round
Offered: Freshman, Sophomore, Junior, Senior, Graduate level
Other Engineering Support Programs: Summer and Fall orientation program for all freshmen and transfer; English, mathematics, and reading placement testing program; career counseling service; international student office; disabled student services office; placement services.

310 Lake Superior State University

INSTITUTION PROFILE
HEAD OF THE INSTITUTION
Robert Arbuckle
Phone: (906)635-2202 Fax: (906)635-2111

GENERAL INFORMATION
[All Students—Fall 1991]
Undergraduate enrollment 2,515
Total institution enrollment 2,515
Type of institution: Public **Calendar:** Semesters
Location: Sault Ste. Marie **Population:** 14,000
Setting: Small Town
Types of engineering degrees: Engineering Technology

ENGINEERING TECHNOLOGY ADMISSIONS
ADMISSIONS OFFICE CONTACT
Bruce Johnson
Lake Superior State University
1000 University Drive
Sault Ste. Marie, MI 49783
Phone: (906)635-2231 Fax: (906)635-2111

TECHNOLOGY COLLEGE ADMISSIONS INFORMATION
Entrance Requirements: ACT, High School courses—Beginning Algebra (1 year), Advanced Algebra (1 year), Science with a laboratory (1 year)
Entrance Recommendations: High School courses—Trigonometry (1 year), Geometry (1 year)

DEPARTMENTAL ADMISSIONS INFORMATION
Admission to the engineering technology department: At the time of admission to the institution.

FINANCIAL INFORMATION
ESTIMATED EXPENSES (ACADEMIC YEAR)
[Expenses are for the 1992-93 nine-month academic year.]

State Residents	Undergraduate	Graduate
Tuition and fees	$ 2,694	$ –
College room and board	$ 3,855	$ –
Books and supplies	$ 500 (C)	$ –
Other expenses	$ – (1)	$ –
Total estimated expenses	**$ 7,049**	**$ –**

Out-of-State Residents	Undergraduate	Graduate
Tuition and fees	$ 5,250	$ –
College room and board	$ 3,855	$ –
Books and supplies	$ 500 (C)	$ –
Other expenses	$ – (1)	$ –
Total estimated expenses	**$ 9,605**	**$ –**

Notes: (C) Estimated data. (1) $12 Lab Fee for Engr. Tech. courses

FINANCIAL AID OFFICE CONTACT
William Munsell, *Director of Financial Aid*
Lake Superior State University
1000 University Dr.
Sault Ste. Marie, MI 49783
Phone: (906)635-1311 Fax: (906)635-2111

GENERAL FINANCIAL AID INFORMATION
Forms accepted/required: FAF, SAR
Additional financial aid information: LSSU distributes over $5 million to 2/3rds of our students. Scholarships include Michigan competitive, board of regents, distinguished scholars, Phil Hart, and many individual departmental scholarships. Employment includes college work study, which is needs based, and department payroll, for which all students are eligible. Loans include Perkins, Stafford, PLUS, and short term for emergencies. Grants include PELL, Institutional, TIP, and BIA (Native American Tuition waiver).

ENGINEERING TECHNOLOGY COLLEGE ADDRESS
Engineering Technology Department
Lake Superior State University
1000 University Drive
Sault Ste. Marie, MI 49783
Phone: (906)632-6841 Fax: (906)635-2111

TECHNOLOGY ENROLLMENTS—BY CLASS
[Numbers are technology baccalaureate enrollments for the fall 1991 term, unless otherwise footnoted.]
1st-year students/Freshmen 67
2nd-year students/Sophomores 88
3rd-year students/Juniors 59
4th-year students/Seniors 64
Total **278**

NUMBER OF DEGREES AWARDED—BY PROGRAM
[Numbers are engineering technology baccalaureate degrees awarded during the 1991-1992 academic year, unless otherwise footnoted. For full details about each engineering technology program, refer to the *Table of Degree Programs*.]
Automated Systems Engineering Technology 11
Computer Engineering Technology .... –
Drafting & Design Engineering Technology .... –
Electrical/Electronic(s) Engineering Technology 30
Mechanical Engineering Technology 14
Total **55**

PERCENTAGE OF DEGREES AWARDED—BY CATEGORY
[Percentages are of all engineering technology baccalaureate degrees awarded during the 1991-1992 academic year, unless otherwise footnoted.]
African Americans .... – %
Asian/Pacific Island Americans .... – %
Hispanic Americans .... – %
Native Americans .... – %
Foreign Citizens .... – %
All Others .... 100 %
Women .... – %
Persons with disabilities .... – %
Students over 25 years of age .... – %

FULL- & PART-TIME FACULTY—BY DEPARTMENT
[Figures are the head count of full-time faculty and the full-time equivalent (FTE) of part-time faculty for each engineering technology department or equivalent.]

Department	Full	Part
Electrical Eng Tech	3	2.0
Mechanical Eng Tech	7	1.0
Automated Systems Eng Tech	2	1.0

TECHNOLOGY COLLEGE DESCRIPTION
LSSU is the smallest public university in the state of Michigan. Student/faculty ratio of approx. 20:1. All engineering technology programs are TAC/ABET accredited. Only four bachelors degree programs in eng. tech. are TAC/ABET accredited in the state of Michigan, three of these programs are at LSSU. The automated systems engineering technology program is the only TAC/ABET accredited program of its kind in the U.S. Co-op programs are available, but not required, for selected students in the beginning of their junior year.

TECHNOLOGY DEGREE PROGRAMS DESCRIPTION
B.S. Electrical Engineering Technology: Junior sequence consists of digital design, analog design, and full year of network analysis. Senior year has two options: Communications and Industrial Controls/Instrumentation. B.S. Mechanical Engineering Technology: Junior sequence consists of kinematics, dynamics, NC/CNC manufacturing processes, quality control, and fluid mechanics. Senior sequence in both thermodynamics/heat transfer and machine design. B.S. Automated Systems Engineering Technology: Capstone curriculum drawing together mechanical, electrical, computer, and manufacturing technology concepts. Junior sequence based on previous background to create well rounded engineering technologist. Common courses include robotics, control systems, programmable logic controllers, computer control concepts. Senior sequence includes sensors, machine vision, automated manufacturing systems, and quality control. All programs include group senior design projects that draw on the experience gained through previous coursework.

TRANSFER INFORMATION
Residency Requirements: Bachelors degree candidates must complete 32 of their final 40 credits and at least fifty percent of their required departmental 300/400 level credits in courses offered by Lake Superior State University.

Transfer via Articulation Agreements
Admission to engineering technology: At the end of the sophomore year
Technology graduates transferred from ... 4-yr: 4 2-yr: 10
Requirements: 2.0 gradepoint and be eligible to return to their previous college unless they have completed their required coursework. Must provide official high school transcript, satisfactory GED scores, or a satisfactory score on one of the tests approved by the U.S. Dept. of Education to meet the terms of the 'Ability to Benefit' regulation.

Transfer without Articulation Agreements
Admission to engineering technology: At any time
Technology graduates transferred from ... 4-yr: – 2-yr: 10
Requirements: Same as for articulation agreements.

GRADUATION REQUIREMENTS
Students must maintain a 2.00 gradepoint in their major and complete a total of at least 35 credits of general education requirements

STUDENT PROGRAMS
PROFESSIONAL AND HONORARY SOCIETIES
[For key to acronyms, see *Introduction*.]
Professional Societies: ASME, IEEE, SAE, SAME, SME
Honorary Societies: Tau Alpha Pi

SUPPORT PROGRAMS
Student Chapter Organizations: Society of Women Engineers
Other Engineering Support Programs: Weekend programs for freshman orientation and transfer student orientation. Tutoring services available through the honor societies and the counseling center. Counseling and career aptitude testing available through the counseling center.

311 Louisiana Tech University

■ INSTITUTION PROFILE
HEAD OF THE INSTITUTION
Daniel D. Reneau
Phone: (318)257-3785   Fax: (318)257-2928

GENERAL INFORMATION
[All Students—Fall 1991]
Undergraduate enrollment 9,283
Graduate enrollment 1,097
Total institution enrollment 10,380
Type of institution: Public   **Calendar:** Quarters
Nearest city: Shreveport   **Population:** 198,525
Miles from main campus: 72   **Setting:** Small Town
Types of engineering degrees: Engineering & Technology
Other degree-granting colleges: Arts & Sciences, Education, Human Ecology, Life Sciences, Administration and Business

■ ENGINEERING TECHNOLOGY ADMISSIONS
ADMISSIONS OFFICE CONTACT
Karen T. Akin
Louisiana Tech University
Box 3178
Wyly Tower 221
Ruston, LA 71272
Phone: (318)257-3036   Fax: (318)257-2499

TECHNOLOGY COLLEGE ADMISSIONS INFORMATION
Admission to the engineering technology college: Good academic standing and at least a 2.2 GPA based on hours attempted in the Freshman Technology Curriculum.
Entrance Requirements: High School courses—English (4 years), Mathematics (3 years), Social Studies, Science (3 years), Electives (4.5 years), For Fall of 1993, an ACT composite of 22, or a 2.0 GPA on admission courses, or rank in the upper 50% of graduating class is required.
Entrance Recommendations: High School courses—English (4 years), Algebra (2 years), Plane Geometry (1 year), Trigonometry, Chemistry, Physics (1 year)
Requirements for foreign students: TOEFL (Score: 500); financial statement

DEPARTMENTAL ADMISSIONS INFORMATION
Admission to the engineering technology department: Good academic standing and at least a 2.2 GPA based on hours attempted in the Freshmen Technology Curriculum.

■ FINANCIAL INFORMATION
ESTIMATED EXPENSES (ACADEMIC YEAR)
[Expenses are for the 1992-93 nine-month academic year.]

State Residents	Undergraduate	Graduate
Tuition and fees	$ 2,118	$ 2,118
College room and board	$ 2,115	$ 2,115
Books and supplies	$ 600	$ 600
Other expenses	$ 111 [1]	$ 111 [1]
Total estimated expenses	**$ 4,944**	**$ 4,944**

Out-of-State Residents	Undergraduate	Graduate
Tuition and fees	$ 3,273	$ 3,273
College room and board	$ 2,115	$ 2,115
Books and supplies	$ 600	$ 600
Other expenses	$ 111 [1]	$ 111 [1]
Total estimated expenses	**$ 6,099**	**$ 6,099**

Notes: (1) Engineering Fee

FINANCIAL AID OFFICE CONTACT
Etienna R. Winzer, *Director, Division of Financial Aid*
Louisiana Tech University
P.O. Box 7925 T.S.
Ruston, LA 71272
Phone: (318)257-2641

GENERAL FINANCIAL AID INFORMATION
Forms accepted/required: SAR, SDF, Institutional
Additional financial aid information: Need-based scholarships; Pell Grant; Supplemental Educational Opportunity Grant; National Direct Student Loan; College Work-Study; Stafford Loan; State Student Incentive Grant; Louisiana College Tuition Plan are all available.

■ TECHNOLOGY COLLEGE INFORMATION
[For additional personnel, refer to the *Appendix*.]

HEAD OF ENGINEERING TECHNOLOGY
Barry A. Benedict
Phone: (318)257-4647   Fax: (318)257-2562
EMail: benedict@engr.latech.edu

ENGINEERING TECHNOLOGY COLLEGE ADDRESS
College of Engineering
Louisiana Tech University
P.O. Box 10348 T.S.
Ruston, LA 71272
Phone: (318)257-4647   Fax: (318)257-2562

TECHNOLOGY ENROLLMENTS—BY CLASS
[Numbers are technology baccalaureate enrollments for the fall 1991 term, unless otherwise footnoted.]
1st-year students/Freshmen 16
2nd-year students/Sophomores 16
3rd-year students/Juniors 17
4th-year students/Seniors 36
Total 85

NUMBER OF DEGREES AWARDED—BY PROGRAM
[Numbers are engineering technology baccalaureate degrees awarded during the 1991-1992 academic year, unless otherwise footnoted. For full details about each engineering technology program, refer to the *Table of Degree Programs*.]
Construction Engineering Technology 4
Electrical Engineering Technology 17
Total 21

PERCENTAGE OF DEGREES AWARDED—BY CATEGORY
[Percentages are of all engineering technology baccalaureate degrees awarded during the 1991-1992 academic year, unless otherwise footnoted.]
African Americans .... 33.0 %
Asian/Pacific Island Americans .... – %
Hispanic Americans .... – %
Native Americans .... – %
Foreign Citizens .... – %
All Others .... 67.0 %
Women .... 0.1 %
Persons with disabilities .... – %
Students over 25 years of age .... – %

ENGINEERING TECHNOLOGY STUDENT DATA
Applicants to the engineering technology college
Number of applicants to engineering technology college ..... – (A)
Percent offered admission .... –% (A)
Matriculated engineering technology students
Percentage in top quartile (25%) of High School class ...... – (A)
Average ACT scores: Math—19, Composite—19.4
Notes: (A) Data not available.

FULL- & PART-TIME FACULTY—BY DEPARTMENT
[Figures are the head count of full-time faculty and the full-time equivalent (FTE) of part-time faculty for each engineering technology department or equivalent.]

Department	Full	Part
Construction Eng Tech	3	–
Electrical Eng Tech	3	–

TECHNOLOGY COLLEGE DESCRIPTION
The college offers a 4-year undergraduate program in engineering as well as graduate training leading to the M.S., D.E., and Ph.D. All engineering students follow a common freshman curriculum. The college consists of 7 departments and 11 programs (including 2 technology programs). All programs are accredited by ABET and each curriculum meets state general education requirements. The curricula provide a strong foundation in engineering fundamentals with liberal arts studies incorporated throughout the program. Development of communication skills is stressed. Computers are thoroughly integrated throughout the curricula. Average number of years required to complete the B.S. degree: 4.7.

TECHNOLOGY DEGREE PROGRAMS DESCRIPTION
The College of Engineering four-year programs in technology are practice-oriented programs which prepare graduates for industry. Both programs lead to the B.S. degree and are accredited by ABET. The programs are less mathematically oriented than their engineering counterparts and instead, include more training in the area of business and project management.

DUAL DEGREE PROGRAMS
Technology baccalaureate graduates with dual degrees 0

TRANSFER INFORMATION
Residency Requirements: Three-fourths of the hours required for graduation must have been completed in college residence.

Transfer via Articulation Agreements
Admission to engineering technology: At any time
Technology graduates transferred from ... 4-yr: –   2-yr: –
Requirements: Students must have an overall GPA of at least 2.0 out of 4.0 in all transfer courses. Transfer students are subject to the same requirements for admission as all other students.

Transfer without Articulation Agreements
Admission to engineering technology: At any time
Technology graduates transferred from ... 4-yr: –   2-yr: –
Requirements: Students must have an overall GPA of at least 2.0 out of 4.0 in all transfer courses. Transfer students are subject to the same requirements for admission as all other students.

GRADUATION REQUIREMENTS
B.S. program in Construction Engineering Technology requires 131 semester credit hours. B.S. program in Electrical Engineering Technology requires 127 semester credit hours. Overall GPA of 2.0 is required for graduation.

■ STUDENT PROGRAMS
PROFESSIONAL AND HONORARY SOCIETIES
[For key to acronyms, see *Introduction*.]
Professional Societies: ASCE, IEEE, Association of Electrical Engineering Technology, Associated General Contractors (AGC)
Honorary Societies: Tau Alpha Pi

SUPPORT PROGRAMS
Student Chapter Organizations: National Society of Black Engineers

Minority Engineering Program
For: Ethnic Minorities   **Available:** Year round
Offered: Prematriculation, Freshman, Sophomore, Junior, Senior, Graduate level

Minority Scholarship Program
For: Ethnic Minorities   **Available:** Year round
Offered: Prematriculation, Freshman, Sophomore, Junior, Senior, Graduate level

Counseling Center
For: Women & Ethnic Minorities   **Available:** Year round
Offered: Prematriculation, Freshman, Sophomore, Junior, Senior, Graduate level

Career Planning and Placement Center
For: Women & Ethnic Minorities   **Available:** Year round
Offered: Prematriculation, Freshman, Sophomore, Junior, Senior, Graduate level

Other Engineering Support Programs: Five summer orientation sessions for all freshmen: Counseling Center offering career, academic, and personal counseling; Career Planning & Placement Center; International Student Office; Financial Aid Office; Services for Handicapped; and Engineering Scholarship program.

312 University of Maine

■ INSTITUTION PROFILE
HEAD OF THE INSTITUTION
Frederick E. Hutchinson
Phone: (207)581-1512   Fax: (207)581-1517
GENERAL INFORMATION
[All Students—Fall 1991]
Undergraduate enrollment 10,727
Graduate enrollment 2,077
Total institution enrollment 12,804
Type of institution: Public   **Calendar:** Semesters
Nearest city: Bangor   **Population:** 35,000
Miles from main campus: 10   **Setting:** Small Town
Types of engineering degrees: Engineering Technology
Other degree-granting colleges: Business Administration, Education, Nursing, Applied Sciences & Agriculture, Social and Behavioral Sciences, Arts and Humanities, Forest Resources, Sciences

■ ENGINEERING TECHNOLOGY ADMISSIONS
ADMISSIONS OFFICE CONTACT
William B. Munsey
University of Maine
115 Chadbourne Hall
Orono, ME 04469-5713
Phone: (207)581-1561   Fax: (207)581-1213
TECHNOLOGY COLLEGE ADMISSIONS INFORMATION
Entrance Requirements: SAT (Score: 900), High School courses—English (4 years), Mathematics through Algebra II and Trigonometry (4 years), Physics (1 year), History (1 year)
Entrance Recommendations: High School courses—Foreign Language (2 years), Computer Sciences (1 year), Fine Arts (1 year), Advanced Mathematics (1 year)
Requirements for foreign students: TOEFL (Score: 450); financial statement
DEPARTMENTAL ADMISSIONS INFORMATION
Admission to the engineering technology department: At the time of admission to the institution.

■ FINANCIAL INFORMATION
ESTIMATED EXPENSES (ACADEMIC YEAR)
[Expenses are for the 1992-93 nine-month academic year.]

State Residents	Undergraduate	Graduate
Tuition and fees	$ 2,937	$ _(A)
College room and board	$ 4,362	$ _(A)
Books and supplies	$ 400	$ _(A)
Other expenses	$ 546 (1)	$ _(A)
Total estimated expenses	$ 8,245	$ –

Out-of-State Residents	Undergraduate	Graduate
Tuition and fees	$ 8,316	$ _(A)
College room and board	$ 4,362	$ _(A)
Books and supplies	$ 400	$ _(A)
Other expenses	$ 546 (1)	$ _(A)
Total estimated expenses	$13,624	$ –

Notes: (A) Data not available. (1) Includes lab fees and comprehensive fees.

FINANCIAL AID OFFICE CONTACT
Peggy L. Crawford, *Director of Student Aid*
University of Maine
Wingate Hall
Orono, ME 04469-5781
Phone: (207)581-1324   Fax: (207)581-3261
GENERAL FINANCIAL AID INFORMATION
Forms accepted/required: FAF
Additional financial aid information: University and college need based scholarships. Work-study jobs available.

■ TECHNOLOGY COLLEGE INFORMATION
[For additional personnel, refer to the *Appendix*.]
HEAD OF ENGINEERING TECHNOLOGY
John J. McDonough
Phone: (207)581-2341   Fax: (207)581-2113
ENGINEERING TECHNOLOGY COLLEGE ADDRESS
School of Engineering Technology
University of Maine
221 East Annex
Orono, ME 04469-5725
Phone: (207)581-2341   Fax: (207)581-2113
TECHNOLOGY ENROLLMENTS—BY CLASS
[Numbers are technology baccalaureate enrollments for the fall 1991 term, unless otherwise footnoted.]
1st-year students/Freshmen 116
2nd-year students/Sophomores 105
3rd-year students/Juniors 83
4th-year students/Seniors 93
Total 397
NUMBER OF DEGREES AWARDED—BY PROGRAM
[Numbers are engineering technology baccalaureate degrees awarded during the 1991-1992 academic year, unless otherwise footnoted. For full details about each engineering technology program, refer to the *Table of Degree Programs*.]
Construction Management Technology 8
Electrical Engineering Technology 21
Mechanical Engineering Technology 25
Total 54
PERCENTAGE OF DEGREES AWARDED—BY CATEGORY
[Percentages are of all engineering technology baccalaureate degrees awarded during the 1991-1992 academic year, unless otherwise footnoted.]
African Americans .... – %
Asian/Pacific Island Americans .... – %
Hispanic Americans .... – %
Native Americans .... 4.0 %
Foreign Citizens .... – %
All Others .... 96.0 %
Women .... 4.0 %
Persons with disabilities .... – %
Students over 25 years of age .... 27.0 %
ENGINEERING TECHNOLOGY STUDENT DATA
Applicants to the engineering technology college
Number of applicants to engineering technology college ... 130 (A)
Percent offered admission .... 70.0 %

Matriculated engineering technology students
Percentage in top quartile (25%) of High School class 13.0
Average SAT scores: Math—500, Verbal—450, Combined—950
Notes: (A) Data not available.
FULL- & PART-TIME FACULTY—BY DEPARTMENT
[Figures are the head count of full-time faculty and the full-time equivalent (FTE) of part-time faculty for each engineering technology department or equivalent.]

Department	Full	Part
Civil/Construction Management Eng Tech	4	–
Electrical/Eng Tech	4	0.5
Mechanical Eng Tech	6	–

TECHNOLOGY COLLEGE DESCRIPTION
Co-op is available to all upper level students, for degree credit. Depending upon the students schedule, no extra time is required to participate in co-op. The BSMET students senior design project involves design, fabrication, and competition of various vehicles for specific handicapped persons. The winning design is donated to that person.
TECHNOLOGY DEGREE PROGRAMS DESCRIPTION
Although all programs are Bachelor's degree programs, the Associate degree is available upon request. The Associate degree described with the Construction Management Technology program is in Civil Engineering Technology.
TRANSFER INFORMATION
Residency Requirements: One full year must be completed in residence for Bachelor's degree. One semester for Associate's degree.
Transfer without Articulation Agreements
Admission to engineering technology: At the end of the freshman year
Technology graduates transferred from ... 4-yr: 10   2-yr: 5
GRADUATION REQUIREMENTS
2.00 GPA in all major courses. 2.00 GPA in all discipline specific courses. Pass all required courses.

■ STUDENT PROGRAMS
PROFESSIONAL AND HONORARY SOCIETIES
[For key to acronyms, see *Introduction*.]
Professional Societies: ACSM, ASCE, ASHRAE, ASME, IEEE, SME, TAPPI
Honorary Societies: Tau Alpha Pi
SUPPORT PROGRAMS
Student Chapter Organizations: Society of Women Engineers
Other Engineering Support Programs: There are no specific Engineering Technology support programs. All support programs are University based for the entire student population. These include minority, handicapped, etc.

313 Mankato State University

INSTITUTION PROFILE
HEAD OF THE INSTITUTION
Richard R. Rush
Phone: (507)389-1111 Fax: (507)389-5859

GENERAL INFORMATION
[All Students—Fall 1991]
Undergraduate enrollment 13,100
Graduate enrollment 1,675
Total institution enrollment **14,775**
Type of institution: Public Calendar: Quarters
Nearest city: Minneapolis-St. Paul Population: 2,500,000
Miles from main campus: 70 Setting: Urban
Types of engineering degrees: Engineering & Technology
Other degree-granting colleges: Education, Nursing, Arts and Humanities, Health and Human Performance, Natural Sciences, Math and Home Economics, Social and Behavioral Sciences, Business

ENGINEERING TECHNOLOGY ADMISSIONS
ADMISSIONS OFFICE CONTACT
Jack Parkins
Mankato State University
Admissions Office
MSU Box 55
Mankato, MN 56002-8400
Phone: (507)389-1822 Fax: (507)389-1040

TECHNOLOGY COLLEGE ADMISSIONS INFORMATION
Entrance Requirements: ACT (Score: 21)
Entrance Recommendations: High School courses—Mathematics (3 years), Physics and Computer Science (1 year), English (4 years), Industrial Arts (4 years)
Requirements for foreign students: TOEFL (Score: 500); financial statement; If from a non-English speaking country, successful completion of Level 109 in an English Language School; or a Michigan English Proficiency Test Score of 80. Past academic work documents must be officially notarized.

DEPARTMENTAL ADMISSIONS INFORMATION
Admission to the engineering technology department: At the time of admission to the institution.

FINANCIAL INFORMATION
ESTIMATED EXPENSES (ACADEMIC YEAR)
[Expenses are for the 1992-93 nine-month academic year.]

State Residents	Undergraduate	Graduate
Tuition and fees	$ 2,392	$ 1,700[1]
College room and board	$ 2,620	$ 2,620
Books and supplies	$ 400	$ 500
Other expenses	$ –	$ –
Total estimated expenses	**$ 5,412**	**$ 4,820**

Out-of-State Residents	Undergraduate	Graduate
Tuition and fees	$ 4,382	$ 2,370[1]
College room and board	$ 2,620	$ 2,620
Books and supplies	$ 400	$ 500
Other expenses	$ –	$ –
Total estimated expenses	**$ 7,402**	**$ 5,490**

Notes: (1) Based on 8 credits per quarter for three quarters

FINANCIAL AID OFFICE CONTACT
Sandra Loerts, *Director, Student Financial Aid Office*
Mankato State University
109 Wigley Administration Building
MSU Box 37
Mankato, MN 56002-8400
Phone: (507)389-1185 Fax: (507)389-5114

GENERAL FINANCIAL AID INFORMATION
Forms accepted/required: FFS
Additional financial aid information: Allis Foundation scholarships, University scholarships, Pell Grants, Minnesota State Grants, Supplemental Educational Opportunity Grants, Stafford Loans, Perkins Loans, and 27 College of Physics, Engineering and Technology academic scholarships.

TECHNOLOGY COLLEGE INFORMATION
[For additional personnel, refer to the *Appendix*.]
HEAD OF ENGINEERING TECHNOLOGY
Robert J. Herickhoff
Phone: (507)389-1521 Fax: (507)389-1095

ENGINEERING TECHNOLOGY COLLEGE ADDRESS
College of Physics, Engineering and Technology
Mankato State University
Trafton Science Center
MSU Box 3
Mankato, MN 56002-8400
Phone: (507)389-1521 Fax: (507)389-1095

TECHNOLOGY ENROLLMENTS—BY CLASS
[Numbers are technology baccalaureate enrollments for the fall 1991 term, unless otherwise footnoted.]
1st-year students/Freshmen 61
2nd-year students/Sophomores 53
3rd-year students/Juniors 61
4th-year students/Seniors 94
Total **269**

NUMBER OF DEGREES AWARDED—BY PROGRAM
[Numbers are engineering technology baccalaureate degrees awarded during the 1991-1992 academic year, unless otherwise footnoted. For full details about each engineering technology program, refer to the *Table of Degree Programs*.]
Automotive Engineering Technology 10
Electronic Engineering Technology 28
Manufacturing Engineering Technology 6
Total **44**

PERCENTAGE OF DEGREES AWARDED—BY CATEGORY
[Percentages are of all engineering technology baccalaureate degrees awarded during the 1991-1992 academic year, unless otherwise footnoted.]
African Americans .... – %
Asian/Pacific Island Americans .... 5.0 %
Hispanic Americans .... – %
Native Americans .... – %
Foreign Citizens .... 1.0 %
All Others .... 94.0 %
Women .... 1.0 %
Persons with disabilities .... 0.5 %
Students over 25 years of age .... 8.0 %

ENGINEERING TECHNOLOGY STUDENT DATA
Applicants to the engineering technology college
Number of applicants to engineering technology college 61
Percent offered admission 100.0%

Matriculated engineering technology students
Percentage in top quartile (25%) of High School class 67.0

FULL- & PART-TIME FACULTY—BY DEPARTMENT
[Figures are the head count of full-time faculty and the full-time equivalent (FTE) of part-time faculty for each engineering technology department or equivalent.]

Department	Full	Part
Electronic Eng Tech	3	–
Manufacturing Eng Tech	3	–
Automotive Eng Tech	2	–

TECHNOLOGY COLLEGE DESCRIPTION
MSU's technology programs represent an outgrowth of earlier program options for science majors. Started in the 1980's, the programs offer fully accredited degrees in EET, MET and AET with options for minors in computer science, business, and mathematics; as well as other areas of science and technology. The programs offer areas of emphasis in each of these disciplines. Students enjoy access to over $9,000,000 in equipment including complete cleanroom facilities, advanced analytical instrumentation, automotive electronics laboratories, machining facilities, robotic capabilities, and state-of-the-art facilities for communications design and analysis. The programs work cooperatively with the Community Colleges and Technical Schools throughout Minnesota to assure ease of transition from these institutions into the MSU environment.

TECHNOLOGY DEGREE PROGRAMS DESCRIPTION
Mankato State University offers B.S. degree programs in 3 engineering technology disciplines: Automotive Engineering Technology; Electronic Engineering Technology; Manufacturing Engineering Technology. These programs are distinguished by excellent state-of-the-art equipment and laboratory facilities. Student projects are encouraged and supported. Such projects have won numerous regional and national awards. Most of the faculty hold Ph.D.s and have had extensive industry experience. Close liaison with industry has greatly benefited the engineering technology programs!

DUAL DEGREE PROGRAMS
Technology baccalaureate graduates with dual degrees 27

TRANSFER INFORMATION
Residency Requirements: Transfer students must complete at least 45 hours at Mankato State University during the last 2 years prior to graduation.

Transfer via Articulation Agreements
Admission to engineering technology: At any time
Technology graduates transferred from ... 4-yr: – 2-yr: 20%
Requirements: There is a joint admission agreement with the Minnesota Community College System.

Transfer without Articulation Agreements
Admission to engineering technology: At any time
Technology graduates transferred from ... 4-yr: – 2-yr: 5%

STUDENT PROGRAMS
PROFESSIONAL AND HONORARY SOCIETIES
[For key to acronyms, see *Introduction*.]
Professional Societies: ACS, AIP, APS, ASME, IEEE, SAE, SME, SPS
Honorary Societies: Phi Kappa Phi

SUPPORT PROGRAMS
Student Chapter Organizations: Society of Women Engineers

Cultural Diversity Program
For: Women & Ethnic Minorities Available: Year round
Offered: Freshman, Sophomore, Junior, Senior, Graduate level

Women's Studies
For: Women Available: Year round
Offered: Freshman, Sophomore, Junior, Senior, Graduate level

Other Engineering Support Programs: MSU offers a variety of support and services to incoming students. At the freshman level, MSU offers an introductory engineering sequence. In addition, the college supports dormitory programs for eng. technology students. These programs include the availability of computer facilities, laboratory facilities, and faculty advisors in selected locations. Several student organizations are active.

314 University of Massachusetts, Dartmouth

■ INSTITUTION PROFILE
HEAD OF THE INSTITUTION
Joseph C. Deck, *Chancellor*
Phone: (508)999-8004　　**Fax:** (508)999-8860

GENERAL INFORMATION
[All Students—Fall 1991]
Undergraduate enrollment 5,049
Graduate enrollment 223
Total institution enrollment 5,272
Type of institution: Public　　**Calendar:** Semesters
Nearest city: New Bedford　　**Population:** 95,000
Miles from main campus: 9　　**Setting:** Suburban
Types of engineering degrees: Engineering & Technology
Other degree-granting colleges: Arts & Sciences, Nursing, Visual & Performing Arts, Business & Industry

■ ENGINEERING TECHNOLOGY ADMISSIONS
ADMISSIONS OFFICE CONTACT
Raymond Barrows
University of Massachusetts, Dartmouth
Old Westport Road
North Dartmouth, MA 02747-2300
Phone: (508)999-8782　　**Fax:** (508)999-8901

TECHNOLOGY COLLEGE ADMISSIONS INFORMATION
Entrance Requirements: High School courses—English (4 years), Foreign language (2 years), Social sciences (2 years), Mathematics (3 years)
Requirements for foreign students: TOEFL

DEPARTMENTAL ADMISSIONS INFORMATION
Admission to the engineering technology department: At the time of admission to the institution.

■ FINANCIAL INFORMATION
ESTIMATED EXPENSES (ACADEMIC YEAR)
[Expenses are for the 1992-93 nine-month academic year.]

State Residents	Undergraduate	Graduate
Tuition and fees	$ 3,453	$ 3,803
College room and board	$ 4,300	$ 4,300
Books and supplies	$ 500	$ 500
Other expenses	$ –	$ –
Total estimated expenses	**$ 8,253**	**$ 8,603**

Out-of-State Residents	Undergraduate	Graduate
Tuition and fees	$ 8,339	$ 8,339
College room and board	$ 4,300	$ 4,300
Books and supplies	$ 500	$ 500
Other expenses	$ –	$ –
Total estimated expenses	**$13,139**	**$13,139**

FINANCIAL AID OFFICE CONTACT
Gerald S. Coutinho, *Director*
University of Massachusetts, Dartmouth
Old Westport Road
North Dartmouth, MA 02747-2300
Phone: (508)999-8632　　**Fax:** (508)999-8901

GENERAL FINANCIAL AID INFORMATION
Forms accepted/required: FAF
Additional financial aid information: The usual range of financial aid is available through federal, state and university sources, as described in the University's Admissions Bulletin.

■ TECHNOLOGY COLLEGE INFORMATION
[For additional personnel, refer to the *Appendix*.]
HEAD OF ENGINEERING TECHNOLOGY
L. Bryce Andersen
Phone: (508)999-8539　　**Fax:** (508)999-8485

ENGINEERING TECHNOLOGY COLLEGE ADDRESS
College of Engineering
University of Massachusetts, Dartmouth
Old Westport Road
North Dartmouth, MA 02747-2300
Phone: (508)999-8000　　**Fax:** (508)999-8485

TECHNOLOGY ENROLLMENTS—BY CLASS
[Numbers are technology baccalaureate enrollments for the fall 1991 term, unless otherwise footnoted.]
1st-year students/Freshmen 80
2nd-year students/Sophomores 61
3rd-year students/Juniors 46
4th-year students/Seniors 41
Total **228**

NUMBER OF DEGREES AWARDED—BY PROGRAM
[Numbers are engineering technology baccalaureate degrees awarded during the 1991-1992 academic year, unless otherwise footnoted. For full details about each engineering technology program, refer to the *Table of Degree Programs*.]
Electrical Engineering Technology 30
Mechanical Engineering Technology 14
Total **44**

PERCENTAGE OF DEGREES AWARDED—BY CATEGORY
[Percentages are of all engineering technology baccalaureate degrees awarded during the 1991-1992 academic year, unless otherwise footnoted.]
African Americans .... – %(A)
Asian/Pacific Island Americans .... – %(A)
Hispanic Americans .... – %(A)
Native Americans .... – %(A)
Foreign Citizens .... – %(A)
All Others .... 100 %
Women .... – %(A)
Persons with disabilities .... – %(A)
Students over 25 years of age .... – %(A)

Notes: (A) Data not available.

ENGINEERING TECHNOLOGY STUDENT DATA
Applicants to the engineering technology college
Number of applicants to engineering technology college 95
Percent offered admission 70.5%
Matriculated engineering technology students
Percentage in top quartile (25%) of High School class .... –(A)
Average SAT scores: Math—482, Verbal—425, Combined—907
Notes: (A) Data not available.

■ FULL- & PART-TIME FACULTY—BY DEPARTMENT
[Figures are the head count of full-time faculty and the full-time equivalent (FTE) of part-time faculty for each engineering technology department or equivalent.]

Department	Full	Part
Mechanical Eng Tech	10	0.3
Electrical Eng Tech	4	3.0

TECHNOLOGY COLLEGE DESCRIPTION
The College of Engineering emphasizes undergraduate education in the setting of a small yet diversified university. The modern 700 acre suburban campus includes residence halls for 220 students. Engineering technology laboratory facilities are modern and equipment is state-of-the-art. Engineering technology freshmen need not declare a major at the time of admission. One of the two majors may be selected at any time during the freshman year. Freshmen with strong high school preparation can complete the B.S. program in 4 years. For a variety of reasons some students choose to extend their programs over 4 1/2 or 5 years. Such an extension provides more time for in-depth study, part-time employment or active participation in student organizations or athletics.

TECHNOLOGY DEGREE PROGRAMS DESCRIPTION
The B.S. programs are applications-oriented. They prepare the graduate for entry-level professional positions.

TRANSFER INFORMATION
Transfer via Articulation Agreements
Admission to engineering technology: At any time
Transfer without Articulation Agreements
Admission to engineering technology: At any time

GRADUATION REQUIREMENTS
B.S.E.E.T., 127 credits; B.S.M.E.T., 129 credits

■ STUDENT PROGRAMS
PROFESSIONAL AND HONORARY SOCIETIES
[For key to acronyms, see *Introduction*.]
Professional Societies: ASME, IEEE, NSPE, SAE, SME
Honorary Societies: Eta Kappa Nu, Tau Alpha Pi

SUPPORT PROGRAMS
Student Chapter Organizations: Society of Women Engineers
College Now
For: Women & Ethnic Minorities　　**Available:** Academic year
Offered: Freshman
Other Engineering Support Programs: Orientations are held for freshmen and transfer students. An optional summer course is available to strengthen mathematics and physics background for freshmen. Tutoring is available for all courses.

315 University of Massachusetts, Lowell

■ INSTITUTION PROFILE
HEAD OF THE INSTITUTION
William Hogan
Phone: (508)934-2201

GENERAL INFORMATION
[All Students—Fall 1991]
Undergraduate enrollment 1,286
Total institution enrollment **1,286**
Type of institution: Public **Calendar:** Semesters
Location: Lowell **Population:** 70,000
Setting: Urban
Types of engineering degrees: Engineering Technology
Other degree-granting colleges: Arts & Sciences, Education, Health Professions, Fine Arts, Management Science

■ ENGINEERING TECHNOLOGY ADMISSIONS
ADMISSIONS OFFICE CONTACT
Sharon L. Quigley
University of Massachusetts, Lowell
One University Ave
Lowell, MA 01854
Phone: (508)934-2570 Fax: (508)452-1445

TECHNOLOGY COLLEGE ADMISSIONS INFORMATION
Admission to the engineering technology college: Proof of high school diploma or a Graduate Equivalency Diploma (DED). A major must be declared upon the completion of 18 semester hours in their major.

DEPARTMENTAL ADMISSIONS INFORMATION
Admission to the engineering technology department: At the time of admission to the institution.

■ FINANCIAL INFORMATION
ESTIMATED EXPENSES (ACADEMIC YEAR)
[Expenses are for the 1992-93 nine-month academic year.]

All Students	Undergraduate	Graduate
Tuition and fees	$ 1,755 [2]	$ 1,390 [1]
College room and board	$ —	$ —
Books and supplies	$ —	$ —
Other expenses	$ —	$ —
Total estimated expenses	**$ 1,755**	**$ 1,390**

Notes: (1) Continuing Education Program, based of $135/credit hour for graduate programs. (2) Continuing Education Program, based on $95/credit hour for undergraduate programs.

FINANCIAL AID OFFICE CONTACT
Walter Costello, *Director of Financial Aid*
University of Massachusetts, Lowell
McGauvran Student Center South Campus
Lowell, MA 01854
Phone: (508)934-4233

GENERAL FINANCIAL AID INFORMATION
Forms accepted/required: FAF, IRS, Institutional

■ TECHNOLOGY COLLEGE INFORMATION
[For additional personnel, refer to the *Appendix*.]
HEAD OF ENGINEERING TECHNOLOGY
Aldo Crugnola
Phone: (508)934-2575 Fax: (508)452-1445

ENGINEERING TECHNOLOGY COLLEGE ADDRESS
James B. Francis College of Engineering
University of Massachusetts, Lowell
One University Ave
Lowell, MA 01854
Phone: (508)934-2570 Fax: (508)452-1445

TECHNOLOGY ENROLLMENTS—BY CLASS
[Numbers are technology baccalaureate enrollments for the fall 1991 term, unless otherwise footnoted.]
1st-year students/Freshmen .... (A)
2nd-year students/Sophomores .... (A)
3rd-year students/Juniors .... (A)
4th-year students/Seniors .... (A)
Total **0**
Notes: (A) Data not available.

NUMBER OF DEGREES AWARDED—BY PROGRAM
[Numbers are engineering technology baccalaureate degrees awarded during the 1991-1992 academic year, unless otherwise footnoted. For full details about each engineering technology program, refer to the *Table of Degree Programs*.]
Civil Engineering Technology 6
Electronic(s) Engineering Technology 29
Mechanical Engineering Technology 21
Total **56**

PERCENTAGE OF DEGREES AWARDED—BY CATEGORY
[Percentages are of all engineering technology baccalaureate degrees awarded during the 1991-1992 academic year, unless otherwise footnoted.]
African Americans .... — % (A)
Asian/Pacific Island Americans .... — % (A)
Hispanic Americans .... — % (A)
Native Americans .... — % (A)
Foreign Citizens .... — % (A)
All Others .... 100 %
Women .... — % (A)
Persons with disabilities .... — % (A)
Students over 25 years of age .... — % (A)
Notes: (A) Data not available.

FULL- & PART-TIME FACULTY—BY DEPARTMENT
[Figures are the head count of full-time faculty and the full-time equivalent (FTE) of part-time faculty for each engineering technology department or equivalent.]

Department	Full	Part
Civil Eng Tech	2	3.0
Electrical/Electronic(s) Eng Tech	2	9.0
Mechanical Eng Tech	2	8.0

TECHNOLOGY COLLEGE DESCRIPTION
The Engineering Technology programs are an evening program administered by the Division of Continuing Education. Advising and course selection is provided by the full-time faculty of the Engineering Technology Department.

TECHNOLOGY DEGREE PROGRAMS DESCRIPTION
The Engineering Technology degree programs are offered cooperatively between the College of Engineering and the Division of Continuing Education. The College of Engineering controls the academic issues of the program such as curriculum content, faculty hiring, academic standards, and official student advising. The B.S. in Engineering Technology is an integrated 8-year evening program. Students may elect to receive the Associate Degree upon completion of a stipulated 4-year program; most Engineering Technology students do elect to receive an Associate Degree. Degrees are offered in Mechanical, Civil, and Electronic Engineering Technology.

TRANSFER INFORMATION
Residency Requirements: Minimum of 30 credits.
Transfer without Articulation Agreements
Admission to engineering technology: At any time
Technology graduates transferred from ... 4-yr: 2 2-yr: 18

GRADUATION REQUIREMENTS
A cum of 2.00 in the major, as well as an overall cum of 2.00. Also a minimum of 30 semester hours in residence and 120 credit hours total for a BS.

■ STUDENT PROGRAMS
PROFESSIONAL AND HONORARY SOCIETIES
[For key to acronyms, see *Introduction*.]
Professional Societies: SME
Honorary Societies: Alpha Sigma Lambda

SUPPORT PROGRAMS
Student Chapter Organizations: Society of Women Engineers, National Society of Black Engineers

Equal Opportunity Program
For: Ethnic Minorities **Available:** Academic year
Offered: Freshman, Sophomore, Junior, Senior

Office of Minorities Services
For: Women & Ethnic Minorities **Available:** Academic year
Offered: Freshman, Sophomore, Junior, Senior, Graduate level

Other Engineering Support Programs: The support services include faculty advisors, a tutoring referral service, and workshops on career development issues, such as self-assessment, decision making, etc. Also, membership in Alpha Sigma Lambda honor society is available for those who maintain a grade point average of 3.2 or higher and are in the highest 10% of the class.

316 Memphis State University

INSTITUTION PROFILE
HEAD OF THE INSTITUTION
V. Lane Rawlins
Phone: (901)678-2234 Fax: (901)678-2000
GENERAL INFORMATION
[All Students—Fall 1991]
Undergraduate enrollment 15,767
Graduate enrollment 3,451
Total institution enrollment 19,218
Type of institution: Public **Calendar:** Semesters
Location: Memphis **Population:** 610,337
Setting: Urban
Types of engineering degrees: Engineering & Technology
Other degree-granting colleges: Arts & Sciences, Business Administration, Communication & Journalism, Education, Fine & Performing Arts, Law, Nursing, University College

ENGINEERING TECHNOLOGY ADMISSIONS
ADMISSIONS OFFICE CONTACT
Carol Ferguson
Memphis State University
Corner of Central and Zach Curlin
Engineering Science Building Room 201
Memphis, TN 38152
Phone: (901)678-2171 Fax: (901)678-4180
TECHNOLOGY COLLEGE ADMISSIONS INFORMATION
Entrance Requirements: ACT (Score: 18), High School courses—English (4 years), Science (4 years), Math (4 years)
Entrance Recommendations: ACT (Score: 18), High School courses—Trigonometry (1 year), Calculus (1 year), Physics (1 year), Chemistry (1 year)
Requirements for foreign students: TOEFL (Score: 550); F-1 and J-1 Student Visa Health certificate
DEPARTMENTAL ADMISSIONS INFORMATION
Admission to the engineering technology department: At the time of admission to the institution.

FINANCIAL INFORMATION
ESTIMATED EXPENSES (ACADEMIC YEAR)
[Expenses are for the 1992-93 nine-month academic year.]

State Residents	Undergraduate	Graduate
Tuition and fees	$ 828	$ 991
College room and board	$ 700	$ 700
Books and supplies	$ 250	$ 250
Other expenses	$ _(A)	$ _(B)
Total estimated expenses	**$ 1,778**	**$ 1,941**

Out-of-State Residents	Undergraduate	Graduate
Tuition and fees	$ 2,490	$ 2,673
College room and board	$ 700	$ 700
Books and supplies	$ 250	$ 250
Other expenses	$ _(A)	$ _(B)
Total estimated expenses	**$ 3,440**	**$ 3,623**

Notes: (A) Data not available. (B) Data not applicable.

GENERAL FINANCIAL AID INFORMATION
Forms accepted/required: ACT
Additional financial aid information: Pell Grants, Supplemental Educational Opportunity Grant, and Tennessee Student Assistance Program.

TECHNOLOGY COLLEGE INFORMATION
[For additional personnel, refer to the *Appendix*.]
HEAD OF ENGINEERING TECHNOLOGY
John D. Ray
Phone: (901)678-2171 Fax: (901)678-4180
ENGINEERING TECHNOLOGY COLLEGE ADDRESS
Engineering Technology
Memphis State University
Corner of Central and Zach Curlin
Engineering Science Building Room 201
Memphis, TN 38152
Phone: (901)678-2171 Fax: (901)678-4180
TECHNOLOGY ENROLLMENTS—BY CLASS
[Numbers are technology baccalaureate enrollments for the fall 1991 term, unless otherwise footnoted.]
1st-year students/Freshmen 49
2nd-year students/Sophomores 54
3rd-year students/Juniors 103
4th-year students/Seniors 217
Total 423
NUMBER OF DEGREES AWARDED—BY PROGRAM
[Numbers are engineering technology baccalaureate degrees awarded during the 1991-1992 academic year, unless otherwise footnoted. For full details about each engineering technology program, refer to the *Table of Degree Programs*.]
Architectural Engineering Technology 15
Computer Engineering Technology 16
Electrical Engineering Technology 27
Manufacturing Engineering Technology 14
Total 72
PERCENTAGE OF DEGREES AWARDED—BY CATEGORY
[Percentages are of all engineering technology baccalaureate degrees awarded during the 1991-1992 academic year, unless otherwise footnoted.]
African Americans 21.0%
Asian/Pacific Island Americans .... _ % (A)
Hispanic Americans .... _ % (A)
Native Americans .... _ % (A)
Foreign Citizens .... _ % (A)
All Others 79.0%
Women 18.0%
Persons with disabilities .... _ % (A)
Students over 25 years of age 12.0%
Notes: (A) Data not available.
GRADUATE ENROLLMENTS & DEGREES AWARDED
Master's enrollment 14
Master's degrees awarded 4
Doctoral enrollment .... _
Doctoral degrees awarded .... _

ENGINEERING TECHNOLOGY STUDENT DATA
Applicants to the engineering technology college
Number of applicants to engineering technology college ..... _ (A)
Percent offered admission .... _% (A)
Matriculated engineering technology students
Percentage in top quartile (25%) of High School class .... _ (A)
Average ACT scores: Math—21.3, Composite—22.1
Notes: (A) Data not available.
FULL- & PART-TIME FACULTY—BY DEPARTMENT
[Figures are the head count of full-time faculty and the full-time equivalent (FTE) of part-time faculty for each engineering technology department or equivalent.]

Department	Full	Part
Electrical/Electronic(s) Eng Tech	4	1.0
Manufacturing Eng Tech	5	1.0
Computer Eng Tech	4	1.0
Architecture Tech	3	1.0

TECHNOLOGY COLLEGE DESCRIPTION
Our goal is to provide a quality education for students pursuing a degree in engineering technology.
TECHNOLOGY DEGREE PROGRAMS DESCRIPTION
Our college strives to give a quality education to students seeking an Engineering Technology Degree.
TRANSFER INFORMATION
Residency Requirements: Thirty of last 33 hours must be completed on campus
Transfer via Articulation Agreements
Admission to engineering technology: At any time
Technology graduates transferred from ... 4-yr: 10 2-yr: 25
Transfer without Articulation Agreements
Admission to engineering technology: At any time
Technology graduates transferred from ... 4-yr: _ 2-yr: _
GRADUATION REQUIREMENTS
2.0 GPA in major 2.0 in all courses. 2.0 GPA in all courses taken at Memphis State University.

STUDENT PROGRAMS
PROFESSIONAL AND HONORARY SOCIETIES
[For key to acronyms, see *Introduction*.]
Professional Societies: ASCE, ASME, IEEE, SAE, SME
Honorary Societies: Pi Tau Sigma, Tau Alpha Pi, Tau Beta Pi
SUPPORT PROGRAMS
Student Chapter Organizations: Society of Women Engineers, National Society of Black Engineers
Early Scholars
For: Ethnic Minorities **Available:** Summer only
Offered: Prematriculation
Other Engineering Support Programs: Summer orientation Program for Incoming Freshman and Transfer students. engineering learning center which provides free tutoring programs for engineering students.

317 Metropolitan State College of Denver

INSTITUTION PROFILE

HEAD OF THE INSTITUTION
Thomas B. Brewer
Phone: (303)556-3022   Fax: (303)556-3912

GENERAL INFORMATION
[All Students—Fall 1991]
Undergraduate enrollment 17,864
Total institution enrollment 17,864
Type of institution: Public   Calendar: Semesters
Location: Denver   Population: 500,000
Setting: Urban
Types of engineering degrees: Engineering Technology
Other degree-granting colleges: Business, Letters, Arts and Science

ENGINEERING TECHNOLOGY ADMISSIONS

ADMISSIONS OFFICE CONTACT
Kenneth Curtis
Metropolitan State College of Denver
P.O. Box 173362, Campus Box 16
Denver, CO 80217-3362
Phone: (303)556-3018   Fax: (303)556-3999

TECHNOLOGY COLLEGE ADMISSIONS INFORMATION
Entrance Requirements: Either ACT or SAT (but not both) required for student under 20 years old.
Entrance Recommendations: SAT, ACT
Requirements for foreign students: TOEFL (Score: 500); financial statement; Will be required to pass English proficiency test.

DEPARTMENTAL ADMISSIONS INFORMATION
Admission to the engineering technology department: At the time of admission to the institution.

FINANCIAL INFORMATION

ESTIMATED EXPENSES (ACADEMIC YEAR)
[Expenses are for the 1992-93 nine-month academic year.]

State Residents	Undergraduate	Graduate
Tuition and fees	$ 1,746	$ —
College room and board	$ 4,620	$ —
Books and supplies	$ 482	$ —
Other expenses	$ 1,096	$ —
Total estimated expenses	**$ 7,944**	**$ —**

Out-of-State Residents	Undergraduate	Graduate
Tuition and fees	$ 5,460	$ —
College room and board	$ 4,620	$ —
Books and supplies	$ 482	$ —
Other expenses	$ 1,096	$ —
Total estimated expenses	**$11,658**	**$ —**

FINANCIAL AID OFFICE CONTACT
Cheryl Judson, *Director of Financial Aid*
Metropolitan State College of Denver
P.O. Box 173362, Campus Box 2
Denver, CO 80217-3362
Phone: (303)556-3043

GENERAL FINANCIAL AID INFORMATION
Forms accepted/required: ACT
Additional financial aid information: Full range of financial aid is available, including loans, grants, and scholarships.

TECHNOLOGY COLLEGE INFORMATION
[For additional personnel, refer to the *Appendix*.]

HEAD OF ENGINEERING TECHNOLOGY
Bill T. Rader
Phone: (303)556-2978   Fax: (303)556-2159

ENGINEERING TECHNOLOGY COLLEGE ADDRESS
School of Professional Studies
Metropolitan State College of Denver
P.O. Box 173362, Campus Box 8
Denver, CO 80217-3362
Phone: (303)556-3018   Fax: (303)556-3999

TECHNOLOGY ENROLLMENTS—BY CLASS
[Numbers are technology baccalaureate enrollments for the fall 1991 term, unless otherwise footnoted.]
1st-year students/Freshmen .... (A)
2nd-year students/Sophomores .... (A)
3rd-year students/Juniors .... (A)
4th-year students/Seniors .... (A)
Total 0

Notes: (A) Data not available.

NUMBER OF DEGREES AWARDED—BY PROGRAM
[Numbers are engineering technology baccalaureate degrees awarded during the 1991-1992 academic year, unless otherwise footnoted. For full details about each engineering technology program, refer to the *Table of Degree Programs*.]
Civil Engineering Technology 11
Electronic(s) Engineering Technology 38
Mechanical Engineering Technology 15
Total 64

PERCENTAGE OF DEGREES AWARDED—BY CATEGORY
[Percentages are of all engineering technology baccalaureate degrees awarded during the 1991-1992 academic year, unless otherwise footnoted.]
African Americans .... — % (A)
Asian/Pacific Island Americans .... — % (A)
Hispanic Americans .... — % (A)
Native Americans .... — % (A)
Foreign Citizens .... — % (A)
All Others .... 100 %
Women .... — % (A)
Persons with disabilities .... — % (A)
Students over 25 years of age .... — % (A)

Notes: (A) Data not available.

FULL- & PART-TIME FACULTY—BY DEPARTMENT
[Figures are the head count of full-time faculty and the full-time equivalent (FTE) of part-time faculty for each engineering technology department or equivalent.]

Department	Full	Part
Civil Eng Tech	6	0.3
Electronics & Mechanical Eng Tech	9	1.7

TECHNOLOGY COLLEGE DESCRIPTION
Metropolitan State College of Denver is a four year state supported college located on a 169 acre campus next to downtown Denver. The campus is shared by the University of Colorado at Denver and the Community College of Denver. The college prides itself on teaching excellence.

TECHNOLOGY DEGREE PROGRAMS DESCRIPTION
Areas of emphasis in CET: environmental, structures, land surveying, & controlling surveying. Areas of emphasis in EET: computers, communications, control systems, & power. Areas of emphasis in MET: manufacturing & mechanical.

TRANSFER INFORMATION
Residency Requirements: Minimum of 30 semester hours at college, including at least 8 hours upper-division in major. Last 12 hours taken for degree must be in residence at college.

Transfer without Articulation Agreements
Admission to engineering technology: At any time

GRADUATION REQUIREMENTS
2.0 minimum GPA in major, and 2.0 overall GPA required for graduation.

STUDENT PROGRAMS

PROFESSIONAL AND HONORARY SOCIETIES
[For key to acronyms, see *Introduction*.]
Professional Societies: ACM, ASCE, ASME, IEEE
Honorary Societies: Tau Alpha Pi, Golden Key Honor Society

SUPPORT PROGRAMS
Other Engineering Support Programs: There is a freshman year program to assist new students. There is an advising center, and a tutoring center.

318 Milwaukee School of Engineering

■ INSTITUTION PROFILE
HEAD OF THE INSTITUTION
Hermann Viets
Phone: (414)277-7100 Fax: (414)277-7468

GENERAL INFORMATION
[All Students—Fall 1991]
Undergraduate enrollment 2,775
Graduate enrollment 391
Total institution enrollment **3,166**
Type of institution: Private **Calendar:** Quarters
Location: Milwaukee **Population:** 1,400,000
Setting: Urban
Types of engineering degrees: Engineering & Technology

■ ENGINEERING TECHNOLOGY ADMISSIONS
ADMISSIONS OFFICE CONTACT
T. O. Smith
Milwaukee School of Engineering
P.O. Box 644
1025 N. Milwaukee Street
Milwaukee, WI 53201-0644
Phone: (800)332-6763 Fax: (414)277-7186

TECHNOLOGY COLLEGE ADMISSIONS INFORMATION
Entrance Requirements: ACT, High School courses—English (3 years), Science (4 years), Mathematics (Algebra 2, Geometry 1) (3 years)
Entrance Recommendations: SAT, ACT, High School courses—Chemistry (1 year), Plane Trigonometry (1 year), Computers (1 year), Mechanical Drawing (1 year)
Requirements for foreign students: TOEFL (Score: 500); financial statement; Translated Transcripts

DEPARTMENTAL ADMISSIONS INFORMATION
Admission to the engineering technology department: At the time of admission to the institution.

■ FINANCIAL INFORMATION
ESTIMATED EXPENSES (ACADEMIC YEAR)
[Expenses are for the 1992-93 nine-month academic year.]

All Students	Undergraduate	Graduate
Tuition and fees	$ 9,810	$ –
College room and board	$ 3,045	$ –
Books and supplies	$ 1,000	$ –
Other expenses	$ 1,920	$ –
Total estimated expenses	**$15,775**	**$ –**

FINANCIAL AID OFFICE CONTACT
Susan Hebert, *Director of Financial Aid*
Milwaukee School of Engineering
1025 N. Broadway
Milwaukee, WI 53202-3109
Phone: (414)277-7224 Fax: (414)277-7450

GENERAL FINANCIAL AID INFORMATION
Forms accepted/required: ACT, AFSA, CSS-FAF, FAT, FSA, FFS, Stafford, SAR, Supplemental, Institutional, As required
Additional financial aid information: Academic and need-based scholarships; Federal and State grant programs; Federal and institutional loan programs; College work-study program.

■ TECHNOLOGY COLLEGE INFORMATION
[For additional personnel, refer to the *Appendix*.]
HEAD OF ENGINEERING TECHNOLOGY
Thomas W. Davis
Phone: (414)277-7324 Fax: (414)277-7477
EMail: davis@kirk.msoe.edu

ENGINEERING TECHNOLOGY COLLEGE ADDRESS
Milwaukee School of Engineering
1025 N. Broadway
Milwaukee, WI 53202-3109
Phone: (414)277-7300 Fax: (414)277-7477

TECHNOLOGY ENROLLMENTS—BY CLASS
[Numbers are technology baccalaureate enrollments for the fall 1991 term, unless otherwise footnoted.]
1st-year students/Freshmen 368
2nd-year students/Sophomores 272
3rd-year students/Juniors 213
4th-year students/Seniors 115
Total **968**

NUMBER OF DEGREES AWARDED—BY PROGRAM
[Numbers are engineering technology baccalaureate degrees awarded during the 1991-1992 academic year, unless otherwise footnoted. For full details about each engineering technology program, refer to the *Table of Degree Programs*.]
Electrical/Electronic(s) Engineering Technology 44
Manufacturing Engineering Technology 7
Mechanical Engineering Technology 30
Total **81**

PERCENTAGE OF DEGREES AWARDED—BY CATEGORY
[Percentages are of all engineering technology baccalaureate degrees awarded during the 1991-1992 academic year, unless otherwise footnoted.]
African Americans .... 2.0 %
Asian/Pacific Island Americans .... 2.0 %
Hispanic Americans .... 1.0 %
Native Americans .... 1.0 %
Foreign Citizens .... 1.0 %
All Others .... 93.0 %
Women .... 10.0 %
Persons with disabilities .... – % (A)
Students over 25 years of age .... – % (A)
Notes: (A) Data not available.

ENGINEERING TECHNOLOGY STUDENT DATA
Applicants to the engineering technology college
Number of applicants to engineering technology college ... 147
Percent offered admission 90.0%

Matriculated engineering technology students
Percentage in top quartile (25%) of High School class 43.0
Average SAT scores: Math—500, Verbal—490, Combined—990
Average ACT scores: Math—22.89, Composite—21.78

FULL- & PART-TIME FACULTY—BY DEPARTMENT
[Figures are the head count of full-time faculty and the full-time equivalent (FTE) of part-time faculty for each engineering technology department or equivalent.]

Department	Full	Part
Electrical Eng Tech	32	11.5
Manufacturing Eng Tech	22	8.0
Mechanical Eng Tech	22	8.0

TECHNOLOGY COLLEGE DESCRIPTION
MSOE offers four year degree programs in three engineering technology disciplines. In each program students complete laboratory intensive courses in their field of specialization plus numerous cross disciplinary courses in other engineering fields, and appropriate course work in humanities, social science, business and communication. There are opportunities for minors in business and management, and technical communications. Undergraduates have the opportunity to gain additional experience by conducting industrially sponsored applied research through the college's Applied Technology Center and research consortia in fluid power, biomedical systems, rapid prototyping and other areas. Average number of years required to actually complete the bachelor's degree: 6.2

TECHNOLOGY DEGREE PROGRAMS DESCRIPTION
The degree programs prepare students for entry-level professional positions and for continued graduate study. Courses and laboratory sessions are taught by qualified faculty with current industrial experience in their field of specialty. Laboratories are industrially sponsored and utilize full size industrial quality equipment. Course sequencing and course development are guided by 18 business and industry advisor committees. Students may pursue advanced study at the Master's level in engineering or engineering management. These programs are industrially oriented and multidisciplinary. Students in the Master of Science degree program in engineering may pursue specialities in materials, fluid power, computers, electronics and controls, manufacturing and imaging.

TRANSFER INFORMATION
Residency Requirements: Students must complete at least 50% of required courses at MSOE for a degree from MSOE.

Transfer via Articulation Agreements
Admission to engineering technology: At any time
Technology graduates transferred from ... 4-yr: – 2-yr: 325
Requirements: Official transcripts of all high school, college and university courses must be submitted. A grade of 'C' or better is required for a course to be considered for transfer. A financial aid transcript from the Financial Aid Office of any previously attended college must be submitted to the Financial Aid Office. Completion of minimum course work as specified in the Academic Catalog.

Transfer without Articulation Agreements
Admission to engineering technology: At any time
Technology graduates transferred from ... 4-yr: 89 2-yr: 98
Requirements: Evaluated on an individual basis.

GRADUATION REQUIREMENTS
Bachelor of engineering technology programs require 203 to 208 quarter credits, depending on major selected. A 2.0 cumulative grade point average and a 2.0 major grade point average is required.

■ STUDENT PROGRAMS
PROFESSIONAL AND HONORARY SOCIETIES
[For key to acronyms, see *Introduction*.]
Professional Societies: ACM, ASHRAE, ASME, DPMA, IEEE, SAE, SME
Honorary Societies: Eta Kappa Nu, Kappa Eta Kappa, Tau Alpha Pi, Tau Omega Mu

SUPPORT PROGRAMS
Student Chapter Organizations: Society of Women Engineers, National Society of Black Engineers, Society of Hispanic Professional Engineers, Society of International Students

Upward Bound
For: Ethnic Minorities **Available:** Summer only
Offered: Prematriculation

Catalyst for Future Success
For: Ethnic Minorities **Available:** Summer only
Offered: Prematriculation

IDEAS Program
For: Women **Available:** Academic year
Offered: Freshman

Other Engineering Support Programs: Fall orientation program for all freshmen and transfer students; New student orientation course (1 quarter) for all freshmen; First year mentoring program; Learning Resource Center; Personal and career counseling services; Foreign student office; Veteran's services; Services for handicapped students; Placement services; Student support services; Student Life activities.

319 Montana State University

INSTITUTION PROFILE

HEAD OF THE INSTITUTION
Michael P. Malone
Phone: (406)994-2341 Fax: (406)994-1893

GENERAL INFORMATION
[All Students—Fall 1991]
Undergraduate enrollment 9,313
Graduate enrollment 798
Total institution enrollment 10,111
Type of institution: Public   Calendar: Semesters
Nearest city: Billings   Population: 80,000
Miles from main campus: 140   Setting: Small Town
Types of engineering degrees: Engineering & Technology
Other degree-granting colleges: Agricultural & Environmental, Architecture, Arts & Sciences, Business Administration, Education, Fine & Performing Arts, Nursing

ENGINEERING TECHNOLOGY ADMISSIONS

ADMISSIONS OFFICE CONTACT
Rhonda Duffus
Montana State University
120 Hamilton Hall
Bozeman, MT 59717
Phone: (406)994-2452 Fax: (406)994-1923

TECHNOLOGY COLLEGE ADMISSIONS INFORMATION
Entrance Requirements: High School courses—English (4 years), Mathematics (3 years), Social Studies (3 years), Laboratory Science (2 years), ACT or SAT
Entrance Recommendations: SAT (Score: 800), ACT (Score: 20), High School courses—Foreign Language (2 years)
Requirements for foreign students: TOEFL (Score: 525); financial statement; Evaluation of academic credentials by ECE

DEPARTMENTAL ADMISSIONS INFORMATION
Admission to the engineering technology department: At the time of admission to the institution.

FINANCIAL INFORMATION

ESTIMATED EXPENSES (ACADEMIC YEAR)
[Expenses are for the 1992-93 nine-month academic year.]

State Residents	Undergraduate	Graduate
Tuition and fees	$ 1,850	$ 1,850
College room and board	$ 3,500	$ 3,500
Books and supplies	$ 550	$ 550
Other expenses	$ 2,300 [1]	$ 2,300 [1]
Total estimated expenses	$ 8,200	$ 8,200

Out-of-State Residents	Undergraduate	Graduate
Tuition and fees	$ 5,500	$ 5,500
College room and board	$ 3,500	$ 3,500
Books and supplies	$ 550	$ 550
Other expenses	$ 2,300 [1]	$ 2,300 [1]
Total estimated expenses	$11,850	$11,850

Notes: (1) Personal/Transportation

FINANCIAL AID OFFICE CONTACT
James R. Craig, *Director, Financial Aid Services*
Montana State University
135 Strand Union Building
Bozeman, MT 59717
Phone: (406)994-2845 Fax: (406)994-5488

GENERAL FINANCIAL AID INFORMATION
Forms accepted/required: AFSA, Stafford, SAR
Additional financial aid information: Students applying for financial assistance are considered for all aid programs for which they are eligible. Assistance is offered in the form of grants, scholarships, long term loans, and work opportunities. Priority consideration is given students who apply by March 1 of each year.

TECHNOLOGY COLLEGE INFORMATION
[For additional personnel, refer to the *Appendix*.]

HEAD OF ENGINEERING TECHNOLOGY
David F. Gibson
Phone: (406)994-2272 Fax: (406)994-6098
EMail: ADEDG@MTSUNIX1.BITNET

ENGINEERING TECHNOLOGY COLLEGE ADDRESS
College of Engineering
Montana State University
212 Roberts Hall
Bozeman, MT 59717-0382
Phone: (406)994-2272 Fax: (406)994-6098

TECHNOLOGY ENROLLMENTS—BY CLASS
[Numbers are technology baccalaureate enrollments for the fall 1991 term, unless otherwise footnoted.]
1st-year students/Freshmen 47
2nd-year students/Sophomores 77
3rd-year students/Juniors 95
4th-year students/Seniors .... 119 [1]
Total 338

Notes: (1) Includes second degree students.

NUMBER OF DEGREES AWARDED—BY PROGRAM
[Numbers are engineering technology baccalaureate degrees awarded during the 1991-1992 academic year, unless otherwise footnoted. For full details about each engineering technology program, refer to the *Table of Degree Programs*.]
Construction Engineering Technology 28
Electrical/Electronic(s) Engineering Technology 13
Mechanical Engineering Technology 32
Total 73

PERCENTAGE OF DEGREES AWARDED—BY CATEGORY
[Percentages are of all engineering technology baccalaureate degrees awarded during the 1991-1992 academic year, unless otherwise footnoted.]
African Americans .... – % (A)
Asian/Pacific Island Americans .... – % (A)
Hispanic Americans .... – % (A)
Native Americans .... – % (A)
Foreign Citizens .... – % (A)
All Others .... 100 %
Women .... 1.3 %
Persons with disabilities .... – % (A)
Students over 25 years of age .... – % (A)

Notes: (A) Data not available.

ENGINEERING TECHNOLOGY STUDENT DATA
Applicants to the engineering technology college
Number of applicants to engineering technology college 74
Percent offered admission 88.1%

Matriculated engineering technology students
Percentage in top quartile (25%) of High School class .... – (A)
Average SAT scores: Math—592 [1], Verbal—495 [1], Combined—1087 [1]
Average ACT scores: Composite—24.5 [1]

Notes: (A) Data not available. (1) Includes engineering and computer science

FULL- & PART-TIME FACULTY—BY DEPARTMENT
[Figures are the head count of full-time faculty and the full-time equivalent (FTE) of part-time faculty for each engineering technology department or equivalent.]

Department	Full	Part
Civil & Agricultural Eng	19	3.7
Electrical Eng	16	1.0
Mechanical Eng	11	1.6

TECHNOLOGY COLLEGE DESCRIPTION
The College of Engineering is the largest college on the university campus. It offers four-year undergraduate programs in six areas of engineering, as well as three engineering technology programs and computer science. Programs in engineering technology emphasize hand-on laboratory experiences and applications-oriented coursework. Intern programs with industrial firms are available in some programs. State-of-the-art computer capabilities exist in both the College and the University. Each program in engineering technology is associated with a student club, and most of these are affiliated with a national professional society. Engineering technology coursework is complemented by a University Core Curriculum to help develop critical thinking. The average number of years required to actually complete the bachelor's degree is approximately 4.5 years.

TECHNOLOGY DEGREE PROGRAMS DESCRIPTION
The College's four year degree programs in engineering technology prepare graduates for entry level positions in construction or industrial firms. The programs emphasize industrial applications of theoretical and mathematical concepts. The engineering technology graduate works closely with the engineer to transform the engineer's designs into completed construction or manufactured product. All three of the College's engineering technology programs are accredited by ABET-TAC.

DUAL DEGREE PROGRAMS
Technology baccalaureate graduates with dual degrees 0

TRANSFER INFORMATION
Residency Requirements: A minimum of 30 residence credits are required and a minimum of 23 of the last 30 credits earned to meet the graduation requirement must be resident credits.

Transfer via Articulation Agreements
Admission to engineering technology: At any time
Technology graduates transferred from ... 4-yr: 25   2-yr: 5
Requirements: In-state residents must be admissible to their former school. Out-of-state residents must have a 2.0 transferable GPA.

GRADUATION REQUIREMENTS
Engineering technology students are required to maintain a 2.0 GPA to be eligible for graduation. In addition, a student must receive a C grade or better in all courses required for graduation.

STUDENT PROGRAMS

PROFESSIONAL AND HONORARY SOCIETIES
[For key to acronyms, see *Introduction*.]
Professional Societies: AGCA, ASCE, ASME, IEEE, NSPE
Honorary Societies: Chi Epsilon, Eta Kappa Nu, Pi Tau Sigma, Tau Beta Pi

SUPPORT PROGRAMS
Student Chapter Organizations: Society of Women Engineers, American Indian Science and Engineering Society

Women's Center
For: Women   Available: Year round
Offered: Freshman, Sophomore, Junior, Senior, Graduate level

Center for Native American Studies
For: Ethnic Minorities   Available: Year round
Offered: Prematriculation, Freshman, Sophomore, Junior, Senior, Graduate level

Other Engineering Support Programs: Programs for support of engineering technology students include freshman/transfer orientation programs; programs for students over the traditional age; programs for students returning to school after an extended period, as well as those needing special academic assistance. Other programs are conducted by Disabled Student Services, Career Services, Veterans Affairs, and International Students.

320 Murray State University

INSTITUTION PROFILE
HEAD OF THE INSTITUTION
Ronald J. Kurth
Phone: (502)762-3763 **Fax:** (502)762-3413

GENERAL INFORMATION
[All Students—Fall 1991]
Undergraduate enrollment 6,794
Graduate enrollment 1,255
Total institution enrollment **8,049**

Type of institution: Public **Calendar:** Semesters
Nearest city: Nashville, TN **Population:** 750,000
Miles from main campus: 120 **Setting:** Small Town
Types of engineering degrees: Engineering Technology
Other degree-granting colleges: Industry and Technology, Business and Public Affairs, Fine Arts and Communications, Science, Humanistic Studies

ENGINEERING TECHNOLOGY ADMISSIONS
ADMISSIONS OFFICE CONTACT
Philip Bryan
Murray State University
112 Sparks Hall
Murray, KY 42071
Phone: (502)762-3741

TECHNOLOGY COLLEGE ADMISSIONS INFORMATION
Entrance Requirements: SAT (Score: 700), ACT (Score: 18)
Requirements for foreign students: TOEFL (Score: 500)

DEPARTMENTAL ADMISSIONS INFORMATION
Admission to the engineering technology department: At the time of admission to the institution.

FINANCIAL INFORMATION
ESTIMATED EXPENSES (ACADEMIC YEAR)
[Expenses are for the 1992-93 nine-month academic year.]

State Residents	Undergraduate	Graduate
Tuition and fees	$ 1,600	$ 1,740
College room and board	$ 1,170	$ 1,170
Books and supplies	$ —(A)	$ —(A)
Other expenses	$ —(A)	$ —(A)
Total estimated expenses	**$ 2,770**	**$ 2,910**

Out-of-State Residents	Undergraduate	Graduate
Tuition and fees	$ 4,280	$ 4,700
College room and board	$ 1,170	$ 1,170
Books and supplies	$ —(A)	$ —(A)
Other expenses	$ —(A)	$ —(A)
Total estimated expenses	**$ 5,450**	**$ 5,870**

Notes: (A) Data not available.

FINANCIAL AID OFFICE CONTACT
Johnny McDougal, *Director*
Murray State University
Office of Student Financial Aid
Murray, KY 42071
Phone: (502)762-2546

GENERAL FINANCIAL AID INFORMATION
Forms accepted/required: ACT, AFSA, FAF, FSA
Additional financial aid information: Various types of financial aid are available: They include grants, loans, scholarships, student employment on campus. A specific listing may be obtained through the Student Financial Aid Office at Murray State.

TECHNOLOGY COLLEGE INFORMATION
[For additional personnel, refer to the *Appendix*.]
HEAD OF ENGINEERING TECHNOLOGY
Thomas B. Auer
Phone: (502)762-3391 **Fax:** (502)762-3631

ENGINEERING TECHNOLOGY COLLEGE ADDRESS
Department of Industrial and Engineering Technology
Murray State University
Martha Layne Collins Center
Murray, KY 42071
Phone: (502)762-3393 **Fax:** (502)762-3653

TECHNOLOGY ENROLLMENTS—BY CLASS
[Numbers are technology baccalaureate enrollments for the fall 1991 term, unless otherwise footnoted.]
1st-year students/Freshmen 122
2nd-year students/Sophomores 77
3rd-year students/Juniors 69
4th-year students/Seniors 96
Total **364**

NUMBER OF DEGREES AWARDED—BY PROGRAM
[Numbers are engineering technology baccalaureate degrees awarded during the 1991-1992 academic year, unless otherwise footnoted. For full details about each engineering technology program, refer to the *Table of Degree Programs*.]
Civil Engineering Technology 10
Civil/Construction Engineering Technology 3
Electrical Engineering Technology 10
Electrical/Computer Engineering Technology 1
Environmental Engineering Technology 6
Manufacturing Engineering Technology 11
Total **41**

PERCENTAGE OF DEGREES AWARDED—BY CATEGORY
[Percentages are of all engineering technology baccalaureate degrees awarded during the 1991-1992 academic year, unless otherwise footnoted.]
African Americans .... — %(A)
Asian/Pacific Island Americans .... — %(A)
Hispanic Americans .... — %(A)
Native Americans .... — %(A)
Foreign Citizens .... — %(A)
All Others .... 100 %
Women .... 5.0 %
Persons with disabilities .... — %(A)
Students over 25 years of age .... — %(A)

Notes: (A) Data not available.

GRADUATE ENROLLMENTS & DEGREES AWARDED
Master's enrollment 51
Master's degrees awarded 3
Doctoral enrollment .... —
Doctoral degrees awarded .... —

FULL- & PART-TIME FACULTY—BY DEPARTMENT
[Figures are the head count of full-time faculty and the full-time equivalent (FTE) of part-time faculty for each engineering technology department or equivalent.]

Department	Full	Part
Civil Eng Tech (General)	2	1.7
Electrical Eng Tech (Power/Communications)	3	—
Manufacturing Eng Tech	3	1.7
Mechanical Eng Tech	3	1.7
Electrical Eng Tech (Computer)	1	—
Civil Eng Tech (Construction)	1	1.7
Environmental Eng Tech	1	—
Mining Management Tech	—	—

TECHNOLOGY COLLEGE DESCRIPTION
Murray State University focuses upon providing pre-eminent undergraduate education. Within Engineering Technology state-of-the-art systems and applications are stressed. Cooperative education experiences are required in Civil, Construction, and Environmental programs and are offered as an option in other programs. The College of Industry and Technology occupied the Martha Layne Collins Center for Industry and Technology on September 7, 1991. The building design and the engineering technology programs housed within are devoted to providing 21st century education. There are thirteen teaching and research laboratories.

TECHNOLOGY DEGREE PROGRAMS DESCRIPTION
The department offers programs at the associate and baccalaureate levels. the department also offers a technical minor and master of science degree.

TRANSFER INFORMATION
Transfer without Articulation Agreements
Admission to engineering technology: At any time
Technology graduates transferred from . . . 4-yr: –   2-yr: –

GRADUATION REQUIREMENTS
A candidate for a baccalaureate degree must complete a minimum of 128 semester hours with at least three/fourths of the credit earned as courses in residence. Candidate must have a scholastic standing of at least 2.0 on a 4.0 scale in all credits completed and in each major, minor or area of study.

STUDENT PROGRAMS
PROFESSIONAL AND HONORARY SOCIETIES
[For key to acronyms, see *Introduction*.]
Professional Societies: AGCA, IEEE, SME
Honorary Societies: Epsilon Pi Tau

SUPPORT PROGRAMS
Other Engineering Support Programs: All entering freshman are required to enroll in Freshman Orientation, 1 credit hour. The class consists of meetings with advisors, departmental personnel, service areas, and campus field trips. Availability of the university resources is stressed with emphasis on personal needs.

321 University of New Hampshire

INSTITUTION PROFILE

HEAD OF THE INSTITUTION
Dale F. Nitzschke
Phone: (603)862-2450   Fax: (603)862-3060

GENERAL INFORMATION
[All Students—Fall 1991]
Undergraduate enrollment 10,704
Graduate enrollment 1,553
Total institution enrollment 12,257
Type of institution: Public **Calendar:** Semesters
Nearest city: Dover, NH **Population:** 26,000
Miles from main campus: 4 **Setting:** Small Town
Types of engineering degrees: Engineering & Technology
Other degree-granting colleges: Liberal Arts, Life Science and Agriculture, Business and Economics, Health and Human Services

ENGINEERING TECHNOLOGY ADMISSIONS

ADMISSIONS OFFICE CONTACT
David W. Kraus
University of New Hampshire
Grant House
Durham, NH 03824
Phone: (603)862-1360

TECHNOLOGY COLLEGE ADMISSIONS INFORMATION

Requirements for foreign students: TOEFL (Score: 550)

DEPARTMENTAL ADMISSIONS INFORMATION
Admission to the engineering technology department: At the end of the second year.
Additional information: Must have graduated from a TAC ABET accredited institution, or provide evidence of the ability to successfully complete the requirements of the program.

FINANCIAL INFORMATION

ESTIMATED EXPENSES (ACADEMIC YEAR)
[Expenses are for the 1992-93 nine-month academic year.]

State Residents	Undergraduate	Graduate
Tuition and fees	$ 3,290	$ 4,057
College room and board	$ 3,600	$ 3,600
Books and supplies	$ 250	$ 250
Other expenses	$ —	$ —
Total estimated expenses	$ 7,140	$ 7,907
Out-of-State Residents	Undergraduate	Graduate
Tuition and fees	$ 9,840 [1]	$ 10,530
College room and board	$ 3,600	$ 3,600
Books and supplies	$ 250	$ 250
Other expenses	$ —	$ —
Total estimated expenses	$13,690	$14,380

Notes: (1) Students from Vermont and Connecticut pay a yearly tuition of $4935.

FINANCIAL AID OFFICE CONTACT
Richard Craig, *Director*
University of New Hampshire
Financial Aid
Stoke Hall
Durham, NH 03824
Phone: (603)862-3600

GENERAL FINANCIAL AID INFORMATION
Forms accepted/required: FAF, IRS
Additional financial aid information: Aid is available from University grants and scholarships, Pell Grant Programs, UNH loan funds, Perkins Loans, Higher Education Act Loans and college work-study programs.

TECHNOLOGY COLLEGE INFORMATION
[For additional personnel, refer to the *Appendix*.]

HEAD OF ENGINEERING TECHNOLOGY
Otis J. Sproul
Phone: (603)862-1781   Fax: (603)862-2486

ENGINEERING TECHNOLOGY COLLEGE ADDRESS
College of Engineering and Physical Sciences
University of New Hampshire
Kingsbury Hall
Durham, NH 03824
Phone: (603)862-1234   Fax: (603)862-2486

TECHNOLOGY ENROLLMENTS—BY CLASS
[Numbers are technology baccalaureate enrollments for the fall 1991 term, unless otherwise footnoted.]
1st-year students/Freshmen .... (B)
2nd-year students/Sophomores .... (B)
3rd-year students/Juniors 43
4th-year students/Seniors 66
Total 109

Notes: (B) Data not applicable.

NUMBER OF DEGREES AWARDED—BY PROGRAM
[Numbers are engineering technology baccalaureate degrees awarded during the 1991-1992 academic year, unless otherwise footnoted. For full details about each engineering technology program, refer to the *Table of Degree Programs*.]
Electrical Engineering Technology 10
Mechanical Engineering Technology 13
Total 23

PERCENTAGE OF DEGREES AWARDED—BY CATEGORY
[Percentages are of all engineering technology baccalaureate degrees awarded during the 1991-1992 academic year, unless otherwise footnoted.]
African Americans .... — %
Asian/Pacific Island Americans .... — %
Hispanic Americans .... — %
Native Americans .... — %
Foreign Citizens .... — %
All Others .... 100.0 %
Women .... 2.2 %
Persons with disabilities .... — %
Students over 25 years of age .... — % (A)

Notes: (A) Data not available.

ENGINEERING TECHNOLOGY STUDENT DATA
Applicants to the engineering technology college
Number of applicants to engineering technology college 29
Percent offered admission 62.1%

Matriculated engineering technology students
Percentage in top quartile (25%) of High School class ...... — (A)

Notes: (A) Data not available.

FULL- & PART-TIME FACULTY—BY DEPARTMENT
[Figures are the head count of full-time faculty and the full-time equivalent (FTE) of part-time faculty for each engineering technology department or equivalent.]

Department	Full	Part
Electrical Eng Tech	2	—
Mechanical Eng Tech	2	—

TECHNOLOGY COLLEGE DESCRIPTION
An outstanding faculty devoted to teaching on a campus that is ideally located in a small town within an hour's drive of the ocean, the mountains and Boston, MA.

TECHNOLOGY DEGREE PROGRAMS DESCRIPTION
The University of New Hampshire offers a two-plus-two type of engineering technology program, curricula in electrical engineering technology and mechanical engineering technology are available. Students may continue study in their fields of specialization, select electives that broaden their educational background and participate in project courses as part of a team where their talents are utilized in solving real industrial design and analysis problems.

TRANSFER INFORMATION
Residency Requirements: The last 32 credits must be taken in residence.

Transfer via Articulation Agreements
Admission to engineering technology: At the end of the sophomore year
Technology graduates transferred from ... 4-yr: — 2-yr: 18
Requirements: An Associate's degree from a TAC ABET accredited institution, or evidence of the ability to successfully complete the requirements of the program.

GRADUATION REQUIREMENTS
Students must acquire a minimum of 128 credits in order to graduate. In addition, they must have a cumulative grade point average of 2.00 for all courses taken at the University of New Hampshire.

STUDENT PROGRAMS

PROFESSIONAL AND HONORARY SOCIETIES
[For key to acronyms, see *Introduction*.]
Professional Societies: ASME, IEEE
Honorary Societies: Phi Kappa Phi

SUPPORT PROGRAMS
Student Chapter Organizations: Society of Women Engineers

Office of Multicultural Student Affairs
For: Ethnic Minorities   **Available:** Year round
Offered: Freshman, Sophomore, Junior, Senior, Graduate level
Other Engineering Support Programs: The University provides orientation programs for all transfer students. There are also facilities that provide aid in learning skills, personal counseling, mathematics, career planning and placement. All students have assigned faculty advisors.

322 New Jersey Institute of Technology

■ INSTITUTION PROFILE
HEAD OF THE INSTITUTION
Saul K. Fenster
Phone: (201)596-3101 **Fax:** (201)624-2541
GENERAL INFORMATION
[All Students—Fall 1991]
Undergraduate enrollment 4,876
Graduate enrollment 2,521
Total institution enrollment **7,397**
Type of institution: Public **Calendar:** Semesters
Location: Newark **Population:** 275,221
Setting: Urban
Types of engineering degrees: Engineering & Technology
Other degree-granting colleges: Science and Liberal Arts, Architecture, Industrial Management

■ ENGINEERING TECHNOLOGY ADMISSIONS
ADMISSIONS OFFICE CONTACT
William Anderson
New Jersey Institute of Technology
University Heights
Newark, NJ 07102-9938
Phone: (201)596-3300 **Fax:** (201)802-1854
TECHNOLOGY COLLEGE ADMISSIONS INFORMATION
Entrance Requirements: Engineering Technology students are upper division transfer students. NJIT does not ask such transfer students for either SAT scores or high school information.
Requirements for foreign students: TOEFL (Score: 520)
DEPARTMENTAL ADMISSIONS INFORMATION
Admission to the engineering technology department: At the end of the second year.

■ FINANCIAL INFORMATION
ESTIMATED EXPENSES (ACADEMIC YEAR)
[Expenses are for the 1992-93 nine-month academic year.]

All Students	Undergraduate	Graduate
Tuition and fees	$ 4,524 [1]	$ — [3]
College room and board	$ 4,980 [2]	$ — [3]
Books and supplies	$ 750	$ — [3]
Other expenses	$ —	$ — [3]
Total estimated expenses	**$10,254**	**—**

Notes: (1) In-state (2) Average (3) Graduate studies not offered in Engineering Technology

FINANCIAL AID OFFICE CONTACT
Mary Hurdle, *Director*
New Jersey Institute of Technology
Financial Aid
University Heights
Newark, NJ 07102-9938
Phone: (201)596-3479
GENERAL FINANCIAL AID INFORMATION
Forms accepted/required: CSS-FAF, FAF, FAT, Stafford, IRS, SAR, Supplemental, Institutional, New Jersey Financial Aid Form (NJFAF)
Additional financial aid information: Financial aid offered includes: Pell grants, SEOG's State grants/scholarships, academic merit scholarships, private scholarships, Stafford loans, Perkins loans, Plus/SLS loans and institutional fund loans. Aid is in the form of scholarships, grants, loans and part-time campus employment.

■ TECHNOLOGY COLLEGE INFORMATION
[For additional personnel, refer to the *Appendix*.]
HEAD OF ENGINEERING TECHNOLOGY
George Pincus
Phone: (201)596-3213 **Fax:** (201)596-2316
EMail: Pincus@Admin1.NJIT.Edu

ENGINEERING TECHNOLOGY COLLEGE ADDRESS
Newark College of Engineering
New Jersey Institute of Technology
University Heights
323 Martin Luther King Blvd.
Newark, NJ 07102-9938
Phone: (201)596-3000 **Fax:** (201)596-2316
TECHNOLOGY ENROLLMENTS—BY CLASS
[Numbers are technology baccalaureate enrollments for the fall 1991 term, unless otherwise footnoted.]
1st-year students/Freshmen .... 0 [1]
2nd-year students/Sophomores .... 0 [1]
3rd-year students/Juniors 291
4th-year students/Seniors 223
Total **514**

Notes: (1) Engineering Technology is an upper division program only.

NUMBER OF DEGREES AWARDED—BY PROGRAM
[Numbers are engineering technology baccalaureate degrees awarded during the 1991-1992 academic year, unless otherwise footnoted. For full details about each engineering technology program, refer to the *Table of Degree Programs*.]
Computer Engineering Technology 6
Construction & Contracting Engineering Technology .. 32
Electrical Engineering Technology 51
Manufacturing Engineering Technology 7
Mechanical Engineering Technology 21
Surveying Engineering Technology 1
Total **118**

PERCENTAGE OF DEGREES AWARDED—BY CATEGORY
[Percentages are of all engineering technology baccalaureate degrees awarded during the 1991-1992 academic year, unless otherwise footnoted.]
African Americans 8.5%
Asian/Pacific Island Americans 7.6%
Hispanic Americans 6.8%
Native Americans .... — %
Foreign Citizens .... 2.5% [2]
All Others 74.6%
Women 3.4%
Persons with disabilities .... — % [A]
Students over 25 years of age 76.3%

Notes: (A) Data not available. (2) International students - not foreign born. NJIT maintains data on a citizen/permanent resident versus non-citizen basis.

ENGINEERING TECHNOLOGY STUDENT DATA
Applicants to the engineering technology college
Number of applicants to engineering technology college ... 208
Percent offered admission 71.1%
Matriculated engineering technology students
Percentage in top quartile (25%) of High School class ..... — [2]

Notes: (2) Engineering Technology is upper division only. NJIT does not usually ask transfer students to provide either SAT scores or high school rank information.

FULL- & PART-TIME FACULTY—BY DEPARTMENT
[Figures are the head count of full-time faculty and the full-time equivalent (FTE) of part-time faculty for each engineering technology department or equivalent.]

Department	Full	Part
Eng Tech	13	4.2

TECHNOLOGY COLLEGE DESCRIPTION
Co-operative education is available to engineering students at NJIT and consists of two separate semesters of full-time work. Participation in Co-op may extend the time required to complete the degree program by one year. The program includes courses in the basic sciences and the humanities and the social sciences. Thus, the overall program provides students with an education designed to permit students to make major contributions in many areas. More than 30 major research and public service centers on campus including Centers for Manufacturing Systems and Hazardous Substance Management.

TECHNOLOGY DEGREE PROGRAMS DESCRIPTION
The BSET program can be completed in two years of full-time day study or four years of part-time evening study (normally three evenings per week), and hence is available to those employed full time in industry. A core curriculum is required of all students. Each curriculum prepares the graduate to have a practical approach to the solution of everyday problems and to work closely with the engineer or scientist as an important member of the engineering scientific team.

DUAL DEGREE PROGRAMS
Technology baccalaureate graduates with dual degrees 0
Enrollment requirements: Written approval to undertake this curriculum must be obtained from the department and the dean of the appropriate college.

TRANSFER INFORMATION
Residency Requirements: Students must complete at least 33 credits approved by the department of their major study.
Transfer via Articulation Agreements
Admission to engineering technology: At the end of the sophomore year
Technology graduates transferred from ... 4-yr: — 2-yr: 113
Requirements: Completion of an AAS degree or its equivalent. Grade point average (GPA) of 2.5 or better recommended. Transcripts of all attempted post secondary school work. A rolling admission procedure is used.
Transfer without Articulation Agreements
Admission to engineering technology: At the end of the sophomore year
Technology graduates transferred from ... 4-yr: 1 2-yr: 4
Requirements: Completion of an AAS degree or its equivalent. 2.5 GPA or better recommended. Transcripts of all attempted post-secondary school work. A rolling admission procedure is used.

GRADUATION REQUIREMENTS
The BSET degree requires 132-137 credits (including 64 transfer credits) depending on major selected. Students must attain a grade point average of 2.0 in all the courses listed in the catalog as being required in the third and fourth years of the appropriate curriculum.

■ STUDENT PROGRAMS
PROFESSIONAL AND HONORARY SOCIETIES
[For key to acronyms, see *Introduction*.]
Professional Societies: ACI, ACSM, AGCA, ASHRAE, ASME, IEEE, ISA, SAE, SME, SPE*
Honorary Societies: Tau Alpha Pi, Sigma Lambda Chi
SUPPORT PROGRAMS
Educational Opportunity Program
For: Ethnic Minorities **Available:** Year round
Offered: Prematriculation, Junior, Senior

Women in Engineering, Science and Technology
For: Women **Available:** Year round
Offered: Prematriculation, Junior, Senior

Women Students Over 25
For: Women **Available:** Academic year
Offered: Junior, Senior

Big Sister/Little Sister
For: Women **Available:** Academic year
Offered: Junior, Senior

Other Engineering Support Programs: Transfer student orientation, personal counseling, academic counseling, career counseling, tutoring program, career services, international student services, health services and services to physically challenged and learning disabled students.

323 New Mexico State University

INSTITUTION PROFILE
HEAD OF THE INSTITUTION
James E. Halligan
Phone: (505)646-2035   Fax: (505)646-6334

GENERAL INFORMATION
[All Students—Fall 1991]
Undergraduate enrollment 13,251
Graduate enrollment 2,094
Total institution enrollment 15,345

Type of institution: Public **Calendar:** Semesters
Nearest city: El Paso, TX **Population:** 875,000
Miles from main campus: 50 **Setting:** Small Town
Types of engineering degrees: Engineering & Technology
Other degree-granting colleges: Arts & Sciences, Education, Business Administration and Economics, Agriculture & Home Economics, Human & Community Services

ENGINEERING TECHNOLOGY ADMISSIONS
ADMISSIONS OFFICE CONTACT
Bill Bruner
New Mexico State University
Box 30001, Dept. 3A
Las Cruces, NM 88003-0001
Phone: (505)646-3121   Fax: (505)646-6330

TECHNOLOGY COLLEGE ADMISSIONS INFORMATION
Admission to the engineering technology college: Admitted regularly if their English ACT score is 16 or greater and their performance on a local math exam is satisfactory. Provisional admission is possible.
Entrance Requirements: High School courses—English (4 years), Science (beyond General Science) (2 years), Mathematics (3 years), Foreign language or fine arts (1 year), Regular admission requires a 2.0 HS GPA & Enhanced ACT of 20 or a HS GPA of 2.5 or an Enhanced ACT of 21.
Entrance Recommendations: ACT (Score: 20), High School courses—Foreign language (2 years), Technical drawing (1 year), Computers (1 year)
Requirements for foreign students: TOEFL (Score: 500); financial statement

DEPARTMENTAL ADMISSIONS INFORMATION
Admission to the engineering technology department: Regular admission requires an Enhanced ACT of 16 or greater and a satisfactory score on a local math exam. Provisional admission is allowed.
Additional information: Students not qualifying for regular admission will be admitted into the General ET program. Such students have two years to complete any deficiencies with a 2.0 GPA.

FINANCIAL INFORMATION
ESTIMATED EXPENSES (ACADEMIC YEAR)
[Expenses are for the 1992-93 nine-month academic year.]

State Residents	Undergraduate	Graduate
Tuition and fees	$ 1,756	$ 1,876
College room and board	$ 2,668 [1]	$ 2,668 [1]
Books and supplies	$ 250	$ 580
Other expenses	$ –	$ –[B]
Total estimated expenses	**$ 4,674**	**$ 5,124**

Out-of-State Residents	Undergraduate	Graduate
Tuition and fees	$ 5,686	$ 5,806
College room and board	$ 2,668 [1]	$ 2,668 [1]
Books and supplies	$ 250	$ 580
Other expenses	$ –	$ –[B]
Total estimated expenses	**$ 8,604**	**$ 9,054**

Notes: (B) Data not applicable. (1) Board cost is for 12 meals per week.

FINANCIAL AID OFFICE CONTACT
Greeley Myers, *Director of Financial Aid*
New Mexico State University
Box 30001, Dept. 5100
Las Cruces, NM 88003-0001
Phone: (505)646-4105   Fax: (505)646-6330

GENERAL FINANCIAL AID INFORMATION
Forms accepted/required: FAT, Stafford, IRS, SAR, Supplemental, United Student Aid Fund Single File Application, New Mexico Supplemental Information Form
Additional financial aid information: Academic scholarships (including special scholarships for out of state students) from the University, the College of Engineering, and Air Force & Army ROTC. Need based financial aid includes Pell Grant, SEOG, Perkins DSL, and work study. Minority students are also eligible for National Action Council for Minorities (NACME) Scholarships. Non-need based work study is also available.

TECHNOLOGY COLLEGE INFORMATION
[For additional personnel, refer to the *Appendix*.]
HEAD OF ENGINEERING TECHNOLOGY
J. Derald Morgan
Phone: (505)646-2911   Fax: (505)646-3549

ENGINEERING TECHNOLOGY COLLEGE ADDRESS
Department of Engineering Technology
New Mexico State University
Box 30001, Dept 3566
Las Cruces, NM 88003-0001
Phone: (505)646-2236   Fax: (505)646-6107

TECHNOLOGY ENROLLMENTS—BY CLASS
[Numbers are technology baccalaureate enrollments for the fall 1991 term, unless otherwise footnoted.]
1st-year students/Freshmen 49
2nd-year students/Sophomores 58
3rd-year students/Juniors 72
4th-year students/Seniors 138
Total 317

NUMBER OF DEGREES AWARDED—BY PROGRAM
[Numbers are engineering technology baccalaureate degrees awarded during the 1991-1992 academic year, unless otherwise footnoted. For full details about each engineering technology program, refer to the *Table of Degree Programs*.]
Civil Engineering Technology 11
Electronic(s) Engineering Technology 55
Mechanical Engineering Technology 13
Total 79

PERCENTAGE OF DEGREES AWARDED—BY CATEGORY
[Percentages are of all engineering technology baccalaureate degrees awarded during the 1991-1992 academic year, unless otherwise footnoted.]
African Americans .... 1.0 %
Asian/Pacific Island Americans .... 2.0 %
Hispanic Americans .... 35.0 %
Native Americans .... 4.0 %
Foreign Citizens .... 2.0 %
All Others .... 56.0 %
Women .... 11.0 %
Persons with disabilities .... 3.0 %
Students over 25 years of age .... 60.0 %

FULL- & PART-TIME FACULTY—BY DEPARTMENT
[Figures are the head count of full-time faculty and the full-time equivalent (FTE) of part-time faculty for each engineering technology department or equivalent.]

Department	Full	Part
Eng Tech	3	–

TECHNOLOGY COLLEGE DESCRIPTION
The College offers four year programs in engineering, engineering technology, and surveying. Freshman students are integrated into the College of Engineering. Students with adequate mathematics skills are immediately enrolled in a FORTRAN course taught by an ET faculty member. Students in all options may complete an Environmental Minor which includes two courses taught via interactive television from NM Tech or the University of New Mexico. Mechanical option students may choose an emphasis in composite materials or manufacturing. Electronic option students may choose a computer emphasis or communications emphasis. Civil students can earn a dual degree in ET and surveying in about three additional semesters.

TECHNOLOGY DEGREE PROGRAMS DESCRIPTION
Students receive degrees in engineering technology with options in civil, electronics, or mechanical technology. Students in all options may complete the requirements for a minor in Waste Management with about one additional semester. The Mechanical option offers interdisciplinary options in Composite Materials and an emphasis in Manufacturing. The Electronics option offers an emphasis in computer technology or emphasis in electronic communications. Finally, students in the Civil option may earn a second BS in Surveying in about three additional semesters. The College of Engineering offers a unique Environmental Minor which requires student to take at least two courses by interactive television from other universities. New Mexico State University has an unusually diverse student body. Students from Indian reservations, farms, ranches, Hispanic communities, metropolitan El Paso and Albuquerque, and Los Alamos study, work and play with students from many other states and foreign countries. New Mexico State University is the only Carnegie I Research Institution which is also a member of the Hispanic Association of Colleges and Universities.

DUAL DEGREE PROGRAMS
Technology baccalaureate graduates with dual degrees 0

TRANSFER INFORMATION
Residency Requirements: Students must complete 30 hours in residence.
Transfer without Articulation Agreements
Admission to engineering technology: Transfer students are placed in the General ET status. Regular admission requires establishing a 2.0 GPA over certain courses.
Technology graduates transferred from ... 4-yr: 10% 2-yr: 30%

GRADUATION REQUIREMENTS
Overall GPA of 2.0. Last 30 hours must be taken at NMSU.

STUDENT PROGRAMS
PROFESSIONAL AND HONORARY SOCIETIES
[For key to acronyms, see *Introduction*.]
Professional Societies: AIAA, AMS, ASAE, ASCE, ASHRAE, ASME, IEEE, IIE, SAE, SME, American Society for Certified Engineering Technicians
Honorary Societies: Tau Alpha Pi, Phi Kappa Phi

SUPPORT PROGRAMS
Student Chapter Organizations: Society of Women Engineers, National Society of Black Engineers, Society of Hispanic Professional Engineers, American Indians in Science and Engineering

American Indian Programs
For: Ethnic Minorities   **Available:** Year round
Offered: Prematriculation, Freshman, Sophomore, Junior, Senior, Graduate level

Black Programs
For: Ethnic Minorities   **Available:** Year round
Offered: Prematriculation, Freshman, Sophomore, Junior, Senior, Graduate level

Chicano Programs
For: Ethnic Minorities   **Available:** Year round
Offered: Prematriculation, Freshman, Sophomore, Junior, Senior, Graduate level

Women's Center
For: Women   **Available:** Year round
Offered: Prematriculation, Freshman, Sophomore, Junior, Senior, Graduate level

Other Engineering Support Programs: Summer orientation programs for new students, fall and spring orientation courses, Center for Counseling and Student Development, Center for Learning Assistance, federally funded student and peer counseling, veterans services, and free student organization sponsored tutoring. The ET Department also has a student study hall.

324 University of North Carolina-Charlotte

■ INSTITUTION PROFILE
HEAD OF THE INSTITUTION
James H. Woodward
Phone: (704)547-2201   Fax: (704)547-2144
GENERAL INFORMATION
[All Students—Fall 1991]
Undergraduate enrollment 13,087
Graduate enrollment 2,276
Total institution enrollment 15,363
Type of institution: Public   Calendar: Semesters
Nearest city: Charlotte   Population: 417,000
Miles from main campus: 1   Setting: Urban
Types of engineering degrees: Engineering & Technology
Other degree-granting colleges: Architecture, Arts & Sciences, Business Administration, Education, Nursing

■ ENGINEERING TECHNOLOGY ADMISSIONS
ADMISSIONS OFFICE CONTACT
Kathi M. Baucom
University of North Carolina-Charlotte
Director of Admissions
Charlotte, NC 28223
Phone: (704)547-2213   Fax: (704)547-2144
TECHNOLOGY COLLEGE ADMISSIONS INFORMATION

Requirements for foreign students: TOEFL (Score: 500); financial statement
DEPARTMENTAL ADMISSIONS INFORMATION
Admission to the engineering technology department: At the end of the second year.
Additional information: All students must have an Associate in Applied Science degree in an Engineering Technology program which would normally include any necessary prerequisite courses.

■ FINANCIAL INFORMATION
ESTIMATED EXPENSES (ACADEMIC YEAR)
[Expenses are for the 1992-93 nine-month academic year.]

State Residents	Undergraduate	Graduate
Tuition and fees	$ 1,191	$ –
College room and board	$ 3,126	$ –
Books and supplies	$ 600	$ –
Other expenses	$ 125	$ –
Total estimated expenses	**$ 5,042**	**$ –**

Out-of-State Residents	Undergraduate	Graduate
Tuition and fees	$ 6,863	$ –
College room and board	$ 3,126	$ –
Books and supplies	$ 600	$ –
Other expenses	$ 125	$ –
Total estimated expenses	**$10,714**	**$ –**

FINANCIAL AID OFFICE CONTACT
Curtis R. Whalen, *Director*
University of North Carolina-Charlotte
Student Financial Aid
King Building
Charlotte, NC 28223
Phone: (704)547-2461   Fax: (704)547-2144
GENERAL FINANCIAL AID INFORMATION
Forms accepted/required: FAF
Additional financial aid information: Scholarships, grants, loans, part-time employment; College Work Study Program. Required forms: Financial Aid Form of the College Scholarship Service.

■ TECHNOLOGY COLLEGE INFORMATION
[For additional personnel, refer to the *Appendix*.]
HEAD OF ENGINEERING TECHNOLOGY
Robert D. Snyder
Phone: (704)547-2301   Fax: (704)547-2352
ENGINEERING TECHNOLOGY COLLEGE ADDRESS
College of Engineering
University of North Carolina-Charlotte
Highway 49
Smith Engineering Building
Charlotte, NC 28223
Phone: (704)547-2305   Fax: (704)547-2352
TECHNOLOGY ENROLLMENTS—BY CLASS
[Numbers are technology baccalaureate enrollments for the fall 1991 term, unless otherwise footnoted.]
1st-year students/Freshmen 0
2nd-year students/Sophomores 0
3rd-year students/Juniors 227
4th-year students/Seniors 183
Total **410**
NUMBER OF DEGREES AWARDED—BY PROGRAM
[Numbers are engineering technology baccalaureate degrees awarded during the 1991-1992 academic year, unless otherwise footnoted. For full details about each engineering technology program, refer to the *Table of Degree Programs*.]
Civil Engineering Technology 20
Electrical Engineering Technology 59
Manufacturing Engineering Technology 6
Mechanical Engineering Technology 27
Total **112**
PERCENTAGE OF DEGREES AWARDED—BY CATEGORY
[Percentages are of all engineering technology baccalaureate degrees awarded during the 1991-1992 academic year, unless otherwise footnoted.]
African Americans 5.2%
Asian/Pacific Island Americans 3.5%
Hispanic Americans 0.9%
Native Americans 0.9%
Foreign Citizens 3.5%
All Others 86.0%
Women 5.2%
Persons with disabilities .... – %
Students over 25 years of age 69.3%
ENGINEERING TECHNOLOGY STUDENT DATA
Applicants to the engineering technology college
Number of applicants to engineering technology college ... 202
Percent offered admission 93.1%
FULL- & PART-TIME FACULTY—BY DEPARTMENT
[Figures are the head count of full-time faculty and the full-time equivalent (FTE) of part-time faculty for each engineering technology department or equivalent.]

Department	Full	Part
Civil Eng Tech	3	1.1
Electrical Eng Tech	7	1.1
Manufacturing Eng Tech	2	1.0
Mechanical Eng Tech	6	1.1

TECHNOLOGY COLLEGE DESCRIPTION
The Engineering Technology programs provide the upper two (i.e. junior and senior) years of a two-plus-two program. All students are admitted as juniors subsequent to earning an Associate in Applied Science degree elsewhere, usually at a two-year community college or technical institute. All programs are of an applied nature and laboratory-intensive. Co-op is optional for all programs. Average number of years required to actually complete the bachelor's degree: Approximately 2.5.

TECHNOLOGY DEGREE PROGRAMS DESCRIPTION
The engineering technology programs provide an opportunity for graduates of two-year associate degree engineering technology programs to continue their education at the baccalaureate level and to pursue careers as engineering technologists. All graduates receive the degree of Bachelor of Science in Engineering Technology.
TRANSFER INFORMATION
Residency Requirements: Students must complete the last 30 semester hours of credit toward the degree in residence as well as the last 12 semester hours of the work in the major field.
Transfer via Articulation Agreements
Admission to engineering technology: At the end of the sophomore year
Technology graduates transferred from ... 4-yr: 8 2-yr: 106
Requirements: Minimum 2.5 GPA for entry into Manufacturing Engineering Technology and 2.2 for all others.
GRADUATION REQUIREMENTS
Bachelor of Science in Engineering Technology programs require 96 quarter credits (64 semester credits) in the two-year A.A.S. degree plus 64 to 66 semester credits, depending on major selected. Co-op students must complete at least three work periods.

■ STUDENT PROGRAMS
PROFESSIONAL AND HONORARY SOCIETIES
[For key to acronyms, see *Introduction*.]
Professional Societies: AGCA, ASCE, ASHRAE, ASME, IEEE, ITE, NSPE, SAE, SME, Society of Women Engineers (SWE)
Honorary Societies: Tau Alpha Pi
SUPPORT PROGRAMS
Student Chapter Organizations: Society of Women Engineers

University Transitional Opportunity Prog (UTOP)
For: Ethnic Minorities   Available: Summer only
Offered: Prematriculation

Student Advising for Freshman Excellence (SAFE)
For: Ethnic Minorities   Available: Year round
Offered: Freshman, Sophomore

Student Support Services (SSS)
For: Ethnic Minorities   Available: Year round
Offered: Junior, Senior

Other Engineering Support Programs: Summer and fall orientation program for all freshmen and transfers, Learning Center, tutorial services, supplemental instruction, Disabled Student Services, Writing Resources Center, Counseling Center, Center for Student Employment and Career Services, Center for International Studies.

325 Northeastern University

INSTITUTION PROFILE

HEAD OF THE INSTITUTION
John A. Curry
Phone: (617)437-2101 Fax: (617)437-5015

GENERAL INFORMATION
[All Students—Fall 1991]
Undergraduate enrollment 23,590
Graduate enrollment 5,292
Total institution enrollment 31,779
Type of institution: Private   Calendar: Quarters
Location: Boston   Population: 578,000
Setting: Urban
Types of engineering degrees: Engineering & Technology
Other degree-granting colleges: Allied Health Sciences, Arts & Sciences, Business Administration, Law, Nursing, Pharmacy, Technology & Applied Sciences, Computer Science, Criminal Justice

ENGINEERING TECHNOLOGY ADMISSIONS

ADMISSIONS OFFICE CONTACT
Michael J. Clifford
Northeastern University
Undergraduate Admissions
139 Richards Hall
Boston, MA 02115
Phone: (617)437-2200 Fax: (617)437-8780

TECHNOLOGY COLLEGE ADMISSIONS INFORMATION
Entrance Requirements: SAT (Score: 850), High School courses—English (4 years), Advanced algebra, geometry, trigonometry (1 year), Laboratory sciences (2 years), Social Sciences (2 years)
Entrance Recommendations: High School courses—Computer science (1 year), Mechanical drawing (1 year)
Requirements for foreign students: TOEFL (Score: 550); financial statement

DEPARTMENTAL ADMISSIONS INFORMATION
Admission to the engineering technology department: At the time of admission to the institution.

FINANCIAL INFORMATION

ESTIMATED EXPENSES (ACADEMIC YEAR)
[Expenses are for the 1992-93 nine-month academic year.]

All Students	Undergraduate	Graduate
Tuition and fees	$11,490 [1]	$ –
College room and board	$ 6,000 [2]	$ –
Books and supplies	$ – [A]	$ –
Other expenses	$ 795 [3]	$ –
Total estimated expenses	$18,285	$ –

Notes: (A) Data not available. (1) Tuition for freshmen for 3 quarters (2) Varies depending on the type of accommodation and meal plan selected (3) Includes required student center and health service fee; others, such as parking, are optional

FINANCIAL AID OFFICE CONTACT
Jean C. Eddy, *Dean of Financial Aid*
Northeastern University
356 Richards Hall
360 Huntington Avenue
Boston, MA 02115
Phone: (617)437-3190 Fax: (617)437-8623

GENERAL FINANCIAL AID INFORMATION
Forms accepted/required: FAF
Additional financial aid information: Financial aid includes state assistance programs, federal programs and University and other private scholarships.

TECHNOLOGY COLLEGE INFORMATION
[For additional personnel, refer to the *Appendix*.]

HEAD OF ENGINEERING TECHNOLOGY
Thomas E. Hulbert
Phone: (617)437-2500 Fax: (617)437-2501

ENGINEERING TECHNOLOGY COLLEGE ADDRESS
School of Engineering Technology
Northeastern University
120 Snell Engineering Center
360 Huntington Avenue
Boston, MA 02115
Phone: (617)437-2500 Fax: (617)437-2501

TECHNOLOGY ENROLLMENTS—BY CLASS
[Numbers are technology baccalaureate enrollments for the fall 1991 term, unless otherwise footnoted.]
1st-year students/Freshmen 134
2nd-year students/Sophomores .... 98 [1]
3rd-year students/Juniors .... 94 [1]
4th-year students/Seniors 94
Total 420

Notes: (1) Program is a five-year program; enrollment in Middler Year (between sophomore and junior) was 103

NUMBER OF DEGREES AWARDED—BY PROGRAM
[Numbers are engineering technology baccalaureate degrees awarded during the 1991-1992 academic year, unless otherwise footnoted. For full details about each engineering technology program, refer to the *Table of Degree Programs*.]
Aerospace Maintenance Engineering Technology 4
Computer Technology 18
Electrical Engineering Technology 76
Manufacturing Engineering Technology 1
Mechanical Engineering Technology 47
Mechanical-Structural Engineering Technology 4
Total 150

PERCENTAGE OF DEGREES AWARDED—BY CATEGORY
[Percentages are of all engineering technology baccalaureate degrees awarded during the 1991-1992 academic year, unless otherwise footnoted.]
African Americans .... –   %
Asian/Pacific Island Americans .... –   %
Hispanic Americans .... –   %
Native Americans .... –   %
Foreign Citizens .... –   %
All Others 100   %
Women .... –   %
Persons with disabilities .... –   %
Students over 25 years of age .... –   %

ENGINEERING TECHNOLOGY STUDENT DATA
Applicants to the engineering technology college
Number of applicants to engineering technology college ... 431
Percent offered admission 97.0%
Average SAT scores: Math—490, Verbal—410, Combined—900

FULL- & PART-TIME FACULTY—BY DEPARTMENT
[Figures are the head count of full-time faculty and the full-time equivalent (FTE) of part-time faculty for each engineering technology department or equivalent.]

Department	Full	Part
Electrical/Electronic(s) Eng Tech	4	2.0
Mechanical Eng Tech	2	3.0
Computer Tech	3	–

TECHNOLOGY COLLEGE DESCRIPTION
The five-year engineering technology programs are laboratory-intensive. Although co-op is required of full-time students, it is not available for part-time students. The school places particular emphasis on seeking students with academic promise who face cultural, economic, or personal barriers and assisting them in reaching their greatest potential. There is a large part-time program for working adults that operates in the evening and on weekends. The part-time associate degrees are four-year programs and the bachelor's degrees are seven-year programs.

TECHNOLOGY DEGREE PROGRAMS DESCRIPTION
The five-year degree programs in engineering technology emphasize industrial application of theoretical concepts to prepare work-ready graduates. All engineering technology programs award the BSET. The school also awards associate degrees part-time.

TRANSFER INFORMATION
Residency Requirements: Students must complete the last full year in residence.

Transfer via Articulation Agreements
Admission to engineering technology: At any time
Technology graduates transferred from . . . 4-yr: –   2-yr: 2
Requirements: The total student record is evaluated and acceptances are based on a rolling admission procedure.

Transfer without Articulation Agreements
Admission to engineering technology: At any time
Technology graduates transferred from . . . 4-yr: 4   2-yr: 12
Requirements: The total student record is evaluated and acceptances are based on a rolling admission procedure.

GRADUATION REQUIREMENTS
Bachelor of science in engineering technology programs require 186-189 quarter hours of study, depending on major selected. Associate in engineering programs require 100-105 quarter hours, depending on major selected.

STUDENT PROGRAMS

PROFESSIONAL AND HONORARY SOCIETIES
[For key to acronyms, see *Introduction*.]
Professional Societies: ASCE, ASME, IEEE, SAE, SME
Honorary Societies: Tau Alpha Pi

SUPPORT PROGRAMS
Student Chapter Organizations: Society of Women Engineers, Black Engineering Student Society

NUPRIME
For: Ethnic Minorities   **Available:** Year round
Offered: Prematriculation, Freshman, Sophomore, Junior, Senior

Women in Engineering
For: Women   **Available:** Year round
Offered: Prematriculation, Freshman, Sophomore, Junior, Senior

Other Engineering Support Programs: Remedial programs: problem solving and college algebra, graphics, English, economics--1st quarter. Math, graphics, English, physics--2nd quarter. Calculus, physics, Foundations of Communication, computer programming--3rd quarter. The Educational Assistance Center focuses on math, English and reading skills and is available to all students. Library tutoring offers assistance in specific subject

326 Oklahoma State University

INSTITUTION PROFILE
HEAD OF THE INSTITUTION
John R. Campbell
Phone: (405)744-6384   Fax: (405)744-6285

GENERAL INFORMATION
[All Students—Fall 1991]
Undergraduate enrollment 14,835
Graduate enrollment 3,810
Total institution enrollment 19,476

Type of institution: Public **Calendar:** Semesters
Nearest city: Oklahoma City **Population:** 403,000
Miles from main campus: 67 **Setting:** Small Town
Types of engineering degrees: Engineering Technology
Other degree-granting colleges: Arts & Sciences, Business Administration, Education, Agriculture, Human and Environmental Sciences, Veterinary Medicine, Graduate

ENGINEERING TECHNOLOGY ADMISSIONS
ADMISSIONS OFFICE CONTACT
Gordon L. Reese
Oklahoma State University
103 Whitehurst
Stillwater, OK 74078-0233
Phone: (405)744-6866

TECHNOLOGY COLLEGE ADMISSIONS INFORMATION
Entrance Requirements: ACT (Score: 22), High School courses—English (3 years), History (including American History) (1 year), Laboratory Science (1 year), Mathematics (2 years)
Entrance Recommendations: High School courses—English (4 years), History (must include American History) (2 years), Laboratory Science (2 years), Mathematics (3 years), SAT equivalent accepted in lieu of ACT score.
Requirements for foreign students: TOEFL (Score: 500); financial statement

DEPARTMENTAL ADMISSIONS INFORMATION
Admission to the engineering technology department: At the time of admission to the institution.

FINANCIAL INFORMATION
ESTIMATED EXPENSES (ACADEMIC YEAR)
[Expenses are for the 1992-93 nine-month academic year.]

State Residents	Undergraduate	Graduate
Tuition and fees	$ 1,644	$ 2,064[1]
College room and board	$ 2,960	$ 2,960
Books and supplies	$ 536	$ 576
Other expenses	$ 2,374	$ 2,374
Total estimated expenses	**$ 7,514**	**$ 7,974**

Out-of-State Residents	Undergraduate	Graduate
Tuition and fees	$ 4,622	$ 5,724[1]
College room and board	$ 2,960	$ 2,960
Books and supplies	$ 536	$ 576
Other expenses	$ 2,374	$ 2,374
Total estimated expenses	**$10,492**	**$11,634**

Notes: (1) Graduate programs not offered in Engineering Technology.

FINANCIAL AID OFFICE CONTACT
Charles W. Bruce, *Director, Financial Aid*
Oklahoma State University
101 Hanner Hall
Stillwater, OK 74078-0233
Phone: (405)744-7440   Fax: (405)744-6438

GENERAL FINANCIAL AID INFORMATION
Forms accepted/required: ACT, AFSA, FAF, FSA, FFS, Stafford, IRS, SAR
Additional financial aid information: All information on financial aids is available from the Office of Financial Aids at the University. A full spectrum of aid is available.

TECHNOLOGY COLLEGE INFORMATION
[For additional personnel, refer to the *Appendix*.]

HEAD OF ENGINEERING TECHNOLOGY
Karl N. Reid
Phone: (405)744-5140   Fax: (405)744-7545

ENGINEERING TECHNOLOGY COLLEGE ADDRESS
Division of Engineering Technology
Oklahoma State University
294 Cordell South
Stillwater, OK 74078-0233
Phone: (405)744-5638   Fax: (405)744-5283

TECHNOLOGY ENROLLMENTS—BY CLASS
[Numbers are technology baccalaureate enrollments for the fall 1991 term, unless otherwise footnoted.]
1st-year students/Freshmen 65
2nd-year students/Sophomores 101
3rd-year students/Juniors 167
4th-year students/Seniors 303
Total **636**

NUMBER OF DEGREES AWARDED—BY PROGRAM
[Numbers are engineering technology baccalaureate degrees awarded during the 1991-1992 academic year, unless otherwise footnoted. For full details about each engineering technology program, refer to the *Table of Degree Programs*.]
Construction Management Technology 8
Electronic(s) & Computer Technology 35
Fire Protection & Safety Technology 27
General Technology 1
Manufacturing Engineering Technology 4
Mechanical Design Technology 29
Mechanical Power Technology 9
Total **113**

PERCENTAGE OF DEGREES AWARDED—BY CATEGORY
[Percentages are of all engineering technology baccalaureate degrees awarded during the 1991-1992 academic year, unless otherwise footnoted.]
African Americans .... – %
Asian/Pacific Island Americans .... 2.3 %
Hispanic Americans .... – %
Native Americans .... 0.8 %
Foreign Citizens .... 14.7 %
All Others .... 82.2 %
Women .... 4.7 %
Persons with disabilities .... – % (A)
Students over 25 years of age .... – % (A)

Notes: (A) Data not available.

ENGINEERING TECHNOLOGY STUDENT DATA
Applicants to the engineering technology college
Number of applicants to engineering technology college ...... – (A)
Percent offered admission .... –% (A)

Matriculated engineering technology students
Percentage in top quartile (25%) of High School class .... – (A)
Average ACT scores: Composite—24.0

Notes: (A) Data not available.

FULL- & PART-TIME FACULTY—BY DEPARTMENT
[Figures are the head count of full-time faculty and the full-time equivalent (FTE) of part-time faculty for each engineering technology department or equivalent.]

Department	Full	Part
Electronics/Computer Tech	7	–
Construction Management Tech	3	–
Mechanical Power Tech	5.5	–
Fire Protection & Safety Tech	4	–
Mechanical Design & Manufacturing Tech	7	–

TECHNOLOGY COLLEGE DESCRIPTION
The Engineering Technology programs are a part of a spectrum of programs in the College of Engineering, Architecture, and Technology that is as broad as offered anywhere, although Associates Degrees have been phased out in the programs that formerly offered them. Co-op education is available as an option to students in the College in the alternating term format that adds about one year to the duration of the program.

TECHNOLOGY DEGREE PROGRAMS DESCRIPTION
The Engineering Technology degree programs are bachelor's level programs. Students will normally take at least one course in the major field each semester, beginning in the freshman year. (Associates degree programs have been discontinued.) No graduate programs are offered, but students may pursue masters degrees in engineering with additional undergraduate preparation.

TRANSFER INFORMATION
Residency Requirements: One half of major requirements must be earned in residence, during at least two semesters, one semester and a summer, or three summers.

Transfer via Articulation Agreements
Admission to engineering technology: At any time

Transfer without Articulation Agreements
Admission to engineering technology: At any time
Technology graduates transferred from ... 4-yr: 36   2-yr: 29

GRADUATION REQUIREMENTS
A minimum of 40 to 42 hours in the major is required in each of the degree plans. A 2.0 overall GPA and a 2.0 in the major are required.

STUDENT PROGRAMS
PROFESSIONAL AND HONORARY SOCIETIES
[For key to acronyms, see *Introduction*.]
Professional Societies: AIDD, IEEE, SAE, SFPE, SME, SPE
Honorary Societies: Tau Alpha Pi

SUPPORT PROGRAMS
Student Chapter Organizations: Society of Women Engineers, National Society of Black Engineers, Society of Hispanic Professional Engineers, American Indian Science and Engineering Society, Society of Asian-American Engineers

HUES (Housing for Underrepresented Eng. Stu.)
For: Ethnic Minorities   **Available:** Academic year
Offered: Prematriculation, Freshman, Sophomore, Junior

ESAAP (Summer Academic Achievement Program)
For: Ethnic Minorities   **Available:** Summer only
Offered: Freshman

Minority Special Orientation
For: Ethnic Minorities   **Available:** Academic year
Offered: Freshman

Other Engineering Support Programs: Special support programs for Technology students at Oklahoma State University are provided through the College of Engineering, Architecture and Technology available to all students in the College.

327 Old Dominion University

INSTITUTION PROFILE

HEAD OF THE INSTITUTION
James V. Koch
Phone: (804)683-3159 Fax: (804)683-5679

GENERAL INFORMATION
[All Students—Fall 1991]
Undergraduate enrollment 11,624
Graduate enrollment 5,062
Total institution enrollment **16,686**
Type of institution: Public Calendar: Semesters
Location: Norfolk Population: 261,229
Setting: Urban
Types of engineering degrees: Engineering & Technology
Other degree-granting colleges: Education, Arts and Letters, Business and Public Administration, Health Sciences, Sciences

ENGINEERING TECHNOLOGY ADMISSIONS

ADMISSIONS OFFICE CONTACT
Angela N. Boyd
Old Dominion University
Office of Admissions
Norfolk, VA 23529-0050
Phone: (804)683-3637 Fax: (804)683-3647

TECHNOLOGY COLLEGE ADMISSIONS INFORMATION
Entrance Requirements: SAT (Score: 850)
Entrance Recommendations: High School courses—English (4 years), Natural Sciences (3 years), Mathematics (3 years), Foreign language (3 years)
Requirements for foreign students: TOEFL (Score: 550); financial statement

DEPARTMENTAL ADMISSIONS INFORMATION
Admission to the engineering technology department: At the time of admission to the institution.

FINANCIAL INFORMATION

ESTIMATED EXPENSES (ACADEMIC YEAR)
[Expenses are for the 1992-93 nine-month academic year.]

State Residents	Undergraduate	Graduate
Tuition and fees	$ 3,978	$ –
College room and board	$ 4,450	$ –
Books and supplies	$ 550	$ –
Other expenses	$ –(A)	$ –
Total estimated expenses	**$ 8,978**	**$ –**

Out-of-State Residents	Undergraduate	Graduate
Tuition and fees	$ 9,792	$ –
College room and board	$ 4,450	$ –
Books and supplies	$ 550	$ –
Other expenses	$ –(A)	$ –
Total estimated expenses	**$14,792**	**$ –**

Notes: (A) Data not available.

FINANCIAL AID OFFICE CONTACT
Helga A. Greenfield, *Director*
Old Dominion University
Financial Aid Office
Norfolk, VA 23529-0052
Phone: (804)683-3683 Fax: (804)683-5357

GENERAL FINANCIAL AID INFORMATION
Forms accepted/required: FAF, Stafford, Institutional, Free Application for Federal Student Financial Aid (FAFSA)
Additional financial aid information: Need based scholarships; merit based scholarships; grants; loans; part-time employment on campus; work study programs.

TECHNOLOGY COLLEGE INFORMATION

[For additional personnel, refer to the *Appendix*.]

HEAD OF ENGINEERING TECHNOLOGY
Ernest J. Cross Jr.
Phone: (804)683-3787 Fax: (804)683-4898

ENGINEERING TECHNOLOGY COLLEGE ADDRESS
College of Engineering and Technology
Old Dominion University
Kaufman-Duckworth Hall, Room 101
Norfolk, VA 23529-0236
Phone: (804)683-3787 Fax: (804)683-4898

TECHNOLOGY ENROLLMENTS—BY CLASS
[Numbers are technology baccalaureate enrollments for the fall 1991 term, unless otherwise footnoted.]
1st-year students/Freshmen .... 68 [1]
2nd-year students/Sophomores .... 69 [1]
3rd-year students/Juniors .... 155 [1]
4th-year students/Seniors .... 276 [1]
Total **568**

Notes: (1) Fall 1991 data

NUMBER OF DEGREES AWARDED—BY PROGRAM
[Numbers are engineering technology baccalaureate degrees awarded during the 1991-1992 academic year, unless otherwise footnoted. For full details about each engineering technology program, refer to the *Table of Degree Programs*.]
Civil Engineering Technology 31
Electrical Engineering Technology 53
Mechanical Engineering Technology 38
Technical Operations 13
Total **135**

PERCENTAGE OF DEGREES AWARDED—BY CATEGORY
[Percentages are of all engineering technology baccalaureate degrees awarded during the 1991-1992 academic year, unless otherwise footnoted.]
African Americans 4.5%
Asian/Pacific Island Americans 7.4%
Hispanic Americans .... – %
Native Americans .... – %
Foreign Citizens 2.2%
All Others 85.9%
Women 12.6%
Persons with disabilities .... – %
Students over 25 years of age .... – % (A)

Notes: (A) Data not available.

ENGINEERING TECHNOLOGY STUDENT DATA
Applicants to the engineering technology college
Number of applicants to engineering technology college ... 107
Percent offered admission 80.0%

Matriculated engineering technology students
Percentage in top quartile (25%) of High School class 20.0
Average SAT scores: Math—510, Verbal—430, Combined—940

FULL- & PART-TIME FACULTY—BY DEPARTMENT
[Figures are the head count of full-time faculty and the full-time equivalent (FTE) of part-time faculty for each engineering technology department or equivalent.]

Department	Full	Part
Eng Tech	19	4.0

TECHNOLOGY COLLEGE DESCRIPTION
The University offers four year undergraduate programs in engineering and engineering technology as well as engineering degrees at the master's and doctoral level. The engineering technology curricula are laboratory intensive and utilize several of the College's new facilities including robotics and CAD labs. Over 60% of the junior and senior classes are made up of junior/community college transfers with associate degrees. Co-operative education is available to students in all programs. Capstone courses are required in each major and students are strongly urged to complete the national standardized Fundamentals of Engineering exam during their last year. The average number of years required to actually complete the bachelor's degree is 4.5.

TECHNOLOGY DEGREE PROGRAMS DESCRIPTION
The University's B.S. degree programs in engineering technology emphasize the operational and applied aspects of engineering related activities and utilization of devices and systems. All degrees are BSET. The only non-accredited program, Technical Operations, which combines technical and business courses is designed primarily for evening students with associate degrees so they might enhance advancement capabilities.

TRANSFER INFORMATION
Residency Requirements: Student must earn a minimum of 30 semester credits at the University.
Transfer via Articulation Agreements
Admission to engineering technology: At any time
Technology graduates transferred from ... 4-yr: – 2-yr: 100
Requirements: Must be in good academic standing at previous institution. Only grades of C or better transfer.
Transfer without Articulation Agreements
Admission to engineering technology: At any time
Requirements: Must be in good academic standing at previous institution. Only grades if C or better transfer.

GRADUATION REQUIREMENTS
B.S. programs in engineering technology require 133-134 semester hours, depending on major; exit writing exam; 2.00 GPA in major and overall.

STUDENT PROGRAMS

PROFESSIONAL AND HONORARY SOCIETIES
[For key to acronyms, see *Introduction*.]
Professional Societies: ACI, AGCA, AIAA, AIPE, ASCE, ASHRAE, ASME, IEEE, SAE, SME
Honorary Societies: Chi Epsilon, Eta Kappa Nu, Pi Tau Sigma, Tau Alpha Pi, Tau Beta Pi, Theta Tau

SUPPORT PROGRAMS
Student Chapter Organizations: Society of Women Engineers, Fellowship of Minority Engineers and Scientists, Affiliated with Nat'l Society of Black Engineers

Under-Represented Minorities in Engr Programs
For: Ethnic Minorities **Available:** Academic year
Offered: Freshman, Sophomore, Junior, Senior

Women's Center
For: Women **Available:** Year round
Offered: Freshman, Sophomore, Junior, Senior, Graduate level
Other Engineering Support Programs: The University conducts a summer orientation program for freshmen and transfers. During the program all testing, advising and registration is completed.

328 Oregon Institute of Technology

■ INSTITUTION PROFILE
HEAD OF THE INSTITUTION
Lawrence J. Wolf
Phone: (503)885-1101 Fax: (503)885-1515

GENERAL INFORMATION
[All Students—Fall 1991]
Undergraduate enrollment 2,666
Total institution enrollment 2,666
Type of institution: Public Calendar: Quarters
Nearest city: Portland, OR Population: 700,000
Miles from main campus: 275 Setting: Small Town
Types of engineering degrees: Engineering Technology

■ ENGINEERING TECHNOLOGY ADMISSIONS
ADMISSIONS OFFICE CONTACT
Russ Lyon
Oregon Institute of Technology
3201 Campus Drive
Klamath Falls, OR 97601-8801
Phone: (503)885-1150

TECHNOLOGY COLLEGE ADMISSIONS INFORMATION
Entrance Requirements: High School courses—English (4 years), Mathematics (3 years), Social Studies (3 years), Science (2 years)
Entrance Recommendations: SAT (Score: 890), ACT (Score: 21)
Requirements for foreign students: TOEFL (Score: 520); financial statement

DEPARTMENTAL ADMISSIONS INFORMATION
Admission to the engineering technology department: At the time of admission to the institution.
Additional information: Laser Optical Engineering Technology -- Placement in College Algebra or higher.

■ FINANCIAL INFORMATION
ESTIMATED EXPENSES (ACADEMIC YEAR)
[Expenses are for the 1992-93 nine-month academic year.]

State Residents	Undergraduate	Graduate
Tuition and fees	$ 2,097	$ –
College room and board	$ 3,390	$ –
Books and supplies	$ 520	$ –
Other expenses	$ 1,840	$ –
Total estimated expenses	**$ 7,847**	**$ –**

Out-of-State Residents	Undergraduate	Graduate
Tuition and fees	$ 6,180	$ –
College room and board	$ 3,390	$ –
Books and supplies	$ 520	$ –
Other expenses	$ 1,840	$ –
Total estimated expenses	**$11,930**	**$ –**

FINANCIAL AID OFFICE CONTACT
John Huntley, *Director of Financial Aid*
Oregon Institute of Technology
3201 Campus Drive
Klamath Falls, OR 97601-8801
Phone: (503)885-1280

GENERAL FINANCIAL AID INFORMATION
Forms accepted/required: USAFund Forms.
Additional financial aid information: All federal and state financial aid programs are available, plus many private need or merit-based scholarships.

■ TECHNOLOGY COLLEGE INFORMATION
[For additional personnel, refer to the *Appendix*.]
HEAD OF ENGINEERING TECHNOLOGY
James McAtee
Phone: (503)885-1240
EMail: mcateej@oit.osshe.edu

ENGINEERING TECHNOLOGY COLLEGE ADDRESS
Division of Engineering Technology
Oregon Institute of Technology
3201 Campus Drive
Klamath Falls, OR 97601-8801
Phone: (503)885-1000 Fax: (503)885-1515

TECHNOLOGY ENROLLMENTS—BY CLASS
[Numbers are technology baccalaureate enrollments for the fall 1991 term, unless otherwise footnoted.]
1st-year students/Freshmen 318
2nd-year students/Sophomores 219
3rd-year students/Juniors 208
4th-year students/Seniors 371
Total 1,116

NUMBER OF DEGREES AWARDED—BY PROGRAM
[Numbers are engineering technology baccalaureate degrees awarded during the 1991-1992 academic year, unless otherwise footnoted. For full details about each engineering technology program, refer to the *Table of Degree Programs*.]
Civil Engineering Technology 30
Computer Systems Technology 20
Electrical/Electronic(s) Engineering Technology 37
Laser Optical Engineering Technology 15
Manufacturing Engineering Technology 14
Mechanical Engineering Technology 41
Surveying 10
Total 167

PERCENTAGE OF DEGREES AWARDED—BY CATEGORY
[Percentages are of all engineering technology baccalaureate degrees awarded during the 1991-1992 academic year, unless otherwise footnoted.]
African Americans .... – %
Asian/Pacific Island Americans 5.5%
Hispanic Americans 1.5%
Native Americans 1.0%
Foreign Citizens 2.0%
All Others 90.0%
Women 5.5%
Persons with disabilities .... – %(A)
Students over 25 years of age .... – %(A)

Notes: (A) Data not available.

FULL- & PART-TIME FACULTY—BY DEPARTMENT
[Figures are the head count of full-time faculty and the full-time equivalent (FTE) of part-time faculty for each engineering technology department or equivalent.]

Department	Full	Part
Civil Eng Tech	7	–
Electrical/Electronic(s) Eng Tech	12	–
Manufacturing Eng Tech	6	–
Mechanical Eng Tech	8	–
Survey	3	–
Laser Optical Eng Tech	3	–
Computer Systems Eng Tech	9	–

TECHNOLOGY COLLEGE DESCRIPTION
Small class size. Hands-on philosophy; lots of lab time. Heavy computerization and computer networking. Optional co-ops. Exchange programs in Japan, China, and Europe. Senior projects.

DUAL DEGREE PROGRAMS
Technology baccalaureate graduates with dual degrees 5

TRANSFER INFORMATION
Residency Requirements: 45 resident quarter hours with last 30 on-campus.
Transfer via Articulation Agreements
Admission to engineering technology: At any time
Requirements: Transfer GPA of 2.0 required.
Transfer without Articulation Agreements
Admission to engineering technology: At any time
Requirements: Transfer GPA of 2.0 required.

■ STUDENT PROGRAMS
PROFESSIONAL AND HONORARY SOCIETIES
[For key to acronyms, see *Introduction*.]
Professional Societies: ACM, ASCE, ASME, IEEE, SME
SUPPORT PROGRAMS
Student Chapter Organizations: Society of Women Engineers
Other Engineering Support Programs: Orientation Programs. Learning Assistance Center. Study Skills Seminars. Academic and Personal Counseling.

329 Pennsylvania State University at Erie, The Behrend College

INSTITUTION PROFILE

HEAD OF THE INSTITUTION
John M. Lilley
Phone: (814)898-6160 Fax: (814)898-6461

GENERAL INFORMATION
[All Students—Fall 1991]
Undergraduate enrollment 2,974
Graduate enrollment 216
Total institution enrollment 3,190
Type of institution: Public Calendar: Semesters
Nearest city: Erie Population: 108,000
Miles from main campus: 2 Setting: Suburban
Types of engineering degrees: Engineering & Technology
Other degree-granting colleges: Business Administration, Communication & Journalism, Humanities & Social Sciences

ENGINEERING TECHNOLOGY ADMISSIONS

ADMISSIONS OFFICE CONTACT
Mary-Ellen Madigan
Pennsylvania State University at Erie, The Behrend College
Station Road
Erie, PA 16563-0105
Phone: (814)898-6100 Fax: (814)898-6461

TECHNOLOGY COLLEGE ADMISSIONS INFORMATION
Entrance Requirements: SAT, ACT, High School courses—Academic Math (2 years), English (4 years), Social Studies/Art (4 years), SAT or ACT scores are considered in combination with high school grades.
Requirements for foreign students: TOEFL (Score: 550); financial statement

DEPARTMENTAL ADMISSIONS INFORMATION
Admission to the engineering technology department: At the time of admission to the institution.
Additional information: Penn State - Behrend associate degree students must apply for readmission to enter the baccalaureate engineering technology programs. Students with a cumulative GPA less than 2.6 will be admitted to these programs on a space available basis with the approval of the School of Engineering and Engineering Technology. Engineering students must complete the requirements for an associate degree in engineering technology before being admitted to the baccalaureate engineering technology programs.

FINANCIAL INFORMATION

ESTIMATED EXPENSES (ACADEMIC YEAR)
[Expenses are for the 1992-93 nine-month academic year.]

State Residents	Undergraduate	Graduate
Tuition and fees	$ 4,618	$ –
College room and board	$ 3,650	$ –
Books and supplies	$ 404	$ –
Other expenses	$ –(1)	$ –
Total estimated expenses	**$ 8,672**	**$ –**

Out-of-State Residents	Undergraduate	Graduate
Tuition and fees	$ 9,644	$ –
College room and board	$ 3,650	$ –
Books and supplies	$ 404	$ –
Other expenses	$ –(1)	$ –
Total estimated expenses	**$13,698**	**$ –**

Notes: (1) Engineering surcharge of $400 for junior and senior students.

FINANCIAL AID OFFICE CONTACT
Kate Delfino, *Assistant Director of Admissions & Financial Aid*
Pennsylvania State University at Erie, The Behrend College
Station Road
Erie, PA 16563-0105
Phone: (814)898-6162 Fax: (814)898-6461

GENERAL FINANCIAL AID INFORMATION
Forms accepted/required: FAT, Stafford, Pennsylvania State Grant, Federal Student Aid
Additional financial aid information: Need based: loans, grants, college work study. Merit based: scholarships.

TECHNOLOGY COLLEGE INFORMATION
[For additional personnel, refer to the *Appendix*.]

HEAD OF ENGINEERING TECHNOLOGY
Richard C. Progelhof
Phone: (814)898-6153 Fax: (814)898-6213
EMail: rcp5@psuvm.psu.edu

ENGINEERING TECHNOLOGY COLLEGE ADDRESS
School of Engineering and Engineering Technology
Pennsylvania State University at Erie, The Behrend College
Station Road
Erie, PA 16563-0203
Phone: (814)898-6153 Fax: (814)898-6213

TECHNOLOGY ENROLLMENTS—BY CLASS
[Numbers are technology baccalaureate enrollments for the fall 1991 term, unless otherwise footnoted.]
1st-year students/Freshmen 77
2nd-year students/Sophomores 105
3rd-year students/Juniors 94
4th-year students/Seniors 102
Total 378

NUMBER OF DEGREES AWARDED—BY PROGRAM
[Numbers are engineering technology baccalaureate degrees awarded during the 1991-1992 academic year, unless otherwise footnoted. For full details about each engineering technology program, refer to the *Table of Degree Programs*.]
Electrical Engineering 15
Mechanical Engineering 21
Plastics Engineering 9
Total 45

PERCENTAGE OF DEGREES AWARDED—BY CATEGORY
[Percentages are of all engineering technology baccalaureate degrees awarded during the 1991-1992 academic year, unless otherwise footnoted.]
African Americans .... – %
Asian/Pacific Island Americans .... – %
Hispanic Americans .... – %
Native Americans .... – %
Foreign Citizens .... – %
All Others .... 100 %
Women .... – %
Persons with disabilities .... – %
Students over 25 years of age .... 81.0 %

ENGINEERING TECHNOLOGY STUDENT DATA

Applicants to the engineering technology college
Number of applicants to engineering technology college 44
Percent offered admission 89.1%

Matriculated engineering technology students
Percentage in top quartile (25%) of High School class .... – (A)

Notes: (A) Data not available.

FULL- & PART-TIME FACULTY—BY DEPARTMENT
[Figures are the head count of full-time faculty and the full-time equivalent (FTE) of part-time faculty for each engineering technology department or equivalent.]

Department	Full	Part
Electrical Eng Tech	5	–
Mechanical Eng Tech	8	1.7
Plastics Eng Tech	3	–

TECHNOLOGY COLLEGE DESCRIPTION
Penn State Erie, The Behrend College, located on a 600-acre tract in Erie, PA, provides a small-university environment within the context of a major university. Enrolling more than 3,000 students, Penn State - Behrend is a comprehensive four-year and graduate college with a full range of supporting services. Students can earn the Bachelor of Arts or Bachelor of Science degree at the four year level or an associate degree in selected two-year programs. Laboratories are equipped with state-of-the-art equipment. Graduates of the baccalaureate engineering technology programs are trained to work side-by-side with engineers, scientists and other skilled workers in the constantly changing field of technology.

TECHNOLOGY DEGREE PROGRAMS DESCRIPTION
Penn State-Behrend awards 2-year Associate degrees in Electrical Engineering Technology and Mechanical Engineering Technology. In addition, Bachelor of Science degrees are offered in the fields of Electrical Engineering Technology, Mechanical Engineering Technology, and Plastics Engineering Technology. Each of the baccalaureate majors is a "2+2" curriculum designed primarily for junior level entry of graduates of TAC-ABET accredited two year programs. The ABET-accredited Electrical Engineering Technology program provides study in the areas of microprocessors, linear systems analysis, filter theory, and control and power systems. The ABET-accredited Mechanical Engineering Technology major emphasizes thermodynamics, fluid dynamics, heat transfer, finite-element analysis, CAD/CAM, and machine design. The Plastics Engineering Technology program is one of only three such programs in the country and introduces students to the plastics industry and plastics materials, including mold design, design of plastics parts and plastics production equipment. All programs make extensive use of commercial grade computer hardware and software and culminate in a senior-level design experience. The vast majority of these "capstone" design projects are done in conjunction with regional industry.

TRANSFER INFORMATION
Transfer without Articulation Agreements
Admission to engineering technology: Engineering students must complete requirements for an associate degree before being admitted to the baccalaureate program.

GRADUATION REQUIREMENTS
Students graduating from Behrend College baccalaureate majors must achieve a minimum grade-point average of at least 2.00; achieve an overall grade-point average of at least a 2.00 within the Prescribed, Additional, and Supporting courses in the Requirements for the major, including chosen Options, if any; and earn a C grade or better in all 300- and 400-level courses within these same Prescribed, Additional, and Supporting courses. If a student receives a grade below C, he/she must repeat that course or a School-Approved alternative, and earn a grade of C or better.

SUPPORT PROGRAMS
Student Chapter Organizations: National Society of Black Engineers, Association of Black Collegians, Asian Study Organization, Organization of Latin Americans

330 Pennsylvania State University - Harrisburg

■ INSTITUTION PROFILE
HEAD OF THE INSTITUTION
Ruth Leventhal
Phone: (717)948-6101
GENERAL INFORMATION
[All Students—Fall 1991]
Undergraduate enrollment 2,100
Graduate enrollment 1,400
Total institution enrollment **3,500**
Type of institution: Public **Calendar:** Semesters
Nearest city: Harrisburg **Population:** 200,000
Miles from main campus: 8 **Setting:** Small Town
Types of engineering degrees: Engineering Technology
Other degree-granting colleges: Allied Health Sciences, Arts & Sciences, Business Administration, Communication & Journalism, Education, Humanities & Social Sciences

■ ENGINEERING TECHNOLOGY ADMISSIONS
ADMISSIONS OFFICE CONTACT
Ed Escalet
Pennsylvania State University - Harrisburg
777 W. Harrisburg Pike
Middletown, PA 17057
Phone: (717)948-6250

TECHNOLOGY COLLEGE ADMISSIONS INFORMATION

Requirements for foreign students: TOEFL (Score: 500); financial statement

DEPARTMENTAL ADMISSIONS INFORMATION
Admission to the engineering technology department: At the end of the second year.
Additional information: A set of courses in Basic Sciences, Math, Humanities and Social Sciences listed in catalog.

■ FINANCIAL INFORMATION
ESTIMATED EXPENSES (ACADEMIC YEAR)
[Expenses are for the 1992-93 nine-month academic year.]

State Residents	Undergraduate	Graduate
Tuition and fees	$ 3,978	$ 4,450
College room and board	$ 3,330	$ 3,330
Books and supplies	$ 400	$ 500
Other expenses	$ –	$ –
Total estimated expenses	**$ 7,708**	**$ 8,280**

Out-of-State Residents	Undergraduate	Graduate
Tuition and fees	$ 8,374	$ 8,890
College room and board	$ 3,330	$ 3,330
Books and supplies	$ 500	$ 500
Other expenses	$ –	$ –
Total estimated expenses	**$12,204**	**$12,720**

■ TECHNOLOGY COLLEGE INFORMATION
[For additional personnel, refer to the *Appendix*.]
HEAD OF ENGINEERING TECHNOLOGY
Gautam Ray
Phone: (717)948-6108 **Fax:** (717)948-6401
EMail: GXR6@PSUADMIN
ENGINEERING TECHNOLOGY COLLEGE ADDRESS
School of Science, Engineering & Technology
Pennsylvania State University - Harrisburg
777 W. Harrisburg Pike
Middletown, PA 17057
Phone: (717)948-6250 **Fax:** (717)948-6401

TECHNOLOGY ENROLLMENTS—BY CLASS
[Numbers are technology baccalaureate enrollments for the fall 1991 term, unless otherwise footnoted.]
1st-year students/Freshmen .... (2)
2nd-year students/Sophomores .... (2)
3rd-year students/Juniors 252
4th-year students/Seniors 325
Total **577**
Notes: (2) This campus offers Junior and Senior courses only.

NUMBER OF DEGREES AWARDED—BY PROGRAM
[Numbers are engineering technology baccalaureate degrees awarded during the 1991-1992 academic year, unless otherwise footnoted. For full details about each engineering technology program, refer to the *Table of Degree Programs*.]
Electrical & Computer Engineering Technology 68
Environmental Engineering Technology 137
Mechanical Engineering Technology 53
Structural Design & Construction Engineering Tech 30
Total **288**

PERCENTAGE OF DEGREES AWARDED—BY CATEGORY
[Percentages are of all engineering technology baccalaureate degrees awarded during the 1991-1992 academic year, unless otherwise footnoted.]
African Americans .... 2.4 %
Asian/Pacific Island Americans .... – %
Hispanic Americans .... – %
Native Americans .... – %
Foreign Citizens .... – %
All Others .... 97.6 %
Women .... 2.0 %
Persons with disabilities .... – %
Students over 25 years of age .... 30.1 %

FULL- & PART-TIME FACULTY—BY DEPARTMENT
[Figures are the head count of full-time faculty and the full-time equivalent (FTE) of part-time faculty for each engineering technology department or equivalent.]

Department	Full	Part
Electrical & Computer Eng Tech	12	2.0
Mechanical Eng Tech	10	2.0
Structural Design & Construction Eng Tech	6	2.0
Environmental Eng Tech	5	4.0

TECHNOLOGY COLLEGE DESCRIPTION
Only Junior and Senior students are admitted to Electrical and Computer, Mechanical, Structural Design and Construction and Environmental Engineering Technology programs. 70 to 72 credit hours are required for graduation.

TRANSFER INFORMATION
Transfer via Articulation Agreements
Admission to engineering technology: At the end of the sophomore year
Technology graduates transferred from ... 4-yr: 60 2-yr: 192
Transfer without Articulation Agreements
Admission to engineering technology: At the end of the sophomore year

■ STUDENT PROGRAMS
PROFESSIONAL AND HONORARY SOCIETIES
[For key to acronyms, see *Introduction*.]
Professional Societies: ACI, ACM, AGCA, ASA*, ASCE, ASME, IEEE

SUPPORT PROGRAMS
Minority Student Support
For: Ethnic Minorities **Available:** Year round
Offered: Junior, Senior, Graduate level

331 Pittsburg State University

■ INSTITUTION PROFILE
HEAD OF THE INSTITUTION
Donald W. Wilson
Phone: (316)235-4103 Fax: (316)232-7515

GENERAL INFORMATION
[All Students—Fall 1991]
Undergraduate enrollment 4,810
Graduate enrollment 1,356
Total institution enrollment 6,166
Type of institution: Public Calendar: Semesters
Nearest city: Greater Kansas City Population: 1,400,000
Miles from main campus: 120 Setting: Small Town
Types of engineering degrees: Engineering Technology
Other degree-granting colleges: Arts & Sciences, Business Administration, Education

■ ENGINEERING TECHNOLOGY ADMISSIONS
ADMISSIONS OFFICE CONTACT
James Taylor
Pittsburg State University
1701 South Broadway
Pittsburg, KS 66762-9987
Phone: (316)231-7000 Fax: (316)232-7515

TECHNOLOGY COLLEGE ADMISSIONS INFORMATION
Entrance Requirements: Traditional freshmen must have ACT score for "assessment" but it is not used for admission.
Entrance Recommendations: High School courses—Physics (1 year), Chemistry (1 year), Computers (1 year), Mathematics (4 years)
Requirements for out-of-state residents: PSU, by law, admits all graduates of Kansas high schools accredited by the State Board. Graduates of accredited high schools outside Kansas are eligible if student in upper 50% of graduating class.
Requirements for foreign students: TOEFL (Score: 550); financial statement

DEPARTMENTAL ADMISSIONS INFORMATION
Admission to the engineering technology department: At the time of admission to the institution.

■ FINANCIAL INFORMATION
ESTIMATED EXPENSES (ACADEMIC YEAR)
[Expenses are for the 1992-93 nine-month academic year.]

State Residents	Undergraduate	Graduate
Tuition and fees	$ 782 [1]	$ –
College room and board	$ 2,334 [1]	$ –
Books and supplies	$ 200 [1]	$ –
Other expenses	$ – [B]	$ –
Total estimated expenses	$ 3,316	$ –

Out-of-State Residents	Undergraduate	Graduate
Tuition and fees	$ 2,222 [1]	$ –
College room and board	$ 3,774 [1]	$ –
Books and supplies	$ 200 [1]	$ –
Other expenses	$ – [A]	$ –
Total estimated expenses	$ 6,196	$ –

Notes: (A) Data not available. (B) Data not applicable. (1) Data is for one semester - double for annual expenses.

FINANCIAL AID OFFICE CONTACT
Ronald G. Hopkins, *Director*
Pittsburg State University
Financial Aid
1701 South Broadway
Pittsburg, KS 66762-9987
Phone: (316)231-7000 Fax: (316)232-7515

GENERAL FINANCIAL AID INFORMATION
Forms accepted/required: AFSA, FAT, FFS, Stafford, IRS, SAR, Supplemental, Scholarship/Student Data Form
Additional financial aid information: The university participates in all federal aid programs including PELL Grants, Supplemental Opportunity Grants, Army ROTC Scholarships, the College Work Study Program, and the Guaranteed Student Loan Programs.

■ TECHNOLOGY COLLEGE INFORMATION
[For additional personnel, refer to the *Appendix*.]
HEAD OF ENGINEERING TECHNOLOGY
Vern C. Goold
Phone: (316)235-4363 Fax: (316)231-4231

ENGINEERING TECHNOLOGY COLLEGE ADDRESS
Engineering Technology Department
Pittsburg State University
1701 South Broadway
Pittsburg, KS 66762-9987
Phone: (316)231-7000 Fax: (316)231-4231

TECHNOLOGY ENROLLMENTS—BY CLASS
[Numbers are technology baccalaureate enrollments for the fall 1991 term, unless otherwise footnoted.]
1st-year students/Freshmen 95
2nd-year students/Sophomores 73
3rd-year students/Juniors 66
4th-year students/Seniors .... 146 [1]
Total 380

Notes: (1) Four students are classified as "special" and are not counted above.

NUMBER OF DEGREES AWARDED—BY PROGRAM
[Numbers are engineering technology baccalaureate degrees awarded during the 1991-1992 academic year, unless otherwise footnoted. For full details about each engineering technology program, refer to the *Table of Degree Programs*.]
Construction Engineering Technology 5
Construction Management Technology 8
Electronic(s) Engineering Technology 21
Manufacturing Engineering Technology 7
Mechanical Engineering Technology 8
Plastics Engineering Technology 28
Total 77

PERCENTAGE OF DEGREES AWARDED—BY CATEGORY
[Percentages are of all engineering technology baccalaureate degrees awarded during the 1991-1992 academic year, unless otherwise footnoted.]
African Americans 1.3%
Asian/Pacific Island Americans 1.3%
Hispanic Americans 1.3%
Native Americans .... – %
Foreign Citizens 11.5%
All Others 84.6%
Women 3.8%
Persons with disabilities .... – % [A]
Students over 25 years of age 33.3%

Notes: (A) Data not available.

ENGINEERING TECHNOLOGY STUDENT DATA
Applicants to the engineering technology college
Number of applicants to engineering technology college ..... – [A]
Percent offered admission .... –% [A]

Matriculated engineering technology students
Percentage in top quartile (25%) of High School class .... – [A]
Average ACT scores: Math—19.76, Composite—20.03

Notes: (A) Data not available.

FULL- & PART-TIME FACULTY—BY DEPARTMENT
[Figures are the head count of full-time faculty and the full-time equivalent (FTE) of part-time faculty for each engineering technology department or equivalent.]

Department	Full	Part
Electrical/Electronic(s) Eng Tech	4	–
Manufacturing Eng Tech	3	–
Mechanical Eng Tech	4	–
Construction Eng Tech	3	–
Plastics Eng Tech	4	–
Construction Management Tech	3	–

TECHNOLOGY COLLEGE DESCRIPTION
PSU's Engineering Technology programs are established as part of the total engineering education spectrum and are committed to excellence in engineering education. The philosophy of this commitment is derived from the motto on PSU's official seal which states, 'By Doing Learn'. The Engineering Technology programs at PSU are comprised of elements of the technological spectrum requiring scientific and engineering knowledge plus the operational methods and skills devoted to achieving practical purpose in support of product producing industries. PSU does not imply it has an 'engineering' program. It has an 'Engineering Technology' program which is an alternative in engineering education.

TECHNOLOGY DEGREE PROGRAMS DESCRIPTION
The educational emphasis of the Engineering Technology program is less theoretical and less mathematical than the engineering counterpart and is more hardware and process oriented. Programs are laboratory intensive. Curricula are characterized as including mathematics courses through calculus and applied science and technology courses which emphasize the application of technical knowledge and methods to problems found in industry. Co-op and summer internships are available. Students should complete the program in four years providing they have the proper educational preparation. Faculty advisors are noted for working with deprived students who show academic promise. Average number of years required to actually complete the bachelor's degree: 4.5.

DUAL DEGREE PROGRAMS
Technology baccalaureate graduates with dual degrees 2

TRANSFER INFORMATION
Residency Requirements: A minimum of 30 semester hours of credit must be earned in residence courses taken from Pittsburg State University with a grade point average of 2.0 for all resident hours attempted. These minimum resident hours must include 8 semester hours of credit in the major department. Community college transfers must complete a minimum of sixty credit hours at an accredited four-year university or college.

Transfer without Articulation Agreements
Admission to engineering technology: At any time
Technology graduates transferred from ... 4-yr: 22 2-yr: 18
Requirements: Transfer students, except international students, with a 2.0 or higher grade point average on a 4 point scale are eligible for admission to the University.

GRADUATION REQUIREMENTS
Students must successfully complete a minimum of 124 semester hours of credit with an earned grade point average of 2.0000 for all hours attempted and included in the GPA computation.

■ STUDENT PROGRAMS
PROFESSIONAL AND HONORARY SOCIETIES
[For key to acronyms, see *Introduction*.]
Professional Societies: AGCA, ASME, IEEE, ISA, SAMPE, SME, SPE*, American Foundrymen's Society (AFS), Society of Women Engineers (SWE)
Honorary Societies: Sigma Lambda Chi

SUPPORT PROGRAMS
Student Chapter Organizations: Society of Women Engineers
Women's Studies
For: Women **Available:** Academic year
Offered: Freshman, Sophomore, Junior, Senior, Graduate level
Equal Opportunity Office
For: Women & Ethnic Minorities **Available:** Year round
Offered: Prematriculation, Freshman, Sophomore, Junior, Senior, Graduate level

Other Engineering Support Programs: Summer orientation program for all freshmen and transfers. College Success Program, career counseling services, Financial Aid Office, international student office, veteran's services, National Student Exchange Program, student advising, Student Health Services.

University of Pittsburgh at Johnstown

■ INSTITUTION PROFILE
HEAD OF THE INSTITUTION
Frank H. Blackington III
Phone: (814)269-2090 **Fax:** (814)269-2096
GENERAL INFORMATION
[All Students—Fall 1991]
Undergraduate enrollment 3,241
Total institution enrollment **3,241**
Type of institution: Public **Calendar:** Semesters
Nearest city: Johnstown, Pennsylvania **Population:** 20,000
Miles from main campus: 8 **Setting:** Small Town
Types of engineering degrees: Engineering Technology
Other degree-granting colleges: Arts & Sciences, Business Administration, Communication & Journalism, Education, Humanities & Social Sciences, Nursing

■ ENGINEERING TECHNOLOGY ADMISSIONS
ADMISSIONS OFFICE CONTACT
Thomas J. Wonders
University of Pittsburgh at Johnstown
133 Biddle Hall
Johnstown, PA 15904
Phone: (814)269-7050 **Fax:** (814)269-2096
TECHNOLOGY COLLEGE ADMISSIONS INFORMATION
Entrance Requirements: SAT (Score: 950), High School courses—English (4 years), Algebra (2 years), Geometry (1 year), Chemistry (1 year)
Entrance Recommendations: High School courses—Solid Geometry, Higher Math (1 year), Math Placement Exam
DEPARTMENTAL ADMISSIONS INFORMATION
Admission to the engineering technology department: At the end of the second year.
Additional information: Successful completion of the first three terms of the Engineering Technology Program.

■ FINANCIAL INFORMATION
ESTIMATED EXPENSES (ACADEMIC YEAR)
[Expenses are for the 1992-93 nine-month academic year.]

State Residents	Undergraduate	Graduate
Tuition and fees	$ 6,320	$ –
College room and board	$ 3,544	$ –
Books and supplies	$ 350(C)	$ –
Other expenses	$ –(A)	$ –
Total estimated expenses	**$10,214**	**$ –**

Out-of-State Residents	Undergraduate	Graduate
Tuition and fees	$13,130	$ –
College room and board	$ 3,544	$ –
Books and supplies	$ 350(C)	$ –
Other expenses	$ –(B)	$ –
Total estimated expenses	**$17,024**	**$ –**

Notes: (A) Data not available. (B) Data not applicable. (C) Estimated data.

FINANCIAL AID OFFICE CONTACT
Julie A. Salem, *Associate Director*
University of Pittsburgh at Johnstown
Admissions and Student Aid
125 Biddle Hall
Johnstown, PA 15904
Phone: (814)269-7045 **Fax:** (814)269-2069
GENERAL FINANCIAL AID INFORMATION
Forms accepted/required: FAT, Stafford, SAR, Free Application for Federal Student Aid, PHEAA Aid Information Request
Additional financial aid information: Need-based grants and scholarships, Loans, Part-time need-based campus jobs.

■ TECHNOLOGY COLLEGE INFORMATION
[For additional personnel, refer to the *Appendix*.]
HEAD OF ENGINEERING TECHNOLOGY
Paul H. Saylor
Phone: (814)269-7250 **Fax:** (814)269-7245
ENGINEERING TECHNOLOGY COLLEGE ADDRESS
Engineering Technology Division
University of Pittsburgh at Johnstown
450 Schoolhouse Road
Johnstown, PA 15904
Phone: (814)269-7250 **Fax:** (814)269-7245
TECHNOLOGY ENROLLMENTS—BY CLASS
[Numbers are technology baccalaureate enrollments for the fall 1991 term, unless otherwise footnoted.]
1st-year students/Freshmen 86
2nd-year students/Sophomores 98
3rd-year students/Juniors 81
4th-year students/Seniors 69
Total **334**
NUMBER OF DEGREES AWARDED—BY PROGRAM
[Numbers are engineering technology baccalaureate degrees awarded during the 1991-1992 academic year, unless otherwise footnoted. For full details about each engineering technology program, refer to the *Table of Degree Programs*.]
Civil Engineering Technology 19
Electrical Engineering Technology 23
Mechanical Engineering Technology 17
Total **59**
PERCENTAGE OF DEGREES AWARDED—BY CATEGORY
[Percentages are of all engineering technology baccalaureate degrees awarded during the 1991-1992 academic year, unless otherwise footnoted.]
African Americans .... – %
Asian/Pacific Island Americans .... – %
Hispanic Americans .... – %
Native Americans .... – %
Foreign Citizens .... – %
All Others .... 100 %
Women .... 10.8 %
Persons with disabilities .... – %
Students over 25 years of age .... – %

■ ENGINEERING TECHNOLOGY STUDENT DATA
Applicants to the engineering technology college
Number of applicants to engineering technology college ... 203
Percent offered admission 77.0%
Matriculated engineering technology students
Percentage in top quartile (25%) of High School class 26.0
FULL- & PART-TIME FACULTY—BY DEPARTMENT
[Figures are the head count of full-time faculty and the full-time equivalent (FTE) of part-time faculty for each engineering technology department or equivalent.]

Department	Full	Part
Civil Eng Tech	6	–
Electrical Eng Tech	6	–
Mechanical Eng Tech	6	–

TECHNOLOGY DEGREE PROGRAMS DESCRIPTION
Students enrolled in the Engineering Technology Program major in Civil, Electrical, or Mechanical Engineering Technology. Within these three majors, a variety of areas of concentration is offered.
TRANSFER INFORMATION
Transfer without Articulation Agreements
Admission to engineering technology: At any time
Technology graduates transferred from ... 4-yr: 8 2-yr: –
Requirements: Minimum 2.50 QPA from an ABET accredited institution.
GRADUATION REQUIREMENTS
(1) Complete required courses with passing grades; (2) Earn total number of credits required by major area; (3) Attain minimum QPA of 2.0 in all courses on University of Pittsburgh record and major area (CET, EET, MET); (4) Complete senior year (30 credits) while registered in the Division of Engineering Technology.

■ STUDENT PROGRAMS
PROFESSIONAL AND HONORARY SOCIETIES
[For key to acronyms, see *Introduction*.]
Professional Societies: ASCE, ASHRAE, ASME, IEEE
Honorary Societies: Chi Lambda Tau Honorary Leadership Society
SUPPORT PROGRAMS
Student Chapter Organizations: Society of Women Engineers
Black Action Society
For: Women & Ethnic Minorities **Available:** Year round
Offered: Freshman, Sophomore, Junior, Senior
Association of Women Students
For: Women & Ethnic Minorities **Available:** Year round
Offered: Freshman, Sophomore, Junior, Senior
Other Engineering Support Programs: Freshman Orientation for new incoming freshmen and transfer students. A campus wide Learning Skills Center is available to all students. Freshman Seminar Engineering Technology Faculty Advisors

333 Point Park College

■ INSTITUTION PROFILE

HEAD OF THE INSTITUTION
J. Matthew Simon
Phone: (412)392-3990 **Fax:** (412)391-1980

GENERAL INFORMATION
[All Students—Fall 1991]
Undergraduate enrollment 2,821
Graduate enrollment 111
Total institution enrollment **2,932**
Type of institution: Private **Calendar:** Semesters
Location: Pittsburgh **Population:** 350,000
Setting: Urban
Types of engineering degrees: Engineering Technology

■ ENGINEERING TECHNOLOGY ADMISSIONS

ADMISSIONS OFFICE CONTACT
Terrance Kizina
Point Park College
201 Wood Street
Pittsburgh, PA 15222
Phone: (412)391-4100 **Fax:** (412)391-1980

TECHNOLOGY COLLEGE ADMISSIONS INFORMATION
Entrance Requirements: High School courses—English (4 years), Math (2 years), Science (3 years), Social Science (4 years), Either the SAT or ACT are required but there is no minimum required score in either case
Entrance Recommendations: Either the SAT or ACT are recommended
Requirements for foreign students: TOEFL (Score: 500); financial statement

DEPARTMENTAL ADMISSIONS INFORMATION
Admission to the engineering technology department: At the time of admission to the institution.

■ FINANCIAL INFORMATION

ESTIMATED EXPENSES (ACADEMIC YEAR)
[Expenses are for the 1992-93 nine-month academic year.]

All Students	Undergraduate	Graduate
Tuition and fees	$ 8,622	$ 4,824
College room and board	$ 4,390 [1]	$ 4,390 [1]
Books and supplies	$ 400	$ 400
Other expenses	$ 500	$ 500
Total estimated expenses	**$13,912**	**$10,114**

Notes: (1) Cost includes the Flexible Meal Plan

FINANCIAL AID OFFICE CONTACT
Deidre Smith, *Director of Financial Aid*
Point Park College
201 Wood Street
Pittsburgh, PA 15222
Phone: (412)391-4100 **Fax:** (412)391-1980

GENERAL FINANCIAL AID INFORMATION
Forms accepted/required: Pennsylvania Higher Education Administration Agency (PHEAA)
Additional financial aid information: Scholarships based on need, merit; departmental scholarships; College Work Study; loans, grants

ENGINEERING TECHNOLOGY COLLEGE ADDRESS
Department of Natural Sciences and Engineering Technology
Point Park College
201 Wood Street
Pittsburgh, PA 15222
Phone: (412)391-4100 **Fax:** (412)391-1980

TECHNOLOGY ENROLLMENTS—BY CLASS
[Numbers are technology baccalaureate enrollments for the fall 1991 term, unless otherwise footnoted.]
1st-year students/Freshmen 154
2nd-year students/Sophomores 42
3rd-year students/Juniors 147
4th-year students/Seniors 222
Total **565**

NUMBER OF DEGREES AWARDED—BY PROGRAM
[Numbers are engineering technology baccalaureate degrees awarded during the 1991-1992 academic year, unless otherwise footnoted. For full details about each engineering technology program, refer to the *Table of Degree Programs*.]
Civil Engineering Technology 12
Electrical Engineering Technology 50
Mechanical Engineering Technology 21
Total **83**

PERCENTAGE OF DEGREES AWARDED—BY CATEGORY
[Percentages are of all engineering technology baccalaureate degrees awarded during the 1991-1992 academic year, unless otherwise footnoted.]
African Americans .... 2.0 %
Asian/Pacific Island Americans .... – %
Hispanic Americans .... – %
Native Americans .... – %
Foreign Citizens .... – % (A)
All Others .... 98.0 %
Women .... 4.4 %
Persons with disabilities .... – % (A)
Students over 25 years of age .... – % (A)

Notes: (A) Data not available.

ENGINEERING TECHNOLOGY STUDENT DATA
Applicants to the engineering technology college
Number of applicants to engineering technology college 15
Percent offered admission 80.0%

Matriculated engineering technology students
Percentage in top quartile (25%) of High School class .... –
Average SAT scores: Math—470, Verbal—380, Combined—850

FULL- & PART-TIME FACULTY—BY DEPARTMENT
[Figures are the head count of full-time faculty and the full-time equivalent (FTE) of part-time faculty for each engineering technology department or equivalent.]

Department	Full	Part
Natural Sciences & Eng Tech	7	2.7
Civil Eng Tech	2	–
Electrical/Electronic(s) Eng Tech	3	1.0
Mechanical Eng Tech	2	1.0

TECHNOLOGY COLLEGE DESCRIPTION
Point Park College provides professionally oriented studies in the arts, business, communications and technology. All engineering technology programs emphasize 'hands on' experience. All three areas include drafting and CAD courses. Computer Integrated Manufacturing (CIM) theory and laboratory activities used in curriculum. Engineering Technology majors may complete their programs without any required co-op or exchange programs, but internships are available and encouraged. Students pursuing BS degrees may make formal application during their final semester for the EIT exam leading to eventual registration as a professional engineer.

TECHNOLOGY DEGREE PROGRAMS DESCRIPTION
Point Park College offers A.S., B.S., and post-baccalaureate programs in civil, electrical and mechanical engineering technology. Students majoring in EET concentrate in: Electronics, Electrical Power or Process Control and Instrumentation. The latter follows the curriculum developed by the education department of the ISA.

DUAL DEGREE PROGRAMS
Technology baccalaureate graduates with dual degrees 1

TRANSFER INFORMATION
Residency Requirements: All transfer students must complete at least 30 credit hours and a minimum of 4 departmental courses in residence and meet specified requirements for each program.
Transfer via Articulation Agreements
Admission to engineering technology: At any time
Transfer without Articulation Agreements
Admission to engineering technology: At any time
Technology graduates transferred from ... 4-yr: 5 2-yr: 31

GRADUATION REQUIREMENTS
All programs require a minimum GPA of 2.00. The B.S. programs in Civil and Mechanical Engineering Technology require 135 credits with 61 credits within the department. Electrical Engineering Technology requires 136 credits with 62 credits within the department.

■ STUDENT PROGRAMS

PROFESSIONAL AND HONORARY SOCIETIES
[For key to acronyms, see *Introduction*.]
Professional Societies: ASME, IEEE, ISA
Honorary Societies: Alpha Chi (Full-time students), Alpha Sigma Lambda (Part-time students)

SUPPORT PROGRAMS
Other Engineering Support Programs: There are orientation programs for all incoming students. The Program for Academic Success (PAS) sponsors summer academic remedial classes for incoming freshmen as well as tutoring, academic counseling etc.

334 Prairie View A&M University

INSTITUTION PROFILE

HEAD OF THE INSTITUTION
Julius W. Becton Jr.
Phone: (409)857-2111 **Fax:** (409)857-3928

GENERAL INFORMATION
[All Students—Fall 1991]
Undergraduate enrollment 5,130
Graduate enrollment 530
Total institution enrollment 5,660
Type of institution: Public **Calendar:** Semesters
Nearest city: Houston **Population:** 2,000,500
Miles from main campus: 45 **Setting:** Small Town
Types of engineering degrees: Engineering Technology
Other degree-granting colleges: Arts & Sciences, Business Administration, Education, Nursing

ENGINEERING TECHNOLOGY ADMISSIONS

ADMISSIONS OFFICE CONTACT
Mary E. Gooch
Prairie View A&M University
P.O. Box 2610
Prairie View, TX 77446
Phone: (409)857-2626 **Fax:** (409)857-2425

TECHNOLOGY COLLEGE ADMISSIONS INFORMATION
Entrance Requirements: SAT, ACT, 1) completed application form; 2) certified final HS transcript or a GED transcript; 3) scores from SAT or ACT; 4) A Counselor Recommendation Form completed by high school counselor.
Entrance Recommendations: SAT (Score: 700), ACT (Score: 17)
Requirements for foreign students: TOEFL (Score: 500); Graduate students whose first language is not English must present a score of 550 on the TOEFL in addition to passing additional instruments required by their departmental major.

DEPARTMENTAL ADMISSIONS INFORMATION
Admission to the engineering technology department: At the time of admission to the institution.

FINANCIAL INFORMATION

ESTIMATED EXPENSES (ACADEMIC YEAR)
[Expenses are for the 1992-93 nine-month academic year.]

State Residents	Undergraduate	Graduate
Tuition and fees	$ 509	$ 421
College room and board	$ 1,700	$ –
Books and supplies	$ 300	$ 150
Other expenses	$ –	$ –
Total estimated expenses	**$ 2,509**	**$ 571**

Out-of-State Residents	Undergraduate	Graduate
Tuition and fees	$ 2,209	$ 1,663
College room and board	$ 1,700	$ –
Books and supplies	$ 300	$ 150
Other expenses	$ –	$ –
Total estimated expenses	**$ 4,209**	**$ 1,813**

FINANCIAL AID OFFICE CONTACT
A. D. James, *Director of Financial Aid*
Prairie View A&M University
P.O. Box C
Prairie View, TX 77446
Phone: (409)857-2424 **Fax:** (409)857-2425

GENERAL FINANCIAL AID INFORMATION
Forms accepted/required: FAF, FSA, Stafford, Supplemental, University Scholarship by programs, Ethnic Recruitment Scholarships
Additional financial aid information: Pell Grant and College Work Study Program

TECHNOLOGY COLLEGE INFORMATION

[For additional personnel, refer to the *Appendix*.]
HEAD OF ENGINEERING TECHNOLOGY
Haku Israni
Phone: (409)857-4122 **Fax:** (409)857-2097

ENGINEERING TECHNOLOGY COLLEGE ADDRESS
College of Applied Sciences and Engineering Technology
Prairie View A&M University
P.O. Box 308
Prairie View, TX 77446
Phone: (409)857-4717 **Fax:** (409)857-2097

TECHNOLOGY ENROLLMENTS—BY CLASS
[Numbers are technology baccalaureate enrollments for the fall 1991 term, unless otherwise footnoted.]
1st-year students/Freshmen 142
2nd-year students/Sophomores 55
3rd-year students/Juniors 24
4th-year students/Seniors 19
Total **240**

NUMBER OF DEGREES AWARDED—BY PROGRAM
[Numbers are engineering technology baccalaureate degrees awarded during the 1991-1992 academic year, unless otherwise footnoted. For full details about each engineering technology program, refer to the *Table of Degree Programs*.]
Computer Engineering Technology .... 14 (C)
Computer Science .... –
Electrical Engineering Technology .... 13 (C)
Mechanical Engineering Technology .... 13 (C)
Total **40**

Notes: (C) Estimated data.

PERCENTAGE OF DEGREES AWARDED—BY CATEGORY
[Percentages are of all engineering technology baccalaureate degrees awarded during the 1991-1992 academic year, unless otherwise footnoted.]
African Americans .... 85.0 %
Asian/Pacific Island Americans .... 2.0 %
Hispanic Americans .... 2.0 %
Native Americans .... 2.0 %
Foreign Citizens .... 9.0 %
All Others .... – %
Women .... 36.6 %
Persons with disabilities .... – %
Students over 25 years of age .... – %

ENGINEERING TECHNOLOGY STUDENT DATA

Applicants to the engineering technology college
Number of applicants to engineering technology college ..... – (A)
Percent offered admission .... –% (A)

Matriculated engineering technology students
Percentage in top quartile (25%) of High School class 30.0
Average SAT scores: Combined—950
Average ACT scores: Composite—19

Notes: (A) Data not available.

FULL- & PART-TIME FACULTY—BY DEPARTMENT
[Figures are the head count of full-time faculty and the full-time equivalent (FTE) of part-time faculty for each engineering technology department or equivalent.]

Department	Full	Part
Electrical Eng Tech	–	3.0
Computer Eng Tech	–	3.0
Mechanical Eng Tech	5.0	3.0

TECHNOLOGY COLLEGE DESCRIPTION
1. 85% of the Engineering Technology faculty have earned M.S. or above in Electrical Engineering or Mechanical Engineering. 2. 83% of the Computer Science faculty have earned M.S. or above in Computer Science. 3. Computer Engineering Technology operates 4 state-of-the-art laboratories. 4. Electrical Engineering Technology operates 6 state-of-the-art laboratories. 5. Computer Science operates 5 state-of-the-art laboratories. 6. Mechanical Engineering Technology operates 4 laboratories. 7. Co-op slots are available in all the engineering technology programs.

TECHNOLOGY DEGREE PROGRAMS DESCRIPTION
B.S. Degree M.S. Degree Joint Ph.D. Degree Prairie View A&M University

TRANSFER INFORMATION
Residency Requirements: Last 30 semester hours must be carried at Prairie View A&M University
Transfer via Articulation Agreements
Admission to engineering technology: At any time
Requirements: Need to present evidence of good standing at the last institution attended and official transcripts of all work completed at other institutions. Transfer students must satisfy all Prairie View A&M University requirements for graduation.

STUDENT PROGRAMS

PROFESSIONAL AND HONORARY SOCIETIES
[For key to acronyms, see *Introduction*.]
Professional Societies: SME
Honorary Societies: Epsilon Pi Tau Honor Society in Technology

SUPPORT PROGRAMS
Student Chapter Organizations: Society of Women Engineers, National Society of Black Engineers, The Computer Engineering Technology Association, Electrical/Electronics Engineering Technology Assn

Texas Alliance for Minorities in Engineering
For: Ethnic Minorities **Available:** Academic year
Offered: Freshman, Sophomore
Other Engineering Support Programs: Texas AMP program for computer science majors. Supported through academic and financial support

335 Purdue University

■ INSTITUTION PROFILE

HEAD OF THE INSTITUTION
Steven C. Beering
Phone: (317)494-9708 Fax: (317)494-7875

GENERAL INFORMATION
[All Students—Fall 1991]
Undergraduate enrollment 31,568
Graduate enrollment 6,500
Total institution enrollment **38,068**

Type of institution: Public Calendar: Semesters
Nearest city: Indianapolis Population: 1,250,000
Miles from main campus: 60 Setting: Urban
Types of engineering degrees: Engineering Technology
Other degree-granting colleges: Health Sciences, School of Management, Agriculture, School of Liberal Arts, Veterinary Medicine, Science, Engineering, Graduate School, Consumer and Family Sciences

TECHNOLOGY COLLEGE ADMISSIONS INFORMATION
Entrance Requirements: High School courses—Algebra (1 year), Geometry (1 year), Laboratory Science (1 year), English (4 years), SAT or ACT.
Entrance Recommendations: High School courses—Math (3 years), Science (3 years)
Requirements for foreign students: TOEFL (Score: 550)

DEPARTMENTAL ADMISSIONS INFORMATION
Admission to the engineering technology department: At the time of admission to the institution.

■ FINANCIAL INFORMATION

ESTIMATED EXPENSES (ACADEMIC YEAR)
[Expenses are for the 1992-93 nine-month academic year.]

State Residents	Undergraduate	Graduate
Tuition and fees	$ 2,520	$ 2,520
College room and board	$ 3,650	$ 3,650
Books and supplies	$ 480	$ 480
Other expenses	$ 1,290 (1)	$ 1,290 (1)
Total estimated expenses	**$ 7,940**	**$ 7,940**

Out-of-State Residents	Undergraduate	Graduate
Tuition and fees	$ 8,192	$ 8,192
College room and board	$ 3,650	$ 3,650
Books and supplies	$ 480	$ 480
Other expenses	$ 1,290 (1)	$ 1,290 (1)
Total estimated expenses	**$13,612**	**$13,612**

Notes: (1) Misc. (laundry, amusement, etc.) and travel home for holidays.

FINANCIAL AID OFFICE CONTACT
Joyce Hall, *Director*
Purdue University
Division of Financial Aid
1102 Schleman Hall Rm. 305
West Lafayette, IN 47907-1102
Phone: (317)494-5050 Fax: (317)494-6707

GENERAL FINANCIAL AID INFORMATION
Forms accepted/required: ACT, AFSA, FAF, FAT, FFS, Stafford, IRS, SAR, Parent Loan Form, PLUS Loan
Additional financial aid information: Need based scholarships, Grants, Industry Scholarships/Grants, Part-time Employment Opportunities, Work Study, Government Loans.

■ TECHNOLOGY COLLEGE INFORMATION
[For additional personnel, refer to the *Appendix*.]

HEAD OF ENGINEERING TECHNOLOGY
Don K. Gentry
Phone: (317)494-2552 Fax: (317)494-0486

ENGINEERING TECHNOLOGY COLLEGE ADDRESS
School of Technology
Purdue University
1410 Knoy Hall of Technology
West Lafayette, IN 47907-1410
Phone: (317)494-2554 Fax: (317)494-0486

TECHNOLOGY ENROLLMENTS—BY CLASS
[Numbers are technology baccalaureate enrollments for the fall 1991 term, unless otherwise footnoted.]
1st-year students/Freshmen 326
2nd-year students/Sophomores 394
3rd-year students/Juniors 180
4th-year students/Seniors 225
Total **1,125**

NUMBER OF DEGREES AWARDED—BY PROGRAM
[Numbers are engineering technology baccalaureate degrees awarded during the 1991-1992 academic year, unless otherwise footnoted. For full details about each engineering technology program, refer to the *Table of Degree Programs*.]
Computer Integrated Manufacturing Technology 25
Electrical Engineering Technology 111
Mechanical Engineering Technology 101
Total **237**

PERCENTAGE OF DEGREES AWARDED—BY CATEGORY
[Percentages are of all engineering technology baccalaureate degrees awarded during the 1991-1992 academic year, unless otherwise footnoted.]
African Americans .... – %
Asian/Pacific Island Americans .... – %
Hispanic Americans .... 1.0 %
Native Americans .... – %
Foreign Citizens .... 1.2 %
All Others .... 97.8 %
Women .... 2.0 %
Persons with disabilities .... – % (A)
Students over 25 years of age .... 12.0 %

Notes: (A) Data not available.

ENGINEERING TECHNOLOGY STUDENT DATA
Applicants to the engineering technology college
Number of applicants to engineering technology college ... 805
Percent offered admission 86.0%

Matriculated engineering technology students
Percentage in top quartile (25%) of High School class 55.0
Average SAT scores: Math—516, Verbal—426, Combined—942
Average ACT scores: Math—23, Composite—23

FULL- & PART-TIME FACULTY—BY DEPARTMENT
[Figures are the head count of full-time faculty and the full-time equivalent (FTE) of part-time faculty for each engineering technology department or equivalent.]

Department	Full	Part
Electrical Eng Tech	24	–
Mechanical Eng Tech	16	–
Computer Integrated Manufacturing Tech	6	–

TECHNOLOGY COLLEGE DESCRIPTION
The School's engineering technology programs are laboratory-intensive. A co-op program is optional for the student. The bachelor of science degree awarded under the School of Technology's 'two-plus-two' education plan is unique. A student who follows this plan earns first the associate degree in two years and then a bachelor of science degree in two more years. Average number of years required to actually complete the bachelor's degree: 4.5

TECHNOLOGY DEGREE PROGRAMS DESCRIPTION
The School of Technology's four-year degree programs in engineering technology focus upon the applications of principles of science, engineering, and techniques to a wide variety of electrical, electronic, mechanical, and manufacturing engineering technology areas, depending upon the program selected. The bachelor of science degree awarded under the School of Technology's 'two-plus-two' education plan is unique. A student who follows this plan earns first the associate degree in two years and then a bachelor of science degree in two more years. **Purdue University is an Equal Access/Equal Opportunity University.**

TRANSFER INFORMATION
Residency Requirements: Resident study at Purdue University for at least two semesters and the enrollment in and completion of at least 32 semester hours of course work required and approved for completion of the degree.

Transfer without Articulation Agreements
Admission to engineering technology: The application deadline is one month prior to the beginning of each semester.
Technology graduates transferred from . . . 4-yr: – 2-yr: –

GRADUATION REQUIREMENTS
The bachelor of science degree in engineering technology programs require 132 total credits. The associate degree requires 69 credits for the electrical engineering technology program and 66 credits for the mechanical engineering technology program.

■ STUDENT PROGRAMS

PROFESSIONAL AND HONORARY SOCIETIES
[For key to acronyms, see *Introduction*.]
Professional Societies: ASHRAE, ASME, IEEE, SAE, SME, Fluid Power Society (FPS), American Society for Quality Control (ASQC)
Honorary Societies: Tau Alpha Pi

SUPPORT PROGRAMS
Student Chapter Organizations: Society of Women Engineers

Minority Technology Association
For: Ethnic Minorities **Available:** Academic year
Offered: Freshman, Sophomore, Junior, Senior

Technology Seventh and Eighth Grade Summer Program
For: Ethnic Minorities **Available:** Summer only
Offered: Prematriculation

Other Engineering Support Programs: Orientation programs for freshmen, transfer, and change of degree objective students, career counseling, academic advising, tutoring program, support services for handicapped students, informational programs for prospective students, and general referral activities.

336 Rochester Institute of Technology

■ INSTITUTION PROFILE
HEAD OF THE INSTITUTION
Albert J. Simone
Phone: (716)475-2394
GENERAL INFORMATION
[All Students—Fall 1991]
Undergraduate enrollment 11,150
Graduate enrollment 1,868
Total institution enrollment **13,018**
Type of institution: Private **Calendar:** Quarters
Nearest city: Rochester **Population:** 250,000
Miles from main campus: 2 **Setting:** Suburban
Types of engineering degrees: Engineering & Technology
Other degree-granting colleges: Business Administration, Continuing Education, Liberal Arts, Science, Imaging Arts & Sciences

■ ENGINEERING TECHNOLOGY ADMISSIONS
ADMISSIONS OFFICE CONTACT
Daniel Shelley
Rochester Institute of Technology
One Lomb Memorial Drive
P.O. Box 9887
Rochester, NY 14623-0887
Phone: (716)475-6631
TECHNOLOGY COLLEGE ADMISSIONS INFORMATION
Entrance Requirements: High School courses—Math (3 years), Physics or Chemistry (1 year)
Entrance Recommendations: SAT (Score: 1000), High School courses—Technology
Requirements for foreign students: TOEFL (Score: 525)
DEPARTMENTAL ADMISSIONS INFORMATION
Admission to the engineering technology department: At the time of admission to the institution.

■ FINANCIAL INFORMATION
ESTIMATED EXPENSES (ACADEMIC YEAR)
[Expenses are for the 1992-93 nine-month academic year.]

All Students	Undergraduate	Graduate
Tuition and fees	$12,525	$ –
College room and board	$ 5,000	$ –
Books and supplies	$ 500	$ –
Other expenses	$ 200 [1]	$ –
Total estimated expenses	**$18,225**	**$ –**

Notes: (1) fees

FINANCIAL AID OFFICE CONTACT
Verna J. Hazen, *Director*
Rochester Institute of Technology
One Lomb Memorial Drive
P.O. Box 9887
Rochester, NY 14623-0887
Phone: (716)475-2186
GENERAL FINANCIAL AID INFORMATION
Forms accepted/required: FAF, FFS
Additional financial aid information: Scholarships, Grants, Entitlements, Loans, Employment

■ TECHNOLOGY COLLEGE INFORMATION
[For additional personnel, refer to the *Appendix*.]
HEAD OF ENGINEERING TECHNOLOGY
W. David Baker
Phone: (716)475-2915 **Fax:** (716)475-5275
EMail: wdbite@ritvax.isc.rit.edu
ENGINEERING TECHNOLOGY COLLEGE ADDRESS
School of Engineering Technology
Rochester Institute of Technology
One Lomb Memorial Drive
P.O. Box 9887
Rochester, NY 14623-0887
Phone: (716)475-2411

TECHNOLOGY ENROLLMENTS—BY CLASS
[Numbers are technology baccalaureate enrollments for the fall 1991 term, unless otherwise footnoted.]
1st-year students/Freshmen 152
2nd-year students/Sophomores 164
3rd-year students/Juniors 617
4th-year students/Seniors 423
Total **1,356**

NUMBER OF DEGREES AWARDED—BY PROGRAM
[Numbers are engineering technology baccalaureate degrees awarded during the 1991-1992 academic year, unless otherwise footnoted. For full details about each engineering technology program, refer to the *Table of Degree Programs*.]
Civil Engineering Technology 52
Computer Engineering Technology 22
Electrical Engineering Technology 112
Manufacturing Engineering Technology 41
Mechanical Engineering Technology 75
Telecommunications Engineering Technology .. 3
Total **305**

PERCENTAGE OF DEGREES AWARDED—BY CATEGORY
[Percentages are of all engineering technology baccalaureate degrees awarded during the 1991-1992 academic year, unless otherwise footnoted.]
African Americans .... 1.0 %
Asian/Pacific Island Americans .... 3.0 %
Hispanic Americans .... 0.1 %
Native Americans .... 0.3 %
Foreign Citizens .... 1.0 %
All Others .... 94.6 %
Women .... 4.0 %
Persons with disabilities .... – %
Students over 25 years of age .... 30.0 %

ENGINEERING TECHNOLOGY STUDENT DATA
Applicants to the engineering technology college
Number of applicants to engineering technology college ... 253
Percent offered admission .... 82.0 %
Average SAT scores: Math—536[1], Verbal—432[1], Combined—968[1]

Notes: (1) Mean of middle 50th percentile.

FULL- & PART-TIME FACULTY—BY DEPARTMENT
[Figures are the head count of full-time faculty and the full-time equivalent (FTE) of part-time faculty for each engineering technology department or equivalent.]

Department	Full	Part
Civil Eng Tech	5	–
Electrical Eng Tech	14	–
Manufacturing Eng Tech	4	–
Mechanical Eng Tech	7	–

TECHNOLOGY COLLEGE DESCRIPTION
All engineering technology programs accept freshman and transfer students. Program emphasis is placed upon applications of current technology, and each program involves extensive laboratory experience. All students participate in paid cooperative work experience during their junior and senior year and thus graduate with over one year of practical work experience.

TECHNOLOGY DEGREE PROGRAMS DESCRIPTION
All programs prepare graduates for entry positions in the engineering field, ranging from field service to design. Electrical Engineering Technology offers specialties in computer design, electronic communications, power systems or microelectronics. Mechanical Engineering Technology offers a special sequence in HVAC energy systems. Elective paths in construction management, environmental controls, structures, water resources, and heavy construction are offered within Civil Engineering Technology.

TRANSFER INFORMATION
Residency Requirements: A minimum of 45 credit hours must be successfully completed in residence at the Institute in the College granting the degree. Forty (40) of these credits must be in the principal field of study.
Transfer via Articulation Agreements
Admission to engineering technology: At the end of the sophomore year
Technology graduates transferred from ... 4-yr: – 2-yr: 225
Requirements: Minimum GPA of 2.30 for transfers with an associate degree. These students are admitted with third year status.
Transfer without Articulation Agreements
Admission to engineering technology: At the end of the sophomore year
Requirements: Same as for those students from schools with articulation agreements.

GRADUATION REQUIREMENTS
Bachelor Degree programs in engineering technology require 190-196 quarter credits, depending on the major selected.

■ STUDENT PROGRAMS
PROFESSIONAL AND HONORARY SOCIETIES
[For key to acronyms, see *Introduction*.]
Professional Societies: ASCE, ASHRAE, ASME, IEEE, SME
Honorary Societies: Tau Alpha Pi
SUPPORT PROGRAMS
Student Chapter Organizations: Society of Women Engineers, National Society of Black Engineers
Other Engineering Support Programs: RIT offers a Summer Orientation Program for all freshmen and transfers, a Learning Assessment Program, a College Anticipation Program, a College Restoration Program, the English Learning Center, as well as personal and career counseling through the Counseling Center.

337 St. Cloud State University

■ INSTITUTION PROFILE

HEAD OF THE INSTITUTION
Robert O. Bess
Phone: (612)255-2122   Fax: (612)654-5139

GENERAL INFORMATION
[All Students—Fall 1991]
Undergraduate enrollment 14,211
Graduate enrollment 1,296
Total institution enrollment 15,507
Type of institution: Public   **Calendar:** Quarters
Nearest city: St. Paul/Minneapolis   **Population:** 2,200,000
Miles from main campus: 70   **Setting:** Small Town
Types of engineering degrees: Engineering & Technology
Other degree-granting colleges: Education, Business, Social Sciences, Fine Arts and Humanities, Science and Technology

■ ENGINEERING TECHNOLOGY ADMISSIONS

ADMISSIONS OFFICE CONTACT
Sherwood Reid
St. Cloud State University
115 Administrative Services Building
St. Cloud, MN 56301-4498
Phone: (612)255-2243   Fax: (612)654-5367

TECHNOLOGY COLLEGE ADMISSIONS INFORMATION
Admission to the engineering technology college: Admission to the institution, transfer for undecided status or from a community/technical college.
Entrance Requirements: PSAT, SAT (Score: 900), ACT (Score: 25)
Entrance Recommendations: High School courses—Mathematics (3 years), Physics (1 year), Chemistry (1 year), Computers (1 year)
Requirements for foreign students: TOEFL (Score: 475); 1. A completed Undergraduate Application for Admission. 2. A $15 application fee. 3. An English translation of all educational transcripts. 4. Proof of English proficiency (TOEFL, Michigan).

DEPARTMENTAL ADMISSIONS INFORMATION
Admission to the engineering technology department: Admission to the institution, transfer from undecided status or from a community/technical college.

■ FINANCIAL INFORMATION

ESTIMATED EXPENSES (ACADEMIC YEAR)
[Expenses are for the 1992-93 nine-month academic year.]

State Residents	Undergraduate	Graduate
Tuition and fees	$ 1,743	$ 3,020
College room and board	$ 2,300	$ _(1)
Books and supplies	$ 720	$ _(1)
Other expenses	$ 675 (2)	$ _(1)
Total estimated expenses	**$ 5,438**	**$ 3,020**

Out-of-State Residents	Undergraduate	Graduate
Tuition and fees	$ 2,990	$ 4,356
College room and board	$ 2,300	$ _(1)
Books and supplies	$ 720	$ _(1)
Other expenses	$ 675 (2)	$ _(1)
Total estimated expenses	**$ 6,685**	**$ 4,356**

Notes: (1) No available (2) Estimated insurance premium and fees per quarter

FINANCIAL AID OFFICE CONTACT
Frank Loncorich, *Director of Financial Aids*
St. Cloud State University
106 Administrative Services Building
St. Cloud, MN 56301-4498
Phone: (612)255-2047   Fax: (612)654-5367

GENERAL FINANCIAL AID INFORMATION
Forms accepted/required: ACT, Institutional
Additional financial aid information: Gift aid: Money that does no have to be repaid. This includes grants and scholarships. Work aid: Money that is earned through on-and off-campus employment. Loan aid: Money that is borrowed in the form of long-and short-term and must be repaid.

■ TECHNOLOGY COLLEGE INFORMATION
[For additional personnel, refer to the *Appendix*.]

HEAD OF ENGINEERING TECHNOLOGY
Richard G. Hogan
Phone: (612)255-2192   Fax: (612)255-4262

ENGINEERING TECHNOLOGY COLLEGE ADDRESS
Department of Technology
St. Cloud State University
720 Fourth Avenue South
216 Headley Hall
St. Cloud, MN 56301-4498
Phone: (612)255-2107   Fax: (612)255-4262

TECHNOLOGY ENROLLMENTS—BY CLASS
[Numbers are technology baccalaureate enrollments for the fall 1991 term, unless otherwise footnoted.]
1st-year students/Freshmen 30
2nd-year students/Sophomores 26
3rd-year students/Juniors 29
4th-year students/Seniors 44
Total **129**

NUMBER OF DEGREES AWARDED—BY PROGRAM
[Numbers are engineering technology baccalaureate degrees awarded during the 1991-1992 academic year, unless otherwise footnoted. For full details about each engineering technology program, refer to the *Table of Degree Programs*.]
Engineering Technology (General) 2
Imaging Engineering Technology 7
Imaging Science 1
Manufacturing Engineering Technology 15
Total **25**

PERCENTAGE OF DEGREES AWARDED—BY CATEGORY
[Percentages are of all engineering technology baccalaureate degrees awarded during the 1991-1992 academic year, unless otherwise footnoted.]
African Americans .... – %
Asian/Pacific Island Americans .... 10.0 %
Hispanic Americans .... – %
Native Americans .... – %
Foreign Citizens .... – %
All Others .... 90.0 %
Women .... 35.0 %
Persons with disabilities .... – %
Students over 25 years of age .... 25.0 %

FULL- & PART-TIME FACULTY—BY DEPARTMENT
[Figures are the head count of full-time faculty and the full-time equivalent (FTE) of part-time faculty for each engineering technology department or equivalent.]

Department	Full	Part
Manufacturing Eng Tech	7	1.7

TECHNOLOGY COLLEGE DESCRIPTION
St. Cloud State University offers a four year plan for full time students. Part time students may choose to be included in the night classes package. The Department of Technology prepares individuals for professional positions in manufacturing, imaging sciences, photofinishing, aviation and other industries. The curricula are based upon recommendations by Industrial Advisory Boards, the American Society for Engineering Education, National Association of Industrial Technology, Photo Marketing Association International, The Society for Imaging Science and Technology, Federal Aviation Administration, University Aviation Association, and upon the trends defined by the Society of Manufacturing Engineers.

TECHNOLOGY DEGREE PROGRAMS DESCRIPTION
The B.S. in Engineering Technology - Manufacturing emphasis is an accredited program by the Technology Accreditation Commission of the Accreditation Board for Engineering and Technology (ABET). St. Cloud State University's curriculum for Manufacturing Engineering Technology Program stresses classroom studies with extensive field applications and state-of-the-art systems and applications. The course study includes core classes in mathematics and sciences, electronics, and computer science. Computer-aided-Design and computer-integrated manufacturing, as well as technical writing courses are also included. Emphasis is made in obtaining hands-on experience in methods engineering, materials and processes, fluid mechanics, safety, quality assurance, classical and the latest developments in production systems, automation and robotics. Many students with senior status are employed as interns by regional manufacturers, acquiring with this, the valuable experience towards starting a career.

TRANSFER INFORMATION
Transfer without Articulation Agreements
Admission to engineering technology: At any time
Technology graduates transferred from ... 4-yr: 10 2-yr: 9

GRADUATION REQUIREMENTS
To be admitted to the Engineering Manufacturing Engineering major, a student must complete the designated pre-major classes, have a 2.25 overall GPA and a minimum of 2.25 grade point in the designated pre-major classes. In addition, the student requires to have satisfied the Institution requirements.

■ STUDENT PROGRAMS

PROFESSIONAL AND HONORARY SOCIETIES
[For key to acronyms, see *Introduction*.]
Professional Societies: AIDD, ASA*, IIE, SME, American Production & Inventory Control Society (APICS), American Society for Quality Control (ASQC)
Honorary Societies: Eta Kappa Nu, Kappa Eta Kappa

SUPPORT PROGRAMS
Student Chapter Organizations: Society of Women Engineers, Mexican American Student Association (MASA), American Indian Club

Minority Student Programs
For: Women & Ethnic Minorities   **Available:** Year round
Offered: Prematriculation, Freshman, Sophomore, Junior, Senior, Graduate level

Other Engineering Support Programs: Orientation programs for all freshmen and transfers. Career Counseling services. Financial Aid office, International Student office, Veteran's services, Student advising, Student Health services. International programs with Denmark, England, Germany, France, Costa Rica, and Japan.

338 South Carolina State University

INSTITUTION PROFILE
HEAD OF THE INSTITUTION
Barbara R. Hatton
Phone: (803)536-7013 Fax: (803)533-3622
GENERAL INFORMATION
[All Students—Fall 1991]
Undergraduate enrollment 4,777
Graduate enrollment 368
Total institution enrollment 5,145
Type of institution: Public **Calendar:** Semesters
Nearest city: Columbia **Population:** 250,000
Miles from main campus: 45 **Setting:** Small Town
Types of engineering degrees: Engineering Technology
Other degree-granting colleges: Business, Home Economics and Human Services, Arts & Sciences, Education, Graduate Studies

ENGINEERING TECHNOLOGY ADMISSIONS
ADMISSIONS OFFICE CONTACT
Benny R. Mayfield
South Carolina State University
P.O. Box 7127
300 College Street
Orangeburg, SC 29117-1627
Phone: (803)536-8411 Fax: (803)536-8420
TECHNOLOGY COLLEGE ADMISSIONS INFORMATION
Entrance Requirements: SAT (Score: 700), High School courses—English (4 years), Math (3 years), Laboratory Science (2 years)
Entrance Recommendations: SAT (Score: 760), High School courses—Algebra (2 years), Trigonometry (1 year), Geometry (1 year), Computer Science (1 year)
Requirements for foreign students: TOEFL (Score: 500); financial statement
DEPARTMENTAL ADMISSIONS INFORMATION
Admission to the engineering technology department: At the end of the first year.
Additional information: Grade of 'C' or above in Engineering Technology courses.

FINANCIAL INFORMATION
ESTIMATED EXPENSES (ACADEMIC YEAR)
[Expenses are for the 1992-93 nine-month academic year.]

State Residents	Undergraduate	Graduate
Tuition and fees	$ 2,050	$ 2,050
College room and board	$ 2,520	$ 2,520
Books and supplies	$ 500	$ 500
Other expenses	$ 120	$ 210
Total estimated expenses	**$ 5,190**	**$ 5,280**

Out-of-State Residents	Undergraduate	Graduate
Tuition and fees	$ 4,080	$ 2,050
College room and board	$ 2,520	$ 2,520
Books and supplies	$ 500	$ 500
Other expenses	$ 210	$ 210
Total estimated expenses	**$ 7,310**	**$ 5,280**

FINANCIAL AID OFFICE CONTACT
Margaret C. Black, *Director of Financial Aid*
South Carolina State University
P.O. Box 1886
300 College Street
Orangeburg, SC 29117-1886
Phone: (803)536-7067 Fax: (803)536-8420
GENERAL FINANCIAL AID INFORMATION
Forms accepted/required: CSS-FAF, FAF, FAT, FSA, Stafford, IRS, SAR, Supplemental, Institutional
Additional financial aid information: Need-Based Financial aid; Long-term loans; need-based scholarships, merit-based scholarships; college student employment program. Federal work-study program, short-term and long-term loans; loans and grants.

TECHNOLOGY COLLEGE INFORMATION
[For additional personnel, refer to the *Appendix*.]
HEAD OF ENGINEERING TECHNOLOGY
Shoi Y. Hwang
Phone: (803)536-8479 Fax: (803)533-3623
ENGINEERING TECHNOLOGY COLLEGE ADDRESS
School of Engineering Technology
South Carolina State University
Box 8063, SCSU
300 College Street, N.E.
Orangeburg, SC 29117
Phone: (803)536-7132 Fax: (803)533-3623
TECHNOLOGY ENROLLMENTS—BY CLASS
[Numbers are technology baccalaureate enrollments for the fall 1991 term, unless otherwise footnoted.]
1st-year students/Freshmen 151
2nd-year students/Sophomores 123
3rd-year students/Juniors 60
4th-year students/Seniors 123
Total **457**
NUMBER OF DEGREES AWARDED—BY PROGRAM
[Numbers are engineering technology baccalaureate degrees awarded during the 1991-1992 academic year, unless otherwise footnoted. For full details about each engineering technology program, refer to the *Table of Degree Programs*.]
Civil Engineering Technology 15
Electrical Engineering Technology 31
Industrial Engineering Technology 7
Mechanical Engineering Technology 7
Total **60**
PERCENTAGE OF DEGREES AWARDED—BY CATEGORY
[Percentages are of all engineering technology baccalaureate degrees awarded during the 1991-1992 academic year, unless otherwise footnoted.]
African Americans 95.2%
Asian/Pacific Island Americans 1.6%
Hispanic Americans 1.6%
Native Americans .... – %
Foreign Citizens 1.6%
All Others .... – %
Women 12.0%
Persons with disabilities .... – %
Students over 25 years of age 21.9%
ENGINEERING TECHNOLOGY STUDENT DATA
Applicants to the engineering technology college
Number of applicants to engineering technology college ... 285
Percent offered admission 66.0%
Matriculated engineering technology students
Percentage in top quartile (25%) of High School class .... – (A)
Average SAT scores: Math—373, Verbal—330, Combined—703
Notes: (A) Data not available.
FULL- & PART-TIME FACULTY—BY DEPARTMENT
[Figures are the head count of full-time faculty and the full-time equivalent (FTE) of part-time faculty for each engineering technology department or equivalent.]

Department	Full	Part
Civil & Mechanical Eng Tech	8	–
Industrial & Electrical Eng Tech	8	–
Civil & Mechanical Eng Tech	8	–
Industrial & Electrical Eng Tech	8	–

TECHNOLOGY COLLEGE DESCRIPTION
The School's four year engineering technology programs are laboratory-intensive. Both co-op and summer intern programs are available. The College also offers an off-campus upper division B.S. degree program in Electrical Engineering Technology at selected South Carolina Technical Colleges: Greenville Technical College, Midlands Technical College, Trident Technical College and Piedmont Technical College. The School places emphasis on assisting the students in reaching their greatest potential and upgrading the caliber of minority engineering technologists graduating from South Carolina State University.

TECHNOLOGY DEGREE PROGRAMS DESCRIPTION
The college's B.S. degree programs in engineering technology emphasize practical application of theories and hands-on experience.
DUAL DEGREE PROGRAMS
Technology baccalaureate graduates with dual degrees 0
Enrollment requirements: Minimum grade of 2.0 on a 4.00 scale in all courses. Note: Only offer dual degrees in Engineering Technology and Physics or Applied Mathematics.
TRANSFER INFORMATION
Residency Requirements: Transfer student must earn the last thirty semester credit hours in residence.
Transfer via Articulation Agreements
Admission to engineering technology: First three years at the other four year institution(s), then transfer to S.C. State University for the last two years.
Technology graduates transferred from   ... 4-yr: 3   2-yr: 6
Requirements: Minimum 2.0 GPA. on a 4.0 scale. Only grades of a 'C' or better are accepted for transfer.
Transfer without Articulation Agreements
Admission to engineering technology: At any time
Technology graduates transferred from   ... 4-yr: 2   2-yr: 10
Requirements: Minimum 2.0 GPA. Only grades of 'C' or better are accepted for transfer.
GRADUATION REQUIREMENTS
One must pass a college-wide English proficiency exam. A grade of 'C' or better must be obtained in every engineering technology course. Also the minimum overall GPA must be 2.0 (on a 4.00 scale).

STUDENT PROGRAMS
PROFESSIONAL AND HONORARY SOCIETIES
[For key to acronyms, see *Introduction*.]
Professional Societies: ASCE, ASME, IEEE, IIE, SME
SUPPORT PROGRAMS
Student Chapter Organizations: National Society of Black Engineers
Study Center
For: Women & Ethnic Minorities   **Available:** Year round
Offered: Freshman, Sophomore, Junior, Senior, Graduate level
Psychometric Center
For: Women & Ethnic Minorities   **Available:** Year round
Offered: Freshman, Sophomore, Junior, Senior, Graduate level
Counseling & Self-Development Center
For: Women & Ethnic Minorities   **Available:** Year round
Offered: Freshman, Sophomore, Junior, Senior, Graduate level
Career Development Center
For: Women & Ethnic Minorities   **Available:** Year round
Offered: Freshman, Sophomore, Junior, Senior, Graduate level
Other Engineering Support Programs: School of Freshmen Studies - Freshmen/Transfer Orientation Program; Psychometric Center - Academic Testing/Counseling and Academic Advisement Program; Counseling & Self-development Center - Provides Advisement to all; Career Development Center - CO-OP Education and Job Placement Program; Engineering Summer Institute - Strengthening future freshmen programs.

339 Southern College of Technology

INSTITUTION PROFILE

HEAD OF THE INSTITUTION
Stephen R. Cheshier
Phone: (404)528-7230　　　　Fax: (404)528-7483

GENERAL INFORMATION
[All Students—Fall 1991]
Undergraduate enrollment 3,603
Graduate enrollment 319
Total institution enrollment 3,922
Type of institution: Public　　**Calendar:** Quarters
Nearest city: Atlanta　　**Population:** 2,000,000
Miles from main campus: 15　　**Setting:** Suburban
Types of engineering degrees: Engineering Technology
Other degree-granting colleges: Architecture, Arts & Sciences, Management

ENGINEERING TECHNOLOGY ADMISSIONS

ADMISSIONS OFFICE CONTACT
Virginia A. Head
Southern College of Technology
Director of Admissions
1100 S. Marietta Parkway
Marietta, GA 30060-2896
Phone: (404)528-7281　　　　Fax: (404)528-7292

TECHNOLOGY COLLEGE ADMISSIONS INFORMATION
Entrance Requirements: High School courses—English (4 years), Both Mathematics and Science for three years (3 years), Social Science (3 years), Foreign Language (2 years), SAT-V 410 and SAT-M 440 or ACT English of 21 and Math of 21 required for regular admission. SAT-V 350 and SAT-M 350 or ACT English of 18 and Math of 16 required for admission to developmental studies.
Entrance Recommendations: High School courses—Calculus (1 year), Physics (1 year), Computer Science (1 year)
Requirements for foreign students: 50 in each part of the TOEFL and a financial affidavit are required for international students

DEPARTMENTAL ADMISSIONS INFORMATION
Admission to the engineering technology department: At the time of admission to the institution.

FINANCIAL INFORMATION

ESTIMATED EXPENSES (ACADEMIC YEAR)
[Expenses are for the 1992-93 nine-month academic year.]

State Residents	Undergraduate	Graduate
Tuition and fees	$ 1,548	$ 1,548
College room and board	$ 3,219	$ 3,219
Books and supplies	$ 465	$ 465
Other expenses	$ 1,557	$ 1,557
Total estimated expenses	$ 6,789	$ 6,789

Out-of-State Residents	Undergraduate	Graduate
Tuition and fees	$ 4,230	$ 4,230
College room and board	$ 3,219	$ 3,219
Books and supplies	$ 465	$ 465
Other expenses	$ 1,557	$ 1,557
Total estimated expenses	$ 9,471	$ 9,471

FINANCIAL AID OFFICE CONTACT
Emerelle McNair, *Director of Financial Aid*
Southern College of Technology
Financial Aid
1100 S. Marietta Parkway
Marietta, GA 30060-2896
Phone: (404)528-7290

GENERAL FINANCIAL AID INFORMATION
Forms accepted/required: FAF
Additional financial aid information: Need-based scholarships and grants; short-term loans; merit-based scholarships; part-time jobs on campus; college work-study program.

TECHNOLOGY COLLEGE INFORMATION

[For additional personnel, refer to the *Appendix*.]

HEAD OF ENGINEERING TECHNOLOGY
William D. Rezak
Phone: (404)528-7234　　　　Fax: (404)528-7454

ENGINEERING TECHNOLOGY COLLEGE ADDRESS
School of Technology
Southern College of Technology
1100 S. Marietta Parkway
Marietta, GA 30060-2896
Phone: (404)528-7230　　　　Fax: (404)528-7483

TECHNOLOGY ENROLLMENTS—BY CLASS
[Numbers are technology baccalaureate enrollments for the fall 1991 term, unless otherwise footnoted.]
1st-year students/Freshmen 701
2nd-year students/Sophomores 600
3rd-year students/Juniors 623
4th-year students/Seniors 635
Total 2,559

NUMBER OF DEGREES AWARDED—BY PROGRAM
[Numbers are engineering technology baccalaureate degrees awarded during the 1991-1992 academic year, unless otherwise footnoted. For full details about each engineering technology program, refer to the *Table of Degree Programs*.]
Apparel Engineering Technology 2
Civil Engineering Technology 46
Computer Engineering Technology 15
Electrical Engineering Technology 118
Industrial Engineering Technology 80
Mechanical Engineering Technology 72
Textile Engineering Technology 4
Total 337

PERCENTAGE OF DEGREES AWARDED—BY CATEGORY
[Percentages are of all engineering technology baccalaureate degrees awarded during the 1991-1992 academic year, unless otherwise footnoted.]
African Americans 16.8%
Asian/Pacific Island Americans 2.9%
Hispanic Americans 1.5%
Native Americans 0.2%
Foreign Citizens 7.5%
All Others 71.1%
Women 13.2%
Persons with disabilities 0.2%
Students over 25 years of age 45.0%

GRADUATE ENROLLMENTS & DEGREES AWARDED
Master's enrollment 53
Master's degrees awarded .... –
Doctoral enrollment .... –
Doctoral degrees awarded .... –

ENGINEERING TECHNOLOGY STUDENT DATA

Applicants to the engineering technology college
Number of applicants to engineering technology college ... 346
Percent offered admission 88.2%

Matriculated engineering technology students
Percentage in top quartile (25%) of High School class .... –(B)
Average SAT scores: Math—508, Verbal—415, Combined—923
Average ACT scores: Math—18, Composite—19

Notes: (B) Data not applicable.

FULL- & PART-TIME FACULTY—BY DEPARTMENT
[Figures are the head count of full-time faculty and the full-time equivalent (FTE) of part-time faculty for each engineering technology department or equivalent.]

Department	Full	Part
Civil Eng Tech	10	0.4
Electrical/Electronic(s) Eng Tech	20	0.7
Mechanical Eng Tech	14	0.4
Apparel & Textile Eng Tech	3	–
Industrial Eng Tech	13	1.3

TECHNOLOGY COLLEGE DESCRIPTION

The four-year engineering technology programs are laboratory intensive. An active co-op program is offered. A developmental studies program is offered to students who are not academically prepared to pursue college work. The college also offers associate degree transfer programs in engineering technology.

TRANSFER INFORMATION
Residency Requirements: Students must earn the final 45 credit hours required for the degree in residence at the college.

Transfer via Articulation Agreements
Admission to engineering technology: At any time
Technology graduates transferred from ... 4-yr: –　2-yr: 30
Requirements: Students should be in good standing at the previous institution. A rolling admission procedure is used.

Transfer without Articulation Agreements
Admission to engineering technology: At any time
Technology graduates transferred from ... 4-yr: 189　2-yr: 25
Requirements: Students should be in good standing at the previous institution. A rolling admission procedure is used.

GRADUATION REQUIREMENTS
An overall GPA of 2.00 is required.

STUDENT PROGRAMS

PROFESSIONAL AND HONORARY SOCIETIES
[For key to acronyms, see *Introduction*.]
Professional Societies: ACM, AIA, ASCE, ASHRAE, ASME, IEEE, IIE, SME, American Society for Quality Control
Honorary Societies: Tau Alpha Pi

SUPPORT PROGRAMS
Student Chapter Organizations: National Society of Black Engineers, Black Men in Unity, Black Student Association, NAACP

Minority Academic Counseling
For: Ethnic Minorities　　**Available:** Year round
Offered: Freshman, Sophomore, Junior, Senior

Minority Tutoring
For: Ethnic Minorities　　**Available:** Year round
Offered: Freshman, Sophomore, Junior, Senior

Other Engineering Support Programs: Orientation program for all freshmen and transfers; counseling services; veterans' services; placement services; services for handicapped

340 University of Southern Colorado

■ INSTITUTION PROFILE
HEAD OF THE INSTITUTION
Robert C. Shirley
Phone: (719)549-2306   **Fax:** (719)549-2219
GENERAL INFORMATION
[All Students—Fall 1991]
Undergraduate enrollment 4,303
Graduate enrollment 185
Total institution enrollment 4,488
Type of institution: Public   **Calendar:** Semesters
Location: Pueblo   **Population:** 100,000
Setting: Urban
Types of engineering degrees: Engineering & Technology
Other degree-granting colleges: Applied Science and Engineering Technology, Science and Mathematics, Humanities and Social Sciences, Business, Center for Teaching and Learning

■ ENGINEERING TECHNOLOGY ADMISSIONS
ADMISSIONS OFFICE CONTACT
Margie Wade
University of Southern Colorado
2200 Bonforte Blvd.
Pueblo, CO 81001-4901
Phone: (719)549-2461   **Fax:** (719)549-2219
TECHNOLOGY COLLEGE ADMISSIONS INFORMATION
Entrance Requirements: SAT, ACT, Admission is based on an index of test scores and high school grade point average.
Entrance Recommendations: High School courses—English (4 years), Algebra (2 years), Trigonometry (1 year), Physics (1 year)
Requirements for foreign students: TOEFL (Score: 500); financial statement
DEPARTMENTAL ADMISSIONS INFORMATION
Admission to the engineering technology department: At the time of admission to the institution.

■ FINANCIAL INFORMATION
ESTIMATED EXPENSES (ACADEMIC YEAR)
[Expenses are for the 1992-93 nine-month academic year.]

All Students	Undergraduate	Graduate
Tuition and fees	$ 1,730	$ 1,730
College room and board	$ 4,288	$ 4,288
Books and supplies	$ 700	$ 700
Other expenses	$ –	$ –
Total estimated expenses	$ 6,718	$ 6,718

FINANCIAL AID OFFICE CONTACT
Gina T. Mestas, *Director of Financial Aid*
University of Southern Colorado
2200 Bonfort Blvd
Pueblo, CO 81001-4901
Phone: (719)549-2753   **Fax:** (719)549-2938
GENERAL FINANCIAL AID INFORMATION
Forms accepted/required: ACT, AFSA, CSS-FAF, FAT, FSA, FFS, Stafford, IRS, SAR, Supplemental, Institutional, Comprehensive Financial Aid Report (CFAR), Financial Aid Form Need Analysis Report (FAFNAR)
Additional financial aid information: Need-based scholarships College Work-Study Program Student Loan Programs Private Scholarships USC President's Scholarships

■ TECHNOLOGY COLLEGE INFORMATION
[For additional personnel, refer to the *Appendix*.]
HEAD OF ENGINEERING TECHNOLOGY
Ray L. Sisson
Phone: (719)549-2696   **Fax:** (719)549-2519
EMail: SISSON@STARBURST.USCOLO.EDU
ENGINEERING TECHNOLOGY COLLEGE ADDRESS
College of Applied Science and Engineering Technology
University of Southern Colorado
2200 Bonforte Blvd.
Pueblo, CO 81001-4901
Phone: (719)549-2100   **Fax:** (719)549-2219
TECHNOLOGY ENROLLMENTS—BY CLASS
[Numbers are technology baccalaureate enrollments for the fall 1991 term, unless otherwise footnoted.]
1st-year students/Freshmen 73
2nd-year students/Sophomores 45
3rd-year students/Juniors 50
4th-year students/Seniors 84
Total 252
NUMBER OF DEGREES AWARDED—BY PROGRAM
[Numbers are engineering technology baccalaureate degrees awarded during the 1991-1992 academic year, unless otherwise footnoted. For full details about each engineering technology program, refer to the *Table of Degree Programs*.]
Civil Engineering Technology 17
Electronic(s) Engineering Technology 12
Mechanical Engineering Technology 15
Total 44
PERCENTAGE OF DEGREES AWARDED—BY CATEGORY
[Percentages are of all engineering technology baccalaureate degrees awarded during the 1991-1992 academic year, unless otherwise footnoted.]
African Americans .... 2.4 %
Asian/Pacific Island Americans .... 2.4 %
Hispanic Americans .... 30.6 %
Native Americans .... 1.9 %
Foreign Citizens .... 7.3 %
All Others .... 55.4 %
Women .... 7.1 %
Persons with disabilities .... – % (A)
Students over 25 years of age .... – % (A)
Notes: (A) Data not available.
ENGINEERING TECHNOLOGY STUDENT DATA
Applicants to the engineering technology college
Number of applicants to engineering technology college 63
Percent offered admission .... 100.0 %
Matriculated engineering technology students
Percentage in top quartile (25%) of High School class .... – (A)
Average SAT scores: Combined—901
Average ACT scores: Composite—20.8
Notes: (A) Data not available.
FULL- & PART-TIME FACULTY—BY DEPARTMENT
[Figures are the head count of full-time faculty and the full-time equivalent (FTE) of part-time faculty for each engineering technology department or equivalent.]

Department	Full	Part
Civil Eng Tech	4	–
Electronics Eng Tech	3.5	–
Mechanical Eng Tech	4	–

TECHNOLOGY COLLEGE DESCRIPTION
USC is a Phase I institution in the Colorado Space Grant Consortium, which seeks to increase the pool of professionals in the space industry through programs in teaching, research and outreach. The Engineering Technology Dept. manages the USC program with funding of $45,000 per year. Currently, a graduate assistantship is supported in an outreach project with an elementary school, and 4 undergraduate students are supported in a rocket project which will result in a launch of a sounding rocket to measure ozone absorption. The Engineering Technology Dept. also directs a NASA funded grant program at Ames Research Center which employs 5 full-time faculty research associates with a funding level of about $500,000 per year. This program provides research opportunities for both students and faculty in the area of software reliability and fault-tolerant computer systems.

TRANSFER INFORMATION
Residency Requirements: A minimum of 30 semester hours of resident instruction as approved by the department of the major must be earned in residence. Of the last 32 semester credits earned immediately preceding graduation, no more than 16 may be completed at other universities.
Transfer via Articulation Agreements
Admission to engineering technology: At any time
Technology graduates transferred from   . . . 4-yr: –   2-yr: 12
Transfer without Articulation Agreements
Admission to engineering technology: At any time
Technology graduates transferred from   . . . 4-yr: 2   2-yr: 17
GRADUATION REQUIREMENTS
An overall GPA of 2.0 on a 4.0 scale is required.

■ STUDENT PROGRAMS
PROFESSIONAL AND HONORARY SOCIETIES
[For key to acronyms, see *Introduction*.]
Professional Societies: AGCA, ASME, IEEE, IIE, SME
Honorary Societies: Tau Alpha Pi
SUPPORT PROGRAMS
Student Chapter Organizations: Society of Women Engineers, Society of Hispanic Professional Engineers

Colorado Space Grant Consortium
For: Women & Ethnic Minorities   **Available:** Year round
Offered: Prematriculation, Freshman, Sophomore, Junior, Senior, Graduate level
Other Engineering Support Programs: Indian Summer Orientation Programs (freshmen and transfers) Counseling and Career Services

341 Southern Illinois University at Carbondale

■ INSTITUTION PROFILE

HEAD OF THE INSTITUTION
John C. Guyon
Phone: (618)453-2341   Fax: (618)453-5362

GENERAL INFORMATION
[All Students—Fall 1991]
Undergraduate enrollment 20,339
Graduate enrollment 4,427
Total institution enrollment **24,766**

Type of institution: Public   **Calendar:** Semesters
Nearest city: St. Louis, Missouri   **Population:** 2,500,000
Miles from main campus: 100   **Setting:** Small Town
Types of engineering degrees: Engineering Technology

Other degree-granting colleges: Business Administration, Education, Law, Medicine, Agriculture, Communications and Fine Arts, Liberal Arts, Science, Technical Careers

■ ENGINEERING TECHNOLOGY ADMISSIONS

ADMISSIONS OFFICE CONTACT
Jerre C. Pfaff
Southern Illinois University at Carbondale
Admissions and Records
Carbondale, IL 62901
Phone: (618)453-2982   Fax: (618)453-3250

TECHNOLOGY COLLEGE ADMISSIONS INFORMATION
Entrance Requirements: High School courses—English (4 years), Mathematics (3 years), Science (3 years), Social Studies (3 years), Students in the upper half of their high school class must have a 18 ACT. Students in the lower half must have a 20 ACT score.
Entrance Recommendations: High School courses—English (4 years), Mathematics (3.5 years), Science (2 years)
Requirements for foreign students: TOEFL (Score: 520); financial statement

DEPARTMENTAL ADMISSIONS INFORMATION
Admission to the engineering technology department: At the time of admission to the institution.

■ FINANCIAL INFORMATION

ESTIMATED EXPENSES (ACADEMIC YEAR)
[Expenses are for the 1992-93 nine-month academic year.]

State Residents	Undergraduate	Graduate
Tuition and fees	$ 3,006	$ 3,006
College room and board	$ 3,024	$ 3,024
Books and supplies	$ 550	$ 650
Other expenses	$ 2,600	$ 3,000
Total estimated expenses	**$ 9,180**	**$ 9,680**

Out-of-State Residents	Undergraduate	Graduate
Tuition and fees	$ 7,406	$ 7,406
College room and board	$ 3,024	$ 3,024
Books and supplies	$ 550	$ 650
Other expenses	$ 2,600	$ 3,000
Total estimated expenses	**$13,580**	**$14,080**

FINANCIAL AID OFFICE CONTACT
Pamela A. Britton, *Director*
Southern Illinois University at Carbondale
Financial Aid
Carbondale, IL 62901
Phone: (618)453-4334

GENERAL FINANCIAL AID INFORMATION
Forms accepted/required: ACT, AFSA, CSS-FAF, FFS, PHEA
Additional financial aid information: Scholarships and awards are based on scholastic achievements of new freshmen and Illinois community college transfers; federal and state grant programs; and on-campus student work jobs.

■ TECHNOLOGY COLLEGE INFORMATION
[For additional personnel, refer to the *Appendix*.]

HEAD OF ENGINEERING TECHNOLOGY
Juh W. Chen
Phone: (618)453-4321   Fax: (618)453-4235

ENGINEERING TECHNOLOGY COLLEGE ADDRESS
College of Engineering
Southern Illinois University at Carbondale
Carbondale, IL 62901-6603
Phone: (618)453-4321   Fax: (618)453-4235

TECHNOLOGY ENROLLMENTS—BY CLASS
[Numbers are technology baccalaureate enrollments for the fall 1991 term, unless otherwise footnoted.]
1st-year students/Freshmen 12
2nd-year students/Sophomores 33
3rd-year students/Juniors 53
4th-year students/Seniors 130
Total **228**

NUMBER OF DEGREES AWARDED—BY PROGRAM
[Numbers are engineering technology baccalaureate degrees awarded during the 1991-1992 academic year, unless otherwise footnoted. For full details about each engineering technology program, refer to the *Table of Degree Programs*.]
Civil Engineering Technology 8
Electrical Engineering Technology 24
Manufacturing Systems .... – (B)
Mechanical Engineering Technology 20
Total **52**

Notes: (B) Data not applicable.

PERCENTAGE OF DEGREES AWARDED—BY CATEGORY
[Percentages are of all engineering technology baccalaureate degrees awarded during the 1991-1992 academic year, unless otherwise footnoted.]
African Americans 5.7%
Asian/Pacific Island Americans 1.9%
Hispanic Americans 1.9%
Native Americans .... – %
Foreign Citizens 9.6%
All Others 80.9%
Women 7.6%
Persons with disabilities .... – % (B)
Students over 25 years of age 40.0%

Notes: (B) Data not applicable.

ENGINEERING TECHNOLOGY STUDENT DATA
Applicants to the engineering technology college
Number of applicants to engineering technology college 31
Percent offered admission 90.0%

Matriculated engineering technology students
Percentage in top quartile (25%) of High School class 57.0
Average ACT scores: Math—23.8, Composite—23.1

FULL- & PART-TIME FACULTY—BY DEPARTMENT
[Figures are the head count of full-time faculty and the full-time equivalent (FTE) of part-time faculty for each engineering technology department or equivalent.]

Department	Full	Part
Tech	9	1.0

TECHNOLOGY COLLEGE DESCRIPTION
The Bachelor of Science Degree programs in Engineering Technology prepare the graduates for work involving technical, operational, and production roles in work assignments which require the application of technical knowledge to such activities as development, design, and construction. Day-to-day operational and support engineering activities are important responsibilities of the graduates of the engineering technology programs. The civil engineering technology program includes courses in structures, structural graphics, construction, strength of materials, soil mechanics, hydraulics, surveying, highways and water treatment. Electrical engineering technology students study electrical circuits, logic design, communications, microprocessors and computers, and power distribution systems. The mechanical engineering technology program includes courses in drafting, mechanics, dynamics, refrigeration and air conditioning, heat, power, pneumatics, machine design and power systems technology

TECHNOLOGY DEGREE PROGRAMS DESCRIPTION
The college's four year B.S. degree programs in engineering technology provide graduates with sound principles of technology and their applications to industrial operations, development, design and construction. Graduates are prepared to deal capably with technical and production problems on the job. An M.S. degree in Manufacturing Systems is available to technology graduates interested in designing and implementing modern manufacturing systems, increasing production and improving product quality.

TRANSFER INFORMATION
Residency Requirements: Complete 90 semester hours of credit at SIU-C or complete sixty senior institution hours and the last thirty consecutive semester hours are taken at SIUC. Thirty of the sixty must be in engineering technology courses.

Transfer via Articulation Agreements
Admission to engineering technology: At any time
Technology graduates transferred from . . . 4-yr: –   2-yr: –
Requirements: Students who have an overall C average (2.0 on a 4.0 scale) and are eligible to continue their enrollment at the last institution of attendance will be eligible for admission to any semester.

Transfer without Articulation Agreements
Admission to engineering technology: At any time
Technology graduates transferred from . . . 4-yr: –   2-yr: –
Requirements: Students who have an overall C average or better and are eligible to continue their enrollment at the last institution of attendance will be eligible for admission to any semester.

GRADUATION REQUIREMENTS
B.S. programs in engineering technology require 34 semester hours of general education; 11 semester hours of basic sciences; 13-14 semester hours of mathematics and 69-70 semester hours of engineering technology major courses, for a total of 128 credits. Graduates must have at least a 2.0 GPA (A=4.0) in their major and overall.

■ STUDENT PROGRAMS

PROFESSIONAL AND HONORARY SOCIETIES
[For key to acronyms, see *Introduction*.]
Professional Societies: AGCA, ASHRAE, IEEE, SME, Illinois Professional Land Surveyors Association, Blacks in Engineering and Allied Technology
Honorary Societies: Tau Alpha Pi

SUPPORT PROGRAMS
Student Chapter Organizations: Society of Women Engineers, National Society of Black Engineers, Society of Hispanic Professional Engineers, Blacks in Engineering and Allied Technology

Minority Engineering Program
For: Ethnic Minorities   **Available:** Summer only
Offered: Prematriculation

Other Engineering Support Programs: Comprehensive university orientation program for all new students and parents; academic advisement provided by trained advisors in the unit; career development center; testing services; counseling center; services for students with disabilities; international programs and services; and university placement center.

342 University of Southern Mississippi

■ INSTITUTION PROFILE

HEAD OF THE INSTITUTION
Aubrey K. Lucas
Phone: (601)266-5001   Fax: (601)266-5756

GENERAL INFORMATION
[All Students—Fall 1991]
Undergraduate enrollment 10,457
Graduate enrollment 1,891
Total institution enrollment 12,348
Type of institution: Public   Calendar: Semesters
Nearest city: Jackson   Population: 300,000
Miles from main campus: 95   Setting: Urban
Types of engineering degrees: Engineering Technology
Other degree-granting colleges: Business Administration, Fine & Performing Arts, The Arts, Science and Technology, Education and Psychology, Liberal Arts, Health and Human Sciences

■ ENGINEERING TECHNOLOGY ADMISSIONS

ADMISSIONS OFFICE CONTACT
Richard W. Pyle
University of Southern Mississippi
Office of Admissions
Southern Station Box 5011
Hattiesburg, MS 39406-5011
Phone: (601)266-5555   Fax: (601)266-5816

TECHNOLOGY COLLEGE ADMISSIONS INFORMATION
Entrance Requirements: ACT (Score: 18), High School courses—English (4 years), Mathematics (3 years), Sciences (3 years), Social Sciences (2.5 years), For more information contact the Admission Office.
Entrance Recommendations: ACT (Score: 18), High School courses—Foreign Language (2 years), Computer Science (1 year), Typing/Keyboarding, For more information contact the Admission Office.
Requirements for foreign students: financial statement; All International Student must apply to the University through International Admissions Office (601) 266-5645.

DEPARTMENTAL ADMISSIONS INFORMATION
Admission to the engineering technology department: At the time of admission to the institution.

■ FINANCIAL INFORMATION

ESTIMATED EXPENSES (ACADEMIC YEAR)
[Expenses are for the 1992-93 nine-month academic year.]

All Students	Undergraduate	Graduate
Tuition and fees	$ 1,936	$ —
College room and board	$ 2,170 [1]	$ —
Books and supplies	$ — [2]	$ —
Other expenses	$ — [2]	$ —
Total estimated expenses	$ 4,106	$ —

Notes: (1) Per semester costs are $630 for room and $455 for 5-day meal plan; $510 for 7-day plan. (2) Varies by student.

FINANCIAL AID OFFICE CONTACT
Vernetta P. Fairley, *Director, Financial Aid*
University of Southern Mississippi
Southern Station, Box 5101
Hattiesburg, MS 39406-5101
Phone: (601)266-4774

GENERAL FINANCIAL AID INFORMATION
Forms accepted/required: ACT, Stafford, Institutional
Additional financial aid information: Contact the Financial Aid office for information.

■ TECHNOLOGY COLLEGE INFORMATION
[For additional personnel, refer to the *Appendix*.]

HEAD OF ENGINEERING TECHNOLOGY
Ruth A. Cade, *Director*
Phone: (601)266-4896   Fax: (610)266-5829

ENGINEERING TECHNOLOGY COLLEGE ADDRESS
School of Engineering Technology
University of Southern Mississippi
Southern Station Box 5137
Hattiesburg, MS 39406-5137
Phone: (601)266-4896   Fax: (601)266-5829

TECHNOLOGY ENROLLMENTS—BY CLASS
[Numbers are technology baccalaureate enrollments for the fall 1991 term, unless otherwise footnoted.]
1st-year students/Freshmen 76
2nd-year students/Sophomores 66
3rd-year students/Juniors 122
4th-year students/Seniors 183
Total **447**

NUMBER OF DEGREES AWARDED—BY PROGRAM
[Numbers are engineering technology baccalaureate degrees awarded during the 1991-1992 academic year, unless otherwise footnoted. For full details about each engineering technology program, refer to the *Table of Degree Programs*.]
Architectural Engineering Technology 13
Computer Engineering Technology 7
Construction Engineering Technology 16
Electronic Engineering Technology 36
Engineering Technology .... — [B]
Industrial Engineering Technology 15
Manufacturing Technology .... — [B]
Mechanical Engineering Technology 8
Total **95**

Notes: (B) Data not applicable.

PERCENTAGE OF DEGREES AWARDED—BY CATEGORY
[Percentages are of all engineering technology baccalaureate degrees awarded during the 1991-1992 academic year, unless otherwise footnoted.]
African Americans 21.9%
Asian/Pacific Island Americans 3.1%
Hispanic Americans 1.1%
Native Americans .... — %
Foreign Citizens .... — % [A]
All Others 73.9%
Women 15.4%
Persons with disabilities .... — % [A]
Students over 25 years of age .... — % [A]

Notes: (A) Data not available.

GRADUATE ENROLLMENTS & DEGREES AWARDED
Master's enrollment 34
Master's degrees awarded .... — [A]
Doctoral enrollment .... —
Doctoral degrees awarded .... —

Notes: (A) Data not available.

ENGINEERING TECHNOLOGY STUDENT DATA
Applicants to the engineering technology college
Number of applicants to engineering technology college ..... — [A]
Percent offered admission .... —% [A]
Matriculated engineering technology students
Percentage in top quartile (25%) of High School class .... — [A]
Average ACT scores: Math—20.1 [2], Composite—21.4 [2]

Notes: (A) Data not available. (2) All first-time freshman Fall 91.

■ FULL- & PART-TIME FACULTY—BY DEPARTMENT
[Figures are the head count of full-time faculty and the full-time equivalent (FTE) of part-time faculty for each engineering technology department or equivalent.]

Department	Full	Part
Construction Eng Tech	3	—
Electronics Eng Tech	5	—
Mechanical Eng Tech	2	—
Master of Science in Eng Tech	2	—
Computer Eng Tech	3	—
Architectural Eng Tech	3	—
Industrial Eng Tech	2	—
Pre-Eng	—	—

TECHNOLOGY COLLEGE DESCRIPTION
The University offers 4 year undergraduate programs in engineering technology as well as a new master's degree in engineering technology. The engineering technology curricula are laboratory intensive where students are exposed to state of the art technology. Over 50% of the junior and senior classes are comprised of junior/community college transfers with associate degrees. Co-operative education is available as well as the opportunity to minor in any of the undergraduate programs.

TECHNOLOGY DEGREE PROGRAMS DESCRIPTION
The University's B.S. degree programs in engineering technology emphasize the operational and applied aspects of engineering and utilization of devices and systems.

TRANSFER INFORMATION
Residency Requirements: Last 32 semester hours in residence

Transfer via Articulation Agreements
Admission to engineering technology: Through the Admissions office pending review of transferring GPA status.
Requirements: All students must be in good standing at time of withdrawal from previous institution. Transfer students must meet freshman admission requirements unless (1) over 21 years of age or (2) in possession of an associate degree.

Transfer without Articulation Agreements
Admission to engineering technology: By Director of School of Engineering Technology
Requirements: All transfers must be in good standing at time of withdrawal from previous institution. Transfer students must meet freshman admission requirements unless (1) over 21 years of age or (2) in possession of associate degree.

GRADUATION REQUIREMENTS
Completion of General Education (core) curriculum (55 sem-hrs) plus major requirements; minimum total program 128 sem-hrs (varies with major). GPA 2.00 overall and 2.00 in major courses required.

■ STUDENT PROGRAMS

PROFESSIONAL AND HONORARY SOCIETIES
[For key to acronyms, see *Introduction*.]
Professional Societies: AGCA, AIA, ASME, IEEE, NAHB, SME, Construction Specifications Institute (CSI)
Honorary Societies: Tau Alpha Pi, Sigma Lambda Chi

SUPPORT PROGRAMS
Student Chapter Organizations: Society of Women Engineers, National Society of Black Engineers
Other Engineering Support Programs: The School of Engineering Technology offers early advisement for all students. For more information on advisement schedule contact the main office (601) 266-4896.

343 State University of New York College of Technology at Farmingdale

INSTITUTION PROFILE

HEAD OF THE INSTITUTION
Frank A. Cipriani
Phone: (516)420-2145   Fax: (516)420-2753

GENERAL INFORMATION
[All Students—Fall 1991]
Undergraduate enrollment 9,684
Total institution enrollment 9,684
Type of institution: Public   **Calendar:** Semesters
Nearest city: New York   **Population:** 7,300,000
Miles from main campus: 30   **Setting:** Suburban
Types of engineering degrees: Engineering Technology

ENGINEERING TECHNOLOGY ADMISSIONS

ADMISSIONS OFFICE CONTACT
Janet Snyder
State University of New York College of Technology at Farmingdale
Route 110
Farmingdale, NY 11735
Phone: (516)420-2457   Fax: (516)420-2633

TECHNOLOGY COLLEGE ADMISSIONS INFORMATION
Admission to the engineering technology college: Upon admission to the institution and into specific curriculum.
Entrance Requirements: High School courses—Sequential Math (2 years), Laboratory Science (1 year)
Entrance Recommendations: High School courses—Physics or Chemistry (1 year)
Requirements for foreign students: TOEFL (Score: 500); financial statement; Some academic requirements

DEPARTMENTAL ADMISSIONS INFORMATION
Admission to the engineering technology department: At the time of admission to the institution.
Additional information: Two years of Sequential Mathematics and one year of Physics or Chemistry with the associated laboratory course.

FINANCIAL INFORMATION

ESTIMATED EXPENSES (ACADEMIC YEAR)
[Expenses are for the 1992-93 nine-month academic year.]

State Residents	Undergraduate	Graduate
Tuition and fees	$ 1,620	$ –
College room and board	$ 2,160	$ –
Books and supplies	$ 600	$ –
Other expenses	$ 1,255 (1)	$ –
Total estimated expenses	**$ 5,635**	**$ –**

Out-of-State Residents	Undergraduate	Graduate
Tuition and fees	$ 3,570	$ –
College room and board	$ 2,160	$ –
Books and supplies	$ 600	$ –
Other expenses	$ 1,255 (1)	$ –
Total estimated expenses	**$ 7,585**	**$ –**

Notes: (1) $755 personal expenses; $500 transportation expenses (dorm student); $1000 transportation expenses (commuter student).

FINANCIAL AID OFFICE CONTACT
Nancy N. Dunnagan, *Financial Aid Director*
State University of New York College of Technology at Farmingdale
Rte 110
Farmingdale, NY 11735
Phone: (516)420-2328   Fax: (516)420-2357

GENERAL FINANCIAL AID INFORMATION
Forms accepted/required: ACT, AFSA, CSS-FAF, FAF, IRS, Institutional, Farmingdale Financial Aid Application
Additional financial aid information: Major sources of financial aid: Pell Grant, Supplemental Education Opportunity Grant, Stafford Loan, Parent Loan for Undergraduate Students, Supplemental Loan for Students, Perkins Loans, College Work Study Program, N.Y. State Tuition Assistance Program (TAP) and Aid for Part Time Study Program (APTS), Vietnam Veterans Tuition Award, N.Y. State Regents Award for Children of Deceased Police Officers or Firefighters, VESID, Aid to Native Americans.

TECHNOLOGY COLLEGE INFORMATION
[For additional personnel, refer to the *Appendix*.]

HEAD OF ENGINEERING TECHNOLOGY
Arthur A. Ezra
Phone: (516)420-2115   Fax: (516)420-2194
EMail: EZRAAA @ SNYFARVA

ENGINEERING TECHNOLOGY COLLEGE ADDRESS
School of Engineering Technologies
State University of New York College of Technology at Farmingdale
Route 110
Farmingdale, NY 11735
Phone: (516)420-2000   Fax: (516)420-2194

TECHNOLOGY ENROLLMENTS—BY CLASS
[Numbers are technology baccalaureate enrollments for the fall 1991 term, unless otherwise footnoted.]

1st-year students/Freshmen 402
2nd-year students/Sophomores 305 (1)
3rd-year students/Juniors 55
4th-year students/Seniors 55 (2)
Total **817**

Notes: (1) Freshman and Sophomore part time students equal 155 (2) Junior and Senior part time students equal 72

NUMBER OF DEGREES AWARDED—BY PROGRAM
[Numbers are engineering technology baccalaureate degrees awarded during the 1991-1992 academic year, unless otherwise footnoted. For full details about each engineering technology program, refer to the *Table of Degree Programs*.]

Aerospace Technology .... –
Aircraft Maintenance Technology .... –
Automotive Engineering Technology .... –
Biomedical Engineering Technology .... –
Civil Engineering Technology .... –
Computing Graphics Technology .... –
Construction/Architectural Engineering Technology ... –
Electrical Engineering Technology 31
Engineering Science .... –
Manufacturing Engineering Technology 7
Mechanical Engineering Technology .... –
Total **38**

PERCENTAGE OF DEGREES AWARDED—BY CATEGORY
[Percentages are of all engineering technology baccalaureate degrees awarded during the 1991-1992 academic year, unless otherwise footnoted.]

African Americans .... – %
Asian/Pacific Island Americans .... – %
Hispanic Americans .... – %
Native Americans .... – %
Foreign Citizens .... – %
All Others .... 100 %
Women .... – %
Persons with disabilities .... – %
Students over 25 years of age .... – %

ENGINEERING TECHNOLOGY STUDENT DATA

Applicants to the engineering technology college
Number of applicants to engineering technology college . 1,034
Percent offered admission 72.0%

Matriculated engineering technology students
Percentage in top quartile (25%) of High School class ...... – (A)

Notes: (A) Data not available.

FULL- & PART-TIME FACULTY—BY DEPARTMENT
[Figures are the head count of full-time faculty and the full-time equivalent (FTE) of part-time faculty for each engineering technology department or equivalent.]

Department	Full	Part
Civil Eng Tech	5	2.0
Electrical/Electronic(s) Eng Tech	13	2.3
Manufacturing Eng Tech	3	–
Mechanical Eng Tech	3	1.8
Automotive Eng Tech	5	–
Biomedical Eng Tech	1	–
Construction/Architectural Eng Tech	5	2.0
Aerospace Tech	4	2.3
Computing Graphics	1.5	–
Eng Science	2	2.0

TECHNOLOGY COLLEGE DESCRIPTION

There is an emphasis on hands-on laboratory experience for the students in each curriculum. Evening courses are provided so that students who work full time can attend part time to obtain a baccalaureate degree in Electrical or Manufacturing Engineering Technology.

TECHNOLOGY DEGREE PROGRAMS DESCRIPTION
The Electrical Engineering Technology and the Manufacturing Engineering Technology programs lead to B.T. degrees. Admission is at the Junior year level.

TRANSFER INFORMATION
Residency Requirements: A minimum of 30 credits must be earned at the college in program of study in which matriculated.

Transfer via Articulation Agreements
Admission to engineering technology: Upon completion of Associate Degree.
Requirements: Associate degree in related program of study with a 2.5 cumulative average or better.

Transfer without Articulation Agreements
Admission to engineering technology: Upon completion of A.S. degree.
Requirements: Associate degree in related program of study with 2.5 cumulative grade point average or better.

GRADUATION REQUIREMENTS
For a B.T degree in Electrical Engineering Technology, 140 undergraduate semester hours are required with a minimum overall grade of 'C' or better in the upper division EET courses. For a B.T. degree in Manufacturing Engineering Technology, 139 undergraduate semester hours are required with a minimum overall grade of 'C' or better in the upper division MET courses.

STUDENT PROGRAMS

PROFESSIONAL AND HONORARY SOCIETIES
[For key to acronyms, see *Introduction*.]
Professional Societies: AIAA, AIDD, ASME, IEEE, NSPE, SAE, SES, SME, Society for Biomedical Technology (SBET), Associated General Contractors (AGC)
Honorary Societies: Alpha Beta Gamma, Phi Theta Kappa

SUPPORT PROGRAMS
Other Engineering Support Programs: There is a freshmen/transfer orientation program combined with student advisement in June preceding the Fall semester. There are summer courses to strengthen a student's background. There is a Pre-Tech program for students who have some deficiencies in their preparation for entering a degree program in any of the Engineering Technologies.

State University of New York Institute of Technology, Utica/Rome

■ INSTITUTION PROFILE

HEAD OF THE INSTITUTION
Peter J. Cayan
Phone: (315)792-7400　　　Fax: (315)792-7222

GENERAL INFORMATION
[All Students—Fall 1991]
Undergraduate enrollment 2,250
Graduate enrollment 258
Total institution enrollment 2,508

Type of institution: Public　　Calendar: Semesters
Nearest city: Utica, New York　　Population: 75,000
Miles from main campus: 1　　Setting: Small Town
Types of engineering degrees: Engineering Technology
Other degree-granting colleges: Arts & Sciences, Business Administration, Nursing, Information Systems & Engineering Technology

■ ENGINEERING TECHNOLOGY ADMISSIONS

ADMISSIONS OFFICE CONTACT
Eileen M. Collins
State University of New York Institute of Technology, Utica/Rome
Office of Admissions
P.O. Box 3050
Utica, NY 13504-3050
Phone: (315)792-7208

TECHNOLOGY COLLEGE ADMISSIONS INFORMATION
Admission to the engineering technology college: Freshmen are not admitted due to the upper division nature of the institution.

DEPARTMENTAL ADMISSIONS INFORMATION
Admission to the engineering technology department: At the time of admission to the institution.
Additional information: 2.5 or higher GPA. and 56 transferable credit hours, generally from parallel associate degree programs.

■ FINANCIAL INFORMATION

ESTIMATED EXPENSES (ACADEMIC YEAR)
[Expenses are for the 1992-93 nine-month academic year.]

State Residents	Undergraduate	Graduate
Tuition and fees	$ 2,938	$ 4,163
College room and board	$ 4,615	$ 4,615
Books and supplies	$ 500	$ 500
Other expenses	$ –	$ –
Total estimated expenses	**$ 8,053**	**$ 9,278**

Out-of-State Residents	Undergraduate	Graduate
Tuition and fees	$ 6,778	$ 7,479
College room and board	$ 4,615	$ 4,615
Books and supplies	$ 500	$ 500
Other expenses	$ –	$ –
Total estimated expenses	**$11,893**	**$12,594**

FINANCIAL AID OFFICE CONTACT
Edward A. Hutchinson, *Director of Financial Aid*
State University of New York Institute of Technology, Utica/Rome
Marcy Campus
P.O. Box 3050
Utica, NY 13504-3050
Phone: (315)792-7210　　　Fax: (315)792-7802

GENERAL FINANCIAL AID INFORMATION
Forms accepted/required: FAF, FAT, IRS, SUNY Institute of Technology aid application., Pell Grant Student Aid Reports
Additional financial aid information: Approximately 75% of all students receive some form of financial aid. The institute participates in almost all federal and state aid programs. College sponsored scholarships are available in many areas.

■ TECHNOLOGY COLLEGE INFORMATION
[For additional personnel, refer to the *Appendix*.]

HEAD OF ENGINEERING TECHNOLOGY
Ronald Sarner
Phone: (315)792-7234　　　Fax: (315)792-7800
EMail: ron@sunyit.edu

ENGINEERING TECHNOLOGY COLLEGE ADDRESS
School of Information Systems and Engineering Technology
State University of New York Institute of Technology, Utica/Rome
P.O. Box 3050
Utica, NY 13504-3050
Phone: (315)792-7234　　　Fax: (315)792-7800

TECHNOLOGY ENROLLMENTS—BY CLASS
[Numbers are technology baccalaureate enrollments for the fall 1991 term, unless otherwise footnoted.]
1st-year students/Freshmen 0
2nd-year students/Sophomores 0
3rd-year students/Juniors 203
4th-year students/Seniors 157
Total 360

NUMBER OF DEGREES AWARDED—BY PROGRAM
[Numbers are engineering technology baccalaureate degrees awarded during the 1991-1992 academic year, unless otherwise footnoted. For full details about each engineering technology program, refer to the *Table of Degree Programs*.]
Computer Systems Technology 36
Electrical Engineering Technology .... 7 [1]
Industrial Engineering Technology 11
Mechanical Engineering Technology 20
Total 74

Notes: (1) Admit upper level students only.

PERCENTAGE OF DEGREES AWARDED—BY CATEGORY
[Percentages are of all engineering technology baccalaureate degrees awarded during the 1991-1992 academic year, unless otherwise footnoted.]
African Americans .... 4.0 %
Asian/Pacific Island Americans .... 3.0 %
Hispanic Americans .... 3.0 %
Native Americans .... – %
Foreign Citizens .... 2.0 %
All Others .... 88.0 %
Women .... 4.0 %
Persons with disabilities .... – % [A]
Students over 25 years of age .... – % [A]

Notes: (A) Data not available.

FULL- & PART-TIME FACULTY—BY DEPARTMENT
[Figures are the head count of full-time faculty and the full-time equivalent (FTE) of part-time faculty for each engineering technology department or equivalent.]

Department	Full	Part
Electrical/Electronic(s) Eng Tech	8	0.3
Mechanical Eng Tech	4	0.6
Industrial Eng Tech	2	0.3
Computer Tech	1	–

TECHNOLOGY COLLEGE DESCRIPTION
In addition to programs in engineering technology, the School of Information Systems and Engineering Technology also offers degree programs in Computer Science, Photonics, and Telecommunications. Both the Telecommunications and Photonics programs were the first of their kind to be offered in New York State. The school is located in new buildings on a new campus. Average number of years required to complete the bachelor's degree: 2.

TECHNOLOGY DEGREE PROGRAMS DESCRIPTION
The Board of Regents and the New York State Education Department has authorized the State University of New York Institute of Technology at Utica/Rome to confer the following undergraduate degrees for the Electrical Engineering Technology, Industrial Engineering Technology and Mechanical Engineering Technology programs: Bachelor of Technology and Bachelor of Science. The Computer Technology program confers the Bachelor of Science degree.

TRANSFER INFORMATION
Residency Requirements: 24 semester credits must be earned in residence.
Transfer via Articulation Agreements
Admission to engineering technology: Minimum of 56 credits of prior work required for admission.
Requirements: A GPA of 2.5 is required.
Transfer without Articulation Agreements
Admission to engineering technology: At any time
Requirements: A GPA of 2.5 or higher is required.

GRADUATION REQUIREMENTS
Satisfactory completion of 124 credit hours with a minimum GPA of 2.00 for all coursework taken at the Institute of Technology is required for graduation in addition to completing all program requirements.

■ STUDENT PROGRAMS

PROFESSIONAL AND HONORARY SOCIETIES
[For key to acronyms, see *Introduction*.]
Professional Societies: ASME, IEEE, IIE, SME
Honorary Societies: Tau Alpha Pi

SUPPORT PROGRAMS
Student Chapter Organizations: Society of Women Engineers

Collegiate Science and Technology Entry Program
For: Ethnic Minorities　　Available: Academic year
Offered: Junior, Senior

Society of Women Engineers
For: Women　　Available: Academic year
Offered: Junior, Senior

Other Engineering Support Programs: Orientation Program for all new students; Learning Center; Career Planning and Placement Office; Veterans' Services; International Student Services; student clubs and chapters of professional societies.

345 Temple University

INSTITUTION PROFILE

HEAD OF THE INSTITUTION
Peter J. Liacouras
Phone: (215)787-7405

GENERAL INFORMATION
[All Students—Fall 1991]
Undergraduate enrollment	22,000
Graduate enrollment	10,000
Total institution enrollment	**32,000**

Type of institution: Public and Private **Calendar:** Semesters
Location: Philadelphia **Population:** 1,600,000
Setting: Urban
Types of engineering degrees: Engineering Technology
Other degree-granting colleges: Allied Health Sciences, Arts & Sciences, Business Administration, Communication & Journalism, Dentistry, Education, Fine & Performing Arts, Law, Medicine, Nursing, Pharmacy, Technology & Applied Sciences, Social Administration, Landscape Architecture, Health, Physical Education, Recreation & Dance

TECHNOLOGY COLLEGE ADMISSIONS INFORMATION
Entrance Requirements: SAT (Score: 900), High School courses—English (4 years), Mathematics (3 years), Foreign Language (2 years), Laboratory Science (2 years)
Entrance Recommendations: High School courses—Computer Science (1 year), Advanced Placement Tests
Requirements for foreign students: TOEFL (Score: 500)

DEPARTMENTAL ADMISSIONS INFORMATION
Admission to the engineering technology department: At the time of admission to the institution.

FINANCIAL INFORMATION

ESTIMATED EXPENSES (ACADEMIC YEAR)
[Expenses are for the 1992-93 nine-month academic year.]

State Residents	Undergraduate	Graduate
Tuition and fees	$ 4,983	$ 4,356[1]
College room and board	$ 5,540	_[A]
Books and supplies	$ 800[C]	_[A]
Other expenses	$ _[A]	$ _[A]
Total estimated expenses	**$11,323**	**$ 4,356**

Out-of-State Residents	Undergraduate	Graduate
Tuition and fees	$ 9,197	$ 5,490[1]
College room and board	$ 5,540	_[A]
Books and supplies	$ 800[C]	_[A]
Other expenses	$ _[A]	$ _[A]
Total estimated expenses	**$15,537**	**$ 5,490**

Notes: (A) Data not available. (C) Estimated data. (1) Based on 9 credit hours per semester.

FINANCIAL AID OFFICE CONTACT
John Morris, *Director*
Temple University
2nd Floor Conwell Hall
Broad & Montgomery Streets
Philadelphia, PA 19130
Phone: (215)787-1492

GENERAL FINANCIAL AID INFORMATION
Forms accepted/required: State PHEAA form
Additional financial aid information: University, state, and federal low-interest loans, awards based on merit, work-study positions, ROTC scholarships.

TECHNOLOGY COLLEGE INFORMATION
[For additional personnel, refer to the *Appendix*.]

HEAD OF ENGINEERING TECHNOLOGY
Charles K. Alexander
Phone: (215)787-7959 Fax: (215)787-6936

ENGINEERING TECHNOLOGY COLLEGE ADDRESS
College of Engineering, Computer Science, and Architecture
Temple University
12th & Norris Streets
Philadelphia, PA 19130
Phone: (215)787-7800 Fax: (215)787-6936

TECHNOLOGY ENROLLMENTS—BY CLASS
[Numbers are technology baccalaureate enrollments for the fall 1991 term, unless otherwise footnoted.]

1st-year students/Freshmen	204
2nd-year students/Sophomores	136
3rd-year students/Juniors	107
4th-year students/Seniors	175
Total	**622**

NUMBER OF DEGREES AWARDED—BY PROGRAM
[Numbers are engineering technology baccalaureate degrees awarded during the 1991-1992 academic year, unless otherwise footnoted. For full details about each engineering technology program, refer to the *Table of Degree Programs*.]

Civil & Construction Engineering	26
Civil/Construction Engineering Technology	–
Electrical Engineering Technology	37
Engineering Technology	2
Environmental Engineering Technology	19
Mechanical Engineering Technology	23
Total	**107**

PERCENTAGE OF DEGREES AWARDED—BY CATEGORY
[Percentages are of all engineering technology baccalaureate degrees awarded during the 1991-1992 academic year, unless otherwise footnoted.]

African Americans	12.0 %
Asian/Pacific Island Americans	7.4 %
Hispanic Americans	2.7 %
Native Americans	– %
Foreign Citizens	8.3 %
All Others	69.6 %
Women	11.1 %
Persons with disabilities	– %[A]
Students over 25 years of age	– %[A]

Notes: (A) Data not available.

ENGINEERING TECHNOLOGY STUDENT DATA
Applicants to the engineering technology college
Number of applicants to engineering technology college ..... –[A]
Percent offered admission .... –%[A]

Matriculated engineering technology students
Percentage in top quartile (25%) of High School class .... 50.0[C]
Average SAT scores: Math—500[C], Verbal—450[C], Combined—950[C]

Notes: (A) Data not available. (C) Estimated data.

FULL- & PART-TIME FACULTY—BY DEPARTMENT
[Figures are the head count of full-time faculty and the full-time equivalent (FTE) of part-time faculty for each engineering technology department or equivalent.]

Department	Full	Part
Civil/Construction Eng Tech	10	3.0
Electrical Eng Tech	16	–
Mechanical Eng Tech	12	3.5
Environmental Eng Tech	10	3.0
General Eng Tech	12	3.5

TECHNOLOGY COLLEGE DESCRIPTION
The college offers a four-year B.S. degree program in civil/construction engineering technology, environmental engineering technology, mechanical engineering technology, electrical engineering technology, and general engineering technology. The first year is dominated by required core courses. Fundamentals of engineering technology are emphasized during the second year, with specialized courses offered during years three and four.

TECHNOLOGY DEGREE PROGRAMS DESCRIPTION
The B.S. degree prepares a student for entry-level positions in engineering technology. The M.B.A. degree is advocated for graduates seeking to further advanced training.

TRANSFER INFORMATION
Residency Requirements: Students must be matriculated at Temple University for the last 30 credits prior to receiving their degree. Community college transfer credits are limited to 64 semester hours.

Transfer via Articulation Agreements
Admission to engineering technology: At any time
Requirements: Transfer students must have completed at least 15 semester hours. with a GPA. of 2.5 from a non-accredited (ABET.) engineering technology program or a non-engineering technology major. A 2.0 GPA. is minimum GPA. from an ABET. approved program.

Transfer without Articulation Agreements
Admission to engineering technology: At any time
Requirements: Same as entrance requirements for units with articulation agreements.

GRADUATION REQUIREMENTS
B.S. degree candidates must successfully complete 124 semester hours of courses and be in good standing.

STUDENT PROGRAMS

PROFESSIONAL AND HONORARY SOCIETIES
[For key to acronyms, see *Introduction*.]
Professional Societies: ACI, ACM, ASCE, ASME, IEEE, SAE, American Society of Metals (ASM), Pennsylvania Society of Land Surveyors (PSLS)
Honorary Societies: Eta Kappa Nu

SUPPORT PROGRAMS
Student Chapter Organizations: Society of Women Engineers, Minority Engineering Student Association (MESA)
Other Engineering Support Programs: Orientation program for freshman and transfer students; free tutoring center in college; career counseling and job placement; veterans, international, and handicapped student offices; academic counseling office; psychological services are university services.

346 University of Tennessee, Martin

■ INSTITUTION PROFILE

HEAD OF THE INSTITUTION
Margaret N. Perry
Phone: (901)587-7500 **Fax:** (901)587-7019

GENERAL INFORMATION
[All Students—Fall 1991]
Undergraduate enrollment 5,212
Graduate enrollment 282
Total institution enrollment **5,494**
Type of institution: Public **Calendar:** Semesters
Nearest city: Memphis **Population:** 650,000
Miles from main campus: 120 **Setting:** Small Town
Types of engineering degrees: Engineering Technology
Other degree-granting colleges: Agricultural & Environmental, Arts & Sciences, Business Administration, Education, Fine & Performing Arts, Agriculture and Home Economics

■ ENGINEERING TECHNOLOGY ADMISSIONS

ADMISSIONS OFFICE CONTACT
Randy L. Perry
University of Tennessee, Martin
Room 113 EPS Building
Martin, TN 38238
Phone: (901)587-7380 **Fax:** (901)587-7375

TECHNOLOGY COLLEGE ADMISSIONS INFORMATION
Entrance Requirements: ACT (Score: 19), High School courses—English (4 years), Social Science, Natural Science, Foreign Language (2 years), Adv.Math (1 unit geometry, calculus, or adv. Geom) (3 years), Fine and Performing Arts (1 year)
Entrance Recommendations: SAT (Score: 750), ACT (Score: 19), High School courses—History, Social Studies (1 year), Mathematics (4 years), Natural/Physical Science (2 years)
Requirements for foreign students: TOEFL (Score: 500); financial statement; All degree-seeking international students must complete an English Placement Test; the student must enroll in the appropriate English course.

DEPARTMENTAL ADMISSIONS INFORMATION
Admission to the engineering technology department: At the time of admission to the institution.

■ FINANCIAL INFORMATION

ESTIMATED EXPENSES (ACADEMIC YEAR)
[Expenses are for the 1992-93 nine-month academic year.]

State Residents	Undergraduate	Graduate
Tuition and fees	$ 2,592	$ 3,255
College room and board	$ 3,585	$ 3,585
Books and supplies	$ 900	$ 1,050
Other expenses	$ –	$ –
Total estimated expenses	**$ 7,077**	**$ 7,890**
Out-of-State Residents	Undergraduate	Graduate
Tuition and fees	$ 7,992	$ 8,655
College room and board	$ 3,585	$ 3,585
Books and supplies	$ 900	$ 1,050
Other expenses	$ –	$ –
Total estimated expenses	**$12,477**	**$ 13,290**

FINANCIAL AID OFFICE CONTACT
Randall D. Hall, *Director of Financial Aid*
University of Tennessee, Martin
205 Administration Building
Martin, TN 38238
Phone: (901)587-7040 **Fax:** (901)587-7019

GENERAL FINANCIAL AID INFORMATION
Forms accepted/required: ACT, AFSA, CSS-FAF, FAT, Stafford, SAR, Supplemental
Additional financial aid information: 1) Federal Work-Study, 2) Federal SEOG, 3) Federal Pell Grant, 4) Tennessee Student Assistance Award, 5) Federal Stafford Loan, 6) Federal Plus/SLS Loan, 7) Federal Perkins Loan.

■ TECHNOLOGY COLLEGE INFORMATION
[For additional personnel, refer to the *Appendix*.]

HEAD OF ENGINEERING TECHNOLOGY
Randy L. Perry
Phone: (901)587-7380 **Fax:** (901)587-7375
EMail: RLPERRY@MARTN.Bitnet

ENGINEERING TECHNOLOGY COLLEGE ADDRESS
School of Engineering Technology and Engineering
University of Tennessee, Martin
Room 113, EPS Building
Martin, TN 38238
Phone: (901)587-7380 **Fax:** (901)587-7375

TECHNOLOGY ENROLLMENTS—BY CLASS
[Numbers are technology baccalaureate enrollments for the fall 1991 term, unless otherwise footnoted.]
1st-year students/Freshmen 73
2nd-year students/Sophomores 32
3rd-year students/Juniors 16
4th-year students/Seniors 34
Total **155**

NUMBER OF DEGREES AWARDED—BY PROGRAM
[Numbers are engineering technology baccalaureate degrees awarded during the 1991-1992 academic year, unless otherwise footnoted. For full details about each engineering technology program, refer to the *Table of Degree Programs*.]
Civil Engineering Technology 2
Electrical Engineering Technology 11
Mechanical Engineering Technology 7
Total **20**

PERCENTAGE OF DEGREES AWARDED—BY CATEGORY
[Percentages are of all engineering technology baccalaureate degrees awarded during the 1991-1992 academic year, unless otherwise footnoted.]
African Americans .... 5.0 %
Asian/Pacific Island Americans .... – %
Hispanic Americans .... – %
Native Americans .... – %
Foreign Citizens .... 15.0 %
All Others .... 80.0 %
Women .... 10.0 %
Persons with disabilities .... – %
Students over 25 years of age .... 25.0 %

ENGINEERING TECHNOLOGY STUDENT DATA
Applicants to the engineering technology college
Number of applicants to engineering technology college 43
Percent offered admission 93.0%
Matriculated engineering technology students
Percentage in top quartile (25%) of High School class .... –(A)
Average ACT scores: Math—21.8, Composite—22.1

Notes: (A) Data not available.

FULL- & PART-TIME FACULTY—BY DEPARTMENT
[Figures are the head count of full-time faculty and the full-time equivalent (FTE) of part-time faculty for each engineering technology department or equivalent.]

Department	Full	Part
Civil Eng Tech	3	–
Electrical/Electronic(s) Eng Tech	3	–
Mechanical Eng Tech	3	–

TECHNOLOGY COLLEGE DESCRIPTION
The School offers the Bachelor of Science degree in Engineering Technology with majors in civil, electrical, and mechanical. This program of study is designed to prepare individuals to participate in engineering activities which require applied solutions to technological problems. The curricula embrace practical orientations to teaching the application of scientific and engineering knowledge and methods, but they do not contain the mathematical and theoretical rigor of baccalaureate degrees in traditional engineering programs. The programs include the fundamental courses such as mathematics, physics, engineering graphics, and basic mechanics as well as required technical courses. The curricula also include courses in management, economics, communications and technical writing, which prepares individuals for managerial type positions. Engineering Technology students may participate in the Cooperative Education program.

TECHNOLOGY DEGREE PROGRAMS DESCRIPTION
The School's four-year degree program in engineering technology with majors in civil, electrical, and mechanical emphasize practical orientations to teaching the application of scientific and engineering knowledge and methods to prepare work-ready graduates. The B.S. degree in Civil Engineering Technology encompasses mapping, building layout, topography control, construction layout and planning, highway layout and design, and sub-division planning. The B.S. in Electrical Engineering Technology trains its graduates to put engineering design into operations and to supervise electrical technicians; trained to direct proper utilization of electrical equipment used in manufacturing or power distribution. The B.S. degree in Mechanical Engineering Technology graduate is developed to organize employees, materials, and equipment and to put into effect engineering design and principles; the program prepares its graduate with practical and theoretical knowledge of manufacturing tools and methods and is able to supervise and train technicians, craftsmen, and others.

TRANSFER INFORMATION
Residency Requirements: Must have minimum of 60 hours at senior institution; last 30 semesters hours at UTM.
Transfer via Articulation Agreements
Admission to engineering technology: At any time
Technology graduates transferred from . . . 4-yr: – 2-yr: 1
Requirements: A transfer student with junior standing (minimum 60 semester hours) at the receiving institution is exempt from meeting the high school unit requirements; one with less than 60 semester hours, who graduated from high school after 1988, is required to have the high school units listed in #13 above. Must have minimum 2.00 cumulative GPA. Must have minimum of one academic year in residence.

Transfer without Articulation Agreements
Admission to engineering technology: At any time
Technology graduates transferred from . . . 4-yr: – 2-yr: 1
Requirements: Same as for articulation agreements.

GRADUATION REQUIREMENTS
Bachelor of Science degree in the School of Engineering Technology requires the successful completion of from 131-134 semester hours and a grade point average of at least 2.00 on all work attempted; must have a minimum of one academic year in residence at UT-Martin.

■ STUDENT PROGRAMS

PROFESSIONAL AND HONORARY SOCIETIES
[For key to acronyms, see *Introduction*.]
Professional Societies: IEEE, SAE, SME, Society of Women Engineers (SWE)

SUPPORT PROGRAMS
Student Chapter Organizations: Society of Women Engineers
Black Student Association
For: Ethnic Minorities **Available:** Year round
Offered: Prematriculation, Freshman, Sophomore, Junior, Senior, Graduate level
Other Engineering Support Programs: FRESHMAN STUDIES-An orientation course, is offered each Fall Semester to freshman students the week prior to (and continuing through) the Fall term. How to Study sessions, placement testing, advising conferences, class registration and orientation to the University environment round out the week-long program; other planned activities specific to their discipline continues throughout the term.

347 Texas A&M University

■ INSTITUTION PROFILE
HEAD OF THE INSTITUTION
William H. Mobley
Phone: (409)845-2217 Fax: (409)845-5027

GENERAL INFORMATION
[All Students—Fall 1991]
Undergraduate enrollment 33,024
Graduate enrollment 7,973
Total institution enrollment 40,997
Type of institution: Public Calendar: Semesters
Nearest city: Houston, TX Population: 2,782,414
Miles from main campus: 97 Setting: Urban
Types of engineering degrees: Engineering Technology
Other degree-granting colleges: Agricultural & Environmental, Architecture, Business Administration, Education, Medicine, Agricultural & Life Sciences, Geosciences, Liberal Arts, Science, Veterinary Medicine

■ ENGINEERING TECHNOLOGY ADMISSIONS
ADMISSIONS OFFICE CONTACT
Billy G. Lay
Texas A&M University
Director of Admissions
College Station, TX 77843-0100
Phone: (409)845-1157 Fax: (409)845-0727

TECHNOLOGY COLLEGE ADMISSIONS INFORMATION
Entrance Requirements: SAT (Score: 1000), ACT (Score: 24), High School courses—English (4 years), Mathematics (3 years), Science (2 years), Social Studies (2.5 years), Figures above are for Priority Admission. Regular Admission minimum SAT of 800 or ACT of 19 if in 1st Qtr. of h.s. class, 950 or 22 in 2nd Qtr., & 1100 or 27 if in 3rd/4th Qtrs.
Entrance Recommendations: High School courses—Foreign Language (2 years), Computer Science (1 year), ECAT and MAT, Level I & Level II
Requirements for out-of-state residents: Non-residents are not eligible for priority admission unless they rank in top 25% of high school class with minimum SAT of 1100 or ACT of 27. No different admission criteria under 'regular admission' policy.
Requirements for foreign students: TOEFL (Score: 550); financial statement; Written & Oral English proficiency exam tested locally before registration.

DEPARTMENTAL ADMISSIONS INFORMATION
Admission to the engineering technology department: At the time of admission to the institution.
Additional information: Students who meet university entrance requirements enter College of Engineering with Lower Division classification. Admission to major degree sequence may be limited by availability of instructional resources. To be considered for admission to major degree sequence student must be in good academic standing & have received credit for specific courses. For engr. tech. they are: CHEM 102 & 112, ENDG 105, ENGL 104, ENGR 109 or CPSC 110, MATH 151, & 6 hrs. of directed electives or equiv.

■ FINANCIAL INFORMATION
ESTIMATED EXPENSES (ACADEMIC YEAR)
[Expenses are for the 1992-93 nine-month academic year.]

State Residents	Undergraduate	Graduate
Tuition and fees	$ 1,544	$ 1,744
College room and board	$ 2,241	$ –(B)
Books and supplies	$ 600	$ –(B)
Other expenses	$ 1,914	$ –(B)
Total estimated expenses	**$ 6,299**	**$ 1,744**

Out-of-State Residents	Undergraduate	Graduate
Tuition and fees	$ 4,856	$ 5,296
College room and board	$ 2,241	$ –(B)
Books and supplies	$ 600	$ –(B)
Other expenses	$ 1,914	$ –(B)
Total estimated expenses	**$ 9,611**	**$ 5,296**

Notes: (B) Data not applicable.

FINANCIAL AID OFFICE CONTACT
Donald L. Engelage, *Director of Student Financial Aid*
Texas A&M University
Mail Stop 1252
College Station, TX 77843-1252
Phone: (409)845-8874 Fax: (409)847-9061

GENERAL FINANCIAL AID INFORMATION
Forms accepted/required: ACT, AFSA, CSS-FAF, FAF, FFS, Note: FAF or FFS required, not both.
Additional financial aid information: Various types of financial aid are available including institutional, state and federal aid. This aid consists of academic and need-based scholarships, state and federal grants, employment opportunities both on and off campus, as well as the College-Work Study Program, and student loans. The Student Financial Aid Office at Texas A&M is very responsive to student needs and its primary purpose is to provide resources to students who would otherwise be unable to pursue a post-secondary education.

■ TECHNOLOGY COLLEGE INFORMATION
[For additional personnel, refer to the *Appendix*.]

HEAD OF ENGINEERING TECHNOLOGY
Kenneth L. Peddicord
Phone: (409)845-7203 Fax: (409)845-8986
EMail: KLP7201@SUMMA.TAMU.EDU

ENGINEERING TECHNOLOGY COLLEGE ADDRESS
Engineering Technology Department
Texas A&M University
MS# 3367
College Station, TX 77843-3367
Phone: (409)845-4951 Fax: (409)847-9396

TECHNOLOGY ENROLLMENTS—BY CLASS
[Numbers are technology baccalaureate enrollments for the fall 1991 term, unless otherwise footnoted.]
1st-year students/Freshmen 34
2nd-year students/Sophomores 74
3rd-year students/Juniors 209
4th-year students/Seniors 482
Total 799

NUMBER OF DEGREES AWARDED—BY PROGRAM
[Numbers are engineering technology baccalaureate degrees awarded during the 1991-1992 academic year, unless otherwise footnoted. For full details about each engineering technology program, refer to the *Table of Degree Programs*.]
Electronic Engineering Technology .... 56 [2]
Engineering Technology 111
Manufacturing Engineering Technology .... 19 [2]
Mechanical Engineering Technology .... 36 [2]
Total 222

Notes: (2) This data is included in the total for Engineering Technology Degree Program.

PERCENTAGE OF DEGREES AWARDED—BY CATEGORY
[Percentages are of all engineering technology baccalaureate degrees awarded during the 1991-1992 academic year, unless otherwise footnoted.]
African Americans 6.3%
Asian/Pacific Island Americans 4.5%
Hispanic Americans 10.0%
Native Americans .... – %
Foreign Citizens .... – %
All Others 79.2%
Women 1.8%
Persons with disabilities .... – % (A)
Students over 25 years of age 41.0%

Notes: (A) Data not available.

ENGINEERING TECHNOLOGY STUDENT DATA
Applicants to the engineering technology college
Number of applicants to engineering technology college 42
Percent offered admission 66.6%

Matriculated engineering technology students
Percentage in top quartile (25%) of High School class ... 100.0
Average SAT scores: Math—570, Verbal—470, Combined—1040
Average ACT scores: Math—29, Composite—29

FULL- & PART-TIME FACULTY—BY DEPARTMENT
[Figures are the head count of full-time faculty and the full-time equivalent (FTE) of part-time faculty for each engineering technology department or equivalent.]

Department	Full	Part
Electronics Eng Tech	9	–
Manufacturing Eng Tech	6	–
Mechanical Eng Tech	7	–
Eng Tech	22	–

TECHNOLOGY COLLEGE DESCRIPTION
Texas A&M's three TAC/ABET accredited ET programs emphasize high technology. The curricula have a strong math/science base using differential and integral calculus and calculus-based physics taken by engineering majors. The technical specialty courses have a well-coordinated prerequisite structure, building on the math/science base and developing strong hand-on laboratory experiences with modern equipment and computing systems. The ET faculty are active in applied research and draw ET students into their research projects to enhance the students' education. The research is funded from many sources including industry and state and Federal agencies. A strong co-op program is available. The Telecommunications Option in Electronics Engineering Technology cooperates with the College of Business in an MBA in Telecommunications Technology Management Program where the advanced technical courses are offered by the EET faculty.

TECHNOLOGY DEGREE PROGRAMS DESCRIPTION
The engineering technology programs offered at Texas A&M University focus upon the latest technologies and make special efforts to infuse creativity and innovation. While directly related, these curricula are distinct from engineering by virtue of the greater focus on hardware, laboratory procedures and instrumentation, and the development of technological skills. Due to the greater emphasis on hands-on laboratory experience and the development of technological skills, there is less theory analysis than is required in the typical engineering curricula. Engineering technology graduates fulfill vital roles in industries as members of the technological team who work closely with engineers as well as with technicians and craftsmen. They fulfill such tasks as improving product design and development, material and product testing for assurance of quality, production management, and operation/service of complex technological systems. Programs are offered in electronics, manufacturing and mechanical engineering technology. The electronics program has an area of specialization in telecommunications. Graduates of these programs are awarded B.S. degrees in engineering technology.

TRANSFER INFORMATION
Transfer without Articulation Agreements
Admission to engineering technology: At any time
Technology graduates transferred from ... 4-yr: – 2-yr: –
Requirements: 30 SCH's or less - 3.0 GPR & meet requirements for entering freshmen. 31 to 45 SCH's - Overall GPR of 3.0 or higher; 3.0 for each of two most recent semesters. 46 to 60 SCH's - Overall GPR of 2.5 or higher; 2.5 ea. of two most recent semesters. 61 or more SCH's - Overall GPR of 2.0 or higher; 2.0 each of two most recent semesters.

GRADUATION REQUIREMENTS
Minimum of 2.0 GPA in major courses and 2.0 overall.

■ STUDENT PROGRAMS
PROFESSIONAL AND HONORARY SOCIETIES
[For key to acronyms, see *Introduction*.]
Professional Societies: ASME, IEEE, SME
Honorary Societies: Tau Alpha Pi, Engineering Scholars Program

SUPPORT PROGRAMS
Student Chapter Organizations: Society of Women Engineers, National Society of Black Engineers, Mexican American Engineers, College of Engineering Minority Scholarship Program
Other Engineering Support Programs: Summer and fall orientation program for all freshmen and transfer students; services for handicapped and veterans; alcohol and drug education and prevention services; international student services; student counseling services; career planning and placement services; and a cooperative education program.

348 Texas Tech University

■ INSTITUTION PROFILE
HEAD OF THE INSTITUTION
Robert W. Lawless
Phone: (806)742-2121 **Fax:** (806)742-2138

GENERAL INFORMATION
[All Students—Fall 1991]
Undergraduate enrollment 202,879
Graduate enrollment 4,420
Total institution enrollment **247,070**
Type of institution: Public **Calendar:** Semesters
Nearest city: Amarillo **Population:** 150,000
Miles from main campus: 120 **Setting:** Urban
Types of engineering degrees: Engineering & Technology
Other degree-granting colleges: Allied Health Sciences, Architecture, Arts & Sciences, Business Administration, Education, Law, Medicine, Nursing, Home Economics, Agricultural Sciences

TECHNOLOGY COLLEGE ADMISSIONS INFORMATION
Entrance Requirements: SAT (Score: 900), ACT (Score: 22), High School courses—English (4 years), Chemistry (1 year), Physics (1 year), Math (Geometry, Trigonometry, & Algebra II) (3 years), Texas Academic Skills Program (TASP) Test
Requirements for foreign students: TOEFL (Score: 550); financial statement; Should apply one year in advance.

DEPARTMENTAL ADMISSIONS INFORMATION
Admission to the engineering technology department: At the time of admission to the institution.
Additional information: A minimum Grade Point Average of 2.0 is required.

■ FINANCIAL INFORMATION
ESTIMATED EXPENSES (ACADEMIC YEAR)
[Expenses are for the 1992-93 nine-month academic year.]

State Residents	Undergraduate	Graduate
Tuition and fees	$ 750	$ –
College room and board	$ 3,452	$ –
Books and supplies	$ 600	$ –
Other expenses	$ 25	$ –
Total estimated expenses	**$ 4,827**	**$ –**

Out-of-State Residents	Undergraduate	Graduate
Tuition and fees	$ 2,810	$ –
College room and board	$ 3,452	$ –
Books and supplies	$ 600	$ –
Other expenses	$ 25	$ –
Total estimated expenses	**$ 6,887**	**$ –**

FINANCIAL AID OFFICE CONTACT
Ronny Barnes
Texas Tech University
Office for Financial Aids for Students
Box 45011
Lubbock, TX 79409-5011
Phone: (806)742-3681

GENERAL FINANCIAL AID INFORMATION
Forms accepted/required: ACT, AFSA, CSS-FAF, FAF, FAT, FAR, FSA, FFS, Stafford, IRS, SAR, SDF
Additional financial aid information: Financial aid available includes need-based and merit-based scholarships, loans, grants, part-time jobs on campus, college work-study programs.

■ TECHNOLOGY COLLEGE INFORMATION
[For additional personnel, refer to the *Appendix*.]
HEAD OF ENGINEERING TECHNOLOGY
Mason H. Somerville
Phone: (806)742-3451 **Fax:** (806)742-3493

ENGINEERING TECHNOLOGY COLLEGE ADDRESS
Department of Engineering Technology
Texas Tech University
Box 43107
University & Broadway
Lubbock, TX 79409-3107
Phone: (806)742-3538 **Fax:** (806)742-1900

TECHNOLOGY ENROLLMENTS—BY CLASS
[Numbers are technology baccalaureate enrollments for the fall 1991 term, unless otherwise footnoted.]
1st-year students/Freshmen 37
2nd-year students/Sophomores 41
3rd-year students/Juniors 44
4th-year students/Seniors 88
Total **210**

NUMBER OF DEGREES AWARDED—BY PROGRAM
[Numbers are engineering technology baccalaureate degrees awarded during the 1991-1992 academic year, unless otherwise footnoted. For full details about each engineering technology program, refer to the *Table of Degree Programs*.]
Construction Engineering Technology 8
Electrical/Electronic(s) Engineering Technology 18
Mechanical Engineering Technology 16
Total **42**

PERCENTAGE OF DEGREES AWARDED—BY CATEGORY
[Percentages are of all engineering technology baccalaureate degrees awarded during the 1991-1992 academic year, unless otherwise footnoted.]
African Americans 4.3%
Asian/Pacific Island Americans 4.3%
Hispanic Americans 12.4%
Native Americans .... – %
Foreign Citizens 1.4%
All Others 77.6%
Women .... – %
Persons with disabilities 6.2%
Students over 25 years of age 36.7%

ENGINEERING TECHNOLOGY STUDENT DATA
Applicants to the engineering technology college
Number of applicants to engineering technology college 38
Percent offered admission 100.0%
Matriculated engineering technology students
Percentage in top quartile (25%) of High School class ... 100.0
Average SAT scores: Combined—1009

FULL- & PART-TIME FACULTY—BY DEPARTMENT
[Figures are the head count of full-time faculty and the full-time equivalent (FTE) of part-time faculty for each engineering technology department or equivalent.]

Department	Full	Part
Eng Tech	10	0.5

TECHNOLOGY COLLEGE DESCRIPTION
The program leading to the degree of B.S. in Engineering Technology supports construction, electrical/electronic, and mechanical engineering technologies. The degree is designed for students whose basic aptitude and interests are in the application of established procedures to the solution of technical problems rather than in the research-oriented professions of engineering. It is considered to be a terminal occupational degree preparing students for technical careers in such fields as applied design, construction, operations, maintenance, quality control and sales. The faculty provide careful guidance and counsel to enable their students to become successful in spite of barriers such as financial, cultural, economic or personal hardships. The integrity of the program is maintained as the students overcome obstacles. Average number of years required to actually complete the bachelor's degree is 4.5.

TECHNOLOGY DEGREE PROGRAMS DESCRIPTION
The degree of Bachelor of Science in Engineering Technology with the construction option prepares the graduate to work in the construction industry through courses in methods, estimation, scheduling, as well as working with knowledge of materials and their properties. The electrical/electronics option keeps the students abreast of the latest electronic developments as related to electrical manufacturing, maintenance, quality control and the like. The mechanical option prepares graduates for career opportunities in the areas of: manufacturing, HVAC design, power production, and mechanical component design.

TRANSFER INFORMATION
Residency Requirements: A student must complete their last 30 hours in residence and are allowed a maximum of 66 transfer hours from a Junior College.

Transfer via Articulation Agreements
Admission to engineering technology: At any time
Requirements: A grade of 'C' or better is required to transfer a course. A student must have an overall GPA of 2.0 to be admitted to degree programs.

Transfer without Articulation Agreements
Admission to engineering technology: At any time
Requirements: A grade of 'C' or better is required to transfer a course. A student must have an overall GPA of 2.0 or higher to be admitted to the program.

GRADUATION REQUIREMENTS
Bachelor of Science degrees in Engineering Technology require 132 semester hours, with a minimum 2.00 GPA.

■ STUDENT PROGRAMS
PROFESSIONAL AND HONORARY SOCIETIES
[For key to acronyms, see *Introduction*.]
Professional Societies: ACI, AGCA, ASCE, ASHRAE, ASME, IEEE, IIE, NSPE, SAE, SPE
Honorary Societies: Alpha Epsilon, Alpha Pi Mu, Chi Epsilon, Eta Kappa Nu, Omega Chi Epsilon, Tau Alpha Pi, Tau Beta Pi, Upsilon Pi Epsilon, Pi Epsilon Tau

SUPPORT PROGRAMS
Student Chapter Organizations: Society of Women Engineers, National Society of Black Engineers, Society of Hispanic Professional Engineers, Minority Engineering Program

Minority Engineering Program
For: Ethnic Minorities **Available:** Year round
Offered: Freshman, Sophomore

Other Engineering Support Programs: The University hosts several Summer Orientation Sessions for both entering and transfer students. Support service available for Engineering Technology students include: Engineering Communications Center, a Student Mentor Program, the Minority Engineering Program, Programs for Academic Support Services (PASS), Inc., Career Planning and Placement Center.

349 Tri-State University

■ INSTITUTION PROFILE

HEAD OF THE INSTITUTION
William G. Meyers
Phone: (219)665-4187   Fax: (219)665-4292

GENERAL INFORMATION
[All Students—Fall 1991]
Undergraduate enrollment 1,060
Total institution enrollment 1,060
Type of institution: Private   Calendar: Quarters
Nearest city: Fort Wayne, Indiana   Population: 200,000
Miles from main campus: 45   Setting: Small Town
Types of engineering degrees: Engineering & Technology
Other degree-granting colleges: Arts & Sciences, Business Administration

■ ENGINEERING TECHNOLOGY ADMISSIONS

ADMISSIONS OFFICE CONTACT
Walter Lilley
Tri-State University
Admissions Office
Angola, IN 46703-0307
Phone: (219)665-4132   Fax: (219)665-4292

TECHNOLOGY COLLEGE ADMISSIONS INFORMATION
Entrance Requirements: SAT, ACT, High School courses—Algebra (1 year), Science and Social Studies (2 years), Geometry, and Trigonometry (1 year), English (4 years), Students are required to take either the SAT or the ACT. Additionally, all freshman are given mathematics and English placement examinations during orientation.
Entrance Recommendations: High School courses—Advanced Algebra and Trigonometry (1 year), Mechanical Drawing (1 year), Computer Fundamentals (1 year), Chemistry and Physics (1 year)
Requirements for foreign students: financial statement; English proficiency verification is required. TOEFL is recommended. Mathematics and English placement examinations are given during orientation.

DEPARTMENTAL ADMISSIONS INFORMATION
Admission to the engineering technology department: At the time of admission to the institution.

■ FINANCIAL INFORMATION

ESTIMATED EXPENSES (ACADEMIC YEAR)
[Expenses are for the 1992-93 nine-month academic year.]

All Students	Undergraduate	Graduate
Tuition and fees	$10,500 [2]	$ — [1]
College room and board	$ 3,900	$ — [1]
Books and supplies	$ 500	$ — [1]
Other expenses	$ 500 [3]	$ — [1]
Total estimated expenses	$15,400	$ —

Notes: (1) No graduate programs (2) Non-engineering full-time tuition is approximately $9476. Tuition for a quarter credit hour is $214 for engineering and $184 for all other schools. (3) Personal Expenses

FINANCIAL AID OFFICE CONTACT
Susan Stroh, Financial Aid Director
Tri-State University
Financial Aid Office
Angola, IN 46703-0307
Phone: (219)665-4174   Fax: (219)665-4292

GENERAL FINANCIAL AID INFORMATION
Forms accepted/required: FAF, Stafford, Supplemental, Institutional, Perkins Loan
Additional financial aid information: Scholarships (need and merit based), Grants (Institutional, State and Federal), Campus Employment, Work Study, Athletic Awards, Loans

■ TECHNOLOGY COLLEGE INFORMATION
[For additional personnel, refer to the Appendix.]

HEAD OF ENGINEERING TECHNOLOGY
William G. Meyers
Phone: (219)665-4187   Fax: (219)665-4292

ENGINEERING TECHNOLOGY COLLEGE ADDRESS
School of Engineering, Technology Division
Tri-State University
300 South Darling Street
Angola, IN 46703-0307
Phone: (219)665-4100   Fax: (219)665-4292

TECHNOLOGY ENROLLMENTS—BY CLASS
[Numbers are technology baccalaureate enrollments for the fall 1991 term, unless otherwise footnoted.]
1st-year students/Freshmen 20
2nd-year students/Sophomores 15
3rd-year students/Juniors 1
4th-year students/Seniors 0
Total 36

NUMBER OF DEGREES AWARDED—BY PROGRAM
[Numbers are engineering technology baccalaureate degrees awarded during the 1991-1992 academic year, unless otherwise footnoted. For full details about each engineering technology program, refer to the Table of Degree Programs.]
Computer Aided Drafting & Design Technology .... — [3]
Drafting & Design Technology .... — [1]
Manufacturing Technology .... — [1]
Total .... —

Notes: (1) Associate degree program (3) The baccalaureate engineering technology program is new and as such has not yet produced a graduate.

PERCENTAGE OF DEGREES AWARDED—BY CATEGORY
[Percentages are of all engineering technology baccalaureate degrees awarded during the 1991-1992 academic year, unless otherwise footnoted.]
African Americans .... — % [1]
Asian/Pacific Island Americans .... — % [1]
Hispanic Americans .... — % [1]
Native Americans .... — % [1]
Foreign Citizens .... — % [1]
All Others .... 100 % [1]
Women .... — % [1]
Persons with disabilities .... — % [1]
Students over 25 years of age .... — % [1]

Notes: (1) The baccalaureate engineering technology program is new and as such has not yet produced a graduate.

ENGINEERING TECHNOLOGY STUDENT DATA
Applicants to the engineering technology college
Number of applicants to engineering technology college 53
Percent offered admission 92.0%
Matriculated engineering technology students
Percentage in top quartile (25%) of High School class 26.0
Average SAT scores: Math—457, Verbal—371, Combined—829
Average ACT scores: Math—20, Composite—19

FULL- & PART-TIME FACULTY—BY DEPARTMENT
[Figures are the head count of full-time faculty and the full-time equivalent (FTE) of part-time faculty for each engineering technology department or equivalent.]

Department	Full	Part
Eng Tech	2	0.5

TECHNOLOGY COLLEGE DESCRIPTION
Tri-State University provides a unique university experience to students through small classes and a faculty committed to quality, exclusively undergraduate, education. The curriculum of the Technology Division of the School of Engineering emphasizes many of the underlying principles of component design and the skills required to communicate with engineers, scientists, and production personnel. The emphasis is on teaching and hands-on laboratory instruction. All classes and laboratories are presented and directed by experienced faculty. Questions are welcomed through the faculty open-door policy.

TECHNOLOGY DEGREE PROGRAMS DESCRIPTION
A TAC/ABET accredited associate degree program in Drafting and Design Technology and an associate degree program in Manufacturing Technology produce productive technicians for engineering teams. The baccalaureate degree in Computer Aided Drafting and Design Technology results in a technologist with technical knowledge and communication skills for supervisory roles.

DUAL DEGREE PROGRAMS
Technology baccalaureate graduates with dual degrees 0

TRANSFER INFORMATION
Residency Requirements: A minimum of 45 quarter credit hours must be earned at Tri-State University to be eligible for a baccalaureate degree. A minimum of 25 quarter credit hours must be earned at Tri-State University to be eligible for an associate degree.
Transfer via Articulation Agreements
Admission to engineering technology: At any time
Requirements: Satisfactory academic performance at the previous institution.
Transfer without Articulation Agreements
Admission to engineering technology: At any time
Requirements: Satisfactory academic performance at the previous institution.

GRADUATION REQUIREMENTS
102 quarter credit hours must be completed with an average overall grade point of at least 2.0/4.0 for an associate degree. 192 quarter credit hours must be completed with an average overall grade point of at least 2.0/4.0 for a baccalaureate degree. Additionally, a student must satisfy program course requirements.

■ STUDENT PROGRAMS

PROFESSIONAL AND HONORARY SOCIETIES
[For key to acronyms, see Introduction.]
Professional Societies: ACM, AIAA, AIChE, AIDD, ASCE, ASME, IEEE, SAE, SME, National Society of Minority Students, Society of Women Engineers
Honorary Societies: Chi Epsilon, Eta Kappa Nu, Omega Chi Epsilon, Pi Tau Sigma, Sigma Gamma Tau, Sigma Phi Delta, Tau Beta Pi, Phi Eta Sigma, Skull and Bones

SUPPORT PROGRAMS
Student Chapter Organizations: Society of Women Engineers, National Society of Black Engineers, National Society of Minority Students, Malaysian Student Association, International Student Association, Muslim Student Association

International Students Office
For: Ethnic Minorities   Available: Year round
Offered: Prematriculation, Freshman, Sophomore, Junior, Senior

English Language Center
For: Ethnic Minorities   Available: Year round
Offered: Prematriculation, Freshman, Sophomore, Junior, Senior
Other Engineering Support Programs: The Charter Counseling Center, a Health Services Center and the Gettig Fitness and Wellness Center are available for students as needed. The Career Center provides assistance in locating co-op and full-time employment. Non-credit preparatory courses in mathematics, chemistry and physics are also available.

350 Virginia State University

INSTITUTION PROFILE
HEAD OF THE INSTITUTION
Nathanael Pollard Jr.
Phone: (804)524-5070　　　　Fax: (804)524-6506
GENERAL INFORMATION
[All Students—Fall 1991]
Undergraduate enrollment 3,861
Graduate enrollment 389
Total institution enrollment **4,250**
Type of institution: Public　　**Calendar:** Semesters
Nearest city: Richmond, VA　　**Population:** 250,000
Miles from main campus: 25　　**Setting:** Small Town
Types of engineering degrees: Engineering Technology
Other degree-granting colleges: Agricultural & Environmental, Allied Health Sciences, Architecture, Arts & Sciences, Business Administration, Communication & Journalism, Education, Fine & Performing Arts, Humanities & Social Sciences

ENGINEERING TECHNOLOGY ADMISSIONS
ADMISSIONS OFFICE CONTACT
Karen R. Winston
Virginia State University
Box 9018, 20708 Fourth Avenue
1 Hayden Drive
Petersburg, VA 23806
Phone: (804)524-5902　　　　Fax: (804)524-5055
TECHNOLOGY COLLEGE ADMISSIONS INFORMATION
Entrance Requirements: High School courses—Math (2 years), Science (2 years), English (4 years)
Entrance Recommendations: SAT (Score: 700), ACT (Score: 20), High School courses—Physics (2 years), Chemistry (2 years)
DEPARTMENTAL ADMISSIONS INFORMATION
Admission to the engineering technology department: At the time of admission to the institution.

FINANCIAL INFORMATION
ESTIMATED EXPENSES (ACADEMIC YEAR)
[Expenses are for the 1992-93 nine-month academic year.]

State Residents	Undergraduate	Graduate
Tuition and fees	$ 2,913 [1]	$ –
College room and board	$ 4,127 [1]	$ –
Books and supplies	$ 300 [1]	$ –
Other expenses	$ –	$ –
Total estimated expenses	**$ 7,340**	**$ –**
Out-of-State Residents	Undergraduate	Graduate
Tuition and fees	$ 6,315	$ –
College room and board	$ 4,126	$ –
Books and supplies	$ 300	$ –
Other expenses	$ –	$ –
Total estimated expenses	**$10,741**	**$ –**

Notes: (1) Per year.

FINANCIAL AID OFFICE CONTACT
Henry Debose, *Director*
Virginia State University
P.O. Box 9031
1 Hayden Drive
Petersburg, VA 23806
Phone: (804)524-5990　　　　Fax: (804)524-6818

GENERAL FINANCIAL AID INFORMATION
Forms accepted/required: FAF, Stafford, IRS, SDF, Supplemental, VSU Application for Financial Assistance
Additional financial aid information: Foundation monies from such companies as Philip Morris, Allied Fibers, etc. Alumni program gives money to students in need. Pell Grant and In-State-Tuition are some of the need-based scholarships. College work study is part-time campus job.

TECHNOLOGY COLLEGE INFORMATION
[For additional personnel, refer to the *Appendix*.]
HEAD OF ENGINEERING TECHNOLOGY
Arthur L. Allen
Phone: (804)524-5631　　　　Fax: (804)524-5950
ENGINEERING TECHNOLOGY COLLEGE ADDRESS
Department of Engineering Technology
Virginia State University
P.O. Box 9032
1 Hayden Drive
Petersburg, VA 23806
Phone: (804)524-5185　　　　Fax: (804)524-5950
TECHNOLOGY ENROLLMENTS—BY CLASS
[Numbers are technology baccalaureate enrollments for the fall 1991 term, unless otherwise footnoted.]
1st-year students/Freshmen 69
2nd-year students/Sophomores 29
3rd-year students/Juniors 17
4th-year students/Seniors 35
Total **150**
NUMBER OF DEGREES AWARDED—BY PROGRAM
[Numbers are engineering technology baccalaureate degrees awarded during the 1991-1992 academic year, unless otherwise footnoted. For full details about each engineering technology program, refer to the *Table of Degree Programs*.]
Electronic Engineering Technology 9
Mechanical Engineering 9
Total **18**
PERCENTAGE OF DEGREES AWARDED—BY CATEGORY
[Percentages are of all engineering technology baccalaureate degrees awarded during the 1991-1992 academic year, unless otherwise footnoted.]
African Americans .... 66.0 %
Asian/Pacific Island Americans .... 5.0 %
Hispanic Americans .... – %
Native Americans .... – %
Foreign Citizens .... 5.0 %
All Others .... 24.0 %
Women .... 22.0 %
Persons with disabilities .... – %
Students over 25 years of age .... 33.0 %
ENGINEERING TECHNOLOGY STUDENT DATA
Applicants to the engineering technology college
Number of applicants to engineering technology college 87
Percent offered admission .... 90.0 %
Matriculated engineering technology students
Percentage in top quartile (25%) of High School class .... –[A]
Average SAT scores: Math—400, Verbal—350, Combined—750
Notes: (A) Data not available.

FULL- & PART-TIME FACULTY—BY DEPARTMENT
[Figures are the head count of full-time faculty and the full-time equivalent (FTE) of part-time faculty for each engineering technology department or equivalent.]

Department	Full	Part
Electronic(s) Eng Tech	3	–
Mechanical Eng Tech	3	–

TECHNOLOGY COLLEGE DESCRIPTION
Majors in Electronics and Mechanical Engineering Technology are offered. Both are integrated four-year B.S. programs. Junior and Senior level courses are offered both day and evening for the convenience of working students. Transfer students with Associate degrees in related fields are accepted; transfer credits are evaluated on a course by course basis. Students may Minor in Mathematics, Physics, or Computer Science. All faculty have earned M.S. or above and 30% are Licensed P.E. The Department operates six state-of-the-art laboratories supported by highly qualified technicians. Summer internships and Co-op programs are available with local governments and industries.

DUAL DEGREE PROGRAMS
Technology baccalaureate graduates with dual degrees 0
Enrollment requirements: Must satisfy departmental requirements and must take minimum of 30 credit hours from the department.

TRANSFER INFORMATION
Transfer via Articulation Agreements
Admission to engineering technology: At any time
Technology graduates transferred from ... 4-yr: 2　2-yr: 5
Transfer without Articulation Agreements
Admission to engineering technology: At any time
Technology graduates transferred from ... 4-yr: 3　2-yr: 5

GRADUATION REQUIREMENTS
A cumulative GPA of 2.0 is required for graduation. Minimum of "C" grade is required for all technical and science courses. 129 semester credit hours is required for graduation.

STUDENT PROGRAMS
PROFESSIONAL AND HONORARY SOCIETIES
[For key to acronyms, see *Introduction*.]
Professional Societies: ASME, Engineering Technology Club
Honorary Societies: Tau Alpha Pi
SUPPORT PROGRAMS
Other Engineering Support Programs: University conducts one-week Summer orientation program for first-time students. Orientation is an academic requirement for all freshmen. Through federal grants, university provides special academic and counseling services to eligible students via Talent Search, Upward Bound, and Special Services Projects.

351 Weber State University

INSTITUTION PROFILE

HEAD OF THE INSTITUTION
Paul H. Thompson
Phone: (801)626-6001   Fax: (801)626-7922

GENERAL INFORMATION
[All Students—Fall 1991]
Undergraduate enrollment 14,371
Graduate enrollment 124
Total institution enrollment 14,495
Type of institution: Public **Calendar:** Quarters
Nearest city: Salt Lake City **Population:** 160,000
Miles from main campus: 35 **Setting:** Urban
Types of engineering degrees: Engineering Technology
Other degree-granting colleges: Education, Health Professions, Arts & Humanities, Business & Economics, Science, Social & Behavioral Sciences

ENGINEERING TECHNOLOGY ADMISSIONS

ADMISSIONS OFFICE CONTACT
L. Winslow Hurst
Weber State University
Ogden, UT 84408-1015
Phone: (801)626-6050   Fax: (801)626-6747

TECHNOLOGY COLLEGE ADMISSIONS INFORMATION
Entrance Requirements: ACT (Score: 12)
Entrance Recommendations: ACT (Score: 12)
Requirements for out-of-state residents: Graduate high school with cumulative GPA at least 2.20 Submit ACT or SAT.
Requirements for foreign students: TOEFL (Score: 500); financial statement

DEPARTMENTAL ADMISSIONS INFORMATION
Admission to the engineering technology department: At the time of admission to the institution.

FINANCIAL INFORMATION

ESTIMATED EXPENSES (ACADEMIC YEAR)
[Expenses are for the 1992-93 nine-month academic year.]

State Residents	Undergraduate	Graduate
Tuition and fees	$ 1,542	$ 1,578
College room and board	$ 2,760	$ 2,760
Books and supplies	$ 522	$ 900
Other expenses	$ —(1)	$ —(1)
Total estimated expenses	$ 4,824	$ 5,238

Out-of-State Residents	Undergraduate	Graduate
Tuition and fees	$ 4,332	$ 4,476
College room and board	$ 2,760	$ 2,760
Books and supplies	$ 522	$ 900
Other expenses	$ —(1)	$ —(1)
Total estimated expenses	$ 7,614	$ 8,136

Notes: (1) For three quarters

FINANCIAL AID OFFICE CONTACT
Richard Effiong, *Director*
Weber State University
Ogden, UT 84408-1017
Phone: (801)626-7131   Fax: (801)626-7408

GENERAL FINANCIAL AID INFORMATION
Forms accepted/required: ACT, AFSA, CSS-FAF, FAT, Stafford, IRS, SAR, Institutional, Information Verification Form
Additional financial aid information: College Work Study, Stafford Loan, Pell Grant, Perkins Loans, Short Term Emergency Loans, State Student Incentive Grants, Supplemental Education Opportunity Grant

TECHNOLOGY COLLEGE INFORMATION

[For additional personnel, refer to the *Appendix*.]

HEAD OF ENGINEERING TECHNOLOGY
Warren R. Hill
Phone: (801)626-6303   Fax: (801)626-7531

ENGINEERING TECHNOLOGY COLLEGE ADDRESS
College of Applied Science and Technology
Weber State University
Ogden, UT 84408-1801
Phone: (801)626-6303   Fax: (801)626-7531

TECHNOLOGY ENROLLMENTS—BY CLASS
[Numbers are technology baccalaureate enrollments for the fall 1991 term, unless otherwise footnoted.]
1st-year students/Freshmen 190
2nd-year students/Sophomores 113
3rd-year students/Juniors 114
4th-year students/Seniors 286
Total **703**

NUMBER OF DEGREES AWARDED—BY PROGRAM
[Numbers are engineering technology baccalaureate degrees awarded during the 1991-1992 academic year, unless otherwise footnoted. For full details about each engineering technology program, refer to the *Table of Degree Programs*.]
Electronic Engineering Technology 87
Manufacturing Engineering Technology 23
Mechanical Engineering Technology 4
Total **114**

PERCENTAGE OF DEGREES AWARDED—BY CATEGORY
[Percentages are of all engineering technology baccalaureate degrees awarded during the 1991-1992 academic year, unless otherwise footnoted.]
African Americans .... 1.0 %
Asian/Pacific Island Americans .... 3.7 %
Hispanic Americans .... 2.0 %
Native Americans .... 0.6 %
Foreign Citizens .... 2.3 %
All Others .... 90.4 %
Women .... 8.1 %
Persons with disabilities .... — % (A)
Students over 25 years of age .... — % (A)

Notes: (A) Data not available.

FULL- & PART-TIME FACULTY—BY DEPARTMENT
[Figures are the head count of full-time faculty and the full-time equivalent (FTE) of part-time faculty for each engineering technology department or equivalent.]

Department	Full	Part
Electronics Eng Tech	11	—
Manufacturing Eng Tech	11	—
Mechanical Eng Tech	5	—

TECHNOLOGY COLLEGE DESCRIPTION
The College of Applied Science and Technology prides itself on providing hands-on experiences for its students taught by faculty with relevant industrial backgrounds. These experiences are further bolstered by appropriate required year-long senior projects where students work in teams, frequently on multi-disciplinary assignments. Co-op placements are available in all programs and co-op classes are treated as elective courses. The College houses the Center for Aerospace Technology which provides opportunities for students to design, build, test, and fly small satellites. Students also develop the experiments which go on board the satellites and operate the ground station used for data collection from the satellites. In addition, the College contains the Technology Assistance Center which provides faculty and students the possibility of working with local businesses and industries to solve their specific technical problems.

TECHNOLOGY DEGREE PROGRAMS DESCRIPTION
All of the engineering technology programs offer both the AAS and BS degrees. All of the programs emphasize the hands-on aspects of the discipline which are strongly reinforced by required year-long team senior projects. The Manufacturing Engineering Technology program is known for its machining capabilities, particularly CNC. The Electronics Engineering Technology program has excellent capabilities in both the communications and digital, particularly microprocessor, areas. The Mechanical Engineering Technology program provides students with a very broad background with a special emphasis in technical writing.

TRANSFER INFORMATION
Residency Requirements: A candidate for the Bachelor of Arts, Bachelor of Science, Bachelor of Fine Arts, or Bachelor of Integrated Studies degree must have a minimum of 45 credits in residence at Weber State and be registered during at least one quarter following the last commencement prior to graduation.

Transfer via Articulation Agreements
Admission to engineering technology: At any time
Requirements: 2.00 cumulative GPA in all previous college work, good standing where currently registered. Transfers with AA or AS within Utah System of Higher Education are considered as having met the Weber State general education requirements.

Transfer without Articulation Agreements
Admission to engineering technology: At any time

GRADUATION REQUIREMENTS
(1) Grade of C or better in all major courses. (2) Overall GPA of 2.0. (3) A year long sequence of senior projects courses.

STUDENT PROGRAMS

PROFESSIONAL AND HONORARY SOCIETIES
[For key to acronyms, see *Introduction*.]
Professional Societies: ASME, IEEE, SAE, SME

SUPPORT PROGRAMS
Other Engineering Support Programs: Academic Advisement, Career Services, Counseling & Psychological Services, Drug & Alcohol Program, Education Resource Programs, Educational Support Services, Literacy Support, Technological Support, Research Support, Veterans/Military Support Services, International Student Center, Learning Center, Services for Physically Challenged Students, Student Health Service, Women's Resource Center

352 Wentworth Institute of Technology

■ INSTITUTION PROFILE
HEAD OF THE INSTITUTION
John F. Van Domelen
Phone: (617)442-9010 Fax: (617)427-2852

GENERAL INFORMATION
[All Students—Fall 1991]
Undergraduate enrollment 3,299
Total institution enrollment 3,299
Type of institution: Private **Calendar:** Semesters
Location: Boston **Population:** 567,000
Setting: Urban
Types of engineering degrees: Engineering & Technology
Other degree-granting colleges: Arts & Sciences, Design and Construction

■ ENGINEERING TECHNOLOGY ADMISSIONS
ADMISSIONS OFFICE CONTACT
Thomas J. McGinn
Wentworth Institute of Technology
550 Huntington Avenue
Boston, MA 02115-5998
Phone: (617)442-9010 Fax: (617)427-2852

TECHNOLOGY COLLEGE ADMISSIONS INFORMATION
Entrance Requirements: High School courses—English (4 years), Laboratory Science (1 year), Mathematics (2 years), SAT or ACT is required for applicants to the Bachelor of Science, Associate in Engineering or Associate in Applied Science degree programs.
Entrance Recommendations: High School courses—Algebra (2 years), Plane Geometry (1 year), Trigonometry or Advanced Mathematics (1 year), College preparatory Physics (1 year), SAT and ACT are recommended for first-year applicants to other degree programs.
Requirements for foreign students: TOEFL (Score: 510); financial statement

DEPARTMENTAL ADMISSIONS INFORMATION
Admission to the engineering technology department: At the time of admission to the institution.
Additional information: Specific program prerequisites must be met.

■ FINANCIAL INFORMATION
ESTIMATED EXPENSES (ACADEMIC YEAR)
[Expenses are for the 1992-93 nine-month academic year.]

All Students	Undergraduate	Graduate
Tuition and fees	$ 8,840 [1]	$ –
College room and board	$ 5,900	$ –
Books and supplies	$ 900	$ –
Other expenses	$ –	$ –
Total estimated expenses	**$15,640**	**$ –**

Notes: (1) Fees include Health Center Fee ($30/semester) and application fee ($30).

FINANCIAL AID OFFICE CONTACT
Carol A. Rubel, *Director of Financial Aid*
Wentworth Institute of Technology
550 Huntington Avenue
Boston, MA 02115-5998
Phone: (617)442-9010 Fax: (617)427-2852

GENERAL FINANCIAL AID INFORMATION
Forms accepted/required: FAF, IRS, Institutional
Additional financial aid information: Need-based scholarships; MELA, TERI loan programs; merit-based scholarships; Wentworth Opportunity Grant; Academic Performance Loan/Scholarship; SEOG, Perkins Loan, GSL, SLS, PLUS, PELL Grant; College Work-Study Program.

■ TECHNOLOGY COLLEGE INFORMATION
[For additional personnel, refer to the *Appendix*.]
HEAD OF ENGINEERING TECHNOLOGY
Alexander W. Avtgis
Phone: (617)442-9010 Fax: (617)427-2852

ENGINEERING TECHNOLOGY COLLEGE ADDRESS
College of Engineering and Technology
Wentworth Institute of Technology
550 Huntington Avenue
Boston, MA 02115-5998
Phone: (617)442-9010 Fax: (617)427-2852

TECHNOLOGY ENROLLMENTS—BY CLASS
[Numbers are technology baccalaureate enrollments for the fall 1991 term, unless otherwise footnoted.]
1st-year students/Freshmen 875
2nd-year students/Sophomores 765
3rd-year students/Juniors .... 567 [1]
4th-year students/Seniors .... 301 [1]
Total **2,508**

Notes: (1) Includes students on cooperative work semester

NUMBER OF DEGREES AWARDED—BY PROGRAM
[Numbers are engineering technology baccalaureate degrees awarded during the 1991-1992 academic year, unless otherwise footnoted. For full details about each engineering technology program, refer to the *Table of Degree Programs*.]
Architectural Engineering Technology 63
Building Construction Technology 42
Civil Engineering Technology 55
Computer Engineering Technology 19
Electronic(s) Engineering Technology 99
Manufacturing Engineering Technology 17
Mechanical Engineering Technology 54
Total **349**

PERCENTAGE OF DEGREES AWARDED—BY CATEGORY
[Percentages are of all engineering technology baccalaureate degrees awarded during the 1991-1992 academic year, unless otherwise footnoted.]
African Americans .... 7.9 %
Asian/Pacific Island Americans .... 5.0 %
Hispanic Americans .... 2.9 %
Native Americans .... 0.2 %
Foreign Citizens .... 6.1 %
All Others .... 77.9 %
Women .... 11.4 %
Persons with disabilities .... – % (A)
Students over 25 years of age .... – % (A)

Notes: (A) Data not available.

ENGINEERING TECHNOLOGY STUDENT DATA
Applicants to the engineering technology college
Number of applicants to engineering technology college . 2,240
Percent offered admission 72.0%

Matriculated engineering technology students
Percentage in top quartile (25%) of High School class ...... – (A)

Notes: (A) Data not available.

FULL- & PART-TIME FACULTY—BY DEPARTMENT
[Figures are the head count of full-time faculty and the full-time equivalent (FTE) of part-time faculty for each engineering technology department or equivalent.]

Department	Full	Part
Civil & Environmental Tech	7	–
Electrical/Electronic(s) Eng Tech	17	–
Mechanical/Manufacturing Eng Tech	17	0.4
Aeronautics	2	–
Architecture	20	6.3
Construction Sciences	8	1.0

TECHNOLOGY COLLEGE DESCRIPTION
Wentworth's four-year engineering technology programs are laboratory intensive. Cooperative work experience is required in the day programs. A five year program is available for those who have the ability to complete the baccalaureate program but lack the educational preparation. Wentworth places particular emphasis on seeking out students with academic promise who face cultural, economic or personal barriers and assisting them in reaching their greatest potential. Wentworth also offers two-year programs in engineering technology.

TECHNOLOGY DEGREE PROGRAMS DESCRIPTION
The Institute's four-year baccalaureate programs in engineering technology emphasize industrial application of theoretical concepts to prepare work-ready graduates. The cooperative work requirement increases their preparation. Wentworth also awards associate (two-year) degrees in all the fields listed in the Table as well as aeronautical, aviation, aircraft maintenance, electrical engineering, electronic maintenance, mechanical, mechanical design and mechanical drafting technology. An associate degree program in engineering technology (undeclared) is offered.

TRANSFER INFORMATION
Residency Requirements: A minimum of 50% of the total semester hours of any degree-granting program (except the one-year professional program, which shall be 100% of the total semester hours) must be completed in residence.

Transfer via Articulation Agreements
Admission to engineering technology: Any level. Transfer programs are articulated with many different academic institutions, including community colleges.
Requirements: Specific program prerequisites must be met.

Transfer without Articulation Agreements
Admission to engineering technology: At any time
Requirements: Specific program prerequisites must be met.

GRADUATION REQUIREMENTS
Students who are candidates for graduation must have shown a sincere desire to meet all requirements of every course in their program and must successfully complete every course listed in the prescribed program in which they are enrolled. Such students must also satisfy the faculty with regard to their attendance and character. The scholastic index for the total of courses taken must not be below 2.00.

■ STUDENT PROGRAMS
PROFESSIONAL AND HONORARY SOCIETIES
[For key to acronyms, see *Introduction*.]
Professional Societies: ACI, ACM, AGCA, AIA, ASCE, ASHRAE, ASME, IEEE, NAHB, SME, Associated Builders and Contractors (ABC), Student Association of Interior Design (SAID)
Honorary Societies: Tau Alpha Pi

SUPPORT PROGRAMS
Student Chapter Organizations: Society of Women Engineers, National Society of Black Engineers

Women's Resource Center
For: Women **Available:** Year round
Offered: Freshman, Sophomore, Junior, Senior
Other Engineering Support Programs: Summer and fall orientation for all freshman and transfers; Faculty Advisors, Freshman Mentor Program; Learning Center and peer tutors; career counseling services; international student counselor; counseling center.

353 West Virginia Institute of Technology

■ INSTITUTION PROFILE
HEAD OF THE INSTITUTION
John P. Carrier
Phone: (304)442-3146   Fax: (304)442-3057

GENERAL INFORMATION
[All Students—Fall 1991]
Undergraduate enrollment 3,036
Graduate enrollment 15
Total institution enrollment 3,051
Type of institution: Public   Calendar: Semesters
Nearest city: Charleston, WV   Population: 80,000
Miles from main campus: 30   Setting: Small Town
Types of engineering degrees: Engineering & Technology
Other degree-granting colleges: Arts & Sciences, Business & Economics

■ ENGINEERING TECHNOLOGY ADMISSIONS
ADMISSIONS OFFICE CONTACT
Robert P. Scholl
West Virginia Institute of Technology
210 Old Main
Montgomery, WV 25136
Phone: (304)442-3151   Fax: (304)442-3057

TECHNOLOGY COLLEGE ADMISSIONS INFORMATION
Admission to the engineering technology college: With required ACT scores, or upon completion of Intermediate Studies (one semester) if prerequisites/ACT minimums not met.
Entrance Requirements: SAT (Score: 690), ACT (Score: 17), High School courses—Algebra (1 year), Plane Geometry (1 year), Trigonometry, Placement in English and Math based on standard test scores. English: ACT 17 or SAT 370; Mathematics: ACT 21 or SAT 470 for normal entry.
Entrance Recommendations: ACT, High School courses—Physics (1 year)
Requirements for foreign students: TOEFL (Score: 550); financial statement

DEPARTMENTAL ADMISSIONS INFORMATION
Admission to the engineering technology department: At the time of admission to the institution.

■ FINANCIAL INFORMATION
ESTIMATED EXPENSES (ACADEMIC YEAR)
[Expenses are for the 1992-93 nine-month academic year.]

State Residents	Undergraduate	Graduate
Tuition and fees	$ 1,784	$ 1,976
College room and board	$ 3,752	$ 3,752
Books and supplies	$ 400	$ 400
Other expenses	$ –	$ –
Total estimated expenses	$ 5,936	$ 6,128

Out-of-State Residents	Undergraduate	Graduate
Tuition and fees	$ 3,954	$ 4,578
College room and board	$ 3,752	$ 3,752
Books and supplies	$ 400	$ 400
Other expenses	$ –	$ –
Total estimated expenses	$ 8,106	$ 8,730

FINANCIAL AID OFFICE CONTACT
Nina M. Morton, *Director of Financial Aid*
West Virginia Institute of Technology
Montgomery, WV 25136
Phone: (304)442-3032   Fax: (304)442-3057

GENERAL FINANCIAL AID INFORMATION
Forms accepted/required: FAF, Institutional
Additional financial aid information: Scholarship, grants, loans, and work-study. The Engineering Technology Division offers several merit-based scholarships each year. The Community and Technical College has monies for special needs students.

■ TECHNOLOGY COLLEGE INFORMATION
[For additional personnel, refer to the *Appendix*.]
HEAD OF ENGINEERING TECHNOLOGY
Frank A. Gourley Jr.
Phone: (304)442-3098   Fax: (304)442-3245

ENGINEERING TECHNOLOGY COLLEGE ADDRESS
Division of Engineering Technology/Industrial Technology
West Virginia Institute of Technology
218 Davis Hall
Montgomery, WV 25136
Phone: (304)442-3338   Fax: (304)442-3245

TECHNOLOGY ENROLLMENTS—BY CLASS
[Numbers are technology baccalaureate enrollments for the fall 1991 term, unless otherwise footnoted.]
1st-year students/Freshmen 124
2nd-year students/Sophomores 177
3rd-year students/Juniors 15
4th-year students/Seniors 53
Total 369

NUMBER OF DEGREES AWARDED—BY PROGRAM
[Numbers are engineering technology baccalaureate degrees awarded during the 1991-1992 academic year, unless otherwise footnoted. For full details about each engineering technology program, refer to the *Table of Degree Programs*.]
Civil/Surveying Engineering Technology 6
Drafting & Design Engineering Technology 18
Electrical Engineering Technology 23
Electronic(s) Engineering Technology 30
Engineering Technology 6
Industrial Technology 16
Mechanical Engineering Technology 12
Surveying Engineering Technology 3
Technical Studies 7
Technology Education 4
Total 125

PERCENTAGE OF DEGREES AWARDED—BY CATEGORY
[Percentages are of all engineering technology baccalaureate degrees awarded during the 1991-1992 academic year, unless otherwise footnoted.]
African Americans 7.0%
Asian/Pacific Island Americans 3.0%
Hispanic Americans 3.0%
Native Americans 3.0%
Foreign Citizens 3.0%
All Others 81.0%
Women 33.0%
Persons with disabilities 1.0%
Students over 25 years of age .... – %

ENGINEERING TECHNOLOGY STUDENT DATA
Applicants to the engineering technology college
Number of applicants to engineering technology college . 1,978
Percent offered admission 95.0%
Average ACT scores: Composite—19.3

FULL- & PART-TIME FACULTY—BY DEPARTMENT
[Figures are the head count of full-time faculty and the full-time equivalent (FTE) of part-time faculty for each engineering technology department or equivalent.]

Department	Full	Part
Civil Eng Tech	2	–
Electrical/Electronic(s) Eng Tech	5	–
Tech Education	2	–
Mechanical Eng Tech	2	–
Drafting & Design/Industrial Eng Tech	3	–

TECHNOLOGY COLLEGE DESCRIPTION
The college offers four associate degree engineering technology programs: Civil/Surveying, Drafting & Design, Electrical, and Mechanical Engineering Technology. Associate degree graduates may enter one of the five Bachelor of Science Degree programs in electronic, surveying, or general engineering technology; industrial technology; or technology education. Tech is a small teaching-oriented institution. It is possible to get help from full-time qualified faculty and from administrators. There is a College of Engineering on campus which allows for interchange and activities of mutual interest. All required courses are normally taught by full-time faculty. Many courses are taught both fall and spring semester. Multiple faculty (5-7) serve most programs.

TECHNOLOGY DEGREE PROGRAMS DESCRIPTION
West Virginia Institute of Technology offers both two-year Associate of Science and 'plus-two' Bachelor of Science Engineering Technology programs through its Community and Technical College. In addition, four-year engineering programs are offered through its College of Engineering. The 'plus-two' programs allow students opportunity for greater breadth and flexibility in developing their background to meet demands of today's job market. The five 'plus-two' Bachelor of Science programs are designed to provide a variety of backgrounds for students transferring with associate degrees. The study of engineering applications and hands-on laboratory experiences are provided by faculty with extensive industrial experience. Graduates of both associate and baccalaureate degree programs are prepared for employment in entry-level technical/professional positions or for continuing their education at an advanced level. Supporting these programs are technical laboratories such as robotics, computer-aided drafting & design, automated manufacturing, process control, digital electronics, programmable logic control, fluid power, PC computers, manufacturing, hot & cold metals, environmental control, soil mechanics, electrical power, materials testing, and surveying.

DUAL DEGREE PROGRAMS
Technology baccalaureate graduates with dual degrees 0

TRANSFER INFORMATION
Residency Requirements: Thirty of the last 36 hours must be taken in residence at West Virginia Institute of Technology.

Transfer via Articulation Agreements
Admission to engineering technology: At any time
Technology graduates transferred from ... 4-yr: –   2-yr: 5
Requirements: Associate degree in closely related program of study. Evaluation of transcript required, as a minimum, with possibility of some additional course work after transfer.

Transfer without Articulation Agreements
Admission to engineering technology: At any time
Technology graduates transferred from ... 4-yr: 5   2-yr: –
Requirements: Students transfer from the college of engineering on campus and from other four-year programs. Evaluation of transcripts is required.

GRADUATION REQUIREMENTS
(1) 30 of last 36 hours in residence. (2) Minimum of 40 semester hours in upper division courses. (3) Overall 2.0 GPA in all courses attempted. (4) Overall 2.0 GPA at West Virginia Tech. (5) Minimum 2.0 GPA in major and minor (if elected). (6) CORE curriculum requirements must be met.

■ STUDENT PROGRAMS
PROFESSIONAL AND HONORARY SOCIETIES
[For key to acronyms, see *Introduction*.]
Professional Societies: ASCE, ASME, IEEE, ISA, American Design and Drafting Association (ADDA)
Honorary Societies: Phi Theta Kappa

SUPPORT PROGRAMS
Peer Sponsor Program
For: Women   **Available:** Academic year
Offered: Freshman

Black Student Union
For: Ethnic Minorities   **Available:** Academic year
Offered: Freshman, Sophomore, Junior, Senior

Association of Women Students
For: Women   **Available:** Academic year
Offered: Freshman, Sophomore, Junior, Senior

Other Engineering Support Programs: New students can pre-register during the summer; all freshman are required to take Technology Orientation; an 'Intermediate Studies' program is available for students with low ACT scores; tutoring services are available through Intermediate Studies and the Learning Center; career counseling services; foreign student advisor; veteran's services; placement services, transfer agreements.

354 Western Carolina University

INSTITUTION PROFILE
HEAD OF THE INSTITUTION
Myron L. Coulter
Phone: (704)227-7100 **Fax:** (704)227-7176

GENERAL INFORMATION
[All Students—Fall 1991]
Undergraduate enrollment 5,439
Graduate enrollment 933
Total institution enrollment **6,372**
Type of institution: Public **Calendar:** Semesters
Nearest city: Asheville **Population:** 100,000
Miles from main campus: 60 **Setting:** Small Town
Types of engineering degrees: Engineering Technology

ENGINEERING TECHNOLOGY ADMISSIONS
ADMISSIONS OFFICE CONTACT
Drumont Bowman
Western Carolina University
Director, Admissions Office
HFR Administration Building
Cullowhee, NC 28723
Phone: (704)227-7317 **Fax:** (704)227-7202

TECHNOLOGY COLLEGE ADMISSIONS INFORMATION
Entrance Requirements: SAT, High School courses—Social Studies (inc. 1 U.S. History) (2 years), English (4 years), Math (1-algebra I,1-Algebra II,1-Geom or/Adv Math) (3 years), Science (1 life/biological, 1 phys sci, 1 lab) (3 years), Will use ACT. Minimum score: 19 composite
Entrance Recommendations: High School courses—Foreign language (2 years), Math (in twelfth grade) (1 year)
Requirements for foreign students: TOEFL (Score: 550); SAT Declaration of Certification of Finances and History of Immunization form.

DEPARTMENTAL ADMISSIONS INFORMATION
Admission to the engineering technology department: At the time of admission to the institution.
Additional information: Regular departmental admission requirements: Students should be in the top 50% of their class and SAT score equal to or higher than 350 on each part of the test.

FINANCIAL INFORMATION
ESTIMATED EXPENSES (ACADEMIC YEAR)
[Expenses are for the 1992-93 nine-month academic year.]

State Residents	Undergraduate	Graduate
Tuition and fees	$ 1,375	$ 1,311
College room and board	$ 2,364	$ 2,364
Books and supplies	$ 225	$ 450
Other expenses	$ 1,332	$ –
Total estimated expenses	**$ 5,296**	**$ 4,125**

Out-of-State Residents	Undergraduate	Graduate
Tuition and fees	$ 7,047	$ 6,983
College room and board	$ 2,364	$ 2,364
Books and supplies	$ 225	$ 450
Other expenses	$ 786	$ –
Total estimated expenses	**$10,422**	**$ 9,797**

FINANCIAL AID OFFICE CONTACT
Glenn Hardesty, *Director, Student Financial Aid*
Western Carolina University
Student Financial Aid
HFR Administration Building
Cullowhee, NC 28723-9023
Phone: (704)227-7290 **Fax:** (704)227-7202

GENERAL FINANCIAL AID INFORMATION
Forms accepted/required: FFS
Additional financial aid information: Need based scholarships, short term loans, merit based scholarships, part-time jobs on campus, work-study program.

TECHNOLOGY COLLEGE INFORMATION
[For additional personnel, refer to the *Appendix*.]
HEAD OF ENGINEERING TECHNOLOGY
J. Dale Pounds
Phone: (704)227-7272 **Fax:** (704)227-7705

ENGINEERING TECHNOLOGY COLLEGE ADDRESS
School of Applied Sciences
Western Carolina University
Belk Building
Cullowhee, NC 28723-9646
Phone: (702)227-7272 **Fax:** (704)227-7705

TECHNOLOGY ENROLLMENTS—BY CLASS
[Numbers are technology baccalaureate enrollments for the fall 1991 term, unless otherwise footnoted.]
1st-year students/Freshmen 20
2nd-year students/Sophomores 15
3rd-year students/Juniors 47
4th-year students/Seniors 50
Total **132**

NUMBER OF DEGREES AWARDED—BY PROGRAM
[Numbers are engineering technology baccalaureate degrees awarded during the 1991-1992 academic year, unless otherwise footnoted. For full details about each engineering technology program, refer to the *Table of Degree Programs*.]
Electronic(s) Engineering Technology 11
Manufacturing Engineering Technology 16
Total **27**

PERCENTAGE OF DEGREES AWARDED—BY CATEGORY
[Percentages are of all engineering technology baccalaureate degrees awarded during the 1991-1992 academic year, unless otherwise footnoted.]
African Americans 3.8%
Asian/Pacific Island Americans 3.0%
Hispanic Americans .... – %
Native Americans .... – %
Foreign Citizens .... – %
All Others 93.2%
Women 7.6%
Persons with disabilities .... – %
Students over 25 years of age .... – %

FULL- & PART-TIME FACULTY—BY DEPARTMENT
[Figures are the head count of full-time faculty and the full-time equivalent (FTE) of part-time faculty for each engineering technology department or equivalent.]

Department	Full	Part
Electrical/Electronic(s) Eng Tech	3	–
Manufacturing Eng Tech	4	–
Industrial Tech	4	–

TECHNOLOGY COLLEGE DESCRIPTION
The School's four-year engineering technology programs are laboratory intensive with strong industrial advisory committees. An optional co-op program is available and students are encouraged to participate. The School also has articulation agreements with a number of North Carolina and South Carolina two-year institutions. Average number of years required to actually complete the bachelor's degree: 4

TECHNOLOGY DEGREE PROGRAMS DESCRIPTION
The School's four-year degree programs in engineering technology emphasize industrial applications of theoretical concepts to practice work-ready graduates. At the completion of 128 semester hours (4 years normally) the graduate will receive a B.S. degree.

TRANSFER INFORMATION
Transfer via Articulation Agreements
Admission to engineering technology: Majority are two-year college graduates, but students may transfer in at any point in the program.
Technology graduates transferred from . . . 4-yr: –   2-yr: 30

GRADUATION REQUIREMENTS
2.0 overall, 2.0 in major. Bachelor of Science of engineering technology programs require 128 semester hour credits.

STUDENT PROGRAMS
PROFESSIONAL AND HONORARY SOCIETIES
[For key to acronyms, see *Introduction*.]
Professional Societies: SME, American Society of Safety Engineers, Professional Association of Industrial Distributors
Honorary Societies: Tau Alpha Pi

SUPPORT PROGRAMS
There are at least 13 student organizations
For: Ethnic Minorities **Available:** Academic year
Other Engineering Support Programs: Academic counseling services are available in the Career and Academic Planning Center. Orientation to engineering technology is included in freshman level engineering graphics.

355 Western Kentucky University

INSTITUTION PROFILE

HEAD OF THE INSTITUTION
Thomas C. Meredith
Phone: (502)745-4492

GENERAL INFORMATION
[All Students—Fall 1991]
Undergraduate enrollment 13,711
Graduate enrollment 2,039
Total institution enrollment **15,750**
Type of institution: Public **Calendar:** Semesters
Nearest city: Nashville, Tennessee **Population:** 532,000
Miles from main campus: 65 **Setting:** Suburban
Types of engineering degrees: Engineering Technology
Other degree-granting colleges: Business Administration, Education, Humanities & Social Sciences, Science, Technology, and Health, Graduate

ENGINEERING TECHNOLOGY ADMISSIONS

ADMISSIONS OFFICE CONTACT
Cheryl C. Chambless
Western Kentucky University
1526 Russellville Road
Bowling Green, KY 42101
Phone: (502)745-5422

TECHNOLOGY COLLEGE ADMISSIONS INFORMATION
Entrance Requirements: ACT (Score: 17), High School courses—English (4 years), Mathematics (Algebra I, Geometry, Algebra II) (3 years), Social Studies (2 years), Science (including Biology I, Phys I, or Chem I) (2 years)
Entrance Recommendations: ACT (Score: 17), High School courses—Foreign Languages (1 year)
Requirements for out-of-state residents: High school GPA 2.2/4.0, ACT composite at least 19 or SAT total at least 720, and rank in top half of class.
Requirements for foreign students: TOEFL (Score: 500)

DEPARTMENTAL ADMISSIONS INFORMATION
Admission to the engineering technology department: At the time of admission to the institution.

FINANCIAL INFORMATION

ESTIMATED EXPENSES (ACADEMIC YEAR)
[Expenses are for the 1992-93 nine-month academic year.]

State Residents	Undergraduate	Graduate
Tuition and fees	$ 1,542	$ 1,684
College room and board	$ 1,160 (1)	$ 1,160 (1)
Books and supplies	$ 300	$ 400
Other expenses	$ –	$ –
Total estimated expenses	**$ 3,002**	**$ 3,244**

Out-of-State Residents	Undergraduate	Graduate
Tuition and fees	$ 4,224	$ 4,644
College room and board	$ 1,160 (1)	$ 1,160 (1)
Books and supplies	$ 300	$ 400
Other expenses	$ –	$ –
Total estimated expenses	**$ 5,684**	**$ 6,204**

Notes: (1) Housing only, Board not estimated

FINANCIAL AID OFFICE CONTACT
Marilyn Clark, *Director, Dept of Student Financial Assistance*
Western Kentucky University
1526 Russellville Road
Bowling Green, KY 42101
Phone: (502)745-2756

GENERAL FINANCIAL AID INFORMATION
Forms accepted/required: SAR, Institutional, Kentucky Financial Aid Form
Additional financial aid information: Grants, loans, student employment

TECHNOLOGY COLLEGE INFORMATION
[For additional personnel, refer to the *Appendix*.]

HEAD OF ENGINEERING TECHNOLOGY
Charles E. Kupchella
Phone: (502)745-4448 **Fax:** (502)745-6471

ENGINEERING TECHNOLOGY COLLEGE ADDRESS
Department of Engineering Technology
Western Kentucky University
1526 Russellville Road
Bowling Green, KY 42101
Phone: (502)745-2461 **Fax:** (502)745-6471

TECHNOLOGY ENROLLMENTS—BY CLASS
[Numbers are technology baccalaureate enrollments for the fall 1991 term, unless otherwise footnoted.]
1st-year students/Freshmen 84
2nd-year students/Sophomores 56
3rd-year students/Juniors 59
4th-year students/Seniors 97
Total **296**

NUMBER OF DEGREES AWARDED—BY PROGRAM
[Numbers are engineering technology baccalaureate degrees awarded during the 1991-1992 academic year, unless otherwise footnoted. For full details about each engineering technology program, refer to the *Table of Degree Programs*.]
Civil Engineering Technology 10
Electrical Engineering Technology 16
Electromechanical Engineering Technology 3
Mechanical Engineering Technology 11
Total **40**

PERCENTAGE OF DEGREES AWARDED—BY CATEGORY
[Percentages are of all engineering technology baccalaureate degrees awarded during the 1991-1992 academic year, unless otherwise footnoted.]
African Americans .... – %
Asian/Pacific Island Americans .... – %
Hispanic Americans .... – %
Native Americans .... – %
Foreign Citizens .... 2.3 %
All Others .... 97.7 %
Women .... 2.3 %
Persons with disabilities .... – %
Students over 25 years of age .... 10.0 %

ENGINEERING TECHNOLOGY STUDENT DATA
Applicants to the engineering technology college
Number of applicants to engineering technology college 76
Percent offered admission 100.0%
Matriculated engineering technology students
Percentage in top quartile (25%) of High School class .... – (A)
Average ACT scores: Composite—20.5

Notes: (A) Data not available.

FULL- & PART-TIME FACULTY—BY DEPARTMENT
[Figures are the head count of full-time faculty and the full-time equivalent (FTE) of part-time faculty for each engineering technology department or equivalent.]

Department	Full	Part
Eng Tech	9	9.5

TECHNOLOGY COLLEGE DESCRIPTION
Co-op programs available.

TRANSFER INFORMATION
Residency Requirements: minimum 36 weeks during which 32 semester hours earned; 18 of the 36 weeks must be in the senior year
Transfer via Articulation Agreements
Admission to engineering technology: At the end of the sophomore year
Technology graduates transferred from . . . 4-yr: – 2-yr: –
Requirements: GPA of 'C' for last semester, cumulative GPA of 'C', good standing at institution transferring from
Transfer without Articulation Agreements
Admission to engineering technology: At any time

GRADUATION REQUIREMENTS
Minimum scholastic standing of 2.0/4.0 in all credits presented for graduation; all credits completed at WKU; overall in the major and minor subjects; in the major and minor subjects completed at WKU. Minimum of 128 semester hours.

STUDENT PROGRAMS

PROFESSIONAL AND HONORARY SOCIETIES
[For key to acronyms, see *Introduction*.]
Professional Societies: ASCE, ASME, IEEE, SME
Honorary Societies: Phi Kappa Phi

SUPPORT PROGRAMS
Office of Black Student Retention
For: Ethnic Minorities **Available:** Year round
Offered: Prematriculation, Freshman, Sophomore, Junior, Senior, Graduate level
Other Engineering Support Programs: Orientation and Registration, Office of Academic Advising, Career Planning and Placement, Cooperative Education.

356 Western Michigan University

INSTITUTION PROFILE
HEAD OF THE INSTITUTION
Diether H. Haenicke
Phone: (616)387-2351 **Fax:** (616)387-2355
GENERAL INFORMATION
[All Students—Fall 1991]
Undergraduate enrollment 20,130
Graduate enrollment 3,928
Total institution enrollment **24,058**
Type of institution: Public **Calendar:** Semesters
Nearest city: Grand Rapids **Population:** 181,843
Miles from main campus: 50 **Setting:** Urban
Types of engineering degrees: Engineering Technology
Other degree-granting colleges: Arts & Sciences, Education, Haworth College of Business, Health & Human Services, Lee Honors College, Fine Arts

ENGINEERING TECHNOLOGY ADMISSIONS
ADMISSIONS OFFICE CONTACT
Stanley E. Henderson
Western Michigan University
2240 Seibert Administration Building
Kalamazoo, MI 49008
Phone: (616)387-2000 **Fax:** (616)387-2355
TECHNOLOGY COLLEGE ADMISSIONS INFORMATION
Entrance Requirements: Admission requirements same as for university.
Entrance Recommendations: High School courses—Math and English (each) (4 years), Physical or Life Sciences (3 years), Social Sciences (2 years), Basic Computer Course (1 year), Same as for university
Requirements for foreign students: TOEFL (Score: 550); MTELP (85%); TOEFL score can be 500 if student takes remedial English class during first semester.
DEPARTMENTAL ADMISSIONS INFORMATION
Admission to the engineering technology department: At the time of admission to the institution.
Additional information: Satisfactory completion ('C' or better) of basic writing, mathematics, and sciences courses is required before enrollment in upper level Engineering Technology (300-400) courses. Students should contact their academic advisor for the list of required courses that applies to their curriculum.

FINANCIAL INFORMATION
ESTIMATED EXPENSES (ACADEMIC YEAR)
[Expenses are for the 1992-93 nine-month academic year.]

State Residents	Undergraduate	Graduate
Tuition and fees	$ 2,905 [(1)]	$ –
College room and board	$ 3,830	$ –
Books and supplies	$ 470	$ –
Other expenses	$ –	$ –
Total estimated expenses	**$ 7,205**	**$ –**
Out-of-State Residents	Undergraduate	Graduate
Tuition and fees	$ 6,795 [(2)]	$ –
College room and board	$ 3,830	$ –
Books and supplies	$ 470	$ –
Other expenses	$ –	$ –
Total estimated expenses	**$11,095**	**$ –**

Notes: (1) Based on 15 hrs per sem as full-time; figure shown is average of tuition/fees for lower (fresh/soph) and upper (jr/sr) level--lower is 2780, upper is 3030. (2) Based on 15 hrs per sem as full-time; figure shown is average of tuition/fees for lower (fresh/soph) and upper (jr/sr) level--lower is 6490, upper is 7100.

FINANCIAL AID OFFICE CONTACT
John A. Kundel, *Director of Student Financial Aid*
Western Michigan University
Student Financial Aid Office
3306 Student Services Building
Kalamazoo, MI 49008
Phone: (616)387-6000 **Fax:** (616)387-6989
GENERAL FINANCIAL AID INFORMATION
Forms accepted/required: FSA, FFS, Institutional
Additional financial aid information: Need-based scholarships and grants; short-term loans; merit-based scholarships; part-time jobs on campus; student employment referral; College Work-Study program.

TECHNOLOGY COLLEGE INFORMATION
[For additional personnel, refer to the *Appendix*.]
HEAD OF ENGINEERING TECHNOLOGY
Leonard R. Lamberson
Phone: (616)387-4017 **Fax:** (616)387-4024
EMail: LAMBERSON@GW.WMICH.EDU
ENGINEERING TECHNOLOGY COLLEGE ADDRESS
College of Engineering and Applied Sciences
Western Michigan University
2022 Kohrman Hall
Kalamazoo, MI 49008-5062
Phone: (616)387-4017 **Fax:** (616)387-4024
TECHNOLOGY ENROLLMENTS—BY CLASS
[Numbers are technology baccalaureate enrollments for the fall 1991 term, unless otherwise footnoted.]
1st-year students/Freshmen 163
2nd-year students/Sophomores 236
3rd-year students/Juniors 253
4th-year students/Seniors 345
Total **997**
NUMBER OF DEGREES AWARDED—BY PROGRAM
[Numbers are engineering technology baccalaureate degrees awarded during the 1991-1992 academic year, unless otherwise footnoted. For full details about each engineering technology program, refer to the *Table of Degree Programs*.]
Aircraft Maintenance Engineering Technology .... –
Automotive Engineering Technology 6
Aviation Technology & Operations 55
Construction Science & Management 21
Engineering Graphics 30
Engineering Metallurgy .... –
Industrial Design 9
Manufacturing Engineering Technology 10
Total **131**
PERCENTAGE OF DEGREES AWARDED—BY CATEGORY
[Percentages are of all engineering technology baccalaureate degrees awarded during the 1991-1992 academic year, unless otherwise footnoted.]
African Americans .... 3.1 %
Asian/Pacific Island Americans .... – %
Hispanic Americans .... – %
Native Americans .... – %
Foreign Citizens .... 3.1 %
All Others .... 93.8 %
Women .... 9.9 %
Persons with disabilities .... – % [(A)]
Students over 25 years of age .... – % [(A)]

Notes: (A) Data not available.

FULL- & PART-TIME FACULTY—BY DEPARTMENT
[Figures are the head count of full-time faculty and the full-time equivalent (FTE) of part-time faculty for each engineering technology department or equivalent.]

Department	Full	Part
Eng Tech	19	10.0

TECHNOLOGY COLLEGE DESCRIPTION
The College offers four-year undergraduate programs in engineering, engineering technology and related technical management and support areas. General education in the humanities and social sciences is distributed throughout the four years. Computer-aided engineering is emphasized at all levels, as are communication skills. Co-op and other field experience opportunities are optional. During the last semester, students complete a senior design project, usually involving a real problem in an area industry.

TECHNOLOGY DEGREE PROGRAMS DESCRIPTION
The B.S. degrees are designed to provide graduates with the background necessary to successfully assume a variety of positions in manufacturing, construction, automotive, and design industries. The combination of specialized and general education is intended to allow employment flexibility, although most graduates are placed in industries closely related to their field of study.

TRANSFER INFORMATION
Residency Requirements: One-half of program must be from a four-year accredited institution; 30 semester hours must be from WMU.
Transfer without Articulation Agreements
Admission to engineering technology: At any time
GRADUATION REQUIREMENTS
Bachelor of Science programs in Engineering Technology require 128 to 141 semester credits, depending on major selected.

STUDENT PROGRAMS
PROFESSIONAL AND HONORARY SOCIETIES
[For key to acronyms, see *Introduction*.]
Professional Societies: NAHB, SAE, SME, SPE*, ASM International, Industrial Designers Society of America (IDSA)
Honorary Societies: Alpha Eta Rho
SUPPORT PROGRAMS
Student Chapter Organizations: Society of Women Engineers, National Society of Black Engineers, Pre-college intro programs to engineering, WMU

Martin Luther King Program
For: Ethnic Minorities **Available:** Year round
Offered: Freshman

Minority Student Services
For: Ethnic Minorities **Available:** Year round
Offered: Prematriculation, Freshman, Sophomore, Junior, Senior, Graduate level

Center for Women's Services
For: Women **Available:** Year round
Offered: Prematriculation, Freshman, Sophomore, Junior, Senior, Graduate level

Other Engineering Support Programs: Aviation alumni and career day, October 10, 1991, featured alumni and Career Planning/Placement staff speaking to current aviation students.

THE JOURNAL OF ENGINEERING EDUCATION

The only publication that offers current reports on today's innovations in engineering education.

Engineering education is rapidly changing. Teaching has become the most critical aspect of an engineering educator's livelihood. Keeping up with current innovations in the field can be difficult.

Let the Journal of Engineering Education make it easier.

The American Society for Engineering Education, with support from the National Science Foundation, is addressing the important issues in engineering education with the publication of a distinguished journal.

In January 1993, ASEE launched the *Journal of Engineering Education,* a quarterly, peer-reviewed journal, that replaced *Engineering Education,* last published in August 1991. A distinguished volunteer editor, assisted by an editorial board will set editorial policy and guide the journals operations.

A Scholarly Journal

The *Journal of Engineering Education* seeks to serve the needs of the engineering education enterprise. It will maintain the highest standards of scholarship while serving as an agent of change.

The content of the *Journal* will be a record of where engineering education is, where it has come from, and where it appears to be moving. Specifically, these thrusts are exemplified by papers that describe:

▲ The present status of engineering education
▲ What we can learn from our history
▲ The future course of current trends
▲ Intra-institutional, interdisciplinary efforts
▲ Inter-institutional efforts
▲ Outreach beyond the college of engineering.

Valuable Global Perspectives

Through its 100 years of service to the engineering education community, ASEE has achieved global recognition and established relationships with university administrators, faculty, corporate representatives, and esteemed authors. The *Journal* will publish scholarly papers from around the world that enunciate educational principles, not simply present superficial analyses of classroom data or experiments.

A benefit of ASEE membership

The *Journal of Engineering Education* will give you the information you need to stay on top of the constantly changing world of engineering education.

To order, please refer to the ASEE Publication order form in the back of this directory or contact ASEE Publication Sales at (202) 986-8528.

Appendixes & Indexes

APPENDIXES & INDEXES
Engineering and Engineering Technology Personnel **527**
Index to Engineering and Engineering Technology Personnel **570**
ABET/CEAB Accredited Degree Programs **585**
Index to ABET/CEAB Accredited Degree Programs **596**

ASEE MEMBERSHIP

Vital to continued professional development in engineering and engineering technology education

Involvement in ASEE is the key to enhancing your professional development through timely publications, meetings, projects, and services.

ENHANCED MEMBERSHIP BENEFITS INCLUDE:

- A subscription to the Society's official magazine, *ASEE PRISM*, valued at $75.

- The *Directory of Engineering and Engineering Technology Undergraduate Programs.*

- The *Directory of Engineering Graduate Studies and Research.*

- The *ASEE Membership Directory.*

- Other publications, including numerous technical division publications, regional newsletters, and special reports of major studies and national surveys.

- Lower, member rates at ASEE conferences.

- Essential professional development and networking opportunities with academic colleagues and key corporate contacts who have a direct interest in engineering and engineering technology education.

ANNUAL MEMBERSHIP DUES

$60 Individual Membership

$20 Associate Membership (open only to graduate students now performing teaching functions and those interested in teaching as a career).

Join ASEE today and begin taking advantage of the benefits immediately!

If you have any questions, please call
ASEE Member Services at (202) 986-8518.

Engineering & Engineering Technology Personnel

ENGINEERING

001 U of Akron
Head of the Institution
Peggy G. Elliott
Phone: (216)972-7074   Fax: (216)972-8652

Head of Engineering
Nicholas D. Sylvester
Phone: (216)972-6978   Fax: (216)972-5162
EMail: DELUCA@ENGINEER@UAKRON

Associate/Assistant Deans
Max S. Willis, *Associate Dean, Research & Graduate Studies*
Phone: (216)972-6978   Fax: (216)972-5162
Graham Kelly, *Associate Dean for Undergraduate Studies*
Phone: (216)972-7817   Fax: (216)972-5162
Paul C. Lam, *Assistant Dean & Director of Minority Engineering Program*
Phone: (216)972-7741   Fax: (216)972-5162
Richard S. Rice, *Assistant Dean & Director of Cooperative Education*
Phone: (216)972-7817   Fax: (216)972-5162

Department Chair(s)
Chemical Eng
Sunggyu Lee
Phone: (216)972-7250   Fax: (216)972-5856
Civil Eng
Paul Chang
Phone: (216)972-7286   Fax: (216)972-6020
EMail: R11PC@VMI.CC@UAKRON.EDU
Electrical Eng
Chiou S. Chen
Phone: (216)972-7648   Fax: (216)972-6487
EMail: R1CSC@VMI.CC@UAKRON.EDU
Biomedical Eng
Daniel B. Sheffer
Phone: (216)972-6650   Fax: (216)972-8834
EMail: R1DBS@VMI.CC@UAKRON.EDU
Mechanical Eng
Benjamin T. Chung
Phone: (216)972-6307   Fax: (216)972-6027
EMail: R1BTC@VMI.CC@UAKRON.EDU

Admissions Office Contact
Martha A. Booth
Phone: (216)972-7100

Financial Aid Office Contact
Robert D. Hahn, *Director*
Phone: (216)972-7032   Fax: (216)972-7139

002 U of Alabama
Head of the Institution
E. Roger Sayers
Phone: (205)348-5100   Fax: (205)348-8377

Head of Engineering
Robert F. Barfield
Phone: (205)348-6405   Fax: (205)348-8573
EMail: BARFIELD@UA1VM.UA.EDU

Associate/Assistant Deans
Karen L. Frair, *Associate Dean for Administration*
Phone: (205)348-6431   Fax: (205)348-8573
Verle N. Schrodt, *Assistant Dean for Research & Graduate Studies*
Phone: (205)348-1591   Fax: (205)348-9455
William K. Rey, *Assistant Dean for Undergraduate Programs*
Phone: (205)348-6408   Fax: (205)348-8573

Department Chair(s)
Chemical Eng
William J. Hatcher
Phone: (205)348-6450   Fax: (205)348-9659
EMail: WHATCHER@UA1VM.UA.EDU
Civil Eng
Daniel S. Turner
Phone: (205)348-6550   Fax: (205)348-8573
EMail: DTURNER@UA1VM.UA.EDU
Electrical Eng
Russell L. Pimmell
Phone: (205)348-6351   Fax: (205)348-6959
EMail: RPIMMELL@UA1VM.UA.EDU
Industrial Eng
Dennis B. Webster
Phone: (205)348-1608   Fax: (205)348-8573
EMail: DWEBSTER@UA1VM.UA.EDU
Mechanical Eng
Donald C. Raney
Phone: (205)348-1616   Fax: (205)348-6419
EMail: DRANEY@UA1VM.UA.EDU
Aerospace Eng
John E. Jackson
Phone: (205)348-7306   Fax: (205)348-8573
EMail: AEDEPT@UA1VM.UA.EDU
Computer Science
David B. Brown
Phone: (205)348-6363   Fax: (205)348-8573
EMail: BROWN@CS.UA.EDU
Eng Mechanics
James L. Hill
Phone: (205)348-1628   Fax: (205)348-8573
EMail: JHILL@UA1VM.UA.EDU
Metallurgical & Materials Eng
John T. Berry
Phone: (205)348-1747   Fax: (205)348-8573
EMail: JBERRY@UA1VM.UA.EDU
Mineral Eng
Lloyd A. Morley
Phone: (205)348-1677   Fax: (205)348-8573
EMail: LMORLEY@UA1VM.UA.EDU
Eng Tech
James L. Keating
Phone: (205)348-6555   Fax: (205)348-8573
EMail: JKEATING@UA1VM.UA.EDU

Admissions Office Contact
Roy C. Smith
Phone: (205)348-5666   Fax: (205)348-9046

Financial Aid Office Contact
Molly M. Lawrence, *Director*
Phone: (205)348-6756   Fax: (205)348-2989

003 U of Alabama at Birmingham
Head of the Institution
Charles A. McCallum
Phone: (205)934-4636

Head of Engineering
Jay Goldman
Phone: (205)934-8400   Fax: (205)934-8437
EMail: JGOLDMAN@ENGSYS.ENG.UAB.EDU

Department Chair(s)
Civil Eng
Edmund P. Segner, Jr.
Phone: (205)934-8430   Fax: (205)934-9855
EMail: ESEGNER@ENGSYS.ENG.UAB.EDU
Electrical & Computer Eng
David A. Commer
Phone: (205)934-8440   Fax: (205)934-8437
EMail: DACONNER@ENGSYS.ENG.UAB.EDU
Mechanical Eng
Terry Wright
Phone: (205)934-8460   Fax: (205)934-8437
EMail: TWRIGHT@ENGSYS.ENG.UAB.EDU
Materials Science & Eng
Barry Andrews
Phone: (205)934-8450   Fax: (205)934-8485
EMail: BANDREWS@ENGSYS.ENG.UAB.EDU
Biomedical Eng
Ernest M. Stokely
Phone: (205)934-8420   Fax: (205)935-4919
EMail: ESTOKELY@ENGSYS.ENG.UAB.EDU

Admissions Office Contact
Norma E. Sorenson
Phone: (205)934-8410   Fax: (205)934-8437

Financial Aid Office Contact
Claude E. McCann, *Director, Financial Aid*
Phone: (205)934-8223

004 U of Alabama at Huntsville
Head of the Institution
Frank A. Franz
Phone: (205)895-6340   Fax: (205)895-6538

Head of Engineering
Lynn D. Russell
Phone: (205)895-6474   Fax: (205)895-6758

Associate/Assistant Deans
Kenneth O. Thompson, *Associate Dean*
Phone: (205)895-6877   Fax: (205)895-6758

Department Chair(s)
Chemical Eng
James E. Smith, Jr.
Phone: (205)895-6349   Fax: (205)895-6758
Civil Eng
Gerald R. Karr
Phone: (205)895-6154   Fax: (205)895-6758
Electrical & Computer Eng
Stephen T. Kowel
Phone: (205)895-6316   Fax: (205)895-6803
Industrial & Systems Eng
Bernard J. Schroer
Phone: (205)895-6256   Fax: (205)895-6733
Mechanical Eng
Gerald R. Karr
Phone: (205)895-6154   Fax: (205)895-6758

Admissions Office Contact
Kenneth O. Thompson
Phone: (205)895-6877   Fax: (205)895-6758

Financial Aid Office Contact
James B. Gibson, *Director*
Phone: (205)895-6241

005 U of Alaska, Fairbanks
Head of the Institution
Joan K. Wadlow
Phone: (907)474-5213   Fax: (907)474-5213

Head of Engineering
Frank F. Williams
Phone: (907)474-7330   Fax: (907)474-6087
EMail: FFFLW@ACAD3.ALASKA.EDU

Associate/Assistant Deans
John P. Zarling, *Associate Dean*
Phone: (907)474-7330   Fax: (907)474-6087

Department Chair(s)
Civil Eng
Robert F. Carlson
Phone: (907)474-7241   Fax: (907)464-6087
EMail: FFRFC1@ACAD3.ALASKA.EDU
Electrical Eng
John Aspnes
Phone: (907)474-7137   Fax: (907)474-6087
Mechanical Eng
Ronald A. Johnson
Phone: (907)474-7209   Fax: (907)474-6087
EMail: FYMEDPT@ACAD3.ALASKA.EDU

Admissions Office Contact
Ann Tremarello
Phone: (907)474-7821   Fax: (907)474-5379

Financial Aid Office Contact
Donald E. Scheaffer, *Director of Financial Aid*
Phone: (907)474-7256   Fax: (907)474-7900

006 U of Alberta
Head of the Institution
Paul T. Davenport
Phone: (403)492-3212   Fax: (403)492-2726

Head of Engineering
Fred D. Otto
Phone: (403)492-3596   Fax: (403)492-0500
EMail: fotto@vm.ualberta.ca

Associate/Assistant Deans
David T. Lynch, *Associate Dean (Planning)*
Phone: (403)492-5398   Fax: (403)492-0500
Loverne R. Plitt, *Associate Dean (Student Services)*
Phone: (403)492-3399   Fax: (403)492-0500
Ken C. Porteous, *Associate Dean (Cooperative Education)*
Phone: (403)492-5152   Fax: (403)492-2732

Department Chair(s)
Chemical Eng
Murray R. Gray
Phone: (403)492-3321   Fax: (403)492-2881
Civil Eng
Daniel W. Smith
Phone: (403)492-4235   Fax: (403)492-0249
Electrical Eng
Clarence E. Capjack
Phone: (403)492-3332   Fax: (403)492-1811
Mechanical Eng
J. Doug Dale
Phone: (403)492-3598   Fax: (403)492-2200
Mining, Metallurgical & Petroleum Eng
Ken Barron
Phone: (403)492-3337

Financial Aid Office Contact
Jiang Liu, *Emergency Aid & Bursary Coordinator*
Phone: (403)492-3483   Fax: (403)492-6701

007 Alfred U
Head of the Institution
Edward G. Coll, Jr.
Phone: (607)871-2101   Fax: (607)871-2339

Head of Engineering
Richard W. Ott
Phone: (607)871-2137   Fax: (607)871-2339
EMail: Ott

Department Chair(s)
Electrical Eng
James Lancaster
Phone: (607)871-2130
EMail: FLancaster
Mechanical Eng
Bruce R. Hollworth
Phone: (607)871-2100
EMail: Hollworth
Ceramic Eng & Sciences
Alastair Cormack
Phone: (607)871-2422   Fax: (607)871-2392
EMail: Cormack

Admissions Office Contact
Daniel L. Meyer
Phone: (800)541-9229   Fax: (607)871-2198

Financial Aid Office Contact
Earl E. Pierce, Jr., *Director of Student Financial Aid*
Phone: (607)871-2159   Fax: (607)871-2198

008 U of Arizona
Head of the Institution
Manuel T. Pacheco
Phone: (602)621-5511   Fax: (602)621-7475

Head of Engineering
Ernest T. Smerdon
Phone: (602)621-6594   Fax: (602)621-2232

Engineering & Engineering Technology Personnel

Associate/Assistant Deans
Vern R. Johnson, *Associate Dean*
Phone: (602)621-6032   Fax: (602)621-9995
William P. Cosart, *Associate Dean*
Phone: (602)659-6596   Fax: (602)621-2232
John R. Sevier, *Associate Dean*
Phone: (602)621-6598   Fax: (602)621-2232
Morris Farr, *Assistant Dean*
Phone: (602)621-6032   Fax: (602)621-9995

Department Chair(s)
Chemical Eng
Thomas W. Peterson
Phone: (602)621-2591   Fax: (602)621-6048
Civil Eng & Eng Mechanics
Dinshaw N. Contractor
Phone: (602)621-2266   Fax: (602)621-2550
Electrical & Computer Eng
Kenneth F. Galloway
Phone: (602)621-6193   Fax: (602)621-8076
Systems & Industrial Eng
Pitu Mirchandani
Phone: (602)621-6551   Fax: (602)621-6646
Aerospace & Mechanical Eng
Parviz Nikravesh
Phone: (602)621-2236   Fax: (602)621-8191
Agricultural & Biosystems Eng
Donald C. Slack
Phone: (602)621-3691   Fax: (602)621-3963
Eng Mathematics
David Lomen
Phone: (602)621-6868   Fax: (602)621-8322
Eng Physics
John D. McCullen
Phone: (602)621-6812   Fax: (602)621-4721
Hydrology & Water Resources
Soroosh Sorooshian
Phone: (602)621-7120   Fax: (602)621-1422
Materials Science & Eng
Donald R. Uhlmann
Phone: (602)322-2960   Fax: (602)621-8059
Mining & Geological Eng
Ben Sternberg
Phone: (602)621-2147   Fax: (602)621-8330
Nuclear & Energy Eng
Morris Farr
Phone: (602)621-2311   Fax: (602)621-8096

Admissions Office Contact
Jerome A. Lucido
Phone: (602)621-3237   Fax: (602)621-9799

Financial Aid Office Contact
Phyliss K. Bolt-Bannister, *Student Financial Aid Director*
Phone: (602)621-1643

009 Arizona State U

Head of the Institution
Lattie F. Coor
Phone: (602)965-5606   Fax: (602)965-0865

Head of Engineering
David C. Chang
Phone: (602)965-1722   Fax: (602)965-2267
EMail: icdcc@asuacad

Associate/Assistant Deans
Charles E. Backus, *Associate Dean*
Phone: (602)965-2825   Fax: (602)965-2267
Richard W. Kelly, *Associate Dean*
Phone: (602)965-3874   Fax: (602)965-5089
William E. Lewis, *Associate Dean*
Phone: (602)965-1730   Fax: (602)965-2267
C. Edward Wallace, *Associate Dean*
Phone: (602)965-1726   Fax: (602)965-2267

Department Chair(s)
Chemical, Bio, & Materials Eng
Joseph D. Henry
Phone: (602)965-3313   Fax: (602)965-0037
EMail: icjdh@asuacad

Civil Eng
Larry W. Mays
Phone: (602)965-3589   Fax: (602)965-0557
EMail: iclwm@asuacad
Electrical Eng
Peter E. Crouch
Phone: (602)965-3424   Fax: (602)965-3837
EMail: iacpec@asuacad
Industrial & Management Systems Eng
Philip M. Wolfe
Phone: (602)965-3185   Fax: (602)965-8692
EMail: iacpmw@asuacad
Mechanical & Aerospace Eng
Don L. Boyer
Phone: (602)965-3291   Fax: (602)965-1384
EMail: iddlb@asuacad
Computer Science & Eng
Ben M. Huey
Phone: (602)965-3190   Fax: (602)965-2751
EMail: iacbmh@asuacad

Admissions Office Contact
Susan Clouse
Phone: (602)965-7788   Fax: (602)965-2120

Financial Aid Office Contact
Paul Barberini, *Director*
Phone: (602)965-4045

010 U of Arkansas

Head of the Institution
Daniel E. Ferritor
Phone: (501)575-4148   Fax: (501)575-7575

Head of Engineering
Neil M. Schmitt
Phone: (501)575-3054   Fax: (501)575-4346
EMail: nms@nschmitt.uark.edu

Associate/Assistant Deans
Robert C. Welch, *Associate Dean of Engineering*
Phone: (501)575-6010   Fax: (501)575-4346
Jim L. Gattis, *Associate Dean of Engineering*
Phone: (505)575-6010   Fax: (501)575-4346
William K. Warnock, *Assistant Dean of Engineering*
Phone: (501)575-6012   Fax: (501)575-4346

Department Chair(s)
Chemical Eng
Robert E. Babcock
Phone: (501)575-4951   Fax: (501)575-7926
EMail: reb@engr.uark.edu
Civil Eng
Robert P. Elliott
Phone: (501)575-4954   Fax: (501)575-7168
EMail: rpe@engr.uark.edu
Electrical Eng
William D. Brown
Phone: (501)575-3005   Fax: (501)575-7967
EMail: wdb@engr.uark.edu
Industrial Eng
Eric M. Malstrom
Phone: (501)575-3156   Fax: (501)575-8431
EMail: emm@engr.uark.edu
Mechanical Eng
William F. Schmidt
Phone: (501)575-3153   Fax: (501)575-6982
EMail: wfs@engr.uark.edu
Biological & Agricultural Eng
Carl L. Griffis
Phone: (501)575-2351   Fax: (501)575-2846
EMail: cgriffis@saturn.uark.edu
Computer Systems Eng
Ronald W. Skeith
Phone: (501)575-6036   Fax: (501)575-5339
EMail: rws@engr.uark.edu

Financial Aid Office Contact
Lenthon Clark, *Director of Financial Aid*
Phone: (501)575-3806   Fax: (501)575-7575

011 Arkansas Tech U

Head of the Institution
Kenneth Kersh
Phone: (501)968-0237

Head of Engineering
Jack Hamm
Phone: (501)968-0353   Fax: (501)968-0677

Department Chair(s)
Engineering
Roy R. Culp
Phone: (501)968-0259   Fax: (501)968-0677

Admissions Office Contact
Tammy Rhodes
Phone: (501)968-0343

Financial Aid Office Contact
Shirley Goines, *Director of Student Aid*
Phone: (501)968-0399

012 Auburn U

Head of the Institution
William V. Muse
Phone: (205)844-4650   Fax: (205)844-6179

Head of Engineering
William F. Walker
Phone: (205)844-4326   Fax: (205)844-2672
EMail: wwalker@eng.auburn.edu

Associate/Assistant Deans
Larry D. Benefield, *Associate Dean for Academics*
Phone: (205)844-4326   Fax: (205)844-2672
M. Dayne Aldridge, *Associate Dean for Cross Disciplinary Programs*
Phone: (205)844-4333   Fax: (205)844-2672
John M. Owens, *Associate Dean for Research*
Phone: (205)844-4326   Fax: (205)844-2672

Department Chair(s)
Chemical Eng
Robert P. Chambers
Phone: (205)844-4827   Fax: (205)844-2063
EMail: chambers@eng.auburn.edu
Civil Eng
Loren D. Lutes
Phone: (205)844-4320   Fax: (205)844-6290
EMail: llutes@eng.auburn.edu
Electrical Eng
J. David Irwin
Phone: (205)844-1800   Fax: (205)844-1809
EMail: jdirwin@eng.auburn.edu
Industrial Eng
Vernon E. Unger
Phone: (205)844-1400   Fax: (205)844-1381
EMail: eunger@eng.auburn.edu
Mechanical Eng
John S. Goodling
Phone: (205)844-3308   Fax: (205)844-3307
EMail: goodling@eng.auburn.edu
Aerospace Eng
John E. Cochran, Jr.
Phone: (205)844-6802   Fax: (205)844-6803
EMail: jcochran@eng.auburn.edu
Agricultural Eng
Paul K. Turnquist
Phone: (205)844-4180   Fax: (205)844-3530
Computer Science & Eng
Stephen B. Seidman
Phone: (205)844-4330   Fax: (205)844-6329
EMail: seidman@eng.auburn.edu
Textile Eng
William K. Walsh
Phone: (205)844-5453   Fax: (205)844-4068
EMail: wwalsh@eng.auburn.edu

Admissions Office Contact
Charles F. Reeder
Phone: (204)844-4080   Fax: (205)844-6436

Financial Aid Office Contact
Clark Aldridge, *Director*
Phone: (205)844-4723   Fax: (205)844-6085

013 Baylor U

Head of the Institution
Herbert H. Reynolds
Phone: (817)755-1311

Head of Engineering
James D. Bargainer
Phone: (817)755-3871   Fax: (817)755-2716
EMail: jim_bargainer@engineering.baylor.edu

Department Chair(s)
Division of Eng
James M. Warren
Phone: (817)755-3871   Fax: (817)755-2716
EMail: jim_warren@engineering.baylor.edu

Admissions Office Contact
Herman D. Thomas
Phone: (817)755-1811

Financial Aid Office Contact
William J. Dube, *Dean for Acad Schlrshps & Student Financial Aid*
Phone: (817)755-2611

014 Boston U

Head of the Institution
John Silber
Phone: (617)353-2200   Fax: (617)353-9764

Head of Engineering
Charles DeLisi
Phone: (617)353-2800   Fax: (617)353-6322
EMail: in%"delisi@buenga.bu.edu"

Associate/Assistant Deans
Thomas J. Kerr, *Associate Dean of Engineering Undergraduate Academic Affairs*
Phone: (617)353-6447   Fax: (617)353-5769
William Taft, *Associate Dean of Graduate Engineering Academic Affairs*
Phone: (617)353-9760

Department Chair(s)
Electrical, Computer & Systems Eng
Thomas G. Kincaid
Phone: (617)353-2806   Fax: (617)353-6322
EMail: in%"tgk@buenga.bu.edu"
Aerospace & Mechanical Eng
John Baillieul
Phone: (617)353-4841   Fax: (617)353-5866
EMail: in%"johnb@buenga.bu.edu"
Biomedical Eng
Herbert Voigt
Phone: (617)353-2833   Fax: (617)353-6322
EMail: in%"hfv@buenga.bu.edu"
Manufacturing Eng
Peter Z. Bulkeley
Phone: (617)353-2837   Fax: (617)353-6322
EMail: in%'pzb@buenga.bu.edu'

Financial Aid Office Contact
Barbara Tornow
Phone: (617)353-2965   Fax: (617)353-7300

015 Bradley U

Head of the Institution
John R. Brazil
Phone: (309)677-3167   Fax: (309)677-2330

Head of Engineering
John E. Francis
Phone: (309)677-2721   Fax: (309)677-3670

Associate/Assistant Deans
James G. Seckler, *Associate Dean*
Phone: (309)677-2720   Fax: (309)677-3670
Robert J. Podlasek, *Assistant Dean*
Phone: (309)677-2714   Fax: (309)677-3670

Department Chair(s)
Civil Eng & Construction
Amir W. Al-Khafaji
Phone: (309)677-2941   Fax: (309)677-3670
Electrical & Computer Eng & Tech
Brian D. Huggins
Phone: (309)677-2732   Fax: (309)677-3670

Engineering & Engineering Technology Personnel

Industrial Eng
Kalikathon S. Krishnamoorthi
Phone: (309)677-2740   Fax: (309)677-3670
Mechanical Eng
Max A. Wessler
Phone: (309)677-2711   Fax: (309)677-3670
Manufacturing
Arnold E. Ness
Phone: (309)677-2938   Fax: (309)677-3670
Admissions Office Contact
Gary Bergman
Phone: (800)447-6460   Fax: (309)677-2797
Financial Aid Office Contact
David Pardieck, *Director of Financial Aid*
Phone: (309)677-3089   Fax: (309)677-2798

016   U of Bridgeport
Head of the Institution
Edward Eigel
Phone: (203)576-4665   Fax: (203)576-4983
Head of Engineering
Bruce C. Skinner
Phone: (203)576-4111   Fax: (203)576-4766
EMail: skinner@cse.bridgeport.edu
Department Chair(s)
Electrical Eng
Wenelin D. Janeff
Phone: (203)576-4296
Mechanical Eng
Richard D. Schile
Phone: (203)576-4343
Computer Science & Eng
Stephen E. Grodzinsky
Phone: (203)576-4145   Fax: (203)576-4766
EMail: grodzinsky@cse.bridgeport.edu
Management Eng
Paul T. Bauer
Phone: (203)576-4379   Fax: (203)576-4766
EMail: bauer@cse.bridgeport.edu
Admissions Office Contact
Barbara Maryak
Phone: (203)576-4560   Fax: (203)576-4941
Financial Aid Office Contact
Bessie Phakias
Phone: (203)576-4568

017   Brigham Young U
Head of the Institution
Rex E. Lee
Phone: (802)378-2521
Head of Engineering
L. Douglas Smoot
Phone: (801)378-4326   Fax: (801)378-5705
Associate/Assistant Deans
Steven E. Benzley, *Associate Dean*
Phone: (801)378-4326   Fax: (801)378-5705
John J. Kunzler, *Associate Dean*
Phone: (801)378-4326   Fax: (801)378-5705
David K. Anthony, *Assistant Dean*
Phone: (801)378-5780   Fax: (801)378-5705
Ronald E. Terry, *Assistant Dean*
Phone: (801)378-4326   Fax: (801)378-5705
Department Chair(s)
Chemical Eng
Richard L. Rowley
Phone: (801)378-2586   Fax: (801)378-7799
Civil Eng
S. Olani Durrant
Phone: (801)378-2811   Fax: (801)378-2478
Electrical & Computer Eng
David J. Comer
Phone: (801)378-4012   Fax: (801)378-6586
EMail: cdav@ee.byu.edu
Mechanical Eng
Geoffrey J. Germane
Phone: (801)378-2625   Fax: (801)378-5037

Manufacturing Eng & Eng Tech
A. Brent Strong
Phone: (801)378-6300   Fax: (801)378-7575
Admissions Office Contact
Tom M. Gourley
Phone: (801)378-4264   Fax: (801)378-2507
Financial Aid Office Contact
Norman B. Finlinson, *Director*
Phone: (801)378-4104

018   Bucknell U
Head of the Institution
Gary A. Sojka
Phone: (717)524-1511   Fax: (717)524-3760
Head of Engineering
Thomas P. Rich
Phone: (717)524-3711   Fax: (717)524-3760
Associate/Assistant Deans
Trudy B. Cunningham, *Associate Dean of Engineering*
Phone: (717)524-3705   Fax: (717)524-3760
Department Chair(s)
Chemical Eng
William E. King
Phone: (717)524-1114   Fax: (717)524-3760
Civil Eng
Jai B. Kim
Phone: (717)524-1112   Fax: (717)524-3760
Electrical Eng
Maurice F. Aburdene
Phone: (717)524-1234   Fax: (717)524-3760
Mechanical Eng
James N. Zaiser
Phone: (717)524-3193   Fax: (717)524-3760
Computer Science
Jerud J. Mead
Phone: (717)524-1394   Fax: (717)524-3760
Admissions Office Contact
Mark D. Davies
Phone: (717)524-1101   Fax: (717)524-3760
Financial Aid Office Contact
Ronald T. Laszewski, *Director of Financial Aid*
Phone: (717)524-1331   Fax: (717)524-3760

019   U of Cal at Berkeley
Head of the Institution
Chang-Lin Tien, *Chancellor*
Phone: (510)642-7464
Head of Engineering
David A. Hodges
Phone: (510)642-5771
Associate/Assistant Deans
William C. Webster, *Associate Dean*
Phone: (510)642-7594   Fax: (510)643-8653
Steven E. Schwarz, *Associate Dean for Undergraduate Matters & Curriculum*
Phone: (510)642-7594   Fax: (510)643-8653
H. Frank Morrison, *Associate Dean for Special Programs*
Phone: (510)642-1734
George Leitmann, *Associate Dean, Research Services*
Phone: (510)642-3984
Edwin R. Lewis, *Associate Dean for Interdisciplinary Studies*
Phone: (510)642-8790
Department Chair(s)
Civil Eng
Keith C. Crandall
Phone: (510)642-3261
Electrical Eng & Computer Sciences
Paul R. Gray
Phone: (510)642-3214
Industrial Eng & Operations Research
Shmuel S. Oren
Phone: (510)642-3424

Mechanical Eng
David B. Bogy
Phone: (510)642-2544
Materials Science & Mineral Eng
Ronald Gronsky
Phone: (510)642-3801
Naval Architecture & Offshore Eng
Ronald W. Yeung
Phone: (510)642-5464
Nuclear Eng
T. Kenneth Fowler
Phone: (510)642-5010
Chemical Eng
Morton Denn
Phone: (510)643-8749

020   U of Cal, Davis
Head of the Institution
Theodore L. Hullar
Phone: (916)752-2066   Fax: (916)752-2400
Head of Engineering
Mohammed S. Ghausi
Phone: (916)752-0554   Fax: (916)752-8058
EMail: msghausi@ucdavis.edu
Associate/Assistant Deans
Billy R. Sanders, *Assistant Dean*
Phone: (916)752-0560   Fax: (916)752-8058
James F. Shackelford, *Associate Dean, Undergraduate Studies*
Phone: (916)752-0556   Fax: (916)752-8058
Benjamin J. McCoy, *Associate Dean, Research*
Phone: (916)752-1435   Fax: (916)752-8058
Zuhair A. Munir, *Associate Dean, Graduate Studies*
Phone: (916)752-0559   Fax: (916)752-8058
Department Chair(s)
Chemical Eng
Brian G. Higgins
Phone: (916)752-0400   Fax: (916)752-1031
EMail: bghiggins@ucdavis.edu
Civil Eng
Melvin R. Ramey
Phone: (916)752-0586   Fax: (916)752-7872
EMail: mrramey@ucdavis.edu
Electrical & Computer Eng
S.L. Hakimi
Phone: (916)752-0583   Fax: (916)752-8428
EMail: slhakimi@ucdavis.edu
Mechanical, Aeronautical & Materials Eng
Allan A. McKillop
Phone: (916)752-0580   Fax: (916)752-4158
EMail: aamckillop@ucdavis.edu
Biological & Agricultural Eng
David Hills
Phone: (916)752-0102   Fax: (916)752-2640
EMail: ddhills@ucdavis.edu
Computer Science & Eng
Peter Linz
Phone: (916)752-7004   Fax: (916)752-4767
EMail: linz@cs.ucdavis.edu
Applied Science
Neville C. Luhmann, Jr.
Phone: (510)422-9787   Fax: (510)422-8681
Admissions Office Contact
Gary Tudor
Phone: (916)752-2971   Fax: (916)752-1280
Financial Aid Office Contact
Ronald W. Johnson, *Director*
Phone: (916)752-2390   Fax: (916)752-7339

021   U of Cal, Irvine
Head of the Institution
Dennis Smith
Phone: (714)856-5111   Fax: (714)725-2087
Head of Engineering
William A. Sirignano
Phone: (714)856-6002   Fax: (714)856-7966

Associate/Assistant Deans
Allen R. Stubberud, *School Administrator, Research & Graduate Studies*
Phone: (714)856-4842   Fax: (714)856-7966
Harry H. Tan, *Associate Dean, Undergraduate Student Affairs*
Phone: (714)856-6737   Fax: (714)856-7966
Department Chair(s)
Biochemical Eng
Henry C. Lim
Phone: (714)856-8290   Fax: (714)725-2541
Civil Eng
Medhat A. Haroun
Phone: (714)856-5333   Fax: (714)725-2117
Electrical & Computer Eng
Leonard A. Ferrari
Phone: (714)856-5689   Fax: (714)856-4152
Mechanical & Aerospace Eng
William E. Schmitendorf
Phone: (714)856-5406   Fax: (714)856-8585
Admissions Office Contact
Ann Williams
Phone: (714)856-6703
Financial Aid Office Contact
Otto W. Reyer, *Director*
Phone: (714)856-6261   Fax: (714)856-4876

022   U of Cal, Los Angeles
Head of the Institution
Charles E. Young, *Chancellor*
Phone: (310)825-2121   Fax: (310)206-6030
Head of Engineering
A.R. Frank Wazzan
Phone: (310)206-8245   Fax: (310)206-4061
Associate/Assistant Deans
Stephen E. Jacobsen, *Associate Dean - Student Affairs*
Phone: (310)825-2941   Fax: (310)206-4061
Alan N. Willson, Jr., *Associate Dean - Faculty Relations*
Phone: (310)206-7875   Fax: (310)206-4061
Alan N. Willson, Jr., *Associate Dean - Planning & Research*
Phone: (310)206-7875   Fax: (310)206-4061
Michael K. Stenstrom, *Assistant Dean - Computing Resources*
Phone: (310)825-1408   Fax: (310)206-4061
Department Chair(s)
Chemical Eng
David T. Allen
Phone: (310)206-0300
Civil Eng
Michael K. Stenstrom
Phone: (310)825-1050
Electrical Eng
Nicolaos G. Alexopoulos
Phone: (310)825-1027
Computer Science Department
Leonard Kleinrock
Phone: (310)825-8878
Materials Science & Eng
Kanji Ono
Phone: (310)825-5233
Mechanical, Aerospace & Nuclear Eng
Jason L. Speyer
Phone: (310)206-4451
Admissions Office Contact
Stephen E. Jacobsen
Phone: (310)825-2941   Fax: (310)825-4061
Financial Aid Office Contact
Lawrence W. Burt
Phone: (310)206-0400   Fax: (310)206-1728

023   U of Cal, San Diego
Head of the Institution
Richard C. Atkinson, *Chancellor*
Phone: (619)534-3135   Fax: (619)534-6523

Head of Engineering
M. Lea Rudee
Phone: (619)534-4575   Fax: (619)534-4771
EMail: rudee@ucsd.edu

Department Chair(s)
Electrical Eng
Manuel Rotenberg
Phone: (619)534-2726   Fax: (619)534-2486
EMail: mrotenberg@ucsd.edu

Applied Mechanics & Eng Sciences
David R. Miller
Phone: (619)534-0113   Fax: (619)534-7078
EMail: kounaves@ames.ucsd.edu

Computer Science & Eng
S. Gill Williamson
Phone: (619)534-1126   Fax: (619)534-7029
EMail: gwilliamson@ucsd.edu

Financial Aid Office Contact
Thomas M. Rutter, *Director*
Phone: (619)534-3800   Fax: (619)534-5459

024   U of Cal, Santa Barbara

Head of the Institution
Barbara S. Uehling
Phone: (805)893-2231   Fax: (805)893-4445

Head of Engineering
Venkatesh Narayanamurti
Phone: (805)893-3141   Fax: (805)893-8124

Associate/Assistant Deans
Roger C. Wood, *Associate Dean for Academic Affairs*
Phone: (805)893-4512   Fax: (805)893-8124
Jacqueline A. Hynes, *Assistant to the Dean for Undergraduate Studies*
Phone: (805)893-3885   Fax: (805)893-8124
W. Henry Weinberg, *Associate Dean for Graduate Programs & Coll Development*
Phone: (805)893-4802   Fax: (805)893-8124

Department Chair(s)
Chemical & Nuclear Eng
L. Gary Leal
Phone: (805)893-8510   Fax: (805)893-4731

Electrical & Computer Eng
Evelyn L. Hu
Phone: (805)893-3821   Fax: (805)893-3262

Mechanical & Environmental Eng
Robert M. McMeeking
Phone: (805)893-8434   Fax: (805)893-8651

Materials Eng
David R. Clarke
Phone: (805)893-8275   Fax: (805)893-8486

Computer Science
John L. Bruno
Phone: (805)893-4321   Fax: (805)893-8553

Admissions Office Contact
William J. Villa
Phone: (805)893-2881

Financial Aid Office Contact
Ron Andrade, *Acting Director of Student Financial Services*
Phone: (805)893-2432

025   U of Cal, Santa Cruz

Head of the Institution
Patrick E. Mantey
Phone: (408)459-2158   Fax: (408)459-4829

Head of Engineering
Patrick E. Mantey
Phone: (408)459-2158   Fax: (408)459-4829
EMail: mantey@cse.ucsc.edu

Associate/Assistant Deans
Glen G. Langdon, *Undergraduate Director*
Phone: (408)459-2212   Fax: (408)459-4829
Tracy Larrabee, *Assistant Undergraduate Director*
Phone: (408)459-3476   Fax: (408)459-4829
Virginia P. Carrillo, *Student Affairs Officer*
Phone: (408)459-2868   Fax: (408)459-4829

Department Chair(s)
Computer Eng
Patrick E. Mantey
Phone: (408)459-2158   Fax: (408)459-4829
EMail: mantey@cse.ucsc.edu

Admissions Office Contact
Patrick E. Mantey
Phone: (408)459-2320   Fax: (408)459-4829

Financial Aid Office Contact
Esperanza L. Nee, *Director of Financial Aid*
Phone: (408)459-2963

026   Cal Inst of Tech

Head of the Institution
Thomas E. Everhart
Phone: (818)356-6301   Fax: (818)449-9374

Head of Engineering
John H. Seinfeld
Phone: (818)356-4100   Fax: (818)585-1729
EMail: john_seinfeld@starbase1.caltech.edu

Department Chair(s)
Chemical Eng
Manfred Morari
Phone: (818)356-4186   Fax: (818)568-8743
EMail: mm@imc.caltech.edu

Civil Eng
James K. Knowles
Phone: (818)356-4135   Fax: (818)568-2719
EMail: jkknepem@cadre2.caltech.edu

Electrical Eng
Robert J. McEliece
Phone: (818)356-3891

Mechanical Eng
Allan J. Acosta
Phone: (818)356-4110   Fax: (818)568-2719

Applied Mathematics
Donald S. Cohen
Phone: (818)356-4559   Fax: (818)449-2677
EMail: dsc@ama-1.caltech.edu

Environmental Eng Science
Norman H. Brooks
Phone: (818)356-4404

Graduate Aeronautical Laboratories
Hans G. Hornung
Phone: (818)356-4551   Fax: (818)449-2677
EMail: hans@shock.caltech.edu

Applied Mechanics
James K. Knowles
Phone: (818)356-4135   Fax: (818)568-2719
EMail: jkknepem@cadre2.caltech.edu

Applied Physics
Noel R. Corngold
Phone: (818)356-4129

Eng Science
Theodore Y. Wu
Phone: (818)356-4230

Materials Science
Brent T. Fultz
Phone: (818)356-4426

Computer Science
Jan J.L. Van de Snepscheut
Phone: (818)356-4269
EMail: jan.vlsi.cs.caltech.edu

Computation & Neural Systems
Demetri Psaltis
Phone: (818)356-4856

Admissions Office Contact
Carol L. Snow
Phone: (818)356-6341   Fax: (818)564-8136

Financial Aid Office Contact
David J. Levy, *Director of Financial Aid*
Phone: (818)356-6280

027   Cal Poly State U

Head of the Institution
Warren J. Baker
Phone: (805)756-6000   Fax: (805)756-1129

Head of Engineering
Peter Y. Lee
Phone: (805)756-2131   Fax: (805)756-6503
EMail: plee@oasis.calpoly.edu

Associate/Assistant Deans
J. Kent Butler, *Student Affairs & Curriculum*
Phone: (805)756-2131   Fax: (805)756-6503
Daniel W. Walsh, *Administration & Planning*
Phone: (805)756-2131   Fax: (805)756-6503

Department Chair(s)
Civil & Environmental Eng
Edward Nowatzki
Phone: (805)756-2137

Electronic & Electrical Eng
Saul Goldberg
Phone: (805)756-2781   Fax: (805)756-1458
EMail: sgoldber@ohm.calpoly.edu

Industrial Eng
Joanne Freeman
Phone: (805)756-2341   Fax: (805)756-5439

Mechanical Eng
Ronald L. Mussulman
Phone: (805)756-1334

Aeronautical Eng
Russell Cummings
Phone: (805)756-2562   Fax: (805)756-2376

Computer Eng
Zane Motteler
Phone: (805)756-2824   Fax: (805)756-2956

Eng Science
Daniel W. Walsh
Phone: (805)756-2131   Fax: (805)756-6503
EMail: dwalsh@zeus.calpoly.edu

Materials Eng
Robert H. Heidersbach
Phone: (805)756-2568   Fax: (805)756-2299

Agricultural Eng
Edgar J. Carnegie
Phone: (805)756-2378   Fax: (805)756-2626

Architectural Eng
Tom Ballew
Phone: (805)756-2281   Fax: (805)756-5986

Financial Aid Office Contact
Diane L. Ryan, *Director of Financial Aid*
Phone: (805)756-2927

028   Cal State U, Chico

Head of the Institution
Robin S. Wilson
Phone: (916)898-5201   Fax: (916)898-5077

Head of Engineering
Gary Z. Watters
Phone: (916)898-5963   Fax: (916)898-5995
EMail: GWATTERS@OAVAXCSUCHICO.EDU

Department Chair(s)
Civil Eng
Thomas C. Ferrara
Phone: (916)898-5342   Fax: (916)898-4576
EMail: TFERRARA@OAVAX.CSUCHICO.EDU

Electrical/Electronic Eng
Orlando Baiocchi
Phone: (916)898-5343   Fax: (916)898-4956
EMail: OBAIOCCHI@OAVAX.CSUCHICO.EDU

Mechanical Eng & Manufacturing
Ramesh Varahamurti
Phone: (916)898-5346   Fax: (916)898-4070
EMail: RVARAHAMURTI@OAVAX.CSUCHICO.EDU

Computer Science & Eng
Orlando S. Madrigal
Phone: (916)898-6442   Fax: (916)898-5995
EMail: OMADRIGAL@OAVAX.CSUCHICO.EDU

Admissions Office Contact
Kenneth C. Edson
Phone: (916)898-6321   Fax: (916)898-6824

Financial Aid Office Contact
David Cook, *Financial Aids Officer*
Phone: (916)898-5065   Fax: (916)898-6824

029   Cal State U, Fresno

Head of the Institution
John D. Welty
Phone: (209)278-2324   Fax: (209)278-4715

Head of Engineering
Elden K. Shaw
Phone: (209)278-2500   Fax: (209)278-7071

Department Chair(s)
Civil & Surveying Eng
Karl Longley
Phone: (209)278-2889   Fax: (209)278-6759

Electrical & Computer Eng
Medhat Ibrahim
Phone: (209)278-2726   Fax: (209)278-6297

Mechanical & Industrial Eng
Walter V. Loscutoff
Phone: (209)278-2328   Fax: (209)278-6759

Computer Science
Henderson Yeung
Phone: (209)278-4638   Fax: (209)278-4197

Admissions Office Contact
Richard Backer
Phone: (209)278-2191   Fax: (209)278-4812

Financial Aid Office Contact
Joseph W. Heuston, *Director of Financial Aid*
Phone: (209)278-6563   Fax: (209)278-7044

030   Cal State U, Long Beach

Head of the Institution
Curtis L. McCray
Phone: (310)985-4121   Fax: (310)985-5584

Head of Engineering
J. Richard Williams
Phone: (310)985-5123   Fax: (310)985-7561

Associate/Assistant Deans
Mihir K. Das, *Associate Dean for Instruction*
Phone: (310)985-8032   Fax: (310)985-7561
Ralph C. Cooper, *Associate Dean for Research*
Phone: (213)985-8029   Fax: (213)985-7561
James Ary, *Associate Dean for Industrial Relations & Development*
Phone: (310)985-8053   Fax: (310)985-7561

Department Chair(s)
Chemical Eng
Lloyd Hile
Phone: (310)985-1508   Fax: (310)985-7561

Civil Eng
Peter Cowan
Phone: (310)985-5135   Fax: (310)985-7561

Electrical Eng
Radhe Das
Phone: (310)985-8048   Fax: (310)985-7561

Mechanical Eng
Ortwin Ohtmer
Phone: (310)985-1518   Fax: (310)985-4408

Aerospace Eng
Cebeci Tuncer
Phone: (310)985-4551   Fax: (310)985-7561

Computer Eng & Computer Science
Michael K. Mahoney
Phone: (310)985-1550   Fax: (310)985-7561

Eng Tech
Tesfai Goitom
Phone: (310)985-8020   Fax: (310)985-7561

Admissions Office Contact
Fay Denny
Phone: (310)985-5471   Fax: (310)985-8887

Financial Aid Office Contact
Gloria J. Kapp, *Director*
Phone: (310)985-8403

031   Cal State U, Los Angeles

Head of the Institution
James M. Rosser
Phone: (213)343-3030   Fax: (213)343-2670

Engineering & Engineering Technology Personnel

Head of Engineering
Raymond B. Landis
Phone: (213)343-4500   Fax: (213)343-4555
EMail: rlandis@calstatela.edu

Associate/Assistant Deans
Don M. Maurizio, *Associate Dean*
Phone: (213)343-4510   Fax: (213)343-4555

Department Chair(s)
Civil Eng
Raymond I. Jeng
Phone: (213)343-4450   Fax: (213)343-4555
EMail: rjeng@calstatela.edu
Electrical Eng
Martin S. Roden
Phone: (213)343-4470   Fax: (213)343-4555
EMail: mroden@calstatela.edu
Mechanical Eng
Neda S. Fabris
Phone: (213)343-4490   Fax: (213)343-4555
EMail: nfabris@calstatela.edu

Admissions Office Contact
Kevin M. Browne
Phone: (213)343-3901   Fax: (213)343-2670

Financial Aid Office Contact
Vincent Deanda, *Director*
Phone: (213)343-3240   Fax: (213)343-2670

032 Cal State U, Northridge

Head of the Institution
Blenda J. Wilson
Phone: (818)885-2121   Fax: (818)885-2254

Head of Engineering
Diane L. Schwartz
Phone: (818)885-4501   Fax: (818)885-2140
EMail: DSCHWARTZ@VAX.CSUN.EDU

Associate/Assistant Deans
Gregg Dixon, *Associate Dean*
Phone: (818)885-2183   Fax: (818)885-2140

Department Chair(s)
Civil & Industrial Eng & Applied Mechanics
Miguel Macias
Phone: (818)885-2166   Fax: (818)885-2140
EMail: CIAM.OFF@VAX.CSUN.EDU
Electrical & Computer Eng
Sharlene Katz
Phone: (818)885-2190   Fax: (818)885-2140
EMail: SKATZ@VAX.CSUN.EDU
Mechanical Eng
William Rivers
Phone: (818)885-2187   Fax: (818)885-2140
EMail: ME.OFF@VAX.CSUN.EDU
Computer Science
Dorothy Miller
Phone: (818)885-3398   Fax: (818)995-2140
EMail: COMPSCI.OFF@VAX.CSUN.EDU

033 Cal State U, Sacramento

Head of the Institution
Donald R. Gerth
Phone: (916)278-7737   Fax: (916)278-6959

Head of Engineering
Donald H. Gillott
Phone: (916)278-6366   Fax: (916)278-5949
EMail: gillottd@ecs.csus.edu

Associate/Assistant Deans
John N. Hester, *Associate Dean, Academic Programs & Research*
Phone: (916)278-6366   Fax: (916)278-5949
John T. Oldenburg, *Associate Dean, Resources & Facilities Planning*
Phone: (916)278-6366   Fax: (916)278-5949
Larry A. Hill, *Assistant Dean*
Phone: (916)278-6366   Fax: (916)278-5949

Department Chair(s)
Electrical & Electronic Eng
Karl E. Stoffers
Phone: (916)278-6873   Fax: (916)278-5949
EMail: Stoffers@ecs.csus.edu

Civil Eng
Vishnu L. Agaskar
Phone: (916)278-6982   Fax: (916)278-5949
EMail: agaskar@ecs.csus.edu
Mechanical Eng
Andrew R. Banta
Phone: (916)278-6624   Fax: (916)278-5949
EMail: Bantaa@ecs.csus.edu
Computer Science
Anne-Louise Radimsky
Phone: (916)278-6834   Fax: (916)278-5949
EMail: Radimsky@ecs.csus.edu
Biomedical Eng Program
John T. Oldenburg
Phone: (916)278-6127   Fax: (916)278-5949
EMail: Oldenbrg@ecs.csus.edu
Construction Management Program
Keith Bisharat
Phone: (916)278-6616   Fax: (916)278-5949
EMail: Bisharak@ecs.csus.edu
Mechanical Eng Tech
Joseph H. Harralson
Phone: (916)278-7081   Fax: (911)278-5949
EMail: Harralsj@ecs.csus.edu
Computer Eng
Ronald W. Becker
Phone: (916)278-6844   Fax: (916)278-5949
EMail: Beckerr@ecs.csus.edu

Admissions Office Contact
Larry Glasmire
Phone: (916)278-7111   Fax: (916)278-5443

Financial Aid Office Contact
Starla Satchell, *Director of Financial Aid*
Phone: (916)278-6554

034 Cal State Poly, Pomona

Head of the Institution
Bob H. Suzuki
Phone: (909)869-2290

Head of Engineering
Edward C. Hohmann
Phone: (909)869-2600   Fax: (909)869-4370

Associate/Assistant Deans
Carl E. Rathmann, *Associate Dean*
Phone: (909)869-2600   Fax: (909)869-4370

Department Chair(s)
Chemical Eng
Julie M. Schoenung
Phone: (909)869-2626   Fax: (909)869-4370
Civil Eng
Ronald L. Carlyle
Phone: (909)869-2488   Fax: (909)869-4370
Electrical & Computer Eng
Richard H. Cockrum
Phone: (909)869-2511   Fax: (909)869-4370
Industrial & Manufacturing Eng
Phillip R. Rosenkrantz
Phone: (909)869-2555   Fax: (909)869-4370
Mechanical Eng
George F. Engelke
Phone: (909)869-2575   Fax: (909)869-4370
Aerospace Eng
Paul A. Lord
Phone: (909)869-2470   Fax: (909)869-4370
Agricultural Eng
Joe Y. Hung
Phone: (909)869-2221   Fax: (909)869-4370

Admissions Office Contact
Joseph C. Marshall
Phone: (909)869-2000

Financial Aid Office Contact
Al Andino, *Director of Financial Aid*
Phone: (714)869-3700

035 Calvin Coll

Head of the Institution
Anthony J. Diekema
Phone: (616)957-6100   Fax: (616)957-8551

Head of Engineering
David Hoekema
Phone: (616)957-6442   Fax: (616)957-8551

Department Chair(s)
Eng
Robert J. Hoeksema
Phone: (616)957-6050   Fax: (616)957-8551
EMail: HOER@CALVIN.EDU

Admissions Office Contact
Thomas E. McWhertor
Phone: (800)688-0122   Fax: (616)957-8551

Financial Aid Office Contact
Wayne K. Hubers, *Director*
Phone: (616)957-6134   Fax: (616)957-8551

036 Carnegie Mellon U

Head of the Institution
Robert Mehrabian
Phone: (412)268-2200   Fax: (412)268-2330

Head of Engineering
Stephen W. Director
Phone: (412)268-2537   Fax: (412)268-6421
EMail: director@orion.ece.cmu.edu

Associate/Assistant Deans
Chris T. Hendrickson, *Associate Dean for Academic Affairs*
Phone: (412)268-2478   Fax: (412)268-6421
Robert P. Kail, *Associate Dean for Undergraduate Studies*
Phone: (412)268-2479   Fax: (412)268-6421
Patricia Laughlin, *Associate Dean for Administration*
Phone: (412)268-5008   Fax: (412)268-6421

Department Chair(s)
Chemical Eng
John L. Anderson
Phone: (412)268-2230   Fax: (412)268-7139
EMail: JA0B@VAXB.CMU.EDU
Civil Eng
Richard G. Luthy
Phone: (412)268-2941   Fax: (412)268-7813
EMail: luthy@ece.cmu.edu
Electrical & Computer Eng
Robert M. White
Phone: (412)268-3545   Fax: (412)268-5787
EMail: white@gauss.ece.cmu.edu
Mechanical Eng
Adnan Akay
Phone: (412)268-2501   Fax: (412)268-3488
EMail: AKAY@ANDREW.CMU.EDU
Eng & Public Policy
Granger Morgan
Phone: (412)268-2672   Fax: (412)268-3757
EMail: GM5D@ANDREW.CMU.EDU
Materials Science & Eng
Paul Wynblatt
Phone: (412)268-2700   Fax: (412)268-7596
EMail: PW01@CMCCVB.CMU.EDU
Manufacturing Eng
Harold Paxton
Phone: (412)268-2947   Fax: (412)268-7596
EMail: HP0L@ANDREW.CMU.EDU

Financial Aid Office Contact
Walter C. Cathie, *Associate V.P. for Financial Resources*
Phone: (412)268-2068   Fax: (412)268-7837

037 Case Western Reserve U

Head of the Institution
Agnar Pytte
Phone: (216)368-4344   Fax: (216)368-5861

Head of Engineering
Thomas P. Kicher
Phone: (216)368-4436   Fax: (216)368-6939
EMail: Internet:tpk@po.cwru.edu

Associate/Assistant Deans
W. Sanford Topham, *Associate Dean*
Phone: (216)368-2928   Fax: (216)368-4718

Department Chair(s)
Chemical Eng
J. Adin Mann
Phone: (216)368-4150   Fax: (368)368-3016
EMail: Internet:jam12@po.cwru.edu
Civil Eng
Adel S. Saada
Phone: (216)368-2952   Fax: (216)368-5229
EMail: Internet:axs31@po.cwru.edu
Electrical Eng & Applied Physics
Donald E. Schuele
Phone: (216)368-4085   Fax: (216)368-2668
EMail: Internet:des3@po.cwru.edu
Systems Eng
Kenneth A. Loparo
Phone: (216)368-4053   Fax: (216)368-3123
EMail: Internet:kal4@po.cwru.edu
Mechanical & Aerospace Eng
Joseph M. Prahlr
Phone: (216)368-2940   Fax: (216)368-6445
EMail: Internet:jmp@po.cwru.edu
Biomedical Eng
Gerald M. Saidel
Phone: (216)368-4094   Fax: (216)368-4969
EMail: Internet:gms3@po.cwru.edu
Computer Eng & Science
Lee J. White
Phone: (216)368-2802   Fax: (216)368-2801
EMail: Internet:ljw@po.cwru.edu
Macromolecular Science
John Blackwell
Phone: (216)368-4172   Fax: (216)368-4202
EMail: Internet:jxb6@po.cwru.edu
Material Science & Eng
Joe H. Payer
Phone: (216)368-4230   Fax: (216)368-8618
EMail: Internet:jmp@po.cwru.edu
Systems Eng
Kenneth A. Loparo
Phone: (216)368-4053   Fax: (216)368-3123
EMail: Internet:kal4@po.cwru.edu

Admissions Office Contact
William T. Conley
Phone: (216)368-4450   Fax: (216)368-5111

Financial Aid Office Contact
Donald Chenelle, *Director of U Financial Aid*
Phone: (216)368-4530   Fax: (216)368-5054

038 The Catholic U of America

Head of the Institution
Patrick Ellis
Phone: (202)319-5100   Fax: (202)319-4441

Head of Engineering
John J. McCoy
Phone: (202)319-5160   Fax: (202)319-4499

Associate/Assistant Deans
Timothy W. Kao, *Associate Dean*
Phone: (202)319-5177   Fax: (202)319-4499
John J. Gilheany, *Assistant Dean*
Phone: (202)319-5170   Fax: (202)319-4499

Department Chair(s)
Civil Eng
Timothy W. Kao
Phone: (202)319-5163   Fax: (202)319-4499
Electrical Eng
Robert Meister
Phone: (202)319-5193   Fax: (202)319-4499
Mechanical Eng
John J. Gilheany
Phone: (201)319-5170   Fax: (202)319-4499

Admissions Office Contact
David R. Gibson
Phone: (202)319-5305   Fax: (202)319-5831

Financial Aid Office Contact
Doris Torosian, *Director of Financial Aid*
Phone: (202)319-5307

Engineering & Engineering Technology Personnel

039 U of Central Florida
Head of the Institution
John C. Hitt
Phone: (407)823-1823   Fax: (407)823-5407
Head of Engineering
Gary E. Whitehouse
Phone: (407)823-2156   Fax: (407)823-5483
EMail: WHITEHSE @ UCF1VM.CC.UCF.EDU
Associate/Assistant Deans
Richard N. Miller, *Associate Dean for Undergraduate Affairs*
Phone: (407)823-2455   Fax: (407)823-5483
Stephen L. Rice, *Associate Dean for Research & Graduate Affairs*
Phone: (407)823-2156   Fax: (407)823-5483
Department Chair(s)
Civil & Environmental Eng
A.E. Radwan
Phone: (407)823-2841   Fax: (407)823-3315
EMail: AERADWAN @ UCF1VM.CC.UCF.EDU
Electrical & Computer Eng
Nicolaos S. Tzannes
Phone: (407)823-2786   Fax: (407)823-5835
EMail: TZANNES @ UCF1VM.CC.UCF.EDU
Industrial Eng
William W. Swart
Phone: (407)823-2204   Fax: (407)823-3413
EMail: SWART @ UCF1VM.CC.UCF.EDU
Mechanical & Aerospace Eng
David W. Nicholson
Phone: (407)823-2416   Fax: (407)823-0208
EMail: NICHOLSN @ UCF1VM.CC.UCF.EDU
Admissions Office Contact
John F. Bush
Phone: (407)823-3000   Fax: (407)823-5652
Financial Aid Office Contact
Mary H. McKinney, *Director of Financial Aid*
Phone: (407)823-2827   Fax: (407)823-5652

040 Central State U
Head of the Institution
Arthur E. Thomas
Phone: (513)376-6332   Fax: (513)376-6530
Head of Engineering
Melvin A. Johnson, Jr.
Phone: (513)376-6324   Fax: (513)376-6530
Department Chair(s)
Manufacturing Eng
William A. Grissom
Phone: (513)376-6435   Fax: (513)376-6598
Admissions Office Contact
Robert E. Johnson
Phone: (513)376-6348   Fax: (513)376-6648
Financial Aid Office Contact
Sunny Terrell, *Executive Director, Financial Aid*
Phone: (513)376-6575   Fax: (513)376-6530

041 Christian Brothers U
Head of the Institution
T. "Brother" Drahmann, FSC
Phone: (901)722-0250   Fax: (901)722-0494
Head of Engineering
Ray W. Brown
Phone: (901)722-0408   Fax: (901)722-0494
Department Chair(s)
Chemical Eng
Henry H. Luttrell
Phone: (901)722-0412   Fax: (901)722-0494
Civil Eng
Siripong Malasri
Phone: (901)722-0419   Fax: (901)722-0494
Electrical Eng
L. "Brother" Althaus
Phone: (901)722-0411   Fax: (901)722-0494
Mechanical Eng
Michael Santi
Phone: (901)722-0572   Fax: (901)722-0494
Eng Management
Craig Blackman
Phone: (901)722-0283   Fax: (901)722-0494
Admissions Office Contact
M. "Brother" Smith, FSC
Phone: (901)722-0205   Fax: (901)722-0494
Financial Aid Office Contact
Sandi Mayo, *Director of Financial Aid*
Phone: (901)722-0306   Fax: (901)722-0494

042 U of Cincinnati
Head of the Institution
Joseph A. Steger
Phone: (513)556-2201   Fax: (513)556-3010
Head of Engineering
Constantine N. Papadakis
Phone: (513)556-2933   Fax: (513)556-3626
EMail: CONSTANTINE.PAPADAKIS@UC.EDU
Associate/Assistant Deans
James F. McDonough, *Associate Dean for Academic Affairs*
Phone: (513)556-5418   Fax: (513)556-3626
Ravi K. Jain, *Associate Dean for Research & International Engineering*
Phone: (513)556-5438   Fax: (513)556-3626
Robert K. Michael, *Assistant Dean (Undergraduate Advising)*
Phone: (513)556-3465   Fax: (513)556-3626
Donald E. Hagedorn, *Assistant Dean (Undergraduate Advising)*
Phone: (513)556-5424   Fax: (513)556-3626
Edward N. Prather, *Assistant Dean for Emerging Ethnic Engineers Program*
Phone: (513)556-1164   Fax: (513)556-6488
Department Chair(s)
Chemical Eng
Sun-Tak Hwang
Phone: (513)556-2761   Fax: (513)556-3473
EMail: SHWANG@UCENG@UC.EDU
Civil & Environmental Eng
Issam A. Minkarah
Phone: (513)556-3645   Fax: (513)556-2599
Electrical & Computer Eng
Vik J. Kapoor
Phone: (513)556-4461   Fax: (513)556-7326
EMail: VKAPOOR@UC.ENG.EDU
Mechanical, Industrial, & Nuclear Eng
Roy E. Eckart
Phone: (513)556-2739   Fax: (513)556-3390
Aerospace Eng & Eng Mechanics
George J. Simitses
Phone: (513)556-3548   Fax: (513)556-5038
Materials Science & Eng
Relva C. Buchanan
Phone: (513)556-3096   Fax: (513)556-2569
Mechanical, Industrial, & Nuclear Eng
Roy E. Eckart
Phone: (513)556-2739   Fax: (513)556-5038
Admissions Office Contact
Rudolph Jones
Phone: (513)556-1100   Fax: (513)556-1105
Financial Aid Office Contact
James Williams, *Director of Student Financial Aid Office*
Phone: (513)556-6982

043 The Citadel
Head of the Institution
Claudius E. Watts, III
Phone: (803)792-5012   Fax: (803)792-6767
Associate/Assistant Deans
Charles Lindbergh, *Associate Dean of Engineering & Program Development*
Phone: (803)792-5083   Fax: (803)792-6328
Department Chair(s)
Civil Eng
Charles Lindbergh
Phone: (803)792-5083   Fax: (803)792-6328
Electrical Eng
Harold W. Askins
Phone: (803)792-5057
Admissions Office Contact
Wallace I. West
Phone: (803)792-5230   Fax: (803)792-7084
Financial Aid Office Contact
Hank M. Fuller, *Director of Financial Aid & Scholarships*
Phone: (803)792-5187   Fax: (803)792-7084

044 CUNY-City Coll
Head of the Institution
Augusta Souza-Kappner, *Acting*
Phone: (212)650-7285   Fax: (212)650-7680
Head of Engineering
Charles B. Watkins
Phone: (212)650-5435   Fax: (212)650-5768
EMail: ENGACBW
Associate/Assistant Deans
Shee-Ming Chen, *Associate Dean, Undergraduate Engineering Studies*
Phone: (212)650-8020
Gerard G. Lowen, *Associate Dean, Engineering Graduate Studies*
Phone: (212)650-8030
Leslie Isaacs, *Assistant Dean for Management & Computing*
Phone: (212)650-7146   Fax: (212)650-5768
Department Chair(s)
Chemical Eng
Robert A. Graff
Phone: (212)650-7232   Fax: (212)650-6660
Civil Eng
Charles A. Miller
Phone: (212)650-8000   Fax: (212)650-6965
EMail: Miller@CE-Mail.ENGR.CCNY.CUNY.EDU
Electrical Eng
Samir Ahmed
Phone: (212)650-7248   Fax: (212)650-8249
Mechanical Eng
Zeev Dagan
Phone: (212)650-5220   Fax: (212)650-8013
EMail: Dagan@MECL2C6.ENGR.CCNY.CUNY.EDU
Computer Science
Gary Bloom
Phone: (212)650-6631   Fax: (212)650-6184
EMail: CSCGSB@CCNYVME
Admissions Office Contact
Nancy P. Campbell
Phone: (212)650-6448   Fax: (212)650-6417
Financial Aid Office Contact
Thelma R. Mason, *Diirector, Financial Aid*
Phone: (212)650-6656   Fax: (212)650-5829

045 CUNY-Staten Island
Head of the Institution
Edmond L. Volpe
Phone: (718)390-7940   Fax: (718)273-0533
Head of Engineering
Fred R. Naider
Phone: (718)390-7925   Fax: (718)273-0533
Department Chair(s)
Applied Sciences
William Monaghan
Phone: (718)390-6521   Fax: (718)273-0533
Admissions Office Contact
Panagiotis Razelos
Phone: (718)390-7972   Fax: (718)273-0533
Financial Aid Office Contact
Sherman Whipkey, *Director of Student Financial Assistance*
Phone: (718)390-7760

046 Clarkson U
Head of the Institution
Richard H. Gallagher
Phone: (315)268-6444   Fax: (315)268-3872
Head of Engineering
William R. Wilcox
Phone: (315)268-6446   Fax: (315)268-3841
EMail: SOE1@CLVM
Department Chair(s)
Chemical Eng
R. Shankar Subramanian
Phone: (315)268-6648   Fax: (315)268-6654
Civil & Environmental Eng
Anthony G. Collins
Phone: (315)268-6529   Fax: (315)268-7985
EMail: COLLINS@SUN.SOE
Electrical & Computer Eng
Paul B. McGrath
Phone: (315)268-6511   Fax: (315)268-7600
EMail: MCGRATH@SUN.SOE
Mechanical & Aeronautical Eng
Goodarz Ahmadi
Phone: (315)268-2322   Fax: (315)268-6438
EMail: AHMADI@SUN.SOE
Admissions Office Contact
Robert A. Croot
Phone: (315)268-6479   Fax: (315)268-5605
Financial Aid Office Contact
Donald T. Mills, *Director of Financial Aid*
Phone: (315)268-6471

047 Clemson U
Head of the Institution
A. Max Lennon
Phone: (803)656-3413   Fax: (803)656-4676
Head of Engineering
Thomas M. Keinath
Phone: (803)656-3202   Fax: (803)656-0859
EMail: tom.keinath@eng.clemson.edu
Associate/Assistant Deans
William B. Barlage, *Associate Dean for Instruction*
Phone: (803)656-3200   Fax: (803)656-0859
A. Wayne Bennett, *Associate Dean for Research & External Affairs*
Phone: (803)656-4236   Fax: (803)656-0859
Walter E. Castro, *Assistant Dean for Undergraduate Affairs*
Phone: (803)656-4440   Fax: (803)656-0859
Department Chair(s)
Chemical Eng
Charles H. Barron
Phone: (803)656-3056   Fax: (803)656-0784
EMail: charles_barron.coe_mail@quick-mail.clemson.edu
Civil Eng
Russell H. Brown
Phone: (803)656-3002   Fax: (803)656-2670
EMail: ce_postmaster.coe_mail@quickmail.clemson.edu
Electrical & Computer Eng
Wilson Pearson
Phone: (803)656-5650   Fax: (803)656-2698
EMail: pearson@coe.eng.a1.clemson.edu
Industrial Eng
Michael S. Leonard
Phone: (803)656-4717   Fax: (803)656-0795
EMail: mike_leonard.coe_mail@quickmail.clemson.edu
Mechanical Eng
Christian E. Przirembel
Phone: (803)656-2482   Fax: (803)656-4435
EMail: chris_przirembel.coe_mail@quick-mail.clemson.edu

Engineering & Engineering Technology Personnel

Agricultural & Biological Eng
Richard O. Hegg
Phone: (803)656-4043   Fax: (803)656-0338
EMail: rhegg@clust1.clemson.edu

Bioengineering
Andreas von Recum
Phone: (803)656-5556   Fax: (803)656-4466
EMail: af_von_recum.coe_mail@quickmail.clemson.edu

Ceramic Eng
Gordon Lewis
Phone: (803)656-3038   Fax: (803)656-1095
EMail: glcre00@clemson.clemson.edu

Environmental Systems Eng
A. Ray Abernathy
Phone: (803)656-5572   Fax: (803)656-0672
EMail: aa@ese.clemson.edu

Admissions Office Contact
Michael R. Heintze
Phone: (803)656-2287   Fax: (803)656-0622

Financial Aid Office Contact
Marvin G. Carmichael, *Director of Financial Aid*
Phone: (803)656-2280   Fax: (803)656-0622

048   Cleveland State U

Head of the Institution
J. Taylor Sims
Phone: (216)687-3544   Fax: (216)687-9333

Head of Engineering
George A. Coulman
Phone: (216)687-2558   Fax: (216)687-9280
EMail: Coulman@CSVAXD.CSUOHIO.EDU (INTERNET)

Associate/Assistant Deans
Edward S. Godleski, *Associate Dean for Undergraduate Affairs*
Phone: (216)687-2557   Fax: (216)687-9280
Chittaranjan Jain, *Associate Dean for Graduate Affairs*
Phone: (216)687-2555   Fax: (216)687-9280

Department Chair(s)
Chemical Eng
E. Earl Graham
Phone: (216)687-2572   Fax: (216)687-9220
EMail: RO174@CSUOHIO

Civil Eng
Paul X. Bellini
Phone: (216)687-2412

Electrical Eng
James H. Burghart
Phone: (216)687-2586

Industrial Eng
Louis A. Tuzi
Phone: (216)687-4664

Mechanical Eng
Edward S. Keshock
Phone: (216)687-9249

Tech
Donald J. Anthan
Phone: (216)687-2559

Admissions Office Contact
Ruth Ann Moyer
Phone: (216)687-3755   Fax: (216)687-9210

Financial Aid Office Contact
William R. Bennett, *Director*
Phone: (216)687-3764   Fax: (216)687-9247

049   U of Colorado at Boulder

Head of the Institution
Judith E. Albino
Phone: (303)492-6201   Fax: (303)492-6772

Head of Engineering
A.R. Seebass
Phone: (303)492-7006   Fax: (303)492-2199

Associate/Assistant Deans
John M. Dunn, *Associate Dean for Academic Affairs*
Phone: (303)492-3646   Fax: (303)492-2199

Lloyd Griffiths, *Associate Dean for Research & Graduate Education*
Phone: (303)492-7427   Fax: (303)492-2199

Department Chair(s)
Chemical Eng
Robert Davis
Phone: (303)492-7314   Fax: (303)492-4341
EMail: DAVIS_R

Civil, Env'l, & Architectural Eng
James P. Heaney
Phone: (303)492-3276   Fax: (303)492-7317
EMail: HEANEY_J

Electrical & Computer Eng
William M. Waite
Phone: (303)492-3511   Fax: (303)492-2758
EMail: WAITE_W

Mechanical Eng
John W. Daily
Phone: (303)492-7110   Fax: (303)492-3498
EMail: DAILY_J

Aerospace Eng Sciences
Robert D. Culp
Phone: (303)492-7974   Fax: (303)492-2825
EMail: CULP_R

Computer Science
Robert B. Schnabel
Phone: (303)492-7554   Fax: (303)492-2844
EMail: SCHNABEL_R

Financial Aid Office Contact
Jerry Sullivan, *Director, Office of Financial Aid*
Phone: (303)492-5091   Fax: (303)492-0838

050   U of Colorado-Col Spr

Head of the Institution
Judith Albino
Phone: (303)492-6201   Fax: (303)492-6772

Head of Engineering
Pieter A. Frick
Phone: (719)593-3226   Fax: (719)593-3542
EMail: pfrick@wetterhorn.uccs.edu

Associate/Assistant Deans
Richard Y.C. Kwor, *Assistant Dean*
Phone: (719)593-3322   Fax: (719)593-3542

Department Chair(s)
Electrical & Computer Eng
Rodger E. Ziemer
Phone: (719)593-3351   Fax: (719)593-3542
EMail: reziemer@happy.uccs.edu

Financial Aid Office Contact
Lee Ingalls, *Director of Financial Aid/Student Employment*
Phone: (719)593-3460   Fax: (719)593-3362

051   U of Colorado at Denver

Head of the Institution
John C. Buechner, *Chancellor*
Phone: (303)556-2643   Fax: (303)556-2164

Head of Engineering
Peter E. Jenkins
Phone: (303)556-2870   Fax: (303)556-2511
EMail: PJenkins@cudnvr.denver.colorado.EDU

Associate/Assistant Deans
Oren G. Strom, *Associate Dean*
Phone: (303)556-2870   Fax: (303)556-2511

Department Chair(s)
Civil Eng
Nien-Yin Chang
Phone: (303)556-2871   Fax: (303)556-2368
EMail: NChang@cudnvr.denver.colorado.EDU

Electrical Eng
Gary Leininger
Phone: (303)556-2872   Fax: (303)556-2511
EMail: GLeininger@cudnvr.denver.colorado.EDU

Mechanical Eng
James Gerdeen
Phone: (303)556-8516   Fax: (303)556-2511
EMail: JGerdeen@cudnvr.denver.colorado.EDU

Computer Science & Eng
John R. Clark
Phone: (303)556-4314   Fax: (303)556-2511
EMail: JClark@cudnvr.denver.colorado.EDU

Admissions Office Contact
Barbara Schneider
Phone: (303)556-3287   Fax: (303)556-4838

Financial Aid Office Contact
Elinore Miller, *Director of Financial Aid*
Phone: (303)556-2886   Fax: (303)556-4822

052   Colorado School of Mines

Head of the Institution
George S. Ansell
Phone: (303)273-3280   Fax: (303)273-3040

Head of Engineering
Franklin D. Schowengerdt
Phone: (303)273-3320   Fax: (303)273-3040
EMail: FSCHOWEN@DEAN

Department Chair(s)
Chemical Eng & Petroleum Refining
Robert M. Baldwin
Phone: (303)273-3720   Fax: (303)273-3730

Eng
Joan P. Gosink
Phone: (303)273-3650   Fax: (303)273-3602

Eng Physics
John U. Trefny
Phone: (303)273-3830   Fax: (303)273-3278

Geology & Geological Eng
Roger Slatt
Phone: (303)273-3800   Fax: (303)273-3278

Geophysical Eng
Phillip R. Romig
Phone: (303)273-3935   Fax: (303)273-3478

Metallurgical & Materials Eng
John J. Moore
Phone: (303)273-3780   Fax: (303)273-3795

Mining Eng
Donald W. Gentry
Phone: (303)273-3701   Fax: (303)273-3719

Petroleum Eng
Craig W. Van Kirk
Phone: (303)273-3740   Fax: (303)273-3278

Admissions Office Contact
William Young
Phone: (303)273-3220   Fax: (303)273-3278

Financial Aid Office Contact
Roger Koester, *Director of Financial Aid*
Phone: (303)273-3301   Fax: (303)273-3278

053   Colorado State U

Head of the Institution
Albert C. Yates
Phone: (303)491-6211   Fax: (303)491-0501

Head of Engineering
Frank A. Kulacki
Phone: (303)491-6603   Fax: (303)491-5569
EMail: fkulacki%dean%engadmin@vines.colostate.edu

Associate/Assistant Deans
Johannes Gessler, *Associate Dean for Undergraduate Studies*
Phone: (303)491-1058   Fax: (303)491-5569
Fred W. Smith, *Associate Dean for Research & Graduate Studies*
Phone: (303)491-8657   Fax: (303)491-8671
David R. Martinez, *Assistant Dean for Minority Affairs*
Phone: (303)491-6220   Fax: (303)491-5569

Department Chair(s)
Agricultural & Chemical Eng
Terry G. Lenz
Phone: (303)491-5252   Fax: (303)491-7369
EMail: lenz@longs.lance.colostate.edu

Civil Eng
Neil S. Grigg
Phone: (303)491-5048   Fax: (303)491-7727
EMail: ngrigg@vines.colostate.edu

Electrical Eng
Jorge I. Aunon
Phone: (303)491-6600   Fax: (303)491-2249
EMail: aunon@longs.lance.colostate.edu

Mechanical Eng
C. Byron Winn
Phone: (303)491-6558   Fax: (303)491-1055
EMail: byron@longs.lance.colostate.edu

Eng Science
Johannes Gessler
Phone: (303)491-1058   Fax: (303)491-5569
EMail: jgessler%ce%engadmin@vines.colostate.edu

Atmospheric Science
Stephen S. Cox
Phone: (303)491-8360   Fax: (303)491-8449
EMail: mtucker@vines.colostate.edu

Financial Aid Office Contact
G.K. Jacks, *Director*
Phone: (303)491-6321   Fax: (303)491-5010

054   Columbia U

Head of the Institution
Michael I. Sovern
Phone: (212)854-2825   Fax: (212)854-6466

Head of Engineering
David H. Auston
Phone: (212)854-2993   Fax: (212)864-0104
EMail: dha3@cunixf.cc.columbia.edu

Associate/Assistant Deans
John R. Kender, *Vice Dean*
Phone: (212)854-2993   Fax: (212)864-0104
Johna Harvey, *Associate Dean*
Phone: (212)854-2996   Fax: (212)864-0104

Department Chair(s)
Chemical Eng & Applied Chemistry
Jordan L. Spencer
Phone: (212)854-4453   Fax: (212)854-8257
EMail: spencer@cunixf.cc.columbia.edu

Civil Eng & Eng Mechanics
Morton B. Friedman
Phone: (212)854-3143
EMail: friedman@cucevx.civil.columbia.edu

Electrical Eng
Thomas E. Stern
Phone: (212)854-3105
EMail: tom@ctr.columbia.edu

Industrial Eng & Operations Research
Donald Goldfarb
Phone: (212)854-2941
EMail: goldfarb@cunixf.cc.columbia.edu

Mechanical Eng
Herbert Deresiewicz
Phone: (212)854-2965

Applied Physics
Gerald A. Navratil
Phone: (212)854-4457
EMail: navratil@cuplvx.apne.columbia.edu

Computer Science
Zvi Galil
Phone: (212)854-2736
EMail: galil@cs.columbia.edu

Henry Krumb School of Mines
Ponisseril Somasundaran
Phone: (212)854-2905
EMail: somasund@cunixf.cc.columbia.edu

Admissions Office Contact
Lawrence J. Momo
Phone: (212)854-2522   Fax: (212)854-1209

Financial Aid Office Contact
Deborah B. Pointer, *Director of Financial Aid*
Phone: (212)854-3711   Fax: (212)854-5353

055 Concordia U

Head of the Institution
Patrick Kenniff
Phone: (514)848-4849   Fax: (514)848-8765

Head of Engineering
M.N.S. Swamy
Phone: (514)848-3060   Fax: (514)848-4509

Associate/Assistant Deans
F. Douglas Hamblin, *Associate Dean*
Phone: (514)848-3063   Fax: (514)848-8646
Terrill Fancott, *Associate Dean*
Phone: (514)848-3074   Fax: (514)848-8646
George D. Xistris, *Assistant Dean, Planning & Priorities*
Phone: (514)848-3141   Fax: (514)848-8646

Department Chair(s)
Civil Eng
Oscar A. Pekau
Phone: (514)848-7809   Fax: (514)848-2809
Electrical & Computer Eng
J. Charles Giguere
Phone: (514)848-3091   Fax: (514)848-8646
Mechanical Eng
M.O.M. Osman
Phone: (514)848-3133   Fax: (514)848-8646
Centre for Building Studies
Paul P. Fazio
Phone: (514)848-3210   Fax: (514)848-7965

Admissions Office Contact
Thomas E. Swift
Phone: (514)848-2668   Fax: (514)848-8631

Financial Aid Office Contact
Roger Cote, *Director, Financial Aid*
Phone: (514)848-3522   Fax: (514)848-3494

056 U of Conn

Head of the Institution
Harry J. Hartley
Phone: (203)486-2337   Fax: (203)486-2627

Head of Engineering
Harold D. Brody
Phone: (203)486-2221   Fax: (203)486-0318

Associate/Assistant Deans
Eric P. Soulsby, *Associate Dean for Undergraduate Programs*
Phone: (203)486-2223   Fax: (203)486-0318
Tuz Chin Ting, *Associate Dean for Research*
Phone: (203)486-5462   Fax: (203)486-0318

Department Chair(s)
Chemical Eng
Thomas F. Anderson
Phone: (203)486-4019   Fax: (203)486-2959
Civil Eng
John W. Leonard
Phone: (203)486-4018   Fax: (203)486-2298
Electrical & Systems Eng
Peter K. Cheo
Phone: (203)486-4816   Fax: (203)486-3789
Mechanical Eng
Roman Solecki
Phone: (203)486-2090   Fax: (203)486-5088
Computer Science & Eng
Keith Barker
Phone: (203)486-3719   Fax: (203)486-4817
Metallurgy
Owen F. Devereux
Phone: (203)486-4620   Fax: (203)486-4745

Admissions Office Contact
Ann L. Huckenbeck
Phone: (203)486-3137   Fax: (203)486-1476

Financial Aid Office Contact
Veronica G. O'Dette, *Director of Student Financial Aid*
Phone: (203)486-2819

057 The Cooper Union

Head of the Institution
John J. Iselin
Phone: (212)353-4240   Fax: (212)353-4244

Head of Engineering
Eleanor Baum
Phone: (212)353-4285   Fax: (212)353-4341
EMail: baum@green.cooper.edu

Associate/Assistant Deans
Art Lucchesi, *Associate Dean of Engineering*
Phone: (212)353-4289   Fax: (212)353-4341

Department Chair(s)
Chemical Eng
Irving Brazinsky
Phone: (212)353-4373   Fax: (212)353-4341
Civil Eng
Jameel Ahmad
Phone: (212)353-4294   Fax: (212)353-4341
Electrical Eng
Melvin Sandler
Phone: (212)353-4336   Fax: (212)353-4341
Mechanical Eng
Jean LeMee
Phone: (212)353-4295   Fax: (212)353-4341
Eng
Art Lucchesi
Phone: (212)353-4289   Fax: (212)353-4341

Admissions Office Contact
Richard Bory
Phone: (212)353-4121   Fax: (212)353-4343

Financial Aid Office Contact
Anne-Marie Wiemer-Sumner, *Director, Financial Aid & Career Counseling*
Phone: (212)353-4111   Fax: (212)353-4343

058 Cornell U

Head of the Institution
Frank H. Rhodes
Phone: (607)255-5201   Fax: (607)255-9412

Head of Engineering
William B. Streett
Phone: (607)255-9679   Fax: (607)255-9606

Associate/Assistant Deans
Gerald E. Rehkugler, *Associate Dean for Undergraduate Programs*
Phone: (607)255-8240   Fax: (607)255-9606
John Hopcroft, *Associate Dean for Coll Relations*
Phone: (607)255-6087   Fax: (607)255-9606
S. Leigh Phoenix, *Associate Dean for Graduate Education*
Phone: (607)255-8818   Fax: (607)255-9606

Department Chair(s)
Chemical Eng
C. Cohen
Phone: (607)255-7292   Fax: (607)255-9166
Electrical Eng
N.C. MacDonald
Phone: (607)255-3388   Fax: (607)255-4565
Agricultural & Biological Eng
R.B. Furry
Phone: (607)255-2270   Fax: (607)255-4080
Applied & Eng Physics
R.A. Buhrman
Phone: (607)255-3732   Fax: (607)255-7658
Civil & Environmental Eng
A.H. Meyburg
Phone: (607)255-3690   Fax: (607)255-9004
Computer Science
J.E. Hartmanis
Phone: (607)255-9208   Fax: (607)255-4428
Geological Sciences
D.E. Karig
Phone: (607)255-3679   Fax: (607)254-4780
Materials Science & Eng
J.M. Blakely
Phone: (607)255-5149   Fax: (607)255-2365
Mechanical & Aerospace Eng
F. Moore
Phone: (607)255-4100   Fax: (607)255-1222
Nuclear Science & Eng
D.D. Clark
Phone: (607)255-5224   Fax: (607)255-9417
Operations Research & Eng
J.A. Muckstadt
Phone: (607)255-9123   Fax: (607)255-9129
Theoretical & Applied Mechanics
J. Jenkins
Phone: (607)255-7185   Fax: (607)255-2011

Admissions Office Contact
Richard Hale
Phone: (607)255-5008   Fax: (607)255-9606

059 Dartmouth Coll

Head of the Institution
James O. Freedman
Phone: (603)646-2222   Fax: (603)646-2266

Head of Engineering
Charles E. Hutchinson
Phone: (603)646-2238   Fax: (603)646-3856

Associate/Assistant Deans
Graham B. Wallis, *Associate Dean*
Phone: (603)646-3844   Fax: (603)646-3856
Carol B. Muller, *Associate Dean*
Phone: (603)646-3058   Fax: (603)646-3856

Department Chair(s)
Eng Sciences
Alvin O. Converse
Phone: (603)646-3677   Fax: (603)646-3856

Admissions Office Contact
Maria Laskaris
Phone: (603)646-2875   Fax: (603)646-1216

Financial Aid Office Contact
Virginia S. Hazen, *Director of Financial Aid*
Phone: (603)646-2451

060 U of Dayton

Head of the Institution
Raymond L. Fitz, S.M.
Phone: (513)229-4122   Fax: (513)229-3433

Head of Engineering
Joseph Lestingi
Phone: (513)229-2736   Fax: (513)229-2756

Associate/Assistant Deans
Norman S. Phillips, *Assistant Dean*
Phone: (513)229-2736   Fax: (513)229-2756
Franklin E. Eastep, *Assoc. Dean*
Phone: (513)229-2241   Fax: (513)229-2471
Carol M. Shaw, *Assoc. Dean*
Phone: (513)229-4632   Fax: (513)229-4666

Department Chair(s)
Chemical Eng
James A. Snide
Phone: (513)229-2627   Fax: (513)229-2756
Civil & Eng Mechanics
Fred K. Bogner
Phone: (513)229-3847   Fax: (513)229-2756
Electrical Eng
Donald L. Moon
Phone: (513)229-3611   Fax: (513)229-2756
Mechanical Eng
John J. Schauer
Phone: (513)229-2835   Fax: (513)229-2756

Admissions Office Contact
Myron H. Achbach
Phone: (513)229-4411   Fax: (513)229-3433

Financial Aid Office Contact
Joyce J. Wilkins, *Director of Financial Aid*
Phone: (513)229-4311   Fax: (513)229-3433

061 U of Delaware

Head of the Institution
David P. Roselle
Phone: (302)831-2111

Head of Engineering
Stuart L. Cooper
Phone: (302)831-8017   Fax: (302)831-6751

Associate/Assistant Deans
Dan L. Boulet, Jr., *Assistant Dean*
Phone: (302)831-8659   Fax: (302)831-8179
Michael L. Vaughan, *Assistant Dean*
Phone: (302)831-6315   Fax: (302)831-8179

Department Chair(s)
Chemical Eng
Michael T. Klein
Phone: (302)831-8155
Civil Eng
Ib A. Svendsen
Phone: (302)831-2441
Electrical Eng
Peter J. Warter
Phone: (302)831-2407
Mechanical Eng
John D. Meakin
Phone: (302)831-1672
Materials Science
Ian W. Hall
Phone: (302)831-2062

Admissions Office Contact
N. Bruce Walker
Phone: (302)831-8123

062 U of Denver

Head of the Institution
Daniel L. Ritchie
Phone: (303)871-2111

Head of Engineering
John Kice
Phone: (303)871-2693   Fax: (303)871-2500

Department Chair(s)
Electrical Eng
Albert J. Rosa
Phone: (303)871-2102   Fax: (303)871-4450
EMail: arosa@ducair.edu
Mechanical Eng
Albert J. Rosa
Phone: (303)871-2102   Fax: (303)871-4450
EMail: arosa@ducair.edu
General Eng
Albert J. Rosa
Phone: (303)871-2102   Fax: (303)871-4450
EMail: arosa@ducair.edu

Admissions Office Contact
Albert J. Rosa
Phone: (303)871-2102   Fax: (303)871-4450

Financial Aid Office Contact
Colleen Hillmeyer, *Director, Financial Aid*
Phone: (303)871-2681   Fax: (303)871-2341

063 U of the District of Columbia

Head of the Institution
Tilden J. LeMelle
Phone: (202)282-7550   Fax: (202)282-3681

Head of Engineering
Philip L. Brach
Phone: (202)282-7427   Fax: (202)282-3677

Associate/Assistant Deans
Alfred O. Taylor, *Associate Dean*
Phone: (202)282-7427   Fax: (202)282-3677
Thedola H. Milligan, *Associate Dean*
Phone: (202)282-7427   Fax: (202)282-3677

Department Chair(s)
Civil Eng
Fred F. Chang
Phone: (202)282-7349   Fax: (202)282-3677

Engineering & Engineering Technology Personnel

Electrical Eng
Samuel Lakeou
Phone: (202)282-7347 Fax: (202)282-3677
Mechanical Eng
Calvin Brooks
Phone: (202)282-3500 Fax: (202)282-3677
Admissions Office Contact
Sandra B. Dolphin
Phone: (202)282-3200 Fax: (202)282-3682
Financial Aid Office Contact
Kenneth Howard, *Director of Financial Aid*
Phone: (202)282-3239 Fax: (202)282-3344

064 Drexel U

Head of the Institution
Richard D. Breslin
Phone: (215)895-2100 Fax: (215)895-1714
Head of Engineering
Yatish T. Shah
Phone: (215)895-2210 Fax: (215)895-4929
Associate/Assistant Deans
James E. Mitchell, *Associate Dean for Undergraduate Affairs*
Phone: (215)895-5927 Fax: (215)895-4929
Nihat M. Bilgutay, *Associate Dean for Research & Graduate Studies*
Phone: (215)895-2210 Fax: (215)895-4929
Department Chair(s)
Chemical Eng
Charles B. Weinberger
Phone: (215)895-2226 Fax: (215)895-5837
EMail: WEINBECB@DUVM.OCS.DREXEL.EDU
Civil & Architectural Eng
Mohamed Elgaaly
Phone: (215)895-2364 Fax: (215)895-1363
Electrical & Computer Eng
Bruce A. Eisenstein
Phone: (215)895-2359 Fax: (215)895-1695
EMail: bruce_eisenstein@cbis.ece.drexel.edu
Mechanical Eng & Mechanics
Shlomo Carmi
Phone: (215)895-2352 Fax: (215)895-1478
EMail: Shlomo.Carmi@CoE.DREXEL.EDU
Materials Eng
Ihab Kamel
Phone: (215)895-2323 Fax: (215)895-6760
EMail: KAMEL@DUVM.OCS.DREXEL.EDU
Admissions Office Contact
John Russel
Phone: (215)895-2400 Fax: (215)895-5939
Financial Aid Office Contact
Nicholas Flocco, *Director*
Phone: (215)895-2537 Fax: (215)895-5939

065 Duke U

Head of the Institution
H. Keith H. Brodie
Phone: (919)684-2424
Head of Engineering
Earl H. Dowell
Phone: (919)660-5389 Fax: (919)684-4860
Associate/Assistant Deans
Marion L. Shepard, *Associate Dean for Academic Affairs*
Phone: (919)660-5387 Fax: (919)684-4860
Department Chair(s)
Civil & Environmental Eng
Henry Petroski
Phone: (919)660-5200 Fax: (919)660-5219
Electrical Eng
H. Craig Casey, Jr.
Phone: (919)660-5252 Fax: (919)660-5293
Mechanical Eng & Materials Science
Robert M. Hochmuth
Phone: (919)660-5310 Fax: (919)660-8963
Biomedical Eng
James H. McElhaney
Phone: (919)660-5131 Fax: (919)684-4488
Admissions Office Contact
Christoph O. Guttentag
Phone: (919)684-3214
Financial Aid Office Contact
James A. Belvin, Jr., *Director*
Phone: (919)684-6225 Fax: (919)660-9811

066 École Polytechnique

Head of the Institution
Jean-Paul Gourdeau
Phone: (514)340-4704 Fax: (514)340-3222
Head of Engineering
André Bazergui
Phone: (514)340-4943 Fax: (514)340-4600
Associate/Assistant Deans
Louis Courville, *Directeur des études de premier cycle*
Phone: (514)340-4707 Fax: (514)340-5836
Gilbert Drouin, *Directeur des études supérieures et de la recherche*
Phone: (514)340-4990 Fax: (514)340-4440
Gabriel Garneau, *Directeur des personnels et des systèmes d'information*
Phone: (514)340-4059 Fax: (514)340-4440
Louise V. Châtillon, *Directeur des services administratifs*
Phone: (514)340-4738 Fax: (514)340-3261
Department Chair(s)
Chemical Eng
Denis Rouleau
Phone: (514)340-4818 Fax: (514)340-4159
Civil Eng
Jules Houde
Phone: (340)514-4257 Fax: (514)340-5918
Electrical Eng
Bernard Lanctôt
Phone: (514)340-4696 Fax: (514)340-4078
Industrial Eng
Michel Normandin
Phone: (514)340-4978 Fax: (514)340-4173
Mechanical Eng
Jean Rousselet
Phone: (514)340-4757 Fax: (514)340-4052
Metallurgical & materials engineering
Jean-Paul Baïlon
Phone: (514)340-4787 Fax: (514)340-4468
Mineral engineering
Guy Valiquette
Phone: (514)340-4777 Fax: (514)340-4191
Eng Physics
Guy Faucher
Phone: (514)340-4768 Fax: (514)340-3218
Applied mathematics
Marc Moore
Phone: (514)340-4825 Fax: (514)340-4463
Energy engineering
Daniel Rozon
Phone: (514)340-4803 Fax: (514)340-4192
Biomedical engineering
Robert A. Leblanc
Phone: (514)340-4895 Fax: (514)340-4611
Admissions Office Contact
Claude Brissette
Phone: (514)340-4724 Fax: (514)340-5836
Financial Aid Office Contact
France Gaudron, *Responsable de l'aide financière*
Phone: (514)340-4842 Fax: (514)340-5836

067 Embry-Riddle-Daytona Beach

Head of the Institution
Steven Sliwa
Phone: (904)226-6200 Fax: (904)226-6299
Head of Engineering
Ray Wimberly
Phone: (904)226-6634 Fax: (904)226-6299
Department Chair(s)
Electrical Eng
James Lyall
Phone: (602)776-3833
Aerospace & Aeronautical Eng
Howard Curtis
Phone: (904)226-6748
Eng Physics
Shiv Aggarwal
Phone: (904)226-6709
Admissions Office Contact
Darryl W. Niemeyer
Phone: (904)226-6100 Fax: (904)226-6299
Financial Aid Office Contact
Phillip C. Ledbetter, *Director of Financial Aid*
Phone: (904)226-6195

068 Embry-Riddle-Western Campus

Head of the Institution
Steven M. Sliwa
Phone: (904)226-6200 Fax: (904)226-6299
Head of Engineering
Paul S. Daly
Phone: (602)776-3800 Fax: (602)776-3827
Department Chair(s)
Electrical Eng
Raymond D. Bellem
Phone: (602)776-3885 Fax: (602)776-3827
Aerospace Eng
Richard F. Felton
Phone: (602)776-3844 Fax: (602)776-3827
Financial Aid Office Contact
Dan Lupin, *Director of Financial Aid*
Phone: (602)776-3762 Fax: (602)445-3184

069 U of Evansville

Head of the Institution
James S. Vinson
Phone: (812)479-2151 Fax: (812)479-2320
Head of Engineering
John R. Tooley
Phone: (812)479-2651 Fax: (812)479-2780
EMail: JACK_TOOLEY%WAYNE-MTS@UM.CC.UMICH.EDU
Department Chair(s)
Mechanical & Civil Eng
Phillip M. Gerhart
Phone: (812)479-2648 Fax: (812)479-2780
Electrical Eng & Computer Science
Dick K. Blandford
Phone: (812)479-2291 Fax: (812)479-2780
Mechanical & Civil Eng
Phillip M. Gerhart
Phone: (812)479-2648 Fax: (812)479-2780
Admissions Office Contact
John Byrd
Phone: (800)423-8633 Fax: (812)479-2320
Financial Aid Office Contact
Tom Stone, *Director of Financial Aid*
Phone: (812)479-2364 Fax: (812)479-2320

070 Ferris State U

Head of the Institution
Helen Popovich
Phone: (616)592-2500
Head of Engineering
Joel Galloway
Phone: (616)592-2890 Fax: (616)592-2946
Associate/Assistant Deans
Vordyn Nelson, *Associate Dean*
Phone: (616)592-2890 Fax: (616)592-2946
Department Chair(s)
Construction Department
Ralph Shields
Phone: (616)592-2360 Fax: (616)592-2946
EMail: Swan 312
Admissions Office Contact
Duncan Sargent
Financial Aid Office Contact
Robert Bopp, *Director*
Phone: (616)592-2110

071 U of Florida

Head of the Institution
John V. Lombardi
Phone: (904)392-1311 Fax: (904)392-9506
Head of Engineering
Winfred M. Phillips
Phone: (904)392-6000 Fax: (904)392-9673
EMail: wphil@engnet.ufl.edu
Associate/Assistant Deans
M. Jack Ohanian, *Associate Dean, Research & Administration*
Phone: (904)392-0946 Fax: (904)392-9673
Warren Viessman, Jr., *Associate Dean, Academic Programs*
Phone: (904)392-0943 Fax: (904)392-9673
Jonathan F.K. Earle, *Assistant Dean, Academic Programs*
Phone: (904)392-0944 Fax: (904)392-9673
Department Chair(s)
Chemical Eng
Timothy J. Anderson
Phone: (904)392-0881 Fax: (904)392-9513
EMail: chemical@engnet.ufl.edu
Civil Eng
Paul Y. Thompson
Phone: (904)392-9537 Fax: (904)392-3394
EMail: pyt@ce.ufl.edu
Electrical Eng
Martin A. Uman
Phone: (904)392-0913 Fax: (904)392-8671
EMail: admin@ee.ufl.edu
Industrial & Systems Eng
D. Jack Elzinga
Phone: (904)392-1464 Fax: (904)392-3537
EMail: isedept@ise.ufl.edu
Mechanical Eng
Robert B. Gaither
Phone: (904)392-0828 Fax: (904)392-1071
EMail: carl@cimar.me.ufl.edu
Aerospace Eng, Mechanics & Eng Science
Martin A. Eisenberg
Phone: (904)392-0961 Fax: (904)392-7303
EMail: ccoop@engnet.ufl.edu
Agricultural Eng
Otto J. Loewer
Phone: (904)392-1864 Fax: (904)392-4092
EMail: ces@agen.ufl.edu
Coastal & Oceanographic Eng
Robert G. Dean
Phone: (904)392-1436 Fax: (904)392-3466
EMail: dean@coed.ufl.edu
Computer & Information Sciences
Stephen S. Yau
Phone: (904)392-1211 Fax: (904)392-1220
EMail: dgt@bikini.cis.ufl.edu
Environmental Eng Sciences
Joseph J. Delfino
Phone: (904)392-0841 Fax: (904)392-3076
EMail: sjame@mailgate.engnet.ufl.edu
Materials Science & Eng
Reza Abbaschian
Phone: (904)392-1453 Fax: (904)392-6359
EMail: phowe@mse.ufl.edu
Nuclear Eng Sciences
James S. Tulenko
Phone: (904)392-1401 Fax: (904)392-3380
EMail: nuceng@pine.circa.ufl.edu

Engineering & Engineering Technology Personnel

072 Florida A&M U

Head of the Institution
Frederick S. Humphries
Phone: (904)599-3223   Fax: (904)561-2152

Head of Engineering
Ching-Jen Chen
Phone: (904)487-6100   Fax: (904)487-6486
EMail: CJChen@EVAX.ENG.FSU.EDU

Associate/Assistant Deans
J.W. Toliver, *Associate Dean*
Phone: (904)487-6423   Fax: (904)487-6486

Department Chair(s)
Chemical Eng
Michael H. Peters
Phone: (904)487-6144   Fax: (904)487-6150
EMail: bitnet peters @FSU
Civil Eng
Soronadi Nnaji
Phone: (904)487-6136   Fax: (904)487-6142
Electrical Eng
Thomas J. Harrison
Phone: (904)487-6457   Fax: (904)487-6479
EMail: Harrison@EVAX.ENG.FSU.EDU
Industrial Eng
S.A. Awoniyi
Phone: (904)487-6354   Fax: (904)487-6342
EMail: Awoniyi@EVAX.ENG.FSU.EDU
Mechanical Eng
Anjaneyulu Krothapalli
Phone: (904)487-6331   Fax: (904)487-6337
EMail: Krothapalli@ENG.FSU.EDU

Admissions Office Contact
Barbara Cox
Phone: (904)599-3797   Fax: (904)561-2248

Financial Aid Office Contact
Alton Royal, *Director of Financial Aid*
Phone: (904)599-3730   Fax: (904)599-3952

073 Florida Atlantic U

Head of the Institution
Anthony J. Catanese
Phone: (407)367-3450   Fax: (407)367-2777

Head of Engineering
Craig S. Hartley
Phone: (407)367-3400   Fax: (407)367-2659
EMail: HARTLEY@ACC.FAU.EDU

Associate/Assistant Deans
Roger A. Messenger, *Associate Dean for Academic Affairs*
Phone: (407)367-3407   Fax: (407)367-2659
Joseph A. Campbell, *Associate Dean for Resource Development*
Phone: (407)367-3484   Fax: (407)367-2659

Department Chair(s)
Electrical Eng
Henry F. Helmken
Phone: (407)367-3410   Fax: (407)367-2336
EMail: HELMKENH@CSE.FAU.EDU
Mechanical Eng
Karl Stevens
Phone: (407)367-3430   Fax: (407)367-2825
EMail: STEVENS@ACC.FAU.EDU
Ocean Eng
Stanley Dunn
Phone: (407)367-3430   Fax: (407)367-3885
EMail: DUNNS@ACC.FAU.EDU
Computer Science & Eng
Neal S. Coulter
Phone: (407)367-3855   Fax: (407)367-2800
EMail: NEAL@CSE.FAU.EDU

Admissions Office Contact
Barbara T. Fincher
Phone: (904)392-1374   Fax: (904)392-3987

Financial Aid Office Contact
Karen L. Fooks, *Director of Financial Aid*
Phone: (904)392-1275   Fax: (904)392-2861

074 Florida Inst of Tech

Head of the Institution
Lynn E. Weaver
Phone: (407)768-8000   Fax: (407)984-8461

Head of Engineering
Robert L. Sullivan
Phone: (407)768-8000   Fax: (407)984-8461

Associate/Assistant Deans
Charles D. Beach, *Assistant Dean*
Phone: (407)768-8000   Fax: (407)984-8461

Department Chair(s)
Chemical Eng
Pat L. Mangonon
Phone: (407)768-8000   Fax: (407)984-8461
Civil & Environmental Eng
Edward H. Kalajian
Phone: (407)768-8000   Fax: (407)984-8461
Electrical & Computer Eng
Thomas J. Sanders
Phone: (407)768-8000   Fax: (407)984-8461
Mechanical & Aerospace Eng
Paavo Sepri
Phone: (407)768-8000   Fax: (407)984-8461
Oceanography/Ocean Eng & Env'l Science
N. Thomas Stephens
Phone: (407)768-8000   Fax: (407)984-8461

Admissions Office Contact
Louis T. Levy
Phone: (407)768-8000   Fax: (407)723-9468

Financial Aid Office Contact
Leonard E. Gude, *Director of Financial Aid & Scholarship*
Phone: (407)768-8000   Fax: (407)984-8461

075 Florida International U

Head of the Institution
Modesto A. Maidique
Phone: (305)348-2111   Fax: (305)348-3660

Head of Engineering
Gordon R. Hopkins
Phone: (305)348-2521   Fax: (305)348-3582
EMail: HOPKINS@SERVAX

Associate/Assistant Deans
Gustavo A. Roig, *Associate Dean for Academic Programs*
Phone: (305)348-3027   Fax: (305)348-3582

Department Chair(s)
Mechanical Eng
M.A. Ebadian
Phone: (305)348-2569   Fax: (305)348-4176
Industrial Eng
Fred W. Swift
Phone: (305)348-2256   Fax: (305)348-3721
Electrical Eng
James R. Story
Phone: (305)348-2808   Fax: (305)348-3707
Civil Eng
L. David Shen
Phone: (305)348-3814   Fax: (305)348-2802

Admissions Office Contact
Carmen Brown
Phone: (305)348-3675   Fax: (305)348-3648

Financial Aid Office Contact
Ana R. Sarasti, *Director, Financial Aid*
Phone: (305)348-2431   Fax: (305)348-2346

076 Florida State U

Head of the Institution
Dale W. Lick
Phone: (904)644-1085   Fax: (904)644-0172

Head of Engineering
Ching-Jen Chen
Phone: (904)487-6100   Fax: (904)487-6486
EMail: CJChen@EVAX.ENG.FSU.EDU

Associate/Assistant Deans
J.W. Toliver, *Associate Dean*
Phone: (904)487-6423   Fax: (904)487-6486

Department Chair(s)
Chemical Eng
Michael H. Peters
Phone: (904)487-6144   Fax: (904)487-6150
EMail: bitnet peters @FSU
Civil Eng
Soronadi Nnaji
Phone: (904)487-6136   Fax: (904)487-6142
Electrical Eng
Thomas J. Harrison
Phone: (904)487-6457   Fax: (904)487-6479
EMail: Harrison@EVAX.ENG.FSU.EDU
Industrial Eng
Samuel A. Awoniyi
Phone: (904)487-6345   Fax: (904)487-6342
EMail: Awoniyi.EVAX0.ENG.FSU.EDU
Mechanical Eng
Anjaneyulu Krothapalli
Phone: (904)487-6331   Fax: (904)487-6337
EMail: Krothapalli@ENG.FSU.EDU

Admissions Office Contact
Peter F. Metarko
Phone: (904)644-6200   Fax: (904)644-6404

Financial Aid Office Contact
Robert McCloud, *Director of Financial Aid*
Phone: (904)644-5871   Fax: (904)644-6404

077 GMI

Head of the Institution
James E. John
Phone: (313)762-9864   Fax: (313)762-9807

Head of Engineering
John D. Lorenz
Phone: (313)762-7949   Fax: (313)762-9836

Associate/Assistant Deans
James T. Luxon, *Associate Dean, Graduate Studies, Extension Services & Research*
Phone: (313)762-7996   Fax: (313)762-9836
Bruce P. Henderson, *Associate Dean for Academic Affairs*
Phone: (313)762-9775   Fax: (313)762-9836
John E. Rolfe, *Assistant Dean, Academic Services*
Phone: (313)762-7882   Fax: (313)762-9836

Department Chair(s)
Electrical & Computer Eng
David J. Leffen
Phone: (313)762-7900   Fax: (313)762-9836
Industrial & Manufacturing Systems Eng
Norman L. Crawford
Phone: (313)762-7940   Fax: (313)762-9924
Mechanical Eng
George T. Kartsounes
Phone: (313)762-7833   Fax: (313)762-7860
Manufacturing Systems Eng
Norman L. Crawford
Phone: (313)762-7940   Fax: (313)762-9924

Admissions Office Contact
Kevin A. Pollock
Phone: (313)762-7865   Fax: (313)762-9837

Financial Aid Office Contact
Mark J. Delorey, *Director of Financial Aid*
Phone: (313)762-7859   Fax: (313)762-9807

078 Gannon U

Head of the Institution
David A. Rubino
Phone: (814)871-5800   Fax: (814)871-7338

Head of Engineering
William D. Gregory
Phone: (814)871-7616   Fax: (814)455-2631

Associate/Assistant Deans
Stanley J. Zagorski, *Associate Dean*
Phone: (814)871-7617   Fax: (814)455-2631
Howard T. Wilson, *Assistant Dean*
Phone: (814)871-7760   Fax: (814)455-2631

Department Chair(s)
Electrical Eng
Samuel L. Hazen
Phone: (814)871-7775   Fax: (814)455-2631
Mechanical Eng
Ludwik A. Medeksza
Phone: (814)871-7620   Fax: (814)455-2631

Admissions Office Contact
Joyce Scheid-Gilman
Phone: (814)871-7407   Fax: (814)455-6277

Financial Aid Office Contact
James A. Treiber, *Director of Financial Aid*
Phone: (814)871-7337   Fax: (814)455-6277

079 George Mason U

Head of the Institution
George W. Johnson
Phone: (703)993-8700   Fax: (703)993-8707

Head of Engineering
Andrew P. Sage
Phone: (703)993-1500   Fax: (703)993-1521
EMail: asage@gmuvax.gmu.edu

Associate/Assistant Deans
James D. Palmer, *Associate Dean for Graduate Affairs*
Phone: (703)993-1507   Fax: (703)993-1521
Peter Denning, *Associate Dean for Computing*
Phone: (703)993-1530   Fax: (703)993-1521
E. Bernard White, *Assistant Dean for Undergraduate Affairs*
Phone: (703)993-1511   Fax: (703)993-1521
George R. Umberger, *Assistant Dean for Administration & Outreach*
Phone: (703)993-1516   Fax: (703)993-1521

Department Chair(s)
Electrical & Computer Eng
Gerald Cook
Phone: (703)993-1569   Fax: (703)993-1521
Systems Eng
Alexander Levis
Phone: (703)993-1645   Fax: (703)993-1521
Urban Systems Eng
Terry Ryan
Phone: (703)993-1500   Fax: (703)993-1521
Information & Software Systems Eng
Larry Kerschberg
Phone: (703)993-1640   Fax: (703)993-1521
Applied & Eng Statistics
Edward J. Wegman
Phone: (703)993-1698   Fax: (703)993-1700
Computer Science
Peter Denning
Phone: (703)993-1530   Fax: (703)993-1521
Operations Research & Eng
Carl M. Harris
Phone: (703)993-1670   Fax: (703)993-1521

Admissions Office Contact
Patricia Riordan
Phone: (703)993-2400   Fax: (703)993-2392

Financial Aid Office Contact
Jennifer Douglas, *Director*
Phone: (703)993-4350

Engineering & Engineering Technology Personnel

080 The George Washington U
Head of the Institution
Stephen J. Trachtenberg
Phone: (202)994-6500   Fax: (202)994-0654
Head of Engineering
Gideon Frieder
Phone: (202)994-6080   Fax: (202)994-4522
EMail: frieder@seas.gwu.edu
Associate/Assistant Deans
Richard M. Soland, *Acting Associate Dean*
Phone: (202)994-7179   Fax: (202)994-4522
Department Chair(s)
Eng Management
E. Lile Murphree, Jr.
Phone: (202)994-3795   Fax: (202)994-4606
EMail: murphree@seas.gwu.edu
Operations Research
James E. Falk
Phone: (202)994-6084   Fax: (202)994-0245
EMail: falk@seas.gwu.edu
Continuing Eng Education
Ronald W. Witzel
Phone: (202)994-6106   Fax: (202)872-0645
EMail: rwitzel@seas.gwu.edu
Civil, Mechanical & Environmental Eng
Ali M. Kiper
Phone: (202)994-6749   Fax: (202)994-0238
EMail: kiper@seas.gwu.edu
Electrical Eng & Computer Science
Robert J. Harrington
Phone: (202)994-6083   Fax: (202)994-4027
EMail: rharring@seas.gwu.edu
Financial Aid Office Contact
Vicki J. Baker
Phone: (202)994-6620   Fax: (202)994-0906

081 U of Georgia
Head of the Institution
Charles B. Knapp
Phone: (706)542-1214   Fax: (706)542-0995
Head of Engineering
William P. Flatt
Phone: (706)542-3924   Fax: (706)542-0803
Associate/Assistant Deans
Chris J.B. Smit, *Associate Dean & Director*
Phone: (706)542-1611   Fax: (706)542-2130
Department Chair(s)
Biological & Agricultural Eng
Ernest D. Threadgill
Phone: (706)542-1653   Fax: (706)542-8806
EMail: BAEATH@UGA
Admissions Office Contact
Claire C. Swann
Phone: (706)542-8776   Fax: (706)542-1466
Financial Aid Office Contact
Ray Tripp, *Director*
Phone: (706)542-8208

082 Georgia Tech
Head of the Institution
John P. Crecine
Phone: (404)894-5051   Fax: (404)853-9163
Head of Engineering
John A. White
Phone: (404)894-3350   Fax: (404)853-0168
EMail: jwhite@gatech.edu
Associate/Assistant Deans
J. Narl Davidson, *Associate Dean*
Phone: (404)894-2972   Fax: (404)894-9809
Jack R. Lohmann, *Associate Dean*
Phone: (404)894-3355   Fax: (404)894-9809
Department Chair(s)
Chemical Eng
Ronald W. Rousseau
Phone: (404)894-2867   Fax: (404)894-2866
EMail: ronald.rousseau@che.gatech.edu
Civil Eng (incl Env'l Eng & Eng Sci Mech)
Jean-Lou A. Chameau
Phone: (404)894-2201   Fax: (404)894-2278
EMail: jeanlou.chameau@ce.gatech.edu
Electrical Eng (incl Computer Eng)
Roger P. Webb
Phone: (404)894-2902   Fax: (404)853-9171
EMail: roger.webb@ee.gatech.edu
Industrial & Systems Eng
John J. Jarvis
Phone: (404)894-2303   Fax: (404)894-2301
EMail: john.jarvis@isye.gatech.edu
Mechanical Eng (incl Nuclear Eng)
Ward O. Winer
Phone: (404)894-3200   Fax: (404)894-8336
EMail: ward.winer@me.gatech.edu
Aerospace Eng
Alvin Pierce
Phone: (404)894-3000   Fax: (404)894-2760
EMail: al.pierce@aerospace.gatech.edu
Materials Science & Eng (incl Ceramics)
Miroslav I. Marek
Phone: (404)894-2816   Fax: (404)853-9140
EMail: miroslav.marek@mse.gatech.edu
Textile & Fiber Eng (incl Polymer & Textile Chem)
Fred L. Cook
Phone: (404)894-2536   Fax: (404)894-8780
EMail: fred.cook@textiles.gatech.edu
Admissions Office Contact
Deborah D. Smith
Phone: (404)894-4154   Fax: (404)894-9511
Financial Aid Office Contact
Curley M. Williams, *Director*
Phone: (404)894-4160   Fax: (404)853-9396

083 Gonzaga U
Head of the Institution
Bernard J. Coughlin
Phone: (509)328-4220   Fax: (509)484-2818
Head of Engineering
Zia A. Yamayee
Phone: (509)328-4220   Fax: (509)484-2818
Associate/Assistant Deans
William P. Ilgen, *Associate Dean*
Phone: (509)328-4220   Fax: (509)484-2818
Department Chair(s)
Civil Eng
William P. Ilgen
Phone: (509)328-4220   Fax: (509)484-2818
Electrical Eng
Raymond A. Birgenheier
Phone: (509)328-4220   Fax: (509)484-2818
Mechanical Eng
John J. Marciniak
Phone: (509)328-4220   Fax: (509)484-2818
Admissions Office Contact
Phillip Ballinger
Phone: (509)328-4220   Fax: (509)484-2818
Financial Aid Office Contact
Nancy E. Ryan, *Associate Director, Financial Aid*
Phone: (509)328-4220   Fax: (509)484-2818

084 Grand Valley State U
Head of the Institution
Arend D. Lubbers
Phone: (616)895-2182   Fax: (616)895-3503
Head of Engineering
Paul D. Plotkowski, *Director*
Phone: (616)771-6750   Fax: (616)771-6642
EMail: plotkowp@gvsu.edu
Department Chair(s)
School of Eng
Paul D. Plotkowski
Phone: (616)771-6750   Fax: (616)771-6642
EMail: plotkowp@gvsu.edu
Admissions Office Contact
Paul D. Plotkowski
Phone: (616)771-6750   Fax: (616)771-6642
Financial Aid Office Contact
Ken Fridsma, *Director of Financial Aid*
Phone: (616)895-3234   Fax: (616)895-3180

085 Grove City Coll
Head of the Institution
Jerry H. Combee
Phone: (412)458-2500   Fax: (412)458-2190
Head of Engineering
Joseph F. Goncz, Jr.
Phone: (412)458-2033   Fax: (412)458-2190
Department Chair(s)
Department of Eng
Joseph F. Goncz, Jr.
Phone: (412)458-2033   Fax: (412)458-2190
Department of Eng
Joseph F. Goncz, Jr.
Phone: (412)458-2033   Fax: (412)458-2190
Admissions Office Contact
Jeffrey C. Mincey
Phone: (412)458-2100   Fax: (412)458-2190
Financial Aid Office Contact
Anne P. Bowne, *Director of Financial Aid*
Phone: (412)458-2163   Fax: (412)458-2190

086 Harvard U
Head of the Institution
Neil L. Rudenstine
Phone: (617)495-1502
Head of Engineering
Paul C. Martin
Phone: (617)495-5829   Fax: (617)495-9837
Associate/Assistant Deans
Albert Gold, *Associate Dean for Administration*
Phone: (617)495-2908   Fax: (617)495-9837
Department Chair(s)
Division of Applied Sciences
Paul C. Martin
Phone: (617)495-5829   Fax: (617)495-9837
Admissions Office Contact
Marlyn M. Lewis
Phone: (617)495-5339
Financial Aid Office Contact
James S. Miller, *Director of Financial Aid*
Phone: (617)495-1580

087 Harvey Mudd Coll
Head of the Institution
Henry E. Riggs
Phone: (714)621-8120
Head of Engineering
John I. Molinder
Phone: (714)621-8019   Fax: (714)621-8465
EMail: JMOLINDER@HMCVAX.CLAREMONT.EDU
Associate/Assistant Deans
J. Richard Phillips, *Director, Engineering Clinic*
Phone: (714)621-8020   Fax: (714)621-8465
Department Chair(s)
Eng
John I. Molinder
Phone: (714)621-8019   Fax: (714)621-8465
EMail: JMOLINDER@HMCVAX.CLAREMONT.EDU
Admissions Office Contact
Patricia Coleman
Phone: (714)621-8011   Fax: (714)621-8360
Financial Aid Office Contact
Noe Ortiz, *Director of Financial Aid*
Phone: (714)621-8055

088 U of Hawaii at Manoa
Head of the Institution
Yuen C. Paul
Phone: (808)956-5280
Head of Engineering
Reginald H. Young
Phone: (808)956-7727   Fax: (808)956-2291
EMail: young
Associate/Assistant Deans
Deane H. Kihara, *Assistant Dean*
Phone: (808)956-8404   Fax: (808)956-2291
Department Chair(s)
Civil Eng
Harold S. Hamada
Phone: (808)956-7550   Fax: (808)956-5014
EMail: hamada
Electrical Eng
Shu Lin
Phone: (808)956-7586   Fax: (808)956-3427
EMail: slin
Mechanical Eng
Ping Cheng
Phone: (808)956-7167   Fax: (808)956-2373
EMail: cheng
Admissions Office Contact
Deane H. Kihara
Phone: (808)956-8404   Fax: (808)956-2291
Financial Aid Office Contact
Annabelle C. Fong, *Director, Financial Aid Office*
Phone: (808)956-7251   Fax: (808)956-5076

089 Hofstra U
Head of the Institution
James M. Shuart
Phone: (516)463-6800   Fax: (516)564-4296
Head of Engineering
David M. Rooney
Phone: (516)463-5545   Fax: (516)564-4296
Department Chair(s)
Department of Eng
David M. Rooney
Phone: (516)463-5545   Fax: (516)564-4296
Admissions Office Contact
Joan I. Mohr
Phone: (516)463-6700   Fax: (516)564-4296
Financial Aid Office Contact
Jean A. Belmont, *Director*
Phone: (516)463-6680   Fax: (516)564-4291

090 U of Houston
Head of the Institution
James H. Pickering
Phone: (713)743-8820   Fax: (713)743-8838
Head of Engineering
Roger Eichhorn
Phone: (713)743-4200   Fax: (713)743-4214
EMail: Eichhorn@uh.edu
Associate/Assistant Deans
Charles Dalton, *Graduate Associate Dean*
Phone: (713)743-4200   Fax: (713)743-4214
David P. Shattuck, *Undergraduate Associate Dean*
Phone: (713)743-4200   Fax: (713)743-4214
Department Chair(s)
Chemical Eng
Dan Luss
Phone: (713)743-4300   Fax: (713)747-6323
EMail: DLUSS@JETSON.UH.EDU
Civil & Environmental Eng
Michael W. O'Neill
Phone: (713)743-4250   Fax: (713)743-4260
EMail: "IN%ONEILL@uh.edu"
Electrical Eng
Stuart A. Long
Phone: (713)743-4400   Fax: (713)743-4444
EMail: Long@uh.edu

Mechanical Eng
Larry C. Witte
Phone: (713)743-4500   Fax: (713)743-4503
EMail: Mechair@Jetson.uh.edu
Industrial Eng
Charles E. Donaghey
Phone: (713)743-4180   Fax: (713)743-4190
EMail: Donaghey@Jetson.uh.edu.
Financial Aid Office Contact
Robert Sheridan, *Director of Scholarships & Financial Aids*
Phone: (713)743-1010

091   Howard U
Head of the Institution
Franklyn G. Jenifer
Phone: (202)806-2500   Fax: (202)806-5960
Head of Engineering
M. Lucius Walker, Jr.
Phone: (202)806-6565   Fax: (202)462-1810
EMail: walker@echo.eng.umd.edu
Associate/Assistant Deans
Carmen Cannon, *Associate Dean*
Phone: (202)806-6638   Fax: (202)462-1810
Department Chair(s)
Chemical Eng
Joseph N. Cannon
Phone: (202)806-6624   Fax: (202)806-5960
Civil Eng
James H. Johnson
Phone: (202)806-6570   Fax: (202)806-5960
Electrical Eng
James A. Momoh
Phone: (202)806-6587   Fax: (202)806-5960
Mechanical Eng
Lewis Thigpen
Phone: (202)806-6600   Fax: (202)806-5960
Systems & Computer Science
Don M. Coleman
Phone: (202)806-6595   Fax: (202)806-5960
Admissions Office Contact
Emmett R. Griffin, Jr.
Phone: (202)806-2700   Fax: (202)806-4365
Financial Aid Office Contact
Adrienne W. Price, *Director, Financial Aid & Student Employment*
Phone: (202)806-2800

092   U of Idaho
Head of the Institution
Elizabeth A. Zinser
Phone: (208)885-6365   Fax: (208)885-6558
Head of Engineering
Richard T. Jacobsen
Phone: (208)885-6479   Fax: (208)885-6645
Associate/Assistant Deans
Weldon T. Tovey, *Associate Dean for Academics*
Phone: (208)885-6479   Fax: (208)885-6645
David M. Woodall, *Associate Dean / Director of Research*
Phone: (208)885-6479   Fax: (208)885-6645
Department Chair(s)
Chemical Eng
Roger A. Korus
Phone: (208)885-6005   Fax: (208)885-7462
Civil Eng
Fred J. Watts
Phone: (208)885-6602   Fax: (208)885-6645
Electrical Eng
Joseph J. Feeley
Phone: (208)885-7482   Fax: (208)885-7579
Mechanical Eng
E.C. Lemmon
Phone: (208)885-7134   Fax: (208)885-6645
Computer Science
John W. Dickinson
Phone: (208)885-7227   Fax: (208)885-6645

Admissions Office Contact
Weldon R. Tovey
Phone: (208)885-6479   Fax: (208)885-6645
Financial Aid Office Contact
Daniel D. Davenport, *Director*
Phone: (208)885-6312

093   Idaho State U
Head of the Institution
Richard L. Bowen
Phone: (208)236-3440   Fax: (208)236-4000
Head of Engineering
V. Charyulu
Phone: (206)236-2902   Fax: (208)236-4538
Associate/Assistant Deans
Richard M. Wabrek, *Associate Dean*
Phone: (208)236-4399   Fax: (208)236-4538
Department Chair(s)
Eng
Richard M. Wabrek
Phone: (208)236-4399   Fax: (208)236-4538
Admissions Office Contact
V. Charyulu
Phone: (208)236-2902   Fax: (208)236-4538
Financial Aid Office Contact
Doug Severs, *Director*
Phone: (208)236-2756   Fax: (208)236-4000

094   U of Illinois at Chicago
Head of the Institution
Stanley O. Ikenberry
Phone: (312)996-8800
Head of Engineering
Paul M. Chung
Phone: (312)996-2400   Fax: (312)996-8664
Associate/Assistant Deans
William DeFotis, *Associate Dean of Undergraduate Administration*
Phone: (312)996-2402   Fax: (312)996-8664
Stephen Szepe, *Associate Dean*
Phone: (312)996-0418   Fax: (312)996-8664
Arlene F. Norsym, *Assistant Dean*
Phone: (312)996-2419   Fax: (312)996-8664
Thomas E. Glenn, *Assistant to the Dean*
Phone: (312)413-7624   Fax: (312)996-8664
Department Chair(s)
Chemical Eng
Irving F. Miller
Phone: (312)996-5711   Fax: (312)996-0808
Civil Eng, Mechanics, & Metallurgy
Chien-Heng Wu
Phone: (312)996-3428   Fax: (312)996-2426
Electrical Eng & Computer Science
Wai-Kai Chen
Phone: (312)996-2462   Fax: (312)413-0024
Bioengineering Program
Irving F. Miller
Phone: (312)996-2335   Fax: (312)996-5921
Mechanical Eng
Piergiorgio Uslenghi
Phone: (312)996-5096   Fax: (312)413-0447

095   Illinois, Urbana-Champaign
Head of the Institution
Stanley O. Ikenberry
Phone: (217)333-3070   Fax: (217)333-3072
Head of Engineering
William R. Schowalter
Phone: (217)333-2150   Fax: (217)244-7705
EMail: schow@ux1.cso.uiuc.edu
Associate/Assistant Deans
Carl J. Altstetter, *Associate Dean for Academic Affairs*
Phone: (217)333-7870   Fax: (217)244-7705

Anthony F. Graziano, *Associate Dean for Administrative Affairs*
Phone: (217)333-2152   Fax: (217)244-7705
Howard L. Wakeland, *Associate Dean for Undergraduate Studies*
Phone: (217)333-2280   Fax: (217)244-4974
Department Chair(s)
Chemical Eng
R.C. Alkire
Phone: (217)333-3640   Fax: (217)244-8068
Civil Eng
N.M. Hawkins
Phone: (217)333-3814   Fax: (217)333-9464
Electrical & Computer Eng
T.N. Trick
Phone: (217)333-2300   Fax: (217)333-7427
EMail: ame@ece.uiuc.edu
Aeronautical & Astronautical Eng
W.C. Solomon
Phone: (217)333-2651   Fax: (217)244-7705
Agricultural Eng
R.L. Pershing
Phone: (217)333-3570   Fax: (217)244-0323
EMail: agrengr@uiucvmd
Computer Science
D.H. Lawrie
Phone: (217)333-4428   Fax: (217)333-3501
EMail: admin@cs.uiucedu
General Eng
T.F. Conry
Phone: (217)333-2730   Fax: (217)244-7705
Materials Science & Eng
J. Economy
Phone: (217)333-1441   Fax: (217)333-2736
Nuclear Eng
B.G. Jones
Phone: (217)333-2295   Fax: (217)333-2906
Physics Eng
D.K. Campbell
Phone: (217)333-3761   Fax: (217)333-9819
EMail: PHYSDEPT@VMD.CSO.UIUC.EDU
Theoretical & Applied Mech. Eng
D.E. Carlson
Phone: (217)333-2322   Fax: (217)244-5707
Mechanical & Industrial Eng
A.L. Addy
Phone: (217)333-1176   Fax: (217)244-6534
Financial Aid Office Contact
Orlo Austin, *Director*
Phone: (217)333-0100

096   Illinois Inst of Tech
Head of the Institution
Lewis Collens
Phone: (312)567-5198   Fax: (312)567-3004
Head of Engineering
Stephen M. Copley
Phone: (312)567-3009   Fax: (312)567-5205
Associate/Assistant Deans
Gerald F. Saletta, *Associate Dean*
Phone: (312)567-3013   Fax: (312)567-5205
Department Chair(s)
Chemical Eng
Hamid Arastoopour
Phone: (312)567-3038   Fax: (312)567-8874
EMail: CHEARASTOOPO
Civil Eng
David Arditi
Phone: (312)567-3540   Fax: (312)567-3519
Electrical & Computer Eng
Henry Stark
Phone: (312)567-3400   Fax: (312)567-8976
EMail: EESTARK
Mechanical & Aerospace Eng
Hassan Nagib
Phone: (312)567-3175   Fax: (312)567-9079
EMail: MENAGIB

Metallurgical & Materials Eng
John S. Kallend
Phone: (312)567-3050   Fax: (312)567-8875
EMail: METMKALLEND
Environmental Eng
Kenneth E. Noll
Phone: (312)567-3535   Fax: (312)567-3548
Admissions Office Contact
William Black
Phone: (312)567-3025   Fax: (312)567-8828
Financial Aid Office Contact
Walter J. O'Neill, *Director of Financial Aid*
Phone: (312)567-3033   Fax: (312)567-3302

097   Indiana-Purdue, Indianapolis
Head of the Institution
Gerald L. Bepko, *Chancellor*
Phone: (317)274-4615   Fax: (317)274-4615
Head of Engineering
R. Bruce Renda
Phone: (317)274-0800   Fax: (317)274-0832
Associate/Assistant Deans
H. Öner Yurtseven, *Associate Dean for Academic Affairs*
Phone: (317)274-9708   Fax: (317)274-4567
Gary Burkart, *Associate Dean*
Phone: (317)266-5601   Fax: (317)266-5615
Christine Fitzpatrick, *Associate Dean for Student Affairs*
Phone: (317)274-0804   Fax: (317)274-4567
Pat Fox, *Assistant Dean for Financial Affairs*
Phone: (317)274-0807   Fax: (317)274-4567
Department Chair(s)
Electrical Eng
A.S.C. Sinha
Phone: (317)274-9721   Fax: (317)274-4493
Mechanical Eng
N. Paydar
Phone: (317)274-9716   Fax: (317)274-9744
Admissions Office Contact
Christine Fitzpatrick
Phone: (317)274-0804   Fax: (317)274-4567
Financial Aid Office Contact
Natala Hart
Phone: (317)274-4162

098   U of Iowa
Head of the Institution
Hunter R. Rawlings, III
Phone: (319)335-3549
Head of Engineering
Richard K. Miller
Phone: (319)335-5766
Department Chair(s)
Chemical & Biochemical Eng
Gregory R. Carmichael
Phone: (319)335-1399   Fax: (319)335-2951
EMail: grcarmichael@icaen.uiowa.edu
Civil & Environmental Eng
Gene F. Parkin
Phone: (319)335-5655   Fax: (319)335-5777
Electrical & Computer Eng
Sudhakar M. Reddy
Phone: (319)335-5196   Fax: (319)335-6028
EMail: reddy@eng.uiowa.edu
Industrial Eng
Andrew Kusiak
Phone: (319)335-5934   Fax: (319)335-5777
EMail: ankusiak@icaen.uiowa.edu
Mechanical Eng
Lea-Der Chen
Phone: (319)335-5673   Fax: (319)335-5669
EMail: ldchen@icaen.uiowa.edu
Biomedical Eng
Vijay K. Goel
Phone: (319)335-5638   Fax: (319)335-5777
EMail: goel@bmevax.eng.uiowa.edu

Engineering & Engineering Technology Personnel

099 Iowa State U
- 104 Lafayette Coll

Admissions Office Contact
Michael Barron
Phone: (319)335-1566

Financial Aid Office Contact
Mark S. Warner, *Director of Student Financial Aid*
Phone: (319)335-1450

099   Iowa State U

Head of the Institution
Martin C. Jischke
Phone: (515)294-2042

Head of Engineering
David T. Kao
Phone: (515)294-5933   Fax: (515)294-9273

Associate/Assistant Deans
Arvid R. Eide, *Associate*
Phone: (515)294-1309   Fax: (515)294-9273
William Lord, *Associate*
Phone: (515)294-6617   Fax: (515)294-9273
George Burnet, *Associate*
Phone: (515)294-2416   Fax: (515)294-9273

Department Chair(s)
Chemical Eng
Terry S. King
Phone: (515)294-7642   Fax: (515)294-2689
Civil & Construction Eng
Lowell F. Greimann
Phone: (515)294-2140   Fax: (515)294-8216
Electrical Eng & Computer Eng
Randall Geiger
Phone: (515)294-2663   Fax: (515)294-8432
Industrial & Manufacturing Systems Eng
Way Kuo
Phone: (515)294-1682   Fax: (515)294-3524
Mechanical Eng
Theodore H. Okiishi
Phone: (515)294-1423   Fax: (515)294-8584
Agricultural & Biosystems Eng
James R. Gilley
Phone: (515)294-0462   Fax: (515)294-9589
Biomedical Eng
Mary H. Greer
Phone: (515)294-6520   Fax: (515)294-8500
Materials Science & Eng
Krishna Vedula
Phone: (515)294-1214   Fax: (515)294-9273
Aerospace Eng & Eng Mechanics
David K. Holger
Phone: (515)294-6240   Fax: (515)294-8584

Admissions Office Contact
Karsten Smedal
Phone: (515)294-0815   Fax: (515)294-1088

Financial Aid Office Contact
Earl Dowling, *Director*
Phone: (515)294-2223   Fax: (515)294-0907

100   The Johns Hopkins U

Head of the Institution
William C. Richardson
Phone: (410)516-8068   Fax: (410)516-7075

Head of Engineering
Don P. Giddens
Phone: (410)516-8350   Fax: (410)516-8627
EMail: dgiddens@jhuvms.bitnet

Associate/Assistant Deans
Ross B. Corotis, *Associate Dean*
Phone: (410)516-7395   Fax: (410)516-8627
Candice V. Dalrymple, *Assistant Dean*
Phone: (410)516-8909   Fax: (410)516-8627

Department Chair(s)
Chemical Eng
Marc D. Donohue
Phone: (410)516-7761   Fax: (410)516-5510
EMail: mdd@mdd.che.jhu.edu
Electrical & Computer Eng
Charles R. Westgate
Phone: (410)516-7033   Fax: (410)516-5566
EMail: eed_wcrw@jhunix.bitnet
Civil Eng
Bruce R. Ellingwood
Phone: (410)516-8443   Fax: (410)516-7473
EMail: ced_wbe@jhuvms.bitnet
Mechanical Eng
William N. Sharpe, Jr.
Phone: (410)516-7132   Fax: (410)516-1111
EMail: wnsharpe@jhuvms.bitnet
Biomedical Eng
Murray Sachs
Phone: (410)955-3131   Fax: (410)955-0549
EMail: msachs@eureka-gold.wbme.jhu.edu
Computer Science
Gerald M. Masson
Phone: (410)516-7013   Fax: (410)516-6134
EMail: masson@jhuvms.bitnet
Geography & Environmental Eng
Charles R. O'Melia
Phone: (410)516-7102   Fax: (410)516-8996
EMail: dog_wcro@jhuvms.bitnet
Materials Science & Eng
Robert E. Green, Jr.
Phone: (410)516-6115   Fax: (410)516-5293
EMail: mse-wreg@jhunix.bitnet
Mathematical Sciences
John C. Wierman
Phone: (410)516-7211   Fax: (410)516-7459
EMail: msc_wjcw@jhuvms.bitnet

Admissions Office Contact
Richard M. Fuller
Phone: (410)516-8171   Fax: (410)516-5200

Financial Aid Office Contact
Ellen Frishberg, *Director, Student Financial Services*
Phone: (410)516-8028   Fax: (410)516-6025

101   U of Kansas

Head of the Institution
Gene A. Budig
Phone: (913)864-3131   Fax: (913)864-4120

Head of Engineering
Carl E. Locke, Jr.
Phone: (913)864-3881   Fax: (913)864-3199

Associate/Assistant Deans
Thomas E. Mulinazzi, *Associate Dean*
Phone: (913)864-3881   Fax: (913)864-3199

Department Chair(s)
Chemical & Petroleum Eng
Paul Willhite
Phone: (913)864-4965   Fax: (913)864-4967
Civil Eng
Stanley T. Rolfe
Phone: (913)864-3766   Fax: (913)864-3199
Electrical & Computer Eng
James A. Roberts
Phone: (913)864-4620   Fax: (913)864-3199
Mechanical Eng
Terry Faddis
Phone: (913)864-3181   Fax: (913)864-5254
Aerospace Eng
David R. Downing
Phone: (913)864-4267   Fax: (913)864-3597
Architectural Eng
Thomas E. Glavinich
Phone: (913)864-3434   Fax: (913)864-5394
Eng Physics
Francis Prosser
Phone: (913)864-4626   Fax: (913)864-5262
Eng Management
David Kraft
Phone: (913)341-4554   Fax: (913)341-3301

Admissions Office Contact
Deborah B. Castrop
Phone: (913)864-3911   Fax: (913)864-5230

Financial Aid Office Contact
Diane DelBuono, *Director of Financial Aid*
Phone: (913)864-4700

102   Kansas State U

Head of the Institution
Jon Wefald
Phone: (913)532-6221   Fax: (913)532-7639

Head of Engineering
Donald E. Rathbone
Phone: (913)532-5590   Fax: (913)532-7810
EMail: DEANENGR@KSUVM.KSU.EDU

Associate/Assistant Deans
Kenneth K. Gowdy, *Associate Dean*
Phone: (913)532-5590   Fax: (913)532-7810
Gale G. Simons, *Associate Dean*
Phone: (913)532-5844   Fax: (913)532-7810
John P. Dollar, *Assistant Dean*
Phone: (913)532-5592   Fax: (913)532-7810
Ray E. Hightower, *Assistant Dean*
Phone: (913)532-5592   Fax: (913)532-7810

Department Chair(s)
Chemical Eng
Liang T. Fan
Phone: (913)532-5584   Fax: (913)532-7372
EMail: FAN@KSUVM.KSU.EDU
Civil Eng
Stuart E. Swartz
Phone: (913)532-5862   Fax: (913)532-7810
EMail: ENDSLEY@KSUVM.KSU.EDU
Electrical & Computer Eng
David L. Soldan
Phone: (913)532-5600   Fax: (913)532-7810
EMail: SOLDAN@KSUVM.KSU.EDU
Industrial Eng
R. Michael Harnett
Phone: (913)532-5606   Fax: (913)532-7810
EMail: MRH@KSUVM.KSU.EDU
Mechanical Eng
Donald L. Fenton
Phone: (913)532-5610   Fax: (913)532-7810
EMail: FENTON@KSUVM.KSU.EDU
Architectural Eng & Construction Science
Charles L. Burton
Phone: (913)532-5964   Fax: (913)532-7810
EMail: CLB@KSUVM.KSU.EDU
Nuclear Eng
N. Dean Eckhoff
Phone: (913)532-5625   Fax: (913)532-6952
EMail: NDE@KSUVM.KSU.EDU
Agricultural Eng
Stanley J. Clark
Phone: (913)532-5580   Fax: (913)532-7810
EMail: SJCLARK@KSUVM.KSU.EDU

Admissions Office Contact
Richard N. Elkins
Phone: (913)532-6250   Fax: (913)532-5632

Financial Aid Office Contact
Larry Moeder, *Director*
Phone: (913)532-6420   Fax: (913)532-5632

103   U of Kentucky

Head of the Institution
Charles T. Wethington, Jr.
Phone: (606)257-1704   Fax: (606)257-5640

Head of Engineering
Thomas W. Lester
Phone: (606)257-1687   Fax: (606)258-4922
EMail: thomas.lester@ukwang.uky.edu

Associate/Assistant Deans
Frederick C. Trutt, *Associate Dean for Administration & Undergraduate Studies*
Phone: (606)257-8827   Fax: (606)258-4922
John N. Walker, *Associate Dean for Graduate Studies, Research & Development*
Phone: (606)257-6262   Fax: (606)258-1035

Department Chair(s)
Chemical Eng
Frederick C. Trutt
Phone: (606)257-4956   Fax: (606)257-3342
EMail: fred.trutt@ukwang.uky.edu
Civil Eng
Donn E. Hancher
Phone: (606)257-4857   Fax: (606)257-3342
EMail: donn.e.hancher@ukwang.uky.edu
Electrical Eng
Syed A. Nasar
Phone: (606)257-8042   Fax: (606)257-3342
EMail: syed.a.nasar@ukwang.uky.edu
Mechanical Eng
Andrew F. Seybert
Phone: (606)257-2809   Fax: (606)257-3342
EMail: andrew.f.seybert@ukwang.uky.edu
Agricultural Eng
I. Joseph Ross
Phone: (606)257-3000   Fax: (606)257-5671
EMail: joe.ross@ukwang.uky.edu
Eng Mechanics
Theodore R. Tauchert
Phone: (606)257-2629   Fax: (606)257-3342
EMail: ted.tauchert@ukwang.uky.edu
Materials Science & Eng
Phillip J. Reucroft
Phone: (606)257-8723   Fax: (606)258-1929
EMail: p.j.reucroft@ukwang.uky.edu
Mining Eng
Lee W. Saperstein
Phone: (606)257-1173   Fax: (606)258-1962
EMail: lee.saperstein@ukwang.uky.edu

Admissions Office Contact
Joseph L. Fink, III
Phone: (606)257-2000   Fax: (606)257-3823

Financial Aid Office Contact
David H. Stockham, *Director*
Phone: (606)257-3172

104   Lafayette Coll

Head of the Institution
Robert I. Rotberg
Phone: (215)250-5200   Fax: (215)250-0157

Head of Engineering
Michael A. Paolino
Phone: (215)250-5403   Fax: (215)250-0351
EMail: PM#0@Lafayacs

Associate/Assistant Deans
Rebecca L. Rosenbauer, *Assistant to the Director*
Phone: (215)250-5400   Fax: (215)250-0351

Department Chair(s)
Chemical Eng
J. Ronald Martin
Phone: (215)250-5430   Fax: (215)250-0351
EMail: MJ#4@Lafayacs
Eng
B. Vincent Viscomi
Phone: (215)250-5438   Fax: (215)250-0351
EMail: VB#1@Lafayacs
Civil Eng
Terence J. McGhee
Phone: (215)250-5439   Fax: (215)250-0351
EMail: MT#0@Lafayacs
Electrical Eng
William A. Hornfeck
Phone: (215)250-5423   Fax: (215)250-0351
EMail: HW#4@Lafayacs
Mechanical Eng
Richard A. Merz
Phone: (215)250-5451   Fax: (215)250-0351
EMail: MR#5@Lafayacs

Admissions Office Contact
G. Gary Ripple
Phone: (215)250-5110   Fax: (215)250-9850

Financial Aid Office Contact
Barry W. McCarty, *Director of Student Financial Aid*
Phone: (215)250-5055   Fax: (215)250-9850

105 Lamar U

Head of the Institution
Brock Brentlinger, *Interim*
Phone: (409)880-8401   Fax: (409)880-8404

Head of Engineering
Fred M. Young
Phone: (409)880-8741   Fax: (409)880-8121
EMail: (fred@lub001.lamar.educ)

Department Chair(s)
Chemical Eng
Jack R. Hopper
Phone: (409)880-8784
Civil Eng
Enno Koehn
Phone: (409)880-8759
Electrical Eng
Bernard J. Maxum
Phone: (409)880-8747   Fax: (409)880-8121
Industrial Eng
Victor Zaloom
Phone: (409)880-8805
Mechanical Eng
William E. Simon
Phone: (409)880-8769
Mathematics
John R. Cannon
Phone: (409)880-8792
Computer Science
Ronald S. King
Phone: (409)880-8775

Admissions Office Contact
James Rush
Phone: (409)880-8353

Financial Aid Office Contact
Ralynn Castete, *Director, Student Financial Aid*
Phone: (409)880-8450

106 Lawrence Tech U

Head of the Institution
Richard E. Marburger
Phone: (313)356-0200   Fax: (313)356-6458

Head of Engineering
Joseph B. Olivieri
Phone: (313)356-0200   Fax: (313)356-6458
EMail: Bitnet: OLIVIERI@LTUVAX

Associate/Assistant Deans
Richard S. Maslowski, *Associate Dean*
Phone: (313)356-0200   Fax: (313)356-6458

Department Chair(s)
Civil Eng
Alan L. Prasuhn
Phone: (313)356-0200   Fax: (313)356-6458
EMail: Bitnet: PRASUHN@LTUVAX
Electrical Eng
Robert Farrah
Phone: (313)356-0200   Fax: (313)356-6458
EMail: Bitnet: FARRAH@LTUVAX
Mechanical Eng
Wayne M. Brehob
Phone: (313)356-0200   Fax: (313)356-6458
EMail: Bitnet: WAYNE@LTUVAX

Admissions Office Contact
Timothy R. Kennedy
Phone: (313)356-0200   Fax: (313)356-6458

Financial Aid Office Contact
Paul F. Kinder, *Director of Financial Aid & Veterans Affairs*
Phone: (313)356-0200   Fax: (313)356-6458

107 LeTourneau U

Head of the Institution
Alvin O. Austin
Phone: (903)753-0231   Fax: (903)237-2730

Head of Engineering
John R. Busch
Phone: (903)753-0231   Fax: (903)237-2732

Department Chair(s)
Electrical Eng
R. William Graff
Phone: (903)753-0231   Fax: (903)237-2732
Mechanical Eng
W.C. Crisman
Phone: (903)753-0231   Fax: (903)237-2732
Welding Eng
William H. Kielhorn
Phone: (903)753-0231   Fax: (903)237-2732

Admissions Office Contact
Howard G. Wilson
Phone: (903)753-0231   Fax: (903)237-2732

Financial Aid Office Contact
Willard Rusk
Phone: (903)753-0231   Fax: (903)237-2732

108 Lehigh U

Head of the Institution
Peter W. Likins
Phone: (215)758-3155   Fax: (215)758-5402

Head of Engineering
Sunder H. Advani
Phone: (215)758-5308   Fax: (215)758-5623

Associate/Assistant Deans
Kenneth N. Sawyers, *Associate Dean for Undergraduate Studies*
Phone: (215)758-4025   Fax: (215)758-5623

Department Chair(s)
Chemical Eng
Dennis W. Hess
Phone: (215)758-4260   Fax: (215)758-5057
Civil Eng
Irwin J. Kugelman
Phone: (215)758-3566   Fax: (215)758-4522
Computer Science & Electrical Eng
Kenneth K. Tzeng
Phone: (215)758-4070   Fax: (215)758-5623
Industrial Eng
Marlin U. Thomas
Phone: (215)758-4050   Fax: (215)758-4886
Mechanical Eng
Robert P. Wei
Phone: (215)758-4102   Fax: (215)758-5623
Materials Science & Eng
David B. Williams
Phone: (215)758-4220   Fax: (215)758-4244

Admissions Office Contact
Patricia G. Boig
Phone: (215)758-3100   Fax: (215)758-4361

Financial Aid Office Contact
William E. Stanford, *Director, Financial Aid*
Phone: (215)758-3181   Fax: (215)758-4361

109 Louisiana State U

Head of the Institution
William E. Davis
Phone: (504)388-6977   Fax: (504)388-5982

Head of Engineering
Edward McLaughlin
Phone: (504)388-5701   Fax: (504)388-5990

Associate/Assistant Deans
Julius P. Langlinais, *Associate Dean for Instruction & Undergraduate Activities*
Phone: (504)388-5703   Fax: (504)388-5990
Arthur M. Sterling, *Associate Dean for Research & Graduate Activities*
Phone: (504)388-5700   Fax: (504)388-5990

Department Chair(s)
Chemical Eng
John R. Collier
Phone: (504)388-1426   Fax: (504)388-1476
Civil Eng
Raymond R. Avent
Phone: (504)388-8442   Fax: (504)388-5990
Electrical & Computer Eng
Alan H. Marshak
Phone: (504)388-5241   Fax: (504)388-5200
Industrial & Manufacturing Systems Eng
L.K. Keys
Phone: (504)388-5112   Fax: (504)388-5990
Mechanical Eng
Victor A. Cundy
Phone: (504)388-5792   Fax: (504)388-5990
Biological & Agricultural Eng
Lalit R. Verma
Phone: (504)388-3153   Fax: (504)388-3492
Petroleum Eng
Zaki A. Bassiouni
Phone: (504)388-5215   Fax: (504)388-5990

Admissions Office Contact
Lisa B. Harris
Phone: (504)388-1175   Fax: (504)388-5991

Financial Aid Office Contact
Ester M. Hill
Phone: (504)388-3103   Fax: (504)388-6300

110 Louisiana Tech U

Head of the Institution
Daniel D. Reneau
Phone: (318)257-3785   Fax: (318)257-2928

Head of Engineering
Barry A. Benedict
Phone: (318)257-4647   Fax: (318)257-2562
EMail: benedict@engr.latech.edu

Associate/Assistant Deans
James D. Nelson, *Associate Dean for Academic Affairs*
Phone: (318)257-2842   Fax: (318)257-2562

Department Chair(s)
Chemical Eng
Houston M. Huckabay
Phone: (318)257-2483   Fax: (318)257-2484
EMail: huckabay@engr.latech.edu
Civil Eng
Leslie K. Guice
Phone: (318)257-4176   Fax: (318)257-2562
EMail: guice@engr.latech.edu
Electrical Eng
Louis E. Roemer
Phone: (318)257-4921   Fax: (318)257-2562
EMail: roemer@engr.latech.edu
Industrial Eng
Dileep R. Sule
Phone: (318)257-3394   Fax: (318)257-2562
EMail: sule@engr.latech.edu
Mechanical Eng
Robert O. Warrington
Phone: (318)257-2357   Fax: (318)257-2562
EMail: row@engr.latech.edu
Biomedical Eng
Paul N. Hale
Phone: (318)257-2645   Fax: (318)255-4175
EMail: phale@engr.latech.edu
Computer Science
Barry L. Kurtz
Phone: (318)257-2436   Fax: (318)257-2562
EMail: kurtz@engr.latech.edu
Petroleum Eng & Geosciences
Robert M. Caruthers
Phone: (318)257-3972   Fax: (318)257-2562
EMail: rmc@engr.latech.edu

Admissions Office Contact
Karen T. Akin
Phone: (318)257-3036   Fax: (318)257-2499

Financial Aid Office Contact
Etienna R. Winzer, *Director, Division of Financial Aid*
Phone: (318)257-2641

111 U of Louisville

Head of the Institution
Donald C. Swain
Phone: (502)588-5420   Fax: (502)588-5682

Head of Engineering
Thomas R. Hanley
Phone: (502)588-6281   Fax: (502)588-7033
EMail: TRHANL01@ULKYVM.LOUISVILLE.EDU

Associate/Assistant Deans
Leo B. Jenkins, *Associate Dean*
Phone: (502)588-6281   Fax: (502)588-7033
Scherrill G. Russman, *Assistant Dean*
Phone: (502)588-6281   Fax: (502)588-7033
Donald L. Cole, *Assistant Dean*
Phone: (502)588-6194   Fax: (502)588-7033

Department Chair(s)
Chemical Eng
James C. Watters
Phone: (502)588-6347   Fax: (502)588-6355
EMail: JCWATT01@ULKYVM.LOUISVILLE.EDU
Civil Eng
Louis F. Cohn
Phone: (502)588-6276   Fax: (502)588-7033
EMail: LFCOHN01@ULKYVM.LOUISVILLE.EDU
Electrical Eng
Darrel L. Chenoweth
Phone: (502)588-6289   Fax: (502)588-6807
EMail: DLCHEN01@ULKYVM.LOUISVILLE.EDU
Industrial Eng
Mickey R. Wilhelm
Phone: (502)588-6342   Fax: (502)588-7033
EMail: MRWILH01@ULKYVM.LOUISVILLE.EDU
Mechanical Eng
Harry G. Schaeffer
Phone: (502)588-6331   Fax: (502)588-6053
EMail: HGSCHA01@ULKYVM.LOUIS-VILLE,EDU
Eng Math & Computer Science
Khaled A. Kamel
Phone: (502)588-6304   Fax: (502)852-4713
EMail: KAKAME01@ULKYVM.LOUISVILLE,EDU

Admissions Office Contact
Donald L. Cole
Phone: (502)588-6194   Fax: (502)588-7033

Financial Aid Office Contact
Gilbert B. Tanner, *Director*
Phone: (588)502-5511   Fax: (502)588-0182

112 Loyola Coll in Maryland

Head of the Institution
Joseph A. Sellenger, S.J.
Phone: (410)617-2201

Head of Engineering
David Roswell
Phone: (410)617-2563

Department Chair(s)
Electrical Eng
Robert D. Shelton
Phone: (410)617-2852
Electrical Eng & Eng Science
Robert D. Shelton
Phone: (410)617-2852   Fax: (410)617-5123
EMail: rds@loyola.edu

Admissions Office Contact
Robert D. Shelton
Phone: (410)617-2852   Fax: (410)617-5123

Financial Aid Office Contact
Mark Lindenmeyer
Phone: (410)617-5000

113 Loyola Marymount U

Head of the Institution
Thomas P. O'Malley
Phone: (310)338-2775   Fax: (310)338-2766

Head of Engineering
Gerald S. Jakubowski
Phone: (310)338-2834   Fax: (310)338-7339

Engineering & Engineering Technology Personnel

114 U of Maine

Associate/Assistant Deans
W. Thomas Calder, *Associate*
Phone: (310)338-2823 Fax: (310)338-7339

Department Chair(s)
Civil Eng
Michael E. Mulvihill
Phone: (310)338-2995 Fax: (310)338-7339
Electrical Eng
John A. Page
Phone: (310)338-7358 Fax: (310)338-7339
Mechanical Eng
Joseph P. Callinan
Phone: (310)338-1875 Fax: (310)338-7339

Admissions Office Contact
Matthew X. Fissinger
Phone: (310)338-2750 Fax: (310)338-2797

Financial Aid Office Contact
Donna Palmer, *Director of Financial Aid*
Phone: (310)338-2753

114 U of Maine

Head of the Institution
Frederick E. Hutchinson
Phone: (207)581-1512 Fax: (207)581-1517

Head of Engineering
Norman Smith
Phone: (207)581-2216 Fax: (207)581-2220

Associate/Assistant Deans
Wayne A. Hamilton, *Associate Dean*
Phone: (207)581-2217 Fax: (207)581-2220
Clinton H. Winne, *Assistant Dean*
Phone: (207)581-2225 Fax: (207)581-2220

Department Chair(s)
Chemical Eng
Joseph M. Genco
Phone: (207)581-2277 Fax: (207)581-2323
Civil Eng
Chet A. Rock
Phone: (207)581-2171 Fax: (207)581-2202
Electrical & Computer Eng
John C. Field
Phone: (207)581-2223 Fax: (207)581-2220
Mechanical Eng
Donald A. Grant
Phone: (207)581-2120 Fax: (207)581-2202
Surveying Eng
David A. Tyler
Phone: (207)581-2188 Fax: (207)581-2206
Eng Physics
Kenneth Brownstein
Phone: (207)581-1016 Fax: (207)581-1039
Bio-Resource Eng
Hayden M. Soule
Phone: (207)581-2715 Fax: (207)581-2725
Forest Eng
Thomas J. Corcoran
Phone: (207)581-2846 Fax: (207)581-2858

Admissions Office Contact
William J. Munsey
Phone: (207)581-1561 Fax: (207)581-1556

Financial Aid Office Contact
Peggy L. Crawford, *Director*
Phone: (207)581-1324

115 U of Manitoba

Head of the Institution
Arnold Naimark
Phone: (204)474-9345 Fax: (204)275-1160

Head of Engineering
Garland E. Laliberte
Phone: (204)474-9806 Fax: (204)275-3773
EMail: BITNET LALIBER@ UOFMCC

Associate/Assistant Deans
Sami Rizkalla, *Associate Dean*
Phone: (204)474-9809 Fax: (204)275-3773

Department Chair(s)
Civil Eng
R. Bruce Pinkney
Phone: (204)474-8212 Fax: (204)261-9534
Agricultural Eng
N. Ross Bulley
Phone: (204)474-9868 Fax: (204)275-0233
Electrical & Computer Eng
Steve Onyshko
Phone: (204)474-9603 Fax: (204)261-4639
Geological Eng
Brian Stimpson
Phone: (204)474-8270 Fax: (204)275-0839
Mechanical & Industrial Eng
Douglas W. Ruth
Phone: (204)474-9803 Fax: (204)275-7507

Admissions Office Contact
Shari Campbell
Phone: (204)474-8813 Fax: (204)275-6534

Financial Aid Office Contact
Peter Dueck
Phone: (204)474-9261 Fax: (204)275-6534

116 Mankato State U

Head of the Institution
Richard R. Rush
Phone: (507)389-1111 Fax: (507)389-5859

Head of Engineering
Robert J. Herickhoff
Phone: (507)389-1521 Fax: (507)389-1095

Department Chair(s)
Electrical Eng
Thomas Hendrickson
Phone: (507)389-5747 Fax: (507)389-1095
Mechanical Eng
Jerzy Fiszdon
Phone: (507)389-6383 Fax: (507)389-1095

Admissions Office Contact
Jack Parkins
Phone: (507)389-1822 Fax: (507)389-1040

Financial Aid Office Contact
Sandra Loerts, *Director, Student Financial Aid Office*
Phone: (507)389-1185 Fax: (507)389-5114

117 Marietta Coll

Head of the Institution
Patrick D. McDonough
Phone: (614)374-4701 Fax: (614)374-4896

Head of Engineering
Robert W. Chase
Phone: (614)374-4776 Fax: (614)374-4896

Department Chair(s)
Industrial Eng
David Cress
Phone: (614)374-4780 Fax: (614)374-4896
Petroleum Eng
Robert W. Chase
Phone: (614)374-4776 Fax: (614)374-4896
Binary Eng Programs
David Cress
Phone: (614)374-4780 Fax: (614)374-4896

Admissions Office Contact
Dennis R. DePerro
Phone: (800)331-7896 Fax: (614)374-4896

Financial Aid Office Contact
James M. Bauer, *Associate Dean & Director of Financial Aid*
Phone: (800)331-2709 Fax: (614)374-4896

118 Marquette U

Head of the Institution
Albert J. DiUlio, S.J.
Phone: (414)288-7223

Head of Engineering
Robert L. Reid
Phone: (414)288-6720 Fax: (414)288-7082

Associate/Assistant Deans
Jon K. Jensen, *Assistant Dean for Academic Affairs*
Phone: (414)288-7080 Fax: (414)288-7082
William E. Brower, *Associate Dean for Research*
Phone: (414)288-7081 Fax: (414)288-7082

Department Chair(s)
Civil & Environmental Eng
Keith F. Faherty
Phone: (414)288-7030 Fax: (414)288-7082
Electrical & Computer Eng
Russell J. Niederjohn
Phone: (414)288-6820 Fax: (414)288-7082
Industrial Eng
G.E.O. Widera
Phone: (414)288-3543 Fax: (414)288-7082
Mechanical Eng
G.E.O. Widera
Phone: (414)288-7259 Fax: (414)288-7082
Biomedical Eng
John H. Linehan
Phone: (414)288-3375 Fax: (414)288-7082

Admissions Office Contact
Michael T. Istwan
Phone: (414)288-7302 Fax: (414)288-3764

Financial Aid Office Contact
Daniel L. Goyette, *Director of Student Financial Aid*
Phone: (414)288-7390

119 Maryland Baltimore Cty

Head of the Institution
Freeman A. Hrabowski, III
Phone: (410)455-2274 Fax: (410)455-1210

Head of Engineering
Duane F. Bruley
Phone: (410)455-3714 Fax: (410)455-3559
EMail: bruley@umbc2.umbc.edu

Department Chair(s)
Chemical & Biochemical Eng
Antonio R. Moreira
Phone: (410)455-3417 Fax: (410)455-1049
Electrical Eng
Gary M. Carter
Phone: (410)455-3509 Fax: (410)455-3559
Mechanical Eng
Akhtar S. Khan
Phone: (410)455-3301 Fax: (410)455-1052
Eng Management
Theodore W. Cadman
Phone: (410)455-3434 Fax: (410)455-3559

Admissions Office Contact
Mindy A. Hand
Phone: (410)455-2291 Fax: (410)455-1094

Financial Aid Office Contact
Thomas R. Taylor, Jr., *Director of Financial Aid*
Phone: (410)455-2387 Fax: (410)455-1094

120 U of Maryland, Coll Park

Head of the Institution
William E. Kirwan
Phone: (301)405-5803

Head of Engineering
George E. Dieter
Phone: (301)405-3868 Fax: (301)314-9867

Associate/Assistant Deans
Herbert Rabin, *Associate Dean for Research*
Phone: (301)405-3904 Fax: (301)314-9867
Marilyn R. Berman, *Associate Dean for Student Affairs*
Phone: (301)405-3871 Fax: (301)314-9867
Horace L. Russell, *Assistant Dean for Undergraduate Studies*
Phone: (301)405-5284 Fax: (301)314-9867

Department Chair(s)
Chemical Eng
Richard V. Calabrese
Phone: (301)405-1939
EMail: RVC@Eng.UMD.EDU
Civil Eng
James Colville
Phone: (301)405-1972 Fax: (301)314-9867
Electrical Eng
William W. Destler
Phone: (301)405-3683
EMail: Destler@eng.umd.edu
Mechanical Eng
Davinder K. Anand
Phone: (301)405-5294
Fire Protection Eng
John L. Bryan
Phone: (301)405-3996
EMail: JBRYAN@UMDACC.UMD.EDU
Materials & Nuclear Eng
Manfred R. Wuttig
Phone: (301)405-5946
EMail: Wuttig@chemserv.eng.umd.edu
Agricultural Eng
Larry E. Stewart
Phone: (301)405-1193
Aerospace Eng
Sung W. Lee
Phone: (301)405-1129
EMail: LEE@Hellcat.UMD.EDU

Admissions Office Contact
Linda Clement
Phone: (301)314-8385

Financial Aid Office Contact
Ulysses Glee, Jr., *Director*
Phone: (301)314-8313

121 U of Mass

Head of the Institution
Michael K. Hooker
Phone: (617)287-7000

Head of Engineering
Keith R. Carver
Phone: (413)545-0300 Fax: (413)545-0724

Associate/Assistant Deans
Duane E. Cromack, *Associate Dean*
Phone: (413)545-0300 Fax: (413)545-0724
Nancy B. Hellman, *Assistant Dean for Undergraduate Affairs*
Phone: (413)545-2035 Fax: (413)545-0724
Melton M. Miller, *Assistant Dean for Transfer Programs*
Phone: (413)545-1569 Fax: (413)545-0724

Department Chair(s)
Chemical Eng
Michael F. Doherty
Phone: (413)545-2359 Fax: (413)545-1647
Civil Eng
William H. Highter
Phone: (413)545-2508 Fax: (413)545-2840
Electrical Eng
Lewis E. Franks
Phone: (413)545-0962 Fax: (413)545-4611
Industrial Eng
Richard J. Giglio
Phone: (413)545-2851 Fax: (413)545-0724
Mechanical Eng
Thomas R. Blake
Phone: (413)545-5904 Fax: (413)545-1027

Admissions Office Contact
Timm R. Rinehart
Phone: (413)545-0222

Financial Aid Office Contact
Burt Batty, *Director*
Phone: (413)545-0801

122 U of Mass, Dartmouth

Head of the Institution
Joseph C. Deck, *Chancellor*
Phone: (508)999-8004   Fax: (508)999-8860

Head of Engineering
L. Bryce Andersen
Phone: (508)999-8539   Fax: (508)999-8485

Department Chair(s)
Civil Eng
Frederick M. Law
Phone: (508)999-8464   Fax: (508)999-8485
Electrical & Computer Eng
Lee E. Estes
Phone: (508)999-8474   Fax: (508)999-8485
Mechanical Eng
Ronald DiPippo
Phone: (508)999-8541   Fax: (508)999-8485

Admissions Office Contact
Raymond Barrows
Phone: (508)999-8605   Fax: (508)999-8901

Financial Aid Office Contact
Gerald S. Coutinho, *Director*
Phone: (508)999-8632   Fax: (508)999-8901

123 U of Mass, Lowell

Head of the Institution
William Hogan
Phone: (508)934-2201

Head of Engineering
Aldo Crugnola
Phone: (508)934-2575   Fax: (508)452-1445

Associate/Assistant Deans
Louis J. Petrovic, *Assistant Dean*
Phone: (508)934-2577   Fax: (508)452-1445

Department Chair(s)
Chemical & Nuclear Eng
Jose Martin
Phone: (508)934-3167   Fax: (508)452-1445
Civil Eng
Donald G. Leitch
Phone: (508)934-2279   Fax: (508)452-1445
Electrical Eng
Bodo Reinisch
Phone: (508)934-3363   Fax: (508)452-1445
Industrial Eng
David Colling
Phone: (508)934-2591   Fax: (508)452-1445
Mechanical Eng
William Kyros
Phone: (508)934-2959   Fax: (508)452-1445
Plastics Eng
Robert Nunn
Phone: (508)934-3435   Fax: (508)452-1445
Work Environment
David Wegman
Phone: (508)934-3265   Fax: (500)452-1445
Eng Tech
Donald Pottle
Phone: (508)934-2597   Fax: (508)452-1445

Admissions Office Contact
Sharon L. Quigley
Phone: (508)934-2570   Fax: (508)452-1445

Financial Aid Office Contact
Walter Costello, *Director of Financial Aid*
Phone: (508)934-4223

124 Mass Inst of Tech

Head of the Institution
Charles M. Vest
Phone: (617)253-0148   Fax: (617)253-3124

Head of Engineering
Joel Moses
Phone: (617)253-3292   Fax: (617)253-8549
EMail: Moses@LArch.Lcs.MIT.Edu

Associate/Assistant Deans
John B. Vander Sande, *Associate Dean of Engineering*
Phone: (617)253-6933   Fax: (617)253-8549
Alfred R. Doig, *Assistant Dean for Resource Development*
Phone: (617)253-2222   Fax: (617)253-8549
Donna R. Savicki, *Assistant Dean of Engineering for Administration*
Phone: (617)253-3294   Fax: (617)253-8549

Department Chair(s)
Chemical Eng
Robert A. Brown
Phone: (617)253-5726   Fax: (617)253-9695
EMail: RAB@ATHENA.MIT.EDU
Civil & Environmental Eng
Rafael L. Bras
Phone: (617)253-2117   Fax: (617)258-6099
EMail: RLBRAS@MIT.EDU
Electrical Eng & Computer Science
Paul Penfield
Phone: (617)253-2506   Fax: (617)258-7354
EMail: PENFIELD@MIT.EDU
Mechanical Eng
Nam P. Suh
Phone: (617)253-2225   Fax: (617)258-6156
Ocean Eng
T. Francis Ogilvie
Phone: (617)253-4330   Fax: (617)253-8125
Aeronautics & Astronautics
Earll M. Murman
Phone: (617)253-3284   Fax: (617)258-7566
EMail: MURMAN@ATHENA.MIT.EDU
Materials Science & Eng
Merton S. Flemings
Phone: (617)253-3233   Fax: (617)258-6886
EMail: FLEMINGS@MIT.EDU
Nuclear Eng
Mujid S. Kazimi
Phone: (617)253-4206   Fax: (617)258-7437

Admissions Office Contact
Michael C. Behnke
Phone: (617)258-5515   Fax: (617)253-8000

Financial Aid Office Contact
Stanley G. Hudson, *Director of Student Financial Aid*
Phone: (617)253-4971   Fax: (617)258-8301

125 McGill U

Head of the Institution
David L. Johnston
Phone: (514)398-4180   Fax: (514)398-7379

Head of Engineering
Pierre R. Bélanger
Phone: (514)398-7251   Fax: (514)398-7379
EMail: pierre@eng1.lan.McGill.CA

Associate/Assistant Deans
James W. Provan, *Associate Dean (Academic)*
Phone: (541)398-7254   Fax: (514)398-7379
Robert D. Japp, *Associate Dean (Student Affairs)*
Phone: (514)398-7258   Fax: (514)398-7379

Department Chair(s)
Chemical Eng
Musa R. Kamal
Phone: (514)398-4262   Fax: (514)398-6678
Civil Eng
John C. Osler
Phone: (514)398-6857   Fax: (514)398-7361
Electrical Eng
Nicolas C. Rumin
Phone: (514)398-7113   Fax: (514)398-4470
EMail: rumin@rosy.lan.McGill.CA
Mechanical Eng
Abdul M. Ahmed
Phone: (514)398-6275   Fax: (514)398-7365
Mining And Metallurical Eng
John Gruzleski
Phone: (514)398-4365   Fax: (514)398-4492

Admissions Office Contact
Mariela Johansen
Phone: (514)398-3672   Fax: (514)398-4193

Financial Aid Office Contact
Judy Stymest, *Director of Student Aid Office*
Phone: (514)398-6013

126 Memphis State U

Head of the Institution
V. Lane Rawlins
Phone: (901)678-2234   Fax: (901)678-2000

Head of Engineering
John D. Ray
Phone: (901)678-2171   Fax: (901)678-4180

Associate/Assistant Deans
Frank J. Claydon, *Associate Dean*
Phone: (901)678-2171   Fax: (901)678-4180

Department Chair(s)
Civil Eng
Martin Lipinski
Phone: (901)678-2746   Fax: (901)678-4180
Electrical Eng
Carl E. Halford
Phone: (901)678-2175   Fax: (901)678-4180
Mechanical Eng
Ed Perry
Phone: (901)678-2173   Fax: (901)678-4180
Biomedical Eng
Michael Yen
Phone: (901)678-3263   Fax: (901)678-4180

Admissions Office Contact
Carol Ferguson
Phone: (901)678-2171   Fax: (901)678-4180

127 Mercer U

Head of the Institution
R. Kirby Godsey
Phone: (912)752-2500   Fax: (912)752-2108

Head of Engineering
Carroll B. Gambrell
Phone: (912)752-2377   Fax: (912)752-2166

Associate/Assistant Deans
John D. Patterson, *Associate Dean*
Phone: (912)752-2343   Fax: (912)752-2166

Department Chair(s)
Electrical & Computer Eng
Edward M. O'Brien
Phone: (912)752-4146   Fax: (912)752-2166
Biomedical & Environmental Eng
Benjamin S. Kelley
Phone: (912)752-4146   Fax: (912)752-2166
Computer & Information Systems
John L. Palmer
Phone: (912)752-2138   Fax: (912)752-2166
Eng Core
John L. Palmer
Phone: (912)752-2138   Fax: (912)752-2166
Industrial & Systems Eng
Allen F. Grum
Phone: (912)752-2453   Fax: (912)752-2166
Mechanical & Aerospace Eng
Lawrence E. Hooks
Phone: (912)752-2138   Fax: (912)752-2166
Technical Communications
Marjorie Davis
Phone: (912)752-2430   Fax: (912)752-2166

Admissions Office Contact
Lea Weissenburger
Phone: (912)752-2650   Fax: (912)752-2828

Financial Aid Office Contact
Pam Anderson, *Office Manager*
Phone: (912)752-2670   Fax: (912)752-2313

128 U of Miami

Head of the Institution
Edward T. Foote
Phone: (305)284-5155   Fax: (305)284-3768

Head of Engineering
Martin Becker
Phone: (305)284-2404   Fax: (305)284-4792

Associate/Assistant Deans
Samuel S. Lee, *Associate Dean*
Phone: (305)284-2408   Fax: (305)284-4792

Department Chair(s)
Civil Eng
David A. Chin
Phone: (305)284-3391   Fax: (305)284-4792
Electrical Eng
Tzay Young
Phone: (305)284-3291   Fax: (305)284-4792
Industrial Eng
Tarek M. Khalil
Phone: (305)284-2344   Fax: (305)284-4040
Mechanical Eng
Sadik Kakac
Phone: (305)284-2571   Fax: (305)284-4792
EMail: KAKAC@COEGLD.ENG.MIAMI.EDU
Biomedical Eng
Peter P. Tarjan
Phone: (305)284-2135   Fax: (305)284-4792
Architectural Eng
David A. Chin
Phone: (305)284-3391   Fax: (305)284-4792

Admissions Office Contact
Martina S. Hahn
Phone: (305)284-2404   Fax: (305)284-4792

Financial Aid Office Contact
Martin J. Carney, *Director, Financial Assistance Services*
Phone: (305)284-5212   Fax: (305)284-4082

129 Miami U

Head of the Institution
Paul G. Risser
Phone: (513)529-2345   Fax: (513)529-2121

Head of Engineering
David C. Haddad
Phone: (513)529-4036   Fax: (513)529-3841

Associate/Assistant Deans
Errol A. Gundler, *Associate Dean*
Phone: (513)529-4036   Fax: (513)529-3841

Department Chair(s)
Manufacturing Eng
Donald L. Byrkett
Phone: (513)529-2650   Fax: (513)529-3841
Paper Science & Eng
William E. Scott
Phone: (513)529-2200   Fax: (513)529-3841
Eng Management
Errol A. Gundler
Phone: (513)529-4036   Fax: (513)529-3841

Admissions Office Contact
Errol A. Gundler
Phone: (513)529-4036   Fax: (513)529-3841

Financial Aid Office Contact
Diane Stemper, *Director of Student Financial Aid*
Phone: (513)529-4734   Fax: (513)529-3841

130 U of Michigan

Head of the Institution
James J. Duderstadt
Phone: (313)764-6270

Head of Engineering
Peter M. Banks
Phone: (313)764-8475   Fax: (313)763-9487
EMail: Peter_Banks@um.cc.umich.edu

Engineering & Engineering Technology Personnel

Associate/Assistant Deans
Erdogan Gulari, *Senior Associate Dean*
Phone: (313)763-5464   Fax: (313)763-9487
George Carignan, *Associate Dean for Graduate Education & Research*
Phone: (313)763-2174   Fax: (313)763-9487
Michael G. Parsons, *Associate Dean for Undergraduate Education*
Phone: (313)936-3045   Fax: (313)763-9487
Lynn Conway, *Associate Dean for Instruction & Instructional Tech*
Phone: (313)763-5509   Fax: (313)763-9487

Department Chair(s)
Chemical Eng
Johannes Schwank
Phone: (313)763-2384   Fax: (313)763-0459
EMail: Johannes_Schwank@ub.cc.umich.edu
Civil & Environmental Eng
E. Benjamin Wylie
Phone: (313)764-8495   Fax: (313)764-4292
EMail: E._B._Wylie@um.cc.umich.edu
Electrical Eng & Computer Science
George I. Haddad
Phone: (313)764-2390   Fax: (313)763-1503
EMail: George_Haddad@um.cc.umich.edu
Industrial & Operations Eng
Chelsea C. White
Phone: (313)764-6473   Fax: (313)764-3451
EMail: Chip_White@um.cc.umich.edu
Mechanical Eng & Applied Mechanics
Panos Papalambros
Phone: (313)764-8464   Fax: (313)747-3170
EMail: Panos_Papalambros@um.cc.umich.edu
Aerospace Eng
N. Harris McClamroch
Phone: (313)764-3310   Fax: (313)763-0578
EMail: N._Harris_McClamroch@um.cc.umich.edu
Atmospheric, Oceanic & Space Sciences
S. Roland Drayson
Phone: (313)764-3335   Fax: (313)764-4585
EMail: Roland_Drayson@um.cc.umich.edu
Materials Science & Eng
Ronald Gibala
Phone: (313)763-4970   Fax: (313)763-4788
EMail: Ronald_Gibala@um.cc.umich.edu
Naval Architecture & Marine Eng
Robert F. Beck
Phone: (313)764-6470   Fax: (313)936-8820
EMail: Bob_Beck@um.cc.umich.edu
Nuclear Eng
William R. Martin
Phone: (313)764-4260   Fax: (313)763-4540
EMail: Bill_Martin@um.cc.umich.edu

Admissions Office Contact
Theodore L. Spencer
Phone: (313)764-7433

131   U of Michigan-Dearborn

Head of the Institution
Bernard W. Klein
Phone: (313)593-5500   Fax: (313)593-5452

Head of Engineering
Subrata Sengupta
Phone: (313)593-5290   Fax: (313)593-9967

Associate/Assistant Deans
Keshav S. Varde, *Associate Dean*
Phone: (313)593-5117   Fax: (313)593-9967

Department Chair(s)
Electrical & Computer Eng
Malayappan Shridhar
Phone: (313)593-5420   Fax: (313)593-9967
EMail: Shridhar@umdsun2.umd.umich.edu
Industrial & Systems Eng
Swatantra Kachhal
Phone: (313)593-5361   Fax: (313)593-9967
EMail: Kachhal@umdsun2.umd.umich.edu
Mechanical Eng
Chi L. Chow
Phone: (313)593-5241   Fax: (313)593-9967

Admissions Office Contact
Carol Mack
Phone: (313)593-5199   Fax: (313)593-5452

Financial Aid Office Contact
John A. Mason, *Director, Office of Financial Aid*
Phone: (313)593-5300   Fax: (313)593-5452

132   Michigan State U

Head of the Institution
Gordon Guyer
Phone: (517)355-6560   Fax: (517)355-4670

Head of Engineering
Theodore A. Bickart
Phone: (517)355-5113   Fax: (517)355-2288
EMail: BICKART@egr.msu.edu

Associate/Assistant Deans
George Van Dusen, *Assistant Dean*
Phone: (517)355-5128   Fax: (517)336-1356
Nicholas Altiero, *Associate Dean*
Phone: (517)355-3522   Fax: (517)355-2288

Department Chair(s)
Agricultural Eng
Robert D. von Bernuth
Phone: (517)353-7268   Fax: (517)353-8982
EMail: vonbern@age.msu.edu
Chemical Eng
Donald K. Anderson
Phone: (517)355-5135   Fax: (517)336-1105
EMail: ANDERSON@che.msu.edu
Civil Eng
William E. Saul
Phone: (517)355-5107   Fax: (517)336-1827
EMail: saul@ce.msu.edu
Computer Science
Anthony S. Wojcik
Phone: (517)355-5218   Fax: (517)336-1061
EMail: WOJCIK@cps.msu.edu
Electrical Eng
Jes Asmussen
Phone: (517)355-4620   Fax: (517)353-1980
EMail: asmussen@ee.msu.edu
Mechanical Eng
Ronald Rosenberg
Phone: (517)355-5131   Fax: (517)353-1750
EMail: Rosenberg@me.msu.edu
Materials Science & Mechanics
Kalinath Mukherjee
Phone: (517)355-5141   Fax: (517)353-9842
EMail: KMMMM@msu.edu

Admissions Office Contact
William Turner
Phone: (517)355-8332   Fax: (517)336-2069

133   Michigan Tech U

Head of the Institution
Curtis J. Tompkins
Phone: (906)487-2200   Fax: (906)487-2935

Head of Engineering
Vernon B. Watwood, *Interim*
Phone: (906)487-2005   Fax: (906)487-2782
EMail: vbwatwoo@mtu.edu

Associate/Assistant Deans
Duane L. Abata, *Associate Dean*
Phone: (906)487-2005   Fax: (906)487-2782
Thomas G. Ellis, *Assistant Dean*
Phone: (906)487-2005   Fax: (906)487-2782

Department Chair(s)
Chemical Eng
Edward R. Fisher
Phone: (906)487-2047   Fax: (906)487-2061
EMail: ERFISHER@MTUS5
Civil & Environmental Eng
C. Robert Baillod
Phone: (906)487-2520   Fax: (906)487-2943
EMail: crbaillo@mtu.edu
Electrical Eng
Jon A. Soper
Phone: (906)487-2757   Fax: (906)487-2949
EMail: jasoper@mtu.edu
Chemistry
John H. Adler
Phone: (906)487-2047   Fax: (906)487-2061
Geological Eng, Geology, & Geophysics
William I. Rose
Phone: (906)487-2531   Fax: (906)487-3371
EMail: raman@mtu.edu (internet) sulfur@mtuS5 (Bitnet)
Mining Eng
Duane L. Abata
Phone: (906)487-2610   Fax: (906)487-2495
EMail: duane@mtu.edu
Metallurgical & Materials Eng
Thomas Courtney
Phone: (906)487-2630   Fax: (906)487-2934
Mechanical Eng
John H. Johnson
Phone: (906)487-2551   Fax: (906)487-2822
EMail: jjohnson@mtu.edu
General Eng
Vernon B. Watwood
Phone: (906)487-3057   Fax: (906)487-2582
EMail: vbwatwoo@mtu.edu

Admissions Office Contact
Joseph Galetto
Phone: (906)487-2335   Fax: (906)487-2245

Financial Aid Office Contact
Tim T. Malette, *Director of Financial Aid*
Phone: (906)487-2622   Fax: (906)487-2398

134   Milwaukee Sch of Eng

Head of the Institution
Hermann Viets
Phone: (414)277-7100   Fax: (414)277-7468

Head of Engineering
Thomas W. Davis
Phone: (414)277-7324   Fax: (414)277-7470
EMail: davis@kirk.msoe.edu

Department Chair(s)
Electrical Eng
Ray W. Palmer
Phone: (414)277-7325   Fax: (414)277-7465
EMail: palmer@kirk.msoe.edu
Industrial Eng
John H. Farrow
Phone: (414)277-7287   Fax: (414)277-7470
Mechanical Eng
John H. Farrow
Phone: (414)277-7287   Fax: (414)277-7470
Computer Eng
Ray W. Palmer
Phone: (414)277-7325   Fax: (414)277-7465
EMail: palmer@kirk.msoe.edu
Biomedical Eng
Ray W. Palmer
Phone: (414)277-7325   Fax: (414)277-7465
EMail: palmer@kirk.msoe.edu
Architectural Eng
Matthew W. Fuchs
Phone: (414)277-7302   Fax: (414)277-7479

Admissions Office Contact
T.O. Smith
Phone: (800)332-6763   Fax: (414)277-7475

Financial Aid Office Contact
Susan Hebert, *Director of Financial Aid*
Phone: (414)277-7224   Fax: (414)277-7450

135   U of Minnesota, Duluth

Head of the Institution
Lawrence A. Ianni
Phone: (218)726-7106   Fax: (218)726-6535

Head of Engineering
Sabra S. Anderson
Phone: (218)726-7201   Fax: (218)726-6360
EMail: SANDERSO@UB.D.UMN.EDU

Associate/Assistant Deans
Timothy B. Holst, *Associate Dean*
Phone: (218)726-7585   Fax: (218)726-6360
Janny B. Walker, *Assistant to the Deans*
Phone: (218)726-7585   Fax: (218)726-6360

Department Chair(s)
Chemical Eng
Dianne Dorland
Phone: (218)726-7126   Fax: (218)726-6360
EMail: DDORLAND@UB.D.UMN.EDU
Industrial Eng
Alden L. Kendall
Phone: (218)726-6161   Fax: (218)726-6360
EMail: LKENDALL@UB.D.UMN.EDU
Computer Eng
Nazmi M. Shehadeh
Phone: (218)726-6147   Fax: (218)726-7267
EMail: NSHEHADE@UB.D.UMN.EDU

Admissions Office Contact
Gerald R. Allen
Phone: (218)726-7500   Fax: (218)726-6389

Financial Aid Office Contact
Nicholas F. Whelihan, *Director*
Phone: (218)726-7500   Fax: (218)726-8787

136   U of Minnesota

Head of the Institution
Nils Hasselmo
Phone: (612)626-1616   Fax: (612)625-3875

Head of Engineering
Gordon S. Beavers
Phone: (612)624-2006   Fax: (612)624-2841
EMail: beavers@mailbox.mail.umn.edu

Associate/Assistant Deans
Sally Gregory Kohlstedt, *Associate Dean for Academic Affairs*
Phone: (612)626-1802   Fax: (612)625-2841
Russell K. Hobbie, *Associate Dean for Student Affairs*
Phone: (612)624-5091   Fax: (612)624-0261
Walter H. Johnson, *Acting Associate Dean*
Phone: (612)624-2006   Fax: (612)624-2841

Department Chair(s)
Chemical Eng & Materials Science
H.T. Davis
Phone: (612)625-1313   Fax: (612)626-7246
EMail: davis@csfa.cs.umn.edu
Civil & Mineral Eng
Steven Crouch
Phone: (612)625-5522   Fax: (612)624-0293
Electrical Eng
Mostafa Kaveh
Phone: (612)625-0720   Fax: (612)625-4583
EMail: kaveh@ee.umn.edu
Mechanical Eng
Richard J. Goldstein
Phone: (612)625-5552   Fax: (612)625-3434
EMail: goldstei@mailbox.mail.umn.edu
Aerospace Eng & Mechanics
William L. Garrard
Phone: (612)625-8000   Fax: (612)626-1558
EMail: garrard@aem.umn.edu
Agricultural Eng
R. Vance Morey
Phone: (612)625-7733   Fax: (612)624-3005
EMail: rvmorey@ux.acs.umn.edu
Computer Science
K.S.P. (Pat) Kumar
Phone: (612)625-4002   Fax: (612)625-0572
EMail: kumar@ee.umn.edu

137 U of Mississippi

Head of the Institution
R. Gerald Turner, *Chancellor*
Phone: (601)232-7111   Fax: (601)232-5935

Head of Engineering
Allie M. Smith
Phone: (601)232-7407   Fax: (601)232-7796
EMail: ENAS@UMSVM

Associate/Assistant Deans
James G. Vaughan, *Associate Dean*
Phone: (601)232-5378   Fax: (601)232-7191
Damon Wall, *Assistant Dean*
Phone: (601)232-5373   Fax: (601)232-7796

Department Chair(s)
Chemical Eng
Peter C. Sukanek
Phone: (601)232-7023   Fax: (601)232-7796
EMail: CMPES@UMSVM

Civil Eng
Robert M. Hackett
Phone: (601)232-7191   Fax: (601)232-7191
EMail: ENAS@UMSVM

Electrical Eng
Charles E. Smith
Phone: (601)232-7231   Fax: (601)232-7010
EMail: EEDEPT@UMSVM

Mechanical Eng
Jeffrey A. Roux
Phone: (601)232-7219   Fax: (601)232-7191
EMail: MEJAR@UMSVM

Geology & Geological Eng
George D. Brunton
Phone: (601)232-7498   Fax: (601)232-7796
EMail: GEBRUNTN@UMSVM

Computer & Information Science
P. Tobin Maginnis
Phone: (601)232-7396   Fax: (601)232-7010
EMail: PTM@CS.OLEMISSEDU

Admissions Office Contact
Allie M. Smith
Phone: (601)232-7407   Fax: (601)232-7796

Financial Aid Office Contact
Thomas G. Hood, *Director of Financial Aid*
Phone: (601)232-7175   Fax: (601)234-8155

138 Mississippi State U

Head of the Institution
Donald W. Zacharias
Phone: (601)325-3221   Fax: (601)325-8028

Head of Engineering
Robert A. Altenkirch
Phone: (601)325-2269   Fax: (601)325-8573
EMail: raa@de.msstate.edu

Associate/Assistant Deans
Clayborne D. Taylor, *Associate Dean for Research & Graduate Studies*
Phone: (601)325-2269   Fax: (601)325 8573
William N. Smyer, *Assistant Dean for Undergraduate Affairs*
Phone: (601)325-2266   Fax: (601)325-8573

Department Chair(s)
Chemical Eng
Donald O. Hill
Phone: (601)325-2480   Fax: (601)325-2482
EMail: hill@che.MsState.Edu

Civil Eng
Joseph H. Sherrard
Phone: (601)325-3050   Fax: (601)325-8573
EMail: sherrard@civil.MsState.Edu

Electrical & Computer Eng
G. Marshall Molen
Phone: (601)325-3912   Fax: (601)325-2298
EMail: molen@ee.MsState.Edu

Industrial Eng
Larry G. Brown
Phone: (601)325-3865   Fax: (601)325-8573
EMail: lgb@dengr.MsState.Edu

Mechanical & Nuclear Eng
W. Glenn Steele
Phone: (601)325-3260   Fax: (601)325-8573
EMail: wgs1@me.MsState.Edu

Agricultural & Biological Eng
R. Kenneth Matthes
Phone: (601)325-3282   Fax: (601)325-3853
EMail: kmatt@abe.MsState.Edu

Aerospace Eng
John C. McWhorter, III
Phone: (601)325-3623   Fax: (601)325-8573
EMail: mcwho@ae.MsState.Edu

Petroleum Eng
Charles T. Carley
Phone: (601)325-3607   Fax: (601)325-7223
EMail: ct@carpenter.MsState.Edu

Admissions Office Contact
Jerry B. Inmon
Phone: (601)325-2224   Fax: (601)325-1846

Financial Aid Office Contact
Audrey S. Lambert, *Director of Student Financial Aid & Scholarships*
Phone: (601)325-2450

139 U of Missouri-Columbia

Head of the Institution
Charles Kiesler
Phone: (314)882-3387

Head of Engineering
Anthony L. Hines
Phone: (314)882-4378   Fax: (314)882-2490

Associate/Assistant Deans
Paul W. Braisted, *Associate*
Phone: (314)882-4375   Fax: (314)882-2490
Elaine M. Charlson, *Associate*
Phone: (314)882-4375   Fax: (314)882-2490

Department Chair(s)
Chemical Eng
Dabir S. Viswanath
Phone: (314)882-4877   Fax: (314)884-4940

Civil Eng
James W. Baldwin
Phone: (314)882-6084   Fax: (314)882-4784

Electrical & Computer Eng
Jon Meese
Phone: (314)882-3379   Fax: (314)882-0397

Industrial Eng
Larry G. David
Phone: (314)882-2692   Fax: (314)882-0397

Mechanical & Aerospace Eng
Richard C. Warder
Phone: (314)882-9569   Fax: (314)882-2490

Nuclear Eng
Jay F. Kunze
Phone: (314)882-3550   Fax: (314)882-0397

Agricultural Eng
James C. Frisby
Phone: (314)882-2369   Fax: (314)882-1115

Civil Eng (Kansas City)
George F. Hauck
Phone: (816)235-1286   Fax: (816)235-1260

Electrical Eng (Kansas City)
Jerome Knopp
Phone: (816)235-5267   Fax: (816)235-1260

Mechanical Eng (Kansas City)
Donald R. Smith
Phone: (816)235-1252   Fax: (816)235-1260

Financial Aid Office Contact
Joseph M. Camille, *Director*
Phone: (314)882-3795

140 U of Missouri-Kansas City

Head of the Institution
George A. Russell
Phone: (314)882-2011   Fax: (314)884-4204

Head of Engineering
Anthony L. Hines
Phone: (314)882-4378   Fax: (314)882-0397

Associate/Assistant Deans
Paul W. Braisted, *Associate Dean*
Phone: (314)882-4377   Fax: (314)882-0397

Department Chair(s)
Civil Eng
George F. Hauck
Phone: (816)235-5268   Fax: (816)235-1260

Electrical & Computer Eng
Jerome Knopp
Phone: (816)235-1277   Fax: (816)235-1260

Mechanical & Aerospace Eng
Donald R. Smith
Phone: (816)235-1461   Fax: (816)235-1260

Admissions Office Contact
Lorraine F. O'Brien
Phone: (816)235-1250   Fax: (816)235-1260

Financial Aid Office Contact
Buford B. Baber, *Director, Financial Aid*
Phone: (816)235-1154

141 U of Missouri-Rolla

Head of the Institution
John T. Park
Phone: (314)341-4114   Fax: (314)341-6306

Head of Engineering
Robert L. Davis
Phone: (314)341-4151   Fax: (314)341-4979

Associate/Assistant Deans
Don L. Warner, *Dean, School of Mines & Metallurgy*
Phone: (314)341-4153   Fax: (314)341-4192
Ronald A. Kohser, *Assistant Dean, Mines & Metallurgy*
Phone: (314)341-4734   Fax: (314)341-4192
Jerry R. Bayless, *Associate Dean, Engineering Undergraduate Affairs*
Phone: (314)341-4151   Fax: (314)341-4979
Gary K. Patterson, *Associate Dean, Engineering, Graduate Affairs & Research*
Phone: (314)341-4151   Fax: (314)341-4979
Nicholas Tsoulfanidis, *Assistant Dean, Mines & Metallurgy*
Phone: (314)341-4745   Fax: (314)341-4192

Department Chair(s)
Chemical Eng
Stephen L. Rosen
Phone: (314)341-4443

Civil Eng
Joseph E. Minor
Phone: (314)341-4461   Fax: (314)341-4729

Electrical Eng
Walter J. Gajda, Jr.
Phone: (314)341-4509   Fax: (314)341-4532

Basic Eng
Ronald D. Fannin
Phone: (314)341-4581

Ceramic Eng
Robert E. Moore
Phone: (314)341-4401

Geological & Petroleum Eng
John D. Rockaway
Phone: (314)341-4867

Metallurgical Eng
John L. Watson
Phone: (314)341-4753

Nuclear Eng
Albert E. Bolon
Phone: (314)341-4720

Eng Management
Yildirim Omurtag
Phone: (314)341-4558

Mechanical & Aerospace Eng & Eng Mechanics
Bassem F. Armaly
Phone: (314)341-4662

Mining Eng
John W. Wilson
Phone: (314)341-4753

Admissions Office Contact
David J. Allen
Phone: (800)522-0938   Fax: (314)341-6308

Financial Aid Office Contact
Robert W. Whites, *Associate Director*
Phone: (314)341-4282   Fax: (314)341-4082

142 Monmouth Coll

Head of the Institution
Samuel H. Magill
Phone: (908)571-3402   Fax: (908)571-3570

Head of Engineering
Richard A. Kuntz
Phone: (908)571-3409   Fax: (908)571-3523
EMail: kuntz@monmouth.edu

Department Chair(s)
Electronic Eng
Richard W. Benjamin
Phone: (908)571-3446   Fax: (908)571-3693
EMail: benjamin@monmouth.edu

Admissions Office Contact
Barry W. Ward
Phone: (908)571-3456   Fax: (908)571-3629

Financial Aid Office Contact
Hank Mackiewicz, *Dean of Financial Aid*
Phone: (908)571-3463   Fax: (908)571-3629

143 Montana Coll of Mineral Sci

Head of the Institution
Lindsay D. Norman
Phone: (406)496-4129   Fax: (406)496-4133

Head of Engineering
Thomas Waring
Phone: (406)496-4127   Fax: (406)496-4133

Associate/Assistant Deans
Dan Bradley, *Associate Dean, Petroleum Engineering & Geoph. Eng.*
Phone: (406)496-4254   Fax: (406)496-4133
H. Peter Knudsen, *Associate Dean, Mining & Minerals Engineering*
Phone: (406)496-4395   Fax: (406)496-4133
Rodney James, *Associate Dean, Environmental Engineering & Natural Sciences*
Phone: (406)496-4446   Fax: (406)496-4133

Department Chair(s)
Petroleum Eng
Dan Bradley
Phone: (406)496-4254   Fax: (406)496-4133

Environmental Eng
Rodney James
Phone: (406)496-4446   Fax: (406)496-4133

Metallurgical Eng
Theodore Jordan
Phone: (406)496-4112   Fax: (406)496-4133

Geophysical Eng
William Sill
Phone: (406)496-4216   Fax: (406)496-4133

Geological Eng
Willis Weight
Phone: (406)496-4329   Fax: (406)496-4133

Mining Eng
H. Peter Knudsen
Phone: (406)496-4395   Fax: (406)496-4133

Admissions Office Contact
Ed Johnson
Phone: (406)496-4178   Fax: (406)496-4133

Financial Aid Office Contact
Frank Kondelis, *Director of Financial Aid*
Phone: (406)496-4212   Fax: (406)496-4133

144 Montana State U

Head of the Institution
Michael P. Malone
Phone: (406)994-2341   Fax: (406)994-1893

Head of Engineering
David F. Gibson
Phone: (406)994-2272 Fax: (406)994-6098
EMail: ADEDG@MTSUNIX1.BITNET

Associate/Assistant Deans
Joseph J. Fedock, *Associate Dean*
Phone: (406)994-2272 Fax: (406)994-6098

Department Chair(s)
Chemical Eng
John T. Sears
Phone: (406)994-2222 Fax: (406)994-6098
EMail: john_s@ercvx1.montana.edu
Civil & Agricultural Eng
Theodore E. Lang
Phone: (406)994-2111 Fax: (406)994-6105
EMail: marla_w@civil.montana.edu
Electrical Eng
Victor Gerez
Phone: (406)994-2505 Fax: (406)994-6098
EMail: victor_g@ece.montana.edu
Industrial & Management Eng
William R. Taylor
Phone: (406)994-3971 Fax: (406)994-6098
EMail: bob_t@ercvx1.montana.edu
Mechanical Eng
Byron Bennett
Phone: (406)994-2203 Fax: (406)994-6292
EMail: ameby@msu.oscs.montana.edu

Admissions Office Contact
Rhonda Duffus
Phone: (406)994-2452 Fax: (406)994-1923

Financial Aid Office Contact
James R. Craig, *Director, Financial Aid Services*
Phone: (406)994-2845 Fax: (406)994-5488

145 Morgan State U

Head of the Institution
Earl S. Richardson
Phone: (410)319-3200 Fax: (410)319-3107

Head of Engineering
Eugene M. Deloatch
Phone: (410)319-3231 Fax: (410)319-3843
EMail: deloatch@echo.eng.umd.edu

Associate/Assistant Deans
Jerome A. Atkins, *Assistant Dean*
Phone: (410)319-3231 Fax: (410)319-3843

Department Chair(s)
Civil Eng
Lewis P. Clopton
Phone: (410)319-3903 Fax: (410)319-3843
EMail: clopton@echo.eng.umd.edu
Electrical Eng
Pamela L. Mack
Phone: (410)319-3073 Fax: (410)319-3843
EMail: mack@echo.eng.umd.edu
Industrial Eng
Irving J. Winters
Phone: (410)319-3129 Fax: (410)319-3843
EMail: winters@echo.eng.umd.edu

Admissions Office Contact
Chelseia Harold-Miller
Phone: (410)319-3000

Financial Aid Office Contact
Reginald T. Cureton, *Director of Financial Aid*
Phone: (410)319-3170 Fax: (410)319-3852

146 U of Nebraska - Lincoln

Head of the Institution
Graham B. Spanier
Phone: (402)472-2116 Fax: (402)472-5110

Head of Engineering
Stanley R. Liberty
Phone: (402)472-3181 Fax: (402)472-7792
EMail: sliberty@unl.edu

Associate/Assistant Deans
Morris H. Schneider, *Associate Dean*
Phone: (402)472-3181 Fax: (402)472-7792

Samy E.G. Elias, *Associate Dean for Research*
Phone: (402)472-3810 Fax: (402)472-7792

Department Chair(s)
Chemical Eng
Richard E. Gilbert
Phone: (402)472-2750 Fax: (402)472-2750
EMail: chrdreg@unl.ed
Civil Eng
William E. Kelly
Phone: (402)472-2371 Fax: (402)472-8934
EMail: cerdwek@unl.ed
Electrical Eng
Rodney J. Soukup
Phone: (402)472-3771 Fax: (402)472-4732
EMail: eerdrjs@unl.edu
Industrial & Management Systems Eng
Michael W. Riley
Phone: (402)472-3495 Fax: (402)472-7792
EMail: ieidrile@unl.edu
Mechanical Eng
Russell C. Nelson
Phone: (402)472-2375 Fax: (402)472-1465
Agricultural Eng
Glenn R. Hoffman
Phone: (402)472-1413 Fax: (402)472-6338
EMail: bsen001@unlvm
Bilogical Systems Eng
Glenn R. Hoffman
Phone: (402)472-1413 Fax: (402)472-6338
EMail: bsen001@unlvm
Computer Eng
Joseph Y-T Leung
Phone: (402)472-3200 Fax: (402)472-7767
EMail: jyl@centaur.unl.edu

Admissions Office Contact
John E. Beacon
Phone: (402)472-3620 Fax: (402)472-3603

Financial Aid Office Contact
John E. Beacon, *Director*
Phone: (402)472-2030 Fax: (402)472-3603

147 U of Nevada, Las Vegas

Head of the Institution
Robert C. Maxson
Phone: (702)895-3201 Fax: (702)895-1088

Head of Engineering
William R. Wells
Phone: (702)895-3699 Fax: (702)895-4059
EMail: wcube@unlv.edu

Associate/Assistant Deans
Walter C. Vodrazka, *Associate Dean*
Phone: (702)895-3699 Fax: (702)895-4059

Department Chair(s)
Civil & Environmental Eng
Edward S. Neumann
Phone: (702)895-3701 Fax: (702)895-3936
Electrical & Computer Eng
William L. Brogan
Phone: (702)597-4183 Fax: (702)895-4075
EMail: eewlb@big.unlv.edu
Mechanical Eng
Robert F. Boehm
Phone: (702)895-1331 Fax: (702)895-3936
EMail: boehm@mg.unlv.edu

Admissions Office Contact
Larry Mason
Phone: (702)895-3443 Fax: (702)895-3850

Financial Aid Office Contact
Judy Belanger, *Director*
Phone: (702)895-3695 Fax: (702)895-1353

148 New England Coll

Head of the Institution
William R. O'Connell, Jr.
Phone: (603)428-2222 Fax: (603)428-7230

Head of Engineering
Donald G. Blanchard
Phone: (603)428-2245 Fax: (603)428-7230

Department Chair(s)
Civil Eng
Donald G. Blanchard
Phone: (603)428-2245 Fax: (603)428-7230

Admissions Office Contact
John F. Spaulding
Phone: (603)428-2223 Fax: (603)428-7230

Financial Aid Office Contact
Sandy Schneider, *Director*
Phone: (603)428-2211 Fax: (603)428-7230

149 U of New Hampshire

Head of the Institution
Dale F. Nitzschke
Phone: (603)862-2450 Fax: (603)862-3060

Head of Engineering
Otis J. Sproul
Phone: (603)862-1781 Fax: (603)862-2486

Associate/Assistant Deans
Donald W. Melvin, *Associate Dean*
Phone: (603)862-3101 Fax: (603)862-2486

Department Chair(s)
Chemical Eng
Stephen S. Fan
Phone: (603)862-3656 Fax: (603)862-3747
Civil Eng
David L. Gress
Phone: (603)862-1428 Fax: (603)862-2364
Electrical & Computer Eng
John L. Pokoski
Phone: (603)862-1355 Fax: (603)862-2486
Mechanical Eng
Kenneth C. Baldwin
Phone: (603)862-1352 Fax: (603)862-2486

Admissions Office Contact
David W. Kraus
Phone: (603)862-1360

Financial Aid Office Contact
Richard Craig, *Director, Financial Aid*
Phone: (603)862-3600

150 U of New Haven

Head of the Institution
Lawrence J. DeNardis
Phone: (203)932-7275 Fax: (203)937-0756

Head of Engineering
M.J. Kenig
Phone: (203)932-7168 Fax: (203)932-7394

Associate/Assistant Deans
John Sarris, *Associate Dean*
Phone: (203)932-7146 Fax: (203)932-7394
B. Badri Saleeby, *Special Assistant to the Dean*
Phone: (203)932-7172 Fax: (203)932-7394
Betsy J. Hogan, *Coordinator of Special Programs*
Phone: (203)932-7728 Fax: (203)932-7394

Department Chair(s)
Chemistry/Chemical Eng
Michael J. Saliby
Phone: (203)932-7169 Fax: (203)932-7394
Civil & Environmental Eng
David J. Wall
Phone: (203)932-7157 Fax: (203)932-7394
Electrical & Computer Eng
Andrew J. Fish
Phone: (203)932-7162 Fax: (203)932-7394
Industrial Eng
M. Ali Montazer
Phone: (203)932-7050 Fax: (203)932-7394
Mechanical Eng
John Sarris
Phone: (203)932-7146 Fax: (203)932-7394

Computer Science
Roger G. Frey
Phone: (203)932-7065 Fax: (203)932-7394

Admissions Office Contact
Steven Briggs
Phone: (203)932-7319 Fax: (203)933-5610

Financial Aid Office Contact
Jane C. Sangeloty, *Director of Financial Aid*
Phone: (203)932-7312

151 New Jersey Inst of Tech

Head of the Institution
Saul K. Fenster
Phone: (201)596-3101 Fax: (201)624-2541

Head of Engineering
George Pincus
Phone: (201)596-3213 Fax: (201)596-2316
EMail: Pincus@Admin1.NJIT.Edu

Associate/Assistant Deans
Richard Parker, *Associate Dean of Engineering*
Phone: (201)596-3223 Fax: (201)596-2316

Department Chair(s)
Chemical Eng, Chemistry & Env'l Science
Gordon Lewandowski
Phone: (201)596-3573 Fax: (201)596-8436
EMail: Lewandows@Admin1.NJIT.Edu
Civil & Environmental Eng
William Spillers
Phone: (201)596-2479 Fax: (201)242-1823
EMail: Spillers@Admin1.NJIT.Edu
Electrical & Computer Eng
Jacob Klapper
Phone: (201)596-3516 Fax: (201)596-5680
EMail: Klapper@Admin1.NJIT.Edu
Mechanical & Industrial Eng
Bernard Koplik
Phone: (201)596-3331 Fax: (201)643-0674
EMail: Koplik@Admin1.NJIT.Edu
Mechanical & Industrial Eng
Bernard Koplik
Phone: (201)596-3331 Fax: (201)643-0674
EMail: Koplik@Admin1.NJIT.Edu
Eng Tech
Robert English
Phone: (201)596-3224 Fax: (201)596-2316
EMail: English_P@Admin1.NJIT.Edu

Admissions Office Contact
William Anderson
Phone: (201)596-3300 Fax: (201)802-1854

Financial Aid Office Contact
Mary Hurdle, *Director of Financial Aid*
Phone: (201)596-3479

152 The U of New Mexico

Head of the Institution
Richard E. Peck
Phone: (505)277-2626 Fax: (505)277-5965

Head of Engineering
James E. Thompson
Phone: (505)277-5521 Fax: (505)277-0813

Associate/Assistant Deans
David Kauffman, *Associate Dean*
Phone: (505)277-5521 Fax: (505)277-0813
Brian T. Smith, *Assistant Dean*
Phone: (505)277-5521 Fax: (505)277-0813

Department Chair(s)
Chemical & Nuclear Eng
Norman F. Roderick
Phone: (505)277-5431 Fax: (505)277-0813
EMail: roderick@unmb
Civil Eng
Jerome W. Hall
Phone: (505)277-2722 Fax: (505)277-1988
Electrical & Computer Eng
Nasir Ahmed
Phone: (505)277-2436 Fax: (505)277-1439
EMail: ahmed@houdini.eece.unm.edu

Mechanical Eng
Joe H. Mullins
Phone: (505)277-2761 Fax: (505)277-0813
Computer Science
Bernard M. Moret
Phone: (505)277-3112 Fax: (505)277-0813
EMail: moret@unmvax.cs.unm.edu

Admissions Office Contact
Cynthia Stuart
Phone: (505)277-2446 Fax: (505)277-6686

Financial Aid Office Contact
John E. Whiteside, *Director*
Phone: (505)277-2041

153 New Mexico Inst of Mining

Head of the Institution
Laurence H. Lattman
Phone: (505)835-5600 Fax: (505)835-6329

Head of Engineering
Carl J. Popp, *Vice President for Academic Affairs*
Phone: (505)835-5227 Fax: (505)835-6329

Associate/Assistant Deans
James L. Corey, *Associate Vice President, Academic Affairs*
Phone: (505)835-5190 Fax: (505)835-6329

Department Chair(s)
Electrical Eng
Bill Rison
Phone: (505)835-5330 Fax: (505)835-5707
Mineral & Environmental Eng
Catherine Aimone-Martin
Phone: (504)835-5345 Fax: (505)835-5252
Materials & Metallurgical Eng
Osman Inal
Phone: (505)835-5229 Fax: (505)835-5626
Petroleum Eng
Robert Bretz
Phone: (505)835-5412 Fax: (505)835-6329

Admissions Office Contact
Louise Chamberlin
Phone: (800)428-8324 Fax: (505)835-6329

Financial Aid Office Contact
Anne Hansen, *Director, Financial Aid*
Phone: (505)835-5333 Fax: (505)835-6329

154 New Mexico State U

Head of the Institution
James E. Halligan
Phone: (505)646-2035 Fax: (505)646-6334

Head of Engineering
J. Derald Morgan
Phone: (505)646-2914 Fax: (505)646-3549

Associate/Assistant Deans
J. Eldon Steelman, *Associate Dean (Email: esteelma@nmsu.EDU)*
Phone: (505)646-2912 Fax: (505)646-3549
Rose Marie Melon, *Assistant to the Dean*
Phone: (505)646-3547 Fax: (505)646-3549

Department Chair(s)
Chemical Eng
James Eakman
Phone: (505)646-1214 Fax: (505)646-4149
Civil Eng
Kenneth R. White
Phone: (505)646-3801 Fax: (505)646-6049
Electrical Eng
M. Don Merrill
Phone: (505)646-3117 Fax: (505)646-1435
Industrial Eng
Satish Kamat
Phone: (505)646-2974 Fax: (505)646-2976
Mechanical Eng
George Mulholland
Phone: (505)646-3501 Fax: (505)646-6111
Surveying
James P. Reilly
Phone: (505)646-5375 Fax: (505)646-3549

Agricultural Eng
J. Phillip King
Phone: (505)646-6103 Fax: (505)646-6049
Geological Eng
Joseph Finney
Phone: (505)646-4140 Fax: (505)644-6049

Admissions Office Contact
Bill Bruner
Phone: (505)646-3121 Fax: (505)646-6330

Financial Aid Office Contact
Greeley Myers, *Director of Financial Aid*
Phone: (505)646-4105 Fax: (505)646-6330

155 U of New Orleans

Head of the Institution
Gregory M. O'Brien
Phone: (504)286-6201 Fax: (504)286-6872

Head of Engineering
John N. Crisp
Phone: (504)286-6327 Fax: (504)286-7413

Associate/Assistant Deans
Richard R. Bishop, *Associate Dean*
Phone: (504)286-6327 Fax: (504)286-7413
W.J. Lannes, *Associate Dean*
Phone: (504)286-6327 Fax: (504)286-7413

Department Chair(s)
Civil Eng
Kenneth L. McManis
Phone: (504)286-6668 Fax: (504)286-7413
Electrical Eng
Rasheed Azzam
Phone: (504)286-6650 Fax: (504)286-7413
Mechanical Eng
Hudy C. Hewitt, Jr.
Phone: (504)286-6652 Fax: (504)286-7413
Naval Architecture & Marine Eng
Robert Latorre
Phone: (504)286-7180 Fax: (504)286-7413

Admissions Office Contact
Roslyn Sheley
Phone: (504)286-6000

Financial Aid Office Contact
Avon Dennis, *Director*
Phone: (504)286-6603 Fax: (504)286-7393

156 New York Inst of Tech

Head of the Institution
Matthew Schure
Phone: (516)686-7650 Fax: (516)686-6830

Head of Engineering
Heskia Heskiaoff
Phone: (516)686-7931 Fax: (516)626-0673

Department Chair(s)
Electrical Eng
Edward Nelson
Phone: (516)686-7523 Fax: (800)626-0419
Mechanical Eng
Gottlieb Koenig
Phone: (516)686-7828 Fax: (800)626-0419
Industrial Eng
Gottlieb Koenig
Phone: (516)686-7828 Fax: (800)626-0419

Admissions Office Contact
Arthur Lambert
Phone: (516)686-7520 Fax: (516)626-0419

Financial Aid Office Contact
Doreen Meyers, *Director of Financial Aid*
Phone: (516)686-7680 Fax: (516)626-2627

157 NY State Coll of Ceramics

Head of the Institution
Edward G. Coll, Jr.
Phone: (607)871-2101 Fax: (607)871-2339

Head of Engineering
Alastair N. Cormack
Phone: (607)871-2422 Fax: (607)871-2305
EMail: Cormack@Ceramics.alfred.edu

Associate/Assistant Deans
James W. McCauley, *Dean of the New York State Coll of Ceramics*
Phone: (607)871-2411 Fax: (607)871-2344

Department Chair(s)
Ceramic Eng & Sciences
Alastair N. Cormack
Phone: (607)871-2422 Fax: (607)871-2305
EMail: Cormack@Ceramics.alfred.edu

Admissions Office Contact
Daniel L. Meyer
Phone: (607)871-2115 Fax: (607)871-2198

Financial Aid Office Contact
Earl E. Pierce, *Director of Financial Aid*
Phone: (607)871-2159 Fax: (607)871-2198

158 UNC-Charlotte

Head of the Institution
James H. Woodward
Phone: (704)547-2201 Fax: (704)547-2144

Head of Engineering
Robert D. Snyder
Phone: (704)547-2301 Fax: (704)547-2352

Associate/Assistant Deans
Harry J. Leamy, *Associate Dean*
Phone: (704)547-4096 Fax: (704)547-3183
Silvia G. Middleton, *Assistant Dean*
Phone: (704)547-2301 Fax: (704)547-2352

Department Chair(s)
Civil Eng
L.E. King
Phone: (704)547-2304 Fax: (704)547-2352
Electrical Eng
Farid M. Tranjan
Phone: (704)547-2302 Fax: (704)547-2352
Mechanical Eng
Paul H. DeHoff
Phone: (704)547-2303 Fax: (704)547-2352
Computer Science
Gyorgy E. Revesz
Phone: (704)547-4880 Fax: (704)547-2352
EMail: Lejk@mosaic.uncc.edu

Admissions Office Contact
Kathi M. Baucom
Phone: (704)547-2213 Fax: (704)547-2144

Financial Aid Office Contact
Curtis R. Whalen, *Director of Student Financial Aid*
Phone: (704)547-2461 Fax: (704)547-2144

159 North Carolina A&T

Head of the Institution
Edward B. Fort
Phone: (919)334-7940

Head of Engineering
Harold L. Martin
Phone: (919)334-7589 Fax: (919)334-7540
EMail: hlm@vanity.ncat.edu

Associate/Assistant Deans
Lonnie Sharpe, *Associate Dean - Undergraduate Programs*
Phone: (919)334-7589 Fax: (919)334-7540
John Kelly, *Associate Dean - Graduate Programs & Research*
Phone: (919)334-7589 Fax: (919)334-7540
Vernal G. Alford, III, *Assistant to the Dean*
Phone: (919)334-7589 Fax: (919)334-7540

Department Chair(s)
Chemical Eng
Franklin G. King
Phone: (919)334-7564 Fax: (919)334-7540

Civil Eng
Kenneth H. Murray
Phone: (919)334-7737 Fax: (919)334-7667
EMail: kmurray@garfield.ncat.edu
Electrical Eng
Tony L. Mitchell
Phone: (919)334-7760 Fax: (919)334-7716
EMail: tlm@garfield.ncat.edu
Industrial Eng
Eui H. Park
Phone: (919)334-7780 Fax: (919)334-7729
EMail: park@vanity.ncat.edu
Mechanical Eng
William J. Craft
Phone: (919)334-7620 Fax: (919)334-7417
EMail: craft@garfield.ncat.edu
Architectural Eng
Peter Rojeski
Phone: (919)334-7575 Fax: (919)334-7126
EMail: projeski@garfield.ncat.edu
Agricultural Eng
Godfrey A. Gayle
Phone: (919)334-7787 Fax: (919)334-7244
Computer Science
Joseph Monroe
Phone: (919)334-7245 Fax: (919)334-7244

Admissions Office Contact
John Smith
Phone: (919)334-7946

Financial Aid Office Contact
Delores S. Davis, *Director of Financial Aid*
Phone: (919)334-7973

160 North Carolina State U

Head of the Institution
Larry K. Monteith
Phone: (919)515-2191

Head of Engineering
Wilbur L. Meier
Phone: (919)515-2311 Fax: (919)515-2463
EMail: wmeier@eos.ncsu.edu

Associate/Assistant Deans
Tildon H. Glisson, *Associate Dean for Academic Affairs*
Phone: (919)515-3693 Fax: (919)515-2463
William E. Isler, *Associate Dean for Research Programs*
Phone: (919)515-2345 Fax: (919)515-2463
Robert M. Turner, *Assistant Dean for Student Services*
Phone: (919)515-3263 Fax: (919)515-2463
Hubert Winston, *Assistant Dean for Academic Affairs*
Phone: (919)515-2315 Fax: (919)515-2463

Department Chair(s)
Chemical Eng
George W. Roberts
Phone: (919)515-7328 Fax: (919)515-3465
EMail: 'roberts@ncsuche'
Civil Eng
E.D. Brill
Phone: (919)515-2352 Fax: (919)515-7908
Electrical & Computer Eng
Ralph K. Cavin
Phone: (919)515-5078 Fax: (919)515-5523
EMail: rkc@eos.ncsu.edu
Industrial Eng
Stephen D. Roberts
Phone: (919)515-2362 Fax: (919)515-5281
EMail: roberts@eos.ncsu.edu
Mechanical & Aerospace Eng
Carl F. Zorowski
Phone: (919)515-2366 Fax: (919)515-7968
EMail: cfz@ncsumae
Biological & Agricultural Eng
David B. Beasley
Phone: (919)515-2694 Fax: (919)515-6772
EMail: beasley@eos.ncsu.edu

Engineering & Engineering Technology Personnel

Computer Science
Alan L. Tharp
Phone: (919)515-7926 Fax: (919)515-7896
EMail: tharp@adm.csc.ncsu.edu
Materials Science & Eng
John J. Hren
Phone: (919)515-3568 Fax: (919)515-7724
Nuclear Eng
Donald J. Dudziak
Phone: (919)515-2301 Fax: (919)515-5115
EMail: dudziak@eos.ncsu.edu
Textile Eng, Chemistry, & Science
Charles D. Livengood
Phone: (919)515-6647 Fax: (919)515-6532
EMail: charles_livengood@ncsu.edu

Admissions Office Contact
George R. Dixon
Phone: (919)515-2434 Fax: (919)515-5039

Financial Aid Office Contact
Julia E. Rice, *Director*
Phone: (919)515-2421

161 U of North Dakota

Head of the Institution
Kendall R. Baker
Phone: (701)777-2121 Fax: (701)777-3866

Head of Engineering
Mogens Henriksen
Phone: (701)777-3412 Fax: (701)777-4838
EMail: henrik@eng.und.nodak.edu

Associate/Assistant Deans
Thomas C. Owens, *Associate Dean*
Phone: (701)777-4244 Fax: (701)777-4838

Department Chair(s)
Chemical Eng
Thomas C. Owens
Phone: (701)777-4244 Fax: (701)777-4838
Civil Eng
Ronald A. Apanian
Phone: (701)777-3562 Fax: (701)777-4838
Electrical Eng
Sastry Kuruganty
Phone: (701)777-4331 Fax: (701)777-4838
Mechanical Eng
Donald P. Naismith
Phone: (701)777-2571 Fax: (701)777-4838
Geology & Geological Eng
Patricia H. Kelley
Phone: (701)777-2248 Fax: (701)777-4838
Eng Management
Don V. Mathsen
Phone: (701)777-5128 Fax: (701)777-3119
Eng Physics
Glenn I. Lykken
Phone: (701)777-2911 Fax: (701)777-3650

Admissions Office Contact
Donna M. Bruce
Phone: (701)777-3821 Fax: (701)777-2696

Financial Aid Office Contact
Mark Brickson, *Director*
Phone: (701)777-3121

162 North Dakota State U

Head of the Institution
J.L. Ozbun
Phone: (701)237-7211 Fax: (701)237-7050

Head of Engineering
Joseph Stanislao
Phone: (701)237-7494 Fax: (701)237-7195

Associate/Assistant Deans
Donald A. Smith, *Associate Dean for the Ph.D. Program*
Phone: (701)237-7994 Fax: (701)237-7195

Department Chair(s)
Civil Eng & Construction
Don Richard
Phone: (701)237-7244 Fax: (701)237-7195

Electrical & Electronics Eng
Edward C. Bertnolli
Phone: (701)237-7019 Fax: (701)237-8677
Industrial Eng
Allen J. Henderson
Phone: (701)237-7287 Fax: (701)237-7195
Mechanical Eng
Karl G. Maurer
Phone: (701)237-8671 Fax: (701)237-7195
Agricultural Eng
Earl C. Stegman
Phone: (701)237-7261 Fax: (701)298-1008
Eng Science
Aurel Carcoana
Phone: (701)237-7105 Fax: (701)237-7195
Aero-Manufacturing Eng Tech
Scott G. Danielson
Phone: (701)237-7121 Fax: (701)237-7195
Aerospace Studies
Brian Stephens
Phone: (701)237-8186 Fax: (701)241-2089
Military Science
Timothy Lenzmeier
Phone: (701)237-7575 Fax: (701)241-2089

Admissions Office Contact
Robert A. Preloger
Phone: (701)237-8643

Financial Aid Office Contact
Wayne K. Tesmer, *Financial Aid Director*
Phone: (701)237-7538

163 Northeastern U

Head of the Institution
John A. Curry
Phone: (617)437-2101 Fax: (617)437-5015

Head of Engineering
Paul H. King
Phone: (617)437-2152 Fax: (617)437-8504

Associate/Assistant Deans
Richard J. Scranton, *Associate Dean for Undergraduate Programs*
Phone: (617)437-2152 Fax: (617)437-8504
David C. Blackman, *Assistant Dean & Director of NUPRIME*
Phone: (617)437-5904 Fax: (617)437-8504
Paula G. Leventman, *Assistant Dean & Director of Women's Programs*
Phone: (617)437-4835 Fax: (617)437-2501

Department Chair(s)
Chemical Eng
Ralph A. Buonopane
Phone: (617)437-2989 Fax: (617)437-2501
Civil Eng
Mishac K. Yegian
Phone: (617)437-2445 Fax: (617)437-4419
Electrical & Computer Eng
John G. Proakis
Phone: (617)437-4159 Fax: (617)437-8970
Industrial Eng & Information Systems
Stuart J. Deutsch
Phone: (617)437-2740 Fax: (617)437-2501
EMail: sdeutsch@lynx.northeastern.edu
Mechanical Eng
John W. Cipolla, Jr.
Phone: (617)437-2982 Fax: (617)437-2921
EMail: jwc@meceng.coe.northeastern.edu

Admissions Office Contact
Michael F. Clifford
Phone: (617)437-2200 Fax: (617)437-8780

Financial Aid Office Contact
Jean C. Eddy, *Dean of Financial Aid*
Phone: (617)437-3190 Fax: (617)437-8623

164 Northern Arizona U

Head of the Institution
Eugene M. Hughes
Phone: (602)523-3232 Fax: (602)523-4230

Head of Engineering
Spencer L. Brinkerhoff
Phone: (602)523-2880 Fax: (602)523-2300

Associate/Assistant Deans
Walter G. Hopkins, III, *Associate Dean*
Phone: (602)523-2050 Fax: (602)523-2300

Department Chair(s)
Civil Eng
Richard A. Mirth
Phone: (602)523-4339 Fax: (602)523-2300
Electrical Eng
Walter G. Hopkins
Phone: (602)523-4359 Fax: (602)523-2300
EMail: wgh@sun.cse.nau.edu
Mechanical Eng
David E. Hartman
Phone: (602)523-3450 Fax: (602)523-2300
Computer Science & Eng
Lanny J. Mullens
Phone: (602)523-3130 Fax: (602)523-2300
EMail: ljm@agassiz.cse.nau.edu

Admissions Office Contact
Molly S. Carder
Phone: (602)523-6002 Fax: (602)523-2220

Financial Aid Office Contact
James D. Pritchard, *Director of Financial Aid*
Phone: (602)523-4951

165 Northern Illinois U

Head of the Institution
John E. LaTourette
Phone: (815)753-9501 Fax: (815)753-8686

Head of Engineering
Romualdas Kasuba
Phone: (815)753-1283 Fax: (815)753-1310
EMail: kasuba@ceet.niu.edu

Associate/Assistant Deans
Joy M. Pauschke, *Associate Dean*
Phone: (815)753-0745 Fax: (815)753-0362

Department Chair(s)
Electrical Eng
Alan P. Genis
Phone: (815)753-9962 Fax: (815)753-1310
EMail: genis@ceet.niu.edu
Industrial Eng
Mohamed I. Dessouky
Phone: (815)753-9980 Fax: (815)753-1310
EMail: dessouky@ceet.niu.edu
Mechanical Eng
Parviz Payvar
Phone: (815)753-9970 Fax: (815)753-1310
EMail: payvar@ceet.niu.edu
Tech
Dennis Stoia
Phone: (815)753-0533 Fax: (815)753-3702
EMail: stoia@ceet.niu.edu

Admissions Office Contact
Joy M. Pauschke
Phone: (815)753-1442 Fax: (815)753-0362

Financial Aid Office Contact
Jerry D. Augsburger, *Director, Student Financial Aid*
Phone: (815)753-1300

166 Northwestern U

Head of the Institution
Arnold R. Weber
Phone: (708)491-7456 Fax: (708)491-8406

Head of Engineering
Jerome B. Cohen
Phone: (708)491-5220 Fax: (708)491-8539
EMail: jbc@ccadmin.tech.nwu.edu

Associate/Assistant Deans
William T. Brazelton, *Associate Dean*
Phone: (708)491-7379 Fax: (708)491-8539
William C. Cohen, *Associate Dean*
Phone: (708)491-7850 Fax: (708)491-8539

Carolyn H. Krulee, *Assistant Dean*
Phone: (708)491-5195 Fax: (708)491-8539
Stephen H. Carr, *Assistant Dean*
Phone: (708)491-4097
Geraldine O. Garner, *Assistant Dean*
Phone: (708)491-3366

Department Chair(s)
Chemical Eng
Julio M. Ottino
Phone: (708)491-3558 Fax: (708)491-3728
Civil Eng
Leon M. Keer
Phone: (708)491-4046 Fax: (708)491-4011
EMail: l-keer@nwu.edu
Electrical Eng & Computer Science
Abraham H. Haddad
Phone: (708)491-3641 Fax: (708)491-4455
EMail: ahaddad@eecs.nwu.edu
Industrial Eng
Robert H. Fourer
Phone: (708)491-3151 Fax: (708)491-8005
EMail: fourer@iems.nwu.edu
Mechanical Eng
Elmer E. Lewis
Phone: (708)491-3579 Fax: (708)491-3915
EMail: lewis@plato.nwu.edu
Biomedical Eng
Andrew E. Kertesz
Phone: (708)491-7672 Fax: (708)491-5299
EMail: kertesz@ccadmin.tech.nwu.edu
Eng Science & Applied Mathematics
Edward Olmstead
Phone: (708)491-3345 Fax: (708)491-2178
EMail: olmstead@ccadmin.tech.nwu.edu
Materials Science & Eng
Yip-Wah Chung
Phone: (708)491-7823 Fax: (708)491-7820
EMail: ywchung@nwu.edu

Admissions Office Contact
Carol Lunkenheimer
Phone: (708)491-7271 Fax: (708)467-7317

Financial Aid Office Contact
Carolyn Lindley, *Director of Financial Aid*
Phone: (708)491-7400

167 Norwich U

Head of the Institution
Richard W. Schneider
Phone: (802)485-2065 Fax: (802)485-2580

Head of Engineering
Eugene Sevi
Phone: (802)485-2275 Fax: (802)485-2580
EMail: Sevi

Associate/Assistant Deans
John B. Stevens, *Associate Dean*
Phone: (802)485-2261 Fax: (802)485-2580

Department Chair(s)
Civil Eng
John B. Stevens
Phone: (802)485-2261 Fax: (802)485-2580
EMail: Stevens
Electrical & Computer Eng
Ronald Lessard
Phone: (802)485-2270 Fax: (802)485-2580
EMail: Lessard
Mechanical Eng
Kenneth Craig
Phone: (802)485-2279 Fax: (802)485-2580
EMail: Craig
Environmental Eng Tech
Wight Greg
Phone: (802)485-2276 Fax: (802)485-2580
EMail: Wight
Architecture
Robert E. Schmidt
Phone: (802)828-8599 Fax: (802)485-2580
EMail: Schmidt

Engineering & Engineering Technology Personnel

Admissions Office Contact
Frank E. Griffis
Phone: (802)485-2000   Fax: (802)485-2580
Financial Aid Office Contact
Karen Waring, *Financial Aid Director*
Phone: (802)485-2015   Fax: (802)485-2580

168   U of Notre Dame
Head of the Institution
Edward A. Malloy
Phone: (219)239-6755   Fax: (219)239-7428
Head of Engineering
Anthony N. Michel
Phone: (219)239-5534   Fax: (219)239-8007
Associate/Assistant Deans
James I. Taylor, *Associate Dean*
Phone: (219)239-5533   Fax: (219)239-8007
Jerry J. Marley, *Associate Dean*
Phone: (219)239-5531   Fax: (219)239-8007
John D. Miles, *Assistant Dean*
Phone: (219)239-5532   Fax: (219)239-8007
Terrence J. Akai, *Assistant Dean*
Phone: (219)239-7446   Fax: (219)239-8007
Department Chair(s)
Chemical Eng
Hsueh-Chia Chang
Phone: (219)239-5697   Fax: (219)239-8366
Civil Eng & Geological Sciences
William G. Gray
Phone: (219)239-5380   Fax: (219)239-8007
Electrical Eng
Daniel J. Costello
Phone: (219)239-5480   Fax: (219)239-8007
Aerospace & Mechanical Eng
Thomas J. Mueller
Phone: (219)239-5433   Fax: (219)239-8007
Computer Science & Eng
Steven C. Bass
Phone: (219)239-8320   Fax: (219)239-8007
School of Architecture
Thomas G. Smith
Phone: (219)239-6137   Fax: (219)239-8007
Admissions Office Contact
Kevin M. Rooney
Phone: (219)239-7505
Financial Aid Office Contact
Joseph A. Russo, *Director of Financial Aid*
Phone: (219)239-6436

169   Oakland U
Head of the Institution
Sandra P. Parckard
Phone: (313)370-3500   Fax: (313)370-3504
Head of Engineering
Howard R. Whitt
Phone: (313)370-2217   Fax: (313)370-4261
EMail: witt@argo.acs.oakland.edu
Associate/Assistant Deans
Bhushan L. Bhatt, *Associate Dean for Administration*
Phone: (313)370-2233   Fax: (313)370-4261
Robert N.K. Loh, *Associate Dean for Research & Development*
Phone: (313)370-2233   Fax: (313)370-4261
Department Chair(s)
Mechanical Eng
Joseph D. Hovanesian
Phone: (313)370-2210   Fax: (313)370-4261
EMail: perria@vela.acs.oakland.edu
Computer Science & Eng
Subramanian Ganesan
Phone: (313)270-2200   Fax: (313)370-4261
EMail: ganesan@vela.oakland.edu
Electrical & Systems Eng
Naim A. Kheir
Phone: (313)370-2177   Fax: (313)370-4261
EMail: dahlmann@vela.acs.oakland.edu

Admissions Office Contact
J.W. Rose
Phone: (313)370-3360   Fax: (313)370-2286
Financial Aid Office Contact
Lee Anderson, *Director of Financial Aid Programs*
Phone: (313)370-3370   Fax: (313)370-2286

170   Ohio U
Head of the Institution
Charles Ping
Phone: (614)593-1804   Fax: (614)593-9196
Head of Engineering
T. Richard Robe
Phone: (614)593-1479   Fax: (614)593-0659
Associate/Assistant Deans
Joseph E. Essman, *Associate Dean*
Phone: (614)593-1482   Fax: (614)593-0659
Pamela Parker, *Assistant Dean for Development*
Phone: (614)593-1488   Fax: (614)593-0659
Department Chair(s)
Chemical Eng
Paul Jepson
Phone: (614)693-1498   Fax: (614)593-4684
Civil Eng
Glenn Hazen
Phone: (614)593-1469   Fax: (614)593-4684
EMail: Ent.OhioU.Edu.GHazen
Electrical Eng
Jerrel Mitchell
Phone: (614)593-1566   Fax: (614)593-0007
EMail: Mitchell@BobCat.Ent.OhioU.Edu
Industrial & Systems Eng
Charles Parks
Phone: (614)593-1554   Fax: (614)593-4684
EMail: CParks@BobCat.Ent.OhioU.Edu
Mechanical Eng
Jay Gunasekera
Phone: (614)593-1555   Fax: (614)593-4684
EMail: Gunaselr@Ouaccvmb
Financial Aid Office Contact
Carolyn Sabatino, *Director of Financial Aid*
Phone: (614)593-4141   Fax: (614)593-4140

171   Ohio Northern U
Head of the Institution
DeBow Freed
Phone: (419)772-2031   Fax: (419)772-1932
Head of Engineering
Bruce E. Burton
Phone: (419)772-2372   Fax: (419)772-2404
EMail: bburton@newton.onu.edu
Department Chair(s)
Civil Eng
Marlin D. Minich
Phone: (419)772-2376   Fax: (419)772-2404
EMail: mminich@newton.onu.edu
Electrical Eng
Fred L. Grismore
Phone: (419)772-1849   Fax: (419)772-2404
EMail: fgrismore@newton.onu.edu
Mechanical Eng
Leo R. Maier
Phone: (419)772-2385   Fax: (419)772-2404
EMail: lmaier@newton.onu.edu
Admissions Office Contact
Karen Condeni
Phone: (419)772-2260   Fax: (419)772-2313
Financial Aid Office Contact
Karen Condeni, *Vice President/Dean, Financial Aid*
Phone: (419)772-2260   Fax: (419)772-2313

172   Ohio State U
Head of the Institution
E. Gordon Gee
Phone: (614)292-2424   Fax: (614)292-1231

Head of Engineering
Jose B. Cruz, Jr.
Phone: (614)292-2836   Fax: (614)292-9021
Associate/Assistant Deans
George L. Smith, Jr., *Associate Dean, Academic Affairs*
Phone: (614)292-2651   Fax: (614)292-9021
John T. Demel, *Associate Dean, Student Services*
Phone: (614)292-2651   Fax: (614)292-9021
Richard D. Frasher, *Assistant Dean, Academic Programs & Services*
Phone: (614)292-8143   Fax: (614)292-9021
Marianne S. Mueller, *Assistant Dean, Engineering Career Services*
Phone: (614)292-6651   Fax: (614)292-4794
Minnie M. McGee, *Assistant Dean, Minority Affairs*
Phone: (614)292-4309   Fax: (614)292-9021
Department Chair(s)
Chemical Eng
Jacques L. Zakin
Phone: (614)292-6986   Fax: (614)292-3769
EMail: Zakin.1@osu.edu
Civil Eng
Tien H. Wu
Phone: (614)292-7338   Fax: (614)292-3780
EMail: Wu.26@osu.edu
Electrical Eng
Daniel B. Hodge
Phone: (614)292-2571   Fax: (614)292-7596
EMail: hodge@eagle.eng.ohio.state.edu
Industrial Eng
George L. Smith, Jr.
Phone: (614)292-6041   Fax: (614)292-7852
EMail: Smith.14@osu.edu
Mechanical Eng
Lawrence A. Kennedy
Phone: (614)292-5782   Fax: (614)292-3163
EMail: Kennedy.15@osu.edu
Aeronautical & Astronautical Eng
Gerald M. Gregorek
Phone: (614)292-2691   Fax: (614)292-9021
Agricultural Eng
Robert J. Gustafson
Phone: (614)292-6131   Fax: (614)292-9448
EMail: Gustafson.4@osu.edu
Aviation
William E. Pippin
Phone: (614)292-5593   Fax: (614)292-5020
Computer & Information Science
Mervin E. Muller
Phone: (614)292-5973   Fax: (614)292-2911
EMail: Muller.3@osu.edu
Eng Mechanics
Carl H. Popelar
Phone: (614)292-2731   Fax: (614)292-9021
EMail: Popelar.2@osu.edu
Eng Physics
Frank C. DeLucia
Phone: (614)292-2653
Materials Science & Eng
Robert H. Wagoner
Phone: (614)292-2491   Fax: (614)292-1537
EMail: Wagoner.2@osu.edu
Geodetic Science & Surveying
Clyde C. Goad
Phone: (614)292-6753   Fax: (614)292-2957
EMail: Goad.1@osu.edu
Welding Eng
William A. Baeslack
Phone: (614)292-6841   Fax: (614)292-6842
Eng Graphics
John T. Demel
Phone: (614)292-2427   Fax: (614)292-9021
EMail: John.T.Demel@osu.edu
Admissions Office Contact
James J. Mager
Phone: (614)292-3980   Fax: (614)292-4818

Financial Aid Office Contact
Mary B. Haldane, *Director of Student Financial Aid*
Phone: (614)292-0300   Fax: (614)292-9264

173   U of Oklahoma
Head of the Institution
Richard L. Van Horn
Phone: (405)325-3916   Fax: (405)325-7605
Head of Engineering
Billy L. Crynes
Phone: (405)325-2621   Fax: (405)325-7508
EMail: crynes@mailhost.ecn.uoknor.edu
Associate/Assistant Deans
Hillel J. Kumin, *Associate Dean for Academic Programs*
Phone: (405)325-2621   Fax: (405)325-7508
Walter D. Ballew, *Interim Associate Dean for Research*
Phone: (405)325-2621   Fax: (405)325-7508
Donald R. Geis, *Assistant Dean for Development*
Phone: (405)325-2621   Fax: (405)325-7508
Department Chair(s)
Chemical Eng
Jeffrey H. Harwell
Phone: (405)325-5811   Fax: (405)325-5813
EMail: harwell@mailhost.ecn.uoknor.edu
Civil Eng & Environmental Science
Ronald L. Sack
Phone: (405)325-5911
EMail: sack@mailhost.ecn.uoknor.edu
Electrical Eng
Theodore E. Batchman
Phone: (405)325-4721
EMail: batchman@mailhost.ecn.uoknor.edu
Industrial Eng
Jerry L. Purswell
Phone: (405)325-3721   Fax: (405)325-7555
EMail: purswell@mailhost.ecn.uoknor.edu
Aerospace & Mechanical Eng
Charles W. Bert
Phone: (405)325-5011   Fax: (405)325-1088
EMail: bert@mailhost.ecn.uoknor.edu
Eng Physics
Helmut J. Fischbeck
Phone: (405)325-3961
EMail: fischbeck@astron.nhn.uoknor.edu
Petroleum & Geological Eng
Ronald D. Evans
Phone: (405)325-2921   Fax: (405)325-7417
EMail: evans@mailhost.ecn.uoknor.edu
Computer Science
John Y. Cheung
Phone: (405)325-4324
EMail: cheung@mailhost.ecn.uoknor.edu
Admissions Office Contact
Marc S. Borish
Phone: (405)325-2251   Fax: (405)325-7047
Financial Aid Office Contact
Mary Mowdy, *Interim Director of Financial Aid*
Phone: (405)325-4521   Fax: (405)325-7608

174   Oklahoma Christian
Head of the Institution
J. Terry Johnson
Phone: (405)425-5100   Fax: (405)425-5316
Head of Engineering
W. Joe Watson
Phone: (405)425-5426   Fax: (405)425-5316
Associate/Assistant Deans
Troy J. Pemberton, *Engineering Department Chair*
Phone: (405)425-5416   Fax: (405)425-5316
Department Chair(s)
Electrical Eng
Lynn S. Nored
Phone: (405)425-5409   Fax: (405)425-5316

Mechanical Eng
James N. Cutbirth
Phone: (405)425-5410 Fax: (405)425-5316
Admissions Office Contact
Bob Rowley
Phone: (405)425-5054 Fax: (405)425-5316
Financial Aid Office Contact
Andy Carpenter, *Director of Financial Aid*
Phone: (405)425-5105 Fax: (405)425-5316

175   Oklahoma State U

Head of the Institution
John R. Campbell
Phone: (405)744-6386 Fax: (405)744-6285
Head of Engineering
Karl N. Reid
Phone: (405)744-5140 Fax: (405)744-7545
EMail: kreid@master.ceat.okstate.edu
Associate/Assistant Deans
David R. Thompson, *Associate Dean for Instruction & Extension*
Phone: (405)744-5140 Fax: (405)744-7545
Allen E. Kelly, *Associate Dean for Research*
Phone: (405)744-5957 Fax: (405)744-7545
Department Chair(s)
Chemical Eng
Robert L. Robertson
Phone: (405)744-5280 Fax: (405)744-6187
Civil Eng
Robert K. Hughes
Phone: (405)744-5190 Fax: (405)744-6187
Electrical & Computer Eng
James E. Baker
Phone: (405)744-5151 Fax: (405)744-6187
Industrial Eng & Management
Timothy J. Greene
Phone: (405)744-6055 Fax: (405)744-6187
Mechanical & Aerospace Eng
Lawrence L. Hoberock
Phone: (405)744-5900 Fax: (405)744-6187
Agricultural Eng
Bill J. Barfield
Phone: (405)744-5431 Fax: (405)744-6059
Architecturaal Eng
James F. Knight
Phone: (405)744-6043 Fax: (405)744-6187
Admissions Office Contact
Larry D. Zirkle
Phone: (405)744-5276 Fax: (495)744-6187
Financial Aid Office Contact
Charles W. Bruce, *Director of Financial Aids*
Phone: (405)744-6604 Fax: (405)744-6438

176   Old Dominion U

Head of the Institution
James V. Koch
Phone: (804)683-3159 Fax: (804)683-5679
Head of Engineering
Ernest J. Cross, Jr.
Phone: (804)683-3787 Fax: (804)683-4898
Associate/Assistant Deans
William A. Drewry, *Associate Dean for Undergraduate Programs & Administration*
Phone: (804)683-3789 Fax: (804)683-4898
Griffith J. McRee, *Associate Dean for Research & Graduate Studies*
Phone: (804)683-4897 Fax: (804)683-4898
Gary R. Crossman, *Associate Dean for Community, Industrial, & Alumni Relations*
Phone: (804)683-3768 Fax: (804)683-4898
Department Chair(s)
Civil & Environmental Eng
William A. Drewry
Phone: (804)683-3765 Fax: (804)683-5354
Electrical & Computer Eng
Roland R. Mielke
Phone: (804)683-3741 Fax: (804)683-3220

Mechanical Eng & Mechanics
Gregory V. Selby
Phone: (804)683-3720 Fax: (804)683-5344
Eng Management
Laurence D. Richards
Phone: (804)683-4558 Fax: (804)683-5640
Admissions Office Contact
Angela N. Boyd
Phone: (804)683-3637 Fax: (804)683-3647
Financial Aid Office Contact
Helga A. Greenfield, *Director*
Phone: (804)683-3683 Fax: (804)683-5357

177   Oregon State U

Head of the Institution
John V. Byrne
Phone: (503)737-4133 Fax: (503)737-3033
Head of Engineering
S John T. Owen
Phone: (503)737-4525 Fax: (503)737-3462
EMail: owensj@ccmail.orst.edu
Associate/Assistant Deans
Tom M. West, *Associate Dean for Administration*
Phone: (503)737-4525 Fax: (503)737-3462
R. Gary Hicks, *Associate Dean for Research & Graduate Studies*
Phone: (503)737-5318 Fax: (503)737-3462
Roy C. Rathja, *Assistant Dean for Undergraduate Studies*
Phone: (503)737-4525 Fax: (503)737-3462
Department Chair(s)
Chemical Eng
W. James Frederick
Phone: (503)737-4791 Fax: (503)737-4600
EMail: frederwj@ccmail.orst.edu
Civil Eng
Wayne M. Huber
Phone: (503)737-4934 Fax: (503)737-3052
EMail: huberw@ccmail.orst.edu
Electrical & Computer Eng
Gabor C. Temes
Phone: (503)737-3617 Fax: (503)737-1300
EMail: temes@ece.orst.edu
Industrial & Manufacturing Eng
Sabah Randhawa
Phone: (503)737-2365 Fax: (503)737-5241
EMail: randhas@ccmail.orst.edu
Mechanical Eng
Gordon M. Reistad
Phone: (503)737-3441 Fax: (503)737-2600
EMail: reistagm@ccmail.orst.edu
Computer Science
Walter Rudd
Phone: (503)737-3273 Fax: (503)737-3014
EMail: rudd@cs.orst.edu
Eng Physics
Kenneth S. Krane
Phone: (503)737-4631 Fax: (503)737-1683
EMail: krane@physics.orst.edu
Nuclear Eng
Alan H. Robinson
Phone: (503)737-2343 Fax: (503)737-0480
EMail: robinsah@ccmail.orst.edu
Admissions Office Contact
Kay Conrad
Phone: (503)737-4411 Fax: (503)737-2400
Financial Aid Office Contact
Keith McCreight, *Director, Financial Aid Office*
Phone: (503)737-2241

178   U of the Pacific

Head of the Institution
Bill Atchley
Phone: (209)946-2222 Fax: (209)946-2652
Head of Engineering
Ashland Brown
Phone: (209)946-3091 Fax: (209)946-3086

Associate/Assistant Deans
Robert E. Hamernik, *Associate Dean*
Phone: (209)946-3060 Fax: (209)946-3086
Thomas Cheney, *Assistant Dean*
Phone: (209)946-3062 Fax: (209)946-3086
Department Chair(s)
Civil Eng
David Q. Fletcher
Phone: (209)946-3076 Fax: (209)946-3086
Electrical & Computer Eng
Richard H. Turpin
Phone: (209)946-3075 Fax: (209)946-3086
Mechanical Eng
Edwin Pejack
Phone: (209)946-3082 Fax: (209)946-3086
Eng Physics
Andres F. Rodriguez
Phone: (209)946-2227 Fax: (209)947-3086
Eng Management
James R. Morgali
Phone: (209)946-3061 Fax: (209)946-3086
Admissions Office Contact
Edward L. Schoenberg
Phone: (209)946-2211
Financial Aid Office Contact
Lynn Fox, *Director*
Phone: (209)946-2421

179   U of Penn

Head of the Institution
Sheldon Hackney
Phone: (215)898-7221 Fax: (215)898-9659
Head of Engineering
Gregory C. Farrington
Phone: (215)898-7244 Fax: (215)573-2018
EMail: farringt@eniac.seas.upenn.edu
Associate/Assistant Deans
John D. Keenan, *Associate Dean for Undergraduate Education*
Phone: (215)898-7246 Fax: (215)898-1130
Dwight L. Jaggard, *Associate Dean for Graduate Education & Research*
Phone: (215)898-8241 Fax: (215)898-1130
Jacob M. Abel, *Associate Dean for Educational Development & Special Programs*
Phone: (215)898-8342 Fax: (215)898-1130
Department Chair(s)
Chemical Eng
Eduardo D. Glandt
Phone: (215)898-8351 Fax: (215)573-2093
EMail: glandt@cheme.seas.upenn.edu
Electrical Eng
Saleem Kassam
Phone: (215)898-5990 Fax: (215)573-2068
EMail: kassam@pender.ee.seas.upenn.edu
Mechanical Eng & Applied Mechanics
Ira M. Cohen
Phone: (215)898-7076 Fax: (215)573-2065
EMail: cohen@eniac.seas.upenn.edu
Materials Science & Eng
David P. Pope
Phone: (215)898-9837 Fax: (215)573-2128
EMail: Pope@eniac.seas.upenn.edu
Bioengineering
Lawrence E. Thibault
Phone: (215)898-8501 Fax: (215)573-2071
Systems
Kenneth A. Fegley
Phone: (215)898-9390 Fax: (215)573-2065
EMail: fegley@eniac.seas.upenn.edu
Computer & Information Science
Norman Badler
Phone: (215)898-8560 Fax: (215)898-0587
EMail: badler@central.cis.seas.upenn.edu
Admissions Office Contact
Willis J. Stetson
Phone: (215)898-7507 Fax: (215)898-9670

Financial Aid Office Contact
William M. Schilling, *Director of Student Financial Aid*
Phone: (215)898-6784 Fax: (215)573-5428

180   Penn State U

Head of the Institution
Joab L. Thomas
Phone: (814)865-7611 Fax: (814)863-4631
Head of Engineering
David N. Wormley
Phone: (814)865-7537 Fax: (814)863-4749
EMail: CHW1@PSUADMIN
Associate/Assistant Deans
George J. McMurtry, *Associate Dean for Administration & Planning*
Phone: (814)865-2151 Fax: (814)863-4749
Michael M. Reischman, *Associate Dean for Graduate Study & Research*
Phone: (814)865-4542 Fax: (814)863-0497
Carl H. Wolgemuth, *Associate Dean for Undergraduate Studies*
Phone: (814)863-3750 Fax: (814)863-4749
Joseph S. DiGregorio, *Assoc Dean for Commonwealth & Continuing Educ & Internat'l Programs*
Phone: (814)865-7644 Fax: (814)863-4749
Department Chair(s)
Chemical Eng
J.L. Duda
Phone: (814)865-2574 Fax: (814)865-7846
EMail: JLD6@PSUADMIN
Civil & Environmental Eng
Chin Y. Kuo
Phone: (814)865-8391 Fax: (814)863-7304
EMail: CYK1@PSUADMIN
Electrical & Computer Eng
Larry C. Burton
Phone: (814)865-7667 Fax: (814)865-7065
EMail: LCB2@PSUADMIN
Industrial & Management Systems Eng
Allen L. Soyster
Phone: (814)865-7601 Fax: (814)863-4745
EMail: ALS7@PSUADMIN
Mechanical Eng
Harold R. Jacobs
Phone: (814)865-2519 Fax: (814)863-4848
EMail: HRJ1@PSUADMIN
Aerospace Eng
Dennis K. McLaughlin
Phone: (814)865-2569 Fax: (814)865-7092
EMail: DKM2@PSUADMIN
Agricultural & Biological Eng
Dennis E. Buffington
Phone: (814)865-7792 Fax: (814)863-1031
EMail: DEB2@PSUADMIN
Architectural Eng
Paul A. Seaburg
Phone: (814)865-6394 Fax: (814)863-4789
EMail: PQS1@PSUADMIN
Eng Science & Mechanics
Richard P. McNitt
Phone: (814)865-4523 Fax: (814)863-7967
EMail: RPM1@PSUADMIN
Nuclear Eng
Edward H. Klevans
Phone: (814)865-1341 Fax: (814)865-8499
EMail: EHK2@PSUADMIN
Mining Eng
Raja V. Ramani
Phone: (814)863-1617 Fax: (814)865-3248
EMail: RVR1@PSUADMIN
Ceramic Science & Eng
David J. Green
Phone: (814)863-2011 Fax: (814)865-2917
EMail: ARS2@PSUADMIN
Metals Science & Eng
Donald A. Koss
Phone: (814)865-5447 Fax: (814)865-2917
EMail: SMH2@PSUADMIN

Engineering & Engineering Technology Personnel

Petroleum & Natural Gas Eng
Turgay Ertekin
Phone: (814)865-6082 Fax: (814)863-1875
EMail: EUR@PSUVM

Bioengineering
Herbert H. Lipowsky
Phone: (814)865-1407 Fax: (814)865-0490
EMail: hhl1@PSUADMIN

Admissions Office Contact
Anna M. Griswold
Phone: (814)865-5471 Fax: (814)863-7590

Financial Aid Office Contact
Anna M. Griswold, *Director, Student Aid*
Phone: (814)865-6301 Fax: (814)863-0322

181 U of Pittsburgh

Head of the Institution
John D. O'Connor
Phone: (412)624-4200 Fax: (412)624-1150

Head of Engineering
Charles A. Sorber
Phone: (412)624-9800 Fax: (412)624-1108

Associate/Assistant Deans
Larry J. Shuman, *Associate Dean*
Phone: (412)624-9815 Fax: (412)624-1108
George E. Klinzing, *Associate Dean*
Phone: (412)624-9814 Fax: (412)624-1108

Department Chair(s)
Civil Eng
Fred Moses
Phone: (412)624-9870 Fax: (412)624-0135

Chemical & Petroleum Eng
Gerald D. Holder
Phone: (412)624-9631 Fax: (412)624-9639

Electrical Eng
Marwan A. Simaan
Phone: (412)624-8002 Fax: (412)624-8003
EMail: KKT@EE.PITT.EDU

Industrial Eng
Harvey Wolfe
Phone: (412)624-9830 Fax: (412)624-9831

Mechanical Eng
Michael J. Kolar
Phone: (412)624-7661 Fax: (412)624-1108

Materials Science & Eng
Nicholas G. Eror
Phone: (412)624-9720 Fax: (412)624-1108
EMail: EROR@PITT.VMS.CIS.PITT.EDU

Freshman Program & Career Development
Sandra L. Bishop
Phone: (412)624-9825 Fax: (412)624-1108
EMail: IN%"JCRESTO@VMS.CIS.PITT.EDU"

Admissions Office Contact
Betsy A. Porter
Phone: (412)624-7488 Fax: (412)648-8815

Financial Aid Office Contact
Betsy A. Porter, *Director of Admissions & Financial Aid*
Phone: (412)624-7488 Fax: (412)648-8815

182 Poly U

Head of the Institution
George Bugliarello
Phone: (718)260-3500 Fax: (718)260-3755

Head of Engineering
Roger P. Roess
Phone: (718)260-3550 Fax: (718)260-3136

Department Chair(s)
Chemical Eng
Allan S. Myerson
Phone: (718)260-3620 Fax: (718)260-3136

Civil Eng
Ilan Juran
Phone: (718)260-3220 Fax: (718)260-3136

Electrical Eng
Henry Bertoni
Phone: (718)260-3590 Fax: (718)260-3136

Industrial Eng
William McShane
Phone: (718)260-3112 Fax: (178)260-3136

Mechanical Eng
William McShane
Phone: (718)260-3160 Fax: (718)260-3136

Aerospace Eng
Pasquale Sforza
Phone: (718)260-3160 Fax: (718)260-3136

Metallurgy & Materials Science
Irving Cadoff
Phone: (718)260-3250 Fax: (718)260-3136

Admissions Office Contact
Ellen Hartigan
Phone: (718)260-3100 Fax: (718)260-3136

Financial Aid Office Contact
Veronica Lukas, *Director of Financial Aid*
Phone: (718)260-3300 Fax: (718)260-3136

183 U of Portland

Head of the Institution
David T. Tyson
Phone: (503)283-7101 Fax: (503)283-7399

Head of Engineering
Thomas J. Nelson
Phone: (503)283-7314 Fax: (503)283-7399

Associate/Assistant Deans
Khalid H. Khan, *Associate Dean*
Phone: (503)283-7276 Fax: (503)283-7399

Department Chair(s)
Civil Eng
Mehmet I. Inan
Phone: (503)283-7151 Fax: (503)283-7399

Electrical Eng
Aziz S. Inan
Phone: (503)283-7429 Fax: (503)283-7399

Mechanical Eng
Miroslav C. Rokos
Phone: (503)283-7254 Fax: (503)283-7399

Admissions Office Contact
Daniel B. Reilly
Phone: (503)283-7147 Fax: (503)283-7399

Financial Aid Office Contact
Rita A. Lambert, *Director of Financial Aid*
Phone: (503)283-7311 Fax: (503)283-7399

184 Portland State U

Head of the Institution
Judith A. Ramaley
Phone: (503)725-4411 Fax: (503)725-4499

Head of Engineering
H. Chik M. Erzurumlu
Phone: (503)725-4631 Fax: (503)725-4298
EMail: bfhe@eas.pdx.edu

Associate/Assistant Deans
Richard D. Morris, *Assistant Dean*
Phone: (503)725-4698 Fax: (503)725-4298

Department Chair(s)
Civil Eng
Franz N. Rad
Phone: (503)725-4282 Fax: (503)725-4298
EMail: franz@eas.pdx.edu

Electrical Eng
Rolf Schaumann
Phone: (503)725-3806 Fax: (503)725-4882
EMail: schaumann@ee.pdx.edu

Mechanical Eng
Graig A. Spolek
Phone: (503)725-4290 Fax: (503)725-4298
EMail: graigs@eas.pdx.edu

Computer Science
Leonard Shapiro
Phone: (503)725-4036 Fax: (503)725-3211
EMail: len@cs.pdx.edu

Eng Management Graduate Program
Dundar F. Kocaoglu
Phone: (503)725-4660 Fax: (503)725-4667
EMail: d6emp@psuorvm.cc.pdx.edu

Admissions Office Contact
H. Chik M. Erzurumlu
Phone: (503)725-4631 Fax: (503)725-4298

Financial Aid Office Contact
John E. Anderson, *Director of Financial Aid*
Phone: (503)725-3461 Fax: (503)725-4882

185 Prairie View A&M U

Head of the Institution
Julius W. Becton
Phone: (409)857-2111 Fax: (409)857-3928

Head of Engineering
John Foster
Phone: (409)857-2212 Fax: (409)857-2222

Associate/Assistant Deans
James O. Morgan, *Associate Dean*
Phone: (409)857-2212 Fax: (409)857-2222

Department Chair(s)
Chemical Eng
Jorge Gabitto
Phone: (409)857-2427 Fax: (409)857-2222
EMail: jgabitto@pvcea.pvam.edu

Civil Eng
Ramalinjam Radha
Phone: (409)857-2418 Fax: (409)857-2222
EMail: rradha@pvcea.pvam.edu

Electrical Eng
John H. Fuller
Phone: (409)857-3923 Fax: (409)857-2222
EMail: jfuller@pvcea.pvam.edu

Mechanical Eng
Shield B. Lin
Phone: (409)857-4023 Fax: (409)857-2222
EMail: slin@pvcea.pvam.edu

Architecture
Simon R. Wiltz
Phone: (409)857-2014 Fax: (409)857-2222
EMail: switz@pvcea.pvam.edu

Graduate Program
James O. Morgan
Phone: (409)857-2211 Fax: (409)857-2222
EMail: jmorgan@pvcea.pvam.edu

Admissions Office Contact
Robert F. Ford
Phone: (409)857-2618 Fax: (409)857-4956

Financial Aid Office Contact
Advergus D. James, *Director*
Phone: (409)857-2424

186 Princeton U

Head of the Institution
Harold T. Shapiro
Phone: (609)258-6100 Fax: (609)258-1294

Head of Engineering
James Wei
Phone: (609)258-2260 Fax: (609)258-6744
EMail: jameswei@pucc

Associate/Assistant Deans
Bradley W. Dickinson, *Associate Dean for Academic Affairs*
Phone: (609)258-2916 Fax: (609)258-6744
Richard L. Golden, *Associate Dean for Operations & Research*
Phone: (609)258-4553 Fax: (609)258-6744
Harold Y. McCulloch, *Assistant Dean for Undergraduate Affairs*
Phone: (609)258-3662 Fax: (609)258-6744

Department Chair(s)
Chemical Eng
William B. Russel
Phone: (609)258-4590 Fax: (609)258-0211
EMail: wbrussel@pucc

Civil Eng & Operations Research
J.H. Prevost
Phone: (609)258-5424 Fax: (609)258-1270
EMail: jean@soil

Electrical Eng
Stuart C. Schwartz
Phone: (609)258-4618 Fax: (609)258-3745
EMail: stuart@princeton

Mechanical & Aerospace Eng
Garry L. Brown
Phone: (609)258-6083 Fax: (609)258-6109
EMail: glb@pucc

Computer Science
Robert Sedgewick
Phone: (609)258-4345 Fax: (609)258-1771
EMail: rs@cs

Financial Aid Office Contact
Don M. Betterton, *Director, Undergraduate Financial Aid*
Phone: (609)258-3330 Fax: (609)258-2853

187 U of Puerto Rico

Head of the Institution
Alejandro M. Ruiz
Phone: (809)265-3878 Fax: (809)834-3031

Head of Engineering
José F. Lluch
Phone: (809)833-1121 Fax: (809)833-1190
EMail: J_Lluch @ Rumac.UPR.CLU.Edu

Associate/Assistant Deans
David Serrano, *Assistant Dean for Research & Graduate School*
Phone: (809)265-3826 Fax: (809)832-0119
Jorgé Ortiz, *Associate Dean for Academic Affairs*
Phone: (809)265-3823 Fax: (809)833-1190
Rafael Fernandez-Sein, *Assistant Dean for Administrative Affairs*
Phone: (809)265-3825 Fax: (809)833-1190

Department Chair(s)
Chemical Eng
Felix Santiago
Phone: (809)265-3818 Fax: (809)834-3655
EMail: J_Briano @ Rumac.UPR.CLU.Edu

Civil Eng
Felipé Luyanda
Phone: (809)265-3815 Fax: (809)833-8260
EMail: S_Luyanda @ RUMAD.UPR.CLU.EDU

Electrical Eng
Samuel Irrizary
Phone: (809)265-3821 Fax: (809)832-0119
EMail: REVE @ RMSE 13.UPR.CLU.EDU

Industrial Eng
Jack Allison
Phone: (809)265-3819 Fax: (809)832-0119
EMail: J_Allison @ RUMAD.UPR.CLU.EDU

Mechanical Eng
Fernando Pla
Phone: (809)265-3817 Fax: (809)265-3817
EMail: D_Serrano @ RUMAC.UPR.CLU.EDU

General Eng
Anand D. Sharma
Phone: (809)265-3816 Fax: (809)832-0119
EMail: L_Pumarada @ RUMAD.UPR.CLU.EDU

Admissions Office Contact
Jorgé Ortiz
Phone: (809)832-4040 Fax: (809)833-1190

Financial Aid Office Contact
Pedro J. Aubret, *Director*
Phone: (809)832-4040

188 Purdue U

Head of the Institution
Steven C. Beering
Phone: (317)494-9708

Head of Engineering
Henry T. Yang
Phone: (317)494-5346

Engineering & Engineering Technology Personnel

189 Rensselaer Poly Inst

Head of the Institution
Roland W. Schmitt
Phone: (518)276-6211   Fax: (518)276-8702

Head of Engineering
James M. Tien
Phone: (518)276-6486   Fax: (518)276-8788
EMail: JMTIEN@RPITSMTS.Bitnet

Associate/Assistant Deans
Robert C. Block, *Associate Dean*
Phone: (518)276-6620   Fax: (518)276-8788
Lester A. Gerhardt, *Associate Dean*
Phone: (518)276-6203   Fax: (518)276-8788
John E. Kolb, *Assistant Dean*
Phone: (518)276-6626   Fax: (518)276-8788

Department Chair(s)
Chemical Eng
Arthur Fontijn
Phone: (518)276-6379   Fax: (518)276-4030
EMail: mulson@rpitsmts
Civil & Environmental Eng
George J. Dvorak
Phone: (518)276-8650   Fax: (518)276-4833
EMail: usergzbh@rpitsmts
Electrical, Computer, & Systems Eng
Arthur C. Sanderson
Phone: (518)276-6316   Fax: (518)276-6261
EMail: acs@ecs.rpi.edu
Decision Sciences & Eng Systems
Madabhushi Raghavachari
Phone: (518)276-6485   Fax: (518)276-8227
EMail: ragavm@aix.rpi.edu
Mechanical Eng, Aeronautical Eng, & Mechanics
Erhard Krempl
Phone: (518)276-6351   Fax: (518)276-2623
EMail: ekrempl@meche.rpi.edu
Biomedical Eng
Rob J. Roy
Phone: (518)276-6548   Fax: (518)276-3035
EMail: userhc7m@rpitsmts
Electric Power Eng
J. Keith Nelson
Phone: (518)276-6329   Fax: (518)276-6226
EMail: userhcr7@rpitsmts
Materials Eng
Robert H. Doremus
Phone: (518)276-6373   Fax: (518)276-8554
EMail: userfnv6@rpitsmts
Nuclear Eng & Eng Physics
Robert C. Block
Phone: (518)276-6110   Fax: (518)276-4832
EMail: blockr@rpi.edu

Admissions Office Contact
Conrad Sharrow
Phone: (518)276-6216   Fax: (518)276-4072

Financial Aid Office Contact
James H. Stevenson, *Director of Financial Aid*
Phone: (518)276-6813   Fax: (518)276-4072

190 U of Rhode Island

Head of the Institution
Robert L. Carothers
Phone: (401)792-2444   Fax: (401)792-7149

Head of Engineering
Thomas J. Kim, *Interim Dean*
Phone: (401)792-2186   Fax: (401)782-1066

Department Chair(s)
Chemical Eng
Stanley M. Barnett
Phone: (401)792-2443   Fax: (401)782-1066
Civil & Environmental Eng
Daniel W. Urish
Phone: (401)792-2267   Fax: (401)782-1066
Electrical Eng
William J. Ohley
Phone: (401)792-2505   Fax: (401)782-1066
Industrial & Manufacturing Eng
Winston A. Knight
Phone: (401)792-2455   Fax: (401)782-1388
Mechanical Eng & Applied Mechanics
Martin H. Sadd
Phone: (401)792-2524   Fax: (401)782-1066
Ocean Eng
Malcolm L. Spaulding
Phone: (401)792-2273   Fax: (401)782-2819

Admissions Office Contact
Catherine L. Zeiser
Phone: (401)792-7100   Fax: (401)792-2002

Financial Aid Office Contact
Horace J. Amaral, Jr., *Assistant Dean of Student Financial Aid*
Phone: (401)792-2314

191 Rice U

Head of the Institution
George E. Rupp
Phone: (713)527-4041   Fax: (713)285-5271

Head of Engineering
Michael M. Carroll
Phone: (713)527-4009   Fax: (713)285-5300
EMail: DENG@ricevm1.rice.edu

Associate/Assistant Deans
Hardy M. Bourland, *Assoc. Dean*
Phone: (713)527-4955   Fax: (713)285-5300

Department Chair(s)
Chemical Eng
Clarence A. Miller
Phone: (713)527-4902   Fax: (713)524-5237
EMail: CENG@ricevm1.rice.edu
Civil Eng
Ronald P. Nordgren
Phone: (713)527-4949   Fax: (713)285-5268
EMail: CIVI@ricevm1.rice.edu
Electrical & Computer Eng
Frank K. Tittel
Phone: (713)527-4020   Fax: (713)524-5237
EMail: ELEC@ricevm1.rice.edu
Mechanical Eng & Materials Science
C.-C. Wang
Phone: (713)527-4906   Fax: (713)285-5423
EMail: MEMS@ricevm1.rice.edu
Computer Science
John E. Dennis
Phone: (713)527-4834   Fax: (713)285-5136
EMail: COMP@ricevm1.rice.edu
Environmental Science & Eng
Philip B. Bedient
Phone: (713)527-4951   Fax: (713)285-5203
EMail: ENVI@ricevm1.rice.edu
Computational & Applied Mathematics
William W. Symes
Phone: (713)527-4805   Fax: (713)285-5318
EMail: MASC@ricevm1.rice.edu
Statistics
David W. Scott
Phone: (713)527-6032   Fax: (713)285-5476
EMail: STAT@ricevm1.rice.edu

Admissions Office Contact
Ron W. Moss
Phone: (713)527-4036   Fax: (713)285-5271

Financial Aid Office Contact
G.D. Hunt, *Director, Office of Financial Aid*
Phone: (723)527-4958   Fax: (713)285-5322

192 U of Rochester

Head of the Institution
G. Dennis O'Brien
Phone: (716)275-8356   Fax: (716)256-2473

Head of Engineering
Bruce W. Arden
Phone: (716)275-4151   Fax: (716)461-4735
EMail: arden@ee.rochester.edu

Associate/Assistant Deans
William E. Kiker, *Associate Dean*
Phone: (716)275-4154   Fax: (716)461-4735
Richard H. Heist, *Associate Dean for Graduate Studies*
Phone: (716)275-4153   Fax: (716)461-4735
Donna Lampen Smith, *Assistant Dean for Undergraduate Affairs*
Phone: (716)275-4155   Fax: (716)461-4735

Department Chair(s)
Chemical Eng
Harvey J. Palmer
Phone: (716)275-4041   Fax: (716)442-6686
EMail: GAV2%UHURA.CC.ROCHESTER.EDV@UORVM
Electrical Eng
Kevin J. Parker
Phone: (716)275-4066   Fax: (716)473-0486
EMail: parker@ee.rochester.edu
Mechanical Eng
Richard C. Benson
Phone: (716)275-4071   Fax: (716)256-2509
EMail: benson@me.rochester.edu
The Inst of Optics
Duncan T. Moore
Phone: (716)275-5248   Fax: (716)473-6745
EMail: moore@moe.optics.rochester.edu

Admissions Office Contact
Wayne A. Locust
Phone: (716)275-3221   Fax: (716)461-4595

Financial Aid Office Contact
Ryan Williams, *Director of Financial Aid*
Phone: (716)275-3226   Fax: (716)461-4595

193 Rochester Inst of Tech

Head of the Institution
Albert J. Simone
Phone: (716)475-2396   Fax: (716)475-5700

Head of Engineering
Paul E. Petersen
Phone: (716)475-2146   Fax: (716)475-6879
EMail: PEPEEE

Associate/Assistant Deans
N. Richard Reeve, *Associate Dean*
Phone: (716)475-7048   Fax: (716)475-6879

Department Chair(s)
Electrical Eng
Raman M. Unnikrishnan
Phone: (716)475-2165   Fax: (716)475-6879
EMail: RXVEEE
Industrial & Manufacturing Eng
Jasper E. Shealy
Phone: (716)475-2598   Fax: (716)475-6879
EMail: JESEIE
Mechanical Eng
Charles W. Haines
Phone: (716)475-2162   Fax: (716)475-6879
EMail: CWHEME
Computer Eng
Roy S. Czernikowski
Phone: (716)475-2987   Fax: (716)475-6879
EMail: RSCEEC
Microelectronic Eng
Lynn F. Fuller
Phone: (716)475-6065   Fax: (716)475-6879
EMail: LFFEEE

Admissions Office Contact
Daniel Shelley
Phone: (716)475-6631   Fax: (716)475-5476

Financial Aid Office Contact
Verna J. Hazen, *Director*
Phone: (716)475-5520   Fax: (716)475-7270

194 Rose-Hulman

Head of the Institution
Samuel F. Hulbert
Phone: (812)877-8202   Fax: (812)877-9925

Head of Engineering
James R. Eifert
Phone: (812)877-8222   Fax: (812)877-9925
EMail: Eifert@NEXTWORK.Rose-Hulman.EDU'

Department Chair(s)
Chemical Eng
Noel E. Moore
Phone: (812)877-8292   Fax: (812)877-3198
EMail: MOORE@CH.ROSE-HULMAN.EDU
Civil Eng
James L. McKinney
Phone: (812)877-8335   Fax: (812)877-3198
EMail: MCKINNEY@CH.ROSE-HULMAN.EDU
Electrical & Computer Eng
Buck F. Brown
Phone: (812)877-8226   Fax: (812)877-3198
EMail: BROWN@EE.ROSE-HULMAN.EDU
Mechanical Eng
Robert Steinhauser
Phone: (812)877-8320   Fax: (812)877-3198
EMail: STEINHAUS@ME.Rose-Hulman.EDU

Admissions Office Contact
Chuck Howard
Phone: (812)877-1511   Fax: (812)877-3198

Financial Aid Office Contact
R. Paul Steward, *Director of Financial Aid*
Phone: (812)877-1511   Fax: (812)877-9925

Associate/Assistant Deans (Purdue — 188)
John F. McLaughlin, *Associate Dean*
Phone: (317)494-5349
Warren H. Stevenson, *Assistant Dean*
Phone: (317)494-5341
Mark S. Lundstrom, *Assistant Dean*
Phone: (317)494-5641

Department Chair(s)
Aeronautics & Astronautics
Alten F. Grandt, Jr.
Phone: (317)494-5117
Agricultural Eng
Larry F. Huggins
Phone: (317)494-1162
Chemical Eng
G.V. Reklaitis
Phone: (317)494-4075
Civil Eng
Vincent P. Drnevich
Phone: (317)494-2159
Construction Eng & Management
Daniel W. Halpin
Phone: (317)494-2240
Electrical Eng
Richard J. Schwartz
Phone: (317)494-3539
Freshman Eng
Phillip C. Wankat
Phone: (317)494-3884
Industrial Eng
Ferdinand F. Leimkuhler
Phone: (317)494-5444
Interdisciplinary Eng Studies
David P. Kessler
Phone: (317)494-7422
Materials Eng
Gerald L. Liedl
Phone: (317)494-4095
Mechanical Eng
Frank P. Incropera
Phone: (317)494-5688
Nuclear Eng
Victor H. Ransom
Phone: (317)494-5741

Financial Aid Office Contact
Joyce Hall, *Director of Financial Aid*
Phone: (317)494-5050

195 Rutgers, The State U

Head of the Institution
Francis L. Lawrence
Phone: (908)932-7454   Fax: (908)932-8060

Head of Engineering
Ellis H. Dill
Phone: (908)932-2214   Fax: (908)932-5313

Associate/Assistant Deans
Fred R. Bernath, *Associate Dean*
Phone: (908)932-2212   Fax: (908)932-5313
Jeffery L. Rankin, *Assistant Dean*
Phone: (908)932-2212   Fax: (908)932-5313
Paul W. Krygar, *Assistant Dean*
Phone: (908)932-2687   Fax: (908)932-5313

Department Chair(s)
Chemical & Biochemical Eng
Burton Davidson
Phone: (908)932-2228   Fax: (908)932-5313
Civil & Environmental Eng
Yong S. Chae
Phone: (908)932-2232   Fax: (908)932-0577
Electrical & Computer Eng
Bogoljub Lalevic
Phone: (908)932-3262   Fax: (908)932-5313
Industrial Eng
Elsayed A. Elsayed
Phone: (908)932-3654   Fax: (908)932-5467
Mechanical & Aerospace Eng
Abdelfattah Zebib
Phone: (908)932-2248   Fax: (908)932-5313
Ceramics
Malcolm G. McLaren
Phone: (908)932-2220   Fax: (908)932-3258
Biomedical Eng
Evangelia Tzanakou
Phone: (908)932-3155   Fax: (908)932-3753
Mechanics & Materials Science
Bernard H. Kear
Phone: (908)932-2245   Fax: (908)932-5977

Admissions Office Contact
M.E. Mitchell
Phone: (908)932-3770   Fax: (908)932-0237

Financial Aid Office Contact
Carl H. Buck, *Director of Financial Aid*
Phone: (908)932-7755

196 Saginaw Valley State U

Head of the Institution
Eric R. Gilbertson
Phone: (517)790-4041   Fax: (517)790-1314

Head of Engineering
Thomas E. Kullgren
Phone: (517)790-4141   Fax: (517)790-2717

Department Chair(s)
Electrical Eng
Tirumale Ramesh
Phone: (517)790-4192   Fax: (517)790-2717
Mechanical Eng
Terry Ishihara
Phone: (517)790-4154   Fax: (517)790-2717

Admissions Office Contact
Thomas E. Kullgren
Phone: (517)790-4144   Fax: (517)790-2717

Financial Aid Office Contact
William L. Healy, *Director, Scholarship & Financial Aid*
Phone: (517)790-4106

197 St. Cloud State U

Head of the Institution
Robert O. Bess
Phone: (612)255-2122

Head of Engineering
G. Richard Hogan
Phone: (612)255-2192   Fax: (612)255-4262

Associate/Assistant Deans
Dale A. Williams, *Assistant Dean*
Phone: (612)255-2192   Fax: (612)255-4262

Department Chair(s)
Electrical Eng
Bruce W. Ellis
Phone: (612)255-3252   Fax: (612)654-5127
EMail: ellis@kanga.stcloud.msus.edu
Manufacturing Eng
Andrew Bekkala
Phone: (612)255-3255   Fax: (612)654-5127
EMail: bekkala@kanga.stcloud.msus.edu

Admissions Office Contact
Bruce W. Ellis
Phone: (612)255-3252   Fax: (612)654-5127

Financial Aid Office Contact
Frank Loncorich, *Director of Financial Aid*
Phone: (612)255-2047

198 Parks Coll of Saint Louis U

Head of the Institution
Lawrence H. Biondi, S.J.
Phone: (314)658-2471   Fax: (314)658-7105

Head of Engineering
Peggy Baty
Phone: (618)337-7575   Fax: (618)332-6802

Associate/Assistant Deans
Dan Hejde, *Acting Associate Dean*
Phone: (618)337-7575   Fax: (618)332-6802

Department Chair(s)
Electrical Eng
Habibur Rahman
Phone: (618)337-7575   Fax: (618)332-6802
Aerospace Eng
John A. George
Phone: (618)337-7575   Fax: (618)332-6802

Admissions Office Contact
Sarah Nandor
Phone: (618)337-7575   Fax: (618)332-6802

Financial Aid Office Contact
Rachel M. Phillipone, *Financial Aid Coordinator*
Phone: (618)337-7575   Fax: (618)332-6802

199 U of San Diego

Head of the Institution
Author E. Hughes
Phone: (619)260-4520   Fax: (619)260-6833

Head of Engineering
Thomas A. Kanneman
Phone: (619)260-4628   Fax: (619)260-4619
EMail: t_kanneman@usdcsd.acusd.edu

Department Chair(s)
Electrical Eng
Thomas A. Kanneman
Phone: (619)260-4628   Fax: (619)260-4619
EMail: t_kanneman@usdcsd.acusd.edu

Admissions Office Contact
Warren W. Muller
Phone: (619)260-4506   Fax: (619)260-6836

Financial Aid Office Contact
Judith Lewis-Logue, *Director, Financial Aid*
Phone: (619)260-4514

200 San Diego State U

Head of the Institution
Thomas B. Day
Phone: (619)594-5201   Fax: (619)594-5642

Head of Engineering
George T. Craig
Phone: (619)594-6061   Fax: (619)594-6005
EMail: gcraig@sciences.sdsu.edu

Associate/Assistant Deans
Nihad A. Hussain, *Associate Dean of Engineering*
Phone: (619)594-6061   Fax: (619)594-6005
Fang H. Chou, *Assistant Dean for Student Affairs*
Phone: (619)594-6061   Fax: (619)594-6005

Department Chair(s)
Civil Eng
Janusz Supernak
Phone: (619)594-6071   Fax: (619)594-6005
EMail: supernak@jcsnext.sdsu.edu
Electrical Eng
Leonard R. Marino
Phone: (619)594-5718   Fax: (619)594-6005
EMail: lmarino@sciences.sdsu.edu
Mechanical Eng
John G. Pinto
Phone: (619)594-6067   Fax: (619)594-6005
EMail: jpinto@sciences.sdsu.edu
Aerospace Eng & Eng Mechanics
Nagy Nosseir
Phone: (619)594-6074   Fax: (619)594-6005
EMail: aeror1009@ucsvax.sdsu.edu
Eng Sciences/Applied Mechanics
Mauro Pierucci
Phone: (619)594-6079   Fax: (619)594-6005
EMail: mpierucci@sciences.sdsu.edu

Financial Aid Office Contact
William D. Boyd, *Director of Financial Aid*
Phone: (619)594-6323   Fax: (619)594-5642

201 San Francisco State U

Head of the Institution
Robert A. Corrigan
Phone: (415)338-1381

Head of Engineering
V.V. Krishnan
Phone: (415)338-1174   Fax: (415)338-0525

Department Chair(s)
Civil Eng
Norman Owen
Phone: (415)338-7740   Fax: (415)338-0525
Electrical Eng
Richard Zimmerman
Phone: (415)338-7742   Fax: (415)338-0525
Mechanical Eng
Anthony Wheeler
Phone: (415)338-1053   Fax: (415)338-0525

Financial Aid Office Contact
Jeffrey S. Baker, *Director, Financial Aids*
Phone: (415)338-2437

202 San Jose State U

Head of the Institution
Handel Evans
Phone: (408)924-2400

Head of Engineering
Jay D. Pinson
Phone: (408)924-3800   Fax: (408)924-3818

Associate/Assistant Deans
James J. Freeman, *Associate Dean*
Phone: (408)924-3800   Fax: (408)924-3818
Kenneth Challenger, *Associate Dean*
Phone: (408)924-3800   Fax: (408)923-3818

Department Chair(s)
Chemical Eng
Mike B. Jennings
Phone: (408)924-3909   Fax: (408)924-3818
Civil Eng
Thailia Anagnos
Phone: (408)924-3900   Fax: (408)924-3818
Electrical Eng
Ray R. Chen
Phone: (408)924-3950   Fax: (408)924-3818
Industrial & Systems Eng
Ernest A. Unwin
Phone: (408)924-4050   Fax: (408)924-3818
Mechanical Eng
Tai-Ran Hsu
Phone: (408)924-3850   Fax: (408)924-3818
Aerospace Eng
Richard D. Desautel
Phone: (408)924-3840   Fax: (408)924-3818
Computer Eng
Nicholas L. Pappas
Phone: (408)924-4100   Fax: (408)924-3818
Materials Eng
K.S. Sree Harsha
Phone: (408)924-4050   Fax: (408)924-3818
General Eng
Nabil Ibrahim
Phone: (408)924-3968   Fax: (408)924-3818

Financial Aid Office Contact
Donald Ryan, *Financial Aid Director*
Phone: (408)924-1676

203 Santa Clara U

Head of the Institution
Paul L. Locatelli, S.J.
Phone: (408)554-4100

Head of Engineering
Terry E. Shoup
Phone: (408)554-4600   Fax: (408)554-5474
EMail: tshoup@scu.bitnet

Associate/Assistant Deans
Mohammad A. Ketabchi, *Associate Dean, Engineering*
Phone: (408)554-2731   Fax: (408)554-5474
Steven C. Chiesa, *Assistant Dean, Undergraduate Services*
Phone: (408)554-4697   Fax: (408)554-5474
George R. Fegan, *Assistant Dean, Graduate Services*
Phone: (408)554-4181   Fax: (408)554-5474

Department Chair(s)
Civil Eng
Sukhmander Singh
Phone: (408)554-6869   Fax: (408)554-5474
Electrical Eng
Timothy J. Healy
Phone: (408)554-5309   Fax: (408)554-5474
Mechanical Eng
Mark D. Ardema
Phone: (408)554-4173   Fax: (408)554-5474
Computer Eng
Daniel W. Lewis
Phone: (408)554-4449   Fax: (408)554-5474

Admissions Office Contact
Daniel J. Saracino
Phone: (408)554-4700   Fax: (408)554-5255

Financial Aid Office Contact
Rita LeBarre, *Director of Financial Aid*
Phone: (408)554-4505   Fax: (408)554-6926

204 Seattle U

Head of the Institution
William J. Sullivan, S.J.
Phone: (206)296-1891   Fax: (206)296-2163

Head of Engineering
Kathleen Mailer
Phone: (206)296-5500   Fax: (206)296-2179
EMail: kmailer@seattleu.edu

Associate/Assistant Deans
Robert J. Smith, *Assistant Dean*
Phone: (206)296-5502   Fax: (206)296-2179
Rolf T. Skrinde, *Associate Dean for Research & Design*
Phone: (206)296-5525   Fax: (206)296-2179

Department Chair(s)
Civil & Environmental Eng
Percy Chien
Phone: (206)296-5520   Fax: (206)296-2179
Electrical Eng
Patricia D. Daniels
Phone: (206)296-5970   Fax: (206)296-2179

Engineering & Engineering Technology Personnel

Mechanical Eng
Dennis W. Wiedemeier
Phone: (206)296-5540   Fax: (206)296-2179
Admissions Office Contact
Lee Gerig
Phone: (206)296-5800
Financial Aid Office Contact
James White, *Director of Financial Aid*
Phone: (206)296-5840   Fax: (206)296-2163

205   U of South Alabama
Head of the Institution
Frederick P. Whiddon
Phone: (205)460-6111   Fax: (205)460-7541
Head of Engineering
David T. Hayhurst
Phone: (205)460-6140   Fax: (205)460-6343
Associate/Assistant Deans
Wilford D. Raburn, *Associate Dean*
Phone: (205)460-7506   Fax: (205)460-6140
Department Chair(s)
Chemical Eng
H. Ted Huddleston, Jr.
Phone: (205)460-6160   Fax: (205)460-6343
Civil Eng
Joseph M. Olsen
Phone: (205)460-6174   Fax: (205)460-6343
Electrical Eng
Russell M. Hayes, Jr.
Phone: (205)460-6117   Fax: (205)460-6343
Mechanical Eng
George W. Douglas
Phone: (205)460-6168   Fax: (205)460-6343
Admissions Office Contact
J. David Stearns
Phone: (205)460-6494
Financial Aid Office Contact
Grady L. Collins, *Director*
Phone: (205)460-6231

206   U of South Carolina
Head of the Institution
John M. Palms
Phone: (803)777-2001   Fax: (803)777-9480
Head of Engineering
W.K. Humphries
Phone: (803)777-4259   Fax: (803)777-9597
Associate/Assistant Deans
Joseph H. Gibbons, *Associate Dean for Undergraduate Studies*
Phone: (803)777-4177   Fax: (803)777-9597
James B. Radziminski, *Associate Dean for Graduate Studies & Research*
Phone: (803)777-4178   Fax: (803)777-9597
Department Chair(s)
Chemical Eng
Ralph E. White
Phone: (803)777-4181   Fax: (803)777-8265
Civil Eng
Joseph H. Bradburn
Phone: (803)777-3614   Fax: (803)777-9597
Electrical Eng
Robert O. Pettus
Phone: (803)777-4195   Fax: (803)777-8045
Mechanical Eng
Walter H. Peters
Phone: (803)777-4185   Fax: (803)777-0106
Financial Aid Office Contact
Denise Wellman
Phone: (803)777-3215   Fax: (803)777-0941

207   South Dakota School of Mines
Head of the Institution
Richard J. Gowen
Phone: (605)394-2411   Fax: (605)394-6131

Head of Engineering
William L. Hughes
Phone: (605)394-2256   Fax: (605)394-6131
Department Chair(s)
Chemical Eng
James M. Munro
Phone: (605)394-2421   Fax: (605)394-6131
Civil Eng
William L. Hughes
Phone: (605)394-2442   Fax: (605)394-6131
Electrical Eng
A.L. Riemenschneider
Phone: (605)394-2451   Fax: (605)394-6131
Industrial Eng
Wayne B. Krause
Phone: (394)394-1270   Fax: (605)394-6131
Mechanical Eng
Wayne B. Krause
Phone: (605)394-1270   Fax: (605)394-6131
Geological Eng
William M. Roggenthen
Phone: (605)394-2461   Fax: (605)394-6131
Metallurgical Eng
Kenneth N. Han
Phone: (605)394-2342   Fax: (605)394-6131
Mining Eng
John D. Erickson
Phone: (605)394-2345   Fax: (605)394-6131
Admissions Office Contact
Gary A. Bjordal
Phone: (605)394-2400   Fax: (605)394-6131
Financial Aid Office Contact
Sharon K. Colombe, *Director of Financial Aid*
Phone: (605)394-2274   Fax: (605)394-6131

208   South Dakota State U
Head of the Institution
Robert T. Wagner
Phone: (605)688-4111   Fax: (605)688-5822
Head of Engineering
Duane E. Sander
Phone: (605)688-4161   Fax: (605)688-5878
EMail: DYERB@ENGSDSTATE.EDU
Department Chair(s)
Civil Eng
Dwayne A. Rollag
Phone: (605)688-5427   Fax: (605)688-5878
Electrical Eng
Virgil G. Ellerbruch
Phone: (605)688-4526   Fax: (605)688-5880
Mechanical Eng
Donell P. Froehlich
Phone: (605)688-5426   Fax: (605)688-5878
Agricultural Eng
Ralph Alcock
Phone: (605)688-5141   Fax: (605)688-6065
Computer Science
Gerald E. Bergum
Phone: (605)688-5719   Fax: (605)688-5822
EMail: CS00@SDSUMUS.BITNET
General Eng
H. Frank Kornbaum
Phone: (605)688-6417   Fax: (605)688-5878
EMail: KORNBAUF@MG.SDSTATE.EDU
Eng Physics
Warren W. Hein
Phone: (605)688-5428   Fax: (605)688-5878
EMail: KRUEGERS@MG.SDSTATE.EUD
Admissions Office Contact
Tracy Welsh
Phone: (605)688-4121   Fax: (605)688-6384
Financial Aid Office Contact
Jay Larsen, *Financial Aid Director*
Phone: (605)688-4695   Fax: (605)688-5822

209   U of South Florida
Head of the Institution
Francis T. Borkowski
Phone: (813)974-2791   Fax: (813)974-5530
Head of Engineering
Michael G. Kovac
Phone: (813)974-3780   Fax: (813)974-5094
EMail: kovac@ec.usf.edu
Associate/Assistant Deans
Melvin W. Anderson, *Associate Dean for Academic Affairs*
Phone: (813)974-3782   Fax: (813)974-5094
John A. Llewellyn, *Associate Dean of Computer Applications*
Phone: (813)974-3788   Fax: (813)974-5094
Andrew J. Barrett, *Assistant Dean*
Phone: (813)974-3783   Fax: (813)447-1501
Department Chair(s)
Chemical Eng
Richard A. Gilbert
Phone: (813)974-3997   Fax: (813)974-3651
EMail: gilbert@ec.usf.edu
Civil Eng
Wayne F. Echelberger
Phone: (813)974-2275   Fax: (813)974-3651
EMail: echelber@ec.usf.edu
Electrical Eng
Elias K. Stefanakos
Phone: (813)974-2369   Fax: (813)974-5250
EMail: stefanak@ec.usf.edu
Industrial Eng
Paul E. Givens
Phone: (813)974-2269   Fax: (813)974-3651
EMail: givens@ec.usf.edu
Mechanical Eng
Ronald H. Howell
Phone: (813)974-2280   Fax: (813)974-3539
EMail: howell@ec.usf.edu
Computer Science & Eng
Abraham Kandel
Phone: (813)974-3652   Fax: (813)974-5456
EMail: kandel@sol.usf.edu
Admissions Office Contact
George R. Card
Phone: (813)974-2684   Fax: (813)974-5094
Financial Aid Office Contact
Gwyndolyn Francis, *Director of Financial Aid*
Phone: (813)974-4700   Fax: (813)974-5144

210   Southern U & A&M Coll
Head of the Institution
Marvin L. Yates
Phone: (504)771-5020   Fax: (504)771-2026
Head of Engineering
Trent V. Montgomery
Phone: (504)771-5290   Fax: (504)771-2072
EMail: trent@subrvm.bitnet
Associate/Assistant Deans
Thomas L. Henderson, *Associate Dean*
Phone: (504)771-3798   Fax: (504)771-2072
Department Chair(s)
Civil Eng
Huey K. Lawson
Phone: (504)771-5870   Fax: (504)771-4320
Electrical Eng
James A. Anderson
Phone: (504)771-5292   Fax: (504)775-9828
Mechanical Eng
Sahib S. Chehl
Phone: (504)771-4701   Fax: (504)771-4877
Division of Tech
Eddie Hildreth, Jr.
Phone: (504)771-4193   Fax: (504)771-2495
Admissions Office Contact
Henry J. Bellaire
Phone: (504)771-5124

Financial Aid Office Contact
Cynthia L. Tarver, *Director*
Phone: (504)771-2796

211   U of Southern Cal
Head of the Institution
Steven B. Sample
Phone: (213)740-2111
Head of Engineering
Leonard M. Silverman
Phone: (213)740-0884   Fax: (213)740-8493
Associate/Assistant Deans
Clarke T. Howatt, *Assoicate Dean for Student Affairs*
Phone: (213)740-4530   Fax: (213)740-8690
Richard K. Miller, *Associate Dean for Academic Affairs*
Phone: (213)740-0621   Fax: (213)740-8493
Lloyd J. Griffiths, *Associate Dean for Research & Academic Affairs*
Phone: (213)740-0877   Fax: (213)740-8493
Thomas Garrow, *Associate Dean for Planning & External Relations*
Phone: (213)740-7099   Fax: (213)740-8493
Department Chair(s)
Chemical Eng
Yanis C. Yortsos
Phone: (213)740-0317   Fax: (213)740-8053
EMail: Yortsos@muskat.usc.edu
Civil Eng
Mihran Agbabian
Phone: (213)740-0610   Fax: (213)744-1426
EMail: agbabian@mizar.usc.edu
Mechanical Eng
Redekopp G. Larry
Phone: (213)740-5369   Fax: (213)740-8071
Industrial Eng
Gerald Nadler
Phone: (213)740-4892
EMail: nadler@mizar.usc.edu
Electrical Eng
Hans H. Kuehl
Phone: (213)740-4700   Fax: (213)740-8677
EMail: Kuehl@mizar.usc.edu
Aerospace Eng
E. Phillip Muntz
Phone: (213)740-5366   Fax: (213)740-4303
EMail: emuntz@chaph.usc.edu
Biomedical Eng
Vasilis Z. Marmarelis
Phone: (213)740-0838
Computer Science
Ellis Horowitz
Phone: (213)740-8056   Fax: (213)740-7285
EMail: horowitz@pollux.usc.edu
Admissions Office Contact
Clarke T. Howatt
Phone: (213)740-4530   Fax: (213)740-4530
Financial Aid Office Contact
Catherine Thomas, *Director, Financial Aid*
Phone: (212)740-5444

212   U of Southern Colorado
Head of the Institution
Robert C. Shirley
Phone: (719)549-2306   Fax: (719)549-2219
Head of Engineering
Ray L. Sisson
Phone: (719)549-2696   Fax: (719)549-2519
EMail: SISSON@COMET.USCOLO.EDU
Department Chair(s)
Industrial & Systems Eng
Nancy L. Mills
Phone: (719)549-2036   Fax: (719)549-2519
EMail: MILLS@STARBURST.USCOLO.EDU
Admissions Office Contact
Margie Wade
Phone: (719)549-2461   Fax: (719)549-2219

213 Southern Illinois-Carbondale

Head of the Institution
John C. Guyon
Phone: (618)453-2341   Fax: (618)453-5362

Head of Engineering
Juh W. Chen
Phone: (618)453-4321   Fax: (618)453-4235

Associate/Assistant Deans
James L. Evers, *Associate Dean*
Phone: (618)453-4321   Fax: (618)453-4235
Echol E. Cook, *Associate Dean*
Phone: (618)453-4321   Fax: (618)453-4235

Department Chair(s)
Civil Eng
Sedat Sami
Phone: (618)536-2368   Fax: (618)453-4235
Electrical Eng
Glafkos D. Galanos
Phone: (618)536-2364   Fax: (618)453-4235
Mechanical Eng
Albert C. Kent
Phone: (618)536-2396   Fax: (618)453-4235
Mining Eng
Yoginder P. Chugh
Phone: (618)536-6637   Fax: (618)453-4235

Admissions Office Contact
Jerre C. Pfaff
Phone: (618)453-4321   Fax: (618)453-3250

Financial Aid Office Contact
Pamela A. Britton, *Director, Financial Aid*
Phone: (618)453-4334

214 Southern Illinois-Edwardsville

Head of the Institution
Earl E. Lazerson
Phone: (618)692-2475   Fax: (618)692-2270

Head of Engineering
Colby V. Ardis
Phone: (618)692-2541   Fax: (618)692-3374

Associate/Assistant Deans
Harlan H. Bengtson, *Assistant Dean*
Phone: (618)692-2534   Fax: (618)692-2555

Department Chair(s)
Civil Eng
Mark P. Rossow
Phone: (618)692-2533   Fax: (618)692-2555
Electrical Eng
Raghupathy Bollini
Phone: (618)692-2524   Fax: (618)692-3374
Industrial Eng
Jacob H. Van Roekel
Phone: (618)692-3389   Fax: (618)692-2555
Mechanical Eng
Chi Loong Chow
Phone: (618)692-3389   Fax: (618)692-2555

Admissions Office Contact
Harlan H. Bengtson
Phone: (618)692-2534   Fax: (618)692-2555

Financial Aid Office Contact
William D. Burns, *Acting Director*
Phone: (618)692-3880

215 U of Southern Maine

Head of the Institution
Richard Pattenaude
Phone: (207)780-4480   Fax: (207)780-4549

Head of Engineering
Brian C. Hodgkin
Phone: (207)780-5582   Fax: (207)780-5129

Associate/Assistant Deans
Richard Carter, *Assist. Dean*
Phone: (207)780-5141   Fax: (207)780-5129

Department Chair(s)
Electrical Eng
James W. Smith
Phone: (207)780-5584   Fax: (207)780-5129

Admissions Office Contact
Daniel Palubniak
Phone: (207)780-4970

Financial Aid Office Contact
Helen F. Parker, *Senior Associate Director*
Phone: (207)780-5250

216 Southern Methodist U

Head of the Institution
A. Kenneth Pye
Phone: (214)768-3300   Fax: (214)768-4138

Head of Engineering
André G. Vacroux
Phone: (214)768-3050   Fax: (214)768-3845
EMail: dorle@seas.smu.edu

Associate/Assistant Deans
James G. Dunham, *Assistant Dean for Undergraduate & Graduate Studies*
Phone: (214)768-3484   Fax: (214)768-3883
E. Douglas Harris, *Assistant Dean for Undergraduate Recruiting*
Phone: (214)768-3041   Fax: (214)768-3883

Department Chair(s)
Electrical Eng
Jerome K. Butler
Phone: (214)768-3113   Fax: (214)768-3883
EMail: jkb@seas.smu.edu
Computer Science & Eng
Jeffery L. Kennington
Phone: (214)768-3278   Fax: (214)768-3085
EMail: jlk@seas.smu.edu
Mechanical Eng
Bijan Mohraz
Phone: (214)768-3128   Fax: (214)678-3883
EMail: mohraz@seas.smu.edu

217 Stanford U

Head of the Institution
Gerhard R. Casper
Phone: (415)723-2481   Fax: (415)725-6847

Head of Engineering
James F. Gibbons
Phone: (415)723-3938   Fax: (415)723-5599
EMail: Gibbons@Sierra.Stanford.Edu

Associate/Assistant Deans
Charles H. Kruger, *Senior Associate Dean for Faculty & Academic Affairs*
Phone: (415)723-3936   Fax: (415)723-5599
John C. Bravman, *Associate Dean for Undergraduate Education*
Phone: (415)723-9106   Fax: (415)723-5599
Gene F. Franklin, *Assistant Dean for Undergraduate Education*
Phone: (415)723-9106   Fax: (415)723-5599
James V. Jucker, *Assistant Dean for Undergraduate Education*
Phone: (415)725-9265   Fax: (415)723-5599
Noe P. Lozano, *Associate Dean for Minority & Affirmative Action Programs*
Phone: (415)723-9107   Fax: (415)723-5599

Department Chair(s)
Chemical Eng
Channing R. Robertson
Phone: (415)723-4986   Fax: (415)725-7294
EMail: Channing@Rio.Stanford.Edu
Civil Eng
Gilbert M. Masters
Phone: (415)723-3921   Fax: (415)725-8662
EMail: Masters@Cive.Stanford.Edu
Electrical Eng
Joseph W. Goodman
Phone: (415)723-5782   Fax: (415)723-1882
EMail: Goodman@Sierra.Stanford.Edu
Industrial Eng & Eng Management
Robert C. Carlson
Phone: (415)723-9279   Fax: (415)725-8799
EMail: Carlson@Sierra.Stanford.Edu
Mechanical Eng
William C. Reynolds
Phone: (415)725-2073   Fax: (415)725-4862
EMail: Reynolds@Sierra.Stanford.Edu
Computer Science
Jeffrey D. Ullman
Phone: (415)723-9745   Fax: (415)725-7411
EMail: Ullman@CS.Stanford.Edu
Materials Science & Eng
William D. Nix
Phone: (415)725-2605   Fax: (415)725-4034
EMail: nix@Sierra.Stanford.Edu
Aeronautics & Astronautics
George S. Springer
Phone: (415)723-4135   Fax: (415)723-0062
EMail: Springer@Sierra.Stanford.Edu
Eng-Economic Systems
James L. Sweeney
Phone: (415)723-2847   Fax: (415)723-1614
EMail: Sweeney@Sierra.Stanford.Edu
Operations Research
Richard W. Cottle
Phone: (415)725-0557   Fax: (415)723-4107
EMail: Cottle@Sierra.Stanford.Edu

Admissions Office Contact
James M. Montoya
Phone: (415)723-2091

Financial Aid Office Contact
Robert P. Huff, *Director of Financial Aids*
Phone: (415)723-3058

218 SUNY-Binghamton

Head of the Institution
Lois B. DeFleur
Phone: (607)777-2131   Fax: (607)777-4000

Head of Engineering
Lyle D. Feisel
Phone: (607)777-2871   Fax: (607)777-4822
EMail: Feisel@BINGTJW

Associate/Assistant Deans
Michael F. McGoff, *Associate Dean for Academic Affairs & Administration*
Phone: (607)777-6204   Fax: (607)777-4822
Douglas M. Green, *Associate Dean for Research & External Affairs*
Phone: (607)777-4555   Fax: (607)777-4822

Department Chair(s)
Electrical Eng
James E. Morris
Phone: (607)777-4856   Fax: (607)777-4822
EMail: JMORRIS@BINGTJW
Mechanical & Industrial Eng
John A. Fillo
Phone: (607)777-4747   Fax: (607)777-4822
EMail: JFillo@BINGTJW
Computer Science
Sudhir Aggarwal
Phone: (607)777-4802   Fax: (607)777-4822
EMail: SAGGARWAL@BINGTJW
Systems Science
George J. Klir
Phone: (607)777-6510   Fax: (607)777-4822
EMail: GKLIR@BINGTJW

Admissions Office Contact
Michael F. McGoff
Phone: (607)777-6203   Fax: (607)777-4822

Financial Aid Office Contact
Christina M. Knickerbocker, *Director*
Phone: (607)777-2428

219 SUNY-Buffalo

Head of the Institution
William R. Greiner
Phone: (716)645-2901   Fax: (716)645-3728

Head of Engineering
George C. Lee
Phone: (716)645-2771   Fax: (716)645-2495
EMail: FEAGCLEE@ubvms.cc.buffalo.edu

Associate/Assistant Deans
Kenneth M. Kiser, *Associate Dean*
Phone: (716)645-2772   Fax: (716)645-2495
Warren H. Thomas, *Associate Dean-Undergraduate Student Services*
Phone: (716)645-2774   Fax: (716)645-2495

Department Chair(s)
Chemical Eng
Ralph T. Yang
Phone: (716)645-2911   Fax: (716)645-3822
EMail: CMEYANG@ubvms.cc.buffalo.edu
Civil Eng
Dale D. Meredith
Phone: (716)645-2114   Fax: (716)645-3733
EMail: CIEDALE@ubvms.cc.buffalo.edu
Electrical & Computer Eng
Wayne K. Anderson
Phone: (716)645-2422   Fax: (716)645-3656
EMail: ELEANDER@ubvms.cc.buffalo.edu
Industrial Eng
Rajan Batta
Phone: (716)645-2357   Fax: (716)645-3302
EMail: INDBATTA@ubvms.cc.buffalo.edu
Mechanical & Aerospace Eng
Andres Soom
Phone: (716)645-2593   Fax: (716)645-3875
EMail: MECDG@ubvms.cc.buffalo.edu

Admissions Office Contact
Kevin Durkin
Phone: (716)829-2333   Fax: (716)829-3902

Financial Aid Office Contact
Michael D. Randall, *Director*
Phone: (716)829-3724   Fax: (716)829-2022

220 SUNY-Env'l Science & Forestry

Head of the Institution
Ross S. Whaley
Phone: (315)470-6681   Fax: (315)470-6932

Head of Engineering
Robert H. Brock, *Director*
Phone: (315)470-6633   Fax: (315)470-6958

Department Chair(s)
Forest Eng
Robert H. Brock
Phone: (315)470-6633   Fax: (315)470-6958
Paper Science & Eng
Leland R. Schroeder
Phone: (315)470-6502   Fax: (315)470-6945
Wood Products Eng
Leonard A. Smith
Phone: (315)470-6880   Fax: (315)470-6879

Admissions Office Contact
Dennis O. Stratton
Phone: (315)470-6600   Fax: (315)470-6933

Financial Aid Office Contact
John E. View, *Director of Financial Aid & EOP Program*
Phone: (315)470-6670   Fax: (315)470-6933

221 SUNY-Maritime

Head of the Institution
Floyd H. Miller
Phone: (718)409-7270   Fax: (718)409-7392

Head of Engineering
Jose Femenia, *Chairman*
Phone: (718)409-7411   Fax: (718)409-7421

Financial Aid Office Contact
Gina T. Mestas, *Director of Financial Aid*
Phone: (719)549-2753   Fax: (719)549-2938

Department Chair(s)
Electrical Eng
Conrad C. Youngren
Phone: (718)409-7424   Fax: (718)409-7421
Mechanical Eng
John L. Mathieson
Phone: (718)409-7418   Fax: (718)409-7421
Marine Eng
John L. Mathieson
Phone: (718)409-7418   Fax: (718)409-7421
Naval Architecture
Robert B. Zubaly
Phone: (718)409-7417   Fax: (718)409-7421
Facilities Eng
Jose Femenia
Phone: (718)409-7411   Fax: (718)409-7421

Admissions Office Contact
Peter Cooney
Phone: (718)409-7222   Fax: (718)409-7392

Financial Aid Office Contact
Howard L. English, *Director of Financial Aid*
Phone: (718)409-7277   Fax: (718)409-7392

222 SUNY-New Paltz

Head of the Institution
Alice Chandler
Phone: (914)257-3288

Head of Engineering
Owen Hill
Phone: (914)257-3720   Fax: (914)257-3009

Department Chair(s)
Electrical Eng
Owen Hill
Phone: (914)257-3720   Fax: (914)257-3009

Admissions Office Contact
Robert J. Seaman
Phone: (914)257-3200

Financial Aid Office Contact
Daniel Sistarenik, *Director of Financial Aid Programs*
Phone: (914)257-3250

223 SUNY-Stony Brook

Head of the Institution
John H. Marburger
Phone: (516)632-6265   Fax: (516)632-6252

Head of Engineering
Yacov Shamash
Phone: (516)632-8380   Fax: (516)632-8205
EMail: yshamash@ccmail.sunysb.edu

Associate/Assistant Deans
Marian Visich, Jr., *Associate Dean*
Phone: (516)632-8380   Fax: (516)632-8205
Joseph S. Hogan, *Associate Dean*
Phone: (516)632-8380   Fax: (516)632-8205
Joan M. Kenny, *Assistant Dean*
Phone: (516)632-8381   Fax: (516)632-8205

Department Chair(s)
Electrical Eng
Kenneth L. Short
Phone: (516)632-8420   Fax: (516)632-8494
Mechanical Eng
James Tasi
Phone: (516)632-8300   Fax: (516)632-8544
Materials Science & Eng
Clive R. Clayton
Phone: (516)632-8484   Fax: (516)632-8052

Admissions Office Contact
GiGi Lamens
Phone: (516)632-6868   Fax: (516)632-9027

Financial Aid Office Contact
Sherwood Johnson, *Director, Financial Aid*
Phone: (516)632-6840

224 Swarthmore Coll

Head of the Institution
Alfred H. Bloom
Phone: (215)328-8314   Fax: (215)328-8673

Head of Engineering
Frederick L. Orthlieb
Phone: (215)328-8080   Fax: (215)328-8082
EMail: forthli1@swarthmore.edu

Department Chair(s)
Eng
Frederick L. Orthlieb
Phone: (215)328-8080   Fax: (215)328-8082
EMail: forthli1@swarthmoreedu

Admissions Office Contact
Robert A. Barr
Phone: (215)328-8308   Fax: (215)328-8673

Financial Aid Office Contact
Laura Talbot, *Director of Financial Aid*
Phone: (215)328-8358   Fax: (215)328-8673

225 Syracuse U

Head of the Institution
Kenneth A. Shaw, *Chancellor*
Phone: (315)443-2235   Fax: (315)443-3503

Head of Engineering
Steven C. Chamberlain
Phone: (315)443-4341   Fax: (315)443-4936
EMail: Steve_Chamberlain@ISR.SYR.EDU

Associate/Assistant Deans
Samuel P. Clemence, *Associate Dean for Research & Graduate Affairs*
Phone: (315)443-4317   Fax: (315)443-4936
Earl J. Kletsky, *Associate Dean for Academic Programs*
Phone: (315)443-2545   Fax: (315)443-4936
Anne L. Shelly, *Associate Dean for Academic Programs*
Phone: (315)443-4442   Fax: (315)443-4936
Richard A. Lisbon, *Assistant Dean for Administration*
Phone: (315)443-4343   Fax: (315)443-4936

Department Chair(s)
Chemical Eng & Materials Science
Cynthia S. Hirtzel
Phone: (315)443-2557   Fax: (315)443-2559
EMail: CHIRTZEL@SUVM.ACS.SYR.EDU
Civil & Environmental Eng
Raymond D. Letterman
Phone: (315)443-2311   Fax: (315)443-1243
EMail: RDLETTER@SUVM.ACS.SYR.EDU
Electrical & Computer Eng
Kamal Jabbour
Phone: (315)443-2655   Fax: (315)443-2583
EMail: JABBOUR@SUVM.ACS.SYR.ACS
Mechanical, Aerospace & Manufacturing Eng
Volker Weiss
Phone: (315)443-2341   Fax: (315)443-9099
EMail: VWEISS@SUNRISE.ACS.SYR.EDU
Bioengineering
Gustav A. Engbretson
Phone: (315)443-1931   Fax: (315)443-1184
EMail: Gus_Engbretson@ISR.SYR.EDU
Aerospace Eng
John E. LaGraff
Phone: (315)443-4366   Fax: (315)443-9099
EMail: JLAGRAFF@SUVM.ACS.SYR.EDU
Computer Eng
Edward Stabler
Phone: (315)443-4370   Fax: (315)443-2583
EMail: HLPEPS@SUVM.ACS.SYR.EDU
Eng Physics
Volker Weiss
Phone: (315)443-3918   Fax: (315)443-9099
EMail: VWEISS@SUNRISE.ACS.SYR.EDU
Environmental Eng
Stephen J. Nix
Phone: (315)443-3347   Fax: (315)443-1243
EMail: SJNIX@SUVM.ACS.SYR.EDU
Manufacturing Eng
John F. Barrows
Phone: (315)443-2826   Fax: (315)443-9099
EMail: WARNERP@SUAIS.BITNET
Mechanical Eng
Edward A. Bogucz
Phone: (315)443-4366   Fax: (315)443-9099
EMail: eabogucz@sunrise.syr.edu

Admissions Office Contact
David C. Smith
Phone: (315)443-3611

Financial Aid Office Contact
Christopher Walsh, *Director of Financial Aid*
Phone: (315)449-1513   Fax: (315)443-1531

226 Technical U of Nova Scotia

Head of the Institution
Peter F. Adams
Phone: (902)420-7644   Fax: (902)420-7551

Head of Engineering
Donald A. Roy
Phone: (902)420-7600   Fax: (902)420-7551

Associate/Assistant Deans
William F. Caley, *Associate Dean, Undergraduate*
Phone: (902)420-7608   Fax: (902)420-7551
Tom A. Gill, *Associate Dean, Graduate Studies & Research*
Phone: (902)420-7759   Fax: (902)420-7551

Department Chair(s)
Chemical Eng
Paul A. Amyotte
Phone: (902)420-7697   Fax: (902)420-7551
Civil Eng
Leslie A. Baikie
Phone: (902)420-7678   Fax: (902)420-7551
Electrical Eng
C. Robert Baird
Phone: (902)420-7717   Fax: (902)420-7551
Industrial Eng
G. Peter Wilson
Phone: (902)420-7588   Fax: (902)420-7551
Mechanical Eng
Adam C. Bell
Phone: (902)420-7788   Fax: (902)420-7551
Agricultural Eng
K. Chris Watts
Phone: (902)420-7578   Fax: (902)420-7551
Mining & Metallurgical Eng
Laszlo A. Adorjan
Phone: (902)420-7708   Fax: (902)420-7551
Applied Mathematics
William J. Phillips
Phone: (902)420-7599   Fax: (902)420-7551
Food Science & Tech
Marvin A. Tung
Phone: (902)420-7735   Fax: (902)420-7551

Admissions Office Contact
Lamont Pelletier
Phone: (902)420-7624   Fax: (902)420-7551

Financial Aid Office Contact
William F. Caley, *Associate Dean, Undergraduate*
Phone: (902)420-7608   Fax: (902)420-7551

227 Temple U

Head of the Institution
Peter J. Liacouras
Phone: (215)787-7405

Head of Engineering
Charles K. Alexander
Phone: (215)787-7959   Fax: (215)787-6936

Associate/Assistant Deans
Thomas J. Ward, *Associate Dean for Administration*
Phone: (215)787-1050   Fax: (215)787-6936
Henry M. Sendaula, *Associate Dean for Research*
Phone: (215)787-7819   Fax: (215)787-6936

Department Chair(s)
Civil Eng
Frederick Schmitt
Phone: (215)787-7831   Fax: (215)787-6936
Electrical Eng
Brian P. Butz
Phone: (215)787-7212   Fax: (215)787-6936
Mechanical Eng
Steven Ridenour
Phone: (215)787-8825   Fax: (215)787-6936

Financial Aid Office Contact
John Morris, *Director*
Phone: (215)787-1492

228 U of Tennessee, Chattanooga

Head of the Institution
Frederick W. Obear
Phone: (615)755-4141   Fax: (615)756-5559

Head of Engineering
Ronald B. Cox
Phone: (615)755-4121   Fax: (615)755-5229

Department Chair(s)
Chemical Eng
William Q. Gurley
Phone: (615)755-4121   Fax: (615)755-5229
Civil Eng
Edwin P. Foster
Phone: (615)755-4121   Fax: (615)755-5229
Electrical Eng
Jack L. Thompson
Phone: (615)755-4349   Fax: (615)755-5229
Industrial/Engr Mgt/Manufacturing
Gregory A. Sedrick
Phone: (615)755-4121   Fax: (615)755-5229
Mechanical Eng
William Q. Gurley
Phone: (615)755-4121   Fax: (615)755-5229

Admissions Office Contact
Patsy Reynolds
Phone: (615)755-4662

Financial Aid Office Contact
Ray P. Fox, *Dean, Admissions & Records*
Phone: (615)755-4511

229 U of Tennessee, Knoxville

Head of the Institution
Joseph E. Johnson
Phone: (615)974-2241   Fax: (615)974-3753

Head of Engineering
Jerry E. Stoneking
Phone: (615)974-5321   Fax: (615)974-2669
EMail: BITNET:Stonekin:UTK VAX

Associate/Assistant Deans
William A. Miller, *Associate Dean*
Phone: (615)975-2454   Fax: (615)974-2669
James T. Pippin, *Assistant Dean*
Phone: (615)974-4457   Fax: (615)974-2669
June Q. Moore, *Assistant Dean*
Phone: (615)974-5323   Fax: (615)974-2669

Department Chair(s)
Chemical Eng
John W. Prados
Phone: (615)974-6053   Fax: (615)974-2669
EMail: INTERNET:Prados:UTKVX1.UTK.EDU
Civil Eng
Gregory D. Reed
Phone: (615)974-2503   Fax: (615)974-1669
EMail: BITNET:GD Reed.UTKVX
Electrical Eng
Joseph M. Googe
Phone: (615)974-5465   Fax: (615)974-2669
EMail: Internet:Googe:UTK.Engr.Edu.
Industrial Eng
Charles H. Aikens
Phone: (615)974-3333   Fax: (615)974-2669
EMail: BITNET:PA82578

Materials Science & Eng
Joseph E. Spruiell
Phone: (615)974-5336   Fax: (615)974-2669
EMail: INTERNET:Spruiell:UTKUXL.UTK.EDU

Eng Science & Mechanics
Thomas G. Carley
Phone: (615)974-2171   Fax: (615)974-2669
EMail: Internet:Carley:UTKVX.UTK.EDU

Nuclear Eng
Thomas W. Kerlin
Phone: (615)974-2525   Fax: (615)974-2669
EMail: BITNET:SALMON:UTKVX

Mechanical & Aerospace Eng
Donald R. Pitts
Phone: (615)974-5115   Fax: (615)974-2669
EMail: BITNET:PA85508

Agricultural Eng
Fred D. Tompkins
Phone: (615)974-7266   Fax: (615)974-4514

Eng Physics
William Bugg
Phone: (615)974-3342   Fax: (615)974-7843
EMail: BITNET:Bugg:SLAC.VM

Admissions Office Contact
Gordon E. Stanley
Phone: (615)974-2184   Fax: (615)974-6435

Financial Aid Office Contact
John E. Mays, *Director of Financial Aid*
Phone: (615)974-3131

230 Tennessee State U

Head of the Institution
James A. Hefner
Phone: (615)320-3432   Fax: (615)320-3376

Head of Engineering
Decatur B. Rogers
Phone: (615)320-3550   Fax: (615)320-3554

Associate/Assistant Deans
Mohan J. Malkani, *Associate Dean*
Phone: (615)320-3550   Fax: (615)320-3554

Department Chair(s)
Civil Eng
Farouk P. Mishu
Phone: (615)320-3277   Fax: (615)320-3554
EMail: MISHU@TSU—Bitnet

Electrical Eng
Satinderpaul S Devgan
Phone: (615)320-3268   Fax: (615)320-3554
EMail: DEVGAN@TSU—Bitnet

Mechanical Eng
Chinyere Onwubiko
Phone: (615)320-3555   Fax: (615)320-3554
EMail: ONWUBIKO@TSU—Bitnet

Architectural Eng
Walter Vincent
Phone: (615)320-3560   Fax: (615)320-3554
EMail: VINCENT@TSU—Bitnet

Admissions Office Contact
Erskine R. Vanderbilt
Phone: (615)320-3420   Fax: (615)320-3114

Financial Aid Office Contact
Wilson Lee, *Director of Financial Aid*
Phone: (615)320-3440   Fax: (615)320-3114

231 Tennessee Tech

Head of the Institution
Angelo A. Volpe
Phone: (615)372-3241   Fax: (615)372-3898

Head of Engineering
George M. Swisher
Phone: (615)372-3172   Fax: (615)372-6172
EMail: GMS8735@TNTECH.EDU

Associate/Assistant Deans
Marie B. Ventrice, *Associate Dean*
Phone: (615)372-3172   Fax: (615)372-6172
John T. Mason, III, *Assistant Dean*
Phone: (615)372-3172   Fax: (615)372-6172

Department Chair(s)
Chemical Eng
David W. Yarbrough
Phone: (615)372-3297   Fax: (615)372-6172
EMail: DWY1460@TNTECH.EDU

Civil Eng
Rafael B. Bustamante
Phone: (615)372-3454   Fax: (615)372-6172

Electrical Eng
P.K. Rajan
Phone: (615)372-3397   Fax: (615)372-6172
EMail: PKR1259@TNTECH.EDU

Industrial Eng
Sidney G. Gilbreath, III
Phone: (615)372-3465   Fax: (615)372-6172
EMail: SGG0491@TNTECH.EDU

Mechanical Eng
Edwin I. Griggs
Phone: (615)372-3254   Fax: (615)372-6172
EMail: EIG5242@TNTECH.EDU

Admissions Office Contact
James C. Perry
Phone: (615)372-3888   Fax: (615)372-3898

Financial Aid Office Contact
Raymond L. Holbrook, *Director, Financial Aid*
Phone: (615)372-3073   Fax: (615)372-6138

232 U of Texas at Arlington

Head of the Institution
Wendell H. Nedderman
Phone: (817)273-2101   Fax: (817)794-5656

Head of Engineering
John H. McElroy
Phone: (817)273-2571   Fax: (817)273-2548

Associate/Assistant Deans
Floyd L. Cash, *Associate Dean for Academic Affairs*
Phone: (817)273-2571   Fax: (817)273-2548
Hal Sobol, *Associate Dean for Research*
Phone: (817)794-5639   Fax: (817)273-2548

Department Chair(s)
Civil Eng
Clinton Parker
Phone: (817)794-5055   Fax: (817)273-2630

Electrical Eng
Owen R. Mitchell
Phone: (817)273-2649   Fax: (817)273-2548

Industrial Eng
Gladstone T. Stevens
Phone: (817)273-3092   Fax: (817)273-3406

Mechanical Eng
Kent L. Lawrence
Phone: (817)273-2561   Fax: (817)273-2952

Aerospace Eng
Charles W. Jiles
Phone: (817)273-2603   Fax: (817)273-5010

Computer Science Eng
Billy D. Carroll
Phone: (817)273-3785   Fax: (817)273-2548

Admissions Office Contact
Zack Prince
Phone: (817)273-3565

Financial Aid Office Contact
Judy Walker, *Director*
Phone: (817)273-3561   Fax: (817)273-5555

233 U of Texas at Austin

Head of the Institution
Robert M. Berdahl
Phone: (512)471-1232   Fax: (512)471-8102

Head of Engineering
Herbert H. Woodson
Phone: (512)471-1136   Fax: (512)471-3955
EMail: H_Woodson.Dean_of_Eng@Engdean-gate.CE.UTexas.EDU

Associate/Assistant Deans
Ned H. Burns, *Associate Dean for Academic Affairs*
Phone: (512)471-7995   Fax: (512)471-3955
Dale E. Klein, *Associate Dean for Research*
Phone: (512)471-4325   Fax: (512)471-3955
Alvin H. Meyer, *Associate Dean for Student Affairs*
Phone: (512)471-4321   Fax: (512)471-3955
Douglas R. Lloyd, *Associate Dean for Graduate Student Recruiting*
Phone: (512)471-1519   Fax: (512)471-3955
John C. Halton, *Assistant Dean for Development*
Phone: (512)471-3395   Fax: (512)471-3955

Department Chair(s)
Chemical Eng
Thomas F. Edgar
Phone: (512)471-3080   Fax: (512)471-7060
EMail: Edgar@CHE.UTexas.EDU

Civil Eng
C.M. Walton
Phone: (512)471-4921   Fax: (512)471-0592
EMail: Mike_Walton.CEMail@CEMail-gate.CE.UTexas.EDU

Electrical & Computer Eng
Mario J. Gonzalez, Jr.
Phone: (512)471-6179   Fax: (512)471-3652

Mechanical Eng
Kenneth R. Diller
Phone: (512)471-0796   Fax: (512)471-8727
EMail: Diller@UTVMS.cc.UTEXAS.edu

Aerospace Eng & Eng Mechanics
Richard W. Miksad
Phone: (512)471-4596   Fax: (512)471-3788
EMail: Miksad@UTXVMS

Petroleum Eng
Larry W. Lake
Phone: (512)471-8233   Fax: (512)471-9605
EMail: Larry_Lake@UTPE.PE.UTEXAS.EDU

Admissions Office Contact
Shirley F. Binder
Phone: (512)471-1711   Fax: (512)471-3529

Financial Aid Office Contact
Patricia S. Harris, *Director*
Phone: (512)471-4001   Fax: (512)475-6296

234 U of Texas at Dallas

Head of the Institution
Robert H. Rutford
Phone: (214)690-2201   Fax: (214)690-2237

Head of Engineering
Blake E. Cherrington
Phone: (214)690-2974   Fax: (214)690-2813
EMail: cher@utdallas.edu

Associate/Assistant Deans
Bernie H. List, *Associate Dean*
Phone: (214)690-2977   Fax: (214)690-2813
William J. Pervin, *Assistant Dean & Coll Master*
Phone: (214)690-2892   Fax: (214)690-2813

Department Chair(s)
Electrical Eng
Grover Wetsel
Phone: (214)690-2474   Fax: (214)690-2813
EMail: gcw@utdallas.edu

Admissions Office Contact
Jackie Beitler
Phone: (214)690-2976   Fax: (214)690-2813

Financial Aid Office Contact
Michael O'Rear, *Director*
Phone: (214)690-2941   Fax: (214)690-2947

235 U of Texas at El Paso

Head of the Institution
Diana S. Natalicio
Phone: (915)747-5555

Head of Engineering
Stephen Riter
Phone: (915)747-5460   Fax: (915)747-5616

Associate/Assistant Deans
Andrew Swift, *Assistant Dean of Research*
Phone: (915)747-6904   Fax: (915)747-5616
Darrell Schroder, *Assistant Dean*
Phone: (915)747-6970   Fax: (915)747-5616

Department Chair(s)
Civil Eng
Charles Turner
Phone: (915)747-6908   Fax: (915)747-5616

Electrical Eng
Michael E. Austin
Phone: (915)747-6966   Fax: (915)747-5616

Industrial Eng
Thomas J. McLean
Phone: (915)747-6903   Fax: (915)747-5616

Mechanical Eng
Thomas J. McLean
Phone: (915)747-6903   Fax: (919)747-5616

Metallurgical & Materials Eng
Larry Murr
Phone: (915)747-6929   Fax: (915)747-5616

Computer Science
Andrew Bernat
Phone: (915)747-6950   Fax: (915)747-5145
EMail: abernat@cs.ep.utexas.edu

Admissions Office Contact
Stephen Riter
Phone: (915)747-5460   Fax: (915)747-5616

Financial Aid Office Contact
Beto Lopez, *Director Recruitment & Scholarships*
Phone: (915)747-5896

236 U of Texas at San Antonio

Head of the Institution
Samuel A. Kirkpatrick
Phone: (210)691-4101   Fax: (210)691-4655

Head of Engineering
James H. Tracey
Phone: (210)691-4450   Fax: (210)691-4445
EMail: jtracey@lonestar

Associate/Assistant Deans
Lawrence R. Williams, *Associate Dean*
Phone: (210)691-4430   Fax: (210)691-4445

Department Chair(s)
Division of Eng
G. V. S. Raju
Phone: (210)691-4490   Fax: (210)691-5589

Civil Eng
Alberto Arroyo
Phone: (210)691-4490   Fax: (210)691-5589

Electrical Eng
John L. Schmalzel
Phone: (210)691-4490   Fax: (210)691-5589

Mechanical Eng
Jahan Eftekhar
Phone: (210)691-4490   Fax: (210)691-5589

237 Texas A&I U

Head of the Institution
Manuel L. Ibanez
Phone: (512)595-3207   Fax: (512)595-3218

Head of Engineering
Phil V. Compton
Phone: (512)595-2001   Fax: (512)595-2106
EMail: KFPVC00@TAIMVS1

Department Chair(s)
Chemical Eng & Natural Gas Eng
P. Walter Pritchett
Phone: (512)595-2002   Fax: (512)595-2106

Civil Eng
John W. Weber
Phone: (512)595-2266   Fax: (512)595-2106

Electrical Eng & Computer Science
Yuan R. Wang
Phone: (512)595-2004   Fax: (512)595-2106

Engineering & Engineering Technology Personnel

Mechanical Eng & Industrial Eng
Robert A. McLauchlan
Phone: (512)595-2003   Fax: (512)595-2106
Environmental Eng
Ray N. Finch
Phone: (512)595-3046   Fax: (512)595-2011
Industrial Tech
Frank Mullen
Phone: (512)595-2608   Fax: (512)595-2011
Admissions Office Contact
Ruth T. Fletcher
Phone: (512)595-2811
Financial Aid Office Contact
Auturo Pecos, *Director, Financial Aid*
Phone: (512)595-3911

238 Texas A&M U
Head of the Institution
William H. Mobley
Phone: (409)845-2217   Fax: (409)845-5027
Head of Engineering
Kenneth L. Peddicord
Phone: (409)845-7203   Fax: (409)845-8986
EMail: KLP7201
Associate/Assistant Deans
Carl A. Erdman, *Executive Associate Dean*
Phone: (409)845-5220   Fax: (409)845-8986
Kenneth R. Hall, *Associate Dean*
Phone: (409)845-1322   Fax: (409)845-8986
W.D. Turner, *Associate Dean*
Phone: (409)845-8699   Fax: (409)847-8654
Karan L. Watson, *Assistant Dean*
Phone: (409)862-4367   Fax: (409)847-8654
Department Chair(s)
Chemical Eng
Ray W. Flumerfelt
Phone: (409)845-9807   Fax: (409)845-6446
EMail: RWF8043
Civil Eng
Jim T.P. Yao
Phone: (409)845-1318   Fax: (409)845-6156
EMail: JTY0735
Electrical Eng
Alton D. Patton
Phone: (409)845-3429   Fax: (409)845-6259
EMail: DEBBIE@EEMIPS.TAMU.EDU
Industrial Eng
Gary Hogg
Phone: (409)845-5502   Fax: (409)847-9005
EMail: GLH9988
Mechanical Eng
Walter L. Bradley
Phone: (409)845-1326   Fax: (409)845-3081
EMail: WLB3621
Aerospace Eng
Walt E. Haisler
Phone: (409)845-1640   Fax: (409)845-6051
EMail: WEH4201
Computer Science
Dick A. Volz
Phone: (409)845-8873   Fax: (409)847-8578
EMail: VOLZCS.TAMU.EDU
Nuclear Eng
John W. Poston
Phone: (409)845-4161   Fax: (409)845-6443
EMail: JWP8890
Petroleum Eng
James E. Russell
Phone: (409)845-2241   Fax: (409)845-1307
EMail: RUSSELL@SPINDLETOP.TAMU.EDU
Eng Tech
John A. Weese
Phone: (409)845-4951   Fax: (409)847-9396
EMail: JAW2833
Agricultural Eng
Donald L. Reddell
Phone: (409)845-3931   Fax: (409)845-3932
EMail: D1R8527@VENUS.TAMU.EDU

Admissions Office Contact
Gary R. Engelgau
Phone: (409)845-1040   Fax: (409)845-0727
Financial Aid Office Contact
Donald L. Engelage, *Director*
Phone: (409)845-3236

239 Texas Tech U
Head of the Institution
Robert W. Lawless
Phone: (806)742-2121   Fax: (806)742-2138
Head of Engineering
Mason H. Somerville
Phone: (806)742-3451   Fax: (806)742-3493
Associate/Assistant Deans
John Borrelli, *Associate Dean for Academic Affairs*
Phone: (806)742-3454   Fax: (806)742-3493
Ernst W. Kiesling, *Associate Dean for Research*
Phone: (806)742-3451   Fax: (806)742-3493
Department Chair(s)
Chemical Eng
Raghu Narayan
Phone: (806)742-3553   Fax: (806)742-3552
Civil Eng
Kent Wray
Phone: (806)742-3523   Fax: (806)742-3488
Electrical Eng
Marion O. Hagler
Phone: (806)742-3533   Fax: (806)742-1245
Industrial Eng
James L. Smith
Phone: (806)742-3543   Fax: (806)742-3411
Mechanical Eng
Edward E. Anderson
Phone: (806)742-3563   Fax: (806)742-3540
Petroleum Eng
John J. Day
Phone: (806)742-3573   Fax: (806)742-3502
Computer Science
William M. Marcy
Phone: (806)742-3527   Fax: (806)742-1900
Eng Tech
John J. Day
Phone: (806)742-3538   Fax: (806)742-1900
Admissions Office Contact
Marty Grassel
Phone: (806)742-1482   Fax: (806)742-2007
Financial Aid Office Contact
Ronny Barnes
Phone: (806)742-3681

240 U of Toledo
Head of the Institution
Frank E. Horton
Phone: (419)537-2211
Head of Engineering
Edward Lumsdaine
Phone: (419)537-2707
Associate/Assistant Deans
Robert A. Bennett, *Associate Dean*
Phone: (419)537-2707
Kenneth J. DeWitt, *Assistant Dean*
Phone: (419)537-2707
Richard Springman, *Assistant Dean*
Phone: (417)537-2707
Department Chair(s)
Chemical Eng
Bruce Poling
Phone: (419)537-2639
Civil Eng
Naser Mostaghel
Phone: (419)537-3558
Electrical Eng
Kai Fong Lee
Phone: (419)537-2580

Computer Science & Eng
Hilda Standley
Phone: (419)537-3558
Industrial Eng
Roger McNichols
Phone: (419)537-2412
Mechanical Eng
Richard Irey
Phone: (419)537-2642
Eng Tech
Edward Stobbe
Phone: (419)537-3586
Eng Physics
John Simon
Phone: (419)537-2648
Admissions Office Contact
R. Eastop
Phone: (419)537-2696
Financial Aid Office Contact
Richard Lasko, *Director of Financial Aid*
Phone: (419)537-2056

241 Tri-State U
Head of the Institution
William G. Meyers
Phone: (219)665-4187   Fax: (219)665-4292
Head of Engineering
William G. Meyers
Phone: (219)665-4187   Fax: (219)665-4292
Associate/Assistant Deans
Thomas J. Enneking, *Assistant Dean*
Phone: (219)665-4216   Fax: (219)665-4292
Department Chair(s)
Chemical Eng
Andrew J. Wilson
Phone: (219)665-4223   Fax: (219)665-4292
Civil Eng
Sandanand Kundapur
Phone: (219)665-4214   Fax: (219)665-4292
Electrical Eng
Alan R. Stoudinger
Phone: (219)665-4189   Fax: (219)665-4292
Mechanical Eng
Frank R. Swenson
Phone: (219)665-4228   Fax: (219)665-4292
Aerospace Eng
Robert K. Wattson
Phone: (219)665-4234   Fax: (219)665-4292
Eng Administration
Thomas J. Enneking
Phone: (219)665-4216   Fax: (219)665-4292
Admissions Office Contact
Walter Lilley
Phone: (219)665-4139   Fax: (219)665-4292
Financial Aid Office Contact
Susan Stroh, *Financial Aid Director*
Phone: (219)665-4174   Fax: (219)665-4292

242 Trinity U
Head of the Institution
Ronald K. Calgaard
Phone: (210)736-8401   Fax: (210)736-8400
Head of Engineering
John S. Dickey
Phone: (210)736-7414   Fax: (210)736-7229
Department Chair(s)
Eng Science
John P. Giolma
Phone: (210)736-7511   Fax: (210)736-7569
EMail: paul@engr.trinity.edu
Admissions Office Contact
Beth Allen
Phone: (210)736-7207   Fax: (210)736-8164
Financial Aid Office Contact
Estelle Frerichs, *Director of Financial Aid*
Phone: (210)736-8315

243 Tufts U
Head of the Institution
John DiBiaggio
Phone: (617)627-3300
Head of Engineering
Frederick C. Nelson
Phone: (617)627-3237   Fax: (617)627-3819
Associate/Assistant Deans
Edward J. Maskalenko, *Associate Dean*
Phone: (617)627-3237   Fax: (617)627-3819
Department Chair(s)
Chemical Eng
Gregory D. Botsaris
Phone: (617)627-3900   Fax: (617)627-3991
Civil Eng
Lewis Edgers
Phone: (617)627-3211   Fax: (617)627-3994
Electrical Eng
Denis W. Fermental
Phone: (617)627-3217   Fax: (617)627-3220
Mechanical Eng
Behrouz Abedian
Phone: (617)627-3239   Fax: (617)627-3819
Admissions Office Contact
David D. Cuttino
Phone: (617)628-5000   Fax: (617)627-3860
Financial Aid Office Contact
William F. Eastwood, *Director of Financial Aid*
Phone: (617)628-5000

244 Tulane U
Head of the Institution
Eamon M. Kelly
Phone: (504)865-5201   Fax: (504)865-5202
Head of Engineering
William C. Van Buskirk
Phone: (504)865-5766   Fax: (504)862-8747
Associate/Assistant Deans
Efstathios E. Michaelides, *Associate Dean for Graduate Studies & Research*
Phone: (504)865-5819   Fax: (504)862-8747
Samuel L. Sullivan, Jr., *Associate Dean for Undergraduate Studies*
Phone: (504)865-5764   Fax: (504)862-8747
Johnette Hassell, *Assistant Dean of Engineering & Computer Science*
Phone: (504)865-5764   Fax: (504)862-8747
Department Chair(s)
Chemical Eng
Raymond V. Bailey
Phone: (504)865-5843   Fax: (504)865-6744
EMail: Ray@che.che.tulane.edu
Civil & Environmental Eng
John L. Niklaus
Phone: (504)865-5778   Fax: (504)862-8747
Electrical Eng
Andrew B. Martinez
Phone: (504)865-5875   Fax: (504)865-5526
EMail: martinez@bourbon.ee.tulane.edu
Mechanical Eng
P. Michael Lynch
Phone: (504)865-5775   Fax: (504)865-5345
EMail: lynch@control.me.tulane.edu
Biomedical Eng
Cedric F. Walker
Phone: (504)865-5897   Fax: (504)862-8779
EMail: cfw@mv3600.bmen.tulane.edu
Computer Science
Johnette Hassell
Phone: (504)865-5840   Fax: (504)862-8747
EMail: hassell@comus.cs.tulane.edu
Financial Aid Office Contact
Thomas P. Lovett, *Director of Financial Aid*
Phone: (504)865-5723

245 U of Tulsa

Head of the Institution
Robert H. Donaldson
Phone: (918)631-2305 Fax: (918)631-2347

Head of Engineering
Lewis M. Duncan
Phone: (918)631-2288 Fax: (918)631-2286
EMail: LDuncan@vax1.utulsa.edu

Associate/Assistant Deans
Steven J. Bellovich, *Associate Dean for Academic Affairs*
Phone: (918)631-2478 Fax: (918)631-2286
Kraemer D. Luks, *Associate Dean for Research*
Phone: (918)631-3264 Fax: (918)631-2286

Department Chair(s)
Chemical Eng
Richard E. Thompson
Phone: (918)631-2227 Fax: (918)631-3268
Electrical Eng
Gerald R. Kane
Phone: (918)631-3280 Fax: (918)631-3220
EMail: grk@ohm.ee.tulsa.edu
Mechanical Eng
Edmund F. Rybicki
Phone: (918)631-2996 Fax: (918)631-2397
Eng Physics
Kenneth A. Kuenhold
Phone: (918)631-3032 Fax: (918)631-2995
Petroleum Eng
Stefan Miska
Phone: (918)631-3035 Fax: (918)631-2059

Admissions Office Contact
John C. Corso
Phone: (918)631-2307

Financial Aid Office Contact
David L. Gruen, *Director*
Phone: (918)631-2526

246 Tuskegee U

Head of the Institution
Benjamin F. Payton
Phone: (205)727-8501 Fax: (205)727-5276

Head of Engineering
Shaik Jeelani
Phone: (205)727-8355 Fax: (205)727-8090

Associate/Assistant Deans
Shaik Jeelani, *Associate Dean*
Phone: (205)727-8946 Fax: (205)727-8090

Department Chair(s)
Chemical Eng
Nadar Vahdat
Phone: (205)727-8798 Fax: (205)727-8090
Electrical Eng
Warren F. Clayton
Phone: (205)727-8298 Fax: (205)727-8090
Mechanical Eng
Pradosh K. Ray
Phone: (205)727-8918 Fax: (205)727-8090
Aerospace Science Eng
Y. C. Yang
Phone: (205)727-8768 Fax: (205)727-8090

Admissions Office Contact
Lee Young
Phone: (205)727-8390

Financial Aid Office Contact
Barbara T. Blair, *Director of Financial Aid*
Phone: (205)727-8201

247 Union Coll

Head of the Institution
Roger H. Hull
Phone: (518)370-6101

Head of Engineering
Lawrence J. Hollander
Phone: (518)370-6038 Fax: (518)370-6789
EMail: HOLLANDL@GAR.UNION.EDU

Department Chair(s)
Civil Eng
Mohammad Mafi
Phone: (518)370-6313 Fax: (518)370-6789
Electrical Eng
Michael Rudko
Phone: (518)370-6316 Fax: (518)370-6789
Mechanical Eng
Frank F. Milillo
Phone: (518)370-6322 Fax: (518)370-6789

Admissions Office Contact
Daniel M. Lundquist
Phone: (518)370-6112

Financial Aid Office Contact
Michael S. Brown, *Director of Financial Aid*
Phone: (518)370-6123

248 US Air Force Academy

Head of the Institution
Bradley C. Hosmer
Phone: (719)472-4140 Fax: (719)472-4146

Head of Engineering
Cary A. Fisher
Phone: (719)472-2531 Fax: (719)472-2944
EMail: dfemmail@falconnet.usafa.af.mil

Department Chair(s)
Civil Eng
James L. Brickell
Phone: (719)472-3150 Fax: (719)472-2944
EMail: dfcemail@falconnet.usafa.af.mil
Electrical Eng
Alan R. Klayton
Phone: (719)472-3190 Fax: (719)472-3135
EMail: dfeebb@falconnet.usafa.af.mil
Eng Mechanics
Cary A. Fisher
Phone: (719)472-2531 Fax: (719)472-2944
EMail: dfemmail@falconnet.usafa.af.mil
Astronautics
Robert B. Giffen
Phone: (719)472-4110 Fax: (719)472-3723
EMail: dfasmail@falconnet.usafa.af.mil
Aeronautics
Michael L. Smith
Phone: (719)472-4010 Fax: (719)472-3135
EMail: dfanmail@falconnet.usafa.af.mil

Admissions Office Contact
Rolland R. Stoneman
Phone: (719)472-2520 Fax: (719)472-3647

249 US Coast Guard Academy

Head of the Institution
Thomas T. Matteson
Phone: (203)444-8285 Fax: (203)444-8288

Head of Engineering
John C. Maxham
Phone: (203)444-8547 Fax: (203)444-8546

Department Chair(s)
Civil Eng
Howard C. Dunn, Jr.
Phone: (203)444-8537 Fax: (203)444-8546
Electrical Eng
Ben B. Peterson
Phone: (203)444-8541 Fax: (203)444-8546
Naval Architecture & Marine Eng
Dwight G. Hutchinson
Phone: (203)444-8525 Fax: (203)444-8546

Admissions Office Contact
Robert W. Thorne
Phone: (203)444-8500 Fax: (203)444-8288

250 US Merchant Marine Academy

Head of the Institution
Paul L. Krinsky
Phone: (516)773-5348 Fax: (516)773-5390

Head of Engineering
James A. Harbach
Phone: (516)773-5471 Fax: (516)773-5534

Financial Aid Office Contact
Frances Ferguson, *Financial Aid Officer*
Phone: (516)773-5391 Fax: (516)773-5509

251 US Military Academy

Head of the Institution
Howard D. Graves
Phone: (914)938-2610 Fax: (914)938-3214

Head of Engineering
Gerald E. Galloway, Jr.
Phone: (914)938-2000 Fax: (914)938-5438

Associate/Assistant Deans
Edward G. Tezak, *Associate Dean*
Phone: (914)938-3615 Fax: (914)938-5438

Department Chair(s)
Civil & Mechanical Eng
Fletcher M. Lamkin
Phone: (914)938-2668 Fax: (914)938-5522
Electrical Eng
Daniel M. Litynski
Phone: (914)938-2201 Fax: (914)938-5956
EMail: DD1408@eecs1.eecs.usma.edu
Geography & Environmental Eng
John H. Grubbs
Phone: (914)938-2300 Fax: (914)938-3339
Systems Eng
James L. Kays
Phone: (914)938-2701 Fax: (914)938-5919
Physics
Raymond J. Winkel, Jr.
Phone: (914)938-3901

Admissions Office Contact
Pierce A. Rushton, Jr.
Phone: (914)938-4041 Fax: (914)938-3021

252 US Naval Academy

Head of the Institution
Thomas C. Lynch
Phone: (410)267-2202 Fax: (410)267-2303

Head of Engineering
Philip F. Grasser
Phone: (410)267-2391 Fax: (410)267-2591
EMail: E05205@n1.usna.navy.mil

Associate/Assistant Deans
Philip F. Grasser, *Director, Division of Engineering & Weapons*
Phone: (410)267-2391 Fax: (410)267-2591
John F. McKernan, *Executive Assistant*
Phone: (410)267-2391 Fax: (410)267-2591

Department Chair(s)
Aerospace Eng
Maido Saarlas
Phone: (410)267-3284 Fax: (410)267-2591
EMail: e01012@n1.usna.navy.mil
Mechanical Eng
John O. Geremia
Phone: (410)267-2792 Fax: (410)267-2591
EMail: e03007@n1.usna.navy.mil
Electrical Eng
Richard L. Martin
Phone: (410)267-2896 Fax: (410)267-3493
EMail: e02031@n1.usna.navy.mil
Weapons & Systems Eng
E. Eugene Mitchell
Phone: (410)267-2586 Fax: (410)267-3493
EMail: e05006@n1.usna.navy.mil
Naval Architecture, Ocean, & Marine Eng
Marshall L. Nuckols
Phone: (410)267-3871 Fax: (410)267-2591
EMail: e04109@n1.usna.navy.mil

Admissions Office Contact
John W. Renard
Phone: (410)267-4361 Fax: (410)267-4348

253 U of Utah

Head of the Institution
Arthur K. Smith
Phone: (801)581-5701 Fax: (801)581-6892

Head of Engineering
David W. Pershing
Phone: (801)581-6911 Fax: (801)581-8692
EMail: deandave@dean.eng.utah.edu

Head of Engineering/Tech
Francis H. Brown, *Dean, Coll of Mines & Earth Sciences*
Phone: (801)581-8767 Fax: (801)581-5560

Associate/Assistant Deans
Francis H. Brown, *Dean, Coll of Mines & Earth Sciences*
Phone: (801)581-8767 Fax: (801)581-5560
Dietrich K. Gehmlich, *Associate Dean for Academic Affairs, Coll of Engineering*
Phone: (801)581-6911 Fax: (801)581-8692
K. Larry DeVries, *Associate Dean for Research, Coll of Engineering*
Phone: (801)581-7101 Fax: (801)581-8692
Peter F. Gerity, *Associate Dean for Community Relations, Coll of Engineering*
Phone: (801)581-8346 Fax: (801)581-8692
Edward M. Trujillo, *Assistant Dean for Minority Affairs, Coll of Engineering*
Phone: (801)581-4460 Fax: (801)581-8692

Department Chair(s)
Chemical & Fuels Eng
A. Lamont Tyler
Phone: (801)581-6920 Fax: (801)581-8692
Civil Eng
Lawrence D. Reavely
Phone: (801)581-7101 Fax: (801)581-8692
Electrical Eng
Om P. Gandhi
Phone: (801)581-3629 Fax: (801)581-5281
EMail: gandhi@ee.utah.edu
Mechanical Eng
Robert B. Roemer
Phone: (801)581-3851 Fax: (801)581-8692
EMail: roemer@me.mech.utah.edu
Computer Science
Thomas C. Henderson
Phone: (801)581-3601 Fax: (801)581-5843
EMail: tch@cs.utah.edu
Metallurgical Eng
J. Gerald Byrne
Phone: (801)581-6386 Fax: (801)581-5560
Mining Eng
M. Kim McCarter
Phone: (801)581-7198 Fax: (801)581-5560
Geological Eng
A.A. (Tony) Ekdale
Phone: (801)581-7162 Fax: (801)581-7065
Materials Science & Eng
Richard H. Boyd
Phone: (801)581-6865 Fax: (801)581-4816
EMail: boyd@polyvax.utah.edu
Bioengineering
Richard A. Normann
Phone: (801)581-8528 Fax: (801)581-8692

Admissions Office Contact
J. Stayner Landward
Phone: (801)581-7281 Fax: (801)585-3034

Financial Aid Office Contact
Harold R. Weight, *Director of Financial Aid*
Phone: (801)581-6211 Fax: (801)585-3034

254 Utah State U

Head of the Institution
Emert George
Phone: (801)750-1157 Fax: (801)750-3880

Head of Engineering
A. Bruce Bishop
Phone: (801)750-2776 Fax: (801)750-1248

Engineering & Engineering Technology Personnel

Associate/Assistant Deans
A P. Moser, *Associate Dean*
Phone: (801)750-3316   Fax: (801)750-1248
Ronald L. Thurgood, *Associate Dean*
Phone: (801)750-2775   Fax: (801)750-1248

Department Chair(s)
Civil & Environmental Eng
Loren R. Anderson
Phone: (801)750-2932   Fax: (801)750-1185
Electrical Eng
Richard W. Harris
Phone: (801)750-2840   Fax: (801)750-3054
Industrial Tech & Education
Maurice G. Thomas
Phone: (801)750-1795   Fax: (801)750-2567
Mechanical & Aerospace Eng
Frank J. Redd
Phone: (801)750-2868   Fax: (801)750-2417
Biological & Irrigation Eng
Wynn R. Walker
Phone: (801)750-2785   Fax: (801)750-1248

Admissions Office Contact
J. Rodney Clark
Phone: (801)750-1106   Fax: (801)750-2644

Financial Aid Office Contact
Vicki L. Atkinson, *Director of Financial Aid*
Phone: (801)750-1022   Fax: (801)750-2571

255   Valparaiso U

Head of the Institution
Alan F. Harre
Phone: (219)464-5115   Fax: (219)464-5381

Head of Engineering
Stuart G. Walesh
Phone: (219)464-5121   Fax: (219)464-5381

Department Chair(s)
Civil Eng
Bradford H. Spring
Phone: (219)464-5220   Fax: (219)464-5381
Electrical & Computer Eng
Demosthenes P. Gelopulos
Phone: (219)464-5203   Fax: (219)464-5381
Mechanical Eng
Robert D. Palumbo
Phone: (219)464-5135   Fax: (219)464-5381

Admissions Office Contact
Karen Foust
Phone: (219)464-5011   Fax: (219)464-5381

Financial Aid Office Contact
Katherine E. Wehling, *Director*
Phone: (219)464-5014   Fax: (219)464-5381

256   Vanderbilt U

Head of the Institution
Joe B. Wyatt
Phone: (615)322-2168

Head of Engineering
Edward A. Parrish
Phone: (615)322-2762   Fax: (615)343-8006
EMail: eap@vuse.vanderbilt.edu

Associate/Assistant Deans
Edward J. White, *Associate Dean*
Phone: (615)322-2762   Fax: (615)343-8006
George E. Cook, *Associate Dean for Research*
Phone: (615)343-5032   Fax: (615)343-8006
John F. Carney, III, *Associate Dean for Graduate Affairs*
Phone: (615)322-3518
Robert E. Stammer, Jr., *Assistant Dean for Student Affairs*
Phone: (615)343-8060
Carolyn Ruth A Williams, *Assistant Dean for Minority Affairs*
Phone: (615)322-2724

Department Chair(s)
Chemical Eng
Tomlinson Fort
Phone: (615)322-2441
Civil & Environmental Eng
Edward L. Thackston
Phone: (615)322-2697
Electrical Eng
David V. Kerns
Phone: (615)322-2771
EMail: kerns@vuse.vanderbilt.edu
Mechanical Eng
Alvin M. Strauss
Phone: (615)322-2413
EMail: ams@vuse.vanderbilt.edu
Computer Science
Patrick C. Fischer
Phone: (615)322-2796
EMail: pcf@vuse.vanderbilt.edu
Materials Science & Eng
George T. Hahn
Phone: (615)343-6868
Biomedical Eng
Thomas R. Harris
Phone: (615)322-3521
EMail: trh@vuse.vanderbilt.edu

Admissions Office Contact
Neill F. Sanders
Phone: (615)322-2561

Financial Aid Office Contact
David Mohning, *Director, Financial Aid*
Phone: (615)322-3591

257   U of Vermont

Head of the Institution
Thomas P. Salmon
Phone: (802)656-3186   Fax: (802)656-8432

Head of Engineering
George F. Pinder
Phone: (802)656-3390   Fax: (802)656-8802
EMail: pinder@uvm.edu

Associate/Assistant Deans
Richard M. Foote, *Associate Dean*
Phone: (802)656-2940   Fax: (802)656-8802
Ann C. Livingston, *Assistant Dean, Student Affairs*
Phone: (802)656-3392   Fax: (802)656-8279

Department Chair(s)
Civil Eng
Jean-Guy L. Beliveau
Phone: (802)656-3800   Fax: (802)656-8446
Electrical Eng
Kenneth I. Golden
Phone: (802)656-3330   Fax: (802)656-0696
EMail: golden@uvmgen.edu
Mechanical Eng
Mahendra S. Hundal
Phone: (802)656-3320   Fax: (802)656-1929
EMail: hundal@uvm.edu

Admissions Office Contact
Susan A. Wertheimer
Phone: (802)656-3370   Fax: (802)656-8611

Financial Aid Office Contact
Donald M. Honeman, *Director*
Phone: (802)656-3156

258   Villanova U

Head of the Institution
Edmund J. Dobbin
Phone: (215)645-4510   Fax: (215)645-4514

Head of Engineering
Robert D. Lynch
Phone: (215)645-4940   Fax: (215)645-4941
EMail: RDLYNCH@VUVAXCOM

Department Chair(s)
Chemical Eng
C. Michael Kelly
Phone: (215)645-4950   Fax: (215)645-4941
EMail: CMKELLY@VUVAXCOM
Civil Eng
Lewis J. Mathers
Phone: (215)645-4960   Fax: (215)645-4941
EMail: MATHERS@VUVAXCOM
Electrical Eng
S.S. Rao
Phone: (215)645-4970   Fax: (215)645-4436
EMail: EEOFFICE@VUVAXCOM
Mechanical Eng
Alan M. Whitman
Phone: (215)245-4980   Fax: (215)645-7312
EMail: WHITMAN@VUVAXCOM
Computer Eng
S.S. Rao
Phone: (215)645-4970   Fax: (215)645-4436
EMail: EEOFFICE@VUVAXCOM

Financial Aid Office Contact
George Walter, *Financial Aid Officer*
Phone: (215)645-4010   Fax: (215)645-7599

259   U of Virginia

Head of the Institution
John T. Casteen, III
Phone: (804)924-3337   Fax: (804)924-3792

Head of Engineering
Edgar A. Starke, Jr.
Phone: (804)924-3593   Fax: (804)982-2734
EMail: EAS1O@virginia.edu

Associate/Assistant Deans
David Morris, *Associate Dean for Academic Programs*
Phone: (804)924-3310   Fax: (804)982-2734
James L. Kelly, *Assistant Dean for Undergraduate Programs*
Phone: (804)924-3164   Fax: (804)982-2734
George L. Cahen, *Assistant Dean for Graduate Programs*
Phone: (804)924-3897   Fax: (804)982-2734
Eugene R. Seeloff, *Assistant Dean for Career Planning & Placement*
Phone: (804)924-3050   Fax: (804)982-2734
William J. Thurneck, *Assistant Dean for Administrative & Academic Affairs*
Phone: (804)924-3155   Fax: (804)982-2734

Department Chair(s)
Chemical Eng
John P. O'Connell
Phone: (804)924-3428   Fax: (804)982-2658
EMail: jpo2x@virginia.EDU
Civil Eng
Furman W. Barton
Phone: (804)924-6361   Fax: (804)982-2951
EMail: fwb@virginia.EDU
Electrical Eng
Robert J. Mattauch
Phone: (804)924-6112   Fax: (804)924-8818
EMail: rjm@virginia.EDU
Mechanical, Aerospace & Nuclear Eng
Paul E. Allaire
Phone: (804)924-6209   Fax: (804)982-2037
EMail: pea@virginia.EDU
Applied Mathematics
James G. Simmonds
Phone: (804)924-1041
EMail: jgs@virginia.EDU
Biomedical Eng
Jen-Shih Lee
Phone: (804)924-5095   Fax: (804)982-3870
EMail: jl@virginia.EDU
Computer Science
Anita K. Jones
Phone: (804)982-2200   Fax: (804)982-2214
EMail: akj7a@virginia.EDU
Humanities
Omar A. Gianniny
Phone: (804)924-6115   Fax: (804)924-6270
EMail: oag@virginia.EDU
Materials Science & Eng
William A. Jesser
Phone: (804)982-5640   Fax: (804)982-5660
EMail: thc2a@virginia.EDU
Systems Eng
James P. Ignizio
Phone: (804)924-6216   Fax: (804)982-2972
EMail: jpi@virginia.EDU

Admissions Office Contact
William J. Thurneck
Phone: (804)924-3155   Fax: (804)982-2734

Financial Aid Office Contact
Wayne M. Sparks, *Director of Financial Aid to Students*
Phone: (804)924-3725

260   Virginia Military Inst

Head of the Institution
John W. Knapp
Phone: (703)464-7311   Fax: (703)464-7660

Head of Engineering
Richard S. Trandel
Phone: (703)464-7308   Fax: (703)464-7663

Department Chair(s)
Civil & Environmental Eng
James R. Groves
Phone: (703)464-7331   Fax: (703)464-7618
Electrical Eng
Richard Skutt
Phone: (703)464-7236   Fax: (703)464-7662
Mechanical Eng
Michael R. Sexton
Phone: (703)464-7308   Fax: (703)464-7663

Admissions Office Contact
Daniel A. Troppoli
Phone: (703)464-7211   Fax: (703)464-7746

Financial Aid Office Contact
Timothy P. Golden, *Director of Financial Aid*
Phone: (703)464-7208

261   Virginia Tech

Head of the Institution
James D. McComas
Phone: (703)231-6231   Fax: (703)231-4264

Head of Engineering
Wayne Clough
Phone: (703)231-6641   Fax: (703)231-7248
EMail: GWC @ VTVM1.CC.VT.EDU

Associate/Assistant Deans
James F. Marchman, *Associate Dean for Academic Affairs*
Phone: (703)231-6643   Fax: (703)231-7248
Walter F. O'Brien, *Associate Dean for Research & Graduate Studies*
Phone: (703)231-9171   Fax: (703)231-7248
Benjamin B. Blanchard, *Assistant Dean for Extension*
Phone: (703)231-5458   Fax: (703)231-7248
Pamela S. Kurstedt, *Assistant Dean for Enrichment Programs*
Phone: (703)231-9764   Fax: (703)231-7248
Joseph G. Tront, *Assistant Dean for Computing*
Phone: (703)231-5067   Fax: (703)231-3362

Department Chair(s)
Chemical Eng
William L. Conger
Phone: (703)231-6632   Fax: (703)231-5022
EMail: WLCONGER @ VTVM1.cc.vt.edu
Civil Eng
David F. Kibler
Phone: (703)231-6635   Fax: (703)231-7532
EMail: KiblerDF @ VTVM1.cc.vt.edu

Electrical Eng/Computer Eng
F.W. Stephenson
Phone: (703)231-6646 Fax: (703)231-3362
EMail: FREQY-Stephenson @ VTVM1.cc.vt.edu
Industrial & Systems Eng
Robert D. Dryden
Phone: (703)231-6656 Fax: (703)231-3322
EMail: Dryden @ VTVM1.cc.vt.edu
Mechanical Eng
Robert A. Comparin
Phone: (703)231-6661 Fax: (703)231-9100
EMail: Comparin @ VTVM1.cc.vt.edu
Aerospace & Ocean Eng
Joseph A. Schetz
Phone: (703)231-6611 Fax: (703)231-9632
EMail: AOE @ VTVM1.cc.vt.edu
Agricultural Eng
John V. Perumpral
Phone: (703)231-6615 Fax: (703)231-3199
EMail: Perump @ VTVM1.cc.vt.edu
Eng Science & Mechanics
Edmund G. Henneke
Phone: (703)231-6651 Fax: (703)231-4574
EMail: Henneke @ VTMV1.cc.vt.edu
Materials Science & Eng
Ronald S. Gordon
Phone: (703)231-6655 Fax: (703)231-8919
EMail: GordonR @ VTVM1.cc.vt.edu
Mining Eng
Michael E. Karmis
Phone: (703)231-6671 Fax: (703)231-4070
EMail: MKarmis @ VTMV1.cc.vt.edu

Admissions Office Contact
David R. Bousquet
Phone: (703)341-6267 Fax: (703)231-3242

Financial Aid Office Contact
Anne H. Clarke, *Director*
Phone: (703)231-5179 Fax: (703)231-9139

262 U of Washington

Head of the Institution
William P. Gerberding
Phone: (206)543-5010 Fax: (206)543-3951

Head of Engineering
J.R. Bowen
Phone: (206)543-0340 Fax: (206)685-0666
EMail: bowen@engr.washington.edu

Associate/Assistant Deans
Ashley F. Emery, *Associate Dean for Academic Affairs*
Phone: (206)543-8590 Fax: (206)685-0666
Keith A. Holsapple, *Associate Dean for Continuing Educ. & Televised Instruction*
Phone: (206)543-6198 Fax: (206)543-0217
Paul R. Young, *Associate Dean, Research, Facilities, & External Affairs*
Phone: (206)543-8388 Fax: (206)543-6264
Gregory L. Zick, *Associate Dean for Computing*
Phone: (206)543-6156 Fax: (206)543-2907

Department Chair(s)
Chemical Eng
Bruce A. Finlayson
Phone: (206)543-2253 Fax: (206)543-3778
EMail: finlayson@u.washington.edu
Civil Eng
John F. Ferguson
Phone: (206)543-2390 Fax: (206)543-1543
EMail: jferg@u.washington.edu
Electrical Eng
Thomas A. Seliga
Phone: (206)543-2150 Fax: (206)543-3842
EMail: seliga@ee.washington.edu
Industrial Eng
Scott C. Iverson
Phone: (206)543-1427 Fax: (206)685-3072
Mechanical Eng
Richard C. Corlett
Phone: (206)543-5090 Fax: (206)685-8047

Aeronautics & Astronautics
Walter H. Christiansen
Phone: (206)543-1950 Fax: (206)543-0217
EMail: walt@aa.washington.edu
Materials Science & Eng
Thomas G. Stoebe
Phone: (206)543-7090 Fax: (206)543-3100
EMail: stoebe@u.washington.edu
Computer Science & Eng
Jean L. Baer
Phone: (206)543-1695 Fax: (206)543-2969
EMail: baer@cs.washington.edu
Technical Communication
Mark P. Haselkorn
Phone: (206)543-2567 Fax: (206)543-8858
EMail: d5878@u.washington.edu

Admissions Office Contact
William J. Heideger
Phone: (206)543-1770 Fax: (206)685-0666

Financial Aid Office Contact
Eric Godfrey, *Assistant Vice President & Director*
Phone: (206)543-6101

263 Washington U

Head of the Institution
William H. Danforth
Phone: (314)935-5100

Head of Engineering
Christopher I. Byrnes
Phone: (314)935-6166 Fax: (314)935-4434

Associate/Assistant Deans
John K. Russell, *Associate Dean for Academic Instruction & Registrar*
Phone: (314)935-6100 Fax: (314)935-4434
William L. Marsden, *Associate Dean*
Phone: (314)935-6130 Fax: (314)935-4434
Michael D. Moll, *Associate Dean & Business Manager*
Phone: (314)935-6111 Fax: (314)935-4434

Department Chair(s)
Chemical Eng
John L. Kardos
Phone: (314)935-6062
Civil Eng
Phillip L. Gould
Phone: (314)935-6383
Electrical Eng
Barry E. Spielman
Phone: (314)935-5565
Mechanical Eng
Salvatore P. Sutera
Phone: (314)935-6047
Computer Science
Seymour V. Pollack
Phone: (314)935-6198
Eng & Policy
William P. Darby
Phone: (314)935-5484
Systems Science & Mathematics
Norman Katz
Phone: (314)935-6083

Admissions Office Contact
William L. Marsden
Phone: (314)935-6130 Fax: (314)935-4434

Financial Aid Office Contact
Dennis J. Martin, *Assistant Provost & Director of Financial Aid*
Phone: (314)935-5900

264 Washington State U

Head of the Institution
Samuel H. Smith
Phone: (509)335-6666 Fax: (509)335-0103

Head of Engineering
Reid C. Miller
Phone: (509)335-5593 Fax: (509)335-9608

Associate/Assistant Deans
Denny C. Davis, *Associate Dean for Instruction*
Phone: (509)335-5593 Fax: (509)335-9608
Richard L. Zollars, *Acting Associate Dean for Research & Extended Studies*
Phone: (509)335-5593 Fax: (509)335-9608

Department Chair(s)
Chemical Eng
William J. Thomson
Phone: (509)335-4332 Fax: (509)335-9608
Civil & Environmental Eng
Rafik Y. Itani
Phone: (509)335-9578 Fax: (509)335-7632
Electrical Eng & Computer Science
John A. Ringo
Phone: (509)335-8148 Fax: (509)335-3818
Mechanical & Materials Eng
Stephen D. Antolovich
Phone: (509)335-8654 Fax: (509)335-4662
Biological Systems Eng
Ralph P. Cavalieri
Phone: (509)335-1578 Fax: (509)335-2722
Architecture
Rafi Samizay
Phone: (509)335-5539 Fax: (509)335-6132

265 Wayne State U

Head of the Institution
David W. Adamany
Phone: (313)577-2230 Fax: (313)577-3200

Head of Engineering
Fred W. Beaufait
Phone: (313)577-3775 Fax: (313)577-5300
EMail: internet: beaufait@dbo.eng.wayne.edu

Associate/Assistant Deans
Donald J. Silversmith, *Associate Dean of Engineering for Research & Graduate Studies*
Phone: (313)577-3861 Fax: (313)577-5300
H. Allan Knappenberger, *Associate Dean for Academic Affairs*
Phone: (313)577-3040 Fax: (313)577-5300
Mark A. Jackson, *Assistant Dean for Student Affairs & Minority Programs*
Phone: (313)577-3780 Fax: (313)577-5300

Department Chair(s)
Chemical Eng
Ralph H. Kummler
Phone: (313)577-3800 Fax: (313)577-3810
EMail: internet: kummler@dbo.eng.wayne.edu
Civil & Environmental Eng
Mumtaz A. Usmen
Phone: (313)577-3789 Fax: (313)577-3881
Electrical & Computer Eng
Michael P. Polis
Phone: (313)577-3920 Fax: (313)577-1101
EMail: internet: mpolis@ece.eng.wayne.edu
Industrial & Manufacturing Eng
Donald R. Falkenburg
Phone: (313)577-3821 Fax: (313)577-8833
EMail: internet: falken@mie.eng.wayne.edu
Mechanical Eng
Kenneth A. Kline
Phone: (313)577-3843 Fax: (313)577-3881
EMail: internet: kkline@nova.wayne.edu
Materials Science & Eng
Ralph H. Kummler
Phone: (313)577-3800 Fax: (313)577-3810
EMail: internet: kummler@dbo.eng.wayne.edu

Admissions Office Contact
H. Allan Knappenberger
Phone: (313)577-3040 Fax: (313)577-5300

Financial Aid Office Contact
Judith L. Florian, *Director*
Phone: (313)577-3378 Fax: (313)577-6648

266 Webb Inst

Head of the Institution
James J. Conti
Phone: (516)671-2277 Fax: (516)674-9838

Head of Engineering
Jacques B. Hadler
Phone: (516)671-2215 Fax: (516)674-9838

Associate/Assistant Deans
Lawrence W. Ward, *Assistant Dean*
Phone: (516)671-2215 Fax: (516)674-9838

Admissions Office Contact
William G. Murray
Phone: (516)671-2213 Fax: (516)674-9838

Financial Aid Office Contact
William G. Murray, *Director of Admissions*
Phone: (516)671-2213 Fax: (516)674-9838

267 West Virginia U

Head of the Institution
Neil S. Bucklew
Phone: (304)293-5531 Fax: (304)293-7417

Head of Engineering
Robert M. Desmond
Phone: (304)293-4821 Fax: (304)293-5024
EMail: DESMOND@FACULTY.COE.WVU.WVNET.EDU

Associate/Assistant Deans
Thomas R. Long, *Associate Dean for Academic Affairs, Finance & Personnel*
Phone: (304)293-4821 Fax: (304)293-5024
John T. Jurewicz, *Associate Dean for Academic Affairs & Research*
Phone: (304)293-4821 Fax: (304)293-5024

Department Chair(s)
Chemical Eng
Eugene V. Cilento
Phone: (304)293-2111 Fax: (304)293-5024
EMail: CILENTO@FACULTY.COE.WVU.WVNET.EDU
Civil Eng
Sam A. Kiger
Phone: (304)293-3031 Fax: (304)293-5024
EMail: KIGER@FACULTY.COE.WVU.WVNET.EDU
Electrical & Computer Eng
Roy S. Nutter
Phone: (304)293-6361 Fax: (304)293-5024
EMail: RSN@A.COE.WVU.WVNET.EDU
Industrial Eng
Ralph W. Plummer
Phone: (304)293-4607 Fax: (304)293-5024
EMail: PLUMMER@IE.COE.WVU.WVNET.EDU
Mechanical & Aerospace Eng
Donald W. Lyons
Phone: (304)293-3111 Fax: (304)293-6689
EMail: LYONS@FACULTY.COE.WVU.WVNET.EDU

Admissions Office Contact
Glenn Carter
Phone: (304)293-2124 Fax: (304)293-3080

Financial Aid Office Contact
Neil E. Bolyard, *Director*
Phone: (304)293-5242 Fax: (304)293-4890

268 West Virginia Inst of Tech

Head of the Institution
John C. Carrier
Phone: (304)442-3146 Fax: (304)442-3059

Head of Engineering
Ernest E. Nester
Phone: (304)442-3161 Fax: (304)442-3307

Department Chair(s)
Chemical Eng
Edward H. Crum
Phone: (304)442-3163 Fax: (304)442-3307

Engineering & Engineering Technology Personnel

Civil Eng
Larry C. Nottingham
Phone: (304)442-3391   Fax: (304)442-3307
Electrical Eng
Gerald S. Mersten
Phone: (304)442-3379   Fax: (304)442-3307
Mechanical Eng
Govindappa Puttaiah
Phone: (304)442-3374   Fax: (304)442-3307
Admissions Office Contact
Robert P. Scholl
Phone: (304)442-3151   Fax: (304)442-3059
Financial Aid Office Contact
Nina M. Morton, *Director of Financial Aid*
Phone: (304)442-3032   Fax: (304)442-3059

269 Western Michigan U
Head of the Institution
Diether H. Haenicke
Phone: (616)387-2351   Fax: (616)387-2355
Head of Engineering
Leonard R. Lamberson
Phone: (616)387-4017   Fax: (616)387-4024
EMail: LAMBERSON@GW.WMICH.EDU
Associate/Assistant Deans
Molly W. Williams, *Associate Dean*
Phone: (616)387-4017   Fax: (616)387-4024
Department Chair(s)
Electrical Eng
Thomas F. Piatkowski
Phone: (616)387-4057   Fax: (616)387-4024
EMail: PIATKOWSKI@GW.WMICH.EDU
Industrial Eng
Richard E. Munsterman
Phone: (616)387-3737   Fax: (616)387-4024
EMail: MUNSTERMAN@GW.WMICH.EDU
Mechanical & Aeronautical Eng
Jerry H. Hamelink
Phone: (616)387-3366   Fax: (616)387-4024
Paper & Printing Science & Eng
Arvon D. Byle
Phone: (616)387-2770   Fax: (616)387-2768
Admissions Office Contact
Stanley E. Henderson
Phone: (616)387-2000   Fax: (616)387-2355
Financial Aid Office Contact
John A. Kundel, *Director of Student Financial Aid*
Phone: (616)387-6000   Fax: (616)387-6989

270 Western New England Coll
Head of the Institution
Beverly W. Miller
Phone: (413)782-1243   Fax: (413)782-1746
Head of Engineering
Stephen C. Crist
Phone: (413)782-1273   Fax: (413)782-1746
Associate/Assistant Deans
Richard A. Grabiec, *Assistant Dean*
Phone: (413)782-1271   Fax: (413)782-1746
Department Chair(s)
Electrical Eng
William G. Bradley
Phone: (413)782-1491   Fax: (413)782-1746
Industrial & Manufacturing Eng
John B. Nelson
Phone: (413)782-1289   Fax: (413)782-1746
Mechanical Eng
Robert C. Azar
Phone: (413)782-1334   Fax: (413)782-1746
Admissions Office Contact
Lori-Ann Paterwic
Phone: (413)782-3111   Fax: (413)782-1746
Financial Aid Office Contact
Kathy M. Chambers, *Director of Finanial Aid*
Phone: (413)782-3111   Fax: (413)796-2008

271 Wichita State U
Head of the Institution
Warren B. Armstrong
Phone: (316)689-3001   Fax: (316)689-3093
Head of Engineering
William J. Wilhelm
Phone: (316)689-3400   Fax: (316)689-3853
Associate/Assistant Deans
Mark T. Jong, *Associate Dean*
Phone: (316)689-3408   Fax: (316)689-3853
Department Chair(s)
Electrical Eng
Roy H. Norris
Phone: (316)689-3415   Fax: (316)689-3853
Industrial Eng
Don L. Hommertzheim
Phone: (316)689-3425   Fax: (316)689-3853
Mechanical Eng
Richard T. Johnson
Phone: (316)689-3402   Fax: (316)689-3853
Aerospace Eng
Bert L. Smith
Phone: (316)389-3410   Fax: (316)689-3853
Admissions Office Contact
Rita Abent
Phone: (800)362-2594   Fax: (316)689-3795
Financial Aid Office Contact
Larry G. Rector, *Director, Student Financial Planning & Assistance*
Phone: (316)689-3430

272 Widener U
Head of the Institution
Robert J. Bruce
Phone: (215)499-4100   Fax: (215)876-9751
Head of Engineering
Thomas G. McWilliams
Phone: (215)499-4036   Fax: (215)499-4059
Associate/Assistant Deans
Richard C. Jones, *Assist. Dean*
Phone: (215)499-4039   Fax: (215)499-4059
David H.T. Chen, *Asst. Dean for Graduate Programs & Research*
Phone: (215)499-4049   Fax: (215)499-4059
Department Chair(s)
Chemical Eng
Gennaro J. Maffia
Phone: (215)499-4089   Fax: (215)499-4059
EMail: PFGJMAFFIA@CYBER.WIDENER.EDUC
Civil Eng
Charles L. Bartholomew
Phone: (215)499-4249   Fax: (215)499-4059
Electrical Eng
Alfred T. Johnson
Phone: (215)499-4053   Fax: (215)499-4059
EMail: PFAT JOHNSON@SY-BER.WIDENER.EDU
Mechanical Eng
Louis T. Hayes
Phone: (215)499-4192   Fax: (215)499-4059
Graduate Programs & Research
David H.T. Chen
Phone: (215)499-4049   Fax: (215)499-4059
Admissions Office Contact
Michael Mahoney
Phone: (215)499-4126   Fax: (215)876-9751
Financial Aid Office Contact
Ethel Desmarais, *Director*
Phone: (215)499-4171   Fax: (215)876-9751

273 Wilkes U
Head of the Institution
Christopher N. Breiseth
Phone: (717)831-4000   Fax: (717)824-2934

Head of Engineering
Umid R. Nejib
Phone: (717)831-4800   Fax: (717)829-2434
EMail: nejib@wilkesl.wilkes.edu
Associate/Assistant Deans
Bing T. Wong, *Associate Dean*
Phone: (717)831-4803   Fax: (717)829-2434
Department Chair(s)
GeoEnvironmental Eng
Dale Bruns
Phone: (717)831-4610   Fax: (717)829-2434
EMail: dbruns@wilkes1.wilkes.edu
Electrical & Computer Eng
Ahmad Armand
Phone: (717)831-4810   Fax: (717)829-2434
EMail: armand@wilkes1.wilkes.edu
Mechanical & Materials Eng
John L. Orehotsky
Phone: (717)831-4810   Fax: (717)829-2434
EMail: jorehotsky@Wilkes1.Wilkes.edu
Admissions Office Contact
Emory P. Guffrovich, Jr.
Phone: (800)945-5378   Fax: (717)829-2434
Financial Aid Office Contact
Rachael L. Lohman, *Director of Financial Aid*
Phone: (717)824-4651   Fax: (717)829-6662

274 U of Wisconsin-Madison
Head of the Institution
Katharine Lyall, *Acting*
Phone: (608)262-4048
Head of Engineering
John G. Bollinger
Phone: (608)262-3482   Fax: (608)262-6400
EMail: boleng@engr.wisc.edu
Associate/Assistant Deans
Thomas W. Chapman, *Associate Dean - International Engineering*
Phone: (608)263-2191   Fax: (608)262-6707
Dietmeyer L. Dietmeyer, *Associate Dean - Academic Affairs*
Phone: (608)262-3484   Fax: (608)262-6400
Hampton Alfred, *Assistant Dean - Minority Affairs*
Phone: (608)262-7764   Fax: (608)262-1064
Donald C. Woolston, *Assistant Dean - Pre-Engineering*
Phone: (608)262-2473   Fax: (608)262-6400
William W. Wuerger, *Associate Dean - Operations*
Phone: (608)265-2001   Fax: (608)262-6400
Department Chair(s)
Chemical Eng
James A. Dumesic
Phone: (608)262-1092   Fax: (608)262-5434
EMail: dumesic@engr.wisc.edu
Civil & Environmental Eng
Peter L. Monkmeyer
Phone: (608)262-3542   Fax: (608)262-5199
EMail: monkmeyer@engr.wisc.edu
Electrical & Computer Eng
Bahaa E. Saleh
Phone: (608)262-6504   Fax: (608)262-1267
EMail: bsaleh@engr.wisc.edu
Industrial Eng
Arne Thesen
Phone: (608)262-9660   Fax: (608)262-8454
EMail: thesen@engr.wisc.edu
Mechanical Eng
Marvin F. DeVries
Phone: (608)262-0666   Fax: (608)265-2316
EMail: devries@engr.wisc.edu
Eng Mechanics
Edward G. Lovell
Phone: (608)262-3990   Fax: (608)262-3735
EMail: lovell@engr.wisc.edu
Nuclear Eng & Eng Physics
Gilbert A. Emmert
Phone: (608)263-1648   Fax: (608)262-6707
EMail: emmert@engr.wisc.edu
Material Science & Eng
Frank J. Worzala
Phone: (608)262-1821   Fax: (608)262-8353
EMail: worzala@engr.wisc.edu
Eng Professional Development
Donald M. Walker
Phone: (608)262-7988   Fax: (608)263-3160
EMail: donald@engr.wisc.edu
Admissions Office Contact
Keith White
Phone: (608)262-3961

275 U of Wisconsin-Milwaukee
Head of the Institution
John H. Schroeder, *Chancellor*
Phone: (414)229-4331   Fax: (414)229-4553
Head of Engineering
Charles F. James, Jr.
Phone: (414)229-4126   Fax: (414)229-6958
Associate/Assistant Deans
Kenneth F. Neusen, *Associate Dean for Administration & Planning*
Phone: (414)229-4272   Fax: (414)229-6958
Fattah A. Shaikh, *Associate Dean for Graduate Studies & Research*
Phone: (414)229-4667   Fax: (414)229-6958
Gilbert L. Roderick, *Associate Dean for Undergraduate Studies*
Phone: (414)229-4667   Fax: (414)229-6958
Robert J. Bonk, *Assistant Dean for Business Affairs*
Phone: (414)229-4171   Fax: (414)229-6958
Department Chair(s)
Civil Eng & Mechanics
Donald R. Sherman
Phone: (414)229-5166   Fax: (414)229-6958
Electrical Eng
Joseph D. McPherson
Phone: (414)229-5362   Fax: (414)229-6958
EMail: joe@ee.uwm.edu
Industrial & Systems Eng
Umesh K. Saxena
Phone: (414)229-4052   Fax: (414)229-6958
Mechanical Eng
Robert T. Balmer
Phone: (414)229-5176   Fax: (414)229-6958
EMail: balmer@watt.cae.uwm.edu
Materials Eng
George S. Baker
Phone: (414)229-6463   Fax: (414)229-6958
Admissions Office Contact
Beth L. Weckmueller
Phone: (414)229-3800   Fax: (414)229-6940
Financial Aid Office Contact
Mary E. Roggeman, *Director, Department of Financial Aid*
Phone: (414)229-4541

276 U of Wisconsin-Platteville
Head of the Institution
Lee A. Halgren, *Interim*
Phone: (608)342-1234
Head of Engineering
Ronald A. Yeske
Phone: (608)342-1561   Fax: (608)342-1566
Associate/Assistant Deans
D. Joanne Wilson, *Assistant Dean*
Phone: (608)342-1686   Fax: (608)342-1566
Department Chair(s)
Civil Eng
Max L. Anderson
Phone: (608)342-1543   Fax: (608)342-1566
Electrical Eng
Richard D. Shultz
Phone: (608)342-1536   Fax: (608)342-1566

Engineering & Engineering Technology Personnel

Industrial Eng
Swaminathan Balachandran
Phone: (608)342-1715   Fax: (608)342-1566

Mechanical Eng
Richard D. Strunk
Phone: (608)342-1721   Fax: (608)342-1566

General Eng
John A. Krogman
Phone: (608)342-1711   Fax: (608)342-1566

Admissions Office Contact
Richard R. Schumacher
Phone: (608)342-1125   Fax: (608)342-1122

Financial Aid Office Contact
Elizabeth Tucker, *Director of Financial Aids*
Phone: (608)342-1836

277   Worcester Poly Inst

Head of the Institution
Jon C. Strauss
Phone: (508)831-5200   Fax: (508)831-5753

Head of Engineering
Francis C. Lutz
Phone: (508)831-5404   Fax: (508)831-5774
EMail: FCLUTZ@JAKE.WPI.EDU

Associate/Assistant Deans
Lance E. Schachterle, *Associate Dean of Undergraduate Studies*
Phone: (508)831-5514   Fax: (508)831-5485

Department Chair(s)

Chemical Eng
Albert Sacco
Phone: (508)831-5250   Fax: (508)831-5483
EMail: ASACCO@JAKE.WPI.EDU

Civil Eng
Robert W. Fitzgerald
Phone: (508)831-5530   Fax: (508)831-5808
EMail: RFITZ@WPI.WPI.EDU

Electrical Eng
John A. Orr
Phone: (508)831-5231   Fax: (508)831-5491
EMail: ORR@WPI.WPI.EDU

Manufacturing Eng
Richard D. Sisson
Phone: (508)831-5335   Fax: (508)831-5680
EMail: SISSON@WPI.WPI.EDU

Mechanical Eng
Mohammad N. Noori
Phone: (508)831-5236   Fax: (508)831-5680
EMail: MNNOORI@WPI.WPI.EDU

Biomedical Eng
Robert A. Peura
Phone: (508)831-5447   Fax: (508)831-5483
EMail: RAPEURA@WPI.WPI.EDU

Fire Protection Eng
David A. Lucht
Phone: (508)831-5104   Fax: (508)831-5680
EMail: DALUCHT@JAKE.WPI.EDU

Materials Eng
Ronald R. Biederman
Phone: (508)831-5453   Fax: (508)831-5680
EMail: RRB@WPI.WPI.EDU

Admissions Office Contact
Robert G. Voss
Phone: (508)831-5286   Fax: (508)831-5483

Financial Aid Office Contact
Michael J. Curley, *Director of Financial Aid*
Phone: (508)831-5469   Fax: (508)831-5753

278   Wright State U

Head of the Institution
Paige E. Mulhollan
Phone: (513)873-2312   Fax: (513)873-2421

Head of Engineering
James E. Brandeberry
Phone: (513)873-5007   Fax: (513)873-5009
EMail: jbrand@matrix.cs.wright.edu

Associate/Assistant Deans
Marc E. Low, *Associate Dean*
Phone: (513)873-5001   Fax: (513)873-5009
Giorgio M. McBeath, *Assistant Dean*
Phone: (513)873-5001   Fax: (513)873-5009
Richard K. Rathbun, *Assistant Dean*
Phone: (513)873-5001   Fax: (513)873-5009

Department Chair(s)

Electrical Eng
Raymond E. Siferd
Phone: (513)873-5037   Fax: (513)873-5009
EMail: rsiferd@valhalla.cs.wright.edu

Mechanical & Materials Eng
Richard J. Bethke
Phone: (513)873-5040   Fax: (513)873-5009
EMail: rbethke@matrix.cs.wright.edu

Biomedical & Human Factors Eng
Anthony J. Cacioppo
Phone: (513)873-5044   Fax: (513)873-5009
EMail: acacio@matrix.cs.wright.edu

Computer Science & Eng
Robert D. Dixon
Phone: (513)873-5131   Fax: (513)873-5009
EMail: rdixon@valhalla.cs.wright.edu

Admissions Office Contact
Kenneth A. Davenport
Phone: (513)873-2211   Fax: (513)873-3301

Financial Aid Office Contact
David R. Darr, *Director of Financial Aid*
Phone: (513)873-2321   Fax: (513)873-3301

279   U of Wyoming

Head of the Institution
Terry P. Roark
Phone: (307)766-4121   Fax: (307)766-2271

Head of Engineering
Samuel D. Hakes
Phone: (307)766-4253   Fax: (307)766-4444

Associate/Assistant Deans
John W. Steadman, *Associate Dean*
Phone: (307)766-2240   Fax: (307)766-4444
David Whitman, *Assistant Dean*
Phone: (307)766-4253   Fax: (307)766-4444
Bruce R. Dewey, *Assistant Dean*
Phone: (307)766-4253   Fax: (307)766-4444

Department Chair(s)

Chemical Eng
Chang-Yul Cha
Phone: (307)766-2837   Fax: (307)766-4444

Civil & Architectural Eng
Arthur P. Boresi
Phone: (307)766-5255   Fax: (307)766-4444

Electrical Eng
John W. Steadman
Phone: (307)766-2240   Fax: (307)766-4444
EMail: JWSTEAD@UWYO

Mechanical Eng
Kynric M. Pell
Phone: (307)766-2213   Fax: (307)766-2695

Petroleum Eng
H. Gordon Harris
Phone: (307)766-5186   Fax: (307)766-4444

Admissions Office Contact
Richard A. Davis
Phone: (307)766-5160   Fax: (307)766-4042

Financial Aid Office Contact
John F. Nutter, *Director*
Phone: (307)766-2118   Fax: (307)766-4042

280   Yale U

Head of the Institution
Howard R. Lamar
Phone: (203)432-2550   Fax: (203)432-7105

Head of Engineering
Gary L. Haller, *Chairman*
Phone: (203)432-4220   Fax: (203)432-2797

Department Chair(s)

Chemical Eng
Csaba G. Horvath
Phone: (203)432-4357   Fax: (203)432-7232

Electrical Eng
Tso-Ping Ma
Phone: (203)432-2212   Fax: (203)432-2797

Mechanical Eng
Robert E. Apfel
Phone: (203)432-4223   Fax: (203)432-2797

Applied Physics
Werner P. Wolf
Phone: (203)432-4282   Fax: (203)432-4283

Admissions Office Contact
Richard H. Shaw, Jr.
Phone: (203)432-1900   Fax: (203)432-7329

Financial Aid Office Contact
Jacqueline Foster, *Director of Financial Aid*
Phone: (203)432-0360

ENGINEERING TECHNOLOGY

281   U of Akron

Head of the Institution
Peggy G. Elliott
Phone: (216)972-7074

Head of Engineering Tech
Frederick J. Sturm
Phone: (216)972-7220   Fax: (216)972-5300

Associate/Assistant Deans
Minnie C. Pritchard, *Associate Dean, Community & Technical Coll*
Phone: (216)972-7220   Fax: (216)972-5300

Department Chair(s)

Surveying & Construction Tech
Ron G. Adams
Phone: (216)972-7050   Fax: (216)972-6990

Electronic Tech
Ron G. Adams
Phone: (216)972-7050   Fax: (216)972-6990

Manufacturing & Automated Manufacturing Tech
Ron G. Adams
Phone: (216)972-7050   Fax: (216)972-6990

Mechanical Tech
Ron G. Adams
Phone: (216)972-7050   Fax: (216)972-6990

Drafting & Computer Drafting Tech
Ron G. Adams
Phone: (216)972-7050   Fax: (216)972-6990

Admissions Office Contact
Martha A. Booth
Phone: (216)972-7100

Financial Aid Office Contact
Robert D. Hahn, *Director*
Phone: (216)972-7032   Fax: (216)972-7139

282   Alabama A&M U

Head of the Institution
David Henson
Phone: (205)851-5230

Head of Engineering Tech
Arthur J. Bond
Phone: (205)851-5560   Fax: (205)851-5580

Department Chair(s)

Eng Tech
Edward L. Bernstein
Phone: (205)851-5581   Fax: (205)851-5580

Admissions Office Contact
James O. Heyward
Phone: (205)851-5245

Financial Aid Office Contact
Percy N. Lanier, *Director*
Phone: (205)851-5400

283   Arizona State U

Head of the Institution
Lattie F. Coor
Phone: (602)965-5606   Fax: (602)965-0865

Head of Engineering Tech
David C. Chang
Phone: (602)965-1722   Fax: (602)965-2267
EMail: icdcc@asuacad

Associate/Assistant Deans
Richard W. Kelly, *Associate Dean & Director School of Tech*
Phone: (602)965-3874   Fax: (602)965-5089
Charles E. Backus, *Associate Dean*
Phone: (602)965-2825   Fax: (602)965-2267
William E. Lewis, *Associate Dean*
Phone: (602)965-1730   Fax: (602)965-2267
C. Edward Wallace, *Associate Dean*
Phone: (602)965-1726   Fax: (602)965-2267

Department Chair(s)
Aeronautical Tech
Robert O. Meitz
Phone: (602)965-7775   Fax: (602)965-5089
EMail: idrom@asuacad

Electronics & Computer Tech
Albert L. McHenry
Phone: (602)965-3137   Fax: (602)965-0723
EMail: iacaxm@asuacad

Manufacturing & Industrial Tech
Donald W. Collins
Phone: (602)965-3791   Fax: (602)965-5089
EMail: aodwc@asuacad

Admissions Office Contact
Susan Clouse
Phone: (602)965-7788   Fax: (602)965-2120

Financial Aid Office Contact
Paul Barberini, *Director*
Phone: (602)965-4045

284   U of Arkansas at Little Rock

Head of the Institution
Alan B. Sugg
Phone: (501)686-2505   Fax: (501)686-2507

Head of Engineering Tech
Charles A. Stevens
Phone: (501)569-3150   Fax: (501)569-8020
EMail: CASTEVENS%EIVAX@UALR.EDU

Associate/Assistant Deans
J W. Wiggins, *Associate Dean for Undergraduate Affairs*
Phone: (501)569-3247   Fax: (501)569-8020
Gaylord M. Northrop, *Associate Dean for Research/Director of GIT*
Phone: (501)569-8211   Fax: (501)569-8020

Department Chair(s)
Electronics Eng Tech
Hirak C. Patangia
Phone: (501)569-8202   Fax: (501)569-8020
EMail: HCPATANGIA@UALR.EDU

Manufacturing Eng Tech
Mamdouh M. Bakr
Phone: (501)569-8228   Fax: (501)569-8020
EMail: MMBAKR@UALR.EDU

Mechanical Eng Tech
Burt Henderson
Phone: (501)569-8224   Fax: (501)569-8020
EMail: BHENDERS@UALRET.UALR.EDU

Computer Eng Tech
Harry A. Tschumi
Phone: (501)569-8227   Fax: (501)569-8020
EMail: PETE@UALRET.UALR.EDU

Construction Eng Tech
Charles B. Conine
Phone: (501)569-8229   Fax: (501)569-8020

Surveying
Julian S. Rouch
Phone: (501)569-8204   Fax: (501)569-8020
EMail: JSROUCH@UALR.EDU

Admissions Office Contact
Sue Pines
Phone: (501)569-3127

285   Auburn U

Head of the Institution
William V. Muse
Phone: (205)844-4650   Fax: (205)844-6179

Head of Engineering Tech
William F. Walker
Phone: (205)844-4326   Fax: (205)844-2672
EMail: wwalker@eng.auburn.edu

Associate/Assistant Deans
Larry D. Benefield, *Associate Dean for Academics*
Phone: (205)844-4326   Fax: (205)844-2672
M. Dayne Aldridge, *Associate Dean for Cross Disciplinary Programs*
Phone: (204)844-4333   Fax: (205)844-2672

John M. Owens, *Associate Dean for Research*
Phone: (205)844-4326   Fax: (205)844-2672

Department Chair(s)
Textile Eng
William K. Walsh
Phone: (205)844-5453   Fax: (205)844-4068
EMail: wwalsh@eng.auburn.edu

Admissions Office Contact
Charles F. Reeder
Phone: (205)844-4080   Fax: (205)844-6436

Financial Aid Office Contact
Clark Aldridge, *Director*
Phone: (205)844-4723   Fax: (205)844-6085

286   Bradley U

Head of the Institution
John R. Brazil
Phone: (309)677-3167   Fax: (309)677-2330

Head of Engineering Tech
John E. Francis
Phone: (309)677-2720   Fax: (309)677-3670

Associate/Assistant Deans
James G. Seckler, *Associate Dean*
Phone: (309)677-2720   Fax: (309)677-3670
Robert J. Podlasek, *Assistant Dean*
Phone: (309)677-2714   Fax: (309)677-3670

Department Chair(s)
Electrical Eng Tech
Brian D. Huggins
Phone: (309)677-2732   Fax: (309)677-3670

Manufacturing Tech
Arnold E. Ness
Phone: (309)677-2938   Fax: (309)677-3670

Admissions Office Contact
Gary Bergman
Phone: (800)447-6460   Fax: (309)677-2797

Financial Aid Office Contact
David Pardieck, *Director of Financial Aid*
Phone: (309)677-3089   Fax: (309)677-2798

287   Brigham Young U

Head of the Institution
Rex E. Lee
Phone: (801)378-2521

Head of Engineering Tech
L. Douglas Smoot
Phone: (801)378-4326   Fax: (801)378-5705

Associate/Assistant Deans
Steven E. Benzley, *Associate Dean*
Phone: (801)378-4326   Fax: (801)378-5705
John J. Kunzler, *Associate Dean*
Phone: (801)378-4326   Fax: (801)378-5705
David K. Anthony, *Assistant Dean*
Phone: (801)378-5780   Fax: (801)378-5075
Ronald E. Terry, *Assistant Dean*
Phone: (801)378-4326   Fax: (801)378-5705

Department Chair(s)
Manufacturing Eng & Eng Tech
John J. Kunzler
Phone: (801)378-6300   Fax: (801)378-7575

Manufacturing Eng & Eng Tech
A. Brent Strong
Phone: (801)378-6300   Fax: (801)378-7575

Manufacturing Eng
A. Brent Strong
Phone: (801)378-6300   Fax: (801)378-7575

Manufacturing Eng & Eng Tech
A. Brent Strong
Phone: (801)378-6300   Fax: (801)378-7575

Admissions Office Contact
Tom M. Gourley
Phone: (801)378-2507   Fax: (801)378-4264

Financial Aid Office Contact
Norman B. Finlinson, *Director*
Phone: (801)378-4104

288   Cal State U, Sacramento

Head of the Institution
Donald R. Gerth
Phone: (916)278-7737   Fax: (916)278-6959

Head of Engineering Tech
Donald H. Gillott
Phone: (916)278-6366   Fax: (916)278-5949

Associate/Assistant Deans
John N. Hester, *Associate Dean, Acaedemic Programs & Research*
Phone: (916)278-6366   Fax: (916)278-5949
John T. Oldenburg, *Associate Dean, Resources & Facilities Planning*
Phone: (917)278-6366   Fax: (916)278-5949
Larry A. Hill, *Assistant Dean*
Phone: (916)278-6366   Fax: (916)278-5949

Department Chair(s)
Mechanical Eng Tech
Joseph H. Harralson
Phone: (916)278-7081   Fax: (916)278-5949
EMail: Harralsj@ecs.csus.edu

Construction Management
Keith Bisharat
Phone: (916)278-6616   Fax: (916)278-5949
EMail: Bisharak@ecs.csus.edu

Admissions Office Contact
Larry Glasmire
Phone: (916)278-7111   Fax: (916)278-5443

Financial Aid Office Contact
Starla Satchell, *Director of Financial Aid*
Phone: (916)278-6554

289   Cal State Poly, Pomona

Head of the Institution
Bob H. Suzuki
Phone: (909)869-2290

Head of Engineering Tech
Edward C. Hohmann
Phone: (909)869-2600   Fax: (909)869-4370

Associate/Assistant Deans
Carl E. Rathmann, *Associate Dean*
Phone: (909)869-2600   Fax: (909)869-4370

Department Chair(s)
Eng Tech
Edward V. Clancy
Phone: (909)869-2492   Fax: (909)869-4370

Admissions Office Contact
Joseph C. Marshall
Phone: (909)869-2000

Financial Aid Office Contact
Al Andino, *Director of Financial Aid*
Phone: (909)869-3700

290   Capitol Coll

Head of the Institution
G. William Troxler
Phone: (301)953-0060   Fax: (301)953-3876

Head of Engineering Tech
Earl E. Gottsman
Phone: (301)953-0060   Fax: (301)953-3876

Department Chair(s)
Electrical/Electronic(s) Eng Tech
Robert L. Weiler
Phone: (301)953-0060   Fax: (301)953-3876
Computer & Mathematics Department
Frank D. Sheng
Phone: (301)953-0060   Fax: (301)953-3876

Admissions Office Contact
Anthony G. Miller
Phone: (301)953-0060   Fax: (301)953-3876

Financial Aid Office Contact
Sheila Sauls-White, *Director*
Phone: (301)953-0060   Fax: (301)953-3876

291   Central Conn State U

Head of the Institution
John W. Shumaker
Phone: (203)827-7203   Fax: (203)827-7033

Head of Engineering Tech
John R. Wright
Phone: (203)827-7378   Fax: (203)827-7982

Associate/Assistant Deans
Andrew W. Baron, *Assistant Dean*
Phone: (203)827-7997   Fax: (203)827-7982

Department Chair(s)
Civil & Construction Eng Tech
Daryll C. Dowty
Phone: (203)827-7370   Fax: (203)827-7982
Manufacturing Eng Tech
Daryll C. Dowty
Phone: (203)827-7370   Fax: (203)827-7982
Industrial Systems Eng Tech
Daryll C. Dowty
Phone: (203)827-7370   Fax: (203)827-7982

Admissions Office Contact
Hakim-Salahu A Din
Phone: (203)827-7543   Fax: (203)827-7220

Financial Aid Office Contact
John Taylor, *Director of Financial Aid*
Fax: (203)827-7200

292   U of Central Florida

Head of the Institution
John C. Hitt
Phone: (407)823-1823   Fax: (407)823-5407

Head of Engineering Tech
Gary E. Whitehouse
Phone: (407)823-2156   Fax: (407)823-5483
EMail: WHITEHSE @ UCF1VM.CC.UCF.EDU

Associate/Assistant Deans
Richard N. Miller, *Associate Dean for Undergraduate Affairs*
Phone: (407)823-2455   Fax: (407)823-5483
Stephen L. Rice, *Associate Dean for Research & Graduate Affairs*
Phone: (407)823-2156   Fax: (407)823-5483

Department Chair(s)
Eng Tech
James D. McBrayer
Phone: (407)823-2269   Fax: (407)823-3413
EMail: MCBRAYER @ UCF1VM.CC.UCF.EDU

Admissions Office Contact
John F. Bush
Phone: (407)823-3000   Fax: (407)823-5652

Financial Aid Office Contact
Mary H. McKinney, *Director of Financial Aid*
Phone: (407)823-2827   Fax: (407)823-5652

293   U of Cincinnati

Head of the Institution
Joseph A. Steger
Phone: (513)556-2201   Fax: (513)556-3010

Head of Engineering Tech
Fritz J. Kryman
Phone: (513)556-6556   Fax: (513)556-4599

Associate/Assistant Deans
Lawrence G. Gilligan, *Associate Dean of Academic & Faculty Affairs*
Phone: (513)556-6580   Fax: (513)556-4599
Cheryll A. Dunn, *Associate Dean, Student Affairs*
Phone: (513)556-6561   Fax: (513)556-4599
Patricia K. Lloyd, *Assistant Dean, Administration*
Phone: (513)556-6578   Fax: (513)556-4599

Department Chair(s)
Construction Science
Forest D. Atkins
Phone: (513)556-6553   Fax: (513)556-4224
EMail: ATKINSFJ@UC.EDU

Engineering & Engineering Technology Personnel

Electrical Eng Tech
Elvin D. Stepp
Phone: (513)556-6558   Fax: (513)556-4224
EMail: STEPPE@UC.EDU

Mechanical Eng Tech
A. Allen Arthur
Phone: (513)556-5305   Fax: (513)556-4015
EMail: ARTHURAA@UC.EDU

Chemical Tech
John C. Spille
Phone: (513)556-6591   Fax: (513)556-4224
EMail: SPILLEJC@UC.EDU

Mathematics, Physics, & Computing Tech
Robert E. Schlemmer
Phone: (513)556-6565   Fax: (513)556-4224
EMail: SCHLEMR@UCBEH

Humanities/Social Sciences/Communication
Sam C. Geonetta
Phone: (513)556-6562   Fax: (513)556-4224
EMail: GEONETSC@UC.EDU

Admissions Office Contact
Cheryll A. Dunn
Phone: (513)556-6561   Fax: (513)556-4599

294 U of Dayton

Head of the Institution
Raymond L. Fitz
Phone: (513)229-4122   Fax: (513)229-4545

Head of Engineering Tech
Robert L. Mott
Phone: (513)229-2333   Fax: (513)229-2756

Department Chair(s)
Electronic Eng Tech
Joseph M. Farren
Phone: (513)229-3614   Fax: (513)229-2756

Industrial Eng Tech
James F. Courtright
Phone: (513)229-4216   Fax: (513)229-2756

Manufacturing Eng Tech
Robert L. Wolff
Phone: (513)229-4216   Fax: (513)229-2756

Eng Tech Service Courses
Robert L. Mott
Phone: (513)229-2333   Fax: (513)229-2756

Mechanical Eng Tech
Philip E. Doepker
Phone: (513)229-4216   Fax: (513)229-2756

Chemical Process Tech
David I. Gross
Phone: (513)229-3627   Fax: (513)229-2756

Environmental Eng Tech
David I. Gross
Phone: (513)229-3627   Fax: (513)229-2756

Admissions Office Contact
Myron H. Achbach
Phone: (513)229-4411   Fax: (513)229-4545

295 DeVry Inst-Chicago

Head of the Institution
E. Arthur Stunard
Phone: (312)929-8500   Fax: (312)348-1780

Head of Engineering Tech
Eugene E. Miller
Phone: (312)929-8500   Fax: (312)348-1780

Department Chair(s)
Electronics Eng Tech
Eugene A. Miller
Phone: (708)953-1300   Fax: (708)953-1236

Admissions Office Contact
Richard L. Yaconis
Phone: (312)929-6550   Fax: (312)348-1780

Financial Aid Office Contact
Betty Glenn, *Director of Financial Aid*
Phone: (312)929-8500   Fax: (312)348-1780

296 DeVry Inst-Columbus

Head of the Institution
Richard A. Czerniak
Phone: (614)253-7291   Fax: (614)252-4108

Head of Engineering Tech
Edward A. Wilson
Phone: (614)253-7291   Fax: (614)252-4108

Department Chair(s)
Electronics Eng Tech
Edward A. Wilson
Phone: (614)253-7291   Fax: (614)252-4108

Admissions Office Contact
Richard Rodman
Phone: (614)253-1525   Fax: (614)252-4108

Financial Aid Office Contact
Cynthia A. Price, *Director of Financial Aid*
Phone: (614)253-7291   Fax: (614)252-4108

297 DeVry Inst-Atlanta

Head of the Institution
Ron Bush
Phone: (404)292-7900   Fax: (404)292-2321

Head of Engineering Tech
Jack Griffin
Phone: (404)292-7900   Fax: (404)292-2321

Department Chair(s)
Electronic Eng Tech
John Dunbar
Phone: (404)292-7900   Fax: (404)292-2321

Admissions Office Contact
Susan Hirst
Phone: (404)292-2645   Fax: (404)292-2321

Financial Aid Office Contact
Loretta Franklin, *Dean of Student Finance*
Phone: (404)292-7900   Fax: (404)292-2321

298 DeVry Inst-DuPage

Head of the Institution
Jerry R. Dill
Phone: (708)953-1300   Fax: (708)953-1236

Head of Engineering Tech
Sylvia Washington
Phone: (708)953-1300   Fax: (708)953-1236

Department Chair(s)
Electronics Eng Tech
Sylvia Washington
Phone: (708)953-1300   Fax: (708)953-1236

Admissions Office Contact
Virginia Mechnig
Phone: (708)953-2000   Fax: (708)953-1236

Financial Aid Office Contact
Diane Battistella, *Director of Financial Aid*
Phone: (708)953-1300   Fax: (708)953-1236

299 DeVry Inst-Dallas

Head of the Institution
Frank Cannon
Phone: (214)258-6767   Fax: (214)659-1748

Head of Engineering Tech
Charles Reck
Phone: (214)258-6767   Fax: (214)659-1748

Department Chair(s)
Electronic Eng Tech
Charles Reck
Phone: (214)258-6767   Fax: (214)659-1748

Admissions Office Contact
Danny Millan
Phone: (214)258-6330   Fax: (214)659-1748

Financial Aid Office Contact
Laura Myers, *Director of Financial Aid*
Phone: (214)258-6767   Fax: (214)659-1748

300 DeVry Inst-Kansas City

Head of the Institution
C. Robert LeValley
Phone: (816)941-0430   Fax: (816)941-0896

Head of Engineering Tech
Paul Stephanchick
Phone: (816)941-0430   Fax: (816)941-0896

Department Chair(s)
Electronics Eng Tech
Paul Stephanchick
Phone: (816)941-0430   Fax: (816)941-0896

Admissions Office Contact
Gale Dykes-Grimmett
Phone: (816)941-2810   Fax: (816)941-0896

Financial Aid Office Contact
Sharon Lamson, *Director of Financial Aid*
Phone: (816)941-0430   Fax: (816)941-0896

301 DeVry Inst-Los Angeles

Head of the Institution
David Moore
Phone: (213)699-9927   Fax: (213)692-6272

Head of Engineering Tech
Iraj Borbor
Phone: (213)699-9927   Fax: (213)692-6272

Department Chair(s)
Electronics Eng Tech
Mohammad Tasooji
Phone: (213)699-9927   Fax: (213)692-6272

Admissions Office Contact
Keith E. Paridy
Phone: (213)692-0551   Fax: (213)692-6272

Financial Aid Office Contact
Lynn Taylor, *Director of Financial Aid*
Phone: (213)699-9927   Fax: (213)692-6272

302 DeVry Inst-Phoenix

Head of the Institution
James A. Dugan
Phone: (602)870-9222   Fax: (602)870-1209

Head of Engineering Tech
Michael Thomas
Phone: (602)870-9222   Fax: (602)870-1209

Department Chair(s)
Electronics Eng Tech
Michael Thomas
Phone: (602)870-9222   Fax: (602)870-1209

Admissions Office Contact
Kim Galetti
Phone: (602)870-9201   Fax: (602)870-1209

Financial Aid Office Contact
Scott Morrison, *Director of Financial Aid*
Phone: (602)870-9222   Fax: (602)870-1209

303 U of the District of Columbia

Head of the Institution
Tilden J. LeMelle
Phone: (202)282-7550   Fax: (202)282-3681

Head of Engineering Tech
Philip L. Brach
Phone: (202)282-7427   Fax: (202)282-3677

Associate/Assistant Deans
Alfred O. Taylor, *Associate Dean*
Phone: (202)282-7427   Fax: (202)282-3677
Thedola H. Milligan, *Associate Dean*
Phone: (202)282-7427   Fax: (202)282-3677

Department Chair(s)
Architectural & Civil Eng Tech
Clarence W. Pearson
Phone: (202)282-7450   Fax: (202)282-3677

Electrical & Mechanical Eng Tech
Edward L. Walker
Phone: (202)282-7425   Fax: (202)282-3677

Admissions Office Contact
Sandra B. Dolphin
Phone: (202)282-7300   Fax: (202)282-3679

Financial Aid Office Contact
Kenneth Howard, *Director of Financial Aid*
Phone: (202)282-3239   Fax: (202)282-3344

304 East Tennessee State U

Head of the Institution
Roy S. Nicks
Phone: (615)929-4211   Fax: (615)929-4004

Head of Engineering Tech
James A. Hales
Phone: (615)929-4465   Fax: (615)929-6435
EMail: A31HALES@ETSU

Associate/Assistant Deans
John S. Vaglia, *Associate Dean, School of Applied Science & Tech*
Phone: (615)929-4299   Fax: (615)929-6435

Department Chair(s)
Electronics Eng Tech
Wayne D. Andrews
Phone: (615)929-4310   Fax: (615)929-6435
EMail: WANDY@ETSUCIM

Manufacturing Eng Tech
Wayne D. Andrews
Phone: (615)929-4310   Fax: (615)929-6435
EMail: WANDY@ETSUCIM

Biomedical Eng Tech
Wayne D. Andrews
Phone: (615)929-4310   Fax: (615)929-6435
EMail: WANDY@ETSUCIM

Construction Eng Tech
Wayne D. Andrews
Phone: (615)929-4310   Fax: (615)929-6435
EMail: WANDY@ETSUCIM

Eng Design Graphics Tech
Wayne D. Andrews
Phone: (615)929-4310   Fax: (615)929-6435
EMail: WANDY@ETSUCIM

Industrial Tech
Wayne D. Andrews
Phone: (615)929-4310   Fax: (615)929-6435
EMail: WANDY@ETSUCIM

Surveying & Mapping
Wayne D. Andrews
Phone: (615)929-4310   Fax: (615)929-6435
EMail: WANDY@ETSUCIM

Admissions Office Contact
Nancy L. Dishner
Phone: (615)929-4213   Fax: (615)929-5770

Financial Aid Office Contact
Linda J. Clemons, *Director, Office of Financial Aid*
Phone: (615)929-4313   Fax: (615)929-5770

305 Embry-Riddle-Daytona Beach

Head of the Institution
Steven Silwa
Phone: (905)226-6200   Fax: (904)226-6299

Head of Engineering Tech
William Martin
Phone: (904)226-6821   Fax: (904)226-6011

Department Chair(s)
Civil Eng
John B. Stevens
Phone: (802)485-2261   Fax: (802)485-2580
EMail: Stevens

Electrical Eng
William Till
Phone: (802)485-2272   Fax: (802)485-2580
EMail: Till

Mechanical Eng
Kenneth Craig
Phone: (802)485-2279   Fax: (802)485-2580
EMail: Craig

Engineering & Engineering Technology Personnel

Environmental Eng Tech
Wight Greg
Phone: (802)485-2276   Fax: (802)485-2580
EMail: Wight

Computer Science & Computer Eng
Michael Murphy
Phone: (802)828-2265   Fax: (802)485-2265
EMail: Murphy

Architecture
Robert E. Schmidt
Phone: (802)828-8599   Fax: (802)485-2580
EMail: Schmidt

Admissions Office Contact
Darryl W. Niemeyer
Phone: (904)226-6100   Fax: (904)226-6299

Financial Aid Office Contact
Phillip C. Ledbetter, *Director of Financial Aid*
Phone: (904)226-6195

306   Ferris State U

Head of the Institution
Helen Popovich
Phone: (616)592-2500

Head of Engineering Tech
Joel Galloway
Phone: (616)592-2898   Fax: (616)592-2946

Associate/Assistant Deans
Vordyn Nelson, *Associate Dean*
Phone: (616)592-2898   Fax: (616)592-2946

Department Chair(s)
Construction Department
Ralph Shields
Phone: (616)592-2360   Fax: (616)592-2946
EMail: Swan 312

Electrical/Electronics Eng Tech
Philip Marcotte
Phone: (616)592-2388   Fax: (616)592-2946
EMail: Swan 405

Manufacturing Eng Technologies
Steve Hickel
Phone: (616)592-2511   Fax: (616)592-2946
EMail: Swan 109

Admissions Office Contact
Duncan Sargent
Phone: (616)592-2100

Financial Aid Office Contact
Robert Bopp, *Director*
Phone: (616)592-2110

307   Franklin U

Head of the Institution
Paul J. Otte
Phone: (614)341-6234   Fax: (614)221-7723

Head of Engineering Tech
Gena R. Proulx
Phone: (614)341-6351   Fax: (614)221-7723

Department Chair(s)
Electronics Eng Tech
Rafic A. Bachnak
Phone: (614)341-6328   Fax: (614)221-7723

Mechanical Eng Tech
Mohamad Qatu
Phone: (614)341-6402   Fax: (614)221-7723

Admissions Office Contact
Linda M. Steele
Phone: (614)224-6237   Fax: (614)224-7723

Financial Aid Office Contact
Katherine M. Faye, *Director of Financial Aid*
Phone: (614)224-6237   Fax: (614)221-7723

308   U of Hartford

Head of the Institution
Humphrey Tonkin
Phone: (203)768-4417   Fax: (203)768-5417

Head of Engineering Tech
Alan J. Hadad
Phone: (203)768-4308   Fax: (203)768-5074

Associate/Assistant Deans
Richard P. Mills, Jr., *Associate Dean*
Phone: (203)768-4795   Fax: (203)768-5074

Department Chair(s)
Electronic & Computer Eng Tech
Everitt K. Smith
Phone: (203)768-4756   Fax: (203)768-5074

Architectural Eng Tech
Allen I. Bernholtz
Phone: (203)768-4755   Fax: (203)768-5074

Mathematics & Technicial Communications
Phyllis S. Katz
Phone: (203)768-4754   Fax: (203)768-5074

Admissions Office Contact
Richard P. Mills, Jr.
Phone: (800)766-4024   Fax: (203)768-5074

Financial Aid Office Contact
Daniel E. Small, *Director of Student Financial Assistance*
Phone: (203)768-4904   Fax: (203)768-4961

309   Indiana-Purdue, Indianapolis

Head of the Institution
Gerald L. Bepko, *Chancellor*
Phone: (317)274-4417   Fax: (317)274-4615

Head of Engineering Tech
R. Bruce Renda
Phone: (317)274-0800   Fax: (317)274-0832

Associate/Assistant Deans
H. Öner Yurtseven, *Associate Dean for Academic Affairs*
Phone: (317)274-9708   Fax: (317)274-4567
Gary Burkart, *Associate Dean*
Phone: (317)226-5601   Fax: (317)226-5615
Christine Fitzpatrick, *Assistant Dean for Student Affairs*
Phone: (317)274-0804   Fax: (317)274-4567
Pat Fox, *Assistant Dean for Financial Affairs*
Phone: (317)274-0807   Fax: (317)274-4567

Department Chair(s)
Construction Tech
Edgar Fleenor
Phone: (317)274-8720   Fax: (317)274-4567

Electrical Eng Tech
Richard Pfile
Phone: (317)274-2756   Fax: (317)274-4567

Manufacturing Tech
Robert E. Peale
Phone: (317)274-3429   Fax: (317)274-4567

Admissions Office Contact
Christine Fitzpatrick
Phone: (317)274-0804   Fax: (317)274-4567

Financial Aid Office Contact
Natala Hart
Phone: (317)274-4162

310   Lake Superior State U

Head of the Institution
Robert Arbuckle
Phone: (906)635-2202   Fax: (906)635-2111

Department Chair(s)
Electrical Eng Tech
Patrick M. Grounds
Phone: (906)635-2207   Fax: (906)635-2111

Mechanical Eng Tech
Patrick M. Grounds
Phone: (906)635-2207   Fax: (906)635-2111

Automated Systems Eng Tech
Patrick M. Grounds
Phone: (906)635-2207   Fax: (906)635-2111

Admissions Office Contact
Bruce Johnson
Phone: (906)635-2231   Fax: (906)635-2111

Financial Aid Office Contact
William Munsell, *Director of Financial Aid*
Phone: (906)635-1311   Fax: (906)635-2111

311   Louisiana Tech U

Head of the Institution
Daniel D. Reneau
Phone: (318)257-3785   Fax: (318)257-2928

Head of Engineering Tech
Barry A. Benedict
Phone: (318)257-4647   Fax: (318)257-2562
EMail: benedict@engr.latech.edu

Associate/Assistant Deans
James D. Nelson, *Associate Dean for Academic Affairs*
Phone: (318)257-2842   Fax: (318)257-2562

Department Chair(s)
Construction Eng Tech
Richard B. Lewis
Phone: (318)257-4723   Fax: (313)257-2562

Electrical Eng Tech
John W. Ray
Phone: (318)257-2262   Fax: (318)257-2562
EMail: ray@engr.latech.edu

Admissions Office Contact
Karen T. Akin
Phone: (318)257-3036   Fax: (318)257-2499

Financial Aid Office Contact
Etienna R. Winzer, *Director, Division of Financial Aid*
Phone: (318)257-2641

312   U of Maine

Head of the Institution
Frederick E. Hutchinson
Phone: (207)581-1512   Fax: (207)581-1517

Head of Engineering Tech
John J. McDonough
Phone: (207)581-2341   Fax: (207)581-2113

Department Chair(s)
Civil/Construction Management Eng Tech
Howard M. Gray
Phone: (207)581-2181   Fax: (208)581-2113

Electrical/Eng Tech
George H. Elliott
Phone: (207)581-2350   Fax: (207)581-2113

Mechanical Eng Tech
Herb L. Crosby
Phone: (207)581-2134   Fax: (207)581-2113

Admissions Office Contact
William B. Munsey
Phone: (207)581-1561   Fax: (207)581-1213

Financial Aid Office Contact
Peggy L. Crawford, *Director of Student Aid*
Phone: (207)581-1324   Fax: (207)581-3261

313   Mankato State U

Head of the Institution
Richard R. Rush
Phone: (507)389-1111   Fax: (507)389-5859

Head of Engineering Tech
Robert J. Herickhoff
Phone: (507)389-1521   Fax: (507)389-1095

Department Chair(s)
Electronic Eng Tech
Thomas Hendrickson
Phone: (507)389-5747   Fax: (507)389-1095

Manufacturing Eng Tech
Kirk L. Ready
Phone: (507)389-6383   Fax: (507)389-1095

Automotive Eng Tech
Kirk L. Ready
Phone: (507)389-6383   Fax: (507)389-1095

Admissions Office Contact
Jack Parkins
Phone: (507)389-1822   Fax: (507)389-1040

Financial Aid Office Contact
Sandra Loerts, *Director, Student Financial Aid Office*
Phone: (507)389-1185   Fax: (507)389-5114

314   U of Mass, Dartmouth

Head of the Institution
Joseph C. Deck, *Chancellor*
Phone: (508)999-8004   Fax: (508)999-8860

Head of Engineering Tech
L. Bryce Andersen
Phone: (508)999-8539   Fax: (508)999-8485

Department Chair(s)
Mechanical Eng Tech
Ronald DiPippo
Phone: (508)999-8541   Fax: (508)999-8485

Electrical Eng Tech
Robert C. Helgeland
Phone: (508)999-8387   Fax: (508)999-8485

Admissions Office Contact
Raymond Barrows
Phone: (508)999-8782   Fax: (508)999-8901

Financial Aid Office Contact
Gerald S. Coutinho, *Director*
Phone: (508)999-8632   Fax: (508)999-8901

315   U of Mass, Lowell

Head of the Institution
William Hogan
Phone: (508)934-2201

Head of Engineering Tech
Aldo Crugnola
Phone: (508)934-2575   Fax: (508)452-1445

Associate/Assistant Deans
Louis J. Petrovic, *Assistant Dean*
Phone: (508)934-2577   Fax: (508)452-1445

Department Chair(s)
Civil Eng Tech
Donald Pottle
Phone: (508)934-2579   Fax: (508)452-1445

Electrical/Electronic(s) Eng Tech
Donald Pottle
Phone: (508)934-2579   Fax: (508)452-1445

Mechanical Eng Tech
Donald Pottle
Phone: (508)934-2579   Fax: (508)452-1445

Admissions Office Contact
Sharon L. Quigley
Phone: (508)934-2570   Fax: (508)452-1445

Financial Aid Office Contact
Walter Costello, *Director of Financial Aid*
Phone: (508)934-4233

316   Memphis State U

Head of the Institution
V. Lane Rawlins
Phone: (901)678-2234   Fax: (901)678-2000

Head of Engineering Tech
John D. Ray
Phone: (901)678-2171   Fax: (901)678-4180

Associate/Assistant Deans
Frank J. Claydon, *Associate Dean*
Phone: (901)678-2171   Fax: (901)678-4180
Weston T. Brooks, *Engineering Tech Department Chair*
Phone: (901)678-2225   Fax: (901)678-4180

Department Chair(s)
Electrical/Electronic(s) Eng Tech
Susan Simons
Phone: (901)678-2225   Fax: (901)678-4180

Manufacturing Eng Tech
Ron Day
Phone: (901)678-2225   Fax: (901)678-4180
Computer Eng Tech
Susan Simons
Phone: (901)678-2225   Fax: (901)678-4180
Architecture Tech
Sherry Bryan-Hagge
Phone: (901)678-2225   Fax: (901)668-4180

Admissions Office Contact
Carol Ferguson
Phone: (901)678-2171   Fax: (901)678-4180

317   Metropolitan State-Denver

Head of the Institution
Thomas B. Brewer
Phone: (303)556-3022   Fax: (303)556-3912

Head of Engineering Tech
Bill T. Rader
Phone: (303)556-2978   Fax: (303)556-2159

Associate/Assistant Deans
Mary A. Miller, *Associate Dean*
Phone: (303)556-2978   Fax: (303)556-2159

Department Chair(s)
Civil Eng Tech
Hugh H. Brown
Phone: (303)556-3227   Fax: (303)556-2159
Electronics & Mechanical Eng Tech
Larry G. Keating
Phone: (303)556-2503   Fax: (303)556-2159
EMail: Keating@ZENO.MSCD.EDU
Electronics & Mechanical Eng Tech
Larry G. Keating
Phone: (303)556-2503   Fax: (303)556-2159
EMail: Keating@ZENO.MSCD.EDU

Admissions Office Contact
Kenneth Curtis
Phone: (303)556-3018   Fax: (303)556-3999

Financial Aid Office Contact
Cheryl Judson, *Director of Financial Aid*
Phone: (303)556-3043

318   Milwaukee Sch of Eng

Head of the Institution
Hermann Viets
Phone: (414)277-7100   Fax: (414)277-7468

Head of Engineering Tech
Thomas W. Davis
Phone: (414)277-7324   Fax: (414)277-7477
EMail: davis@kirk.msoe.edu

Department Chair(s)
Electrical Eng Tech
Ray W. Palmer
Phone: (414)277-7325   Fax: (414)277-7465
EMail: palmer@kirk.msoe.edu
Manufacturing Eng Tech
John H. Farrow
Phone: (414)277-7287   Fax: (414)277-7470
Mechanical Eng Tech
John H. Farrow
Phone: (414)277-7287   Fax: (414)277-7470

Admissions Office Contact
T.O. Smith
Phone: (800)332-6763   Fax: (414)277-7186

Financial Aid Office Contact
Susan Hebert, *Director of Financial Aid*
Phone: (414)277-7224   Fax: (414)277-7450

319   Montana State U

Head of the Institution
Michael P. Malone
Phone: (406)994-2341   Fax: (406)994-1893

Head of Engineering Tech
David F. Gibson
Phone: (406)994-2272   Fax: (406)994-6098
EMail: ADEDG@MTSUNIX1.BITNET

Associate/Assistant Deans
Joseph J. Fedock, *Associate Dean*
Phone: (406)994-2272   Fax: (406)994-6098

Department Chair(s)
Civil & Agricultural Eng
Theodore E. Lang
Phone: (406)994-2111   Fax: (406)994-6105
Electrical Eng
Victor Gerez
Phone: (406)994-2505   Fax: (406)994-6098
Mechanical Eng
Byron Bennett
Phone: (406)994-2203   Fax: (406)994-6292

Admissions Office Contact
Rhonda Duffus
Phone: (406)994-2452   Fax: (406)994-1923

Financial Aid Office Contact
James R. Craig, *Director, Financial Aid Services*
Phone: (406)994-2845   Fax: (406)994-5488

320   Murray State U

Head of the Institution
Ronald J. Kurth
Phone: (502)762-3763   Fax: (502)762-3413

Head of Engineering Tech
Thomas B. Auer
Phone: (502)762-3391   Fax: (502)762-3631

Associate/Assistant Deans
James T. Vaughan, *Assistant Dean: Coll of Industry & Tech*
Phone: (502)762-3391   Fax: (502)762-3631

Department Chair(s)
Civil Eng Tech (General)
Paul R. McNeary
Phone: (502)762-3393   Fax: (502)762-3653
Electrical Eng Tech (Power/Communications)
Paul R. McNeary
Phone: (502)762-3393   Fax: (502)762-3653
Manufacturing Eng Tech
Paul R. McNeary
Phone: (502)762-3393   Fax: (502)762-3653
Mechanical Eng Tech
Paul R. McNeary
Phone: (502)762-3393   Fax: (502)762-3653
Electrical Eng Tech (Computer)
Paul R. McNeary
Phone: (502)762-3393   Fax: (502)762-3653
Civil Eng Tech (Construction)
Paul R. McNeary
Phone: (502)762-3393   Fax: (502)762-3653
Environmental Eng Tech
Paul R. McNeary
Phone: (502)762-3393   Fax: (502)762-3653
Mining Management Tech
Paul R. McNeary
Phone: (502)762-3393   Fax: (502)762-3653

Admissions Office Contact
Philip Bryan
Phone: (502)762-3741

Financial Aid Office Contact
Johnny McDougal, *Director*
Phone: (502)762-2546

321   U of New Hampshire

Head of the Institution
Dale F. Nitzschke
Phone: (603)862-2450   Fax: (603)862-3060

Head of Engineering Tech
Otis J. Sproul
Phone: (603)862-1781   Fax: (603)862-2486

Associate/Assistant Deans
Donald W. Melvin, *Associate Dean*
Phone: (603)862-3101   Fax: (603)862-2486

Department Chair(s)
Electrical Eng Tech
David A. Forest
Phone: (603)862-1827   Fax: (603)862-1856
Mechanical Eng Tech
David A. Forest
Phone: (603)862-1827   Fax: (603)862-1856

Admissions Office Contact
David W. Kraus
Phone: (603)862-1360

Financial Aid Office Contact
Richard Craig, *Director*
Phone: (603)862-3600

322   New Jersey Inst of Tech

Head of the Institution
Saul K. Fenster
Phone: (201)596-3101   Fax: (201)624-2541

Head of Engineering Tech
George Pincus
Phone: (201)596-3213   Fax: (201)596-2316
EMail: Pincus@Admin1.NJIT.Edu

Associate/Assistant Deans
Richard Parker, *Associate Dean of Engineering*
Phone: (201)596-3223   Fax: (201)596-2316

Department Chair(s)
Eng Tech
Robert English
Phone: (201)596-3224   Fax: (201)596-2316
EMail: English@Admin1.NJIT.Edu
Eng Tech
Robert English
Phone: (201)596-3224   Fax: (201)596-2316
EMail: English@Admin1.NJIT.Edu
Eng Tech
Robert English
Phone: (201)596-3224   Fax: (201)596-2316
EMail: English@Admin1.NJIT.Edu
Eng Tech
Robert English
Phone: (201)596-3224   Fax: (201)596-2316
EMail: English@Admin1.NJIT.Edu

Admissions Office Contact
William Anderson
Phone: (201)596-3300   Fax: (201)802-1854

Financial Aid Office Contact
Mary Hurdle, *Director*
Phone: (201)596-3479

323   New Mexico State U

Head of the Institution
James E. Halligan
Phone: (505)646-2035   Fax: (505)646-6334

Head of Engineering Tech
J. Derald Morgan
Phone: (505)646-2911   Fax: (505)646-3549

Associate/Assistant Deans
George D. Alexander, *Department Head, Engineering Tech*
Phone: (505)646-2236   Fax: (505)646-6107
J. Eldon Steelman, *Associate Dean of Engineering*
Phone: (505)646-2912   Fax: (505)646-3549
Joe L. Creed, *Assistant Dean*
Phone: (505)646-2913   Fax: (505)646-3549
Rose Marie Melon, *Assistant to the Dean*
Phone: (505)646-3547   Fax: (505)646-3549

Department Chair(s)
Eng Tech
H. James Skidmore
Phone: (505)646-2236   Fax: (505)646-6107
Eng Tech
Al Romero
Phone: (505)646-2236   Fax: (505)646-6107
Eng Tech
William S. Fleming
Phone: (505)646-2236   Fax: (505)646-6107

Admissions Office Contact
Bill Bruner
Phone: (505)646-3121   Fax: (505)646-6330

Financial Aid Office Contact
Greeley Myers, *Director of Financial Aid*
Phone: (505)646-4105   Fax: (505)646-6330

324   UNC-Charlotte

Head of the Institution
James H. Woodward
Phone: (704)547-2201   Fax: (704)547-2144

Head of Engineering Tech
Robert D. Snyder
Phone: (704)547-2301   Fax: (704)547-2352

Associate/Assistant Deans
Harry J. Leamy, *Associate Dean*
Phone: (704)547-4096   Fax: (704)547-3183
Silvia G. Middleton, *Assistant Dean*
Phone: (704)547-2301   Fax: (704)547-2352

Department Chair(s)
Civil Eng Tech
Edwin R. Braun
Phone: (704)547-2305   Fax: (704)547-2352
Electrical Eng Tech
Edwin R. Braun
Phone: (704)547-2305   Fax: (704)547-2352
Manufacturing Eng Tech
Edwin R. Braun
Phone: (704)547-2305   Fax: (704)547-2352
Mechanical Eng Tech
Edwin R. Braun
Phone: (704)547-2305   Fax: (704)547-2352

Admissions Office Contact
Kathi M. Baucom
Phone: (704)547-2213   Fax: (704)547-2144

Financial Aid Office Contact
Curtis R. Whalen, *Director*
Phone: (704)547-2461   Fax: (704)547-2144

325   Northeastern U

Head of the Institution
John A. Curry
Phone: (617)437-2101   Fax: (617)437-5015

Head of Engineering Tech
Thomas E. Hulbert
Phone: (617)437-2500   Fax: (617)437-2501

Department Chair(s)
Electrical/Electronic(s) Eng Tech
Ronald Scott
Phone: (617)437-3807   Fax: (617)437-2501
Mechanical Eng Tech
George Kent
Phone: (617)437-5240   Fax: (617)437-2501
Computer Tech
Nonna Lehmkuhl
Phone: (617)437-3500   Fax: (617)437-2501

Admissions Office Contact
Michael J. Clifford
Phone: (617)437-2200   Fax: (617)437-8780

Financial Aid Office Contact
Jean C. Eddy, *Dean of Financial Aid*
Phone: (617)437-3190   Fax: (617)437-8623

326   Oklahoma State U

Head of the Institution
John R. Campbell
Phone: (405)744-6384   Fax: (405)744-6285

Head of Engineering Tech
Karl N. Reid
Phone: (405)744-5140   Fax: (405)744-7545

Associate/Assistant Deans
David R. Thompson, *Associate Dean for Instruction & Extension*
Phone: (405)744-5140   Fax: (405)744-7545

James E. Bose, *Director, Division of Engineering Tech*
Phone: (405)744-5638   Fax: (405)744-6187

Department Chair(s)
Electronics/Computer Tech
Thomas G. Bertenshaw
Phone: (405)744-5716   Fax: (405)744-5283
Construction Management Tech
Charles A. Rich
Phone: (405)744-5712   Fax: (405)744-5283
Mechanical Power Tech
Marvin D. Smith
Phone: (405)744-5711   Fax: (405)744-5283
Fire Protection & Safety Tech
Marvin D. Smith
Phone: (405)744-8771   Fax: (405)744-5283
Mechanical Design & Manufacturing Tech
Gerald McClain
Phone: (405)744-5710   Fax: (405)744-6187

Admissions Office Contact
Gordon L. Reese
Phone: (405)744-6866

Financial Aid Office Contact
Charles W. Bruce, *Director, Financial Aid*
Phone: (405)744-7440   Fax: (405)744-6438

327   Old Dominion U

Head of the Institution
James V. Koch
Phone: (804)683-3159   Fax: (804)683-5679

Head of Engineering Tech
Ernest J. Cross, Jr.
Phone: (804)683-3787   Fax: (804)683-4898

Associate/Assistant Deans
William A. Drewry, *Associate Dean for Undergraduate Programs & Administration*
Phone: (804)683-3789   Fax: (804)683-4898
Gary R. Crossman, *Associate Dean of Community, Industrial, & Alumni Relations*
Phone: (804)683-3768   Fax: (804)683-4898

Department Chair(s)
Eng Tech
William D. Stanley
Phone: (804)683-3775   Fax: (804)683-5655

Admissions Office Contact
Angela N. Boyd
Phone: (804)683-3637   Fax: (804)683-3647

Financial Aid Office Contact
Helga A. Greenfield, *Director*
Phone: (804)683-3683   Fax: (804)683-5357

328   Oregon Inst of Tech

Head of the Institution
Lawrence J. Wolf
Phone: (503)885-1101   Fax: (503)885-1515

Head of Engineering Tech
James McAtee
Phone: (503)885-1240
EMail: mcateej@oit.osshe.edu

Department Chair(s)
Civil Eng Tech
Richard Zbinden
Phone: (503)885-1510
Electrical/Electronic(s) Eng Tech
Mel Turner
Phone: (503)885-1550
Manufacturing Eng Tech
Don Skudstad
Phone: (503)885-1540
Mechanical Eng Tech
James Fenner
Phone: (503)885-1993
Survey
Jack Walker
Phone: (503)885-1511

Laser Optical Eng Tech
Nancy Kincheloe
Phone: (503)885-1570
Computer Systems Eng Tech
Don Phillips
Phone: (503)885-1595

Admissions Office Contact
Russ Lyon
Phone: (503)885-1150

Financial Aid Office Contact
John Huntley, *Director of Financial Aid*
Phone: (503)885-1280

329   Penn State-Erie

Head of the Institution
John M. Lilley
Phone: (814)898-6160   Fax: (814)898-6461

Head of Engineering Tech
Richard C. Progelhof
Phone: (814)898-6153   Fax: (814)898-6213
EMail: rcp5@psuvm.psu.edu

Associate/Assistant Deans
John N. Grode, *Assistant Director*
Phone: (814)898-6153   Fax: (814)898-6213

Department Chair(s)
Electrical Eng Tech
Ronald P. Krahe
Phone: (814)898-6346   Fax: (814)898-6213
EMail: rpk3@psuvm.psu.edu
Mechanical Eng Tech
Kenneth J. Fisher
Phone: (814)898-6387   Fax: (814)898-6213
Plastics Eng Tech
Paul E. Koch
Phone: (814)898-6345   Fax: (814)898-6006
EMail: pek3@psuvm.psu.edu

Admissions Office Contact
Mary-Ellen Madigan
Phone: (814)898-6100   Fax: (814)898-6461

Financial Aid Office Contact
Kate Delfino, *Assistant Director of Admissions & Financial Aid*
Phone: (814)898-6162   Fax: (814)898-6461

330   Penn State U - Harrisburg

Head of the Institution
Ruth Leventhal
Phone: (717)948-6101

Head of Engineering Tech
Gautam Ray
Phone: (717)948-6108   Fax: (717)948-6401
EMail: GXR6@PSUADMIN

Department Chair(s)
Electrical & Computer Eng Tech
Jerry Shoup
Phone: (717)948-6114
Mechanical Eng Tech
William Aungst
Phone: (717)948-6118
Structural Design & Construction Eng Tech
Joseph Cecere
Phone: (717)948-6135
Environmental Eng Tech
Charles Cole
Phone: (717)948-6133

Admissions Office Contact
Ed Escalet
Phone: (717)948-6250

331   Pittsburg State U

Head of the Institution
Donald W. Wilson
Phone: (316)235-4103   Fax: (316)232-7515

Head of Engineering Tech
Vern C. Goold
Phone: (316)235-4363   Fax: (316)231-4231

Associate/Assistant Deans
James L. Otter, *Associate Chairperson*
Phone: (316)235-4350   Fax: (316)231-4231

Department Chair(s)
Electrical/Electronic(s) Eng Tech
Steve M. Hefley
Phone: (316)235-4357   Fax: (316)231-4231
Manufacturing Eng Tech
William L. Williamson
Phone: (316)235-4355   Fax: (316)231-4231
Mechanical Eng Tech
Timothy E. Thomas
Phone: (316)235-4353   Fax: (316)231-4231
Construction Eng Tech
James L. Otter
Phone: (316)235-4358   Fax: (316)231-4231
Plastics Eng Tech
George W. Graham
Phone: (316)231-4356   Fax: (316)231-4231
Construction Management Tech
James L. Otter
Phone: (316)235-4358   Fax: (316)231-4231

Admissions Office Contact
James Taylor
Phone: (316)231-7000   Fax: (316)232-7515

Financial Aid Office Contact
Ronald G. Hopkins, *Director*
Phone: (316)231-7000   Fax: (316)232-7515

332   U of Pittsburgh at Johnstown

Head of the Institution
Frank H. Blackington, III
Phone: (814)269-2090   Fax: (814)269-2096

Head of Engineering Tech
Paul H. Saylor
Phone: (814)269-7250   Fax: (814)269-7245

Associate/Assistant Deans
James L. Hales, *Associate Director*
Phone: (814)269-7250   Fax: (814)269-7245

Department Chair(s)
Civil Eng Tech
Harry R. Feller
Phone: (814)269-7247   Fax: (814)269-7245
Electrical Eng Tech
Richard M. Bender
Phone: (814)269-7246   Fax: (814)269-7245
Mechanical Eng Tech
John G. Klavuhn
Phone: (814)269-7248   Fax: (814)269-7245

Admissions Office Contact
Thomas J. Wonders
Phone: (814)269-7050   Fax: (814)269-2096

Financial Aid Office Contact
Julie A. Salem, *Associate Director*
Phone: (814)269-7045   Fax: (814)269-2069

333   Point Park Coll

Head of the Institution
J. Matthew Simon
Phone: (412)392-3990   Fax: (412)391-1980

Department Chair(s)
Natural Sciences & Eng Tech
Mark O. Farrell
Phone: (412)392-3879   Fax: (412)391-1980
Civil Eng Tech
Daniel J. Reed
Phone: (412)392-3868   Fax: (412)391-1980
Electrical/Electronic(s) Eng Tech
Khalil Thanaa
Phone: (412)392-3898   Fax: (412)392-1980
Mechanical Eng Tech
Aloysius E. Dapprich
Phone: (412)392-3875   Fax: (412)392-1980

Admissions Office Contact
Terrance Kizina
Phone: (412)391-4100   Fax: (412)391-1980

Financial Aid Office Contact
Deidre Smith, *Director of Financial Aid*
Phone: (412)391-4100   Fax: (412)391-1980

334   Prairie View A&M U

Head of the Institution
Julius W. Becton, Jr.
Phone: (409)857-2111   Fax: (409)857-3928

Head of Engineering Tech
Haku Israni
Phone: (409)857-4122   Fax: (409)857-2097

Associate/Assistant Deans
Charles T. Edwards, Jr., *Associate Dean*
Phone: (409)857-4518   Fax: (409)857-2097

Department Chair(s)
Electrical Eng Tech
David A. Kirkpatrick
Phone: (409)857-4717   Fax: (409)857-2097
Computer Eng Tech
David A. Kirkpatrick
Phone: (409)857-4717   Fax: (409)857-2097
Mechanical Eng Tech
David A. Kirkpatrick
Phone: (409)857-4717   Fax: (409)857-2097

Admissions Office Contact
Mary E. Gooch
Phone: (409)857-2626   Fax: (409)857-2425

Financial Aid Office Contact
A.D. James, *Director of Financial Aid*
Phone: (409)857-2424   Fax: (409)857-2425

335   Purdue U

Head of the Institution
Steven C. Beering
Phone: (317)494-9708   Fax: (317)494-7875

Head of Engineering Tech
Don K. Gentry
Phone: (317)494-2552   Fax: (317)494-0486

Associate/Assistant Deans
Fred W. Emshousen, *Associate Dean*
Phone: (317)494-2554   Fax: (317)494-0486

Department Chair(s)
Electrical Eng Tech
Larry D. Hoffman
Phone: (317)494-7483   Fax: (317)494-0486
Mechanical Eng Tech
William K. Dalton
Phone: (317)494-7514   Fax: (317)494-0486
Computer Integrated Manufacturing Tech
William K. Dalton
Phone: (317)494-7514   Fax: (317)494-0486

Financial Aid Office Contact
Joyce Hall, *Director*
Phone: (317)494-5050   Fax: (317)494-6707

336   Rochester Inst of Tech

Head of the Institution
Albert J. Simone
Phone: (716)475-2394

Head of Engineering Tech
W. David Baker
Phone: (716)475-2915   Fax: (716)475-5275
EMail: wdbite@ritvax.isc.rit.edu

Associate/Assistant Deans
James Scudder, *Associate Director*
Phone: (716)475-5190   Fax: (716)475-5275

Department Chair(s)
Civil Eng Tech
Robert Easton
Phone: (716)475-2183   Fax: (716)475-5275
Electrical Eng Tech
Thomas Young
Phone: (716)475-2179   Fax: (716)475-5275

Manufacturing Eng Tech
V. Raju
Phone: (716)475-2270   Fax: (716)475-5275
Mechanical Eng Tech
Robert Merrill
Phone: (716)475-6174   Fax: (716)475-5275
Admissions Office Contact
Daniel Shelley
Phone: (716)475-6631
Financial Aid Office Contact
Verna J. Hazen, *Director*
Phone: (716)475-2186

337   St. Cloud State U
Head of the Institution
Robert O. Bess
Phone: (612)255-2122   Fax: (612)654-5139
Head of Engineering Tech
Richard G. Hogan
Phone: (612)255-2192   Fax: (612)255-4262
Associate/Assistant Deans
Dale A. Williams, *Assistant Dean*
Phone: (612)255-2193   Fax: (612)255-4262
Department Chair(s)
Manufacturing Eng Tech
Juan J. Diaz
Phone: (612)255-2107   Fax: (612)255-4262
EMail: Diaz@Tigger; Diaz@Tigger.StCloud.MSUS.EDU
Admissions Office Contact
Sherwood Reid
Phone: (612)255-2243   Fax: (612)654-5367
Financial Aid Office Contact
Frank Loncorich, *Director of Financial Aids*
Phone: (612)255-2047   Fax: (612)654-5367

338   South Carolina State U
Head of the Institution
Barbara R. Hatton
Phone: (803)536-7013   Fax: (803)533-3622
Head of Engineering Tech
Shoi Y. Hwang
Phone: (803)536-8479   Fax: (803)533-3623
Department Chair(s)
Civil & Mechanical Eng Tech
Stanley N. Ihekweazu
Phone: (803)536-8392   Fax: (803)533-3623
Industrial & Electrical Eng Tech
Ramachandra R. Sandrapaty
Phone: (803)536-8476   Fax: (803)533-3623
Civil & Mechanical Eng Tech
Stanley N. Ihekweazu
Phone: (803)536-7117   Fax: (803)533-3623
Industrial & Electrical Eng Tech
Ramachandra R. Sandrapaty
Phone: (803)536-8476   Fax: (803)533-3623
Admissions Office Contact
Benny R. Mayfield
Phone: (803)536-8411   Fax: (803)536-8420
Financial Aid Office Contact
Margaret C. Black, *Director of Financial Aid*
Phone: (803)536-7067   Fax: (803)536-8420

339   Southern Coll of Tech
Head of the Institution
Stephen R. Cheshier
Phone: (404)528-7230   Fax: (404)528-7483
Head of Engineering Tech
William D. Rezak
Phone: (404)528-7234   Fax: (404)528-7454
Department Chair(s)
Civil Eng Tech
Boyce D. Tate
Phone: (404)528-7261

Electrical/Electronic(s) Eng Tech
Julian A. Wilson
Phone: (404)528-7246
Mechanical Eng Tech
Britt K. Pearce
Phone: (404)528-7274
Apparel & Textile Eng Tech
Lawrence T. Haddock
Phone: (404)528-7272
Industrial Eng Tech
Thomas H. Carmichael
Phone: (404)528-7243
Admissions Office Contact
Virginia A. Head
Phone: (404)528-7281   Fax: (404)528-7292
Financial Aid Office Contact
Emerelle McNair, *Director of Financial Aid*
Phone: (404)528-7290

340   U of Southern Colorado
Head of the Institution
Robert C. Shirley
Phone: (719)549-2306   Fax: (719)549-2219
Head of Engineering Tech
Ray L. Sisson
Phone: (719)549-2696   Fax: (719)549-2519
EMail: SISSON@STARBURST.USCOLO.EDU
Associate/Assistant Deans
Perry R. McNeill, *Chair Engineering Tech*
Phone: (719)549-2703   Fax: (719)549-2519
Department Chair(s)
Civil Eng Tech
Larry O. Womack
Phone: (719)549-2877   Fax: (719)549-2519
EMail: PERKINS@STARBURST.USCOLO.EDU
Electronics Eng Tech
Peter C. Burton
Phone: (719)549-2877   Fax: (719)549-2519
EMail: PERKINS@STARBURST.USCOLO.EDU
Mechanical Eng Tech
Perry R. McNeill
Phone: (719)549-2877   Fax: (719)549-2519
EMail: PERKINS@STARBURST.USCOLO.EDU
Admissions Office Contact
Margie Wade
Phone: (719)549-2461   Fax: (719)549-2219
Financial Aid Office Contact
Gina T. Mestas, *Director of Financial Aid*
Phone: (719)549-2753   Fax: (719)549-2938

341   Southern Illinois-Carbondale
Head of the Institution
John C. Guyon
Phone: (618)453-2341   Fax: (618)453-5362
Head of Engineering Tech
Juh W. Chen
Phone: (618)453-4321   Fax: (618)453-4235
Associate/Assistant Deans
James L. Evers, *Associate Dean*
Phone: (618)453-4321   Fax: (618)453-4235
Echol E. Cook, *Associate Dean*
Phone: (618)453-4321   Fax: (618)453-4235
Department Chair(s)
Tech
Gary J. Butson
Phone: (618)536-3396   Fax: (618)453-7455
Admissions Office Contact
Jerre C. Pfaff
Phone: (618)453-2982   Fax: (618)453-3250
Financial Aid Office Contact
Pamela A. Britton, *Director*
Phone: (618)453-4334

342   U of Southern Mississippi
Head of the Institution
Aubrey K. Lucas
Phone: (601)266-5001   Fax: (601)266-5756
Head of Engineering Tech
Ruth A. Cade, *Director*
Phone: (601)266-4896   Fax: (610)266-5829
Associate/Assistant Deans
Shelton L. Houston, *Assistant Director*
Phone: (601)266-4896   Fax: (601)266-5829
Department Chair(s)
Construction Eng Tech
George L. Mathis
Phone: (601)266-4894   Fax: (601)266-5829
Electronics Eng Tech
Cecil A. Harrison
Phone: (601)266-5628   Fax: (601)266-5829
Mechanical Eng Tech
Shri K. Vajpayee
Phone: (601)266-4727   Fax: (601)266-5829
Master of Science in Eng Tech
Ruth A. Cade
Phone: (601)266-4896   Fax: (601)266-5829
Computer Eng Tech
Cecil A. Harrison
Phone: (601)266-5628   Fax: (601)266-5829
Architectural Eng Tech
George L. Mathis
Phone: (601)266-4894   Fax: (601)266-5829
Industrial Eng Tech
Shri K. Vajpayee
Phone: (601)266-4727   Fax: (601)266-5829
Pre-Eng
Cecil A. Harrison
Phone: (601)266-5628   Fax: (601)266-5829
Admissions Office Contact
Richard W. Pyle
Phone: (601)266-5555   Fax: (601)266-5816
Financial Aid Office Contact
Vernetta P. Fairley, *Director, Financial Aid*
Phone: (601)266-4774

343   SUNY-Farmingdale
Head of the Institution
Frank A. Cipriani
Phone: (516)420-2145   Fax: (516)420-2753
Head of Engineering Tech
Arthur A. Ezra
Phone: (516)420-2115   Fax: (516)420-2194
EMail: EZRAAA @ SNYFARVA
Associate/Assistant Deans
Terry L. Smith, *Assoc. Dean*
Phone: (516)420-2310   Fax: (516)420-2194
Louis A. Scala, *Asst. Dean*
Phone: (516)420-2039   Fax: (516)420-2194
Department Chair(s)
Civil Eng Tech
Amit Bandyopadhyay
Phone: (516)420-2378   Fax: (516)420-2194
EMail: BANDYOA @ SNYFARVA
Electrical/Electronic(s) Eng Tech
Socrates Thanasas
Phone: (516)420-2084   Fax: (516)420-2194
Manufacturing Eng Tech
John Tiedemann
Phone: (516)420-2326   Fax: (516)420-2194
EMail: TIEDEMJE @ SNYFARVA
Mechanical Eng Tech
John Tiedemann
Phone: (516)420-2326   Fax: (516)420-2194
EMail: TIEDEMJE @ SNYFARVA
Automotive Eng Tech
Robert J. Lagnese
Phone: (516)420-2117   Fax: (516)420-2194
Biomedical Eng Tech
Bruce J. Morgan
Phone: (516)420-2140   Fax: (516)420-2194

Construction/Architectural Eng Tech
Amit Bandyopadhyay
Phone: (516)420-2378   Fax: (516)420-2194
EMail: BANDYOA @ SNYFARVA
Aerospace Tech
Paul Baumann
Phone: (516)420-2314   Fax: (516)420-2194
Computing Graphics
Mahendra Shah
Phone: (516)420-2311   Fax: (516)420-2194
Eng Science
Edward Garcia
Phone: (516)420-2187   Fax: (516)420-2197
Admissions Office Contact
Janet Snyder
Phone: (516)420-2457   Fax: (516)420-2633
Financial Aid Office Contact
Nancy N. Dunnagan, *Financial Aid Director*
Phone: (516)420-2328   Fax: (516)420-2357

344   SUNY-Utica/Rome
Head of the Institution
Peter J. Cayan
Phone: (315)792-7400   Fax: (315)792-7222
Head of Engineering Tech
Ronald Sarner
Phone: (315)792-7234   Fax: (315)792-7800
EMail: ron@sunyit.edu
Department Chair(s)
Electrical/Electronic(s) Eng Tech
Windsor S. Thomas
Phone: (315)792-7357   Fax: (315)792-7222
Mechanical Eng Tech
Digendra K. Das
Phone: (315)792-7421   Fax: (315)792-7800
Industrial Eng Tech
Atlas Hsie
Phone: (315)792-7122   Fax: (315)792-7800
Computer Tech
Windsor S. Thomas
Phone: (315)792-7357   Fax: (315)792-7222
Admissions Office Contact
Eileen M. Collins
Phone: (315)792-7208
Financial Aid Office Contact
Edward A. Hutchinson, *Director of Financial Aid*
Phone: (315)792-7210   Fax: (315)792-7802

345   Temple U
Head of the Institution
Peter J. Liacouras
Phone: (215)787-7405
Head of Engineering Tech
Charles K. Alexander
Phone: (215)787-7959   Fax: (215)787-6936
Associate/Assistant Deans
Thomas J. Ward, *Associate Dean for Administration*
Phone: (215)787-1050   Fax: (215)787-6936
Henry M. Sendaula, *Associate Dean for Research*
Phone: (215)787-7819   Fax: (215)787-6936
Department Chair(s)
Civil/Construction Eng Tech
Frederick Schmitt
Phone: (215)787-7831   Fax: (215)787-6936
Electrical Eng Tech
Brian P. Butz
Phone: (215)787-7212   Fax: (215)787-6936
Mechanical Eng Tech
Steven Ridenour
Phone: (215)787-8825   Fax: (215)787-6936
Environmental Eng Tech
Frederick Schmitt
Phone: (215)787-7831   Fax: (215)787-6936
General Eng Tech
Steven Ridenour
Phone: (215)787-8825   Fax: (215)787-6936

Engineering & Engineering Technology Personnel

Financial Aid Office Contact
John Morris, *Director*
Phone: (215)787-1492

346 U of Tennessee, Martin

Head of the Institution
Margaret N. Perry
Phone: (901)587-7500   Fax: (901)587-7019

Head of Engineering Tech
Randy L. Perry
Phone: (901)587-7380   Fax: (901)587-7375
EMail: RLPERRY@MARTN.Bitnet

Admissions Office Contact
Randy L. Perry
Phone: (901)587-7380   Fax: (901)587-7375

Financial Aid Office Contact
Randall D. Hall, *Director of Financial Aid*
Phone: (901)587-7040   Fax: (901)587-7019

347 Texas A&M U

Head of the Institution
William H. Mobley
Phone: (409)845-2217   Fax: (409)845-5027

Head of Engineering Tech
Kenneth L. Peddicord
Phone: (409)845-7203   Fax: (409)845-8986
EMail: KLP7201@SUMMA.TAMU.EDU

Associate/Assistant Deans
Carl A. Erdman, *Executive Associate Dean*
Phone: (409)845-5220   Fax: (409)845-8986
William D. Turner, *Associate Dean, Undergraduate*
Phone: (409)845-8699   Fax: (409)847-8654
Kenneth R. Hall, *Associate Dean*
Phone: (409)845-1322   Fax: (409)845-8986
G.K. Bennett, *Associate Dean*
Phone: (409)845-1722   Fax: (409)845-8986
Karan L. Watson, *Assistant Dean*
Phone: (409)845-6920   Fax: (409)847-8654

Department Chair(s)
Electronics Eng Tech
Joseph A. Morgan
Phone: (409)845-5966   Fax: (409)847-9396
Manufacturing Eng Tech
John E. Mayer
Phone: (409)845-4957   Fax: (409)847-9396
Mechanical Eng Tech
Swaminadham Midturi
Phone: (409)845-3276   Fax: (409)847-9396
Eng Tech
John A. Weese
Phone: (409)845-4951   Fax: (409)847-9396
EMail: JAW2833@SUMMA.TAMU.EDU

Admissions Office Contact
Billy G. Lay
Phone: (409)845-1157   Fax: (409)845-0727

Financial Aid Office Contact
Donald L. Engelage, *Director of Student Financial Aid*
Phone: (409)845-8874   Fax: (409)847-9061

348 Texas Tech U

Head of the Institution
Robert W. Lawless
Phone: (806)742-2121   Fax: (806)742-2138

Head of Engineering Tech
Mason H. Somerville
Phone: (806)742-3451   Fax: (806)742-3493

Associate/Assistant Deans
John Borrelli, *Associate Dean for Academic Affairs*
Phone: (806)742-3454   Fax: (806)742-3493

Department Chair(s)
Eng Tech
Ronald Pigott
Phone: (806)742-3538   Fax: (806)742-1900

Financial Aid Office Contact
Ronny Barnes
Phone: (806)742-3681

349 Tri-State U

Head of the Institution
William G. Meyers
Phone: (219)665-4187   Fax: (219)665-4292

Head of Engineering Tech
William G. Meyers
Phone: (219)665-4187   Fax: (219)665-4292

Associate/Assistant Deans
Thomas J. Enneking, *Assistant Dean*
Phone: (219)665-4216   Fax: (219)665-4292

Department Chair(s)
Eng Tech
Edward J. Nagle
Phone: (219)665-4262   Fax: (219)665-4292

Admissions Office Contact
Walter Lilley
Phone: (219)665-4132   Fax: (219)665-4292

Financial Aid Office Contact
Susan Stroh, *Financial Aid Director*
Phone: (219)665-4174   Fax: (219)665-4292

350 Virginia State U

Head of the Institution
Nathanael Pollard, Jr.
Phone: (804)524-5070   Fax: (804)524-6506

Head of Engineering Tech
Arthur L. Allen
Phone: (804)524-5631   Fax: (804)524-5950

Department Chair(s)
Electronic(s) Eng Tech
Cevat Kardan
Phone: (804)524-5408   Fax: (804)524-5950
Mechanical Eng Tech
Cevat Kardan
Phone: (804)524-5408   Fax: (804)524-5950

Admissions Office Contact
Karen R. Winston
Phone: (804)524-5902   Fax: (804)524-5055

Financial Aid Office Contact
Henry Debose, *Director*
Phone: (804)524-5990   Fax: (804)524-6818

351 Weber State U

Head of the Institution
Paul H. Thompson
Phone: (801)626-6001   Fax: (801)626-7922

Head of Engineering Tech
Warren R. Hill
Phone: (801)626-6303   Fax: (801)626-7531

Department Chair(s)
Electronics Eng Tech
Wayne E. Andrews
Phone: (801)626-6898   Fax: (801)626-7578
Manufacturing Eng Tech
Robert E. Wallentine
Phone: (801)626-6305   Fax: (801)626-7531
Mechanical Eng Tech
Robert E. Wallentine
Phone: (801)626-6305   Fax: (801)626-7531

Admissions Office Contact
L. Winslow Hurst
Phone: (801)626-6050   Fax: (801)626-6747

Financial Aid Office Contact
Richard Effiong, *Director*
Phone: (801)626-7131   Fax: (801)626-7408

352 Wentworth Inst of Tech

Head of the Institution
John F. Van Domelen
Phone: (617)442-9010   Fax: (617)427-2852

Head of Engineering Tech
Alexander W. Avtgis
Phone: (617)442-9010   Fax: (617)427-2852

Associate/Assistant Deans
Wilfred J. Caissie, *Assistant Dean*
Phone: (617)442-9010   Fax: (617)427-2852

Department Chair(s)
Civil & Environmental Tech
Michael Kupferman
Phone: (617)442-9010   Fax: (617)427-2852
Electrical/Electronic(s) Eng Tech
Jerome L. Krasner
Phone: (617)442-9010   Fax: (617)427-2852
Mechanical/Manufacturing Eng Tech
James M. Knowlton
Phone: (617)442-9010   Fax: (617)427-2852
Aeronautics
Wilfred J. Caissie
Phone: (617)442-9010   Fax: (617)427-2852
Architecture
Terrence G. Heinlein
Phone: (617)442-9010   Fax: (617)427-2852
Construction Sciences
Frederick E. Gould
Phone: (617)442-9010   Fax: (617)427-2852

Admissions Office Contact
Thomas J. McGinn
Phone: (617)442-9010   Fax: (617)427-2852

Financial Aid Office Contact
Carol A. Rubel, *Director of Financial Aid*
Phone: (617)442-9010   Fax: (617)427-2852

353 West Virginia Inst of Tech

Head of the Institution
John P. Carrier
Phone: (304)442-3146   Fax: (304)442-3057

Head of Engineering Tech
Frank A. Gourley, Jr.
Phone: (304)442-3098   Fax: (304)442-3245

Associate/Assistant Deans
Martha K. Shouldis, *Dean, Community & Technical Coll*
Phone: (304)442-3226   Fax: (304)442-3245

Department Chair(s)
Civil Eng Tech
James E. Cook
Phone: (304)442-3300   Fax: (304)442-3245
Electrical/Electronic(s) Eng Tech
James W. Piercy
Phone: (304)442-3348   Fax: (304)442-3245
Tech Education
Thearn H. Ellis
Phone: (304)442-3079   Fax: (304)442-3245
Mechanical Eng Tech
Reuben K. Ward
Phone: (304)442-3169   Fax: (304)442-3245
Drafting & Design/Industrial Eng Tech
James R. Blevins
Phone: (304)442-3339   Fax: (304)442-3245

Admissions Office Contact
Robert P. Scholl
Phone: (304)442-3151   Fax: (304)442-3057

Financial Aid Office Contact
Nina M. Morton, *Director of Financial Aid*
Phone: (304)442-3032   Fax: (304)442-3057

354 Western Carolina U

Head of the Institution
Myron L. Coulter
Phone: (704)227-7100   Fax: (704)227-7176

Head of Engineering Tech
J. Dale Pounds
Phone: (704)227-7272   Fax: (704)227-7705

Associate/Assistant Deans
Noelle Kehrberg, *Associate Dean*
Phone: (704)227-7272   Fax: (704)227-7705

Department Chair(s)
Electrical/Electronic(s) Eng Tech
George DeSain
Phone: (704)227-7272   Fax: (704)227-7705
Manufacturing Eng Tech
George DeSain
Phone: (704)227-7272   Fax: (704)227-7705
Industrial Tech
George DeSain
Phone: (704)227-7272   Fax: (704)227-7705

Admissions Office Contact
Drumont Bowman
Phone: (704)227-7317   Fax: (704)227-7202

Financial Aid Office Contact
Glenn Hardesty, *Director, Student Financial Aid*
Phone: (704)227-7290   Fax: (704)227-7202

355 Western Kentucky U

Head of the Institution
Thomas C. Meredith
Phone: (502)745-4492

Head of Engineering Tech
Charles E. Kupchella
Phone: (502)745-4448   Fax: (502)745-6471

Associate/Assistant Deans
Martin R. Houston, *Associate Dean Ogden Coll of Science, Tech, & Health*
Phone: (502)745-4448   Fax: (502)745-6471
Franklin D. Conley, *Assistant Dean Ogden Coll of Science, Tech, & Health*
Phone: (502)745-4448   Fax: (502)745-6471

Department Chair(s)
Eng Tech
John P. Russell
Phone: (502)745-6394   Fax: (502)745-6471

Admissions Office Contact
Cheryl C. Chambless
Phone: (502)745-5422

Financial Aid Office Contact
Marilyn Clark, *Director, Dept of Student Financial Assistance*
Phone: (502)745-2756

356 Western Michigan U

Head of the Institution
Diether H. Haenicke
Phone: (616)387-2351   Fax: (616)387-2355

Head of Engineering Tech
Leonard R. Lamberson
Phone: (616)387-4017   Fax: (616)387-4024
EMail: LAMBERSON@GW.WMICH.EDU

Associate/Assistant Deans
Molly W. Williams, *Associate Dean*
Phone: (616)387-4017   Fax: (616)387-4024

Department Chair(s)
Eng Tech
Pnina Ari-Gur
Phone: (616)387-6515   Fax: (616)387-4024
EMail: ARIGURP@GW.WMICH.EDU

Admissions Office Contact
Stanley E. Henderson
Phone: (616)387-2000   Fax: (616)387-2355

Financial Aid Office Contact
John A. Kundel, *Director of Student Financial Aid*
Phone: (616)387-6000   Fax: (616)387-6989

Index to Engineering & Technology Personnel

A

Name	Institution	Page
Abata, Duane L.	Michigan Tech U	133
Abbaschian, Reza	U of Florida	071
Abedian, Behrouz	Tufts U	243
Abel, Jacob M.	U of Penn	179
Abent, Rita	Wichita State U	271
Abernathy, A. Ray	Clemson U	047
Aburdene, Maurice F.	Bucknell U	018
Achbach, Myron H.	U of Dayton 060	294
Acosta, Allan J.	Cal Inst of Tech	026
Adamany, David W.	Wayne State U	265
Adams, Peter F.	Technical U of Nova Scotia	226
Adams, Ron G.	U of Akron	281
Addy, A.L.	Illinois, Urbana-Champaign	095
Adler, John H.	Michigan Tech U	133
Adorjan, Laszlo A.	Technical U of Nova Scotia	226
Advani, Sunder H.	Lehigh U	108
Agaskar, Vishnu L.	Cal State U, Sacramento	033
Agbabian, Mihran	U of Southern Cal	211
Aggarwal, Shiv	Embry-Riddle-Daytona Beach	067
Aggarwal, Sudhir	SUNY-Binghamton	218
Ahmad, Jameel	The Cooper Union	057
Ahmadi, Goodarz	Clarkson U	046
Ahmed, Abdul M.	McGill U	125
Ahmed, Nasir	The U of New Mexico	152
Ahmed, Samir	CUNY-City Coll	044
Aikens, Charles H.	U of Tennessee, Knoxville	229
Aimone-Martin, Catherine	New Mexico Inst of Mining	153
Akai, Terrence J.	U of Notre Dame	168
Akay, Adnan	Carnegie Mellon U	036
Akin, Karen T.	Louisiana Tech U 110	311
Albino, Judith	U of Colorado-Col Spr	050
Albino, Judith E.	U of Colorado at Boulder	049
Alcock, Ralph	South Dakota State U	208
Aldridge, Clark	Auburn U 012	285
Aldridge, M. Dayne	Auburn U 012	285
Alexander, Charles K.	Temple U 227	345
Alexander, George D.	New Mexico State U	323
Alexopoulos, Nicolaos G.	U of Cal, Los Angeles	022
Alford, Vernal G., III	North Carolina A&T	159
Alfred, Hampton	U of Wisconsin-Madison	274
Al-Khafaji, Amir W.	Bradley U	015
Alkire, R.C.	Illinois, Urbana-Champaign	095
Allaire, Paul E.	U of Virginia	259
Allen, Arthur L.	Virginia State U	350
Allen, Beth	Trinity U	242
Allen, David J.	U of Missouri-Rolla	141
Allen, David T.	U of Cal, Los Angeles	022
Allen, Gerald R.	U of Minnesota, Duluth	135
Allison, Jack	U of Puerto Rico	187
Altenkirch, Robert A.	Mississippi State U	138
Althaus, L. "Brother"	Christian Brothers U	041
Altiero, Nicholas	Michigan State U	132
Altstetter, Carl J.	Illinois, Urbana-Champaign	095
Amaral, Horace J., Jr.	U of Rhode Island	190
Amyotte, Paul A.	Technical U of Nova Scotia	226
Anagnos, Thailia	San Jose State U	202
Anand, Davinder K.	U of Maryland, Coll Park	120
Andersen, L. Bryce	U of Mass, Dartmouth 122	314
Anderson, Donald K.	Michigan State U	132
Anderson, Edward E.	Texas Tech U	239
Anderson, James A.	Southern U & A&M Coll	210
Anderson, John E.	Portland State U	184
Anderson, John L.	Carnegie Mellon U	036
Anderson, Lee	Oakland U	169
Anderson, Loren R.	Utah State U	254
Anderson, Max L.	U of Wisconsin-Platteville	276
Anderson, Melvin W.	U of South Florida	209
Anderson, Pam	Mercer U	127
Anderson, Sabra S.	U of Minnesota, Duluth	135
Anderson, Thomas F.	U of Conn	056
Anderson, Timothy J.	U of Florida	071
Anderson, Wayne K.	SUNY-Buffalo	219
Anderson, William	New Jersey Inst of Tech 151	322
Andino, Al	Cal State Poly, Pomona 034	289
Andrade, Ron	U of Cal, Santa Barbara	024
Andrews, Barry	U of Alabama at Birmingham	003
Andrews, Wayne D.	East Tennessee State U	304
Andrews, Wayne E.	Weber State U	351
Ansell, George S.	Colorado School of Mines	052
Anthan, Donald J.	Cleveland State U	048
Anthony, David K.	Brigham Young U 017	287
Antolovich, Stephen D.	Washington State U	264
Apanian, Ronald A.	U of North Dakota	161
Apfel, Robert E.	Yale U	280
Arastoopour, Hamid	Illinois Inst of Tech	096
Arbuckle, Robert	Lake Superior State U	310
Ardema, Mark D.	Santa Clara U	203
Arden, Bruce W.	U of Rochester	192
Ardis, Colby V.	Southern Illinois-Edwardsville	214
Arditi, David	Illinois Inst of Tech	096
Ari-Gur, Pnina	Western Michigan U	356
Armaly, Bassem F.	U of Missouri-Rolla	141
Armand, Ahmad	Wilkes U	273
Armstrong, Warren B.	Wichita State U	271
Arroyo, Alberto	U of Texas at San Antonio	236
Arthur, A. Allen	U of Cincinnati	293
Ary, James	Cal State U, Long Beach	030
Askins, Harold W.	The Citadel	043
Asmussen, Jes	Michigan State U	132
Aspnes, John	U of Alaska, Fairbanks	005
Atchley, Bill	U of the Pacific	178
Atkins, Forest D.	U of Cincinnati	293
Atkins, Jerome A.	Morgan State U	145
Atkinson, Richard C.	U of Cal, San Diego	023
Atkinson, Vicki L.	Utah State U	254
Aubret, Pedro J.	U of Puerto Rico	187
Auer, Thomas B.	Murray State U	320
Augsburger, Jerry D.	Northern Illinois U	165
Aungst, William	Penn State U - Harrisburg	330
Aunon, Jorge I.	Colorado State U	053
Austin, Alvin O.	LeTourneau U	107
Austin, Michael E.	U of Texas at El Paso	235
Austin, Orlo	Illinois, Urbana-Champaign	095
Auston, David H.	Columbia U	054
Avent, Raymond R.	Louisiana State U	109
Avtgis, Alexander W.	Wentworth Inst of Tech	352
Awoniyi, S.A.	Florida A&M U	072
Awoniyi, Samuel A.	Florida State U	076
Azar, Robert C.	Western New England Coll	270
Azzam, Rasheed	U of New Orleans	155

B

Name	Institution	Page
Babcock, Robert E.	U of Arkansas	010
Baber, Buford B.	U of Missouri-Kansas City	140
Bachnak, Rafic A.	Franklin U	307
Backer, Richard	Cal State U, Fresno	029
Backus, Charles E.	Arizona State U 009	283
Badler, Norman	U of Penn	179
Baer, Jean L.	U of Washington	262
Baeslack, William A.	Ohio State U	172
Baikie, Leslie A.	Technical U of Nova Scotia	226
Bailey, Raymond V.	Tulane U	244
Baillieul, John	Boston U	014
Baillod, C. Robert	Michigan Tech U	133
Baiocchi, Orlando	Cal State U, Chico	028
Baird, C. Robert	Technical U of Nova Scotia	226
Baker, George S.	U of Wisconsin-Milwaukee	275
Baker, James E.	Oklahoma State U	175
Baker, Jeffrey S.	San Francisco State U	201
Baker, Kendall R.	U of North Dakota	161
Baker, Vicki J.	The George Washington U	080
Baker, W. David	Rochester Inst of Tech	336
Baker, Warren J.	Cal Poly State U	027
Bakr, Mamdouh M.	U of Arkansas at Little Rock	284
Balachandran, Swaminathan	U of Wisconsin-Platteville	276
Baldwin, James W.	U of Missouri-Columbia	139
Baldwin, Kenneth C.	U of New Hampshire	149
Baldwin, Robert M.	Colorado School of Mines	052
Ballew, Tom	Cal Poly State U	027
Ballew, Walter D.	U of Oklahoma	173
Ballinger, Phillip	Gonzaga U	083
Balmer, Robert T.	U of Wisconsin-Milwaukee	275
Bailon, Jean-Paul	École Polytechnique	066
Bandyopadhyay, Amit	SUNY-Farmingdale	343
Banks, Peter M.	U of Michigan	130
Banta, Andrew R.	Cal State U, Sacramento	033
Barberini, Paul	Arizona State U 009	283
Barfield, Bill J.	Oklahoma State U	175
Barfield, Robert F.	U of Alabama	002
Bargainer, James D.	Baylor U	013
Barker, Keith	U of Conn	056
Barlage, William B.	Clemson U	047
Barnes, Ronny	Texas Tech U 239	348
Barnett, Stanley M.	U of Rhode Island	190
Baron, Andrew W.	Central Conn State U	291
Barr, Robert A.	Swarthmore Coll	224
Barrett, Andrew J.	U of South Florida	209
Barron, Charles H.	Clemson U	047
Barron, Ken	U of Alberta	006
Barron, Michael	U of Iowa	098
Barrows, John F.	Syracuse U	225
Barrows, Raymond	U of Mass, Dartmouth 122	314
Bartholomew, Charles L.	Widener U	272
Barton, Furman W.	U of Virginia	259
Bass, Steven C.	U of Notre Dame	168
Bassiouni, Zaki A.	Louisiana State U	109
Batchman, Theodore E.	U of Oklahoma	173
Batta, Rajan	SUNY-Buffalo	219
Battistella, Diane	DeVry Inst-DuPage	298
Batty, Burt	U of Mass	121
Baty, Peggy	Parks Coll of Saint Louis U	198
Baucom, Kathi M.	UNC-Charlotte 158	324
Bauer, James M.	Marietta Coll	117
Bauer, Paul T.	U of Bridgeport	016
Baum, Eleanor	The Cooper Union	057
Baumann, Paul	SUNY-Farmingdale	343
Bayless, Jerry R.	U of Missouri-Rolla	141
Bazergui, André	École Polytechnique	066
Beach, Charles D.	Florida Inst of Tech	074
Beacon, John E.	U of Nebraska - Lincoln	146
Beasley, David B.	North Carolina State U	160
Beaufait, Fred W.	Wayne State U	265
Beavers, Gordon S.	U of Minnesota	136
Beck, Robert F.	U of Michigan	130
Becker, Martin	U of Miami	128
Becker, Ronald W.	Cal State U, Sacramento	033
Becton, Julius W., Jr.	Prairie View A&M U 185	334
Bedient, Philip B.	Rice U	191
Beering, Steven C.	Purdue U 188	335
Behnke, Michael C.	Mass Inst of Tech	124
Beitler, Jackie	U of Texas at Dallas	234
Bekkala, Andrew	St. Cloud State U	197
Belanger, Judy	U of Nevada, Las Vegas	147
Beliveau, Jean-Guy L.	U of Vermont	257
Bell, Adam C.	Technical U of Nova Scotia	226
Bellaire, Henry J.	Southern U & A&M Coll	210
Bellem, Raymond D.	Embry-Riddle-Western Campus	068
Bellini, Paul X.	Cleveland State U	048
Bellovich, Steven J.	U of Tulsa	245
Belmont, Jean A.	Hofstra U	089
Belvin, James A., Jr.	Duke U	065
Bender, Richard M.	U of Pittsburgh at Johnstown	332
Benedict, Barry A.	Louisiana Tech U 110	311
Benefield, Larry D.	Auburn U 012	285
Bengtson, Harlan F.	Southern Illinois-Edwardsville	214
Benjamin, Richard W.	Monmouth Coll	142
Bennett, A. Wayne	Clemson U	047
Bennett, Byron	Montana State U 144	319
Bennett, G.K.	Texas A&M U	347
Bennett, Robert A.	U of Toledo	241
Bennett, William R.	Cleveland State U	048
Benson, Richard C.	U of Rochester	192
Benzley, Steven E.	Brigham Young U 017	287
Bepko, Gerald L.	Indiana-Purdue, Indianapolis 097	309
Berdahl, Robert M.	U of Texas at Austin	233
Bergman, Gary	Bradley U 015	286
Bergum, Gerald E.	South Dakota State U	208
Berman, Marilyn R.	U of Maryland, Coll Park	120
Bernat, Andrew	U of Texas at El Paso	235
Bernath, Fred R.	Rutgers, The State U	195
Bernholtz, Allen I.	U of Hartford	308
Bernstein, Edward L.	Alabama A&M U	282
Berry, John T.	U of Alabama	002

570 ASEE 1993 Directory of Engineering and Engineering Technology Undergraduate Programs

Bert, Charles W. ... U of Oklahoma 173
Bertenshaw, Thomas G. ... Oklahoma State U 326
Bertnolli, Edward C. ... North Dakota State U 162
Bertoni, Henry ... Poly U 182
Bess, Robert O. ... St. Cloud State U 197 337
Bethke, Richard J. ... Wright State U 278
Betterton, Don M. ... Princeton U 186
Bhatt, Bhushan L. ... Oakland U 169
Bickart, Theodore A. ... Michigan State U 132
Biederman, Ronald R. ... Worcester Poly Inst 277
Bilgutay, Nihat M. ... Drexel U 064
Binder, Shirley F. ... U of Texas at Austin 233
Biondi, Lawrence H., S.J. ... Parks Coll of Saint Louis U 198
Birgenheier, Raymond A. ... Gonzaga U 083
Bisharat, Keith ... Cal State U, Sacramento 033 288
Bishop, A. Bruce ... Utah State U 254
Bishop, Richard R. ... U of New Orleans 155
Bishop, Sandra L. ... U of Pittsburgh 181
Bjordal, Gary A. ... South Dakota School of Mines 207
Black, Margaret C. ... South Carolina State U 338
Black, William ... Illinois Inst of Tech 096
Blackington, Frank H., III ... U of Pittsburgh at Johnstown 332
Blackman, Craig ... Christian Brothers U 041
Blackman, David C. ... Northeastern U 163
Blackwell, John ... Case Western Reserve U 037
Blair, Barbara T. ... Tuskegee U 246
Blake, Thomas R. ... U of Mass 121
Blakely, J.M. ... Cornell U 058
Blanchard, Benjamin B. ... Virginia Tech 261
Blanchard, Donald G. ... New England Coll 148
Blandford, Dick K. ... U of Evansville 069
Bélanger, Pierre R. ... McGill U 125
Blevins, James R. ... West Virginia Inst of Tech 353
Block, Robert C. ... Rensselaer Poly Inst 189
Bloom, Alfred H. ... Swarthmore Coll 224
Bloom, Gary ... CUNY-City Coll 044
Boehm, Robert F. ... U of Nevada, Las Vegas 147
Bogner, Fred K. ... U of Dayton 060
Bogucz, Edward A. ... Syracuse U 225
Bogy, David B. ... U of Cal at Berkeley 019
Boig, Patricia G. ... Lehigh U 108
Bollinger, John G. ... U of Wisconsin-Madison 274
Bollini, Raghupathy ... Southern Illinois-Edwardsville 214
Bolon, Albert E. ... U of Missouri-Rolla 141
Bolt-Bannister, Phyliss K. ... U of Arizona 008
Bolyard, Neil E. ... West Virginia U 267
Bond, Arthur J. ... Alabama A&M U 282
Bonk, Robert J. ... U of Wisconsin-Milwaukee 275
Booth, Martha A. ... U of Akron 001 281
Bopp, Robert ... Ferris State U 070 306
Borbor, Iraj ... DeVry Inst-Los Angeles 301
Boresi, Arthur P. ... U of Wyoming 279
Borish, Marc S. ... U of Oklahoma 173
Borkowski, Francis T. ... U of South Florida 209
Borrelli, John ... Texas Tech U 239 348
Bory, Richard ... The Cooper Union 057
Bose, James E. ... Oklahoma State U 326
Botsaris, Gregory D. ... Tufts U 243
Boulet, Dan L., Jr. ... U of Delaware 061
Bourland, Hardy M. ... Rice U 191
Bousquet, David R. ... Virginia Tech 261
Bowen, J.R. ... U of Washington 262
Bowen, Richard L. ... Idaho State U 093
Bowman, Drumont ... Western Carolina U 354
Bowne, Anne P. ... Grove City Coll 085
Boyd, Angela N. ... Old Dominion U 176 327
Boyd, Richard H. ... U of Utah 253
Boyd, William D. ... San Diego State U 200
Boyer, Don L. ... Arizona State U 009
Brach, Philip L. ... U of the District of Columbia 063 303
Bradburn, Joseph H. ... U of South Carolina 206
Bradley, Dan ... Montana Coll of Mineral Sci 143
Bradley, Walter L. ... Texas A&M U 238
Bradley, William G. ... Western New England Coll 270
Braisted, Paul W. ... U of Missouri-Columbia 139
... U of Missouri-Kansas City 140
Brandeberry, James E. ... Wright State U 278
Bras, Rafael L. ... Mass Inst of Tech 124
Braun, Edwin R. ... UNC-Charlotte 324
Bravman, John C. ... Stanford U 217
Brazelton, William T. ... Northwestern U 166
Brazil, John R. ... Bradley U 015 286
Brazinsky, Irving ... The Cooper Union 057
Brehob, Wayne M. ... Lawrence Tech U 106
Breiseth, Christopher N. ... Wilkes U 273
Brentlinger, Brock ... Lamar U 105
Breslin, Richard D. ... Drexel U 064
Bretz, Robert ... New Mexico Inst of Mining 153
Brewer, Thomas B. ... Metropolitan State-Denver 317
Brickell, James L. ... US Air Force Academy 248
Brickson, Mark ... U of North Dakota 161
Briggs, Steven ... U of New Haven 150
Brill, E.D. ... North Carolina State U 160
Brinkerhoff, Spencer L. ... Northern Arizona U 164
Brissette, Claude ... École Polytechnique 066
Britton, Pamela A. ... Southern Illinois-Carbondale 213 341
Brock, Robert H. ... SUNY-Env'l Science & Forestry 220
Brodie, H. Keith H. ... Duke U 065
Brody, Harold D. ... U of Conn 056
Brogan, William L. ... U of Nevada, Las Vegas 147
Brooks, Calvin ... U of the District of Columbia 063
Brooks, Norman H. ... Cal Inst of Tech 026
Brooks, Weston T. ... Memphis State U 316
Brower, William E. ... Marquette U 118
Brown, Ashland ... U of the Pacific 178
Brown, Buck F. ... Rose-Hulman 194
Brown, Carmen ... Florida International U 075
Brown, David B. ... U of Alabama 002
Brown, Francis H. ... U of Utah 253
Brown, Garry L. ... Princeton U 186
Brown, Hugh H. ... Metropolitan State-Denver 317
Brown, Larry G. ... Mississippi State U 138
Brown, Michael S. ... Union Coll 247
Brown, Ray W. ... Christian Brothers U 041
Brown, Robert A. ... Mass Inst of Tech 124
Brown, Russell H. ... Clemson U 047
Brown, William D. ... U of Arkansas 010
Browne, Kevin M. ... Cal State U, Los Angeles 031
Brownstein, Kenneth ... U of Maine 114
Bruce, Charles W. ... Oklahoma State U 175 326
Bruce, Donna M. ... U of North Dakota 161
Bruce, Robert J. ... Widener U 272
Bruley, Duane F. ... Maryland Baltimore Cty 119
Bruner, Bill ... New Mexico State U 154 323
Bruno, John L. ... U of Cal, Santa Barbara 024
Bruns, Dale ... Wilkes U 273
Brunton, George D. ... U of Mississippi 137
Bryan, John L. ... U of Maryland, Coll Park 120
Bryan, Philip ... Murray State U 320
Bryan-Hagge, Sherry ... Memphis State U 316
Buchanan, Relva C. ... U of Cincinnati 042
Buck, Carl H. ... Rutgers, The State U 195
Bucklew, Neil S. ... West Virginia U 267
Budig, Gene A. ... U of Kansas 101
Buechner, John C. ... U of Colorado at Denver 051
Buffington, Dennis E. ... Penn State U 180
Bugg, William ... U of Tennessee, Knoxville 229
Bugliarello, George ... Poly U 182
Buhrman, R.A. ... Cornell U 058
Bulkeley, Peter Z. ... Boston U 014
Bulley, N. Ross ... U of Manitoba 115
Buonopane, Ralph A. ... Northeastern U 163
Burghart, James H. ... Cleveland State U 048
Burkart, Gary ... Indiana-Purdue, Indianapolis 097 309
Burnet, George ... Iowa State U 099
Burns, Ned H. ... U of Texas at Austin 233
Burns, William D. ... Southern Illinois-Edwardsville 214
Burt, Lawrence W. ... U of Cal, Los Angeles 022
Burton, Bruce E. ... Ohio Northern U 171
Burton, Charles L. ... Kansas State U 102
Burton, Larry C. ... Penn State U 180
Burton, Peter C. ... U of Southern Colorado 340
Busch, John R. ... LeTourneau U 107
Bush, John F. ... U of Central Florida 039 292
Bush, Ron ... DeVry Inst-Atlanta 297
Bustamante, Rafael B. ... Tennessee Tech 231
Butler, J. Kent ... Cal Poly State U 027
Butler, Jerome K. ... Southern Methodist U 216
Butson, Gary J. ... Southern Illinois-Carbondale 341
Butz, Brian P. ... Temple U 227 345
Byle, Arvon D. ... Western Michigan U 269
Byrd, John ... U of Evansville 069
Byrkett, Donald L. ... Miami U 129
Byrne, J. Gerald ... U of Utah 253
Byrne, John V. ... Oregon State U 177
Byrnes, Christopher I. ... Washington U 263

C

Cacioppo, Anthony J. ... Wright State U 278
Cade, Ruth A. ... U of Southern Mississippi 342
Cadman, Theodore W. ... Maryland Baltimore Cty 119
Cadoff, Irving ... Poly U 182
Cahen, George L. ... U of Virginia 259
Caissie, Wilfred J. ... Wentworth Inst of Tech 352
Calabrese, Richard V. ... U of Maryland, Coll Park 120
Calder, W. Thomas ... Loyola Marymount U 113
Caley, William F. ... Technical U of Nova Scotia 226
Calgaard, Ronald K. ... Trinity U 242
Callinan, Joseph P. ... Loyola Marymount U 113
Camille, Joseph M. ... U of Missouri-Columbia 139
Campbell, D.K. ... Illinois, Urbana-Champaign 095
Campbell, John R. ... Oklahoma State U 175 326
Campbell, Joseph A. ... Florida Atlantic U 073
Campbell, Nancy P. ... CUNY-City Coll 044
Campbell, Shari ... U of Manitoba 115
Cannon, Carmen ... Howard U 091
Cannon, Frank ... DeVry Inst-Dallas 299
Cannon, John R. ... Lamar U 105
Cannon, Joseph N. ... Howard U 091
Capjack, Clarence E. ... U of Alberta 006
Carcoana, Aurel ... North Dakota State U 162
Card, George R. ... U of South Florida 209
Carder, Molly S. ... Northern Arizona U 164
Carignan, George ... U of Michigan 130
Carley, Charles T. ... Mississippi State U 138
Carley, Thomas G. ... U of Tennessee, Knoxville 229
Carlson, D.E. ... Illinois, Urbana-Champaign 095
Carlson, Robert C. ... Stanford U 217
Carlson, Robert F. ... U of Alaska, Fairbanks 005
Carlyle, Ronald L. ... Cal State Poly, Pomona 034
Carmi, Shlomo ... Drexel U 064
Carmichael, Gregory R. ... U of Iowa 098
Carmichael, Marvin G. ... Clemson U 047
Carmichael, Thomas H. ... Southern Coll of Tech 339
Carnegie, Edgar J. ... Cal Poly State U 027
Carney, John F., III ... Vanderbilt U 256
Carney, Martin J. ... U of Miami 128
Carothers, Robert L. ... U of Rhode Island 190
Carpenter, Andy ... Oklahoma Christian 174
Carr, Stephen H. ... Northwestern U 166
Carrier, John C. ... West Virginia Inst of Tech 268
Carrier, John P. ... West Virginia Inst of Tech 353
Carrillo, Virginia P. ... U of Cal, Santa Cruz 025
Carroll, Billy D. ... U of Texas at Arlington 232
Carroll, Michael M. ... Rice U 191
Carter, Gary M. ... Maryland Baltimore Cty 119
Carter, Glenn ... West Virginia U 267
Carter, Richard ... U of Southern Maine 215
Caruthers, Robert M. ... Louisiana Tech U 110
Carver, Keith R. ... U of Mass 121
Casey, H. Craig, Jr. ... Duke U 065
Cash, Floyd L. ... U of Texas at Arlington 232
Casper, Gerhard R. ... Stanford U 217
Casteen, John T., III ... U of Virginia 259
Castete, Ralynn ... Lamar U 105
Castro, Walter E. ... Clemson U 047
Castrop, Deborah B. ... U of Kansas 101
Catanese, Anthony J. ... Florida Atlantic U 073
Cathie, Walter C. ... Carnegie Mellon U 036
Cavalieri, Ralph P. ... Washington State U 264
Cavin, Ralph K. ... North Carolina State U 160
Cayan, Peter J. ... SUNY-Utica/Rome 344
Cecere, Joseph ... Penn State U - Harrisburg 330
Cha, Chang-Yul ... U of Wyoming 279
Chae, Yong S. ... Rutgers, The State U 195

Index to Engineering & Technology Personnel

Name	Institution	Page
Challenger, Kenneth	San Jose State U	202
Chamberlain, Steven C.	Syracuse U	225
Chamberlin, Louise	New Mexico Inst of Mining	153
Chambers, Kathy M.	Western New England Coll	270
Chambers, Robert P.	Auburn U	012
Chambless, Cheryl C.	Western Kentucky U	355
Chameau, Jean-Lou A.	Georgia Tech	082
Chandler, Alice	SUNY-New Paltz	222
Chang, David C.	Arizona State U 009	283
Chang, Fred F.	U of the District of Columbia	063
Chang, Hsueh-Chia	U of Notre Dame	168
Chang, Nien-Yin	U of Colorado at Denver	051
Chang, Paul	U of Akron	001
Chapman, Thomas W.	U of Wisconsin-Madison	274
Charlson, Elaine M.	U of Missouri-Columbia	139
Charyulu, V.	Idaho State U	093
Chase, Robert W.	Marietta Coll	117
Chehl, Sahib S.	Southern U & A&M Coll	210
Chen, Ching-Jen	Florida A&M U	072
	Florida State U	076
Chen, Chiou S.	U of Akron	001
Chen, David H.T.	Widener U	272
Chen, Juh W.	Southern Illinois-Carbondale 213	341
Chen, Lea-Der	U of Iowa	098
Chen, Ray R.	San Jose State U	202
Chen, Shee-Ming	CUNY-City Coll	044
Chen, Wai-Kai	U of Illinois at Chicago	094
Chenelle, Donald	Case Western Reserve U	037
Cheney, Thomas	U of the Pacific	178
Cheng, Ping	U of Hawaii at Manoa	088
Chenoweth, Darrel L.	U of Louisville	111
Cheo, Peter K.	U of Conn	056
Cherrington, Blake E.	U of Texas at Dallas	234
Cheshier, Stephen R.	Southern Coll of Tech	339
Cheung, John Y.	U of Oklahoma	173
Chien, Percy	Seattle U	204
Chiesa, Steven C.	Santa Clara U	203
Chin, David A.	U of Miami	128
Chou, Fang H.	San Diego State U	200
Chow, Chi L.	U of Michigan-Dearborn	131
Chow, Chi Loong	Southern Illinois-Edwardsville	214
Christiansen, Walter H.	U of Washington	262
Châtillon, Louise V.	École Polytechnique	066
Chugh, Yoginder P.	Southern Illinois-Carbondale	213
Chung, Benjamin T.	U of Akron	001
Chung, Paul M.	U of Illinois at Chicago	094
Chung, Yip-Wah	Northwestern U	166
Cilento, Eugene V.	West Virginia U	267
Cipolla, John W., Jr.	Northeastern U	163
Cipriani, Frank A.	SUNY-Farmingdale	343
Clancy, Edward V.	Cal State Poly, Pomona	289
Clark, D.D.	Cornell U	058
Clark, J. Rodney	Utah State U	254
Clark, John R.	U of Colorado at Denver	051
Clark, Lenthon	U of Arkansas	010
Clark, Marilyn	Western Kentucky U	355
Clark, Stanley J.	Kansas State U	102
Clarke, Anne H.	Virginia Tech	261
Clarke, David R.	U of Cal, Santa Barbara	024
Claydon, Frank J.	Memphis State U 126	316
Clayton, Clive R.	SUNY-Stony Brook	223
Clayton, Warren F.	Tuskegee U	246
Clemence, Samuel P.	Syracuse U	225
Clement, Linda	U of Maryland, Coll Park	120
Clemons, Linda J.	East Tennessee State U	304
Clifford, Michael F.	Northeastern U	163
Clifford, Michael J.	Northeastern U	325
Clopton, Lewis P.	Morgan State U	145
Clough, Wayne	Virginia Tech	261
Clouse, Susan	Arizona State U 009	283
Cochran, John E., Jr.	Auburn U	012
Cockrum, Richard H.	Cal State Poly, Pomona	034
Cohen, C.	Cornell U	058
Cohen, Donald S.	Cal Inst of Tech	026
Cohen, Ira M.	U of Penn	179
Cohen, Jerome B.	Northwestern U	166
Cohen, William C.	Northwestern U	166
Cohn, Louis F.	U of Louisville	111
Cole, Charles	Penn State U - Harrisburg	330
Cole, Donald L.	U of Louisville	111
Coleman, Don M.	Howard U	091
Coleman, Patricia	Harvey Mudd Coll	087
Coll, Edward G., Jr.	NY State Coll of Ceramics	157
Coll, Edward G., Jr.	Alfred U	007
Collens, Lewis	Illinois Inst of Tech	096
Collier, John R.	Louisiana State U	109
Colling, David	U of Mass, Lowell	123
Collins, Anthony G.	Clarkson U	046
Collins, Donald W.	Arizona State U	283
Collins, Eileen M.	SUNY-Utica/Rome	344
Collins, Grady L.	U of South Alabama	205
Colombe, Sharon K.	South Dakota School of Mines	207
Colville, James	U of Maryland, Coll Park	120
Combee, Jerry H.	Grove City Coll	085
Comer, David J.	Brigham Young U	017
Commer, David A.	U of Alabama at Birmingham	003
Comparin, Robert A.	Virginia Tech	261
Compton, Phil V.	Texas A&I U	237
Condeni, Karen	Ohio Northern U	171
Conger, William L.	Virginia Tech	261
Conine, Charles B.	U of Arkansas at Little Rock	284
Conley, Franklin D.	Western Kentucky U	355
Conley, William T.	Case Western Reserve U	037
Conrad, Kay	Oregon State U	177
Conry, T.F.	Illinois, Urbana-Champaign	095
Conti, James J.	Webb Inst	266
Contractor, Dinshaw N.	U of Arizona	008
Converse, Alvin O.	Dartmouth Coll	059
Conway, Lynn	U of Michigan	130
Cook, David	Cal State U, Chico	028
Cook, Echol E.	Southern Illinois-Carbondale	341
Cook, Fred L.	Georgia Tech	082
Cook, George E.	Vanderbilt U	256
Cook, Gerald	George Mason U	079
Cook, James E.	West Virginia Inst of Tech	353
Cooney, Peter	SUNY-Maritime	221
Cooper, Ralph C.	Cal State U, Long Beach	030
Cooper, Stuart L.	U of Delaware	061
Coor, Lattie F.	Arizona State U 009	283
Copley, Stephen M.	Illinois Inst of Tech	096
Corcoran, Thomas J.	U of Maine	114
Corey, James L.	New Mexico Inst of Mining	153
Corlett, Richard C.	U of Washington	262
Cormack, Alastair	Alfred U	007
Cormack, Alastair N.	NY State Coll of Ceramics	157
Corngold, Noel R.	Cal Inst of Tech	026
Corotis, Ross B.	The Johns Hopkins U	100
Corrigan, Robert A.	San Francisco State U	201
Corso, John C.	U of Tulsa	245
Cosart, William P.	U of Arizona	008
Costello, Daniel J.	U of Notre Dame	168
Costello, Walter	U of Mass, Lowell 123	315
Cote, Roger	Concordia U	055
Cottle, Richard W.	Stanford U	217
Coughlin, Bernard J.	Gonzaga U	083
Coulman, George A.	Cleveland State U	048
Coulter, Myron L.	Western Carolina U	354
Coulter, Neal S.	Florida Atlantic U	073
Courtney, Thomas	Michigan Tech U	133
Courtright, James F.	U of Dayton	294
Courville, Louis	École Polytechnique	066
Coutinho, Gerald S.	U of Mass, Dartmouth 122	314
Cowan, Peter	Cal State U, Long Beach	030
Cox, Barbara	Florida A&M U	072
Cox, Ronald B.	U of Tennessee, Chattanooga	228
Cox, Stephen S.	Colorado State U	053
Craft, William J.	North Carolina A&T	159
Craig, George T.	San Diego State U	200
Craig, James R.	Montana James U 144	319
Craig, Kenneth	Norwich U	167
	Embry-Riddle-Daytona Beach	305
Craig, Richard	U of New Hampshire 149	321
Crandall, Keith C.	U of Cal at Berkeley	019
Crawford, Norman L.	GMI	077
Crawford, Peggy L.	U of Maine 114	312
Crecine, John P.	Georgia Tech	082
Creed, Joe L.	New Mexico State U	323
Cress, David	Marietta Coll	117
Crisman, W.C.	LeTourneau U	107
Crisp, John N.	U of New Orleans	155
Crist, Stephen C.	Western New England Coll	270
Cromack, Duane E.	U of Mass	121
Croot, Robert A.	Clarkson U	046
Crosby, Herb L.	U of Maine	312
Cross, Ernest J., Jr.	Old Dominion 176	327
Crossman, Gary R.	Old Dominion 176	327
Crouch, Peter E.	Arizona State U	009
Crouch, Steven	U of Minnesota	136
Crugnola, Aldo	U of Mass, Lowell 123	315
Crum, Edward H.	West Virginia Inst of Tech	268
Cruz, Jose B., Jr.	Ohio State U	172
Crynes, Billy L.	U of Oklahoma	173
Culp, Robert D.	U of Colorado at Boulder	049
Culp, Roy R.	Arkansas Tech U	011
Cummings, Russell	Cal Poly State U	027
Cundy, Victor A.	Louisiana State U	109
Cunningham, Trudy B.	Bucknell U	018
Cureton, Reginald T.	Morgan State U	145
Curley, Michael J.	Worcester Poly Inst	277
Curry, John A.	Northeastern U 163	325
Curtis, Howard	Embry-Riddle-Daytona Beach	067
Curtis, Kenneth	Metropolitan State-Denver	317
Cutbirth, James N.	Oklahoma Christian	174
Cuttino, David D.	Tufts U	243
Czerniak, Richard A.	DeVry Inst-Columbus	296
Czernikowski, Roy S.	Rochester Inst of Tech	193

D

Name	Institution	Page
Dagan, Zeev	CUNY-City Coll	044
Daily, John W.	U of Colorado at Boulder	049
Dale, J. Doug	U of Alberta	006
Dalrymple, Candice V.	The Johns Hopkins U	100
Dalton, Charles	U of Houston	090
Dalton, William K.	Purdue U	335
Daly, Paul S.	Embry-Riddle-Western Campus	068
Danforth, William H.	Washington U	263
Daniels, Patricia D.	Seattle U	204
Danielson, Scott G.	North Dakota State U	162
Dapprich, Aloysius E.	Point Park Coll	333
Darby, William P.	Washington U	263
Darr, David R.	Wright State U	278
Das, Digendra K.	SUNY-Utica/Rome	344
Das, Mihir K.	Cal State U, Long Beach	030
Das, Radhe	Cal State U, Long Beach	030
Davenport, Daniel D.	U of Idaho	092
Davenport, Kenneth A.	Wright State U	278
Davenport, Paul T.	U of Alberta	006
David, Larry G.	U of Missouri-Columbia	139
Davidson, Burton	Rutgers, The State U	195
Davidson, J. Narl	Georgia Tech	082
Davies, Mark D.	Bucknell U	018
Davis, Delores S.	North Carolina A&T	159
Davis, Denny C.	Washington State U	264
Davis, H.T.	U of Minnesota	136
Davis, Marjorie	Mercer U	127
Davis, Richard A.	U of Wyoming	279
Davis, Robert	U of Colorado at Boulder	049
Davis, Robert L.	U of Missouri-Rolla	141
Davis, Thomas W.	Milwaukee Sch of Eng 134	318
Davis, William E.	Louisiana State U	109
Day, John J.	Texas Tech U	239
Day, Ron	Memphis State U	316
Day, Thomas B.	San Diego State U	200
Dean, Robert G.	U of Florida	071
Deanda, Vincent	Cal State U, Los Angeles	031
Debose, Henry	Virginia State U	350
Deck, Joseph C.	U of Mass, Dartmouth 122	314
DeFleur, Lois B.	SUNY-Binghamton	218
DeFotis, William	U of Illinois at Chicago	094
DeHoff, Paul H.	UNC-Charlotte	158
DelBuono, Diane	U of Kansas	101
Delfino, Joseph J.	U of Florida	071
Delfino, Kate	Penn State-Erie	329
DeLisi, Charles	Boston U	014
Deloatch, Eugene M.	Morgan State U	145
Delorey, Mark J.	GMI	077
DeLucia, Frank C.	Ohio State U	172
Demel, John T.	Ohio State U	172
DeNardis, Lawrence J.	U of New Haven	150
Denn, Morton	U of Cal at Berkeley	019
Denning, Peter	George Mason U	079
Dennis, Avon	U of New Orleans	155
Dennis, John E.	Rice U	191
Denny, Fay	Cal State U, Long Beach	030
DePerro, Dennis R.	Marietta Coll	117
Deresiewicz, Herbert	Columbia U	054
DeSain, George	Western Carolina U	354

Name	Institution	Page
Desautel, Richard D.	San Jose State U	202
Desmarais, Ethel	Widener U	272
Desmond, Robert M.	West Virginia U	267
Dessouky, Mohamed I.	Northern Illinois U	165
Destler, William W.	U of Maryland, Coll Park	120
Deutsch, Stuart J.	Northeastern U	163
Devereux, Owen F.	U of Conn	056
Devgan, Satinderpaul S	Tennessee State U	230
DeVries, K. Larry	U of Utah	253
DeVries, Marvin F.	U of Wisconsin-Madison	274
Dewey, Bruce R.	U of Wyoming	279
DeWitt, Kenneth J.	U of Toledo	240
Diaz, Juan J.	St. Cloud State U	337
DiBiaggio, John	Tufts U	243
Dickey, John S.	Trinity U	242
Dickinson, Bradley W.	Princeton U	186
Dickinson, John W.	U of Idaho	092
Diekema, Anthony J.	Calvin Coll	035
Dieter, George E.	U of Maryland, Coll Park	120
Dietmeyer, Dietmeyer L.	U of Wisconsin-Madison	274
DiGregorio, Joseph S.	Penn State U	180
Dill, Ellis H.	Rutgers, The State U	195
Dill, Jerry R.	DeVry Inst-DuPage	298
Diller, Kenneth R.	U of Texas at Austin	233
Din, Hakim-Salahu A	Central Conn State U	291
DiPippo, Ronald	U of Mass, Dartmouth 122	314
Director, Stephen W.	Carnegie Mellon U	036
Dishner, Nancy L.	East Tennessee State U	304
DiUlio, Albert J., S.J.	Marquette U	118
Dixon, George R.	North Carolina State U	160
Dixon, Gregg	Cal State U, Northridge	032
Dixon, Robert D.	Wright State U	278
Dobbin, Edmund J.	Villanova U	258
Doepker, Philip E.	U of Dayton	294
Doherty, Michael F.	U of Mass	121
Doig, Alfred R.	Mass Inst of Tech	124
Dollar, John P.	Kansas State U	102
Dolphin, Sandra B.	U of the District of Columbia 063	303
Donaghey, Charles E.	U of Houston	090
Donaldson, Robert H.	U of Tulsa	245
Donohue, Marc D.	The Johns Hopkins U	100
Doremus, Robert H.	Rensselaer Poly Inst	189
Dorland, Dianne	U of Minnesota, Duluth	135
Douglas, George W.	U of South Alabama	205
Douglas, Jennifer	George Mason U	079
Dowell, Earl H.	Duke U	065
Dowling, Earl	Iowa State U	099
Downing, David R.	U of Kansas	101
Dowty, Daryll C.	Central Conn State U	291
Drahmann, T. "Brother", FSC	Christian Brothers U	041
Drayson, S. Roland	U of Michigan	130
Drewry, William A.	Old Dominion U 176	327
Drnevich, Vincent P.	Purdue U	188
Drouin, Gilbert	École Polytechnique	066
Dryden, Robert D.	Virginia Tech	261
Dube, William J.	Baylor U	013
Duda, J.L.	Penn State U	180
Duderstadt, James A.	U of Michigan	130
Dudziak, Donald J.	North Carolina State U	160
Dueck, Peter	U of Manitoba	115
Duffus, Rhonda	Montana State U 144	319
Dugan, James A.	DeVry Inst-Phoenix	302
Dumesic, James A.	U of Wisconsin-Madison	274
Dunbar, John	DeVry Inst-Atlanta	297
Duncan, Lewis M.	U of Tulsa	245
Dunham, James G.	Southern Methodist U	216
Dunn, Cheryll A.	U of Cincinnati	293
Dunn, Howard C., Jr.	US Coast Guard Academy	249
Dunn, John M.	U of Colorado at Boulder	049
Dunn, Stanley	Florida Atlantic U	073
Dunnagan, Nancy N.	SUNY-Farmingdale	343
Durkin, Kevin	SUNY-Buffalo	219
Durrant, S. Olani	Brigham Young U	017
Dvorak, George J.	Rensselaer Poly Inst	189
Dykes-Grimmett, Gale	DeVry Inst-Kansas City	300

E

Name	Institution	Page
Eakman, James	New Mexico State U	154
Earle, Jonathan F.K.	U of Florida	071
Eastep, Franklin E.	U of Dayton	060
Easton, Robert	Rochester Inst of Tech	336
Eastop, R.	U of Toledo	240
Eastwood, William F.	Tufts U	243
Ebadian, M.A.	Florida International U	075
Echelberger, Wayne F.	U of South Florida	209
Eckart, Roy E.	U of Cincinnati	042
Eckhoff, N. Dean	Kansas State U	102
Economy, J.	Illinois, Urbana-Champaign	095
Eddy, Jean C.	Northeastern U 163	325
Edgar, Thomas F.	U of Texas at Austin	233
Edgers, Lewis	Tufts U	243
Edson, Kenneth C.	Cal State U, Chico	028
Edwards, Charles T., Jr.	Prairie View A&M U	334
Effiong, Richard	Weber State U	351
Eftekhar, Jahan	U of Texas at San Antonio	236
Eichhorn, Roger	U of Houston	090
Eide, Arvid R.	Iowa State U	099
Eifert, James R.	Rose-Hulman	194
Eigel, Edward	U of Bridgeport	016
Eisenberg, Martin A.	U of Florida	071
Eisenstein, Bruce A.	Drexel U	064
Ekdale, A.A. (Tony)	U of Utah	253
Elgaaly, Mohamed	Drexel U	064
Elias, Samy E.G.	U of Nebraska - Lincoln	146
Elkins, Richard N.	Kansas State U	102
Ellerbruch, Virgil G.	South Dakota State U	208
Ellingwood, Bruce R.	The Johns Hopkins U	100
Elliott, George H.	U of Maine	312
Elliott, Peggy N.	U of Akron 001	281
Elliott, Robert P.	U of Arkansas	010
Ellis, Bruce W.	St. Cloud State U	197
Ellis, Patrick	The Catholic U of America	038
Ellis, Thearn H.	West Virginia Inst of Tech	353
Ellis, Thomas G.	Michigan Tech U	133
Elsayed, Elsayed A.	Rutgers, The State U	195
Elzinga, D. Jack	U of Florida	071
Emery, Ashley F.	U of Washington	262
Emmert, Gilbert A.	U of Wisconsin-Madison	274
Emshousen, Fred W.	Purdue U	335
Engbretson, Gustav A.	Syracuse U	225
Engelage, Donald L.	Texas A&M U 238	347
Engelgau, Gary R.	Texas A&M U	238
Engelke, George F.	Cal State Poly, Pomona	034
English, Howard L.	SUNY-Maritime	221
English, Robert	New Jersey Inst of Tech 151	322
Enneking, Thomas J.	Tri-State U 241	349
Erdman, Carl A.	Texas A&M U 238	347
Erickson, John D.	South Dakota School of Mines	207
Eror, Nicholas G.	U of Pittsburgh	181
Ertekin, Turgay	Penn State U	180
Erzurumlu, H. Chik M.	Portland State U	184
Escalet, Ed	Penn State U - Harrisburg	330
Essman, Joseph E.	Ohio U	170
Estes, Lee E.	U of Mass, Dartmouth	122
Evans, Handel	San Jose State U	202
Evans, Ronald D.	U of Oklahoma	173
Everhart, Thomas E.	Cal Inst of Tech	026
Evers, James L.	Southern Illinois-Carbondale 213	341
Ezra, Arthur A.	SUNY-Farmingdale	343

F

Name	Institution	Page
Fabris, Neda S.	Cal State U, Los Angeles	031
Faddis, Terry	U of Kansas	101
Faherty, Keith F.	Marquette U	118
Fairley, Vernetta P.	U of Southern Mississippi	342
Falk, James E.	The George Washington U	080
Falkenburg, Donald R.	Wayne State U	265
Fan, Liang T.	Kansas State U	102
Fan, Stephen S.	U of New Hampshire	149
Fancott, Terrill	Concordia U	055
Fannin, Ronald D.	U of Missouri-Rolla	141
Farr, Morris	U of Arizona	008
Farrah, Robert	Lawrence Tech U	106
Farrell, Mark O.	Point Park Coll	333
Farren, Joseph M.	U of Dayton	294
Farrington, Gregory C.	U of Penn	179
Farrow, John H.	Milwaukee Sch of Eng 134	318
Faucher, Guy	École Polytechnique	066
Faye, Katherine M.	Franklin U	307
Fazio, Paul P.	Concordia U	055
Fedock, Joseph J.	Montana State U 144	319
Feeley, Joseph J.	U of Idaho	092
Fegan, George R.	Santa Clara U	203
Fegley, Kenneth A.	U of Penn	179
Feisel, Lyle D.	SUNY-Binghamton	218
Feller, Harry R.	U of Pittsburgh at Johnstown	332
Felton, Richard F.	Embry-Riddle-Western Campus	068
Femenia, Jose	SUNY-Maritime	221
Fenner, James	Oregon Inst of Tech	328
Fenster, Saul K.	New Jersey Inst of Tech 151	322
Fenton, Donald L.	Kansas State U	102
Ferguson, Carol	Memphis State U 126	316
Ferguson, Frances	US Merchant Marine Academy	250
Ferguson, John F.	U of Washington	262
Fermental, Denis W.	Tufts U	243
Fernandez-Sein, Rafael	U of Puerto Rico	187
Ferrara, Thomas C.	Cal State U, Chico	028
Ferrari, Leonard A.	U of Cal, Irvine	021
Ferritor, Daniel E.	U of Arkansas	010
Field, John C.	U of Maine	114
Fillo, John A.	SUNY-Binghamton	218
Finch, Ray N.	Texas A&I U	237
Fincher, Barbara T.	U of Florida	071
Fink, Joseph L., III	U of Kentucky	103
Finlayson, Bruce A.	U of Washington	262
Finlinson, Norman B.	Brigham Young U 017	287
Finney, Joseph	New Mexico State U	154
Fischbeck, Helmut J.	U of Oklahoma	173
Fischer, Patrick C.	Vanderbilt U	256
Fish, Andrew J.	U of New Haven	150
Fisher, Cary A.	US Air Force Academy	248
Fisher, Edward R.	Michigan Tech U	133
Fisher, Kenneth J.	Penn State-Erie	329
Fissinger, Matthew X.	Loyola Marymount U	113
Fiszdon, Jerzy	Mankato State U	116
Fitz, Raymond L., S.M.	U of Dayton 060	294
Fitzgerald, Robert W.	Worcester Poly Inst	277
Fitzpatrick, Christine	Indiana-Purdue, Indianapolis 097	309
Flatt, William P.	U of Georgia	081
Fleenor, Edgar	Indiana-Purdue, Indianapolis	309
Fleming, William S.	New Mexico State U	323
Flemings, Merton S.	Mass Inst of Tech	124
Fletcher, David Q.	U of the Pacific	178
Fletcher, Ruth T.	Texas A&I U	237
Flocco, Nicholas	Drexel U	064
Florian, Judith L.	Wayne State U	265
Flumerfelt, Ray W.	Texas A&M U	238
Fong, Annabelle C.	U of Hawaii at Manoa	088
Fontijn, Arthur	Rensselaer Poly Inst	189
Fooks, Karen L.	U of Florida	071
Foote, Edward T.	U of Miami	128
Foote, Richard M.	U of Vermont	257
Ford, Robert F.	Prairie View A&M U	185
Forest, David A.	U of New Hampshire	321
Fort, Edward B.	North Carolina A&T	159
Fort, Tomlinson	Vanderbilt U	256
Foster, Edwin P.	U of Tennessee, Chattanooga	228
Foster, Jacqueline	Yale U	280
Foster, John	Prairie View A&M U	185
Fourer, Robert H.	Northwestern U	166
Foust, Karen	Valparaiso U	255
Fowler, T. Kenneth	U of Cal at Berkeley	019
Fox, Lynn	U of the Pacific	178
Fox, Pat	Indiana-Purdue, Indianapolis 097	309
Fox, Ray P.	U of Tennessee, Chattanooga	228
Frair, Karen L.	U of Alabama	002
Francis, Gwyndolyn	U of South Florida	209
Francis, John E.	Bradley U 015	286
Franklin, Gene F.	Stanford U	217
Franklin, Loretta	DeVry Inst-Atlanta	297
Franks, Lewis E.	U of Mass	121
Franz, Frank A.	U of Alabama at Huntsville	004
Frasher, Richard D.	Ohio State U	172
Frederick, W. James	Oregon State U	177
Freed, DeBow	Ohio Northern U	171
Freedman, James O.	Dartmouth Coll	059
Freeman, James J.	San Jose State U	202
Freeman, Joanne	Cal Poly State U	027
Frerichs, Estelle	Trinity U	242
Frey, Roger G.	U of New Haven	150
Frick, Pieter A.	U of Colorado-Col Spr	050
Fridsma, Ken	Grand Valley State U	084
Frieder, Gideon	The George Washington U	080
Friedman, Morton B.	Columbia U	054
Frisby, James C.	U of Missouri-Columbia	139

Index to Engineering & Technology Personnel

G

Name	Institution	Page
Frishberg, Ellen	The Johns Hopkins U	100
Froehlich, Donell P.	South Dakota State U	208
Fuchs, Matthew W.	Milwaukee Sch of Eng	134
Fuller, Hank M.	The Citadel	043
Fuller, John H.	Prairie View A&M U	185
Fuller, Lynn F.	Rochester Inst of Tech	193
Fuller, Richard M.	The Johns Hopkins U	100
Fultz, Brent T.	Cal Inst of Tech	026
Furry, R.B.	Cornell U	058
Gabitto, Jorge	Prairie View A&M U	185
Gaither, Robert B.	U of Florida	071
Gajda, Walter J., Jr.	U of Missouri-Rolla	141
Galanos, Glafkos D.	Southern Illinois-Carbondale	213
Galetti, Kim	DeVry Inst-Phoenix	302
Galetto, Joseph	Michigan Tech U	133
Galil, Zvi	Columbia U	054
Gallagher, Richard H.	Clarkson U	046
Galloway, Gerald E., Jr.	US Military Academy	251
Galloway, Joel	Ferris State U 070	306
Galloway, Kenneth F.	U of Arizona	008
Gambrell, Carroll B.	Mercer U	127
Gandhi, Om P.	U of Utah	253
Ganesan, Subramanian	Oakland U	169
Garcia, Edward	SUNY-Farmingdale	343
Garneau, Gabriel	École Polytechnique	066
Garner, Geraldine O.	Northwestern U	166
Garrard, William L.	U of Minnesota	136
Garrow, Thomas	U of Southern Cal	211
Gattis, Jim L.	U of Arkansas	010
Gaudron, France	École Polytechnique	066
Gayle, Godfrey A.	North Carolina A&T	159
Gee, E. Gordon	Ohio State U	172
Gehmlich, Dietrich K.	U of Utah	253
Geiger, Randall	Iowa State U	099
Geis, Donald R.	U of Oklahoma	173
Gelopulos, Demosthenes P.	Valparaiso U	255
Genco, Joseph M.	U of Maine	114
Genis, Alan P.	Northern Illinois U	165
Gentry, Don K.	Purdue U	335
Gentry, Donald W.	Colorado School of Mines	052
Geonetta, Sam C.	U of Cincinnati	293
George, Emert	Utah State U	254
George, John A.	Parks Coll of Saint Louis	198
Gerberding, William P.	U of Washington	262
Gerdeen, James	U of Colorado at Denver	051
Geremia, John O.	US Naval Academy	252
Gerez, Victor	Montana State U 144	319
Gerhardt, Lester A.	Rensselaer Poly Inst	189
Gerhart, Phillip M.	U of Evansville	069
Gerig, Lee	Seattle U	204
Gerity, Peter F.	U of Utah	253
Germane, Geoffrey J.	Brigham Young U	017
Gerth, Donald R.	Cal State U, Sacramento 033	288
Gessler, Johannes	Colorado State U	053
Ghausi, Mohammed S.	U of Cal, Davis	020
Gianniny, Omar A.	U of Virginia	259
Gibala, Ronald	U of Michigan	130
Gibbons, James F.	Stanford U	217
Gibbons, Joseph H.	U of South Carolina	206
Gibson, David F.	Montana State U 144	319
Gibson, David R.	The Catholic U of America	038
Gibson, James B.	U of Alabama at Huntsville	004
Giddens, Don P.	The Johns Hopkins U	100
Giffen, Robert B.	US Air Force Academy	248
Giglio, Richard J.	U of Mass	121
Giguere, J. Charles	Concordia U	055
Gilbert, Richard A.	U of South Florida	209
Gilbert, Richard E.	U of Nebraska - Lincoln	146
Gilbertson, Eric R.	Saginaw Valley State U	196
Gilbreath, Sidney G., III	Tennessee Tech	231
Gilheany, John J.	The Catholic U of America	038
Gill, Tom A.	Technical U of Nova Scotia	226
Gilley, James R.	Iowa State U	099
Gilligan, Lawrence G.	U of Cincinnati	293
Gillott, Donald H.	Cal State U, Sacramento 033	288
Giolma, John P.	Trinity U	242
Givens, Paul E.	U of South Florida	209
Glandt, Eduardo D.	U of Penn	179
Glasmire, Larry	Cal State U, Sacramento 033	288
Glavinich, Thomas E.	U of Kansas	101
Glee, Ulysses, Jr.	U of Maryland, Coll Park	120
Glenn, Betty	DeVry Inst-Chicago	295
Glenn, Thomas E.	U of Illinois at Chicago	094
Glisson, Tildon H.	North Carolina State U	160
Goad, Clyde C.	Ohio State U	172
Godfrey, Eric	U of Washington	262
Godleski, Edward S.	Cleveland State U	048
Godsey, R. Kirby	Mercer U	127
Goel, Vijay K.	U of Iowa	098
Goines, Shirley	Arkansas Tech U	011
Goitom, Tesfai	Cal State U, Long Beach	030
Gold, Albert	Harvard U	086
Goldberg, Saul	Cal Poly State U	027
Golden, Kenneth I.	U of Vermont	257
Golden, Richard L.	Princeton U	186
Golden, Timothy P.	Virginia Military Inst	260
Goldfarb, Donald	Columbia U	054
Goldman, Jay	U of Alabama at Birmingham	003
Goldstein, Richard J.	U of Minnesota	136
Goncz, Joseph F., Jr.	Grove City Coll	085
Gonzalez, Mario J., Jr.	U of Texas at Austin	233
Gooch, Mary E.	Prairie View A&M U	334
Goodling, John S.	Auburn U	012
Goodman, Joseph W.	Stanford U	217
Googe, Joseph M.	U of Tennessee, Knoxville	229
Goold, Vern C.	Pittsburg State U	331
Gordon, Ronald S.	Virginia Tech	261
Gosink, Joan P.	Colorado School of Mines	052
Gottsman, Earl E.	Capitol Coll	290
Gould, Frederick E.	Wentworth Inst of Tech	352
Gould, Phillip L.	Washington U	263
Gourdeau, Jean-Paul	École Polytechnique	066
Gourley, Frank A., Jr.	West Virginia Inst of Tech	353
Gourley, Tom M.	Brigham Young U 017	287
Gowdy, Kenneth K.	Kansas State U	102
Gowen, Richard J.	South Dakota School of Mines	207
Goyette, Daniel L.	Marquette U	118
Grabiec, Richard A.	Western New England Coll	270
Graff, R. William	LeTourneau U	107
Graff, Robert A.	CUNY-City Coll	044
Graham, E. Earl	Cleveland State U	048
Graham, George W.	Pittsburg State U	331
Grandt, Alten F., Jr.	Purdue U	188
Grant, Donald A.	U of Maine	114
Grassel, Marty	Texas Tech U	239
Grasser, Philip F.	US Naval Academy	252
Graves, Howard D.	US Military Academy	251
Gray, Howard M.	U of Maine	312
Gray, Murray R.	U of Alberta	006
Gray, Paul R.	U of Cal at Berkeley	019
Gray, William G.	U of Notre Dame	168
Graziano, Anthony F.	Illinois, Urbana-Champaign	095
Green, David J.	Penn State U	180
Green, Douglas M.	SUNY-Binghamton	218
Green, Robert E., Jr.	The Johns Hopkins U	100
Greene, Timothy J.	Oklahoma State U	175
Greenfield, Helga A.	Old Dominion U 176	327
Greer, Mary H.	Iowa State U	099
Greg, Wight	Norwich U	167
	Embry-Riddle-Daytona Beach	305
Gregorek, Gerald M.	Ohio State U	172
Gregory, William D.	Gannon U	078
Gregory Kohlstedt, Sally	U of Minnesota	136
Greimann, Lowell F.	Iowa State U	099
Greiner, William R.	SUNY-Buffalo	219
Gress, David L.	U of New Hampshire	149
Griffin, Emmett R., Jr.	Howard U	091
Griffin, Jack	DeVry Inst-Atlanta	297
Griffis, Carl L.	U of Arkansas	010
Griffis, Frank E.	Norwich U	167
Griffiths, Lloyd	U of Colorado at Boulder	049
Griffiths, Lloyd J.	U of Southern Cal	211
Grigg, Neil S.	Colorado State U	053
Griggs, Edwin I.	Tennessee Tech	231
Grismore, Fred L.	Ohio Northern U	171
Grissom, William A.	Central State U	040
Griswold, Anna M.	Penn State U	180
Grode, John N.	Penn State-Erie	329
Grodzinsky, Stephen E.	U of Bridgeport	016
Gronsky, Ronald	U of Cal at Berkeley	019
Gross, David I.	U of Dayton	294
Grounds, Patrick M.	Lake Superior State U	310

H

Name	Institution	Page
Groves, James R.	Virginia Military Inst	260
Grubbs, John H.	US Military Academy	251
Gruen, David L.	U of Tulsa	245
Grum, Allen F.	Mercer U	127
Gruzleski, John	McGill U	125
Gude, Leonard E.	Florida Inst of Tech	074
Guffrovich, Emory P., Jr.	Wilkes U	273
Guice, Leslie K.	Louisiana Tech U	110
Gulari, Erdogan	U of Michigan	130
Gunasekera, Jay	Ohio U	170
Gundler, Errol A.	Miami U	129
Gurley, William Q.	U of Tennessee, Chattanooga	228
Gustafson, Robert J.	Ohio State U	172
Guttentag, Christoph O.	Duke U	065
Guyer, Gordon	Michigan State U	132
Guyon, John C.	Southern Illinois-Carbondale 213	341
Hackett, Robert M.	U of Mississippi	137
Hackney, Sheldon	U of Penn	179
Hadad, Alan J.	U of Hartford	308
Haddad, Abraham H.	Northwestern U	166
Haddad, David C.	Miami U	129
Haddad, George I.	U of Michigan	130
Haddock, Lawrence T.	Southern Coll of Tech	339
Hadler, Jacques B.	Webb Inst	266
Haenicke, Diether H.	Western Michigan U 269	356
Hagedorn, Donald E.	U of Cincinnati	042
Hagler, Marion O.	Texas Tech U	239
Hahn, George T.	Vanderbilt U	256
Hahn, Martina S.	U of Miami	128
Hahn, Robert D.	U of Akron 001	281
Haines, Charles W.	Rochester Inst of Tech	193
Haisler, Walt E.	Texas A&M U	238
Hakes, Samuel D.	U of Wyoming	279
Hakimi, S.L.	U of Cal, Davis	020
Haldane, Mary B.	Ohio State U	172
Hale, Paul N.	Louisiana Tech U	110
Hale, Richard	Cornell U	058
Hales, James A.	East Tennessee State U	304
Hales, James L.	U of Pittsburgh at Johnstown	332
Halford, Carl E.	Memphis State U	126
Halgren, Lee A.	U of Wisconsin-Platteville	276
Hall, Ian W.	U of Delaware	061
Hall, Jerome W.	The U of New Mexico	152
Hall, Joyce	Purdue U 188	335
Hall, Kenneth R.	Texas A&M U 238	347
Hall, Randall D.	U of Tennessee, Martin	346
Haller, Gary L.	Yale U	280
Halligan, James E.	New Mexico State U 154	323
Halpin, Daniel W.	Purdue U	188
Halton, John C.	U of Texas at Austin	233
Hamada, Harold S.	U of Hawaii at Manoa	088
Hamblin, F. Douglas	Concordia U	055
Hamelink, Jerry H.	Western Michigan U	269
Hamernik, Robert E.	U of the Pacific	178
Hamilton, Wayne A.	U of Maine	114
Hamm, Jack	Arkansas Tech U	011
Han, Kenneth N.	South Dakota School of Mines	207
Hancher, Donn E.	U of Kentucky	103
Hand, Mindy A.	Maryland Baltimore Cty	119
Hanley, Thomas R.	U of Louisville	111
Hansen, Anne	New Mexico Inst of Mining	153
Harbach, James A.	US Merchant Marine Academy	250
Hardesty, Glenn	Western Carolina U	354
Harnett, R. Michael	Kansas State U	102
Harold-Miller, Chelseia	Morgan State U	145
Haroun, Medhat A.	U of Cal, Irvine	021
Harralson, Joseph H.	Cal State U, Sacramento 033	288
Harre, Alan F.	Valparaiso U	255
Harrington, Robert J.	The George Washington U	080
Harris, Carl M.	George Mason U	079
Harris, E. Douglas	Southern Methodist U	216
Harris, H. Gordon	U of Wyoming	279
Harris, Lisa B.	Louisiana State U	109
Harris, Patricia S.	U of Texas at Austin	233
Harris, Richard W.	Utah State U	254
Harris, Thomas A.	Vanderbilt U	256
Harrison, Cecil A.	U of Southern Mississippi	342
Harrison, Thomas J.	Florida A&M U	072
	Florida State U	076

Index to Engineering & Technology Personnel

Name	Institution	Page
Hart, Natala	Indiana-Purdue, Indianapolis 097	309
Hartigan, Ellen	Poly U	182
Hartley, Craig S.	Florida Atlantic U	073
Hartley, Harry J.	U of Conn	056
Hartman, David E.	Northern Arizona U	164
Hartmanis, J.E.	Cornell U	058
Harvey, Johna	Columbia U	054
Harwell, Jeffrey H.	U of Oklahoma	173
Haselkorn, Mark P.	U of Washington	262
Hassell, Johnette	Tulane U	244
Hasselmo, Nils	U of Minnesota	136
Hatcher, William J.	U of Alabama	002
Hatton, Barbara R.	South Carolina State U	338
Hauck, George F.	U of Missouri-Columbia	139
	U of Missouri-Kansas City	140
Hawkins, N.M.	Illinois, Urbana-Champaign	095
Hayes, Louis T.	Widener U	272
Hayes, Russell M., Jr.	U of South Alabama	205
Hayhurst, David T.	U of South Alabama	205
Hazen, Glenn	Ohio U	170
Hazen, Samuel L.	Gannon U	078
Hazen, Verna J.	Rochester Inst of Tech 193	336
Hazen, Virginia S.	Dartmouth Coll	059
Head, Virginia A.	Southern Coll of Tech	339
Healy, Timothy J.	Santa Clara U	203
Healy, William L.	Saginaw Valley State U	196
Heaney, James P.	U of Colorado at Boulder	049
Hebert, Susan	Milwaukee Sch of Eng 134	318
Hefley, Steve M.	Pittsburg State U	331
Hefner, James A.	Tennessee State U	230
Hegg, Richard O.	Clemson U	047
Heideger, William A.	U of Washington	262
Heidersbach, Robert H.	Cal Poly State U	027
Hein, Warren W.	South Dakota State U	208
Heinlein, Terrence G.	Wentworth Inst of Tech	352
Heintze, Michael R.	Clemson U	047
Heist, Richard H.	U of Rochester	192
Hejde, Dan	Parks Coll of Saint Louis U	198
Helgeland, Robert C.	U of Mass, Dartmouth	314
Hellman, Nancy B.	U of Mass	121
Helmken, Henry F.	Florida Atlantic U	073
Henderson, Allen J.	North Dakota State U	162
Henderson, Bruce P.	GMI	077
Henderson, Burt	U of Arkansas at Little Rock	284
Henderson, Stanley E.	Western Michigan U 269	356
Henderson, Thomas C.	U of Utah	253
Henderson, Thomas L.	Southern U & A&M Coll	210
Hendrickson, Chris T.	Carnegie Mellon U	036
Hendrickson, Thomas	Mankato State U 116	313
Henneke, Edmund G.	Virginia Tech	261
Henriksen, Mogens	U of North Dakota	161
Henry, Joseph D.	Arizona State U	009
Henson, David	Alabama A&M U	282
Herickhoff, Robert J.	Mankato State U 116	313
Heskiaoff, Heskia	New York Inst of Tech	156
Hess, Dennis W.	Lehigh U	108
Hester, John N.	Cal State U, Sacramento 033	288
Heuston, Joseph W.	Cal State U, Fresno	029
Hewitt, Hudy C., Jr.	U of New Orleans	155
Heyward, James O.	Alabama A&M U	282
Hickel, Steve	Ferris State U	306
Hicks, R. Gary	Oregon State U	177
Higgins, Brian G.	U of Cal, Davis	020
Highter, William H.	U of Mass	121
Hightower, Ray E.	Kansas State U	102
Hildreth, Eddie, Jr.	Southern U & A&M Coll	210
Hile, Lloyd	Cal State U, Long Beach	030
Hill, Donald O.	Mississippi State U	138
Hill, Ester M.	Louisiana State U	109
Hill, James L.	U of Alabama	002
Hill, Larry A.	Cal State U, Sacramento 033	288
Hill, Owen	SUNY-New Paltz	222
Hill, Warren R.	Weber State U	351
Hillmeyer, Colleen	U of Denver	062
Hills, David	U of Cal, Davis	020
Hines, Anthony L.	U of Missouri-Columbia	139
	U of Missouri-Kansas City	140
Hirst, Susan	DeVry Inst-Atlanta	297
Hirtzel, Cynthia S.	Syracuse U	225
Hitt, John C.	U of Central Florida 039	292
Hobbie, Russell K.	U of Minnesota	136
Hoberock, Lawrence L.	Oklahoma State U	175
Hochmuth, Robert M.	Duke U	065

Name	Institution	Page
Hodge, Daniel B.	Ohio State U	172
Hodges, David A.	U of Cal at Berkeley	019
Hodgkin, Brian C.	U of Southern Maine	215
Hoekema, David	Calvin Coll	035
Hoeksema, Robert J.	Calvin Coll	035
Hoffman, Glenn R.	U of Nebraska - Lincoln	146
Hoffman, Larry D.	Purdue U	335
Hogan, Betsy J.	U of New Haven	150
Hogan, G. Richard	St. Cloud State U	197
Hogan, Joseph S.	SUNY-Stony Brook	223
Hogan, Richard G.	St. Cloud State U	337
Hogan, William	U of Mass, Lowell 123	315
Hogg, Gary	Texas A&M U	238
Hohmann, Edward C.	Cal State Poly, Pomona 034	289
Holbrook, Raymond L.	Tennessee Tech	231
Holder, Gerald D.	U of Pittsburgh	181
Holger, David K.	Iowa State U	099
Hollander, Lawrence J.	Union Coll	247
Hollworth, Bruce R.	Alfred U	007
Holsapple, Keith A.	U of Washington	262
Holst, Timothy B.	U of Minnesota, Duluth	135
Hommertzheim, Don L.	Wichita State U	271
Honeman, Donald M.	U of Vermont	257
Hood, Thomas G.	U of Mississippi	137
Hooker, Michael K.	U of Mass	121
Hooks, Lawrence E.	Mercer U	127
Hopcroft, John	Cornell U	058
Hopkins, Gordon R.	Florida International U	075
Hopkins, Ronald G.	Pittsburg State U	331
Hopkins, Walter G., III	Northern Arizona U	164
Hopper, Jack R.	Lamar U	105
Hornfeck, William A.	Lafayette Coll	104
Hornung, Hans G.	Cal Inst of Tech	026
Horowitz, Ellis	U of Southern Cal	211
Horton, Frank E.	U of Toledo	240
Horvath, Csaba G.	Yale U	280
Hosmer, Bradley C.	US Air Force Academy	248
Houde, Jules	École Polytechnique	066
Houston, Martin R.	Western Kentucky U	355
Houston, Shelton L.	U of Southern Mississippi	342
Hovanesian, Joseph D.	Oakland U	169
Howard, Chuck	Rose-Hulman	194
Howard, Kenneth I.	U of the District of Columbia 063	303
Howatt, Clarke T.	U of Southern Cal	211
Howell, Ronald H.	U of South Florida	209
Hrabowski, Freeman A., III	Maryland Baltimore Cty	119
Hren, John J.	North Carolina State U	160
Hsie, Atlas	SUNY-Utica/Rome	344
Hsu, Tai-Ran	San Jose State U	202
Hu, Evelyn L.	U of Cal, Santa Barbara	024
Huber, Wayne M.	Oregon State U	177
Hubers, Wayne K.	Calvin Coll	035
Huckabay, Houston M.	Louisiana Tech U	110
Huckenbeck, Ann L.	U of Conn	056
Huddleston, H. Ted, Jr.	U of South Alabama	205
Hudson, Stanley G.	Mass Inst of Tech	124
Huey, Ben M.	Arizona State U	009
Huff, Robert P.	Stanford U	217
Huggins, Brian D.	Bradley U 015	286
Huggins, Larry F.	Purdue U	188
Hughes, Author E.	U of San Diego	199
Hughes, Eugene M.	Northern Arizona U	164
Hughes, Robert K.	Oklahoma State U	175
Hughes, William L.	South Dakota School of Mines	207
Hulbert, Samuel F.	Rose-Hulman	194
Hulbert, Thomas E.	Northeastern U	325
Hull, Roger H.	Union Coll	247
Hullar, Theodore L.	U of Cal, Davis	020
Humphries, Frederick S.	Florida A&M U	072
Humphries, W.K.	U of South Carolina	206
Hundal, Mahendra S.	U of Vermont	257
Hung, Joe Y.	Cal State Poly, Pomona	034
Hunt, G.D.	Rice U	191
Huntley, John	Oregon Inst of Tech	328
Hurdle, Mary	New Jersey Inst of Tech 151	322
Hurst, L. Winslow	Weber State U	351
Hussain, Nihad A.	San Diego State U	200
Hutchinson, Charles E.	Dartmouth Coll	059
Hutchinson, Dwight E.	US Coast Guard Academy	249
Hutchinson, Edward A.	SUNY-Utica/Rome	344
Hutchinson, Frederick E.	U of Maine 114	312
Hwang, Shoi Y.	South Carolina State U	338
Hwang, Sun-Tak	U of Cincinnati	042

Name	Institution	Page
Hynes, Jacqueline A.	U of Cal, Santa Barbara	024

I

Name	Institution	Page
Ianni, Lawrence A.	U of Minnesota, Duluth	135
Ibanez, Manuel L.	Texas A&I U	237
Ibrahim, Medhat	Cal State U, Fresno	029
Ibrahim, Nabil	San Jose State U	202
Ignizio, James P.	U of Virginia	259
Ihekweazu, Stanley N.	South Carolina State U	338
Ikenberry, Stanley O.	U of Illinois at Chicago	094
	Illinois, Urbana-Champaign	095
Ilgen, William P.	Gonzaga U	083
Inal, Osman	New Mexico Inst of Mining	153
Inan, Aziz S.	U of Portland	183
Inan, Mehmet I.	U of Portland	183
Incropera, Frank P.	Purdue U	188
Ingalls, Lee	U of Colorado-Col Spr	050
Inmon, Jerry B.	Mississippi State U	138
Irey, Richard	U of Toledo	240
Irrizary, Samuel	U of Puerto Rico	187
Irwin, J. David	Auburn U	012
Isaacs, Leslie	CUNY-City Coll	044
Iselin, John J.	The Cooper Union	057
Ishihara, Terry	Saginaw Valley State U	196
Isler, William E.	North Carolina State U	160
Israni, Haku	Prairie View A&M U	334
Istwan, Michael T.	Marquette U	118
Itani, Rafik Y.	Washington State U	264
Iverson, Scott C.	U of Washington	262

J

Name	Institution	Page
Jabbour, Kamal	Syracuse U	225
Jacks, G.K.	Colorado State U	053
Jackson, John E.	U of Alabama	002
Jackson, Mark A.	Wayne State U	265
Jacobs, Harold R.	Penn State U	180
Jacobsen, Richard T.	U of Idaho	092
Jacobsen, Stephen E.	U of Cal, Los Angeles	022
Jaggard, Dwight L.	U of Penn	179
Jain, Chittaranjan	Cleveland State U	048
Jain, Ravi K.	U of Cincinnati	042
Jakubowski, Gerald S.	Loyola Marymount U	113
James, A.D.	Prairie View A&M U	334
James, Advergus D.	Prairie View A&M U	185
James, Charles F., Jr.	U of Wisconsin-Milwaukee	275
James, Rodney	Montana Coll of Mineral Sci	143
Janeff, Wenelin D.	U of Bridgeport	016
Japp, Robert D.	McGill U	125
Jarvis, John J.	Georgia Tech	082
Jeelani, Shaik	Tuskegee U	246
Jeng, Raymond I.	Cal State U, Los Angeles	031
Jenifer, Franklyn G.	Howard U	091
Jenkins, J.	Cornell U	058
Jenkins, Leo B.	U of Louisville	111
Jenkins, Peter E.	U of Colorado at Denver	051
Jennings, Mike B.	San Jose State U	202
Jensen, Jon K.	Marquette U	118
Jepson, Paul	Ohio U	170
Jesser, William A.	U of Virginia	259
Jiles, Charles W.	U of Texas at Arlington	232
Jischke, Martin C.	Iowa State U	099
Johansen, Mariela	McGill U	125
John, James E.	GMI	077
Johnson, Alfred T.	Widener U	272
Johnson, Bruce	Lake Superior State U	310
Johnson, Ed	Montana Coll of Mineral Sci	143
Johnson, George W.	George Mason U	079
Johnson, J. Terry	Oklahoma Christian	174
Johnson, James H.	Howard U	091
Johnson, John H.	Michigan Tech U	133
Johnson, Joseph E.	U of Tennessee, Knoxville	229
Johnson, Melvin A., Jr.	Central State U	040
Johnson, Richard T.	Wichita State U	271
Johnson, Robert E.	Central State U	040
Johnson, Ronald G.	U of Alaska, Fairbanks	005
Johnson, Ronald W.	U of Cal, Davis	020
Johnson, Sherwood	SUNY-Stony Brook	223
Johnson, Vern R.	U of Arizona	008
Johnson, Walter H.	U of Minnesota	136

Name	Institution	Page
Johnston, David L.	McGill U	125
Jones, Anita K.	U of Virginia	259
Jones, B.G.	Illinois, Urbana-Champaign	095
Jones, Richard C.	Widener U	272
Jones, Rudolph	U of Cincinnati	042
Jong, Mark T.	Wichita State U	271
Jordan, Theodore	Montana Coll of Mineral Sci	143
Jucker, James V.	Stanford U	217
Judson, Cheryl	Metropolitan State-Denver	317
Juran, Ilan	Poly U	182
Jurewicz, John T.	West Virginia U	267

K

Name	Institution	Page
Kachhal, Swatantra	U of Michigan-Dearborn	131
Kail, Robert P.	Carnegie Mellon U	036
Kakac, Sadik	U of Miami	128
Kalajian, Edward H.	Florida Inst of Tech	074
Kallend, John S.	Illinois Inst of Tech	096
Kamal, Musa R.	McGill U	125
Kamat, Satish	New Mexico State U	154
Kamel, Ihab	Drexel U	064
Kamel, Khaled A.	U of Louisville	111
Kandel, Abraham	U of South Florida	209
Kane, Gerald R.	U of Tulsa	245
Kanneman, Thomas A.	U of San Diego	199
Kao, David T.	Iowa State U	099
Kao, Timothy W.	The Catholic U of America	038
Kapoor, Vik J.	U of Cincinnati	042
Kapp, Gloria J.	Cal State U, Long Beach	030
Kardan, Cevat	Virginia State U	350
Kardos, John L.	Washington U	263
Karig, D.E.	Cornell U	058
Karmis, Michael E.	Virginia Tech	261
Karr, Gerald R.	U of Alabama at Huntsville	004
Kartsounes, George T.	GMI	077
Kassam, Saleem	U of Penn	179
Kasuba, Romualdas	Northern Illinois U	165
Katz, Norman	Washington U	263
Katz, Phyllis S.	U of Hartford	308
Katz, Sharlene	Cal State U, Northridge	032
Kauffman, David	The U of New Mexico	152
Kaveh, Mostafa	U of Minnesota	136
Kays, James L.	US Military Academy	251
Kazimi, Mujid S.	Mass Inst of Tech	124
Kear, Bernard H.	Rutgers, The State U	195
Keating, James L.	U of Alabama	002
Keating, Larry G.	Metropolitan State-Denver	317
Keenan, John D.	U of Penn	179
Keer, Leon M.	Northwestern U	166
Kehrberg, Noelle	Western Carolina U	354
Keinath, Thomas M.	Clemson U	047
Kelley, Benjamin S.	Mercer U	127
Kelley, Patricia H.	U of North Dakota	161
Kelly, Allen E.	Oklahoma State U	175
Kelly, C. Michael	Villanova U	258
Kelly, Eamon M.	Tulane U	244
Kelly, Graham	U of Akron	001
Kelly, James L.	U of Virginia	259
Kelly, John	North Carolina A&T	159
Kelly, Richard W.	Arizona State U	009 283
Kelly, William E.	U of Nebraska - Lincoln	146
Kendall, Alden L.	U of Minnesota, Duluth	135
Kender, John R.	Columbia U	054
Kenig, M.J.	U of New Haven	150
Kennedy, Lawrence A.	Ohio State U	172
Kennedy, Timothy R.	Lawrence Tech U	106
Kenniff, Patrick	Concordia U	055
Kennington, Jeffery L.	Southern Methodist U	216
Kenny, Joan M.	SUNY-Stony Brook	223
Kent, Albert C.	Southern Illinois-Carbondale	213
Kent, George	Northeastern U	325
Kerlin, Thomas W.	U of Tennessee, Knoxville	229
Kerns, David V.	Vanderbilt U	256
Kerr, Thomas J.	Boston U	014
Kerschberg, Larry	George Mason U	079
Kersh, Kenneth	Arkansas Tech U	011
Kertesz, Andrew E.	Northwestern U	166
Keshock, Edward J.	Cleveland State U	048
Kessler, David P.	Purdue U	188
Ketabchi, Mohammad A.	Santa Clara U	203
Keys, L.K.	Louisiana State U	109
Khalil, Tarek M.	U of Miami	128
Khan, Akhtar S.	Maryland Baltimore Cty	119
Khan, Khalid H.	U of Portland	183
Kheir, Naim A.	Oakland U	169
Kibler, David F.	Virginia Tech	261
Kice, John	U of Denver	062
Kicher, Thomas P.	Case Western Reserve U	037
Kielhorn, William H.	LeTourneau U	107
Kiesler, Charles	U of Missouri-Columbia	139
Kiesling, Ernst W.	Texas Tech U	239
Kiger, Sam A.	West Virginia U	267
Kihara, Deane H.	U of Hawaii at Manoa	088
Kiker, William E.	U of Rochester	192
Kim, Jai B.	Bucknell U	018
Kim, Thomas J.	U of Rhode Island	190
Kincaid, Thomas G.	Boston U	014
Kincheloe, Nancy	Oregon Inst of Tech	328
Kinder, Paul F.	Lawrence Tech U	106
King, Franklin G.	North Carolina A&T	159
King, J. Phillip	New Mexico State U	154
King, L.E.	UNC-Charlotte	158
King, Paul H.	Northeastern U	163
King, Ronald S.	Lamar U	105
King, Terry S.	Iowa State U	099
King, William E.	Bucknell U	018
Kiper, Ali M.	The George Washington U	080
Kirkpatrick, David A.	Prairie View A&M U	334
Kirkpatrick, Samuel A.	U of Texas at San Antonio	236
Kirwan, William E.	U of Maryland, Coll Park	120
Kiser, Kenneth M.	SUNY-Buffalo	219
Kizina, Terrance	Point Park Coll	333
Klapper, Jacob	New Jersey Inst of Tech	151
Klavuhn, John G.	U of Pittsburgh at Johnstown	332
Klayton, Alan R.	US Air Force Academy	248
Klein, Bernard W.	U of Michigan-Dearborn	131
Klein, Dale E.	U of Texas at Austin	233
Klein, Michael T.	U of Delaware	061
Kleinrock, Leonard	U of Cal, Los Angeles	022
Kletsky, Earl J.	Syracuse U	225
Klevans, Edward H.	Penn State U	180
Kline, Kenneth A.	Wayne State U	265
Klinzing, George E.	U of Pittsburgh	181
Klir, George J.	SUNY-Binghamton	218
Knapp, Charles B.	U of Georgia	081
Knapp, John W.	Virginia Military Inst	260
Knappenberger, H. Allan	Wayne State U	265
Knickerbocker, Christina M.	SUNY-Binghamton	218
Knight, James F.	Oklahoma State U	175
Knight, Winston A.	U of Rhode Island	190
Knopp, Jerome	U of Missouri-Columbia	139
	U of Missouri-Kansas City	140
Knowles, James K.	Cal Inst of Tech	026
Knowlton, James M.	Wentworth Inst of Tech	352
Knudsen, H. Peter	Montana Coll of Mineral Sci	143
Kocaoglu, Dundar F.	Portland State U	184
Koch, James V.	Old Dominion U	176 327
Koch, Paul E.	Penn State-Erie	329
Koehn, Enno	Lamar U	105
Koenig, Gottlieb	New York Inst of Tech	156
Koester, Roger	Colorado School of Mines	052
Kohser, Ronald A.	U of Missouri-Rolla	141
Kolar, Michael J.	U of Pittsburgh	181
Kolb, John E.	Rensselaer Poly Inst	189
Kondelis, Frank	Montana Coll of Mineral Sci	143
Koplik, Bernard	New Jersey Inst of Tech	151
Kornbaum, H. Frank	South Dakota State U	208
Korus, Roger A.	U of Idaho	092
Koss, Donald A.	Penn State U	180
Kovac, Michael G.	U of South Florida	209
Kowel, Stephen T.	U of Alabama at Huntsville	004
Kraft, David	U of Kansas	101
Krahe, Ronald P.	Penn State-Erie	329
Krane, Kenneth S.	Oregon State U	177
Krasner, Jerome L.	Wentworth Inst of Tech	352
Kraus, David W.	U of New Hampshire	149 321
Krause, Wayne B.	South Dakota School of Mines	207
Krempl, Erhard	Rensselaer Poly Inst	189
Krinsky, Paul L.	US Merchant Marine Academy	250
Krishnamoorthi, Kalikathon S.	Bradley U	015
Krishnan, V.V.	San Francisco State U	201
Krogman, John A.	U of Wisconsin-Platteville	276
Krothapalli, Anjaneyulu	Florida A&M U	072
	Florida State U	076
Kruger, Charles H.	Stanford U	217
Krulee, Carolyn H.	Northwestern U	166
Krygar, Paul W.	Rutgers, The State U	195
Kryman, Fritz J.	U of Cincinnati	293
Kuehl, Hans H.	U of Southern Cal	211
Kuenhold, Kenneth A.	U of Tulsa	245
Kugelman, Irwin J.	Lehigh U	108
Kulacki, Frank A.	Colorado State U	053
Kullgren, Thomas E.	Saginaw Valley State U	196
Kumar, K.S.P. (Pat)	U of Minnesota	136
Kumin, Hillel J.	U of Oklahoma	173
Kummler, Ralph H.	Wayne State U	265
Kundapur, Sandanand	Tri-State U	241
Kundel, John A.	Western Michigan U	269 356
Kuntz, Richard A.	Monmouth Coll	142
Kunze, Jay F.	U of Missouri-Columbia	139
Kunzler, John J.	Brigham Young U	017 287
Kuo, Chin Y.	Penn State U	180
Kuo, Way	Iowa State U	099
Kupchella, Charles E.	Western Kentucky U	355
Kupferman, Michael	Wentworth Inst of Tech	352
Kurstedt, Pamela S.	Virginia Tech	261
Kurth, Ronald J.	Murray State U	320
Kurtz, Barry L.	Louisiana Tech U	110
Kuruganty, Sastry	U of North Dakota	161
Kusiak, Andrew	U of Iowa	098
Kwor, Richard Y.C.	U of Colorado-Col Spr	050
Kyros, William	U of Mass, Lowell	123

L

Name	Institution	Page
Lagnese, Robert J.	SUNY-Farmingdale	343
LaGraff, John E.	Syracuse U	225
Lake, Larry W.	U of Texas at Austin	233
Lakeou, Samuel	U of the District of Columbia	063
Lalevic, Bogoljub	Rutgers, The State U	195
Laliberte, Garland E.	U of Manitoba	115
Lam, Paul C.	U of Akron	001
Lamar, Howard R.	Yale U	280
Lamberson, Leonard R.	Western Michigan U	269 356
Lambert, Arthur	New York Inst of Tech	156
Lambert, Audrey S.	Mississippi State U	138
Lambert, Rita A.	U of Portland	183
Lamens, GiGi	SUNY-Stony Brook	223
Lamkin, Fletcher M.	US Military Academy	251
Lampen Smith, Donna	U of Rochester	192
Lamson, Sharon	DeVry Inst-Kansas City	300
Lancaster, James	Alfred U	007
Lanctôt, Bernard	École Polytechnique	066
Landis, Raymond B.	Cal State U, Los Angeles	031
Landward, J. Stayner	U of Utah	253
Lang, Theodore E.	Montana State U	144 319
Langdon, Glen G.	U of Cal, Santa Cruz	025
Langlinais, Julius P.	Louisiana State U	109
Lanier, Percy N.	Alabama A&M U	282
Lannes, W.J.	U of New Orleans	155
Larrabee, Tracy	U of Cal, Santa Cruz	025
Larry, Redekopp G.	U of Southern Cal	211
Larsen, Jay	South Dakota State U	208
Laskaris, Maria	Dartmouth Coll	059
Lasko, Richard	U of Toledo	242
Laszewski, Ronald T.	Bucknell U	018
Latorre, Robert	U of New Orleans	155
LaTourette, John E.	Northern Illinois U	165
Lattman, Laurence H.	New Mexico Inst of Mining	153
Laughlin, Patricia	Carnegie Mellon U	036
Law, Frederick M.	U of Mass, Dartmouth	122
Lawless, Robert W.	Texas Tech U	239 348
Lawrence, Francis L.	Rutgers, The State U	195
Lawrence, Kent L.	U of Texas at Arlington	232
Lawrence, Molly M.	U of Alabama	002
Lawrie, D.H.	Illinois, Urbana-Champaign	095
Lawson, Huey K.	Southern U & A&M Coll	210
Lay, Billy G.	Texas A&M U	347
Lazerson, Earl E.	Southern Illinois-Edwardsville	214
Leal, L. Gary	U of Cal, Santa Barbara	024
Leamy, Harry J.	UNC-Charlotte	158 324
LeBarre, Rita	Santa Clara U	203
Leblanc, Robert A.	École Polytechnique	066
Ledbetter, Phillip C.	Embry-Riddle-Daytona Beach	067 305
Lee, George C.	SUNY-Buffalo	219
Lee, Jen-Shih	U of Virginia	259

Index to Engineering & Technology Personnel

Name	Institution	Page
Lee, Kai Fong	U of Toledo	240
Lee, Peter Y.	Cal Poly State U	027
Lee, Rex E.	Brigham Young U 017	287
Lee, Samuel S.	U of Miami	128
Lee, Sung W.	U of Maryland, Coll Park	120
Lee, Sunggyu	U of Akron	001
Lee, Wilson	Tennessee State U	230
Leffen, David J.	GMI	077
Lehmkuhl, Nonna	Northeastern U	325
Leimkuhler, Ferdinand F.	Purdue U	188
Leininger, Gary	U of Colorado at Denver	051
Leitch, Donald G.	U of Mass, Lowell	123
Leitmann, George	U of Cal at Berkeley	019
LeMee, Jean	The Cooper Union	057
LeMelle, Tilden J.	U of the District of Columbia 063	303
Lemmon, E.C.	U of Idaho	092
Lennon, A. Max	Clemson U	047
Lenz, Terry G.	Colorado State U	053
Lenzmeier, Timothy	North Dakota State U	162
Leonard, John W.	U of Conn	056
Leonard, Michael S.	Clemson U	047
Lessard, Ronald	Norwich U	167
Lester, Thomas W.	U of Kentucky	103
Lestingi, Joseph	U of Dayton	060
Letterman, Raymond D.	Syracuse U	225
Leung, Joseph Y-T	U of Nebraska - Lincoln	146
LeValley, C. Robert	DeVry Inst-Kansas City	300
Leventhal, Ruth	Penn State U - Harrisburg	330
Leventman, Paula G.	Northeastern U	163
Levin-Stankevich, Brian	Florida Atlantic U	073
Levis, Alexander	George Mason U	079
Levy, David J.	Cal Inst of Tech	026
Levy, Louis F.	Florida Inst of Tech	074
Lewandowski, Gordon	New Jersey Inst of Tech	151
Lewis, Daniel W.	Santa Clara U	203
Lewis, Edwin R.	U of Cal at Berkeley	019
Lewis, Elmer E.	Northwestern U	166
Lewis, Gordon	Clemson U	047
Lewis, Marlyn M.	Harvard U	086
Lewis, Richard B.	Louisiana Tech U	311
Lewis, William E.	Arizona State U 009	283
Lewis-Logue, Judith	U of San Diego	199
Liacouras, Peter J.	Temple U 227	345
Liberty, Stanley R.	U of Nebraska - Lincoln	146
Lick, Dale W.	Florida State U	076
Liedl, Gerald L.	Purdue U	188
Likins, Peter W.	Lehigh U	108
Lilley, John M.	Penn State-Erie	329
Lilley, Walter	Tri-State U 241	349
Lim, Henry C.	U of Cal, Irvine	021
Lin, Shield B.	Prairie View A&M U	185
Lin, Shu	U of Hawaii at Manoa	088
Lindbergh, Charles	The Citadel	043
Lindenmeyer, Mark	Loyola Coll in Maryland	112
Lindley, Carolyn	Northwestern U	166
Linehan, John H.	Marquette U	118
Linz, Peter	U of Cal, Davis	020
Lipinski, Martin	Memphis State U	125
Lipowsky, Herbert H.	Penn State U	180
Lisbon, Richard A.	Syracuse U	225
List, Bernie H.	U of Texas at Dallas	234
Litynski, Daniel M.	US Military Academy	251
Liu, Jiang	U of Alberta	006
Livengood, Charles D.	North Carolina State U	160
Livingston, Ann C.	U of Vermont	257
Llowellyn, John A.	U of South Florida	209
Lloyd, Douglas R.	U of Texas at Austin	233
Lloyd, Patricia K.	U of Cincinnati	293
Lluch, José F.	U of Puerto Rico	187
Locatelli, Paul L., S.J.	Santa Clara U	203
Locke, Carl E., Jr.	U of Kansas	101
Locust, Wayne A.	U of Rochester	192
Loerts, Sandra	Mankato State U 116	313
Loewer, Otto J.	U of Florida	071
Loh, Robert N.K.	Oakland U	169
Lohman, Rachael L.	Wilkes U	273
Lohmann, Jack R.	Georgia Tech	082
Lombardi, John V.	U of Florida	071
Lomen, David	U of Arizona	008
Loncorich, Frank	St. Cloud State U 197	337
Long, Stuart A.	U of Houston	090
Long, Thomas R.	West Virginia U	267
Longley, Karl	Cal State U, Fresno	029
Loparo, Kenneth A.	Case Western Reserve U	037
Lopez, Beto	U of Texas at El Paso	235
Lord, Paul A.	Cal State Poly, Pomona	034
Lord, William	Iowa State U	099
Lorenz, John D.	GMI	077
Loscutoff, Walter V.	Cal State U, Fresno	029
Lovell, Edward G.	U of Wisconsin-Madison	274
Lovett, Thomas P.	Tulane U	244
Low, Marc E.	Wright State U	278
Lowen, Gerard G.	CUNY-City Coll	044
Lozano, Noe P.	Stanford U	217
Lubbers, Arend D.	Grand Valley State U	084
Lucas, Aubrey K.	U of Southern Mississippi	342
Lucchesi, Art	The Cooper Union	057
Lucht, David A.	Worcester Poly Inst	277
Lucido, Jerome A.	U of Arizona	008
Luhmann, Neville C., Jr.	U of Cal, Davis	020
Lukas, Veronica	Poly U	182
Luks, Kraemer D.	U of Tulsa	245
Lumsdaine, Edward	U of Toledo	240
Lundquist, Daniel M.	Union Coll	247
Lundstrom, Mark S.	Purdue U	188
Lunkenheimer, Carol	Northwestern U	166
Lupin, Dan	Embry-Riddle-Western Campus	068
Luss, Dan	U of Houston	090
Lutes, Loren D.	Auburn U	012
Luthy, Richard G.	Carnegie Mellon U	036
Luttrell, Henry H.	Christian Brothers U	041
Lutz, Francis C.	Worcester Poly Inst	277
Luxon, James T.	GMI	077
Luyanda, Felipé	U of Puerto Rico	187
Lyall, James	Embry-Riddle-Daytona Beach	067
Lyall, Katharine	U of Wisconsin-Madison	274
Lykken, Glenn I.	U of North Dakota	161
Lynch, David T.	U of Alberta	006
Lynch, P. Michael	Tulane U	244
Lynch, Robert D.	Villanova U	258
Lynch, Thomas C.	US Naval Academy	252
Lyon, Russ	Oregon Inst of Tech	328
Lyons, Donald W.	West Virginia U	267

M

Name	Institution	Page
Ma, Tso-Ping	Yale U	280
MacDonald, N.C.	Cornell U	058
Macias, Miguel	Cal State U, Northridge	032
Mack, Carol	U of Michigan-Dearborn	131
Mack, Pamela L.	Morgan State U	145
Mackiewicz, Hank	Monmouth Coll	142
Madigan, Mary-Ellen	Penn State-Erie	329
Madrigal, Orlando S.	Cal State U, Chico	028
Maffia, Gennaro J.	Widener U	272
Mafi, Mohammad	Union Coll	247
Mager, James J.	Ohio State U	172
Magill, Samuel H.	Monmouth Coll	142
Maginnis, P. Tobin	U of Mississippi	137
Mahoney, Michael	Widener U	272
Mahoney, Michael K.	Cal State U, Long Beach	030
Maidique, Modesto A.	Florida International U	075
Maier, Leo R.	Ohio Northern U	171
Mailer, Kathleen	Seattle U	204
Malasri, Siripong	Christian Brothers U	041
Malette, Tim T.	Michigan Tech U	133
Malkani, Mohan J.	Tennessee State U	230
Malloy, Edward A.	U of Notre Dame	168
Malone, Michael P.	Montana State U 144	319
Malstrom, Eric M.	U of Arkansas	010
Mangonon, Pat L.	Florida Inst of Tech	074
Mann, J. Adin	Case Western Reserve U	037
Mantey, Patrick E.	U of Cal, Santa Cruz	025
Marburger, John H.	SUNY-Stony Brook	223
Marburger, Richard E.	Lawrence Tech U	106
Marchman, James F.	Virginia Tech	261
Marciniak, John J.	Gonzaga U	083
Marcotte, Philip	Ferris State U	306
Marcy, William M.	Texas Tech U	239
Marek, Miroslav I.	Georgia Tech	082
Marino, Leonard R.	San Diego State U	200
Marley, Jerry J.	U of Notre Dame	168
Marmarelis, Vasilis Z.	U of Southern Cal	211
Marsden, William L.	Washington U	263
Marshak, Alan H.	Louisiana State U	109
Marshall, Joseph C.	Cal State Poly, Pomona 034	289
Martin, Dennis J.	Washington U	263
Martin, Harold L.	North Carolina A&T	159
Martin, J. Ronald	Lafayette Coll	104
Martin, Jose	U of Mass, Lowell	123
Martin, Paul C.	Harvard U	086
Martin, Richard L.	US Naval Academy	252
Martin, William	Embry-Riddle-Daytona Beach	305
Martin, William R.	U of Michigan	130
Martinez, Andrew B.	Tulane U	244
Martinez, David R.	Colorado State U	053
Maryak, Barbara	U of Bridgeport	016
Maskalenko, Edward J.	Tufts U	243
Maslowski, Richard S.	Lawrence Tech U	106
Mason, John A.	U of Michigan-Dearborn	131
Mason, John T., III	Tennessee Tech	231
Mason, Larry	U of Nevada, Las Vegas	147
Mason, Thelma R.	CUNY-City Coll	044
Masson, Gerald M.	The Johns Hopkins U	100
Masters, Gilbert M.	Stanford U	217
Mathers, Lewis J.	Villanova U	258
Mathieson, John L.	SUNY-Maritime	221
Mathis, George L.	U of Southern Mississippi	342
Mathsen, Don V.	U of North Dakota	161
Mattauch, Robert J.	U of Virginia	259
Matteson, Thomas T.	US Coast Guard Academy	249
Matthes, R. Kenneth	Mississippi State U	138
Maurer, Karl G.	North Dakota State U	162
Maurizio, Don M.	Cal State U, Los Angeles	031
Maxham, John C.	US Coast Guard Academy	249
Maxson, Robert C.	U of Nevada, Las Vegas	147
Maxum, Bernard J.	Lamar U	105
Mayer, John E.	Texas A&M U	347
Mayfield, Benny R.	South Carolina State U	338
Mayo, Sandi	Christian Brothers U	041
Mays, John E.	U of Tennessee, Knoxville	229
Mays, Larry W.	Arizona State U	009
McAtee, James	Oregon Inst of Tech	328
McBeath, Giorgio M.	Wright State U	278
McBrayer, James D.	U of Central Florida	292
McCallum, Charles A.	U of Alabama at Birmingham	003
McCann, Claude E.	U of Alabama at Birmingham	003
McCarter, M. Kim	U of Utah	253
McCarty, Barry W.	Lafayette Coll	104
McCauley, James W.	NY State Coll of Ceramics	157
McClain, Gerald	Oklahoma State U	326
McClamroch, N. Harris	U of Michigan	130
McCloud, Robert	Florida State U	076
McComas, James D.	Virginia Tech	261
McCoy, Benjamin J.	U of Cal, Davis	020
McCoy, John J.	The Catholic U of America	038
McCray, Curtis L.	Cal State U, Long Beach	030
McCreight, Keith	Oregon State U	177
McCullen, John D.	U of Arizona	008
McCulloch, Harold Y.	Princeton U	186
McDonough, James F.	U of Cincinnati	042
McDonough, John J.	U of Maine	312
McDonough, Patrick D.	Marietta Coll	117
McDougal, Johnny	Murray State U	320
McElhaney, James H.	Duke U	065
McEliece, Robert J.	Cal Inst of Tech	026
McElroy, John H.	U of Texas at Arlington	232
McGee, Minnie M.	Ohio State U	172
McGhee, Terence J.	Lafayette Coll	104
McGinn, Thomas J.	Wentworth Inst of Tech	352
McGoff, Michael F.	SUNY-Binghamton	218
McGrath, Paul B.	Clarkson U	046
McHenry, Albert L.	Arizona State U	283
McKernan, John F.	US Naval Academy	252
McKillop, Allan A.	U of Cal, Davis	020
McKinney, James L.	Rose-Hulman	194
McKinney, Mary H.	U of Central Florida 039	292
McLaren, Malcolm G.	Rutgers, The State U	195
McLauchlan, Robert A.	Texas A&I U	237
McLaughlin, Dennis K.	Penn State U	180
McLaughlin, Edward	Louisiana State U	109
McLaughlin, John F.	Purdue U	188
McLean, Thomas J.	U of Texas at El Paso	235
McManis, Kenneth L.	U of New Orleans	155
McMeeking, Robert M.	U of Cal, Santa Barbara	024
McMurtry, George J.	Penn State U	180
McNair, Emerelle	Southern Coll of Tech	339
McNeary, Paul R.	Murray State U	320

Name	Institution	Page
McNeill, Perry R.	U of Southern Colorado	340
McNichols, Roger	U of Toledo	240
McNitt, Richard P.	Penn State U	180
McPherson, Joseph D.	U of Wisconsin-Milwaukee	275
McRee, Griffith J.	Old Dominion U	176
McShane, William	Poly U	182
McWhertor, Thomas E.	Calvin Coll	035
McWhorter, John C., III	Mississippi State U	138
McWilliams, Thomas G.	Widener U	272
Mead, Jerud J.	Bucknell U	018
Meakin, John D.	U of Delaware	061
Mechnig, Virginia	DeVry Inst-DuPage	298
Medeksza, Ludwik A.	Gannon U	078
Meese, Jon	U of Missouri-Columbia	139
Mehrabian, Robert	Carnegie Mellon U	036
Meier, Wilbur L.	North Carolina State U	160
Meister, Robert	The Catholic U of America	038
Meitz, Robert O.	Arizona State U	283
Melon, Rose Marie	New Mexico State U 154	323
Melvin, Donald W.	U of New Hampshire 149	321
Meredith, Dale D.	SUNY-Buffalo	219
Meredith, Thomas C.	Western Kentucky U	355
Merrill, M. Don	New Mexico State U	154
Merrill, Robert	Rochester Inst of Tech	336
Mersten, Gerald S.	West Virginia Inst of Tech	268
Merz, Richard A.	Lafayette Coll	104
Messenger, Roger A.	Florida Atlantic U	073
Mestas, Gina T.	U of Southern Colorado 212	340
Metarko, Peter F.	Florida State U	076
Meyburg, A.H.	Cornell U	058
Meyer, Alvin H.	U of Texas at Austin	233
Meyer, Daniel L.	NY State Coll of Ceramics	157
	Alfred U	007
Meyers, Doreen	New York Inst of Tech	156
Meyers, William G.	Tri-State U 241	349
Michael, Robert K.	U of Cincinnati	042
Michaelides, Efstathios E.	Tulane U	244
Michel, Anthony N.	U of Notre Dame	168
Middleton, Silvia G.	UNC-Charlotte 158	324
Midturi, Swaminadham	Texas A&M U	347
Mielke, Roland R.	Old Dominion U	176
Miksad, Richard W.	U of Texas at Austin	233
Miles, John D.	U of Notre Dame	168
Mililio, Frank F.	Union Coll	247
Millan, Danny	DeVry Inst-Dallas	299
Miller, Anthony G.	Capitol Coll	290
Miller, Beverly W.	Western New England Coll	270
Miller, Charles A.	CUNY-City Coll	044
Miller, Clarence A.	Rice U	191
Miller, David R.	U of Cal, San Diego	023
Miller, Dorothy	Cal State U, Northridge	032
Miller, Elinore	U of Colorado at Denver	051
Miller, Eugene L.	DeVry Inst-Chicago	295
Miller, Eugene E.	DeVry Inst-Chicago	295
Miller, Floyd H.	SUNY-Maritime	221
Miller, Irving F.	U of Illinois at Chicago	094
Miller, James S.	Harvard U	086
Miller, Mary A.	Metropolitan State-Denver	317
Miller, Melton M.	U of Mass	121
Miller, Reid C.	Washington State U	264
Miller, Richard K.	U of Iowa	098
	U of Southern Cal	211
Miller, Richard N.	U of Central Florida 039	292
Miller, William A.	U of Tennessee, Knoxville	229
Milligan, Thedola H.	U of the District of Columbia 063	303
Mills, Donald T.	Clarkson U	046
Mills, Nancy L.	U of Southern Colorado	212
Mills, Richard P., Jr.	U of Hartford	308
Mincey, Jeffrey C.	Grove City Coll	085
Minich, Marlin D.	Ohio Northern U	171
Minkarah, Issam A.	U of Cincinnati	042
Minor, Joseph E.	U of Missouri-Rolla	141
Mirchandani, Pitu	U of Arizona	008
Mirth, Richard A.	Northern Arizona U	164
Mishu, Farouk P.	Tennessee State U	230
Miska, Stefan	U of Tulsa	245
Mitchell, E. Eugene	US Naval Academy	252
Mitchell, James E.	Drexel U	064
Mitchell, Jerrel	Ohio U	170
Mitchell, M.E.	Rutgers, The State U	195
Mitchell, Owen R.	U of Texas at Arlington	232
Mitchell, Tony L.	North Carolina A&T	159
Moas, Olga V.	Florida Atlantic U	073
Mobley, William H.	Texas A&M U 238	347
Moeder, Larry	Kansas State U	102
Mohning, David	Vanderbilt U	256
Mohr, Joan I.	Hofstra U	089
Mohraz, Bijan	Southern Methodist U	216
Molen, G. Marshall	Mississippi State U	138
Molinder, John I.	Harvey Mudd Coll	087
Moll, Michael D.	Washington U	263
Momo, Lawrence J.	Columbia U	054
Momoh, James A.	Howard U	091
Monaghan, William	CUNY-Staten Island	045
Monkmeyer, Peter L.	U of Wisconsin-Madison	274
Monroe, Joseph	North Carolina A&T	159
Montazer, M. Ali	U of New Haven	150
Monteith, Larry K.	North Carolina State U	160
Montgomery, Trent V.	Southern U & A&M Coll	210
Montoya, James M.	Stanford U	217
Moon, Donald L.	U of Dayton	060
Moore, David	DeVry Inst-Los Angeles	301
Moore, Duncan T.	U of Rochester	192
Moore, F.	Cornell U	058
Moore, John J.	Colorado School of Mines	052
Moore, June Q.	U of Tennessee, Knoxville	229
Moore, Marc	École Polytechnique	066
Moore, Noel E.	Rose-Hulman	194
Moore, Robert E.	U of Missouri-Rolla	141
Morari, Manfred	Cal Inst of Tech	026
Moreira, Antonio R.	Maryland Baltimore Cty	119
Moret, Bernard M.	The U of New Mexico	152
Morey, R. Vance	U of Minnesota	136
Morgali, James R.	U of the Pacific	178
Morgan, Bruce J.	SUNY-Farmingdale	343
Morgan, Granger	Carnegie Mellon U	036
Morgan, J. Derald	New Mexico State U	154
Morgan, James O.	Prairie View A&M U	185
Morgan, Joseph A.	Texas A&M U	347
Morley, Lloyd A.	U of Alabama	002
Morris, David	U of Virginia	259
Morris, James E.	SUNY-Binghamton	218
Morris, John	Temple U 227	345
Morris, Richard D.	Portland State U	184
Morrison, H. Frank	U of Cal at Berkeley	019
Morrison, Scott	DeVry Inst-Phoenix	302
Morton, Nina M.	West Virginia Inst of Tech 268	353
Moser, A P.	Utah State U	254
Moses, Fred	U of Pittsburgh	181
Moses, Joel	Mass Inst of Tech	124
Moss, Ron W.	Rice U	191
Mostaghel, Naser	U of Toledo	240
Mott, Robert L.	U of Dayton	294
Motteler, Zane	Cal Poly State U	027
Mowdy, Mary	U of Oklahoma	173
Moyer, Ruth Ann	Cleveland State U	048
Muckstadt, J.A.	Cornell U	058
Mueller, Marianne S.	Ohio State U	172
Mueller, Thomas J.	U of Notre Dame	168
Mukherjee, Kalinath	Michigan State U	132
Mulhollan, Paige E.	Wright State U	278
Mulholland, George	New Mexico State U	154
Mulinazzi, Thomas E.	U of Kansas	101
Mullen, Frank	Texas A&I U	237
Mullens, Lanny J.	Northern Arizona U	164
Muller, Carol B.	Dartmouth Coll	059
Muller, Mervin E.	Ohio State U	172
Muller, Warren W.	U of San Diego	199
Mullins, Joe H.	The U of New Mexico	152
Mulvihill, Michael E.	Loyola Marymount U	113
Munir, Zuhair A.	U of Cal, Davis	020
Munro, James M.	South Dakota School of Mines	207
Munsell, William	Lake Superior State U	310
Munsey, William B.	U of Maine	312
Munsey, William J.	U of Maine	114
Munsterman, Richard E.	Western Michigan U	269
Muntz, E. Phillip	U of Southern Cal	211
Murman, Earll M.	Mass Inst of Tech	124
Murphree, E. Lile, Jr.	The George Washington U	080
Murphy, Michael	Embry-Riddle-Daytona Beach	305
Murr, Larry	U of Texas at El Paso	235
Murray, Kenneth H.	North Carolina A&T	159
Murray, William G.	Webb Inst	266
Muse, William V.	Auburn U 012	285
Mussulman, Ronald L.	Cal Poly State U	027
Myers, Greeley	New Mexico State U 154	323
Myers, Laura	DeVry Inst-Dallas	299
Myerson, Allan S.	Poly U	182

N

Name	Institution	Page
Nadler, Gerald	U of Southern Cal	211
Nagib, Hassan	Illinois Inst of Tech	096
Nagle, Edward J.	Tri-State U	349
Naider, Fred R.	CUNY-Staten Island	045
Naimark, Arnold	U of Manitoba	115
Naismith, Donald P.	U of North Dakota	161
Nandor, Sarah	Parks Coll of Saint Louis U	198
Narayan, Raghu	Texas Tech U	239
Narayanamurti, Venkatesh	U of Cal, Santa Barbara	024
Nasar, Syed A.	U of Kentucky	103
Natalicio, Diana S.	U of Texas at El Paso	235
Navratil, Gerald A.	Columbia U	054
Nedderman, Wendell H.	U of Texas at Arlington	232
Nee, Esperanza L.	U of Cal, Santa Cruz	025
Nejib, Umid R.	Wilkes U	273
Nelson, Edward	New York Inst of Tech	156
Nelson, Frederick C.	Tufts U	243
Nelson, J. Keith	Rensselaer Poly Inst	189
Nelson, James E.	Louisiana Tech U 110	311
Nelson, John B.	Western New England Coll	270
Nelson, Russell C.	U of Nebraska - Lincoln	146
Nelson, Thomas J.	U of Portland	183
Nelson, Vordyn	Ferris State U 070	306
Ness, Arnold E.	Bradley U 015	286
Nester, Ernest E.	West Virginia Inst of Tech	268
Neumann, Edward S.	U of Nevada, Las Vegas	147
Neusen, Kenneth F.	U of Wisconsin-Milwaukee	275
Nicholson, David W.	U of Central Florida	039
Nicks, Roy S.	East Tennessee State U	304
Niederjohn, Russell J.	Marquette U	118
Niemeyer, Darryl W.	Embry-Riddle-Daytona Beach 067	305
Niklaus, John I.	Tulane U	244
Nikravesh, Parviz	U of Arizona	008
Nitzschke, Dale F.	U of New Hampshire 149	321
Nix, Stephen J.	Syracuse U	225
Nix, William D.	Stanford U	217
Nnaji, Soronadi	Florida A&M U	072
	Florida State U	076
Noll, Kenneth E.	Illinois Inst of Tech	096
Noori, Mohammad N.	Worcester Poly Inst	277
Nordgren, Ronald P.	Rice U	191
Nored, Lynn S.	Oklahoma Christian	174
Norman, Lindsay D.	Montana Coll of Mineral Sci	143
Normandin, Michel	École Polytechnique	066
Normann, Richard A.	U of Utah	253
Norris, Roy H.	Wichita State U	271
Norsym, Arlene F.	U of Illinois at Chicago	094
Northrop, Gaylord M.	U of Arkansas at Little Rock	284
Nosseir, Nagy	San Diego State U	200
Nottingham, Larry C.	West Virginia Inst of Tech	268
Nowatzki, Edward	Cal Poly State U	027
Nuckols, Marshall L.	US Naval Academy	252
Nunn, Robert	U of Mass, Lowell	123
Nutter, John F.	U of Wyoming	279
Nutter, Roy S.	West Virginia U	267

O

Name	Institution	Page
Obear, Frederick W.	U of Tennessee, Chattanooga	228
O'Brien, Edward M.	Mercer U	127
O'Brien, G. Dennis	U of Rochester	192
O'Brien, Gregory M.	U of New Orleans	155
O'Brien, Lorraine F.	U of Missouri-Kansas City	140
O'Brien, Walter F.	Virginia Tech	261
O'Connell, John P.	U of Virginia	259
O'Connell, William R., Jr.	New England Coll	148
O'Connor, John D.	U of Pittsburgh	181
O'Dette, Veronica G.	U of Conn	056
Ogilvie, T. Francis	Mass Inst of Tech	124
Ohanian, M. Jack	U of Florida	071
Ohley, William J.	U of Rhode Island	190
Ohtmer, Ortwin	Cal State U, Long Beach	030
Okiishi, Theodore H.	Iowa State U	099
Oldenburg, John T.	Cal State U, Sacramento 033	288
Olivieri, Joseph B.	Lawrence Tech U	106
Olmstead, Edward	Northwestern U	166

Olsen, Joseph M.	U of South Alabama	205
O'Malley, Thomas P.	Loyola Marymount U	113
O'Melia, Charles R.	The Johns Hopkins U	100
Omurtag, Yildirim	U of Missouri-Rolla	141
O'Neill, Michael W.	U of Houston	090
O'Neill, Walter J.	Illinois Inst of Tech	096
Ono, Kanji	U of Cal, Los Angeles	022
Onwubiko, Chinyere	Tennessee State U	230
Onyshko, Steve	U of Manitoba	115
O'Rear, Michael	U of Texas at Dallas	234
Orehotsky, John L.	Wilkes U	273
Oren, Shmuel S.	U of Cal at Berkeley	019
Orr, John A.	Worcester Poly Inst	277
Orthlieb, Frederick L.	Swarthmore Coll	224
Ortiz, Jorge	U of Puerto Rico	187
Ortiz, Noe	Harvey Mudd Coll	087
Osler, John C.	McGill U	125
Osman, M.O.M.	Concordia U	055
Ott, Richard W.	Alfred U	007
Otte, Paul J.	Franklin U	307
Otter, James L.	Pittsburg State U	331
Ottino, Julio M.	Northwestern U	166
Otto, Fred D.	U of Alberta	006
Owen, Norman	San Francisco State U	201
Owen, S John T.	Oregon State U	177
Owens, John M.	Auburn U 012	285
Owens, Thomas C.	U of North Dakota	161
Ozbun, J.L.	North Dakota State U	162

P

Pacheco, Manuel T.	U of Arizona	008
Page, John A.	Loyola Marymount U	113
Palmer, Donna	Loyola Marymount U	113
Palmer, Harvey J.	U of Rochester	192
Palmer, James D.	George Mason U	079
Palmer, John L.	Mercer U	127
Palmer, Ray W.	Milwaukee Sch of Eng 134	318
Palms, John M.	U of South Carolina	206
Palubniak, Daniel	U of Southern Maine	215
Palumbo, Robert D.	Valparaiso U	255
Paolino, Michael A.	Lafayette Coll	104
Papadakis, Constantine N.	U of Cincinnati	042
Papalambros, Panos	U of Michigan	130
Pappas, Nicholas L.	San Jose State U	202
Parckard, Sandra P.	Oakland U	169
Pardieck, David	Bradley U 015	286
Paridy, Keith E.	DeVry Inst-Los Angeles	301
Park, Eui H.	North Carolina A&T	159
Park, John T.	U of Missouri-Rolla	141
Parker, Clinton	U of Texas at Arlington	232
Parker, Helen F.	U of Southern Maine	215
Parker, Kevin J.	U of Rochester	192
Parker, Pamela	Ohio U	170
Parker, Richard	New Jersey Inst of Tech 151	322
Parkin, Gene F.	U of Iowa	098
Parkins, Jack	Mankato State U 116	313
Parks, Charles	Ohio U	170
Parrish, Edward A.	Vanderbilt U	256
Parsons, Michael G.	U of Michigan	130
Patangia, Hirak C.	U of Arkansas at Little Rock	284
Paterwic, Lori-Ann	Western New England Coll	270
Pattenaude, Richard	U of Southern Maine	215
Patterson, Gary K.	U of Missouri-Rolla	141
Patterson, John D.	Mercer U	127
Patton, Alton D.	Texas A&M U	238
Paul, Yuen C.	U of Hawaii at Manoa	088
Pauschke, Joy M.	Northern Illinois U	165
Paxton, Harold	Carnegie Mellon U	036
Paydar, N.	Indiana-Purdue, Indianapolis	097
Payer, Joe H.	Case Western Reserve U	037
Payton, Benjamin F.	Tuskegee U	246
Payvar, Parviz	Northern Illinois U	165
Peale, Robert E.	Indiana-Purdue, Indianapolis	309
Pearce, Britt K.	Southern Coll of Tech	339
Pearson, Clarence W.	U of the District of Columbia	303
Pearson, Wilson	Clemson U	047
Peck, Richard E.	The U of New Mexico	152
Pecos, Auturo	Texas A&I U	237
Peddicord, Kenneth L.	Texas A&M U 238	347
Pejack, Edwin	U of the Pacific	178
Pekau, Oscar A.	Concordia U	055

Pell, Kynric M.	U of Wyoming	279
Pelletier, Lamont	Technical U of Nova Scotia	226
Pemberton, Troy J.	Oklahoma Christian	174
Penfield, Paul	Mass Inst of Tech	124
Perry, Ed	Memphis State U	126
Perry, James C.	Tennessee Tech	231
Perry, Margaret N.	U of Tennessee, Martin	346
Perry, Randy L.	U of Tennessee, Martin	346
Pershing, David W.	U of Utah	253
Pershing, R.L.	Illinois, Urbana-Champaign	095
Perumpral, John V.	Virginia Tech	261
Pervin, William J.	U of Texas at Dallas	234
Peters, Michael H.	Florida A&M U	072
	Florida State U	076
Peters, Walter H.	U of South Carolina	206
Petersen, Paul E.	Rochester Inst of Tech	193
Peterson, Ben B.	US Coast Guard Academy	249
Peterson, Thomas W.	U of Arizona	008
Petroski, Henry	Duke U	065
Petrovic, Louis J.	U of Mass, Lowell 123	315
Pettus, Robert O.	U of South Carolina	206
Peura, Robert A.	Worcester Poly Inst	277
Pfaff, Jerre C.	Southern Illinois-Carbondale 213	341
Pfile, Richard	Indiana-Purdue, Indianapolis	309
Phakias, Bessie	U of Bridgeport	025
Phillipone, Rachel M.	Parks Coll of Saint Louis U	198
Phillips, Don	Oregon Inst of Tech	328
Phillips, J. Richard	Harvey Mudd Coll	087
Phillips, Norman S.	U of Dayton	060
Phillips, William J.	Technical U of Nova Scotia	226
Phillips, Winfred M.	U of Florida	071
Phoenix, S. Leigh	Cornell U	058
Piatkowski, Thomas F.	Western Michigan U	269
Pickering, James H.	U of Houston	090
Pierce, Alvin	Georgia Tech	082
Pierce, Earl E.	NY State Coll of Ceramics	157
	Alfred U	007
Piercy, James W.	West Virginia Inst of Tech	353
Pierucci, Mauro	San Diego State U	200
Pigott, Ronald	Texas Tech U	348
Pimmell, Russell L.	U of Alabama	002
Pincus, George	New Jersey Inst of Tech 151	322
Pinder, George F.	U of Vermont	257
Pines, Sue	U of Arkansas at Little Rock	284
Ping, Charles	Ohio U	170
Pinkney, R. Bruce	U of Manitoba	115
Pinson, Jay D.	San Jose State U	202
Pinto, John G.	San Diego State U	200
Pippin, James T.	U of Tennessee, Knoxville	229
Pippin, William E.	Ohio State U	172
Pitts, Donald R.	U of Tennessee, Knoxville	229
Pla, Fernando	U of Puerto Rico	187
Plitt, Loverne R.	U of Alberta	006
Plotkowski, Paul D.	Grand Valley State U	084
Plummer, Ralph W.	West Virginia U	267
Podlasek, Robert J.	Bradley U 015	286
Pointer, Deborah B.	Columbia U	054
Pokoski, John L.	U of New Hampshire	149
Poling, Bruce	U of Toledo	240
Polis, Michael P.	Wayne State U	265
Pollack, Seymour V.	Washington U	263
Pollard, Nathanael, Jr.	Virginia State U	234
Pollock, Kevin A.	GMI	077
Pope, David P.	U of Penn	179
Popelar, Carl H.	Ohio State U	172
Popovich, Helen	Ferris State U 070	306
Popp, Carl J.	New Mexico Inst of Mining	153
Porteous, Ken C.	U of Alberta	006
Porter, Betsy A.	U of Pittsburgh	181
Poston, John W.	Texas A&M U	238
Pottle, Donald	U of Mass, Lowell 123	315
Pounds, J. Dale	Western Carolina U	354
Prados, John W.	U of Tennessee, Knoxville	229
Prahlr, Joseph M.	Case Western Reserve U	037
Prasuhn, Alan L.	Lawrence Tech U	106
Prather, Edward N.	U of Cincinnati	042
Preloger, Robert A.	North Dakota State U	162
Prevost, J.H.	Princeton U	186
Price, Adrienne W.	Howard U	091
Price, Cynthia A.	DeVry Inst-Columbus	296
Prince, Zack	U of Texas at Arlington	232
Pritchard, James D.	Northern Arizona U	164
Pritchard, Minnie C.	U of Akron	281

Pritchett, P. Walter	Texas A&I U	237
Proakis, John G.	Northeastern U	163
Progelhof, Richard C.	Penn State-Erie	329
Prosser, Francis	U of Kansas	101
Proulx, Gena R.	Franklin U	307
Provan, James W.	McGill U	125
Przirembel, Christian E.	Clemson U	047
Psaltis, Demetri	Cal Inst of Tech	026
Purswell, Jerry L.	U of Oklahoma	173
Puttaiah, Govindappa	West Virginia Inst of Tech	268
Pye, A. Kenneth	Southern Methodist U	216
Pyle, Richard W.	U of Southern Mississippi	342
Pytte, Agnar	Case Western Reserve U	037

Q

Qatu, Mohamad	Franklin U	307
Quigley, Sharon L.	U of Mass, Lowell 123	315

R

Rabin, Herbert	U of Maryland, Coll Park	120
Raburn, Wilford D.	U of South Alabama	205
Rad, Franz N.	Portland State U	184
Rader, Bill T.	Metropolitan State-Denver	317
Radha, Ramalinjam	Prairie View A&M U	185
Radimsky, Anne-Louise	Cal State U, Sacramento	033
Radwan, A.E.	U of Central Florida	039
Radziminski, James B.	U of South Carolina	206
Raghavachari, Madabhushi	Rensselaer Poly Inst	189
Rahman, Habibur	Parks Coll of Saint Louis U	198
Rajan, P.K.	Tennessee Tech	231
Raju, G. V. S.	U of Texas at San Antonio	236
Raju, V.	Rochester Inst of Tech	336
Ramaley, Judith A.	Portland State U	184
Ramani, Raja V.	Penn State U	180
Ramesh, Tirumale	Saginaw Valley State U	196
Ramey, Melvin R.	U of Cal, Davis	020
Randall, Michael D.	SUNY-Buffalo	219
Randhawa, Sabah	Oregon State U	177
Raney, Donald C.	U of Alabama	002
Rankin, Jeffery L.	Rutgers, The State U	195
Ransom, Victor H.	Purdue U	188
Rao, S.S.	Villanova U	258
Rathbone, Donald E.	Kansas State U	102
Rathbun, Richard K.	Wright State U	278
Rathja, Roy C.	Oregon State U	177
Rathmann, Carl E.	Cal State Poly, Pomona 034	289
Rawlings, Hunter R., III	U of Iowa	098
Rawlins, V. Lane	Memphis State U 126	316
Ray, Gautam	Penn State U - Harrisburg	330
Ray, John D.	Memphis State U 126	316
Ray, John W.	Louisiana Tech U	311
Ray, Pradosh K.	Tuskegee U	246
Razelos, Panagiotis	CUNY-Staten Island	045
Ready, Kirk L.	Mankato State U	313
Reavely, Lawrence D.	U of Utah	253
Reck, Charles	DeVry Inst-Dallas	299
Rector, Larry G.	Wichita State U	271
Redd, Frank J.	Utah State U	254
Reddell, Donald L.	Texas A&M U	238
Reddy, Sudhakar M.	U of Iowa	098
Reed, Daniel J.	Point Park Coll	333
Reed, Gregory D.	U of Tennessee, Knoxville	229
Reeder, Charles F.	Auburn U 012	285
Reese, Gordon L.	Oklahoma State U	326
Reeve, N. Richard	Rochester Inst of Tech	193
Rehkugler, Gerald E.	Cornell U	058
Reid, Karl N.	Oklahoma State U 175	326
Reid, Robert L.	Marquette U	118
Reid, Sherwood	St. Cloud State U	337
Reilly, Daniel B.	U of Portland	183
Reilly, James P.	New Mexico State U	154
Reinisch, Bodo	U of Mass, Lowell	123
Reischman, Michael M.	Penn State U	180
Reistad, Gordon M.	Oregon State U	177
Reklaitis, G.V.	Purdue U	188
Renard, John W.	US Naval Academy	252
Renda, R. Bruce	Indiana-Purdue, Indianapolis 097	309
Reneau, Daniel D.	Louisiana Tech U 110	311
Reucroft, Phillip J.	U of Kentucky	103

Name	Institution	Page
Revesz, Gyorgy E.	UNC-Charlotte	158
Rey, William K.	U of Alabama	002
Reyer, Otto W.	U of Cal, Irvine	021
Reynolds, Herbert H.	Baylor U	013
Reynolds, Patsy	U of Tennessee, Chattanooga	228
Reynolds, William C.	Stanford U	217
Rezak, William D.	Southern Coll of Tech	339
Rhodes, Frank H.	Cornell U	058
Rhodes, Tammy	Arkansas Tech U	011
Rice, Julia E.	North Carolina State U	160
Rice, Richard S.	U of Akron	001
Rice, Stephen L.	U of Central Florida 039	292
Rich, Charles A.	Oklahoma State U	326
Rich, Thomas P.	Bucknell U	018
Richard, Don	North Dakota State U	162
Richards, Laurence D.	Old Dominion U	176
Richardson, Earl S.	Morgan State U	145
Richardson, William C.	The Johns Hopkins U	100
Ridenour, Steven	Temple U 227	345
Riemenschneider, A.L.	South Dakota School of Mines	207
Riggs, Henry E.	Harvey Mudd Coll	087
Riley, Michael W.	U of Nebraska - Lincoln	146
Rinehart, Timm R.	U of Mass	121
Ringo, John A.	Washington State U	264
Riordan, Patricia	George Mason U	079
Ripple, G. Gary	Lafayette Coll	104
Rison, Bill	New Mexico Inst of Mining	153
Risser, Paul G.	Miami U	129
Ritchie, Daniel L.	U of Denver	062
Riter, Stephen	U of Texas at El Paso	235
Rivers, William	Cal State U, Northridge	032
Rizkalla, Sami	U of Manitoba	115
Roark, Terry P.	U of Wyoming	279
Robe, T. Richard	Ohio U	170
Roberts, George W.	North Carolina State U	160
Roberts, James A.	U of Kansas	101
Roberts, Stephen D.	North Carolina State U	160
Robertson, Channing R.	Stanford U	217
Robertson, Robert L.	Oklahoma State U	175
Robinson, Alan H.	Oregon State U	177
Rock, Chet A.	U of Maine	114
Rockaway, John D.	U of Missouri-Rolla	141
Roden, Martin S.	Cal State U, Los Angeles	031
Roderick, Gilbert L.	U of Wisconsin-Milwaukee	275
Roderick, Norman F.	The U of New Mexico	152
Rodman, Richard	DeVry Inst-Columbus	296
Rodriguez, Andres F.	U of the Pacific	178
Roemer, Louis E.	Louisiana Tech U	110
Roemer, Robert B.	U of Utah	253
Roess, Roger P.	Poly U	182
Rogers, Decatur B.	Tennessee State U	230
Roggeman, Mary E.	U of Wisconsin-Milwaukee	275
Roggenthen, William M.	South Dakota School of Mines	207
Roig, Gustavo A.	Florida International U	075
Rojeski, Peter	North Carolina A&T	159
Rokos, Miroslav C.	U of Portland	183
Rolfe, John E.	GMI	077
Rolfe, Stanley T.	U of Kansas	101
Rollag, Dwayne A.	South Dakota State U	208
Romero, Al	New Mexico State U	323
Romig, Phillip R.	Colorado School of Mines	052
Rooney, David M.	Hofstra U	089
Rooney, Kevin M.	U of Notre Dame	168
Rosa, Albert J.	U of Denver	062
Rose, J.W.	Oakland U	169
Rose, William I.	Michigan Tech U	133
Roselle, David P.	U of Delaware	061
Rosen, Stephen L.	U of Missouri-Rolla	141
Rosenbauer, Rebecca L.	Lafayette Coll	104
Rosenberg, Ronald	Michigan State U	132
Rosenkrantz, Phillip R.	Cal State Poly, Pomona	034
Ross, I. Joseph	U of Kentucky	103
Rosser, James M.	Cal State U, Los Angeles	031
Rossow, Mark P.	Southern Illinois-Edwardsville	214
Roswell, David	Loyola Coll in Maryland	112
Rotberg, Robert I.	Lafayette Coll	104
Rotenberg, Manuel	U of Cal, San Diego	023
Rouch, Julian S.	U of Arkansas at Little Rock	284
Rouleau, Dennis	École Polytechnique	066
Rousseau, Ronald W.	Georgia Tech	082
Rousselet, Jean	École Polytechnique	066
Roux, Jeffrey A.	U of Mississippi	137
Rowley, Bob	Oklahoma Christian	174
Rowley, Richard L.	Brigham Young U	017
Roy, Donald A.	Technical U of Nova Scotia	226
Roy, Rob J.	Rensselaer Poly Inst	189
Royal, Alton	Florida A&M U	072
Rozon, Daniel	École Polytechnique	066
Rubel, Carol A.	Wentworth Inst of Tech	352
Rubino, David A.	Gannon U	078
Rudd, Walter	Oregon State U	177
Rudee, M. Lea	U of Cal, San Diego	023
Rudenstine, Neil L.	Harvard U	086
Rudko, Michael	Union Coll	247
Ruiz, Alejandro M.	U of Puerto Rico	187
Rumin, Nicolas C.	McGill U	125
Rupp, George E.	Rice U	191
Rush, James	Lamar U	105
Rush, Richard R.	Mankato State U 116	313
Rushton, Pierce A., Jr.	US Military Academy	251
Rusk, Willard	LeTourneau U	107
Russel, John	Drexel U	064
Russel, William B.	Princeton U	186
Russell, George A.	U of Missouri-Kansas City	140
Russell, Horace L.	U of Maryland, Coll Park	120
Russell, James E.	Texas A&M U	238
Russell, John K.	Washington U	263
Russell, John P.	Western Kentucky U	355
Russell, Lynn P.	U of Alabama at Huntsville	004
Russman, Scherrill G.	U of Louisville	111
Russo, Joseph A.	U of Notre Dame	168
Rutford, Robert H.	U of Texas at Dallas	234
Ruth, Douglas W.	U of Manitoba	115
Rutter, Thomas M.	U of Cal, San Diego	023
Ryan, Diane L.	Cal Poly State U	027
Ryan, Donald	San Jose State U	202
Ryan, Nancy E.	Gonzaga U	083
Ryan, Terry	George Mason U	079
Rybicki, Edmund F.	U of Tulsa	245

S

Name	Institution	Page
Saada, Adel S.	Case Western Reserve U	037
Saarlas, Maido	US Naval Academy	252
Sabatino, Carolyn	Ohio U	170
Sacco, Albert	Worcester Poly Inst	277
Sachs, Murray	The Johns Hopkins U	100
Sack, Ronald L.	U of Oklahoma	173
Sadd, Martin H.	U of Rhode Island	190
Sage, Andrew P.	George Mason U	079
Saidel, Gerald M.	Case Western Reserve U	037
Saleeby, B. Badri	U of New Haven	150
Saleh, Bahaa A.	U of Wisconsin-Madison	274
Salem, Julie A.	U of Pittsburgh at Johnstown	332
Saletta, Gerald F.	Illinois Inst of Tech	096
Saliby, Michael J.	U of New Haven	150
Salmon, Thomas P.	U of Vermont	257
Sami, Sedat	Southern Illinois-Carbondale	213
Samizay, Rafi	Washington State U	264
Sample, Steven B.	U of Southern Cal	211
Sander, Duane E.	South Dakota State U	208
Sanders, Billy R.	U of Cal, Davis	020
Sanders, Neill F.	Vanderbilt U	256
Sanders, Thomas J.	Florida Inst of Tech	074
Sanderson, Arthur C.	Rensselaer Poly Inst	189
Sandler, Melvin	The Cooper Union	057
Sandrapaty, Ramachandra R.	South Carolina State U	338
Sangeloty, Jane C.	U of New Haven	150
Santi, Michael	Christian Brothers U	041
Santiago, Felix	U of Puerto Rico	187
Saperstein, Lee W.	U of Kentucky	103
Saracino, Daniel J.	Santa Clara U	203
Sarasti, Ana R.	Florida International U	075
Sargent, Duncan	Ferris State U 070	306
Sarner, Ronald	SUNY-Utica/Rome	344
Sarris, John	U of New Haven	150
Satchell, Starla	Cal State U, Sacramento 033	288
Saul, William E.	Michigan State U	132
Sauls-White, Sheila	Capitol Coll	290
Savicki, Donna R.	Mass Inst of Tech	124
Sawyers, Kenneth N.	Lehigh U	108
Saxena, Umesh K.	U of Wisconsin-Milwaukee	275
Sayers, E. Roger	U of Alabama	002
Saylor, Paul H.	U of Pittsburgh at Johnstown	332
Scala, Louis A.	SUNY-Farmingdale	343
Schachterle, Lance E.	Worcester Poly Inst	277
Schaeffer, Harry G.	U of Louisville	111
Schauer, John J.	U of Dayton	060
Schaumann, Rolf	Portland State U	184
Scheaffer, Donald E.	U of Alaska, Fairbanks	005
Scheid-Gilman, Joyce	Gannon U	078
Schetz, Joseph A.	Virginia Tech	261
Schile, Richard D.	U of Bridgeport	016
Schilling, William M.	U of Penn	179
Schlemmer, Robert E.	U of Cincinnati	293
Schmalzel, John L.	U of Texas at San Antonio	236
Schmidt, Robert E.	Norwich U	167
	Embry-Riddle-Daytona Beach	305
Schmidt, William F.	U of Arkansas	010
Schmitendorf, William E.	U of Cal, Irvine	021
Schmitt, Frederick	Temple U 227	345
Schmitt, Neil M.	U of Arkansas	010
Schmitt, Roland W.	Rensselaer Poly Inst	189
Schnabel, Robert B.	U of Colorado at Boulder	049
Schneider, Barbara	U of Colorado at Denver	051
Schneider, Morris H.	U of Nebraska - Lincoln	146
Schneider, Richard W.	Norwich U	167
Schneider, Sandy	New England Coll	148
Schoenberg, Edward L.	U of the Pacific	178
Schoenung, Julie M.	Cal State Poly, Pomona	034
Scholl, Robert P.	West Virginia Inst of Tech 268	353
Schowalter, William R.	Illinois, Urbana-Champaign	095
Schowengerdt, Franklin D.	Colorado School of Mines	052
Schroder, Darrell	U of Texas at El Paso	235
Schrodt, Verle N.	U of Alabama	002
Schroeder, John H.	U of Wisconsin-Milwaukee	275
Schroeder, Leland R.	SUNY-Env'l Science & Forestry	220
Schroer, Bernard J.	U of Alabama at Huntsville	004
Schuele, Donald E.	Case Western Reserve U	037
Schumacher, Richard R.	U of Wisconsin-Platteville	276
Schure, Matthew	New York Inst of Tech	156
Schwank, Johannes	U of Michigan	130
Schwartz, Diane L.	Cal State U, Northridge	032
Schwartz, Richard J.	Purdue U	188
Schwartz, Stuart C.	Princeton U	186
Schwarz, Steven E.	U of Cal at Berkeley	019
Scott, David W.	Rice U	191
Scott, Ronald	Northeastern U	325
Scott, William E.	Miami U	129
Scranton, Richard J.	Northeastern U	163
Scudder, James	Rochester Inst of Tech	336
Seaburg, Paul A.	Penn State U	180
Seaman, Robert J.	SUNY-New Paltz	222
Sears, John T.	Montana State U	144
Seckler, James G.	Bradley U 015	286
Sedgewick, Robert	Princeton U	186
Sedrick, Gregory A.	U of Tennessee, Chattanooga	228
Seebass, A.R.	U of Colorado at Boulder	049
Seeloff, Eugene R.	U of Virginia	259
Segner, Edmund P., Jr.	U of Alabama at Birmingham	003
Seidman, Stephen B.	Auburn U	012
Seinfeld, John H.	Cal Inst of Tech	026
Selby, Gregory V.	Old Dominion U	176
Seliga, Thomas A.	U of Washington	262
Sellenger, Joseph A., S.J.	Loyola Coll in Maryland	112
Sendaula, Henry M.	Temple U 227	345
Sengupta, Subrata	U of Michigan-Dearborn	131
Sepri, Paavo	Florida Inst of Tech	074
Serrano, David	U of Puerto Rico	187
Severs, Doug	Idaho State U	093
Sevi, Eugene	Norwich U	167
Sevier, John R.	U of Arizona	008
Sexton, Michael R.	Virginia Military Inst	260
Seybert, Andrew F.	U of Kentucky	103
Sforza, Pasquale	Poly U	182
Shackelford, James F.	U of Cal, Davis	020
Shah, Mahendra	SUNY-Farmingdale	343
Shah, Yatish T.	Drexel U	064
Shaikh, Fattah A.	U of Wisconsin-Milwaukee	275
Shamash, Yacov	SUNY-Stony Brook	223
Shapiro, Harold T.	Princeton U	186
Shapiro, Leonard	Portland State U	184
Sharma, Anand D.	U of Puerto Rico	187
Sharpe, Lonnie	North Carolina A&T	159
Sharpe, William N., Jr.	The Johns Hopkins U	100
Sharrow, Conrad	Rensselaer Poly Inst	189
Shattuck, David P.	U of Houston	090
Shaw, Carol M.	U of Dayton	060

Index to Engineering & Technology Personnel

Name	Institution	Page
Shaw, Elden K.	Cal State U, Fresno	029
Shaw, Kenneth A.	Syracuse U	225
Shaw, Richard H., Jr.	Yale U	280
Shealy, Jasper E.	Rochester Inst of Tech	193
Sheffer, Daniel B.	U of Akron	001
Shehadeh, Nazmi M.	U of Minnesota, Duluth	135
Sheley, Roslyn	U of New Orleans	155
Shelley, Daniel	Rochester Inst of Tech 193	336
Shelly, Anne L.	Syracuse U	225
Shelton, Robert D.	Loyola Coll in Maryland	112
Shen, L. David	Florida International U	075
Sheng, Frank D.	Capitol Coll	290
Shepard, Marion L.	Duke U	065
Sheridan, Robert	U of Houston	090
Sherman, Donald R.	U of Wisconsin-Milwaukee	275
Sherrard, Joseph H.	Mississippi State U	138
Shields, Ralph	Ferris State U 070	306
Shirley, Robert C.	U of Southern Colorado 212	340
Short, Kenneth L.	SUNY-Stony Brook	223
Shouldis, Martha K.	West Virginia Inst of Tech	353
Shoup, Jerry	Penn State U - Harrisburg	330
Shoup, Terry E.	Santa Clara U	203
Shridhar, Malayappan	U of Michigan-Dearborn	131
Shuart, James M.	Hofstra U	089
Shultz, Richard D.	U of Wisconsin-Platteville	276
Shumaker, John W.	Central Conn State U	291
Shuman, Larry J.	U of Pittsburgh	181
Siferd, Raymond E.	Wright State U	278
Silber, John	Boston U	014
Sill, William	Montana Coll of Mineral Sci	143
Silverman, Leonard M.	U of Southern Cal	211
Silversmith, Donald J.	Wayne State U	265
Silwa, Steven	Embry-Riddle-Daytona Beach	305
Simaan, Marwan A.	U of Pittsburgh	181
Simitses, George J.	U of Cincinnati	042
Simmonds, James G.	U of Virginia	259
Simon, J. Matthew	Point Park Coll	333
Simon, John	U of Toledo	240
Simon, William E.	Lamar U	105
Simone, Albert J.	Rochester Inst of Tech 193	336
Simons, Gale G.	Kansas State U	102
Simons, Susan	Memphis State U	316
Sims, J. Taylor	Cleveland State U	048
Singh, Sukhmander	Santa Clara U	203
Sinha, A.S.C.	Indiana-Purdue, Indianapolis	097
Sirignano, William A.	U of Cal, Irvine	021
Sisson, Ray L.	U of Southern Colorado 212	340
Sisson, Richard D.	Worcester Poly Inst	277
Sistarenik, Daniel	SUNY-New Paltz	222
Skeith, Ronald W.	U of Arkansas	010
Skidmore, H. James	New Mexico State U	323
Skinner, Bruce C.	U of Bridgeport	016
Skrinde, Rolf T.	Seattle U	204
Skudstad, Don	Oregon Inst of Tech	328
Skutt, Richard	Virginia Military Inst	260
Slack, Donald C.	U of Arizona	008
Slatt, Roger	Colorado School of Mines	052
Sliwa, Steven	Embry-Riddle-Daytona Beach	305
Sliwa, Steven M.	Embry-Riddle-Western Campus	068
Small, Daniel E.	U of Hartford	308
Smedal, Karsten	Iowa State U	099
Smerdon, Ernest T.	U of Arizona	008
Smit, Chris J.B.	U of Georgia	081
Smith, Allie M.	U of Mississippi	137
Smith, Arthur K.	U of Utah	253
Smith, Bert L.	Wichita State U	271
Smith, Brian T.	The U of New Mexico	152
Smith, Charles E.	U of Mississippi	137
Smith, Daniel W.	U of Alberta	006
Smith, David C.	Syracuse U	225
Smith, Deborah D.	Georgia Tech	082
Smith, Deidre	Point Park Coll	333
Smith, Dennis	U of Cal, Irvine	021
Smith, Donald A.	North Dakota State U	162
Smith, Donald R.	U of Missouri-Columbia	139
	U of Missouri-Kansas City	140
Smith, Everitt K.	U of Hartford	308
Smith, Fred W.	Colorado State U	053
Smith, George L., Jr.	Ohio State U	172
Smith, James E., Jr.	U of Alabama at Huntsville	004
Smith, James L.	Texas Tech U	239
Smith, James W.	U of Southern Maine	215
Smith, John	North Carolina A&T	159
Smith, Leonard A.	SUNY-Env'l Science & Forestry	220
Smith, M. "Brother", FSC	Christian Brothers U	041
Smith, Marvin D.	Oklahoma State U	326
Smith, Michael L.	US Air Force Academy	248
Smith, Norman	U of Maine	114
Smith, Robert J.	Seattle U	204
Smith, Roy C.	U of Alabama	002
Smith, Samuel H.	Washington State U	264
Smith, T.O.	Milwaukee Sch of Eng 134	318
Smith, Terry L.	SUNY-Farmingdale	343
Smith, Thomas G.	U of Notre Dame	168
Smoot, L. Douglas	Brigham Young U 017	287
Smyer, William N.	Mississippi State U	138
Snide, James A.	U of Dayton	060
Snow, Carol L.	Cal Inst of Tech	026
Snyder, Janet	SUNY-Farmingdale	343
Snyder, Robert D.	UNC-Charlotte 158	324
Sobol, Hal	U of Texas at Arlington	232
Sojka, Gary A.	Bucknell U	018
Soland, Richard M.	The George Washington U	080
Soldan, David L.	Kansas State U	102
Solecki, Roman	U of Conn	056
Solomon, W.C.	Illinois, Urbana-Champaign	095
Somasundaran, Ponisseril	Columbia U	054
Somerville, Mason H.	Texas Tech U 239	348
Soom, Andres	SUNY-Buffalo	219
Soper, Jon A.	Michigan Tech U	133
Sorber, Charles A.	U of Pittsburgh	181
Sorenson, Norma E.	U of Alabama at Birmingham	003
Sorooshian, Soroosh	U of Arizona	008
Soukup, Rodney J.	U of Nebraska - Lincoln	146
Soule, Hayden M.	U of Maine	114
Soulsby, Eric P.	U of Conn	056
Souza-Kappner, Augusta	CUNY-City Coll	044
Sovern, Michael I.	Columbia U	054
Soyster, Allen L.	Penn State U	180
Spanier, Graham B.	U of Nebraska - Lincoln	146
Sparks, Wayne M.	U of Virginia	259
Spaulding, John F.	New England Coll	148
Spaulding, Malcolm L.	U of Rhode Island	190
Spencer, Jordan L.	Columbia U	054
Spencer, Theodore L.	U of Michigan	130
Speyer, Jason L.	U of Cal, Los Angeles	022
Spielman, Barry E.	Washington U	263
Spille, John C.	U of Cincinnati	293
Spillers, William	New Jersey Inst of Tech	151
Spolek, Graig A.	Portland State U	184
Spring, Bradford H.	Valparaiso U	255
Springer, George S.	Stanford U	217
Springman, Richard	U of Arkansas	010
Sproul, Otis J.	U of New Hampshire 149	321
Spruiell, Joseph E.	U of Tennessee, Knoxville	229
Sree Harsha, K.S.	San Jose State U	202
Stabler, Edward	Syracuse U	225
Stammer, Robert E., Jr.	Vanderbilt U	256
Standley, Hilda	U of Toledo	240
Stanford, William E.	Lehigh U	108
Stanislao, Joseph	North Dakota State U	162
Stanley, Gordon E.	U of Tennessee, Knoxville	229
Stanley, William D.	Old Dominion U	327
Stark, Henry	Illinois Inst of Tech	096
Starke, Edgar A., Jr.	U of Virginia	259
Steadman, John W.	U of Wyoming	279
Stearns, J. David	U of South Alabama	205
Steele, Linda M.	Franklin U	307
Steele, W. Glenn	Mississippi State U	138
Steelman, J. Eldon	New Mexico State U 154	323
Stefanakos, Elias K.	U of South Florida	209
Steger, Joseph A.	U of Cincinnati 042	293
Stegman, Earl C.	North Dakota State U	162
Steinhauser, Robert	Rose-Hulman	194
Stemper, Diane	Miami U	129
Stenstrom, Michael K.	U of Cal, Los Angeles	022
Stephanchick, Paul	DeVry Inst-Kansas City	300
Stephens, Brian	North Dakota State U	162
Stephens, N. Thomas	Florida Inst of Tech	074
Stephenson, F.W.	Virginia Tech	261
Stepp, Elvin D.	U of Cincinnati	293
Sterling, Arthur M.	Louisiana State U	109
Stern, Thomas E.	Columbia U	054
Sternberg, Ben	U of Arizona	008
Stetson, Willis J.	U of Penn	179
Stevens, Charles A.	U of Arkansas at Little Rock	284
Stevens, Gladstone T.	U of Texas at Arlington	232
Stevens, John B.	Norwich U	167
	Embry-Riddle-Daytona Beach	305
Stevens, Karl	Florida Atlantic U	073
Stevenson, James H.	Rensselaer Poly Inst	189
Stevenson, Warren H.	Purdue U	188
Steward, R. Paul	Rose-Hulman	194
Stewart, Larry E.	U of Maryland, Coll Park	120
Stimpson, Brian	U of Manitoba	115
Stobbe, Edward	U of Toledo	240
Stockham, David H.	U of Kentucky	103
Stoebe, Thomas G.	U of Washington	262
Stoffers, Karl E.	Cal State U, Sacramento	033
Stoia, Dennis	Northern Illinois U	165
Stokely, Ernest M.	U of Alabama at Birmingham	003
Stone, Tom	U of Evansville	069
Stoneking, Jerry E.	U of Tennessee, Knoxville	229
Stoneman, Rolland R.	US Air Force Academy	248
Story, James R.	Florida International U	075
Stoudinger, Alan R.	Tri-State U	241
Stratton, Dennis O.	SUNY-Env'l Science & Forestry	220
Strauss, Alvin M.	Vanderbilt U	256
Strauss, Jon C.	Worcester Poly Inst	277
Streett, William B.	Cornell U	058
Stroh, Susan	Tri-State U 241	349
Strom, Oren G.	U of Colorado at Denver	051
Strong, A. Brent	Brigham Young U 017	287
Strunk, Richard D.	U of Wisconsin-Platteville	276
Stuart, Cynthia	The U of New Mexico	152
Stubberud, Allen R.	U of Cal, Irvine	021
Stunard, E. Arthur	DeVry Inst-Chicago	295
Sturm, Frederick J.	U of Akron	281
Stymest, Judy	McGill U	125
Subramanian, R. Shankar	Clarkson U	046
Sugg, Alan B.	U of Arkansas at Little Rock	284
Suh, Nam P.	Mass Inst of Tech	124
Sukanek, Peter C.	U of Mississippi	137
Sule, Dileep R.	Louisiana Tech U	110
Sullivan, Jerry	U of Colorado at Boulder	049
Sullivan, Robert L.	Florida Inst of Tech	074
Sullivan, Samuel L., Jr.	Tulane U	244
Sullivan, William J., S.J.	Seattle U	204
Supernak, Janusz	San Diego State U	200
Sutera, Salvatore P.	Washington U	263
Suzuki, Bob H.	Cal State Poly, Pomona 034	289
Svendsen, Ib A.	U of Delaware	061
Swain, Donald C.	U of Louisville	111
Swamy, M.N.S.	Concordia U	055
Swann, Claire C.	U of Georgia	081
Swart, William W.	U of Central Florida	039
Swartz, Stuart E.	Kansas State U	102
Sweeney, James L.	Stanford U	217
Swenson, Frank R.	Tri-State U	241
Swift, Andrew	U of Texas at El Paso	235
Swift, Fred W.	Florida International U	075
Swift, Thomas E.	Concordia U	055
Swisher, George M.	Tennessee Tech	231
Sylvester, Nicholas D.	U of Akron	001
Symes, William W.	Rice U	191
Szepe, Stephen	U of Illinois at Chicago	094

T

Name	Institution	Page
Taft, William	Boston U	014
Talbot, Laura	Swarthmore Coll	224
Tan, Harry H.	U of Cal, Irvine	021
Tanner, Gilbert B.	U of Louisville	111
Tarjan, Peter P.	U of Miami	128
Tarver, Cynthia L.	Southern U & A&M Coll	210
Tasi, James	SUNY-Stony Brook	223
Tasooji, Mohammad	DeVry Inst-Los Angeles	301
Tate, Boyce D.	Southern Coll of Tech	339
Tauchert, Theodore R.	U of Kentucky	103
Taylor, Alfred O.	U of the District of Columbia 063	303
Taylor, Clayborne D.	Mississippi State U	138
Taylor, James	Pittsburg State U	331
Taylor, James I.	U of Notre Dame	168
Taylor, John	Central Conn State U	291
Taylor, Lynn	DeVry Inst-Los Angeles	301
Taylor, Thomas R., Jr.	Maryland Baltimore Cty	119
Taylor, William R.	Montana State U	144
Temes, Gabor C.	Oregon State U	177

Index to Engineering & Technology Personnel

Name	Institution	Page
Terrell, Sunny	Central State U	040
Terry, Ronald E.	Brigham Young U 017	287
Tesmer, Wayne K.	North Dakota State U	162
Tezak, Edward G.	US Military Academy	251
Thackston, Edward L.	Vanderbilt U	256
Thanaa, Khalil	Point Park Coll	333
Thanasas, Socrates	SUNY-Farmingdale	343
Tharp, Alan L.	North Carolina State U	160
Thesen, Arne	U of Wisconsin-Madison	274
Thibault, Lawrence E.	U of Penn	179
Thigpen, Lewis	Howard U	091
Thomas, Arthur E.	Central State U	040
Thomas, Catherine	U of Southern Cal	211
Thomas, Herman D.	Baylor U	013
Thomas, Joab L.	Penn State U	180
Thomas, Marlin U.	Lehigh U	108
Thomas, Maurice G.	Utah State U	254
Thomas, Michael	DeVry Inst-Phoenix	302
Thomas, Timothy E.	Pittsburg State U	331
Thomas, Warren H.	SUNY-Buffalo	219
Thomas, Windsor S.	SUNY-Utica/Rome	344
Thompson, David R.	Oklahoma State U 175	326
Thompson, Jack L.	U of Tennessee, Chattanooga	228
Thompson, James E.	The U of New Mexico	152
Thompson, Kenneth O.	U of Alabama at Huntsville	004
Thompson, Paul H.	Weber State U	351
Thompson, Paul Y.	U of Florida	071
Thompson, Richard E.	U of Tulsa	245
Thomson, William J.	Washington State U	264
Thorne, Robert W.	US Coast Guard Academy	249
Threadgill, Ernest D.	U of Georgia	081
Thurgood, Ronald L.	Utah State U	254
Thurneck, William J.	U of Virginia	259
Tiedemann, John	SUNY-Farmingdale	343
Tien, Chang-Lin	U of Cal at Berkeley	019
Tien, James M.	Rensselaer Poly Inst	189
Till, William	Embry-Riddle-Daytona Beach	305
Ting, Tuz Chin	U of Conn	056
Tittel, Frank K.	Rice U	191
Toliver, J.W.	Florida A&M U	072
	Florida State U	076
Tompkins, Curtis J.	Michigan Tech U	133
Tompkins, Fred D.	U of Tennessee, Knoxville	229
Tonkin, Humphrey	U of Hartford	308
Tooley, John R.	U of Evansville	069
Topham, W. Sanford	Case Western Reserve U	037
Tornow, Barbara	Boston U	014
Torosian, Doris	The Catholic U of America	038
Tovey, Weldon R.	U of Idaho	092
Tovey, Weldon R.	U of Idaho	092
Tracey, James H.	U of Texas at San Antonio	236
Trachtenberg, Stephen J.	The George Washington U	080
Trandel, Richard S.	Virginia Military Inst	260
Tranjan, Farid M.	UNC-Charlotte	
Trefny, John U.	Colorado School of Mines	052
Treiber, James A.	Gannon U	078
Tremarello, Ann	U of Alaska, Fairbanks	005
Trick, T.N.	Illinois, Urbana-Champaign	095
Tripp, Ray	U of Georgia	081
Tront, Joseph G.	Virginia Tech	261
Troppoli, Daniel A.	Virginia Military Inst	260
Troxler, G. William	Capitol Coll	290
Trujillo, Edward M.	U of Utah	253
Trutt, Frederick C.	U of Kentucky	103
Tschumi, Harry A.	U of Arkansas at Little Rock	284
Tsoulfanidis, Nicholas	U of Missouri-Rolla	141
Tucker, Elizabeth	U of Wisconsin-Platteville	276
Tudor, Gary	U of Cal, Davis	020
Tulenko, James S.	U of Florida	071
Tuncer, Cebeci	Cal State U, Long Beach	030
Tung, Marvin A.	Technical U of Nova Scotia	226
Turner, Charles	U of Texas at El Paso	235
Turner, Daniel S.	U of Alabama	002
Turner, Mel	Oregon Inst of Tech	328
Turner, R. Gerald	U of Mississippi	137
Turner, Robert M.	North Carolina State U	160
Turner, W.D.	Texas A&M U	238
Turner, William	Michigan State U	132
Turner, William D.	Texas A&M U	347
Turnquist, Paul K.	Auburn U	012
Turpin, Richard H.	U of the Pacific	178
Tuzi, Louis A.	Cleveland State U	048
Tyler, A. Lamont	U of Utah	253
Tyler, David A.	U of Maine	114
Tyson, David T.	U of Portland	183
Tzanakou, Evangelia	Rutgers, The State U	195
Tzannes, Nicolaos S.	U of Central Florida	039
Tzeng, Kenneth K.	Lehigh U	108

U

Name	Institution	Page
Uehling, Barbara S.	U of Cal, Santa Barbara	024
Uhlmann, Donald R.	U of Arizona	008
Ullman, Jeffrey D.	Stanford U	217
Uman, Martin A.	U of Florida	071
Umberger, George R.	George Mason U	079
Unger, Vernon E.	Auburn U	012
Unnikrishnan, Raman M.	Rochester Inst of Tech	193
Unwin, Ernest A.	San Jose State U	202
Urish, Daniel W.	U of Rhode Island	190
Uslenghi, Piergiorgio	U of Illinois at Chicago	094
Usmen, Mumtaz A.	Wayne State U	265

V

Name	Institution	Page
Vacroux, André G.	Southern Methodist U	216
Vaglia, John S.	East Tennessee State U	304
Vahdat, Nadar	Tuskegee U	246
Vajpayee, Shri K.	U of Southern Mississippi	342
Valiquette, Guy	École Polytechnique	066
Van Buskirk, William C.	Tulane U	244
Van de Snepscheut, Jan J.L.	Cal Inst of Tech	026
Van Domelen, John F.	Wentworth Inst of Tech	352
Van Dusen, George	Michigan State U	132
Van Horn, Richard L.	U of Oklahoma	173
Van Kirk, Craig W.	Colorado School of Mines	052
Van Roekel, Jacob H.	Southern Illinois-Edwardsville	214
Vander Sande, John B.	Mass Inst of Tech	124
Vanderbilt, Erskine R.	Tennessee State U	230
Varahamurti, Ramesh	Cal State U, Chico	028
Varde, Keshav S.	U of Michigan-Dearborn	131
Vaughan, James G.	U of Mississippi	137
Vaughan, James T.	Murray State U	320
Vaughan, Michael L.	U of Delaware	061
Vedula, Krishna	Iowa State U	099
Ventrice, Marie B.	Tennessee Tech	231
Verma, Lalit R.	Louisiana State U	109
Vest, Charles M.	Mass Inst of Tech	124
Viessman, Warren, Jr.	U of Florida	071
Viets, Hermann	Milwaukee Sch of Eng 134	318
View, John E.	SUNY-Env'l Science & Forestry	220
Villa, William J.	U of Cal, Santa Barbara	024
Vincent, Walter	Tennessee State U	230
Vinson, James S.	U of Evansville	069
Viscomi, B. Vincent	Lafayette Coll	104
Visich, Marian, Jr.	SUNY-Stony Brook	223
Viswanath, Dabir S.	U of Missouri-Columbia	139
Vodrazka, Walter C.	U of Nevada, Las Vegas	147
Voigt, Herbert	Boston U	014
Volpe, Angelo A.	Tennessee Tech	231
Volpe, Edmond L.	CUNY-Staten Island	045
Volz, Dick A.	Texas A&M U	238
von Recum, Andreas	Clemson U	047
von Bernuth, Robert D.	Michigan State U	132
Voss, Robert G.	Worcester Poly Inst	277

W

Name	Institution	Page
Wabrek, Richard M.	Idaho State U	093
Wade, Margie	U of Southern Colorado 212	340
Wadlow, Joan K.	U of Alaska, Fairbanks	005
Wagner, Robert T.	South Dakota State U	208
Wagoner, Robert H.	Ohio State U	172
Waite, William M.	U of Colorado at Boulder	049
Wakeland, Howard L.	Illinois, Urbana-Champaign	095
Walesh, Stuart G.	Valparaiso U	255
Walker, Cedric F.	Tulane U	244
Walker, Donald M.	U of Wisconsin-Madison	274
Walker, Edward L.	U of the District of Columbia	303
Walker, Jack	Oregon Inst of Tech	328
Walker, Janny B.	U of Minnesota, Duluth	135
Walker, John N.	U of Kentucky	103
Walker, Judy	U of Texas at Arlington	232
Walker, M. Lucius, Jr.	Howard U	091
Walker, N. Bruce	U of Delaware	061
Walker, William F.	Auburn U 012	285
Walker, Wynn R.	Utah State U	254
Wall, Damon	U of Mississippi	137
Wall, David J.	U of New Haven	150
Wallace, C. Edward	Arizona State U 009	283
Wallentine, Robert E.	Weber State U	351
Wallis, Graham B.	Dartmouth Coll	059
Walsh, Christopher	Syracuse U	225
Walsh, Daniel W.	Cal Poly State U	027
Walsh, William K.	Auburn U 012	285
Walter, George	Villanova U	258
Walton, C.M.	U of Texas at Austin	233
Wang, C.-C.	Rice U	191
Wang, Yuan R.	Texas A&I U	237
Wankat, Phillip C.	Purdue U	188
Ward, Barry W.	Monmouth Coll	142
Ward, Lawrence W.	Webb Inst	266
Ward, Reuben K.	West Virginia Inst of Tech	353
Ward, Thomas J.	Temple U 227	345
Warder, Richard C.	U of Missouri-Columbia	139
Waring, Karen	Norwich U	167
Waring, Thomas	Montana Coll of Mineral Sci	143
Warner, Don L.	U of Missouri-Rolla	141
Warner, Mark S.	U of Iowa	098
Warnock, William K.	U of Arkansas	010
Warren, James M.	Baylor U	013
Warrington, Robert O.	Louisiana Tech U	110
Warter, Peter J.	U of Delaware	061
Washington, Sylvia	DeVry Inst-DuPage	298
Watkins, Charles B.	CUNY-City Coll	044
Watson, John L.	U of Missouri-Rolla	141
Watson, Karan L.	Texas A&M U 238	347
Watson, W. Joe	Oklahoma Christian	174
Watters, Gary Z.	Cal State U, Chico	028
Watters, James C.	U of Louisville	111
Watts, Claudius E., III	The Citadel	043
Watts, Fred J.	U of Idaho	092
Watts, K. Chris	Technical U of Nova Scotia	226
Wattson, Robert K.	Tri-State U	241
Watwood, Vernon B.	Michigan Tech U	133
Wazzan, A.R. Frank	U of Cal, Los Angeles	022
Weaver, Lynn E.	Florida Inst of Tech	074
Webb, Roger P.	Georgia Tech	082
Weber, Arnold R.	Northwestern U	166
Weber, John W.	Texas A&I U	237
Webster, Dennis B.	U of Alabama	002
Webster, William C.	U of Cal at Berkeley	019
Weckmueller, Beth L.	U of Wisconsin-Milwaukee	275
Weese, John A.	Texas A&M U 238	347
Wefald, Jon	Kansas State U	102
Wegman, David	U of Mass, Lowell	123
Wegman, Edward J.	George Mason U	079
Wehling, Katherine E.	Valparaiso U	255
Wei, James	Princeton U	186
Wei, Robert P.	Lehigh U	108
Weight, Harold R.	U of Utah	253
Weight, Willis	Montana Coll of Mineral Sci	143
Weiler, Robert L.	Capitol Coll	290
Weinberg, W. Henry	U of Cal, Santa Barbara	024
Weinberger, Charles B.	Drexel U	064
Weiss, Volker	Syracuse U	225
Weissenburger, Lea	Mercer U	127
Welch, Robert C.	U of Arkansas	010
Wellman, Denise	U of South Carolina	206
Wells, William R.	U of Nevada, Las Vegas	147
Welsh, Tracy	South Dakota State U	208
Welty, John D.	Cal State U, Fresno	029
Wertheimer, Susan A.	U of Vermont	257
Wessler, Max A.	Bradley U	015
West, Tom M.	Oregon State U	177
West, Wallace I.	The Citadel	043
Westgate, Charles R.	The Johns Hopkins U	100
Wethington, Charles T., Jr.	U of Kentucky	103
Wetsel, Grover	U of Texas at Dallas	234
Whalen, Curtis R.	UNC-Charlotte 158	324
Whaley, Ross S.	SUNY-Env'l Science & Forestry	220
Wheeler, Anthony	San Francisco State U	201
Whelihan, Nicholas F.	U of Minnesota, Duluth	135
Whiddon, Frederick P.	U of South Alabama	205
Whipkey, Sherman	CUNY-Staten Island	045
White, Chelsea C.	U of Michigan	130

Name	Institution	Page
White, E. Bernard	George Mason U	079
White, Edward J.	Vanderbilt U	256
White, James	Seattle U	204
White, John A.	Georgia Tech	082
White, Keith	U of Wisconsin-Madison	274
White, Kenneth R.	New Mexico State U	154
White, Lee J.	Case Western Reserve U	037
White, Ralph E.	U of South Carolina	206
White, Robert M.	Carnegie Mellon U	036
Whitehouse, Gary E.	U of Central Florida	039 292
Whites, Robert W.	U of Missouri-Rolla	141
Whiteside, John E.	The U of New Mexico	152
Whitman, Alan M.	Villanova U	258
Whitman, David	U of Wyoming	279
Whitt, Howard R.	Oakland U	169
Widera, G.E.O.	Marquette U	118
Wiedemeier, Dennis W.	Seattle U	204
Wiemer-Sumner, Anne-Marie	The Cooper Union	057
Wierman, John C.	The Johns Hopkins U	100
Wiggins, J W.	U of Arkansas at Little Rock	284
Wilcox, William R.	Clarkson U	046
Wilhelm, Mickey R.	U of Louisville	111
Wilhelm, William J.	Wichita State U	271
Wilkins, Joyce J.	U of Dayton	060
Willhite, Paul	U of Kansas	101
Williams, Ann	U of Cal, Irvine	021
Williams, Carolyn Ruth A	Vanderbilt U	256
Williams, Curley M.	Georgia Tech	082
Williams, Dale A.	St. Cloud State U	197 337
Williams, David B.	Lehigh U	108
Williams, Frank F.	U of Alaska, Fairbanks	005
Williams, J. Richard	Cal State U, Long Beach	030
Williams, James	U of Cincinnati	042
Williams, Lawrence R.	U of Texas at San Antonio	236
Williams, Molly W.	Western Michigan U	269 356
Williams, Ryan	U of Rochester	192
Williamson, S. Gill	U of Cal, San Diego	023
Williamson, William L.	Pittsburg State U	331
Willis, Max S.	U of Akron	001
Willson, Alan N., Jr.	U of Cal, Los Angeles	022
Wilson, Andrew J.	Tri-State U	241
Wilson, Blenda J.	Cal State U, Northridge	032
Wilson, D. Joanne	U of Wisconsin-Platteville	276
Wilson, Donald W.	Pittsburg State U	331
Wilson, Edward A.	DeVry Inst-Columbus	296
Wilson, G. Peter	Technical U of Nova Scotia	226
Wilson, Howard G.	LeTourneau U	107
Wilson, Howard T.	Gannon U	078
Wilson, John W.	U of Missouri-Rolla	141
Wilson, Julian A.	Southern Coll of Tech	339
Wilson, Robin S.	Cal State U, Chico	028
Wiltz, Simon R.	Prairie View A&M U	185
Wimberly, Ray	Embry-Riddle-Daytona Beach	067
Winer, Ward O.	Georgia Tech	082
Winkel, Raymond J., Jr.	US Military Academy	251
Winn, C. Byron	Colorado State U	053
Winne, Clinton H.	U of Maine	114
Winston, Hubert	North Carolina State U	160
Winston, Karen R.	Virginia State U	350
Winters, Irving J.	Morgan State U	145
Winzer, Etienna R.	Louisiana Tech U	110 311
Witte, Larry C.	U of Houston	090
Witzel, Ronald W.	The George Washington U	080
Wojcik, Anthony S.	Michigan State U	132
Wolf, Lawrence J.	Oregon Inst of Tech	328
Wolf, Werner P.	Yale U	280
Wolfe, Harvey	U of Pittsburgh	181
Wolfe, Philip M.	Arizona State U	009
Wolff, Robert L.	U of Dayton	294
Wolgemuth, Carl H.	Penn State U	180
Womack, Larry O.	U of Southern Colorado	340
Wonders, Thomas J.	U of Pittsburgh at Johnstown	332
Wong, Bing T.	Wilkes U	273
Wood, Roger C.	U of Cal, Santa Barbara	024
Woodall, David M.	U of Idaho	092
Woodson, Herbert H.	U of Texas at Austin	233
Woodward, James H.	UNC-Charlotte	158 324
Woolston, Donald C.	U of Wisconsin-Madison	274
Wormley, David N.	Penn State U	180
Worzala, Frank J.	U of Wisconsin-Madison	274
Wray, Kent	Texas Tech U	239
Wright, John R.	Central Conn State U	291
Wright, Terry	U of Alabama at Birmingham	003
Wu, Chien-Heng	U of Illinois at Chicago	094
Wu, Theodore Y.	Cal Inst of Tech	026
Wu, Tien H.	Ohio State U	172
Wuerger, William W.	U of Wisconsin-Madison	274
Wuttig, Manfred R.	U of Maryland, Coll Park	120
Wyatt, Joe B.	Vanderbilt U	256
Wylie, E. Benjamin	U of Michigan	130
Wynblatt, Paul	Carnegie Mellon U	036

X

Name	Institution	Page
Xistris, George D.	Concordia U	055

Y

Name	Institution	Page
Yaconis, Richard L.	DeVry Inst-Chicago	295
Yamayee, Zia A.	Gonzaga U	083
Yang, Henry T.	Purdue U	188
Yang, Ralph T.	SUNY-Buffalo	219
Yang, Y. C.	Tuskegee U	246
Yao, Jim T.P.	Texas A&M U	238
Yarbrough, David W.	Tennessee Tech	231
Yates, Albert C.	Colorado State U	053
Yates, Marvin L.	Southern U & A&M Coll	210
Yau, Stephen S.	U of Florida	071
Yegian, Mishac K.	Northeastern U	163
Yen, Michael	Memphis State U	126
Yeske, Ronald A.	U of Wisconsin-Platteville	276
Yeung, Henderson	Cal State U, Fresno	029
Yeung, Ronald W.	U of Cal at Berkeley	019
Yortsos, Yanis C.	U of Southern Cal	211
Young, Charles E.	U of Cal, Los Angeles	022
Young, Fred M.	Lamar U	105
Young, Lee	Tuskegee U	246
Young, Paul R.	U of Washington	262
Young, Reginald H.	U of Hawaii at Manoa	088
Young, Thomas	Rochester Inst of Tech	336
Young, Tzay	U of Miami	128
Young, William	Colorado School of Mines	052
Youngren, Conrad C.	SUNY-Maritime	221
Yurtseven, H. Öner	Indiana-Purdue, Indianapolis	097 309

Z

Name	Institution	Page
Zacharias, Donald W.	Mississippi State U	138
Zagorski, Stanley J.	Gannon U	078
Zaiser, James N.	Bucknell U	018
Zakin, Jacques L.	Ohio State U	172
Zaloom, Victor	Lamar U	105
Zarling, John P.	U of Alaska, Fairbanks	005
Zbinden, Richard	Oregon Inst of Tech	328
Zebib, Abdelfattah	Rutgers, The State U	195
Zeiser, Catherine L.	U of Rhode Island	190
Zick, Gregory L.	U of Washington	262
Ziemer, Rodger E.	U of Colorado-Col Spr	050
Zimmerman, Richard	San Francisco State U	201
Zinser, Elizabeth A.	U of Idaho	092
Zirkle, Larry D.	Oklahoma State U	175
Zollars, Richard L.	Washington State U	264
Zorowski, Carl F.	North Carolina State U	160
Zubaly, Robert B.	SUNY-Maritime	221

ENGINEERING IN JAPAN

Education, Practice, Future Outlook
Robert S. Cutler, Editor

Japan's recent advancements in high technology demonstrate a science and engineering system that differs in significant ways from that of the United States. Moreover, unique features of Japanese-style engineering are not well understood by American scientists, engineers, and policy makers.

Robert S. Cutler of Georgetown University has compiled this 139-page book of papers presented at a symposium at the 1991 annual meeting of the American Association for the Advancement of Science in Washington, D.C. Ten Japanese and American science policy observers from academe and industry examine key elements of present and future "engineering systems" in Japan.

The symposium addressed current problems, opportunities, and future plans for Japanese engineering education and industrial practice. Included are a discussion of the basic concepts of "knowledge management" and technology development that differ between the two cultures; the plans for reforms in graduate school education for engineers at the University of Tokyo; and a description of a successful 30-year, cross-cultural engineering collaboration between a major Japanese and a leading American company.

Engineering in Japan: Education, Practice, Future Outlook **is a must for your library.** As a practicing engineer or an engineering educator, it is imperative to stay abreast of the issues of engineering in Japan. Simply send payment and the completed order form to receive this vital collection of papers. ($19.95 plus $1.50 shipping)

To order, use the ASEE Publications order form in the back of this directory or call the publications coordinator at (202) 986-8528.

ABET/CEAB Accredited Degree Programs

This appendix is divided into two parts: engineering programs and engineering technology programs. Each is organized alphabetically by school, with degree programs also listed alphabetically for each school. Each section includes all baccalaureate (usually four-year bachelor's) degree programs accredited by the Accreditation Board for Engineering and Technology (ABET). The engineering section also includes programs accredited by the Canadian Engineering Accreditation Board (CEAB). After the school name, the city and state or province of the main campus are given in parentheses. Next, in brackets, are references to the section of the directory ("E" for Engineering and/or "T" for Engineering Technology) and the institution profile reference number of a participating school's full-page profile(s). Following this appendix is an alphabetical index of accredited engineering and engineering technology degree programs.

The Accreditation Board for Engineering and Technology (ABET) is the official accrediting authority for engineering and technology programs in the United States. In Canada, the official accrediting body for engineering programs is the Canadian Engineering Accreditation Board (CEAB). These organizations accredit individual engineering and engineering technology programs, not universities, colleges, schools or degrees. For complete information on the accreditation process, you should contact: ABET, 345 East 47th Street, New York, NY 10017-2397; (212) 705-7685. For information on accreditation in Canada, contact CEAB, 601-116 Albert Street, Ottawa, ON, K1P 5G3; (613) 232-2474.

ENGINEERING

A

University of Akron
(Akron, OH) [E-001]
Chemical Engineering
Civil Engineering
Electrical Engineering
Mechanical Engineering

University of Alabama
(Tuscaloosa, AL) [E-002]
Aerospace Engineering
Chemical Engineering
Civil Engineering
Electrical Engineering
Industrial Engineering
Mechanical Engineering
Metallurgical Engineering
Mineral Engineering

University of Alabama at Birmingham
(Birmingham, AL) [E-003]
Civil Engineering
Electrical Engineering
Materials Engineering
Mechanical Engineering

University of Alabama at Huntsville
(Huntsville, AL) [E-004]
Chemical Engineering
Civil Engineering
Computer Engineering
Electrical Engineering
Industrial & Systems Engineering
Mechanical Engineering

University of Alaska-Anchorage
(Anchorage, AK)
Civil Engineering

University of Alaska, Fairbanks
(Fairbanks, AK) [E-005]
Civil Engineering
Electrical Engineering
Geological Engineering
Mechanical Engineering
Mining Engineering

University of Alberta
(Edmonton, AB-Canada) [E-006]
Agricultural Engineering
Chemical Engineering
Civil Engineering
Computer Engineering
Electrical Engineering
Engineering Physics
Mechanical Engineering
Metallurgical Engineering
Mineral Engineering
Mineral Process Engineering
Mining Engineering
Petroleum Engineering

Alfred University
(Alfred, NY) [E-007]
Electrical Engineering
Industrial Engineering
Mechanical Engineering

University of Arizona
(Tucson, AZ) [E-008]
Aerospace Engineering
Agricultural & Biosystems Engineering
Chemical Engineering
Civil Engineering
Computer Engineering
Electrical Engineering
Geological Engineering
Industrial Engineering
Materials Science & Engineering
Mechanical Engineering
Mining Engineering
Nuclear Engineering
Systems Engineering

Arizona State University
(Tempe, AZ) [E-009]
Aerospace Engineering
Bioengineering
Chemical Engineering
Civil Engineering
Computer Systems Engineering
Electrical Engineering
Industrial Engineering
Interdisciplinary Studies Engineering
Mechanical Engineering
Special Studies Engineering

University of Arkansas
(Fayetteville, AR) [E-010]
Biological & Agricultural Engineering
Chemical Engineering
Civil Engineering
Computer Systems Engineering
Electrical Engineering
Industrial Engineering
Mechanical Engineering

Arkansas Tech University
(Russellville, AR) [E-011]
Engineering

Auburn University
(Auburn University, AL) [E-012]
Aerospace Engineering
Agricultural Engineering
Chemical Engineering
Civil Engineering
Computer Engineering
Electrical Engineering
Industrial Engineering
Materials Engineering
Mechanical Engineering

B

Baylor University
(Waco, TX) [E-013]
Engineering

Boston University
(Boston, MA) [E-014]
Aerospace Engineering
Biomedical Engineering
Computer Engineering
Electrical Engineering
Manufacturing Engineering
Mechanical Engineering
Systems Engineering

Bradley University
(Peoria, IL) [E-015]
Civil Engineering
Electrical Engineering
Industrial Engineering
Manufacturing Engineering
Mechanical Engineering

University of Bridgeport
(Bridgeport, CT) [E-016]
Computer Engineering
Electrical Engineering
Mechanical Engineering

Brigham Young University
(Provo, UT) [E-017]
Chemical Engineering
Civil Engineering
Electrical Engineering
Mechanical Engineering

University of British Columbia
(Vancouver, BC-Canada)
Agricultural Engineering
Bio-Resource Engineering
Chemical Engineering
Civil Engineering
Electrical Engineering
Engineering Physics
Geological Engineering
Mechanical Engineering
Metallurgical Engineering
Metals & Materials Engineering
Mineral Engineering
Mining & Mineral Process Engineering

Bucknell University
(Lewisburg, PA) [E-018]
Chemical Engineering
Civil Engineering
Electrical Engineering
Mechanical Engineering

C

University of Calgary
(Calgary, AB-Canada)
Chemical Engineering
Civil Engineering
Electrical Engineering
Mechanical Engineering
Surveying Engineering

University of California at Berkeley
(Berkeley, CA) [E-019]
Chemical Engineering
Civil Engineering
Computer Science
Electrical Engineering
Industrial Engineering
Mechanical Engineering
Mineral Engineering
Naval Architecture
Nuclear Engineering

University of California, Davis
(Davis, CA) [E-020]
Aeronautical Science & Engineering
Agricultural Engineering
Chemical Engineering
Civil Engineering
Computer Science & Engineering
Electrical Engineering
Materials Science & Engineering
Mechanical Engineering

University of California, Irvine
(Irvine, CA) [E-021]
Civil Engineering
Electrical Engineering
Mechanical Engineering

University of California, Los Angeles
(Los Angeles, CA) [E-022]
Aerospace Engineering
Chemical Engineering
Civil Engineering
Computer Science & Engineering
Electrical Engineering
Materials Engineering
Mechanical Engineering

University of California, San Diego
(La Jolla, CA) [E-023]
Bioengineering
Chemical Engineering
Electrical Engineering
Mechanical Engineering
Structural Engineering
Systems & Control Engineering

University of California, Santa Barbara
(Santa Barbara, CA) [E-024]
Chemical Engineering
Electrical Engineering
Mechanical Engineering
Nuclear Engineering

University of California, Santa Cruz
(Santa Cruz, CA) [E-025]
Computer Engineering

California Institute of Technology
(Pasadena, CA) [E-026]
Chemical Engineering
Engineering & Applied Sciences

California Polytechnic State University
(San Luis Obispo, CA) [E-027]
Aeronautical Engineering
Agricultural Engineering
Architectural Engineering
Civil Engineering
Electrical Engineering
Electronic(s) Engineering
Environmental Engineering
Industrial Engineering
Mechanical Engineering
Metallurgical & Materials Engineering

California State University, Chico
(Chico, CA) [E-028]
Civil Engineering
Computer Engineering
Electrical/Electronic Engineering
Mechanical Engineering

ABET/CEAB Accredited Degree Programs

California State University, Fresno
(Fresno, CA) [E-029]
Civil Engineering
Electrical Engineering
Industrial Engineering
Mechanical Engineering
Surveying Engineering

California State University, Long Beach
(Long Beach, CA) [E-030]
Chemical Engineering
Civil Engineering
Computer Science & Engineering
Electrical Engineering
Mechanical Engineering

California State University, Los Angeles
(Los Angeles, CA) [E-031]
Civil Engineering
Electrical Engineering
Mechanical Engineering

California State University, Northridge
(Northridge, CA) [E-032]
Engineering

California State University, Sacramento
(Sacramento, CA) [E-033]
Civil Engineering
Computer Engineering
Electrical/Electronic Engineering
Mechanical Engineering

California State Polytechnic University, Pomona
(Pomona, CA) [E-034]
Aerospace Engineering
Agricultural Engineering
Chemical Engineering
Civil Engineering
Electrical Engineering
Industrial Engineering
Manufacturing Engineering
Mechanical Engineering

Calvin College
(Grand Rapids, MI) [E-035]
Engineering

Carleton University
(Ottawa, ON)
Aerospace Engineering
Civil Engineering
Computer Systems Engineering
Electrical Engineering
Mechanical Engineering

Carnegie Mellon University
(Pittsburgh, PA) [E-036]
Chemical Engineering
Civil Engineering
Computer Engineering
Electrical Engineering
Engineering & Public Policy
Mechanical Engineering
Metallurgical Engineering & Materials Science

Case Western Reserve University
(Cleveland, OH) [E-037]
Biomedical Engineering
Chemical Engineering
Civil Engineering
Computer Engineering
Electrical Engineering
Fluid & Thermal Engineering Science
Materials Science & Engineering
Mechanical Engineering
Polymer Science & Engineering
Systems & Control Engineering

The Catholic University of America
(Washington, DC) [E-038]
Biomedical Engineering
Civil Engineering
Electrical Engineering
Mechanical Engineering

University of Central Florida
(Orlando, FL) [E-039]
Aerospace Engineering
Civil Engineering
Computer Engineering
Electrical Engineering
Environmental Engineering
Industrial Engineering
Mechanical Engineering

Central State University
(Wilberforce, OH) [E-040]
Manufacturing Engineering

Christian Brothers University
(Memphis, TN) [E-041]
Chemical Engineering
Civil Engineering
Electrical Engineering
Mechanical Engineering

University of Cincinnati
(Cincinnati, OH) [E-042]
Aerospace Engineering
Chemical Engineering
Civil Engineering
Computer Engineering
Electrical Engineering
Engineering Mechanics
Environmental Engineering
Industrial Engineering
Materials Engineering
Mechanical Engineering
Nuclear & Power Engineering

The Citadel
(Charleston, SC) [E-043]
Civil Engineering
Electrical Engineering

City College of the City University of New York
(New York, NY) [E-044]
Chemical Engineering
Civil Engineering
Electrical Engineering
Mechanical Engineering

City University of New York, College of Staten Island
(Staten Island, NY) [E-045]
Engineering Science

Clarkson University
(Potsdam, NY) [E-046]
Chemical Engineering
Civil Engineering
Computer Engineering
Electrical Engineering
Mechanical Engineering

Clemson University
(Clemson, SC) [E-047]
Agricultural Engineering
Ceramic Engineering
Chemical Engineering
Civil Engineering
Computer Engineering
Electrical Engineering
Industrial Engineering
Mechanical Engineering

Cleveland State University
(Cleveland, OH) [E-048]
Chemical Engineering
Civil Engineering
Electrical Engineering
Industrial Engineering
Mechanical Engineering

University of Colorado at Boulder
(Boulder, CO) [E-049]
Aerospace Engineering Sciences
Architectural Engineering
Chemical Engineering
Civil Engineering
Electrical & Computer Engineering
Electrical Engineering
Mechanical Engineering

University of Colorado at Colorado Springs
(Colorado Springs, CO) [E-050]
Electrical Engineering

University of Colorado at Denver
(Denver, CO) [E-051]
Civil Engineering
Electrical Engineering
Mechanical Engineering

Colorado School of Mines
(Golden, CO) [E-052]
Chemical & Petroleum-Refining Engineering
Engineering
Engineering Physics
Geological Engineering
Geophysical Engineering
Metallurgical Engineering
Mining Engineering
Petroleum Engineering

Colorado State University
(Fort Collins, CO) [E-053]
Agricultural Engineering
Chemical Engineering
Civil Engineering
Electrical Engineering
Engineering Science(s)
Mechanical Engineering

Columbia University
(New York, NY) [E-054]
Chemical Engineering
Civil Engineering
Electrical Engineering
Industrial Engineering
Mechanical Engineering
Metallurgical Engineering
Mining Engineering

Concordia University
(Montreal, PQ-Canada) [E-055]
Building Engineering
Civil Engineering
Computer Engineering
Electrical Engineering
Mechanical Engineering

University of Connecticut
(Storrs, CT) [E-056]
Chemical Engineering
Civil Engineering
Computer Science & Engineering
Electrical Engineering
Mechanical Engineering

The Cooper Union
(New York, NY) [E-057]
Chemical Engineering
Civil Engineering
Electrical Engineering
Mechanical Engineering

Cornell University
(Ithaca, NY) [E-058]
Agricultural Engineering
Chemical Engineering
Civil Engineering
Electrical Engineering
Engineering Physics
Materials Science & Engineering
Mechanical Engineering
Operations Research & Engineering

D

Dartmouth College
(Hanover, NH) [E-059]
Engineering

University of Dayton
(Dayton, OH) [E-060]
Chemical Engineering
Civil Engineering
Electrical Engineering
Mechanical Engineering

University of Delaware
(Newark, DE) [E-061]
Chemical Engineering
Civil Engineering
Electrical Engineering
Mechanical Engineering

University of Denver
(Denver, CO) [E-062]
Electrical Engineering
Mechanical Engineering

University of Detroit Mercy
(Detroit, MI)
Chemical Engineering
Civil Engineering
Electrical Engineering
Mechanical Engineering

University of the District of Columbia
(Washington, DC) [E-063]
Civil Engineering
Electrical Engineering
Mechanical Engineering

Dordt College
(Sioux Center, IA)
Engineering

Drexel University
(Philadelphia, PA) [E-064]
Architectural Engineering
Chemical Engineering
Civil Engineering
Electrical Engineering
Materials Engineering
Mechanical Engineering

Duke University
(Durham, NC) [E-065]
Biomedical Engineering
Civil Engineering
Electrical Engineering
Mechanical Engineering

E

École Polytechnique de Montréal
(Montréal, PQ-Canada) [E-066]
Génie chimique
Génie civil
Génie des matériaux
Génie des mines
Génie géologique
Génie industriel
Génie informatique
Génie électrique
Génie méchanique
Génie minier
Génie métallurgique
Génie physique

École de Technologie Supérieure
(Montréal, PQ-Canada)
Génie de la production automatisée
Génie et gestion de la construction
Génie électrique
Génie méchanique

Embry-Riddle Aeronautical University - Daytona Beach
(Daytona Beach, FL) [E-067]
Aerospace Engineering

Embry-Riddle Aeronautical University, Western Campus
(Prescott, AZ) [E-068]
Aeronautical Engineering

ABET/CEAB Accredited Degree Programs

University of Evansville
(Evansville, IN) [E-069]
Electrical Engineering
Mechanical Engineering

F

Fairleigh Dickinson University
(Teaneck, NJ)
Electrical Engineering

Ferris State University
(Big Rapids, MI) [E-070]
Surveying Engineering

University of Florida
(Gainesville, FL) [E-071]
Aerospace Engineering
Agricultural Engineering
Chemical Engineering
Civil Engineering
Computer & Information Engineering Science
Electrical Engineering
Engineering Science(s)
Environmental Engineering
Industrial & Systems Engineering
Materials Science & Engineering
Mechanical Engineering
Nuclear Engineering

Florida Agricultural and Mechanical University
(Tallahassee, FL) [E-072]
Chemical Engineering
Civil Engineering
Electrical Engineering
Industrial Engineering
Mechanical Engineering

Florida Atlantic University
(Boca Raton, FL) [E-073]
Electrical Engineering
Mechanical Engineering
Ocean Engineering

Florida Institute of Technology
(Melbourne, FL) [E-074]
Aerospace Engineering
Chemical Engineering
Civil Engineering
Computer Engineering
Electrical Engineering
Mechanical Engineering
Ocean Engineering

Florida International University
(Miami, FL) [E-075]
Civil Engineering
Electrical Engineering
Industrial Engineering
Mechanical Engineering

G

Gannon University
(Erie, PA) [E-078]
Electrical Engineering
Mechanical Engineering

George Mason University
(Fairfax, VA) [E-079]
Electronic(s) Engineering

The George Washington University
(Washington, DC) [E-080]
Civil Engineering
Computer Engineering
Electrical Engineering
Mechanical Engineering
Systems Analysis & Engineering

University of Georgia
(Athens, GA) [E-081]
Agricultural Engineering

Georgia Institute of Technology
(Atlanta, GA) [E-082]
Aerospace Engineering
Ceramic Engineering
Chemical Engineering
Civil Engineering
Computer Engineering
Electrical Engineering
Engineering Science & Mechanics
Industrial & Systems Engineering
Materials Engineering
Mechanical Engineering
Nuclear Engineering
Textile Engineering

GMI Engineering & Management Institute
(Flint, MI) [E-077]
Electrical Engineering
Industrial Engineering
Manufacturing Systems Engineering
Mechanical Engineering

Gonzaga University
(Spokane, WA) [E-083]
Civil Engineering
Electrical Engineering
Mechanical Engineering

Grand Valley State University
(Grand Rapids, MI) [E-084]
Engineering

Grove City College
(Grove City, PA) [E-085]
Electrical Engineering
Mechanical Engineering

University of Guelph
(Guelph, ON)
Agricultural Engineering
Biological Engineering
Water Resources Engineering

H

Hampton University
(Hampton, VA)
Chemical Engineering
Electrical Engineering

University of Hartford
(West Hartford, CT)
Civil Engineering
Electrical Engineering
Mechanical Engineering

Harvard University
(Cambridge, MA) [E-086]
Engineering Sciences

Harvey Mudd College
(Claremont, CA) [E-087]
Engineering

University of Hawaii at Manoa
(Honolulu, HI) [E-088]
Civil Engineering
Electrical Engineering
Mechanical Engineering

Hofstra University
(Hempstead, NY) [E-089]
Electrical Engineering
Engineering Science(s)
Mechanical Engineering

University of Houston
(Houston, TX) [E-090]
Chemical Engineering
Civil Engineering
Electrical Engineering
Industrial Engineering
Mechanical Engineering

Howard University
(Washington, DC) [E-091]
Chemical Engineering
Civil Engineering
Electrical Engineering
Mechanical Engineering

Humboldt State University
(Arcata, CA)
Environmental Resources Engineering

I

University of Idaho
(Moscow, ID) [E-092]
Agricultural Engineering
Chemical Engineering
Civil Engineering
Electrical Engineering
Geological Engineering
Mechanical Engineering
Metallurgical Engineering
Mining Engineering

Idaho State University
(Pocatello, ID) [E-093]
Engineering

University of Illinois at Chicago
(Chicago, IL) [E-094]
Bioengineering
Chemical Engineering
Civil Engineering
Computer Engineering
Electrical Engineering
Industrial Engineering
Mechanical Engineering
Metallurgical Engineering

University of Illinois at Urbana-Champaign
(Urbana, IL) [E-095]
Aeronautical & Astronautical Engineering
Agricultural Engineering
Ceramic Engineering
Chemical Engineering
Civil Engineering
Computer Engineering
Electrical Engineering
Engineering Mechanics
General Engineering
Industrial Engineering
Mechanical Engineering
Metallurgical Engineering
Nuclear Engineering

Illinois Institute of Technology
(Chicago, IL) [E-096]
Aerospace Engineering
Chemical Engineering
Civil Engineering
Electrical Engineering
Mechanical Engineering
Metallurgical Engineering

Indiana University-Purdue University at Indianapolis
(Indianapolis, IN) [E-097]
Electrical Engineering
Mechanical Engineering

University of Iowa
(Iowa City, IA) [E-098]
Biomedical Engineering
Chemical Engineering
Civil Engineering
Electrical Engineering
Industrial Engineering
Mechanical Engineering

Iowa State University
(Ames, IA) [E-099]
Aerospace Engineering
Agricultural Engineering
Ceramic Engineering
Chemical Engineering
Civil Engineering
Computer Engineering
Construction Engineering
Electrical Engineering
Engineering Science(s)
Industrial Engineering
Mechanical Engineering
Metallurgical Engineering
Nuclear Engineering

J

The Johns Hopkins University
(Baltimore, MD) [E-100]
Biomedical Engineering
Chemical Engineering
Civil Engineering
Electrical Engineering
Engineering Mechanics
Materials Science & Engineering
Mechanical Engineering

K

University of Kansas
(Lawrence, KS) [E-101]
Aerospace Engineering
Architectural Engineering
Chemical Engineering
Civil Engineering
Computer Engineering
Electrical Engineering
Engineering Physics
Mechanical Engineering
Petroleum Engineering

Kansas State University
(Manhattan, KS) [E-102]
Agricultural Engineering
Architectural Engineering
Chemical Engineering
Civil Engineering
Computer Engineering
Electrical Engineering
Industrial Engineering
Manufacturing Systems Engineering
Mechanical Engineering
Nuclear Engineering

University of Kentucky
(Lexington, KY) [E-103]
Agricultural Engineering
Chemical Engineering
Civil Engineering
Electrical Engineering
Mechanical Engineering
Metallurgical Engineering
Mining Engineering

L

Lafayette College
(Easton, PA) [E-104]
Agricultural Engineering
Chemical Engineering
Civil Engineering
Electrical Engineering
Engineering
Mechanical Engineering

Lakehead University
(Thunder Bay, ON-Canada)
Chemical Engineering
Civil Engineering
Electrical Engineering
Mechanical Engineering

Lamar University
(Beaumont, TX) [E-105]
Chemical Engineering
Civil Engineering
Electrical Engineering
Industrial Engineering
Mechanical Engineering

Laurentian University
(Sudbury, ON-Canada)
Extractive Metallurgical Engineering
Extractive Metallurgy
Mining Engineering

Université Laval
(Québec, PQ-Canada)
Génie chimique
Génie civil
Génie des matériaux et de la métallurgie
Génie des mines et de la minéralurgie
Génie géologique
Génie électrique
Génie méchanique
Génie minier
Génie métallurgique
Génie physique
Génie rural

Lawrence Technological University
(Southfield, MI) [E-106]
Construction Engineering
Electrical Engineering
Mechanical Engineering

Lehigh University
(Bethlehem, PA) [E-108]
Chemical Engineering
Civil Engineering
Computer Engineering
Electrical Engineering
Industrial Engineering
Materials Science & Engineering
Mechanical Engineering

LeTourneau University
(Longview, TX) [E-107]
Engineering

Louisiana State University
(Baton Rouge, LA) [E-109]
Biological & Agricultural Engineering
Chemical Engineering
Civil Engineering
Computer Engineering
Electrical Engineering
Industrial Engineering
Mechanical Engineering
Petroleum Engineering

Louisiana Tech University
(Ruston, LA) [E-110]
Biomedical Engineering
Chemical Engineering
Civil Engineering
Electrical Engineering
Industrial Engineering
Mechanical Engineering
Petroleum Engineering

Loyola College in Maryland
(Baltimore, MD) [E-112]
Engineering Science

Loyola Marymount University
(Los Angeles, CA) [E-113]
Civil Engineering
Electrical Engineering
Mechanical Engineering

M

University of Maine
(Orono, ME) [E-114]
Bio-Resource Engineering
Chemical Engineering
Civil Engineering
Electrical Engineering
Engineering Physics
Forest Engineering
Mechanical Engineering
Surveying Engineering

Manhattan College
(Riverdale, NY)
Chemical Engineering
Civil Engineering
Electrical Engineering
Environmental Engineering
Mechanical Engineering

University of Manitoba
(Winnipeg, MB-Canada) [E-115]
Agricultural Engineering
Civil Engineering
Computer Engineering
Electrical Engineering
Geological Engineering
Industrial Engineering
Mechanical Engineering

Mankato State University
(Mankato, MN) [E-116]
Electrical Engineering

Marietta College
(Marietta, OH) [E-117]
Petroleum Engineering

Marquette University
(Milwaukee, WI) [E-118]
Biomedical Engineering
Civil Engineering
Electrical Engineering
Industrial Engineering
Mechanical Engineering

University of Maryland Baltimore County
(Baltimore, MD) [E-119]
Chemical Engineering
Mechanical Engineering

University of Maryland, College Park
(College Park, MD) [E-120]
Aerospace Engineering
Agricultural Engineering
Chemical Engineering
Civil Engineering
Electrical Engineering
Engineering
Fire Protection Engineering
Mechanical Engineering
Nuclear Engineering

University of Massachusetts
(Amherst, MA) [E-121]
Chemical Engineering
Civil Engineering
Computer Systems Engineering
Electrical Engineering
Industrial Engineering & Operations Research
Mechanical Engineering

University of Massachusetts, Dartmouth
(North Dartmouth, MA) [E-122]
Civil Engineering
Computer Engineering
Electrical Engineering
Mechanical Engineering

University of Massachusetts, Lowell
(Lowell, MA) [E-123]
Chemical Engineering
Civil Engineering
Electrical Engineering
Mechanical Engineering
Nuclear Engineering
Plastics Engineering

Massachusetts Institute of Technology
(Cambridge, MA) [E-124]
Aeronautics & Astronautics Engineering
Chemical Engineering
Civil Engineering
Computer Science & Engineering
Electrical Science & Engineering
Materials Science & Engineering
Mechanical Engineering
Nuclear Engineering
Ocean Engineering

McGill University
(Montréal, PQ-Canada) [E-125]
Agricultural Engineering
Chemical Engineering
Civil Engineering
Electrical Engineering
Mechanical Engineering
Metallurgical Engineering
Mining Engineering

McMaster University
(Hamilton, ON-Canada)
Ceramic Engineering
Ceramic Engineering & Management
Chemical Engineering
Chemical Engineering & Management
Civil Engineering
Civil Engineering & Engineering Mechanics
Civil Engineering & Management
Computer Engineering
Computer Engineering & Management
Electrical Engineering
Electrical Engineering & Management
Engineering & Computer Systems
Engineering Physics
Engineering Physics & Management
Manufacturing Engineering
Materials Engineering
Materials Engineering & Management
Mechanical Engineering
Mechanical Engineering & Management
Metallurgical Engineering
Metallurgical Engineering & Management

McNeese State University
(Lake Charles, LA)
Engineering

Memorial University of Newfoundland
(St. John's, NF-Canada)
Civil Engineering
Electrical Engineering
Mechanical Engineering
Naval Architectural Engineering
Shipbuilding Engineering

Memphis State University
(Memphis, TN) [E-126]
Civil Engineering
Electrical Engineering
Mechanical Engineering

Mercer University
(Macon, GA) [E-127]
Engineering

Merrimack College
(North Andover, MA)
Civil Engineering
Electrical/Computer Engineering

University of Miami
(Coral Gables, FL) [E-128]
Architectural Engineering
Civil Engineering
Computer Engineering
Electrical Engineering
Industrial Engineering
Mechanical Engineering

Miami University
(Oxford, OH) [E-129]
Manufacturing Engineering

University of Michigan
(Ann Arbor, MI) [E-130]
Aerospace Engineering
Chemical Engineering
Civil Engineering
Computer Engineering
Electrical Engineering
Industrial & Operations Engineering
Materials Science & Engineering
Mechanical Engineering
Naval Architecture & Marine Engineering
Nuclear Engineering

University of Michigan-Dearborn
(Dearborn, MI) [E-131]
Electrical Engineering
Industrial & Systems Engineering
Mechanical Engineering

Michigan State University
(East Lansing, MI) [E-132]
Agricultural Engineering
Chemical Engineering
Civil Engineering
Electrical Engineering
Materials Science & Engineering
Mechanical Engineering

Michigan Technological University
(Houghton, MI) [E-133]
Chemical Engineering
Civil Engineering
Electrical Engineering
Engineering
Environmental Engineering
Geological Engineering
Materials Science & Engineering Option in Metallurgical Engineering
Mechanical Engineering Option in Metallurgical Engineering
Mineral Processing Engineering
Mining Engineering

Milwaukee School of Engineering
(Milwaukee, WI) [E-134]
Architectural Engineering
Biomedical Engineering
Computer Engineering
Electrical Engineering
Industrial Engineering
Mechanical Engineering

University of Minnesota, Duluth
(Duluth, MN) [E-135]
Chemical Engineering
Computer Engineering
Industrial Engineering

University of Minnesota
(Minneapolis, MN) [E-136]
Aerospace Engineering & Mechanics
Agricultural Engineering
Chemical Engineering
Civil Engineering
Electrical Engineering
Extractive Metallurgical Engineering
Geological Engineering
Materials Science & Engineering
Mechanical Engineering

University of Mississippi
(University, MS) [E-137]
Chemical Engineering
Civil Engineering
Electrical Engineering
Geological Engineering
Mechanical Engineering

Mississippi State University
(Mississippi State, MS) [E-138]
Aerospace Engineering
Agricultural Engineering
Biological Engineering
Chemical Engineering
Civil Engineering
Computer Engineering
Electrical Engineering
Industrial Engineering
Mechanical Engineering
Nuclear Engineering
Petroleum Engineering

ABET/CEAB Accredited Degree Programs

University of Missouri-Columbia
(Columbia, MO) [E-139]
Agricultural Engineering
Chemical Engineering
Civil Engineering
Computer Engineering
Electrical Engineering
Industrial Engineering
Mechanical Engineering

University of Missouri-Kansas City
(Independence, MO) [E-140]
Agricultural Engineering
Chemical Engineering
Civil Engineering
Computer Engineering
Electrical Engineering
Industrial Engineering
Mechanical Engineering

University of Missouri-Rolla
(Rolla, MO) [E-141]
Aerospace Engineering
Ceramic Engineering
Chemical Engineering
Civil Engineering
Electrical Engineering
Engineering Management
Geological Engineering
Mechanical Engineering
Metallurgical Engineering
Mining Engineering
Nuclear Engineering
Petroleum Engineering

Université de Moncton
(Moncton, NB-Canada)
Génie civil
Génie industriel
Génie mécanique

Monmouth College
(West Long Branch, NJ) [E-142]
Electronic(s) Engineering

Montana College of Mineral Science and Technology
(Butte, MT) [E-143]
Engineering Science(s)
Environmental Engineering
Geological Engineering
Geophysical Engineering
Metallurgical Engineering
Mining Engineering
Petroleum Engineering

Montana State University
(Bozeman, MT) [E-144]
Bio-Resources Option in Civil Engineering
Chemical Engineering
Civil Engineering
Electrical Engineering
Industrial & Management Engineering
Mechanical Engineering

Morgan State University
(Baltimore, MD) [E-145]
Civil Engineering
Electrical Engineering
Industrial Engineering

N

University of Nebraska - Lincoln
(Lincoln, NE) [E-146]
Agricultural Engineering
Chemical Engineering
Civil Engineering
Electrical Engineering
Industrial Engineering
Mechanical Engineering

University of Nebraska-Omaha
(Omaha, NE)
Civil Engineering

University of Nevada, Las Vegas
(Las Vegas, NV) [E-147]
Civil Engineering
Electrical Engineering
Mechanical Engineering

University of Nevada-Reno
(Reno, NV)
Chemical Engineering
Civil Engineering
Electrical Engineering
Geological Engineering
Mechanical Engineering
Metallurgical Engineering
Mining Engineering

University of New Brunswick
(Fredericton, NB-Canada)
Chemical Engineering
Civil Engineering
Electrical Engineering
Forest Engineering
Geological Engineering
Mechanical Engineering
Surveying Engineering

New England College
(Henniker, NH) [E-148]
Civil Engineering

University of New Hampshire
(Durham, NH) [E-149]
Chemical Engineering
Civil Engineering
Electrical Engineering
Mechanical Engineering

University of New Haven
(West Haven, CT) [E-150]
Civil Engineering
Electrical Engineering
Industrial Engineering
Mechanical Engineering

New Jersey Institute of Technology
(Newark, NJ) [E-151]
Chemical Engineering
Civil Engineering
Electrical Engineering
Industrial Engineering
Mechanical Engineering

The University of New Mexico
(Albuquerque, NM) [E-152]
Chemical Engineering
Civil Engineering
Computer Engineering
Construction Engineering
Electrical Engineering
Mechanical Engineering
Nuclear Engineering

New Mexico Institute of Mining and Technology
(Socorro, NM) [E-153]
Petroleum Engineering

New Mexico State University
(Las Cruces, NM) [E-154]
Agricultural Engineering
Chemical Engineering
Civil Engineering
Electrical Engineering
Geological Engineering
Industrial Engineering
Mechanical Engineering

University of New Orleans
(New Orleans, LA) [E-155]
Civil Engineering
Electrical Engineering
Mechanical Engineering
Naval Architecture & Marine Engineering

New York Institute of Technology
(Old Westbury, NY) [E-156]
Electrical Engineering
Mechanical Engineering

New York State College of Ceramics at Alfred University
(Alfred, NY) [E-157]
Ceramic Engineering
Ceramic Engineering Science
Glass Engineering Science

University of North Carolina-Charlotte
(Charlotte, NC) [E-158]
Civil Engineering
Electrical Engineering
Mechanical Engineering

North Carolina Agricultural and Technical State University
(Greensboro, NC) [E-159]
Agricultural Engineering
Architectural Engineering
Chemical Engineering
Civil Engineering
Electrical Engineering
Industrial Engineering
Mechanical Engineering

North Carolina State University
(Raleigh, NC) [E-160]
Aerospace Engineering
Biological & Agricultural Engineering
Chemical Engineering
Civil Engineering
Computer Engineering
Construction Option in Civil Engineering
Electrical Engineering
Industrial Engineering
Materials Science & Engineering
Mechanical Engineering
Nuclear Engineering
Textile Engineering

University of North Dakota
(Grand Forks, ND) [E-161]
Chemical Engineering
Civil Engineering
Electrical Engineering
Geological Engineering
Mechanical Engineering

North Dakota State University
(Fargo, ND) [E-162]
Agricultural Engineering
Civil Engineering
Construction Engineering
Electrical/Electronic(s) Engineering
Industrial Engineering
Mechanical Engineering

Northeastern University
(Boston, MA) [E-163]
Chemical Engineering
Civil Engineering
Electrical Engineering
Industrial Engineering
Mechanical Engineering

Northern Arizona University
(Flagstaff, AZ) [E-164]
Civil Engineering
Computer Science & Engineering
Electrical Engineering
Mechanical Engineering

Northern Illinois University
(DeKalb, IL) [E-165]
Electrical Engineering
Industrial Engineering
Mechanical Engineering

Northwestern University
(Evanston, IL) [E-166]
Biomedical Engineering
Chemical Engineering
Civil Engineering
Electrical Engineering
Environmental Engineering
Industrial Engineering
Materials Science & Engineering
Mechanical Engineering

Norwich University
(Northfield, VT) [E-167]
Civil Engineering
Electrical Engineering
Mechanical Engineering

University of Notre Dame
(Notre Dame, IN) [E-168]
Aerospace Engineering
Chemical Engineering
Civil Engineering
Electrical Engineering
Mechanical Engineering
Metallurgical Engineering

O

Oakland University
(Rochester, MI) [E-169]
Computer Engineering
Electrical Engineering
Mechanical Engineering
Systems Engineering

Ohio University
(Athens, OH) [E-170]
Chemical Engineering
Civil Engineering
Electrical Engineering
Industrial & Systems Engineering
Mechanical Engineering

Ohio Northern University
(Ada, OH) [E-171]
Civil Engineering
Electrical Engineering
Mechanical Engineering

Ohio State University
(Columbus, OH) [E-172]
Aeronautical & Astronautical Engineering
Agricultural Engineering
Ceramic Engineering
Chemical Engineering
Civil Engineering
Electrical Engineering
Industrial & Systems Engineering
Mechanical Engineering
Metallurgical Engineering
Welding Engineering

University of Oklahoma
(Norman, OK) [E-173]
Aerospace Engineering
Chemical Engineering
Civil Engineering
Electrical Engineering
Engineering
Engineering Physics
Industrial Engineering
Mechanical Engineering
Petroleum Engineering

Oklahoma Christian University of Science and Arts
(Oklahoma City, OK) [E-174]
Electrical Engineering
Mechanical Engineering

Oklahoma State University
(Stillwater, OK) [E-175]
Aerospace Option in Mechanical Engineering
Agricultural Engineering
Architectural Engineering
Chemical Engineering
Civil Engineering
Electrical Engineering

General Engineering
Industrial Engineering & Management
Mechanical Engineering

Old Dominion University
(Norfolk, VA) [E-176]
Civil Engineering
Computer Engineering
Electrical Engineering
Mechanical Engineering

Oregon State University
(Corvallis, OR) [E-177]
Chemical Engineering
Civil Engineering
Computer Engineering
Electrical & Electronics Engineering
Industrial Engineering
Manufacturing Engineering Option in Industrial Engineering
Mechanical Engineering
Nuclear Engineering

University of Ottawa
(Ottawa, ON-Canada)
Chemical Engineering
Civil Engineering
Computer Engineering
Electrical Engineering
Mechanical Engineering

P

University of the Pacific
(Stockton, CA) [E-178]
Civil Engineering
Computer Engineering
Electrical Engineering
Engineering Physics
Mechanical Engineering

University of Pennsylvania
(Philadelphia, PA) [E-179]
Bioengineering
Chemical Engineering
Civil Engineering Systems
Electrical Engineering
Materials Science & Engineering
Mechanical Engineering & Applied Mechanics
Systems Science & Engineering

Pennsylvania State University
(University Park, PA) [E-180]
Aerospace Engineering
Agricultural Engineering
Architectural Engineering
Ceramic Science & Engineering
Chemical Engineering
Civil Engineering
Computer Engineering
Electrical Engineering
Engineering Science
Industrial Engineering
Mechanical Engineering
Metals Science & Engineering
Mining Engineering (Mineral Processing Option)
Mining Engineering (Mining Option)
Nuclear Engineering
Petroleum & Natural Gas Engineering

University of Pittsburgh
(Pittsburgh, PA) [E-181]
Chemical Engineering
Civil Engineering
Electrical Engineering
Industrial Engineering
Materials Science & Engineering
Mechanical Engineering
Metallurgical Engineering

Polytechnic University
(Brooklyn, NY) [E-182]
Aerospace Engineering
Chemical Engineering
Civil Engineering
Computer Engineering

Electrical Engineering
Industrial Engineering
Mechanical Engineering
Metallurgical Engineering (Materials Science & Engineering)

University of Portland
(Portland, OR) [E-183]
Civil Engineering
Electrical Engineering
Mechanical Engineering

Portland State University
(Portland, OR) [E-184]
Civil Engineering
Electrical Engineering
Mechanical Engineering

Prairie View A&M University
(Prairie View, TX) [E-185]
Civil Engineering
Electrical Engineering
Mechanical Engineering

Princeton University
(Princeton, NJ) [E-186]
Aerospace Engineering
Chemical Engineering
Civil Engineering
Electrical Engineering
Engineering Physics
Geological Engineering
Mechanical Engineering

University of Puerto Rico, Mayagüez Campus
(Mayagüez, PR) [E-187]
Chemical Engineering
Civil Engineering
Electrical Engineering
Industrial Engineering
Mechanical Engineering

Purdue University
(West Lafayette, IN) [E-188]
Aeronautical & Astronautical Engineering
Agricultural Engineering
Chemical Engineering
Civil Engineering
Computer & Electrical Engineering
Construction Engineering & Management
Electrical Engineering
Food Process Engineering
Industrial Engineering
Mechanical Engineering
Metallurgical Engineering
Nuclear Engineering
Surveying Engineering

Purdue University-Calumet
(Hammond, IN)
Electrical Engineering
Mechanical Engineering

Q

Québec a Chicoutimi, Université de
(Chicoutimi, PQ-Canada)
Génie géologique
Génie informatique
Génie unifié

Québec à Trois-Rivières, Université de
(Trois-Rivières, PQ-Canada)
Génie chimique
Génie industriel
Génie électrique
Génie méchanique manufacturier

Queen's University
(Kingston, ON-Canada)
Chemical Engineering
Civil Engineering
Electrical Engineering
Engineering Chemistry

Engineering Physics
Geological Engineering
Materials & Metallurgical Engineering
Mathematics & Engineering
Mechanical Engineering
Metallurgical Engineering
Mining Engineering

R

University of Regina
(Regina, SK-Canada)
Electronic Information Systems Engineering
Industrial Systems Engineering
Regional Environmental Systems Engineering
Regional Systems Engineering
Systems Engineering

Rensselaer Polytechnic Institute
(Troy, NY) [E-189]
Aeronautical Engineering
Biomedical Engineering
Chemical Engineering
Civil Engineering
Computer & Systems Engineering
Electric Power Engineering
Electrical Engineering
Environmental Engineering
Industrial & Management Engineering
Materials Engineering
Mechanical Engineering
Nuclear Engineering

University of Rhode Island
(Kingston, RI) [E-190]
Chemical Engineering
Civil Engineering
Computer Engineering
Electrical Engineering
Industrial Engineering
Mechanical Engineering

William Marsh Rice University
(Houston, TX) [E-191]
Chemical Engineering
Civil Engineering
Electrical Engineering
Materials Science & Engineering
Mechanical Engineering

University of Rochester
(Rochester, NY) [E-192]
Chemical Engineering
Electrical Engineering
Mechanical Engineering

Rochester Institute of Technology
(Rochester, NY) [E-193]
Computer Engineering
Electrical Engineering
Industrial Engineering
Mechanical Engineering
Microelectronic Engineering

Rose-Hulman Institute of Technology
(Terre Haute, IN) [E-194]
Chemical Engineering
Civil Engineering
Electrical Engineering
Mechanical Engineering

Royal Military College of Canada
(Kingston, ON-Canada)
Chemical & Materials Engineering
Chemical Engineering
Civil Engineering
Computer Engineering
Electrical Engineering
Engineering & Management
Engineering Physics
Fuels & Materials Engineering
Mechanical Engineering

Rutgers, The State University
(Piscataway, NJ) [E-195]
Bioresource Engineering

Ceramic Engineering
Chemical Engineering
Civil Engineering
Electrical Engineering
Industrial Engineering
Mechanical Engineering

Ryerson Polytechnical Institute
(Toronto, ON-Canada)
Aerospace Engineering
Chemical Engineering
Civil Engineering
Electrical Engineering
Industrial Engineering
Mechanical Engineering

S

Saginaw Valley State University
(University Center, MI) [E-196]
Electrical Engineering
Mechanical Engineering

St. Cloud State University
(St. Cloud, MN) [E-197]
Electrical Engineering

Parks College of Saint Louis University
(Cahokia, IL) [E-198]
Aerospace Engineering
Electrical Engineering

Saint Martin's College
(Lacey, WA)
Civil Engineering

Saint Mary's University
(San Antonio, TX)
Electrical Engineering
Industrial Engineering

University of San Diego
(San Diego, CA) [E-199]
Electrical Engineering

San Diego State University
(San Diego, CA) [E-200]
Aerospace Engineering
Civil Engineering
Electrical Engineering
Mechanical Engineering

San Francisco State University
(San Francisco, CA) [E-201]
Civil Engineering
Electrical Engineering
Mechanical Engineering

San Jose State University
(San Jose, CA) [E-202]
Aerospace Engineering
Chemical Engineering
Civil Engineering
Computer Engineering
Electrical Engineering
Industrial & Systems Engineering
Materials Engineering
Mechanical Engineering

Santa Clara University
(Santa Clara, CA) [E-203]
Civil Engineering
Computer Engineering
Electrical Engineering
Mechanical Engineering

University of Saskatchewan
(Saskatoon, SK-Canada)
Agricultural Engineering
Chemical Engineering
Civil Engineering
Electrical Engineering
Engineering Physics
Geological Engineering
Geological Engineering (Geophysics)

ABET/CEAB Accredited Degree Programs

Mechanical Engineering
Mining Engineering

Seattle University
(Seattle, WA) [E-204]
Civil Engineering
Electrical Engineering
Mechanical Engineering

Seattle Pacific University
(Seattle, WA)
Electrical Engineering

Université de Sherbrooke
(Sherbrooke, PQ-Canada)
Génie chimique
Génie civil
Génie électrique
Génie méchanique

Simon Fraser University
(Burnaby, BC-Canada)
Engineering Science

University of South Alabama
(Mobile, AL) [E-205]
Chemical Engineering
Civil Engineering
Electrical Engineering
Mechanical Engineering

University of South Carolina
(Columbia, SC) [E-206]
Chemical Engineering
Civil Engineering
Electrical Engineering
Mechanical Engineering

South Dakota School of Mines and Technology
(Rapid City, SD) [E-207]
Chemical Engineering
Civil Engineering
Electrical Engineering
Geological Engineering
Mechanical Engineering
Metallurgical Engineering
Mining Engineering

South Dakota State University
(Brookings, SD) [E-208]
Agricultural Engineering
Civil Engineering
Electrical Engineering
Mechanical Engineering

University of South Florida
(Tampa, FL) [E-209]
Chemical Engineering
Civil Engineering
Computer Engineering
Electrical Engineering
Industrial Engineering
Mechanical Engineering

Southern University and A&M College
(Baton Rouge, LA) [E-210]
Civil Engineering
Electrical Engineering
Mechanical Engineering

University of Southern California
(Los Angeles, CA) [E-211]
Aerospace Engineering
Chemical Engineering
Civil Engineering
Electrical Engineering
Industrial & Systems Engineering
Mechanical Engineering

University of Southern Colorado
(Pueblo, CO) [E-212]
Industrial Engineering

Southern Illinois University at Carbondale
(Carbondale, IL) [E-213]
Civil Engineering
Electrical Engineering
Mechanical Engineering
Mining Engineering

Southern Illinois University at Edwardsville
(Edwardsville, IL) [E-214]
Civil Engineering
Electrical Engineering
Industrial Engineering

University of Southern Maine
(Gorham, ME) [E-215]
Electrical Engineering

Southern Methodist University
(Dallas, TX) [E-216]
Civil Engineering
Computer Engineering
Electrical Engineering
Mechanical Engineering

University of Southwestern Louisiana
(Lafayette, LA)
Chemical Engineering
Civil Engineering
Electrical Engineering
Mechanical Engineering
Petroleum Engineering

Stanford University
(Stanford, CA) [E-217]
Chemical Engineering
Civil Engineering
Electrical Engineering
Industrial Engineering
Mechanical Engineering
Petroleum Engineering

State University of New York at Binghamton
(Binghamton, NY) [E-218]
Electrical Engineering
Mechanical Engineering

State University of New York at Buffalo
(Buffalo, NY) [E-219]
Aerospace Engineering
Chemical Engineering
Civil Engineering
Electrical Engineering
Industrial Engineering
Mechanical Engineering

State University of New York, College of Environmental Science and Forestry
(Syracuse, NY) [E-220]
Forest Engineering

State University of New York Maritime College
(Throgs Neck Station, NY) [E-221]
Electrical Engineering
Marine Engineering
Naval Architecture

State University of New York, College at New Paltz
(New Paltz, NY) [E-222]
Electrical Engineering

State University of New York at Stony Brook
(Stony Brook, NY) [E-223]
Computer Engineering (Option in Electrical Engineering)
Electrical Engineering
Engineering Science
Mechanical Engineering

Stevens Institute of Technology
(Hoboken, NJ)
Chemical Engineering
Civil Engineering
Computer Engineering
Electrical Engineering
Engineering
Engineering Management
Engineering Physics
Materials & Metallurgical Engineering
Mechanical Engineering

Swarthmore College
(Swarthmore, PA) [E-224]
Engineering

Syracuse University
(Syracuse, NY) [E-225]
Aerospace Engineering
Bioengineering
Chemical Engineering
Civil Engineering
Computer Engineering
Electrical Engineering
Mechanical Engineering

T

Technical University of Nova Scotia
(Halifax, NS-Canada) [E-226]
Agricultural Engineering
Chemical Engineering
Civil Engineering
Electrical Engineering
Engineering Physics
Industrial Engineering
Mechanical Engineering
Metallurgical Engineering
Mining Engineering

Temple University
(Philadelphia, PA) [E-227]
Civil Engineering
Electrical Engineering
Mechanical Engineering

University of Tennessee, Chattanooga
(Chattanooga, TN) [E-228]
Engineering

University of Tennessee, Knoxville
(Knoxville, TN) [E-229]
Aerospace Engineering
Agricultural Engineering
Chemical Engineering
Civil Engineering
Electrical Engineering
Engineering Science
Industrial Engineering
Materials Science & Engineering
Mechanical Engineering
Nuclear Engineering

Tennessee State University
(Nashville, TN) [E-230]
Architectural Engineering
Civil Engineering
Electrical Engineering
Mechanical Engineering

Tennessee Technological University
(Cookeville, TN) [E-231]
Chemical Engineering
Civil Engineering
Electrical Engineering
Industrial Engineering
Mechanical Engineering

University of Texas at Arlington
(Arlington, TX) [E-232]
Aerospace Engineering
Civil Engineering
Computer Science & Engineering

Electrical Engineering
Industrial Engineering
Mechanical Engineering

University of Texas at Austin
(Austin, TX) [E-233]
Aerospace Engineering
Architectural Engineering
Chemical Engineering
Civil Engineering
Computer Engineering
Electrical Engineering
Mechanical Engineering
Petroleum Engineering

University of Texas at Dallas
(Richardson, TX) [E-234]
Electrical Engineering

University of Texas at El Paso
(El Paso, TX) [E-235]
Civil Engineering
Electrical Engineering
Industrial Engineering
Mechanical Engineering
Metallurgical Engineering

University of Texas at San Antonio
(San Antonio, TX) [E-236]
Civil Engineering
Electrical Engineering
Mechanical Engineering

Texas A&I University
(Kingsville, TX) [E-237]
Chemical Engineering
Civil Engineering
Electrical Engineering
Mechanical Engineering
Natural Gas Engineering

Texas A&M University
(College Station, TX) [E-238]
Aerospace Engineering
Agricultural Engineering
Bioengineering
Chemical Engineering
Civil Engineering
Electrical Engineering
Industrial Engineering
Mechanical Engineering
Nuclear Engineering
Ocean Engineering
Petroleum Engineering
Radiological Health Engineering

Texas A&M University-Galveston
(Galveston, TX)
Marine Engineering

Texas Tech University
(Lubbock, TX) [E-239]
Agricultural Engineering
Chemical Engineering
Civil Engineering
Electrical Engineering
Engineering Physics
Industrial Engineering
Mechanical Engineering
Petroleum Engineering

University of Toledo
(Toledo, OH) [E-240]
Chemical Engineering
Civil Engineering
Computer Science & Engineering
Electrical Engineering
Engineering Physics
Industrial Engineering
Mechanical Engineering

University of Toronto
(Toronto, ON-Canada)
Chemical Engineering
Civil Engineering
Electrical Engineering
Engineering Science

ABET/CEAB Accredited Degree Programs

Geo-Engineering
Geological & Mineral Engineering
Geological Engineering
Geological Engineering & Applied Earth Science
Industrial Engineering
Mechanical Engineering
Metallurgical Engineering & Materials Science
Metallurgy & Materials Science

Tri-State University
(Angola, IN) [E-241]
Aerospace Engineering
Chemical Engineering
Civil Engineering
Electrical Engineering
Mechanical Engineering

Trinity University
(San Antonio, TX) [E-242]
Engineering Science(s)

Tufts University
(Medford, MA) [E-243]
Chemical Engineering
Civil Engineering
Computer Engineering (Option in Electrical Engineering)
Electrical Engineering
Mechanical Engineering

Tulane University
(New Orleans, LA) [E-244]
Biomedical Engineering
Chemical Engineering
Civil Engineering
Electrical Engineering
Mechanical Engineering

University of Tulsa
(Tulsa, OK) [E-245]
Chemical Engineering
Electrical Engineering
Engineering Physics
Mechanical Engineering
Petroleum Engineering

Tuskegee University
(Tuskegee, AL) [E-246]
Aerospace Science Engineering
Chemical Engineering
Electrical Engineering
Mechanical Engineering

U

Union College
(Schenectady, NY) [E-247]
Civil Engineering
Electrical Engineering
Mechanical Engineering

United States Air Force Academy
(USAF Academy, CO) [E-248]
Aeronautical Engineering
Astronautical Engineering
Civil Engineering
Electrical Engineering
Engineering Mechanics
Engineering Science(s)
Mechanical Engineering (Option in Engineering Science(s))

United States Coast Guard Academy
(New London, CT) [E-249]
Civil Engineering
Electrical Engineering
Naval Architecture & Marine Engineering

United States International University
(San Diego, CA)
Civil Engineering

United States Merchant Marine Academy
(Kings Point, NY) [E-250]
Marine Engineering Systems

United States Military Academy
(West Point, NY) [E-251]
Civil Engineering
Electrical Engineering
Engineering Management
Mechanical Engineering

United States Naval Academy
(Annapolis, MD) [E-252]
Aerospace Engineering
Electrical Engineering
Marine Engineering
Mechanical Engineering
Naval Engineering
Ocean Engineering
Systems Engineering

University of Utah
(Salt Lake City, UT) [E-253]
Biomedical Engineering
Chemical Engineering
Civil Engineering
Electrical Engineering
Geological Engineering
Materials Engineering
Materials Science & Engineering
Mechanical Engineering
Metallurgical Engineering
Mining Engineering

Utah State University
(Logan, UT) [E-254]
Agricultural & Irrigation Engineering
Civil Engineering
Electrical Engineering
Manufacturing Engineering (Option in Mechanical Engineering)
Mechanical Engineering

V

Valparaiso University
(Valparaiso, IN) [E-255]
Civil Engineering
Computer Engineering
Electrical Engineering
Mechanical Engineering

Vanderbilt University
(Nashville, TN) [E-256]
Biomedical Engineering
Chemical Engineering
Civil Engineering
Electrical Engineering
Mechanical Engineering

University of Vermont
(Burlington, VT) [E-257]
Civil Engineering
Electrical Engineering
Mechanical Engineering

University of Victoria
(Victoria, BC-Canada)
Computer Engineering
Electrical Engineering
Mechanical Engineering

Villanova University
(Villanova, PA) [E-258]
Chemical Engineering
Civil Engineering
Electrical Engineering
Mechanical Engineering

University of Virginia
(Charlottesville, VA) [E-259]
Aerospace Engineering
Chemical Engineering
Civil Engineering
Electrical Engineering
Mechanical Engineering
Nuclear Engineering
Systems Engineering

Virginia Military Institute
(Lexington, VA) [E-260]
Civil Engineering
Electrical Engineering
Mechanical Engineering

Virginia Polytechnic Institute and State University
(Blacksburg, VA) [E-261]
Aerospace Engineering
Agricultural Engineering
Chemical Engineering
Civil Engineering
Computer Engineering
Electrical Engineering
Engineering Science & Mechanics
Industrial Engineering & Operations Research
Materials Engineering
Mechanical Engineering
Mining Engineering
Ocean Engineering

W

Walla Walla College
(College Place, WA)
Engineering

University of Washington
(Seattle, WA) [E-262]
Aeronautics & Astronautics
Ceramic Engineering
Chemical Engineering
Civil Engineering
Computer Engineering
Electrical Engineering
Industrial Engineering
Mechanical Engineering
Metallurgical Engineering

Washington University
(St. Louis, MO) [E-263]
Chemical Engineering
Civil Engineering
Computer Science & Engineering
Electrical Engineering
Engineering & Public Policy
Mechanical Engineering
Systems Science & Engineering

Washington State University
(Pullman, WA) [E-264]
Agricultural Engineering
Chemical Engineering
Civil Engineering
Electrical Engineering
Geological Engineering
Materials Science & Engineering
Mechanical Engineering

University of Waterloo
(Waterloo, ON-Canada)
Chemical Engineering
Civil Engineering
Computer Engineering
Electrical Engineering
Geological Engineering
Mechanical Engineering
Systems Design Engineering

Wayne State University
(Detroit, MI) [E-265]
Chemical Engineering
Civil Engineering
Electrical Engineering
Industrial Engineering
Materials Science & Engineering
Mechanical Engineering

Webb Institute of Naval Architecture
(Glen Cove, NY) [E-266]
Naval Architecture & Marine Engineering

University of West Florida
(Pensacola, FL)
Systems & Control Engineering

West Virginia University
(Morgantown, WV) [E-267]
Aerospace Engineering
Chemical Engineering
Civil Engineering
Computer Engineering
Electrical Engineering
Engineering of Mines
Industrial Engineering
Mechanical Engineering
Petroleum & Natural Gas Engineering

West Virginia Institute of Technology
(Montgomery, WV) [E-268]
Chemical Engineering
Civil Engineering
Electrical Engineering
Mechanical Engineering

Western Michigan University
(Kalamazoo, MI) [E-269]
Computer Systems Engineering
Electrical Engineering
Industrial Engineering
Mechanical Engineering

Western New England College
(Springfield, MA) [E-270]
Electrical Engineering
Industrial Engineering
Mechanical Engineering

University of Western Ontario
(London, ON-Canada)
Chemical & Biochemical Engineering
Chemical Engineering
Civil Engineering
Electrical Engineering
Materials Engineering
Mechanical Engineering

Wichita State University
(Wichita, KS) [E-271]
Aerospace Engineering
Electrical Engineering
Industrial Engineering
Mechanical Engineering

Widener University
(Chester, PA) [E-272]
Chemical Engineering
Civil Engineering
Electrical Engineering
Engineering
Mechanical Engineering

Wilkes University
(Wilkes-Barre, PA) [E-273]
Electrical Engineering
Materials Engineering

University of Windsor
(Windsor, ON-Canada)
Chemical Engineering
Civil Engineering
Electrical Engineering
Engineering Materials
Geological Engineering
Industrial Engineering
Mechanical Engineering

University of Wisconsin-Madison
(Madison, WI) [E-274]
Agricultural Engineering
Chemical Engineering
Civil Engineering
Electrical Engineering
Engineering Mechanics
Industrial Engineering
Mechanical Engineering

Metallurgical Engineering
Nuclear Engineering
Surveying Option in Civil & Environmental Engineering

University of Wisconsin-Milwaukee
(Milwaukee, WI) [E-275]
Civil Engineering
Electrical Engineering
Industrial Engineering
Materials Engineering
Mechanical Engineering

University of Wisconsin-Platteville
(Platteville, WI) [E-276]
Civil Engineering
Electrical Engineering
Industrial Engineering
Mechanical Engineering

Worcester Polytechnic Institute
(Worcester, MA) [E-277]
Chemical Engineering
Civil Engineering
Electrical Engineering
Manufacturing Systems Engineering
Mechanical Engineering

Wright State University
(Dayton, OH) [E-278]
Biomedical Engineering
Computer Engineering
Electrical Engineering
Engineering Physics
Materials Science & Engineering
Mechanical Engineering

University of Wyoming
(Laramie, WY) [E-279]
Agricultural Engineering
Architectural Engineering
Chemical Engineering
Civil Engineering
Electrical Engineering
Mechanical Engineering
Petroleum Engineering

Y

Yale University
(New Haven, CT) [E-280]
Chemical Engineering
Electrical Engineering
Mechanical Engineering

Youngstown State University
(Youngstown, OH)
Chemical Engineering
Civil Engineering
Electrical Engineering
Industrial Engineering
Mechanical Engineering

ENGINEERING TECHNOLOGY

A

University of Akron
(Akron, OH) [T-281]
Construction Technology
Electronic(s) Technology
Mechanical Technology

University of Alabama
(Tuscaloosa, AL)
Civil Engineering Technology
Electrical Engineering Technology

Alabama A&M University
(Normal, AL) [T-282]
Civil Engineering Technology
Electrical Engineering Technology
Mechanical Drafting & Design Technology
Mechanical Engineering Technology

American Technical Institute
(Brunswick, TN)
Nuclear Engineering Technology

Arizona State University
(Tempe, AZ) [T-283]
Aeronautical Engineering Technology
Electronics Engineering Technology
Manufacturing Engineering Technology

University of Arkansas at Little Rock
(Little Rock, AR) [T-284]
Computer Engineering Technology
Construction Engineering Technology
Electronic(s) Engineering Technology
Manufacturing Engineering Technology
Mechanical Engineering Technology

Auburn University
(Auburn University, AL) [T-285]
Textile Management and Technology

B

Bluefield State College
(Bluefield, WV)
Architectural Engineering Technology
Civil Engineering Technology
Electrical Engineering Technology
Mining Engineering Technology

Bradley University
(Peoria, IL) [T-286]
Electrical Engineering Technology
Mechanical Design Option in Manufacturing Technology
Productions Operations Option in Manufacturing Technology

Brigham Young University
(Provo, UT) [T-287]
Design Engineering Technology
Electronic(s) Engineering Technology
Manufacturing Engineering Technology

C

California Maritime Academy
(Vallejo, CA)
Marine Engineering Technology

California Polytechnic State University
(San Luis Obispo, CA)
Air Conditioning & Refrigeration Technology
Electronic(s) Technology
Manufacturing Processes Technology
Mechanical Technology
Welding Technology

California State University, Sacramento
(Sacramento, CA) [T-288]
Construction Management Option in Engineering Technology
Mechanical Engineering Technology Option in Engineering Technology

California State Polytechnic University, Pomona
(Pomona, CA) [T-289]
Engineering Technology

Capitol College
(Laurel, MD) [T-290]
Computer Engineering Technology
Electronic(s) Engineering Technology

Central Connecticut State University
(New Britain, CT) [T-291]
Engineering Technology-Construction
Engineering Technology-Manufacturing

Central Washington University
(Ellensburg, WA)
Electronic(s) Engineering Technology

University of Cincinnati
(Cincinnati, OH) [T-293]
Architectural Engineering Technology
Electrical Engineering Technology
Mechanical Engineering Technology

CUNY - City College
(New York, NY)
Electromechanical Engineering Technology

Cogswell Polytechnic College North
(Kirkland, WA)
Electronic(s) Engineering Technology
Mechanical Engineering Technology

Cogswell Polytechnic College-Cupertino
(Cupertino, CA)
Electronic(s) Engineering Technology
Mechanical Engineering Technology

Colorado Technical College
(Colorado Springs, CO)
Electronic(s) Engineering Technology

D

University of Dayton
(Dayton, OH) [T-294]
Electronic(s) Engineering Technology
Industrial Engineering Technology
Manufacturing Engineering Technology
Mechanical Engineering Technology

University of Delaware
(Newark, DE)
Agricultural Engineering Technology
Engineering Technology & Technical Management

DeVry Institute of Technology, Chicago
(Chicago, IL) [T-295]
Electronic(s) Engineering Technology

DeVry Institute of Technology, Columbus
(Columbus, OH) [T-296]
Electronic(s) Engineering Technology

DeVry Institute of Technology, Atlanta
(Decatur, GA) [T-297]
Electronic(s) Engineering Technology

DeVry Institute of Technology, DuPage
(Addison, IL) [T-298]
Electronic(s) Engineering Technology

DeVry Institute of Technology, Dallas
(Irving, TX) [T-299]
Electronic(s) Engineering Technology

DeVry Institute of Technology, Kansas City
(Kansas City, MO) [T-300]
Electronic(s) Engineering Technology

DeVry Institute of Technology, Los Angeles
(City of Industry, CA) [T-301]
Electronic(s) Engineering Technology

DeVry Institute of Technology, Phoenix
(Phoenix, AZ) [T-302]
Electronic(s) Engineering Technology

University of the District of Columbia
(Washington, DC) [T-303]
Construction Engineering Technology
Electromechanical Systems Engineering Technology

E

East Tennessee State University
(Johnson City, TN) [T-304]
Construction Technology
Electronic(s) Engineering Technology
Manufacturing Engineering Technology

Eastern Washington University
(Cheney, WA)
Computer Option in Computer Engineering Technology
Mechanical Engineering Technology

Embry-Riddle Aeronautical University - Daytona Beach
(Daytona Beach, FL) [T-305]
Aircraft Engineering Technology

F

Fairmont State College
(Fairmont, WV)
Civil Engineering Technology
Electronic(s) Engineering Technology
Mechanical Engineering Technology

Florida A&M University
(Tallahassee, FL)
Civil Engineering Technology
Construction Engineering Technology
Electronic(s) Engineering Technology

Fort Valley State College
(Fort Valley, GA)
Electronic(s) Engineering Technology

Franklin University
(Columbus, OH) [T-307]
Electronic(s) Engineering Technology
Mechanical Engineering Technology

G

Georgia Southern University
(Statesboro, GA)
Civil Engineering Technology
Electrical Engineering Technology
Industrial Engineering Technology
Mechanical Engineering Technology

H

University of Hartford
(West Hartford, CT) [T-308]
Electronics Engineering Technology

University of Houston
(Houston, TX)
Civil Technology
Computer Engineering Technology
Electrical Technology
Mechanical Technology

University of Houston-Downtown
(Houston, TX)
Process & Piping Design Engineering Technology Option in Engineering Tech

I

Indiana University-Purdue University at Indianapolis
(Indianapolis, IN) [T-309]
Electrical Technology
Mechanical Engineering Technology

K

Kansas State University
(Manhattan, KS)
Electronic Technology Option in Engineering Technology
Mechanical Engineering Technology Option in Engineering Technology

L

Lake Superior State University
(Sault Ste. Marie, MI) [T-310]
Automated Systems Engineering Technology
Electrical Engineering Technology
Mechanical Engineering Technology

Louisiana Tech University
(Ruston, LA) [T-311]
Construction Engineering Technology
Electrical Engineering Technology

M

University of Maine
(Orono, ME) [T-312]
Construction Management Technology
Electrical Engineering Technology
Mechanical Engineering Technology

Maine Maritime Academy
(Castine, ME)
Marine Engineering Technology

Mankato State University
(Mankato, MN) [T-313]
Electronic(s) Engineering Technology
Manufacturing Engineering Technology

University of Massachusetts, Dartmouth
(North Dartmouth, MA) [T-314]
Electrical Engineering Technology
Mechanical Engineering Technology

University of Massachusetts, Lowell
(Lowell, MA) [T-315]
Civil Engineering Technology
Electronic Engineering Technology
Mechanical Engineering Technology

Memphis State University
(Memphis, TN) [T-316]
Architectural Technology
Computer Engineering Technology
Electronic(s) Engineering Technology
Manufacturing Engineering Technology

Metropolitan State College of Denver
(Denver, CO) [T-317]
Civil Engineering Technology
Electronic(s) Engineering Technology
Mechanical Engineering Technology

Midwestern State University
(Wichita Falls, TX)
Manufacturing Engineering Technology

Milwaukee School of Engineering
(Milwaukee, WI) [T-318]
Electrical Engineering Technology
Mechanical Engineering Technology

Missouri Western State College
(St. Joseph, MO)
Construction Engineering Technology
Electronics Engineering Technology

Montana State University
(Bozeman, MT) [T-319]
Construction Engineering Technology
Electrical & Electronics Engineering Technology
Mechanical Engineering Technology

Murray State University
(Murray, KY) [T-320]
Civil Engineering Technology
Construction Technology
Electrical Engineering Technology
Manufacturing Engineering Technology

N

University of Nebraska-Omaha
(Omaha, NE)
Construction Engineering Technology
Electronic(s) Engineering Technology
Manufacturing Engineering Technology

University of New Hampshire
(Durham, NH) [T-321]
Electrical Engineering Technology
Mechanical Engineering Technology

New Jersey Institute of Technology
(Newark, NJ) [T-322]
Construction & Contracting Option in Engineering Technology
Electrical Option in Engineering Technology
Manufacturing Technology Option in Engineering Technology
Mechanical Option in Engineering Technology

New Mexico Highlands University
(Las Vegas, NM)
Electronic(s) Engineering Technology

New Mexico State University
(Las Cruces, NM) [T-323]
Civil Engineering Technology Option in Engineering Technology
Electronic(s) Engineering Technology Option in Engineering Technology
Mechanical Engineering Technology Option in Engineering Technology

New York Institute of Technology
(New York, NY)
Electrical Engineering Technology
(Old Westbury, NY)
Electrical Engineering Technology

University of North Carolina-Charlotte
(Charlotte, NC) [T-324]
Civil Engineering Technology
Electrical Engineering Technology
Mechanical Engineering Technology

Northeastern University
(Boston, MA) [T-325]
Electrical Engineering Technology
Mechanical Engineering Technology
Mechanical/Structural Engineering Technology

Northern Arizona University
(Flagstaff, AZ)
Electrical Engineering Technology
Mechanical Engineering Technology

Northrop-Rice Aviation Institute of Technology
(Inglewood, CA)
Aircraft Maintenance Engineering Technology

Norwich University
(Northfield, VT)
Environmental Engineering Technology

O

Oklahoma State University
(Stillwater, OK) [T-326]
Construction Management Technology
Electronics Technology
Fire Protection & Safety Technology
Manufacturing Technology
Mechanical Design Technology
Mechanical Power Technology

Old Dominion University
(Norfolk, VA) [T-327]
Civil Engineering Technology
Electrical Engineering Technology
Mechanical Engineering Technology

Oregon Institute of Technology
(Klamath Falls, OR) [T-328]
Civil Engineering Technology
Computer Engineering Technology
Electronic(s) Engineering Technology
LASER Optical Engineering Technology
Manufacturing Engineering Technology
Mechanical Engineering Technology
Software Engineering Technology

P

Pennsylvania State University at Erie, The Behrend College
(Erie, PA) [T-329]
Electrical Engineering Technology
Mechanical Engineering Technology

Pennsylvania State University - Harrisburg
(Middletown, PA) [T-330]
Electrical Engineering Technology
Environmental Engineering Technology
Mechanical Engineering Technology
Structural Design & Construction Engineering Technology

Pittsburg State University
(Pittsburg, KS) [T-331]
Construction Engineering Technology
Electronic(s) Engineering Technology
Manufacturing Engineering Technology
Mechanical Engineering Technology
Plastics Engineering Technology

University of Pittsburgh at Johnstown
(Johnstown, PA) [T-332]
Civil Engineering Technology
Electrical Engineering Technology
Mechanical Engineering Technology

Point Park College
(Pittsburgh, PA) [T-333]
Civil Engineering Technology
Electrical Engineering Technology
Mechanical Engineering Technology

Prairie View A&M University
(Prairie View, TX) [T-334]
Computer Engineering Technology
Electrical Engineering Technology

Purdue University
(West Lafayette, IN) [T-335]
Electrical Technology
Mechanical Technology

Purdue University-Calumet
(Hammond, IN)
Construction Technology
Electrical Engineering Technology
Industrial Engineering Technology
Mechanical Engineering Technology

Purdue University - Kokomo
(West Lafayette, IN)
Electrical Engineering Technology

Purdue University - North Central
(Westville, IN)
Mechanical Engineering Technology

R

Rochester Institute of Technology
(Rochester, NY) [T-336]
Civil Engineering Technology
Computer Technology
Electrical Engineering Technology
Energy Engineering Technology
Manufacturing Engineering Technology
Mechanical Engineering Technology

Roger Williams College
(Bristol, RI)
Electrical Engineering Technology
Mechanical Engineering Technology

S

St. Cloud State University
(St. Cloud, MN) [T-337]
Engineering Technology Option in Manufacturing Engineering Technology

Savannah State College
(Savannah, GA)
Civil Engineering Technology
Electronic(s) Engineering Technology
Mechanical Engineering Technology

South Carolina State University
(Orangeburg, SC) [T-338]
Civil Engineering Technology
Electrical Engineering Technology
Industrial Engineering Technology
Mechanical Engineering Technology

Southern College of Technology
(Marietta, GA) [T-339]
Apparel Engineering Technology
Architectural Engineering Technology
Civil Engineering Technology
Computer Engineering Technology
Electrical Engineering Technology
Industrial Engineering Technology
Mechanical Engineering Technology
Textile Engineering Technology

University of Southern Colorado
(Pueblo, CO) [T-340]
Civil Engineering Technology
Electronic(s) Engineering Technology
Mechanical Engineering Technology

Southern Illinois University at Carbondale
(Carbondale, IL) [T-341]
Civil Engineering Technology
Electrical Engineering Technology
Mechanical Engineering Technology

University of Southern Indiana
(Evansville, IN)
Civil Engineering Technology
Electrical Engineering Technology
Mechanical Engineering Technology

University of Southern Mississippi
(Hattiesburg, MS) [T-342]
Architectural Engineering Technology
Computer Engineering Technology
Construction Engineering Technology
Electronic(s) Engineering Technology
Industrial Engineering Technology
Mechanical Engineering Technology

SUNY-Binghamton
(Binghamton, NY)
Electrical Engineering Technology
Electromechanical Engineering Technology
Mechanical Engineering Technology

SUNY-Buffalo
(Buffalo, NY)
Electrical Engineering Technology
Mechanical Engineering Technology

State University of New York College of Technology at Farmingdale
(Farmingdale, NY) [T-343]
Electrical Technology
Manufacturing Technology

State University of New York Institute of Technology, Utica/Rome
(Utica, NY) [T-344]
Computer Technology
Electrical Engineering Technology
Industrial Engineering Technology
Mechanical Engineering Technology

T

Temple University
(Philadelphia, PA) [T-345]
Civil & Construction Engineering Technology
Electrical Engineering Technology
Mechanical Engineering Technology

University of Tennessee, Martin
(Martin, TN) [T-346]
Civil Engineering Technology
Electrical Engineering Technology
Mechanical Engineering Technology

Texas A&M University
(College Station, TX) [T-347]
Electronic Engineering Technology
Manufacturing Engineering Technology
Mechanical Engineering Technology

Texas Tech University
(Lubbock, TX) [T-348]
Construction Technology
Electrical/Electronics Technology
Mechanical Technology

University of Toledo
(Toledo, OH)
Electronic(s) Engineering Technology
Mechanical Engineering Technology

Trenton State College
(Trenton, NJ)
Industrial Engineering Technology
Mechanical Engineering Technology

V

Virginia State University
(Petersburg, VA) [T-350]
Electronic Engineering Technology
Mechanical Engineering Technology

W

Weber State University
(Ogden, UT) [T-351]
Automotive Engineering Technology
Electronic(s) Engineering Technology
Manufacturing Engineering Technology
Mechanical Engineering Technology

Wentworth Institute of Technology
(Boston, MA) [T-352]
Architectural Engineering Technology
Civil Engineering Technology
Computer Engineering Technology
Electronic Engineering Technology
Manufacturing Engineering Technology
Mechanical Engineering Technology

West Virginia Institute of Technology
(Montgomery, WV) [T-353]
Electronic(s) Engineering Technology

Western Carolina University
(Cullowhee, NC) [T-354]
Manufacturing Engineering Technology

Western Kentucky University
(Bowling Green, KY) [T-355]
Civil Engineering Technology
Electrical Engineering Technology
Mechanical Engineering Technology

Western Michigan University
(Kalamazoo, MI) [T-356]
Manufacturing Engineering Technology

Western Washington University
(Bellingham, WA)
Electronic(s) Engineering Technology
Manufacturing Engineering Technology

Y

Youngstown State University
(Youngstown, OH)
Civil Engineering Technology
Electrical Engineering Technology
Mechanical Engineering Technology

Index to Accredited Degree Programs

U.S. ENGINEERING
(ABET-ACCREDITED PROGRAMS)

A

Aeronautical & Astronautical Engineering
University of Illinois at Urbana-Champaign [E-095]
Ohio State University [E-172]
Purdue University [E-188]

Aeronautical Engineering
California Polytechnic State University [E-027]
Embry-Riddle Aeronautical University, Western Campus [E-068]
Rensselaer Polytechnic Institute [E-189]
United States Air Force Academy [E-248]

Aeronautical Science & Engineering
University of California, Davis [E-020]

Aeronautics & Astronautics
University of Washington [E-262]

Aeronautics & Astronautics Engineering
Massachusetts Institute of Technology [E-124]

Aerospace Engineering
University of Alabama [E-002]
University of Arizona [E-008]
Arizona State University [E-009]
Auburn University [E-012]
Boston University [E-014]
University of California, Los Angeles [E-022]
California State Polytechnic University, Pomona [E-034]
Carleton University
University of Central Florida [E-039]
University of Cincinnati [E-042]
Embry-Riddle Aeronautical University - Daytona Beach [E-067]
University of Florida [E-071]
Florida Institute of Technology [E-074]
Georgia Institute of Technology [E-082]
Illinois Institute of Technology [E-096]
Iowa State University [E-099]
University of Kansas [E-101]
University of Maryland, College Park [E-120]
University of Michigan [E-130]
Mississippi State University [E-138]
University of Missouri-Rolla [E-141]
North Carolina State University [E-160]
University of Notre Dame [E-168]
University of Oklahoma [E-173]
Pennsylvania State University [E-180]
Polytechnic University [E-182]
Princeton University [E-186]
Parks College of Saint Louis University [E-198]
San Diego State University [E-200]
San Jose State University [E-202]
University of Southern California [E-211]
State University of New York at Buffalo [E-219]
Syracuse University [E-225]
University of Tennessee, Knoxville [E-229]
University of Texas at Arlington [E-232]
University of Texas at Austin [E-233]
Texas A&M University [E-238]
Tri-State University [E-241]
United States Naval Academy [E-252]
University of Virginia [E-259]
Virginia Polytechnic Institute and State University [E-261]
West Virginia University [E-267]
Wichita State University [E-271]

Aerospace Engineering & Mechanics
University of Minnesota [E-136]

Aerospace Engineering Sciences
University of Colorado at Boulder [E-049]

Aerospace Option in Mechanical Engineering
Oklahoma State University [E-175]

Aerospace Science Engineering
Tuskegee University [E-246]

Agricultural & Biosystems Engineering
University of Arizona [E-008]

Agricultural & Irrigation Engineering
Utah State University [E-254]

Agricultural Engineering
Auburn University [E-012]
University of California, Davis [E-020]
California Polytechnic State University [E-027]
California State Polytechnic University, Pomona [E-034]
Clemson University [E-047]
Colorado State University [E-053]
Cornell University [E-058]
University of Florida [E-071]
University of Georgia [E-081]
University of Idaho [E-092]
University of Illinois at Urbana-Champaign [E-095]
Iowa State University [E-099]
Kansas State University [E-102]
University of Kentucky [E-103]
Lafayette College [E-104]
University of Maryland, College Park [E-120]
Michigan State University [E-132]
University of Minnesota [E-136]
Mississippi State University [E-138]
University of Missouri-Columbia [E-139]
University of Missouri-Kansas City [E-140]
University of Nebraska - Lincoln [E-146]
New Mexico State University [E-154]
North Carolina Agricultural and Technical State University [E-159]
North Dakota State University [E-162]
Ohio State University [E-172]
Oklahoma State University [E-175]
Pennsylvania State University [E-180]
Purdue University [E-188]
South Dakota State University [E-208]
University of Tennessee, Knoxville [E-229]
Texas A&M University [E-238]
Texas Tech University [E-239]
Virginia Polytechnic Institute and State University [E-261]
Washington State University [E-264]
University of Wisconsin-Madison [E-274]
University of Wyoming [E-279]

Architectural Engineering
California Polytechnic State University [E-027]
University of Colorado at Boulder [E-049]
Drexel University [E-064]
University of Kansas [E-101]
Kansas State University [E-102]
University of Miami [E-128]
Milwaukee School of Engineering [E-134]
North Carolina Agricultural and Technical State University [E-159]
Oklahoma State University [E-175]
Pennsylvania State University [E-180]
Tennessee State University [E-230]
University of Texas at Austin [E-233]
University of Wyoming [E-279]

Astronautical Engineering
United States Air Force Academy [E-248]

B

Bioengineering
Arizona State University [E-009]
University of California, San Diego [E-023]
University of Illinois at Chicago [E-094]
University of Pennsylvania [E-179]
Syracuse University [E-225]
Texas A&M University [E-238]

Biological & Agricultural Engineering
University of Arkansas [E-010]
Louisiana State University [E-109]
North Carolina State University [E-160]

Biological Engineering
Mississippi State University [E-138]

Biomedical Engineering
Boston University [E-014]
Case Western Reserve University [E-037]
The Catholic University of America [E-038]
Duke University [E-065]
University of Iowa [E-098]
The Johns Hopkins University [E-100]
Louisiana Tech University [E-110]
Marquette University [E-118]
Milwaukee School of Engineering [E-134]
Northwestern University [E-166]
Rensselaer Polytechnic Institute [E-189]
Tulane University [E-244]
University of Utah [E-253]
Vanderbilt University [E-256]
Wright State University [E-278]

Bio-Resource Engineering
University of Maine [E-114]

Bioresource Engineering
Rutgers, The State University [E-195]

Bio-Resources Option in Civil Engineering
Montana State University [E-144]

C

Ceramic Engineering
Clemson University [E-047]
Georgia Institute of Technology [E-082]
University of Illinois at Urbana-Champaign [E-095]
Iowa State University [E-099]
University of Missouri-Rolla [E-141]
New York State College of Ceramics at Alfred University [E-157]
Ohio State University [E-172]
Rutgers, The State University [E-195]
University of Washington [E-262]

Ceramic Engineering Science
New York State College of Ceramics at Alfred University [E-157]

Ceramic Science & Engineering
Pennsylvania State University [E-180]

Chemical & Petroleum-Refining Engineering
Colorado School of Mines [E-052]

Chemical Engineering
University of Akron [E-001]
University of Alabama [E-002]
University of Alabama at Huntsville [E-004]
University of Arizona [E-008]
Arizona State University [E-009]
University of Arkansas [E-010]
Auburn University [E-012]
Brigham Young University [E-017]
Bucknell University [E-018]
University of California at Berkeley [E-019]
University of California, Davis [E-020]
University of California, Los Angeles [E-022]
University of California, San Diego [E-023]
University of California, Santa Barbara [E-024]
California Institute of Technology [E-026]
California State University, Long Beach [E-030]
California State Polytechnic University, Pomona [E-034]
Carnegie Mellon University [E-036]
Case Western Reserve University [E-037]
Christian Brothers University [E-041]
University of Cincinnati [E-042]
City College of the City University of New York [E-044]
Clarkson University [E-046]
Clemson University [E-047]
Cleveland State University [E-048]
University of Colorado at Boulder [E-049]
Colorado State University [E-053]
Columbia University [E-054]
University of Connecticut [E-056]
The Cooper Union [E-057]

Cornell University [E-058]
University of Dayton [E-060]
University of Delaware [E-061]
University of Detroit Mercy
Drexel University [E-064]
University of Florida [E-071]
Florida Agricultural and Mechanical University [E-072]
Florida Institute of Technology [E-074]
Georgia Institute of Technology [E-082]
Hampton University
University of Houston [E-090]
Howard University [E-091]
University of Idaho [E-092]
University of Illinois at Chicago [E-094]
University of Illinois at Urbana-Champaign [E-095]
Illinois Institute of Technology [E-096]
University of Iowa [E-098]
Iowa State University [E-099]
The Johns Hopkins University [E-100]
University of Kansas [E-101]
Kansas State University [E-102]
University of Kentucky [E-103]
Lafayette College [E-104]
Lamar University [E-105]
Lehigh University [E-108]
Louisiana State University [E-109]
Louisiana Tech University [E-110]
University of Maine [E-114]
Manhattan College
University of Maryland Baltimore County [E-119]
University of Maryland, College Park [E-120]
University of Massachusetts [E-121]
University of Massachusetts, Lowell [E-123]
Massachusetts Institute of Technology [E-124]
University of Michigan [E-130]
Michigan State University [E-132]
Michigan Technological University [E-133]
University of Minnesota, Duluth [E-135]
University of Minnesota [E-136]
University of Mississippi [E-137]
Mississippi State University [E-138]
University of Missouri-Columbia [E-139]
University of Missouri-Kansas City [E-140]
University of Missouri-Rolla [E-141]
Montana State University [E-144]
University of Nebraska - Lincoln [E-146]
University of Nevada-Reno
University of New Hampshire [E-149]
New Jersey Institute of Technology [E-151]
The University of New Mexico [E-152]
New Mexico State University [E-154]
North Carolina Agricultural and Technical State University [E-159]
North Carolina State University [E-160]
University of North Dakota [E-161]
Northeastern University [E-163]
Northwestern University [E-166]
University of Notre Dame [E-168]
Ohio University [E-170]
Ohio State University [E-172]
University of Oklahoma [E-173]
Oklahoma State University [E-175]
Oregon State University [E-177]
University of Pennsylvania [E-179]
Pennsylvania State University [E-180]
University of Pittsburgh [E-181]
Polytechnic University [E-182]
Princeton University [E-186]
University of Puerto Rico, Mayagüez Campus [E-187]
Purdue University [E-188]
Rensselaer Polytechnic Institute [E-189]
University of Rhode Island [E-190]
William Marsh Rice University [E-191]
University of Rochester [E-192]
Rose-Hulman Institute of Technology [E-194]
Rutgers, The State University [E-195]
San Jose State University [E-202]
University of South Alabama [E-205]
University of South Carolina [E-206]
South Dakota School of Mines and Technology [E-207]
University of South Florida [E-209]
University of Southern California [E-211]
University of Southwestern Louisiana
Stanford University [E-217]

State University of New York at Buffalo [E-219]
Stevens Institute of Technology
Syracuse University [E-225]
University of Tennessee, Knoxville [E-229]
Tennessee Technological University [E-231]
University of Texas at Austin [E-233]
Texas A&I University [E-237]
Texas A&M University [E-238]
Texas Tech University [E-239]
University of Toledo [E-240]
Tri-State University [E-241]
Tufts University [E-243]
Tulane University [E-244]
University of Tulsa [E-245]
Tuskegee University [E-246]
University of Utah [E-253]
Vanderbilt University [E-256]
Villanova University [E-258]
University of Virginia [E-259]
Virginia Polytechnic Institute and State University [E-261]
University of Washington [E-262]
Washington University [E-263]
Washington State University [E-264]
Wayne State University [E-265]
West Virginia University [E-267]
West Virginia Institute of Technology [E-268]
Widener University [E-272]
University of Wisconsin-Madison [E-274]
Worcester Polytechnic Institute [E-277]
University of Wyoming [E-279]
Yale University [E-280]
Youngstown State University

Civil Engineering
University of Akron [E-001]
University of Alabama [E-002]
University of Alabama at Birmingham [E-003]
University of Alabama at Huntsville [E-004]
University of Alaska-Anchorage
University of Alaska, Fairbanks [E-005]
University of Arizona [E-008]
Arizona State University [E-009]
University of Arkansas [E-010]
Auburn University [E-012]
Bradley University [E-015]
Brigham Young University [E-017]
Bucknell University [E-018]
University of California at Berkeley [E-019]
University of California, Davis [E-020]
University of California, Irvine [E-021]
University of California, Los Angeles [E-022]
California Polytechnic State University [E-027]
California State University, Chico [E-028]
California State University, Fresno [E-029]
California State University, Long Beach [E-030]
California State University, Los Angeles [E-031]
California State University, Sacramento [E-033]
California State Polytechnic University, Pomona [E-034]
Carleton University
Carnegie Mellon University [E-036]
Case Western Reserve University [E-037]
The Catholic University of America [E-038]
University of Central Florida [E-039]
Christian Brothers University [E-041]
University of Cincinnati [E-042]
The Citadel [E-043]
City College of the City University of New York [E-044]
Clarkson University [E-046]
Clemson University [E-047]
Cleveland State University [E-048]
University of Colorado at Boulder [E-049]
University of Colorado at Denver [E-051]
Colorado State University [E-053]
Columbia University [E-054]
University of Connecticut [E-056]
The Cooper Union [E-057]
Cornell University [E-058]
University of Dayton [E-060]
University of Delaware [E-061]
University of Detroit Mercy
University of the District of Columbia [E-063]
Drexel University [E-064]
Duke University [E-065]

University of Florida [E-071]
Florida Agricultural and Mechanical University [E-072]
Florida Institute of Technology [E-074]
Florida International University [E-075]
The George Washington University [E-080]
Georgia Institute of Technology [E-082]
Gonzaga University [E-083]
University of Hartford
University of Hawaii at Manoa [E-088]
University of Houston [E-090]
Howard University [E-091]
University of Idaho [E-092]
University of Illinois at Chicago [E-094]
University of Illinois at Urbana-Champaign [E-095]
Illinois Institute of Technology [E-096]
University of Iowa [E-098]
Iowa State University [E-099]
The Johns Hopkins University [E-100]
University of Kansas [E-101]
Kansas State University [E-102]
University of Kentucky [E-103]
Lafayette College [E-104]
Lamar University [E-105]
Lehigh University [E-108]
Louisiana State University [E-109]
Louisiana Tech University [E-110]
Loyola Marymount University [E-113]
University of Maine [E-114]
Manhattan College
Marquette University [E-118]
University of Maryland, College Park [E-120]
University of Massachusetts [E-121]
University of Massachusetts, Dartmouth [E-122]
University of Massachusetts, Lowell [E-123]
Massachusetts Institute of Technology [E-124]
Memphis State University [E-126]
Merrimack College
University of Miami [E-128]
University of Michigan [E-130]
Michigan State University [E-132]
Michigan Technological University [E-133]
University of Minnesota [E-136]
University of Mississippi [E-137]
Mississippi State University [E-138]
University of Missouri-Columbia [E-139]
University of Missouri-Kansas City [E-140]
University of Missouri-Rolla [E-141]
Montana State University [E-144]
Morgan State University [E-145]
University of Nebraska - Lincoln [E-146]
University of Nebraska-Omaha
University of Nevada, Las Vegas [E-147]
University of Nevada-Reno
New England College [E-148]
University of New Hampshire [E-149]
University of New Haven [E-150]
New Jersey Institute of Technology [E-151]
The University of New Mexico [E-152]
New Mexico State University [E-154]
University of New Orleans [E-155]
University of North Carolina-Charlotte [E-158]
North Carolina Agricultural and Technical State University [E-159]
North Carolina State University [E-160]
University of North Dakota [E-161]
North Dakota State University [E-162]
Northeastern University [E-163]
Northern Arizona University [E-164]
Northwestern University [E-166]
Norwich University [E-167]
University of Notre Dame [E-168]
Ohio University [E-170]
Ohio Northern University [E-171]
Ohio State University [E-172]
University of Oklahoma [E-173]
Oklahoma State University [E-175]
Old Dominion University [E-176]
Oregon State University [E-177]
University of the Pacific [E-178]
Pennsylvania State University [E-180]
University of Pittsburgh [E-181]
Polytechnic University [E-182]
University of Portland [E-183]
Portland State University [E-184]

Index to Accredited Degree Programs

Prairie View A&M University [E-185]
Princeton University [E-186]
University of Puerto Rico, Mayagüez Campus [E-187]
Purdue University [E-188]
Rensselaer Polytechnic Institute [E-189]
University of Rhode Island [E-190]
William Marsh Rice University [E-191]
Rose-Hulman Institute of Technology [E-194]
Rutgers, The State University [E-195]
Saint Martin's College
San Diego State University [E-200]
San Francisco State University [E-201]
San Jose State University [E-202]
Santa Clara University [E-203]
Seattle University [E-204]
University of South Alabama [E-205]
University of South Carolina [E-206]
South Dakota School of Mines and Technology [E-207]
South Dakota State University [E-208]
University of South Florida [E-209]
Southern University and A&M College [E-210]
University of Southern California [E-211]
Southern Illinois University at Carbondale [E-213]
Southern Illinois University at Edwardsville [E-214]
Southern Methodist University [E-216]
University of Southwestern Louisiana
Stanford University [E-217]
State University of New York at Buffalo [E-219]
Stevens Institute of Technology
Syracuse University [E-225]
Temple University [E-227]
University of Tennessee, Knoxville [E-229]
Tennessee State University [E-230]
Tennessee Technological University [E-231]
University of Texas at Arlington [E-232]
University of Texas at Austin [E-233]
University of Texas at El Paso [E-235]
University of Texas at San Antonio [E-236]
Texas A&I University [E-237]
Texas A&M University [E-238]
Texas Tech University [E-239]
University of Toledo [E-240]
Tri-State University [E-241]
Tufts University [E-243]
Tulane University [E-244]
Union College [E-247]
United States Air Force Academy [E-248]
United States Coast Guard Academy [E-249]
United States International University
United States Military Academy [E-251]
University of Utah [E-253]
Utah State University [E-254]
Valparaiso University [E-255]
Vanderbilt University [E-256]
University of Vermont [E-257]
Villanova University [E-258]
University of Virginia [E-259]
Virginia Military Institute [E-260]
Virginia Polytechnic Institute and State University [E-261]
University of Washington [E-262]
Washington University [E-263]
Washington State University [E-264]
Wayne State University [E-265]
West Virginia University [E-267]
West Virginia Institute of Technology [E-268]
Widener University [E-272]
University of Wisconsin-Madison [E-274]
University of Wisconsin-Milwaukee [E-275]
University of Wisconsin-Platteville [E-276]
Worcester Polytechnic Institute [E-277]
University of Wyoming [E-279]
Youngstown State University

Civil Engineering Systems
University of Pennsylvania [E-179]

Computer & Electrical Engineering
Purdue University [E-188]

Computer & Information Engineering Science
University of Florida [E-071]

Computer & Systems Engineering
Rensselaer Polytechnic Institute [E-189]

Computer Engineering
University of Alabama at Huntsville [E-004]
University of Arizona [E-008]
Auburn University [E-012]
Boston University [E-014]
University of Bridgeport [E-016]
University of California, Santa Cruz [E-025]
California State University, Chico [E-028]
California State University, Sacramento [E-033]
Carnegie Mellon University [E-036]
Case Western Reserve University [E-037]
University of Central Florida [E-039]
University of Cincinnati [E-042]
Clarkson University [E-046]
Clemson University [E-047]
Florida Institute of Technology [E-074]
The George Washington University [E-080]
Georgia Institute of Technology [E-082]
University of Illinois at Chicago [E-094]
University of Illinois at Urbana-Champaign [E-095]
Iowa State University [E-099]
University of Kansas [E-101]
Kansas State University [E-102]
Lehigh University [E-108]
Louisiana State University [E-109]
University of Massachusetts, Dartmouth [E-122]
University of Miami [E-128]
University of Michigan [E-130]
Milwaukee School of Engineering [E-134]
University of Minnesota, Duluth [E-135]
Mississippi State University [E-138]
University of Missouri-Columbia [E-139]
University of Missouri-Kansas City [E-140]
The University of New Mexico [E-152]
North Carolina State University [E-160]
Oakland University [E-169]
Old Dominion University [E-176]
Oregon State University [E-177]
University of the Pacific [E-178]
Pennsylvania State University [E-180]
Polytechnic University [E-182]
University of Rhode Island [E-190]
Rochester Institute of Technology [E-193]
San Jose State University [E-202]
Santa Clara University [E-203]
University of South Florida [E-209]
Southern Methodist University [E-216]
Stevens Institute of Technology
Syracuse University [E-225]
University of Texas at Austin [E-233]
Valparaiso University [E-255]
Virginia Polytechnic Institute and State University [E-261]
University of Washington [E-262]
West Virginia University [E-267]
Wright State University [E-278]

Computer Engineering (Option in Electrical Engineering)
State University of New York at Stony Brook [E-223]
Tufts University [E-243]

Computer Science
University of California at Berkeley [E-019]

Computer Science & Engineering
University of California, Davis [E-020]
University of California, Los Angeles [E-022]
California State University, Long Beach [E-030]
University of Connecticut [E-056]
Massachusetts Institute of Technology [E-124]
Northern Arizona University [E-164]
University of Texas at Arlington [E-232]
University of Toledo [E-240]
Washington University [E-263]

Computer Systems Engineering
Arizona State University [E-009]
University of Arkansas [E-010]
Carleton University
University of Massachusetts [E-121]
Western Michigan University [E-269]

Construction Engineering
Iowa State University [E-099]
Lawrence Technological University [E-106]
The University of New Mexico [E-152]
North Dakota State University [E-162]

Construction Engineering & Management
Purdue University [E-188]

Construction Option in Civil Engineering
North Carolina State University [E-160]

E

Electric Power Engineering
Rensselaer Polytechnic Institute [E-189]

Electrical & Computer Engineering
University of Colorado at Boulder [E-049]

Electrical & Electronics Engineering
Oregon State University [E-177]

Electrical Engineering
University of Akron [E-001]
University of Alabama [E-002]
University of Alabama at Birmingham [E-003]
University of Alabama at Huntsville [E-004]
University of Alaska, Fairbanks [E-005]
Alfred University [E-007]
University of Arizona [E-008]
Arizona State University [E-009]
University of Arkansas [E-010]
Auburn University [E-012]
Boston University [E-014]
Bradley University [E-015]
University of Bridgeport [E-016]
Brigham Young University [E-017]
Bucknell University [E-018]
University of California at Berkeley [E-019]
University of California, Davis [E-020]
University of California, Irvine [E-021]
University of California, Los Angeles [E-022]
University of California, San Diego [E-023]
University of California, Santa Barbara [E-024]
California Polytechnic State University [E-027]
California State University, Fresno [E-029]
California State University, Long Beach [E-030]
California State University, Los Angeles [E-031]
California State Polytechnic University, Pomona [E-034]
Carleton University
Carnegie Mellon University [E-036]
Case Western Reserve University [E-037]
The Catholic University of America [E-038]
University of Central Florida [E-039]
Christian Brothers University [E-041]
University of Cincinnati [E-042]
The Citadel [E-043]
City College of the City University of New York [E-044]
Clarkson University [E-046]
Clemson University [E-047]
Cleveland State University [E-048]
University of Colorado at Boulder [E-049]
University of Colorado at Colorado Springs [E-050]
University of Colorado at Denver [E-051]
Colorado State University [E-053]
Columbia University [E-054]
University of Connecticut [E-056]
The Cooper Union [E-057]
Cornell University [E-058]
University of Dayton [E-060]
University of Delaware [E-061]
University of Denver [E-062]
University of Detroit Mercy
University of the District of Columbia [E-063]
Drexel University [E-064]
Duke University [E-065]
University of Evansville [E-069]
Fairleigh Dickinson University
University of Florida [E-071]
Florida Agricultural and Mechanical University [E-072]
Florida Atlantic University [E-073]
Florida Institute of Technology [E-074]
Florida International University [E-075]
Gannon University [E-078]
The George Washington University [E-080]
Georgia Institute of Technology [E-082]
GMI Engineering & Management Institute [E-077]
Gonzaga University [E-083]
Grove City College [E-085]
Hampton University
University of Hartford
University of Hawaii at Manoa [E-088]
Hofstra University [E-089]
University of Houston [E-090]

Index to Accredited Degree Programs

Howard University [E-091]
University of Idaho [E-092]
University of Illinois at Chicago [E-094]
University of Illinois at Urbana-Champaign [E-095]
Illinois Institute of Technology [E-096]
Indiana University-Purdue University at Indianapolis [E-097]
University of Iowa [E-098]
Iowa State University [E-099]
The Johns Hopkins University [E-100]
University of Kansas [E-101]
Kansas State University [E-102]
University of Kentucky [E-103]
Lafayette College [E-104]
Lamar University [E-105]
Lawrence Technological University [E-106]
Lehigh University [E-108]
Louisiana State University [E-109]
Louisiana Tech University [E-110]
Loyola Marymount University [E-113]
University of Maine [E-114]
Manhattan College
Mankato State University [E-116]
Marquette University [E-118]
University of Maryland, College Park [E-120]
University of Massachusetts [E-121]
University of Massachusetts, Dartmouth [E-122]
University of Massachusetts, Lowell [E-123]
Memphis State University [E-126]
University of Miami [E-128]
University of Michigan [E-130]
University of Michigan-Dearborn [E-131]
Michigan State University [E-132]
Michigan Technological University [E-133]
Milwaukee School of Engineering [E-134]
University of Minnesota [E-136]
University of Mississippi [E-137]
Mississippi State University [E-138]
University of Missouri-Columbia [E-139]
University of Missouri-Kansas City [E-140]
University of Missouri-Rolla [E-141]
Montana State University [E-144]
Morgan State University [E-145]
University of Nebraska - Lincoln [E-146]
University of Nevada, Las Vegas [E-147]
University of Nevada-Reno
University of New Hampshire [E-149]
University of New Haven [E-150]
New Jersey Institute of Technology [E-151]
The University of New Mexico [E-152]
New Mexico State University [E-154]
University of New Orleans [E-155]
New York Institute of Technology [E-156]
University of North Carolina-Charlotte [E-158]
North Carolina Agricultural and Technical State University [E-159]
North Carolina State University [E-160]
University of North Dakota [E-161]
Northeastern University [E-163]
Northern Arizona University [E-164]
Northern Illinois University [E-165]
Northwestern University [E-166]
Norwich University [E-167]
University of Notre Dame [E-168]
Oakland University [E-169]
Ohio University [E-170]
Ohio Northern University [E-171]
Ohio State University [E-172]
University of Oklahoma [E-173]
Oklahoma Christian University of Science and Arts [E-174]
Oklahoma State University [E-175]
Old Dominion University [E-176]
University of the Pacific [E-178]
University of Pennsylvania [E-179]
Pennsylvania State University [E-180]
University of Pittsburgh [E-181]
Polytechnic University [E-182]
University of Portland [E-183]
Portland State University [E-184]
Prairie View A&M University [E-185]
Princeton University [E-186]
University of Puerto Rico, Mayagüez Campus [E-187]
Purdue University [E-188]
Purdue University-Calumet
Rensselaer Polytechnic Institute [E-189]
University of Rhode Island [E-190]
William Marsh Rice University [E-191]
University of Rochester [E-192]
Rochester Institute of Technology [E-193]

Rose-Hulman Institute of Technology [E-194]
Rutgers, The State University [E-195]
Saginaw Valley State University [E-196]
St. Cloud State University [E-197]
Parks College of Saint Louis University [E-198]
Saint Mary's University
University of San Diego [E-199]
San Diego State University [E-200]
San Francisco State University [E-201]
San Jose State University [E-202]
Santa Clara University [E-203]
Seattle University [E-204]
Seattle Pacific University
University of South Alabama [E-205]
University of South Carolina [E-206]
South Dakota School of Mines and Technology [E-207]
South Dakota State University [E-208]
University of South Florida [E-209]
Southern University and A&M College [E-210]
University of Southern California [E-211]
Southern Illinois University at Carbondale [E-213]
Southern Illinois University at Edwardsville [E-214]
University of Southern Maine [E-215]
Southern Methodist University [E-216]
University of Southwestern Louisiana
Stanford University [E-217]
State University of New York at Binghamton [E-218]
State University of New York at Buffalo [E-219]
State University of New York Maritime College [E-221]
State University of New York, College at New Paltz [E-222]
State University of New York at Stony Brook [E-223]
Stevens Institute of Technology
Syracuse University [E-225]
Temple University [E-227]
University of Tennessee, Knoxville [E-229]
Tennessee State University [E-230]
Tennessee Technological University [E-231]
University of Texas at Arlington [E-232]
University of Texas at Austin [E-233]
University of Texas at Dallas [E-234]
University of Texas at El Paso [E-235]
University of Texas at San Antonio [E-236]
Texas A&I University [E-237]
Texas A&M University [E-238]
Texas Tech University [E-239]
University of Toledo [E-240]
Tri-State University [E-241]
Tufts University [E-243]
Tulane University [E-244]
University of Tulsa [E-245]
Tuskegee University [E-246]
Union College [E-247]
United States Air Force Academy [E-248]
United States Coast Guard Academy [E-249]
United States Military Academy [E-251]
United States Naval Academy [E-252]
University of Utah [E-253]
Utah State University [E-254]
Valparaiso University [E-255]
Vanderbilt University [E-256]
University of Vermont [E-257]
Villanova University [E-258]
University of Virginia [E-259]
Virginia Military Institute [E-260]
Virginia Polytechnic Institute and State University [E-261]
University of Washington [E-262]
Washington University [E-263]
Washington State University [E-264]
Wayne State University [E-265]
West Virginia University [E-267]
West Virginia Institute of Technology [E-268]
Western Michigan University [E-269]
Western New England College [E-270]
Wichita State University [E-271]
Widener University [E-272]
Wilkes University [E-273]
University of Wisconsin-Madison [E-274]
University of Wisconsin-Milwaukee [E-275]
University of Wisconsin-Platteville [E-276]
Worcester Polytechnic Institute [E-277]
Wright State University [E-278]
University of Wyoming [E-279]
Yale University [E-280]
Youngstown State University

Electrical Science & Engineering
Massachusetts Institute of Technology [E-124]

Electrical/Computer Engineering
Merrimack College

Electrical/Electronic Engineering
California State University, Chico [E-028]
California State University, Sacramento [E-033]

Electrical/Electronic(s) Engineering
North Dakota State University [E-162]

Electronic(s) Engineering
California Polytechnic State University [E-027]
George Mason University [E-079]
Monmouth College [E-142]

Engineering
Arkansas Tech University [E-011]
Baylor University [E-013]
California State University, Northridge [E-032]
Calvin College [E-035]
Colorado School of Mines [E-052]
Dartmouth College [E-059]
Dordt College
Grand Valley State University [E-084]
Harvey Mudd College [E-087]
Idaho State University [E-093]
Lafayette College [E-104]
LeTourneau University [E-107]
University of Maryland, College Park [E-120]
McNeese State University
Mercer University [E-127]
Michigan Technological University [E-133]
University of Oklahoma [E-173]
Stevens Institute of Technology
Swarthmore College [E-224]
University of Tennessee, Chattanooga [E-228]
Walla Walla College
Widener University [E-272]

Engineering & Applied Sciences
California Institute of Technology [E-026]

Engineering & Public Policy
Carnegie Mellon University [E-036]
Washington University [E-263]

Engineering Management
University of Missouri-Rolla [E-141]
Stevens Institute of Technology
United States Military Academy [E-251]

Engineering Mechanics
University of Cincinnati [E-042]
University of Illinois at Urbana-Champaign [E-095]
The Johns Hopkins University [E-100]
United States Air Force Academy [E-248]
University of Wisconsin-Madison [E-274]

Engineering of Mines
West Virginia University [E-267]

Engineering Physics
Colorado School of Mines [E-052]
Cornell University [E-058]
University of Kansas [E-101]
University of Maine [E-114]
University of Oklahoma [E-173]
University of the Pacific [E-178]
Princeton University [E-186]
Stevens Institute of Technology
Texas Tech University [E-239]
University of Toledo [E-240]
University of Tulsa [E-245]
Wright State University [E-278]

Engineering Science
City University of New York, College of Staten Island [E-045]
Loyola College in Maryland [E-112]
Pennsylvania State University [E-180]
State University of New York at Stony Brook [E-223]
University of Tennessee, Knoxville [E-229]

Engineering Science & Mechanics
Georgia Institute of Technology [E-082]
Virginia Polytechnic Institute and State University [E-261]

Engineering Science(s)
Colorado State University [E-053]
University of Florida [E-071]

Engineering Sciences
Harvard University [E-086]

Engineering Science(s)
Hofstra University [E-089]
Iowa State University [E-099]
Montana College of Mineral Science and Technology [E-143]
Trinity University [E-242]
United States Air Force Academy [E-248]

Environmental Engineering
California Polytechnic State University [E-027]
University of Central Florida [E-039]
University of Cincinnati [E-042]
University of Florida [E-071]
Manhattan College
Michigan Technological University [E-133]
Montana College of Mineral Science and Technology [E-143]
Northwestern University [E-166]
Rensselaer Polytechnic Institute [E-189]

Environmental Resources Engineering
Humboldt State University

Extractive Metallurgical Engineering
University of Minnesota [E-136]

F

Fire Protection Engineering
University of Maryland, College Park [E-120]

Fluid & Thermal Engineering Science
Case Western Reserve University [E-037]

Food Process Engineering
Purdue University [E-188]

Forest Engineering
University of Maine [E-114]
State University of New York, College of Environmental Science and Forestry [E-220]

G

General Engineering
University of Illinois at Urbana-Champaign [E-095]
Oklahoma State University [E-175]

Geological Engineering
University of Alaska, Fairbanks [E-005]
University of Arizona [E-008]
Colorado School of Mines [E-052]
University of Idaho [E-092]
Michigan Technological University [E-133]
University of Minnesota [E-136]
University of Mississippi [E-137]
University of Missouri-Rolla [E-141]
Montana College of Mineral Science and Technology [E-143]
University of Nevada-Reno
New Mexico State University [E-154]
University of North Dakota [E-161]
Princeton University [E-186]
South Dakota School of Mines and Technology [E-207]
University of Utah [E-253]
Washington State University [E-264]

Geophysical Engineering
Colorado School of Mines [E-052]
Montana College of Mineral Science and Technology [E-143]

Glass Engineering Science
New York State College of Ceramics at Alfred University [E-157]

I

Industrial & Management Engineering
Montana State University [E-144]
Rensselaer Polytechnic Institute [E-189]

Industrial & Operations Engineering
University of Michigan [E-130]

Industrial & Systems Engineering
University of Alabama at Huntsville [E-004]
University of Florida [E-071]
Georgia Institute of Technology [E-082]
University of Michigan-Dearborn [E-131]
Ohio University [E-170]
Ohio State University [E-172]
San Jose State University [E-202]
University of Southern California [E-211]

Industrial Engineering
University of Alabama [E-002]
Alfred University [E-007]
University of Arizona [E-008]
Arizona State University [E-009]
University of Arkansas [E-010]
Auburn University [E-012]
Bradley University [E-015]
University of California at Berkeley [E-019]
California Polytechnic State University [E-027]
California State University, Fresno [E-029]
California State Polytechnic University, Pomona [E-034]
University of Central Florida [E-039]
University of Cincinnati [E-042]
Clemson University [E-047]
Cleveland State University [E-048]
Columbia University [E-054]
Florida Agricultural and Mechanical University [E-072]
Florida International University [E-075]
GMI Engineering & Management Institute [E-077]
University of Houston [E-090]
University of Illinois at Chicago [E-094]
University of Illinois at Urbana-Champaign [E-095]
University of Iowa [E-098]
Iowa State University [E-099]
Kansas State University [E-102]
Lamar University [E-105]
Lehigh University [E-108]
Louisiana State University [E-109]
Louisiana Tech University [E-110]
Marquette University [E-118]
University of Miami [E-128]
Milwaukee School of Engineering [E-134]
University of Minnesota, Duluth [E-135]
Mississippi State University [E-138]
University of Missouri-Columbia [E-139]
University of Missouri-Kansas City [E-140]
Morgan State University [E-145]
University of Nebraska - Lincoln [E-146]
University of New Haven [E-150]
New Jersey Institute of Technology [E-151]
New Mexico State University [E-154]
North Carolina Agricultural and Technical State University [E-159]
North Carolina State University [E-160]
North Dakota State University [E-162]
Northeastern University [E-163]
Northern Illinois University [E-165]
Northwestern University [E-166]
University of Oklahoma [E-173]
Oregon State University [E-177]
Pennsylvania State University [E-180]
University of Pittsburgh [E-181]
Polytechnic University [E-182]
University of Puerto Rico, Mayagüez Campus [E-187]
Purdue University [E-188]
University of Rhode Island [E-190]
Rochester Institute of Technology [E-193]
Rutgers, The State University [E-195]
Saint Mary's University
University of South Florida [E-209]
University of Southern Colorado [E-212]
Southern Illinois University at Edwardsville [E-214]
Stanford University [E-217]
State University of New York at Buffalo [E-219]
University of Tennessee, Knoxville [E-229]
Tennessee Technological University [E-231]
University of Texas at Arlington [E-232]
University of Texas at El Paso [E-235]
Texas A&M University [E-238]
Texas Tech University [E-239]
University of Toledo [E-240]
University of Washington [E-262]
Wayne State University [E-265]
West Virginia University [E-267]
Western Michigan University [E-269]
Western New England College [E-270]
Wichita State University [E-271]
University of Wisconsin-Madison [E-274]
University of Wisconsin-Milwaukee [E-275]
University of Wisconsin-Platteville [E-276]
Youngstown State University

Industrial Engineering & Management
Oklahoma State University [E-175]

Industrial Engineering & Operations Research
University of Massachusetts [E-121]
Virginia Polytechnic Institute and State University [E-261]

Interdisciplinary Studies Engineering
Arizona State University [E-009]

M

Manufacturing Engineering
Boston University [E-014]
Bradley University [E-015]
California State Polytechnic University, Pomona [E-034]
Central State University [E-040]
Miami University [E-129]

Manufacturing Engineering Option in Industrial Engineering
Oregon State University [E-177]

Manufacturing Engineering (Option in Mechanical Engineering)
Utah State University [E-254]

Manufacturing Systems Engineering
GMI Engineering & Management Institute [E-077]
Kansas State University [E-102]
Worcester Polytechnic Institute [E-277]

Marine Engineering
State University of New York Maritime College [E-221]
Texas A&M University-Galveston
United States Naval Academy [E-252]

Marine Engineering Systems
United States Merchant Marine Academy [E-250]

Materials & Metallurgical Engineering
Stevens Institute of Technology

Materials Engineering
University of Alabama at Birmingham [E-003]
Auburn University [E-012]
University of California, Los Angeles [E-022]
University of Cincinnati [E-042]
Drexel University [E-064]
Georgia Institute of Technology [E-082]
Rensselaer Polytechnic Institute [E-189]
San Jose State University [E-202]
University of Utah [E-253]
Virginia Polytechnic Institute and State University [E-261]
Wilkes University [E-273]
University of Wisconsin-Milwaukee [E-275]

Materials Science & Engineering
University of Arizona [E-008]
University of California, Davis [E-020]
Case Western Reserve University [E-037]
Cornell University [E-058]
University of Florida [E-071]
The Johns Hopkins University [E-100]
Lehigh University [E-108]
Massachusetts Institute of Technology [E-124]
University of Michigan [E-130]
Michigan State University [E-132]
University of Minnesota [E-136]
North Carolina State University [E-160]
Northwestern University [E-166]
University of Pennsylvania [E-179]
University of Pittsburgh [E-181]
William Marsh Rice University [E-191]
University of Tennessee, Knoxville [E-229]
University of Utah [E-253]
Washington State University [E-264]
Wayne State University [E-265]
Wright State University [E-278]

Materials Science & Engineering
(Option in Metallurgical Engineering)
Michigan Technological University [E-133]

Mechanical Engineering
University of Akron [E-001]
University of Alabama [E-002]
University of Alabama at Birmingham [E-003]
University of Alabama at Huntsville [E-004]
University of Alaska, Fairbanks [E-005]
Alfred University [E-007]
University of Arizona [E-008]
Arizona State University [E-009]
University of Arkansas [E-010]
Auburn University [E-012]
Boston University [E-014]
Bradley University [E-015]
University of Bridgeport [E-016]
Brigham Young University [E-017]
Bucknell University [E-018]
University of California at Berkeley [E-019]
University of California, Davis [E-020]
University of California, Irvine [E-021]
University of California, Los Angeles [E-022]
University of California, San Diego [E-023]
University of California, Santa Barbara [E-024]
California Polytechnic State University [E-027]
California State University, Chico [E-028]
California State University, Fresno [E-029]
California State University, Long Beach [E-030]
California State University, Los Angeles [E-031]
California State University, Sacramento [E-033]
California State Polytechnic University, Pomona [E-034]
Carleton University
Carnegie Mellon University [E-036]
Case Western Reserve University [E-037]
The Catholic University of America [E-038]
University of Central Florida [E-039]
Christian Brothers University [E-041]
University of Cincinnati [E-042]
City College of the City University of New York [E-044]
Clarkson University [E-046]
Clemson University [E-047]
Cleveland State University [E-048]
University of Colorado at Boulder [E-049]
University of Colorado at Denver [E-051]
Colorado State University [E-053]
Columbia University [E-054]
University of Connecticut [E-056]
The Cooper Union [E-057]
Cornell University [E-058]
University of Dayton [E-060]
University of Delaware [E-061]
University of Denver [E-062]
University of Detroit Mercy
University of the District of Columbia [E-063]
Drexel University [E-064]
Duke University [E-065]
University of Evansville [E-069]
University of Florida [E-071]
Florida Agricultural and Mechanical University [E-072]
Florida Atlantic University [E-073]
Florida Institute of Technology [E-074]
Florida International University [E-075]
Gannon University [E-078]
The George Washington University [E-080]
Georgia Institute of Technology [E-082]
GMI Engineering & Management Institute [E-077]
Gonzaga University [E-083]
Grove City College [E-085]
University of Hartford
University of Hawaii at Manoa [E-088]
Hofstra University [E-089]
University of Houston [E-090]
Howard University [E-091]
University of Idaho [E-092]
University of Illinois at Chicago [E-094]
University of Illinois at Urbana-Champaign [E-095]
Illinois Institute of Technology [E-096]
Indiana University-Purdue University at Indianapolis [E-097]
University of Iowa [E-098]
Iowa State University [E-099]
The Johns Hopkins University [E-100]
University of Kansas [E-101]
Kansas State University [E-102]
University of Kentucky [E-103]
Lafayette College [E-104]
Lamar University [E-105]
Lawrence Technological University [E-106]
Lehigh University [E-108]
Louisiana State University [E-109]
Louisiana Tech University [E-110]
Loyola Marymount University [E-113]
University of Maine [E-114]
Manhattan College
Marquette University [E-118]
University of Maryland Baltimore County [E-119]
University of Maryland, College Park [E-120]
University of Massachusetts [E-121]
University of Massachusetts, Dartmouth [E-122]
University of Massachusetts, Lowell [E-123]
Massachusetts Institute of Technology [E-124]
Memphis State University [E-126]
University of Miami [E-128]
University of Michigan [E-130]
University of Michigan-Dearborn [E-131]
Michigan State University [E-132]
Milwaukee School of Engineering [E-134]
University of Minnesota [E-136]
University of Mississippi [E-137]
Mississippi State University [E-138]
University of Missouri-Columbia [E-139]
University of Missouri-Kansas City [E-140]
University of Missouri-Rolla [E-141]
Montana State University [E-144]
University of Nebraska - Lincoln [E-146]
University of Nevada, Las Vegas [E-147]
University of Nevada-Reno
University of New Hampshire [E-149]
University of New Haven [E-150]
New Jersey Institute of Technology [E-151]
The University of New Mexico [E-152]
New Mexico State University [E-154]
University of New Orleans [E-155]
New York Institute of Technology [E-156]
University of North Carolina-Charlotte [E-158]
North Carolina Agricultural and Technical State University [E-159]
North Carolina State University [E-160]
University of North Dakota [E-161]
North Dakota State University [E-162]
Northeastern University [E-163]
Northern Arizona University [E-164]
Northern Illinois University [E-165]
Northwestern University [E-166]
Norwich University [E-167]
University of Notre Dame [E-168]
Oakland University [E-169]
Ohio University [E-170]
Ohio Northern University [E-171]
Ohio State University [E-172]
University of Oklahoma [E-173]
Oklahoma Christian University of Science and Arts [E-174]
Oklahoma State University [E-175]
Old Dominion University [E-176]
Oregon State University [E-177]
University of the Pacific [E-178]
Pennsylvania State University [E-180]
University of Pittsburgh [E-181]
Polytechnic University [E-182]
University of Portland [E-183]
Portland State University [E-184]
Prairie View A&M University [E-185]
Princeton University [E-186]
University of Puerto Rico, Mayagüez Campus [E-187]
Purdue University [E-188]
Purdue University-Calumet
Rensselaer Polytechnic Institute [E-189]
University of Rhode Island [E-190]
William Marsh Rice University [E-191]
University of Rochester [E-192]
Rochester Institute of Technology [E-193]
Rose-Hulman Institute of Technology [E-194]
Rutgers, The State University [E-195]
Saginaw Valley State University [E-196]
San Diego State University [E-200]
San Francisco State University [E-201]
San Jose State University [E-202]
Santa Clara University [E-203]
Seattle University [E-204]
University of South Alabama [E-205]
University of South Carolina [E-206]
South Dakota School of Mines and Technology [E-207]
South Dakota State University [E-208]
University of South Florida [E-209]
Southern University and A&M College [E-210]
University of Southern California [E-211]
Southern Illinois University at Carbondale [E-213]
Southern Methodist University [E-216]
University of Southwestern Louisiana
Stanford University [E-217]
State University of New York at Binghamton [E-218]
State University of New York at Buffalo [E-219]
State University of New York at Stony Brook [E-223]
Stevens Institute of Technology
Syracuse University [E-225]
Temple University [E-227]
University of Tennessee, Knoxville [E-229]
Tennessee State University [E-230]
Tennessee Technological University [E-231]
University of Texas at Arlington [E-232]
University of Texas at Austin [E-233]
University of Texas at El Paso [E-235]
University of Texas at San Antonio [E-236]
Texas A&I University [E-237]
Texas A&M University [E-238]
Texas Tech University [E-239]
University of Toledo [E-240]
Tri-State University [E-241]
Tufts University [E-243]
Tulane University [E-244]
University of Tulsa [E-245]
Tuskegee University [E-246]
Union College [E-247]
United States Military Academy [E-251]
United States Naval Academy [E-252]
University of Utah [E-253]
Utah State University [E-254]
Valparaiso University [E-255]
Vanderbilt University [E-256]
University of Vermont [E-257]
Villanova University [E-258]
University of Virginia [E-259]
Virginia Military Institute [E-260]
Virginia Polytechnic Institute and State University [E-261]
University of Washington [E-262]
Washington University [E-263]
Washington State University [E-264]
Wayne State University [E-265]
West Virginia University [E-267]
West Virginia Institute of Technology [E-268]
Western Michigan University [E-269]
Western New England College [E-270]
Wichita State University [E-271]
Widener University [E-272]
University of Wisconsin-Madison [E-274]
University of Wisconsin-Milwaukee [E-275]
University of Wisconsin-Platteville [E-276]
Worcester Polytechnic Institute [E-277]
Wright State University [E-278]
University of Wyoming [E-279]
Yale University [E-280]
Youngstown State University

Mechanical Engineering & Applied Mechanics
University of Pennsylvania [E-179]

Mechanical Engineering (Option in Engineering Science(s))
United States Air Force Academy [E-248]

Mechanical Engineering Option in Metallurgical Engineering
Michigan Technological University [E-133]

Metallurgical & Materials Engineering
California Polytechnic State University [E-027]

Metallurgical Engineering
University of Alabama [E-002]
Colorado School of Mines [E-052]
Columbia University [E-054]
University of Idaho [E-092]
University of Illinois at Chicago [E-094]
University of Illinois at Urbana-Champaign [E-095]
Illinois Institute of Technology [E-096]
Iowa State University [E-099]
University of Kentucky [E-103]
University of Missouri-Rolla [E-141]
Montana College of Mineral Science and Technology [E-143]
University of Nevada-Reno
University of Notre Dame [E-168]

Ohio State University [E-172]
University of Pittsburgh [E-181]
Purdue University [E-188]
South Dakota School of Mines and Technology [E-207]
University of Texas at El Paso [E-235]
University of Utah [E-253]
University of Washington [E-262]
University of Wisconsin-Madison [E-274]

Metallurgical Engineering & Materials Science
Carnegie Mellon University [E-036]

Metallurgical Engineering (Materials Science & Engineering)
Polytechnic University [E-182]

Metals Science & Engineering
Pennsylvania State University [E-180]

Microelectronic Engineering
Rochester Institute of Technology [E-193]

Mineral Engineering
University of Alabama [E-002]
University of California at Berkeley [E-019]

Mineral Processing Engineering
Michigan Technological University [E-133]

Mining Engineering
University of Alaska, Fairbanks [E-005]
University of Arizona [E-008]
Colorado School of Mines [E-052]
Columbia University [E-054]
University of Idaho [E-092]
University of Kentucky [E-103]
Michigan Technological University [E-133]
University of Missouri-Rolla [E-141]
Montana College of Mineral Science and Technology [E-143]
University of Nevada-Reno
South Dakota School of Mines and Technology [E-207]
Southern Illinois University at Carbondale [E-213]
University of Utah [E-253]
Virginia Polytechnic Institute and State University [E-261]

Mining Engineering (Mineral Processing Option)
Pennsylvania State University [E-180]

Mining Engineering (Mining Option)
Pennsylvania State University [E-180]

N

Natural Gas Engineering
Texas A&I University [E-237]

Naval Architecture
University of California at Berkeley [E-019]
State University of New York Maritime College [E-221]

Naval Architecture & Marine Engineering
University of Michigan [E-130]
University of New Orleans [E-155]
United States Coast Guard Academy [E-249]
Webb Institute of Naval Architecture [E-266]

Naval Engineering
United States Naval Academy [E-252]

Nuclear & Power Engineering
University of Cincinnati [E-042]

Nuclear Engineering
University of Arizona [E-008]
University of California at Berkeley [E-019]
University of California, Santa Barbara [E-024]
University of Florida [E-071]
Georgia Institute of Technology [E-082]
University of Illinois at Urbana-Champaign [E-095]
Iowa State University [E-099]
Kansas State University [E-102]
University of Maryland, College Park [E-120]
University of Massachusetts, Lowell [E-123]
Massachusetts Institute of Technology [E-124]
University of Michigan [E-130]
Mississippi State University [E-138]
University of Missouri-Rolla [E-141]
The University of New Mexico [E-152]

North Carolina State University [E-160]
Oregon State University [E-177]
Pennsylvania State University [E-180]
Purdue University [E-188]
Rensselaer Polytechnic Institute [E-189]
University of Tennessee, Knoxville [E-229]
Texas A&M University [E-238]
University of Virginia [E-259]
University of Wisconsin-Madison [E-274]

O

Ocean Engineering
Florida Atlantic University [E-073]
Florida Institute of Technology [E-074]
Massachusetts Institute of Technology [E-124]
Texas A&M University [E-238]
United States Naval Academy [E-252]
Virginia Polytechnic Institute and State University [E-261]

Operations Research & Engineering
Cornell University [E-058]

P

Petroleum & Natural Gas Engineering
Pennsylvania State University [E-180]
West Virginia University [E-267]

Petroleum Engineering
Colorado School of Mines [E-052]
University of Kansas [E-101]
Louisiana State University [E-109]
Louisiana Tech University [E-110]
Marietta College [E-117]
Mississippi State University [E-138]
University of Missouri-Rolla [E-141]
Montana College of Mineral Science and Technology [E-143]
New Mexico Institute of Mining and Technology [E-153]
University of Oklahoma [E-173]
University of Southwestern Louisiana
Stanford University [E-217]
University of Texas at Austin [E-233]
Texas A&M University [E-238]
Texas Tech University [E-239]
University of Tulsa [E-245]
University of Wyoming [E-279]

Plastics Engineering
University of Massachusetts, Lowell [E-123]

Polymer Science & Engineering
Case Western Reserve University [E-037]

R

Radiological Health Engineering
Texas A&M University [E-238]

S

Special Studies Engineering
Arizona State University [E-009]

Structural Engineering
University of California, San Diego [E-023]

Surveying Engineering
California State University, Fresno [E-029]
Ferris State University [E-070]
University of Maine [E-114]
Purdue University [E-188]

Surveying Option in Civil & Environmental Engineering
University of Wisconsin-Madison [E-274]

Systems & Control Engineering
University of California, San Diego [E-023]
Case Western Reserve University [E-037]
University of West Florida

Systems Analysis & Engineering
The George Washington University [E-080]

Systems Engineering
University of Arizona [E-008]
Boston University [E-014]
Oakland University [E-169]
United States Naval Academy [E-252]
University of Virginia [E-259]

Systems Science & Engineering
University of Pennsylvania [E-179]
Washington University [E-263]

T

Textile Engineering
Georgia Institute of Technology [E-082]
North Carolina State University [E-160]

W

Welding Engineering
Ohio State University [E-172]

CANADIAN ENGINEERING
(CEAB-ACCREDITED PROGRAMS)

A

Agricultural Engineering
University of Alberta [E-006]
University of British Columbia
University of Guelph
University of Manitoba [E-115]
McGill University [E-125]
Ryerson Polytechnical Institute
University of Saskatchewan
Technical University of Nova Scotia [E-226]

B

Bio-Resource Engineering
University of British Columbia

Biological Engineering
University of Guelph

Building Engineering
Concordia University [E-055]

C

Ceramic Engineering
McMaster University

Ceramic Engineering & Management
McMaster University

Chemical & Biochemical Engineering
University of Western Ontario

Chemical & Materials Engineering
Royal Military College of Canada

Chemical Engineering
University of Alberta [E-006]
University of British Columbia
University of Calgary
Lakehead University
McGill University [E-125]
McMaster University
University of New Brunswick
University of Ottawa
Queen's University
Royal Military College of Canada
Ryerson Polytechnical Institute
University of Saskatchewan
Technical University of Nova Scotia [E-226]
University of Toronto
University of Waterloo
University of Western Ontario
University of Windsor

Chemical Engineering & Management
McMaster University

Génie chimique
École Polytechnique de Montréal [E-066]
Université Laval
Québec à Trois-Rivières, Université de
Université de Sherbrooke

Civil Engineering
University of Alberta [E-006]
University of British Columbia
University of Calgary
Concordia University [E-055]
Lakehead University
University of Manitoba [E-115]
McGill University [E-125]
McMaster University
Memorial University of Newfoundland
University of New Brunswick
University of Ottawa
Queen's University
Royal Military College of Canada
Ryerson Polytechnical Institute
University of Saskatchewan
Technical University of Nova Scotia [E-226]
University of Toronto
University of Waterloo
University of Western Ontario
University of Windsor

Génie civil
École Polytechnique de Montréal [E-066]
Université Laval
Université de Moncton
Université de Sherbrooke

Civil Engineering & Engineering Mechanics
McMaster University

Civil Engineering & Management
McMaster University

Computer Engineering
University of Alberta [E-006]
Concordia University [E-055]
University of Manitoba [E-115]
McMaster University
University of Ottawa
Royal Military College of Canada
University of Victoria
University of Waterloo

Computer Engineering & Management
McMaster University

Génie et gestion de la construction
École de Technologie Supérieure

E

Electrical Engineering
University of Alberta [E-006]
University of British Columbia
University of Calgary
Concordia University [E-055]
Lakehead University
University of Manitoba [E-115]
McGill University [E-125]
McMaster University
Memorial University of Newfoundland
University of New Brunswick
University of Ottawa
Queen's University
Royal Military College of Canada
Ryerson Polytechnical Institute
University of Saskatchewan
Technical University of Nova Scotia [E-226]
University of Toronto
University of Waterloo
University of Western Ontario
University of Windsor

Electrical Engineering & Management
McMaster University

Génie électrique
École Polytechnique de Montréal [E-066]
École de Technologie Supérieure
Université Laval
Québec à Trois-Rivières, Université de
Université de Sherbrooke

Electronic Information Systems Engineering
University of Regina

Engineering & Computer Systems
McMaster University

Engineering & Management
Royal Military College of Canada

Engineering Chemistry
Queen's University

Engineering Materials
University of Windsor

Engineering Physics
University of Alberta [E-006]
University of British Columbia
McMaster University
Queen's University
Royal Military College of Canada
University of Saskatchewan
Technical University of Nova Scotia [E-226]

Engineering Physics & Management
McMaster University

Engineering Science
Simon Fraser University
University of Toronto

Extractive Metallurgical Engineering
Laurentian University

Extractive Metallurgy
Laurentian University

F

Forest Engineering
University of New Brunswick

Fuels & Materials Engineering
Royal Military College of Canada

G

Geo-Engineering
University of Toronto

Geological & Mineral Engineering
University of Toronto

Geological Engineering
University of British Columbia
University of Manitoba [E-115]
University of New Brunswick
Queen's University
University of Saskatchewan
University of Toronto
University of Waterloo
University of Windsor

Geological Engineering (Geophysics)
University of Saskatchewan

Geological Engineering & Applied Earth Science
University of Toronto

Génie géologique
École Polytechnique de Montréal [E-066]
Université Laval
Québec a Chicoutimi, Université de

I

Industrial Engineering
University of Manitoba [E-115]
Ryerson Polytechnical Institute
Technical University of Nova Scotia [E-226]
University of Toronto
University of Windsor

Industrial Systems Engineering
University of Regina

Génie industriel
École Polytechnique de Montréal [E-066]
Université de Moncton
Québec à Trois-Rivières, Université de

Génie informatique
École Polytechnique de Montréal [E-066]
Québec a Chicoutimi, Université de

M

Manufacturing Engineering
McMaster University

Materials & Metallurgical Engineering
Queen's University

Materials Engineering
McMaster University
University of Western Ontario

Materials Engineering & Management
McMaster University

Génie des matériaux
École Polytechnique de Montréal [E-066]

Génie des matériaux et de la métallurgie
Université Laval

Mathematics & Engineering
Queen's University

Mechanical Engineering
University of Alberta [E-006]
University of British Columbia
University of Calgary
Concordia University [E-055]
Lakehead University
University of Manitoba [E-115]
McGill University [E-125]
McMaster University
Memorial University of Newfoundland
University of New Brunswick
University of Ottawa
Queen's University
Royal Military College of Canada
Ryerson Polytechnical Institute
University of Saskatchewan
Technical University of Nova Scotia [E-226]
University of Toronto
University of Victoria
University of Waterloo
University of Western Ontario
University of Windsor

Génie méchanique
École Polytechnique de Montréal [E-066]
École de Technologie Supérieure
Université Laval
Université de Moncton
Université de Sherbrooke

Génie méchanique manufacturier
Québec à Trois-Rivières, Université de

Mechanical Engineering & Management
McMaster University

Metallurgical Engineering & Materials Science
University of Toronto

Metallurgical Engineering
University of Alberta [E-006]
University of British Columbia
McGill University [E-125]
McMaster University
Queen's University
Technical University of Nova Scotia [E-226]

Génie métallurgique
École Polytechnique de Montréal [E-066]
Université Laval

Metallurgy & Materials Science
University of Toronto

Metals & Materials Engineering
University of British Columbia

Mineral Engineering
University of Alberta [E-006]
University of British Columbia

Mineral Process Engineering
University of Alberta [E-006]

Génie des mines
École Polytechnique de Montréal [E-066]

Génie des mines et de la minéralurgie
Université Laval

Génie minier
École Polytechnique de Montréal [E-066]
Université Laval

Mining Engineering
University of Alberta [E-006]
Laurentian University
McGill University [E-125]
Queen's University
University of Saskatchewan
Technical University of Nova Scotia [E-226]

Mining & Mineral Process Engineering
University of British Columbia

N

Naval Architectural Engineering
Memorial University of Newfoundland

P

Petroleum Engineering
University of Alberta [E-006]

Génie physique
École Polytechnique de Montréal [E-066]
Université Laval

Génie de la production automatisée
École de Technologie Supérieure

R

Regional Environmental Systems Engineering
University of Regina

Regional Systems Engineering
University of Regina

Génie rural
Université Laval

S

Shipbuilding Engineering
Memorial University of Newfoundland

Systems Engineering
University of Regina
University of Calgary

Systems Design Engineering
University of Waterloo

U

Génie unifié
Québec a Chicoutimi, Université de

W

Water Resources Engineering
University of Guelph

ENGINEERING TECHNOLOGY

A

Aeronautical Engineering Technology
Arizona State University [T-283]

Agricultural Engineering Technology
University of Delaware

Air Conditioning & Refrigeration Technology
California Polytechnic State University

Aircraft Engineering Technology
Embry-Riddle Aeronautical University - Daytona Beach [T-305]

Aircraft Maintenance Engineering Technology
Northrop-Rice Aviation Institute of Technology

Apparel Engineering Technology
Southern College of Technology [T-339]

Architectural Engineering Technology
Bluefield State College
University of Cincinnati [T-293]
Southern College of Technology [T-339]
University of Southern Mississippi [T-342]
Wentworth Institute of Technology [T-352]

Architectural Technology
Memphis State University [T-316]

Automated Systems Engineering Technology
Lake Superior State University [T-310]

Automotive Engineering Technology
Weber State University [T-351]

C

Civil & Construction Engineering Technology
Temple University [T-345]

Civil Engineering Technology
University of Alabama
Alabama A&M University [T-282]
Bluefield State College
Fairmont State College
Florida A&M University
Georgia Southern University
University of Massachusetts, Lowell [T-315]
Metropolitan State College of Denver [T-317]
Murray State University [T-320]
University of North Carolina-Charlotte [T-324]
Old Dominion University [T-327]
Oregon Institute of Technology [T-328]
University of Pittsburgh at Johnstown [T-332]
Point Park College [T-333]
Rochester Institute of Technology [T-336]
Savannah State College
South Carolina State University [T-338]
Southern College of Technology [T-339]
University of Southern Colorado [T-340]
Southern Illinois University at Carbondale [T-341]
University of Southern Indiana
University of Tennessee, Martin [T-346]
Wentworth Institute of Technology [T-352]
Western Kentucky University [T-355]
Youngstown State University

Civil Engineering Technology Option in Engineering Technology
New Mexico State University [T-323]

Civil Technology
University of Houston

Computer Engineering Technology
University of Arkansas at Little Rock [T-284]
Capitol College [T-290]
University of Houston
Memphis State University [T-316]
Oregon Institute of Technology [T-328]
Prairie View A&M University [T-334]
Southern College of Technology [T-339]

Index to Accredited Degree Programs

University of Southern Mississippi [T-342]
Wentworth Institute of Technology [T-352]

Computer Option in Computer Engineering Technology
Eastern Washington University

Computer Technology
Rochester Institute of Technology [T-336]
State University of New York Institute of Technology, Utica/Rome [T-344]

Construction & Contracting Option in Engineering Technology
New Jersey Institute of Technology [T-322]

Construction Engineering Technology
University of Arkansas at Little Rock [T-284]
University of the District of Columbia [T-303]
Florida A&M University
Louisiana Tech University [T-311]
Missouri Western State College
Montana State University [T-319]
University of Nebraska-Omaha
Pittsburg State University [T-331]
University of Southern Mississippi [T-342]

Construction Management Option in Engineering Technology
California State University, Sacramento [T-288]

Construction Management Technology
University of Maine [T-312]
Oklahoma State University [T-326]

Construction Technology
University of Akron [T-281]
East Tennessee State University [T-304]
Murray State University [T-320]
Purdue University-Calumet
Texas Tech University [T-348]

D

Design Engineering Technology
Brigham Young University [T-287]

E

Electrical & Electronics Engineering Technology
Montana State University [T-319]

Electrical Engineering Technology
University of Alabama
Alabama A&M University [T-282]
Bluefield State College
Bradley University [T-286]
University of Cincinnati [T-293]
Georgia Southern University
Lake Superior State University [T-310]
Louisiana Tech University [T-311]
University of Maine [T-312]
University of Massachusetts, Dartmouth [T-314]
Milwaukee School of Engineering [T-318]
Murray State University [T-320]
University of New Hampshire [T-321]
New York Institute of Technology
New York Institute of Technology
University of North Carolina-Charlotte [T-324]
Northeastern University [T-325]
Northern Arizona University
Old Dominion University [T-327]
Pennsylvania State University at Erie, The Behrend College [T-329]
Pennsylvania State University - Harrisburg [T-330]
University of Pittsburgh at Johnstown [T-332]
Point Park College [T-333]
Prairie View A&M University [T-334]
Purdue University-Calumet
Purdue University - Kokomo
Rochester Institute of Technology [T-336]
Roger Williams College
South Carolina State University [T-338]
Southern College of Technology [T-339]
Southern Illinois University at Carbondale [T-341]
University of Southern Indiana
SUNY-Binghamton
SUNY-Buffalo
State University of New York Institute of Technology, Utica/Rome [T-344]
Temple University [T-345]
University of Tennessee, Martin [T-346]
Western Kentucky University [T-355]
Youngstown State University

Electrical Option in Engineering Technology
New Jersey Institute of Technology [T-322]

Electrical Technology
University of Houston
Indiana University-Purdue University at Indianapolis [T-309]
Purdue University [T-335]
State University of New York College of Technology at Farmingdale [T-343]

Electrical/Electronics Technology
Texas Tech University [T-348]

Electromechanical Engineering Technology
CUNY - City College
SUNY-Binghamton

Electromechanical Systems Engineering Technology
University of the District of Columbia [T-303]

Electronic Engineering Technology
University of Massachusetts, Lowell [T-315]
Texas A&M University [T-347]
Virginia State University [T-350]
Wentworth Institute of Technology [T-352]

Electronic Technology Option in Engineering Technology
Kansas State University

Electronics Engineering Technology
Arizona State University [T-283]

Electronic(s) Engineering Technology
University of Arkansas at Little Rock [T-284]
Brigham Young University [T-287]
Capitol College [T-290]
Central Washington University
Cogswell Polytechnic College North
Cogswell Polytechnic College-Cupertino
Colorado Technical College
University of Dayton [T-294]
DeVry Institute of Technology, Chicago [T-295]
DeVry Institute of Technology, Columbus [T-296]
DeVry Institute of Technology, Atlanta [T-297]
DeVry Institute of Technology, DuPage [T-298]
DeVry Institute of Technology, Dallas [T-299]
DeVry Institute of Technology, Kansas City [T-300]
DeVry Institute of Technology, Los Angeles [T-301]
DeVry Institute of Technology, Phoenix [T-302]
East Tennessee State University [T-304]
Fairmont State College
Florida A&M University
Fort Valley State College
Franklin University [T-307]

Electronics Engineering Technology
University of Hartford [T-308]

Electronic(s) Engineering Technology
Mankato State University [T-313]
Memphis State University [T-316]
Metropolitan State College of Denver [T-317]

Electronics Engineering Technology
Missouri Western State College

Electronic(s) Engineering Technology
University of Nebraska-Omaha
New Mexico Highlands University
Oregon Institute of Technology [T-328]
Pittsburg State University [T-331]
Savannah State College
University of Southern Colorado [T-340]
University of Southern Mississippi [T-342]
University of Toledo
Weber State University [T-351]
West Virginia Institute of Technology [T-353]
Western Washington University

Electronic(s) Engineering Technology Option in Engineering Technology
New Mexico State University [T-323]

Electronic(s) Technology
University of Akron [T-281]
California Polytechnic State University

Electronics Technology
Oklahoma State University [T-326]

Energy Engineering Technology
Rochester Institute of Technology [T-336]

Engineering Technology
California State Polytechnic University, Pomona [T-289]

Engineering Technology & Technical Management
University of Delaware

Engineering Technology Option in Manufacturing Engineering Technology
St. Cloud State University [T-337]

Engineering Technology-Construction
Central Connecticut State University [T-291]

Engineering Technology-Manufacturing
Central Connecticut State University [T-291]

Environmental Engineering Technology
Norwich University
Pennsylvania State University - Harrisburg [T-330]

F

Fire Protection & Safety Technology
Oklahoma State University [T-326]

I

Industrial Engineering Technology
University of Dayton [T-294]

Industrial Engineering Technology
Purdue University-Calumet
South Carolina State University [T-338]
Southern College of Technology [T-339]
University of Southern Mississippi [T-342]
State University of New York Institute of Technology, Utica/Rome [T-344]
Trenton State College

Industrial Engineering Technology
Georgia Southern University

L

LASER Optical Engineering Technology
Oregon Institute of Technology [T-328]

M

Manufacturing Engineering Technology
Arizona State University [T-283]
University of Arkansas at Little Rock [T-284]
Brigham Young University [T-287]
University of Dayton [T-294]
East Tennessee State University [T-304]
Mankato State University [T-313]
Memphis State University [T-316]
Midwestern State University
Murray State University [T-320]
University of Nebraska-Omaha
Oregon Institute of Technology [T-328]
Pittsburg State University [T-331]
Rochester Institute of Technology [T-336]
Texas A&M University [T-347]
Weber State University [T-351]
Wentworth Institute of Technology [T-352]
Western Carolina University [T-354]
Western Michigan University [T-356]
Western Washington University

Manufacturing Processes Technology
California Polytechnic State University

Manufacturing Technology
Oklahoma State University [T-326]
State University of New York College of Technology at Farmingdale [T-343]

Manufacturing Technology Option in Engineering Technology
New Jersey Institute of Technology [T-322]

Marine Engineering Technology
California Maritime Academy
Maine Maritime Academy

Mechanical Design Option in Manufacturing Technology
Bradley University [T-286]

Mechanical Design Technology
Oklahoma State University [T-326]

Mechanical Drafting & Design Technology
Alabama A&M University [T-282]

Mechanical Engineering Technology
Alabama A&M University [T-282]
University of Arkansas at Little Rock [T-284]
University of Cincinnati [T-293]
Cogswell Polytechnic College North
Cogswell Polytechnic College-Cupertino
University of Dayton [T-294]
Eastern Washington University
Fairmont State College
Franklin University [T-307]
Georgia Southern University
Indiana University-Purdue University at Indianapolis [T-309]
Lake Superior State University [T-310]
University of Maine [T-312]
University of Massachusetts, Dartmouth [T-314]
University of Massachusetts, Lowell [T-315]
Metropolitan State College of Denver [T-317]
Milwaukee School of Engineering [T-318]
Montana State University [T-319]
University of New Hampshire [T-321]
University of North Carolina-Charlotte [T-324]
Northeastern University [T-325]
Northern Arizona University
Old Dominion University [T-327]
Oregon Institute of Technology [T-328]
Pennsylvania State University at Erie, The Behrend College [T-329]
Pennsylvania State University - Harrisburg [T-330]
Pittsburg State University [T-331]
University of Pittsburgh at Johnstown [T-332]
Point Park College [T-333]
Purdue University-Calumet
Purdue University - North Central
Rochester Institute of Technology [T-336]
Roger Williams College
Savannah State College
South Carolina State University [T-338]
Southern College of Technology [T-339]
University of Southern Colorado [T-340]
Southern Illinois University at Carbondale [T-341]
University of Southern Indiana
University of Southern Mississippi [T-342]
SUNY-Binghamton
SUNY-Buffalo
State University of New York Institute of Technology, Utica/Rome [T-344]
Temple University [T-345]
University of Tennessee, Martin [T-346]
Texas A&M University [T-347]
University of Toledo
Trenton State College
Virginia State University [T-350]
Weber State University [T-351]
Wentworth Institute of Technology [T-352]
Western Kentucky University [T-355]
Youngstown State University

Mechanical Engineering Technology Option in Engineering Technology
California State University, Sacramento [T-288]
Kansas State University
New Mexico State University [T-323]

Mechanical Option in Engineering Technology
New Jersey Institute of Technology [T-322]

Mechanical Power Technology
Oklahoma State University [T-326]

Mechanical Technology
University of Akron [T-281]
California Polytechnic State University
University of Houston
Purdue University [T-335]
Texas Tech University [T-348]

Mechanical/Structural Engineering Technology
Northeastern University [T-325]

Mining Engineering Technology
Bluefield State College

N

Nuclear Engineering Technology
American Technical Institute

P

Plastics Engineering Technology
Pittsburg State University [T-331]

Process & Piping Design Engineering Technology Option in Engineering Tech
University of Houston-Downtown

Productions Operations Option in Manufacturing Technology
Bradley University [T-286]

S

Software Engineering Technology
Oregon Institute of Technology [T-328]

Structural Design & Construction Engineering Technology
Pennsylvania State University - Harrisburg [T-330]

T

Textile Engineering Technology
Southern College of Technology [T-339]

Textile Management and Technology
Auburn University [T-285]

W

Welding Technology
California Polytechnic State University

ASEE Products

ASEE IS PROUD TO ANNOUNCE OFFICIAL ASEE PRODUCTS
AVAILABLE TO MEMBERS AND NONMEMBERS...

ASEE Coffee Mug

This mug is a must for every coffee drinker. Keep it at your desk and proudly display your ASEE affiliation with this white ceramicware mug with a 2–color logo.

ASEE member price: $5.00
Non-member price: $7.50
ASEE Order # 4183-PU-06-131

ASEE Sports Cap

Enjoy all types of sports or simply relax in this cotton twill sports cap. Featuring a glare-proof, padded visor, it is available in white with a 2–color embroidered logo. One size fits all.

ASEE member price: $8.00
Non-member price: $12.00
ASEE order # 4183-PU-06-132

ASEE Sports Tee

You will be the envy of all your colleagues in this 100% preshrunk cotton T-shirt. Soft and absorbent, this shirt is machine washable and carries the ASEE 2–color logo. Sizes: Large and Extra-Large. Color: White.

ASEE member price: $10.00
Non-member price: $15.00
ASEE order # 4183-PU-06-133

When ordering, please refer to the ASEE Publications order form in the back of the directory. For information on other ASEE publications, please call the ASEE Publications Coordinator at (202) 986-8528.

AVAILABLE RESOURCES

The best resources for today's engineering educators
Publications available from the American Society for Engineering Education

The latest on international trends in engineering education

Engineering in Japan—Education, Practice, and Future Outlook
Robert S. Cutler (ed.), 1991; 96 pp.
Members/Non-members: $19.95
ASEE order #: 4183-PU-06-014

Discusses the key elements of a symposium held in February 1991 as a part of the annual meeting of the American Association for the Advancement of Science. Topics include technology transfer, "symmetrical access," the present and future elements of "engineering systems" in Japan, and the informative first-hand experiences of AAAS colleagues recently returned from assignment in Japanese laboratories.

Periodicals

ASEE PRISM
Free to members
Individual Subscriptions: $75.00   ASEE order #: 4182-113
Domestic Library Subscriptions: $125.00   ASEE order #: 4182-111
Foreign Library Subscriptions: $135.00   ASEE order #: 4182-112
Single Copies: $7.50   ASEE order #: 4183-101

ASEE PRISM reflects the many facets of engineering and engineering technology education—people, institutions, issues, problems, challenges, and diversity. Published monthly (except July and August), it highlights government and industry activities, profiles outstanding faculty, and carries classified advertising of academic and corporate openings.

Journal of Engineering Education
Free to members
Individual Subscriptions: $75.00   ASEE order #: 4182-116
Domestic Library Subscriptions: $150.00   ASEE order #: 4182-114
Foreign Library Subscriptions: $160.00   ASEE order #: 4182-115
Single Copies: $25.00   ASEE order #: 4183-102

The *Journal of Engineering Education* is a scholarly, quarterly journal recording the progress of engineering education, discussing current trends and innovations, and offering commentary on new directions.

Directories

1993 Directory of Engineering and Engineering Technology Undergraduate Programs
Free to members, additional copies: $24.95
Students: $24.95   Non-members: $49.95
ASEE order #: 4183-PU-06-037

This directory profiles 356 undergraduate engineering and engineering technology colleges in the U.S. and Canada. Detailed information is provided on degree programs, minority support programs, admission requirements, financial aid, and tuition. It features a College Locator Disk—a custom program to help students find the schools that meet their needs.

1992–93 Directory of Engineering Graduate Studies and Research
Free to members, additional copies: $34.95
Students: $34.95   Non-members: $69.95
ASEE order #: 4183-PU-06-035

Engineering Education in Europe, 1992 Directory
European Society of Engineering Education
Members: $70.00   Non-members: $77.00
ASEE order #: 4183-PU-06-036

This 540-page directory describes 450 colleges of engineering and universities of technology and more than 1,800 study programs in 17 European countries. (In French, English, Spanish, and Italian.)

Directory of Engineering Technology Institutions and Programs
Frank A. Gourley (ed.), June 1990; 80 pp.
Members: $6.80   Non-members: $8.00
ASEE order #: 4183-PU-06-031

Using data supplied by the Engineering Manpower Commission, this directory lists almost 400 institutions and 1,700 programs. It focuses on two-year, four-year, and graduate engineering technology programs in the U.S. Sorted by state and curriculum category, the listings contain the program name and level, ABET-accreditation status, and general contact information for each institution.

Valuable teaching aids and classroom materials

Definition of the Engineering Method
Billy V. Koen, 1985, 5th printing, 1991; 80 pp.
Members: $7.00   Non-members: $8.00
ASEE order #: 4183-PU-06-008

Intended both for engineers, who will appreciate the distinctions drawn between engineering and science, and for others, this provocative essay leads the reader to a fresh appreciation of how engineers change the world. A popular item in university bookstores, Koen's book has been used by engineering faculty across the country.

The Teaching of Elementary Problem Solving in Engineering and Related Fields
James Lubkin (ed.), 1980, 2nd printing, 1985; 198 pp.
Members: $8.50   Non-members: $10.00
ASEE order #: 4183-PU-06-006

This book presents a variety of approaches used by expert teachers of problem solving and encourages professors to experiment in their courses.

Instructional Objectives: A Guide to Effective Teaching
Richard Leuba, 1980; 14 pp.
Members: $2.10   Non-members: $2.50
ASEE order #: 4183-PU-06-004

A concise guide for faculty who want to establish or revise objectives for their courses.

Career information

An Academic Career: It Could Be for You
Raymond Landis, December 1989; 12 pp.
Sold in increments of 25
Members: $.50 each   Non-members: $.77 each (75+, $.45 each)
ASEE order #: 4183-PU-06-012

This useful booklet explains why students should consider an engineering faculty career, describes the personal and professional rewards, and gives tips on getting a Ph.D.

Thinking of an Academic Career
New Engineering Educators Committee, December 1989; 4 pp.
Sold in increments of 25
Members: $.45 each   Non-members: $.60 each (75+, $.45 each)
ASEE order #: 4183-PU-06-013

This guide for prospective engineering faculty members offers employment trends—geographically and by engineering discipline—salary levels, and useful suggestions for job hunting and interview questions.

ASEE Conference Proceedings

ASEE Annual Conference Proceedings
Members: $55.00   Non-members: $70.00
1993 edition, available 6/93—ASEE order #: 4183-PU-06-093
1992 edition—ASEE order #: 4183-PU-06-087
1991 edition—ASEE order #: 4183-PU-06-084

1992 Annual College–Industry Education Conference Proceedings
Members: $24.00   Non-members: $30.00
ASEE order #: 4183-6799

Engineering Foundation Conference Proceedings

Engineering Education: Curriculum Innovation and Integration
Win Aung and Shlomo Carmi (eds.), 1992; 450 pp.
Members: $24.95   Non-members: $29.95
ASEE order #: 4183-PU-06-089

For engineering education at the baccalaureate level, the conclusions are that a structural change is needed both with respect to the curriculum and teaching of engineering and the recruitment and retention of students. The results of some of the recent efforts made in those areas were presented at this conference.

New Approaches to Undergraduate Engineering Education III
1991; 348 pp.
Members: $19.95   Non-members: $24.95
ASEE order #: 4183-PU-06-088

The Engineering Foundation, in cooperation with the National Science Foundation, sponsored the third conference in a series on "New Approaches to Undergraduate Engineering Education." Conference co-chairs were Patricia D. Daniels of Seattle University and M.E. Van Valkenburg of the University of Illinois at Urbana-Champaign.

When ordering, please refer to the ASEE Publications order form in the back of the directory.
If you have any questions about ASEE publications or products, please call (202) 986-8528.